LAND MAMMALS OF OREGON

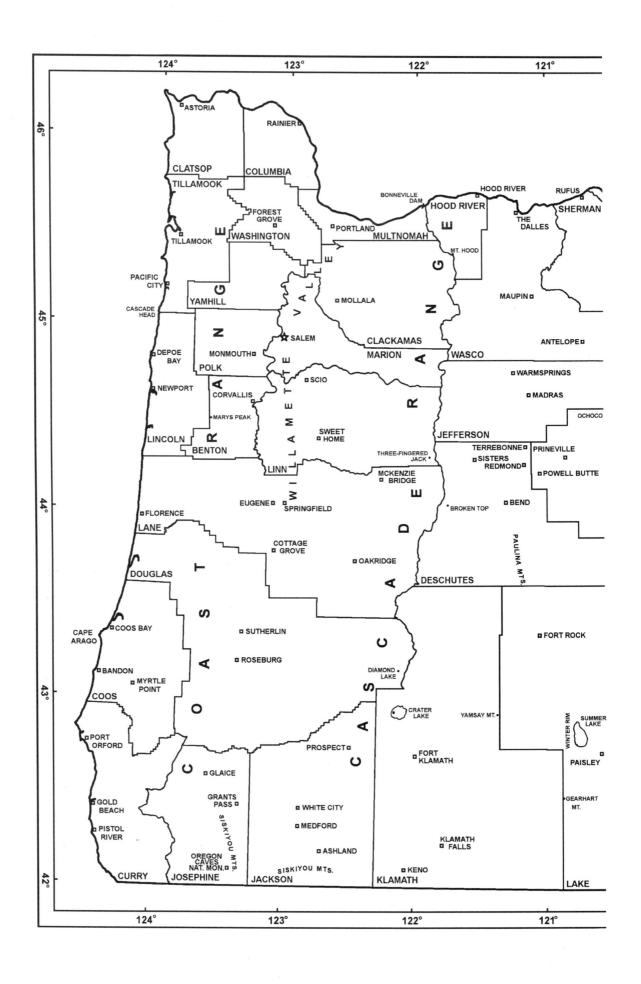

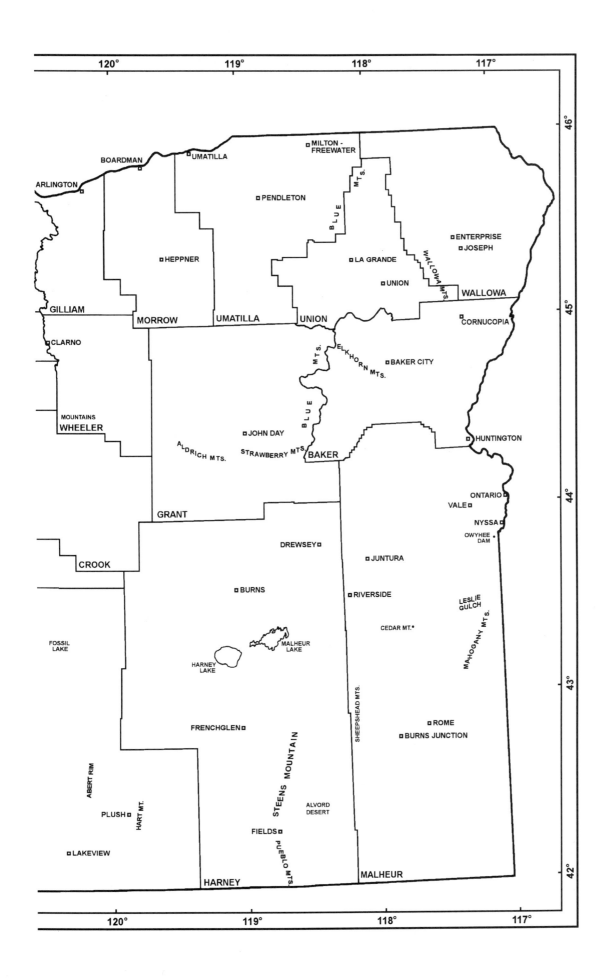

LAND MAMMALS

OF OREGON

B. J. Verts
Leslie N. Carraway

University of California Press
Berkeley / Los Angeles / London

The publisher gratefully acknowledges the generous contributions provided by:

Randal A. Aaberg; Margaret P. Banis; Merlin S. and Elsie K. Eltzroth; Olive B. and Russell D. Hodgkins; Richard F. Hoyer; Dorothy and Arthur Sevigny; J. Mary Taylor, Donna P. and John Theobald; Lois Richardson Verts; Rachel Jane Verts; The Reverend William C. Wright; U.S. Department of the Interior, Bureau of Land Management; Oregon Department of Fish and Wildlife, Game Management Program; Oregon Department of Fish and Wildlife, Wildlife Diversity Program; Thomas G. Scott Achievement Fund, Department of Fisheries and Wildlife, Oregon State University; Oregon Chapter of The Wildlife Society.

University of California Press
Berkeley and Los Angeles, California

University of California Press, Ltd.
London, England

Library of Congress Cataloging-in-Publication Data

Verts, B. J., 1927–
    Land mammals of Oregon / B.J. Verts, Leslie N. Carraway.
        p.   cm.
    Includes bibliographical references and index.
    ISBN 0-520-21199-5 (alk. paper)
    1. Mammals—Oregon.   I. Carraway, Leslie N.   II. Title.
    QL719.07V474   1998
    599´.09795—dc21

                                    97-27039
                                    CIP

Printed in Canada

9 8 7 6 5 4 3 2 1

The paper used in this publication meets the minimum requirements of American National Standards for Information Sciences—Permanence of Paper for Printed Library Materials, ANSI Z39.48-1984.

This book is dedicated to the memory of

Vernon Orlando Bailey

21 June 1864–20 April 1942

Naturalist and Author

The Mammals and Life Zones of Oregon

1936

CONTENTS

PREFACE / xv
   ACKNOWLEDGMENTS / xv

CHAPTER 1. INTRODUCTION / 1
   WHAT IS A MAMMAL? / 1
   BASIC MORPHOLOGY AND MORPHOMETRY OF MAMMALS / 2
   WHENCE CAME MAMMALS? / 2
   NAMING MAMMALS / 6

CHAPTER 2. A BRIEF HISTORY OF MAMMALOGY IN OREGON / 8
   MAMMAL COLLECTIONS / 8
   PUBLICATIONS ON MAMMALS / 11
   FUTURE NEEDS / 13

CHAPTER 3. CHECKLIST OF LAND MAMMALS OF OREGON / 14

CHAPTER 4. PHYSIOGNOMY AND MAMMAL FAUNAS OF OREGON / 16
   THE FORMATION OF OREGON / 16
   EVOLUTION OF MAMMAL FAUNAS OF OREGON / 19
   PHYSIOGRAPHIC PROVINCES AND PRESENT-DAY MAMMALIAN FAUNAS / 22
   BIOGEOGRAPHICAL AFFINITIES OF PRESENT-DAY MAMMALS OF OREGON / 33

CHAPTER 5. ZOONOSES / 35
   VIRAL DISEASES / 35
   BACTERIAL DISEASES / 36
   RICHETTSIAL DISEASES / 37

CHAPTER 6. KEY TO THE ORDERS OF LAND MAMMALS OF OREGON / 39

CHAPTER 7. ORDER DIDELPHIMORPHIA—OPOSSUMS / 40
   FAMILY DIDELPHIDAE—Opossums / 40
     Didelphis virginiana, Virginia opossum / 40

CHAPTER 8. ORDER INSECTIVORA—SHREWS, MOLES, AND SHREW-MOLE / 45
   KEY TO THE INSECTIVORA OF OREGON / 45
   FAMILY SORICIDAE—Shrews / 46
     Sorex bairdi, Baird's Shrew / 46
     Sorex bendirii, Pacific Water or Marsh Shrew / 49
     Sorex merriami, Merriam's Shrew / 51
     Sorex monticolus, Montane or Dusky Shrew / 52
     Sorex pacificus, Pacific Shrew / 54
     Sorex palustris, Water Shrew / 55
     Sorex preblei, Preble's Shrew / 57
     Sorex sonomae, Fog Shrew / 58
     Sorex trowbridgii, Trowbridge's Shrew / 60
     Sorex vagrans, Vagrant Shrew / 62

FAMILY TALPIDAE—Shrew-mole and Moles / 65
  Neurotrichus gibbsii, Shrew-mole / 65
  Scapanus latimanus, Broad-footed Mole / 68
  Scapanus orarius, Coast Mole / 69
  Scapanus townsendii, Townsend's Mole / 71

CHAPTER 9. ORDER CHIROPTERA—BATS / 74
  KEY TO THE CHIROPTERA OF OREGON / 75
  FAMILY VESPERTILIONIDAE—Evening Bats / 76
    Myotis californicus, California Myotis / 79
    Myotis ciliolabrum, Western small-footed Myotis / 82
    Myotis evotis, Long-eared Myotis / 84
    Myotis lucifugus, Little Brown Myotis / 87
    Myotis thysanodes, Fringed Myotis / 90
    Myotis volans, Long-legged Myotis / 93
    Myotis yumanensis, Yuma Myotis / 95
    Lasiurus cinereus, Hoary Bat / 98
    Lasionycteris noctivagans, Silver-haired Bat / 100
    Pipistrellus hesperus, Western Pipistrelle / 105
    Eptesicus fuscus, Big Brown Bat / 107
    Euderma maculatum, Spotted Bat / 110
    Corynorhinus townsendii, Townsend's Big-eared Bat / 113
    Antrozous pallidus, Pallid Bat / 116
  FAMILY MOLOSSIDAE—Free-tailed Bats / 119
    Tadarida brasiliensis, Brazilian Free-tailed Bat / 119

CHAPTER 10. ORDER LAGOMORPHA—RABBITS, HARES, AND PIKAS / 123
  KEY TO THE LAGOMORPHA OF OREGON / 123
  FAMILY OCHOTONIDAE—Pikas / 124
    Ochotona princeps, American Pika / 124
  FAMILY LEPORIDAE—Rabbits and Hares / 128
    Brachylagus idahoensis, Pygmy Rabbit / 128
    Sylvilagus bachmani, Brush Rabbit / 131
    Sylvilagus floridanus, Eastern Cottontail / 134
    Sylvilagus nuttallii, Mountain Cottontail / 138
    Lepus americanus, Snowshoe Hare / 141
    Lepus californicus, Black-tailed Jackrabbit / 145
    Lepus townsendii, White-tailed Jackrabbit / 149

CHAPTER 11. ORDER RODENTIA—GNAWING MAMMALS / 153
  KEY TO THE FAMILIES OF RODENTIA OF OREGON / 153
  FAMILY APLODONTIDAE—Aplodontids / 154
    Aplodontia rufa, Mountain Beaver / 154
  FAMILY SCIURIDAE—Squirrels, Chipmunks, and Marmots / 157
    SOME TAXONOMIC CONSIDERATIONS AT THE GENERIC LEVEL / 158
    Key to the Sciuridae of Oregon / 158
    Tamias amoenus, Yellow-pine Chipmunk / 159
    Tamias minimus, Least Chipmunk / 163
    The *Tamias townsendii* Complex / 166
    Tamias senex, Allen's Chipmunk / 166
    Tamias siskiyou, Siskiyou Chipmunk / 168
    Tamias townsendii, Townsend's Chipmunk / 169

Marmota flaviventris, Yellow-bellied Marmot / 173
Ammospermophilus leucurus, White-tailed Antelope Squirrel / 178
Spermophilus beecheyi, California Ground Squirrel / 181
Spermophilus beldingi, Belding's Ground Squirrel / 184
Spermophilus columbianus, Columbian Ground Squirrel / 187
Spermophilus elegans, Wyoming Ground Squirrel / 190
Spermophilus lateralis, Golden-mantled Ground Squirrel / 193
The *Spermophilus "townsendii"* Complex / 196
Spermophilus canus, Merriam's Ground Squirrel / 196
Spermophilus mollis, Piute Ground Squirrel / 198
Spermophilus washingtoni, Washington Ground Squirrel / 201
Sciurus carolinensis, Eastern Gray Squirrel / 203
Sciurus griseus, Western Gray Squirrel / 207
Sciurus niger, Eastern Fox Squirrel / 210
Tamiasciurus douglasii, Douglas' Squirrel / 212
Tamiasciurus hudsonicus, Red Squirrel / 216
Glaucomys sabrinus, Northern Flying Squirrel / 219

FAMILY GEOMYIDAE—Pocket Gophers / 223
Key to the Geomyidae of Oregon / 223
Thomomys bottae, Botta's Pocket Gopher / 224
Thomomys bulbivorus, Camas Pocket Gopher / 229
Thomomys mazama, Western Pocket Gopher / 231
Thomomys talpoides, Northern Pocket Gopher / 234
Thomomys townsendii, Townsend's Pocket Gopher / 238

FAMILY HETEROMYIDAE—Pocket Mice, Kangaroo Rats, and Kangaroo Mouse / 239
Key to the Heteromyidae of Oregon / 240
Perognathus longimembris, Little Pocket Mouse / 240
Perognathus parvus, Great Basin Pocket Mouse / 245
Microdipodops megacephalus, Dark Kangaroo Mouse / 248
Dipodomys californicus, California Kangaroo Rat / 249
Dipodomys microps, Chisel-toothed Kangaroo Rat / 251
Dipodomys ordii, Ord's Kangaroo Rat / 254

FAMILY CASTORIDAE—Beavers / 256
Castor canadensis, American Beaver / 256

FAMILY MURIDAE—Rats, Mice, Voles, and Muskrats / 262
Key to the Subfamilies of Muridae of Oregon / 262
Subfamily Sigmodontinae—New World Rats and Mice / 262
Key to the Sigmodontinae of Oregon / 263
Reithrodontomys megalotis, Western Harvest Mouse / 263
Peromyscus crinitus, Canyon Mouse / 267
Peromyscus maniculatus, Deer Mouse / 270
Peromyscus truei, Piñon Mouse / 273
Onychomys leucogaster, Northern Grasshopper Mouse / 276
Neotoma cinerea, Bushy-tailed Woodrat / 280
Neotoma fuscipes, Dusky-footed Woodrat / 283
Neotoma lepida, Desert Woodrat / 286
Subfamily Murinae—Old World Rats and Mice / 288
Key to the Murinae of Oregon / 289
Rattus norvegicus, Norway Rat / 289
Rattus rattus, Black Rat / 291
Mus musculus, House Mouse / 293

Subfamily Arvicolinae—Voles and Muskrats / 296
  Key to the Arvicolinae of Oregon / 296
    Clethrionomys californicus, Western Red-backed Vole / 298
    Clethrionomys gapperi, Southern Red-backed Vole / 303
    Phenacomys albipes, White-footed Vole / 304
    Phenacomys intermedius, Heather Vole / 307
    Phenacomys longicaudus, Red Tree Vole / 309
    Microtus californicus, California Vole / 311
    Microtus canicaudus, Gray-tailed Vole / 317
    Microtus longicaudus, Long-tailed Vole / 320
    Microtus montanus, Montane Vole / 322
    Microtus oregoni, Creeping Vole / 325
    Microtus richardsoni, Water Vole / 328
    Microtus townsendii, Townsend's Vole / 330
    Lemmiscus curtatus, Sagebrush Vole / 332
    Ondatra zibethicus, Common Muskrat / 335
FAMILY DIPODIDAE—Jumping Mice / 339
  Key to the Dipodidae of Oregon / 340
    Zapus princeps, Western Jumping Mouse / 340
    Zapus trinotatus, Pacific Jumping Mouse / 343
FAMILY ERETHIZONTIDAE—Porcupines / 345
    Erethizon dorsatum, Common Porcupine / 345
FAMILY MYOCASTORIDAE—Nutria / 348
    Myocastor coypus, Nutria / 348

CHAPTER 12. ORDER CARNIVORA—FLESH EATERS / 354
KEY TO THE FAMILIES OF CARNIVORA OF OREGON / 354
  FAMILY CANIDAE—Doglike Mammals / 354
    Key to the Canidae of Oregon / 355
      Canis latrans, Coyote / 355
      Canis lupus, Gray Wolf / 360
      Urocyon cinereoargenteus, Common Gray Fox / 363
      Vulpes velox, Kit Fox / 366
      Vulpes vulpes, Red Fox / 369
  FAMILY URSIDAE—Bears / 373
    Key to the Ursidae of Oregon / 373
      Ursus americanus, Black Bear / 374
      Ursus arctos, Grizzly or Brown Bear / 378
  FAMILY OTARIIDAE—Eared Seals / 381
    Key to the Otariidae of Oregon / 381
      Callorhinus ursinus, Northern Fur Seal / 381
      Eumetopias jubatus, Northern Sea Lion / 385
      Zalophus californianus, California Sea Lion / 388
  FAMILY PHOCIDAE—Hair Seals / 390
    Key to the Phocidae of Oregon / 391
      Mirounga angustirostris, Northern Elephant Seal / 391
      Phoca vitulina, Harbor Seal / 394
  FAMILY PROCYONIDAE—Ringtail and Raccoon / 397
    Key to the Procyonidae of Oregon / 397
      Bassariscus astutus, Ringtail / 397
      Procyon lotor, Common Raccoon / 400

FAMILY MUSTELIDAE—Mustelids / 405
Key to the Mustelidae of Oregon / 405
Martes americana, American Marten / 406
Martes pennanti, Fisher / 411
Mustela erminea, Ermine / 415
Mustela frenata, Long-tailed Weasel / 418
Mustela vison, Mink / 422
Gulo gulo, Wolverine / 426
Taxidea taxus, American Badger / 428
Lutra canadensis, River Otter / 433
Enhydra lutris, Sea Otter / 437
FAMILY MEPHITIDAE—Skunks / 440
Key to the Mephitidae of Oregon / 440
Spilogale gracilis, Western Spotted Skunk / 441
Mephitis mephitis, Striped Skunk / 445
FAMILY FELIDAE—Cats / 450
Key to the Felidae of Oregon / 450
Puma concolor, Mountain Lion / 450
Lynx canadensis, Lynx / 455
Lynx rufus, Bobcat / 458

CHAPTER 13. ORDER ARTIODACTYLA—EVEN-TOED UNGULATES / 463
KEY TO THE ARTIODACTYLA OF OREGON / 463
FAMILY CERVIDAE—Deer, Elk, and Moose / 463
Cervus elaphus, Elk or Wapiti / 464
Odocoileus hemionus, Mule Deer–Black-tailed Deer / 469
Odocoileus hemionus hemionus, Mule Deer / 469
Odocoileus hemionus columbianus, Columbian Black-tailed Deer / 474
Odocoileus virginianus, White-tailed Deer / 479
FAMILY ANTILOCAPRIDAE—Pronghorn / 484
Antilocapra americana, Pronghorn / 484
FAMILY BOVIDAE—Sheep, Goat, and Bison / 490
Bos bison, Bison / 491
Oreamnos americanus, Mountain Goat / 493
Ovis canadensis, Mountain or Bighorn Sheep / 497

CHAPTER 14. SPECIES OF POSSIBLE OCCURRENCE IN OREGON / 503

GLOSSARY / 505

APPENDIX. SPECIMENS EXAMINED / 511

LITERATURE CITED / 581

INDEX / 653

PREFACE

It is difficult to identify precisely when this work was begun; one of us (BJV) conceived the idea of a new book on the mammals of Oregon shortly after accepting a position in Oregon in 1965 and commenced to accumulate pertinent information shortly thereafter. On several occasions along the way, some writing on the project was accomplished, mostly as a means of improving teaching materials and to ascertain precisely how parts of the work should be organized. Then, both of us were sidetracked, one to further graduate education, the other to serve as an editor; both activities contributed significantly to this work, because, without those experiences, neither of us was sufficiently mature professionally to undertake such an enormous task. Now, at the twilight of one career and shortly after the dawning of another, we decided to put most other work aside and complete this task.

We undertook the task of assembling the information and writing this book not as a means of promoting ourselves or our personal knowledge of Oregon's mammals, but rather to describe Oregon's magnificent natural textbook consisting of 136 nominal species of "land" mammals, about which relatively little is known even after 190 years of work. We and other mammalogists and naturalists, both past and present, who have studied Oregon's mammals have only begun to learn to pose questions for which answers can be found in this "textbook." So, in writing our book, we often have been forced to rely on research conducted outside Oregon, to produce accounts of unequal depth or breadth, and to write simply "We do not know" about fundamental facets of the life history of various species. Thus, we do not contend that our book is the "final word" on Oregon's mammals, as some have treated Bailey's (1936) *Mammals and life zones of Oregon*, which served so well as the starting point for many of us who sought additional knowledge of the fauna. If neither beginning nor end, then ours must be between; we would be presumptuous to suggest that we have progressed more than even a tiny fraction of the distance to true understanding of this "magnificent natural textbook." We prefer to think of our work as "a regrouping," a compilation that we hope will stimulate bright young minds to attempt to "read" more from the "natural textbook," both to further knowledge and to correct what has been misread or misinterpreted. We hasten to emphasize that we do not contend that all interpretations and conclusions presented herein, based on both our own observations and those of others, are correct. We "think" that they are correct, at least in light of our knowledge of topics that impinge on the interpretations and conclusions presented.

We think that "regrouping" is an appropriate descriptor for our work from another point of view also. Too often in the past in our haste to seek new concepts about organisms, investigators have extrapolated to a whole group some bit of new information discovered about one or two species within that group. Such a practice is particularly common among groups of small, difficult-to-study species. Not only does this practice promote error, it also tends to stifle research to ascertain how widespread are relationships observed in only a few members. It gives neophytes the false impression that "everything is known already." Yet, frequently when someone "takes another look" at a phenomenon purported to be universal, a diversity is found within the group as great as that presently known to occur within the entire Class Mammalia. For example, that vagrant shrews (*Sorex vagrans*) are able to perceive some features of their environment by a crude system of ultrasonic echolocation is not cause to think that all congeners possess the ability to echolocate, nor is it cause to think that the system is not more or less crude in congeners. But, the impression given by the common statement that "shrews can echolocate" is that a shrew is just a shrew. We think nothing could be farther from the truth, and in our regrouping we advocate testing for various phenomena in other members of groups. We would like to believe that within the foreseeable future such activity would be given equal priority with that considered "the cutting edge."

ACKNOWLEDGMENTS

Numerous personnel of the Oregon Department of Fish and Wildlife willingly contributed wildlife photographs, data in their files, publications, specimens, or answers to our many questions during the last 6 years. In particular, we thank C. R. Bruce, D. Budeau, V. Coggins, J. Collins, T. Dufour, T. P. Farrell, R. L. Green, J. Greer, M. Henjum, W. R. Humphrey, P. E. Matthews, T. O'Neil, C. Puchy, C. Trainer, and W. Van Dyke.

R. M. Timm allowed us to deposit research specimens in the University of Kansas, Museum of Natural History collection; permitted extensive access to the collection; and assisted with housing arrangements during our many visits to Lawrence, KS. D. F. Markle allowed extensive use of the camera lucida in his laboratory. C. Donohoe assisted with development of the shadings used in various maps throughout the book. T. P. Farrell, G. A. Feldhamer, and J. M. Taylor provided unpublished data for our use. M. P. Banis assisted LNC during numerous trips to museum collections. L. F. Alexander helped prepare specimens on a collecting trip. E. Shvarts verified the dialectal translation of the name of a Russian explorer. R. W. Woods, of *Computer Express*, provided technical support for our computer software and hardware.

Several librarians were of particular assistance: M. P. Kinch and J. M. Perry (Valley Library, Oregon State University) and S. E. Haines, B. Mitchell, and R. Clement (Department of Special Collections, Kenneth Spencer Research Library of the University of Kansas).

Photographs, other than those taken by the authors, are acknowledged in the respective captions. Reproductions of sketches of fossil mammals in CHAPTER 3 are from Scott (1937) and were drawn by R. Bruce Horsfall. Sketches of the opossum, carnivores, rabbits, squirrels, and deer at ends of chapters and of skulls and specimens in Figs. 1-1, 1-2, and 1-3 were drawn by Chris Wall Moore; she also drew the flying squirrel on the jacket. The sketch of the northern pocket gopher at the end of the LITERATURE CITED is from Moore and Reid (1951) and was drawn by Bob Hines; it was reproduced with approval of the Publishing and Information Service, United States Department of

Agriculture. Insofar as possible, we used vernacular names advocated by Jones et al. (1992) in their *Revised checklist of North American mammals north of Mexico, 1991.* Scientific and vernacular names of plants are those advocated by Hitchcock and Cronquist (1973) or Hickman (1993).

For loan of or access to specimens in their care, we thank J. J. Beatty and R. T. Mason (Department of Zoology, Oregon State University, OSMNH), B. J. Betts (Eastern Oregon University, EOSC), E. C. Birney (James Ford Bell Museum, University of Minnesota, JFBM), M. A. Bogan (specimens formerly at Denver Wildlife Research Center, Fort Collins, Colorado, BS/FC; now housed at the University of New Mexico, Museum of Southwestern Biology), J. R. Choate (Sternberg Museum of Natural History, Fort Hayes State University, FHSU), S. P. Cross (Southern Oregon University, SOSC), D. F. Markle (Oregon State University, Department of Fisheries and Wildlife, OSUFW), P. Endzweig (State Museum of Anthropology, University of Oregon, UOMNH), R. D. Fisher (National Museum of Natural History, USNM), R. B. Forbes (Portland State University, PSU), W. K.-H. Fuchs (American Museum of Natural History, AMNH), J. E. Heyning (Natural History Museum of Los Angeles County, LACM), R. E. Johnson (Charles R. Conner Museum, Washington State University, CRCM), M. L. Kennedy (University of Memphis, MSUMZ), G. L. Kirkland, Jr. (Vertebrate Museum, Shippensberg University, SUVM), T. E. Lawlor (Humboldt State University, HSU), E. L. Orr and W. N. Orr (Department of Geology, University of Oregon, UO), E. A. Rickart (Utah Museum of Natural History, Vertebrate Collections, UU), J. Rozdilsky (The Burke Museum, University of Washington, UW), G. Shugart and D. Paulson (James R. Slater Museum of Natural History, University of Puget Sound, PSM), L. S. Skjelstad (Horner Museum, Oregon State University, HM), B. R. Stein (Museum of Vertebrate Zoology, University of California, Berkeley, MVZ), R. M. Timm (Museum of Natural History, University of Kansas, KU), K. T. Wilkins (Strecker Museum, Baylor University, SM), T. L. Yates (Museum of Southwestern Biology, University of New Mexico, MSB), and E. Yensen (Museum of Natural History, Albertson College, CIMNH). The following sent information regarding specimens collected in Oregon in their collections: J. S. Arnold (The University of Michigan, Museum of Zoology, UMMZ), R. J. Baker (The Museum, Texas Tech University, TTU), M. Hafner (Museum of Natural Science, Louisiana State University, LSUMZ), S. B. McLaren (Carnegie Museum of Natural History, CMNH), T. R. Mullican (Dakota Wesleyan University, DWU), and the H. and A. Oliver Museum (John Day, Oregon, HAOM).

Work at the National Museum of Natural History was supported by a short-term visitor grant from the Office of Fellowships and Grants, Smithsonian Institution to LNC. Travel to other museum collections was supported, in part, by a grant from the Oregon Department of Fish and Wildlife to LNC. Research was supported, in part, by Oregon Agricultural Experiment Station project 902. E. Yensen commented on an early draft of some chapters.

Publication of this book was supported, in part, by the U.S. Department of the Interior, Bureau of Land Management; Oregon Department of Fish and Wildlife, Game Management Program; Oregon Department of Fish and Wildlife, Wildlife Diversity Program; Thomas G. Scott Achievement Fund, Department of Fisheries and Wildlife, Oregon State University; and Oregon Chapter of The Wildlife Society. The following persons also contributed support for the publication of this book: Randal A. Aaberg, Margaret P. Banis, Merlin S. and Elsie K. Eltzroth, Olive B. and Russell Hodgkins, Richard F. Hoyer, Dorothy and Arthur Sevigny, J. Mary Taylor, Donna P. and John Theobald, Lois Richardson Verts, Rachel Jane Verts, and The Reverend William C. Wright.

B. J. Verts
Leslie N. Carraway

CHAPTER 1

INTRODUCTION

Mammals are the animals with which humans are most familiar. We are mammals; we commonly keep mammals as household pets; we give our youngest children stuffed toys that resemble (to a degree) mammals; we produce movies featuring cartoon characters representing mammals; we raise mammals as sources of milk, meat, and other products; we use laboratory mammals for testing the safety of products and for medical and basic research; we stalk and kill mammals for sport and food; we keep exotic mammals in captivity for amusement and public education; we spend thousands of dollars and devote millions of hours afield photographing or just viewing mammals; and some of us are fortunate enough to be able to study mammals as a vocation or an avocation. From studies of mammals we gain insight into our own origin and evolution, and, to some degree, into our own behavioral and ecological interrelationships.

Although it is the large species of mammals that possess charisma for most people, they constitute a relatively insignificant part of the diversity of mammals. In Oregon, only 20 (14.7%) species, (including three extirpated species and five pinnipeds) regularly exceed 10 kg; 93 (68.4%) species do not exceed 1 kg. Thus, it is the small, secretive, and largely nocturnal mammals that form most of the mammalian fauna. These small mammals glean seeds, dig fungi, catch insects, and mow grass, and they convert these materials to flesh that serves, in turn, as prey for carnivorous mammals and raptorial birds. Collectively, they outweigh the entire large-mammal fauna many times. Despite their abundance and diversity, small mammals are much less well known by the public, sportsman, and wildlife biologists. However, it is from studies of small mammals that present-day understanding of mammal biology largely was derived.

WHAT IS A MAMMAL?

Mammals are vertebrates: they possess an internal bony skeleton with a spinal column. This eliminates all invertebrates such as sponges, jellyfish, insects, worms, arthropods, and molluscs. However, other characters must be used to differentiate mammals from birds, fishes, frogs, snakes, turtles, and all the other vertebrates. Some characters useful for distinguishing mammals from other vertebrates are:

1. Hair.— Hair may be stiff sensory structures such as vibrissae, defensive structures such as quills, or protective structures such as bristles and awns (guard hairs), all of which overlie the insulative underhair composed of wool, fur, or vellus (Noback, 1951). All these structures arise from roots in the dermis and consist of keratinized dead epidermal cells. Hairs are composed of an outer scale-like layer, the cuticle; a layer of packed cells, the cortex; and a core of squarish air cells, the medulla (Noback, 1951). Contrary to some published accounts (e.g., Bowyer and Curry, 1983), patterns of cuticular scales on guard hairs are not sufficiently unique to species to be diagnostic (Short, 1978). Pigments in the medulla and cortex give hair its color.

All mammals have hair at least sometime in their lives.

The cetaceans (whales, dolphins, and porpoises) commonly have only a few hairs; these are lost at or shortly after birth. At the other extreme, the rabbits and hares have a dense pelage that, except for the nose pad, lips, and anus, covers the entire body, including the toe pads. Mammals usually undergo one or two molts annually. In some species, the hair is lost and replaced in a regular pattern; thus, a distinct molt line may demark the point between the old, worn pelage and the growth of new hair. Many species molt in spring, replacing a heavy winter pelage with a pelage of shorter and less dense fur, and in late summer or early autumn to add insulation for the forthcoming winter.

2. Three auditory ossicles.— In the middle ear of mammals, three bones mechanically transmit vibrations of the tympanic membrane to the oval window. From the oval window outward, these bones are the stapes, incus, and malleus; they were derived from the reptilian columella, quadrate, and articular, respectively. The quadrate, attached to the lower rear of the skull, and the articular, forming part of the mandible, compose the jaw articulation in reptiles.

3. Dentary-squamosal jaw articulation.— With modification of the reptilian quadrate and articular for transmission of sound, the mandible of mammals is composed of only one bone, the dentary, and the articulation of the jaw is between the dentary and the squamosal. The dentary-squamosal jaw articulation is commonly considered the character that separates reptiles from mammals. However, some fossil reptiles and some early mammals possessed jaws with both dentary-squamosal and articular-quadrate articulations. Among mammal embryos, development of the jaw and middle ear parallels the sequence observed in the fossil record leading from reptiles to mammals (Crompton and Jenkins, 1979).

4. Viviparity.— Young are born after prenatal development in the uterus with nourishment of the embryo through a placenta. Embryonic nourishment by means of a placenta is not unique to mammals. However, among mammals, all but three extant species—all of which are restricted to Australia, Tasmania, and New Guinea—are viviparous.

5. Mammae.— These specialized skin glands, unique to mammals, are developed in the female to produce milk for nourishment of neonates during their period of early growth. These glands are stimulated to production shortly after birth by maternal hormones and by the act of nursing by neonates. Even the three species of mammals that lay eggs have mammary glands and produce milk for nourishment of their young.

6. Strongly heterodont dentition.— In mammals, the teeth are set in the premaxillary, maxillary, and dentary, and, in most, their structure is strongly modified for specialized functions:

Incisors—simple; usually unicuspid; uppers set in the premaxillary, lowers in the dentary; and usually functioning as nipping or pinching teeth.

Canines—simple; unicuspid, often elongate; uppers set in the maxillary, lowers in the dentary; and usually functioning as stabbing teeth.

Premolars—often complex, but some commonly simple; often with many cusps; uppers set in the maxillary, lowers in the dentary; and usually functioning as shearing (secodont), crushing (bunodont), or grinding (lophodont and selenodont) teeth.

Molars—usually complex; often with many cusps; uppers set in the maxillary, lowers in the dentary; usually functioning as crushing or grinding teeth.

Incisors, canines, and premolars are usually diphyodont, whereas molars are always monophyodont. In many species the deciduous teeth (or milk teeth) are present in developing young and are replaced by permanent teeth after a period of growth. However, in some species the deciduous teeth are resorbed or shed by the developing embryo and the permanent teeth compose the only functional dentition.

Primitive eutherians were considered to have 44 teeth: three incisors, one canine, four premolars, and three molars in upper and lower jaws on each side. The subsequent evolution and dental homologies among the various groups of mammals are not understood completely. Numbers of each of the types of teeth often are expressed as dental formulas, fraction-like arrangements in which numbers of teeth on one side of the upper jaw are written as a numerator and those on one side of the lower jaw as a denominator. Thus, for the primitive eutherian, represented in the wild in Oregon only by the moles (genus *Scapanus*), the dental formula would be:

I3/3, C1/1, P4/4, M3/3 = 44

or in the abbreviated form used in this book:

3/3, 1/1, 4/4, 3/3 = 44

For a jumping mouse, *Zapus*, that has lost the second and third incisors, the canine, and the first three premolars in the upper jaw, and the second and third incisors, the canine, and all four premolars in the lower jaw, the dental formula is:

1/1, 0/0, 1/0, 3/3 = 18

The loss of premolars is from anterior to posterior; thus, the remaining premolar in the upper jaw of *Zapus* is the fourth premolar. However, molars are lost from posterior to anterior; thus, the remaining molar in the upper jaw of the striped skunk (*Mephitis mephitis*) as indicated by the dental formula:

3/3, 1/1, 3/3, 1/2 = 34

is the first molar and the remaining two molars in the lower jaw are the first and second. Of course, the missing premolar in both jaws is the first. The incisors (except the first), canine, and anterior premolars in shrews (Soricidae) and all six anteriormost teeth in moles (Talpidae) are unicuspid and commonly are referred to as U1–U5 and U1–U6 in the two families, respectively. Dental formulas for the genera of mammals occurring in Oregon are provided in Table 1-1.

7. Enucleated erythrocytes.— The red blood cells extrude their nuclei at maturity probably as a means of enhancing their capacity to bind gasses. In other vertebrates the nucleus is retained.

8. Seven cervical vertebrae.— Except for the manatee (*Trichechus manatus*) which has six; two-toed sloths (*Choloepus*) which have five to seven, or sometimes even eight; and three-toed sloths (*Bradypus*) which have eight or nine, the number of neck vertebrae is seven. Thus, all Oregon mammals have seven cervical vertebrae although in a few species some of the vertebrae are fused.

9. Muscular diaphragm.— In mammals, the thoracic and abdominal cavities are separated by a muscular sheet that when contracted bows posteriorly, thereby increasing the volume of the thoracic cavity and causing the lungs to fill with air. A complete muscular diaphragm is unique to mammals.

10. Enlarged brain.— In comparison with other vertebrates, the mammalian brain is greatly enlarged mostly through the great expansion of the cerebral hemispheres. The neopallium is expanded to cover the midbrain and overlie the cerebellum, those portions of the brain that dominate in other vertebrates. This expanded element dominates other portions of the brain and serves to improve memory and other cognitive skills and as the origin for much motor activity.

11. Simplified skull.— The number of bony elements in the skull of mammals has been greatly reduced from that in other vertebrates. The cranium is greatly enlarged to accommodate the much larger brain.

BASIC MORPHOLOGY AND MORPHOMETRY OF MAMMALS

Use of the keys provided herein to identify the species of mammals in Oregon requires familiarity with the morphology of the mammalian skull (Fig. 1-1) and with commonly measured dimensions of the skull (Fig. 1-2) and body (Fig. 1-3) used to describe the species. With few exceptions, the bones and processes of the skull, irrespective of size or shape, occupy the same relative positions in all species of mammals. Thus, bones and processes of the mammalian skull as depicted for one species (Fig. 1-1) can be used as landmarks in locating diagnostic characters in the remaining species. We strongly suggest that anyone unfamiliar with the morphology of the mammalian skull compare the bones and processes of the juvenile elk (*Cervus elaphus*) labeled in Fig. 1-1 with those of a cleaned skull of a juvenile coyote (*Canis latrans*) or some other relatively large species. Juveniles should be used because sutures between some bones become obliterated in older mammals. Descriptions of the skull bones and processes with their relative positions, and descriptions of dimensions commonly measured, are included in the GLOSSARY.

WHENCE CAME MAMMALS?

Mammals had their origin in the reptilian subclass †Synapsida, an extinct (†) group that arose in the Pennsylvanian, 150 million years before a tiny, homoiothermic, hairy beast inhabited the earth. The synapsids included the orders †Pelycosauria and †Therapsida. The pelycosaurs—the most familiar of which likely is †*Dimetrodon*, the reptilian carnivore with the enormously elongate neural spines that supported a sail-like appendage on the dorsum—possessed but one characteristic to indicate their relationship to mammals: a lateral temporal opening in the skull. †*Haptodus*, a Pennsylvanian-Permian pelycosaur with less charismatic neural spines, probably was closer to the lineage that gave rise to the order †Therapsida.

The therapsids included both herbivorous and carnivorous genera, and graded from a primitive reptilian morphology to reptiles that exhibited many characters that lean strongly toward those of mammals. An advanced carnivo-

Table 1-1.—*Dental formulas for genera of Oregon mammals.*

Dental formula	Genera	Dental formula	Genera
5/4 – 1/1 – 3/3 – 4/4 = 50	*Didelphis*	3/2 – 1/1 – 3/3 – 1/2 = 32	*Enhydra*
3/3 – 1/1 – 4/4 – 3/1 = 44	*Scapanus*[a]	0/3 – 0/1 – 3/3 – 3/3 = 32	*Odocoileus, Antilocapra, Bos, Oreamnos, Ovis*
3/3 – 1/1 – 4/4 – 2/2 = 42	*Canis, Urocyon, Vulpes, Ursus*[b]	3/3 – 1/1 – 3/2 – 3/1 = 30	*Puma*
3/3 – 1/1 – 4/4 – 2/2 = 40	*Bassariscus, Procyon*	2/1 – 1/1 – 4/4 – 1/1 = 30	*Mirounga*
2/3 – 1/1 – 3/3 – 3/3 = 38	*Myotis*	1/2 – 1/1 – 1/2 – 3/3 = 28	*Antrozous*
3/3 – 1/1 – 4/4 – 1/2 = 38	*Martes, Gulo*	3/3 – 1/1 – 2/2 – 1/1 = 28	*Lynx*
3/3 – 1/1 – 2/2 – 3/3 = 36	*Neurotrichus*	2/1 – 0/0 – 3/2 – 3/3 = 28	*Brachylagus, Sylvilagus, Lepus*
2/2 – 1/1 – 2/3 – 3/3 = 36	*Lasionycteris, Corynorhinus*	2/1 – 0/0 – 3/2 – 2/3 = 26	*Ochotona*
3/2 – 1/1 – 4/4 – 2/1 = 36	*Callorhinus*	1/1 – 0/0 – 2/1 – 3/3 = 22	*Aplodontia, Tamias, Marmota, Ammospermophilus, Spermophilus, Tamiasciurus, Glaucomys, Sciurus griseus,* and some *S. carolinensis*
3/3 – 1/1 – 4/3 – 1/2 = 36	*Lutra*		
2/3 – 1/1 – 2/2 – 3/3 = 34	*Pipistrellus, Euderma*	1/1 – 0/0 – 1/1 – 3/3 = 20	*Thomomys, Perognathus, Microdopodops, Dipodomys, Erethizon, Castor, Myocastor, Sciurus niger,* and some *S. carolinensis*
3/2 – 1/1 – 4/4 – 1/1 = 34	*Eumetopias, Zalophus*[c], *Phoca*		
3/1 – 1/3 – 3/2 – 1/2 = 34	*Mustela, Taxidea, Spilogale, Mephitis*	1/1 – 0/0 – 1/0 – 3/3 = 18	*Zapus*
0/3 – 1/1 – 3/3 – 3/3 = 34	*Cervus*	1/1 – 0/0 – 0/0 – 3/3 = 16	*Reithrodontomys, Peromyscus, Onychomys, Neotoma, Rattus, Mus, Clethrionomys, Phenacomys, Microtus, Lemmiscus, Ondatra*
3/1 – 1/1 – 3/1 – 3/3 = 32	*Sorex*[d]		
2/3 – 1/1 – 1/2 – 3/3 = 32	*Eptesicus*		
1/3 – 1/1 – 2/2 – 3/3 = 32	*Lasiurus*[e], *Tadarida*[f]		

[a] In upper jaw, incisors, canine, and second and third premolars commonly referred to as unicuspid teeth and denoted by U1–U6.
[b] One or more premolars frequently missing.
[c] Sometimes two upper molars present.
[d] In upper jaw, second and third incisors, canine, and second and third premolars commonly referred to as unicuspid teeth and denoted by U1–U5.
[e] Third upper premolar frequently missing.
[f] One lower incisor sometimes missing.

rous branch of therapsids, the suborder †Theriodonta included a variety of mammal-like reptiles, one group of which, the infraorder †Cynodontia, usually is conceded to have given rise to the mammals. Some mammalian features of cynodonts included two occipital condyles, a secondary bony palate, heterodont dentition, a mandible composed largely of the dentary bone, wide zygoma for attachment of muscles to close the jaw, and, in †*Probainognathus,* a new jaw joint between the surangular and the squamosal in addition to the usual reptilian quadrate-articular articulation.

With the rise of prototherians of the order †Triconodonta and the therians of the order †Symmetrodonta, a new class of vertebrates, Mammalia, had its beginning in the late Triassic about 200 million years ago. These early mammals were tiny—an order of magnitude smaller than any known mammal-like reptile—but exhibited characteristics considerably advanced over those considered near the main line of their progenitors: a considerably larger brain in relation to body mass, a dentary-squamosal jaw articulation in addition to a quadrate-articular articulation, and a loss of bony elements in the skull.

During the remainder of the Mesozoic (200–65 million years ago)—the first two-thirds of mammalian history—the fossil record for mammals is fragmented with long hiatuses.

All three of the major groups of mammals—monotremes, marsupials, and eutherians—arose in the Mesozoic, but some of the early groups did not persist: the †Triconodonta only from the late Triassic to the early Cretaceous (200–130 million years ago) and the †Docodonta only from the middle and late Jurassic (180–130 million years ago). However, the †Multituberculata, the first herbivorous mammals, were in terms of persistence the most successful group of mammals—ever; their record extends from the late Jurassic to the Eocene (165–45 million years ago). Nevertheless, throughout the Mesozoic, mammals remained a relatively insignificant vertebrate component of the reptile-dominated world.

During the Cretaceous (130–65 million years ago), several events occurred that set the stage for the radiation of mammals: the radiation of flowering plants, termites, beetles, moths, and butterflies created new food niches for mammals; Pangaea fragmented, thereby isolating various lineages and subjecting them to changing climes; and the demise of the dinosaurs and many other reptilian groups dramatically reduced competition. Also during the Cretaceous, the mammals diversified into several groups that formed the foundation for the forthcoming radiation. At the beginning of the Cenozoic (65 million years ago), mammals,

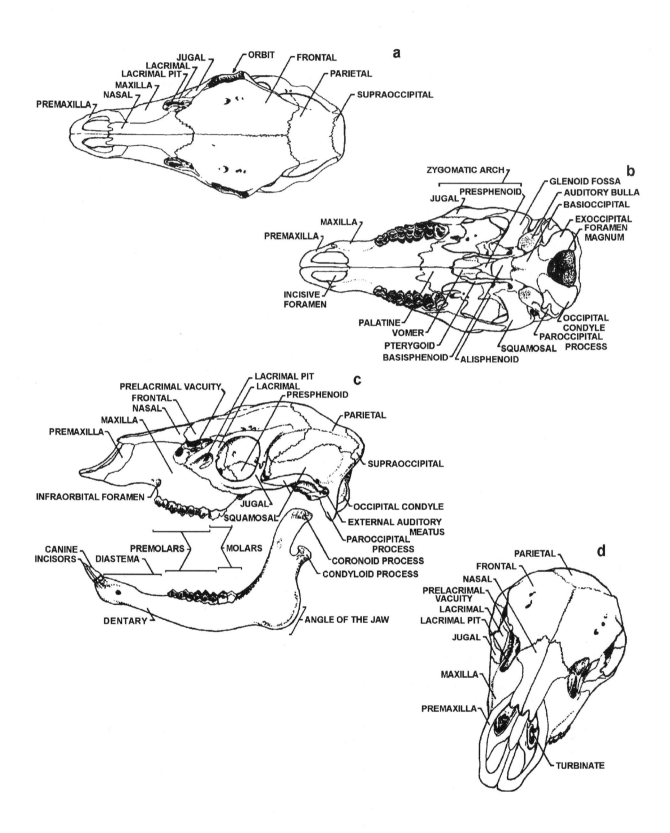

Fig. 1-1. Four views of the skull of a juvenile elk (*Cervus elaphus*) with bones and processes labeled: **a**. dorsal; **b**. vental; **c**. lateral; **d**. anterior-oblique.

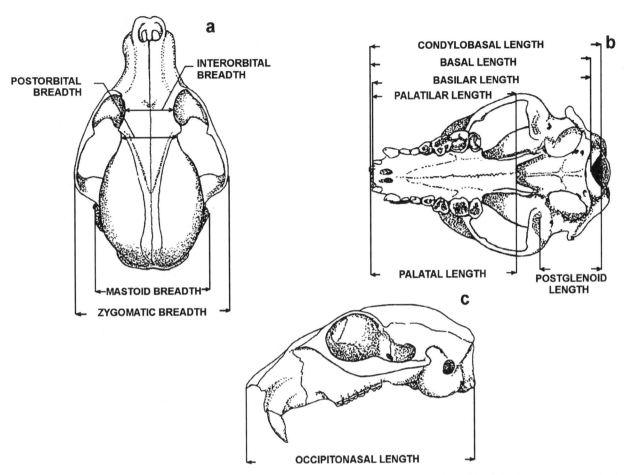

Fig. 1-2. **a**. Dorsal and **b**. ventral views of the skull of the raccoon (*Procyon lotor*) and **c**. the lateral view of the skull of a western gray squirrel (*Sciurus griseus*) illustrating points for measuring commonly used skull dimensions.

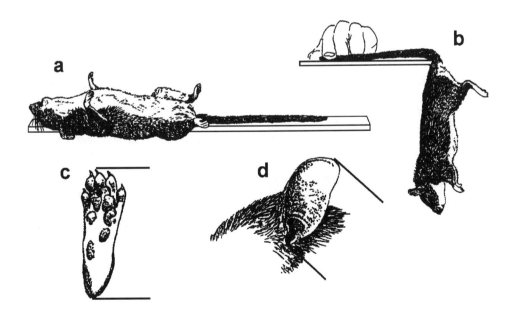

Fig. 1-3. Illustrations of standard dimensions being measured on a bushy-tailed woodrat (*Neotoma cinerea*): **a**. total length; **b**. length of tail; **c**. length of hind foot; **d**. length of ear from notch. Precise points for dimensions are described in the GLOSSARY.

especially the eutherians, began a spectacular radiation. The 13 late Cretaceous families of mammals increased to 41 by the end of the Paleocene (55 million years ago), expanded to 91 by the end of the Eocene ($\cong$38 million years ago), declined to 77 by the late Oligocene (25 million years ago), but increased to 95 in the early Miocene ($\cong$20 million years ago) when mammals attained their greatest diversity (Kurtén, 1972). Although mammal diversity suffered a serious setback with the Pleistocene extinction of many larger forms, mammals remain the dominant form of life, and with the rise of humankind $\cong$3 million years ago came to have the greatest impact on the planet, exerted by any group of organisms, in its 4.5-billion-year history.

NAMING MAMMALS

For most of us, some of the first words we spoke were names of mammals: cat, bear, dog, and cow. These, the vernacular names, serve us well until we recognize that "cat" or "bear" can be applied to a wide array of mammals that have some characteristics in common. Often, a second name can be added to those we first uttered to produce more precise names such as "bobcat" and "black bear." These names, and those that have no modifiers such as "coyote" and "raccoon," serve amply the needs of most of us to identify mammals or to communicate about specific kinds of mammals. However, regional differences in vernacular names commonly are confounding to those who travel widely or to those who attempt to communicate to a wide audience. For example, a mountain beaver (*Aplodontia rufa*) is quickly identified as a "boomer" by Oregonians, whereas, in the mountains of western North Carolina, the red squirrel (*Tamiasciurus hudsonicus*), a species that also occurs in Oregon, is known as a "boomer." As an extreme example, to an Oregonian a pocket gopher of the genus *Thomomys* would be called a "gopher," but in Minnesota a "gopher" is a thirteen-lined ground squirrel (*Spermophilus tridecemlineatus*), and in Florida the name "gopher" is applied to a burrowing land tortoise, *Gopherus polyphemus*. For the mammalogist, wildlife biologist, or zoologist who cannot afford to be misunderstood, vernacular names are useful for communicating about specific kinds of mammals only after those attempting to communicate are certain that all concerned are apprised of the identity of the mammals under discussion. To further this end, a system of naming known as binominal nomenclature is employed.

Binominal nomenclature, the system employed in the naming of all organisms, was developed by the Swedish taxonomist Carolus Linnaeus, or, after he was ennobled in 1761, Carl von Linné. Following this system, two Latinized names are applied to each species of animal. The first is the generic name and is always capitalized; the second is the specific name and is always lower case. Both names are always underlined in typescript, italicized in print. In formal publications on the taxonomy of mammals, the name of the author of the original description and the year of publication are added embellishments. Thus, the name of the pocket gopher endemic to the Willamette Valley would be written *Thomomys bulbivorus* (Richardson, 1829a). The parentheses enclosing the author's name and date alert the reader to the fact that Richardson, in applying the binomial, placed the pocket

gopher in a different genus, *Diplostoma*, that did not reflect the relationship of this species with other pocket gophers. Later, this pocket gopher was recognized to be a close relative of other pocket gophers in western United States (genus *Thomomys*); the name *Diplostoma* was first applied to species now classified in the genus *Geomys*. Absence of parentheses indicates that the generic name first applied has not changed.

During the last part of the 19th century and early part of the 20th, as the continent was explored and new and different organisms discovered, the naming, describing, and classifying of animals constituted a significant part of biology. As more and more specimens of mammals from more and more localities were collected and became available for study it became apparent that individuals with characters intermediate between those of organisms formerly thought to exhibit dramatically different characteristics were fairly common. These intermediates usually were from localities that geographically lay somewhere between the localities of specimens with dramatically different characteristics.

For example, Merriam (1901b) described a deep rufous-colored pocket gopher from Tillamook, Tillamook Co., and named it *Thomomys hesperus*; on the next page, he described an entirely slate black one from Mapleton, Lane Co., and named it *Thomomys niger*. Subsequently, it was learned that intermediates occurred in an area between the known geographic ranges of the two strikingly different-colored pocket gophers (Walker, 1955). The geographic range of *T. hesperus* abutted that of the similar, albeit hazel-colored, *T. douglasii* that, in turn, abutted the range of the russet-brown *T. mazama*, a species described and named 4 years earlier by Merriam (1897a). Because of their similar morphologies and because of the evidence that all small montane pocket gophers in western Oregon probably formed a continuous interbreeding population (at least at intervals in the recent past), the idea that these gophers formed a polytypic species was adopted. Because the name *T. mazama* was the oldest applied to a portion of the species, it became the binomial applied to all. However, in recognition of the geographic variation in color that occurred within this species, a third name (subspecies) was added to *Thomomys mazama* to form a trinomial for each colormorph with a distinct geographic range. Thus, *T. hesperus* became *T. mazama hesperus*, *T. niger* became *T. mazama niger*, and so on. Many species of mammals in Oregon have similar nomenclatural histories: mammals with easily recognizable morphological differences first were described as different species and given distinct binomials only later to be found (or presumed) to form continuous interbreeding populations, whereupon trinomials were applied to recognize the various geographically separate morphotypes.

The problem with recognizing and naming geographic variants lies in the need to delimit their geographic ranges. For example, moderate-sized deer in western Oregon with tails of medium size (mostly black dorsally and white ventrally), moderately spreading antlers, and metatarsal glands about 75 mm long are known as *Odocoileus hemionus columbianus*. Large deer east of the Cascade Range with small white tails that expand toward their black tips, widely spreading antlers, and metatarsal glands about 105 mm long are known as *Odocoileus hemionus*

hemionus. On summer range at higher elevations in the Cascade Range, deer seemingly with all possible combinations and all possible degrees of intermediacy of these characters are common. Where do we draw the line between the geographic ranges of *O. h. columbianus* and *O. h. hemionus*? To which group do we assign the intermediates? And how closely must an intermediate resemble the "typical" characteristics of one subspecies to be classified as a member of that subspecies? These are critical but unanswerable questions—at least unanswerable objectively. Because of demographic differences and differences in hunting pressure on the two groups, it has been deemed necessary to manage them differently. So, for management purposes, the Oregon Department of Fish and Wildlife has taken the only practical course of action: ignore intermediates that indicate a continuous breeding population and arbitrarily establish the crest of the Cascade Range as the line separating the geographical races. Thus, for the purpose of establishing and enforcing hunting regulations those west of the crest of the Cascade Range are black-tailed deer (*O. h. columbianus*) and those east of the crest are mule deer (*O. h. hemionus*), irrespective of actual appearance. Although this is a workable solution to a very sticky problem, it is biological nonsense. The problem of delimiting ranges for races within polytypic species extends from deer to deer mice (*Peromyscus maniculatus*) and from pocket gophers to pronghorns (*Antilocapra americana*). This difficulty caused Ernst Mayr (1965:348) to comment, "The better the geographic variation of a species is known, the more difficult it becomes to delimit subspecies and the more obvious it becomes that many such delimitations are quite arbitrary." Thus, the perpetuation of biological nonsense has extended to biologists.

In this book, trinomials applied by others are provided to alert readers to the possibility of considerable geographic variation within species. Also, where named geographic variants occur, each subspecies is described in general terms, but no attempt is made to define the limits of the ranges of subspecies on range maps for the species considered. However, data in tables of skull and body measurements are grouped by geographic region as defined by others to include one of the nominal races. For tables of mensural data, our goal was to include samples of 30 adult males and 30 adult females from each defined geographic region, but in many instances we were unable to locate sufficient specimens. Only mean and range for dimensions are provided when sample size is <10.

With few exceptions, range maps for the 136 species treated in this book are based on the 55,266 museum specimens from Oregon that we examined. For some species for which few museum specimens were found, we included maps based on published records, reported sight records, or reported records of harvest to better illustrate their geographic distributions. We recognize the potential hazard of this approach, thus we caution readers to use care in interpreting these maps. We included a verbal description of the geographic distribution and a list of specimens examined (APPENDIX) from each locality for each species to supplement the range maps. In lists of specimens examined, those localities for which county was known but were not plotted on range maps for the respective species are indicated by an asterisk (*). Specimens labeled with a place name that we could not locate are included in "additional records." Acronyms for museums containing specimens examined are listed in Yates et al. (1987) and with curators of various collections in the Acknowledgments.

Keys for identification of the orders, families, and species of mammals in Oregon are provided. A key to the orders (CHAPTER 6) directs the reader to keys to families (and in some instances to species) included within each order, which in turn direct the reader to keys to species within each family. Size of drawings of skulls is approximately proportional to the largest member of each group.

In the synonymy included with each species account we listed the original name, names applied to Oregon races of that species, and usually the first occurrence of the currently used name combination. In the latter instance, the name of the species is separated from the name of the author with a colon, as use of a comma or parentheses is specifically excluded by the International Code of Zoological Nomenclature, Third Edition.

CHAPTER 2

A BRIEF HISTORY OF MAMMALOGY IN OREGON

MAMMAL COLLECTIONS

The first collection of mammals in Oregon was made by members of the expedition led by Meriwether Lewis and William Clark. After leaving St. Louis, Missouri, on 14 May 1804, the group traveled up the Missouri River, then down the Clearwater and Columbia rivers, arriving at the Pacific Ocean on 15 November 1805 (Burroughs, 1961). The party remained at Ft. Clatsop, Clatsop Co., until the following spring, when they traveled up the Columbia River on their return to the East. Detailed journals kept by Lewis and Clark, and other members of the expedition, chronicled events and resources observed on the trip, including detailed descriptions of numerous species of mammals and their habitats. Their descriptions were sufficiently detailed that species identification often was possible without the specimens. From Oregon, they characterized or identified the black bear (*Ursus americanus*), raccoon (*Procyon lotor*), badger (*Taxidea taxus*), mink (*Mustela vison*), river otter (*Lutra canadensis*), sea otter (*Enhydra lutris*), striped skunk (*Mephitis mephitis*), gray wolf (*Canis lupus*), coyote (*C. latrans*), gray fox (*Urocyon cinereoargenteus*), bobcat (*Lynx rufus*), Townsend's chipmunk (*Tamias townsendii*), western gray squirrel (*Sciurus griseus*), Douglas' squirrel (*Tamiasciurus douglasii*), beaver (*Castor canadensis*), mountain beaver (*Aplodontia rufa*), Columbian white-tailed deer (*Odocoileus virginianus leucurus*), Columbian black-tailed deer (*O. hemionus columbianus*), and elk (*Cervus elaphus*—Burroughs, 1961). Although several of these and other species were collected and described for the first time, Lewis and Clark did not receive credit for describing the species. Their failure to provide scientific names, the absence of specimens to compliment many of their descriptions, and the fact that they were not "trained naturalists" probably contributed to their work being slighted by scientists of the day (Burroughs, 1961).

Almost all of the specimens collected west of the Rocky Mountains during the trip were lost to insect damage or weather. However, specimens that survived the trip eastward were deposited in Charles Willson Peale's museum in Philadelphia in 1806. There, they served as type specimens when George Ord, Thomas Say, and others applied names and provided descriptions (Gunderson, 1976). When Peale's Museum was disbanded in 1820, the mammal collection was sent to individuals who had storage room for it. In 1850, the entire collection was sold at public auction, half (including all the mammals) to P. T. Barnum for his "American Museum" in New York and half to Moses Kimball, who deposited his purchase in the Boston Museum. In 1865, Barnum's museum burned destroying specimens collected by Lewis and Clark that he had purchased (Burroughs, 1961). Specimens purchased by Kimball, after several transfers and considerable deterioration, ultimately found their way to the Museum of Comparative Zoology at Harvard University; only one specimen, that of a bird, "could be definitely attributed to the expedition" (Burroughs, 1961:x).

In 1834, John Kirk Townsend, an ornithologist accompanied by the botanist Thomas Nuttall, joined the expedition of Captain John B. Weyth to Oregon (Alden and Ifft, 1943). He traveled down the Columbia River to Ft. Vancouver, Washington, and up the Willamette River to the falls (present-day site of Oregon City, Clackamas Co.). In addition to his collection of birds, Townsend collected the first specimens of several species of mammals, at least two of which (*Tamias townsendii* and *Microtus townsendii*) were from Oregon. These were later named by The Reverend John J. Bachman (1839b) and all were given the specific name *townsendii* in honor of the collector. Later (1846–1854), in conjunction with John Audubon, Bachman produced the three-volume work titled *The viviparous quadrupeds of North America* that constituted the first significant report on the mammals of this continent. It included accounts of 197 species (excluding varieties), most of which were figured by Audubon (Hamilton, 1955).

Fig. 2-1. Photograph of Clinton Hart Merriam (left) and Vernon Orlando Bailey in the field in 1891. Merriam and several field assistants, including Bailey, collected mammals and related information on Mt. Mazama (Crater Lake) in August–September 1896. (Courtesy National Archives, photograph no. 22-WB-1929.)

In 1853, Spencer F. Baird, first assistant secretary (later secretary) of the Smithsonian Institution, with the assistance of his father-in-law, Brigadier General Sylvester Churchill, convinced the War Department (in which his father-in-law served as inspector general) to send physician-naturalists with crews surveying routes for railroads to the West. Four survey parties explored Oregon from October 1854 to September 1855, one each along the Columbia River, the Oregon-California border, the Cascade Range, and the Pacific Coast. The physician-naturalists accompanying the crews were Drs. George Suckley, J. S. Newberry, and J. G. Cooper, and Lieutenant W. P. Trowbridge, respectively (Baird, 1858). In his volume on mammals, published as part of the report of the railroad surveys, Baird (1858) listed 61 specimens representing 27 species as having been collected by the physician-naturalists in Oregon. These included the black bear, gray wolf, red fox (*Vulpes vulpes*), fisher (*Martes pennanti*), marten (*M. americana*), river otter, western gray squirrel, Douglas' squirrel, California ground squirrel (*Spermophilus beecheyi*), yellow-bellied marmot (*Marmota flaviventris*), least chipmunk (*Tamias minimus*), Townsend's chipmunk, mountain beaver, Trowbridge's shrew (*Sorex trowbridgii*), vagrant shrew (*S. vagrans*), Townsend's mole (*Scapanus townsendii*), white-tailed jackrabbit (*Lepus townsendii*), black-tailed jackrabbit (*L. californicus*), mountain cottontail (*Sylvilagus nuttallii*), northern pocket gopher (*Thomomys talpoides*), deer mouse (*Peromyscus maniculatus*), montane vole (*Microtus montanus*), Townsend's vole (*M. townsendii*), bushy-tailed woodrat (*Neotoma cinerea*), Norway rat (*Rattus norvegicus*), and Pacific jumping mouse (*Zapus trinotatus*).

As the United States expanded into the western frontier, farmers and ranchers commenced to demand information about mammals and birds that plagued their crops. At the urging of the American Ornithologists' Union, in 1885 the federal government created the Branch of Economic Ornithology within the Division of Entomology under the Commissioner of Agriculture. Clinton Hart Merriam (Fig. 2-1), a physician who had produced major works on both mammals (*Mammals of the Adirondaks*) and birds (*Birds of Connecticut*), was appointed ornithologist (Hamilton, 1955). Within 3 years the new agency became the Division of Economic Ornithology and Mammalogy, in 1896 it became the Division of Biological Survey, and in 1905 it became the Bureau of Biological Survey with Merriam still at the head.

Originally, Congress envisioned an agency to solve individual local problems involving birds and mammals, but Merriam believed that "fire-fighting type" research was not likely to result in development of good solutions for fundamental questions at issue. He believed that common sense dictated that the species present, their distributions, and factors responsible for their distributions were needed to provide a basis for assisting farmers and ranchers. In the late 1880s, the cyclone snap trap (*Journal of Mammalogy*, 78[1]:following p. 269) and museum special trap were invented, allowing for the first time relatively easy capture of many species of small mammals. Previously, understanding of the biology of many small, nocturnal creatures was from specimens thieved from house cats or found dead in buckets and water troughs (Sterling, 1974).

Earlier, Vernon Bailey, a Minnesota farm boy, began to send well-prepared specimens of small mammals to Merriam.

Merriam recognized his talents, and instructed and encouraged him, then in 1887 hired him as a special field agent to collect and prepare specimens. However, Merriam had the ulterior motive of training Bailey as a field naturalist. With a trained field naturalist and newly designed traps, Merriam began to consider the potential and value of long-term, large-scale field studies (Storer, 1969).

As part of Merriam's goal of conducting a nationwide biogeographical survey (Hamilton, 1955), field parties were sent to the West to collect specimens and obtain information on faunal and floral relationships. The first of these in Oregon, a cursory survey west of the Cascade Range and along the Columbia River east to The Dalles, Wasco Co., was conducted 23 September–18 December 1893 by C. P. Streator (Fig. 2-2a). Streator preserved 239 specimens representing 30 species. The next and most comprehensive set of surveys was conducted in 1896 when three collecting parties traveled extensively over central and eastern Oregon in wagons. The parties were led by Vernon Bailey (Fig. 2-2b), C. P. Streator (Fig. 2-2c), and Edward A. Preble (Fig. 2-2d). Facilities were primitive and food and water often were scarce; some of the groups commonly resorted to eating the carcasses of the animals they skinned in preparing specimens. Bailey's party obtained 430 specimens representing 50 species, Streator's group acquired 161 specimens representing 23 species, and Preble's group prepared 719 specimens representing 45 species. Among these specimens, 18 were bats of seven species, the first chiropterans to be collected in Oregon.

In April–June 1913, the Oregon naturalist Alex Walker (Fig. 2-3), accompanied by his father and a friend, collected mammals and birds over much of central and western Oregon (Fig. 2-4), traveling between sites by sternwheeler and horse-drawn wagon. Later, the Oregon Game Department (a precursor to the Oregon Department of Fish and Wildlife) purchased these specimens as an addition to its established mammal and bird collections; in 1914 the department hired Walker to conduct further biological studies and to photograph wildlife (Smith, 1961). In 1914, Vernon Bailey led Stanley Jewett and Luther Goldman (Bureau of Biological Survey), Alfred C. Shelton (University of Oregon), Morton E. Peck (botanist from Willamette University), and Don Lancefield (Reed College) on a cooperative survey up the McKenzie River to the north base of Three Sisters mountains (Fig. 2-5a). Alex Walker joined the party in June at McKenzie Bridge (Fig. 2-6). Later Goldman led a group through western and central Oregon (Fig. 2-5b). Goldman's group preserved 719 specimens representing 59 species and Bailey's party preserved 118 specimens representing 30 species. The final survey accomplished under the jurisdiction of the Bureau of Biological Survey was in 1915 when Edward A. Preble collected extensively in Malheur Co. (Fig. 2-7), where he obtained 259 specimens representing 38 species.

Stanley Jewett, a career naturalist with the Bureau of Biological Survey, developed a private collection of 1,169 specimens representing 97 species of Oregon mammals during his travels in the state between 1910 and 1955 (Fig. 2-8). Upon his death, his collection was deposited in the San Diego Museum of Natural History. Because Jewett exchanged specimens with many other naturalists, small series of specimens that he collected and preserved can be found in many of the major collections in the United States.

In the late 1930s, the Oregon Game Department

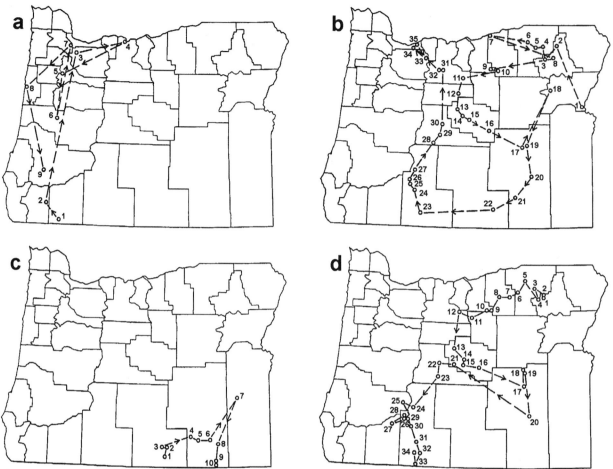

Fig. 2-2. Sequence of collection sites used during early surveys conducted by field parties of the U.S. Bureau of Biological Survey obtained from localities and collection dates of mammal specimens in the National Museum of Natural History: **a.** C. P. Streator in 1893: (1) 23 Sept.–9 Oct., (2) 11–12 Oct., (3) 14–24 Oct., (4) 29 Oct.–9 Nov., (5) 11–18 Nov., (6) 21–23 Nov., (7) 26–29 Nov., (8) 7 Dec., and (9) 15–18 Dec. **b.** Vernon Bailey in 1896: (1) 18 May, (2) 22–30 May, (3) 31 May–2 June, (4) 2 June, (5) 3 June, (6) 3–4 June, (7) 6–10 June, (8) 11 June, (9) 16–17 June, (10) 18–20 June, (11) 21–23 June, (12) 24 June, (13) 25–27 June, (14) 28–29 June, (15) 30 June–1 July, (16) 2–4 July, (17) 6–8 July, (18) 13–14 July, (19) 18–21 July, (20) 22 July, (21) 28 July (22) 31 July–4 Aug., (23) 11 Aug., (24) 12–13 Aug., (25) 13 Aug., (26) 14–25 Aug., (27) 25–26 Aug., (28) 27 Aug., (29) 29 Aug., (30) 30 Aug., (31) 4 Sept., (32) 5 Sept., (33) 6–12 Sept. and 21 Sept., (34) 12–13 Sept., and (35) 18 Sept. **c.** C. P. Streator in 1896: (1) 20–24 June, (2) 27 June–2 July, (3) 3 July, (4) 13–15 July and 18 July, (5) 17–27 July, (6) 30 July, (7) 10 Aug., (8) 13 Aug., (9) 14–20 Aug., and (10) 25–28 Aug. **d.** Edward A. Preble in 1896: (1) 22–30 May, (2) 31 May–1 June, (3) 2–3 June, (4) 3 June, (5) 4–6 June, (6) 8 June, (7) 10 June, (8) 10–15 June, (9) 16–17 June, (10) 17 June, (11) 20 June, (12) 20–22 June, (13) 24–27 June, (14) 28–29 June, (15) 30 June–1 July, (16) 2 July, (17) 6–7 July, (18) 10–16 July, (19) 18–21 July, (20) 23–27 July, (21) 31 July, (22) 2–4 Aug., (23) 6–7 Aug., (24) 9 Aug., (25) 10–13 Aug., (26) 15 Aug. and 19–24 Aug., (27) 27–31 Aug., (28) 17–24 Aug. and 3–6 Sept., (29) 15 Aug. and 2–6 Sept., (30) 7–14 Sept. and 18 Sept., (31) 17–19 Sept., (32) 20–25 Sept., (33) 27 Sept., and (34) 28 Sept. Lines do not represent actual routes of travel.

donated its collection of 977 specimens representing 77 species of mammals to what is now the Department of Fisheries and Wildlife, Oregon State University. At that time, the department had no space for the collection, so it was housed in the Museum of Natural History, Department of Zoology, Oregon State University. In 1970, responsibility for the Oregon Game Department collection was returned to the Department of Fisheries and Wildlife.

Alex Walker continued to collect and preserve mammal specimens for his private collection until the late 1960s. About 1975, his collection of 698 specimens representing 72 species of Oregon mammals was deposited in the Museum of Natural History, Department of Zoology, Oregon State University. In March 1979, all of that collection and most of the mammal specimens in the Department of Zoology's collection (2,085 specimens representing 97 species) were transferred to the Department of Fisheries and Wildlife, Oregon State University. The combined collections of mammals housed in Nash Hall on the Oregon State University campus

now total approximately 10,000 specimens. All but one species (*Euderma maculatum*) of Oregon land mammals are represented in the collection.

Since the surveys conducted in 1896–1915, and with the passing of the era of naturalists who maintained private collections, collection of mammals in Oregon has been sporadic. Small collections of Oregon mammals, each developed by a professor and a cadre of students, are housed in departments of biology at Southern Oregon University, Eastern Oregon University, and Portland State University. In the late 1960s and early 1970s, Chris Maser collected extensively along the Oregon Coast and in the central Cascade Range; his specimens were deposited in the James R. Slater Museum of Natural History, University of Puget Sound, Tacoma, Washington. Also, in recent years, Richard Johnson, John Rozdilski, the two of us, and several others have salvaged large numbers of specimens collected by researchers studying small-mammal community ecology in Oregon. Many of these specimens are on deposit at

Fig. 2-3. Photograph of Alex Walker near McKenzie Bridge, Lane Co., in 1914. (Oregon Historical Society photograph no. 86844.)

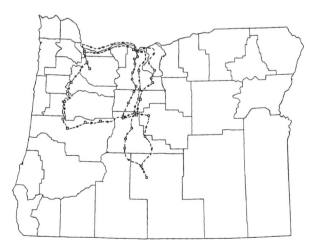

Fig. 2-4. Routes traveled by Alex Walker in collecting mammals in 1913 (circles) and 1914 (squares). Part of the 1914 trip was as a member of the cooperative party led by Vernon Bailey. (Redrawn after Smith, 1961.)

the Charles R. Conner Museum of Natural History, Washington State University; Burke Memorial Washington State Museum, University of Washington; Museum of Natural History, University of Kansas; Department of Fisheries and Wildlife, Oregon State University; and National Museum of Natural History. However, no collection made since those by personnel of the Bureau of Biological Survey represents an organized effort to further knowledge of present-day distribution of mammals in the state as a whole.

PUBLICATIONS ON MAMMALS

From 12 August to 15 September 1896, C. Hart Merriam, with field assistants Vernon Bailey, Edward A. Preble, and Cleveland Allen, camped in tents on Mt. Mazama near Crater Lake while collecting specimens and related information. Based on these studies and specimens he had acquired earlier from Major Charles E. Bendire,

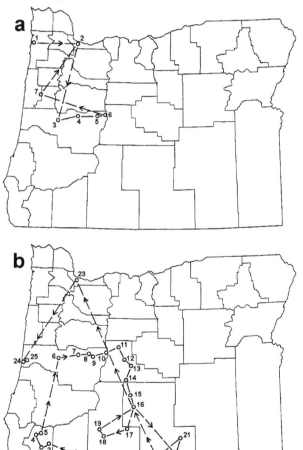

Fig. 2-5. Sequence of collection sites of field parties led by members of the Bureau of Biolgical Survey in 1914 obtained from localities and collection dates of mammal specimens in the National Museum of Natural History. **a**. Group led by Vernon Bailey: (1) 8 June, (2) 11 June and 11–12 Oct., (3) 16–21 June, (4) 22–27 June, (5) 29 June–7 July, (6) 12–19 July, and (7) 25 July; **b**. Group led by Luther Goldman: (1) 4–17 May, (2) 22 May–2 June, (3) 4–5 June, (4) 7–10 June, (5) 12 June, (6) 19–21 June, (7) 22–26 June, (8) 27–28 June, (9) 4–10 July, (10) 13–19 July, (11) 22–26 July, (12) 28 July– Aug., (13) 5–7 Aug., (14) 13–18 Aug., (15) 21–27 Aug., (16) 29 Aug.–6 Sept. and 14–16 Sept., (17) 8 Sept., (18) 10 Sept., (19) 11–12 Sept., (20) 20 Sept., (21) 21–30 Sept., (22) 1–4 Oct., (23) 13–15 Oct., (24) 20 Oct., and (25) 22–25 Oct. Lines do not represent actual routes of travel. Note that the parties were together during late June and early July.

Fig. 2-6. Photograph of the 1914 cooperative party led by Vernon Bailey at Frog Camp near McKenzie Pass, Lane Co. Standing from left to right are Stanley Jewett, Alex Walker, Vernon Bailey, Luther Goldman, Alfred C. Shelton, Morton E. Peck, and Don Lancefield. Seated are R. Bruce Horsefall (artist) and Jack Frye (camp cook) and his dog. (Oregon Historical Society photograph no. 86846.)

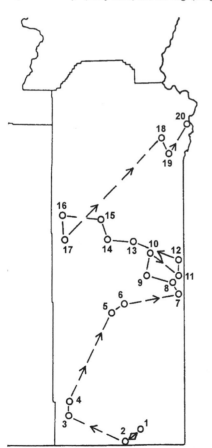

Fig. 2-7. Sequence of collection sites of field party led by Edward A. Preble in 1915 in Malheur Co. obtained from localities and collection dates of mammal specimens in the National Museum of Natural History. Lines do not represent actual routes of travel.

Dr. J. C. Merrill, and a soldier, Samuel Parker, Merriam (1896a) produced the first faunal list specific to a region in Oregon, a 27-page article titled *The mammals of Mount Mazama, Oregon.* The list contained 61 species (59 by present-day taxonomy).

While still an undergraduate, Harold E. Anthony, a native Oregonian who became curator of mammals at the American Museum of Natural History in New York, surveyed mammals along Willow Creek near Ironside, Malheur Co., in August–September 1912. His faunal list (Anthony, 1913), containing accounts of 38 species, and 2 plates, was published under the title *Mammals of northern Malheur County, Oregon.*

Based largely on the mammal collections by the Bureau of Biological Survey and mammals deposited in the National Museum of Natural History, Vernon Bailey produced the first comprehensive treatment of the mammals of Oregon. This was published in 1936 under the title *The mammals and life zones of Oregon* in the *North American Fauna* series, established nearly half century earlier by Merriam to provide an appropriate outlet for scientific findings resulting from the surveys. Other than descriptions and distributions of the taxa, much of the information on other aspects of their biology contained in Bailey's (1936) volume was either anecdotal or based on hearsay. Nevertheless, his work has stood as the authoritative source of information on mammals of the state.

In 1981, Chris Maser and colleagues (Maser et al., 1981b), in cooperation with the U.S. Department of Agriculture, Forest Service and the U.S. Department of the Interior, Bureau of Land Management, produced a volume titled *Natural history of Oregon Coast mammals.* Much of the information contained in the species accounts is based on Maser's observations and collections.

In 1982, we produced a *A bibliography of Oregon mam-*

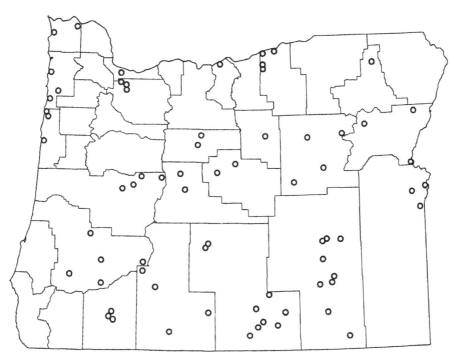

Fig. 2-8. Localities at which Stanley Jewett, Bureau of Biological Survey, collected specimens of mammals between 1910 and 1955. Based on specimens deposited in the San Diego Museum of Natural History.

malogy in which we attempted to list and index all references published between 1890 and 1980 that contained information regarding Oregon mammals (Carraway and Verts, 1982). Searches of numerous books and 107 serial publications by LNC revealed 1,924 articles that contained mention of mammals in Oregon; subsequently, 19 articles were found to have been overlooked. Many of these publications were agency reports or articles produced for the general public. We continued to collect articles on Oregon mammals published since 1980, although neither of us has conducted the page-by-page searches of journals and reports used in producing the bibliography. Many of the more recent papers are cited in this volume to make available in a single source the most up-to-date information on mammals in the state and to credit those who have contributed to the knowledge of mammals in Oregon.

FUTURE NEEDS

With the passing of >100 years since the first systematic surveys of mammals in Oregon and in consideration of the extent of exploitation of the state's forest and range resources during the interval, surveys of the current distribution and status of many of the less common species and species known to be restricted to small areas in the state are warranted. The recent records extending the range of the piñon mouse (*Peromyscus truei*—Carraway et al., 1994) and the apparent shrinking of the range of the pygmy rabbit (*Brachylagus idahoensis*—Weiss and Verts, 1984) indicate that geographic ranges and habitat affinities of some Oregon species are imperfectly known. Statewide surveys

of the type conducted by Bailey, Goldman, Preble, and others are neither practical nor sufficiently detailed and precise to delineate remnant populations of some of these species. Such surveys, if conducted inappropriately, might in themselves be overly exploitive and subject to public criticism. However, concentration of efforts on one species at a time and use of techniques that involve removal of only a few individuals for vouchers, as was done with the pygmy rabbit (Weiss and Verts, 1984), could be employed.

We believe that no wild mammal should be taken for any scientific purpose that does not include preservation of appropriate material and information for deposit in a systematics collection. A variety of readily available sources (Anderson, 1965; Burt, 1948; Hall, 1962, 1981; Hall and Kelson, 1959; Ingles, 1965; Wobeser et al., 1980) contain instructions for preservation of skins and skeletal material. Some repositories now have facilities for preservation of tissues for genetic studies; curators of collections (Yates et al., 1987) should be contacted regarding preservation and shipment of tissues. In addition, we believe that it should be the responsiblity of those who take mammals for various studies to allot sufficient resources and time to acquire relevant information and to prepare each specimen. We believe that the agency that issues permits for the taking of mammals for scientific purposes should monitor those permits to ensure that arrangements were made to preserve specimens taken under those permits. Unfortunately, and despite our efforts to alter the course, the trend seems to be in a direction opposite our philosophies.

CHAPTER 3

CHECKLIST OF LAND MAMMALS OF OREGON

ORDER DIDELPHIMORPHIA
FAMILY DIDELPHIDAE
Didelphis virginiana, Virginia Opossum

ORDER INSECTIVORA
FAMILY SORICIDAE
Sorex bairdi, Baird's Shrew
Sorex bendirii, Pacific Water or Marsh Shrew
Sorex merriami, Merriam's Shrew
Sorex monticolus, Montane or Dusky Shrew
Sorex pacificus, Pacific Shrew
Sorex palustris, Water Shrew
Sorex preblei, Preble's Shrew
Sorex sonomae, Fog Shrew
Sorex trowbridgii, Trowbridge's Shrew
Sorex vagrans, Vagrant Shrew
FAMILY TALPIDAE
Neurotrichus gibbsii, Shrew-mole
Scapanus latimanus, Broad-footed Mole
Scapanus orarius, Coast Mole
Scapanus townsendii, Townsend's Mole

ORDER CHIROPTERA
FAMILY VESPERTILIONIDAE
Myotis californicus, California Myotis
Myotis ciliolabrum, Western Small-footed Myotis
Myotis evotis, Long-eared Myotis
Myotis lucifugus, Little Brown Myotis
Myotis thysanodes, Fringed Myotis
Myotis volans, Long-legged Myotis
Myotis yumanensis, Yuma Myotis
Lasiurus cinereus, Hoary Bat
Lasionycteris noctivagans, Silver-haired Bat
Pipistrellus hesperus, Western Pipistrelle
Eptesicus fuscus, Big Brown Bat
Euderma maculatum, Spotted Bat
Corynorhinus townsendii, Townsend's Big-eared Bat
Antrozous pallidus, Pallid Bat
FAMILY MOLOSSIDAE
Tadarida brasiliensis, Brazilian Free-tailed Bat

ORDER LAGOMORPHA
FAMILY OCHOTONIDAE
Ochotona princeps, American Pika
FAMILY LEPORIDAE
Brachylagus idahoensis, Pygmy Rabbit
Sylvilagus bachmani, Brush Rabbit
Sylvilagus floridanus, Eastern Cottontail
Sylvilagus nuttallii, Mountain Cottontail
Lepus americanus, Snowshoe Hare
Lepus californicus, Black-tailed Jackrabbit
Lepus townsendii, White-tailed Jackrabbit

ORDER RODENTIA
FAMILY APLODONTIDAE
Aplodontia rufa, Mountain Beaver
FAMILY SCIURIDAE
Tamias amoenus, Yellow-pine Chipmunk
Tamias minimus, Least Chipmunk
Tamias senex, Allen's Chipmunk
Tamias siskiyou, Siskiyou Chipmunk
Tamias townsendii, Townsend's Chipmunk
Marmota flaviventris, Yellow-bellied Marmot
Ammospermophilus leucurus, White-tailed Antelope Squirrel
Spermophilus beecheyi, California Ground Squirrel
Spermophilus beldingi, Belding's Ground Squirrel
Spermophilus canus, Merriam's Ground Squirrel
Spermophilus columbianus, Columbian Ground Squirrel
Spermophilus elegans, Wyoming Ground Squirrel
Spermophilus lateralis, Golden-mantled Ground Squirrel
Spermophilus mollis, Piute Ground Squirrel
Spermophilus washingtoni, Washington Ground Squirrel
Sciurus carolinensis, Eastern Gray Squirrel
Sciurus griseus, Western Gray Squirrel
Sciurus niger, Eastern Fox Squirrel
Tamiasciurus douglasii, Douglas' Squirrel
Tamiasciurus hudsonicus, Red Squirrel
Glaucomys sabrinus, Northern Flying Squirrel
FAMILY GEOMYIDAE
Thomomys bottae, Botta's Pocket Gopher
Thomomys bulbivorus, Camas Pocket Gopher
Thomomys mazama, Western Pocket Gopher
Thomomys talpoides, Northern Pocket Gopher
Thomomys townsendii, Townsend's Pocket Gopher
FAMILY HETEROMYIDAE
Perognathus longimembris, Little Pocket Mouse
Perognathus parvus, Great Basin Pocket Mouse
Microdipodops megacephalus, Dark Kangaroo Mouse
Dipodomys californicus, California Kangaroo Rat
Dipodomys microps, Chisel-toothed Kangaroo Rat
Dipodomys ordii, Ord's Kangaroo Rat
FAMILY CASTORIDAE
Castor canadensis, American Beaver
FAMILY MURIDAE
Subfamily Sigmodontinae
Reithrodontomys megalotis, Western Harvest Mouse
Peromyscus crinitus, Canyon Mouse
Peromyscus maniculatus, Deer Mouse
Peromyscus truei, Piñon Mouse
Onychomys leucogaster, Northern Grasshopper Mouse
Neotoma cinerea, Bushy-tailed Woodrat
Neotoma fuscipes, Dusky-footed Woodrat
Neotoma lepida, Desert Woodrat

Subfamily Murinae
Rattus norvegicus, Norway Rat
Rattus rattus, Black Rat
Mus musculus, House Mouse
Subfamily Arvicolinae
Clethrionomys californicus, Western Red-backed Vole
Clethrionomys gapperi, Southern Red-backed Vole
Phenacomys albipes, White-footed Vole
Phenacomys intermedius, Heather Vole
Phenacomys longicaudus, Red Tree Vole
Microtus californicus, California Vole
Microtus canicaudus, Gray-tailed Vole
Microtus longicaudus, Long-tailed Vole
Microtus montanus, Montane Vole
Microtus oregoni, Creeping Vole
Microtus richardsoni, Water Vole
Microtus townsendii, Townsend's Vole
Lemmiscus curtatus, Sagebrush Vole
Ondatra zibethicus, Common Muskrat
FAMILY DIPODIDAE
Zapus princeps, Western Jumping Mouse
Zapus trinotatus, Pacific Jumping Mouse
FAMILY ERETHIZONTIDAE
Erethizon dorsatum, Common Porcupine
FAMILY MYOCASTORIDAE
Myocastor coypus, Nutria
ORDER CARNIVORA
FAMILY CANIDAE
Canis latrans, Coyote
Canis lupus, Gray Wolf
Urocyon cinereoargenteus, Common Gray Fox
Vulpes velox, Kit Fox
Vulpes vulpes, Red Fox
FAMILY URSIDAE
Ursus americanus, Black Bear
Ursus arctos, Grizzly or Brown Bear

FAMILY OTARIIDAE
Callorhinus ursinus, Northern Fur Seal
Eumetopias jubatus, Northern Sea Lion
Zalophus californianus, California Sea Lion
FAMILY PHOCIDAE
Mirounga angustirostris, Northern Elephant Seal
Phoca vitulina, Harbor Seal
FAMILY PROCYONIDAE
Bassariscus astutus, Ringtail
Procyon lotor, Common Raccoon
FAMILY MUSTELIDAE
Martes americana, American Marten
Martes pennanti, Fisher
Mustela erminea, Ermine
Mustela frenata, Long-tailed Weasel
Mustela vison, Mink
Gulo gulo, Wolverine
Taxidea taxus, American Badger
Lutra canadensis, River Otter
Enhydra lutris, Sea Otter
FAMILY MEPHITIDAE
Spilogale gracilis, Western Spotted Skunk
Mephitis mephitis, Striped Skunk
FAMILY FELIDAE
Puma concolor, Mountain Lion
Lynx canadensis, Lynx
Lynx rufus, Bobcat
ORDER ARTIODACTYLA
FAMILY CERVIDAE
Cervus elaphus, Elk or Wapiti
Odocoileus hemionus, Black-tailed and Mule Deer
Odocoileus virginianus, White-tailed Deer
FAMILY ANTILOCAPRIDAE
Antilocapra americana, Pronghorn
FAMILY BOVIDAE
Bos bison, Bison
Oreamnos americanus, Mountain Goat
Ovis canadensis, Mountain Sheep

CHAPTER 4

PHYSIOGNOMY AND MAMMAL FAUNAS OF OREGON

THE FORMATION OF OREGON

Fundamentally, Oregon owes its existence to the fact that it lies at the western edge of the North American continent that for >200 million years has collided with the floor of the Pacific Ocean. Oregon, like the remainder of North America, sits upon one of the many crustal plates (lithosphere) that compose the outermost layer of the earth's surface. These plates, some with land masses atop, "float" on the intensely hot, but unmelted yet slightly plastic, material below. The plates are not stationary but can slide past one another, spread apart (rift) with the space filling with molten lava from below, and override one another with one plunging deep into the plastic layer (subduction). The combination of the latter two processes creates new ocean floor and new mountain ranges (Fig. 4-1). The sediments that originally eroded from the land and compose the surface of the ocean floor are scraped off as the plate slides beneath the leading edge of the landmass. This material, combined with volcanic island chains carried on the plates (terrane), becomes attached (accreted) to the landmass to form a new coastline (Fig. 4-1). As much as three-fourths of present-day Oregon may be composed of materials formed elsewhere and accreted to former coastal regions of the state (Orr et al., 1992). The remainder of the subducting plate, consisting of heavier basalt and some sediments, melts as it dives into the intensely hot material on which the plates float. Some of the newly melted mixture rises to the surface ≅160 km inland in upside-down, teardrop-shaped globs (plutons). This is the material of which volcanos are made (Fig. 4-1). Most of the molten mixture cools below the surface to form great masses of granite that serve as foundations for the volcanos.

The processes just described, with modifications and interruptions, and at various intensities and degrees of complexity, played a major role in the formation of the physiographic features of present-day Oregon. These features affected the climate and vegetation that, in turn, influenced the evolution and ontogeny of various faunas, including those of mammals.

Many of the pieces that ultimately became part of Oregon were formed elsewhere in the world beginning ≅400 million years ago, but not until the early Triassic (≅200 million years ago) did the pieces of this jigsaw-puzzle state commence to come together. At that time, much of what is now Oregon was beneath the sea; only the southeastern corner was dry land. North America separated from Europe; possibly the westward movement of the continent initiated the chain of events that added the remainder of the state. Based on the fossil record, the Klamath and Blue mountains, then low islands, probably were accreted close together near the Idaho border. Still, about two-thirds of the state remained beneath the ocean. Near the beginning of the Cretaceous (≅80 million years ago), space between the recently accreted terranes and the mainland filled with volcanic ash and sediments eroded from the continent, creating the first locally produced fossil fauna. Most fossils were of marine origin; none was a mammal (Orr et al., 1992).

By the end of the Cretaceous (≅65 million years ago), the shoreline had moved westward to central Oregon. During the early Eocene (≅55 million years ago), a north-south row of volcanic islands commenced to form offshore and subsequently were accreted to the westward-moving continent to add ≅80 km to the east-west dimension of the state (Orr et al., 1992). The newly acquired material became the base for the Coast Range; subsequently, a new subduction zone formed seaward of the island chain (Fig. 4-1). Somewhat later in the Eocene (≅50 million years ago), the Coast Range began a 51° clockwise rotation to assume its present position (Orr et al., 1992). As the Coast Range rotated, the crust to the east thinned and in the middle Eocene (≅44–32 million years ago) permitted enormous eruptions to cover central Oregon with volcanic debris that

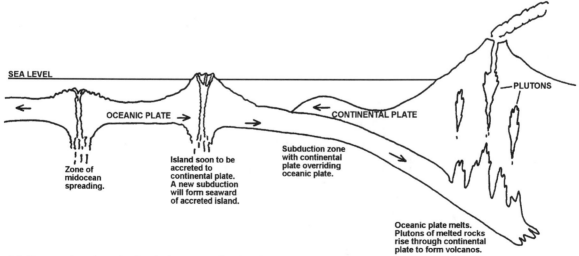

Fig. 4-1. Cross-section schematic of geologic processes involved in movements of crustal plates with resulting processes that add landmass to the continent. Much of Oregon was formed and continues to be added to by these processes.

buried, among other items, what were to become the earliest known fossil mammals in the state (Orr et al., 1992).

Commencing in the late Eocene (≅40 million years ago), further westward movement of the continent and overriding of the ocean plate produced the volcanic eruptions that formed the base of the Western Cascades. Then, in the early Miocene (25–20 million years ago), the crust of central Oregon, stretched thin by movements of the plates, faulted extensively, allowing enormous extrusive floods of lava to flow out and cover vast areas with layer upon layer of basalt. Extensive faulting occurred in the region to produce the block mountains of south-central Oregon. The Western Cascades erupted again with the lava forming a base for the High Cascades. Commencing in the Pliocene (≅4 million years ago), the Western Cascades tilted to the west, permitting outpouring of lava that formed and continues to form a north-south series of volcanos (High Cascades). The tilting and uplifting of the Western Cascades produced a strong rainshadow over much of eastern Oregon. Following the eruptions that produced the High Cascades, chambers that once contained lava collapsed, allowing blocks 15–30 km wide and nearly 500 km long between faults to drop as much as 600 m, thereby lowering the mountain range. More recently, commencing at the beginning of the Pleistocene (≅2 million years ago), volcanism was accompanied by glaciation that further carved the Cascade Range. The High Cascades continue to dominate the present-day topography of Oregon and are responsible for much of the variation in soils, climate, natural vegetation, and faunas of the state (Franklin and Dyrness, 1988; Orr et al., 1992).

Even more recently, the montane glaciers of the ice ages created great lakes in southeastern Oregon and elsewhere when they melted. Ice dams responsible for forming the lakes broke and produced cataclysmic floods on the Snake and Columbia rivers and in the Willamette Valley. And, finally,

the explosion of Mt. Mazama ≅6,000 years ago ejected an estimated 42–50 km³ of tephra that forms the pumice soils in much of eastern Oregon (Alt and Hyndman, 1978).

Even in modern times, cataclysmic events have impacted the flora and fauna of large regions of the state. Wildfires that became known as the Tillamook Burn devastated >130,000 ha of old-growth forest in 1933 and burned over again in 1945, the Oxbow Burn destroyed >18,000 ha of forest lands in Douglas Co. in 1968, and the Swamp Creek fire charred nearly 20,000 ha of rangeland in Malheur Co. in 1994. Scattered among these disastrous fires, thousands of smaller wildfires have reshaped the vegetation and the fauna that it supports on areas totalling hundreds of thousands of hectares.

Finally, perhaps the most cataclysmic event of all was the coming of European settlers with their firearms, axes, plows, and livestock, and most deadly of all, their insatiable quest for material wealth and their unprecedented ability to procreate. No other event has had, and will continue to have, as great an impact on the flora and fauna of the state of Oregon including the distribution and abundance of mammals. In the 150 years since exploration of Oregon by Europeans began, the Wyoming ground squirrel (*Spermophilus elegans*), grizzly bear (*Ursus arctos*), gray wolf (*Canis lupus*), sea otter (*Enhydra lutris*), and bison (*Bos bison*) were extirpated; some species such as the mule and black-tailed deer (*Odocoileus hemionus*) have become many times more abundant; and others such as the Townsend's big-eared bat (*Corynorhinus townsendii*), fisher (*Martes pennanti*), and pygmy rabbit (*Brachylagus idahoensis*) have become much reduced in numbers. Other species such as the mountain sheep (*Ovis canadensis*), beaver (*Castor canadensis*), and pronghorn (*Antilocapra americana*) were reduced to near extirpation, then protected and managed until their populations were vigorous and

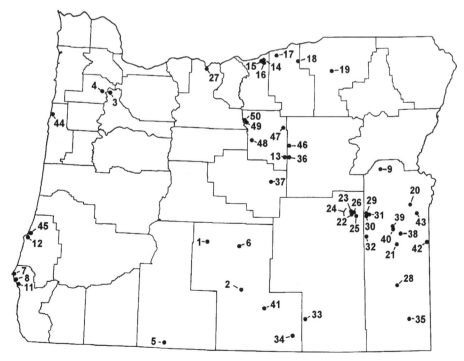

Fig. 4-2. Localities at which mammalian fossils were collected in Oregon. Numbers correspond to sites listed in Table 4-1 and to numbers following names of mammals in text. Locality 10 was too vague to plot.

Table 4-1.—*Names, ages, and legal descriptions of fossil-mammal sites in Oregon with authorities providing faunal lists for each.*

No.[a]	Name of site	Age	County	Legal description	Authorities
1	Fort Rock	Early Holocene	Lake	T25S, R14E, sec. 29[b]	Grayson, 1977, 1979
2	Lake Chewaucan	Late Pleistocene	Lake	T29–36S, R16–22E	Allison, 1982
3	Mill Creek	Late Pleistocene	Marion	T5S, R1W, sec. 17[b]	Packard, 1952
4	Palmer Creek	Late Pleistocene	Yamhill	T4S, R3W, sec. 16	Packard, 1952
5	Enrico Ranch	Blancan	Klamath	T40S, R9E, sec 28	Hutchinson, 1968
6	Fossil Lake	Pleistocene	Lake	T24–29S, R13–21E	Allison, 1966, 1979; Cope, 1883; Elftman, 1931; Matthew, 1902; McCornack, 1914, 1920; Packard, 1952
7	Cape Blanco	Early Pleistocene	Curry	T31S, R16W, sec. 3[b]	Allison et al., 1962; J. Merriam, 1913; Packard, 1947a
8	Elk River formation	Early Pleistocene	Curry	T32S, R15W, sec. 19[b]	Leffler, 1964; Packard, 1947a
9	Ironside	Pliocene	Malheur	T14S, R39E, sec. 28[b]	Merriam, 1916
10	Wilson Springs	Pliocene	"Crooked River"[c]		Colbert, 1938
11	Port Orford	Middle Pliocene	Curry	T33S, R15W, sec. 4	Allison et al., 1962; Barnes and Mitchell, 1975
12	Empire Beds	Late Miocene–early Pliocene	Coos	T25S, R13W, sec. 20[b]	True, 1905, 1909
13	Rattlesnake	Hemphillian	Wheeler	T12S, R25E, sec. 36	Colbert, 1938; Furlong, 1932; Merriam, 1917; Merriam et al., 1916, 1925; Stirton, 1935; Wilson, 1938b
14	Arlington	Late Hemphillian	Gilliam	T3N, R22E, jct. sec. 10, 11, 14, 15	Black, 1963; Shotwell, 1958a, 1963
15	Krebs Ranch I	Late Hemphillian	Gilliam	T3N, R22E, sec. 30	Hutchinson, 1968; Shotwell, 1958b
16	Krebs Ranch II	Hemphillian	Gilliam	T3N, R22E, sec. 30	Shotwell, 1958b, 1967a
17	Boardman	Hemphillian	Morrow	T4N, R24E, sec. 26[b]	Shotwell, 1958b
18	Westend Blowout	Hemphillian	Morrow	T3N, R27E, sec. 7–8	Black, 1963; Hutchinson, 1968; Shotwell, 1958b
19	McKay Reservoir	Hemphillian	Umatilla	T1N, R32E, sec. 2, and T2N, R32E, sec. 35	Black, 1963; Hutchinson, 1968; Shotwell, 1955, 1956, 1958a, 1958b, 1967a, 1967b
20	Little Valley II	Hemphillian	Malheur	T20S, R43E, sec. 33	Hutchinson, 1968; Shotwell, 1970
21	Juniper Creek Canyon	Hemphillian	Malheur	T25S, R41E, sec. 22[b]	Shotwell, 1967a, 1970
22	Bartlett Mountain	Hemphillian	Harney	T21S, R35E, sec. 34	Hutchinson, 1968; Shotwell, 1958a, 1967a, 1967b, 1970; Wilson, 1938b
23	Bartlett Mountain II	Hemphillian	Harney	T21S, R35E, sec. 22	Hutchinson, 1968; Shotwell, 1963
24	Drewsey	Hemphillian	Harney	Composite of sites 22, 23, and 26	Black, 1963; Shotwell, 1958a, 1963
25	Drinkwater	Hemphillian	Harney	T20S, R36E, sec. 34[b]	Shotwell, 1963, 1970
26	Otis Basin	Hemphillian	Harney	T20S, R36E, sec. 18	Hutchinson, 1968; Shotwell, 1963, 1967a, 1970
27	The Dalles	Hemphillian	Wasco	T1N, R13E, sec. 3[b]	Buwalda and Moore, 1930; McCornack, 1914
28	Rome	Hemphillian	Malheur	T32S, R41E, sec. 7	Furlong, 1932; Hutchinson, 1968; Shotwell, 1955, 1958a; Wilson, 1935, 1938a, 1938b
29	Black Butte I	Clarendonian	Malheur	T21S, R37E, sec. 11	Hutchinson, 1968; Shotwell, 1970; Shotwell and Russell, 1963
30	Black Butte II	Clarendonian	Malheur	T21S, R37E, jct. sec. 10, 11, 14, 15	Hutchinson, 1968; Shotwell, 1967a, 1967b, 1970
31	Juntura Basin	Clarendonian	Malheur	T21S, R37E, sec. 11–12[d]	Black, 1963; Shotwell, 1958a, 1963
32	West of Riverside	Clarendonian	Malheur	T24S, R37E, sec. 9	Hutchinson, 1968
33	Beatys Butte	Barstovian	Harney	T36S, R29E, sec. 14	Black, 1963; Hutchinson, 1968; Shotwell, 1958a; Wallace, 1946
34	Guano Ranch	Barstovian	Lake	T38S, R27E, sec. 34	Hutchinson, 1968; Mawby, 1960
35	M1040	Barstovian	Malheur	T35S, R42E, sec. 25	Hutchinson, 1968
36	Mascall	Barstovian	Grant	T12S, R26E, sec. 18[b]	Black, 1963; Downs, 1956; Gazin, 1932; Maxson, 1928; Merriam et al., 1925; Packard, 1952; Shotwell, 1967b; Stock, 1930
37	Paulina Creek	Barstovian	Crook	T16S, R23E[b]	Shotwell, 1958a
38	Quartz Basin	Barstovian	Malheur	T24S, R43E, sec. 33	Hutchinson, 1966, 1968; Shotwell, 1967a, 1967b, 1968
39	Red Basin	Barstovian	Malheur	T23S, R40E, jct. sec. 10, 11, 14, 15	Hutchinson, 1968; Shotwell, 1967a, 1968
40	Skull Springs	Barstovian	Malheur	T23S, R40E, jct. sec. 10, 11, 14, 15	Black, 1963; Gazin, 1932; Hall, 1931a; Hutchinson, 1966; Shotwell, 1958a, 1967a; Wallace, 1946
41	Snyder Creek	Barstovian	Lake	T35S, R23E, sec. 32	Hutchinson, 1968
42	Sucker Creek	Barstovian	Malheur	T25S, R46E, sec. 33[b]	Scharf, 1935; Shotwell, 1958a; Wallace, 1946
43	Oxbow Basin	Late Miocene	Malheur	T22S, R44E[b]	Shotwell, 1968
44	Astoria formation	Middle Miocene	Lincoln	T7S, R11W, sec. 27[b]	Packard, 1947b
45	Fossil Point	Miocene	Coos	T26S, R14W, sec. 3[b]	Shotwell, 1951
46	Upper John Day	Early Miocene	Grant	T11S, R26E[b]	Black, 1963; Colbert, 1938; Dawson, 1958; Dice, 1917; Dougherty, 1940; McCornack, 1920; McGrew, 1941; Shotwell, 1958a; Stirton, 1935; Wood, 1936
47	Haystack Creek	Arikareean	Wheeler	T8S, R25E, sec. 21	Shotwell, 1958a
48	Bridge Creek	Late Oligocene	Wheeler	T10S, R21E, sec. 31[b]	Orr and Orr, 1981
49	Clarno Mammal Quarry	Early Oligocene	Wheeler	T7S, R19E, sec. 32[b]	Orr and Orr, 1981
50	Clarno Fauna	Late Eocene	Wheeler	T7S, R19E, sec. 32[b]	Orr and Orr, 1981

[a]Numbers correspond to those in Fig. 4-2.
[b]Derived from published maps.
[c]No more precise locality available.
[d]Composite of sites 29 and 30.

healthy. The sea otter was reintroduced, but failed to become reestablished. Thus, humans have the potential of maintaining much if not all of the original mammalian fauna of Oregon if they are inclined to do so.

EVOLUTION OF MAMMAL FAUNAS OF OREGON

Although mammals arose in the late Triassic (≅200 million years ago) elsewhere in the world, deposits containing fossil mammals were not laid down in Oregon until the late Eocene (≅41–43 million years ago). Fossil remains of mammals deposited since that time were recovered at numerous sites in Oregon (Fig. 4-2), but each site provides only a limited temporal and spatial sample of the faunas that must have resided in the state. Table 4-1 contains the age of deposits and references in support of taxa collected at each site. Approximate ages of deposits are from Orr and Orr (1981). Superscript numbers after scientific names in the following discussion of the mammal faunas leading to the present-day mammal fauna indicate the site or sites at which fossils of that taxon were collected in Oregon. Names preceded by a dagger (†) are of mammal taxa now extinct.

In the Eocene, the climate was subtropical with coal-forming swamps along what was the Pacific coast and "palms, figs, avocados, pecans, and walnuts" growing in what is now central Oregon (Baldwin, 1981:3). Only one deposit in Oregon contains remains of late Eocene mammals and only a few genera represent the ancient faunas. Notable among these are a cursorial rhinoceros with large tusks (†*Amynodon*[50]), a huge rhino-like perissodactyl with a forked protuberance (not horns) on its snout (†*Brontotherium*[50]), a horned pocket gopher-like mylagaulid rodent (†*Epigaulus*[50]), a clawed horse-like chalicothere (†*Moropus*[50]), an intermediate tapir-rhinoceros perissodactyl (†*Hyrachyus*[50]), a wolf-sized carnivorous creodont (†*Hyaenodon*[50]), and a six-"horned" rhinoceros-sized dinocerat (†*Uintatherium*[50]).

The Oligocene also had a warm, temperate climate with "[†]*Metasequoia*, maple, sycamore, ginkgo, and katsura trees" abundant throughout those portions of the state not covered by shallow seas (Baldwin, 1981:3). The early Oligocene mammal fauna of Oregon included the large carnivorous hyaenodontid creodont (†*Hemipsalodon grandis*[49]). The late Oligocene fauna included an unidentified bat (Chiroptera) and a tapiroid perissodactyl (†*Colodon*[48]).

The early Miocene climate was mild and humid. Oregon was vegetated with extensive forests of †*Metasequoia* (Baldwin, 1981). During the middle Miocene, offshore weather systems were the dominant climatic forces as only the relatively low Western Cascade Range and Klamath Mountains were developed. The flora of the Coast Range was typical of a humid coastal coniferous forest (Axelrod, 1950), an environment not conducive to development of fossiliferous beds. Thus, there are no fossil terrestrial mammals from western Oregon of this age, but specimens of pinniped carnivores (†*Desmatophoca*[44] and †*Pontolias*[7, 12, 45]), indicators of a cooler climate than currently present (Allison et al., 1962), were found at sites along the coast.

East of the Western Cascades, the environment was ideal for development of fossiliferous beds. The early Miocene (Arikareean and Hemingfordian; ≅19–25 million years ago) flora in which mammals occurred was a widespread piñon (*Pinus*)-juniper (*Juniperus*) forest interspersed with meadows, savannas, and grasslands indicative of low relief and

Diceratherium

Amebelodon

Mylodon

uniform climate with more than twice the 38.1 cm/year present-day annual rainfall (Chaney, 1925; Downs, 1956). Primitive mountain beavers (†*Allomys*[46], †*Meniscomys*[46], and †*Sewelleladon*[46, 47]), a variety of squirrels (†*Miosciurus*[46], †*Protosciurus*[46], and †*Protospermophilus*[46]), and a camel (†*Paratylops*[46]) occupied woodlands. Archaic rabbits (†*Archaeolagus*[46] and †*Paleolagus*[46]), early pocket gophers (†*Grangerimus*[46], †*Entoptychus*[46], and †*Pleurolicus*[46]), and gopher-like mylagualids (†*Meniscomys*[46] and †*Mylagaulodon*[46]) occurred outside woodlands. The mylagaulodons were equipped with horns. A terrestrial beaver (†*Paleocastor*[46]) lived in pond banks and dug spiral tunnels. A little later (≅15.4–18.5 million years ago), an antilocaprid with branching deer-like horns (†*Merycodus*[33, 38–40,42]), three types of horses (†*Hypohippus*[33, 36, 39, 40, 42], †*Merychippus*[33, 36, 38–40, 42, 43], and †*Parahippus*[33, 36, 39, 40, 42]), a peccary (†*Prosthennops*[13, 15–19, 21, 26, 33, 36, 38, 39, 42, 43]), and a mastodon (†*Mammut*[16–18, 22, 26, 29, 36, 40, 42]) commenced lines that lasted for millions of years. Another horned gopher-like mylagaulid rodent (†*Mylagaulus*[19, 22, 23, 26, 28, 29, 33, 36, 37, 42]) succeeded †*Meniscomys* and †*Mylagaulodon*, and continued throughout the Miocene. A short-lived race of oreodonts with high-crowned teeth (†*Ticholeptus*[36, 39, 40, 42]), deer-like palaeomerycids with tiny antlers atop enormous bony pedicels (†*Dromomeryx*[33, 36, 38–40, 42]), and unidentified camels and rhinoceroses also were part of the early Miocene mammal fauna.

By the middle Miocene (Barstovian; ≅15 million years ago), the western mountain ranges were beginning to form, resulting in a drier inland climate. Grasslands and savannas expanded and woodlands commenced to lose plant species less tolerant of xeric conditions. In response, the archaic rabbits (†*Archaeolagus* and †*Paleolagus*) of savannas and grasslands were replaced by another group of early rabbits (†*Hypolagus*[13, 15, 16, 18–23, 26, 28, 33, 36, 38]) and were joined by the still-extant hares (*Lepus*[1, 6, 13, 36]). As grasslands expanded, the widely tolerant horned gophers (†*Mylagaulus*) increased in size, underwent great modification of their teeth, and expanded eastward across the Great Plains (Shotwell, 1958a). †*Ticholeptus*, †*Dromomeryx*, †*Mammut*, and the early Miocene horses were joined by †*Archaeohippus*[36], two additional species of †*Merychippus,* two additional oreodonts (†*Merycoidodon*[36] and †*Merycochoerus*[36]), other rhinoceroses (†*Diceratherium*[36]), additional camels (†*Miolabis*[36] and †*Alticamelus*[13, 36]), several canids (†*Amphicyon*[13, 33, 36, 39, 40], *Canis*[6, 13, 15–21, 36], and †*Tephroncyon*[36, 39]), mustelids (†*Leptarctus*[36] and †*Lutrictis*[36]), several heteromyid rodents (†*Prodipodomys*[36, 38, 39], †*Diprionomys*[20–22, 30, 36, 40], and †*Peridionomys*[36, 38, 39]), sciurids (†*Arctomyoides*[36] and †*Protospermophilus*[33, 36, 39, 40, 46]), and the now ubiquitous deer mice (*Peromyscus*[19, 20, 22, 29, 30, 36, 38, 39]).

Slightly more recent Barstovian (≅15 million years ago) faunas included several canids (†*Euplocyon*[39, 40], †*Tomarctus*[39, 40], †*Pliocyon*[39, 40], and †*Hemicyon*[38–40]) in addition to some older forms (†*Tephroncyon* and †*Amphicyon*), mustelids (†*Brachypsalis*[38, 39] and *Martes*[39, 40]), cats (†*Pseudaeluras*[21, 29, 39]), a variety of shrews (†*Alluvisorex*[33, 38, 39, 40], †*Heterosorex*[34, 39, 40], †*Paradomnina*[39, 40], and †*Limnoecus*[40]), moles (†*Mystipterus*[30, 34, 38, 39], †*Scalopoides*[19, 20, 30, 34, 38, 39], †*Scapanoscapter*[35, 39], and *Scapanus*[5, 15, 16, 18, 19, 22, 28, 30, 32, 41]), an aplodontid (†*Liodontia*[19, 22, 23, 26, 31, 33, 39, 40]), a hornless rhinoceros (†*Aphelops*[17, 19, 21, 33, 38, 39]), another deer-like palaeomerycid (†*Rakomeryx*[39]), and a chalicothere (†*Chalicotherium*[39, 40]). A beaver (†*Monosaulax*[38, 39]), an early hedgehog relative (Erinaceidae: †*Lantanotherium*[39]), and representatives of the extinct pocket gopher-like family †Eomyidae (†*Pseudotheridomys*[39]) also were part of the fauna. Sciurids diversified dramatically as they expanded into the grasslands from the woodlands; included, in addition to tree squirrels (*Sciurus*[33]), were early chipmunks (*Tamias*[22, 30, 39]) and ground squirrels (†*Protospermophilus*[33, 36, 39, 40, 46] and *Spermophilus*[19, 20, 22, 26, 29, 38, 39]). The horned gophers (†*Mylagaulus*), oreodonts (†*Ticholeptus*), deer-like palaeomercids (†*Dromomeryx*), heteromyid rodents (†*Diprionomys*, †*Prodipodomys*, and †*Peridipodomys*), the early horses (†*Merychippus*, †*Parahippus*, and †*Hypohippus*), mastodons (†*Mammut*), and deer-like antilocaprids (†*Merycodus*), in addition to various other camelids and rhinoceroids, persisted.

In the late Barstovian (≅15 million years ago), †*Merycodus*, †*Dromomeryx*, †*Prosthennops*, †*Hemicyon*, †*Brachypsalis*, †*Alluvisorex*, †*Scalopoides*, †*Mystipterus*, †*Hypolagus*, †*Merychippus*, †*Aphelops*, †*Monosaulax*, *Peromyscus*, †*Pseudotheridomys,* †*Prodipodomys*, †*Peridomys*, and *Spermophilus* were joined by an unidentified opossum (Didelphidae[38]), an extant genus of mustelids (*Mustela*[13, 38]), a ringtail (*Bassariscus*[17, 18, 38]), other types of shrews (†*Ingentisorex*[38]) and moles (†*Achyoscapter*[38]),

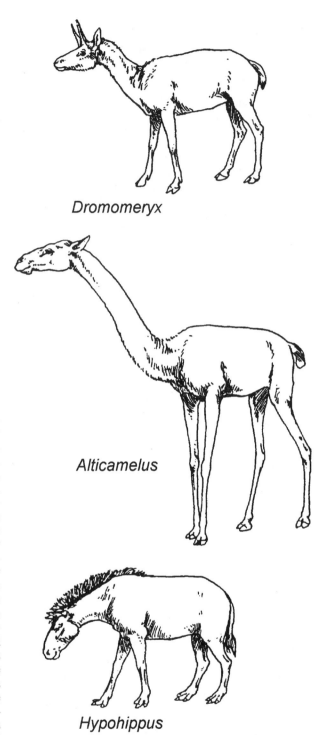

Dromomeryx

Alticamelus

Hypohippus

another eoymid (†*Adjidaumo*[38]), and an early jumping mouse (†*Macrognathomys*[22, 30, 38]).

In the Clarendonian (early late-Miocene; ≅11.3 million years ago), the previously represented extant genera *Peromyscus*, *Martes*, *Scapanus*, *Spermophilus*, and *Tamias* occurred in southeastern Oregon faunas. The extant genera of pocket mice (*Perognathus*[16, 19, 20, 22, 30]) and antelope ground squirrels (*Ammospermophilus*[31]) were new to Oregon faunas. Other genera that occurred in earlier faunas in Oregon were †*Epigaulus*, †*Mylagaulus*, †*Mystipterus*,

†*Scalopoides*, †*Diprionomys*, †*Macrognathomys*, †*Liodontia*, †*Procamelus*, †*Prosthennops*, †*Mammut*, †*Aphelops*, and †*Pseudaeluras*. A new genus of camels (†*Pliauchenia*[6, 13, 20–23, 25, 29, 31]) that continued to the Pleistocene, three new genera of canids (†*Aelurodon*[29], †*Osteoborus*[17, 19, 24, 29, 30], and *Vulpes*[6, 29]), another aplodontid (†*Tardontia*[22, 29–31]), two new beavers (†*Eucastor*[29, 30] and †*Histricops*[22, 23, 29]), a new genus of pocket gophers (†*Pliosaccomys*[19, 20, 22, 26, 30]), and a new gopher-like eomyid (†*Leptodontomys*[19, 22, 30]) became part of the Clarendonian fauna of Oregon. Also added to the fauna were a "shovel-tusked" gomptotherid proboscidean (†*Platybelodon*[29]), three mustelids (†*Pliotaxidea*[18, 19, 22, 23, 29], †*Sthenictis*[29, 30], and †*Eomellivora*[29, 30]), a shrew (†*Hesperosorex*[29]), a three-toed horse (†*Hipparion*[9, 13, 19, 23, 26, 27, 29]), an oreodont (†*Ustatochoerus*[29]), and a heteromyid rodent (†*Cupidinomys*[29]).

Slightly later (≅10 million years ago), new genera of antilocaprids (†*Sphenophalos*[13, 15, 17, 20, 22, 23, 28]), mustelids (†*Plionictus*[28], †*Lutravus*[28], and *Lutra*[28]), single-toed horses (†*Pliohippus*[13, 16, 18, 20, 21, 23, 28]), rhinoceroses (†*Teleoceras*[13, 16, 17, 21, 23, 25, 28]), beavers (†*Dipoides*[13, 15, 16, 19–23, 28] and *Castor*[6, 18, 19, 28]), and muroid rodents (†*Microtoscoptes*[21–23, 28, 30]) were added to the Miocene mammal fauna of southeastern Oregon. Some of these genera remained part of mammal faunas in Oregon until the Pleistocene; *Lutra* and *Castor* are still extant.

In the Hemphillian (≅8.5 million years ago) mammal faunas of eastern Oregon continued to include the antilocaprid with bifurcate horn-cores (†*Sphenophalos*), camels (†*Pliauchenia* and †*Procamelus*), mustelids (†*Pliotaxidea*), moles (†*Scalopoides* and *Scapanus*), rabbits (†*Hypolagus*), mastodons (†*Mammut*), horses (†*Hipparion* and †*Pliohippus*), peccaries (†*Prosthennops*), canids (†*Osteoborus*), rhinoceroses (†*Teleoceras*), beavers (†*Dipoides* and †*Histricops*), ancestral mountain beavers (†*Liodontia* and †*Tardontia*), muroid rodents (†*Microtoscoptes* and *Peromyscus*), early pocket gophers (†*Pliosaccomys*), horned gopher-like mylagaulids (†*Mylagaulus*), ground squirrels (*Spermophilus*), chipmunks (*Tamias*), primitive jumping mice (†*Macrognathomys*), pocket mice (*Perognathus*), and other heteromyids (†*Diprionomys*). These were joined by the enormous shovel-tusked gomptotherian †*Amebelodon*[23], an unidentified bear (Ursidae[6, 13, 17, 18, 21, 22]), an unidentified ground sloth (†Megalonychidae[13, 17, 18, 26]), and a member of the extant genus of kangaroo rats (*Dipodomys*[16, 19, 22]). Slightly later faunas in southeastern Oregon consisted largely of genera also found in these deposits and included a saber-toothed cat (†*Machairodus*[17, 21]) and a bear (†*Indarctos*[13, 21]).

North-central Oregon faunas of about the same, or slightly younger, age included most of the same genera, although often different species than faunas in southeastern Oregon. Also included were several genera not occurring among those of southeastern Oregon faunas. Notable were a new camel (†*Paracamelus*[19]), modern cats (*Felis*[6, 13, 19]), early wolverines (†*Plesiogulo*[19]), a new mole (†*Hydrosacpheus*[19]), a shrew-mole (*Neurotrichus*[19]), a pika (*Ochotona*[1, 19]), a three-toed horse (†*Neohipparion*[17]), a new muroid rodent (†*Prosomys*[19]), a new pocket gopher (†*Parapliosaccomys*[19]), and a marmot (*Marmota*[19]).

Still younger Hemphillian faunas (≅6.4 million years ago) of the same region included the antilocaprid †*Sphenophalos,* the giraffe-like camel (†*Alticamelus*) and other camels (†*Procamelus* and †*Pliauchenia*), peccaries (†*Prosthennops*), canids (†*Amphicyon* and *Canis*), cats

Uintatherium

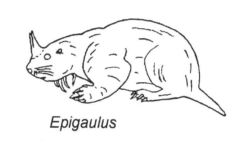

Epigaulus

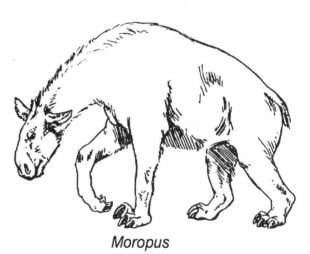

Moropus

(*Felis*), bears (†*Indarctos*), ground sloths (†Megalonychidae), rabbits (†*Hypolagus*) and hares (*Lepus*), horses (†*Hipparion* and †*Pliohippus*), rhinoceroses (†*Teleoceras*), beavers (†*Dipoides*), and ground squirrels (*Spermophilus*). A new shovel-tusked gomptotherian (†*Gomptotherium* [*Tetrabelodon*][9, 13]) and a new antilocaprid with spiraled horns (†*Ilingoseros*[13]) were added to the fauna.

The Pliocene (≅3–5 million years ago) fossil-mammal fauna of Oregon was severely restricted in volume, but certainly not in variety. Species identified in Pliocene-age deposits were the broad-footed mole (*Scapanus* cf. *latimanus*[5]), the harbor seal (*Phoca* cf. *vitulina*[11]), the otariid seal (†*Pontolias magnus*[12]), the peccary (†*Prosthennops* cf. *rex*[10]), the one-toed horse (†*Hipparion anthonyi*[9]), the gomptotherian †*Gomptotherium* (*Tetrabelodon*), and an unidentified rhinocerotid[9]. Similarly, known Pleistocene (since ≅2 million years ago) fossil mammals from Oregon are of limited abundance. The sea otter (*Enhydra*[8]), two tapirs (*Tapirus* cf. †*californicus*[7, 8] and *T.* †*haysii*[7]), two otariid seals (†*Pithanotharia starri*[7] and †*Pontolias magnus*[7]), and an unidentified otariid seal[8] are known from early Pleistocene deposits, and a bison (*Bos*[21]), a camel (†*Camelops vitakerianus*[21]), a peccary (†*Platygonus*[21]), and a horse (*Equus*[21]) are known from late Pleistocene deposits. A ground sloth (†*Mylodon harlani*[3, 4, 11]) is known from mid- and late-Pleistocene deposits in western Oregon. Fossil hares (*Lepus californicus*[1] and *L. townsendii*[1]), the mountain cottontail (*Sylvilagus nuttallii*[1]), and the pika (*Ochotona princeps*[1]) are known from early Holocene deposits near Fort Rock, Lake Co.

From the foregoing, it is obvious that human activities are not the only forces responsible for extinction. However, it must be remembered that the numerous extinctions indicated by the daggers (†) occurred over a period of >40 million years, whereas extirpation of the five species referred to in the previous section occurred during the past 200 years.

PHYSIOGRAPHIC PROVINCES AND PRESENT-DAY MAMMALIAN FAUNAS

Present-day Oregon can be divided into nine provinces based upon the physiognomic variation within the state (Fig. 4-3). Although fairly homogeneous, the provinces are not entirely discrete as transitions between them may be gradual (Franklin and Dyrness, 1988). Several identifiable vegetative zones occur within each province (Fig. 4-4).

An analysis of relationships of physiographic provinces and mammal faunas was conducted by overlaying range maps for each species (except marine species and those introduced by humans) with a transparent map of the physiographic provinces (Franklin and Dyrness, 1988) and counting the number of collection localities in each province. The proportion of collection localities in each province was calculated (Table 4-2) and the resemblance of mammalian faunas among the provinces determined (Table 4-3). Species for which >10 localities were available and for which ≥40% of the localities were within a province were considered characteristic of the mammalian fauna of that province. Lack of precision in plotting of collection localities and in delineating provinces likely contributed to unexpected results for some species. For example, *Thomomys bulbivorus* is endemic to the Willamette Valley Province, but the analysis (Table 4-2) indicates that the species also occurs in both Coast Range and Cascade Range provinces. However, use of a relatively large percentage of localities (40%) was believed to have reduced the impact of this imprecision in characterizing the mammalian faunas of the provinces.

Coast Range Province.—The Coast Range Province (Fig. 4-5) is delimited by the Columbia River to the north, the Klamath Mountains to the south, the Pacific Ocean to the west, and the Willamette Valley to the east (Fig. 4-3). In the southern part of the province, the slopes are sharp, but less so in the northern part. Elevations of the primary ridges range from 450 to 750 m, with the highest peak at 1,249 m (Marys Peak, Benton Co.). Soils are varied, but include reddish clay loams derived from sandstone and reddish-brown silt loams derived from basalt. Sandy soils, with some dune areas, are common in the coastal region.

Mammut

The climate is wet and mild; minimum January temperatures average ≥0°C, maximum July temperatures average 20–28°C, and precipitation ranges from 120 to 300 cm annually. Heavy fogs are common and, especially in summer, may add significantly to precipitation by condensing and dripping from trees.

The vegetation of the province includes parts of the Sitka spruce (*Picea sitchensis*) zone at elevations mostly <160 m and within a few kilometers of the coast and the western hemlock (*Tsuga heterophylla*) zone that occupies the remainder of the province (Fig. 4-4). The most common trees in addition to Sitka spruce in the former zone are western hemlock, western red cedar (*Thuja plicata*), and Douglas-fir (*Pseudotsuga menziesii*), but lodgepole pine (*Pinus contorta*) occurs abundantly along the ocean. Red alder (*Alnus rubra*) is common in riparian zones. Since settlement, forests in most of the zone have been removed by fires or logging, and naturally or artificially propagated forests consist almost entirely of Douglas-fir (Franklin and Dyrness, 1988).

Sixty-three species of nonmarine native mammals occur in the Coast Range Province, the third lowest species diversity among the provinces. Eleven species are considered to characterize (Table 4-2) the mammalian fauna of the province: *Sorex bendirii*, *S. pacificus*, *S. sonomae*, *S. trowbridgii*, *Scapanus orarius*, *Aplodontia rufa*, *Phenacomys albipes*, *P. longicaudus*, *Microtus oregoni*, *M. townsendii*, and *Mustela erminea*. The fauna of the province resembles most closely that of the Willamette Valley and Klamath Mountains (Table 4-3), provinces that adjoin the Coast Range.

Willamette Valley Province.—The ≅200-km-long by

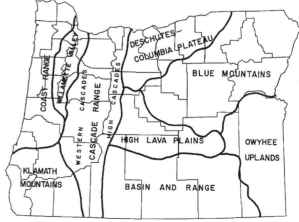

Fig. 4-3. The nine physiographic provinces in Oregon (redrawn after Franklin and Dyrness, 1988 and Orr et al., 1992).

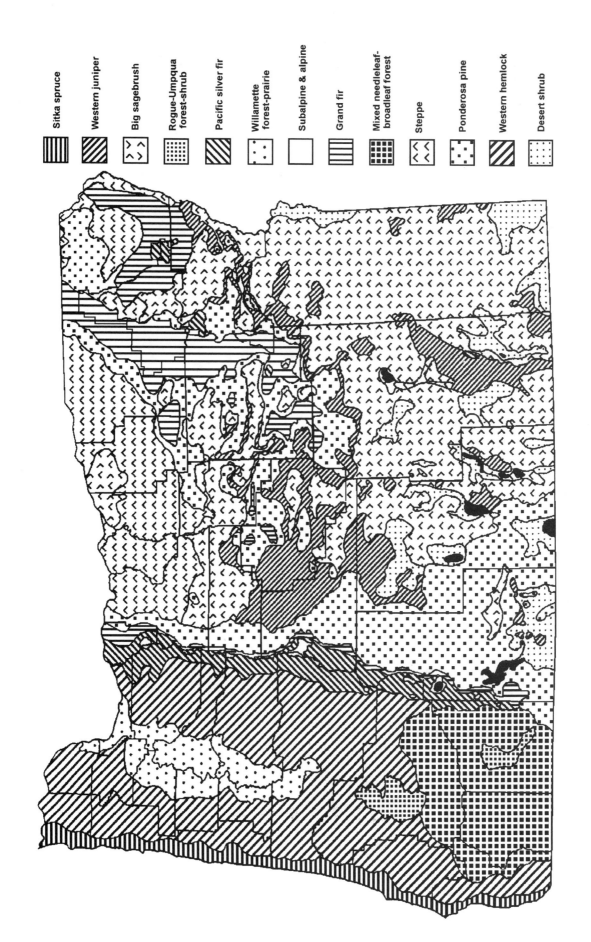

Fig. 4-4. The major vegetative zones in Oregon (redrawn after Loy, 1976).

Legend:

- Sitka spruce
- Western juniper
- Big sagebrush
- Rogue–Umpqua forest-shrub
- Pacific silver fir
- Willamette forest-prairie
- Subalpine & alpine
- Grand fir
- Mixed needleleaf-broadleaf forest
- Steppe
- Ponderosa pine
- Western hemlock
- Desert shrub

Table 4-2.—*Proportion of collection sites (listed as percentages) of land mammals (exclusive of introduced species and species associated primarily with the marine environment) in each of the nine physiographic regions of Oregon (Fig. 4–3).*

Taxa	Localities considered (*n*)	Coast Range	Klamath Mountains	Willamette Valley	Cascade Range	Deschutes-Columbia Plateau	High Lava Plains	Basin and Range	Blue Mountains	Owyhee Uplands
Insectivora										
Sorex bairdi	71	39.4		8.5	52.1					
Sorex bendirii	113	41.6	8.8	8.8	38.9			1.8		
Sorex merriami	11						18.2	54.5	27.3	
Sorex monticolus	69	13.0			20.3			17.4	49.3	
Sorex pacificus	77	46.8	3.9	2.6	46.8					
Sorex palustris	64				40.6	3.1		17.2	39.1	
Sorex preblei	18						22.2	44.4	27.8	5.6
Sorex sonomae	161	41.0	24.8	1.2	32.9					
Sorex trowbridgii	262	40.8	15.6	6.5	35.9			1.1		
Sorex vagrans	327	27.5	7.6	15.6	23.5	0.3	2.1	4.6	18.3	0.3
Neurotrichus gibbsii	167	31.1	15.0	18.0	35.9					
Scapanus latimanus	25		48.0		20.0		8.0	24.0		
Scapanus orarius	139	40.3	14.4	8.6	27.3	4.3	0.7		4.3	
Scapanus townsendii	128	35.2	16.4	43.8	4.7					
Chiroptera										
Myotis californicus	103	24.3	26.2	11.7	15.5		3.9	5.8	10.7	1.9
Myotis ciliolabrum	32				9.4	12.5	3.1	21.9	21.9	31.3
Myotis evotis	103	9.7	12.6	9.7	17.5	2.9	3.9	15.5	24.3	3.9
Myotis lucifugus	113	14.2	8.0	14.2	15.9	5.3	7.1	17.7	10.6	7.1
Myotis thysanodes	24	20.8	29.2	4.2	25.0				20.8	
Myotis volans	69	7.2	10.1	4.3	21.7		2.9	17.4	34.8	1.4
Myotis yumanensis	89	19.1	13.5	11.2	21.3	3.4	4.5	10.1	12.4	4.5
Lasiurus cinereus	39	10.3	23.1	7.7	5.1	2.6	2.6	7.7	38.5	2.6
Lasionycteris noctivagans	89	5.6	14.6	4.5	22.5	1.1	2.2	12.4	33.7	3.4
Pipistrellus hesperus	20						10.0	30.0	10.0	50.0
Eptesicus fuscus	124	9.7	15.3	12.9	19.4	3.2	5.6	12.9	18.5	2.4
Euderma maculatum	2					50.0		50.0		
Corynorhinus townsendii	67	11.9	19.4	13.4	13.4	3.0	7.5	9.0	14.9	7.5
Antrozous pallidus	34	2.9	23.5	2.9	8.8	14.7		20.6	11.8	14.7
Tadarida brasiliensis	7		85.7					14.3		
Lagomorpha										
Ochotona princeps	95		1.1		58.9		1.1	10.5	28.4	
Brachylagus idahoensis	39						23.1	30.8	10.3	35.9
Sylvilagus bachmani	68	23.5	25.0	44.1	7.4					
Sylvilagus nuttalli	101		2.0		3.0	24.8	21.8	14.9	17.8	15.8
Lepus americanus	77	23.4	3.9		40.3		3.9	3.9	24.7	
Lepus californicus	95	3.3	10.5	7.4	1.1	13.7	10.5	20.0	12.6	21.1
Lepus townsendii	25				4.0	28.0	4.0	32.0	24.0	8.0
Rodentia										
Aplodontia rufa	115	47.8	15.7	16.5	20.0					
Tamias amoenus	298		5.0		23.2	4.0	14.4	20.1	31.5	1.7
Tamias minimus	129				4.7	1.6	23.3	35.7	11.6	23.3
Tamias senex	57				61.4	1.8	21.1	15.8		
Tamias siskiyou	78		50.0		50.0					
Tamias townsendii	279	33.0	12.5	19.0	34.4	1.1				
Marmota flaviventris	80				18.8	3.8	8.8	27.5	22.5	18.8
Ammospermophilus leucurus	27							66.7		33.3
Spermophilus beecheyi	137	21.9	18.2	23.4	19.7	11.7		4.4	0.7	
Spermophilus beldingi	181				9.9	7.7	9.4	19.3	40.9	12.7
Spermophilus canus	57					17.5	21.1	8.8	12.3	40.4
Spermophilus columbianus	71								98.6	1.4
Spermophilus elegans	2							50.0		50.0
Spermophilus lateralis	197		5.6		28.9	8.6	7.1	15.2	29.4	5.1
Spermophilus mollis	4							75.0		25.0
Spermophilus washingtoni	13					92.3			7.7	
Sciurus griseus	100	11.0	31.0	26.0	24.0	4.0	2.0	2.0		
Tamiasciurus douglasii	338	24.6	8.6	12.4	27.5	1.2	4.1	7.1	14.5	
Tamiasciurus hudsonicus	100								99.0	1.0
Glaucomys sabrinus	149	24.8	8.7	13.4	35.6		0.7	2.0	14.8	
Thomomys bottae	53	3.8	77.4		17.0			1.9		
Thomomys bulbivorus	58	12.1		81.0	6.9					
Thomomys mazama	208	33.7	8.7	9.1	44.7			3.8		

Table 4–2.—Continued.

Taxa	Localities considered (n)	Coast Range	Klamath Mountains	Willamette Valley	Cascade Range	Deschutes-Columbia Plateau	High Lava Plains	Basin and Range	Blue Mountains	Owyhee Uplands
Thomomys talpoides	226				2.7	9.3	10.6	23.9	42.9	10.6
Thomomys townsendii	34						32.4	20.6		47.1
Perognathus longimembris	9							77.8		22.2
Perognathus parvus	162				3.1	18.5	14.8	36.4	11.1	16.0
Microdipodops megacephalus	35						34.3	62.9		2.9
Dipodomys californicus	27		40.7		3.7			55.6		
Dipodomys microps	35						28.6	62.9		8.6
Dipodomys ordii	149				1.3	16.8	25.5	31.5	4.0	20.8
Castor canadensis	89	29.2	7.9	5.6	10.1	6.7	7.9	18.0	9.0	5.6
Reithrodontomys megalotis	82		11.0		1.2	25.6	12.2	20.7	12.2	17.1
Peromyscus crinitus	50					8.0	12.0	42.0	8.0	30.0
Peromyscus maniculatus	722	18.3	11.8	9.7	19.4	5.8	7.8	10.5	13.6	3.2
Peromyscus truei	60		45.0		8.3	6.7	25.0	11.7	3.3	
Onychomys leucogaster	56					19.6	23.2	33.9	1.8	21.4
Neotoma cinerea	259	20.8	5.0	3.5	17.0	6.6	8.5	10.8	17.4	10.4
Neotoma fuscipes	97	4.1	39.2	29.9	10.3			16.5		
Neotoma lepida	35						25.7	40.0	2.9	31.4
Clethrionomys californicus	223	26.9	17.9	2.2	52.9					
Clethrionomys gapperi	58								98.3	1.7
Phenacomys albipes	55	74.5	9.1	7.3	9.1					
Phenacomys intermedius	36				61.1			5.6	33.3	
Phenacomys longicaudus	109	42.2	13.8	22.9	21.1					
Microtus californicus	55	12.7	76.4	1.8	9.1					
Microtus canicaudus	58	12.1		86.2	1.7					
Microtus longicaudus	202	13.9	6.4	2.5	13.9	2.0	6.9	23.3	24.3	6.9
Microtus montanus	221		2.7		8.1	7.2	10.0	26.7	33.5	11.8
Microtus oregoni	155	41.3	12.3	14.8	31.6					
Microtus richardsoni	60				65.0				35.0	
Microtus townsendii	128	42.2	9.4	35.2	13.3					
Lemmiscus curtatus	55					10.9	20.0	36.4	10.9	21.8
Ondatra zibethicus	92	22.8	6.5	21.7	4.3	6.5	7.6	9.8	16.3	4.3
Zapus princeps	79		7.6		2.5		1.3	24.1	62.0	2.5
Zapus trinotatus	163	38.0	8.0	9.8	44.2					
Erethizon dorsatum	89		6.7	1.1	5.6	14.6	9.0	28.1	28.1	6.7
Carnivora										
Canis latrans	107	1.9	9.3	1.9	7.5	20.6	12.1	17.8	16.8	12.1
Canis lupus	34		23.5		67.6		2.9	2.9		2.9
Urocyon cinereoargenteus	11	18.2	63.6	9.1			9.1			
Vulpes velox	4				25.0		25.0		25.0	25.0
Vulpes vulpes	25			40.0	16.0	4.0	20.0	8.0	12.0	
Ursus americanus	67	13.4	13.4	6.0	34.3	4.5	9.0	1.5	17.9	
Ursus arctos	4						50.0		50.0	
Bassariscus astutus	14		50.0		50.0					
Procyon lotor	78	17.9	23.1	19.2	11.5	6.4		2.6	10.3	9.0
Martes americana	34	23.5	5.9		38.2			2.9	29.4	
Martes pennanti	2		50.0		50.0					
Mustela erminea	94	40.4	8.5	19.1	23.4	1.1		3.2	3.2	1.1
Mustela frenata	137	15.3	8.8	11.7	16.8	8.0	4.4	4.4	23.4	7.3
Mustela vison	79	30.4	11.4	13.9	11.4	3.8	7.6	6.3	13.9	1.3
Gulo gulo	2						50.0		50.0	
Taxidea taxus	68				2.9	13.2	16.2	22.1	29.4	16.2
Spilogale gracilis	82	31.7	18.3	7.3	13.4	6.1	4.9	8.5	4.9	4.9
Mephitis mephitis	60	5.0	21.7	10.0	6.7	10.0	10.0	18.3	15.0	3.3
Lutra canadensis	33	39.4	21.2	12.1	12.1		3.0	6.1	3.0	3.0
Puma concolor	60	10.0	46.7	1.7	21.7	1.7			18.3	
Lynx canadensis	8			12.5	25.0	12.5		12.5	25.0	12.5
Lynx rufus	164	9.8	9.8	1.8	21.3	3.7	3.7	19.5	22.0	8.5
Artiodactyla										
Cervus elaphus	5	20.0		20.0		20.0			40.0	
Odocoileus hemionus	43	20.9	11.6	11.6	20.9	4.7		20.9	9.3	
Odocoileus virginianus	27	7.4	51.9	3.7	11.1				3.7	22.2
Antilocapra americana	9							100.0		
Bos bison	8						12.5		62.5	25.0
Ovis canadensis	7				14.3		14.3	57.1	14.3	

Fig. 4-5. Photograph of typical physiography and vegetation in the Coast Range Province. View toward the southwest from high point near Buxton Creek (T3N, R4W, NE¼ sec. 35), Washington Co.

30–50-km-wide Willamette Valley is a north-south depression between the Coast and Cascade ranges (Fig. 4-3). Most of the valley is flat; it slopes gently to the north, with the elevation declining from ≅130 m at Eugene to <25 m at Portland, and is drained by the sluggish Willamette River. Rounded hills and buttes, a few attaining 300 m elevation, are scattered throughout the length of the valley (Fig. 4-6). Thick sedimentary deposits underlie much of the valley; some were laid down >1.5 million years ago, others ≅0.5 million years ago, and still others during the Bretz floods about 13,000 years ago (Allen et al., 1986; Franklin and Dryness, 1988). Soils are deep sandy loams and silty clay loams without a clay subsoil in well-drained areas; on more poorly drained sites, soils are light-colored silt loams over clay. The climate is moderate with ≅100 cm of precipitation annually and average minimum temperature ≅0.5°C in January and average maximum temperature ≅28°C in

July. Annual snowfall is 15–20 cm.

At European settlement, much of the Willamette Valley was vegetated by grassland communities generated and maintained by annual burning by Indians (Franklin and Dyrness, 1988; Johannessen et al., 1971). Original grasslands probably were dominated by danthonia (*Danthonia californica*) and needlegrasses (*Stipa*), but many present-day grasses (including all annuals) are introduced species (Franklin and Dyrness, 1988). Former grasslands presently support much of the agriculture and human population of the state. Under protection from fire much of valley becomes vegetated initially by oaks (*Quercus garryana*) and maples (*Acer macrophyllum*) replaced in time by Douglas-fir (Johannessen et al., 1971). Common shrubs in both woodlands and grasslands include Pacific poison oak (*Rhus diversiloba*) and sweetbriar rose (*Rosa eglanteria*). Black cottonwood (*Populus trichocarpa*), Oregon ash (*Fraxinus*

Table 4-3.—*Faunal resemblances[a] based on physiographic provinces (Franklin and Dyrness, 1988) and mammalian distribution in Oregon.*

	Coast Range	Klamath Mountains	Willamette Valley	Cascade Range	Deschutes-Columbia Plateau	High Lava Plains	Basin and Range	Blue Mountains	Owyhee Uplands
Coast Range	**63**	0.847	0.944	0.787	0.500	0.462	0.547	0.542	0.409
Klamath Mountains	58	**74**	0.824	0.711	0.563	0.567	0.658	0.594	0.493
Willamette Valley	59	56	**62**	0.554	0.569	0.465	0.523	0.545	0.429
Cascade Range	61	72	43	**92**	0.667	0.642	0.749	0.717	0.577
Deschutes-Columbia Plateau	31	38	35	51	**61**	0.734	0.851	0.803	0.704
High Lava Plains	30	40	30	51	47	**67**	0.805	0.901	0.794
Basin and Range	41	53	39	67	63	62	**87**	0.786	0.781
Blue Mountains	39	46	39	62	57	59	66	**81**	0.772
Owyhee Uplands	26	34	27	45	44	52	59	56	**64**

[a]Presented as the number (below the diagonal) and proportion (above the diagonal) of species common to both provinces. Proportion calculated by doubling the number of species common to both provinces and dividing by the total number of species in both provinces. Bold values along the diagonal are the number of species that occur in that province.

Fig. 4-6. Photograph of typical physiography and vegetation in the Willamette Valley Province. View toward the south, south-east from ridge 0.8 km N, 1.6 km E Lewisburg (T11S, R4W, SW¼ sec. 6), Benton Co.

Fig. 4-7. Photograph of typical physiography and vegetation in the Klamath Mountains Province. View toward the southwest from high point near Siskiyou Summit (T40S, R2E, NE¼ sec. 29), Jackson Co.

latifolia), and red alder mixed with Douglas-fir are common in riparian zones (Franklin and Dyrness, 1988).

Sixty-two species of native mammals have been recorded in the Willamette Valley Province, the second lowest species diversity among the nine provinces. Five species can be considered to characterize the mammalian fauna of the province (Table 4-2): *Scapanus townsendii*, *Sylvilagus bachmani*, *Thomomys bulbivorus*, *Microtus canicaudus*,

and *Vulpes vulpes*. The mammalian fauna of the province resembles most closely those of the Coast Range and Klamath Mountains (Table 4-3).

Klamath Mountains Province.—The Klamath Mountains Province (Fig. 4-7) in Oregon extends from the Oregon-California border northward along the coast almost to Coos Bay, then eastward to the mouth of the North Umpqua River, then south-southeastward to include the

Bear Creek valley near Medford that serves as the boundary with the Western Cascades (Fig. 4-3). The province is delimited by the extent of pre-Tertiary rocks (Franklin and Dyrness, 1988) and includes the Siskiyou Mountains. The topography is deeply folded and faulted; elevations mostly range from 600 to 1,200 m with Mt. Ashland, Jackson Co., the highest at 2,297 m (Franklin and Dyrness, 1988). The area is drained largely by the Rogue, Illinois, Applegate, and Chetco rivers. The area is formed of Triassic–Jurassic-age terranes composed of granitic rocks with late Triassic volcanic and metamorphosed sedimentary deposits and Jurassic unaltered sedimentary deposits. Soils in the western half of the province are dark-colored silt loams or silty clay loams over silty clay subsoils, whereas those in the eastern half are reddish-brown loams with clay loam subsoils (Franklin and Dyrness, 1988). The climate is cool and moist along the coast, but hot and dry in the interior valleys. Precipitation is 60–170 cm annually with <20% falling during the growing season.

Southwestern Oregon is extremely complex floristically with forest-shrub and mixed needle-leaved/broad-leaved sclerophyll vegetative zones (Fig. 4-4) in addition to extensions of the Sitka spruce and western hemlock zones in coastal areas (Franklin and Dyrness, 1988). Most of the province either is forested or has the potential of becoming forested in the absence of disturbance. The Rogue and Umpqua valleys are the most arid areas west of the Cascade Range and are vegetated by a mosaic of oaks (*Quercus kellogii* and *Q. garryana*) with an understory of grasses and shrubs (especially deerbrush, *Ceanothus integerrimus*; white-leaved manzanita, *Arctostaphylos viscida*; and birch-leaf mountain-mahogany, *Cercocarpus montanus*) on driest sites and open stands of Douglas-fir, ponderosa pine (*Pinus ponderosa*), and incense-cedar (*Calocedrus decurrens*) with an understory of oak on more mesic sites. The mixed-evergreen zone includes tanoak (*Lithocarpus densiflorus*), canyon live oak (*Quercus chrysolepis*), madrone (*Arbutus menziesii*), and Douglas-fir. Sugar pine (*Pinus lambertiana*) and Jeffrey pine (*P. jeffreyi*) are characteristic of some of the valleys (Franklin and Dyrness, 1988).

Seventy-four species of native mammals occur in the Klamath Mountains Province, a species diversity slightly above the median for the nine provinces (Table 4-2). Ten species characterize the mammalian fauna of the province: *Scapanus latimanus*, *Tamias siskiyou*, *Thomomys bottae*, *Dipodomys californicus*, *Peromyscus truei*, *Microtus californicus*, *Urocyon cinereoargenteus*, *Bassariscus astutus*, *Puma concolor*, and *Odocoileus virginianus*. The mammalian fauna of the Klamath Mountains has the greatest resemblance to those of the Coast Range and Willamette Valley provinces (Table 4-3).

Cascade Range Province.—The Cascade Range Province (Fig. 4-8) extends the full north-south dimension of the state between the Willamette Valley and Klamath Mountains provinces on the west and the Deschutes-Columbia Plateau, High Lava Plains, and Basin and Range provinces on the east (Fig. 4-3). It consists of two more-or-less distinct subprovinces: the Western Cascades and the High Cascades (Fig. 4-3). The Western Cascades include the Oligocene- and Miocene-aged volcanic deposits that form the west slope of the range. The topography is rugged in the east, but slopes becomes more gentle in the west. Elevations of peaks are as great as 1,800 m, but many of the ridges are ≅1,500 m (Franklin and Dyrness, 1988). The

northern two-thirds of the area is drained largely by the Middle Fork of the Willamette River and tributaries to the Willamette River (McKenzie, Santiam, Mollala, and Clackamas rivers) and the southern third is drained by the Umpqua and Rogue rivers and their tributaries. Bedrock is mostly basalt and andesite; glacial deposits are widespread. Soils mostly are either brown clay loams with silty clay and silty clay loam subsoils formed from pyroclastic materials or brown gravelly loam or sandy loam containing volcanic ash with clay loam or clay subsoils formed from igneous materials. The former type is subject to slumping if poorly drained; the latter type often is deep and well developed on gentle slopes (Franklin and Dyrness, 1988).

The High Cascades consists of much younger rolling terrain at elevations of 1,500–1,800 m with a series of major volcanic peaks (Mt. Hood, Mt. Jefferson, Three Sisters, Broken Top, Bachelor Butte, Mt. Thielsen, Mt. Scott, Mt. McLoughlin) extending to elevations of 2,721–3,427 m and numerous lower peaks and cinder cones. The area is much younger than the Western Cascades; some deposits are only a few hundred years old. Most soils were formed from glacial deposits and are immature with stony, gravelly, or sandy surface layers and stony loam subsoils (Franklin and Dyrness, 1988).

The climate of the Cascade Range Province is cool and wet with both temperature and precipitation strongly related to elevation and latitude. The ranges of average minimum temperatures in January are 2–4°C on lower slopes and –2–0°C on upper slopes; ranges of average maximum temperatures in July are 16–18°C at lower elevations and 14–16°C at higher elevations. Precipitation ranges from ≅150 to >250 cm in the north and from ≅100 to ≅150 cm in the south; much of the precipitation falls as snow with accumulated snowpacks as great as 750 cm on some peaks during some winters. Small glaciers persist on some of the higher peaks, mostly above 2,500 m.

Most of the Western Cascades south to the Klamath Mountains Province below ≅1,000 m is within the western hemlock zone (Figs. 4-3 and 4-4) with western hemlock and western red cedar as climax dominants and Douglas-fir as a subclimax dominant. In the southern part of the zone, incense-cedar, sugar pine, and ponderosa pine may occur. At higher elevations (1,000–2,500 m), the western hemlock zone grades into the subalpine fir zone (Fig. 4-4). At elevations of 1,000–1,500 m, Pacific silver fir (*Abies amabilis*), noble fir (*A. procera*), and western white pine (*Pinus monticola*) become more common. At elevations of 1,500–2,000 m, mountain hemlock (*Tsuga mertensiana*) and Pacific silver fir are dominants and subalpine fir (*Abies lasiocarpa*) and lodgepole pine are subdominants. In the northern part of the Cascade Range Province, Pacific silver fir is a common associate, but disappears in the southern part of the province where Shasta red fir (*Abies magnifica* var. *shastensis*) becomes a major component (Franklin and Dyrness, 1988). At ≅1,800 m, the zone grades into alpine parkland. Timberline areas are characterized by a mosaic of rock outcrops, patches of snow, meadows, and groups of trees. As temperatures moderate and precipitation declines on the east slope of the High Cascades at elevations of 1,100–1,500 m the grand fir (*Abies grandis*) zone forms a narrow band the full length of the province (Figs. 4-3 and 4-4). Grand fir, ponderosa pine, western larch (*Larix occidentalis*), or Douglas-fir may dominate seral communities, with Oregon boxwood (*Pachistima*

Fig. 4-8. Photograph of typical physiography and vegetation in the Cascade Range Province. View toward the south, south-east from Dee Wright Observatory (T15S, R8E) near Belknap Crater, Linn Co.

Fig. 4-9. Photograph of typical physiography and vegetation in the Deschutes-Columbia Plateau Province. View toward the east from high point 4 km N, 3.2 km W Shaniko (T6S, R16E, E½ sec. 28), Wasco Co.

myrsinites), big huckleberry (*Vaccinium membranaceum*), and a variety of shrubs and herbs or pinegrass (*Calamagrostis rubescens*) as major associates. Engelmann spruce (*Picea engelmannii*), subalpine fir, western red cedar, sugar pine, western white pine, and mountain hemlock may occur in local areas (Franklin and Dyrness, 1988). At slightly lower elevations, a narrow ponderosa pine zone occupies the east slope along the northern portion of the High Cascades, but widens further south to include west-

ern portions of other provinces (Figs. 4-3 and 4-4).

Ninety-two species of native mammals occur in the Cascade Range Province, more than in any other physiographic province (Table 4-3). Likely the great species diversity is a reflection of the montane region serving to separate faunas of the mesic region to the west and the arid region to the east, thereby containing elements of both faunas. Fourteen species characterize the mammalian fauna of the province: *Sorex bairdi, S. pacificus, S. palustris, Ochotona princeps,*

Lepus americanus, Tamias senex, T. siskiyou, Thomomys mazama, Clethrionomys californicus, Phenacomys intermedius, Microtus richardsoni, Zapus trinotatus, Canis lupus (extirpated), and *Bassariscus astutus*. Resemblance of the mammalian fauna of the Cascade Range Province was greatest with those of the Coast Range and Basin and Range provinces (Table 4-3).

Deschutes-Columbia Plateau Province.—The Deschutes-Columbia Plateau Province (Fig. 4-9) is bounded on the north by the Columbia River, on the west by the High Cascades Province, and on the south and east by the High Lava Plains and Blue Mountains provinces (Fig. 4-3). Nearly all of the province is underlain by numerous Miocene-age 8–30-m-thick lava flows totalling 600–1,500 m in thickness. The topography is gently rolling to hilly, but rivers have cut steep-sided canyons into the lava. Elevations range from 300 to 600 m over much of the area, but along the Columbia River decline to <150 m (Franklin and Dyrness, 1988). Soils are varied, but in general reflect differences in rainfall; most were formed under grasslands. Most soils are loams or silt loams with subsoils containing calcium carbonate accumulations. The climate is arid to semiarid (precipitation <25–60 cm annually), with hot, dry summers (average maximum temperature in July 30–34°C) and cold winters (average minimum temperature in January –2.5 to –4.3°C).

The natural vegetation over most of the province is steppe and shrub-steppe types (Fig. 4-4) with big sagebrush (*Artemisia tridentata*) and bitter-brush (*Purshia tridentata*) the dominant shrubs, and bluebunch wheatgrass (*Agropyron spicatum*), Idaho fescue (*Festuca idahoensis*), giant wildrye (*Elymus cinereus*), and Thurber needlegrass (*Stipa thurberiana*) the primary bunchgrasses. Most of the area is either under cultivation for dry-land small grains or grazed by livestock, and much of the area has been subjected to range fires. The shrubs are eliminated by fire and the bunchgrasses are intolerant of overgrazing. Thus, non-native grasses such as cheat grass (*Bromus tectorum*), Kentucky bluegrass (*Poa pratensis*), and medusahead wildrye (*Elymus caput-medusae*) have invaded and dominate the uncultivated areas.

The Deschutes-Columbia Plateau Province has the most depauperate mammalian fauna among the nine physiograpic provinces; only 61 native species occur in the province, 39 of which are represented by ≤10% of the total collection localities for the species (Table 4-2). The mammalian fauna of the plateau (Table 4-2) is characterized by only one species: *Spermophilus washingtoni*. Resemblance of the mammalian fauna of the province is greatest with those of the Basin and Range and Blue Mountains provinces (Table 4-3).

High Lava Plains Province.—The High Lava Plains Province (Fig. 4-10) lies south of the Blue Mountains Province and north of the Basin and Range Province (Fig. 4-3). It is an area of young (Pliocene–Pleistocene) lava flows with scattered buttes and cinder cones (Franklin and Dyrness, 1988). Some of the area is covered with shallow temporary lakes or formerly was covered by such lakes. Elevations range from ≅1,245 to 2,434 m with the highest point at Paulina Peak (Newberry Crater, Deschutes Co.). Many of the soils are derived from pumice from eruptions of Mt. Mazama and Paulina Peak, and are sandy loam with loamy subsoils. Soils formed in the lake basins often are silty clay loam over clay subsoils and frequently contain heavy saline concentrations. The climate is decidedly continental with average minimum temperature in January of –4 to –6°C and average maximum temperature in July of 14–22°C. Precipitation averages <25–50 cm annually, most of which falls in winter.

Differences in vegetation in various parts of the province largely reflect levels of precipitation and soil moisture. The ponderosa pine zone along the lower slopes of the High Cascade Province widens to extend >50 km into the more mesic southwestern portion of the province (Figs. 4-3 and 4-4). All of the northwestern corner and a 50-km-wide belt to the east of the remainder of the ponderosa pine zone is occupied by a xeric open savanna western juniper (*Juniperus occidentalis*) zone. The western juniper zone has a shrub understory mostly of big sagebrush with some bitter-brush; bluebunch wheatgrass and Idaho fescue are the predominant grasses. When western juniper and bitter-brush are killed by fire, they are replaced by sagebrush and cheat grass. Much of the remainder of the province is occupied by a shrub-steppe zone somewhat similar to that of the Deschutes-Columbia Plateau Province (Figs. 4-3 and 4-4). Big sagebrush may be replaced by low sagebrush (*Artemisia arbuscula*); rabbit-brush (*Chrysothamnus*) is common. Along the southern border of the province, salt-desert shrub communities dominated by shadscale (*Atriplex confertifolia*), salt sage (*A. nuttallii*), and greasewood (*Sarcobatus vermiculatus*) occur in a scattered zone (Figs. 4-3 and 4-4).

Sixty-seven native species of mammals occur in the High Lava Plains Province, the median among the nine provinces (Table 4-3). The number of collection localities did not total as much as 40% of the total number of localities for any of the species. Thus, despite the moderate species diversity, the mammalian fauna is not characterized by a preponderance of any species. The fauna resembles that of the Blue Mountains and Basin and Range provinces (especially the latter) most closely (Table 4-3).

Basin and Range Province.—The Basin and Range Province occupies most of the south-central portion of the state (Fig. 4-3). It is within the region stretched thin by the clockwise rotation of the coastal mountains that resulted in extensive volcanic activity that covered much of the area with lava and ash often many hundreds of meters thick. The volcanism was followed by extensive faulting that produced the characteristic topography of depressions interspersed with fault-block mountains such as Steens Mountain, Hart Mountain, Abert Rim, and Winter Rim. Much of the region is at an elevation of ≅1,200 m, but peaks rise 300–1,500 m higher with Steens Mountain at 2,852 m the highest (Fig. 4-11). Steens Mountain exhibits evidence of extensive glaciation with Little Blitzen Canyon and Kiger Gorge typical U-shaped glaciated valleys. The faulted region tilts downward to the north and drainage is internal; thus, during the Pleistocene Ice Ages, much of the area was covered with huge lakes of which only remnants remain. Soils in the western portion of the province were derived from Mt. Mazama pumice over gravel or sand and were formed beneath forest vegetation. In the east, soils formed from basalt are stony loam over clay or stony loam subsoils (Franklin and Dyrness, 1988). The climate is arid (annual precipitation <25–50 cm) with cold winters (average minimum January temperature 0 to –6°C) and warm summers (average maximum July temperature <14–22°C).

The vegetation of the province, like that of the province to the north, is related to precipitation and soil moisture. Much of the western third is in the ponderosa pine

Fig. 4-10. Photograph of typical physiography and vegetation in the High Lava Plains Province. View toward the south from 6.8 km E Riley (T23S, R27E, NE¼ sec. 36), Harney Co.

Fig. 4-11. Photograph of typical physiography and vegetation in the Basin and Range Province. View toward the south-southwest from 20 km N Alkali Lake (T28S, R23E, E½ sec. 11), Lake Co.

zone (that includes areas of lodgepole pine) with small scattered areas of western juniper. In the eastern two-thirds of the province, shrub-steppe covers most of the land with desert shrubs in the old lake beds and western juniper on the major fault-block mountains (Figs. 4-3 and 4-4). In mesic microhabitats near streams, communities of quaking aspen (*Populus tremuloides*) occur where forests meet the shrub-steppe and on some of the fault-block moun-

tains (Franklin and Dyrness, 1988).

The Basin and Range Province is second in mammal species diversity with 87 native species represented (Table 4-3). Eight species are considered to characterize (Table 4-2) the mammalian fauna of the province: *Sorex merriami, S. preblei, Ammospermophilus leucurus, Microdipodops megacephalus, Dipodomys californicus, D. microps, Peromyscus crinitus,* and *Neotoma lepida.* Resemblance

Fig. 4-12. Photograph of typical physiography and vegetation in the Owyhee Uplands Province. View toward the south from the summit of Drinkwater Pass (T20S, R36E, SE¼ sec. 34), Harney Co.

Fig. 4-13. Photograph of typical physiography and vegetation in the Blue Mountains Province. View toward the south near Buford Creek (T6N, R44E), Wallowa Co.

of the mammalian fauna of the Basin and Range is greatest with those of the Deschutes-Columbia Plateau and High Lava Plains provinces (Table 4-3).

Owyhee Uplands Province.—This southeastern Oregon province (Fig. 4-3) is similar to the Basin and Range Province except that it is a flat plateau deeply dissected by the Owyhee River and its mostly intermittent tributaries (Fig. 4-

12). Thus, it is unlike the previous province in that its drainage is not internal, there is little faulting except for the Mahogany and Owyhee mountains, and the highest point is <2,000 m (Mahogany Mountain, Malheur Co.). The climate is characterized by hot summers (average maximum temperature in July 14 to >26°C), cold winters (average minimum temperature in January 0 to −6°C), and little precipitation

(<25 cm annually). Vegetation in the province consists largely of the shrub-steppe zone, but relatively small, scattered areas at the periphery of the province are in western juniper and salt-desert shrub zones (Figs. 4-3 and 4-4).

Sixty-four species of mammals occur in the Owyhee Uplands Province, a number exceeded in five of the nine provinces (Table 4-3). Three species characterize the mammalian fauna (Table 4-2) within the province: *Pipistrellus hesperus*, *Spermophilus canus*, and *Thomomys townsendii*. Resemblance of the Owyhee Upland fauna is greatest with those of the High Lava Plains and the Basin and Range provinces (Table 4-3).

Blue Mountains Province.—The Blue Mountains Province (Fig. 4-3) occupies the northeastern corner of the state, extending from the High Lava Plains to the spectacular Hells Canyon on the Snake River and including the Ochoco, Blue, and Wallowa mountains and the Strawberry, Elkhorn, and Greenhorn ranges (Fig. 4-13). The Blue and Ochoco mountains have moderate slopes with elevations to 2,100 m, whereas the Wallowa Mountains are rugged and steep with elevations to 2,900 m; the latter are extensively glaciated. Valleys and basins separate the ranges. In the Ochoco Mountains, valley elevations are ≅750 m, but in the Wallowa Mountains they are ≅900 m. Soils at lower elevations were formed under grassland or shrub-steppe vegetation; in the western half of the province soils are clay loam over clay subsoils, whereas in the eastern half they are silt loam with clay loam subsoils. On ridge tops and north slopes, soils were formed from volcanic ash under forest vegetation; they are deep fine sandy loam or silt loam. Transition soils at forest-grassland ecotones are dark silt loams over silty clay loam to clay subsoils (Franklin and Dyrness, 1988). The climate of the province is variable with broad areas of cold winters (average minimum temperature in January 0 to –6°C), moderate summers (average maximum temperature in July <14–22°C), and precipitation increasing with elevation (≅50–150 cm annually).

Forested areas at lower elevations in the Blue Mountains Province are within the ponderosa pine zone (includes some western juniper). At higher elevations, the grand fir zone predominates with grand fir-white fir (*Abies concolor*) the climax species (big huckleberry is a major understory shrub) and western larch, lodgepole pine, and Douglas-fir the major seral species (Franklin and Dyrness, 1988). Englemann spruce, western white pine, and ponderosa pine may occur in lesser abundance. At still higher elevations and in pockets of cold, the subalpine zone with subalpine fir-Englemann spruce forests is characteristic. Whitebark pine (*Pinus albicaulis*) occurs at timberline. Shrub steppe and steppe zones characterize valleys at lower and higher elevations, respectively (Fig. 4-4).

Eighty species of mammals occur in the Blue Mountains Province; thus, the province has the third most diverse mammal fauna in the state (Table 4-3). Characteristic species are *Sorex monticolus*, *Spermophilus beldingi*, *S. columbianus*, *Tamiasciurus hudsonicus*, *Thomomys talpoides*, *Clethrionomys gapperi*, and *Zapus princeps* (Table 4-2). Resemblance of the mammalian fauna of the Blue Mountains is greatest with those of the High Lava Plains and the Deschutes-Columbia Plateau provinces (Table 4-3).

BIOGEOGRAPHICAL AFFINITIES OF PRESENT-DAY MAMMALS OF OREGON

From the foregoing sections, it is obvious that Oregon had a cataclysmic origin and a dynamic history of development, and that the mammal faunas changed radically with those events. Although Oregon sits more-or-less astride the boundary of northern and southern elements that influence the flora and fauna of the state, the feature that impacts the mammalian fauna the greatest is the great north-south chain of mountains that bisects the state. General distributions and biogeographical affinities of the 136 present-day species of land mammals in Oregon are shown in Table 4-4.

Four species of mammals are endemic to Oregon, major portions of the geographic ranges of six species occur within the state, and six species are marine with distributions in Oregon restricted to beaches and estuaries. Nine species were introduced into Oregon by humans since settlement.

The 15 species that occur entirely or mostly west of the Cascade Range in Oregon are distributed mostly in coastal regions throughout their geographic ranges; the 47 species that occur entirely or mostly east of the Cascade Range in Oregon are distributed mostly in inland regions throughout their geographic ranges. Of the 15 western Oregon species, six have southern affinities, three have northern affinities, and six have affinities more-or-less centered on Oregon. Of the 47 eastern Oregon species, 11 have northern affinities (many boreal species with montane populations in Oregon), 22 have southern affinities (many Great Basin species with northern extensions into Oregon), and 14 have inland affinities more-or-less centered on Oregon. Of the 43 species with essentially statewide distributions, 21 occur continent-wide, 7 have southern affinities, 6 have northern affinities, and 9 have affinities more-or-less centered on Oregon. Six species are unique in having strong southern affinities with geographic ranges including a small portion of southwestern Oregon. Five species have been extirpated since settlement (Table 4-4) and several others are considered endangered (Marshall 1992; Olterman and Verts, 1972).

Table 4-4.—*The 136 species of mammals in Oregon categorized by distribution and biogeographical affinity.*

Species endemic to Oregon—4
  Sorex bairdi, Baird's shrew
  Sorex pacificus, Pacific shrew
  Thomomys bulbivorus, Camas pocket gopher
  Phenacomys longicaudus, Red tree vole
Species for which most of the geographic range occurs in Oregon—6
  Thomomys mazama, Western pocket gopher
  Tamias siskiyou, Siskiyou chipmunk
  Spermophilus canus, Merriam's ground squirrel
  Phenacomys albipes, White-footed vole
  Clethrionomys californicus, Western red-backed vole
  Microtus canicaudus, Gray-tailed vole

Species introduced into Oregon—9
  Didelphis virginiana, Virginia opossum
  Sylvilagus floridanus, Eastern cottontail
  Sciurus carolinensis, Eastern gray squirrel
  Sciurus niger, Eastern fox squirrel
  Rattus norvegicus, Norway rat
  Rattus rattus, Black rat
  Mus musculus, House mouse
  Myocastor coypus, Nutria
  Oreamnos americanus, Mountain goat
Marine species restricted to beaches and estuaries in Oregon—6
  Callorhinus ursinus, Northern fur seal
  Eumetopias jubatus, Northern sea lion

Table 4–4.—Continued.

Zalophus californianus, California sea lion
Mirounga angustirostris, Northern elephant seal
Phoca vitulina, Harbor seal
Enhydra lutris, Sea otter[a]

Species with continent-wide distributions—21

Myotis lucifugus, Little brown myotis
Lasiurus cinereus, Hoary bat
Lasionycteris noctivagans, Silver-haired bat
Eptesicus fuscus, Big brown bat
Castor canadensis, American beaver
Peromyscus maniculatus, Deer mouse
Ondatra zibethicus, Common muskrat
Erethizon dorsatum, Common porcupine
Canis latrans, Coyote
Canis lupus, Gray wolf [a]
Vulpes vulpes, Red fox
Ursus americanus, Black bear
Procyon lotor, Common raccoon
Mustela frenata, Long-tailed weasel
Mustela vison, Mink
Mephitis mephitis, Striped skunk
Lutra canadensis, River otter
Puma concolor, Mountain lion
Lynx rufus, Bobcat
Cervus elaphus, Elk or wapiti
Odocoileus virginianus, White-tailed deer

Species with southern affinities occurring west of the Cascade Range in Oregon—6

Sorex sonomae, Fog shrew
Sylvilagus bachmani, Brush rabbit
Spermophilus beecheyi, California ground squirrel
Sciurus griseus, Western gray squirrel
Neotoma fuscipes, Dusky-footed woodrat
Urocyon cinereoargenteus, Common gray fox

Species with northern affinities distributed west of the Cascade Range in Oregon—3

Sorex bendirii, Pacific water or Marsh shrew
Scapanus townsendii, Townsend's mole
Tamias townsendii, Townsend's chipmunk

Species with nearly equal north-south affinities distributed west of the Cascade Range in Oregon—6

Sorex trowbridgii, Trowbridge's shrew
Neurotrichus gibbsii, Shrew-mole
Aplodontia rufa, Mountain beaver
Microtus oregoni, Creeping vole
Microtus townsendii, Townsend's vole
Zapus trinotatus, Pacific jumping mouse

Species with northern affinities distributed mostly or wholly east of the Cascade Range in Oregon—11

Sorex palustris, Water shrew
Tamias minimus, Least chipmunk
Tamias amoenus, Yellow-pine chipmunk
Spermophilus washingtoni, Washington ground squirrel
Spermophilus columbianus, Columbian ground squirrel
Tamiasciurus hudsonicus, Red squirrel
Thomomys talpoides, Northern pocket gopher
Clethrionomys gapperi, Southern red-backed vole
Phenacomys intermedius, Heather vole
Microtus richardsoni, Water vole
Gulo gulo, Wolverine

Species with southern affinities distributed mostly or wholly east of the Cascade Range in Oregon—22

Sorex merriami, Merriam's shrew
Pipistrellus hesperus, Western pipistrellus
Myotis ciliolabrum, Western small-footed myotis
Euderma maculatum, Spotted bat
Brachylagus idahoensis, Pygmy rabbit
Tamias senex, Allen's chipmunk

Ammospermophilus leucurus, White-tailed antelope squirrel
Spermophilus beldingi, Belding's ground squirrel
Spermophilus mollis, Piute ground squirrel
homomys townsendii, Townsend's pocket gopher
Perognathus longimembris, Little pocket mouse
Perognathus parvus, Great Basin pocket mouse
Microdipodops megacephalus, Dark kangaroo mouse
Dipodomys ordii, Ord's kangaroo rat
Dipodomys microps, Chisel-toothed kangaroo rat
Reithrodontomys megalotis, Western harvest mouse
Peromyscus crinitus, Canyon mouse
Peromyscus truei, Piñon mouse
Onychomys leucogaster, Northern grasshopper mouse
Neotoma lepida, Desert woodrat
Vulpes velox, Kit fox
Antilocapra americana, Pronghorn

Species with nearly equal north-south affinities distributed east of the Cascade Range in Oregon—14

Sorex monticolus, Montane or Dusky shrew
Sorex preblei, Preble's shrew
Ochotona princeps, American pika
Sylvilagus nuttallii, Mountain cottontail
Lepus townsendii, White-tailed jackrabbit
Marmota flaviventris, Yellow-bellied marmot
Spermophilus elegans, Wyoming ground squirrel[a]
Spermophilus lateralis, Golden-mantled ground squirrel
Microtus montanus, Montane vole
Lemmiscus curtatus, Sagebrush vole
Zapus princeps, Western jumping mouse
Taxidea taxus, American badger
Ovis canadensis, Mountain sheep
Bos bison, Bison[a]

Species with northern affinities distributed statewide or nearly so in Oregon—6

Lepus americanus, Snowshoe hare
Glaucomys sabrinus, Northern Flying squirrel
Martes americana, American marten
Martes pennanti, Fisher
Mustela erminea, Ermine
Lynx canadensis, Lynx

Species with southern affinities distributed statewide or nearly so in Oregon—7

Myotis californicus, California myotis
Myotis thysanodes, Fringed myotis
Myotis yumanensis, Yuma myotis
Corynorhinus townsendii, Townsend's big-eared bat
Antrozous pallidus, Pallid bat
Lepus californicus, Black-tailed jackrabbit
Spilogale gracilis, Western spotted skunk

Species with nearly equal north-south affinities distributed statewide or nearly so in Oregon—9

Sorex vagrans, Vagrant shrew
Scapanus orarius, Coast mole
Myotis evotis, Long-eared myotis
Myotis volans, Long-legged myotis
Tamiasciurus douglasii, Douglas' squirrel
Neotoma cinerea, Bushy-tailed woodrat
Microtus longicaudus, Long-tailed vole
Ursus arctos, Grizzly or Brown bear[a]
Odocoileus hemionus, Black-tailed and Mule deer

Species with southern affinities distributed in southwestern Oregon—6

Scapanus latimanus, Broad-footed mole
Tadarida brasiliensis, Brazilian free-tailed bat
Thomomys bottae, Botta's pocket gopher
Dipodomys californicus, California kangaroo rat
Microtus californicus, California vole
Bassariscus astutus, Ringtail

[a]Extirpated.

CHAPTER 5

ZOONOSES

Mammals are the most common reservoirs of zoonoses (animal diseases transmissible to humans) and, worldwide, they reservoir a wide variety of viral, rickettsial, and bacterial pathogens (Childs, 1995). Although some mammal-borne zoonoses are among the most serious and dreaded diseases, the human incidence of most of them in North American tends to be low. A detailed discussion of all zoonoses that might be encountered in Oregon is beyond the scope of this chapter, but we believe that mammalogists, wildlife biologists, ecologists, and others who regularly work with mammals, and the general public, should be apprised of the potential hazard of the most serious and most common ones. In addition, all should be aware of methods available to prevent or avoid these zoonoses yet permit continued study and enjoyment of the mammalian fauna of the state. Thus, it is not our objective to frighten, but to caution.

VIRAL DISEASES

Hantavirus Pulmonary Syndrome.—Hantaviruses cause several diseases in Europe and Asia commonly referred to as hemorrhagic fever. In 1993, a severe outbreak of hantavirus pulmonary syndrome occurred in southwestern United States. It is characterized by fever, muscle aches, headache, nausea, vomiting, coughing, and severe respiratory distress leading to death in >50% of recorded cases (Childs et al., 1995). The Sin Nombre virus isolated from deer mice (*Peromyscus maniculatus*) in the region was considered the causative agent. Several other hantaviruses, some of which are somewhat less virulent human pathogens, have been isolated from other rodents elsewhere in the United States (Childs et al., 1995). In deer mice, after a short period of viremia, the virus becomes established in either the kidneys or the lungs or both and is shed in the urine or saliva for an extended period. However, in other rodents, no persistent infection follows the 6–8-day period of viremia; thus the possibility of transmitting the virus is short-lived or nonexistent in these species (A. Fairbrother, U. S. Environmental Protection Agency, in litt., 3 December 1993).

These viruses are believed to be transmitted to humans by inhalation of particles containing excretions or secretions of rodents. Transmission among rodents and from rodents to humans through bites is known to occur, but arthropods, although they may become infected by feeding on an infected host, are not believed to be common vectors of the disease organism (Childs et al., 1995).

In Benton Co., three of 17 gray-tailed voles (*Microtus canicaudus*), but none of nine deer mice and three house mice (*Mus musculus*) exhibited positive antibody response to Sin Nombre virus (A. Fairbrother, in litt., 3 December 1993). At least one human death has been attributed to hantavirus pulmonary syndrome in Oregon, a 16-year-old male from The Dalles, Wasco Co., who died 4 July 1994 (J. Senior, *The Oregonian*, 26 December 1993, p. B2).

Mammalogists probably have more contact with small rodents, especially the ubiquitous deer mouse, than any other group of workers. At the 75th annual meeting of the American Society of Mammalogists in Washington, D.C.,

researchers from the Communicable Disease Center, Atlanta, Georgia, drew blood to test for antibodies to hantaviruses from 288 mammalogists. Only nine samples were reactive for hantaviruses; eight of the reactions were equivocal (D. E. Wilson, in litt., 25 August 1994). Thus, despite 50 human cases of the disease (30 deaths) in the United States in 1990–1993 (J. Senior, *The Oregonian*, 26 December 1993, p. B2), the risk of contracting the virus seems low. Avoiding handling deer mice or their excreta should further reduce the risk of the disease. Because of aerosol transmission of the virus and the high rate of mortality, elaborate precautionary measures are recommended for researchers who work with potentially infected wild mammals (Mills et al., 1995).

Rabies.—Rabies, sometimes called hydrophobia, was a disease known and feared in antiquity. It is caused by a *Lyssavirus* transmitted through the saliva, most commonly by the bite of an infected mammal. Other methods of transmission are believed possible (Constantine, 1967), although extremely rare (Brass, 1994). Upon gaining entry, the virus may enter nerve endings directly or replicate in muscle then enter nerve endings. It is transported along the nerves at 8–12 mm/day to the central nervous system where it produces an encephalitis or encephalomyelitis (Brass, 1994). Once symptoms develop, the disease is nearly always fatal. Treatment involves techniques for preventing the virus from reaching the central nervous system.

In the United States, rabies is confined mostly to wild mammals because of required vaccination programs for domestic dogs; house cats currently are the most commonly infected domestic species (Krebs et al., 1995). Among wild mammals in the United States, striped skunks (*Mephitis mephitis*), raccoons (*Procyon lotor*), red foxes (*Vulpes vulpes*), coyotes (*Canis latrans*), and bats (Chiroptera) are the most common reservoir-vectors of rabies virus. Among nonvolants, each serves as the dominant reservoir species in one or more geographic regions in the United States and each is associated with a specific variant of the rabies virus (Krebs et al., 1995).

Sylvatic rabies in Oregon is restricted largely to bats; of the 242 cases of the infection reported in wild mammals from 1960 to 1992, 231 (95.5%) were in bats, 7 (2.9%) were in foxes (presumably *Vulpes vulpes*), and 4 (1.7%) were in skunks (presumably *Mephitis mephitis*). Species identification was not available for the bats, but *Lasionycteris noctivagans* reportedly is diagnosed with rabies frequently in the Pacific Northwest and three other species that occur in Oregon (*Eptesicus fuscus, Lasiurus cinereus*, and *Tadarida brasiliensis*) are commonly infected in neighboring regions (Krebs et al., 1995). Rabid bats were reported from 31 of the 36 counties in the state; all seven foxes and three of four skunks were from west of the Cascade Range. Three human deaths were attributed to the disease during the 32-year period (L. P. Williams, Jr., Oregon Public Health Veterinarian, in litt.).

For the general public, the hazard of rabies can be reduced significantly by simple measures to avoid exposure. DO NOT HANDLE BATS! DO NOT APPROACH ANY

CARNIVORE THAT APPEARS TO BEHAVE ABNOR-MALLY. Adopt the philosophy of "if it does not back off from you, you back off from it!" Do not pick up young skunks, foxes, or other carnivores that appear to have been abandoned or any mammal that appears to be sick or dazed.

For those in high-risk categories, such as spelunkers and those who routinely handle carnivores or bats, a preexposure vaccine is available. It should be administered sufficiently in advance of possible exposure that an immune response can be demonstrated; booster vaccinations should be obtained at regular intervals.

Persons bitten by a potentially rabid mammal should wash the wound with soap and water, treat the wound with alcohol or iodine solution, and seek counsel of a physician immediately (Brass, 1994). Postexposure vaccine and human or equine rabies immunoglobulin usually will be administered and nearly always are effective. Nevertheless, from personal experience of one of us (BJV), considerable anxiety is involved as the incubation period of rabies in humans may be lengthy.

Control of rabies in wild mammals during epizootics by reducing populations usually is neither practical nor effective and commonly is socially and scientifically unacceptable (Rosatte, 1987). Recent use of oral vaccines in baits to control epizootics of rabies in terrestrial mammals seemingly has been effective (Rosatte, 1987) and merits further investigation (Brass, 1994).

BACTERIAL DISEASES

Plague.—Plague is caused by *Yersinia pestis*, a bacterium hosted mostly by rodents and transmitted by fleas (Siphonaptera) or by direct contact among mammals. Bubonic, septicemic, and pneumonic plague are the most common forms of the disease. Symptoms include fever, headache, chills, muscle aches, general malaise, and prostration, and in the bubonic form, swollen and painful lymph glands near the bacterium's point of entry. Pneumonic plague may develop from septicemic plague or directly through inhaling respiratory droplets containing the bacterium; its symptoms include high fever, respiratory difficulties, and bloody sputum (Gage et al., 1995). Human-to-human transmission becomes possible with the pneumonic form. Mortality among untreated humans with bubonic plague is 50–60%; among those with pneumonic plague it is 100% (Gage et al., 1995). Plague is treated with streptomycin and other antibiotics. A vaccine is available for those who work with these rodents, but its effectiveness in preventing aerosol transmission of the disease is unknown. Some recommend prophylactic antibiotics for high-risk workers (Gage et al., 1995).

In the Middle Ages, when hygiene and sanitation were minimal, commensal rats (*Rattus*) served as reservoirs and their fleas as vectors of the disease organism to humans. Some humans developed the pneumonic form of the disease and transmitted the disease to other humans through respiratory droplets; about one-third of the human population of Europe died of the disease during the last half of the 14th century (Langer, 1964). The pandemic was followed by numerous outbreaks of the disease in London, Venice, and other major cities of Europe during the 15th and 16th centuries (Langer, 1964). At the present time, commensal rat-borne plague is not a serious threat in the United States.

In the United States, >75 species of mammals have been recorded with the disease. In the West, where the disease is enzootic with occasional epizootics, ground squirrels (*Spermophilus*), chipmunks (*Tamias*), woodrats (*Neotoma*), deer mice, and California voles (*Microtus californicus*) are mammals that occur in Oregon and are frequently infected with plague (Gage et al., 1995). Mortality among rodents in some populations may be low with survivors possibly maintaining the bacterium in their blood and serving as reservoirs of the infection in fleas that feed on them (Gage et al., 1995).

Despite the potential for transmission of plague to humans from rodent populations where the disease is enzootic, incidence of the disease in humans in western United States remains low. In the 24 years before 1994, only 334 cases of plague were reported in humans in the 13 western states, most in the Southwest. In Oregon, 12 human cases were reported in 1970–1995, one of which was believed to have been acquired elsewhere. However, the 12 cases resulted in four deaths (L. P. Williams, Jr., Oregon Public Health Veterinarian, in litt.). Contraction of most of the infections was attributed to flea bites. Thus, avoidance of wild rodents and their fleas seems to be the only reliable preventive measure available to the general public.

Tularemia.—Tularemia is caused by *Francisella tularensis*, a gram-negative bacterium that can be transmitted to humans by handling infected mammals, by bites of infected arthropods (especially ticks), and through water contaminated with urine of infected mammals (Gage et al., 1995). Although >100 species of mammals are known to have been infected with the etiologic agent of tularemia, lagomorphs undoubtedly serve as host-vectors of the disease in most human cases, particularly where rabbit hunting is a major sport. Ticks and biting flies are largely responsible for transmitting the disease among wild mammals; 20 (55.6%) of the 36 human cases reported in Oregon in 1979–1990 were attributed to insect bites, all of which were acquired east of the Cascade Range (L. P. Williams, Jr., Oregon Public Health Veterinarian, in litt.). Aquatic species such as muskrats (*Ondatra zibethicus*) and beavers (*Castor canadensis*) may acquire and transmit the bacterium through water (Gage et al., 1995).

Symptoms of the disease often mimic those of some other zoonoses; fever, chills, general malaise, fatigue, and swollen lymph glands are common (Gage et al., 1995). Involvement of the respiratory system also is common and other syndromes of the disease are known. Those who develop symptoms should seek council of a physician; the physician should be advised of possible contact with rabbits or rodents and of insect bites. The disease organism is sensitive to antibiotic treatment, with streptomycin the most effective. Mortality in untreated cases is 5–7% (Gage et al., 1995).

In Illinois, hunting eastern cottontails (*Sylvilagus floridanus*) is a major sport. The incidence of human tularemia was high and was strongly correlated with density of cottontail populations. Also involved was the relationship between the average date of the first 10 frosts and the opening of the cottontail hunting season. Cold weather was believed to cause ticks to hibernate and to kill infected cottontails, thereby greatly reducing the hazard of handling the meat (Yeatter and Thompson, 1952). Thus, those who hunt rabbits in Oregon might reduce the possibility of acquiring the disease simply by waiting for cold weather before pursuing their sport.

Lyme Disease.—Lyme disease, caused by the spirocete bacterium *Borrelia burgdorferi*, is characterized by fever, muscle aches, and redness of the skin spreading from the point of admission to the body. Later, arthritis, inflammation of the heart, and neurologic disorders develop (Gage et al., 1995). Lyme disease was first recognized as a separate disease in 1975 from an unusual concentration of cases of childhood arthritis following skin rashes in the vicinity of Lyme, Connecticut. Subsequently, the etiologic agent was identified and ticks in the genus *Ixodes* were found to transmit it. Mammals and other terrestrial vertebrates serve as reservoirs of the spirochete; usually a small mammal serves as host to larval ticks and deer (*Odocoileus*) serve as hosts to adult ticks. Relatively few vertebrate host species for the ticks are good reservoirs for the agent causing Lyme disease.

Although the reporting of Lyme disease in Oregon was optional until 1 December 1994, 40 cases of the disease were recorded in the state in 1979–1985, and 158 cases in 1986–1995. Lyme disease was reported in humans in 17 and 26 counties during the two periods, respectively. Although human cases of the disease were distributed widely in Oregon, ≥30% of the cases were in Josephine and Jackson counties. Gage et al. (1995) claimed that the dusky-footed woodrat (*Neotoma fuscipes*) was the primary reservoir of the causative agent of Lyme disease in western United States. However, in Oregon, that species is much more restricted geographically than are cases of Lyme disease.

Preventative measures usually are directed toward avoidance of tick bites through either protective clothing or tick repellents. Removal of attached ticks promptly may reduce risk of acquiring Lyme disease. During early stages of the disease, treatment with antibiotics usually is effective, but sometimes it is not during later stages of the disease. A vaccine for humans has been developed and is being tried experimentally (Gage et al., 1995).

Tick-borne Relapsing Fever.—Other species in the genus *Borellia* are responsible for causing several types of relapsing fever. In the Pacific states, *B. hermsii* is the causative agent of relapsing fever, the tick *Ornithodoros hermsi* is the vector, and chipmunks (*Tamias amoenus*) and red squirrels (*Tamiasciurus hudsonicus*) are considered to be the reservoir species. The spirochetes do not appear in the blood of some other small mammals; thus, they probably are not capable of infecting the ticks that feed on them. Humans usually acquire the disease in cabins containing nests of rodents with infected ticks.

Relapsing fever is characterized by alternating periods of fever (3–7-day bouts) and nonfebrile periods (Gage et al., 1995). Relapses usually are progressively milder. Tetracyclines and penicillin are used to treat the disease; mortality is <5%. Avoiding rodent-infested cabins and other such areas that might be infested with ticks, and rodent-proofing and treating cabins with appropriate pesticides to kill ticks are recommended preventative measures (Gage et al., 1995).

In Oregon, 31 cases of relapsing fever in humans were reported in 1982–1992; the source of infection in 19 (61.3%) was traced to cabins in the central Cascades (L. P. Williams, Jr., Oregon Public Health Veterinarian, in litt.). One of the 31 infected persons died.

Leptospirosis.—Leptospirosis is a zoonotic disease caused by spirochetes in the genus *Leptospira* of which numerous antigenically distinct serotypes have been identified. Although several species of mammals may be infected by a serotype and humans may be infected by several of the serotypes, each serotype usually is associated with a specific mammal or group of mammals. For example, before 1950, most leptospiral infections in humans in the United States were from *Leptospira icterohemorrhagiae*, a serotype usually associated with rats (*Rattus*); since that time most human infections were from *L. canicola*, a serotype associated with domestic dogs (Kaufmann, 1976).

Leptospires are waterborne, enter the body through mucous membranes or by ingestion, and are excreted in the urine. Optimal conditions for survival of the organisms are moist, slightly alkaline or neutral, substrates and temperatures ≥22°C. Flu-like symptoms are characteristic of leptospirosis, but renal damage and involvement of the central nervous system are known to occur. Response to antibiotic therapy usually is good. In Oregon, leptospirosis was reported in 26 humans in 1978–1991; 13 had waded or bathed in lakes or creeks, 6 had contact with domestic mammals, and 5 had contact with wild mammals.

In a survey conducted on two areas in Benton Co., leptospires were not isolated by culture of urine and kidney tissues from 202 individuals of 15 species of mammals. However, microscopic agglutination-lysis tests revealed that 12 (13.6%) of 88 individuals had significant titers to antigens of *L. canicola* or *L. grippotyphosa*. Sera of three of 38 opossums (*Didelphis virginiana*), seven of 18 striped skunks (*Mephitis mephitis*), and one of 14 eastern cottontails had significant titers (≥1:100) to *L. grippotyphosa*, and one each of the opossums, skunks, and cottontails, and the lone red fox (*Vulpes vulpes*), had significant titers to *L. canicola*. None reacted to antigens of *L. pomona*, *L. hardjo*, *L. icterohemorrhagiae*, *L. tarassovi*, *L. alexi*, *L. autumnalis*, or *L. szwajizak* (Stewart, 1979). However, more recently, nine of 20 gray-tailed voles, but none of three house mice from one of the same areas, had titers of ≥1:100 to antigens of *L. grippotyphosa*. One of 41 gray-tailed voles, two of 23 house mice, and none of 41 deer mice from other areas in the Willamette Valley had titers of 1:100 to *L. ballum*, but sera from none of the animals reacted to *L. grippotyphosa* (A. Fairbrother, in litt., 3 December 1993).

Avoiding wading or bathing in lakes, ponds, or creeks and not drinking water therefrom are recommended preventative measures. Researchers and others who handle small mammals should wear waterproof gloves or otherwise handle mammals in a manner to avoid contamination with urine.

RICKETTSIAL DISEASES

Rocky Mountain Spotted Fever.—Rocky Mountain spotted fever is caused by *Rickettsia rickettsii*. The organism is transmitted to humans and among a variety of mammal hosts by ticks, and among ticks transovarially and between larval stages. However, the organism likely cannot be maintained by transmission among ticks alone as reproduction and survival in infected ticks is reduced (Gage et al., 1995). Thus, involvement of mammal hosts is necessary to maintain the disease. In western United States, the tick *Dermacentor andersoni*, a species that occurs in Oregon, is the primary vector of the disease.

Symptoms of the disease in humans include fever, headache, dermal rash, muscle aches, and loss of appetite, with a variety of serious cardiovascular, renal, and central nervous system complications possible (Gage et al., 1995). Tetracyclines, chloramphenicol, and some other antibiotics are used to treat the disease. Before antibiotics were available mortality was 23–70%, but it has declined to only 3–5% since the advent of appropriate treatment (Gage et al., 1995).

Avoidance of areas where ticks are abundant; wearing light-colored clothing tightly fitting at ankles, wrists, and neck to exclude ticks; and use of tick repellents are the most frequently used means of avoiding tick bites. Removal of feeding ticks immediately reduces the risk of acquiring the disease as the rickettsiae do not leave the tick until it has fed for some time (Gage et al., 1995).

Q fever.—Q fever is a rickettsial disease caused by *Coxiella burnetti* transmitted among a wide variety of wild and domestic mammal hosts by several species of ticks. However, transmission from other mammals to humans is mostly by inhalation of materials from mammal tissues and products containing the organism or, less frequently, from unpasteurized milk. Ticks are not believed to play a role in transmission of the organism to humans (Gage et al., 1995). Sheepherders, sheep shearers, and those who work in slaughterhouses are most often infected, but the organism can be acquired from rodents, artiodactyls, and a variety of other wild mammals. In Oregon, 31 cases of Q fever were diagnosed in humans in 1974–1990, with only two likely acquired from wild mammals (L. P. Williams, Jr., Oregon Public Health Veterinarian, in litt.).

The disease in mammals other than humans is usually asymptomatic, although pneumonitis has been recorded. In humans, the rickettsial organism produces an acute febrile disease with sudden onset of headache, chills, sore muscles, swollen lymph glands, conjunctivitis, nausea, diarrhea, and sore pharynx, and sometimes respiratory symptoms (Gage et al., 1995). The disease in humans can be treated successfully with tetracyclines, chloramphenicol, erythromycin, and several other antibiotics. However, it can become chronic with involvement of the endocardium requiring years of treatment with antibiotics (Gage et al., 1995). Mortality is ≅1%. Avoiding areas contaminated with animal tissues or products and wearing a respirator when working in contaminated areas are preventative measures. A vaccine has been developed in Australia (Gage et al., 1995).

Chapter 6

KEY TO THE ORDERS OF LAND MAMMALS OF OREGON

1a. Cranium small; teeth 50; incisors 5/4; epipubic bones present; hallux clawless and opposable; penis bifurcate; females with a marsupium enclosing nipples...
..................................**ORDER DIDELPHIMORPHIA** (p. 40).

1b. Cranium relatively large; teeth 44 or fewer; incisors ≤3/3, epipubic bones absent; hallux, if present, equipped with a claw and not opposable; penis not bifurcate; marsupium enclosing nipples absent...................................2

  2a. Forearm, hand, and fingers modified into membrane-covered wing; membrane between rear limbs incorporates all or most of tail; hind limbs rotated so knees bend backward; ears long with well-developed tragi...............................**ORDER CHIROPTERA** (p. 74).

  2b. Forearm, hand, and fingers not modified into wing; no membrane between rear limbs; tail free if present; knees bend forward; ears without well-developed tragi 3

3a. Feet with hooves; unguligrade; upper incisors absent; ruminant; at least males equipped with horns or antlers...
.....................................**ORDER ARTIODACTYLA** (p. 463).

3b. Feet with claws; plantigrade or digitigrade, but never unguligrade; upper incisors present; stomach simple; horns and antlers lacking...4

  4a. Incisors chisel-like, modified for gnawing; canines absent; a broad diastema between incisors and premolars..5

  4b. Incisors not chisel-like; canines present; no diastema between incisors and premolars.............................6

5a. Incisors 2/1; ears longer than tail...........................
.....................................**ORDER LAGOMORPHA** (p. 123).

5b. Incisors 1/1; ears usually shorter than tail...............
....**ORDER RODENTIA** (p. 153).

  6a. Canines smaller than incisors; zygoma weak or absent; eyes minute or lacking external openings........
....**ORDER INSECTIVORA** (p. 45).

  6b. Canines larger than incisors; zygoma strongly developed; eyes large and always with external openings.......................**ORDER CARNIVORA** (p. 354).

CHAPTER 7

ORDER DIDELPHIMORPHIA—OPOSSUMS

For many years, the metatherian mammals that included New World opossums with related Australian and South American forms were classified in the order Marsupialia. Although the group contained <7% as many species as the eutherian orders, it had undergone remarkable functional radiation; only volant and marine forms were not represented. Consequently, several systems for classifying the group to illustrate better the within-group relationships of metatherians were proposed (Anderson and Jones, 1984). Although each system had its proponents, none was entirely satisfactory; consequently, Marsupialia remained in use at the ordinal level with 16 or 17 modern families recognized.

Marshall et al. (1990) developed a new classification from phylogenetic studies of morphology of modern and fossil forms. In this classification, the New World opossums are ascribed to the order Didelphimorphia, one of 10 orders of metatherians in the supercohort Marsupialia. Thus, it remains appropriate to refer to all mammals in this group as "marsupials," an approach we intend to follow herein. The Didelphimorphia, as defined by Marshall et al. (1990), includes only one extant family, Didelphidae, presently restricted to North America and South America, plus the Argentine fossil family †Sparassocynidae.

The marsupials often are referred to as "pouched mammals" because about 50% of them have a marsupium in which the mammae are located (Vaughan, 1986). With or without a marsupium, marsupials have extremely short gestation periods, are born in an exceedingly underdeveloped state, become attached to a mammae through their own efforts, and undergo an obligate period of extrauterine development that, combined with the gestation period, approximates the gestation period plus the period before weaning of eutherian mammals of equal size. Marsupials have either a choriovitelline placenta or a chorioallantoic placenta lacking villi, and most embryonic nourishment is by "uterine milk," a substance secreted by the uterine wall. By the common, but in our minds regretable, practice of referring to eutherian mammals as "placental mammals," the presence of these features in marsupials is ignored.

FAMILY DIDELPHIDAE—OPOSSUMS

Sixty-three species in 15 genera are included in the family; only one species occurs north of Mexico (Wilson and Reeder, 1993a). Fossil didelphids are known from North America, South America, Antarctica, Europe, western Asia, and northern Africa (Dawson and Kirshtalka, 1984; Marshall et al., 1990; Simons and Bown, 1984). Although thought by some to have originated in North America in the Cretaceous ≅100 million years ago (Dawson and Krishtalka, 1984; Marshall et al., 1990), marsupials currently are represented in Oregon by only an introduced species considered to have invaded North America from South America in the mid-Pleistocene (Marshall, 1984).

Didelphis virginiana Kerr

Virginia Opossum

1792. *Didelphis virginiana* Kerr, p. 193.

Description.—The Virginia opossum is a cat-sized mammal with a pointed nose; unfurred, black, leathery ears with white edges; beady eyes; a hind foot with an opposable hallux; and a naked, scaly, and prehensile tail (Plate I). Males usually average larger ($\overline{X} = 3.4$ kg, $n = 52$) than females ($\overline{X} = 2.4$ kg, $n = 18$) in Oregon (Hopkins and Forbes, 1979) and elsewhere (Hamilton, 1958b); the largest male in Oregon weighed 5.3 kg (Hopkins and Forbes, 1979), the largest female 4.4 kg (OSUFW 7064). Means and ranges of external and cranial measurements for Oregon opossums are presented in Table 7-1. The skull (Fig. 7-1) is characterized by its small braincase, 50 teeth (more than any other North American mammal; Table 1-1), posteriorly flared nasal bones, and inturned (inflected) angle of the jaw. Older individuals have prominent and rugose sagittal and lambdoidal crests.

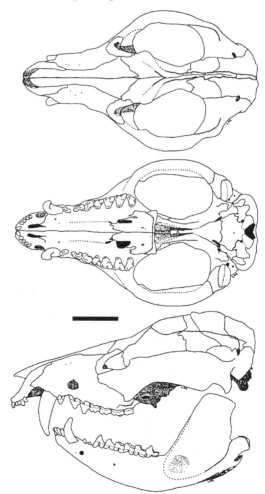

Fig. 7-1. Dorsal, ventral, and lateral views of the cranium, and lateral view of the dentary of the Virginia opossum, *Didelphis virginiana* (OSUFW 1293). Scale bar equals 25 mm.

Table 7-1.—*Means and ranges (in parentheses) of measurements of skin and skull characters for female and male Virginia opossums* (Didelphis virginiana) *from Oregon. Skin characters were recorded from specimen tags; skull characters were measured to the nearest 0.01 mm.*

Sex	n	Total length	Tail length	Hindfoot length	Ear length	Mass (Kg)	Occipitonasal length	Nasal length	Length of maxillary toothrow	Zygomatic breadth	Breadth of braincase	Breadth of molariform toothrow	Skull depth	Length of mandible
♀	7	741[a]	277[a]	64[a]	56[a]	2.00[b]	112.65	53.96	43.60	62.03	31.89	34.25	28.67	94.40
		(684–769)	(195–325)	(54–72)	(54–58)	(1.30–2.70)	(100.76–125.87)	(51.24–59.59)	(39.51–46.81)	(55.30–66.16)	(26.36–34.76)	(31.55–36.89)	(26.97–31.20)	(83.32–104.75)
♂	6	774[c]	295[c]	69[c]	48[c]	3.70[c]	116.87	56.59	41.36	66.95	32.98	32.66	30.24	96.82
		(749–800)	(254–330)	(59–75)	(47–50)	(2.70–5.10)	(101.82–138.49)	(49.99–66.11)	(35.59–47.89)	(52.20–83.37)	(24.27–42.27)	(26.06–38.18)	(26.25–35.39)	(84.22–113.61)

[a]Sample size reduced by 2. [b]Sample size reduced by 5. [c]Sample size reduced by 3.

The pelage consists of woolly underfur overlain sparsely with coarse guard hairs. The common color is light grayish over the head and body with nearly black legs, shoulders, hips, and base of tail; black, cinnamon, and white opossums are known to occur (Hartman, 1952). The female has an anteriorly opening abdominal pouch (marsupium) containing 9–17 (commonly 12–13) mammae (Hamilton, 1958*b*). In males, the scrotum is anterior to the bifurcate (forked) penis; in females, the reproductive tract is bifid (vagina and uterus double). Both sexes have an epipubic bone projecting anteriorly from each pubis; the function of these bones is debated (White, 1984).

Distribution.—The Virginia opossum is broadly distributed in North America east of the Rocky Mountains and in Central America (Fig. 7-2); it was introduced into western North America. The first introduction in Oregon was near

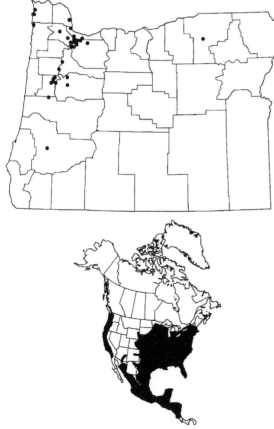

Fig. 7-2. Localities in Oregon from which museum specimens of the Virginia opossum, *Didelphis virginiana*, were examined (listed in the APPENDIX, p. 516), and the distribution of the Virginia opossum in North America (range map redrawn from Hall, 1981:7, map 2, with modifications.)

McKay Creek, Umatilla Co., between 1910 and 1921 (Jewett and Dobyns, 1929). A population apparently was established there as trappers reported catching ≥12 animals during the 1927–1928 trapping season and a specimen collected in the area in 1928 is on deposit at the National Museum of Natural History (USNM 248510). Bailey (1936:393) considered the species "well established and thriving" in that area. However, the population in Umatilla Co. apparently was extirpated sometime during the 1930s (Kebbe, 1955*b*).

Populations of *D. virginiana* were established in northwestern Oregon apparently from releases of animals brought to the state as pets or novelties (Kebbe, 1955*b*; Oregon State Game Commission, 1946*b*). By 1946, the species was established along the lower Columbia, Nehalem, and Young's rivers (Oregon State Game Commission, 1946*b*). Although a Virginia opossum from an unexplained source was killed near Salem (Marion Co.) in 1948, Clatsop Co. and nearby regions in adjoining counties were considered the "only area in the state now known to have opossums" (Kebbe, 1955*b*:42). Hammer (1966:8) reported the species south to the Benton-Lane county line in 1965 and east to "nine miles east of Scio, Linn County, in 1963." Populations now are established throughout the Willamette Valley and other interior valleys, and along the entire Pacific Coast (Fig. 7-2). E. W. Hammer (pers. comm., 24 March 1968) reported examining a specimen killed on U. S. Route 26 in Wasco Co. about 21 km north of the Wasco-Jefferson county line. Also, since the claimed extirpation of the population in Umatilla Co., numerous additional opossums from east of the Cascade Range were reported taken for fur (Table 7-2).

Geographic Variation.—No geographic variation is recognized among populations in Oregon; all are considered *D. v. virginiana* (Gardner, 1973).

Habitat and Density.—In its native range, *D. virginiana* attains its greatest density in wooded areas dissected by small streams and interspersed with cultivated fields (Fitch and Sandidge, 1953; Hartman, 1952; Reynolds, 1945). In these situations, densities range from 2–4/km² where most of the land is under cultivation (Verts, 1963; Wiseman and Hendrickson, 1950) to >100/km² in rich bottomland woods (Holmes and Sanderson, 1965; Sanderson, 1961) and tropical second-growth forests (Fleming, 1972); the greatest density recorded was 235/km² (Holmes and Sanderson, 1965). Small streams, forest communities in various seral stages, and agricultural lands planted to a variety of crops are typical of many habitats occupied by Virginia opossums in Oregon. Thus, the species probably occupies habitats in Oregon similar to those occupied in the native range. In Corvallis, Benton Co., the Virginia opossum used low-density residential and riverfront areas more than expected on the basis of availability; high-density residential, commercial, and industrial areas less than expected;

Table 7-2.—Reported annual harvest of opossums (Didelphis virginiana) in Oregon by county, 1969–1996. Data provided provided by Oregon Department of Fish and Wildlife.

County	Trapping season																										
	1969–1970	1970–1971	1971–1972	1972–1973	1973–1974	1974–1975	1975–1976	1976–1977	1977–1978	1978–1979	1979–1980	1980–1981	1981–1982	1982–1983	1983–1984	1984–1985	1985–1986	1986–1987	1987–1988	1988–1989	1989–1990	1990–1991	1991–1992	1992–1993	1993–1994	1994–1995	1995–1996
Baker	1																										
Benton	43	8	32	81	103	73	176	171	304	237	395	701	251	363	277	364	325	189	164	76	63	108	171	81	68	148	9
Clackamas	34	74	45	124	112	30	57	33	456	336	433	572	361	377	452	295	211	164	150	48	17	17	35	39	84	45	6
Clatsop	32	26	4	7	9	16	28	54	50	83	70	118	67	62	59	32	15	72	18	3	9		168	8		8	
Columbia		51	9	20	97	121	118	133	298	288	208	149	182	211	217	146	303	107	163	119	115	4	163	7	39	5	3
Coos							2	16	41	38	102	41		174	85	150	77	37	46	28	11	11	26	15	11	40	
Crook																											
Curry													3	2		5	2	10	14	6			4	2	4	1	
Deschutes													14		2	2			10								
Douglas	1	2				5	22	30	47	243	377	618	511	561	881	616	364	359	110	122	156		234	135	71	99	45
Gilliam																											
Grant																											
Harney																											
Hood River				5	7	14	3		27	19		40		15				11	19	34	5		10	12	30	95	
Jackson													9		7	7	9	19	56	72	57	85	105	98	66	56	12
Jefferson									1	1																	
Josephine														6	14	42	43	81	207	131	76	151	128	103	147	89	6
Klamath									8														3	3	1	1	
Lake													4														
Lane		13	23	30	137	171	341	496	637	947	1,192	1,273	678	968	1,176	735	668	369	305	196	138	108	267	190	122	85	77
Lincoln		18	11	41	58	61	130	103	223	160	235	110	39	99	57	24	96	85	106	40	16	7	18	5	2	8	3
Linn	131	24	108	175	334	441	454	588	750	635	775	628	457	570	746	597	505	492	369	220	145	63	123	177	157	146	7
Malheur				5																							
Marion	75	67	99	72	64	232	305	511	741	742	707	445	692	558	753	290	324	323	69	46	140	136	237	125	81		25
Morrow																											
Multnomah	39	38	23	29	19	32	24	39	106	77	165	114	101	110	60	78	58	39	53	26	3	17	9	21	2	13	
Polk	17	1	4	13	23	57	56	101	192	157	338	263	75	202	80	415	190	175	316	56	28	30	12	24	7	4	6
Sherman														1													
Tillamook	13	26	63	77	66	133	130	178	149	201	157	86	47	113	72	47	50	82	23	9	18	15	37	17	20		
Umatilla														4	1				18	10	5	4	9	11	10		
Union																											
Wallowa																											
Wasco						2	6	2	1	2				40	8	19		11	11	2	1	2	18	8	6	15	4
Washington	152	7	55	14	70	156	170	202	252	323	258	256	243	621	286	248	167	81	93	14	22	15	39	25	8	32	
Wheeler																											
Yamhill	72	44	30	61	73	42	38	75	324	354	333	285	203	202		578	191	184	231	44	22	10	42	16	31	27	
Totals	610	399	383	766	1,197	1,340	1,972	2,477	4,356	4,584	5,557	5,900	3,962	5,282	4,962	5,453	3,822	2,866	3,143	1,350	909	956	1,473	1,252	1,009	1,028	207
Trappers reporting[a]		721	773	847	1,102	1,350	1,377	1,755	2,746	2,899	3,216	3,373	4,030	5,069	4,271	3,104	3,575	3,397	3,833	1,985	1,536	1,228	1,390	1,454	1,293	1,464	687
Average price/pelt	$0.32	$0.61	$0.49	$0.50	$1.59	$1.53	$1.65	$2.14	$2.06	$2.96	$4.08	$2.17	$2.39	$2.27	$2.02	$1.69	$1.16	$2.22	$1.21	$2.69	$0.96	$1.50	$1.57	$1.06	$2.45	$1.42	$2.47

[a]Not all sought opossums and not all were successful.

and medium-density residential and agricultural and park areas about equal to that expected (Meier, 1983). Although no quantitative estimates of density for the Virginia opossum in Oregon are available, we suspect that greatest densities are attained in urban areas.

Diet.—The Virginia opossum is a "highly opportunistic omnivore" both in Oregon (Hopkins and Forbes, 1980:26) and in its native range (McManus, 1974). In Portland, Multnomah Co., foods of animal origin composed 52.9% of the stomach contents and occurred in 96.9% of 64 stomachs containing food remains; there was relatively little seasonal variation. Mammals (especially opossum flesh), slugs and snails (Gastropoda), and earthworms (Oligochaeta) contributed most of the foods of animal origin (Hopkins and Forbes, 1980). Foods of plant origin composed 29.1% of the stomach contents and occurred in all 64 stomachs containing food remains; however, plant material constituted more of the diet in summer and autumn (Hopkins and Forbes, 1980). Occurrence in the diet of onions, apples, grapes, and other vegetables and fruits, in addition to pilfered pet food and numerous inert items doubtlessly is related to the urban setting. However, the relatively small contribution (9% by volume) of garbage is surprising. Virginia opossums are too slow (maximum speed 7.6 km/h—McManus, 1970) to be efficient predators; most foods of mammalian and avian origin likely are consumed as carrion or consist of nestlings. However, the diet of the Virginia opossum in rural areas of Oregon has not been evaluated; hence the impact that this immigrant species has on populations of ground-nesting birds and mammals in this state is unknown.

Reproduction, Ontogeny, and Mortality.—Adult females produce an average of about 22 young/litter (Hartman, 1952) after a gestation period of only 12.5–13 days (McCrady, 1938). The blind, honeybee-sized ($\overline{X} = 0.16$ g), embryo-like young move from the vaginal opening to the pouch entirely by their own efforts (Hartman, 1952); after entering the pouch, those successful in locating an unoccupied mammae begin to suckle and continue development. The nipple swells within the mouth of the suckling young preventing detachment during the 52–74-day ($\overline{X} = 62$ day) development period (Hartman, 1928). A full complement of young in the pouch is rare as most authorities report an average of 6.8–8.9 pouch young (McManus, 1974). In Portland, Multnomah Co., five females had an average of 8.2 (range, 5–11) young in their pouches (Hopkins and Forbes, 1979) and in Corvallis 11 females had an average of 8.0 (range, 6–9) pouch young (Meier, 1983). Young are weaned at 95–105 days after birth although some solid food is eaten before weaning; dispersal of young occurs about 4 months after birth (Hartman, 1952). Litters are produced in two peaks in most regions (McManus, 1974), but three peaks of breeding were observed in Portland, Multnomah Co.; however, based on ranges in estimated dates of conception, breeding in the region may occur in any month of the year (Hopkins and Forbes, 1979).

Maximum reported longevity is ≥7 years (McManus, 1974), but the average life expectancy of 1.33 years (with population turnover in 4.8 years) in Ohio (Petrides, 1949) was considered "a little higher" than in Maryland (Llewellyn and Dale, 1964:121). No comparable information is available for populations in Oregon. Young individuals may be preyed upon by raptors (Fitch and Shirer, 1970) and the larger carnivores must take opossums occasionally, but few predators seem to prey on opossums consistently (McManus, 1974).

Virginia opossums commonly fall victim to automobiles and, from our casual observations, we suggest that, upon becoming independent, this is the greatest cause of mortality in Oregon. Virginia opossums contract a variety of zoonotic diseases such as histoplasmosis, tularemia, leptospirosis, murine typhus, and Rocky Mountain spotted fever where those diseases are endemic. However, they seem nearly refractory to equine encephalomyelitis and rabies by inoculation (Barr, 1963), but they readily contracted rabies by aerosol transmission in bat caves (Constantine, 1967).

Habits.—Virginia opossums seem to be rather dull witted; they amble about their affairs audibly sniffing to detect food, mates, or potential enemies. Although sufficiently agile to travel along twigs as small as 1 cm in diameter, they possess neither the slyness nor quickness to catch squirrels (*Tamiasciurus hudsonicus*) and chipmunks (*Tamias striatus*) housed in the same cage. Almost all intraspecific encounters are either agonistic or sexual (McManus, 1970); except for adult male–adult female pairs in which individuals seem to become tolerant of each other through mutual avoidance (except when the female becomes receptive), Virginia opossums caged together kill and devour the weakest until only one remains.

When approached too closely, Virginia opossums may freeze or cower, or they may crouch, bare their teeth, hiss, growl, screech, and secrete a greenish liquid from anal glands, flatulate, and defecate in attempts to dissuade the attacker (Francq, 1969; McManus, 1970). If the attacker persists, particularly if the opossum is grabbed and shaken, it may fall on its side, arch its body, open its mouth, allowing its tongue to loll out, and otherwise appear to fall into a catatonic trance. After several minutes, it slowly regains its feet and attempts to escape by slowly increasing the speed of its movements. The precise nature of feigned death in opossums is not understood fully, but there are no changes in patterns of electrical potentials produced either by the heart (electrocardiograph—Francq, 1970) or by the brain (electroencephalograph—Norton et al., 1964). McManus (1970) postulated that the catatonia evolved as a mechanism to reduce mortality during interspecific encounters.

Virginia opossums are active nocturnally; in Corvallis they became active for 2–3 h after sunset and remained active for 3–8.8 h with longer active periods during winter (Meier, 1983). However, they were not active at temperatures ≤3°C or during exceptionally heavy rains. During warm periods following intense cold, Virginia opossums commonly abandon their nocturnal habits and forage during midday (Fitch and Shirer, 1970). Nevertheless, they maintain a relatively constant body temperature (Dills and Manganiello, 1973) and do not hibernate (McManus, 1969).

Virginia opossums more commonly use underground dens when they are available, particularly those in rocky outcrops (Reynolds, 1945; Sandidge, 1953; Wiseman and Hendrickson, 1950; Yeager, 1936), but hollow logs and trees, brush piles, squirrel dreys, bramble tangles, straw piles, barns and outbuildings, and even the frozen carcasses of livestock and big game animals may be used. Nesting material may be transported to dens securely tethered in the prehensile tail (Layne, 1951; Pray, 1921; Smith, 1941). In Corvallis both natural den sites and man-made structures were used by Virginia opossums (Meier, 1983).

In many regions, Virginia opossums seem to have relatively poorly defined and temporally unstable home ranges

(Fitch and Sandidge, 1953; Fitch and Shirer, 1970; Gillette, 1980; Reynolds, 1945) that tend to be 2.7–2.9 times longer than wide (Hunsaker, 1977). Home-range areas calculated by a variety of methods ranged from 4.7 to 78.6 ha (Meier, 1983), with the larger areas thought to be in poorer-quality habitat and to include individuals radiotracked or recaptured for longer periods of time. In Corvallis, home-range areas averaged 21.1 ha for males and 4.9 ha for females (Meier, 1983). Home ranges in the urban environment were much less elongate (1.7 times longer than wide) and tended to be more stable (Meier, 1983); the former characteristic may be related to the more homogeneous habitat of the urban setting and the latter to the large proportion (nine of 17) of individuals radiotracked for relatively short periods.

Remarks.—Apparently because of the unique morphology of the reproductive organs, the presence of a marsupium, and the attachment of young to the nipple during a long period of postnatal development, a volume of mythology and folklore about the Virginia opossum has arisen (Hartman, 1952). Some of the most persistent myths are:

The bifurcate penis seems to be the source of the misconception that the female is bred through the nose and that she blows the spermatozoa either into her vaginal orifice or into her pouch, where fertilization and all development occur.

Because the young are attached to the swollen nipple so tightly, a common belief in some regions is that the young arise by "budding" off the female in a manner similar to the asexual reproduction in some coelenterates.

In regions where sexual reproduction and birth of young are recognized in Virginia opossums, a frequent misconception involves the mode by which young get to the pouch. Some find it incomprehensible that such tiny, poorly developed young could make such an extensive excursion without assistance. Thus, some believe the female picks up the honeybee-sized young with her teeth and attaches them, one-by-one, to her nipples.

Young opossums able to leave the pouch, but not yet independent of maternal care, may ride on the female's back. A novel idea, the origin of which is obscure, is that the female holds her tail over her back and the young grasp it with their own prehensile tails for balance. Although artwork depicting this mother-young interaction is reproduced in several 19th-century books on natural history, no photographs or firsthand accounts of the behavior occur in the scientific literature. The myth persists in Oregon; we encountered a preparator arranging mounted opossums in this posture for a natural-history exhibit.

CHAPTER 8

ORDER INSECTIVORA—SHREWS, MOLES, AND SHREW-MOLE

The insectivores, represented in Oregon only by shrews (Soricidae) and moles (Talpidae), elsewhere in the world include tenrecs (Tenricidae), hedgehogs (Erinaceidae), golden moles (Chrysochloridae), and solenodons (Solenodontidae). The order formerly included the elephant shrews (Macroscelididae) and tree shrews (Tupaiidae), now each placed in a separate order, Macroscelidea and Scandentia, respectively (Wilson and Reeder, 1993*a*). The group includes 428 species and is distributed worldwide except for Australia and most of South America. Fossils considered to be the earliest insectivores are from the late Cretaceous of North America. Both shrews and moles are thought to have arisen from a fossil group of mid-Paleocene–late Eocene age in North America; the earliest fossil shrews are from the Eocene of North America and the earliest moles are from the late Eocene of the Isle of Wight (Dawson and Krishtalka, 1984). Insectivores currently are represented in North America by 33 species of shrews and seven species of moles (Wilson and Reeder, 1993*a*). Ten species of shrews and four species of moles currently are recognized in Oregon.

KEY TO THE INSECTIVORA OF OREGON

1a. Pinnae present; zygomata absent; I1 bicuspid and longer than wide in ventral view; condyloid process double faceted; anterior teeth tipped with red; 32 teeth..
....2 (**Soricidae**)
1b. Pinnae absent; zygomata weak but present; I1 unicuspid and of equal length and witdth in ventral view; condyloid process single faceted; teeth white; ≥36 teeth..
....11 (**Talpidae**)
2a. Pelage usually blackish (some with frosting of silver hairs); U1 and U2 longer than wide in ventral view..
....3
2b. Pelage usually brownish; U1 and U2 not longer than wide in ventral view (usually squarish)..........................5
3a. Venter nearly as dark as dorsum except tail white on underside (sharply bicolored); hind foot <18 mm long; total length usually <145 mm; medial edge of I1s greatly curved; I1s broadly divergent; (least interorbital breadth ≥3.5 mm separates it from *S. vagrans*)..............................
...............*Sorex trowbridgii*, Trowbridge's shrew (p. 60).
3b. Venter either as dark as dorsum with unicolored tail or lighter than dorsum with sharply bicolored tail; hind foot >18 mm long; total length usually >145 mm; medial edge of I1s not greatly curved; I1s appressed or slightly divergent..4
4a. Venter as dark as dorsum; tail uniformly dark; rostrum distinctly downcurved; two posterolingually directed ridges on occlusal surface of p4; length of c1 greater than length of p4.....................................*Sorex bendirii*, Pacific water shrew or marsh shrew (p. 49).
4b. Venter lighter than dorsum; tail sharply bicolored; rostrum not downcurved; one posterolingually directed ridge on occlusal surface of p4; length of c1 less than or equal to length of p4..
.................................*Sorex palustris*, water shrew (p. 55).

5a. U3 equal to or larger than U4.......................................6
5b. U3 distinctly smaller than U4......................................7
6a. Tine present on medial edge of I1; condylobasal length <14.6 mm; maxillary breadth ≤4.2 mm; length of dentary ≤6.5 mm; height of coronoid process ≤3.2 mm; length of c1–m3 ≤4.1 mm; pigment in two–three sections on i1; i1 set at an angle ≤8° from horizontal ramus of the dentary..
.............................*Sorex preblei*, Preble's shrew (p. 57).
6b. No tine on medial edge of I1; condylobasal length >15.8 mm; maxillary breadth ≥5.0 mm; length of dentary ≥6.5 mm; height of coronoid process ≥3.8 mm; length of c1–m3 ≥4.3 mm; pigment in one section on i1; i1 set at an angle ≥12° from horizontal ramus of the dentary..
.........................*Sorex merrami*, Merriam's shrew (p. 51).
7a. No tine or other projection on medial edge of I1; medial edge of I1s abutting for most or all of length of pigmented portion; U5 and P4 not overlapping; zygomatic process of maxillary usually rounded..................
...............................*Sorex sonomae*, fog shrew (p. 58).
7b. Tine or ridge present on medial edge of I1; pigmented portion of I1s separated or divergent; P4 overlapping U5; zygomatic process of maxillary pointed..............8
8a. Medial edges of pigmented portions of I1s parallel but separated by posteriormedial ridge.......................
.........................*Sorex pacificus*, Pacific shrew (p. 54).
8b. Medial edges of pigmented portions of I1s slightly divergent; tine present on medial edge of I................9
9a. Pigmentation on I1 not extending above tine; four or fewer friction pads on second–fourth digits of hind feet; (least interorbital breadth <3.5 mm separates it from *S. trowbridgii*)......*Sorex vagrans*, vagrant shrew (p. 62).
9b. Pigmentation on I1 extending well above tine; five or more friction pads on second–fourth digits of hind feet..10
10a. Pelage dark brown; tine on I1 usually small; height of coronoid process ≥4.3 mm; skull breadth at U4–U4 (1.78388) + greatest length of skull (0.70523) + height of coronoid process (2.55808)–26.9517 is <1.15.....................*Sorex bairdi*, Baird's shrew (p. 46).
10b. Pelage gray-brown (dusty); tine on I1 usually large and robust; height of coronoid process ≤4.2 mm; skull breadth at U4–U4 (1.78388) + greatest length of skull (0.70523) + height of coronoid process (2.55808) –26.9517 is >0.51...
.................*Sorex monticolus*, montane shrew (p. 52).
11a. Total length <130 mm; tail >25% of total length; palm of forefoot longer than wide; hind foot <18 mm long; interior basal projection of M1 and M2 bilobed; 36 teeth..............*Neurotrichus gibbsii*, shrew-mole (p. 65).
11b. Total length >130 mm; tail <25% of total length; palm of forefoot at least as wide as long; hind foot >18 mm long; interior basal projection of M1 and M2 simple; 44 teeth..12
12a. Pelage grayish with a coppery wash; U5 and U6 crowded (usually appressed)..

.........*Scapanus latimanus*, broad-footed mole (p. 68).
12b. Pelage blackish to purplish black (often with fine indistinct whitish spots and occasionally with large orangish splotches); U5 and U6 evenly spaced as other unicuspids...13
13a. Total length usually >200 mm; hind foot >24 mm; total length of skull ≥40 mm; sublacrimal ridge prominent... ....*Scapanus townsendii*, Townsend's mole (p. 71).
13b. Total length usually <200 mm; hind foot <24 mm; total length of skull <40 mm; sublacrimal ridge indistinct................*Scapanus orarius*, coast mole (p. 69).

FAMILY SORICIDAE—SHREWS

Soricids are distributed worldwide except for Australia and most of South America. They are represented in Oregon by 10 species of long-tailed shrews of the genus *Sorex;* two species are endemic to Oregon. All are small (less than half the size of most species of mice), blackish or brownish mammals with minute eyes; 32 red-tipped teeth; bicuspid first upper incisors; first upper incisors longer (anteroposterior) than wide; first lower incisors extending almost straight forward; remaining upper incisors, canines, and premolars (except the fourth) unicuspid (commonly denoted by "U"); dilambdodont (W-shaped) upper molars, the posteriormost the smallest; and double-faceted condyloid processes. They also have a pointed, somewhat proboscis-like nose; a subterminal mouth; and a tail >50% of the length of the head and body. The delicate, jewel-like skull has no zygomata and the tympanic bone is a simple ring (Fig. 8-1). All are pentadactyl and plantigrade. Adult males possess large, externally visible flank glands sparsely covered with short bristles; those of females are not visible externally and the glands are not developed in immatures of either sex.

Shrews commonly enter and become entrapped in discarded soft-drink and beer bottles (Gerard and Feldhamer, 1990; Pagels and French, 1987), there to die of starvation. Sometimes skeletons of several shrews are found per bottle. Pitfalls, constructed by burying coffee cans or plastic buckets and connecting two or more sites with leads of hardware cloth or plastic sheeting, probably are the most effective devices for capturing shrews.

Morphometric data for Oregon shrews are presented in Table 8-1 and reproductive data for species of shrews occurring in the state are presented in Table 8-2.

Sorex bairdi Merriam
Baird's Shrew

1895*a. Sorex bairdi* Merriam, 10:77.
1918. *Sorex obscurus bairdi* Jackson, 31:127.
1918. *Sorex obscurus permiliensis* Jackson, 31:128.
1955. *Sorex vagrans bairdi:* Findley, 9:35.
1955. *Sorex vagrans permiliensis:* Findley, 9:36.
1977. *Sorex monticolus bairdi:* Hennings and Hoffmann, 68:12.
1977. *Sorex monticolus permiliensis:* Hennings and Hoffmann, 68:14.
1990. *Sorex bairdii bairdii:* Carraway, 32:39.
1990. *Sorex bairdii permiliensis:* Carraway, 32:40.

Description.—*Sorex bairdi* is a medium-sized shrew (Table 8-1) in which the third unicuspid is smaller than the fourth (Fig. 8-1) and the level of pigmentation on the anterior face of the first upper incisor extends above the level of the small anteromedial median tine. The tine is usually smaller and more delicate than that on sympatric

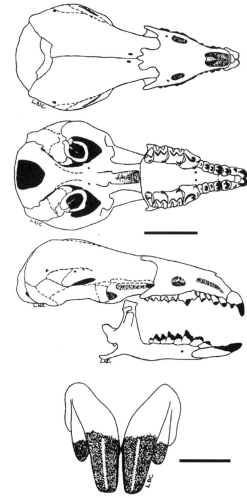

Fig. 8-1. Dorsal, ventral, and lateral views of the cranium (scale bar equals 6 mm); lateral view of the dentary; and anterior view of I1 (scale bar equals 1 mm) of Baird's shrew, *Sorex bairdi* (OSUFW 4832).

and parapatric *S. monticolus* (Carraway, 1990). Alexander (1996) was able to separate 90.6% of sympatric and parapatric *S. bairdi* and *S. monticolus* by use of the following formula:

Score = breadth at U4–U4 (1.78388) + greatest length of skull (0.70523) + height of coronoid process (2.55808) – 26.9517

where, if score <0.51 = *S. monticolus* or if score >1.15 = *S. bairdi*. Individuals with scores between these values could not be assigned to either taxon. By use of a more complex formula involving 12 cranial characters 94.3% were classified correctly (Alexander, 1996).

Pelage color is discussed by subspecies.

Distribution.—*Sorex bairdi* is endemic to Oregon. It occurs in the Coast Range from the Pacific Ocean east to Portland, Multnomah Co., and south of the Columbia River and south to Corvallis, Benton Co., and Reed, Lane Co. It also occurs along the west slope of the Cascade Range from the Columbia River south to central Lane Co. (Fig. 8-2).

Geographic Variation.—Two subspecies are recognized currently (Carraway, 1990): *S. b. bairdi* in the Coast Range and *S. b. permiliensis* on the west slope of the Cascade Range. The former is light medium-brown to dark medium-brown on the dorsum and pinkish gray on the venter; bases of hairs are neutral dark-gray on the dorsum and neutral very-

Table 8-1.—*Means (±SE), ranges (in parentheses), and CVs of measurements (in mm) of skull and skin[a] characters for female and male Soricidae from regions in Oregon. Skin characters were recorded from specimen tags; skull characters were measured to the nearest 0.01 mm. SE and CV not provided if* n <10.

Species and region	Sex	n	Total length	Tail length	Hindfoot length	Mass (g)	Greatest length of skull	Maxillary breadth	Least interorbital breadth	Cranial breadth	Length of maxillary toothrow	Length of mandible	Length of mandibular toothrow
Sorex bairdi													
Coast Range	♀	22	126 ± 1.2 (112–136) 4.6	54 ± 0.8 (48–60) 6.8	15 ± 0.2 (13–18) 6.7	8.34 ± 0.49[b] (5.9–12.3) 23.85	18.49 ± 0.09 (17.63–19.72) 2.52	5.39 ± 0.08 (3.86–5.96) 7.32	3.75 ± 0.04 (3.41–4.19) 5.13	9.09 ± 0.05 (8.51–9.57) 2.37	6.69 ± 0.06 (6.16–7.27) 3.89	8.59 ± 0.06 (7.98–9.05) 3.14	5.35 ± 0.04 (5.11–5.69) 3.15
	♂	29	125 ± 1.4 (103–135) 6.2	54 ± 0.9 (44–67) 9.3	15 ± 0.2 (12–17) 8.7	8.56 ± 0.25[c] (6.9–11.5) 12.64	18.46 ± 0.14 (17.56–20.72) 4.01	5.35 ± 0.07 (3.77–6.14) 7.58	3.76 ± 0.03[d] (3.51–4.28) 4.76	9.29 ± 0.11 (8.64–12.00) 6.60	6.67 ± 0.07 (6.12–7.77) 5.33	8.62 ± 0.19 (7.52–13.50) 11.89	5.36 ± 0.04[d] (5.10–6.04) 4.07
Cascade Range	♀	30	117 ± 0.9 (107–128) 4.1	51 ± 0.5 (42–56) 5.7	14 ± 0.1 (13–15) 3.9		18.85 ± 0.06 (18.23–19.40) 1.65	5.46 ± 0.03 (5.16–5.89) 2.90	3.74 ± 0.02 (3.40–4.05) 3.38	9.24 ± 0.03 (8.94–9.44) 1.56	4.68 ± 0.03 (4.30–4.95) 3.06	8.14 ± 0.04 (7.53–8.73) 3.02	5.38 ± 0.03 (5.02–5.73) 2.55
	♂	30	118 ± 0.9 (102–125) 4.2	51 ± 0.6 (42–57) 6.1	15 ± 0.1[d] (13–15) 3.8	5.80	18.97 ± 0.07 (18.18–19.57) 1.94	5.45 ± 0.03 (5.09–5.95) 3.04	3.73 ± 0.02 (3.50–4.00) 3.21	9.27 ± 0.04 (8.94–9.98) 2.58	4.79 ± 0.06 (4.54–6.56) 7.31	8.13 ± 0.04 (7.82–8.75) 2.65	5.40 ± 0.02 (5.11–5.61) 2.41
Sorex bendirii													
East slope Cascade Range	♂	1	155	71	20		20.86	6.40	4.23	10.49		9.63	5.93
West of crest of Cascade Range	♀	30	159 ± 1.7 (143–180) 5.8	70 ± 1.0 (59–81) 8.1	21 ± 0.1 (20–22) 2.9	17.25 ± 1.17[e] (11.7–23.2) 21.47	23.45 ± 0.09 (22.45–24.23) 2.07	7.06 ± 0.05 (6.50–7.58) 3.92	4.56 ± 0.04 (4.22–5.14) 4.97	11.19 ± 0.05 (10.55–11.78) 2.57	5.45 ± 0.02[b] (5.18–5.60) 2.20	10.42 ± 0.08 (9.60–11.30) 4.42	6.61 ± 0.04 (6.05–7.09) 3.38
	♂	30	165 ± 1.7 (146–181) 5.5	73 ± 0.9 (63–81) 6.6	21 ± 0.2 (19–23) 4.4	18.91 ± 1.06[f] (11.2–23.9) 20.18	23.57 ± 0.08 (22.29–24.22) 1.88	7.03 ± 0.05 (6.45–7.55) 3.90	4.59 ± 0.06 (3.96–5.34) 6.61	11.31 ± 0.06 (10.56–12.04) 2.93	5.45 ± 0.03[c] (5.10–5.66) 2.80	10.56 ± 0.09 (9.74–11.96) 4.97	6.57 ± 0.04 (6.10–6.88) 2.94
Sorex merriami	♀	3	89 (85–94)	34 (30–37)	12 (11–12)	4.01[d]	16.06 (15.80–16.33)	5.25 (5.04–5.57)	3.66 (3.40–4.01)	8.33 (8.25–8.44)	5.91 (5.44–6.20)	7.16 (7.05–7.26)	4.71 (4.69–4.73)
	♂	9	89[d] (82–94)	34[d] (32–36)	11 (11–12)	3.19 (2.4–4.3)	15.72 (15.33–16.26)	4.98 (4.40–5.27)	3.65 (3.36–3.84)	8.14 (7.71–8.56)	5.79 (5.43–6.13)	7.06 (6.46–7.60)	4.62 (4.41–4.75)
Sorex monticolus													
Hood River County	♀	30	112 ± 1.2 (102–127) 5.8	49 ± 0.6 (41–55) 6.3	14 ± 0.3 (13–16) 4.9	6.17 ± 0.21 (3.6–8.8) 18.33	17.77 ± 0.06 (17.21–18.52) 1.83	5.06 ± 0.03 (4.77–5.41) 2.97	3.73 ± 0.03 (3.38–4.15) 4.48	8.86 ± 0.03 (8.47–9.30) 2.04	6.56 ± 0.04 (6.09–7.00) 3.29	7.65 ± 0.06 (7.16–8.26) 3.99	5.14 ± 0.02 (4.93–5.42) 2.29
	♂	31	113 ± 0.8 (105–120) 3.9	49 ± 0.4 (45–54) 4.97	14 ± 0.1 (12–15) 4.8	5.95 ± 0.17 (4.9–9.4) 15.09	17.58 ± 0.06 (17.25–18.61) 1.97	5.04 ± 0.03 (4.61–5.40) 3.70	3.75 ± 0.04 (3.44–4.28) 5.32	8.87 ± 0.03 (8.50–9.17) 1.89	6.61 ± 0.03 (6.20–6.87) 2.95	7.64 ± 0.05 (7.10–8.07) 3.67	5.12 ± 0.02 (4.87–5.33) 2.19
Eastern Oregon	♀	21	112 ± 1.2 (102–127) 4.1	41 ± 0.5 (35–45) 5.7	12 ± 0.2 (10–13) 6.5	3.59 ± 0.13 (2.8–5.2) 16.67	16.26 ± 0.08 (15.61–16.81) 2.19	4.62 ± 0.04 (4.31–5.00) 3.63	3.23 ± 0.03 (3.01–3.44) 4.39	8.18 ± 0.05 (7.85–8.51) 2.55	5.85 ± 0.06 (5.33–6.28) 4.42	7.25 ± 0.04 (6.84–7.53) 2.69	4.82 ± 0.04 (4.43–5.22) 3.67
	♂	30	98 ± 0.9 (85–106) 5.2	40 ± 0.5 (36–44) 6.4	11 ± 0.1 (10–13) 6.8	3.81 ± 0.18[d] (2.4–6.1) 25.30	16.32 ± 0.06 (15.55–16.93) 2.03	4.64 ± 0.03 (4.31–4.97) 3.18	3.30 ± 0.06 (2.95–4.75) 9.25	8.15 ± 0.04 (7.75–8.55) 2.51	5.88 ± 0.04 (5.34–6.28) 4.05	7.28 ± 0.04 (6.97–7.69) 2.76	4.77 ± 0.04 (4.04–5.12) 4.49
Sorex pacificus													
Coast Range	♀	14	132 ± 1.4 (120–138) 3.9	59 ± 1.0 (52–66) 6.5	16 ± 0.2 (15–18) 4.9	7.68 ± 0.53[g] (5.8–12.8) 24.09	19.84 ± 0.09 (19.07–20.34) 1.63	5.89 ± 0.03 (5.68–6.05) 2.12	4.01 ± 0.04 (3.78–4.32) 4.13	9.58 ± 0.04 (9.32–9.97) 1.62	6.46 ± 0.33 (4.69–7.61) 19.30	9.01 ± 0.10 (8.16–9.68) 4.29	5.83 ± 0.04 (5.56–6.01) 2.30
	♂	19	131 ± 1.5 (118–141) 4.9	60 ± 0.8 (54–69) 6.1	16 ± 0.2 (15–17) 5.1	7.46 ± 0.47[h] (4.7–11.6) 24.49	19.82 ± 0.14 (17.92–20.69) 2.99	5.96 ± 0.04 (5.52–6.29) 3.32	4.07 ± 0.03 (3.75–4.27) 3.19	9.69 ± 0.05 (9.23–10.06) 2.36	6.95 ± 0.25 (4.76–8.00) 15.94	9.10 ± 0.09 (8.14–9.60) 4.46	5.86 ± 0.04 (5.60–6.24) 3.07
Cascade Range	♀	30	118 ± 1.2 (103–130) 5.4	51 ± 0.5 (44–55) 5.5	14 ± 0.1 (13–15) 4.9	5.74[i] (4.7–7.4)	19.05 ± 0.07 (18.23–19.65) 1.95	5.56 ± 0.03 (5.28–5.84) 3.04	3.78 ± 0.03 (3.56–4.10) 3.71	9.38 ± 0.04 (8.92–9.75) 2.23	5.07 ± 0.15 (4.41–7.19) 16.28	8.29 ± 0.05 (7.64–9.07) 3.59	5.43 ± 0.02 (5.24–5.66) 2.11
	♂	30	118 ± 0.9 (104–125) 4.4	51 ± 0.6 (44–58) 6.3	14 ± 0.1 (13–15) 3.8	5.79[i] (5.0–6.9)	18.93 ± 0.08 (17.86–19.91) 2.35	5.55 ± 0.03 (5.20–5.91) 2.99	3.76 ± 0.03 (3.47–4.01) 3.87	9.34 ± 0.03 (8.94–9.75) 2.05	4.84 ± 0.09 (4.47–6.79) 10.57	8.19 ± 0.05 (7.50–8.93) 3.11	5.36 ± 0.02 (5.10–5.66) 2.26

Table 8-1.—*Continued.*

Species and region	Sex	n	Total length	Tail length	Hindfoot length	Mass (g)	Greatest length of skull	Maxillary breadth	Least interorbital breadth	Cranial breadth	Length of maxillary toothrow	Length of mandible	Length of mandibular toothrow
Sorex palustris	♀	21	150 ± 1.7 (130–167) 5.1	70 ± 0.9 (65–79) 6.0	19 ± 0.2 (18–20) 3.7	11.14 ± 1.00[j] (6.5–15.5) 28.48	19.58 ± 0.09 (18.86–20.77) 2.24	5.78 ± 0.05 (5.31–6.15) 3.70	3.75 ± 0.02 (3.57–4.01) 2.91	9.74 ± 0.05 (9.07–10.08) 2.32	7.09 ± 0.05 (6.54–7.79) 3.51	9.05 ± 0.06 (8.53–9.45) 2.82	5.67 ± 0.04 (5.19–5.83) 3.21
	♂	17	152 ± 2.5 (131–170) 6.8	74 ± 0.9 (65–79) 5.8	20 ± 0.2 (19–22) 3.7	12.33 ± 0.95[k] (7.5–17.2) 26.67	19.97 ± 0.09 (19.27–20.68) 2.06	5.88 ± 0.05 (5.55–6.35) 3.52	3.89 ± 0.03 (3.66–4.16) 3.69	9.97 ± 0.05 (9.49–10.34) 2.05	7.17 ± 0.08 (6.73–7.66) 4.53	9.14 ± 0.07 (8.55–9.72) 3.37	5.48 ± 0.08 (4.75–5.79) 6.15
Sorex preblei	♀	9	83 (75–92)	35 (32–37)	11 (10–12)	2.23[k] (2.1–2.3)	14.55 (14.31–14.74)	3.89 (3.70–4.04)	2.73 (2.49–2.89)	7.13 (7.01–7.40)	5.08 (4.84–5.35)	6.39 (6.05–6.64)	4.01 (3.89–4.16)
	♂	15	91 ± 1.9 (75–99) 8.2	36 ± 0.8 (29–40) 8.8	11 ± 0.1 (10–11) 4.8	2.41[l] (1.9–2.75)	14.44 ± 0.13 (13.17–14.86) 3.44	4.01 ± 0.03 (3.79–4.21) 2.69	2.76 ± 0.04 (2.54–3.03) 5.11	7.31 ± 0.08 (6.98–8.30) 4.15	5.09 ± 0.05 (4.72–5.43) 3.64	6.22 ± 0.07 (5.65–6.75) 4.19	3.98 ± 0.03 (3.77–4.13) 2.55
Sorex sonomae Coast Range	♀	30	135 ± 2.1 (105–152) 8.4	61 ± 0.9 (47–68) 8.0	17 ± 0.2 (15–19) 6.1	8.50[m] (6.7–9.7)	21.34 ± 0.14 (19.74–22.46) 3.64	6.51 ± 0.07 (5.68–7.02) 5.57	4.21 ± 0.03 (3.82–4.54) 4.38	10.23 ± 0.07 (9.38–10.76) 3.74	5.35 ± 0.04 (4.76–5.72) 4.36	9.24 ± 0.12 (6.41–10.0) 6.84	6.48 ± 0.15 (5.72–9.85) 12.89
	♂	30	135 ± 2.2 (107–153) 9.1	59 ± 1.1 (45–69) 10.2	17 ± 0.2 (15–18) 6.0	9.37 ± 0.52[n] (6.3–12.8) 21.58	21.54 ± 0.17 (19.34–23.01) 4.25	6.52 ± 0.06 (5.89–7.11) 4.96	4.28 ± 0.04 (3.87–4.64) 5.28	10.40 ± 0.07 (9.49–11.04) 3.48	5.32 ± 0.05 (4.81–5.90) 5.05	9.44 ± 0.08 (8.39–10.41) 4.80	6.19 ± 0.06 (5.43–6.75) 5.69
Cascade Range	♀	30	124 ± 1.3 (112–143) 5.8	53 ± 0.8 (44–66) 8.3	14 ± 0.2 (12–16) 7.0	7.07 ± 0.33[o] (5.0–9.8) 19.07	19.09 ± 0.13 (16.50–20.19) 3.69	5.70 ± 0.04 (5.34–6.09) 3.61	3.89 ± 0.03 (3.53–4.29) 4.59	9.48 ± 0.04 (8.91–9.99) 2.58	4.80 ± 0.04 (4.47–5.52) 4.71	8.36 ± 0.06 (7.64–9.24) 3.79	5.53 ± 0.04 (5.25–6.01) 3.48
	♂	30	123 ± 1.6 (106–145) 7.3	52 ± 0.9 (41–64) 9.5	14 ± 0.2 (12–17) 8.4	8.20 ± 0.46[b] (5.2–12.7) 27.27	19.34 ± 0.09 (18.64–20.20) 2.72	5.72 ± 0.04 (5.18–6.12) 3.99	3.82 ± 0.04 (3.14–4.08) 5.92	9.50 ± 0.05 (8.71–10.00) 2.95	4.81 ± 0.08 (4.16–6.87) 9.17	8.38 ± 0.05 (7.67–8.92) 3.59	5.48 ± 0.04 (4.88–5.79) 3.88
Sorex trowbridgii Lake County	♀	1	105	45	14		17.92	5.06	3.72	8.65	6.28	8.18	5.29
	♂	1	113	48	15		18.31	5.36	3.80	9.12	6.33	8.28	5.34
Remainder of Oregon	♀	30	110 ± 1.2 (100–132) 5.9	51 ± 0.5 (46–58) 5.7	13 ± 0.1 (12–14) 4.9	4.53 ± 0.19 (2.9–6.3) 22.78	17.22 ± 0.06 (16.48–17.95) 1.93	4.83 ± 0.02 (4.63–5.18) 2.34	3.69 ± 0.02 (3.54–3.89) 2.23	8.70 ± 0.04 (8.10–9.09) 2.67	5.47 ± 0.18 (4.19–6.79) 18.18	7.58 ± 0.05 (6.60–8.02) 3.62	5.02 ± 0.02 (4.80–5.60) 2.69
	♂	30	113 ± 1.4 (99–126) 6.6	54 ± 0.6 (48–60) 5.6	13 ± 0.1 (11–14) 5.4	4.57 ± 0.19 (3.0–6.7) 22.48	17.39 ± 0.06 (16.81–18.05) 1.94	4.85 ± 0.02 (4.45–5.02) 2.79	3.71 ± 0.02 (3.45–3.96) 3.31	8.75 ± 0.03 (8.13–9.02) 2.13	5.44 ± 0.17 (4.21–6.47) 17.49	7.70 ± 0.04 (7.20–8.15) 3.09	5.00 ± 0.02 (4.76–5.23) 2.45
Sorex vagrans West of Cascade Range	♀	30	96 ± 1.3 (83–109) 7.2	37 ± 0.4 (32–40) 5.9	12 ± 0.1 (10–12) 5.3	5.14 ± 0.24[h] (3.4–7.3) 24.30	16.36 ± 0.67 (15.60–16.92) 2.23	4.74 ± 0.03 (4.31–5.20) 3.86	3.06 ± 0.02 (2.81–3.30) 4.01	8.11 ± 0.04 (7.67–8.58) 2.77	4.10 ± 0.05 (3.77–5.49) 6.90	7.04 ± 0.05 (6.46–7.43) 3.61	4.63 ± 0.03 (4.22–4.80) 2.99
	♂	31	97 ± 1.1 (86–112) 6.4	39 ± 0.5 (31–44) 7.6	12 ± 0.1 (10–13) 6.7	5.00 ± 0.25[l] (3.6–7.8) 23.65	16.39 ± 0.07 (15.46–17.08) 2.42	4.71 ± 0.04 (4.21–5.20) 4.48	3.13 ± 0.03 (2.81–3.45) 4.97	8.19 ± 0.03 (7.97–8.52) 1.83	4.05 ± 0.04 (3.55–5.07) 6.04	7.04 ± 0.04 (6.68–7.51) 3.20	4.59 ± 0.03 (4.27–4.82) 3.31
East of Cascade Range	♀	30	97 ± 1.1 (85–110) 6.4	39 ± 0.5 (33–44) 6.7	11 ± 0.2 (9–13) 8.9	3.69 ± 0.17[d] (2.3–6.0) 24.19	16.14 ± 0.05 (15.68–16.85) 1.76	4.54 ± 0.01 (4.39–4.70) 1.72	3.12 ± 0.02 (2.90–3.47) 4.23	7.92 ± 0.04 (7.36–8.33) 2.70	5.67 ± 0.04 (5.12–6.19) 4.33	6.97 ± 0.03 (6.53–7.42) 2.75	4.62 ± 0.04 (4.25–5.21) 4.19
	♂	30	99 ± 0.9 (89–107) 4.9	39 ± 0.4 (33–42) 5.7	11 ± 0.2 (9–13) 7.1	4.04 ± 0.18[d] (2.1–6.3) 24.02	16.19 ± 0.03 (15.89–16.58) 1.00	4.54 ± 0.02 (4.25–4.83) 2.79	3.12 ± 0.03 (2.80–3.42) 4.86	7.99 ± 0.03 (7.74–8.25) 1.81	5.69 ± 0.04 (5.29–6.20) 3.96	7.07 ± 0.04 (6.61–7.55) 2.77	4.66 ± 0.02 (4.44–4.95) 2.84

[a] Although measures of tail and ear length were recorded, the data were not included because of extreme variation believed to stem from differences in techniques used to measure these characters.
[b] Sample size reduced by 6. [c] Sample size reduced by 8. [d] Sample size reduced by 1. [e] Sample size reduced by 20. [f] Sample size reduced by 17. [g] Sample size reduced by 2. [h] Sample size reduced by 4. [i] Sample size reduced by 22. [j] Sample size reduced by 11. [k] Sample size reduced by 5. [l] Sample size reduced by 9. [m] Sample size reduced by 21. [n] Sample size reduced by 15. [o] Sample size reduced by 13.

dark-gray on the venter. The tail is indistinctly bicolored, dark brown dorsally and white ventrally. *S. b. permiliensis* is medium brown to dark brown on the dorsum and pinkish light-brown on the venter; bases of hairs are neutral very dark-gray on the dorsum and neutral dark-gray on the venter. The tail is indistinctly bicolored, medium brown to dark brown above and pinkish white below (Carraway, 1990).

Habitat and Density.—In April 1986, we collected two specimens in 48 museum special traps set for 4 days in an open Douglas-fir (*Pseudotsuga menziesii*) forested area with numerous rotting logs in Polk Co. We found no published information on habitats or density that we could assign with certainty to *S. bairdi*.

Diet.—We found no published information on the diet

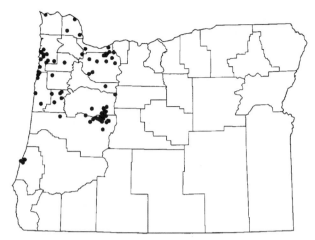

Fig. 8-2. Localities in Oregon from which museum specimens of Baird's shrew, *Sorex bairdi*, were examined (listed in the APPENDIX, p. 551). The species is endemic to Oregon (Carraway, 1990).

of *S. bairdi*. From a slightly higher-than-expected relative bite force, Carraway and Verts (1994*a*) predicted that the species fed on moderately hard prey.

Reproduction, Ontogeny, and Mortality.—We found no published information on these topics concerning *S. bairdi*. We obtained information from 10 pregnant females suggesting that average litter size was midway among those of similar-sized congeners (Table 8-2). It seems likely that those reproducing in July–August were born earlier in the calendar year.

Habits.—*Sorex bairdi* is active diurnally. At about 1430 h, 17 April 1986, while one of us (BJV) baited two other traps to be set at a station, a Baird's shrew darted from beneath a log to lick liver-paste bait from the treadle of a museum special trap just set. It responded to the slightest move-

ment of hands or feet of the observer by dashing back beneath the log, only to return a few seconds later. After several such trips, the other two traps were set; the animal was captured while traps at the next station were being baited. Other species captured on the same trapline with *S. bairdi* were *Cleithronomys californicus* and *Peromyscus maniculatus*.

Remarks.—*Sorex bairdi* was recognized originally as a separate species, then relegated first to a subspecies of *S. obscurus* (=*S. monticolus*) by Jackson (1918) then to one of *S. vagrans* (Findley, 1955), and finally recognized as a separate species on the basis of a multivariate morphometric analysis (Carraway, 1990). The original name applied by Merriam was spelled with one terminal *i*. However, when shrews forming the *obscurus* group were separated from *S. vagrans* and the name *Sorex monticolus* applied to them, Hennings and Hoffmann (1977) were inconsistent in spelling *bairdi*, sometimes using only one and sometimes two terminal *i*'s. When Carraway (1990) reelevated *bairdi* to species level, she mistakenly used two terminal *i*'s. Alexander (1996) corrected the error.

A detailed life-history study of the dark brown woodland shrews in northwestern Oregon is deserved.

Sorex bendirii (Merriam)
Pacific Water or Marsh Shrew

1884*a*. *Atophyrax bendirii* Merriam, 2:217.
1890. *Sorex bendirii*: Dobson, pt. 3, fasc. 1, pl. 23, fig. 17, and explanation.
1895*a*. *Sorex (Atophyrax) bendirii palmeri* Merriam, 10:97.

Description.—*Sorex bendirii* (Plate I) is the largest member of the genus in North America (Table 8-1; Jackson, 1928). It may be distinguished from congeners in Oregon by a blackish pelage on both dorsum and venter and a unicolored tail. The hind foot is fimbriated in young individuals. There is a series of "small, fleshy projections along

Table 8-2.—*Reproduction in some Oregon soricids.*

Species	Number of litters	$\overline{X}$	Range	n	Reproductive season	Basis	Authority	State or province
Sorex bairdi		4.8	4–7	10	Apr.–May; Jul.–Aug.	Various[a]	This study	Oregon
Sorex bendirii		6.0	5–7	2	May	Various[a]	This study	Oregon
		4		1	Mar.	Nestlings	Pattie, 1969	Washington
		7		1	May	Embryos	This study	Oregon
Sorex merriami		6.0	5–7	3	Apr.–May; Jul.	Embryos	Johnson and Clanton, 1954	Washington
Sorex monticolus		7		1	Jul.	Nestlings	Slipp, 1942	Washington
		4		1	Jul.	Embryos	Dalquest, 1948	Washington
		4.5	2–6	6	Jun.–Jul.; Sep.	Embryos	This study	Oregon
Sorex pacificus		4.5	2–7	6	May–Jun.; Aug.	Embryos	This study	Oregon
		4		1		Implant sites	This study	Oregon
Sorex palustris	2–3	5.5	5–6	6	Feb.–Jun.	Embryos	Conaway, 1952	Montana
		5.3	1–8	8	Feb.–Apr.; Jun.–Jul.	Various[b]	Conaway, 1952	Montana
Sorex sonomae		4.1	3–5	12	Mar.; May–Sep.	Embryos	Carraway, 1988	California[c]
		4.3	3–5	8	Apr.–Aug.; Nov.	Embryos	Carraway, 1988	Oregon[c]
		3.2	1–5	26		Implant sites	This study	Oregon
Sorex trowbridgii		5.0	3–6	8	Feb.–May	Embryos	Jameson, 1955	California
		3.9	3–5	28	Feb.–Jun.; Sep.–Oct.	Embryos	Gashwiler, 1976*a*	Oregon
		3.8	3–4	13		Implant sites	Gashwiler, 1976*a*	Oregon
		3.9	3–5	34		Corpora lutea	Gashwiler, 1976*a*	Oregon
		3.8	2–5	9	Mar.–Jul.	Embryos	This study	Oregon
Sorex vagrans	2–3	6.0	2–8	18	Feb.–Jul.; Sep.–Nov.	Embryos	Hooven et al., 1975	Oregon
		5.9	2–8	15	Feb.–Jul.; Sep.–Nov.	Neonates	Hooven et al., 1975	Oregon
		5.4	1–8	34	Mar.–Aug.; Oct.	Embryos	This study	Oregon

[a] Counts of embryos and pigmented sites of implantation.
[b] Counts of blastocysts and pigmented sites of implantation.
[c] Identified in original publication as *Sorex pacificus*, but condylobasal length =20 mm indicates that all are *S. sonomae*.

the outer edge of each nostril" (Maser, 1975:158). The skull is heavy and rugose; the rostrum is distinctly downcurved (Fig. 8-3). Among Oregon shrews, *S. bendirii* is unique in having two posterolingually directed ridges on the occlusal surface of p4 and in having an anteroposterior length of c1 greater than that of p4 (Carraway, 1995).

Distribution.—*Sorex bendirii* occurs (Fig. 8-4) from extreme southwestern British Columbia south through Washington west of the Cascade Range to Oregon, where a branch continues southward along the Pacific Coast into California as far south as the Mendocino-Sonoma county line. In Oregon, it occurs in the northern Cascade Range in Clackamas, Hood River, and Multnomah counties, then west in Clatsop, Columbia, and Washington counties along the Columbia River, and southeasterly from Newport, Lincoln Co., through Benton, Lane, Linn, Jackson, and Kla-

math counties. The species does not occur in the interior valleys of western Oregon.

Geographic Variation.—In Oregon, two geographic races are recognized currently (Hall, 1981; Pattie, 1973): the larger and darker race with the heavier skull (Bailey, 1936) that occurs in the Coast Range is considered *S. b. palmeri*, whereas the lighter and slightly smaller race that occurs on the east and west slopes of the Cascade Range is considered *S. b. bendirii*.

Habitat and Density.—Habitats occupied by *S. bendirii* include alder (*Alnus*) in riparian zones; skunk cabbage (*Lysichitum americanum*) marshes (Maser et al., 1981b); deep, dark, red cedar (*Thuja plicata*) swamps; floating mats of yellow cress (*Rorippa*—Dalquest, 1948), and muddy places in both forests and forest edges (Bailey, 1936). In the Cascade Range, Anthony et al. (1987) captured *S. bendirii* in riparian zones in all three stages (old-growth, mature, and young) of the forest sere sampled. However, *S. bendirii* represented <1% of the total number of small mammals captured and only 6.3% of the shrews captured in May and 3.0% of the shrews captured in August–September. We know of no estimates of density of populations of *S. bendirii*.

Diet.—In Oregon, 24 marsh shrews ate 26 of 72 food items identified in stomach contents of 269 shrews of five species; ≥25% of the food was of aquatic origin. Insect larvae, slugs and snails (Gastropoda), mayfly (Ephemeroptera) naiads,

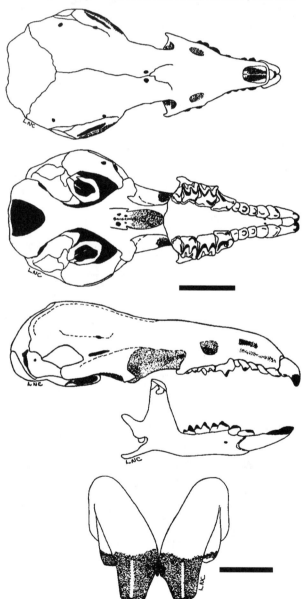

Fig. 8-3. Dorsal, ventral, and lateral views of the cranium (scale bar equals 6 mm); lateral view of the dentary; and anterior view of I1 (scale bar equals 1 mm) of the Pacific water or Marsh shrew, *Sorex bendirii* (OSUFW 6072).

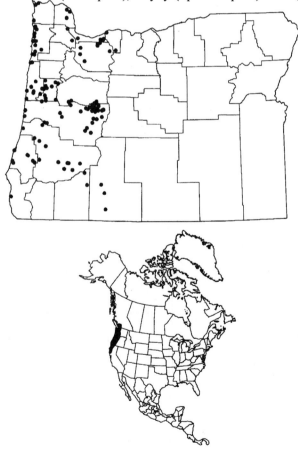

Fig. 8-4. Localities in Oregon from which museum specimens of the Pacific water or Marsh shrew, *Sorex bendirii*, were examined (listed in the APPENDIX, p.551), and the distribution of the Pacific water or Marsh shrew in North America. (Range map redrawn from Hall, 1981:44, map 23, with modifications.)

Table 8-3.—*Percent volume of foods consumed by five species of* Sorex *in western Oregon as determined by analysis of stomach contents. Data from Whitaker and Maser (1976).*

Taxon or category	Sorex trowbridgii	Sorex vagrans	Sorex sonomae[a]	Sorex pacificus[a]	Sorex bendirii
Coleoptera (beetles)	6.5	15.0	4.2	6.0	2.0
Diptera (flies)	3.5	1.9	0.7	0.4	1.2
Hemiptera (true bugs)	6.3	1.3	0.9	1.7	2.6
Hymenoptera (bees & ants)	2.6	2.5		4.1	0.8
Homoptera (leafhoppers)		2.4			
Insect larvae	9.9	28.9	14.8	28.7	41.3
Araneae (spiders) and Phalangida (daddy longlegs)	14.9	5.2	4.0	4.6	8.0
Chilopoda (centipedes)	15.4	0.6	10.7	0.2	1.7
Gastropoda (slugs & snails)	9.4	14.0	28.0	7.0	14.9
Oligochaeta (earthworms)	1.7	3.8	1.8	0.2	7.9
Unidentified invertebrate material	18.6	14.3	6.9	33.8	12.3
Fungi	1.2	6.7	3.7	1.5	0.6
Other material[b]	10.4	6.0	24.0	12.1	6.8
Totals	100.4	100.1	99.7	100.3	100.1

[a] Reclassified according to Carraway, 1990.

[b] Includes other insects, plant materials, insect eggs, and amphibian flesh.

unidentified invertebrates, and earthworms (Oligochaeta) were prey fed upon in greatest amounts (Table 8-3). Relative bite force in *S. bendirii* was lower than expected, causing Carraway and Verts (1994*a*) to predict a relatively soft diet for the species. Pattie (1969) indicated that captive *S. bendirii* usually refused beetles (Coleoptera) and crayfish (Astacidea), but quickly took soft-bodied forms. He reported that earthworms, termites (Isoptera), centipedes (Chilopoda), spiders (Araneae), and sowbugs (Isopoda) were consumed eagerly.

Reproduction, Ontogeny, and Mortality.—We found little information on reproduction (Table 8-2), none on development of young or on mortality.

Habits.—*Sorex bendirii* is truly a "water shrew" despite Bailey's (1936) comment to the contrary. It swims easily both on the surface and while submerged, mostly by alternate strokes of the hind feet; when leaving the water, it literally springs from the surface (Pattie, 1969). It can run on the surface for nearly 5 s. During dives for as long as 3.5 min, *S. bendirii* searched for prey on the substrate largely by use of tactile senses (Pattie, 1969). Lampman (1947) observed a shrew, identified by S. G. Jewett as *S. bendirii* on the basis of the locality, to dive into the water and capture fish (Teleostei). *S. bendirii* leaves the water to eat and to cache excess prey.

In captivity, when disturbed or handled, or when scuffling with a conspecific, *S. bendirii* made a "shrill twittering call" (Pattie, 1969:31). It showed little concern for visual cliffs, commonly falling from elevated surfaces.

Remarks.—Based on distributions of museum specimens in Oregon it appears that published range maps for *S. bendirii* (Hall 1981; Pattie, 1973) do not depict the geographic range of the species realistically. We consider *S. bendirii* a prime candidate for a morphologic, morphometric, and colorimetric analysis of geographic variation, in addition to studies of its ecology and reproduction.

Sorex merriami Dobson
Merriam's Shrew

1890. *Sorex merriami* Dobson, pt. 3, fasc. 1, pl. 23, fig. 6.

Description.—*Sorex merriami*, except for *S. preblei*, is the smallest shrew in Oregon (Table 8-1). This brownish shrew possesses a short, truncate skull (Fig. 8-5); the maxillary breadth is >4.9 mm, the third unicuspid tooth is equal to or larger than the fourth, and the first upper incisors are without tines and are appressed for most of the length of the pigmented area. *S. merriami* is medium dark-brown on the dorsum, pinkish white on the venter; the tail is sharply bicolored in the same tones as the body with dark and light portions about equal. Johnson and Clanton (1954) reported clear-cut summer and winter pelages in Washington, more grayish in winter and more brownish in summer. They found Merriam's shrews undergoing molts to summer pelage in April and to winter pelage in October and November.

Distribution.—The geographic range of *S. merriami* (Fig. 8-6) extends from north-central Washington south through Oregon east of the Cascade Range into eastern California thence southeasterly to southeastern Arizona, and northeastward to central New Mexico, north-central Colorado, extreme western Nebraska and North Dakota, and the eastern half of Montana (Armstrong and Jones, 1971). The only fossil record is for extreme southeastern New Mexico (Kurtén and Anderson, 1980). In Oregon, museum specimens have been collected in Grant, Harney, Lake, and Wasco counties (Fig. 8-6). Gashwiler (1976*b*) reported that one was collected in Deschutes Co., but the specimen was destroyed in a fire.

Geographic Variation.—No geographic variation is recognized in Merriam's shrews in Oregon; all are considered *S. m. merriami* (Armstrong and Jones, 1971).

Habitat and Density.—*Sorex merriami* occupies drier habitats than most congeners. It is reported to be associated with sagebrush (*Artemisia*)-bunchgrass habitats in Washington (Hudson and Bacon, 1956; James, 1953; Johnson and Clanton, 1954), California (Hoffmann, 1955), Colorado (Starrett and Starrett, 1956), Montana (Hooper, 1944), Utah (Osgood, 1909*b*), and Wyoming (Long and Kerfoot, 1963), and in wet meadows in Nebraska (McDaniel, 1967), in mountain-mahogany (*Cercocarpus*) in Wyoming (Brown, 1967*b*) and Colorado (Spencer and Pettus, 1966), and conifer woodlands in Colorado and Arizona (Hall, 1933*b*; Hoffmeister, 1955, 1956). In Oregon, *S. merriami* has been taken most often on slopes vegetated by low sagebrush (*Artemisia arbuscula*) on Steens Mountain west and north of Fish Lake. One specimen was taken in the dunes along the south shore of Harney Lake.

Nowhere does *S. merriami* seem to be abundant; in areas in which it is known to occur, several hundred trap-nights are

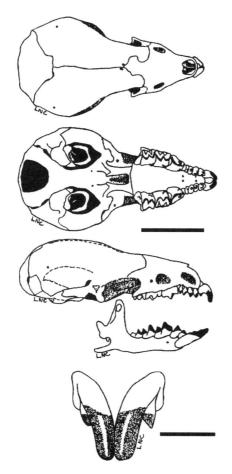

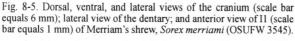

Fig. 8-5. Dorsal, ventral, and lateral views of the cranium (scale bar equals 6 mm); lateral view of the dentary; and anterior view of I1 (scale bar equals 1 mm) of Merriam's shrew, *Sorex merriami* (OSUFW 3545).

required to capture an animal (Johnson and Clanton, 1954).

Diet.—In Washington, spiders (Araneae) contributed 8–60% of the diet, caterpillars (Lepidoptera) 30–90%, beetles (Coleoptera) 5–100%, cave crickets (*Ceutophilus*) 10–85%, ichneumon flies (Ichneumonidae) 10–70%, insect eggs ≤5%, and unidentified insects 1–2% (Johnson and Clanton, 1954). *S. merriami* had the highest relative bite force of 12 taxa of western shrews, causing Carraway and Verts (1994*a*) to suggest that it was adapted for foraging on relatively large, hard-bodied prey.

Reproduction, Ontogeny, and Mortality.—In Washington, Johnson and Clanton (1954) found females with enlarged uteri in March; embryos in April, May, and July; and enlarged mammary glands or evidence of nursing in March, July, and October. Mean litter size is similar to that for most other small shrews in Oregon (Table 8-2). We found no other information regarding reproduction in *S. merriami* in Oregon or concerning development of young. Hoffmann (1955) reported finding skulls of *S. merriami* in pellets regurgitated by owls (species not specified).

Habits.—*Sorex merriami* commonly is associated with sagebrush voles (*Lemmiscus curtatus*) as many have been caught in unbaited snap traps or pitfall traps set in runways made by this species (Brown, 1967*b*; Johnson and Clanton, 1954). Johnson and Clanton (1954) presented evidence that *S. merriami* may be active in daylight hours and that it may not be active when rain is falling.

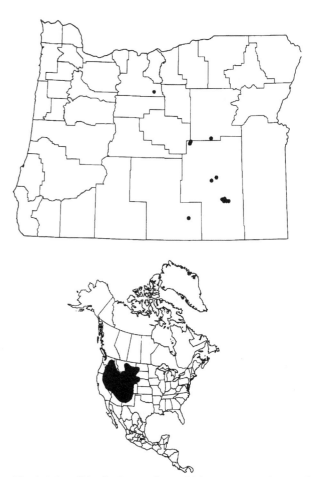

Fig. 8-6. Localities in Oregon from which museum specimens of Merriam's shrew, *Sorex merriami*, were examined (listed in the APPENDIX, p. 552), and the distribution of Merriam's shrew in North America. (Range map redrawn from Hall, 1981:49, map 28, with modifications.)

Remarks.—We believe that *S. merriami* may be more widely distributed in Oregon than indicated by localities at which museum specimens were taken (Fig. 8-6) and that the paucity of records likely is an artifact of sampling effort. Certainly, the species is worthy of a detailed and dedicated research effort, particularly considering that the type of habitat occupied is one often grazed by livestock.

Sorex monticolus Merriam
Montane or Dusky Shrew

1890*a*. *Sorex monticolus* Merriam, 3:43.
1891. *Sorex vagrans similis* Merriam, 5:34.
1895*a*. *Sorex obscurus* Merriam, 10:72.
1899. *Sorex setosus* Elliot, 1:274.
1955. *Sorex vagrans setosus*: Findley, 9:35.
1955. *Sorex vagrans obscurus*: Findley, 9:43.
1977. *Sorex monticolus obscurus*: Hennings and Hoffmann, 68:14.
1977. *Sorex monticolus setosus*: Hennings and Hoffmann, 68:14.

Description.—*Sorex monticolus* is slightly larger and possesses a slightly longer tail than *Sorex vagrans* (Table 8-1), a species with which at one time it was considered conspecific (Findley, 1955; Hennings and Hoffmann, 1977). *S. monticolus* is distinctive in that the third unicuspid is smaller than the fourth (Fig. 8-7), the red pigment on the anterior face of the first upper incisor extends to or beyond the base of the median tine (Hennings and Hoffmann, 1977), the median tine is large and robust

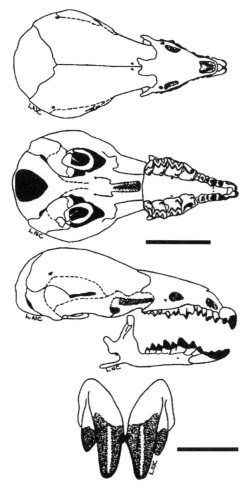

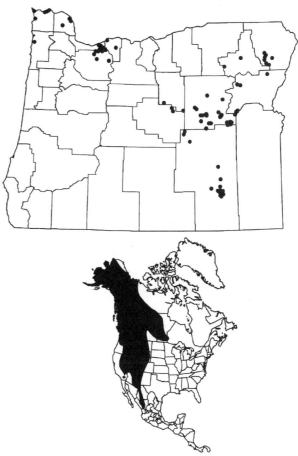

Fig. 8-7. Dorsal, ventral, and lateral views of the cranium (scale bar equals 6 mm); lateral view of the dentary; and anterior view of I1 (scale bar equals 1 mm) of the montane or dusky shrew, *Sorex monticolus* (OSUFW 4810).

Fig. 8-8. Localities in Oregon from which museum specimens of the montane or dusky shrew, *Sorex monticolus*, were examined (listed in the APPENDIX, p. 552), and the distribution of the montane shrew in North America. (Range map redrawn from Alexander, 1996:37, fig. 17.)

(Carraway, 1990), and there are five or six friction pads on the second through fourth toes on the hind foot (van Zyll de Jong, 1982). In northwestern Oregon, separation of *S. monticolus* and *S. bairdi* requires use of a score produced by a complex formula presented in the account of the latter species. Information on molts is not available; color is discussed by geographic race.

Distribution.—The geographic range of *S. monticolus* (Fig. 8-8) is from just south of the Columbia River in northwestern Oregon northward and westward through Washington, British Columbia, Yukon, and Alaska (at least to the Brooks Range—Bee and Hall, 1956), thence southeastward through Alberta and most of Saskatchewan to central Manitoba, and southward through the Rocky Mountains in Idaho, Montana, Wyoming, Utah, Colorado, and in other montane regions as extensions from this range or as disjunct populations in Oregon, California, Nevada, Arizona, New Mexico, Sonora, Chihuahua, Sinaloa, and Durango (Alexander, 1996; Hennings and Hoffmann, 1977; Junge and Hoffmann, 1981). In Oregon, it occurs as a small population near the Columbia River and along the Pacific Coast in Clackamas, Clatsop, Columbia, Hood River, Multnomah, and Tillamook counties (Alexander, 1996; Carraway, 1990) and as a series of populations at elevations above ≅1,200 m in Baker, Crook, Grant, Harney, Union, Wheeler, and Wallowa

counties (Fig. 8-8).

Geographic Variation.—Two geographic races are recognized in Oregon: *S. m. setosus* in northwestern Oregon and *S. m. obscurus* in the northeastern part of the state (Alexander, 1996; Carraway, 1990; Hennings and Hoffmann, 1977). *S. m. setosus* is blackish brown on the dorsum and light brown on the venter; the bases of the hairs are neutral very dark-gray. The tail is distinctly bicolored, blackish brown above, pinkish gray below; the flank glands are covered with pale brown bristles. The dorsal pelage of *S. m. obscurus* has yellow overtones—medium brown to medium yellowish-brown; the venter is white. The tail is indistinctly bicolored, medium brown above grading to white below; the flank glands have white bristles (Carraway, 1990). In both Oregon and Washington, the skull of *S. m. setosus* averages larger than that of *S. m. obscurus* (Alexander, 1996; Carraway, 1990).

Habitat and Density.—In a study conducted in Washington, the greatest number of *S. monticolus* was collected in 40-year-old Douglas-fir (*Pseudotsuga menziesii*)–salal (*Gaultheria shallon*) habitat, but some were collected in six of the nine other habitats sampled (Terry, 1981). Of 30 habitat variables measured, the abundance of these shrews was significantly correlated positively only with presence of dead wood. In a montane area in Utah, *S. monticolus* shifted use of microhabitats seasonally. Immediately after snowmelt and before growth of herbaceous vegetation, these shrews used habitats that included fallen logs and shrubs; in midsummer, new growth of herbaceous material covered the entire area and the shrews used available habitats randomly; at the end of summer, herbaceous material declined and the shrews shifted toward mature quaking aspen (*Populus tremuloides*) stands (Belk et al., 1990).

In California, Ingles (1961*a*) captured 11 montane shrews 143 times on a 0.51-ha study plot encompassing three habitats; 95% of the captures were in the 0.28-ha wet willow (*Salix*)-sedge (*Scirpus microcarpus*) core; 5% were in the surrounding 0.14-ha wet meadow, and none was in the 0.09-ha dry litter-covered forest (*Abies magnifica*, *A. concolor*, and *Pinus jeffreyi*). He attributed occupancy of the willow-sedge core to the abundance of invertebrates that inhabited the thick mat of dead vegetation and that served as food for these shrews. The 39.3/ha (11 shrews) that occupied the willow-sedge area seems to be the only estimate of density available.

In Oregon, only one of 342 small mammals caught in "riparian fringe" and none of 257 caught at streamside was *S. monticolus* (Anthony et al., 1987).

Diet.—Gunther et al. (1983) indicated that fragments of food in stomachs of 17 *S. monticolus* collected on an unburned clear-cutting in July in Washington consisted of 12% lichens, 26% conifer seeds, 4% grass, 51% invertebrates, and 6% unidentified materials. Percentages for four individuals from 110-year-old forest were similar. For four individuals from a burned clear-cutting, lichens, conifer seeds, and invertebrates composed 1%, 25%, and 74%, respectively, of the diet. Rhoades (1986) reported that digestive tracts of *S. monticolus* contained spores, suggesting that fungal sporocarps were eaten, but he did not indicate which of the species of fungi were consumed.

Ingles (1960) indicated that *S. monticolus* foraged on wet wood termites (*Zootermopsis nevadensis*) when they emerge at evening-time in July and August at high elevations in California, accounting for the first major nocturnal peak in activity. Later at night, the species foraged on insects, spiders, and phalangids as lower temperatures immobilized them, accounting for the second major peak in activity.

Reproduction, Ontogeny, and Mortality.—Bailey (1936:361) indicated that pregnant female *S. monticolus* "are found in June, July, and August" and that the number of "embryos vary from 4 to 8 in number, 6 apparently being normal for adults." He also indicated that the species probably produced only one litter annually. Although some of this information may not be reliable, we have few supportive data to refute his statements (Table 8-2).

Slipp (1942:212) described seven nestlings he considered to be two-thirds as long ($\bar{X} = 82.4$ mm) and one-third as heavy ($\bar{X} = 2.4$ g) as adults as "appearing superficially like adults" with closed, undeveloped eyes; nonfunctional, unerupted teeth; and short, stiff pelage, white ventrally and pale tobacco brown dorsally. The young ingested sweetened milk from cotton swabs, but refused insects. The nestlings were obtained from a fist-sized nest of dry grass within a rotten log.

From a mark-recapture study in British Columbia, M. L. Hawes (1975) reported that for the first two litter cohorts combined, survival to the following summer was only 4%. She suggested that the inability to find sufficient food possibly contributed to mortality in young shrews without territories. As for other soricids, raptorial birds may prey on *S. monticolus* and mammalian predators may kill but often

not consume individuals.

We found no information regarding mortality factors or longevity; we presume that both are similar to those of other species of shrews.

Habits.—In a 37-day session, two of three *S. monticolus* maintained in captivity in California were active in all 144 10-min periods of the 24-h day; the three shrews were active an average of 30.0–30.4% of the 72 10-min periods at 0700–1700 h and in an average of 37.5–50.4% of the 10-min periods at 1700–0700 h (Ingles, 1960). The three shrews averaged being 1.38 times as active at night as during the daytime. There were two reasonably well-defined peaks of nocturnal activity. Activity periods averaged 42 min in the daytime, 58 min at night; maximums were 200 and 280 min, respectively. Rest periods averaged 45 min in the daytime, 38 min at night; maximums were 180 and 120 min, respectively (Ingles, 1960).

Remarks.—In this account, we considered shrews identified as the wandering shrew (*Sorex vagrans*) and the dusky shrew (*Sorex vagrans obscurus*) from 2,135 m at Huntington Lake, Fresno Co., California, studied by Ingles (1960, 1961*a*) to be *Sorex monticolus*.

Sorex pacificus Coues
Pacific Shrew

1877. *Sorex pacificus* Coues, 3:650.
1918. *Sorex yaquinae* Jackson, 31:127.
1936. *Sorex pacificus yaquinae*: Bailey, 55:364.
1955. *Sorex vagrans pacificus*: Findley, 9:34.
1955. *Sorex vagrans yaquinae*: Findley, 9:34.
1990. *Sorex pacificus pacificus*: Carraway, 32:54.
1990. *Sorex pacificus cascadensis* Carraway, 32:55.

Description.—*Sorex pacificus* is the only shrew in Oregon without a tine on the anteromedial surface of the first upper incisor (Fig. 8-9) but with a "posteriomedial [*sic*] ridge visible in anterior view through the gap between the incisors" (Carraway, 1990:54). It also is a large brown shrew (Table 8-1) with the third unicuspid smaller than the fourth (Fig. 8-9) and the zygomatic process of the maxillary pointed, not rounded as in *S. sonomae*. Pelage color is discussed by geographic race.

Distribution.—*Sorex pacificus* is endemic to Oregon (Fig. 8-10). It is distributed as two disjunct populations: one in the Coast Range from Cascade Head, Tillamook Co., south to Coos Bay, Coos Co., and eastward to Philomath, Benton Co., and Sutherlin, Douglas Co; the other in the Cascade Range from northeastern Linn Co. to southern Jackson Co. (Carraway, 1990).

Geographic Variation.—Each of the two disjunct populations of *S. pacificus* is recognized as a nominal subspecies: *S. p. pacificus* in the Coast Range and *S. p. cascadensis* in the Cascade Range. The pelage of *S. p. pacificus* is dark brown on the dorsum, pinkish light-brown on the venter; bases of all hairs are neutral dark-gray. The tail is indistinctly bicolored, dark brown dorsally and white ventrally. *Sorex p. cascadensis* is medium reddish-brown dorsally, light brown ventrally; the tail also is indistinctly bicolored, reddish brown above and light brown below (Carraway, 1990). Although there is considerable overlap in the ranges of most dimensions of the two races, *S. p. pacificus* averages larger than *S. p. cascadensis* (Carraway, 1990).

Habitat and Density.—Maser et al. (1981*b*) provided a description of the habitat of *S. pacificus* identical with

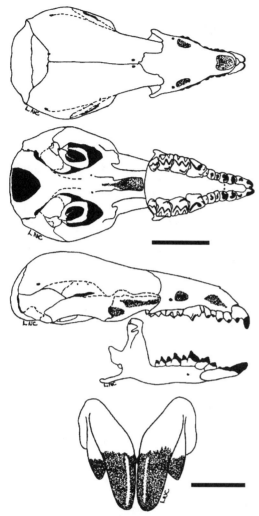

Fig. 8-9. Dorsal, ventral, and lateral views of the cranium (scale bar equals 6 mm); lateral view of the dentary; and anterior view of I1 (scale bar equals 1 mm) of the Pacific shrew, *Sorex pacificus* (OSUFW 8306).

that of *S. sonomae.* Bailey (1936) indicated that the species is often found in moist wooded areas with fallen decaying logs and brushy vegetation. In a quantitative study of 15 habitat variables, Morrison and Anthony (1989:807) found "No differences. . . between capture and noncapture sites for *S. pacificus.*" However, with *S. trowbridgii, S. pacificus* tended to avoid grass habitats occupied by *S. vagrans,* but *S. pacificus* tended to be captured more frequently than *S. trowbridgii* in habitats with a less dense forb component (Morrison and Anthony, 1989). No estimates of density are available; however, only 2% of 829 individuals of 10 species of small mammals captured in live traps were *S. pacificus* (Morrison and Anthony, 1989).

Diet.—Stomachs of 27 *S. pacificus* contained 30 of 72 food items ingested by 269 shrews of five species; internal organs of insects composed 28.6% by volume of the diet. Other foods that contributed substantially to the diet of *S. pacificus* are unidentified insect larvae, slugs and snails (Gastropoda), beetle larvae (Coleoptera), and unidentified invertebrates (Table 8-3). Vegetation composed only 2.4% of the diet, fungi 1.5% (Whitaker and Maser, 1976). Numerous dead specimens of *Omus audouini* (Coleoptera: Cicindelidae) found under logs were considered to have been cached there by *S. pacificus* (Maser, 1973). This diet, combined with a

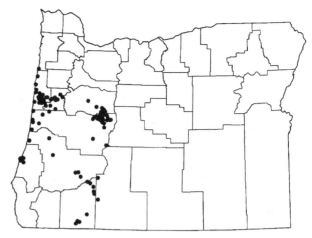

Fig. 8-10. Localities in Oregon from which museum specimens of the Pacific shrew, *Sorex pacificus*, were examined (listed in the APPENDIX, p. 553). The species is endemic to Oregon (Carraway, 1990).

greater-than-average relative bite force and parallel, slightly separated first upper incisors, suggests that *S. pacificus* is adapted for capturing, killing, and eviscerating hard-bodied insects (Carraway and Verts, 1994*a*).

Reproduction, Ontogeny, and Mortality.—Maser et al. (1981*b*) indicated that reproductively active female Pacific shrews were caught in March, April, June, and August and that litter size ranged from three to four. Counts of embryos recorded on tags of a small sample of museum specimens averaged 0.5 embryo greater (Table 8-2) than the maximum reported by Maser et al. (1981*b*). We found no description of developing young and no information regarding mortality factors; we presume both are similar to those of other shrews.

Habits.—Maser et al. (1981*b*) and Whitaker and Maser (1976) indicated that activity in *S. pacificus* is nocturnal. Although we cannot refute this information, we suspect that, as in several other Oregon shrews (Ingles, 1960; Newman, 1976; Rust, 1978; Terry, 1978), activity in *S. pacificus* is merely depressed in the daytime. Relatively little is known of the habits of *S. pacificus*; we presume that other aspects of the behavior are similar to those recorded for other congeners in Oregon.

Remarks.—Shrews formerly referred to as *S. yaquinae* and *S. pacificus yaquinae* are considered herein to be *S. pacificus* (*sensu* Carraway, 1990).

Sorex palustris Richardson
Water Shrew

1828. *Sorex palustris* Richardson, 3:517.
1858. *Neosorex navigator* Baird, 8:11.
1895*a*. *Sorex (Neosorex) palustris navigator*: Merriam, 10:92.

Description.—*Sorex palustris* (Plate I) is a large shrew (almost as large as *S. bendirii*; Table 8-1) with a very dark gray to black dorsal pelage, a white venter and throat, and a sharply bicolored tail. The broad dorsal portion of the tail is colored the same as the body; the white ventral stripe is much narrower. The summer pelage is slightly brownish (Jackson, 1928). The molt to summer pelage commences on the rostrum in February and progresses posteriorly; it is essentially complete by the end of May (Conaway, 1952). The molt to winter pelage is mostly in July and August; it commences on the flanks or middorsal region and progresses anteriorly and posteriorly (Conaway, 1952).

The hind foot is strongly fimbriated. The skull is large (Table 8-1) and angular, but the rostrum is not deflected downward as in *S. bendirii* (Fig. 8-11). Some individuals lack an anteromedial tine on the first upper incisor. In *S. palustris,* as in all other shrews in Oregon except *S. bendirii,* the anteroposterior dimension of c1 is less than that of p4.

Distribution.—The geographic range of *S. palustris* (Fig. 8-12) extends westward from the Atlantic Coast of Newfoundland, New Brunswick, Nova Scotia, and the New England states to the Pacific Coast of southeastern Alaska, British Columbia, and northern Washington, thence south from northern Quebec; James Bay, Ontario; Churchill, Manitoba; Great Slave Lake, Northwest Territories; and Anchorage, Alaska, through the Appalachian Mountains to Great Smoky Mountains National Park, North Carolina, and to Michigan, Wisconsin, Minnesota, and extreme northeastern South Dakota, and through the montane areas of the West south into California, Nevada, Arizona, and New Mexico. In Oregon, *S. palustris* occurs as disjunct populations in the Wallowa, Blue, Ochoco, Strawberry, Steens, and Hart mountains and in the High Cascade Range west and downslope to McKenzie Bridge, Lane Co.; it is absent from the Deschutes-Columbia

Plateau and Basin and Range provinces and from most of the High Lava Plains Province (Fig. 8-12).

Geographic Variation.—No geographic variation is recognized in water shrews in Oregon; all are considered *S. p. navigator* (Jackson, 1928).

Habitat and Density.—*Sorex palustris* is almost always found close to water, a relationship exploited in capturing specimens to the extent that of 101 captured by Conaway (1952) none was taken >17 cm from water. Undercut banks, exposed tree roots, and boulder-strewn streamsides vegetated by willow (*Salix*)-sedge, willow-grass, or willow-alder (*Alnus*) associations seem to be prime habitat for *S. palustris* (Brown, 1967*b;* Conaway, 1952). Of 130 *S. palustris* captured in Manitoba, 92% were from hydric habitats (mostly grass-sedge marshes and willow-alder bordered streams), 8% from mesic habitats (mostly coniferous forests), and none from xeric habitats (Wrigley et al., 1979). Occasionally, *S. palustris* may be captured farther from water and in other habitats (Beneski and Stinson, 1987). *S. palustris* commonly is associated with the beaver (*Castor canadensis*—Wrigley et al., 1979).

We found no published estimates of density, but Conaway (1952) indicated that he caught one *S. palustris*/30 trap-nights in carefully set mouse traps. Wrigley et al. (1979) captured 1,236 *S. cinereus* in the same effort required to capture 130 *S. palustris*. Most authorities suggest that numbers of *S. palustris* are low (Beneski and Stinson, 1987).

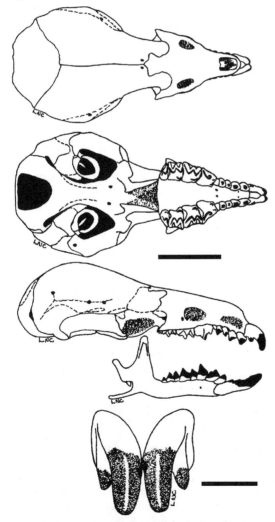

Fig. 8-11. Dorsal, ventral, and lateral views of the cranium (scale bar equals 6 mm); lateral view of the dentary; and anterior view of I1 (scale bar equals 1 mm) of the water shrew, *Sorex palustris* (OSUFW 4853).

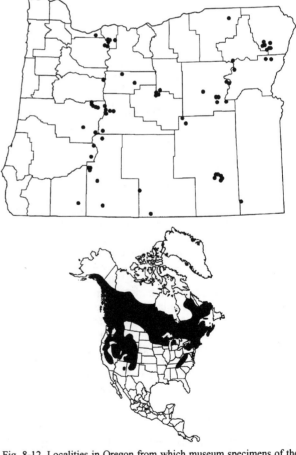

Fig. 8-12. Localities in Oregon from which museum specimens of the water shrew, *Sorex palustris*, were examined (listed in the APPENDIX, p. 554), and the distribution of the water shrew in North America. (Range map redrawn from Hall, 1981:42, map 22, with modifications.)

Diet.—Although some investigators consider *S. palustris* to be insectivorous (Conaway, 1952; Conaway and Pfitzer, 1952; Sorenson, 1962), others report that earthworms (Oligochaeta) and slugs and snails (Gastropoda) contribute significantly to the diet (Whitaker and French, 1984; Whitaker and Schmeltz, 1973) and that mice (Muridae), fish (Teleostei), and salamanders (*Dicamptodon ensatus*) are consumed at least occasionally (Buckner, 1970; Dalquest, 1948; Nussbaum and Maser, 1969; Svihla, 1934). Insects and other invertebrates consumed include stonefly nymphs (Plecoptera), flies (Diptera), caddisflies (Tricoptera), mayflies (Ephemeroptera), and crickets (Gryllidae), spiders (Araneae), and leeches (Hirundinea—Dalquest, 1948; Hamilton, 1930; Sorenson, 1962; Whitaker and Schmeltz, 1973).

Bite force in *S. palustris* is only 94% of that expected from condylobasal length (Carraway and Verts, 1994a). Strangely, the shortened condyloid-coronoid length and the more acute angle between muscle and resistant-moment arms would produce an even lower relative bite force were it not for a shorter-than-expected mandible (Carraway and Verts, 1994a). The common occurrence of soft-bodied prey such as earthworms and aquatic-insect larvae is concordant with the relatively weaker bite force of this shrew.

In captivity, *S. palustris* cached surplus food and maintained a larder of high-caloric, slow-deteriorating food items (Sorenson, 1962). Although members of the genus *Sorex* are not considered to possess toxic saliva as present in some shrews in eastern North America and in Europe (Pearson, 1942; Pucek, 1959), Nussbaum and Maser (1969) reported that a sculpin (*Cottus*) and a salamander appeared to be paralyzed when bitten by a water shrew.

Reproduction, Ontogeny, and Mortality.—All male and most female *S. palustris* do not become sexually active in the year of their birth, but, in Montana, all females captured in late January–August exhibited some stage of reproductive activity and most males captured during the same period contained spermatozoa in their testes (Conaway, 1952). The first litter was born before 19 March and some females were still lactating in mid-July. Females may have a postpartum estrus and at least some mate, but, because of the possibility of resorption of embryos, it is unknown if young are produced from such matings. Two or three litters annually are thought to be produced (Conaway, 1952). Litter size, based on counts of embryos, ranged from three to 10 (Beneski and Stinson, 1987), but by other methods may be even more variable (Table 8-2).

We found no published description of the development of neonates. However, Conaway (1952) indicated that mean weights of both males (9.7 ± 1.1 g) and females (9.7 ± 0.6 g) in their 1st year were significantly less than those of 2nd-year animals of corresponding sex (15.4 ± 1.3 and 12.3 ± 1.6 g, respectively). He also indicated that maximum longevity for *S. palustris* was about 18 months.

Because of its aquatic habits, *S. palustris* is vulnerable to predatory fish in addition to being preyed upon by carnivorous mammals and raptorial birds when on land (Beneski and Stinson, 1987).

Habits.—Like *S. bendirii*, *S. palustris* is an accomplished swimmer and diver, and can walk on the surface of the water supported by air bubbles in the fibrillae of the hind feet (Jackson, 1928). Propulsion in the water is by alternate strokes of the feet (Jackson, 1928; Svihla, 1934), not froglike as suggested by A. B. Howell (1924). When underwater, *Sorex palustris* needs only to cease to paddle to rise to the surface and float, supported by air trapped in the grooved hair; however, the fur rapidly becomes wetted, necessitating frequent dressing with the hind feet (Svihla, 1934). *S. palustris* forages underwater by probing the substrate with its proboscis (Svihla, 1934); it also chases and captures small fish (Nussbaum and Maser, 1969). However, the means by which it detects prey is uncertain; covering eyes and ears with wax and removal of the vibrissae did not affect prey detection (Sorenson, 1962). Hearing, especially of higher frequencies audible to humans, is acute (Sorenson, 1962).

Bouts of activity lasting an average of 30 min are alternated with periods of inactivity averaging 57 min; peaks in activity are at 0400 – 0500 and 1900–2300 h. *S. palustris* is active <15% of the time (Sorenson, 1962). Water shrews held in large cages usually were solitary and most interactions with conspecifics were antagonistic; most such interactions resulted in displacement of one individual, but occasionally two individuals fought viciously, sometimes resulting in the death of one of them (Sorenson, 1962). Nests were not defended, but cached food was marked with feces that caused other individuals to avoid the cache temporarily (Sorenson, 1962).

Remarks.—Observations by Nussbaum and Maser (1969) of *S. palustris* and Pattie (1969) of *S. bendirii* suggest that the toxicity of the saliva of these and other shrews of the genus *Sorex* should be reevaluated, particularly to invertebrates and heterothermic vertebrates that seem to be less resistant than homoiothermic vertebrates to the effects of numerous man-made toxicants.

Sorex preblei Jackson
Preble's Shrew
1922. *Sorex preblei* Jackson, 12:263.

Description.—In Oregon, *S. preblei* may be distinguished from congeners by U3 being larger than U4 and the presence of anteromedial tines on I1 (Fig. 8-13). The anteromedial tine on I1 is long, acutely pointed, and set within the pigmented area. This is the smallest shrew in Oregon (Table 8-1); adults commonly weigh less than a dime (≅3.2 g). The skull is delicate and jewel-like; it is ≤14.8 mm long with a mandibular toothrow <4.2 mm long and a maxillary breadth ≤4.2 mm. The mandible is <6.6 mm long and is deeper than the height of M1s that exhibit even moderate wear. The pelage is medium dark-brown to very dark-gray on the dorsum and silvery gray on the venter; the venter color extends high on the sides. The tail is bicolored, medium dark-brown on the dorsal surface, white on the ventral surface, and darkening toward the tip.

Distribution.—*Sorex preblei* occurs (Fig. 8-14) from extreme southeastern Washington and western Idaho south through the eastern half of Oregon to extreme northeastern California, thence easterly through extreme northern Nevada, the northern third of Utah, southwestern Colorado, and extreme southeastern Wyoming, thence north through western Wyoming and the southwestern three-fourths of Montana (Harris and Carraway, 1993). In Oregon, *S. preblei* has been collected in Deschutes, Grant, Klamath, Lake, Harney, Malheur, and Wallowa counties (Fig. 8-14—Bailey, 1936; C. Hansen, 1956; Hoffmann and Fisher, 1978; Jackson, 1922, 1928; Verts, 1975). Extralimital Pleistocene records are for caves in extreme southern New Mexico (Harris and Carraway, 1993).

Geographic Variation.—No geographic variation is

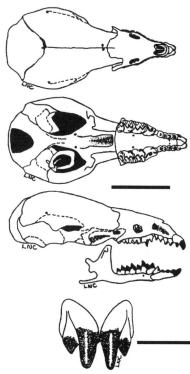

Fig. 8-13. Dorsal, ventral, and lateral views of the cranium (scale bar equals 6 mm); lateral view of the dentary; and anterior view of I1 (scale bar equals 1 mm) of Preble's shrew, *Sorex preblei* (OSUFW 4539). Note: right otic is missing.

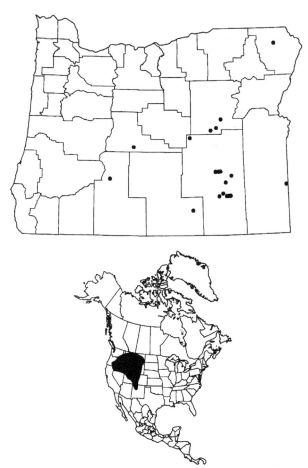

Fig. 8-14. Localities in Oregon from which museum specimens of Preble's shrew, *Sorex preblei*, were examined (listed in the APPENDIX, p. 554), and the distribution of Preble's shrew in North America. (Range map drawn from localities provided by Harris and Carraway, 1993:58, fig 2.)

recognized currently in *S. preblei* (Cornely et al., 1992).

Habitat and Density.—Throughout its range, *S. preblei* has been captured in a wide variety of habitats, including marshes (Bailey, 1936), along streams (Bailey, 1936; Ports and George, 1990), dry bunchgrass (C. Hansen, 1956), and wet, alkaline habitat (Tomasi and Hoffmann, 1984). Grasses and sagebrush (*Artemisia*) were common to most habitats (C. Hansen, 1956; Hoffmann and Fisher, 1978; Hoffmann and Pattie, 1968; Hoffmann et al., 1969; Ports and George, 1990; Williams, 1984). In Oregon in the past 20 years, *S. preblei* was taken in sagebrush-bunchgrass habitat on Steens Mountain, in big sagebrush (*A. tridentata*)–bitter-brush (*Purshia tridentata*) habitat on Hart Mountain, and in a bathtub in a ranch house near the headquarters of Malheur National Wildlife Refuge. Most authorities mentioned the rarity of the species (Armstrong, 1957; Bailey, 1936; Hoffmann and Fisher, 1978; Hoffmann and Pattie, 1968; Junge and Hoffmann, 1981; Ports and George, 1990; Tomasi and Hoffmann, 1984; Verts, 1975); however, on Steens Mountain, in June–July 1953–1954, Charles G. Hansen (1956) took eight specimens in snap traps and pitfalls and in June-September 1978, Alex Flecker captured 24 in pitfalls. We captured four in 12 pitfalls operated for 6 days in June 1992 in sagebrush at the edge of a quaking aspen (*Populus tremuloides*) grove on Steens Mountain, Harney Co.

Remarks.—Essentially no information is available for *S. preblei* regarding diet; reproduction, ontogeny, and mortality; or habits (Cornely et al., 1992). Relatively low bite force in *S. preblei* suggests that it feeds on soft-bodied prey (Carraway and Verts, 1994*a*). We found museum specimens of adult females collected in June–July with the hair surrounding the nipples worn off, indicating recent nursing of young.

Because of the paucity of information, we recommend

that specimens of all small brown shrews captured east of the Cascade Range in Oregon be preserved in such a manner that competent investigators can assess reproductive status and diet, and deposit the skins and skeletons in permanent repositories for future research.

We suspect that the "rarity" of *S. preblei* is largely an artifact of sampling effort. Also, the wide variety of habitat types occupied suggests that the requirements of *S. preblei* may be more specific than those described by the dominant vegetation or soil moisture. Like *S. merriami, S. preblei* is deserving of a detailed and dedicated research effort; in fact, studies of the two species might be combined as the two species commonly are syntopic. For a start, we recommend an assessment of the habitat similar to that conducted by Morrison and Anthony (1989) for small mammals in western Oregon.

Sorex sonomae Jackson

Fog Shrew

1921. *Sorex pacificus sonomae* Jackson, 2:162.
1955. *Sorex vagrans sonomae*: Findley, 9:32.
1990. *Sorex sonomae sonomae*: Carraway, 32:29.
1990. *Sorex sonomae tenelliodus* Carraway, 32:33.

Description.—*Sorex sonomae* (Plate I) is the largest of the brown shrews in Oregon (Table 8-1). It may be distinguished from congeners in Oregon by the third unicuspid being smaller than the fourth combined with the absence of tines or other projections on the first upper incisors (Fig. 8-15), permitting them to be appressed for most or all of the length of the pigmented portion. The fifth upper unicuspid is usually rectangular and is separated from the fourth premolar by a gap. The zygomatic process of the maxillary usually is rounded. *S. sonomae* has four friction pads on the second through fourth toes of the hind foot (Carraway, 1990). Pelage color is discussed by geographic races.

Distribution.—The range of *S. sonomae* (Fig. 8-16) extends from Taft, Lincoln Co., east to near the eastern boundaries of Linn and Lane counties (except it is absent in the Willamette Valley) and southward along the west slope of the Cascade Range and in the Coast Range and Siskiyou Mountains to Hilt, Siskiyou Co., California, then along the Pacific Coast to Crescent City, Del Norte Co., California, and Inverness, Marin Co., California (Carraway, 1990).

Geographic Variation.—Two geographic races are recognized (Carraway, 1990); both occur in Oregon. *S. s. sonomae* occurs throughout the range of the species in the

state, except *S. s. tenelliodus* occupies a narrow band extending along the northern and eastern edges of the range from Taft and Newport, Lincoln Co., to the eastern borders of Linn and Lane counties and along the west slope of the Cascade Range. *S. s. sonomae* is larger and more robust; *S. s. tenelliodus* is smaller and more delicate (Carraway, 1990). Adults of *S. s. sonomae* are light dark-brown on the dorsum and dark brown on the venter; bases of dorsal hairs are neutral black, neutral very dark-gray on the venter. Subadults are slightly lighter. All surfaces of the tail are the same color as the dorsum. Adults of *S. s. tenelliodus* are medium brown to dark gray on the dorsum, pinkish white on the venter; the tail is indistinctly bicolored, medium brown above and pinkish white below (Carraway, 1990).

Habitat and Density.—Maser et al. (1981b:53) described areas occupied by fog shrews as "alder/salmonberry, riparian alder, and skunkcabbage marsh habitats; they are less often found in the mature conifer and immature conifer habitats." Bailey (1936:363) indicated that *S. sonomae* commonly was associated with decaying logs in marshy and brushy areas in Oregon and in "redwood and dense spruce forests as well as in . . . marshes and swamps" in northern California. No information regarding density is available.

Diet.—Bite force and morphology of I1 suggest that *S. sonomae* feeds on organisms with hard exoskeletons or tough skins (Carraway and Verts, 1994a), and, in the wild, snails and slugs (Gastropoda) constitute the primary foods (Whitaker

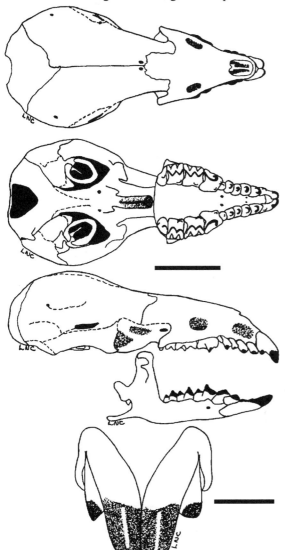

Fig. 8-15. Dorsal, ventral, and lateral views of the cranium (scale bar equals 6 mm); lateral view of the dentary; and anterior view of I1 (scale bar equals 1 mm) of the fog shrew, *Sorex sonomae* (OSUFW 4602).

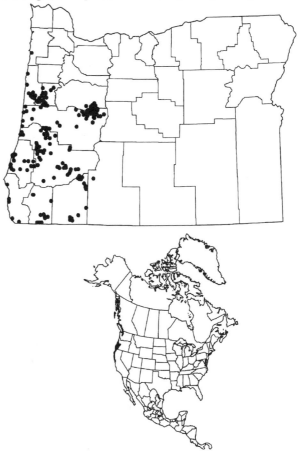

Fig. 8-16. Localities in Oregon from which museum specimens of the fog shrew, *Sorex sonomae*, were examined (listed in the APPENDIX, p. 554), and the distribution of the fog shrew in North America. (Range map drawn from localities provided by Carraway, 1990:34, fig. 16.)

and Maser, 1976). Other items that contribute substantially to the diet are centipedes (Chilopoda), unidentified insect larvae, amphibian flesh, and unidentified invertebrates (Table 8-3); 29 of 72 food items consumed by 269 individuals of five species of *Sorex* occurred in the diet of 30 *S. sonomae*. *S. sonomae* was considered the primary mammalian predator of *Omus dejeani* (Coleoptera: Cicindelidae) in southwestern Oregon (Maser, 1973). Vegetation (mostly seeds) composed 4.1% of the diet, fungi 3.7%. However, in captivity, these shrews consumed a wide variety of vertebrate and invertebrate prey offered (Maser and Hooven, 1974).

Reproduction, Ontogeny, and Mortality.— Other than records of breeding dates and litter size (Table 8-2) obtained from tags on museum specimens and from a small sample of animals from northern California (Carraway, 1988), we have no information on reproduction in this species. No description of offspring at any stage in development is available and no information on mortality is available. We presume that fog shrews are subject to mortality factors similar to those encountered by other forest-dwelling shrews and that they have lifespans no longer than those of other species of *Sorex*.

Habits.—Activity in *S. sonomae* may be more nocturnal than that in other Oregon *Sorex* (Maser and Hooven, 1974; Whitaker and Maser, 1976); however, no quantitative evaluation of time of activity is available for the species. In captivity, *S. sonomae* built nests of natural materials, groomed frequently, slept most of the day except for feeding bouts, established specific sites for defecation and urination, spent considerable time licking the sides of the cage (possibly to reingest feces), and exhibited typical cage-behavior by jumping against the sides of the cage (Maser and Hooven, 1974). Nonflying prey apparently was hunted by olfaction, but flying prey was hunted by sight; *S. sonomae* sometimes caught the flying prey in midair. Prey was either eaten immediately, carried to the nest and eaten, or cached near the nest; most prey was not killed, but immobilized. Hazardous prey (stinging invertebrates) was killed immediately (Maser and Hooven, 1974). Maser and Hooven (1974) provided detailed anecdotal accounts of the behavior of *S. sonomae* capturing, killing, immobilizing, caching, and eating a variety of food items.

Remarks.—Because of descriptions of the animals and localities provided by Bailey (1936), Carraway (1988), Maser and Hooven (1974), Maser et al. (1981*b*), and Whitaker and Maser (1976), we are certain that shrews they identified as *Sorex pacificus* are *Sorex sonomae*.

Carraway (1990) elevated *Sorex pacificus sonomae* to species level because holotypes of *Sorex pacificus pacificus* and *Sorex pacificus yaquinae* possess the same morphological pattern on I1 that with a multivariate morphometric analysis identified the latter taxon as specifically separate from the large dark-brown shrews.

Sorex trowbridgii Baird, 1858
Trowbridge's Shrew

1858. *Sorex trowbridgii* Baird, 8:13.
1913*a*. *Sorex montereyensis mariposae* Grinnell, 10:189.
1923. *Sorex trowbridgii mariposae* Grinnell, 21:314.

Description.—Trowbridge's shrew (Plate II) is another medium-sized shrew (Table 8-1). It can be distinguished from other Oregon shrews by its dark-brown or grayish-black pelage on both dorsum and venter, and its sharply bicolored tail, white below and dark brown or grayish black

above (Jackson, 1928). In addition, it possesses a slightly pigmented ridge that extends from the apex of the upper unicuspid teeth lingually toward the cingulum, but is separated from it by an anteroposterior groove (Carraway, 1987; Hall, 1981). The red pigment on the anterior face of the first upper incisor does not extend above the anteromedial tine, the first upper incisors are widely divergent, the third unicuspid is distinctly smaller than the fourth, and the least interorbital breadth is ≥3.5 mm (Fig. 8-17; Carraway, 1987). Most individuals possess a postmandibular foramen (Junge and Hoffmann, 1981).

Sorex trowbridgii molts from the grayish-black winter pelage to the more brownish summer pelage in May–June, but some in coastal areas may molt as early as April (Jackson, 1928). However, Jameson (1955) found this molt to occur in June–August in the northern Sierra Nevada, California. Few adults live sufficiently long to molt into the winter pelage a second time.

Distribution.—The geographic range of *S. trowbridgii* (Fig. 8-18) extends from southwestern British Columbia south through western Washington, Oregon, and northern California, where it splits, with one branch extending southeast along the Sierra Nevada Range as far as Kern Co., the other along the Coast Range south to Santa Barbara Co. (Hall, 1981). In Oregon, its range is west and south of a

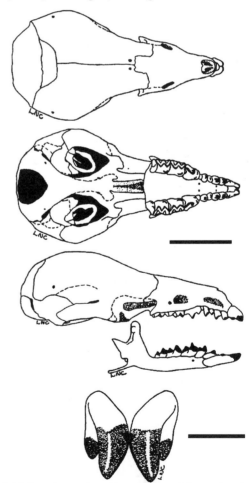

Fig. 8-17. Dorsal, ventral, and lateral views of the cranium (scale bar equals 6 mm); lateral view of the dentary (OSUFW 8267); and anterior view of I1 (scale bar equals 1 mm) of Trowbridge's shrew, *Sorex trowbridgii* (OSUFW 4843).

line connecting Parkdale, Hood River Co.; Pine Grove, Wasco Co.; Alder Spring, Lane Co.; Diamond Lake, Douglas Co.; Gearhart Mountain, Lake Co.; and Lakeview, Lake Co. (Fig. 8-18). The only Pleistocene record is from an asphalt bed near the southern limit of its distribution in Santa Barbara Co., California (Kurtén and Anderson, 1980).

Geographic Variation.—Two geographic races are recognized in Oregon: *S. t. mariposae* east and south of Diamond Lake, Douglas Co., and *S. t. trowbridgii* throughout the remainder of the range of the species within the state (Hall, 1981). *S. t. mariposae* is brownish and with a relatively wide skull, whereas *S. t. trowbridgii* is more grayish and with a relatively narrow skull (Jackson, 1928).

Habitat and Density.—*Sorex trowbridgii* occurs in all stages of the coniferous forest sere from old growth to recent clear-cuttings, including clear-cuttings on which the slash was burned or herbaceous plants were treated with herbicides (Anthony et al., 1987; Black and Hooven, 1974; Borrecco et al., 1979; Gashwiler, 1959, 1970a; Gunther et al., 1983; Hooven, 1969; Hooven and Black, 1976; Jameson, 1955; Terry, 1981; Williams, 1991). From studies in which population densities on unlogged areas are compared with those on clear-cuttings, numbers of *S. trowbridgii* may decline after logging (Hooven, 1969) or they may increase (Gunther et al., 1983), suggesting that factors other than dominant vegetation may be responsible

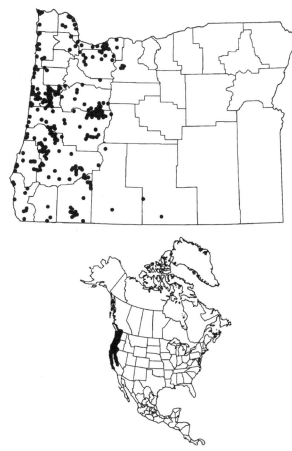

Fig. 8-18. Localities in Oregon from which museum specimens of Trowbridge's shrew, *Sorex trowbridgii*, were examined (listed in the Appendix, p. 556), and the distribution of Trowbridge's shrew in North America. (Range map redrawn from Hall, 1981:48, map 27, with modifications.)

for regulating abundance of the species. Precisely what constitutes prime habitat for the species is somewhat enigmatic. According to Dalquest (1941b:171) it is the most common shrew in "dry fir woods" in western Washington, but it may be "more numerous" in marshy areas "than in dry woods" and where "*Sorex vagrans* is present, *Sorex trowbridgii* is almost never found"; thus, it does not occur in "small sphagnum bogs, surrounded by fir woods" until *S. vagrans* is removed therefrom. Also, presence of *S. trowbridgii* was correlated positively with factors that indicate more dense Douglas-fir (*Pseudotsuga menziesii*) forests and negatively with factors that indicate that the soil is heavily compacted (Terry, 1981). Where *S. trowbridgii* is sympatric with *S. monticolus* the two are thought to be separated by microhabitat differences; the former is a digger and forages in the top layers of soil, whereas the latter forages in the duff (Terry, 1981). In California, numbers of *S. trowbridgii* caught per unit effort in pitfalls were 18–32 times greater in mixed conifer (*Pinus ponderosa-P. lambertiana-Abies concolor-Pseudotsuga menziesii*) habitat than in red fir (*A. magnifica*)-lodgepole (*P. contorta*) and ponderosa pine (*P. ponderosa*) habitats, respectively; none was captured in blue oak (*Quercus douglasii*)-digger pine (*Pinus sabiniana*) habitat (Williams, 1991).

Published estimates of density for *S. trowbridgii* are not available, but where the species occurs, it is commonly the most frequently captured shrew and may be the most frequently captured small mammal (Hooven and Black, 1976).

Diet.—Bite force and morphology of the first upper incisor suggest that *S. trowbridgii* is a food generalist, perhaps more so than *S. vagrans* (Carraway and Verts, 1994a). In concordance, in Oregon, 158 *S. trowbridgii* consumed 47 of 72 food items identified in stomachs of 269 shrews of five species, likely a reflection of the wide variety of habitats occupied and the adaptability of *S. trowbridgii*. The five items consumed in greatest amounts, in order by volume, were centipedes (Chilopoda), spiders (Araneae), internal organs of invertebrates, slugs and snails (Gastropoda), and unidentified invertebrate material. Among insects, Coleoptera, Hemiptera, Diptera, and Hymenoptera were consumed in declining order (Table 8-3). In feeding trials, captive *S. trowbridgii*, like captive *S. vagrans*, selected seeds of Douglas-fir and western white pine (*Pinus monticola*) over those of true firs (*Abies*), but did not discriminate between them (Terry, 1978). Captive *S. trowbridgii* consumed considerable quantities of five of 10 species of mushrooms offered (Terry, 1978). Greatest food consumption is during the reproductive season (Rust, 1978).

Reproduction, Ontogeny, and Mortality.—In the Sierra Nevada Range, California, male *S. trowbridgii* attain reproductive competency in February, females about 2 weeks later; the dates may be as much as a month earlier in brush fields and on south slopes (Jameson, 1955). Body mass increases rapidly from about 4.0–5.5 g to about 6.0–7.5 g in males, with gravid females becoming considerably heavier; testis length increases from 1–3 mm to 4.7–6 mm. Most of eight gravid females were captured in April–May, but one was caught as early as 25 February (Jameson, 1955). Four of the eight pregnant females also were lactating, indicating that at least some individuals produced a minimum of two litters annually; Jameson (1955) speculated that as many as three litters were produced in the short reproductive season that he considered to end by early

June. In Oregon, Gashwiler (1976*a*) found pregnant females in February–June and September–October, with the highest percent pregnant in samples collected in April–May. Whether or not those breeding in late summer–early autumn were young-of-the-year was not stated. Breeding males and females also were heavier (4.5–6.2 g and 3.7–6.2 g, respectively) than nonbreeding counterparts (2.5–5.0 g and 3.0–4.5 g, respectively—Gashwiler, 1976*b*). In April, 22% (*n* = 9) were pregnant and lactating, in May 12% (*n* = 33), and in June 13% (*n* = 16), also suggesting that more than one litter is born annually. Litter size averages five in California, but less than four in Oregon (Table 8-2). Young of *S. trowbridgii* have not been described.

Jameson (1955) recognized two age-classes: juveniles (young-of-the-year) and adults (those in the 2nd calendar year of life); at most times of the year one age-class or the other predominated. The age-classes can be recognized by differences in the degree of erosion of the teeth; juveniles exhibit relatively light wear, whereas individuals ≥1 year older have heavily worn teeth. Rarely does *S. trowbridgii* live for >18 months (Jameson, 1955).

Reported predators of *S. trowbridgii* are the Pacific giant salamander (*Dicamptodon ensatus*—Maser et al., 1981*b*) and the barred owl (*Strix varia*), but a variety of snakes (Serpentes) and raptorial birds probably prey on this species. House cats kill *S. trowbridgii* but usually do not eat it (Maser et al., 1981*b*) presumably because of the odor produced by the flank glands; other carnivorous mammals may behave similarly upon encountering *S. trowbridgii*.

Habits.—*Sorex trowbridgii* tends to be a "digger" in that it forages in the soil beneath the duff (Terry, 1981). In captivity it constructed complex mazes of tunnels unless the soil was extremely wet. Nevertheless, *S. trowbridgii* reportedly climbs trees (Carraway and Verts, 1988) and we saw photographs of *S. trowbridgii* "captured" by automatic camera "traps" set in trees to monitor activities of red tree voles (*Phenacomys longicaudus*) by Brian Biswell, a former graduate student in the Department of Fisheries and Wildlife, Oregon State University.

Sorex trowbridgii also is accused of depredations on seeds of commercially valuable forest trees (Moore, 1942). In feeding trials, *S. trowbridgii* removed about half as many seeds as *S. vagrans* in 1-h tests, but as many as or more than *S. vagrans* in 12-h tests (Terry, 1978). In some instances, *S. trowbridgii* cached seeds; however, whether any of the seeds were recovered at a later date was not stated (Terry, 1978).

Although Whitaker and Maser (1976:104) stated unequivocally that *S. trowbridgii* is "strictly nocturnal," captive animals, although more active at night, are active throughout the diel cycle (Rust, 1978; Terry, 1978). Activity in *S. trowbridgii* was described as occurring in short bouts at regular intervals of about 1 h. The bouts of activity total about 39% of the diel cycle (Rust, 1978). Greatest activity tends to be associated with the reproductive season (Rust, 1978).

Remarks.—*Sorex trowbridgii* would be an interesting and appropriate subject on which to conduct research on such topics as morphological, cytologic, and genetic variation; descriptive ontogeny; long-term population dynamics; and intra- and interspecific behavior. Its broad habitat affinities and apparent abundance should facilitate such research. The species is purported by some to be difficult to maintain in captivity (Dalquest, 1948; Maser et al., 1981*b*), but others have been successful in doing so (Rust, 1978; Terry, 1978); the former authors did not describe the protocol used and the latter authors provided only a brief description.

Sorex vagrans Baird
Vagrant Shrew

1858. *Sorex vagrans* Baird, 8:15.
1858. *Sorex suckleyi* Baird, 8:18.
1891. *Sorex dobsoni* Merriam, 5:33.
1895*a*. *Sorex amoenus* Merriam, 10:69.
1895*a*. *Sorex nevadensi* Merriam, 10:71.
1899*b*. *Sorex shastensis* Merriam, 16:87.
1922. *Sorex trigonirostris* Jackson, 12:264.
1936. *Sorex ornatus trigonirostris*: Bailey, 55:366.

Description.—The vagrant shrew (Plate II) can be distinguished from all other congeners in Oregon by the combination of the upper unicuspids wider than long in ventral view, an anteromedial tine on the first upper incisor with red pigment not extending above its dorsal level (Fig. 8-19; the tine and body of the tooth commonly are separated by a white- or pale-colored area), and the third unicuspid distinctly smaller than the fourth (Carraway, 1990). There are four or fewer friction pads on the second to fourth toes of the hind feet (van Zyll de Jong, 1982). The least interorbital breadth is <3.5 mm and the first upper incisors are straight or barely divergent (Carraway, 1987). The vagrant shrew is medium-sized among Oregon shrews (Table 8-1).

The vagrant shrew is light medium-brown on the dorsum, light pinkish-gray on the sides, and white on the venter; bases of hairs on all three areas are neutral very dark-gray. The tail is weakly bicolored (light medium-brown over white) in adults, sharply bicolored (dark brown over white) in juveniles. Bristles on the flank glands are dark pinkish-gray (Carraway, 1990). The winter pelage of *S. vagrans* is darker than the summer pelage. Spring molts occur March–May, autumn molts in October; some old individuals may molt to a second summer pelage in midsummer (Dalquest, 1944; M. L. Hawes, 1975).

Distribution.—The geographic range of the vagrant shrew (Fig. 8-20) extends from southern British Columbia (including the eastern half of Vancouver Island) and southern Alberta south through Washington, Oregon, and coastal and montane regions of California to Monterey Bay and Yosemite National Park, eastward through much of the northern half of Nevada, most of Idaho, extreme western Wyoming, east and north of Great Salt Lake in Utah, and in northwestern Montana; also, populations occur in Mexico from Michoacan to Puebla and from Tlaxcala to Veracruz (Hall, 1981; Hennings and Hoffmann, 1977; Smith, 1988). In Oregon, the vagrant shrew occurs throughout the state except in the Columbia Basin (Fig. 8-20). Fossils have been recovered from caves in Arizona, Texas, and Arkansas (Kurtén and Anderson, 1980).

Geographic Variation.—No geographic variation is recognized in vagrant shrews in Oregon; all are considered *S. v. vagrans* (Carraway, 1990; Hennings and Hoffmann, 1977).

Habitat and Density.—*Sorex vagrans* tends to be more of a generalist than most Oregon shrews in terms of habitat affinities; nevertheless, it usually is captured in greatest numbers in moist grassy areas and more open areas with patches of shrubs and deciduous trees (Hawes, 1977; Hoffman, 1960; Terry, 1981; Whitaker and Maser, 1976). In a quantitative

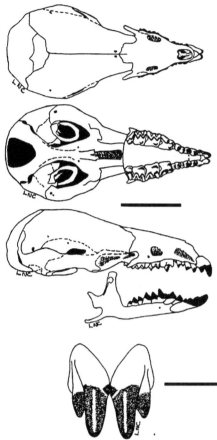

Fig. 8-19. Dorsal, ventral, and lateral views of the cranium (scale bar equals 6 mm); lateral view of the dentary; and anterior view of I1 (scale bar equals 1 mm) of the vagrant shrew, *Sorex vagrans* (OSUFW 8510).

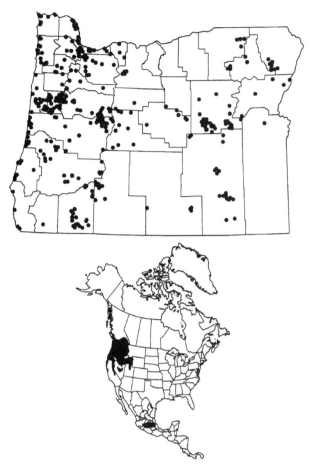

Fig. 8-20. Localities in Oregon from which museum specimens of the vagrant shrew, *Sorex vagrans*, were examined (listed in the APPENDIX, p. 558), and the distribution of the vagrant shrew in North America. (Range map redrawn from Hennings and Hoffmann, 1977:6–7, fig. 3, with modifications.)

analysis of habitat associations of four species of insectivores in western Washington, *S. vagrans* was most abundant in red alder (*Alnus rubra*)-common swordfern (*Polystichum munitum*) habitats and meadows in which the organic matter above the mineral soil was the deepest; it was least abundant or absent in Douglas-fir (*Pseudotsuga menziesii*) forests (Terry, 1981). In the Willamette Valley, *S. vagrans* reportedly is comparatively rare in conifer forests other than in riparian zones; it was most common in grassy habitats below 240 m and decreased in abundance with greater elevation and precipitation (Hooven et al., 1975). In Harney Co., *S. vagrans* is restricted to the "marsh-grassland physiognomic type" (Feldhamer, 1979*b*:217).

Several investigators of effects of forest-management practices on small-mammal populations (Black and Hooven, 1974; Borrecco et al., 1979; Gashwiler, 1959, 1970*a*; Hooven, 1969; Hooven and Black, 1976) indicated that *S. vagrans* was present, sometimes abundant, in forested areas and in clear-cuttings, especially before burning or treatment with herbicides. However, in view that several taxa now considered to differ at the species level were once included in *S. vagrans* (Carraway, 1990; Findley, 1955) and in the absence of voucher specimens, we advise caution in including this species among those considered to occur frequently in forested areas of the Coast and Cascade ranges.

Sorex vagrans is most abundant in summer when population densities as high as 58.2/ha are recorded; lowest densities are in autumn (Hawes, 1977; Newman, 1976). Although comparable quantitative data regarding density for popula-

tions of the species in Oregon are not available, *S. vagrans* often composes 30–35% of the small mammals captured in live traps in the Willamette Valley (Hooven et al., 1975).

Diet.—Bite force and morphology of the first upper incisor suggest that *S. vagrans* is a "food generalist" (Carraway and Verts, 1994*a*), and, indeed, some investigators of its diet have so described it (Terry, 1978). In western Oregon, 30 of 72 items identified in stomachs of 269 shrews of five species of *Sorex* also were eaten by the 30 *S. vagrans* examined. Food items consumed, in order by volume, were insect larvae, slugs and snails (Gastropoda), unidentified invertebrates, fungi (*Endogone*), and spiders (Araneae; Table 8-3). In eastern Oregon, earthworms (Oligochaeta) and spiders contributed most to the diet of *S. vagrans* in an ungrazed mountain meadow, but caterpillars, moths (Lepidoptera), beetle (Coleoptera) larvae, and unidentified flies (Diptera) contributed most in grazed meadows (Whitaker et al., 1983).

In captivity, *S. vagrans* consumes a wide variety of food items of plant and animal origin, including apples; rolled oats; seeds of a variety of forbs and shrubs; a variety of mushrooms; fresh and cooked liver, hamburger, and sausage; canned sardines; canned dog food; tree frogs (*Hyla*); salamanders (*Plethodon*); freshly killed rats (*Rattus*) and mice (*Mus, Peromyscus*); earthworms; grasshoppers (Orthroptera); and even conspecifics. A captive *S. vagrans* reportedly was unable to pierce shells of small snails and the integument of large slugs (Broadbooks, 1939), but other captives consumed slugs and their eggs (Terry, 1978). Terry (1978) suggested that some fruits were not eaten (although seeds removed therefrom were consumed) because they could not be pierced by *S. vagrans*.

Sorex vagrans commonly includes a small amount of vegetable material in its diet and, with other species of shrews and mice, may be involved in depredations on seeds of Douglas-fir and other commercially valuable trees (Moore, 1942; Terry, 1978). In feeding trials with captive animals, *S. vagrans* did not discriminate between seeds of Douglas-fir and western white pine (*Pinus monticola*), but selected seeds of those species over seeds of true firs (*Abies*—Terry, 1978). Maser (1966c) reported that an individual that he observed in captivity fed on seeds of Douglas-fir only the first time they were offered. Both Maser (1966c) and Terry (1978) reported that many seeds offered were cached; if all cached seeds are not recovered at a later time, depredations possibly are offset by shrews placing seeds at sites more favorable for their germination and establishment of seedlings. Broadbooks (1939) indicated that some cached food items (earthworms) were recovered.

Because of the high surface-area:body-mass ratio and a metabolic rate higher than predicted on the basis of body mass in soricine shrews (Eisenberg, 1981), food consumption by shrews is amazingly high. A captive *S. vagrans* weighing 6.5 g consumed an average of 8.36 g (range, 0.65–12.3 g) of food per day during a 12-day period (Broadbooks, 1939); estimated consumption by other captives was even greater (Hooven et al., 1975).

Reproduction, Ontogeny, and Mortality.—Female vagrant shrews that overwinter become receptive in March immediately after completing the spring molt and produce two, sometimes three, litters before succumbing (M. L. Hawes, 1975, 1977). Although some reproduction has been recorded from March–September in Washington (Newman, 1976) and February–July and September–mid-November in the Willamette Valley (Hooven et al., 1975), most offspring are produced in April–June. In the Willamette Valley, 36% were pregnant or lactating in March, 90% in April and May, and 100% in June or July; comparable data were not presented for the autumn reproductive period (Hooven et al., 1975). Hooven et al. (1975) suggested that at least some spring-born individuals attained reproductive competence in autumn as few old adults were captured during this season. Litter size averages about six (Table 8-2).

Young are naked, blind, and dark pinkish-red at birth; they weigh about 0.35 g. By 1 week they weigh about 1.9 g, but are still naked and blind; the sexes can be differentiated by 14 days; by 19 days they weigh 4–5 g; and, at 3 weeks, their eyes open. Young were weaned at about 25 days and attained adult mass by about 41 days (Hooven et al., 1975). Weaned young often remain associated and remain together in the nest (Eisenberg, 1964; Hooven et al., 1975). *S. vagrans* exhibits nearly complete cohort turnover annually in Washington (Newman, 1976) and Oregon (Hooven et al., 1975); the average life expectancy of newly identified individuals was 6.6 months; 8% survived >15 months and only 3% survived to 24 months of age (Newman, 1976). *S. vagrans* is preyed upon by a variety of snakes (Serpentes) and raptorial birds (Hooven et al., 1975); it may be killed by carnivorous mammals, but is

rarely recorded as an item in their diets, presumably because of the strong odor produced by the flank glands (Bee et al., 1980). Older individuals have extensively eroded teeth, suggesting that declining ability to capture, kill, and masticate prey may be the proximal cause of death in old age. Nevertheless, shrews of this and other species with heavily worn teeth that we examined appeared in good physical condition, and in the appropriate season, usually were reproductively active (Carraway et al., 1996).

Habits.—*Sorex vagrans* tends to follow paths at the interface of the mineral soil and the overlying duff (Terry, 1981). In the open, it travels in short bursts of movement usually from one sheltered area to another (Eisenberg, 1964). Thus, in the wild, it is rarely observed except as it streaks along an opened microtine runway or as it explores among the leaves. When moving about, its nose is constantly twitching up and down and from side to side; it may stop occasionally to scratch, yawn, or engage in some other comfort behavior (Eisenberg, 1964; Maser, 1966c). Much time is spent digging (Eisenberg, 1964). *S. vagrans* produces broadband (16–60 mHz) double pulses of sound by which it locates barriers to its movements and detects openings as small as 4 by 4 cm at distances of ≤30 cm; the echolocation system is crude in comparison with that of microchiropterans (Buchler, 1976). It is an agile climber and can negotiate surprisingly thin twigs by use of its "semi-prehensile" tail (Maser, 1966c:52). *S. vagrans* exhibits some behavior that tends to indicate it does not perceive much that transpires in its microcosm; it does not respond to low-frequency sounds; it continues to travel to its nest despite having dropped nesting material or food items it was transporting, then returns to retrieve the items; it often touches or almost touches food items before seemingly recognizing them as such; and it is easily captured in pitfalls (Broadbooks, 1939; Eisenberg, 1964; Maser, 1966c). Nevertheless, it can learn to avoid obstacles (Buchler, 1976). In captivity it tames quickly and can be handled without danger of being bitten; it seems to recognize its usual caretaker (Maser, 1966c).

Familiar conspecifics can be housed together; apparently they communicate by squeaks (when fighting), by chemical odors produced by flank glands, and by touch during naso-nasal or naso-anal encounter postures (Eisenberg, 1964; Hawes, 1976). In staged encounters with both familiar and strange conspecifics, avoidance by moving away is the most common response; fighting is rare, but more common between subadults and between unfamiliar individuals (Eisenberg, 1964). Nests usually are defended (Eisenberg, 1964). Because of different habitat affinities, *S. vagrans* usually avoids interaction with sympatric congeners (Hawes, 1977; Hooven et al., 1975; Terry, 1981). *S. vagrans* sometimes attacks other mammals larger than itself (Broadbooks, 1939).

Nests for both sleeping and rearing young are constructed by grasping a piece of moss (*Eurynchium*), grass, or other dry and insulative material in the mouth and by thrusting ("stitching") it into place with rapid movements of the head and by constantly turning the body (Eisenberg, 1964); nests usually are constructed beneath a piece of bark or similar shielding material (Eisenberg, 1964; Jewett, 1931; Maser, 1966c; Terry, 1981). *S. vagrans* sleeps curled in the nest; after becoming acclimated to captivity, the sleeping animal may be exposed by removal of the overlying material without arousing it (Maser, 1966c).

In captivity and in the wild, *S. vagrans* is considered to be active at all hours of the day and night (Eisenberg, 1964; Hawes, 1977; Hooven et al., 1975). In captivity, 5–10-min bouts of activity were followed by slightly longer rest periods in the nest (Eisenberg, 1964). However, activity as indicated by captures/h in traps operated throughout the 24-h period was mostly during the hours of darkness at all seasons; most captures during daylight hours were in spring (Newman, 1976).

In British Columbia, home-range areas calculated by an elliptical model (Koeppl et al., 1975) for individuals captured ≥10 times ranged from 510 to 1,986 m² for nonbreeding females, from 553 to 1,540 m² for nonbreeding males, from 732 to 5,261 m² for the same females in breeding condition, and from 1,908 to 4,460 m² for the same males in breeding condition (Hawes, 1977). Nonbreeding individuals tended to have nonoverlapping or only slightly overlapping home ranges (thus, were considered to be territories), whereas home ranges of breeding individuals overlapped with those of both breeding and nonbreeding individuals (Hawes, 1977). This territorial behavior tends to corroborate, in part, the purported hierarchy with subadults dominant over adults and adult females dominant over adult males (Hooven et al., 1975).

Remarks.—Because *S. vagrans* is superficially similar to several other shrews in western North America that at one time were considered synonyms of it (Carraway, 1990; Findley, 1955) and because it was difficult to distinguish from *S. monticolus* (=*S. obscurus*) over a broad area and remains so in some regions (George and Smith, 1991), we limited references used for specific life-history events to those that we are reasonably certain are for *Sorex vagrans* (*sensu* Carraway, 1990; Hennings and Hoffmann, 1977). Also, we specifically excluded references to races other than *S. v. vagrans*.

Sorex trigonirostris Jackson, formerly considered a distinct species restricted to Grizzly Peak and near Ashland, Jackson Co., Oregon, was subsumed into *S. vagrans* by Hennings and Hoffmann (1977); Carraway's (1990) multivariate morphometric analysis affirmed Hennings and Hoffmann's (1977) view. Because only 12 specimens were identified as *S. trigonirostris*, the status of the species was of some concern (Olterman and Verts, 1972). Some recent publications (e.g., Nowak, 1991) and agency reports reflect neither the taxonomic changes nor the concomittant change of status.

FAMILY TALPIDAE—SHREW-MOLE AND MOLES

The family Talpidae is distributed throughout all of Europe and Asia south of ≅65° north latitude except for areas south of the Black and Caspian seas and Himalaya Mountains; representatives also occur in Japan, and in southern Canada and the United States except for the western Great Plains, Rocky Mountains, and Great Basin (Yates, 1984). The family includes 42 species in 17 genera (Wilson and Reeder, 1993a).

The talpids are represented in Oregon by four species: three species of moles of the genus *Scapanus* and the shrew-mole, genus *Neurotrichus*. All are dark gray or blackish animals with fusiform bodies; short appendages; manus, forelimbs, and pectoral girdle modified for burrowing; manus rotated so palm faces outward; entirely white teeth; unicuspid first upper incisors; first upper incisors as wide

as long; dilambdodont (W-shaped) upper molars; and odor-producing skin glands. The incisors, canine, and first two premolars are not strongly differentiated and are commonly referred to as unicuspid (abbreviated "U"). All are plantigrade and pentadactyl. The skull is compressed dorsoventrally with a long rostrum, complete auditory bullae, and complete but weak zygomata; there is no jugal bone. Moles are the size of small rats, robustly built with elongated snouts, superior nostrils, short limbs, forefeet widened by a falciforme bone and highly modified for digging, extremely small eyes, no pinnae, velvety pelage lacking guard hairs, 44 teeth, and scaly tails covered with short bristles. The shrew-mole is superficially shrew-like; however, the nostrils are directed laterally, the forelimbs are slightly modified for digging, there are 36 white teeth, and guard hairs are present in the fur.

Reproductive data for species of moles occurring in the state are presented in Table 8-4 and morphometric data for Oregon moles are presented in Table 8-5.

Neurotrichus gibbsii (Baird)
Shrew-mole

1858. *Urotrichus gibbsii* Baird, 8:76.
1880. *Neurotrichus gibbsii*: Günther, pl. 42.
1899b. *Neurotrichus gibbsi major* Merriam, 16:88.

Description.—The shrew-mole (Plate II) is the smallest talpid in Oregon (Table 8-5). The pelage is black; the eyes are rudimentary; the external auditory meatus is slitlike; the nose is elongate, flattened dorsoventrally, opens laterally, and possesses a fringe of bristles; and the mouth is inferior. The feet are pentadactyl, digitigrade, and scaly; the forefeet are about 4 mm wide, 5.4 mm long, and 2 mm deep; tuberculate on the soles; and equipped with long curved claws (Dalquest and Orcutt, 1942). The tail is about 50% of the length of the head and body, fat, sparsely haired, blunt ended, covered with transverse annular rows of scales, and tufted (Jackson, 1915).

Neurotrichus gibbsii can be distinguished from other talpids in Oregon by its smaller size, 36 instead of 44 teeth (Table 1-1), bilobed basal projections on M1 and M2, thin and only slightly bowed zygomata (Fig. 8-21), and forefeet only moderately modified for digging (Carraway and Verts, 1991b). It may be distinguished from shrews of similar size and color by the absence of red pigment on the teeth. *N. gibbsii* is the only known mammal with a pigmented layer covering the anterior surface of the lens of the eye (Lewis, 1983) and the only talpid with a pair of ampullary glands and a lobate prostate gland (Eadie, 1951).

Distribution.—The geographic distribution of *N. gibbsii* (Fig. 8-22) includes the region west of the Cascade and Sierra Nevada ranges from the Frazier River, British Columbia, south through Washington, Oregon, and California as far as Fremont Peak, Monterey Co. It also occurs on Destruction Island, Washington (Carraway and Verts, 1991b). In Oregon, the species occurs as far east as Brooks Meadows, Hood River Co.; Indian Ford Campground, Deschutes Co.; and Fort Klamath, Klamath Co. No museum specimens have been collected in much of the northern Coast Range or in the interior valleys (Fig. 8-22); we do not know if these are unoccupied or unsampled areas. No Pleistocene fossils are known (Kurtén and Anderson, 1980), although fossil congeners are known from Poland

Table 8-4.—*Reproduction in some Oregon talpids.*

Species	Number of litters	Litter size $\bar{X}$	Litter size Range	Litter size n	Reproductive season	Basis	Authority	State or province
Neurotrichus gibbsii		2.8	1–4	5	Feb.–Sep.	Embryos	Dalquest and Orcutt, 1942	Washington
Scapanus orarius	1	3.0[a]			Jan.–Apr.	Embryos	Glendenning, 1959	British Columbia
		3.5	3–4	2		Implant sites	This study	Oregon
Scapanus townsendii	1	2.5	1–3	18	Jan.–Apr.	Embryos	Pedersen, 1963	Oregon
	1	2.8	1–4	43	Jan.–Apr.	Nestlings	Pedersen, 1963	Oregon
		3.1	2–6	9	Feb.–Apr.	Embryos	Moore, 1939	Oregon
		3.3	3–4	7		Implant sites	Moore, 1939	Oregon

[a] Reported as age-specific with yearlings with two embryos, 2-year-olds with three, and older adults with four.

(Storch and Qui, 1983).

Geographic Variation.—No geographic variation in *N. gibbsii* is recognized in Oregon (Carraway and Verts, 1991*b*); all are considered *N. g. gibbsii.*

Habitat and Density.—*Neurotrichus gibbsii* is most abundant in moist sod-free ravines with deep, black-silt soils with high humus content and covered with a layer of dead leaves and twigs. Dominant vegetation in these areas is big-leaf maple (*Acer macrophyllum*), vine maple (*A. circinatum*), red alder (*Alnus rubra*), flowering dogwood (*Cornus nuttallii*), various shrubs and brambles (*Sambuchus racemosa, Rubus spectabilis, R. parviflorus, R. ursinus, Berberis nervosa*), common sword-fern (*Polystichum munitum*), mosses, and, in muddy areas, skunk cabbage (*Lysichitum americanum*). Less frequently, *N. gibbsii* occurs in willow (*Salix*) thickets, conifer forests, riparian hardwoods, pasturelands, headland prairie and shrub habitats, dry woods, burned-over land, and where the soil is dry, hard, or stony (Dalquest, 1941*b*; Dalquest and Orcutt, 1942; Maser et al., 1981*b*; Terry, 1981).

In favorable habitats, 12–15 individuals/ha is the average population size, but after removal of other small mammals, populations as high as 247/ha have been recorded (Dalquest and Orcutt, 1942).

Diet.—Like that of the larger talpids, the diet of *N. gibbsii* consists mostly of earthworms (Oligochaeta); in Oregon, stomachs of 81.8% of 11 individuals contained earthworms composing 48.5% of the volume of the contents (Whitaker et al., 1979). Other items in the diet included 4.3% centipedes (Chilopoda), 4.1% slugs (Gastropoda), 9.9% flies (Diptera), 11.4% beetles (Coleoptera), 13.6% unidentified insects, and 0.6% vegetation. However, in Washington, lichens and seeds of conifers contributed significantly (32% and 36%, respectively) to the diet in July, although by September 75–88% of the diet consisted of invertebrates (Gunther et al., 1983). Digestion is extraordinarily rapid; feces containing residues of ingested foods are passed within 35–40 min (Banfield, 1974).

Reproduction, Ontogeny, and Mortality.—In Wash-

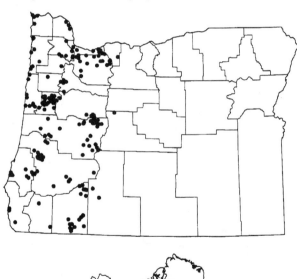

Fig. 8-22. Localities in Oregon from which museum specimens of the shrew-mole, *Neurotrichus gibbsii*, were examined (listed in the APPENDIX, p. 536), and the distribution of the shrew-mole in North America. (Range map redrawn from Hall, 1981:67, map 39, with modifications.)

Fig. 8-21. Dorsal, ventral, and lateral views of the cranium and lateral view of the dentary of the shrew-mole, *Neurotrichus gibbsii* (OSUFW 7376). Scale bar equals 5 mm.

Table 8-5.—*Means (±SE), ranges (in parentheses) and CVs of measurements of skull and skin characters for female and male Talpidae from regions in Oregon. Skin characters were recorded from specimen tags; skull characters were measured to the nearest 0.01 mm. SE and CV not provided if* n *<10.*

Species and region	Sex	n	Total length	Tail length	Hindfoot length	Mass (g)	Greatest length of skull	Maxillary breadth	Least interorbital breadth	Cranial breadth	Length of maxillary toothrow	Length of mandible	Length of mandibular toothrow
Neurotrichus gibbsii	♀	30	108 ± 1.4 (92–121) 7.1	38 ± 0.7 (31–50) 9.5	15 ± 0.2 (12–17) 6.2	8.96 ± 0.36[a] (6–13) 20.62	21.78 ± 0.11 (20.69–22.95) 2.62	5.89 ± 0.05 (5.51–5.63) 4.89	5.23 ± 0.03 (4.93–5.62) 3.07	10.14 ± 0.05 (9.51–10.75) 2.50	7.72 ± 0.05 (7.07–8.25) 3.51	13.82 ± 0.09 (12.82–14.64) 3.31	7.62 ± 0.05 (6.97–8.16) 3.26
	♂	30	108 ± 1.7 (92–121) 8.4	37 ± 0.7 (30–44) 10.2	15 ± 0.3 (11–18) 9.5	8.76 ± 0.28[b] (6.2–13) 16.38	21.78 ± 0.09 (20.68–23.0) 2.48	5.88 ± 0.07 (5.24–7.12) 6.97	5.19 ± 0.03 (4.72–5.61) 3.78	10.16 ± 0.04 (9.56–10.62) 2.32	7.76 ± 0.05 (7.11–8.35) 3.62	13.76 ± 0.08 (12.63–14.52) 3.26	7.59 ± 0.07 (6.82–8.21) 4.84
Scapanus latimaus	♀	8	173[c] (150–183)	39 (30–55)	22[c] (21–23)	73.23[d] (66.7–78)	35.36 (34.37–35.86)	10.45 (10.19–11.06)	7.77 (7.49–8.14)	17.24 (16.89–17.91)	11.79 (11.46–12.02)	22.86 (22.09–23.47)	11.67 (11.35–11.93)
	♂	12	170 ± 3.4 (150–192) 7.0	33 ± 1.5[c] (25–41) 15.0	21 ± 0.8 (16–25) 12.5	65.52[e] (55.6–73.9)	35.41 ± 0.34 (33.57–38.13) 3.33	10.31 ± 0.11 (9.73–10.90) 3.73	7.69 ± 0.09 (7.36–8.16) 3.93	17.02 ± 0.15 (15.98–17.93) 3.05	11.96 ± 0.12 (11.50–13.09) 3.39	22.88 ± 0.25 (21.19–22.0) 3.81	11.77 ± 0.14 (11.07–12.94) 4.27
Scapanus orarius West of Cascade Range	♀	31	161 ± 1.4 (148–179) 4.7	33 ± 0.5 (27–39) 8.5	21 ± 0.2 (19–23) 4.9	56.38 ± 3.33[f] (41–78.8) 18.65	33.48 ± 0.14 (32.62–35.5) 2.31	9.28 ± 0.07 (8.70–10.50) 4.44	7.68 ± 0.04 (7.22–8.14) 3.16	16.14 ± 0.08 (15.15–17.19) 2.65	11.52 ± 0.06 (10.26–12.03) 3.05	21.38 ± 0.13 (20.22–22.93) 3.47	11.37 ± 0.05 (10.90–11.88) 2.59
	♂	30	159 ± 1.8 (138–179) 6.1	34 ± 0.7 (25–41) 11.8	21 ± 0.2 (19–22) 4.8	55.19 ± 2.63[g] (38–73.4) 17.60	33.81 ± 0.14 (32.18–36.04) 2.26	9.28 ± 0.05 (8.83–10.07) 2.85	7.70 ± 0.04 (7.10–8.08) 2.79	16.27 ± 0.05 (15.67–16.87) 1.81	11.58 ± 0.05 (11.11–12.18) 2.39	21.53 ± 0.09 (20.59–22.84) 2.28	11.37 ± 0.05 (10.89–12.0) 2.47
East of Cascade Range	♀	7	156[c] (140–166)	34[c] (28–37)	21[c] (19–23)	50.37[d] (45.73–55)	33.09 (32.08–33.86)	9.58 (8.55–10.11)	7.52 (6.65–8.01)	15.84 (14.78–16.40)	11.49 (10.35–12.04)	21.21 (20.02–22.0)	11.25 (10.17–11.73)
	♂	4	155[a] (149–161)	34[a] (32–34)	21[a] (20–21)	43.0	33.03 (32.19–34.29)	9.79 (9.37–10.03)	7.51 (6.92–7.91)	16.27 (15.83–16.26)	11.49 (11.15–11.96)	21.12 (20.73–21.90)	11.19 (10.80–11.76)
Scapanus townsendii	♀	17	206 ± 1.2 (201–218) 2.3	41 ± 1.0 (33–47) 10.2	25 ± 0.4 (21–28) 6.7	118.8[h] (87.4–176.4)	40.84 ± 0.21 (39.46–42.62) 2.15	11.62 ± 0.08 (11.21–12.44) 2.97	8.30 ± 0.07 (7.85–8.74) 3.28	19.66 ± 0.09 (18.90–20.15) 2.09	14.22 ± 0.09 (13.42–14.96) 2.68	27.43 ± 0.18 (26.39–29.10) 2.71	14.21 ± 0.08 (13.46–14.82) 2.34
	♂	30	210 ± 1.4 (200–230) 3.6	41 ± 0.9 (30–51) 12.2	26 ± 0.3 (22–29) 6.5	142.4 ± 5.32[i] (109–231) 19.04	41.31 ± 0.16 (39.66–43.58) 2.16	11.84 ± 0.07 (11.11–12.73) 3.24	8.36 ± 0.07 (7.63–9.12) 4.44	19.83 ± 0.07 (19.05–20.55) 2.06	14.23 ± 0.08 (13.65–15.21) 3.02	28.04 ± 0.13 (27.11–29.68) 2.59	14.14 ± 0.08 (13.34–15.15) 2.99

[a] Sample size reduced by 2. [b] Sample size reduced by 3. [c] Sample size reduced by 1. [d] Sample size reduced by 5. [e] Sample size reduced by 7. [f] Sample size reduced by 21. [g] Sample size reduced by 16. [h] Sample size reduced by 8. [i] Sample size reduced by 4.

ington, animals in breeding condition were captured from February to September. Although most breeding is March–mid-May, only about 5% of animals captured during this peak period were in breeding condition (Dalquest and Orcutt, 1942). Average litter size is three (Table 8-4). The gestation period is not known. Reproduction in Oregon has not been evaluated.

At birth, *N. gibbsii* is pink and naked, and has closed auditory meatuses; a 1-h-old neonate weighed 0.67 g and had a total length of 26.0 mm, a tail length of 5.0 mm, and a hind-foot length of 3.6 mm (Kritzman, 1972).

Shrew-moles are known to be preyed upon by the barn owl (*Tyto alba*), great horned owl (*Bubo virginianus*), saw-whet owl (*Aegolius acadicus*), long-eared owl (*Asio otis*), red-tailed hawk (*Buteo jamaicensis*), coyote (*Canis latrans*), raccoon (*Procyon lotor*), and garter snake (*Thamnophis ordinoides*—Cowan, 1942; Dalquest and Orcutt, 1942; Forsman and Maser, 1970; Giger, 1965; Maser and Brodie, 1966; Toweill and Anthony, 1988*b*; von Bloeker, 1937).

Habits.—Shrew-moles construct and forage along networks of runways about 40 mm wide and 20 mm deep in the soft soil beneath the duff. Sometimes ≤28-mm-diameter burrows are dug usually within 12.7 cm of the surface but occasionally as deep as 30.5 cm (Dalquest and Orcutt, 1942; Racey, 1929). Sleeping chambers are constructed beneath small openings to the surface. *N. gibbsii* is active for intervals of 2–18 min at all hours; periods of activity are followed by rest or sleep for 1–8 min (Dalquest and Orcutt, 1942). When startled on the surface, shrew-moles scurry to the nearest cover and remain motionless for ≤1 min. They can climb and swim, but are not known to jump.

Because shrew-moles are completely sightless, they rely on their prehensile nose to locate prey; prey is located, then the ground is tapped with the nose as the shrew-mole slowly advances until the prey is touched (Maser et al., 1981*b*). Earthworms are immobilized or cut into sections before being eaten by shrew-moles; more mobile prey may be tipped over with the nose before being pounced upon (Dalquest and Orcutt, 1942).

Shrew-moles are thought to be gregarious and travel in bands of as many as 11 individuals; they move into an area for a few hours or days then move on (Dalquest and Orcutt, 1942). No information is available on the size or stability of home ranges, or on whether portions thereof are defended as territories at some seasons.

Remarks.—Shrew-moles are not known to cause economic damage in any habitat in which they occur.

Because of several unusual morphological features, the anatomy of the shrew-mole has been studied extensively (Carraway and Verts, 1991*b*). However, little is known regarding the physiology, reproduction, and movements of the species. A morphometric study of geographic variation of the species, with special attention to newly discovered populations east of the Cascade Range in Oregon, is warranted.

Scapanus latimanus (Bachman)
Broad-footed Mole

1842. *Scalopus latimanus* Bachman, 4:34.
1894. *Scapanus dilatus* True, p. 2 (preprint of Proc. USNM, 17:242).
1897i. *Scapanus alpinus* Merriam, 11:102.
1913b. *Scapanus latimanus dilatus*: Grinnell, 3:269.

Description.—*Scapanus latimanus* is intermediate in size among Oregon moles (Table 8-5). It can be distinguished from congeners by its lighter gray pelage with a coppery wash and by U5 and U6 being appressed (Fig. 8-23). Occasionally the latter character is exhibited on only one side. Colormorphs similar to those described and depicted for *S. townsendii* (Carraway and Verts, 1991a) have been reported (Palmer, 1937).

Distribution.—The geographic distribution of *S. latimanus* (Fig. 8-24) extends from Fort Rock, Lake Co., Oregon, south through California (except for a narrow coast strip in the northwest corner of the state, the western portion of the Sacramento and San Joaquin valleys, and the Colorado and Mohave deserts) and extreme western Nevada south to the East Walker River, into Baja California as far as La Grulla (Hall, 1981; Palmer, 1937). In Oregon,

S. latimanus occurs south of a line connecting Hugo, Josephine Co.; Prospect, Jackson Co.; Crater Lake, Klamath Co.; and Fort Rock and Goose Lake, Lake Co. (Fig. 8-24). Fossils of *S. latimanus* are known from California within the present-day range (Kurtén and Anderson, 1980).

Geographic Variation.—Although all specimens collected in Oregon were assigned to *S. l. dilatus*, specimens collected in California within 0.4 km of the Oregon border were assigned to *S. l. caurinus* (Hall, 1981; Palmer, 1937). Animals from southern Josephine Co. might be referable to the latter race or might be intergrades between the two races. *S. l. caurinus* is larger and has a darker pelage than *S. l. dilatus* (Palmer, 1937).

Diet.—In captivity, four *S. latimanus* ate an average of 70.2% (range, 53.4–81.0%) of their body mass of dog food (95% horse meat) daily; at least one individual gained mass (Grim, 1958). Beach hoppers (*Orchestoidea californiana*) apparently are eaten by individuals that extend surface tunnels onto the beach in California (McCully, 1967). Stomach contents of one individual captured in Oregon consisted of 80% Gastropoda and 20% Coleoptera (Whitaker et al., 1979).

Remarks.—*Scapanus latimanus* is another Oregon species about which essentially nothing is known regarding its ecology, reproduction, or behavior. Palmer (1937) considered soil moisture limiting for *S. latimanus* and offered as evidence the observation that the distribution in Oregon ex-

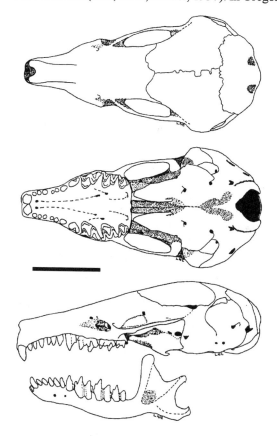

Fig. 8-23. Dorsal, ventral, and lateral views of the cranium and lateral view of the dentary of the broad-footed mole, *Scapanus latimanus* (OSUFW 2157). Scale bar equals 10 mm.

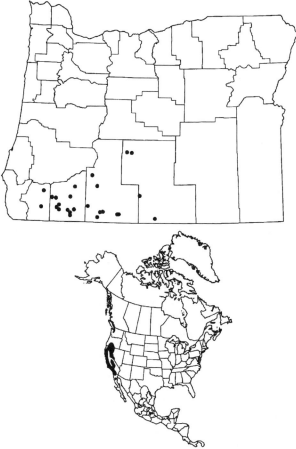

Fig. 8-24. Localities in Oregon from which museum specimens of the broad-footed mole, *Scapanus latimanus*, were examined (listed in the APPENDIX, p. 548), and the distribution of the broad-footed mole in North America. (Range map redrawn from Hall, 1981:69, map 41, with modifications.)

tended farther northward east of the Cascade Range than on the west. He further suggested that the species was limited by arid, baked soils that prevented digging and by looser soils that held fewer invertebrates fed on by this mole.

Relatively large numbers of specimens of *S. latimanus* are held in several collections; Palmer (1937) examined 518 specimens and we recorded 52 from Oregon alone. This suggests that individuals are not especially difficult to trap, although Grinnell and Swarth (1912) complained of having difficulty capturing individuals of this species. We are convinced that a hard-working, energetic, ingenious graduate student such as those that worked on *S. townsendii* (Giger, 1965, 1973; Pedersen, 1963, 1966), using modern techniques, could capture sufficient numbers of broad-footed moles throughout the year to produce the first in-depth treatment of the natural history of the species.

Scapanus orarius True
Coast Mole

1896. *Scapanus orarius* True, 9:52.
1915. *Scapanus orarius schefferi* Jackson, 38:63.

Description.—*Scapanus orarius* (Plate II) is the smallest of the moles in Oregon (Table 8-5). It can be distinguished from *S. latimanus* by its blackish or dark grayish pelage without a coppery wash and evenly spaced unicuspid teeth, and from *S. townsendii* by a total length <200 mm, a total length of the skull <40 mm, and a sublacrimal ridge (Fig. 8-25) not strongly developed (Hartman and Yates, 1985; Jackson, 1915).

Distribution.—The coast mole occurs from the extreme southwestern corner of mainland British Columbia south along the Pacific Coast through Washington and Oregon to Mendocino, California. East of the Cascade Range the distribution extends to the Idaho border in Washington, across northern Oregon, and into extreme west-central Idaho (Fig. 8-26; Hartman and Yates, 1985). In Oregon, the species occurs in Baker, Umatilla, Grant, Crook, Union, Sherman, and Wasco counties east of the Cascade Range and throughout most of the area west of the Cascade Range, except it is absent from much of the Willamette Valley (Fig. 8-26).

No fossils of *S. orarius* are known (Hartman and Yates, 1985).

Geographic Variation.—Two geographic races are recognized in Oregon: those along the McKenzie River in Lane Co. and those in eastern Oregon were considered *S. o. schefferi* and those in the remainder of the state were considered *S. o. orarius* by Hall (1981); no zone of intergradation was depicted. However, Hartman and Yates (1985) indicated a zone of intergradation between eastern and western races in the vicinity of Arlington, Gilliam Co., and Boardman, Morrow Co., a region from which we found no museum specimens (Fig. 8-26). *S. o. schefferi* is larger and has a darker pelage than *S. o. orarius*.

Habitat and Density.—Coast moles inhabit a variety of vegetational and soil types ranging from coastal-dune, pastureland, and meadow habitats to several deciduous- and coniferous-forest habitats, and to sagebrush (*Artemisia*)-grass associations (Hartman and Yates, 1985), but they avoid beach, coastal lake, and tideland river areas (Maser et al., 1981*b*). Although the consensus is that *S. orarius* tends to occupy streamside and forested habitats with light, well-

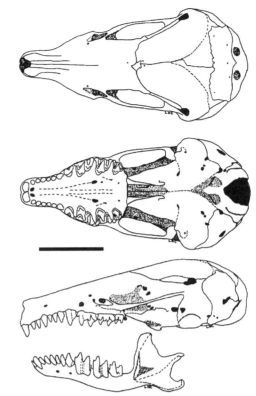

Fig. 8-25. Dorsal, ventral, and lateral views of the cranium and lateral view of the dentary of the coast mole, *Scapanus orarius* (OSUFW 3850). Scale bar equals 10 mm.

drained soils, whereas *S. townsendii* tends to occupy more open habitats with heavier soils with high water content (Hartman and Yates, 1985), Schaefer and Sadleir (1981:37) concluded that *S. orarius* "dug mainly in areas with light, wet soil." Nevertheless, both frequently inhabit the same field where the species are sympatric. Because coast moles occur in such a variety of habitats, vegetation types, and soil densities other factors must be limiting. Glendenning (1959) described a strong positive relationship between density of moles and abundance of earthworms (Oligochaeta); heavily manured soil supported as many as 10 individuals/ha, but populations as low as one individual/14 ha were recorded.

Diet.—Earthworms (Oligochaeta) may contribute more to the diet of *S. orarius* than to that of congeners; Glendenning (1959) reported that 93% of the stomach contents of 108 coast moles consisted of earthworms, 2% of slugs (Gastropoda), 2% of insect larvae, and 1% each of adult insects, earthworm ova, and unidentified material. However, Whitaker et al. (1979) reported that 92% of 25 contained earthworms in their stomachs, but earthworms composed only 56.2% of the volume of the contents. In western Oregon the diet seems much more eclectic as none of 27 other items recorded in the stomach contents contributed >6.3% (centipedes, Chilopoda); Formicidae (ants and termites) contributed 6.1% (plus an additional 3.9% as pupae), small mammals 6.0%, and vegetation 4.1%. Glendenning (1959:40) reported as unique among moles the refusal by *S. orarius* to consume "meat scraps or any dead food"; however, he claimed (p. 39) that an individual "will consume part of its vanquished opponent after a fight with another mole," and Dalquest (1948) reported that a vole (*Microtus oregoni*) caught in a snap trap was partly

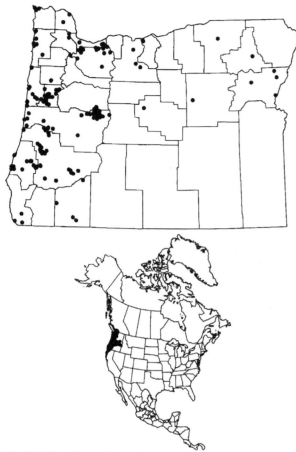

Fig. 8-26. Localities in Oregon from which museum specimens of the coast mole, *Scapanus orarius*, were examined (listed in the APPENDIX, p. 549), and the distribution of the coast mole in North America. (Range map redrawn from Hall, 1981:68, map 40, with modifications.)

eaten by a coast mole.

Reproduction, Ontogeny, and Mortality.—In British Columbia, mating occurs from January to early March but apparently is not synchronous as females with similar-sized embryos were collected a month apart in March and April (Glendenning, 1959). Litter size reportedly is strongly age-specific with yearlings producing two young, 2-year-olds three, and older adults four; sample sizes were not provided and variation in number of offspring within age-classes was not indicated (Table 8-4). Females caught in May "had given birth to young, teats were developed, and the ovaries were shrunken" (Glendenning, 1959:39). No studies of reproduction in *S. orarius* have been conducted in Oregon and productivity records were obtained from only two individuals (Table 8-4). Seemingly, only one litter is produced annually, probably because the testes commence to return to the inactive state in March (Glendenning, 1959). The length of the gestation period is not known. Nests containing two 13–15-g and four 40-g young were discovered in April and May, respectively. The smaller young were nearly naked, measured 5.5 cm long, and possessed well-formed feet and tails; the larger were considered to be about half grown (Glendenning, 1959). Three 32-g young were considered not to be weaned (Glendenning, 1959).

Raptors, snakes (Serpentes), and carnivorous pets are known to take coast moles, but the latter predators seldom

eat them (Hartman and Yates, 1985). In Oregon, remains of twice as many *S. orarius* (28) as *S. townsendii* (14) occurred in regurgitated pellets of barn owls (*Tyto alba*—Giger, 1965). Although remains of *S. orarius* occurred during 6 months as opposed to 3 months for *S. townsendii,* periods with the most records were similar (May–July). The owls also left large numbers of moles uneaten in their roosts (Giger, 1965).

Longevity is unknown, but Glendenning (1959) reported that 6% of 940 coast moles captured in British Columbia were >3 years old; 40% were >1 year old.

Habits.—Coast moles, like Townsend's moles, are largely fossorial; surface activity as indicated by the occurrence of remains in owl pellets is mostly by young-of-the-year during dispersal (Giger, 1965). Greatest activity is in winter when temperatures are lower and soil is more moist (Schaefer and Sadleir, 1981). From midautumn to early spring, individuals produce exclusive networks of 5-cm-diameter tunnels ("encampments") 15–20 cm below the surface and as long as 450 m; 8- by 10-cm chambers are constructed at intervals along the tunnels or at the junction of tunnels. Soil from these excavations is pushed out of voiding holes to form 200–400 mounds, each about 30 cm in diameter and 15 cm high (Glendenning, 1959). Construction of mounds requires an average of about 33 min (Schaefer, 1982). During the mating season the encampments often are connected by long runways marked by large, widely separated mounds; apparently these runways are produced by males as females are not caught along them (Glendenning, 1959). In contrast to Townsend's moles, coast moles do not produce nesting mounds or "fortresses." Young moles, after leaving the nest, forage by pushing up surface tunnels (no mounds are produced), sometimes >90 m/night (Glendenning, 1959).

Although Maser et al. (1981b) claimed that coast moles were active primarily at night, an individual of unknown sex monitored for 50.8 h by radioisotope tracking mostly during the daylight hours "did not confine its activity to any specific part of the day" (Schaefer, 1982:481). Commonly, the individual was active for 4–5 h followed by a 3–4-h period of inactivity. Most activity was in the vicinity of the nest; most foraging was in the vicinity of mounds (Schaefer, 1982). The home range of the same coast mole occupied an area of about 39 by 39 m (Schaefer, 1982); 15 others for which home ranges were described from evidence of activity restricted their movements to areas about 30 by 40 m (Schaefer, 1981).

Remarks.—Considering that *S. orarius* is associated more frequently with forest habitats, its burrowing activities are less likely to be undesirable than those of *S. townsendii.* However, control may be warranted in some situations; trapping with scissor-jaw traps is the most effective and economical method of control (Glendenning, 1959). Glendenning (1959) suggested that in pristine ecosystems the burrowing activities of the coast mole might be beneficial, but that in agricultural lands benefits derived from aeration, drainage, and mixing of the soil and from consumption of noxious invertebrates were outweighed by the deleterious effects of tunneling and mound construction on plants and machinery. We believe that the economic contributions remain open to evaluation.

In view that we found only 14 specimens of *S. orarius* from east of the Cascade Range in Oregon, and relatively little published information on the coastal race, much more extensive investigation of the status, ecology, and behavior of the species is warranted. However, of foremost importance is a quantitative analysis of geographic variation in *S. orarius* accompanied by an appropriate taxonomic revision; knowledge of the affinities of populations in the McKenzie River drainage and the relationship of populations separated by the Cascade Range is particularly critical. From our cursory observations of coast mole activity east of the Cascade Range, we suspect that the locality records based on museum specimens in that region (Fig. 8-26) represent widely disjunct populations.

Scapanus townsendii (Bachman)
Townsend's Mole

1839*b. Scalopus Townsendii* Bachman, 8:58.
1848.   *Scapanus Tow[n]sendii*: Pomel, 9:247.
1853.   *Scalops aeneus* Cassin, 1853:299.

Description.—Townsend's mole (Fig. 8-27) is the largest talpid in Oregon (Table 8-5). It can be distinguished from *S. latimanus* by its almost black pelage, absence of a coppery wash on the fur, and even-spaced unicuspids, and from *S. orarius* by a total length >200 mm, length of hind foot >24 mm, total length of the skull usually ≥40 mm, and prominent sublacrimal ridges (Fig. 8-28; Carraway et al., 1993; Jackson, 1915). A variety of colormorphs—including some with small white spots or splotches of reddish yellow on both grayish and normally black pelages; some with mottled pelages of gray, brown, grayish brown, and light yellow on both dorsum and venter; some entirely white shading to yellow; some entirely reddish yellow; and some with a variety of these colors in patterns—have been recorded (Carraway and Verts, 1991*a*). Males possess a tiny, almost transparent baculum averaging 0.42 mm long and 0.27 mm wide (Maser and Brown, 1972).

Fig. 8-27. Photograph of Townsend's mole, *Scapanus townsendii*. (Reprinted from Carraway et al., 1993, with permission of the American Society of Mammalogists.)

Distribution.—*Scapanus townsendii* occurs from Huntington, British Columbia, southward through western Washington, Oregon, and the coastal region of California to Ferndale (Fig. 8-29; Carraway et al., 1993). In Oregon, it is restricted to the interior valleys and coastal regions west of the foothills of the Cascade Range (Fig. 8-29). No fossils of *S. townsendii* are known.

Geographic Variation.—No geographic variation is recognized in Townsend's moles in Oregon; all are con-

sidered *S. t. townsendii* (Carraway et al., 1993).

Habitat and Density.—In Oregon and most areas within its geographic range, *S. townsendii* occupies pastures, prairies, and shrub habitats in lowlands and river flood plains (Pedersen, 1963; Yates and Pedersen, 1982). In Washington, it was reported to occur in fir (*Abies*) forests (Dalquest, 1948).

Highest densities (12.4 moles/ha) were recorded in pasture areas in Tillamook Co., but, in the same region, densities as low as 0.42 moles/ha were recorded in poorly drained areas with few earthworms (Oligochaeta). Removal of 1.9 moles/ha and 1.0 moles/ha from two large fields resulted in cessation of mole activity (Giger, 1973). Also, eight nest mounds were recorded in a 6.1-ha field in the same region (Kuhn et al., 1966).

Diet.—The primary food of Townsend's mole is earthworms and insects; stomachs of 308 *S. townsendii* from the Willamette Valley contained an average of 72.5% earthworms with a frequency of occurrence of 85.6% earthworms, 30.7% earthworm cocoons, 31.0% insect larvae and pupae, 13.0% centipedes (Chilopoda), 2.6% each for slugs (Gastropoda) and adult insects, and 2.6% miscellaneous items that included a wild bulb, grain, and unidentified starch granules (Wight, 1928). Stomachs of 106 male *S. townsendii* from Tillamook Co. contained 63.7% earthworms and 36.3% grass roots by volume and those of 76 females contained

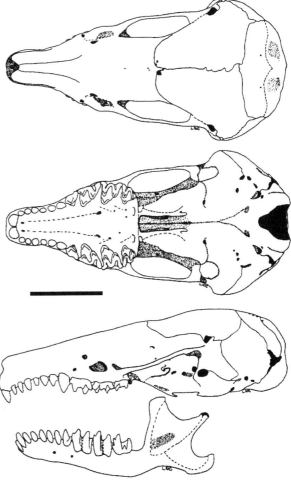

Fig. 8-28. Dorsal, ventral, and lateral views of the cranium and lateral view of the dentary of Townsend's mole, *Scapanus townsendii* (OSUFW 1005). Scale bar equals 10 mm.

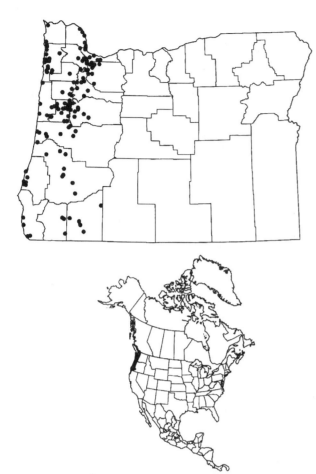

Fig. 8-29. Localities in Oregon from which museum specimens of Townsend's mole, *Scapanus townsendii*, were examined (listed in the APPENDIX, p. 550), and the distribution of Townsend's mole in North America. (Range map redrawn from Hall, 1981:69, map 41, with modifications.)

an average of an 65.8% earthworms and 39.2% grass roots (Whitaker et al., 1979). Samples collected elsewhere in Oregon also contained a preponderance of earthworms (Moore, 1933; Whitaker et al., 1979). Small amounts of vegetation including a variety of garden vegetables were recorded (Moore, 1933), but this material may be eaten largely to obtain water (Wight, 1928). Food consumption averages one-third to two-thirds of body mass/day or about 18 kg/year (Cahalane, 1947; Wick and Landforce, 1962).

Reproduction, Ontogeny, and Mortality.—Onset of the breeding season, as indicated by males with enlarged testes and females with vaginal plugs, commences in February in Tillamook Co. (Moore, 1939). However, the first obviously pregnant female was not taken until 13 March, and none of 14 males caught between 17 March and 20 April was considered in breeding condition. In another study in the same area, a specimen collected on 22 February exhibited the earliest evidence of pregnancy, and a specimen with well-developed embryos was collected on 8 March (Pedersen, 1963). Moore (1939) caught no pregnant females after 12 April, but Pedersen (1963) caught one with tiny embryos on 4 April. The gestation period is unknown. It seems likely that males are not in breeding condition sufficiently long for more than one litter to be produced annually (Carraway et al., 1993).

Litter size, as measured by various methods, seems to

average slightly less than three with a range of one to four (Table 8-4). At birth, the naked, pinkish young weigh about 5 g. They have no teeth; the eyes are not distinguishable; and the claws are soft (Kuhn et al., 1966; Pedersen, 1963). By 10 days, the young are grayish; by 15–20 days, pelage has begun to develop; and by 30 days, they are completely furred. They weigh 60–80 g by 1 month of age (Kuhn et al., 1966). Reproductive competence is attained at about 10 months.

Of 180 nestlings marked in April–May on a 150-ha study area near Tillamook, Tillamook Co., Giger (1973) found 58 (32.2%) dead in the nests in June, 44 (24.4%) were recaptured after mid-September, and 78 (43.3%) were not accounted for. Because of its fossorial habit, *S. townsendii* largely avoids most predation to which terrestrial small mammals are subjected (Giger, 1973). Nevertheless, raptorial birds, snakes (Serpentes), and carnivorous mammals are known to prey on Townsend's moles; most of those preyed upon are juveniles (Giger, 1965). Domestic dogs and house cats commonly kill moles but usually do not eat them (Maser et al., 1981b; Silver, 1933). Although Townsend's moles are excellent swimmers, flooding can have a catastrophic affect on their populations; apparently because of poor eyesight they cannot find higher ground (Giger, 1973).

Habits.—Townsend's moles are almost entirely fossorial; surface activity seems restricted to times when their tunnels are flooded and to juveniles during dispersal as indicated by the occurrence of their remains in barn owl (*Tyto alba*) pellets during May–June (Giger, 1965). They construct extensive permanent tunnel systems, relatively shallow in the open, but deeper under the protection of fencerows, foundations, and roadbeds (Silver, 1933; Yates and Pedersen, 1982). Spoil resulting from tunneling is pushed to the surface to form mounds of characteristic shape on the surface (Fig. 8-30); as many as 805 mounds/ha were counted (Yates and Pedersen, 1982) and one male produced 302 mounds in 77 days (Nowak and Paradiso, 1983). Interconnecting tunnel systems produced by individuals with adjacent home ranges may be used during dispersal, but the absence of overlapping movements tends to indicate intraspecific agonistic behavior. As might be expected of a fossorial mammal, movements tend to be extremely restricted; the average ($\pm SD$) distance between points of capture for 24 adults caught twice was 23 ± 15.5 m and for 14 caught three or more times it was 40.5 ± 28 m (Giger, 1973). Of 32 individuals displaced 34 – 455 m (some across canals, roadways, or rivers) 14 successfully returned to their

Fig. 8-30. Photograph of mounds produced by a Townsend's mole, *Scapanus townsendii*. (Reprinted from Carraway et al., 1993, with permission of the American Society of Mammalogists.)

former home ranges (Giger, 1973). Of marked nestlings recaptured, 86.3% had dispersed <152 m and 61.4% had dispersed <305 m; 6.8% had crossed paved roads.

Nest mounds produced from spoil from construction of nest cavities may be 0.75–1.25 m in diameter or may take the form of a series of small mounds within a 1.8–3.1-m-diameter area (Kuhn et al., 1966). Nests, consisting of a core of fine dry grass and an outer layer of green and damp vegetation, usually are constructed close to the surface so that neonates are warmed by the heat of the sun and heat produced by the fermenting outer layer of vegetation (Kuhn et al., 1966; Pedersen, 1966).

Remarks.—Because of the large number of mounds of earth they throw up, Townsend's moles are considered pests when they inhabit lawns, golf courses, cemeteries, and pasture lands. In relatively small areas, control of damage can be accomplished by removal trapping with scissor-jaw, Cinch, squeeze, diamond-jaw, or choker-loop traps. Pamphlets containing diagrams and descriptions of techniques for setting these devices for efficient capture of moles and limiting their hazard for other species are available from county agents. However, trapping is not economically feasible in pasture lands and other large expanses. Most toxic baits are no longer registered for use, and, because of the largely insectivorous diet of the species, commonly are not effective for moles, but often kill more individuals of other small-mammal species (Yates and Pedersen, 1982). Excavation of breeding nests and destruction of young may be more effective (Kuhn et al., 1966).

Moles are significant contributors to ecosystems of which they are a part; their tunneling and mound-building activities aerate and mix soil layers and provide drainage. Moles also eat large numbers of insects, insect larvae, and other invertebrate pests (Kuhn and Edge, 1990).

Because of its fossorial habits and because of the difficulty of capturing and observing individuals, much information available for terrestrial species of similar size, abundance, and distribution is not available for *S. townsendii.* Thus, the species is deserving of study of a wide variety of biological phenomena (Yates and Pedersen, 1982).

CHAPTER 9

ORDER CHIROPTERA—BATS

Bats are mammals that fly. Thus, the most distinctive features of bats are those associated with flight. The arm, hand, fingers, legs, and, in some species, tail are modified to support membranes that serve as wings and other aerodynamic surfaces (Fig. 9-1), yet are of sufficiently low mass to permit the wings to provide the necessary lift, thrust, and control for flight. The bones of the arm are modified to allow the flight membranes to be folded when the bat is at rest, but to maintain a rigid wing when in flight (Vaughan, 1986). Bats with long, tapering wings tend to be more rapid flyers than those with broad wings, but ≅80 km/h is maximum for even the fastest flyers in level flight (Vaughan, 1986). By comparison, the bats in Oregon are relatively slow flyers (Table 9-1).

Bats possess all of the attributes considered diagnostic of mammals, but none of the sinister features and behaviors attributed to them in myth and sorcery. In general, they are small, long-lived, have long dependency as infants, low fecundity, and low mortality; unlike most small mammals, they do not attain sexual maturity early in life and usually are members of stable communities in equilibrium (K-strategists—Findley, 1993). Also, as a group they exhibit relatively little morphological differentiation, probably because of adaptations necessary for flight. As a result, bats are some of the most difficult mammals to identify to species (Findley, 1993).

The order Chiroptera is extremely diverse; its 925 named species, classified in 17 families and 177 genera, compose 20% of the recognized species of mammals in the world. It is exceeded in diversity only by the Rodentia (Wilson and Reeder, 1993a). The families are grouped into suborders: Megachiroptera (Old World fruit bats) with one family and Microchiroptera with the remaining 16 families. The Megachiroptera have large eyes and, except for three or four species, depend upon them for orientation and navigation.

The Microchiroptera have small eyes, but produce ultrasonic vocalizations, the echos of which provide information to the individual bat regarding location of prey and obstacles, and other features of the environment. Translocated bats frequently return to their original roosts, sometimes from extraordinary distances. Within areas of familiarity homing likely is based on memory, vision, and echolocation, but bats translocated beyond those areas usually do not return to their home roosts more frequently than expected by random encounter with areas of familiarity (Perkins, 1977).

Bats are distributed worldwide as far north as trees grow and all are crepuscular or nocturnal. As a group, bats have exploited a wide variety of food resources, nearly as diverse as that exploited by all other mammals. Sanguivory is widely attributed to bats in general, but actually engaged in only by three species (≅0.3% of all species) in the New World tropics. Various other species of bats are insectivorous (≅70%), frugivorous (≅23%), nectarivorous (≅5%), carnivorous (≅0.7%), or piscivorous (≅0.6%). A few, at times, may eat a variety of foods so can be considered omnivorous. None is truly herbivorous in the sense of consuming largely leafy material (Hill and Smith, 1984).

The oldest fossil bats date from the early Eocene (≅60 million years ago) of Europe and North America, but because known fossils are remarkably similar to modern bats, the group likely originated much earlier, perhaps as early as the mid-Cretaceous (100 million years ago—Hill and Smith, 1984). Some of the earliest fossils are extremely well preserved, even to including the remains of moths (Lepidoptera) consumed by the bat just before it died.

The bats in Oregon are represented by 15 species in the families Molossidae (free-tailed bats) with one species and Vespertilionidae (evening bats) with the remaining 14 spe-

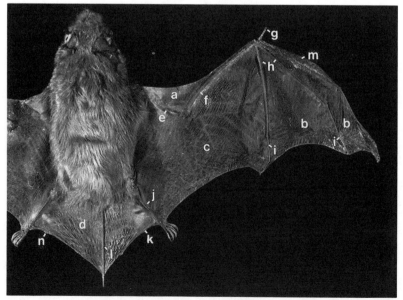

Fig. 9-1. Flight membranes and supporting structures of the big brown bat, *Eptesicus fuscus*: **a**. propatagium; **b**. chiropatagium; **c**. plagiopatagium; **d**. uropatagium; **e**. humerus; **f**. radius and ulna (forearm); **g**. pollex; **h**. metacarpals; **i**. digits; **j**. tibia and fibula; **k**. calcar; **l**. tail; **m**. dactylopatagium, and **n**. keel (epiblema).

Table 9-1.—*Flight speed of some species of bats that occur in Oregon.*[a]

Species	Minimum (km/h)	Maximum (km/h)	Average (km/h)	Individual bats (n)	Total trials (n)
Myotis californicus	9.7	16.4	13.0	3	10
Myotis thysanodes	11.9	15.8	14.0	3	11
Myotis volans	15.0	17.1	15.9	1	6
Myotis yumanensis	10.6	14.3	12.4[b]	3	15[b]
Eptesicus fuscus	12.9	24.9	16.7[b]	4	16[b]
Pipistrellus hesperus	7.7	9.9	8.7	1	2
Corynorhinus townsendii	10.3	19.8	13.2	4	20
Lasionycteris noctivagans	17.2	18.7	18.0	2	6
Lasiurus cinereus	15.8	21.2	18.2	1	14
Antrozous pallidus	12.1	17.1	16.6	6	25
Tadarida brasiliensis	11.1	17.1	14.5	6	25

[a]Recalculated from Hayward and Davis, 1964.
[b]Estimated; original data incomplete.

cies. All are essentially entirely insectivorous; one species gleans insects from the ground, one gleans insects from leaves of trees or bushes, and the remainder forage on flying insects (Nowak, 1991). Freeman (1981) reported a significant relationship between hardness of the diet and robustness of the skulls of 41 species of insectivorous bats. We reanalyzed her data for the 14 Oregon species that she included in her study; similarly, a significant relationship between hardness of the diet and robustness of the skulls was improved with the omission of *Lasiurus* (Fig. 9-2).

Bats are feared by many people. Perhaps some fear is justified because some bats can serve as reservoirs and vectors of the virus that causes rabies, but most likely is unjustified because it has a basis in ignorance, folklore, or superstition. Bats often are considered obnoxious by those whose attics are selected as roosts by colonies, but few homeowners are willing to invest the effort required to exclude bats from a house in which a roost has become established. Most just want to kill the bats on the false assumption that their "bat problem" will be solved without thinking that other bats likely also will find their attics at-

tractive sites in which to establish roosts. Few recognize the contribution of bats to the control of insects and welcome their presence by spreading a sheet of polyethylene beneath the roost in their attic on which to collect the guano to prevent possible staining of ceilings. An annual mid-winter trip to the attic to replace the polyethylene sheet to prevent a build-up of unpleasant odors is a small price to pay for the guano for fertilizing flower beds or garden and for avoiding use of pesticides to control insects.

In general, bat populations in North America have declined precipitously at least since the mid-1940s (Humphrey, 1982; van Zyll de Jong, 1985). A variety of causes were cited as responsible for the declines, including insecticides (directly and through impact on food resources), disturbance of cave roosts by spelunkers and vandals, and destruction of roosting habitat of forest-dwelling species by harvest of timber. Possibly all contributed, but considering that bats are considered to be K-strategists, thus dependent upon communities in equilibrium (Findley, 1993), almost any alteration of the environment might be expected to have a deleterious effect on them.

Morphometric data for Oregon Chiroptera are presented in Table 9-2.

KEY TO THE CHIROPTERA OF OREGON

1a. About one-third of tail projecting beyond posterior margin of uropatagium; fibula robust, half the diameter of tibia; lower incisors bifid; a combination of 32 teeth including one upper incisor and rostrum narrower than braincase..............................***Tadarida brasiliensis***, Brazilian free-tailed bat (**Molossidae**) (p. 119).

1b. No more than a few millimeters of tail protruding beyond posterior margin of uropatagium; fibula slender or rudimentary; lower incisors trifid; number of teeth differ, but if 32, then two upper incisors and rostrum as wide as braincase............................2 (**Vespertilionidae**)

2a. Ear ≥28 mm long..3

2b. Ear ≤26 mm long..5

3a. Pelage black with three large white spots on dorsum; zygomata widen abruptly in the middle; 34 teeth.. ....***Euderma maculatum***, spotted bat (p. 110).

3b. Pelage various colors, but lacking large white spots on dorsum; zygomata of more or less uniform width; number of teeth differ, but never 34....................................4

4a. Pelage light brown to gray on dorsum; individual hairs dark at base; fleshy protuberance projecting dorsally

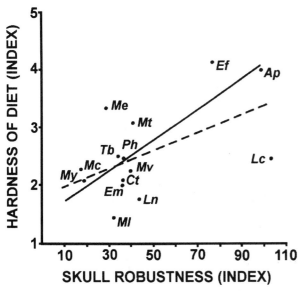

Fig. 9-2. Relationship between hardness of the diet and robustness of the skull of 14 species of insectivorous bats from Oregon (dashed line; r^2 = 29.76, F = 5.085, P = 0.044). Solid line (r^2 = 51.43, F = 11.645, P = 0.006) represents relationship with *Lasiurus cinereus* excluded. Data from Freeman (1981) restructured and reanalyzed. Species designated by first letter of generic and specific names; no data available for *Myotis ciliolabrum*.

and medially from each side of snout (may be obscure in dried specimens); zygomatic breadth about equal to width of braincase; 36 teeth.....*Corynorhinus townsendii*, Townsend's big-eared bat (p. 113).
4b. Pelage yellowish, light tan, or cream colored on dorsum; individual hairs light at base; no fleshy protuberances at sides of snout; zygomatic breadth much greater than width of braincase; 28 teeth.................
............................*Antrozous pallidus*, pallid bat (p. 116).
5a. Dorsal pelage red or mahogany with silver-tipped hairs on the dorsum and extending to the posterior edge of the uropatagium; only one upper incisor; (rostrum as wide as braincase separates skull from that of *Tadarida*)..
............................*Lasiurus cinereus*, hoary bat (p. 98).
5b. Dorsal pelage not red or mahogany and not extending over >75% of the anterior portion of the uropagatium; two upper incisors.................6
6a. Pelage black with many silver-tipped hairs on dorsum; 36 teeth; (zygomatic breadth slightly greater than width of braincase separates skull from that of *Corynorhinus*)............................
............*Lasionycteris noctivagans*, silver-haired bat (p. 100).
6b. Pelage variously colored, but never with silver-tipped hairs on dorsum; 32, 34, or 38 teeth.....................7
7a. Length of head and body >60 mm; forearm >40 mm long; 32 teeth.................
....................*Eptesicus fuscus*, big brown bat (p. 107).
7b. Length of head and body <60 mm; forearm <40 mm long; 34 or 38 teeth.................8
8a. Tragus curved anteriorly and blunt, <5 mm long; total length of skull ≤11 mm; 34 teeth.................
....*Pipistrellus hesperus*, western pipistrelle (p. 105).
8b. Tragus straight and pointed, >6 mm long; total length of skull ≥12 mm; 38 teeth.................9
9a. Edge of uropatagium between calcaria with a conspicuous fringe of hairs.................
....................*Myotis thysanodes*, fringed myotis (p. 90).
9b. Uropatagium lacking a conspicuous fringe of hairs on edge.................10
10a. Ear >16 mm long and extending >5 mm beyond tip of nose when gently laid forward.................
....................*Myotis evotis*, long-eared myotis (p. 84).
10b. Ear <16 mm long and extending ≤2 mm beyond tip of nose when gently laid forward.................11
11a. Calcar with a keel.................12
11b. Calcar without a keel.................14
12a. Pelage of dorsum extending onto anterior 25% of uropatagium; pelage of venter extending onto plagiopagatium to a line from the elbow to the knee..........*Myotis volans*, long-legged myotis (p. 93).
12b. Pelage of dorsum not extending onto uropatagium; pelage of venter not extending onto plagiopatagium...
.................13
13a. Pelage on dorsum dull; skull sloped steeply between rostrum and braincase; third metacarpal usually >30.5 mm long; height of coronoid process <2.75 mm.........
...............*Myotis californicus*, California myotis (p. 79).
13b. Pelage on dorsum glossy; skull sloped gradually between rostrum and braincase; third metacarpal usually <30 mm long; height of coronoid process ≥2.75 mm...
.....*Myotis ciliolabrum*, western small-footed myotis (p. 82).

14a. Pelage on dorsum dull; total length of skull usually <14 mm; forearm usually <35 mm; skull sloped steeply from rostrum to braincase.................
.................*Myotis yumanensis*, Yuma myotis (p. 95).
14b. Pelage on dorsum glossy; total length of skull usually >14 mm; forearm usually ≥35 mm; skull sloped gradually from rostrum to braincase.................
.................*Myotis lucifugus*, little brown myotis (p. 87).

FAMILY VESPERTILIONIDAE—EVENING BATS

Except for most islands in the western Pacific Ocean, vespertilionid bats occur worldwide from ≅65° north latitude southward (Koopman, 1984). The family is the most diverse in the order Chiroptera with 318 named species (Wilson and Reeder, 1993a). Thus, among mammals it is the second most diverse family, being exceeded only by the Muridae (Rodentia). Fourteen of the 15 species of bats that occur in Oregon are evening bats.

These small- to medium-sized bats are characterized by a long tail that extends to the trailing edge of a wide uropatagium but does not extend beyond it by >4–5 mm. Females commonly average larger than males (Williams and Findley, 1979). The lips are simple and not thickened and the eyes are diminutive. The incisors are separated medially. The upper molars have well-developed W-shaped ectolophs. These bats produce echolocation calls through the mouth or nose by which obstacles and prey are located; most produce cries at ultrasonic frequencies, but *Euderma maculatum* and *Lasiurus cinereus* produce echolocation calls within the range of detection by the human ear (Barbour and Davis, 1969; Thomas et al., 1987).

Most vespertilionids are insectivorous, but the family includes fishing bats and two Oregon species may consume other vertebrates occasionally (Bell, 1982; Engler, 1943; Koopman, 1984). Most of the insects are caught in flight, commonly in the uropatagium or wing tip; one Oregon species gleans insects from plants and the ground, and we observed another Oregon species to glean moths (Lepidoptera) from the wall of our garage. Small insects caught in flight are consumed in flight, but bats may rest briefly while consuming large insects. Although much dietary information available for this group in Oregon is based on analysis of stomach or fecal contents (Henny et al., 1982; Whitaker et al., 1977, 1981a), many of these species feed opportunistically on swarms of insects (Warner, 1985). Thus, reported differences in diet among some species may be largely a function of prey availability rather than based on selectivity by the various species.

Vespertilionids are active nocturnally or crepuscularly. During the daylight hours, most seek refuge in caves, but lava tubes, crevices in rimrock and cliffs, mine shafts, abandoned tunnels, dry cisterns, barns, attics of houses, hollow trees, loose bark on trees, tile roofs, and thick foliage of shrubs and trees may serve as daytime roosts (Nowak, 1991). Night roosts in which bats rest between foraging bouts usually are in places other than those used as day roosts and commonly provide less protection. For example, a narrow alcove separating exterior doors between our garage and the living quarters of our house has served as a night roost for one–three evening bats each summer and autumn for many years. In caves, tunnels, barns, and other

Table 9-2.—*Means (±SE), ranges (in parentheses), and CVs of measurements of skull and skin[a] characters for female and male Chiroptera from regions in Oregon. Skin characters were recorded from specimen tags; skull characters were measured to the nearest 0.01 mm. SE and CV not provided if n < 10.*

Species and region	Sex	n	Total length	Tail length	Ear length	Forearm length	Mass (g)	Condylobasal length	Least interorbital breadth	Zygomatic breadth	Breadth of braincase	Mastoid breadth	Length of mandible
Antrozous pallidus West of Cascade Range	♀	6	115 (111–118)	43 (39–46)	32 (30–34)	58.87 (57.20–61.00)	23.95 (21.5–28.0)	22.36 (22.05–22.73)	4.44 (4.18–4.73)	13.15 (12.96–13.31)	10.04 (9.70–10.56)	10.58 (10.33–10.78)	14.96 (14.73–15.36)
	♂	14	112 ± 1.6[b] (98–122) 5.2	41 ± 1.3[b] (29–46) 10.9	31 ± 0.8[b] (26–35) 8.7	57.57 ± 0.62[b] (53.00–61.00) 4.07	22.28 ± 0.82[c] (16.9–27.0) 12.79	21.84 ± 0.16 (20.73–22.72) 2.73	4.31 ± 0.04 (3.98–4.49) 3.72	12.89 ± 0.08 (12.02–13.34) 2.42	9.79 ± 0.07 (9.17–10.11) 2.49	10.49 ± 0.08 (10.03–10.90) 2.78	14.66 ± 0.11 (13.92–15.49) 2.92
East of Cascade Range	♀	13	116 ± 1.5 (108–125) 4.7	45 ± 1.6 (37–55) 12.7	30[d] (26–34)	53.83 ± 0.68 (48.07–58.00) 4.58	22.11[e] (19.0–26.0)	21.34 ± 0.17 (19.79–22.17) 2.88	4.26 ± 0.05 (3.90–4.52) 4.27	12.54 ± 0.07 (12.14–12.85) 1.94	9.59 ± 0.07 (9.11–9.96) 2.55	10.09 ± 0.09 (9.62–10.61) 3.11	14.45 ± 0.11 (13.75–15.06) 2.79
	♂	16	112 ± 1.4 (103–120) 4.9	42 ± 1.3 (35–55) 12.1	29 ± 0.8[f] (25–36) 10.1	53.29 ± 0.62 (46.69–57.00) 4.62	17.99 ± 0.75[e] (14.6–21.0) 13.19	21.09 ± 0.12 (20.16–21.76) 2.19	4.28 ± 0.04 (3.95–4.55) 4.16	12.40 ± 0.08 (11.61–12.94) 2.47	9.54 ± 0.08 (9.03–10.42) 3.41	10.02 ± 0.07 (9.56–10.65) 2.78	14.11 ± 0.08 (13.66–14.71) 2.19
Corynorhinus townsendii West of Cascade Range	♀	19	106 ± 0.8 (100–112) 3.2	51 ± 0.8 (46–57) 7.1	36 ± 0.8[c] (34–40) 6.1	43.49 ± 0.41 (39.54–46.00) 4.11	11.2 ± 0.54[g] (6.8–12.5) 15.14	16.87 ± 0.05 (16.58–17.29) 1.18	3.69 ± 0.02 (3.49–3.82) 2.87	8.91 ± 0.05 (8.67–9.58) 2.48	8.45 ± 0.04 (8.16–8.89) 2.32	9.28 ± 0.03 (8.97–9.49) 1.59	10.51 ± 0.03 (10.21–10.77) 1.39
	♂	9	103 (93–118)	47 (41–54)	34[c] (30–38)	43.11 (34.01–55.60)	12.0[h] (10.0–19.0)	17.10 (16.21–20.50)	3.76 (3.51–4.35)	9.26 (8.52–12.17)	8.53 (8.18–9.17)	9.32 (9.04–9.74)	10.79 (10.17–13.74)
East of Cascade Range	♀	7	101 (95–110)	50 (47–52)	37 (34–40)	43.64 (42.00–45.72)	10.5[d] (10.0–11.0)	16.91 (16.36–17.13)	3.70 (3.46–4.09)	8.93 (8.55–9.37)	8.55 (8.06–8.73)	9.33 (9.14–9.47)	10.55 (10.29–10.66)
	♂	8	101 (80–116)	43 (33–51)	33 (26–39)	46.26 (41.00–52.40)	10.9[f] (6.1–14.0)	18.13 (16.20–21.70)	3.81 (3.38–4.33)	10.13 (8.76–12.71)	8.89 (8.07–10.25)	9.40 (8.92–10.41)	11.78 (10.16–14.74)
Eptesicus fuscus	♀	26	119 ± 1.9 (105–156) 8.1	45 ± 0.9 (35–54) 9.9	16 ± 0.7[i] (12–20) 15.8	46.68 ± 0.52 (37.75–49.67) 5.66	14.84[j] (12.7–19.0)	19.43 ± 0.09 (18.48–20.34) 2.26	4.37 ± 0.03 (3.87–4.67) 3.95	12.91 ± 0.07 (12.03–13.69) 2.93	9.58 ±0.06 (9.01–10.22) 3.44	10.09 ± 0.07 (9.43–10.86) 4.43	14.14 ± 0.07 (13.32–14.67) 2.57
	♂	30	113 ± 1.0 (102–122) 5.3	44 ± 0.8 (37–51) 9.5	17 ± 0.3[d] (12–19) 10.3	45.88 ± 0.38 (40.03–49.05) 4.52	15.57 ± 0.65[k] (10.5–20.0) 18.64	18.89 ± 0.09 (17.52–20.13) 2.86	4.45 ± 0.04 (4.12–5.00) 4.61	12.40 ± 0.08 (11.74–13.18) 3.34	9.37 ± 0.05 (8.83–9.82) 3.05	9.74 ± 0.05 (9.02–10.30) 2.94	13.64 ± 0.07 (12.82–14.58) 2.89
Euderma maculatum[l]	♀	3	119 (115–124)	47 (45–49)	42 (40–46)	52.72 (52.00–54.00)	17.0[b] (16.0–18.0)	19.15 (19.08–19.28)	4.16 (4.15–4.18)	10.37 (10.23–10.50)	9.98 (9.87–10.15)	10.08 (9.95–10.27)	12.17 (11.87–12.52)
	♂	3	115[b] (112–117)	49[b] (46–52)	42[b]	50.03 (49.69–50.43)	12.6[b] (10.0–15.2)	18.84 (18.58–19.03)	4.17 (5.94–6.03)	10.17 (10.06–10.24)	9.66 (9.61–9.72)	9.91 (9.80–9.99)	11.82 (11.75–11.93)
Lasionycteris noctivagans	♀	22	103 ± 1.1 (95–115) 5.0	41 ± 0.6 (35–44) 7.1	15 ± 0.6[f] (10–18) 15.9	41.43 ± 0.39 (37.60–47.00) 4.39	11.9 ± 0.48[d] (8.1–16.3) 16.50	16.43 ± 0.06 (16.07–17.03) 1.66	4.25 ± 0.04 (3.83–4.53) 3.99	9.83 ± 0.06 (9.17–10.56) 2.88	8.15 ± 0.04 (7.91–8.63) 2.06	8.61 ± 0.07 (8.20–9.65) 3.68	11.60 ± 0.06 (11.16–12.38) 2.29
	♂	30	101 ± 1.4 (80–115) 7.5	39 ± 0.6 (31–45) 8.6	15 ± 0.4[g] (12–18) 13.9	40.73 ± 0.30 (35.16–43.00) 4.08	10.3 ± 0.35[m] (7.8–13.0) 14.84	16.27 ± 0.05 (15.49–16.69) 1.62	4.36 ± 0.02 (4.07–4.62) 3.09	9.87 ± 0.04 (9.29–10.38) 2.17	8.17 ± 0.03 (7.87–8.45) 2.27	8.61 ± 0.03 (8.27–8.98) 1.89	11.50 ± 0.05 (10.93–12.16) 2.36
Lasiurus cinereus	♀	3	136 (133–139)	59 (53–65)	17	54.86 (54.11–54.47)	25.4	17.56 (17.82–17.88)	5.08 (4.92–5.19)	12.69 (12.33–13.14)	10.03 (9.99–10.07)	10.23 (10.12–10.33)	13.27 (13.14–13.39)
	♂	17	133 ± 1.5 (120–143) 4.7	57 ± 1.1 (48–64) 8.1	16 ± 0.7[c] (12–20) 15.9	51.73 ± 0.69 (41.27–54.08) 5.56	24.1 ± 0.84[h] (19.3–28.4) 12.49	17.22 ± 0.07 (16.43–17.66) 1.79	5.28 ± 0.03 (5.10–5.49) 2.12	12.25 ± 0.05 (11.70–12.59) 1.79	9.78 ± 0.04 (9.50–10.08) 1.76	10.09 ± 0.04 (9.82–10.40) 1.66	12.72 ± 0.06 (12.15–13.16) 1.86
Myotis californicus West of Cascade Range	♀	20	82 ± 0.8 (75–90) 4.5	36 ± 0.6 (32–44) 8.0	14 ± 0.3 (11–15) 9.0	32.42 ± 0.38 (29.78–38.00) 5.18	4.98 ± 0.27 (4.0–6.5) 18.92	13.23 ± 0.07 (12.55–14.14) 2.47	3.25 ± 0.03 (3.09-3.49) 3.64	7.85 ± 0.06 (7.37–8.60) 2.58	6.50 ± 0.04 (6.20–6.84) 2.50	6.85 ± 0.04 (6.42–7.18) 2.68	9.00 ± 0.07 (8.55–9.86) 3.27
	♂	18	78 ± 1.6 (60–91) 8.6	34 ± 0.6 (29–37) 7.4	14 ± 0.4 (10–17) 12.5	31.15 ± 0.36 (26.42–33.00) 4.93	4.48 ± 0.19 (4.0–6.7) 16.69	13.10 ± 0.05 (12.56–13.44) 1.62	3.30 ± 0.03 (3.15–3.70) 4.09	7.82 ± 0.07 (7.22–8.38) 3.89	6.47 ± 0.03 (6.18–6.68) 2.04	6.87 ± 0.04 (6.43–7.20) 2.42	8.95 ± 0.07 (8.28–9.44) 3.08
East of Cascade Range	♀	18	84 ± 1.2 (75–92) 5.8	38 ± 0.9 (30–47) 10.9	14 ± 0.3 (10–15) 8.4	32.47 ± 0.29 (29.75–34.00) 3.91	4.97 ± 0.17 (4.2–6.5) 13.32	13.44 ± 0.09 (12.51–14.00) 3.09	3.22 ± 0.02 (3.01–3.37) 2.95	8.00 ± 0.08 (6.93–8.44) 4.24	6.52 ± 0.04 (6.11–6.78) 2.51	6.88 ± 0.04 (6.43–7.14) 2.69	9.16 ± 0.08 (8.56–9.68) 3.63
	♂	21	83 ± 0.9 (76–90) 4.8	39 ± 0.9 (31–46) 10.2	14 ± 0.2 (12–16) 6.4	32.70 ± 0.40 (30.00–37.00) 5.67	4.69 ± 0.26 (2.5–9.0) 24.99	13.18 ± 0.06 (12.72–13.72) 2.09	3.28 ± 0.03 (3.04–3.53) 4.15	7.87 ± 0.05 (7.45–8.16) 2.66	6.47 ± 0.02 (6.28–6.64) 1.73	6.83 ± 0.04 (6.39–7.07) 2.68	9.03 ± 0.05 (8.75–9.49) 2.31
Myotis ciliolabrum	♀	10	83 ± 1.2 (78–91) 4.7	39 ± 0.9 (35–45) 7.2	14 ± 0.5 (12–16) 9.9	31.93 ± 0.35 (29.57–33.64) 3.50	4.13 ± 0.44 (2.8–6.0) 25.95	13.47 ± 0.09 (13.07–13.94) 2.14	3.15 ± 0.03 (3.04–3.30) 2.69	8.12 ± 0.07 (7.80–8.49) 2.81	6.53 ± 0.05 (6.30–6.76) 2.60	6.83 ± 0.06 (6.51–7.17) 2.89	9.29 ± 0.09 (8.94–9.74) 3.12
	♂	4	79 (75–86)	39 (33–46)	14 (12–16)	29.72 (23.41–33.00)	4.07 (3.0–4.7)	13.49 (13.16–13.77)	3.21 (3.15–3.27)	8.17 (7.91–8.43)	6.59 (6.42–6.80)	6.80 (6.71–6.97)	9.29 (9.18–9.48)
Myotis evotis West of Cascade Range	♀	14	91 ± 1.7 (80–99) 7.1	41 ± 1.1 (31–46) 9.8	21 ± 0.6 (15–24) 11.2	36.61 ± 0.62 (32.00–40.00) 2.48	5.68 ± 0.19 (4.4–7.2) 12.42	15.70 ± 0.10 (14.96–16.49) 2.48	3.83 ± 0.06 (3.05–4.04) 6.34	9.22 ± 0.11 (8.13–9.80) 4.38	7.52 ± 0.05 (7.19–7.85) 2.64	7.73 ± 0.09 (6.78–8.12) 4.34	10.89 ± 0.06 (10.56–11.25) 2.17
	♂	15	87 ± 1.7 (76–98)	39 ± 0.9 (34–46)	21 ± 0.4 (18–24)	36.68 ± 0.38 (34.46–39.00)	5.34 ± 0.29 (4.1–7.0)	15.47 ± 0.11 (14.44–15.99)	3.79 ± 0.03 (3.61–3.98)	9.17 ± 0.05 (8.81–9.38)	7.35 ± 0.04 (7.12–7.63)	7.68 ± 0.04 (7.42–7.86)	10.76 ± 0.07 (10.24–11.17)

Table 9-2.—*Continued.*

Species and region	Sex	n	Total length	Tail length	Ear length	Forearm length	Mass (g)	Condylobasal length	Least interorbital breadth	Zygomatic breadth	Breadth of braincase	Mastoid breadth	Length of mandible
	♀		7.7	8.9	7.4	3.97	17.98	2.79	3.00	2.18	2.07	1.81	2.49
East of Cascade Range		31	87 ± 0.7	38 ± 0.5	19 ± 0.2	37.18 ± 0.26	5.74 ± 0.16	15.40 ± 0.05	3.74 ± 0.02	8.92 ± 0.05	7.27 ± 0.03	7.60 ± 0.03	10.68 ± 0.04
			(80–94)	(34–45)	(16–22)	(32.01–39.08)	(4.7–8.5)	(14.40–15.87)	(3.51–3.91)	(8.23–9.50)	(6.96–7.63)	(7.29–7.93)	(10.12–11.09)
	♂		4.2	7.6	6.3	3.87	14.71	1.89	2.55	2.97	2.19	1.93	2.10
		30	87 ± 0.8	39 ± 0.6	19 ± 0.3	36.60 ± 0.25	5.53 ± 0.11	15.48 ± 0.06	3.78 ± 0.02	8.95 ± 0.08	7.25 ± 0.03	7.58 ± 0.03	10.69 ± 0.05
			(79–98)	(35–52)	(17–23)	(33.66–40.00)	(4.5–6.6)	(14.64–16.18)	(3.56–3.96)	(7.08–9.52)	(6.89–7.60)	(7.16–8.03)	(10.14–11.24)
Myotis lucifugus			5.0	8.9	8.5	3.79	10.02	2.19	2.69	4.75	2.46	2.39	2.62
West of Cascade Range	♀	30	90 ± 0.9	38 ± 0.4	15 ± 0.3	36.07 ± 0.29	6.77 ± 0.24	14.15 ± 0.06	3.87 ± 0.03	8.67 ± 0.05	7.12 ± 0.03	7.45 ± 0.04	9.79 ± 0.06
			(79–98)	(33–41)	(10–17)	(33.00–40.00)	(4.1–9.5)	(13.53–14.74)	(3.65–4.19)	(8.22–9.44)	(6.81–7.37)	(7.11–7.86)	(9.27–10.47)
	♂		5.2	5.5	10.6	4.51	19.11	2.21	4.10	3.36	2.39	2.76	3.17
		22	87 ± 1.1	37 ± 0.4	14 ± 0.6	35.14 ± 0.32	6.09 ± 0.31	14.13 ± 0.09	3.87 ± 0.03	8.66 ± 0.06	7.08 ± 0.04	7.43 ± 0.05	9.72 ± 0.08
			(76–96)	(34–41)	(7–16)	(31.00–37.00)	(4.0–9.0)	(13.43–14.75)	(3.58–4.04)	(7.99–9.12)	(6.64–7.49)	(6.95–7.82)	(8.92–10.21)
	♀		5.7	4.4	19.4	4.29	19.97	3.07	3.26	3.52	2.96	3.00	3.82
East of Cascade Range		32	86 ± 1.1	37 ± 0.7	13 ± 0.4	35.80 ± 0.28	6.59 ± 0.28	14.16 ± 0.06	3.83 ± 0.03	8.58 ± 0.05	7.07 ± 0.04	7.46 ± 0.04	9.84 ± 0.05
			(65–96)	(25–48)	(10–17)	(31.96–39.00)	(5.0–9.8)	(13.57–14.80)	(3.51–4.22)	(8.01–9.25)	(6.59–7.61)	(6.95–8.02)	(9.17–10.34)
	♂		7.5	10.7	14.9	4.38	20.73	2.48	4.70	3.43	3.19	3.15	2.94
		28	82 ± 1.1	35 ± 0.6	13 ± 0.3	35.04 ± 0.27	5.75 ± 0.14	14.06 ± 0.06	3.80 ± 0.03	8.59 ± 0.04	7.04 ± 0.03	7.38 ± 0.04	9.69 ± 0.05
			(70–91)	(25–41)	(10–16)	(31.00–37.00)	(4.0–7.0)	(13.52–14.67)	(3.45–4.18)	(8.04–9.05)	(6.78–7.46)	(7.10–7.83)	(9.17–10.26)
	♀		7.2	9.1	12.6	4.05	11.54	2.16	4.15	2.49	2.09	2.51	2.94
Northeastern Oregon		12	92 ± 2.0	37 ± 0.8	15 ± 0.6	36.86 ± 0.26	7.00	14.25 ± 0.07	3.76 ± 0.06	8.70 ± 0.06	7.08 ± 0.03	7.44 ± 0.04	9.84 ± 0.04
			(79–100)	(32–42)	(14–16)	(34.81–38.00)		(13.93–14.83)	(3.36–4.06)	(8.45–9.09)	(6.90–7.29)	(7.24–7.75)	(9.60–10.06)
	♂		7.6	7.4	6.7	2.45		1.67	5.71	2.58	1.60	1.66	1.48
		6	84	35	14	35.69	6.75	14.21	3.85	8.72	7.14	7.43	9.77
			(80–88)	(31–39)	(13–15)	(33.42–38.00)	(6.5–7.0)	(13.78–14.55)	(3.63–4.04)	(8.26–9.09)	(6.92–7.46)	(7.32–7.73)	(9.23–10.09)
Myotis thysanodes	♀	3	85	38	19	38.78	7.40	15.75	3.92	9.36	7.57	7.85	11.02
			(81–92)	(35–43)		(36.45–41.91)		(15.37–16.42)	(3.83–4.07)	(9.15–9.74)	(7.29–7.71)	(7.53–8.08)	(10.62–11.31)
	♂	9	87	38	17	39.59	6.17	16.06	3.88	9.59	7.65	7.84	11.23
			(80–93)	(33–41)	(13–20)	(37.49–42.55)	(4.7–7.0)	(15.15–16.70)	(3.64–4.03)	(9.21–10.0)	(7.18–8.08)	(7.25–8.35)	(10.71–11.60)
Myotis volans	♀												
West of Cascade Range		8	92	42	11	38.05	8.00	13.98	3.89	8.48	7.11	7.58	9.75
			(87–101)	(39–45)	(10–14)	(36.35–39.62)	(6.5–9.0)	(13.66–14.26)	(3.71–4.11)	(7.98–8.75)	(6.82–7.33)	(7.22–7.92)	(9.26–10.10)
	♂	8	86	37	12	36.76	5.61	13.81	3.90	8.28	6.97	7.43	9.62
			(68–97)	(30–44)	(8–16)	(32.35–40.00)	(3.1–7.9)	(13.56–14.42)	(3.66–4.14)	(8.00–8.62)	(6.75–7.08)	(7.20–7.67)	(9.11–10.15)
	♀												
East of Cascade Range		30	97 ± 0.7	43 ± 0.5	13 ± 0.3	39.34 ± 0.18	8.28 ± 0.20	14.14 ± 0.04	3.98 ± 0.02	8.71 ± 0.04	7.25 ± 0.03	7.74 ± 0.03	9.94 ± 0.04
			(88–103)	(36–48)	(10–15)	(36.09–40.69)	(6.5–11.0)	(13.74–14.63)	(3.83–4.20)	(8.29–9.04)	(7.00–7.58)	(7.42–7.99)	(9.57–10.37)
	♂		4.2	6.8	10.8	2.45	12.32	1.58	2.77	2.26	2.17	2.01	2.07
		30	99 ± 1.0	44 ± 0.6	14 ± 0.2	39.07 ± 0.41	7.38 ± 0.15	14.05 ± 0.05	4.01 ± 0.02	8.69 ± 0.03	7.23 ± 0.03	7.69 ± 0.03	9.89 ± 0.04
			(84–110)	(37–50)	(10–15)	(35.31–48.50)	(5.5–8.5)	(13.54–14.49)	(3.78–4.25)	(8.19–8.99)	(6.90–7.50)	(7.44–7.98)	(9.49–10.36)
Myotis yumanensis	♀		5.6	7.5	8.4	5.76	11.04	1.79	2.70	2.20	2.08	1.90	2.37
West of Cascade Range		19	85 ± 1.1	36 ± 0.4	15 ± 0.1	33.82 ± 0.49	5.71 ± 0.26	13.93 ± 0.04	3.76 ± 0.02	8.29 ± 0.05	6.84 ± 0.03	7.16 ± 0.03	9.51 ± 0.04
			(72–91)	(34–40)	(14–15)	(25.63–36.00)	(5.0–7.0)	(13.50–14.22)	(3.61–3.95)	(7.92–8.80)	(6.57–7.31)	(6.96–7.42)	(9.07–9.75)
	♂		5.6	4.6	3.6	6.35	12.23	1.38	2.49	2.47	2.20	1.53	1.77
		15	86 ± 1.3	37 ± 0.6	15 ± 0.4	33.92 ± 0.58	5.81 ± 0.24	13.95 ± 0.09	3.82 ± 0.03	8.31 ± 0.08	6.90 ± 0.03	7.20 ± 0.03	9.49 ± 0.09
			(79–99)	(31–41)	(11–16)	(26.82–36.07)	(5.0–7.0)	(13.36–14.58)	(3.65–4.11)	(7.66–8.82)	(6.69–7.17)	(7.07–7.53)	(9.01–10.24)
	♀		5.8	6.8	9.5	6.58	13.89	2.57	3.50	3.83	1.74	1.77	3.52
East of Cascade Range		30	85 ± 0.5	34 ± 0.4	14 ± 0.3	34.69 ± 0.17	6.04 ± 0.25	13.89 ± 0.04	3.81 ± 0.02	8.35 ± 0.04	6.94 ± 0.02	7.25 ± 0.02	9.58 ± 0.04
			(77–90)	(30–38)	(10–15)	(31.66–36.67)	(3.8–8.3)	(13.37–14.38)	(3.59–4.02)	(8.03–9.06)	(6.71–7.17)	(6.99–7.52)	(9.09–10.06)
	♂		3.5	6.8	12.2	2.69	20.29	1.65	2.69	2.49	1.75	1.54	2.28
		16	83 ± 0.9	33 ± 0.6	14 ± 0.3	34.08 ± 0.18	5.34 ± 0.31	13.78 ± 0.10	3.81 ± 0.03	8.29 ± 0.09	6.89 ± 0.05	7.24 ± 0.05	9.49 ± 0.07
			(77–89)	(30–38)	(12–16)	(32.60–35.10)	(4.0–9.0)	(13.20–14.89)	(3.58–4.06)	(7.78–9.12)	(6.61–7.48)	(6.96–7.89)	(9.02–10.35)
			4.4	6.9	8.7	2.09	22.45	2.92	3.51	4.80	2.98	3.04	3.14
Pipistrellus hesperus	♀	15	77 ± 0.6	30 ± 0.7	12 ± 0.2[f]	29.97 ± 0.17	5.3 ± 0.21[f]	12.01 ± 0.07	3.32 ± 0.03	7.55 ± 0.04	6.23 ± 0.03	6.41 ± 0.03	8.04 ± 0.05
			(73–80)	(23–34)	(11–13)	(28.92–31.00)	(4.0–6.5)	(11.47–12.48)	(3.18–3.63)	(7.28–7.95)	(6.02–6.46)	(6.21–6.57)	(7.62–8.30)
			2.9	9.1	5.1	2.21	13.46	2.20	3.43	2.20	1.97	1.85	2.24
	♂	12	72 ± 0.9	29 ± 0.7	11 ± 0.2[f]	28.39 ± 0.30	3.9 ± 0.26[f]	11.78 ± 0.07	3.25 ± 0.03	7.32 ± 0.07	6.08 ± 0.04	6.24 ± 0.04	7.84 ± 0.07
			(68–76)	(25–33)	(11–12)	(26.95–30.00)	(3.0–5.0)	(11.33–12.25)	(3.01–3.41)	(6.82–7.57)	(5.84–6.33)	(5.99–6.44)	(7.39–8.15)
			4.3	7.8	4.6	3.71	19.46	1.97	3.52	3.54	2.33	2.33	3.00
Tadarida brasiliensis	♀	6	94	34	16[b]	42.85	9.84[b]	16.64	4.02	9.64	8.54	9.33	11.32
			(89–104)	(28–40)	(13–20)	(42.26–43.90)	(8.0–11.4)	(16.34–17.11)	(3.88–4.13)	(9.25–9.79)	(8.38–8.70)	(9.13–9.60)	(11.09–11.65)
	♂	12	93 ± 1.9	33 ± 1.0	16 ± 0.6[b]	42.17 ± 0.31	10.4 ± 0.61[b]	16.93 ± 0.08	3.97 ± 0.02	9.79 ± 0.06	8.46 ± 0.03	9.36 ± 0.06	11.45 ± 0.06
			(81–106)	(27–39)	(13–20)	(40.62–45.00)	(7.7–15.0)	(16.53–17.43)	(3.85–4.11)	(9.43–10.13)	(8.24–8.67)	(9.08–9.69)	(11.09–11.81)
			6.9	10.9	11.7	2.55	19.35	1.66	2.11	2.17	1.35	2.07	1.69

[a] Although measures of hind foot and tragus length were recorded, the data were not included because of extreme variation believed to stem from differences in techniques used to measure these characters.

[b] Sample size reduced by 1. [c] Sample size reduced by 2. [d] Sample size reduced by 5. [e] Sample size reduced by 6. [f] Sample size reduced by 3. [g] Sample size reduced by 9. [h] Sample size reduced by 4. [i] Sample size reduced by 12. [j] Sample size reduced by 18. [k] Sample size reduced by 10.

[l] Includes only one female from Oregon.

[m] Sample size reduced by 11.

open areas, evening bats alight on the ceiling by flying up-ward and grasping the surface with their thumbs and toes; upon gripping the ceiling firmly with the toes, they release the hold with the thumbs and hang head downward (Nowak, 1991). Taking flight is simply a matter of the bats releasing their hold on the ceiling and opening their wings.

In the active season, bats enter a semitorpor during the daylight hours except during the late stages of pregnancy (Studier and O'Farrell, 1972) and during spermatogenesis (Kurta and Kunz, 1988). In late autumn, after accumulating stores of body fat, bats seek roosts with stable environ-ments and spend the winter in torpor, during which physi-ologic processes are greatly reduced. Hibernation may continue for 2–3 months without arousal and as long as 8 months with arousal for brief intervals (Humphrey, 1982).

Females of at least some species breed in autumn and store sperm until the following spring when they ovulate and become pregnant. Although the testes cease to pro-duce spermatozoa during winter, the male accessory glands remain functional and the epididymides are filled with mature spermatozoa. Some animals were observed to copu-late in winter when they were aroused from hibernation (Wimsatt, 1945). All species produce only one litter annu-ally and many give birth to only one offspring that nearly always is implanted in the right horn of the uterus. However, four Oregon species are known to produce more than one young frequently (Bailey, 1929; Koford and Koford, 1948; Provost and Kirkpatrick, 1952; Storer, 1931). In some spe-cies, the maternal female sometimes carries neonates during flights, but probably not while foraging (Fenton, 1969). Young vespertilionid bats commence to fly at 3 weeks (Davis et al., 1968) to 2 months of age (Koopman, 1984).

Some species are colonial at all times, whereas others live in colonies only in winter; colonies may consist of only one species or of two or more species in separate clusters (Koopman, 1984). In some species, females are solitary during the birthing and rearing seasons; others form nurs-ery colonies. Males of species that form nursery colonies roost separately; some form all-male colonies. During win-ter, some species hibernate in the same refuges in which they cared for young during summer; other species hiber-nate in caves not used in summer; and still other species (tree dwellers) migrate long distances to wintering areas. Winter roosts of some species have not been discovered (Humphrey, 1982).

Bats are preyed upon by many carnivorous mammals and birds; nevertheless, they tend to be exceptionally long lived for their size (Paradiso and Greenhall, 1967). Marked individuals 30 years of age have been recorded (Keen and Hitchcock, 1980).

Vespertilionids, like bats in some other families, can be asymptomatic reservoirs and vectors of rabies; however, few individuals examined are infected. Nevertheless, all bats should be handled with appropriate precautions and any bat on the ground or flying during daylight hours should be treated as if it were infected with the virus.

Myotis californicus (Audubon and Bachman)
California Myotis

1842. *Vespertilio californicus* Audubon and Bachman, 8:285.
1897a. *Myotis californicus*: Miller, 13:69.
1897a. *Myotis californicus caurinus* Miller, 13:72.

Description.—In *Myotis californicus* (Plate III), the face and ears are brown, the dorsal pelage is not glossy or

burnished, the braincase is rounded, the rostrum is short and separated from the braincase by an abrupt step (Fig. 9-3), the third metacarpal is >30.5 mm (as long as the forearm—Miller and Allen, 1928), and the calcar has a prominent keel. The hind feet are small and weak. In dorsal view in fresh specimens, the anterior-posterior length of the naked portion of the snout is about equal to the width of the nostrils, whereas in *M. ciliolabrum* the length is ≅1.5 times the width of the nostrils (van Zyll de Jong, 1985). See Remarks for additional discussion related to separat-ing the two species.

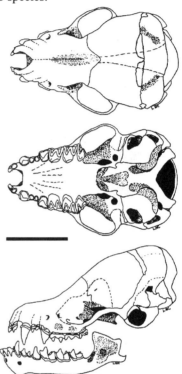

Fig. 9-3. Dorsal, ventral, and lateral views of the cranium, and lateral view of the dentary of the California myotis, *Myotis californicus* (MVZ 94197). Scale bar equals 3 mm.

Distribution.—*Myotis californicus* occurs from the central coastal region of British Columbia south to the tip of Baja California Sur and central Chiapas, Mexico, and eastward to western Montana, western Colorado, central New Mexico, and western Texas (Fig. 9-4; Simpson, 1993). In Oregon, the species occurs throughout the state except for the Columbia Basin (Fig. 9-4).

Remains of *M. californicus* were reported from Ho-locene (8,000 year old) deposits in a cave in Texas (Kurtén and Anderson, 1980). The site is extralimital.

Geographic Variation.—Two of the four recognized subspecies occur in Oregon: the bright orangish-brown *M. c. californicus* east of the Cascade Range and the dark brown *M. c. caurinus* west of the Cascade Range (Bailey, 1936; Simpson, 1993).

Habitat and Density.—*Myotis californicus* occupies a variety of habitats throughout its range; these include such diverse communities as shrub-steppe, shrub desert, ponde-rosa pine (*Pinus ponderosa*) forest, juniper (*Juniperus occidentalis*)-sagebrush (*Artemisia tridentata*), grand fir (*Abies grandis*) and Douglas-fir (*Pseudotsuga menziesii*) forests, and humid coastal forests (van Zyll de Jong, 1985; Whitaker et al., 1981a). Although a wide

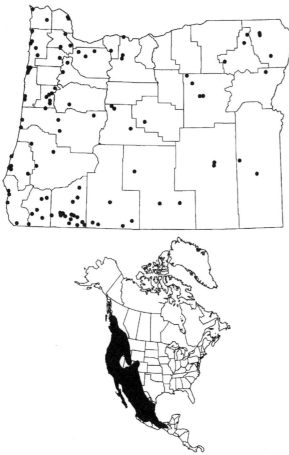

Fig. 9-4. Localities in Oregon from which museum specimens of the California myotis, *Myotis californicus*, were examined (listed in the Appendix, p. 531), and the distribution of the Calfornia myotis in North America. (Range map redrawn from Hall, 1981:186, map 146, with modifications.)

variety of habitats may be used, some are used at a much greater rate than others. For example, in Arizona, Bell (1980) recorded echolocation calls of *M. californicus* for 5-min periods throughout the night at 10 sites in each of three habitat types. Within the same hour-long intervals between 2000 and 0500 h, the average number of calls/5-min period along riparian forests was 3.4–26.0 times greater than in desert areas and 2.1–21.8 times greater than in juniper (*Juniperus*)-scrub. In desert areas in southern California, the species occurs most frequently near water either in rocky canyons or on sand flats. West of the desert region, it occurs in riparian groves of cottonwood (*Populus*), sycamore (*Platanus*), and willow (*Salix*—Krutzsch, 1954b). In British Columbia, 36 (54.5%) of 66 were captured at the mouth of a canyon along a fast-flowing creek bordered by alders (*Alnus*) and ponderosa pine. Habitats at other sites that produced nine (13.6%) and seven (10.6%) individuals were described as willow and alder along the bank of the Okanagan River and a "stony creek with poplar border" (Fenton et al., 1980:417, fig. 1).

Krutzsch (1954b) classified *M. californicus* as a crevice dweller as most roost sites used by the species were narrow and crevice-like. He suggested (p. 539) that the species roosted "in the most available site when their evening foraging is completed" and offered a series of observations of these bats temporarily roosting at sites

unavailable earlier. These included a fold in a canvas shelter, beneath a rag hanging in an unused building, and behind sign boards in addition to more commonly used bat roosts in buildings, mine tunnels, hollow trees, and rock crevices. Many of these sites did not exhibit characteristic urine stains and accumulations of guano; thus, they likely were used irregularly.

Perkins et al. (1990) recorded one *M. californicus* hibernating in a cave in Malheur Co. in February and another in a cave in Deschutes Co. in November. Hall (1946) claimed that these bats hibernated in mine tunnels and suspected that they also used caves.

Night roosts of *M. californicus* in southern Nevada and northern Arizona, as determined by observing upon release individuals captured in nets and fitted with gelatin capsules containing a chemiluminescent cold-light compound, were primarily in desert shrubs and trees (Hirshfeld et al., 1977). Hirshfeld et al. (1977) believed that during periods of greater moonlight these and other species of bats sought night roosts that provided greater shelter, possibly as a means of avoiding being preyed upon.

Whitaker et al. (1981a:285) reported that *M. californicus* was "widely distributed throughout eastern Oregon, [but] was nowhere easily collected." We are uncertain if this is a reference to its abundance or to its behavior. However, for western Oregon, these bats were considered "abundant and generally distributed" and were "easily obtained" from roosts beneath "tarpaper siding on buildings" (Whitaker et al., 1977:50).

Perkins (1987) captured 1,090 bats of 13 species in mist nets set at 267 isolated water sites and at entrances to caves and mines in Oregon during summer 1982–1986. *M. californicus* was captured at 19 sites; only 48 individuals (4.4%) were of this species. We know of no other quantitative reports of abundance.

Diet.—In eastern Oregon, moths (Lepidoptera) composed 52–84% of the material identified in fecal droppings and stomach contents, whereas in western Oregon they composed <15% of the material in stomachs of *M. californicus* (Table 9-3). However, flies (Diptera) composed only 10–18% in eastern Oregon, but >60% in western Oregon. Beetles (Coleoptera), lacewings (Neuroptera), and leafhoppers (Homoptera) composed a moderate portion of some samples from eastern Oregon, but occurred in tiny amounts or not at all in stomach samples from western Oregon (Table 9-3). The reverse was true for termites (Isoptera) and caddisflies (Trichoptera), with nearly 5% of stomach contents from western Oregon composed of each order; only one sample from eastern Oregon contained a small amount of termite remains and none contained remains of caddisflies (Table 9-3).

In Arizona, *M. californicus* sometimes took moths, but commonly fed on beetles and flies (Fenton and Bell, 1979). Counts of moth scales/g in fecal samples for *M. californicus*-*M. leibii* (=*M. ciliolabrum*) combined were higher than those for any other *Myotis*; Black (1974:149) suggested that one species might be a "moth strategist," whereas the other was a "beetle strategist." Such was not borne out by analysis of stomach samples of the two species in eastern Oregon (Whitaker et al., 1981a:283, table 1). In a later study conducted in Arizona, *M. californicus*, in addition to several other species of insectivorous bats, responded opportunistically to patches of prey produced

Table 9-3.—Percent volume of foods consumed by five species of Myotis in eastern and western Oregon as determined by analysis of stomach contents and feces. Data from Whitaker et al. (1977, 1981a) and Henry et al. (1982).

| Taxon or category | Myotis lucifugus | | | Myotis yumanensis | | | Myotis evotis | | | | Myotis californicus | | | | | Myotis volans | | |
| | Western | Eastern | Eastern | Western | Eastern | Eastern | Western | Eastern | Eastern | Eastern | Western | Eastern | Eastern | Eastern | Western | Eastern | Eastern | Eastern |
	Stomach	Stomach	Feces	Stomach	Stomach	Feces	Stomach	Stomach	Feces	Stomach	Stomach	Stomach	Feces	Stomach	Stomach	Stomach	Feces	Stomach
n	67	28	3	25	7	4	13	121	69	93	31	28	64	11	25	90	139	67
Lepidoptera (moths and larvae)	5.1	21.8	73.3	14.8	22.9	33.8	46.2	53.7	58.6	50.4	14.4	84.3	52.0	73.2	78.2	58.6	77.9	47.7
Isoptera (termites)	8.9	3.2		18.8	12.9		3.1	0.1	0.3	0.1	4.8		0.7		7.6		0.1	
Diptera (flies)	51.7	36.8	11.7	52.9	10.0	21.3	12.3	8.7	11.2	7.4	60.2	10.4	18.4	18.3	1.6	8.4	6.9	9.9
Hymenoptera (bees and ants)	2.7	3.9				2.5	4.2	2.0	1.5	2.1	0.6		1.2			0.4	2.1	0.5
Hemiptera (true bugs)	1.5	7.3	3.3				3.8	4.5	4.9	4.8	1.6	1.8	1.8	4.5	1.6	7.8	1.2	6.9
Orthoptera (crickets)	0.6							0.2		0.2	1.1				4.0		0.1	
Trichoptera (caddisflies)	8.4				13.6	12.5	18.0	18.1	6.1	21.2	4.7							
Coleoptera (beetles)	1.9	14.3		1.6	13.6	7.5	1.2	5.9	2.0	6.2	0.3	3.0	7.8	4.1		9.1	4.7	10.6
Homoptera (leafhoppers)	1.8	3.3	5.0	0.3	13.6	5.0		2.8	4.9	2.3	0.6		4.7		1.8	13.2	1.0	21.0
Neuroptera (lacewings)	0.4	0.7	6.7					2.7	6.2	3.1			9.5				3.5	
Unidentified insects	16.9	3.6		10.0	13.6	17.6	1.5	1.6			4.2	0.7	3.9		0.4	2.4	1.9	3.2
Araneae (spiders)				0.8			9.6	0.1	1.3	2.1	7.3				4.8	0.1	0.1	
Chilopoda (centipedes)																		
Vegetation									2.9								0.7	
Unidentified material		3.6																
Totals	99.9	98.5	100.0	100.0	100.2	100.2	99.9	100.4[a]	99.9[b]	99.9	99.8	100.2	100.0	100.1	100.0	99.9	100.2	99.8

[a]Total given as 100 in original.
[b]Total given as 99.8 in original.

experimentally by use of ultraviolet light, but exhibited no evidence of partitioning food resources by space or time (Bell, 1980).

As many flies consumed by M. californicus, especially midges (Chironomidae), are associated with water, variation in diet of the species as revealed in the two regions of Oregon (Table 9-3) likely is the result of differences in availability of prey at the sites and seasons that the bats or their feces were collected. M. californicus forages "along margins of tree clumps, around the edge of the tree canopy, over water, and well above ground in open country" (Simpson, 1993:3). Where sympatric with M. ciliolabrum, M. californicus seems to exclude it from foraging over water (Simpson, 1993), but we know of no similar interspecific interactions that might account for the observed differences in diet in M. californicus in eastern and western Oregon. Although possible, we seriously doubt that selectivity on the part of the bats was involved in the regional differences in diet observed.

The diet of M. californicus was classified as moderately hard by Freeman (1981), but her index to skull robustness for the species was the least for species that occur in Oregon (Fig. 9-2). However, references on which her index to hardness was based included research on diet conducted in western Oregon in which flies (rated hardness no. 1) predominated, but not that in eastern Oregon in which moths (rated hardness no. 2) predominated. Thus, the index to hardness of the diet may have been underestimated.

Reproduction, Ontogeny, and Mortality.—Relatively little is known regarding these topics for M. californicus. Cockrum (1955), by assembling published records, found one female of six examined contained two embryos; the remainder contained only one. Krutzsch (1954b) reported that M. californicus in southern California had only one young per litter. He provided accounts of his observations of several pregnant females, each containing one embryo, and several lactating females, each with one neonate, as evidence. He also provided evidence of a wide range of parturition dates, but none after mid-June. However, Hall (1946) recorded a pregnant female in southwestern Nevada on 4 July, but one each taken on 3 June and 23 June was not pregnant.

Duke et al. (1979) recorded a longevity of 15 years for a banded individual.

Habits.—In California, M. californicus commences to forage before dark and continues to do so until the early morning hours. Numbers captured and retained at a night roost by 30-min intervals increased rapidly from 1930 to 2300 h, then declined precipitously during the following 30 min and more gradually thereafter to 0200 h (Krutzsch, 1954b:543, fig. 1). Such data likely indicate at least two peaks of foraging with an intervening rest at a night roost. In Nevada, numbers captured in a net placed over a water tank at a desert spring in fall and winter were greatest during the first 1.5 h after sunset, but some individuals were captured during each 30-min interval for the first 4.5 h after sunset (O'Farrell et al., 1967). One individual each was captured during the 8.5–9-h and the 11–11.5-h intervals, and two individuals were captured during the 11.5–12-h interval. M. californicus was captured when ambient temperatures were as low as 2°C; 71% were captured at temperatures <10°C (O'Farrell et al., 1967). In a report of an extension of the previously mentioned study, O'Farrell and Bradley (1970) reported highest rates of capture in January–March and June–July, but some individuals were captured during all months. They captured more females than males at all seasons; overall, they captured approximately twice as many females as males. O'Farrell and Bradley (1970) believed that some *M. californicus* in the region remained in sustained hibernation for several months, but others hibernated only intermittently. *M. californicus* is capable of flying as a partial ectotherm (O'Farrell and Bradley, 1970).

In Arizona, in riparian forest, activity evaluated by monitoring echolocation calls was greatest during the 1st hour of darkness, followed by lower levels of activity during the next 3 h, an increase at midnight, and a peak at 0100–0200 h, followed by a gradual decline to daylight. Activity of *M. californicus* in other habitats peaked earlier in the evening and was more frequent during the night (Bell, 1980:1878, table 1).

In Arizona, where open water was scarce, California myotis usually foraged "along the margins of clumps of trees or around the edges of tree canopies," but they foraged over water where it was present (Fenton and Bell, 1979:1272). Their foraging behavior was similar to that of *M. lucifugus* in that their flight was slow, they made frequent abrupt changes in direction, and they often attempted to catch several insects within a short distance (Fenton and Bell, 1979). The same researchers did not observe the species to forage in groups, to engage in intraspecific aggression suggestive of territoriality, or to attempt to catch other than flying insects. Also, they indicated that *M. californicus* located concentrations of insect prey quickly. In Nevada, *M. californicus* usually foraged ≤3 m above the ground (O'Farrell et al., 1967).

In desert regions, *M. californicus* frequently flew low over water to drink by dragging its lower jaw through the surface. Krutzsch (1954b) reported one individual drinking along a 1.8-m-long pool six times before leaving the area and, on another occasion, six individuals drinking simultaneously.

Remarks.—*Myotis californicus* and *M. ciliolabrum* likely are more similar than any two other species of bats in western North America (Bogan, 1974). Consequently, considerable effort has been directed to separating the two species (Bogan, 1974; van Zyll de Jong, 1984). In a bivariate plot of cranial depth against rostral breadth for specimens from New Mexico, values for *M. ciliolabrum* fell to the right of a line originating at 4.7 mm and directed upward 52° from the horizontal, whereas those for *M. californicus* fell to the left of the line (Bogan, 1974:52, fig. 1). In a morphometric analysis of skull characters of two subspecies of *M. californicus* from British Columbia and Arizona, two subspecies of *M. ciliolabrum* from broad areas of their geographic ranges, and *M. leibii* from eastern and midwestern North America, van Zyll de Jong (1984) found height of the coronoid process separated the two species; in *M. californicus* the dimension was <2.75 mm, whereas in *M. ciliolabrum* it was >2.75 mm. Also, van Zyll de Jong (1984:2523, fig. 4) claimed that a line tangential to the slope of the forehead crossed the posterior edge of the canine in *M. californicus*, but the anterior edge in *M. ciliolabrum*. Finally, the top of the skull over the cerebellar fossa was purported to exhibit less flattening in *M. californicus* than in *M. ciliolabrum*.

Table 9-4.—*Number of* Myotis californicus *and* M. ciliolabrum *from east of the Cascade Range in Oregon, as identified by collectors and curators of systematics collections, that possess combinations of two of four characteristics considered diagnostic for each of the species (Bogan, 1974; van Zyll de Jong, 1984). Combinations above the diagonal are for 34* M. californicus *and those below the diagonal are for 14* M. ciliolabrum.

Characteristic	A	B	C	D
Cranial depth to rostral breadth[a]	—	15	20	11
Height of coronoid process[b]	3	—	14	11
Slope of forehead[c]	1	5	—	9
Shape of skull over cerebellar fossa[d]	1	4	2	—

[a]Values to the left of a line originating at 4.7 mm and directed upward 52° from the horizontal considered *M. californicus*.
[b]Values <2.75 mm considered *M. californicus*.
[c]Line tangential to slope of forehead passing posterior to midline of canine at alveolus considered *M. californicus*.
[d]Top of skull rounded over cerebellar fossa considered *M. californicus*.

We found little agreement in classification by use of the foregoing four characters, of *Myotis* possessing a keel and without pelage extending onto the flight membranes from east of the Cascade Range, where both species are purported to occur (Table 9-4). Only 1 (2.9%) of 34 classified by collectors and curators as *M. californicus* possessed all four characteristics ascribed to that species and only 2 (14.3%) of 14 possessed all four characteristics ascribed to *M. ciliolabrum*. Cluster analysis of 16 skull characters for 98 Oregon specimens of these bats from the entire state and for 48 specimens from the region east of the Cascade Range failed to produce groups significantly correlated with any of the foregoing characters. Although we are unwilling to synonymize *C. californicus* and *M. ciliolabrum* on the basis of our analysis of relatively small numbers of specimens from a limited geographic region, we believe that a critical analysis of variation between and within the two nominal species in western North America is essential.

Considering our inability to separate the two species by cluster analysis of cranial characters, by characters used by others, or to find new characters, we resorted to relying on identifications of others in preparing the range maps and specimens-examined lists. We find this regrettable, but we prefer to admit that "we do not know" than to claim otherwise.

Myotis ciliolabrum (Merriam)
Western Small-footed Myotis

1886. *Vespertilio ciliolabrum* Merriam, 4:2.
1890a. *Vespertilio melanorhinus* Merriam, 3:46.
1928. *Myotis subulatus melanorhinus*: Miller and Allen, 144:169.
1968. *Myotis leibii ciliolabrum*: Glass and Baker, 81:259.
1984. *Myotis ciliolabrum*: van Zyll de Jong, 62:2525.
1984. *Myotis ciliolabrum melanorhinus*: van Zyll de Jong, 62:2526.

Description.—*Myotis ciliolabrum* (Plate III) is among the smaller members of the genus in Oregon (Table 9-2). The face and ears are black, the dorsal pelage is a rich yellowish-brown and is glossy or burnished, the ventral pelage is buff, the braincase is flattened, the rostrum is separated from the braincase by a gradual slope (Fig. 9-5), and the third metacarpal is <30.5 mm (shorter than the forearm—Miller and Allen, 1928; van Zyll de Jong, 1985). The calcar is keeled. *M. ciliolabrum* and *M. californicus* are extremely similar; see Remarks for *M. californicus* regarding separation of the two species.

Distribution.—*Myotis ciliolabrum* occurs from southern British Columbia, Alberta, and Saskatchewan south

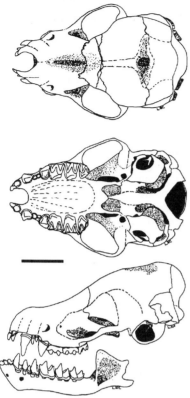

Fig. 9-5. Dorsal, ventral, and lateral views of the cranium, and lateral view of the dentary of the western small-footed myotis, *Myotis ciliolabrum* (PSM 20800). Scale bar equals 3 mm.

through the United States from the Great Plains westward (except humid coastal regions) and into Mexico as far south as central Nuevo León and southern Zacatecas (Fig. 9-6; Hall, 1981; van Zyll de Jong, 1985). In Oregon, the species occurs east of the Cascade Range (Fig. 9-6).

Fossil remains of *M. ciliolabrum* are known from late Pleistocene cave deposits in Wyoming (Kurtén and Anderson, 1980).

Geographic Variation.—Only the more richly colored of the two recognized subspecies occurs in Oregon: *M. c. melanorhinus* (Hall, 1981; van Zyll de Jong, 1985).

Habitat and Density.—Relatively little is known regarding these topics for *M. ciliolabrum* (Hoffmeister, 1986), almost nothing from Oregon. Whitaker et al. (1981*a*) reported taking six in nets in Grant Co., where they indicated the species was associated with arid rangelands. There, it foraged back and forth along the face of cliffs. Perkins et al. (1990) recorded a few individuals hibernating in each of several caves in Deschutes, Grant, Malheur, and Wallowa counties in November–February.

Much of the information available regarding habitats used by *M. ciliolabrum* is largely anecdotal or of a general nature. For example, of 73 specimens taken by Hall (1946) in Nevada, three were found flying in houses and the remainder were shot over ponds or streams. In southern California, a maternity colony of four adult females and one young was found between loose wallpaper and the wooden wall of an abandoned house; 19 days later, ≥21 adults and 16 young were taken in a similar situation in another room of the house (Koford and Koford, 1948). Hoffmeister (1986) believed that *M. ciliolabrum* in Arizona roosted in crevices and cavities in cliffs or rocks or perhaps in caves

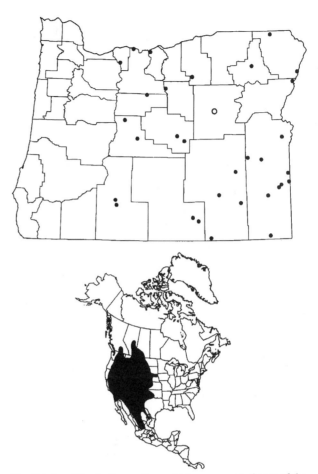

Fig. 9-6. Localities in Oregon from which museum specimens of the western small-footed myotis, *Myotis ciliolabrum*, were examined (listed in the Appendix, p. 532), and the distribution of the western small-footed myotis in North America. (Range map redrawn from Hall, 1981:188, map 147, and van Zyll de Jong, 1985:113, unnumbered figure.) Open symbol indicates record for county.

or mines. In South Dakota, Tuttle and Heaney (1974) found lactating females and young in 12 of 35 holes and cracks in rocks and along faces of vertical banks at which guano was found. In Colorado, Svoboda and Choate (1987) caught *M. ciliolabrum* at the entrance to an abandoned iron mine that, from mid-June to October, housed a colony of ≅100,000 predominantly male *Tadarida brasiliensis* and eight other species of bats. Douglas (1967) collected seven as they flew over a gravel road through piñon-juniper woodland interspersed with small stands of sagebrush (*Artemisia tridentata*) in the Mesa Verde National Park, Colorado. Pearson et al. (1952) found *M. ciliolabrum* in the same night roosts and hibernacula occupied by *Corynorhinus townsendii*; in one instance, two small-footed bats were in the same cluster with 15 hibernating western big-eared bats. In British Columbia, 33 (62.2%) of 53 *M. ciliolabrum* were captured in nets and traps set on a "talus slope with scrub" and at the "mouth of [a] canyon, [with a] fast-flowing creek bordered by alders and Ponderosa Pine" (Fenton et al., 1980:417, fig. 1).

Bailey (1936) believed that this species was relatively abundant in eastern Oregon. However, Perkins et al. (1990) never recorded more than six individuals hibernating in any of 13 caves occupied by *M. ciliolabrum*. Among 1,090 bats of 13 species captured at 164 (61.4%) of 267 water sites and entrances to caves and mines in Oregon at which mist nets were set in summer 1982–1986, only 17 *M. ciliolabrum* were captured; they occurred at only seven sites (Perkins, 1987). In maternity colonies in the Badlands, South Dakota, Tuttle and Heaney (1974) found only one adult female in nine of 11 roosts; the other two roosts contained two and four adult females. They also found a lone adult male occupying a horizontal crack in a rock on a hillside. Thus, the species either may not be as abundant as formerly or individuals are so widely dispersed among suitable sites that overall abundance seems to be low.

Diet.—Stomachs of six *M. ciliolabrum* from eastern Oregon contained 65.0% moths (Lepidoptera), 15.8% true bugs (Hemiptera), 12.5% flies (Diptera), 2.5% beetles (Coleoptera), 2.5% leafhoppers (Homoptera), 0.8% bees and wasps (Hymenoptera), and 0.8% unidentified insects (Whitaker et al., 1981*a*). In fecal samples from eight individuals from Arizona, moths occurred in 100%, beetles in 88%, flies in 63%, true bugs in 38%, and bees and wasps and lacewings (Neuroptera) each in 25% (Warner, 1985).

Reproduction, Ontogeny, and Mortality.—Koford and Koford (1948) reported 19 adult females (plus two or more of unknown sex that escaped) and 16 nonvolant young in a maternity colony. They also reported observations of E. McMillan who collected four females at the same site 3 weeks earlier; one had an attached neonate and another gave birth to a stillborn offspring 2 days later. Although these observations suggest that the species produces only one young, Tuttle and Heaney (1974) found five nursing young with four lactating females, suggesting that two young may be produced at times. In British Columbia, Fenton et al. (1980) collected lactating females on 13 and 23 June and pregnant females on 21 and 23 June. Such suggests that parturition may extend over a 2–3-week period.

Habits.—In some situations, *M. ciliolabrum* is considered an early flyer, whereas in others, it is thought to become active later than other species of *Myotis* (Hoffmeister, 1986; Whitaker et al., 1977). In British Columbia, the species is reported to commence to forage shortly after sunset and to exhibit peaks in activity at 2200–2300 h and at 0100–0200 h (van Zyll de Jong, 1985). Also, in areas where *M. californicus* and *M. ciliolabrum* are sympatric in British Columbia, the former forages over water and the latter forages along rocky bluffs, but *M. ciliolabrum* may forage over water when not associated with *M. californicus* (Simpson, 1993). The two species were suggested to coexist by spatially partitioning food resources (van Zyll de Jong, 1985).

Remarks.—*Myotis subulatus* (Say, 1823*b*), the name by which this species was known earlier (Bailey, 1936; Hall, 1981; Hall and Kelson, 1959), was suppressed to permit retention of the name of *Myotis yumanensis* H. Allen, 1864 by which the Yuma myotis had been known for >100 years (Glass and Baker, 1965). The oldest species name available for the small-footed myotis was *leibii* published as *Vespertilio leibii* Audubon and Bachman, 1842, thus forming the basis for establishing the name *Myotis leibii* with one subspecies (*M. l. leibii*) in eastern North America and two subspecies (*M. l. ciliolabrum* and *M. l. melanorhinus*) in the western part of the continent (Glass and Baker, 1965). Western forms of *Myotis leibii* were shown not to intergrade with the eastern race based on multivariate analysis of morphometric characters of the skull; thus they were placed in a separate species with *ciliolabrum* elevated to specific rank (van Zyll de Jong, 1984). Jones et al. (1986, 1992) recognized separation of *M.*

ciliolabrum as a distinct species, but Koopman (1993) did not, retaining the name *M. leibii.*

Myotis evotis (H. Allen)
Long-eared Myotis

1864. *Vespertilio evotis* Allen, 7(publ. 165):48.
1896. *Vespertilio chrysonotus* Allen, 8:240.
1897*a. Myotis evotis*: Miller, 13:77.
1943*b. Myotis evotis pacificus* Dalquest, 56:2.

Description.—*Myotis evotis* (Fig. 9-7), like its near relative *M. thysanodes* (Reduker et al., 1983), has long ears (Table 9-2) and a calcar without a keel, but is easily distinguished from it by the absence of a conspicuous fringe of stiff hairs on the trailing edge of the uropatagium and a forearm usually <40 mm long. Some *M. evotis*, especially in northeastern Oregon, may have some minute hairs on the edge of the uropatagium (Whitaker et al., 1981*a*), but they do not form such a conspicuous fringe as those of *M. thysanodes*. The ears, when gently laid forward, extend >5 mm beyond the nose (Manning and Jones, 1989). Overall, *M. evotis* has slightly larger body measurements than *M. thysanodes*, but not as great a body mass or cranial dimensions (Table 9-2).

The dorsal pelage is a glossy pale brown (eastern Oregon) to golden brown (western Oregon); the venter is whitish to buff. The bases of the hairs are black. The ears, wings, and uropatagium are blackish.

The skull is narrow with an oval braincase (Fig. 9-8). The rostrum grades gradually to the braincase without a distinct step.

Distribution.—*Myotis evotis* occurs from central British Columbia southeastward to western North Dakota then southward along the Pacific coast to the southern tip of Baja California Sur and in the Great Basin and Rocky Mountain regions south to southern Nevada and Utah and to central Arizona and New Mexico (Fig. 9-9; Manning and Jones, 1989). In Oregon, *M. evotis* occurs throughout the state (Fig. 9-8).

Late Pleistocene (Wisconsinan) fossils have been collected in caves in Wyoming, Arizona, and Texas, and Holocene (8,000 years old) remains were obtained from a cave in Texas (Kurtén and Anderson, 1980). The Texas and Arizona records are extralimital.

Geographic Variation.—Both of the named subspecies occur in Oregon: the lighter-colored *M. e. evotis* east of the Cascade Range and the darker-colored *M. e. pacificus* west of the Cascade Range (Manning and Jones, 1989).

Habitat and Density.—*Myotis evotis* is primarily a bat of coniferous forests in much of Oregon (Albright, 1959; Bailey, 1936; Maser et al., 1981*b*; Whitaker et al., 1977, 1981*a*) and elsewhere (Hoffmann and Pattie, 1968; Ingles, 1949; Vaughan, 1954), but may occur far from trees in shrub-steppe regions of the state (Bailey, 1936). It forages in openings in dense forest, between the trees beneath the canopy in ponderosa pine (*Pinus ponderosa*), and over willow (*Salix*)-bordered creeks (Bailey, 1936). The species is known to enter dwellings and other buildings through open windows and doors, and to forage on moths (Lepidoptera) therein.

Myotis evotis often day roosts in buildings, but may use many other natural and man-made structures, including caves and mines, bridges, hollow trees and loose bark on trees, and fissures in rock outcrops (Manning and Jones, 1989). Although a large number of *M. evotis* used a cave

Fig. 9-7. Photograph of the long-eared myotis, *Myotis evotis*.

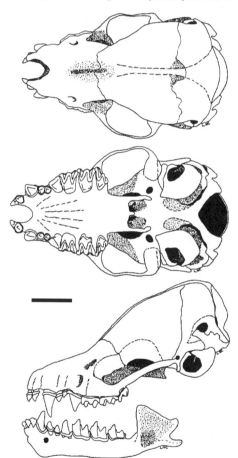

Fig. 9-8. Dorsal, ventral, and lateral views of the cranium, and lateral view of the dentary of the long-eared myotis, *Myotis evotis* (OSUFW 3057). Scale bar equals 3 mm.

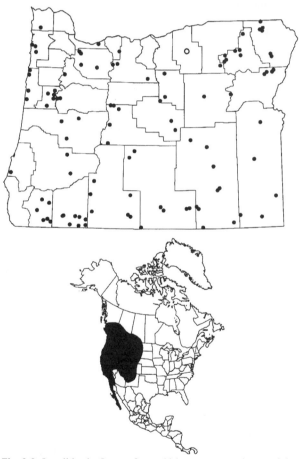

Fig. 9-9. Localities in Oregon from which museum specimens of the long-eared myotis, *Myotis evotis*, were examined (listed in the APPEN-DIX, p. 532), and the distribution of the long-eared myotis in North America. (Range map redrawn from Hall, 1981:207, map 164, with modifications.) Open symbol indicates record for county.

as a day roost during summer in Idaho (Davis, 1939), Barbour and Davis (1969) claimed the species seldom used caves for this purpose.

Whitaker et al. (1981a:285) considered *M. evotis* to be "the most abundant bat in northeastern Oregon forests." Of 280 *Myotis* they collected for analysis of stomach con-

tents, 121 (43.2%) were *M. evotis*. Some specimens were collected by shooting. The relatively slow and straight flight of *M. evotis* likely makes it more vulnerable to being collected by shooting than some congeners that have more erratic flight. Whitaker et al. (1977) reported that "these bats were not abundant" in western Oregon although they were widely distributed in the region. Only 13 (7.9%) of 165 *Myotis* they collected in western Oregon were *M. evotis.* Of 383 (total reported [381] does not correspond with total of numbers reported for each species) bats banded in July–September at Oregon Caves National Monument, Josephine Co., by Albright (1959), 213 (55.6%) were *M. evotis;* 86.9% were males. Thomas (1988), in his study of echolocation calls, did not distinguish among several species in a group of *Myotis* that included *M. evotis;* thus relative frequency of detection of its echolocation calls in the Coast Range from that source is not available.

Diet.—Long-eared myotis forage by "picking their prey from the surface of foliage, tree trunks, rocks, or even from the ground" and are classified as "hovering gleaners" (Findley, 1987:44–45). Whitaker et al. (1977, 1981*a*) recorded spiders (Araneae) and centipedes (Chilopoda) in the diet of *M. evotis;* these groups are nonvolant.

In both eastern and western Oregon, moths composed approximately one-half of the volume of food in the stomachs and fecal material of *M. evotis* (Table 9-3). Beetles (Coleoptera) composed 18–21% of the stomach contents in both eastern and western Oregon, but only about one-third as much in fecal material of the species collected in eastern Oregon. Flies (Diptera) ranked third in stomach contents, second in fecal samples. Spiders, leafhoppers (Homoptera), and true bugs (Hemiptera) contributed moderately to the diet (Table 9-3).

Freeman (1981) indicated that *M. evotis* consumed a harder diet than that of any congener in Oregon (Fig. 9-2), but that the robustness of its skull was exceeded by that of several *Myotis,* including *M. volans.* However, Reduker (1983), in a detailed comparison of masticatory musculature and jaw mechanics of *M. evotis* and *M. volans* (an aerial insectivorous species), concluded that *M. evotis* has a more forceful bite than *M. volans,* but only when the gape is <60°. He further postulated that a speedy and forceful bite would be an advantage in picking insects from substrates and subduing them with quick, strong nipping motions.

In New Mexico, in areas where *M. evotis* and a closely related congener (*M. auriculus,* formerly considered a conspecific—Genoways and Jones, 1969) are allopatric, a large proportion of the diet of both species consisted of moths, with males eating significantly more moths than females. However, where sympatric with *M. auriculus, M. evotis* shifted its diet to include more beetles; the diets of the two sexes are not significantly different (Husar, 1976). Husar (1976) noted no differences in foraging times for the two species and relative abundance of moths available to the bats could not be used to explain the shift in diet of *M. evotis.* Husar (1976) believed that her observations represented an example of either niche partitioning or character displacement in foraging behavior to reduce competitive interactions. *M. auriculus* does not occur in Oregon and as *M. evotis* is distributed statewide, the possibility of a similar comparison with other congeners here is precluded.

Reproduction, Ontogeny, and Mortality.—Essentially all information related to these topics is anecdotal (Manning and Jones, 1989). Cockrum (1955) listed records for nine *M. evotis* from scattered localities throughout the range of the species each with one embryo or one young. Manning and Jones (1989) listed records for four additional females each with one fetus. Pregnant females reportedly were collected at various localities in the geographic range

from 19 May to 15 July (Manning and Jones, 1989).

Testes of five males taken on 30 June–9 July in Montana averaged 4.5 mm (range, 3–6 mm) long and those of seven taken in late July averaged 4.2 mm (range, 3–5 mm—Jones et al., 1973*b*). Males taken by various researchers in South Dakota had testes 2–4 mm long in May and 6.0–7.5 mm long in late July–early August, whereas those taken in North Dakota had testes 5 mm long in late June (Manning and Jones, 1989). Two males that we took in Benton Co. in late July–early August had testes 2.5 by 4.5 mm and 3 by 5 mm.

Although decidedly circumstantial evidence, dates when pregnant females and males with enlarged testes were taken suggest a breeding season commencing in late summer–early autumn with sperm storage during winter and ovulation and fertilization in spring. Several other *Myotis* follow a similar reproductive sequence.

Four neonates born in captivity on 16 July to females taken in western Washington 1 day earlier weighed 1.08–1.36 g and were 40–47 mm long. The young were naked, but some of their deciduous teeth had erupted.

Essentially nothing is known regarding population structure and dynamics, predators, or other inimical factors for *M. evotis* (Manning and Jones, 1989; van Zyll de Jong, 1985). Tuttle and Stevenson (1982) recorded a longevity of 22 years.

Habits.—Some published information regarding onset of foraging activity appears to be contradictory (Manning and Jones, 1989). For example, Vaughan (1954) reported *M. evotis* foraged for ≅2.25 h commencing 30 min after sunset in California, Whitaker et al. (1977) indicated that these bats commenced to forage 10–40 min after full darkness, and Dalquest (1948) believed the species foraged late at night, sometimes after midnight. Albright (1959) observed these bats to enter Oregon Caves, Josephine Co., at 2200–0200 h where 213 of 383 bats he captured for marking were of this species. However, he noted relatively few bats in the cave during the day and presumed that they were "using secluded recesses" during the day (p. 27). Albright's (1959) records, and our observations of night-roosting *M. evotis* at 2300–2400 h and of flying bats at 0300 h, likely indicate an early foraging period shortly after dark followed by a period in a night roost and another bout of foraging before dawn. We suggest that the bats that Albright (1959) observed may have used the cave as a night roost and day roosted elsewhere instead of using some secluded recess. Nevertheless, Marcot (1984) recorded observations of others that indicate that the cave is used as a day roost by small numbers of these bats.

Males and nonpregnant females use separate roosts from postpartum females and their offspring in maternity roosts. Maternity roosts often are small, commonly containing as few as five (Maser et al., 1981*b*) or 12–30 individuals (Cowan and Guiguet, 1965), but may contain as many as "dozens" (Davis, 1939:114). Relatively little is known regarding migration and use of hibernacula, but Manning and Jones (1989:3) surmised that "these bats probably migrate short distances between summer haunts and winter retreats." They also suggested that the species probably uses caves or mines as hibernacula. In Oregon, Perkins et al. (1990) recorded one *M. evotis* hibernating in a cave in Clackamas Co. in February and two others in a cave in Douglas Co. in November.

Remarks.—During July-September most years since 1971, one to three *Myotis evotis* have foraged on alfalfa looper moths (Noctuidae: *Autographa californica*) that alight on the walls in our garage. Also, these bats occasionally used the alcove between the doors to our house and garage as a night roost. Over the years, we collected eight of these (OSUFW 1481, 1492, 1494, 4608–4610, 8769, 9008) either in a gloved hand on the night roost (2300–2400 h) or by trapping them in the garage by quickly closing the doors and waiting until they lit on the walls. Both males and females were involved. Most often we initially detected their presence by their droppings and the wings of the moths affixed to our automobile in the garage or in our gardening shoes in the alcove each morning. We often saw them flying in the garage in the headlights of our automobile when we came home after dark, and once, when awakened at about 0300 h, one of us (BJV) observed one to hover momentarily at a screened, open window. We observed only one individual to forage in the garage at a time and several days elapsed after one was captured until evidence of use of the garage by another was observed. On one occasion, we observed a long-eared myotis to hover for a second, then pluck an alfalfa looper moth from the inside wall of our garage; thus, we can attest to their gleaning mode of foraging.

In July 1995, wings of moths and fecal droppings of bats again commenced to appear on our automobile during the night. On 18 July, the overhead door had been closed earlier in the evening, but the access door remained open. At 2215 h when the lights were turned on, a *M. evotis* was found hanging in a corner inside the garage. We closed the access door to observe our visitor, but shortly it commenced to fly about in the garage. When it twice crashed into the window in the access door, we opened the door and allowed it to escape. The following afternoon, we washed our automobile to observe how long the garage remained bat-free after the disturbance. New droppings and moth fragments decorated our vehicle the morning of 20 July; thus, either our disturbance did not deter the bat from entering our garage or more than one individual foraged in our garage.

Myotis lucifugus (Le Conte)
Little Brown Myotis

1831.   *V[espertilio]. lucifugus* Le Conte, 1(App.):431.
1897a. *Myotis lucifugus alascensis* Miller, 13:63.
1904.   *Myotis (Leuconoe) carissima* Thomas, 13:383.
1917.   *Myotis lucifugus carissima*: Cary, 42:43.

Description.—*Myotis lucifugus* (Fig. 9-10) averages slightly larger than other members of the genus in the state except for *M. volans* and *M. evotis* (Table 9-2). The ≅17-mm-long calcar usually lacks a keel and hairs on the hind foot extend beyond the toes. The ears extend <2 mm beyond the nose when laid forward; the tragus is blunt. The dorsal pelage is usually a glossy golden brown to dark brown, whereas that of the venter is whitish to buff. The hairs on the venter have dark bases and are not glossy. The flight membranes and ears are dark brown.

The greatest length of the skull is usually >14 mm, the rostrum is relatively short, the forehead forms a gradual slope to the braincase (Fig. 9-11), and the braincase is flattened (Barbour and Davis, 1969; Fenton and Barclay, 1980). See Remarks for additional discussion regarding identification of *M. lucifugus*.

Distribution.—*Myotis lucifugus* occurs from central

Fig. 9-10. Photograph of the little brown myotis, *Myotis lucifugus*. (Photograph courtesy R. B. Hayward.)

Alaska south and east across the southern half of Canada south through most of the United States (except for the southern Great Plains) into northern Chihuahua, Mexico (Fig. 9-12). The population at high elevations in south-central Mexico possibly is disjunct (Barbour and Davis, 1969; Hall, 1981). The little brown myotis occurs throughout Oregon (Fig. 9-12).

Pleistocene-age (Wisconsinan) fossils from cave deposits were recorded for Pennsylvania, Virginia, Missouri, Texas, and Wyoming (Kurtén and Anderson, 1980).

Geographic Variation.—Two of the six subspecies currently recognized occur in Oregon: the lighter-colored, brassy *M. l. carissima* in most of the region east of the Cascade Range and the darker-colored, nearly bronze *M. l. alascensis* west of the Cascade Range and in the extreme northeastern corner of the state (Dalquest, 1948; Hall, 1981).

Habitat and Density.—In summer, *M. lucifugus* occurs in a wide variety of habitats, but seems especially prone to establish residence near a lake, pond, or stream. In western Oregon, the little brown myotis reportedly is associated with deciduous and coniferous forested areas where it forages "among scattered trees or along edges of dense timber" (Whitaker et al., 1977:47). In eastern Oregon, where it occurs in timbered areas, it forages at the edges of openings, but in shrub-grassland areas it forages mostly over water (Whitaker et al., 1981a).

Buildings—especially churches, barns, garages, pumphouses, cabins, and other structures subject to relatively little disturbance by humans—seem to be the primary sites for maternity colonies. Such colonies also are known to occur with lesser frequency in caves, beneath bridges, under shingles, in hollow trees, and in other similar refuges (Barbour and Davis, 1969; Fenton and Barclay, 1980). Maternity sites must be sufficiently warm to promote rapid pre- and postnatal growth; thus, dimly lit sites with relatively high ambient temperatures are requisite. *M. lucifugus* can withstand temperatures in excess of 50°C in maternity colonies. We have no information regarding size of maternity colonies for Oregon, but elsewhere they commonly include 300–800 individuals; much smaller and much larger colonies are known to occur (Barbour and Davis, 1969).

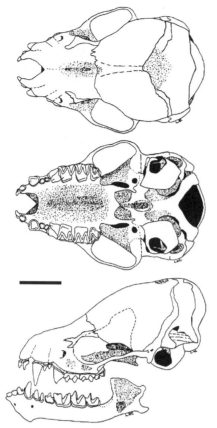

Fig. 9-11. Dorsal, ventral, and lateral views of the cranium, and lateral view of the dentary of the little brown myotis, *Myotis lucifugus* (OSUFW 2969). Scale bar equals 3 mm.

Fig. 9-12. Localities in Oregon from which museum specimens of the little brown myotis, *Myotis lucifugus*, were examined (listed in the APPENDIX, p. 532), and the distribution of the little brown myotis in North America. (Range map redrawn from Hall, 1981:192, map 149, with modifications.)

Day roosts for males and nonreproductive females usually are at some distance from maternity roosts. Only one to a few individuals occupy these sites; the sites tend to be cooler than maternity roosts and bats occupying them commonly are torpid (Fenton and Barclay, 1980). Upon dispersal from maternity colonies, young bats may spend the day in a variety of locations, as on trunks of trees and sides of buildings, and in other exposed places.

Night roosts at sites different from those used during the day commonly are occupied by numerous individuals; the bats pack themselves into confined spaces after the first foraging bout of the night. Although the function of night roosts separate from day roosts is not clear, ambient temperatures may be involved as indicated by the tight congregations of bats and the fact that night roosts are not used when temperatures exceed 15°C (Fenton and Barclay, 1980). Night roosts are used by pregnant females, but not by lactating females.

Although Fenton and Barclay (1980) found no record of the species hibernating in buildings, a little brown myotis was reported hibernating in a barn in Washington Co. in December (Perkins et al., 1990). Four specimens also were found in a mine in Wallowa Co. in January (Perkins et al., 1990). Several caves in Oregon are known to be used as hibernacula by *M. lucifugus* (J. M. Perkins, in litt.).

Myotis lucifugus possibly is the most abundant bat in Oregon. In the Coast Range, 48.8% of 4,470 echolocation calls recorded by Thomas (1988) were those of either *M. lucifugus* or *M. yumanensis;* echolocation calls were considered too similar to discriminate between the species.

Whitaker et al. (1977) collected more of this species than any other for their dietary study of bats in western Oregon; for a similar study in eastern Oregon, numbers of little brown myotis were outranked by those of several species (Whitaker et al., 1981*a*). Perkins (1987) reported collecting more *Lasionycteris noctivagans* than *M. lucifugus* in mist nets in all regions of Oregon, but *M. lucifugus* ranked second in number of captures in the state. However, we suspect that effort directed toward capture of the various species was not equal. Nevertheless, its frequent association with human habitation may make its abundance more apparent than real as "populations of *M. lucifigus* have drastically declined in many parts of its range" (Fenton and Barclay, 1980:4). Also, because of its association with human habitation, the species often is subjected to various measures to eliminate colonies from buildings (Barclay et al., 1980).

Diet.—*Myotis lucifugus* in Oregon, as elsewhere, feeds heavily on flies (Diptera), especially those of the family Chironomidae (Table 9-3; Anthony and Kunz, 1977; Belwood and Fenton, 1976; Herd and Fenton, 1983). In stomach samples, dipterans ranked first among orders of invertebrates consumed by little brown myotis in both western and eastern Oregon (Table 9-3). However, in fecal samples from eastern Oregon, dipterans ranked second, far behind moths (Lepidoptera); Whitaker et al. (1981*a*) considered differential digestion responsible for the discrepancy in volume of these two orders consumed by *M. lucifugus*. In western Oregon, termites (Isoptera) and caddisflies (Trichoptera) ranked second and third, respectively, whereas, in eastern Oregon, moths and beetles (Coleoptera) filled those slots. Lacewings (Neuroptera) ranked third in fecal samples from eastern Oregon. Other orders of invertebrates contributed ≤5% of the diet of *M. lucifugus* in both eastern and western Oregon (Table 9-3).

Little brown bats likely fill and empty their digestive tracts two or more times each night (Buchler, 1975). This, and differential availability of various insect prey temporally as well as seasonally, may serve to produce unrepresentative dietary information if materials were from bats collected in nets or by shooting only at dusk and within 2–3 h thereafter.

Based on the analysis of Freeman (1981), *M. lucifugus* has the softest diet of any of the bats in Oregon, but the index to the robustness of the skull, although low, is greater than that of three congeners (Fig. 9-2). Bailey (1936) estimated that a very fat, 8.8-g female with a stomach full of insects weighing 2.4 g likely consumed at least half its body mass each night. He also surmised that less obese individuals may consume a mass of insects approximately equal to their own mass each night.

Reproduction, Ontogeny, and Mortality.—Essentially nothing is known regarding these topics for *M. lucifugus* in Oregon. We examined one pregnant female containing a 16-mm (crown-rump) embryo collected on 28 June in Lake Co. Maser et al. (1981*b*) recorded pregnant females with 2–5-mm embryos on 3 May in Coos Co. and two with 16- and 23-mm embryos on 4 July in Columbia Co.

Elsewhere, copulation commences about mid-August, peaks, then in hibernacula, continues with a lesser frequency throughout the winter. Whether the frequency of natural mating in hibernacula is equal to that of hibernating individuals disturbed by the presence and activities of human observers is not known (Fenton and Barclay, 1980). Mating is random and promiscuous; males mate with as many females as possible and females may mate more than once. All ages of females mate, but young-of-the-year males are not sexually active (Fenton and Barclay, 1980). Herd and Fenton (1983) described two nonexclusive reproductive strategies for this species: in northern parts of the range, parturition is later and pregnancy in young of the previous year is infrequent, whereas, in the southern parts of the range, parturition is earlier and young of the previous year typically bear young. Which of these strategies is followed by the species in Oregon is unknown.

Ovulation and fertilization occur in spring when females leave hibernacula; embryonic development requires 50–60 days (Wimsatt, 1945), depending upon roost temperatures and frequency and duration of torpor (Racey, 1973). Parturient females assume a head-up position and young are born into the cupped uropatagium. Only one young usually is produced during each pregnancy (Cockrum, 1955; Fenton and Barclay, 1980), but twins and twin fetuses have been recorded occasionally (Barbour and Davis, 1969). Presentation is breech (Wimsatt, 1945). The maternal female often severs the umbilical cord by chewing and sometimes eats the placenta, but occasionally neonates with attached placentae are observed (Wimsatt, 1945).

At birth, neonates weigh 1.5–1.9 g; are pinkish beige except for the nearly black wings, ears, legs, tail, and uropatagium; and are covered with fine silky hairs (Wimsatt, 1945). Twenty of the 22 deciduous teeth are erupted at birth; permanent teeth commence to erupt when young are ≅58 mm long (Fenton, 1970). Lower incisors are the first permanent teeth to erupt followed by the upper canines; P4, m1, m2, c1, M1, and M2 erupt simultaneously; and M3, p3, and P3 are last to erupt (Fenton, 1970). Young grow rapidly; by 3 weeks of age their forearms and ears are adult-size, and they are weaned and can fly well (Fenton and Barclay, 1980; van Zyll de Jong, 1985). Newly volant young forage by waiting for prey to approach while they sit on a perch (Fenton and Barclay, 1980).

Although individuals ≥10 years of age are common and maximum recorded longevity exceeds 31 years, mortality during the 1st winter of life often is high. Estimated survival rates were 1.55 years for males and 1.17–2.15 years for females (Fenton and Barclay, 1980). A variety of predatory vertebrates take *M. lucifugus*; most predation on the species occurs when an individual predator encounters a concentration of the bats. Accidents—including becoming entangled on spiny plants (Verts, 1988) and impaled on barbed-wire fences, and drowning in hibernacula during floods—contribute to mortality (Fenton and Barclay, 1980). Pesticides, applied (now in contradiction to registration) directly to bats to eliminate them from buildings or acquired by the bats indirectly through consumption of insect prey for which the toxicants were applied, also take a toll (Kunz et al., 1977).

Habits.—In spring and summer, *M. lucifugus* leaves the day roost and commences to forage at late dusk. It commonly forages low over water or in openings at forest edges. In Michigan, the median distances above a stream and from the shoreline at which *M. lucifugus* foraged were 1.5 m and 0.9 m, respectively; the values were significantly less than for syntopically foraging *Eptesicus fuscus* (Kurta, 1982). It is an adept swimmer in the event that it falls into the water (Patten and Patten, 1956).

Although some individuals fill their stomachs quickly, others forage for 2–3 h after emerging. These bats, except for lactating females, then congregate at a night roost; later they forage again, then return to their day roost. Lactating females return to the day roost to nurse their young between bouts of foraging.

Although most prey is taken while in flight, some is gleaned from the surface of the water. Apparently, prey is detected by echolocation at short range (≤1 m), but individual bats, as indicated by abrupt turns, often attempt to capture several prey items per minute (Fenton and Bell, 1979). In the laboratory, individual *M. lucifugus* captured as many as 20 fruit flies (*Drosophila*) in 1 min and on three occasions two flies were caught in 0.5 s (Griffin et al., 1960). Some small, slow-flying insects are caught in the mouth, but most are captured in a pouch formed by cupping the uropatagium; prey then is transferred to the mouth in flight. When the flight path of the bat is to one side of that of the prey item, the prey may be caught in the plagiopatagium or flipped with the plagiopatagium into position to be caught in the forward-curved uropatagium (Webster and Griffin, 1962).

Fenton and Bell (1979) observed >30 individuals of this species foraging in an area of only 3 by 10 m over a lake, but observed no social interactions other than occasional "honking" by an individual on a collision course with another. They also found no evidence that bats from one colony were the exclusive users of a feeding area or that a feeding area was defended by bats from one colony.

Echolocation calls produced by *M. lucifugus* are 1–5 ms in duration and sweep downward from ≅80 to ≅40 kHz. A cruising bat produces ≅20 pulses/s, whereas one chasing a flying prey item or approaching an obstacle produces >50 pulses/s (Fenton and Bell, 1979). *M. lucifugus* also uses the echolocation calls of conspecifics to locate roost sites. Males at mating sites within hibernacula also produce echolocation calls (Fenton and Barclay, 1980). Although it uses echolocation calls for detecting prey and obstacles, *M. lucifugus* depends upon visual cues for orientation (Mueller, 1968).

In midsummer, males and nonbreeding females commence to congregate at hibernacula, usually caves or abandoned mines. Females and their now-independent offspring follow shortly thereafter, so by August populations are composed of both sexes. These bats respond to the calls of conspecifics and swarm at entrances to hibernacula. Although the bats involved in swarming at a hibernaculum may actually hibernate elsewhere, the swarming behavior is believed to familiarize young bats with the locations of hibernacula, serve to mix populations, and be a prenuptial activity (Fenton and Barclay, 1980). Mating activity commences at this time.

Remarks.—*Myotis lucifugus* is similar to *M. yumanensis* (van Zyll de Jong, 1985). Many workers claim that within a region all but a small number of specimens usually can be distinguished readily by use of morphological characters (Harris, 1974; Herd and Fenton, 1983). The glossy dorsal guard hairs of *M. lucifugus* commonly are sufficient to distinguish it from *M. yumanensis*, which possesses dull dorsal guard hairs. However, in British Columbia, this character was not considered useful as the pelage of *M. lucifugus* there often is not glossy, especially in nulliparous females (Herd and Fenton, 1983). Of morphometric characters, both Harris (1974) in the Southwest and Herd and Fenton (1983) in British Columbia listed mastoid breadth as the most useful for distinguishing the two species, even though ranges in Utah (*M. lucifugus* = 7.4–8.0 mm, $\overline{X}$ = 7.66 mm; *M. yumanensis* = 7.0–7.5 mm, $\overline{X}$ = 7.26) and British Columbia (*M. lucifugus* = 7.15–7.75 mm, $\overline{X}$ = 7.44; *M. yumanensis* = 6.90–7.45, $\overline{X}$ = 7.23) overlapped somewhat. Ranges in mastoid breadth for the two species in Oregon overlapped almost completely (Table 9-2). Parkinson (1979) reported that 93% of 72 specimens with forearm lengths ≥37 mm and 94% of 225 specimens with forearm lengths ≤35 mm were classified correctly into their a priori groups by discriminant analysis. She referred the former group to *M. lucifugus* and the latter to *M. yumanensis*. Of specimens taken in Oregon, 100 (83.3%) of 120 identified as *M. lucifugus* on the basis of glossy pelage and forehead gradually tapering from rostrum to braincase had forearm lengths ≥35 mm, whereas 57 (71.3%) of 80 of those identified as *M. yumanensis* by dull, nonglossy pelage and a rostrum separated from the braincase by a moderately sharp step had forearm lengths <35 mm.

Specimens morphologically intermediate between the two species taken in Washington, Oregon, California, Utah, and New Mexico were suspected of being hybrids (Barbour and Davis, 1969; Harris, 1974; Harris and Findley, 1962;

Herd and Fenton, 1983; Parkinson, 1979). Herd and Fenton (1983) found no evidence of hybridization by use of electrophoretic variation at 31 protein loci even though, based on discriminant analysis of morphometric characters, 14.0% of the 93 bats were considered intermediate and 7.5% could have been designated hybrids.

Myotis lucifugus is sort of the "flying *Peromyscus maniculatus*" of North American Chiroptera in that it survives well in captivity and has been the subject of extensive laboratory investigations, particularly those dealing with physiology (Fenton and Barclay, 1980). For example, heart rate ranges from 0.5 beats/min at a body temperature of –3°C to 20/min at 7°C, 110/min at 25°C, and 210/min at 36°C in quiescent bats. The maximum recorded was 1,368 beats/min (Kallen, 1977). Information concerning kidney function, production of hormones, brown fat, torpor, olfaction, vision, and other characteristics also is available (Fenton and Barclay, 1980).

Myotis thysanodes Miller
Fringed Myotis

1897a. *Myotis thysanodes* Miller, 13:80.

Description.—The fringed myotis (Fig. 9-13), in comparison with other members of the genus, has long ears (Table 9-2) and a large calcar without a keel. However, the species is best distinguished by a conspicuous fringe of stiff hairs on the trailing edge of the uropatagium (Fig. 9-14) and a forearm usually >40 mm long. Head and body length of females is significantly greater than among males (Williams and Findley, 1979).

The skull is slender and has a well-developed sagittal crest; it is the most robust among the species of *Myotis* that occur in Oregon (Fig. 9-15). The molars are much simplified in comparison with those of other members of the genus in Oregon (O'Farrell and Studier, 1980). The baculum is dumbbell shaped with a ventral groove; it averages 0.77 mm long (Krutzsch and Vaughan, 1955).

The pelage is yellowish brown to almost brassy; the bases of the hairs are nearly black. The venter is considerably lighter in color. The wing membranes and uropatagium are dark brown and are resistant to puncture (O'Farrell and Studier, 1980).

Fig. 9-13. Photograph of the fringed myotis, *Myotis thysanodes*. (Photograph courtesy R. B. Hayward.)

Fig. 9-14. Photograph of the uropatagium of a fringed myotis, *Myotis thysanodes* in flight; note fringe. (Photograph courtesy R. B. Hayward.)

Distribution.—The fringed myotis occurs from south-central British Columbia south through Washington, Oregon, and California to extreme northern Baja California Norte, then eastward through southeastern Idaho and southwestern Wyoming to central Colorado and south through New Mexico to Veracruz and Chiapas (Fig. 9-16; O'Farrell and Studier, 1980). A disjunct population, considered a distinct subspecies, occurs in the Black Hills of Wyoming and South Dakota (Fig. 9-16; Jones and Choate, 1978). In Oregon, *M. thysanodes* occurs in the Coast Range from Jackson Co. to Clatsop Co. and in the northeastern corner of the state (Fig. 9-16).

There is no fossil record (O'Farrell and Studier, 1980).

Geographic Variation.—One of the three named subspecies occurs in Oregon: *M. t. thysanodes* (Hall, 1981; O'Farrell and Studier, 1980).

Habitat and Density.—Miller and Allen (1928) considered *M. thysanodes* a cave-dwelling bat, even though most of the specimens they examined were from buildings. Several other studies were conducted on colonies occupying buildings (Musser and Durrant, 1960; O'Farrell and Studier, 1973, 1975).

The species is known to occur in the caverns at Oregon Caves National Monument, Josephine Co., in late summer. Albright (1959) captured 29 fringed myotis at night by blocking the opening of the minelike entrance tunnel to the cave. He believed that these bats were returning to the caverns presumably after foraging. He observed few bats in the cave during the daylight hours, thus (p. 27) presumed that roosts were in "secluded recesses" of the cave. The possibility that the bats were using the cave as a night roost apparently was not considered, although the cave is now considered a night roost during the autumn breeding season (Cross, 1979). The entrance to this cave is at an elevation of ≅1,220 m (Albright, 1959) and is surrounded largely by old-growth Douglas-fir (*Pseudotsuga menziesii*). Nearby streams are bordered by western yew (*Taxus brevifolia*), Port Orford cedar (*Chamaecyparis lawsoniana*), and big-leaf maple (*Acer macrophyllum*—Roest, 1951). Maser et al. (1981*b*) reported collecting three specimens in Lincoln Co. from a day roost under loose flashing around a chimney in alder (*Alnus rubra*)-salmonberry (*Rubus spectabilis*) habitat near immature conifers. In Washington, Williams (1968) captured two females along a willow (*Salix*)-bordered creek in an area in which the predominant vegetation was sagebrush (*Artemisia tridentata*). Elsewhere, oak (*Quercus*) and piñon (*Pinus*) seemingly are the "most commonly used

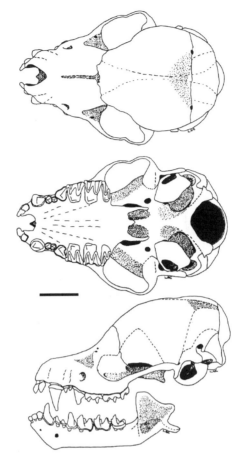

Fig. 9-15. Dorsal, ventral, and lateral views of the cranium, and lateral view of the dentary of the fringed myotis, *Myotis thysanodes* (OSUFW 5989). Scale bar equals 3 mm.

vegetative associations" (O'Farrell and Studier, 1980:3). Thus, it seems that the fringed myotis is adapted to live in areas with diverse vegetative substrates.

We found few quantitative or comparative data related to abundance of *M. thysanodes* in Oregon. In the Coast Range, Thomas (1988) identified 228 echolocation calls of *M. thysanodes* and *Eptesicus fuscus* combined among 4,470 calls detected. He detected 2.5–5.9 times as many calls of the two species per hour in old-growth Douglas-fir-western hemlock (*Tsuga heterophylla*) as in younger stands. Olterman and Verts (1972) reported that S. P. Cross at Southern Oregon University had related to them that ≅3% of the bats that he captured in mist nets in southwestern Oregon were *M. thysanodes*. However, they were unable to assemble sufficient information to assess the status of the species in the state.

Diet.—The diet of *M. thysanodes* is considered one of the hardest among the species of *Myotis* that occur in Oregon (Fig. 9-2). Black (1974:149), in assessing a bat community in New Mexico, considered the fringed myotis to be a "beetle strategist," as 73% of 11 fecal samples contained beetles (Coleoptera) and only 36% contained remains of moths (Lepidoptera). Findley (1987:44) classified *M. thysanodes* as one of the "hovering gleaners" that picks its prey from various substrates while in flight.

Stomach contents of four fringed myotis from western Oregon contained 46.2% moths, 31.2% spiders (Araneae) and harvestmen (Phalangida), 16.3% beetles, and 6.3% flies

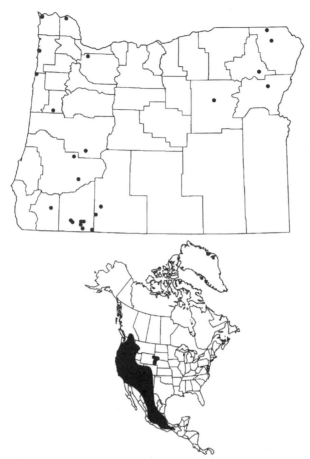

Fig. 9-16. Localities in Oregon from which museum specimens of the fringed myotis, *Myotis thysanodes*, were examined (listed in the APPENDIX, p. 533), and the distribution of the fringed myotis in North America. (Range map redrawn from Hall, 1981:204, map 161, with modifications.)

(Diptera—Whitaker et al., 1977). Fecal droppings of four individuals from eastern Oregon contained 42.5% moths, 17.5% leafhoppers (Homoptera), 7.5% lacewings (Neuroptera), 7.5% flies, 2.5% true bugs (Hemiptera), and 22.5% unidentified insects (Whitaker et al., 1981*b*). Thus, Black's (1974) categorization of *M. thysanodes* as primarily a beetle-eater is not supported by dietary analyses in Oregon. However, in Arizona, insect prey available to an 11-species bat community that included *M. thysanodes* was "extremely patchy both in space and time" (Bell, 1980:1880). Often, a prey species was abundant on 1 night, but did not occur in samples during the remainder of the summer or was present in small numbers for only a few nights after the peak (Bell, 1980). Consequently, consumption of various taxa of prey by these and other species of insectivorous bats may be more a function of prey availability than one of selection of specific taxa by the bats.

Reproduction, Ontogeny, and Mortality.—No information regarding these topics is available for *M. thysanodes* in Oregon.

In New Mexico, where most major studies of the species were conducted, *M. thysanodes* did not copulate before leaving the summer roosts in September, as did *M. lucifugus*. *M. thysanodes* commenced to return to summer roosts in the interval 10–18 April, but did not become abundant until 1–8 May. Uteri of females collected in April contained spermatozoa, but the females did not ovulate until after 18 April. By

15 May, all were pregnant; females produce only one young (Cockrum, 1955). Postimplantation development requires 50–60 days. Parturition is fairly synchronous with young born in the 2-week period 25 June–7 July. We speculate that dates may be somewhat later in Oregon.

At birth, neonates are about half as long and weigh about one-fourth as much as the maternal female (O'Farrell and Studier, 1973). At birth, the eyes are closed and the pinnae folded, but the eyes open and the pinnae unfold during the 1st day of life. The skin is pink, but commences to become pigmented and covered with dark gray-brown fur within 1 week. The young are tolerant of each other, in contradiction to the squabbling noted among adults (O'Farrell and Studier, 1973). O'Farrell and Studier (1973) found that while most adults were foraging, two–10 females remained with the cluster of young, but Baker (1962) reported that young were left unattended. Young that fall from the roost produce a distress call and are retrieved by a female alighting nearby and allowing the juvenile to attach itself for a return flight to the roost. O'Farrell and Studier (1973) believed that females retrieving young were not those that gave birth to them. When young <15 days old were handled, they constantly produced distress calls; females remaining with the cluster of young commonly alighted upon the investigator's hand holding the young bat, but never attempted to bite the investigator (O'Farrell and Studier, 1973). Females returning from foraging were believed to retrieve their own young and nurse them during the day. Although pregnant females were shy and fled when approached, those with nursing young were not; they commonly allowed investigators to touch them without taking flight.

Prenatal growth is extremely rapid, producing a logarithmic function with preparturition interval; however, growth likely is affected by climatic variables (O'Farrell and Studier, 1973).

When 16.5 days old, young fringed bats are capable of flying in a straight line for a short distance, but are unable to gain altitude. Five days later their flight is similar to that of adults (O'Farrell and Studier, 1973). Juvenile males are first to leave maternity roosts, followed by juvenile females, then adult females. Juvenile males are not believed to become reproductively active in the autumn of their birth (O'Farrell and Studier, 1973).

In New Mexico, mortality among neonates was only 1% during the developmental period (O'Farrell and Studier, 1973). Recorded maximum longevity is 18.3 years (Tuttle and Stevenson, 1982).

Habits.—Almost nothing is known regarding *M. thysanodes* in Oregon. Although several authorities indicated that the species migrated in winter (O'Farrell and Studier, 1980), evidence for migrations seems largely circumstantial. Fringed myotis return to summer roosts in spring and leave in late summer, but neither the magnitude nor the direction of movements is known. Arrival at the maternity roosts in New Mexico occurred during a short interval; the population there remained relatively stable until September then declined rapidly (O'Farrell and Studier, 1973). O'Farrell and Studier (1975) suggested that such was indicative that the summer population remained a coherent group during winter. O'Farrell and Studier (1975, 1980) also believed that the species possibly was active periodically throughout winter in New Mexico. However, Hoffmeister (1970) indicated that the winter range of the species in Arizona, except for two banding records, was restricted to the extreme southeastern portion of the state. O'Farrell and Studier (1980) suggested that migrations were of short distances or to more southern areas.

Winter records of the species in Oregon are few. Perkins et al. (1990) recorded one in a mine in Baker Co. in December and another in a mine in Lane Co. in November. We examined a specimen (USNM 265389) taken in a cave in Clackamas Co. in January.

Elsewhere, males and females form separate colonies in summer, although an occasional male may be found in a maternity colony. In one instance, however, female *M. thysanodes* associated in the same cluster with males of another species of *Myotis* in summer (Baker, 1962). Maternity colonies of as many as several hundred female *M. thysanodes* commonly are established in caves and in buildings in southwestern United States (Baker, 1962; Barbour and Davis, 1969; Musser and Durrant, 1960; O'Farrell and Studier, 1973). However, some caves seem to be used in summer only as night roosts (Albright, 1959; Easterla, 1966). Most of the fringed myotis (26 of 29) taken at Oregon Caves National Monument, Josephine Co., in summer were males (Albright, 1959).

Flight speed of *M. thysanodes* is moderate in comparison with that of other species of Oregon bats (Table 9-1). Fringed myotis were reported to produce echolocation calls at 49–31 kHz with two harmonics and a maximum duration of 8 ms (Fenton and Bell, 1981). Thomas et al. (1987) provided a range of 60–30 kHz and a duration of only 1–2 ms.

Remarks.—Bailey (1936) did not include *M. thysanodes* in his treatment of the mammal fauna of Oregon. He obviously was aware of the species and its characteristics as he included it in his monograph on the mammals of New Mexico (Bailey, 1931). However, Alex Walker (1942) reported that he had collected one in 1928; apparently the specimen was not identified as a fringed myotis until several years after Bailey (1936) published his work on Oregon mammals. Other Oregon specimens that we examined were collected in 1939 or later.

Myotis volans (H. Allen)
Long-legged Myotis

1866. *V[espertilio]. volans* Allen, 18:282.
1886. *Vespertilio longicrus* True, 8:588.
1911. *Myotis altifrons* Hollister, 56(26):3.
1914. *Myotis longicrus interior* Miller, 27:211.
1928. *Myotis volans longicrus*: Miller and Allen, 144:140.
1928. *Myotis volans interior*: Miller and Allen, 144:142.

Description.—*Myotis volans* (Fig. 9-17) is a relatively large member of the genus (Table 9-2) with short, round ears; small hind feet; a well-developed keel on the calcar; and fur on the ventral side of the plagiopatagium extending from the body to a line connecting the elbow and the knee. The fur of the dorsum extends onto the anterior 25% of the uropatagium (Barbour and Davis, 1969). The dorsal pelage is ocherous buff east of the Cascade Range and dark smoky-brown west of the Cascade Range (Hall, 1981). The venter is a lighter hue. The ears and flight membranes are blackish.

The skull (Fig. 9-18) is small with a short rostrum and a globose braincase; there is a sharp break between the rostrum and braincase. The occiput is elevated. The skull and forearm are significantly longer in females than in males, but the length of the head and body is not significantly

different between the sexes (Williams and Findley, 1979).

Distribution.—*Myotis volans* is a western species (Fig. 9-19) that occurs from the panhandle of Alaska south to the tip of Baja California Sur and eastward through the southern parts of British Columbia and Alberta to north-central North Dakota thence south through the Great Plains into Mexico to southern Durango. A disjunct population occurs in central Mexico from Jalisco to Veracruz. The species occurs throughout Oregon except the Columbia Basin (Fig. 9-19).

The only Pleistocene record of *M. volans* is from a cave site in Wyoming (Kurtén and Anderson, 1980). Roth (1972) recorded an extralimital Holocene specimen from central Texas.

Geographic Variation.—Two of the four currently recognized geographic races occur in Oregon: the darker *M. v. longicrus* west of the Cascade Range and the lighter *M. v. interior* east of the Cascade Range (Warner and Czaplewski, 1984).

Fig. 9-17. Photograph of the long-legged myotis, *Myotis volans*. (Photograph courtesy R. B. Hayward.)

Habitat and Density.—In general, *M. volans* is a species associated with montane coniferous forests, but it also occurs in some desert and riparian habitats (Barbour and Davis, 1969). Hoffmeister (1970) indicated that the species shifted its range seasonally in Arizona. Maternity colonies are established in buildings, hollow trees, and crevices in rock outcrops (Barbour and Davis, 1969). Caves and mines are used as hibernacula and night roosts (Barbour and Davis, 1969; Schowalter, 1980). In several caves in Deschutes Co., Perkins et al. (1990) recorded two–64 individuals in November. We examined one specimen (PSM 1953) taken in December in a cave in Clackamas Co.

In the Coast Range, Thomas (1988) detected echolocation calls of *M. volans* only 1.4 times more frequently in old-growth (>200 years old) stands of Douglas-fir (*Pseudo-*

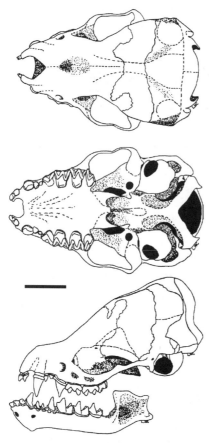

Fig. 9-18. Dorsal, ventral, and lateral views of the cranium, and lateral view of the dentary of the long-legged myotis, *Myotis volans* (OSUFW 4586). Scale bar equals 3 mm.

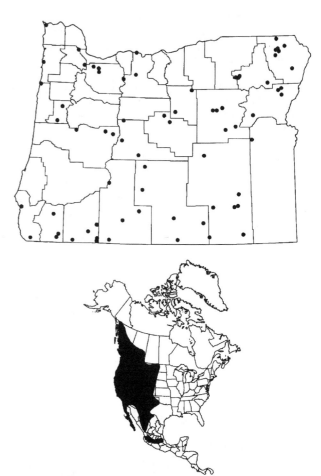

Fig. 9-19. Localities in Oregon from which museum specimens of the long-legged myotis, *Myotis volans*, were examined (listed in the Appendix, p. 533), and the distribution of the long-legged myotis in North America. (Range map redrawn from Hall, 1981:198, map 154, with modifications.)

tsuga menziesii) and western hemlock (*Tsuga heterophylla*) than in mature (100–165 years old) stands, but 20.9 times more frequently than in young (<75 years old) stands. Also, he detected echolocation calls of *M. volans* at a frequency equal to that of *M. californicus*, *M. ciliolabrum*, and *M. evotis* combined, and 1.2 times more frequently than *Lasionycteris noctivagans*, 3.8 times more than *Lasiurus cinereus*, 7.7 times more than *Corynorhinus townsendii*, and 2.1 times more than *Eptesicus fuscus* and *M. thysanodes* combined. However, a combination of *M. lucifugus* and *M. yumanensis* was detected 4.5 times more frequently than *M. volans*. Despite its apparent abundance, at least during the summer months, relatively little is known regarding populations of *M. volans* or other aspects of its biology.

Diet.—*Myotis volans* feeds heavily on moths (Lepidoptera). In western Oregon, more than three-fourths of contents of stomachs of the species consists of moths, but in eastern Oregon only about one-half of the material in stomachs consists of Lepidoptera (Table 9-3). In eastern Oregon, however, more than three-fourths of fecal material of long-legged myotis consists of remains of moths (Table 9-3). In western Oregon, termites (Isoptera) and spiders (Araneae) ranked second and third, respectively, among foods eaten, whereas in eastern Oregon, beetles (Coleoptera) and leafhoppers (Homoptera) occupied those ranks (Table 9-3). In Arizona, moths were eaten by all specimens of *M. volans* examined, but remains of beetles, leafhoppers, termites, and spiders occurred in feces of few or none (Warner, 1985).

Although the robustness of the skull of *M. volans* is exceeded by that of only one other species in the genus (*M. thysanodes*) in Oregon, the relative hardness of the diet is only moderate (Fig. 9-2). Reduker (1983) found that the masticatory apparatus of *M. volans* permitted a forceful closure of the jaw from a wide gape. He postulated that such was adaptive for removing and immobilizing insects caught in the uropatagium or plagiopatagium. The long legs, fur on the ventral side of the flight membranes, and short rostrum are characteristic of species that subdue and temporarily immobilize flying insects in their flight membranes (Findley, 1972).

Reproduction, Ontogeny, and Mortality.—Essentially nothing is known regarding these topics for *M. volans* in Oregon. Bailey (1936) concluded from collection in mid- to late August of four males at one site and four females at another site by Luther Goldman that the sexes were segregated during the breeding season. Because pregnant females were captured elsewhere at this latitude as late as late July and mid-August (Czaplewski et al., 1979; Findley, 1954), we believe that he intended to convey that the sexes were segregated during the littering and rearing seasons. Elsewhere, adult males either do not occur in maternity colonies or occur only in small numbers (Dalquest and Ramage, 1946; Davis and Barbour, 1970).

In New Mexico, spermatogenesis peaked in late July,

copulation commenced in late August, ovulation occurred in March–May, and parturition extended from May through August (Druecker, 1972). In southern California, parturition was in early June (Dalquest and Ramage, 1946), but not until late July in Wyoming (Findley, 1954). In Alberta, all adult and most juvenile males taken in August–September were reproductively active; those taken on 17 August had large testes, but small dark epididymides, and those taken on 21 September had regressed testes, but large and pale epididymides. A female taken on 21 September had spermatozoa in her uterus (Schowalter, 1980). Because lengths of testes measured in Alberta were similar to those measured in New Mexico at the same season (Druecker, 1972), Schowalter (1980) concluded that the timing of breeding was similar throughout the north-south extent of the species' range.

Each female produces only one ovum (Druecker, 1972) and one young annually (Dalquest and Ramage, 1946; Hall, 1946). Although the ovum may be produced by either ovary, implantation is always in the right horn of the uterus (Druecker, 1972). Druecker (1972) believed that the left ovary was more active in young-of-the-year females, accounting for his finding either a corpus luteum or a mature Graafian follicle in the left ovary of 11 of 15 examined. Druecker (1972) also believed that a single copulation in autumn did not ensure the presence of spermatozoa for fertilization in spring.

Seemingly, nothing is known regarding postpartum development and survival in *M. volans*. Maximum longevity recorded for the species is 21 years (Warner and Czaplewski, 1984).

Habits.—Along the Pacific Coast, Maser et al. (1981*b*) reported *M. volans* to commence to fly as early as 2000 h on warm, overcast evenings in August, but not until 45 min later when the overcast was light and apparently not at all on cold, clear evenings. Based on time of capture in a montane area in Arizona, this species commenced to fly within the first 30 min after sunset; capture rates remained steady for 1.5 h, then declined gradually during the following 2 h (Cockrum and Cross, 1964–1965). Based on the frequency of detection of echolocation calls in summer in the same region, *M. volans* exhibited a peak of activity at 2000–2300 h and another at 0300–0400 h (Bell, 1980). Nevertheless, at least some individuals are active at all hours of the night (Warner and Czaplewski, 1984). In Alberta in September, *M. volans* was active on nights when ambient temperatures were as low as –8°C (Schowalter, 1980).

The flight of *M. volans* is direct and faster than that of its congeners in Oregon (Table 9-1). Prey is pursued over long distances; it is capable of detecting prey as far away as 5 m, perhaps 10 m (Fenton and Bell, 1979). The species forages at treetop level early in the evening, but later near the ground beneath the forest canopy. It forages exclusively on flying insects, commonly making complex maneuvers in the last stages of pursuit of a prey item (Fenton and Bell, 1979). One individual seems to use the same foraging route each evening (Grinnell, 1918) and one individual was observed possibly to exhibit territorial behavior by chasing another individual (Maser et al., 1981*b*). However, feeding territories, if established, are not well defined (Maser et al., 1981*b*). Fenton and Bell (1979) observed no evidence of cooperative feeding or territorial interactions; however, they suggested that patchiness of insect prey in areas in which they conducted their study might be less conducive to territorial defense than to mutual avoidance.

In late summer and early autumn, long-legged myotis sometimes swarm at cave entrances (Schowalter, 1980). Three-fourths of those swarming at two caves in Alberta were males (Schowalter, 1980). Davis and Hitchcock (1965) believed that swarming was associated with selection of a suitable site for hibernation. However, Bradbury (1977) considered swarming possibly associated with some form of mating competition. Although *M. volans* is known to hibernate in some caves in Oregon (Perkins et al., 1990), we do not know whether it swarms at entrances to these caves in late summer or early autumn.

Myotis volans produces frequency-modulated echolocation calls with a minimum frequency of 40–35 kHz and a maximum frequency of 89–80 kHz with one-three harmonics. The pulses of sound last 2–8 ms and contain a constant-frequency component (Fenton and Bell, 1979; Thomas et al., 1987). The constant-frequency component is responsible for the ability of *M. volans* to detect prey at relatively long distances (Fenton and Bell, 1979). Echolocation calls also are produced at a high rate as a bat approaches a landing site. During foraging, when on a collision course with another individual, long-legged myotis lower the frequency of the terminal portion of the call to produce a "honk" (Fenton and Bell, 1979).

Residues and metabolites of DDT in tissues of *M. volans* 1 year after spraying large areas in northeastern Oregon with the insecticide to control the tussock moth (*Orgyia pseudotsugata*) averaged 6.21 ppm (range, 2.0–11.0 ppm), a level exceeded only in *M. californicus* (Henny et al., 1982). After 2 years, the average residues and metabolites had declined to 3.96 ppm (range, 1.8–5.3 ppm) and after 3 years to 1.17 ppm (range, 0.26–3.8 ppm). The species also contained residues of several other organochlorine insecticides. Henny et al. (1982) concluded that information regarding prey of the different species was not adequate to attribute differences in levels of remains of the toxicants among species of bats to differences in their diets.

Myotis yumanensis (H. Allen)
Yuma Myotis

1864. *Vespertilio yumanensis* Allen, 7(publ. 165):58.
1897*a*. *Myotis yumanensis saturatus* Miller, 13:68.
1914. *Myotis yumanensis sociabilis* Grinnell, 12:318.

Description.—*Myotis yumanensis* (Plate III) is similar to *M. lucifugus* except that it is slightly smaller (Table 9-2), has short and dull (not glossy) dorsal pelage, and the rostrum is separated from the braincase by a moderately sharp step (Fig. 9-20). The greatest length of the skull is usually <14 mm. Herd and Fenton (1983:2030) claimed to have used a "simple antibody test" to distinguish *M. yumanensis* and *M. lucifugus*. They did not describe the process, but only referenced Herd's unpublished dissertation as the source.

The dorsal pelage ranges from an olive tan to dark brown; the venter is whitish (especially the throat) to buffy or tan. The calcar lacks a keel. The feet are large and the ears are short. The ears and nose pad are paler than those of *M. lucifugus* (Herd and Fenton, 1983). The wings and uropatagium are dark brown.

Mensural characters used elsewhere to distinguish the

two species are described and an evaluation of their reliability for separating Oregon specimens is included in the Remarks of the account on *M. lucifugus*. Also, a discussion of putative hybridization of the two species is included in the account on the same species.

Distribution.—*Myotis yumanensis* occurs from southern British Columbia south along the Pacific Coast to the southern tip of Baja California Sur and eastward to western Montana, western Idaho, and western Nevada; all of Arizona and New Mexico with a northward extension to northeastern Utah; and western Texas. The range also extends into Mexico south to Hidalgo and Michoacán (Fig. 9-21; Hall, 1981). In Oregon, the species occurs throughout the state (Fig. 9-21). We found no published report of fossils of *M. yumanensis*.

Geographic Variation.—Two of the six recognized subspecies occur in Oregon: the darker *M. y. saturatus* west of the Cascade Range and the lighter *M. y. sociabilis* east of the Cascade Range (Dalquest, 1948; Hall, 1981).

Habitat and Density.—*Myotis yumanensis* is associated more closely with water than any other North American species of bat, even *M. lucifugus* (Barbour and Davis, 1969; van Zyll de Jong, 1985). Whitaker et al. (1977:47) indicated that the species was associated with "large streams, rivers, ponds, or lakes" in western Oregon, but Whitaker et al. (1981b) found too few in eastern Oregon to comment on the habitat associations of the species.

In British Columbia, 139 (88.5%) of 157 *M. yumanensis* were captured in willow (*Salix*) and alder (*Alnus*) along the bank of a river, at the mouth of a canyon with a fast-flowing creek bordered by alders and ponderosa pine (*Pinus*

ponderosa), or in a narrow strip of desert near a lake (Fenton et al., 1980).

Dalquest (1947b:231) suggested that *M. yumanensis* probably was a cave dweller originally, but had become completely adapted to "man-made structures." He described summer colonies in the attic and belfry of a church, in an adobe building, in a power house with a corrugated-iron roof, in the attic of a ranch store, in a barn, on the porch of a high school, in a warehouse, and beneath tiles on the roof of a university building. He also suggested that hollows and cracks in trees and loose pieces of bark were used as retreats. With few exceptions, sites occupied by *M. yumanensis* were close to water, near trees, dimly lit, contained crevices, and were rarely visited by humans (Dalquest, 1947b). In British Columbia, Herd and Fenton (1983) found maternity colonies of *M. yumanensis* in cabins.

Perkins (1987), in capturing 1,090 bats at 164 (61.4%) of 267 sites at which nets were set in summer 1982–1986, caught only 16 *M. yumanensis*; this species was captured at only six sites. Whitaker et al. (1981a:282) considered the species "exceedingly scarce" in eastern Oregon. They obtained only seven individuals among 413 of 12 species of vespertilionids collected for analysis of stomach contents; all were from Grant Co. In western Oregon, Whitaker et al. (1977) captured 25 among 239 of 11 species obtained for the same purpose.

Diet.—In western Oregon, flies (Diptera) composed

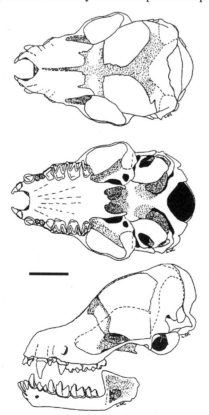

Fig. 9-20. Dorsal, ventral, and lateral views of the cranium, and lateral view of the dentary of the Yuma myotis, *Myotis yumanensis* (OSUFW 36). Scale bar equals 3 mm.

Fig. 9-21. Localities in Oregon from which museum specimens of the Yuma myotis, *Myotis yumanensis*, were examined (listed in the APPENDIX, p. 533), and the distribution of the Yuma myotis in North America. (Range map redrawn from Hall, 1981:190, map 148, with modifications.)

slightly more than half of the stomach contents of *M. yumanensis* and this item combined with moths (Lepidoptera) and termites (Isoptera) composed 86.5% of the diet (Table 9-3). In eastern Oregon, moths contributed almost 25% and flies composed only 10% of the stomach contents, but five other categories each contributed ≥12.9% of the diet (Table 9-3). Analyses of fecal samples (Table 9-3) suggested that *M. yumanensis* in eastern Oregon was more of a dietary generalist.

In British Columbia, *M. yumanensis* fed entirely on flies (particularly Chironomidae) in early May, but fed mostly on caddisflies (Trichoptera) and mayflies (Ephemeroptera) in July. In August, 90% of the diet of adults was caddisflies and mayflies, whereas 55% of the diet of juveniles was flies and 44% was caddisflies and mayflies (Herd and Fenton, 1983).

According to our reanalysis of Freeman's (1981) data, indices to hardness of the diet and skull robustness for *M. yumanensis* were near those expected on the basis of analysis of 14 species of Oregon bats (Fig. 9-2). Although the index to hardness of the diet was more than twice that of *M. lucifugus,* the index to skull robustness was less than half the estimated value for that species (Fig. 9-2).

Reproduction, Ontogeny, and Mortality.—Relatively little is known regarding these topics in *M. yumanensis* either in Oregon or elsewhere. Maser et al. (1981*b*) recorded pregnancies as early as mid-April along the Oregon coast. Dalquest (1947*b*) reported that pregnant females were collected on 20 April in California; of 70 females examined from 20 April to 19 June, 63 were pregnant and each carried only one embryo. Bailey (1936) and Maser et al. (1981*b*) agreed that only one young was produced.

Dalquest (1947*b*) reported full-term embryos on 3–19 June; he believed that most young were born in California before mid-June and that those in desert areas produced young no earlier that those in temperate areas. However, two specimens taken in late June in Washington each carried one embryo (Dalquest, 1948) and Bailey (1936) reported pregnant females as late as 7 July in western Montana. In western Texas, Easterla (1973) reported catching pregnant females on 12 June and lactating females on 10 June. In British Columbia, Herd and Fenton (1983) caught the first lactating female of the year on 19 June and the last pregnant female on 28 June. Thus, it seems that birth dates may extend over a 3–4-week period north of Mexico, but that no north-south cline in birth dates is evident.

Myotis yumanensis follows the reproductive strategy of early parturition and production of offspring by females in their 1st year of life. Herd and Fenton (1983) suggested that this strategy was more characteristic of bats in lower latitudes and in *M. yumanensis* possibly reflected relatively short-term sympatry with *M. lucifugus* in British Columbia. Pregnancy and lactation in *M. yumanensis* occurred ≅2 weeks earlier than in *M. lucifugus* and male *M. yumanensis* attained breeding condition earlier than most male *M. lucifugus* (Herd and Fenton, 1983).

Male Yuma myotis attain sexual competency in early August (Maser et al., 1981*b*) and copulation probably occurs in late summer-early autumn with sperm storage in the uterus during winter (Dalquest, 1947*b*).

Dalquest (1947*b*) claimed that maternal females do not carry their young while foraging. However, in New Mexico, Commissaris (1959) reported capturing eight females in flight in a storage building four of which had a young ≤4 days old

clinging to their venters. Bailey (1936) provided secondhand information that some females collected by shooting in western Montana had small young clinging to them.

Known inimical factors for *M. yumanensis* are few. Maser et al. (1981*b*) reported a house cat that became proficient at leaping into the air and catching these bats as they flew into a garage. It did not eat the bats. Dalquest (1947*b*), in noting the low-flying behavior of the species, suggested that several agile predatory mammals might be in a position to capture one from time to time. Cockrum (1973) recorded a maximum longevity of 8.8 years.

Habits.—*Myotis yumanensis* is an early flyer, sometimes becoming active before sunset. These bats have been taken at night roosts 15 min after dusk with stomachs packed with insects; they may use night roosts several times during the night (Dalquest, 1947*b*). Although Dalquest (1947*b*) claimed that fecal droppings are not found on the floors of buildings used by *M. yumanensis* as night roosts, Maser et al. (1981*b*) reported that night roosts frequented by the species along the Oregon coast exhibited ample evidence of use, including accumulations of guano. Maser et al. (1981*b*) reported night roosts established under bridges, on the exposed chimney of a house beneath an overhanging roof, and on the ceiling beams in an open garage.

Females may form large maternity colonies, but males commonly spend the day alone (Dalquest, 1947*b*). *M. yumanensis* shared summer day roosts with *M. lucifugus* at Ft. Klamath and Olene, Klamath Co., and at two localities in northern California. The two species also were collected as they foraged together at Tule Lake, California (Parkinson, 1979). However, Herd and Fenton (1983) observed that individuals of the two species held temporarily in bags tended to associate in clusters with conspecifics. They suggested that within roosts the two species may form separate aggregations. They also observed that marked bats of the two species exhibited significantly different patterns of habitat use and that the different patterns corresponded with observed differences in their diets. In California, *M. yumanensis* was observed to occupy the same roosts with *Tadarida brasiliensis* and several species of vespertilionids; in some instances, they associated in the same clusters with the other species (Dalquest, 1947*b*).

Day roosts in attics of buildings may become exceedingly hot during late afternoons on sunny days. Licht and Leitner (1967) reported that *M. yumanensis* in a nursery colony roosted on a beam supporting the roof of a barn. In the morning, when the temperature 5 mm above the surface of the beam was 30–32°C, the bats clustered tightly near the ceiling, but as the temperature at this point reached 40°C, the bats moved away from the ceiling and became distributed over the beam. When the temperature near the surface of the beam at the ceiling reached 48.6°C, the few bats that had not moved off the beam to roost in a cool stairwell of the barn had moved as far as possible from the ceiling to a point at which the temperature was 41.5°C. Nevertheless, no panting or other cooling behavior was noted.

Dalquest (1947*b*) described the flight of *M. yumanensis* as a series of curves; some individuals upon release flew in circles as small as 0.3 m in diameter, but most flew in circles or spirals 1.2–3.7 m in diameter. The bats flew with a steady wing beat, uninterrupted by glides. A landing site is approached from ≅0.3 m below; the bat pulls upward at ≅45°, half closes its wings, and touches down head upward with legs widely spread and uropatagium expanded. The bat then turns to hang head down; such turns usually are counterclockwise (Dalquest, 1947b).

In British Columbia, foraging activity of both M. yumanensis and M. lucifugus in 1982 before 26 July was greater before midnight than after, but after that date foraging was more evenly distributed during the night (Herd and Fenton, 1983). Both before and after that date, the two species usually foraged at different sites. In one region after 26 July, the two species foraged over different water areas <100 m apart. Although both species foraged over similar habitats at times, M. yumanensis foraged in fewer habitat types than M. lucifugus. M. lucifugus was observed to forage in all habitats sampled, whereas M. yumanensis foraged only over water (Herd and Fenton, 1983).

Along the Oregon coast as elsewhere, M. yumanensis forages low over the water, usually <0.3 m above the surface (Dalquest, 1947b; Herd and Fenton, 1983; Maser et al., 1981b). Individuals usually fly regular routes while foraging, circles or "S's" over ponds and lakes and straight back-and-forth patterns over streams (Dalquest, 1947b; Maser et al., 1981b). In British Columbia on windy nights, M. yumanensis foraged beneath the tree canopy (Fenton et al., 1980).

Winter habits of M. yumanensis are largely unknown (Dalquest, 1947b; Maser et al., 1981b). The species is reported to be active at temperatures when some other species of bats are in torpor. Dalquest (1947b) suggested that M. yumanensis probably hibernates where temperatures are relatively low. However, hibernacula are unknown. Perkins et al. (1990) did not list M. yumanensis among bats found in >650 caves and mines searched during winter in Oregon and Washington.

Remarks.—Myotis yumanensis seems particularly prone to abandon roosts when disturbed by humans. In view of the paucity of recent records of the species, especially in eastern Oregon, every effort should be made to avoid disturbing nursery colonies of the species.

Lasiurus cinereus (Palisot de Beauvois)
Hoary Bat

1796. Vespertilio cinereus (misspelled linereus) Palisot de Beauvois, p. 18.
1864. Lasiurus cinereus Allen, 7(publ. 165):21.
1912. Nycteris cinerea: Miller, 79:64.

Description.—Lasiurus cinereus (Plate III) is the largest (Table 9-2) and most distinctively colored bat in Oregon. Adults weigh 20–35 g (Shump and Shump, 1982) with females averaging 3.9% larger than males (Williams and Findley, 1979). The ears are short and rounded, the tragus short and broad, and the calcar long and narrowly keeled. The wingspread ranges from 34 to 41 cm. The uropatagium is entirely covered with heavy fur. In flight the species can be identified by its large size; long, pointed wings; slow, regular wingbeat; and fast, direct flight (Grinnell, 1918; Merriam, 1884b; Provost and Kirkpatrick, 1952).

The head of the hoary bat is light yellowish-brown blending to dark brown over the eyes and around the mouth; the chin patch is yellow. The ears are yellowish brown with black edges. Hairs forming the fur on the dorsum are banded; their bases are blackish brown followed in order by bands of yellowish brown, blackish brown, and white. The bands produce a frosty or hoary effect. Patches of fur

at the wrist and elbow are white or cream colored, contrasting with the black and brown wing membranes (Shump and Shump, 1982; van Zyll de Jong, 1985). The venter is white on the belly and pale brown on the chest (Hall, 1981).

The skull is short, broad, and in profile tapers upward in a smooth curve from rostrum to rear of braincase; there is no step between rostrum and braincase (Fig. 9-22). The nasal openings are broad and the zygomata extend well beyond the braincase (Shump and Shump, 1982). There usually are 32 teeth, but one or both of the tiny P3's sometimes are missing.

Distribution.—In North America, Lasiurus cinereus occurs from southeastern British Columbia, northeastward to Great Slave Lake, Northwest Territories, then southeastward across the southern provinces and south through all of the United States (except the southern tip of Florida) and most of Mexico to Guatemala (Fig. 9-23). Other populations occur in South America and in Hawaii. Extralimital records of presumed wanderers are from Southampton Island, Northwest Territories; Iceland; Bermuda; Hispaniola; and the Orkney Islands off Scotland (Shump and Shump, 1982). It is the most widely distributed of North American bats (Barbour and Davis, 1969).

In Oregon, hoary bats are represented by museum specimens collected at scattered localities over most of the region west of the Cascade Range and in montane regions east of the Cascade Range. No specimens have been collected in desert areas or the Columbia Basin area (Fig. 9-23).

Late Pleistocene fossils are known from Kansas, New Mexico, Texas, and Nuevo León, Mexico (Shump and Shump, 1982). Lasiurus †fossilis is considered ancestral to L. cinereus (Kurtén and Anderson, 1980).

Geographic Variation.—Only one of the three recognized subspecies occurs in North America: L. c. cinereus (Shump and Shump, 1982).

Habitat and Density.—Hoary bats usually are associated with montane boreal forests (Bailey, 1936). Whitaker et al. (1981b) indicated that during spring and autumn migrations they may be found in arid shrub-steppe; however, they seem rarely to be collected in such habitat in Oregon (Fig. 9-23). They commonly are referred to as "tree bats" as they usually day roost in foliage 3–10 m above the ground (van Zyll de Jong, 1985). Two individuals reportedly were shot in western Oregon as they flew in mixed coniferous and deciduous forest (Whitaker et al., 1977). Hoary bats have been found in a variety of other roosts, including squirrel nests, cavities in trees excavated by birds, and on the sides of buildings (Shump and Shump, 1982). Hoary bats sometimes enter caves as evidenced by skulls, skeletons, mummified remains, and occasionally live specimens found therein (Beer, 1954; Myers, 1960). However, they are not believed to leave the caves, but die therein (Myers, 1960). Hoary bats often forage over water or roads through forests and over openings in the forest (Whitaker et al., 1981b; Zinn and Baker, 1979).

Whitaker et al. (1977) considered hoary bats uncommon in western Oregon. Thomas (1988) recorded echolocation calls of L. cinereus in the Coast Range of Oregon, but they were too infrequent to use in analyses of relative detection rates by age of forest stands. However, calls were recorded twice as frequently as those of Corynorhinus townsendii, but 3.1 times less frequently than those of Lasionycteris noctivagans, 3.8 times less than those of

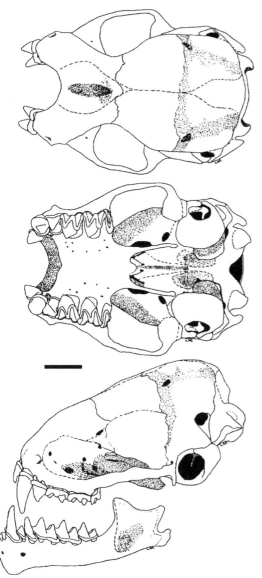

Fig. 9-22. Dorsal, ventral, and lateral views of the cranium, and lateral view of the dentary of the hoary bat, *Lasiurus cinereus* (OSUFW 6000). Scale bar equals 3 mm.

Fig. 9-23. Localities in Oregon from which museum specimens of the hoary bat, *Lasiurus cinereus*, were examined (listed in the APPENDIX, p. 520), and the distribution of the hoary bat in North America. (Range map redrawn from Hall, 1981:226, map 176, with modifications.)

Myotis volans and a combination of three other species of *Myotis,* and 1.8 times less than those of a combination of *Eptesicus fuscus* and *Myotis thysanodes.* In western Washington, only one of 3,029 echolocation calls recorded was that of *L. cinereus.*

Diet.—Although hoary bats are known to prey on other orders of insects (Ross, 1967; Zinn and Baker, 1979), they commonly are considered to be specialists on moths (Lepidoptera—Black, 1972; Shump and Shump, 1982). Fecal droppings of six *L. cinereus* collected in eastern Oregon contained 98.3% moths, 0.8% flies (Diptera), 0.3% unidentified insects, and 0.5% vegetation by volume (Whitaker et al., 1981a). In western Oregon, the stomach of one individual contained entirely moths and another contained entirely culicid flies (Whitaker et al., 1977). Hoary bats reportedly approach moths (Lepidoptera) from the rear and, upon capture, bite off the abdomen and thorax, allowing the head and wings to fall away (Shump and Shump, 1982).

A hoary bat was observed to pursue a smaller bat (Orr,

1950) and another was found consuming one it apparently had caught (Bishop, 1947). The frequency that other bats are included in the diet of the hoary bat is unknown.

Reproduction, Ontogeny, and Mortality.—Some information regarding reproduction in hoary bats seemingly is based on supposition and much of the remainder is derived from scattered records. Shump and Shump (1982) suggested that copulation likely occurs in autumn during migration or in wintering areas, but referenced no publication in support of their contention. As most temperate-climate, hibernating vespertilionids (Nowak, 1991), including congeners of hoary bats (Layne, 1958a; Stuewer, 1948), copulate (at least for the first time each year) in late summer–early autumn, their assessment of the mating season probably is correct. Females with implanted embryos reportedly were taken at scattered locations throughout the United States from 14 May to 20 June and females with young were taken from 14 June to 17 July (Cockrum, 1955).

Litter size based on counts of embryos and young/female commonly is two (Table 9-5; Cockrum, 1955); Barbour and Davis (1969:144) indicated that "in every case litter size is two." However, Shump and Shump (1982) listed the range as one to four, with two being most common. Bogan (1972) reported that one of 10 pregnant females maintained in captivity had only one embryo; the remainder had two. We found no records for litter size in Oregon.

Bogan (1972:612) described neonates as "covered with fine, silvery-gray hair on the dorsum of the head, shoulders, uropatagium, and feet; otherwise they are naked." The haired areas were pigmented as were the edges of the pinnae and the fingers. The eyes and ears were closed and the pinnae were closely appressed to the head. At 2 days of age, the young had developed more hair on the dorsum and were able to make high-pitched chirps; at 3 days, the ears were open and the pinnae were erect; at 7 days, the thorax was sparsely covered with hairs and the young could hang alone; by 11 days, hairs were becoming denser on the thorax and pinnae; by 12 days, the eyes were open; by 18 days, the white patches at wrists and elbows were present and the venter was furred. By day 22, young were colored much as adults, but a distinct molt was not observed; by day 27, young could glide; by day 34, young could fly and were weaned. Although captive females were able to carry their young in flight until the young were 6–7 days old (Bogan, 1972), females in the wild usually leave their young hanging in trees while they forage (Barbour and Davis, 1969).

We know of no records of causes or magnitude of mortality in hoary bats in Oregon. Elsewhere, they are known to be taken at least occasionally by raptors and snakes (Serpentes), and in some regions they frequently are infected with rabies (Shump and Shump, 1982).

Habits.—*Lasiurus cinereus* is noted for its fast flight (Merriam, 1884*b;* Provost and Kirkpatrick, 1952; Shump and Shump, 1982); in timed trials it averaged faster than all other bats tested (Table 9-1). It usually flies in straight-line paths in open areas (Barclay, 1985). It produces "single harmonic search-approach calls that are low (20–17 kHz), essentially constant frequency signals" particularly suited to detecting large insects at long range (Barclay, 1986:2700). Such a foraging strategy reduces the availability and probability of small insects as prey (Barclay, 1985). Hoary bats also make audible calls (Barbour and Davis, 1969).

Hoary bats usually are late flyers, becoming active after most other species have begun to fly. Merriam (1884*b*) claimed that onset of activity was in response to the degree of darkness, but later, in the same publication, indicated that he had not observed the species to fly when temperatures were above 15°C. There are records, however, of hoary bats flying even before sunset and before some species of birds had gone to roost (Dalquest, 1943*a;* Provost and Kirkpatrick, 1952). Provost and Kirkpatrick (1952) questioned whether either light or temperature were key factors in onset of activity in hoary bats; only two of eight they collected were flying when the ambient temperature was <15.6°C and one was taken when the ambient temperature was 22.8°C.

The hoary bat has long been considered to be migratory (Merriam, 1884*b;* Grinnell, 1918) and, indeed, considerable circumstantial evidence has been presented to support some seasonal movements (Findley and Jones, 1964). In the Pacific Northwest, plots of localities at which museum specimens were collected in August–October were interpreted by Dalquest (1943*a:*20, fig. 1) to indicate progressive southward migration of the species. He indicated that there were no October records for British Columbia (p. 23), yet his map (p. 20, fig. 1) contains a symbol (triangle) for that month in the province. Considering that Dalquest (1943*a*) provided records of the species in British Columbia, Washington, Oregon, and California during

all 3 months, we find his evidence of a fall migration to be weak. This, combined with our more recent records of the species being collected in Oregon in December and February (other records extend from April to October), suggests either that some individuals are residents or that migrations are overlapping north-south shifts. Hoary bats are known to assemble and fly in large groups in spring and autumn when migrations would be expected to occur (Barbour and Davis, 1969; Findley and Jones, 1964), but migration routes are unknown and wintering areas are not well documented (Shump and Shump, 1982). Dalquest (1943*a*) considered hoary bats in the Pacific Northwest to winter in California from San Francisco to the Mexican border; he believed that the northward migration began in April and by late May the bats were in northern California. However, Vaughan and Krutzsch (1954) found hoary bats in southern California in all months except July and August. They suggested that the sexes segregated in late spring with males moving to the foothills and females remaining in the lowlands and coastal valleys. Thus, migrations may be elevational as well as latitudinal, possibly accounting, at least in part, for the seemingly contradictory evidence for latitudinal migrations.

During most seasons, hoary bats are solitary. They usually react aggressively when handled. However, during late pregnancy females in captivity become more tractable and can be handled without gloves (Bogan, 1972). The aggressiveness returns at weaning or earlier if the young die. Bogan (1972) suggested that the temporary docility permitted toleration of contact by young with the usually solitary female and that a female with young likely would remain hidden at the approach of a potential predator at a time that it could not easily flee or fight.

Remarks.—The paucity of direct evidence of seasonal migrations (Shump and Shump, 1982) and even lesser understanding of wintering areas (Findley and Jones, 1964) suggest a fruitful avenue for research. The species, although not particularly abundant, is fairly easy to capture in nets, at least seasonally (Findley and Jones, 1964). It is sufficiently large to carry miniature beacon transmitters (Wai-Ping and Fenton, 1989), thereby permitting application of modern radiotelemetry techniques to accomplish such an investigation.

Lasionycteris noctivagans (Le Conte)
Silver-haired Bat

1831. *V* [*espertilio*]. *noctivagans* Le Conte, 1(App.):431.
1866. *Lasionycteris noctivagans:* Peters, 1865:648.

Description.—In the silver-haired bat (Fig. 9-24) the wings, ears, and uropatagium are black, and the fur, except for the silvery-white tips of the hairs on the dorsum, is dark brown to black. The anterior half of the uropatagium is furred on the upper side. The ears are short, rounded, and naked; the tragus is broad and blunt. The two sexes are nearly identical in size (Williams and Findley, 1979).

The skull is flattened; there is no step between the rostrum and braincase (Fig. 9-25). The rostrum and interorbital region are broad and the rostrum has a depressed area on each side. The zygomata extend markedly beyond the widest point of the braincase. There are 36 teeth, two fewer than occur in other dark but slightly smaller bats in Oregon (genus *Myotis*).

Distribution.—*Lasionycteris noctivagans* occurs from

Table 9-5.—*Reproduction in some Oregon bats.*

Species	Number of litters	Litter size $\overline{X}$	Range	n	Reproductive season[a]	Basis	Authority	State or province
Antrozous pallidus	1	1.8	1–3	28	Oct.–Nov.	Variable[b]	Orr, 1954	California
	1	1.8		88		Variable[b]	Cockrum, 1955[c]	Range-wide
Corynorhinus townsendii	1	1.0	1[d]	470	Oct.–Feb.[e]	Variable[f]	Pearson et al., 1952	California
Eptesicus fuscus	1	1.0	1	4		Embryos	This study	Oregon
	1	1.0	1	15		Embryos	Hall, 1946	Nevada
Euderma maculatum	1	1.0		3		Neonates	Easterla, 1971, 1976	Texas
Lasionycteris noctivagans	1	1.9	1–2	7		Variable[g]	Parsons et al., 1986	Ontario
Lasiurus cinereus	1	2.0	2	16		Embryos	Cockrum, 1955[c]	Various[h]
Pipistrellus hesperus	1	1.9	1–2	31		Embryos	Cockrum, 1955[c]	Various

[a]Breeding season; sperm storage delays parturition several months more than required by embryonic development.
[b]Includes counts of embryos and young born in captivity.
[c]Based on a variety of published and unpublished sources.
[d]Hall (1946) reported one of 10 females with two embryos in Nevada.
[e]All females were inseminated in October, but evidence of additional copulations during hibernation was noted.
[f]Based on counts of embryos and corpora lutea, and females with young.
[g]Combination of embryos and nonvolant young in maternity colony.
[h]Data compiled from the literature and museum-specimen labels; none from the Pacific Northwest.

southeastern Alaska across the southern provinces of Canada south through all of the United States except for Florida and the southern half of the southeastern states and the southern two-thirds of California, parts of Arizona, and west Texas (Fig. 9-26; Kunz, 1982; Parsons et al., 1986). The range extends into northeastern Mexico. The species occurs statewide in Oregon except for most of the Columbia Basin (Fig. 9-26).

Wisconsinan-age fossils are known from caves in Wyoming (Kurtén and Anderson, 1980).

Geographic Variation.—No subspecies are recognized (Kunz, 1982).

Habitat and Density.—Whitaker et al. (1977, 1981a) indicated that *L. noctivagans* was primarily associated with coniferous forests in Oregon, including the juniper (*Juniperus*) woodlands in the southeastern portion of the state. They also reported that it sometimes occurs in mixed deciduous-coniferous forest, and during migrations in May and September it occurs in rangelands where it forages along small streams.

By sampling with mist nets placed over water at 62 sites in three types of forest stands in Oregon on 105 nights in June-September, Perkins and Cross (1988) found average capture rates for *L. noctivagans* were 3.7–49.5 times greater in stands >200 years old than in younger stands (Fig. 9-27). However, they were unable to locate suitable sampling sites within younger stands of two forest types (Fig. 9-27). Within the >200-year-old stands, average capture rates were 2.3 times greater in Douglas-fir (*Pseudotsuga menziesii*)–western hemlock (*Tsuga heterophylla*) forests than in ponderosa pine (*Pinus ponderosa*)-sugar pine (*P. lambertiana*) forests and 2.8 times greater than in grand fir (*Abies grandis*)-white fir (*A. concolor*) forests (Fig. 9-27). Perkins and Cross (1988) believed that older forest stands, particularly those with Douglas-fir, contained more crevices for roosting by silver-haired bats. Differences in abundance of suitable crevices result from natural features of the bark of different tree species and changes related to age of the trees, including an increase in lightning and wind damage, activities of cavity-nesting birds, exfoliation of bark, and shedding of lower limbs.

By sampling forest stands in the Oregon Coast Range with ultrasonic bat detectors in mid-July–early September,

Fig. 9-24. Photograph of the silver-haired bat, *Lasionycteris noctivagans*. (Photograph courtesy R. B. Hayward.)

Thomas (1988) obtained similar results. Echolocation calls were detected in old-growth (>200 years old) Douglas-fir–western hemlock stands 68.3 times more frequently than in mature stands (100–165 years old) but only 5.4 times more frequently than in young (<75 years old) stands. Silver-haired bats were detected much less frequently in the Cascade Range in Washington and differences in detection rates among stands of different age-classes also were much less. They were detected only 1.7 and 1.8 times more frequently in old-growth stands than in mature and young stands, respectively (Thomas, 1988).

In Manitoba, visual searches of trees on a ridge at the south edge of Lake Manitoba revealed that most *L. noctivagans* (89.8% were females) used day roosts <3 m above the ground in peach-leaved willow (*Salix amygdaloides*) trees averaging 1.62 m in circumference (Barclay et al., 1988). Smaller willows and other species of trees in the area had smoother bark, and fewer cracks and splits in which silver-haired bats could roost.

In addition to roosting in trees in summer, silver-haired bats, in parts of the range at least, are known to use mines,

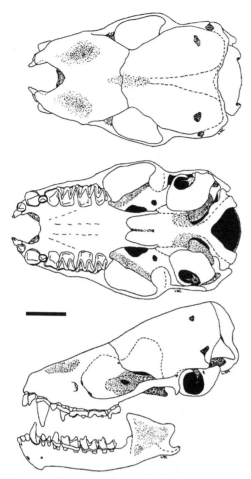

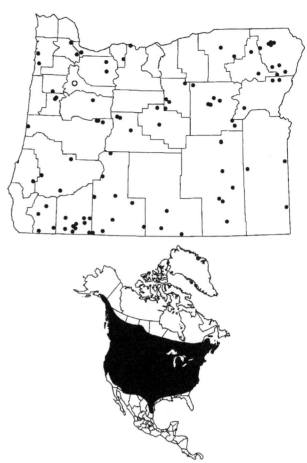

Fig. 9-25. Dorsal, ventral, and lateral views of the cranium, and lateral view of the dentary of the silver-haired bat, *Lasionycteris noctivagans* (OSUFW 6015). Scale bar equals 3 mm.

Fig. 9-26. Localities in Oregon from which museum specimens of the silver-haired bat, *Lasionycteris noctivagans*, were examined (listed in the Appendix, p. 519), and the distribution of the silver-haired bat in North America. (Range map redrawn from Hall, 1981:210, map 166, with modifications.) Open symbol indicates record for county.

caves, rock crevices, trees, and buildings for winter hibernacula (Beer, 1956; Cowan, 1933; Frum, 1953; Gosling, 1977; Layne, 1958a; Pearson, 1962). Some of those in mines were roosting in crevices (Layne, 1958a). However, records of *L. noctivagans* wintering north of the −6.5°C mean daily minimum isotherm for January are sparse (Izor, 1979). Western Oregon and the Deschutes-Columbia Plateau Province are not within the area from which *L. noctivagans* is thought to be excluded by low temperatures. However, Izor (1979) provided no winter records of the species in Oregon, and of 79 museum specimens taken in Oregon for which we recorded dates of collection, none was collected between 7 November and 2 April. Perkins (1987) did not list *L. noctivagans* among seven species noted in 478 potential hibernacula searched for *Corynorhinus townsendii* in winter 1982–1986. However, *L. noctivagans* may remain in British Columbia in winter (Cowan, 1933; Schowalter et al., 1978). Nevertheless, *L. noctivagans* must be rare in Oregon in winter if, indeed, it occurs here at all. Consequently, suitable sites to be used as hibernacula in Oregon in winter are not considered critical to the species' continued existence in the state.

Relatively little is known regarding the abundance of *L. noctivagans* in Oregon or elsewhere (Kunz, 1982). Barbour and Davis (1969:106) considered the species to be "erratic in abundance," but scarce over most of its range. Others indicated that the species ranked second to fourth in abun-

dance among bat species at various localities (Kunz, 1982). In the Oregon Coast Range, echolocation calls of *L. noctivagans* were detected 3.1 times more frequently than those of *Lasiurus cinereus* and 6.2 times more frequently than those of *Corynorhinus townsendii*, but only 0.8 times as frequently as those of *Myotis volans* and those of a combination of *M. californicus*, *M. ciliolabrum*, and *M. evotis*. Also, *L. noctivagans* was detected only 0.2 times as frequently as a combination of *M. lucifugus* and *M. yumanensis* and 1.7 times as frequently as a combination of *Eptesicus fuscus* and *M. thysanodes* (Thomas, 1988).

Diet.—Kunz (1982), in summarizing published information regarding the diet of the species, concluded that *L. noctivagans* was an opportunistic feeder that took a variety of insects. Published studies on the diet of the species in Oregon (Table 9-6) tend to corroborate that assessment. Both east and west of the Cascade Range, moths (Lepidoptera) contribute most to the diet, but prey in nine additional orders of insects plus spiders (Araneae) also were reported to be consumed in both regions (Table 9-6).

The diet of *L. noctivagans* east of the Cascade Range differed considerably depending on whether contents of fecal droppings or of stomachs were analyzed (Table 9-6; Whitaker et al., 1981b). However, based on the most extensive analyses of stomach contents, the diet of *L. noctivagans* east and west of the Cascade Range was similar (Table 9-6). East of

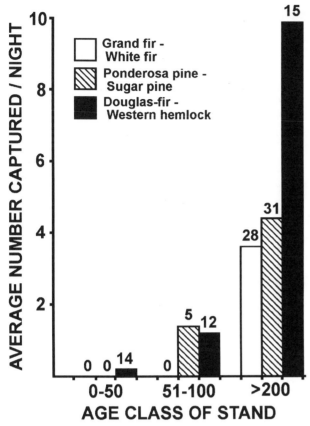

Fig. 9-27. Average number of silver-haired bats, *Lasionycteris noctivagans*, captured in mist nets/night by age-class of three types of forest in Oregon. Numerals above bars indicate number of nights sampling was conducted in each age-class of each habitat type. (Data from Perkins and Cross, 1988:23, table 2.)

the Cascade Range the volume of moth remains was more than twice as great and lacewings (Neuroptera) were >12 times greater in fecal droppings than in stomachs, but the volume of bees and ants (Hymenoptera), beetles (Coleoptera), true bugs (Hemiptera), and hoppers (Homoptera) was 3–15 times greater in stomach contents (Table 9-6). Because these bats were collected over a broad area of eastern Oregon, Whitaker et al. (1981*b*) suggested that differential availability of prey likely was responsible for differences in volumes of various prey taxa determined by the two methods. In samples collected in Grant Co., percent volumes of moths and flies (Diptera) obtained by analysis of stomachs and feces were similar although the samples were not collected at the same time. However, the similarity in findings based on stomach contents in eastern and western Oregon (Table 9-6) suggests that differences obtained by examining stomach contents and fecal droppings may not be entirely geographical or temporal. Henny et al. (1982) presented dietary information for 18 *L. noctivagans* from Wallowa Co. that differed greatly from that presented by others for Oregon silver-haired bats (Table 9-6). We are not certain, but it seems likely that these 18 individuals constitute a subsample of the 41 from eastern Oregon reported on by Whitaker et al. (1981*b*).

Reproduction, Ontogeny, and Mortality.—Relatively little is known regarding reproduction in the silver-haired bat and essentially none of what is known regarding the subject was derived from studies in the Pacific Northwest. Layne (1958*a*) found two males and two females hibernating in a mine in southern Illinois on 12 February; epididymides of one male examined contained relatively few spermatozoa, but spermatozoa were found in oviducts of both females. Kunz (1971) caught pregnant females in Iowa as late as 21 June and lactating females between 21 June and 24 July. He estimated the median date of parturition was 16 June. He caught flying young commencing 22 July. Parsons et al. (1986) reported a female with two well-developed embryos taken on 7 August in northern Saskatchewan was the latest record of pregnancy for the species.

Silver-haired bats are polytocous, with two young apparently the norm (Cockrum, 1955; Kunz, 1971; Kurta and Stewart, 1990). However, in Ontario, seven females had a total of 13 young (Table 9-5; Parsons et al., 1986). Kunz (1971) reported 1.7 volant young per adult female in Iowa, suggesting some mortality between birth and independence. Reports of the existence of large maternity colonies are mostly discounted (Barbour and Davis, 1969; Kunz, 1982), but records of small maternity colonies seem reliable. In Ontario, eight adult females (one pregnant, seven lactating) and 12 nonvolant young were captured when a dead section of a tree containing cavities excavated by flickers (*Colaptes auratus*) was cut. Three other adults, one carrying two young, were seen to escape (Parsons et al., 1986). Parsons et al. (1986) also recorded descriptions of preserved specimens from another small maternity colony in Saskatchewan.

A parturient captive female assumed an upright position with uropatagium curled ventrally and upward to form a pocket into which the young were born. Births of both young were breech presentations; the first, a female, commenced to use the hind legs as soon as they were free to aid in the birth process. The adult female began to lick the young before the head and shoulders were expelled from the birth canal and continued to do so after the young was free. Both young were free of the amnion. Before the umbilical cord was severed, the female neonate moved to one of the nipples and commenced to suckle. The second, a male, was born 21 min later; it did not nurse. Both young were active; they could crawl across the cage floor, climb its walls, and hang vertically. The adult female ate the placentae (Kurta and Stewart, 1990).

Kunz (1971) described neonates as pink and hairless with closed eyes and folded ears; the male and female composing the litter weighed 1.9 g and 1.8 g, respectively. Kurta and Stewart (1990) described neonates as pink with mottled tan and black wings, and black, folded pinnae. The pinnae including the tragi, on one individual became erect within 2 h of birth. The male and female in the litter each weighed 2.5 g at birth.

Because silver-haired bats rarely occur in groups (Barclay et al., 1988), except perhaps in small maternity colonies (Parsons et al., 1986) and do not often use caves extensively (Layne, 1958*a*; Pearson, 1962), they probably are not as subject to harassment and vandalism as the colonial bat species. For similar reasons, observations of predation on the species are rare and often noteworthy. Thus, descriptions by Byre (1990) of teams of juvenile peregrine falcons (*Falco peregrinus*) hunting migrating *L. noctivagans* over Lake Michigan likely are unique. J. M. Perkins (pers. comm., 17 April 1995) provided a maximum longevity record of 4.5 years for an individual banded in Oregon as an adult.

Habits.—Bailey (1936) and several others who investigated the species (Kunz, 1982) indicated that *L. noctivagans* in Oregon and elsewhere is a late flyer, emerging well after other species commence to feed. However, Whitaker et al. (1977) claimed that the species in western Oregon was an early flyer, emerging 15–45 min before other species began to forage. Cowan and Guiguet (1965) in British Columbia and Barclay (1984) in Manitoba also reported that the species was an early flyer. Explanation for the apparent contradiction may lie in the presence and relative abundance of other species of bats using the same foraging area as *L. noctivagans*. Reith (1980) noted that *L. noctivagans* in Oregon foraged later in the evening when *Eptesicus fuscus* was more abundant at water holes over which both species foraged. However, where *E. fuscus* was absent or outnumbered *L. noctivagans* by ≤3 times, *L. noctivagans* foraged earlier in the evening. In other regions, *L. noctivagans* is known to forage in the same areas as a wide variety of other species of vespertilionid bats (Kunz, 1982).

Although *L. noctivagans* and *Lasiurus cinereus* also are reported to forage at different times in some regions (Kunz, 1973), the two species are known to use different foraging strategies (Barclay, 1985, 1986), presumably to avoid significant niche overlap. *L. noctivagans* is more maneuverable, and produces "multiharmonic search-approach calls with an initial frequency sweep and a constant frequency tail" suitable for "bats foraging in open but near objects, and pursuing prey detected at relatively close range" (Barclay, 1986:2700). It is adapted to feed on a wide range of swarming insects (Barclay, 1985, 1986).

Although the species has been observed to fly at temperatures as low as –2°C (Jones, 1965), Barclay et al. (1988) found undisturbed individuals in torpor in day roosts with temperatures as high as 20°C. Body temperatures of undisturbed individuals in day roosts were within 1–2°C of ambient temperatures to 20°C. Thus, *L. noctivagans* routinely enters torpor during the day (Barclay et al., 1988).

In Manitoba, of 177 individuals located in day roosts in summer, 59.9% roosted singly, 16.9% in pairs, and 23.2% in eight groups of three–six. A few of the roosts were used consistently within and among years. In captivity, *L. noctivagans* did not use roosts previously used by other individuals more frequently than those not previously used, suggesting that the bats did not depend on olfactory cues to locate suitable roosts (Barclay et al., 1988).

Lasionycteris noctivagans is considered to be a migratory species, but information regarding movement patterns is derived mostly from observations of seasonal distributions. In Alberta, peaks in numbers collected were in spring (almost all adult females) and in late summer–early autumn (almost all adult females and juveniles). In British Columbia, except in the northeastern portion where the climate is more like that in Alberta, both sexes occur in more equal numbers and the species is reported to hibernate there (Cowan, 1933; Schowalter et al., 1978). In June–July on a study area in Manitoba, of 68 individuals found in dayroosts and marked with bands, none was found roosting in another location on the study area, suggesting that they had moved northward (Barclay et al., 1988). Females are believed to move farther than males (Kunz, 1982).

Remarks.—Carcasses of *L. noctivagans* collected at stock ponds in ponderosa pine forests in Wallowa Co. where 0.85 kg/ha DDT was applied to control the Douglas-fir tussock moth (*Orgyia pseudotsugata*) 1–3 years earlier contained 0.56–6.5 ppm of DDT residues, whereas those collected on a control area contained levels from less than detectable to 0.8 ppm (Henny et al., 1982). Levels in brain tissue of *L. noctivagans* were as high as 85 ppm, the highest level recorded among five species of bats tested. No relationship between residue levels and insect taxa in the diet of the five species was established (Henny et al., 1982).

Lasionycteris noctivagans commences to arrive in Oregon in early spring; we found specimens collected as early as

Table 9-6.—*Percent volume of foods consumed by three species of vespertilionid bats in eastern and western Oregon as determined by analysis of stomach contents and feces.*

Taxon or category	Lasionycteris noctivagans				Eptesicus fuscus				Pipistrellus hesperus[a]
	Western Stomach[b]	Eastern			Western Stomach[b]	Eastern			Eastern Stomach[e]
		Stomach[c]	Feces[e]	Stomach[d]		Stomach[e]	Feces[e]	Stomach[d]	
n	15	41	124	18	30	60	177	13	26
Lepidoptera (moths)	32.0	29.7	67.7	8.6	21.3	13.8	24.1	16.9	15.9
Isoptera (termites)	14.0		0.4		12.7	0.3	0.6		
Diptera (flies)	18.9	11.0	15.1	1.8	15.9	1.7	8.0		31.6
Hymenoptera (bees and ants)	7.0	12.2	0.7	25.3	5.5	9.2	2.5	42.3	22.6
Hemiptera (true bugs)	6.6	7.9	2.5	8.6	1.6	6.3	6.4	12.6	10.9
Orthoptera (crickets)	3.8		1.2		2.3	0.9			
Trichoptera (caddisflies)	2.7	1.1				0.5	0.8		
Coleoptera (beetles)	2.7	10.7	1.9	3.4	34.4	52.6	37.8	20.3	4.3
Homoptera (leafhoppers)	3.4	24.5	1.5	49.9	1.5	10.2	9.8	5.6	5.2
Neuroptera (lacewings)	0.3	0.5	6.4	0.3		1.3	5.5	0.4	1.2
Unidentified insects	7.0	2.4	1.4	2.2	1.8	2.8	4.0		9.0
Araneae (spiders)	1.4		tr		2.8	0.4	tr		
Vegetation			1.1				0.6		
Unidentified material			0.2						
Totals	99.8	100.0	100.1	100.1	99.8	100.0	100.1	100.0	100.7[f]

[a]Species does not occur west of the Cascade Range.
[b]Data from Whitaker et al. (1977).
[c]Data from Whitaker et al. (1981*b*).
[d]Data from Henny et al. (1982).
[e]Data from Whitaker et al. (1981*b*).
[f]Total given as 99.9 in original.

2–3 April. However, nighttime temperatures commonly fall below freezing even at low elevations and rain (or sometimes snow) is frequent at this season. Barclay et al. (1988) believed that *L. noctivagans* encountering these conditions simply entered torpor and waited until the weather improved. However, torpor retards embryonic development (Racey, 1973), thereby delaying birth of young and providing less time for the maternal female and the young to accumulate fat reserves for migration, hibernation, or both. Studies of migration and hibernation in the species certainly would improve knowledge of the biology of *L. noctivagans* and likely would provide insight into mechanisms by which individuals maximize their fitness in variable environments. The species is reasonably abundant and easily captured in nets as evidenced by various studies referenced herein; modern radiotelemetry techniques are sufficiently advanced (Wai-Ping and Fenton, 1989) to permit such investigations.

Pipistrellus hesperus (H. Allen)
Western Pipistrelle

1864. *Scotophilus hesperus* Allen, 7(publ. 165):43.
1897a. *Pipistrellus hesperus*: Miller, 13:88.

Description.—The western pipistrelle (Fig. 9-28) is the smallest bat in Oregon (Table 9-2) and the smallest bat north of Mexico. Some individuals possess skulls <1 cm long. Males are 2–6% smaller than females (Findley and Traut, 1970).

The pelage is light yellowish gray, slightly lighter on the venter than on the dorsum. The bases of the hairs are black. The ears and nose are nearly black; the wings and uropatagium are slightly lighter. The ears are short and broad; the tragus is blunt and bent forward at the tip. The calcar is keeled.

The skull (Fig. 9-29) is without a distinct step between rostrum and braincase, although the interorbital area is slightly depressed. Of the upper incisors the inner one is simple, but the outer one possesses an accessory cusp. There are 34 teeth, the same as in *Euderma maculatum*, but size alone suffices to separate the two species.

Distribution.—*Pipistrellus hesperus* occurs in southeastern Washington south through shrub-steppe areas of eastern Oregon and southwestern Idaho to the southern tip of Baja California Sur and eastward through parts of Utah and Colorado, western Texas, and extreme southwestern Oklahoma south through Mexico (except for the central highlands) to Morelos and Guerrero in the west and to Hidalgo in the east (Fig. 9-30; Hall, 1981). In Oregon, the species occurs along the lower Deschutes and John Day rivers in north-central Oregon and in Malheur and eastern Harney counties (Fig. 9-30).

Holocene-age deposits in Texas contain the only fossil remains of *P. hesperus* (Kurtén and Anderson, 1980).

Geographic Variation.—One of the two recognized subspecies occurs in Oregon: *P. h. hesperus* (Hall, 1981).

Habitat and Density.—Western pipistrelles are denizens of the desert and shrub-steppe regions of Oregon where they find daytime retreats in crevices in canyon walls, cliffs, and rimrocks. They sometimes use caves and mines that have crevices in their walls and ceilings; occasionally they enter buildings (Barbour and Davis, 1969; Whitaker et al., 1981b). In Arizona, individual *P. hesperus* roosted alone in crevices

in large cliffs. Although individuals exhibited little fidelity to specific crevices, they commonly used the same roosting area (Cross, 1965). These dayroosts were not used at night by western pipistrelles, but individuals were observed to alight on rocks near water holes and rest for several minutes (Cross, 1965). Cross (1965) also reported an observation by P. H. Krutzsch of western pipistrelles roosting at night in a building in southern California.

Diet.—In Oregon, stomachs of *P. hesperus* contained mostly remains of flies (Diptera) and bees and wasps (Hymenoptera), with moths (Lepidoptera) and true bugs (Hemiptera) ranking third and fourth, respectively (Table 9-6). Leafhoppers (Homoptera), beetles (Coleoptera), lacewings (Neuroptera), and unidentified insects each contributed <10% of the volume of stomach contents (Table 9-6). Thus, the diet of the western pipistrelle was considerably different from that of most larger species of bats (Tables 9-3 and 9-6).

In Arizona, leafhoppers, moths, and flying ants (Hymenoptera) composed most of the prey consumed, but the contribution of each of these prey types differed seasonally. Moths were consumed in greatest amount in winter, least in summer; leafhoppers were eaten most in spring and autumn, least in summer and winter; and flying ants contributed most to the diet in summer, but were not represented at other seasons (Ross, 1967). Ross (1967) also reported that the diet varied greatly among four sites within a season (August). The

Fig. 9-28. Photograph of the western pipistrelle, *Pipistrellus hesperus*. (Photograph courtesy R. B. Hayward.)

likelihood of such seasonal and site differences in Oregon makes interpretation of the diet of the pipistrelle (Table 9-6) difficult, as Whitaker et al. (1981b) collected samples in six eastern Oregon counties and did not report the season(s) at which their samples were collected.

Reproduction, Ontogeny, and Mortality.—Seemingly, little is known regarding these topics. Bailey (1936) reported that E. A. Preble had taken a female containing two embryos on 19 July; he further indicated that there were many records of females with two young. Koford and Koford (1948) found three lactating females each with two young on 15 June in southern California. Based on records

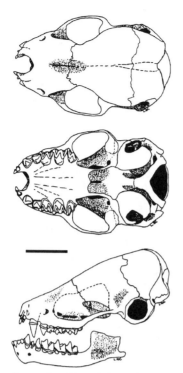

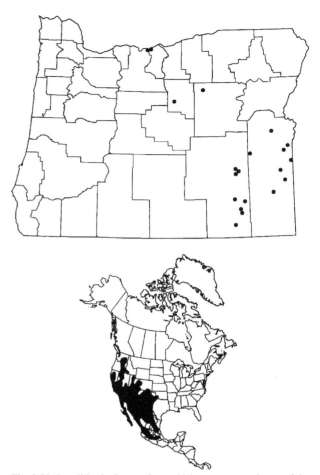

Fig. 9-29. Dorsal, ventral, and lateral views of the cranium, and lateral view of the dentary of the western pipistrelle, *Pipistrellus hesperus* (OSUFW 5655). Scale bar equals 3 mm.

Fig. 9-30. Localities in Oregon from which museum specimens of the western pipistrelle, *Pipistrellus hesperus*, were examined (listed in the APPENDIX, p. 547), and the distribution of the western pipistrelle in North America. (Range map redrawn from Hall, 1981:211, map 167, with modifications.)

from the literature (including those in the previous two references) and from museum specimen tags, 28 of 31 pregnant females from throughout the range contained two embryos and only three contained one embryo (Table 9-5; Cockrum, 1955).

In southern Nevada, the species composed 74.9% of the 865 bats of eight species captured during 70 nights over a desert spring (O'Farrell and Bradley, 1970). Recorded maximum longevity is 4 years (Paradiso and Greenhall, 1967).

Habits.— The western pipistrelle likely is the earliest-flying bat in Oregon. They were observed to fly in full daylight well before sunset (Barbour and Davis, 1969) and they also are known to fly after dawn (Van Gelder and Goodpaster, 1952). They were noted to forage with violet-green swallows (*Tachycineta thalassina*) and to compete with them for flying insects (Van Gelder and Goodpaster, 1952). The peak of activity is between 30 and 59 min after sunset (Cockrum and Cross, 1964–1965).

The flight of the western pipistrelle is described as "fluttery" (Barbour and Davis, 1969:112). They are slow and weak flyers; they cannot fly in even a moderate breeze and often are handicapped by a slight breeze. O'Farrell and Bradley (1970), in capturing 648 *P. hesperus*, took only one individual when wind speeds exceeded 14.5 km/h.

In southern Nevada, *Pipistrellus hesperus* was active throughout the year, but was captured in greatest abundance in summer (O'Farrell and Bradley, 1970). Based on numbers of museum specimens collected each month, at least some males are active throughout the year, but females are active March–September (Findley and Traut, 1970). Although *P. hesperus* can be induced to hibernate in the laboratory (O'Farrell and Bradley, 1970), there is some evidence that it may not spend long periods in torpor, at least in southwestern United States. It is a partial ectotherm; it is known to fly with body temperatures of 22–38°C de-

pending upon ambient temperatures, and it has been known to fly when ambient temperatures were –8°C (O'Farrell and Bradley, 1970). Activity at lower body temperatures allows less expenditure of energy, permitting these bats to supplement their stored fat with a few insects captured at dusk in winter; such likely contributes to their winter survival (O'Farrell and Bradley, 1970). Cross (1965) indicated that only males were active in winter, but O'Farrell and Bradley (1970) captured both sexes at all seasons, although fewer females in winter.

O'Farrell and Bradley (1970) believed that rock crevices of the type used as day roosts in summer would not be adequate hibernacula because of widely fluctuating temperatures and low humidity. Perkins et al. (1990) did not list *P. hesperus* among species found in >650 caves and mines and 70 buildings searched for hibernating bats during winter in Oregon and Washington. Findley and Traut (1970:741) proclaimed that "no migrations are known to occur." Whether western pipistrelles in Oregon hibernate, migrate, or engage in intermittent torpor combined with winter foraging as described by O'Farrell and Bradley (1970) is unknown. Thus, we are not certain that Bailey's (1936:384) assertion that "hibernation carries them through the period of cold and scarcity of food" is entirely applicable to the species in Oregon, but we cannot offer a well-documented alternative.

The species forages "in canyons, along cliffs, and over lava flows" (Whitaker et al., 1981*a*:290). Ross (1967:222) believed that pipistrelles, upon detecting swarms of insects, flew through the cloud open mouthed and engulfed prey by "pure chance" rather than by employing echolocation to locate individual prey. His observation of an individual with four flies in its mouth when it was collected was offered as evidence of "filter feeding" by western pipistrelles. However, Black (1974:152) indicated that *P. hesperus* was among those bats that "probably echolocate individual insects."

Bailey (1936) described western pipistrelles coming from their day roosts in crevices in rocks to the Owyhee River to drink before they commenced to search for food. Nevertheless, these bats are highly adapted to the desert environment and, under experimental conditions, were able to maintain a positive water balance when drinking water was withheld for 1.5 weeks, a feat not duplicated by eight other species tested (Geluso, 1978). Pipistrelles produced the most concentrated urine of the species in the test.

Western pipistrelles were reported to produce echolocation calls at 91–53 kHz with two harmonics and a duration of 4 ms (Fenton and Bell, 1981). However, the species likely exhibits some variability in characteristics of echolocation calls as Thomas et al. (1987) reported a frequency range of 100–60 kHz and a duration of 1–2 ms.

Eptesicus fuscus (Palisot de Beauvois)
Big Brown Bat

1796. *Vespertilio fuscus* Palisot de Beauvois, p. 18.
1900. *Eptesicus fuscus*: Méhely, p. 206.
1902a. *Eptesicus fuscus bernardinus* Rhoads, 53:619.

Description.—*Eptesicus fuscus* (Fig. 9-31) is a medium-sized (Table 9-2) bat with a large head; short, broad wings; thick, round ears; large, bright eyes; and a heavy body. The calcar is keeled and the tail extends ≅3 mm beyond the uropatagium (Kurta and Baker, 1990). The baculum is 0.8 mm long and arrow shaped (Hamilton, 1949).

The pelage is long, lax, and glossy; it is bright brownish-tan on the dorsum and slightly lighter on the venter. The wings, ears, nose, feet, and uropatagium are blackish. The uropatagium has a light sprinkling of hairs on the proximal one-fourth (Bailey, 1936; Kurta and Baker, 1990).

The skull (Fig. 9-32) is large, somewhat angular, and with relatively thin zygomata extending well beyond the braincase. The rostrum is broad, rounded, and slightly upturned; it is separated from the braincase by a shallow step. *E. fuscus* has 32 teeth; it is the only bat in Oregon with a combination of two upper incisors and four upper molariform teeth (Table 1-1).

Distribution.—*Eptesicus fuscus* occurs from northern Alberta south through the southern one-third of the southern tier of Canadian provinces, through the entire United States, most of Mexico, the central highland region of the Central American states, into Colombia and Venezuela. It also occurs in Cuba, Hispaniola, Jamaica, Puerto Rico, and the Bahamas (Fig. 9-33; Kurta and Baker, 1990). In Oregon, the big brown bat occurs throughout the state (Fig. 9-33).

Fossil big brown bats are known from >30 Pleistocene sites in the United States and from sites in Mexico, Puerto Rico, and the Bahamas. Some sites in midwestern and eastern United States are 600,000 years old (Kurta and Baker, 1990).

Geographic Variation.—Only one of the 11 named subspecies occurs in Oregon: *E. f. bernardinus* (Hall, 1981; Kurta and Baker, 1990). Hall (1981:216, map 169) depicted the range of *E. f. pallidus* as extending into southeastern Oregon, but listed no specimen from the state.

Habitat and Density.—Whitaker et al. (1981*a*) believed that big brown bats were widely distributed in eastern Oregon, but that they were more abundant in forested areas than in shrub-steppe regions. In eastern Oregon, big brown bats were reported to forage "over the forest canopy, along roads through the trees, along the forest edge, over forest clearings, and along cliffs and canyon streams" (Whitaker et al., 1981*a*:290). In western Oregon, the species "was usually associated with coniferous and deciduous forest" (Whitaker et al., 1977:51).

Summer roosts commonly are in buildings, beneath bridges, and occasionally in hollow trees; winter roosts (hibernacula) sometimes are in buildings, but more commonly in rock crevices, caves, mines, and tunnels. In summer, females tend to relocate to warmer quarters to raise their young, whereas males may continue to reside in tunnels and other cooler sites. However, females may abandon roosts temporarily if temperatures therein rise above ≅33–35°C (Davis et al., 1968). In winter, when temperatures in hibernacula fall below freezing, big brown bats arouse and seek warmer quarters (Barbour and Davis, 1969).

Eptesicus fuscus is considered to be one of the more abundant species of bats (Maser et al., 1981*b*). However, in the Coast Range of Oregon, detection of echolocation calls of *E. fuscus* and *Myotis thysanodes* combined was the least among five categories of bats in old-growth stands, but increased in rank in mature (fourth) and young (third) stands of Douglas-fir (*Pseudotsuga menziesii*)–western hemlock (*Tsuga heterophylla*—Thomas, 1988). The species is considered less abundant in coniferous forest regions than in deciduous forest regions (Kurta and Baker, 1990).

Diet.—Although *E. fuscus* tends to be a generalist in selecting foraging habitat and consumes insects in several

Fig. 9-31. Photograph of the big brown bat, *Eptesicus fuscus*. (Photograph courtesy R. B. Hayward.)

orders, it tends to specialize in feeding on small beetles (Coleoptera—Kurta and Baker, 1990). In both eastern and western Oregon, and by analysis of both stomach contents and feces, beetles contributed greatly to the diet of the big brown bat (Table 9-6). However, moths (Lepidoptera) also contributed considerably (Table 9-6). Several researchers elsewhere found moths to contribute much less (or not at all) to the diet of big brown bats (Hamilton, 1933a; Ross, 1967; Whitaker, 1972). Whitaker et al. (1981a) believed that the relatively large volume of moths in the diet of big brown bats likely was a reflection of the abundance of moths in eastern Oregon, but that the difference noted in volume of moths between stomach and fecal samples may have been the result of differential digestion. The volume of bees and ants (Hymenoptera) in stomachs of a small sample of these bats from eastern Oregon was 4.6-16.9-fold greater than that reported by other investigators (Table 9-6). Such likely represents a combination of opportunistic feeding by *E. fuscus*, a sample collected mostly at the same local-

ity and within a short interval, and small sample size.

Reproduction, Ontogeny, and Mortality.—We found few published records regarding these topics for Oregon. In eastern United States, production of spermatozoa peaked in August but declined in September; the cauda epididymides were packed with spermatozoa in specimens collected in October (Christian, 1956). In Indiana, records of *E. fuscus* copulating were obtained each month from November to March (Mumford, 1958). Phillips (1966) found spermatozoa in reproductive tracts of females taken in October in Kansas. In some regions at least, two–five ova are produced; where twinning occurs, one blastocyst implants in each horn of the uterus, but where only one young is produced nearly three-fourths of the embryos implant in the right horn (Schowalter and Gunson, 1979). In Maryland, ovulation occurs about 1 April and parturition occurs about 1 June, giving a postfertilization period of development of ≅60 days (Barbour and Davis, 1969). However, Christian (1956) believed that ovulation in *E. fuscus* in Pennsylvania occurred about the 1st week of April with young born from 15 May to 22 June. Thus, embryonic development varied from ≅45 days to ≅83 days. Such likely represents rates of development being related to temperatures of roosts.

Bailey (1936) collected four near-term pregnant females on 21 June in Eugene, Lane Co., and two with fully

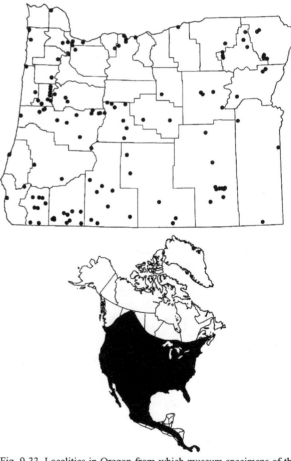

Fig. 9-32. Dorsal, ventral, and lateral views of the cranium, and lateral view of the dentary of the big brown bat, *Eptesicus fuscus* (OSUFW 3020). Scale bar equals 3 mm.

Fig. 9-33. Localities in Oregon from which museum specimens of the big brown bat, *Eptesicus fuscus*, were examined (listed in the APPENDIX, p. 517), and the distribution of the big brown bat in North America. (Range map redrawn from Hall, 1981:216, map 169, and van Zyll de Jong, 1985:162, unnumbered figure, with modifications.) The species also occurs on some Caribbean islands and in parts of South America.

developed fetuses on 26 June near Vida, Lane Co.; all carried only one embryo. He believed that young were born in Oregon in late June. We examined three pregnant females collected from the same maternal colony in Benton Co. on 17 June; the range in crown-rump lengths of embryos was 38–59 mm, suggesting minor variation in future dates of parturition. We also have a record of a pregnant female with a 24-mm embryo taken 11 July in Klamath Co. by one of us (BJV). Thus, dates of parturition in Oregon may be as much as a month later than in southern California (Krutzsch, 1946) and comparable to dates of parturition in Alberta (Schowalter and Gunson, 1979).

In western North America, litter size is almost always one, whereas in the Great Plains and eastward the species commonly has two young (Barbour and Davis, 1969; Cockrum, 1955). Christian (1956) found 16 pregnant females each with two embryos and one female with one embryo in Pennsylvania; he predicted that 83% of females in the area had two young. Brenner (1968) indicated that, during a 3-year period, young per female in a colony in Ohio varied between one and two, and females in a colony in Pennsylvania had at least two young. Gates (1937) recorded 23 young produced by 11 wild-caught pregnant females (with 16 known to have been produced by seven females) in Louisiana, indicating that at least some eastern females at times have three young, perhaps four (Phillips, 1966). Harper (1929) reported finding a female with four embryos in Georgia. In Nevada, Hall (1946) reported all of 15 pregnant females carried one embryo (included in Cockrum's, 1955 summary) and each of the four pregnant Oregon females that we examined carried only one embryo (Table 9-5). Krutzsch (1946) recorded an instance of two young in southern California; Cockrum (1955) reported two other instances of two young produced by females in western United States. In Alberta, 15 (13.0%) of 115 pregnant females carried twin fetuses; the remainder had only one fetus (Schowalter and Gunson, 1979). Thus, it seems that the average number of young per female may be less, but not as variable in the West.

In Ohio, reproductive success, calculated as the number of volant young per postpartum female counted during 15 July–15 August, decreased significantly with increases in density (Mills et al., 1975). Mills et al. (1975) suggested that the density-dependent factor might be availability of food resources to juvenile bats after they commence to forage. During the summer with the lowest temperatures, hence likely low or irregular emergence of insects, efficiency of rearing young was lowest, suggesting a possible mechanism for the density-dependent relationship.

Neonates are naked and almost immobile at birth; the eyes and ears are closed but open within a few hours (Kurta and Baker, 1990). Neonates weighed 3.1 g (range, 2.7–3.6 g; $n = 5$) in Kansas (Kunz, 1974), 3.3 g ($n = 16$) in Massachusetts (Burnett and Kunz, 1982), and 4.0 g (range, 3.4–4.5 g; $n = 10$) in Kentucky (Davis et al., 1968). We know of no body-mass data for neonate E. fuscus in Oregon or of an evaluation of possible differences in average birth mass between young produced in litters of one and those of two.

Growth of young is rapid; length of the forearm approaches that of adults in 15 days, but body mass, although more variable, continues to increase at least to 2 months of age (Burnett and Kunz, 1982). Young commence to fly at 18–35 days of age (Kurta and Baker, 1990), but are not weaned until ≅40 days old (Kunz, 1974). Males become sexually mature in the year of their birth, but not all females breed their first autumn (Schowalter and Gunson, 1979).

In Minnesota and Wisconsin, annual mortality for the 1st year after banding was estimated to be 62%, 33% the 2nd year, and 23% the 3rd year. Populations in the region were believed to have declined ≅60% during the period of study (Beer, 1955). A 20-year study of an E. fuscus population wintering in a storm sewer in the same region revealed that <50% of the bats were alive 2 years after banding (Goehring, 1972). In Ohio, mortality in one population was 68.1% the 1st year after banding, 28.7% the 2nd year, and 72.0% the 3rd year; in another population, the values were 89.5%, 30.0%, and 42.9% for the 3 years, respectively (Mills et al., 1975). Recorded maximum longevity for the species is 19 years (Paradiso and Greenhall, 1967), but both Beer (1955) and Goehring (1972) indicated that few individuals attained even half that age. Mortality factors include lack of adequate energy reserves in winter, predation by a variety of raptors and carnivores, and accidents (Kurta and Baker, 1990). Predation by mammals and birds likely is infrequent and probably has little impact on populations (van Zyll de Jong, 1985). Widespread use of certain insecticides poses a potential hazard (Luckens and Davis, 1964). However, average residue levels in E. fuscus from forested areas sprayed with DDT in northeastern Oregon ranged from 1.93 (1.5–≅2.8, $n = 3$) to 0.58 (below detectable levels–2.7) ppm 1–3 years later; 13 specimens from unsprayed areas averaged 0.43 (0.31–0.92) ppm. Residue levels in E. fuscus were penultimate among five species tested (Henny et al., 1982). Rabies is enzootic in E. fuscus in the United States with occasional, short-lived epizootics (Kurta and Baker, 1990).

Sex ratios at birth are only slightly biased toward males; in Alberta, 52.1% were males (Schowalter and Gunson, 1979). Many (but not all—Mills et al., 1975) populations of adults in caves also are biased toward males (Beer, 1955; Goehring, 1972). Beer (1955) presented evidence that differential mortality after bats were banded was not responsible for bias in sex ratios. He suggested that segregation of the sexes occurred in hibernacula as well as in maternity roosts; thus sex ratios of big brown bats in winter samples obtained in caves were not representative.

Habits.—In coastal Oregon, big brown bats leave their roosts and commence to forage 30–40 min before dark (Maser et al., 1981b), probably about the same time as recorded in Kansas (Phillips, 1966). Foraging areas often are 1–2 km from roosts. Early in the evening they commonly forage above the forest >50 m above the ground, but later descend to ≅15 m above the ground or to even as low as 7–9 m when foraging over a road (Whitaker et al., 1977). Over streams, adults forage higher above the water than juveniles (Kurta, 1982). Foraging is not continuous, but totals ≅1.7 h/night with the remainder of the time spent in night roosts usually separate from day roosts (Kurta and Baker, 1990). Big brown bats are capable of detecting low-frequency sounds such as those produced by groups of insects at distances as great as 600 m (Buchler and Childs, 1981). Echolocation calls permit detection of small prey at close range; however, Brigham et al. (1989) found significant inter- and intraindividual variation in certain characteristics of the calls. In addition, E. fuscus produces audible chatter while in flight apparently as a means of communicating with nearby conspecifics (Barbour and Davis, 1969).

In captivity, *E. fuscus* consumed an average of 4.4 g of insects/day and excreted 10–13% of the mass of the food ingested (Coutts et al., 1973). Elytra and legs of beetles and wings and antennae of moths, especially on larger species, were removed before the remainder was consumed. Also, the abdomens of moths containing eggs and the tips of the abdomens of some beetles were rejected by the bats (Coutts et al., 1973). Undesirable parts of prey commonly accumulate with guano beneath roosts. In the wild, 10 juveniles collected after known periods of feeding had obtained insects at an average rate of 1.2 g/h, whereas one adult accumulated insects at a rate of 2.7 g/h (Gould, 1955). Big brown bats can fill their stomachs in <20 min under favorable conditions (van Zyll de Jong, 1985).

In August, big brown bats commence to store fat; they may nearly double their body mass (to 30 g) before hibernation. In cold regions, males form tight clusters in mines or caves whereas females roost singly (Barbour and Davis, 1969).

After an initial bout of feeding, big brown bats commonly retire to a night roost to rest. Postpartum females instead return to the site of the maternal colony to find and nurse their young; even the smallest young are left in clusters in nursery colonies while maternal females forage (Davis et al., 1968). Hall (1946) reported collecting adult females carrying young as they left a small cave to forage in Nevada. Although other young attempt to nurse any female that arrives in the colony, females do not permit nursing by other than their own young; identification of young is verified by the female licking its face and lips (Barbour and Davis, 1969). Young bats that become dislodged and fall to the floor of the roost site make squeaking sounds and usually are retrieved by the maternal female. Young may be brooded by the female holding it close beneath the wing membranes (Barbour and Davis, 1969).

Big brown bats are surprisingly sedentary and exhibit remarkable fidelity for natal sites. In Minnesota and Wisconsin, of 992 banded individuals recaptured during winter only 13 were taken at sites other than their original points of capture. Average distance from original point of capture for these 13 was 11.8 km (range, 0.8–98.1 km). All 12 of those recovered in summer were taken at sites other than where they were banded, but the average distance moved was only 12.2 km (range, 1.0–53.1 km). Of those recaptured ≥1 year from time of banding, only seven were retaken at sites other than their original points of capture (Beer, 1955). In Ohio, Mills et al. (1975) recorded a movement of 288 km, but only seven of 1,677 females and none of 265 males were recaptured at sites other than where they were banded. Big brown bats apparently use the glow in the western sky after sunset as a point of reference in orientation to return to colonies after foraging (Buchler and Childs, 1982).

Maternity colonies are not considered random aggregations of individuals but are cohesive social units. Quality of available roost sites differs; big brown bats choose sites at which reproductive output is enhanced (Brigham and Fenton, 1986).

Big brown bats seem particularly adaptable to captivity; such likely is responsible for the numerous published reports on various aspects of the physiology of the species (Kurta and Baker, 1990). They can be maintained in captivity on artificial diets that include mealworms (Tenebrionidae), banana, cottage cheese, and vitamin supplements (Rasweiler, 1977).

Remarks.—Because big brown bats commonly roost in buildings, they often are considered uninvited and undesirable guests by occupants or owners. Exclusion is the most reliable method of permanently evicting these bats (Barclay et al., 1980), but requires that all cracks and openings sufficiently large to admit a bat be closed permanently. Such an undertaking should be accomplished in two stages: first, all holes and cracks *except* the entrance-exit commonly used by the bats should be closed and second, when *all* the bats have left the building to forage their last access to the building should be closed. Special care must be taken not to conduct the exclusion operation at a season (June-August) when nonflying young are left in the roost while adults forage. Unless the operation is conducted carefully and all cracks and holes are sealed, success in evicting the bats is problematical as bats commonly find previously unused access holes (Brigham and Fenton, 1986). However, the evicted bats likely will establish new roosts nearby (Brigham and Fenton, 1986). Thus, nearby erection of a structure specifically designed for roosting bats ("bat house"—Greenhall, 1982) may minimize the impact of exclusion on the bats and reduce the possibility of the bat colony becoming established in a nearby residence or other building.

Euderma maculatum (J. A. Allen)
Spotted Bat

1891. *Histiotus maculatus* Allen, 3:195.
1894. *Euderma maculata*: H. Allen, 43:61.
1897a. *Euderma maculatum* Miller, 13:46.

Description.—The spotted bat (Plate III) is distinctive in possessing enormous pinkish-gray ears joined at their bases over the forehead, a black dorsal pelage with ≅15-mm-diameter white spots on each shoulder and on the rump, a smaller white spot behind each ear, and a ≅10-mm-diameter bare area on the throat (van Zyll de Jong, 1985). The latter character is hidden from view until the throat region is extended by tipping the head backward (Watkins, 1977). The ventral pelage is white, but the hairs have black bases. The flight membranes are pinkish red (Easterla, 1965) to gray brown (van Zyll de Jong, 1985). When asleep at room temperature or during cold-induced torpor, the ears are curled laterally and downward over the forearms (Constantine, 1961a). The forearm is significantly longer in females than in males, the only character among 16 examined by Best (1988) to exhibit sexual dimorphism. The calcar is not keeled.

The skull has an elongate braincase; in dorsal view the width of the braincase and the zygomatic breadth are essentially equal; thus the two sides of the skull are parallel (Fig. 9-34). The zygomata are heavy with a postorbital expansion near the middle. There is no sagittal crest. There are 34 teeth; P1 and p1 are minute (Watkins, 1977).

Distribution.—Based on museum specimens examined by Best (1988), the spotted bat occurs from extreme southern British Columbia south through eastern Oregon, western Nevada, southern California, southwestern Arizona, and southern Chihuahua and Queretaro, Mexico, and eastward to the Big Bend area of Texas, central New Mexico, southwestern Colorado, central Wyoming, and south-central Montana (Fig. 9-35). In Oregon, one specimen was collected near Mickey Spring, Harney Co. (McMahon et al., 1981) and one along the John Day River near Clarno,

Wheeler Co. (Barss and Forbes, 1984; Fig. 9-35).

No fossils of *E. maculatum* are known (Watkins, 1977).

Geographic Variation.—The species is considered monotypic. Best (1988) found geographic variation in 10 of 16 morphometric characters that he analyzed. However, he was reluctant to invoke population isolation as a possible explanation for the observed variation because specimens available were too few and from too many habitat types.

Habitat and Density.—Fenton et al. (1987:142–143, fig. 1) detected spotted bat echolocation calls at 10 of 80 areas (79 illustrated) in western United States and British Columbia; they concluded that "there was no obvious association of spotted bat activity with any particular habitat conditions." Best (1988) indicated that the species occurred in many habitat types, but most often in dry, rough, desert terrain. Many of these spotted bats preserved as museum specimens were captured in nets set at desert water holes (Easterla, 1971, 1976; McMahon et al., 1981; Pouché and Bailie, 1974) or at livestock watering tanks in montane meadows in ponderosa pine (*Pinus ponderosa*) forests (Constantine, 1961*a;* Findley and Jones, 1965; Jones, 1961). In the Okanagan Valley, British Columbia, spotted bats were discovered to forage in openings in a montane ponderosa pine woodland (Leonard and Fenton, 1983; Woodsworth et al., 1981). Irrespective of foraging area, spotted bats roost in crevices in cliffs or canyon walls (Easterla, 1976; Leonard and Fenton, 1983; Reynolds, 1981; Woodsworth et al., 1981). One specimen taken in Oregon was at a desert water hole (McMahon et al., 1981); the other was found dead on "the floor of a small crevice in the base of a west-facing andesitic cliff" near areas of shrub-steppe and fields of alfalfa (Barss and Forbes, 1984). Seemingly, critical habitat requirements are cliffs with crevices for roosting sites and low vegetation or openings in forests for foraging areas. The frequency of capture near water and the considerable intake of water in captivity (Vorhies, 1935) suggest that proximity to water may further restrict foraging areas. The species has been captured from 57 m below to 3,230 m above sea level (Best, 1988; Reynolds, 1981).

The spotted bat has been referred to as America's rarest mammal (Nowak, 1991). In 1985, Best (1988) sent queries to 506 individuals and institutions seeking museum specimens for his study of geographic variation in the species; he found 73 specimens and was able to use 67 of them. A few other specimens also may be preserved in a fashion unsuitable for some studies (e.g., Mickey, 1961). Studies of habitat use by spotted bats based on visual observation and on detection of echolocation calls within the frequency range of the human ear (Leonard and Fenton, 1983; Woodsworth et al., 1981) suggest that capture of specimens reveals little information regarding either abundance or distribution of the species. From detection of echolocation calls during widespread sampling within the range of the species, however, Fenton et al. (1987) concluded that capture data provided a reasonable indication of the distribution of the species. Nevertheless, they failed to detect calls at 18 widely scattered areas from which spotted bats had been reported; thus, such a comparison seems without basis. In addition, in Oregon at least, their searches did not include the major areas of montane ponderosa pine habitat, a type in which the species is known to forage in British Columbia and New Mexico.

Diet.—Pouché and Bailie (1974:256) reported that,

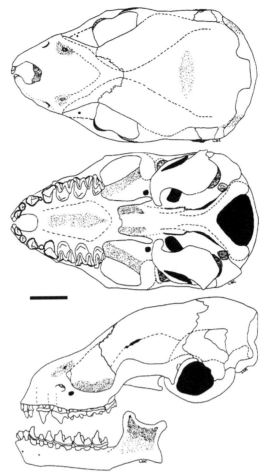

Fig. 9-34. Dorsal, ventral, and lateral views of the cranium, and lateral view of the dentary of the spotted bat, *Euderma maculatum* (MSB 24999). Scale bar equals 3 mm.

upon release, an individual they had captured in a net flew a short distance, hovered for a "split-second," dropped to the ground, captured a grasshopper (Orthoptera), and within 10 s was in flight again. From this observation, spotted bats, like some other long-eared species, were believed to glean insects from the ground. However, examination of contents of stomachs of spotted bats from Utah ($n = 2$) and Texas ($n = 15$), and fecal pellets from New Mexico ($n = 18$) revealed mostly remains of moths (Lepidoptera—Easterla, 1965; Easterla and Whitaker, 1972; Ross, 1961). Two individuals from Texas had eaten June beetles (Coleoptera: Scarabaeidae—Easterla and Whitaker, 1972). Rarely were heads, wings, or legs of the prey included among the remains, suggesting that spotted bats discard those parts before ingesting the remainder.

Of >71,000 attempts by spotted bats to capture prey, all were attempts to capture flying insects (Leonard and Fenton 1983). From their low-frequency echolocation calls; their high-level, fast, and maneuverable flight; and their frequent steep dives, tympanate moths were believed to constitute primary prey (Leonard and Fenton, 1983; Woodsworth et al., 1981). Tympanate moths likely are unable to detect low-frequency ecolocation calls until spotted bats are 0.1–2.0 m away (Woodsworth et al., 1981).

Reproduction, Ontogeny, and Mortality.—So few spotted bats have been examined that essentially all information regarding reproduction is anecdotal. Three

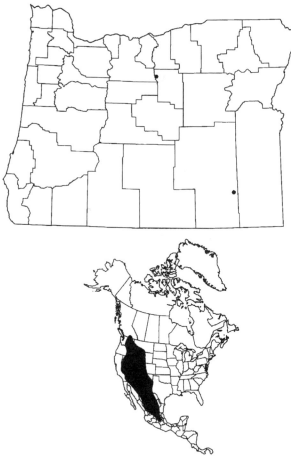

Fig. 9-35. Localities in Oregon from which museum specimens of the spotted bat, *Euderma maculatum*, were examined (listed in the APPENDIX, p. 518), and the distribution of the spotted bat in North America. (Range map redrawn from Hall, 1981:231, map 181, and van Zyll de Jong, 1985:155, unnumbered figure, with modifications.)

females captured 5–11 June in Texas and one captured on 16 June in British Columbia were pregnant (Easterla, 1971, 1976; van Zyll de Jong, 1985); the three captured in Texas gave birth in captivity. Lactating females were captured 12 June–18 August in Texas, Utah, and New Mexico (Easterla, 1976; Findley and Jones, 1965; Jones, 1961). The right uterine horn of both lactating females caught 30 June–1 July in New Mexico was enlarged, but only one contained a pigmented implantation site (Findley and Jones, 1965). Epididymides of two males captured on 23 June did not contain mature spermatozoa (Jones, 1961). The position of the testes of males captured 19 August and 17 September in New Mexico could not be determined (Constantine, 1961*a*), but a male captured 21 August in Utah had testes 3 by 7 mm (Easterla, 1965). Based on the limited observations of spotted bats, litter size is one, parturition is in June, and lactation is in June–August; however, too few data are available to establish the breeding season.

Easterla (1971, 1976) observed three neonates born in captivity; all died within 5 days of birth. Shortly after birth they weighed 2.6, 3.25, and 4.0 g. The largest of the three nursed nearly continuously for the first 48 h after birth. The youngster often clung to its mother when the female was permitted to fly around a room daily. When the female fed, the young was left hanging on the wall of the cage; upon its return, the female pulled the offspring to

it and allowed it to nurse (Easterla, 1971).

Spotted bats released during daylight hours are subject to being preyed upon by raptors (Black, 1976; Watkins, 1977). Some specimens were obtained after the bats drowned, but they may have struck nets set for them (Easterla, 1965).

Habits.—Because several were captured after midnight, spotted bats were believed to be late flyers (Easterla, 1976; McMahon et al., 1981; Pouché and Bailie, 1974). However, Leonard and Fenton (1983) observed spotted bats or detected their echolocation calls throughout the night. These authors also indicated that cloud cover, wind, precipitation, or phase of the lunar cycle had no effect on the pattern of activity of the bats. In May–June in British Columbia, four radiotracked individuals left their roosts an average of 13.3 ± 8.5 min after sunset; time of sunset, but not temperature, influenced departures significantly (Wai-Ping and Fenton, 1989).

In British Columbia, early in the season, spotted bats tended to forage for 3–5 min in each of several clearings sequentially and repeatedly, but later in the year they often flew for as long as an hour in one clearing. In both foraging patterns, neighboring spotted bats seemed to avoid each other (Woodsworth et al., 1981), but otherwise used overlapping foraging areas. Leonard and Fenton (1983) suggested that the low-frequency echolocation calls may serve to space individuals on feeding grounds. Foraging flights usually were 10–30 m above the ground and ≥20 m from the edge of vegetation. The bats flew continuously from leaving the roost to returning thereto; they did not use night roosts (Wai-Ping and Fenton, 1989).

Echolocation calls produced by spotted bats may be classified as search, approach, and terminal phases. Search and approach calls may consist of one or two notes with multiharmonics, whereas the terminal phase is a single pulse without harmonics. In all echolocation calls the fundamental frequency sweeps downward from ≅12 to ≅6 kHz, frequencies much lower than echolocation calls produced by most other species of bats. The low frequencies "are less directional and less subject to atmospheric attenuation," but they restrict detection of objects to those ≥10 mm in diameter (Leonard and Fenton, 1984:125). Such features seem correlated with size and predator-avoidance capabilities of species in the diet of spotted bats. Calls with these characteristics also are suitable for long-range communication and maintenance of individual spacing either by an individual monitoring the intrusion of a conspecific into a foraging area or by an individual advertising its presence in an area much in the same manner as a singing bird. However, low-frequency calls are more conspicuous than ultrasonic calls made by most other species of bats (Leonard and Fenton, 1984). Spotted bats responded positively to playback of recorded echolocation calls by approaching the speaker and producing buzzing vocalizations, behaviors considered to be aggressive (Leonard and Fenton, 1984). The frequencies used for echolocation calls produced by spotted bats are within the range of frequencies detectable by the human ear; echolocation calls of spotted bats reportedly can be heard by humans at a distance of 250 m or more (Easterla, 1970; Woodsworth et al., 1981).

Several spotted bats caught in mist nets have had broken bones or tears in the skin. Easterla (1965, 1970) suggested that the high flight speed at which the bats hit the net was responsible for the injuries as the species seemed docile when being removed from nets and in captivity. Constantine (1961b:95) reported that a male that he caught on 19 August "had a gentle disposition, did not seem to resent handling, and made no effort to bite or escape" but another caught 17 September "had an unpleasant disposition, biting and fighting desperately when handled." From reports regarding caged individuals, spotted bats do not seem to survive well in captivity (Constantine, 1961a; Durrant, 1935; Easterla, 1965, 1970, 1971, 1976).

Remarks.—Considering the abundance of seemingly suitable roosting and foraging habitats in Oregon, we would not be surprised to learn that an energetic student with excellent hearing had detected the echolocation calls of spotted bats at several additional locations in the state. We suggest that anyone who might undertake such a mission might improve the possibility of detecting spotted bats by playing recordings of echolocation calls with appropriate equipment. We encourage someone to try.

Despite its rarity not only in Oregon, but everywhere, Marshall (1992) did not include the spotted bat among those species listed in his *Sensitive vertebrates of Oregon*. Even if Oregon does not support resident populations and the known specimens strayed into the state, spotted bats deserve protection while in Oregon. If our prognosis is correct that discovery of additional spotted bats in the state may be only a matter of looking (and listening), then measures to protect the species here are even more appropriate.

Corynorhinus townsendii (Cooper)
Townsend's Big-eared Bat

1837. *Plecotus townsendii* Cooper, 4:73.
1865. *Corynorhinus townsendi* Allen, 17:175.
1897a. *Corynorhinus macrotis pallescens* Miller, 13:52.
1914. *Corynorhinus macrotis intermedius* Grinnell, 12:320.
1936. *Corynorhinus rafinesquii townsendii*: Bailey, 55:386.
1936. *Corynorhinus rafinesquii pallescens*: Bailey, 55:388.
1936. *Corynorhinus rafinesquii intermedius*: Bailey, 55:389.
1959. *Plecotus townsendii pallescens*: Handley, 110:190.

Description.—*Corynorhinus townsendii* (Plate IV) is a large bat with exceptionally long ears (Table 9-2) connected at their bases across the forehead. Each side of the muzzle between the nostril and eye is adorned with large sebaceous and sudoriferous (pararhinal) glands, producing a lump-nosed appearance (Quay, 1970). The third–fifth digits on the manus are subequal. The pelage is light brown to dark chocolate on the dorsum and light beige to brown on the venter. The light-colored hairs have dark bases. The calcar is not keeled. The wing and tail membranes are extremely thin (Dalquest, 1947a).

The skull (Fig. 9-36) is slender with zygomata of nearly uniform thickness and of approximately the same breadth as the braincase. The short rostrum joins the braincase by a moderate step. *C. townsendii* has 36 teeth, more than any of the Oregon bats with exceptionally long ears.

Distribution.—Townsend's big-eared bat occurs from southern British Columbia south to extreme northern Baja California Norte and most of Sonora then eastward to central South Dakota, eastern Colorado, western Oklahoma, central Texas, and in the central highlands of Mexico south to the Isthmus of Tehuantepec, Mexico. Disjunct populations occur in the limestone-cave regions of Missouri-Arkansas-Oklahoma and Kentucky-Virginia-West Virginia (Fig. 9-37; Barbour and Davis, 1969; Hall, 1981; Kunz

and Martin, 1982).

In Oregon, *C. townsendii* has been collected throughout most of the state except in parts of the Blue Mountains Province and in the western part of the Basin and Range Province (Fig. 9-37).

Pleistocene fossils of extinct species of *Corynorhinus* were discovered in cave deposits in Mexico and Maryland. Fossil *C. townsendii* of Wisconsinan age were found in Arizona, New Mexico, and Virginia, and fossils of *Corynorhinus* of uncertain species were obtained from cave deposits in Pennsylvania, Tennessee, Missouri, and West Virginia (Kurtén and Anderson, 1980).

Geographic Variation.—Two of the five currently recognized subspecies occur in Oregon: the darker *C. t. townsendii* west of the Cascade Range and the lighter *C. t. pallescens* east of the Cascade Range (Dalquest, 1948; Hall, 1981). Bailey (1936) speculated that the range of a dull-colored race collected near the Oregon border in California extended into the Klamath Valley, Oregon. However, bats in this population were considered to exhibit various grades of intermediacy between *C. t. townsendii* and *C. t. pallescens*, and the race was placed in synonymy in part with *C. t. pallescens* and in part with *C. t. townsendii* by Handley (1959).

Habitat and Density.—Townsend's big-eared bat is

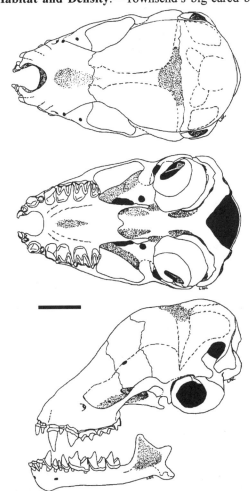

Fig. 9-36. Dorsal, ventral, and lateral views of the cranium, and lateral view of the dentary of Townsend's big-eared bat, *Corynorhinus townsendii* (OSUFW 5002). Scale bar equals 3 mm.

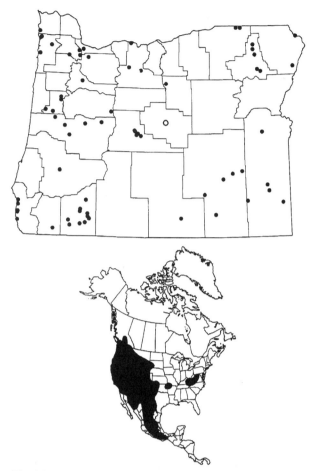

Fig. 9-37. Localities in Oregon from which museum specimens of Townsend's big-eared bat, *Corynorhinus townsendii*, were examined (listed in the APPENDIX, p. 516), and the distribution of Townsend's big-eared bat in North America. (Range map redrawn from Hall, 1981:235, map 184, with modifications.) Open symbol indicates record for county.

associated with a wide variety of habitats within its range, but commonly occurs in areas vegetated by "desert scrub, piñon-juniper, and pine forest." It does not occur in extremely arid desert regions (Barbour and Davis, 1969:165). In western Oregon, the species is associated with coniferous forests (Whitaker et al., 1977). In ≥40-ha stands of <75- to >200-year-old coniferous forest (Douglas-fir, *Pseudotsuga menziesii*–western hemlock, *Tsuga heterophylla* sere) in the Oregon Coast Range, Thomas (1988) detected echolocation calls of 4.2 *C. townsendii*/100 h of observation with handheld ultrasonic detectors. In comparison, detection rates were 32.3/100 h for *Myotis volans,* 26.1/100 h for *Lasionycteris noctivagans,* and 8.5/100 h for *Lasiurus cinereus.* In similar habitats in the western Cascade Range in Washington, Thomas (1988) detected only 0.3 echolocation calls of *C. townsendii*/100 h of observation.

Townsend's big-eared bat commonly is considered a cave-dwelling species (Bailey, 1936; Barbour and Davis, 1969), and caves and abandoned mine tunnels are considered critical habitat for the species (Perkins, 1987). Nevertheless, buildings frequently are used as night roosts, and at higher elevations in western parts of the state, *C. townsendii* may use buildings as day roosts (Barbour and Davis, 1969). In California, Dalquest (1947a) reported *C.*

townsendii to occupy buildings in summer in which he found temperatures uncomfortably high, but hibernacula must be cold although temperatures must not fall much below freezing.

In Oregon, Perkins (1987) found hibernating *C. townsendii* in 34 (7.1%) of 478 caves, mines, buildings, and tunnels searched in winter from 1982 to 1986. Only two sites contained ≥200 individuals, most sites contained ≥50, and half the sites contained <5. He counted 997 Townsend's big-eared bats at the sites. In summer, 31 (4.0%) of 772 potential day or night roosts searched contained 1,013 members of the species. Two nursery colonies (≅80 individuals) and eight roost sites (11 individuals) were found west of the Cascade Range and four nursery colonies (≅450 individuals) and four roost sites (eight individuals) were located in central Oregon (Perkins, 1987). Six individuals were captured in mist nets at two of 267 sites sampled. However, *Corynorhinus townsendii* is noted for its exceptional ability to detect and avoid mist nets (Kunz and Martin, 1982; Whitaker et al., 1981a). Perkins (1987) estimated that 2,300–2,600 Townsend's big-eared bats remained in the state, of which 800–900 were in southwestern Oregon, 300–400 in the northwest, 500–600 in central Oregon, 100 in the northeast, and 500–600 in southeastern Oregon. He did not reveal the method used to obtain the estimates. However, in an unpublished report of a survey conducted in winter 1989–1990, J. M. Perkins (in litt.) indicated that he had found 947 hibernating *C. townsendii* in Oregon.

Diet.—Whitaker et al. (1977, 1981a) analyzed contents of stomachs of 16 *C. townsendii* from western Oregon and one from eastern Oregon. Sixteen of the 17 contained only moths (Lepidoptera); one from western Oregon contained 95% moths and 5% true bugs (Hemiptera). These investigators considered the diet to reflect selectivity on the part of the bats rather than differential availability of types of prey. Although Howell (1920d:171) indicated that *Corynorhinus* was known to "pick insects from leaves and other resting places," likely most prey is taken during aerial feeding (Kunz and Martin, 1982).

Reproduction, Ontogeny, and Mortality.—In California, in August–October, testes of male *C. townsendii* as young a 2–3 months commence to enlarge and exhibit evidence of spermatogenesis. Some young males may produce a few spermatozoa and store them in their epididymides, but it is unlikely that they are fertile during their first winter because of small numbers of spermatozoa and small size of accessory glands. No spermatozoa appear in the epididymides of young-of-the-year after November (Pearson et al., 1952). Among adult males, spermatogenesis commences in mid-August and progresses rapidly with mature spermatozoa produced in ≅3 weeks and stored in the epididymides a few days later. By mid-September the testes commence to regress and are devoid of spermatozoa by early November. Motile spermatozoa are present in the epididymides of almost all adult males from November to late April (Pearson et al., 1952).

In California, some females may have copulated before they enter hibernacula, but all females collected, including young-of-the-year, were inseminated during the first 3 weeks of October. Nevertheless, the peak in copulations was in winter roosts in November-February, and females with vaginae from which spermatozoa could be expressed were taken as late as the end of February (Pearson et al.,

1952). Multiple copulations are known to occur and at least some copulations occur while the female is torpid. Several ovarian follicles enlarge in late September-early October; the follicles that will rupture in spring become even larger after copulation, but do not continue to enlarge throughout winter. Both ovaries are about equally functional, but 98% of implantations occur in the right horn of the uterus (Pearson et al., 1952). Pearson et al. (1952) speculated that Bailey's (1936:390) report of a female taken on 13 October with an embryo "the size of a no. 8 shot" was based on his misinterpretation of a uterine horn that had not regressed to its original size.

Timing of ovulation and duration of embryonic development are related to age of the maternal female, temperature of roost areas, elevation, and latitude. In California, the earliest estimated date of parturition was 19 April at an elevation of 91 m and the latest was 22 July at 1,770 m (Pearson et al., 1952). Scheffer (1930a) recorded births in Washington in the 2nd week of July. Most females have one young (Table 9-5). Studies conducted at widely separated portions of the geographic range indicate that natality in C. townsendii is >90% (Humphrey and Kunz, 1976).

Neonate Townsend's big-eared bats average 2.4 g (Pearson et al., 1952), ≅25% of that of the maternal female (Kunz and Martin, 1982). They are naked with a pinkish-gray body, dark-gray ears, and a pale-gray uropatagium grading to nearly black at the trailing edge (Dalquest, 1947a). A gray fuzz appears by 3 days of age (Pearson et al., 1952). The ears cover the unopened eyes, but become erect by 1 week; the eyes open by 9 days (Pearson et al., 1952). Growth is extremely rapid; the young bats are able to fly at 2.5–3 weeks of age, attain near adult size by 4 weeks, and are weaned by 6–8 weeks (Barbour and Davis, 1969; Kunz and Martin, 1982). We know of no published account of an investigation of reproduction or development in C. townsendii in Oregon.

In California, survival, as measured by return of females to nursery colonies, was 73–81% for adults and 38–54% for young of the previous year. Of those in the latter category, 75% returned as 2-year olds, and of those, 80% returned as 3-year olds (Pearson et al., 1952). Perkins (in litt.) and Barbour and Davis (1969) indicated that the greatest inimical factors for populations of Townsend's big-eared bats probably are vandalism and other disturbance of hibernacula and roosts by humans. Pearson et al. (1952:317) suggested that abundance of C. townsendii may depend upon "the number of suitable winter roosting sites and the number of summer roosting sites surrounded by adequate feeding territory." They discounted predation and disease as serious inimical factors. Paradiso and Greenhall (1967) reported a longevity record for C. townsendii of 16 years 5 months, and J. M. Perkins (pers. comm., 17 April 1995) advised us of a record of 21 years 2 months for the species.

Habits.—Corynorhinus townsendii does not roost in crevices, but commonly hangs in shallow depressions in the roof in dark or dimly lit areas in caves or mine tunnels. It does not crawl after alighting (Pearson et al., 1952). In spring and summer, females assemble in nursery colonies; males are solitary. Warmer roosting areas are used at these seasons and the animals remain active; they may become torpid if temperatures decline. In winter, most C. townsendii roost singly with fur fluffed and wings wrapping the body and interlocked, but some form clusters of two–three to several dozen. In clusters, the bats do not wrap themselves in their wings, but fold the wings alongside their bodies (Humphrey and Kunz, 1976). Corynorhinus townsendii at rest often curls its ears "back and down tightly against the side of the neck like a ram's horn" (Handley, 1959:178), or occasionally the ears may be "curled backward under the wings, the tragus protruding outward" (Alcorn, 1944:309).

Corynorhinus townsendii tends to be a late-flying species (Cockrum and Cross, 1964–1965; Pearson et al., 1952); it is difficult to collect by shooting because it usually flies when it is too dark to be seen (Bailey, 1936; Whitaker et al., 1981a). It commonly flies about inside the roost for a considerable period before emerging (Pearson et al., 1952); apparently, light intensity outside the roost is being sampled to ascertain the appropriate level for emergence. It is a moderately fast flyer (Table 9-1) and it is capable of flight at relatively low body temperatures (Kunz and Martin, 1982). Upon emergence from the day roost, C. townsendii forages for a time then retires to a night roost often shared with other species of bats (Dalquest, 1947a; Pearson et al., 1952). It does not return to its day roost until shortly before sunrise (Pearson et al., 1952); thus it may engage in a second bout of foraging before retiring for the day.

In flight, the external ears are directed forward; Kunz and Martin (1982) suggested that they might be involved aerodynamically by providing lift. Nevertheless, the primary function of the pinnae is related to hearing and echolocation. C. townsendii produces low-intensity sounds from mouth or nose with nearly equal effectiveness and can discriminate echolocation sounds from background noise adeptly (Kunz and Martin, 1982).

During hibernation, Townsend's big-eared bats arouse frequently and may change position within the cave or building or shift among nearby hibernacula. Members of this species, males particularly, are prone to fly about in hibernacula even in midwinter (Pearson et al., 1952). More than half of the autumn body mass may be lost during the period of hibernation, more during the early part of the period (Humphrey and Kunz, 1976). These bats tend to seek relatively cold sites for hibernation. In California, but not elsewhere, females were reported to hibernate in colder sites than those used by males (Humphrey and Kunz, 1976; Pearson et al., 1952).

Corynorhinus townsendii tends to be relatively sedentary; it does not engage in long-distance migrations. In California, the maximum recorded movement was 32.2 km (Pearson et al., 1952). In the southern Great Plains, movements between maternity roosts and hibernacula did not exceed 39.7 km (Humphrey and Kunz, 1976). Females exhibit considerable fidelity to nursery colonies, returning annually to the same site (Pearson et al., 1952).

Townsend's big-eared bats seem particularly sensitive to almost any type of disturbance. An entire cluster commonly takes flight when placed in the beam of a flashlight; they are especially difficult to maintain in captivity as many cannot be taught to feed; pregnant females held captive usually resorb or abort their embryos; and when placed in a cage they pile on top of one another so tightly that some often smother (Barbour and Davis, 1969; Pearson et al., 1952). Roosts commonly are deserted and the bats move to alternate, frequently less-desirable roosts in response to the mere presence of humans in occupied caverns (Humphrey and Kunz, 1976). Although the species is less sensitive to disturbance by humans in occupied caverns in winter, any added activity caused by disturbance in winter may contribute to depletion of energy reserves (Humphrey and Kunz, 1976). Young-of-the-year may be particularly vulnerable to losses related to exhaustion of stored energy reserves (Humphrey and Kunz, 1976; Pearson et al., 1952).

Attachment of bands to wings for recognition of individuals commonly causes inflammation of tissues and sometimes abnormal bone growth, especially in females. These bats commonly chew bands attached to them, obliterating identifying numerals and letters (Humphrey and Kunz, 1976).

Remarks.—In describing Townsend's big-eared bat, Cooper (1837) placed it with the Old World big-eared bats in the genus *Plecotus*. Allen (1865) separated the Old World and New World forms and erected the genus *Corynorhinus* for the latter group, a generic name used for Townsend's big-eared bat in the previous treatment of Oregon mammals (Bailey, 1936). Handley (1959) returned the New World big-eared bats to the genus *Plecotus*, retaining *Corynorhinus* as a subgenus. This taxonomy was followed by most North American mammalogists until Frost and Timm's (1992) phylogenetic treatment of plecotine bats based on 25 morphological and 11 karyological characters. Because this analysis showed that New World species formerly placed in the genus *Plecotus* formed a separate clade, these authors reelevated *Corynorhinus* to generic level.

We concur wholeheartedly with the recommendations of Perkins (1987) regarding protection of *C. townsendii* populations in Oregon by closure to entry by humans of caves and abandoned mines known to have been occupied by the species. However, we disagree equally strongly with his recommendation that populations be monitored annually or biennially unless such can be done without humans entering the caverns.

Antrozous pallidus (Le Conte)
Pallid Bat

1856. *V[espertilio]. pallidus* Le Conte, 7:437.
1864. *Antrozous pallidus*: Allen, 7(publ. 165):68.
1897a. *Antrozous pallidus pacificus* Merriam, 11:180.
1936. *Antrozous pallidus cantwelli* Bailey, 55:391.

Description.—The pallid bat (Plate IV) is a large (Table 9-2) bat with a cream to light-brown dorsum and a paler venter; the pelage is woolly. The ears are large and, in comparison with other Oregon bats, the eyes also are large. The ears are not joined at the base as in *Euderma* and *Corynorhinus*, the other large-eared Oregon bats. The tragus is long and pointed, more than half the length of the ear, and serrate on the outer edge (Orr, 1954). The face is adorned with several wartlike pararhinal glands (Walton and Siegel, 1966). The glands produce a material with a skunk-like odor, especially when individuals are disturbed; the odoriferous material is thought to be involved in defense (Orr, 1954). The muzzle is blunt and squarish; the nostrils open forward beneath a ridge at the front of the muzzle.

The skull (Fig. 9-38) has a high braincase and the length of the rostrum is greater than half the length of the skull (Hall, 1981). The zygomata are much wider than the braincase. There are only 28 teeth, the fewest among Oregon bats (Table 1-1).

Distribution.—The pallid bat occurs in the Okanagan Valley, British Columbia, south through eastern Washington and Oregon, and in western Oregon, from Lane Co.

southward. It also occurs in the area circumscribed by California; Baja California Sur, Sonora, Sinaloa, Nayarit, Jalisco, Queretaro, and Nuevo León, Mexico; and western Texas, Oklahoma, southern Kansas, southern Wyoming, and southern Idaho (Fig. 9-39). A disjunct population also occurs on the island of Cuba (Koopman, 1993).

Pallid bats occur in the interior valleys of western Oregon from near the Lane-Benton county line south to the California border and throughout the state east of the Cascade Range except for the Blue Mountain region (Fig. 9-39).

Pleistocene-Holocene fossils are known from California, New Mexico, and Arizona (Kurtén and Anderson, 1980).

Geographic Variation.—Two of the six currently recognized subspecies occur in Oregon: the larger and darker *A. p. pacificus* in the interior valleys west of the Cascade Range and the smaller and lighter-colored *A. p. pallidus* east of the Cascade Range (Martin and Schmidly, 1982). Pallid bats in southeastern Washington, eastern Oregon, and northern Nevada were designated as the separate subspecies *A. p. cantwelli* by Bailey (1936). This name was placed in synonymy with *A. p. pallidus* by Martin and Schmidly (1982).

Habitat and Density.—Pallid bats usually are associated with desert areas (Hermanson and O'Shea, 1983). In Oregon, the distribution (Fig. 9-39) reflects this association as the geographic range of *Antrozous* east of the Cascade Range does not include most of the forested Blue Mountains Province and west of the Cascade Range is restricted to the drier interior valleys of the southern portion of the state. The species usually is found in brushy, rocky terrain, but has been observed at edges of coniferous and deciduous woods and in open farmland (Orr, 1954). In central Oregon, pallid bats occupied a typical semiarid desert area vegetated by grass (mostly *Poa sandbergii*), sagebrush (*Artemisia tridentata*), and juniper (*Juniperus*—Lewis, 1994). Areas used by *Antrozous* commonly include sources of water (Orr, 1954), but water does not seem to be requisite (Findley et al., 1975).

Narrow crevices in caves, mine shafts, and buildings serve as day roosts in summer for most pallid bats, but the species has been known to roost in a variety of other places, including rock piles, piles of burlap sacks, and hollow trees (Hermanson and O'Shea, 1983). Day roosts used by pallid bats are characterized by semidarkness and protection from above (Orr, 1954). Night roosts used by pallid bats to rest between foraging bouts or in which to consume particularly large prey usually are near, but separated from, day roosts. In central Oregon, pallid bats used deserted buildings, rock overhangs, and space under bridges as night roosts. Suitable bridges were those constructed of wood or cement with I-beam or box-type support; bridges constructed of steel or those with flat undersides were not used (Lewis, 1994). Warmer temperatures may play a significant role in the selection of night roosts (O'Shea and Vaughan, 1977). Also, darkness, protection from precipitation and wind, and being sufficiently open to allow free flight when the bats are disturbed are other criteria of suitable night roosts (Lewis, 1994).

Winter roosts of pallid bats are not well known. All of the 50 Oregon specimens for which we obtained dates of capture were taken between 11 April and 24 September.

Orr (1954) provided evidence of one–four wintering individuals in scattered locations, but none at sites used as summer roosts. Hall (1946) found two males in separate crevices in a mine tunnel in December in Nevada. Twente (1955) found *Antrozous* hibernating in a 3.8–7.6-mm-wide crack in the ceiling of a cave in southern Kansas; ≅20 individuals were present one winter, >100 the next.

We know of no estimates of abundance for pallid bats in Oregon or elsewhere. We located only 81 specimens from the state; compared with *Eptesicus* and several *Myotis* the species seems relatively scarce. However, *Antrozous* occupies narrow crevices, likely making it less accessible to collectors than the species for which more specimens are available.

Diet.—Observation of pallid bats feeding around an ultraviolet light and examination of their stomach contents and fecal pellets revealed that the species feeds mostly on

beetles (Coleoptera) and moths (Lepidoptera) by gleaning them from the ground (Bell, 1982; Easterla and Whitaker, 1972; Hermanson and O'Shea, 1983). O'Shea and Vaughan (1977) characterized insects identified from parts falling to the ground beneath a night roost used by pallid bats in Arizona as ground-dwellers, weak flyers, and strong flyers that rest on vegetation frequently; all were large, commonly 20–70 mm long (Hermanson and O'Shea, 1983). The most numerous discarded insect parts beneath a night roost in Washington were those of Jerusalem crickets (*Stenopelmatus*) and June beetles (*Polyphylla crinta*), both ground-dwelling species (McNeil, 1956). Hermanson and O'Shea (1983) listed arachnid prey as including scorpions (Scorpiones) and solpugids (Solpugida), and, among insects, crickets (Gryllacrididae), beetles (Tenebrionidae, Carabidae, Silphidae, Cerambycidae), grasshoppers (Acrididae), cicadas (Cicadidae), katydids (Tettigoniidae), and praying mantids (Mantidae). O'Shea and Vaughan (1977) included all of these and added hawkmoths (Sphingidae); Herreid (1961) added long-horned grasshoppers (Tettigoniidae), dragonflies (Libellulidae), bugs (Pentatomidae), alderflies (Sialidae), aphis lions (Chrysopidae), caddisflies (Trichoptera), moths (Noctuidae, Saturniidae), and wasps (Scoliidae). Bell (1980, 1982) observed that pallid bats ignored the great numbers of

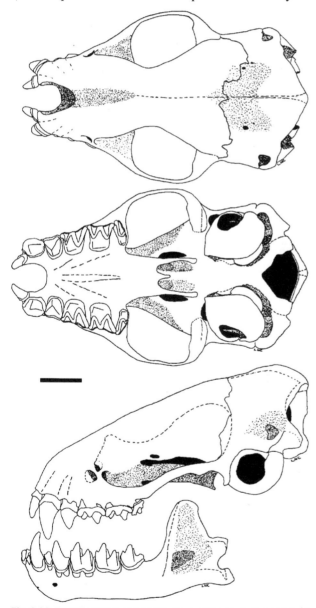

Fig. 9-38. Dorsal, ventral, and lateral views of the cranium, and lateral view of the dentary of the pallid bat, *Antrozous pallidus* (OSUFW 2785). Scale bar equals 3 mm.

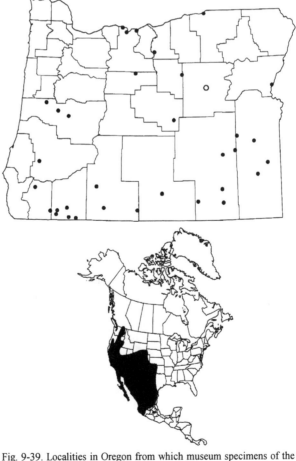

Fig. 9-39. Localities in Oregon from which museum specimens of the pallid bat, *Antrozous pallidus*, were examined (listed in the Appendix, p. 511), and the distribution of the pallid bat in North America. (Range map redrawn from Hall, 1981:237, map 185, with modifications based on Clark and Stromberg, 1987, and Martin and Schmidly, 1982.) A disjunct population occurs in Cuba (Koopman, 1993). Open symbol indicates record for county.

flying insects attracted to ultraviolet lights, but preyed only upon those that alighted on the ground. In addition to insects, pallid bats are known to prey on lizards (Squamata) and mice occasionally (Bell, 1982; O'Shea and Vaughan, 1977) and are suspected of consuming smaller bats (Engler, 1943).

We know of no published evaluation of the dietary habits of pallid bats in either eastern or western Oregon. Although taxa preyed upon may differ between these regions, and between both of these regions and the nearest regions in which the diet has been evaluated (Bell, 1982; McNeil, 1956; Orr, 1954; O'Shea and Vaughan, 1977), we do not suspect that size, habitat, or behavior of prey consumed in Oregon differs significantly.

Reproduction, Ontogeny, and Mortality.—In central California, the testes of pallid bats increase in size in August, remain large in September, and commence to regress by mid-October. Pallid bats in captivity copulated most frequently in October–November, but some copulations were observed in January–February (Orr, 1954).

In captivity, females apparently ovulate and become pregnant in response to increases in ambient temperature as captives were induced to give birth in February, ≅83 days after being placed in a warm cage. Dates of parturition in wild and wild-caught near-term captives were in May–June in central California, with some variation among years and colonies (Orr, 1954). However, Cockrum (1955) recorded as pregnant one taken on 4 July. Near Clarno, Wasco Co., only two (5.1%) of 39 females caught in 1990 were not pregnant or lactating, whereas in the cooler spring of 1991, eight (14.0%) of 57 were neither pregnant nor lactating (Lewis, 1993). Also, in 1990, the first lactating pallid bats were caught on 17 June and most caught after 20 June were lactating, whereas in 1991, the first lactating bat was caught on 14 July and some had not yet given birth "well into August that year" (Lewis, 1993:1431). Lactating females weighed significantly less in 1991 than in 1990. In the year with the cooler spring (1991), female bats may have spent more time in torpor, accounting for the lower body mass and the retarded embryonic development (Lewis, 1993).

Two young per litter is most common, but litters of three and one have been recorded; the average litter size is usually less than two (Table 9-5). Yearling females usually bear only one young (Davis, 1969).

Young are altricial; they are born with eyes closed, ears folded tightly against the head, and are pink and naked except for a few hairs that can be seen only with magnification (Orr, 1954). The young are carried by the maternal female on foraging flights for only 1–2 days after birth (Twente, 1955). In Arizona, among free-living pallid bats, the eyes opened at 2–5 days, the ears were open at 5–11 days, the dorsum was fully haired by 11–17 days, the venter was fully haired by 14–20 days, and the young bats made their first flights at 33–36 days (Davis, 1969). Among captives in California, the eyes and ears opened at 8–10 days, the dorsum was fully haired at 18 days, the venter was fully haired at 24 days, and the young bats first flew at 49 days (Orr, 1954). The early pelage of young is darker than that of adults because the lighter basal portion of the hairs has not yet grown out. Davis (1969) indicated that young could be separated from adults by forearm length through the first autumn of life. Postpartum females lactate for 2–3 months; they do not nurse other than their own young (Davis, 1969). Recurved cusps on the deciduous

teeth, especially those on the incisors, enhance the grasp of the young on a nipple of the maternal female, thereby preventing detachment in flight. Manually detaching young pallid bats from nipples without damage to either young or adult is difficult (Orr, 1954).

Pallid bats may survive 8–9 years in captivity, but mortality rates are not available for the species (Hermanson and O'Shea, 1983). Their habit of feeding on the ground makes them particularly vulnerable to debilitating injury and predation (Hermanson and O'Shea, 1983). They are known to be preyed upon by snakes (Serpentes) and raptors (Tytonidae, Strigidae, Accipitridae). Young bats may be particularly vulnerable to avian predation (O'Shea and Vaughan, 1977). However, prey items consumed in Oregon by some of the raptors known to have taken pallid bats elsewhere do not include the species (Maser et al., 1970, 1980; Maser and Hammer, 1972). Recorded maximum longevity is 9.1 years (Cockrum, 1973).

Habits.—Pallid bats fly more slowly and lack the maneuverability of smaller bats, but they crawl and climb extremely well (Orr, 1954) and they use a variety of "gaits, speeds and positions" in "walking" (Dietz, 1973:790). In straight flight they make 10–11 wingbeats/s, but increase to ≅13.5 wingbeats/s in making turns (Orr, 1954).

Pallid bats leave day roosts later than some other species of bats with which they cohabit, but time of emergence varies seasonally (Orr, 1954). In Kansas, they commonly commenced to fly around inside a barn (day roost) shortly after sunset, but did not emerge from the barn until the light intensity was ≅10 lux (Twente, 1955). In summer, both sexes of pallid bats may roost together in colonies during the day (Orr, 1954; Storer, 1931) or may segregate into one-sex colonies (Hall, 1946). O'Shea and Vaughan (1977) presented evidence that adult males joined colonies of females and their offspring only after the offspring commenced to forage. In most situations, *A. pallidus* roosts in clusters of 20–200 individuals, but may occur singly or in small groups occasionally. *Tadarida brasiliensis* and *Myotis yumanensis* sometimes roost within clusters of *A. pallidus*, but more commonly the clusters do not include other species (Hermanson and O'Shea, 1983).

Emergence from day roosts commences about 2100–2200 h at the summer solstice and becomes progressively earlier thereafter (Lewis, 1994:222, fig. 1). Upon emergence, pallid bats usually forage within ≅3 km of day roosts at moderate speeds and at elevations of 0.1–10 m with frequent swoops close to the ground (Bell, 1982; O'Shea and Vaughan, 1977). They may become nearly stationary in flight by riding the wind, and they are capable of hovering for a few wingbeats. Prey seems to be located from as far as 10 m and pallid bats may approach potential prey at elevations of 0.2–1.0 m several times before alighting atop or near the prey to capture it. Prey is taken from the ground, or sometimes gleaned from plants (Hermanson and O'Shea, 1983). However, pallid bats do not produce echolocation calls at this phase of the attack, so likely locate prey by the rustling sounds made by the prey (Bell, 1982). Pallid bats do not respond to species of insects that call from protection of cover. Prey usually is subdued and the bat airborne within 10 s; small prey is consumed in flight, whereas pallid bats carry large prey to night roosts where wings, legs, and heads of insects are removed before the remainder of the prey is consumed (Bell, 1982). In night roosts, individuals consuming prey remain separate from clusters.

After the initial bout of foraging, pallid bats retire to a night roost and form clusters; they exchange vocalizations by which they locate one another. Throughout the summer in central Oregon, pallid bats commenced to arrive in night roosts within 10–15 min after the first bats emerged from day roosts (Lewis, 1994). Pallid bats spend 40–75% of the time away from day roosts in night roosts (O'Shea and Vaughan, 1977). In addition to providing places for rest, to digest food, and to reduce loss of energy to thermoregulation, night roosts are believed to serve a social function. The short interval between emergence from day roosts and arrival at night roosts, the arrival of pallid bats in groups, and the mixing of bats from different day roosts tend to suggest that sociality is involved in use of night roosts (Lewis, 1994). Females with young left at the day roost return to the day roost rather than spending the remainder of the night at a night roost (Twente, 1955). Pallid bats engage in another bout of foraging before returning to their day roosts 15 min to 2 h before sunrise (O'Shea and Vaughan, 1977). As these bats approach the day roost, they rally outside the entrance. These swarms of bats produce distinctive calls at intervals of several seconds; the swarming bats examine the entrance to the roost repeatedly for 15–45 min. Finally, one bat enters the roost and continues to call from within; the remainder of the rallying bats enter within a few minutes (Vaughan and O'Shea, 1976).

Maternal females continue an association with their offspring for 12 months or more after the young commence to fly. At this season, pallid bats returning to day roosts commonly fly in groups of two or three including one or two smaller, less agile flyers (O'Shea and Vaughan, 1977). Also at this season, a different roost is used nearly each day; thus, young pallid bats become accustomed to seeking the visual and vocal activity of conspecifics rallying near the entrance to a day roost rather than becoming attached to a specific roost site (O'Shea and Vaughan, 1977).

In summer, pallid bats seek locations within day roosts at which they can maintain a body temperature of ≅30°C passively. They enter torpor, but may shift about to form loose clusters in cooler parts of the roost when the ambient temperature rises or to cluster more tightly in warmer sites when the ambient temperature declines (Vaughan and O'Shea, 1976). Pallid bats also form clusters and enter torpor in night roosts for 2–3 h in summer and 4–5 h in autumn (O'Shea and Vaughan, 1977). In some areas pallid bats enter torpor and subsist only on body reserves in winter (Twente, 1955), but in Nevada, individuals were caught in nets at a desert spring in November and February when ambient temperatures were 2–3°C (O'Farrell et al., 1967).

FAMILY MOLOSSIDAE—FREE-TAILED BATS

Free-tailed bats are widely distributed in warmer areas of the world; they occur in southern parts of Europe; Asia north to Korea and Japan; all of Malaysia, Australia, and Africa; South America to 45° south; and in North America to the upper Midwest and British Columbia (Koopman, 1984; Nowak, 1991). At present, this family includes 12 genera and 80 species (Koopman, 1993) only one of which, Tadarida brasiliensis, occurs in Oregon. Except for one specimen of Nyctinomops macrotis (=Tadarida molossa) collected in British Columbia in 1938 (considered an accidental occurrence—Nagorsen, 1990), the several colonies

of Tadarida brasiliensis in southern Oregon constitute the northernmost extent of the range of the family in North America.

Free-tailed bats are characterized by long, narrow wings; a narrow uropatagium with the tail projecting far beyond the trailing edge; short, velvetlike pelage; thick, leathery ears; small eyes; large lips; short, strong legs with broad feet; and a fringe of stiff hairs on both sides of the hind feet. These bats produce a strong, penetrating, and characteristic odor that remains for a considerable period.

Molossids produce echolocation calls to locate prey and avoid obstacles; feed largely on flying insects; seek shelter in caves, buildings, and hollow trees; and form colonies numbering from a few to millions. They are rapid flyers. In general, free-tailed bats are active throughout the year, but in northern areas often are less active in winter. They do not enter true hibernation, but may enter torpor when placed in a cold environment and may survive there for a short period. Some make extensive seasonal migrations. Species in other families of bats often roost with molossids.

Tadarida brasiliensis (I. Geoffroy)
Brazilian Free-tailed Bat

1824. *Nyctinomus brasiliensis* I. Geoffroy St.-Hilaire, 1:343.
1860. *M[olossus]. mexicanus* Saussure, 12:283.
1955. *Tadarida brasiliensis mexicana*: Schwartz, 36:108.

Description.—*Tadarida brasiliensis* (Plate IV) is a small bat (Table 9-2) with approximately one-half of the tail extending beyond the uropatagium; vertical grooves in the upper lip; large, round, and forward-directed ears with tiny papillae on the anterior edge; and a calcar without a keel. The wings are long and narrow, and have relatively low camber (Wilkins, 1989).

The dorsal pelage is short (2–3 mm) and dull dark-brown, whereas that of the venter is longer (3–4 mm) and slightly paler. Juveniles are gray, but acquire the brown pelage in autumn (LaVal, 1973). The dorsal hairs are uniformly colored, whereas the ventral hairs have whitish tips. The dark brown flight membranes, although seemingly unfurred, are covered with extremely short hairs (Wilkins, 1989). The sides of the feet have a fringe of stiff hairs; long, curved, stiff hairs on the toes extend far beyond the nails.

Tadarida brasiliensis is covered with sebaceous glands that produce highly odoriferous compounds that coat whatever substrate a bat rests upon. These compounds may persist for years and mark caves or buildings with distinctive odors that can be detected by "even the insensitive human nose" from ≥2 km downwind (Glass, 1982:132).

The skull (Fig. 9-40) is robust, somewhat triangular, and with zygomata extending slightly beyond the braincase. One of the three tiny lower incisors sometimes is missing, p4 is the tallest tooth in the lower jaw except for the canine, the premolars are not molariform, M1 and M2 have W-shaped occlusal patterns, and M3 has a Z-shaped pattern (Barbour and Davis, 1969; Wilkins, 1989).

Distribution.—*Tadarida brasiliensis* has one of the most extensive distributions of any North American bat (Wilkins, 1989). It extends from a line connecting southern Oregon, Utah, Nebraska, Arkansas, Mississippi, and southern North Carolina south through Central America and into South America (except for much of Amazon and Orinoco basins, the caatingas of Brazil, and the east slope of the northern Andes) as far south as Península Valdés,

Argentina. The species also occurs widely in the Caribbean islands. The distribution in South America is poorly known (Wilkins, 1989). *T. brasiliensis* attains the northern limit of its range in Oregon, where it occurs in Douglas, Jackson, and Klamath counties (Fig. 9-41).

Pleistocene fossils of this species are known from Arizona, Florida, Texas, New Mexico, and Kentucky (extralimital). Fossils also are known from Puerto Rico, Cuba, and Antigua (Kurtén and Anderson, 1980; Wilkins, 1989).

Geographic Variation.—Of the nine recognized subspecies of *T. brasiliensis* only one occurs in Oregon: *T. b. mexicana* (Hall, 1981).

Habitat and Density.—In southern Oregon, *T. brasiliensis* occurs largely in buildings (Cross, 1979), but in Texas, New Mexico, Arizona, and Mexico, colonies formerly numbering millions of bats resided in each of several caves (Davis et al., 1962; Geluso et al., 1976; Villa-R., 1967; Wilkins, 1989). Wilkins (1989) believed that these bats were largely cave dwellers until buildings were constructed, but indicated that in Florida, the species roosted in hollow trees. Caves occupied by *T. brasiliensis* in Texas had large rooms (≥18 by 18 by 10 m high) and large entrances (≅9 by 5 m) or vertical shafts (≅9 by 9 m—Davis et al., 1962). During summer when these caves are occupied by bats, they are extremely hot and humid, and have high concentrations of ammonia from decomposing urine and feces (Barbour and Davis, 1969). Buildings used by these bats in winter usually are heated and in regions in which winter temperatures permit insect activity periodically (Cross, 1979; Davis et al., 1962). *T. brasiliensis* in some regions makes extensive migrations (≤1,840 km—Glass, 1982), but those in Oregon and northern California remain in the region year-round (Wilkins, 1989).

Until about 1960, *T. brasiliensis* was known from only a few specimens in Oregon (see Remarks), and indeed may have been present in the state in low numbers. During the 1960s, at least 10 additional specimens were deposited in systematics collections, all from Jackson Co. Olterman and Verts (1972) listed 15 museum records and considered the species to be rare in Oregon despite a personal communication from S. P. Cross that colonies were known to occur in a barn near Ashland and under the tiles in the roof of a building on the Southern Oregon University campus in Ashland, Jackson Co. They also reported that Cross deemed the species fairly common in the Ashland area. Cross (1979) reported that one colony of *T. brasiliensis* in southern Oregon contained several thousand individuals.

Diet.—We know of no published report on the diet of *T. brasiliensis* in Oregon. However, in California and elsewhere small moths (Lepidoptera) composed the primary prey taken by this species (Ross, 1961; Storer, 1926). Also consumed in lesser amounts were leafhoppers (Homoptera), beetles (Coleoptera), flies (Diptera), true bugs (Hemiptera), and ants (Hymenoptera). Campbell (1925) claimed that these bats provided a valuable service by preying largely upon anopheline mosquitoes (Diptera: Culicidae). This notion was dispelled by Storer (1926), who was unable to find remains of mosquitoes in samples of feces from colonies that Campbell (1925) had established for mosquito control.

Reproduction, Ontogeny, and Mortality.—We know of no information on these topics obtained in Oregon. In Florida, the seminiferous tubules commence to enlarge in September, but no spermatozoa were found in the testes before 20 January. No spermatozoa were found in the epididymides of specimens collected in January, but they were abundant in both testes and epididymides in February and March (Sherman, 1937). None was present in either organ in mid-April.

In a maternity roost in a church in California, embryos were first detected on 28 April, although females taken on 3 April had swollen uteri. Only lactating females and neonates were observed on 9 July and all young were believed to have been born within a 3–4-day period. Krutzsch (1955) suggested that young were in utero for ≅100 days. However, in Florida, Sherman (1937) found females with spermatozoa in their uteri as early as 13 February, 2-mm embryos on 23 March, and the first evidence of parturition on 31 May. Most young were born from ≅31 May to 11 June during one year and from ≅13 June to 30 June the next. Sherman (1937) believed the gestation period to be ≅11–12 weeks and Davis et al. (1962:345) placed it at "a few days in excess of 90 days." In Louisiana, LaVal (1973) found embryos as early as 11 March and many neonates with placentas still attached on 4 June. In Texas, >90% of young were born within 15 days of the mid-June average date of birth (Davis et al., 1962).

In Texas and Florida, *T. brasiliensis* was reported to give birth to only one young (Davis et al., 1962; Sherman, 1937); in California, most females had one young, but several with two

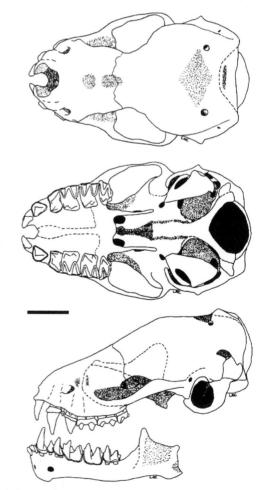

Fig. 9-40. Dorsal, ventral, and lateral views of the cranium, and lateral view of the dentary of the Brazilian free-tailed bat, *Tadarida brasiliensis* (OSUFW 8869). Scale bar equals 3 mm.

Fig. 9-41. Localities in Oregon from which museum specimens of the Brazilian free-tailed bat, *Tadarida brasiliensis*, were examined (listed in the APPENDIX, p. 565), and the distribution of the Brazilian free-tailed bat in North America. (Range map redrawn from Hall, 1981:242, map 188, with modifications.) The species also occurs on some Caribbean islands and in parts of South America; extralimital specimens are known from several sites north of the depicted range (Wilkins, 1989).

embryos were collected (Krutzsch, 1955). However, of 26 pregnant females from a colony in Alabama, 11 carried one embryo, 12 had two, and 3 had three (Di Salvo et al., 1969).

Growth and development of young *T. brasiliensis* are rapid. In slightly more than a month, the naked, flightless neonates become fully furred, almost equal in size to adults, and able to fly sufficiently well to forage for themselves (Davis et al., 1962; Pagels and Jones, 1974). Females become sexually mature when about 9 months old, but males likely do not become reproductively active until their 2nd year (Sherman, 1937).

Populations of *T. brasiliensis* occupying caves in southwestern United States formerly numbered in the millions, but have declined dramatically during the last half of the 20th century (Geluso et al., 1976). Organochlorine insecticides, and their residues and metabolites, were implicated in the declines by some (Geluso et al., 1976, 1981; Humphrey, 1975), but not by others (Clark et al., 1975). Seemingly, the disparity in findings was related to different levels of organochlorine residues among bats from several caves and to the fact that the compounds acquired during nursing are stored in fat during early life. The young bats died during their first migration when fat containing these compounds was metabolized for energy (Geluso et al., 1976).

Disturbance of free-tailed bats by humans involved in economic, recreational, or scientific endeavors also may have contributed to the decline in numbers in caves.

Snakes (Serpentes) and raptorial birds are known to prey on *T. brasiliensis*; instances of predation by other vertebrates are uncommon (Davis et al., 1962; Krutzsch, 1955). Paradiso and Greenhall (1967) recorded a maximum longevity for the species as 4 years and 6 months, but Glass (1982) reported recovery of one marked individual in its 8th year of life and several in their 5th year or older. LaVal (1973) also recorded maximum longevity for the species of 8 years, but Davis et al. (1962) estimated maximum longevity for the species at 15 years.

Habits.—Average ($\pm SE$) time of emergence of Brazilian free-tailed bats from day roosts in several Texas caves was 11 ± 2.51 min ($n = 98$) after sunset (Herreid and Davis, 1966). Before the exodus began, activity and vocalization within the cave increased (Twente, 1955). Herreid and Davis (1966) described two basic types of flight patterns: more commonly bats exited the caves, spread out in several directions, and commenced to forage individually; occasionally, commencing before sunset, a stream of bats poured forth and maintained its integrity as a great serpentine column across the sky for many kilometers. Barbour and Davis (1969:199) described the sound of wings of such flights as "like the roar of a white-water river." Clusters of several hundred bats then broke from the column and moved away. Areas used for foraging by *T. brasiliensis* often are ≥ 50 km from day roosts; flights to foraging areas are direct and swift (Wilkins, 1989). Individual bats averaged 3 h and 48 min away from the caves (Herreid and Davis, 1966). As *T. brasiliensis* in Oregon is not a cave dweller, such spectacular exodus flights are not expected.

Free-tailed bats that live in buildings often exit in groups, but if they leave single file, they commonly form groups as they spiral upward 15–30 m before flying off toward a foraging area (Davis et al., 1962). *T. brasiliensis* usually does not use night roosts. The first group of bats to return to the day roost arrives $\cong 1$–1.5 h after the first bats leave. Groups return to day roosts throughout the night; as dawn approaches extensive "traffic jams" develop at entrances to roosts. Bats often must make numerous "passes" at the entrance before gaining access to the day roost; those that fail to gain access seek roosts elsewhere, commonly roosting alone. When these bats were released in groups during daylight hours they spiraled upward until they could no longer be seen with binoculars; Davis et al. (1962) estimated that they climbed to >3,000 m. Williams et al. (1973) tracked groups of flying *T. brasiliensis* by radar at altitudes in excess of 3,000 m and at speeds of >40 km/h. They concluded that from that altitude both echolocation and vision likely were inadequate for the bats to orient and navigate in areas without prominent landmarks, such as in Texas where they worked. They suggested that these bats possibly oriented by use of the positions of celestial bodies, but the hypothesis had not been tested.

Davis et al. (1962) reported that females returning to nursery roosts with full mammaries permitted nursing by the first two young encountered. However, a more recent investigation revealed that maternal females correctly identified and permitted nursing by their own offspring 83% of the time (McCracken, 1984). The 17% of maternal females that permitted random nursing likely were unable to locate their own offspring among "creches" containing as many as 40 young/100 cm² before being mobbed by scores of ravenous young (McCracken, 1984).

Based on the weight of the contents of stomachs, Davis et al. (1962) estimated that *T. brasiliensis* ingested about 1.0 g of insects per night. From this estimate and estimates of the number of bats in Texas they calculated that the species consumed about 6 million kg of insects per year in the state. Barbour and Davis (1969) believed that the estimate should be at least 3 times as great. Digestion of these insects results in production of enormous deposits of guano in day roosts. Guano is of considerable economic value as fertilizer and large quantities of the material were removed from some caves in southwestern United States and northern Mexico for that purpose. Also, in the past, guano was used as a source of potassium nitrate for the manufacture of gunpowder (Wilkins, 1989).

Remarks.—The early history of the occurrence of *T. brasiliensis* in Oregon is convoluted. Shamel (1931) listed a specimen from "Fort Dalles" (=The Dalles, Wasco Co.). However, Stager (1945) questioned the validity of the record because he was unable to locate the specimen in systematics collections in which Shamel (1931) claimed to have examined specimens. Bailey (1936) did not include the species in his treatment of the mammal fauna of Oregon, presumably because he also was unable to locate the specimen. We, too, failed to locate the specimen from "Fort Dalles" in either of the collections mentioned by Shamel (1931) or in any of the other collections we visited.

Thus, the first valid published record of the species in the state vouched by a specimen was that of Stager (1945:196), who reported a *T. brasiliensis* was collected "under the roof of the laundry of the Sacred Hearts Hospital in Medford, Oregon" on 1 October 1940 by Dennis Constantine. Stager (1945) claimed that the specimen was deposited at the Department of Zoology at our institution, but Jewett (1955) indicated that the specimen was in his private collection (no. 2846). At the present time it is in the mammal collection at the University of Puget Sound (PSM 7707). We do not know the circumstances of its transfer, but we know that a large portion of S. G. Jewett's collection is now at the San Diego Museum of Natural History. Jewett (1955) reported receiving three specimens taken from a large colony in a building in Ashland, Jackson Co., by Mr. W. W. Wells on 6 February 1954. Three specimens (PSM 7708–7710) at the University of Puget Sound bear that date, locality, and collector's name.

A moribund female was found in a hanger at Kingsley Air Force Base, Klamath Falls, Klamath Co., on 7 September 1960, but its disposition was not reported (Constantine, 1961*b*). We did not find a specimen from that locality or one collected on that date.

Tadarida brasiliensis was the subject of the most cruel and inhumane use of animals of which we are aware. During World War II, the U.S. Navy initiated Project X-ray to investigate the use of these bats to carry small incendiary bombs. Thousands of the bomb-carrying bats were to be released from airplanes at appropriate sites in anticipation of their seeking refuge in enemy buildings. The bomb was designed to produce a 50-cm flame for 8 min, thereby incinerating the bat and igniting the building. During a field test, several bomb-laden bats escaped, entered the headquarters of Project X-ray, and set it ablaze (Yalden and Morris, 1975). The project was canceled!

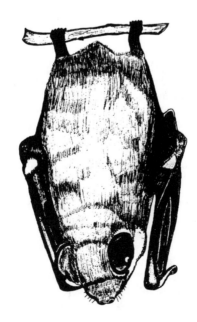

CHAPTER 10

ORDER LAGOMORPHA—RABBITS, HARES, AND PIKAS

The lagomorphs include 79 Recent species (25 species of pikas all in one genus and 54 species of rabbits and hares in 11 genera—Hoffmann, 1993; Wilson and Reeder, 1993a). The order is distributed nearly worldwide; the European rabbit, *Oryctolagus cuniculus*, was introduced into Australia, New Zealand, the southern part of South America, and some oceanic islands, areas not occupied naturally by lagomorphs. In Oregon, the lagomorphs are represented by a pika (*Ochotona princeps*), four species of rabbits (*Brachylagus* and *Sylvilagus*), and three species of hares (*Lepus*). One species, *Sylvilagus floridanus*, was introduced into the state from the Midwest.

The relationship of lagomorphs to other groups of mammals is uncertain; at various times, triconodonts, marsupials, insectivores, primates, and artiodactyls have been considered ancestral to lagomorphs (Diersing, 1984). Anatomically and serologically they are most similar to the artiodactyls; other evidence suggests that the group has been separated from other eutherian mammals for a long time. The earliest known lagomorphs were from Paleocene deposits and the two families diverged in the Oligocene. Lagomorphs were long thought to be closely allied with the rodents, but the idea was abandoned nearly a century ago. However, the concept of a Paleocene divergence of lagomorphs and rodents from a common ancestor has been resurrected to some extent by discovery of fossil forms classified as a separate order †Mixodontia (Dawson and Krishtalka, 1984).

The lagomorphs are characterized by a fenestrated maxilla; two upper incisors, the second directly behind the first; a broad diastema between incisors and premolars; ears longer than the tail; and haired soles of feet (Diersing, 1984). None weigh >10 kg; most weigh <2 kg. Females usually average heavier than males.

KEY TO THE LAGOMORPHA OF OREGON

1a. Ear rounded, not longer than wide; ear edged with white; no external tail; length of front and rear legs subequal; hind foot <40 mm long; toe pad naked; frontal without supraorbital process; dental formula 2/1, 0/0, 3/2, 2/3 = 26....*Ochotona princeps*, pika (**Ochotonidae**; p. 124).

1b. Ear at least twice as long as wide; edge of ear at least as dark as remainder of ear; external tail short but present; hind leg longer than front leg; hind foot >60 mm long; toe pad furred; frontal bone with supraorbital process; dental formula 2/1, 0/0, 3/2, 3/3 = 28........2 (**Leporidae**)

2a. Hind foot ≤100 mm long; interparietal separated from parietals by a distinct suture; young altricial................3

2b. Hind foot ≥105 mm long; interparietal fused to and indistinguishable from parietals; young precocial.......6

3a. Ear <60 mm long; anterior and posterior projections of supraorbital process subequal; one reentrant angle on anterior margin of P2; margin of reentrant angle on P3 straight; auditory bulla inflated.......................................
........*Brachylagus idahoensis*, pygmy rabbit (p. 128).

3b. Ear ≥60 mm long; anterior projection of supraorbital

process less than half as long as posterior projection; two or three (usually three) reentrant angles on anterior margin of P2; margin of reentrant angle on P3 crenate; auditory bulla not greatly inflated.....................4

4a. Dorsal pelage uniformly dark brown; hairs on venter lead colored with buffy tips; tip of ear not noticeably darker than other portions; ventral surface of tail not cottony white; hind foot ≤85 mm long; combination of supraoccipital shield squarish and posterior projection of supraorbital process free of (at most barely touching) squamosal; thin loop of enamel separating anterior and posterior basins on P4, M1, and M2 not crenate...
..............*Sylvilagus bachmani*, brush rabbit (p. 131).

4b. Dorsal pelage light brownish-gray with rump lighter than remainder of dorsum; hairs on venter lead colored with white tips; tip of ear black; ventral surface of tail cottony white; hind foot usually ≥85 mm long; either rounded (or angular) supraoccipital shield or posterior projection of supraorbital process fused to squamosal; thin loop of enamel separating anterior and posterior basins on P4, M1, and M2 crenate.................................5

5a. Pelage on rump more grayish than brownish; supraoccipital shield usually angular or rounded (shield shaped) with a median spine; opening of bony external auditory meatus larger than the circle formed by the zygomatic process of the squamosal and the posterior portion of the jugal; posterior projection of supraorbital process usually free of (at most barely touching) squamosal...
........*Sylvilagus nuttallii*, Mountain cottontail (p. 138).

5b. Pelage on rump more brownish than grayish; supraoccipital shield usually squarish, often with a median notch, but occasionally with median spine; opening of bony external auditory meatus equal to or smaller than the circle formed by the zygomatic process of the squamosal and the posterior projection of the jugal; posterior projection of supraorbital process overlapping and fused to squamosal...
........*Sylvilagus floridanus*, eastern cottontail (p. 134).

6a. Ear ≤80 mm long; anterior projection of supraorbital process lacking; occipitonasal length <75 mm; (brownish in summer, commonly white in winter, except in the Coast Range where pelage is always brownish)...
..............*Lepus americanus*, snowshoe hare (p. 141).

6b. Ear >80 mm long; anterior projection of supraorbital process present; occipitonasal length >75 mm............7

7a. Tail black on dorsal surface, buffy or brownish on ventral surface; rostrum narrow and tapering; posterior projection of supraorbital process usually long and tapering, sometimes fused to squamosal[1]............................
...*Lepus californicus*, black-tailed jackrabbit (p. 145).

7b. Tail white except occasionally with dusky dorsal stripe; rostrum broad and with nearly parallel sides; posterior projection of supraorbital process usually short and obtusely angled, rarely fused to the squamosal[1]...
.....*Lepus townsendii*, white-tailed jackrabbit (p. 149).

---

[1]*Lepus californicus* and *L. townsendii* often are not distinguishable by skulls alone (Hoffmann and Pattie, 1968).

Table 10-1.—*Means (±SE), ranges (in parentheses), and* CVs *of measurements of skull and skin[a] characters for female and male pikas* (Ochotona princeps) *from regions in Oregon. Skin characters were recorded from specimen tags; skull characters were measured to the nearest 0.01 mm.* SE *and* CV *not provided if* n <10.

Region	Sex	n	Total length	Hind foot length	Mass (g)	Basilar length	Length of nasals	Length of maxillary toothrow	Length of bulla
Jefferson, Deschutes, southern Wasco, and eastern Clackamas, Marion, and Linn counties	♀	11	196 ± 1.7[b] (190–208) 2.7	31 ± 0.3 (30–33) 3.1	182.2[c] (159.3–210)	38.54 ± 0.36 (36.02–40.28) 3.09	14.03 ± 0.17 (13.15–15.28) 4.03	8.57 ± 0.09 (8.12–9.11) 3.81	9.91 ± 0.04 (9.28–10.76) 4.82
	♂	17	191 ± 2.3 (175–207) 4.9	32 ± 0.35 (29–34) 4.6	157.6[d] (138–194)	38.78 ± 0.47 (32.54–40.53) 5.03	14.28 ± 0.17 (13.30–15.56) 5.01	8.55 ± 0.10 (7.17–9.24) 4.99	10.04 ± 0.12 (9.23–11.16) 5.01
Union, Wallowa, northern Baker and Grant, and eastern Umatilla counties	♀	16	189 ± 1.9 (170–200) 4.1	31 ± 0.3 (28–32) 4.2	173.9[e] (150.9–193.8)	38.28 ± 0.45 (32.77–40.05) 4.71	13.72 ± 0.16 (12.66–14.77) 4.64	8.30 ± 0.07 (7.81–8.66) 3.28	10.40 ± 0.09 (9.72–10.91) 3.73
	♂	18	190 ± 2.8 (162–206) 6.4	30 ± 0.3[b] (28–32) 4.1	156.4 ± 4.8[d] (136.1–184.1) 9.73	38.31 ± 0.57 (32.83–40.82) 6.29	13.73 ± 0.12 (12.94–14.92) 3.76	8.33 ± 0.07 (7.72–8.81) 3.73	10.54 ± 0.13 (9.86–12.17) 5.12
West slope of the Cascade Range	♀	6	198 (190–205)	32 (31–33)	182.9[f] (161.2–206.7)	39.68 (37.86–40.88)	13.93 (13.11–4.46)	8.99 (8.18–9.42)	10.92 (9.84–12.29)
	♂	18	189 ± 3.2 (162–211) 7.1	31 ± 0.4 (28–34) 5.5	161.6[g] (130–183)	39.93 ± 0.50 (33.87–41.77) 5.48	13.95 ± 0.20 (11.83–15.03) 6.19	8.73 ± 0.09 (7.78–9.22) 4.53	10.49 ± 0.20 (9.55–11.86) 8.19
Eastern Jackson, southern Klamath, Lake, and Harney counties	♀	4	183 (180–186)	28 (27–29)		35.79 (31.54–37.82)	13.17 (13.01–13.31)	7.74 (7.67–7.8)	10.28 (9.67–10.56)
	♂	9	177[b] (148–190)	28[b] (26–30)		35.66 (32.09–37.93)	13.15 (12.43–14.25)	7.84 (7.59–8.23)	10.34 (8.89–10.78)

[a]Although measures of tail and ear length were recorded, the data were not included because of extreme variation believed to stem from differences in techniques used to measure these characters.
[b]Sample size reduced by 1. [c]Sample size reduced by 5. [d]Sample size reduced by 8. [e]Sample size reduced by 9. [f]Sample size reduced by 2. [g]Sample size reduced by 11.

FAMILY OCHOTONIDAE—PIKAS

The 25 Recent species of pikas all are classified in the genus *Ochotona*; two species occur in North America (one of which occurs in Oregon) and the remainder are distributed through most of Asia north of the Himalaya Mountains (Diersing, 1984; Hall, 1981). Many of the Asian species are poorly known because their habitats are relatively inaccessible (Corbet, 1978); however, some are considered pests in regions in which small grains and tree fruits are grown (Diersing, 1984).

Twenty-four extinct genera are recognized; the classification of some is questionable (Diersing, 1984). The earliest fossil record for the family was of Oligocene age in Europe; one fossil form (†*Prolagus*) may have survived to the last half of the 18th century (Kurtén, 1968). Pikas entered North America in the Pleistocene. Some authorities consider the two North American pikas to be conspecific.

All pikas are small, compact, short-eared, externally tailless mammals with small eyes, soft fur, 26 teeth, and a single fenestra on the frontals. Females have two–three pairs of mammae; males have no scrotum. Most are active diurnally, herbivorous, and coprophagous (Diersing, 1984).

Ochotona princeps (Richardson)
American Pika

1828. *Lepus* (*Lagomys*) *princeps* Richardson, 3:520.
1912. *Ochotona taylori* Grinnell, 25:129.
1919. *Ochotona fenisex brunnescens* Howell, 32:108.
1919. *Ochotona fenisex fumosa* Howell, 32:109.
1919. *Ochotona schisticeps jewetti* Howell, 32:109.
1924. *Ochotona princeps brunnescens* A. H. Howell, 47:31.
1924. *Ochotona princeps fumosa* A. H. Howell, 47:33.
1951b. *Ochotona princeps jewetti*: Hall, 5:130.
1951b. *Ochotona princeps taylori*: Hall, 5:133.

Description.—The pika (Plate V) is a rat-sized lagomorph (Table 10-1) characterized by rounded ears, no external tail, bare plantar pads, and hind feet scarcely longer than the front feet. The frontal bones lack supraorbital processes (Fig. 10-1). There are 26 teeth (Table 1-1). *O. princeps* is brownish black on the dorsum blending to ash gray on the sides and to buffy or whitish on the venter; the ears have a narrow buffy or whitish border.

Distribution.—The geographic distribution of *O. princeps* (Fig. 10-2) extends from central British Columbia south through the Cascade and Sierra Nevada ranges to Inyo Co., California, on the west and through the Rocky Mountains south to southern Utah and northern New

Table 10-1. *Extended.*

Breadth of braincase	Parietal breadth	Zygomatic breadth	Nasal breadth	Depth of skull	Length of mandible	Length of mandibular toothrow	Depth of mandibular ramus
19.93 ± 0.13 (19.24–20.46) 2.19	7.75 ± 0.23 (6.59–8.98) 9.78	21.47 ± 0.11 (20.94–22.0) 1.72	4.09 ± 0.09 (3.57–4.57) 7.55	16.35 ± 0.13 (15.69–16.87) 2.63	32.21 ± 0.24 (30.82–33.93) 2.47	8.39 ± 0.05 (8.08–8.62) 2.16	5.61 ± 0.07 (5.3–5.97) 3.87
20.13 ± 0.11 (19.15–20.95) 2.29	7.98 ± 0.22 (6.27–10.28) 11.27	21.61 ± 0.15 (20.54–22.84) 2.80	4.29 ± 0.07 (3.84–4.82) 6.86	16.39 ± 0.09 (15.66–17.1) 2.34	32.63 ± 0.25 (30.25–34.01) 3.09	8.40 ± 0.05 (8.04–8.81) 2.32	5.65 ± 0.07 (4.85–6.14) 5.37
19.79 ± 0.12 (18.94–20.58) 2.51	8.12 ± 0.31 (6.76–11.7) 15.13	21.26 ± 0.14 (20.39–22.17) 2.61	4.08 ± 0.07 (3.56–4.56) 7.11	16.35 ± 0.10 (15.54–17.41) 2.51	31.95 ± 0.39 (27.69–33.66) 4.87	8.06 ± 0.07 (7.69–8.55) 3.23	5.43 ± 0.06 (4.94–5.78) 4.56
19.93 ± 0.14 (19.26–21.28) 3.08	8.77 ± 0.45 (6.74–12.64) 21.61	21.47 ± 0.15 (20.7–22.77) 2.89	4.29 ± 0.08 (3.75–4.94) 8.42	16.45 ± 0.06 (15.79–16.87) 1.63	31.84 ± 0.48 (27.54–34.34) 6.46	8.17 ± 0.07 (7.78–8.86) 3.76	5.56 ± 0.07 (4.99–5.95) 5.18
19.99 (18.63–2.27)	7.72 (6.89–8.62)	21.80 (20.62–22.74)	4.62 (4.3–4.92)	16.81 (16.13–17.44)	32.88 (30.86–34.21)	8.67 (8.26–9.44)	5.67 (5.05–6.19)
19.66 ± 0.27 (17.37–21.26) 5.77	8.17 ± 0.15 (7.3–9.71) 7.78	21.49 ± 0.18 (20.16–22.68) 3.5	4.39 ± 0.07 (3.74–4.91) 7.19	16.41 ± 0.14 (15.05–17.12) 3.61	32.29 ± 0.49 (27.26–34.63) 6.47	8.44 ± 0.06 (7.87–8.83) 3.11	5.67 ± 0.08 (4.88–6.17) 5.82
19.84 (19.34–20.36)	8.31 (7.47–9.65)	20.54 (20.22–20.79)	3.95 (3.76–4.13)	16.03 (15.7–16.45)	29.67 (27.03–30.97)	7.77 (7.35–8.27)	5.11 (4.95–5.26)
19.45 (18.98–20.75)	8.45 (6.94–11.37)	20.43 (19.5–21.43)	3.86 (3.53–4.46)	15.91 (15.11–16.66)	29.22 (26.7–31.54)	7.68 (7.2–7.91)	5.16 (4.65–5.53)

Mexico to the east; several disjunct populations occupy isolated ranges between the major ranges (Hall, 1981). In Oregon the species is limited to suitable habitats in the Cascade Range and the Wallowa, Blue, Strawberry, Steens, Hart, and Warner mountains, and at Newberry Crater in Deschutes Co. and Grizzly and Cougar peaks in western Lake Co. (Fig. 10-2). Apparently, *O. princeps* was much more widely distributed in Oregon 7,000–11,000 years ago (Grayson, 1977) and occupied habitats no longer suitable for the species (Mead, 1987).

Mid-Pleistocene fossils of *O. princeps* are known from Maryland and West Virginia far to the east of the present range, and late-Pleistocene fossils are known from Wyoming, Colorado, Idaho, and Nevada within the present-day range of the species (Kurtén and Anderson, 1980).

Geographic Variation.—Four nominal geographic races occur in Oregon: *O. p. brunnescens* (brownish with buffy underparts) in some of the western Cascade Range, *O. p. fumosa* (less brownish and more blackish) in the central Cascade Range to Newberry Crater, *O. p. taylori* (smaller) from Lower Klamath Lake to Steens Mountain, and *O. p. jewetti* (larger and more cinnamon) in the Wallowa, Blue, and Strawberry mountains (Hall, 1981; A. H. Howell, 1924).

Habitat and Density.—The pika usually is associated with talus containing boulders 0.2–1 m in diameter and immediately adjacent to meadows containing suitable forage plants. Sometimes forested areas, including recently cut-over lands, are occupied (Gashwiler, 1959; Roest, 1953). Man-made habitats consisting of road fill, mining spoils, and piles of lumber or other refuse are sometimes occupied (Broadbooks, 1965; Roest, 1953; Smith, 1974). It commonly is considered a montane mammal and, east of the Cascade Range, it usually occurs above 1,500 m; however, west of the Cascade Range the pika occurs as low as about 40 m at Multnomah Falls, Multnomah Co. (Roest, 1953) and 138 m near Estacada, Clackamas Co. Thus, elevation per se is not the critical component of the habitat as often implied. However, temperature plays a significant role in determining suitability of habitats. Pikas seemingly cannot remain active at lower elevations in summer except during the morning and late afternoon when temperatures are lower. During warmer periods of the day they retreat to burrows beneath the boulders (Smith, 1974). Thus, low-elevation habitats at Multnomah Falls, Multnomah Co., and Estacada, Clackamas Co., probably are suitable because they are near water and on north-facing slopes where temperatures tend to be much lower.

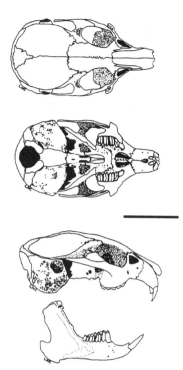

Fig. 10-1. Dorsal, ventral, and lateral views of the cranium, and lateral view of the dentary of the pika, *Ochotona princeps* (OSUFW 6122). Scale bar equals 15 mm.

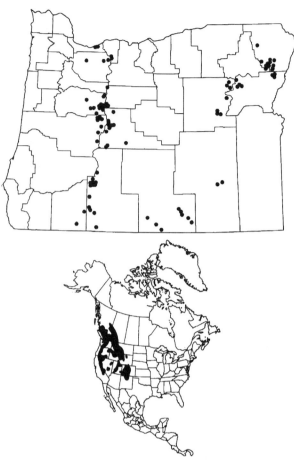

Fig. 10-2. Localities in Oregon from which museum specimens of the pika, *Ochotona princeps*, were examined (listed in the APPENDIX, p. 537), and the distribution of the pika in North America. (Range map redrawn from Hall, 1981:288, map 217, with modifications.)

Diet.—Pikas are generalist herbivores (Smith and Weston, 1990). Because they cache vegetation for winter use, habitats occupied and winter diet are, to some degree, a reflection of the stored materials. However, while grazing (on-site consumption) pikas select more grasses than forbs, but while haying they select more forbs than grasses and include more of the coarser parts of forbs in caches (Huntly et al., 1986). Species cached, of course, depend on their abundance and phenology, but pikas tend to harvest plants that have the highest nutritional value (Smith and Weston, 1990). Because pikas seemingly are not absolutely dependent on food caches (Millar and Zwickel, 1972) they must forage extensively along snow tunnels near the meadow-talus interface in winter (Huntly et al., 1986).

At a high-elevation site in Washington, caches contained ≥23 species of plants, but usually were composed mostly of three or four species common in nearby meadows (Broadbooks, 1965). The composite *Luina,* lupine (*Lupinus*), peavine (*Lathyrus*), and huckleberry (*Vaccinium*) were common plants in most caches, but reed grass (*Calamagrostis*), other grasses, aspen (*Populus*), arnica (*Arnica*), buckwheats (*Eriogonum*), yarrow (*Achillea*), phacelia (*Phacelia*), ocean-spray (*Holodiscus*), balsamroot (*Balsamorhiza*), snowberry (*Symphoricarpos*), elderberry (*Sambucus*), sedge (*Carex*), gilia (*Gilia*), sagebrush (*Artemisia*), and juniper (*Juniperus*) were represented in small amounts in a few caches (Broadbooks, 1965).

Maser et al. (1978*a*) found three pikas collected in the Cascade Range of Oregon had consumed fungi; both hypogeous and epigeous fungi were represented.

Pikas, like the leporids, are coprophagous, ingesting cecal pellets directly; unlike leporids, they sometimes store cecal pellets in food caches for future use (Smith and Weston, 1990). Seasonality in the degree of coprophagy practiced by pikas depends on their nutritional demands

(Johnson and Maxell, 1966).

Reproduction, Ontogeny, and Mortality.—Relatively little information on these topics for *O. princeps* in Oregon is available (Millar, 1973). Throughout the range of the pika the reproductive season differs considerably with latitude and elevation, but the breeding season for pikas commonly commences about a month before snowmelt so that the energy-demanding lactation period coincides with spring "green-up" (Millar, 1972); in most regions the breeding season is May–July. The gestation period is 30 days (Severaid, 1950). Litter size depends considerably on the method used to estimate it (Table 10-2), but the number born always is considerably less than the potential (Markham and Whicker, 1973; Millar, 1973, 1974). Prenatal losses are related directly to litter size (Millar, 1973). In addition, losses between birth and weaning are high; whereas 77% of females were successful in rearing their first litters to weaning, only 8% were successful in weaning their second litters (Millar, 1974). Overall, from a potential of 2.6 at ovulation, only 1.8 offspring were reared to weaning; rarely do females rear more than two young, although 41% of 80 litters in Alberta contained three young at birth (Millar, 1973).

In northeastern Oregon, "characteristics of the habitat were strongly related to the number of offspring that female pikas were able to wean in a year" (Brandt, 1989:122). Of the features investigated, number of entrances to dens accounted for 49% of the variation in number of offspring weaned /female; females occupying dens with only one entrance were most successful. Maternal age and body mass were not related to reproductive success. However, females mating with heavier males weaned more offspring/year, but were 15% more likely to succumb during the summer than those mating with lighter-weight males. Also, females nesting near vegetation that provided suitable food weaned more offspring and exhibited slightly higher survival, presumably by reducing exposure to predation while traveling between den and food source (Brandt, 1989).

In Alberta, average body mass of neonates was 10 g; the smallest individual trapped weighed 32 g (Millar and Tapper, 1973). Juveniles attained the average body mass of adults captured May–August (133 g) in slightly >80 days (Millar and Tapper, 1973).

Habits.—*Ochotona princeps* is active year-round despite much of its range covered by a meter or more of snow during winter months. During such times pikas are able to travel about in snow-free areas beneath boulders and through tunnels dug in the snow. The seasonality of various activities engaged in by pikas depends largely on elevation and climate (Smith, 1974). After the breeding season, pikas (adult males first, then adult females, then juveniles) begin to cut and store vegetation in piles commonly, but not always, beneath overhanging rocks. Contrary to earlier reports (Seton, 1929*b*) vegetation is not spread for drying then stored, but is placed directly in haypiles (Smith and Weston, 1990). Strangely, haypiles do not seem to be critical as winter survival was not affected

when they were removed (Millar and Zwickel, 1972).

Pikas are active above ground about a third of the daylight hours and spend about half of that time sitting on a prominent boulder surveying their domain. Other aboveground activities include feeding, haying, vocalizing, and territory establishment and defense (Smith and Weston, 1990). Pikas frequently vocalize; short calls are used as an alarm call when predators are sighted or in an attempt to intimidate conspecifics that trespass on occupied territories. Long calls are produced usually, but not exclusively, by males during the breeding season (Smith and Weston, 1990). Territories are centered on haypiles; territorial defense includes chasing and fighting, but such interactions are uncommon, usually occur among individuals of like sex, and most often are between unfamiliar individuals (Smith and Weston, 1990). Social hierarchies may be established with males, larger individuals, and adults dominant over females, smaller individuals, and juveniles, respectively (Kawamichi, 1976). Although a tendency for pikas of opposite sex to have adjacent home ranges has been reported (Smith and Weston, 1990), Brown et al. (1989) found patterns during 5 years that did not differ from random in Colorado. Essentially all territories are filled continuously; thus, juveniles face a low probability of finding a vacant territory, the hazards of dispersing, or parental aggression if they remain on the natal territory.

Female pikas do not select their mates randomly. In northeastern Oregon, females mated more frequently with males holding single-entrance dens. Males that held single-entrance dens during the mating period produced more

Table 10-2.—*Reproduction in Oregon lagomorphs.*

Species	Number of litters	Litter size $\bar{X}$	Range	n	Reproductive season	Basis	Authority	State or Province
Brachylagus idahoensis		5.9	4–8	14	Feb.–Apr.	Embryos	Janson, 1946	Utah
Lepus americanus	2–3	3.2	2–5	5	Apr.–Aug.	Embryos	Black, 1965	Oregon
	2.9	2.8	1-5	41	Mar.–Jul.	Embryos[a]	Adams, 1959	Montana
	2–4	3.0	1–6	157	Mar.–Aug.	Live births[b]	Severaid, 1945	Maine
Lepus californicus		3.4		245	Jan.–Jul.	Embryos	French et al., 1965	Idaho
	2.5	4.9	2–7	22		Embryos	Feldhamer, 1979*a*	Idaho
	4.3	2.4[c]		85	Jan.–Aug.[d]	Embryos	Lechleitner, 1959	California
		3.4	1–7	17		Embryos	Hall, 1946	Nevada
Lepus townsendii	3–4	4.9[b]	1–9	128	Feb.–Sep.	Embryos	James and Seabloom, 1969	North Dakota
	3–4	3.6	1–5	15	Feb.–Jul.	Embryos	Kline, 1963	Iowa
	3	4.8[e]	2–9	104	Feb.–Jul.	Embryos	Rogowitz, 1992	Wyoming
Ochotona princeps	2	2.6	2–4	175	May–Jul.	Corpora lutea	Millar, 1973, 1974	Alberta
		2.3	1–4	80		Embryos	Millar, 1973	Alberta
	2	3.2	1–6	53	May–Aug.	Embryos	Markham and Whicker, 1973	Colorado
		3.6	1–6	33		Implant sites	Markham and Whicker, 1973	Colorado
		2.9	1–6	21		Captive birth[f]	Markham and Whicker, 1973	Colorado
Sylvilagus bachmani	5–6	2.9	1–4	15	Feb.–Aug.	Embryos	Chapman and Harman, 1972	Oregon
		3.3	3–5	9	Feb.–Aug.	Implant sites	Chapman and Harman, 1972	Oregon
Sylvilagus floridanus	5–8	5.6[g]	3–8	106	Jan.–Sep.	Embryos	Trethewey and Verts, 1971	Oregon
Sylvilagus nuttallii	4–5	5.0	3–8	31	Feb.–Jul.	Corpora lutea	Powers and Verts, 1971	Oregon
		4.6[h]	1–6	31		Embryos	Powers and Verts, 1971	Oregon
	3	7.7	7–8	3	Mar.–May	Corpora lutea	Green and Flinders, 1980*a*	Idaho

[a] Counted by palpation.
[b] Animals bred in captivity.
[c] Includes an average of 0.2 embryos lost per litter.
[d] Sporadic breeding throughout the year.
[e] Average litter size varied with litter; $\bar{X}$ = 4.4 for first litter, 6.2 for second, and 4.0 for third.
[f] Animals bred in the wild and held captive until parturition.
[g] Includes an average of 0.5 resorbing embryos per litter.
[h] Includes an average of 0.3 resorbing embryos per litter.

calls/min (Brandt, 1989).

Remarks.—Low-elevation populations in Oregon, possibly the lowest anywhere, that likely are snow-free during a large portion of the winter, deserve study.

FAMILY LEPORIDAE—RABBITS AND HARES

The leporids are represented in Oregon by a pygmy rabbit of the genus *Brachylagus*, three species of rabbits (two cottontails) of the genus *Sylvilagus*, and three hares (two jackrabbits) of the genus *Lepus*. Members of the family are characterized by ears at least twice as long as wide; a visible, but short external tail; hind legs longer than the front legs; and supraorbital processes on the frontal bones. All are relatively small as none exceeds 0.75 m in total length. All have 28 teeth (Table 1-1). All are brownish or grayish with white, buff, or light gray underparts; some populations have entirely white pelages in winter. The uterus is duplex with each "horn" having a separate cervical canal; thus, transuterine migration of ova, a phenomenon known to occur (Lechleitner, 1959), must be from one ostium to the other through the body cavity.

In Oregon, leporids are classified as "predatory animals" (ORS 610.002) thus, are not regulated by hunting seasons (Oregon Department of Fish and Wildlife, 1993–1994).

Brachylagus idahoensis (Merriam)

Pygmy Rabbit

1891. *Lepus idahoensis* Merriam, 5:76.
1904. *Brachylagus idahoensis* Lyon, 1904:411.
1930. *Sylvilagus idahoensis*: Grinnell, Dixon, and Linsdale, 35:553.

Description.—The pygmy rabbit (Plate V) is the smallest leporid in Oregon (Table 10-3). The ears are short (<60 mm long), rounded, and covered with long silky hairs inside and out; the tail is small and covered with hairs possessing wide buffy bands with narrow blackish tips above and below; and the feet are short, densely furred below, and colored a light orangy-buff. The orangy-buff nape patch is more conspicuous in winter than in summer. The dorsal pelage is much lighter in autumn and winter than in spring and summer. In spring, dorsal hairs have a long basal band of dark smoky-gray, a narrow band of buffy white, and a slightly longer band of blackish brown at the tip; in autumn the dorsal hairs have a long basal band of dark smoky-gray, a wide band of buff, and a short band of blackish brown at the tip (somewhat longer on rump). The hairs on the venter have slate-gray bases and white or buffy tips, giving an overall appearance of a white venter splotched with gray.

The skull of *B. idahoensis* (Fig. 10-3) is distinctive in that the anterior projection of the supraorbital process is subequal to the posterior projection, comparatively much longer than in *Sylvilagus*. Also, the anterior face of P2 has only one reentrant angle and the margin of the reentrant angle on P3 is straight (Fig. 10-4). The auditory bullae are inflated.

Distribution.—The geographic distribution of the pygmy rabbit extends from extreme southwestern Montana south through southern Idaho and most of western Utah west through the northern half of Nevada to extreme eastern California and north to include the southeastern third of Oregon. A disjunct population occurs in southeastern Washington (Fig. 10-5; Green and Flinders, 1980a; Hall, 1981).

In Oregon, Olterman and Verts (1972) indicated that

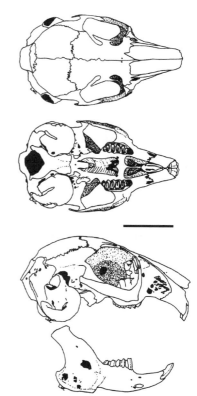

Fig. 10-3. Dorsal, ventral, and lateral views of the cranium, and lateral view of the dentary of the pygmy rabbit, *Brachylagus idahoensis* (OSUFW 6116). Scale bar equals 15 mm.

pygmy rabbits had been collected at 37 sites east and south of a line connecting Klamath Falls, Klamath Co.; Fremont, Lake Co.; Redmond, Deschutes Co.; and Baker City, Baker Co. Museum specimens that we examined were collected at 41 sites within the same region (Fig. 10-5). However, based on sightings and physical evidence, Weiss and Verts (1984) found only 51 occupied sites among 211 potentially suitable sites examined within this area. The 51 sites were circumscribed by a line connecting Lakeview and Fort Rock, Lake Co.; Millican, Deschutes Co.; Paulina, Crook Co.; Seneca, Grant Co.; Burns Junction, Malheur Co.; and Fields, Harney Co. Thus, the geographic range has shrunk considerably in historic times.

Geographic Variation.—*Brachylagus idahoensis* is monotypic (Green and Flinders, 1980a).

Habitat and Density.—*Brachylagus idahoensis* is closely tied to habitats dominated by big sagebrush (*Arte-*

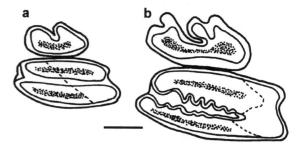

Fig. 10-4. Occlusal view of P2 and P3 of **a**. *Brachylagus idahoensis* showing only one reentrant angle on the anterior face of P2 and a reentrant angle on P3 without crenations, and **b**. *Sylvilagus nuttallii* showing three reentrant angles on the anterior face of P2 and a reentrant angle on P3 with crenate margins. Scale bar equals 1 mm.

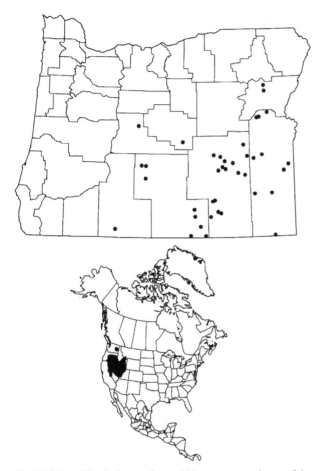

Fig. 10-5. Localities in Oregon from which museum specimens of the pygmy rabbit, *Brachylagus idahoensis*, were examined (listed in the APPENDIX, p. 512), and the distribution of the pygmy rabbit in North America. (Range map redrawn from Green and Flinders, 1980*a*:3, fig. 2, with modifications.)

misia tridentata) on deep, friable soils (Green and Flinders, 1980*b*; Orr, 1940; Weiss and Verts, 1984). In Oregon, soil and vegetation components of habitat sampled at 15 sites occupied by *B. idahoensis* indicated that mean (±*SE*) soil depth (51.0 ± 2.3 cm), soil strength in surface (0.8 ± 0.2 kg/cm²) and subsurface (3.8 ± 0.3 kg/cm²) horizons, shrub height (84.4 ± 5.8 cm), and shrub cover (28.8 ± 1.4%) were significantly greater than at 21 adjacent unoccupied sites (Weiss and Verts, 1984). Mean (±*SE*) basal area of perennial grasses (3.7 ± 0.9%), density of annual grasses (5.2 ± 2.1/1,000 cm²), density of forbs (3.4 ± 0.6/1,000 cm²), and cryptogam cover (2.4 ± 0.5%) were not significantly different between occupied and unoccupied sites. Occupancy of such areas likely was related to "availability of forage, security from predation, and ease of burrow construction" (Weiss and Verts, 1984:563). In Idaho, shrub cover and shrub height also were significantly greater on occupied sites than on unoccupied sites (Green and Flinders, 1980*b*).

Although densities of populations may fluctuate dramatically among years (Green and Flinders, 1980*a*; Weiss and Verts, 1984), evidence for multiannual cycles is lacking (Green and Flinders, 1980*a*). Janson (1946) estimated densities of 4.3 and 8.6/ha on two small areas in Utah. No estimates of density are available for *B. idahoensis* in Oregon, but we do not suspect that densities greater than those estimated in Utah are common.

Diet.—The principal food of *B. idahoensis* is big sagebrush, even where other tall shrubs such as bitter-brush (*Purshia tridentata*) are common (Green and Flinders, 1980*b*; Orr, 1940). On an annual basis, shrubs (essentially entirely big sagebrush) composed 67% of the diet, grasses 26%, and forbs only 6% (Green and Flinders, 1980*b*). In spring, summer, and autumn, grasses (*Agropyron* and *Poa*) composed 30–40% of the diet, but in winter composed only 1% (Green and Flinders, 1980*b*).

Reproduction, Ontogeny, and Mortality.—In Utah, testis size increased from December to April (Janson, 1946); in Idaho, testes were scrotal in January and contained abundant spermatozoa by March (Green and Flinders, 1980*a*). Pregnant females were collected in February–April in Utah (Janson, 1946) and March–May in Idaho (Green and Flinders, 1980*a*). In Oregon, Dice (1926) reported seeing several small young in Baker Co. in late June. Litter size averages larger in *Brachylagus* than among *Sylvilagus*, but the reported number of litters is smaller (Table 10-2). Although postpartum estrus has not been reported in the species, the abbreviated breeding season seemingly would not allow time for more than exceedingly brief anestrous periods.

No information on development of young is available.

Carnivores and raptors known to prey on *B. idahoensis* are weasels (*Mustela*), bobcats (*Lynx rufus*), coyotes (*Canis latrans*), red foxes (*Vulpes vulpes*), northern harriers (*Circus cyaneus*), and long-eared owls (*Asio otus*— Gashwiler et al., 1960; Green and Flinders, 1980*b*; Janson, 1946; Orr, 1940). Brodie and Maser (1967) listed *B. idahoensis* as composing 1.3% of 696 identified food items in regurgitated pellets of great horned owls (*Bubo virginianus*) in Deschutes Co. The relatively small areas of habitats occupiable by pygmy rabbits, the relatively large number of young produced per litter, and at least the occasional dramatic decline in abundance suggest that mortality may be high in most populations (Weiss and Verts, 1984). Janson (1946) suspected that weaned young attempting to locate suitable home ranges are particularly vulnerable to predators.

Habits.—Pygmy rabbits are unique among Oregon leporids in that they live in burrows that they construct themselves or in those modifed from burrows dug by other mammals (Green and Flinders, 1980*a*; Janson, 1946). Natural cavities are sometimes used; N. T. Weiss, a former graduate student, found a "colony" of *B. idahoensis* occupying spaces beneath a group of abandoned automobiles. Burrows usually are simple, commonly with two entrances (but as many as 10 have been recorded), and located at the base of a sagebrush (Janson, 1946). Burrows do not contain nesting material, but may contain a few sprigs of uneaten food (Orr, 1940). In addition to burrows, pygmy rabbits use forms similar to those used by *Sylvilagus* as daytime retreats (Janson, 1946). Pygmy rabbits may not use burrows as escape cover when frightened (Anthony, 1913; Janson, 1946; Orr, 1940); however, a technique for capture of pygmy rabbits with nooses in winter developed by Green and Flinders (1979*a*) was based on the animals retreating to their burrows. Seemingly, nests of *B. idahoensis* are unknown; excavated burrows in the vicinity of capture sites of lactating females contained no young (Janson, 1946).

Pygmy rabbits may be active at any time of day, but usually they feed early in the morning and late in the evening (Janson, 1946; Orr, 1940). Sometimes they feed by climb-

Table 10-3.—*Means (±SE), ranges (in parentheses), and CVs of measurements of skull and skin[a] characters for female and male* Brachylagus *and* Sylvilagus *from regions in Oregon. Skin characters were recorded from specimen tags; skull characters were measured to the nearest 0.01 mm. SE and CV not provided if* n *<10.*

Species and region	Sex	n	Total length	Hind foot length	Mass (g)	Basilar length	Length of nasals	Length of maxillary toothrow	Length of bulla
Brachylagus idahoensis	♀	22	272 ± 4.5 (220–301) 7.8	70 ± 1.3[b] (50–65) 8.4	312.1[c] (266.3–357.8)	44.85 ± 0.53 (39.48–50.24) 5.54	18.50 ± 0.29 (16.18–21.14) 7.41	9.09 ± 0.08 (7.94–9.72) 4.55	11.35 ± 0.23 (9.37–12.85) 9.59
	♂	30	275 ± 2.9 (230–300) 5.8	71 ± 0.5 (66–76) 3.9		45.62 ± 0.33 (41.52–48.79) 3.92	18.39 ± 0.23 (15.41–20.11) 6.82	9.18 ± 0.07 (8.44–9.76) 4.21	11.51 ± 0.18 (9.35–12.91) 8.69
Sylvilagus bachmani West of Cascade Range except Jackson County	♀	30	338 ± 4.0 (285–369) 6.5	83 ± 0.9 (75–94) 6.1	735.9 ± 21.3[b] (483.0–987.1) 15.86	55.67 ± 0.31 (56.23–61.37) 2.86	28.89 ± 0.23 (25.95–30.72) 4.42	13.23 ± 0.14 (11.88–14.85) 5.63	10.10 ± 0.24 (7.95–11.65) 12.92
	♂	24	323 ± 4.2 (265–355) 6.3	85 ± 0.9 (76–92) 5.3	683.8 ± 14.4 (541.5–827.2) 11.51	58.30 ± 0.25 (55.44–61.16) 2.14	28.65 ± 0.18 (26.87–30.27) 3.04	12.89 ± 0.13 (11.91–14.26) 4.21	10.33 ± 0.26 (8.40–11.88) 9.59
Jackson and Klamath counties	♀	1	331	78		59.12	27.35	13.09	9.39
Sylvilagus floridanus	♀	30	408 ± 4.5 (339–445) 6.0	93 ± 0.9 (82–104) 5.7	1266.5 ± 30.3 (899.4–1599.4) 13.09	63.55 ± 0.31 (59.35–66.66) 2.67	31.76 ± 0.18 (29.69–33.64) 3.09	14.39 ± 0.07 (13.42–15.32) 2.81	10.53 ± 0.08 (9.49–11.49) 4.36
	♂	30	400 ± 3.5 (340–430) 4.8	93 ± 0.9 (83–100) 5.1	1081.1 ± 20.4 (755.6–1319.6) 10.36	63.39 ± 0.27 (60.29–66.95) 2.30	31.51 ± 0.21 (29.37–33.52) 3.59	14.28 ± 0.09 (13.20–15.22) 3.71	10.54 ± 0.09 (9.25–11.95) 4.66
Sylvilagus nuttallii	♀	30	337 ± 4.0 (285–369) 6.5	85 ± 0.9 (75–94) 6.2	745.2 ± 24.7 (420.1–1065.8) 17.54	58.56 ± 0.33 (54.91–61.41) 3.12	28.89 ± 0.21 (26.71–30.84) 4.01	13.24 ± 0.13 (11.74–14.56) 5.54	9.73 ± 0.13 (8.53–11.17) 7.25
	♂	30	324 ± 4.6 (265–375) 7.8	85 ± 0.8 (76–92) 5.4	707.3 ± 15.1 (541.5–830.6) 10.43	57.82 ± 0.34 (54.25–61.51) 3.19	28.36 ± 0.22 (26.11–30.83) 4.19	12.83 ± 0.13 (11.75–14.26) 5.53	9.92 ± 0.15 (8.47–11.71) 8.24

[a]Although measures of tail and ear length were recorded, the data were not included because of extreme variation believed to stem from differences in techniques used to measure these characters.
[b]Sample size reduced by 2. [c]Sample size reduced by 20. [d]Sample size reduced by 1.

ing into tops of sagebrush (Janson, 1946). At other times they rest in forms beneath sagebrush or in the entrances to burrows. *B. idahoensis* travels with a "low scampering gait" at speeds of ≅24 km/h (Janson, 1946:74).

Home ranges of *B. idahoensis* tend to be small at least in winter; Janson (1946) indicated that activity was concentrated within 27 m of burrows, but Green and Flinders (1979*b*) recorded one-way movements as great as 450 m. A juvenile that escaped from a holding pen was recaptured 211 days later within 200 m of its original capture site and 2.5 km from the point at which it escaped (Green and Flinders, 1979*b*). The latter authors (p. 88) suggested that this record might be indicative of the "dispersal and pioneering capabilities of the pygmy rabbit." Nevertheless, one of these authors (J. T. Flinders) indicated that pygmy rabbits were reluctant to cross open spaces and roads (Weiss and Verts, 1984); thus, dispersal from occupied sites to unoccupied sites or to formerly occupied sites from which the species was extirpated might not be as easy as implied by the 2.5-km movement record.

Remarks.—The historical path leading to the name *Brachylagus idahoensis* for the brush rabbit is tortuous (Green and Flinders, 1980*a*). Originally named *Lepus*

idahoensis by Merriam, the species was assigned to its own monotypic genus (*Brachylagus*) 12 years later where it remained for 26 years. Then, *Brachylagus* became a subgenus within the genus *Sylvilagus* and remained so classified for 33 years whereupon it was returned to the generic rank where it remains (Green and Flinders, 1980*a*). Some workers did not agree with the elevation of the subgenus *Brachylagus* to generic level (Hall, 1981).

Because of its rigid habitat requirements and because of removal of sagebrush in connection with range improvement and other agricultural practices, Olterman and Verts (1972) expressed some concern regarding the retention of the pygmy rabbit as a part of the mammal fauna of Oregon. In addition to loss of habitat, Weiss and Verts (1984) were concerned about fragmentation of habitat and loss of avenues of dispersal that might prevent reestablishment of populations extirpated during dramatic declines in abundance such as occurred in 1983. These concerns, plus the shrinking of the range in historic times in Oregon, suggest to us that the setting aside and protection of some of the occupied habitats would be a wise course of action for management agencies to consider. The wildlife management agency in Washington has proceeded in this endeavor (McAllister, 1995).

Table 10-3. *Extended.*

Breadth of braincase	Parietal breadth	Zygomatic breadth	Nasal breadth	Depth of skull	Length of mandible	Length of mandibular toothrow	Depth of mandibular ramus
21.46 ± 0.23 (18.58–23.59) 5.13	16.42 ± 0.18 (15.21–18.02) 5.20	26.93 ± 0.23 (24.8–28.95) 3.95	8.88 ± 0.11 (7.87–9.96) 5.79	23.95 ± 0.19 (22.68–26.74) 3.63	36.38 ± 0.45[d] (31.48–40.39) 5.64	9.08 ± 0.10 (7.93–9.82) 5.19	7.24 ± 0.09 (6.04–8.15) 6.13
21.27 ± 0.21 (19.05–23.40) 5.43	17.09 ± 0.17 (15.49–18.84) 5.51	27.28 ± 0.14 (25.41–28.49) 2.75	9.21 ± 0.11 (7.82–10.32) 6.30	24.25 ± 0.14 (23.07–26.74) 3.06	36.89 ± 0.33 (33.22–39.77) 4.89	9.01 ± 0.10 (7.38–10.12) 6.25	7.14 ± 0.09 (6.45–8.23) 6.03
23.06 ± 0.22 (21.34–25.09) 5.12	18.24 ± 0.31 (15.54–21.38) 9.29	32.07 ± 0.69 (12.13–33.87) 11.88	13.01 ± 0.13 (11.25–14.56) 5.53	28.74 ± 0.19 (26.99–30.67) 3.67	51.08 ± 0.29 (48.69–54.56) 3.17	13.23 ± 0.09 (12.28–14.32) 4.13	9.89 ± 0.1 (8.91–11.16) 6.38
22.75 ± 0.23 (21.10–24.93) 4.91	18.99 ± 0.36 (15.08–22.45) 9.30	32.67 ± 0.10 (31.35–33.34) 2.75	13.36 ± 0.18 (11.67–14.9) 6.50	29.17 ± 0.22 (27.34–30.90) 3.74	50.44 ± 0.27 (48.05–53.58) 2.61	12.93 ± 0.12 (11.87–14.32) 4.54	9.74 ± 0.09 (9.16–10.98) 4.41
23.31	16.91	31.87	13.40	27.98	47.68	13.09	9.33
24.35 ± 0.11 (22.75–25.70) 2.55	20.80 ± 0.17 (18.89–23.01) 4.55	5.64 ± 0.19 (33.55–37.49) 2.88	14.41 ± 0.14 (12.64–15.83) 5.47	31.04 ± 0.14 (29.81–32.46) 2.47	56.76 ± 0.31 (53.55–59.16) 3.02	14.56 ± 0.09 (12.92–15.48) 3.32	11.00 ± 0.72) (9.87–11.72) 4.21
24.24 ± 0.13 (23.35–25.82) 2.96	21.00 ± 0.17 (19.38–23.11) 4.54	35.62 ± 0.17 (34.04–38.55) 2.61	14.80 ± 0.15 (13.01–16.74) 5.49	31.39 ± 0.13 (29.97–32.72) 2.35	56.15 ± 0.35 (53.20–60.10) 3.45	14.49 ± 0.09 (13.33–15.43) 3.31	11.15 ± 0.43) (10.07–12.43) 4.95
23.15 ± 0.13 (21.47–24.46) 3.19	18.27 ± 0.27 (15.15–21.25) 7.98	32.81 ± 0.16 (31.66–36.16) 2.67	13.01 ± 0.13 (11.25–14.13) 5.67	28.79 ± 0.19 (26.87–30.48) 3.78	50.85 ± 0.32 (47.85–53.83) 3.42	13.15 ± 0.11 (12.0–14.16) 4.67	9.81 ± 0.25) (8.89–11.25) 5.92
23.09 ± 0.14 (21.74–24.26) 3.39	18.88 ± 0.24 (15.74–21.1) 7.04	32.64 ± 0.10 (31.13–33.85) 1.69	13.25 ± 0.17 (11.45–14.78) 6.94	28.93 ± 0.19 (26.65–30.47) 3.56	49.74 ± 0.32 (45.59–53.83) 3.73	12.81 ± 0.11 (11.86–14.14) 4.77	9.75 ± 0.86) (9.02–10.86) 3.90

Sylvilagus bachmani (Waterhouse)
Brush Rabbit

1839. *Lepus bachmani* Waterhouse, 1838:103.
1899. *Lepus bachmani ubericolor* Miller, p. 383.
1904. *Sylvilagus (Microlagus) bachmani*: Lyon, 45:336.
1904. *Sylvilagus (Microlagus) bachmani ubericolor*: Lyon, 45:337.
1935. *Sylvilagus bachmani tehamae* Orr, 48:27.

Description.—The brush rabbit (Plate V) is the small-est member of the genus in Oregon (Table 10-3). The general color of the dorsal pelage is a dark, somewhat mottled, orangy buff produced by hairs banded steel gray, black, orangy buff, and black from base to tip. Hairs on the rump and pate have longer dark tips, producing a darker pelage in those areas. The light-colored band on hairs on the sides lacks the orangy tone of those on the dorsum. Hairs on the venter are gray with whitish or buffy tips, producing a mottled appearance. The nape is orangy—the hairs have no black band or tip. The feet are creamy with scattered black-tipped hairs. The ears are dark grayish, at most, barely darker at the tip; the hairs are slightly tipped with orangy buff. The muzzle and sides of the face are grayish, a color produced by blackish hairs with whitish tips. The tail in comparison with that of the cottontails is much abbrevi-ated; on the dorsal surface the hairs are dark gray with orangy-buff tips and on the ventral surface the hairs are lighter gray with short white tips. Fleeing individuals do not show a "cottony" tail as do congeners. Overall, the color of *S. bachmani* is much darker and the hairs are much finer than in *S. floridanus*, but the pelage tends to be sleek (but dull, not shiny) as in *S. floridanus*, not fluffy as in *S. nuttallii*.

The skull in *S. bachmani* is distinctive in that the posterior extension of the supraorbital process never touches the squamosal (Fig. 10-6). The supraoccipital shield is squarish (Fig. 10-7). As in *Brachylagus* and all *Sylvilagus*, the interparietal bone is separated from the parietal and supraoccipital bones by distinct sutures. As in all *Sylvilagus*, but not *Brachylagus*, the anterior projections of the supraorbital processes are less than half the length of the posterior projections, P2 has three reentrant angles on its anterior face, and the margins of the reentrant angle on P3 (but not on P4, M1, M2, or M3) are crenate (Fig. 10-4). Females are not significantly larger than males (Chapman, 1971a).

Distribution.—The geographic range of *S. bachmani* (Fig. 10-8) extends from the Columbia River southward to the tip of Baja California west of the Cascade and Sierra Nevada ranges and desert areas of eastern California

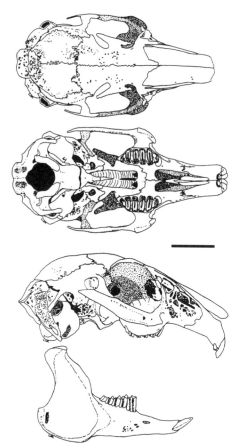

Fig. 10-6. Dorsal, ventral, and lateral views of the cranium, and lateral view of the dentary of the brush rabbit, *Sylvilagus bachmani* (OSUFW 2668). Scale bar equals 15 mm.

(Chapman, 1974; Hall, 1981). The species does not occur in the Central Valley, California.

In Oregon, *S. bachmani* occurs in the Willamette Valley and other interior valleys, in coastal areas, and in valleys along coastal streams from the Columbia River south, and from the foothills of the Cascade Range west (Fig. 10-8).

Late Pleistocene fossils are known from California (Kurtén and Anderson, 1980).

Geographic Variation.—Two nominal geographic races occur in Oregon: *S. b. ubericolor* and *S. b. tehamae*. *S. b. ubericolor* is darker colored and has shorter ears, longer hind

feet, and smaller auditory bullae than *S. b. tehamae*; it occupies most of the range of the species in Oregon. *S. b. tehamae* occurs mostly in California, but its range extends into southern Oregon northward to Prospect, Jackson Co. (Orr, 1935).

Habitat and Density.—Throughout its range, the brush rabbit is a denizen of clumps of extremely dense brush interspersed among grassy areas (Chapman, 1971*b*; Connell, 1954; Orr, 1940; Shields, 1960). In the Willamette Valley, disjunct clumps of brambles (especially *Rubus discolor*, *R. laciniatus*, and *Rosa eglanteria*) ≥460 m² scattered among forbs and grasses (e.g., *Daucus carota*, *Dipsacus sylvestris*, *Poa*, *Festuca*, and *Phalaris arundinacea*) are the principal habitats occupied (Chapman, 1971*b*). Hinschberger (1978) recorded *S. bachmani* as occurring in communities dominated by beachgrass (*Ammophila arenaria*), sitka spruce (*Picea sitchensis*), black cottonwood (*Populus trichocarpa*), or willow (*Salix*) along the Columbia River. In coastal regions, we observed brush rabbits in bush lupine (*Lupinus arboreus*).

No estimates of density are available for Oregon populations of *S. bachmani*, but in California, Orr (1940) counted 26 inhabiting an area of about 0.6 ha. For an introduced population on Año Nuevo Island, California, Zoloth (1969) estimated densities as great as 73/ha on a 1.6-ha study area and 125/ha on a 0.8-ha study area. From casual observations we believe that Oregon population densities are considerably less.

Diet.—The food habits of *S. bachmani* have not been studied in Oregon. Orr (1940) indicated that in California grasses contributed most to the diet, but that other species of plants were consumed, particularly in autumn and winter. In California, *S. bachmani* is known to eat congeners of several plants that occur in Oregon. Among the grasses are *Eragrostris*, *Bromus*, and *Avena*; among the forbs are *Chenopodium*, *Cirsium*, *Sonchus*, *Juncus*, *Conium*, *Dipsacus*, *Potentilla*, *Trifolium*, and *Hydrocotyle*; and among the shrubs and brambles are *Rosa*, *Rubus*, *Ceanothus*, *Rhamnus*, and *Symphoricarpos*. In Oregon, Chapman (1974) observed brush rabbits feeding on blackberries (*Rubus*) and we saw them with purplish-black stains around their mouths, presumably from ingesting blackberries.

Reproduction, Ontogeny, and Mortality.—Male *S. bachmani* attain reproductive competence in January, possibly as early as December (Chapman and Harman, 1972). First conceptions occur in February; from April to July nearly all adult females are pregnant, lactating, or both preg-

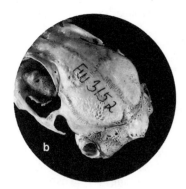
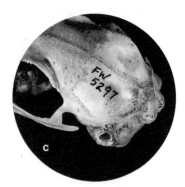

Fig. 10-7. Photographs of the left ventral portions of skulls of the **a**. brush rabbit, *Sylvilagus bachmani*; **b**. Nuttall's cottontail, *Sylvilagus nuttallii*; and **c**. eastern cottontail, *Sylvilagus floridanus*. Note shapes of supraoccipital shields, lengths of posterior projections of supraorbital processes, and diameters of external auditory meatuses.

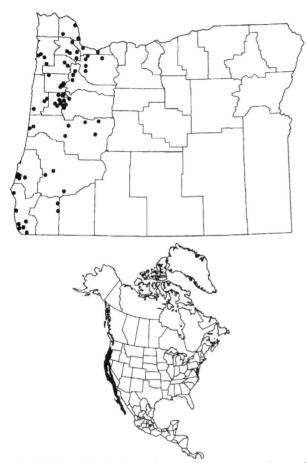

Fig. 10-8. Localities in Oregon from which museum specimens of the brush rabbit, *Sylvilagus bachmani*, were examined (listed in the APPENDIX, p. 564), and the distribution of the brush rabbit in North America. (Range map redrawn from Hall, 1981:298, map 221, with modifications.)

nant and lactating. The last parturitions are about mid-August (Chapman and Harman, 1972; Verts, 1967*a*). Both onset and cessation of breeding are later in Oregon than in California (Mossman, 1955). Whether or not breeding is synchronous as in Oregon congeners is unknown.

Mean litter size based on counts of embryos was somewhat smaller than that based on counts of pigmented sites of implantation (Table 10-2), a difference probably accounted for by 16.3% of the ova failing to implant and 15.5% of those implanting being resorbed during pregnancy (Chapman and Harman, 1972). Mean litter size based on counts of embryos was smaller in Oregon (Table 10-2) than reported for the species in California (Mossman, 1955; Orr, 1940). Mean litter sizes for first and subsequent pregnancies were not significantly different (Chapman and Harman, 1972).

Based on a gestation period of 27 days (Chapman and Harman, 1972; Mossman, 1955), a breeding season of 170 days, a pregnancy rate of 81.1%, and a mean litter size of 2.87, Chapman and Harman (1972) estimated that the average female brush rabbit produced about 15 offspring/year. This is considerably fewer than produced by Oregon congeners (Powers and Verts, 1971; Trethewey and Verts, 1971). Breeding by juveniles in the year of their birth is not known to occur.

Five nestlings estimated to be 3 days old averaged 25.5 g (range, 23.8–27.3 g), the dorsum appeared black from tips of incoming hairs, the venter was pink, the ears lay flat to the head, and the feet were naked (Orr, 1942). The young were able to crawl feebly. The eyes opened on the 10th and 11th days and the young were able to hop. The average body mass of the three individuals remaining in the nest on the 13th day was 75.8 g (Orr, 1942). By the 14th day all had abandoned the nest (Orr, 1942).

Brush rabbits are known to be preyed upon by coyotes (*Canis latrans*), gray foxes (*Urocyon cinereoargenteus*), bobcats (*Lynx rufus*), long-tailed weasels (*Mustela frenata*), red-tailed hawks (*Buteo jamaicensis*), Cooper's hawks (*Accipiter cooperi*), barn owls (*Tyto alba*), great horned owls (*Bubo virginianus*), California jays (*Aphelocoma coerulescens*), rattlesnakes (*Crotalus virdis*), and gopher snakes (*Pituophis catenifer*) in California (Orr, 1940). No information on longevity or on age structure of populations of brush rabbits is available. However, based on their relatively lower productivity and their secretive nature, we suspect that brush rabbits, upon attaining adulthood, are subject to lower rates of mortality than their congeners in Oregon.

Habits.—Brush rabbits rarely venture more than a few meters from the extremely dense brush that constitutes the requisite component of their habitat. Runways, resembling those constructed by *Microtus* except for size (they are 11–12 cm wide), are interlaced through brushy clumps and may be discernible 2–3 m into surrounding grassy areas. These trails are kept clear of vegetation and permit brush rabbits easy and quick access to and from foraging areas. Brush rabbits construct forms along runways within the edges of brushy clumps from which they view surrounding areas before emerging to feed (Orr, 1940). Upon emerging from brushy cover, brush rabbits are extremely cautious; strange sights, sounds, or scents may be cause for them to freeze momentarily then run for cover or to bolt directly down the nearest runway leading to brushy cover. Brush rabbits will not leave their brushy sanctuaries even when pursued by dogs; possibly they climb upward among the brambles and allow the dogs to course below them (Chapman, 1974). If the observer remains quiet after startling a brush rabbit into a clump of brambles it commonly will make thumping sounds (Orr, 1940). On Año Nuevo Island, California, where predators are absent, animals in a population established from stock introduced 18 years earlier behaved much less cautiously (Zoloth, 1969).

Brush rabbits are considered crepuscular, but activity at other times has been recorded (Orr, 1940; Zoloth, 1969). During midmorning and late afternoon, brush rabbits spend considerable time basking, especially after rainy periods (Orr, 1940); Chapman (1974) indicated that in captivity individuals climbed into the branches of small trees to rest.

Brush rabbits seem behaviorally subordinate to eastern cottontails and in large outdoor pens were sometimes killed by them (Chapman and Verts, 1969). Brush rabbits are known to hybridize with eastern cottontails (Verts and Carraway, 1980).

Home ranges of brush rabbits are small (Chapman, 1971*b*; Connell, 1954; Orr, 1940; Shields, 1960) and tend to "conform to the size and shape of the brushy habitat available" (Chapman, 1971*b*:689). In the Willamette Valley, the mean standard diameter (root mean square of radio positions from the calculated center of activity—Harrison, 1958) was 37.3 ± 14.0 m (range, 19.4–72.3 m) for 12 adult males and 33.1 ± 16.9 m (range, 9.3–71.2 m) for 24 adult females, a significant difference. The mean standard diameter was significantly larger for juvenile males and slightly smaller for juvenile females than for adults of the same sex (Chapman, 1971*b*). Of 24 animals displaced ≥165 m from their centers of activity, 15 successfully returned to their home ranges, but none of five displaced 195–352 m returned to its home range. No rabbit displaced >7.9 standard diameters from its center of activity homed successfully (Chapman, 1971*b*). Homing ability was not related to sex, age, or compass direction displaced, but 15 of 16 individuals that homed did so on clear nights; some individuals displaced on a cloudy nights waited for clear nights to return to their home ranges.

Remarks.—Brush rabbits are accused of depredations on tree seedlings planted for forest regeneration. Most of the damage is in close proximity to dense brush, thus probably can be reduced by planting seedlings ≥6 m from such features (Canutt, 1969; Mitchell, 1950). In view of information available on diet, the close association of brush rabbits with dense brush and brambles, and forest-management practices designed to reduce brush and slash we question whether the depredations are extensive.

The fecundity of *S. bachmani* × *S. floridanus* hybrids and the extent of introgression of genetic material into brush rabbit populations from eastern cottontails (and vice versa) need to be evaluated. Hybridization of the two species is thought to be rare (Verts and Carraway, 1980), but F$_1$s likely are not often distinguished from eastern cottontails. Whether F$_1$s are cross fertile with individuals of both, one, or neither of the parental species would be useful information on which to base management decisions. However, we consider the statement that through competition and hybridization the eastern cottontail is "well on the way to eliminating the brush rabbit from the valley floor and lower foothills" (*The Chat*, 27[3]:21, November 1997) to be totally without scientific basis.

Sylvilagus floridanus (J. A. Allen)
Eastern Cottontail

1890*c. Lepus sylvaticus floridanus* Allen, 3:160.
1894. *Sylvilagus sylvaticus mearnsii* J. A. Allen, 6:171.
1904. *Sylvilagus (Sylvilagus) floridanus mearnsi:* Lyon, 45:336.

Description.—*Sylvilagus floridanus* (Plate V) is the largest member of the genus in Oregon (Table 10-3). Overall the dorsal pelage is brownish, becoming darker (almost black) on the rump and lighter buffy brown on the flanks; the nape patch is orangy brown without black hairs. The head is the same color as the dorsum. The hairs on the dorsum have steel gray bases followed by bands of brownish black, buff, and black. Occasional specimens have pelages in which dorsal hairs lack brownish-black and black bands. The venter hairs are white with gray bases, giving a splotched appearance when the hairs are spread. The tail hairs are white to the base. The feet are whitish splotched with light tan and are >80 mm but <105 mm long. The pelage is much coarser and tends to be sleeker than that of *S. nuttallii*.

Skull characteristics useful for separating *S. floridanus* from congeners are supraorbital processes with a posterior projection that does not taper greatly and that overlaps the squamosal at the point at which they abut; an external auditory meatus decidedly smaller than the partial circle formed by the zygomatic process of the squamosal and the

posterior extension of the jugal (Fig. 10-7); and a squarish supraoccipital shield, often with a median notch (Fig. 10-7). As in all congeners and in *Brachylagus*, the interparietal bone is separated from the parietal and superoccipital bones by distinct sutures. As in all congeners, but not *Brachylagus*, the anterior projections of the supraorbital processes are less than half the length of the posterior projections (Fig. 10-9), P2 has three reentrant angles on its anterior face, and P3, P4, M1, and M2 have a reentrant angle with crenate margins (Fig. 10-4).

Distribution.—The eastern cottontail ranges throughout most of eastern United States except New England west to about the 105th parallel, then west into New Mexico and Arizona and south through much of Central America and into extreme northern South America (Fig. 10-10; Chapman et al., 1980). The species has been introduced into Oregon, Washington, and British Columbia (Chapman et al., 1980; Cowan and Guiguet, 1965).

Sylvilagus floridanus was introduced into Benton Co. in 1937 and into Linn Co. in 1941 from Ohio and Illinois, respectively (Fig. 10-10; Graf, 1955). From these sites, eastern cottontails have spread at least through the mid-Willamette Valley (Verts and Carraway, 1981; Verts et al., 1972). The source of animals in Portland, Multnomah Co., and environs is unknown, but may be from Missouri stock introduced near Battle Ground, Washington, in 1933

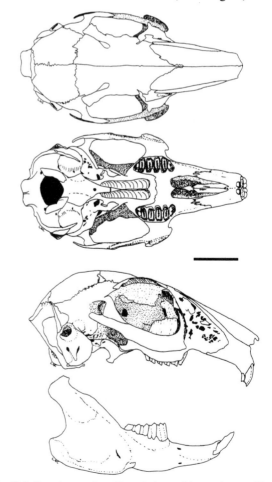

Fig. 10-9. Dorsal, ventral, and lateral views of the cranium, and lateral view of the dentary of the eastern cottontail, *Sylvilagus floridanus* (OSUFW 5353). Scale bar equals 15 mm.

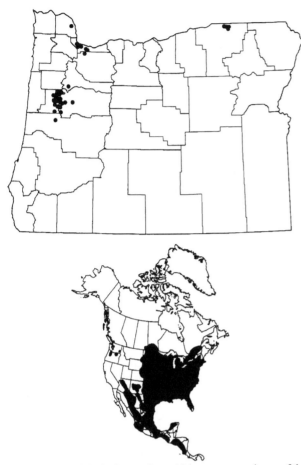

Fig. 10-10. Localities in Oregon from which museum specimens of the eastern cottontail, *Sylvilagus floridanus*, were examined (listed in the APPENDIX, p. 564), and the distribution of the eastern cottontail in North America. (Range map redrawn from Hall, 1981:303, map 223, with modifications.)

(Dalquest, 1941*a*). Whether or not the entire Willamette Valley is occupied is unknown. From casual observations and from data collected during two surveys (Verts and Carraway, 1981; Verts et al., 1972) we believe that the range of eastern cottontails does not extend into the foothills of either the Cascade or Coast ranges. *S. floridanus* also occurs in Umatilla Co. and possibly elsewhere in northeastern Oregon (Fig. 10-10); we suspect that these animals are from stock introduced into Washington in 1926–1927 from Kansas (Dalquest, 1941*a*). A former graduate student, Mark Rhodes, indicated that he had observed eastern cottontails along the central Oregon coast, but we neither observed them nor examined specimens from the coastal region.

Geographic Variation.—Nagorsen (1990) referenced an unpublished dissertation that indicated that *S. floridanus* in Oregon, Washington, and mainland British Columbia belonged to the race *S. f. alacer*. However, animals introduced into Benton and Linn counties came from regions in which eastern cottontails are considered referable to *S. f. mearnsii* (Hall, 1981). Graf (1955) indicated that cottontails introduced into Linn Co. probably were *S. f. mearnsii*. Those introduced into Washington from Missouri were from a region in which the cottontails are considered referable to *S. f. alacer*, whereas those from Kansas might be referable to *S. f. alacer*, *S. f. mearnsii*, *S. f. similis*, or *S. f. llanensis* depending upon the locality from which the stock was obtained (Hall, 1981).

Considering that in their native range cottontails frequently were transplanted within and among states in the 1920s and 1930s, races may have been mixed before the species was introduced into western North America.

Habitat and Density.—In the Willamette Valley, coverts occupied by *S. floridanus* commonly contain large clumps of blackberries (*Rubus discolor* and *R. laciniatus*) interlaced around small Oregon white oak (*Quercus garryana*), Oregon ash (*Fraxinus latifolia*), and black cottonwood (*Populus trichocarpa*) trees and interspersed among grasses and forbs (Verts and Carraway, 1981). Fencerows, railroad rights-of-way, and riparian zones containing these plants are thought to serve as avenues of dispersal for eastern cottontails (Verts and Carraway, 1981). The large fields of grass (mostly *Lolium perenne* and *L. multiflorum*) and small grains are not occupied by eastern cottontails, except perhaps for a few meters at their borders. Near Umapine, Umatilla Co., *S. floridanus* occupies brushy fencerows, riparian zones, and low sagebrush (*Artemisia*)-covered ridges adjacent to alfalfa fields; at night in winter, eastern cottontails venture 50–75 m or more into the fields to feed. One of us (BJV) and a crew of students occasionally netted cottontails in alfalfa fields while night-lighting pheasants (*Phasianus colchicus*).

No direct measures of density are available, but some indirect measures provide an indication of the density. On several occasions, eastern cottontails were collected on the E. E. Wilson Wildlife Area in Benton Co., probably as near prime cottontail habitat as is available in the Willamette Valley. In 18 months in 1968–1969, 486 *S. floridanus* were collected on the 708-ha area without a perceptible increase in effort required to collect animals (Trethewey and Verts, 1971). In 1973, 78 were collected on the same area in a 5–week period (Sullins et al., 1976); usually eight–10 individuals were collected in a 1–2-h period by spotlighting and shooting with a shotgun. In 1979, 60 were collected in a 30-day period on the same area by use of the same techniques; commonly only 45 min–1 h was required to collect 10 individuals. In 30 days in 1991, 32 more were collected, but considerable more effort was required; 2–3 h were required to collect two–11 individuals.

Diet.—Like all lagomorphs in Oregon, *S. floridanus* is an obligate herbivore. Although the diet of *S. floridanus* has not been studied in Oregon, extensive studies of the diet and food selection conducted elsewhere provide some indication of what likely is eaten in Oregon. From direct observations of feeding cottontails, Dalke (1942) indicated that grasses and grasslike plants were eaten more frequently except during winter months when shrubs, vines, and trees were foraged upon extensively; forbs were rarely consumed in winter and somewhat less than grasses during the entire year. Klimstra and Corder (1957) also indicated that perennial grasses were used extensively as food by cottontails in all seasons. However, Hendrickson (1938) indicated that grasses were little fed upon in winter. Bailey (1969*a*), in a laboratory study, found that juvenile cottontails consistently gained weight when fed alfalfa and some wild forbs, but usually lost weight and commonly did not survive the 10-day experimental period when fed only grasses. In feeding trials, Bailey and Siglin (1966) found that dandelion (*Taraxacum officinale*) and prickly lettuce (*Lactuca serriola*) were selected over eight other species of forbs and one species of grass by juvenile cottontails; both species of forbs occur in the Willamette Valley.

Reproduction, Ontogeny, and Mortality.—In western Oregon, male eastern cottontails are in breeding condition from December to August, and adult females commence to breed in January or February depending on the severity of climatic conditions (Trethewey and Verts, 1971). Onset of the breeding season is earlier in western Oregon than in most areas of the native range. Breeding tends to be synchronous and, because of postpartum estrus among many females, synchrony tends to be maintained during more than half of the breeding season. Breeding synchrony may be improved by warm periods following inclement weather at the onset of the breeding season (Trethewey and Verts, 1971).

Gestation in the eastern cottontail usually is considered to be 28 days, although periods of 25–27 days were recorded (Rongstad, 1969). Near-term fetuses and neonates usually weigh 25–30 g and have hind feet 19–20 mm long (Beule and Studholme, 1942; Rongstad, 1969). Fetuses in smaller litters may weigh slightly more and those in larger litters slightly less, and within litters, males usually average slightly heavier than females (Verts et al., 1997). At birth, the skin is pink; the eyes and ears are sealed; the incisors are barely erupted; and the molariform teeth are not visible. The dorsum darkens and the entire body surface becomes sparsely covered with light-colored guard hairs within 24 h (Beule and Studholme, 1942). The eyes of the young open in about 1 week and the young commonly leave the nest before the end of the 2nd week of life (Beule and Studholme, 1942).

R. Hoyer, a local falconer, brought us two exceptionally large and well-developed embryos that constituted the potential litter of a female his hawk had caught. They weighed 42.6 and 48.3 g, had hind feet 23 mm long, had erupted upper and lower incisors, and had molariform teeth easily visible through the thin gums; the skin of the dorsum was nearly black and was well covered with 1.5–2.5-mm-long hairs (dark on the dorsum, light on the venter). Development had progressed at least 2–3 days beyond that of near-term fetuses and neonates nestlings described by Beule and Studholme (1942). We are unsure of the cause of the apparent extended development, but suspect that it may be related to the small size of the litter.

Essentially all (132 of 135; 97.8%) adult females collected during the breeding season are pregnant or lactating or both pregnant and lactating. Females, lactating but not pregnant, may have ceased to breed before the end of the breeding season or failed to breed during one of the synchronous breeding periods (Trethewey and Verts, 1971). In addition to adults, some juvenile females born in the first and second litter cohorts produced one or more litters ($\overline{X}$ = 3.4 young, n = 5) of their own during the breeding season in which they were born (Trethewey and Verts, 1971). Juvenile females both pregnant and lactating produce a minimum of two litters, but a method to detect more than two litters in juveniles is lacking.

Although we present average litter size for the entire breeding season (Table 10-2), we consider it a rather meaningless statistic because average sizes of each of the sequential litter cohorts may differ by as much as 2.5 potential offspring (Trethewey and Verts, 1971). Thus, unless all samples of adult females pregnant in all litter cohorts are identical, the overall average will be biased by sample

size. First litters produced during the breeding season were the smallest ($\overline{X}$ = 3.8, n = 4, 1968; $\overline{X}$ = 4.35, n = 12, 1969) and second litters were the largest ($\overline{X}$ = 6.3, n = 16, 1968; $\overline{X}$ = 6.3, n = 9, 1969); most of the remaining litters averaged <0.5 potential offspring less than the largest average litter (Trethewey and Verts, 1971). Intrauterine losses averaged nearly 0.5 potential offspring per litter, nearly twice the rate of loss reported for the species in its native range (Trethewey and Verts, 1971).

Potentially, productivity of eastern cottontail populations in western Oregon is enormous, so much so that the adage "to breed like rabbits" in reference to the species is most appropriate. An adult female that bred at the onset of the breeding season and at each postpartum estrus would produce about 39 young if it produced the average number of offspring for each litter (Trethewey and Verts, 1971). Even those that ceased to breed in late June or July might produce 24–35 offspring. And, to this, must be added production by juvenile females in the first (and possibly the second) litter that survived to reproduce in the breeding season in which they were born.

With such potential for increase, cottontail populations would explode were it not for the equally enormous mortality. Rain that drowns nestlings, low temperatures that chill neonates, abandonment of young, predators, diseases, accidents, and a myriad of other factors take a continuous toll of cottontails. In a sample of 42 adults (closed epiphyseal cartilages) collected in August–September 1973, 23 (54.8%) were in their 2nd calendar year of life, 16 (38.1%) in their 3rd year, and only 3 (7.1%) in their 4th year (Sullins et al., 1976). Thus, even after reaching adulthood, life for the eastern cottontail is so precarious that few survive to attain only a third of the ≅10-year maximum longevity (Lord, 1961a).

Habits.—When inactive, eastern cottontails spend most of their time "hunched in forms roofed over by adjacent vegetation" (Marsden and Holler, 1964:9). During the inactive period, eastern cottontails reingest cecal pellets for 3–6.5 h, twice as long in winter as in summer. The soft, elongate, and mucus-covered pellets are taken into the mouth directly from the anus and swallowed without mastication (Heisinger, 1962).

Lord (1961b) believed that cottontails in Illinois became active in late afternoon or early evening about the same time and ceased activity in the morning about the same time, irrespective of the season. Mech et al. (1966) found that onset and cessation of activity in a small sample of eastern cottontails followed by radiotelemetry in Wisconsin were related to sunset and sunrise, respectively. However, Holler and Marsden (1970) found that both authorities were correct because early in the year, when Mech et al. (1966) conducted their study, onset of activity is related to sunset, but later in spring and summer, when Lord (1961b) conducted his study, onset activity is unrelated to sunset. In the native range, rainfall and snow depth had positive effects on the amount of roadside activity, whereas moonlight, changes in temperature, and wind velocity had negative effects (Fafarman and Whyte, 1979; Kline, 1965).

Eastern cottontails tend to be rather sedentary. In the Willamette Valley, mean distance moved between points of capture on an 11.7-ha study area was significantly greater for juvenile males ($\overline{X}$ = 54.0 m; range, 4.3–222.0 m; n = 25 captures) than for adult males ($\overline{X}$ = 44.4 m; range, 7.9–

86.0 m; $n = 11$ captures), juvenile females ($\overline{X} = 42.4$ m; range, 0–106.1 m; $n = 33$ captures), and adult females ($\overline{X} = 39.8$ m; range, 13.4–164.6 m; $n = 34$ captures). Only one of 67 marked individuals was known to have left the area (Chapman and Trethewey, 1972a).

Male and female eastern cottontails are equally susceptible to capture in live traps at first capture, but females are about twice as susceptible as males to recapture (Chapman and Trethewey, 1972b). Juveniles are easier to capture than adults. Susceptibility to capture is correlated positively with barometric pressure and negatively with temperature. Cottontails are captured with greatest ease in winter and with greatest difficulty in summer (Chapman and Trethewey, 1972b).

Nonsocial behavior includes exploration (such as searching for receptive females), dusting and loafing (doglike rolling in dust, presumably to remove parasites), feeding and drinking, grooming (face washing, licking, scratching, and biting), and displacement activities (activities in social situations that seem out of context). Occasionally, rabbits involved in feeding or grooming sit upright and wave their paws up and down rapidly; the significance of this behavior is unknown (Marsden and Holler, 1964).

As for S. nuttallii, most observed social behavior is involved with reproduction (Marsden and Holler, 1964). Three basic postures exhibited by cottontails in social situations are recognized: male alert, female threat, and submissive postures. In the complete repertoire of reproductive interactions the dominant male approaches an estrous female and assumes an alert posture; the female, in turn, assumes a threat posture, resulting in a "face-off" that is a prelude to further interactions (Marsden and Holler, 1964). If the male continues to approach, the female may respond by boxing with the forepaws or occasionally by charging at the male; if the male retreats, the female may give chase. However, no physical contact is involved. The male responds by dashing past the female, often emitting a jet of urine in the direction of the female; the female responds by shaking and grooming, presumably to remove the urine. The male may continue to dash by the female and emit urine as many as 20 times. Sometimes, the female jumps into the air and emits a jet of urine at the male one or more times. After two or three jumps the male commonly loses interest in the female; thus, the jump sequence is thought to be a form of appeasement for a highly sexually aroused male (Marsden and Holler, 1964). As the female moves away, the male follows closely behind, making periodic attempts to mount. When the female adopts a submissive posture and presents, the male quickly mounts and thrusts rapidly, one of the pair squeals (it is unknown which sex squeals), and the male jumps backward at completion of coitus.

Other social interactions include avoidance of dominant individuals by those lower in the social hierarchy, dislodgement of a subordinate by a dominant individual, chasing by a dominant animal, and fighting (including biting, striking, and kicking). Cottontails exhibit linear dominance hierarchies; in males, this structure tends to prevent fighting (Marsden and Holler, 1964). Territoriality is not exhibited although nests containing young may be defended by females. Young cottontails <2 months old engage in play that includes dashing close by one another and reciprocal chasing. Lone youngsters sometimes dash repeatedly up and down a pathway (Marsden and Holler, 1964).

Chapman and Verts (1969) provided evidence that eastern cottontails were behaviorally dominant over native brush rabbits (Sylvilagus bachmani) and behaved aggressively toward them in thickets occupied by both species. In a 9- by 18-m pen, an adult female and juvenile male cottontail killed a juvenile female brush rabbit. However, Verts and Carraway (1980) provided evidence that eastern cottontails and brush rabbits hybridize under natural conditions at least occasionally.

Remarks.—Chapman and Trethewey (1972a) indicated that the ecologic and economic impacts of the unsanctioned introduction of eastern cottontails into Oregon were unknown. Except for the potentially deleterious effect of the possible introgression of their genetic material into populations of brush rabbits and the interspecific social strife, we see little ecologic impact from the >50-year-old "experiment." Also, we know of no serious negative economic effects, although we know firsthand of the destructive effects of S. floridanus in a vegetable garden. However, we believe that hunting of cottontails, a sport at which hunting colleagues scoffed in the mid-1960s, is gradually increasing in popularity in Oregon. Since 1982, data on hunting effort and harvest of cottontails on the E. E. Wilson Wildlife Area, Benton Co., gathered by personnel of the Oregon Department of Fish and Wildlife (Table 10-4), indicate that both have tended to increase. The harvest statistics also provide another index to density, although we admit that the E. E. Wilson Area likely is much better cottontail habitat than most of the Willamette Valley.

The introduction of the alien cottontail provided an opportunity to test hypotheses related to "Gloger's rule" (Rhodes, 1982). This "rule" is that animals in areas with cool, moist climates tend to be darker in color than conspecifics or congeners in areas with warm, dry climates. For example, the brush rabbit, a species restricted to the mesic region west of the Cascade Range, is much darker than its nearest native relative, the Mountain cottontail, restricted to arid regions east of the Cascade Range. If selective forces responsible for the differences in color are still operating, then eastern cottontails in the Willamette Valley should gradually become darker; if the selective forces are sufficiently strong and have operated sufficiently long then it might be possible to detect a difference in color between cottontails introduced into the Willamette Valley and those in the area from which the Oregon population originated. Rhodes (1982) tested the

Table 10-4.—*Number harvested, number of hunters attempting to harvest, and number harvested/100 hunter hours for eastern cottontails* (Sylvilagus floridanus) *during 15 October–28 February, 1983–1996, on the E. E. Wilson Wildlife Area, Benton Co., Oregon.*

Hunting season	Number harvested	Number hunters	Number harvested/ 100 hunter hours
1983–1984	411	180	83.09
1984–1985	291	267	44.84
1985–1986	391	340	35.02
1986–1987	564	343	46.64
1987–1988	999	491	58.52
1988–1989	413	374	32.02
1989–1990	853	481	55.54
1990–1991[a]	703	464	46.21
1991–1992	793	765	29.24
1992–1993	873	754	32.35
1993–1994	856	750	32.17
1994–1995	1,093	722	42.31
1995–1996	686	689	31.03

[a]No hunting permitted in October.

hypothesis of no difference with samples from Benton Co. and Ohio. Surprisingly, not only was it impossible to reject the hypothesis, but eastern cottontails from Oregon actually averaged a bit lighter in color than those from Ohio. Several possible circumstances may explain the reversal, but obviously those selective forces responsible for "Gloger's rule" are not among them. The clever student should be able to design other experiments to take advantage of this introduced species to test other hypotheses.

Sylvilagus nuttallii (Bachman)
Mountain Cottontail

1837. *Lepus nuttallii* Bachman, 7:345.
1839a. *Lepus artemisia* Bachman, 8:94.
1904. *Sylvilagus nuttallii*: Lyon, 45:323.

Description.—The mountain cottontail (Plate V) is intermediate in size among congeners in Oregon (Table 10-3). The pelage is light grayish-brown on the dorsum shading to light buffy-gray on the sides and to gray on the rump; the nape is rusty. The head is a light grayish-buff. The hairs on the dorsum have wide steel-gray basal bands followed by a narrow blackish band, a wide buffy-brown band, and a short blackish tip. The venter hairs are white with gray bases, giving a splotched appearance when hairs are spread. The tail hairs are white to the base. The leading edge of the distal portion of the ears is black. The hind feet, whitish splotched with light tan, are >80 mm but <100 mm long. The pelage is finer and tends to be fluffier than that of *S. floridanus*. Externally, *S. nuttallii* can be distinguished from other leporids in Oregon by a combination of its decidedly grayish rump; white tail; and relatively broad, rounded ears.

Destinctive skull characters include a supraorbital process with a posterior projection at most barely touching the squamosal, an external auditory meatus with a diameter equal to or larger than the incomplete circle formed by the zygomatic process of the squamosal and the posterior extension of the jugal, and a supraoccipital shield usually angular or rounded (shield shaped) with a median spine (Figs. 10-7 and 10-11). As in all congeners and *Brachylagus*, the interparietal bone is separated from the parietal and supraoccipital bones by distinct sutures. As in all congeners, but not *Brachylagus*, the anterior projections of the supraorbital processes are less than half the length of the posterior projections (Fig. 10-7 and 10-11), P2 has three reentrant angles on its anterior face, and P3, P4, M1, and M2 have a reentrant angle with crenate margins (Fig. 10-4).

Distribution.—*Sylvilagus nuttallii* occurs throughout the intermountain West (Fig. 10-12) from extreme southern British Columbia south along the east edge of the Cascade and Sierra Nevada ranges to the Death Valley region, Inyo Co., California, thence eastward through Nevada, northern Arizona, and New Mexico to the east edge of the Rocky Mountains in Colorado and Wyoming with eastern extensions into Nebraska and North Dakota and northward extensions into Saskatchewan and Manitoba (Hall, 1981). In Oregon, *S. nuttallii* occurs throughout the state east of the Cascade Range with a western extension into Josephine Co. (Fig. 10-12).

Late Pleistocene fossils of *S. nuttallii* are known from cave deposits in Idaho and New Mexico (Kurtén and Anderson, 1980).

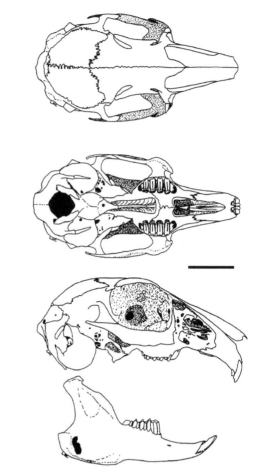

Fig. 10-11. Dorsal, ventral, and lateral views of the cranium, and lateral view of the dentary of the mountain cottontail, *Sylvilagus nuttallii* (OSUFW 481). Scale bar equals 15 mm.

Geographic Variation.—No geographic variation is recognized in *S. nuttallii* in Oregon; all are considered *S. n. nuttallii* (Hall, 1981).

Habitat and Density.—In Oregon and elsewhere, *S. nuttallii* usually is associated with rocky outcrops with nearby dominant vegetation consisting of big sagebrush (*Artemisia tridentata*), bitter-brush (*Purshia tridentata*), rabbit-brush (*Chrysothamnus*), western juniper (*Juniperus occidentalis*), and mountain-mahogany (*Cercocarpus ledifolius*) at lower elevations, and piñon pine (*Pinus edulis*), Douglas-fir (*Pseudotsuga menziesii*), and quaking aspen (*Populus tremuloides*) at higher elevations (Ingles, 1965; Janson, 1946; Orr, 1940). Sagebrush seems to be the common component of habitats with which the species usually is associated (McKay and Verts, 1978a; Orr, 1940; Sullivan et al., 1989). It sometimes occurs in willows (*Salix*) near springs or along streams (Orr, 1940). It avoids cultivated land (Sullivan et al., 1989).

On a study area in Deschutes Co. vegetated by a sagebrush-juniper scabland community, *S. nuttallii* was considered least abundant in a flat habitat with fine pumice soils and relatively sparse vegetation consisting of relatively few species of herbaceous plants. It was considered most abundant in a habitat consisting of a "mosaic of small lava hummocks between narrow valleys with deep pumice soils" with a herbaceous vegetation consisting largely of cheat grass (*Bromus tectorum*) and a

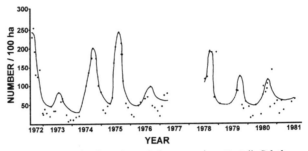

Fig. 10-13. Estimated numbers of the mountain cottontail, *Sylvilagus nuttallii*, at approximately monthly intervals each year 1972–1981 on an 87-ha study area near Terrebonne, Deschutes Co. Population density in 1977 was too low to estimate. Parts of graph from McKay and Verts (1978*b*) and Skalski and Verts (1981).

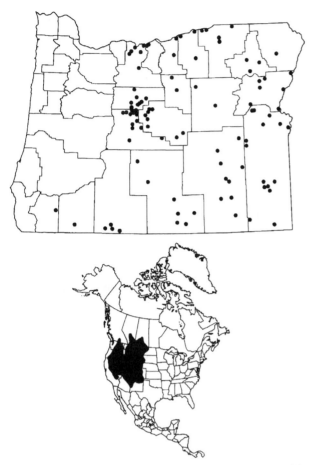

Fig. 10-12. Localities in Oregon from which museum specimens of the mountain cottontail, *Sylvilagus nuttallii*, were examined (listed in the APPENDIX, p. 565), and the distribution of the mountain cottontail in North America. (Range map redrawn from Hall, 1981:303, map 223, with modifications.)

relatively large number of forbs and other grasses (McKay and Verts, 1978*a*:364). However, the two methods used to evaluate dispersion gave conflicting results, suggesting that time of day, response of cottontails to traps, and the relative ease of sighting flushed cottontails contributed as much as differences in habitat to the observed dispersions (McKay and Verts, 1978*a*).

Density of mountain cottontails varies dramatically with season and among years (McKay and Verts, 1978*b*; Skalski and Verts, 1981). Populations as low as 6.6/100 ha and as high as 254.4/100 ha were estimated for a study area in Deschutes Co. (McKay and Verts, 1978*b*); however, in 1977 the density was so low that previously used estimators could not be employed (Fig. 10-13). Without doubt, water, either available as free water or in succulent foods, is the critical and limiting factor in regulating density of mountain cottontails (Verts et al., 1984). Densities in British Columbia ranged from 22.5 to 42.5/100 ha; density was lower during drier summers (Sullivan et al., 1989).

Diet.—In central Oregon, during spring and early summer *S. nuttallii* fed mostly on grasses, especially cheat grass. As cheat grass matured, *S. nuttallii* fed more on forbs and later-maturing grasses; then, as these forage plants became desiccated in midsummer, *S. nuttallii* increased consumption of juniper until 25% of the diet was composed of the leaves of this tree in August (Verts et al., 1984). About

75% of the diet consisted of newly sprouted *Bromus* and *Agropyron* after late-summer rains (Verts et al., 1984). Because some junipers exude water from the tips of boughs in early morning during summer and because juniper is relatively low in nutrients and contains terpenoid compounds known to interfere with digestion of cellulose by gut microflora, Verts et al. (1984) postulated that juniper was ingested largely for its water content rather than its nutritional content.

In Idaho, fecal pellets of *S. nuttallii* selected to represent long-term diets of the species contained >60% residues of grasses, mostly wheatgrass, needle-and-thread grass (*Stipa comata*), bluegrass (*Poa*), and cheat grass (Johnson and Hansen, 1979*b*).

A *S. nuttallii* from southeastern Oregon had eaten a hypogeous ascomycete, suggesting to Maser et al. (1978*a*) that the fungus had been obtained by digging.

The mountain cottontail, like other Oregon leporids, is coprophagous. Individuals collected in the afternoon after they have been inactive for most of the day usually have stomachs packed with reingested cecal pellets being coalesced into an amorphous mass. Passage of material through the intestinal tract a second time permits extraction of additional nutrients made available by cellulolytic activity of cecal microflora (Bailey, 1969*b*) and of certain vitamins (Kulwich et al., 1953).

Reproduction, Ontogeny, and Mortality.—In central Oregon, estimated earliest dates of breeding for *S. nuttallii* were 22 February in 1969, 5 March in 1972, 27 February in 1973, and 12 February in 1981; corresponding estimated latest dates of parturition were 30 July, 18 June, 19 May, and 5 July, respectively (Gehman, 1984; McKay, 1975; Powers and Verts, 1971). Commonly, young are produced in four, sometimes five, synchronously conceived litter cohorts; however, during especially dry years, fewer litter cohorts may be produced (McKay and Verts, 1978*b*; Powers and Verts, 1971). In 1969, 31 adult females shed an average of 5.0 ± 0.2 (range, 3–8) ova, implanted an average of 4.6 ± 0.2 (range, 2–6) blastocysts, and contained an average of 4.3 ± 0.1 (range, 1–6) viable embryos when collected; means for each increased between first and second litters then declined for third and fourth litters (Table 10-2). Apparently, juvenile females rarely breed later during the breeding season in which they were born (McKay, 1975; Powers and Verts, 1971).

No description of neonates is available. Orr (1940) found individuals as small as 40 g that had abandoned their nest and were moving about; he speculated that the nest may have been disturbed. One of us (BJV) captured two young in Sherman live traps; the one preserved weighed 112 g.

Mortality among juvenile litter cohorts from birth to 31 August usually ranged between 40 and 80% and seemed related largely to precipitation within a few weeks before and after the estimated date of conception (Hundertmark, 1982; Verts et al., 1984). In late summer and early autumn, mortality remained high, probably because of losses among individuals with low body mass in later-born litter cohorts when temperatures begin to fall below freezing. During most of October–November, mortality was low, probably because surviving individuals had relatively high body masses, and food and moisture resources were abundant during autumn "green-up" following precipitation in late summer or early autumn. However, as many as 80% of the cottontails surviving to late autumn may succumb during short periods of intense cold (McKay and Verts, 1978b). Late-winter mortality is relatively low, probably because of a more favorable resource:density ratio resulting from high mortality in early winter (McKay and Verts, 1978b).

Precipitation, seemingly responsible for high mortality among juveniles when amounts are low, is believed to serve as a selective force that alters the genetic composition of populations at least at one locus (Skalski and Verts, 1981). Transferrin, a beta globulin in blood, is polymorphic in S. nuttallii and encoded by three codominant alleles distinguished by their relative mobilities when serum is subjected to starch-gel electrophoresis: A, the fastest-migrating allele; B, the next fastest allele; and C, the slowest allele. Potentially, these could form six genotypes (phenotypes), but no individuals with AA or AC genotypes were found, the AB genotype occurred infrequently and disappeared from the area during the study, and CC genotype never exceeded 10% of the population. Thus, BB and BC genotypes predominated. In dry years, juvenile survival was low and the BB genotype was favored; in years with greater moisture, juvenile survival was higher and the BC genotype was favored. Skalski and Verts (1981) suggested that the amount of precipitation falling in successive 2-3-year periods was sufficiently variable to avoid fixation by elimination of either the B or C allele by selection.

Predation accounts for some losses; during cruises of a 116-ha study area in 1981, Gehman (1984) observed five instances of predators (two coyotes, Canis latrans; a red-tailed hawk, Buteo jamaicensis; a golden eagle, Aquila chrysaetos; and an unidentified falcon) feeding on mountain cottontails. He also observed five encounters of S. nuttallii with great horned owls (Bubo virginianus) and one encounter with a red-tailed hawk, none of which resulted in the capture of the cottontail. Brodie and Maser (1967) reported that 3.2% of 696 food items identified in pellets regurgitated by great horned owls consisted of mountain cottontails.

Of 21 S. nuttallii with closed epiphyseal cartilages, 3 contained no adhesion lines in the periosteal zone, 13 had one line, 4 had two lines, and 1 had three lines. Thus, 3 were in the year of their birth, 13 in the 2nd calendar year of life, 4 in the 3rd calendar year, and 1 in the 4th calendar year. A former graduate student, Kris Hundertmark, caught two cottontails that a previous graduate student, John

Skalski, had marked as juveniles on the study area near Terrebonne, Deschutes Co., 5 years previously.

Habits.—Mountain cottontails tend to be solitary (Orr, 1940) and, during the day, spend a large proportion of the time sitting in forms or beneath shrubs or hidden in crevices in rocks or burrows dug by other animals. Activity during the daylight hours tends to be bimodal with peaks in the the 2nd hour after sunrise and the hour before sunset (Verts and Gehman, 1991). During periods of activity, S. nuttallii spent most of the time feeding and moving from one place to another. Grooming accounted for only 1.6% of the time that cottontails were in view (Verts and Gehman, 1991). Only 7.3% of 1,192 behavioral acts were classified as social, and, of those, 94.6% were reproductive interactions. Because sampling was conducted during a large part of the breeding season, the paucity of reproductive interactions suggests that most of these energy-demanding activities occur at night (Verts and Gehman, 1991).

In Deschutes Co., mountain cottontails regularly climb trees during the early morning hours in July–August, a rather unusual behavior for a leporid (Verts and Gehman, 1991; Verts et al., 1984). Although S. nuttallii is sometimes observed to bask on horizontal branches or atop rocks (Plate V), acquisition of water is believed to be the primary function of climbing in juniper trees. Tree-climbing; restriction of most reproductive activity to the hours of darkness when humidity is high; and the solitary life-style that produces uniform dispersion and, in turn, reduces competition for forage resources (especially those containing much water) are thought to be behavioral adaptations that permit S. nuttallii to occupy regions in the intense rain shadow of the Cascade Range at the periphery of its geographic range (Verts and Gehman, 1991). Whether or not S. nuttallii in mesic habitats climbs trees is unknown.

Remarks.—Although Sullivan et al. (1989) indicated that S. nuttallii in British Columbia caused sufficient damage to horticultural crops to require implementation of control measures in the 1950s and 1960s, we know of no situation in Oregon where control is required. S. nuttallii is not classified as a game animal in Oregon.

A survey, conducted by a former undergraduate student, Mark Henjum, in 1973, of about 1% of the licensed hunters indicated that an estimated 139,410 cottontails (both species combined) were harvested in Oregon. Thus, cottontails were the third-ranking upland game animals in the state that year. Nevertheless, 85% of the respondents indicated that they did not hunt leporids; nearly half did not because they preferred to hunt big game, but a large proportion of the remainder indicated that they did not hunt them because of a lack of familiarity with leporids as a harvestable resource. In addition to being locally abundant, mountain cottontails possess some excellent sporting qualities and are a superb, low-cholesterol meat.

The role of morphological differences in the auditory meatus and bulla in hearing at high elevations where transmission of sound is attenuated is as intriguing today as it was >50 years ago (Orr, 1940). A pilot project conducted by an undergraduate student, Meg Eden, tended to indicate that S. nuttallii possessed larger auditory meatuses at higher elevations, but samples were too small for statistical testing.

Lepus americanus Erxleben
Snowshoe Hare

1777. [*Lepus*] *americanus* Erxleben, 1:330.
1855. *Lepus washingtoni* Baird, 7:333.
1869. *Lepus bairdii* Hayden, 3:115.
1875. [*Lepus americanus*] var. *Washingtoni*: Allen, 17:431.
1899*b*. *Lepus klamathensis* Merriam, 16:100.
1934. *Lepus bairdii oregonus* Orr, 15:152.
1936. *Lepus americanus klamathensis*: Bailey, 55:95.
1942. *Lepus americanus oregonus*: Dalquest, 23:179.

Description.—*Lepus americanus* (Plate VI) is the smallest member of the genus in Oregon (Table 10-5), but except for the much shorter ears, is much like its congeners in overall conformation. The hind feet are >105 mm, but the ears are ≤80 mm.

The interparietal bone in *L. americanus* is not separated from the parietal bones by distinguishable sutures. Occipitonasal length is <75 mm and the supraorbital process lacks an anterior projection (Fig. 10-14). There are eight mammae (Bittner and Rongstad, 1982).

The snowshoe hare in eastern parts of its geographic range undergoes three molts (Lyman, 1943): a molt from white to brown in spring, a molt from brown to brown in autumn, and a molt from brown to white in winter. In Oregon, individuals in populations east of the Cascade Range, and some individuals in some populations in the Cascade Range, become white in winter and are brown in summer. For individuals brown in winter, whether the brown autumn pelage is retained over winter or molted and replaced by a new brown winter pelt is unknown (Nagorsen, 1983).

For snowshoe hares white in winter, the pelage is formed by hairs white only at the tip. The base of these hairs is grayish (about half the length in the Cascade Range, but <10% of the length in the Wallowa Mountains) followed by a subterminal band of tannish buff to light pinkish-cinnamon and a terminal band of white. A few hairs may have blackish tips. The ears have a narrow band of black with some brown hairs extending part way down the ear. For those in summer pelage and those brown in winter, the pelage consists of hairs with light-gray bases, a band of dark brown, a band of tannish buff, and a black tip on the dorsum. On the sides and throat, most hairs lack the black tip. The venter is white with pure white hairs along the midline and with hairs with gray bases and long white tips toward the sides. The ears are edged with white inside, black outside.

Distribution.—The geographic distribution of *L. americanus* (Fig. 10-15) extends from Alaska to Newfoundland south through the Appalachian Range to North Carolina, the Great Lakes states, and the northern prairie states; through the Rocky Mountains to northern New Mexico and southern Utah; through the Cascade and Sierra Nevada ranges to central California and Nevada; and through the

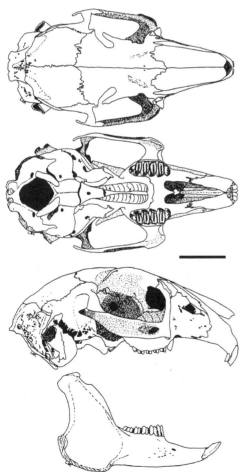

Fig. 10-14. Dorsal, ventral, and lateral views of the cranium, and lateral view of the dentary of the snowshoe hare, *Lepus americanus* (OSUFW 1406). Scale bar equals 15 mm.

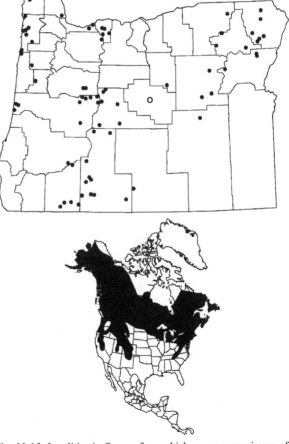

Fig. 10-15. Localities in Oregon from which museum specimens of the snowshoe hare, *Lepus americanus*, were examined (listed in the APPENDIX, p. 520), and the distribution of the snowshoe hare in North America. (Range map redrawn from Hall, 1981:317, map 227, with modifications.) Open symbol indicates record for county.

Table 10-5.—*Means (±SE), ranges (in parentheses), and CVs of measurements of skull and skin[a] characters for female and male Lepus from regions in Oregon. Skin characters were recorded from specimen tags; skull characters were measured to the nearest 0.01 mm. SE and CV not provided if* n <10.

Species and region	Sex	n	Total length	Hind foot length	Mass (g)	Basilar length	Length of nasals	Length of maxillary toothrow	Length of bulla
Lepus americanus									
West of crest of Cascade Range and Hood River, and western Jefferson and Wasco counties	♀	16	412 ± 8.3 (324–470) 8.1	119 ± 1.5 (111–131) 4.9	1,230 ± 42.4[b] (1,020–1,428) 11.43	65.55 ± 0.50 (62.81–69.04) 3.05	30.37 ± 13.79 (27.13–32.17) 4.56	13.79 ± 0.16 (12.82–15.00) 4.57	9.69 ± 0.18 (8.61–11.06) 7.28
	♂	10	397 ± 7.4 (340–420) 5.9	113[c] (105–125)	1,069[d] (950–1,416)	65.30 ± 0.48 (63.09–67.00) 2.31	30.44 ± 0.69 (25.96–33.12) 7.18	13.89 ± 0.18 (12.97–14.68) 4.09	9.31 ± 0.24 (8.54–10.49) 8.32
Deschutes, southwestern Crook, eastern Klamath, and western two-thirds Lake counties	♀	4	431 (410–447)	126 (120–132)		66.29 (64.48–67.50)	30.38 (28.15–32.42)	13.94 (13.36–14.27)	10.04 (9.70–10.31)
	♂	2	426 (410–441)	136 (131–141)		65.54 (65.15–65.93)	30.00 (28.24–31.76)	13.69 (13.63–13.76)	10.19 (9.85–10.52)
Umatilla, Union, Wallowa, Baker, Grant, and northern Harney and Malheur counties	♀	5	451 (425–466)	137 (134–142)	1,408[d] (1,310–1,506)	68.44 (67.72–71.08)	31.46 (30.61–33.04)	14.26 (13.72–15.05)	10.68 (10.23–11.14)
	♂	7	410 (370–440)	135 (128–146)	1,069[e] (1,027–1,134)	65.38 (62.35–68.97)	30.35 (28.27–32.36)	13.57 (13.09–14.07)	10.34 (9.61–11.24)
Lepus californicus									
West of Cascade Range	♀	5	534[c] (457–611)	132[c] (110–143)	2,366[f] (1,461.5–3,310)	82.61 (71.72–89.94)	39.66 (29.81–43.99)	15.58 (13.81–16.69)	11.01 (10.68–11.60)
	♂	6	570 (535–625)	129 (105–147)	2,143[f] (2,011–2,299)	83.03 (79.08–85.79)	40.79 (37.73–43.05)	16.10 (15.32–16.75)	11.53 (11.04–12.14)
East of Cascade Range	♀	25	535 ± 8.0[g] (475–609) 6.5	128 ± 2.3[g] (64–155) 7.8	2,610	78.49 ± 0.52 (73.68–85.02) 3.28	35.36 ± 0.56 (31.59–44.86) 7.91	15.66 ± 0.16 (14.09–17.51) 4.99	12.52 ± 0.13 (11.39–14.00) 5.25
	♂	25	530 ± 7.2[h] (472–580) 5.5	125 ± 1.2[h] (114–132) 3.9	1,795[i] (1,557–2,042)	77.51 ± 0.46 (73.23–81.99) 2.96	34.74 ± 0.37 (30.63–38.17) 5.29	15.38 ± 0.13 (13.64–16.31) 4.32	12.63 ± 0.14 (11.52–14.46) 5.71
Lepus townsendii	♂	7	570 (530–598)	143[c] (130–150)	2,550[b] (2,540–2,560)	79.37 (77.63–82.63)	36.64 (34.30–39.63)	16.15 (15.73–16.89)	11.78 (10.44–12.59)

[a]Although measures of tail and ear length were recorded, the data were not included because of extreme variation believed to stem from differences in techniques used to measure these characters. [b]Sample sizes reduced by 5. [c]Sample sizes reduced by 1. [d]Sample sizes reduced by 3. [e]Sample sizes reduced by 4. [f]Sample sizes reduced by 2. [g]Sample sizes reduced by 6. [h]Sample sizes reduced by 9. [i]Sample sizes reduced by 22.

Olympic and Coast ranges to the central Oregon Coast (Hall, 1981). In Oregon, *L. americanus* occurs in the Coast Range south to Florence, Lane Co.; on both slopes of the Cascade Range; and in the Ochoco, Strawberry, Blue, and Wallowa mountains (Fig. 10-15).

Pleistocene fossils are known from Virginia, Maryland, Pennsylvania, Kentucky, Missouri, Arkansas, Idaho, and California; those in Missouri, Arkansas, and Kentucky are extralimital (Kurtén and Anderson, 1980).

Geographic Variation.—Three of the 15 nominal subspecies are purported to occur in Oregon: *L. a. washingtoni* in the Coast Range, northern Cascade Range, and on the west slope of the southern Cascade Range; *L. a. klamathensis* on the east slope of the southern Cascade Range; and *L. a. oregonus* in the northeastern mountains (Hall, 1981). However, Nagorsen (1985:577) concluded from a morphometric study of cranial variation throughout the range of the species that "there is little merit in recognizing subspecies of *L. americanus*."

Habitat and Density.—*Lepus americanus* is considered to be associated with dense thickets of young conifers, especially those with lower branches touching the ground and especially firs (presumably *Abies*) and western larch (*Larix occidentalis*) interspersed with small clearings vegetated by grasses and forbs such as *Collinsia parviflora*, *Castilleja*, and *Lupinus* in eastern Oregon (Orr, 1934, 1940). In western Oregon, it is associated with Douglas-fir (*Pseudotsuga menziesii*) and western hemlock (*Tsuga heterophylla*) with an understory of Indian basket grass (*Xerophyllum tenax*) and bracken fern (*Pteridium aquilinum*—Black, 1965). In a measure of habitat use by *L. americanus* in western Oregon, Black (1965) claimed no significant relationship between mean number of fecal pellets per 4.1-m^2 circular plot and the degree of closure of the canopy. However, our reanalysis of his data indicated that the hypothesis of equal distribution of pellets among canopy-density classes must be rejected ($\chi^2 = 25.64$, *d.f.* = 4, $P < 0.01$). Indeed, greater use was made of portions of the habitat in which canopy cover was greatest.

On a 259-ha study area in Alberta, snowshoe hares pri-

Table 10-5.—*Extended.*

Breadth of braincase	Parietal breadth	Zygomatic breadth	Nasal breadth	Depth of skull	Length of mandible	Length of mandibular toothrow	Depth of mandibular ramus
23.97 ± 0.25 (22.08–25.83) 4.10	19.13 ± 0.25 (17.37–20.60) 5.14	36.20 ± 0.17 (35.09–37.34) 1.89	15.19 ± 0.18 (14.21–16.75) 4.69	30.19 ± 0.17 (29.07–31.60) 2.25	55.42 ± 0.51 (52.24–58.74) 3.68	14.10 ± 0.18 (13.25–15.71) 5.03	10.62 ± 0.11 (9.78–11.22) 4.29
23.42 ± 0.18 (22.42–24.44) 2.42	19.85 ± 0.33 (18.25–21.46) 5.27	36.04 ± 0.15 (35.21–36.93) 1.34	15.94 ± 0.29 (14.28–17.75) 5.84	30.68 ± 0.20 (29.36–32.31) 2.04	54.99 ± 0.32 (53.68–57.04) 1.86	14.19 ± 0.11 (13.49–14.63) 2.36	10.42 ± 0.14 (9.89–11.07) 4.18
22.92 (21.79–23.82)	20.11 (19.35–21.36)	36.52 (35.40–37.43)	15.11 (14.71–16.01)	30.29 (27.94–32.01)	55.21 (53.31–56.41)	14.17 (13.65–14.93)	10.64 (10.26–10.84)
23.24 (22.07–24.40)	18.97 (17.67–20.26)	36.48 (35.44–37.51)	14.76 (14.35–15.16)	29.86 (29.81–29.91)	54.73 (54.01–55.44)	14.15 (13.95–14.35)	10.67 (10.42–10.92)
24.24 (23.57–24.59)	20.06 (19.14–21.67)	37.38 (36.83–38.27)	16.18 (15.25–17.27)	31.87 (31.06–32.48)	57.57 (55.28–60.48)	14.64 (14.05–15.47)	11.02 (10.61–11.49)
24.16 (23.07–25.18)	19.91 (18.20–21.36)	36.91 (36.14–37.66)	15.22 (13.72–16.07)	31.33 (29.26–33.02)	55.19 (52.14–57.38)	14.05 (13.31–14.48)	10.56 (9.77–11.16)
26.39 (25.09–27.22)	24.95 (22.37–26.38)	43.57 (40.97–45.78)	19.58 (16.23–21.27)	36.21 (31.75–38.00)	68.64 (58.05–75.79)	16.29 (14.24–17.63)	13.08 (11.78–14.12)
26.39 (24.00–27.10)	24.84 (24.20–25.55)	42.87 (41.71–44.46)	20.56 (18.14–21.93)	38.29 (36.08–41.01)	68.01 (63.24–70.04)	16.77 (15.93–17.52)	12.74 (12.03–13.20)
24.48 ± 0.22 (22.02–27.75) 4.52	23.21 ± 0.28 (20.71–26.07) 5.95	40.69 ± 0.29 (38.57–43.83) 3.51	17.17 ± 0.35 (15.07–21.52) 10.16	35.27 ± 0.28 (32.82–39.19) 3.91	66.19 ± 0.53 (61.60–73.71) 3.97	16.45 ± 0.16 (14.50–18.05) 5.00	13.39 ± 0.14 (11.69–14.68) 5.33
24.62 ± 0.17 (23.24–26.42) 3.48	22.91 ± 0.25 (20.59–25.70) 5.43	40.39 ± 0.26 (37.33–42.62) 3.22	17.00 ± 0.18 (14.73–18.35) 5.33	34.97 ± 0.17 (33.36–36.49) 2.49	65.31 ± 0.45 (60.37–69.31) 3.43	16.43 ± 0.14 (14.90–18.15) 4.32	13.37 ± 0.11 (12.17–14.62) 3.99
26.84 (25.72–28.43)	25.74 (23.71–27.48)	43.15 (41.24–43.98)	19.52 (17.96–21.98)	38.48 (37.40–39.33)	67.04 (64.97–68.55)	16.86 (15.82–17.93)	13.26 (12.24–14.18)

marily occupied "a tangle of deadfalls and red raspberry (*Rubus idaeus*) canes, with a dense regrowth of quaking aspen (*Populus tremuloides*) and willows (*Salix* spp.)" (Keith, 1966:829). Also occupied were thickets at edges of bogs consisting of black spruce (*Picea mariana*), alder (*Alnus*), and hazelnut (*Corylus cornuta*). As density declined, habitats with the greatest brushy cover remained occupied whereas those with the least brushy cover became depopulated; movements of individuals were primarily responsible for changes in distribution (Keith, 1966).

Densities of snowshoe hare populations follow a 9–10-year cycle of peak numbers in the boreal region of North America (Elton and Nicholson, 1942b; Keith, 1963; MacLulich, 1937). In the montane regions of British Columbia south to California, snowshoe hare populations are not considered to cycle (Wolff, 1982), or if they cycle the amplitude of the fluctuations is much less than farther north (Adams, 1959). Where populations cycle, peak densities in spring were 2.1/ha in Minnesota (Green and Evans, 1940a), 1.5–8.9/ha in Alberta (Keith and Windberg, 1978),

and 2.1–3.5/ha in Yukon Territory (Krebs et al., 1986b). Minimum densities in spring (more difficult to estimate— Krebs et al., 1986b) were about 6.7%, 2.0–5.9%, and 0.5–0.9% of peak densities in spring for the three regions, respectively. In general, from peak densities, the decline phase of the cycle requires 2 (Krebs et al., 1986b), 3 (Green and Evans, 1940a), or 4 years (Keith and Windberg, 1978). Changes in fecundity, survival, recruitment, and body mass at different stages of the cycle seem equally inconsistent (Keith et al., 1984; Krebs et al., 1986b).

Keith (1974) hypothesized that the decline in density was initiated in response to a shortage of food, followed by an increase in predation. Wolff (1980, 1982) hypothesized that in northern regions low densities of predators permitted hares to disperse into marginal habitats during the increase phase of the cycle, but that when the carrying capacity was exceeded starvation and greater predation eliminated hares in marginal habitats and restricted remanent populations to refugia in the more favorable habitats remaining. Wolff (1982) further explained that, in the north, high survival of initial dispersers that moved into essentially continuous occupiable habitat permitted buildup of density, whereas in the south patchiness of occupiable habitat greatly inhibited survival of dispersers that moved into the less-than-optimal habitat. Also, predators in the south are facultative, resident, and of greater variety; thus, may be involved in preventing the buildup of numbers. The two hypotheses are compatible; however, on study areas in Yukon Territory, addition of food (either rabbit chow or natural browse) did not prevent the decline (Krebs et al., 1986a, 1986b). Also, Keith et al. (1984) found food shortage on only two of three study areas, but high levels of predation on all three areas during the decline phase in Alberta. Although starvation and predation likely play significant roles, we are confident that complete understanding of population cycles in snowshoe hares has not yet been attained.

The only estimates of density in Oregon are those by Black (1965) of 4.0/ha in spring and of approximately twice that at the end of the breeding season, and that by Hooven (1966) of 7.4/ha. Whether fluctuations in numbers of snowshoe hares are of sufficient amplitude or regularity in any part of the species range in Oregon to permit them to be recognized as cycles is unknown. In north-central Washington, numbers of fecal pellets counted in 10 3.1-m² circular plots along each of 24 100-m transects indicated that snowshoe hare densities were 1.2–15.8 times greater in ≤25-year-old lodgepole pine (Pinus contorta) forest than in older forests and declined by 5.7–59.1% during a 4-year period. Koehler (1990) believed that the observed magnitude of change was insufficient to indicate that populations cycled in southern parts of the range of the species.

Diet.—In western Oregon, primary foods of *L. americanus* are western hemlock, oval-leaved huckleberry (*Vaccinium ovalifolium*), red blueberry (*V. parvifolium*), willow (*Salix sitchensis*), and Douglas-fir (Black, 1965). Heavily used browse plants in Montana were Douglas-fir and barberry (*Berberis repens*) in winter; balsamroot (*Balsamorhiza*), arnica (*Arnica*), and shiny-leaf spiraea (*Spiraea betulifolia*) in summer; and pale fawn-lily (*Erythronium grandiflorum*) in spring (Adams, 1959). Lesser-used species included ponderosa pine (*Pinus ponderosa*) and Saskatoon serviceberry (*Amelanchier alnifolia*) in winter and bearberry (*Arctostaphylos uva-ursi*) in summer. The ingestion of sand by snowshoe hares was reported, but no explanation for the seemingly aberrant behavior was provided (Adams, 1959).

In Alberta, snowshoe hares require a maximum of about 300 g/day of "small twigs, stems, and branches of various woody shrubs and saplings" (Keith et al., 1984:26). Stems ≤3–4 mm in diameter contain the greatest nutrient content and are most digestible by snowshoe hares (Pease et al., 1979). Larger-diameter browse may slow rates of weight loss in winter when smaller-diameter browse is not sufficient to provide the daily requirement (Keith et al., 1984).

Reproduction, Ontogeny, and Mortality.—In Montana, the testes begin to enlarge by mid-February and most are scrotal by mid-March. Females become pregnant about 2 weeks later and are continuously pregnant or lactating (or both) until October (Adams, 1959). The latest observation of scrotal testes and the latest pregnancy were in the first half of July; lactation may last as long as 56 days after the last parturition of the year (Adams, 1959). In Michigan, males were fertile from February to August and females were determined to be pregnant by palpation from

mid-April to the 1st week in September (Bookhout, 1965). In Newfoundland, males were fertile from March through August (Dodds, 1965). Females produced an average of 2.9 litters in Montana (Adams, 1959) and 2.8 in Alberta (Rowan and Keith, 1956).

Gestation in captives averaged 37.2 (range, 36–40) days (Severaid, 1945). Developing young can be detected with certainty by palpation 10 days after copulation (Bookhout, 1964). Average mass (in g) and crown-rump length (in mm) or forehead-rump length (after 18 days of gestation) for embryos (n = 1–6) from females for which dates of breeding were known were: 9 days, 0.01, 3.5; 10 days, 0.02, 5.5; 11 days,—, 6.8; 12 days 0.14, 9.3; 14 days, 0.35, 14.3; 15 days, 0.46, 17.0; 18 days, 1.6, 25.2; 20 days, 2.9, 35.3; 22 days, 4.4, 40.7; 24 days, 8.4, 45.5; 25 days, 12.4, 60.8; 26.5 days, 21.0, 75.0; 28 days, 19.8, 67.6; 30 days, 37.9, 91.5; 31 days, 42.2, 94.0; 33 days, 42.0, 97.0; 34 days, 65.9, 110.7 (Bookhout, 1964). Use of age-weight and age-length growth curves did not produce discrepancies in estimated dates of breeding of >1 day. Also, developmental landmarks such as presence of digits (≥15 days of age), nails (≥19 days of age), fat deposits (≥21 days of age), vibrissae (≥23 days of age), and fur (29 days of age) are reliable criteria of length of gestation (Bookhout, 1964).

Based on a compilation of reports by several authorities, body mass of young ranged from 39.6 to 85.0 g at birth (Adams, 1959). A male born in captivity and estimated to weigh 85 g at birth doubled its mass by 8 days of age, again by 16 days, a third time by 38 days, and a fourth time by 67 days; it weighed 1,518.0 g at 120 days of age (Grange, 1932). Neonates are precocial; they are fully furred with open eyes and ears, and capable of moving about, but not hopping. Development is rapid; by 8 days of age they begin to take solid food (Nice et al., 1956) and by 9 days of age they are difficult to catch by hand in the field (Grange, 1932).

Young remain together from 1 to 4 days after birth, then scatter in the area near the birth site; 18 m may separate siblings at 8 days, 30 m by 12 days, and by 14 days they may range over 0.6–1.0 ha (Rongstad and Tester, 1971). The young assemble each evening, the maternal female joins the litter, and the young nurse for 5–10 min, whereupon the female leaves and spends the remainder of the 24-h period as much as 275 m away; thus, young are nursed only once each day. The assembly site is not fixed, but usually within 9–15 m of previous sites. The time of nursing is remarkably constant; it seems to be related to the end of nautical twilight (time the sun is 12° below the horizon). Young usually are nursed for 25–28 days, but those in the last litter of the season may be nursed for twice that long (Rongstad and Tester, 1971).

Average litter size differs by month, by litter period, by latitude, by year, and by region (Bookhout, 1965; Dodds, 1965; Green and Evans, 1940b; Rowan and Keith, 1956; Severaid, 1945). Adams (1959) reported that average litter size, as recorded by several workers, ranged from 2.88 to 4.00. As common among leporids, the first litter is smallest, the second largest, and later litters intermediate (Severaid, 1945); thus, annual average litter sizes (Table 10-2) may be biased by the number of individuals for each of the several litter cohorts.

Although a relationship between average litter size and stage of the cycle is denied by some workers (Adams, 1959; Green and Evans, 1940b), Rowan and Keith (1956:275) considered that the trend from an average of 4.11/litter (n

= 27) at peak density to 2.20/litter (n = 5) at the population low "strongly suggests that reduced litter size accompanies the cyclic decline." They considered the difference too great to be accounted for by including larger numbers of first litters when the population is low.

Survival of snowshoe hares from birth to the following breeding season was 18% in western Oregon (Black, 1965). Survival of telemetered snowshoe hares in Alberta from birth to 1 year of age was 16%; for adults, annual survival was 58% from February 1971 to January 1972, and 33% from August 1972 to July 1973 (Brand et al., 1975). Predators were responsible for deaths of nine (64%) of 14 juveniles and all 12 adults; four kills were by lynxes (*Lynx canadensis*), three by coyotes (*Canis latrans*), two by weasels (*Mustela*), three by great horned owls (*Bubo virginianus*), two by goshawks (*Accipiter gentilis*), and six by unknown predators. Although Brand et al. (1975) misused chi-square with a 2 by 2 contingency table containing a zero, their claimed significant difference in predator mortalities among age-classes was corroborated by our reanalysis of their data with Fisher's exact test (P = 0.03). In Montana, survival from birth to the following February was 9% for juveniles; for adults survival from the reproductive period to the following February was 31% (Adams, 1959).

Habits.—As measured by traffic on snowshoe hare trails in Alberta in summer, <15% of the total daily activity was between 0330 h and 1530 h (Keith, 1964). Activity during the remaining hours produced a nearly symmetrical curve peaking at 2300 h. The proportion of activity between 2030 h and 0130 h increased with the seasonal increase in hours of darkness (Keith, 1964).

Agonistic behavior in snowshoe hares ranges from raising or lowering the ears to chases to actual striking or biting of one individual by another (Graf, 1985); except for physical contact, most of the aggressive interactions are similar in form and context to those described by Marsden and Holler (1964) for eastern cottontails (*Sylvilagus floridanus*). Snowshoe hares exhibit linear dominance hierarchies with females dominant in summer and males dominant in winter (Graf, 1985). In tests with captive pairs, FitzGerald and Keith (1990) found adults to dominate juveniles and females to dominate males when both are in breeding condition, but were unable to demonstrate conclusively that heavier individuals dominate lighter individuals of the same sex, age, and breeding condition.

Occasionally, in heterogeneous habitats, the sexes segregated by habitat characteristics: males were associated with sites at which cover provided by both overstory and understory was greatest, whereas females were associated with sites having the most forage (Litvaitis, 1990). Different roles in reproduction may explain differences in use of microhabitats, as seasonal shifts in dominance between sexes do not correspond with the dominant sex occupying the densest habitat where survival is greatest (Litvaitis, 1990).

Snowshoe hares make grunting (growling), screaming, whining, and chirping or clicking sounds. The context of some of the vocalizations is known to differ; hence, their significance is not understood completely (Forcum, 1966; Trapp and Trapp, 1965). Snowshoe hares reingest soft fecal pellets in much the same manner as that observed in numerous other leporids (Adams, 1959). Forms used by snowshoe hares in western Oregon commonly are in beargrass (*Xerophyllum tenax*—Black, 1965).

The mean radii of circular bivariate home ranges in western Oregon were 114.3 m for males and 97.6 m for females (Black, 1965); circles with these radii cover slightly >4 ha and slightly <3 ha for the two sexes, respectively. In Alaska, average home-range areas estimated by the exclusive boundary method (Stickel, 1954) for individuals captured four or more times were 5.9 ± 0.5 ha for males and 5.8 ± 1.0 ha for females; 95% of recaptures were within 336 m of the last capture (O'Farrell, 1965). In another study in Alaska, 50% of activity was in <1 ha (Wolff, 1980).

"Trap sickness" or "shock disease," characterized by lack of wariness, greatly increased hunger, opisthotonic seizures, and death, is common in snowshoe hares caught in live traps; symptoms are more common in adults in January–April and in young hares (Adams, 1959). Stress related to confinement, cold or wet conditions, or high temperatures resulting in acute hypoglycemia are responsible for the symptoms (Keith et al., 1968). Intramuscular or intraperitoneal injections of 1–3 ml 50% dextrose plus warming in a 50– 65°C oven for 5–15 min usually results in full recovery (Keith et al., 1968).

Remarks.—Snowshoe hares may cause serious depredations upon nursery stock used in reforestation (Mitchell, 1950), although naturally regenerated seedlings seem to be affected much less (Moore, 1940). Odors from planter's hands, cultural methods including fertilization of nursery stock, and differences in succulence were rejected as possible causes of differences in depredations on seedlings. In an often-referenced experiment, 38% of Douglas-fir seedlings from nurseries were browsed by *L. americanus* within 1 month; within 1 year 94% were browsed and 35% were "heavily, and usually fatally" browsed (Moore, 1940:19). Reduction of population densities of snowshoe hares by poisoning, snaring, and shooting; screening placed around seedlings; and chemical repellents applied to seedlings have been used in attempts to protect new forest plantings (Hooven, 1966). None of the methods provided adequate protection of seedlings.

The snowshoe hare has been neglected as a research animal in Oregon; few research reports have been published on the species (Carraway and Verts, 1982). A relatively simple project on the species in western Oregon would be to mark individuals with dyes in summer and autumn to ascertain if individuals that do not become white in winter have a winter molt and develop a new brown winter coat or have only two annual molts and retain the brown autumn coat over winter. A long-term study of density in the area in which the brown coat was retained, with special emphasis on stages of community development, might be enlightening. An investigation of genetic differences in relation to coat color in populations in which some individuals develop a white pelage in winter and some do not possibly would produce interesting results.

Lepus californicus Gray
Black-tailed Jackrabbit

1837. *Lepus californicus* Gray, 1:586.
1896. *Lepus texianus deserticola* Mearns, 18:564.
1904*a. Lepus texianus wallawalla* Merriam, 17:137.
1909. *Lepus californicus wallawalla*: Nelson, 29:132.
1909. *Lepus californicus deserticola*: Nelson, 29:137.
1926. *Lepus californicus vigilax* Dice, 166:11.
1932. *Lepus californicus depressus* Hall and Witlow, 45:71.

Description.—*Lepus californicus* (Plate VI) is slightly smaller than *L. townsendii* but considerably larger than *L.*

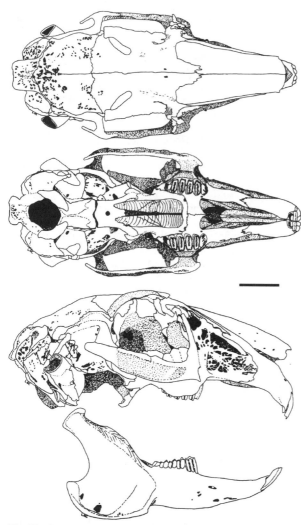

Fig. 10-16. Dorsal, ventral, and lateral views of the cranium, and lateral view of the dentary of the black-tailed jackrabbit, *Lepus californicus* (OSUFW 5165). Scale bar equals 15 mm.

americanus (Table 10-5). In conformation *L. californicus* is much like *L. townsendii*; the hind feet are >105 mm and the ears are >80 mm. There are six mammae (Lechleitner, 1959).

Skulls of *L. californicus* (Fig. 10-16), like those of congeners, have an interparietal bone fused to and indistinguishable from the parietal bones. The rostrum is narrow, long, and tapering; the occipitonasal length is >75 mm; the supraorbital process has an anterior projection; the posterior projection of the supraorbital process usually is long, tapering, and commonly fused to the cranium. However, there are no characteristics of the skull that can be used to distinguish *L. townsendii* from *L. californicus* with certainty (Hoffmann and Pattie, 1968).

West of the Cascade Range the dorsal hairs of *L. californicus* have gray blending to dark-brown or blackish bases followed by a narrow band of buff and a black tip. Hairs on the throat, sides, and rump have gray bases blending to buff with short black tips. On the venter, hairs are white with light pinkish-buff tips. The tail is black on the dorsum and dark buff on the venter. The ears are dark buff peppered with black and blending to black at the tip. The feet are mostly white with splotches of buff dorsally. Overall, the coloration is dark buff shading to black on the dorsum. In summer, the pelage is much lighter; black bands on hairs are shorter.

East of the Cascade Range the light-colored bands on the dorsal hairs are almost white instead of buff and the dark-colored bands are much narrower. The overall appearance of the pelage in winter is similar to that of the summer pelage west of the Cascade Range, but more grizzled. The feet are much darker and the dorsal surface of the tail has more black.

Distribution.—*Lepus californicus* occurs from southern Washington, southwestern Montana, southern Nebraska, and southwestern Missouri south to the tip of Baja California Sur, and central Sonora, Querétaro, and Hidalgo (Fig. 10-17; Hall, 1981). In Oregon, west of the Cascade Range, *L. californicus* is restricted to the Rogue, Umpqua, and Willamette valleys; east of the Cascade Range it occurs throughout the sagebrush (*Artemisia*) regions (Fig. 10-17). Gaps in the illustrated distribution in unforested regions east of the Cascade Range, except for high mountains, likely are largely a reflection of unequal sampling effort.

Pleistocene fossils are known from numerous sites in California, Nevada, Arizona, and Texas, and from Mexico (Kurtén and Anderson, 1980).

Geographic Variation.—Two nominal subspecies occur in Oregon: *L. c. californicus* west of the Cascade Range

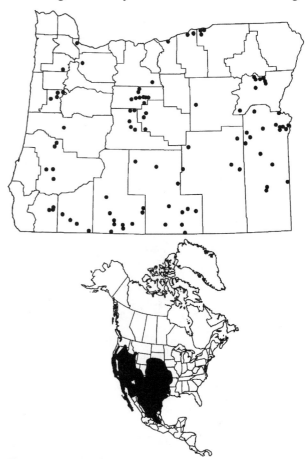

Fig. 10-17. Localities in Oregon from which museum specimens of the black-tailed jackrabbit, *Lepus californicus*, were examined (listed in the APPENDIX, p. 521), and the distribution of the black-tailed jackrabbit in North America. (Range map redrawn from Hall, 1981:327, map 230, with modifications.)

and *L. c. wallawalla* east of the Cascade Range. Color differences of the two races are presented in the Description section. A published range map indicates that *L. c. deserticola* occurs in eastern Malheur Co., but no locality records for the form are listed for Oregon (Hall, 1981:327, map 230).

Habitat and Density.—*Lepus californicus* uses a wide variety of habitats within its range (Johnson and Anderson, 1984) and findings regarding habitat use often are inconsistent or contradictory. In Idaho, highest densities were associated with high proportions of grass cover. However, *L. californicus* rarely was seen or flushed in grassy areas in the daytime, but commonly was observed in or flushed from shrub cover (Johnson and Anderson, 1984). Thus, daytime association with shrub habitats may be to obtain protection from predators and from the sun (Lechleitner, 1959). However, in another study conducted in Idaho, greatest densities of black-tailed jackrabbits as indexed by spotlight censuses at night usually occurred in habitats dominated by black greasewood (*Sarcobatus vermiculatus*), followed, in order, by those dominated by big sagebrush (*Artemisia tridentata*), by big sagebrush and winterfat (*Eurotia lanata*), and by shadscale (*Atriplex confertifolia*) and winterfat (Smith and Nydegger, 1985). In a third study conducted in Idaho, densities were significantly higher near cultivated lands than in isolated rangelands (Fagerstone et al., 1980). Black-tailed jackrabbits reportedly replace white-tailed jackrabbits (*L. townsendii*) on ranges where grasses have been replaced by shrubs (Anthony, 1913; Bear and Hansen, 1966; Couch, 1927; Dalquest, 1948; Grayson, 1977), and the former species is most abundant in pastures subject to light or moderate grazing in summer and in which cover is heaviest. Where these two jackrabbit species are sympatric, black-tailed jackrabbits use the more densely vegetated areas (Flinders and Hansen, 1975).

Most published densities for *L. californicus* in various parts of its geographic range are 0.5–2.7/ha (Lechleitner, 1958b), but Palmer (1897) reported that about 8,000 black-tailed jackrabbits were killed during drives conducted during 9 days on 259 ha in California ($\cong$31/ha). In Utah, residents claimed that populations of *L. californicus* are cyclic with 7-year intervals between peaks (Gross et al., 1974); we heard similar claims for black-tailed jackrabbit populations in southeastern Oregon. Gross et al. (1974) provided evidence of three peaks in density in the Curlew Valley, Utah, between 1951 and 1970: in the early 1950s, early 1960s, and late 1960s or early 1970s. The density in 1970 was about 9 times that at the same season in 1967, an order of magnitude less than the difference observed in snowshoe hare (*L. americanus*) populations in some regions. Gross et al. (1974) also indicated that peaks in density were not synchronized among local areas. Much of the variation in density among years may be explained by changes in mortality; these rates are correlated with indices to coyote (*Canis latrans*) abundance (Wagner and Stoddart, 1972) and are thought to be related, in part, to predation by coyotes (Clark, 1972).

Diet.—Black-tailed jackrabbits are generalist herbivores. Because they commonly are associated with a variety of habitats, their diets are extremely variable; for example, in Idaho, 43 of 66 plant species known to be present on a study area were consumed by this jackrabbit (Johnson and Anderson, 1984). In general, grasses dominate in the diet in spring and early summer; forbs increase in late summer

and early autumn; and shrubs increase in late autumn and winter (Currie and Goodwin, 1966; Fagerstone et al., 1980; Johnson and Anderson, 1984; Sparks, 1968a). In June–July in Idaho, grasses composed ≥50% of the diet on most sites; the only other plant taxa selected by *L. californicus* were winterfat, starvation cholla (*Opuntia polyacantha*), and members of Boraginaceae and Leguminosae (Johnson and Anderson, 1984). Also in Idaho, *L. californicus* near cultivated lands selected barley and crested wheatgrass (*Agropyron cristatum*), and ate some alfalfa in summer, but did not eat large quantities of potato leaves (Fagerstone et al., 1980). There is some discrepancy in species reportedly eaten by *L. californicus*; some species such as cheat grass (*Bromus tectorum*) may be consumed in large quantities in some areas and little used or unused in other areas (Johnson and Anderson, 1984).

Black-tailed jackrabbits select forage plants for their succulence (Hansen and Flinders, 1969). Westoby (1980) indicated that in July–August in the Curlew Valley, Utah, most plants did not contain sufficient water for *L. californicus* to avoid water stress. Some plants that contain high levels of water and are eaten in summer also contain high concentrations of oxalate; presumably *L. californicus* concentrates urine to eliminate the salts (Westoby, 1980).

Some cultivars of grasses used in reclamation of rangelands are selected over others by black-tailed jackrabbits. This caused Ganskopp et al. (1993) to suggest that the less palatable types be planted in critical areas to reduce damage when populations of *L. californicus* are high and to reduce competition between *L. californicus* and livestock.

Reproduction, Ontogeny, and Mortality.—No information regarding reproduction, development, and survival rates is available for *L. californicus* in Oregon. In California, the breeding season is essentially year-round, but most reproduction is confined to January–August (Lechleitner, 1959). In Idaho, onset and cessation of breeding differ somewhat seasonally among years but the length of the breeding season remains relatively stable ($\overline{X} = 128$ days; range, 124–135 days—French et al., 1965).

Gestation was reported to average 43 days (range, 41–47 days—Haskell and Reynolds, 1947), but Goodwin and Currie (1965) reported a captive female producing three litters with intervals of 38 and 40 days between litters. Gross et al. (1974) used 40 days as an estimate of gestation for their study. Female black-tailed jackrabbits undergo postpartum estrus; thus the interval between litters may not exceed the gestation period (Lechleitner, 1959). A large proportion of females in a population may be both pregnant and lactating during much of the breeding season (Lechleitner, 1959). Gross et al. (1974) indicated that breeding (and subsequent parturition) occurred such that four or five litter cohorts were produced. Although several investigators indicated that breeding by young-of-the-year was unusual or did not occur (Feldhamer, 1979a; French et al., 1965; Lechleitner, 1959; Tiemeier, 1965), Gross et al. (1974) found that an average of 27% (range, 0–63%) of juveniles produced during the first litter of the year in Utah themselves bred and produced a litter during 7 of 9 years.

Average litter size increases to a peak then declines: when ordered by month, the peak occurs in April ($\overline{X} \cong 4$) in California (Lechleitner, 1959) and in May ($\overline{X}$s = 3.7–6.3) in Idaho (French et al., 1965); when ordered by litter cohort, either the second or third litter cohort is largest (Gross et al., 1974). Pre- and postimplantation losses were 6.7

and 6.2%, respectively, in California, but totaled 46.3% of ova shed in Idaho (Feldhamer, 1979a) and from 1 to 31% in different years in Utah (Gross et al., 1974). Lechleitner (1959) recognized that overall litter size was related to size of samples of each litter cohort, but reported that overall litter size in California declined by an average of only 0.2 young/litter when weighted by sample size. Litter size increases as population density declines and decreases as density rises (French et al., 1965). French et al. (1965) also suggested that litter size was correlated inversely with length of the breeding season (and latitude), a relationship discernible from data presented herein (Table 10-2). Estimates of total annual production of offspring were 9.8/female in California (Lechleitner, 1959), 10.2/female in Idaho (Feldhamer, 1979a), 12.3/female in Kansas (Tiemeier, 1965) and 13–14/female among captives in Arizona (Haskell and Reynolds, 1947).

At birth, young *L. californicus* have open eyes and ears, and are fully furred and capable of moving about weakly (Orr, 1940). By 2–3 days of age young begin to exhibit well-coordinated hopping (Haskell and Reynolds, 1947); within a week they are capable of crossing a road.

Young may or may not be deposited in depressions with no nest preparation; they commonly leave the nest area soon after birth; and littermates may stay together a week or more after leaving the birth area (Tiemeier, 1965). Within 24 h of birth, neonates average 3% of the body mass of adult females and average lengths of ear, hind foot, and body were 22%, 17%, and 29%, respectively, of the same measurements for adults (Goodwin and Currie, 1965). At 1 month of age, body length was 50% of that of adults, but body mass was only 13% of adult mass; by 2.5 months of age, body length, length of hind foot, and ear length were 90% of those of adults, but body mass still was only 65% of adult body mass (Goodwin and Currie, 1965). Females were reported to nurse offspring for only 17–20 days (Tiemeier, 1965), but Sparks (1968b) found milk in stomachs of individuals estimated to be 4 and 12–13 weeks old. Lechleitner (1957) found young rabbits with both milk curds and reingested soft fecal pellets in their stomachs, suggesting that young animals reingest soft feces of the maternal female to inoculate their digestive tracts with cellulolytic microflora. Before juveniles began to produce fecal pellets, maternal females reportedly "sucked the anuses of the young," presumably to extract feces as a measure to reduce detection by predators (Haskell and Reynolds, 1947:130).

Survival rates of only 9% at 1 year, 6% at 2 years, and 2% at 3 years were calculated for a population in Idaho (Feldhamer, 1979a); another study in the same state indicated that 3.5% survived 1 year (French et al., 1965). In Kansas, average life span was 1.8 years (range, 1.5–2.1 years—Tiemeier, 1965); 22% survived ≥1.5 years, 9% survived ≥2 years, and <1% survived ≥3 years (Tiemeier and Plenert, 1964). In California, average life span was 1.4 years and turnover of the population required ≅5 years (Lechleitner, 1959).

Coyotes (*Canis latrans*), badgers (*Taxidea taxus*), bobcats (*Lynx rufus*), golden eagles (*Aquila chrysaetos*), several species of hawks (*Buteo lagopus*, *B. swainsoni*, *Circus cyaneus*, *Falco peregrinus*) and owls (*Tyto alba* and *Bubo virginianus*), scrub jays (*Aphelocoma coerulescens*), rattlesnakes (*Crotalus viridis*), and gopher snakes

(*Pituophis melanoleucus*) are known to prey on *L. californicus* (Orr, 1940; Tiemeier, 1965). In Idaho, during a severe storm (temperatures ≅ –30°C; wind velocity = 80–95 km/h; snowfall = 36 cm) in January–February 1982, 37 of 59 telemetered *L. californicus* seemingly died of weather-induced causes (Stoddart, 1985). A storm of similar severity in Kansas in March 1957 also killed large numbers of black-tailed jackrabbits (Tiemeier, 1965).

Habits.—Commencing about daylight, black-tailed jackrabbits leave their foraging areas and return to daytime resting areas in denser and taller vegetation (Tiemeier, 1965). They spend the daytime hours in forms (never underground) reingesting soft fecal pellets and cleaning themselves (Lechleitner, 1957, 1958a). Forms may be depressions several centimeters deep dug in the soil at the base of a shrub or clump of vegetation or they may be a space pressed into grasses and forbs sufficiently large to conceal a jackrabbit. The animal sits facing outward with its rump toward the rear of the form at the base of a plant; forms provide some protection from the weather and predators. Sometimes animals rest hunched with ears laid back in the open or stretched on their sides basking in the sunshine (Lechleitner, 1958a).

Hearing is acute; movements, but not distant still objects, are detected by sight; and olfaction is used to identify foods and detect conspecifics (Lechleitner, 1958a). At the approach of danger, *L. californicus* becomes alert by raising its ears, and, if the threat is still distant, by standing on the hind legs, with ears erect and nose twitching. When an individual becomes alert, others in the vicinity also become alert; the mode of intraspecific communication is not known. When the threat gets near the animals either flee or freeze (Lechleitner, 1958a). If the danger is at a distance and the jackrabbit is in dense vegetation, it may lay its ears back and attempt to sneak away, then become nearly invisible by remaining motionless. If the potential threat approaches rapidly or when approached in relatively thin cover, black-tailed jackrabbits commonly flee, taking long leaps with an occasional longer and higher leap (Lechleitner, 1958a). *L. californicus* can attain speeds of 43–61 km/h (Cottam and Williams, 1943; Lechleitner, 1958a). When fleeing at high speed, black-tailed jackrabbits often collide with conspecifics, other animals, and fences (Lechleitner, 1958a).

"The major feeding period is at dusk and probably extends well into the night with another surge of feeding activity near dawn" (Lechleitner, 1957:482). However, wind, falling snow, and fog, and sometimes sharply falling temperatures, reduce or prevent nighttime activity (Tiemeier, 1965). Often large numbers of individuals are observed to feed in the same field, but these aggregations seem to be a response to quality and availability of food rather than to a social organization (Lechleitner, 1958a).

The only social interactions other than those related to reproduction seem to be between maternal females and their offspring, and these last only to weaning (Lechleitner, 1958a). Relatively little is known regarding behavior of neonates or of neonates and maternal females, but Stoddart (1984) hypothesized that, although no nest is prepared, neonates remain grouped and exhibit strong site fidelity for at least a week. At the approach of danger, neonates remain motionless except when the threat of an attack is obvious, whereupon they move away independently. Adult females stay away from the litter except when nursing, but they are thought to regroup a litter that disperses in response to disturbance (Stoddart, 1984).

Lechleitner (1958a) described the "sexual hunt" by male *L. californicus* as consisting of random running, usually along trail systems with the nose close to the ground. Males made brief pauses to sniff at vegetation and other objects during coursing that commonly took them over an area of 14 ha and required as long as 1 h. Upon encountering another male, the approaching male stretched its neck and continued to sniff with ears erect. Usually, encounters ended with the males separating after sniffing each other, but sometimes they engaged in reciprocal short aggressive chases. Upon encountering a female, the female faced the male and lowered its ears; if the male continued to approach slowly, the female charged and boxed the male; if the male persisted, the female jumped into the air as the male ran under and discharged urine. Sometimes the male jumped and the female ran under it. If, at the onset, the female retreated, a chase ensued; sometimes other individuals joined the chase. Males attempted to mount whenever the female was caught, but were rarely successful. Frequently, capture led to male-female fights. Fighting consisted of the pair standing on their hind feet or leaping into the air and boxing with the forefeet. Although combatants were not observed to bite, they commonly spat fur after the engagements. Copulation sometimes followed chases and boxing. This behavior is essentially identical to that described in eastern cottontails (*Sylvilagus floridanus*) by Marsden and Holler (1964). Sometimes the sequence was interrupted suddenly with individuals engaging in displacement behavior, usually feeding or digging and rolling in the dirt (Lechleitner, 1958a).

In summer, home ranges of adult males in Kansas averaged 17.2 ha whereas those of adult females averaged 17.1 ha; range for both sexes combined was 5.3–78.5 ha (Tiemeier, 1965). In winter, black-tailed jackrabbits moved to areas supporting succulent vegetation, sometimes >3 km from summer ranges. In California, home ranges were well defined, usually <20 ha, and overlapping; the size was thought to be determined by the "pattern of food, water, and cover in the environment" (Lechleitner, 1958b:383). When driven out by a flood, several individuals returned to their home ranges even though most landmarks were obliterated (Lechleitner, 1958b).

Remarks.—Black-tailed jackrabbits often are blamed for deterioration of grazing lands, especially because densities of this species increase where lands are heavily grazed. However, considering that jackrabbits and other native grazing mammals were part of a dynamic, but more stable, ecosystem before the advent of livestock grazing, it seems more likely that under moderate grazing jackrabbits exert forces favoring community development. On rangelands sufficiently deteriorated that weeds dominate over grasses, indeed, jackrabbits may contribute toward further deterioration (Bond, 1945; Hansen and Flinders, 1969).

Because of the extreme variation in results for studies of habitat use, diet, reproduction, and short-term density, we believe that replication in Oregon likely will not contribute significantly to understanding the overall biology of the species. A more productive approach might be a long-term ($\cong$20 years) and fine-scale ($\leq$5 km) study of density over a broad ($\cong$625 km^2) area. Even if based on

once-a-year (prerecruitment) indices, such a study might provide insight into temporal and spatial patterns in fluctuations in abundance of *L. californicus*. Such a project, of course, would be a career commitment.

We are somewhat concerned about the status of *L. californicus* populations west of the Cascade Range, particularly those in the Willamette Valley with which we are most familiar. In 1980, we (Verts and Carraway, 1981) drove 3,047.2 km on secondary roads in a 1,501.9-km^2 area in Linn Co. during early morning or late evening hours and saw 96 eastern cottontails, but only one black-tailed jackrabbit. It was a road kill. In 1973, 1979, and 1991, we made collections totalling 172 eastern cottontails at night by use of a spotlight on the Oregon Department of Fish and Wildlife, E. E. Wilson Wildlife Area in northeastern Benton Co.; we saw no *L. californicus*, although Don Trethewey collected three specimens there in 1968. Our only other observations of *L. californicus* in the Willamette Valley since 1979 were a road kill (OSUFW 7220) near Amity, Yamhill Co., and a live individual at the corner of Walnut Boulevard and Highland Drive in Corvallis, Benton Co. We suspect that *L. californicus* may be associated with dense blackberry (*Rubus*) brambles during daylight hours, thus difficult to observe or flush. However, if population densities were even a 10th of the minimum reported (0.5/ha), seemingly we should have observed some during our nighttime collecting trips.

Lepus townsendii Bachman
White-tailed Jackrabbit

1839a. *Lepus townsendii* Bachman, 8:90, pl. 2.
1904c. *Lepus campestris sierrae* Merriam, 17:132.

Description.—*Lepus townsendii* (Plate VI) is the largest of the lagomorphs in Oregon (Table 10-5). Summer and winter pelages of white-tailed jackrabbit are drastically different. In summer, the hairs are light gray at the base followed by a narrow band of brown, a wide band of buff, and a narrow tip of black; the overall appearance is grizzled dark grayish with overtones of pinkish buff, blending lighter on the sides. The venter is white except for throat and chest regions colored similar to that of the sides. The ears are slightly darker than the dorsum except for a white strip on the posterior lateral margin and a black tip. The feet are mottled white and buff. The tail is entirely white. In winter, the hairs are white at the base followed by a band of gray, a narrow band of white, and a black tip; the overall appearance is the color of dirty snow, shading lighter on the sides and snowy white on the venter. The chest has a few dark hairs. The ears are white with a few partly grayish hairs. The feet are white with buffy soles. The tail is white except for a sooty band on the dorsal portion. In Colorado, the winter pelage varied with some individuals nearly pure white and others with grades of darkness; differences were considered not to be the result of selection resulting from differential predation (Hansen and Bear, 1963). The hind feet are >105 mm and the ears are >80 mm. There are six–10 mammae (Seton, 1929b).

In *L. townsendii*, the occipitonasal length is >75 mm; the supraorbital process has an anterior projection; the posterior projection of the supraorbital process usually is short and obtusely angled, rarely fused to the cranium; and the rostrum is broad and with nearly parallel sides (Fig.

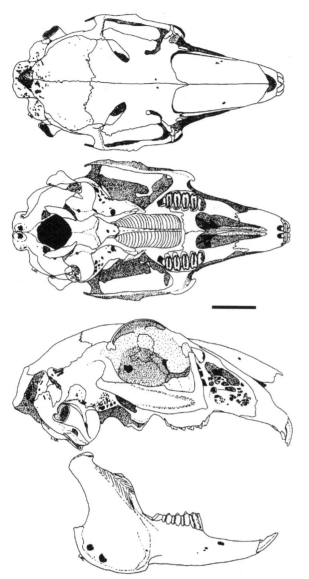

Fig. 10-18. Dorsal, ventral, and lateral views of the cranium, and lateral view of the dentary of the white-tailed jackrabbit, *Lepus townsendii* (OSUFW 1489). Scale bar equals 15 mm.

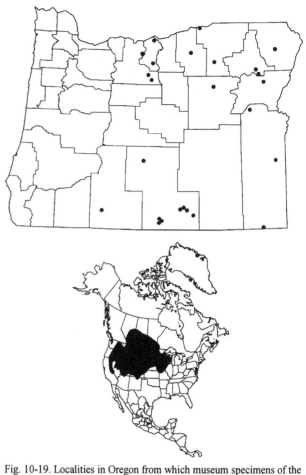

Fig. 10-19. Localities in Oregon from which museum specimens of the white-tailed jackrabbit, *Lepus townsendii*, were examined (listed in the APPENDIX, p. 521), and the distribution of the white-tailed jackrabbit in North America. (Range map redrawn from Lim, 1987:2, fig. 3, with modifications.) The range has expanded eastward and northward since settlement (Lim, 1987).

10-18). As with other *Lepus* the interparietal bone is fused to and indistinguishable from the parietal bones. There are no characteristics of the skull that can be used to distinguish *L. townsendii* from *L. californicus* with absolute assurance (Hoffmann and Pattie, 1968).

Distribution.—The geographic range of *L. townsendii* (Fig. 10-19) extends from extreme southern British Columbia and east of the Cascade Range in Washington and Oregon south to Mono Co., California, then eastward to Lake Michigan in Wisconsin, extreme northwestern Illinois, and central Missouri. Latitudinally, the range extends from about 53°N (central Saskatchewan) to about 36°N (northern New Mexico). The eastern and northern portions of what is now the current range were invaded in historic times in response to conversion of forests to agricultural and park lands (Lim, 1987). In Oregon, the white-tailed jackrabbit occurs east and south of a line connecting Rufus, Sherman Co.; Maupin and Antelope, Wasco Co.; John Day, Grant Co.; Juntura, Malheur Co.; Fields, Harney

Co.; Fort Rock, Lake Co.; and Fort Klamath and Klamath Falls, Klamath Co. (Fig. 10-19).

Pleistocene fossils tentatively identified as *L. townsendii* are known from Alberta, Idaho, Wyoming, New Mexico, and Texas; sites in the latter two states are extralimital (Kurtén and Anderson, 1980).

Geographic Variation.—Only one of the two nominal subspecies, *L. t. townsendii*, occurs in Oregon (Lim, 1987).

Habitat and Density.—*Lepus townsendii* usually is associated with bunchgrass habitats in the western part of its range (Anthony, 1913; Bear and Hansen, 1966; Couch, 1927; Dalquest, 1948; Flinders and Hansen, 1973; Orr, 1940). In many areas in which grasses have been replaced by shrubs the white-tailed jackrabbit has been replaced by the black-tailed jackrabbit (*Lepus californicus*—Anthony, 1913; Bear and Hansen, 1966; Couch, 1927; Dalquest, 1948; Grayson, 1977). In Colorado, the shift from one species to the other was thought to occur in the late 1800s (Flinders and Hansen, 1972); in Oregon, the shift commenced 5,000–7,000 years ago (Grayson, 1977). In winter in Utah, white-tailed jackrabbits were abundant in hilly areas vegetated with big sagebrush (*Artemisia tridentata*), bitter-brush (*Purshia tridentata*), and giant wildrye (*Elymus cinereus*). In summer they were common on north slopes supporting patches of chokecherry (*Prunus virginiana*), snowberry (*Symphoricarpos rotundifolus*), and mountain myrtle (*Pachistima myrsinites*—Janson, 1946). Other habitats occupied in Utah also were described as containing shrubs (Janson, 1946). *L. townsendii* reportedly moves to lower altitudes in winter (Orr, 1940) and, in severe weather, may move into the edges of forested areas (Bailey, 1927; Kline, 1963). In Colorado, *L. townsendii* occurs in alpine areas above the Krummholz in all seasons (Braun and Streeter, 1968).

Although no estimates of density are available for *L. townsendii* in Oregon, estimates were 1.4/km² in northeastern Colorado (Flinders and Hansen, 1973), 3.9– 44.0/km² in Iowa (Kline, 1963), and 3.5–32.0/km² in Minnesota (Mohr and Mohr, 1936). Rogowitz and Wolfe (1991) reported March and October estimates of density of 7.0 and 7.2/km², 6.2 and 6.4/km², and 6.4 and 9.0/km², respectively, during 3-consecutive years in Wyoming. Mohr and Mohr (1936) suggested a negative relationship between density and precipitation during the growing season. Olterman and Verts (1972), based on conversations with wildlife biologists and naturalists, believed that *L. townsendii* was relatively common in Oregon although occupied habitats were scattered.

Diet.—In Colorado, the summer diet of *L. townsendii* consisted of 70% forbs, 19% grasses and grasslike plants, and 7% shrubs; in autumn, grasses and grasslike plants composed 43% of the diet, forbs 34%, and shrubs 14%; in winter, 76% of the diet consisted of shrubs, 12% forbs, and 5% grasses; and in spring, 87% of the diet was shrubs, 4% forbs, and 4% grasses (Bear and Hansen, 1966). Major species eaten were shrubs—Parry's rabbit-brush (*Chrysothamnus parryi*), fringed sagebrush (*Artemisia frigida*), and winterfat (*Eurotia lanata*); grasslike plants—bent sedge (*Carex obtusata*); and forbs—clover (*Trifolium*), common dandelion (*Taraxacum officinale*), goosefoot (*Chenopodium*), and Indian-paintbrush (*Castilleja integra*). Only the woody portion of some shrubs was eaten, but the entire plant of sagebrush was eaten (Bear and Hansen, 1966).

In Utah, grasses were a principal component of the diet in winter (both dried grasses and new growth); hay, waste grain, and new growth of winter grain also were consumed in large amounts (Janson, 1946). When snow covered grasses, sagebrush, rabbit-brush, and snakeweed (*Gutierrezia sarothrae*) were eaten. In some areas in Utah in late winter, prickly pear cactus (*Opuntia*) and juniper (*Juniperus*) berries were eaten. In summer, the diet consisted mostly of green grasses, sedges, forbs, and a small amount of sagebrush. Grain and alfalfa were consumed in a strip about 9 m wide adjacent to cover (Janson, 1946).

White-tailed jackrabbits are coprophagous; most animals collected in midday have material identifiable as reingested pellets in their stomachs (Spencer, 1955).

Reproduction, Ontogeny, and Mortality.—In North Dakota, the testes were flaccid and averaged <2 g in August–December; they became more turgid and increased in mass in January; and they attained a peak mass ($\overline{X} \cong 11$ g) in March, but remained >8 g in February–June. In July, testis mass averaged >4 g, but varied greatly, possibly a result of including masses of reproductively active individuals collected early in the month with those of reproductively inactive individuals collected late in the month (James and Seabloom, 1969). In Wyoming, *L. townsendii* commenced to breed from late February to mid-March and

ceased to breed in July; late snow melt retarded onset of breeding (Rogowitz, 1992).

Breeding appeared to be synchronous with immediate postpartum estrus (Kline, 1963; Rogowitz, 1992); thus, females, except a few that failed to breed during postpartum estrus, were continuously pregnant from onset of the breeding season at least through birth of the third litter. The gestation period usually is assumed to be 42 days (Kline, 1963; Rogowitz, 1992). Only 29% of those examined during August–September (period of gestation for the fourth litter) were pregnant (James and Seabloom, 1969). Juveniles do not breed in the year of their birth (Bear and Hansen, 1966; Kline, 1963; Rogowitz, 1992).

As for the eastern cottontail (*Sylvilagus floridanus*), an overall average litter size may be a nonsensical statistic (Table 10-2); variation in average litter size among litter cohorts is such that an overall average would be biased by examining different numbers of females pregnant with each of the synchronous litters. In North Dakota and Wyoming, second litters were largest, first litters next in size, and third or fourth litters smallest (James and Seabloom, 1969; Rogowitz, 1992). Others reported production of only one annual litter (Bailey, 1927; Gunderson and Beer, 1953).

Preimplantation losses were high, averaging 0.9 ova or zygotes/pregnancy in North Dakota; postimplantation losses averaged 0.3 embryos/pregnancy (James and Seabloom, 1969). In Iowa, preimplantation losses averaged 36.6% (Kline, 1963). In Wyoming, an average of 6.6–7.1 ova/female was shed during each of the first two litter periods, but 36% and 5–9% interuterine mortality reduced the average number of viable fetuses to 4.2– 4.5 and 6.2 for first and second litters, respectively (Rogowitz, 1992). Production of 5.3–6.0 ova/female in the third litter period during 2 years and high (up to 86%) interuterine mortality during another year reduced the average number of viable fetuses to 3.5–4.1 for the third litter (Rogowitz, 1992). Average annual production by adult females was estimated to be 15 young in North Dakota (James and Seabloom, 1969).

Young white-tailed jackrabbits are highly precocial; they are fully furred, their eyes and ears are open, and their incisors erupt before birth (Bear and Hansen, 1966; Davis, 1939). Near term, fetuses average about 90 g, and the smallest young caught in traps also weigh about 90 g (Bear and Hansen, 1966). By 2 weeks of age, young average about 280 g, by 4 weeks about 765 g, and by 6 weeks about 1,160 g; by 24 weeks young average 76–92% of adult body mass (Bear and Hansen, 1966).

Coyotes (*Canis latrans*), bobcats (*Lynx rufus*), and eagles (*Aquila chrysaetos* and *Haliaeetus leucocephalus*) are the primary predators of white-tailed jackrabbits in Washington (Dalquest, 1948). We suspect that a variety of mammalian, avian, and reptilian predators take young white-tailed jackrabbits. In Wyoming, of 245 individuals >6 months old, only 26% were >12 months old and only 12% were >18 months old. Annual survival rates for both sexes combined were 9% and 20% during 2 consecutive years; modal longevity was estimated to be <1 year (Rogowitz and Wolfe, 1991).

Habits.—White-tailed jackrabbits usually are active at night; in the daylight hours they remain in their forms, even until almost being stepped upon. Forms commonly are beneath sagebrush (Bear and Hansen, 1966; Janson, 1946). In winter, these jackrabbits commonly use cavities at the end of ≅1-m-long tunnels in the snow (Bear and Hansen, 1966). To escape danger or when wounded, they may enter dens dug by other species (Janson, 1946). Females usually are solitary, whereas males often form groups of two to five individuals (Bear and Hansen, 1966); however, large concentrations have been observed occasionally in the northern portion of the range both during the day and night (Brunton, 1981; Lahrman, 1980).

Anthony (1913:18) described the gait of the white-tailed jackrabbit as "a halting, one-sided lope." When pursued, these jackrabbits may attain a speed of 55 km/h (Cottam and Williams, 1943).

Blackburn (1973) described the mating behavior of the black-tailed jackrabbit as lasting 5–20 min and involving circling, male and female approaches, long chases, and urine emission followed by copulation. He indicated that courtship behavior of the white-tailed jackrabbit was similar but that jumping was more pronounced.

Remarks.—Much of the published information available was based on studies of the eastern subspecies (*L. t. campanius*),

thus may not provide a reliable assessment of the natural history of *L. townsendii* in Oregon. Exceptions are the more recent studies in Wyoming (Rogowitz, 1990, 1992; Rogowitz and Gessaman, 1990; Rogowitz and Wolfe, 1991).

We are not certain that any of the populations in Oregon could withstand collections of the magnitude of those made by Bear and Hansen (1966), Flinders and Hansen (1972), James and Seabloom (1969), or Rogowitz (1992) without seriously affecting the results of a study. Nevertheless, several aspects of the biology of the species could be evaluated in Oregon without large collections of specimens. We believe that foremost among such studies should be an evaluation of the extent and nature of occupied habitats, and estimates of abundance. Because the white-tailed jackrabbit is associated with widely scattered, late-seral grassland communities frequently disturbed by agricultural and grazing practices, we believe that the species could become seriously depleted or extirpated locally, despite assurances by biologists and naturalists that it was common in Oregon (Olterman and Verts, 1972).

PLATE I

Didelphis virginiana, Virginia opossum—Lee W. Kuhn

Sorex bendirii, Pacific water or Marsh shrew—Ronn Altig

Sorex palustris, Water shrew—Richard B. Forbes

Sorex sonomae, Fog shrew—Robert B. Smith

PLATE II

Sorex trowbridgii, Trowbridge's shrew—Ronn Altig

Sorex vagrans, Vagrant shrew—Robert M. Storm

Neurotrichus gibbsii, Shrew-mole—Ronn Altig

Scapanus orarius, Coast mole—Richard B. Forbes

PLATE III

Myotis yumanensis, Yuma myotis—Merlin D. Tuttle, Bat Conservation International

Myotis ciliolabrum, Western small-footed myotis—Richard B. Forbes

Myotis californicus, California myotis—John P. Hayes

Lasiurus cinereus, Hoary bat—Richard B. Forbes

Euderma maculatum, Spotted bat—
Merlin D. Tuttle, Bat Conservation International

PLATE IV

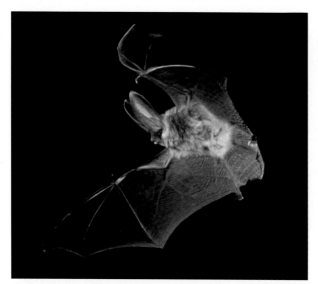

Corynorhinus townsendii, Townsend's big-eared bat—
Merlin D. Tuttle, Bat Conservation International

Antrozous pallidus, Pallid bat—
James Collins, Oregon Department of Fish and Wildlife

Tadarida brasiliensis in flight—Merlin D. Tuttle, Bat Conservation International

Tadarida brasiliensis, Brazilian free-tailed bat—
Merlin D. Tuttle, Bat Conservation International

PLATE V

Ochotona princeps, American pika—
Chuck Stelling, Oregon Department of Fish and Wildlife

Brachylagus idahoensis, Pygmy rabbit—
Jeffrey S. Green and Jerran T. Flinders

Sylvilagus nuttallii, Mountain cottontail—
Marguerite M. Smith

Sylvilagus floridanus, Eastern cottontail—Ronn Altig

Sylvilagus bachmani, Brush rabbit—Ronn Altig

Sylvilagus nuttallii in a juniper tree—Steven D. Gehman

PLATE VI

Lepus townsendii in winter pelage—Jerran T. Flinders

Lepus americanus, Snowshoe hare—Ronn Altig

Lepus californicus, Black-tailed jackrabbit

Lepus townsendii, White-tailed jackrabbit, in summer pelage—Jerran T. Flinders

PLATE VII

Aplodontia rufa, Mountain beaver—Ronn Altig

Tamias senex, Allen's chipmunk—Richard B. Forbes

Tamias townsendii, Townsend's chipmunk—Ronn Altig

Tamias amoenus, Yellow-pine chipmunk—Richard B. Forbes

Tamias minimus, Least chipmunk—Ronn Altig

Tamias siskiyou, Siskiyou chipmunk—Richard B. Forbes

PLATE VIII

Ammospermophilus leucurus,
White-tailed antelope squirrel—Ronn Altig

Marmota flaviventris, Yellow-bellied marmot—
James Collins, Oregon Department of Fish and Wildlife

Spermophilus beecheyi, California ground squirrel—Winston P. Smith

Spermophilus beldingi, Belding's ground squirrel—Robert M. Storm

Spermophilus beldingi juveniles—Richard B. Forbes

PLATE IX

Spermophilus columbianus, Columbian ground squirrel

Spermophilus elegans,
Wyoming ground squirrel—Eric Yensen

Spermophilus washingtoni,
Washington ground squirrel—Eric Yensen

Spermophilus canus, Merriam's ground squirrel—Ronn Altig

Spermophilus lateralis, Golden-mantled ground squirrel—Ronn Altig

PLATE X

Sciurus carolinensis, Eastern gray squirrel—Richard B. Forbes

Sciurus niger, Eastern fox squirrel—Richard B. Forbes

Sciurus griseus, Western gray squirrel—M. S. (Elzy) Eltzroth

PLATE XI

Tamiasciurus hudsonicus, Red squirrel—
Mike Kemp, Oregon Department of Fish and Wildlife

Tamiasciurus douglasii, Douglas' squirrel—Ronn Altig

Glaucomys sabrinus, Northern flying squirrel—James Collins, Oregon Department of Fish and Wildlife

PLATE XII

Thomomys talpoides, Northern pocket gopher—Ronn Altig

Thomomys mazama burrow casts

Thomomys bulbivorus, Camas pocket gopher—Kenneth L. Gordon

Thomomys bulbivorus excavating earth—David Budeau

PLATE XIII

Perognathus parvus, Great Basin pocket mouse—Ronn Altig

Microdipodops megacephalus, Dark kangaroo mouse

Dipodomys ordii, Ord's kangaroo rat—Ronn Altig

Dipodomys microps, Chisel-toothed kangaroo rat—Ronn Altig

Dipodomys californicus, California kangaroo rat—Stephen Cross

PLATE XIV

Castor canadensis, American beaver—Ronn Altig

Onychomys leucogaster, Northern grasshopper mouse—
Ronn Altig

Peromyscus maniculatus, Deer mouse—Ronn Altig

Peromyscus truei, Piñon mouse

PLATE XV

Neotoma fuscipes, Dusky-footed woodrat—Ronn Altig

Neotoma cinerea nest in juniper tree

Neotoma cinerea, Bushy-tailed woodrat—Ronn Altig

Neotoma lepida, Desert woodrat—Ronn Altig

PLATE XVI

Rattus norvegicus, Norway rat—Richard B. Forbes

Mus musculus, House mouse—Ronn Altig

PLATE XVII

Clethrionomys gapperi, Southern red-backed vole—Ronn Altig

Microtus californicus, California vole

Phenacomys longicaudus, Red tree vole—Ronn Altig

Microtus longicaudus, Long-tailed vole—Ronn Altig

PLATE XVIII

Microtus richardsoni, Water vole

Microtus montanus, Montane vole—Richard B. Forbes

Ondatra zibethicus, muskrat lodge—
Oregon Department of Fish and Wildlife

Lemmiscus curtatus, Sagebrush vole—
Murray L. Johnson

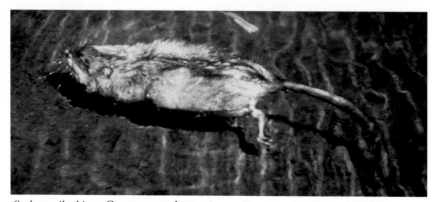

Ondatra zibethicus, Common muskrat—Oregon Department of Fish and Wildlife

PLATE XIX

Zapus princeps, Western jumping mouse—Ronn Altig

Erethizon dorsatum, Common porcupine—Robert M. Storm

Zapus trinotatus, Pacific jumping mouse—
Richard B. Forbes

Myocastor coypus, Nutria—Ron Romig, *Albany Democrat-Herald*

PLATE XX

Canis latrans, Coyote—James Collins, Oregon Department of Fish and Wildlife

Canis lupus, Gray wolf—Richard B. Forbes

PLATE XXI

Urocyon cinereoargenteus, Common gray fox—Ronn Altig

Vulpes velox, Kit fox—Cal Geisler, Oregon Department of Fish and Wildlife

Vulpes vulpes, Red fox—Ronn Altig

PLATE XXII

Ursus americanus, Black bear—Ronn Altig

Ursus arctos, Grizzly or Brown bear—Richard B. Forbes

PLATE XXIII

Zalophus californianus, California sea lion—National Marine Fisheries Service

Eumetopias jubatus, Northern sea lion—
National Marine Fisheries Service

Callorhinus ursinus, Northern fur seal—
Thomas R. Loughlin, National Marine Fisheries Service

Eumetopias jubatus rookery—Thomas R. Loughlin, National Marine Fisheries Service

PLATE XXIV

Phoca vitulina, Harbor seal—David J. Rugh, National Marine Fisheries Service

Mirounga angustirostris, Northern elephant seal—Richard B. Forbes

PLATE XXV

Procyon lotor, Common raccoon—James Collins, Oregon Department of Fish and Wildlife

Martes americana, American marten—Richard B. Forbes

PLATE XXVI

Mustela erminea, Ermine in summer pelage—Ronn Altig

Mustela vison, Mink—Richard B. Forbes

Mustela erminea in winter pelage—Matthew G. Hunter

Mustela frenata, Long-tailed weasel—Ronn Altig

PLATE XXVII

Taxidea taxus, American badger—Robert M. Storm

Gulo gulo, Wolverine—Ellis Mason, Oregon Department of Fish and Wildlife

PLATE XXVIII

Enhydra lutris, Sea otter—Richard B. Forbes

Lutra canadensis, River otter—Richard B. Forbes

Mephitis mephitis, Striped skunk—Ronn Altig

Spilogale gracilis, Western spotted skunk—Ronn Altig

PLATE XXIX

Puma concolor, Mountain lion—Richard B. Forbes

Lynx rufus, Bobcat—Richard B. Forbes

Lynx canadensis, Lynx—Richard B. Forbes

PLATE XXX

Odocoileus hemionus hemionus, female and male Mule deer—Robert M. Storm

Odocoileus hemionus columbianus, female Black-tailed deer—
Oregon Department of Fish and Wildlife

Odocoileus hemionus hemionus,
male Mule deer—Aaron Drew

Odocoileus virginianus, female White-tailed deer—Richard B. Forbes

PLATE XXXI

Cervus elaphus, male Elk—Robert M. Storm

Cervus elaphus, female Elk—
Richard B. Forbes

Bos bison, male Bison—Richard B. Forbes

Bos bison, female Bison—Richard B. Forbes

PLATE XXXII

Antilocapra americana, male Pronghorn—Richard B. Forbes

Oreamnos americanus, female with young Mountain goat—
Oregon Department of Fish and Wildlife

Oreamnos americanus, male Mountain goat—Richard B. Forbes

Ovis canadensis, male Mountain sheep—Ronn Altig

Ovis canadensis, female Mountain sheep—
Oregon Department of Fish and Wildlife

CHAPTER 11

ORDER RODENTIA—GNAWING MAMMALS

Rodents are distributed worldwide. Originally they did not occur in Antarctica, New Zealand, and some oceanic islands, but species commensal with humans are well established everywhere. They are the most successful mammal group, not only in terms of distribution, but also in diversity as >2,000 living species are known (Wilson and Reeder, 1993a). Fossil rodents date from the late Paleocene of North America (family †Ischyromyidae), but the group probably had an ancestry in common with the lagomorphs in the order †Anagalida of eastern Asia (Carroll, 1988). Most of the rodents are small; the largest in North America is the beaver (*Castor canadensis*). However, some of the extinct species were huge; the largest, †*Eumegamys*, a South American caviomorph rodent, had a skull 60 cm long (Carroll, 1988).

The rodents are characterized by a pair of large, evergrowing incisors in both upper and lower jaws; the incisors have enamel only on their anterior surfaces; thus, by differential wear, they remain sharpened chisels. The canines are absent, producing a broad diastema between incisors and molariform teeth. The molariform teeth often are reduced in number, either brachydont or hypsodont, and with occlusal surfaces ranging from tuberculate to prismatic to complexly laminate (Carleton, 1984).

The skull ranges from smooth and streamlined to angular and rugose with prominent sagittal, lambdoidal, or temporal ridges. The glenoid fossa is large and shallow, permitting lateral and anterior-posterior motion of the jaw. The infraorbital foramen in some is small and serves as passage only for blood vessels and nerves, but in others may be keyhole shaped or broadly oval, permitting passage of part of the masseter muscle that originates on the rostrum. The masseter provides most of the force to close the jaw as the temporalis muscle is relatively small. There are usually 13 thoracic and six lumbar vertebrae. A gallbladder and a baculum (males) are present in most species.

In Oregon, the rodents are represented by 63 species, 46.3% of the land mammals in the state. These are divided into nine families and 27 genera, and include arboreal, fossorial, terrestrial, and aquatic forms. Some are adapted for gliding, running, leaping, swimming, climbing, or digging. Most are herbivorous, but some are omnivorous and much of the diet of a few at times is largely of animal origin. Many have high reproductive rates and correspondingly high rates of mortality. They serve as prey for many carnivorous birds and mammals.

KEY TO THE FAMILIES OF RODENTIA IN OREGON

1a. Tail extremely short; pelage uniformly dark brown with white spot below ear; upper molariform teeth with labial spinelike projections; lower molariform teeth with lingual spinelike projections; auditory bulla flask shaped....................................**Aplodontidae** (p. 154).

1b. Tail readily visible; pelage variously colored, but if white spot below ear, then pelage not uniformly colored; molariform teeth lacking labial or lingual spine-like projections; auditory bulla not flask shaped........2

2a. Infraorbital foramen round and equal to or larger than foramen magnum; either a combination of hind feet webbed and tail round and scaly or hind feet not webbed and tail equipped with quills.........................3

2b. Infraorbital foramen oval, round, or keyhole shaped, but never as large as foramen magnum; if hind feet webbed, then tail not round; if hind feet not webbed, then tail not equipped with quills.................................4

3a. Four digits on front feet, five on hind feet; feet not webbed; many hairs on dorsum and tail modified as barbed quills; paroccipital process does not extend ventrally as far as auditory bulla; posterior margin of incisive foramen anterior to zygomatic plate........... ....**Erethizontidae** (p. 345).

3b. Pentadactyl; hind feet webbed; no quills; paroccipital process extends ventrally far beyond auditory bulla; posterior margin of incisive foramen even with anterior margin of zygomatic plate.................................... ....**Myocastoridae** (p. 348).

4a. Tail spatulate, scaly, and nearly hairless; hind feet webbed with web extending beyond proximal edge of toenail; basal length >75 mm........................... ....**Castoridae** (p. 256).

4b. Tail not spatulate, if flattened dorsoventrally, then well haired; feet not webbed, or if webbed, then entire distal phalanges extend beyond web; basal length <75 mm...5

5a. Cheek pouches open on either side of the mouth; infraorbital foramen on side of rostrum anterior to zygomatic plate..6

5b. No cheek pouches with external openings (some have internal membranous cheek pouches); infraorbital foramen opening on or near zygomatic plate..............7

6a. Tail longer than length of head and body; hind legs longer than front legs; infraorbital foramina connected through opening on side of rostrum.......................... ....**Heteromyidae** (p. 239).

6b. Tail shorter than length of head and body; front legs not noticeably shorter than hind legs; infraorbital foramina not connected through opening on side of rostrum....................................**Geomyidae** (p. 223).

7a. Tail usually bushy, often flattened dorsoventrally; skull with prominent postorbital processes on frontal bones; four molariform teeth in lower jaw....**Sciuridae** (p. 157).

7b. Tail not bushy or flattened dorsoventrally (except in *Neotoma cinerea*); no postorbital processes on frontal bones; three molariform teeth in lower jaw....................8

8a. Tail ≅150% of length of head and body; hind feet modified for saltatorial locomotion; four upper molariform teeth; yellow wash on sides of body; dental battery consisting of 18 teeth..................... ....**Dipodidae** (p. 339).

8b. Tail <150% of length of head and body; hind feet not greatly modified for saltatorial locomotion; three upper molariform teeth; sides usually not washed with yellow; dental battery consisting of 16 teeth........... ....**Muridae** (p. 262).

FAMILY APLODONTIDAE—APLODONTIDS

The family is represented by only one extant species.

Aplodontia rufa (Rafinesque)
Mountain Beaver

1817a. *Anisonyx? rufa* Rafinesque, 2:45.
1829b. *Aplodontia* Richardson, 4:334.
1899a. *Aplodontia pacifica* Merriam, 13:19.
1914. *Aplodontia rufa chryseola* Kellogg, 12:295.
1918. *Aplodontia rufa pacifica*: Taylor, 17:467.

Description.—*Aplodontia rufa* (Plate VII) is a medium-sized (Table 11-1) muskrat-like (*Ondatra zibethicus*) rodent often lacking a visible tail. However, the mountain beaver has an extremely short, fur-covered tail, and otherwise differs from the muskrat by possessing five-toed feet without fimbriae, a partially opposable pollex, and a one-piece bifurcate baculum. The pelage is dark brown dorsally, slightly lighter ventrally, and there is a small white spot at the base of each ear.

The skull is flattened dorsoventrally and widened posteriorly; the nasals are arched slightly (Fig. 11-1). The cheekteeth lack complex folds of enamel; each consists of a puddle of dentine surrounded by enamel. In adults, each cheektooth possesses a spinelike projection, laterally in the upper jaw, medially in the lower jaw (Fig. 11-1). The auditory bulla is flask shaped.

Distribution.—*Aplodontia rufa* occurs west of the Cascade Range from southern British Columbia south through western Washington, Oregon, and extreme northwestern California (Fig. 11-2). Disjunct populations occur along the Sierra Nevada Range of eastern California and extreme western Nevada, and as tiny populations in each Mendocino and Marin counties, California (Dalquest and Scheffer, 1945; Hall, 1981). In Oregon, *A. rufa* occurs in forested areas on the west slope of the Cascade Range west to the Pacific Ocean; it is absent from the interior valleys (Fig. 11-2).

Despite a somewhat diverse fossil lineage extending from the early-middle Oligocene (Rensberger, 1975) and perhaps from the late Eocene (Savage and Russell, 1983), fossils of *Aplodontia* are known from only three Pleistocene-age sites, all in northern California and all within the present-day range of the genus (Kurtén and Anderson,

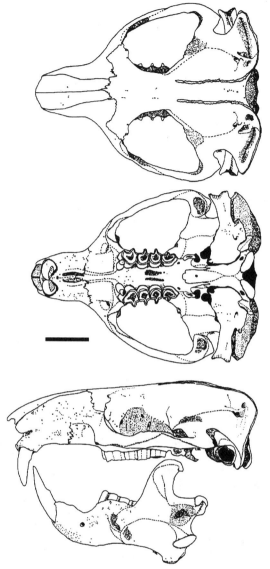

Fig. 11-1. Dorsal, ventral, and lateral views of the cranium, and lateral view of the dentary of the mountain beaver, *Aplodontia rufa* (OSUFW 7404). Scale bar equals 15 mm.

Table 11-1.—*Means (±SE), ranges (in parentheses), and* CVs *of measurements of skin and skull characters for female and male mountain beavers* (Aplodontia rufa) *from regions in Oregon. Skin characters were recorded from specimen tags; skull characters were measured to the nearest 0.01 mm.*

Region	Sex	n	Total length	Tail length	Hind foot length	Ear length	Mass (kg)	Occipitonasal length	Nasal length	Length of maxillary toothrow	Zygomatic breadth
West slope Cascade Range	♀	12	336 ± 4.6 (316–360) 4.7	29 ± 1.7 (21–43) 20.1	56 ± 0.7 (50–60) 4.3	28 ± 2.0[a] (26–30) 10.1		72.15 ± 0.45 (68.52–74.11) 2.17	25.26 ± 0.33 (23.24–27.03) 4.53	18.63 ± 0.22 (17.48–0.22) 4.08	54.39 ± 0.78 (49.84–58.45) 4.96
	♂	21	339 ± 4.3 (294–385) 5.8	29 ± 1.8 (17–52) 28.0	57 ± 0.9 (49–63) 7.3	26 ± 2.0[b] (22–31) 15.8	0.99 ± 0.15[c] (0.78–1.29) 0.26	72.16 ± 0.66 (65.80–77.96) 4.22	24.68 ± 0.39 (21.80–28.95) 7.35	18.53 ± 0.19 (16.75–20.36) 4.79	54.92 ± 0.69 (49.46–61.29) 5.84
Coast Range	♀	34	312 ± 5.9 (270–445) 11.2	19 ± 1.5 (6–42) 45.3	51 ± 0.5 (43–59) 7.1	23 ± 0.5[d] (16–28) 11.4	0.84 ± 0.03[e] (0.40–1.12) 0.20	69.73 ± 0.44 (65.33–75.04) 3.72	25.03 ± 0.26 (22.58–28.84) 6.11	18.82 ± 0.09 (17.70–20.01) 2.83	51.79 ± 0.43 (47.91–58.46) 4.84
	♂	37	311 ± 4.5 (205–380) 8.8	25 ± 1.6 (13–57) 39.8	49 ± 0.5 (45–59) 6.1	22 ± 0.6[f] (12–28) 14.9	0.85 ± 0.03[g] (0.57–1.11) 0.17	70.09 ± 0.37 (65.71–75.71) 3.22	25.51 ± 0.24 (22.80–29.71) 5.68	18.56 ± 0.10 (17.26–19.83) 4.85	52.56 ± 0.39 (48.24–58.60) 4.54

[a]Sample size reduced by 10. [b]Sample size reduced by 17. [c]Sample size reduced by 18. [d]Sample size reduced by 1. [e]Sample size reduced by 5. [f]Sample size reduced by 6. [g]Sample size reduced by 11.

1980). Fossils of aplodontoid rodents have been recovered from north-central and southeastern Oregon (Carraway and Verts, 1993).

Geographic Variation.—Two of the seven named subspecies occur in Oregon: the larger and more brownish *A. r. rufa* in the Cascade Range and the smaller and more blackish *A. r. pacifica* in the Coast Range (Taylor, 1918).

Habitat and Density.—Within its range, *A. rufa* occupies all stages of the forest sere and occurs at all elevations below treeline (Carraway and Verts, 1993). However, it is most abundant in early to midseral stages with dense tangles of Douglas-fir (*Pseudotsuga menziesii*), western hemlock (*Tsuga heterophylla*), red alder (*Alnus rubra*), vine maple (*Acer circinatum*), hazelnut (*Corylus cornuta*), salal (*Gaultheria shallon*), huckleberry (*Vaccinium parvifolium*), salmonberry (*Rubus spectabilis*), Pacific blackberry (*R. ursinus*), thimbleberry (*R. parviflorus*), bracken fern (*Pteridium aquilinum*), and common sword-fern (*Polystichum munitum*—Hacker, 1991; Lovejoy and Black, 1979a, 1979b).

Although estimated densities as high as 15–20/ha were reported (Hacker, 1991; Hooven, 1977), most estimates are no more than one-half as great. Voth (1968) removed 21 individuals from a 2-ha area in western Oregon. Also in western Oregon, estimates of density by direct enumeration on a 10-ha area (5.5-ha trapping grid plus a 40-m-wide boundary representing the average distance between captures) during a 20-month period ranged from 4.1 to 5.4/ha (Lovejoy and Black, 1979a, 1979b). By adding the same boundary strip to the area sampled by Voth (1968), Lovejoy and Black (1979b) indicated that the estimated density of that population was essentially identical with their minimum estimate. However, based on a comparison with an estimate by use of the Lincoln Index, Lovejoy and Black (1979b) suggested that densities by direct enumeration probably were underestimated. Also in the Coast Range, Hacker (1991) estimated densities of 2.3–18.2/ha based on removal of animals from 12 areas ranging from 1.3 to 6.0 ha to which she added a 10-m-wide border (about one-half the diameter of burrow systems); densities did not differ significantly among stands 1, 4–5, and 40–60 years since clear-cutting. Densities were related significantly to counts of tunnels (presumably burrow entrances) along 10 50- by 1-m transects in each area (Hacker, 1991).

Table 11-1.—*Extended.*

Breadth of braincase	Breadth of molariform toothrow	Skull depth	Length of mandible
33.29 ± 0.87	14.89 ± 0.46	20.78 ± 0.18	46.19 ± 0.62
(27.72–36.24)	(13.13–17.97)	(19.80–21.75)	(41.96–49.22)
9.00	10.61	3.00	4.65
34.95 ± 0.47	14.79 ± 0.20	20.56 ± 0.20	46.36 ± 0.47
(28.47–38.92)	(13.35–17.34)	(18.90–22.63)	(42.90–50.82)
6.23	6.25	4.47	4.75
26.53 ± 0.41	16.64 ± 0.12	20.38 ± 0.12	46.21 ± 0.32
(24.04–34.21)	(15.25–18.04)	(19.32–22.22)	(42.85–50.42)
8.96	4.33	3.43	3.99
27.57 ± 0.58	16.04 ± 0.15	20.49 ± 0.12	46.03 ± 0.32
(23.99–35.02)	(13.51–17.72)	(18.43–22.71)	(42.51–51.32)
12.69	5.74	3.59	4.22

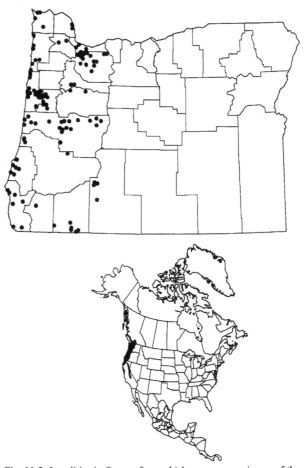

Fig. 11-2. Localities in Oregon from which museum specimens of the mountain beaver, *Aplodontia rufa*, were examined (listed in APPENDIX, p. 511), and the distribution of the mountain beaver in North America. (Range map redrawn from Hall, 1981:335, map 233, with modifications.)

Following removal of *A. rufa* from 12 areas in western Oregon, characteristics of 84 colonized sample plots differed significantly from those of 79 sample plots not recolonized (Hacker and Coblentz, 1993). Populated plots contained more sword-fern, shrubs, forbs, downed wood, forage-plant cover, and uprooted stumps, but less grass, and were more likely to be in drainages, on softer soils, on steeper slopes, and on north slopes (Hacker and Coblentz, 1993).

Diet.—As determined by identification of epidermal fragments of plants in the feces, the diet of *A. rufa* in the Coast Range consists largely of ferns (*Polystichum munitum* and *Pteridium aquilinum*) at all seasons and for all sex-age-classes (Voth, 1968). Although 31 other species were eaten by adults (mostly males) throughout the year and 14 other species were eaten by juveniles during the first 4 months after weaning, none of the species contributed as much as 10% of the overall diet. However, among lactating females, 18.4% of the diet consisted of grasses and 33.9% consisted of conifers, mostly Douglas-fir (Voth, 1968). The two ferns also dominated food caches located outside tunnel systems and in food chambers within tunnel systems (Voth, 1968). Most seasonal differences in diet and differences in diet among sex-age groups possibly were related to protein requirements and availability. Adults consumed red alder during autumn, the season during which its protein content is highest; also, some grasses and forbs were consumed in spring by lactating females and by juveniles during the season of growth (Voth, 1968). Strangely, *P. aquilinum* and some of the other plants consumed by *A. rufa* are toxic to many vertebrates; Voth (1968) suggested that *A. rufa* specialized in foraging on the usually toxic plants. A variety of other species reportedly eaten by *A. rufa* elsewhere (Carraway and Verts, 1993) did not occur, or occurred only in extremely small amounts, in fecal samples analyzed by Voth (1968). Crouch (1968a) reported that mountain beavers clipped 22% of 900 trees and shrubs on a 138-ha study area in western Oregon; clipped most frequently were vine maple, big-leaf maple (*Acer macrophyllum*), huckleberry, red alder, cascara (*Rhamnus purshiana*), and hazelnut.

Reproduction, Ontogeny, and Mortality.—Reproductive activity in mountain beavers commences in November or December with a rapid increase in size of testes, prostate, and bulbourethral glands, and the testes becoming semiscrotal (males do not possess a true scrotum—Hubbard, 1922; Lovejoy et al., 1978; Pfeiffer, 1956b). Although estrus is fairly synchronous within a population, onset of estrus differs with locality and may occur in February–May (Pfeiffer, 1958); in the Coast Range of Oregon, most females are pregnant in March (Hacker, 1991). Mountain beavers produce only one litter per year after a 28–30-day gestation period (Pfeiffer, 1958; Scheffer, 1929).

Some workers indicated that female mountain beavers do not attain reproductive competency in the year following birth (Lovejoy et al., 1978; Pfeiffer, 1958), but 50% of females from 1-year-old clear-cuttings and 44% on 4–5-year-old clear-cuttings bred in the year following their birth (Hacker, 1991). However, none bred in the year following the year of their birth on 40–60-year-old forest stands. In the 2nd year following birth, 91% bred and among older females 100% bred regardless of stand type (Hacker, 1991). Males exhibited a similar pattern; in the year following birth, 62.5% of males became reproductively mature on 1-year-old clear-cuttings whereas only 28.6% and 11.1% matured in 4–5-year-old and 40–60-year-old stands, respectively (Hacker, 1991). All males ≥2 years old became reproductively active in all three stand types (Hacker, 1991).

Overall, the average litter size (Table 11-2) seems low for an animal as small as a mountain beaver. Hacker (1991) presented evidence that litter size tended to increase with maternal age and to decrease with the advance of community development. The values for litter size were based on counts of embryos and counts of pigmented sites of implantation combined; she did not justify statistically the combining of data sets. However, overall values (Table 11-2) are similar to those for *A. rufa* based on various methodologies in other parts of its range (Carraway and Verts, 1993).

Young may be born either head or tail first (Cramblet and Ridenhour, 1956), and are pink, naked, blind, and helpless (Lovejoy et al., 1978). Vibrissae are present and the anterior toes are clawed, but eyes and auditory meatuses are closed. At 2 days of age, two young averaged 19.8 g and 66 mm in total length (Lovejoy et al., 1978), smaller than the 27 g and 87 mm averages for three 1-day-old young described by Cramblet and Ridenhour (1956). At 3 days, the young squeaked; by 5 days, the dorsum and pate were pigmented; by 7 days, the skin was wrinkled and baggy; by 9 days, fine hairs covered the body sparsely; by 10 days, the pinnae unfolded; by 14 days, the neonates were covered with hair; by 21 days, the auditory meatuses opened; by 28 days, the lower incisors erupted; by 33 days, the upper incisors erupted; by 45 days, the eyes opened; and by 49 days, solid food was accepted although young still nursed (Lovejoy et al., 1978). The increase in mass was exponential through the 8th week when the two averaged 347 g (Lovejoy et al., 1978).

From studies conducted in the Coast and Cascade ranges, both bobcats (*Lynx rufus*) and coyotes (*Canis latrans*) prey heavily on the mountain beaver in some regions (Nussbaum and Maser, 1975; Toweill and Anthony, 1988a, 1988b; Witmer and deCalesta, 1986). Mountain lions (*Puma concolor*) and golden eagles (*Aquila chrysaetos*) are known to prey on mountain beavers at least occasionally (Toweill and Maser, 1985); smaller carnivores are thought to take young individuals (Hooven, 1977). Because of their interference with community development in the early forest sere (Crouch, 1968a; Neal and Borrecco, 1981), depredations by mountain beavers sometimes are subject to control by reducing food and cover plants by burning, use of herbicides, or mechanical means, or by reducing populations by use of toxic baits and trapping (Hooven, 1977).

Habits.—In summer, mountain beavers forage above ground five or six times during each 24-h period; they are

Table 11-2.—*Reproduction in four rodent species in families monotypic in Oregon.*

Species	Number of litters	Litter size $\bar{X}$	Range	n	Reproductive season	Basis	Authority	State or province
Aplodontia rufa	1	3.4	2–6	9		Implant sites	This study	Oregon
	1	3.0	1–4	5	Feb.–Apr.	Embryos	This study	Oregon
	1	2.8[a]		82		Embryos and implant sites[b]	Hacker, 1991	Oregon
Castor canadensis	1	3.5	3–4	6	Feb.–Apr.	Embryos	Scheffer, 1925	Oregon
Erethizon dorsatum	1	1.0	1[c]		Sep.–Nov.[d]	Unknown	Dodge, 1982	Unknown[e]
Myocastor coypus	2–3?	4.9[f]	1–11	60	All year	Embryos	Peloquin, 1969	Oregon

[a]Based on samples from all age classes and three stages of the forest sere.
[b]Counts were by two methods combined.
[c]Rare instances of twins have been discussed but never authenticated.
[d]The usual breeding season; occasional births in August indicate that breeding may occur as late as January.
[e]Authority conducted research on porcupine in Washington and Massachusetts, perhaps elsewhere.
[f]Living embryos only; an additional 0.6 embryos were undergoing resorption.

50–60% more active during the hours of darkness than during daylight hours (Ingles, 1959). The greater activity at night caused some workers to indicate that mountain beavers are nocturnal (Carraway and Verts, 1993). Although aboveground activity nearly ceases in winter, mountain beavers do not hibernate (Ingles, 1959).

Mountain beavers ingest 25–33% of their body mass of water daily as free water or as water in succulent food plants (Nungesser et al., 1960); they also produce copious amounts (267 ± 25.7 ml/kg daily) of hypotonic (220 ± 14 mOsm/l during daytime and 314 ± 34 mOsm/l at night) urine. The high demand for water was suggested as being responsible for the geographic range of the species being restricted to wet areas of the Pacific Northwest (Nungesser and Pfeiffer, 1965). In Oregon, high summer temperatures and annual summer droughts likely are responsible for the absence of mountain beavers from some of the interior valleys (Fig. 11-2).

Mountain beavers are hindgut fermenters, and as such reingest their fecal material. Both hard and soft fecal pellets are produced and are intermixed in the large intestine. While defecating, the mountain beaver sits on its rump with tail and partially everted anus directed upward. Hard pellets are taken into the mouth and, with a flick of the head, are tossed into a latrine; the 4-times-larger soft pellets are reingested (Ingles, 1961b).

Mountain beavers construct extensive burrow systems that tend to radiate from a nest site (Martin, 1971). Burrows are 15–25 cm wide, 13–18 cm high (Hubbard, 1922), and only 15–46 cm below the surface (Camp, 1918). The shallow burrows may be so extensive in some areas (Scheffer, 1929) that walking becomes difficult; Camp (1918) found ≥100 entrances in a 0.46-ha area. Earth excavated in tunneling is pushed into spoils at only a few burrow entrances; some entrances are plugged, others only concealed by vegetation (Camp, 1918). Sometimes burrow entrances are covered with tentlike structures constructed of sticks and succulent vegetation (Johnson, 1975). Aspect selected for burrow systems seemingly is to take advantage of moderate microclimates; Johnson (1971) found that average temperatures of burrows in Oregon did not vary >3°C in any month and the greatest diel variation was 4°C. As part of burrow systems, nest chambers ≅30 cm in diameter are 0.3–1.5 m underground (Martin, 1971); they may contain ≤35 l of vegetation or dried leaves trampled into a pad (Camp, 1918; Martin, 1971). In earth-plugged chambers along tunnel walls, mountain beavers store roots, stems, and leaves.

Commencing in midsummer, piles of wilted and partly dried vegetation commonly are found in or near entrances to mountain beaver burrow systems. Some authorities claimed or implied the piles were hay being cured and the material was moved into burrows where it was eaten, some indicated the dried vegetation was used as nesting material, and some claimed the material was used for both food and nesting (Carraway and Verts, 1993). A similar controversy involves sounds produced by mountain beavers. A variety of whistles and booming sounds have been attributed to the mountain beaver (hence the vernacular name "boomer" applied to the species by some people), but more recent workers indicated that the only sounds produced are soft whines or sobs, a grinding of the teeth, and a high-pitched squeal (Carraway and Verts, 1993); other sounds attributed to *A. rufa* likely are produced by birds or other mammals (Scheffer, 1929).

Mountain beavers tend to be asocial except during the reproductive and rearing seasons. Even though home ranges overlap and the species can withstand considerable crowding, nests and burrows are defended (Martin, 1971). The strong musty odor produced by mountain beavers may be a mode of interspecific communication (Seton, 1929b) or a means of enhancing the territorial confidence of a resident animal (Nolte et al., 1993).

In western Oregon, home ranges (length times width) of mountain beavers captured seven or more times at five or more trap stations averaged ($\pm SE$) 0.17 ± 0.05 ha (range, 0.05–0.40 ha; $n = 7$) for adult females and 0.32 ± 0.05 (range, 0.09–0.79; $n = 14$) ha for adult males (Lovejoy and Black, 1979a). In Washington, minimum-area home ranges for 10 adults of both sexes radiotracked for 3–19 months averaged 0.12 ha (range, 0.03–0.20 ha—Martin, 1971). Individuals with maximum and minimum home-range areas moved a maximum of 42.6 and 36.6 m from their nests during a 12-month period, respectively (Martin, 1971). Maximum dispersal distance was 564 m for a female and 379 m for a male (Martin, 1971). Recolonization of 1.3–6.0-ha areas from which mountain beavers were removed by trapping to reduce depredations on forest plantings required <1 year (Hacker, 1991), as might be expected from dispersal distances recorded by Martin (1971).

Remarks.—*Aplodontia rufa* commonly is referred to as the most primitive living rodent (Kurtén and Anderson, 1980) presumably because it is the only living rodent in which the origin of the masseter is limited to the ventral surface of the zygoma (Kurtén and Anderson, 1980; Rensberger, 1975). However, the karyotype is considered to be advanced because of the absence of acrocentric autosomes (McMillin and Sutton, 1972).

FAMILY SCIURIDAE—SQUIRRELS, CHIPMUNKS, AND MARMOTS

The squirrels first appeared in the fossil record in the middle Oligocene of North America then spread to Europe in the late Oligocene (Kurtén and Anderson, 1980), to Asia in the Miocene, and to South America and Africa in the Pleistocene (McLaughlin, 1984). Currently, sciurids occur worldwide except for the Australian region, Madagascar, southern South America, and some Old World desert regions (Kurtén and Anderson, 1980; McLaughlin, 1984). Currently, 273 species in 50 genera are recognized (Wilson and Reeder, 1993a).

Members of the family are characterized by the presence of a postorbital process on the frontal bones, the fourth toe the longest on all feet, a bushy tail often flattened dorsoventrally, and a small infraorbital foramen. The zygomatic plate is broad and forms the origin of the lateral masseter muscle on the anterior surface. There are four cheekteeth in the lower jaw, P3 is reduced or absent, and the molariform teeth feature transverse ridges. There are four toes on the forefeet, five on the rear; the limb bones are unspecialized; and the freedom of movement of the wrist, ankle, and elbow is not restricted (Kurtén and Anderson, 1980; McLaughlin, 1984). A baculum is present in all species, but much reduced in *Tamias* and *Tamiasciurus*.

Except for the murids, the squirrels are the most diverse

family of mammals in Oregon; 21 species grouped into six genera are represented in the state. The family includes a marmot, nine ground squirrels, five chipmunks, five tree squirrels, and a flying squirrel. The marmot is characterized by its large size and a dark bar across the nose. The ground squirrels can be recognized by the absence of stripes on the face and twisted, dorsoventrally flattened zygomata that converge anteriorly. Tree squirrels also lack facial stripes, but the zygomata are untwisted, flattened laterally, and nearly parallel. Chipmunks have stripes on the face and incisors with tiny longitudinal grooves on their anterior face, and the infraorbital foramen is visible when the skull is viewed from the ventral surface. The flying squirrel is unique in possessing a fur-covered membrane between the fore- and hind limbs, a strongly distichous tail, and a notch in each frontal bone through which the maxillary is visible when the skull is viewed from above.

Sciurids are primarily herbivorous, with tree squirrels, chipmunks, and flying squirrels feeding primarily on nuts, fruits, buds, and fungi, and ground squirrels and marmots feeding primarily on leaves of grasses and forbs. Ground squirrels and chipmunks have internal cheek pouches in which they transport food to winter caches. Ground squirrels, chipmunks, and marmots tend to be associated with early successional stages whereas tree squirrels and flying squirrels tend to be members of more mature communities. The marmots and most ground squirrels spend less favorable seasons in torpor in underground dens; chipmunks may enter torpor during inclement periods, but become active on mild days; and tree squirrels may only forgo activity during storms. Flying squirrels are primarily nocturnal, but other squirrels in Oregon are primarily diurnal.

SOME TAXONOMIC CONSIDERATIONS AT THE GENERIC LEVEL

Genus *Tamias* Illiger, 1811

Chipmunks in western North America possess a striping pattern on the pelage, upper premolar count, baculum shape, and other morphological characters somewhat different than found in the chipmunk (*Tamias striatus*) in eastern North America (White, 1953). Thus, traditionally, western chipmunks were placed in a separate genus (*Eutamias*). Ellerman (1940) and Bryant (1945) did not consider some of these characters as conservative and placed all chipmunks in the genus *Tamias*. Nevertheless, many authorities, including those who published checklists of North American mammals (Jones et al., 1973a, 1975, 1979), did not recognize the change. On the basis of karyology (Nadler et al., 1977) and protein biochemistry and immunology (Hight et al., 1974) that supported long separation of the chipmunk lineage from that of other sciurids, Nadler et al. (1977) recommended that all North American chipmunks be referred to *Tamias*. Ellis and Maxson (1979) added immunological evidence, and claimed that, combined with other evidence derived from the work of Nadler et al. (1977), continued recognition of two genera was valid. From the work of Nadler et al. (1977), Hight et al. (1974), and additional information provided by Levenson et al. (1985), Jones et al. (1982, 1986) recognized *Tamias* as the genus for all North American chipmunks and have continued to do so "because no one has published data convincing us that it is incorrect to do so" (Jones et al., 1992:4).

Nevertheless, the latter authors recognized (p. 4) that "controversy remains" regarding the clarity of chipmunk affinities above the species level (Patterson and Heaney, 1987).

Genus *Spermophilus* F. Cuvier, 1825

Citellus Oken, 1816, the long-used generic name of a group of ground squirrels (*Spermophilus*), was declared "not available" in Opinion 417, International Commission on Zoological Nomenclature, 1956. Nevertheless, *Citellus* and the subgeneric names *Callospermophilus* and *Otospermophilus* still are used occasionally as generic names in reference to some of these spermophiles. Thus, during searches for information regarding a species of *Spermophilus* it is prudent to extend the search to include the previously used generic names.

KEY TO THE SCIURIDAE OF OREGON

1a. Pelage on face with a dark bar across nose followed by a white patch in front of eyes; zygomatic breadth >48 mm; occipitonasal length >70 mm...................***Marmota flaviventris***, yellow-bellied marmot (p. 173).

1b. Pelage on face lacking dark bar and white patch; zygomatic breadth <48 mm; occipitonasal length <70mm....2

2a. Pelage on face striped; infraorbital foramen through the zygomatic plate, the opening visible in ventral aspect; incisors with numerous minute longitudinal grooves in the enamel...3

2b. Pelage on face not striped; infraorbital foramen a canal between the zygomatic plate and the rostrum, the opening visible only from the anterior; incisors lacking longitudinal grooves.......................................7

3a. Occipitonasal length >35 mm.....................................4

3b. Occipitonasal length <35 mm.....................................6

4a. Three to five dorsal dark stripes usually black; dorsal light stripes usually cinnamon buff to tawny with outermost pair about the same color as innermost pair; flanks also cinnamon buff to tawny; diameter of baculum <20% of its length.......................***Tamias townsendii***, Townsend's chipmunk (p. 169).

4b. Only middorsal dark stripe black (or at least darker than other four); dorsal light stripes gray to whitish; flanks ocherous; diameter of baculum >20% of its length...5

5a. Tip of baculum longer than shaft; inner pair of dorsal light stripes usually light gray; outer pair of light stripes much brighter than inner pair; shoulders and rump usually washed with brown.................................... ....***Tamias siskiyou***, Siskiyou chipmunk (p. 168).

5b. Tip of baculum less than one-half length of shaft; inner pair of dorsal light stripes grayish mixed with some brown; both pairs of light stripes with much white; shoulders and rump usually washed with gray........ ....***Tamias senex***, Allen's chipmunk (p. 166).

6a. Pelage on rump more grayish than brownish; flanks sometimes lightly washed with ochre; pelage on underside of tail usually lemon yellow; length of bent tip of baculum <28% of length of shaft; palatal length (1.45163) + condylonasal length (0.41994) + breadth of braincase (0.32907) is <39.8728.............................. ....***Tamias minimus***, least chipmunk (p. 163).

6b. Pelage on rump more brownish than grayish; flanks

usually brightly washed with ochre; pelage on underside of tail usually fulvous; length of bent tip of baculum >28% of length of shaft; palatal length (1.45163) + condylonasal length (0.41994) + breadth of braincase (0.32907) >40.3238.............................
....***Tamias amoenus***, yellow pine chipmunk (p. 159).
7a. Zygomata untwisted and compressed laterally; zygomatic arches parallel...8
7b. Zygomata twisted and compressed dorsoventrally; zygomatic arches strongly converging anteriorly...........13
8a. Pelage soft and fluffy; membrane between fore- and hind limbs, a modification for gliding; pelage of tail strongly compressed dorsoventrally; maxillary bones visible through notches of interorbital constriction when skull viewed in dorsal aspect.............***Glaucomys sabrinus***, northern flying squirrel (p. 219).
8b. Guard hairs coarse, giving pelage a sleek appearance; no membrane between fore- and hind limbs; pelage of tail compressed dorsoventrally, but not excessively so; maxillary bones not visible through notch of interorbital constriction when skull viewed in dorsal aspect...9
9a. Occipitonasal length <55 mm; anterior margin of orbit opposite P4 when skull viewed from ventral aspect; baculum vestigial..10
9b. Occipitonasal length >55 mm; anterior margin of orbit opposite M1 when skull viewed from ventral aspect; baculum well developed..11
10a. Pelage on dorsum reddish gray to reddish olive; pelage on venter white; pelage on tail usually with black-tipped hairs..
.......***Tamiasciurus hudsonicus***, red squirrel (p. 216).
10b. Pelage on dorsum dusky olive to grayish olive; pelage on venter light yellowish to deep orange; pelage on tail usually with yellowish or white-tipped hairs...
Tamiasciurus douglasii, Douglas' squirrel (p. 212).
11a. Pelage on dorsum rusty red; pelage on venter orange reddish; P3 lacking; skull and other bones reddish......
...........................***Sciurus niger***, fox squirrel (p. 210).
11b. Pelage on dorsum grayish or silvery gray; pelage on venter white; P3 usually present; skull and other bones whitish..12
12a. Hind foot >70 mm long; P4 wider than long; pelage silvery on dorsum...
.........***Sciurus griseus***, western gray squirrel (p. 207).
12b. Hind foot <70 mm long; P4 longer than wide; pelage gray, but not silvery, on dorsum.........................
..............***Sciurus carolinensis***, gray squirrel (p. 203).
13a. Pelage marked only with light stripe on sides; infraorbital foramen a narrow oval with a spinelike projection directly ventral to the opening........***Ammospermophilus leucurus***, white-tailed antelope squirrel (p. 178).
13b. Pelage lacking light stripe on sides, or if light stripe present it is sandwiched between two dark stripes; infraorbital foramen triangular or broadly oval with a large projection ventral and lateral to the opening.....14
14a. Pelage on side of dorsum with light stripe bordered with black; pelage of head and shoulders a deep golden yellow; temporal ridges absent or obscure.....
....***Spermophilus lateralis***, golden-mantled ground squirrel (p. 193).
14b. Pelage lacking stripes on side of dorsum; pelage of head and shoulders brownish or grayish (sometimes

with yellowish wash), never a deep golden yellow; V-shaped or U-shaped temporal ridges prominent (may be obscure in young individuals)......................15
15a. Pelage with large dark patch (nearly black) on nape; tail >128 mm long; parastyle ridge on M1 and M2 straight..***Spermophilus beecheyi***, California ground squirrel (p. 181).
15b. Pelage lacking large dark patch on nape; tail <128 mm long; parastyle ridge on M1 and M2 angled..........16
16a. Pelage on dorsum buffy gray with small whitish spots; pelage on face, legs, and tail distinctly reddish; hind foot >49 mm long; temporal ridges acutely V-shaped; greatest length of skull >49 mm........***Spermophilus columbianus***, Columbian ground squirrel (p. 187).
16b. Pelage on dorsum not spotted, or if spotted the spots >5 mm in diameter; pelage on face, legs, and tail not reddish; hind foot <49 mm long; temporal ridges distinctly U-shaped; greatest length of skull <49 mm...17
17a. Pelage on dorsum mottled with whitish splotches >5 mm in diameter..................................***Spermophilus washingtoni***, Washington ground squirrel (p. 201).
17b. Pelage on dorsum uniformly colored..........................18
18a. Dark pelage on dorsum sharply separated from markedly lighter pelage on venter; lighter ventral pelage extends high on sides; hind foot <39 mm long.........19
18b. Dark pelage on dorsum gradually blending into slightly lighter pelage on venter; lighter ventral pelage does not extend high on sides; hind foot >39 mm long..20
19a. Discriminant value >0.85[1]................***Spermophilus mollis***, Piute ground squirrel (p. 198).
19b. Discriminant value <0.85[1]...................***Spermophilus canus***, Merriam's ground squirrel (p. 196).
20a. Pelage on dorsum usually with a moderately well-defined broad reddish or brownish band; pelage on ventral surface of tail cinnamon colored; distal tail hairs with three distinct color bands.......***Spermophilus beldingi***, Belding's ground squirrel (p. 184).
20b. Pelage on dorsum lacking broad reddish or brownish band; pelage on ventral surface of tail buff colored; distal tail hairs with five to seven alternate light and dark color bands.....................***Spermophilus elegans***, Wyoming ground squirrel (p. 190).

---

[1]Discriminant value=occipitonasal length (0.87257) + nasal length (−0.75297) + length of maxillary toothrow (0.81479) + zygomatic breadth (−0.94707) + breadth of braincase (−1.13649) + breadth across molar toothrows (2.15414) + skull depth (−0.81291) + length of mandible (0.76948) − 13.2225.

Tamias amoenus J. A. Allen
Yellow-pine Chipmunk

1890a. *Tamias amoenus* Allen, 3:90.
1897f. *Eutamias amoenus*: Merriam, 11:194.
1911.   *Eutamias ludibundus* Hollister, 56(26):1.
1913.   *Eutamias amoenus propinquus* Anthony, 32:6.
1922.   *Eutamias amoenus ludibundus*: Howell, 3:184.
1925.   *Eutamias amoenus ochraceus* Howell, 6:54.
1947.   *Eutamias amoenus albiventris* Booth, 28:7.

Description.—*Tamias amoenus* (Plate VII) is penultimate in the size range of Oregon chipmunks (Table 11-3); only *T. minimus* is smaller and it is only slightly so. It is similar to congeners in that the face and dorsum are marked with a dark stripe; the middorsal dark stripe is followed laterally by two alternate light and dark stripes. The outermost light stripe is nearly white, whereas the innermost light stripe is a mixture of gray, ochre, and white hairs. The sides usually are washed with ochre, the feet usually are buff, and the tail hairs commonly have ocherous tips. The underside of the tail commonly is cinnamon. *T. amoenus* usually is more brightly colored than *T. minimus*, but some populations contain individuals colored similar to the latter species (Anderson, 1974). The skull (Fig. 11-3) also is similar to that of *T. minimus*.

According to Anderson (1974) the most reliable means of distinguishing *T. amoenus* from *T. minimus* is by summing three skull dimensions. We developed an equation based on skull dimensions of adult specimens from Oregon to separate *T. amoenus* and *T. minimus* as follows:

Value = palatal length (1.45163) + condylonasal
length (0.41994) + breadth of braincase (0.32907)

Adult specimens of small chipmunks for which the calculated value is >40.3238 are *T. amoenus*. Only nine (1.7%) of a sample of 543 small chipmunks from east of the Cascade Range could not be classified by use of the equation (Carraway and Verts, 1996).

Distribution.—The yellow-pine chipmunk occurs from central British Columbia south in the Cascade and Sierra Nevada ranges to Mono Co., California, and east in montane regions through Idaho, and western portions of Wyoming, Montana, and Alberta. The species also has been recorded from extreme northern portions of eastern Nevada and western Utah, in coastal areas of southwestern British Columbia, and in isolated populations in northwestern Nevada and on the Olympic Peninsula, Washington (Fig. 11-4; Hall, 1981). In Oregon, *T. amoenus* occurs on the east slope of the Western Cascades and eastward through most of the remainder of Oregon, except it is absent from most of the Columbia Basin and much of southern Harney, eastern Malheur, and southern Baker counties. Its range also extends westward in the Siskiyou Mountains in southern Jackson and Josephine counties (Fig. 11-4).

Remains of chipmunks have been found in cave deposits of Pleistocene and Holocene ages in California, Idaho, and Wyoming, but seemingly none has been identified as those of *T. amoenus* (Kurtén and Anderson, 1980).

Geographic Variation.—Four of the 14 currently recognized subspecies occur in Oregon: the larger, darker, and more grayish *T. a. albiventris* in the Wallowa Mountains; the smaller and smoky *T. a. amoenus* over most of the basin and range country east of the Cascade Range; the smaller and more grayish *T. a. ludibundus* in the northern portion of the Cascade Range; and the larger and more brownish *T. a. ochraceus* in the Siskiyou Mountains (Booth, 1947; Hall, 1981; A. H. Howell, 1924, 1929).

Habitat and Density.—In central Oregon, habitats in which *T. amoenus* was studied were described as open juniper (*Juniperus occidentalis*)-ponderosa pine (*Pinus ponderosa*) forests with understories of big sagebrush (*Artemisia tridentata*) and rabbit-brush (*Chrysothamnus nauseosus*); ponderosa pine with scattered lodgepole pine (*Pinus contorta*) forests with understories of bitter-brush (*Purshia tridentata*); and open ponderosa pine forests with dense patches of pine regeneration and some replacement of ponderosa pine by Douglas-fir (*Pseudotsuga menziesii*) and with understories of tobacco-brush (*Ceanothus velutinus*) and manzanita (*Arctostaphylos patula*—States, 1976). Both in this region

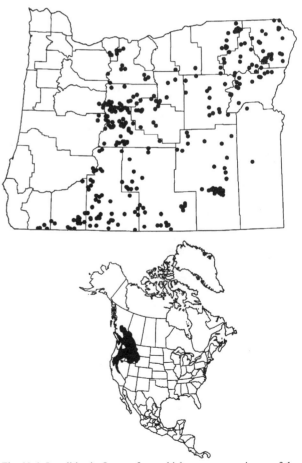

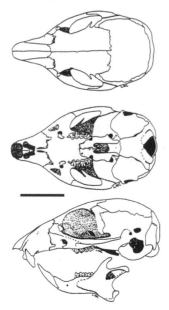

Fig. 11-3. Dorsal, ventral, and lateral views of the cranium, and lateral view of the dentary of the yellow-pine chipmunk, *Tamias amoenus* (OSUFW 1572). Scale bar equals 10 mm.

Fig. 11-4. Localities in Oregon from which museum specimens of the yellow-pine chipmunk, *Tamias amoenus*, were examined (listed in APPENDIX, p. 565), and the distribution of the yellow-pine chipmunk in North America. (Range map redrawn from Hall, 1981:349, map 237.)

Table 11-3.—*Means (±SE), ranges (in parentheses), and CVs of measurements of skull and skin[a] characters for female and male chipmunks (Tamias) from regions in Oregon. Skin characters were recorded from specimen tags; skull characters were measured to the nearest 0.01 mm. SE and CV not provided if n <10.*

Species and region	Sex	n	Total length	Tail length	Hind foot length	Mass (g)	Greatest length of skull	Length of maxillary toothrow	Zygomatic breadth	Breadth of braincase	Breadth of molariform toothrow	Skull length	Length of mandible
Tamias amoenus Cascade Range	♀	10	207 ± 4.4 (171–221) 6.8	89 ± 3.8 (60–102) 13.2	32 ± 0.4 (30–34) 3.9		33.26 ± 0.21 (32.39–34.40) 2.02	5.55 ± 0.11 (4.99–5.90) 6.22	18.49 ± 0.14 (17.87–19.14) 2.39	15.79 ± 0.11 (15.24–16.39) 2.19	7.94 ± 0.05 (7.61–8.12) 2.14	13.41 ± 0.06 (13.14–13.63) 1.39	17.78 ± 0.12 (17.23–18.36) 2.06
	♂	9	201 (176–210)	88 (64–100)	32 (30–33)	47.90[b] (47.7–48.1)	32.72 (32.29–33.39)	5.37 (5.13–5.86)	17.96 (17.53–18.55)	15.80 (15.50–16.15)	7.81 (7.54–8.07)	13.27 (12.99–13.44)	17.53 (16.85–18.09)
Wallowa Mountains	♀	30	214 ± 1.3 (195–230) 3.2	94 ± 1.3 (80–106) 7.3	32 ± 0.2 (28–34) 4.2	53.20[c] (45.7–59.5)	33.52 ± 0.12 (32.01–34.62) 1.97	5.61 ± 0.04 (5.12–6.13) 4.13	18.59 ± 0.07 (17.71–19.37) 2.01	16.11 ± 0.06 (15.57–16.83) 1.91	8.09 ± 0.04 (7.75–8.80) 2.59	13.40 ± 0.05 (13.00–14.07) 2.13	17.73 ± 0.09 (16.78–18.81) 2.90
	♂	33	209 ± 1.9 (166–227) 5.3	91 ± 1.3 (70–105) 8.3	32 ± 0.4 (26–40) 6.9	43.85[d] (38.0–47.5)	33.15 ± 0.12 (31.53–34.36) 2.09	5.54 ± 0.04 (4.93–6.06) 4.09	18.33 ± 0.07 (17.77–19.09) 2.06	16.12 ± 0.06 (15.34–16.69) 2.22	7.95 ± 0.04 (7.58–8.38) 2.82	13.66 ± 0.08 (13.13–15.56) 3.35	17.39 ± 0.08 (16.64–18.21) 2.74
East of Cascade Range except Wallowa Mountains	♀	30	203 ± 1.9 (188–229) 4.9	86 ± 1.8 (65–113) 11.2	30 ± 0.6 (20–33) 10.7	51.21 ± 2.11[b] (39.6–87.4) 19.79	32.98 ± 0.12 (31.66–34.41) 2.04	5.39 ± 0.05 (4.98–5.87) 4.83	18.38 ± 0.08 (17.56–19.31) 2.30	15.75 ± 0.07 (14.94–16.56) 2.28	7.93 ± 0.05 (7.53–8.51) 3.25	13.32 ± 0.06 (12.79–14.06) 2.46	17.66 ± 0.09 (16.57–18.60) 2.65
	♂	30	195 ± 1.9 (170–223) 5.5	81 ± 1.8 (54–98) 11.8	31 ± 0.3 (24–33) 5.7	45.43 ± 0.88 (36.0–57.0) 10.64	32.70 ± 0.14 (31.36–34.75) 2.31	5.24 ± 0.04 (4.85–5.77) 3.89	18.30 ± 0.08 (17.55–19.47) 2.32	15.89 ± 0.07 (15.29–16.84) 2.55	7.89 ± 0.04 (7.55–8.38) 2.58	13.33 ± 0.06 (12.77–14.09) 2.29	17.45 ± 0.08 (16.49–18.52) 2.43
Jackson and Josephine counties	♀	6	217 (204–244)	95 (91–105)	33 (32–35)	38.0	33.22 (32.08–36.34)	5.68 (5.26–6.76)	18.59 (18.19–19.64)	16.29 (15.85–17.00)	8.08 (7.73–8.73)	13.84 (13.05–14.99)	17.89 (16.89–19.23)
	♂	5	204 (196–215)	91 (86–98)	32 (30–34)		32.73 (32.01–33.20)	5.36 (4.94–5.73)	18.17 (17.80–18.64)	15.91 (15.63–16.26)	7.85 (7.56–8.21)	13.34 (12.86–13.81)	16.96 (16.34–17.49)
Tamias minimus	♀	30	195 ± 2.0 (157–222) 5.7	86 ± 1.8 (51–98) 11.1	29 ± 0.3 (27–34) 5.0	39.58[e] (32.4–54.4)	29.72 ± 0.18 (27.80–33.18) 3.35	4.99 ± 0.04 (4.49–5.46) 4.08	16.70 ± 0.13 (15.50–19.24) 4.34	14.92 ± 0.10 (14.04–16.37) 3.74	7.25 ± 0.06 (6.77–8.14) 4.34	12.38 ± 0.08 (11.85–13.76) 3.36	15.66 ± 0.14 (14.66–18.49) 4.92
	♂	30	191 ± 2.7 (160–250) 7.7	86 ± 1.7 (60–98) 10.5	29 ± 0.3 (24–31) 5.0	38.45[f] (31.0–41.0)	29.37 ± 0.15 (27.45–31.24) 2.82	4.93 ± 0.03 (4.59–5.16) 3.15	16.41 ± 0.11 (15.34–18.25) 3.64	14.87 ± 0.08 (13.82–15.68) 3.08	7.09 ± 0.04 (6.68–7.78) 3.23	12.45 ± 0.06 (11.82–13.31) 2.59	15.48 ± 0.12 (14.36–17.88) 4.31
Tamias senex	♀	30	241 ± 1.5 (225–252) 3.3	101 ± 1.2 (87–115) 6.3	35 ± 0.2 (33–38) 3.1	84.62 ± 2.73[g] (73.8–100.0) 10.69	37.92 ± 0.13 (36.23–39.14) 1.89	6.45 ± 0.05 (5.92–7.00) 4.48	20.89 ± 0.09 (19.42–21.71) 2.27	17.27 ± 0.08 (16.27–18.27) 2.55	9.35 ± 0.06 (8.69–10.23) 3.55	14.98 ± 0.06 (14.03–15.91) 2.30	20.48 ± 0.11 (19.40–21.83) 2.94
	♂	30	238 ± 1.3 (223–252) 2.9	98 ± 1.8 (78–112) 9.9	35 ± 0.3 (31–38) 5.4	85.71 ± 1.38[h] (72.4–103.1) 8.06	38.09 ± 0.14 (36.44–39.73) 1.99	6.32 ± 0.04 (5.73–6.84) 3.68	21.03 ± 0.08 (20.09–21.92) 1.96	17.47 ± 0.08 (16.79–18.40) 2.36	9.38 ± 0.03 (9.05–9.71) 1.86	15.17 ± 0.06 (14.38–15.77) 2.15	20.40 ± 0.09 (19.27–21.24) 2.43
Tamias siskiyou	♀	30	251 ± 2.3[i] (213–276) 4.9	105 ± 1.3[i] (92–123) 6.5	36 ± 0.3[i] (31–38) 4.23	93.60 ± 3.97[j] (68.3–112.0) 13.40	38.57 ± 0.23 (33.45–40.48) 3.20	6.61 ± 0.07 (5.74–7.34) 5.66	21.19 ± 0.10 (19.05–22.13) 2.63	17.33 ± 0.08 (16.02–18.09) 2.64	9.37 ± 0.05 (8.31–9.76) 3.07	15.02 ± 0.07 (13.43–15.59) 2.73	20.57 ± 0.15 (17.43–21.58) 3.92
	♂	30	243 ± 1.4 (227–260) 3.2	104 ± 1.4 (88–117) 7.2	36 ± 0.4 (27–38) 5.81	95.50[k] (81.0–120.0)	38.12 ± 0.14 (36.63–39.89) 2.02	6.63 ± 0.06 (5.89–7.18) 4.84	21.02 ± 0.10 (20.03–22.06) 2.65	17.44 ± 0.09 (16.16–18.30) 2.86	9.32 ± 0.04 (8.99–9.78) 2.12	15.12 ± 0.06 (14.47–15.67) 2.14	20.51 ± 0.13 (18.58–22.06) 3.50
Tamias townsendii Coast Range	♀	30	256 ± 2.4 (229–294) 5.2	111 ± 2.0 (90–145) 9.9	36 ± 0.7 (20–40) 10.13	85.53 ± 3.07[l] (63.7–113.0) 15.22	39.03 ± 0.15 (37.36–41.43) 2.16	6.45 ± 0.05 (5.97–6.99) 4.41	21.74 ± 0.07 (20.90–22.47) 1.83	17.77 ± 0.16 (16.78–22.14) 4.99	9.71 ± 0.04 (9.41–10.21) 2.13	15.39 ± 0.06 (14.91–16.07) 2.04	21.26 ± 0.11 (20.43–22.80) 2.81
	♂	30	252 ± 3.8 (209–298) 8.2	108 ± 1.9 (80–135) 9.6	36 ± 0.8 (29–55) 11.6	79.05 ± 3.23[m] (41.0–115.0) 18.74	38.57 ± 0.16 (37.02–40.06) 2.21	6.38 ± 0.05 (5.85–6.97) 4.69	21.46 ± 0.09 (20.46–22.45) 2.35	17.57 ± 0.08 (16.88–18.50) 2.46	9.53 ± 0.06 (8.98–10.70) 3.23	15.46 ± 0.07 (14.51–16.22) 2.33	20.91 ± 0.12 (19.60–22.13) 3.19
West slope of Cascade Range	♀	30	247 ± 2.7 (216–285) 5.9	102 ± 1.7 (80–121) 9.1	36 ± 0.4 (31–45) 6.8	86.22 ± 2.84[h] (56.7–109.0) 16.44	38.21 ± 0.18 (36.42–40.49) 2.59	6.39 ± 0.05 (5.65–7.00) 4.62	21.29 ± 0.08 (20.22–22.03) 2.17	17.44 ± 0.07 (16.86–18.22) 2.07	9.52 ± 0.06 (8.99–10.32) 3.25	15.24 ± 0.07 (14.55–16.38) 2.46	20.88 ± 0.12 (19.26–22.01) 3.08
	♂	31	235 ± 2.5 (205–270) 5.9	99 ± 1.8 (65–116) 9.9	35 ± 0.5 (27–40) 7.5	79.93 ± 1.87 (59.6–101.7) 13.06	37.79 ± 0.11 (36.85–38.99) 1.62	6.31 ± 0.04 (5.97–6.88) 3.79	20.89 ± 0.06 (20.22–21.63) 1.83	17.35 ± 0.08 (16.50–18.10) 2.49	9.29 ± 0.04 (8.97–9.94) 2.21	15.22 ± 0.04 (14.59–15.68) 1.53	20.49 ± 0.09 (19.46–21.46) 2.39

[a]Although measures of ear length were recorded, the data were not included because of extreme variation believed to stem from differences in techniques used to measure these characters.
[b]Sample size reduced by 7. [c]Sample size reduced by 24. [d]Sample size reduced by 29. [e]Sample size reduced by 21. [f]Sample size reduced by 25. [g]Sample size reduced by 19. [h]Sample size reduced by 5. [i]Sample size reduced by 1. [j]Sample size reduced by 20. [k]Sample size reduced by 26. [l]Sample size reduced by 12. [m]Sample size reduced by 9.

and elsewhere in the state, ponderosa pine seems to be the common factor in habitats occupied by the species (Bailey, 1936). Nevertheless, individual *T. amoenus* occasionally range short distances from the forest edge into shrub-steppe communities; at one locality, 30% of captures in shrub-steppe were >130 m from the forest edge, but at another site, none was as far as 130 m (States, 1976).

In central Oregon, estimates of spring density ranged from 1.7 to 10.82/ha, whereas those in late summer after recruitment of young ranged from 2.63 to 12.83/ha (States, 1976). Densities tended to be low in areas where *T. amoenus* ranged into shrub-steppe to forage and increase progressively in areas toward the middle of the "ponderosa pine transition zone," but declined where ponderosa pine was being succeeded by Douglas-fir (States, 1976:228). In Washington, average (±SD) densities were 1.50 ± 0.43 and 1.66 ± 0.55/ha on control and experimental grids, respectively, before manipulation, but 1.48 ± 0.26

and 2.23 ± 0.38/ha, respectively, after *T. townsendii* was removed from the experimental grid (Trombulak, 1985).

We frequently observed and occasionally captured *T. amoenus* on an area near Terrebonne, Deschutes Co., where studies of the mountain cottontail (*Sylvilagus nuttallii*) were conducted (McKay and Verts, 1978*a*, 1978*b*), but the species never seemed as abundant there as in ponderosa pine forests farther west. The vegetation was shrub-steppe with widely spaced junipers in pumice soil with many rock outcrops (McKay and Verts, 1978*a*).

Diet.—Based on observations of individuals while they forage, seeds, fruits, bulbs and tubers, insects, birds' eggs, berries, flowers, green foliage, roots, grains, and tender buds of woody plants commonly are listed as foods of *T. amoenus* (Maser and Maser, 1987). In Washington, *T. amoenus* spent half its observed feeding time consuming seeds, particularly those of grasses and pines (*Pinus*); half the remaining time was spent eating fruits, especially those of queen's cup (*Clintonia uniflora*) and broadpetal strawberry (*Fragaria virginiana*). Fungi and miscellaneous items were consumed during the remaining time (Trombulak, 1985).

In Oregon, fungi constituted 50% of the contents of *T. amoenus* stomachs examined and occurred in 82% of specimens examined (Maser et al., 1978*a*). Ten genera of hypogeous fungi were found in stomach contents of 135 *T. amoenus* collected in Grant Co. in July, a season when availability of fungi should be relatively low. *Rhizopogon* was the most common genus of fungus eaten; 60% of the males and 59% of the females had consumed only *Rhizopogon* (Maser and Maser, 1987). In California, fungi also constituted a large proportion of the diet (Tevis, 1952). By volume, fungi ranked first in stomach contents during summer and second during spring and autumn; seeds ranked first during spring and autumn (Tevis, 1953). Arthropods ranked third at all seasons (Tevis, 1953).

Reproduction, Ontogeny, and Mortality.—In the Cascade Range in Washington, *T. amoenus* possessed enlarged uteri upon emergence from winter torpor. Essentially all females bred, most during a week-long period near the end of March. Most young were born during the last week of April; young emerged near mid-June; and lactation ceased the 1st week of July (Kenagy and Barnes, 1988). All males had enlarged seminal vesicles and spermatozoa in their cauda epididymides when they emerged from winter torpor in the 3rd week of March; after the 1st week of May, males rarely possessed spermatozoa and none possessed more than a few (Kenagy and Barnes, 1988). In Alberta, emergence and breeding were nearly a month later than in Washington; some males possessed spermatozoa in their epididymides until late June (Sheppard, 1969). During a 4-year period, 18–71% of yearling males and 14–67% of yearling females bred; presumably, the remaining proportion first breed at ≅23 months of age (Sheppard, 1969). Most authorities indicate that *T. amoenus* produces only one litter annually (Table 11-4), but based on one female simultaneously lactating and pregnant collected in July, Negus and Findley (1959) suggested that a second litter possibly was produced in Wyoming.

Kenagy and Barnes (1988) believed that litter size (Table 11-4) in *T. amoenus* varied with latitude and altitude with those at higher latitudes and elevations averaging smaller. In support, they compared average litter size of a sample collected at 600–675 m in Washington with that of a sample collected by Tevis (1955) at 1,370–1,525 m in California. The Washington average was 1.1 young/litter less than that

Table 11-4.—*Reproduction in some Oregon chipmunks* (Tamias), *tree squirrels* (Sciurus *and* Tamiasciurus), *and the northern flying squirrel* (Glaucomys sabrinus).

Species	Number of litters	$\overline{X}$	Range	n	Reproductive season	Basis	Authority	State or province
Glaucomys sabrinus	1?	5.0	3–6	3		Implant sites	This study	Oregon
Sciurus griseus	1	2.2	2–3	13	Dec.–Sep.[b]	Embryos	Fletcher, 1963	California
	1	2.7	2–4	9	Jan.–Jun.	Embryos	Ingles, 1947	California
	1	2.8	2–4	10		Embryos	Swift, 1977	California
	1	3.0	2–4	25	Feb.–Jul.	Embryos	W. C. Asserson, III, in litt.	California
Tamias amoenus	1	5.0	3–7	43	Mar.–Apr.	Embryos	Kenagy and Barnes, 1988	Washington
	1	5.8	5–7	7	Apr.–Jun.	Embryos	Broadbooks, 1958	Washington
	1	6.1	4–8	17		Embryos	Tevis, 1955	California
		5.0	2–7	5	May	Implant sites and embryos	This study	Oregon
	1	4.6		42	Apr.–Jun.	Embryos	Sheppard, 1969	Alberta
		4.7		59		Implant sites	Sheppard, 1969	Alberta
Tamias minimus	1	5.8	4–7	7		Embryos	Tevis, 1955	California
		4.8	2–6	4	Apr.–May	Implant sites and embryos	This study	Oregon
	1	6.4	5–9	11	Apr.–Jun.	Embryos	Skryja, 1974	Wyoming
		5.4	3–8	24		Implant sites	Skryja, 1974	Wyoming
	1	4.2		21	May–Jun.	Embryos	Sheppard, 1969	Alberta
		4.0		32		Implant sites	Sheppard, 1969	Alberta
Tamias senex	1	4.5	3–5	6		Embryos	Tevis, 1955	California
Tamias townsendii	1	4.2	3–5	17	Apr.–May[a]	Implant sites	Gashwiler, 1976*c*	Oregon
		4.0		5	Apr.–May[a]	Embryos	Gashwiler, 1976*c*	Oregon
	1	3.8	3–4	12	Apr.–May	Embryos	Kenagy and Barnes, 1988	Washington
Tamiasciurus douglasii		3.8	1–8	12		Implant sites	This study	Oregon

[a]Period of fecundity (pregnancy and lactation) extended from April to September, but lactating females were collected as early as June, indicating that many if not most females probably gave birth in May.
[b]Most reproductive efforts complete by May.

observed in California, but identical with that of a small sample collected at ≅1,700 m in Oregon (Table 11-4). In Alberta, based on counts of corpora lutea, yearling females produced significantly fewer ova ($\bar{X}$ = 4.68 ± 0.12, n = 44) than older adults ($\bar{X}$ = 5.42 ± 0.11, n = 71; Sheppard, 1969).

In Oregon, four neonates averaged 1.7 g; body measurements (in mm) averaged: total length, 44; length of tail, 12; length of hind foot, 7 (Booth, 1942). The young were bright pink and completely without hair; the eyes were closed. The young emitted faint squeaks when handled (Booth, 1942). In Washington, young averaged 2.7 g at birth, about 5.3% of the body mass of their mothers; at first capture they averaged 21 g or 41% of adult mass (Kenagy and Barnes, 1988). Young in another Washington litter also averaged 2.7 g at birth; they gained 0.5 g/day, their ears unfolded at 6–7 days, I1s erupted at 22–29 days, and the eyes opened at 30–33 days (Broadbooks, 1958). The light and dark lateral stripes appeared at 10 days and the middorsal black strip appeared at 14 days (Broadbooks, 1958).

Recorded maximum longevity for free-living *T. amoenus* is 5 years and 2 months (Broadbooks, 1958). The primary predators of yellow-pine chipmunks probably are long-tailed weasels (*Mustela frenata*), coyotes (*Canis latrans*), badgers (*Taxidea taxus*), bobcats (*Lynx rufus*), goshawks (*Accipiter gentilis*), and rattlesnakes (*Crotalus*—Broadbooks, 1958).

Habits.—Yellow-pine chipmunks are active during the daylight hours, with activity commonly commencing slightly before sunrise, then, following a midday break (0900–1500 h) during hot days, continuing until slightly after sunset. On cloudy days, chipmunks may continue to be active throughout the day, even during light rain (Broadbooks, 1958). During periods of strong winds, *T. amoenus* is observed infrequently, but trapping records reveal that they are active; they probably avoid winds by staying near or under cover (Broadbooks, 1958).

Yellow-pine chipmunks are "sprightly and active. They seem always to be moving restlessly about, running, investigating for food, and watching for enemies" (Dalquest, 1948:254). Movements tend to be nervous and jerky; even arboreal activity is in spurts of climbing for 1–2 m with intervening stops. Relative brain size in *T. amoenus* is slightly greater than that of *T. minimus,* but considerably less than that of other Oregon chipmunks (Budeau and Verts, 1986). This supports the hypothesis that species of mammals that climb in or otherwise negotiate multilayered habitats have evolved greater brain capacity to process the more complex information than closely related species that live in simpler habitats (Eisenberg and Wilson, 1978, 1981).

Although these chipmunks sometimes can be approached to within a few meters, more commonly the initial sighting is of an individual, often with tail held perpendicular to the body, scampering off into a tangle of logs and understory vegetation then suddenly reappearing 20 m or so away. Broadbooks (1968) described 10 distinct calls produced by *T. amoenus,* several of which served to alert conspecifics to some perceived threat.

Tamias amoenus remains in its burrow for ≅4 months, but arouses from torpor at intervals of about 5–7 days to feed upon materials stored in the burrow system (Geiser and Kenagy, 1987; A. Svihla, 1936). It does not store fat. Three caches obtained from burrow systems excavated in

November contained an average of 35,400 seeds. In Washington, burrows averaged 68.6 cm long (n = 14) and terminated in nests that averaged 16.5 cm in diameter. Grasses, pappi, feathers, lichens, and cotton (pilfered from traps) were used to construct and line the nests.

Yellow-pine chipmunks, in the absence of seasonal cues in captivity, spontaneously exhibit reproductive and hibernatory cycles similar to those of chipmunks in nature, but at intervals <1 year. Torpor is promoted by cold, absent at warm (23°C) temperatures, and partially inhibited by long (16L:8D) photoperiods (Kenagy, 1981). Even at 23°C, activity, body mass, and water consumption declined. Thus, the rhythms seem endogenous although probably synchronized by environmental cues (Kenagy, 1981). Animals fed a highly unsaturated lipid diet exhibited longer bouts of torpor, lower minimum body temperatures, and lower metabolic rates than animals on a saturated lipid diet (Geiser and Kenagy, 1987).

In Washington, home-range size calculated by an exclusive boundary-strip method averaged 1.0 ± 0.1 ha (n = 45) for females and 1.6 ± 0.1 ha (n = 28) for males. Home ranges tended to be linear (about twice as long as wide) and tended to be in an east-west orientation (Broadbooks, 1970). Breeding females often have somewhat overlapping home ranges. Although chases are observed occasionally, more often territorial behavior is not exhibited when one individual encounters another (Broadbooks, 1970); the area in the immediate vicinity of the den may be defended (Gordon, 1943). Conspecifics, when meeting, kiss (touch noses), smell the sides of the face and neck, then sniff the anus of each other, all within 5–6 s (Broadbooks, 1958). Although *T. amoenus* commonly dust bathes, such activity does not seem to be a means of marking territories.

The yellow-pine chipmunk is behaviorally dominant to the least chipmunk, and, where sympatric, likely restricts it to less productive habitats (e.g., shrub-steppe) and occupies adjacent more productive habitats (e.g., ponderosa pine forests—Heller, 1971; Sheppard, 1971). A more detailed discussion of interspecific aggressive interactions is included with the account on *T. minimus.*

Remarks.—We have included the findings of States (1976) as he reported them. In view of the difficulty experienced by others in separating *T. amoenus* and *T. minimus* (Anderson, 1974; Hall, 1981; Howell, 1929; Ingles, 1965; Sutton and Nadler, 1969), however, we cannot refrain from denoting that he did not mention precisely how he distinguished the two species in areas where he considered them to occupy the same or adjacent habitats. States (1976) did not mention depositing voucher specimens; we found none among specimens we examined.

Tamias minimus Bachman
Least Chipmunk

1839*b*. *Tamias minimus* Bachman, 8:71.
1890*a*. *Tamias minimus pictus* Allen, 3:115.
1934. *Eutamias minimus scrutator* Hall and Hatfield, 40:321.

Description.—*Tamias minimus* (Plate VII) is the smallest chipmunk in Oregon, but it is only slightly smaller than *T. amoenus* (Table 11-3). As in all chipmunks, the face is marked with a dark stripe through the eye bordered on each side with a light- and a dark-colored stripe. The middorsum is marked with a dark-colored stripe followed on each side by two alternating light- and dark-colored stripes. The innermost light-colored stripe is grayish, whereas the outermost light-colored stripe is white grading to grayish anteriorly and posteriorly. Contrary to some published descriptions (Clark and Stromberg, 1987:88), the outermost side stripe is dark. The hips, shoulders, and feet are gray and the flanks of some individuals are lightly washed with ochre. The underside of the tail of most individuals tends to be more yellowish than orangish. Although *T. minimus* tends to be less brightly colored (more gray and less orangish) than *T. amoenus*, color markings of *T. minimus* differ considerably even among individuals within the same population. Thus, *T. minimus* cannot be distinguished reliably from *T. amoenus* by color markings alone (Anderson, 1974). Also, the skull of *T. minimus* (Fig. 11-5) is similar to that of *T. amoenus*.

According to Anderson (1974) the most reliable means of separating the two small chipmunks in Oregon is by use of a combination of three skull dimensions. We developed an equation based on skull dimensions of adult specimens from Oregon to separate *T. mimimus* and *T. amoenus* as follows:

Value = palatal length (1.45163) + condylonasal length (0.41994) + breadth of braincase (0.32907).

Adult specimens of small chipmunks for which the calculated value is <39.8728 are *T. minimus*. Only nine (1.7%) of a sample of 543 small chipmunks from east of the Cascade Range could not be classified by use of the equation (Carraway and Verts, 1996).

Distribution.—The least chipmunk has the largest geographic range of any chipmunk (Sullivan, 1985). It occurs in Canada from west-central Yukon Territory southeastward to Great Slave Lake and along the west side of Hudson Bay and James Bay to southeastern Ontario, then southward through northeastern British Columbia and most of Alberta, Saskatchewan, Manitoba, and Ontario. In the United States, *T. minimus* occurs in northern North Dakota, Minnesota, Wisconsin, and the upper peninsula of Michigan; and from western North Dakota, South Dakota,

and Nebraska, and central Colorado, northern New Mexico, and northern Arizona westward to eastern Washington, Oregon, and California (Fig. 11-6; Hall, 1981). In Oregon, *T. minimus* occurs east of the Cascade Range except in the Columbia Basin and most of the Blue Mountains (Fig. 11-6).

Remains of least chipmunks have been identified in Pleistocene and Holocene deposits in caves in Idaho, Wyoming, and Utah; also, remains of the species have been found in cave deposits of similar age in Tennessee and Virginia (Kurtén and Anderson, 1980).

Geographic Variation.—Only one of the 19 recognized subspecies of *T. minimus* occurs in Oregon: *T. m. scrutator* (Hall, 1981).

Habitat and Density.—In general, *T. minimus* in Oregon is the chipmunk of the shrub-steppe, but sometimes it occurs at the edges of open forests. In central Oregon, 2–13% of captures of *T. minimus* on adjacent grids that straddled the border between ponderosa pine (*Pinus ponderosa*) forest and shrub-steppe were "in shrubby areas inside tree line" (States, 1976:227). A few individuals were believed to reside within the forest edge. On Steens Mountain, *T. minimus* was collected from Blitzen Valley to near the crest; from ≅1,770 m to ≅2,745 m *T. minimus* reportedly was associated with *T. amoenus*, but at elevations <1,770 m it was the only chipmunk (C. Hansen, 1956). *T.*

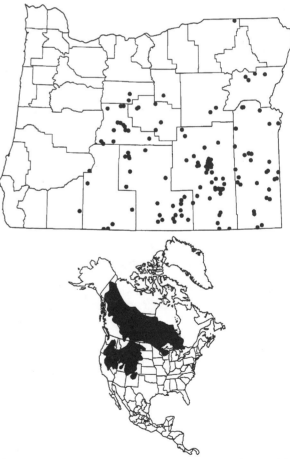

Fig. 11-6. Localities in Oregon from which museum specimens of the least chipmunk, *Tamias minimus*, in Oregon, were examined (listed in APPENDIX, p. 567), and the distribution of the least chipmunk in North America. (Range map redrawn from Hall, 1981:345, map 236, with modifications.)

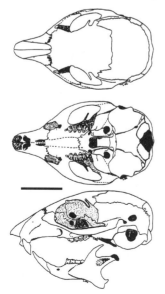

Fig. 11-5. Dorsal, ventral, and lateral views of the cranium, and lateral view of the dentary of the least chipmunk, *Tamias minimus* (OSUFW 2738). Scale bar equals 10 mm.

minimus occurred most frequently in aspen (*Populus tremuloides*) groves and in rocky situations (C. Hansen, 1956).

Also, at lower elevations in southeastern Oregon, *T. minimus* was captured on sample grids established in sagebrush (*Artemisia*) and greasewood (*Sarcobatus vermiculatus*) habitats, but not in grassland or marsh habitats (Feldhamer, 1979b). On the shrub areas, average (±SD) estimates of density ranged from 2.0 ± 0.6 to 13.9 ± 1.0/ha in sagebrush and from 1.7 ± 0.7 to 16.9 ± 1.7/ha in greasewood. Density was correlated positively with soil depth and strength, and proportion of clay in the soil. These factors likely were related to the stability of burrows dug by chipmunks; seemingly chipmunks are better able to dig in heavy soils than other small mammals in the region (Feldhamer, 1979b). Chipmunks also seemed to select for areas with sparse cover in the first 15 cm above the surface of the soil, but with dense cover above 15 cm. Such cover likely affords concealment from predators without interfering with rapid escape movements of *T. minimus* (Feldhamer, 1977b). In Colorado, estimated densities ranged from 11.4 to 22.2/ha during a 3-year period (Vaughan, 1974). Among seven shrub communities in Utah, *T. minimus* occurred only in the sagebrush community; estimated density was 3.0/ha (Fautin, 1946). In Minnesota, Erlien and Tester (1984) suggested that populations of *T. minimus* possibly were cyclic with peaks at intervals of 5–7 years; however, <20% of variation among annual estimates of density was attributable to cycles.

Diet.—*Tamias minimus* commonly is classified as an omnivore (Carleton, 1966; Schreiber, 1979; Vaughan, 1974). In Colorado, 77.0–93.5% of the diet was plant materials, with seeds composing ≅50% of the diet in summer; seeds of sedge (*Carex*), bitterroot (*Lewisia*), collomia (*Collomia*), and violet (*Viola*) contributed most. Seeds also were a highly sought food item during and immediately after snow melt despite their low availability at that season. Flowers, leaves, fruits, and fungi composed 24–36% of the diet (Vaughan, 1974). Arthropods contributed most to the diet in early summer. In another Colorado study, dandelion (*Taraxacum*) seed heads composed a large proportion of the diet where the plant species was abundant, but contributed only a small proportion of the diet in other areas (Carleton, 1966). Schreiber (1979) found that *T. minimus* consumed a diet higher in energy content in autumn than in summer, but was less efficient in extracting energy from the oil-rich seeds consumed in autumn.

Reproduction, Ontogeny, and Mortality.—In Wyoming, testes of *T. minimus* weighed the most in March and declined to a minimum in July; spermatozoa were present in the epididymides in March and April, but few (or none) were present later in the year (Skryja, 1974). The earliest mating was on 8 April and the latest date that a pregnant female was captured was 16 June. At various localities within the range of the species, about 10–40% of females in the populations may not mate. Litter size varies not only with locality but also with method of estimation (Table 11-4).

In Alberta, based on counts of corpora lutea, yearling female *T. minimus* shed significantly fewer ova (4.36 ± 0.24, n = 11; thus presumably produced smaller litters) than older females (4.59 ± 0.14, n = 41; Sheppard, 1969). Sheppard (1969) believed that lower reproductive rates (smaller litters and later breeding) put *T. minimus* at a competitive disadvantage with *T. amoenus* in adjacent forest

habitats. He also believed that the best explanation for the lower reproductive rate in *T. minimus* lay in the lower availability of food in the relatively poorer-quality habitat above treeline where the species lived.

At birth, young in a litter of five observed by Forbes (1966) were naked, pink, and somewhat wrinkled, and had closed eyes and auditory meatuses. At 27 days, the incisors were erupted, the average mass was 11.7 g, the eyes and auditory meatuses of some individuals were commencing to open, and the young were able to climb on wire mesh; by 31 days, the eyes and ears were completely open on all surviving individuals. At 36 days, young were beginning to eat solid food and by 40 days m1 and M1 had erupted; the young were "alert, active, and quick" (Forbes, 1966:160).

In Wyoming, the principal predators are "long-tailed weasels, mink, red foxes, marsh hawks, and red-tailed hawks" (Clark and Stromberg, 1987:89). We suspect that an equal or greater variety of raptorial birds and carnivorous mammals take least chipmunks in Oregon.

Habits.—Least chipmunks are active during the daylight hours; we know of no study regarding peaks of activity. Like its congeners, *T. minimus* does not fatten in autumn (Bailey, 1936), but presumably enters bouts of torpor with interspersed periods of feeding on foods cached in dens. Like *T. amoenus*, *T. minimus* moves swiftly through its habitat, disappearing when pursued too closely, but reappearing a little further away if the observer remains motionless for a few moments.

Least chipmunks often can be seen sitting in the tops of shrubs, presumably to obtain a better view of their surroundings (Bailey, 1936). Nevertheless, of four species of chipmunks in Oregon, *T. minimus* has the smallest brain in relation to body size, reflecting the absence of arboreality and the relative simplicity of the shrub-steppe habitat in which it occurs (Budeau and Verts, 1986).

In Montana, home ranges of six individuals (three adults, three young-of-the-year when observations commenced) whose movements were followed for 2 consecutive years ranged from 0.7 to 1.8 ha; no pattern of home-range size in relation to age was evident, but for five individuals, the size was similar between years (Martinsen, 1968).

From a laboratory study with four species of California chipmunks, Heller (1971:315) reported that *T. minimus* and *T. amoenus* did not interact aggressively "in a significant number of encounters, but when aggression does occur *amoenus* is dominant." Most of the observed aggression was between females; males of the two species exhibited little aggression when paired. Nevertheless, Heller (1971) concluded that the upper limit of the altitudinal range of *T. minimus* possibly was determined by aggression of *T. amoenus*. In a somewhat similar study with two chipmunk species in Alberta, *T. amoenus* was highly successful in encounters with *T. minimus*; the latter species tended to exhibit avoidance behavior (Sheppard, 1971). Despite the reversal of altitudes occupied, *T. minimus* also seemingly was competitively excluded from more productive habitats in forested valleys and on lower slopes by *T. amoenus*, but because of its smaller size, *T. minimus* was able to survive on the limited food available on the high ridges above treeline (Sheppard, 1971). Also, *T. minimus* seems to be behaviorally subordinate to the golden-mantled ground squirrel (*Spermophilus lateralis*), with which it may be syntopic in some areas (Carey, 1978). We suspect that interspecific aggression plays a role in *T. minimus* occupying the least productive chipmunk habitats in Oregon.

Remarks.—We were surprised by the relative paucity of quantitative life-history information available for *T. minimus*, not only for Oregon, but throughout its geographic range. To us, the species seemingly has several attributes that would make it an excellent model for studies of population and other ecological phenomena.

We believe that a study of interactions of *T. minimus* and *T. amoenus* on Steens Mountain to ascertain if, indeed, the two species commingle in the same habitats at elevations >1,770 m, as suggested by C. Hansen (1956), is warranted. If the two species actually are syntopic over a broad area, the mechanism by which they coexist may be unique.

The *Tamias townsendii* Complex

Originally, this complex consisted of six subspecies of *T. townsendii* plus *T. alleni*, *T. quadrimaculatus*, *T. merriami*, and *T. dorsalis* (Howell, 1929), based largely upon the large body size and white or buff edging on the tail (Sutton, 1993). Of these taxa, *T. t. townsendii*, *T. t. cooperi*, *T. t. ochrogenys*, *T. t. senex*, and *T. t. siskiyou* initially were considered to occur in Oregon (Howell, 1929). On the basis of differences in body and skull morphometrics and in morphology of the baculum, Sutton and Nadler (1974) elevated three of these taxa to species level: *T. ochrogenys*, *T. senex*, and *T. siskiyou*. They also restricted the distribution of *T. ochrogenys* to California, leaving the two subspecies of *T. townsendii*, and *T. senex* and *T. siskiyou* as occurring in Oregon. Because Sutton and Nadler (1974) used only four body and skull characters in their analyses and because morphology of the baculum may not be particularly conservative, Levenson and Hoffmann (1984) reevaluated morphological variation by use of 253 specimens and 14 body and cranial characters. They also produced a cladogram based on electrophoretic data for 20 proteins in 60 individuals of four taxa of the *T. townsendii* group. Because of extensive overlap in morphometric characters, the latter authors (p. 166) concluded that "the taxa are not easily or consistently distinguishable by these criteria, and that there apparently are not sharp boundaries or clear zones of intergradation." They also claimed (p. 166) that "results of our cladistic analyses of electrophoretic data are extremely different from those of Sutton and Nadler's [1974] analyses of genital bone morphology." Later, Levenson et al. (1985) treated the Oregon taxa of these large chipmunks as subspecies of *T. townsendii*. In response, Sutton (1987) made pair-wise comparisons of 14 body and cranial dimensions for *T. ochrogenys*, *T. senex*, *T. siskiyou*, and *T. townsendii* from coastal regions of southwestern Oregon and north western California. He concluded (p. 376) that "these taxa [were] separated geographically by river boundaries and readily separated in analysis of the morphological characters studied." Gannon and Lawlor (1989) contended that discriminant analysis of six characters of the alarm calls produced by the four taxa supported separation of these chipmunks as distinct species. They also indicated that calls produced by chipmunks at species' boundaries were more distinctive and less variable than those produced by individuals near the centers of species' ranges. Authors of the most recent checklist of North American mammals (Jones et al., 1992)

acknowledged these findings by listing *T. townsendii*, *T. senex*, and *T. siskiyou* as distinct species.

Although considerable work on the problem of speciation and distribution of this group of chipmunks was conducted in extreme southwestern Oregon and southward in California, these topics have been largely ignored in the Cascade Range. Only ten individuals from this region were included in the morphometric analysis by Sutton and Nadler (1974): four males listed as *T. t. cooperi* and one male and one female listed as *T. t. siskiyou* from McKenzie Bridge, Lane Co., and one female listed as *T. t. senex* and one male and two females listed as *T. t. siskiyou* from Crater Lake National Park, Klamath Co. Sutton and Nadler (1974:204) claimed "no intergradation of characters apparent in the specimens from either location," a statement obviously not well supported by such small numbers of specimens even though morphological differences among specimens of different taxa at the same location may have been considerable. Nevertheless, distributions of *T. t. cooperi*, *T. senex*, and *T. siskiyou* in the Cascade Range were depicted (Sutton and Nadler, 1974:202, fig. 1) much as Howell (1929) and Hall and Kelson (1959) had depicted ranges based largely upon pelage color when all were considered subspecies of *T. townsendii*. Levenson and Hoffmann (1984) did not provide localities of specimens used in their analyses, and Sutton (1987) and Gannon and Lawlor (1989) limited their studies to areas other than the southern Cascade Range in Oregon.

Because coastal rivers become extremely low in summer, Levenson and Hoffmann (1984) were critical of the suggestion that they act as barriers to interchange among the nominal species of chipmunks. Thus, Sutton (1993) admitted that there was reason to investigate the taxonomic status of the group further.

Herein, we have treated *T. townsendii*, *T. senex*, and *T. siskiyou* as separate species. We were forced to rely upon pelage color and markings to distinguish the species in the southern Cascade Range because the baculum was not preserved or it remains in the dried phallis of most male museum specimens. Because several investigators failed to deposit voucher specimens, we often have relied upon localities of study sites in considering some ecological information obtained by those who considered all of these chipmunks to be *T. townsendii*. Because some have claimed that at least two of the species occur at the same site, such an approach obviously may be subject to error.

Tamias senex J. A. Allen
Allen's Chipmunk

1890a. *Tamias senex* Allen, 3:83.
1897f. *Eutamias senex*: Merriam, 11:196.

Description.—*Tamias senex* (Plate VII) is the smallest chipmunk in the *townsendii* complex in Oregon (Table 11-3). The pelage is marked with five dark and four light stripes on the dorsum; the middorsal strip is usually black and nearly always darker than the other four dark stripes. The inner pair of dark stripes is usually grayish mixed with some brown; both pairs of light stripes contain much white. The shoulders and rump are usually washed with gray. The flanks are ocherous. The sides of the face are marked with three brown and two light gray stripes; a patch behind the ear is light gray. The tail is blackish frosted with ochre dorsally and rusty brown ventrally. The venter commonly is white or cream.

The diameter of the baculum is >20% of its length and the tip is less than one-half the length of the shaft, characters useful in separating *T. senex* from other species in the *townsendii* complex. The skull (Fig. 11-7) is similar to that of other species in the *townsendii* complex. Pelage color and markings seem useful in separating *T. senex* from *T. townsendii* and *T. siskiyou*, but may not be absolutely reliable. However, we were unable to verify our identifications by comparison with baculum morphology.

Allen's chipmunk produces a call of a rapid series of three–four to as many as 10 syllables (Gannon and Lawlor, 1989).

Distribution.—In Oregon, Allen's chipmunk occurs in forested areas along the eastern portion of the Cascade Range from Ollalie Forest Camp and Warm Springs in Jefferson Co. south to 22.5 km (14 mi.) E Ashland, Jackson Co., and Lakeview, Lake Co. (Fig. 11-8). The species also occurs along the Sierra Nevada Range in California south to Madera Co. and in extreme western Nevada (Hall, 1981; Sutton and Nadler, 1974) and in California west to the Pacific Coast between the Klamath and Eel rivers (Fig. 11-8; Gannon and Lawlor, 1989).

Geographic Variation.—The species is monotypic (Hall, 1981).

Habitat and Density.—These chipmunks are forest dwellers that, in addition to climbing in trees, frequent dense thickets of buckbrush (*Ceanothus*), "shrubby oak chapparal," and rocky outcrops (Bailey, 1936:131). Tevis (1955:71) indicated that *T. senex* in northeastern California "inhabits forests that are relatively dense, also brushy areas that are relatively moist." Of the four species of chipmunks that he caught, *T. senex* was the least frequently in association with the golden-mantled squirrel (*Spermophilus lateralis*).

In the 1970s and early 1980s, we collected (by shooting) a series of sciurids on the east slope of Black Butte, Deschutes Co., in September each year. The habitat was open ponderosa pine (*Pinus ponderosa*) with an understory of thick clumps of manzanita (*Arctostaphylos*) and an occasional bitter-brush (*Purshia tridentata*). Numerous rocky outcrops were scattered throughout the area. *T. senex* was collected less frequently than either *T. amoenus* or *S. lateralis*, but more frequently than *Tamiasciurus douglasii*. We are unsure if *T. senex* was numerically inferior or simply less easily collected than the ground squirrel and other chipmunk. In northeastern California, with equal effort at 18 sites, Tevis (1955) caught 2.6–4.6 times as many *S. lateralis* and three other species of *Tamias* as of *T. senex*. Tevis (1955:72) related that from "all appearances, trees, shrubs, forbs, ground litter, volcanic soil, lava rock, degree of slope, elevation, and food supply did not differ" at the sites. We know of no more detailed treatments of habitat or density for *T. senex*.

Diet.—Bailey (1936:132) listed "acorns, chinquapins, hazelnuts, seeds of pines, spruces, and hemlocks, cherries, berries, seeds of grasses, and a great variety of plants, roots, tubers, green vegetation, flowers, insects and small animal life" among items used as food by *T. senex*. He further indicated that seeds and fruit of the little bitter cherry (*Prunus emarginata*) were particularly relished and that "manzanita berries, currants, gooseberries, raspberries,

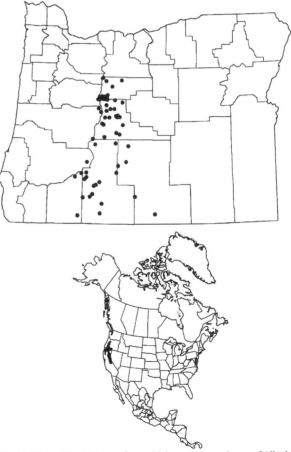

Fig. 11-8. Localities in Oregon from which museum specimens of Allen's chipmunk, *Tamias senex*, in Oregon, were examined (listed in APPENDIX, p. 568), and the distribution of Allen's chipmunk in North America. (Range map redrawn from Hall, 1981:351, map 238, with modifications.)

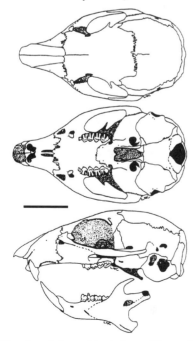

Fig. 11-7. Dorsal, ventral, and lateral views of the cranium, and lateral view of the dentary of Allen's chipmunk, *Tamias senex* (OSUFW 3373). Scale bar equals 10 mm.

thimbleberries, blueberries, and strawberries" were eaten. The species also consumes fungi.

Reproduction, Ontogeny, and Mortality.—Bailey (1936:132) claimed that *T. senex* has "4 to 6 young, born in May or June and appearing aboveground in July." Tevis (1955) indicated that *T. senex* in California bred about 1 month after emergence from hibernation. He recorded one individual active on 19 March, but most did not emerge until mid-April, suggesting that the breeding season was in May. In addition to the records of Tevis (1955) regarding litter size (Table 11-4), we recorded five pigmented sites of implantation in each of two females that we caught. One of these was taken at Suttle Lake, Jefferson Co., on 5 May, suggesting that it had given birth before that date.

Habits.—Bailey (1936) claimed that *T. senex* was less secretive and more like congeners that occupy more open habitats than like other members of the *townsendii* complex. He described *T. senex* (p. 131) as "quicker, more dependent on escape in open places" than similar species. These chipmunks are superb climbers, an ability to which we can attest. While collecting sciurids, one of us (BJV) flushed an Allen's chipmunk that in <30 s climbed to the crown of a nearby ponderosa pine estimated to be >40 m tall. Although the chipmunk was completely exposed, several shots were required at the maximum range of the light shotgun the investigator used to collect sciurids to obtain a "hit" with a pellet.

Bailey (1936) indicated that *T. senex* did not become fat in autumn; he believed that these chipmunks aroused periodically from torpor and survived by consuming food stored in their dens. Tevis (1955) reported that *T. senex* in northeastern California gained about 20% in body mass, but did not add fat until shortly before immergence. He also reported that more than half of the animals still had a remnant of fat when they emerged in spring.

Tamias siskiyou (A. H. Howell)
Siskiyou Chipmunk

1922. *Eutamias townsendii siskiyou* Howell, 3:180.
1974. *Eutamias siskiyou*: Sutton and Nadler, 19:211.

Description.—*Tamias siskiyou* (Plate VII) is slightly smaller than *T. townsendii* and slightly larger than *T. senex* (Table 11-3). The pelage is marked with five dark and four light stripes on the dorsum; the middorsal strip is usually black and nearly always darker than the other four dark stripes. The inner pair of light stripes usually is light gray and the outer pair usually is much brighter than the inner pair. The shoulders and rump are usually washed with brown. The flanks are ocherous. The sides of the face are marked with three brown and two light gray stripes; a patch behind the ear is light gray. The tail is blackish frosted with ochre dorsally and rusty brown ventrally. The undersides commonly are ocherous.

The diameter of the baculum is >20% of its length and the tip is longer than the shaft, characters useful in separating *T. siskiyou* from other species in the *townsendii* complex. The skull (Fig. 11-9) is similar to that of other species in the *townsendii* complex. Pelage color and markings seem useful in separating *T. siskiyou* from *T. townsendii* and *T. senex*, but may not be absolutely reliable. However, we were unable to verify our identifications by comparison with baculum morphology.

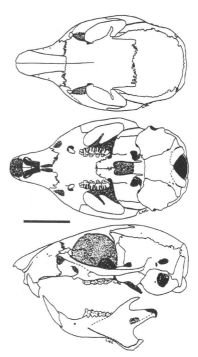

Fig. 11-9. Dorsal, ventral, and lateral views of the cranium, and lateral view of the dentary of the Siskiyou chipmunk, *Tamias siskiyou* (OSUFW 4045). Scale bar equals 10 mm.

Siskiyou chipmunks produce calls of "a single syllable, greater frequency range, greater intensity, and longer duration between calls" than other species in the *T. townsendii* complex (Gannon and Lawlor, 1989:740).

Distribution.—In Oregon, *T. siskiyou* occurs in Curry, Josephine, and Jackson counties south of the Rogue River; in extreme western Klamath Co.; in extreme eastern Douglas and Lane counties; and in extreme southeastern Linn Co. (Fig. 11-10). The remainder of the species range is in northwestern California north of the Klamath River (Fig. 11-10; Hall, 1981; Sutton and Nadler, 1974).

Geographic Variation.—The species is monotypic (Hall, 1981).

Habitat and Density.—In Jackson Co., McIntire (1980) studied populations of *T. siskiyou* in areas selectively logged 7–9 years earlier. The dominant species were white fir (*Abies concolor*), Douglas-fir (*Pseudotsuga menziesii*), western white pine (*Pinus monticola*), and Pacific yew (*Taxus brevifolia*) with an understory of huckleberry (*Vaccinium membranaceum*), common snowberry (*Symphoricarpus albus*), and wood rose (*Rosa gymnocarpa*). Herbaceous species included inside-out flower (*Vancouveria hexandra*), windflower (*Anemone deltoidea*), and starflower (*Trientalis latifolia*). One 2.56-ha trapping grid was established at a site at which slash was piled and burned 1–2 years after logging, and another was established at a site at which slash was not manipulated. Only five of 27 measured habitat variables differed significantly between the two areas; all significant characteristics were related to the density of woody vegetation. Numbers of individual *T. siskiyou* captured per unit effort were similar on the two grids: treated grid = 9.6/100 trap-days in 1978, 5.1/100 trap-days in 1979; untreated grid = 8.9/100 trap-days in 1978, 4.0/100 trap-days in 1979. The total number of individuals captured in the 2 years

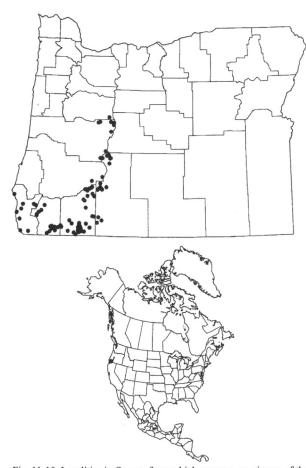

Fig. 11-10. Localities in Oregon from which museum specimens of the Siskiyou chipmunk, *Tamias siskiyou*, were examined (listed in APPENDIX, p. 568), and the distribution of the Siskiyou chipmunk in North America. (Range map redrawn from Hall, 1981:351, map 238, with modifications.)

combined (effort was essentially identical on both grids) did not differ significantly from an expected equal number on both grids (McIntire, 1980:46, table 8).

Diet.—Bailey (1936) recorded information provided by Ned Hollister regarding *T. siskiyou* feeding on little bitter cherries (*Prunus emarginata*) in September and on acorns (*Quercus*) and pine (*Pinus*) nuts the following month. He also commented on the large capacity of cheek pouches that these chipmunks use to transport seeds to their winter caches. Individuals carried as many as 5 acorns, 19 seeds of sugar pine (*Pinus lambertiana*), or 65 seeds of other pines.

In a thinned forest stand in Jackson Co., 99.1% of 224 and 96.8% of 250 fecal samples from sites on which slash was piled and burned or unmanipulated, respectively, contained spores of one or more types of hypogeous fungi (McIntire, 1980). A minimum of 15 genera of fungi in the classes Basidiomycetes, Ascomycetes, and Phycomycetes were represented. *Rhizopogon* (false truffles) and indistinguishable related genera were identified most frequently; *Endogone*, commonly eaten by other rodents (Maser et al., 1978a), was among the least frequently identified fungi in feces of *T. siskiyou* (McIntire, 1980). Obviously, the sporocarps of fungi contribute significantly to the diet of these chipmunks. McIntire (1980) noted evidence of these chipmunks foraging heavily upon common thistle (*Cirsium vulgare*).

Reproduction, Ontogeny, and Mortality.—We know

of no published information regarding these topics. We located one record of a litter size of six based on pigmented implantation sites (OSUFW 8028).

Habits.—Siskiyou chipmunks on an area in Jackson Co. exhibited three diurnal peaks of activity: an early morning peak, a midday peak, and the maximum peak in early evening (McIntire, 1980). On a site on which slash was burned, the early morning activity was greater but peaked about an hour later, the midday activity was much less extensive, and the evening activity was more variable than on a site at which slash was not burned. McIntire (1980) believed that the reduced cover on the slash-burned site likely was responsible for differences in time of activity on the two areas.

The activity budget (proportion of time spent in various activities) of *T. siskiyou* on sites in Jackson Co. where slash was burned and unburned (in parentheses) was: foraging, 39.8% (42.7%); social interactions, 1.3% (2.2%); food handling, 43.3% (27.9%); maintenance, 4.9% (19.5%); and protection, 10.6% (7.7%). The last three were significantly different (*P* < 0.01) between sites (McIntire, 1980). The greater time engaged in protective behavior on the slash-burned site was believed to be a response to the loss of ground cover.

Habitat objects used most frequently on the site where slash was burned were: burnpiles, 29.6%; open ground, 24.8%; slash, 17.7%; and Pacific yew trees, 10.4%. The extensive use of burnpiles was largely a result of chipmunks foraging on thistles that grew on them. On the unburned site, habitat objects used most frequently were: slash, 51.5%; open ground, 19.4%; rocks, 8.5%; and overstory trees, 6.6% (McIntire, 1980).

Home-range areas for adults calculated by the inclusive boundary-strip method (Stickel, 1954) averaged 0.52 ha (range, 0.43–0.59 ha) for six males, and was 0.34 ha for one female on a forest site on which slash was burned. The areas averaged 0.47 ha (range, 0.31–0.59 ha) for four males and 0.53 ha (range, 0.44–0.78 ha) for six females where slash was not burned (McIntire, 1980). McIntire (1980:95) suggested that the similarity of home-range areas for males on the two areas was evidence that the slash-burned site provided the necessities "in a similar manner" to the unmanipulated site. Although direct observations tended to corroborate use of similar-sized home-range areas (McIntire, 1980), we believe it appropriate to point out that home-range areas calculated by the inclusive boundary-strip method usually are approximately twice as large as those produced by the more commonly used minimum-area method. McIntire (1980) found considerable overlap in home ranges among neighboring chipmunks, but noted evidence that small areas (presumably near dens) were defended from conspecifics.

Tamias townsendii Bachman
Townsend's Chipmunk

1839b. *Tamias Townsendii* Bachman, 8:68.
1855. *Tamias cooperi* Baird, 7:334.
1897f. *E.[utamias] townsendi*: Merriam, 11:192.
1903b. *Tamias townsendii littoralis* Elliot, 3(10):153.
1919. *Eutamias townsendii cooperi*: Taylor, 9:110.

Description.—Townsend's chipmunk (Plate VII) is the largest member of the genus in Oregon (Table 11-3). The pelage of *T. townsendii* is dark and dull in comparison with that of congeners, but as in other Oregon chipmunks there is a dark brown to blackish middorsal stripe with alternate light and dark stripes laterally, a total of five dark and four light stripes. Both pairs of light dorsal stripes either are cinnamon brown to tawny (Coast Range and western valleys) or may contain dull white or cream (Cascade Range). Alternate dark (three) and light (two) stripes adorn the sides of the face. The throat, belly, and a patch behind the ear are white. The tail is black on the tip and the margins are frosted above with buff- or white-tipped hairs; the underside of the tail is rusty brown. The feet and sides are cinnamon buff.

Tamias townsendii may be separated from *T. amoenus* and *T. minimus* by an occipitonasal length >35 mm and from other congeners in Oregon by a baculum with a diameter <20% of its length. The skull (Fig. 11-11) is similar to that of other species in the *townsendii* complex. Pelage color and markings seem useful in separating *T. townsendii* from *T. siskiyou* and *T. senex*, but may not be absolutely reliable (see Remarks). However, we were unable to verify our identifications by comparison with baculum morphology.

Tamias townsendii and other chipmunks produce "chip" calls that "typically consist of a rapid series of high-intensity syllables normally directed toward intruders such as conspecifics, predators, or other trespassers" (Gannon and Lawlor, 1989:742). Except for a call recorded in Washington Co., the number of syllables per call for the species at four localities in Oregon averaged 2.4–2.8 (Warner, 1971) and 2.8 (range, 2.0–3.6—Gannon and Lawlor, 1989). These averages are twice that of calls produced by *T. senex*, but only ≅80% of calls produced by *T. siskiyou*. Gannon and Lawlor (1989) were able to separate the three species by discriminant analysis of six characteristics of chip calls produced by 227 chipmunks; overall classification success was 91%.

Distribution.—In Oregon, *T. townsendii* occurs in the Coast Range and the western interior valleys south to the Rogue River and on both slopes of the Cascade Range south to Mt. Jefferson, then at progressively lower altitudes in eastern Linn, Lane, and Douglas counties (Fig. 11-12; see Remarks). The species also occurs in western Washington and extreme southwestern British Columbia, including the southern tip of Vancouver Island (Fig. 11-12; Hall, 1981).

Geographic Variation.—Both of the currently recognized subspecies occur in Oregon: the all dark *T. t. townsendii* in the Coast Range south to the Rogue River and in the western valleys, and *T. t. cooperi*, with white or cream outermost dorsal light stripes, in the Cascade Range (Hall, 1981).

Habitat and Density.—Gashwiler (1976c) described habitats in which he captured *T. t. cooperi* on the west slope of the Cascade Range in Clackamas, Linn, and Lane counties as old-growth forests and clear-cuttings. The former were vegetated largely by Douglas-fir (*Pseudotsuga menziesii*), with some western hemlock (*Tsuga heterophylla*) and western red cedar (*Thuja plicata*), and subdominant Pacific yew (*Taxus brevifolia*) and vine maple (*Acer circinatum*) with an understory of salal (*Gaultheria shallon*), sword-fern (*Polystichum munitum*), Cascade Oregongrape (*Berberis nervosa*), twinflower (*Linnaea borealis*), and goldthread (*Coptis laciniata*). Clear-cuttings after 1–2 years were described as vegetated by sprouts and seedlings of vine maple, big-leaf maple (*Acer*

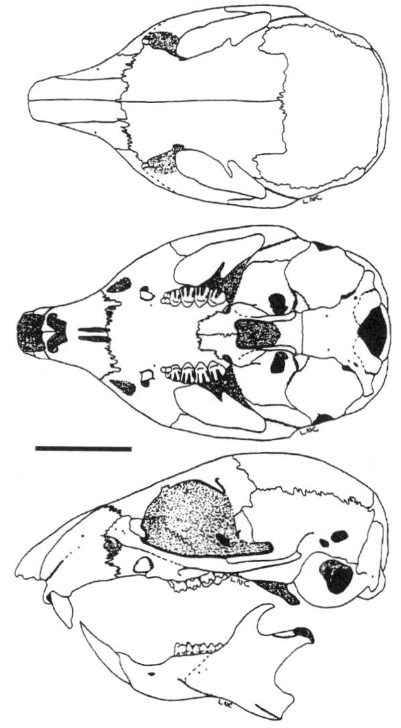

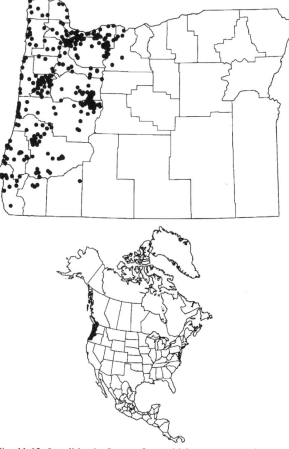

Fig. 11-12. Localities in Oregon from which museum specimens of Townsend's chipmunk, *Tamias townsendii*, were examined (listed in Appendix, p. 569), and the distribution of Townsend's chipmunk in North America. (Range map redrawn from Hall, 1981:351, map 238, with modifications; see Remarks.)

Fig. 11-11. Dorsal, ventral, and lateral views of the cranium, and lateral view of the dentary of Townsend's chipmunk, *Tamias townsendii* (OSUFW 8064). Scale bar equals 10 mm.

macrophyllum), flowering dogwood (*Cornus nuttallii*), tobacco-brush (*Ceanothus velutinus*), blackberry (*Rubus ursinus*), and some of the conifers. Also, fireweed (*Epilobium angustifolium*), horseweed (*Conyza canadensis*), and other herbaceous plants invaded.

In a comparison of small-mammal abundance in four paired riparian- and upland-forest sites on the west slope of the Cascade Range in Lane Co., Doyle (1990) found that chipmunks were more abundant in upland than in riparian habitats. Upland habitats were vegetated similarly to those described by Gashwiler (1976c), but riparian habitats included considerable big-leaf maple, vine maple, and alder (*Alnus rubra*). She reported that number of captures was significantly correlated positively with the extent of decay of downed logs; density of snags and percent cover of lichens, evergreen herbs, evergreen shrubs, and evergreen trees; and correlated negatively with percent cover of deciduous herbs and deciduous trees.

Rosenberg and Anthony (1993a) compared populations of chipmunks in five old-growth (>400 years old) forest stands with those in five second-growth (30–60 years old) forest stands on the west slope of the Cascade Range. Both old-growth and second-growth stands were dominated by Douglas-fir and western hemlock; western red cedar was common in the older stands whereas incense cedar (*Calocedrus decurrens*) was common in developing stands. Other plant species in the understory included those listed by Gashwiler (1976c) plus California hazel (*Corylus cornuta*), huckleberry (*Vaccinium*), Pacific rhododendron (*Rhododendron macrophyllum*), and Oregon oxalis (*Oxalis oregana*). Average densities of chipmunks in second growth during their 3-year study were 2.2/ha (range, 1.9–2.5/ha) in autumn and 3.3/ha (range, 1.9–4.6/ha) in spring, whereas in old growth average densities were 4.5/ha (range, 3.6–5.3/ha) in autumn and 5.7/ha (range, 4.7–7.7/ha) in spring. In old-growth, but not in second-growth communities, density of chipmunks was related significantly to density of large (≥50 cm-diameter) snags.

Some specimens used in the three previously described analyses of chipmunk habitats on the west slope of the Cascade Range were collected in the vicinity of McKenzie Bridge, Lane Co., a site at which Sutton and Nadler (1974) reported the occurrence of both *T. townsendii* and *T. siskiyou*. Although we classified the five voucher specimens that we found for these studies as *T. townsendii* (OSUFW 9117, KU 145170–145173, all from old-growth forest stands deposited by Rosenberg), we believe that the values and relationships reported for these studies should be viewed as suspect at least until such time as the questions regarding species and distribution of chipmunks in the region are resolved.

Along the Columbia River, Hinschberger (1978) captured *T. townsendii* in alder habitats (0.2/100 trap-nights) near the mouth (river miles 0–12) and in willow (*Salix*; 0.2/100 trap-nights) and in Douglas-fir and maple (1.6/100 trap-nights) opposite the Bonneville Pool (river miles 145–192). He observed these chipmunks in tidal Sitka spruce (*Picea sitchensis*) and black cottonwood (*Populus trichocarpa*) habitats, but did not capture them at river miles 12–79.

In a 3-year comparison of chipmunk populations in a 125-year-old Douglas-fir stand with those on burned and unburned clear-cuttings in the western foothills of the Cascade Range, Hooven and Black (1976) obtained somewhat

conflicting results. With equal effort, number of captures of chipmunks (animals were released) during the 3 years averaged 39.0, 28.5, and 9.5/ha on the 125-year-old stand, whereas they averaged 7.5, 24.8, and 8.5/ha on the unburned plot and 24.8, 36.5, and 9.3/ha on the burned plot. The dramatic decline in number of captures per unit area during the 3rd year and their similarity among treatments suggests that factors other than the treatments being tested were influencing chipmunk populations greatly.

For ≤7-year-old clear-cuttings in the central Coast Range (Lane and Lincoln counties), Anthony and Morrison (1985) recorded estimates of ≅4–8/ha on two control grids and ≅3–5/ha on two grids before and 1 year after herbicide treatment to suppress shrubs. However, in the 2nd year after treatment, density on one treated grid increased to nearly 10/ha, whereas that on the other declined to 0/ha. Vegetation on the clear-cuttings was dominated by salmonberry (*Rubus spectabilis*), thimbleberry (*R. parviflorus*), and vine maple with an understory of bracken fern, tansy ragwort (*Senecio jacobaea*), foxglove (*Digitalis purpurea*), pearly-everlasting (*Anaphalis margaritacea*), Oregon oxalis, and several grasses. Herbicide treatment defoliated >50% of the salmonberry and thimbleberry, but few of the shrubs were killed, so by 2 years postspray the shrub layer resembled that before treatment. Hayes et al. (1995) estimated densities ranging from 0 to 23.8/ha in three coniferous stands >140 years old and nine 10–35-year-old Douglas-fir plantations in the Coast Range. Densities did not differ significantly between young and mature stands. However, Hayes et al. (1995) reported a significant relationship between density of *T. townsendii* and percentage of salal in the understory. Carey and Witt (1991) reported densities of 0.72–3.06/ha on three sites in the Coast Range; densities did not correspond with indices to abundance obtained from track-counts on smoked sheets of aluminum.

In British Columbia, at the onset of a study, an experimental population of *T. townsendii* was only two-thirds as dense as the population on a control area. When provided with supplementary food (sunflower seeds and whole oats) the experimental population increased and maintained a density ≅1.4 times that on the control area (Sullivan et al., 1983). Juvenile survival, body mass, growth rate, and possibly litter size were greater, and range length was less on the experimental area than on the control area. These findings caused Sullivan et al. (1983) to conclude that food supply limited population density of Townsend's chipmunk populations. Although a seasonal cycle in population density was evident with peaks at or near the time of maximum recruitment and nadirs usually just before onset of recruitment (Sullivan et al., 1983:746, fig. 2), there is no evidence of a multiannual cycle in density (Gashwiler, 1970a; Sullivan et al., 1983) as suggested for some other chipmunk species (Tryon and Snyder, 1973).

In six habitat types in a subalpine area in the Washington Cascades, *T. townsendii* was captured more frequently than expected in traps set in tree islands (dominated by subalpine fir, *Abies lasiocarpa*; or by a combination of subalpine fir and mountain hemlock, *Tsuga mertensiana*; with some whitebark pine, *Pinus albicaulis*) and talus slides and less frequently than expected in traps set in meadow communities and in ecotones (Meredith, 1972).

Diet.—Bailey (1936) commented on the variety of food items consumed by these chipmunks. He indicated (p. 128) that the diet included "nuts, acorns, seeds, berries, and other fruits, roots, bulbs, green vegetation, insects, and other small animal life." He further observed them to eat "raspberries, thimbleberries, blackberries, salal berries, seeds of spruce and hemlock cones, and a great variety of other seeds." He provided second-hand information that they ate "salmon-berries, red and black elderberries, gooseberries, crab apples, plums, and prunes; also the seeds of maple, boxelder, rose, thistles, grasses, and grains."

Hooven and Black (1976) reported that, in season, blackberries and western raspberries (R. leucodermis) were common items in the diet of T. townsendii as revealed by purplish stains on mouths of captured individuals and by remains of the berries in feces left in traps. They also recorded consumption of seeds of the common thistle (Cirsium vulgare) as revealed by piles of thistle achenes. Chipmunks caught in traps selected Douglas-fir seeds over other seeds in the mix used as trap bait; often they "cached" Douglas-fir seeds in one corner of the trap (Hooven and Black, 1976).

Tamias townsendii is an avid mycophagist; stomachs of 93% of 14 individuals examined contained fungi; fungi composed 77% of the volume of material in the stomachs (Maser et al., 1978a). These chipmunks contained an average of 3.6 (range, 1–8) taxa of hypogeous fungi; Basidiomycetes, Ascomycetes, and Endogonaceae composed 73%, 20%, and 7% of the taxa identified, respectively.

Gashwiler (1963) found seeds of tobacco-brush and vine maple, in addition to oats and wheat used as trap bait, in pouches of chipmunks that he caught in live traps. One chipmunk had 1,091 tobacco-brush seeds and another had a total of 6.0 ml of seeds of several species in its pouches. Gashwiler (1963) referred to these chipmunks as T. t. cooperi, but the locality provided for their capture was not sufficiently precise to exclude the possibility of another taxon being involved.

Reproduction, Ontogeny, and Mortality.—On the west slope of the Cascade Range, 37.5–100% of monthly samples of male chipmunks collected by Gashwiler (1976c) from soon after emergence in March to July were in breeding condition. However, males in samples collected in August to the end of the annual period of activity in October were not. Females were believed to attain breeding condition in April about 1 month later than males. Some females (14.3–100%) were either pregnant or lactating when collected in April-September with the peak proportion in May. Average litter size based on counts of pigmented sites of implantation was slightly greater than that based on counts of embryos (Table 11-4). Gashwiler (1976c) claimed that the average based on counts of corpora lutea was slightly larger than the average by either of the other methods. Because he listed 21 individuals with corpora lutea, but only five with embryos, we are skeptical that collections made throughout the breeding and lactating seasons would produce such a disproportion of individuals in the preimplantation stage of pregnancy. Therefore, we suspect that Gashwiler (1976c) mistook some developing follicles that ultimately would become atretic for corpora lutea.

At an elevation of 600–675 m on the east slope of the Cascade Range in Washington, T. townsendii emerged from a 4.5-month hibernation in late March. Females bred in late April and early May, gave birth in late May and early June,

and lactated from late May to the end of July (Kenagy and Barnes, 1988). Average litter size was slightly smaller than reported for the species in Oregon (Table 11-4) and the smallest among the three co-occurring sciurids examined in Washington. Many males did not emerge for ≥2 weeks after the first males emerged; late-emerging males did not attain reproductive condition (Kenagy and Barnes, 1988).

A natal nest of T. townsendii was described as a sphere lined with short pieces of sedge (Carex spectabilis) followed by a layer of "gray moss" (Usnea) and covered by broad blades of sedge. The nest contained three nestlings that averaged 4.8 g (range, 4.7–5.0 g) and were estimated to be 2–3 days old. They were flesh-colored, toothless, blind, hairless except for vibrissae, and without stripes, but they possessed well-developed claws (Shaw, 1944).

Forbes and Turner (1972) described development of young in two litters of chipmunks born in captivity to females caught at Lost Prairie Campground in eastern Linn Co. (T13S, R6E, NE¼ Sec. 34). Both females gave birth at night, one to a litter of five (one dead when discovered) on 17–18 June, the other to a litter of four on 23–24 June. At birth, body mass of the young ranged from 3.2 to 3.9 g and total length ranged from 55 to 60 mm. Young in the second-born litter averaged 3.8 g on day 1, 9.4 g on day 10, 19.2 g on day 20, 28.5 g on day 29, and 33.7 g on day 34. Vibrissae were present on day 1; tips of dark hairs and the position of dorsal stripes were visible on day 4; hair extended onto the tail, shoulders, and flanks with the adult color pattern visible by day 10; and the young were essentially fully furred by day 16 (Forbes and Turner, 1972). The incisors erupted on days 20–22, the auditory meatuses opened on days 24–25, the eyes opened on days 27–29, and dP3 and dP4 erupted on day 40. Young attain the adult dental complement at about 90 days; they attain reproductive competence in spring of the 2nd calendar year of life (Forbes and Turner, 1972).

Dalquest (1948:261) claimed that weasels (Mustela frenata), minks (M. vison), and other predatory mammals were the "greatest menace" to these chipmunks in Washington.

Habits.—Townsend's chipmunks tend to be more secretive than most congeners. Although active only during the daylight hours, their "chuck…, chuck…, chuck…, chuck" calls are heard more often than the animals are seen. When active, they tend to stay in the shadows or hidden by thick vegetation. When observed, they commonly are sitting on a stump, log, or low branch. Although they tend not to move in the quick, jerky spurts characteristic of smaller congeners, they can move quickly to escape by climbing or by diving into burrows. During rapid escape movements on the ground, the tail is held erect. Brain size relative to body mass is significantly greater for T. townsendii than for T. minimus, T. amoenus, or T. senex (Budeau and Verts, 1986). Such supports the premise of Eisenberg and Wilson (1978, 1981) that species of mammals that climb or fly through complex, multilayered habitats tend to have larger brains to process the more complex information.

In a study of captive groups of four individuals, 59.9% of 10,739 encounters were judged agonistic with 58.5% of the agonistic encounters consisting of chases (Sherman, 1973). Although dominance hierarchies were established quickly in the groups, they did not seem to reduce the rate of agonistic encounters. Individual differences in activity and aggressiveness were deemed to contribute more than sex or body size to rank in hierarchies. Although dominance did not reduce time and energy lost through chases, it reduced serious confrontations and increased tolerance among individuals (Sherman, 1973).

Juveniles born in captivity, reared with siblings or cross-fostered, and subjected to pair-wise tests at ≅2 months of age exhibited behaviors with frequencies that varied with relatedness and familiarity of pairs. Fuller and Blaustein (1990) speculated that kin recognition in *T. townsendii* possibly enabled individuals to direct nepotistic behavior toward kin selectively or promoted outbreeding.

To prepare for winter, in September and October these chipmunks may be more visible as they scurry about gathering foods to store as fat or to cache for winter use. In winter, Townsend's chipmunks become torpid for varying periods depending upon the elevation and the severity of weather. In the Coast Range, we have heard their calls on moderate days throughout winter, but rarely on days with significant rainfall or with near-freezing temperatures.

Gashwiler (1965) reported the home-range area calculated by the exclusive boundary-strip method for a male captured 59 times between 28 August 1955 and 24 May 1962 was 0.69 ha. Although this specimen was classified as *T. t. cooperi* it was captured in a region where *T. siskiyou* is reported to occur (Sutton and Nadler, 1974). For three female and seven male *T. townsendii* captured 3–16 times each in Washington (Meredith, 1972), we calculated minimum-area home ranges from graphic representations of their points of capture. Those of females averaged ≅0.14 ha (range, 0.09 – 0.21 ha) and those of males averaged ≅0.51 ha (range, 0.02–1.14 ha).

Remarks.—Because pelage in chipmunks seemingly is selected to favor adaptations to local ecological conditions, variation in color and markings may not reflect interspecies differences (Gannon and Lawlor, 1989). Consequently, we were not surprised to find that some specimens that we identified as *T. townsendii* were within the range of *T. siskiyou* (Figs. 11-10 and 11-12) established by others on the basis of baculum morphology and alarm calls.

A comparative study of skull and baculum morphometrics in *Tamias* in the southern Cascade Range in Oregon is essential for better understanding of the differentiation and distribution within the group. In the interim, we beg ecologists who study small-mammal populations and communities in that region to deposit series of voucher specimens from each of their study plots in some permanent repository.

Although Townsend's chipmunks sometimes are considered to influence reforestation negatively because of the seeds they consume, caching of seeds may have a positive effect (Bailey, 1936). In addition, the spread of spores of hypogeous fungi may have the greatest positive impact of all (Maser et al., 1978b).

Marmota flaviventris (Audubon and Bachman)
Yellow-bellied Marmot

1841. *Arctomys flaviventer* Audubon and Bachman, 1:99.
1899. *Arctomys flaviventer avarus* Bangs, 1:68.
1904. [*Marmota*] *flaviventer*: Trouessart, Suppl., p. 344.
1904. [*Marmota flaviventer*] *avarus*: Trouessart, Suppl., p. 344.

Description.—The yellow-bellied marmot (Plate VIII) is the largest sciurid in Oregon (Table 11-5). This large

and heavy squirrel has short legs, a short and bushy tail, and ears short and covered with fur. The feet are plantigrade and pentadactyl; the pollex is rudimentary. Except for the pollex, which has a nail, all digits are equipped with strong claws. The sole of the foot has five pads, three at the base of the digits; the heel pad on the hind foot is ovate (Frase and Hoffmann, 1980). There are five pairs of mammae, two pectoral, two abdominal, and one inguinal. Marmots also have cheek and anal glands.

The pelage consists of a dense, woolly underfur covered by long, coarse guard hairs and is distinctively colored and marked (Frase and Hoffmann, 1980). The overall color is yellowish brown to tannish brown frosted with buffy white over the back and sides, a buffy to ocherous venter, tannish brown legs and feet, and a whitish chin and nose. A horizontal stripe of buffy white in front of the eyes is separated from the white of the chin and nose by a bar of brown.

The skull (Fig. 11-13) is heavy, strong, and nearly flat dorsally in profile. The interorbital region is considerably

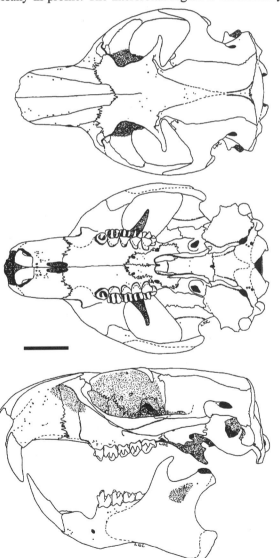

Fig. 11-13. Dorsal, ventral, and lateral views of the cranium, and lateral view of the dentary of the yellow-bellied marmot, *Marmota flaviventris* (PSM 14757). Scale bar equals 15 mm.

wider than the postorbital constriction. Its large size (zygomatic breadth >48 mm, occipitonasal length >70 mm) is sufficient to separate it from that of other sciurids in Oregon.

Distribution.—The yellow-bellied marmot occurs from southern British Columbia south through the Cascade and Sierra Nevada ranges to south-central California; eastward through montane and basin regions to southern Alberta, eastern Montana, Wyoming, and Colorado; and into northern New Mexico (Fig. 11-14; Hall, 1981). A disjunct population lies on the border between Wyoming and South Dakota (Frase and Hoffmann, 1980). In the northern portion of its range *M. flaviventris* occurs in warm, arid habitats, whereas in the southern portion it is restricted to montane "islands" (Frase and Hoffmann, 1980). In Oregon, *M. flaviventris* occurs in suitable habitats east of a line connecting Mt. Hood, Hood River Co., and Mt. Mazama, Klamath Co., except for the Columbia Basin (Fig. 11-14).

Pleistocene and sub-Recent fossils of *M. flaviventris* are known from numerous sites within the present range and south through southern California, Arizona, New Mexico, and extreme western Texas into Mexico to central Nuevo León (Frase and Hoffmann, 1980; Kurtén and Anderson, 1980).

Geographic Variation.—Two of the 11 currently recognized subspecies occur in Oregon: *M. f. flaviventris* along the east slope of the Cascade Range and the paler and slightly smaller *M. f. avara* through the remainder of the

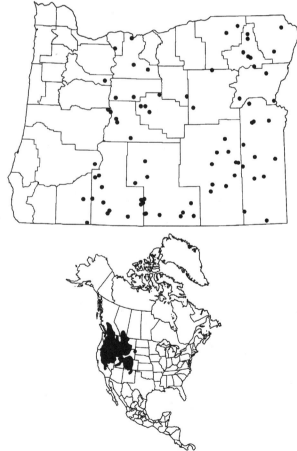

Fig. 11-14. Localities in Oregon from which museum specimens of the yellow-bellied marmot, *Marmota flaviventris*, were examined (listed in APPENDIX, p. 522), and the distribution of the yellow-bellied marmot in North America. (Range map redrawn from Hall, 1981:373, map 251, with modifications.)

state east of the Cascade Range (Hall, 1981; Howell, 1915).

Habitat and Density.—The primary requisites of suitable habitats for *M. flaviventris* are boulders or piles of rocks (e.g., talus slopes, lava fields, rimrocks and ledges, and stone fences) and an abundance of succulent vegetation in close proximity thereto. Occasionally, an abandoned building or a pile of logs serves as a substitute for rocks. Absence of one or the other of these requisites likely is responsible for the spotty distribution (Fig. 11-14; Couch, 1930). Van Vuren and Armitage (1991:1757) considered that time of snow melt, because it influenced the length of the growing season, was "a major determinant of habitat quality for yellow-bellied marmots" in alpine and subalpine environments.

S. Thompson (1979) studied marmot populations on a 110-ha study area in central Oregon part of which was an irrigated meadow planted to alfalfa, bluegrass (*Poa pratensis*), and orchard grass (*Dactylis glomerata*); a larger portion was dryland with vegetation dominated by rabbit-brush (*Chrysothamnus nauseosus* and *C. viscidiflorus*), bitter-brush (*Purshia tridentata*), green-leaf manzanita (*Arctostaphylos patula*), sagebrush (*Artemisia tridentata*) with scattered grasses and forbs. The area was nearly surrounded by juniper (*Juniperus occidentalis*)-pine (*Pinus ponderosa*) forest. We have observed marmots in similar habitats throughout the range in Oregon.

Because areas with the requisite combination of rock piles and forage tend to be scattered widely, estimates of absolute density likely would be extremely difficult, if not impossible, and possibly of questionable value. Thus, economic density, the number per unit area of habitat actually inhabited and used by a colony, is a more realistic measure of marmot abundance. In central Oregon habitats in which S. Thompson (1979) studied marmots, population density at recruitment (May–June) was 0.82, 0.25, 0.16, and 0.55/ha during 1975–1978, respectively. In 3 of 4 years, peak density occurred at recruitment, but in 1 year, density was highest during April, a month before young emerged from natal dens. In most years, dispersal did not play a significant role in the density of marmots on the study area at 1,100 m elevation in Deschutes Co., but in 1 year, 34 (70.8%) of 48 adult-sized marmots left the study area (S. Thompson, 1979).

Diet.—Yellow-bellied marmots are essentially entirely herbivorous (Frase and Hoffmann, 1980) although Bailey (1936) reported that an individual with nearly a full stomach of greenstuff also had consumed 55 black and yellow larvae of the sphynx moth (*Deilephila lineata*). Grasses, clovers (*Trifolium*), stonecrops (*Sedum*), and a variety of other plants are eaten, and in late summer marmots consume more flowers and seeds (Bailey, 1936). Couch (1930) claimed that *M. flaviventris* ate all native grasses near their burrow systems other than ryegrass (*Elymus*); he also claimed that alfalfa was particularly relished, a characteristic doubtlessly well known among eastern Oregon ranchers (S. Thompson, 1979). Occasionally, yellow-bellied marmots consume tree fruits and bark of fruit trees (Couch, 1930).

Despite depredations on crops that may appear more extensive, yellow-bellied marmots consume only 2.0–6.4% of net primary production available to them (Armitage, 1979). In a test of food selectivity with captives, Armitage (1979) found that plants or parts of plants that contained secondary compounds (alkaloids) were avoided or

Table 11-5.—*Means (±SE), ranges (in parentheses), and CVs of measurements of skull and skin* characters for female and male ground squirrels (*Ammospermophilus* and *Spermophilus*) and the yellow-bellied marmot (*Marmota flaviventris*) from regions in Oregon. Skin characters were recorded from specimen tags; skull characters were measured to the nearest 0.01 mm. SE and CV not provided if* n < 10.

Species and region	Sex	n	Total length	Hind foot length	Mass (g)	Greatest length of skull	Length of maxillary toothrow	Zygomatic breadth	Breadth of braincase	Breadth of molariform toothrow	Skull depth	Length of mandible
Ammospermophilus lecurus	♀	15	206 ± 2.2 (192–226) 4.0	38 ± 0.5 (34–40) 4.9	95.2[b] (81.4–105.0)	38.32 ± 0.39 (36.56–43.14) 4.02	7.06 ± 0.17 (6.10–8.88) 9.07	21.85 ± 0.29 (20.94–25.72) 5.26	18.78 ± 0.14 (17.91–20.05) 2.99	9.66 ± 0.13 (9.16–11.20) 5.06	16.75 ± 0.12 (16.05–17.65) 2.72	21.11 ± 0.15 (20.32–22.12) 2.68
	♂	8	215 (205–230)	38 (32–40)	93.1[c] (89.2–97.0)	38.71 (37.01–40.86)	6.86 (6.55–7.19)	22.05 (21.09–23.45)	19.12 (18.36–20.03)	9.47 (9.20–9.95)	16.97 (16.75–17.23)	21.39 (20.21–22.48)
Marmota flaviventris East slope of Cascade Range	♀	4	490	65	1,464.0	82.25 (79.75–86.22)	19.41 (33.35–37.02)	55.18 (18.61–20.24)	31.55 (30.75–33.05)	33.49 (22.03–25.01)	31.43 (30.88–31.82)	55.44 (51.95–58.14)
East of base of Cascade Range	♀	8	504 (428–585)	69 (60–75)	2,046.0[d] (1,288–3,050)	77.94 (73.80–81.99)	19.13 (18.14–20.55)	50.44 (47.18–52.41)	28.87 (27.82–30.49)	22.15 (21.25–23.19)	29.72 (27.09–31.53)	50.40 (46.62–54.01)
	♂	8	554[e] (450–640)	79[e] (73–85)	3,430.8[e] (3,042–3,742)	84.14 (77.81–92.97)	19.86 (19.21–20.52)	54.29 (49.20–61.71)	32.00 (29.77–33.66)	23.11 (21.90–24.11)	30.76 (29.36–33.66)	53.94 (48.21–61.16)
Spermophilus beecheyi	♀	27	445 ± 4.2[e] (415–489) 4.9	57 ± 0.5[f] (52–63) 5.0	583.9[g] (247–703)	58.45 ± 0.43 (54.89–63.64) 3.92	11.74 ± 0.12 (10.15–12.77) 5.54	35.99 ± 0.35 (33.29–40.69) 5.28	23.68 ± 0.13 (22.25–25.07) 3.03	16.17 ± 0.10 (14.99–17.26) 3.47	22.60 ± 0.14 (21.46–24.37) 3.38	35.20 ± 0.33 (31.76–38.96) 5.08
	♂	27	462 ± 4.9[h] (394–510) 5.5	59 ± 0.9[e] (42–66) 8.9	670.4[i] (466.0–843.1)	60.30 ± 0.49 (54.65–65.35) 4.42	11.98 ± 0.09 (11.17–13.11) 4.45	37.18 ± 0.38 (33.14–42.10) 5.65	24.19 ± 0.10 (22.98–25.79) 2.27	15.98 ± 0.10 (14.96–17.38) 3.44	23.16 ± 0.16 (21.84–25.05) 3.89	36.40 ± 0.34 (32.94–39.73) 5.08
Spermophilus beldingi Southeastern Harney and Malheur counties	♀	7	279 (255–302)	44 (41–45)	344.7[e] (241.3–429.6)	45.33 (42.75–47.66)	10.11 (9.63–10.47)	29.46 (27.82–31.09)	20.05 (19.61–20.39)	12.69 (12.17–13.19)	18.88 (17.84–19.49)	28.21 (26.52–29.64)
	♂	13	285 ± 3.9 (255–300) 4.9	43 ± 0.6 (39–47) 4.8	323.4[j] (265.0–365.3)	45.62 ± 0.35 (44.05–49.09) 2.73	10.12 ± 0.08 (9.72–10.68) 2.97	29.11 ± 0.29 (27.19–30.92) 3.55	20.23 ± 0.14 (19.11–20.96) 2.51	12.58 ± 0.08 (12.18–13.07) 2.25	19.06 ± 0.14 (18.31–20.05) 2.68	28.02 ± 0.26 (27.12–30.34) 3.39
Elsewhere in Oregon	♀	34	268 ± 2.9 (234–338) 6.3	42 ± 1.0 (37–73) 13.9	269.6 ± 6.97[e] (174.6–346.0) 14.62	43.79 ± 0.25 (41.64–49.22) 3.36	9.74 ± 0.06 (9.22–11.07) 3.33	28.55 ± 0.19 (26.01–31.61) 4.05	19.55 ± 0.10 (18.42–21.18) 3.04	12.29 ± 0.06 (11.87–13.37) 2.80	18.52 ± 0.09 (17.35–19.98) 2.99	26.26 ± 0.18 (24.53–30.22) 3.98
	♂	29	269 ± 3.3 (221–298) 6.5	42 ± 0.5 (35–47) 6.2	283.2 ± 12.94[j] (182.1–455.2) 20.93	44.95 ± 0.24 (41.64–47.30) 2.86	9.64 ± 0.07 (8.79–10.38) 4.05	29.16 ± 0.18 (25.98–30.96) 3.37	20.03 ± 0.12 (18.70–21.38) 3.09	12.49 ± 0.07 (11.53–13.17) 3.06	18.89 ± 0.11 (17.49–20.01) 3.25	27.53 ± 0.23 (25.41–30.07) 4.51
Spermophilus canus Northern Malheur County	♀	12	217 ± 3.1 (191–231) 4.9	33 ± 0.4 (30–34) 3.8	171.7[c] (143.8–210.0)	37.22 ± 0.31 (34.65–38.32) 2.88	8.20 ± 0.15 (7.46–9.54) 6.29	24.77 ± 0.22 (23.31–25.74) 3.08	17.79 ± 0.15 (17.03–18.67) 2.96	10.36 ± 0.12 (9.72–11.00) 3.87	16.03 ± 0.12 (15.16–16.54) 2.63	22.79 ± 0.31 (20.56–24.35) 4.77
	♂	18	221 ± 3.3 (200–257) 6.4	34 ± 0.3 (31–37) 4.0	196.6[k] (146.0–300.2)	38.18 ± 0.33 (34.81–40.56) 3.71	8.32 ± 0.05 (7.79–8.61) 2.44	25.17 ± 0.28 (22.00–26.69) 4.78	18.04 ± 0.09 (17.42–18.66) 2.02	10.50 ± 0.07 (9.79–11.01) 2.91	16.69 ± 0.15 (15.59–17.88) 3.71	23.19 ± 0.28 (20.16–25.16) 5.14
Remainder of Oregon	♀	14	205 ± 2.8 (188–218) 5.1	31 ± 0.3 (29–33) 3.5		35.83 ± 0.24 (34.30–37.51) 2.51	7.65 ± 0.07 (7.20–7.94) 3.27	23.32 ± 0.18 (22.33–24.43) 2.89	17.27 ± 0.15 (16.16–18.10) 3.19	10.12 ± 0.04 (9.83–10.36) 1.63	15.36 ± 0.11 (14.74–16.33) 2.56	21.46 ± 0.14 (20.88–22.69) 2.39
	♂	21	203 ± 1.7 (189–216) 3.8	32 ± 0.3 (29–34) 4.2	134.7	36.18 ± 0.19 (34.69–37.66) 2.39	7.73 ± 0.05 (7.32–8.24) 3.09	23.28 ± 0.14 (22.34–25.12) 2.81	17.38 ± 0.09 (16.85–18.38) 2.33	9.89 ± 0.05 (9.33–10.31) 2.39	15.51 ± 0.11 (14.64–16.26) 3.17	21.75 ± 0.11 (20.93–22.64) 2.32
Spermophilus columbianus Harney, Malheur, and southern Grant counties	♀	9	342 (265–370)	50 (43–53)	490.1[e] (376.8–646.0)	52.86 (51.35–54.21)	11.99 (11.53–12.58)	32.66 (31.33–33.71)	20.94 (19.85–22.47)	14.42 (14.03–14.82)	21.09 (20.49–21.94)	31.97 (31.24–32.76)
	♂	6	352 (296–380)	52 (49–55)	716.1[l] (678.0–754.1)	52.88 (46.08–55.40)	12.17 (11.80–12.76)	32.41 (26.61–35.29)	21.21 (19.45–22.03)	14.28 (13.49–15.52)	21.29 (19.44–22.22)	32.11 (29.59–33.87)
Northeastern Oregon	♀	11	354 ± 2.6[f] (340–366) 2.3	50 ± 0.9[f] (45–55) 6.1	498.1[d] (396–711)	52.91 ± 0.39 (50.46–54.85) 2.48	11.61 ± 0.09 (11.02–12.16) 2.79	32.54 ± 0.31 (31.12–33.89) 3.21	21.73 ± 0.28 (20.38–23.86) 4.26	14.36 ± 0.14 (13.64–15.11) 3.28	21.31 ± 0.16 (20.42–21.96) 2.45	32.10 ± 0.25 (30.52–33.11) 2.54
	♂	25	360 ± 5.5 (290–410) 7.6	53 ± 0.6 (46–61) 5.6	488.9[m] (382.7–583.4)	53.59 ± 0.33 (50.38–56.93) 3.08	11.88 ± 0.13 (10.97–13.17) 5.38	33.49 ± 0.27 (30.09–35.58) 3.99	21.71 ± 0.10 (20.49–22.51) 2.31	14.42 ± 0.09 (13.59–15.20) 2.99	21.90 ± 0.11 (20.90–23.33) 2.52	32.00 ± 0.47 (22.14–35.68) 7.34
Spermophilus elegans	♀	2	293 (290–295)	43 (42–44)	453.6	44.77 (44.76–44.78)	9.78 (9.61–9.94)	29.15 (28.96–29.33)	19.99 (19.93–20.06)	13.30 (13.21–13.39)	18.83 (18.64–19.02)	27.64 (27.53–27.74)
	♂	1	315	48		45.34	10.33	28.92	19.99	13.07	18.50	27.40
Spermophilus lateralis Josephine and Curry counties	♀	7	252 (241–265)	39 (34–43)	138.6	41.55 (40.20–42.76)	8.34 (7.77–8.96)	25.14 (25.78–28.19)	18.92 (18.55–19.29)	10.97 (10.52–11.59)	17.09 (16.74–17.72)	24.50 (23.01–26.25)
	♂	6	271 (240–300)	40 (38–42)		44.39 (40.36–49.62)	8.31 (7.94–8.63)	26.32 (24.59–27.99)	19.47 (18.41–20.69)	10.91 (10.39–11.62)	18.12 (16.97–22.06)	25.31 (23.93–27.70)
Cascade Range and Columbia Basin	♀	30	261 ± 4.0 (225–336) 8.5	38 ± 0.6[f] (31–46) 7.7	163.1 ± 4.97[c] (116.5–221.0) 14.92	41.95 ± 0.25 (39.90–45.13) 3.29	8.42 ± 0.07 (7.72–9.26) 4.65	25.49 ± 0.15 (23.87–26.63) 3.14	19.45 ± 0.08 (18.69–20.33) 2.17	11.08 ± 0.06 (10.43–11.94) 3.15	17.26 ± 0.08 (16.45–18.66) 2.43	24.61 ± 0.18 (23.06–27.43) 3.98
	♂	30	258 ± 4.3 (225–356) 9.2	39 ± 0.6 (33–47) 8.1	174.0 ± 5.99 (104.4–230.4) 18.86	42.48 ± 0.29 (36.17–44.87) 3.86	8.31 ± 0.09 (6.70–9.34) 5.99	25.96 ± 0.22 (21.83–28.01) 4.54	19.59 ± 0.11 (18.00–20.76) 2.99	11.13 ± 0.07 (9.62–11.52) 3.32	17.32 ± 0.07 (16.64–18.38) 2.24	24.81 ± 0.14 (23.31–25.98) 3.13

Table 11-5.—Continued.

Species and region	Sex	n	Total length	Hind foot length	Mass (g)	Greatest length of skull	Length of maxillary toothrow	Zygomatic breadth	Breadth of braincase	Breadth of molariform toothrow	Skull depth	Length of mandible
Harney, Malheur, and southern Baker counties	♀	14	273 ± 4.3 (249–305) 5.8	41 ± 0.5 (37–44) 4.2	206.5[n] (186.0–240.0)	42.40 ± 0.22 (41.03–43.87) 1.91	8.48 ± 0.10 (7.69–9.10) 4.54	26.24 ± 0.23 (25.15–28.22) 3.22	19.92 ± 0.17 (18.73–21.02) 3.11	11.46 ± 0.07 (10.96–11.85) 2.36	17.29 ± 0.11 (16.41–17.91) 2.45	24.92 ± 0.20 (23.58–26.21) 3.06
	♂	15	274 ± 2.7 (255–293) 3.8	41 ± 0.3 (39–43) 3.2	210.7[o] (149–283)	42.88 ± 0.15 (41.76–43.64) 1.37	8.65 ± 0.07 (8.13–9.15) 3.34	26.49 ± 0.23 (24.89–28.36) 3.31	19.96 ± 0.13 (19.22–20.88) 2.47	11.45 ± 0.11 (10.81–12.19) 3.77	17.41 ± 0.09 (16.75–18.04) 2.14	24.95 ± 0.12 (24.12–25.73) 1.93
Wallowa Mountains	♀	11	270 ± 2.9 (255–285) 3.6	41 ± 0.5[f] (38–44) 3.9	181.1[j] (157.0–201.8)	42.28 ± 0.30 (40.09–43.79) 2.37	8.86 ± 0.13 (8.30–9.84) 4.98	25.73 ± 0.28 (24.49–26.98) 3.66	19.42 ± 0.13 (18.89–20.05) 2.18	11.29 ± 0.09 (10.80–11.76) 2.83	17.17 ± 0.14 (16.36–17.99) 2.61	24.48 ± 0.16 (23.66–25.90) 2.22
	♂	10	273 ± 4.1 (260–296) 4.7	41 ± 0.6 (38–43.5) 4.7	184.5[c] (172.2–198.3)	42.77 ± 0.39 (41.06–44.65) 2.92	8.65 ± 0.11 (7.86–9.12) 4.16	26.64 ± 0.29 (25.78–28.19) 3.39	19.89 ± 0.22 (18.70–20.91) 3.53	11.46 ± 0.08 (10.96–11.69) 2.10	17.44 ± 0.12 (16.81–17.84) 1.85	25.45 ± 0.26 (24.41–26.75) 3.17
Spermophilus mollis	♀	5	202 (190–212)	33 (29–36)	165.4[h] (132.2–198.6)	35.56 (34.46–36.76)	7.92 (7.42–8.23)	22.36 (21.48–23.94)	17.26 (16.73–17.70)	9.94 (9.65–10.35)	15.43 (14.64–15.75)	21.41 (21.04–21.89)
Spermophilus washingtoni	♀	11	215 ± 2.9 (201–233) 4.5	34 ± 0.3 (32–35) 2.8		38.57 ± 0.24 (37.54–40.08) 2.09	8.57 ± 0.06 (8.27–8.88) 2.25	24.42 ± 0.21 (23.79–25.83) 2.83	17.55 ± 0.11 (17.06–18.12) 2.13	10.96 ± 0.05 (10.72–11.23) 1.44	16.59 ± 0.13 (16.11–17.68) 2.51	23.67 ± 0.20 (22.44–24.62) 2.83
	♂	19	224 ± 2.3 (205–242) 4.4	35 ± 0.3 (32–38) 3.9	203.1	39.46 ± 0.19 (38.23–41.20) 2.09	8.50 ± 0.07 (7.98–9.03) 3.71	25.39 ± 0.19 (23.43–27.03) 3.23	17.74 ± 0.12 (16.83–18.38) 2.99	10.73 ± 0.13 (8.72–11.45) 5.37	16.68 ± 0.07 (16.21–17.38) 1.75	24.03 ± 0.13 (23.07–25.22) 2.45

[a] Although measures of tail and ear length were recorded, the data were not included because of extreme variation believed to stem from differences in techniques used to measure these characters. [b] Sample size reduced by 10. [c] Sample size reduced by 6. [d] Sample size reduced by 5. [e] Sample size reduced by 2. [f] Sample size reduced by 1. [g] Sample size reduced by 21. [h] Sample size reduced by 3. [i] Sample size reduced by 23. [j] Sample size reduced by 8. [l] Sample size reduced by 4. [m] Sample size reduced by 16. [o] Sample size reduced by 12. [n] Sample size reduced by 7. [k] Sample size reduced by 13.

consumed in relatively small amounts, but those without the toxic compounds were consumed readily. He suggested that populations of *M. flaviventris* might be limited or otherwise affected where plants containing secondary compounds were abundant. Many of the plant species or congeners thereof that Armitage (1979) tested occur in Oregon.

Reproduction, Ontogeny, and Mortality.—In California, *M. flaviventris* emerges from hibernation in reproductive condition and breeds within 2 weeks; emergence, hence onset of reproduction, is dependent upon temperature. At elevations of 1,675–2,165 m this was in April (Nee, 1969). In Washington, parturition reportedly is 15 March–15 April, depending on elevation (Couch, 1930); thus, based on a 30-day gestation period (Armitage, 1962), onset of reproduction probably was about mid-February to mid-March. In addition to the method used for estimates, locality, elevation, year-to-year differences, and age of females affect average litter size (Table 11-6; Frase and Hoffmann, 1980; Nee, 1969); however, Howell's (1915) claim of three to eight young per litter seems to be a reasonable range. In alpine and subalpine habitats, time of snow melt explained significant proportions of differences in litter size, body mass of young-of-the-year, and frequency that females reproduced, but not growth rate of young (Van Vuren and Armitage, 1991). Late-born young suffer relatively high mortality, probably because they do not have time to accumulate sufficient fat reserves to overwinter and cannot compensate by increasing their rate of growth.

Young appear above ground about 30 days after birth when they are about the size of an adult Columbian ground squirrel (*Spermophilus columbianus* — Couch, 1930). At ≅1,100 m elevation in central Oregon, young-of-the-year commenced to emerge from natal dens in early May, about 60–70 days after adults first emerged from their hibernacula (S. Thompson, 1979). In 4 consecutive years, 14 of 20, 2 of 19, 1 of 10, and 10 of 10 resident adult females produced litters that emerged from natal dens. Survival of emergent young to hibernation was 57 of 60 in 1975, 0 of 9 in 1976, 4 of 4 in 1977, and ≤10 of 45 in 1978

(S. Thompson, 1979).

In central Oregon, other than three animals that died in traps, all marmot mortalities on a 110-ha area during a 4-year period were attributable to predators. Ten incidences of predation were observed and >50 marmots were believed to have been preyed upon (S. Thompson, 1979). Although several species of raptors were present on the area no losses were attributed to them; coyotes (*Canis latrans*) and badgers (*Taxidea taxus*) were responsible for all losses. Badgers dug extensively at the man-made rock piles at which many young had hibernated but failed to reappear the following spring (S. Thompson, 1979). In addition, badgers were thought to have been responsible for the extensive loss of young between emergence and immergence in 1978. In Colorado, ages of 32 marmots preyed upon by coyotes were determined from remains in coyote fecal droppings; 17 were adults, 13 were yearlings, and only 2 were juveniles. Juveniles were thought not to be especially vulnerable to predation by coyotes because they tended to remain close to the natal den, into which they could escape easily at the approach of a coyote (Van Vuren, 1991). Bailey (1936) found remains of juvenile marmots beneath nests of Swainson's (*Buteo swainsoni*) and red-tailed hawks (*B. jamaicensis*) and great horned owls (*Bubo virginianus*) on Steens Mountain. In Colorado, only two instances of predation were witnessed in >5,000 h of observation during a 20-year period; in both, coyotes were the predator species (Armitage, 1982). In the same area in 1984–1989, however, 17% of 395 coyote fecal droppings contained remains of marmots (Van Vuren, 1991). In central Oregon, losses in hibernacula not attributable to predation were believed not to occur (S. Thompson, 1979), but in Colorado losses during the period of hibernation were thought to be extensive (Armitage and Downhower, 1974).

Habits.—Marmots hibernate; body temperature, heart rate, respiration, and physiological processes decline to extremely low levels. Captives that hibernated in dens in outdoor cages under near-natural conditions in a region in central Oregon with relatively mild winters lost 21.7–35.7%

Table 11-6.—*Reproduction in some Oregon ground squirrels* (Ammospermophilus *and* Spermophilus) *and the yellow-bellied marmot* (Marmota flaviventris).

Species	Number of litters	Litter size $\overline{X}$	Range	n	Reproductive season	Basis	Authority	State or province
Ammospermophilus leucurus		7.8	5–11	10	Mar.–Jun.	Embryos	Hall, 1946	Nevada
Marmota flaviventris	1	4.8	3–6	5	Apr.–May	Embryos	Nee, 1969	California
	1	4.4	≤5	27		Emergent young	Thompson, 1979	Oregon
Spermophilus beecheyi		6.4[a]		9		Implant sites	Chapman and Lind, 1973	Oregon
		5.5[a]		13		Implant sites	Chapman and Lind, 1973	Oregon
	1	5.1[b]	3–7	8	Mar.–May	Embryos	Edge, 1931	Oregon
Spermophilus beldingi		7.4		175		Embryos	Sullins and Verts, 1978	Oregon[c]
		5.2		109[d]		Embryos	Costain and Verts, 1982	Oregon[e]
		5.4		120[d]		Implant sites	Costain and Verts, 1982	Oregon[e]
		8.1		302[d]		Implant sites	Costain and Verts, 1982	Oregon[f]
		7.1	3–10	37		Embryos	Tevis, 1955	California
Spermophilus columbianus	1	3.7	2–5	3		Embryos and implant sites	This study	Oregon
	1	4.6	2–7	5	Mar.–Apr.	Neonates[g]	Shaw, 1925b	Washington
	1	3.3	1–5	28		Embryos	Murie et al., 1980	Alberta
Spermophilus elegans	1	4.3		31[h]		Emergent young	Pfeifer, 1982	Colorado
	1	6.1	5–9	40	Mar.–May	Implant sites	Clark, 1970a	Wyoming
	1	5.3	4–9	16	Mar.–May	Embryos	Clark, 1970a	Wyoming
Spermophilus lateralis	1	5.6	3–8	43		Embryos	Tevis, 1955	California
		5.0	3–8	36	Mar.–Jun.	Embryos	McKeever, 1965	California
		5.2	2–7	88	Mar.–Jun.	Implant sites	McKeever, 1964a	California
		5.2	3–7	5		Implant sites and embryos	This study	Oregon
	1	7.4	5–11	43	Mar.–Apr.	Embryos	Kenagy and Bartholomew, 1985	California
Spermophilus mollis	1	—[i]	—[i]			Embryos and implant sites[j]	Smith and Johnson, 1985	Idaho
	1	9.3	5–15	146	Mar.–May	Embryos	Alcorn, 1940	Nevada
		9.1[h]	5–16	157		Embryos	Rickart, 1988	Utah
		6.6[h]	5–10	38		Embryos	Rickart, 1988	Utah
Spermophilus washingtoni	1	8.0	5–11	26	Feb.–Apr.	Embryos and	T. Scheffer, 1941	Washington

[a]Litter size inversely related to latitude and average annual rainfall and positively related to average altitude, temperature, and freeze-free period.
[b]Includes an average of 0.4 embryos undergoing resorption.
[c]Near Monument, Grant Co., ≅600 m elevation.
[d]Values recalculated from significantly different age-specific values.
[e]At Harris Ranch, Grant Co., ≅1,700 m elevation.
[f]Near Izee, Grant Co., ≅1,300 m elevation.
[g]Young born in captivity.
[h]Recalculated average for yearlings and adult combined.
[i]Average litter size ranged from 7.2 to 9.1 ($n = 36$–40) in years during which females produced young. During a drought year, females produced no young.
[j]Average litter sizes based on counts of embryos and counts of implantation sites were not significantly different so were combined.

of their preemergence body mass overwinter (S. Thompson, 1979).

In central Oregon at 1,100 m, marmots emerged from hibernation the last week of February or the 1st week of March, adult males first, followed in order by adult females, yearling females, and yearling males (S. Thompson, 1979). Adults remained active for 135–150 days. Except for juveniles, marmots entered hibernation by the end of July; juveniles remained active to mid-August. In 3 of 4 years, adult females were first to immerge; yearling females followed, then adult and yearling males, and finally the juveniles. In the 4th year, females that produced young and juveniles were last to immerge, the sequence reported to occur in Colorado (Armitage, 1962). At another central

Oregon site only 30 km away but at 500 m greater elevation, the annual cycle of emergence, reproduction, fattening, and immergence was shifted ≅2 months later in the year. However, the length of the active period was approximately the same at both sites (S. Thompson, 1979). Marmots translocated from the higher to the lower elevation shifted their period of activity to correspond with that of the new locale.

Upon emerging in spring, many marmots on the 110-ha study area in central Oregon moved ≅1.25 km (range, 0.73–2.61 km) from winter hibernacula in a 20- by 13-m rock outcrop to summer burrows in man-made rock piles adjacent to alfalfa fields; they returned to the rock outcrop in autumn to hibernate. However, about half the young born on the area hibernated in dens in man-made rock piles near the edges of fields; survival was signficantly greater among those that migrated to the rock outcrop before hibernating (S. Thompson, 1979). S. Thompson (1979) believed that marmots elsewhere in central and eastern Oregon probably made similar seasonal movements of several hundred meters from rimrocks and rock outcrops to burrows near croplands and lush vegetation in riparian zones.

After emergence from hibernation, marmots do not feed for 10–20 days. In central Oregon, marmots first fed in dryland sites, then 10–15 days later commenced to feed in hay fields. In April, the peak of feeding activity was near midday, in mid-May peaks occurred in midmorning and midafternoon, and in June marmots fed largely in the crepuscular hours and were inactive during midday (S. Thompson, 1979). Although aboveground activities of *M. flaviventris* usually are considered diurnal (Couch, 1930) and bimodal with peaks in morning and late afternoon (Armitage, 1962; S. Thompson, 1979), under pressure from hunters they may "become partially nocturnal" (Dalquest, 1948:264).

Upon emergence, some adult males establish territories; during 1 year, S. Thompson (1979) found that two males on his study area in central Oregon each established narrow territories ≅900 m long and separated by ≅300 m on the crest of a rocky ridge. In 2 other years, one male used the entire ridge. Territories were patrolled regularly for a few days to 2–3 weeks; males traveled an average of >2 km/day, stopping to mark prominent rocks with products of their cheek glands and to advertise their presence by tail flagging. Other males usually avoided those with territories, but those that did not were chased by territorial males. Subordinate males either lived in rock piles at the periphery of the area or dispersed. Females were greeted with tail flagging and by nose-to-nose contact; the male smelled the female's anogenital region and attempted to mount. Copulation rarely was observed, but males exhibited little interest beyond the initial greeting during subsequent encounters with the same female, suggesting that copulation had occurred out of sight (S. Thompson, 1979). After the breeding season, territorial males may move to a summer residence and become tolerant of other males; those that remain on the territory maintain amicable relations with females, but remain dominant to other males.

Marmota flaviventris produces three vocalizations: a whistle, a scream, and a chattering of the teeth. Six variations of the whistle are used that function as alert, alarm, and threat calls (Waring, 1966). The scream indicates fear or excitement and the tooth chatter is a threat. In addition to other marmots, individuals of some other species of mammals often respond to marmot alarm whistles (Waring, 1966).

Remarks.—Bailey (1936) and Couch (1930) commented on the use of the yellow-bellied marmot as human food; both considered them excellent fare.

Ammospermophilus leucurus (Merriam)
White-tailed Antelope Squirrel

1889c. *Tamias leucurus* Merriam, 2:20.
1892. *Ammospermophilus* Merriam, 7:27.
1907. *Ammospermophilus leucurus* [*leucurus*]: Mearns, 56:viii, 299.
1903d. *Citellus l*[*eucurus*]. *vinnulus* Elliot, 3:241.

Description.—*Ammospermophilus leucurus* (Plate VIII) is the smallest species of ground squirrel in Oregon (Table 11-5). Although rather cryptically colored, it is marked distinctively: the grayish-brown dorsum and sides are separated by a white stripe on each side. The stripes are tapered at both ends and extend from behind the ear to near the base of the tail. A white line encircles each eye, the underside of the tail is white grading to grayish near the tip, and the venter is white (hairs of the venter have lead-colored bases).

The skull of *A. leucurus* (Fig. 11-15) is similar to that of ground squirrels of the genus *Spermophilus* in that the zygomata are twisted, flattened dorsoventrally, and converge anteriorly. It can be separated from those of other ground squirrels in Oregon by a narrow oval infraorbital foramen with a spinelike projection directly ventral to the opening.

Distribution.—*Ammospermophilus leucurus* occurs from southeastern Oregon and southwestern Idaho south through eastern California and Nevada to the southern tip of Baja California Sur and eastward through Nevada and Utah to western Colorado and into northern Arizona and northwestern New Mexico (Fig. 11-16). In Oregon, *A. leucurus* occurs south and east of a line connecting Vale, Malheur Co.; Harney Lake, Harney Co.; and Paisley, Lake Co. (Fig. 11-16).

Miocene and Pliocene fossils of *Ammospermophilus*, some of which are similar to *A. leucurus*, are known from California, Oregon, and Washington. Pleistocene and Holocene fossils of *A. leucurus* are known from Texas (extralimital), Utah, Arizona, Nevada, and California (Belk and Smith, 1991).

Geographic Variation.—Only one of the nine currently recognized subspecies occurs in Oregon: *A. l. leucurus* (Hall, 1981).

Habitat and Density.—In Utah, *A. leucurus* occurs most frequently in the narrow ecotone between rocky situations and desert flats (Armstrong, 1979). In the same state, the species was captured most often on sandy soils where vegetation consisted of tall, relatively dense shrubs with

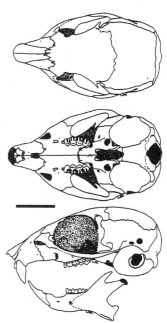

Fig. 11-15. Dorsal, ventral, and lateral views of the cranium, and lateral view of the dentary of the white-tailed antelope squirrel, *Ammospermophilus leucurus* (OSUFW 6725). Scale bar equals 10 mm.

relatively few annuals (Robey et al., 1987). In north-central Nevada, *A. leucurus* was caught only on plots vegetated by shadscale (*Atriplex confertifolia*), big sage-brush (*Artemisia tridentata*), and shadscale-big sagebrush associations (O'Farrell and Clark, 1984). In southern Nevada, white-tailed antelope squirrels are associated with blackbrush (*Coleogyne ramosissima*) and joshua trees (*Yucca brevifolia*—Bradley, 1967), species that do not occur in Oregon (Hitchcock and Cronquist, 1973). Grinnell and Dixon (1918) indicated that vegetation at sites occupied by *A. leucurus* in California was variable, but substrates of hard-surfaced gravelly soils on slopes at occupied sites were common. Most specimens that we took near Fields, Harney Co., were in shadscale-sagebrush associations on moderately soft soils, but one taken at the city landfill was on extremely hard soil.

Bradley (1967) recorded densities of 6.4/km² in summer and 34.7/km² in autumn in southern Nevada.

Diet.—Seeds commonly are considered to be the staple of the diet of *A. leucurus* (Bradley, 1968; Grinnell and Dixon, 1918; Hall, 1946; Howell, 1938), but it cannot sur-vive on seeds alone (Hudson, 1962). Green vegetation, insects, and vertebrates constitute significant portions of the diet during some seasons (Bradley, 1968); thus, *A. leucurus* is an omnivore (Belk and Smith, 1991). In south-ern Nevada, *A. leucurus* consumed >50% green-plant

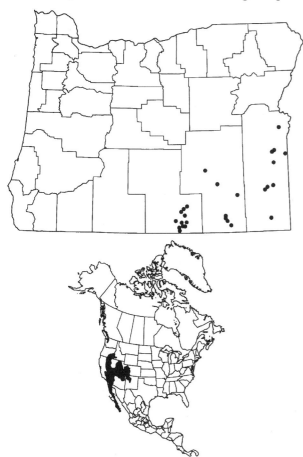

Fig. 11-16. Localities in Oregon from which museum specimens of the white-tailed antelope squirrel, *Ammospermophilus leucurus*, were ex-amined (listed in Appendix, p. 511), and the distribution of the white-tailed antelope squirrel in North America, (Range map redrawn from Hall, 1981:379, map 254).

material and <20% animal material during March–May, but in summer it ate <30% green-plant material and >25% animal material (Bradley, 1968). Because seeds and green plants are masticated so thoroughly, identification of much of the stomach contents of *A. leucurus* is impossible. How-ever, *A. leucurus* transports much food in its cheek pouches (Bradley, 1967); contents of cheek pouches provide samples of potentially consumed materials more easily identified. Many of the plant taxa identified among seeds in cheek pouches of these ground squirrels in southern Nevada (Bra-dley, 1968) are not represented among the flora of Oregon (Hitchcock and Cronquist, 1973). Because *A. leucurus* tends to be a food generalist elsewhere (Bradley, 1968), we suspect that the common seed-producing plants within its range in Oregon serve as the source of the staple diet of the species here.

Although *A. leucurus* makes use of free water when it is available (Grinnell and Dixon, 1918), free water is uncom-mon in broad areas within the region occupied by the species. *A. leucurus*, unlike the heteromyids, cannot survive on metabolic water and the small proportion of free water in seeds (Hudson, 1962). Consequently, seasonal changes in consumption of green-plant material, insects, and small vertebrates may be as much responses to demands for water as for specific nutrients (Bradley, 1968). The increase in consumption of green-plant material in March-May probably is related to demands for water for lactation, and consumption of more insects and vertebrates in summer probably is in response to desiccation of green plants (Bradley, 1968). Much of the vertebrate flesh consumed is as carrion (Bradley, 1968), but some is obtained by predation (Mogart, 1985).

Reproduction, Ontogeny, and Mortality.—We know of no quantitative information published about *A. leucurus* in Oregon on these topics (Carraway and Verts, 1982). Based on a survey of research findings at scattered locali-ties throughout the range, both males and females attain reproductive competence at 1 year of age, only one litter is believed to be produced annually, the gestation period ranges from 30 to 35 days, and average litter size and aver-age birth mass are relatively large among ground-dwelling sciurids (Belk and Smith, 1991).

In California, the testes of *A. leucurus* commence to in-crease in mass in November; spermatogenesis is initiated in December; and the peak mass of testes is attained in February, a month before the peak mass of epididymides and seminal vesicles (Kenagy and Bartholomew, 1985). Mature spermatozoa occur in the epididymides for ≅5 months/year, but the accessory glands are undeveloped for most of that period, thereby further restricting the period of reproductive competence. There is only one breeding period each year and many males are infertile by the time offspring are born about a month later (Kenagy and Bartholomew, 1985). In Nevada, most pregnant females were taken in March, but others were taken during the fol-lowing 3 months (Hall, 1946); thus, males at some sites must remain fertile longer or those at different sites must be fertile at different times.

Most mating in southern California was in the first half of March and nearly all young were born in the first half of April. The average litter size (Table 11-6) in that area depends upon climatic conditions; in a year in which pre-cipitation was only 26% of normal, the average litter size was 1.1 less than the average for 3 years and 2.4 less than for the highest of the 3 years (Kenagy and Bartholomew, 1985). Grinnell and Dixon (1918) recorded two pregnant individuals with 13 and 14 embryos collected in California. Belk and Smith (1991) indicated that the overall average litter size reported by several investigators was about eight.

At birth, *A. leucurus* averages 2.9 g; at 5 days, 4.6 g; at 10 days, 6.1 g; at 15 days, 7.6 g; at 20 days, 9.0 g; at 25 days, 13.2 g; at 30 days, 18.0 g; and at 45 days, 33.1 g (Maxwell and Morton, 1975). Although body mass at birth is large (3.2% of adult body mass) in comparison with that of *Spermophilus beldingi* (2.9% of adult body mass), growth is slow. *A. leucurus* attains only 36.1% of adult body mass by 45 days of age, whereas *S. beldingi* attains 73.2% of adult body mass by that age (Maxwell and Morton, 1975). Also, as another example of its relatively slow postnatal development, *A. leucurus* is unable to thermoregulate completely until 45 days of age, whereas *S. beldingi* displays limited ability to thermoregulate at 10 days of age and attains complete thermoregulatory abilities by 25 days of age (Maxwell and Morton, 1975).

Habits.—*Ammospermophilus leucurus* is active at all seasons; it neither hibernates nor estivates (Bradley, 1967; Fisler, 1976; Grinnell and Dixon, 1918; Hall, 1946; Hudson, 1962; Kramm, 1972). Consequently, there is no annual cycle of fat deposition and metabolism (Kenagy and Bartholomew, 1985). Torpor can occur and arousal is possible at ambient temperatures >20° C; many do not survive torpor at ambient temperatures <10° C (Kramm, 1972).

The white-tailed antelope squirrel is active diurnally; there are two peaks of daily activity in summer, but only one in winter (Chappell and Bartholomew, 1981). In north-central Nevada, *A. leucurus* commonly was observed along roadways during crepuscular hours (O'Farrell and Clark, 1984). Bradley (1967) once saw one at night when moonlight was exceptionally bright.

Burrows occupied by *A. leucurus* may be under shrubs or in the open; commonly burrows abandoned by *Dipodomys* are appropriated. A female captured 61 times entered seven different burrows upon release (Bradley, 1967). In addition to the nest burrow, several burrows scattered in the area used by one individual are used for escape from potential predators or as a cooler place to unload heat acquired during aboveground activity (Bradley, 1967; Chappell and Bartholomew, 1981). When active above ground, white-tailed antelope squirrels move from one shady spot to the next extremely rapidly with the tail held over the back as a sunshade; less heat is acquired by running than would be acquired by walking in the sunshine (Chappell and Bartholomew, 1981). Occasionally, they climb into shrubs or stand on boulders or other objects in "picket-pin" posture to view their surroundings (Bradley, 1967; Grinnell and Dixon, 1918) or to lose heat in the wind by convection (Chappell and Bartholomew, 1981).

The body temperature of *A. leucurus* is extremely labile at all seasons; body temperatures of 38.0–40.2°C in summer and 33.7–39.1°C in winter are common in free-living individuals (Chappell and Bartholomew, 1981). Within the range of the species, environmental temperatures commonly exceed the upper critical temperature of *A. leucurus* for most of the daylight hours in summer. *A. leucurus* becomes hyperthermic to permit longer periods of activity at such

temperatures; excess heat is dissipated rapidly (0.8°C of body temperature/min) by conduction within burrow systems (Chappell and Bartholomew, 1981). By periodically dissipating excess heat, *A. leucurus* can remain active at ambient temperatures of at least 43.6°C. When critical body temperatures are exceeded, as might occur when caught in traps, white-tailed antelope squirrels salvate profusely (Hudson, 1962) and smear the saliva over their heads to lower body temperature by evaporation; such obviously is a last-resort measure.

In winter, *A. leucurus* minimizes loss of energy by allowing the body temperature to fall to 32–33°C at night. The energy cost of rewarming is estimated to be about 15% of the energy saved by nighttime hypothermia and as much as 25% of that may be recovered by basking (Chappell and Bartholomew, 1981).

In southern Nevada, home-range areas calculated by the minimum-area method for five males and four females, each captured ≥20 times, averaged 6.0 ha; greatest distances between points of capture for the same individuals averaged 343.6 m (Bradley, 1967). In the same region, greatest distances between points of capture for *A. leucurus* captured in a variety of habitats averaged 117.1 m for males and 101.2 m for females; the maximum distances between points of capture were 515.2 m for a male and 322.0 m for a female (Allred and Beck, 1963b). Averages were considerably different among sites with similar plant communities; however, average movements on sites disturbed by nuclear blasts were nearly double those on undisturbed sites (Allred and Beck, 1963b).

Ammospermophilus leucurus exhibits a linear social hierarchy without defense of a territory throughout the entire year (Fisler, 1976). The hierarchical system is maintained by visual and tactile signals seemingly understood by all adults in the population. Olfaction is not considered a significant means of communication (Fisler, 1976). Upon encountering one another, individuals greet by a nasonasal or nasooral "kiss"; actual contact may or may not occur (Fisler, 1976). If dominance were established previously, the subordinate individual retreats and interaction ceases; however, if dominance is uncertain between approaching individuals, one or both individuals may stretch their bodies low to the ground while ≅0.5 m apart. Sometimes individuals in stretched posture kiss. The subordinate retreats, usually without being pursued (Fisler, 1976). If dominance is not established or recognized during greeting, additional and increasing agonistic behavior ensues. Hierarchical rank may change over time among interacting individuals. Body mass seems to play a role in the rank held by an individual; heavier individuals usually are dominant (Fisler, 1976). Juveniles, upon emerging from natal dens, do not use or acknowledge the various signals for 2–3 weeks, even after being knocked off their feet when they do not respond appropriately to a signal by a ranking individual (Fisler, 1976).

Remarks.—We can offer little regarding the present status of *A. leucurus* in Oregon beyond the >25-year-old conclusion of Olterman and Verts (1972), based on hearsay accounts of agency personnel, that within its range the species was common. To 1980, we were unable to locate a primary research report devoted largely to any aspect of the biology of *A. leucurus* in Oregon (Carraway and Verts, 1982); we know of none published subsequently.

Spermophilus beecheyi (Richardson)
California Ground Squirrel

1829*a. Arctomys*? (*Spermophilus*?) *Douglasii* Richardson, 1:172.
1913*b. Citellus beecheyi douglasi*: Grinnell, 3:345.

Description.—*Spermophilus beecheyi* (Plate VIII) is a large, long-tailed (Table 11-5) gray squirrel with a large, nearly black, triangular patch between light-gray shoulder patches. The gray dorsal pelage is speckled with buffy-white spots about 5 mm in diameter; the venter and feet are buffy. The tail is bushy, but not so full and spreading as those of tree squirrels (Bailey, 1936), and >128 mm long (Hall, 1981). In the field, the long tail, light-gray shoulders, and dark patch on the foreback are distinctive.

The skull of *S. beecheyi* (Fig. 11-17) is similar to that of congeners in that the zygomata are twisted, flattened dorsoventrally, and converge anteriorly. The temporal ridges do not have a distinct V-shape. The parastyle ridges on M1 and M2 are straight, not angled. The upper incisors are stout and distinctly recurved (Hall, 1981).

Distribution.—California ground squirrels are distributed from south-central Washington south through western Oregon and most of California (except for southeastern desert areas) to central Baja California Norte (Fig. 11-18; Hall, 1981). In Oregon, *S. beecheyi* occurs throughout the area west of the Cascade Range and eastward through southern Klamath and Lake counties to Lakeview, Lake Co. Also, an arm of the range extends from the Columbia River in Hood River, Wasco, and Sherman counties southward through Jefferson and Crook counties to Prineville, Crook Co. (Fig. 11-18).

Wisconsinan-age fossils have been found in California (Kurtén and Anderson, 1980).

Geographic Variation.—Only one of the eight named races of *S. beecheyi* occurs in Oregon: *S. b. douglasii* (Hall, 1981).

Habitat and Density.—*Spermophilus beecheyi* is considered among the most generalized of the ground squirrels; it inhabits a variety of habitats (Tomich, 1962). Edge (1934) described areas containing concentrations of burrows of *S. beecheyi* both east and west of the Cascade Range as pastureland. Southern slopes seem to be favored over northern slopes for establishing burrows (Edge, 1934).

On the west slope of the Cascade Range, a population of *S. beecheyi* became established in a clear-cutting at ≅1,000 m elevation within 3 years of harvest of the Douglas-fir (*Pseudotsuga menziesii*, 87%)-western hemlock (*Tsuga heterophylla*, 12%) forest (Gashwiler, 1970*a*). Density was estimated at 0.83–1.23/ha 8–10 years after harvest. A. W. Moore, a longtime employee of the U.S. Fish and Wildlife Service involved with control of rodent pests, reported never having caught a California ground squirrel in old-growth forest; he believed that clear-cuttings were colonized by ground squirrels moving along

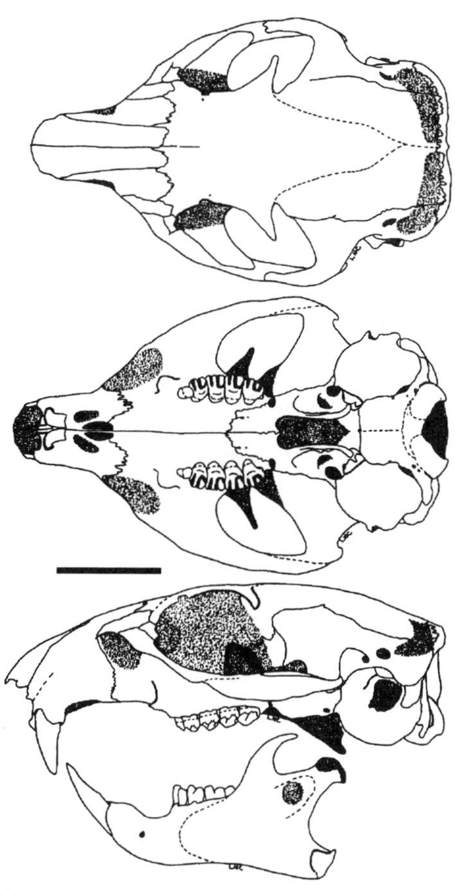

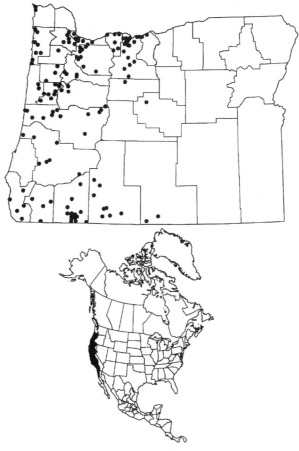

Fig. 11-18. Localities in Oregon from which museum specimens of the California ground squirrel, *Spermophilus beecheyi*, were examined (listed in APPENDIX, p. 560), and the distribution of the California ground squirrel in North America. (Range map redrawn from Hall, 1981:402, map 269, with modifications).

Fig. 11-17. Dorsal, ventral, and lateral views of the cranium, and lateral view of the dentary of the California ground squirrel, *Spermophilus beecheyi* (OSUFW 3905). Scale bar equals 15 mm.

logging roads (Gashwiler, 1970a). In California, densities ranging from 1.2 to 6.8/ha were reported (Evans and Holdenried, 1943).

Diet.—In northern California, *S. beecheyi* consumed seeds of field lupine (*Lupinus micranthus*), milk-thistle (*Silybum marianum*), buckhorn plantain (*Plantago lanceolata*), oak (*Quercus*), barley, bur-clover (*Medicago*), mountain brome (*Bromus carinatus*), tocalote (*Centaurea melitensis*), and manzanita (*Arctostaphlos manzanita*—Grinnell and Dixon, 1918). Evans and Holdenried (1943) listed three species of brome (*Bromus*), two of fescue (*Festuca*), two of thistle (*Centaurea*), two of clover (*Medicago*), and one each of barley (*Hordeum*), oats (*Avena*), sweet-clover (*Melilotus*), alfilaria (*Erodium*), torilis (*Torilis*), popcorn flower (*Plagiobothrys*), and thistle (*Silybum*) as most commonly eaten in the San Francisco Bay region of California. They also reported that acorns (*Quercus*), buckeyes (*Aesculus*), cascara (*Rhamnus*), elderberries (*Sambucus*), and some insects were eaten. Stanton (1944:158) indicated that "green vegetation, fruits and seeds, of which about one-fourth came from cultivated crops" composed the principal foods in stomachs of 204 *S. beecheyi* collected in western Oregon. He indicated that animal material composed 1.5% of the contents and occurred in 13.7% of the stomachs. To our knowledge, no more detailed studies of the diet of *S. beecheyi* have been conducted in Oregon.

Reproduction, Ontogeny, and Mortality.—Edge (1931) reported that California ground squirrels in western Oregon commenced to breed in late March and that most young were born about mid-April; however, these dates of breeding and parturition are not congruent with the 25–35 day gestation period that he also reported for the species or for the ≅30-day gestation period reported by Grinnell and Dixon (1918). Bickford (1980) reported breeding activity among *S. beecheyi* in western Oregon from early February to late March with most or all females pregnant by late March. Females produced young and nursed them in May–June, but became aggressive toward juveniles by mid-July (Bickford, 1980). In California, estimated birth dates ranged from early April to late June in 1 year, but from mid-April to late July the next year (Tomich, 1962). At approximately the same latitude but further west, Evans and Holdenried (1943) recorded the last pregnant female on 23 April. In southern California, the species breeds throughout the year (Storer, 1930). Thus, it appears that both onset and termination of the breeding season may vary among years and at different localities.

Females breed for the first time as they approach 1 year of age and essentially all females breed each year (Tomich, 1962). However, aggressive males may prevent some males from establishing breeding territories, thus excluding them from the breeding population (Bickford, 1980). The sex ratio usually is biased in favor of females, often 1:1.2–1.5 (Bickford, 1980; Evans and Holdenried, 1943; Tomich, 1962).

Chapman and Lind (1973) reported a significant (r^2 = 0.762, n = 10) inverse relationship between average litter size and latitude, the reverse of the relationship found in many species of mammals (Blus, 1966; Lord, 1960; Rowan and Keith, 1956). Because of a significant relationship between average litter size and the average frost-free period, they hypothesized that the southward increase in

average litter size was a response to greater mortality resulting from greater opportunity for activity and correspondingly greater predation on *S. beecheyi* in the warmer climates. Average litter sizes in western Oregon (Table 11-6) are near the low end of the range of values reported from throughout the range (Chapman and Lind, 1973).

Based on a litter of five born to a female captured in western Oregon, average mass of young was 11.8 g at 1 day of age, 19.5 g at 1 week, 24.0 g at 2 weeks, 35.2 g at 4 weeks, 47.0 g at 5 weeks, 53.5 g at 6 weeks, 64.5 g at 7 weeks, and 73.0 g at 8 weeks (Edge, 1931). The average mass of a much larger sample of young in California was 16.5 g, 30.5 g, 45.9 g, 64.2 g, 88.1 g, 116.6 g, and 163.4 g at weeks 1–7, respectively (Tomich, 1962), suggesting that growth of the Oregon litter both before and after young commenced to eat solid food was impaired. Such impairment possibly was the result of an inadequate diet provided the maternal female, as cannibalism of one young was noted (Edge, 1931).

At birth, young are hairless, reddish, and wrinkled; the eyes are closed and movements are poorly coordinated (Edge, 1931). By 13 days, hair appears on the head, by 20 days hair appears on the dorsum, and by 27 days the full pelage of short hair is complete (Tomich, 1962). However, at 5–6 weeks the tail is still not bushy. By 24–28 days, the auditory meatuses open, but the eyes do not open until 33–38 days (Tomich, 1962). The lower incisors erupt at ≅21 days, but the upper incisors do not erupt until ≅28 days.

Young are able to crawl slightly by swimming motions of the front limbs at 6 days; at 3 weeks mobility is the result of coordinated movements of both front and hind limbs; by 5 weeks young are able to leave and return to the nest; and by 6.5 weeks young are able to "frisk" about the cage (Tomich, 1962). Young do not cling to nipples when the maternal female leaves the nest.

Edge (1934) estimated the maximum longevity of *S. beecheyi* to be 5–6 years, whereas Tomich (1962) estimated it to be 3–4 years. Evans and Holdenried (1943) reported survival of 40.2% of 102 young from June to April; 15.7% remained the following April. However, the next year, of 135 young marked only 20.7% survived from June to the following year. Evans and Holdenried (1943) did not consider the life span to be >2 years.

Nussbaum and Maser (1975) reported that 2.8 and 2.9% of fecal droppings of bobcats (*Lynx rufus*) in the Coast and Cascade ranges, respectively, contained remains of *S. beecheyi*. Witmer and deCalesta (1986) indicated that remains of *S. beecheyi* occurred in droppings of both coyotes (*Canis latrans*) and bobcats in the Coast Range of southwestern Oregon at all seasons; the greatest frequency of occurrence in both carnivores was in summer and autumn. Other species known to prey on *S. beecheyi* in Oregon include red foxes (*Vulpes vulpes*—Livezey and Evenden, 1943), gray foxes (*Urocyon cinereoargentus*—Wilcomb, 1948), and great horned owls (*Bubo virginianus*—Maser et al., 1970). In California, a variety of reptilian, avian, and mammalian species are known to prey on *S. beecheyi* (Evans and Holdenried, 1943), and they probably also prey on the species in Oregon.

Habits.—In western Oregon, California ground squirrels usually enter hibernacula in November and emerge therefrom in late February or early March (Edge, 1931). A similar period of torpor was reported for the race in northern California (Grinnell and Dixon, 1918). *S. beecheyi* is one of only a few species of ground squirrels in which females are reported by some authorities to emerge before males (Michener, 1984); however, Linsdale (1946) and Bickford (1980) caught males earlier in the year than females. Bickford (1980) claimed that *S. beecheyi* in western Oregon immerged into their hibernacula in September or early October, but we commonly have seen individuals active above ground in November. The occurrence of remains of *S. beecheyi* in fecal droppings of carnivores at all seasons (Witmer and deCalesta, 1986) suggests that activity in these ground squirrels in southern Oregon may be similar to that reported for the species in central California; although no individual was active throughout the year in California some activity was evident at all seasons (Tomich, 1962).

Burrows are constructed on slopes or other well-drained sites, sometimes in considerable density; Edge (1934) recorded >36 burrows/ha. Burrow entrances commonly are located adjacent to a log or rock. Mounds at burrow entrances commonly are flattened and hard packed; occupied burrow systems have well-worn trails leading from them. Of 14 burrows excavated by Edge (1934), two exceeded 1.2 m in greatest depth below the surface, whereas three extended only to ≅0.3 m. Most burrows had only one entrance, but some that interconnected with other burrow systems had more than one. Burrows commonly are branched and have a chamber containing grasses, leaves, and moss for a nest; other chambers contain refuse and feces or are filled with earth from later excavations. Food storage was mostly in passageways near the nest chamber (Edge, 1934). Edge (1934) considered vertical shafts dug near nest chambers to be drains. At least some burrow systems are occupied by several *S. beecheyi* without conflict; however, squirrels that escape pursuit by entering the nearest available burrow often are ejected by the resident (Evans and Holdenried, 1943).

Although *S. beecheyi* is a ground-dwelling squirrel it has a strong propensity to climb (Grinnell and Dixon, 1918). It commonly is observed to sit on fence posts, stumps, brush piles, foundations of razed buildings, boulders, or other objects that extend above the grade, and occasionally up in trees as high as 10 m. Grinnell and Dixon (1918) considered the long tail an adaptation for climbing.

Of 106 individuals caught in 46 traps set on a 3.3-ha area in California, 23 adults and 32 young occupied home ranges with greatest diameters of ≤137 m; 4 adults and 4 young shifted to new home ranges ≤137 m from previously occupied home ranges; 3 adults and 12 young made sallies of ≥274 m from their home ranges; 1 adult and 12 young shifted to home ranges 274–1,097 m from previously occupied home ranges; and 13 young maintained home ranges with greatest diameters of ≥274 m (then as they matured 8 reduced the size of the area used and the other 5 disappeared—Evans and Holdenried, 1943). Bickford (1980) indicated that home ranges of *S. beecheyi* in Oregon overlapped extensively.

Spermophilus beecheyi commonly rubs the cheeks, top of head, and dorsum on various substrates by twisting or dirt-bathing motions; presumably these are scent-marking behaviors as areas so rubbed receive the attention of conspecifics as they commonly sniff the sites (Bickford, 1980). Also, significantly more male *S. beecheyi* are captured in

traps containing scent of male conspecifics, but females seemingly do not respond to scent of either male or female conspecifics. Males may avoid traps containing scent of females (Salmon and Marsh, 1989).

California ground squirrels upon meeting a conspecific commonly engage in mutual anogenital sniffing; also males engage in "paw-pushing" as a prelude to courtship or other interactions (Bickford, 1980). The paw-push is produced by a male when facing another individual; the right front foot is placed against the right shoulder of the conspecific (or the left front foot against the left shoulder). Recipients of the paw-push frequently turn and raise the tail to engage in anogenital sniffing which, in turn, commonly precipitates chases (Bickford, 1980). In California, the meeting behavior was described as "foot-to-cheek," but was variable in that a male sometimes grasps both cheeks of a female with both front paws and sometimes just touches the cheek with its nose (Owings et al., 1977). Adult males generally are more aggressive than females or juveniles; they sometimes tolerate weanlings in their proximity, but as the young mature, they are chased away (Bickford, 1980; Evans and Holdenried, 1943). Of 55 adult males captured in April–June in California, 21 had scars on their heads and bodies presumably resulting from intraspecific conflicts (Evans and Holdenried, 1943). Dobson (1983) indicated that both male and female *S. beecheyi* exhibit agonistic and territorial behaviors toward members of their own sex during prebreeding, breeding, and early pregnancy periods; males usually are dominant over females. He suggested that males were territorial to obtain and maintain mates and that females were territorial to protect environmental resources necessary for survival of offspring.

Some adult female *S. beecheyi* engage in infanticide; they actively stalk newly emerged young, and kill and consume them. Trulio et al. (1986) believed that some individuals developed the habit of killing young and may have done so to obtain a high-protein food. The possibility that reduction of competition for scarce resources is responsible for infanticide cannot be ruled out. However, some individuals also kill and consume moles (*Scapanus latimanus*) and lizards (Squamata). Females whose young are killed usually move surviving young to other burrow systems; they also commonly move young after encounters with snakes (Serpentes).

Upon detecting raptors flying overhead, *S. beecheyi* runs to its burrow and emits a whistle, but upon sighting a predaceous mammal it assumes an alert posture and chatters (Owings et al., 1977). Upon detecting snakes, *S. beecheyi* harasses them, sometimes tooth chattering and kicking sand at them; in the process, it makes a variety of flagging motions with the tail. Tail flagging attracts other squirrels that also may harass the snake and, in turn, tail flag themselves. Conspecifics that observe tail flagging can infer from variations therein the degree of hazard and the type of harassing behavior about to occur (Hennessy et al., 1981). Upon encountering snakes or other predators in burrows, California ground squirrels plug the burrows with earth (Coss and Owings, 1978).

Remarks.—In Oregon, the California ground squirrel commonly is referred to as the Douglas ground squirrel, digger squirrel, or gray digger.

California ground squirrels are a potential and sometimes a realized source of plague in humans. The peak in prevalence of infected *S. beecheyi* occurs soon after young emerge from natal burrows, and the young are considered a particular hazard because they commonly bleed from the nose (epistaxus) during bacteremia (Williams and Cavanaugh, 1983). Infected adults seemingly do not exhibit epistaxus.

The implication that *S. beecheyi* is a significant predator on eggs in nests of game birds (Stanton, 1944) is largely unsupported by the findings on which that conclusion seemingly was based. No "eggless" dummy nests were established as controls. Thus, capture of 70 *S. beecheyi* at dummy nests containing eggs of game birds is of little significance considering that only three had eggshells in their stomachs and that most captives refused to eat unbroken eggs.

Spermophilus beldingi Merriam
Belding's Ground Squirrel

1888a. *Spermophilus beldingi* Merriam, 4:317.
1898a. *Spermophilus oregonus* Merriam, 12:69.
1938. *Citellus beldingi oregonus:* Howell, 56:83.
1940. *Citellus beldingi crebrus* Hall, 21:59.

Description.—*Spermophilus beldingi* (Plate VIII) is a medium-sized (Table 11-5) ground squirrel without spots, stripes, or splotches. The pelage is smoky gray with some pinkish on the face, feet, and venter, and with a more or less well-defined reddish or brownish band in the middorsal region. The tail is cinnamon on the ventral surface. Distal tail hairs have three distinct color bands (Davis, 1939). The hind foot usually is >39 mm long.

The skull of *S. beldingi* (Fig. 11-19) is similar to that of other ground squirrels in that the zygomata are twisted, flattened dorsoventrally, and converge anteriorly. It can be separated from that of *S. columbianus* by having U-shaped temporal ridges and from *S. beecheyi* by an angled parastyle ridge on M1 and M2. However, characters of skulls of *S. canus, S. mollis, S. elegans, S. lateralis,* and *S. washingtoni* overlap those of *S. beldingi* sufficiently to be of little value in distinguishing the latter species.

Spermophilus beldingi has a diploid number of 30 chromosomes, less than in any other species of stripeless ground squirrel in Oregon (Nadler, 1966).

Distribution.—Hall (1981) depicted the geographic range of *S. beldingi* as extending from montane regions of northeastern Oregon south to the central Sierra Nevada Range, California, and south of the Snake River in western Idaho into central Nevada, and east into extreme northwestern Utah. However, our records extend the range westward to the Deschutes River and Broken Top Mountain in Deschutes Co., Diamond Lake in Douglas Co., and Fish Lake in Jackson Co. Durrant and Hansen (1954) indicated that the species had been taken north of the Snake River in Jerome Co., Idaho. We have made these and other minor adjustments in Hall's (1981:388, map 261) range map for the species (Fig. 11-20). In Oregon (Fig. 11-20), *S. beldingi* occurs south and east of a line connecting Enterprise, Wallowa Co.; Heppner, Morrow Co.; Maupin, Wasco Co.; Sisters, Deschutes Co.; Diamond Lake, Douglas Co.; and Fish Lake, Jackson Co. Altitudinally, *S. beldingi* occurs from <550 to >2,400 m in Oregon.

Spermophilus beldingi has not been found in any Pleistocene fauna (Jenkins and Eshelman, 1984).

Geographic Variation.—Two of the three named races

of *S. beldingi* occur in Oregon: the darker *S. b. oregonus* throughout the range in the state except for the extreme southeastern corner occupied by the lighter *S. b. creber* (Hall, 1940, 1981).

Habitat and Density.—Belding's ground squirrels mostly occur in steppe and shrub-steppe areas, particularly in meadows; sagebrush (*Artemisia*) flats; and small-grain, pasture, and hay-crop fields, and sometimes in openings in woodlands (Bailey, 1936; Grinnell and Dixon, 1918; Turner, 1972a, 1972b). Vegetation in much of the area occupied by Belding's ground squirrels in Oregon is characterized by sagebrush (mostly *Artemisia tridentata*), rabbit-brush (*Chrysothamnus nauseosus* and *C. viscidiflorus*), bitter-brush (*Purshia tridentata*), bluebunch wheatgrass (*Agropyron spicatum*), Idaho fescue (*Festuca idahoensis*), cheat grass (*Bromus tectorum*), Sandberg bluegrass (*Poa sandbergii*), giant wildrye (*Elymus cinereus*) with scattered western junipers (*Juniperus occidentalis*), and, in moist areas, some Ponderosa pine (*Pinus ponderosa* — Costain, 1978; Sullins, 1976; Turner, 1972a).

In drier areas, *S. beldingi* usually occurs in the vicinity of free water; in more mesic regions, habitats remote from water are occupied. However, rocky areas and areas subject to flooding are avoided (Turner, 1972a). Where *S. beldingi* occurs in sympatry with *S. columbianus* the former is associated with habitats more xeric that those occupied by the latter, in contradiction to the prediction of Durrant and Hansen (1954). However, in regions in which *S. columbianus* is absent, *S. beldingi* occupies mesic habitats similar to those occupied by *S. columbianus* where the two species are sympatric (Turner, 1972b). Where sympatric with *S. mollis* in Nevada, *S. beldingi* occupies the more mesic habitats (Hall, 1946); from our casual observations a similar relationship is evident in Oregon. Heavy clay soils

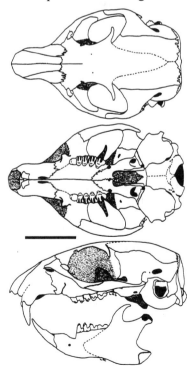

Fig. 11-19. Dorsal, ventral, and lateral views of the cranium, and lateral view of the dentary of Belding's ground squirrel, *Spermophilus beldingi* (OSUFW 4443). Scale bar equals 15 mm.

and very fine sandy soils tend to be avoided; intermediate loams and sandy loams are favored (Turner, 1972a).

To our knowledge the highest estimate of density of Belding's ground squirrels in Oregon was the 466 individuals counted on 0.81 ha (575/ha) in Klamath Co. by I. N. Gabrielson and recorded by Howell (1938). Although densities in central Oregon sufficient to elicit complaints regarding crop depredations were common in the 1970s, estimates of densities of some of those populations often were an order of magnitude less than that recorded by Howell (1938). In Grant Co., Turner (1972a) reported observing 13 adults on 0.40 ha (32.5/ha) at ≅1,000 m elevation and 120 adults on 1.6 ha (75/ha) at ≅1,500 m. Costain (1978), a few kilometers to the southwest at ≅1,700 m, recorded maximum densities of adults of 22.9–37.4/ha and maximum densities of 25.4–79.0/ha in June after emergence of young. In the northwestern part of the county at 550–1,200 m, Sullins and Verts (1978) reported removing 20–89 adults from 1-ha circular plots in 2-day periods.

Diet.—*Spermophilus beldingi* is primarily an opportunistic herbivore, but insects and other animal material (including conspecifics) may be consumed on occasion (Jenkins and Eshelman, 1984). Overall, grasses contribute most to the diet, but as the season progresses, *S. beldingi* first shifts to forbs such as clover (*Trifolium*), alfalfa, common dandelion (*Taraxacum officinale*), yarrow (*Achillea*

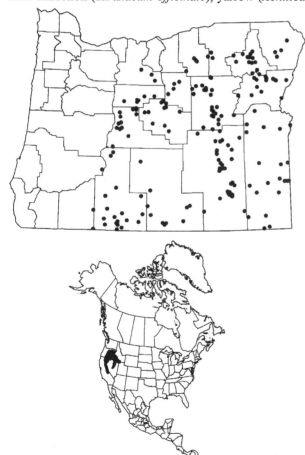

Fig. 11-20. Localities in Oregon from which museum specimens of Belding's ground squirrel, *Spermophilus beldingi*, were examined (listed in Appendix, p. 561), and the distribution of Belding's ground squirrel in North America. (Range map redrawn from Hall, 1981:388, map 261, with modifications.)

millefolium), and filaree (*Erodium cicutarium*), then later to seeds of cheat grass, meadow barley, and filaree (Morton, 1975; Turner, 1972a). Arthropods and miscellaneous items constitute <20% of the diet at all seasons (Morton, 1975). Road-killed conspecifics seem to be particularly relished (Turner, 1972a).

In a field test of potential baits for control of populations of *S. beldingi*, succulent materials (chopped apple or chopped lettuce) ranked first (and commonly first and second) among 12 items offered in both tests conducted in each month February–July (Sullins and Verts, 1978). Consumption of grain baits (especially those composed of oats) increased to a peak in May, but, as for the succulent baits, declined in June and July, likely in response to the great increase in biomass of agricultural crops available at that season and perhaps fewer individuals active above ground.

Belding's ground squirrels consume enormous amounts of food. McKeever (1963) reported that stomachs and their contents of adult male and female *S. beldingi*, respectively, weighed an average of 9.2 and 11.4 g in April, 14.2 and 18.3 g in June, and 30.0 and 30.0 g in August. Turner (1972a) reported a mass of >50 g for the stomach and its contents for an August-caught adult female; this constituted >20% of the total mass of the individual. In Klamath Co., uncontrolled populations of *S. beldingi* removed 36.1–53.8% of the first cutting of alfalfa (Kalinowski and deCalesta, 1981). Grinnell and Dixon (1918) estimated that during the growing season 750 *S. beldingi* ate as much as one steer. Food is not stored (Bailey, 1936).

Reproduction, Ontogeny, and Mortality.—Female Belding's ground squirrels attain reproductive competence as yearlings, whereas males delay until the 2nd year of life (Michener, 1984; Morton and Gallup, 1975). Courtship and mating commence about 1 week after emergence of females and continue for about 1 month (Turner, 1972a; McKeever, 1966). Thus, these activities are related to elevation and to year-to-year variation in temperatures.

Females are receptive for an average (±*SD*) of only 4.7 ± 0.3 h (Hanken and Sherman, 1981). In Grant Co. at ≅1,700 m, pregnant and lactating females were captured from April through June (Costain, 1978); thus, mating and fertilization probably commenced in March at this elevation as the gestation period is 23–28 days (Jenkins and Eshelman, 1984). Onset of breeding was about 2 weeks earlier in 1976 than in 1975; the difference was attributed to earlier emergence in response to warmer temperatures earlier in spring 1976 (Costain, 1978). In Grant Co., at ≅1,700 m, most young were born in May (Costain, 1978). A delay in reproductive activities for a few days or weeks after emergence by females born the previous year likely is responsible for the 4–6-week spread in mating and birth dates (Turner, 1972a). Among adults, the strong sex bias in favor of females (Verts and Costain, 1988) and the low proportion of nonbreeding females (Costain and Verts, 1982) indicate a polygynous mating system. In California, multiple matings of females with several males were common and multiple paternity of young occurred in 78% of litters for which paternity could be established (Hanken and Sherman, 1981).

Only one litter is produced annually at all elevations, but overall average litter size differs considerably with elevation (Table 11-6). In Grant Co., females in older age-classes produced significantly more offspring (Costain and Verts, 1982). In California, litter size increased with maternal age to 4 years then declined (Sherman and Morton, 1984); however, in Grant Co., based on counts of periosteal zonations, no individual attained 4 years of age (Costain and Verts, 1982). Because of age-specific differences, at least some of the variation in average litter size at sites at different elevations (Table 11-6) may be the result of differences in proportions of populations composed of females in various age cohorts. In addition to maternal age, quantity and quality of food may affect litter size (Morton and Sherman, 1978).

Eighteen (16.5%) of 109 pregnant females from ≅1,700 m in Grant Co. contained one or more resorbing embryos; intrauterine losses totaled 4.4% of the 570 embryos (Costain and Verts, 1982).

Young are born hairless except for 1-mm-long vibrissae, the skin is pink and so thin that viscera are visible, the digits are fused, and the pinnae do not protrude. Neonates can crawl weakly, right themselves with some effort, and squeak (Morton and Tung, 1971). By 5 days of age, the dorsum, forehead, and eye region are pigmented, but remain hairless; the distal one-third of the digits are free and the claws are pigmented; the vibrissae are ≅4 mm long; and the pinnae protrude ≅1 mm. By 10 days of age, fine hairs cover the entire body, gray on the dorsum and reddish on the head; the toes are free for about one-half their length; the vibrissae are ≅7 mm long and the pinnae are about 4 mm. The young can crawl by use of the forelimbs. By 15 days of age, fur on the dorsum is brown, white on the venter; the toes are completely separated; and the eyes and ears remain closed. The eyes of some young open by day 18. By day 20, the eyes of most young are open; the pinnae are 5–6 mm long and the auditory meatuses are open; upper incisors have erupted; and anal glands can be everted. By day 25, the pelage is similar to that of the adult. The young are weaned and attempt to defend themselves when handled (Morton and Tung, 1971). In captivity, offspring of females from California weighed 6.9 g at birth, 15.1 g by day 5, 26.6 g by day 10, 41.0 g by day 15, 54.4 g by day 20, and 69.0 g by day 25; by 45 days of age, some young weighed as much as some adults (Morton and Tung, 1971).

Great horned owls (*Bubo virginianus*), burrowing owls (*Athene cunicularia*), Swainson's hawks (*Buteo swainsoni*), red-tailed hawks (*B. jamaicensis*), prairie falcons (*Falco mexicanus*), goshawks (*Accipiter gentilis*), badgers (*Taxidea taxus*), weasels (*Mustela frenata*), coyotes (*Canis latrans*), martens (*Martes americana*), gopher snakes (*Pituophis melanoleucus*), and rattlesnakes (*Crotalis viridus*) are known to prey on *S. beldingi* (Bailey, 1936; Brodie and Maser, 1967; Grinnell and Dixon, 1918; Maser et al., 1971; Robinson, 1980; Sherman, 1977; Turner, 1972a). Near highways, many, especially males, may be killed by automobiles (Sherman and Morton, 1984). Nevertheless, in a California population, the most hazardous period was winter, when 35.5% of adult females and 38.5% of adult males captured ≤30 days before they entered torpor were not captured after emergence the following spring (Sherman and Morton, 1984). Rates of over-winter disappearance were 66.7% for juvenile females and 70.2% for juvenile males; for yearlings, the rates were 29.6% and 37.1% for females and males, respectively (Sherman and Morton, 1984). Life expectancies at birth in the California

population were 1.33 and 1.07 years for males and females, respectively (Jenkins and Eshelman, 1984). In California, infanticide accounted for ≥29% of juvenile mortality before immergence (Sherman, 1981).

In Grant Co., mortality among males and females in the different age-classes may be considerably different from that for the California population. Sex ratios for juveniles were equal and only moderately biased in favor of females (1.4–1.6:1) for yearlings; however, among adults, sex ratios were biased as much as 6.0:1 in favor of females (Verts and Costain, 1988). These changes were believed to indicate that the greatest mortality occurred among yearling males between immergence in autumn and emergence as adults in spring (Verts and Costain, 1988). Conversely, in California, male-biased dispersal among juveniles (Holekamp, 1984) may have been at least partly responsible for sex ratios of both yearling and adult cohorts being biased in favor of females.

Habits.—Belding's ground squirrels spend 6–8 months in torpor; duration of torpor and dates of immergence and emergence vary with elevation and with age-class of the squirrels. Emergence at low elevation in Umatilla Co. was recorded as early as 22 January (Howell, 1938). In Grant Co. at 550–1,200 m, sufficient numbers were above ground that bait testing could commence in February, but testing was suspended in mid-July because individuals had commenced to enter torpor (Sullins, 1976). At ≅1,700 m, some were still active on 3 September (Costain, 1978). From seasonal changes in sex ratios, adult males immerge first followed in order by adult females, yearling females, juvenile females, juvenile males, and yearling males (Costain, 1978; Verts and Costain, 1988). Among ground squirrels, extended activity by males in the year before attaining reproductive competence (yearling *S. beldingi*) is requisite (Michener, 1984). In baiting trials, a larger proportion of males in samples taken in June and July had consumed bait (Sullins and Verts, 1978), probably because consumption of baits by females declined as females immerged, a relationship misinterpreted by Jenkins and Eshelman (1984) to indicate that males had immerged first. Adult males emerge first in spring, commonly by burrowing through a meter or more of snow.

In preparation for torpor, Belding's ground squirrels become obese as they must rely on body fats for energy during dormancy and, after emergence, until suitable forage becomes available. The increase in mass may be extremely rapid; the body mass of a juvenile female in western Nevada increased from 180 g to 357 g in only 8 days (Loehr and Risser, 1977). In California, immediately before immergence, the mass of lipids was 126.4 and 122.1% of the body mass of males and females, respectively, after fats and water were extracted (Morton, 1975). In spring, the values were 40.1 and 41.3% for the two sexes, respectively. However, lowest values were 9.1% in late June for males and 7.6% in mid-July for females (Morton, 1975). Shortly after attaining maximum body mass, Belding's ground squirrels become dormant, even when available food is abundant (Morton, 1975). Thus, dormancy probably is not induced directly by temperature, desiccation of vegetation, photoperiod, or other environmental factors (Turner, 1972a). Although *S. beldingi* does not remain in torpor continuously for >18 days (McKeever,

1966), individuals, during the brief periods of arousal, do not leave their hibernacula.

During the season that *S. beldingi* is active, individuals leave their burrows shortly after sunlight strikes the entrances to their burrows (Loehr and Risser, 1977). Initially, individuals stay close to the entrances to their burrows, but as more individuals emerge they gradually extend their movements. By midday, activity declines appreciably. In some populations, a second, but less intense, period of activity occurs in late afternoon (Loehr and Risser, 1977), but not in others (Turner, 1972*a*).

Feeding occupies most of the morning hours, ≅20% of the total time spent above ground in spring and ≅40% in summer (Loehr and Risser, 1977). While feeding, individuals sit upright and manipulate foodstuffs with the forefeet. Basking, grooming, running, digging, and "picketing" (presumably sitting upright in alert posture) are other common activities (Loehr and Risser, 1977).

Until young are weaned, parous females defend 100–1,600-m² territories. Territories are defined by scent-marking; defense is by chasing and fighting. Aggression toward yearling males and alien adult females is especially vigorous, probably because they engage in infanticide; the former eat those killed, but the latter do not (Sherman, 1981). Individuals in other sex, age, and reproductive classes defend only areas in the immediate vicinity of their burrows.

A burrow described and depicted by Grinnell and Dixon (1918) had two entrances (10.0 by 9.0 cm and 7.5 by 6.0 cm), was 20.1 m long, and reached a maximum depth of 1.1 m. Burrows may be classified as short (0.15–4.0 m long) with one entrance or as long with two entrances (Turner, 1972*a*, 1973). Burrows are rarely used for >1 year, but burrows of other species such as northern pocket gophers (*Thomomys talpoides*) may be appropriated and modified. The different types of burrows are used for different purposes; one-entrance burrows apparently are emergency escape burrows, as they are entered more frequently in response to the approach of an aerial predator, whereas two-entrance burrows are more likely entered at the approach of a terrestrial predator (Turner, 1973).

Belding's ground squirrels produce predator-specific alarm calls upon sighting potential danger; a short vocalization or "chirp" is given in response to the appearance of an aerial predator or the sudden and close appearance of a terrestrial threat. A longer vocalization or "churr" is the usual response to the appearance of a terrestrial predator (Turner, 1973). Other individuals respond differently to the two calls. Only those nearby respond to the "churr" and those responding usually go to two-entrance burrows although one-entrance burrows often are closer. The "chirp" elicits widespread and immediate response with squirrels going to or entering the nearest burrow; one-hole and two-hole burrows are used about equally (Turner, 1973). The "churr" call also is produced in agonistic encounters, upon discovery of strange objects such as traps, and during courtship (Turner, 1972*a*). At the dissolution of the family unit (female and young), juvenile males disperse, whereas most juvenile females remain within the boundaries of the home range of the maternal female (Holekamp, 1984). Because alarm calls produced by females tend to alert their close kin, the production of alarm calls by *S. beldingi* is considered an expression of nepotism (Sherman, 1977).

Spermophilus columbianus (Ord)
Columbian Ground Squirrel

1815. *Arctomys Columbianus* Ord, 2:292 (described on p. 303).
1817*a*. *Anisonyx brachiura* Rafinesque, 2:45.
1891. *Spermophilus columbianus*: Merriam, 5:39.
1928. *Citellus columbianus ruficaudus* A. H. Howell, 41:212.

Description.—*Spermophilus columbianus* (Plate IX) is the larger of the two short-tailed, spotted ground squirrels in Oregon (Table 11-5). It can be distinguished from other short-tailed ground squirrels by hind feet ≥49 mm long. As the hind foot of juveniles attains ≥92% of adult size at a time that body mass is ≤33% of that of adults (Elliott and Flinders, 1980), this character should be useful for identifying all but the smallest individuals. Also, the temporal ridges are acutely V-shaped (Fig. 11-21); all other short-tailed ground squirrels in Oregon have moderately to distinctly U-shaped temporal ridges.

The pelage of *S. columbianus* is strikingly different from that of other ground squirrels in Oregon. The dorsum is brownish to dark gray with orangy-tan spots; the face, feet, and legs are a bright orangy rust; and the venter is yellowish rust. The head and neck are grayish. The tail is grayish or rusty.

Distribution.—Columbian ground squirrels occur in montane regions of eastern British Columbia and western Alberta south through eastern Washington and northeastern Oregon east through the northern two-thirds of Idaho

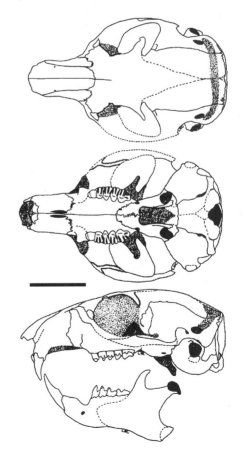

Fig. 11-21. Dorsal, ventral, and lateral views of the cranium, and lateral view of the dentary of the Columbian ground squirrel, *Spermophilus columbianus* (OSUFW 290). Scale bar equals 15 mm.

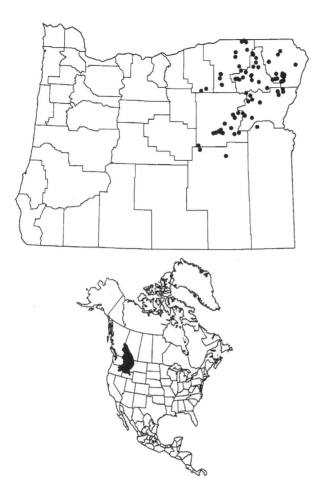

Fig. 11-22. Localities in Oregon from which museum specimens of the Columbian ground squirrel, *Spermophilus columbianus*, were examined (listed in APPENDIX, p. 562), and the distribution of the Columbian ground squirrel in North America. (Range map redrawn from Hall, 1981:389, map 262, with modifications.)

and the western one-fourth of Montana (Fig. 11-22; Hall, 1981). In Oregon, the species occurs in the Wallowa and Blue mountains (Fig. 11-22).

Fossil Columbian ground squirrels are known from late-Pleistocene-age cave deposits in Idaho (Kurtén and Anderson, 1980).

Geographic Variation.—Both of the two nominal subspecies occur in Oregon: the smaller and lighter *S. c. columbianus* throughout most of the Wallowa and Blue mountains and the larger and darker *S. c. ruficaudus* only along the southern edge of the range in northern Malheur and Harney counties and southern Grant Co. (Hall, 1981; Howell, 1938). Range maps indicating the absence of *S. c. columbianus* in Oregon or *S. c. ruficaudus* as an endemic to Oregon (Elliott and Flinders, 1984, 1991) apparently were lapsus.

Habitat and Density.—In Oregon, *S. columbianus* occurs in "small openings and meadows in forested areas"; most of these areas flood each spring so the ground squirrels are restricted to the edges of meadows or to

mounds within them (Turner, 1972*b*:914). The species is most abundant in the grand fir (*Abies grandis*) zone, but occurs in the upper part of the ponderosa pine (*Pinus ponderosa*) zone and the lower part of the subalpine fir (*Abies lasiocarpa*) zone. Within these zones, *S. columbianus* commonly is associated with early seral quaking aspen (*Populus tremuloides*) and lodgepole pine (*Pinus contorta*—Turner, 1972*a*). In many of these areas, the native grasses such as Idaho fescue (*Festuca idahoensis*) have been replaced by Kentucky and Canada bluegrasses (*Poa pratensis* and *P. compressa*) as a result of heavy grazing pressure (Turner, 1972*a*). Sedges (*Carex*) are characteristic of the moist meadows occupied by *S. columbianus*.

In Montana, *S. columbianus* occupied 5-, 10-, and 15-year-old clear-cuttings in the subalpine fir-Queen's cup (*Clintonia uniflora*) association, in which larch (*Larix occidentalis*) was the early seral dominant, and open subalpine basins in 15-year-old clear-cuttings in the subalpine-woodrush (*Luzula hitchcockii*) association. Densities were not significantly different among clear-cuttings (Ramirez and Hornocker, 1981). However, these ground squirrels were absent in uncut fir-Queen's cup associations and less than half as abundant in uncut fir-woodrush associations (Ramirez and Hornocker, 1981).

Where sympatric with *S. beldingi*, *S. columbianus* occupies more mesic habitats at higher elevation and enters torpor later in the season. In portions of its range in which *S. beldingi* is absent, *S. columbianus* occurs in both xeric and mesic habitats (Turner, 1972*a*). Both species enter torpor earlier at low elevations where forage becomes desiccated earlier in the season.

To our knowledge, no studies that included estimates of density of populations of *S. columbianus* have been conducted in Oregon. In Washington, Shaw (1920) reported densities of 25.7/ha in wheatfields and 61.7/ha on bottomland. In Alberta, Festa-Bianchet (1981) estimated 31.1 and 70.3/ha adults and yearlings at emergence in 2 successive years on a 0.74-ha meadow; emerging juveniles numbered 54.1 and 64.9/ha in the same 2 years. Festa-Bianchet (1981:1034) considered the habitat at this meadow to be of "particularly good quality." In another Alberta study, the density of adults and yearlings ranged from 11.6 to 16.1/ha during a 7-year period; the density of juveniles ranged from 4.6 to 20.7/ha (Boag and Murie, 1981). Sex ratios were biased in favor of females among adults and yearlings, but were balanced among juveniles (Boag and Murie, 1981). Betts (1991) studied a population consisting of five–seven adult females, three adult males, six–seven yearling females and three–four yearling males that occupied a 0.6-ha grassy meadow in Montana (density = 28.3–35.0/ha).

Diet.—In Washington, of 43 stomachs from *S. columbianus,* all contained vegetative material and 86% contained only vegetative material; 13.9% contained some insect parts and 2% contained hair of other species of mammals (Shaw, 1925*a*). After emergence in spring when

the ground was partly covered with snow, adult males ate yarrow (*Achillea*) and dug roots of wild onions (*Allium*), then later consumed buttercup (*Ranunculus*) blossoms and wild oats (*Avena fatua*). Columbian ground squirrels did not eat bunchgrass (*Agropyron*) even when the leaves were green and succulent, but avidly consumed bluegrass (*Poa*). Wild lettuce (*Lactuca*) was devoured eagerly (Shaw, 1925*a*). Captive animals fed clover and apple consumed ≅135 g/day, ≅33% of their body mass. Shaw (1925*a*) considered the species to cause serious depredations on wheat.

Some Columbian ground squirrels cache food in part of their burrow systems; 21 of 41 burrows examined contained a larder (Shaw, 1926). For 13 burrows, the sex of the occupant was known; 12 were those of males. Most of the stored material was crop plants: potatoes were stored in eight burrows, wheat in five, oats in three, barley in two, and apple in one. Tonella (*Tonella tenella*) occurred in nine burrows (Shaw, 1926). Shaw (1926) presented some evidence that males ate stored food after they emerged and before snowmelt made other foods available.

Reproduction, Ontogeny, and Mortality.—Testes of male *S. columbianus* are at maximum size at emergence from dormancy (Shaw, 1926); mating commences when females emerge and continues for ≅3 weeks. Yearlings in many populations do not breed (Betts, 1976; Michener, 1977; Murie and Harris, 1982; Murie et al., 1980; Moore, 1937), but breeding by yearlings has been reported in some populations (Festa-Bianchet, 1981). Festa-Bianchet (1981) considered high-quality habitat and favorable climatic conditions responsible for early sexual development in yearling females in Alberta. Adult females that fail to breed during the first estrous period may become receptive at about 2-week intervals for two or three times (Shaw, 1925*b*). Gestation is 24 days (Shaw, 1925*b*).

No formal study of reproduction of Columbian ground squirrels has been conducted in Oregon and reproductive information is recorded on labels of only a few specimens (Table 11-6). In general, productivity as measured by means of counts of embryos, corpora lutea, or pigmented sites of implantation is relatively low for a member of the genus *Spermophilus* (Table 11-6; Murie et al., 1980). However, recorded average litter sizes range from 2.7 to 7.0 (Elliott and Flinders, 1991). These averages vary inversely with elevation (Murie et al., 1980) and directly with population density (Festa-Bianchet, 1981). The apparent incongruity of the latter relationship may have been the result of a common response of both density and productivity to the high quality of the habitat (Festa-Bianchet, 1981). In an analysis that included data from several studies, litter size was correlated negatively with latitude and positively with elevation (Wroot et al., 1987).

At birth, young are naked, pink, and blind, and the auditory meatuses are closed (Shaw, 1925*b*). In Washington, body mass of young averaged 8.6 g (*n* = 23), but average mass at birth may range from 6.8 to 11.4 g (Wroot et al., 1987). The average (range in parentheses) interval from

birth to healing of the umbilicus is 5.5 (2–7) days, to eruption of i1 is 16.2 (13–17) days, to eruption of I1 is 19.8 (17–21) days, to opening of the auditory meatuses is 22.3 (20–23) days, and to opening of the eyes is 20.1 (18–23) days (Ferron, 1981). The average interval from birth to developing the ability to crawl forward is 2.7 (1–6) days, to walk on all four legs is 15.1 (14–17) days, to climb is 17.9 (13–20) days, to leave the nest at will is 25.5 (23–26) days, to drink is 28.6 (26–34) days, and to eat solid food is 23.8 (21–26) days (Ferron, 1981).

Growth rates obtained from several published studies were not correlated with latitude or elevation singly, but were correlated with latitude adjusted by 5° for each 1,000 m of elevation (Wroot et al., 1987).

In Alberta, only 40% of females and 35% of males survived to 1 year of age; however, annual survival for adults was 50–64% among females and 37–54% among males (Boag and Murie, 1981). Columbian ground squirrels have been reported to be preyed upon by brown bears (*Ursus arctos*), coyotes (*Canis latrans*), martens (*Martes americana*), badgers (*Taxidea taxus*), lynxes (*Lynx canadensis*), mountain lions (*Puma concolor*), weasels (*Mustela*), golden eagles (*Aquila chrysaetos*), red-tailed hawks (*Buteo jamaicensis*), and goshawks (*Accipiter gentilis*—Elliott and Flinders, 1991).

Habits.—In Alberta, Columbian ground squirrels spend an average of 245–255 days in torpor and an average of only 69–94 days active; much of the variation is related to climatic conditions and to differences in times of immergence and emergence of the various sex and age groups (Michener, 1977). The sequence of immergence is adult males, adult females, yearlings, and juveniles (Betts, 1976). The sequence of emergence is adult males, adult females, and yearlings (Michener, 1977; Murie and Harris, 1982). Dates of emergence are correlated positively with altitude and latitude (Murie and Harris, 1982). Duration of daily aboveground activity differs among various sex and age-classes, but tends to decrease with season and is believed to be dependent upon temperature during the diel cycle (Betts, 1976; Michener, 1977).

While above ground, Columbian ground squirrels spend more time alert than in any other activity; time in alert posture and time spent feeding were related inversely (Betts, 1976). Adult males engaged in aggressive behavior more frequently than other sex and age-classes, and tended to exhibit aggression more frequently during the early part of the active season (Betts, 1976). Activities related to body care commonly were the next most frequent.

Spermophilus columbianus engages in a greeting behavior that resembles "kissing." Upon meeting, the two squirrels touch mouth and nasal areas usually for 1–5 s; kissing commonly occurs before other social behavior, including sexual interactions. The behavior seems relatively common in *S. columbianus* and there is abundant literature on the function and evolution of the behavior pattern in this species (Betts, 1976).

Columbian ground squirrels communicate by sound, posture, touch, and probably scent. Betts (1976) indicated that *S. columbianus* had a repertoire of seven different sounds: shrill chirp, soft chirp, churr, tooth chatter, squeal, squawk, and growl; he provided sonograms for the first six. The churr and both chirps function as alarm calls, the tooth chatter and growl as threats, and the squeal and squawk as expressions of fear or pain. Information also is transmitted by postures and tail-flicking both during aggressive encounters and "outside the context of direct social interaction" (Betts, 1976:671). Several of the more common postures of *S. columbianus* are illustrated across the bottom of the previous 2 pages (redrawn from Betts, 1976:655): arched postures and head-down, rear-up postures are threats and upright postures indicate alarm. Tail-flicking can occur in any posture and may indicate excitation or serve to alert distant squirrels. Communication by scent is a less well-established behavior; kissing may permit exchange of scent by which participants identify each other, and certain rolling, rubbing, and stretching activities may serve to deposit scent (Betts, 1976). Scent also may be deposited passively as individuals enter burrows or touch other objects. Allogrooming, kissing, and inguinal nosing, in addition to other functions, may function to communicate by touch.

In Alberta, male Columbian ground squirrels have overlapping home ranges averaging 4,200 m²; core areas averaging 460 m² within these home ranges are defended more vigorously (Murie and Harris, 1978). Males within their core areas are dominant, but may become subordinate when outside their core areas. Most 2–3-year-old males are subordinate, whereas all males ≥4 years old are dominant (Murie and Harris, 1978). Adult females have home ranges averaging 200–800 m²; home-range size varies among periods within years and among years (Festa-Bianchet and Boag, 1982). Part of the home range is defended, especially vigorously during early gestation and during lactation (Festa-Bianchet and Boag, 1982). Territoriality in adult females is believed to be related mainly to protection of young (Festa-Bianchet and Boag, 1982). Betts (1991) reported that overlap of home-range areas of sex-age-classes other than adult females remained relatively constant through the reproductive period, but overlap in home ranges of adult females was significantly less during lactation than before young were born or after they were weaned. Also, adult females chased other adult females in relation to the distance between the approaching female and the territorial female's burrow, but not in relation to the distance between the two females. Adult females chased individuals in other sex-age-classes in relation to the distance between them, not in relation to their proximity to the adult female's burrow. Betts (1991) interpreted these behaviors to indicate that the greatest threat of infanticide was that from another adult female, not that from adult males as predicted earlier.

Juvenile males tend to disperse (sometimes as far as 6.7 km—Boag and Murie, 1981), whereas juvenile females exhibit fidelity for natal areas and often "inherit" natal burrows from their mothers (Harris and Murie, 1984).

Remarks.—In Oregon, the Columbian ground squirrel often is referred to as the "red digger" because of its striking orangy-rust face, feet, and legs, and its burrowing activities.

Spermophilus elegans Kennicott
Wyoming Ground Squirrel

1863. *Spermophilus richardsonii elegans* Kennicott, 15:158.
1928. *Citellus elegans nevadensis* A. H. Howell, 41:211.
1938. *Citellus richardsonii nevadensis* Howell, 56:77.
1971. *Spermophilus elegans nevadensis*: Nadler, Hoffmann, and Greer, 20:303.

Description.—*Spermophilus elegans* (Plate IX) is a medium-sized ground squirrel (Table 11-5) without spots, stripes, or splotches. The dorsal pelage is grayish with a slight buffy wash; the venter is ocherous buff. The underside of the tail is buffy. Distal tail hairs have five to seven alternate light and dark color bands (Davis, 1939). In the field, *S. elegans* appears distinctly more yellowish than sympatric congeners with which it might be confused. Also, the tail usually is longer, both actually and relatively, than in sympatric unstriped spermophiles (Hall, 1946).

The skull of *S. elegans* (Fig. 11-23) is similar to that of other ground squirrels in that the zygomata are twisted, flattened dorsoventrally, and converge anteriorly. Characters of skulls of *S. elegans* overlap those of sympatric congeners such that they are of little value in identifying the species.

Distribution.—The geographic distribution of *S. elegans* includes three disjunct populations: one in southeastern Oregon (extirpated), southwestern Idaho, and northern Nevada; another in southwestern Montana and eastern Idaho; and a third largely in the southern half of Wyoming and northwestern third of Colorado, but extending into Utah and Nebraska (Fig. 11-24). In Oregon, *S. elegans* is known from seven specimens collected at two localities in Malheur Co. near McDermitt, Nevada (Fig. 11-24).

Fossils of the "*S. richardsonii* complex," in addition to records at sites within the present-day range of *S. richardsonii* in Alberta, are known from specimens from Kansas, Texas,

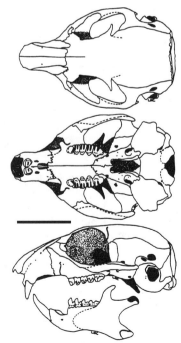

Fig. 11-23. Dorsal, ventral, and lateral views of the cranium, and lateral view of the dentary of the Wyoming ground squirrel, *Spermophilus elegans* (OSUFW 3911). Scale bar equals 15 mm.

and Oklahoma that likely represent *S. elegans* (Zegers, 1984). *S. richardsonii* and *S. elegans* may have diverged from a common ancestor as recently as 11,000 years ago. Several biogeographically based hypotheses to explain the divergence of *S. richardsonii* and the three populations of *S. elegans* have been proposed (Zegers, 1984).

Geographic Variation.—Specimens taken in Oregon belong to only one of the three named races: *S. e. nevadensis* (Hall, 1981).

Habitat and Density.—In northern Nevada, *S. elegans* more commonly occurs in "meadows or meadowlike habitats," but may occur, at least occasionally, in sagebrush (*Artemisia*) or bottomlands (Hall, 1946:305). We captured specimens in this region in a sagebrush-dominated right-of-way along a short section of abandoned highway; other plants in the area included cheat grass (*Bromus tectorum*), rabbit-brush (*Chrysothamnus*), and an unidentified mustard (Cruciferae). In Wyoming, several studies of *S. elegans* were conducted in short-grass prairie that included wheatgrass (*Agropyron*), junegrass (*Koeleria*), bluegrass (*Poa*), needlegrass (*Stipa*), rabbit-brush, greasewood (*Sarcobatus*), and saltbush (*Atriplex*—Clark, 1970*a*). In Oregon, E. A. Preble captured four specimens in a meadow that also supported a population of *S. beldingi* (Bailey, 1936).

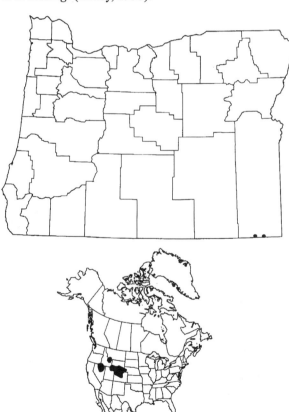

Fig. 11-24. Localities in Oregon from which museum specimens of the Wyoming ground squirrel, *Spermophilus elegans*, were examined (listed in APPENDIX, p. 562), and the distribution of the Wyoming ground squirrel in North America. (Range map redrawn from Hall, 1981:386, map 259, with modifications.)

In Wyoming, densities of 0.2/ha in March and 1.2/ha in June were reported (Clark, 1970*a*). In Colorado, estimates of density were 20/ha in April, 14/ha in May, 48/ha in June, and 20/ha in July; the increase in June was related to recruitment and the decline in July was related to adults entering torpor and juvenile dispersal, mortality, and torpor (Zegers, 1981).

Diet.—In Oregon, the only report of plants eaten by *S. elegans* is the oft-cited record of Bailey (1936) of 1,160 capsules of collomia (*Collomia*) and a few seeds of collinsia (*Collinsia*) in the cheek pouches of an adult male taken near the headwaters of Rattlesnake Creek, Malheur Co. In Wyoming, green-plant material composed most of the food eaten in April–June, but seeds contributed an increasingly greater proportion of the diet in July–August (Clark, 1968). Stomachs ($n = 82$) of *S. elegans* contained seeds of bluestem wheatgrass (*Agropyron smithii*), blue gramma (*Bouteloua gracilis*), needle-and-thread grass (*Stipa comata*), Indian ricegrass (*Oryzopsis hymenoides*), junegrass (*Koeleria*), and common peppergrass (*Lepidum densiflorum*), and flowers of thistle (*Cirsium*) and prickly-pear cactus (*Opuntia polycantha*); most of these plant species occur in Oregon. *S. elegans* also was observed to feed on road-killed conspecifics and on the feces of cattle (Clark, 1968). Clark (1968) indicated that the diet of *S. elegans* was governed largely by availability.

Reproduction, Ontogeny, and Mortality.—In Colorado, females emerged 16–18 days after males and bred within 1–4 days of emergence; the gestation period is 22–23 days (Fagerstone, 1988). Both yearlings and adults of both sexes bred during most years, but most yearling females failed to breed during a year that they emerged late and at low body mass in response to low temperatures and delayed snowmelt (Fagerstone, 1988). The testes are enlarged at emergence, but regress to one-tenth that mass within 2 weeks; however, males may remain fertile for a short time because of spermatozoa stored in the epididymides (Fagerstone, 1988). Recrudescence commences in June, a month before onset of torpor (Clark, 1970*a*; Fagerstone, 1988).

Also in Colorado, average (±*SE*) litter size at emergence of young was less for yearling females (3.6 ± 0.8, $n = 12$) than for adult females (4.8 ± 0.5, $n = 19$), but not significantly so (Table 11-6; Pfeifer, 1982). However, females that occupied frequently used (83% occupancy) burrow systems produced significantly larger litters (5.2 ± 0.4, $n = 21$; $t = 4.66$) than females that occupied less frequently used (59% occupancy) burrow systems (2.7 ± 0.2, $n = 15$). Use of specific burrow systems during natal periods seemed related to snowmelt (Pfeifer, 1982). Litter size reported for *S. elegans* in Wyoming was considerably larger (Table 11-6). However, Stanton et al. (1992) reported that litter size based on counts of implantation sites differed among sites, age-classes of females, and years. They believed that variation was related to quantity and quality of food available.

On the day of birth, seven captive neonates weighed an average of 5.6 g and averaged 46 mm long (Denniston, 1957). In another captive litter, neonates averaged 5.9 g and 53 mm at 2 days of age, 11.7 g and 78 mm at 8 days, 21.0 g and 103 mm at 14 days, 48.9 g and 165 mm at 28 days, and 125.4 g and 225 mm at 42 days (Clark, 1970*b*).

Juveniles attained 100% of adult size for the hind foot by 42 days of age, for the tail by 56 days, and for total length by 63 days; 100% of adult body mass was not attained until 100 days of age (Clark, 1970b).

At 2 days of age, the skin of *S. elegans* is reddish, smooth, and translucent; the viscera are visible through the skin of the abdomen (Clark, 1970b). The digits are fused and the pinnae are not separated from the skin of the head, but the vibrissae are visible. By 5 days of age, grayish claws are visible on the still-fused digits, the eyes are darkly pigmented, and the dorsum and sides are grayish, but the venter is still pinkish. The young can right themselves and move about with some difficulty. By 2 weeks, young are covered with a pelage of fine hairs, the digits are separated, and they can right themselves and move about readily. By 4 weeks, the upper incisors have erupted, the eyes and auditory meatuses are open, and the color of the pelage is that of adults. By 5 weeks, captive young were eating solid food (Clark, 1970b).

In Wyoming, eight (66.7%) of 12 marked in July survived to the following April and six (75.0%) of those eight were alive in July; thus, annual survival for the sample was 50% (Clark, 1970a). In Colorado, overall survival was 54 and 31% during 2 consecutive summers and 65 and 100% during the following winters, respectively. During the 1st year, survival of juveniles was approximately half that of adults, but within an age-class it was about equal between the sexes; during the 2nd year, survival was about equal among the sex and age-classes (Zegers and Williams, 1979). Thus, the difference in survival between years was largely in adult cohorts.

Habits.—In Wyoming, the first individuals emerged from hibernacula in early March in each of 3 years; however, individuals continued to emerge during the following 2–3 weeks. Males emerged before females (Clark, 1970a). Juveniles were first captured in late May. By late July, adults had entered torpor (males about a week before females) and by late August all aboveground activity had ceased. Although ground squirrels were active during a period of about 6 months, no individual was active for the entire period (Clark, 1970a). Adult males and adult females gained an average of 29 g (from 279 ± 53.6 to 308 ± 60.9 g) and 84 g (from 203 ± 32.1 to 287 ± 39.5 g), respectively, between March and May, and they added an average of 92 g (from 308 ± 60.9 to 400 ± 20.0 g) and 58 g (from 287 ± 39.5 to 345 ± 51.0 g), respectively, before entering torpor (Clark, 1970a).

In Colorado, males emerged in early March during 1 year and in late March during 2 years; females emerged 16–18 days later. Yearlings emerged about the same time as females (Fagerstone, 1988). Juveniles emerged in late June. Although temperature and snowcover were shown to influence emergence of Wyoming ground squirrels, they did not emerge when temperatures and snowcover were appropriate in February, suggesting that endogenous factors in addition to exogenous factors influence the timing of emergence. Yearling and adult males entered hibernacula in late July and early August; yearling and adult females immerged about 1.5 weeks earlier than males. Juveniles did not immerge until September with females immerging before males (Fagerstone, 1988). However, radio-equipped squirrels continued to move for as long as 2 weeks after entering hibernacula, indicating that they did not enter torpor immediately upon immerging. Although juveniles immerged in response to declining temperatures and food shortages, immergence of adults occurred when food plants were still green and lush, and before temperatures had commenced to decline, suggesting other cues for initiation of torpor. Fagerstone (1988) believed that immergence of adults was in response to excessive obesity, and the concomitant difficulty in running and entering burrows; it simply became energetically more efficient to hibernate than to attempt to maintain optimal mass while avoiding predators in a less-agile condition.

In Alberta, males of a closely related species, *S. richardsonii*, spend considerably less time in torpor than females before emergence. To compensate, most males, but no females, cache food in their hibernacula. Michener (1993) found 1,295 g of stored grain in the hibernaculum of one male. Michener (1992:397) also suggested that food caches permitted male *S. richardsonii* to "terminate torpor several weeks in advance of emergence, during which time they recouped mass and developed sperm in preparation for the forthcoming mating season." Considering that some species of ground squirrels that occur in Oregon also are known to cache food (Shaw, 1926), we suspect that *S. elegans* may exhibit similar behavior.

During the active season, individuals usually became active within an hour after sunrise. Early in the season, activity was unimodal with the peak occurring about 1330 h (Clark and Denniston, 1970). Later in the season, most activity ceased in midday as the ambient temperature approached 25°C, although an individual might emerge for a brief period when temperatures exceeded this level (Clark, 1970a). Captive animals exhibited evidence of heat stress when subjected to temperatures above 25°C. In late afternoon after temperatures declined, these ground squirrels emerged from their burrows to forage; they returned to their burrows for the night (Clark, 1970a). Snow, heavy rain, and hail storms caused *S. elegans* to cease aboveground activity, but light showers did not affect activity (Clark, 1970a).

In Idaho, 42 burrows of *S. elegans* averaged (±*SE*) 7.8 ± 0.3 cm (range, 2.4–12.3 cm) wide by 5.6 ± 0.2 cm (range, 1.4–7.3 cm) high (Laundré, 1989).

The approach of raptorial birds or humans caused *S. elegans* to assume one of the rigid alert postures similar to those observed in congeners or, more commonly, the unique upright posture with an arched back and occasionally the unique prone posture with head up (Clark and Denniston, 1970). The latter posture was used most often by juveniles in the first few days immediately after they emerged from their natal burrows. Wyoming ground squirrels also responded to the alert posture of nearby individuals of other ground-dwelling sciurids by assuming an alert posture themselves. Except during the reproductive season, *S. elegans* in the field exhibited little agonistic behavior. However, captive individuals challenged other adults introduced into the chamber containing their artificial burrow system by vocalizing, whistling, and growling, and by making warning rushes toward the intruder (Clark and Denniston, 1970). During flight from an animal that approached too closely or was in pursuit, *S. elegans* extends the hairs on the tail to give the appearance of a "bottle brush" (Clark and

Denniston, 1970). Domestic livestock was ignored.

In Wyoming, the home-range area calculated by the minimum-area method for nine males was 0.21 ha (range, 0.04–0.49 ha); juveniles had larger home ranges ($\bar{X}$ = 0.32 ha) than yearlings ($\bar{X}$ = 0.15 ha) or adults of unknown age ($\bar{X}$ = 0.21 ha—Clark, 1970a). No comparable data are available for females. Some individuals transplanted 0.4 km established home ranges where released, some returned to their original home range, and some disappeared (Clark, 1970a).

Remarks.—The Wyoming ground squirrel was described originally as a distinct species, *Spermophilus elegans*, but was synonymized with *S. richardsonii* by Howell (1938). We know of no publication in which the Wyoming ground squirrel formally was reelevated to specific status. Jones et al. (1982) cited Koeppl and Hoffmann's (1981) comparative study of four species of ground squirrels as the basis for again recognizing *S. elegans* as a distinct species. However, bacula, cranial morphology, blood proteins, karyotypes, and vocalizations are sufficiently different between *S. elegans* and *S. richardsonii* to cause others to suggest that separation at the species level was warranted (Burt, 1960; Fagerstone, 1987b; Koeppl et al., 1978; Nadler et al., 1971, 1974; Robinson and Hoffmann, 1975).

The status of *S. elegans* in Oregon is uncertain. E. A. Preble collected four specimens on 2–3 June 1915 in Malheur Co. just north of McDermitt, Nevada, and Vernon Bailey collected three on 1–2 July 1927 between the headwaters of the Quinn River (mostly in Nevada) and Rattlesnake Creek. To our knowledge, none has been collected subsequently. Olterman and Verts (1972:11), based on the time since the last record and on information related by L. W. Turner, considered the species "probably endangered if present in the state." On 25 June 1982, we observed large numbers of *S. elegans* along the wide right-of-way along U.S. 95 in Humboldt Co., Nevada, to within a kilometer of the Oregon border; we collected two specimens in Nevada 43 km north of Winnemucca (OSUFW 6878 and OSUFW Study-46). The right-of-way north of the Oregon-Nevada line was only one-fourth as wide and we saw no *S. elegans*. Also, we searched the area near the headwaters of Rattlesnake Creek on 25–26 June 1982 but, like Turner, we saw only *S. beldingi*. On 19 June 1992, we again traveled U.S. 95 from Winnemucca to 53 km north of McDermitt, Nevada, but saw no Wyoming ground squirrels and little evidence of recent burrowing activity. A few kilometers north of McDermitt we saw numerous *S. beldingi* in hayfields, but no *S. elegans*.

One specimen of the same race that occurred in Oregon was collected at Riddle, Idaho (Davis, 1939; Howell, 1938), but none was collected in that state since those reports were published (Zegers, 1984). The absence of recent records at the northern periphery of the range of this race tends to support the hypothesis of Durrant and Hansen (1954) that the race is a relict and possibly is being replaced by *S. mollis* in xeric habitats and by *S. beldingi* in mesic habitats.

The paucity of investigations of *S. e. nevadensis* has caused us to rely essentially entirely upon studies of the other races conducted in Wyoming and Colorado for information on the life history and ecology of the species.

Spermophilus lateralis (Say)
Golden-mantled Ground Squirrel

1823b. *S[ciurus]. lateralis* Say, 2:46.
1831. *Spermophilus lateralis*: Cuvier, 1:335.
1890b. *Tamias chrysodeirus* Merriam, 4:19.
1901c. *Callospermophilus chrysodeirus trinitatis* Merriam, 14:126.
1910. *Callospermophilus trepidus* Taylor, 5(6):283.
1931. *Callospermophilus chrysodeirus connectens* Howell, 12:161.
1938. *Citellus lateralis chrysodeirus*: Howell, 56:203.
1938. *Citellus lateralis connectens* Howell, 56:205.
1938. *Citellus lateralis trepidus*: Howell, 56:206.
1938. *Citellus lateralis trinitatis*: Howell, 56:211.

Description.—*Spermophilus lateralis* (Plate IX) doubtlessly is the most distinctively marked ground squirrel in Oregon; a white stripe bordered on both sides by a black stripe extends from the shoulder to the hip (Hall, 1981). From nose to nape above the eye, the head is russet. The back between the stripes is grizzled dark grayish-brown becoming less grizzled on the rump; lateral to the stripes the color grades to a light buffy-gold on the venter. The face, shoulders, front legs, and feet are a bright orangish-gold; the hind feet are buffy. The tail is blackish with scattered orangish-gold hairs above and orangish rust hairs below. A buffy-white line encircles the eye. *S. lateralis* is easily distinguishable from the chipmunks (*Tamias*), with which it is sometimes confused, by the absence of stripes on the face. *S. lateralis* is among the smaller members of the genus in Oregon (Table 11-5). The tail usually is <128 mm.

The skull (Fig. 11-25) is similar to that of other ground squirrels in that the zygomata are twisted, flattened dorsoventrally, and converge anteriorly. The temporal ridges are absent or obscure and the parastyle ridges on M1 and M2 are straight. The upper incisors are slender and straight.

Distribution.—*Spermophilus lateralis* occurs in the Rocky Mountains from east-central British Columbia and west-central Alberta southeastward through western Montana, Wyoming, Colorado, New Mexico, and Arizona

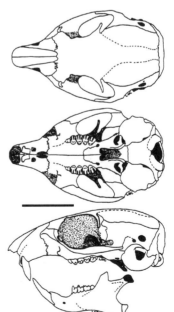

Fig. 11-25. Dorsal, ventral, and lateral views of the cranium, and lateral view of the dentary of the golden-mantled ground squirrel, *Spermophilus lateralis* (OSUFW 1590). Scale bar equals 15 mm.

westward in montane regions of Idaho, Utah, Nevada, southeastern Washington, Oregon, and California (Fig. 11-26; Hall, 1981). *S. lateralis* occurs throughout Oregon east and south of a line connecting Mt. Hood, Hood River Co.; McKenzie Bridge, Lane Co.; Diamond Lake, Douglas Co.; Butte Falls, Jackson Co.; and Galice, Josephine Co. (Fig. 11-26).

Pleistocene fossils of *S. lateralis* are known from cave deposits in Wyoming, Colorado, Idaho, and Arizona (Kurtén and Anderson, 1980).

Geographic Variation.—Four of the 16 nominal species of *S. lateralis* occur in Oregon: the short-tailed, relatively dark *S. l. chrysodeirus* occupies the east slope of the Cascade Range and most of central Oregon; the still darker and browner *S. l. trinitatis* occurs in the Siskiyou Mountains; the dark *S. l. connectens*, with a light underside of the tail, occurs in the Wallowa Mountains; and the longer-tailed, lighter-colored *S. l. trepidus* occurs on Steens Mountain and southeastward south of Huntington, Baker Co. (Hall, 1981; Howell, 1938).

Habitat and Density.—Most authorities list open timber with an abundance of rocks, fallen logs, and stumps as the prime habitat of *S. lateralis* (Bailey, 1936; Gordon, 1943; Grinnell and Dixon, 1918; Hatt, 1927; Howell, 1938; McKeever, 1964a). Indeed, *S. lateralis* often is associated

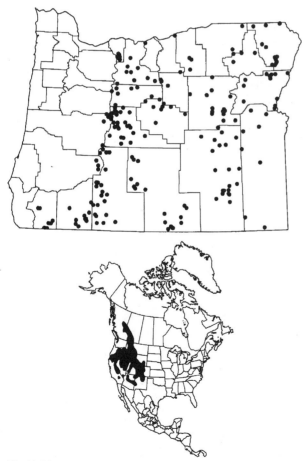

Fig. 11-26. Localities in Oregon from which museum specimens of the golden-mantled ground squirrel, *Spermophilus lateralis*, were examined (listed in Appendix, p. 563), and the distribution of the golden-mantled ground squirrel in North America. (Range map redrawn after Hall, 1981:409, map 274, with modifications.)

with open forests, but it is not restricted thereto. After having collected numerous specimens in a ponderosa pine (*Pinus ponderosa*)-manzanita (*Arcostaphylos*) community in western Deschutes Co., we were a bit amazed to encounter *S. lateralis* on a nearly barren talus slope in southern Malheur Co. A few scattered sagebrush (*Artemisia*), rabbit-brush (*Chrysothamnus*), and mustard (Cruciferae) plants grew among the rocks; there were no trees. Gordon (1943) indicated that the deep shade of dense forests was avoided, but that the species ranged down from open forest into sagebrush and even into meadows where rocks were abundant.

Gordon (1938, 1943) estimated 9.3 adults/ha on a 1.6-ha site in Wheeler Co. and 6.2–12.4/ha at other sites; however, he described no method used for estimating density. McKeever (1964a) related that he captured 83 specimens at three inspections of 180 traps set for 2 days. He also reported relative abundances of 5.14 in ponderosa pine, 0.78 in lodgepole pine (*Pinus contorta*), and 0.25 in red cedar (*Thuja plicata*) forests for comparisons of habitats.

Diet.—The diet of *S. lateralis* differs dramatically seasonally. In northeastern California, grasses and forbs (especially mules ear, *Wyethis mollis*—McKeever, 1964a) were the principal foods eaten in early spring, but hypogeous fungi composed the major portion of foods eaten during the remainder of the season (McKeever, 1964a; Tevis, 1953). In a sample of 12 *S. lateralis* from Oregon, 58% contained fungi and fungi composed 44% of the stomach contents (Maser et al., 1978a). In California in late summer, these ground squirrels ate seeds of conifers and buckbrush (*Ceanothus*); however, captives refused conifer seeds when fungi with their higher-protein content (McKeever, 1964a) were available (Tevis, 1953). Howell (1938) listed seeds of oak (*Quercus*), Douglas-fir (*Pseudotusga menziesii*), ponderosa pine, rose (*Rosa*), lupine (*Lupinus*), stoneseed (*Lithospermum*), alfilaria (*Erodium*), clover (*Trifolium*), smartweed (*Polygonum*), shepherd's-purse (*Capsella*), milk-vetch (*Astragalus*), willow-weed (*Epilobium*), penstemon (*Penstemon*), bitter-brush (*Purshia tridentata*), gilia (*Gilia*), violet (*Viola*), and frasera (*Frasera*), and fruits of serviceberry (*Amelanchier*), currant (*Ribes*), and blackberry (*Rubus*) as eaten by *S. lateralis*. Also, *S. lateralis* reportedly feeds on bulbs of wild onion (*Allium*) in autumn (McKeever, 1964a). During an outbreak of the California tortoise-shell butterfly (*Nymphalis californica*), 60% of the diet of *S. lateralis* consisted of caterpillars of the species; however, it returned to eating fungi when the outbreak abated (Tevis, 1953). *S. lateralis* also feeds on the flesh of mammals (including that of conspecifics) and birds when available (McKeever, 1964a; Tevis, 1953).

Reproduction, Ontogeny, and Mortality.—Testicular recrudescence commences in autumn before males enter torpor and continues during the period of dormancy (Tevis, 1955). The testes and seminal vesicles attain maximum size in late March and April, respectively (McKeever, 1965); testes double in length, prostate and epididymis widths triple, and the diameter of the bulbo-urethral gland increases by five times (Tevis, 1955). The scrotum blackens and becomes pendulous. Most *S. lateralis* mate in late March; McKeever (1964a) reported 83% of females collected in northeastern California in April were pregnant and 17% had enlarged uteri; in May, 35% were pregnant and 65%

were lactating. The gestation period is 27–28 days (Cameron, 1967). Litter size averages slightly more than five (Table 11-6); only one litter is produced annually (Tevis, 1955). After the breeding season, regression of the male gonads is rapid with minimum size reached in June (McKeever, 1965).

At birth, young are pink and, except for 2-mm-long vibrissae and minute hairs on the pate, hairless; there is no stripe pattern on the skin, the eyes and ears are closed, and the toes with barely discernible claws are fused (McKeever, 1964a). At 1 day of age, individuals in a litter of four averaged 3.6 g, whereas those in a litter of three averaged 6.6 g (McKeever, 1964a). The lateral stripes become evident as pigmented areas in the skin by 4 days of age; hair becomes visible on the anterior portion of the dorsum by 8 days; the entire dorsum is covered with hair, but the tail remains sparsely furred by 15 days; the tail hair is well developed, but without pattern by 18 days; and the typical adult pattern appears by 40 days (McKeever, 1964a).

The average (range in parentheses) interval from birth to healing of the umbilicus is 5.7 (3–10) days, to eruption of i1 is 17.5 (15–19) days, to eruption of I1 is 21.2 (20–24) days, to opening of the auditory meatuses is 18.3 (17–20) days, and to opening of the eyes is 25.3 (22–29) days (Ferron, 1981). The average interval from birth to developing the ability to crawl forward is 2.1 (1–4) days, to walk on all four legs is 20.1 (16–23) days, to climb is 31.4 (30–35) days, to leave the nest at will is 36.6 (36–37) days, to eat solid food is 29.9 (28–32) days, and to drink is 35.5 (35–36) days (Ferron, 1981).

In more severe environments with abbreviated seasons of activity, fecundity is less; however, rates of growth and development are related inversely to litter size (Phillips, 1981). Thus, in more severe environments, production of small litters allows young to become prepared for torpor at an earlier age and maternal females can commence to prepare for torpor earlier in the season (Phillips, 1981). M. T. Bronson (1979) reported a decrease in length of the active season, an increase in survivorship of adult females (but no change in that of males or juveniles), a decrease in litter size, and fewer 1- and 2-year-olds of both sexes breeding with an increase in elevation.

We examined a specimen (OSUFW 6607) collected 4 September 1948 and maintained in captivity by Alex Walker until it died on 14 August 1956.

Habits.—*Spermophilus lateralis* is active diurnally from late winter–early spring to midautumn. Most of its activity is confined to the ground where it travels rapidly among stumps, logs, or other slightly elevated prominences. Burrows usually are constructed under bushes, trees, stumps, rocks, or logs (Gordon, 1943), probably accounting for the paucity of descriptions of excavated burrow systems. *S. lateralis* can climb and sometimes may be seen in shrubs or occasionally as high as 10 m in trees (Howell, 1938). When not moving, it rarely assumes the "picket-pin" posture characteristic of many species of ground squirrels, but sits crouched and alert (Howell, 1938).

In late summer, *S. lateralis* is highly visible as it scurries about gleaning seeds. Grinnell and Dixon (1918) indicated that *S. lateralis* stored large amounts of food that it transported in well-developed internal cheek pouches. They provided an anecdotal account of a semitame individual scatter-hoarding fruit it was given. Howell (1938)

and Gordon 1943) claimed that the species stored food in underground chambers. However, McKeever (1964a) found no evidence of food stored in burrows by the species in the wild, and Hatt (1927) reported no food in the burrow that he excavated. In the Cascade Range in Oregon, about 50% of the bitter-brush (*Purshia tridentata*) shrubs and about 15% of the ponderosa pine seedlings were the result of scatter-hoarding by *S. lateralis* and chipmunks (West, 1968).

Spermophilus lateralis undergoes an endogenously regulated annual cycle of fattening, torpor, and reproduction; for captives from Oregon, the interval between peaks in body mass was about 51 weeks for females and 46 weeks for males, suggesting an annual resetting of the cycles by the environment to prevent them from becoming out of phase (Blake, 1972). Captive animals from Oregon, where elevations are lower and climates milder, stored proportionately less fat, but averaged longer cycles than those from Colorado (Blake, 1972). In one study, captives maintained at 20°C became sluggish, but did not enter torpor, whereas those maintained at 4°C became torpid if they had accumulated sufficient fat (Jameson, 1965). Thus, cold, in itself, was not believed to be sufficient to induce torpor. However, in another study, *S. lateralis* maintained at 0°C and 21°C became torpid, but those kept at 35–38°C did not (Pengelley and Fisher, 1963). Torpor is not continuous, as individuals aroused periodically; at 0°C, no animal remained in torpor for >16 days continuously (Pengelley and Fisher, 1961). Arousals are more frequent near the beginning and end of the season of torpor. In captivity, during arousal small amounts of food are eaten (Jameson, 1965). Because McKeever (1964a) found no evidence of food stored in hibernacula, feeding during arousal may be a behavior restricted to captives.

Also during arousal, these squirrels defecate and urinate, groom, rearrange their bedding, and explore their cages, but most of the time they are in a curled position with dorsum upright and tail extending beneath and beyond the head (Torke and Twente, 1977). Captives from Oregon were torpid less frequently than those from Colorado (Blake, 1972).

At 1,400–1,500 m in northeastern California, adult *S. lateralis* commenced to enter torpor by the end of August; most young did not disappear until early October and the last one was seen on 22 October (Tevis, 1955). In the same region, *S. lateralis* emerged in mid-March (McKeever, 1965).

Because urine is formed only when individuals are in the active homoiothermic state, Pengelley and Fisher (1961) suggested that periodic arousal in *S. lateralis* was induced by the metabolic products accumulated during torpor. However, Muchlinski and Carlisle (1982) presented evidence that the opposite may be true; they found the osmolarity of urine produced during arousal was not significantly different (actually a bit less) from that produced during the summer active period. They suggested that water produced during the metabolism of fat while torpid may dilute body fluids. This, in turn, may induce arousal.

Spermophilus lateralis is territorial and exhibits a linear social hierarchy (Gordon, 1936). Gordon (1943) recorded movements of marked individuals of 122 m and 274 m from dens they occupied; because he provided food for these ground squirrels at bait stations the reported movements may have been affected by the baits. He also indicated that home-range areas for *S. lateralis* ranged from 0.4 ha to 4.1 ha, presumably based on observations of marked individuals; however, no method for estimating the home-range area was described.

Gordon (1943) conducted a variety of experiments with *S. lateralis* in the field by use of mazes and other puzzle devices that required the squirrels to unlock doors to obtain food. The objectives of the research were unclear, the samples were small, and conclusions were limited to indicating that the species was amenable to the method of testing.

Remarks.—We were surprised by the paucity of readily available published information on population density and movements. Initially, we thought that a highly visible, easily captured species such as *S. lateralis* would be an excellent subject species for studies of this type. However, the vigor and persistence exhibited by *S. lateralis* in attempting to enter a food store after the entrance was blocked (Gordon, 1943) suggest that individuals can become habituated easily to a specific site (as a trap) if bait is provided. The examples of long-distance foraging by individual ground squirrels (Gordon, 1943) may indicate that fidelity to home areas is not strong and that baited traps may lure individuals for considerable distances from outside a trapping area.

The *Spermophilus* "*townsendii*" Complex

Of Oregon mammals, this group doubtlessly has the most complex and confusing taxonomic history (Howell, 1938; Scheffer, 1946). A specimen collected in 1836 by J. K. Townsend "on the Columbia River, about 300 miles from the mouth, in July" (Townsend, 1839:316) was deposited in the collection at the Philadelphia Academy of Natural Sciences and was given the name *Spermophilus townsendii* by Bachman (1839b). The mounted specimen with skull inside was long thought to represent the spotted ground squirrels that occur east of the Columbia River in Washington and south of the river in Oregon (Bailey, 1936). The July-caught specimen, probably exceedingly fat in preparation for dormancy, became covered with grease and dirt. Howell (1938) cleaned the specimen by placing it in petroleum ether for a few days. Spotting was not evident on the cleaned skin, so the specimen was declared to be representative of the unspotted ground squirrel that occurs only west of the Columbia River in Washington and the name *Citellus* (=*Spermophilus*) *townsendii* was reallocated to this group (Howell, 1938). The resulting unnamed spotted ground squirrels that occur east of the Columbia River in Washington and in Umatilla, Gilliam, and Morrow counties, Oregon, became known as *Spermophilus washingtoni*. Scheffer (1946) was much upset by the arrangement, seemingly because of the possibility that the specimen was bleached by washing and exposure to light, and because a description by Audubon and Bachman (1854) might be construed to indicate the specimen possessed spots. Nevertheless, the change in taxonomy continues to be recognized (Hoffmann et al., 1993), although to our knowledge, the skull in the specimen collected by Townsend has not been compared with those of spotted and unspotted ground squirrels by a modern nondestructive imaging technique.

In 1863, Kennicott applied the name *Spermophilus* *mollis* to a small, gray, unspotted ground squirrel in Utah that now is known to occur in southern Idaho, extreme southeastern Oregon, extreme eastern California, most of Nevada, and the western one-third of Utah. Thirty-five years later, Merriam (1898a) added *canus* as a subspecies of *S. mollis*; the name was applied to the small, gray unspotted ground squirrels that occur over most of the remainder of the sagebrush desert regions south of the Columbia Basin and Blue Mountains in Oregon and extreme northwestern Nevada. Fifteen years still later, C. H. Merriam (1913) named *vigilis* as a subspecies of *canus,* a taxon apparently elevated to species level at the same time. This taxon was restricted to ground squirrels in a relatively small area near Huntington, Baker Co., and Vale and Ontario, Malheur Co. Although he used the now-invalid generic name *Citellus*, this was the species-level taxonomy for the small, gray unspotted ground squirrels that Bailey (1936) used in his treatment of Oregon mammals.

Two years following Bailey's (1936) publication, Howell (1938) designated *canus*, *mollis*, and *vigilis*, and a taxon restricted to north of the Snake River in Idaho (*artemisiae*) as subspecies of *townsendii*, the species name applied to the small, gray unspotted ground squirrels west of the Columbia River in Washington. Howell (1938) continued to recognize another taxon (*idahoensis*) restricted to an area north of the Snake River in Idaho and named by C. H. Merriam (1913) at the specific level. However, Davis (1939) reduced these squirrels to subspecific level within *mollis*. Hall and Kelson (1959) recognized *townsendii*, *canus*, *mollis*, *vigilis*, *artemisiae*, and *idahoensis* as subspecies of *Spermophilus townsendii*. On the basis of differences in karyotype, Nadler (1968) separated the *S. townsendii* in Washington east and north of the Yakima River (2n = 38, FN = 66) from those west and south of the river (2n = 36, FN = 68); he designated the former a new subspecies (*S. t. nancyae*). Nadler (1968) also showed that karyotypes of some races differed dramatically: *S. t. canus*, 2n = 46, FN = 68; *S. t. vigilis*, 2n = 46, FN = 66; *S. t. mollis* and *S. t. idahoensis*, 2n = 38; FN = 66. Rickart et al. (1985) added the karyotype of *S. t. artemisiae*: 2n = 38, FN = 66. The latter authors found no evidence of intergradation between the chromosomal forms within the *S. townsendii* complex. Nevertheless, Rickart (1987), in his species account, treated the group as one species, although giving consideration to each taxon in instances that published information differed.

Hoffmann et al. (1993) considered the complex to consist of three species: *S. mollis*, including *artemisiae*, *idahoensis*, *mollis*, and *nancyae* (only *mollis* occurs in Oregon); *S. canus*, including *canus* and *vigilis* (both occur in Oregon); and *S. townsendii* (restricted to Washington). We have followed this species-level taxonomy in our treatment of Oregon ground squirrels.

Spermophilus canus Merriam
Merriam's Ground Squirrel
1913. *Citellus canus vigilis* C. H. Merriam, 26:137.

Description.—*Spermophilus canus* (Plate IX) is one of the two small gray ground squirrels without stripes or spots (Table 11-5). As in *S. mollis*, it has a hind foot <39 mm long and the darker dorsal pelage is sharply separated from lighter-colored pelage of the venter high on the sides. The pinnae are inconspicuous. Also as in *S. mollis*, the braincase (Fig. 11-27) is relatively broad and the rostrum is relatively narrow. The tail is short; Howell (1938) claimed that it ranged from 37 to 42 mm for specimens from Oregon and used that character to separate *S. canus* and *S. mollis*. We found the range recorded from labels attached to museum specimens to be somewhat greater (30–61, *n* = 65) and to overlap that of *S. mollis*. Variation in techniques used by collectors to measure tail length may be responsible for the wider range.

Spermophilus canus possesses a karyotype (2n = 46, FN = 68) considerably different from that of *S. mollis* (2n = 38, FN = 66). The two species exhibit no evidence of intergradation (Nadler, 1968; Rickart et al., 1985).

Distribution.—Most of the geographic range of *S. canus* is in Oregon, where it occurs south and east of a line connecting Huntington, Baker Co.; North Powder, Union Co.; Squaw Butte, Wheeler Co.; Maupin, Wasco Co.; Warm Springs, Jefferson Co.; Bend, Deschutes Co.; and Fort Rock, Summer Lake, and Plush, Lake Co. (Fig. 11-28). The species does not occur south of the North Fork Owyhee River in Malheur Co. (Fig. 11-28; Rickart et al., 1985). The species also occurs in extreme northwestern Nevada and in the Snake River plain in west-central Idaho south of the Snake River (Fig. 11-28).

Geographic Variation.—Both subspecies occur in Oregon: the grayer *S. c. canus* over most of the range and the browner *S. c. vigilis* in the Malheur and Snake river plains along the Oregon-Idaho border (Howell, 1938).

Habitat and Density.—Bailey (1936:154) indicated that these ground squirrels occupied the "Upper Sonoran sagebrush plains" on the high ridges between the Deschutes and John Day rivers and southward in the valleys into Nevada. He also stated that the species occurred in sagebrush and in grain and alfalfa fields in the Malheur and Owyhee river valleys along the Oregon-Idaho border. On Malheur National Wildlife Refuge, Harney Co., Feldhamer (1977*b*)

reported catching *S. canus* on grids dominated by big sagebrush (*Artemisia tridentata*) and greasewood (*Sarcobatus vermiculatus*), but not on those classified as marsh and grassland.

West of Terrebonne, Deschutes Co., we observed *S. canus* in pastures, grassland areas, and occasionally on the big sagebrush-western juniper (*Juniperus occidentalis*) scabland area on which populations of *Sylvilagus nuttallii* were studied (McKay and Verts, 1978*a*). One specimen was collected on the scabland area on 28 May 1975 (OSUFW 3125). Also, one of us (BJV) observed a colony of *S. canus* occupying a 1–2-m-high rimrock area east of Christmas Lake, Lake Co. Nearby vegetation was mostly big sagebrush and cheat grass (*Bromus tectorum*); one individual was collected (OSUFW 573).

Diet.—Bailey (1936) claimed that *S. canus* consumed a variety of grasses and forbs, and was especially attracted to green crops such as alfalfa and clover. He also reported that the species consumed ripening grains and insects. Near Huntington, Baker Co., these ground squirrels were observed to feed on flower heads of sunflowers (presumably *Helianthus*), alfilaria (*Erodium*), and various legumes and other composites. Nevertheless, Bailey (1936:156), in referring to *S. c. canus,* indicated that these "squirrels do not come in touch with agriculture" over most of their range,

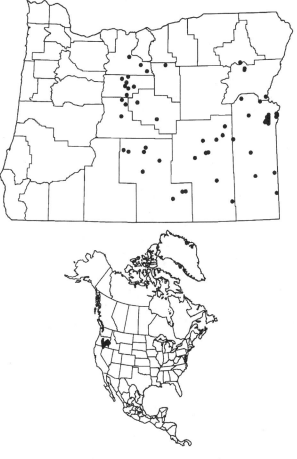

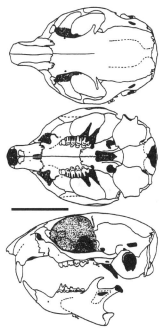

Fig. 11-27. Dorsal, ventral, and lateral views of the cranium, and lateral view of the dentary of Merriam's ground squirrel, *Spermophilus canus* (OSUFW 3125). Scale bar equals 15 mm.

Fig. 11-28. Localities in Oregon from which museum specimens of Merriam's ground squirrel, *Spermophilus canus*, were examined (listed in APPENDIXX, p. 562), and the distribution of Merriam's ground squirrel in North America. (Range map redrawn from Rickart et al., 1985:101, fig. 5, with modifications.)

but reported that *S. c. vigilis* commonly invaded irrigated and cultivated land and caused extensive damage to crops. We know of no more-recent or detailed analysis of the diet of the species.

Maser and Shaver (1976) reported that a *S. canus* with severely maloccluded incisors fed by biting off long blades of grass with its molars.

Reproduction, Ontogeny, and Mortality.—The short period of aboveground activity precludes the possibility of more than one litter annually (Bailey, 1936). Bailey (1936) believed that the litter was born in late April or early May. George Feldhamer, a former graduate student, collected a pregnant female near Harney Lake, Harney Co., on 28 April 1975 (OSUFW 2996) that contained nine embryos with crown-rump lengths of 30 mm. We know of no evaluation of the potential productivity of this species.

Remains of six *S. canus* occurred among 1,112 food items identified in 438 pellets regurgitated by barn owls (*Tyto alba*) near Vale, Malheur Co. (Maser et al., 1980), and one occurred among 187 prey items of great horned owls (*Bubo virginianus*) on the Crooked River National Grasslands, Jefferson Co. (Maser et al., 1970). Likely a variety of raptorial birds and carnivorous mammals prey on *S. canus* (Bailey, 1936).

Habits.—Bailey (1936) reported that these ground squirrels emerged in early March, bred, reared their young, became exceedingly fat, and immerged into their hibernacula by early August. He reported that most had entered their hibernacula by 16 June at Riverside, Malheur Co.

Remarks.—Nadler (1968) and Rickart et al. (1985) examined cytologically specimens collected within areas depicted by Hall and Kelson (1959) and Hall (1981) as ranges of the various described subspecies of *Spermophilus townsendii*. Their findings were the basis for elevating several subspecies to species level (see p. 196).

Originally, we considered specimens collected within the geographic area from which Nadler (1968) obtained specimens with 2n = 46 karyotypes to be *S. canus* and those examined by Rickart et al. (1985) from the region in which they obtained specimens with 2n = 38 karyotypes to be *S. mollis*. We applied discriminant analysis to eight skull characters (Table 11-5) for 65 specimens from the former area and 35 specimens from the latter (six from Oregon, the remainder from Nevada, Idaho, and Utah). One from Idaho and 19 from Utah had been identified cytologically as *S. mollis* (2n = 38). Seven specimens from each a priori group were classified into the opposite group, an overall 86% correct classification. Four of the six a priori Oregon specimens of *S. mollis* grouped with *S. canus*; all were collected 18–58 km from a site at which Rickart et al. (1985) obtained specimens with 2n = 38 karyotypes. The four a priori specimens of *S. mollis* classified as *S. canus* by discriminate analysis were grouped with the latter species for reanalysis, but the a priori specimens of *S. canus* were not regrouped with *S. mollis* as six were from localities at which Nadler (1968) reported collecting specimens with 2n = 46 and one was from a site >270 km from the nearest known specimen with 2n = 38.

By reanalysis of the newly formed groups, we developed an equation that produced 95% correct classification based on skull dimensions of adult specimens to separate *S. canus* and *S. mollis* as follows:

Value = occipitonasal length (0.87257) + nasal length (−0.75297) + length of maxillary toothrow (0.81479) + zygomatic breadth (−0.94707) + breadth of braincase (−1.13649) + breadth of molar toothrow (2.15414) + skull depth (−0.81291) + length of mandible (0.76948) − 13.2225

Adult specimens for which calculated values are <0.85 are *S. canus* and those for which calculated values are >0.85 are *S. mollis*. Despite the high percentage correctly classified, we recommend that some caution be employed in use of the equation, especially with specimens from near Vale and Ontario, Malheur Co.

The paucity of primary research reports dealing with aspects of the biology of this species is appalling. Thus, for a student interested in quantitative natural history, *S. canus* should be a prime candidate for study.

We deemed the vernacular name used for this taxon in the past too likely to be confused with those applied to other Oregon ground squirrels. Therefore, we have coined a new vernacular name to honor the describer of the taxon, Clinton Hart Merriam.

Spermophilus mollis Kennicott
Piute Ground Squirrel

1863. *Spermophilus mollis* Kennicott, 15:157.
1898a. *Spermophilus mollis stephensi* Merriam, 12:69.

Description.—*Spermophilus mollis* is the other of the two small gray ground squirrels without stripes or spots (Table 11-5). As in *S. canus* the hind foot is <39 mm long and the darker dorsal pelage is separated sharply from the lighter ventral pelage high on the sides. The pinnae are inconspicuous. Also as in *S. canus,* the braincase (Fig. 11-29) is relatively broad and the rostrum is relatively narrow. The tail is relatively long. Howell (1938) claimed that it ranged from 44 to 61 mm and used that character to separate *S. mollis* and *S. canus.* We found the range recorded from labels on museum specimens to be somewhat greater

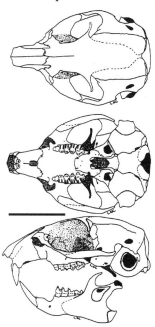

Fig. 11-29. Dorsal, ventral, and lateral views of the cranium, and lateral view of the dentary of the Piute ground squirrel, *Spermophilus mollis* (UU 28285). Scale bar equals 15 mm.

(33–53, $n = 5$) and to overlap that of *S. canus*. Variation in methods used by collectors to measure tail length may be responsible for the wider range.

We know of no external or cranial character that will separate *S. mollis* and *S. canus* reliably. However, karyotypes of the two species differ greatly; *S. mollis* has $2n = 38$, $FN = 66$, whereas *S. canus* has $2n = 46$, $FN = 68$ (Nadler, 1968; Rickart et al., 1985). No individuals with intermediate karyotypes are known. An equation to distinguish between skulls of the two species is presented in Remarks in the account of *S. canus* (see p. 198).

Distribution.—*Spermophilus mollis* occurs in southern Idaho, southeastern Oregon, extreme western California, all but the southern tip of Nevada, and western Utah (Fig. 11-30; Hoffmann et al., 1993). In Oregon, the species occurs south of Sheepshead and Cedar mountains, Malheur Co. (Fig. 11-30; Rickart et al., 1985).

Geographic Variation.—One of the four recognized subspecies occurs in Oregon: *S. m. mollis* (Hall, 1981; Hoffmann et al., 1993; Rickart et al., 1985).

Habitat and Density.—*Spermophilus mollis* is a species of the high desert and commonly occurs in habitats in which the dominant shrub is big sagebrush (*Artemisia tridentata*), saltbush (*Atriplex confertifolia*), or greasewood (*Sarcobatus vermiculatus*—Rickart, 1987, 1988). Cheat

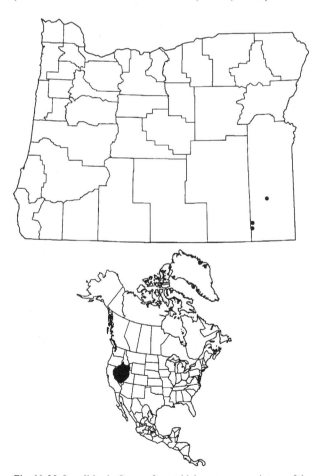

Fig. 11-30. Localities in Oregon from which museum specimens of the Piute ground squirrel, *Spermophilus mollis*, were examined (listed in APPENDIX, p. 564, and the distribution of the Piute ground squirrel in North America. (Range map redrawn from Rickart et al., 1985:101, fig. 5, with modifications.)

grass (*Bromus tectorum*), Sandberg bluegrass (*Poa sandbergii*), rabbit-brush (*Chrysothamnus nauseosus*), Indian ricegrass (*Oryzopsis hymenoides*), and needle-and-thread grass (*Stipa comata*) are common understory plants (Rickart, 1987, 1988).

The species is commonly associated with rocky outcrops, levees, railroad embankments, ditchbanks, and sand dunes (Bailey, 1936; Davis, 1939). Some occur in fencerows and edges of alfalfa and small-grain fields. Where sympatric with *S. beldingi*, *S. mollis* is restricted to the more arid habitats (Durrant and Hansen, 1954).

In southwestern Idaho, densities of adults ranged from 11.7 ± 1.8 to 15.9 ± 1.3/ha during a 4-year study (Smith and Johnson, 1985). For the same data set, mark-recapture estimates (with size of the effective study area adjusted by use of the average and maximum distances moved) ranged from 4.4 ± 0.2 to 7.6 ± 0.4/ha (Johnson et al., 1987). In Nevada, following control operations to reduce ground squirrel populations, Alcorn (1940) counted 120 (296/ha) and 134 (331/ha) dead adults and juveniles on two 0.4-ha plots considered to contain representative populations; however, immigration from surrounding areas may have inflated the estimates of density (Rickart, 1987).

Densities of these ground squirrels are known to undergo extraordinary fluctuations. In desert areas of Utah, populations so dense that before 1936 "they were the most conspicuous mammal observed" were reduced to "a few scattered individuals . . . from scattered localities" in 1937 (R. Hansen, 1956:123). Populations had not recovered 18 years later. Similar die-offs were recorded in Nevada between 1938 and 1939 (Alcorn, 1940) and in California between 1916 and 1917 (Grinnell and Dixon, 1918). In Idaho, numbers of active burrow entrances on study plots declined from 128 in 1982 to 68 in 1988; numbers and stability of burrows were greatest in areas vegetated by winterfat (*Eurotia lanata*) and Sandberg bluegrass, moderate in areas of big sagebrush, and least in shadscale communities (Yensen et al., 1992). Numbers of occupied burrows were correlated negatively with density of cheat grass and other exotic annuals. Changes in vegetation resulting from fire were believed responsible for the decline in numbers of occupied burrows in this region (Yensen et al., 1992).

The following anecdotal account suggests that similar perturbations occur in populations in Oregon. In southern Malheur Co., on 26 June 1982, we observed numerous *S. mollis* and many active burrows along the then unimproved White Horse Ranch road; we collected one specimen (FW 6858). On 19 June 1992, in traveling the same route along the now improved (graveled and ditched) road, we observed none and found no active burrows during several off-road excursions on foot.

Diet.—Based on studies conducted in Idaho (Smith and Johnson, 1985) and Utah (Rickart, 1982), the diet of *S. mollis* consists primarily of green vegetation (mostly grasses) in spring with a shift to seeds of both grasses and forbs later in the season. Insects constitute only a minor component of the diet. In western Idaho, on the Snake River Birds of Prey Area north of the Snake River, 60% of the food ingested consisted of grasses, 21% forbs, 16% leaves of shrubs, and 2% insects; cheat grass provided most of the food eaten (Smith and Johnson, 1985). Also, at four sites in Idaho, diets of *S. mollis* evaluated by microhistological examination of stomach contents consisted of 7–87% grasses (Yensen and Quinney, 1992). Depending on the site, native species (mostly Sandberg bluegrass, winterfat, big sagebrush, and anns, *Vulpia octoflora*) composed 7–96% of the diet and exotic species (mostly cheat grass; Russian thistle, *Salsola iberica*; and tansy, *Descurainia*) composed 4 – 68%. At each site, only two–four species composed >90% of the diet (Yensen and Quinney, 1992).

Results of these studies corroborate the observations of Bailey (1936), who claimed that similar stomach contents probably contained 50–75% water, sufficient to supply the needs of this desert rodent. He also claimed that where this species resides near the edges of alfalfa and small-grain fields, it does considerable damage to crops.

Reproduction, Ontogeny, and Mortality.—Among adult male *S. mollis*, the testes are scrotal upon emergence; females emerge 1–2 weeks after males and commonly breed within 1 week (Rickart, 1982). In 1938 in Nevada, males first emerged on 14 February, females on 3 March; breeding commenced in the 5–20 March interval (Alcorn, 1940). In some populations, most males attain reproductive competence as yearlings, whereas in others few have scrotal testes until they approach 2 years of age (Rickart, 1987; Smith and Johnson, 1985). Among females, 49–92% of yearlings and 97–100% of adults in Idaho (Smith and Johnson, 1985) and 94–95% of yearlings and 100% of adults in Utah (Rickart, 1987, 1988) bred. In Nevada, 90% of all females bred when the population was at a peak, but only 50% during a population decline (Alcorn, 1940). However, during a severe drought (1977), all females, except those residing at the edges of irrigated fields, failed to become reproductively active (Smith and Johnson, 1985). Gestation was estimated at 23–24 days (Alcorn, 1940).

Litter size, based on counts of embryos, may be as great as 16 (Alcorn, 1940; Rickart, 1988); as many as 17 neonates were found in one nest (Alcorn, 1940). Average litter sizes, based on counts of embryos and pigmented sites of implantation, commonly approximated half the maximum number (Table 11-6). Rickart (1988) indicated that yearling females averaged producing 15–19% fewer young than adult females and that average litter size was greater in more mesic habitats. Rickart (1987) also indicated that interuterine losses ranged from 0 to 21% during a 4-year period in Utah.

Neonates are pink, blind, and naked, but have long, sharp claws on the forefeet. In Utah, neonates averaged 3.9 g (range, 2.7–5.2 g; *n* = 21—Rickart, 1986). Growth of young, in comparison with that of other members of the genus, is comparatively slow; more than 40 days are required for young to attain 50% of adult mass (Rickart, 1986). Until weaned, average growth of young is limited by milk supply available to individuals, which, in turn, is related inversely to litter size (Rickart, 1986). However, after weaning and commencing aboveground activity, juvenile *S. mollis* can attain predormancy body mass in ≤2 months (Rickart, 1986).

Survival of juveniles and adult females was greater than that of adult males in Idaho, probably accounting for the shift in sex ratios from favoring males among juveniles to favoring females among yearlings and adults (Smith and Johnson, 1985). Overall, annual survival was 22% for males and 35% for females, but ranged from 16 to 41% on five

study areas (Smith and Johnson, 1985). In two populations in Utah, survivorship was less and fecundity was greater than in the Idaho population (Rickart, 1988). Maximum life span, based on numbers of periosteal adhesion lines in the mandible, was 5 years in Utah (Rickart, 1987, 1988).

On the Snake River Birds of Prey Area in Idaho, *S. mollis* constituted the major prey item of rattlesnakes (*Crotalis viridus*), prairie falcons (*Falco mexicanus*), and badgers (*Taxidea taxus*). Many also were taken by gopher snakes (*Pituophis melanoleucus*), red-tailed hawks (*Buteo jamaicensis*), ferrunginous hawks (*Buteo regalis*), and ravens (*Corvus corax*), and some were taken by weasels (*Mustela frenata*) and coyotes (*Canis latrans*—Smith and Johnson, 1985). Swainson's hawks (*Buteo swainsoni*) and rough-legged hawks (*Buteo lagopus*) also are known to take *S. mollis* (Alcorn, 1940). In Nevada, American Indians used *S. mollis* for food (Alcorn, 1940).

Habits.—These ground squirrels usually are active for 3–5 months each year (Alcorn, 1940; Rickart, 1982); they emerge from their hibernacula in February–early March (Rickart, 1982). Ambient temperatures at emergence commonly are below freezing at night and reach only 4–10°C during the day (Alcorn, 1940). Adult males emerge 1–3 weeks before adult and yearling females (Alcorn, 1940; Rickart, 1982, 1987; Smith and Johnson, 1985); in regions where the testes of yearling males do not become scrotal they emerge after the females have bred (Smith and Johnson, 1985), but where they emerge with adult males they engage in reproductive activities (Rickart, 1982, 1987). Young-of-the-year first emerge from their burrows from late March to early May (Rickart, 1987; Smith and Johnson, 1985). Immergence usually is in June or early July, but emergence for a short period in autumn is known to occur during some years (Alcorn, 1940). The sequence of immergence is adult males, adult females, juvenile females, and juvenile males (Rickart, 1987).

In Utah, adult males commenced to store body fat about 6 weeks after emergence; females commenced to fatten 2–3 weeks later after young were weaned. Stored lipids composed 60–71% of pre-immergent body mass among adults and ≅67% of pre-immergent body mass in juveniles. While dormant an average of 270 days, adult males lost 51% of pre-emergent body mass, adult females 46%. Among juveniles dormant for an average of 240 days, males lost an average of 46% of pre-emergent body mass, females 50% (Rickart, 1982). However, at emergence, fat composed ≅9% of the body mass of adult males, but 20–29% of that of adult females. As an adaptation for the arid environment in which it lives, the species has the greatest urine-concentrating capacity among sciurid rodents (Rickart, 1989).

Although the species is diurnal, most activity is restricted to the cooler morning hours when winds are calm. Much of the active period is devoted to feeding (Alcorn, 1940). *S. mollis* climbs into shrubs to feed on the leaves or to obtain a better view of the surrounding area (Alcorn, 1940). It also can swim well (Davis, 1939).

Burrows of *S. mollis* were classified as those of adults, those of juveniles, and "axillary" burrows (Alcorn, 1940). Burrows occupied by adults were 3–17 m long, 7.5–15.0 cm in diameter, extended to a depth of 60–147 cm, usually had two or three openings, and contained a nest chamber 18–25 cm in diameter. Burrows occupied by juveniles were much less elaborate in that they were 2–5 m long, 7–8 cm in diameter, extended to a depth of 59–79 cm, had only one opening, and had nest chambers 17–20 cm in diameter. "Axillary" burrows were least complex; they were <1 m long, 7–10 cm in diameter, extended only 30–45 cm below ground, possessed only one opening, and had no nest chamber. The latter burrows usually were near good feeding areas and were connected to primary burrows by well-worn trails; they serve as emergency escape tunnels for both adults and juveniles. In response to a potential threat, *S. mollis* "whistles," runs quickly to an axillary burrow, and stands upright on the hind feet (Bailey, 1936; hence, the vernacular name "picket-pin" commonly used to refer to the species) to resurvey the situation; other individuals nearby respond similarly (Alcorn, 1940). Laundré (1989) and Reynolds and Wakkinen (1987) measured characteristics of series of burrows, but did not classify burrows by sex and age of individuals occupying them; the latter authors noted a bimodal distribution of dimensions, suggesting more than one type of burrow. Soils used by *S. mollis* for burrowing tend to be characterized by low percentages of clay; percentages of sand and silt tend to be more variable (Laundré, 1989; Reynolds and Wakkinen, 1987). Burrows of these ground squirrels permit as much as 34% more precipitation to infiltrate the soil and also permit deeper penetration of moisture, thereby reducing evaporative losses (Laundre, 1993).

Although results of detailed studies of social behavior in *S. mollis* have not been published (Rickart, 1987), sufficient observations have been recorded to know that adults establish and defend territories, den alone, engage in direct combat with conspecifics, and engage in infanticide and cannibalism (Alcorn, 1940). These squirrels avoid traps in which individuals in other sex and age groups have been caught (Nydegger and Johnson, 1989). *S. mollis* produces whistlelike alarm calls (Alcorn, 1940), but because its sociality has not been determined precisely (Michener, 1983) it is not known whether calling is an expression of nepotism, as it seemingly is in *S. beldingi* (Sherman, 1977).

In Idaho, home-range areas for nine males and five females captured ≥10 times each averaged 1,357 ± 189.7 m² (Smith and Johnson, 1985). On the same area, the average distance between successive captures (an index to home-range area) was 28.6 ± 1.3 m (*n* = 181 intervals) for adult males, 26.9 ± 4.4 m (*n* = 55) for yearling males, 23.0 ± 2.4 m (*n* = 58) for adult females, and 25.3 ± 2.0 m (*n* = 70) for yearling females; the index for adult males was significantly greater than for other sex and age groups (Nydegger and Johnson, 1989). Overlap of home ranges was significantly more frequent among members of different sex and age groups than among members of the same group; the overlap was extensive between adult males and adult females (Nydegger and Johnson, 1989).

A population estimated at 74/ha reduced to <3/ha by applications of poisoned bait during a 10-day period, increased to about 20/ha 10 days after control measures ceased. Some of the immigrants traveled as far as 400 m (Alcorn, 1940). Rickart (1982, 1988) considered *S. mollis* difficult to catch in live traps; however, Nydegger and Johnson (1989) considered the species easy to livetrap.

Remarks.—Remarks for *Spermophilus canus* contains a discussion of our analyses of morphometric differences between *S. mollis* and *S. canus,* and an equation that seems to be useful in distinguishing between the two species. Of

primary concern to future researchers and conservationists should be the identification of populations with 2n = 38, FN = 66 karyotypes. If the distribution of *S. mollis* based on these investigations remains approximately as presently known, populations should be monitored regularly and appropriate measures taken to ensure that the species remains a component of the mammalian fauna of the state.

Spermophilus washingtoni (Howell)
Washington Ground Squirrel
1938. *Citellus washingtoni* Howell, 56:69.

Description.—*Spermophilus washingtoni* (Plate IX) is the smaller of the two short-tailed, spotted ground squirrels in Oregon (Table 11-5). The dorsum has squarish grayish-white spots about 4 mm across on a background of pale smoky-gray with a pinkish wash to brownish gray. The venter is grayish white and, somewhat similar to the markings of *S. canus* and *S. mollis,* extends high on the sides and is separated sharply from the dorsal pelage.

Of the short-tailed ground squirrels, only *S. columbianus* and *S. beldingi* are sympatric or parapatric with *S. washingtoni. S. washingtoni* may be separated from the former by a hind foot <43 mm long and from the latter by the spotted dorsum (Rickart and Yensen, 1991). The skull of *S. washingtoni* (Fig. 11-31) is similar to that of *S. canus* but tends to be longer and narrower.

Distribution.—The Washington ground squirrel is endemic to the Deschutes-Columbia Plateau Province (Orr et al., 1992) east and south of the Columbia River and east of the John Day River (Hall, 1981; Rickart and Yensen, 1991). The known range (Fig. 11-32), probably contiguous when the region was settled, now consists of three disjunct populations, two in Washington and one in Oregon (Betts, 1990:30, fig. 1). In Oregon, museum speci-

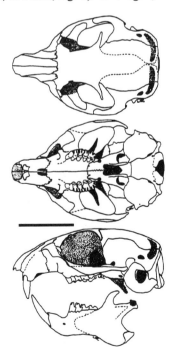

Fig. 11-31. Dorsal, ventral, and lateral views of the cranium (OSUFW 3810), and lateral view of the dentary (OSUFW 3806) of the Washington ground squirrel, *Spermophilus washingtoni.* Scale bar equals 15 mm.

mens have been collected in Gilliam, Morrow, and Umatilla counties from the Columbia River south to Heppner, Morrow Co. (Fig. 11-32). However, in historic times, the geographic range occupied by *S. washingtoni* in Oregon has been reduced considerably (Betts, 1990).

Geographic Variation.—*Spermophilus washingtoni* currently is considered to be monotypic. Dalquest (1948), on finding no difference in size, synonymized the two subspecies, *S. w. washingtoni* and *S. w. loringi*, described by Howell (1938). Subsequent unpublished studies support monotypy (Rickart and Yensen, 1991).

Habitat and Density.—Originally, several of the widespread vegetative associations in the shrub-steppe region occupied by *S. washingtoni* were dominated by big sagebrush (*Artemisia tridentata*) or bluebunch wheatgrass (*Agropyron spicatum*) or both (Franklin and Dyrness, 1969). Grazing, fire, cultivation, and irrigation have removed or altered much of the vegetation of the region. On nonagricultural lands, rabbit-brush (*Chrysothamnus viscidiflorus* and *C. nauseosus*) and cheat grass (*Bromus tectorum*) have replaced much of the original flora (Franklin and Dyrness, 1969).

In an analysis of 23 vegetative and edaphic features in paired occupied and unoccupied *S. washingtoni* colonies at 13 localities, Betts (1990) found significantly greater

Fig. 11-32. Localities in Oregon from which museum specimens of the Washington ground squirrel, *Spermophilus washingtoni*, were examined (listed in Appendix, p. 564), and the distribution of the Washington ground squirrel in North America. (Range map redrawn from Rickart and Yensen, 1991:3, fig. 3.)

values for percent cover of annual grasses, total grasses, and forbs and grasses on occupied sites. Also, soil depth was significantly greater, and soil strength and percent clay were significantly less on occupied sites than on unoccupied sites. The greater depth and less strength of soils at sites occupied by *S. washingtoni* likely is related to requirements for excavation of burrows (Betts, 1990).

On a 3-ha study site ≅24 km south of Boardman, Morrow Co., captures per unit effort (listed in declining order) were greatest in *Agropyron-Artemisia*, *Artemisia-Bromus*, and *Artemisia-Stipa* associations and least in *Artemisia*, *Stipa-Bromus*, and *Agropyron-Bromus* associations (Carlson et al., in litt.). The apparent contribution of shrubs in the shrub-grass associations producing the greatest number of captures per unit effort was not supported by comparison of occupied and unoccupied sites by Betts (1990). However, the moderate number of captures per unit effort in some grass associations, such as (listed in declining order) *Agropyron*, *Bromus*, *Agropyron-Stipa*, and *Stipa* associations (Carlson et al., in litt.), may indicate that the critical features of habitats supporting *S. washingtoni* were not identified in either study.

From early April to the 1st week in July, Carlson et al. (in litt.) captured 86 individual *S. washingtoni* (15 adult males, 11 adult females, 32 juvenile males, and 28 juvenile females) on a 3-ha (64-trap) grid. They provided an estimate of 90 individuals (30/ha) but did not indicate the model used to produce the estimate. Bailey (1936) estimated 124–247/ha and Dalquest (1948) estimated ≥124/ha, but neither author provided a description of methods used to obtain their estimates. Betts (1990) examined 56 sites in Oregon indicated by others to have supported colonies of *S. washingtoni* previously and found evidence of occupancy at only 36; 10 exhibiting no evidence of *S. washingtoni* were occupied 10 years earlier (Carlson et al., in litt.). Betts (1990) classified 23 colonies that he located in Oregon as small, 7 as moderate in size, and 6 as large; however, he provided no quantification for the size categories.

Diet.—In spring, *S. washingtoni* was observed to feed on bluebunch wheatgrass, needle-and-thread grass (*Stipa comata*), Sandberg bluegrass (*Poa sandbergii*), and cheat grass, but in June, as plants became desiccated, the ground squirrels shifted to seeds of the grasses (Carlson et al., in litt.). In Washington, *S. washingtoni* was observed to feed on globemallow (*Sphaeralcea*), wheatgrass (*Agropyron*), plantain (*Plantago*), cheat grass, Indian ricegrass (*Oryzopsis*), tumblemustard (*Sisymbrium*), alfalfa, oats, and wheat (Howell, 1938; T. Scheffer, 1941). Shaw (1921) observed *S. washingtoni* to seek alfilaria (*Erodium*) during the predormant fattening period. Some insects are eaten (Carlson et al., in litt.). Howell (1938) considered *S. washingtoni* to be a serious agricultural pest primarily from its consumption of crops.

Reproduction, Ontogeny, and Mortality.—In Washington, females had already bred and some had given birth by late February; thus, *S. washingtoni* probably breeds in late January or early February (T. Scheffer, 1941). T. Scheffer (1941) claimed that *S. washingtoni* was polygamous. He also claimed that he had never taken a virgin female; thus, females likely first breed as yearlings. The short interval that *S. washingtoni* is active above ground precludes production of more than one litter annually.

Production of young seems to be similar to that recorded for *S. mollis* (Table 11-6).

In Washington, 12 young averaging 38.3 g (range, 24–44 g) taken in late March were active above ground and feeding on green-plant material; they probably were weaned as their stomachs contained no milk curds (T. Scheffer, 1941). In late April, eight young averaged 116 g (range, 89–139 g), and in late May, two young males weighed 175 and 205 g and three young females weighed 147, 159, and 193 g (T. Scheffer, 1941). In Oregon, young first appeared above ground on 5 April (Carlson et al., in litt.).

Remains of *S. washingtoni* were found in nests of golden eagles (*Aquila chrysaetos*), red-tailed hawks (*Buteo jamaicensis*), ferruginous hawks (*Buteo regalis*), and Swainson's hawks (*Buteo swainsoni*) and in stomachs of gopher snakes (*Pituophis melanoleucus*) and rattlesnakes (*Crotalis virdis*). Marsh hawks (*Circus cyaneus*), rough-legged hawks (*Buteo lagopus*), prairie falcons (*Falco mexicanus*), and short-eared owls (*Asio flammeus*) were seen to hunt over colonies of *S. washingtoni* (Carlson et al., in litt.). Badgers (*Taxidea taxus*) excavate large numbers of Washington ground squirrel burrows (Bailey, 1936; Carlson et al., in litt.; T. Scheffer, 1941), thus probably are among the top predators of the species. Coyotes (*Canis latrans*), weasels (*Mustela frenata*), and burrowing owls (*Athene cunicularia*) were observed in or over colonies of *S. washingtoni* and are considered potential predators of the species (Carlson et al., in litt.).

Habits.—*Spermophilus washingtoni* emerges from dormancy in January–early March depending on elevation (Shaw, 1921); males emerge before females (Bailey, 1936; T. Scheffer, 1941). After reproducing and fattening, adults commence to immerge in late May or early June; juveniles remain above ground for another month (Carlson et al., in litt.). A captive individual remained dormant for 244 days (Shaw, 1921). In 1979 in Oregon, the last adult male was captured on 26 May, the last adult female on 1 June, and only one juvenile was captured on 5 July, the last day that trapping was conducted (Carlson et al., in litt.). Vegetation on grazed and weedy colony sites became desiccated 3–4 weeks earlier than at a colony site on ungrazed grassland, prompting early immergence and possibly contributing to lower survival (Carlson et al., in litt.). All activity is during daylight hours and most is during the morning hours (T. Scheffer, 1941).

Although no studies of social behavior are available for *S. washingtoni* (Rickart and Yensen, 1991), the species is colonial (Bailey, 1936). At the approach of a potential threat, *S. washingtoni* produces "a soft, lisping whistle"; other members of the colony respond by standing upright, repeating the whistle, and quickly retiring to their burrows (Howell, 1938:7).

Among juveniles on a 3-ha trapping grid in Morrow Co., 31% of marked males but only 18% of marked females were never recaptured, suggesting greater dispersal among juvenile males, as commonly observed in some other ground squirrel species. However, sex ratios for both adults and juveniles captured on the grid were slightly, but not significantly, male biased (Carlson et al., in litt.).

Recapture rates (average number of recaptures/marked individual) were 1.7 for adult males, 4.25 for juvenile males, 3.9 for adult females, and 4.1 for juvenile females (Carlson et al., in litt.). Averages of greatest distance moved between captures were 239 m for adult males, 159 m for adult females, and 131 m for juveniles of both sexes (Carlson et al., in litt.).

Remarks.—Until 1938, the Washington ground squirrel was known to the scientific community as *Spermophilus* [*Citellus*] *t. townsendii* (Bachman, 1839b; Howell, 1938). See the section titled "The *Spermophilus* 'townsendii' complex" (p. 196) for further details regarding classification of this species.

Some of the more detailed and quantified life-history information on the Washington ground squirrel is contained in an unpublished report based on a study supported by the Student-Oriented Studies Program of the National Science Foundation and conducted on the U.S. Naval Weapons Systems Training Facility near Boardman, Morrow Co., in 1979. The report, dated 12 September 1980, was prepared by L. Carlson, G. Geupel, J. Kjelmyr, J. MacIvor, M. Morton, and N. Shishido of Lewis and Clark College, Portland, Oregon. We made extensive use of the information contained in this report, and in accordance with the style followed herein, we acknowledged information obtained from the report by referencing "Carlson et al., in litt."

Bailey (1936) and Howell (1938) considered *S. washingtoni* to be extremely abundant within its range and a serious agricultural pest. Olterman and Verts (1972) located 35 museum specimens collected in Oregon before 1935, but none during the following 35 years. They also described an unsuccessful search for colonies of *S. washingtoni* in Umatilla and Morrow counties by L. W. Turner in 1971. This information apparently stimulated interest in the species as Rohweder et al. (1979) reported observing and collecting voucher specimens on the U.S. Naval Weapons Systems Training Facility near Boardman, Morrow Co.; these were the first specimens collected in Oregon in >35 years. Subsequently, state and private agencies have expressed some concern about the status of the species. The geographic range of the species has been reduced in historic times in both Washington and Oregon and many colonies in both states are considered moderately or highly vulnerable to extirpation (Betts, 1990). A 2,095-ha Research Natural Area established on the U.S. Naval Weapons Systems Training Facility in 1978 probably is the most secure refuge for the species in the state (Betts, 1990). Because of the possibility of a catastrophe befalling a lone colony, greater security for other colonies is imperative if the likelihood of extirpation is to be reduced.

Sciurus carolinensis Gmelin
Eastern Gray Squirrel

1788. [*Sciurus*] *carolinensis* Gmelin, 1:148.
1815. *Sciurus Pennsylvanica* Ord, 2:292.
1894e. *Sciurus pennsylvanicus*: Rhoads, p. 19.

Description.—*Sciurus carolinensis* (Plate X) is the smallest member of the genus in Oregon (Table 11-7). A specimen from Salem, Marion Co., is yellowish brown to rusty on the head, forelegs, anterior two-thirds of the dorsum, and hind legs. The flanks, hips, and proximal one-third of the tail are steel gray with some scattered yellowish-brown hairs. The distal two-thirds of the tail also has scattered yellowish-brown hairs; the entire tail is bordered with white-tipped hairs; and the underside of the tail is slightly lighter than the dorsal surface. The tail is much narrower and shorter than that of congeners in Oregon. The hairs of the venter are white to the base; the white venter and gray flanks are separated by a 10–20-mm-wide band of yellowish brown. Entirely black and entirely white individuals are known to occur in some parts of the native range; 86% of an introduced population in Vancouver, British Columbia, is the black phase (Robinson and Cowan, 1954). Colormorphs may occur occasionally in Oregon.

The skull (Fig. 11-33) is characteristic of tree squirrels in that the zygomata are untwisted, compressed laterally, and parallel (not converging anteriorly). The skull (and other bones) is whitish, distinguishing it from the skull of *S. niger,* and a peglike P3 usually is present, separating it from skulls of congeners in Oregon. The posterior margin of the nasals does not extend beyond the premaxillary-frontal suture, the hind foot is <70 mm long, and P4 usually is longer than wide, characters that distinguish the species from *S. griseus.* Like congeners, *S. carolinensis* has four toes on the forefeet and five toes on the hind feet; all are equipped with strong, recurved claws.

Distribution.—The native range of the eastern gray squirrel was from southern Saskatchewan southward east of the Great Plains to eastern Texas then eastward to Florida and through parts of southern Canada to Maine (Fig. 11-34; Hall, 1981). *S. carolinensis* was introduced and populations have been established in several western states and provinces, and in Great Britain, South Africa, and Australia (Flyger and Gates, 1982). *S. carolinensis* was introduced into Oregon. We have seen or received reports of the species in parks, campuses, and residential areas in Salem, Marion Co.; Portland and Milwaukie, Multnomah Co.; and Vale, Malheur Co. Individuals were transplanted from one area to another in Salem and from Salem to Sweet Home, Linn Co., in 1971 (Oregon State Game Commission, 1973); we do not know if the transplant to Sweet Home resulted in a viable population.

Geographic Variation.—The population in Salem reportedly originated from stock within the region considered to be occupied by *S. c. pennsylvanicus* (Hall, 1981). We do not know whether all populations in Oregon originated from this stock.

Habitat and Density.—In its native range, *S. carolinensis* tends to occur in, or be more abundant than *S. niger* in, larger woodlots in which community development is more advanced. In Missouri, the former species is more common in bottomland woods, whereas the latter is more abundant in upland woodlots (Schwartz and Schwartz, 1981).

All populations of *S. carolinensis* in Oregon of which we are aware are in urban areas; introduced populations of *S. carolinensis* also did not expand beyond urban areas in Idaho (Larrison and Johnson, 1981) and in British Columbia (Robinson and Cowan, 1954). In an urban-park area in Vancouver, British Columbia, individuals in an introduced

Fig. 11-34. Localities in Oregon from which museum specimens of the eastern gray squirrel, *Sciurus carolinensis,* were examined (listed in Appendix, p. 550), and the distribution of the eastern gray squirrel in North America. (Range map redrawn from Hall, 1981:418, map 279, with modifications.) Introduced populations occur in several western states (Flyger and Gates, 1982).

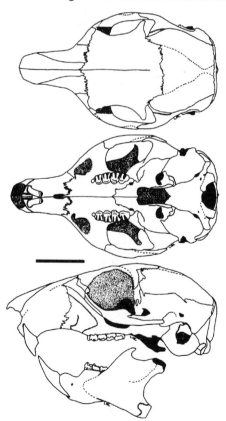

Fig. 11-33. Dorsal, ventral, and lateral views of the cranium, and lateral view of the dentary of the eastern gray squirrel, *Sciurus carolinensis* (OSUFW 2191). Scale bar equals 15 mm.

Table 11-7.—Means (±SE), ranges (in parentheses) and CVs of measurements of skull and skin characters for female and male tree squirrels (Sciurus and Tamiasciurus) from regions in Oregon. Skin characters were recorded from specimen tags; skull characters were measured to the nearest 0.01 mm. SE and CV not provided if n <10.

Species and region	Sex	n	Total length	Tail length	Hind foot length	Ear length	Mass (g)	Greatest length of skull	Nasal length	Length of maxillary toothrow	Zygomatic breadth	Breadth of braincase	Breadth of molariform toothrow	Skull depth	Length of mandible
Sciurus carolinensis	♀	1	413	145	61	32	561.5	61.61	22.35	11.15	34.65	24.55	13.91	26.11	34.8
Sciurus griseus	♀	17	569 ± 4.2[a] (526–594) 2.9	279 ± 2.5[a] (264–301) 3.6	81 ± 1.0[a] (72–92) 5.1	38 ± 0.7[b] (35–41) 5.6	766.5 ± 24.72[b] (635–907) 10.19	66.69 ± 0.28 (64.21–68.60) 1.72	23.79 ± 0.17 (22.40–24.89) 3.01	12.42 ± 0.12 (11.51–13.11) 3.83	37.93 ± 0.23 (35.87–40.15) 2.50	26.95 ± 0.11 (26.33–27.84) 1.68	15.78 ± 0.11 (15.01–16.59) 2.82	28.53 ± 0.11 (27.41–29.33) 1.63	39.80 ± 0.25 (37.95–41.19) 2.56
	♂	30	578 ± 6.98 (490–680) 6.5	275 ± 6.9 (187–400) 13.7	77 ± 1.2 (51–84) 8.2	34 ± 1.1[c] (24–40) 14.6	825.2 ± 20.86[d] (707.3–1032) 10.72	67.12 ± 0.24 (63.13–70.39) 1.98	24.21 ± 0.13 (22.41–25.35) 2.84	12.49 ± 0.09 (11.59–13.53) 3.85	37.94 ± 0.17 (35.88–40.62) 2.50	27.18 ± 0.14 (25.09–28.31) 2.74	15.83 ± 0.08 (14.87–17.07) 2.61	28.99 ± 0.14 (26.68–30.28) 2.59	39.94 ± 0.19 (37.75–42.30) 2.69
Sciurus niger	♀	7	477 (411–565)	225 (198–261)	70 (62–81)	31 (27–39)	654.4[a] (429–870.1)	61.78 (59.03–63.93)	21.03 (19.46–22.12)	11.29 (10.35–12.10)	34.69 (33.12–36.19)	25.05 (24.37–25.86)	14.97 (14.22–15.58)	25.72 (24.71–26.35)	34.45 (33.07–35.83)
	♂	13	482 ± 5.7 (440–507) 4.3	214 ± 3.7 (190–238) 6.2	65 ± 1.7 (50–72) 9.3	29 ± 1.5 (13–33) 17.9	618.3 ± 22.81 (496.8–748) 13.29	61.60 ± 0.51 (58.88–64.60) 3.01	20.75 ± 0.19 (19.80–22.08) 3.32	10.91 ± 0.13 (10.23–11.49) 4.26	34.96 ± 0.19 (34.03–36.35) 1.92	25.12 ± 0.18 (23.94–26.23) 2.54	14.64 ± 0.15 (13.61–15.68) 3.77	25.94 ± 0.19 (25.09–27.53) 2.72	34.77 ± 0.35 (33.07–36.89) 3.65
Tamiasciurus douglasii	♂	30	339 ± 2.4 (306–371) 3.9	135 ± 2.0 (105–170) 8.2	52 ± 0.5[a] (48–57) 4.6	26 ± 0.5[e] (22–29) 7.4	247.9 ± 7.75[f] (200.3–294) 11.27	48.87 ± 0.16 (47.06–50.32) 1.77	15.44 ± 0.10 (14.53–16.50) 3.55	8.79 ± 0.05 (8.10–9.24) 3.22	28.15 ± 0.15 (26.56–29.63) 3.00	21.17 ± 0.07 (20.49–22.07) 1.82	11.18 ± 0.06 (10.52–11.87) 2.71	21.42 ± 0.07 (20.63–22.39) 1.72	27.97 ± 0.12 (26.37–28.95) 2.26
Clatsop to Multnomah and Coos counties	♀	30	323 ± 3.9[a] (287–375) 4.2	131 ± 2.2[a] (100–151) 8.7	51 ± 1.0[g] (45–76) 10.8	22 ± 0.9[h] (13–28) 16.8	229.8 ± 8.15[i] (195.3–314.7) 13.27	47.41 ± 0.18 (45.66–49.39) 2.09	14.46 ± 0.12 (13.19–15.61) 4.54	8.43 ± 0.06 (7.78–8.99) 3.70	27.10 ± 0.13 (25.15–28.40) 2.69	20.41 ± 0.08 (19.75–21.45) 2.02	10.92 ± 0.04 (10.53–11.58) 1.84	21.18 ± 0.08 (20.33–22.40) 1.95	27.11 ± 0.14 (25.34–28.78) 2.84
	♂	30	315 ± 2.7 (285–343) 4.7	126 ± 1.4 (115–142) 6.0	51 ± 0.7 (45–63) 7.0	23 ± 0.6 (17–27) 11.5	213.1 ± 7.05[j] (178.2–261) 11.93	47.27 ± 0.18 (45.20–49.00) 2.11	14.45 ± 0.14 (13.20–16.01) 5.22	8.59 ± 0.06 (7.98–9.31) 3.69	26.87 ± 0.12 (25.84–28.01) 2.44	20.43 ± 0.09 (19.63–21.82) 2.42	10.83 ± 0.07 (10.06–11.55) 3.50	21.31 ± 0.07 (20.59–22.01) 1.89	26.73 ± 0.13 (24.86–28.17) 2.69
A band from Curry to Hood River counties	♀	16	319 ± 4.1 (288–358) 5.2	132 ± 2.2[a] (120–143) 6.3	50 ± 0.6 (47–55) 4.8	24[b] (19–28)	194.8[i] (110–291)	46.59 ± 0.24 (44.47–47.94) 2.03	14.18 ± 0.23 (12.05–15.65) 6.48	8.53 ± 0.13 (7.35–9.50) 5.87	26.62 ± 0.29 (24.04–28.67) 4.31	20.40 ± 0.15 (19.27–21.63) 2.97	10.96 ± 0.09 (10.45–11.71) 3.48	21.20 ± 0.08 (20.50–21.64) 1.58	26.99 ± 0.22 (25.43–28.74) 3.22
	♂	11	326 ± 4.4[a] (303–345) 4.3	133 ± 3.2[a] (120–147) 7.6	52 ± 0.6 (49–55) 3.9	22[j] (19–23)	222.8	48.08 ± 0.34 (46.22–49.71) 2.32	14.83 ± 0.29 (12.95–16.24) 6.41	8.72 ± 0.11 (8.25–9.35) 4.29	27.48 ± 0.27 (26.07–28.90) 3.28	20.62 ± 0.19 (19.48–21.75) 3.09	11.07 ± 0.07 (10.70–11.43) 2.14	21.34 ± 0.14 (20.63–22.03) 2.21	27.45 ± 0.25 (25.95–29.04) 3.06
East of Cascade Range	♀	32	332 ± 2.3 (295–358) 3.9	129 ± 1.6[a] (112–148) 6.7	52 ± 0.3 (46–55) 3.6	27 ± 0.8[k] (22–32) 11.7	246.7 ± 5.83[k] (213–287) 8.18	48.25 ± 0.13 (46.84–49.47) 1.55	14.60 ± 0.11 (13.64–15.89) 4.14	8.47 ± 0.04 (8.13–9.07) 2.49	27.79 ± 0.09 (26.76–29.21) 1.82	20.74 ± 0.07 (19.95–21.67) 2.04	11.01 ± 0.05 (10.52–11.48) 2.56	21.40 ± 0.07 (20.63–22.16) 1.72	27.73 ± 0.11 (26.55–28.77) 2.17
	♂	30	333 ± 2.4 (297–355) 3.9	133 ± 1.6 (111–150) 6.6	52 ± 0.4 (46–57) 3.9	27 ± 0.6 (20–33) 13.2	242.5 ± 2.75[l] (227.1–278) 5.21	48.46 ± 0.17 (45.35–49.82) 1.91	14.81 ± 0.16 (12.77–16.70) 5.79	8.56 ± 0.05 (8.01–9.29) 3.48	27.75 ± 0.17 (24.28–29.80) 3.43	20.89 ± 0.09 (19.81–22.35) 2.37	10.98 ± 0.06 (10.44–11.54) 2.81	21.59 ± 0.06 (20.89–22.15) 1.52	27.93 ± 0.16 (25.68–29.28) 3.15
Tamiasciurus hudsonicus	♀	30	329 ± 2.0 (305–354) 3.3	132 ± 1.8 (117–165) 7.5	51 ± 0.6 (38–55) 6.2	24[m] (15–28)	255.7[n] (236.9–295.5)	48.23 ± 0.22 (45.46–51.34) 2.46	15.04 ± 0.12 (13.78–16.37) 4.24	8.77 ± 0.04 (8.31–9.25) 2.51	27.99 ± 0.16 (25.01–29.15) 3.10	21.01 ± 0.08 (20.15–21.99) 2.19	11.17 ± 0.06 (10.41–11.71) 2.76	21.21 ± 0.09 (19.43–22.23) 2.30	27.88 ± 0.14 (26.04–29.31) 2.68

[a]Sample size reduced by 1. [b]Sample size reduced by 7. [c]Sample size reduced by 12. [d]Sample size reduced by 10. [e]Sample size reduced by 16. [f]Sample size reduced by 17. [g]Sample size reduced by 14. [h]Sample size reduced by 2. [i]Sample size reduced by 13. [j]Sample size reduced by 8. [k]Sample size reduced by 20. [l]Sample size reduced by 9. [m]Sample size reduced by 22. [n]Sample size reduced by 25.

population of *S. carolinensis* were associated largely with deciduous and mixed-forest areas; except where supported by unnatural food sources, they avoided coniferous forests (Robinson and Cowan, 1954). Densities on a 24-ha study area ranged from 1.7/ha in spring to 2.2/ha in autumn.

Diet.—In their native range, *S. carolinensis* and *S. niger*, especially where they are syntopic, consume essentially identical food items, with nuts, acorns, fungi, samaras, berries, buds, leaves, bark, and a few insects composing most of the diet (Nixon et al., 1968). Both species rarely consume more than three or four food items at a time (Brown and Yeager, 1945; Nixon et al., 1968), but shift their diet to take advantage of seasonal abundances of various foods. In the native ranges of both species, food-producing plants (especially nut-producing trees) commonly do not produce crops that can be used by squirrels each year. Thus, gray squirrels tend to be opportunists, taking advantage of available nuts, acorns, and other foods even though those being eaten might not be selected if other species were available (Nixon et al., 1968).

Eastern gray squirrels in an introduced population in Vancouver, British Columbia, foraged primarily on the buds, leaves, flowers, and samaras of vine maple (*Acer circinatum*) and big-leaf maple (*A. macrophyllum*), with vegetative and reproductive parts of a variety of shrubs and trees contributing a small amount to the diet. Unnatural foods fed to squirrels by humans contributed significantly to their diet (Robinson and Cowan, 1954).

Reproduction, Ontogeny, and Mortality.—In its native range, *S. carolinensis* commonly produces two litters annually; the first breeding season usually occurs in January–March, the second in July–October depending on latitude. The gestation period averages ≅44 days (Hayssen et al., 1993). Litter size, based on number of embryos, averaged from 2.5 to 4.0 for several studies (Hayssen et al., 1993); the maximum reported was eight young found in a nest (Barkalow, 1967). In an urban-park situation in Vancouver, British Columbia, *S. carolinensis* produced two litters (born 9 June–6 July and late July to late August) averaging 1.5 weanling young/litter (*n* = 11—Robinson and Cowan, 1954). We have no information regarding reproduction in *S. carolinensis* in Oregon.

Neonates weigh 13–17 g, and at 1 day of age have body lengths of 78–80 mm. At 1 week they weigh 26–35 g; at 2 weeks, 36–45 g; at 3 weeks, 46–55 g; at 4 weeks 56–80 g; at 5 weeks, 81–108 g; at 6 weeks, 108–130 g; and at 7 weeks, 131–150 g (Uhlig, 1955). Young are weaned at 7–10 weeks of age. A few individuals attain sexual maturity at 4–6 months of age, but most mate for the first time at 10–14 months (Hayssen et al., 1993).

Habits.—Eastern gray squirrels are active throughout the year, but may remain in dreys during extremely cold periods and may spend many hours with belly flat on a limb and with legs and tail draped alongside during extremely warm periods. Activity is restricted to daylight periods; in Illinois, activity by eastern gray squirrels may commence as early as 0400 h. Activity is greatest in the morning, peaking at 0600–0700 h; there is a second period of activity in the afternoon, peaking about 1700–1800 h (Brown and Yeager, 1945). In their native range, we have observed eastern gray squirrels to cross busy thoroughfares by climbing utility poles and "wire-walking" the thin cable to avoid the hazards of automobile traffic.

Like its congeners, *S. carolinensis* occupies cavity nests in trees and dreys; commonly, two or more nests or dreys are used concurrently (Brown and Yeager, 1945). In Vancouver, British Columbia, 20 (76.9%) of 26 dreys were built in large western hemlock (*Tsuga heterophylla*), 4 (15.4%) in cedar (*Thuja plicata*), and 1 (3.8%) each in big-leaf maple and Douglas-fir (*Pseudotsuga menziesii*) trees (Robinson and Cowan, 1954). These squirrels were believed to build fewer dreys than those in the native range, perhaps because of the abundance of tree dens.

In native regions, home ranges of eastern gray squirrels usually average <1 ha, rarely as much as 2 ha (Flyger and Gates, 1982); however, home-range size in an introduced population in Vancouver, British Columbia, was 20–22 ha for males and 2–6 ha for females (Robinson and Cowan, 1954). Adult female eastern gray squirrels often share their home ranges with their offspring (Nixon et al., 1986). Interactions among juvenile eastern gray squirrels are affected by their kinship; amicable behavior essentially is always among littermates, whereas interactions among nonlittermates usually are agonistic (Koprowski, 1993*c*).

Both sexes and all age-classes of free-living eastern gray squirrels engage in scent-marking by cheek-rubbing on limbs as they move through the trees or more deliberately at traditional marking places on the underside of limbs or at the base of trees. Typically at traditional marking places, the bark is gnawed without ingestion of particles, then the cheeks and chin are rubbed over the area vigorously; sometimes the area is urinated upon (Koprowski, 1993*a*). Gray squirrels do not mark as frequently as fox squirrels.

During the 1 day of estrus, a female eastern gray squirrel is surrounded by four–18 males; some males fight to be close to the female whereas other males simply congregate within the home range of the estrous female and are attentive to her activities (Koprowski, 1993*d*). Gray squirrels usually copulate in trees; other pursuers and satellite males commonly attack a copulating pair, biting and sometimes knocking both animals from the trees. Females copulate with one to eight males during an estrous period, but rarely with the same male twice (Koprowski, 1993*d*). Active pursuit is the tactic of older, dominant males whereas younger, subordinate males tend to be satellites; some males 2.75–3.25 years old use both tactics within a breeding season, suggesting that this age is the switching point (Koprowski, 1993*d*). Active pursuit is the most successful mating tactic, but females also copulate with satellite males (Koprowski, 1993*d*).

Remarks.—Governor Ben Olcott (1919–1923) reportedly was responsible for having 48 *S. carolinensis* shipped from Pennsylvania in November 1919; about 20 survived the trip and were released on the capitol grounds in Salem (Hattan, 1976; Oregon State Game Commission, 1973; S. L. Auman, *Oregon Statesman*, Salem, Oregon, 1 April 1970). In 1971, 21 nest boxes were erected on the capitol grounds and nearby areas; later the same year all boxes exhibited evidence of having been used, with 10 containing nests (Oregon State Game Commission, 1973).

A few years ago, an employee of the City of Salem Parks Department brought us an oak (*Quercus*) branch from which much of the bark was stripped from the terminal portion by eastern gray squirrels. He indicated that extensive damage to oaks by these squirrels was common in summer in parks. Although bark-stripping to mark territories cannot be ruled out, much evidence supports the hypothesis that food shortage is responsible for bark-stripping. Most bark-stripping occurs in late spring and summer after buds, shoots, and flowers are no longer available and before nuts and acorns become available in autumn. Also, squirrels enter live traps readily during periods when trees are stripped, but are less easily trapped when food is abundant; tree-stripping is particularly prevalent in poorer habitats (Kenward, 1983).

Although this alien species seems to have remained within urban boundaries in Oregon, and both the damage for which it may be responsible and its interspecific interactions with native species do not seem particularly severe, we find introduced species aesthetically unpalatable. We have difficulty condoning efforts to enhance or to extend populations of introduced species, such as transplanting individuals or providing nest boxes and food for eastern gray squirrels, when native species such as the western gray squirrel (*Sciurus griseus*) are given little attention.

Sciurus griseus Ord
Western Gray Squirrel

1818. *Sciurus griseus* Ord, 87:152.
1841. *Sciurus leporinus* Audubon and Bachman, 1:101.
1848. *Sciurus fossor* Peale, 8:55.

Description.—*Sciurus griseus* (Plate X) is the largest tree squirrel in the state (Table 11-7). It has the typical characteristics of sciurids and those of the genus, but it is distinguishable from congeners by a hind foot >70 mm long, large ears without tufts, a bright silvery-gray pelage on the dorsum and a pure white venter, and a long, plumose tail (Carraway and Verts, 1994*b*). It has plantigrade, pentadactyl feet with heavy, strongly curved claws on the four functional toes of the forefeet and the five toes of the hind feet; the pollex is rudimentary and equipped with a nail.

The skull (Fig. 11-35) is large (usually 65–70 mm long—Ingles, 1965), but dimensions overlap those of congeners. The skull can be distinguished from that of *S. niger* by the absence of red pigments in the bone and from that of *S. carolinensis* by P4 usually being wider than long and the posterior margin of the nasals extending well beyond the premaxillary-frontal suture. P3 is present.

Distribution.—The western gray squirrel occurs from Lake Chelan and the south edge of Puget Sound, Washington, south in the Cascade Range of Oregon and Sierra Nevada and minor mountain ranges of California to near the United States-Mexico border, then west to the foothills of the Coast Range in Oregon and, except for the Central Valley, to the Pacific Ocean in California (Fig. 11-36; Hall, 1981). In Oregon, *S. griseus* occurs from central Wasco, Jefferson, Deschutes, and Klamath counties west, except for unforested portions of the Willamette Valley, to central Washington, Benton, Lane, Douglas, Coos, and Curry counties (Fig. 11-36). We have unverified reports of the species occurring in the Freemont National Forest in western Lake Co.

A Pleistocene fossil is known from California (Kellogg, 1912).

Geographic Variation.—Only one of the three nominal subspecies of the western gray squirrel occurs in Oregon: *S. g. griseus* (Hall, 1981).

Habitat and Density.—The western gray squirrel west of the Cascade Range commonly is associated with mixed-forest communities in which Oregon white oak (*Quercus garryana*) is a major feature. Dalquest (1948) considered the geographic distribution of *S. griseus* in Washington to be determined largely by the distribution of this oak. However, Bailey (1936:117) considered that *S. griseus* was "most common in the interior valleys with the oaks, maples, yellow pines, and sugar pines." In Deschutes and Jefferson counties we have observed these squirrels in ponderosa pine (*Pinus ponderosa*)-manzanita (*Arctostaphylos*) communities. In California, in addition to several species of oaks, black walnut (*Juglans hindsii*) commonly seems to be a predominant feature of habitats supporting western gray squirrels (Ingles, 1947).

In southern Oregon, Cross (1969) recorded 12–13 *S. griseus* on a 1.8-ha study area during 2 of 3 years that sampling was conducted repeatedly; during the 3rd year, only four of the eight individuals caught thereon used the area regularly. Densities as high as 4.3/ha were recorded in California (Ingles, 1947), and unquantified eruptions were reported in the Sierra Nevada Range several times during the 20th century (Carraway and Verts, 1994*b*).

Population density of *S. griseus* in Oregon seems to fluctuate dramatically. In southern Oregon, an index to density (number seen/distance traveled) obtained along established routes ranged from 0.05 to 0.44/km with three peaks and

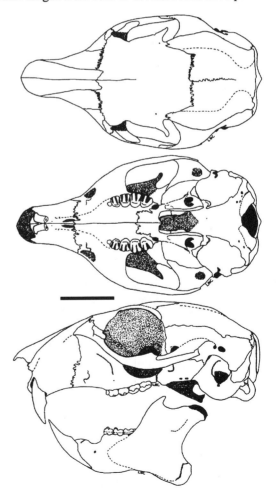

Fig. 11-35. Dorsal, ventral, and lateral views of the cranium, and lateral view of the dentary of the western gray squirrel, *Sciurus griseus* (OSUFW 2138). Scale bar equals 15 mm.

three lows during 1960–1968 (Cross, 1969). A similar trend is evident in the harvest/unit effort, with four peaks and three lows during 1969–1981 (Table 11-8). Dramatic changes in density in California usually were attributed to outbreaks of scabies or mange mites (*Notoedres*) following eruptions (Carraway and Verts, 1994*b*). Cross (1969) suggested that some *S. griseus* with skin lesions may be infected with ringworm (*Trichophyton rubrum*) instead of mange mites.

Diet.—Based on two studies in California (Stienecker, 1977; Stienecker and Browning, 1970) and one in Oregon (Cross, 1969) involving 689 specimens, hypogeous fungi (especially *Rhizopogon*) contributed most to the diet of *S. griseus*. Nevertheless, the same authorities considered mast (fruits of California bay, *Umbellularia californica*; acorns; and pine nuts) to be the critical food for the species. Other foods known to be eaten by *S. griseus* in Oregon include berries, buds, bark, sap, and nuts, and seeds of conifers (Douglas-fir, *Pseudotsuga menziesii*; subalpine fir, *Abies lasiocarpa*; sitka spruce, *Picea sitchensis*; hemlock, *Tsuga*; and other pines, *Pinus*—Bailey, 1936). In Washington in spring, *S. griseus* in parks commonly strips bark from Douglas-fir, ponderosa pine, and maple trees (presumably *Acer macrophyllum*) to obtain cambium (Bowles, 1921; Scheffer, 1952). Removal of bark commonly is sufficient to girdle tops or limbs of trees, thereby resulting in their death.

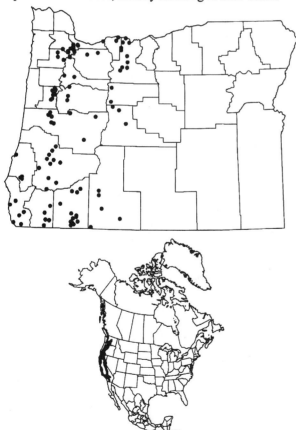

Fig. 11-36. Localities in Oregon from which museum specimens of the western gray squirrel, *Sciurus griseus*, were examined (listed in APPENDIX, p. 550), and the distribution of the western gray squirrel in North America. (Range map redrawn from Hall, 1981:434, map 288, with modifications.)

In southern Oregon, stomach contents of 36 of 59 individuals contained fungi, fungi were eaten in 10 of 12 months, and fungi composed 12–62% of the stomach contents of squirrels collected during those 10 months (Cross, 1969). Also, nuts and acorns occurred in stomach contents of squirrels collected during all months and composed 38–100% of the contents. In general, fungi constituted a greater proportion of the stomach contents and were eaten by a larger proportion of squirrels collected in April, May, and June; conifer nuts ranked highest in July–August; and acorns were high in September–October (Cross, 1969). Insect parts were found in one individual (Cross, 1969); in California, *S. griseus* is known to eat aphids (*Aphis*) that cause leaf-roll in Oregon ash (*Fraxinus latifolia*—Ingles, 1947); thus, insectivory may not be entirely coincidental to ingestion of plant materials.

Reproduction, Ontogeny, and Mortality.—Little information regarding these topics is available for *S. griseus* in Oregon. In southern Oregon, Cross (1969) collected pregnant females in late March and recorded the birth of young in mid-March to a female held captive for 1 month. We collected a pregnant female on 31 March with three 16-mm-long embryos and examined a road-killed female obtained on 16 November with two dark and three light pigmented implantation sites; both were from Benton Co.

In California, young are born from February to July after a gestation of >43 days (Table 11-4); age structure of young-of-the-year commonly indicates two peaks of breeding (W. C. Asserson III, in litt.; Fletcher, 1963; Swift, 1977). Although collection of lactating females in October suggests the possibility of two litters annually, there are no published records of second pregnancies in *S. griseus* unrelated to intrauterine loss of an earlier litter (Carraway and Verts, 1994*b*).

At four sites in California, average litter size based on counts of embryos ranged from 2.2 to 3.0 (Table 11-4); individual litters usually range from two to four (Carraway and Verts, 1994*b*). There is some indication that younger females produce fewer offspring (Swift, 1977).

Young are thought to be born in tree-hole nests, but are moved to dreys when space limitations or ectoparasites become intolerable (Ingles, 1947). Young ≅1 week old weighed 74.6 g and 80.4 g, were 205 mm and 225 mm long, and had tails 93 and 100 mm long. The eyes were not open, the skin was loose and wrinkled, and hair was present only on the dorsum (Storer, 1922). Another individual, also thought to be ≅1 week old, weighed 55 g and had 1–3-mm-long fur on the dorsum, head, and tail (Swift, 1977).

House cats, red-tailed hawks (*Buteo jamaicensis*), golden eagles (*Aguila chrysaetos*), and coyotes (*Canis latrans*) are known predators of *S. griseus*, and goshawks (*Accipiter gentilis*), great horned owls (*Bubo virginianus*), gray foxes (*Urocyon cinereoargenteus*), martens (*Martes americana*), and fishers (*M. pennanti*) are thought to prey upon it (Carraway and Verts, 1994*b*). In urban and populated rural areas, accidents involving automobiles contribute considerably to mortality in local populations of western gray squirrels.

In 1969, the Oregon State Game Commission (*Game Division Annual Report*, 1969) declared that "Although the silver gray squirrel is classified as a game species, there is little interest in hunting the species." However, at that time an estimated 5,400 hunters killed an estimated 21,760 gray squirrels; most were considered to have been harvested by hunters whose primary quarry was game birds or big game. By 1981, the last year for which data are available, more than twice as many hunters killed nearly 2.5 times as many squirrels and the area considered to be inhabited by *S. griseus* declined by >28% (Table 11-8).

Table 11-8.—*Harvest statistics[a] for the western gray squirrel* (Sciurus griseus) *in Oregon, 1969–1981.*

Year[b]	Inhabited area (km²)	Number hunters	Days hunted ($\overline{X}$)	Number harvested	Number harvested /km²	Number harvested/ 100 hunter days
1969	40,575	5,400	4.34	21,760	0.536	92.8
1970	40,575	6,200	5.05	26,730	0.659	85.4
1971	40,575	3,990	4.36	12,460	0.307	71.6
1972	40,575	5,820	3.41	19,852	0.489	100.0
1973	38,575	4,448	5.28	19,282	0.500	82.1
1974	38,575	4,892	4.34	13,173	0.341	62.0
1975	38,575	4,933	6.55	15,864	0.411	49.1
1976	38,575	6,931	4.71	27,846	0.722	82.3
1977	38,575	7,364	5.88	38,470	0.997	88.2
1979	29,091	10,065	5.59	41,610	1.430	74.0
1980	29,091	8,084	6.18	26,812	0.922	53.7
1981	29,091	11,964	5.24	50,524	1.737	80.6

[a]Data obtained from Oregon State Game Commission, *Game Division Annual Reports.*
[b]No data available for 1978.

Habits.—Western gray squirrels are largely aboreal, although overland travel between trees and foraging on the ground are common. They are wary and secretive, but curious; they examine new objects placed in their environment but, once captured, they often avoid live traps (Cross, 1969). However, the western gray squirrel never seems to become as street-wise as its eastern urban relatives by using utility poles and cables to cross busy thoroughfares.

Dreys constructed and used by *S. griseus* are multilayered with the largest sticks on the outside and the finer and softer material as lining for the chamber (Cross, 1969; Merriam, 1930). In southern Oregon, dreys were observed only in coniferous trees (Cross, 1969), but in California, they are constructed in both coniferous and deciduous trees (Grinnell and Storer, 1924; Ingles, 1947; Merriam, 1930; Sumner and Dixon, 1953). Although dreys are used by *S. griseus* for rearing young, cavities created by woodpeckers (Picidae) commonly are used for sleeping at other seasons (Ingles, 1947).

Sciurus griseus is active at all seasons and exhibits diurnal activity almost exclusively. Activity commences near civil sunrise, peaks ≅1–2 h later, declines by midday when only 60% are active at any time, remains relatively stable the remainder of the afternoon, and declines again commencing ≅3 h before civil sunset; by sunset, >90% have retired to dreys or cavities (Cross, 1969). Upon becoming active, *S. griseus* usually grooms for 3–15 min, then commences to explore, either by making short hops along the ground or by traveling rapidly through the trees (Cross, 1969). While abroad, western gray squirrels commonly spend long periods resting; they may sit on a limb with the tail in an S-curve over the back, lie on a limb with the head elevated and the tail in an S-curve over the back, or lie with belly and chin flattened on a limb and legs and tail hanging alongside the limb (Cross, 1969; Ingles, 1947).

Activity tends to be greatest on cloudy days when wind velocity is low. Although rain does not seem to reduce activity (Ingles, 1947), western gray squirrels may remain in their nests on stormy days (Grinnell and Storer, 1924).

While foraging, western gray squirrels hop along with nose close to the ground; they frequently stop to dig at the surface. They also cut cones and allow them to fall to the ground, then recover and carry them to a log or tree to remove the seeds (Grinnell and Storer, 1924). Acorns and seeds not eaten immediately are scatter-hoarded, then relocated at a later date by olfaction (Cross, 1969; Ingles, 1947); defective seeds may be detected by touch, then discarded (Cross, 1969).

In southern Oregon, convex-polygon home ranges based on livetrapping, direct observation, and radiotelemetered locations for four males and four females in February–April covered 0.79–3.49 ha, whereas for 12 individuals of mixed sex and age in July–August they covered 1.75–5.56 ha (Cross, 1969). Cross (1969) claimed that home-range areas differed significantly within sex and age groups, but Carraway and Verts (1994b) suggested that the data were not appropriate for analyses necessary to support these conclusions.

Linear-right social hierarchies (not absolutely fixed or stable straight-line pecking order) with few reversals tend to become established in populations of *S. griseus*; within age groups, males are dominant to females (Cross, 1969). Although chases are the most commonly observed agonistic behavior, threat postures, foot-stamping, tail-flicking, and tooth-chattering also are used in intraspecific interactions. Threat postures are characterized by an extended head and neck, elevated rump, raised dorsal pelage, and tail held parallel to the dorsum. Actual combat is rare, but when it occurs, it may be sufficiently fierce to produce injury (Cross, 1969; Ingles, 1947). Subordination may be characterized by running, resumption of feeding, and use of alternate routes of travel (Cross, 1969). Subordinate individuals may be forced to wait to feed at good food sources until dominant individuals have fed (Cross, 1969).

In California, *S. griseus* must compete with California ground squirrels (*Spermophilus beecheyi*) and introduced tree squirrels (*Sciurus niger* and *S. carolinensis*) for food (Ingles, 1947). All three species occur within the range of *S. griseus* in Oregon; interspecific interactions with the ground squirrel were thought to be responsible for compensating inverse changes in density of populations of the two species (Cross, 1969). Also, in 1969, a former student, V. P. McCrow, indicated (pers. comm.) that *S. niger* was replacing *S. griseus* as the primary sciurid in nut orchards where he hunted squirrels in Washington Co. Interspecific interactions between *S. griseus* and acorn woodpeckers (*Balanosphyra formicivora*), scrub jays (*Aphelocoma coerulescens*), and Steller's jays (*Cyanocitta stelleri*) are known to occur, with acorn woodpeckers commonly driving the squirrels away from food stores (Cross, 1969; Ingles, 1947).

Remarks.—The paucity of information regarding the biology of the western gray squirrel in Oregon—a species so highly visible, so easily attracted with food, and so subject to live capture, and a species with a regulated annual sport harvest—is inexcusable. On two occasions failure to obtain funding prevented one of us (BJV) from initiating studies of the population biology of *S. griseus* in Oregon.

Sciurus niger Linnaeus
Eastern Fox Squirrel

1758. [*Sciurus*] *niger* Linnaeus, 1:64.

Description.—The eastern fox squirrel (Plate X) is among the larger of the tree squirrels in Oregon (Table 11-7). Its body is the typical squirrel shape. The ears are short, rounded, and without tufts; the tail is flattened dorsoventrally and somewhat fluffy; the claws on all feet are strong and recurved. The feet are plantigrade with four toes on the forefeet and five toes on the hind feet. There are four pairs of mammae. Although all-black individuals and grayish individuals with black heads occur in some parts of the natural range, all museum specimens from Oregon and all those that we have seen in the state have a grizzled rusty-brown dorsum with an orangish venter.

The skull of *S. niger* (Fig. 11-37) can be distinguished from that of *S. griseus* and most *S. carolinensis* by the absence of P3 and from that of *S. carolinensis* by the posterior margin of the nasals usually extending well beyond the premaxillary-frontal suture.

Sciurus niger is unique in that its bones are red or pink and fluoresce bright red under ultraviolet light (Turner, 1937) because of presence of uroporphyrin I resulting from a deficiency of the enzyme uroporphyrinogen III cosynthetase in the blood and tissues (Levin and Flyger, 1971). Thus, the skull is easily separable from those of congeners by color alone. Even the urine is red at times (Turner, 1937).

Distribution.—*Sciurus niger* is native to the region from extreme southern Manitoba, Canada, to northern Coahuila and from south-central Montana, eastern Colorado, and west-central Texas eastward through the United States except for the New England region (Fig. 11-38; Hall, 1981). Introduced populations are established in several western states (Flyger and Gates, 1982).

We examined museum specimens from Multnomah, Washington, Marion, Lane, Union, Clackamas, Yamhill, and Baker counties, Oregon (Fig. 11-38) and have unverified reports of the species occurring in Polk Co.

Geographic Variation.—We were unable to locate a written record of the source of stock or the site of first introduction into Oregon, or to find present or former wildlife-agency personnel with knowledge thereof.

Habitat and Density.—All populations of *S. niger* in Oregon of which we are aware occur in urban areas or in association with nut orchards. In California, where the species also has been introduced into urban areas, *S. niger* usually remained near the site of introduction, but sometimes invaded the countryside where it nested in eucalyptus (*Eucalyptus*) windbreaks (Wolf and Roest, 1971). In these windbreaks, the number of eastern fox squirrels

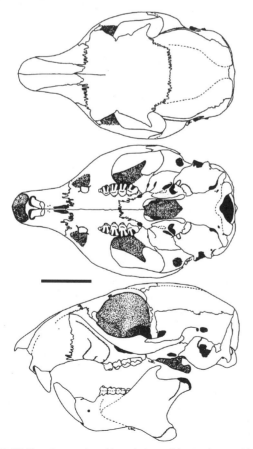

Fig. 11-37. Dorsal, ventral, and lateral views of the cranium, and lateral view of the dentary of the eastern fox squirrel, *Sciurus niger* (OSUFW 663). Scale bar equals 15 mm.

Fig. 11-38. Localities in Oregon from which museum specimens of the eastern fox squirrel, *Sciurus niger*, were examined (listed in APPENDIX, p. 551), and the distribution of the eastern fox squirrel in North America. (Range map redrawn from Hall, 1981:428, map 285, with modifications.) Introduced populations occur in several western states (Flyger and Gates, 1982).

observed averaged 0.93/100 m (range, 0.1–2.2/100 m). No indices or estimates of density are available for fox squirrel populations in Oregon. Wolf and Roest (1971) believed that the fox squirrel in California may occupy oak (*Quercus*) woodland and riparian woods in the future; however, neither Ingles (1965) nor Jameson and Peeters (1988) indicated that such habitats had been invaded.

In its native range, the fox squirrel usually is associated with small woodlots of hardwood (particularly oak and hickory, *Carya*), riparian hardwood zones, and sometimes mixed pine (*Pinus*) and hardwood woodlands (Flyger and Gates, 1982). Fox squirrels can occupy areas with remarkably few trees, particularly where other sources of food are available; on a 2.6-km² area in northwestern Illinois on which about half the land was planted to corn and that supported only two trees other than the several in each of the lawns of three farmsteads, one of us (BJV) commonly observed fox squirrels. Thus, following settlement with the opening of the eastern deciduous forest for agriculture, fox squirrels increased in abundance; with settlement of the prairie and planting of Osage orange (*Maclura pomifera*) fencerows the fox squirrel extended its range far to the west. Land permitted to revert to forest and protected from grazing soon becomes unsuitable for fox squirrels; they are replaced by *S. carolinensis* in such situations (Flyger and Gates, 1982). However, removal of hardwoods and reforestation with pure stands of pine produces habitat unsuitable for *S. niger*.

Diet.—No study of the diet of the eastern fox squirrel in Oregon has been conducted, but from complaints of depredations we know that it consumes filberts, walnuts, and the bark of terminal branches of some trees. In California, introduced fox squirrels consume walnuts, oranges, avocados, strawberries, and tomatoes where these crops are grown near eucalyptus windbreaks (Wolf and Roest, 1971). We have no knowledge of depredations on truck crops in Oregon. In its native range, staples in the fox squirrel diet are nuts, acorns, corn, and soybeans, but buds, sap, seeds, berries, leaves, insect larvae, and many other items are consumed (Flyger and Gates, 1982). However, within a local area, squirrels usually are dependent upon fruit (nuts, seeds, and berries) of a relatively few species; when these species fail to produce crops of fruit, squirrel populations often are subject to severe mortality (Koprowski, 1991*a*).

Reproduction, Ontogeny, and Mortality.—In its native range, *S. niger* produces one or two litters annually; peaks of breeding usually are in January–March and June–August depending on latitude, with corresponding peaks in births after a 44–45-day gestation period. Average litter size reported in several studies ranged from 2.2 to 3.2 depending on the method of evaluation, age of squirrels, and season; range within studies usually was one–four (Hayssen et al., 1993) with a maximum number of embryos reported as seven (Hoover, 1954). We have two records of litter size in Oregon: a specimen from Union Co. with three pigmented sites of implantation and one from Marion Co. with nine implantation sites. We suspect the latter number is the result of multiple litters. No other information regarding reproduction is available for fox squirrels in Oregon.

Neonates are 50–60 mm long and weigh 14–17 g. Young first emerge and commence to eat solid food at about 7 weeks of age and are weaned when 8–12 weeks old (Hayssen et al., 1993). Sexual maturity is attained at 10–12 months of age (Hayssen et al., 1993).

Habits.—The eastern fox squirrel is active throughout the year; during periods of intense cold or extreme heat it may become lethargic, but it does not become torpid. All activity is during daylight; greatest activity is during the early morning and late afternoon hours (Brown and Yeager, 1945). Although highly adapted for arboreal life and commonly traveling through the forest by climbing and jumping, fox squirrels spend considerable time foraging on the ground. During autumn, they are exceptionally active on the ground, scatter-hoarding nuts and acorns as they mature and ripen. In addition to stripping bark to obtain nutrients, fox squirrels remove bark, then rub the spot with their oral glands and sometimes urinate on the area as a means of scent-marking (Koprowski, 1991*b*). All sex and age groups engage in scent-marking, but males mark more than females; the function of scent-marking behavior is not known (Koprowski, 1993*a*).

Fox squirrels seek shelter in tree cavities or in dreys. An individual commonly uses two or more nest cavities or dreys concurrently. In some areas where food is abundant, fox squirrels inhabit woodlands that contain few or no trees with cavities suitable for nests (Allen, 1942). Dreys may be constructed as low as 3 m from the ground, but usually are placed considerably higher in trees, sometimes in the uppermost branches (Allen, 1942).

Within a season, individual fox squirrels usually confine their activities to 4–5 ha, but during a year may range over 16–17 ha or more. During late summer or early autumn, juveniles disperse, sometimes nearly 25 km, but more often only 1–3 km. However, mass migrations of fox squirrels, as sometimes observed in populations of *S. carolinensis*, are not known to occur (Brown and Yeager, 1945).

Although adult females may exhibit a degree of mutual avoidance and possibly defend small areas around nest sites or sites where food was scatter-hoarded, neither females nor males maintain nonoverlapping home ranges (Nixon et al., 1986). Unlike *S. carolinensis*, female fox squirrels do not share home ranges with offspring during winter (Nixon et al., 1986).

During estrus, a female fox squirrel commonly is pursued by four–seven males; dominant males actively pursue the female, and subordinate males sit or forage within the female's home range (Koprowski, 1993*b*). Females copulate (mostly in trees, but occasionally on the ground) as many as seven ($\bar{X} = 2.5 \pm 2.1$, $n = 12$) times during an estrous period, rarely with the same male twice. Active pursuit is the most successful mating tactic employed by males, but satellite males account for nearly 40% of copulations, mostly when females escape from dominant males (Koprowski, 1993*b*).

Remarks.—D. B. Marshall (pers. comm., 3 August 1993), formerly of the U.S. Fish and Wildlife Service, recalled seeing an eastern fox squirrel in southeastern Portland near the beginning of World War II. In 1969, V. P. McCrow, a former graduate student, related that he had hunted fox squirrels in nut orchards near Scholls, Washington Co., as a boy and from his observations, believed that populations of the western gray squirrel (*S. griseus*) declined as those of the fox squirrel increased. N. R. TenEyck (pers. comm., 2 August 1993), District Biologist, Oregon Department of Fish and Wildlife, indicated that fox squirrels had dispersed south and west along the Willamette and Yamhill rivers and had reached Rickreall and Dallas, Polk Co., during the previous 15 years. His observations corroborated those of McCrow regarding the displacement of *S. griseus* by *S. niger.* Specimens on deposit in the Department of Fisheries and Wildlife, Oregon State University mammal collection were obtained near Scholls, Washington Co., in 1969; near Waconda, Marion Co., in 1973 and 1979; in Eugene, Lane Co., in 1973; and in Union, Union Co., in 1971.

W. R. Humphreys, District Biologist, Oregon Department of Fish and Wildlife, indicated that fox squirrels were introduced into Baker City, Baker Co., in the early 1950s by members of the Junior Chamber of Commerce. The stock was obtained from an introduced population in Boise, Idaho. He suspected that populations in other cities and towns in eastern Oregon probably originated from the resulting population in Baker City.

The fox squirrel in its native range is a valuable game species; more than a million are killed annually for sport and food in several midwestern and eastern states. Nevertheless, even in that region, it can be a pest in some situations; it sometimes enters attics in homes, chews insulation from electric wires, strips bark from branches of shade trees, and consumes garden crops. Considering that it is associated largely with urban habitats in Oregon, there is the potential for similar problems. More importantly, the suspected competition with or other deleterious impacts on native species are sufficient for us to advocate prohibition of additional transplants. Also, we believe that various agencies and organizations should not promote practices, such as erection of nest boxes and feeders, designed to enhance established populations that, in turn, may be responsible for the invasion of currently unoccupied areas by fox squirrels. The potential for competition with native fauna and for damage to property outweigh any aesthetic value of the introduced species.

Tamiasciurus douglasii (Bachman)

Douglas' Squirrel

1839*d. Sciurus Douglasii* Bachman, 1838:99.
1841. *Sciurus molli-pilosus* Audubon and Bachman, 1:102.
1842. *Sciurus Belcheri* Gray, 10:263.
1890*b. Sciurus hudsonicus californicus* Allen, 3:165.
1898*b. Sciurus douglasii cascadensis* Allen, 10:277.
1898*a. Sciurus douglasii albolimbatus* Allen, 10:453.
1940. *Tamiasciurus douglasii albolimbatus*: Hayman and Holt, 1:347.
1940. *Tamiasciurus douglasii douglasii*: Hayman and Holt, 1:347.
1940. *Tamiasciurus douglasii mollipilosus*: Hayman and Holt, 1:347.

Description.—*Tamiasciurus douglasii* (Plate XI) is one of the smaller tree squirrels in Oregon (Table 11-7). The color and markings of *T. douglasii* differ "individually, geographically, and seasonally" (Hall and Kelson, 1959:403). Typically, the dorsum is dusky olive to brownish gray with an indistinct band of reddish brown; the venter is deep orange, yellowish, or dirty white; and the tail hairs are much like those on the dorsum for the proximal two-thirds, but have a black band followed by a buff or white tip. A blackish band along the flanks separates the ventral and dorsal pelages. In winter, the band may become lighter or may nearly disappear and the feet change from orange to grayish brown. The ears are tuffed with black and the eyes are ringed with yellowish-white or orange hairs.

The feet are pentadactyl and plantigrade. The hind toes and four functional toes on the forefeet are equipped with claws; the pollex is rudimentary and equipped with a nail. Contrary to some published statements, a tiny baculum is present at the tip of the glans penis; Maser (1969) measured one at 0.33 mm long and 0.17 mm wide. Like those of other tree squirrels, the zygomata of *T. douglasii* are untwisted, compressed laterally, and parallel (Fig. 11-39). In Oregon, the skull may be separated from those of members of the genus *Sciurus* by size (the occipitonasal length is <55 mm in *Tamiasciurus*) and from those of the genus *Glaucomys* by the maxillae being completely obscured by the frontals in dorsal view. Although *T. douglasii* from west of the Ochoco Mountains usually can be separated from *T. hudsonicus* from northeastern Oregon by pelage characters, we know of no skull or pelage characters that can be used to separate the two species in Grant and southwestern Baker counties with a high degree of reliability (see Remarks).

Distribution.—Douglas' squirrels occur from coastal southwestern British Columbia south to San Francisco Bay and to near Bakersfield in California; the range extends eastward through the Cascade Range to the Columbia River in Washington, to the Ochoco Mountains in Crook Co., Oregon, and to near Lake Tahoe in Nevada (Fig. 11-40). A disjunct population occurs in Baja California Norte (Hall, 1981). In Oregon, *T. douglasii* occurs in coniferous forests from the Pacific Coast probably as far east as western Baker Co. (Fig. 11-40). In parts of Grant and Baker counties, *T. douglasii* may be sympatric with its nearly indistinguishable congener *T. hudsonicus* (Hatton and Hoffmann, 1979; Lindsay, 1982; Smith, 1981); whether the two species hybridize is unknown.

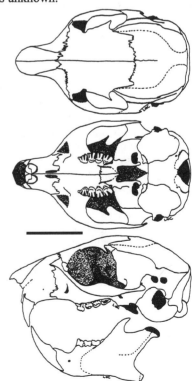

Fig. 11-39. Dorsal, ventral, and lateral views of the cranium, and lateral view of the dentary of Douglas' squirrel, *Tamiasciurus douglasii* (OSUFW 778). Scale bar equals 15 mm.

Late Pleistocene fossils are known from cave deposits at two sites in Nevada (Kurtén and Anderson, 1980).

Geographic Variation.—Three of the four nominal subspecies occur in Oregon: *T. d. douglasii* with the rich orange venter occurs in the Coast Range north of Coos Bay, Coos Co.; *T. d. mollipilosus* with the pale-yellow venter occurs in a narrow diagonal band on the west slope of the Cascade Range as far south as Sweet Home, Linn Co., then in the Coast Range south to the California border; and *T. d. albolimbatus* with the dirty-white venter occurs through the remaining forested areas of the Cascade Range eastward at least to the Ochoco Mountains, Crook Co. (Hall, 1981).

Habitat and Density.—Douglas' squirrels are denizens of coniferous forests where, for food and shelter, they depend largely upon western hemlock (*Tsuga heterophylla*), Douglas-fir (*Pseudotsuga menziesii*), grand fir (*Abies grandis*), Sitka spruce (*Picea sitchensis*), western red cedar (*Thuja plicata*), and western white pine (*Pinus monticola)* west of the Cascade Range. East of the Cascade Range they depend on ponderosa pine (*Pinus ponderosa*), lodgepole pine (*Pinus contorta*), Engelmann spruce (*Picea engelmannii*), subalpine fir (*Abies lasiocarpa*), and Douglas-fir (Smith, 1981). Although ranging from more advanced forest communities into pole- and sapling-sized forest plantings occasionally (Fisch and Dimock, 1978), *T.*

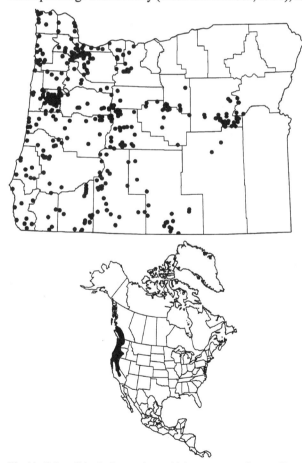

Fig. 11-40. Localities in Oregon from which museum specimens of Douglas' squirrel, *Tamiasciurus douglasii*, were examined (listed in APPENDIX, p. 571), and the distribution of Douglas' squirrel in North America. (Range map redrawn from Hall, 1981:447, map 296, with modifications.) See Remarks regarding classification of the species in Grant and southeastern Baker counties.

douglasii seems dependent upon forests with trees of cone-bearing age. Gashwiler (1959:138), based on his study of a 19.4-ha clear-cutting in a forest in the Cascade Range that formerly supported Douglas' squirrels, opined that "logging apparently ruined the area as a habitat for them."

Where the geographic range of *T. douglasii* is believed to be sympatric with that of *T. hudsonicus*, *T. douglasii* reportedly is restricted to habitats free of lodgepole pine, whereas *T. hudsonicus* occurs in forest areas in which lodgepole pine is a dominant component (Hatton and Hoffmann, 1979; Lindsay, 1982).

Densities of populations of Douglas' squirrels "fluctuate dramatically in response to seasonal or annual variations in food supply" (Buchanan et al., 1990:577). West of the Cascade Range, production of cones by Douglas-fir and western hemlock ranges from near absence to abundant at intervals of $\cong$3–4 years. In the Gifford-Pinchot National Forest just across the Columbia River in Washington, during a year of high production of cones, ranges in estimated densities of Douglas' squirrel populations based on transect counts in winter were 40–144/100 ha in four old-growth (375 years old) stands and 9–76/100 ha in four younger (42–140 years old) stands (Buchanan et al., 1990). The highest estimate among the four younger stands was believed to be related to the proximity of the stand to one that produced an abundance of cones; thus, in most instances, densities were 2.1–16.4 times greater in old-growth stands than in younger stands. Numbers of squirrels detected along transects during a year following 2 successive years of poor production of cones were about one-fifth to one-half those detected during the year with high production of cones (Buchanan et al., 1990). In northern California, populations of *T. douglasii* in a forest of incense cedar (*Calocedrus decurrens*), white fir (*Abies concolor*), Douglas-fir, black oak (*Quercus kelloggii*), ponderosa pine, and sugar pine (*P. lambertiana*), ranged from 15 to 22/100 ha in successive years (Koford, 1982). In British Columbia, the range in densities was 22–89/100 ha on a 9-ha trapping grid in a hemlock–western red cedar–Douglas-fir forest (Sullivan and Sullivan, 1982). Addition of food material (sunflower seeds and whole oats) on nearby areas was believed to have caused the 5–10-fold increases in density noted, suggesting that densities of Douglas' squirrel populations are regulated by food supply (Sullivan and Sullivan, 1982).

Diet.—Based on an analysis of stomach contents, *T. douglasii*, in forests of ponderosa pine, jeffery pine (*Pinus jeffreyi*), lodgepole pine, red fir (*Abies magnifica*), and white fir in northeastern California, consumed mostly tree seeds and fungi. Based on volume, fungi contributed 33–91% of the diet during all months, whereas tree seeds contributed 3–64%; tree seeds contributed most in late summer and autumn (McKeever, 1964*b*). Cambium was eaten in winter and spring (25–63%), but rarely at other seasons. Leaves contributed 56% of the diet in June, but ≤7% during other months. Some flowers were eaten in May and June (4–12% of the diet) and arthropods contributed 2% in September (McKeever, 1964*b*).

Smith (1970) and Buchanan et al. (1990) indicated that population densities of *T. douglasii* fluctuated with the crop of cones. Such suggests that the seeds of conifers, especially those of Douglas-fir and ponderosa pine, provide much of the energy required by *T. douglasii* in autumn and winter. C. Smith (1968) provided lists of foods eaten by *T. douglasii* and the energy content of many of those food items; foods commonly eaten, in general, contained the highest levels of stored energy. The consumption of cambium (a food with less energy content than conifer seeds) in winter and spring suggests that the availability of conifer seeds was not adequate during the year of McKeever's (1964*b*) study.

In young stands of Douglas-fir, especially those near older forest stands, Douglas' squirrels commonly clip terminal and lateral shoots from the topmost whorl of branches, then consume the vegetative buds. Such activity usually occurs in winter or early spring and may involve as much as 38% of the pole- and sapling-sized trees (Fisch and Dimock, 1978); such also may indicate a deficient supply of conifer seeds. Recovery of damaged trees is rapid (Fisch and Dimock, 1978).

Smith (1970) hypothesized that differences in bite force between populations of *T. douglasii* and *T. hudsonicus* west and east of the Cascade Range, respectively, in British Columbia were related to hardness of cones and number of seeds in cones. These differences, in turn, were related to selective pressures for hard cones and few seeds, that, in turn, were produced by *T. hudsonicus* foraging largely on the harder, closed cones of lodgepole pine and *T. douglasii* foraging largely on the softer, open cones of Douglas-fir.

Reproduction, Ontogeny, and Mortality.—Productivity of populations of *T. douglasii* as measured by age at sexual maturity, number of litters produced annually, and number of young per litter seems to be related strongly to the cone crop (Smith, 1981). Several authorities indicate that Douglas' squirrels may produce one or two litters annually, one in June–July and the other in September–October (Cowan and Guiguet, 1965; Dalquest, 1948; Jameson and Peeters, 1988; Koford, 1982; Smith, 1981). However, records of the same individual producing two litters in the same year seem to be relatively uncommon (Koford, 1982; Smith, 1981). Koford (1982) recorded an instance of a female producing a litter later in the year of its birth and Smith (1981:356) suggested that females may "delay or postpone reproduction when food is scarce." Although two peaks in breeding and recruitment of young may be evident, all females likely are not productive during each period.

Litter size, both within and among years, appears to be variable (Smith, 1981), but at least some of the variation might be attributable to combining counts based on different types of evidence. The wide variation in litter size in Oregon based on counts of pigmented sites of implantation (Table 11-4) may be related, in part, to specimens being collected from different population areas and during different years. Thus, average litter size for the Douglas' squirrel throughout its range in the state, as for some other species, may not be a particularly valuable statistic.

Koford (1982) assumed a gestation period of 40 days for *T. douglasii* based on a similar period reported by Hamilton (1939) for *T. hudsonicus* in New York. Ingles (1965) believed the gestation period to be between 36 and 40 days. Because the gestation period for four free-living *T. hudsonicus* in Quebec was between 31 days, 20 h and 35 days, 3 h (Lair, 1985), it seems likely that the gestation period in *T. douglasii* may be shorter than heretofore believed.

Neonates were estimated to weigh 5 g at birth; on the basis of this estimate, three young gained 1.67–2.35 g/day (Koford, 1982). Only the maternal female provides care for young, but no food is brought to the nest cavity or drey for the young (C. Smith, 1968).

Habits.—Douglas' squirrels are active during the daylight hours year-round; although they may remain in their nests or tree dens for a day or two during inclement weather, they do not become torpid. During the nonbreeding season, much of the time is spent foraging and loafing; Douglas's squirrels commonly sit at the axil of a branch with their tails curled in a tight S-shape over the back. Sometimes they lie along a branch on their bellies with legs dangling alongside. Douglas' squirrels commonly use abandoned woodpecker (Picidae) nest cavities, but sometimes build dreys (Maser et al., 1981*b*). During late summer and autumn, Douglas' squirrels spend considerable time and effort cutting cones from trees and allowing them to fall to the ground. Cones then are gathered and stored, often hundreds in one cache. Sometimes cones are cached near springs or in boggy areas; moist cones remain closed, thus protecting seeds they contain (Shaw, 1936).

Douglas' squirrels are strongly territorial. Both sexes maintain nonoverlapping, contiguous territories, and defend them throughout the year by vocalizing and occasionally chasing invaders, but rarely engaging in fights (C. Smith, 1968). Individuals chasing invaders occasionally overrun their territory boundaries and, in turn, become the "chasee." Some vagrant squirrels wander in search of unoccupied territories or sometimes drive territory owners from part or all of their territory (C. Smith, 1968).

Douglas' squirrels produce four vocal calls (rattle, screech, growl, and buzz) associated with territorial behavior and one call (chirp) in response to predators (Smith, 1978). The chirp call may be produced continuously for >30 min in the presence of goshawks (*Accipiter gentilis*) and humans (Smith, 1978); the chirp also may be produced in response to the calls of some birds and mammals. Chirp calls, especially those at the beginning of a bout of calling, are difficult for an observer (and, presumably, a potential predator) to locate (Smith, 1978). Rattles and screeches alone or in combination are used in defense of territories; growls are used to intimidate, thereby avoiding fights; and buzz calls are used by individuals as they approach another squirrel, but not in an attempt to steal its territory (Smith, 1978).

Koford (1982) depicted convex-polygon home ranges (territories) that included 75% of captures or observed locations for nine females and six males during the breeding season and for three other females and three other males after the breeding season. Although size statistics for these home ranges were not presented, from scales provided, most home ranges were <0.6 ha and some were as small as ≅0.05 ha. Koford (1982) indicated that home-range size decreased as population density increased.

During the breeding season, Douglas' squirrels were within the home range of another squirrel 36% of the time, but after the breeding season, they rarely entered the 75% home range of other individuals (Koford, 1982). Invasions of 75% home ranges of members of the opposite sex were significantly more frequent than invasions of home ranges of the same sex by both males and females (Koford, 1982). On the 1 day that a female is receptive, several males enter its territory, move about, and produce rattle calls. The female runs about or sits and sometimes eats; one male usually stays within a few meters, gives rattle calls, and chases other males that approach the female too closely. Females are not receptive during most attempts by males to mount; they growl and move away when a male approaches too closely. When receptive, the female does not move away when approached by a male; mounting usually lasts 1–25 min, is initiated on the ground, and the female with mounted male sometimes climbs the trunk of a tree (Koford, 1982). Although females are accompanied by dominant males 95% of the time, they commonly mate with subordinate males; however, the dominant male in the area commonly is first to mate with an estrous female.

Remarks.—*Tamiasciurus douglasii* and *T. hudsonicus* are allopatric throughout most of their geographic ranges, but Lindsay (1982) considered them to be sympatric in narrow zones at points in British Columbia, Washington, and east-central Oregon. Earlier, Bailey (1936) indicated a gap of ≅97 km between ranges of the two species in Oregon, whereas Hall and Kelson (1959) depicted the gap at ≅40 km. Hatton and Hoffmann (1979) noted that the vegetation in the depicted gaps was forest of a type occupied elsewhere by the two species; they collected and examined 125 specimens from the area formerly considered not to be occupied by either species and 178 specimens from other nearby regions long known to be within the range of one species or the other. Nevertheless, they were unable to establish the taxonomic affinity of those collected within the contact zone in eastern Oregon on the basis of color of the venter and tail fringe. Because some individuals resembled one or the other of the species and some exhibited a variety of intermediate characters, Hatton and Hoffmann (1979) suggested the possibility of hybridization of the two species within the contact zone. However, C. C. Smith communicated to Hatton and Hoffmann (1979) that he had conducted an analysis of calls of tree squirrels in the region and that all those south and west of John Day, Grant Co., gave calls characteristic of those of *T. douglasii*, whereas those east of that point gave calls characteristic of those of *T. hudsonicus*. C. C. Smith further indicated that all those considered to be *T. hudsonicus* were in forests dominated by lodgepole pine and most of those considered to be *T. douglasii* were in areas free of lodgepole pine (Hatton and Hoffmann, 1979). From this, Hatton and Hoffmann (1979) suggested that instead of hybridization, character convergence might be responsible for the presence of individuals with intermediate characters within the contact zone. Hall (1981), although not making the requisite taxonomic changes, suggested that morphological intermediates possibly indicated an absence of a species barrier between *T. douglasii* and *T. hudsonicus.*

Based on a multivariate analysis of 20 skull characters of squirrels from 38 localities throughout the Pacific Northwest, Lindsay (1982) concluded that the two species were clearly separated morphometrically. Although he included supposed hybrids from Washington and British Columbia that indeed clustered with one taxon or the other, he did

not include specimens from the contact zone in Oregon. Both species attain the large ends of clines in size in east-central Oregon, prompting Lindsay (1982) to suggest that his morphometric findings were congruent with Hatton and Hoffmann's (1979) explanation (character convergence) for intermediate colormorphs in the zone of contact.

In British Columbia, Smith (1970) found that east of the Cascade Range (the primary range of *T. hudsonicus* in that region), cones of lodgepole pine were exceptionally hard and did not open for several years, and that cones of Douglas-fir contained relatively few seeds. West of the Cascade Range (the range of *T. douglasii* in British Columbia) lodgepole pines were rare and did not have closed cones, and Douglas-fir cones contained many seeds. Thus, when the Douglas-fir cone crop failed east of the Cascade Range, *T. hudsonicus* could shift to a diet of lodgepole pine seeds, but similar crop failures in Douglas-fir west of the Cascade Range were responsible for steep declines in population density of *T. douglasii*. Smith (1970) hypothesized that the frequency of forest fires east of the Cascade Range in British Columbia led to conifers producing harder cones that did not open for several years and cones with fewer seeds. These hard cones, in turn, selected for large *Tamiasciurus* with jaws with greater static bite force and stronger jaw musculature that led to larger skulls in *T. hudsonicus* east of the Cascade Range than in the 16% smaller *T. douglasii* west of the Cascade Range.

In eastern Oregon, a significantly greater proportion of *T. hudsonicus* than *T. douglasii* had a "sharp, knife-edged, sagittal crest . . . [that] indicates a greater development of the temporal muscles" (Smith, 1970:354). This was considered to parallel differences observed in jaw musculature of the two species in British Columbia and to occur in a zone in the Blue Mountains of Oregon where lodgepole pines with closed cones grade into those whose cones open. The reference that Smith (1970) cited in support of the change in type of cone was Trappe and Harris (1958), who actually wrote that most lodgepole pine trees in the Blue Mountains of Oregon produced cones that did not remain closed for long periods and only an occasional tree bore cones that remained closed.

Thus, the following contradictions are presented in the available literature on *T. douglasii* and *T. hudsonicus* in Oregon:

1. The two species reportedly are clearly separable morphometrically (Lindsay, 1982), but exhibit morphological character convergence in a contact zone (Hatton and Hoffmann, 1979; Smith, 1970);

2. They have cranial morphometrics that parallel pelage differences typically used to distinguish the two species, but no specimens from the contact zone were measured (Lindsay, 1982); and

3. They have differences in jaw mechanics and musculature, and skull architecture that reflect selection for greater strength to open closed cones produced by lodgepole pine (Smith, 1970), but most lodgepole pines in eastern Oregon produce cones that do not remain closed (Trappe and Harris, 1958).

Dissatisfied with the contradictions, we attempted a morphometric approach to separating the two species in Oregon. We were able to separate, by use of discriminant analysis of five skull dimensions, only 78% of *T. douglasii*

(n = 143) from *T. hudsonicus* (n = 94) from areas east of the Cascade Range other than where the two species were considered to be sympatric (Lindsay, 1982). We used the five dimensions (greatest length of skull, skull height, least width of M3–M3, length of foramen magnum, and width between infraorbital foramina) that reportedly account for overall differences in size and shape of skulls of the two species (Lindsay, 1982). Despite the poor degree of classification in these areas, we used these data to create diagnosis files to separate the *Tamiasciurus* from Grant and southwestern Baker counties, where the two species are considered to be sympatric. Only those individuals for which Geisser classification probabilities were ≥75% were classified to species for purposes of producing distribution maps; the remaining specimens from Grant and southwestern Baker counties that we examined are listed as "*Tamiasciurus* unclassified" (Appendix). We believe that a definitive treatment of the systematic relationships of *T. douglasii* and *T. hudsonicus* east of the Cascade Range in Oregon awaits application of molecular techniques followed by morphometric analyses.

Tamiasciurus douglasii, in addition to being referred to as Douglas' squirrel, often is called the chickaree.

Tamiasciurus hudsonicus (Erxleben)
Red Squirrel

1777. [*Sciurus vulgaris*] *hudsonicus* Erxleben, 1:416.
1839d. *Sciurus Richardsoni* Bachman, 1838:100.
1939. *Tamiasciurus hudsonicus richardsoni*: Davis, p. 227.

Description.—*Tamiasciurus hudsonicus* (Plate XI) also is a small tree squirrel; it is only slightly larger than its congener *T. douglasii* (Table 11-7). Typically, the dorsal pelage is grayish olive with a broad, but indistinct, rusty middorsal band; a black band along the flanks separates the dark dorsum from the grayish-white venter. The ears have short black tufts; the eyes are ringed with creamy white; the feet are mottled grayish tan; and the tail has a rusty base grading into an all-black distal half. In some individuals, the dark tail hairs are tipped with ochre.

Like its congener, the red squirrel has five toes on each foot; except for the rudimentary pollex equipped with a nail, the toes are equipped with strong, recurved claws. Also, in male *T. hudsonicus* a baculum is present (Maser, 1969) and in females an os clitoridis is present; the bones are tiny in both sexes (Layne, 1952).

The zygomata are untwisted, compressed laterally, and nearly parallel (Fig. 11-41). The skull may be separated from those of other tree squirrels except *T. douglasii* and *Glaucomys sabrinus* by size (greatest length of the skull <55 mm in *Tamiasciurus*). It is distinguishable from those of the latter species by the maxillae being completely obscured by the frontals when the skull is viewed from above. Lindsay (1982) reported that skulls of *T. hudsonicus* and *T. douglasii* were separable in multivariate space, but in an analysis of specimens from areas other than where the two species are considered sympatric, we were able to separate only 78% of large samples of skulls of the two species (see Remarks in the *T. douglasii* account). Although *T. hudsonicus* from northeastern Oregon usually can be separated from *T. douglasii* from west of the Ochoco Mountains by pelage characters, we know of no skull or pelage characters that can be used to separate the two species in

Grant and southwestern Baker counties with a high degree of reliability.

Distribution.—Red squirrels occur from Seward Peninsula, Alaska; Great Bear Lake, Northwest Territories; Hudson Bay and Ugava Bay, Quebec; and coastal Labrador south throughout the boreal forest region of Canada and northern United States south in montane regions to Arizona and New Mexico in the West and to western South Carolina in the East (Fig. 11-42; Hall, 1981). In Oregon, red squirrels occur in the montane forested portions of Wallowa, Union, Umatilla, Morrow, Malheur, Baker, and Grant counties (Fig. 11-42). In parts of Grant and Baker counties, *T. hudsonicus* may be sympatric with its nearly indistinguishable congener *T. douglasii* (Hatton and Hoffmann, 1979; Lindsay, 1982; Smith, 1981); whether the two species hybridize is unknown.

Red squirrel fossils dating to the Irvingtonian (early to mid-Pleistocene) are known from several sites in the Midwest and eastern United States (Kurtén and Anderson, 1980). Some of the sites are south of the present-day range of the species.

Geographic Variation.—Only one (*T. h. richardsoni*) of the 25 currently recognized subspecies occurs in Oregon (Hall, 1981).

Habitat and Density.—Like its congener, *T. hudsonicus* is a largely arboreal, forest-dwelling species and, although often occupying areas vegetated by other conifers (ponderosa pine, *Pinus ponderosa*; Douglas-fir, *Pseudotsuga menziesii*; grand fir, *Abies grandis*; subalpine fir, *A. lasiocarpa*; and Engelmann spruce, *Picea engelmannii*), it is associated most frequently with lodgepole pine (*Pinus contorta*—Smith, 1981). In mixed stands in British Columbia, during years that other conifers failed to produce cones, *T. hudsonicus* left areas supporting other conifers

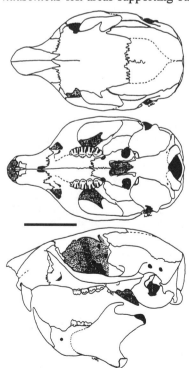

Fig. 11-41. Dorsal, ventral, and lateral views of the cranium, and lateral view of the dentary of the red squirrel, *Tamiasciurus hudsonicus* (OSUFW 1852). Scale bar equals 15 mm.

and immigrated to stands of lodgepole pine (Smith, 1981). In Oregon, in a narrow contact zone between ranges of *T. hudsonicus* and *T. douglasii*, the former, as identified by their calls, reportedly occupied stands of lodgepole pine nearly exclusively (Hatton and Hoffmann, 1979).

Ranges in densities of red squirrel populations were 20.8–62.5/100 ha in a white spruce (*Picea glauca*) forest in Alaska (M. Smith, 1968), 100.0–192.2 and 74.4–122.2/100 ha on two areas in white spruce-lodgepole pine-subalpine fir forest in British Columbia (Sullivan, 1990). In Alberta, densities of 31.8–39.1/100 ha in aspen (*Populus tremuloides*)–balsam poplar (*Populus balsamifera*)-dominated forest and 53.8–58.5/100 ha in jack pine (*Pinus banksiana*) with an understory of alder (*Alnus crispa*) were recorded (Kemp and Keith, 1970). In the spruce-fir forest in British Columbia, addition of food (sunflower seeds) to 9-ha trapping grids increased densities from 20.0 and 42.2/100 ha to levels as great as 588.9 and 464.4/100 ha, respectively (Sullivan, 1990). The latter study was criticized by Koford (1992), who suggested that red squirrels with home ranges outside the 9-ha grids may have extended their ranges to take advantage of the supplemental food and were counted as residents. Sullivan and Klenner (1992) countered by showing that deletion of individuals captured 1–2

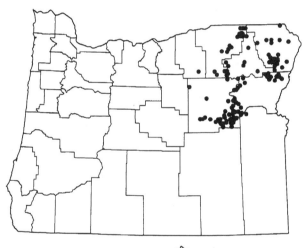

Fig. 11-42. Localities in Oregon from which museum specimens of the red squirrel, *Tamiasciurus hudsonicus*, were examined (listed in APPENDIX, p. 572), and the distribution of the red squirrel in North America. (Range map redrawn from Hall, 1981:444, map 295, with modifications.) See Remarks in *T. douglasii* account regarding classification of the species in Grant and southeastern Baker counties.

times (thus, presumably transients) did not alter their reported results. That food, especially in the form of cached cones, is the proximate regulator of density of populations of red squirrels seemingly is a widely held view among those who have studied the species (Kemp and Keith, 1970; Rusch and Reeder, 1978; Sullivan, 1990; Sullivan and Klenner, 1992; Sullivan and Sullivan, 1982).

Diet.—Red squirrels consume a variety of hypogeous and epigeous fungi; seeds, buds, or berries of conifers, some deciduous trees, and a few shrubs; and fruit or leaves of a few shrubs and herbaceous plants (Kemp and Keith, 1970; Maser et al., 1978a; Rusch and Reeder, 1978; C. Smith, 1968). Occasionally, bones, egg shells, insects, and miscellaneous animal material are eaten (C. Smith, 1968). However, most food eaten "consists of the reproductive parts of fungi, conifers, and angiosperms and of pine cambium" (C. Smith, 1968:48). When feeding on fungi, red squirrels consume primarily the gleba of truffles (Ascomycetes) and gills of mushrooms (Basidomycetes), thereby increasing the energy per unit volume eaten by $\cong$20%. Similarly, they discard the seed coats of conifer seeds to eliminate material with relatively low-energy content (C. Smith, 1968).

Red squirrels have larger masseter muscles and more advantageous jaw mechanics than Douglas' squirrels; thus, red squirrels can open the closed cones of lodgepole pine with ease. In British Columbia, lodgepole pines form a more reliable food source for *T. hudsonicus* than species of conifers whose cones open but that do not produce a crop of cones each year (Smith, 1970). In eastern Oregon, however, cones of most lodgepole pines open spontaneously (Trappe and Harris, 1958). Nevertheless, red squirrels in Oregon usually are associated with stands of lodgepole pine; cut, gather, and cache cones; and attain the largest size for the species in western North America (Hatton and Hoffmann, 1979; Lindsay, 1982).

Reproduction, Ontogeny, and Mortality.—In British Columbia, male red squirrels were reproductively functional March-September in 1966, but January–March in 1967. Breeding commenced in early May 1966, but in February 1967; severe weather and unusually deep snow were believed to have retarded the breeding season in 1966 (Millar, 1970a). No males attained sexual maturity in the year of their birth; all males matured as yearlings, some as young as 5 months of age. Also, no females bred in the year of their birth in British Columbia (Millar, 1970a) as they sometimes do in the eastern part of the species' range (Layne, 1954). All yearling and adult females in British Columbia produced two litters in 1966; nevertheless, Millar (1970a) considered one litter annually to be the rule. In 1967, 65% ($n = 72$) of yearlings failed to breed (Millar, 1970a). In Alberta, litter size and proportion of females breeding differed considerably between years ($\bar{X} = 3.4$; range, 1–5; $n = 43$; 67% breeding in 1967 and $\bar{X} = 4.3$; range, 2–8; $n = 39$; 88% breeding in 1968—Kemp and Keith, 1970). In British Columbia, the ovulation rate as indicated by counts of corpora lutea did not differ between yearlings and adults at two sites or between years, but was significantly greater on Vancouver Island in 1966 ($\bar{X} = 5.1 \pm 0.7$, range 4–7, $n = 40$) than on the island in 1967 ($\bar{X} = 4.1 \pm 0.3$, range 4–5, $n = 9$) or on the mainland in 1967 ($\bar{X} = 3.6 \pm 0.7$, range 2–6, $n = 34$—Millar, 1970b). Also in British Columbia, preimplantation losses averaged 22.7% ($n = 9$) and 7.4% ($n = 25$) and postimplantation losses averaged 0% ($n = 6$) and 9.4% ($n = 17$) for yearlings and adults, respectively (Millar, 1970b). The gestation period for four free-living females followed by radiotelemetry in Quebec ranged from 31 days, 20 h to 35 days, 3 h, considerably shorter than the ≅40 days previously thought for the species (Lair, 1985).

We examined a female with three pigmented implantation sites from Wallowa Co. (KU 148149), and recorded a pregnant female with four embryos taken in Umatilla Co. on 21 May (USNM 274364).

Development in juvenile *T. hudsonicus* is less rapid than in the Columbian ground squirrel (*Spermophilus columbianus*) and golden-mantled ground squirrel (*S. lateralis*), but more rapid than in the northern flying squirrel (*Glaucomys sabrinus*—Ferron, 1981). Average age (in days) for appearance of certain characteristics and behaviors is: healed umbilicus, 10.2; eruption of i1, 17.7; eruption of I1, 27.9; opened ears, 20.1; opened eyes, 31.4; crawl forward, 2.0; regularly walk on all four feet, 26.5; climb, 31.4; out of nest, 38.3; drink, 38.7; and eat solid food, 38.8 (Ferron, 1981).

In Alberta, annual mortality was 67% among postweaning juveniles, 34% among yearlings, and 61% among adults (Kemp and Keith, 1970). Kemp and Keith (1970) indicated that populations in vast regions of Canada were synchronized, with peaks occurring at 2.5–2.8-year intervals. They termed the populations "cyclic" because peak densities occurred at intervals more regularly than expected on the basis of chance. They speculated that peaks in density were associated with production of cones by white spruce, which in turn possibly was associated with weather, especially precipitation. Kemp and Keith (1970) hypothesized that in the spring of years preceding a good cone crop, red squirrels shifted their diet to flower buds that stimulated reproduction. Thus, more young were produced in anticipation of the good cone crop, which maximized the ability of red squirrels to use a fluctuating food supply.

Habits.—Activity patterns of red squirrels are similar to those of Douglas' squirrels: diurnal, year-round, and without periods of torpor. And, indeed, much activity is directed toward foraging, gathering and caching cones, feeding, and resting. Like the Douglas' squirrel, *T. hudsonicus* uses cavities created and abandoned by woodpeckers (Picidae); it also constructs dreys.

Tamiasciurus hudsonicus also is strongly territorial; territories are maintained for exclusive use by the owner, but trespass by neighboring red squirrels is common. At least in the northern part of the species' range the productivity of the territory in terms of number of cones and the number of seeds contained therein determines to a great extent the probability of survival for a red squirrel. Whether territoriality is as critical to survival in Oregon, especially at lower elevations where snow cover often is absent permitting red squirrels to forage for fungi and other items in winter, is unknown. On two areas in British Columbia, minimum-area home ranges averaged 0.63 and 0.88 ha (Sullivan, 1990). Because red squirrels sometimes trespass on territories of neighbors, these areas may be larger than those actually defended.

In the Yukon Territory, Canada, 30% of maternal female red squirrels "bequeath" their territories to their offspring and disperse in search of new territories, a rare and unusual behavior (Price and Boutin, 1993). Females that breed early are less likely to disperse than those that breed late in the season, presumably because their offspring are more likely to find a suitable territory independently. Thus, in some complex manner, bequeathal of a territory must enhance survival of offspring sufficiently to offset the likely reduced probability of the dispersing maternal female finding a suitable territory and surviving to produce additional offspring the following year. Bequeathal was not related to body condition or the quality of the territory of the maternal female, although females in poor condition and with small caches of cones were likely to lose their territories to invaders (Price and Boutin, 1993). Boutin et al. (1993) found that by removing squirrels with territories during the period that juveniles became independent, maternal females, not juveniles, moved to the experimentally vacated territories. Seemingly, such behavior permits more offspring to remain on natal areas, thereby enhancing juvenile survival. Whether or not the phenomenon of bequeathal of natal territories to offspring occurs in red squirrel populations farther south is unknown; because "well-stocked territories [are] less crucial to survival" a hypothesis of no bequeathal of territories at lower latitudes would be appropriate to test in Oregon (Price and Boutin, 1993:149).

Like their congeners, red squirrels produce four calls associated with territorial behavior (rattle, screech, growl, and buzz) and one call in response to predators (chirp—Smith, 1978). The four territorial calls of *T. hudsonicus* differ in frequency and length from those of *T. douglasii*, but seemingly function similarly. The chirp call often is initiated with a low-amplitude, high-frequency "peep" added at the approach of a questionable predator (Smith, 1978). Unlike the chirp calls of *T. douglasii*, those of *T. hudsonicus* are relatively easy to locate. Differences in calls of the two closely related species may be related to differences in the ability of forest types in which the two species evolved to reflect sound. The hemlock forests in which *T. hudsonicus* is thought to have evolved tend to reflect echos of the peep-chirp calls that give the impression that two squirrels in different locations, one higher than the other, are calling (Smith, 1978). Smith (1978) considered the chirp call of *Tamiasciurus* possibly to be altruistic; however, such behavior may increase the survival of close kin, as offspring often establish territories adjacent to those of their mother where these studies were conducted in British Columbia.

Remarks.—A discussion of the problem of distinguishing between the two species of *Tamiasciurus* in eastern Oregon and a description of attempts to distinguish between the species where they are sympatric are presented in Remarks in the account of *T. douglasii* (see pp. 215–216).

The paucity of published information on *T. hudsonicus* in Oregon might be attributable to its relatively small and comparatively remote geographic range in the state were it not that >300 specimens from the state are on deposit in various systematics collections. We suspect that ease of collection and of preparing attractive skins has outweighed the concentrated effort necessary to quantify attributes of the species and its populations. Such is unfortunate, as we have been required to depend largely upon life-history information gathered elsewhere even though some attributes of the species have been shown to vary spatially in addition to temporally.

Glaucomys sabrinus (Shaw)

Northern Flying Squirrel

1801. *Sciurus Sabrinus* Shaw, 2:157.
1839c. *Pteromys oregonensis* Bachman, 8:101.
1897c. *Sciuropterus alpinus klamathensis* Merriam, 11:225.
1898a. *Sciuropterus alpinus bangsi* Rhoads, 49:321.
1898a. *Sciuropterus alpinus fuliginosus* Rhoads, 49:321.
1918. *Glaucomys sabrinus bangsi*: Howell, 44:38.
1918. *Glaucomys sabrinus oregonensis*: Howell, 44:44.
1918. *Glaucomys sabrinus fuliginosus*: Howell, 44:47.
1918. *Glaucomys sabrinus klamathensis*: Howell, 44:52.

Description.—*Glaucomys sabrinus* (Plate XI) is the smallest arboreal squirrel in Oregon (Table 11-9). It is typically squirrel-like, except that the fore- and hind legs are connected by a furred patagium that extends from the ankle to a thin, cartilaginous, styliform process that articulates with the bones of the wrist (Wells-Gosling and Heaney, 1984). The styliform process increases the area of the patagium, which, with the long (≅80% of the length of the head and body), distichously haired, dorsoventrally compressed tail, serves as an airfoil whereby the animals make gliding flights.

The hairs of the dorsum are lead colored with buffy-brown to brown tips shortening toward the edge of the patagium; the underfur of the tail is buffy, but the guard hairs are black tipped, more so distally; the venter is much lighter overall with a light buffy wash to whitish-tipped hairs with lead-colored bases. The underside of the tail is buffy darkening to nearly black distally. The color of individuals east of the Cascade Range tends to be lighter.

The zygomata are untwisted, but neither as parallel as those of other arboreal squirrels nor as convergent as in the ground-dwelling squirrels (Fig. 11-43). However, the skull is unique among Oregon arboreal squirrels in that the maxilla is visible through a notch at the anterior base of each postorbital process when the skull is viewed in dorsal aspect (Fig. 11-43).

Distribution.—*Glaucomys sabrinus* occurs in the bo-

real forest region from southern Alaska to coastal New Brunswick and southward in the Appalachian Mountains to western North Carolina and eastern Tennessee; in the Rocky Mountains and associated ranges south to southern Utah; and in the Coast, Cascade, and Sierra Nevada ranges to southern California (Fig. 11-44; Hall, 1981). In Oregon, *G. sabrinus* occurs in forested areas west of the Cascade Range and eastward to near Lakeview, Lake Co., and Paulina Lake, Deschutes Co.; the species also occurs in the Blue, Ochoco, and Wallowa mountains (Fig. 11-44).

Late Pleistocene (Wisconsinan) and Recent fossils of *G. sabrinus* are known from Tennessee, Pennsylvania, Virginia, West Virginia, Arkansas, and California (Kurtén and Anderson, 1980).

Geographic Variation.—Four of the 25 named races of the northern flying squirrel occur in Oregon: the darkest and smallest race in Oregon is *G. s. oregonensis* in the Coast Range and Willamette Valley; the slightly larger and browner *G. s. fuliginosus* occurs in the Siskiyou and Cascade ranges; the paler *G. s. klamathensis* occurs on the east slope of the southern Cascade Range; and the grayer *G. s. bangsi* occurs in the Wallowa and Blue mountains (Bailey, 1936; Hall, 1981).

Habitat and Density.—Throughout its range, *G. sabrinus* is associated with both coniferous and deciduous forests (Wells-Gosling and Heaney, 1984). In Oregon, it

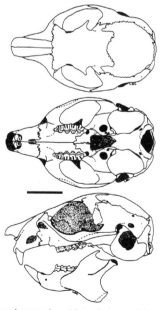

Fig. 11-43. Dorsal, ventral, and lateral views of the cranium, and lateral view of the dentary of the northern flying squirrel, *Glaucomys sabrinus* (OSUFW 3748). Scale bar equals 10 mm.

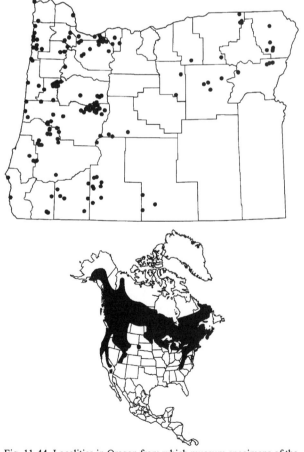

Fig. 11-44. Localities in Oregon from which museum specimens of the northern flying squirrel, *Glaucomys sabrinus*, were examined (listed in APPENDIX, p. 519), and the distribution of the northern flying squirrel in North America. (Range map redrawn from Hall, 1981:452, map 298, with modifications.)

Table 11-9.—*Means (±SE), ranges (in parentheses), and* CVs *of measurements of skull and skin characters for female and male flying squirrels* (Glaucomys sabrinus) *from regions in Oregon. Skin characters were recorded from specimen tags; skull characters were measured to the nearest 0.01 mm.* SE *and* CV *not provided if* n *<10.*

Region	Sex	n	Total length	Tail length	Hind foot length	Ear length	Mass (g)	Greatest length of skull	Nasal length	Length of maxillary toothrow	Zygomatic breadth
Coast Range	♀	24	299 ± 2.6 (264–328) 4.2	130 ± 2.2[a] (102–157) 8.1	39 ± 0.4 (35–42) 4.4	25 ± 0.5[b] (22–30) 8.5	138.87 ± 5.32[b] (100.6–196.9) 15.79	40.24 ± 0.22 (38.27–41.98) 2.67	12.65 ± 0.13 (11.10–13.63) 5.23	8.70 ± 0.05 (8.28–9.25) 2.79	23.56 ± 0.13 (22.33–24.74) 2.73
	♂	30	302 ± 1.7 (279–324) 3.1	134 ± 1.4 (120–151) 5.6	39 ± 0.3 (37–43) 4.1	25 ± 0.5[c] (23–29) 7.3	129.37 ± 3.69[c] (109.6–153.2) 9.87	40.38 ± 0.16 (38.89–42.62) 2.13	12.57 ± 0.11 (11.60–13.93) 4.75	8.44 ± 0.05 (7.97–9.21) 3.38	23.43 ± 0.11 (21.93–24.26) 2.62
A band from southern Josephine County to Wasco County	♀	23	304 ± 2.7 (273–341) 4.3	133 ± 2.0 (102–147) 7.3	39 ± 0.5 (32–41) 5.8	25 ± 0.3[d] (22–27) 5.1	126.36 ± 5.89[d] (91.8–179.5) 19.25	40.32 ± 0.19 (38.36–42.29) 2.29	12.61 ± 0.12 (10.91–13.55) 4.51	8.49 ± 0.11 (7.63–9.34) 6.32	23.68 ± 0.12 (22.19–24.71) 2.51
	♂	20	297 ± 3.3 (267–326) 5.0	129 ± 1.6 (117–144) 5.5	39 ± 0.5 (35–43) 5.9	24 ± 0.5[e] (21–27) 7.8	125.76 ± 3.65[f] (98.7–147.2) 11.60	40.47 ± 0.15 (39.52–42.05) 1.71	12.62 ± 0.08 (11.80–13.13) 2.85	8.47 ± 0.09 (7.71–8.97) 4.62	23.19 ± 0.12 (21.85–24.24) 2.32
East slope Cascade Range	♀	5	321 (310–338)	147 (135–159)	41 (40–44)			41.43 (40.55–42.81)	13.15 (12.39–13.94)	8.89 (8.51–9.15)	24.05 (22.46–25.02)
	♂	5	317 (306–332)	141 (128–151)	42 (40–46)			41.43 (40.92–42.17)	13.15 (12.38–13.68)	8.41 (7.92–8.73)	23.89 (23.60–24.40)
Northeastern Oregon	♀	13	326 ± 3.5[a] (303–341) 3.7	160 ± 11.7 (135–299) 26.4	45 ± 2.7[g] (39–71) 19.6	32[h] (27–42)	207.57[i] (199.0–214.7)	41.16 ± 0.45 (39.40–44.19) 3.92	12.92 ± 0.19 (11.85–14.06) 5.46	8.65 ± 0.13 (8.01–9.36) 5.37	24.96 ± 0.17 (24.00–25.80) 2.42
	♂	7	316 (300–344)	151 (145–162)	43 (40–47)	29[g] (24–32)	183.00[g] (128.0–213.0)	41.62 (38.95–43.49)	13.01 (11.72–13.94)	8.69 (7.89–9.39)	24.80 (23.13–25.81)

[a]Sample size reduced by 1. [b]Sample size reduced by 7. [c]Sample size reduced by 18. [d]Sample size reduced by 6. [e]Sample size reduced by 3. [f]Sample size reduced by 4. [g]Sample size reduced by 2. [h]Sample size reduced by 8. [i]Sample size reduced by 10.

occurs in Douglas-fir (*Pseudotsuga menziesii*), lodgepole pine (*Pinus contorta*), and mixed coniferous forests of grand fir (*Abies grandis*), Engelmann spruce (*Picea engelmannii*), and western larch (*Larix occidentalis*); Douglas-fir and ponderosa pine (*Pinus ponderosa*); Douglas-fir, western hemlock (*Tsuga heterophylla*), and red cedar (*Thuja plicata*) or incense-cedar (*Calocedrus decurrens*—Maser et al., 1985; Rosenberg and Anthony, 1992). We also know that *G. sabrinus* occupied nest boxes erected for wood ducks (*Aix sponsa*) in riparian zones in which the common trees were red alder (*Alnus rubra*), Oregon white oak (*Quercus garryana*), hawthorn (*Crataegus*), big-leaf maple (*Acer macrophyllum*), Oregon ash (*Fraxinus latifolia*), and Douglas-fir (Morse and Wight, 1969).

In the western Cascade Range, estimates of density of populations of *G. sabrinus* were slightly, but not significantly, less in five 30–50-year-old stands of Douglas-fir (range, 1.1–2.5/ha) than in five stands >400 years old (range, 1.4–2.9/ha). The similar densities are contrary to earlier suggestions by some workers that the species is more abundant in old-growth forests (Rosenberg and Anthony, 1992; Witt, 1992). Rosenberg and Anthony (1992) found lowest densities in stands <500 m from nests of spotted owls (*Strix occidentalis*). Based on an estimate of 25% of the northern flying squirrels within the foraging territory of a spotted owl being preyed upon, they concluded that predation may have a strong regulatory effect on densities of *G. sabrinus*. They suggested that the similar densities of *G. sabrinus* in second-growth and old-growth forest stands might be the result of differential rates of predation by owls.

The more dense and homogeneous canopies of second-growth stands may interfere with flight and the ability of owls to capture arboreal prey.

In the Coast Range (Douglas Co.) in 1981–1983, three estimates of density averaged 1.2 ± 0.5/ha (range, 0.7–1.7/ha) at one site in old-growth Douglas-fir forest and nine estimates averaged 0.9 ± 0.4/ha (range, 0.6–1.5/ha) at another site in old growth (Witt, 1992). Seven estimates at five sites in second growth (45–65 years old) based on a different method of estimating density averaged 0.1 ± 0.1/ha (range, 0.0–0.7/ha), overall about 6 times less than in old-growth stands (Witt, 1992). Witt (1992) claimed that estimates produced by the two methods were not significantly different. However, he indicated that estimates of home-range size incorporated into calculations of density were larger on second-growth sites than on old-growth sites, possibly making differences between densities in the two habitat types greater than indicated by calculated values. In nearby areas (the Miner Creek area possibly was the same as that used by Witt, 1992) in 1985–1989, average estimated density of *G. sabrinus* reportedly was 1.9 ± 0.1/ha in old-growth forest and 0.9 ± 0.2/ha in young managed forests, a significant difference (Carey et al., 1992). However, methods, statistical tests, and numerous assumptions employed in this study were challenged (Rosenberg et al., 1994) and the challenge rebutted (Carey, 1995).

Both Carey et al. (1992) and Witt (1992) contended that their findings supported the commonly held view that northern flying squirrels are more abundant in old-growth forests than in second-growth forests. Both also suggested that a microclimate more favorable to fungi on which *G.*

Table 11-9.—*Extended.*

Breadth of braincase	Breadth of molariform toothrow	Skull depth	Length of mandible
18.61 ± 0.08 (17.76–19.37) 2.15	10.05 ± 0.07 (9.37–10.66) 3.49	17.93 ± 0.09 (17.21–18.94) 2.45	23.02 ± 0.12 (21.65–23.81) 2.50
18.26 ± 0.09 (17.11–19.46) 2.64	10.01 ± 0.06 (9.41–10.55) 3.12	17.64 ± 0.09 (16.67–18.49) 2.65	22.99 ± 0.11 (21.94–24.24) 2.58
18.56 ± 0.09 (17.61–19.30) 2.48	10.15 ± 0.05 (9.54–10.54) 2.35	17.95 ± 0.10 (17.01–18.81) 2.69	23.05 ± 0.14 (21.90–24.59) 2.99
18.32 ± 0.07 (17.48–18.74) 1.82	9.98 ± 0.09 (9.36–10.92) 3.89	17.83 ± 0.07 (17.30–18.56) 1.73	22.99 ± 0.14 (22.06–24.05) 2.71
18.72 (18.32–19.24)	10.17 (9.85–10.57)	17.97 (17.59–18.38)	23.57 (22.68–24.17)
18.49 (18.33–18.66)	10.27 (9.94–10.58)	18.22 (17.58–19.12)	23.63 (22.49–24.37)
19.11 ± 0.12 (18.55–19.76) 2.22	10.50 ± 0.11 (9.82–11.05) 3.91	18.33 ± 0.18 (17.41–19.24) 3.47	23.54 ± 0.29 (22.02–25.07) 4.49
19.37 (17.94–20.02)	10.39 (9.36–11.09)	18.39 (17.48–19.46)	24.07 (22.36–27.18)

sabrinus feeds and the greater availability and variety of den sites were factors responsible for the greater density of northern flying squirrels in old-growth stands. However, the disparity between findings of these studies and those of Rosenberg and Anthony (1992) suggests that differences in density between second-growth and old-growth forests await demonstration and, if they exist, that factors responsible for those differences await definition.

Diet.—In northwestern Oregon, essentially 100% of the volume of the stomach contents of northern flying squirrels (n = 21) consisted of fungi; in northeastern Oregon (n = 61) ≥90% consisted of fungi or lichens. No lichens occurred in the stomachs of *G. sabrinus* in northwestern Oregon (Maser et al., 1985). In northeastern Oregon, lichens contributed most to the diet in winter–spring, but fungi contributed most in summer–autumn. Other items in the stomach contents, all recorded in trace amounts, were starch, lipid, leaves, seeds, hair, and invertebrates. In northeastern Oregon, staminate flowers of conifers were consumed in spring. The diet did not seem to be appreciably different between the sexes (Maser et al., 1985).

In northwestern and northeastern Oregon combined, 20 genera of fungal sporocarps were eaten. Of these 20, the percentage of sporocarps of *Rhizopogon* (combined with other genera with similar spores) was greatest (19–28%) for both sexes in both eastern and western Oregon (Maser et al., 1985). Of sporocarps identified in northwestern Oregon, *Gaultieria* and *Leucogaster* composed 14–19%, and in northeastern Oregon, lichens composed 25%. In southwestern Oregon, these three genera plus *Hysterangium* and *Geopora* were identified in fecal samples taken in all

months; 15 other genera were identified seasonally (Maser et al., 1986). In northwestern Oregon, the number of genera of fungi recorded in digestive tracts of individual *G. sabrinus* averaged 5.0 (range, 3–7) in males and 5.3 (range, 1–10) in females; in northeastern Oregon, the averages were 3.2 and 3.3 (ranges, 1–6 for both) for males and females, respectively (Maser et al., 1985).

In other regions, especially in the eastern part of its range, *G. sabrinus* reportedly consumes acorns, nuts, buds, catkins, fruits, insects, tree sap, roosting birds, eggs, and flesh of other vertebrates (Wells-Gosling and Heaney, 1984). Because of its tendency to feed opportunistically on various fungi (Maser et al., 1985), and because captives selected food items on the basis of protein content (Laurance and Reynolds, 1984), we suspect that *G. sabrinus* occasionally may forage opportunistically and take advantage of locally abundant foods in addition to feeding on fungi and lichens.

Reproduction, Ontogeny, and Mortality.—Although some observers reported that *G. sabrinus* produced two or three litters annually, others indicated that only one is produced (Wells-Gosling and Heaney, 1984). Cowan (1936*b*) was emphatic in his claim of only one litter annually in British Columbia, but left open the possibility of more than one in more southern latitudes. Most young are born in May–July, but occasionally a litter may be born as late as mid-November (Raphael, 1984; Witt, 1991). The observed gestation period was 37 days (Muul, 1969); thus mating probably occurs mostly from late March to late May.

Relatively little information regarding average litter size is available for *G. sabrinus* in the western part of its range (Table 11-4). Rust (1946) indicated that in northern Idaho one to three young, with two the most common, were found in "grass nests," but that litters of three were most common in tree cavities. However, these values were based on counts of nestlings subject to postnatal mortality; in addition, sample sizes were not presented. Cowan (1936*b*) listed two nests with three young each and a pregnant female with two embryos for British Columbia.

In three neonates born in captivity, the skin was bright pink, completely without pigment; the eyes were closed and the mouth, genitalia, and anus were covered with skin. The patagium was evident as a loose fold of skin and the toes were webbed (Booth, 1946). Vibrissae appeared and toes separated by day 6; pigments became visible in the skin of the head and shoulders and a few hairs commenced to appear by day 8; the back and sides became pigmented and fine hair grew over most of the body by day 11; the young could crawl by day 18; the lower incisors erupted by day 26; the young commenced to walk and the tail was well furred by day 41; and the young ate solid food by day 47 and were thought to be weaned by day 65 although some nursing was observed to day 85 (Booth, 1946). Ferron (1981) provided somewhat different average (range in parentheses) dates of development for captives: crawl foward, 1.1 (1–2) days; umbilicus healed, 13.4 (9–21) days; eruption of i1 19.9 (18–23) days; eruption of I1, 30.5 (28–33) days; auditory meatuses open, 23.7 (20–26) days; eyes open, 36.9 (33–42) days; regularly walk on all four feet, 32.1 (31–34) days; climb, 32.3 (30–33) days; out of nest at will, 49.0 (48–50) days; drink, 53.3 (52–54) days; and eat solid food, 51.1 (49–55) days.

Cowan (1936*b*) indicated that tails of three 160-mm-long young were covered with fine, short fur, but there was no evidence of dorsoventral flattening of the tail; however, Booth (1946) claimed that the tail was well furred and flattened in three individuals 41 days old for which total length ranged from 146 to 167 mm. The young commenced to glide by 3 months of age.

A male at 1 day of age weighed 6.0 g and its total length was 70 mm; at 5 days, it weighed 9.5 g, whereas its two female siblings weighed 8.2 and 6.8 g; at 15 days the three weighed 17.1, 15.0, and 11.6 g; at 25 days they weighed 23.6, 20.6, and 17.3 g; at 35 days they weighed 27.1, 24.7, and 21.1 g; and at 56 days they weighed 32.0, 28.9, and 27.8 g (Booth, 1946). At 1 year of age the disparity in mass of the three individuals was still apparent: 104.0, 84.0, and 76.0 g. Their wild-caught mother weighed 139 g, suggesting that growth of the young in captivity may have been stunted.

In the Cascade Range of western Oregon, 139 (30.3%) of 459 *G. sabrinus* were recaptured during the year following their initial capture and 36 (19.0%) of 189 were recaptured during the 2nd year following their initial capture (Rosenberg, 1991). Although differences were not significant, a greater proportion of recaptures were of females in second-growth than in old-growth forest stands, and more females than males were recaptured in both second-growth and old-growth forest stands (Rosenberg, 1991).

Forsman et al. (1984) indicated that *G. sabrinus* was the primary prey item in the diet of spotted owls in western Oregon. Other known predators of northern flying squirrels include a variety of raptors, martens (*Martes americana*), house cats, weasels (*Mustela*), and foxes (*Urocyon* and *Vulpes*—Wells-Gosling and Heaney, 1984). Occasionally, the patagium becomes impaled upon sharp objects such as barbed wire, resulting in death of the flying squirrel (Findley, 1945).

Habits.—Activity in *G. sabrinus* is mostly nocturnal although individuals may be observed abroad during daylight hours on rare occasions (Tanner, 1927). Northern flying squirrels usually are active for ≅2 h immediately after sunset, then after a sojourn in the nest are active for ≅1.25 h in the few hours before sunrise (Weigl and Osgood, 1974). Cloud cover advanced, but inclement weather delayed, onset of activity. *G. sabrinus* avoided gliding during rain, high wind, or fog (Weigl and Osgood, 1974). Activity continues throughout the year, even at low temperatures and in snow; the species is not known to enter torpor.

In northeastern Oregon, most northern flying squirrels use cavities in trees for denning, especially in winter (Maser et al., 1981a). In addition to cavities, dreys are constructed and used in areas with more moderate climates, including western Oregon (Bailey, 1936; Cowan, 1936b). Nest boxes erected for use by other species sometimes are appropriated by northern flying squirrels, but natural cavities in trees probably provide better-insulated den sites (Maser et al., 1981a).

Wing loading in *G. sabrinus* is about 50 Newtons/m², considerably greater than that of bats (Chiroptera), but about the same as that recommended for hang-gliders (Thorington and Heaney, 1981). The glide ratio is about 2:1 to 3:1, that is, a glide of 40 m would require a drop of 13–20 m (Sumner and Dixon, 1953; Thorington and Heaney, 1981). In practice, the initial part of the glide is steep so that the squirrel attains optimal velocity quickly; the squirrel levels the flight path to increase glide distance, then, near the end of the glide, in-

creases the angle of attack to slow the landing speed and to regain part of the altitude used initially to attain speed (Thorington and Heaney, 1981).

In the Cascade Range, home-range size of *G. sabrinus* as indexed by average maximum distance moved between points of capture averaged (±SE) 86.6 ± 6.1 m in second-growth stands and 83.5 ± 5.5 m in old-growth stands; the differences were not significant (Rosenberg and Anthony, 1992). On some of the same areas in the Cascade Range, average (±SE) home-range areas calculated by the adaptive kernel method (Whorton, 1989) for *G. sabrinus* were 5.9 ± 0.8 ha (n = 20) for males and 3.9 ± 0.4 ha (n = 19) for females; the difference was significant (P < 0.05—Martin, 1994). However, home-range areas for flying squirrels occupying second-growth stands ≅40 years old and old-growth stands >400 years old were not statistically different (Martin, 1994). Directly comparable data on movements are not available for populations in the Coast Range. However, Witt (1992) indicated that home-range areas calculated from points of capture by inclusive boundary-strip and minimum convex-polygon methods averaged (±SE) 3.9 ± 0.08 and 2.3 ± 0.08 ha, respectively, in an old-growth stand. Based on both points of capture and locations acquired by radiotelemetry, home-range areas averaged 4.2 ± 0.3 ha by the minimum convex-polygon method. Witt (1992) suggested that home-range areas based on capture points alone might be more representative of foraging areas as three of five individuals that he radiotracked had den sites outside home-range areas based on capture points. Comparable data were not available for *G. sabrinus* in second-growth stands.

Northern flying squirrels are known to occur in aggregations of as many as four individuals in a nest (Maser et al., 1981a; Osgood, 1935). Maser et al. (1981a) claimed that northern flying squirrels were segregated by sex in 30 (97%) of 31 nest boxes erected for American kestrels (*Falco sparverius*) they occupied. Although neither aggregation nor separation by sex is questioned, support for the claim is misleading because a minimum of 17 of the boxes were occupied by only one individual.

Witt (1991) found that live traps placed on the side of ≥38-cm-diameter trees ≅1.5 m from the base produced 3 times the number of individual flying squirrels captured and >10 times as many captures/trap-night than traps set on the ground. Greatest trapping success was in April–June and October–December. Rosenberg and Anthony (1993b) observed differential mortality among sex and age-classes during 16–21-day trapping sessions; rates were 32.3% for juvenile females, 11.1% for juvenile males, 5.1% for adult females, and 4.1% for adult males.

Remarks.—Interest in the biology of the northern flying squirrel has escalated in recent years because of the role it plays in the forest ecosystem by dispersing spores of mycorrhizal fungi that contribute to nutrient cycling, productivity, and plant succession (Maser et al., 1986), and because it is the principal constituent in the diet of the spotted owl, which currently is considered threatened (Forsman et al., 1984). We applaud the effort to add to the knowledge of the species, but we think that *G. sabrinus* is of sufficient interest as a member of the mammalian fauna of Oregon to deserve research effort for its own sake. Were it not for fewer than half-a-dozen recently published papers, knowledge of the species in the state would be based largely on anecdotal accounts. As it stands, knowledge of some topics is infinitesimal; for example, published records of average litter size in the western part of the range all seem to be based on three or fewer litters or gravida (Wells-Gosling and Heaney, 1984), and number of litters produced annually does not seem to have been established with certainty (Wells-Gosling and Heaney, 1984).

Arbogast (1992) observed mitochondrial-DNA sequence divergence between northern flying squirrels in the Pacific states and those in the remainder of the range nearly as great as that between the two recognized species of *Glaucomys* in North America. Such suggests that the flying squirrels in the Pacific states may be taxonomically distinct at the species level from those in the remainder of North America.

FAMILY GEOMYIDAE—POCKET GOPHERS

The pocket gophers are New World endemics for which the fossil history commences in the late Miocene (Kurtén and Anderson, 1980). They are closely related to the pocket mice, kangaroo rats, and kangaroo mice (Heteromyidae) and are classified with them in the superfamily Geomyoidea (McLaughlin, 1984). The family includes five Recent genera and 35 species (Wilson and Reeder, 1993a).

In Oregon, the geomyids are represented by five species of pocket gophers of the genus *Thomomys*. Members of the family are characterized by short, nearly naked tails; external, fur-lined cheek pouches opening laterally to the mouth; and small ears, beady eyes, and lips that close behind the incisors. The shoulders are powerfully built, the hips slim, the feet pentadactyl and plantigrade, the front feet equipped with long claws for digging, and the fingers and palms bordered by fringes of stiff bristles. The skull is heavy, flattened dorsoventrally, and strongly angular with narrow infraorbital foramina, and small bullae. The premolars are 8-shaped, each portion almost appearing as a separate tooth. All teeth are rootless and ever-growing, with upper and lower incisors growing 0.62 and 0.99 mm/day, respectively, in one species (Howard and Smith, 1952).

Pocket gophers are nearly contiguously allopatric everywhere, although, in Oregon, *T. bulbivorus* and *T. mazama* are reported to occur within "one hundred yards of each other" (Walker, 1955:122). A few instances of sympatry are recorded in other states; where it was recorded, it was between two morphologically divergent types (Thaeler, 1968a). Andersen and MacMahon (1981) suggested that the parapatry in members of the family is the result of differences among the species in burrowing efficiency in different soils, a view presented earlier by Thaeler (1968a). Pocket gophers are strongly territorial; the only social groupings are formed during the breeding season (Hansen and Miller, 1959). At all other times pocket gophers probably are solitary.

Pocket gophers are fossorial mammals that excavate extensive tunnel systems; individuals in adjacent territories probably have interconnecting burrows, but probably keep them separated with plugs of earth most of the time. Soil is loosened from the faces of new tunnels with incisors and front feet, then pushed out of tunnel openings bulldozer fashion with the front feet and head (Plate XII). In summer, excavated soil is pushed into fan-shaped mounds at burrow entrances, whereas in winter, when deep snow covers the ground, excavated soil is packed into tunnels formed above ground in the snow. When the snow melts, earth casts of the tunnels lie snake-like on the surface of the ground (Plate XII). Pocket gophers usually plug burrow openings when not actively excavating earth and plug breaks in actively used portions of their burrows, a behavior taken advantage of in trapping these species as opening a tunnel creates all the "lure" required.

Pocket gophers are among the most morphologically variable mammals. Intraspecific variation frequently is as great as or greater than interspecific variation. Thus, separation of some parapatric species by external morphological characters or skull characters often is difficult.

Because of their depredations on agricultural crops and seedlings planted for reforestation, and because of the mounds they form in lawns and fields, a variety of control measures have been devised to reduce pocket gopher populations. Where soils, terrain, and field size permit its use, the burrow-builder, a device pulled with a tractor that constructs an artificial burrow and places toxic baits at intervals along the tunnel, probably is the most effective and probably is least hazardous to nontarget species. Traps (Macabee and Victor for small species, Cinch for large species) can be used to reduce numbers in small areas; however, tunnel systems may be reoccupied quickly by dispersers (Tunberg et al., 1984). Several live traps have been designed (Baker and Williams, 1972; Hansen, 1962; Howard, 1952), but none is available commercially.

Dr. Walter E. Howard, University of California–Davis, tested the efficacy of several of the "electronic" devices commonly advertised to rid the lawn and garden of gophers and moles; all were found completely ineffective. Although we know of no controlled test of its efficacy, we suspect that the locally advocated measure of placing a specific brand and flavor of chewing gum in burrow systems as a control for pocket gophers is equally ineffective. We have considered conducting tests of gum, but have never been able to ascertain if it were to be left in the wrapper, unwrapped, or prechewed.

Despite immediate economic losses attributable to them, pocket gophers play vital roles in the ecosystems of which they are a component. Burrowing activities of pocket gophers bring soil to the surface, thereby returning leached nutrients to the root zone; the tunnels thus formed provide channels for surface drainage with deep penetration of water and reduction of runoff and erosion. On depleted grasslands, elimination of grazing by sheep caused an increase in porosity and a decrease in bulk density of soils; additionally, when pocket gophers were present, soil tilth and fertility increased, primary production increased, and the structure of the plant community changed (Laycock and Richardson, 1975). In some regions, pocket gophers are a dominant influence on ecosystems, not only controlling the character of the vegetation, but strongly influencing the distribution and abundance of other small mammals and even the movement patterns of some migrating birds (Vaughan, 1974).

KEY TO THE GEOMYIDAE OF OREGON

1a. Sphenoidal fissure present...2
1b. Sphenoidal fissure absent...4
   2a. Pelage strongly ocherous (pumpkin-colored); hind foot

<33 mm long...
...*Thomomys bottae*, Botta's pocket gopher (p. 224).
2b. Pelage grayish or dark brownish, never ocherous; hind
foot ≥34 mm long...3
3a. Pelage on dorsum usually grayish or light grayish
brown, venter slightly lighter and without white throat
patch; hind foot <40 mm long in males, <38 mm long
in females; pterygoids straight; exoccipital lacking
grooves..*Thomomys*
townsendii, Townsend's pocket gopher (p. 238).
3b. Pelage on dorsum usually dark sooty brown, venter
lead colored with irregular white throat patch; hind foot
>40 mm long in males, >38 mm long in females; ptery-
goids concave on inner surfaces, convex on outer surfaces; exoccipital with a pair of grooves..........
...*Thomomys bulbivorus*, camas pocket gopher (p. 229).
4a. Pelage on dorsum black, blackish with reddish dorsal
stripe, or reddish brown; baculum 22–31 mm long in
adult males; mandible with wide flange (≥1.5 mm)
on posteriolateral edge...........................*Thomomys*
mazama, western pocket gopher (p. 231).
4b. Pelage on dorsum yellowish gray to yellowish brown;
baculum 12–17 mm long in adult males; mandible
with narrow flange (≅1 mm) on posteriolateral
edge...*Thomomys*
talpoides, northern pocket gopher (p. 234).

Thomomys bottae (Eydoux and Gervais)
Botta's Pocket Gopher

1836. *Oryctomys (Saccophorus) bottae* Eydoux and Gervais, 6:23.
1855. *Thomomys laticeps* Baird, 7:335.
1897d. *Thomomys leucodon* Merriam, 11:215.
1935. *Thomomys bottae detumidus* Grinnell, 40:405.

Description.—*Thomomys bottae* is a member of the
heavy-rostrum group (subgenus *Megascapheus*—Thaeler,
1980) of pocket gophers, intermediate in size between the
T. mazama-*T. talpoides* gophers and the *T. bulbivorus*-*T.
townsendii* gophers (Table 11-10). *T. bottae* possesses a
sphenoidal fissure (Fig. 11-45), a character that with body
size serves to distinguish the species from all other gophers in the state. The hind foot is <33 mm long.

The pelage of *T. bottae* tends to be dark rusty-ocherous
on the dorsum and light buffy-ocherous on the venter; the
nose is dusky and the feet and chin are white. Often the
cheeks and venter are spotted white (Bailey, 1915). The
incisors have predominantly white faces in inland populations, but those in the coastal population have yellow faces
(Grinnell, 1935).

Distribution.—The distribution of Botta's pocket gopher (Fig. 11-46) extends from west-central Oregon into
Baja California east through central Nevada and Utah
into southern Colorado, the western three-fourths of
New Mexico, and into Texas west of the Pecos River with
southward projections into Sonora, Chihuahua, Coahuila,
and Nuevo León (Patton and Smith, 1990). In Oregon, *T.
bottae* occurs as disjunct populations: one in Curry Co.
only, the other in southwestern Klamath Co., southern
Jackson Co., and western Josephine Co. with an extension through central Douglas Co. to Cottage Grove, Lane
Co. (Fig.11-46).

Pleistocene fossils are known from California, Nevada,
New Mexico, and Texas; the latter site is extralimital

and the oldest record (103,000 years ago—Kurtén and
Anderson, 1980).

Geographic Variation.—Three of the nearly 150 named
subspecies of *T. bottae* are reported to occur in Oregon:
T. b. laticeps and *T. b. detumidus* in Curry Co., and *T. b.
leucodon* in the remainder of the range (Hall and Kelson,
1959). We do not consider specimens previously referred
to *T. b. detumidus* to be sufficiently different from those
considered to be *T. b. laticeps* to warrant a separate trinomial (see Remarks).

Habitat and Density.—We know of no published information on *T. bottae* regarding these topics in Oregon.
However, we have records of *T. bottae* taken in a 3-year-
old ponderosa pine (*Pinus ponderosa*) plantation with an
understory of a light growth of bristleweed (*Haplopappus*),
ceanothus (*Ceanothus*), and cheat grass (*Bromus tectorum*)
on heavy clay soils and in an oak (*Quercus*)-grass meadow
in Klamath Co. We also have examined burrow systems of
T. bottae in weedy lawns, pastures, and highway rights-of-
way on hard-packed clay soils in Lane Co. In addition to
our observations, Bailey (1936) indicated an association
between *T. bottae* and clay soils; however, we also have
taken the species in dark-colored soils containing considerable sand in Curry Co., and the species has been reported
to occur in soft sands and friable loams in addition to hard
clays in other states (Ghiselin, 1965; Thaeler, 1968a). The
latter authority considered *T. bottae* in northern California
to occupy the broadest spectrum of environmental conditions of five forms studied; grass and riparian associations
were occupied at low elevations and montane meadows and

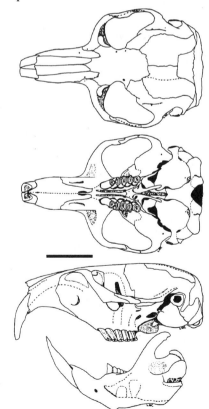

Fig. 11-45. Dorsal, ventral, and lateral views of the cranium, and lateral
view of the dentary of Botta's pocket gopher, *Thomomys bottae* (OSUFW
7519). Note sphenoidal fissure in lateral view of cranium. Scale bar equals
10 mm.

open pine forests at high elevations. *T. bottae* is able to exclude other species of pocket gophers from hard soils (Thaeler, 1968*a*). However, in Colorado, *T. bottae* has broader niche requirements than *T. talpoides* (Miller, 1964), but contrary to expectations, *T. talpoides* is more aggressive than *T. bottae* which, in part, may be responsible for the competitive interference that separates the two species spatially (Baker, 1974). Like *T. talpoides*, *T. bottae* rarely occurs in heavily forested areas (Bailey, 1936; Thaeler, 1968*a*).

In California, densities of adults were 3–84/ha with highest values in alfalfa fields, intermediate values in grasslands, and lowest values in desert areas (Patton and Smith, 1990). Sex ratios in desert areas are near 1:1, but in more productive areas, they are significantly biased in favor of females (Patton and Smith, 1990).

Diet.—In California, on serpentine soils (characterized by high magnesium:calcium ratios and extremely low fertility) mildly toxic to many species of plants, but supporting cluster lilies (*Brodiaea puchella*, *B. coronaria*, and *B. laxa*), two of five *T. bottae* contained corms of these plants in their cheek pouches and all five contained largely masticated corms in their stomachs (Proctor and Whitten, 1971). Congeners of these plants and serpentine soils are common in the range of *T. bottae* in Oregon (Franklin and Dyrness, 1969). During a study of the response to fertiliza-

tion of a plant community on serpentine soil, *T. bottae* located many of the fenced 50- by 50-cm plots and consumed much of the plant material in the new communities (Hobbs et al., 1988). In montane regions, quaking aspen (*Populus tremuloides*) may be restricted to rocky outcrops by *T. bottae* feeding on belowground parts of these trees in meadows (Cantor and Whitman, 1989). We suspect that *T. bottae* in Oregon, like its conspecifics in California, forages on essentially all palatable plants within a body-length radius of feeding holes (Howard and Childs, 1959).

Reproduction, Ontogeny, and Mortality.—For Oregon, we have records of only one pregnant female (with five embryos in May) and four individuals with pigmented implantation sites (Table 11-11). In California, average litter size ranged from 4.1 to 5.6 and breeding seasons ranged from 4 to ≅10 months with lowest values in desert areas, intermediate values in grasslands, and highest values in alfalfa fields (Patton and Smith, 1990). Multiple annual litters among gophers were common in alfalfa fields (38–46% of breeding females), less so in grasslands (rare to 20%), and nonexistent in deserts (Patton and Smith, 1990).

Gestation is 18–19 days (Schramm, 1961). Neonates are extremely altricial, blind, and hairless except for a few fine, colorless sensory tufts, and have a ≅36-mm crown-rump length. The claws are hooflike. The tail is constricted at the base and ends in a strange peglike structure (Hill, 1934). In California, growth in wild juveniles caught repeatedly averaged ($\pm SE$) 1.17 ± 0.37 g/day ($n = 11$) for males and 0.74 ± 0.41 g/day ($n = 14$) for females (Daly and Patton, 1986). Growth continues throughout life in males, but not in females (Daly and Patton, 1986; Howard and Childs, 1959).

In California, average life expectancy for *T. bottae* taken after 1 September (when most individuals are about 6 months old) was 7.6 months for males and 12.3 months for females; longevity was ≤30 months for males and ≤54 months for females (Howard and Childs, 1959). Predators of *T. bottae* include gopher snakes (*Pituophis catenifer*), rattlesnakes (*Crotalus viridis*), red-tailed hawks (*Buteo jamaicensis*), great horned owls (*Bubo virginianus*), barn owls (*Tyto alba*), and coyotes (*Canis latrans*—Howard and Childs, 1959).

Habits.—In June–September in California, *T. bottae* was active ≅40% of the time from 0400 to 1600 h, ≅60% of the time from 1600 to 2000 h, and ≤20% of the time from 2000 to 0400 h. About 15 h/day were spent in nest chambers; bouts of activity averaged ($\pm SE$) 25.9 ± 5.5 min and rest periods averaged 39.5 ± 6.8 min. Neither timing nor location of activity within the burrow system was influenced directly by environmental factors (Gettinger, 1984).

Movements of *T. bottae* introduced into an area from which original resident pocket gophers were removed ranged from 0 to ≅275 m ($\overline{X} = 60.4$ m, $n = 18$); 33% were recaptured at the site of release, 51% were recaught in the same burrow system, and 60% had moved <61 m (Vaughan, 1963). Thus, the ability or the pressure to disperse seemed to be much less for *T. bottae* than for *T. talpoides* in the same region (Vaughan, 1963).

Like its congeners, *T. bottae* is territorial; only during the breeding season can more than one individual be captured in the same burrow system. Even then, it is not known for certain if multiple captures indicate cohabitation or sequential invasion of undefended burrow systems by individuals residing nearby (Howard and Childs, 1959).

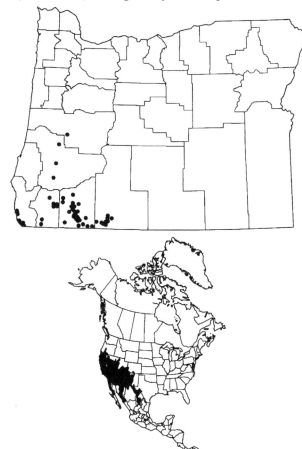

Fig. 11-46. Localities in Oregon from which museum specimens of Botta's pocket gopher, *Thomomys bottae*, were examined (listed in APPENDIX, p. 574), and the distribution of Botta's pocket gopher in North America. (Range map redrawn from Patton and Smith, 1990:4, fig. 1.1, with modifications.)

Table 11-10.—Means (±SE), ranges (in parentheses), and CVs of measurements (in mm) of skull and skin characters for female and male pocket gophers (Thomomys) from regions in Oregon. Skin characters were recorded from specimen tags; skull characters were measured to the nearest 0.01 mm. SE and CV not provided is n <10.

Species and region	Sex	n	Total length	Tail length	Hindfoot length	Mass (g)	Basilar length	Zygomatic breadth	Mastoid breadth	Least interorbital breadth	Length of maxillary toothrow	Skull depth	Length of mandible	Length of mandibular toothrow
Thomomys bottae														
Northern Curry County	♀	6	219 (211–223)	68 (59–71)	29 (28–30)	130.0[b]	32.64 (31.53–33.43)	24.06 (23.03–25.39)	19.82 (19.08–20.64)	6.56 (6.32–6.88)	8.54 (8.11–8.83)	14.87 (13.63–15.60)	0.30 (22.02–24.08)	9.07 (8.64–9.65)
	♂	7	240 (224–263)	75 (68–82)	31 (30–33)	156.0[c]	35.34 (34.33–37.66)	25.69 (24.43–27.44)	21.35 (20.58–22.72)	6.39 (5.81–6.60)	9.26 (8.88–9.52)	15.17 (14.51–15.95)	24.95 (23.41–27.18)	9.28 (8.84–9.77)
Southern Curry County	♀	2	215 (212–217)	67 (65–69)	29 (28–29)		31.65 (30.74–32.56)	23.18 (22.21–24.15)	19.40 (19.14–19.66)	6.81 (6.69–6.93)	8.71 (8.49–8.93)	14.61 (13.94–15.27)	22.33 (21.58–23.08)	8.48 (8.02–8.94)
Western interior valleys and Jackson and Klamath counties	♀	22	206 ± 4.6 (185–280) 10.5	59 ± 1.6 (43–75) 12.8	27 ± 0.4 (22–30) 7.6	132.8[d] (93.3–168.6)	32.31 ± 0.34 (29.49–35.10) 4.87	24.45 ± 0.19 (22.63–26.67) 3.59	19.74 ± 0.19 (17.83–21.18) 4.44	6.50 ± 0.06 (6.02–6.93) 4.03	8.03 ± 0.12 (7.22–9.11) 6.79	14.09 ± 0.17 (12.48–15.53) 5.78	22.46 ± 0.26 (20.25–24.66) 5.48	8.04 ± 0.13 (7.07–9.19) 7.86
	♂	17	224 ± 3.3 (198–246) 6.2	67 ± 1.7 (58–81) 10.6	29 ± 0.4 (27–33) 5.2	147.7[e] (112.4–183)	5.41 ± 0.49 (31.02–38.68) 5.78	27.19 ± 0.33 (24.74–29.19) 5.04	21.43 ± 0.31 (19.13–23.56) 5.92	6.38 ± 0.09 (5.80–7.06) 5.64	8.08 ± 0.11 (7.10–8.66) 5.62	15.19 ± 0.18 (14.34–16.76) 4.76	24.56 ± 0.36 (21.62–27.02) 6.02	8.09 ± 0.12 (7.34–9.40) 6.32
Thomomys bulbivorus														
North of Chehalem Mountains	♀	17	279 ± 2.6 (258–301) 3.8	74 ± 1.3 (61–82) 7.1	39 ± 0.4 (36–41) 3.9	359.9 ± 15.86[e] (261–439) 15.27	45.78 ± 0.39 (43.14–48.61) 3.54	35.41 ± 0.29 (33.66–37.93) 3.44	29.29 ± 0.27 (27.71–31.59) 3.83	7.25 ± 0.06 (6.83–7.89) 3.60	11.02 ± 0.16 (10.00–12.25) 5.87	20.81 ± 0.23 (19.28–22.63) 4.50	32.84 ± 0.28 (30.67–34.89) 3.49	11.34 ± 0.15 (10.47–12.30) 5.42
	♂	16	289 ± 5.2 (234–320) 7.2	79 ± 3.3 (39–95) 16.7	42 ± 0.4 (38–44) 3.4	443.9[f] (378.2–544.7)	48.53 ± 0.30 (46.05–50.98) 2.51	37.42 ± 0.42 (33.65–40.16) 4.54	30.75 ± 0.29 (28.71–33.08) 3.76	7.29 ± 0.09 (6.61–7.81) 5.05	11.00 ± 0.19 (9.58–12.96) 6.97	22.56 ± 0.24 (20.76–24.28) 4.27	35.06 ± 0.25 (33.03–36.63) 2.83	11.41 ± 0.19 (10.11–13.04) 6.61
South of Chehalem Mountains	♀	34	272 ± 2.6 (236–303) 5.6	74 ± 1.2 (61–90) 9.1	38 ± 0.4[g] (33–42) 5.3	345.2 ± 11.46[b] (215.8–449.3) 18.19	45.41 ± 0.34 (39.22–49.00) 4.43	34.65 ± 0.27 (30.64–37.29) 4.61	28.59 ± 0.28 (24.81–32.00) 5.69	7.06 ± 0.07 (6.14–8.25) 5.88	11.06 ± 0.11 (9.39–12.28) 5.67	21.01 ± 0.18 (18.65–22.58) 5.08	32.51 ± 0.27 (27.59–35.07) 4.87	11.46 ± 0.10 (10.50–12.76) 5.14
	♂	30	291 ± 2.9 (266–327) 5.4	79 ± 1.4 (63–99) 9.8	41 ± 0.5 (36–47) 6.8	428.5 ± 17.06[b] (289.1–629.4) 21.07	48.59 ± 0.51 (43.94–54.43) 5.77	37.60 ± 0.40 (33.82–43.12) 5.85	30.94 ± 0.33 (28.17–35.05) 5.85	7.21 ± 0.09 (6.35–8.15) 6.59	11.17 ± 0.13 (10.10–13.56) 6.13	22.44 ± 0.31 (19.86–26.06) 7.55	34.69 ± 0.38 (31.00–38.88) 5.94	11.51 ± 0.09 (10.75–13.42) 4.52
Thomomys mazama														
Northeast coastal	♀	30	212 ± 1.8 (192–232) 4.6	67 ± 1.0 (51–82) 8.3	28 ± 0.2 (25–30) 4.0	91.8 ± 2.31[i] (72.0–107.0) 9.41	30.23 ± 0.18 (28.56–32.07) 3.32	20.65 ± 0.17 (18.95–22.54) 4.44	17.19 ± 0.10 (15.80–18.41) 3.28	6.08 ± 0.06 (5.54–6.67) 5.62	8.09 ± 0.07 (7.41–8.93) 4.75	13.03 ± 0.06 (12.28–13.59) 2.59	21.23 ± 0.15 (19.51–22.79) 3.80	7.95 ± 0.06 (6.91–8.59) 4.12
	♂	30	222 ± 2.9 (191–252) 7.3	70 ± 1.2 (55–83) 9.7	29 ± 0.3 (25–32) 5.7	102.0[j] (83.0–115.0)	31.60 ± 0.24 (29.06–35.51) 4.24	21.84 ± 0.18 (19.56–25.18) 4.56	17.99 ± 0.14 (16.50–19.93) 4.24	6.16 ± 0.07 (5.31–6.96) 6.63	8.34 ± 0.09 (7.14–9.28) 5.61	13.30 ± 0.10 (12.27–15.00) 4.28	22.36 ± 0.19 (20.11–25.52) 4.84	8.05 ± 0.08 (7.08–9.15) 5.26
Eastern Coast Range and Columbia River	♀	35	194 ± 2.0 (175–220) 6.1	56 ± 1.1 (40–73) 12.0	27 ± 0.3 (24–30) 5.9	88.7[k] (80.0–100.0)	29.38 ± 0.19 (27.39–31.49) 3.92	20.82 ± 0.18 (19.29–22.72) 5.09	17.05 ± 0.17 (15.55–19.22) 5.91	6.45 ± 0.04 (6.01–7.06) 3.62	7.71 ± 0.06 (7.04–8.43) 4.57	12.79 ± 0.08 (11.85–13.85) 3.67	20.85 ± 0.13 (19.41–22.86) 3.59	7.69 ± 0.05 (6.93–8.17) 4.00
	♂	35	204 ± 1.8 (179–225) 5.3	59 ± 1.1 (43–70) 11.2	27 ± 0.3 (24–30) 5.6	98.5 ± 3.17[l] (82.0–110.0) 10.67	30.65 ± 0.23 (28.50–33.97) 4.53	21.66 ± 0.19 (19.65–24.44) 5.44	17.64 ± 0.18 (16.25–20.65) 6.08	6.54 ± 0.05 (5.79–6.99) 4.12	7.87 ± 0.05 (7.23–8.52) 3.84	13.08 ± 0.09 (11.79–14.09) 4.04	21.72 ± 0.15 (19.82–23.71) 4.18	7.79 ± 0.05 (7.10–8.31) 3.61
Central coastal	♀	31	214 ± 1.2 (203–228) 3.2	68 ± 0.9 (57–78) 7.8	28 ± 0.3 (25–30) 5.3		30.72 ± 0.13 (29.11–32.03) 2.44	20.95 ± 0.13 (19.49–22.41) 3.37	17.31 ± 0.09 (16.38–18.20) 2.83	6.04 ± 0.03 (5.59–6.39) 2.91	8.45 ± 0.05 (7.99–9.07) 3.53	13.00 ± 0.06 (12.29–13.85) 2.54	22.31 ± 0.13 (20.79–23.73) 3.23	8.12 ± 0.04 (7.71–8.66) 2.77
	♂	21	221 ± 2.5 (196–243) 5.3	69 ± 1.7 (52–83) 10.9	29 ± 0.3 (26–31) 3.9		32.21 ± 0.31 (29.06–34.82) 4.36	22.55 ± 0.26 (20.07–24.33) 5.24	18.22 ± 0.19 (16.76–20.46) 4.89	6.19 ± 0.08 (5.56–6.85) 5.75	8.75 ± 0.09 (8.00–9.52) 4.57	13.59 ± 0.12 (12.40–14.57) 4.17	23.27 ± 0.24 (21.30–25.29) 4.81	8.41 ± 0.07 (7.96–9.04) 3.61

Region	Sex	n												
Southeast coastal	♀	7	204 (195–207)	64 (58–70)	28 (27–28)	89.0	29.73 (28.89–30.38)	20.35 (19.55–20.87)	17.18 (16.51–17.67)	6.30 (5.86–6.77)	7.84 (7.56–8.07)	12.77 (12.33–13.22)	21.25 (20.45–21.92)	7.65 (7.29–8.02)
	♂	8	213 (207–218)	67 (61–70)	28 (26–29)	112.5[m] (102.0–123.0)	30.57 (29.13–32.28)	21.21 (20.08–22.98)	17.72 (16.64–18.65)	6.27 (5.91–6.56)	8.12 (7.79–8.49)	12.92 (12.49–13.48)	21.77 (20.94–23.57)	7.95 (7.74–8.26)
Western Cascade Range extending to central Coast Range	♀	41	199 ± 2.0 (160–219) 6.4	61 ± 1.0 (45–73) 10.5	27 ± 0.3 (22–30) 6.2	92.2 ± 2.76[o] (67.0–123.0) 14.65	29.71 ± 0.16 (27.54–32.05) 3.49	20.38 ± 0.20 (18.08–23.70) 6.37	17.01 ± 0.16 (15.47–19.61) 5.86	6.30 ± 0.06 (5.52–7.54) 5.80	7.94 ± 0.08 (6.88–8.99) 6.56	12.69 ± 0.09 (11.57–14.14) 4.67	21.38 ± 0.15 (19.31–24.02) 4.61	7.75 ± 0.09 (5.90–8.95) 8.09
	♂	33	208 ± 2.4 (179–237) 6.5	65 ± 1.6 (45–90) 13.8	27 ± 0.4 (20–30) 7.5	97.8 ± 3.34[m] (72.0–135.0) 17.72	30.77 ± 0.24 (28.69–33.56) 4.55	20.56 ± 0.21 (18.58–23.73) 5.99	17.32 ± 0.17 (15.62–19.76) 5.67	6.14 ± 0.04 (5.61–6.63) 3.78	7.92 ± 0.07 (7.11–9.01) 5.09	12.89 ± 0.09 (11.88–14.12) 4.19	21.75 ± 0.19 (19.67–23.92) 5.04	7.69 ± 0.07 (6.97–8.21) 4.96
East slope central Cascade Range	♀	19	202 ± 2.1 (191–223) 4.4	63 ± 1.3 (53–76) 8.9	27 ± 0.3 (25–30) 4.7	82.5 ± 1.57[f] (76.0–88.3) 6.30	29.25 ± 0.25 (27.27–31.50) 3.74	19.46 ± 0.19 (18.36–21.19) 4.23	16.50 ± 0.17 (15.34–18.13) 4.49	6.28 ± 0.07 (5.76–6.77) 4.56	7.78 ± 0.08 (7.17–8.50) 4.24	12.32 ± 0.09 (11.44–13.15) 3.53	20.66 ± 0.20 (18.96–22.36) 4.25	7.51 ± 0.07 (7.02–7.97) 3.85
	♂	21	212 ± 2.1 (197–234) 4.5	63 ± 1.2 (56–72) 8.4	28 ± 0.3 (25–30) 4.6	98.9[o] (80.0–123.7)	30.92 ± 0.36 (27.80–33.82) 5.32	20.30 ± 0.26 (18.05–22.95) 5.95	17.09 ± 0.16 (15.57–18.39) 4.29	6.19 ± 0.06 (5.83–6.71) 4.67	7.89 ± 0.07 (7.43–8.49) 3.87	12.79 ± 0.13 (11.63–13.58) 4.52	21.31 ± 0.26 (19.04–23.27) 5.56	7.65 ± 0.06 (7.23–8.52) 3.67
Thomomys talpoides														
Eastern Columbia Basin	♀	14	197 ± 2.4[b] (184–213) 4.3	59 ± 1.1[b] (55–65) 6.2	27 ± 0.3[b] (24–28) 4.4	77.8 ± 2.75[b] (75–80.5) 5.00	30.69 ± 0.43 (27.94–33.33) 5.20	20.53 ± 0.27 (18.85–21.77) 4.91	17.84 ± 0.18 (16.81–18.87) 3.79	6.49 ± 0.09 (6.02–6.93) 5.05	7.77 ± 0.11 (7.17–8.55) 5.29	12.86 ± 0.16 (12.00–13.70) 4.65	21.06 ± 0.31 (19.21–22.59) 5.45	7.56 ± 0.11 (6.77–8.04) 5.37
	♂	12	212 ± 2.8[b] (198–223) 4.2	62 ± 1.5[b] (54–70) 7.8	28 ± 0.5[b] (25–30) 5.3	95.5	34.04 ± 0.57 (29.21–36.32) 5.81	22.28 ± 0.32 (19.76–24.05) 4.97	19.39 ± 0.23 (17.52–20.60) 4.13	6.72 ± 0.11 (6.01–7.25) 5.56	7.71 ± 0.11 (7.15–8.38) 4.86	13.93 ± 0.17 (12.69–14.78) 4.27	22.82 ± 0.33 (20.52–24.30) 4.98	7.68 ± 0.09 (7.20–8.07) 3.91
Wallowa and Blue Mountains	♀	30	185 ± 1.8 (160–205) 5.5	55 ± 1.2 (43–67) 11.6	25 ± 0.3 (20–27) 5.6	81.2 ± 3.75[b] (64.3–99.7) 14.62	28.55 ± 0.19 (26.35–30.44) 3.76	19.73 ± 0.31 (12.04–22.29) 8.66	16.82 ± 0.13 (15.28–18.06) 4.23	6.30 ± 0.07 (5.67–7.05) 5.87	7.21 ± 0.07 (6.50–8.06) 4.95	12.30 ± 0.09 (11.14–13.38) 4.20	19.73 ± 0.16 (17.99–21.49) 4.44	7.12 ± 0.05 (6.53–7.58) 3.54
	♂	30	194 ± 2.0 (175–213) 5.8	59 ± 1.4 (45–70) 12.9	26 ± 0.3 (23–28) 5.6	89.2 ± 5.28[g] (64.9–117) 19.63	29.94 ± 0.27 (27.33–32.62) 4.93	20.99 ± 0.19 (18.73–23.07) 4.95	17.67 ± 0.17 (15.11–19.36) 5.20	6.41 ± 0.08 (5.40–7.66) 7.15	7.33 ± 0.06 (6.14–7.76) 4.80	12.62 ± 0.09 (11.31–13.56) 4.31	20.56 ± 0.17 (18.77–22.65) 4.55	7.29 ± 0.07 (6.49–8.25) 5.38
Remainder of eastern Oregon	♀	30	188 ± 1.8 (166–211) 5.2	56 ± 0.9 (43–65) 8.9	26 ± 0.3[g] (22–29) 5.9	82.0[r] (65.7–101)	29.12 ± 0.23 (27.15–32.83) 4.38	20.49 ± 0.19 (18.60–22.75) 5.19	17.23 ± 0.12 (15.90–18.62) 3.86	6.36 ± 0.07 (5.41–7.18) 6.39	7.15 ± 0.08 (6.38–7.97) 5.96	12.59 ± 0.11 (11.44–14.11) 4.88	20.28 ± 0.18 (18.85–23.06) 4.80	7.13 ± 0.07 (6.33–7.84) 5.29
	♂	30	203 ± 2.4 (178–237) 6.4	61 ± 1.4 (48–80) 12.7	27 ± 0.3[g] (23–29) 5.4	121.5[s] (68.2–172)	31.19 ± 0.37 (27.28–34.57) 6.52	22.03 ± 0.29 (18.82–24.87) 7.24	18.53 ± 0.22 (16.50–20.73) 6.54	6.42 ± 0.08 (5.88–7.40) 6.72	7.49 ± 0.09 (6.60–8.71) 6.43	13.19 ± 0.16 (11.51–14.45) 6.44	21.45 ± 0.28 (18.88–24.14) 7.07	7.31 ± 0.08 (6.39–8.19) 5.82
Thomomys townsendii														
Harney and southeastern Malheur counties	♀	18	246 ± 3.7 (222–289) 6.4	72 ± 1.8 (62–90) 10.3	34 ± 0.6 (27–39) 7.58	245.1[t] (190–356.3)	38.72 ± 0.54 (36.05–45.61) 5.86	29.11 ± 0.33 (26.89–33.57) 4.85	24.38 ± 0.34 (22.78–29.24) 5.94	7.02 ± 0.08 (6.46–7.62) 4.73 5.01	9.94 ± 0.12 (9.07–11.04) 4.06	15.54 ± 0.17 (16.55–19.90) 6.33	28.48 ± 0.42 (26.03–34.09) 3.83	9.87 ± 0.09 (9.00–10.36) 5.39
	♂	25	261 ± 3.3[a] (235–288) 6.2	76 ± 1.5[a] (63–95) 9.6	36 ± 0.4[a] (32–42) 5.9	281.6[f] (203.2–378)	40.81 ± 0.44 (36.58–44.92) 5.39	30.49 ± 0.26 (27.90–32.95) 4.30	25.67 ± 0.28 (23.42–28.93) 5.42	7.02 ± 0.09 (6.11–7.95) 6.87	10.19 ± 0.07 (9.56–10.93) 3.64	18.21 ± 0.16 (16.74–19.85) 4.42	30.03 ± 0.38 (26.54–33.61) 6.30	10.00 ± 0.08 (9.37–10.85) 4.24
Vale–Ontario area, Malheur County	♀	18	271 ± 2.9 (248–288) 4.5	86 ± 1.2 (75–98) 5.8	37 ± 0.7[g] (33–47) 7.9	277.2[u] (265–288)	42.40 ± 0.27 (39.63–43.71) 2.72	31.22 ± 0.25 (28.93–32.54) 3.35	25.49 ± 0.23 (23.90–26.61) 3.77	7.41 ± 0.06 (6.94–7.81) 3.44	9.89 ± 0.16 (8.52–10.93) 6.79	18.29 ± 0.17 (16.35–19.86) 4.02	30.31 ± 0.31 (27.71–32.39) 4.33	10.30 ± 0.11 (9.42–11.08) 4.49
	♂	19	291 ± 3.9 (258–340) 5.9	92 ± 2.2 (78–113) 10.5	38 ± 0.4 (36–42) 4.3	346.6[t] (291–380)	45.21 ± 0.53 (41.20–50.16) 5.13	33.21 ± 0.42 (30.32–36.92) 5.48	27.28 ± 0.36 (24.69–30.33) 5.81	7.37 ± 0.05 (6.77–7.83) 3.09	10.07 ± 0.12 (9.29–10.92) 5.39	19.36 ± 0.23 (17.95–21.55) 5.17	32.83 ± 0.45 (29.14–36.39) 5.96	10.37 ± 0.08 (9.73–10.88) 3.56

[a] Although measures of ear length were recorded, the data were not included because of extreme variation believed to stem from differences in techniques used to measure these characters.
[b] Sample size reduced by 4; [c] Sample size reduced by 5; [d] Sample size reduced by 14; [e] Sample size reduced by 15; [f] Sample size reduced by 1; [g] Sample size reduced by 2; [h] Sample size reduced by 16; [i] Sample size reduced by 22; [j] Sample size reduced by 26; [k] Sample size reduced by 24; [m] Sample size reduced by 6; [n] Sample size reduced by 17; [o] Sample size reduced by 12; [p] Sample size reduced by 20; [q] Sample size reduced by 27; [r] Sample size reduced by 23; [s] Sample size reduced by 11; [u] Sample size reduced by 13.

Table 11-11.—*Reproduction in Oregon geomids.*

Species	Number of litters	Litter size			Reproductive season	Basis	Authority	State or province
		$\overline{X}$	Range	n				
Thomomys bottae		4.5	3–6	4		Implant sites	This study	Oregon
Thomomys bulbivorus	1[a]	4.7	2–8	59	Mar.[b]–Sept.	Implant sites	Verts and Carraway, 1991	Oregon
Thomomys mazama		3.9	1–7	7		Implant sites	This study	Oregon
Thomomys talpoides	2[c]	6.3[d]	3–9	25		Embryos	Wight, 1930	Oregon
	1	4.9	3–8	12		Implant sites	Andersen and MacMahon, 1981	Utah
Thomomys townsendii	2[e]	6.8	3–10	12		Embryos	Horn, 1923	Oregon
	1[f]	4.6	2–9	106[g]		Embryos and implant sites	Howard and Childs, 1959	California

[a]May produce a second litter in irrigated fields.
[b]Wight (1918).
[c]Based on faulty assumption, probably in error.
[d]Reported as 6.5, but data provided indicate that 25 females contained 158 embryos.
[e]Based on individuals simultaneously pregnant and lactating.
[f]Some individuals may have had two.
[g]Estimated from a graphic presentation of data.

Burrow systems of *T. bottae* in Colorado do not cross each other, but spacing within and between burrow systems is remarkably uniform regardless of sex or reproductive condition of the occupants. However, burrow length and linearity were greater for reproductive males than for other gophers (Reichman et al., 1982). Burrow length was almost exactly half that on a site with double the plant resources (Reichman et al., 1982). Extensions of burrow systems by *T. bottae* seemingly are random; at least there is no indication that the gophers detect food sources, then burrow in the direction thereof (Vleck, 1981).

In Mima moundfields (see account of *T. talpoides* for a description of Mima mounds and their presumed formation), *T. bottae* tends to move soil moundward and upward, especially where the soil is shallowest and most poorly drained. Iron pellets incorporated into soil plugs and relocated with a metal detector 1 year later were moved moundward an average of 41.3 cm and upward an average of 4.9 cm (Cox and Allen, 1987). Such findings were interpreted to add credibility to the hypothesis that pocket gophers are responsible for constructing Mima mounds (Cox and Allen, 1987).

Foraging on crops and forest plantings, covering crops with excavated soil, and burrowing through dams and levees are negative economic impacts attributed to *T. bottae* (Howard and Childs, 1959). Although some authorities (Howard and Childs, 1959) find few, if any, redeeming qualities in *T. bottae*, we suspect that its contribution to ecosystems of which it is a component is similar to that of congeners in other areas of the state.

Remarks.—Hall (1981) interpreted the hybridization of *T. bottae* and *T. umbrinus* reported to occur in a zone of sympatry in Arizona to be evidence of conspecificity (Hoffmeister, 1969). He synonymized the named forms under the specific epithet *T. umbrinus*. Other taxonomists have considered the two species to be reproductively isolated, therefore deserving of distinct species names (Jones et al., 1973a).

Bailey (1936) considered the white-toothed race (*T. b. leucodon*) to have acquired the white-faced incisors by erosion of the yellow enamel through its use of the incisors to dig through heavy-clay soils. Considering that congeners that live in areas with heavy-clay soils have yellow-faced incisors highly procumbent as an adaptation for digging (e.g., *T. bulbivorus*), we think that the claimed source of the white-faced incisors in *T. bottae* should be challenged. Because incisors of pocket gophers grow at a phenomenal rate, maintaining an individual or two on light soils or in artificial burrows in captivity for a few weeks should indicate if the white face on the incisors is a result of erosion of yellow enamel.

We suspect that the political boundary between states has such little effect that information on the biology of *T. bottae* derived in California is largely applicable to the species in Oregon. Nevertheless, information on populations of *T. bottae* in the long projections at the northern periphery of its range (Fig. 11-46) for comparison with that obtained elsewhere (Howard and Childs, 1959; Patton and Smith, 1990) might provide new insight into the ecology, genetics, and evolution of the species.

The disjunct population at the mouth of the Pistol River, Curry Co., heretofore referred to *T. b. detumidus* is considered by some to be of special concern. The taxon reportedly was known only from the type locality (Hall, 1981), although we examined specimens taken at other localities nearby. However, we know of no specimens taken since 1949 from any of these localities. Grinnell (1935) distinguished these pocket gophers as a separate taxon because female specimens available to him were slightly smaller; a bit lighter in color with more white on the face, chest, and venter; possessed a weaker rostrum; and had more forward-projecting incisors. Because interpopulation differences in these and other characters among pocket gophers from localities only a few kilometers apart often are equal to or greater than those described between *T. b. detumidus* and *T. b. laticeps*, we consider the specimens from the vicinity of Pistol River to represent merely a population of *T. b. laticeps*. On 23 June 1984, we found considerable pocket gopher activity at Harris Beach State Park north of Brookings and collected one specimen (OSUFW 7390), but we found no sign of pocket gophers farther north at the Pistol River.

Thomomys bulbivorus (Richardson)
Camas Pocket Gopher

1829*a. Diplostoma bulbivorum* Richardson, 1:206.
1855. *Thomomys bulbivora*: Brandt, 9(pt. 2):188.
1858. *Thomomys bulbivorus*: Baird, 8(pt. 1):389.

Description.—*Thomomys bulbivorus* (Plate XII) is the largest member of the genus (Table 11-10), although large size is not a distinctive feature of the species (Thaeler, 1980). The largest individual previously reported was 326 mm long and weighed 542.7 g (Verts and Carraway, 1987*b*); we caught a male (OSUFW 9231) that was 321 mm long and weighed 633.8 g. The length of the hind foot is ≥40 mm in males, ≥38 mm in females. The claws of the forefeet are shortest relative to body size within the genus (Bailey, 1915).

The pelage is a dark-sooty brown on the dorsum; the ears and nose are blackish. The venter is lead colored except for an irregular- and variable-shaped patch of white on the throat (Bailey, 1915). The winter pelage is long and furry; the summer pelage is short and coarse (Bailey, 1915).

The skull (Fig. 11-47) is typical for the genus except that the incisors of *T. bulbivorus* are strongly procumbent, medial grooves on the anterior face of incisors are absent or obscure, exoccipitals are grooved, and pterygoids are concave on their medial surfaces. A sphenoidal fissure is present. *T. bulbivorus* is a member of the heavy-rostrum group (subgenus *Megascapheus*) in which the base of P4 is deflected anteriorly and I1 originates between the bases of P4 and M1 (Thaeler, 1980).

Distribution.—*Thomomys bulbivorus* is endemic to the Willamette Valley, Oregon. Specimens have been collected from the vicinity of Hillsboro, Washington Co., south to Eugene, Lane Co. (Fig. 11-48). We know of only one locality at which the species was collected with an elevation >125 m above sea level (the northeast slope of the Chehalem Mountains, Washington Co.).

The geographic range of *T. bulbivorus* matches almost exactly the extent (≅122 m above sea level) that the Bretz Flood of ≅13,000 years ago inundated the Willamette Valley (Carraway and Kennedy, 1993:953, fig. 1).

Geographic Variation.—*Thomomys bulbivorus* is monotypic. However, Carraway and Kennedy (1993) indicated that individuals from the vicinity of the Chehalem Mountains were distinguishable genetically from those in the remainder of the range.

Habitat and Density.—*Thomomys bulbivorus* is associated with early seral plant communities or cultivated croplands that mimic early seral situations. We have taken these pocket gophers in alfalfa, wheat, and oat fields; in filbert orchards; and in weedy lawns and waste-ground areas. In most occupied sites, soils are heavy clays. *T. bulbivorus* usually does not reside in wetlands or grass fields in which drainage is poor. In addition to adequate drainage, we suspect that the most critical factor in determining the suitability of a site for *T. bulbivorus* is the presence of plants with reserves stored in bulbs or tap roots palatable to the species.

The only estimates of density for *T. bulbivorus* of which we are aware are our nearly complete censuses on small areas. In late summer and early autumn 1987–1990, we trapped 10–16 individuals in 4–24-h periods on 0.5–1.0-

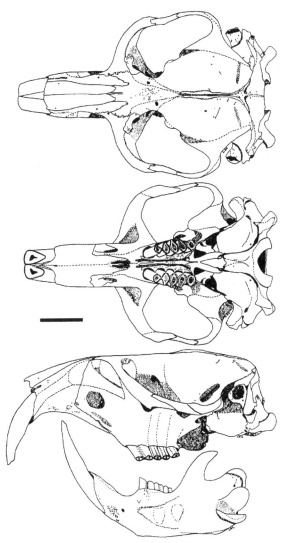

Fig. 11-47. Dorsal, ventral, and lateral views of the cranium, and lateral view of the dentary of the camas pocket gopher, *Thomomys bulbivorus* (KU 147726). Note sphenoidal fissure in lateral view of cranium. Scale bar equals 10 mm.

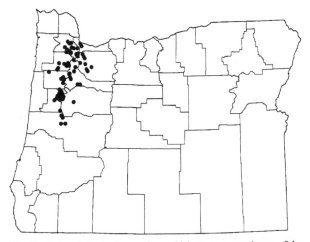

Fig. 11-48. Localities in Oregon from which museum specimens of the camas pocket gopher, *Thomomys bulbivorus*, were examined (listed in APPENDIX, p. 574). The species is endemic to Oregon (Verts and Carraway, 1987*b*).

ha sites in Washington, Yamhill, Polk, and Benton counties; subsequent mound building by those that remained indicated that the species was not eliminated at those sites. Also, by jogging behind a tractor and plow for 4 h, the two of us collected 40 individuals thrown from their burrows in a 2.6-ha alfalfa field in Yamhill Co. We are certain that we failed to capture at least 10 other individuals that quickly reentered partly opened burrow systems in the dead furrow. Thus, densities near the peak of the annual cycle at our sample sites probably ranged from ≅10 to ≅32/ha.

Diet.—*Thomomys bulbivorus* cuts and stores roots of spotted cats-ear (*Hypochaeris radicata*), vetch (*Vicia*), roots of fruit and nut trees, root crops (e.g., carrots, parsnips, potatoes), plantains (*Plantago*), and roots of grasses. Many that we have handled smelled of wild onion (*Allium amplectens*); sometimes the odor was sufficiently strong to bring tears to the eyes while preparing specimens. Bulbs of camus (*Camassia*) were thought to be scarce because *T. bulbivorus* fed upon them (Bailey, 1936), but Wight (1918) did not believe the bulbs composed much of the diet of the species. Occasionally, *T. bulbivorus* forages on the surface near mounds; we have found freshly cut vetch plants pulled partway into tunnel systems. In captivity, *T. bulbivorus* selected false dandelion in 39 of 47 trials; other foods were selected five times and no food was selected over another in three trials (Wight, 1918). Foodstuffs may be eaten immediately or cut into pieces and transported in the cheek pouches to store piles. We have excavated caches of roots of false dandelions and alfalfa, and seed heads of wheat, in tunnel systems we opened in the process of setting traps for *T. bulbivorus*.

Reproduction, Ontogeny, and Mortality.—Although Scheffer (1930*b*, 1938*a*) considered the breeding season for *T. bulbivorus* to extend only from early April to early June, Wight (1918, 1922) reported taking a pregnant female in late March and another on 22 July. Verts and Carraway (1991) extended the breeding season to the 1st week in September when they collected females in early stages of pregnancy in mid-August. They further concluded that the extended breeding season was related to early August irrigation of the filbert orchard in which the females were collected, possibly in response to some secondary compound in the rapidly growing plants stimulated by the added moisture.

Average litter size for *T. bulbivorus* was reported as 4.2 (Scheffer, 1938*a*:224) based on "actual count of embryos in pregnant females." Seemingly, the reported value actually was based on a combination of counts of embryos and of pigmented sites of implantation (Verts and Carraway, 1987*b*). Wight (1918, 1922) reported the range in litter size as four to nine, but listed no litter less than five (Verts and Carraway, 1987*b*). Average litter size based on implantation sites in 59 postpartum females collected at 10 sites along a transect extending the length of the Willamette Valley (Table 11-11) was not significantly different among localities or years (Verts and Carraway, 1991). For individuals with implantation sites of two intensities of pigmentation, the average of the lighter set was 2.9, significantly smaller than that of the darker set; however, the difference was attributed to factors other than an increase in litter size with maternal age (Verts and Carraway, 1991).

Sexual maturity was assumed to be attained by the breeding season following birth in *T. bulbivorus* (Verts and Carraway, 1987*b*); however, three of the four females exhibiting evidence of pregnancy in mid-August were young-of-the-year (Verts and Carraway, 1991). Nistler et al. (1993) confirmed that these three individuals were captured in the year of their birth. They also provided evidence that two females from unirrigated fields became pregnant during the year of their birth; however, the absence of changes in the pelvis typical of postpartum females suggested that the pregnancies terminated before parturition. Scheffer (1930*b*) discounted the possibility of more than one litter annually. However, the other of the four late-breeding females examined by Verts and Carraway (1991) was an adult exhibiting evidence of a recent pregnancy, presumably earlier in the year.

Growth and development of young *T. bulbivorus* are rapid. Neonates are about 50 mm long and weigh about 6.1 g; they lack hair, teeth, and pockets (Wight, 1918). Weight and length are about 23 g and 9 mm at 2 weeks, 36 g and 108 mm at 3 weeks, 54 g and 124 mm at 4 weeks, 71 g and 153 mm at 5 weeks, and 86 g and 164 mm at 6 weeks. By 2 weeks, hair is developing; by 3 weeks, they crawl and begin to eat solid food; by 4 weeks, they develop pockets; by 5 weeks, they open their eyes; and by 6 weeks, they are weaned (Wight, 1918). In a sample of males collected throughout the range of the species, 30% of young-of-the-year weighed more than the average older individual, but among females only 9.4% of young weighed as much as the average older female. Such suggests that final adult body mass is attained well before reproductive competence in males, but not in females (Nistler et al., 1993).

We know of no published information on mortality of *T. bulbivorus*. However, from our experience in trapping *T. bulbivorus*, populations seem to recover rapidly, probably a result of surplus animals from nearby areas dispersing to reoccupy burrow systems from which we removed animals.

Remains of four *T. bulbivorus* were recovered from 621 regurgitated pellets of great horned owls (*Bubo virginianus*), but none was recovered from 140 pellets of long-eared owls (*Asio otus*), 277 of barn owls (*Tyto alba*), and 115 of owls of unknown species (Maser and Brodie, 1966). Other raptorial birds and carnivorous mammals probably prey on *T. bulbivorus* (Bailey, 1936), but additional information on the species as prey is not available.

Habits.—*Thomomys bulbivorus* is described as "morose and savage" (Bailey, 1936:250) and "one of the most vicious animals known for its size" (Wight, 1918:16). Nevertheless, these pocket gophers flee from potential predators whenever the opportunity arises (Wight, 1918). In captivity, they are tamed readily.

Activity in *T. bulbivorus* is mostly subterranean, although we once intercepted a juvenile as it crossed a macadam road late at night in autumn. The occurrence of remains of *T. bulbivorus* in pellets of nocturnal-foraging birds of prey also suggests some surface activity; however, in excavating soil from tunnels, *T. bulbivorus* is exposed to predation for brief periods. Most activity and expenditure of energy are in construction of a complex tunnel system sometimes exceeding 240 m in total length (Wight, 1918). Usually, several tunnels 51–127 mm in diameter are constructed parallel to the surface of the ground at depths of 0.08–0.91 m; tunnels tend to be deeper on hillsides in drier soils (Wight, 1918). Because *T. bulbivorus* has relatively weak front feet and claws, most excavation at the advancing face of the tunnel is with the procumbent incisors (Wight, 1918). In heavy clays, we observed mounds constructed of fragments of soil that bear the distinct curve formed as the earth was peeled from the tunnel face with the incisors. The excavated soil is pushed into an unused portion of the tunnel system or pushed out of an open lateral tunnel to form a fan-shaped mound (Fig. 11-49); open tunnels are plugged with several decimeters of soil and a new lateral opening to the surface excavated when transportation of soil to the old mound becomes energetically excessive.

Individuals with burrow systems in fields adjacent to roadways commonly construct tunnels that open on cutbanks above roadside ditches, possibly because of the ease of dumping excavated soil. In winter, after periods of heavy rainfall, we observed streams of water pouring from the tunnels as the fields drained. We have not excavated such tunnel systems, but we speculate that rapid drainage afforded by such tunnels may permit occupancy of portions of fields that otherwise might not be habitable by *T. bulbivorus*.

Apparently unique among pocket gophers is the construction of "chimney" mounds by *T. bulbivorus* in late winter or early spring when soils are saturated. These mounds, commonly constructed of mud, rise straight upward 15–25 cm with an unplugged opening centered in the top. Wight (1918) postulated that these mounds were constructed to remove mud from the main tunnels, to admit air to dry the tunnel system, and to deter entrance by conspecifics. Although we have never observed a pocket gopher to enter or exit a chimney mound, we are at a loss to explain how an open chimney mound would deter entrance by an alien pocket gopher. We have not observed more than one chimney mound in a burrow system. We believe that a chimney mound likely functions as a fluid-dynamic pump using Bernoulli's principle to withdraw air from the tunnel system (Vogel, 1994:471, fig. 9); ventilation of the tunnel system is provided by air entering an open tunnel at the opposite end of the system.

Burrow systems of several pocket gophers may communicate although tunnels of adjacent individuals may be exclusive territories; nevertheless, it is commonly possible to distinguish the activities of individuals where several live in close proximity, especially in late summer and early autumn. Several newly constructed mounds in a group usually indicate activities of one gopher.

Thomomys bulbivorus produces chattering and grinding sounds with the teeth, and crooning and purrs when males and females are put together. Young make twittering sounds (Wight, 1918).

Remarks.—The large size of *T. bulbivorus* has necessitated the design of a special trap for its capture for scientific purposes or pest control. The Cinch trap kills even the largest *T. bulbivorus* instantly, whereas the Victor and Macabee traps usually produce horrible wounds.

Genetic variation in *T. bulbivorus* is as great as that in congeners with geographic ranges nearly 200 times greater; appears decoupled from the apparent homogeneity in gross morphology of the species; and reflects cataclysmic historical events to which the species was subjected, which, in turn, affected random genetic drift, effective population size, and gene flow (Carraway and Kennedy, 1993).

Thomomys mazama Merriam
Western Pocket Gopher

1897*d.* *Thomomys mazama* Merriam, 11:214.
1897*d.* *Thomomys nasicus* Merriam, 11:216.
1901*b.* *Thomomys oregonus* Merriam, 14:115.
1901*b.* *Thomomys hesperus* Merriam, 14:116.
1901*b.* *Thomomys niger* Merriam, 14:117.
1903*b.* *Thomomys helleri* Elliot, 3:165.

Description.—*Thomomys mazama* is one of the two small pocket gophers (Table 11-10) in Oregon without a sphenoidal fissure (Fig. 11-50). It also is a member of the slender-rostrum group (subgenus *Thomomys*) of species in which the bases of P4 and M1 are not as divergent and the base of I1 does not penetrate the space between them

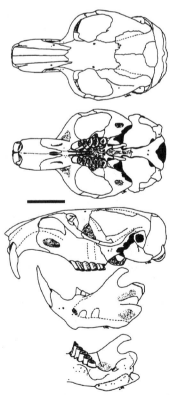

Fig. 11-50. Dorsal, ventral, and lateral views of the cranium; lateral western view of the dentary; and medial-oblique view of the posterior portion of the dentary of the mazama pocket gopher, *Thomomys mazama* (OSUFW 2518). Note absence of sphenoidal fissure in lateral view of the cranium and the wide flange on the dentary. Scale bar equals 10 mm.

Fig. 11-49. Photograph of a typical fan-shaped surface mound produced by the camas pocket gopher, *Thomomys bulbivorus*, in excavating burrows. Small rounded area at upper edge of mound is earth plug in burrow.

(Thaeler, 1980).

Depending upon the region, the pelage ranges from pure black with purplish and greenish overtones to brown to hazel to yellowish hazel on the dorsum and from lead colored to buff to ocherous on the venter (Bailey, 1915). The face is lead colored to black, the postauricular patch is relatively large and black, and the feet and distal portion of the tail are usually whitish. Of the all-black race, 38% have white splotches on the venter, or on both the venter and face (Walker, 1949). Overall, pelage color in *T. mazama* tends to be richer and darker than that of *T. talpoides*, but differences are not sufficient to distinguish the species reliably.

Thomomys mazama can be separated from all congeners in Oregon except *T. talpoides* by the absence of a sphenoidal fissure (Fig. 11-50). The baculum in adult males that have attained reproductive competency is 22–31 mm in *T. mazama* (only 12–17 mm in *T. talpoides*—Johnson and Benson, 1960). Because the baculum in *T. mazama* is relatively much thinner, juveniles also can be identified by use of the baculum; Johnson and Benson (1960) indicated that the shortest baculum for *T. mazama* they examined was 13.3 mm long with a base 0.65 mm in diameter. The baculum of a juvenile *T. talpoides* with a base of this diameter would be <10 mm long (Johnson and Benson, 1960). Use of such a character requires the assumption that females captured in company with adult males are of the same species. Considering that sympatry is uncommon and syntopy even rarer in geomyids, the assumption that females captured with adult male *T. mazama* are of the same species probably is valid. We found characters such as the serrate or smooth interparieto-parietal sutures, length-width ratios of the interparietal, and degree of parallelism in temporal ridges (Ingles, 1965) to be of little value in separating these two species. However, the width of the flange extending posteriorly and ventrally at the angle of the mandible (Fig. 11-50; Thaeler, 1980) seems to be a useful character for separating *T. mazama* (wide, usually ≥1.5 mm) from *T. talpoides* (narrow, usually ≅1 mm).

Distribution.—*Thomomys mazama* occurs from the Columbia River to the California-Oregon border in the Coast Range and along the coast (with a hiatus in Coos and northern Curry counties and interior Lane and Douglas counties) and in the Cascade Range, with eastward extensions to Mosier, Wasco Co., Paulina Lake, Deschutes Co., and Merrill, Klamath Co., south beyond our limits to Tehama Co., California (Fig. 11-51). The species also occurs as disjunct populations in the Olympic Mountains and in the vicinity of Puget Sound in Washington (Fig. 11-51; Hall, 1981). Along the Pacific Coast in Curry Co., *T. mazama* appears to occur in sympatry with *T. bottae* (Figs. 11-46 and 11-51).

Geographic Variation.—Six of the 15 nominal subspecies of *T. mazama* reportedly occur in Oregon: the dark brown *T. m. helleri* at the mouth of the Rogue River, Curry Co.; the shiny black *T. m. niger* between the Siuslaw and Alsea rivers; the dark auburn (sometimes black with brownish venter) *T. m. hesperus* from within the Nehalem River drainage south to near the Alsea River (specimens referable to *T. m. hesperus* and to *T. m. niger* occur on Mary's Peak, Benton Co.); the light hazel *T. m. oregonus* on the east slope of the Coast Range from Scapoose, Columbia Co., to Pedee, Polk Co., and east to Parkdale, Hood River Co.; the reddish brown *T. m. mazama* in the Cascade Range

(mostly on the west slope) from Mosier, Wasco Co., south to the Oregon-California line at the Klamath River and west to the Siskiyou Mountains; and the yellowish brown *T. m. nasicus* on the east slope of the Cascade Range from Bend, Deschutes Co., south to Ft. Klamath, Klamath Co., and east to Paulina Lake, Deschutes Co., and Merrill, Klamath Co. Thaeler (1980) reported diploid chromosome numbers for Oregon specimens of the various subspecies as follows: *T. m. helleri*, 42; *T. m. hesperus*, 44; *T. m. mazama*, 58; *T. m. nasicus*, 58; *T. m. niger*, 44; and *T. m. oregonus*, 40.

Habitat and Density.—In general, *T. mazama* occupies a narrower spectrum of environments than *T. talpoides* despite a broad overlap in habitats and food requirements of the two species (Thaeler, 1968a). In northern California, the two species occur within a few decimeters of each other, but usually no overlap was found (Thaeler, 1968a).

In Jackson Co., the habitat was described as "meadows and open, parklike habitats . . . [and] the more open timbered areas common to the ponderosa pine region" (Hooven, 1971b:347). Walker (1949) found *T. m. niger* on well-drained benchlands above small streams, on abandoned hillside farms where early seral communities were maintained by grazing, and in recent forest plantings. We captured *T. mazama* in Tombstone Prairie and Lost Prairie just east of the summit of Tombstone Pass, Jefferson Co. (U.S. Highway 20), and a short distance into the forest at

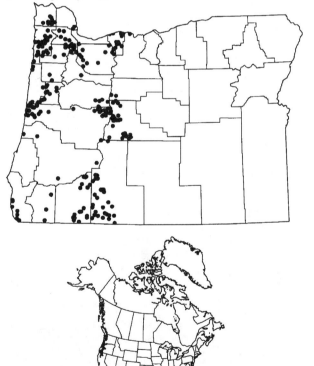

Fig. 11-51. Localities in Oregon from which museum specimens of the western pocket gopher, *Thomomys mazama*, were examined (listed in Appendix, p. 575), and the distribution of the western pocket gopher in North America. (Range map redrawn from Hall, 1981:467, map 301, with modifications.)

the edge of the prairies. We believe that *T. mazama*, like *T. talpoides*, does not occupy dense forest areas.

Common thistle (*Cirsium vulgare*), Canada thistle (*C. arvense*), miner's lettuce (*Montia perfoliata*), pussypaws (*Spraguea umbellata*), pink microsteris (*Microsteris gracilis*), false Solomon's seal (*Smilacina stellata*), common mullein (*Verbascum thapsus*), common dandelion (*Taraxacum officinale*), coastal strawberry (*Fragaria chiloensis*), vetch (*Vicia*), collinsia (*Collinsia*), bleeding heart (*Dicentra oregana*), phacelia (*Phacelia*), red sorrel (*Rumex acetosella*), squaw currant (*Ribes cereum*), and grasses (especially *Bromus*) were common plant species in early seral communities occupied by *T. mazama* in Jackson Co. (Hooven, 1971*b*). Pocket gopher populations on these areas were sufficient to cause considerable damage to ponderosa pine (*Pinus ponderosa*) seedlings planted on the area formerly dominated by Douglas-fir (*Pseudotsuga menziesii*) and white fir (*Abies concolor*).

Vegetation on the Klamath Co. area where Burton and Black (1978) studied dietary habits of *T. mazama* was a ponderosa pine–bitter-brush (*Purshia tridentata*)–needlegrass (*Stipa occidentalis*) community. After a forest fire, the area was planted to ponderosa pine seedlings "which failed largely because of depredations by pocket gophers, frost heaving, and drought" (Burton and Black, 1978:383). After the fire, vegetation consisted of annual and perennial grasses and annual forbs with scattered shrubs. Common species were cheat grass (*Bromus tectorum*), mountain brome (*B. carinatus*), spreading groundsmoke (*Gayophytum diffusum*), autumn willow-weed (*Epilobium paniculatum*), mountain knotweed (*Polygonum douglasii*), bitter-brush, squaw currant, and rabbitbrush goldenweed (*Haplopappus bloomeri*).

Walker (1949) estimated that 2,000–5,000 *T. m. niger* occurred in the ≅1,792-km² range of that race, but admitted that the estimate might be greatly in error.

Diet.—In Klamath Co., even though the cover provided by grasses was never >10%, the diet of *T. mazama* consisted of 16.5–60.5% grasses, with the highest values in January, the lowest in July. Mountain brome, cheat grass, and western needlegrass composed most of the grasses available and all grasses eaten (Burton and Black, 1978). Forbs produced 0.0–19.8% of plant cover and composed 4.1–60.4% of the diet with highest values in July and lowest in January. Of the forbs eaten, groundsmoke in July and knotweed in September were consumed in greatest amounts among annuals and velvet lupine (*Lupinus leucophyllus*) in July and goosefoot violet (*Viola purpurea*) in July among biennials and perennials. Woody plants composed a trace to 6.4% of the diet with the greatest amount in January (Burton and Black, 1978). Plant species identified in stomach contents of *T. mazama* ranged from five to 23 in bimonthly samples; 67% of the plant species available were consumed by *T. mazama* (Burton and Black, 1978). Ponderosa pine constituted 1.5–6.2% of the diet in samples collected in January–May and in November (Burton and Black, 1978).

In captivity, *T. mazama* was observed to reingest its fecal pellets after chewing them thoroughly, thus explaining the absence of distinct pellets in stomach contents of pocket gophers as commonly are found in lagomorphs (Wilks, 1962).

Reproduction, Ontogeny, and Mortality.—We examined one pregnant and seven postpartum females from the southern Cascade Range (Table 11-11). The pregnant individual contained six 12-mm embryos and was collected in May. Walker (1949) collected a female on 21 March that exhibited no evidence of reproductive activity, but one collected on 10 April had greatly thickened and enlarged vagina and uterus. He also collected a female pregnant with three embryos on 3 July. We found no published information on reproduction or development.

In a sample of 127 *T. mazama* from Klamath Co., Livezey and Verts (1979) found 94.5% concordance between adult:juvenile ratios determined by counts of periosteal zonations and other methods of estimating ages of pocket gophers (open pubic symphyses and evidence of parousness in females, and bacula >20 mm or <20 mm long in males). However, based on counts of periosteal zonations none was ≥3 years old and only six (4.7%) were ≥2 years old, suggesting a rapid turnover in the populations or a sample not representative of the age structure of the populations.

Habits.—Western pocket gophers construct burrows ≅3.8–4.4 cm in diameter ≅10–15 cm beneath the surface with a vertical tunnel connecting deeper burrows that include the nest (Walker, 1949). A system of foraging tunnels excavated by Walker (1949) occupied an area of < 20.5 m².

In California, Tunberg et al. (1984) reported that *T. mazama* was quick to invade and use nearby unoccupied or unused burrow systems of conspecifics. Invasion of vacant burrow systems after removal of the original occupant (*n* = 8) required an average of 8.4 (range, 1–16) days, invasion after removal of first invaders (*n* = 4) required an average of 6.5 (range, 4–8) days, and invasion by a third invader (*n* = 1) required 6 days. Two of the invaders were adults, seven were subadults, and four were of unknown age. Based on this behavior, Tunberg et al. (1984) recommended placing long-lasting toxic baits consisting of wheat, paraffin, and a toxicant in burrow systems for most efficient reduction of populations.

Remarks.—The taxonomy of the forms currently recognized as races of *T. mazama* has a tortuous history. Five (*helleri*, *hesperus*, *mazama*, *nasicus*, and *niger*) were named as distinct species; three of these (*helleri*, *mazama*, and *nasicus*) were reduced to subspecies of *T. monticola* but the other two were retained as distinct species by Bailey (1915). Bailey (1936) used the same classification in his treatment of Oregon mammals. However, Goldman (1943*a*) also subsumed *hesperus* and *niger* into *T. monticola*, producing a classification recognized by Hall and Kelson (1959). The sixth form, *oregonus*, originally was named *T. douglasi oregonus* by Merriam (1901*b*), a classification retained by Bailey (1915, 1936). Goldman (1939) considered *douglasi* a subspecies of *T. talpoides*, then classified *oregonus* as a subspecies of *T. monticola* (Goldman, 1943*a*). Johnson and Benson (1960) arranged all six Oregon forms listed herein plus eight forms in Washington and one in California as subspecies of *T. mazama*, leaving *T. monticola* in montane regions of eastern California and extreme western Nevada as a monotypic species (Hall, 1981).

In the National Museum of Natural History we examined a series of specimens with reddish-brown pelages collected 5 miles N Noti, Lane Co., and 2 miles E Scottsburg, Douglas Co., by K. M. Walker. These specimens were within the geographic range of *T. m. niger*

(Bailey, 1936) and were so cataloged. Walker originally had considered the specimens to be *T. m. oregonus,* but to us they appear identical with specimens of *T. m. mazama* from the Cascade Range directly east of these localities. Karyotypes of specimens from these localities would be interesting, perhaps enlightening. The locality near Scottsburg was listed by Bailey (1936) as within the range of *T. m. niger,* but we were unable to locate a black specimen from this locality. Similarly, we were unable to locate the specimen of *T. m. niger* described by Bailey (1936) as being an albino.

A band, fluorescent under ultraviolet light, appeared in vibrissae and claws of individual *T. mazama* 7–15 days after being given 30 mg/kg Rhodamine B. The marker lasted 5–6 weeks (Lindsey, 1983). Such a marker likely would be useful in evaluating various baits and baiting techniques to be used in control measures for *T. mazama* and other pocket gophers (Sullins and Verts, 1978).

Again we are struck by the paucity of information available on a species purported to be involved so frequently in depredations on forest plantings.

Thomomys talpoides (Richardson)
Northern Pocket Gopher

1828. *Cricetus talpoides* Richardson, 3:518.
1897d. *Thomomys quadratus* Merriam, 11:214.
1914. *Thomomys fuscus columbianus* Bailey, 27:117.
1933. *Thomomys quadratus wallowa* Hall and Orr, 46:41.
1939. *Thomomys talpoides columbianus:* Goldman, 20:234.
1939. *Thomomys talpoides quadratus:* Goldman, 20:234.
1939. *Thomomys talpoides wallowa:* Goldman, 20:234.

Description.—*Thomomys talpoides* (Plate XII) is the other small pocket gopher in Oregon (Table 11-10). Its dorsal pelage ranges from a rich brown to yellowish brown or buffy gray; the venter is lighter and usually washed with buff. The tail is usually brown dorsally and the feet are usually white. The skull (Fig. 11-52) is strongly angular and lacks a sphenoidal fissure. *T. talpoides* is a member of the slender-rostrum group (subgenus *Thomomys*) of pocket gophers in which the base of M1 is less inclined posteriorly and the base of I1 does not invade the space between P4 and M1 (Thaeler, 1980).

The baculum is 12–17 mm long in adult males and does not overlap with the 22–31-mm baculum of *T. mazama,* thus serves to distinguish the two species (Johnson and Benson, 1960). Also, width of the flange extending posteriorly and ventrally at the angle of the mandible (Fig. 11-50; Thaeler, 1980) seems to be a useful character for separating *T. talpoides* (narrow, usually ≅1 mm) from *T. mazama* (wide, usually ≥1.5 mm). A more detailed discussion of means of distinguishing the two pocket gophers that do not possess sphenoidal fissures is included with the account of *T. mazama.*

Distribution.—Northern pocket gophers occur from near Kamaloops, British Columbia; Edmonton, Alberta; Prince Albert, Saskatchewan; and Winnepeg, Manitoba; south to Mono Co., California; Nye Co., Nevada; Garfield Co., Utah; Valencia Co., New Mexico; and east to Kit Carson Co., Colorado; Garden Co., Nebraska; Roberts Co., South Dakota; Traill Co., North Dakota; and Kittson Co., Minnesota (Fig. 11-53; Hall, 1981). *T. talpoides* occurs throughout Oregon east of the Cascade Range (Fig. 11-53).

Pleistocene fossil specimens are known from cave depos-

its in Idaho, Wyoming, Utah, and New Mexico, and extralimital specimens tentatively identified as *T. talpoides* were found in Texas and Missouri (Kurtén and Anderson, 1980).

Geographic Variation.—Three of the 56 currently recognized subspecies occur in Oregon: the relatively large and light-colored *T. t. columbianus* near the Columbia River in northern Umatilla, Morrow, and Gilliam counties; the small and dark *T. t. wallowa* in the Blue and Wallowa mountains; and the small and bright-colored *T. t. quadratus* in the remainder of the state east of the Cascade Range (Hall, 1981; Patton, 1993). Davis (1938) reported that *T. talpoides* decreased in size with an increase in stoniness of soils and with elevation from 1,372 to 2,134 m in Idaho, whereas Tryon and Cunningham (1968) reported that it increased in size with deeper and more friable soils, and with elevation from 2,256 to 3,100 m in Wyoming. Hansen and Bear (1964) found size differences at different altitudes no greater than year-to-year differences in size in Colorado. Thus, it seems to us that nongeographic variation in size probably is equal to or greater than geographic variation. Also, we suggest that interpopulation variation in color may be of a magnitude similar to that of nominal subspecies.

Habitat and Density.—*Thomomys talpoides* occurs in prairie areas, mountain meadows, agricultural fields, and some forest areas, especially recently cutover or thinned areas; it usually does not occur in densely forested areas (Moore and Reid, 1951). We observed mounds produced by the species in soft soils derived from volcanic ash, in hard crusty soils containing considerable gravel, and in heavy soils in stream valleys. In southern Oregon, areas

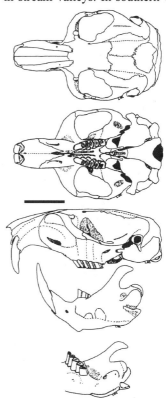

Fig. 11-52. Dorsal, ventral, and lateral views of the cranium; lateral view of the dentary; and medial-oblique view of the posterior portion of the dentary of the northern pocket gopher, *Thomomys talpoides* (OSUFW 3475). Note absence of sphenoidal fissure in lateral view of the cranium and the narrow flange on the dentary. Scale bar equals 10 mm.

with "deeper soils, more forb cover, less shrub cover, and aspect that ranged from south to northwest" had a higher probability of being inhabited by *T. talpoides* (Anderson, 1977:29). Thaeler (1968*a*) indicated that *T. talpoides* occupied a broader spectrum of environments than *T. mazama*, but in the few places that *T. talpoides* and *T. bottae* were sympatric the latter species was able to limit the former to soft tractable soils.

Dalquest (1948:306) reported that alfalfa plants near mounds produced by *T. talpoides* in Washington invariably were largest; he further indicated that the "commensal relation between the gopher and alfafa was understood by many farmers" to the extent that he was denied permission to collect specimens in their fields. There seems little question that *T. talpoides* usually is associated with the more fertile soils, but the species is considered "an indicator of good soil fertility rather than its cause" (Hansen and Morris, 1968:397–398).

When *T. talpoides* encounters large rocks and logs it undermines them, causing them to sink into the earth (Dalquest, 1948); however, gravel-sized stones are brought to the surface (Hansen and Morris, 1968). These habits, engaged in by many generations of *T. talpoides*, may speed decomposition of the rocks, tend to deepen soils, and possibly are responsible for some montane soils containing

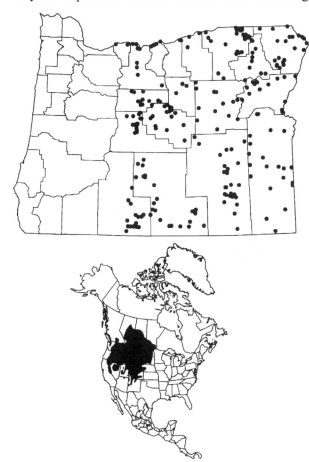

Fig. 11-53. Localities in Oregon from which museum specimens of the northern pocket gopher, *Thomomys talpoides*, were examined (listed in APPENDIX, p. 576), and the distribution of the northern pocket gopher in North America. (Range map redrawn from Hall, 1981:460–461, maps 299–300, with modifications.)

large rocks, but no gravel (Hansen and Morris, 1968).

Throughout much of its range, *T. talpoides* occupies grasslands on mounds that commonly attain 2 m in height, 20 m in diameter (some >30 m), and densities of 25–50/ha (Cox and Hunt, 1990). These mounds, called Mima mounds, are thought to have been formed by the "gradual translocation of soil . . . toward the deep well-drained microsites" that result from "the backward displacement that accompanies outward tunneling from centers of activity" by *T. talpoides* (Cox, 1989:77). The size of the mounds and the number of pocket gopher territories thereon are strongly correlated ($r = 0.935$—Cox and Hunt, 1990). Small Mima mounds occupied by only one gopher tend to be a relatively smooth dome shape and tall in relation to diameter; however, large mounds occupied by more than one individual tend to be flatter and irregular in cross section (Cox and Hunt, 1990). Intermound areas have thin soils marked by polygonal networks of beds of sorted stones thought to have been produced by the "mining" of soil by *T. talpoides* during the formation of Mima mounds (Cox, 1989; Cox and Hunt, 1990). In Wasco Co., vegetation on Mima mounds is dominated by Idaho fescue (*Festuca idahoensis*) and bluebunch wheatgrass (*Agropyron spicatum*), whereas intermound areas are vegetated predominantly by stiff sagebrush (*Artemisia rigida*), pine bluegrass (*Poa scabrella*), biscuit-root (*Lomatium*), and bitterroot (*Lewisia rediviva*—Cox, 1989). In Colorado, Mima mounds were considered to provide "high ground" during snowmelt; survival of *T. talpoides* thereon is enhanced. Because Mima mounds tend to be occupied by adults and intermound areas by juveniles, the differential survival on mounds was thought to be responsible for higher rates of overwinter survival of adults than juveniles (Hansen, 1962).

Many ranchers and land administrators contend that *T. talpoides* not only competes with livestock for forage, but it alters rangeland vegetation deleteriously (Keith et al., 1959), a view supported by investigations such as that of Moore and Reid (1951). Others contend that *T. talpoides*, in the absence of grazing by livestock, has little effect on soils or vegetation on rangelands (Ellison, 1946; Ellison and Aldous, 1952). Seemingly, the debt and asset columns for the economics of *T. talpoides* have not been totaled; thus its net value to agriculture remains to be determined.

Thomomys talpoides survived in areas that received ≤25 cm volcanic ash and debris from the eruption of Mount St. Helens, Washington, in May 1980; its burrowing activities led to modification of the physical structure of the volcanic deposits and to pre-eruption soils transposed to the surface of the newly deposited volcanic materials. Gopher mounds had a richer residual plant community and supported more seedlings than undisturbed volcanic debris, and survival of seedlings on mounds was enhanced (Andersen and MacMahon, 1985*b*).

Estimated densities of *T. talpoides* in Colorado in late summer and early autumn ranged from 7.4 to 84.0/ha (Reid et al., 1966) and in Utah at the same season from 30.9 to 97.1/ha (Richens, 1965). In Wyoming, densities of *T. talpoides* fluctuated in synchrony at three sites in an altitudinal transect extending from 2,256 to 3,100 m (Tryon and Cunningham, 1968). In Utah in four stages of an Engelmann spruce (*Picea engelmannii*)-climax sere, late summer estimates ranged from 4.4 to 54.4/ha with densities tending to decrease as succession progressed

(Andersen et al., 1980). Also in Utah, in a spruce-fir sere, belowground energy density of plants palatable to *T. talpoides* was such that burrowing time to obtain minimum daily energy requirements was 2 h in meadows, 3.9 h in quaking aspen (*Populus tremuloides*), 5.2 h in grand fir (*Abies grandis*), and 30.0 h in Engelmann spruce. Such explains population densities of 6.2–62.5, 2.1–33.3, 0–10.4, and 0–2.1/ha in the four seral stages, respectively (Andersen and MacMahon, 1981). However, in four successional stages of spruce-fir forest in Idaho, *T. talpoides* was more abundant on 1–10-year-old sites and >80-year-old sites than on 11–39-year-old or 40–79-year-old sites (Scrivner and Smith, 1981). Production of perennial forbs declined by 83% and production of grasses increased by 30% on rangelands sprayed with the herbicide 2,4-D (2,4-dichlorophenoxyacetic acid) in Colorado; populations of *T. talpoides* were reduced by 87%, indicating the association of these gophers with the forb component of communities (Keith et al., 1959). We know of no comparable population or community data for Oregon.

In Utah, the annual energy requirement of *T. talpoides* is 50–75 MJ depending on sex, activity, and reproductive status of the individual. To meet this requirement, populations at high density ($\cong$62.5/ha) can remove $\geq$17% of the annual primary production of montane meadows, >30% of the net annual primary production of belowground plant parts (Andersen and MacMahon, 1981).

Reid et al. (1966) found that counts of new mounds formed in a 2-day period in late summer or early autumn (after dispersal of young) on 0.4-ha plots were significantly related ($r = 0.97$) positively and curvilinearly to the number of pocket gophers residing thereon with a standard error of the estimate of four gophers. To produce an estimate within 20% of the mean population size required counting sign on 55 0.4-ha plots when the population was 27/ha, but on only seven such plots when the population was 54/ha.

Diet.—*Thomomys talpoides* is herbivorous, feeding on both aboveground parts and roots of most plant species within its range (Moore and Reid, 1951). In summer, the diet consists largely of leaves and stems of plants, whereas in winter, belowground parts are eaten more frequently (Cox, 1989). Forbs seem to be selected over woody species and grasses, but *T. talpoides* will eat small roots and cambium from large roots and boles of trees, and is known to girdle trees 2–3 m above ground by burrowing through the snowpack in montane regions. In an orchard in Hood River Co., *T. talpoides* fed largely on alfilaria (*Erodium*), black nightshade (*Solanum nigrum*), and cambium of apple and cherry trees (Wight, 1930). In mountain meadows in the Ochoco National Forest, common dandelion (*Taraxacum officinale*), bluegrass (*Poa*), little oniongrass (*Melica fugax*), Gairdner's yampah (*Perideridia gairdneri*), snakeweed (*Polygonum bistortoides*), agoseris (*Agoseris*), wild onion (*Allium*), yellow fritillary (*Fritillaria pudica*), pale fawn-lily (*Erythronium grandiflorum*), and brodiaea (*Brodiaea*) were among the most common species consumed (Moore and Reid, 1951).

On a Mima-mound area in Wasco Co., stomach contents of 15 *T. talpoides* captured in June consisted of 20.7% roots and 79.1% aboveground parts; of aboveground parts, 97.0% was from forbs and 2.4% from grasses. Despite the low percentage of belowground parts, root material was found in 12 of the 15 stomachs examined.

Aboveground parts of *Lupinus* composed 56.1% of the stomach contents and aboveground parts of *Erigonum* composed 12.5% (Cox, 1989). Arthropod fragments were found in the stomach of one specimen. In a cafeteria-type test conducted in the field, *T. talpoides* consumed species from both mound and intermound areas, but species dominant on intermound areas were taken significantly more often than those dominant on the mounds (Cox, 1989).

In mountainous areas of Utah, dandelion, Rydberg's penstemon (*Penstemon rydbergii*), sweet sage (*Artemisia michauxiana*), yarrow (*Achillea lanulosa*), Fendler's meadowrue (*Thalictrum fendleri*), and slender wheatgrass (*Agropyron caninum*) composed 72% of the diet of *T. talpoides* (Aldous, 1951); all these species or congeners thereof occur in Oregon.

In Colorado, on areas with unaltered vegetation, 82% of materials eaten by *T. talpoides* was forbs, 18% grasses; on areas on which forbs were reduced by application of the herbicide 2,4-D, the diet was composed equally of the two types of plants; and on areas treated 2 years consecutively, the diet was 91% grasses (Keith et al., 1959). Whether gophers are able to subsist on a diet composed largely of grasses is unknown (Keith et al., 1959).

In Oregon, *T. talpoides* on ungrazed plots indirectly was responsible for lack of reproduction of ponderosa pine (*Pinus ponderosa*); deer mice (*Peromyscus maniculatus*) use pocket gopher burrows as shelter while gleaning pine seeds (Moore, 1943b). *T. talpoides* rejected pine seeds as food and pine regeneration occurred on plots from which gophers were removed and their tunnels collapsed, thereby eliminating shelter for the mice.

Reproduction, Ontogeny, and Mortality.—Wight (1930:44) reported that of 112 mature females taken 9 March–11 April in Hood River Co., 59 (52.6%) were pregnant, 28 (25.0%) were nursing young, and 25 (22.3%) were "normal" (presumably indicating no evidence of present or past reproduction). Of the 59 pregnant females, 21 (35.6%) were accompanied by young, suggesting to Wight (1930) that at least some females have a minimum of two litters in quick succession. Also, of the 59 pregnant females, 31 (52.5%) were accompanied by males, suggesting that a second estrus might occur immediately postpartum. However, Hansen (1960) provided evidence that capture of pregnant females and young in the same tunnel system was not a reliable indication of multiple pregnancies. Scheffer (1938a) reported no evidence of more than one litter in *T. talpoides* in Washington.

Although Wight's (1930) observations delineated a peak in the breeding season, the extent of the breeding season of *T. talpoides* in Oregon remains unknown. Upriver and on the Washington side, pregnant females were taken from mid-February to late April (Scheffer, 1938a). Gestation, based on observed copulation and parturition, is 18 days (Andersen, 1978).

Average litter size for *T. talpoides* in Oregon (Table 11-11) is similar to that reported for the species in Washington (Scheffer, 1938a) and at one site in Colorado (Hansen, 1960), but 1.4–3.1 more young per litter than those reported for other areas of the Rocky Mountains (Hansen, 1960).

Wight (1930) trapped 220 young *T. talpoides* from 79 separate tunnel systems ($\bar{X} = 2.8$), considerably fewer than expected on the basis of mean numbers of embryos per pregnant female (Table 11-11). Although he admitted that some of the difference might be attributed to mortality, he suggested that most likely was related to dispersal of young from natal burrow systems.

At birth, average mass of siblings in a litter of six ($\overline{X}$ = 2.8 g) was significantly less than that of young in four litters of five ($\overline{X}$ = 3.6 g; Andersen, 1978). The young are hairless, with eyes visible as dark spots under the skin, and the pinnae are pinhead-sized buds (Andersen, 1978). By 9 days of age, the dorsum is covered with gray-black fur, the venter is sparsely covered with white fur, and the upper and lower incisors are separated by a gap. By day 16, the space between the incisors is closed, the pinnae protrude from the head, and the young move about actively. Solid food is eaten the next day. On day 20, the pockets are visible but closed and the auditory meatuses are closed. By day 26, the eyes and ears open. By 30 days, agonistic behavior among siblings is sufficient to affect growth, as indicated by the inflection point in the sigmoid growth curve at that age (Andersen, 1978). By 100 days, the young can attain the size of an adult female. The rapid growth in T. talpoides might result from selection for ability to compete with conspecifics rather than from greater reproductive capacity or less susceptibility to predation (Andersen, 1978).

In Colorado, of 78 individuals marked, 14 were recaptured 1 year later, and only two survived for 3 years (Hansen, 1962). In Utah, the adult survival annually was ≥28%, 18%, 23%, and 70% for 1975–1976, 1976–1977, 1977–1978, and 1978–1979, respectively (Andersen and MacMahon, 1981).

In Wasco and Wheeler counties, T. talpoides was preyed upon differentially by great horned owls (Bubo virginianus), barn owls (Tyto alba), and long-eared owls (Asio otus) on the basis of body mass (Janes and Barss, 1985). However, the three owl species responded differently than predicted to seasonal changes in body mass of T. talpoides; when predation rates were compared by predicted sizes within the T. talpoides population, B. virginianus selected T. talpoides randomly with respect to size, whereas A. otus consumed smaller individuals. However, attacks on T. talpoides by A. otus were thought to be random, but captures were size-related. Barn owls consumed intermediate-sized T. talpoides, and the seasonal presence of this owl corresponded to the period that T. talpoides of the appropriate size occurred on the study area (Janes and Barss, 1985).

Repopulation after 5.5 months of 4-ha areas from which T. talpoides was removed was retarded significantly by treating burrows with clay pellets to which anal-gland compounds from stoats (Mustela erminea) and ferrets (M. putorius) were applied. However, similar treatment of occupied burrows with anal-gland compounds did not reduce numbers of gophers on 4-ha areas (Sullivan et al., 1990).

Habits.—Thomomys talpoides is considered to be largely fossorial although it engages in surface activity sufficiently frequently that its remains often occur in pellets regurgitated by raptorial birds (Anderson, 1955; Janes and Barss, 1985; Moore and Reid, 1951). Some aboveground exposure undoubtedly occurs while depositing earth on the surface during the construction of burrows. However, we captured seven subadults and one adult in six two-pitfall arrays on the west slope of Steens Mountain, Harney Co., during 8 nights in June 1992. Such suggests that surface activity by T. talpoides may be considerably more extensive than that associated with burrow construction. Most

authorities agree that activity is greater at night than during the daylight hours; however, Andersen and MacMahon (1981) found that activity of only one of four individuals followed by radiotelemetry deviated significantly from an equal distribution throughout the 24-h period. Periods of activity or rest usually were <1 h, rarely >2 h. T. talpoides, in Utah, averaged spending 52% of each 24-h period in some kind of activity (Andersen and MacMahon, 1981).

Extensive burrow systems are constructed; these consist of near-surface runways within ≅25 cm of the surface and deep runways 1.2–1.5 m below the surface. The two runways are connected by a vertical shaft with the ≅25-cm-diameter nest cavity commonly situated in the deep runway about 3 m from the vertical shaft (Wight, 1930). Blind tunnels radiate from near the nest; one type is used for food storage, the other as a latrine.

The capacity of T. talpoides to excavate tunnels is phenomenal; one individual constructed 0.5 m of tunnel in compact clay-loam soil in 15 min, 0.8 m in <1 h, 32 m in 8 days, 119 m in <3 months, and 146 m in 5 months (Richens, 1966). In this endeavor, the gopher produced 161 mounds averaging 3.0 kg (range, 1.4–5.3 kg); thus, at this rate, a population of 74/ha would excavate soil at a rate of >34,500 kg/ha each year (Richens, 1966). Average burrowing rate in captivity was 1.47 cm/min (Andersen and MacMahon, 1981).

Of 112 mature females captured in Hood River Co. in March–April, 52 (46.4%) were accompanied by males, but only two (1.8%) were accompanied by other females. Of 133 mature males captured, 36 (27.1%) were solitary and 38 (28.6%) were with young; 12 (9.0%) were captured in company with other males, but 11 males possibly invaded the tunnel system after the first animal was removed (Wight, 1930). Thus, it seems that intraspecific intolerance is largely intrasex specific. However, 95 (43.8%) of 217 (112 males, 105 females) young were solitary, indicating that intolerance by parents came early in the life of offspring and was not sex-related. In Colorado, young gophers occurred together (10.8% of plural captures) much less frequently than in Oregon (38.9% of plural captures), and adult females, but not adult males, occurred together (10.8% female-female, 9.5% male-male) more commonly than in Oregon (1.7% female-female; 9.0% male-male—Hansen and Miller, 1959). Plural occupancy of burrow systems in Colorado was strongly seasonal, with territories relaxed during the breeding season (Hansen and Miller, 1959). The strong territoriality probably sets the upper limit for density of populations even though density-independent factors probably are responsible for regulating numbers below that limit (Andersen and MacMahon, 1981).

Because T. talpoides has broader niche requirements than those of T. bottae and some other geomyids with which it is parapatric, Miller (1964) suggested that T. talpoides should be an inferior competitor to T. bottae. However, contrary to expectations, interspecific paired encounters indicated that T. talpoides was more aggressive than the larger T. bottae (Baker, 1974).

Average movements of marked T. talpoides recaptured in Colorado were 28 m for juvenile males, 18 m for juvenile females, and 11 m for both adult males and adult females. Maximum movements were 82 m for adult females and 101 m for adult males; maximum movement in 24 h was 64 m by a male (Hansen, 1962). Movements of individuals introduced into an area from which resident pocket gophers were removed averaged 239 m (range, 15–790 m; $n = 13$), much greater than those of *T. bottae* in the same region (Vaughan, 1963). In Utah, all movements by marked and recaptured adults were <40 m annually, and the maximum distance was by a male; several immatures moved ≅100 m over winter (Andersen and MacMahon, 1981).

Remarks.—Hansen (1962) described an ingenious method of livetrapping *T. talpoides* by use of large glass jars commonly used in home canning of fruits and vegetables. Lids were modified to serve as one-way doors that permitted gophers to enter the jars, but prevented their egress. Because of the low cost of materials, small amount of labor involved in constructing the trap door, and high success rate reported, we are surprised to have not found subsequent reports of use of the device.

The wide distribution and abundance of *T. talpoides*, its apparent ease of capture alive, and the several hypotheses developed through sampling by kill-trapping seemingly should make *T. talpoides* a prime candidate for study of movements, activity, and social interactions by use of implanted (Virchow, 1977) or collar-type (Tunberg et al., 1984) radio transmitters. Again, we suspect the culprit involved in the absence of such a study is the common practice of applying knowledge of one species to all related forms that share a common life-style. We caution that all pocket gophers are not alike; we suspect that variation in ecology and behavior among species, and even among populations within species, may equal the morphological variation found among species and populations.

Thomomys townsendii (Bachman)
Townsend's Pocket Gopher

1839c. *Geomys Townsendii* Bachman, 8:105.
1897d. *Thomomys nevadensis* Merriam, 11:213.
1914. *Thomomys nevadensis atrogriseus* Bailey, 27:118.
1937. *Thomomys townsendii bachmani* Davis, 18:150.

Description.—*Thomomys townsendii* is a large pocket gopher (Table 11-10), nearly as large as *T. bulbivorus*, and like *T. bulbivorus* it possesses a sphenoidal fissure. It is a member of the heavy-rostrum group (subgenus *Megascapheus*—Thaeler, 1980). Unlike *T. bulbivorus*, the pterygoids are straight (Fig. 11-54). The pelage is dark buffy-gray to sooty gray on the dorsum and a rich buff on the venter; the face, nose, and auricular patch are black; the feet are dirty gray; and the chin is white (Bailey, 1915). Melanistic individuals have been collected in other states, but we know of none from Oregon.

Distribution.—*Thomomys townsendii* is a relict species whose present range (Fig. 11-55) consists of disjunct populations in southern Idaho, northern Nevada, northeastern California, and southeastern Oregon (Rogers, 1991b). In Oregon, the species occurs on small areas in Malheur and Harney counties (Fig. 11-55).

Extralimital Pleistocene fossil specimens are known from Idaho and Fossil Lake, Lake Co., Oregon (Kurtén and Anderson, 1980).

Geographic Variation.—Both of the two currently recognized subspecies occur in Oregon: *T. t. townsendii* along the Malheur and Owyhee river valleys in Malheur Co., and *T. t. nevadensis* in southern Harney Co. (Rogers, 1991a, 1991b). See Remarks for additional details.

Habitat and Density.—In California, *T. townsendii* occupies alkaline soils in sagebrush (*Artemisia*) desert to fertile loamy soils of forest-bordered meadows (Thaeler, 1968b). Despite this apparent disparity, the distribution of the species is considered to be limited to deep soils associated with Pleistocene lakes (Davis, 1937; Rogers, 1991a; Thaeler, 1968b). We have taken *T. townsendii* in alkaline soils along the south shore of Harney Lake, and in deep, fertile soils near Sod House School near the headquarters of the Malheur National Wildlife Refuge and at 1,980 m in Stergen Meadows in the Pueblo Mountains, Harney Co. Also, we have seen mounds produced by the species in loamy soils along roadsides near Vale, Malheur Co.

No information regarding population density is available.

Diet.—We know of no information regarding the diet of *T. townsendii.*

Reproduction, Ontogeny, and Mortality.—Horn (1923) collected 20 *T. townsendii* near Vale, Malheur Co., on 27 March–1 April, of which 18 exhibited evidence of having bred (Table 11-11). Some individuals were both pregnant and lactating, indicating that two or more litters were produced annually. One pregnant female that also was lactating was captured from a burrow system from which

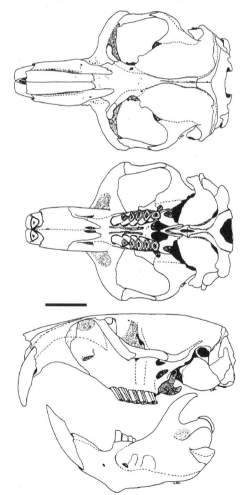

Fig. 11-54. Dorsal, ventral, and lateral views of the cranium, and lateral view of the dentary of Townsend's pocket gopher, *Thomomys townsendii* (OSUFW 1646). Note presence of sphenoidal fissure in lateral view of the cranium. Scale bar equals 10 mm.

five young of "identical size" (one was 180 mm long) were captured during the next 24 h (Horn, 1923:37–38). None of 16 individuals collected nearby in Idaho from 28 May to 30 August was pregnant (Davis, 1939). If *T. townsendii* has an 18–19-day gestation period as do *T. talpoides* (Andersen, 1978) and *T. bottae* (Schramm, 1961), if it weans its young at 6 weeks as does *T. bulbivorus* (Wight, 1918), and if Horn's (1923) specimens were pregnant for half the gestation period when captured, then onset of the breeding season for *T. townsendii* would be about 7 February or earlier.

Townsend's pocket gophers composed 5.3% of the prey identified in regurgitated pellets of a pair of barn owls (*Tyto alba*) roosting and nesting in a cliff near Vale, Malheur Co. (Maser et al., 1980).

Habits.—We found no information on activity, movements, or social interactions specific to *T. townsendii*.

Remarks.—Thaeler (1968*b*) found morphological evidence of hybridization between *T. townsendii* and *T. bottae* in restricted areas in northern California. This, and similar hybridization of *T. bottae* and *T. umbrinus* also reported by Thaeler (1968*b*), led Hall (1981) to consider the hybrids evidence of intergradation and to subsume the three forms as subspecies under the epithet *T. umbrinus*. However, based on electrophretic evidence in addition to that

Fig. 11-55. Localities in Oregon from which museum specimens of Townsend's pocket gopher, *Thomomys townsendii*, were examined (listed in Appendix, p. 577), and the distribution of Townsend's pocket gopher in North America. (Range map redrawn from Rogers, 1991*b*:110, fig. 1.)

derived from cranial morphology and color, Patton et al. (1984) reported that *T. townsendii* hybridized with *T. bottae* in a narrow zone in Honey Lake Valley along Gold Run Creek in Lassen Co., California. Both F$_1$ and backcrosses were identified, but only 12 hybrids were found among 104 gophers examined. Introgression of genetic material could not be detected even 1.6 km from the hybrid zone, causing Patton et al. (1984) to conclude that *T. townsendii* and *T. bottae* indeed were reproductively isolated and should be classified as separate species, a practice followed by some mammalogists since Thaeler's (1968*b*) original publication (Jones et al., 1973*a*).

Davis (1937) recognized seven distinct centers of differentiation, each fairly homogeneous and distinct from the others; he recognized each as a separate subspecies. Based on morphometric and colorimetric analysis, Rogers (1991*a*) concluded that the seven subspecies of Davis (1937) were not supported. She indicated that populations along the Snake River, Idaho (*T. t. owyhensis* and *T. t. similis*) and *T. t. townsendii* along the Malheur and Owyhee rivers in Oregon were differentiated from Humboldt River, Nevada–Honey Lake Valley, California populations (including *T. t. nevadensis*, *T. t. elkoensis*, and *T. t. bachmani* from northern Nevada and extreme southeastern Oregon near Malheur Lake and Alvord Lake, Harney Co., and McDermitt, Nevada; and *T. t. relictus* from Honey Lake Valley, California). Rogers (1991*b*) also found fixed differences at the polymorphic glutamate-oxaloacetate transaminase locus (GOT-1); the Snake River populations all possessed the same allele at the locus, whereas the Humboldt River–Honey Lake Valley populations all possessed a different allele at that locus. Thus, Rogers (1991*b*) recognized only two subspecies, with *T. t. townsendii* having name priority among the four Snake River populations and *T. t. nevadensis* having name priority among the three Humboldt River–Honey Lake Valley populations (Rogers, 1991*b*).

FAMILY HETEROMYIDAE—POCKET MICE, KANGAROO RATS, AND KANGAROO MOUSE

The heteromyids are New World endemics for which the fossil record began in the Early Oligocene in North America. The family includes six Recent genera (59 Recent species) and 13 extinct genera. The heteromyids are closely related to the Geomyidae and are classified with them in the superfamily Geomyoidea (McLaughlin, 1984).

In Oregon, the heteromyids are represented by two species of pocket mice, three species of kangaroo rats, and one species of kangaroo mouse. All six have fur-lined cheek pouches that open alongside the mouth; long hind feet and tails; soft, silky fur, usually buffy gray or tan on the dorsum; infraorbital foramina connected through the rostrum; and weak zygomata. The auditory bullae are enormously inflated (especially in the kangaroo rats and kangaroo mice), permitting the detection of low-frequency sounds to avoid predators (Webster and Webster, 1971, 1972, 1975), but restricting jaw gape and possibly limiting food size (Nikolai and Bramble, 1983). Some of the cervical vertebrae are fused in kangaroo rats and the kangaroo mouse, but not in pocket mice (Hatt, 1932). The upper incisors have a longitudinal groove on their anterior faces. The molariform teeth of kangaroo rats are ever-growing, whereas those of the pocket mice and kangaroo mouse are not.

Heteromyids are granivorous or at least granivorous at some seasons; they glean seeds carried by the wind into depressions in the surface of the ground. The seeds are transported in the cheek pouches and stored either in underground tunnels or in shallow pits dug near the entrances to their burrows. Kangaroo rats and the kangaroo mouse are bipedal, but pocket mice are quadrapedal, except they assume a bipedal stance when foraging.

Heteromyids in Oregon are mammals of shrub deserts; where the understory of grasses becomes too dense these species are excluded. They are highly adapted for life in arid lands; many such adaptations are concerned with acquisition and conservation of water. They excrete highly concentrated urine; they spend the daylight hours in humid subterranean burrows that they keep plugged to prevent escape of moisture; when at rest they place their noses in the inguinal region so moisture-laden air can be rebreathed; their feces are essentially dry pellets; and they are active nocturnally when the ambient temperature is lower and humidity higher. Some species can exist solely on the tiny amount of preformed water contained in seeds and water derived in the process of converting starches to sugars in their digestive systems.

Some heteromyids undergo periods of torpor in winter and some become torpid for intervals during summer. They usually are captured easily in either live traps or snap traps baited with rolled oats, sunflower seeds, or other grains. Heteromyids, especially kangaroo rats, usually are docile when captured; they usually do not bite when handled.

KEY TO THE HETEROMYIDAE OF OREGON

1a. Hind feet ≥34 mm long; anterior base of jugal at attachment to frontal and maxillary broad (usually >6 mm wide)..2
1b. Hind feet <27 mm long; anterior base of jugal at attachment to frontal and maxillary narrow (usually <5 mm wide)..4
2a. Pelage on dorsum dark blackish-brown; tail usually tipped with white; tail ≅150% of length of head and body; four toes on hind foot; occipitonasal length ≥38 mm..***Dipodomys californicus***, California kangaroo rat (p. 249).
2b. Pelage on dorsum light yellowish-tan to brown; tail not tipped with white; tail ≤140% of length of head and body; five toes on hind foot (including inconspicuous claw on inside of foot about midway between four toes and heel); occipitonasal length usually <38 mm..3
3a. Dark ventral tail stripe uniform width to tip of tail; tail >130% of length of head and body; lower incisors flat across anterior faces and with squarish cutting edges..***Dipodomys microps***, chisel-toothed kangaroo rat (p. 251).
3b. Dark ventral tail stripe tapering to point anterior to tip of tail; <130% of length of head and body; lower incisors rounded on anterior surfaces and with awl-shaped cutting edges..***Dipodomys ordii***, Ord's kangaroo rat (p. 254).
4a. Tail thickest in the middle, tapering to both ends; hind legs and feet greatly modified for bipedal, saltatorial locomotion; auditory bullae enormously expanded....

....***Microdipodops megacephalus***, dark kangaroo mouse (p. 248).
4b. Tail not noticeably thicker in the middle; hind legs and feet not greatly modified for bipedal, saltatorial locomotion; auditory bullae only moderately inflated..5
5a. Antitragus lobed; hind foot >20 mm long; occipitonasal length >24 mm; interparietal about 5.4 mm wide; baculum about 7.5 mm long..***Perognathus parvus***, Great Basin pocket mouse (p. 245).
5b. Antitragus not lobed; hind foot <20 mm long; occipitonasal length <24 mm; interparietal 3.5–3.8 mm wide; baculum <4.7 mm long....................***Perognathus longimembris***, little pocket mouse (p. 240).

Perognathus longimembris (Coues)
Little Pocket Mouse

1875.   O[*tognosis*]. *longimembris* Coues, 27:305.
1894a. *Perognathus nevadensis* Merriam, 46:264.
1933.   *Perognathus longimembris nevadensis*: Grinnell, 40:147.

Description.—*Perognathus longimembris* is one of the smallest rodents in North America (Bartholomew and Cade, 1957) and certainly the smallest in Oregon (Table 11-12). As in other members of the genus, the auditory bullae are not greatly inflated; the cheekteeth are hypsodont, rooted, and tuberculate; and p3 is larger than p4 (Fig. 11-56). It can be distinguished from its only congener in Oregon by an unlobed antitragus, a hind foot <20 mm long, and an occipitonasal length usually <24 mm (Hall, 1981).

The dorsal pelage is pinkish buff to ocherous buff with overlying blackish hairs; the venter is buff. The tail is bicolored.

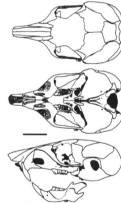

Fig. 11-56. Dorsal, ventral, and lateral views of the cranium, and lateral view of the dentary of the little pocket mouse, *Perognathus longimembris* (OSUFW 6822). Scale bar equals 5 mm.

Distribution.—The geographic range of *P. longimembris* (Fig. 11-57) extends from southeastern Oregon and northwestern Utah south through the desert areas of Utah, Nevada, California, and western Arizona into Baja California Norte and Sonora (Hall, 1981). In Oregon, *P. longimembris* is known from 20 specimens collected at nine localities in Harney and Malheur counties (Fig. 11-57).

Geographic Variation.—Only one, *P. l. nevadensis,* of the 17 currently recognized races (Hall, 1981; Hoffmeister, 1986) occurs in Oregon.

Habitat and Density.—We took two *P. longimembris* in Oregon: one near the crest of a sand dune, the other in a flat area vegetated largely by sagebrush (*Artemisia*); both were in the vicinity of Fields, Harney Co. Hall (1946) indicated that *P. longimembris* in Nevada sometimes was found in soft sands on valley floors, but was usually more abundant on firm sands overlain with pebbles and vegetated by widely spaced shrubs. He further indicated an association between the presence of *P. longimembris* and the presence of a rose-colored mallow (*Sidalcea*). In California, Kenagy (1973) found saltbush (*Atriplex*) growing on hillocks produced by *P. longimembris* in burrowing.

On three sites in northwestern Nevada, *P. longimembris*, in company with other heteromyids, tended to occupy microhabitats characterized by shrubs and by soils with a high percentage of sand (Bowers, 1986). However, on one site, *P. longimembris* responded to reduction of populations of all heteromyids by increasing use of open microhabitats. Lemen and Freeman (1987) also found that removal of *Dipodomys* caused *P. longimembris* to shift its activity to more open space; the reverse was not observed. In California, Thompson (1982*a*) observed a similar response by *P. longimembris* from trapping records, but not from direct observation of little pocket mice equipped with ß-lights. However, addition of artificial shrubs (cardboard shelters) to reduce the intershrub distance caused *P. longimembris* to forage more in open spaces between shrubs (Thompson, 1982*b*). Thus, at least at some sites, *P. longimembris*

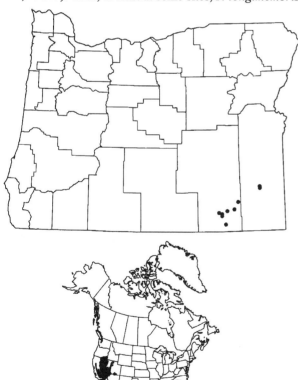

Fig. 11-57. Localities in Oregon from which museum specimens of the little pocket mouse, *Perognathus longimembris*, were examined (listed in APPENDIX, p. 539), and the distribution of the little pocket mouse in North America. (Range map redrawn from Hall, 1981:538, map 325, with modifications.)

may be limited to using specific microhabitats by interference competition of other species (Bowers, 1986) or by greater risk of predation (Thompson, 1982*b*).

Hall (1946) indicated that in some situations during some years in Nevada *P. longimembris* was the most abundant small mammal, attaining densities of nearly 1,000/ha. However, he further reported that population density was subject to marked fluctuations; where the species was extremely abundant at one sampling period it was not taken during subsequent sampling and its sign (tracks and diggings in the sand) was no longer found. Fautin (1946) reported densities ≤1.3/ha in western Utah.

Diet.—*Perognathus longimembris* is considered a true granivore because in captivity it can subsist on commercial bird seed alone; succulent vegetative material is refused (Hoffmeister, 1986). Indeed, in Nevada, Hall (1946) caught an animal with cheek pouches filled with seeds of lambsquarters (*Chenopodium album*) and Burt (1934) caught individuals that carried seeds of burweed (*Ambrosia dumosa*), plantain (*Plantago*), and fescue (*Festuca*). Hall (1946) also caught an individual in June with flowers of buckwheat (*Eriogonum gracillimum*) in its cheek pouches, suggesting that some individuals may consume plant parts other than seeds at least at some seasons. Specimens kept in captivity refused commonly used trap baits such as rolled oats, rolled oats and peanut butter, and peanuts (Hoffmeister, 1986).

Numerous investigators have addressed the problem of avoidance of competition for food resources (resource partitioning) in small-mammal communities, many of which included *P. longimembris*. Selection of seed on the basis of size, an oft-cited means of resource partitioning in desert rodents suggested by Brown and Lieberman (1973), is contradicted by evidence that seeds selected by different species do not differ in mass (Lemen, 1978) or by species (Reichman, 1975). However, the dispersion of seeds may influence use by various heteromyid species (Price, 1978); *P. longimembris* tends to forage on scattered seeds rather than those clumped by action of the wind or some other physical phenomenon. The mechanisms by which food resources are partitioned by heteromyids likely are complex, involving gait, speed of travel, search time, period of foraging stops, handling time, and resource density (Bowers, 1982).

Reproduction, Ontogeny, and Mortality.—In California, recrudescence of the testes began before males emerged from their hibernacula and attained peak size (≅5 times inactive size) shortly after emergence in late March. Sperm production began ≅5 weeks before mating occurred in late April or early May; however, the seminal vesicles and epididymides did not mature until >1 month after emergence (Kenagy and Bartholomew, 1985).

In Nevada, pregnant females were collected only in May (29 of 149) and June (one of 75—Hall, 1946).

In California, litter size in *P. longimembris* (Table 11-13) varied by ≤0.5 young/pregnancy (Kenagy and Bartholomew, 1985). At birth, body mass of young, conceived and born in captivity, averaged 0.5 g (2.4 g/litter or ≅33% of maternal mass) and at emergence it averaged 4.0 g (19.0 g/litter or ≅260% of maternal mass). Lactation ceased by mid-June and young-of-the-year first appeared in live traps at that time (Kenagy and Bartholomew, 1985). However, in another study, young averaged ≅1.3 g at birth,

Table 11-12.—*Means (±SE), ranges (in parentheses), and CVs of measurements of skull and skin[a] characters for female and male Heteromyidae from regions in Oregon. Skin characters were recorded from specimen tags; skull characters were measured to the nearest 0.01 mm. SE and CV not provided if n <10.*

Species and region	Sex	n	Total length	Tail length	Hind foot length	Mass (g)	Greatest length of skull	Occipitonasal length	Length of maxillary toothrow	Length of bulla	Least interorbital breadth	Parietal length
Dipodomys californicus	♀	19	302 ± 4.3 (257–338) 6.2	183 ± 2.7 (152–204) 6.5	45.2 ± 0.6 (40–49) 5.3	84.96[b] (76.5–93.1)	39.52 ± 0.36 (35.25–41.60) 3.99	37.96 ± 0.33 (34.17–39.78) 3.82	5.29 ± 0.05 (4.99–5.83) 4.34	13.78 ± 0.13 (12.19–14.55) 4.05	12.57 ± 0.10 (11.55–13.16) 3.47	2.71 ± 0.08 (2.24–3.70) 12.39
	♂	26	304 ± 3.7[c] (254–337) 6.0	184 ± 2.9[c] (152–204) 7.9	45.4 ± 0.4[d] (42–49) 4.5	76.71 ± 3.59[e] (56.6–96.6) 17.52	38.89 ± 0.33 (34.01–42.13) 4.32	37.30 ± 0.31 (32.91–40.28) 4.25	5.17 ± 0.06 (4.68–6.10) 6.02	14.06 ± 0.11 (12.85–14.71) 3.92	12.47 ± 0.09 (11.39–3.65) 4.07	268 ± 0.06 (2.12–3.77) 12.51
Dipodomys microps	♀	19	265 ± 4.2 (227–299) 6.9	154 ± 2.9 (131–182) 8.1	41.5 ± 0.3 (39–44) 3.4	55.16 ± 2.23[f] (36.0–68.1) 14.54	35.67 ± 0.28 (32.63–37.80) 3.39	34.07 ± 0.30 (30.96–36.57) 3.89	4.58 ± 0.03 (4.35–4.87) 3.20	14.29 ± 0.13 (12.88–5.37) 3.96	11.84 ± 0.09 (11.23–12.59) 3.19	2.01 ± 0.07 (1.52–2.48) 15.38
	♂	15	270 ± 1.8 (257–282) 2.6	160 ± 3.3 (142–197) 7.9	41.7 ± 0.5 (38–45) 4.8	60.96 ± 2.49[g] (48.2–77.5) 13.56	35.99 ± 0.26 (34.60–37.99) 2.81	34.33 ± 0.25 (32.86–36.4) 2.82	4.67 ± 0.04 (4.40–4.90) 3.50	14.27 ± 0.11 (13.66–15.22) 2.99	11.82 ± 0.11 (11.14–12.75) 3.46	2.15 ± 0.09 (1.43–2.71) 16.17
Dipodomys ordii	♀	30	242 ± 2.7 (217–280) 6.1	134 ± 1.6 (116–151) 6.6	39.6 ± 0.3 (36–43) 4.3	50.28 ± 1.38 (36.8–79.0) 15.0	36.85 ± 0.14 (34.92–38.63) 2.4	34.87 ± 0.51 (33.24–36.88) 2.28	4.89 ± 0.05 (4.50–5.42) 5.21	14.90 ± 0.07 (14.16–15.78) 2.64	11.91 ± 0.05 (11.25–12.50) 2.42	2.80 ± 0.06 (2.26–3.59) 12.59
	♂	31	239 ± 2.9 (205–274) 6.8	133 ± 1.9 (100–150) 8.1	40.2 ± 0.3 (37–43) 3.7	51.35 ± 1.29[c] (31.0–66.2) 13.82	37.30 ± 0.4 (35.61–38.49) 2.10	35.26 ± 0.14 (33.70–36.69) 2.19	4.94 ± 0.04 (4.45–5.50) 4.67	14.86 ± 0.07 (14.04–15.68) 2.74	12.04 ± 0.07 (11.27–12.72) 3.03	2.73 ± 0.05 (2.13–3.34) 10.94
Microdipodops megacepahlus	♀	30	152 ± 1.1 (140–165) 3.9	84 ± 0.9 (74–94) 6.2	24.8 ± 0.2 (20–26) 5.1	13.05 ± 0.33[h] (10.2–15.1) 11.74	27.69 ± 0.14 (25.20–28.75) 2.68	25.17 ± 0.12 (23.03–26.32) 2.52	3.39 ± 0.04 (3.04–3.70) 5.89	13.64 ± 0.09 (12.61–14.43) 3.65	6.74 ± 0.03 (6.50–7.08) 2.18	0.76 ± 0.03[h] (0.61–1.07) 15.82
	♂	30	155 ± 1.6 (130–170) 5.7	87 ± 1.4 (57–97) 8.7	25.0 ± 0.2 (23–27) 3.3	13.32 ± 0.35[h] (10.5–16.7) 11.88	27.93 ± 0.12[c] (26.47–29.28) 2.37	25.49 ± 0.17 (23.88–28.39) 3.61	3.37 ± 0.04 (2.66–3.84) 8.44	13.67 ± 0.09 (12.86–4.81) 3.74	6.77 ± 0.05 (6.24–7.52) 3.99	0.74 ± 0.05[j] (0.34–1.35) 32.17
Perognathus longimembris	♀	6	129 (125–136)	72 (68–76)	18.3 (17–19)	6.83[i] (6.4–7.0)	20.99 (20.24–21.46)	20.47 (19.95–20.80)	3.02 (2.78–3.17)	7.68 (7.49–7.90)	5.00 (4.85–5.12)	4.04 (3.62–4.64)
	♂	6	127 (111–140)	66 (51–80)	17.8 (17–19)	6.43[i] (5.3–7.2)	21.68 (20.60–22.61)	21.01 (20.37–21.46)	3.01 (2.87–3.11)	7.91 (7.40–8.36)	5.17 (4.94–5.51)	4.38 (4.19–4.83)
Perognathus parvus Jackson, Klamath, and western Lake counties	♀	5	143 (138–147)	72 (65–74)	21.6 (19–24)	9.5	23.89 (23.50–24.22)	23.90 (23.53–24.37)	3.51 (3.33–3.79)	7.65 (7.47–7.86)	5.74 (5.49–5.91)	5.42 (4.97–5.80)
	♂	12	182 ± 2.2 (162–205) 5.7	88 ± 2.2 (76–97) 8.3	23.5 ± 0.2 (22–24) 2.8	17.40[h] (16.1–20.0)	25.57 ± 0.17 (24.44–26.66) 2.34	25.57 ± 0.18 (24.46–26.51) 2.37	3.85 ± 0.09 (3.34–4.43) 8.77	8.10 ± 0.11 (7.49–8.63) 4.54	6.11 ± 0.08 (5.52–6.56) 4.69	5.51 ± 0.12 (4.87–6.29) 7.37
Harney, Malheur, and eastern Lake counties	♀	12	175 ± 4.7 (142–205) 9.3	91 ± 4.1 (53–110) 15.7	23.0 ± 0.3 (21–25) 3.7	22.00[j] (16.4–29.5)	25.71 ± 0.29 (24.18–27.63) 3.98	25.62 ± 0.29 (24.27–27.56) 3.88	3.99 ± 0.06 (3.65–4.30) 5.32	8.06 ± 0.09 (7.71–8.54) 3.79	5.97 ± 0.06 (5.71–6.39) 3.48	5.52 ± 0.10 (4.89–5.96) 6.52
	♂	32	167 ± 2.9 (154–180) 6.7	95 ± 1.6 (70–115) 9.4	23.9 ± 0.2 (21–27) 5.6	25.27 ± 1.68[k] (14–41) 27.47	26.74 ± 0.22 (24.46–28.98) 4.67	26.62 ± 0.22 (24.32–29.02) 4.65	4.02 ± 0.03 (3.61–4.46) 4.53	8.29 ± 0.06 (7.68–8.85) 3.98	6.14 ± 0.05 (5.69–6.81) 4.99	5.69 ± 0.09 (4.75–6.92) 9.28
Central Oregon	♀	23	161 ± 1.7 (141–173) 5.2	84 ± 1.3 (68–95) 7.7	21.6 ± 0.2 (19–24) 4.9	16.17[l] (14.1–19.5)	24.69 ± 0.13 (23.20–25.69) 2.52	24.68 ± 0.2 (23.39–25.55) 3.45	3.88 ± 0.04 (3.67–4.23) 4.88	7.93 ± 0.05 (7.50–8.53) 3.12	5.74 ± 0.03 (5.31–6.03) 2.87	5.01 ± 0.07 (4.21–5.56) 6.97
	♂	33	165 ± 1.5 (143–178) 5.2	88 ± 1.2 (64–98) 8.1	22.2 ± 0.2 (20–24) 4.6	18.16[m] (14.9–23.0)	25.54 ± 0.13 (23.77–27.31) 2.73	25.45 ± 0.12 (23.74–26.86) 2.59	3.99 ± 0.04 (3.58–4.81) 6.42	8.25 ± 0.06 (7.68–9.11) 3.88	5.88 ± 0.06 (5.28–6.66) 5.59	5.23 ± 0.11 (4.06–7.82) 11.89

[a]Although measures of ear length were recorded, the data were not included because of extreme variation believed to stem from differences in techniques used to measure these characters. [b]Sample size reduced by 10. [c]Sample size reduced by 1. [d]Sample size reduced by 3. [e]Sample size reduced by 12. [f]Sample size reduced by 6. [g]Sample size reduced by 4. [h]Sample size reduced by 9. [i]Sample size reduced by 2. [j]Sample size reduced by 7. [k]Sample size reduced by 15. [l]Sample size reduced by 16. [m]Sample size reduced by 24.

had nearly transparent skin, and were naked, and the eyes and ears were sealed (Hayden and Gambino, 1966). No cheek pouches were evident at birth, but by day 3 they had begun to evaginate. By day 5, the pinnae had unfolded but the meatuses were still sealed; incisors first penetrated the gums at that time. By day 8, sparse hair was evident, and by day 10 one individual ate carrot. By day 14, the juvenile pelage was complete, the eyes were open, the sex could be distinguished by differences in external genitalia, and the pouches were functional. By day 18, young were self-sufficient, but a litter whose maternal parent died on day 14 survived. By day 21, young had attained 65% of adult body mass and the hind foot had attained approximately adult size (Hayden and Gambino, 1966).

Perognathus longimembris is extremely prone to reproductive failure, much more so than the related kangaroo rats (*Dipodomys*); this proneness is believed to be related to the small size and high mass-specific energy demands of the little pocket mouse (Kenagy and Bartholomew, 1985). In California, during a year in which precipitation was only 63% of average, *P. longimembris* experienced complete reproductive failure, possibly explaining the great year-to-year fluctuations in density (Kenagy and Bartholomew, 1985).

Orr (1939) reported that a little pocket mouse in captivity lived 7.5 years. Based on relative toothwear observed

Table 11-12.—*Extended.*

Zygomatic breadth	Skull depth	Length of mandible	Length of mandibular toothrow
20.39 ± 0.18	13.43 ± 0.10	17.16 ± 0.24	5.31 ± 0.08
(18.72–21.94)	(12.60–14.17)	(14.72–19.23)	(4.78–6.55)
3.93	3.26	6.10	7.33
20.23 ± 0.21	13.06 ± 0.09	16.77 ± 0.15	5.28 ± 0.06
(17.31–22.06)	(11.91–14.12)	(14.92–18.27)	(4.65–6.15)
5.42	3.87	4.53	5.89
17.60 ± 0.09	12.32 ± 0.12	16.01 ± 0.13	4.81 ± 0.06
(16.86–18.37)	(11.31–13.29)	(15.24–17.28)	(4.33–5.35)
2.34	4.32	3.44	5.62
17.87 ± 0.19	12.36 ± 0.15	16.09 ± 0.14	4.89 ± 0.06
(16.45–18.89)	(11.17–13.06)	(15.35–17.36)	(4.53–5.22)
4.18	4.77	3.49	4.37
17.02 ± 0.08	12.49 ± 0.06	15.08 ± 0.09	4.53 ± 0.05
(15.98–18.03)	(11.71–13.11)	(13.91–15.88)	(3.96–5.09)
2.52	2.79	3.29	5.78
16.99 ± 0.08[c]	12.54 ± 0.06	15.08 ± 0.08	4.67 ± 0.04
(16.03–17.69)	(11.88–13.07)	(13.99–15.94)	(3.99–5.10)
2.54	2.49	3.11	5.04
11.55 ± 0.07	7.52 ± 0.06	10.16 ± 0.05	3.15 ± 0.04
(10.91–12.24)	(7.33–7.75)	(9.57–10.61)	(2.77–3.71)
3.15	5.29	2.71	7.52
11.60 ± 0.07	7.54 ± 0.17	10.21 ± 0.06	3.16 ± 0.05
(10.80–12.28)	(6.73–7.86)	(9.34–10.85)	(2.79–3.96)
3.51	6.12	3.45	8.39
10.84	8.63	8.78	2.89
(10.29–11.24)	(8.36–8.78)	(8.56–9.10)	(2.69–3.21)
11.19	8.83	9.14	3.02
(10.85–11.97)	(8.24–9.14)	(8.72–9.72)	(2.93–3.13)
12.30	8.63	9.98	3.22
(11.91–12.62)	(8.36–8.78)	(9.67–10.51)	(3.14–3.27)
12.86 ± 0.09	8.83 ± 0.07	10.88 ± 0.15	3.5 ± 0.09
(12.24–13.32)	(8.24–9.14)	(10.22–12.19)	(3.17–4.23)
2.39	2.79	4.71	9.13
12.79 ± 0.10	8.71 ± 0.08	11.42 ± 0.10	3.65 ± 0.06
(12.21–13.32)	(8.20–9.14)	(11.01–12.00)	(3.25–3.94)
2.72	3.25	3.12	5.94
13.21 ± 0.08	8.79 ± 0.06	11.73 ± 0.11	3.72 ± 0.06
(12.11–14.22)	(7.86–9.48)	(10.58–12.85)	(3.25–4.40)
3.53	4.03	5.62	8.73
12.46 ± 0.08	8.51 ± 0.04	10.92 ± 0.07	3.61 ± 0.05
(11.17–12.95)	(8.10–8.93)	(10.25–11.48)	(3.20–4.16)
3.08	2.41	3.11	6.78
12.85 ± 0.06	8.65 ± 0.04	11.09 ± 0.06	3.63 ± 0.05
(12.15–13.60)	(8.35–9.25)	(10.33–11.85)	(3.22–4.12)
2.82	2.72	3.32	7.43

in wild-caught specimens, he suggested that some individuals may attain comparable age in the wild. Hayden and Lindberg (1976) provided a survival curve for a cohort of 216 *P. longimembris* born and raised in the laboratory; median survival was 33 months and maximum survival was 8.3 years. They further reported that the mortality schedule of *P. longimembris* was considerably different from that reported for other rodents in that a moderate rate of mortality was maintained between days 500 and 2,000. They also cited an unpublished reference of a similar study in which 50 individuals had a median survival of 45 months and a maximum survival of 5 years. French et al. (1967) estimated average life-spans of 3.8–4.4 months in fenced

populations and 1.9 months in an unfenced population. However, they recorded life-spans of 32–61 months in 25 individuals. In California, Chew and Butterworth (1964) recaptured 19 (30.6%) of 62 individuals marked a year earlier. The exceptional longevity in *P. longimembris* seems to be related to adverse growing conditions and the resultant reduction in productivity of plants (French et al., 1967; Hayden and Lindberg, 1976). Low precipitation results in poor production of seeds; the reduced food supply, in turn, causes greater incidence and duration of torpor, which results in less exposure to predation, and the lower body temperature results in a lower rate of senescence at the cellular level (Hayden and Lindberg, 1976).

Habits.—Little pocket mice in Nevada are active above ground from April to November with the greatest number of individuals captured in May–July (Hall, 1946; O'Farrell, 1974). In spring–autumn, both sexes of *P. longimembris* exhibit greatest activity (as measured by numbers captured) within 2 h of sunset with activity declining rapidly 4–6 h after sunset (O'Farrell, 1974). Of environmental factors, moonlight seemed to have the greatest effect on frequency of capture of *P. longimembris*; activity increased perceptibly even when a cloud passed in front of the moon (O'Farrell, 1974). When active on the surface, *P. longimembris* spends a large proportion of the time searching beneath shrubs for seeds. Items are identified as seeds by touch (Lawhon and Hafner, 1981) and placed in the pouches by use of the forefeet.

In California, the adjusted range-length (maximum distance between captures + the distance between traps) used as an index of home-range size averaged 64.4 m (range, 38.7–84.4 m) with the greatest value recorded when the population density of *Dipodomys* was least (Chew and Butterworth, 1964).

Little pocket mice spend a large portion of their lives in burrows that may extend 1–2 m below the surface of the ground (French, 1976, 1977). In winter, they remain underground continuously; much of the time they are torpid (Kenagy, 1973). They do not subsist on stored fat as do hibernating sciurids, but rely on stored seeds for energy at this time. In captivity, the animals exhibit a circannual rhythm in belowground and aboveground activity and are thought to cue on soil temperature to synchronize the cycles (French, 1977; Kenagy, 1973).

Burrows constructed by *P. longimembris* are 1.5–2.0 cm in diameter and lead to a nest chamber about 5 cm high and 8 cm wide 1–2 m below the surface (Kenagy, 1973). The nest chamber usually contains a loose bed of roots and leaves, and is used when the animal is in torpor. Other tunnels are constructed at different levels below the surface, some as shallow as 1 cm below the surface. Shallow tunnels are used in relation to the temperature of the soil: in spring during the morning hours, *P. longimembris* rests 30–40 cm below the surface when ambient temperatures are 12–14°C; when temperatures rise to 29°C in midmorning, *P. longimembris* moves to tunnels ≅1 cm below the surface. Also, individuals move to the west side of shrubs when shallow tunnels used in the morning become shaded (Kenagy, 1973). Even in spring and summer they enter torpor during the last part of the night.

In captivity, *P. longimembris* had a rectal temperature of 37°C when maintained at ambient temperatures of 19–23°C and 35.6°C at 2.5–3.0°C. Even at a constant ambient

Table 11-13.—*Reproduction in Oregon heteromyids.*

Species	Number of litters	Litter size $\overline{X}$	Range	n	Reproductive season	Basis	Authority	State or province
Dipodomys californicus	2	2.7	2–4	20	Feb.–Sep.[a]	Embryos	Dale, 1939	California
Dipodomys microps		2.6[b]	2–4	11		Neonates	Daly et al., 1984	California
		2.4[c]	1–4	38		Neonates	Daly et al., 1984	California
		2.3	1–4	39	Apr.–Jul.[d]	Embryos	Hall, 1946	Nevada
	1	2.4	1–3	65	Jan.–May[e]	Embryos	Kenagy and Bartholomew, 1985	California
Dipodomys ordii		3.5	2–7	11		Implant sites	This study	Oregon
		3.5	1–6	80	Feb.–Jul.	Embryos	Hall, 1946	Nevada
Microdipodops megacephalus		3.9	2–7	54	Apr.–Sep.	Embryos	Hall, 1946	Nevada
Perognathus longimembris	1	4.8	4–6	16	Apr.–Jun.	Embryos	Kenagy and Bartholomew, 1985	California
	1	4.3	2–8	30	May–Jun.	Embryos	Hall, 1946	Nevada
Perognathus parvus	2[f]	4.9	4–6	23	May–Aug.	Embryos	Iverson, 1967	British Columbia
	2	5.2	2–8	132	May–Aug.	Embryos	Scheffer, 1938b	Washington
		5.5	3–8	33	May–Jul.	Embryos	Hall, 1946	Nevada
		6.1	5–8	7		Implant sites	This study	Oregon

[a] No specimens examined in July–August.

[b] Born in captivity of wild-bred females.

[c] Captive-bred wild-caught females.

[d] No animals examined before April.

[e] Parturition usually complete by end of April, but did not cease until 3rd week of May in year following autumn precipitation.

[f] At elevations below 305 m a sufficient number of females had three litters to raise the average to 2.06 litters/year; above ≅670 m the average was 1.3 litters/year.

temperature (24°C) variation in body temperature ranged from ≅34.8 to ≅37.4°C within a 24-h period; it commonly shifted as much as 1°C in 10 min (Bartholomew and Cade, 1957). At temperatures of 2–9°C, little pocket mice were active continuously—feeding, digging, filling cheek pouches—and were never seen to sleep. At temperatures of 20–30°C, they were active intermittently: they dug, groomed, and transported seeds only a few minutes at a time, but none exhibited a diurnal cycle of activity that paralleled its body temperature. At ambient temperatures that reached or exceeded their body temperature, little pocket mice exhibited intermittent periods of hyperactivity; they alternately ate, dug, or groomed vigorously for a few minutes, then became completely quiet for a few minutes. They did not pant or drool as observed in some other small mammals exposed to high temperatures (Bartholomew and Cade, 1957).

As in some other small mammals, *P. longimembris* may enter torpor when captured in live traps. However, in captivity, little pocket mice may enter torpor spontaneously even with food available and without stress of low temperatures; they can be induced to enter torpor by removing food for 24–36 h (Bartholomew and Cade, 1957). Between 2 and 20°C, body temperature followed ambient temperature when in torpor. Even in torpor, with body temperatures as low as 25°C, little pocket mice may shift position, take a few shaky steps, slowly scratch, and respond to being handled. Arousal required about 40 min, was accompanied by shivering until the body temperature reached ≅30°C, and was complete when the body temperature reached 30–32°C. In captivity, torpor may last from a few hours to a day or two; in the wild, torpor is believed to last as long as a week at a time (Bartholomew and Cade, 1957). *P. longimembris* lost less

mass at 8°C than at 21°C because torpor began sooner, the metabolic rate was lower, and arousal was less frequent (Bartholomew and Cade, 1957).

The ability of little pocket mice to live in extremely arid environments and to exploit the meager resources thereof may be related to a large extent to their small size. Their large surface:mass ratio is conducive to a high rate of heat loss, which, in turn, increases demands for energy (carbohydrates), resulting in the production of greater amounts of metabolic water from digestion of starches. Demand for foods high in carbohydrates also results in less urea being produced, which reduces demand for water for production of urine. Smaller mammals tend to have greater relative thickness of the medulla of the kidney, permitting greater urine-concentrating ability (Lawler and Geluso, 1986). On the other side of the ledger, small size increases evaporative water loss, especially because of the higher metabolic rate, which requires a greater respiration rate, which, in turn, results in more water being lost from their respiratory tracts (Lawler and Geluso, 1986).

Despite the differential in size, *P. longimembris* was dominant to *Microdipodops megacephalus* in 11 of 12 staged interspecific encounters (Blaustein and Risser, 1974a). This dominance, combined with O'Farrell's (1974) findings that *P. longimembris* tended to be active at seasons that *M. megacephalus* was inactive and vice versa, may be the mechanism by which the two species coexist (Blaustein and Risser, 1974a). Irrespective of sex of pairs, the larger *P. parvus* initiated fighting and was clearly dominant in 13 of 15 staged encounters with *P. longimembris*; in one encounter, dominance was not established and in another a male *P. longimembris* was dominant over a female *P. parvus* (Blaustein and Risser, 1974b). The dominance exhibited by *P. parvus* may be responsible for the absence of syntopy where ranges of the two species overlap (Blaustein and Risser, 1974*b*).

Remarks.—From dusty footprints, these tiny mice seem to have been frequent visitors to our live traps. However, we have caught only two, one in combination with a Jerusalem cricket (*Stenopelmatus fuscus*) that causes us to think that these extremely small mammals frequently entered our traps without being captured. Only the combined mass of the cricket and the mouse was sufficient to spring the trap.

The paucity of information concerning *P. longimembris* in Oregon doubtlessly results from the infrequency that the species is collected here, which, in turn, likely results from a lack of effort and possibly some specimens' being mistaken for juvenile *P. parvus*. In view of the widespread occurrence of shrub-vegetated sandy soils in southeastern Oregon, we believe that the species likely occurs at more than the few areas from which it currently is known.

Perognathus parvus (Peale)
Great Basin Pocket Mouse

1848. *Cricetodipus parvus* Peale, 8:53.
1858. *Perognathus parvus*: Cassin, 8:48.
1858. *Perognathus monticola* Baird, 8(pt. 1):422.
1868. *Abromys lordi* Gray, p. 202.
1875. *P*[*erognathus*]. *mollipilosus* Coues, 27:296.
1889a. *Perognathus olivaceous* Merriam, 1:15.
1900. *Perognathus parvus mollipilosus*: Osgood, 18:36.
1900. *Perognathus parvus olivaceous*: Osgood, 18:37.
1939. *Perognathus parvus lordi*: Davis, p. 266.

Description.—*Perognathus parvus* (Plate XIII) is the larger member of the genus in Oregon (Table 11-12). As in its congeners, the auditory bullae are not greatly inflated and they meet or nearly meet anteriorly (Fig. 11-58). The molars are hypsodont, tuberculate, and rooted; p4 is smaller than p3 (Osgood, 1900). The tail is slender, not constricted proximally, and neither crested nor conspicuously tufted. *P. parvus* may be distinguished from the only congener that occurs in the state (*P. longimembris*) by a lobed antitragus, a hind foot >20 mm long, an occipitonasal length >24 mm, an interparietal ≅5.4 mm wide, and a baculum ≅7.5 mm long.

The dorsal pelage is pinkish buff or ocherous buff overlain with black hairs; the venter is white to buffy. A lateral line, usually somewhat olive colored, separates the dorsal and ventral pelages. The tail is distinctly bicolored. There are two annual molts, one in February–March averaging 35 days (range, 15–63 days), the other in June–August averaging 31 days (range, 11–90 days—Speth, 1969). A small sebaceous gland is present on the underside of the proximal third of the tail (Quay, 1965).

Distribution.—The geographic range of *Perognathus parvus* (Fig. 11-59) extends from Kamloops, British Columbia, south through most of eastern Washington; Oregon east of the Cascade Range, except for the Wallowa and Blue mountains; eastern California; Nevada, except for the southern tip; the northwest corner of Arizona north of the Grand Canyon; Utah, except for the central montane region and the eastern edge; and southern Idaho, with extensions into southwestern Montana and southwestern Wyoming (Hall, 1981; Verts and Kirkland, 1988). In Oregon, *P. parvus* occurs throughout east of the Cascade Range, except it does not occur in the Wallowa and Blue mountains (Fig. 11-59).

Pleistocene fossils of *P. parvus* are known from cave deposits in Idaho (Kurtén and Anderson, 1980) and fossils thought to be *P. parvus* were recovered from a cave in Nevada (Miller, 1979). A similar fossil species, *P. †stevei*,

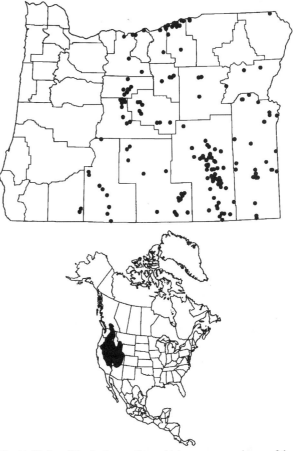

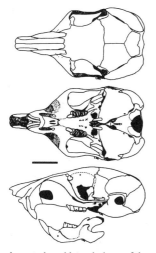

Fig. 11-58. Dorsal, ventral, and lateral views of the cranium, and lateral view of the dentary of the Great Basin pocket mouse, *Perognathus parvus* (OSUFW 7038). Scale bar equals 5 mm.

Fig. 11-59. Localities in Oregon from which museum specimens of the Great Basin pocket mouse, *Perognathus parvus*, were examined (listed in APPENDIX, p. 539), and the distribution of the Great Basin pocket mouse in North America. (Range map redrawn from Hall, 1981:532, map 323, with modifications.)

was found in Oregon at McKay Reservoir, Umatilla Co. (Martin, 1984). All known fossil sites are within the current distribution of the species (Verts and Kirkland, 1988).

Geographic Variation.—Three of the 11 subspecies are considered to occur in Oregon: *P. p. lordi*, *P. p. mollipilosus*, and *P. p. parvus* (Hall, 1981). Hall (1981) depicted a zone of intergradation between *P. p. olivaceus* and *P. p. parvus* in southeastern Oregon, but listed no specimens of the former race in Oregon. However, in one collection we found specimens from that region labeled *P. p. olivaceus*. Because of described color phases and considerable differences in size within named races, and relatively minor differences between the named forms, we believe that reevaluation of geographic variation in *P. parvus* is essential. Considering that Williams (1978) found major differences in karyotypes and Ferrell (1995) found at least two divergent lineages by analysis of sequence divergences in mitochrondrial DNA between specimens from the Columbia Plateau and the Great Basin, we suspect that what is considered *P. parvus* herein actually is composed of at least two species. Separate treatment awaits morphologic and morphometric distinction sufficient to provide adequate descriptions and delineation of geographic ranges.

Habitat and Density.—*Perognathus parvus* is a species commonly associated with light sandy soils in arid and semiarid regions (Verts and Kirkland, 1988). In southeastern Oregon, it was taken almost exclusively in habitats dominated by big sagebrush (*Artemisia tridentata*) or black greasewood (*Sarcobatus vermiculatus*) with understories primarily of cheat grass (*Bromus tectorum*—Feldhamer, 1979*b*). In Morrow Co., *P. parvus* occurred in considerable abundance in a habitat consisting largely of green rabbit-brush (*Chrysothamnus nauseosus*) with an understory of cheat grass, Sandberg bluegrass (*Poa sandbergii*), and snakeweed (*Gutierrezia sarothrae*—Small and Verts, 1983). We also have taken *P. parvus* in juniper (*Juniperus occidentalis*)-sagebrush scablands in Deschutes Co. and in rimrock and adjacent talus in Harney Co.

Feldhamer (1979*b*) caught fewer *P. parvus* than expected on study plots with <40% ground cover and more than expected on plots with >40% ground cover. On plots on which fire had killed the sagebrush and bitter-brush in southern Washington, *P. parvus* was only one-third as abundant as on unburned plots (Gano and Rickard, 1982).

Perognathus parvus commonly is reported to be the most abundant small mammal in communities in which it occurs, often composing >90% of the small-mammal community. Its abundance is strongly and positively correlated with precipitation falling in October–April (Verts and Kirkland, 1988). Densities of ≥80/ha were estimated in Washington (Gray, 1943; O'Farrell et al., 1975); 80/ha is considered the maximum that can be supported by the average annual seed crop (Schreiber, 1978).

In Oregon, 137 Great Basin pocket mice were marked on a 1.8-ha grid; eight died in handling, leaving 129 marked animals, of which 104 were removed in a 3-day trapping period. In the 8 days subsequent to onset of removal trapping, 88 unmarked individuals were captured (Small and Verts, 1983). Later in a similar experiment on the same grid, Verts and Carraway (1986) showed that two-thirds of the unmarked pocket mice captured on the grid after removal trapping probably originated on the grid. Thus, in the study conducted by Small and Verts (1983) a minimum

of 196 individuals probably occupied the grid (109/ha).

Diet.—The diet of *P. parvus* consists mostly of seeds of a wide variety of grasses, forbs, and shrubs (Bailey, 1936; Hall, 1946; Kritzman, 1970; Scheffer, 1938*b*; Smith, 1942). Seeds of cheat grass, nightshade (*Solanum*), goosefoot (*Chenopodium*), pigweed (*Amaranthus*), tumblemustard (*Sisymbrium altissimum*), Russian thistle (*Salsola iberica*), and buckwheat (*Eriogonum*) commonly are represented among materials transported in the cheek pouches. Contrary to the "strict vegetarian" classification of *P. parvus* by Scheffer (1938*b*:11), insects are fed upon by members of the species before grass seeds ripen (Jameson, 1954; Kritzman, 1970; O'Farrell et al., 1975). Iverson (1967) reported that *P. parvus* in British Columbia fed on 50% seeds, 25% vegetation, and 25% animal material at lower elevations but 40% seeds, 40% vegetation, and 20% animal material at higher elevations. Vegetative material and insects sometimes are found among materials in the cheek pouches (Kritzman, 1970) and sometimes not (Iverson, 1967).

Reproduction, Ontogeny, and Mortality.—In southeastern Oregon, *P. parvus* males were in reproductive condition from May to early August, with the peak in June. Males were reproductively active about a month before females (Feldhamer, 1979*b*). Most pregnant females were trapped in June and recruitment was greatest in July. No reproductively active individuals were captured after 29 August (Feldhamer, 1979*b*).

In southern Washington, males with scrotal testes were captured first in February–March, attained peak numbers in April–May, declined in numbers in June–July, and were not captured after August (O'Farrell et al., 1975). Young-of-the-year males developed scrotal testes during 1 of 4 years (O'Farrell et al., 1975).

Length of the breeding season, number of litters produced, and breeding by young-of-the-year in *P. parvus* are dependent largely on precipitation and the resulting production of winter annuals (Verts and Kirkland, 1988). In dry years, *P. parvus* in Washington and Idaho produced a maximum of one litter; a large proportion of females failed to reproduce (O'Farrell et al., 1975; Speth et al., 1968). However, in years of above-average precipitation, all females produced at least one litter, many produced two, and a few produced three (Iverson, 1967; O'Farrell et al., 1975). Female young-of-the-year produced offspring only during the wet year, the same year that young-of-the-year males developed scrotal testes (O'Farrell et al., 1975). The breeding season in Washington ranged from 3 to 6 months, with 5 months recorded in most years (O'Farrell et al., 1975). First pregnancies were noted in April during 2 years and in May during 4 years; the first lactating females were captured in May each year, but lactation was last detected in July during 1 year, September during 3 years, and October during 1 year, and females were not examined after May during 1 year (O'Farrell et al., 1975). Some seasonal and altitudinal variations were noted in reproductive and lactational periods in nearby areas (Iverson, 1967; Scheffer, 1930*b*; Speth et al., 1968). Gestation was estimated at 21–28 days (Scheffer, 1938*b*) and 21–25 days (Iverson, 1967).

Not all females are pregnant simultaneously. In Washington, 11.0% were pregnant in May, 32.5% in June, 39.5% in July, and 13.1% in August (Scheffer, 1938*b*); in Nevada, the percentages were 18.6, 27.6, and 9.2 for May–

July, respectively (Hall, 1946).

In Washington, average litter size was 5.4 (n = 18) in 1921, 5.0 (n = 77) in 1923, and 5.3 (n = 37) in 1924 (Scheffer, 1938b). In 1923, when samples were collected at approximately monthly intervals, average litter size was 3.8 (n = 12) in May, 5.0 (n = 25) in June, 5.7 (n = 32) in July, and 4.6 in August (Scheffer, 1938b:10, table 3). Thus, annual and seasonal differences and the small sample likely were responsible for the slightly greater average litter size in Oregon (Table 11-13).

Information on postnatal development is extremely limited (Verts and Kirkland, 1988). Juveniles at 3 days of age weighed 2.2 g (Speth et al., 1968). In Washington, young were first captured in June, but the average date of first capture for young was 31 July (O'Farrell et al., 1975).

In Washington, survival from weaning to the breeding season during the following year ranged from 56 to 80% with the highest values "during the year of lowest precipitation, poorest food supply, and reduced reproductive rate" (O'Farrell et al., 1975:20). Only 17–19% survived to the 3rd year, 2–3% to the 4th year. Survivorship did not meet the equal-probability assumption required for use of several models for estimating density (O'Farrell et al., 1975).

Reported predators of *P. parvus* include rattlesnakes (*Crotalus*), burrowing owls (*Athene cunicularia*), short-eared owls (*Asio flammeus*), coyotes (*Canis latrans*), weasels (*Mustela frenata* and *M. erminea*), striped skunks (*Mephitis mephitis*), badgers (*Taxidea taxus*), foxes (*Vulpes vulpes*, *V. velox*, and *Urocyon cinereoargenteus*), northern grasshopper mice (*Onychomys leucogaster*), and deer mice (*Peromyscus maniculatus*—Banfield, 1974; Scheffer, 1938b).

Habits.—*Perognathus parvus* constructs burrows ≅25 mm in diameter that extend 0.13–1 m below the surface. Permanent burrows contain granaries for storage of foodstuff, a nest cavity ≅8 cm in diameter, and several entrances (Iverson, 1967; Scheffer, 1938b). Nests consist of fine bits of plant material (Scheffer, 1938b). Escape burrows are simple, shallow (20–30 cm), lack nests and granaries, and have two or more entrances (Iverson, 1967). Burrows may be situated at the base of shrubs (Scheffer, 1938b) or in the open (Kritzman, 1970). Mounds of excavated earth appear as miniatures of those constructed by pocket gophers (Fig. 11-49), including a plugged entrance (Scheffer, 1938b).

Perognathus parvus commonly becomes torpid for various periods at all seasons; even in summer, individuals caught in live traps often enter torpor. During live-trapping studies, we commonly carried a pocket mouse in a shirt pocket inside a down vest for an hour or more until it aroused sufficiently to be released. *P. parvus* may remain in torpor for ≅90% of the winter and, by doing so, reduce food requirements from 300 g to only 50 g of seeds (Meehan, 1978). In addition to conserving energy through torpor, *P. parvus*, like other heteromyids, is a great conserver of water; *P. parvus* can maintain a positive water balance from preformed water in seeds and metabolic water produced through conversion of starches to sugars even at temperatures as high as 29°C (Rubin, 1979). Such may permit *P. parvus* to forage earlier in the night than other desert granivores, thereby avoiding direct competition with them (Rubin, 1979).

Great Basin pocket mice spend much of their lives in their burrows; even during the active season they usually are above ground only late in the evening or at night, and not even then when the moon is bright (Verts and Kirkland, 1988). Adult males emerge from a 3–4-month winter torpor from late March to late April and adult females emerge from mid-April to early May; soil temperature may be the cue for emergence (O'Farrell et al., 1975). Although some Great Basin pocket mice may be captured during all months except perhaps December and January (O'Farrell et al., 1975:10, fig. 4), individual adults average being active for only 60–90 days.

Perognathus parvus sandbathes, especially after being handled; sometimes "wallows" are formed by repeated sandbathing in the same spot (Kritzman, 1970). Some authorities consider *P. parvus* relatively docile and not prone to bite when handled, whereas others report contrary experiences (Verts and Kirkland, 1988).

Great Basin pocket mice have extremely small home ranges. In Washington, maximum distance between captures for five males captured 4–10 times each ranged from 60 to 297 m ($\overline{X}$ = 148 m—Broadbooks, 1961). In Harney Co., 28 home-range areas calculated by the standard-diameter method for males and females averaged 288.4 and 277.7 m², respectively, in greasewood habitats and 336.9 and 267.2 m², respectively, in sagebrush habitats (Feldhamer, 1979b). The small size of home ranges also is illustrated by data obtained during removal-replacement studies in which we were involved (Small and Verts, 1983; Verts and Carraway, 1986). On a 10- by 10-trap grid with 15 m between trap stations, only 12 (10.2%) of 118 pocket mice caught four or more times were caught in four different traps; only one was caught in five. Whether home ranges of *P. parvus* overlap significantly is disputed (Verts and Kirkland, 1988).

Small and Verts (1983) concluded that immigrants from nearby regions were responsible for the rapid replacement of pocket mice removed from a 1.8-ha grid in Morrow Co. However, in a similar study in which potential immigrants in areas surrounding the same grid were marked, Verts and Carraway (1986) observed a similar replacement by unmarked individuals of animals removed. Because few individuals marked in the area surrounding the grid were captured, they suggested that much of the apparent recovery of the population on the grid was by behaviorally subordinate mice commencing to enter traps after dominant individuals were removed.

Although social structure seems to play a significant role in the population ecology of *P. parvus*, the species is not considered to be particularly social. In the wild, individuals are thought to occupy separate nests, and caged pairs engage in vigorous sparring and occasionally bizarre behavior (Verts and Kirkland, 1988). *P. parvus* is strongly agonistic toward and usually dominant over congeners and other mice (Ambrose and Meehan, 1977; Blaustein, 1972; Blaustein and Risser, 1974b; Huey, 1959; Kritzman, 1974).

Remarks.—Their occurrence in some habitats as nearly single-species small-mammal communities, their granivorous diet and gleaning foraging pattern, their ease of capture and recapture, and the seemingly unique social structure in their populations (O'Farrell et al., 1975; Small and Verts, 1983; Verts and Carraway, 1986) make Great Basin pocket mice superb subjects for study of population phenomena and behavioral ecology. Perhaps even their long period of torpor, their apparent aversion to moonlight, and the positive response of their numbers to winter precipitation may be used to the advantage of the investigator in some studies of *P. parvus*. We suspect that much could be learned about this species through use of trap-timing devices that record the time that individuals were captured (Barry et al., 1989).

Microdipodops megacephalus Merriam
Dark Kangaroo Mouse

1891. *Microdipodops* Merriam, 5:115.
1901a. *Microdipodops megacephalus oregonus* Merriam, 14:127.

Description.—*Microdipodops megacephalus* (Plate XIII) is a small (Table 11-12), bipedal heteromyid rodent with enormously expanded auditory bullae (Fig. 11-60), rooted molariform teeth, and a tail thickest in the middle and tapering to both ends. Hind feet <27 mm long distinguishes the species from *Dipodomys*, the only other group in Oregon with greatly inflated auditory bullae.

The pelage of *M. megacephalus* is grayish or brownish dorsally; the venter hairs are white with lead-colored bases. The tail is light colored proximally grading darker to the color of the dorsum distally.

Distribution.—The geographic range of *M. megacephalus* includes Upper Sonoran sagebrush (*Artemisia*) desert regions of southeastern Oregon, extreme eastern California, much of northwestern and eastern Nevada, and a small portion of west-central Utah (Fig. 11-61; Hall, 1981). In Oregon, *M. megacephalus* occurs within the area circumscribed by a line connecting Denio, Nevada; Fort Rock, Lake Co.; Powell Butte, Crook Co.; Malheur National Wildlife Refuge Headquarters, Harney Co.; and Burns Junction, Malheur Co. (Fig. 11-61).

There is no fossil record for the species (O'Farrell and Blaustein, 1974).

Geographic Variation.—Of the 12 subspecies recognized by Hall (1981), only one, *M. m. oregonus,* occurs in Oregon.

Habitat and Density.—*Microdipodops megacephalus* is a species of the Upper Sonoran Desert. In Nevada, especially where it is sympatric with its only congener, *M. megacephalus* commonly is associated with gravelly soils (Ghiselin, 1970; Hall, 1946). However, near the periphery of its range in California and Utah, it is reported to be found on substrates of fine sand (Hall, 1946). Fautin (1946) estimated densities of 0.2–3.8/ha in communities dominated by shadscale (*Atriplex confertifolia*), winterfat (*Eurotia lanata*), greasewood (*Sarcobatus vermiculatus*), and horsebrush (*Tetradymia*). In Oregon, we took *M. megacephalus* at the edge of dunes of fine sand vegetated by greasewood and shadscale along the northeastern edge of Harney Lake, Harney Co. Feldhamer (1977b) captured three on Malheur National Wildlife Refuge in a habitat of 77.2% big sagebrush (*Artemisia tridentata*), 16.9% saltbush (*Atriplex*), and 5.4% greasewood on deep sandy loam; total vegetative cover was 38.3%.

Diet.—Hall (1946) concluded from contents of cheek pouches of trapped individuals that the diet of *M. megacephalus* in Nevada probably consisted mostly of seeds. However, he captured several specimens that contained animal material in their cheek pouches, including a roach (*Arenivaga erratica*), a tenebrionid beetle (*Eusattus*),

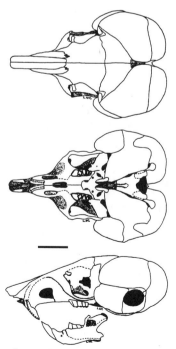

Fig. 11-60. Dorsal, ventral, and lateral views of the cranium, and lateral view of the dentary of the dark kangaroo mouse, *Microdipodops megacephalus* (OSUFW 2940). Scale bar equals 5 mm.

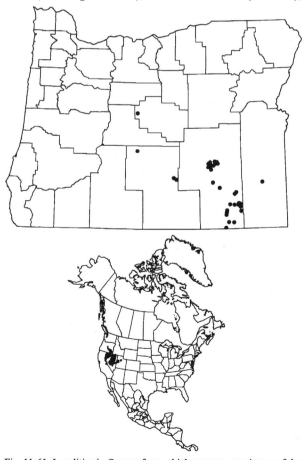

Fig. 11-61. Localities in Oregon from which museum specimens of the dark kangaroo mouse, *Microdipodops megacephalus*, were examined (listed in APPENDIX, p. 523), and the distribution of the dark kangaroo mouse in North America. (Range map redrawn from Hall, 1981:560, map 340, with modifications.)

and fragments of a lizard (*Crotaphytus wislizenii*). Based on examination of fecal pellets in the same state, arthropods formed the greatest proportion of the diet of *M. megacephalus* in most years (Harris, 1986). Arthropod prey was mostly Lepidoptera larvae, beetles (Coleoptera), and Jerusalem crickets (*Stenopelmatus fuscus*) in spring; the same plus a small June beetle (*Polyphylla*?) in summer; and all these plus spiders (Araneae) and solifugids (Solifugae) in autumn (Harris, 1986). In one autumn, a true bug (Lygaeidae) occurred in the diet frequently (Harris, 1986). However, some green vegetation was consumed (mostly in spring) and seeds were used in summer and autumn (Harris, 1986). In summer, seeds used most were those of ricegrass (*Oryzopsis*) and milk-vetch (*Astragalus*); in autumn, most were those of rabbit-brush (*Chrysothamnus*).

In captivity, we maintained an individual for 10 months on a diet of sunflower seeds and an occasional piece of carrot. Blaustein and Risser (1974*a*) fed captives sunflower seeds, rolled oats, and millet with a weekly supplement of lettuce.

In laboratory experiments, *M. megacephalus* harvested more seed from scattered distributions than from clumped distributions whether presented separately or simultaneously (Harris, 1984). In the field, based on the incidence of tracks, *M. megacephalus* visited significantly more plots in the open than those near shrubs whether or not the plots were baited (Harris, 1984). Although *M. megacephalus* occurred more frequently in the open, it tended to concentrate its foraging activity on scattered seed dispersions; *Dipodomys* used both intershrub and near-shrub microhabitats, thus forcing *M. megacephalus* to use plots with scattered seed more than plots with clumped seed (Harris, 1984). This suggests that foraging by *M. megacephalus* incorporates aspects of the foraging behavior of both *Dipodomys* (open habitats) and *Perognathus* (scattered seed), which, in turn, suggests that use of open habitats by desert rodents is related more to gait (ricochetal) than to body size. The enormous auditory bullae and bipedalism (and ricochetal gait) are probably antipredator mechanisms that permit use of open areas for foraging by *M. megacephalus*, but the species is excluded from harvesting clumps of seed by the larger *Dipodomys*, which harvests clumps more efficiently (Harris, 1984).

Reproduction, Ontogeny, and Mortality.—In Nevada, 5 of 14 adult females were pregnant in April, 8 of 91 in May, 34 of 102 in June, 5 of 198 in July, 0 of 26 in August, 2 of 45 in September, and 0 of 2 in October (Hall, 1946). The long breeding season and relatively small average litter size (Table 11-13) suggest that more than one litter may be produced annually; nevertheless, we know of no direct evidence of multiple litters.

In California, only three (13.6%) of 22 individuals marked in 1980 on two 0.25-ha trap grids were recaptured in spring 1981, but 11 (37.9%) of 29 marked on the same areas in 1979 were recaptured in spring 1980 (Harris, 1987). Survival was related directly to the circumference of the tail, thus, to the caudal-fat reserves (Harris, 1987). Tail circumference was considered an index to overall condition, as the size of the fat reserve in the tail was too small to be the sole store of energy for winter.

Longevity in captivity of 5 years and 5 months has been recorded for a wild-caught individual (Egoscue et al., 1970).

Habits.—In Nevada, on a daily basis, activity began

with an intense burst during the first 2 h after sunset, declined to almost nil at 6 h after sunset, then, especially in summer, increased again before sunrise (O'Farrell, 1974). Seasonally, *M. megacephalus* first appeared in traps during the 2nd week of March; activity increased to mid-April, declined and remained low until August, increased again, then ceased at the end of October (O'Farrell, 1974). During the period of activity, the intensity thereof seemed antipodal to that of *Perognathus longimembris* in the same area (O'Farrell, 1974).

In 11 of 12 staged encounters, *M. megacephalus* was subordinate to *P. longimembris* despite being significantly heavier (Blaustein and Risser, 1974*a*). The latter authors considered their findings combined with those of O'Farrell (1974) to indicate that coexistence of the two species is related to *M. megacephalus*, despite being the inferior competitor, having the broadest temporal niche.

O'Farrell and Blaustein (1974) referenced their dissertation and thesis, respectively, in reporting that *M. megacephalus* constructed elaborate nests, cached seeds in its burrow system, and slept on its back. An individual that we maintained in captivity from 15 August 1973 to 11 June 1974 also frequently slept on its back.

Hall (1946) indicated that he had followed the tracks of an individual *M. megacephalus* to a point 21.3 m from their origin; as a radius, this would produce a 1-night circular home-range area of 1,425.3 m². O'Farrell and Blaustein (1974) cited the former's dissertation in reporting annual circular home-range areas of 6,613 m² for males and 3,932 m² for females.

Remarks.—In our bibliographic compendium on Oregon mammals (Carraway and Verts, 1982) we did not list even one article devoted substantially to the biology of *M. megacephalus*. Searches of bibliographic sources published subsequently also failed to reveal publications on the species in Oregon.

Blaustein and Risser (1974*a*) provide an interesting hypothesis that could be tested by removing *P. longimembris* and monitoring activity of *M. megacephalus*. Activity of *M. megacephalus* should not decline in summer if *P. longimembris* is removed. Such a study, if suitable syntopic populations could be located, likely would produce considerable new information on other aspects of the ecology and behavior of both species in Oregon.

Dipodomys californicus Merriam
California Kangaroo Rat

1890*d. Dipodomys californicus* Merriam, 4:49.
1921. *Dipodomys heermanni californicus* Grinnell, 2:95.
1925. *Dipodomys heermanni gabrielsoni* Goldman, 38:33–34.

Description.—*Dipodomys californicus* (Plate XIII) is the largest kangaroo rat in Oregon (Table 11-12), but is only of moderate size within the genus. It has a moderately broad face, relatively large ears, and awl-shaped lower incisors. There is no hallux on the hindfoot. The maxillary arms of the zygomata are relatively broad (Fig. 11-62). The occipitonasal length is ≥38 mm. The tail is ≅150% of the length of the head and body.

Overall, *D. californicus* is the darkest-colored kangaroo rat in Oregon. The pelage of the dorsum is composed of hairs with dark-gray bases, a narrow dark-buff band, and a black terminus. The buffy band becomes longer and the gray base becomes white with increasing distance from the midline, producing lighter-colored flanks. The venter; feet; upper lip; and base, sides, and tip of tail are white; there is a white spot above each eye and behind each ear. A white stripe across each hind leg isolates a patch of dorsum-colored fur. A black moustache; black vibrissae, eyelids, and dorsal and ventral tail stripes (the latter tapering to the tip); and white and black splotched ears produce a striking attire for this handsome rodent. The fur on the terminal one-half of the tail rises at increasingly acute angles distally to produce a crest.

Distribution.—*Dipodomys californicus* occurs between the humid coastal regions and the foothills of the Cascade and Sierra Nevada ranges from a line connecting Brownsboro, Jackson Co., and Gerber Reservoir, Klamath Co., Oregon, south to a line connecting Suisun Bay and Lake Tahoe, California (Fig. 11-63; Kelt, 1988). In Oregon, *D. californicus* occurs only in the southern portions of Jackson, Klamath, and Lake counties (Fig. 11-63).

Geographic Variation.—Of the three nominal subspecies of *D. californicus*, only *D. c. californicus* occurs in Oregon (Kelt, 1988).

Habitat and Density.—*Dipodomys californicus* is a fairly common inhabitant of brushy hillsides on the east and north sides of the Rogue River Valley (Gabrielson, 1931) and other valleys within the range of the species. In Klamath Co., *D. californicus* occurred in an early seral community after a wildfire and subsequent salvage-logging removed the ponderosa pine (*Pinus ponderosa*)–bitterbrush (*Purshia tridentata*)–needlegrass (*Stipa occidentalis*) community (Burton and Black, 1978). Squaw currant (*Ribes cereum*) was a common shrub; rabbitbrush

goldenweed (*Haplopappus bloomeri*), autumn willow-weed (*Epilobium paniculatum*), spreading groundsmoke (*Gayophytum diffusum*), and mountain knotweed (*Polygonum douglasii*) were common forbs. Ground cover was 24–33%, of which 26–41% was grasses and 59–70% was forbs (Burton and Black, 1978).

In northern California, optimum habitat was on shallow (15–30 cm), well-drained, rocky or stony soils supporting scattered common buckbrush (*Ceanothus cuneatus*), but no trees. Scattered boulders and underlying rock projected through in many places. Lush vegetation was avoided (Dale, 1939). Soils in winter often were wet, causing *D. californicus* to excavate mud from its burrows (Dale, 1939). However, Grinnell and Linsdale (1929) indicated that soils inhabited by *D. californicus* must be well drained, especially in the winter months.

Diet.—Green vegetation and bulbs are the primary food materials found in cheek pouches of *D. californicus* in spring and summer (Dale, 1939; Grinnell et al., 1930), and may be the source of water during the dry season (Bailey, 1936). Seeds and berries of green-leaf manzanita (*Arctostaphylos patula*) and seeds of buckbrush, rabbitbrush (*Chrysothamnus*), lupine (*Lupinus*), bur-clover (*Medicago*), slender oats (*Avena barbata*), and foxtail (*Alopecurus*) commonly are found in the cheek pouches (Bailey 1936; Smith, 1942). *D. californicus* may scatter-

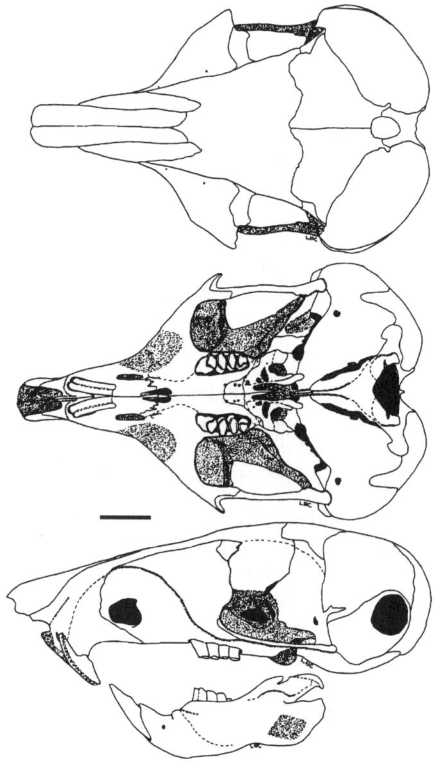

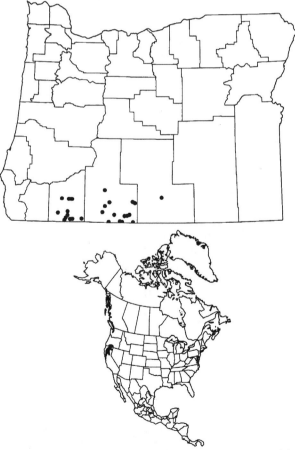

Fig. 11-63. Localities in Oregon from which museum specimens of the California kangaroo rat, *Dipodomys californicus*, were examined (listed in APPENDIX, p. 516), and the distribution of the California kangaroo rat in North America. (Range map redrawn from Kelt, 1988:2, fig. 3, with modifications.)

Fig. 11-62. Dorsal, ventral, and lateral views of the cranium, and lateral view of the dentary of the California kangaroo rat, *Dipodomys californicus* (OSUFW 2684). Scale bar equals 5 mm.

hoard as large food caches are not maintained in burrows (Dale, 1939).

Reproduction, Ontogeny, and Mortality.—In northern California, pregnant females were collected February–September although none was examined in July–August (Dale, 1939). In Oregon, Bailey (1936) examined one pregnant individual each in April and September; the former contained two embryos, the latter four, the range in numbers of embryos in a sample of 20 individuals from California (Table 11-13).

Other than changes in skull morphology related to age of a sample for which skull length was ≅87% of that of adults, no information is available regarding development in *D. californicus* (Dale, 1939).

Bailey (1936) implied that mammalian and avian predators that commonly prey on small nocturnal mammals may serve as a natural check on populations of this species. Dale (1939) indicated that predators may serve to suppress growth of populations when the density is low. Dale (1939) further suggested that local climatic conditions tended to exert a greater force in the regulation of numbers of *D. californicus*. We are less skeptical of the latter claim.

Habits.—*Dipodomys californicus* does not hibernate, but may be active at temperatures as low as –11°C even when snow covers the ground or, at warmer temperatures, when it is raining (Dale, 1939; Grinnell et al., 1930). Activity is nocturnal, commencing shortly after dark (Dale, 1939).

Like its congeners, *D. californicus* constructs burrows as daytime retreats. Burrows commonly are on little knolls, but where soils are thin and rocky, they often are constructed beneath boulders (Dale, 1939). In entering and leaving burrows, *D. californicus* usually follows the same routes; thus travel pathways radiate from entrances, but fade as the distance from entrances increases. Black fecal pellets often are scattered along the pathways (Dale, 1939). Locomotion is described as a "scamper" rather than richocetal as in congeners (Dale, 1939).

Seemingly, nothing is known regarding the social life of *D. californicus*.

Remarks.—For many years, based on the existence of a few individuals with a rudimentary hallux, the four-toed kangaroo rats in southern Oregon and northern California were considered conspecific with the otherwise superficially similar five-toed kangaroo rats that occur mostly farther south (Grinnell, 1922; Patton et al., 1976). All were classified as *Dipodomys heermanni*. However, Dale (1939) suggested, on morphological grounds, that *D. californicus* and *D. heermanni* deserved specific status. Fashing (1973) found that the northern four-toed groups had 52 chromosomes (FN = 96), whereas southern five-toed groups had 64 (FN = 90 or 94). Genetically, the four- and five-toed groups are more different from each other than either is from an allopatric five-toed congener, and the four-toed group is no more closely related to other four-toed congeners than to five-toed congeners (Patton et al., 1976). The concordance of morphological, electrophoretic, and karyological data caused Patton et al. (1976) to reclassify the northern three four-toed races of *D. heermanni* as subspecies of *D. californicus*.

Bailey (1936) claimed that *D. californicus* might become sufficiently numerous to be a pest locally and indicated that he had obtained specimens killed by poisoned grain

distributed as a means of controlling rodents. We opine that depredations on cultivated crops by the species, if indeed they occur, are of little economic consequence.

Dipodomys californicus seems to be another species that researchers largely have ignored. We suspect that this may be related, in part, to its uncertain taxonomic status until the works of Fashing (1973) and Patton et al. (1976) were published. Now that the taxonomy seems more firmly established, its limited distribution in Oregon and its somewhat restricted habitat elsewhere (Dale, 1939) suggest that this handsome rodent deserves study to ensure that it continues to be part of the fauna of the state.

Dipodomys microps (Merriam)
Chisel-toothed Kangaroo Rat

1904*b. Perodipus microps* Merriam, 17:145.
1921. *Perodipus microps preblei* Goldman, 2:233.
1939. *Dipodomys microps preblei*: Hall and Dale, 4:54.

Description.—*Dipodomys microps* (Plate XIII) is an intermediate-sized kangaroo rat (Table 11-12) with a narrow face, small ears, and flat-faced, nearly square-edged lower incisors shaped like miniature chisels. A hallux (little more than an exposed nail) is present. Among kangaroo rats, the cheek pouches in relation to the size of the head are exceptionally large, likely an adaptation for transporting the lightweight materials eaten during a large part of the year (Nikolai and Bramble, 1983). The maxillary arms of the zygomata are relatively narrow (Fig. 11-64). The tail is about 130–140% as long as the head and body.

The pelage on the dorsum is composed of hairs with medium-gray bases, a buffy band, and a tiny blackish tip; on the sides, the bases are white instead of gray and the proportion of hairs with black tips is reduced. Overall the

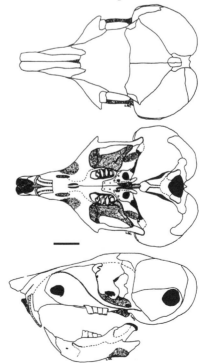

Fig. 11-64. Dorsal, ventral, and lateral views of the cranium, and lateral view of the dentary of the chisel-toothed kangaroo rat, *Dipodomys microps* (OSUFW 6881). Scale bar equals 5 mm.

dorsum is a dirty-sand color. The venter, upper lip, feet, and base and sides of the tail are white and there is a white spot above each eye and behind each ear; a white stripe across each hind leg isolates a patch of dorsum-colored fur. The tail is ringed with white at its base and the tip is grayish brown; the fur on the distal one-third is longer, producing a crest or flag. A black moustache, vibrissae, eyelids, ears, and dorsal and ventral tail stripes (the latter extending undiminished to the tip) complete the pelage of this strikingly marked small mammal. In contradiction to a published report regarding the species (Hayssen, 1991), the moustache in Oregon specimens of *D. microps* seems no more prominent than in Oregon specimens of *D. ordii*.

Distribution.—*Dipodomys microps* occurs from near Mountain Home and Murphy, Idaho, and Malheur National Wildlife Refuge, Harney Co., and Summer Lake, Lake Co., Oregon, southward through extreme eastern California and most of Nevada to San Bernardino Co., California, and the Colorado River in northwestern Arizona; another arm extends west of Great Salt Lake, Utah, north into south-central Idaho south of the Snake River (Fig. 11-65; Hall, 1981; Larrison and Johnson, 1981). Hardy (1949) indicated that *D. microps* crossed the Colorado River on a man-made bridge in northwestern Arizona. In Oregon, *D. microps* occurs in Lake, Harney, and Malheur counties south of a line

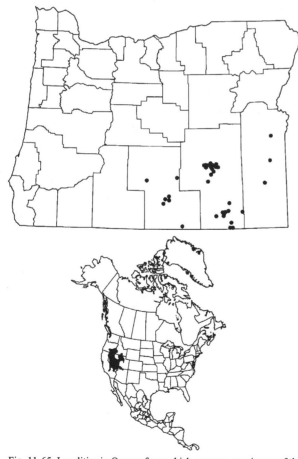

Fig. 11-65. Localities in Oregon from which museum specimens of the chisel-toothed kangaroo rat, *Dipodomys microps*, were examined (listed in APPENDIX, p. 516), and the distribution of the chisel-toothed kangaroo rat in North America. (Range map redrawn from Hall, 1981:571, map 343, with modifications provided by Larrison and Johnson, 1981:66, unnumbered figure.)

connecting Summer Lake, Malheur Lake, and Watson in the three counties, respectively (Fig. 11-65).

Pleistocene fossils of *D. microps* were found in cave deposits in Utah (Heaton, 1985).

Geographic Variation.—Only one, *D. m. preblei*, of the 13 nominal subspecies of *D. microps* occurs in Oregon (Hall, 1981). The range map provided by Hall (1981:571, map 343) shows *D. m. idahoensis* occurring in Oregon, but he listed no specimens of the race from the state.

Habitat and Density.—Throughout its range, *D. microps* is associated with shrub-dominated habitats (Allred et al., 1963; Beatley, 1976; Csuti, 1979; Fautin, 1946; Jorgensen and Hayward, 1965). In more southern parts of its range, *D. microps* is abundant in shadscale (*Atriplex confertifolia*) communities in desert valleys, in upland-desert communities that contain blackbush (*Coleogyne ramosissima*—Csuti, 1979), and in hopsage (*Atriplex*)–desert-thorn (*Lycium*) communities (Allred and Beck, 1963a). It was least abundant in creosotebush (*Larrea tridentata*)-dominated habitats (Allred and Beck, 1963a). In Utah, it was most abundant in saltbush and horse-brush (*Tetradymia*) communities, least abundant in sagebrush (*Artemisia*) and greasewood (*Sarcobatus*) communities.

In Harney Co., Oregon, we captured *D. microps* in a greasewood and shadscale community at the edge of the dunes northeast of Harney Lake, in a sagebrush flat west of the Buena Vista guard station, and on the sand dunes immediately south of Fields and those just north of Denio, Nevada. Feldhamer (1977b) captured *D. microps* in sagebrush and greasewood habitats on the Malheur National Wildlife Refuge; however, one plot in each of the habitat types contained shadscale.

In Nevada, percent cover by shrubs and precipitation-temperature ratios were primary determinants for the abundance of *D. microps*; at the 20 sites (of 58 undisturbed sites sampled) with the highest densities, the former averaged 33.0% (range, 17.4–50.4%), the latter 14.5% (range, 11.5–22.9%). The relatively high turnover rates of body water (Mullen, 1971) and the low urine-concentrating ability (Kenagy, 1973) likely are responsible for restricting *D. microps* to areas with high precipitation-temperature ratios (Beatley, 1976). Edaphic factors were secondary controls (Beatley, 1976); however, in southwestern Utah, *D. microps* was most abundant on older, moderately shallow, saline, and less-gritty soils with few or no large rocks, but occasionally with some gravel (Hardy, 1945). In the same state, Fautin (1946) indicated that the greatest abundance of *D. microps* occurred where the soil texture was coarse gravel and large rocks or medium to fine gravel.

In Nevada, the number of individuals captured on a 49-trap grid during a 2-day period at each of 20 sites averaged 23.2/ha (range, 11.4–50.9/ha). On most undisturbed sites, *D. microps* unquestionably was the dominant *Dipodomys*, but it essentially disappeared from disturbed sites and was replaced by congeners, often by numbers exceeding those of original populations of *D. microps* (Beatley, 1976). In Idaho, population densities in "healthy" shadscale communities were 2–3 times those in "depleted stands" (Larrison and Johnson, 1973). In Utah, population densities of 8.5–34.0/ha were recorded in various shrub habitats (Fautin, 1946). In Nevada, Vollmer et al. (1976) reported population densities of only 0.5–1.0/ha. No comparable data are available for Oregon, but from our limited experience in trapping specifically for the species, we suggest that population densities are near the minimum reported in these studies.

Diet.—*Dipodomys microps* feeds mostly on leaves of shadscale; these kangaroo rats climb into the shrubs, harvest the leaves, and carry them in their cheek pouches to burrow systems for storage (Kenagy, 1972). However, in Nevada, only eight of 30 burrows excavated contained food caches; there is some evidence that *D. microps* scatter-hoards (Anderson and Allred, 1964).

In spring when the water content is ≅80%, the *Atriplex* leaves are eaten whole, but in the remainder of the year when the water content is ≅50%, the leaves are peeled of the hypersaline epidermis and the vesicular trichome layers and the remaining parenchyma, vascular tissue, and mesophyll are eaten. The shavings, removed by drawing the forefoot-held leaves over the cutting edge of the chisel-shaped lower incisors about 10 times on each side, are discarded (Kenagy, 1972). In California, Csuti (1979) captured individuals with leaves of greasewood and desert-thorn in addition to those of shadscale in their cheek pouches.

Some seeds also are eaten in late summer in Idaho (Johnson, 1961), Nevada (Anderson and Allred, 1964), and Utah (Warnock and Grundmann, (1963). The latter authors also found bones in stomach and intestinal contents of 14% of 29 *D. microps* examined; identity of the bones was not provided. In southern Utah, caches in burrow systems contained seeds exclusively (Hardy, 1945, 1949). In northwestern Arizona, Csuti (1979) captured these kangaroo rats with blackbush leaves and seeds of foxtail chess (*Bromus rubens*) in their cheek pouches. Caches in burrow systems in the same area contained seeds of foxtail chess, barley, fescue (*Festuca*), filaree (*Erodium cicutarium*), deervetch (*Lotus*), peppergrass (*Lepidium femontii*), blackbush, and turpentine broom (*Thamnosma montana*). In captivity, blackbush leaves were eaten whole without being peeled (Csuti, 1979). In southern Idaho, leaves and seeds of shadscale, halogeton (*Halogeton glomeratus*), clasping peppergrass (*L. perfoliatum*), and saltsage (*Atriplex nuttallii*), and seeds of tansy mustard (*Descurainia pinnata*), and grasses were found in cheek pouches of *D. microps* (Johnson, 1961).

From a series of feeding trials in captivity, Csuti (1979) concluded that *D. microps*, from an area in which the frequently deciduous blackbush is the only plant whose leaves are eaten and from a shadscale-dominated community, actually is more of a generalist than indicated by its food niche in either area. At least some individuals from the former area could be induced to peel leaves of shadscale for food (and water), but animals from neither area survived when provided dry seeds alone. Seemingly, *D. microps* is capable of "shifting its food niche and associated physiology and behavior when placed in a condition simulating other habitats . . ., as long as the shift is not too abrupt" (Csuti, 1979:50). However, animals from the shadscale-dominated community were more efficient at browsing on shadscale, but were not as able to cope with moisture stress as those from the blackbush area (Csuti, 1979).

In southeastern Oregon, a trace of unidentified fungi was eaten by *D. microps* (Maser et al., 1988).

Reproduction, Ontogeny, and Mortality.—The ges-

tation period in *D. microps* is ≅31 days (Daly et al., 1984). In captivity, there is no postpartum estrus, but females whose litters were removed at 25 days of age became estrous 29–39 days postpartum (Daly et al., 1984). Average litter size for females bred in captivity was slightly, but not significantly, smaller than that for captive females bred in the wild (Table 11-13).

In southern California, males commence to produce spermatozoa in November and mating begins in January or February when testes and epididymides attain maximum mass. In a year following precipitation in September, the season of parturition was extended ≅3 weeks, but lactation ceased too early for survival of some young born in late litters (Kenagy and Bartholomew, 1985). The usual season of parturition coincides with the peak in water content in *Atriplex*. Curtailment of the breeding season may be related to declining water content in *Atriplex*; congeners with access to irrigated fields produced litters in autumn, whereas syntopic *D. microps* did not. Juvenile males do not become reproductively competent in the year of their birth, but some juvenile females became pregnant in a year in which the breeding season was prolonged (Kenagy and Bartholomew, 1985).

Average mass of neonates was 4.0 g; at emergence, young weighed an average of 21.0 g (Kenagy and Bartholomew, 1985).

French et al. (1967) estimated an average life span of 4.9 months in unmanipulated populations on fenced plots and only 1.6 months on an unfenced plot.

Habits.—In Oregon, specimens have been collected only in May–August, likely a reflection of nonrandom effort, as in desert areas of southern California, *D. microps* is active throughout the year, even on nights when temperatures are as low as –19°C and as high as 30°C (Kenagy, 1973). In Nevada, *D. microps* was captured throughout the year in some plant communities, but only during part of the year in others (Allred et al., 1963).

In southern California, activity began at twilight at light intensities 2– 4 times that of full moonlight and continued at a declining rate to a time of similar light intensity in the morning twilight (Kenagy, 1973, 1976). Kenagy (1976) postulated that the early evening emergence of *D. microps*, despite the hazard of exposure to predation by day-active raptors, was related to the need to maintain exclusive "territories" in the vicinity of the home burrow. Those emerging too early were selected against by day-active predators, whereas those emerging too late were selected against by failing to find an eligible mate. Although activity of *D. microps* declined during the four successive quarters of the night, activity of its most abundant congener increased (Kenagy, 1976). Nighttime movements usually were ≤50 m (Kenagy, 1973). However, in Nevada, maximum distance between points of capture averaged 79.5 m for males and 68.6 m for females; 79% of the males and 87% of the females moved <122 m (Allred and Beck, 1963b). On the same area, home-range areas ranged from 1.1 to 2.5 ha (Jorgensen and Hayward, 1965). Maximum dispersal distance recorded was 411.6 m for a male and 434.5 m for a female (Allred and Beck, 1963b).

Dipodomys microps constructs multientrance burrows; 25 spoil mounds (often referred to as "hillocks") in southern California averaged (±*SE*) 67 ± 5 cm high and 2.5 ± 0.1 m wide. Greasewood was the dominant shrub on

76% of the hillocks; average shrub height at mound sites was 85 ± 5 cm (Kenagy, 1973). Tunnels often were 6–8 cm in diameter, but where soils were firm and moist, they sometimes were as small as 5 cm in diameter. Nest chambers averaged 40 cm below the surface, but some were as shallow as 8 cm (Kenagy, 1973). Most nest chambers were on the southern aspect of hillocks. Nests were composed mostly of ricegrass (*Oryzopsis*). In Nevada, depth of burrows averaged 23.4–39.9 cm beneath the surface; shallowest burrows were in *Atriplex-Kochia* (red sage) habitats and deepest were in *Coleogyne* habitats (Anderson and Allred, 1964). Burrows had a maximum of three–six entrances ($\bar{X}$ = 3.2 in *Atriplex-Lycium*, 1.8 in *Atriplex-Kochia*). The most complex burrows were in *Salsola* and *Atriplex-Lycium* habitats (Anderson and Allred, 1964).

Dipodomys microps is aggressive toward conspecifics regardless of sex and cannot be maintained in captivity as mated pairs (Daly et al., 1984). To be successful, attempts to mate *D. microps* in captivity must be in a neutral arena; after mating, animals must be separated quickly to avoid males being killed (Daly et al., 1984).

Remarks.—Our records of few specimens from so few localities, combined with the findings of Beatley (1976) and Larrison and Johnson (1973) that disturbance of shrub communities drastically reduced or essentially eliminated *D. microps,* suggest that rangeland renovation projects and widespread grazing of shrub communities may have placed the species in jeopardy in Oregon. Such may explain our capturing *D. microps* most commonly on sand dunes—areas unlikely to be grazed or included in renovation projects. Further investigation of the distribution, abundance, and habitat associations of the species with the goal of ensuring the continued occurrence of the species in Oregon is warranted.

Dipodomys ordii Woodhouse

Ord's Kangaroo Rat

1853. D[*ipodomys*]. *ordii* Woodhouse, 6:224.
1894*b. Perodipus ordii columbianus* Merriam, 9:115.
1921. *Dipodomys ordii columbianus*: Grinnell, 2:96.

Description.—*Dipodomys ordii* (Plate XIII) is a medium-sized (Table 11-12), narrow-faced, small-eared kangaroo rat with awl-shaped lower incisors. A hallux (little more than an exposed nail) is present. The maxillary arms of the zygomata are relatively broad (Fig. 11-66). The tail is 120–130% of the length of the head and body.

The hairs of the dorsum have broad gray bases, a narrow buffy band, and tiny black tips; along the sides, the bases are white instead of gray and the black tips are present on only a few of the hairs. *D. ordii* is the lightest-colored kangaroo rat in Oregon; overall the dorsum is a rich buff with gray overtones. The venter, upper lip, feet, and sides of the tail are white; there is a white spot above each eye and behind each ear, and a white stripe crosses each thigh. The tail is ringed with white at its base and the tip is grayish brown. The moustache, eyelids, and ears are blackish. Dorsal and ventral tail stripes are brownish, the latter tapering to a point proximal to the tip of the tail. Hairs on the distal one-third of the tail are progressively longer distally, producing a crest or flag.

Distribution.—*Dipodomys ordii* occurs from southern Saskatchewan and Alberta to southern Hidalgo, and, ex-

cept for some montane regions, from central Oregon, eastern California, and western Arizona east to central South Dakota, eastern Nebraska and Kansas, central Oklahoma, and western Texas (Fig. 11-67; Garrison and Best, 1990). In Oregon, *D. ordii* occurs east of a line connecting The Dalles, Hood River Co.; Sisters, Deschutes Co.; and Lakeview, Lake Co., except none has been taken in the Ochoco, Blue, and Wallowa mountains (Fig. 11-67).

Pleistocene fossils referable to *D. ordii* have been collected at numerous sites, but none in Oregon (Garrison and Best, 1990; Kurtén and Anderson, 1980).

Geographic Variation.—Only one, *D. o. columbianus*, of the 34 currently recognized subspecies occurs in Oregon (Garrison and Best, 1990).

Habitat and Density.—In Morrow Co., *D. ordii* did not occur on sampling plots established in areas vegetated primarily by cheat grass (*Bromus tectorum*) or needle-and-thread grass (*Stipa comata*), but composed 51% of the small mammals caught on a big sagebrush (*Artemisia tridentata*)-western juniper (*Juniperus occidentalis*) plot (Rogers and Hedlund, 1980). In the same county, we captured a few individuals on an area vegetated largely by gray rabbit-brush (*Chrysothamnus nauseosus*) and cheat grass (Small and Verts, 1983; Verts and Carraway, 1986), but only in traps set near a dirt road that traffic and wind had eroded 20–30 cm into the soil; burrows of *D. ordii* were common in the road banks, but nowhere else in the area. In Harney Co., *D. ordii* was more abundant in sagebrush habitats than in black greasewood (*Sarcobatus vermiculatus*) habitats (Feldhamer, 1979*b*). We have captured *D. ordii* in sagebrush-juniper scablands in Deschutes Co. where soils were deep and firm, but not on hummocks with shallow soils containing pebbles and stones or on flats of fine pumice

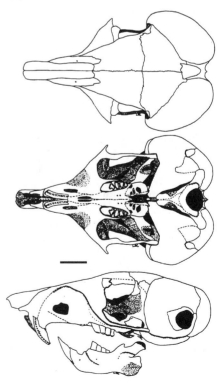

Fig. 11-66. Dorsal, ventral, and lateral views of the cranium, and lateral view of the dentary of Ord's kangaroo rat, *Dipodomys ordii* (OSUFW 3365). Scale bar equals 5 mm.

soils (habitats described by McKay and Verts, 1978*a*).

In a quantitative comparison in Idaho, *D. ordii* was least abundant in *Artemisia-Chrysothamnus*-grass, *Elymus*-forb, and *Juniperus* associations and most abundant in *Juniperus-Chrysothamnus-Eurotia* and *Oryzopsis-Stipa* associations. *D. ordii* was the most abundant small mammal in the latter two associations (Allred, 1973). In Utah, *D. ordii* was more abundant in greasewood- or sagebrush-dominated communities than in other shrub communities; estimated densities ranged from 2.8 to 13.5/ha (Fautin, 1946).

Diet.—We know of no quantitative evaluation of the diet of *D. ordii* in Oregon, but elsewhere the species is primarily a seed-eater. In Idaho in April, stomachs of all individuals contained seeds of halogeton (*Halogeton glomeratus*) and 3% contained seeds of shadscale (*Atriplex confertifolia*); later in the season, a wider variety of seeds was eaten (Johnson, 1961). However, the percent frequency of occurrence of leaves followed a similar pattern of increase during spring and summer (Johnson, 1961). In Utah, however, contents of stomachs and intestines of all 34 *D. ordii* collected in spring–autumn contained vegetative parts of plants, but only four (11.8%) contained seeds. Strangely, five (14.7%) contained bones, but no further description was provided (Warnock and Grundmann, 1963). Insects may compose a small part of the diet (Johnson, 1961; Warnock and Grundmann, 1963).

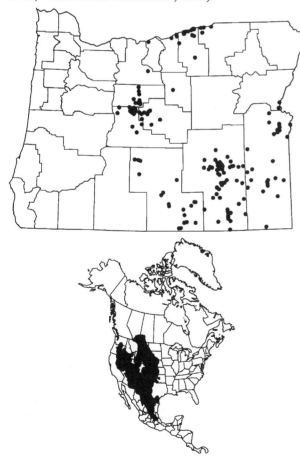

Fig. 11-67. Localities in Oregon from which museum specimens of Ord's kangaroo rat, *Dipodomys ordii*, were examined (listed in APPENDIX, p. 517), and the distribution of Ord's kangaroo rat in North America. (Range map redrawn from Garrison and Best, 1990:2, fig. 3, with modifications.)

In New Mexico, *D. ordii* consumed 29 of 30 species of seeds available in the desert soil; three species (cryptantha, *Cryptantha crassisepala*; Russian thistle, *Salsola kali*; and Indian ricegrass, *Oryzopsis hymenoides*) composed >50% of the diet. Green vegetation occurred in 24% of 134 individuals examined; invertebrates contributed only 2% of the diet (Henderson, 1990). *D. ordii* selected seeds of large size, high in protein and low in saponins and nonprotein amino acids; seed length rather than biomass was a better indicator of seed selection, suggesting that ease of detection was the reason that a species of seed was chosen. Predation by *D. ordii*, and perhaps other desert rodents, may be the selective force responsible for many desert plants having tiny seeds (Henderson, 1990).

Reproduction, Ontogeny, and Mortality.—During proestrus as the ovarian follicles develop, the vaginal orifice in *D. ordii* swells, reaching a maximum diameter of nearly 11 mm; such enlargement may persist for long periods during growth and atresia of follicles. After ovulation, the vulva regresses rapidly and remains small during anestrus (Pfeiffer, 1960). Injection of gonadotropins in anestrous females caused swelling of the vulva, but in proestrus it caused involution of the swollen vulva (Pfeiffer, 1960). Ovulation occurs when follices are 560–600 µm (Duke, 1944) and is considered spontaneous (Pfeiffer, 1956*a*).

Litter size in *D. ordii* seems greater than that of congeners in Oregon (Table 11-13). Two litters are thought to be produced annually; the breeding periods are January–March and September – October in Utah (Duke, 1944). In Nevada, approximately three-fourths of a sample of females was pregnant in February and nearly half of a sample collected in March–April was pregnant, suggesting production of multiple litters (Alcorn, 1941). In Oklahoma and Texas, *D. ordii* bred from late summer–early autumn through winter; reproduction was considered to be influenced greatly by precipitation, food supply, and population density (McCulloch and Inglis, 1961). We suspect that breeding in Oregon is coordinated with primary production following winter rains, and that breeding in summer and autumn ceases except during rare years with extensive precipitation during those periods.

The gestation period is 28–32 days (Duke, 1944). In *D. o. columbianus* in Utah, embryos 22 mm long were considered near term (Duke, 1944); however, we recorded embryos with a crown-rump length of 28 mm. Body mass during the first 3 days of life averaged 7.0 g; body measurements (in mm) were total length, 74.4; tail length, 23.3; length of ear, 2.7; length of hind foot, 16.6 (Smith et al., 1978). With proper environmental conditions, female *D. ordii* may become sexually mature at 2 months of age (McCulloch and Inglis, 1961).

Of 1,099 small-mammal remains identified in 438 barn owl (*Tyto alba*) pellets from Malheur Co. (where *D. ordii* is abundant), only 23 (2.1%) were those of *D. ordii* (Maser et al., 1980). Also, of 5,559 pellets produced by the burrowing owl (*Athene cunicularia*) from the Columbia Basin and of 3,592 containing remains of vertebrate prey only 136 contained remains of *D. ordii* (Green et al., 1993). The paucity of remains in the regurgitated pellets of owls may be a reflection of the ability of kangaroo rats to detect low-frequency sounds produced by owl wings (Webster and Webster, 1971). However, in Idaho, 21.7–23.7% of the biomass consumed by burrowing owls consisted of *D. ordii* (Gleason and Johnson, 1985), suggesting lesser ability to detect the smaller owl. Other reported predators of *D. ordii* include coyotes (*Canis latrans*), kit foxes (*Vulpes velox*), great horned owls (*Bubo virginianus*), and long-eared owls (*Asio otus*—Egoscue, 1956; Johnson and Hansen, 1979*a;* Marti, 1969; Rickart, 1972).

Recorded maximum longevity in captivity is 7 years and 5 months; we have kept several wild-caught individuals for >3 years.

Habits.—Although occasionally seen abroad during daylight hours (Moore, 1929*b*), *D. ordii* is active mostly at night; however, nighttime activity is affected by moonlight, temperature, and inclement weather (O'Farrell, 1974). Nighttime activity may be limited to areas under overhead cover when moonlight is bright (Kaufman and Kaufman, 1982). Daylight hours usually are spent in burrows.

Burrows commonly are constructed beneath desert shrubs, and excavated earth thrown up often buries the lower branches. Randomly selected horizontal and vertical diameters of nine burrows averaged (±*SE*) 7.6 ± 0.8 cm (range, 5.7–13.2 cm) and 4.7 ± 0.4 cm (range, 3.7–7.3 cm), respectively; the two means were significantly different, but neither mean diameter was significantly different from those of burrows of *Peromyscus maniculatus, Spermophilus mollis,* or *S. elegans* (Laundré, 1989). Burrows usually are kept plugged by residents to prevent escape of moisture and entrance of predators. Entrances of active burrows can be located easily in early morning hours when moist plugs contrast against the surrounding dry soil.

Based on anecdotal accounts, *D. ordii* may be aggressive toward conspecifics, especially during the reproductive season (Allan, 1944, 1946). Often opponents kick with the hind feet, bite, and, more rarely, grapple with the forefeet. Captive animals exhibit evidence of being strongly territorial as individuals first introduced into an arena tend to be dominant and individual cages are defended (Allan, 1946). Placing individuals of the opposite sex into small cages commonly results in the death of one individual (Booth, 1943). During mating, chases of either sex by the other seemingly are a prelude to copulation (Allan, 1946). In our experience, most newly captured individuals are extremely docile; usually they can be dumped from a live trap onto the outstretched palm without struggle or attempted flight. However, an attempt to touch the dorsum results in a sideways jump, and an attempt to grasp the individual may provide the stimulus for being bitten. Individuals commonly sandbathe (groom) by sliding and rolling in soft sandy soil, especially after being handled.

Published accounts to the contrary notwithstanding (Davis, 1939; Grinnell, 1922), *D. ordii* can swim (Stock, 1972). It also drinks water, but by the unique system of scooping water into the mouth with the forefeet (Allan, 1946).

Dipodomys ordii possesses a specialized sebaceous gland on the dorsum that becomes most active in both sexes during the molt. The function of the secretion is debated because glandular activity does not seem to be related to the same factors among all species in the genus (Quay, 1953). A scent-marking function is probable (Quay, 1953).

Remarks.—Ord's kangaroo rats in captivity can be maintained on sunflower seeds and a twice-weekly piece of carrot; however, a soft sandy substrate is essential for

sandbathing to prevent the pelage from becoming oily and matted.

FAMILY CASTORIDAE

This Holarctic family includes one genus and two species, one Nearctic and one Palearctic; some authorities consider the two species conspecific and the family monotypic.

Castor canadensis Kuhl
American Beaver

1820. *Castor canadensis* Kuhl, abth. 1, p. 64.
1869. *Castor canadensis leucodonta* Gray, 4:293.
1898*c. Castor canadensis pacificus* Rhoads, 19:422.
1916. *Castor subauratus shastensis* Taylor, 12:433.
1927. *Castor canadensis baileyi* Nelson, 40:125.
1933. *Castor canadensis shastensis:* Grinnell, 40:166.
1940. *Castor canadensis idoneus* Jewett and Hall, 21:87.

Description.—The beaver (Plate XIV), the largest rodent in North America (Table 11-14), commonly weighs in excess of 25 kg. The beaver is highly modified for aquatic life: the body is compact; the legs are short with naked, pentadactyl, and unguiculate feet; the hind feet are webbed and the claws on the first and second toes are split and function in grooming; the ears and eyes are small; and the ears and nostrils are valvular. Underwater, nictitating membranes cover the eyes. The paddlelike tail is broad, scaly, and nearly without hairs; the distal caudal vertebrae are flattened. The baculum is a thin cylindrical bone with a bulbous proximal end that enlarges with age at least through the 4th year (Taber, 1969). The reproductive and digestive tracts open into a common cloaca. Paired anal glands open to the outside at the cloacal opening and paired castor glands open into the urethra; there is no sexual dimorphism in either type of gland although different compounds are secreted by the castor glands of the two sexes (Novak, 1987). The size of both types of gland is strongly correlated with body mass (Svendsen, 1978). The pharnyx and tongue are modified (intranarial epiglottis) to permit breathing while the open mouth is submerged, but breathing through the mouth is prevented (Cole, 1970). Females possess four pectoral mammae.

The thick underfur is overlain with coarse guard hairs; overall, the pelage is dark brown dorsally shading to a lighter brown ventrally. The tail and feet are blackish brown. There is a single annual molt.

The skull is extremely heavy and angular (Fig. 11-68); the infraorbital foramen is small, but the zygomatic plate is massive and contains a large concavity on the anteroventral face to accommodate a large masseter muscle. The basioccipital possesses a pitlike depression. The cheekteeth are strongly hypsodont, but not ever-growing; however, the complexly folded occlusal patterns of enamel and dentine do not change greatly during the life-span of individuals and the cheekteeth rarely, if ever, erode sufficiently to become nonfunctional. The braincase is narrow, the rostrum is broad and deep, and the auditory bulla possesses a neck that extends to the meatus (Fig. 11-68).

Distribution.—The beaver occurs throughout North America except for northern Alaska and northern Canada, southern Nevada, and coastal and southern California; the range extends south into northern Mexico (Fig. 11-69; Hall, 1981). American beavers were introduced successfully into parts of Europe, Asia, and South America (Novak, 1987). In Oregon, *C. canadensis* occurs in suitable habitats throughout the state (Fig. 11-69).

The fossil record of the family Castoridae extends from the early Oligocene of North America and is composed of a variety of fossil species, including the bear-sized giant beaver (†*Castorides ohioensis*). Numerous fossil deposits of *C. canadensis* occur throughout much of North America commencing with a fossil from California dating from about 2 million years ago (Kurtén and Anderson, 1980).

Geographic Variation.—Four of the 24 nominal sub-

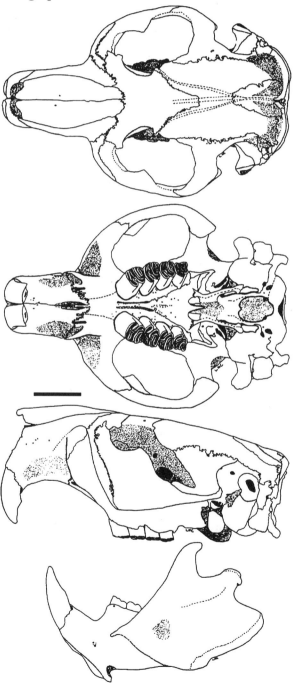

species occur in Oregon: the small, dark *C. c. idoneus* in northwestern Oregon; the large, chestnut-brown *C. c. leucodontus* in the northern two-thirds of the state east of the Cascade Range; the small, dull, and pale *C. c. baileyi* in the Harney Basin drainage; and the bright chestnut-brown *C. c. shastensis* in south-central and southwestern Oregon (Bailey, 1936; Hall, 1981; Jewett and Hall, 1940). To alleviate damage or to enhance populations, beavers have been transplanted on numerous occasions (Kebbe, 1960; Meyers, 1946; V. Scheffer, 1941); subspecies involved at either capture or release points doubtfully was a matter of concern.

Habitat and Density.—The beaver almost always is associated with riparian or lacustrine habitats bordered by a zone of trees, especially cottonwood and aspen (*Populus*), willow (*Salix*), alder (*Alnus*), and maple (*Acer*). Small streams with a constant flow of water that meander through relatively flat terrain in fertile valleys and are subject to being dammed seem especially productive of beavers (Hill, 1982). Streams with rocky bottoms through steep terrain and more subject to wide fluctuations in water levels are less suitable to beavers. In large lakes with broad expanses subject to extensive wave action, beavers usually are restricted to protected inlets.

Because beavers associate in family groups, it is common to report densities in terms of groups (colonies)

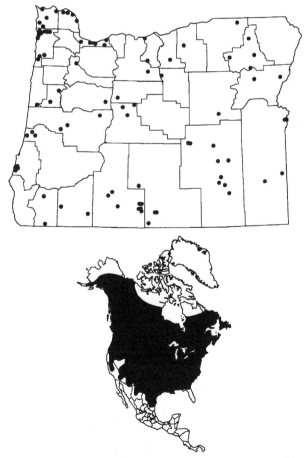

Fig. 11-68. Dorsal, ventral, and lateral views of the cranium, and lateral view of the dentary of of the beaver, *Castor canadensis* (OSUFW 8769). Scale bar equals 25 mm.

Fig. 11-69. Localities in Oregon from which museum specimens of the beaver, *Castor canadensis*, were examined (listed in APPENDIX, p. 513), and the distribution of the beaver in North America. (Range map redrawn from Hall, 1981:604, map 357.)

Table 11-14.—*Means and ranges (in parentheses) of measurements of skin and skull characters for female and male beavers* (Castor canadensis) *from regions in Oregon. Skin characters were recorded from specimen tags; skull characters were measured to the nearest 0.01 mm.*

Region	Sex	n	Total length	Tail length	Hind foot length	Ear length	Mass (kg)	Occipitonasal length	Nasal length	Length of maxillary toothrow	Zygomatic breadth
Northern Coast Range	♀	7						132.96 (124.67–146.15)	48.78 (43.50–53.91)	30.66 (29.65–32.42)	94.07 (88.99–103.87)
	♂	8						137.09 (131.93–141.37)	50.94 (45.92–56.12)	31.70 (29.88–33.55)	96.48 (92.37–101.35)
North-central and eastern Oregon	♀	5					13.50 (12.00–15.00)	129.25 (119.18–139.01)	47.05 (42.47–51.79)	29.92 (27.84–32.07)	92.19 (82.37–100.12)
Southwestern Oregon	♀	3	984[a] (914–1,053)	279[a] (260–299)	178[a] (165–184)		19.00	125.73 (117.02–134.21)	45.76 (40.70–49.33)	31.24 (29.17–33.42)	90.46 (84.98–93.35)
	♂	2	1,090	310	171	30	18.20	136.10 (135.81–136.39)	49.41 (49.06–49.77)	31.48 (31.41–31.55)	90.83 (90.04–91.61)
South-central Oregon	♀	6	1,137	318	191		25.50[b] (24.00–27.00)	135.87 (126.59–142.59)	50.15 (45.07–53.92)	30.61 (27.91–33.29)	94.42 (87.22–101.99)
	♂	6	1,068[c] (1,026–1,118)	287[c] (264–305)	201[c] (172–278)		21.00 (17.00–24.00)	133.52 (128.53–140.73)	48.75 (43.98–51.70)	29.96 (28.66–32.11)	94.82 (90.49–98.58)

[a]Sample size reduced by 1. [b]Sample size reduced by 4. [c]Sample size reduced by 2.

per unit length of streams, but sometimes they are reported as numbers per unit area (Hill, 1982). For Oregon, we found neither type of estimate. Novak (1987) contended that harvest records probably reflected the status of beaver populations, suggesting to us that relative abundances by county could be based on long-term harvest per unit area. On this basis, density of beaver populations is greatest in the northwestern corner of the state (Fig. 11-70; Table 11-15).

Diet.—Beavers are herbivorous. In summer, a variety of green herbaceous vegetation, especially aquatic species, is eaten (Jenkins and Busher, 1979; Svendsen, 1980). Although no dietary studies for beavers have been conducted in Oregon, herbaceous plants common in Oregon and frequently eaten by beavers elsewhere include sword-fern (*Polystichum*), pondweed (*Potamogeton*), horsetail (*Equisetum*), waterweed (*Elodea*), and yellow water-lily (*Nuphar*—Svendsen, 1980). In autumn and winter as green herbaceous vegetation disappears, beavers shift their diet

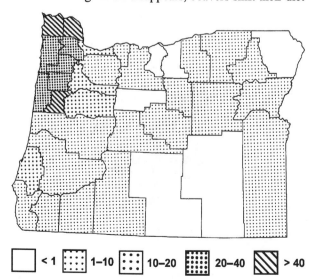

Fig. 11-70. Reported total harvest of beavers, *Castor canadensis*, 1969–1992/1,000 ha in Oregon by county. Data obtained from Oregon Department of Fish and Wildlife annual reports.

to stems, leaves, twigs, and bark of many of the woody species that grow near the water. Bulbous roots of aquatic species also may be eaten in winter (Beer, 1942). Beavers cut mostly deciduous trees such as cottonwood, willow, alder, maple, and birch (*Betula*), but in some regions, coniferous species may be used (Jenkins, 1979).

Despite the seemingly generalist nature of the diet, beavers may be highly selective of species on which they feed. For example, Jenkins and Busher (1979) indicated that although 16 of 17 genera of trees available were cut by beavers, 90% of >900 trees cut consisted of only six genera. Also, of 882 trees available to one family group of beavers, nine were black cherry (*Prunus virginiana*); during 20 months of observation, six of the nine were cut by beavers (Jenkins, 1979). Tree species selected by beavers may change seasonally and from year to year. Also, oaks (*Quercus*) were cut more frequently in the year that they did not produce acorns than in years that acorns were produced (Jenkins, 1979). Jenkins (1979) noted that beavers cut small (<25 cm²) samples from many trees but did not make further use of them. He suggested that seasonal or year-to-year differences in nutritional composition of some trees may be responsible for their selection or rejection by beavers. The distance that trees grow from the shore also may affect their use as food by beavers; trees of a wider range of diameters are cut near the shore and the diameter of trees cut decreases farther from the shore (Jenkins, 1980). These relationships may be related to greater susceptibility of beavers to predation farther from the shore or to the need to expend greater amounts of energy to transport trees cut farther from shore (Jenkins, 1980). The quantity of trees cut by a family group of beavers often is enormous. In eastern Washington, one family group of seven beavers cached 13.2 m³ of woody material before freeze-up; this composed about 65% of the winter food for the group (Beer, 1942). Beavers in the climatically moderate Willamette Valley and coastal regions do not cache food, but east of the Cascade Range they do so (Hill, 1982).

In southeastern Oregon, riparian-zone trees have been reduced or eliminated in many areas by browsing herbivores. However, comparison of growth of red willow (*Salix lasiandra*) in an area inaccessible to cattle, but occupied

Table 11-14.—*Extended.*

Breadth of braincase	Breadth of molariform toothrow	Skull depth	Length of mandible
43.01 (39.33–44.88)	31.24 (29.13–35.84)	55.45 (47.16–61.54)	91.23 (85.47–100.54)
43.26 (41.44–45.96)	34.37 (31.80–37.69)	54.02 (45.19–60.56)	94.29 (90.38–98.13)
45.45 (43.72–47.56)	31.71 (27.26–37.82)	51.99 (46.44–59.21)	88.66 (78.95–100.48)
42.72 (42.09–43.13)	33.29 (30.50–37.43)	52.11 (48.83–56.06)	89.48 (81.91–94.73)
43.32 (43.00–43.63)	35.09 (34.46–35.73)	46.98 (45.19–48.77)	89.95 (88.53–91.37)
44.16 (41.85–45.75)	31.25 (29.00–34.27)	56.86 (52.29–61.43)	92.48 (86.99–98.14)
44.32 (42.92–46.57)	31.33 (28.72–33.65)	56.48 (53.13–61.30)	91.73 (87.98–95.06)

by beavers with that in an area inaccessible to both cattle and beavers indicated that beavers were not responsible for the deterioration. Although beavers harvested 82% of available stems annually, they cut them at a season after growth was completed and reserves were translocated to roots. Subsequent growth of cut willows increased exponentially in relation to the proportion of the stems cut by beavers (Kindschy, 1985).

Reproduction, Ontogeny, and Mortality.—Reproductive performance in *C. canadensis* populations is related to quality of the habitat, nutrition, degree of exploitation of populations, age at which females attain reproductive competence, age of breeding females, and availability of mates, but not to latitude or temperature (Hill, 1982). Both sexes may attain sexual maturity at ≅21 months of age, but usually <50% (sometimes <10–15%) of a population becomes sexually mature as yearlings; there are no records of females breeding at <12 months of age (Novak, 1987). Copulation is in the water or in the lodge; behavior formerly considered to be venter-to-venter copulation is now believed to be a form of play (Tevis, 1950). Most breeding is believed to be in January–February, but estimated dates of birth based on growth and development indicated that breeding may occur as early as November or as late as April.

For a small sample of pregnant females from Oregon (Table 11-2), the average number of embryos was almost exactly midway among those reported in 26 studies conducted throughout the range of the species (Hill, 1982:261, table 14.2) and was only 0.1 embryo more than that reported for 244 pregnant females from Idaho (Leege and Williams, 1967). Because many factors affect litter size so strongly, the value of such averages—unless accompanied by estimates of age, population attributes, and relative quality of the habitat—is questionable.

Most recent authorities consider the gestation period to be 105–107 days (Novak, 1987). Young are born in the lodge. Contrary to early reports that the maternal female excludes the male from the lodge during parturition, more recent observations revealed the adult male and other group members in attendance at parturition (Novak, 1987).

The ≅0.5-kg neonates are precocial; at birth, the incisors are erupted, the eyes are open, the body is completely furred, and the young can walk and swim. Young beavers may nibble on leaves brought to the lodge by the adult male at 4 days of age; they may begin to groom themselves by 7 days of age; and they may leave the lodge for the first time by 2 weeks of age. Neonates are cared for by older members of the group for 4–5 weeks. Nevertheless, they nurse until 2–3 months of age and are dependent upon older members for maintaining dams and lodges and for storing the winter food supply (Novak, 1987).

Estimated annual rate of increase was 49% for an Ohio population in which 74% of females >1.5 years old had ovulated, 40% of females <1.5 years old had ovulated, number of pigmented sites of implantation averaged 3.8, 62% of females contained implantation sites, prenatal mortality ranged from 11 to 16%, and 1st-year trapping and juvenile mortality was 42% (Henry and Bookhout, 1969). Henry and Bookhout (1969) suggested that the population could be maintained at a stationary level with an annual harvest of 32%.

Information obtained elsewhere indicates that the primary predator on beavers was the wolf (*Canis lupus*), now extirpated from Oregon. Common Oregon predators known to prey occasionally on young beavers elsewhere include coyotes (*Canis latrans*), red foxes (*Vulpes vulpes*), minks (*Mustela vison*), and river otters (*Lutra canadensis*—Novak, 1987).

Habits.—Beavers are powerful swimmers; propulsion is largely by use of the webbed hind feet and tail as the front feet are held close under the chin. The tail also is used to maneuver in the water, as a balance organ, as a fat reservoir, and as a heat-exchanger. On land, the beaver is a clumsy waddler.

Of all the beaver's behavioral characteristics, none, doubtlessly, is as legendary as its ability to cut trees for food or construction of dams and lodges. Although Hatt (1944) chronicled beavers cutting 1.5- and 1.7-m-diameter black cottonwoods (*Populus trichocarpa*) in British Columbia, most trees cut are much smaller; only a few are >20 cm in diameter and most may be <5 cm in diameter (Hall, 1960; Jenkins, 1980).

Beavers live in colonies composed of family groups usually consisting of a mated pair of adults, their yearlings, and their young-of-the-year, although occasionally groups may contain individuals >24 months old other than the mated pair (Novak, 1987). Pairing commonly is considered monogamous and long term, as reproduction usually is suppressed in subordinate-female group members of reproductive age. However, in eight groups composing a high-density population in Nevada, Busher et al. (1983) found seven contained more than one adult female and three contained more than one adult male. Two lactating females were captured from three of four colonies 1 year and from all four colonies the following year. Only one example (an adult male) of intercolony activity was noted. Busher et al. (1983) suggested that the social organization of beaver colonies, under some circumstances, may be more plastic than previously supposed.

On small streams, beavers construct dams to maintain more constant water levels; within the flooded area, an islandlike lodge of logs, sticks, and mud is constructed as living quarters. On larger streams with flows too great to be dammed, lodges are built against a bank into which tunnels are dug (Novak, 1987). In addition to the main lodge,

Table 11-15.—*Reported annual harvest of beavers (Castor canadensis) in Oregon by county, 1969–1996. Data provided by Oregon Department of Fish and Wildlife.*

County	1969-1970	1970-1971	1971-1972	1972-1973	1973-1974	1974-1975	1975-1976	1976-1977	1977-1978	1978-1979	1979-1980	1980-1981	1981-1982	1982-1983	1983-1984	1984-1985	1985-1986	1986-1987	1987-1988	1988-1989	1989-1990	1990-1991	1991-1992	1992-1993	1993-1994	1994-1995	1995-1996
Baker	196	84	112	135	159	143	62	54	138	78	140	209	62	57	91	120	70	168	122	85	50	22	99	5	38	38	17
Benton	399	160	227	415	244	363	194	415	348	279	407	660	186	383	270	243	358	360	446	146	243	234	188	225	288	205	
Clackamas	417	232	246	247	195	312	119	289	215	183	360	229	211	140	129	169	207	157	163	48	84	98	93	89	191	72	14
Clatsop	827	313	654	693	873	671	539	822	496	435	629	733	361	389	508	648	642	535	573	573	608	527	582	212	625	821	
Columbia	423	229	176	169	163	432	430	478	383	262	333	331	361	203	249	460	474	436	466	325	502	419	358	197	66	196	
Coos	674	409	636	750	439	551	305	511	321	203	243	301	136	177	137	367	282	182	322	197	305	269	269	19	209	126	
Crook	220	49	138	262	99	111	67	43	109	55	55	81	33	55	48	53	23	83	53	24	13	56	40	50	44	29	
Curry	103	73	99	154	140	97	51	177	64	31	83	97	37	78	36	53	51	13	22	7	8	7	50	3	3		
Deschutes	35	12	30	19	46	54	27	38	18	18	45	45	43	53	49	41	70	68	66	10	31	63	34	34	51	31	
Douglas	1,358	730	997	1,227	782	1,006	593	1,422	734	644	1,440	1,196	302	516	419	619	500	435	614	283	229	190	172	274	264	264	4
Gilliam	14	12	15	13	8	8	20	2	1	21	37	4	4	4	1	13	18	33	8	9	6	9				18	
Grant	268	109	118	238	185	198	237	266	198	89	417	314	115	178	107	98	137	143	205	153	51	32	38	24	68	18	
Harney	134	167	75	166	164	121	57	132	100	32	175	151	43	65	80	34	52	75	69	92	178	51	4	13	122	35	
Hood River	34	24	48	24	45	20	44	31	27	72	66	28	66	60	34	31	19	28	66	65	40	40	21	24	18	22	9
Jackson	125	65	44	95	127	110	78	246	75	149	273	201	53	162	145	100	205	211	211	92	63	68	70	29	53	63	
Jefferson	3	18	17	8	22	33	19	21	8	10	22	17	14	21	1	10	27	18	6	6	21	5	12	7	31	4	
Josephine	43	20	64	45	61	97	43	117	47	69	199	124	41	119	100	78	115	165	115	57	57	36	36	22	66	73	1
Klamath	238	153	162	155	108	108	86	150	130	85	245	217	65	139	130	107	178	249	225	66	153	124	124	22	109	73	6
Lake	160	24	7	10	102	80	35	63	31	11	91	64	6	64	97	24	92	42	47	11	69	27	21	21	9	3	
Lane	2,036	1,016	989	1,618	1,292	1,639	864	1,806	974	1,058	1,871	1,416	586	913	692	1,000	1,140	1,101	1,178	495	601	607	398	625	626	626	2
Lincoln	681	197	304	439	474	655	300	414	363	427	675	526	251	396	339	386	577	658	385	142	127	286	247	330	330	266	6
Linn	708	366	329	497	381	382	311	674	468	285	499	391	189	298	250	354	537	629	478	339	408	173	235	261	261	228	
Malheur	142	23	21	223	97	88	156	434	145	488	398	70	118	154	305	134	454	185	123	134	54	84	318	121	121	4	
Marion	322	235	203	243	229	305	209	364	205	257	270	281	122	238	157	273	191	296	208	88	210	242	120	111	239		
Morrow	47	34	81	60	12	39	15	63	45	37	97	31	10	19	20	11	20	14	105	28	77	155	79	23	23		
Multnomah	135	56	55	69	69	70	48	101	96	104	95	51	40	46	56	17	59	105	69	37	34	48	47	48	2		
Polk	428	67	67	336	90	153	106	246	207	161	311	431	64	211	304	135	274	341	268	172	269	109	122	64	9		1
Sherman	9	2	8	2	5	8	22	5	22	4	16	1	1	7	4	14	8	5	7								
Tillamook	280	208	275	635	248	422	303	591	271	406	509	395	197	234	282	263	425	437	445	120	70	199	121	181	45	45	
Umatilla	173	105	89	189	158	150	47	257	183	106	220	119	174	62	113	151	118	243	285	103	82	85	65	91	27	27	
Union	117	41	38	82	90	73	41	50	56	36	45	66	29	28	10	15	8	35	66	18	23	3	13	35			
Wallowa	108	15	23	41	27	24	6	28	18	35	61	56	24	15	17	23	15	37	31	28	25	19	19	5	5	50	
Wasco	84	28	39	42	42	109	75	88	77	41	110	85	54	43	60	96	73	102	80	97	62	86	33	70	24	24	
Washington	277	87	14	140	145	130	168	298	117	171	251	250	142	147	24	249	295	346	300	145	117	174	227	105	10	10	
Wheeler	31	37	68	49	43	55	17	53	31	10	39	34	11	19	7	15	33	122	62	31	12	6	15	23	32	32	
Yamhill	426	88	122	203	145	259	64	278	197	189	361	181	78	235	247	173	245	299	274	68	101	106	89	104	82	82	3
Totals	11,675	5,490	6,539	9,499	7,674	9,090	5,656	10,784	7,252	6,153	11,148	9,812	3,999	5,815	5,517	6,714	7,656	8,663	8,295	4,262	4,346	4,562	2,075	4,689	4,538	4,538	61
Trappers reporting[a]	721	773	731	847	1,102	1,350	1,377	1,755	2,746	2,899	3,216	3,373	3,069	3,980	3,293	3,104	3,575	3,397	3,833	1,985	1,536	1,228	1,390	1,454	1,293	1,464	687
Average price/pelt[b]	$11.59	$9.52	$13.90	$16.67	$16.22	$13.28	$13.61	$18.04	$13.06	$18.13	$28.11	$19.15	$15.23	$11.38	$12.40	$14.77	$17.60	$20.93	$12.69	$11.79	$11.07	$5.65	$9.94	$7.00	$21.00	$12.00	$20.00

[a] Not all sought beavers and not all were successful.
[b] Rounded to the nearest dollar after the 1991–1992 season.

some groups construct secondary lodges (Tevis, 1950). Much of the unused portion of trees cut for food is used in construction of dams and lodges. Although observers disagree regarding which members of family groups contribute to construction of dams and lodges, they agree that the efforts are independent, not cooperative (Novak, 1987). Living quarters within lodges may be as large as 2.3 by 1.2 by 0.6 m (1.66 m^3), but more commonly range from 0.33 to 0.56 m^3 (Novak, 1987). During periods of high water, beavers work on dams, but when water levels are low they work on lodges to ensure that entrances remain below water. Dam building apparently is stimulated by the sound of running water (Novak, 1987). Beavers also may construct canals by which to travel to food supplies and to transport materials cut for food. Although each is viewed by ethologists largely as a distinct series of fixed action patterns, construction of dams, lodges, and canals commonly is viewed by the public as involving extraordinary engineering feats.

Within lodges, members of family groups interact by mutual grooming; most such activity is concentrated in regions of the body in which selfgrooming is difficult. Mutual grooming does not seem to promote social bonding. During most activities outside of lodges, group members act independently of each other. When a feeding beaver is approached by another group member, it usually abandons the item on which it fed and moves away (Tevis, 1950). When alarmed, beavers slap the water with their tails to produce a resounding noise; such behavior may be mostly to warn other members of possible danger as young beavers usually flee to the lodge and older beavers move to deeper water and may dive (Novak, 1987). Other suggested functions of tail-slapping include frightening of potential foes, causing a foe to reveal its position, and play (Novak, 1987). Tail-slapping is engaged in more frequently by older beavers, and more frequently by females than males.

Beavers establish scent mounds usually near the edge of the water, usually in late winter and spring, and usually where activity is greatest such as on dams, lodges, or trails. Scent mounds consist of mud or vegetation raked into piles on which products of the castor glands (castoreum) and possibly those of the anal glands are deposited periodically. Members of the family group discriminate between castoreum produced by group members and that of strangers. Transients and neighbors usually avoid areas with scent mounds (Aleksiuk, 1968). This behavior, combined with the correspondence of seasons of scent-mound production and the dispersal of near-2-year-olds, suggests that information regarding sex and age of group members may be transmitted to transients (Novak, 1987). In addition to its territorial function, such information may aid groups that have lost a reproductive male or female to acquire a replacement. In some regions where the water does not freeze, beavers do not construct scent mounds (Novak, 1987).

Beavers are most active in the evening or at night, but it is possible to observe beavers engaged in various activities at any hour. Daily activity usually lasts from 8 to 13 h; activity in winter commences later and extends for a shorter period. Beavers do not hibernate, although activity during winter may not be conspicuous because it is largely beneath the ice, especially when temperatures are below −10°C (Novak, 1987).

Small ponds such as a 4-ha one described by Tevis (1950) support only one family group and all activities are restricted to the pond and its environs. On streams, beavers in one colony commonly restrict their activities to 0.6–1.3 km, but colonies are not contiguous (Novak, 1987). In the Sierra Nevada Range of Nevada and California, distances between beaver colonies occupying small streams ranged from 0.76 to 1.55 km (Busher et al., 1983). Dispersal movements as great as 21 km along streams and 11.7 km overland have been recorded (Novak, 1987).

Beavers, like many other rodents, are coprophagous. In captivity, reingestion of feces commenced at 10 days of age, >5 weeks before the young were weaned, but only 2 days after the first observed ingestion of aspen bark and leaves (Buech, 1984). Among older individuals, a period of reingestion commenced in late morning, peaked at midday, and declined in early afternoon, then was repeated in early evening. Marking of ingesta with a magnetic fluorescent material and periodic collection of the material from feces with a magnet revealed that reingested feces and newly ingested food remained largely separated in the digestive tract (Buech, 1984).

Beavers, because of their ability to fell trees, dam streams (and irrigation ditches), dig canals, and tunnel into banks, and because of their taste for certain crops, doubtlessly have the greatest potential of any wild mammal in the state to affect the environment. Their economic value, both positive and negative, can be enormous, depending largely upon the point of view of those affected. However, the more subtle contributions—such as to flood control, to maintenance of water flows, to fisheries management, and to soil conservation—resulting from their activities, in the long term, may have the greatest economic value.

Remarks.—*Castor canadensis* is considered specifically distinct from the Eurasian beaver (*C. fiber*) based on karyology and cranial morphology (Hill, 1982; Wilson, 1993). However, in recent reviews of the biology of *C. canadensis* (Hill, 1982; Jenkins and Busher, 1979; Novak, 1987), information obtained for both species commonly was homogenized such that inspection of supporting references was necessary to ensure that it applied to the beaver in North America. We have endeavored to limit our account to *C. canadensis*.

Of Oregon's diverse mammal fauna, probably no species has been the source of more misconceptions regarding its life history and behavior. Many misconceptions doubtlessly arose partly from misinterpretations of events actually observed, but some are so far-fetched that they defy a logical explanation for their existence (Warren, 1927). For example, one of the items listed as "Things that a beaver does not do" by Warren (1927:166) is that they do not "suck the air from wood in order to make it sink." Although underwater caching of food stores likely was the source of the idea, the mechanism by which beavers could suck air from wood is difficult to imagine!

The beaver played a key role in exploration and settlement of the West as it was the primary target of fur trappers and fur traders who were among the first to visit and establish outposts in Oregon country. Because they reside in colonies, produce an abundance of highly visible sign, are strongly territorial, respond to the odor of castoreum produced by noncolony members, and have predictable routes of travel into lodges and along canals and trails, beavers are extremely vulnerable to trapping. The saga of the exploitation, near demise, and partial recovery of beaver populations is similar to that for many natural resources, and does not need retelling in its entirety. However, the following synopsis of Kebbe's (1960) chronicle regarding transplants of beavers and changes in regulations governing harvest of beavers in Oregon, combined with added notes about beaver management from other sources, recapitulates the fumbling attempts to manage many natural resources that continue little abated:

1893—Before this date, the entire state was open to trapping beavers without restriction; however, in this year, trapping of beavers was prohibited in Baker and Malheur counties.

1899—Trapping of beavers prohibited statewide.

1917—Killing of beavers permitted at all times of the year in Benton and Marion counties.

1923—Beavers permitted to be trapped in November–February statewide except in national forests and in Coos, Curry, Douglas, Jackson, and Josephine counties.

1931—Beavers protected statewide except in Clatsop, Columbia, Multnomah, Marion, and western Douglas counties.

1932—Beavers protected statewide. A beaver-relocation program was instigated; during the following 6 years, 962 nuisance beavers were moved in attempts to reestablish populations where beavers had been extirpated.

1937—Beavers continued to be protected statewide, but nuisance animals could be removed by representatives of the Oregon State Game Commission.

1938—The Federal Aid in Wildlife Restoration Act provided funds on a 3:1 matching basis, some of which were used to reestablish beaver populations through transplants. During the following 6 years, 1,384 beavers were transplanted (Oregon State Game Commission, 1946a). The success of attempts to establish beaver colonies in national forests east of the Cascade Range during the previous 6 years was evaluated 1 year later at 187 sites; colonies were established at or near the release points at 114 (61.0%) of the sites (V. Scheffer, 1941).

1945—The emphasis shifted from transplanting to killing nuisance beavers; >2,000 removed during this year. Also, cooperative agreements with landowners with persistent nuisance-beaver problems were established that permitted Game Commission trappers to harvest beavers. During the following 5 years, 3,000–6,000/year were removed and sold for fur.

1950—Removal of 6,000 beavers failed to reduce complaints of beaver damage. A 3-month (November–January) open season on beavers in agricultural areas was established. Trappers were required to purchase a $2.00 tag to be affixed to the pelt of each beaver. Trappers were permitted to purchase a maximum of 200 tags; unused tags could be returned for refunds.

1951—Regulations altered such that trappers could purchase only 50 tags, but price of tags reduced to $1.00.

1952—Regulations altered such that trappers could purchase 100 tags; price of tags remained unchanged.

1959—During the previous 8 years the annual harvest of beavers declined from 15,257 to 9,786.

The annual harvest from 1969–1992 of beavers in Oregon fluctuated between ≅4,000 to >11,000. During 1993–1995 it declined in response to low fur prices and flooding (Table 11-15). From 1981 to 1991, 56.2% of the 10,719 complaints of property or crop damage received by the Oregon Department of Fish and Wildlife concerning damage by furbearers involved beavers.

FAMILY MURIDAE—RATS, MICE, VOLES, AND MUSKRATS

These rodents compose the most diverse family of mammals; 1,326 (28.6%) of the 4,629 currently recognized species of mammals in the world (Musser and Carleton, 1993) and 25 (18.3%) of the 136 species of mammals (39.7% of 63 species of rodents) in Oregon are members of the family Muridae. These species are arranged in 17 subfamilies, three of which occur in Oregon: Sigmodontinae, Murinae, and Arvicolinae. Each of the three subfamilies will be treated separately.

Formerly, this worldwide family was considered to encompass only the Old World rats and mice now classified in the subfamily Murinae. The Sigmodontinae (formerly part of Cricetinae) and the Arvicolinae (formerly Microtinae) were classified in the family Cricetidae (Arata, 1967).

KEY TO THE SUBFAMILIES OF MURIDAE IN OREGON

1a. Tail scaly; occlusal surfaces of upper molars tuberculate with tubercules arranged serially in three longitudinal rows..........**Subfamily Murinae** (p. 288).
1b. Tail not scaly; occlusal surfaces of upper molars tuberculate or prismatic, or if tuberculate, then tubercules arranged serially in two longitudinal rows....................2
2a. Ears short and mostly concealed by hairs that originate in front of them; anterior margin of braincase tapering abruptly to postorbital constriction; occlusal surfaces of molars with two rows of triangular puddles of dentine surrounded by enamel (prisms) and only terminal prisms on M1 and M2 extending the full width of the teeth..**Subfamily Arvicolinae** (p. 296).
2b. Ears long, membranous, and not concealed by hairs that originate in front of them; anterior margin of braincase tapering gradually to postorbital constriction; occlusal surfaces of upper molars tuberculate, or if prismatic, then prisms other than terminal ones on M1 and M2 extending the full width of the teeth..........................**Subfamily Sigmodontinae** (p. 262).

SUBFAMILY SIGMODONTINAE—NEW WORLD RATS AND MICE

There are 423 species of New World rats and mice in 79 genera (Wilson and Reeder, 1993a). Those in Oregon are characterized by occlusal surfaces of the molars either tuberculate with tubercules arranged in two longitudinal rows or prismatic with prisms extending the full width of the teeth (in Oregon, *Neotoma* only); membranous, nearly naked ears; beady, protruding eyes; a smooth and gradual tapering of the braincase to the postorbital constriction; and keyhole-shaped infraorbital foramina.

Several of these rodent species occupy a diversity of habitats in Oregon; one, the deer mouse (*Peromyscus maniculatus*), is nearly ubiquitous. Others may be restricted to specific habitats; for example, the dusky-footed woodrat

(*Neotoma fuscipes*) seems to be restricted largely to habitats containing poison oak (*Rhus diversiloba*). These rats and mice usually do not cohabit with humans although occasionally they may invade a little-used cabin or summer home. Seeds, buds, insects, and fungi, or, for some, an occasional relative, form the major part of the diet. All are nocturnal and active throughout the year, although they venture forth sparingly during inclement weather and some exhibit diurnal torpor.

Breeding and productivity are variable among the species. Number of litters per year is a difficult reproductive characteristic to determine in some species of New World rats and mice. Some produce one or two litters in a spring breeding period and another one or two in autumn or early winter (Dunmire, 1961; Smith, 1936) and all females do not breed in synchrony. In some species, females attain reproductive competence in the year of their birth, whereas others breed for the first time in the year following their birth.

Although population density of New World rats and mice may fluctuate from year to year, dramatic perturbations at more or less regular multiannual intervals as noted in some species of Arvicolinae are not reported among Oregon species of Sigmodontinae.

KEY TO THE SIGMODONTINAE OF OREGON

1a. Rat-sized; hind foot >28 mm long; occlusal surfaces of molars prismatic with middle prisms of M1 and M2 extending completely across tooth................................2
1b. Mouse-sized; hind foot <28 mm long; occlusal surfaces of molars with two longitudinal rows of tubercles...4
2a. Tail distichous, squirrel-like; sphenopalatine vacuities usually absent, or if present, only a hairline crack; M1 with two reentrant angles on lingual side..............
......*Neotoma cinerea*, bushy-tailed woodrat (p. 280).
2b. Tail round, not squirrel-like; sphenopalatine vacuities present; M1 with one reentrant angle on lingual side (juveniles may have two)..3
3a. Hairs on venter white to base; tail >150 mm long; basal length of skull usually >40 mm; posterior reentrant angle of M3 angled..
......*Neotoma fuscipes*, dusky-footed woodrat (p. 283).
3b. Hairs on venter with lead-colored bases; tail <140 mm long; basal length of skull usually <40 mm; posterior reentrant angle of M3 slightly curved............................
....................*Neotoma lepida*, desert woodrat (p. 286).
4a. Basal length of skull <17.5 mm; anterior face of I1 with deep longitudinal groove......***Reithrodontomys megalotis***, western harvest mouse (p. 263).
4b. Basal length of skull >18.5 mm; anterior face of I1 not grooved 5
5a. Tail <60% of length of head and body; coronoid process extending higher than condyloid process; lateral margins of nasals converging posteriorly...***Onychomys leucogaster***, northern grasshopper mouse (p. 276).
5b. Tail >70% of length of head and body; condyloid process extending higher than coronoid process; lateral margins of nasals nearly parallel................................6
6a. Females with four mammae, pectoral mammae absent; accessory cusps usually not present in primary folds of labial side of M1 and M2, or if present, usually rudimentary; enamel of M1 lacking indentation

on anterior margin (may be slightly indented in young individuals)..
...........***Peromyscus crinitus***, canyon mouse (p. 267).
6b. Females with six mammae, one pair of pectoral mammae present; primary folds on labial side of M1 and M2 usually with well-developed accessory cusps; enamel of M1 indented on anterior margin..................7
7a. Ear >22 mm long, ≥75% of the length of the hind foot.............***Peromyscus truei***, piñon mouse (p. 273).
7b. Ear <22 mm long, <75% of the length of the hind foot...
..........***Peromyscus maniculatus***, deer mouse (p. 270).

Reithrodontomys megalotis (Baird)
Western Harvest Mouse

1858. *Reithrodon megalotis* Baird, 8(pt. 1):451.
1858. *Reithrodon longicauda* Baird, 8(pt. 1):451.
1893a. *Reithrodontomys megalotis*: Allen, 5:79.
1899b. *Reithrodontomys klamathensis* Merriam, 16:93.
1913b. *Reithrodontomys megalotis longicauda*: Grinnell, 3:303.

Description.—*Reithrodontomys megalotis* (Fig. 11-71) is the smallest sigmodontine rodent in Oregon (Table 11-16) and it is the only member of the subfamily with a longitudinal groove in the anterior surface of each upper incisor. The incisive foramina are relatively large, but terminate anterior to the molar toothrows (Fig. 11-72).

Hairs on the dorsum have a sooty base (about one-half the length) followed by a light buffy band and a dark brownish-black tip. The length of the dark tip becomes progressively shorter laterally. Hairs on the venter have dark-gray bases and white tips. The tail is bicolored, but the dark dorsum is not sharply demarked from the whitish ventral portion. The feet are whitish, the ears are approximately the same color as the dorsum, and the mystacial region is black.

Distribution.—*Reithrodontomys megalotis* occurs from extreme southern British Columbia, Alberta, and Saskatchewan south to southern Mexico and, except for the Rocky Mountain region, from California eastward to Indiana, northeastern Arkansas, and western Oklahoma and Texas (Fig. 11-73; Webster and Jones, 1982). In Oregon, *R. megalotis* occurs east of the Cascade Range, except it is absent from the Blue and Wallowa mountains; it also occurs in the southern parts of Josephine and Jackson counties west of the Cascades (Fig. 11-73).

Pleistocene fossils are known from California, Missouri, New Mexico, Texas, Kansas, and Nuevo León within the present-day range and from sites in Texas east of the modern range (Kurtén and Anderson, 1980; Webster and Jones, 1982).

Fig. 11-71. Photograph of the western harvest mouse, *Reithrodontomys megalotis*.

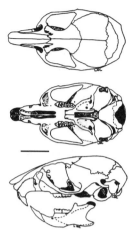

Fig. 11-72. Dorsal, ventral, and lateral views of the cranium, and lateral view of the dentary of the western harvest mouse, *Reithrodontomys megalotis* (OSUFW 6285). Scale bar equals 5 mm.

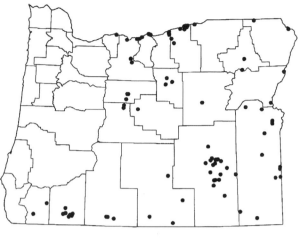

Fig. 11-73. Localities in Oregon from which museum specimens of the western harvest mouse, *Reithrodontomys megalotis*, were examined (listed in APPENDIX, p. 548), and the distribution of the western harvest mouse in North America. (Range map redrawn after Hall, 1981:639, map 377, with modifications.)

Geographic Variation.—Of the 16 subspecies currently recognized (Webster and Jones, 1982) two are reported to occur in Oregon: the buffy-brown *R. m. longicaudus* in Josephine and Jackson counties and the buffy-gray *R. m. megalotis* in the remainder of the occupied portion of the state.

Habitat and Density.—In Harney Co., *R. megalotis* occurred in one or more 1.1-ha plots in big sagebrush (*Artemisia tridentata*), black greasewood (*Sarcobatus vermiculatus*), marsh, and grassland (meadow). The greatest number of captures and the greatest relative abundance were in marsh, the lowest in both categories in sagebrush (Feldhamer, 1979*b*). On one heavily grazed marsh grid, *R. megalotis* was the only species captured; however, it was captured more commonly with *Microtus montanus* (G. A. Feldhamer, in litt.). In a study conducted on both sides of the Columbia River in Washington and Oregon, Hinschberger (1978) also found the greatest relative abundance of *R. megalotis* in marsh habitats. Nevertheless, he collected the species (in order of increasing relative abundance) in habitats dominated by sagebrush, bitter-brush (*Purshia tridentata*), rabbit-brush (*Chrysothamnus*), Russian olive (*Elaeagnus angustifolia*), and cottonwood (*Populus*)-willow (*Salix*). He collected a few individuals in sparsely vegetated rock riprapping along the river bank. Thus, habitat associations in Oregon seem to be essentially identical with those reportedly occupied by *R. megalotis* elsewhere (Webster and Jones, 1982).

In Washington, a density of 11.9/ha was estimated for a population occupying an area dominated by sagebrush, greasewood, and cheat grass (*Bromus tectorum*—Gray, 1943). In New Mexico, population densities of *R. megalotis* remained <5/ha for 3 of 4 years on a playa vegetated mostly by grasses, but during the 4th year the density suddenly peaked at >60/ha (Whitford, 1976). During the peak, individuals moved onto a bajada dominated by creosotebush (*Larrea tridentata*). In Utah, estimates of density as high as 9.5/ha were recorded in greasewood communities (Fautin, 1946). We have no comparable data for Oregon, but G. A. Feldhamer (in litt.) caught 20 individuals on a 49-trap grid (1.1-ha sampling area) in grass habitat during a 10-day trapping period.

Diet.—*Reithrodontomys megalotis* tends to feed largely on plant material, but some arthropods are consumed. In California, flowers, leaves, and seeds of California buckwheat (*Eriogonum fasciculatum*), lemonadeberry (*Rhus integrifolia*), and grasses, and insects and insect larvae (especially Lepidoptera) constituted staples in the diet throughout the year (Meserve, 1976). Seasonally, leaves of California sagebrush (*Artemisia californica*) and birdfoot-trefoil (*Lotus scoparius*) contributed significantly to the diet. Although these plant species do not occur in Oregon, congeners thereof do and might be expected to be eaten by *R. megalotis*. However, in captivity, *R. megalotis* refused to eat, or consumed only in small amounts, several of these species, but tended to select grasses and forbs over shrubs (Meserve, 1976:305, table 2). These harvest mice usually ignored leaves and stems of shrubs and forbs, eating only the flowering parts (Meserve, 1976).

Captives fed wild bird seed containing 10% sunflower seeds, 40% white proso millet, 35% red fortune millet, 10% grain sorghum, 3% hulled oats, and 2% red spring wheat were unusually fecund (Bancroft, 1967) and were considered to have been maintained in "optimal captive conditions" (Webster and Jones, 1982:3).

Reproduction, Ontogeny, and Mortality.—*Reithrodontomys megalotis* has an enormous potential to produce offspring; in captivity, two females each produced

Table 11-16.—*Means (±SE), ranges (in parentheesees) and CVs of measurements of skull and skin* characters for female and male sigmodontine rodents other than* Peromyscus *from regions in Oregon. Skin characters were recorded from specimen tags; skull characters were measured to the nearest 0.01 mm. SE and CV not provided if* n <10.

Species and region	Sex	n	Total length	Tail length	Hind foot length	Mass (g)	Greatest length of skull	Basal length	Length of maxillary toothrow	Zygomatic breadth	Breadth of braincase	Least interorbital breadth	Skull depth
Neotoma cinerea													
Northwestern Oregon	♀	12	413 ± 7.3	195 ± 5.2	39 ± 0.7	262.5 ± 11.43[b]	51.28 ± 0.49	45.68 ± 0.51	10.37 ± 0.13	25.79 ± 0.23	18.80 ± 0.18	5.99 ± 0.09	15.53 ± 0.11
			(351–449)	(158–219)	(39–48)	(191.0–319.0)	(48.12–54.19)	(43.06–49.11)	(9.84–11.28)	(24.76–27.17)	(18.01–19.60)	(5.69–6.57)	(14.89–16.17)
			6.1	9.2	5.4	13.77	3.33	3.83	4.36	3.12	3.35	5.02	2.35
	♂	15	423 ± 9.1[c]	197 ± 5.3	47 ± 1.0	329.8 ± 20.82[d]	51.89 ± 0.59	46.43 ± 0.58	10.71 ± 0.10	26.34 ± 0.34	19.13 ± 0.18	5.96 ± 0.08	15.73 ± 0.09
			(355–458)	(147–222)	(37–52)	(194.0–429.3)	(48.15–55.62)	(42.42–50.08)	(10.23–11.66)	(23.95–28.43)	(18.20–20.39)	(5.62–6.92)	(15.18–16.31)
			8.0	10.5	8.5	20.94	4.42	4.84	3.71	5.01	3.59	5.02	2.22
Cascade Range	♀	4	377	162	41	264.7	49.36	44.28	9.79	26.04	18.75	5.82	15.41
			(341–395)	(135–178)	(37–43)	(174.5–361.0)	(47.04–52.10)	(42.00–46.25)	(9.59–10.03)	(24.95–28.15)	(18.15–19.60)	(5.52–6.08)	(15.02–16.10)
	♂	4	384	175	46	279.8	49.13	44.37	9.92	25.63	18.65	5.59	15.51
			(354–410)	(150–210)	(43–49)	(247.7–302.0)	(47.48–50.57)	(42.22–46.83)	(9.64–10.15)	(24.30–27.23)	(18.50–18.88)	(5.25–6.01)	(15.00–16.79)
Southern coastal and south-central Oregon	♀	7	379	177	43		47.90	19.36	9.52	24.27	17.71	5.73	15.42
			(347–417)	(165–197)	(41–47)		(45.46–50.83)	(39.48–45.28)	(9.11–10.39)	(22.68–26.12)	(17.00–19.12)	(5.30–6.03)	(14.90–15.94)
	♂	8	390	183	45	466.0	49.00	43.78	9.81	24.66	17.91	5.94	15.76
			(364–415)	(166–199)	(42–47)		(45.67–54.25)	(39.91–50.43)	(9.31–10.24)	(23.42–27.92)	(17.48–18.51)	(5.57–6.21)	(15.11–16.47)
Eastern Oregon	♀	32	376 ± 6.8[c]	168 ± 2.5[b]	43 ± 0.4[c]	252.9 ± 16.49[e]	48.86 ± 0.32	44.02 ± 0.32	9.95 ± 0.07	25.51 ± 0.19	17.96 ± 0.15	5.92 ± 0.03	15.65 ± 0.09
			(336–552)	(147–205)	(37–48)	(117.5–362.0)	(45.09–52.73)	(40.31–46.93)	(8.88–10.66)	(22.35–27.46)	(15.48–20.50)	(5.32–6.28)	(14.26–16.55)
			10.1	8.3	5.2	24.39	3.72	4.06	4.19	4.37	4.69	3.09	3.34
	♂	29	397 ± 5.4	172 ± 2.9	45 ± 0.6[c]	303.9 ± 22.73[e]	51.04 ± 0.46	46.33 ± 0.50	10.13 ± 0.07	26.45 ± 0.26	18.31 ± 0.13	5.84 ± 0.05	15.98 ± 0.10
			(344–460)	(145–205)	(39–51)	(199.6–452.0)	(45.90–55.26)	(41.02–52.03)	(9.27–10.81)	(23.94–29.29)	(16.65–19.44)	(5.40–6.38)	(14.76–17.00)
			7.3	9.1	6.5	24.81	4.85	5.86	3.73	5.39	3.82	4.57	3.49
Neotoma fuscipes	♀	30	419 ± 3.9	206 ± 2.5	41 ± 0.3	265.4 ± 10.24[b]	49.68 ± 0.34	43.14 ± 0.37	9.76 ± 0.05	25.30 ± 0.21	18.60 ± 0.14	6.26 ± 0.07	15.75 ± 0.08
			(355–469)	(183–238)	(37–44)	(164.9–430.0)	(43.98–53.28)	(36.33–46.55)	(9.21–10.37)	(22.87–27.02)	(16.90–20.14)	(5.45–6.96)	(14.89–16.44)
			5.2	6.7	3.7	20.41	3.69	4.66	3.01	4.44	4.04	6.08	2.90
	♂	30	431 ± 5.2	206 ± 3.4[c]	42 ± 0.5	306.7 ± 13.56[f]	51.37 ± 0.26[c]	45.13 ± 0.31	9.87 ± 0.06	25.99 ± 0.19	18.92 ± 0.14	6.29 ± 0.05	15.99 ± 0.09
			(359–475)	(171–238)	(32–47)	(203.1–415.5)	(48.03–54.25)	(40.53–48.51)	(9.13–10.49)	(23.35–27.74)	(17.37–20.44)	(5.77–6.97)	(15.09–17.05)
			6.6	8.9	6.6	20.74	2.78	3.73	3.55	3.99	4.10	4.73	3.12
Neotoma lepida	♀	7	273	116	30	104.2[b]	38.82	34.20	8.14	20.27	15.89	5.15	13.64
			(246–298)	(99–133)	(29–32)	(76.0–122.0)	(36.01–41.20)	(31.17–36.32)	(7.90–8.67)	(18.75–21.77)	(15.44–16.60)	(4.91–5.44)	(13.00–14.04)
	♂	25	282 ± 4.2	118 ± 1.6[c]	31 ± 0.3	128.7 ± 12.19[g]	39.35 ± 0.48	34.75 ± 0.45	8.31 ± 0.07	20.62 ± 0.24	16.41 ± 0.11	5.27 ± 0.06	14.15 ± 0.11
			(246–329)	(99–134)	(28–34)	(73.6–243.0)	(33.33–46.04)	(31.31–40.83)	(7.61–9.08)	(18.89–23.76)	(15.37–17.48)	(4.92–5.90)	(13.32–15.67)
			7.4	6.9	4.1	36.69	6.14	6.52	4.21	5.86	3.22	5.52	3.99
Onychomys leucogaster													
Southeastern Oregon	♀	14	138 ± 2.2	38 ± 1.2	19 ± 0.2	27.0[h]	26.38 ± 0.22	22.26 ± 0.25	3.97 ± 0.06	13.93 ± 0.18	12.49 ± 0.10	4.80 ± 0.04	9.55 ± 0.06
			(120–152)	(30–45)	(19–21)	(21.8–34.0)	(25.05–27.87)	(20.70–23.94)	(3.45–4.29)	(12.71–15.09)	(11.64–12.93)	(4.54–5.06)	(9.17–10.01)
			5.9	11.4	3.6		3.09	4.15	5.77	4.74	3.03	3.35	2.33
	♂	13	136 ± 2.3	38 ± 1.4	19 ± 0.2	24.9[f]	26.64 ± 0.18	22.53 ± 0.18	3.99 ± 0.05	14.10 ± 0.19	12.65 ± 0.06	4.83 ± 0.03	9.62 ± 0.07
			(115–145)	(30–49)	(19–21)	(20.5–28.0)	(25.28–27.56)	(20.99–23.65)	(3.72–4.25)	(12.97–15.43)	(12.33–13.05)	(4.65–5.10)	(9.17–10.07)
			6.1	13.4	3.8		2.37	2.93	4.26	5.04	1.79	2.54	2.44
Northeastern Oregon	♀	5	130	35	19	27.3	25.64	21.49	4.13	13.27	12.26	4.65	9.43
			(126–139)	(34–37)	(18–19)		(25.26–26.39)	(20.72–22.20)	(3.91–4.29)	(12.91–13.73)	(12.08–12.48)	(4.56–4.77)	(9.18–9.56)
	♂	3	133	38	19	26.2[c]	26.35	21.96	3.93	13.89	12.49	4.80	9.54
			(130–136)	(34–41)	(18–19)	(25.2–27.2)	(25.85–26.84)	(21.43–22.25)	(3.87–3.97)	(13.14–14.29)	(12.13–12.68)	(4.67–4.89)	(9.41–9.62)
Reithrodontomys megalotis													
Josephine and Curry counties	♀	3	140	66	17	11.0	20.83	16.75	2.96	10.93	10.34	3.28	7.76
			(129–151)	(62–71)	(15–19)		(20.17–21.38)	(15.99–17.18)	(2.92–2.99)	(10.54–11.26)	(10.07–10.58)	(3.26–3.29)	(7.54–7.99)
	♂	1					20.81	17.00	3.01	10.72	10.44	3.36	7.77
Remainder of Oregon	♀	30	139 ± 1.4	68 ± 0.9	17 ± 0.2[c]	12.6 ± 0.47[i]	20.40 ± 0.14	16.36 ± 0.12	3.13 ± 0.02	10.42 ± 0.06	9.97 ± 0.05	3.22 ± 0.03	7.64 ± 0.05
			(126–153)	(57–79)	(16–19)	(11.0–15.5)	(18.84–21.90)	(15.24–17.42)	(2.88–3.42)	(9.71–11.07)	(9.37–10.39)	(2.90–4.06)	(7.03–8.03)
			5.6	7.7	4.7	11.85	3.88	3.88	4.14	3.25	2.51	5.89	3.26
	♂	30	138 ± 1.1	68 ± 1.1	17 ± 0.3	10.4 ± 0.34[j]	20.62 ± 0.09	16.53 ± 0.09	3.13 ± 0.02	10.47 ± 0.05	9.99 ± 0.04	3.21 ± 0.02	7.67 ± 0.04
			(126–151)	(56–79)	(12–19)	(7.10–13.0)	(19.48–21.39)	(15.42–17.38)	(2.97–3.40)	(9.88–10.96)	(9.57–10.31)	(2.92–3.46)	(7.19–8.06)
			4.4	9.1	8.9	12.75	2.58	3.15	2.88	2.55	1.98	3.95	3.17

[a] Although measures of ear length were recorded, the data were not included because of extreme variation believed to stem from differences in techniques used to measure these characters.
[b] Sample size reduced by 2. [c] Sample size reduced by 1. [d] Sample size reduced by 4. [e] Sample size reduced by 18. [f] Sample size reduced by 8. [g] Sample size reduced by 10. [h] Sample size reduced by 6. [i] Sample size reduced by 20. [j] Sample size reduced by 15.

14 litters in a 1-year period (Bancroft, 1967). With average litters of 4.07 and 4.14, one produced 57 young during the year, the other 58. The average interval between parturitions was 28.8 and 26.4 days; the minimum interval between birth of any two litters was 22 days (Bancroft, 1967).

Fisler (1971) indicated that 3% of female *R. megalotis* in California were pregnant or lactating in January–February, 16% in March–April, 15% in May–June, 37% in July–August, 17% in September–October, and 12% in November–December. Although *R. megalotis* is known to breed throughout the year in some parts of its range, Smith (1936) considered that most reproduction was confined to early spring and late autumn. In Nevada, Hall (1946) found pregnant females only in May–August. Thus, productivity usually is much less in wild populations than among captives (Table 11-17).

Neonates weighed 1.0–1.2 g and had a crown-rump length of ≅21 mm. They were naked (except for vibrissae), the skin was smooth and pink, and the eyes and ears were closed. They were capable of making a shrill screech

Table 11-17.—*Reproduction in Oregon sigmodontine rodents.*

Species	Number of litters	Litter size $\bar{X}$	Range	n	Reproductive season	Basis	Authority	State or province
Neotoma cinerea	2	3.6	1–6	7		Embryos	Finley, 1958	Colorado
	>1	3.9	3–6	7	Apr.–Jul.	Embryos	Warren, 1926	Colorado
	3[a]	3.4[e]	1–6	34	Feb.–Aug.	Live births	Egoscue, 1962a	Utah[c]
		4.4	2–8	11		Implant sites	This study	Oregon
Neotoma fuscipes		3.3	1–5	15		Implant sites	This study	Oregon
		2.8	2–3	10	Feb.–May	Embryos and young	English, 1923	Oregon
Neotoma lepida	2?	3.0	1–5	25	Feb.–Jun.	Embryos	Hall, 1946	Nevada
		2.6	1–5	19		Live births	Cameron, 1973	California[d]
		2.3	1–5[b]	100	All year	Live births	Egoscue, 1957	Utah[c]
		2.7	2–4	14	Oct.–Apr.	Live births	Schwartz and Bleich, 1975	California[a]
Onychomys leucogaster	5.1	3.6	1–6	205	All Year	Young produced	Egoscue, 1960	Utah[c]
		3.7	1–6	153	All year	Young produced	Pinter, 1970	Colorado[c]
Peromyscus crinitus	≤2	3.0	2–4	4	Apr.–Jun.	Embryos	This study	Oregon
	≤8	3.1	1–5	130	All year	Young produced	Egoscue, 1964	Utah[c]
		2.9	?–6	206	All year	Young produced	Drickamer and Vestal, 1973	California[f]
Peromyscus maniculatus		4.9[g]	1–8[h]	55		Implant sites	This study	Oregon
	≤2	4.8[g]	1–8	38	May–Nov.	Embryos	This study	Oregon
	2.7[i]	4.4	1–9	427	All Year[j]	Embryos	Gashwiler, 1979	Oregon
Peromyscus truei		4.1	3–9	12	May–Nov.	Embryos	This study	Oregon
		4.0	3–6	13	Jun.–Sept.	Embryos	Douglas, 1969	Colorado
	6.8	3.6	1–6	110		Young produced	Egoscue, 1960	Utah[c]
	3.4	3.5	1–7	27	Apr.–Oct.	Embryos	McCabe and Blanchard, 1950	California
		4.3	3–5	16	May–Jul.	Embryos	Hall, 1946	Nevada
Reithrodontomys megalotis	1	2.6	1–4	18	Mar.–Aug.	Young[k]	Vestal, 1938	California
		4.0	1–6	22	May–Aug.	Embryos	Hall, 1946	Nevada

[a]Born in captivity to wild-caught pregnant females.
[b]Only 8% of litters contained more than three offspring.
[c]Captive colony; source of stock.
[d]Average; maximum number recorded was seven.
[e]Recalculated from data provided; Egoscue (1962a) gave mean as 3.5.
[f]According to King et al., 1968.
[g]Mean (±SE) values for neither implant sites (4.875 ± 0.854, n = 8 and 4.936 ± 0.245, n = 47) nor embryos (4.818 ± 0.313, n = 22 and 4.750 ± 0.382, n = 16) east and west of the Cascade Range, respectively, were significantly different (t = 0.069 and t = 0.138); thus, values east and west of the Cascade Range for both implant sites and for embryos were combined.
[h]Records of implant sites greater than the maximum number of embryos recorded were deleted because more than one litter likely was represented.
[i]Calculated average number of litters produced by females each year (Gashwiler, 1972) ranged from 1.7 to 3.2 during an 8-year study; both maximum and minimum occurred during years when seed production by conifers was poor.
[j]No pregnant female was collected during November–March of years when seed production of conifers was poor, but at least a few were collected in all months during years when seed production was good.
[k]Juveniles found in woodrat houses.

(Smith, 1936). At 1 day of age the skin was darker and more wrinkled, and the young had begun to crawl. At 2 days, the pinnae were unfolded, but the auditory meatuses remained closed. At 3 days, a line at which the eyelids will separate was visible, the skin was nearly black, and hair was evident. At 4 days, the mandibular incisors erupted; the young could stand, but not walk; and the hair color was visible. At 10 days, young could walk, respond to noise, and use the tail to maintain balance; the auditory meatuses were open. At 11 days, the eyes opened and the young struggled when picked up by the maternal female. By 21 days, young were fully weaned (Smith, 1936).

In California, most individuals probably do not survive >6 months after becoming susceptible to capture in traps; the oldest recorded was 18 months of age (Fisler, 1971).

Thus, annual turnover in populations seems the rule, but attrition seems to be gradual rather than at a specific time, at least in areas where recruitment occurs year-round (Fisler, 1971).

In Oregon, *R. megalotis* is known to be preyed upon by great horned owls (*Bubo virginianus*), short-eared owls (*Asio flammeus*), long-eared owls (*Asio otus*), barn owls (*Tyto alba*), and burrowing owls (*Athene cunicularia*—Brodie and Maser, 1967; Maser et al., 1970, 1971, 1980). Likely, canids, felids, mustelids, and snakes (Serpentes) prey on these tiny mammals occasionally (Webster and Jones, 1982). Hall (1946) recounted the observation by Henry Fitch of a scorpion (Scorpiones) feeding on a freshly killed *R. megalotis* in Nevada.

Habits.—*Reithrodontomys megalotis* commonly uses

Microtus runways; it is active nocturnally, with greatest activity on moonless and rainy nights (Pearson, 1960*a*). Irrespective of season, almost all activity was between sunset and sunrise; in late autumn, activity was relatively uniform throughout the night, but in winter and spring activity was greatest immediately after sunset with some indication of three peaks (Pearson, 1960*a*).

Although *R. megalotis* is capable of torpor in the laboratory (Coulombe, 1970; Fisler, 1965) and hibernation was suggested as a possible explanation of winter absence and spring appearance of the species in traps during a year-long sampling in Nevada (O'Farrell, 1974), specimens have been collected in all 12 months in Oregon. In a California population, activity of *R. megalotis* monitored by automatic cameras set in *Microtus* runways was year-round, but greatest in autumn and winter and least in late spring and summer (Pearson, 1960*a*:59, fig. 1). Meserve (1977) found 30.1–37.5% of activity was above the ground on shrubs.

Harvest mice spend periods of inactivity in ball-shaped nests about 75–125 mm in diameter constructed on the surface of the ground; nests sometimes are placed beneath shrubs or debris. Nests are entered through a ≅1.1-cm-diameter opening lined with the same material used to line the nests; the opening is on the side of the nest (Hall, 1946; Smith, 1936). *R. megalotis* is known to enter burrows, but there is no evidence that burrows are excavated by harvest mice.

Home ranges of *R. megalotis* tend to be small; Pearson (1960*a*) reported that only two of 407 records were of individuals that passed both automatic cameras placed 14–69 m apart, but 68 of 286 records were of individuals that passed both cameras placed 1.8–9.5 m apart. Meserve (1977) calculated three-dimensional home ranges for *R. megalotis*; they averaged 3,257 m³.

In comparison with many small rodents, *R. megalotis* seems unusually tolerant of conspecifics. Individuals introduced into cages simultaneously or in sequence seemed amicable regardless of sex (Smith, 1936). However, males of mated pairs placed in the same cage fought, and the females in such cages usually failed to produce offspring (Bancroft, 1967). *R. megalotis* tends to avoid traps in which *Microtus californicus* was caught previously, but only when both species are reproductively active (Heske and Repp, 1986).

Remarks.—Our experience with *R. megalotis* is limited to capture of a few individuals incidental to studies of other species or during attempts directed toward capture of other species. An intriguing aspect of the natural history of *R. megalotis* is the occasional capture of an individual in a desert or other situation far from the marsh or meadow habitat usually occupied by members of the species (Hall, 1946).

Peromyscus crinitus (Merriam)
Canyon Mouse

1891. *Hesperomys crinitus* Merriam, 5:53.
1899. *Peromyscus crinitus*: Bangs, 1:67.

Description.—*Peromyscus crinitus* is a moderate- to small-sized mouse-like rodent (Table 11-18) with a long tail; nearly naked ears equal in length to the hind feet; and a long, lax, and silky pelage (Johnson and Armstrong, 1987). The dorsal pelage consists of hairs the basal two-thirds of which are dusty gray to dark steel-gray; hairs are banded with light buff or ocherous buff and tipped with

black. In the same month, but in different years, we took richly colored specimens on dark lava boulders at Wright's Point south of Burns, Harney Co., and light buffy-colored specimens on ashy-colored soil at the base of rimrock south of Fields in the same county. In addition to the difference in the color of the light band, the black tip is longer and the basal portion of the hair is darker on specimens from Wright's Point. Thus, some correlation between pelage color and color of the substrate seems evident although not as dramatic as that reported elsewhere (Hardy, 1945). The basal portion of the hairs on the venter is similar to that of the dorsal hairs; however, the tips of hairs on the venter are white.

These peromyscine mice usually can be distinguished from congeners by their soft, silky pelage. More diagnostic characters are the two pairs of mammae (both inguinal), the usual absence of accessory cusps on M1 and M2, and the usual absence of an indentation in the enamel on the anterior margin of M1 (except a slight indentation may be present in young individuals; Fig. 11-74).

Distribution.—*Peromyscus crinitus* occurs from north-central Oregon and southwestern Idaho southward to northern Sonora and Baja California Norte and from eastern California eastward to western Colorado and northwestern New Mexico (Fig. 11-75). The species is absent from montane areas of central Utah (Johnson and Armstrong, 1987). In Oregon, most localities at which *P. crinitus* has been taken are in Harney and Malheur counties, but the distribution includes a group of localities in Crook, Jefferson, Wasco, and Wheeler counties and another group in Lake Co. (Fig. 11-75). There also are records of the species in extreme southeastern Baker Co.

Fossil remains of *P. crinitus* are known from a late Pleistocene deposit in California and a Holocene deposit in New Mexico; both sites are extralimital (Johnson and Armstrong, 1987).

Geographic Variation.—Of the eight currently recognized subspecies, only the nominate race, *P. c. crinitus*, occurs in Oregon (Hall, 1981).

Habitat and Density.—Rock, particularly rimrock and talus slopes, seems to be the key component of habitats

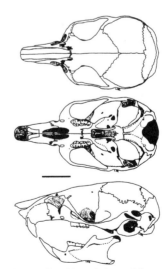

Fig. 11-74. Dorsal, ventral, and lateral views of the cranium, and lateral view of the dentary of the canyon mouse, *Peromyscus crinitus* (OSUFW 7438). Scale bar equals 5 mm.

Table 11-18.—*Means (±SE), ranges, and CVs of measurements of skull and skin*[a] *characters for female and male* Peromyscus *from regions in Oregon. Skin characters were recorded from specimen tags; skull characters were measured to the nearest 0.01 mm.* SE *and* CV *not provided if* n <10.

Species and region	Sex	n	Total length	Tail length	Hind foot length	Mass (g)	Greatest length of skull	Basal length	Length of maxillary toothrow	Zygomatic breadth	Breadth of braincase	Least interorbital breadth	Skull depth
Peromyscus crinitus	♀	30	177 ± 1.3 (160–189) 4.0	93 ± 1.2 (79–101) 7.0	20 ± 0.1 (18–21) 3.8	20.1 ± 1.10[b] (14.0–30.0) 23.91	24.71 ± 0.06 (23.97–25.42) 1.33	19.69 ± 0.08 (18.84–21.20) 2.29	3.58 ± 0.04 (3.11–4.03) 5.79	12.28 ± 0.06 (11.49–13.01) 2.59	11.89 ± 0.04 (11.52–12.37) 1.65	4.22 ± 0.04 (3.89–4.49) 3.17	8.80 ± 0.04 (8.23–9.38) 2.66
	♂	30	174 ± 0.8 (164–180) 2.4	92 ± 0.7 (81–97) 4.3	20 ± 0.1 (19–21) 3.3	16.9 ± 0.80[c] (10.0–24.5) 20.13	24.73 ± 0.08 (23.95–25.98) 1.69	19.65 ± 0.09 (18.96–21.79) 2.69	3.62 ± 0.03 (3.38–3.89) 3.80	12.26 ± 0.06 (11.77–13.55) 2.50	11.82 ± 0.03 (11.55–12.21) 1.49	4.19 ± 0.02 (3.78–4.48) 2.89	8.67 ± 0.04 (8.27–9.04) 2.50
Peromyscus maniculatus West of Cascade Range	♀	30	172 ± 2.3 (146–192) 7.3	84 ± 1.8 (67–105) 11.8	20 ± 0.3 (17–24) 7.3	23.1 ± 0.92 (14.4–33.0) 21.71	25.93 ± 0.14 (23.88–27.93) 2.96	21.19 ± 0.12 (19.48–22.76) 3.09	3.67 ± 0.03 (3.20–3.88) 3.89	13.13 ± 0.08 (12.06–13.98) 3.39	11.79 ± 0.05 (11.26–12.63) 2.31	4.00 ± 0.03 (3.77–4.40) 3.82	9.04 ± 0.05 (8.11–9.46) 3.02
	♂	31	170 ± 2.2 (148–194) 7.2	85 ± 1.7 (67–110) 11.1	21 ± 0.3 (16–23) 7.4	20.2 ± 0.46[d] (15.1–24.0) 11.69	25.55 ± 0.14 (23.49–26.59) 3.01	20.83 ± 0.14 (18.61–21.98) 3.68	3.66 ± 0.04 (3.27–3.99) 5.36	12.91 ± 0.08 (11.58–13.84) 3.65	11.69 ± 0.06 (11.04–12.61) 2.94	3.96 ± 0.03 (3.61–4.46) 4.30	8.98 ± 0.05 (8.34–9.49) 3.26
Harney and Malheur counties	♀	30	160 ± 1.8 (142–179) 6.2	71 ± 0.9 (61–85) 7.4	20 ± 0.2 (17–22) 5.4	21.7 ± 0.15[e] (13.0–34.2) 26.09	25.31 ± 0.12 (23.73–26.42) 2.66	20.70 ± 0.12 (19.30–21.69) 3.19	3.61 ± 0.03 (3.34–3.98) 4.87	12.72 ± 0.09 (11.45–13.72) 4.19	11.59 ± 0.06 (10.76–12.09) 2.69	3.98 ± 0.03 (3.61–4.28) 4.34	8.77 ± 0.05 (8.29–9.47) 3.39
	♂	30	154 ± 2.4 (120–178) 8.5	71 ± 1.3 (55–90) 10.3	19 ± 0.3 (15–24) 8.7	19.3 ± 0.63[f] (15.0–27.0) 16.19	25.07 ± 0.13 (24.00–27.07) 2.89	20.51 ± 0.15 (19.10–22.45) 3.98	3.67 ± 0.04 (3.37–4.14) 5.59	12.79 ± 0.11 (11.79–14.53) 4.69	11.60 ± 0.09 (10.80–13.19) 4.36	3.94 ± 0.05 (3.20–4.61) 6.57	8.79 ± 0.06 (8.22–9.86) 3.89
Northestern Oregon	♀	26	166 ± 1.7 (144–191) 5.2	74 ± 1.2 (62–87) 8.3	20 ± 0.1 (18–22) 3.6	22.9[g] (14.8–28.0)	25.31 ± 0.09 (24.65–26.52) 1.95	20.79 ± 0.10 (19.81–21.65) 2.51	3.74 ± 0.03 (3.43–4.10) 4.20	12.86 ± 0.09 (11.87–13.96) 3.59	11.65 ± 0.06 (10.97–12.26) 2.55	3.93 ± 0.03 (3.64–4.17) 3.74	8.76 ± 0.04 (8.40–9.26) 2.12
	♂	29	163 ± 1.2 (150–175) 3.9	74 ± 1.1 (62–86) 8.1	20 ± 0.1 (19–21) 3.4	20.7[h] (19.0–22.0)	25.40 ± 0.12 (23.98–26.77) 2.53	20.77 ± 0.11 (19.45–21.86) 2.97	3.75 ± 0.03 (3.39–4.10) 4.79	12.94 ± 0.07 (12.06–13.86) 2.90	11.77 ± 0.06 (11.25–12.47) 2.62	4.04 ± 0.03 (3.78–4.39) 4.17	8.82 ± 0.06 (8.19–9.43) 3.77
Remainder of Oregon	♀	23	166 ± 2.1 (147–184) 6.1	73 ± 1.1 (65–87) 7.3	20 ± 0.2 (18–21) 3.8	20.3[i] (12.0–32.6)	25.44 ± 0.13 (23.90–26.67) 2.45	20.85 ± 0.13 (19.40–21.94) 2.96	3.75 ± 0.03 (3.49–3.94) 3.53	12.90 ± 0.09 (11.89–13.52) 3.51	11.68 ± 0.08 (10.94–12.32) 3.09	3.99 ± 0.03 (3.71–4.30) 4.00	8.81 ± 0.03 (8.53–9.22) 1.89
	♂	28	159 ± 2.0 (138–180) 6.7	71 ± 1.3 (59–84) 9.9	21 ± 0.1 (19–22) 3.9	14.6 ± 1.04[j] (11.0–20.0) 23.71	25.36 ± 0.12 (24.35–26.52) 2.53	20.57 ± 0.10 (19.51–21.60) 2.64	3.18 ± 0.04 (3.45–4.26) 5.23	12.96 ± 0.08 (12.20–13.75) 3.23	11.79 ± 0.05 (11.23–12.30) 2.46	4.03 ± 0.03 (3.76–4.46) 3.86	8.88 ± 0.05 (8.40–9.37) 2.85
Peromyscus truei Curry and Josephine counties	♀	2	218 (209–226)	106 (101–111)	23 (22–24)	26.6	29.85 (29.41–30.28)	24.08 (23.76–24.39)	4.43 (4.28–4.58)	14.06 (13.94–14.18)	13.24 (13.16–13.31)	4.55 (4.43–4.67)	10.00 (9.67–10.32)
	♂	3	201 (191–211)	100 (96–104)	24 (23–24)		28.99 (28.60–29.52)	23.46 (23.04–23.84)	4.45 (4.29–4.69)	14.17 (13.95–14.33)	13.60 (13.52–13.66)	4.56 (4.47–4.69)	10.44 (10.26–10.64)
Jackson and Klamath counties	♀	6	192[k] (180–205)	97[k] (85–110)	23[k] (21–24)	30.6[k] (28.0–36.6)	28.57 (26.59–30.17)	23.37 (21.99–24.45)	4.23 (3.74–4.55)	13.89 (13.15–14.63)	12.98 (12.32–13.52)	4.49 (4.19–4.63)	9.77 (8.85–10.26)
	♂	11	197[l] (181–216)	99[l] (96–104)	24[l] (23–25)	28.8[d] (25.5–32.0)	28.62 ± 0.26 (27.23–29.97) 2.97	23.13 ± 0.21 (22.10–24.21) 3.08	4.24 ± 0.04 (3.97–4.44) 3.51	14.00 ± 0.08 (13.58–14.42) 1.98	13.24 ± 0.05 (13.06–13.49) 1.14	4.49 ± 0.03 (4.35–4.69) 2.41	10.26 ± 0.08 (9.78–10.74) 2.74
Remainder of eastern Oregon	♀	26	182 ± 2.1 (161–199) 5.9	89 ± 1.1 (80–101) 6.2	22 ± 0.4 (17–27) 8.0	25.2 ± 1.11[m] (17.9–38.7) 21.13	28.14 ± 0.16 (26.22–29.82) 2.95	22.77 ± 0.13 (21.45–24.07) 2.96	4.07 ± 0.04 (3.76–4.45) 4.56	13.98 ± 0.07 (13.21–14.65) 2.63	13.10 ± 0.06 (12.64–14.05) 2.34	4.56 ± 0.03 (4.33–5.00) 3.19	10.15 ± 0.06 (9.28–10.75) 3.16
	♂	34	178 ± 1.7 (158–202) 5.7	86 ± 1.1 (71–100) 7.7	23 ± 0.2 (21–25) 4.0	23.8 ± 0.86[d] (15.4–39.3) 19.88	27.86 ± 0.12 (26.04–29.43) 2.58	22.38 ± 0.15 (19.27–24.05) 3.81	4.09 ± 0.03 (3.68–4.38) 3.96	13.81 ± 0.09 (12.08–14.92) 4.01	13.01 ± 0.05 (11.97–13.62) 2.31	4.50 ± 0.03 (4.18–4.99) 3.85	10.09 ± 0.05 (9.18–10.47) 2.76

[a] Although measures of ear length were recorded, the data were not included because of extreme variation believed to stem from differences in techniques used to measure these characters.
[b] Sample size reduced by 11. [c] Sample size reduced by 12. [d] Sample size reduced by 4. [e] Sample size reduced by 6. [f] Sample size reduced by 5. [g] Sample size reduced by 19. [h] Sample size reduced by 22. [i] Sample size reduced by 15. [j] Sample size reduced by 17. [k] Sample size reduced by 1. [l] Sample size reduced by 2. [m] Sample size reduced by 3.

occupied by *P. crinitus*, as the species essentially is found nowhere else. We collected a series of specimens along Wright's Point south of Burns, Harney Co.; a line of traps set halfway up the side of and parallel to the former lava-filled riverbed produced *P. crinitus* where vegetation was sparse or nonexistent and *P. maniculatus* in vegetated zones. A similar trapline set west of Frenchglen, Harney Co., produced similar results. Such seems an example of *P. crinitus* being relegated to less productive habitats by a congener.

Johnson and Armstrong (1987) listed published estimates of density for *P. crinitus* in other states ranging as high as 42.9/ha. We know of no comparable data for Oregon, but, in our experience, relative abundance of *P. crinitus* is almost always less than for adjacent populations of *P. maniculatus*. For example, 10 traps set across a talus slope west of Frenchglen, Harney Co., in June 1992 produced two *P. crinitus*, two *Neotoma lepida*, and three *P. maniculatus*. However, in the only habitat supporting populations of both *P. crinitus* and *P. maniculatus* in western Utah (black sage, *Salvia mellifera*), the former was nearly 4 times as abundant as the latter (Fautin, 1946).

Diet.—Morton (1979) classified *P. crinitus* as omnivo-

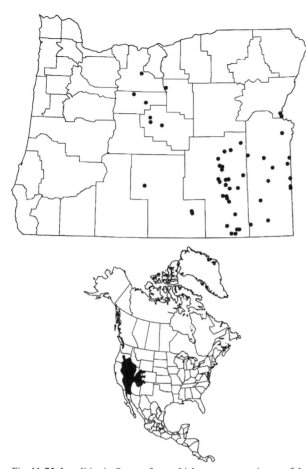

Fig. 11-75. Localities in Oregon from which museum specimens of the canyon mouse, *Peromyscus crinitus*, were examined (listed in APPENDIX, p. 540), and the distribution of the canyon mouse in North America. (Range map redrawn after Johnson and Armstrong, 1987:2, fig. 3, with modifications.)

rous. However, based on unusually high and uniform urine concentrations, MacMillen (1972) suggested that *P. crinitus* may subsist on a diet of insects and other animal material rich in both protein and water. Later, on the basis of the response of body mass to water deprivation, *P. crinitus* was suggested to have the option of subsisting as a granivore or shifting to a more succulent diet of animal material or green vegetation (MacMillen and Christopher, 1974); from the exceptionally high concentrations of urea excreted in winter the same authors predicted a shift to a primarily animal (insect) diet at that season. We know of no list of specific food items consumed by *P. crinitus*. However, Oregon specimens collected during the nonfruiting season of fungi contained traces of both epigeous and hypogeous species (Maser et al., 1978a, 1988).

Reproduction, Ontogeny, and Mortality.—Canyon mice reportedly are seasonally polyestrous, but produce young throughout the year in the laboratory (Egoscue, 1964). The interval between litters was 27–31 days, but gestation in the absence of lactation may be only 24–25 days; thus, gestation may be 3–5 days longer than in congeners (Egoscue, 1964). We collected simultaneously pregnant and lactating females in Harney Co. in April and June, suggesting that at least some individuals may undergo postpartum estrus. Females may produce as many as eight litters annually in the laboratory, but average only two

(Egoscue, 1964). Litter size averages smaller than that of Oregon congeners (Table 11-17), but the average tends to increase with the number of litters produced (Drickamer and Vestal, 1973). One captive female that gave birth to its first litter at 145 days of age produced 25 litters during the following 45 months; all were sired by the same male (Egoscue, 1964). In captivity, parents abandoned or killed 26.9% of litters in one study (Drickamer and Vestal, 1973) and 52% in another (Rood, 1966).

At birth, young weigh 1.8–2.6 g ($\bar{X}$ = 2.2 g) and developmentally are 2.5–3.0 days advanced over newborn *P. maniculatus* (Egoscue, 1964). Neonates are slightly pigmented on the dorsum, have vibrissae, and the pinnae are folded over the auditory meatuses. At 7 days, they weigh 5.0–6.0 g, have short fur with scattered guard hairs, have vibrissae 10 mm long, and the venter has a few light-colored hairs. At 15–17 days, the eyes open (Egoscue, 1964). At 23 days, the young are weaned (Drickamer and Vestal, 1973). At 28 days, young weighed 13.2–15.0 g. Pairs bred as early as 70 days of age, but most were 4–6 months old when they commenced to breed (Egoscue, 1964).

Habits.—Canyon mice usually are active nocturnally, but in captivity they may leave their nests for short periods during daylight hours to eat, drink, or eliminate. However, they tend to be active at light intensities that include full moonlight in open habitats (Kavanau and Havenhill, 1976), considerably lighter than most other desert-dwelling small mammals. Although active throughout the year, *P. crinitus* enters torpor diurnally when deprived of food (Morhardt and Hudson, 1966) and water (MacMillen, 1983). Torpor, weight loss, and a urine-concentrating ability that exceeds that of some of the ablest heteromyids are water-conservation measures that permit *P. crinitus* to subsist without preformed water (MacMillen, 1972).

Canyon mice are unusually agile peromyscines; they usually carry their long tails arched over their backs, commonly squeeze into narrow crevices, and frequently make jumps as long as 20 cm. When pursued, they sometimes travel by a series of hops (Egoscue, 1964). In a laboratory comparison with several conspecifics, *P. crinitus* exhibited excellent abilities as a runner and climber, moderate abilities as a digger and gnawer, and poor swimming abilities (King et al., 1968). During brief stops while foraging, canyon mice commonly place one forefoot or hind foot on a nearby vertical object, apparently as a means of making a quick start if pursued. Those whose tails have been lost during encounters with conspecifics (Eisenberg, 1963) are handicapped in their ability to escape (Egoscue, 1964). *P. crinitus* is more quarrelsome than its congeners; such is especially true of parturient females, perhaps explaining the high mortality among neonates (Egoscue, 1964). Females often defend their poorly constructed nests from conspecifics (Egoscue, 1964; Eisenberg, 1963).

Johnson and Armstrong (1987), based on the former's masters thesis, indicated that home-range size for *P. crinitus* in Utah calculated by the adjusted convex-polygon method averaged ($\pm SE$) 0.35 ± 0.03 ha for males and 0.38 ± 0.08 ha for females. The differences are not statistically significant.

Remarks.—Earlier, we listed no reference directed specifically to any aspect of the natural history of *P. crinitus* in Oregon (Carraway and Verts, 1982), and we know of none published subsequently.

Potential factors responsible for the unusual distribution pattern of *P. crinitus* in Oregon might be an interesting and enlightening research topic. However, that the observed distribution pattern is no more than a product of inadequate and unsystematic sampling remains a strong possibility.

Peromyscus maniculatus (Wagner)
Deer Mouse

1845. *Hesperomys maniculatus* Wagner, 1:148.
1853. *Hesp[eromys]. sonoriensis* Le Conte, 6:413.
1858. *Hesperomys gambelii* Baird, 8(pt. 1):464.
1894c. *Sitomys americanus artemisiae* Rhoads, 46:260.
1898b. *Peromyscus maniculatus* Bangs, 32:496.
1901. *Peromyscus oreas rubidus* Osgood, 14:193.
1903b. *Peromyscus perimekurus* Elliot, 3:156.
1909a. *Peromyscus maniculatus artemisiae*: Osgood, 28:58.
1909a. *Peromyscus maniculatus rubidus*: Osgood, 28:65.
1909a. *Peromyscus maniculatus gambeli*: Osgood, 28:67.
1909a. *Peromyscus maniculatus sonoriensis*: Osgood, 28:89.

Description.—*Peromyscus maniculatus* (Plate XIV) is a moderately small mouse-like rodent (Table 11-18) that in Oregon, as elsewhere, exhibits considerable variation in color, tail length (both actual and relative), and markings (especially relative proportions of dark and light on the bicolored tail). In general, the pelage of adult deer mice is buff to dark brown on the dorsum and white on the venter; among juveniles the pelage is steel gray to lead colored. At a given locality, the pelage averages 13.8% more hairs per unit area in winter than in summer and the average proportions of black overhairs, large banded hairs, and underfur hairs shifts from 16.3, 8.9, and 74.8%, respectively, to 15.8, 7.2, and 76.9%, respectively (Huestis, 1931).

The ears are moderately long, essentially naked, and usually held erect and directed forward. The eyes are black and beady. The vibrissae are long and usually are in constant motion. The feet are pentadactyl (the hallux is vestigial) and plantigrade; the digits are slender and equipped with delicate claws. The tail is semiprehensile and longer than the head and body in the semiarboreal races, but shorter than the head and body in races that climb little. Flex-tail (tail bent at right angle—Huestis and Barto, 1936; Meyer, 1947), albinistic, melanistic, and hairless specimens, and specimens with a variety of other anomalies, are known to occur.

The skull of *P. maniculatus* (Fig. 11-76) is thin and delicate. M1 and M2 usually possess well-developed accessory cusps; M1 usually is indented on the anterior face. The molars are tuberculated with two rows of tubercules discernible when the occlusal surfaces are viewed nearly parallel to the anterior-posterior toothrows.

Distribution.—The deer mouse has the broadest distribution of any species within the genus. It occurs from central Yukon and Northwest Territories east through the southern half of the prairie provinces, most of Ontario, the southern portion of Quebec, all of New Brunswick and Nova Scotia, and most of Newfoundland south through the United States (except for the southeastern states and coastal regions of states south of New Hampshire), and along the central highlands of Mexico and throughout Baja California (Fig. 11-77; Hall, 1981). *P. maniculatus* occurs throughout Oregon (Fig. 11-77).

Fossil remains of *P. maniculatus* have been recovered from >25 late Pleistocene (Wisconsinan) sites scattered throughout the United States (Kurtén and Anderson, 1980).

Geographic Variation.—Four of the approximately 70 recognized subspecies occur in Oregon: the dark-brown, long-tailed *P. m. rubidus* west of the Cascade Range; the bright cinnamon-brown, short-tailed *P. m. gambelii* through much of Oregon east of the Cascade Range; the pale buff, long-furred, short-tailed *P. m. sonoriensis* in southeastern Oregon; and the medium brown, moderately long-tailed *P. m. artemisiae* in northeastern Oregon (Bailey, 1936; Hall, 1981).

Habitat and Density.—*Peromyscus maniculatus* is nearly ubiquitous in Oregon. Below treeline, it occurs as part of essentially all communities in all stages of the sere: from recent clear-cuttings to old-growth forests; from sagebrush (*Artemisia*) steppe to renovated grasslands; and from rowcrops and pastures to whatever preceded agriculture. Outside of urban and metropolitan areas, and perhaps recently tilled fields, it truly would be difficult to set a line of 100 mouse (Sherman or museum special) traps baited with peanut butter and not catch at least one deer mouse during a night or two. Summaries of the following studies conducted in Oregon provide insight into the diversity of habitats occupied by *P. maniculatus* and the relative abundance of the species in each.

During a 12-year study on the west slope of the Cascade Range, estimated densities of *P. maniculatus* averaged 4.2/ha in spring and 3.7/ha in autumn (maximum 9.9/ha) on virgin Douglas-fir (*Pseudotsuga menziesii*)–western hemlock (*Tsuga heterophylla*)–western red cedar (*Thuja plicata*) forest. On a burned-over clear-cutting nearby, estimated densities averaged 7.7/ha in spring and 16.5/ha in autumn; they ranged from 2.2 to 31.6/ha and "fluctuated widely and irregularly" (Gashwiler, 1970a:1023). The deer mouse population on the clear-cutting commenced to increase within 1 year after the slash was burned when only 2% of the area was covered with vegetation. On another clear-cutting in the same region, estimated densities ranged from 1.0 to 10.6/ha in spring and from 5.9 to 21.2/ha in autumn (Gashwiler, 1970b). Population densities in autumn were not correlated with abundance of seeds of Douglas-fir and western hemlock on the ground. The spring-to-autumn increase (6.2 to 15.3/ha) in density on a 19.4-ha

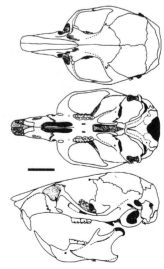

Fig. 11-76. Dorsal, ventral, and lateral views of the cranium, and lateral view of the dentary of the deer mouse, *Peromyscus maniculatus* (OSUFW 7349). Scale bar equals 5 mm.

study site continued unabated despite clear-cut logging operations in June–August and the slash being burned in October (Gashwiler, 1959). During another long-term (1964–1974) study on a clear-cutting in the same region, estimated densities of *P. maniculatus* ranged from 3.0 to 67.7/ha and tended to be highest during years following good crops of cones on conifers (Hooven, 1976).

In forest clear-cuttings in western Oregon, deer mice were more abundant on five of six plots treated with herbicides to reduce competition by forbs and grasses with Douglas-fir seedlings than on untreated plots (Borrecco et al., 1979). These findings suggest that peak population densities of *P. maniculatus* occur in early seral communities more advanced than the grass-forb stage. Nevertheless, the increase in density may commence immediately after removal of the old-growth forest (Gashwiler, 1959). Thus, it is not surprising that relative abundance of *P. maniculatus* in the Coast Range was nearly 3 times greater in 5–10-year-old conifer-shrub habitats than in deciduous (red alder, *Alnus rubra*-bigleaf maple, *Acer macrophyllum*) forests, nearly 9 times greater in 20–35-year-old sapling-pole conifer or 110–200-year-old sawtimber conifer stands, and 12 times greater than in 200+-year-old conifer forests (Gomez, 1992). *P. maniculatus* was only slightly less abundant at streamside sites than at sites 200 m upslope (Gomez, 1992).

Along the Columbia River, *P. maniculatus* occurred in

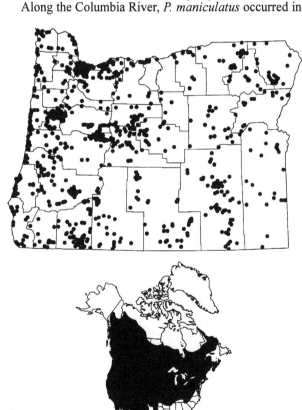

Fig. 11-77. Localities in Oregon from which museum specimens of the deer mouse, *Peromyscus maniculatus*, were examined (listed in APPENDIX, p. 540), and the distribution of the deer mouse in North America. (Range map redrawn after Hall, 1981:674–676, maps 392–394.)

all six segments sampled from the mouth to McNary Dam, Umatilla Co. It was least abundant in beach and dune segments near the mouth of the river and most abundant in segments that contained tree, shrub, and grassland habitats dominated by deciduous trees (Hinschberger, 1978).

In the Columbia Basin, *P. maniculatus* was uncommon (captured in 1 of 12 months; 7/432 trap-nights) in a cheat grass (*Bromus tectorum*) community, moderately common (captured in 8 of 10 months; 3–11/432 trap-nights) in a needle-and-thread grass (*Stipa comata*) community, and common (captured in each of 10 months; 5–27/432 trap-nights) in a sagebrush-juniper community. From an earlier study in similar habitats in Washington (O'Farrell et al., 1975), the observed absence of *P. maniculatus* in grassland communities in midsummer was expected (Rogers and Hedlund, 1980).

In the Harney Basin, *P. maniculatus* was most abundant in habitats dominated by greasewood (*Sarcobatus vermiculatus*), 37.5% less abundant in big sagebrush (*Artemisia tridentata*) habitats, and 79.5% less abundant in marsh (hardstem bullrush, *Scirpus acutus*; cat-tail, *Typha latifolia*; baltic rush, *Juncus balticus*; and sedges, *Carex*) habitats. Only one individual was captured in grasslands (mostly Sandberg bluegrass, *Poa sandbergii*; saltgrass, *Distichlis stricta*; and bluestem wheatgrass, *Agropyron smithii*—Feldhamer, 1979*b*). In northeastern Oregon, female *P. maniculatus* in sagebrush-greasewood and ponderosa pine (*Pinus ponderosa*) habitats, but not in winterfat (*Eurotia lanata*)-saltbush (*Atriplex*) habitats, are larger and tend to occur in more mesic microhabitats than males (Bowers and Smith, 1979).

Where they are sympatric, *Microtus townsendii* may exclude *P. maniculatus* from some grassland habitats (Redfield et al., 1977). Montane voles (*Microtus montanus*) possibly were responsible for the low relative abundance of *P. maniculatus* in grassland and marsh habitats reported by Feldhamer (1979*b*).

Diet.—*Peromyscus maniculatus* is omnivorous; it commonly adjusts its food intake to exploit particularly abundant or available food resources. In many situations, particularly in spring and summer and in sparsely vegetated habitats, *P. maniculatus* may consume more animal material than plant material (Jameson, 1952; Osborne and Sheppe, 1971). Although both hypogeous and epigeous fungi are consumed, most are of the former type; 62% of 21 *P. maniculatus* from southeastern Oregon and northwestern Nevada and 73% of 62 from western British Columbia and western Oregon contained fungi in their stomachs (Maser et al., 1978*a*). In autumn and winter, *P. maniculatus* in forest habitats often shifts to conifer seeds during years that production is high (Gashwiler, 1970*b*, 1979). Thus, the diet of *P. maniculatus* may vary with season and from year to year (Jameson, 1952).

In the northern Sierra Nevada Range, California, the overall diet (by volume) of *P. maniculatus* based on 275 specimens collected throughout 1 year consisted of 53% seeds, 4% fruits, 32% arthropods, 7% leaves, 1% fungi, and 3% miscellaneous items in forested areas and 35% seeds, 22% fruits, 31% arthropods, 7% leaves, <1% fungi, and 5% miscellaneous items in brushfields. The following year, the overall diet based on 503 specimens consisted of 51% seeds, 0% fruits, 26% arthropods, 7% leaves, 7% fungi, and 9% miscellaneous items in forested areas and

34% seeds, 6% fruits, 36% arthropods, 12% leaves, 2% fungi, and 10% miscellaneous items in brushfields (Jameson, 1952). Some of the miscellaneous items were other vertebrates, mostly birds. In a coastal-dune habitat in northwestern California, the spring diet (by volume) of 30 *P. maniculatus* consisted of 18.5% plant material; the remaining animal material was composed largely of arthropods. The stomachs of seven individuals contained remains of the amphipod *Orchestoidea californiana* and one individual contained a trace of the dactylozooids of the hydrozoan coelenterate *Velella velella*, many of which had washed onto the beach (Osborne and Sheppe, 1971). The latter food item, rather unusual for a terrestrial small mammal, illustrates the ability of *P. maniculatus* to take advantage of nearly all resources. Likely such ability is partly responsible for the wide geographic distribution of the species and the nearly universal occurrence of the species in all seral stages of all vegetative associations within its range.

Reproduction, Ontogeny, and Mortality.—In some studies, some reproductive characteristics in *P. maniculatus* were shown to vary with environmental or population characteristics, but in other instances investigators reported an absence of the same relationships (Millar, 1989). First litters average 0.9 fewer young than second litters, but third litters average only 0.1 fewer young than second litters. Some workers reported positive or negative relationships of litter size with temperature, season, rainfall, and size of the maternal female, but not with age of the maternal female, size of seed crops, type of habitat, or length of the breeding season (Millar, 1989). Inconsistent relationships were reported between litter size and elevation or latitude, but elevation and latitude combined explained a significant amount of the variation in litter size. Based on published information, Millar (1989) devised formulas to predict litter size and duration of the breeding season from latitude and longitude. With these formulas, predicted values for these reproductive characteristics at Salem, Marion Co., were 5.1 young/litter and 26.8 weeks. Predicted litter size was slightly greater than that obtained from counts of embryos or implantation sites, and predicted length of the breeding season was less than that obtained from records of pregnancies in Oregon (Table 11-17). The gestation period is 22–23 days; ectopic pregnancies are known to occur (Moore, 1929a).

In forested areas along the west slope of the Cascade Range in Oregon, the proportion of adult females pregnant in autumn and winter during years that production of seeds by conifers was good (39 of 219, 17.8%; ≥2.66 kg seed/ha) was significantly greater than during years when seed production was poor (11 of 484, 2.3%; ≤0.45 kg seed/ha). However, during spring and summer, the reverse was true (31 of 161, 19.3% and 108 of 284, 38.0% during years of good and poor seed crops, respectively—Gashwiler, 1979). Proportions of females pregnant in years following good and poor seed crops were not significantly different (Gashwiler, 1979). In the same study, overall average litter size (Table 11-17) was not related to the size of the seed crop; however, monthly average litter sizes ranged from a low of 3.5 in January to a high of 6.5 in November; in general, litter size averaged smaller in spring than in summer. Gashwiler (1979) suggested that the greater proportion of primiparous maternal females and females nearing the end of their reproductive lifespan in populations in spring may account for the lower averages at that season. Although seasonal differences may account for the large ($\bar{X} = 6.0$, $n = 10$—Beer, 1959) average litter size reported for *P. maniculatus* taken in August in western Oregon, they cannot be invoked to explain the small ($\bar{X} = 3.4$, $n = 65$—Tevis, 1956) average litter sizes of specimens taken in May–October in northern California.

Body mass of neonates usually ranges from 1.3 to 2.2 g ($\cong$8% of adult body mass) and varies with geographic race, size of the litter, and size of the maternal female (Layne, 1968); many of those <1.3 g do not survive (Millar, 1989). One-half adult body mass is attained by 23–27 days of age, one-half adult total length by 10 days, one-half adult body length by 6 days, one-half adult tail length by 15 days, one-half adult foot length by 5 days, and one-half adult ear length in 10–12 days (Layne, 1968). In captivity, average age at first estrus in 189 females (source of stock not reported) was 48.7 days; the youngest was 28 days old (Clark, 1938). One female conceived at 34 days of age. Some males may attain reproductive competence by 40 days of age, most by 60 days of age (Clark, 1938). However, at least in some races of *P. maniculatus*, presence of a parent retards or prevents attainment of reproductive competence in young (Wolff, 1989).

At birth, the digits are fused; the ears are folded forward, closing the auditory meatuses; and the eyes are visible through sealed eyelids. Neonates are naked except for vibrissae and the skin is without pigment. The ears open in 1–2 days and become erect by the 3rd day. Young may be weaned as early as 19 days of age (McCabe and Blanchard, 1950).

A wide variety of snakes (Serpentes), carnivorous mammals, and raptorial birds prey on *P. maniculatus* (Maser et al., 1981b). The following species are recorded as preying on *P. maniculatus* in Oregon: saw-whet owl (*Aegolius acadicus*—Forsman and Maser, 1970), long-eared owl (*Asio otus*), barn owl (*Tyto alba*), great horned owl (*Bubo virginianus*—Maser and Brodie, 1966; Maser et al., 1970, 1980), short-eared owl (*Asio flammeus*—Maser et al., 1970), burrowing owl (*Athene cunicularia*—Maser et al., 1971), coyote (*Canis latrans*), bobcats (*Lynx rufus*—Witmer and deCalesta, 1986), red fox (*Vulpes vulpes*), gray fox (*Urocyon cinereoargenteus*—Wilcomb, 1948), spotted skunk (*Spilogale gracilis*), and ermine (*Mustela erminea*—Maser et al., 1981b). Maser et al. (1970) concluded that the most intense predator pressure on *P. maniculatus* by three species of owls was in winter; however, the deer mouse ranked first in the diet of only one species, second for the other two. In summer, the deer mouse ranked second for two species, sixth for the third. Green et al. (1993) reported that only 1.6% of 5,559 pellets regurgitated by burrowing owls in Oregon contained remains of deer mice.

Habits.—*Peromyscus maniculatus* is active nocturnally; time of onset of activity is cued by light and is remarkably precise, but does not seem to be influenced by clouds (Falls, 1968). In captivity under red light, deer mice from western Oregon exhibited three peaks of activity: 2100–2200, $\cong$2400, and 0200–0500 h (Laban, 1966). For captives obtained in California, 75% of activity was between 1800 and 2400 h, 15% was between 2400 and 0600 h, and 10% was during the 12 h of light; there was no definite pattern to feeding or drinking (French, 1956). Detection of seeds, even buried ones, is by olfaction (Howard and Cole, 1967).

Peromyscus maniculatus is among the several species in the genus known to be able to enter torpor to conserve energy; deprivation of food, negative water balance (possibly by suppressing appetite), and reduction of the oxygen concentration in burrow systems are known to stimulate entry into torpor (Hill, 1983). Some individuals apparently do not enter torpor; also, the frequency and duration of torpor may vary widely among individuals that enter torpor. Typically, torpor is restricted to the daylight hours, lasts for only part of the day, and recurs on successive days (Hill, 1983). Body temperature of torpid individuals may fall to 13–17°C, but below that level individuals usually do not survive even with artificial rewarming. However, body temperature during torpor is dependent upon the individual, ambient temperature, food supply, and possibly other factors (Hill, 1983).

Insulation provided by nesting material also is involved in the conservation of energy. *P. maniculatus* nests in trees, burrows in the ground, crevices in rocks, woodrat (*Neotoma*) houses, and a variety of other places. Communal nesting is observed frequently; huddling also is a means by which *P. maniculatus* conserves energy. Among the various races of *P. maniculatus,* nest sites selected, in general, reflect the type of habitat; woodland forms commonly, but not always, nest in trees, whereas grassland and shrub-steppe forms usually nest in burrows (Wolff, 1989). Several studies of various aspects of the ecology of the species have been conducted by use of artificial nest boxes; rather elaborate nest boxes have been designed (Kaufman and Kaufman, 1989).

In the Harney Basin, standard-diameter home ranges (2δ) averaged 372.1 m^2 (range, 108.5–603.3 m^2; $n = 29$) for males and 298.6 m^2 (range, 179.2–640.5 m^2; $n = 18$) for females in greasewood habitat and 371.7 m^2 (range, 126.7–571.8 m^2; $n = 17$) for males and 392.0 m^2 (range, 173.2–573.3 m^2; $n = 8$) for females in sagebrush habitat (Feldhamer, 1979c). Home-range areas were significantly larger for males than for females in greasewood habitat; in sagebrush habitat home-range areas for females were slightly, but not significantly, larger than for males. Home-range areas were significantly inversely related to population density among males in greasewood habitat, and for both sexes combined in sagebrush habitat. In greasewood habitat, reproductively active males were claimed to have smaller home ranges than nonreproductive males, but the calculated t value (1.72) actually was not significant (Feldhamer, 1979c:53).

In western Oregon, 71 (24%) of 297 deer mice marked on a grid established in a 19.4-ha burned-over clear-cutting also were captured during the same month in traps on lines established in old-growth forest ≥91.5 m on opposite sides of the grid (Gashwiler, 1971). A greater proportion (52%) of the mice moved from the clear-cutting to the old-growth forest during the month after the clear-cutting was burned; neither seed production by conifers in the old-growth forest nor sex and age of mice seemed to influence movements between habitat types. The mice moved an average of 212.8 m and a maximum of ≥425 m (Gashwiler, 1971). Four deer mice displaced ≅800 m were not recaptured at the site at which they were obtained originally. However, following 13 releases of five individuals

≅400 m from points of capture, two mice were recaptured at their home site, one each from opposite directions (Broadbooks, 1961). Such suggests that in Gashwiler's (1971) study both clear-cutting and old-growth habitats may have been within the range of familiarity of the deer mice, if not within their home ranges.

Social interactions have been studied in some races of *P. maniculatus*, but whether the observed behaviors are applicable to any or all of the races that occur in Oregon is unknown. At higher densities, home ranges of *P. maniculatus* overlap 14–37%, and the species is intraspecifically territorial by aggressive interactions and may be interspecifically territorial with syntopic congeners; at lower densities, home ranges overlap 7% and aggressive interactions are uncommon (Wolff, 1989). Peripheral portions of territories are defended less frequently than central portions, and strangers are met with greater aggression than neighbors. Territories of males, in addition to serving as locales for feeding and nesting, may function to provide access to females; however, when density of populations is low, males may abandon their territories in their search for reproductive females (Wolff, 1989). Territories of females, in addition to serving as locales for feeding and nesting, function in the rearing of young and possibly in protecting young from infanticide; conspecific females are excluded from territories by aggressive behavior because they frequently kill unprotected young (Wolff, 1989).

Although *P. maniculatus* may exhibit some characteristics of a monogamous mating system (family groups and overlapping home ranges), based on a series of criteria selected to indicate monogamy, it ranked last among 11 species of *Peromyscus*. Based on an analysis of blood proteins, 19–43% of litters produced by *P. maniculatus* were sired by more than one male (Wolff, 1989). Polyandry was suggested as a mechanism by which females reduce the probability of infanticide by males that sometimes kill unrelated young.

Remarks.—Doubtlessly, no genus of North American mammals has been researched more intensively than *Peromyscus*, and within that genus no species has received more attention than *P. maniculatus*. Two books that summarize and synthesize the voluminous information available on the biology of the genus (much of which is on *P. maniculatus*) have been published (King, 1968; Kirkland and Layne, 1989). These volumes are an appropriate starting place for anyone wishing to research almost any aspect of the biology of this or other species of *Peromyscus*. Nevertheless, by recommending these volumes, we certainly do not intend to give the impression that everything is known about the deer mouse.

Peromyscus truei (Shufeldt)
Piñon Mouse

1885. *Hesperomys truei* Shufeldt, 8:407.
1893b. *Sitomys gilberti* Allen, 5:188.
1894. *P[eromyscus]. Truei*: Thomas, 14:365.
1909a. *Peromyscus truei gilberti*: Osgood, 28:169.
1936. *Peromyscus truei preblei* Bailey, 55:188.
1941. *Peromyscus truei sequoiensis* Hoffmeister, 54:129.

Description.—*Peromyscus truei* (Plate XIV) is the largest member of the genus in Oregon (Table 11-18). It is characterized by its enormous ears; the ear from notch to tip is almost always longer than the hind foot, with the latter dimension usually >22 mm (Hoffmeister, 1981). So disproportionately large are the ears in some populations that, in the field, we affectionately call specimens of *P. truei* "dumbos" after Walt Disney's mythical flying elephant that used its ears for wings. There are three pairs of mammae, two inguinal and one pectoral.

The pelage is long and somewhat lax, but not so much so as in *P. crinitus*. Hair color varies geographically; overall, *P. truei* in central and eastern Oregon is a pale-buff color with a wash of black on the dorsum, whereas specimens from southwestern Oregon are bright ocherous on the shoulders and flanks and a dirty brown on the dorsum. Dorsal hairs on the former group are black at the base with a light-tan subterminal band and a black tip; hairs on the venter have black bases and white tips. For the southwestern group, the dorsal hairs are black at the base, have a narrow ocherous band, and a short black or brownish tip; hairs on the flanks have no black tip; and the hairs of the venter have black bases and creamy white tips. The tail is sharply bicolored, with the black dorsal stripe covering about one-third of the circumference of the tail.

The skull is large (Fig. 11-78) and the auditory bulla is large and inflated. Prominent accessory cusps are present on M1 and M2, and the enamel on the anterior margin of M1 usually is indented.

Distribution.—*Peromyscus truei* occurs from north-central Oregon and southern Idaho (Larrison and Johnson, 1981) south through montane and coastal regions of California into Baja California Norte and through montane regions of Nevada, Utah, and Colorado south through Arizona and New Mexico, and Chihuahua and other central highland states of Mexico to central Oaxaca (Fig. 11-79; Hall, 1981). In Oregon, *P. truei* has been taken in Josephine, Jackson, Klamath, Lake, Deschutes, Jefferson, Grant, Crook, and Harney counties (Fig. 11-79; Carraway et al., 1994). We located a catalog entry for a *P. truei* (AMNH 30782) collected 25 June 1910 by J. T. Nichols at

"Cloud Cap Inn" (surely not Hood River Co.), but no specimen with that number or from that locality was found.

Although Pleistocene fossils tentatively identified as *P. truei* were found in New Mexico, no fossils of the species were reported from Oregon (Kurtén and Anderson, 1980).

Geographic Variation.—Published range maps (Hall, 1981; Hoffmeister, 1951, 1981) depict four nominal subspecies occurring in Oregon: the small *P. t. preblei* with buffy-gray dorsal pelage in Crook, Deschutes, and Jefferson counties; the medium-sized *P. t. gilberti* with pinkish-cinnamon dorsal pelage overlain with black in Jackson and eastern Josephine counties; the extremely large *P. t. sequoiensis* with blackish and dark-reddish dorsal pelage in western Josephine and Curry counties; and *P. t. truei* along the east slope of the southern Cascade Range. However, we found no specimens from Curry Co. and no specimens from Oregon have been referred to *P. t. truei* (Fig. 11-79). Recently collected specimens from Lake, Harney, and Grant counties (Carraway et al., 1994) here are referred tentatively to *P. t. preblei*.

Habitat and Density.—East of the Cascade Range, *P. truei* is almost always associated with western juniper (*Juniperus occidentalis*) in rimrocks (Carraway et al., 1994); in southwestern Oregon, it is associated with rocky outcrops vegetated by oaks (*Quercus kelloggii* and *Q. garryana*—Fisher, 1976); and farther south beyond our

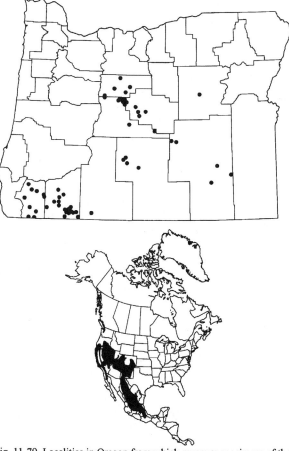

Fig. 11-79. Localities in Oregon from which museum specimens of the piñon mouse, *Peromyscus truei*, were examined (listed in Appendix, p. 545), and the distribution of the piñon mouse in North America. (Range map redrawn from Hoffmeister, 1981:3, fig. 5, with modifications provided by Carraway et al., 1994, and Gufar et al., 1980.)

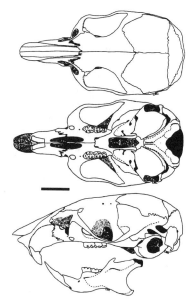

Fig. 11-78. Dorsal, ventral, and lateral views of the cranium, and lateral view of the dentary of the piñon mouse, *Peromyscus truei* (OSUFW 4187). Scale bar equals 5 mm.

limits, it commonly is associated with rocky areas and piñon pines (*Pinus edulis*—Douglas, 1969; Hall, 1946; Hoffmeister, 1951, 1981). However, in some areas in Nevada and Utah, *P. truei* was captured in sagebrush (*Artemisia*) remote from junipers or piñon pines (Armstrong, 1979; O'Farrell and Clark, 1984). In Harney Co. in June 1992, we were able to capture *P. truei* only by setting traps well beneath juniper trees growing in low (1–2 m tall) rimrocks (Carraway et al., 1994); traps set in the same rimrock <1 m beyond the drip-line of a juniper captured only *P. maniculatus*. In Deschutes Co., on the study area described by McKay and Verts (1978a) and in similar habitats nearby, we captured *P. truei* on and at the base of low rimrocks and on lava scabs that supported junipers. However, at these sites it was not necessary to place traps directly beneath junipers, but traps placed 2–3 m from either junipers or rocks produced only *P. maniculatus*. In southwestern Oregon, Fisher (1976) interpreted collection of 90 *P. truei* and eight *P. maniculatus* on one study site to indicate separation of the two species, but he did not indicate if habitats occupied by the two species on the site were distinguishable.

Although he did not estimate density per se, Fisher (1976) captured and followed the movements of 17 individuals on a 0.7-ha study area (24.3/ha) between 17 September and 1 March in southwestern Oregon.

Diet.—We know of no study of the diet of *P. truei* in Oregon, although Fisher (1976) reported that captive *P. truei* from southwestern Oregon selected acorns (*Quercus*) over sunflower seeds. In Colorado, stomachs of *P. truei* contained various parts of sagebrush, penstemon (*Penstemon*), muttongrass (*Poa fendleriana*), buckwheat (*Eriogonum*), snakeweed (*Gutierrezia sarothrae*), yucca (*Yucca*), chitin, and feathers (Douglas, 1969). Of the plants, the same species or congeners thereof occur in Oregon. Douglas (1969) further suggested that winter food must consist largely of juniper berries that remain on the trees all year. In New Mexico, Smartt (1978) reported that 30% of the diet of *P. truei* consisted of arthropods, 31% of juniper fruit, 12% of juniper pollen cones, 5% of prickly pear (*Opuntia*), and 7% of mistletoe (?*Phoradendron*). In California, in midsummer Bradford (1974) found remains of insects (≅62%), foliage (≅16%), fungi (≅16%), and seeds (≅4%) in stomach contents of *P. truei*. In late summer, ≅75% of the diet consisted of acorn mast, ≅18% of insects, and ≅1% of foliage.

Reproduction, Ontogeny, and Mortality.—In the laboratory, females attain first estrus at an average age of 50.1 ± 2.4 days; the youngest was at 28 days (Clark, 1938). Gestation in nonlactating females required 25–27 days, 40 days in one lactating female (Svihla, 1932).

Over the geographic range of the species, litter size seems to vary considerably (Table 11-17). Douglas (1969) suggested that *P. truei* does not have more than six young per litter; however, we found one female each with seven and eight pigmented sites of implantation and another with nine embryos. The Oregon sample, although relatively small, equals the highest average number of embryos reported to date (Table 11-17). These data may indicate a north-south cline of decrease in average litter size as described in some other species of rodents (Chapman and Lind, 1973).

Similarly, the breeding season varies over the geographic range of the species. We have records of pregnant individuals in Oregon in May, October, and November, but our effort was decidedly seasonal. Fisher (1976:56) indicated that breeding on a study area in southwestern Oregon "seemed to be year round," but later (p. 96) stated that "Signs of breeding were noted in all seasons except late summer." He further suggested (p. 96) that restrictions in the supply of food, especially a poor acorn crop, in late summer were "significant in limiting the breeding season of these mice." In Colorado, the breeding season was June–September (Douglas, 1969); in California, March–October (McCabe and Blanchard, 1950); in Nevada, May–July (Hall, 1946); in Arizona, February–November (Hoffmeister, 1986); and in New Mexico, all months except October and December (Holdenreid and Morlan, 1956).

At birth, *P. truei* averages 2.3 g (*n* = 41), nearly twice the average of *P. maniculatus* neonates (McCabe and Blanchard, 1950; Svihla, 1932). Although growth is rapid—with young averaging 5.1 g at 1 week, 8.3 g at 2 weeks, 10.5 g at 3 weeks, 13.9 g at 4 weeks, 16.7 g at 5 weeks, 19.4 g at 6 weeks, and 24.1 g (adult weight) at 10 weeks—development in form and color is decidedly behind that of *P. maniculatus* (McCabe and Blanchard, 1950). At 1 week, the eyelids are indicated by a groove, the ears are open, but flat and fleshy (ears unfold at 3 days—Svhila, 1932), and the dorsum still appears bare without magnification. At 2 weeks young still have a squarish face, the ears are still flat and fleshy, the skin is dark, and the pelage is growing. The eyes open at 15–20 days of age (Svihla, 1932). By 3 weeks of age, development of form and color in *P. truei* is still behind that of *P. maniculatus*, but development of strength, fitness, speed, and combativeness is ahead of that of *P. maniculatus* (McCabe and Blanchard, 1950). By 4 weeks, young have entered violent play, bite viciously if frightened, and still cling to the female's nipples although no milk is produced. By 12 weeks, the postjuvenile molt that commenced at 7 weeks is complete or nearing completion (McCabe and Blanchard, 1950).

We found no mention of *P. truei* occurring in the diet of any raptorial bird or carnivorous mammal in Oregon. Fisher (1976) listed several potential predators of *P. truei* in southwestern Oregon and suggested that it probably was most susceptible to predation by snakes (Serpentes) because of the habitat in which it lives. He also indicated that piñon mice whose movements were being tracked by use of tiny battery-powered lights attached to their heads retreated to cover of rocks at the first sound produced by one of the three owls (*Bubo virginianus*) that occupied his study area. Elsewhere, a few are known to be eaten by coyotes (*Canis latrans*) and red foxes (*Vulpes vulpes*—Douglas, 1969). We suspect that *P. truei*, because of its association with rocky habitats, may be less vulnerable than open-habitat congeners to predation by raptorial birds and large carnivores.

Habits.—Piñon mice are largely nocturnal. In the laboratory, they commonly enter daily torpor when subjected to severe food or water stress (Bradford, 1974). *P. truei* has an ability to concentrate urine about equal to that of other *Peromyscus*, but far less than that of strongly xeric-adapted species of rodents (Bradford, 1974). Most water is obtained from food.

In Colorado, nests usually are associated with juniper trees, with hollow trees being favored (Douglas, 1969). However, in southwestern Oregon, nests usually are in crevices in rocks (Fisher, 1976). A hole in an oak tree ≅1.5 m above the ground was used as a refuge or feeding station at night, but not as a daytime nest (Fisher, 1976).

In Colorado, home ranges calculated by the exclusive boundary-strip method averaged (±*SD*) 0.32 ± 0.12 ha (range, 0.13–0.53 ha; *n* = 16) for males and 0.25 ± 0.08 ha (range, 0.12–0.46 ha; *n* = 22) for females (Douglas, 1969). In southwestern Oregon, minimum-area home ranges for three adult males captured ≥10 times averaged 0.12 ha (range, 0.10–0.16 ha) and the home range for one adult female was 0.04 ha (Fisher, 1976). Minimum-area home ranges obtained by continuously monitoring movements of individuals to which battery-powered lights were attached averaged 0.04 ha (range, 0.02–0.06 ha; *n* = 5) for adult females and 0.14 ha (range, 0.06–0.23 ha; *n* = 6) for adult males (Fisher, 1976).

In captivity, conspecifics usually are tolerant of each other even in close confinement; however, a slight shortage of food or water often produces cannibalistic tendencies in those held together in cages (Douglas, 1969). Paired encounters of *P. truei* with *P. maniculatus* revealed little evidence of heterospecific avoidance or interspecific fighting; during opposite-species encounters the mice spent most of the periods sitting close together (Fisher, 1976).

When frightened, piñon mice commonly run in semisaltatorial fashion, often jumping objects 0.5 m tall. In Colorado, frightened individuals usually ran to the nearest tree and climbed to heights of 3–6 m (Douglas, 1969), whereas in southwestern Oregon, those followed upon release usually entered crevices in rocks. However, during late summer, before acorns had fallen, arboreal activity by *P. truei* as revealed by light-tracking was relatively common (Fisher, 1976).

Remarks.—In reference to *P. truei* in Nevada, Hall (1946:520) stated "Piñons without rocks and rocks without piñons seem not to provide habitats suitable to the mice." In Oregon, western juniper (or oaks in the southwestern part of the state) seemingly assumes the role played in the ecology of *P. truei* by piñon pines in Nevada. Such possibly explains the failure of *P. truei* to become more widely distributed in Oregon, and tends to indicate that recent discovery of the species in Lake and Harney counties was not the result of a recent invasion. Western juniper, based on dendrochronological studies, is a climax species in rimrock areas, with many individuals 185–365 years old, whereas downslope on alluvial soils it is a seral species and a recent invader, with essentially all individuals 33–88 years old (Burkhardt and Tisdale, 1969; Young and Evans, 1981). Consequently, *P. truei* probably is a climax species and likely many, if not most, *P. truei* populations in eastern and central Oregon are relicts. Thus, with the recent state-sanctioned and supported harvest of western juniper for wood products (*Corvallis Gazette-Times*, 10 September 1995, p. B8), central and eastern Oregon populations may be in some jeopardy.

We suggest that a study of genetic variation within and among several of these disjunct populations might prove interesting, perhaps enlightening.

Onychomys leucogaster (Wied-Neuwied)
Northern Grasshopper Mouse

1841. *Hypudaeus leucogaster* Wied-Neuwied, Reise in das innere Nord-America . . ., 2:99.
1858. *Onychomys leucogaster*: Baird, 8(pt. 1):459.
1890a. *Onychomys fuliginosus* Merriam, 3:59.
1891. *Onychomys leucogaster brevicaudus* Merriam, 5:52.
1913. *O[nychomys]. l[eucogaster]. fuliginosus*: Hollister, 26:216.
1986. *Onychomys leucogaster durranti* Riddle and Choate, 67:248.

Description.—*Onychomys leucogaster* (Plate XIV) is a stockily built mouse with a thick tail <60% of the length of the head and body (Table 11-16). The dorsal pelage is dark sepia along the midline grading moderately sharply to a light tannish along the sides. The ventral pelage is white, extends onto the flanks and ribs, and is sharply demarked from the dorsal pelage; the white hairs have dark bases except on the throat, chin, face, and front legs. A dark sepia line extends across the face between the eyes. The ears are thinly furred, sepia on the external surface and light grayish-tan on the internal surface. The feet and tip of the tail are white.

The skull (Fig. 11-80) is similar to that of the peromyscine rodents except it is heavier with a relatively shorter rostrum and wider braincase. The nasals are wedge shaped with converging lateral margins rather than elongate with nearly parallel lateral margins as in *Peromyscus*.

Distribution.—*Onychomys leucogaster* occurs on sandy soils in prairie, intermontane, and desert regions from southwestern Alberta, central Saskatchewan, and southeastern Manitoba south through western Minnesota, northwestern Iowa, eastern Nebraska and Kansas, and central Oklahoma, and Texas to the northern tier of states in Mexico west to Sonora, and in the United States, west to Arizona, Nevada, Oregon, and southeastern Washington (Fig. 11-81; Hall, 1981). In Oregon, northern grasshopper mice have been collected in grassland and desert areas in Crook, Deschutes, Gilliam, Harney, Jefferson, Klamath, Lake, Malheur, Morrow, Sherman, Umatilla, and Wheeler counties (Fig. 11-81).

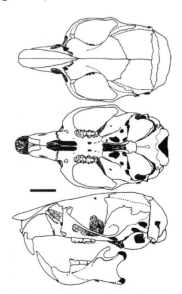

Fig. 11-80. Dorsal, ventral, and lateral views of the cranium, and lateral view of the dentary of the northern grasshopper mouse, *Onychomys leucogaster* (OSUFW 7430). Scale bar equals 5 mm.

Fossils of *O. leucogaster* are widespread in late Pleistocene deposits in the eastern and southern portions of the present range and eastward beyond the present range. *O. †fossilis*, a late Pliocene species from Kansas, is thought to be ancestral to *O. leucogaster* (Kurtén and Anderson, 1980).

Geographic Variation.—Two of the 11 currently recognized subspecies occur in Oregon: the smaller *O. l. durranti* in Jefferson, Crook, Morrow, Umatilla, Gilliam, Sherman, and Wheeler counties, and the larger *O. l. brevicuadus* through the remainder of the range within the state (Engstrom and Choate, 1979; Hall, 1981; Riddle and Choate, 1986).

Earlier, all northern grasshopper mice taken in Oregon were considered to be members of the subspecies *O. l. fuscogriseus* although Hall (1981:729, map 421) indicated that the range of *O. l. brevicuadus* extended into extreme southeastern Oregon. He did not, however, list Oregon localities for the latter subspecies. Riddle and Choate (1986) erected a new subspecies (*O. l. durranti*) for the grasshopper mice in southeastern Washington and north and west of the Blue and Wallowa mountains in Oregon. They synonymized *O. l. fuscogriseus* in the remaining portion of its range with *O. l. brevicaudus*.

Habitat and Density.—In Morrow Co., we captured *O. leucogaster* on an area vegetated largely by gray rabbit-

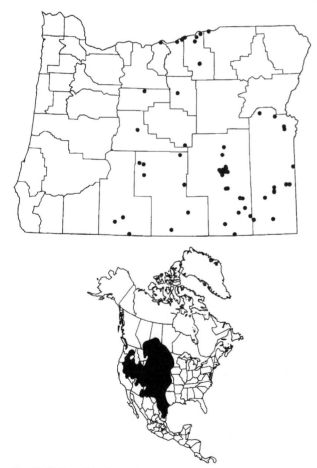

Fig. 11-81. Localities in Oregon from which museum specimens of the northern grasshopper mouse, *Onychomys leucogaster*, were examined (listed in APPENDIX, p. 538), and the distribution of the northern grasshopper mouse in North America. (Range map redrawn after Hall, 1981:729, map 421, with modifications.)

brush (*Chrysothamnus nauseosus*) with scattered cheat grass (*Bromus tectorum*), buckwheat (*Eriogonum*), Sandberg bluegrass (*Poa sandbergii*), and snakeweed (*Gutierrezia sarothrae*—Small and Verts, 1983). The soil there is sandy. In Harney Co., G. A. Feldhamer (in litt.) caught these mice on two of five study plots dominated by sagebrush (*Artemisia tridentata*), but on none of four plots each in marsh and grassland or five plots in black greasewood (*Sarcobatus vermiculatus*).

In Utah, Egoscue (1960:101) noted that *O. leucogaster* tended to avoid "marshy areas, extremely rocky situations, precipitous hillsides, the extensive but generally featureless shadscale (*Atriplex confertifolia*) flats with their alkaline soils, and the pickleweed (*Allenrolfea occidentalis*) hummocks." However, he noted no association with a particular habitat, but indicated that the need for the species to sandbathe might be responsible for its being most common on dunes of fine sand where the vegetation was composed of saltbush (*Atriplex canescens*), rabbit-brush, buckwheat (*Eriogonum dubium*), horsebrush (*Tetradymia*), greasewood, Indian rice grass (*Oryzopsis hymenoides*), and other grasses and forbs. Somewhat in contradiction, Gennaro (1968) indicated that *O. leucogaster* in the Rio Grande Valley, New Mexico, occupied shrub-grassland areas with compact soils.

We know of no estimates of density of the northern grasshopper mouse in Oregon or elsewhere. However, in most comparative studies, population density in relation to that of many other syntopic species of small mammals seems to be low (McCarty, 1978). For example, Egoscue (1960) reported catching two *O. leucogaster* and 36 specimens of four other species in 50 traps set for 5 days. In Morrow Co., we caught two *O. leucogaster* in 2,984 trap-nights while capturing 297 *Perognathus parvus* and seven *Dipodomys ordii* (Verts and Carraway, 1986). In the same region, Green (1983) caught four during trapping that produced 425 individuals of three other species. In Harney Co., Feldhamer (1979*b*) captured seven *O. leucogaster* in 26,460 trap-nights that produced 1,573 other mammals of 14 species. Thus, we are skeptical of the suggestion that populations may become sufficiently dense to "have a decided influence on the abundance of animal life that furnishes their food" (Bailey and Sperry, 1929:3).

Diet.—*Onychomys leucogaster* is largely carnivorous or insectivorous. Bailey and Sperry (1929) indicated that foods of animal origin composed 88.9% of the materials in stomachs of 96 specimens collected in 13 states and in every month of the year. Grasshoppers and crickets (Orthoptera) were found in 86.7% of the animals collected in June–October and composed 38.8% of the food eaten. Beetles (Coleoptera) composed 20.7% of the diet, moths and their larvae and pupae (Lepidoptera) 17.0%, small mammals 3.1%, and miscellaneous arthropods 3.3%. Foods of vegetable origin consisted largely of cultivated grains, particularly wheat and barley. Other grasses included "foxtail and feathergrass" and forbs included "climbing buckwheat, lamb's-quarters, geranium, and spurge" (Bailey and Sperry, 1929:17). In captivity, a wide variety of invertebrates and vertebrates were consumed, but several invertebrates such as millipedes (Myriapoda), blister beetles (*Epicauta*), ants (Formicidae:Hymenoptera), slugs (Limacidae:Gastropoda), and some of the woolly caterpillars (Lepidoptera) with defensive mechanisms were rejected. However, *O. leucogaster* relished scorpions (Scorpiones) and darkling beetles (Tenebrionidae), both of which have powerful defensive weapons. When given a choice of tenebrionids with and without defensive secretions and crickets (*Acheta domesticus*), *O. leucogaster* selected prey in conformance with optimal-foraging theory; tenebrionids with defensive secretions provided the least energy per unit of handling time for both juveniles and adults (Slobodchikoff et al., 1987). For juveniles, crickets provided more energy per unit of handling time than tenebrionids without defensive secretions, but for adults, crickets and tenebrionids without defensive secretions provided equal energy per unit of handling time (Slobodchikoff et al., 1987). Bailey and Sperry (1929) also remarked that captive grasshopper mice ate beef, lamb, and poultry; a family group that we maintained in captivity seemed particularly excited when we provided a few morsels of leftover turkey that we roasted for Thanksgiving.

In captivity, *O. leucogaster* quickly kills, then consumes, other small rodents, even those larger than itself (Bailey and Sperry, 1929; Egoscue, 1960; Ruffer, 1968). Commonly, a small amount of skin and bone is all that remains of prey. Egoscue (1960) reported capturing specimens with bloodstains on their throats and chests, suggesting that wild *O. leucogaster* preys on vertebrates.

Captives allowed to feed by selecting from an array of semipurified and purified nutrients thrived as well as those fed commercial laboratory chow. From this, Harriman (1973) concluded that appetites reflected physiological needs of *O. leucogaster* and were not influenced by palatability.

Reproduction, Ontogeny, and Mortality.— *Onychomys leucogaster* readily breeds in captivity; this, combined with the seemingly low density of its populations, likely is responsible for most information related to reproduction in the species being derived from captive colonies. In addition to the major studies (Table 11-17), R. Svihla (1936) reported an average of 3.1 young per litter for seven litters born in captivity and McCarty (1978) referenced reports of average litter sizes of 3.2 and 3.8.

In captivity, *O. leucogaster* breeds throughout the year, but most litters are produced in spring and summer (Egoscue, 1960; Pinter, 1970). Egoscue (1960) reported that 88.8% of 116 litters were born during February–August, whereas Pinter (1970) reported that only 65.3% of 153 litters were born during that period. None of the captive females maintained by Egoscue (1960) became sexually mature until ≅160 days of age, but Pinter (1970) reported that some females gave birth when ≅120 days old, although most were 150–180 days old when they produced their first litter. R. Svihla (1936) recorded a pregnancy in a female only 95 days old. Females in captivity produce as many as 10 (commonly three to six) litters per year (Egoscue, 1960) and average about half that number (Table 11-17). One wild-caught female produced 11 litters totalling 42 young during the first 27 months in captivity, but no captive-born female produced young after 19 months of age (Pinter, 1970). Pinter (1970) also reported that males as young as 4 months sired litters.

We found records of two pregnant wild females in Oregon; one, collected in May, contained five embryos and the other, collected in June, contained four. Bailey (1936) recorded one with four embryos from Oregon in August.

R. Svihla (1936) recorded gestation periods of 32 and 46 days for females producing first litters and periods of 33, 39, and 47 days for three lactating females. Egoscue (1960) reported ranges (*n* not reported) in gestation periods of 32–38 days and 29–32 days for lactating and nonlactating females, respectively. Horner (1968) recorded a 27-day gestation period. Obviously, the gestation period is variable, but factors that influence its length are unclear.

On 18–19 August 1982, we captured a male and female in Morrow Co. and maintained them in separate cages. On 28 August, we presented the pair to J. Mary Taylor, who immediately placed the two mice together in a large terrarium. The pair produced a litter on 24 September and during the following year produced 10 additional litters ($\overline{X}$ = 4.2/litter) with interbirth intervals averaging 36.1 days (range, 28–43 days). The female was euthanized on 17 September 1985 because of a large scapular tumor and posterior paralysis; during the 37 months in captivity the pair produced 83 young in 23 litters ($\overline{X}$ = 3.6/litter).

At birth, body mass of four *O. l. brevicaudus* neonates averaged 2.8 g; by 5 days they averaged 4.8 g, by 10 days 7.5 g, by 15 days 9.6 g, and by 23 days 13.1 g. Neonates are naked and pinkish and have folded pinnae at birth; they commence to show pigments in the skin in 4.5–7.0 days, the pinnae commence to unfold by day 3, and the pelage first becomes apparent at day 6 (Horner, 1968). The lower incisors commence to erupt by day 8; uppers are erupted by day 11. The digits on the forefeet were separated and the young washed their faces and ears by day 13; the first adult-type call was heard at this time, but none of the other individuals responded (Horner, 1968). The eyes opened by day 20 and the young continued to nurse through day 23, although they also ate solid food (Horner, 1968). Occurrence of developmental events seems to vary somewhat among the subspecies (Horner, 1968; Ruffer, 1965b).

Onychomys leucogaster is rarely identified as a food item of raptors and carnivores in areas where it occurs in Oregon. Maser et al. (1970) did not record it in 675 regurgitated pellets of three species of owls (Strigidae) in central Oregon. Green (1983) reported that combined with several other species of vertebrates it composed only 0.1% of the biomass of prey in 5,559 burrowing owl (*Athene cunicularis*) pellets from the Columbia Basin. Maser et al. (1971) did not identify *O. leucogaster* among the remains of 577 small mammals in 347 burrowing owl pellets from central Oregon. Toweill (1982) did not record it among food remains in 102 digestive tracts of eastern Oregon bobcats (*Lynx rufus*). Elsewhere, it also seems rare as a prey item (McCarty, 1978).

Habits.—Northern grasshopper mice are strongly nocturnal, with peak activity at the new moon and minimum activity at the full moon. Light rain produced an increase in activity, whereas heavy rain suppressed activity (Jahoda, 1970). *O. leucogaster* is not known to enter torpor (McCarty, 1978).

Bailey (1936) claimed never to have found a burrow that he could attribute to the digging abilities of *O. leucogaster;* thus he concluded that burrows used by the species had been appropriated from other species of small mammals. As noted by Ruffer (1965a), 5 years earlier Bailey (1931) indicated that these mice probably made burrows for themselves, but 2 years before that he indicated that they were rarely taken at burrows that they could have made themselves (Bailey and Sperry, 1929). Captive grasshopper mice provided with appropriate substrates dug four types of burrows: a nest burrow (the largest), retreat burrows (for escape), cache burrows (for storing excess food), and miscellaneous burrows (for defecation and for marking—Ruffer, 1965a). Nest burrows provided a more constant temperature and possibly a more stable humidity. Captives placed in enclosures following 3-day occupancy by heteromyids used the burrows constructed by the heteromyids the 1st night, but constructed and used their own burrow systems thereafter. Vaughan (1961) indicated that wild *O. leucogaster* in Colorado was known to use tunnels constructed by geomyids.

Onychomys leucogaster probably is best known for its predatory habits, especially those involving other species of small mammals. In both interspecific and intraspecific encounters, the aggressor pursues and repeatedly pounces upon the intended victim until it secures a hold with the forefeet and bites through the rear portion of its skull (Ruffer, 1968). If the victim assumes a defensive posture, the *Onychomys* bites at the tail and legs of the prey until it turns to flee, whereupon the *Onychomys* pounces on it and attempts to seize it. A conspecific could delay the skull bite by assuming a submissive posture, but ultimately the attacks resumed irrespective of whether or not a submissive posture was assumed (Ruffer, 1968).

Clark (1962) claimed that when northern grasshopper mice with no previous experience at fighting were paired with an individual from another species it "acted out" the characteristic attack, but did not kill the other animal. He further indicated that such a different-species pair could live together amicably, even share a nest. However, the *Onychomys* could be induced to attack the alien species by painful stimuli, changing the environment, vigorous activity by the alien, confinement of the two in a small space, or conditioning the *Onychomys* to fear. Clark (1962) believed that the attacks tended to be reinforced by success of previous attacks; however, he provided no quantification to support his assertions. Ruffer (1968) believed that wild *O. leucogaster* delineate (by scent-marking during sandbathing) and defend large territories and exclude all conspecifics therefrom, but when maintained in small cages in captivity, territoriality disappears and the individuals live amicably. Thus, he believed that Clark's (1962) observations regarding amicable associations might have been the result of small cage size.

Despite these complex and somewhat contradictory records of interspecific and intraspecific interactions, some believe that *O. leucogaster* in the wild associate in male-female pairs (Egoscue, 1960; Ruffer, 1965b). Egoscue (1960) recorded capture of six male-female pairs, each within an area of <0.8 ha. On the two occasions that we caught an adult *O. leucogaster* on a 1.8-ha trapping grid (Small and Verts, 1983; Verts and Carraway, 1986), we caught an adult of the opposite sex within 2 days on the same grid, but no more individuals during the several days remaining of a 15-day trapping period.

Ranges in monthly home-range areas calculated by the exclusive boundary-strip method on a grid with irregularly spaced traps were 1.54–1.94 ha for an adult male (2 months; nine and 16 captures), 3.24–4.62 ha for a subadult male (3 months; eight, eight, and seven captures), and 0.61–0.85 ha (2 months; six and six captures—Blair, 1943). These home-range areas are considerably larger than for most small mammals of similar size possibly because of the predatory nature of *O. leucogaster* (Blair, 1953; Ruffer, 1968).

Probably no aspect of the biology of *O. leucogaster* is more intriguing or more romanticized than the vocalizations it produces. Bailey (1931, 1936), Bailey and Sperry (1929), and R. Svihla (1936) claimed that one adult call was like the howl of a miniature wolf (*Canis lupus*), probably because it is sometimes produced while the caller stands upright on its hind legs and tail with nose pointed upward, eyes partly closed, and mouth wide open. Other postures are at least equally common during calling (Hafner and Hafner, 1979; Ruffer, 1966). Egoscue (1960) ignored the wolf analogy, and Hildebrand (1961) somewhat scoffingly described the differences between the calls of the northern grasshopper mouse and the wolf. Egoscue (1960), Hildebrand (1961), and Ruffer (1966), among others, described some of the various types of calls produced, the postures of calling individuals, and some of the circumstances and contexts in which the calls are produced. Finally, Hafner and Hafner (1979) described quantitatively the vocal repertoire of neonates and adults, and analyzed frequency characteristics for calls within and among individuals, geographic areas, and species of *Onychomys*. They recognized two calls produced by neonates (a series of four–six chirps per second at ≅8 kHz in response to disturbance and a 0.5-s-long pure tone at 7–9 kHz produced while nursing). Adults produce four calls: a single "chit" produced in agonistic contexts, a soft chirping produced by the male during copulation, a low-intensity 0.8-s-long pure tone at 8–9 kHz produced by both sexes when in contact with a mate, and a loud, piercing pure tone of 0.7–1.2-s duration at 9.5–13.5 kHz. The latter call, considered the "wolf call" by others, is produced by both sexes, although mature males in reproductive condition predominate among those producing this call; females that produced the call were in various reproductive conditions. Ruffer (1966) considered this call a means by which widely dispersed conspecifics locate each other; Hafner and Hafner (1979) agreed and suggested that the low frequency at which the call is produced enhances long-distance transmission of the sound. Such sounds can be detected by humans at distances as great as 100 m. Further, Hafner and Hafner (1979) found by discriminant analysis of six length and frequency characters of this call that the calls of each individual were unique, that calls produced by the two sexes could be distinguished, and that calls of *O. leucogaster* could be differentiated from those of its congener *O. torridus*. They suggested that the individual differences might be used in the application of a technique to remote-census populations of grasshopper mice.

Courtship in *O. leucogaster* is considered to be the most complex within the order Rodentia (Ewer, 1968). It involves a 10-step pattern (seven of which are essential to successful copulation) that includes circling, reciprocal anogenital sniffing, nasonasal "kissing," following, backward somersaulting, grooming, neck-nibbling, and finally mounting (Ruffer, 1965b). The sequence often is interrupted at several steps by the male sitting on its haunches and placing the nose at the base of the tail. Sometimes 3 h are required for completion of the sequence. In the final stages of courtship, the female lies on her side before assuming the more typical lordotic position. The male approaches from behind and places his forefeet on the female's shoulders. After intromission, the pair roll onto their sides before breaking the copulatory lock. McCarty (1978) recorded observations of others indicating that multiple copulatory locks enhanced transfer of spermatozoa and fertilization.

Remarks.—Once, when we were preparing to air-freight a group of live grasshopper mice to a colleague at another university, we kept a group of half-dozen or so individuals in a cage in our garage for a time. Each night as we drove into our garage following a 2–3-h stint in our offices, one of the mice rushed to the corner of the cage nearest to our automobile, stood on its hind legs, and howled. We were never certain if the howl was a greeting or a challenge.

Neotoma cinerea (Ord)
Bushy-tailed Woodrat

1815. *Mus cinereus* Ord, 2:292.
1855. *Neotoma occidentalis* Baird, 7:335.
1858. *Neotoma cinerea*: Baird, 8(pt. 1):499.
1891. *Neotoma cinerea occidentalis*: Merriam, 5:58.
1894. *Neotoma occidentalis fusca* True, p. 2 (preprint of Proc. USNM 17:354).
1897. [*Neotoma cinerea*] *fusca*: Trouessart, p. 544.
1903*b*. *Neotoma fuscus apicalis* Elliot, 3:160.
1940. *Neotoma cinerea alticola* Hooper, 42:409.
1940. *Neotoma cinerea pulla* Hooper, 42:411.

Description.—*Neotoma cinerea* (Plate XV) is a large rat-like mammal (Table 11-16); its somewhat distichous squirrel-like tail, unique among members of the genus, is gray above and whitish below and ≅75% as long as the head and body. The dorsum is buffy gray to dark brownish-black and the venter is white to buff depending on the geographic race. The sole of the foot from heel to posteriormost tubercle is completely furred. Males possess a 0.5–1.5- by 5.0-cm holocrine sebaceous gland on the midline of the venter that secretes a brownish, waxy, musky material that stains much of the venter brownish or yellowish (Escherich, 1981). Some females develop a similar, but smaller, gland while nursing.

The skull (Fig. 11-82) is heavy and angular; the braincase is relatively short. The sphenopalatine vacuities usually absent, but if present it is no more than a hairline crack. The occlusal surface of teeth of woodrats differs from that of teeth of other members of the subfamily in consisting of puddles of dentine, each nearly surrounded by a dam of enamel (Fig. 11-82). The teeth differ from those of arvicoline rodents in that each puddle extends the full width of the tooth. M1 has two reentrant angles on the lingual side. Males may be as much as 15% larger than females; coastal races exhibit less size dimorphism (Escherich, 1981).

Distribution.—*Neotoma cinerea* occurs from southeastern Yukon and southwestern Northwest Territories south to northern and eastern California, southern Nevada, northern Arizona, and northern New Mexico, then eastward to central North Dakota, western South Dakota and Nebraska, and central Colorado (Fig. 11-83; Hall, 1981). In Oregon, *N. cinerea* occurs statewide (Fig. 11-83).

Late Pleistocene (Rancholabrean) fossils of *N. cinerea* are known from numerous sites within the present geographic range of the species, although none is listed from Oregon (Kurtén and Anderson, 1980).

Geographic Variation.—Four of the 13 currently recognized subspecies occur in Oregon: the large, blackish-brown *N. c. fusca* with cinnamon chest and sides of venter in the northern three-fourths of the Coast Range and western Willamette Valley; the large, light buffy-gray *N. c. alticola* from east of the Cascade Range; the large, blackish-brown *N. c. occidentalis* with white venter in the northern three-fourths of the Cascade Range and the eastern Rogue River Valley; and the small *N. c. pulla* with brownish-cinnamon dorsal pelage along the southern portion of the Coast and Cascade ranges (Hall, 1981; Hooper, 1940).

Habitat and Density.—*Neotoma cinerea* occurs in a wide variety of habitats. Along the Oregon Coast, Maser et al. (1981*b*:190) indicated that the species could be found in "mature conifer (Douglas-fir variant), immature conifer, alder/salmonberry, lodgepole pine/rhododendron, and Sitka spruce/salal habitats"; they also indicated that it probably occurs in other habitats. In central Oregon, we trapped numerous specimens in western juniper (*Juniperus occidentalis*)-big sagebrush (*Artemisia tridentata*) scabland, an area described and pictured by McKay and Verts (1978*a*). In Lincoln Co., one of us (BJV) took a specimen in a trap set beneath a large log at the edge of an old-growth stand of Douglas-fir (*Pseudotsuga mensiezii*). We also took specimens along a rim in Harney Co. that supported large junipers and populations of *Peromyscus truei* and *P. maniculatus*.

Grayson and Livingston (1989) reported records of *N. cinerea* at elevations as high as 4,342 m and described occupied sites at lower elevations on White Mountain, California; occupied sites were in scattered bouldery outcrops with much of the intervening areas devoid of veg-

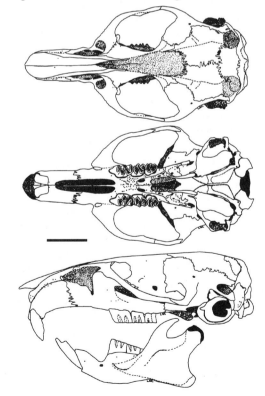

Fig. 11-82. Dorsal, ventral, and lateral views of the cranium, and lateral view of the dentary of the bushy-tailed woodrat, *Neotoma cinerea* (OSUFW 4204). Note absence of sphenopalatine vacuities in ventral view of cranium. Scale bar equals 10 mm.

etation. Sagebrush, currants (*Ribes cereum*), and scattered grasses characterized occupied sites, but all sites with such characteristics were not occupied.

We know of no estimates of density in any habitat in Oregon or elsewhere. Finley (1958) believed that population density was limited by the number of den shelters available as dens in Colorado were nearly always placed in vertical fissures in the faces of cliffs or in caves. Escherich (1981) indicated that the low population estimates derived from counts of urine marks were confirmed by the small number of woodrats caught in traps; nevertheless, he provided no estimates of population density.

Diet.—Bushy-tailed woodrats are largely, if not entirely, herbivorous. Finley (1958) considered foods of *N. cinerea* to consist largely of foliage rather than flowers, fruits, stems, or woody parts of plants; he also indicated that forbs were selected over grasses and that sagebrush tended to be avoided.

In Idaho, 54.3% of identified particles in feces were those of prickly pear (*Opuntia polyacantha*), 17.3% were Munro globemallow (*Sphaeralcea munroana*), 9.9% were tansy mustard (*Descurainia pinnata*), 9.3% were vetches (*Astragalus* and *Vicia*), and six other species each were represented by <2% of the particles. Also, 1.2% of the plant particles were not identified and 1.2% were those of arthropods (Johnson and Hansen, 1979*b*). In Colorado,

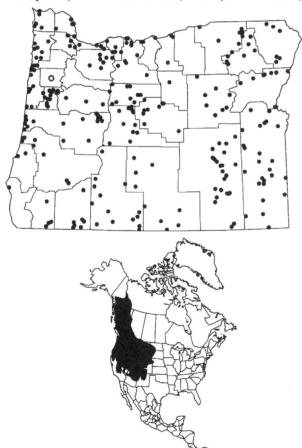

Fig. 11-83. Localities in Oregon from which museum specimens of the bushy-tailed woodrat, *Neotoma cinerea*, were examined (listed in Appendix, p. 534), and the distribution of the bushy-tailed woodrat in North America. (Range map redrawn after Hall, 1981:770, map 440.) Open symbol indicates record for county.

none of eight fecal samples contained arthropods or any other animal material (Haufler and Nagy, 1984); also missing were grasses, lichens, mosses, and seeds. Among forbs, milk-vetch (*Astragalus*), buckwheat (*Erigeron*), and globemallow (*Sphaeralcea*) were represented, and among woody plants, snowberry (*Symphoricarpos*), pine (*Pinus*), juniper (*Juniperus*), and bitter-brush (*Purshia tridentata*) were represented; 65.5% of the material identified consisted of the latter species (Haufler and Nagy, 1984).

Stomachs of seven of 10 *N. cinerea*, most of which were obtained in lodgepole pine (*Pinus contorta*) and grand fir (*Abies grandis*) forests of northeastern Oregon, contained fungi; both hypogeous and epigeous fungi were represented, but most were of the former type (Maser et al., 1978*a*).

Reproduction, Ontogeny, and Mortality.—Information regarding reproduction in free-living *N. cinerea* is meager at best, but Egoscue (1962*a*) described in detail reproduction and productivity in a captive colony established from Utah stock (Table 11-17). Except where noted, discussion of these characteristics is based on his findings.

Females are polyestrous and most that produce more than one litter per breeding season have a postpartum estrus 24–48 h after parturition and produce litters at intervals only slightly longer than the 27–32-day gestation period. Others that do not have an immediate postpartum estrus breed and produce young at intervals of about 2 months. Some captive females produced as many as seven litters in the February–September breeding season, but averaged only 3.0 (Egoscue, 1962*a*). Although a breeding season sufficiently long to produce several litters is suggested by our records of pregnant females having been collected in Oregon in April, May, and July, we seriously doubt that many individuals in the wild produce more than two litters annually.

Although onset of breeding is later among yearlings than among older individuals, the critical factor controlling onset of breeding is body mass (Hickling et al., 1991). Daily food consumption by pregnant females during the 2 weeks before parturition averaged 24% greater than in nonbreeding females; during lactation, food consumption averaged 93% greater. Body fat of females was greatest during pregnancy, but least during lactation (Hickling et al., 1991).

Litter size based on counts of pigmented sites of implantation for a small sample from Oregon averaged ≥0.5 offspring greater than reported for the species elsewhere, including that of a captive colony (Table 11-17). Although a few litters of five and six were recorded (seven of 34), most (26 of 34) ranged from two to four (Egoscue, 1962*a*). Therefore, counts of implantation sites of seven and eight that we recorded likely represent multiple litters. Thus, our data should not be interpreted to indicate that *N. cinerea* in Oregon is more fecund than elsewhere.

Body mass of 28 captive-born neonates of Utah stock averaged 13.5 g (range, 12.0–17.3 g); differences in mass among siblings ranged from ≤1 to 3.3 g (Egoscue, 1962*a*). At birth, the pinnae were folded in most, but not all young; some unfolded within 3 h and all had unfolded by 72 h. By 8 days of age, body mass ranged from 28.1 to 36.9 g. By 14–15 days, the eyes of most young were open and young attempted to consume solid food (Egoscue, 1962*a*). Hickling et al. (1991) indicated that nestlings in captivity commenced to eat solid food provided for the maternal female at 3 weeks of age. At 3 weeks, males averaged 93.2 g and females averaged 78.1 g; young were weaned and commenced to store foodstuffs at 26–30 days of age (Egoscue, 1962a).

In British Columbia, variation in size among individuals in a free-living litter from the 5th to the 26th day was remarkable; by the 15th day, "the largest young was about one third greater than the smallest" (Hováth, 1966:7). Development also differed among individuals; for example, new slightly brownish-gray hair appeared on the dorsum of the largest young on the 9th day of life; on the second largest on the 10th day, on the third largest on the 12th day, and on the smallest on the 15th day. When the maternal female flees, young, from shortly after birth, cling to a nipple and are dragged along (Hováth, 1966). Products of the ventral scent gland developed by maternal females during lactation presumably are used to mark nestlings so they can be located if they become detached from the female's nipples (Escherich, 1981).

In a captive colony, one or more young died before weaning age in two of four litters of five and in all three litters of six born in captivity (Egoscue, 1962a). Egoscue (1962a) claimed that mortality among litters of four or fewer young was less than among larger litters, but presented no supporting evidence. Although he did not speculate directly on a causal relationship, he noted that the greatest mortality was in litters in which the number of young exceeded the number (four) of nipples available for nursing and transporting young.

In the Coast Range, remains of N. cinerea occurred in <10% of feces of coyotes (Canis latrans) and bobcats (Lynx rufus) collected in summer-winter, but none was found in either species in spring (Witmer and deCalesta, 1986). In the Cascade Range, N. cinerea occurred in 11 (2.2%) of 494 fecal droppings of bobcats collected throughout the year (Toweill and Anthony, 1988a). Although Toweill and Anthony (1988b) reported that coyotes in the same area included 19 species of rodents in their diet, they listed only those species that occurred most frequently; however, they seemed not to have identified N. cinerea in the 844 fecal passages examined. In eastern Oregon, N. cinerea occurred in 21% of 117 digestive tracts of bobcats collected in winter (Toweill, 1982), and in the Cascade Range it occurred in two (5.9%) of 34 droppings collected throughout the year (Nussbaum and Maser, 1975). In central Oregon, remains of N. cinerea were found in 12 (1.4%) of 883 regurgitated pellets of great horned owls (Bubo virginianus), but none was found in 378 pellets regurgitated by short-eared owls (Asio flammeus) or 110 pellets of long-eared owls (Asio otus—Brodie and Maser, 1967; Maser et al., 1970). Only one of 1,112 food items identified in 438 barn owl (Tyto alba) pellets collected in Malheur Co. was N. cinerea (Maser et al., 1980). From these records, it seems that N. cinerea, although taken by several carnivorous mammals and raptorial birds, does not constitute a significant part of the diet of any of them. Such may be a reflection of relatively low numbers of these woodrats and, because of the type of shelters in which these woodrats nest, their relatively low availability to predators.

Habits.—Neotoma cinerea is active nocturnally. Sometimes, upon leaving the protection of its den, the woodrat may lie motionless on a rock or limb, watching for a long period before moving forth. N. cinerea runs with alternate movements of fore- and hind feet on opposite sides; when pressed it shifts to saltatorial locomotion with leaps as high as 0.5 m (Escherich, 1981).

We found no estimates of home-range size for any population. However, Escherich (1981) depicted minimum-area home ranges for several individuals based on radiotelemetry and live-capture locations. Most were long and narrow; the largest was ≅15 m wide and ≅135 m long.

Bushy-tailed woodrats often are known by the vernacular name "pack rats" because of their habit of moving almost any item small enough for them to carry. Sticks as large as 5.1–7.6 cm in diameter and 61.0 cm long can be carried 2 m up a vertical wall (Escherich, 1981). In a juniper-sagebrush community in central Oregon along the Deschutes River, N. cinerea commonly lives in juniper trees, particularly those that have been damaged by lightning. Sticks, antlers, bits of plastic, dried cow manure, stones, empty shotgun shells, tin-can lids, and a host of miscellaneous items are crammed into hollows and splits in the trees to form a barrier to potential predators (Plate XV). Bases of the occupied trees are nearly surrounded by middens consisting of several centimeters of amorphous fecal pellets. Commonly, well-worn paths lead away from the trees in several directions to disappear among the sagebrush. At times, the woodrats cut small boughs from junipers; many freshly cut boughs litter the ground. Whether these boughs are cut for food (or water) as reported for a congener (N. stephensi—Vaughan, 1982) is unknown. Similarly, whether the boughs simply are dropped accidentally while being handled or are discarded deliberately (possibly because of excessive terpenoid content—Verts et al., 1984) is unknown. In western Oregon, N. cinerea occasionally has been observed to construct houses high (15.2 m) in Douglas-fir and Sitka spruce (Picea sitchensis) trees (Maser, 1965, 1966a), but tree shelters west of the Cascade Range seem much less common than we observed in junipers in central Oregon.

In other areas, bushy-tailed woodrats fortify abandoned buildings, hollow logs, and crevices in rimrocks with piles of sticks and other items. Nests within shelters are constructed of grass, sagebrush bark, and other fibrous materials; nests usually are cup shaped and rarely are built with roofs (Eschrich, 1981; Finley, 1958). One maternal female enlarged holes in the ground and lined them with insulation material for her developing young; young were moved several times, seemingly in response to infestations of fleas (Siphonaptera—Hováth, 1966).

In northwestern California (Murray and Barnes, 1969) and in Klamath and Lake counties, Oregon (Hammer and Maser, 1973), N. cinerea and N. fuscipes are sympatric, and in some areas syntopic. In such situations, the former species usually occupies houses in rimrocks, whereas the latter nests in lodges in shrubs and junipers downslope from the base of the rim (Murray and Barnes, 1969). Hansen (1956) indicated that N. cinerea on Steens Mountain, Harney Co., was restricted nearly entirely to talus slopes and rimrocks; N. lepida in the region tended to occupy areas at lower elevations and often built nests in trees, sometimes at a distance from rimrocks.

Bushy-tailed woodrats on 20 30–350-m-long rock outcrops separated by 100–2,000 m in Alberta exhibited considerable fidelity for natal areas; of 37 yearling females, 28 (76%) bred on their natal rock outcrop whereas 10 (29%)

of 34 yearling males did so (Moses and Millar, 1992). Paired encounters of wild-caught bushy-tailed woodrats whose maternal relationships were known from unique radioisotopes acquired during intrauterine development and lactation revealed that interactions among adult females and unrelated individuals were strongly agonistic, whereas adult females and their offspring interacted amicably (Moses and Millar, 1992). Thus, close kin tend to maintain a stable, cohesive social structure despite the likelihood of strong competition for resources, particularly den sites. Nevertheless, the competition does not seem to suppress reproduction or delay maturity of offspring. Moses and Millar (1992) suggested that amicable relations between the maternal female and its yearling offspring may provide a means by which the female promotes access to critical resources by its offspring, thereby improving over-winter survival of offspring that remain in the natal area.

In captivity, bushy-tailed woodrats habituate to the presence of people poorly. Those handled from birth become recalcitrant at about 3–4 weeks of age and remain so for several months. However, these individuals ultimately become the most tractable (Escherich, 1981). Captives usually select elevated sites for construction of nests. They produce a variety of vocalizations, many of which are used in special situations, such as a "buzz" made during copulation (Escherich, 1981).

Neotoma cinerea, like other members of the genus in Oregon, is coprophagous. Fecal pellets are taken between the incisors from the anus. Those ingested are swallowed whole; those rejected are flipped into the midden with a quick motion of the head (Escherich, 1981).

Neotoma cinerea sometimes damages fruit trees by removing the bark (Dice, 1925).

Remarks.—Bailey (1936) proclaimed that *N. cinerea* usually deposited urine on the edges of certain protruding rocks along the faces of rimrocks, where, upon evaporation, it formed white encrustations. It is now understood that the woodrats, especially males, mark their territories with urine in addition to the products of the holocrine sebaceous gland on their venters. Food ingested by *N. cinerea* contains large amounts of calcium oxalate; microbial agents in the gut metabolize the oxalate and the calcium is absorbed, then excreted in the cloudy urine of the woodrat. The organic products of urine voided on rocks supposedly are leached by precipitation, leaving the white calcareous material; deposits in protected sites remain colored (Escherich, 1981). Although admitting that such deposits often have the appearance of geologic formations, Bailey (1936) proclaimed that they invariably constituted evidence of the presence of the species. Escherich (1981) further indicated that calcareous deposits containing fresh urine could be used to estimate the abundance of woodrats on a rimrock. We have misgivings about the validity of some of Bailey's (1936) and Escherich's (1981) assertions because we have found the white encrustations on rocky areas that exhibit no evidence of having been inhabited by woodrats. Even if the encrustations were formed by urine deposited by woodrats that inhabited the area in the far-distant past, these areas would be expected to include the requisite fissures and caves, which in turn should contain remnants of houses. Further, some of the desert areas occupied by *N. cinerea* in Oregon receive <20 cm of precipitation annually; thus, the leaching story seems an all-too-convenient explanation for the absence of organic materials in the calcareous encrustations. From odor and taste, we are satisfied that many of the encrustations have not been marked by woodrats recently. Consequently, we shall remain skeptical of the assertion that all white encrustations on rimrocks in eastern Oregon are evidence of occupancy (or past occupancy) by woodrats until analyses of deposits in areas exhibiting no evidence of having been inhabited by woodrats reveal materials that could have been produced only by woodrats. We do not mean to imply that we disbelieve that woodrats repeatedly urine-mark on the same substrate or that calcareous encrustations result therefrom. We just find it difficult to accept that all such encrustations were the work of woodrats.

Neotoma fuscipes Baird
Dusky-footed Woodrat

1858. *Neotoma fuscipes* Baird, 8(pt. 1):495.
1893. *Neotoma macrotis* Thomas, 12:234.
1895. *Neotoma splendens* True, 17:353.
1894a. *Neotoma monochroura* Rhoads, 28:67.

Description.—*Neotoma fuscipes* (Plate XV) is a medium-sized rat-like form (Table 11-16) with large, nearly naked ears, protruding eyes, and a long tail. The longer hind feet have five toes and six plantar tubercles, whereas the shorter forefeet have four toes and five plantar tubercles (Vestal, 1938). Among adults, sphenopalatine vacuities are present, the greatest length of skull usually is >42 mm, M1 has one reentrant angle on the lingual side, and the posterior reentrant angle on M3 is angled (Fig. 11-84); these are characters useful for separating *N. fuscipes* from Oregon congeners.

The dorsal pelage consists of hairs with steel gray bases, a band of ocherous buff, and a tip of black. Toward the venter, the black is omitted or a white tip is added to the gray, buff, and black sequence on progressively more hairs. Hairs on the throat and belly are white to their bases. Although variable in extent, a dusky splotch occurs on the dorsal surface of the white feet. The long dorsal vibrissae are black, but those originating more ventrally are shorter and gray or white.

Distribution.—*Neotoma fuscipes* occurs in western Oregon (except for the Coast Range), California west of the Sierra Nevada Range and southeastern desert (except none occurs in the Central Valley), and northern Baja California Norte (Fig. 11-85; Carraway and Verts, 1991c). In Oregon, *N. fuscipes* occurs from the California border northward along the coast to near Bandon, Coos Co.; northward inland through the Willamette Valley and other interior valleys to near Mollala, Clackamas Co., and Monmouth, Polk Co.; and northward to Brownsboro, Jackson Co., the Sprague River, Klamath Co., and Lake Abert, Lake Co., south of the Cascade Range (Fig. 11-85). Young (1962) reported a specimen taken at Burlington, Multnomah Co., but we found no specimen from that locality. Hammer and Maser (1973) thought that the extension of the range into Klamath and Lake counties was a relatively recent event. A specimen collected near Keno, Klamath Co., on 21 June 1966 was the earliest record that we could find for the species in that area.

Fossil remains of *N. fuscipes* were found at several sites in California, but none is reported from Oregon (Kurtén and Anderson, 1980).

Geographic Variation.—Two of the 11 currently recognized geographic races occur in Oregon: the cinnamon-buff *N. f. fuscipes* with bicolored tail occurs in southern Lake and Klamath counties and southeastern Jackson Co., and the dark cinnamon *N. f. monochroura* with unicolored tail is distributed throughout the remainder of the range of the species in Oregon (Hooper, 1938).

Habitat and Density.—*Neotoma fuscipes* occupies dense thickets and tangles of small trees, shrubs, vines, brambles, and forbs. In California, it reportedly is most numerous where tree cover is ≥90% and has not been disturbed by fire, and least abundant in open areas and in arid areas (Carraway and Verts, 1991*c*). The nearly complete overlap of the geographic distributions of *N. fuscipes* and poison oak (*Rhus diversiloba*) suggests a possible symbiotic relationship between the two species (Carraway and Verts, 1991*c*).

In western Oregon, big-leaf maple (*Acer macrophyllum*), Oregon white oak (*Quercus garryana*), Douglas-fir (*Pseudotsuga menziesii*), red willow (*Salix lasiandra*), poison oak, blackberry (*Rubus*), nettles (*Urtica*), and rose (*Rosa*) are common plants in habitats occupied by *N. fuscipes* (English, 1923). East of the Cascade Range, juniper (*Juniperus*) seems to be the requisite component of the habitat, but additional components include curlyleaf mountain-mahogany (*Cercocarpus ledifolius*), bitter-brush (*Purshia tridentata*), ponderosa pine (*Pinus ponderosa*), Saskatoon serviceberry (*Amelanchier alnifolia*), willow, buttercups (*Ranunculus*), and false Solomon's seal (*Smilacina*—Hammer and Maser, 1973).

Near Corvallis, Benton Co., densities for *N. fuscipes* of

1.5–3.9/ha and 2.7–5.2/ha were reported for 2.8-ha and 3.2-ha study areas, respectively (Hooven, 1959). Lowest densities were in December and peaks were in June and August on the two areas. In California, estimates ranged from 5.1 to 37.1/ha depending upon the openness of the habitat (Fitch, 1947). More recent estimates of average density in five seral stages of Douglas-fir–tanoak (*Lithocarpus densiflora*) forest in northwestern California ranged from 0/ha in small and large sawtimber to 0.41/ha in old-growth stands to 0.97/ha in seedling-shrub stands to 81.12/ha in sapling-brushy poletimber (Sakai and Noon, 1993). Others have reported densities of 39.5/ha (Wallen, 1982) and 45.2/ha (Chew et al., 1959).

Diet.—Most of the information regarding the diet of *N. fuscipes* is based on identification of cached food items (Carraway and Verts, 1991*c*). In the Willamette Valley, 37 species of plants were identified in caches in food chambers of houses occupied by *N. fuscipes*; as many as 16 species were found in a single house (English, 1923). Species occurring in ≥10 of the 63 houses were (in order of frequency of occurrence): apple, vine maple (*Acer circinatum*) seed, crabapple (*Pyrus fusca*), snowberry (*Symphoricarpos albus*), willow, rose, Douglas-fir, red currant (*Ribes sanguineum*), and Oregon white oak. In the same region, Hooven (1959) found madrone (*Arbutus menziesii*), blackberry, and common sword-fern (*Polystichum*

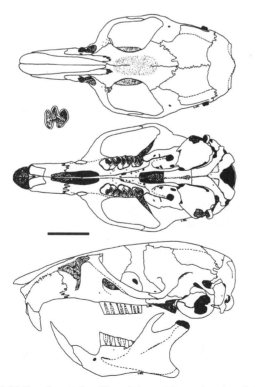

Fig. 11-84. Dorsal, ventral, and lateral views of the cranium; lateral view of the dentary; and occlusal view of M3 of the dusky-footed woodrat, *Neotoma fuscipes* (OSUFW 1566). Note open sphenopalatine vacuity in ventral view of cranium and angled posterior reentrant angle in M3. Scale bar equals 10 mm.

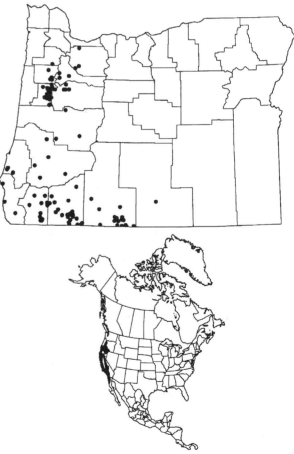

Fig. 11-85. Localities in Oregon from which museum specimens of the dusky-footed woodrat, *Neotoma fuscipes*, were examined (listed in APPENDIX, p. 535), and the distribution of the dusky-footed woodrat in North America. (Range map redrawn after Hall, 1981:767, map 439, with modifications.)

munitum) packed into houses, but when offered to caged individuals these items were not eaten. Captive *N. fuscipes* stripped bark from Douglas-fir boughs for nesting material, but did not eat it (Hooven, 1959). In Klamath and Lake counties, juniper serves as the principal item in the diet, but mountain-mahogany, bitter-brush, ponderosa pine, serviceberry, willow, buttercups, and false Solomon's seal also were eaten (Hammer and Maser, 1973).

The diet of individual dusky-footed woodrats often consists of only two–three plant species and commonly is dominated by a single species (Atsatt and Ingram, 1983; Hammer and Maser, 1973). Many of the plants on which *N. fuscipes* feeds are of low quality and contain relatively high levels of tannins and related phenolic compounds. Concentration of some of these compounds declines with storage when food items are cached in nests, but storage of some items reduces their nutritional value. The uniformity in diet may increase the efficiency of gut microflora, permitting *N. fuscipes* to ingest more food thereby maintaining nitrogen balance from high-fiber, low-nutrient foods (Atsatt and Ingram, 1983). Reingestion of fecal pellets also permits *N. fuscipes* to subsist on low-quality foods (Atsatt and Ingram, 1983).

Reproduction, Ontogeny, and Mortality.—An index to testicular volume (width squared times length) averaged 243 in October, 359 in November, 1,038 in December, and 5,070 in February. Thus, reproductive activity in males seems to commence in December in Oregon, corresponding with the reported onset in California (Vestal, 1938). We do not know when regression commences, and the only information concerning the breeding season in females in Oregon (Table 11-17) is that of English (1923). However, in California, the breeding season is considered to extend from February through September (Vestal, 1938), although pregnant individuals were collected in December (Gander, 1929). Despite the relatively long breeding season and small average litters (Table 11-17), Vestal (1938) reported that *N. fuscipes* produced only one litter annually. We examined three females from Benton and Polk counties collected 1 February, 6 February, and 14 March that contained both darkly and lightly pigmented implantation sites, but none of eight females collected before 15 March was pregnant, suggesting that the parous females produced two litters the previous year. Gander (1929) also speculated that more than one litter was born annually. The gestation period is 33 days (Wood, 1935).

Neonates weigh 12.5–13.8 g, with females averaging slightly heavier than males (Wood, 1935). They are dull reddish and naked except for vibrissae. The eyes open on days 11–16 (Wood, 1935). When the female parent is flushed, suckling young cling to the nipples even when she jumps; upper and lower incisors are divergent, thereby permitting suckling young to grasp the female's nipples viselike between the opposing "V's" (Vestal, 1938). Dislodged young usually squeak, whereupon the maternal female usually retrieves them within 2–4 h (Vestal, 1938).

In Oregon, *N. fuscipes* reportedly is preyed upon by bobcats (*Lynx rufus*—Nussbaum and Maser, 1975), great horned owls (*Bubo virginianus*—Maser and Brodie, 1966), and spotted owls (*Strix occidentalis*—Forsman et al., 1984). Elsewhere, a variety of mammalian, avian, and reptilian species have been reported to prey on this woodrat (Carraway and Verts, 1991c); we suspect that the same species prey on *N. fuscipes* in Oregon. Richard Hoyer, Corvallis, Benton Co., provided us with many specimens obtained by falconing.

Habits.—*Neotoma fuscipes* constructs houses of sticks and other debris (bolts, wire, bits of glass, bottles, bones, shotgun shells, cloth, metal, feathers, shovel handles, socks, mouse traps, surveyor's stakes, tin cans, soap, nails, and a variety of human foods including even a jellyroll—Carraway and Verts, 1991c) in trees or on the ground; commonly trees, faces of rocky bluffs, or fences are used as support for houses, although some houses are simply conical piles of material on the ground. Stick houses average ≅1.5 m in diameter and ≅1.2 m high (Vestal, 1938) and are perforated by openings that lead to large central chambers through a maze of passageways. Sleeping nests usually are near the outer walls and have no opening to the outside; a central chamber is used for food storage; and some houses have a latrine.

Commonly, adjacent stick houses are connected by paths between ground-level openings (Gander, 1929). *N. fuscipes* sometimes travels through mountain beaver (*Aplodontia rufa*) burrows (Pfeiffer, 1953). Houses often are the work of many generations of woodrats and may be occupied continuously for >13 years (Linsdale and Tevis, 1956). The average (±*SE*) period of occupancy by an individual was 5.0 ± 3.5 months in California (M'Closkey, 1972).

Neotoma fuscipes is more active at night, but may be abroad during daylight hours (English, 1923). Most activity is restricted to areas of dense cover (Cranford, 1977), although we occasionally saw them on macadam roads at night on the E. E. Wilson Wildlife Area near Corvallis, Benton Co., while we spotlighted cottontails (*Sylvilagus floridanus*). The spotlight caused the woodrats to make a frantic dash for cover.

In California, home ranges averaged 1,924 and 2,289 m² for adult females and males, respectively (Cranford, 1977). Overlap of home ranges was 25% for females, 15% for males, and 28% for members of the opposite sex except during the breeding season when overlap was 57%. Juveniles establishing independence commonly have home ranges that overlap those of maternal females by as much as 62% (Cranford, 1977).

When disturbed within houses or when flushed from houses, *N. fuscipes* may "tail-rattle"; the tail is vibrated rapidly against vegetation or debris to produce a characteristic clattering sound (Vestal, 1938). Upon capture, young may squeal, but adults usually are silent (Gander, 1929). However, adults may produce squeals, clicks, or grating sounds during intraspecific encounters (Vestal, 1938). Wild-caught dusky-footed woodrats often have torn pinnae and body scars, indicating aggressive interactions (Donat, 1933); in captivity, most animals become calm and tractable, but exhibit agonistic displays (Cameron, 1971) or fight viciously when placed together (Wood, 1935). In the wild, *N. fuscipes* usually is dominant in small-mammal communities of which it is a member (MacMillen, 1964).

Remarks.—Stick houses constructed by *N. fuscipes* constitute a critical element of ecosystems of which the species is a component; numerous other species of mammals, and a variety of species of amphibians, reptiles, insects, and other invertebrates are dependent upon houses constructed by this woodrat (Gander, 1929; Vestal, 1938; Walters and Roth, 1950; Wood, 1935). Fecal pellets produced by *N. fuscipes* also contribute significantly to the ecosystems; one female produced an average of 144 pellets/day, which, by expansion, was used to produce an estimate of 76.2 m³ annually for the woodrats occupying 516 houses on a 10-ha area (Vestal, 1938).

Dusky-footed woodrats may damage Douglas-fir saplings that grow in thickets in hardwood brush, but not in large pure stands or after trees attain a height of 7.5–9.1 m (Hooven, 1959). Nowhere is damage widespread.

During our research we were disappointed but, in view of the association of the species with poison oak in this region, not surprised by the paucity of studies conducted on the species in Oregon. Otherwise, this species seemingly would make a near ideal subject for studies of density, mortality, recruitment, relatedness, dispersal, and other population attributes. It is relatively easy to capture; it usually becomes calm and tractable in captivity; and it is sufficiently small to be handled easily, but sufficiently large to be fitted or implanted with radio transmitters.

Neotoma lepida Thomas
Desert Woodrat

1893. *Neotoma lepida* Thomas, 12:235.
1910. *Neotoma nevadensis* Taylor, 5:289.
1946. *Neotoma lepida nevadensis*: Hall, p. 530.

Description.—*Neotoma lepida* (Plate XV) is the smallest member of the genus that occurs in Oregon (Table 11-16). The tail is round, short (<140 mm), and bicolored; the vibrissae are long; and the ears are long and lightly furred. The pelage is long and soft, a mixture of buff and black dorsally and whitish or light buff ventrally; except for the chest, all body hairs have lead-colored bases. The feet and underside of the tail are white.

The skull of *N. lepida* (Fig. 11-86) is less rugose and angular than that of Oregon congeners. Among adults, sphenopalatine vacuities are present, greatest length of the skull usually is <42 mm, M1 has only one reentrant angle on the lingual side, and the posterior reentrant angle on M3 is straight.

Distribution.—*Neotoma lepida* occurs from southeastern Oregon and southwestern Idaho south through desert areas of Nevada, southern California, western Utah, and western Arizona, and northwestern Sonora to the tip of Baja California Sur; the range also extends along the Colorado, Green, and White rivers through eastern Utah into western Colorado (Fig. 11-87). In Oregon *N. lepida* occurs in Malheur, Harney, and southern Lake counties (Fig. 11-87).

Fossils of *N. lepida* are known from late Pleistocene (Rancholabrean) deposits in California, Nevada, and New Mexico; the latter site is extralimital (Kurtén and Anderson, 1980).

Geographic Variation.—Only one of the 30 subspecies currently recognized occurs in Oregon: the dark-colored *N. l. nevadensis* (Hall, 1946, 1981; Mascarello, 1978).

Habitat and Density.—Throughout its range, *N. lepida* tends to be a habitat generalist (Meserve, 1974), but in Oregon it usually occurs in sagebrush (*Artemisia*) habitats, especially those associated with rimrocks or other rocky outcrops. In Harney Co., Oregon, Feldhamer (1979*b*) captured two individuals on study plots in sagebrush, but none on plots in black greasewood (*Sarcobatus*

vermiculatus), marsh, or grassland. All those we have captured were associated with crevices in rimrocks or with talus at the base of rimrocks. In California, *N. lepida* used microhabitats that provided "shelter from predation or climatic factors such as moonlight, extreme temperatures or extreme humidities" for >95% of all activities; small bolders or talus that provided few refuges and open areas devoid of rocks were rarely used (Thompson, 1982*c*:574). Brown et al. (1972) concluded that the extent to which habitats could afford protection from predators determined their capacity to support desert woodrats.

On a 9.7-ha study plot in an area vegetated by juniper (*Juniperus osteosperma*) and big sagebrush (*Artemisia tridentata*) in Utah, estimated density of *N. lepida* was 4.4–7.7/ha (Stones and Hayward, 1968). Desert woodrats on this area build houses in and under juniper trees, thus likely are much more uniformly distributed than might be expected of woodrats that erect houses in crevices in rimrocks.

Diet.—Desert woodrats are herbivorous; they eat berries, fruits, seeds, leaves, and stems of desert plants. In captivity, pigweed (*Amaranthus*), saltbush (*Atriplex*), seablite (*Suaeda*), greasewood, smartweed (*Polygonum*), nettles (*Urtica*), dock (*Rumex*), and a variety of cultivated grains and vegetables were consumed; meat, both cooked and raw, was refused (Bailey, 1936). In California, their diet differed among various habitats (Cameron and Rainey, 1972); however, within habitats, *N. lepida* tends to specialize on relatively few food items, probably on the basis of their water content or of toxic compounds they may contain (Meserve, 1974). Thus, that many of the plants eaten in California have no close relatives in habitats occupied by *N. lepida* in Oregon should not be surprising. *N. lepida*, at least in some parts of its range, seems remarkably able to consume plants containing toxic compounds (Meserve, 1974).

Reproduction, Ontogeny, and Mortality.—Bailey

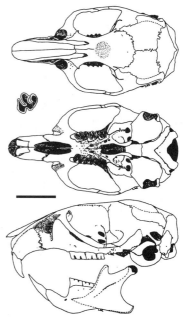

Fig. 11-86. Dorsal, ventral, and lateral views of the cranium; lateral view of the dentary; and occlusal view of M3 of the desert woodrat, *Neotoma lepida* (OSUFW 5621). Note sphenopalatine vacuities in ventral view of cranium and slightly curved posterior reentrant angle in M3. Scale bar equals 10 mm.

(1936) claimed that *N. lepida* gave birth from February to May, sometimes as late as July, and that two–four young were more common, but five had been recorded. These values compare favorably with those for wild-caught and captive-bred females (Cameron, 1973; Egoscue, 1957). We found few records of reproduction in *N. lepida* in Oregon; pregnant females were taken in April and June, each with four embryos (one embryo resorbing in April). A parous female taken in October exhibited six implantation sites.

In a captive colony, young were produced throughout the year; one pair produced as many as five litters and 14 offspring in 1 year (Egoscue, 1957). In the wild, females simultaneously pregnant and lactating suggest that at least two litters may be produced annually (Hall, 1946). Litter size in *N. lepida* averages smaller than that of other species of woodrats associated with more mesic habitats (Table 11-17); also, litter size of *N. lepida* in more arid habitats averages smaller ($\bar{X} = 2.1$, $n = 15$) than that of the species in mesic habitats ($\bar{X} = 4.5$, $n = 4$—Cameron, 1973). The gestation period is 30–36 days (Egoscue, 1957). Offspring as young as 60 days of age may become sexually mature in captive colonies (Egoscue, 1957).

For young born in captivity to wild-caught pregnant females, average mass at birth (6.3–10.7 g) and average growth rate (0.7–1.6 g/day) were related inversely to litter size; average age that young first ate solid food (13–27 days) and average age at weaning (16–42 days) were re-

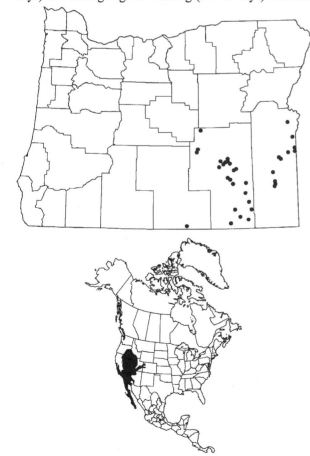

Fig. 11-87. Localities in Oregon from which museum specimens of the desert woodrat, *Neotoma lepida*, were examined (listed in APPENDIX, p. 536), and the distribution of the desert woodrat in North America. (Range map redrawn after Hall, 1981:759, map 435, with modifications.)

lated directly to litter size. Although survival of offspring in litters of one, two, and three was 100%, survival in litters of four was 75%, but only 50% for males and 34% for females in litters of five (Cameron, 1973). However, average mass at weaning (32.0–35.5 g) and litter size were not correlated significantly (Cameron, 1973). Finally, the only losses among maternal females occurred after weaning among those that produced litters of five. These several relationships combined with those related to litter size and aridity of habitats prompted Cameron (1973:492) to conclude that selection was "toward a smaller family size within the nutritional capabilities of the female" and that variation in litter size was "maintained in the gene pool by responses to long-term environmental heterogeneity."

Habits.—Desert woodrats are active nearly exclusively at night. Because of their low resting metabolism and their ability to maintain homeothermy only between 15 and 35°C, they avoid temperature extremes by foraging nocturnally and by resting in well-insulated nests (Nelson and Yousef, 1979).

In new environments, desert woodrats move about cautiously, but when pathways between feeding sites or den sites are established, travel becomes rapid. Activity of desert woodrats followed by radiotelemetry was predictable in that individuals used the same routes to travel between feeding sites and den sites located near the periphery of their home ranges. Consumption of food was in rock shelters that protected the woodrats from potential predators and probably had more favorable levels of temperature and humidity (Thompson, 1982c).

In southern California, home-range areas of desert woodrats estimated by the exclusive boundary-strip method (Stickel, 1954) averaged 371 m² for males ($n = 7$) and 433 m² for females ($n = 3$; Bleich and Schwartz, 1975). Although foraging areas were depicted (with a scale), Thompson (1982c:571, fig. 1) did not estimate home-range areas; nevertheless, he indicated that home ranges that incorporated a larger proportion of open habitats or those with few refuges were larger. In Utah, maximum movements recorded for a male and a female were 147.9 and 195.7 m, respectively; one male traveled 125.6 m in 24 h (Stones and Hayward, 1968).

Desert woodrats construct houses of sticks, dried cow dung, and a variety of miscellaneous items ranging from shed antlers, shotgun shells, and beer cans to bits of colored string, bottle caps, and plastic bags. Usually, houses serve to guard entrances to burrows appropriated from other small mammals or to fortify crevices in rocks; Bailey (1936) considered houses to rarely contain nests, but studies of desert woodrats conducted elsewhere indicate that certain types of houses often contain nests (Cameron and Rainey, 1972; Stones and Hayward, 1968). Houses that guard entrances to burrows usually consist of a bushel (35.2 l) or so of material, but some of those in crevices in rimrocks are composed of enormous quantities of material. In Harney Co. we caught specimens at the base of a rimrock in which a crevice nearly 1 m wide and tapering upward 12–15 m was packed to the top with material gathered by desert woodrats. We do not know how deeply into the crevice the material was packed, but to support the column, the base may have been 2–3 m or more thick. In captivity, individual woodrats selected different proportions of materials provided to construct their houses; for example, one chose 75.0% sticks, 12.5% cholla (*Opuntia*) joints, and 12.5% jar caps, whereas another chose 28.8% sticks, 41.4% cholla joints, and 29.8% jar caps (Bonaccorso and Brown, 1972). One individual added 359 units of these materials in only 1 night. Despite the disposition for transporting various items, Thompson (1982*c*), in his study of the use of space by desert woodrats, found no evidence of food caching. However, Cameron and Rainey (1972) and Stones and Hayward (1968) found cached materials in stick houses of *N. lepida*, but the former authors did not find caches in houses constructed largely of cholla joints.

During mating, male desert woodrats make a rasping vocalization to which females respond by assuming a receptive pose (lordosis); females temporarily deafened or paired with males that cannot vocalize do not respond as frequently (White and Fleming, 1987). Normal sexual behavior in male *N. lepida* requires that the female not only possess the attractive odors associated with estrus, but also engage in appetitive precopulatory behavior (Fleming et al., 1981). Copulation in *N. lepida* is unique among *Neotoma* in that intromission is of short duration ($\bar{X} = 4.9$ s; range, 2–8 s; $n = 42$ copulations) and no locking is involved. No intravaginal thrusting, ejaculation on a single insertion, and multiple ejaculations also were observed (Estep and Dewsbury, 1976).

In some regions of California (but not of Oregon—Hall, 1981), *N. lepida* is sympatric with *N. fuscipes*. Cameron (1971) considered *N. fuscipes* to be behaviorally dominant to *N. lepida*, but Meserve (1974), based on 12 laboratory encounters, considered the two species to be mutually intolerant. Cameron (1971) suggested that in some habitats *N. fuscipes* was able to prevent *N. lepida* from using some food plants that composed a large part of its diet in habitats where *N. fuscipes* was absent. Meserve (1974), on the basis of analysis of fecal droppings and food-selection trials in the laboratory, concluded that the diet of *N. lepida* was less diverse than that of *N. fuscipes*. However, he believed that *N. lepida*, by specializing on food species, sacrificed diversity of food resources for more dependable food plant species that had become widespread by evolving toxins and other mechanisms to deter predators. The mechanisms by which these deterrents are circumvented to permit *N. lepida* to use these species as food are unknown (Meserve, 1974).

Remarks.—The geographic ranges of *N. lepida* and *N. cinerea* overlap in southeastern Oregon (as well as in some other areas—Hall, 1981). We captured one species or the other in houses constructed in crevices in rimrocks, but never both species along the same rimrock. Because rimrocks at which *N. cinerea* was captured invariably supported some juniper trees, whereas junipers were absent on rimrocks at which *N. lepida* was captured, we suspect that moisture may be the ultimate factor determining which species can occupy a specific habitat. Nevertheless, the mechanisms by which one species excludes the other remain to be determined.

SUBFAMILY MURINAE—OLD WORLD RATS AND MICE

The Old World rats and mice are extremely diverse; the most recently published checklist includes 529 species, >11% of all the mammal species in the world (Musser and Carleton, 1993). Three of these species, all commensals of humans, came to North America with the early voyagers to this continent; all three are established in Oregon. These murine rodents are characterized by their relatively small size; scaly, nearly hairless tails; small infraorbital foramina; tuberculate occlusal surfaces on the molars with the tubercles arranged serially in three longitudinal rows; and long, narrow rostrums. All are pentadactyl, but the pollex is rudimentary.

These three introduced murid rodents are omnivorous; they eat grain and mixed feeds, stored succulent fruits and vegetables, eggs, and the carcasses of other animals, even their own kind. They are exceedingly destructive; they consume or foul with their excrement ≥10% of the food stored by humans annually and they cause enormous damage to packaging and warehouses. In addition, they are reservoirs of several serious zoonotic diseases; in the Middle Ages, as much as a third of the human population in Europe died of rat-borne plague. Nevertheless, laboratory strains of the same three species are invaluable in medical research and diagnosis.

The three Oregon murine rodents are exceptionally fecund. In barns, warehouses, and other situations where abundant foodstuffs are available, these species are capable of bringing forth litters during any month. Gestation is 19–22 days (slightly longer in lactating females), and female offspring of some species may attain sexual maturity and breed for the first time when 5–7 weeks old (Nowak, 1991). Litter size usually ranges from 3 to 12 (19 embryos have been recorded), but 5–6 is more common in house mice (*Mus musculus*) with 5–10 litters (maximum 14) produced annually under suitable conditions (Nowak, 1991). In rats (*Rattus*), litter size averages about eight (range, 1–11) with as many as 12 litters produced per year; sexual maturity is attained at ≅80 days of age (Nowak, 1991). Thus, these rats and mice have enormous potential to increase in abundance. Population density can become astronomical; 205,000 house mice/ha were reported at a site in California and 1,600 rats/ha were reported on a farm in Iowa (Nowak, 1991). Feral populations of house mice, as indicated by occasional records of the species in samples from small-mammal communities (Anderson, 1955; Black and Hooven, 1974), occur in Oregon, but such populations likely are temporary and limited to warmer and drier seasons.

Control of populations by use of toxic baits, drugs that cause internal hemorrhage, traps, introduction of predatory mammals, ultrasonic noise, and reproductive inhibitors has been attempted; none of the systems is entirely satisfactory. Probably the most successful method of controlling losses caused by these rodents is through exclusion with concrete footings and floors, and metal siding and roofing, in combination with removal of trash and garbage in a wide area. However, exclusion often is difficult as adult house mice are able to negotiate holes as small as 1 cm in diameter (Berry, 1981*b*) and rats often chew through wood and sometimes make inroads on concrete. Also, constant vigilance is required to prevent introduction of these commensals in packages and crates brought for storage in the rodent-proof structures. In trapping these rats and mice, baits that differ markedly from foodstuffs available usually are most effective; for example, in a granary, succulent fruit or peanut butter likely would be unusually attractive to these pests.

KEY TO THE MURINAE OF OREGON

1a. Length of head and body <100 mm; occipitonasal length <35 mm; length of M1 more than half the length of molar row; upper incisors notched on posterior surface..............***Mus musculus***, house mouse (p. 293).
1b. Length of head and body >100 mm; occipitonasal length >35 mm; length of M1 less than half the length of molar row; upper incisors not notched on posterior surface...2
2a. Tail more than length of head and body; temporal ridges not parallel; length of parietal along temporal ridges less than distance between ridges..............
.................................***Rattus rattus***, black rat (p. 291).
2b. Tail less than length of head and body; temporal ridges parallel; length of parietal along temporal ridges more than distance between ridges...................................
.......................***Rattus norvegicus***, Norway rat (p. 289).

Rattus norvegicus (Berkenhout)
Norway Rat
1769. [*Mus*] *norvegicus* Berkenhout, 1:5.
1932. *Rattus norvegicus* Cabrera, 57:264.

Description.—*Rattus norvegicus* (Plate XVI) is the largest member of the subfamily in Oregon (Table 11-19). This heavy-bodied rat has a scantily haired, scaly tail shorter than the length of the head and body. The ears are membranous and lightly furred. The pelage is coarse, a grizzled brownish or rusty grayish dorsally and dirty white to yellowish gray ventrally. Albino, melanistic, and spotted specimens are known to occur in free-living populations.

The skull (Fig. 11-88) is broad and heavy. Dorsally, prominent ridges extend from the interorbital constriction to the anterior edge of the supraoccipital; along the parietals these ridges are essentially parallel. The length of the parietals along the ridges is greater than the maximum distance between the ridges. The molars are tuberculate with tubercles arranged serially in three longitudinal rows. The longitudinal rows are most easily viewed by holding the ventral surface of the skull at an acute angle to the line of sight.

Distribution.—The Norway rat is distributed throughout the world in association with humans. For Oregon, museum specimens are on deposit from most of the counties west of the Cascade Range and from some counties along the Columbia River east of the Cascade Range (Fig. 11-89). Doubtlessly, this commensal species also occurs wherever humans have established permanent residence or industry.

The Norway rat is a native of China and Siberia; it spread to Europe as a commensal of humans in postglacial times (Kurtén, 1968). Banfield (1974) claimed that the species reached western Europe about 1700 AD, England about 30 years later, and North America during the Revolutionary War about 45 years still later. The first record of the species in Oregon was a specimen collected at Astoria by Lt. Trowbridge in 1855 (Baird, 1858). In northern parts of invaded territory in the New World, it displaced its congener, *Rattus rattus,* which earlier had dispersed similarly from southeastern Asia.

Geographic Variation.—*Rattus norvegicus* is monotypic (Lowery, 1974).

Habitat and Density.—*Rattus norvegicus* nearly always resides near human activity. In urban areas, rats occupy houses, warehouses, stores, sewers, garbage dumps, and any other place that provides adequate shelter and a nearby source of food. In rural areas, they reside in houses, barns, sheds, poultry coops, stables, granaries, silos, greenhouses, haystacks, woodpiles, refuse piles, and almost any other type of structure in which livestock feed or foodstuffs for human consumption are stored or are available nearby. In California, noncommensal populations of Norway rats were found to occupy riparian woodland near fallow agricul-

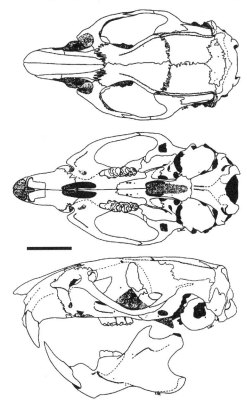

Fig. 11-88. Dorsal, ventral, and lateral views of the cranium, and lateral view of the dentary of the Norway rat, *Rattus norvegicus* (OSUFW 3519). Scale bar equals 10 mm.

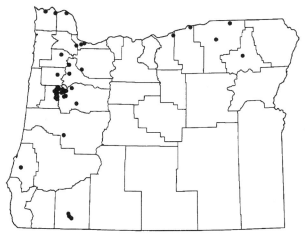

Fig. 11-89. Localities in Oregon from which museum specimens of the Norway rat, *Rattus norvegicus*, were examined (listed in APPENDIX, p. 548). The species occurs worldwide in association with humans.

Table 11-19.—*Means (±SE), ranges (in parentheses), and CVs of measurements of skull and skin[a] characters for female and male Old World mice (Mus) and rats (Rattus) from Oregon. Skin characters were recorded from specimen tags; skull characters were measured to the nearest 0.01 mm. SE and CV not provided if n <10.*

Species and region	Sex	n	Total length	Tail length	Hind foot length	Mass (g)	Greatest length of skull	Basal length	Length of maxillary toothrow	Zygomatic breadth	Breadth of braincase	Least interorbital breadth	Skull depth
Mus musculus	♀	9	156 (148–170)	76 (64–84)	17 (11–20)	17.18[b] (13.5–23.8)	21.38 (20.28–22.15)	18.31 (17.06–19.19)	3.66 (3.42–3.89)	11.08 (10.41–11.66)	9.78 (9.45–10.12)	3.65 (3.47–3.90)	7.67 (7.31–7.99)
	♂	19	155 ± 3.0 (131–182) 8.5	74 ± 1.6 (61–87) 9.3	18 ± 0.4 (12–20) 9.5	16.27 ± 0.77[c] (10.6–24.0) 19.50	21.36 ± 0.19[b] (19.53–22.64) 3.86	18.19 ± 0.19 (16.32–19.62) 4.67	3.58 ± 0.05 (3.20–4.06) 5.82	10.84 ± 0.12 (9.92–11.80) 4.94	9.64 ± 0.06 (9.20–10.02) 2.51	3.62 ± 0.03 (3.40–3.84) 3.68	7.58 ± 0.06 (7.25–8.09) 3.19
Rattus norvegicus	♀	7	405 (370–446)	185 (159–203)	41 (40–42)	277.23[b] (194.8–401.0)	46.82 (44.16–50.00)	42.08 (39.39–45.60)	7.78 (7.21–8.17)	22.93 (21.19–24.85)	16.91 (15.75–18.25)	6.66 (6.15–7.41)	14.31 (13.84–15.02)
	♂	14	405 ± 7.0 (358–445) 6.5	185 ± 4.2 (162–219) 8.6	42 ± 0.7 (38–46) 6.1	326.39 ± 23.43[b] (209.9–463.0) 25.89	46.83 ± 0.66 (43.00–50.29) 5.29	42.36 ± 0.59 (38.81–44.99) 5.25	7.79 ± 0.09 (7.30–8.36) 4.33	23.84 ± 0.48 (20.52–25.96) 7.51	17.34 ± 0.14 (16.39–18.31) 2.98	6.88 ± 0.07 (6.50–7.48) 3.81	14.67 ± 0.16 (13.56–15.69) 3.97
Rattus rattus	♀	1	358	188	37	119.8	39.32	35.04	7.22	18.12	16.95	5.92	13.58

[a]Although measures of ear length were recorded, the data were not included because of extreme variation believed to stem from differences in techniques used to measure these characters.
[b]Sample size reduced by 1. [c]Sample size reduced by 2.

tural land several kilometers from human habitation (Stroud, 1982). *Rattus norvegicus* tends to be a ground-level-dwelling species (Nowak, 1991) and to restrict its congener *R. rattus* to trees, roofs, or other elevated sites, but occasionally the two species occur together (Stroud, 1982). Frequently, *R. norvegicus* even lives partly below ground as it burrows beneath the floors of buildings, under piles of lumber or debris, or into ditch banks and dirt piles.

Diet.—Rats eat anything that humans or livestock eat plus some seemingly unpalatable items, such as soap, manure, hides, paper, and beeswax (Kingdon, 1974). Likely, the source of food for most rats in urban areas is garbage, and in rural areas, stored grain and livestock feed. It is difficult to list a digestible item not laced with some toxicant or chemical repellent that rats are likely to refuse to eat. Although they have specialized in omnivory (Barnett, 1975) and survive easily in captivity as herbivores, Norway rats function as carnivores opportunistically. They attack and feed on mice, other Norway rats, chickens, ducks, lambs, pigs, and sometimes even larger animals and humans; they are not particular about their prey being dead when they feed on it. Norway rats also are adept at catching fish (Nowak, 1991).

Reproduction, Ontogeny, and Mortality.—We know of no information on these topics derived from studies conducted in Oregon.

The reproductive potential of populations in an eastern urban area was compared with that of a population occupying a breeding farm for horses close-by. The city rats were significantly larger, attained reproductive competence at a smaller size, exhibited a greater incidence of pregnancy, and produced significantly larger litters ($\bar{X}$ = 10.1 city; $\bar{X}$ = 8.2 farm—Davis, 1951*b*). In another study conducted in the same urban area, rats in an increasing population had a higher prevalence of pregnancy than those in stationary or declining populations, but the number of young produced and weaned was approximately equal in all three types of populations (Davis, 1951*a*). Mean litter size was 10.34 in stationary populations (*n* = 86), 10.29 in increasing populations (*n* = 73), and 9.93 in declining populations (*n* = 45). Survival of weaned young in increasing populations was higher than in stationary or decreasing populations (Davis, 1951*a*). Davis (1953) reported about 400,000 Norway rats occupied the city of Baltimore, Maryland, in 1944, and, after extensive clean-up campaigns,

about 165,000 in 1947 and 65,000 in 1949. In 1944, females averaged 9.9 embryos in each of 4.7 litters (46.5 young annually), and in 1947–1949, 10.1 embryos in each of 4.9 litters (49.5 young annually).

In a western urban area, the prevalence of pregnancy among 12,962 females with total lengths ≥356 mm was 21.3% and the number of embryos per litter for the 2,768 pregnant females averaged 9.0 (Storer and Davis, 1953). Monthly prevalence of pregnancy exhibited minor peaks in spring (April–May) and in late summer–early autumn (September–October) in each of 3 years, but monthly average litter size did not vary greatly (*CV* = 6.5, *n* = 36 months). Among larger females, prevalence of pregnancy was greater, but litters did not average larger (Storer and Davis, 1953).

In the eastern urban area, growth rates of Norway rats were significantly greater during the cool season (November–April) than during the warm season (May–October), but the reverse was true in nearby parkland (Glass et al., 1988). In Hawaii, both sexes of offspring of wild-caught female Norway rats reared in the laboratory weighed ≅15 g at 1 week of age, but by the 2nd week, males outweighed females 28.4 g to 27.8 g. By 5 weeks, the disparity was 88.2 g to 76.8 g; by 10 weeks, 186.3 g to 123.0 g; by 15 weeks, 240.7 g to 150.6 g; and by 20 weeks, 301.2 g to 190.7 g (Hirata and Nass, 1974).

About 5% of rats that reach the weaning stage live 1 year (Davis, 1948). Of 100 rats alive on a farm in early spring 74 were alive in late spring, 60 in summer, 54 in late summer, 27 in early autumn, 15 in late autumn, and only 5 in winter (Davis, 1948). House cats likely are the most serious predator on Norway rats in both urban and rural areas, but the impact of their predation on density of rat populations is considered to be negligible.

Habits.—Much of the information regarding intraspecific social behavior of Norway rats is derived from studies conducted in enclosures. One such study was initiated with five rats of each sex placed in a 0.09-ha predator-proof enclosure containing nine sunken nest boxes in each corner, an internal fence partly separating each corner, and a central feeding area; food and water were available continously (Calhoun, 1963). During the 27-month study, the population increased from 10 to 170 rats. Rats in the enclosure segregated into social classes; rats in the dominant class were avoided by low-ranking rats and rats in the low-ranking class often were chased by those in the high-ranking class. Rats in the high-ranking class excluded other rats from one corner of the enclosure throughout the study; dominant males in this class established a territory around burrows of several high-ranking, reproductively active females. In this area, the frequency that subordinate males attempted to mate with females in estrus was much reduced, and reproductive and rearing success was enhanced. Low-ranking rats, especially males, migrated to another corner of the enclosure where males were strongly predominant, no territoriality was evident, numerous males chased estrous females and engaged in hundreds of copulations with each of them, females often did not construct separate burrows or nests before giving birth, and few young were raised. Reverse migrations did not occur; thus, as the population increased a larger proportion of the rats were in the low-ranking group, suggesting that a population-regulating mechanism involved sociality. In addition, growth is affected negatively by the frequency of social interactions as the number of rats with which a weanling is associated increases.

Even in free-living populations, Norway rats commonly restrict their movements to extremely small areas. In an eastern urban area and on a nearby farm, 89% of 325 rats and 99% of 757 rats, respectively, moved ≤18 m between successive captures (Davis, 1953). In a noncommensal population occupying a riparian woodland, average distances moved were 50.9 m for males ($n = 9$) and 43.0 m for females ($n = 8$; Stroud, 1982). Movements of rats are largely along a maze of trails constructed between burrows and objects or sites that provide some needed resource. Wherever possible, trails are constructed to take advantage of the cover provided by some vertical object. The trails are only wide enough for one individual; both individuals meeting on a trail usually step slightly to the side, allowing passage, but a juvenile meeting an adult is walked over by the adult (Calhoun, 1963). Nevertheless, adults are rarely overtly aggressive toward juveniles, although threats are common.

Of 36 burrows excavated in a poultry yard, 31 were ≤92 cm long and the longest was 198 cm; most were <30 cm below the surface of the ground. The burrows were 5–6 cm in diameter, had one or more exits in addition to the main entrance, and commonly included short blind branches, some of which were packed with debris. Some of the auxiliary exits were plugged with earth or were dug sufficiently close to the surface that a rat could push through easily. Each burrow system included one or two 12–15-cm-diameter dens floored with leaves and stems of grasses and one or two feeding stations containing a pile of debris (Pisano and Storer, 1948).

The spatial isolation provided by social structure is furthered by temporal isolation. Members of the socially dominant segment of the population tend to visit food stores twice each night, whereas those of low social rank feed before, after, and between the periods used by the bimodally feeding dominant individuals (Calhoun, 1963). In undisturbed situations, rats may be abroad during daylight hours (Pisano and Storer, 1948). Most rats cache food, especially lactating females and rats of low social rank; much of the stored food is not used (Calhoun, 1963).

Remarks.—Norway rats are extremely destructive. In addition to the human foodstuffs and livestock feed they consume or foul with their excrement, they are accomplished gnawers and frequently cut holes in floors, doors, and walls of buildings they occupy. They make direct attacks on about 14,000 humans annually in the United States (Nowak, 1991). Also, they frequently serve as reservoirs of several potentially pandemic zoonotic diseases. Their only redeeming attribute lies in the contribution made to science and medicine by specially inbred lines of the species (Lowery, 1974).

Rattus rattus (Linnaeus)
Black Rat

1758. [*Mus*] *rattus* Linnaeus, 1:61.
1803. *Mus alexandrinus* Geoffroy, p. 192.
1916. *Rattus rattus*: Hollister, 29:126.

Description.—The black rat is slightly smaller than the Norway rat (*Rattus norvegicus*), but much larger than the house mouse (*Mus musculus*), the only other members of the subfamily in Oregon (Table 11-19). The black rat is similar to other members of the subfamily in that it possesses a scaly, scantily haired tail; membranous, nearly naked ears; and a soft pelage. It differs from the Norway rat in having a tail much longer than the length of its head and body and commonly having a darker-colored pelage. The pelage ranges from black to sandy brown dorsally and from lead colored to nearly white ventrally.

The skull (Fig. 11-90) is long and slender, with somewhat pointed nasals. Prominent ridges extend from the interorbital constriction to the anterior edge of the supraoccipital. Along the lateral edges of the parietals the ridges are bowed apart. The maximum distance between the ridges is greater than the length of the ridges along the parietals. As in other members of the subfamily in Oregon, the molars are tuberculate with the tubercles arranged serially in three longitudinal rows. The longitudinal rows are most easily viewed by holding the ventral surface of the skull at an acute angle to the line of sight.

Distribution.—The black rat occurs nearly worldwide.

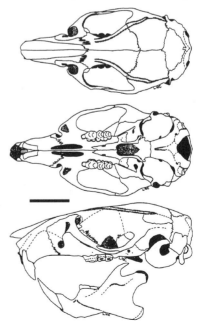

Fig. 11-90. Dorsal, ventral, and lateral views of the cranium, and lateral view of the dentary of the black rat, *Rattus rattus* (PSM 2699). Scale bar equals 10 mm.

In North America it occurs mostly in coastal and seaport cities in British Columbia southward, in urban areas of eastern and southern United States, and in most of Mexico and Central America. In Oregon, specimens from Portland, Salem, and several smaller cities along the Coast are on deposit in museum collections (Fig. 11-91). Pimentel (1949) reported taking specimens 11.2 km south of Florence, Lane Co., and on the northwest shore of Tahkenich Lake, Douglas Co.; we did not locate the specimens.

The black rat, a native of southeastern Asia (Kurtén, 1968), became a commensal of humans and spread throughout Eurasia. It is believed to have arrived in the New World with the "Spanish conquistadores in the early 1500s and to have come ashore at Jamestown with the English colonists in 1607" (Lowery, 1974:283).

Geographic Variation.—At least three subspecies were introduced into North America (Hall, 1981): the all dark-gray *R. r. rattus*, *R. r. alexandrinus* with brownish dorsum and dark-gray venter, and *R. r. frugivorus* with brownish dorsum and white venter. Results of cross-breeding of wild-caught *R. rattus* typifying these races caused Caslick (1956) to suggest that the "subspecies" were no more than color phases similar to those reported for red foxes (*Vulpes vulpes*), black bears (*Ursus americanus*), and eastern fox squirrels (*Sciurus niger*). However, Lowery (1974) claimed that he and other mammalogists were not convinced by the evidence presented. If, indeed, the colormorphs represent Old World subspecies, mixing since introduction makes recognition of individuals in North America as belonging to one or the other of the named subspecies a futile exercise.

Habitat and Density.—Black rats are restricted largely to port cities and coastal regions in Oregon, where they are associated with human enterprises. Wharves; piers; dockside warehouses; piles of crates, nets, traps, and debris; barns; houses; and waterfront stores all provide domiciles for black rats. Feral populations of the species also inhabit wet coastal forests in some areas (Banfield, 1974). In Oregon, Pimentel (1949:52) captured black rats "in dank undergrowth beneath lodgepole pines and . . . near standing water in a mixture of Sitka spruce and alder" in coastal Lane Co. and "in marsh vegetation and . . . in the willow border of the marsh" in Douglas Co. In Washington,

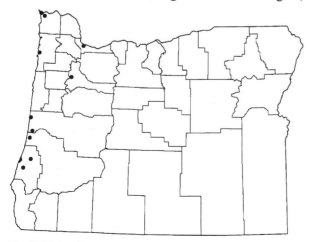

Fig. 11-91. Localities in Oregon from which museum specimens of the black rat, *Rattus rattus*, were examined (listed in APPENDIX, p. 548). The species occurs worldwide in association with humans.

Dalquest (1948:364) claimed that the species seemed to be "widely spread and to live in the wild" in the "coniferous forests on the west slope of the Cascades." In foothill areas in California, black rats are common in blackberry (*Rubus*) thickets (Jameson and Peeters, 1988).

Diet.—Like its congener, the Norway rat, the black rat consumes essentially anything edible and some items of doubtful savor. In association with humans, stored food, livestock feed, food-processing wastes, and garbage probably supply most of the materials ingested by black rats. Feral populations in British Columbia live "on seeds, nuts, insects, and small birds and mammals" (Banfield, 1974:221).

Reproduction, Ontogeny, and Mortality.—The length of the estrous cycle (day of disappearance to the following day of disappearance of leucocytes in vaginal smears) averaged ($\pm SE$) 4.5 ± 0.1 days ($n = 1,077$—Dewsberry et al., 1977). Gestation lasts 21–22 days in primiparous females and 23–29 days in lactating females (Ewer, 1971). There is a postpartum estrus within 24 h of parturition; one female is known to have given birth three times in a 54-day period (Ewer, 1971).

In a western urban area, prevalence of pregnancy was 26.9% among 3,633 females ≥381 mm in total length collected throughout the year. Although the monthly prevalence of pregnancy for 36 months in a 4-year period ranged from 12.1 to 38.3%, no seasonal peaks were discernible. Average litter size for the 978 pregnant females was 7.3 (Storer and Davis, 1953). Variation in monthly average litter size for black rats ($CV = 9.9$, $n = 36$ months) was somewhat greater than for Norway rats. The prevalence of pregnancy was slightly greater and the litter size somewhat smaller than in Norway rats, but the estimated average annual production of young by females was essentially the same, ≅40 offspring (Storer and Davis, 1953).

On Oahu, Hawaii, females and males attained sexual maturity at an average body mass of 76.7 g (range, 70.9–82.9 g) and 107.9 g (range, 101.2–115.0 g), respectively. Of 172 mature females, only 19 (11.0%) were pregnant, 10 for the first time; the average number of embryos per litter was 5.2 (Tamarin and Malecha, 1972). Females pregnant for the first time had significantly fewer (4.5) living embryos than those females that had produced one or more previous litters (5.7). The estimated average number of litters was 2.3 with an estimated annual production of offspring of only 11.8 (Tamarin and Malecha, 1972).

Differences in productivity of populations studied by Storer and Davis (1953) and Tamarin and Malecha (1972) indicate that black rats can exist in habitats with disparate resources. Tamarin and Malecha (1971) reported that survival in a free-living population of black rats was density dependent.

In Hawaii, male offspring of wild-caught female black rats reared in the laboratory to 1 week of age averaged weighing 11.1 g, 16.5% less (13.3 g) than female offspring. Average body mass of males continued to be less than that of females until the 9th week of age. At 5 weeks the disparity was 55.9 g to 62.2 g in favor of females; by 10 weeks, 123.0 g to 115.2 g in favor of males; by 15 weeks, 157.1 g to 130.8 g; and by 20 weeks, 184.8 g to 150.2 g (Hirata and Nass, 1974).

Neonates have darkly pigmented heads at 2 days of age, are fully pigmented at 5 days, have fine fur on the head and shoulders at 7 days, and are fully furred with fine hair at 8 days; the eyes open at 10–16 days. The young first nibble solid food at 15–18 days and commence to explore outside the nest at 17–23 days (Ewer, 1971). At birth, the tail is shorter than the length of the head and body, but after an extremely rapid growth period it is longer than the head and body by the time the young leave the nest. By the 2nd week of life, young retreat from a precipitous edge and exhibit the clinging response, both adaptations to avoid falls from the nest (Ewer, 1971).

In cooler regions of North America, R. rattus was replaced by R. norvegicus after it arrived nearly 2 centuries later. In warmer regions, the less aggressive R. rattus was able to survive through its greater ability to climb; it moved into attics, walls, and tile roofs of houses, and even into trees. In these areas, the vernacular name "roof rat" commonly is used. Where multilevel habitat was not available, R. rattus commonly was replaced by R. norvegicus (Ecke, 1954). Barnett and Spencer (1951) believed that R. norvegicus appropriates nest sites used by R. rattus, thereby evicting the latter species from a region.

Habits.—The social structure of a free-living population observed at a feeding site consisted of a dominant male, two or three high-ranking females subordinate to the top-ranking male, but dominant to others in the group, and sometimes a lower male hierarchy. Within-group attacks always are directed downward along the social scale. When attacked by a higher-ranking individual, subordinates either flee or attempt to appease the attacker by mouth-to-mouth contact at the angle of the jaw, holding the hindquarters low, extending the neck, and holding the ears back alongside the head. Appeasement permits high-ranking animals to maintain their social status without hurting subordinates (Ewer, 1971). In fights, individuals may jump vertically, box, kick, and bite; bites on the tail are common.

Black rats defend a "group" feeding territory against invasion. Territories are marked by rubbing the cheeks or venter on surfaces. Interlopers are chased by both sexes, but exclusion of females is largely by females within the group as males are inhibited from attacking females. Occasionally, an especially determined individual forces itself in and becomes a member of the group (Ewer, 1971).

Black rats exhibit several behavior patterns in response to specific stimuli (Ewer, 1971). Tooth-gnashing is a response to an unfamiliar stimulus, not in itself a threat; it also may be an alarm signal. Piloerection combined with extension of the hind legs is the usual threat pattern. Tail-vibrating may be used to thwart a mating attempt. Social grooming and sitting with bodies in contact are the most common amicable behaviors.

Black rats are agile climbers and accomplished wirewalkers; they can run along a 1.6-mm-diameter wire (Ewer, 1971). Individuals in feral populations build nests of "twigs and leaves in a log or stump, or in the crotch of a tree" (Banfield, 1974:221).

As in several other species of rodents (Bourlière, 1954; Layne, 1972), the tail of the black rat has structural specializations allowing loss of the distal portion of the sheath (tail autotomy), thereby permitting an animal grasped near the end of the tail to escape (Michener, 1976). Black rats grasped by the tail gyrate vigorously to cause the tail sheath to break; they also void urine and feces in an attempt to startle an attacker. The frequency of damage to tails differs among colormorphs, suggesting a selective advantage among morphs with the lowest proportion of tail autotomy (Michener, 1976).

Remarks.—Other than the note by Pimentel (1949) regarding the occurrence and habitats of two noncommensal populations, we know of no information available on R. rattus in Oregon.

Rattus rattus consumes stored food and grain, and fouls much that remains with its excrement. It can serve as a reservoir of several extremely contagious and serious zoonotic diseases. It, and the fleas that parasitize it, were responsible for the pandemic of plague that killed one-fourth to one-third of the human population of Europe in the Middle Ages (Langer, 1964). It also was involved in outbreaks of the same disease in India, Africa, and several major cities of the United States during the first 3 decades of the 20th century (Nowak, 1991). On Hawaiian and Midway islands, it is responsible for the extirpation or extinction of several species of birds (Tomich, 1986). Campbell (1968) recorded predation on eggs of ancient murrelets (Synthliboramphus antiquum) by black rats in the Queen Charlotte Islands, British Columbia. R. rattus is used less frequently in medical and scientific research than R. norvegicus; thus, it has even fewer redeeming qualities than its congener (Lowery, 1974).

Mus musculus Linnaeus, 1758
House Mouse
1758. [Mus] musculus Linnaeus, 1:62.

Description.—The house mouse (Plate XVI) is the smallest member of the subfamily in Oregon (Table 11-19) and smaller than most other members of the family. The ears are large and membranous; the tail is long, tapering, and scaly; and the soles of the feet are naked. The pelage is short, harsh, and usually grayish brown to nearly black dorsally, lighter brown or buffy ventrally. Albino, spotted, and other colormorphs are known to occur.

The skull (Fig. 11-92) is flattened with a short rostrum. The occipitonasal length is <35 mm, M1 contributes more than half the length of the upper molar row, and I1 is notched posteriorly.

Distribution.—House mice occur throughout the world in association with humans. In Oregon, specimens from 20 of the 36 counties are deposited in museum collections (Fig. 11-93), but we have little doubt that the species is ubiquitous in association with humans.

Geographic Variation.—Several of the many named forms likely were introduced into North America. We consider it impossible to designate subspecies in Oregon.

Habitat and Density.—Human dwellings, barns, sheds, warehouses, grocery and feed stores, and any other structures that supply shelter and contain a supply of items that can be appropriated for food are the primary habitats occupied by house mice. In some regions, populations are established in fields, but in Oregon, the paucity of reports of house mice as components of small-mammal communities leads us to believe that feral populations are not common here. Anderson (1955) reported catching one among 29 mammals captured in short grass and sagebrush (Artemisia) in Gilliam Co., and Black and Hooven (1974) reported two among 1,146 small mammals captured on three clear-cuttings in western Oregon. The disparity and

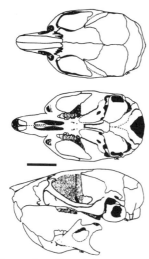

Fig. 11-92. Dorsal, ventral, and lateral views of the cranium, and lateral view of the dentary of the house mouse, *Mus musculus* (OSUFW 5109). Scale bar equals 5 mm.

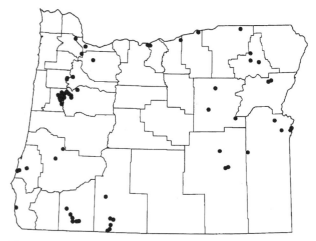

Fig. 11-93. Localities in Oregon from which museum specimens of the house mouse, *Mus musculus*, were examined (listed in APPENDIX, p. 529). The species occurs worldwide in association with humans.

commonplaceness of these habitats suggest that the infrequency of reports of capture of the species in the field is not related to habitat specificity. However, these studies of small-mammal communities and those listing *M. musculus* among species consumed by raptors (e.g., Maser and Brodie, 1966; Maser et al., 1980) usually do not include descriptions of the proximity of human structures from which the captured individuals possibly dispersed. Several specimens on deposit in museum collections were caught in meadows or dense grass; however, the proximity of structures was not recorded on specimen tags. We have captured only one house mouse in a field: a specimen taken in late summer within 100 m of human habitation. We suspect that most populations in open fields in Oregon are temporary, near human structures, and limited to warmer and drier seasons. However, feral populations were reported to occur in extremely wet and cold environments on sub-Antarctic islands (Berry et al., 1979), suggesting that these environmental factors may not be limiting. Conversely, Berry (1981b) claimed that the greatest hazard to feral house mice was the cold. Alternatively, the apparent failure of feral populations of house mice to become established and persist in Oregon may be because voles (*Microtus*) are interfering with their reproduction (Blaustein, 1980), recruitment (DeLong, 1966), or some combination of population processes (Lidicker, 1966) as *Microtus californicus* is known to do in California. We know of no estimates of density for either feral or commensal populations of house mice in Oregon.

The potential of house mice to increase is exemplified in an experimental population initiated with 10 males and 24 females in 16 nest boxes with interconnecting hardware-cloth runways. After an initial loss of two males and four females to fighting, the population increased to ≅400 individuals in 250 days with population growth following a classic sigmoid curve (Reimer and Petras, 1967). Free-living populations may attain astronomical densities; Hall (1927) reported that in California a population of 205,000/ha occurred on a dry lake bed. Hall (1927) also reported that >1,800 kg of house mice were killed in one granary in 1 day. Nevertheless, these numbers pale by comparison with those recorded in Australia, where house mice destroyed

in a relatively small area during a 4-month period were measured in hundreds of tons (Osborne, 1932). Such spectacular outbreaks of mice invariably are related to a greater-than-usual number of potential breeders surviving a usually unfavorable period, such as might occur during a particularly mild winter (Berry, 1981b).

Diet.—*Mus musculus* is truly an omnivore. It consumes essentially anything with caloric content with the potential of being digested. Bailey (1936:170) listed "grains, seeds, nuts, and fruits . . . meat, fat, butter, cheese, milk, cream, bread, cake, vegetables, and any cooked or uncooked food" among items consumed by house mice. F. Bronson (1979) described a colony of mice that lived in a frozen-meat locker at −10°C and consumed only frozen meat. Hall (1927) described feral mice cannibalizing individuals killed on a highway and themselves falling victim to traffic, then to other hungry individuals. Elton (1936:553) described a population in a 550-m-deep coal mine in England that subsisted on "feed of pit ponies and on crusts thrown away by the miners."

We were plagued by house mice entering our garage and establishing residence in the open bag of dog food until we bought a metal garbage can to protect our springers' dinners. Three snap traps baited with peanut butter and set a centimeter from the wall behind the furnace and hot-water tank rid the garage of our personal commensals.

Reproduction, Ontogeny, and Mortality.—*Mus musculus* can breed successfully at ambient temperatures from −20 to +32°C, a wider range than for any other mammal (Pelikán, 1981). In general, commensal populations residing within human structures commonly breed during all months, whereas in feral populations breeding may be seasonal. Litter size usually ranges from four to eight with a mode of six, but litters as large as 12–14 are known. Litter size usually peaks in summer; litter size and maternal body size usually are directly correlated (Pelikán, 1981). Gestation requires 18–21 days (Nowak, 1991).

For Oregon, we have records of litters of eight and 10 based on counts of embryos and of three, five, and seven based on counts of implantation sites; all females were taken in human structures in November–December.

The naked neonates become furred by 10 days of age and their eyes open by day 14. By day 21 the young are weaned and ready to disperse; by 35 days of age they are capable of breeding (Nowak, 1991). Maximum recorded longevity is 6 years for captives, but doubtlessly few attain 1 year of age in free-living populations.

House mice composed 2.0% of prey items in the diet of a pair of barn owls (*Tyto alba*) roosting and nesting in a cliff near Vale, Malheur Co. (Maser et al., 1980) and 0.5% of items eaten by two great horned owls (*Bubo virginianus*) near Monmouth, Polk Co. (Maser and Brodie, 1966). One of 13 opossums (*Didelphis virginiana*) collected in summer in the vicinity of Portland, Multnomah Co., contained remains of a house mouse in its stomach, but none of 51 collected at other seasons contained remains of the species (Hopkins and Forbes, 1980).

Habits.—House mice usually are more active at night, but within barns, warehouses, and other buildings with subdued light may be active at any time. Even when active, these mice usually spend most of their time behind or beneath some object and dash between the items that provide cover. Most of these rapid movements between cover-objects are along walls or some other vertical structure that provides partial cover.

House mice exhibit extreme flexibility in requirements for space. Where food resources are relatively sparse and interspecific competitive pressures light or nonexistent, house mice are able to move sufficiently for their populations to flourish, but not when faced with sparse resources and interspecific competition. For example, in California near San Francisco Bay, feral mice in a field occupied by an expanding population of voles (*Microtus californicus*) had an average home-range area of 122.3 m², but the mouse population disappeared 2½ months after the study commenced. In a nearby field where voles were rare, house mice had an average home-range area of 364.6 m² and the density of the population increased by one-third in 3 weeks (Quadagno, 1968). Where food resources are abundant house mice are able to compress their activities into an extremely small area. For example, for house mice occupying a corn crib in Ontario, the maximum distance moved was 3.0 m for adult males and 0.9 m for adult females (Reimer and Petras, 1968). In an urban building in Wisconsin, 92% of 934 recaptures of marked house mice were within 9.1 m of the previous capture site of the same individual; the average distance from previous captures was only 3.6 m (Young et al., 1950).

House mice, at least in experimental populations, form territories occupied by a dominant male, two–five females, and as many as three subordinate males (Crowcroft, 1955; Reimer and Petras, 1967). Females as well as males in these breeding units (demes) engage in territory defense; however, young females sometimes move between demes. Demes may be stable over several generations even though density may become exceedingly high. Mice in small, well-established demes may defend their territory from invasion by mice in an adjacent, but many times larger, deme (Reimer and Petras, 1967). Such leads to inbreeding and to random distribution of allele frequencies between neighboring colonies (Berry, 1981*b*).

When given a choice, individual house mice of both sexes respond to the odors of opposite-sex mice more frequently than to their own odor, and males respond to the odor of other males more frequently than to their own odor. Such suggests that odors play a role in identification of kin and that response by males to odors of other males may be related to latent aggressive behavior (Rowe, 1970). Territoriality in house mice is not pronounced in feral populations (Nowak, 1991).

Remarks.—The house mouse and humans have had a long association. The earliest record of the association of these mice with an urban community is at a neolithic archaeological site in Turkey dating about 6500–5650 BC. Thousands of years earlier, remains of another member of the genus (*M. †petteri*) were deposited at the early Pleistocene Olduvai hominid locality in Africa (Brothwell, 1981). The house mouse had invaded Europe and was well distributed on that continent by the time that early voyagers commenced to travel to the Far East and New World. Thus, beginning with these early travelers, humans became responsible for the worldwide distribution of the species. Although we know of no record of house mice taken in Oregon before a specimen taken at Elkhead, Douglas Co., on 5 May 1880 (USNM 13414/36945), we suspect that the species arrived here much earlier, as the Norway rat (*Rattus norvegicus*), another introduced commensal, was known to occur here as early as 1855 (Bailey, 1936; Baird, 1858).

Mus musculus has played a significant and ever-increasing role in science and medicine. The 16th–17th-century physician-anatomist William Harvey and the 18th-century clergyman-chemist Joseph Priestley used mice in their researches. Also, a mouse of unstated source was used in a laboratory experiment on the effects of air pressure by Robert Hooke in 1664 (Berry, 1981*b*). Mice were used in early genetics experiments, and indeed, Gregor Mendel may have used mice to develop his laws of segregation before he conducted his "ecclesiastically correct" research with peas (Berry, 1981*b*). From these timid beginnings, use of mice in research likely has followed a curve similar to a growth curve of an irrupting population of these mice. Nowak (1991) recorded that 30 years ago the National Institutes of Health in Bethesda, Maryland, used 800,000 mice per year. We cannot help but believe that use of mice in research and medicine in the interim has increased manyfold despite efforts by some to curtail use of vertebrate animals for such purposes. In addition to their contributions to science, historically house mice have played roles in religion, witchcraft, fortune-telling, and other aspects of human culture, including the town mouse–country mouse story by Horace in 65 AD (Berry, 1981*b*). They even are maintained as pets and for show (Berry, 1981*b*).

The house mouse is extremely variable morphologically, biochemically, and karyologically (Berry, 1981*b*). It is a mammal "weed," able to reproduce rapidly, live in a wide range of conditions, and adjust quickly to environmental changes (Berry, 1981*b*). Its success has been attributed to its genetic variation, ability to breed seasonally or nonseasonally and in a broad range of temperatures, independence of plant secondary compounds to initiate breeding, ability to adapt reproductive processes to local conditions, dispersal at high density, and the ability of young females to become pregnant shortly after dispersing (F. Bronson, 1979).

Likely from the long association with humans, more is known about *M. musculus* than about any other mammal except perhaps humans (Berry, 1981*a*). Despite the impossibility of producing a synoptic account of the species in a few pages because of the voluminous information available, knowledge of the species derived from within the state of Oregon is virtually nonexistent.

SUBFAMILY ARVICOLINAE—VOLES AND MUSKRATS

The arvicolines, consisting of 143 species in 26 genera, have a Holarctic distribution extending southward to Central America, northern Africa, northern India, and southwestern China. In Oregon, the subfamily is represented by 13 species of voles (two *Clethrionomys*, three *Phenacomys*, seven *Microtus*, and one *Lemmiscus*) and the muskrat (*Ondatra*).

Voles and the muskrat are characterized by prismatic molars with only the terminal prisms extending the full width of the teeth (Fig. 11-94), ears furred and partially hidden by the long hairs that originate in front of them, small beady eyes, and a shelflike protrusion of the squamosal, causing the braincase to taper abruptly to the postorbital constriction. Except for the muskrat, all are small (usually <150 g) and all have tails <65% of the length of the head and body.

Many species of voles in Oregon are associated with grasslands, but a few are associated with forest areas and one (possibly two) is arboreal; one of the latter lives almost exclusively in Douglas-fir (*Pseudotsuga menziesii*) and a few other species of coniferous trees. The muskrat is semiaquatic.

Reproduction in voles and the muskrat is highly variable and is influenced by a wide variety of environmental conditions. In addition to effects by environmental factors known to influence reproduction in many species of mammals (Sadleir, 1969), reproduction in voles seems to be sensitive to minute environmental cues. For example, onset of breeding in some species seems to be triggered by compounds in growing plants (Negus and Berger, 1977), but other compounds in plants may inhibit reproduction (Berger et al., 1977). Also, at least in the laboratory, removal of the male within 24 h of mating or subsequent substitution of another male usually results in abortion of young by the female (Berger and Negus, 1982; Stehn and Richmond, 1975), and familiarity as acquired by being reared together reduces fecundity (Boyd and Blaustein, 1985) regardless of genetic relatedness. These results suggest that social structure of populations influences reproduction in voles dramatically. Maternal age also affects productivity as older females commonly have larger litters (Foster, 1961; Gilston, 1976; Tyser, 1975).

Population density of some voles undergoes multiannual perturbations with peaks usually occurring at 3–4-year intervals. Peak densities of 10,000 voles/ha have been recorded for montane voles (*Microtus montanus*) in Oregon (Vertrees, 1959); between peaks, populations often become so low that the species seems to have been extirpated. Precise mechanisms responsible for these fluctuations in numbers have not been defined, although numerous hypotheses have been offered (Krebs et al., 1973; Lidicker, 1988).

Several species of voles whose populations exhibit multiannual perturbations undergo major changes in size depending on the phase of the cycle. In the increase phase, voles often are as much as 20% larger than those in the decrease phase. Consequently, we question the propriety of recognizing subspecies distinguished largely on the basis of size.

Despite some species of voles or different populations of the same species of vole sometimes exhibiting similar traits under similar conditions, we do not intend to imply that all species or populations of voles always behave similarly. Knowledge regarding the biology of the diverse species in Oregon is not equal; for example, the effects of familiarity and relatedness on fecundity, known for captive gray-tailed voles (*Microtus canicaudus*), may or may not apply to other species of voles or even to wild populations of gray-tailed voles.

KEY TO THE ARVICOLINAE OF OREGON

1a. Tail >125 mm long, laterally compressed with strong dorsal and ventral keels; hind foot >50 mm long, partially webbed, and fimbriated; basal length of skull >50 mm; length of upper molar row >10 mm; natatorial... ....***Ondatra zibethicus***, muskrat (p. 335).

1b. Tail <125 mm long, rounded, not keeled; hind foot <50 mm long, not webbed or fimbriated; basal length of skull <50 mm; length of upper molar row <10 mm; terrestrial, semifossorial, or arboreal............................2

2a. Pelage on dorsum light grayish-tan; tail not exceeding length of hind foot by >10 mm; auditory bullae inflated, extending beyond occiput when skull viewed from dorsal aspect.. ....***Lemmiscus curtatus***, sagebrush vole (p. 332).

2b. Pelage on dorsum dark grayish, brownish, or reddish; tail usually exceeding length of extended hind foot by >10 mm; auditory bullae not inflated, scarcely visible when skull is viewed from dorsal aspect.......3

3a. Three loops of enamel on lingual side of M3.................4

3b. Four loops of enamel on lingual side of M3.................7

4a. Eye tiny, 2–4 mm in diameter; molars not rooted and ever-growing in adults; reentrant angles on lingual and labial sides of lower molars about equal; (six plantar tubercles on hind feet).. ....***Microtus oregoni***, creeping vole (p. 325).

4b. Eye ≥4 mm in diameter; molars rooted and not ever-growing in adults; reentrant angles on lingual side of lower molars about twice as deep as those on labial side...5

5a. Tail <50 mm long, <50% of length of head and body; hind foot usually <18 mm long; incisors protrude slightly beyond nasals when skull viewed from dorsal aspect; incisors not strongly recurved; M3 and m3 long and narrow.. ....***Phenacomys intermedius***, heather vole (p. 307).

5b. Tail >50 mm long, >50% of length of head and body; hind foot usually >19 mm long; incisors not protruding beyond nasals when skull viewed from dorsal aspect; incisors strongly recurved; M3 and m3 short and wide..6

6a. Rich dark brown on dorsum; tail slender, scantily haired, and distinctly bicolored; incisive foramina narrow.. ....***Phenacomys albipes***, white-footed vole (p. 304).

6b. Reddish orange to cinnamon on dorsum; tail thick, well haired, and not strongly bicolored; incisive foramina wide.. .....***Phenacomys longicaudus***, red tree vole (p. 309).

7a. Pelage on dorsum with indistinctly outlined reddish median stripe; molars rooted and not ever-growing in adults; most loops of enamel on upper molars rounded...8

7b. Pelage on dorsum lacking reddish median stripe; molars unrooted and evergrowing; most loops of enamel

on upper molars sharply angled.................................9
8a. Posterior margin of palate with a spine..........
....*Clethrionomys*
californicus, western red-backed vole (p. 298).
8b. Posterior margin of palate lacking a spine...........
....*Clethrionomys*
gapperi, Gapper's red-backed vole (p. 303).
9a. Five plantar tubercles on hind feet (may be obscure in dried specimens); posterior one-half of incisive foramina abruptly constricted (posterior portion may be closed in older individuals), forming narrow and pointed slits; condylobasal length >35 mm; incisors protrude greatly beyond nasals when skull viewed in dorsal aspect..
..................*Microtus richardsoni*, water vole (p. 328).
9b. Six plantar tubercles on hind feet (may be obscure in dried specimens); incisive foramina may be constricted posteriorly, but do not form narrow and pointed slits; condylobasal length <35 mm; incisors may protrude beyond nasals, but not greatly so..................................10
10a. Incisors usually completely obscured when skull viewed from dorsal aspect; incisive foramina broadly rounded or tapering gradually posteriorly...............11
10b. Incisors usually barely protrude beyond nasals when skull viewed from dorsal aspect; incisive foramina abruptly constricted posteriorly..................................12
11a. Tail >50% of length of head and body; incisive

foramina widest in the anterior half...................................
.......*Microtus longicaudus*, long-tailed vole (p. 320).
11b. Tail <50% of length of head and body; incisive foramina widest in the posterior half.............................
.........*Microtus californicus*, California vole (p. 311).
12a. Tail nearly uniformly blackish; tail usually >33% of length of head and body; incisive foramina usually >6 mm long; (where sympatric with *Microtus canicaudus,* distinguishable by margin of palate being U-shaped)..
......*Microtus townsendii*, Townsend's vole (p. 330).
12b. Tail usually bicolored, brownish or blackish dorsally and grayish or whitish ventrally; tail usually <33% of length of head and body; incisive foramina usually <5 mm long..13
13a. Tail grayish, usually with sharp brownish stripe; (where sympatric with *Microtus townsendii*, distinguishable by margin of palate being V-shaped).................................
..........*Microtus canicaudus*, gray-tailed vole[1] (p. 317).
13b. Tail bicolored, but usually not sharply so..................
...............*Microtus montanus*, montane vole[1] (p. 322).

[1]*Microtus canicaudus* and *M. montanus* are karyotypically different (Modi, 1986), and apparently reproductively and geographically isolated (Tyser, 1975). However, external characteristics and skull morphology are so variable and overlap so frequently that separation of specimens of the two species without use of geographic localities of capture is difficult.

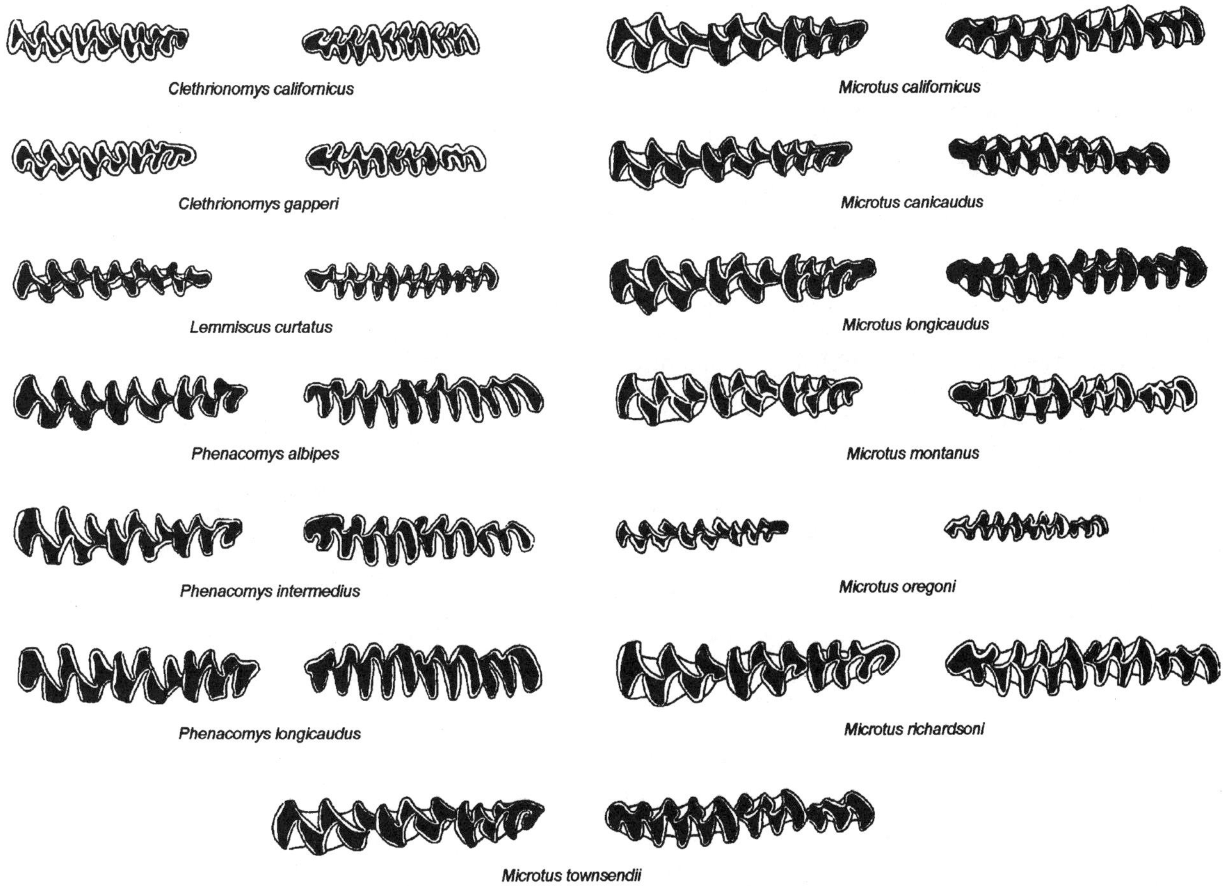

Fig. 11-94. Occlusal views of upper (left) and lower (right) molar toothrows of species of *Clethrionomys, Lemmiscus, Phenacomys,* and *Microtus* that occur in Oregon. Anterior is to the left, labial to the top. Specimens the same as those listed for skull plates.

Clethrionomys californicus (Merriam)
Western Red-backed Vole

1890e. *Evotomys californicus* Merriam, 4:26.
1897e. *Evotomys mazama* Merriam, 11:71.
1897e. *Evotomys obscurus* Merriam, 11:72.
1936. *Clethrionomys californicus mazama*: Bailey, 55:192.
1936. *Clethrionomys californicus obscurus*: Bailey, 55:192.

Description.—*Clethrionomys californicus* (Fig. 11-95) is among the smaller of the voles in Oregon (Table 11-20). It can be distinguished from sympatric arvicoline rodents as follows: from *Microtus oregoni* by eyes >4 mm in diameter,

Fig. 11-95. Photograph of the western red-backed vole, *Clethrionomys californicus*. (Reprinted from Alexander and Verts, 1992, with permission of the American Society of Mammalogists.)

four loops of enamel on the lingual side of M3, and rooted molars in adults; from other *Microtus* by molars rooted in adults and most loops of enamel on upper molars rounded (Fig. 11-94); and from *Phenacomys* by four loops of enamel on the lingual side of M3 and lingual reentrant angles on lower molars less than half the width of the teeth (Hall, 1981; Hall and Cockrum, 1953; Ingles, 1965). *C. californicus* can be distinguished from its only congener in the state, *C. gapperi*, by the presence of a thin, triangular shelf or median spine on the posterior margin of the palate (Fig. 11-96; Maser and Storm, 1970).

The pelage of *C. californicus* consists of a vaguely demarked reddish-brown or chestnut-brown stripe on the dorsum grading to buffy gray to dark gray on the sides and venter; the tail is indistinctly bicolored, light grayish-brown above and whitish below (Bailey, 1936; Maser and Storm, 1970; Merriam, 1890e). The median dorsal stripe is less distinct than in *C. gapperi*.

The tiny Y chromosome in male *C. californicus* is unique among the genus (Modi, 1985) and supports Johnson and Ostenson's (1959) separation of *C. californicus* and *C. gapperi* at the Columbia River.

Distribution.—*Clethrionomys californicus* occurs from the Columbia River south through the Coast Range to about 100 km N San Francisco Bay and through the Cascade and Sierra Nevada ranges to Quincy, Plumas Co., California (Fig. 11-97). It does not occur in the Willamette Valley or other interior valleys west of the Cascade Mountains (Fig. 11-97). There is no fossil record for the species (Alexander and Verts, 1992).

Geographic Variation.—Three geographic races are recognized; all occur in Oregon: *C. c. californicus*, the darkest-colored race, occurs in the Coast Range; *C. c. mazama*, the lightest-colored race, occurs in the higher elevations of the Cascade Range; and *C. c. obscurus*, of intermediate color, occurs in the lower elevations of the Cascade Range

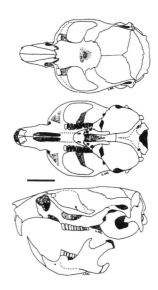

Fig. 11-96. Dorsal, ventral, and lateral views of the cranium, and lateral view of the dentary of the western red-backed vole, *Clethrionomys californicus* (OSUFW 7359). Scale bar equals 5 mm.

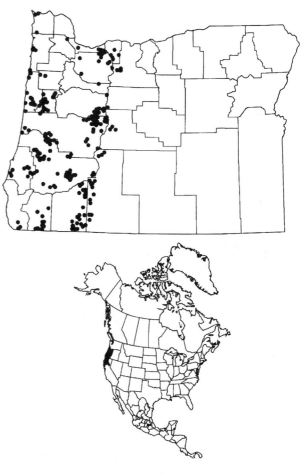

Fig. 11-97. Localities in Oregon from which museum specimens of the western red-backed vole, *Clethrionomys californicus*, were examined (listed in APPENDIX, p. 513), and the distribution of the western red-backed vole in North America. (Range map redrawn after Alexander and Verts, 1992:3, fig. 2, with modifications.)

(Hall, 1981). Maser and Maser (1988) recognized only *C. c. californicus* and *C. c. mazama*, but did not explain their failure to recognize *C. c. obscurus*.

Habitat and Density.—*Clethrionomys californicus* is a denizen of forest habitats, but tends to be most abundant in closed-canopy old-growth forests containing an abundance of fallen logs (Doyle, 1987; Gashwiler, 1959; Maser and Maser, 1988; Rosenberg et al., 1994; Tevis, 1956). In the Coast Range, *C. californicus* occurs in coniferous forests dominated by lodgepole pine (*Pinus contorta*)-Pacific rhododendron (*Rhododendron macrophyllum*), Sitka spruce (*Picea sitchensis*)-salal (*Gaultheria shallon*), and red cedar (*Thuja plicata*)-swamp communities (Maser et al., 1981b). The species also may occur in habitats in which the vegetation is composed of big-leaf maple (*Acer macrophyllum*), red alder (*Alnus rubra*), oval-leafed huckleberry (*Vaccinium ovalifolium*), braken-fern (*Pteridium aquilinum*), salmonberry (*Rubus spectabilis*), and various other shrubs and forbs (MacNab and Dirks, 1941). However, in the Cascade Range, abundance of *C. californicus* was related negatively to percentage cover by deciduous trees (Doyle, 1987). On the west slope of the Cascade Range, the greatest abundance of *C. californicus* was in near-climax Douglas-fir (*Pseudotsuga menziesii*)-dominated forests containing western hemlock (*Tsuga heterophylla*) and cedar with an understory of Cascade Oregongrape (*Berberis nervosa*), vine maple (*Acer circinatum*), various brambles (*Rubus*), rhododendron, salal, and various other shrubs and forbs (Gashwiler, 1959, 1970a). *C. californicus* also occupies areas vegetated by silver fir (*Abies amabilis*), white fir (*A. concolor*), grand fir (*A. grandis*), or ponderosa pine (*Pinus ponderosa*—Doyle, 1987; Hayes and Cross, 1987). Doyle (1987) captured almost 2.5 times as many *C. californicus* in upland habitats as in riparian zones and 3.9 times as many in 250-year-old forest stands as in 100-year-old stands.

Irrespective of the floral composition or its density, the presence (Doyle, 1987; Gashwiler, 1959; Tevis, 1956) and size (Hayes and Cross, 1987) of downed logs are key factors in habitats occupied by *C. californicus*; removal or burning of the logs usually renders the area unsuitable for the species (Gashwiler, 1959; Tevis, 1956). The size and "overhang" areas of logs are correlated positively with the frequency of their use by *C. californicus* (Hayes and Cross, 1987). However, there was some debate regarding the soundness of logs and the abundance of red-backed voles (Doyle, 1987; Hayes and Cross, 1987). Doyle (1987) attributed her negative relationship to the greater abundance of food (mycorrhizae) in rotting logs, whereas Hayes and Cross (1987) attributed their positive relationship to the use of sound logs by voles primarily for external travel lanes and rotting logs for foraging, nesting, or internal travel routes. More recently, Tallmon and Mills (1994) found that radio-tagged red-backed voles in southwestern Oregon in August–September were located significantly more frequently in logs in later stages of decay. The closed-canopy and downed-log requirements of *C. californicus* may be derived from the same physical features of the habitat: the reduced light and cooler temperatures provided by the shade of the overstory or of the logs may be the critical components of the environment (Gashwiler, 1970a; Maser et al., 1981b). To these, Tallmon and Mills (1994) added moisture in and under rotting logs as involved in

production of fungal sporocarps (truffles) required by the voles as food. Clarkson and Mills (1994) captured *C. californicus* within 15 m of 78% of 36 plots containing truffles, but at only 46% of 81 plots containing no truffles. They were essentially absent from recent clear-cuttings, but relatively abundant in forest remnants, especially under logs. In southwestern Oregon, density of *C. californicus* residing on 0.6–2.5-ha mature and old-growth forest remnants increased as both size of remnants and the distance from the edge of the remnants increased; the abundance of sporocarps of hypogeous fungi also was less near the edges of forest remnants (Mills, 1995).

As early seral communities (grass-forb-shrub stage) are replaced by young forest communities, abundance of *Microtus oregoni* decreases and abundance of *C. californicus* increases. Maser et al. (1978b) believed that the two vole species were able to coexist in intermediate stages because *M. oregoni* fed on the persistent grass-forb vegetation, whereas *C. californicus* fed largely on hypogeous fungi in the invading forest.

In eastern Linn and Lane counties, population densities estimated by use of the Lincoln Index on a 19.4-ha old-growth Douglas-fir forest area scheduled for clear-cutting ranged from 1.2 to 7.9/ha in 1954. In May 1955, the density was 4.9/ha; in August, 2 months after logging, 5.4/ha; in early October 12.4/ha; in late October after the slash was burned 4.7/ha; in November 7.4/ha; and in May 1956 zero (Gashwiler, 1959). Six additional samples from the logged and burned area in 1956 produced no red-backed voles, whereas the density in an old-growth stand used as a control was similar to the estimate before logging (Gashwiler, 1959). Black and Hooven (1974) did not capture *C. californicus* on a logged, but unburned portion of a 17,500-ha area devastated by wildfire for 7 years after the fire. Mills (1995) considered this vole "exceptionally rare" on clear-cuttings in southwestern Oregon; he captured three in 1,404 trap-nights on clear-cuttings, but 135 in 7,332 trap-nights on forest remnants.

Diet.—The diet of *C. californicus* consists largely of fungal sporocarps and lichens, more of the former at lower elevations and more of the latter at higher elevations (Hayes et al., 1986; Ure and Maser, 1982). Other food items are small amounts of insect larvae and seeds of conifers (Ure and Maser, 1982).

The most common fungal spores in feces of *C. californicus* in southern Oregon were those of *Rhizopogon* and *Gaultieria*; these taxa and *Hysterangium* and *Leucogaster* were present in feces nearly throughout the year (Hayes et al., 1986). Seven additional genera of fungal spores occurred for periods of 3–6 months at different times of the year and eight other genera occurred infrequently (Hayes et al., 1986). *Azospirillum*, a nitrogen-fixing bacterium, also has been detected in the feces of *C. californicus* (Li et al., 1986). Thus, because of its role in disseminating fungal spores and bacteria, *C. californicus* possibly is a critical component of forest ecosystems of which it is a part (Alexander and Verts, 1992).

Reproduction, Ontogeny, and Mortality.—On the west slope of the Cascade Mountains, some *C. californicus* are in breeding condition during February–November; all males are in breeding condition during March–August and most females are in breeding condition during April–August (Gashwiler, 1977). Young born in early litters

Table 11-20.—Means (±SE), ranges, and CVs of measurements of skull and skin* characters for female and male arvicoline rodents other than Microtus from regions in Oregon. Skin characters were recorded from specimen tags; skull characters were measured to the nearest 0.01 mm. SE and CV not provided if n <10.

Species and region	Sex	n	Total length	Tail length	Hind foot length	Mass (g)	Occipitonasal length	Basal length	Nasal length	Length of incisive foramen	Length of maxillary toothrow	Zygomatic breadth	Breadth of braincase	Least interorbital breadth	Skull depth
Clethrionomys californicus															
Coast Range	♀	30	151 ± 1.8 (128–168) 6.6	49 ± 0.8 (41–55) 8.9	19 ± 0.2[a] (16–21) 6.3	23.6 ± 0.94 (14.3–31.0) 18.15	23.93 ± 0.17 (22.13–27.01) 3.91	21.79 ± 0.15 (20.14–24.03) 3.69	7.09 ± 0.08 (6.38–8.36) 6.42	4.89 ± 0.05 (4.28–5.40) 5.85	5.19 ± 0.05 (4.71–5.81) 5.18	13.15 ± 0.09 (12.32–14.04) 3.99	10.76 ± 0.09 (9.73–11.88) 4.45	4.15 ± 0.03 (3.88–4.53) 3.86	9.33 ± 0.06 (8.82–9.86) 3.27
	♂	30	146 ± 1.9 (117–163) 6.9	49 ± 1.1 (28–63) 12.4	19 ± 0.2[a] (15–20) 7.1	22.6 ± 0.66 (15.4–31.0) 14.90	23.73 ± 0.13 (21.97–25.17) 3.10	21.54 ± 0.13 (19.70–22.94) 3.31	7.06 ± 0.06 (6.16–7.75) 5.00	4.88 ± 0.05 (4.16–5.36) 5.58	5.26 ± 0.04 (4.65–5.65) 3.95	13.07 ± 0.07 (12.37–13.96) 2.99	10.84 ± 0.09 (10.12–12.12) 4.42	4.15 ± 0.03 (3.87–4.57) 3.91	9.38 ± 0.05 (8.91–9.93) 2.78
West slope Cascade Range	♀	30	144 ± 1.6 (128–160) 6.2	46 ± 0.9 (34–54) 10.5	18 ± 0.2[a] (16–20) 5.7	20.9 ± 0.88[a] (15.6–27.0) 18.33	23.58 ± 0.14 (21.90–24.83) 3.09	21.40 ± 0.14 (19.74–22.73) 3.65	7.09 ± 0.09 (6.09–8.20) 7.53	4.89 ± 0.05 (4.18–5.45) 5.24	5.15 ± 0.04 (4.71–5.46) 3.61	12.98 ± 0.09 (11.60–13.77) 3.85	10.92 ± 0.08 (9.83–11.74) 3.86	4.14 ± 0.03 (3.73–4.42) 3.89	9.17 ± 0.05 (8.57–9.63) 3.12
	♂	30	148 ± 1.6 (135–167) 5.8	48 ± 0.8 (40–57) 9.1	18 ± 0.1[a] (16–20) 4.5	21.6 ± 0.91[a] (16.0–29.0) 18.32	23.56 ± 0.14 (22.13–24.93) 3.12	21.38 ± 0.15 (19.78–22.87) 3.72	7.06 ± 0.07 (6.23–8.07) 5.69	4.93 ± 0.05 (4.36–5.44) 5.88	5.11 ± 0.05 (4.16–5.62) 5.53	13.10 ± 0.09 (12.42–14.26) 3.76	10.86 ± 0.09 (9.68–11.93) 5.04	4.11 ± 0.04 (3.55–4.70) 5.46	9.14 ± 0.05 (8.60–9.63) 3.05
High Cascade Range	♀	21	152 ± 1.8 (139–169) 5.4	48 ± 0.9 (42–60) 9.2	18[c] (17–19)	23.8 ± 0.86 (20.8–25.4) 8.09	23.79 ± 0.11 (22.79–24.82) 2.21	21.78 ± 0.12 (20.91–22.82) 2.57	7.08 ± 0.07 (6.45–7.69) 4.43	5.05 ± 0.05 (4.44–5.37) 4.47	5.20 ± 0.05 (4.77–5.70) 4.43	13.37 ± 0.06 (12.85–13.84) 2.05	10.61 ± 0.11 (9.40–11.59) 4.65	4.06 ± 0.03 (3.74–4.38) 3.65	9.22 ± 0.06 (8.73–9.71) 2.93
	♂	25	152 ± 1.8 (137–174) 6.1	49 ± 1.0 (42–62) 10.5	18[a] (13–20)	22.9 ± 3.28 (11.8–30.9) 31.99	24.12 ± 0.11 (22.54–25.12) 2.38	22.09 ± 0.11 (20.36–22.95) 2.57	7.29 ± 0.07 (6.74–8.26) 4.92	5.05 ± 0.04 (4.45–5.33) 4.33	5.22 ± 0.04 (4.76–5.75) 4.26	13.49 ± 0.07 (12.89–14.26) 2.75	10.49 ± 0.07 (9.82–11.23) 3.47	4.09 ± 0.03 (3.88–4.43) 3.54	9.31 ± 0.07 (8.60–10.18) 3.54
Clethrionomys gapperi	♀	31	142 ± 1.3 (125–160) 5.2	43 ± 0.6 (35–48) 7.3	18[b] (16–20)	24.5 ± 1.56 (19.5–34.3) 19.19	23.74 ± 0.14 (22.41–25.64) 3.28	21.79 ± 0.12 (20.34–23.14) 3.17	7.04 ± 0.08 (6.33–8.28) 6.55	4.83 ± 0.04 (4.37–5.24) 4.85	5.45 ± 0.04 (4.82–5.80) 3.80	12.97 ± 0.06 (12.36–13.81) 2.71	10.35 ± 0.08 (9.67–11.42) 4.04	3.89 ± 0.02 (3.71–4.42) 3.38	9.17 ± 0.04 (8.86–9.71) 2.34
	♂	33	141 ± 1.1 (125–150) 4.6	42 ± 0.7 (29–50) 9.4	19[c] (17–20)	22.1 ± 1.34 (16.7–26.8) 17.19	23.89 ± 0.11 (22.55–24.92) 2.62	21.87 ± 0.11 (20.59–22.72) 2.98	7.09 ± 0.06 (6.23–7.64) 4.86	4.84 ± 0.03 (4.28–5.25) 4.05	5.43 ± 0.07 (4.88–7.18) 7.13	13.22 ± 0.08 (11.95–14.40) 3.88	10.59 ± 0.07 (9.90–11.58) 4.06	3.98 ± 0.03 (3.70–4.50) 4.57	9.28 ± 0.04 (8.69–9.77) 2.45
Lemmiscus curtatus	♀	30	125 ± 1.6 (105–141) 7.1	27 ± 0.7 (17–32) 13.8	16 ± 0.1 (15–17) 3.2	24.0 ± 1.63[j] (15.3–32.5) 23.54	23.13 ± 0.13 (21.53–24.09) 3.13	21.87 ± 0.15 (20.17–23.10) 3.65	6.50 ± 0.08 (5.76–7.46) 6.88	4.33 ± 0.06 (3.76–5.34) 8.07	5.95 ± 0.06 (5.30–6.53) 5.49	13.99 ± 0.09 (12.98–14.87) 3.44	10.58 ± 0.17 (9.11–14.94) 8.84	3.26 ± 0.03 (3.06–3.68) 4.58	8.74 ± 0.06 (8.06–9.56) 3.76
	♂	17	126 ± 2.1 (108–138) 7.0	26 ± 0.9 (19–31) 14.1	17 ± 0.1 (16–18) 3.5	23.7[k] (18.5–30.5)	23.67 ± 0.15 (22.62–25.03) 2.66	22.45 ± 0.18 (21.10–24.26) 3.31	6.68 ± 0.10 (6.03–7.67) 6.28	4.43 ± 0.09 (3.89–5.29) 8.06	5.85 ± 0.08 (5.26–6.75) 5.99	14.16 ± 0.16 (13.44–15.35) 4.59	10.56 ± 0.13 (9.56–11.57) 4.99	3.36 ± 0.03 (3.16–3.58) 3.59	8.81 ± 0.07 (8.27–9.32) 3.22
Ondatra zibethicus															
Northern Coast Range	♀	12	574 ± 7.3[a] (545–620) 4.2	272 ± 4.2[a] (249–292) 5.2	82 ± 1.2[a] (73–88) 4.9	1,212.6[j] (908–1,646)	62.69 ± 0.63 (59.84–67.14) 3.48	61.37 ± 0.74 (57.69–66.36) 4.17	21.79 ± 0.28 (20.68–23.57) 4.52	12.64 ± 0.22 (11.31–13.75) 6.10	16.07 ± 0.19 (14.86–17.07) 4.17	39.32 ± 0.52 (35.91–42.54) 4.62	20.94 ± 0.46 (17.96–22.69) 7.64	6.11 ± 0.14 (5.30–7.15) 8.04	21.32 ± 0.28 (19.85–22.61) 4.57
	♂	25	577 ± 6.6[m] (504–620) 5.5	273 ± 6.0[m] (210–321) 10.6	83 ± 0.9[a] (74–90) 5.4	1,126.3[a] (894.2–1,450)	62.78 ± 0.53 (58.81–68.05) 4.25	61.26 ± 0.54 (57.37–66.07) 4.41	21.49 ± 0.25 (19.40–24.44) 5.83	12.69 ± 0.16 (11.05–14.59) 6.17	15.83 ± 0.13 (14.79–17.28) 4.03	38.84 ± 0.36 (34.71–42.72) 4.60	21.45 ± 0.27 (18.22–22.73) 6.32	6.06 ± 0.06 (5.54–6.82) 5.24	21.82 ± 0.20 (20.17–24.07) 4.69

Locality	Sex	N	C1	C2	C3	C4	C5	C6	C7	C8	C9	C10	C11	C12	C13
Klamath and Lake counties	♀	6	526 (488–581)	241 (220–270)	78 (76–80)	1,042.3 (808.7–1,287.0)	58.44 (56.23–62.64)	57.39 (54.76–61.39)	19.45 (18.21–21.25)	12.46 (11.64–13.49)	15.09 (14.38–15.78)	36.84 (34.73–39.80)	21.66 (21.16–21.86)	6.14 (5.88–6.68)	21.22 (20.18–2.44)
	♂	5	560 (549–575) 5.4	256 (250–272)	77 (75–79)	1,153.8 (831–1,704)	61.88 (59.30–64.20)	60.97 (58.57–63.99)	20.99 (19.15–22.90)	12.76 (11.49–13.57)	15.25 (14.50–15.73)	39.65 (37.08–43.04)	22.57 (21.75–23.98)	6.11 (5.19–7.16)	22.29 (21.27–23.79)
Remainder of eastern Oregon	♀	25	532[j] (465–571) 5.7	249[j] (215–290) 8.7	78[j] (71–83) 6.7	1,173[j] (1,042–1,304) 14.77	59.39 ± 0.44 (55.88–63.07) 3.74	58.41 ± 0.47 (54.62–62.76) 4.05	20.57 ± 0.25 (18.56–22.93) 6.05	11.88 ± 0.13 (10.78–13.97) 5.39	15.09 ± 0.10 (14.29–16.35) 3.46	37.26 ± 0.37 (33.62–40.37) 5.02	18.82 ± 0.22 (17.47–21.93) 5.73	5.62 ± 0.06 (4.95–6.02) 5.43	20.06 ± 0.16 (18.89–21.65) 4.08
	♂	33	540 ± 5.8[k] (495–573) 3.9	245 ± 4.6[k] (210–270) 6.8	80 ± 0.9[k] (73–85) 3.9		60.54 ± 0.32 (56.41–64.74) 3.05	59.79 ± 0.36 (55.43–64.59) 3.43	21.25 ± 0.16 (19.54–23.67) 4.31	12.04 ± 0.11 (10.66–13.60) 5.28	15.30 ± 0.07 (14.66–16.36) 2.68	38.42 ± 0.25 (35.28–41.50) 3.68	18.48 ± 0.12 (16.70–19.64) 3.68	5.55 ± 0.09 (4.49–6.76) 8.87	20.54 ± 0.14 (19.11–22.68) 4.05
Phenacomys albipes	♀	17	168 ± 2.2 (155–191) 5.4	66 ± 1.7 (53–86) 10.6	19 ± 0.5 (14–23) 11.6	24.2 ± 0.59[m] (18.9–28.5) 9.24	25.30 ± 0.19 (24.17–27.07) 3.21	22.57 ± 0.16 (21.49–23.96) 2.98	7.66 ± 0.14 (6.99–9.39) 7.42	4.79 ± 0.05 (4.54–5.15) 4.21	6.51 ± 0.12 (4.84–7.15) 7.66	13.48 ± 0.08 (13.00–14.08) 2.37	10.23 ± 0.14 (9.50–11.45) 5.49	3.59 ± 0.07 (3.20–4.46) 7.49	8.89 ± 0.06 (8.46–9.39) 2.99
	♂	30	163 ± 1.7 (149–182) 5.7	64 ± 1.0 (52–75) 8.7	20 ± 0.2 (14–21) 6.7	21.7 ± 0.63[c] (15.3–28.0) 14.77	25.34 ± 0.13 (23.13–26.67) 2.89	22.79 ± 0.10 (21.52–23.73) 2.46	7.47 ± 0.09 (6.43–8.77) 6.31	4.71 ± 0.03 (4.29–5.00) 3.75	6.63 ± 0.04 (6.11–7.05) 3.27	13.46 ± 0.08 (12.58–14.28) 3.31	10.55 ± 0.09 (9.60–11.37) 4.58	3.61 ± 0.03 (3.32–3.96) 3.99	8.95 ± 0.04 (8.51–9.35) 2.61
Phenacomys intermedius Cascade Range	♀	13	133 ± 3.8 (104–149) 10.2	35 ± 1.5 (27–42) 15.3	18 ± 0.3 (15–19) 6.8	30.4[i] (24.0–37.2)	23.94 ± 0.37[e] (22.03–25.55) 5.42	22.54 ± 0.36 (20.09–24.35) 5.77	7.09 ± 0.17 (5.60–7.96) 8.61	4.46 ± 0.12 (3.91–5.28) 9.54	6.33 ± 0.08 (5.83–6.75) 4.56	13.78 ± 0.29 (12.37–15.28) 7.75	10.18 ± 0.21 (8.95–11.03) 7.42	3.66 ± 0.04 (3.44–3.84) 3.78	8.79 ± 0.09 (8.13–9.28) 3.94
	♂	10	135 ± 3.6 (112–146) 8.5	31 ± 1.5 (22–37) 15.3	18 ± 0.2 (17–19) 4.3	25.1[c] (21.6–30.3)	24.94[m] (23.10–26.35)	23.69 ± 0.50 (20.68–25.82) 6.68	7.52 ± 0.21 (5.80–8.08) 8.86	4.48 ± 0.13 (3.59–4.95) 9.22	6.37 ± 0.06 (6.15–6.71) 2.93	13.99 ± 0.24 (12.68–14.74) 5.48	10.39 ± 0.18 (9.28–11.11) 5.49	3.71 ± 0.03 (3.57–3.82) 2.39	9.15 ± 0.10 (8.56–9.46) 3.60
Wallowa and Blue Mountains	♀	4	143 (132–160)	34 (28–39)	18 (17–20)	26.4[m] (22.9–29.9)	24.85 (24.21–25.63)	22.96 (22.05–24.12)	7.30 (7.12–7.63)	4.37 (4.18–4.57)	6.50 (5.63–7.17)	14.79 (14.01–15.23)	9.99 (9.25–10.59)	3.81 (3.48–4.04)	8.86 (8.29–9.20)
	♂	7	149 (144–152)	38 (35–44)	18 (16–19)	32.9[a] (31.2–34.0)	25.96 (24.97–26.75)	24.43 (23.61–24.79)	8.19 (7.64–9.08)	4.66 (4.43–4.90)	6.67 (6.31–7.33)	15.40 (14.51–16.03)	10.48 (10.11–11.58)	3.81 (3.58–4.02)	9.02 (8.59–9.45)
Phenacomys longicaudus North of, and including, Vida, Lane County	♀	31	182 ± 2.3 (150–202) 7.1	78 ± 1.7 (58–94) 12.2	21 ± 0.3 (18–25) 6.5	29.9 ± 1.41[f] (20.0–44.0) 22.98	25.16 ± 0.18 (22.65–26.37) 3.94	22.81 ± 0.16 (20.21–24.51) 4.01	7.53 ± 0.11 (6.35–9.96) 8.43	4.66 ± 0.05 (3.96–5.18) 5.87	6.24 ± 0.08 (4.83–6.81) 6.69	14.20 ± 0.10 (12.98–15.24) 3.93	10.08 ± 0.09 (9.32–11.36) 4.74	3.50 ± 0.04 (2.94–3.92) 5.76	8.81 ± 0.04 (8.09–9.34) 2.83
	♂	44	171 ± 0.9 (162–181) 3.4	71 ± 0.7 (61–81) 6.6	21 ± 0.2 (18–24) 6.2	24.7 ± 1.10[a] (17.0–37.7) 20.43	24.27 ± 0.10 (23.02–26.02) 2.83	21.84 ± 0.10 (20.42–23.79) 3.16	6.95 ± 0.06 (6.22–7.92) 5.42	4.28 ± 0.04 (3.83–5.00) 6.24	6.15 ± 0.03 (5.79–6.56) 2.87	13.57 ± 0.06 (12.80–14.58) 2.93	9.77 ± 0.06 (8.72–10.56) 4.12	3.59 ± 0.02 (3.11–3.89) 4.56	8.66 ± 0.04 (8.09–9.25) 2.69
South of Vida, Lane County	♀	12	172 ± 4.8 (150–206) 9.6	74 ± 3.3 (60–97) 15.7	20 ± 0.6 (15–22) 9.9	22.3[j] (18.8–28.5)	24.44 ± 0.34 (22.86–26.87) 4.81	21.81 ± 0.34 (20.35–24.19) 5.37	7.15 ± 0.15 (6.53–8.27) 7.41	4.53 ± 0.07 (4.14–5.01) 5.38	6.23 ± 0.13 (5.73–7.09) 7.01	13.61 ± 0.23 (12.61–15.07) 5.88	9.86 ± 0.12 (9.48–10.57) 4.09	3.48 ± 0.04 (3.31–3.69) 3.94	8.69 ± 0.08 (8.29–9.26) 3.00
	♂	29	171 ± 1.1 (159–179) 3.6	71 ± 0.6 (65–77) 4.8	21 ± 0.2 (20–23) 3.9	26.4 ± 0.81[o] (19.5–30.2) 13.37	24.27 ± 0.12 (23.03–25.41) 2.72	21.64 ± 0.11 (20.52–22.56) 2.72	6.93 ± 0.08 (6.15–7.80) 6.02	4.37 ± 0.04 (3.83–4.71) 4.98	6.09 ± 0.04 (5.76–6.47) 3.38	13.69 ± 0.08 (12.99–14.84) 3.20	9.84 ± 0.06 (9.28–10.48) 3.22	3.54 ± 0.03 (3.30–3.82) 4.11	8.81 ± 0.06 (8.15–9.33) 3.41

[a] Although measures of ear length were recorded, the data were not included because of extreme variation believed to stem from differences in techniques used to measure these characters
[b] Sample size reduced by 9. [c] Sample size reduced by 4. [d] Sample size reduced by 11. [e] Sample size reduced by 16. [f] Sample size reduced by 20. [g] Sample size reduced by 22. [h] Sample size reduced by 25. [i] Sample size reduced by 23. [j] Sample size reduced by 2. [k] Sample size reduced by 10. [l] Sample size reduced by 5. [m] Sample size reduced by 7.
[n] Sample size reduced by 12. [o] Sample size reduced by 6. [p] Sample size reduced by 1. [q] Sample size reduced by 3. [r] Sample size reduced by 18.

become reproductively active later during the breeding season (Gashwiler, 1959, 1970a); the proportion becoming reproductively active is greater in riparian zones (Doyle, 1990).

Mean litter size in *C. californicus* in the Cascade Range (Table 11-21) was 2.63 by counts of embryos, 2.86 by counts of corpora lutea, and 2.91 by counts of pigmented sites of implantation (Gashwiler, 1977). Gashwiler (1977) estimated an average of 3.1 litters/year was produced by *C. californicus*; however, he considered the estimate low because no pregnant females were caught in July–August. By extrapolating data collected in June to July and those collected in September to August, Gashwiler (1977) estimated 4.6 litters/year.

The gestation period ranges from 17 to 21 days (Maser et al., 1981b). Postpartum estrus seemingly occurs in at least 17% of females (Gashwiler, 1977).

Clethrionomys californicus is preyed upon by saw-whet owls (*Aegolius acadicus*), northern spotted owls (*Strix occidentalis*), bobcats (*Lynx rufus*), coyotes (*Canis latrans*), martens (*Martes americana*), ermines (*Mustela erminea*), long-tailed weasels (*M. frenata*), spotted skunks (*Spilogale gracilis*), feral house cats, and probably other syntopic carnivores and raptors (Forsman, 1975, 1980; Forsman and Maser, 1970; Maser et al., 1981b; Nussbaum and Maser, 1975; Witmer and deCalesta, 1986). Remains of *C.*

californicus constituted 5.5%, 9.2%, and 9.5% of identified items in pellets of the northern spotted owl in the Coast and Cascade ranges, and in the Klamath Mountains, respectively (Forsman et al., 1984).

Habits.—Maser et al. (1981b) indicated that *Clethrionomys* (presumably implying all taxa) was active throughout the 24-h day (p. 195), but later (p. 198) indicated that *C. c. mazama* was active throughout the 24-h day, but *C. c. californicus* was active only at night. Weather may influence activity (Maser et al., 1981b). Most activity is below ground, where *C. californicus* forages on subterranean fungi (Maser et al., 1981b). Nests are built underground, under logs, or beneath the duff (Stephens, 1906).

Hooven (1971a) reported 95% circular home ranges averaged 4.6 ha for males ($n = 12$) and 0.8 ha for females ($n = 8$). No other quantitative information on behavior of *C. californicus* is available (Alexander and Verts, 1992).

Remarks.—A species that associates as closely with a sole feature of the habitat as *C. californicus* does with downed logs should provide numerous opportunities for fundamental research on physical features involved in habitat associations. We wonder if a population might be maintained in a logged area on which slash was burned by providing artificial shade. We also wonder if captive western redbacked voles would associate with "artificial" logs because of the shade they cast and the "security" they provide.

Table 11-21.—*Reproduction in Oregon arvicoline rodents other than* Microtus.

Species	Number of litters	Litter size $\overline{X}$	Range	n	Reproductive season	Basis	Authority	State or province
Clethrionomys californicus	3.1[a]	2.6	1–4	24	Feb.–Nov.	Embryos	Gashwiler, 1977	Oregon
		2.9	2–7	81		Corpora lutea	Gashwiler, 1977	Oregon
		2.9	1–7	57		Implant sites	Gashwiler, 1977	Oregon
Clethrionomys gapperi		6.5	5–8	7	Mar.–Oct.	Embryos	Merritt and Meritt, 1978a	Colorado
		6.0		51	Jun.–Aug.	Embryos	Mihok, 1979	Northwest Territories
		5.6		46		Embryos	Mihok, 1979	Northwest Territories
Lemmiscus curtatus		5.9	4–11	52		Embryos	Johnson et al., 1948	Washington
		5.3	1–11	281	All year	Embryos	James and Booth, 1954	Washington
	3–8[b]	6.1	3–10	66	All year[c]	Embryos + implant sites	Moore, 1943a	Oregon
		4.4	2–13	42	All year[c]	Embryos	Maser et al., 1974	Oregon
Ondatra zibethicus	1.6	7.0	2–9	35	Apr.–Aug.	Nestlings[d]	Reeves and Williams, 1956	Idaho[e]
	2.4	8.0	5–10	31		Nestlings[d]	Reeves and Williams, 1956	Idaho[e]
Phenacomys albipes		3.0	2–4	9		Unknown	Johnson and Maser, 1982	Oregon
Phenacomys intermedius		4.8[b,f]	2–8	43	Jun.–Sep.	Embryos and neonates	Foster, 1961	Manitoba
	2.3	4.2[b,f]	2–7	59	May–Sep.	Neonates	Innes and Millar, 1982	Alberta
		4.8	3–6	4	Jun.–Sep.	Embryos	Vaughan, 1969	Colorado
		5.3	2–9	11		Implant sites	Vaughan, 1969	Colorado
Phenacomys longicaudus		2.9	2–4	7	Dec.–May	Embryos	This study	Oregon

[a]An estimate of 4.6 litters/year was obtained by extrapolating data for summer months when no pregnant females were caught.
[b]In captivity.
[c]Not based on collection of pregnant females throughout the year.
[d]Mean litter size decreased from 8.0/litter for young 3/41 week old to 7.1/litter for those ≥2 weeks old.
[e]First listed data set from Gray's Lake in 1949; second listed data set from Dingle Swamp in 1953.
[f]Pooled sample; litter size related to age.

In view of the potential contribution of western red-backed voles to the health and vigor of coniferous forests, we question whether the red-backed vole–fungi–tree and red-backed vole–bacteria–tree mutualisms are given sufficient consideration in forest-management decisions in western Oregon. If populations of western red-backed voles can be maintained in clear-cuttings on which noncommercial-grade logs are allowed to remain, then assets derived through removal of slash must exceed not only the cost of losing *C. californicus* as prey for carnivores and raptors, as a predator of insects, and as a tunneler and soil mixer, but also those assets associated with its role in scattering disseminules of nitrogen-fixing organisms.

The scientific name applied to the western red-backed vole has differed at various times since it was recognized as a distinct species (Alexander and Verts, 1992; Johnson and Ostenson, 1959). *Clethrionomys occidentalis* was applied to the red-backed vole in western Oregon as recently as 1979 (Jones et al., 1979).

Clethrionomys gapperi (Vigors)
Southern Red-backed Vole

1830. *Arvicola gapperi* Vigors, 5:204.
1891. *Evotomys idahoensis* Merriam, 5:66.
1933. *Clethrionomys gapperi idahoensis*: Whitlow and Hall, 40:265.

Description.—*Clethrionomys gapperi* (Plate XVII) is among the smaller voles in Oregon (Table 11-20). *C. gapperi*, like its congener *C. californicus*, can be distinguished from *Ondatra* by size alone; from *Lemmiscus*, *Phenacomys*, and *Microtus oregoni* by four loops of enamel on the lingual side of M3; and from other *Microtus* by rooted molars in adults and most loops of enamel on the upper molars rounded (Fig. 11-94). It may be distinguished from *C. californicus* by the posterior margin of the palate being truncate, lacking a spinelike shelf. The skull (Fig. 11-98) is relatively smooth in contrast with the angular and rugose skulls of some other arvicolines.

The pelage of *C. gapperi* is silvery on the venter and grayish with an ocherous wash on the sides, and with a rusty or reddish stripe on the dorsum; the tail is dark brown or black above and whitish or grayish below (Merritt, 1981). The summer pelage of adults and the pelage of juveniles tend to be darker.

Distribution.—*Clethrionomys gapperi* occurs (Fig. 11-99) through most of Canada south of Great Slave Lake, Northwest Territories, and Ungava Bay, Quebec, and in the contiguous United States south along the Appalachian Mountains to northern Georgia, in the Midwest to northern Iowa and northern Nebraska, along the Rocky Mountains and associated mountains to southern Arizona and New Mexico, and in the Cascade and Coast ranges south to the Columbia River (Merritt, 1981). In Oregon, *C. gapperi* occurs in the Ochoco, Blue, and Wallowa mountains (Fig. 11-99) southwestward to central Crook Co. and northern Harney Co.

Pleistocene fossils of *C. gapperi* have been found at numerous sites, many of which are extralimital (Kurtén and Anderson, 1980).

Geographic Variation.—Of 29 subspecies currently recognized, only *C. g. idahoensis* occurs in Oregon (Merritt, 1981).

Habitat and Density.—In southeastern Washington and eastern Idaho, of 20 climax vegetative associations sampled,

C. gapperi was captured in only eight (number/1,000 trap-nights in parentheses): *Pinus/Physiocarpus* (7.4), *Pseudotsuga/Physiocarpus* (5.2), *Thuja/Pachistima* (16.3), *Picea-Abies/Pachistima* (3.2), *Picea-Abies/Menziesia* (5.6), *Picea-Abies/Xerophyllum* (5.8), *Thuja-Tsuga/Oplopanax* (7.4), *Alnus sinuata* (22.2—Hoffman, 1960; Rickard, 1960). Both authorities commented regarding the requirement of free water by *C. gapperi* and the absence of the species in more xeric habitats. Downed logs also seem to be a critical component of habitats occupied by *C. gapperi* (Sturges, 1957). Because of the frequency of fungal sporocarps in the diet of *C. gapperi* elsewhere (Maser and Maser, 1988), we suspect that habitat associations of the species are similar to those of its congener *C. californicus*.

In Northwest Territories, density of populations of *C. gapperi* peaked in September at about 21/ha in 1976 and at about 5/ha in 1977 (Mihok, 1979). In Colorado, the density peaked in November at about 48/ha (Merritt and Merritt, 1978*a*). Multiannual cycles of abundance do not seem to occur in *C. gapperi* (Merritt, 1981).

In Ontario, densities of *C. gapperi* in response to clear-cutting in a black spruce (*Picea mariana*) forest were 10.5–14.1/ha after 1 year (the predominant small mammal), 9.6/ha after 2 years, 1.0–2.1/ha after 2 years and 2 months, and 0.3–0.7/ha after 3 years (Martell and Radvanyi, 1977). The paucity of cover was considered to make clear-cuttings unsuitable for *C. gapperi*.

Diet.—In Colorado, *C. gapperi* ate mostly fungi in spring, summer, and autumn; *Endogone* formed a staple item in the diet (Merritt and Merritt, 1978*a*). It shifted its diet to conifer seeds in winter. In Oregon, of five *C. gapperi* examined all contained fungi; three to seven taxa of fungi were represented in the specimens (Maser et al., 1978*a*). Other foods ingested by this omnivorous, opportunistic feeder included green plants, nuts, seeds, berries, mosses, lichens, ferns, and insects (Merritt, 1981).

Reproduction, Ontogeny, and Mortality.—*Clethrionomys gapperi* is polyestrous with immediate postpartum estrus (A. Svihla, 1931). Mating is considered

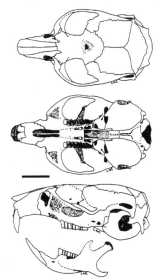

Fig. 11-98. Dorsal, ventral, and lateral views of the cranium, and lateral view of the dentary of the southern red-backed vole, *Clethrionomys gapperi* (OSUFW 7343). Scale bar equals 5 mm.

promiscuous (Mihok, 1979). Gestation requires 17–19 days (A. Svihla, 1930, 1931). In western Washington, litter size in late summer averaged 3.3 (*n* = 3—A. Svihla, 1931); in Northwest Territories and in Colorado, litter size averaged about twice as large (Table 11-21).

Young at birth are hairless except for vibrissae, have closed eyes, and are flesh colored. Within 1 week they are covered with dark gray hair; the reddish tinge to dorsal hairs appears gradually. By 2 weeks, the eyes are open, the young start to eat solid food, and weaning commences. Nestlings are capable of producing a high-pitched squeak when disturbed (A. Svihla, 1931). Young are about 30 days of age when first captured (Mihok, 1979).

In northeastern Oregon, *C. gapperi* composed 13.9% of 4,546 food items identified in pellets of great gray owls (*Strix nebulosa*—Bull et al., 1989). Likely, other raptorial birds and carnivorous mammals that prey on small mammals also take this species.

Habits.—Findings of an investigation in which multiple-capture traps were used provide some insight into the behavior of *C. gapperi* in the wild (Mihok, 1979). Multiple captures were more likely to be of the opposite sex; about half as many were male-male and the fewest were female-female. For example, one female caught 60 times was caught 15 times with one of eight other conspecifics,

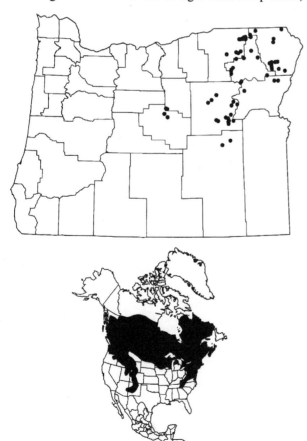

Fig. 11-99. Localities in Oregon from which museum specimens of the southern red-backed vole, *Clethrionomys gapperi*, were examined (listed in APPENDIX, p. 515), and the distribution of the southern red-backed vole in North America. (Range map redrawn after Hall, 1981:782, map 446, with modifications.)

all males. A male caught 77 times was involved in 20 multiple captures with 11 other red-backed voles, but males and females were represented about equally. Mature females were considered to maintain exclusive territories; there was some overlap in space, but none in time. Wounding was low in multiple captures, indicating that *C. gapperi* is amicable and tolerant. Few adult females and juveniles were captured simultaneously, tending to indicate that mother-infant relationships end at weaning. There was no evidence of group structure in populations of *C. gapperi* and there was no evidence that adults suppressed sexual maturity in juveniles, as reported for congeners elsewhere in the world (Mihok, 1979).

Home-range size in Colorado, estimated by the inclusive boundary-strip method, averaged 0.02 ± 0.17 ha (range, 0.003–0.05 ha) for 40 males and 0.02 ± 0.18 ha (range, 0.004–0.05 ha) for 22 females in subnivean environments in winter and 0.01 ± 0.29 ha (range, 0.002–0.06 ha) for 19 males and 0.01 ± 0.17 ha (range, 0.002–0.03 ha) for 19 females in summer (Merritt and Merritt, 1978b). With one exception, home-range area for other studies summarized for comparison averaged 0.02–0.16 ha; the largest estimate was 0.50 ha.

In staged encounters in a neutral arena, relationships generally were amicable, although females exhibited dominance in almost all instances; however, most females tested were in breeding condition. Other evidence suggested a reversal of dominance status between sexes when females were not in breeding condition (Mihok, 1976). Older females or those in later stages of pregnancy were dominant over younger females or those in earlier stages of pregnancy. Aggressive behavior, although relatively uncommon, was more common in female-female pairings (Mihok, 1976).

Remarks.—Other than collection localities, essentially nothing is known regarding *C. gapperi* in Oregon; no published research report is devoted entirely to any aspect of the biology of the species in the state (Carraway and Verts, 1982).

The ingenious use by Mihok (1979) of multiple-capture traps to elucidate complex behavioral interactions calls to mind other trap designs (Barry et al., 1989) and arrangements (Verts and Carraway, 1986), and other techniques used in conjunction with live trapping (Small and Verts, 1983) to obtain information concerning specific aspects of behavior. Time of capture, location of capture, associates with which captured, and response of individuals in certain categories when individuals in the same or other categories are removed temporarily or permanently from a population either singly or in combination can provide, to paraphrase Mihok (1979), stepping-stones to aspects of the social structure of populations of small nocturnal rodents. *C. gapperi* (especially populations west of the Cascade Range north of the Columbia River) and its congener, *C. californicus,* seemingly are prime candidates for a comparative study of behavior by use of such techniques.

Phenacomys albipes Merriam
White-footed Vole

1901c. *Phenacomys albipes* Merriam, 14:125.

Description.—*Phenacomys albipes* (Fig. 11-100) is among the smaller voles (Table 11-20); the pelage consists of dark bluish-gray hairs tipped with a rich brown or black on the dorsum grading lighter on the sides, thence to light

Fig. 11-100. Photograph of the white-footed vole, *Phenacomys alpibes*. (Reprinted from Verts and Carraway, 1995, with permission of the American Society of Mammalogists.)

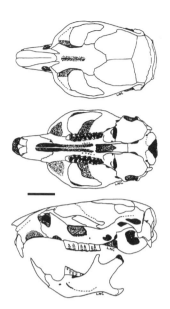

Fig. 11-101. Dorsal, ventral, and lateral views of the cranium, and lateral view of the dentary of the white-footed vole, *Phenacomys albipes* (OSUFW 3113). Scale bar equals 5 mm.

gray to pinkish buff on the venter (Howell, 1926*b*). The tail is distinctly bicolored: white or buff on the ventral surface, brownish black on the dorsal surface. The dark dorsal stripe is much wider than the lighter ventral stripe. The feet are light grayish, but commonly no more so than in other *Phenacomys* (Howell, 1920*a*, 1926*b*).

Phenacomys albipes is separable from other voles by the combination of reentrant angles on the lingual side of the lower molars ≥2 times as deep as those on the labial side (Fig. 11-94); a slender, scantily haired, and sharply bicolored tail >50% of the length of the head and body; and narrow incisive foramina. It also has three loops of enamel on the lingual side of M3 (Fig. 11-94), strongly recurved incisors, a short and wide M3 and m3, a posteriorly sloping septum between the internal nares (Fig. 11-101), and hind feet usually >19 mm long. Superficially, *P. albipes* resembles *Microtus longicaudus* (Howell, 1926*b*).

Distribution.—*Phenacomys albipes* occurs from Old Fort Clatsop, Clatsop Co., and near Rainier, Columbia Co., south through the Coast Range to Arcata, Humboldt Co., California (Hall, 1981). In Lane Co., the range extends eastward on the west slope of the Cascade Range to the H. J. Andrews Experimental Forest (Fig. 11-102).

Geographic Variation.—No geographic variation is recognized currently in *P. albipes*.

Habitat and Density.—*Phenacomys albipes* is captured so infrequently that much of the published information on habitats it occupies is anecdotal. Various Oregon specimens have been caught "among mossy stones along a small stream that flows through a heavy forest of Douglas fir" with an understory of "sword fern, moss, and a tangle of vine maple" (Jewett, 1920:167), in and under "a salmonberry thicket" (Jewett, 1920:167; Sherrell, 1969:39), in an area where "the climax forest tree is noble fir (*Abies nobilis* [*procera*]) with some western hemlock (*Tsuga heterophylla*)," in "a steep sided canyon with some fallen logs, brushy forest and stream sides," "in a grassy spot on a partly logged ridge top," and in a mountain beaver (*Aplodontia rufa*) runway "in dense vegetation" of horsetail (*Equisetum*), common sword-fern (*Polystichum munitum*), deer fern (*Blechnum spicant*), miner's lettuce (*Montia perfoliata*), nettle (*Lamium*), foxglove (*Digitalis purpurea*), Canada thistle (*Cirsium arvense*), salmonberry

(*Rubus spectabilis*), thimbleberry (*R. parviflorus*), vine maple (*Acer circinatum*), and red alder (*Alnus rubra*—Maser and Johnson, 1967:25–26). Maser and Hooven (1969) reported catching a specimen in an early seral plant community resulting from the burning by wildfire of a 3-year-old clear-cutting 2 years earlier. We located museum records of several *P. albipes* captured in clear-cuttings. Voth et al. (1983:1) characterized habitats at which they captured 21 specimens as the "Riparian Alder/Small Stream, *Alnus rubra*/*Rubus spectabilis* (red alder/salmon berry) union."

Howell (1928:153–154) indicated that 11 of the 15 specimens known at that time had been caught on "a gravelly bar at the margin of a little forest stream." Maser and Johnson (1967:26), from a summary of published accounts, information recorded on specimen tags, and personal communications from those who had captured specimens, concluded only that *P. albipes* was "associated with small streams." Voth et al. (1983:6) attributed to Maser and Hooven (1969) the claim that *P. albipes* "may be found several hundred meters from a streamside"; however, the latter authors stated (p. 22) that the individual reported on "was caught 100 to 150 feet from the stream." Voth et al. (1983) suggested that the association of *P. albipes* with streamsides was a result of its foraging on leaves of riparian tree species.

More recent studies have revealed that *P. albipes* is associated with much of the forest sere (Anthony et al., 1987; Gomez, 1992; McComb et al., 1993), but tends to be more abundant in deciduous forest followed in order by conifer-shrub, sapling-pole conifer, old-growth conifer forest, and sawtimber conifer forest (Gomez, 1992). *P. albipes* was captured 5.3 times more frequently ≤25 m of a stream than 200 m upslope (Gomez, 1992).

Maser and Johnson (1967:24) claimed *P. albipes* was the "rarest microtine rodent in North America"; later, this was modified to "one of the rarest microtine rodents north of Mexico" (Voth et al., 1983:1). In the first 29 years after the type specimen was taken in 1899, only 14 additional

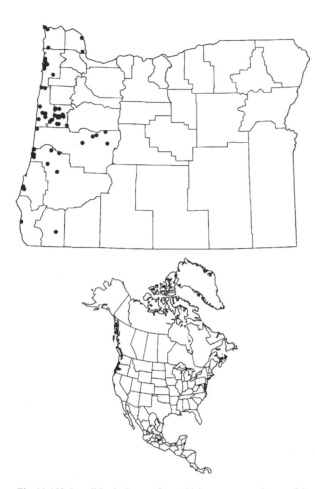

Fig. 11-102. Localities in Oregon from which museum specimens of the white-footed vole, *Phenacomys albipes*, were examined (listed in Appendix, p. 545), and the distribution of the white-footed vole in North America. (Range map based on locality records provided Verts and Carraway, 1995:2, fig. 4.)

specimens were recorded (Howell, 1928). Maser and Johnson (1967) were able to locate only 28 specimens from throughout the range of the species, Olterman and Verts (1972) examined 45 from Oregon, Johnson and Maser (1982) recorded 61 from Oregon and California, and we examined 102 museum specimens from Oregon. Verts and Carraway (1995) recorded 73 additional specimens taken in other studies in 1983–1991, but not all these specimens have been deposited in museums. Although these data are not all directly comparable, they illustrate that the acquisition of study material is accelerating.

Diet.—Howell (1928:154) indicated that specimens from Humboldt Co., California, had eaten "finely triturated foods of some herbaceous plant." Maser and Johnson (1967:26) found "green material in several stomachs, indicating the vegetative portion of plants" was eaten. Voth et al. (1983:6), based on microhistological analysis of stomach contents and colonic fecal pellets, characterized white-footed voles as "obligate browsers." Leaves of hardwood trees, especially red alder and red willow (*Salix lasiandra*), composed 57% of food residues; 15% were residues of leaves of shrubs, principally Cascade Oregongrape (*Berberis nervosa*) and thimbleberry; 23% were residues of forbs, mostly white clover (*Trifolium*

repens) and wood groundsel (*Senecio sylvaticus*); and <2% were residues of grasses and ferns. Residues of seeds, fruits, fungi, and animal matter were not found. Forbs contributed more and hardwoods contributed less in spring than at other seasons (Voth et al., 1983).

Several authorities indicated that they had caught *P. albipes* in traps baited with rolled oats (Howell, 1920*a*; Jewett, 1920; Maser and Johnson, 1967), but whether rolled oats serve to attract *P. albipes* is unknown. In captivity, *P. albipes* has been maintained on a diet of leaves and stems of Pacific blackberry (*Rubus ursinus*—Verts and Carraway, 1995).

Reproduction, Ontogeny, and Mortality.—Maser (1966*b*) reported three females containing three embryos each; one was taken in April, two in July. We recorded a recently parturient female taken in April (KU 145695) and a pregnant female taken in August (KU 145696); the latter individual exhibited evidence of having been nursed recently. The only estimate of litter size available is similar to that of *P. longicaudus* (Table 11-21). An adult female and three 2.4-g young were found in a ground nest (Verts and Carraway, 1995). We know of no information on postnatal development. Maser et al. (1981*b*:201) listed "owls, weasels, mink, spotted skunks, and domestic cats" as potential predators.

Habits.—Because *P. albipes* was taken in traps set on the ground, most workers considered the species to be terrestrial (Jewett, 1920; Maser and Johnson, 1967). However, based on a study of the diet that indicated the species foraged in the tree canopy, Voth et al. (1983:6) claimed that the spatial niche of *P. albipes* extended "from ground level into the hardwood canopy."

Many white-footed voles have been captured in snap traps, although some have been taken in Sherman live traps (Maser and Hooven, 1969). However, the most successful trapping method seems to be with pitfalls (Verts and Carraway, 1995).

Remarks.—The history of nomenclature applied to this species and the basis for our continued use of the generic name *Phenacomys* are outlined in Remarks for the red-tree vole (*Phenacomys longicaudus*).

Phenacomys albipes is listed as a "sensitive species" by the Oregon Department of Fish and Wildlife and is under consideration by the U.S. Fish and Wildlife Service as a threatened and endangered species (Verts and Carraway, 1995).

Despite the rarity of white-footed voles in systematic collections, it is difficult for us to accept that any population of a vole-like mammal has attributes such as low natality, high survival, or low density, or that any vole species exhibits average longevity appreciably greater than that of vole species for which average longevity has been measured. We believe that the key features of life history that will permit frequent capture, necessary for study of population dynamics, of the white-footed vole remain to be discovered. The association of *P. albipes* with hardwood canopies and the large proportion of males (73% of 67) in collections may be clues to some of those features.

Based on experience with only a few specimens, we found that skins of *P. albipes* become extremely fragile when the animals are frozen, even when double-wrapped and frozen for only 3–4 days.

Phenacomys intermedius Merriam

Heather Vole

1889*b*. *Phenacomys intermedius* Merriam, 2:32.
1895*b*. *Phenacomys oramontis* Rhoads, 29:941.
1899. *Phenacomys olympicus* Elliot, 1:225.
1942. *Phenacomys intermedius oramontis*: Anderson, 56:59.

Description.—*Phenacomys intermedius* is among the smaller voles in Oregon (Table 11-20). It can be distinguished from all other small arvicolines except *Lemmiscus*, *Microtus oregoni*, and congeners by three loops of enamel on the lingual side of M3 and from all but congeners by the reentrant angles on the lingual side of the lower molars being twice as deep as those on the labial side (Fig. 11-94). It is distinguishable from other *Phenacomys* by the tail being <50% of the length of the head and body (<42 mm long—McAllister and Hoffmann, 1988), and both M3 and m3 being long and narrow (Johnson and Maser, 1982). Johnson (1967:17) indicated that "the presence of numbers of short, moderately stiff, orange-colored hairs in the peripheral half of the inside of the ears" was useful in distinguishing *P. intermedius* from similar arvicolines.

In external appearance, *P. intermedius* strongly resembles *Microtus montanus*, with which it is sympatric in much of western United States (McAllister and Hoffmann, 1988). It is gray to brownish on the dorsum, whitish to grayish on the venter. There are eight mammae, the nasals protrude only slightly beyond the anterior faces of the incisors (Fig. 11-103), the hind foot usually is <18 mm long, and the molars become rooted when the root openings close.

Distribution.—*Phenacomys intermedius* occurs over most of boreal North America (Fig. 11-104) southward in montane regions to central California, southern Utah, and northern New Mexico. In eastern North America, the range of *P. intermedius* does not extend southward beyond the Great Lakes and the Saint Lawrence River (McAllister and Hoffmann, 1988). In Oregon, the distribution is disjunct, with one population in the Cascade Range and in the vicinity of Gearhart Mountain, Lake Co., and another population in the Blue and Wallowa mountains (Fig. 11-104).

Extralimital fossils, referable to *P. intermedius*, from numerous sites indicate that the Pleistocene range extended south to central Nevada, central Kansas, southern Tennessee, and northern North Carolina (McAllister and Hoffmann, 1988).

Geographic Variation.—Two of the nine nominal subspecies are purported to occur in Oregon: the lighter-colored *P. i. intermedius* in the Blue and Wallowa mountains of northeastern Oregon and the darker *P. i. oramontis* in the Cascade Range.

Habitat and Density.—Inconsistency seemingly is characteristic of descriptions of habitats occupied by *P. intermedius* (Edwards, 1955; Foster, 1961), no doubt a response to the wide variety of habitats occupied by the species. For example, Corn and Bury (1988) reported capturing *P. intermedius* in Oregon in a 5-year-old clear-cutting and a 195-year-old stand of Douglas-fir (*Pseudotsuga menziesii*) on a dry slope. We took it in montane meadows and along streams near forest edges. In Yukon Territory, *P. intermedius* was captured in 10 of 21 habitat types (Krebs and Wingate, 1976) and in five of 18 habitat types when habitats were classified differently (Krebs and Wingate, 1985). In Wyoming, Brown (1967*a*) caught heather voles in only

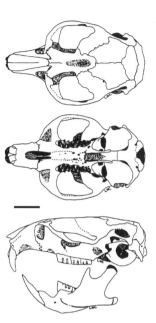

Fig. 11-103. Dorsal, ventral, and lateral views of the cranium, and lateral view of the dentary of the heather vole, *Phenacomys intermedius* (OSUFW 5053). Scale bar equals 5 mm.

three of 12 habitats sampled: subalpine meadow, Engelmann spruce (*Picea engelmannii*)-fir (*Abies lasiocarpa*), and alpine tundra. The distribution of capture sites was not related to the proximity of water or density of ground cover. In a similar 3-year study in Alberta, Millar et al. (1985) captured *P. intermedius* in 12 of 14 habitats sampled; it was most abundant in coniferous riparian areas, but was temporarily abundant in open spruce forest 1 year and in subalpine meadows another. In another Alberta study, *P. intermedius* was caught in mixed stands of quaking aspen (*Populus tremuloides*), lodgepole pine (*P. contorta*), and spruce (*Picea*) and in pure stands of lodgepole pine and on talus slopes at 1,450 m elevation, and in dense stands of spruce and larch (*Larix lyallii*) at 2,240 m (Innes and Millar, 1982). In Ontario, Naylor et al. (1985) caught most *P. intermedius* in jack pine (*P. banksiana*) monocultures where the dense understory of ericaceous (heath) plants provided food, protection from predators and conspecifics, and a favorable microclimate; most were juveniles, thus such habitats were considered dispersal sinks. Foster (1961) analyzed 43 descriptions of habitat in the literature; 80% included mention of dryness, 29% mentioned proximity to water, 20% wet habitats, 32% spruce and pine, 23% treeline, 16% grass, and 11% sagebrush (*Artemisia*) and "heather." Corn and Bury (1988) suggested that the original habitats of *P. intermedius* in the Cascade Range of Oregon were above timberline and in montane meadows, and that, with clear-cut logging, the species extended its range along the higher ridges between west-flowing rivers.

Few estimates of density for *P. intermedius* are available; most authorities refer to heather voles as rare or uncommon, but in a few instances they seem fairly abundant locally for short periods. Captures per 100 trap-nights (trap-nights in parentheses) in Manitoba for the years 1951–1957 were 0 (610), 0.08 (2,427), 1.04 (575), 2.96 (2,300), 2.21 (8,700), 1.46 (5,611), and 0.08 (1,214—Foster, 1961). Similar data for Yukon Territory in 1973–1977 were 0.18

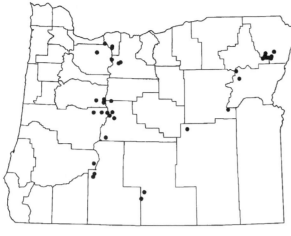

Fig. 11-104. Localities in Oregon from which museum specimens of the heather vole, *Phenacomys intermedius*, were examined (listed in APPENDIX, p. 546), and the distribution of the heather vole in North America. (Range map redrawn from Hall, 1981:787, map 448, with modifications.)

(2,745), 0.17 (4,023), 0.04 (4,509), 0.05 (5,481), and 0.03 (3,519—Krebs and Wingate, 1985). Earlier, Krebs and Wingate (1976) indicated relative densities of 0.05–0.37/ 100 trap-nights. Innes and Millar (1982) provided midsummer estimates of minimum number known alive of 4.3 and 3.0/ha on grids at 1,450 and 2,240 m, respectively, in Alberta.

Diet.—*Phenacomys intermedius* is strictly herbivorous. In Washington, *P. intermedius* ate bark from twigs of moss heather (*Cassiope*), beargrass (*Xerophyllum*), stems of lousewort (*Pedicularis*), and huckleberry (*Vaccinium*— Shaw, 1924). Later, Racey (1928) added hawkweed (*Hieracium alpinum*), western swamp laurel (*Kalmia occidentalis*), and fan-leaf cinquefoil (*Potentilla flabellifolia*) to the list of plants known to occur in food piles of *P. intermedius* in Washington. In British Columbia, bearberry (*Arctostaphylos uva-ursi*) constituted 85–90% of the diet of *P. intermedius,* soapberry (*Shepherdia canadensis*) 6%, and roses (*Rosa*), vetch (*Vicia*), and others 5–10% (Cowan and Guiguet, 1965). In Manitoba, bark of willows (*Salix*), scrub birch (*Betula glandulosa*), and many species of Ericaceae were eaten in winter; in summer, blueberry (*Vaccinium*), bearberry, and cinquefoil (*Potentilla tridentata*) were cached. Smaller amounts of crowberry (*Empetrum nigrum*), fireweed

(*Epilobium angustifolium*), bog-rosemary (*Andromeda polifolia*), mountain cranberry (*Vaccinium vitis-idaea*), and cloudberry (*Rubus chamaemorus*) also were cached by *P. intermedius* (Foster, 1961). In a jack pine monoculture in Ontario, food caches of *P. intermedius* consisted of western swamp laurel, a species usually considered toxic to vertebrates (Naylor et al., 1985).

Reproduction, Ontogeny, and Mortality.—The heather vole is a seasonally polyestrous breeder (Vaughan, 1969) with the breeding season extending from May to early September (Table 11-21). Gestation is 19–24 days (Foster, 1961; Innes and Millar, 1982). The length of the breeding season may vary with elevation (McAllister and Hoffmann, 1988). In Alberta, four litters annually are theoretically possible, but three was the maximum produced and the average was only 2.3 (Innes and Millar, 1982). Young born early in the breeding season breed during the year of their birth, but those born in later litters postpone reproductive maturity until the following breeding season (Vaughan, 1969). Females 4–6 weeks old commence to breed but they produce significantly smaller litters (average 0.79– 2.10 fewer offspring) than overwintered females (Foster, 1961; Innes and Millar, 1982). However, among adults, average litter sizes of primiparous ($\bar{X} = 4.2$, $n = 27$) and multiparous ($\bar{X} = 4.3$, $n = 32$) females were not significantly different (Innes and Millar, 1982). There is some evidence that litter size may be related inversely to latitude (McAllister and Hoffmann, 1988). To our knowledge comparable data for Oregon are not available.

In Alberta, neonates reportedly weighed 1.9 g and to 28 days of age grew at a rate described by $Y = 0.79 + 0.43X$, where $Y = $ mass (g) and $X = $ age (days—Innes and Millar, 1982). Nestlings opened their eyes by 15.5 ± 0.6 days and most commenced to eat solid food by 18 days of age (Innes and Millar, 1982). Both males and females matured in the year of their birth. In Manitoba, average body mass of neonates was 2.4 g (range, 2.0–2.7 g); neonates measured (crown-rump) $\cong 30$ mm long (Foster, 1961). At birth they are pink, wrinkled, hairless, and blind, but by the end of day 1 the dorsum has become slightly blue-black, the vibrissae are visible, and the skin is less wrinkled (Foster, 1961). On day 2, the young can vocalize, the dorsum is darker, and the vibrissae are $\cong 3$ mm long. The dorsum darkens on day 3, by day 5 the rump is slightly furred, and by day 6 the whole body is covered with fine fur. Although movements on days 8–9 remain jerky, coordination has improved. Incisors irrupt on day 9, young begin to walk on day 11, and the eyes open and green food is eaten on day 14. Weaning occurs and young may leave the nest by day 19 (Foster, 1961).

In Alberta, the minimum survival rate for adult female *P. intermedius* caught three or more times in a mark-recapture study was 0.89/2-week period in 1979 and 0.77/ 2-week period in 1980. For juvenile females the minimum survival rate was 0.97/2-week period (Innes and Millar, 1982). Similar estimates could not be calculated for males. Of 32 marked individuals of both sexes only two females were recaptured the following year. Mortality in live traps averaged 35% during the 2-year study (Innes and Millar, 1982).

Bull et al. (1989) reported that 0.8% of 4,546 food items identified in pellets of great gray owls (*Strix nebulosa*) from northeastern Oregon were *P. intermedius*. Remains of *P. intermedius* were found in pellets of the spotted owl (*Strix occidentalis*) in the Cascade Range, but it composed such a small proportion of the diet that its contribution was not reported separately (Forsman et al., 1984). No doubt other raptorial birds and carnivorous mammals syntopic with *P. intermedius* prey on it. Peterson (1956) found remains of *P. intermedius* in the stomachs of eastern brook trout (*Salvelinus fontinalis*).

Habits.—Heather voles are rather docile creatures; they usually make little attempt to flee or to defend themselves when live-captured and handled. Except for females with young, captives reportedly did not behave aggressively toward conspecifics or other species of voles (Foster, 1961). During winter, captives interact amicably, even to the point of several using the same nest simultaneously, but during the breeding season captive males fight viciously. However, wild-caught males do not exhibit wounds (Foster, 1961).

Foster (1961) indicated that *P. intermedius* was active during daylight hours more frequently than was *Clethrionomys gapperi*. Further, he claimed that *P. intermedius* was less active during the day than at twilight or at night; he also indicated that *P. intermedius* in captivity used "activity wheels" more at night than during the day. However, in no instance were supporting data presented for comparison.

In Manitoba in winter, *P. intermedius* constructs nests about 15 cm in diameter of leaves, twigs, grass, and other vegetation commonly at the base of a shrub, stump, or rock (Foster, 1961); in Washington, winter nests were constructed of beargrass lined with snowgrass (*Luzula*—Shaw, 1924). In summer, smaller nests ($\cong$10 cm in diameter) are constructed either <20 cm below ground or under some object such as a log or rock; materials used in construction of summer nests mostly are grasses, mosses, and leaves (Foster, 1961). A tunnel <1 m long connects the nest with a latrine and the surface (Foster, 1961); tunnels about 6 cm in diameter and about 5–15 cm below ground open to the surface through holes about 2.5 cm in diameter (Racey, 1928). Racey (1928) found that all underground tunnels ultimately led to water. Short pathways lead from the openings to the tunnel. Seemingly, *P. intermedius* does not construct and maintain a system of runways similar to those used by some *Microtus*; however, it is known to use runways built by other arvicolines (Foster, 1961).

Phenacomys intermedius exhibits a low susceptibility to capture, males more so than females (Innes and Millar, 1982). Trapping immediately in front of holes that exhibit sign of use by voles improves the rate of capture (Foster, 1961).

Remarks.—Again, mammalogists, including ourselves, have neglected to conduct field research on another member of the small-mammal fauna of Oregon. Comparatively little is known about the life history of *P. intermedius* throughout its continent-wide distribution and often what is learned at one locality contradicts that learned at another. We suspect that *P. intermedius* is much less scarce than might be inferred from its association with boreal habitats and its infrequent appearance in traps set for small mammals. Also, it may have been avoided as a research species because of its low susceptibility to recapture and its high rate of trap mortality. Perhaps the recent emphasis on preserving wild populations of forest-dwelling raptorial birds that, no doubt, prey on *P. intermedius* will stimulate interest in research on the biology of the species.

Phenacomys longicaudus True
Red Tree Vole

1890. *Phenacomys longicaudus* True, 13:303.
1921. *Phenacomys silvicola* Howell, 2:98.

Description.—*Phenacomys longicaudus* (Plate XVII) is a medium-sized vole (Table 11-20). It is uniquely colored among North American voles: bright orangish-red to cinnamon on the dorsum, silvery gray (often with some light orangish hairs) on the venter, and a tail pale orangish on the venter grading to black on the dorsum and distal one-fourth. The pelage tends to be long and soft. There are no hip glands, the molars are rooted, and the toes and claws are long (Howell, 1921, 1926a).

Like its congeners, *P. longicaudus* has lower molars with reentrant angles on the lingual side ≥ 2 times as deep as those on the labial side, and like *Lemmiscus*, *Microtus oregoni*, and other *Phenacomys*, M3 has three loops of enamel on the lingual side (Fig. 11-94). Its tail (>50 mm long) is >50% of the length of the head and body; the nasal bones protrude well beyond the anterior faces of the incisors when the skull is viewed from the ventral aspect; the incisors are strongly recurved; the septum between the internal nares slopes posteriorly; both M3 and m3 are short and wide (Fig. 11-105); and the hind feet are usually >19 mm long. These characters separate *P. longicaudus* from *P. intermedius* but not from *P. albipes*. A well-furred but not strongly bicolored tail, feet orangish on the dorsal surface, and wide incisive foramina separate *P. longicaudus* from *P. albipes*.

Distribution.—*Phenacomys longicaudus* is endemic to western Oregon (Fig. 11-106). It occurs at moderate elevations (<1,040 m—Corn and Bury, 1986) on the west slope of the Cascade Range southward as far as the Douglas-Jackson county line and in the Coast Range to the Oregon-California border. Also, specimens have been collected on some buttes and low mountains (e.g. Chehalem

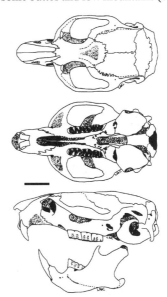

Fig. 11-105. Dorsal, ventral, and lateral views of the cranium, and lateral view of the dentary of the red tree vole, *Phenacomys longicaudus* (OSUFW 6491). Scale bar equals 5 mm.

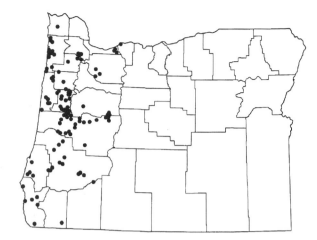

Fig. 11-106. Localities in Oregon from which museum specimens of the red tree vole, *Phenacomys longicaudus*, were examined (listed in APPENDIX, p. 546). The species is considered endemic to Oregon (Johnson and George, 1991).

Mountains, Washington Co.) between the major ranges (Johnson and George, 1991).

One possible fossil record exists: 6,000-year-old material from Gatecliff Shelter, Nye Co., Nevada (Repenning and Grady, 1988).

Geographic Variation.—Two subspecies are recognized currently: the darker-colored *P. l. silvicola* on the west slope of the Coast Range in Tillamook and Lincoln counties and the brighter-colored *P. l. longicaudus* in the remainder of the range (Johnson and George, 1991).

Habitat and Density.—*Phenacomys longicaudus* inhabits a variety of mesic forest habitats dominated by Douglas-fir (*Pseudotsuga menziesii*), grand fir (*Abies grandis*), western hemlock (*Tsuga heterophylla*), and Sitka spruce (*Picea sitchensis*—Johnson and George, 1991). Although *P. longicaudus* may occur in younger stands, old-growth forests seemingly provide optimum habitat. For example, Corn and Bury (1986) caught 17 *P. longicaudus* in pitfall traps; 12 were caught in stands 195–450 years old, whereas only five were captured in stands ≤150 years old. Number captured during six 30-day periods was strongly correlated ($r = 0.87$, $P < 0.05$) with the number of days/30-day period with precipitation ≥1.0 mm.

Gomez (1992) captured *P. longicaudus* in pitfalls set in old-growth stands 4.1 times more frequently than in stands of large saw timber, 7.6 times more frequently than in stands of shrubs, and 17.7 times more frequently than in stands of pole-size trees. He did not capture the species in deciduous stands. Red tree voles were captured slightly more frequently upslope (200 m from streamside) than ≤25 m from a stream (Gomez, 1992). Although Gillesberg and Carey (1991:786) claimed that *P. longicaudus* occurs "in all seral stages of Douglas-fir forest," we believe that we have the only record of an individual caught in a pitfall in a recent clear-cutting (OSUFW 3234). It was a juvenile as evidenced by its mottled grayish and orangish pelage; we think it likely was a disperser.

Diet.—Insofar as known, the only dietary information for *P. longicaudus* obtained by examination of stomach contents is that provided by Wight (1925); he found finely pulverized fir needles, bark, and lichens. From larger twigs, the bark may be removed and only the cambium consumed

(Maser, 1966*b*). Doubtlessly, for most *P. longicaudus,* needles of Douglas-fir compose the bulk of the diet (Maser, 1966*b*). Walker (1930*b*) suggested that young reared in fir, spruce, or hemlock trees might develop a predilection for the leaves of those species and at independence select nest-trees of the same species as their natal trees; the hypothesis remains untested.

Maser (1966*b*) found seeds of bitter cherry (presumably *Prunus virginiana*) and a small piece of Scotch broom (*Cytisus scoparius*) in nests of *P. longicaudus*, suggesting that foods other than foilage of conifers are eaten.

Reproduction, Ontogeny, and Mortality.—Museum records that we examined indicate that pregnant females were collected in December, February, March, April, and May. In addition, Brown (1964) reported collecting and holding temporarily a captive wild-bred female that would not have given birth until June. Maser (1966*b*) and Maser et al. (1981*b*) considered the species to breed throughout the year; however, the claim was not supported by data.

Average litter size is near the smallest among Oregon arvicolines (Table 11-21). Postpartum estrus and pregnancy while lactating are known to occur in *P. longicaudus* (Brown, 1964), but their frequency and the average number of litters produced by females each year are unknown.

Red tree voles may avoid numerous predators because of their life-style; the only arboreal mammalian predator within their range is the marten (*Martes americana*) and it tends not to be common. However, some raptorial birds frequently feed in trees. For example, remains of *P. longicaudus* occurred in 19.1% of 1,214 pellets regurgitated by spotted owls (*Strix occidentalis*) and composed 6.0% of the biomass consumed by the raptor in the Coast Range; comparable values for the Cascade Range were 13.3% and 4.2% ($n = 817$ pellets), respectively (Forsman et al., 1984). Forsman and Maser (1970) found remains of three red tree voles among 42 items identified in 36 pellets of saw-whet owls (*Aegolius acadicus*). However, some ground-feeding raptors reportedly take *P. longicaudus*; Reynolds (1970) found remains of one red tree vole among 153 items identified from 104 pellets regurgitated by long-eared owls (*Asio otus*).

Habits.—*Phenacomys longicaudus* is largely arboreal and builds nests on suitable foundations (commonly nests of birds or other arboreal mammals) in Douglas-fir, western hemlock, grand fir, and Sitka spruce, species that serve as a source of food. Nests constructed of resin ducts from conifer needles (51%), lichens (19%), feces (13%), conifer needles (9%), and fine twigs (9%) are placed in the foliage of large conifers (Gillesberg and Carey, 1991). Height, width, and length of six nests occupied by lone males averaged 25 by 28 by 32 cm; they ranged from 20 by 20 by 30 to 38 by 41 by 46 cm. The same dimensions for 21 nests occupied by lone females or females with young averaged 28 by 37 by 44 cm (excluding three exceptionally large nests). Nests occupied by females ranged from 10 by 15 by 15 to 76 by 61 by 76 cm (Maser, 1966*b*).

Of 117 nests located during searches of 50 felled Douglas-fir trees in Douglas Co., 9% were on the first limb, 42% in the lower third of the canopy, 29% in the middle third, and 19% in the upper third; two (2%) nests were in cavities. Of the 50 trees, 35 contained one or two nests but only two trees contained more than five nests (Gillesberg and Carey, 1991). Most nests are placed against the bole of trees, but some are scattered among the branches. When lower limbs bearing nests die, the nests are abandoned and presumably the voles move upward to be close to food sources (Maser, 1966b). Disturbed nests commonly are abandoned, but in one instance a nest was rebuilt at a site within a tree from which a nest was removed a year earlier (Maser, 1966b). Diameter of trees containing nests of *P. longicaudus* was not significantly larger than expected on the basis of random distribution (B. L. Biswell, pers. comm., 14 May 1994).

Red tree voles flushed from their nests move slowly along the branches and may move between trees along paths formed by interdigitating limbs. In moving among the branches, *P. longicaudus* uses its long tail to counterbalance its weight; in some situations the tail may be used as a semiprehensile grasping organ to avoid falling (Maser, 1966b). Many individuals flushed from nests move down the trunk of their nest-tree head first; some may jump from heights of 20 m or more, spread their legs to slow their fall (Maser, 1966b), and land uninjured after bouncing ≥0.5 m when they hit the thick duff on the forest floor. On the forest floor, the cryptic coloration of red tree voles serves to hide them; nevertheless, they usually quickly scamper beneath some object.

Red tree voles are docile when captured and, except for females with young, usually can be handled without serious risk of being bitten. Except for females with nestlings, they are solitary. However, as evidenced by the occurrence of several nests in some trees (Gillesberg and Carey, 1991), there seem to be some colonial tendencies exhibited by *P. longicaudus.*

Red tree voles commonly clip short boughs and carry them back to the nest to feed. New tender needles are eaten whole, but the lateral resin ducts of older needles are dissected free and discarded before the remainder of the needle is eaten (Maser, 1966b). From discarded resin ducts and feces deposited on tops of nests, it has been inferred that nest tops commonly are used as feeding platforms; however, *P. longicaudus* is known to feed elsewhere in trees at times (Maser, 1966b). Discarded resin ducts also are used to line the spherical nest cavities (Taylor, 1915).

Occasionally, *P. longicaudus* comes to the ground, as evidenced by a few being captured in pitfalls (Corn and Bury, 1986). It is not attracted to baits commonly used in trapping small mammals, hence it rarely is caught in traps.

Remarks.—Taylor (1915) erected the subgenus *Arborimus* to separate *P. longicaudus* from other species within the genus *Phenacomys.* Howell (1920a:243) commented that he could "see nothing to be gained by adopting the subgenus *Arborimus* Taylor" and 6 years later (Howell, 1926b) synonymized the name with *Phenacomys.* Johnson (1973:243) elevated *Arborimus* to generic rank for *longicaudus* based on the number and diversity of differences equaling or exceeding those "currently accepted as valid at the generic level" among "microtine genera." Johnson and Maser (1982) extended the genus *Arborimus* to include *albipes.* Hall (1981:788) indicated that the degree of difference between *Arborimus* and *Phenacomys* was "about the same as between some subgenera in the genus *Microtus*" and retained them as subgenera in the genus *Phenacomys.* Repenning and Grady (1988), within the genus *Phenacomys*, retained the subgenera *Arborimus* (for *longicaudus* and the then-unnamed species *pomo*—

Johnson and George, 1991) and *Phenacomys* (for *intermedius* and the fossil species †*deeringensis*) and erected a new subgenus (*Paraphenacomys*) for *albipes* and two fossil species (†*brachyodus* and †*gryci*). We followed the lead of Repenning and Grady (1988) in use of the generic name *Phenacomys.*

Johnson and George (1991) described the red tree voles with 2n = 40–42 karyotypes that occur south of the Oregon-California border as a separate species, *Arborimus pomo.* This caused the red tree vole in Oregon, which has a karyotype of 2n = 48 or 52 and is herein considered *Phenacomys longicaudus*, to become an endemic species. Consequently, in preparing this account we limited references to those concerning specimens from Oregon.

We find it incredulous that an endemic as unique and charismatic as the red tree vole—a species with an extremely small geographic range, a species that possibly requires habitats currently undergoing fragmentation if not complete destruction through forest-management practices, and a species that serves as prey for an endangered species—has not received greater attention by researchers. Only in the past few years have habitat requirements been quantified; much of the remaining information on the species is anecdotal or supported by evidence based on small samples. Even questions as simple as "Do red tree voles ever eat insects?" which could be answered by examining some fecal pellets from tops of nests for fragments of chitin, have not been addressed. A complete analysis of diet by use of microhistological techniques (Holechek, 1982) seems feasible. Research on the effects of "bottlenecking" of populations on genetic variability resulting from fragmentation of habitats by clear-cutting forests seems critical. Minimum patch size of suitable habitat to maintain a colony seems requisite information to ensure continued existence of the species. Also, the dispersal capabilities and tendencies of the species deserve evaluation.

Microtus californicus (Peale)
California Vole

1848. *Arvicola californica* Peale, 8:46.
1897. [*Microtus*] *californicus*: Trouessart, p. 563.
1918. *Microtus californicus eximius* Kellogg, 21:12.

Description.—*Microtus californicus* (Plate XVII) is a medium-sized vole (Table 11-22) with a cinnamon-brown to tawny-olive dorsum overlain by dark brown or black hairs. The underparts are medium gray sometimes washed with buff; the anal area is white. The vibrissae are light grayish and the tail is bicolored, black above and gray below. The feet are grayish.

For identification of *M. californicus,* size alone excludes *Ondatra* and four loops of enamel on the lingual side of M3 (Fig. 11-94) exclude *Phenacomys, Lemmiscus,* and *Microtus oregoni.* A tail <50% of the length of the head and body (<32% of total length); incisive foramina widest in the posterior one-half (Fig. 11-107), not constricted as in other medium and large *Microtus;* or sharply angled loops of enamel on unrooted, ever-growing molars, will separate *M. californicus* from the remaining arvicolines in Oregon (Kellogg, 1918).

Distribution.—*Microtus californicus* occurs from near Eugene, Lane Co., south through the interior valleys of Oregon and most of California (except the northwest fog

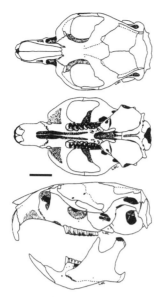

Fig. 11-107. Dorsal, ventral, and lateral views of the cranium, and lateral view of the dentary of the California vole, *Microtus californicus* (OSUFW 3499). Scale bar equals 5 mm.

belt, northeast sagebrush region, and southeast desert area) to El Rosario, Baja California Norte (Fig. 11-108). In Oregon, *M. californicus* occurs in the drainages of the upper Willamette, Umpqua, and Rogue rivers in Lane, Douglas, Jackson, and Josephine counties (Fig. 11-108).

Fossils of *M. californicus* were recovered from Irvingtonian and Rancholabrean deposits in California; one specimen from Vallecito Creek may be the oldest remains of *Microtus* (Zakrzewski, 1985). Records from two localities in Nevada are extralimital (Kurtén and Anderson, 1980). No fossils are known from Oregon.

Geographic Variation.—Of the 17 recognized subspecies, only *M. c. eximius* occurs in Oregon (Hall, 1981).

Habitat and Density.—South of Cottage Grove, Lane Co., Borrecco and Hooven (1972) collected specimens in an 8-year-old clear-cutting planted to Douglas-fir (*Pseudotsuga menziesii*) after the slash was burned. The early seral community consisted largely of common velvet grass (*Holcus lanatus*) and blue wildrye (*Elymus glaucus*) with some hazelnut (*Corylus cornuta*), salal (*Gaultheria shallon*), tall Oregongrape (*Berberis aquifolium*), oxeye daisy (*Chrysanthemum leucanthemum*), false dandelion (*Hypochaeris radicata*), blackberry (*Rubus*), and ceanothus (*Ceanothus*—Borrecco and Hooven, 1972).

In California, investigations into the population biology of *M. californicus* were conducted in habitats in which dominant vegetation was Italian ryegrass (*Lolium multiflorum*), wild oats (*Avena fatua*), ripgut grass (*Bromus rigidus*), and giant wildrye (*Elymus cinereus*); forbs composed ≤30% of the standing crop in summer (Batzli and Pitelka, 1971). In heterogeneous grass habitats, however, higher densities of *M. californicus* were associated with the perennial wildrye than with ripgut grass, the dominant annual (Batzli, 1974). The patches of wildrye serve as refugia when populations are low (Batzli, 1974). Individual California voles tend to restrict their activity to a specific microhabitat type (Ostfeld and Klosterman, 1986).

Some populations of *M. californicus* exhibit multiannual cycles of abundance (Krebs, 1966) and some do not

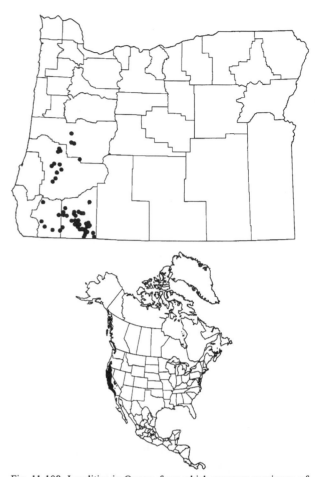

Fig. 11-108. Localities in Oregon from which museum specimens of the California vole, *Microtus californicus*, were examined (listed in APPENDIX, p. 524), and the distribution of the California vole in North America. (Range map redrawn from Hall, 1981:802, map 454, with modifications.)

(Ostfeld and Klosterman, 1986). For example, minimum numbers alive in one California population ranged from 64 males and 122 females/ha at the onset of a study in November to a peak of 387 males and 482 females/ha the following March, then declined to 5 males/ha and 15 females/ha in August (Krebs, 1966). Also, Pearson (1964, 1966, 1971) estimated that densities on a nearby 14-ha area ranged from peaks of 314, 540, and >25/ha to a low of <2/ha. In another California study, densities ranged from 76 to 117/ha during a 27-month period (Ostfeld and Klosterman, 1986). In the latter study, average densities in different microhabitats differed considerably: 127/ha in *Agrostis,* 119/ha in *Holcus,* 58/ha in *Bromus-Lolium,* 47/ha in *Ammophila,* and 14/ha in *Lupinus. M. californicus* has figured prominently in models that involve heterogeneous habitats, social interactions, predation, food resources, and dispersal, singly or in combination, developed to explain multiannual cycles of abundance in arvicolines (Lidicker, 1988).

Diet.—In coastal prairie near San Francisco Bay, California, *M. californicus* ate mostly grasses, especially Italian ryegrass, wild oats, and ripgut grass. When populations of *M. californicus* were high, the standing crop and seed production of these grasses were reduced severely (Batzli and Pitelka, 1971). In laboratory tests, amounts of various plants consumed by *M. californicus* were not correlated with amounts of each plant species provided (Gill, 1977). Grasses were selected over forbs; among grasses, perennial ryegrass (*Lolium perenne*), soft chess (*Bromus mollis*), and rescue grass (*B. catharticus*) were selected over other grasses including wild oats; and seed-head portions of grasses were selected over stem portions (Gill, 1977). Among forbs, stems and flowers of buckhorn plantain (*Plantago lanceolata*) and ox-tongue (*Picris echioides*) were selected over sorrel (*Rumex*), stems and flowers of California figwort (*Scrophularia californica*), stems and flowers of black mustard (*Brassica nigra*), and yarrow (*Achellea borealis*—Gill, 1977).

Reproduction, Ontogeny, and Mortality.—No quantitative information regarding reproduction in Oregon is available. In Marin Co., California, most breeding occurs in late winter and spring with a small peak in autumn and sporadic breeding year-round (Greenwald, 1956). Among captives, development of reproductive organs of females is contingent upon cohabitation with an intact male. Males reared on clean bedding stimulated development of reproductive organs of females, whereas those reared on bedding from their home cage exhibited delayed development and were not able to stimulate development of females (Rissman and Johnston, 1985). The delayed puberty in males in their natal areas may function in avoidance of incestuous matings (Rissman and Johnston, 1985).

Reported average litter sizes are exceptionally variable (Table 11-23), considering the size of samples involved. Litter size also is affected by prior reproductive experience; in California, 75 multiparous females contained an average ($\pm SE$) of 4.45 ± 0.16 embryos, whereas 60 primiparous females contained an average of 3.91 ± 0.18 (Greenwald, 1957). Differences also occur among years (Greenwald, 1957) and among seasons within years (Hoffmann, 1958). Of 302 embryos of ≥ 4 mm crown-rump length, 22 (7.3%) were resorbing; there was a tendency for females to resorb the embryo nearest the cervix (Greenwald, 1957).

Ovulation is induced and occurs <15 h after coitus; however, all follicles do not rupture simultaneously (Greenwald, 1956). Corpora lutea increase in size throughout pregnancy and may attain 2.8 mm in diameter. Significantly more corpora lutea occur in the right ovary and significantly more implantations occur in the right horn of the uterus (Greenwald, 1957). Embryos become visible at about 7 days postcoitus (Greenwald, 1956). The gestation period is 21 days (Hatfield, 1935); parturition is followed by estrus (Greenwald, 1956).

Extra corpora lutea are common in *M. californicus*; ovaries of 82 pregnant females contained a total of 762 corpora lutea, but the females contained only 356 embryos (Greenwald, 1957). Also, extraordinarily large numbers of corpora lutea develop in some females; as many as 29 in both ovaries combined were found in one individual (Greenwald, 1956). The large numbers occur in primiparous and multiparous females, they are not related to duration of pregnancy, and they are not the result of more than a single pregnancy (Greenwald, 1956). However, most large counts occur during the peak of the breeding season. The large counts can be attributed in part to the presence of accessory corpora lutea identified by the retention of an ovum and the remains of the zona pellucida (Greenwald,

1956). Hoffmann (1958) reported that the average number of corpora lutea in primiparous females ($\bar{X} = 12.2$, $n = 55$) did not differ significantly from the average number in multiparous females ($\bar{X} = 12.4$, $n = 43$).

Some females <30 g undergo a sterile cycle at puberty; corpora albicantia are present, but pigmented sites of implantation are absent and mammary tissue has not undergone development (Greenwald, 1956). Biovular, and even triovular, follicles may occur in females of <20 g; however, all such follicles show evidence of atresia. Breeding by females may occur as early as 14 days of age (15.1 g), but breeding at that age likely is exceptional (Greenwald, 1956). Hoffmann (1958) reported that sexual maturity was attained in females at 25–30 g and males at 35–40 g.

Greenwald (1957) estimated that *M. californicus* in a California population produced an average of 4.4 litters in 1952, but produced 5.6 litters in 1953. However, total production remained at about 18 young/female as average ($\pm SE$) litter size declined from 4.93 ± 0.20 to 4.44 ± 0.15 in multiparous females and from 4.50 ± 0.32 to 4.00 ± 0.21 in primiparous females. Hoffmann (1958) reported that the number of litters produced ranged from 9.3 to 11.22 during the 255–320-day breeding season that at the maximum extended from October to August.

Because of accessory corpora lutea, preimplantion losses cannot be estimated; in California, postimplantation mortality was 3.9%, nestling mortality was 32%, and estimates of 0–82% for weanling and juvenile mortality probably were distorted (Hoffmann, 1958).

At birth, *M. californicus* is blind, deaf, and without hair (except vibrissae) and pigment. The top of the head becomes slate blue about 8 h after birth; by 36 h, the anterior half of the dorsum is pigmented and short, downy hair appears on the pate (Hatfield, 1935). By 72 h of age, the entire dorsum including the dorsal surface of the tail is pigmented; the hair on the head is about 2 mm long. The pigment begins to disappear about days 9–10 and finally disappears completely, leaving the skin a light flesh color (Hatfield, 1935). The juvenile pelage, complete by the 5th day, is replaced during the 3rd week by the postjuvenile pelage, which, in turn, is replaced by the adult pelage in the 8th or 9th week. No molt lines are discernible in the pelage, but the extent of the pigment indicates the points at which hair is being replaced (Hatfield, 1935). In captivity, growth and reproduction in young voles of both sexes were suppressed when they were housed with siblings; both also were suppressed in young females by nonsiblings (Batzli et al., 1977). Growth is nearly 1 g/day during the 1st month of life (Greenwald, 1957). Survival and growth were greater in patches of wildrye than in dominant ripgut grass (Batzli, 1974).

In California, both avian and mammalian predators take an enormous toll of *M. californicus*; $>88\%$ of one peak population was eaten by predators before the next breeding season (Pearson, 1964, 1966, 1971). Feral house cats, raccoons (*Procyon lotor*), striped skunks (*Mephitis mephitis*), spotted skunks (*Spilogale gracilis*), and gray foxes (*Urocyon cinereoargenteus*), in order of abundance, were primary mammalian predators on a 14-ha study area. Common avian predators are gulls (Larinae), crows (*Corvus brachyrhynchos*—Gordon, 1977), ravens (*C. corax*—Hall, 1927), black-shouldered (formerly white-tailed) kites (*Elanus caeruleus*—Pearson, 1985), great horned owls

Table 11-22.—*Means (±SE), ranges (in parentheses), and CVs of measurements of skull and skin* characters for female and male Microtus from regions in Oregon. Skin characters were recorded from specimen tags; skull characters were measured to the nearest 0.01 mm. SE and CV not provided if n <10.*

Species and region	Sex	n	Total length	Tail length	Hind foot length	Mass (g)	Occipitonasal length	Basal length	Nasal length	Length of incisive foramen	Length of maxillary toothrow	Zygomatic breadth	Breadth of braincase	Least interorbital length	Skull depth
Microtus californicus	♀	21	167 ± 2.0 (149–182) 5.7	45 ± 0.9 (38–53) 9.1	22 ± 0.3 (18–23) 5.7	43.4 ± 1.79[b] (36.0–55.2) 13.77	28.13 ± 0.23 (26.42–30.19) 3.79	26.39 ± 0.22 (24.62–28.21) 3.82	9.14 ± 0.11 (8.12–9.97) 5.59	5.61 ± 0.06 (4.96–6.11) 5.00	7.24 ± 0.07 (6.47–7.62) 4.22	16.54 ± 0.17 (15.24–18.28) 4.72	11.32 ± 0.13 (10.04–12.09) 5.13	3.66 ± 0.05 (3.35–4.11) 5.81	10.31 ± 0.07 (9.77–11.02) 2.94
	♂	19	174 ± 2.9 (152–196) 7.3	49 ± 1.2 (42–58) 10.3	22 ± 0.2 (21–24) 3.8	47.1[c] (41.4–54.5)	28.49 ± 0.27 (26.18–30.72) 4.19	26.95 ± 0.35 (23.94–29.87) 5.69	9.31 ± 0.16 (8.11–10.40) 7.47	5.61 ± 0.11 (4.89–6.65) 8.29	7.24 ± 0.09 (6.73–7.90) 5.32	16.78 ± 0.18 (15.26–18.00) 4.66	11.38 ± 0.11 (10.64–12.59) 4.33	3.72 ± 0.04 (3.41–4.01) 4.41	10.58 ± 0.07 (10.10–11.07) 2.75
Microtus canicaudus	♀	30	142 ± 1.5 (129–155) 5.7	36 ± 1.3 (25–56) 20.0	19 ± 0.3 (17–24) 7.8	35.3 ± 1.50[d] (24.5–52.0) 20.38	25.97 ± 0.18 (24.37–28.62) 3.81	24.87 ± 0.17 (22.91–27.63) 3.76	7.66 ± 0.09 (6.75–8.86) 6.15	5.19 ± 0.05 (4.46–5.76) 5.78	6.59 ± 0.07 (5.96–8.04) 5.89	15.10 ± 0.09 (14.20–16.21) 3.39	11.40 ± 0.09 (10.27–12.32) 4.28	3.58 ± 0.03 (3.19–4.09) 5.16	10.07 ± 0.06 (9.49–10.90) 3.31
	♂	24	148 ± 2.5 (126–180) 7.9	38 ± 0.9 (29–49) 12.1	19 ± 0.4 (12–22) 9.8	37.6 ± 1.69[e] (27.8–51.6) 20.13	26.24 ± 0.22 (24.54–28.11) 4.12	25.26 ± 0.21 (23.09–27.10) 4.01	7.67 ± 0.09 (6.88–8.56) 5.68	5.10 ± 0.07 (4.61–5.79) 6.32	6.68 ± 0.07 (6.15–7.70) 5.24	15.27 ± 0.16 (14.21–16.98) 5.22	11.44 ± 0.11 (10.24–12.32) 4.92	3.65 ± 0.06 (3.29–4.87) 8.63	10.14 ± 0.08 (9.61–11.18) 4.11
Microtus longicaudus Tillamook County	♀	6	210 (188–224) 6.5	74 (57–85) 9.0	26 (25–27) 3.3	56.2	30.05 (28.19–31.16)	27.95 (25.77–29.71)	9.03 (8.09–10.22)	5.95 (5.44–6.38)	7.51 (7.17–7.98)	16.42 (15.47–17.35)	12.03 (11.72–12.39)	3.97 (3.69–4.16)	10.66 (10.20–11.15)
	♂	8	202 (175–225)	73 (55–85)	25 (22–28)	63.4[f] (46.7–71.0)	28.81 (27.36–30.41)	27.02 (25.04–28.95)	8.37 (7.42–9.88)	5.44 (4.68–6.48)	7.33 (6.89–7.62)	15.87 (14.89–16.73)	11.97 (11.30–12.74)	3.99 (3.68–4.16)	10.49 (9.99–10.84)
Coast Range south of Florence, Lane County	♀	11	192 ± 11.3 (152–258) 19.6	73 ± 6.6 (51–110) 30.1	23 ± 0.9 (20–28) 13.2	48.3[g] (27.1–62.0)	27.89 ± 0.88 (24.99–31.63) 10.45	26.07 ± 0.90 (22.95–30.12) 11.48	8.26 ± 0.42 (7.05–10.46) 16.72	5.14 ± 0.19 (4.48–6.03) 12.22	7.29 ± 0.26 (6.41–8.58) 12.03	15.48 ± 0.46 (13.84–17.91) 9.96	11.30 ± 0.30 (10.21–12.77) 8.87	3.89 ± 0.07 (3.57–4.18) 5.68	10.10 ± 0.18 (9.36–10.91) 5.85
	♂	15	189 ± 8.3 (155–242) 16.9	71 ± 5.1 (50–102) 28.1	23 ± 0.8 (20–29) 13.3	58.1[d] (29.8–79.8)	28.22 ± 0.83 (24.56–33.85) 11.33	26.24 ± 0.79 (22.88–31.54) 11.79	8.25 ± 0.30 (6.72–9.95) 14.22	5.31 ± 0.21 (4.36–6.77) 15.22	7.24 ± 0.21 (6.19–8.53) 11.35	15.42 ± 0.42 (13.59–17.77) 10.58	11.42 ± 0.23 (10.18–12.72) 7.95	3.94 ± 0.07 (3.52–4.33) 6.68	10.33 ± 0.13 (9.53–11.18) 4.96
Wallowa Mountains	♀	14	170 ± 2.9 (149–187) 6.3	59 ± 1.4 (51–68) 9.0	21 ± 0.2 (19–22) 3.3	44.5[h] (32.5–58.1)	25.86 ± 0.23 (24.29–27.47) 3.32	24.27 ± 0.28 (22.21–25.86) 4.28	7.45 ± 0.12 (6.71–8.30) 6.21	4.88 ± 0.07 (4.46–5.26) 5.46	6.48 ± 0.08 (6.02–6.99) 4.54	14.46 ± 0.15 (13.56–15.21) 3.96	10.83 ± 0.11 (10.35–11.75) 3.65	3.75 ± 0.05 (3.44–4.19) 4.95	9.70 ± 0.08 (9.28–10.43) 3.23
	♂	19	173 ± 3.1 (141–198) 7.9	59 ± 1.9 (40–79) 14.6	20 ± 0.6 (11–24) 12.5	40.7[e] (32.5–48.0)	26.53 ± 0.23 (24.50–28.08) 3.84	24.57 ± 0.29 (22.15–26.29) 5.12	7.70 ± 0.12 (6.42–8.39) 6.91	5.14 ± 0.08 (4.46–5.75) 6.74	6.50 ± 0.07 (5.85–7.09) 4.75	14.42 ± 0.16 (13.13–15.53) 4.80	10.92 ± 0.09 (10.12–11.72) 3.71	3.74 ± 0.04 (3.51–4.07) 4.08	9.88 ± 0.06 (9.25–10.29) 2.77
Remainder of eastern Oregon	♀	30	174 ± 1.9 (146–191) 6.3	59 ± 1.2 (47–71) 10.7	21 ± 0.1 (19–22) 3.3	39.5 ± 1.59[i] (31.0–45.7) 12.74	26.49 ± 0.17 (24.49–28.06) 3.44	24.63 ± 0.19 (22.08–26.17) 4.34	7.67 ± 0.08 (6.92–8.39) 5.54	5.03 ± 0.07 (4.11–5.96) 7.10	6.66 ± 0.06 (6.13–7.57) 5.10	14.74 ± 0.12 (13.61–16.00) 4.55	11.33 ± 0.09 (10.38–12.34) 4.36	3.74 ± 0.02 (3.50–4.01) 2.46	9.89 ± 0.06 (9.34–10.78) 3.09
	♂	30	174 ± 1.9 (147–194) 6.1	61 ± 1.1 (52–76) 9.9	21 ± 0.2 (19–23) 4.6	40.6 ± 2.57[j] (20.2–55.9) 25.28	26.89 ± 0.25 (24.91–31.01) 5.14	25.19 ± 0.26 (22.76–28.55) 5.59	7.85 ± 0.11 (6.71–9.33) 7.78	5.07 ± 0.08 (4.20–6.23) 8.49	6.75 ± 0.08 (6.14–7.75) 6.24	15.02 ± 0.16 (13.44–17.25) 5.87	11.32 ± 0.09 (10.41–12.74) 4.24	3.82 ± 0.03 (3.49–4.04) 3.89	10.07 ± 0.04 (9.61–10.61) 2.37
Microtus montanus From Burns to Diamond Lake, Harney County southward	♀	30	159 ± 2.0 (143–189) 7.0	39 ± 0.9 (31–52) 11.8	19 ± 0.2 (18–22) 4.5	41.0 ± 1.69 (27.5–53.0) 16.54	26.23 ± 0.19 (24.32–28.61) 4.03	25.59 ± 0.20 (23.86–27.68) 4.30	7.47 ± 0.10 (6.46–8.54) 7.40	5.39 ± 0.05 (4.89–6.04) 4.88	6.75 ± 0.07 (6.20–7.66) 5.46	15.35 ± 0.15 (14.19–17.56) 5.27	10.64 ± 0.08 (9.73–11.42) 4.35	3.59 ± 0.03 (3.20–3.85) 4.34	9.97 ± 0.07 (9.33–10.91) 3.89
	♂	30	161 ± 2.5 (138–186) 8.4	42 ± 0.9 (35–52) 11.8	20 ± 0.3 (17–23) 6.7	41.9 ± 2.49[k] (28.7–64.7) 25.15	26.68 ± 0.22 (24.19–28.31) 4.43	26.16 ± 0.25 (23.36–28.35) 5.27	7.52 ± 0.11 (5.67–8.53) 8.01	5.45 ± 0.07 (4.69–6.12) 7.43	6.73 ± 0.06 (5.99–7.27) 4.57	15.56 ± 0.17 (13.86–17.25) 5.83	10.75 ± 0.05 (10.25–11.33) 2.75	3.65 ± 0.04 (2.71–3.97) 6.22	10.04 ± 0.08 (9.47–11.06) 4.33

Taxon / Locality	Sex	n													
McDermitt, Malheur County	♀	22	154 ± 1.6 (139–165) 4.9	42 ± 0.7 (36–46) 7.4	19 ± 0.2 (17–21) 5.7	42.6 ± 1.20[d] (32.5–50.7) 10.94	25.76 ± 0.12 (24.75–27.23) 2.18	25.16 ± 0.17 (23.68–27.46) 3.09	7.29 ± 0.09 (6.42–8.28) 6.13	5.16 ± 0.06 (4.70–5.88) 5.33	6.69 ± 0.06 (6.21–7.32) 4.27	15.05 ± 0.10 (13.93–15.91) 3.14	10.72 ± 0.08 (9.79–11.43) 3.44	3.46 ± 0.03 (3.23–3.76) 3.96	9.87 ± 0.06 (9.57–10.46) 2.66
	♂	30	159 ± 2.3 (135–178) 7.8	44 ± 1.6 (26–61) 19.4	20 ± 0.3 (17–23) 6.7	46.7 ± 2.45[l] (24.6–61.0) 19.61	26.08 ± 0.22 (23.68–27.83) 4.57	25.64 ± 0.25 (22.24–27.80) 5.42	7.42 ± 0.09 (6.28–8.30) 7.36	5.34 ± 0.07 (4.42–5.96) 7.43	6.77 ± 0.07 (5.94–7.33) 5.98	15.28 ± 0.15 (13.44–16.74) 5.44	10.73 ± 0.09 (9.70–11.67) 4.78	3.63 ± 0.03 (3.26–3.93) 4.67	10.02 ± 0.07 (9.20–10.86) 4.04
Remainder of Oregon	♀	2	155 (152–158)	40	19		24.77 (24.44–25.10)	23.82 (23.25–24.38)	6.57 (6.47–6.67)	5.43 (5.32–5.53)	6.20 (5.97–6.43)	14.41 (14.20–14.62)	10.97 (10.53–11.41)	3.56 (3.42–3.69)	10.04 (9.84–10.23)
	♂	2	167 (160–174)	45 (42–48)	19 (19–20)		26.67 (25.31–28.02)	26.03 (24.63–27.42)	7.49 (7.22–7.76)	5.27 (5.24–5.29)	6.59 (6.58–6.60)	15.69 (14.86–16.52)	10.93 (10.72–11.13)	3.95 (3.61–4.28)	10.38 (9.94–10.81)
Microtus oregoni Coast Range and northern Cascade Range	♀	30	127 ± 1.6 (108–143) 7.1	33 ± 0.6 (28–42) 9.6	17 ± 0.2 (15–19) 5.8	19.3 ± 0.73[l] (12.6–23.5) 14.07	22.57 ± 0.09 (21.73–23.81) 2.15	21.33 ± 0.12 (19.84–22.58) 3.17	6.68 ± 0.08 (5.88–7.62) 6.38	4.25 ± 0.05 (3.57–4.73) 6.02	5.79 ± 0.03 (5.35–6.10) 3.28	13.26 ± 0.08 (12.53–14.39) 3.37	10.49 ± 0.07 (9.63–11.17) 3.73	3.80 ± 0.03 (3.53–4.03) 3.64	9.43 ± 0.06 (7.92–9.36) 3.65
	♂	30	130 ± 1.7 (112–146) 7.3	34 ± 0.7 (26–43) 10.5	18 ± 0.4 (15–27) 11.8	21.7 ± 1.44[m] (13.4–32.8) 25.74	22.79 ± 0.13 (21.50–24.62) 3.01	21.56 ± 0.15 (20.18–23.11) 3.69	6.69 ± 0.09 (5.72–7.69) 7.54	4.25 ± 0.06 (3.66–4.76) 7.43	5.79 ± 0.04 (5.46–6.29) 3.71	13.59 ± 0.12 (12.61–14.83) 4.79	10.43 ± 0.08 (9.51–11.34) 4.08	3.80 ± 0.03 (3.47–4.23) 4.40	8.55 ± 0.05 (8.11–9.08) 3.09
South of Crater Lake, Klamath County	♂	1	130	32	17		22.95	21.44	5.99	4.28	5.60	12.72	9.79	3.59	8.19
Microtus richardsoni Cascade Range	♀	22	231 ± 3.0 (192–263) 7.2	75 ± 1.9 (59–100) 12.6	28 ± 0.3 (22–30) 5.9	106.2 ± 3.05[k] (88.0–120.7) 9.09	32.27 ± 0.29 (29.29–35.20) 4.23	33.35 ± 0.38 (29.47–37.01) 5.38	9.71 ± 0.16 (8.04–11.14) 7.96	6.17 ± 0.14 (4.69–7.04) 10.85	8.39 ± 0.11 (7.22–9.58) 5.95	20.36 ± 0.26 (18.09–22.36) 5.99	12.64 ± 0.21 (11.33–14.35) 7.66	4.91 ± 0.03 (4.63–5.27) 2.95	11.58 ± 0.09 (10.84–12.23) 3.62
	♂	22	234 ± 3.5 (187–264) 7.5	78 ± 2.2 (59–107) 14.4	28 ± 0.3 (26–30) 4.4	104.1 ± 4.69[b] (71.2–126.0) 14.93	32.81 ± 0.31 (30.57–34.93) 4.39	33.89 ± 0.44 (30.44–36.78) 6.02	9.87 ± 0.18 (8.34–11.96) 8.44	6.19 ± 0.10 (5.33–7.29) 7.82	8.47 ± 0.08 (7.73–9.31) 4.46	20.68 ± 0.27 (18.52–23.34) 6.11	12.96 ± 0.18 (11.82–14.37) 6.37	5.01 ± 0.03 (4.75–5.39) 3.25	11.76 ± 0.09 (11.04–12.52) 3.49
Wallowa and Blue mountains	♀	10	214 ± 4.8 (192–237) 7.1	66 ± 1.8 (60–78) 8.9	26 ± 0.3 (24–27) 3.2	85.7 ± [f] (64.0–106.2)	30.76 ± 0.43 (29.23–32.57) 4.45	31.58 ± 0.74 (28.46–34.23) 7.42	8.80 ± 0.21 (7.68–9.61) 7.61	6.23 ± 0.10 (5.81–6.83) 5.17	8.09 ± 0.19 (7.01–8.76) 7.40	19.27 ± 0.45 (17.37–21.68) 7.37	12.61 ± 0.15 (11.83–13.41) 3.82	4.89 ± 0.03 (4.66–5.00) 2.14	11.06 ± 0.15 (10.46–11.85) 4.38
	♂	7	224 (199–253)	74 (66–86)	26 (25–28)	91.1[a] (76.7–122.0)	31.53 (30.12–33.20)	32.83 (30.44–34.76)	9.24 (8.57–10.42)	6.51 (6.00–7.14)	8.08 (7.29–8.65)	19.53 (18.21–20.50)	12.77 (11.96–14.22)	5.07 (4.92–5.21)	11.14 (10.56–11.65)
Microtus townsendii	♀	30	181 ± 3.0 (151–220) 9.2	52 ± 1.3 (38–62) 13.4	24 ± 0.3 (22–28) 7.2	54.8 ± 2.82[i] (37.2–70.7) 19.29	28.95 ± 0.19 (27.02–31.28) 3.66	27.91 ± 0.23 (25.52–30.62) 4.57	8.38 ± 0.09 (7.49–9.34) 5.75	5.93 ± 0.06 (5.32–6.47) 5.56	7.64 ± 0.06 (6.98–8.27) 3.97	16.58 ± 0.13 (15.00–17.94) 4.31	11.58 ± 0.08 (10.54–12.31) 3.96	4.01 ± 0.04 (3.54–4.47) 5.52	10.93 ± 0.07 (10.39–11.77) 3.27
	♂	30	188 ± 2.5 (162–226) 7.3	58 ± 1.6 (42–84) 15.3	24 ± 0.6 (15–36) 12.7	59.3 ± 3.38[k] (33.0–82.0) 24.17	29.54 ± 0.19 (28.00–31.18) 3.76	28.59 ± 0.19 (26.40–30.44) 3.16	8.54 ± 0.08 (7.60–9.25) 4.85	6.11 ± 0.06 (5.45–6.78) 5.32	7.59 ± 0.05 (6.81–8.24) 3.85	16.85 ± 0.13 (15.01–18.48) 4.27	11.61 ± 0.10 (10.51–12.74) 4.76	4.09 ± 0.03 (3.72–4.37) 4.02	10.98 ± 0.04 (10.49–11.37) 1.95

[a]Although measures of ear length were recorded, the data were not included because of extreme variation believed to stem from differences in techniques used to measure these characters.

[b]Sample size reduced by 11. [c]Sample size reduced by 10. [d]Sample size reduced by 8. [e]Sample size reduced by 7. [f]Sample size reduced by 4. [g]Sample size reduced by 5. [h]Sample size reduced by 9. [i]Sample size reduced by 20. [j]Sample size reduced by 14. [k]Sample size reduced by 15. [l]Sample size reduced by 16. [m]Sample size reduced by 2.

Table 11-23.—*Reproduction in Oregon* Microtus.

Species	Number of litters	Litter size $\overline{X}$	Range	n	Reproductive season	Basis	Authority	State or province
Microtus californicus	4.4[a]	4.2	1–9	154	All year[b]	Embryos	Greenwald, 1957	California
	9–11	6.5	2–10	109	Oct.–Aug.	Embryos	Hoffmann, 1958	California
Microtus canicaudus		4.7	3–7	6	Oct.–Dec.[c]	Embryos	This study	Oregon
		4.6[d]		63	Captives	Live birth	Hagen and Forslund, 1979	Oregon
		4.4[e]		78	Mar.–Dec.	Embryos + implant sites	Wolff et al., 1994	Oregon
Microtus longicaudus		5.0	4–7	4	.	Embryos	This study	Oregon
		4.8	4–6	5		Embryos	Borell and Ellis, 1934	Nevada
		5.6	2–8	39	May–Oct.	Embryos	Hall, 1946	Nevada
	≤2[f]	5.0		26	May–Sep.	Embryos	Van Horne, 1982	Alaska
Microtus montanus	4–5[g]	6.8	1–13	194	Apr.–Oct.	Corpora lutea	Hoffmann, 1958	California
		6.5	2–10	109		Embryos	Hoffmann, 1958	California
		6.0		75		Implant sites	Hoffmann, 1958	California
		4.5	3–6	?	Mar.–Nov.[h]	Neonates	Seidel and Booth, 1960	Washington
		4.2[h]	1–10	254		Neonates	Negus and Pinter, 1965	Wyoming[i]
		6.3	2–8	54	Apr.–Nov.	Embryos	Hall, 1946	Nevada
		5.8	2–10	46		Embryos	Vaughan, 1969	Colorado
Microtus oregoni	4.8[j]	3.1		18	Mar.–Sep.	Embryos	Gashwiler, 1972	Oregon
		3.4		34		Implant sites	Gashwiler, 1972	Oregon
		4.0	2–6	21[k]	Mar.–Sep.[l]	Embryos[k]	Hooven, 1973b	Oregon
Microtus richardsoni		4.0	3–6	5[m]		Embryos	This study	Oregon
	≤2	5.0	2–9	8	May–Sep.	Embryos	Ludwig, 1988	Alberta
		5.3	3–8	9	May–Sep.	Implant sites	Ludwig, 1988	Alberta
	2	7.9	6–10	26	Jun.–Jul.	Embryos	Brown, 1977	Montana
Microtus townsendii		5.9	3–9	10	Apr.–Nov.[c]	Embryos	This study	Oregon
		5.9	3–10	9		Implant sites	This study	Oregon
		5.1[n]		278	Feb.–Nov.	Embryos	Anderson and Boonstra, 1979	British Columbia

[a]Datum is estimate for 1952; in 1953, estimate was 5.6.
[b]Most breeding is in late winter and spring with a minor peak in autumn; sporadic breeding occurs year-round.
[c]Earliest and latest records of pregnancy; collections not continuous throughout the year.
[d]Pooled sample of first litters only; litter size significantly related to age of female.
[e]Mean not significantly different from that of litters ($\overline{X} = 4.6$, $n = 198$) conceived in the laboratory.
[f]Excluding possible fall and winter litters.
[g]Estimated number of litters ranged from 4.4 to 5.2.
[h]In captivity.
[i]Stock transported to Louisiana.
[j]An average calculated on the basis of a 23.8-day gestation period and the proportion of adult females pregnant each month.
[k]Plus counts of young in an unstated number of litters born in live traps.
[l]Indicated that breeding may be throughout the year.
[m]Four specimens collected in July–August may have been young-of-the-year, possibly accounting for the small mean litter size (Ludwig, 1988).
[n]Pooled sample; litter size significantly related to season.

(*Bubo virginianus*—Fitch, 1947), barn owls (*Tyto alba*—Fitch, 1947; Selleck and Glading, 1943), red-tailed hawks (*Buteo jamaicensis*—Fitch, 1947; Fitch et al., 1946), and marsh hawks (*Circus cyaneus*—Selleck and Glading, 1943).

In California, predator populations tend to track those of cycling populations of *M. californicus,* but lag behind them; intensity of predation was least as populations of *M. californicus* increased and was greatest late in the decline phase (Pearson, 1971). Carnivore densities of about 1% of those of vole densities likely prevent growth of maximally reproducing vole populations; those 0.01–0.5% likely slow growth of breeding populations and cause declines in nonbreeding populations; and those <0.01% likely do not prevent increases in breeding populations (Pearson, 1971).

Habits.—*Microtus californicus* spends most of the time underground in its burrows or in those of other species; it makes only brief aboveground excursions (Pearson, 1960b). Activity (188 round trips, mostly by females) monitored at the entrance to a tunnel from which led only one runway indicated that 79% of aboveground activity was in the daytime and that aboveground sojourns were >3 times longer in the daytime ($\overline{X} = 7.4$ min) than at night. However, the temporal distribution of activity, as measured by the number of passages along runways, did not deviate significantly from uniform in autumn, winter, and spring, but, in summer, most activity was in the 4 h after sunrise and activity essentially ceased at midday (Pearson, 1960b). *M. californicus* was significantly more active in the daytime when the moon was full than when there was no moonlight (Pearson, 1960b). Although more activity was recorded on rainy days than on rainless days, the difference was not significant. Photographs of wild California voles as they traversed their runways showed that all sex and age groups commonly carried pieces of grasses (e.g., wild oats) or forbs (e.g., *Rumex*), that all sex and age groups shared runways, and that total activity was not necessarily related to the number of different voles that used a runway (Pearson, 1960*b*).

In staged encounters in a neutral arena, captive *M. californicus* used mutual avoidance to maintain individual distances between males (Colvin, 1973). *M. californicus* also avoided *M. montanus* in interspecific encounters. Also, *M. californicus* was dominant to *Mus musculus* and *Reithrodontomys megalotis* (Blaustein, 1980). Blaustein (1980) found that population density of *M. californicus* was correlated inversely with reproductive success in *R. megalotis* and *M. musculus*; when densities of *M. californicus* were high the latter two species were forced into suboptimal habitat. Also, *R. megalotis* was caught significantly more often in clean traps than traps soiled by *M. californicus*, but only when both species were reproductively active (Heske and Repp, 1986).

Home-range size of *M. californicus* in coastal prairie near San Francisco Bay, California, calculated by the 95% harmonic-mean ($\pm SE$) method was 85.3 ± 13.3 and 128.6 ± 20.9 m^2 for males and females, respectively, when the population density was about 374/ha and 116.5 ± 21.2 and 231.0 ± 31.2 m^2 the following year when the density was about 45/ha (Ostfeld, 1986). Addition of food (carrots) to the area did not affect home ranges of males; they remained largely nonoverlapping. However, added food induced females previously not occupying overlapping home ranges to overlap home ranges extensively.

During the breeding season, adult California voles are captured disproportionately in traps baited with the odor of conspecifics; large males and all adult females were attracted to odors of both males and females (Heske, 1987*a*). However, smaller males were attracted only to odors of females. Juveniles were not attracted to odors of conspecifics and no voles were attracted to odors of conspecifics in the nonbreeding season. These findings suggest that intrasexual interactions are stronger among males than among females (Heske, 1987*a*) and support the idea of Ostfeld (1986) that males are strongly territorial and females are nonterritorial or selectively territorial. In captivity, pregnancy was interrupted by replacing the stud male with a strange male, a phenomenon that may be common to most arvicolines (Heske, 1987*b*).

Remarks.—As Heske's (1987*a*) work with *M. californicus* suggested, previous occupancy of a live trap may affect its attractiveness to animals that subsequently approach the trap. This, in turn, may violate the assumption that all individuals in a population are equally susceptible to being captured, an assumption inherent in applying many models for estimating density. Because olfaction contributes significantly to perception of the environment by many, if not most, wild mammals, the principle of odor affecting susceptiblity to capture may extend to many other small mammals. Although they may reduce the overall rate of capture and add significantly to the labor involved in field studies, clean, odor-free traps likely will enhance the reliability of studies of population dynamics of small mammals.

We emphasized earlier that it might be inappropriate to infer that closely related species share ecologies or behaviors when one species has been investigated, but one or more congeners have not been studied. An account of

Microtus likely is an appropriate place to comment that the reverse also may be true; our failure to record a specific item does not necessarily indicate that a species does not interact with its environment or behave in a manner similar to that of its congeners. More likely it indicates either a paucity of available information or that we failed to find information on that topic for a particular species.

Microtus canicaudus Miller
Gray-tailed Vole

1897*b*. *Microtus canicaudus* Miller, 11:67.

Description.—*Microtus canicaudus* (Fig. 11-109) is a medium-sized vole (Table 11-22) with yellowish-brown or yellowish-gray dorsal pelage and a short tail, blackish or brownish above and light gray below. It can be distinguished from sympatric congeners as follows: from *M. townsendii* by a lighter-colored pelage, a tail <50% of the length of

Fig. 11-109. Photograph of the gray-tailed vole, *Microtus canicaudus*. (Reprinted from Verts and Carraway, 1987*a*, with permission of the American Society of Mammalogists.)

the head and body, and a V-shaped, not U-shaped, notch at the posterior edge of the palate (Fig. 11-110); from *M. oregoni* by its more robustly built body, an eye ≥4 mm in diameter, and a third upper molar with four loops of enamel on the lingual side (Fig. 11-94). It can be distinguished from *M. montanus* (of which it was formerly considered a geographic race) by incisive foramina not markedly constricted posteriorly (Fig. 11-110). Also, *M. canicaudus* and *M. montanus* have karyotypes that "lack homology in six of 22 autosomal arms, differ in the amount and location of autosomal and X chromosome heterochromatin, and differ in the location of one Ag-NOR site" (Modi, 1986:164). Finally, electrophoresis of serum proteins reveals that the secondary globulin bands are farther apart and the hemoglobin molecule migrates more slowly in *M. canicaudus* than in *M. montanus* (Johnson, 1968).

Distribution.—*Microtus canicaudus* is endemic to the Willamette Valley, Oregon, and Clark Co., Washington. In Oregon, the species occurs from near Scapoose, Columbia Co., and Gresham, Multnomah Co., south through the Willamette Valley to near Eugene, Lane Co. (Bailey, 1936; Modi, 1986; Fig. 11-111). We consider all specimens examined from east of the Cascade Range labeled *M. canicaudus* to have been misidentified.

Microtus canicaudus and *M. montanus* are allopatric

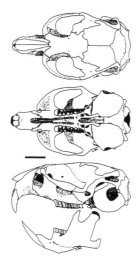

Fig. 11-110. Dorsal, ventral, and lateral views of the cranium, and lateral view of the dentary of the gray-tailed vole, *Microtus canicaudus* (OSUFW 2864). Scale bar equals 5 mm.

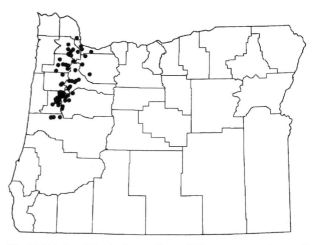

Fig. 11-111. Localities in Oregon from which museum specimens of the gray-tailed vole, *Microtus canicaudus*, were examined (listed in APPENDIX, p. 524). The species also occurs north of the Columbia River in Clark Co., Washington (Verts and Carraway, 1987a).

(Anderson, 1959; Johnson, 1968), but not contiguously so as claimed by Modi (1986).

Geographic Variation.—*Microtus canicaudus* is monotypic.

Habitat and Density.—Presently, gray-tailed voles are associated largely with agricultural lands, particularly small grains, grasses grown for seed, permanent pastures and hayfields, and waste places vegetated with grasses (Bailey, 1936; Goertz, 1964). Formerly, the species likely was associated with the large prairies resulting from annual burning by Indians (Johannessen et al., 1971). To some extent, gray-tailed voles have invaded formerly forested areas now converted to agriculture (Goertz, 1964; Maser and Storm, 1970).

These voles occasionally become so abundant that control measures are necessary (Goertz, 1964). However, few quantitative data are available regarding density. Pearson (1972) estimated the density of a study population at 141 ± 62/ha but used obscure methods. Densities in enclosures as high 2,880/ha were recorded (Wolff et al., 1997). Verts and Carraway (1987a) speculated that fluctuations in density are cyclic. One of us (BJV) observed gray-tailed voles so abundant in November 1974 that, in walking across a nearly denuded field near Corvallis, Benton Co., with two dogs, voles were continuously in view as they scampered for holes leading to belowground retreats. In February 1975, no fresh fecal droppings or cuttings were found and 100 snap traps set across 3-cm-deep runways worn in the soil produced no captures of gray-tailed voles in the field. At peak densities, gray-tailed voles occasionally are killed by automobiles as they disperse across macadam secondary roads. Also at such times, they make runways from lodged pasture grasses to garden plots through several meters of mowed lawn grass.

Diet.—Food habits of gray-tailed voles have not been studied directly. However, Maser and Storm (1970) considered grasses, clover (*Trifolium*), wild onion (*Allium amplectens*), and false dandelion (*Hypochaeris radicata*) to be common items in the diet. We maintained gray-tailed voles in captivity on white clover (*T. repens*), apples, bluegrass (*Poa*), and Italian ryegrass (*Lolium multiflorum*—Verts and Carraway, 1987a). Gile and Gillett (1979) claimed

that gray-tailed voles were omnivorous, but presented no evidence or references to support their contention.

Reproduction, Ontogeny, and Mortality.—Captive female gray-tailed voles as young as 18 days and that weigh only 12.5 g are capable of mating and producing viable offspring (Hagen and Forslund, 1979). Mean litter size for first litters was 5.2 ± 0.3, mean mass of offspring at birth was 2.4 ± 0.03 g, and survival to 18 days was 0.30 (Hagen and Forslund, 1979). Females 28, 42, and 70 days old when first mated produced significantly smaller litters (4.8 ± 0.3, 4.2 ± 0.4, and 4.1 ± 0.2, respectively), significantly heavier young at birth (2.5 ± 0.04, 2.5 ± 0.06, and 2.5 ± 0.03 g, respectively), and young that exhibited significantly greater survival to 18 days (0.86, 0.97, and 0.92, respectively). Litter size continued to decline as age of females at first mating increased, a relationship possibly responsible for lower mean mass of offspring produced by younger females. Females 18 days old when first mated rarely remated immediately after parturition, thus did not produce as many total young as those first mated at ≥28 days of age (Hagen and Forslund, 1979). Wolff et al. (1994) reported that litter size among voles bred in captivity did not differ significantly with age or parity, or from those of wild-caught females based on counts of embryos or implantation sites (Table 11-23). They also reported that 78–92% of adult females (those ≥30 g) caught in live traps in May–October were in reproductive condition; the proportion declined to 68% in November, 18% in December, and 0% in January and February, then increased to 38% in March. Reproductive condition among males, as indexed by testis size, followed a similar seasonal pattern. Males exhibited a greater tendency to move away from their natal home ranges before attaining sexual maturity than did females. Such movements were sufficient to separate voles from opposite-sex relatives or offspring bred only after presumed parents were no longer present (Wolff et al., 1994).

Boyd and Blaustein (1985) found that familiarity, acquired by partners being reared together, suppressed reproduction regardless of their genetic relatedness. Litter size, age of reproductive competency, and survival of offspring were not affected, but number of litters produced was reduced significantly by familiarity of partners. A role of familiarity in population cycles was postulated (Boyd and Blaustein, 1985).

Six specimens on deposit in the Oregon State University, Department of Fisheries and Wildlife mammal collection were pregnant when captured; number of embryos averaged 4.7 (range, 3–7). Five others (seven pregnancies) exhibited pigmented sites of implantation ($\bar{X}$ = 4.9, range, 1–8). Based on differences in intensity of pigmentation, two had two sets; in each instance the lighter was a set of four and the darker a set of six, suggesting that initial litters may be smaller than subsequent litters. Maser and Storm (1970) suggested that gray-tailed voles breed throughout the year; all our records are for October–December, certainly reflecting the strong bias in collecting effort.

Gestation is 21–23 days (Hagen and Forslund, 1979), more commonly 21 days (Tyser, 1975). Animals fed restricted diets weighed significantly less than those fed the same diet ad lib., but did not produce young with sex ratios biased toward females (Goldenberg, 1980) as hypothesized (Trivers and Willard, 1973).

Interspecific matings of *M. canicaudus* and *M. montanus* produced more pregnancies than intraspecific matings of *M. canicaudus* and approximately the same number of pregnancies as intraspecific matings of *M. montanus* (Tyser, 1975). However, mean litter size and survival were significantly less among hybrids than for offspring of intraspecific matings.

At 1 week of age, captive males and females weighed an average of 6.0 ± 0.2 and 5.5 ± 0.1 g, respectively, and were 68.7 ± 1.1 and 66.0 ± 0.8 mm in total length, respectively (Tyser, 1975). By 8 weeks of age, the two sexes weighed an average of 28.8 ± 1.0 and 22.8 ± 0.5 g, respectively, and were 146.9 ± 1.4 and 134.3 ± 0.8 mm long, respectively (Tyser, 1975).

No information regarding survival or population turnover is available; however, we speculate that average life expectancy at birth is extremely short, possibly no more than 1 month. In Benton Co., 104 pellets regurgitated by long-eared owls (*Asio otus*) contained remains of 51 *M. canicaudus* (Reynolds, 1970) and 28 pellets regurgitated by saw-whet owls (*Aegolius acadicus*) contained remains of 14 *M. canicaudus* (Forsman and Maser, 1970). We commonly found remains of gray-tailed voles in pellets of barn owls (*Tyto alba*) from the Willamette Valley used in a classroom exercise. Other predators of *M. canicaudus* include hawks (Falconidae), foxes (*Vulpes vulpes* and *Urocyon cinereoargenteus*), skunks (*Mephitis mephitis*), and house cats (Maser and Storm, 1970). We also observed great blue herons (*Ardea herodias*) and juvenile red-tailed hawks (*Buteo jamaicensis*) to prey on gray-tailed voles in a ryegrass field across the road from our home.

Habits.—*Microtus canicaudus* constructs a complex interlaced system of runways and subterranean burrows; it frequently uses burrows of other species such as those of the camas pocket gopher (*Thomomys bulbivorus*—Maser and Storm, 1970). Belowground tunnels are occupied even when fields flood in winter; apparently air trapped in chambers and tunnels is sufficient for the voles to breathe and for preventing some parts of the burrow system from flooding. Gray-tailed voles swim through flooded tunnels to reach unflooded chambers or sections of tunnels (Maser

and Storm, 1970), but when the entire system floods they abandon the tunnels and seek "high ground" (Verts and Carraway, 1987a). Surface runways are constructed beneath lodged grasses, thus resemble tunnels. Middens are established at intersections of some runways; they may be 8–15 cm long, 3–5 cm wide, and 8–10 cm deep (Maser and Storm, 1970). Nests are constructed beneath the surface of the ground or on the surface beneath bales of hay, discarded boards, and other debris.

In 45- by 45-m square enclosures, minimum-area home ranges of males averaged (94.0 ± 54.3 m^2, $n = 35$) 1.7 times larger than those of females (56.0 ± 30.1 m^2; $n = 24$—J. O. Wolff, pers. comm.; Wolff et al., 1994). Home ranges of males overlapped those of an average ($\pm SD$) of 1.8 ± 1.5 other males by an average of $47.1 \pm 1.5\%$. Initially, females seemed to be territorial and occupied exclusive home ranges, but as density increased in the enclosures home ranges overlapped those of other females by $39.5 \pm 34.4\%$.

Weil (1975) arranged intraspecific and interspecific encounters with laboratory-reared *M. canicaudus* in a neutral arena, the former between pairs of the same and opposite sex, the latter with the sympatric *M. townsendii* and *M. oregoni*. Behavior was scored as dominant-subordinate, mutual avoidance, fraternal, and "no decision" for both intraspecific and interspecific encounters. Intraspecific male-male pairings ($n = 28$) produced behavior scored as 60%, 21%, 3%, and 14% in each of the four categories, respectively; female-female pairings ($n = 16$) produced scores of 50%, 31%, 6%, and 13%; and opposite-sex pairings ($n = 16$) produced scores of 50%, 31%, 0%, and 13%. For male-male pairings scores were 56%, 28%, 3%, and 12% for *M. canicaudus-M. oregoni* encounters and 43%, 53%, 0%, and 3% for *M. canicaudus-M. townsendii* encounters. For female-female pairings of the same species scores were 62%, 0%, 0%, and 37%, and 50%, 37%, 0%, and 13%, respectively. In interspecific encounters with opposite-sex pairings, *M. oregoni* was more often dominant to *M. canicaudus* and *M. canicaudus* was more often dominant to *M. townsendii* (Weil, 1975).

Remarks.—*Microtus canicaudus* does not enter solid metal live traps readily (Maser, 1967), possibly explaining the paucity of information available on the biology of the species in the wild (Verts and Carraway, 1987a). Live traps constructed of wire mesh and set so that voles enter them as they proceed along established runways are most effective (Maser, 1967; Pearson, 1972), but in our experience even these traps capture relatively few gray-tailed voles per unit effort. Even with prebaiting, capture rates of free-living *M. canicaudus* tend to be low. However, *M. canicaudus* in enclosed populations established from wild-caught or laboratory-reared stock seemingly does not exhibit the same reluctance to enter traps (Edge et al., 1995, 1996; Manning et al., 1995; Wolff et al., 1994). *M. canicaudus* seems particularly sensitive to some pesticides (Meyers and Wolff, 1994).

Modi (1986) postulated that mid- to late-Pleistocene montane glaciers separated populations of the widespread ancestor to *M. canicaudus* and *M. montanus*; those west of the Cascade Range diverged to become *M. canicaudus* and those east of the Cascade Range became *M. montanus*. He further suggested that the two species have occupied ranges much the same as the present-day geographic ranges since the speciation event. We cannot support or refute Modi's

(1986) contentions, but we long have been intrigued by the possible effect that the cataclysmic Bretz floods of about 13,000 years ago had on the evolution of species. When an ice-dam broke, water from Pleistocene Lake Missoula cascaded down the Columbia River and backed up the Willamette Valley to approximately the 125-m contour (Allen et al., 1986), an area that matches almost exactly the known distribution of the gray-tailed vole in Oregon. Surely, populations of *M. canicaudus* were severely fragmented and bottlenecked. Also, if *M. canicaudus* and *M. montanus* ever were parapatric as Modi (1986) contended, they certainly became allopatric at that time and likely have been since that era.

Anderson (1959) indicated that M. L. Johnson collected the first specimens of *M. canicaudus* (then considered a subspecies of *M. montanus*) north of the Columbia River in Clark Co., Washington, in 1957. Although skulls of the Washington specimens differed in four characters from those of *M. canicaudus* from the Willamette Valley, Oregon, he considered (p. 454) the Washington specimens in "cranial characters and in color of pelage" to resemble *M. canicaudus* rather than *M. montanus canescens* or *M. m. nanus.* He considered (p. 454) the 13 Washington specimens on deposit at the James R. Slater Museum of Natural History, University of Puget Sound to represent a "variant population of *canicaudus.*"

Both of us examined the 13 specimens plus one other labeled *Microtus canicaudus* from Clark Co., Washington, on deposit at PSM and compared them with specimens of *M. canicaudus* from the Willamette Valley, Oregon, and specimens of *M. m. canescens* and *M. m. nanus* from Washington. The deep V-shaped notch at the posterior edge of the palate, the gradually tapering incisive foramina posteriorly, and the narrow medium-brown dorsal stripe on the tail are considered diagnostic characteristics of *M. canicaudus* in the Willamette Valley (Verts and Carraway, 1987*a*). Although most of the Washington specimens labeled *M. canicaudus* do not possess a well-developed V-shaped notch on the posterior edge of the palate, the incisive foramina and tail stripe of all 14 are more like those of *M. canicaudus* from Oregon than those of *M. m. nanus* from Washington. In addition, the tail in the 14 Washington *M. canicaudus* is much shorter than in *M. m. canescens.* In the absence of karyological or electrophoretic evidence (which seemingly is more definitive than morphological characters—Johnson, 1968; Modi, 1986) to the contrary, we also consider the specimens from Clark Co., Washington, to be *M. canicaudus.* Thus, *M. canicaudus* is not restricted to Oregon as claimed by Wolff et al. (1994).

Microtus longicaudus (Merriam)
Long-tailed Vole

1888*b.* *Arvicola* (*Mynomes*) *longicaudus* Merriam, 22:934.
1891. *Arvicola* (*Mynomes*) *mordax* Merriam, 5:61.
1895. *Microtus* (*Mynomes*) *longicaudus*: Allen, 7:266.
1898. *Microtus angusticeps* Bailey, 12:86.
1923. *Microtus mordax abditus* Howell, 4:36.
1931*a.* *Microtus mordax angustus* Hall, 37:13.
1938. *Microtus longicaudus abditus*: Goldman, 19:491.
1938. *Microtus longicaudus angusticeps*: Goldman, 19:491.
1941. *Microtus mordax halli* Hayman and Holt, 2:603.
1948. *Microtus longicaudus halli*: Dalquest, 2:353.

Description.—*Microtus longicaudus* (Plate XVII) usually is considered to be a medium-sized vole (Table

11-22), although the size of an exceptionally large male may rival that of the largest *M. townsendii.* For identification of *M. longicaudus*, size alone excludes *Ondatra;* four loops of enamel on the lingual side of M3 (Fig. 11-94) exclude *Phenacomys, Lemmiscus,* and *Microtus oregoni.* A combination of the following excludes *M. canicaudus, M. townsendii, M. montanus, M. californicus,* and *M. richardsoni*: tail longer than 50% of the length of the head and body, incisive foramina widest in the anterior half and tapering gradually posteriorly, incisors obscured by nasals in dorsal view (Fig. 11-112), condylobasal length <35 mm, and length of the hind foot ≤36% of the tail length. Unrooted ever-growing molars with sharply angled loops of enamel exclude *Clethrionomys.*

The color of the dorsal pelage "ranges from dull grayish through brownish gray to dark sepia brown" (Smolen and Keller, 1987:1); those west of the Cascade Range have the darkest- and richest-colored pelages. The sides are lighter and more grayish and the venter is grayish white to dull buff (Hall and Cockrum, 1953). The eyes and ears are large. There are four pairs of mammae.

Occasional dental anomalies in *M. longicaudus* have been recorded: grooved upper incisors (Jones, 1978), flattened or blunt incisors, and incisors worn at various angles (Maser and Hooven, 1970*a*).

Distribution.—The geographic range of *M. longicaudus* (Fig. 11-113) extends from northeastern Alaska and extreme northwestern Northwest Territories southeast near the Yukon-Northwest Territories, British Columbia-Alberta, and Alberta-Montana boundaries to northeastern Montana, western South Dakota, and southern New Mexico, then westward through northern Arizona, and eastern and northern California (Hall, 1981). The circumscribed area contains numerous hiatuses, and several of the marginal populations may be relics from a wider Pleistocene distribution (Smolen and Keller, 1987). In Oregon, *M. longicaudus* occurs in 30 of the 36 counties (Fig. 11-113). It is absent from the Willamette Valley and other interior valleys west of the Cascade Range.

Fossils of *M. longicaudus* were found in Wisconsinan-

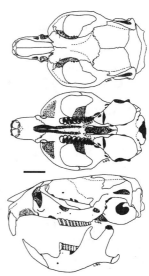

Fig. 11-112. Dorsal, ventral, and lateral views of the cranium, and lateral view of the dentary of the long-tailed vole, *Microtus longicaudus* (OSUFW 3132). Scale bar equals 5 mm.

age cave deposits in Idaho, Wyoming, Colorado, and New Mexico, sites in or near the present-day range (Kurtén and Anderson, 1980).

Geographic Variation.—Three of the 15 named subspecies occur in Oregon: the dark, long-tailed *M. l. abditus* in the northern Coast Range, the brownish *M. l. angusticeps* with creamy underparts in the extreme southwestern corner of the state, and the lighter brownish-gray *M. l. longicaudus* east of the Cascade Range (Bailey, 1936; Hall, 1981). Hall (1981:808, map 457) illustrated the range of *M. l. halli* as occurring in northeastern Oregon; however, he listed no published records for the taxon in the state. Hall (1933a) named *M. mordax angustus*, which later became *M. l. halli*, based largely on its small body size and narrow skull. Because the proportion of large and small individuals in a population is related to the stage of the cycle in some *Microtus* (Boonstra and Krebs, 1979), we question recognition of small specimens from the Wallowa Mountains as a separate taxon.

Habitat and Density.—Within its range, *M. longicaudus* occupies a diversity of habitats: coniferous forest, thickets, forest-meadow ecotones, riparian zones, marshes, and grassy or sagebrush areas (Smolen and Keller, 1987). In eastern Washington, *M. longicaudus* was common in a shrub habitat and in a mixed grass-shrub-

woods habitat during some years, but rarely occurred in grass habitat (Randall and Johnson, 1979). However, these authors suggested that interspecific competition with *M. montanus* played a role in habitat use by *M. longicaudus*. From a study in the same region, Rickard (1960) suggested a close relationship between the presence of *M. longicaudus* and the distribution of a snowberry (*Symphoricarpos*) union (consisting of dwarf shrubs including snowberry; rose, *Rosa*; and cherry, *Prunus*) in which <20% of the light reached the ground. In eastern Oregon, Feldhamer (1979b) found *M. longicaudus* only in marshes vegetated by hardstem bulrush (*Scirpus acutus*), common cat-tail (*Typha latifolia*), Baltic rush (*Juncus balticus*), and sedges (*Carex*). In western Oregon, Black and Hooven (1974) found *M. longicaudus* on 1–10-year-old clear-cuttings burned over by wildfire within 2 years of the fire, but the voles disappeared by 5 years after the fire and were never present on control plots outside the area of the burn. They also found that treatments with herbicides to kill grasses or broad-leaved plants or both did not consistently reduce population levels of long-tailed voles. In the Ruby Mountains, Nevada, Borell and Ellis (1934) found *M. longicaudus* in sagebrush (*Artemisia*) as much as 1.6 km from water, in grass or brush along streams, in meadows, on hillsides covered with chaparral or grass, in willow (*Salix*) thickets, and on rock slides. Occasionally, *M. longicaudus* is taken in the same microhabitat as *Lemmiscus curtatus* (Borell and Ellis, 1934).

In Wyoming, Brown (1967a) found the greatest density of *M. longicaudus* in willow-aspen (*Alnus incana*) with sedge understory, quaking aspen (*Populus tremuloides*) forest with understory of bluegrass (*Poa*) and wheatgrass (*Agropyron*), and subalpine meadows (sedge plus a variety of forbs, including fleabane, *Erigeron*; clover, *Trifolium*; wild sweet-william, *Phlox*; and columbine, *Aquilegia*). In the same state, Clark (1973) recorded *M. longicaudus* in shrub-sedge-grass-savanna (willow; brome, *Bromus*; bluegrass; reedgrass, *Calamagrostis*; aster, *Aster*; and other forbs) with <50% shrub cover and in aspen forest with an understory of aster, little-sunflower (*Helianthella*), and lupine (*Lupinus*). He considered that *M. longicaudus* usually occupied wetter, structurally more complex habitats than other arvicolines in the same region.

We are unaware of studies involving estimates of density of *M. longicaudus* in Oregon; in studies of small-mammal communities, *M. longicaudus* usually is recorded as occurring in relatively low numbers (Black and Hooven, 1974; Feldhamer, 1979b). However, densities as great as 120/ha (Conley, 1976) and 53/ha (Van Horne, 1982) were reported in New Mexico and Alaska, respectively. In California, densities ranged from 4.9 to 16.0/ha (Jenkins, 1948). Although citing unpublished results to the contrary, Randall and Johnson (1979) suggested that densities of *M. longicaudus* populations cycled with 3 years between peaks. However, their evidence was derived by combining data from dissimilar habitats among which populations did not seem to be synchronous. Conley (1976) discounted extrinsic factors being involved in an interannual decline in density. Van Horne (1982) indicated that her study provided no evidence of a multiannual cycle in abundance of *M. longicaudus* in Alaska, but that density was related to the abundance of forbs and berries that declined with seral development.

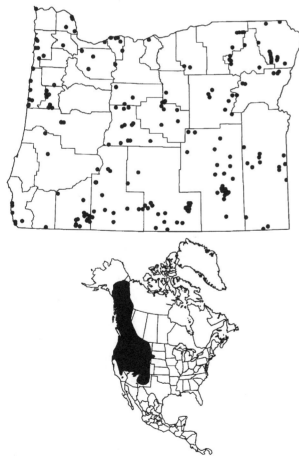

Fig. 11-113. Localities in Oregon from which museum specimens of the long-tailed vole, *Microtus longicaudus*, were examined (listed in APPENDIX, p. 524), and the distribution of the long-tailed vole in North America. (Range map redrawn from Hall, 1981:808, map 457, with modifications.)

Diet.—Few studies of the diet of this vole have been conducted. In Alaska, dicots composed 21–26% of the diet; monocots, 4–6%; stems, 4–13%; fruits and seeds, 52–69%; arthropods, 0–3%; and unidentified materials, 1–5% (Van Horne, 1982). In Wyoming, 38% of stomach contents of *M. longicaudus* was green-plant material, 1% seed fragments, and 61% unidentified materials. In northeastern Oregon, 28% of material in stomachs of four *M. longicaudus* was of one–four taxa of fungi (Maser et al., 1978b); in Colorado, the fungus *Endogone* was consumed by 27.6% of 29 long-tailed voles examined (Williams and Finney, 1964). Linsdale (1938) reported that bark and leaves of sagebrush were eaten.

Reproduction, Ontogeny, and Mortality.—We know of no published studies of reproduction in *M. longicaudus* in Oregon. Four pregnant females on deposit in the Oregon State University, Department of Fisheries and Wildlife mammal collection were collected in June–July; one collected on 6 June was both pregnant and lactating. Also, one collected on 1 July contained pigmented implantation sites and two collected on 20–21 July were lactating but not pregnant. Smolen and Keller (1987) referenced several studies in which mean litter sizes ranging from 4.7 to 6.0 were presented; those in Oregon and Nevada were intermediate (Table 11-23). The breeding season is slightly longer in Nevada than in Alaska (Table 11-23).

Information on gestation, growth, development, and puberty is wanting (Smolen and Keller, 1987).

Predators observed to prey or suspected of preying on *M. longicaudus* include barn owls (*Tyto alba*), great horned owls (*Bubo virginianus*), short-eared owls (*Asio flammeus*), long-eared owls (*Asio otus*), prairie falcons (*Falco mexicanus*), ermines (*Mustela erminea*), long-tailed weasels (*M. frenata*), and pine martens (*Martes americana*); however, the magnitude of losses to this cause has not been reported (Fitzner and Fitzner, 1975; Hayward, 1949; Marti and Braun, 1975; Maser and Brodie, 1966; Quick, 1951; Roth and Powers, 1979; Seidensticker, 1968; Van Horne, 1982; Zielinski et al., 1983).

Van Horne (1982) reported a lower survival rate among adult males than among adult females. Maximum life span of a marked individual in the wild was 13.5 months (Van Horne, 1982).

Habits.—Ordinarily, *M. longicaudus* does not construct runways similar to those made by *M. montanus* and many other species of voles (Hall, 1946; A. B. Howell, 1924); an exception is in grassy areas (Borell and Ellis, 1934). In areas vegetated by sagebrush or chaparral it leaves little or no evidence of its presence (Borell and Ellis, 1934).

Microtus longicaudus tends to be more docile and timid than most congeners with which it is sympatric (Colvin, 1973; Conley, 1976). *M. longicaudus* commonly is displaced by the more aggressive *M. montanus* (Randall, 1978). Colvin (1973) suggested that *M. longicaudus* in Colorado avoided agonistic interactions with *M. montanus* by occupying rocky stream banks. Its tendency to avoid interactions with various congeners may be responsible for the diversity in habitat types it reportedly occupies.

Home-range areas estimated by 95% probability distribution of movements between trap stations reportedly were significantly greater for males than females; however, tabular data presented (presumably a transposition) were 2,360 ± 576 m² for females and 1,615 ± 584 m² for males (Van Horne, 1982). Most activity is nocturnal (Van Horne, 1982). Jenkins (1948) found mean distances between multiple capture sites of 59 m for males and 49 m for females.

Remarks.—The paucity of information in Oregon, and elsewhere, regarding life-history attributes of *M. longicaudus* is appalling. Undoubtedly, the low density of *M. longicaudus* in many situations (Van Horne, 1982), the retiring disposition of the species (Conley, 1976; Colvin, 1973; Randall, 1978), and the somewhat unpredictable nature of occupied habitats contribute to this gap in knowledge of this widely distributed member of the mammalian fauna of the West. We strongly advocate studies of the species, especially west of the Cascade Range.

Microtus longicaudus was referred to as *Microtus mordax* by Bailey (1936).

Microtus montanus (Peale)
Montane Vole

1848. *Arvicola montana* Peale, 8:44.
1891. *Arvicola* (*Mynomes*) *nanus* Merriam, 5:62.
1897. [*Microtus*] *montanus*: Trouessart, p. 563.
1938. *Microtus montanus nanus*: Hall, 51:133.

Description.—*Microtus montanus* (Plate XVIII) is a medium-sized vole (Table 11-22). It can be distinguished from *Ondatra* by size alone; from *Phenacomys*, *Lemmiscus*, and *Microtus oregoni* by four loops of enamel on the lingual side of M3 (Fig. 11-94); and from *M. canicaudus*, *M. townsendii*, *M. californicus*, and *M. richardsoni* by a combination of tail <50% of the length of the head and body, incisive foramina ≤5 mm long and abruptly constricted posteriorly, incisors barely protruding beyond nasals in dorsal view (Fig. 11-114), condylobasal length <35 mm, and length of the hind foot >40% of the tail length (Hall, 1981). Unrooted ever-growing molars with sharply angled loops of enamel exclude *Clethrionomys*. The skull of *M. montanus* tends to be smooth except in older individuals (Bailey, 1900).

The pelage is ashy gray mixed with brown and black on the dorsum shading to light gray or whitish on the venter. The feet are lead colored and the tail is bicolored, black or dark gray above and light gray to whitish below (Bailey,

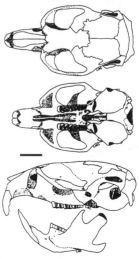

Fig. 11-114. Dorsal, ventral, and lateral views of the cranium, and lateral view of the dentary of the montane vole, *Microtus montanus* (OSUFW 1638). Scale bar equals 5 mm.

1900). Maser et al. (1969) found 21 of 74 caught in Jefferson Co. with splotches of different shades of yellow on the nose, venter, and dorsal surfaces of the feet and tail; the markings occurred only on adults. Jewett (1955) identified melanistic voles originating near Joseph, Wallowa Co., as *M. montanus*.

Distribution.—*Microtus montanus* is distributed throughout most montane and intermontane areas of the West from Williams Lake, British Columbia, south to Inyo Co., California, and east to near Lewistown, Montana; Cheyenne, Wyoming; Hopewell, New Mexico; and Springerville, Arizona (Fig. 11-115). *M. montanus* occurs throughout that portion of Oregon east of the crest of the Cascade Range, and at Diamond Lake, Douglas Co., and several localities in Jackson Co. west of the crest (Fig. 11-115).

Late Pleistocene fossils are known from sites in Colorado, Idaho, Wyoming, and Oregon (Kurtén and Anderson, 1980); like sites in other states, the Fossil Lake site in Oregon, contrary to the statement of Kurtén and Anderson (1980), is within the present-day range of the species.

Geographic Variation.—Of the 15 nominal subspecies, *M. m. montanus* and *M. m. nanus* are purported to occur in Oregon; although *M. m. micropus* is depicted as occurring in extreme southeastern Oregon, the nearest specimen

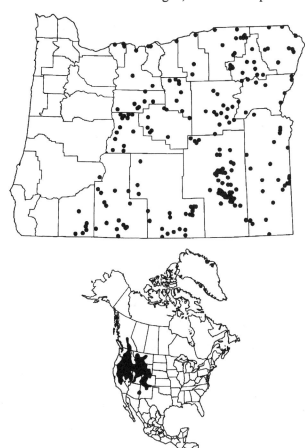

Fig. 11-115. Localities in Oregon from which museum specimens of the montane vole, *Microtus montanus*, were examined (listed in Appendix, p. 526), and the distribution of the montane vole in North America. (Range map redrawn from Hall, 1981:798, map 452, with modifications.)

locality is McDermitt, Nevada (Hall, 1981:798, map 452). The smaller *M. m. nanus* occurs in the northern two-thirds and along the eastern border of the species range in Oregon; the larger *M. m. montanus* occurs in the remainder of the species range in the state (Bailey, 1936; Hall, 1981). Again, we question the validity of recognizing geographic races of arvicoline rodents on the basis of body size, considering that the proportion of large and small individuals in a population of some species varies with the stage of the cycle of abundance (Boonstra and Krebs, 1979).

Habitat and Density.—*Microtus montanus* is essentially always associated with grassy areas, especially in mesic habitats along streams, in mountains, and in marshes (Anderson, 1959). For example, numbers of *M. montanus* captured, adjusted for unequal sampling effort, in four community types on the Malheur National Wildlife Refuge, Harney Co., were sagebrush, 0; greasewood, 7; marsh, 478; and grassland, 171 (Feldhamer, 1979*b*). Catch per unit effort was greatest where the vegetative cover was >80% (Feldhamer, 1979*b*). In Klamath Co., dominant vegetation was meadow foxtail (*Alopecurus pratensis*), bentgrass (*Agrostis alba*), and Kentucky bluegrass (*Poa pratensis*) on a study area used as a source of animals for behavioral studies (Belden, 1965). In the Ruby Mountains, Nevada, *M. montanus* never occupied habitats far from water in summer, but in winter under the snow it moved into areas too dry and too open to be occupied in summer (Borell and Ellis, 1934). In Montana, *M. montanus* was most abundant where vegetative cover was 60–70% in dry-soil habitats dominated by smooth brome (*Bromus inermis*), Idaho fescue (*Festuca idahoensis*), alfalfa, or Canada thistle (*Cirsium arvense*) and where big sagebrush (*Artemisia tridentata*) provided canopy; it was not commonly captured in areas vegetated by sedge (*Carex nebrascensis*) or white clover (*Trifolium repens*—Hodgson, 1972). However, in eastern Washington, plants common in areas occupied by *M. montanus* were bulbous bluegrass (*Poa bulbosa*), cheat grass (*Bromus tectorum*), white sweetclover (*Melilotus alba*), clasping peppergrass (*Lepidium perfoliatum*), and field pennycress (*Thlaspi arvense*—Seidel and Booth, 1960). In parallel studies of the small mammals in different sets of climax communities in the same region, Rickard (1960) caught montane voles only in fescue-snowberry (*Symphoricarpos*) communities and Hoffman (1960) caught them only in Sitka alder (*Alnus sinuata*) communities.

Where *M. montanus* and *M. longicaudus* are sympatric, the former is restricted to grass habitats whereas the latter is restricted to shrub habitats; however, when provided a choice in the absence of the other species, both species selected grass habitats irrespective of prior experience (Randall, 1978). *M. montanus* males are larger and dominate interspecific encounters largely because *M. longicaudus* withdraws to avoid agonistic interactions. Randall (1978) believed that *M. montanus* could exclude *M. longicaudus* from grass habitats. In shrub habitats, densities of the two species fluctuated in opposition (Randall and Johnson, 1979).

Populations of *M. montanus* are irruptive with peak densities occurring several times during the 20th century, and attaining plague proportions in Klamath, Deschutes, Jefferson, Lake, and Crook counties in 1957–1958. Densities as great as 25,000 voles/ha were reported in newspaper accounts, but peak densities probably were 5,000–7,500/ha locally and probably much less generally (Spencer, 1959). On a plot on which mark-recapture estimates of density were <2,000 voles/ha, entrances to subterranean burrows numbered nearly 70,000/ha (Spencer, 1959).

Randall and Johnson (1979) indicated that populations of *M. montanus* cycled with 3 years between peaks. However, their evidence (p. 217, table 1) was derived by combining data from dissimilar habitats among which populations did not seem to be synchronous.

Diet.—In Colorado in summer, leaves of forbs constituted 85% and grasses 9% of the diet of *M. montanus*; five species of plants composed nearly 70% of the diet: small evening-primrose (*Oenothera minor*), American vetch (*Vicia americana*), arnica (*Arnica*), agoseris (*Agoseris glauca*), and narrow-leafed collomia (*Collomia linearis*—Vaughan, 1974). The first three of these species were consumed with much greater frequency than they occurred in plant communities occupied by *M. montanus*. Smooth bromegrass (*Bromus inermis*) and slender wheatgrass (*Agropyron caninum*) were the only grasses regularly eaten by *M. montanus* (Vaughan, 1974). Although some of these species do not occur in Oregon, close relatives occupy similar communities and might be expected to be consumed by *M. montanus*.

In Klamath Co., during the population irruption in 1957–1958, *M. montanus* ate roots and crowns of alsike clover, red clover, alfalfa, fruit trees, and various pasture grasses; heads of grains; and potatoes (Vertrees, 1959). In winter, the voles seemed to move into rangeland near agricultural fields; there they ate "bitter-brush, rabbitbrush, wild cherry, wild plum, and greasewood" (Vertrees, 1959:7).

Captive *M. montanus* ate grasses, carrots, rolled oats, buds of trees, shoots of willow (*Salix*), and roots of some plants. Foods with high water content seemed to be relished (Seidel and Booth, 1960). However, most refused to eat hamburger and dead voles, thereby establishing the strength of herbivory.

Reproduction, Ontogeny, and Mortality.—*Microtus montanus* is polyestrous and exhibits postpartum estrus (Hoffmann, 1958; Negus and Pinter, 1965; Seidel and Booth, 1960). In captivity, as many as 13 consecutive litters may be produced, most at intervals of exactly 21 days (Negus and Pinter, 1965; Seidel and Booth, 1960). Young of both sexes born early in the breeding season become reproductively mature in the year of their birth; in July and August, most litters probably are sired by young males (Vaughan, 1969).

In California, almost all female *M. montanus* that weighed >33 g were reproductively mature. However, females tended to mature at a lower body mass in spring (22–26 g) than in autumn (29–32 g); the lightest reproductively mature female weighed only 19.1 g (Hoffmann, 1958). Males matured at ≅35 g. Average litter size based on counts of corpora lutea was greatest, on embryos next, and on pigmented implantation sites least (Table 11-21). There were no significant differences in average litter size between primiparous and multiparous females based on any of the three counts (Hoffmann, 1958). However, Negus and Pinter (1965) found that the average size of first litters produced in captivity by wild-caught nulliparous females was 3.6 (range, 1–5) and that average litter size increased to 4.9 (range, 2–10) by the fifth litter then declined. Those mated

as adults (presumably parous) produced first litters in captivity averaging 4.2 (range, 2–6); average litter size also reached a peak at the fifth litter ($\bar{X}$ = 5.8; range, 3–10) then declined. Thus, there is some question whether reproductive performance of a wild population is a function of age structure of a population or of a sample from a population. It may be that an overall average litter size for *M. montanus* and perhaps other arvicolines (Tables 11-21 and 11-23) is as biased as that of lagomorphs. Vaughan (1969) indicated that young females produced smaller litters, but their contribution to the population was greater than that of adult females because they were more abundant than adults late in the breeding season.

In captivity, virgin *M. montanus* subjected to an 18-h photoperiod produced larger litters when mated than those subjected to a 6-h photoperiod irrespective of diet; however, those in subgroups receiving a dietary supplement of sprouted wheat at 3-day intervals produced significantly more litters than those receiving the supplement at 15-day intervals (Pinter and Negus, 1965). Plant "estrogens" were postulated to have been responsible for the difference. Negus and Berger (1977) were able to initiate breeding in December and in January in nonbreeding wild populations of *M. montanus* by providing sprouted wheat for 2 weeks. They concluded that *M. montanus* cues on chemical signals in plant foods to initiate reproduction. Timing of breeding to correspond with the production of abundant and high-quality forage during lactation and growth doubtlessly is advantageous.

In wild *M. montanus* in California, the length of the breeding season ranged from 146 to 171 days (Hoffmann, 1958) with the maximum extending from April to October. In captivity, wild-caught females from southeastern Washington produced litters after 20.5–21.0-day gestation periods and remated within 24 h except from mid-November to late February or early March (Seidel and Booth, 1960). Overall, natality and density were considered to be inversely related (Hoffmann, 1958).

Preimplantation mortality was estimated at 5.9% and postimplantation mortality at 2.6–3.9% in montane meadows in California (Hoffmann, 1958). Loss of nestlings ranged from 5.0 to 48.6% and loss of weanlings and juveniles ranged from 7.4 to 45.1%; both losses varied by season and among years (Hoffmann, 1958).

Neonates are pink with light blue-gray pigmentation on the dorsum; the blue-gray extends down the sides by day 3 and onto the venter by day 7. The pigmentation fades as the fur grows, and ultimately disappears except in the perineal region of the male (Seidel and Booth, 1960). At birth, young also are blind, deaf, toothless, and hairless except for vibrissae; they weigh about 2.2 g. The toes are fused and the ears are folded over the auditory meatuses. Young double their body mass by day 5–6, then add ≅0.5 g/day until after weaning; at about day 14, the daily increment added increases to about 0.9–1.0 g/day until day 28, when the daily gain declines to about 0.5–0.6 g/day (Seidel and Booth, 1960). The incisors appear by day 5–6, the young can support their own mass by day 8–9, the eyes open on day 10, the ears unfold on day 10–12, and weaning commences on day 12 and is complete by day 15 (Seidel and Booth, 1960). Seidel and Booth (1960) noted that males begin to lick the genitals of females by day 33, but believed that males probably did not attain reproductive competency until at least day 45.

During the population irruption in 1957–1958 in south-central Oregon, large numbers of raptorial birds (including gulls, *Larus*; shrikes, *Lanius*; and magpies, *Pica*) invaded the Klamath Valley and fed largely upon both live and dead *M. montanus* (Craighead, 1959). In 918 regurgitated pellets from eight species of hawks and owls, 94.4% of the prey items were *M. montanus*; 2,772 voles were represented. In 272 pellets regurgitated by great horned owls (*Bubo virginianus*), remains of 1,164 *M. montanus* were identified, and in 138 pellets of gulls, remains of 361 *M. montanus* were counted (Craighead, 1959).

In Jefferson and Gilliam counties, Maser et al. (1970) found remains of *M. montanus* in 13.4% of 187 pellets regurgitated by great horned owls, in 3.6% of 110 pellets of long-eared owls (*Asio otus*), and in 9.8% of 378 pellets of short-eared owls (*Asio flammeus*). In Deschutes Co., remains of *M. montanus* composed 37.8% of 696 identifiable food items in approximately 205 pellets of great horned owls (Brodie and Maser, 1967). In Jefferson Co., only nine of 1,151 food items identified in pellets of burrowing owls (*Athene cunicularia*) were remains of *M. montanus* (Maser et al., 1971), and in Morrow, Gilliam, and Umatilla counties only 0.3% of 5,559 pellets of the species contained remains of *M. montanus* (Green et al., 1993).

Habits.—In grassy habitats, *M. montanus* constructs 30–60-mm wide runways under the cover of the grasses; runways may be worn as much as 1 cm deep in the soil by the passage of the voles (Seidel and Booth, 1960). The runways usually extend from burrow to burrow with many intersections to form complex webs that may include space under logs, rocks, or boards. In clover and alfalfa habitats, *M. montanus* does not make distinct runways. A ball of dried grasses 120–240 mm in diameter with cavities 35–60 mm in diameter, often lined with finer materials, serves as a nest (Seidel and Booth, 1960). Nests commonly are placed beneath the protection of some object on the surface of the ground. One of us (BJV) found an elongated nest with five near-weanling young beneath a ≅15-cm-diameter fence post lying in a roadside ditch.

During May–November in southeastern Washington, activity of montane voles, based on those caught in live traps equipped with timers designed to record time of capture, occurred at all hours, but during 4 months it was significantly greater than expected during daylight hours. In summer, activity during the hottest part of the day was depressed, producing a diurnally bimodal pattern (Drabek, 1994). Dalquest (1948) claimed that trapping records indicated that *M. montanus* is active both at night and during daylight hours.

At high densities, montane voles exhibit greater intraspecific agonistic behavior, especially when the sex ratio is biased in favor of males (Belden, 1965). Also, dispersal reduces agonistic behavior, an increase in fighting time reduces reproductive output, and juvenile survival declines with increasing density. As density declines, threat display increases and contact fighting decreases (Belden, 1965).

Introduction of a strange male *M. montanus* into a small enclosure with a mated pair resulted in a fight between males, but introduction of a strange female resulted in attacks by both male and female residents (Neider, 1963). In an evaluation of multiple captures, Feldhamer (1977a) recorded 14 male-male captures, 2 male-female captures, and no female-female captures. From the sex and age structure of voles involved in multiple captures, he suggested that juvenile male *M. montanus* may forage and disperse together in late summer and autumn.

In captivity, *M. montanus* males reared by natural parents tended to associate with conspecific females, whereas those cross-fostered and reared by *M. canicaudus* parents did not associate with conspecific females in deference to *M. canicaudus* females. However, *M. montanus* females reared by natural parents did not discriminate between males of the two species, but those reared by *M. canicaudus* parents tended to associate with *M. canicaudus* males (McDonald and Forslund, 1978).

We were unable to locate information on home range and movements of *M. montanus*; we suspect that they might not be greatly different from those of other species of *Microtus*.

Remarks.—During the population irruption in 1957–1958 in south-central Oregon, the causative organisms of both plague and tularemia were isolated from large numbers of *M. montanus* in Klamath Co. (Kartman et al., 1959). We found *M. montanus* not to be particularly difficult to trap with either Sherman live traps or museum special kill traps, as noted for some other species of *Microtus* in Oregon.

The paucity or absence of information regarding several topics indicates that research on *M. montanus* in Oregon has been largely of a "problem-solving" nature rather than of a "knowledge-producing" nature. Unfortunately, when the problems, such as the 1957–1958 irruption, disappeared, interest in and funds for research on the species waned, a most shortsighted method of administering both funds and researchers.

Microtus oregoni (Bachman)
Creeping Vole

1839b. *Arvicola oregoni* Bachman, 8:60.
1896. *Microtus oregoni*: Miller, 12:9.
1897j. *Microtus bairdii* Merriam, 11:75.
1908. *Microtus oregoni adocetus* Merriam, 21:145.

Description.—*Microtus oregoni* (Fig. 11-116) is the smallest vole in Oregon (Table 11-22). It can be distinguished from sympatric and contiguously allopatric voles by its exceedingly short tail (<30% of total length), tiny

Fig. 11-116. Photograph of the creeping vole, *Microtus oregoni*. (Reprinted from Carraway and Verts, 1985, with permission of the American Society of Mammalogists.)

eyes (≤2 mm in diameter), unrooted molars, approximately equal depth of labial and lingual reentrant angles on lower molars, M3 with three reentrant angles on the lingual side (Fig. 11-94), and five plantar tubercles on the hind feet (Hall, 1981). The skull (Fig. 11-117) is low and flat, and the incisive foramina are short and wide (4 by 1 mm) and often constricted slightly posteriorly. The zygomatic arches are relatively thin. The feet are plantigrade, pentadactyl (pollex reduced and clawless), and furred on the sole to the level of the tubercles (Miller, 1896).

The short, dense fur on the dorsum ranges from sooty gray to dark brown or black, with scattered yellowish hairs. Hair on the venter is dusky washed with buff or white; the tail is blackish, slightly darker above than below; and the fur on the ears is black.

Distribution.—The geographic range of *M. oregoni* (Fig. 11-118) extends from Port Moody, British Columbia, south to Mendocino City, California, and east to Lake Chelan, Mt. Aix, and Signal Peak in Washington; Brooks Meadows, Hood River Co., the north base of Three Sisters, and Crater Lake in Oregon; and Beswick and South Yolla Bolly Mountain in California (Hall, 1981). In Oregon, *M. oregoni* ranges (Fig. 11-118) from Brooks Meadows and Crater Lake, Klamath Co., west to the Pacific Ocean. The species has been taken at some localities in the Willamette Valley. However, hiatuses in the distribution occur in the northern Coast Range and on the west slope of the Cascade Range in Clackamas, Marion, and northern Linn counties and in southern Lane, Douglas, and northern Jackson counties. We suspect that the hiatuses largely represent unequal sampling effort.

Geographic Variation.—Three of the four nominal races of *M. oregoni* occur in Oregon: the large, pale reddish-brown *M. o. adocetus* from the vicinity of Medford, Jackson Co.; the yellowish-brown *M. o. bairdii* from Glacier Peak, 7,800 ft, Crater Lake, Klamath Co. (type locality); and the small, dusky-brown *M. o. oregoni* throughout the remainder of the range of the species in the state (Bailey, 1936; Carraway and Verts, 1985; Merriam, 1908).

Habitat and Density.—*Microtus oregoni* is associated with all stages of the moist coniferous forest sere, especially early seral communities in xeric microsites supporting stands of short grasses (Carraway and Verts, 1985). Clear-cuttings and areas on which slash was burned

support 3–8 times as many individuals as virgin forest (Gashwiler, 1972; Goertz, 1964; D. Hawes, 1975; Hooven and Black, 1976). Hooven (1973b) considered that *M. oregoni* required ground cover or mulch, but grass sods were avoided; however, populations were reduced or eliminated by applications of herbicides that selectively reduced grasses and forbs (Black and Hooven, 1974; Borrecco et al., 1979). Herbaceous-dominated sites support higher densities than woody-dominated sites (Gashwiler, 1970a).

Densities as great as 138/ha were estimated in British Columbia (D. Hawes, 1975), but highest reported densities in Oregon are an order of magnitude less (Gashwiler, 1972). Hooven (1973b) reported that the number of creeping voles caught in spring during a 3-year study always exceeded the number caught the previous autumn; however, in another study, this pattern occurred in only 2 of 10 years (Gashwiler, 1972). The latter author considered that densities of populations of creeping voles in early seral communities were higher, but less stable than those in later seral stages.

Although autumn densities fluctuate by an order of magnitude (Gashwiler, 1972), the occurrence of multiannual cycles is questionable (Carraway and Verts, 1985). Sullivan (1980) suggested a 3–4-year cycle in abundance, but Gashwiler (1972) indicated peaks in abundance

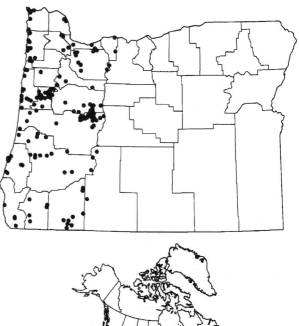

Fig. 11-118. Localities in Oregon from which museum specimens of the creeping vole, *Microtus oregoni*, were examined (listed in APPENDIX, p. 527), and the distribution of the creeping vole in North America. (Range map redrawn from Carraway and Verts, 1985:3, fig. 3, with modifications.)

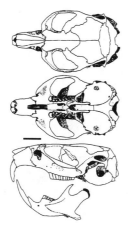

Fig. 11-117. Dorsal, ventral, and lateral views of the cranium, and lateral view of the dentary of the creeping vole, *Microtus oregoni* (OSUFW 7693). Scale bar equals 5 mm.

at consecutive intervals of 5, 3, and possibly 3 years in clear-cuttings and of 5, 1, 2, and possibly 3 years in virgin forest. Sullivan and Krebs (1981) indicated that an increase in body mass of adult voles, sex ratios biased in favor of females, relatively stable habitats, and peaks in density of voles on different grids in overlapping studies were evidence of cycles. They suggested that *M. oregoni* in old-field habitats and possibly early seral forest communities without *M. townsendii* exhibited a 3–4-year cycle. However, D. Hawes (1975) concluded that *M. oregoni* in the same area did not undergo population cycles of the type observed in some other voles.

Diet.—Green vegetation (presumably both forbs and grasses) is considered the primary food of *M. oregoni* (Hooven, 1973b; Maser et al., 1978b, 1981b), although fungi composed 30% of the stomach contents of a sample of 30 (Maser et al., 1978b). Maser et al. (1978b) believed that replacement of most *M. oregoni* by *Clethrionomys californicus* in mature forests was related to the near exclusion of forbs and grasses by shade and the abundance of hypogeous fungi fed on by the latter species.

We maintained captive *M. oregoni* successfully on a diet of bluegrass (*Poa*), clover (*Trifolium*), and apples.

Reproduction, Ontogeny, and, Mortality.—Hooven (1973b:258) indicated that *M. oregoni* "may breed continuously throughout the year"; however, other workers indicate that breeding is limited mostly to March–September (Table 11-21). Gestation averages 23 days 20 h (range, 23 days 8 h–24 days 12 h—Cowan and Arsenault, 1954). Mean litter size (Table 11-23) increases with successive litters at least through the first four litters, but later litters may be smaller (Cowan and Arsenault, 1954); therefore, calculated mean litter size for samples may be influenced by the number of females pregnant with each of the litter cohorts.

Females likely produce four–five litters annually (Cowan and Arsenault, 1954); Gashwiler (1972) estimated an average of 4.8 litters per female. Thus, a total of about 15 young may be produced annually. The first three litters of the year are sufficiently synchronous that three age cohorts can be discerned among young caught before June.

At birth, young *M. oregoni* appear similar to neonates of most small rodents: they are naked and pink and have closed eyes and pinnae folded to close the meatuses (Cowan and Arsenault, 1954); they weigh about 1.6 g (Hooven, 1973b). Dorsal fur and pinnae appear by 1 day of age; pinnae begin to unfold by 2.5 days; incisors are visible beneath the gums, sparse fur is visible on the venter, and the dorsal fur is colored by 3.5 days; and the incisors erupt and the young can crawl weakly by 5.5 days. Ears and eyes open and young begin to eat green food by 10–11.5 days; young are weaned at 13–15 days (Cowan and Arsenault, 1954; Hooven, 1973b). Instantaneous growth rates (K) for 41 young born in captivity were 0.162, 0.063, 0.033, and 0.019 at 0–9, 9–20, 20–30, and 30–38 days, respectively (Cowan and Arsenault, 1954). Rates of gain were similar for males and females to 34–35 days, after which males gained faster. Body mass of males stabilized after sexual maturity until paired with females; paired males added 5–10 g and maintained the higher body mass until the end of the breeding season or until separated from the females (Cowan and Arsenault, 1954).

In British Columbia in summer, minimum survival rates (percent surviving the 14-day interval between trappings) for adults were 70–84% in old fields, 40–75% in grasslands, and 63–91% in brushlands; in winter, combined rates for all age-classes were 55–80%, 70–77%, and 65–78% for the three habitats, respectively (Sullivan and Krebs, 1981). In Oregon, of 713 *M. oregoni* captured 1,758 times, 49% were not recaptured after 1 month, 90% after 5 months, and 99% after 12 months (Gashwiler, 1972), corroborating Cowan and Arsenault's (1954) claim of complete population turnover annually. Maximum longevity of *M. oregoni* is >320 days in captivity (Cowan and Arsenault, 1954) and 480 days in the wild (Gashwiler, 1972).

Coyotes (*Canis latrans*), bobcats (*Lynx rufus*), ermines (*Mustela erminea*), barn owls (*Tyto alba*), saw-whet owls (*Aegolius acadicus*), long-eared owls (*Asio otus*), and great horned owls (*Bubo virginianus*) are known to prey on *M. oregoni* (D. S. deCalesta, in litt.; Forsman and Maser, 1970; Giger, 1965; Maser and Brodie, 1966; Nussbaum and Maser, 1975; Sullivan and Sullivan, 1980). Giger (1965) found 22% of 2,886 skulls of small mammals recovered from 724 pellets regurgitated by barn owls in Tillamook Co. were those of *M. oregoni*.

Habits.—Some workers claim that *M. oregoni* is a burrower that constructs subterranean tunnels in moist, mellow forest soils, sometimes pushing up ridges of soil similar to those made by some moles (Bailey, 1936; Cowan and Arsenault, 1954; Maser et al., 1981b). Ingles (1965) indicated that *M. oregoni* seldom left the tunnels, but Hooven (1973b) was unable to find burrows or tunnels attributable to creeping voles. Bailey (1936) suggested that *M. oregoni* was difficult to collect except by placing traps in tunnels or runways.

In British Columbia, *M. townsendii* is believed to exclude *M. oregoni* from wet habitats, where the latter species is unable to construct burrows. In drier areas, *M. oregoni* may avoid competition from *M. townsendii* by constructing tunnels too small for the latter species to enter (D. Hawes, 1975). However, in encounters of captives in neutral arenas, *M. oregoni* was dominant to *M. townsendii* in all interspecific pairings except those in which both individuals were males (Weil, 1975). In intraspecific encounters, dominance was exhibited in about half of male-male and female-female pairings, but females were dominant to males in >80% of male-female pairings (Weil, 1975).

We collected *M. oregoni* in live and snap traps set near stumps, trees, and logs where tunnels or runways were not conspicuous. Our success in trapping *M. oregoni* without considering trap placement in relation to visible sign and the frequency with which remains of creeping voles are found in regurgitated pellets of raptors (Giger, 1965) suggests to us that activity of creeping voles is not restricted largely to tunnels or covered runways.

Ranges of average home-range areas estimated for *M. oregoni* by the exclusive boundary-strip method were 0.05–0.12 ha for males and 0.04–0.06 ha for females in Oregon. Home-range areas were greatest for both sexes in spring, and least in summer for males and in autumn for females (Gashwiler, 1972). Hooven (1973b), based on work in the same region, reported home ranges of 0.31 and 0.38 ha for males and 0.18 and 0.23 ha for females in burned and unburned clear-cuttings, respectively.

Remarks.—*Microtus oregoni* possibly is unique among

North American mammals in that gonadal and somatic cells have different chromosomal complements (Ohno et al., 1963). The diploid number (2n) is 17 in females and 18 in males; there are 14 metacentric and two submetacentric chromosomes, and the sex chromosomes consist of a large submetacentric X and an acrocentric Y (Hsu and Benirschke, 1969). The somatic cells are XO in the female and XY in the male (Hsu and Benirschke, 1969; Matthey, 1958; Ohno, 1967; Ohno et al., 1963, 1966). A 9% difference in DNA content between X-bearing and Y-bearing spermatozoa allows flow-sorting of sex-specific elements of semen (Pinkel et al., 1982), a technique that has led to attempts to increase the probability of producing offspring of a predetermined sex in humans and domestic animals (Johnson et al., 1989).

We consider an investigation of geographic variation in *M. oregoni* paramount among needs for information regarding the species. The currently recognized subspecies are based on samples of seven or fewer specimens (Merriam, 1897*j*, 1908) and are based on characters (pelage color and tail length) for which individual variation often is great. We question the validity of recognizing subspecies in Oregon.

Microtus richardsoni (DeKay)
Water Vole

1842. *A*[*rvicola*]. *richardsoni* DeKay, part 1:91.
1891. *Arvicola (Mynomes) macropus* Merriam, 5:60.
1894*d. Aulacomys arvicoloides* Rhoads, 28:182.
1897. [*Microtus*] *richardsoni*: Trouessart, p. 565.
1900. *Microtus richardsoni macropus*: Bailey, 17:61.
1900. *Microtus richardsoni arvicoloides*: Bailey, 17:62.

Description.—Other than the muskrat (*Ondatra zibethicus*), *Microtus richardsoni* (Plate XVIII) is the largest arvicoline in Oregon (Tables 11-20 and 11-22). Externally, its large size, hind feet ≥23 mm long, conspicuous flank glands in adults of both sexes, and five plantar tubercles serve to distinguish it from other arvicolines in the state. The skull (Fig. 11-119) is large and angular; the upper incisors are procumbent, extending far anterior of the nasals; the incisive foramina are abruptly constricted, forming a thin slit posteriorly; M3 has four reentrant angles on the lingual side (Fig. 11-94); the frontal shelf is conspicuous; and the paroccipital process is well developed (Ludwig, 1984*a*).

Distribution.—*Microtus richardsoni* is distributed (Fig. 11-120) in two disjunct north-south bands corresponding roughly to the Cascade Range in western British Columbia, Washington, and Oregon and to the Rocky Mountains and nearby montane regions in eastern British Columbia, Alberta, Washington, Oregon, Idaho, Montana, Wyoming, and Utah (Hall, 1981). In Oregon, *M. richardsoni* occurs in the Cascade Range from Mt. Hood to Mt. Mazama and in the Wallowa and Blue mountains east and north of a line connecting Langdon Lake, Umatilla Co.; the North Fork Malheur River, Grant Co.; and Cornucopia, Baker Co. (Fig. 11-120).

Holocene and Pleistocene fossils are known from Montana (Rasmussen, 1974), Alberta (Burns, 1982), and Wyoming (Ludwig, 1984*a*).

Geographic Variation.—Two of the four nominal subspecies occur in Oregon: the larger *M. r. arvicoloides* in the Cascade Range and the smaller *M. r. macropus* in the

Blue and Wallowa mountains (Hall, 1981). Bailey (1936:212) contended that the two races were "at best barely distinguishable." Although average values for most dimensions for specimens collected in the Cascade Range are greater than for those from northeastern Oregon, coefficients of variation are extremely large (Table 11-23). Thus, we remain skeptical of differentiating races of arvicolines on the basis of size.

Habitat and Density.—*Microtus richardsoni* is considered an inhabitant of alpine and subalpine streamsides; in Alberta, 87% of 1,078 captures of *M. richardsoni* were within 5 m of a streambank ($\bar{X} = 2.3$ m—Ludwig, 1984*b*). Others reported similar close association with streamsides (Anderson et al., 1976; Hooven, 1973*a*); however, Anderson et al. (1976) obtained evidence that *M. richardsoni* fed on plant species that did not grow in streamside zones in which it usually is captured in traps.

Occupancy of streamsides is patchy; some stretches are occupied continuously, whereas other stretches were occupied rarely or never (Anderson et al., 1976; Hooven, 1973*a*; Ludwig, 1984*b*). Occupied stretches are more level, have a deeper soil layer, and have more openings in stream banks than unoccupied sites (Ludwig, 1984*b*). In terms of physical features, habitat requirements of *M. richardsoni* are extremely specialized; thus, occupied sites may be separated by broad areas of montane coniferous forests that seemingly serve as barriers to dispersal movements of *M. richardsoni* (Ludwig, 1984*b*).

Maser and Hooven (1970*b*) obtained specimens from a 4-year-old clear-cutting at about 915 m in the Cascade Range in Oregon. The vegetation was dense and consisted of vine maple (*Acer circinatum*), Pacific rhododendron (*Rhododendron macrophyllum*), and redstem ceanothus (*Ceanothus sanquineus*). Although the specimens were

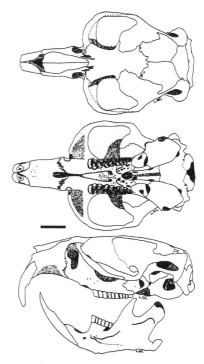

Fig. 11-119. Dorsal, ventral, and lateral views of the cranium, and lateral view of the dentary of the water vole, *Microtus richardsoni* (OSUFW 5677). Scale bar equals 5 mm.

considered a low-altitude record, Hooven (1973*a*) later reported collecting specimens at about 670 m. Woody vegetation in the latter area was similar to that of the former; streamside vegetation consisted of sedges (*Carex*), mints (*Mentha*), horsetail (*Equisetum*), velvet-grass (*Holcus lanatus*), bluegrass (*Poa*), Pacific blackberry (*Rubus ursinus*), and thimbleberry (*R. parviflorus*). Hooven (1973*a*) did not catch *M. richardsoni* in Douglas-fir (*Pseudotsuga menziesii*) forest before clear-cutting, but these voles were captured in two areas after the timber was harvested.

Anderson et al. (1976) recorded August densities of 17.6–32.5 water voles/ha in Alberta.

Diet.—Like many arvicoline rodents, *M. richardsoni* drops fragments of grasses and forbs along its runways through the vegetation. Anderson et al. (1976) assumed that fragments of various species of plants occurred in the runways in the same proportion that they appeared in the diet of *M. richardsoni*. In July, globeflower (*Trollius laxus*), elephant's head (*Pedicularis groenlandica*), paintcup (*Castilleja miniata*), bearberry honeysuckle (*Lonicera involucrata*), fleabane (*Erigeron*), and veiny meadowrue (*Thalictrum venulosum*) were primary items found in runways. In August on another area, grasses and willow (*Salix*) composed 86% of material in runways; other items included aster (*Aster foliaceus*), elephant's head, and unidentified

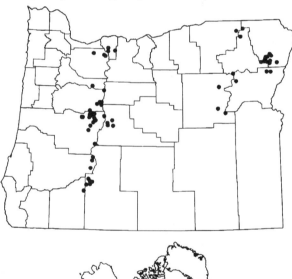

Fig. 11-120. Localities in Oregon from which museum specimens of the water vole, *Microtus richardsoni*, were examined (listed in APPENDIX, p. 528), and the distribution of the water vole in North America. (Range map redrawn from Hall, 1981:821, map 466, with modifications.)

forbs. In August at a different site, arrowleaf groundsel (*Senecio triangularis*) and aster composed 67.5% of cuttings found in runways; 10 other species and unidentified forbs composed the remainder (Anderson et al., 1976). Anderson et al. (1976) considered grasses to play a minor role in the diet of *M. richardsoni*.

Reproduction, Ontogeny, and Mortality.—In Alberta, on a site where snow drifts persisted to late July, the mating season for *M. richardsoni* extended from late May or early June to late August or early September. Populations occupying south-facing slopes bred earliest (Ludwig, 1988). From June to September, the monthly proportion of reproductively active overwintered adults ranged from 84.6 to 100% for females and from 75.0 to 100% for males (Ludwig, 1988). Some young bred during the year of their birth commencing in July; for July–September combined, the proportion of reproductively active young was 25.7% for females (*n* = 105) and 26.3% for males (*n* = 114). In Montana, none of 12 adult females was pregnant in August (Brown, 1977). Individuals that bred during the year of their birth and overwintered also bred as adults (Ludwig, 1988).

Five pregnant females on deposit in the Oregon State University, Department of Fisheries and Wildlife mammal collection were collected in Oregon between 17 May and 24 August. Also, four 15.7–21.3-g juveniles were collected on 14–19 May, ≅2 months earlier than the first young were captured in Alberta (Ludwig, 1988). We also captured half-grown young in October. Thus, the breeding season in Oregon may be considerably longer than in Alberta and Montana.

Gestation is 22 days (Jannett et al., 1979), not 42 days as reported earlier (Brown, 1977). Jannett et al. (1979) obtained evidence to support induced estrus and induced ovulation, but could not discount the possibility that both were spontaneous. Brown (1977) and Jannett et al. (1979) indicated that postpartum estrus and lactational pregnancies were common. Jannett et al. (1979) suggested that lactation did not extend gestation.

Ludwig (1988), based on records accompanying museum specimens, indicated that the range-wide litter size averaged (±*SE*) 5.62 ± 0.18 by counts of embryos (*n* = 86 maternal females) and 6.11 ± 0.43 by counts of implantation sites (*n* = 26). However, Brown (1977) reported a much larger litter size at 1,524 m in northwestern Montana (Table 11-23). Records for Oregon indicated a lower average litter size (Table 11-23) than those of Ludwig (1988) and Brown (1977), but may have included young-of-the-year females; in Alberta, young females averaged significantly fewer (1.35 fewer young/litter) offspring than overwintered females (Ludwig, 1988).

Two litters annually seem to be the maximum produced by *M. richardsoni* (Brown, 1977; Ludwig, 1988). Based on body-mass classes, Anderson et al. (1976) reported three litters were produced by *M. richardsoni* in the same area that Ludwig (1988) conducted his study. However, Anderson et al. (1976) seemingly did not consider the possibility that the lightest cohort might have been produced by juvenile females.

Ludwig (1984*b*, 1988) suggested that the somewhat atypical arvicoline characteristics of moderate litter size, short breeding seasons with one or two litters produced by overwintered females, and the low fecundity of the juvenile female cohort were life-history traits that evolved in response to patchy occupiable habitats. Reduced fecundity may serve to enhance maternal fitness (through production of high-quality emigrants with high survival potential to repopulate streamside patches subject to catastrophes) and to minimize the impact of the species on their food supply of perennial plants (Ludwig, 1984*b*, 1988). However, Brown (1977:281) interpreted the large litters produced in Montana to be "evolutionarily efficient."

In Alberta, annual mortality (including losses through emigration) for all water voles known to be alive July–September was 81.3–100%; for young-of-the-year it was 71.4–100% (Ludwig, 1988). Survival for three consecutive 28-day periods in summer was 95.1%, 65.8%, and 39.0% for males and 95.2%, 73.0%, and 39.4% for females (Ludwig, 1988). Ermines (*Mustela erminea*), martens (*Martes americana*), and possibly accipters (Accipitrinae) are known to prey on *M. richardsoni*; however, the species is not known to be a major prey species of any predator (Ludwig, 1984*a*).

Habits.—*Microtus richardsoni* is active throughout the diel cycle; however, activity is least from dawn to midday, rises during the afternoon, and peaks at night, probably after midnight (Anderson et al., 1976). It is a good swimmer and can swim against the current in small streams both on and below the surface; it commonly takes to water upon release from traps (Anderson et al., 1976).

Adult females tend to maintain exclusive nonoverlapping home-range areas, whereas adult males maintain home ranges near groups of adult females. However, adult males occasionally make extensive exploratory movements, possibly to permit them to locate other adult males and groups of adult females (Ludwig, 1984*b*). In summer, telemetered 72-h home ranges averaged ($\pm SE$) 221.6 ± 75.9 m² for adult females ($n = 16$) and 770.3 ± 358.8 m² for adult males ($n = 9$). In another study in the same region, maximum distances between captures averaged 167.0 m ($n = 2$) for adult males and 38.1 m ($n = 5$) for adult females (Anderson et al., 1976).

In staged encounters in which an adult male in a protective wire basket was placed in the cage of another adult male or a mated pair, the resident male "drum-marked" by alternately raking the sole of one hind foot, then the other across the 10–12-mm long flank gland on the same side, then "stomping" that foot on the floor of the cage several times (Jannett and Jannett, 1974:230). In one instance the male exhibited 16 bouts of drum-marking in 25s; drum-marking usually waned after 5–8 min. Sometimes the resident male raised its tail to a 45° angle and shook it laterally. The heaviest male marked the most; females did not drum-mark. However, the resident female became excited when the resident male confronted a stranger.

Remarks.—The disparity in average litter size among populations in Alberta, Montana, and Oregon, combined with the low-altitude records of the species in Oregon, serves as the basis for interesting questions regarding life-history tactics of *M. richardsoni* in the state. We advocate research to ascertain if adult female *M. richardsoni* in the southern part of the species' range and at altitudes <1,000 m, where the breeding season should not be as restricted as in Alberta and Montana, produce more than two litters annually and relatively small litters, and if some juvenile females breed.

Microtus townsendii (Bachman)
Townsend's Vole

1839*b*. *Arvicola townsendii* Bachman, part 1, 8:60.
1896. *M*[*icrotus*]. *townsendii*: Miller, 12:66.

Description.—*Microtus townsendii* (Fig. 11-121) is a large (Table 11-22) vole with large ears that extend above the fur, a long brownish or blackish tail, and brownish or blackish feet equipped with brown claws. The skull is robust and angular, with heavy ridges; the incisive foramina are long, narrow, and constricted posteriorly (Fig. 11-122).

Fig. 11-121. Photograph of Townsend's vole, *Microtus townsendii*. (Reprinted from Cornely and Verts, 1988, with permission of the American Society of Mammalogists.)

Townsend's vole in Oregon may be separated from sympatric congeners as follows: from *M. oregoni* by an eye >4 mm in diameter, four loops of enamel on the lingual side of M3 (Fig. 11-94), and a tail ≥50% of the length of the head and body; from *M. richardsoni* by six plantar tubercles, incisive foramina not tapering to narrow slits posteriorly, and incisors not greatly procumbent; from *M. canicaudus* by a nearly uniformly blackish or brownish tail, incisive foramina >6 mm long, and a U-shaped or squarish margin of the palate; and from *M. californicus* by a nearly uniformly blackish or brownish tail, incisive foramina widest in the anterior half, and incisors not entirely obscured by the nasals in dorsal view.

Distribution.—*Microtus townsendii* occurs (Fig. 11-123) in extreme southwestern British Columbia (including Vancouver Island), western Washington and Oregon, and south to Humboldt Bay, California (Hall, 1981). In Oregon, specimens have been taken along the Pacific Coast south to southern Coos Co., throughout the Willamette Valley with eastward extensions along the Columbia River to near Hood River, Hood River Co.; the McKenzie River to McKenzie Bridge, Lane Co.; the Umpqua River to Drew, Douglas Co.; and the Rogue River to Prospect, Jackson Co. (Fig. 11-123).

There is no fossil record for the species (Cornely and Verts, 1988).

Geographic Variation.—Of the six named subspecies, only one, *M. t. townsendii*, occurs in Oregon (Hall, 1981).

Habitat and Density.—With few exceptions, *M. townsendii* is associated with moist habitats: meadows, lowland pastures, riparian zones, boggy lands, marshes, and irrigated fields densely vegetated with grasses and sedges (Bailey, 1936; Goertz, 1964; Howell, 1920*b*, 1923; Maser et al., 1981*b*). Occasionally, these voles invade alder (*Alnus*) streamsides and other wooded areas near meadows or other grassy areas (Maser et al., 1981*b*). At

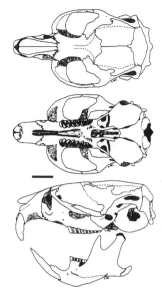

Fig. 11-122. Dorsal, ventral, and lateral views of the cranium, and lateral view of the dentary of Townsend's vole, *Microtus townsendii* (OSUFW 1509). Scale bar equals 5 mm.

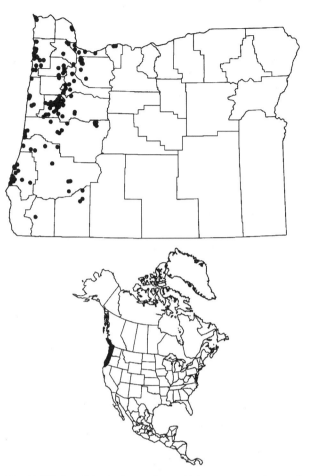

Fig. 11-123. Localities in Oregon from which museum specimens of Townsend's vole, *Microtus townsendii*, were examined (listed in APPENDIX, p. 528), and the distribution of Townsend's vole in North America. (Range map redrawn from Hall, 1981:804, map 455, with modifications.)

times, we took them in late summer in the Willamette Valley in heavily shaded dry upland areas where grasses remained green. Lowland habitats occupied by *M. townsendii* frequently are flooded in winter, sometimes for most of the year (Maser and Storm, 1970)

Although population ecology of *M. townsendii* has not been studied extensively in Oregon, from casual observations we suggest that its populations are subject to fluctuations, resulting in multiannual peaks in density. Despite the difficulty in live capturing *M. townsendii* (Hillborn, 1982) and the biased samples obtained by various methods of live trapping (Beacham, 1982; Boonstra and Krebs, 1978), the population biology of the species has been studied extensively and intensively in an effort to elucidate factors responsible for multiannual cycles. *M. townsendii* in British Columbia undergoes population cycles with peak densities averaging 697 voles/ha (range, 525–800 voles/ha), the highest average attained by any North American *Microtus* (Taitt and Krebs, 1985).

Graf (1947) recorded the number of voles per shock during harvest of hay in fields near Peoria, Linn Co. When populations peaked, he observed one vole/3–8 shocks, but during low populations, he usually observed one vole/40–50 shocks; he recorded one vole/110 shocks during 1 year. He recorded peaks in the population in 1932, 1936, 1939, 1942, and 1943. Given the habitat (hay fields), the region (central Willamette Valley), and the season of haying (summer), we suggest that the voles possibly were not *M. townsendii* as claimed. In our experience, voles occupying hay fields on upland areas in the Willamette Valley are more likely *M. canicaudus*.

Although raptors and carnivorous mammals may take 5.9–30.3% of ear-tagged *M. townsendii* during the spring decline, they are not considered essential for initiation or maintenance of declines in density (Boonstra, 1977). Protection of populations from avian predation with netting, addition of supplemental cover, addition of supplemental food, and disruption of dispersal (fencing) are known to alter various attributes of populations and home-range size

of these voles (Beacham, 1979a, 1980; Boonstra, 1977; Taitt and Krebs, 1981, 1983; Taitt et al., 1981). Dramatic increases in density of fenced populations are the best evidence of the role of dispersal in regulation of numbers of *M. townsendii* (Lidicker, 1985).

Diet.—*Microtus townsendii* forages on the vegetation in which it lives: rushes (*Juncus*), bulrush (*Scirpus*), horsetail (*Equisetum arvense*), clover (*Trifolium*), alfalfa, blue-eyed grass (*Sisyrinchium angustifolium*), common velvet-grass (*Holcus lanatus*), and other grasses (Bailey, 1936; Dalquest, 1948; Goertz, 1964; Maser and Storm, 1970). Bulbs of mint (*Mentha arvensis*) sometimes are dug and cached, occasionally in large amounts (Couch, 1925). *M. townsendii* sometimes is responsible for damaging conifer trees by removing bark and cambium (Harper and Harestad, 1986).

Reproduction, Ontogeny, and Mortality.—In British Columbia, pregnant Townsend's voles were collected in February–November (Anderson and Boonstra, 1979); we did not record pregnant females in Oregon at other than those months (Table 11-23). Although *M. townsendii* is an induced ovulator (MacFarlane and Taylor, 1982a) and can undergo postpartum estrus (Hoffmann, 1958; MacFarlane and Taylor, 1982b), not all postpubertal females are pregnant continuously throughout the breeding season; Anderson and Boonstra (1979) indicated that most were pregnant in spring and summer, fewer than half were pregnant in autumn, and none was pregnant in winter. Gestation periods of 21, 23, and 24 days were recorded, but the latter pregnancy was in a female nursing a litter of five (MacFarlane and Taylor, 1982b). With such a short gestation period and such a long breeding season, the potential number of litters is great; however, few individuals likely live sufficiently long to attain the potential for more than a part of a breeding season (Beacham, 1979b).

Litter size in M. townsendii (Table 11-23) is known to vary with parity, season, phase of the cycle, and maternal body mass (Anderson and Boonstra, 1979), and whether it breeds in the wild or in captivity (MacFarlane and Taylor, 1982b). Success of pregnancy (proportion of pregnant females lactating 2 weeks later) was 46.4% and correlated positively with population growth in British Columbia (Boonstra, 1980). Of 44 unsuccessful pregnancies, 68.2% were lost late in pregnancy or at parturition (Anderson and Boonstra, 1979). Litters born in live traps were almost always unsuccessful.

Offspring are weaned at 15–17 days in captivity, although they may commence to eat solid food a few days earlier; they average 15.1 g (range, 11.4–20.0 g) at weaning (MacFarlane, 1977). Growth is related to density (Boonstra, 1978), sex, season, size, and age (Beacham, 1980).

Short-eared owls (*Asio flammeus*), great horned owls (*Bubo virginianus*), barn owls (*Tyto alba*), snowy owls (*Nyctea scandia*), northern harriers (*Circus cyaneus*), rough-legged hawks (*Buteo lagopus*), red-tailed hawks (*Buteo jamaicensis*), northern shrikes (*Lanius excubitor*), great blue herons (*Ardea herodius*), raccoons (*Procyon lotor*), feral house cats, bobcats (*Lynx rufus*), coyotes (*Canis latrans*), weasels (*Mustela*), minks (*M. vison*), foxes (*Vulpes vulpes* and *Urocyon cinereoargenteus*), skunks (*Spilogale gracilis* and *Mephitis mephitis*), and snakes (Serpentes) are known to prey or suspected of preying on *M. townsendii* (Boonstra, 1977; Cowan and Guiguet, 1965; Maser and Storm, 1970; Maser et al., 1981b).

Habits.—*Microtus townsendii* is active throughout the year and throughout the diel cycle (Bailey, 1936; Dalquest, 1948; Maser and Storm, 1970). It readily enters the water and is a good swimmer and diver (Maser et al., 1981b); it has been captured in traps set underwater (Elliot, 1903a). These voles construct and use a complex system of runways; long-used runways may be worn 50 mm or more into the soil (Maser and Storm, 1970). Middens established at some intersecting runways may contain several liters of fecal pellets. Feeding platforms are established at intervals beneath clumps of grass that overhang nearby runways. Nests of grasses are constructed above ground on hummocks in winter but below ground in summer (Cornely and Verts, 1988).

Bivariate home ranges average ≅900 m² for males and ≅500 m² for females (Madison, 1985). Males tend to move more in the breeding season and have more wounds (especially in the vicinity of the hip glands—MacIsaac, 1977) in the breeding season than earlier in spring. Females move less while lactating than earlier in spring (Taitt and Krebs, 1983). Immigrants recolonizing areas from which *M. townsendii* was removed tend to be more subordinate behaviorally than residents on control areas. However, the immigrants established a breeding population (Krebs et al.,

1978).

In interspecific encounters in neutral arenas, males of *M. townsendii* were dominated by males of the smaller sympatric congeners *M. oregoni* and *M. canicaudus* (Weil, 1975). Mutual avoidance was common when *M. townsendii* was involved. However, Pearson (1972) considered *M. townsendii* possibly to be dominant over *M. canicaudus* where they are syntopic, and D. Hawes (1975) suggested that *M. townsendii* interfered competitively with *M. oregoni* such that the latter avoided the competition by burrowing. When *M. townsendii* was introduced into an area in which *M. oregoni* was the only *Microtus*, the density of *M. oregoni* and the range of habitats it used decreased (D. Hawes, 1975).

Juvenile females have a much stronger tendency to remain in the vicinity of their natal sites than males; in British Columbia, 33% of females, but only 9% of males, entered the breeding population on the grid where they were born (Lambin, 1994). However, the proportion changed seasonally; those born early in the year were more likely to reproduce on the natal grid, in part at least because reproductive maturation is suppressed by opposite-sex adults. Juvenile females were believed to avoid competition with breeding females for resources by dispersing, whereas avoidance of inbreeding influenced dispersal among males (Lambin, 1994).

Remarks.—Doubtlessly, the paucity of information regarding this species in Oregon is related to the difficulty of capturing specimens in live traps and differential susceptibility to capture of various cohorts (Hillborn, 1982; Maser and Storm, 1970). However, specimens are kill-trapped relatively easily with museum special traps set across runways.

Considering that litter size in many species of mammals is correlated positively with latitude (Spencer and Steinhoff, 1968), the possibility of a negative relationship in *M. townsendii* (Table 11-23) deserves further study. Such a study should be designed to control as many as possible of the population and environmental factors that affect reproduction in these voles.

Lemmiscus curtatus (Cooper)
Sagebrush Vole

1868. *Arvicola pauperrima* Cooper, 2:535.
1868. *Arvicola curtata* Cope, 1868:2.
1912. *Lagurus (Lemmiscus) curtatus* Thomas, 9:401.
1913. *Microtus (Lagurus) curtatus artemisiae* Anthony, 32:14.
1939. *Lemmiscus pauperrimus*: Davis, p. 327.
1946. [*Lagurus curtatus*] *pauperrimus*: Hall, p. 560.

Description.—*Lemmiscus curtatus* (Plate XVIII) is among the smaller voles in the state (Table 11-20). The tail does not exceed the length of the hind feet by >10 mm, the auditory bullae extend beyond the occiput (Fig. 11-124), the molars are unrooted and ever-growing, M3 has three loops of enamel on the lingual side (Fig. 11-94), and m3 consists of four prisms. The long, soft, and dense dorsal pelage is grayish tan; the bases of the hairs are lead colored and the tips are black. The ventral pelage is pale buff, the feet light gray, and the tail slightly bicolored. The soles of the feet are densely furred posteriorly. Most adults possess six plantar tubercles, but a few adults and many juveniles possess only five (Johnson et al., 1948). Females have four pairs of mammae. The skull (Fig. 11-124) is typically vole-like, but with strong ridges. Hip glands are present in males; hairs usually are shorter over the hip glands and the glands are characterized by a black circle on the flesh side of the skin (James and Booth, 1954).

Distribution.—*Lemmiscus curtatus* occurs in the Columbia Basin region of Washington, the Great Basin, and the northern Great Plains into Alberta and Saskatchewan (Fig. 11-125; Carroll and Genoways, 1980; Hall, 1981). In Oregon, *L. curtatus* occurs mostly east of a line connecting The Dalles, Hood River Co.; Bend, Deschutes Co.; and Klamath Falls, Klamath Co., except it is absent from the Columbia Basin and most of the Blue and Wallowa mountains (Fig. 11-125).

Fossils of *L. curtatus* were found in Wisconsinan and Holocene deposits in Idaho and in Wisconsinan deposits in Wyoming, Montana, and New Mexico; those from Dry Cave, New Mexico, are the only extralimital fossils (Kurtén and Anderson, 1980).

Geographic Variation.—Of the six nominal geographic races only *L. c. pauperrimus* is considered to occur in Oregon (Hall, 1981). Range maps provided by both Hall (1981:822, map 467) and Carroll and Genoways (1980:2, fig. 4) indicate that *L. c. intermedius* occurs along the Oregon-Nevada border in the extreme southeastern corner of the state. However, Hall (1981) lists no localities for specimens of the latter taxon in Oregon and we know of no specimens collected in that region.

Habitat and Density.—In Harney Co., Moore (1943a) found colonies in dead sagebrush (*Artemisia*), stone walls, rubble piles, and roadside ridges in an area vegetated largely by sagebrush (*Artemisia*).

In Jefferson Co., colonies of *L. curtatus* were found under cow-chips (dried bovine feces), fallen fence posts, old boards, and large pieces of paper, or in abandoned pocket gopher (*Thomomys talpoides*) burrows on loose rocky or sandy soils vegetated mostly by crested wheatgrass (*Agropyron cristatum*), cheat grass (*Bromus tectorum*), green rabbit-brush (*Chrysothamnus viscidiflorus*), gray rabbit-brush (*C. nauseosus*), and big sagebrush (*A. tridentata*), with scattered western junipers (*Juniperus occidentalis*) and bitter-brush (*Purshia tridentata*). Although Maser et al. (1974:195) considered

the area to be "an unusual habitat for the species," they did not specify precisely how the area differed from that expected other than to indicate that much of the area previously was dryland farmed.

In Washington, *L. curtatus* was rarely captured on sandy-loam soil in an *Artemisia-Poa* association at elevations <305 m, but was captured most frequently on stony soils in an *Artemisia-Agropyron* association at 305–610 m and >915 m (O'Farrell and Blaustein, 1972). Similarly, Rickard (1960) captured *L. curtatus* in an *Artemisia-Agropyron* association, but not in an *Agropyron spicatum-Poa sandbergii* association in Washington.

Although we found no estimates of density for *L. curtatus,* several authorities suggested that the abundance of the species may fluctuate dramatically from year to year (Banfield, 1974; Hall, 1928; Soper, 1946). Moore (1943a) provided evidence to suggest that populations might be irruptive.

Diet.—*Lemmiscus curtatus* is strictly herbivorous; in Harney Co., Sandberg bluegrass (*Poa sandbergii*), bottlebrush (*Sitanion hystrix*), and littleflower collinsia (*Collinsia parviflora*) were identified as foods of *L. curtatus* (Moore, 1943a). To this list, Maser et al. (1974) added 18 species of plants consumed by *L. curtatus* in Jefferson Co., Oregon: cheat grass, bulbous bluegrass (*Poa bulbosa*), com-

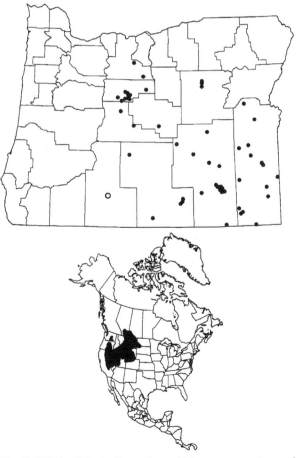

Fig. 11-125. Localities in Oregon from which museum specimens of the sagebrush vole, *Lemmiscus curtatus*, were examined (listed in Appendix, p. 520), and the distribution of the sagebrush vole in North America. (Range map redrawn from Carroll and Genoways, 1980:2, fig. 4, with modifications.) Open symbol indicates record for county.

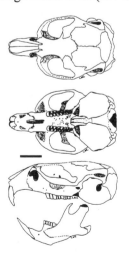

Fig. 11-124. Dorsal, ventral, and lateral views of the cranium, and lateral view of the dentary of the sagebrush vole, *Lemmiscus curtatus* (OSUFW 7407). Scale bar equals 5 mm.

mon velvet-grass (*Holcus lanatus*), crested wheatgrass, medusahead wildrye (*Elymus caput-medusae*), porcupine-grass (*Stipa spartea*), strict erigonum (*Erigonum strictum*) tumblemustard (*Sisymbrium altissimum*), pale alyssum (*Alyssum alyssoides*), cutleaved thelypody (*Thelypodium lasiophyllum*), Wyeth's lupine (*Lupinus wyethii*), alfalfa, Cusick's milk-vetch (*Astragalus cusickii*), filaree (*Erodium cicutarium*), Nevada desert-parsley (*Lomatium nevadense*), green rabbit-brush, big sagebrush, and yarrow (*Achillea millefolium*). James and Booth (1954) indicated that *L. curtatus* in Washington ate big sagebrush, wild onion (*Allium*), cheat grass, bluebunch wheatgrass (*Agropyron spicatum*), and tansy mustard (*Descurainia pinnata*).

The tender culms of newly sprouted leaves of grasses are clipped by *L. curtatus,* thereby stimulating resprouting and prolonging the season during which such foods are available. Seed heads of grasses, except for those of crested wheatgrass in the soft-dough stage, are not eaten by *L. curtatus* (Maser et al., 1974). Stalks of some maturing forbs are cut, seed pods opened, and the seeds eaten by sagebrush voles (Maser et al., 1974). Clipping big sagebrush for food or nesting material by *L. curtatus* may not be as extensive as once supposed; although Hall (1928), James and Booth (1954), and Johnson et al. (1948) indicated that sagebrush was cut by *L. curtatus*, Maser et al. (1974) reported that the sagebrush eaten by *L. curtatus* was clipped by deer mice (*Peromyscus maniculatus*). *L. curtatus* is known to consume *Endogone* (Dowding, 1955) and *Glomus fasciculatus* (Maser et al., 1978a); the latter hypogeous fungus is a symbiont on roots of sheep fescue (*Festuca ovina*), a grass used as food and cover by sagebrush voles (Maser and Strickler, 1978). In captivity, sagebrush voles prospered on a diet of carrots and dandelion (presumably *Taraxacum*—James and Booth, 1954). Although seasonal differences in foods eaten were recognized (Maser et al., 1974), we know of no quantitative evaluation of the diet of *L. curtatus* throughout the year.

Reproduction, Ontogeny, and Mortality.—Although female sagebrush voles were reported to attain reproductive competency as early as 47 days of age (James and Booth, 1954), Maser et al. (1974) indicated that occasionally estrous females as young as 30 days of age were captured. Most authorities claim that *L. curtatus* breeds year-round (Table 11-21); Maser et al. (1974) indicated that pregnant females were detected in all months except September, a month during which James and Booth (1954) reported capturing pregnant females. Gestation is 25 ± 1 days and postpartum mating occurs within 24 h; one captive female gave birth to eight consecutive litters at 25-day intervals (James and Booth, 1954).

At birth, young sagebrush voles are naked, blind, and helpless; the skin is pink and wrinkled, and sufficiently translucent that the ribs and milk in the stomach are visible. The pinnae are folded flat, but the external auditory meatuses open on the 3rd day (James and Booth, 1954). Hair is visible on the 4th day and neonates are fully covered with hair on the 5th day. The eyes open between the 9th and 13th days. By the 15th day the young grasp the mammae sufficiently securely that they are dragged along when the female is flushed from the nest. By the 21st day, young are weaned (James and Booth, 1954).

Neonates weigh 1.5–1.6 g, are 32–33 mm long, and have tails 5 mm long. Growth of neonates seemingly is strongly affected by the number of siblings; a female in a one-young litter weighed 12.1 g and was 82 mm long at 3 weeks, whereas three young in a litter averaged 5.1 g and 58 mm, and four young in another litter averaged 4.5 g and 58 mm (James and Booth, 1954).

In Washington, *L. curtatus* commonly is involved in epizootics of plague (Johnson et al., 1948). In Harney Co., Moore (1943a) found large numbers of sagebrush voles dead in their burrows; they composed 6% of the catch of small mammals in July, a month after they composed 35% of the total catch.

Sagebrush voles are preyed upon by long-tailed weasels (*Mustela frenata*—Maser et al., 1974), bobcats (*Lynx rufus*—Hall, 1946), and probably a variety of wild and domestic carnivores (Maser et al., 1974). On the Crooked River National Grassland, Jefferson Co., 25.6% of 1,151 prey items and 51.1% of 577 mammal prey of the burrowing owl (*Athene cunicularia*) were *L. curtatus;* 295 sagebrush voles were found in 347 owl pellets (0.85/pellet—Maser et al., 1971). In Morrow, Gilliam, and Umatilla counties, however, remains of *L. curtatus* were not identified in 5,559 pellets of burrowing owls (Green et al., 1993), supporting the contention that the species is absent from the Columbia Basin in Oregon. In Gilliam and Jefferson counties, 3.2% of 187 prey items consumed by great horned owls (*Bubo virginianus*), 3.6% of 110 consumed by long-eared owls (*Asio otus*), and 3.7% of 378 consumed by short-eared owls (*A. flammeus*) were *L. curtatus* (Maser et al., 1970). In Deschutes Co., *L. curtatus* composed only 0.3% of 696 food items identified in approximately 205 pellets regurgitated by great horned owls (Brodie and Maser, 1967). Maser et al. (1974) listed 11 other species of raptorial birds that potentially preyed on *L. curtatus* in the Crooked River National Grassland, Jefferson Co.

Habits.—Although most authorities agree that *L. curtatus* may be captured at any time during the diel cycle (Maser et al., 1974), they are in considerable disagreement regarding whether activity is greatest during the daylight hours or at night. Bailey (1936) contended that more were caught at night; James and Booth (1954) agreed that nighttime activity was greatest in winter, but not in summer; Johnson et al. (1948) reported that greatest activity was on bright, sunny days; Moore (1943a) thought that activity on bright, sunny days was limited to brief excursions between points offering shelter; and Maser et al. (1974) indicated that *L. curtatus* was most active from 2–3 h before sunset until 2–3 h after full darkness and from 1–2 h before daylight to 1–2 h after sunrise. Maser et al. (1974) stated that wind had a strong negative effect on the activity of *L. curtatus*. Despite the abundant rhetoric on the subject, no author presented quantitative evidence to substantiate these contentions.

Clustering of burrows tends to indicate that *L. curtatus* resides in colonies (James and Booth, 1954; Johnson et al., 1948; Maser et al., 1974). Maser et al. (1974) reported that sagebrush voles periodically abandoned burrow systems and moved to unused burrow systems as nearby food sources dwindled; thus, burrow systems were occupied, abandoned, and reoccupied sequentially. Nest chambers are 10–13 cm high by 10–18 cm in diameter in open areas and as large as 20 cm high by 25 cm in diameter under partly buried rocks (Maser et al., 1974). Most nests have two or three entrances, but some have as many as six. Nesting material may consist of shredded bark of sagebrush (Johnson et al., 1948; Moore, 1943*a*) and shredded stems, leaves, and seed heads of grasses (Maser et al., 1974; Moore, 1943*a*). Short, blind burrows serve as escape tunnels (Maser et al., 1974).

Hammer and Maser (1969) reported that sagebrush voles on the Crooked River National Grassland, Jefferson Co., commonly excavated the dried feces of domestic cattle; the cavities so constructed were used as feeding stations by *L. curtatus*, but not as nesting sites. In the excavation of the cavity at least some of the fecal material was consumed by the voles.

Uncovered trails 6–8 cm wide are formed between burrow systems, escape tunnels, feeding stations, and food sources; these usually are fairly distinct beneath shrubs, but may be obliterated in the open. Although these commonly are referred to as runways, they are not the well-cleared runways typical of those of some species of *Microtus*, and may be obstructed by growing plants (Hall, 1946). Where *Microtus montanus* used runways of *L. curtatus*, they quickly removed obstructing vegetation (Maser et al., 1974).

From the extent of wounding, most aggressive interactions are among breeding males (Maser et al., 1974). Aggressive interactions among captives occur between a strange male and one or more nonreceptive females and between a strange female and a mated female (James and Booth, 1954).

We know of no published information on home-range size of *L. curtatus*.

Remarks.—The sagebrush vole formerly was classified in the genus *Lagurus* that now is reserved for the Old World steppe lemmings. *Lemmiscus* differs from *Lagurus* in having a m3 consisting of four instead of five prisms, cementum in reentrant angles, ears >50% of length of the hind foot, and a dorsum without a black stripe or yellowish color. Carleton and Musser (1984) considered these characters, combined with the stomach morphology, related *Lemmiscus* more closely to *Microtus* than to *Lagurus*; hence, they agreed with the elevation of subgenus *Lemmiscus* to generic level by Davis (1939).

Several of the published works on the biology of *L. curtatus* in Oregon are anecdotal accounts seemingly based on general impressions with little or no quantification.

Although *L. curtatus* is difficult to capture in conventional live traps for small mammals, it can be live-captured in a specially designed trap (Maser, 1967). Seemingly, some inventive graduate student could combine Maser's (1967) trap with the timing mechanism described by Barry et al. (1989) to produce a piece of equipment by which the daily and annual activity schedules of *L. curtatus* and factors affecting them finally could be evaluated quantitatively.

Ondatra zibethicus (Linnaeus)
Common Muskrat

1766. [*Castor*] *zibethicus* Linnaeus, 1(pt. 1):79.
1863. *Fiber osoyoosensis* Lord, 1863:97.
1903*b*. *Fiber occipitalis* Elliot, 3:162.
1910. *Fiber zibethicus mergens* Hollister, 23:1.
1912. *Ondatra zibethica occipitalis*: Miller, 79:231.
1912. *Ondatra zibethica osoyoosensis*: Miller, 79:231.

Description.—*Ondatra zibethicus* (Plate XVIII) is the largest arvicoline rodent in Oregon (Table 11-22). The body

is heavy, rounded, and, except for the tail and feet, heavily furred. The eyes are beady; the ears are rounded and almost covered with fur; the tail is flattened laterally, scaly, keeled, and naked except for a few hairs on the keel (Willner et al., 1980). The forefeet are relatively small, but the hind feet are large, partly webbed, and strongly fimbriated. Paired perineal musk glands (hence the vernacular name) are present in both sexes. Female muskrats usually have three pairs of mammae, although additional pairs have been recorded (Svihla and Svihla, 1931). The mouth is valvular, permitting closure of the lips and gnawing while under water. The skull (Fig. 11-126), although larger and more rugose, is vole-like; the postorbital shelflike structure of

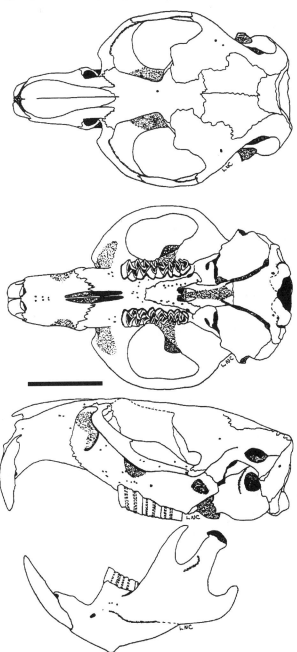

Fig. 11-126. Dorsal, ventral, and lateral views of the cranium, and lateral view of the dentary of the muskrat, *Ondatra zibethicus* (OSUFW 5537). Scale bar equals 15 mm.

the squamosal is exaggerated (Fig. 11-126).

The pelage consists of an exceedingly dense underfur usually overlain by glossy dark-brown guard hairs; the underfur is waterproof (Errington, 1963). Reddish, blackish, silvery, and white individuals have been recorded (Willner et al., 1980). An individual, entirely white except for a few hairs, was recorded from Baker Co. (Kebbe, 1956).

Distribution.—Muskrats occur throughout most of North America north of Mexico (Fig. 11-127; Hall, 1981) and have been introduced widely in Eurasia (Willner et al., 1980). In Oregon, muskrats originally were distributed throughout the Willamette Valley and coastal regions as far south as Coquille, Coos Co., and east of the Cascade Range east of a line from The Dalles, Hood River Co., to Shirk [Home Creek Ranch], Harney Co. (Bailey, 1936; Hall, 1981). High mountain regions and the Klamath, Summer, Abert, and Warner valleys were unoccupied (Bailey, 1936). However, introductions of O. zibethicus from several regions to unpopulated lakes and marshes in Lake, Klamath, and Curry counties and to waterways in northern California that communicate with those in southern Oregon have been chronicled (Hansen, 1965; Twining and Hensley, 1943; Wood, 1974; Yocom, 1970). These regions produce the greatest numbers of muskrat pelts (Table 11-24). Although a plot of collection localities for specimens on deposit in museums depicts a somewhat narrow distribution for the species (Fig. 11-127), fur trappers reportedly took muskrats in every county in the state (Table 11-24).

Fossils of O. zibethicus were recovered from Pleistocene and Recent deposits throughout much of the range of the species, including Oregon (Kurtén and Anderson, 1980; Willner et al., 1980).

Geographic Variation.—Two of the 16 nominal subspecies of O. zibethicus reportedly occur in Oregon: the pale-reddish O. z. occipitalis from Portland, Multnomah Co., west to the Pacific Coast and south to Coquille, Coos Co., and the dark-brown O. z. osoyoosensis north and east of a line from The Dalles, Hood River Co., to Shirk, Harney Co. (Bailey, 1936; Hall, 1981). As the sources of some animals introduced into the region are known (Twining and Hensley, 1943) and some were from defunct fur farms (Willner et al., 1980), muskrats in formerly unoccupied regions of Oregon likely are a heterogeneous mixture of genetic stocks.

Habitat and Density.—Muskrats are highly adapted for the aquatic environment and, although they occasionally make extensive overland treks (Errington, 1963; Kebbe, 1955a), they usually occur in the vicinity of lakes, ponds, sloughs, swamps, marshes, rivers, and creeks. Muskrats usually do not inhabit mountain streams, but may occur in mountain lakes that support appropriate emergent vegetation (Errington, 1963). At Malheur Lake, Harney Co., muskrats live in shallow marshes in which cat-tails (Typha), bulrushes (Scirpus), and rushes (Juncus) are dominant (Errington, 1963).

Muskrat populations undergo dramatic fluctuations in density (Errington, 1963); in some instances the periodicity was sufficiently regular that cycles of 10 years (Elton and Nicholson, 1942a; Errington, 1963), 6 years (Butler, 1962), and 5 years (McLeod, 1948, 1950) were proposed. Although density of populations in Oregon undergoes fluctuations (Errington, 1963), data compiled by the Oregon

Fig. 11-127. Localities in Oregon from which museum specimens of the muskrat, *Ondatra zibethicus*, were examined (listed in APPENDIX, p. 538), and the distribution of the muskrat in North America. (Range map redrawn from Willner et al., 1980:2, fig. 3, with modifications.) Open symbol indicates record for county.

Department of Fish and Wildlife suggest that in 1969–1990 the annual harvests of muskrats never varied by a factor as great as 3 and that, within, among, or over all political subdivisions, regularity in peaks in the harvest was lacking (Table 11-24). We are cognizant of biases in fur-harvest data related to differences in prices paid for pelts, changes in harvest regulations, and differences in social pressures related to the harvest of furbearers. Nevertheless, we find little either in these data, in the several anecdotal accounts (Bailey, 1936; Errington, 1963), or in the short-term studies of density (Sather, 1958) to support the concept of regular cycles in abundance in Oregon similar to those described earlier in this work (see pp. 143-144) for snowshoe hares (*Lepus americanus*) in other regions. We do not question that changes in behavior, natality, and food resources of muskrats, and density of their predators, observed in Oregon and elsewhere, affect numbers of muskrats (Errington, 1963); we merely believe that fluctuations are not sufficiently regular to be considered other than stochastic events.

Diet.—Errington (1963:13) claimed that "the best all-around food" for muskrats was cultivated ear corn, a commodity that muskrats no doubt find in short supply in Oregon. Other food plants that occur in greater abundance and are more widespread in Oregon, and are commonly

Table 11-24.—*Reported annual harvest of muskrats (Ondatra zibethicus) in Oregon by county, 1969–1996. Data provided by Oregon Department of Fish and Wildlife.*

County	Trapping season																										
	1969–1970	1970–1971	1971–1972	1972–1973	1973–1974	1974–1975	1975–1976	1976–1977	1977–1978	1978–1979	1979–1980	1980–1981	1981–1982	1982–1983	1983–1984	1984–1985	1985–1986	1986–1987	1987–1988	1988–1989	1989–1990	1990–1991	1991–1992	1992–1993	1993–1994	1994–1995	1995–1996
Baker	487	483	226	623	359	698	2,498	1,557	1,310	1,144	2,596	2,577	2,431	2,562	1,917	830	720	1,290	1,172	388	613	119	1,357	64	230	171	
Benton	30	58	41	98	43	126	201	451	223	288	393	645	237	371	411	154	128	71	91	36	37	45	24	74	104	81	
Clackamas	207	119	102	140	141	82	251	534	378	283	224	479	145	92	166	128	85	152	207	30	15	8	49	32	88	29	
Clatsop	1,108	1,094	806	502	714	1,090	1,312	1,188	1,519	829	1,597	2,561	1,797	1,149	1,387	1,353	358	428	524	111	94	76	58	15	20	15	
Columbia	368	186	163	297	192	598	644	977	661	739	841	1,220	1,315	814	561	689	376	305	627	320	117	85	67	14	36	155	
Coos	283	131	140	196	234	152	343	455	399	141	328	515	1,440	323	171	392	200	157	127	56	47	11	26	1	5	99	
Crook	67	29	57	67	59	145	213	538	441	97	375	546	270	239	163	86	150	74	324	56	21	10	54	129		13	
Curry	80	68	32	82	43	63	42	42	121	160	62	103	17	52	53	34	13		5								
Deschutes	127	55	20	3	33	55	71	45	22	26	20	84	27	49	55	17	41	71	9	10	21	13	3	2	2		
Douglas	275	100	56	249	262	844	323	506	289	330	418	501	197	302	280	261	164	101	123	100	24	46	37	9	28	35	
Gilliam	1	3						88	2		2							1	3	9			1		8	1	
Grant	25	299	257	203	190	450	321	596	240	103	391	583	340	323	299	145	220	209	352	199	55	14	11	1		20	
Harney	8	3	9,449	71	85	20	12	350	84	68	82	307	173	430	1,302	246	51	141	94	134	323	27	1	37	45	49	
Hood River	23	44	71	55	91	107	211	144	792	169	1,126	326	259	186	190	73	55	48	131	20	84	10	18	2	45	9	
Jackson	1,017	902	856	1,633	1,096	1,847	2,382	1,862	1,258	2,695	1,413	2,437	1,573	1,341	950	894	441	894	586	450	148	186	126	50	182	263	
Jefferson	73	40	194	199	337	387	587	769	543	9	39	888	637	103	34	33	18	44	63	138	19	57	20	8	33	19	
Josephine	691	262	335	428	269	637	420	719	857	1,167	1,337	2,373	1,062	1,063	1,019	961	812	937	1,153	538	141	59	62	48	142	159	
Klamath	13,308	9,912	11,098		14,843	25,812	30,762	26,030	16,115	13,996	14,014	20,941	16,293	19,293	30,234	14,494	16,333	22,710	17,557	8,164	6,573	4,105	6,216	3,748	7,695	8,831	
Lake	78	21	55	72	94	447	732	1,838	283	265	1,627	1,302	2,062	1,237	3,480	1,468	1,938	4,387	5,874	3,080	4,105	2,854	994	3,748	2,558	1,873	186
Lane	1,237	539	602	687	942	1,111	1,989	1,644	1,788	1,144	1,726	992	1,265	1,126	992	1,010	632	816	482	341	235	256	108	112	109		1
Lincoln	277	198	268	224	221	276	489	598	412	630	533	300	215	174	255	240	339	113	170	35	38	12	21	8	5	2	
Linn	73	40	194	199	337	387	587	769	543	448	615	706	637	501	174	438	339	476	425	200	138	57	20	31	42	52	4
Malheur	3,579	3,026	5,058	3,113	3,173	6,497	5,619	6,113	4,827	4,398	8,635	9,256	4,846	3,539	3,721	2,751	1,227	382	2,556	2,373	992	289	608	323	504	625	10
Marion	103	114	164	319	207	588	759	717	628	769	1,307	658	790	610	575	95	382	440	632	175	125	57	133	51	95	52	
Morrow	69	9	55	16	92	71	28	560	199	199	167	23	39	109	95	3	177	189	97	45	8	7	3	2	2	15	
Multnomah	510	392	450	221	352	470	738	528	475	533	808	564	209	256	178	228	177	361	356	111	30	45	52	50	103		
Polk	59	11	10	9	47	145	383	221	84	114	518	268	178	214	172	153	199	32	37	32	11	8	42	103	7		
Sherman								44					1	2				24	27	32	25						
Tillamook	630	376	589	404	624	284	523	423	627	652	574	657	235	231	968	693	652	501	480	54	95	50	160	44	136	103	
Umatilla	994	852	398	1,266	688	892	1,161	1,810	663	1,920	1,590	2,035	1,474	706	804	958	268	987	981	320	121	197	192	309	407	558	1
Union	1,887	1,420	768	846	503	719	2,159	1,857	676	804	1,267	2,582	1,259	1,875	1,494	1,026	682	547	630	325	150	129	181	138	360	179	
Wallowa	607	182	373	1,165	908	1,030	1,332	1,164	675	635	448	731	1,085	1,044	1,574	1,044	800	1,077	1,346	1,241	1,343	714	399	174	138	360	314
Wasco	119	1	182	28	1	48	53	93	165	76	126	279	30	104	91	97	109	174	151	26	20	30	25	61	133	42	
Washington	463	267	190	132	145	441	591	853	782	778	489	1,375	701	5,011	738	680	415	363	588	93	125	32	69	61	79	85	
Wheeler			12				1	11		36		22	16	13	13	2	22	30	50	9	3		8	10	10	30	
Yamhill	81	20	36	46	45	67	123	252	234	330	213	394	169	225	266	160	108	230	184	44	52	36	19	6	13	12	
Totals	29,211	21,215	23,584	37,440	22,911	46,412	56,200	56,050	37,838	36,004	45,041	60,833	41,714	41,735	55,688	32,540	28,324	41,056	38,191	18,685	14,749	9,432	12,101	6,462	13,363	14,110	202
Trappers reporting[a]	721	773	731	847	1,102	1,350	1,377	1,755	2,746	2,899	3,216	3,373	3,069	3,980	3,293	3,104	3,575	3,397	3,833	1,985	1,536	1,228	1,390	1,454	1,293	1,464	687
Average price/pelt[b]	$1.01	$0.91	$1.27	$2.04	$2.22	$2.27	$2.93	$4.38	$4.07	$4.07	$5.58	$5.76	$3.43	$2.58	$3.45	$2.89	$2.47	$3.62	$3.17	$1.44	$0.90	$1.00	$2.00	$1.00	$2.00	$2.00	$3.00

[a] Not all sought muskrats and not all were successful.
[b] Rounded to the nearest dollar after 1989–1990 season.

consumed by *O. zibethicus*, include cat-tails, bulrushes, horsetails (*Equisetum*), bur-reeds (*Sparganium*), arrowheads (*Sagittaria*), willows (*Salix*), sedges (*Carex*), smartweeds (*Polygonum*), and yellow water-lilies (*Nuphar*—Errington, 1963). In spring, new growth of sedges is sought (Sather, 1958). Seemingly, individual tastes determine the diet at least to some extent, as some captive muskrats refuse to eat carrots and apples (Svihla and Svihla, 1931), foods so attractive to wild muskrats that muskrat trappers often use them for bait. Although vegetation composes most of its diet, *O. zibethicus* is not a strict herbivore. Occasionally, fishes (Teleostei), frogs (Anura), invertebrates, and even other muskrats are consumed, usually when food resources are low or a particular prey animal is unusually abundant (Errington, 1963).

Muskrats are voracious feeders; captives consumed 25–30% of their body masses daily depending upon foods offered. In the wild, when populations are high, muskrats literally mow the vegetation (Svihla and Svihla, 1931). Damage to the habitat may be so extensive that it is termed an "eat-out" (Errington, 1963); dramatic die-offs and overland dispersal of muskrats commonly ensue.

Reproduction, Ontogeny, and Mortality.—Although extensive published information regarding reproduction in muskrats in other regions of North America is available (Willner et al., 1980), none is available for populations east or west of the Cascade Range in Oregon. Therefore, we chose to base this portion of our account on information derived from studies conducted close-by on one of the geographic races that occur in our state rather than attempt a summary and synthesis of all available information. Such a summary is available (Willner et al., 1980).

In southeastern Idaho, parturition in muskrats occurred in a 100-day period from early May to late August (Reeves and Williams, 1956); addition of a 30-day gestation period suggests a breeding season of ≅130 days (Table 11-21). Litter size based on number of nestlings per house (Table 11-21) differed by site, year, and age of young. At Gray's Lake, Idaho, one of 16 adult females captured in autumn was barren; the remaining 15 adults contained an average of 10.4 (range, 5–17) pigmented sites of implantation. At Dingle Swamp, Idaho, 18 of 19 autumn-trapped adult females were parous, with an average of 18.0 (range, 7–25) pigmented sites of implantation (Reeves and Williams, 1956). Doubtlessly, the large number of implantation sites represent multiple litters and likely are more representative of overall reproductive output by females than of litter size.

Some authorities claim that young are blind, naked, and helpless at birth (Sather, 1958), whereas others claim that young are born with hair (Svihla and Svihla, 1931). Total length of newborn muskrats in southeastern Idaho averaged 93.5 mm for males and 91.6 mm for females (Reeves and Williams, 1956). In Iowa, neonates (*n* = 16), presumably of both sexes, averaged 96.3 mm long, with tails 26.3 mm long, and weighed 22.0 g; the same dimensions, at 5 and 14 days, respectively, averaged 125.5 mm, 36.4 mm, and 26.6 g, and 199.0 mm, 83.3 mm, and 91.0 g (Sather, 1958). The eyes usually open at 14–16 days and young can swim and dive well at that age (Sather, 1958). Among animals taken during a spring trapping season, body mass averaged 909 g for males and 837 g for females at Gray's Lake, Idaho, and 843 g for males and 830 g for females at Dingle Swamp, Idaho. These animals were considered among the smallest in the United States, considerably smaller than those of the same geographic race in Utah (Reeves and Williams, 1956). Bailey (1936:215) considered eastern Oregon muskrats, also classified as *O. z. osoyoosensis*, to be "large"; he provided "approximate" body masses similar to those listed by Reeves and Williams (1956) for Utah. Thus, size, as indexed by body mass, may differ greatly among muskrats by site within a geographic region; these differences likely are related to quality of the habitat.

Among nestlings, males outnumbered females 1.18:1 (*n* = 277) at Gray's Lake, Idaho, and 1.23:1 (*n* = 356) at Dingle Swamp, Idaho. Among adults taken during a spring trapping season, males outnumbered females 1.18:1 (*n* = 582) and 1.78:1 (*n* = 1,593) at the two sites, respectively (Reeves and Williams, 1956). Similarly biased sex ratios occur elsewhere in much larger samples of all age-classes (Beer and Truax, 1950).

Minks (*Mustela vison*), red foxes (*Vulpes vulpes*), coyotes (*Canis latrans*), raccoons (*Procyon lotor*), badgers (*Taxidea taxus*), bobcats (*Lynx rufus*), fishers (*Martes pennanti*), great horned owls (*Bubo virginianus*), bald eagles (*Haliaeetus leucocephalus*), and ferruginous hawks (*Buteo regalis*) are reported to take muskrats (Dunstan and Harper, 1975; Errington, 1963; Giuliano et al., 1989; Litvaitis et al., 1986; Lokemoen and Duebbert, 1976). Remains of juvenile *O. zibethicus* were recovered from pellets of great horned owls in Deschutes Co. (Brodie and Maser, 1967). Nevertheless, humans, in trapping muskrats for their fur, are the greatest predator of the species.

Habits.—In summer, wild muskrats are active during most of the day with peaks at 1600–1700 h and 2200–2300 h irrespective of weather conditions (Stewart and Bider, 1977). Of various environmental variables considered, rainfall had the most dramatic effect on activity during daylight hours; on rainy days, diurnal activity was evident at all hours, even during 1000–1300 h, a time when activity was extremely low or undetected on rainless days (Stewart and Bider, 1977:493, fig. 3).

Muskrats are powerful swimmers; they can attain 5 km/h, swim backwards (Peterson, 1950), and stay submerged as long as 20 min (Errington, 1961). In a laboratory study, during unrestrained dives the heart rate declined from 310 ± 3 to 54 ± 3 beats/min within 1–2 s of submergence; one individual reduced its heart rate before voluntary submergence, indicating a conditioned response of the heart (Drummond and Jones, 1979). The bradycardia was considered an expression of input from nasal, lung, and carotid chemoreceptors, but precisely how they interact to produce a decline in heart rate is not understood (Drummond and Jones, 1979).

Propulsion through the water is by alternate strokes with the hind feet (Errington, 1963) and skulling by lateral undulations of the tail (Fish, 1982). The tail also prevents yawing, thereby reducing drag (Fish, 1982). The tail also functions as a heat sink (Johansen, 1962). Despite their swimming and diving abilities, muskrats frequently drown when water levels rise rapidly, forcing them to flee their tunnels, or when subjected to high waves (Errington, 1937).

Depending upon the environment, muskrats reside in burrows they dig in banks or in lodges they construct of emergent vegetation and bottom detritis (MacArthur and Aleksiuk, 1979). Lodges (Plate XVIII) are constructed on some type of solid foundation: slightly submerged or slightly emergent rafts of vegetation, mud bars, rock piles, stumps, logs, ice, collapsed and decaying lodges, or debris of any sort that can be used as a "sitting place" by muskrats (Errington, 1963). Dwelling lodges in Manitoba ranged from 0.8 to 2.5 m long, 0.5 to 2.4 m wide, and 0.3 to 1.0 m high; in summer, they were slightly smaller than in winter and contained a single chamber, whereas winter lodges sometimes contained two or three chambers (MacArthur and Aleksiuk, 1979). Errington (1963) indicated that big lodges may be 1.8 m high, 2.4 m in diameter, contain a ring of chambers around a solid center, and have two or three "plunge holes" by which muskrats enter the water. In winter, small (about half-size) lodges are used only as feeding shelters; several of these commonly surround large dwelling lodges (MacArthur, 1978). Burrows may originate from short tunnels with underwater entrances dug by dispersing muskrats seeking a secure dwelling for a few hours or a few days (Errington, 1963). However, complex burrow systems consisting of a maze of tunnels may have been constructed during a period of many years and by many generations of muskrats. Along streams, sloughs, and channels away from marsh environments, muskrats reside in burrows throughout the year, but burrows may not be occupied in winter where lodges can be constructed (MacArthur and Aleksiuk, 1979). Where ambient temperatures ranged from −39 to +34°C, temperatures within lodges and burrows ranged only from −9 to +30°C (MacArthur and Aleksiuk, 1979), approximating the thermoneutral zone of muskrats in air (McEwan et al., 1974).

Muskrats are considered to have relatively small home ranges. Based on recaptures of marked individuals, movements, except during the spring dispersal, usually were <61.0 m (Sather, 1958). In winter, ≥50% of all radio-positions obtained for 11 individuals were ≤15 m from a major dwelling lodge and only one of the 11 moved more than 150 m from a dwelling lodge (MacArthur, 1978). The one individual moved 366 m to another area, then restricted subsequent movements in the same manner as the other 10.

Muskrats are fierce fighters when cornered and can inflict serious wounds with their sharp incisors (Svihla and Svihla, 1931); much of the fighting among muskrats is territorial (Errington, 1963). Some fighting occurs at all seasons, but is most common in spring and early summer, although it may be more frequent when populations attain exceptionally high densities or resources become inadequate (Beer and Meyer, 1951). Fighting is more common among males; onset of severe fighting in spring is believed to be related to an increase in testis mass, but the reduction in fighting precedes the decline in testis mass in autumn (Beer and Meyer, 1951).

Muskrats are considered to be monogamous (Errington, 1963; Sather, 1958). Adult females are thought to overwinter with surviving offspring of the last litter and one or more adult males; first-litter young often establish home ranges within 90 m of the maternal female's home range (Errington, 1963). In spring, young animals disperse whereas older females and perhaps older males remain in their established home ranges (Errington, 1963). Many of the dispersers exhibit evidence of fights (wounding); mortality at this season is high (Beer and Meyer, 1951). The sex ratio among dispersers is about the same as that of the population in general (Beer and Meyer, 1951). Kebbe (1955a) reported counting 287 dead muskrats on approximately 43 km of highway in April 1955 in Klamath Co.; on two sections >12/km were recorded. The muskrats seemed to be moving from Klamath Marsh 8–24 km east of the highway toward the Cascade Range through stands of lodgepole pine (*Pinus contorta*) and ponderosa pine (*Pinus ponderosa*). Overland treks of this nature probably serve to establish (or reestablish) populations in formerly unoccupied water areas; however, few probably survive the journey.

Remarks.—The binomial *Ondatra zibethicus* commonly is misspelled *Ondatra zibethica* (see titles in Literature Cited). The Latin words forming scientific names must agree in case and gender. Despite the terminal "a" that would appear otherwise, *Ondatra* is masculine in gender (Davis and Lowery, 1940); *zibethica* is feminine, thus inappropriate.

The muskrat is a valuable furbearer (Table 11-24); numbers harvested in Oregon commonly exceed those of any other species of furbearer by several times, and in some years the market value of muskrat pelts exceeds that for any other furbearer despite the modest price per pelt paid for muskrats.

By their burrowing activities, muskrats cause damage to river banks, pond dams, and levees sufficient to require control measures (Erickson, 1966). Riprapping banks with stone is the most effective means of reducing damage by burrowing muskrats (Erickson, 1966).

Muskrats commonly are involved in epizootics of hemorrhagic disease, tularemia, leptospirosis, pseudotuberculosis, ringworm, Tyzzer's disease, and other diseases (Willner et al., 1980). Hemorrhagic disease (Errington's disease)—characterized by necrotic foci on the liver, intestinal hemorrhages, and lung hemorrhages (Lord et al., 1956a)—is particularly deadly to muskrats and often is responsible for extensive die-offs at scattered foci in a marsh (Errington, 1963). Lord et al. (1956b) isolated a gram-positive, spore-forming bacterium they identified as a *Clostridium* from muskrats exhibiting characteristic symptoms, and they used material from the naturally infected animals to reproduce the symptoms in captive (albeit wild-caught) muskrats. Errington (1962) suggested that the pneumonic form, characterized by lung lesions, and the enterohepatic form might be separate entities. Wobeser et al. (1978) suggested that the enterohepatic form was the same as Tyzzer's disease and provided additional arguments in support of the premise that Errington's disease actually was two diseases. Tyzzer's disease is produced by *Bacillus piliformis* (Wobeser et al., 1978), an organism similar to the bacterium isolated from muskrats and described by Lord et al. (1956b). Despite decades of work, much seems to remain to be learned about these irruptive and decimating wildlife diseases.

FAMILY DIPODIDAE—JUMPING MICE

The family includes four species of jumping mice in North America in two genera (*Zapus* and *Napaeozapus*), and the Chinese jumping mice (*Eozapus*), birch mice (*Sicista*), and jerboas (10 or 11 genera) in Asia and North Africa (Holden, 1993; Klingener, 1984). The North American jumping mice and Chinese jumping mice are classified in the subfamily

Zapodinae (formerly considered the family Zapodidae). In North America, the fossil record for the subfamily Zapodinae extends to the Miocene (Nowak, 1991).

All North American representatives are small; possess exceptionally long hind legs and feet, hindquarters modified for jumping, and exceptionally long, thin tails; and have a large infraorbital foramen through which part of the anterior medial masseter is transmitted (Klingener, 1984). However, this group does not possess fused neck vertebrae, enlarged auditory bullae, or fused leg bones, characteristic of some other saltatorial rodents or some Old World members of the family. The long tails serve to stabilize the mice during ricochetal locomotion; loss of the tail causes the mice to somersault in midair, but they soon learn to balance without the appendage (Miller, 1900).

In Oregon, the family is represented by two species in the genus *Zapus*. In addition to the familial characteristics, these species possess four upper cheek teeth, upper incisors deeply grooved on their anterior faces, and a yellowish or orangish wash extending high on the sides. Contrary to some reports (Maser et al., 1981*b*), members of the genus do not possess internal cheek pouches (Klingener, 1971). When flushed, jumping mice make two or three bounds then sit motionless near a shrub, fallen log, or clump of grass. At first glance, a jumping mouse might be mistaken for a frog (Anura) as it bounds through tall grasses.

KEY TO THE DIPODIDAE OF OREGON

1a. Pelage on sides usually washed with lemon yellow; coronoid process of mandible short and wide, not diverging greatly from condyloid process; baculum with lance-shaped tip <0.43 mm wide...................................
.........*Zapus princeps*, western jumping mouse (p. 340).
1b. Pelage on sides usually washed with deep orange; coronoid process of mandible long and slender, diverging greatly from condyloid process; baculum with spade-shaped tip >0.43 mm wide...
........*Zapus trinotatus*, Pacific jumping mouse (p. 343).

Zapus princeps J. A. Allen
Western Jumping Mouse

1893*a. Zapus princeps* Allen, 5:71.
1897*h. Zapus pacificus* Merriam, 11:104.
1899. *Zapus princeps oregonus* Preble, 15:24.
1899. *Zapus major* Preble, 15:24.
1954*a. Zapus princeps pacificus*: Krutzsch, 7:412.

Description.—*Zapus princeps* (Plate XIX) is a long-tailed, sleek-bodied mouse (Table 11-25) distinguishable from its allopatric congener, *Z. trinotatus*, by smaller premolars, a narrower skull, and a shorter baculum (Hall, 1981). The pelage is coarse. The yellowish-gray to ocherous sides are separated sharply from the pale-brown to grayish-brown dorsum; the venter is white, often washed with ochre. The tail is bicolored, but not markedly so. The hind legs and feet are strongly modified for saltatorial locomotion.

The coronoid process is short and wide and diverges from the condyloid process at a relatively narrow angle (Fig. 11-128). P4 averages 0.55 mm long by 0.50 mm wide and commonly has only a simple, shallow fold on the occlusal surface (Krutzsch, 1954*a*). The baculum is shorter with a much less expanded tip (<0.43 mm wide) than that

of *Z. trinotatus* (Hall, 1981; Krutzsch, 1954*a*). Hall (1981) reported that the skull was narrow and shallow in comparison with that of *Z. trinotatus*. Nevertheless, our discriminant analyses based on the skull dimensions for samples listed in Table 11-25 for the two species failed to classify 11% of the females and 19% of the males correctly. Thus, we are unable to provide an equation for separating skulls of *Z. princeps* and *Z. trinotatus* as we have for some other difficult-to-separate species.

Distribution.—*Zapus princeps* occurs from southern Yukon south through most of British Columbia, in extreme eastern Washington, in Oregon east of the Cascade Range and in the Rogue River Valley, and in the Sierra Nevada Range in California. The range extends eastward through southern Saskatchewan and Alberta, southeastern Manitoba, much of North Dakota, and extreme northeastern South Dakota. The range also extends southward in montane regions to central Nevada, central Utah, and Arizona and New Mexico (Fig. 11-129). In Oregon, the species occurs throughout most of the area east of the Cascade Range except for the Columbia Basin-Deschutes Plateau region. It also occurs in southern Jackson Co. (Fig. 11-129).

The only known fossil of this species is from a 10,000-year-old cave deposit in Texas, an extralimital record (Kurtén and Anderson, 1980).

Geographic Variation.—Two of the currently recognized 11 subspecies occur in Oregon (Hall, 1981): *Z. p. pacificus* with reddish-brown sides in the Rogue River Valley and *Z. p. oregonus* with yellowish-buff sides throughout the state east of the Cascade Range (Krutzsch, 1954*a*).

Habitat and Density.—*Zapus princeps*, throughout its range, is a denizen of mountain meadows, particularly those with small streams (Bailey, 1936; Borell and Ellis, 1934; Brown, 1970; Clark, 1971; Cranford, 1978; Davis, 1939; R. Svihla, 1931). In some areas, individuals rarely range >30 m from streamside even though the more distant habitat may appear similar (Brown, 1970). Thus, the riparian vegetation where this species resides consists largely of willows (*Salix*), aspen (*Populus tremuloides*), alder (*Alnus*), sedges (*Carex*), wetland forbs (*Polygonum, Epilobium*,

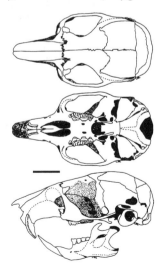

Fig. 11-128. Dorsal, ventral, and lateral views of the cranium, and lateral view of the dentary of the western jumping mouse, *Zapus princeps* (OSUFW 6107). Scale bar equals 5 mm.

Fragaria, Viola, Antennaria) and grasses (especially *Poa*—Brown, 1967a, 1970; Vaughan and Weil, 1980). Some minor, but fairly abrupt, differences in topography seem requisite in habitats occupied by *Z. princeps,* as hibernacula characteristically are located in mounds or banks 25–95 cm higher than adjacent ground level (Cranford, 1978).

In Grand Teton National Park, Wyoming, Clark (1971) sampled six habitats with both live traps and snap traps. No *Z. princeps* was captured in sedge-meadow and sedge-grass communities with either type of trap; 0.9/100 trap-nights and 0.5/100 trap-nights were captured in shrub-swamp with the two traps, respectively; 3.0/100 trap-nights were caught in shrub-sedge-grass-savanna with live traps, but none was caught in the community with snap traps; and 3.2/100 trap-nights were livetrapped and 2.1/100 trap/nights were snap-trapped in lowland aspen. Two individuals caught in live traps in big sagebrush (*Artemisia tridentata*) were believed to have been transients. Thus, few or no western jumping mice were caught in wettest (sedge-meadow, sedge-grass, and shrub-swamp) and driest (big sagebrush) habitats; no mice were caught >100 m from standing water (Clark, 1971).

In a similar study conducted in the Medicine Bow Range, Wyoming, Brown (1967a) caught no *Z. princeps* in mountain-mahogany (*Cercocarpus montanus*), sage-

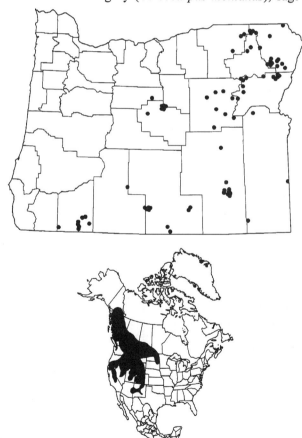

Fig. 11-129. Localities in Oregon from which museum specimens of the western jumping mouse, *Zapus princeps,* were examined (listed in Appendix, p. 578), and the distribution of the western jumping mouse in North America. (Range map redrawn from Hall, 1981:485, map 475, with modifications.)

brush, or alpine tundra communities, 4.6–5.0/100 trap-nights in aspen forest, subalpine meadow, and willow (*Salix*)-alder (*Alnus incana*) communities, and 0.3 – 0.5/100 trap-nights in lodgepole pine (*Pinus contorta*) and spruce (*Picea engelmannii*)-fir (*Abies lasiocarpa*) forest communities. Only 0.4 jumping mice/100 trap-nights were taken >183 m from water (Brown, 1967a).

We know of no study of habitat or density of this species conducted in Oregon. At most, *Z. princeps* is listed among the less common species captured during studies of small-mammal communities, consequently often is considered to contribute relatively little to community structure or dynamics. Nevertheless, the species may be fairly common in small mountain meadows, but appropriate habitats are limited in extent. In one such 4.0-ha mountain meadow in Wyoming, Brown (1970) found an average density following recruitment in August of 4.1/ha (range, 3.7– 4.4/ha) and an average prerecruitment density in June of 2.5/ha (range, 2.2–3.0/ha). Brown (1970:653) commented that the density was "remarkably stable" during his 4-year study. Earlier, Krutzsch (1954a:397) claimed that "its numbers seem to vary considerably from year to year as well as seasonally," but presented no supporting evidence.

Diet.—Seeds, followed by small amounts of green vegetation and arthropods, are listed as the primary foods eaten in most quantitative and anecdotal accounts of the diet of *Z. princeps* (Clark, 1971; Jones et al., 1978). In a study involving western jumping mice from four states or provinces, grass seeds composed 9.5–62.1% of the volume of the stomach contents and unidentified seeds composed an additional 5.6–41.4% (Jones et al., 1978). Some investigators reported stomachs of these mice to contain white starchy material, sometimes in large amounts (Bailey, 1930; Jones et al., 1978); possibly, this also was the heavily masticated remains of seeds. Insects composed 0.8–13.7% of the stomach contents of *Z. princeps* from three regions sampled by Jones et al. (1978), but composed 45.1% in a fourth region (Utah). Vaughan and Weil (1980) reported that insects contributed 63.2% of the diet of *Z. princeps* in June before seeds had matured, whereas they contributed only 15.6–19.7% in July and August after seeds had commenced to mature. The latter authors agreed (p. 123) with Cranford (1978) by stating that "seeds are a vital high-energy food allowing rapid fat deposition" in anticipation of hibernation, but they considered insects a critical source of energy when seeds were not available. In Washington, Dalquest (1948) reported examining *Z. princeps* with chins stained a deep purple from eating the fruits of blackberry (*Rubus ursinus*).

Sporocarps of the fungi *Endogone* composed the entire stomach contents of one specimen taken in Wyoming and occurred in stomachs of all three individuals taken in Colorado (Williams and Finney, 1964) and one individual in Alberta (Dowding, 1955), but did not occur among stomach contents of one taken in North Dakota (Whitaker, 1962). Jones et al. (1978) listed *Endogone* as contributing 1.4–17.7% of stomach contents from four regions within the range of the species.

Reproduction, Ontogeny, and Mortality.—Changes in the reproductive tracts of male *Z. princeps* commence in September with onset of hibernation as an increase in testes mass that continues steadily throughout the period of dormancy. In spring (with timing dependent upon eleva-

Table 11-25.—*Means (±SE), ranges (in parentheses), and CVs of measurements of skull and skin[a] characters for female and male jumping mice* (Zapus) *from regions in Oregon. Skin characters were recorded from specimen tags; skull characters were measured to the nearest 0.01 mm.* SE *and* CV *not provided if* n <10.

Species and region	Sex	n	Total length	Tail length	Hind foot length	Mass (g)	Occipitonasal length	Basal length	Length of maxillary toothrow	Zygomatic breadth	Breadth of braincase	Least interorbital breadth	Skull depth
Zapus princeps													
East of Cascade Range	♀	30	232 ± 1.8 (209–254) 4.3	139 ± 1.2 (127–155) 4.8	32 ± 0.2 (30–35) 3.7	31.97[b] (20.0–44.2)	24.54 ± 0.17 (22.83–25.88) 3.70	19.11 ± 0.14 (17.74–20.59) 4.06	4.33 ± 0.03 (4.10–4.70) 3.63	12.68 ± 0.08 (11.55–13.31) 3.27	11.24 ± 0.05 (10.48–11.87) 2.67	4.64 ± 0.03 (4.30–4.96) 4.09	8.92 ± 0.06 (8.09–9.79) 3.98
	♂	30	237 ± 1.7 (220–256) 3.9	143 ± 1.1 (131–155) 4.3	33 ± 0.3 (30–38) 4.9	28.47 ± 1.05[c] (23.0–35.3) 13.38	24.68 ± 0.14 (23.32–26.19) 3.09	19.19 ± 0.11 (18.33–20.39) 3.19	4.36 ± 0.03 (4.00–4.80) 4.16	12.64 ± 0.07 (11.77–13.38) 2.93	11.19 ± 0.04 (10.69–11.79) 2.06	4.75 ± 0.04 (4.41–5.24) 4.46	8.99 ± 0.05 (8.39–9.55) 2.79
Jackson and Josephine counties	♂	3	220 (218–222)	137 (131–141)	32 (31–32)		23.47 (23.12–23.92)	18.18 (17.80–18.55)	3.97 (3.94–4.03)	12.20 (11.90–12.52)	11.26 (10.94–11.69)	4.42 (4.29–4.55)	8.60 (8.41–8.77)
Zapus trinotatus													
Coast Range	♀	30	231 ± 2.5 (193–262) 5.9	139 ± 2.1 (116–159) 8.2	32 ± 0.3 (29–34) 4.4	25.96 ± 1.02[c] (18.7–34.0) 14.70	24.31 ± 0.14 (22.21–25.50) 3.07	19.12 ± 0.12 (17.64–20.51) 3.49	4.07 ± 0.03 (3.73–4.45) 4.66	12.56 ± 0.06 (11.74–13.06) 2.54	11.06 ± 0.07 (10.45–12.35) 3.25	4.53 ± 0.04 (3.99–4.84) 4.45	8.98 ± 0.04 (8.45–9.41) 2.47
	♂	30	229 ± 1.9 (210–250) 4.5	139 ± 1.6 (118–160) 6.3	32 ± 0.2 (28–34) 4.2	24.07 ± 0.91[d] (19.0–35.4) 17.28	24.31 ± 0.11 (23.19–25.32) 2.48	19.07 ± 0.10 (17.89–19.93) 3.00	4.08 ± 0.03 (3.76–4.43) 3.68	12.28 ± 0.07 (11.62–13.00) 3.00	10.95 ± 0.04 (10.48–11.40) 2.24	4.59 ± 0.03 (4.18–4.96) 4.16	9.00 ± 0.06 (8.43–9.59) 3.48
Cascade Range	♀	12	227 ± 3.8 (206–261) 5.8	133 ± 2.1 (115–144) 5.5	30 ± 0.5 (27–33) 5.7	23.73[e] (17.5–32.0)	24.07 ± 0.18 (23.23–25.63) 2.65	18.89 ± 0.19 (17.83–20.43) 3.49	4.04 ± 0.04 (3.74–4.22) 3.66	12.47 ± 0.09 (12.03–13.20) 2.64	10.97 ± 0.07 (10.54–11.39) 2.37	4.43 ± 0.05 (4.24–4.80) 3.83	8.87 ± 0.08 (8.34–9.32) 3.01
	♂	30	228 ± 1.6 (205–244) 3.9	137 ± 1.2 (123–149) 4.8	31 ± 0.4 (29–39) 6.4	23.18 ± 0.98[f] (18.3–37.9) 19.26	24.20 ± 0.11 (23.14–25.54) 2.59	18.89 ± 0.09 (18.00–19.70) 3.39	4.13 ± 0.04 (3.71–4.54) 5.28	12.44 ± 0.05 (11.97–13.04) 2.18	11.07 ± 0.04 (10.71–11.62) 2.10	4.57 ± 0.04 (4.19–5.27) 4.71	8.93 ± 0.06 (8.56–9.78) 3.39

[a]Although measures of ear length were recorded, the data were not included because of extreme variation believed to stem from differences in techniques used to measure these characters.

[b]Sample size reduced by 21. [c]Sample size reduced by 16. [d]Sample size reduced by 9. [e]Sample size reduced by 8. [f]Sample size reduced by 6.

tion), while still in torpor, development of spermatogonia and primary spermatocytes commences, and a month later a few secondary spermatocytes appear (Brown, 1967c). Males exhibit active spermatogenesis when they arouse from torpor, but mature spermatozoa do not appear until 7–12 days later, and peak testes mass is not reached until 2–3 weeks following attainment of fertility (Brown, 1967c; Clark, 1971). There is some evidence that sexual maturity in yearling males is delayed a few days (Brown, 1967c). By midsummer, just before hibernation, testes mass reaches the annual nadir (Clark, 1971).

In Wyoming, female *Z. princeps* bred <1 week after emergence from hibernation (Brown, 1967c), with the date of breeding dependent upon elevation and age of female. Yearling females (those weighing 21–25 g) bred ≥1 week later than older females (those weighing >26 g). The date of breeding was delayed ≅2 weeks for each 300-m increase in elevation (Brown, 1967c). Reported litter size in Wyoming averaged 5.4 (range, 4–8; $n = 33$—Brown, 1967c) and 5.1 (range, 3–7; $n = 16$—Clark, 1971) based on counts of embryos and 4.7 (range, 3–7; $n = 24$) based on counts of pigmented sites of implantation (Clark, 1971). Brown (1967c) found counts of embryos among yearling females (4.8) averaged less than among older females (6.7). Although Brown (1967c) noted evidence of 2.3% preimplantation losses and no postimplantation mortality among embryos, Clark (1971) reported that only 55% of females exhibited evidence of having bred. As only one litter is produced annually, Brown (1967c) commented that reproductive potential in *Z. princeps* may be the lowest among native mice.

We examined four museum specimens taken 30 June–18 July in Wallowa Co. for which the collector recorded "no embryos, lactating." Pigmented sites of implantation were not mentioned. As gestation is 18 days (Brown, 1967c), these mice may have emerged from hibernation

during the 1st week of June and bred as early as 12 June.

In Wyoming, over-winter disappearance of marked young-of-the-year during a 4-year study averaged 55.6% (range, 44.5–65.5%), whereas only 16.5% (range, 0–28.6%) of adults disappeared (Brown, 1970). Brown (1970) did not believe that predation on hibernating young *Z. princeps* was responsible for the difference, but presented evidence that some young failed to accumulate sufficient fat reserves to sustain them during the long period of dormancy. Rates of disappearance for young of the previous year ($\overline{X} = 30.0\%$; range, 25.0–40.0%) and older adults (27.3%; range, 20.0–33.3%) during the active season (June–August) were similar.

The estimated average life span for marked *Z. princeps* in Wyoming was 16.5 months ($n = 28$) with maximum longevity of ≅4 years. Brown (1970:657) suggested that population turnover of ≅4 years was "a by-product of hibernation" and likely was "highly beneficial to an animal able to produce only one litter per season." He listed several species of terrestrial carnivores as present on his study area, but believed that predation on hibernating jumping mice was negligible. Remains of *Z. princeps* were not listed among those in pellets regurgitated by four species of owls (great horned owl, *Bubo virginianus*; long-eared owl, *Asio otus*; short-eared owl, *A. flammeus*; and burrowing owl, *Athene cunicularia*) in summer in central Oregon (Maser et al., 1970, 1971), two species (screech owl, *Otus asio*, and burrowing owl) in southeastern Oregon (Brown et al., 1986), or the great gray owl (*Strix nebulosa*) in northeastern Oregon (Bull et al., 1989), but the foraging area of the owls may not have included appropriate habitats for jumping mice.

Habits.—Without doubt, "the most unique feature of the life cycle of *Zapus princeps* is its long period of hibernation" (Brown, 1970:653). Clark (1971) indicated that individual mice were active for only 2–2.5 months, although, as a population, some mice were active 2.5–3 months at ≅2,060 m elevation in Grand Teton National Park, Wyoming. Earlier Brown (1967c) reported that the period of activity in the Medicine Bow Range, Wyoming, ranged from a maximum of ≅4 months at ≅2,600 m to a minimum of ≅2.5 months at ≅3,200 m for males and a maximum of 3.5 months at ≅2,600 m to a minimum of ≅2.25 months at ≅3,200 m for females.

In Utah, 26 hibernacula averaged ≅60 cm below the surface with a chamber ≅14.5 cm in diameter; a tunnel averaging 4.3 cm in diameter and 104 cm long led from the surface to the chamber (Cranford, 1978). Tunnels leading to occupied hibernacula were plugged from the ground surface 15–27 cm toward the chamber. No food caches were found in hibernacula. In some, a portion of the tunnel had been used as a latrine for urination (Cranford, 1978). Stems, leaves, needles, and bark of plants were used to construct spherical nests within the chambers; nests were constructed with the finest material on the inside. Typically during hibernation, western jumping mice curl themselves into a ball, then wrap their tails around the ball.

In the laboratory, Z. princeps commences to accumulate fat reserves for the next period of hibernation ≅50 days after emergence from the previous period of torpor. During the ensuing 40 days, the proportion of dry body mass composed of fat increases from ≅20 to ≅64% (Cranford, 1978). Jumping mice from three elevations were maintained in constant darkness at an average temperature of 4°C; all entered hibernation within 5 days of when the field population from the same elevation immerged. The temperature was raised 0.25°C/day in spring; all aroused from torpor within 4 days of the temperature exceeding 8.5°C. In the field, the mice emerged when soil temperature at hibernacula depth reached 8.3–8.8°C (Cranford, 1978). At low elevations, females and males emerged within 1–2 days of each other, but at 2,500 m and above, males emerged 3–9 days before females (Cranford, 1978). During hibernation, animals in a field population lost an average of 0.058 g/day during a 292-day period of torpor.

In the Laramie Mountain Range, Wyoming, home ranges of Z. princeps tended to be long and narrow, and were parallel and adjacent to small streams (Brown, 1970). Many (25–40%) males, but not females, tended to shift their home ranges between June and August sampling periods.

Zapus trinotatus Rhoads
Pacific Jumping Mouse

1895a. Zapus trinotatus Rhoads, 47:421.
1897h. Zapus trinotatus montanus Merriam, 11:104.

Description.—*Zapus trinotatus* (Plate XIX) is a long-tailed, sleek-bodied mouse (Table 11-25) distinguishable from its allopatric congener, *Z. princeps*, by larger premolars, a broader skull, and a longer baculum (Hall, 1981). The pelage is coarse. The ocherous sides are separated sharply from the tawny dorsum; the venter is white, often washed with ochre. The tail is bicolored, but not markedly so. The hind feet are modified for saltatorial locomotion. Gannon (1988) claimed that the ear was fringed with the color of the dorsum, distinguishing it from the white-fringed ear of *Z. princeps*. He attributed the character to Krutzsch (1954a), but the latter author indicated that the ears of one race of each species that occurs in Oregon

were fringed with the same color as the dorsum.

The coronoid process is thin and diverges from the condyloid process at a comparatively wide angle (Fig. 11-130). P4 averages 0.70 mm long by 0.75 mm wide and commonly has a crescent-shaped fold on the occlusal surface. The baculum is spade-shaped with a tip >0.43 mm wide (Hall, 1981). Hall (1981) reported that the skull was broad and deep in comparison with that of *Z. princeps*. Nevertheless, our discriminant analyses based on the skull dimensions for samples listed in Table 11-25 for the two species failed to classify 15% of the females and 20% of the males correctly. Thus, we are unable to provide an equation for separating skulls of *Z. trinotatus* and *Z. princeps* as we have for some other difficult-to-separate species.

Distribution.—*Zapus trinotatus* occurs from southern British Columbia south along the Pacific Coast to San Francisco Bay, California, and along the Cascade Range through Washington and Oregon south to Crater Lake (Fig. 11-131). In Oregon, the species occurs in the Cascade Range and westward to the Pacific Ocean except for southern Jackson Co. (Fig. 11-131).

No fossils are known (Gannon, 1988).

Geographic Variation.—Two of the four nominal subspecies occur in Oregon: the slightly larger and darker *Z. t. trinotatus* in coastal regions and the slightly smaller and lighter *Z. t. montanus* in the Cascade Range (Bailey, 1936; Hall, 1981). Some earlier range maps depicted a broad hiatus between the two races in the southern interior valleys (Gannon, 1988; Hall, 1981), an area from which Bailey (1936) reported specimens. We also found specimens from scattered localities in the region (Fig. 11-131).

Habitat and Density.—Although we know of no quantified evaluation of habitats specifically directed to description of those occupied by *Z. trinotatus*, most authorities agree with Bailey (1936:232) that these mice "are largely marsh and meadow dwellers but are often caught along the creek banks or under ferns and weeds in the woods." In Washington, Dalquest (1941b) caught two or three each year (1937–1940) in a wooded ravine on the campus of the University of Washington. He described the ravine

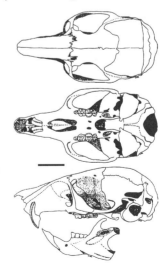

Fig. 11-130. Dorsal, ventral, and lateral views of the cranium, and lateral view of the dentary of the Pacific jumping mouse, *Zapus trinotatus* (OSUFW 5714). Scale bar equals 5 mm.

(p. 170) as vegetated by "broad leaf maple, red cedar, and alder" with an understory of "salmonberry, red elderberry, and devils club" and a ground cover of "grasses of several species, sword ferns, horsetail, and skunk cabbage."

Along the Columbia River, Hinschberger (1978) captured four in 3,009 trap-nights. All were downstream from river mile 79; three were in tidal marsh and one was in alder (*Alnus*), but none was caught in tidal shrub willow (*Salix*), tidal sitka spruce (*Picea sitchensis*), black cottonwood (*Populus trichocarpa*), willow, cottonwood-willow, or Reed canarygrass (*Phalaris arundinacea*) communities.

Borrecco et al. (1979:97) sampled small-mammal populations on three 12–28-ha clear-cuttings "similar in age, past treatment, stocking, cover, wildlife use, and physiographic features" in western Oregon in 1971–1972. The areas were logged and burned 8–12 years earlier, and were "in the late weed stage of secondary succession, but approaching a shrub-dominated period." Although *Z. trinotatus* was captured on all three areas, the species did not compose >9% of the small mammals captured on two areas in either year, but composed 12–29% on one area. The area supporting the greatest abundance of *Z. trinotatus* also supported the greatest abundance of *Microtus oregoni* and the least abundance of *Sorex trowbridgii* and

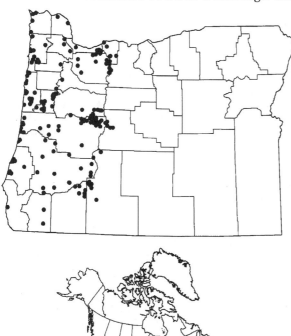

Fig. 11-131. Localities in Oregon from which museum specimens of the Pacific jumping mouse, *Zapus trinotatus*, were examined (listed in APPENDIX, p. 579), and the distribution of the Pacific jumping mouse in North America. (Range map redrawn from Gannon, 1988:2, fig 4, with modifications.)

Peromyscus maniculatus. Hooven (1973*a*) indicated that *Z. trinotatus* was associated closely with *Sorex bendirii* and *Microtus richardsoni* in eastern Lane Co. He indicated that presence of the jumping mice probably was a reflection of their consumption of grasslike plants that grew in the marshy areas shared with the other two species. During a 10-year study in eastern Lane Co., Gashwiler (1970*b*) captured 10 *Z. trinotatus* in 12,000 trap-nights on a 10- by 10-trap grid (15-m trap interval) in a burned-over clear-cutting. By comparison, the estimated density of deer mice (*Peromyscus maniculatus*) on the grid ranged from 1.0 to 21/ha.

Relative abundance of *Z. trinotatus* was much less on clear-cuttings treated with herbicides to suppress grasses and forbs than on untreated clear-cuttings (Borrecco et al., 1979). However, within 2 years, the suppressed vegetation had recovered and the relative abundance of these jumping mice was approximately equal on treated and untreated clear-cuttings (Black and Hooven, 1974).

In a comparison of small-mammal communities in riparian zones with upslope communities in different-aged forest stands in the central Coast Range, Gomez (1992) captured 702 individuals in 100,800 trap-nights with pitfalls. *Z. trinotatus* was significantly more abundant in shrub-conifer communities (2.2/100 trap-nights) than in communities of sapling-pole conifers (0.3/100 trap-nights), large saw-timber conifers (0.3/100 trap-nights), old-growth conifers (0.2/100 trap-nights), and deciduous forest species (0.5/100 trap-nights). The species was significantly more abundant along riparian transects (1.4/100 trap-nights) than along upslope transects (0.4/100 trap-nights).

Diet.—Although foods eaten by *Z. trinotatus* were listed in general terms by several authors of regional works on mammals (Bailey, 1936; Dalquest, 1948; Maser and Franklin, 1974), we know of only one quantitative study of the diet of the species (Jones et al., 1978). The latter study tends to corroborate the general impression of earlier workers. Jones et al. (1978) reported the primary foods eaten by *Z. trinotatus* in Oregon (*n* = 43) and California (*n* = 3) were grass seeds (37.5% by volume), fleshy fruits of huckleberry (*Vaccinium*) and blackberry (*Rubus*; 19.3% by volume), insects (6.3% by volume), and sporocarps of the hypogeous fungi *Endogone* (6.1% by volume). Seeds of other plants contributed 14.9% of the volume of foods eaten. Maser et al. (1978*a*) reported that half of 14 *Z. trinotatus* stomachs contained sporocarps of Endogonaceae, including *Endogone lactiflua* and at least four species of *Glomus*. Nevertheless, the sporocarps composed only 10% of the stomach contents.

Reproduction, Ontogeny, and Mortality.—Bailey (1936) reported that *Z. trinotatus* produced one annual litter consisting of four–eight young. We recorded pregnant females taken in June and lactating females taken in July. We question a record of a pregnant female taken in September (CRCM 89-1315) as most authorities consider this the season that many members of the species enter torpor (Bailey, 1936; Dalquest, 1948; Svihla and Svihla, 1933). The likelihood of young produced from such a late pregnancy accumulating sufficient fat reserves to survive hibernation seemingly would be extremely low. A combination of counts of embryos and pigmented sites of implantation from tags of Oregon-caught museum specimens (*n* = 4) averaged 4.8 (range, 3–6).

Six neonates born to a female captured in a live trap in

Washington averaged 0.8 g (range, 0.7–0.9 g); they were pink, hairless (even lacking vibrissae), and with closed eyes and ears (Svihla and Svihla, 1933). Young are weaned at ≅1 month and they attain sexual maturity the year following birth (Maser et al., 1981*b*).

Although natal nests woven of grasses and other materials by *Z. trinotatus* usually are on the surface of the ground or slightly elevated on stems of grasses, one natal nest was found in a tree (Johnstone, 1979).

Maser and Brodie (1966) identified remains of one *Z. trinotatus* among 277 barn owl (*Tyto alba*) pellets from Polk Co., but none among 116 pellets from Benton Co. They also reported remains of four individuals in 621 pellets regurgitated by great horned owls (*Bubo virginianus*) in Polk Co. Reynolds (1970) found remains of three individuals among 153 prey items in 104 regurgitated pellets of long-eared owls (*Asio otus*) in Benton Co. In the Coast Range, nine of 227 prey items identified in bobcat (*Lynx rufus*) fecal droppings were *Z. trinotatus*, but in the Cascade Range only one of 61 prey items was of that species (Nussbaum and Maser, 1975). However, Toweill and Anthony (1988*a*) reported 10 (2%) of 494 fecal droppings of bobcats on the west slope of the Cascade Range contained remains of *Z. trinotatus*. Other common predators of small mammals probably take *Z. trinotatus* (Maser et al., 1981*b*).

Habits.—In September and October, these jumping mice become exceedingly fat; in captivity, they become progressively less active (Svihla and Svihla, 1933). In their hibernacula, they curl themselves into a tight ball, then wrap their long tails around their curled bodies. Hibernacula commonly are dug in a mound or bank and are ≅75 cm below the surface of the ground and ≅13 cm in diameter with an entrance ≅4 cm in diameter. The nesting chamber is lined with grasses or other fibers; well-shredded newspaper is known to have been used as insulation (Flahaut, 1939). These mice remain in their hibernacula for ≅7 months; if warmed they arouse, but become torpid again upon cooling (Svihla and Svihla, 1933). Torpor is known to occur at ambient temperatures of 8–11°C and as late as 20 April in northern coastal Washington (Edson, 1932). Andersen and MacMahon (1985*a*) believed that occurrence of *Z. trinotatus* in blowdown and tephrafall zones 1–2 years following eruption of Mount St. Helens in Washington in May 1980 was related to individuals being in their hibernacula during the eruption.

When flushed, these mice commonly make three or four 1–2-m-long bounds, changing direction each time they leap, then remain motionless at the base of a tree or clump of grass. They tend to be unafraid and if approached slowly often can be captured by hand; if handled gently, they remain docile and do not bite (Bailey, 1936). Most activity is nocturnal (Bailey, 1936), but we saw them abroad in the late afternoon in June when daylight hours were long.

Remarks.—Other than the taxonomic treatments of part or all of the North American members of the subfamily Zapodinae (Howell, 1920*c*; Krutzsch, 1954*a*) and the quantitative evaluation of the diet of western *Zapus* by Jones et al. (1978), we know of no published article that includes a significant contribution to the biology of *Z. trinotatus*. Contributions for *Z. trinotatus* comparable with those of Brown (1967*a*, 1967*c*, 1970) for *Z. princeps* await an energetic, dedicated, and clever modern naturalist.

FAMILY ERETHIZONTIDAE—PORCUPINES

This New World family contains four genera and 10 species; only one species occurs north of Mexico.

Erethizon dorsatum (Linnaeus)
Common Porcupine

1758. [*Hystrix*] *dorsata* Linnaeus, 1:57.
1835. *Erethizon epixanthus* Brandt, ser. 6, 3:390.
1885. *Erethrizon* [*sic*] *dorsatus epixanthus*: True, 7(app., circ. 29):600.

Description.—The porcupine (Plate XIX) is a large (Table 11-26), short-legged rodent with >30,000 barbed-tipped quills (modified hairs) covering the upper parts of the body and the dorsal and lateral surfaces of the tail. The quills are 2–3 mm in diameter and ≤75 mm long and are scattered among much longer, coarse guard hairs; the underfur is woolly. The quills are arranged in rows across the body; the longest quills are on the rump, the shortest on the face. Quills used in defense are replaced commencing about 10–42 days after loss; quills grow at ≅0.5 mm/day (Woods, 1973). The quills have whitish or yellowish bases and blackish or brownish tips; the overall color of the porcupine is dark brown or blackish. The front feet have four toes, the rear feet five; all have strong, curved claws. The soles are naked. Females have two pairs of mammae.

The skull is heavy, angular, and flattened dorsally (Fig. 11-132); the infraorbital foramen, like that of the nutria (*Myocastor coypus*, the only other histricomorph rodent in North America north of Mexico), is large and oval, and serves as passage for part of the masseter muscle. Also, the molariform toothrows converge anteriorly, the pterygoid fossa is open, and the postcondyloid process of the mandible is prominent. The skull differs from that of the nutria by possessing short paroccipital processes that do not extend ventrally beyond the bullae and incisive foramina that do not extend posteriorly as far as the molariform toothrow. The cheekteeth have complexly folded enamel on their occlusal surfaces. The incisors have strongly pigmented (yellow or orange) enamel on their anterior surface.

Distribution.—The porcupine occurs throughout much of North America from northern Alaska and northern Northwest Territories south to northern Sonora and Chihuahua, the Texas Panhandle, western Oklahoma, Iowa, southern Wisconsin, eastern Tennessee, and southern Virginia (Fig. 11-133; Hall, 1981; Woods, 1973). In Oregon, we examined museum specimens of the porcupine from throughout most of the state east of the Cascade Range (Fig. 11-133). West of the Cascade Range, we found specimens from only a few scattered localities, although we saw live or road-killed porcupines along the Pacific Coast, in the Coast Range, and in the Willamette Valley.

Erethizon dorsatum first appeared in the fossil record in the late Pleistocene. Fossils have been recovered from numerous sites from Oregon east to Pennsylvania and Virginia, and from Alaska south to California and Texas. Several sites are beyond the limits of the Recent distribution (Kurtén and Anderson, 1980).

Geographic Variation.—Only one of the seven named subspecies occurs in Oregon: *E. d. epixanthum* (Hall, 1981).

Habitat and Density.—Although the porcupine reportedly attains its greatest abundance in mixed coniferous and

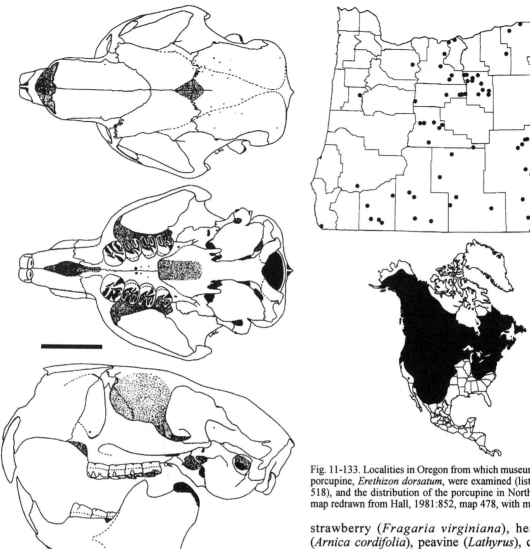

Fig. 11-132. Dorsal, ventral, and lateral views of the cranium, and lateral view of the dentary of the porcupine, *Erethizon dorsatum* (OSUFW 2447). Scale bar equals 25 mm.

Fig. 11-133. Localities in Oregon from which museum specimens of the porcupine, *Erethizon dorsatum*, were examined (listed in APPENDIX, p. 518), and the distribution of the porcupine in North America. (Range map redrawn from Hall, 1981:852, map 478, with modifications.)

hardwood forest, it commonly occurs in deserts, plains, and tundra regions where it is usually associated with riparian areas (Woods, 1973). In forested areas, densities of porcupine populations are greatest in early seral communities (Smith, 1977).

On Mount McLaughlin (=Mt. Pitt) in Jackson Co., porcupines occupied lava fields vegetated by buckbrush (*Ceanothus*) and manzanita (*Arctostaphylos*) that graded into ponderosa pine (*Pinus ponderosa*) at ≅915 m, then to lodgepole pine (*Pinus contorta*) and Engelmann spruce (*Picea engelmannii*) at ≅1,525 m; density was judged greatest in the ponderosa pine zone (Gabrielson, 1928). Near the juncture of Grant, Morrow, and Wheeler counties, Smith (1979) captured 118 porcupines 231 times on a 7.6-km route traversed 68 times in summer 1973 and 1974. The vegetation was predominantly ponderosa pine and Douglas-fir (*Pseudotsuga menziesii*) with an understory of Sandberg bluegrass (*Poa sandbergii*), broadpetal

strawberry (*Fragaria virginiana*), heartleaf arnica (*Arnica cordifolia*), peavine (*Lathyrus*), common snowberry (*Symphoricarpos albus*), and sedge (*Carex*—Smith, 1979). Use of the seven plant communities on the area was not uniform: of 249 porcupines observed, 125 (50.2%) occurred in mixed conifer (either ponderosa pine or Douglas-fir)–pinegrass (mixed grasses and forbs), whereas, on the basis of area occupied by the community, only 75 (30.1%) were expected (Smith, 1982). Also, the ponderosa pine–Douglas-fir–elk sedge (*Carex geyeri*) community was avoided; only 44 (17.7%) of the 249 observations were in the community, whereas 68 (27.3%) were expected (Smith, 1982). An estimated 12.6 porcupines/km² occurred on the study area (Smith, 1977). Similar densities were reported in several other states, but densities as high as 37.2/km² and as low as 3.9/km² also were reported (Smith, 1977).

Based on responses to a questionnaire sent to forest managers and on catch records of U.S. Fish and Wildlife Service trappers in western Oregon, Dodge and Canutt (1969) reported that porcupines were abundant along the west slope of the Cascade Range and in Jackson, Josephine, and Douglas counties, but relatively uncommon in the Coast Range in Coos Co. and northward.

Diet.—In general, porcupines feed primarily on leaves, bark, and cambium of conifers in winter and on succulent

Table 11-26.—*Means (±SE), ranges (in parentheses), and CVs of measurements of skin and skull characters for female and male porcupines* (Erethizon dorsatum) *from Oregon. Skin characters were recorded from specimen tags; skull characters were measured to the nearest 0.01 mm.*

Sex	n	Total length	Tail length	Hind foot length	Ear length	Occipitonasal length	Nasal length	Length of maxillary toothrow	Zygomatic breadth	Breadth of braincase	Breadth of molariform toothrow	Skull depth	Length of mandible
♀	30	631[a]	175[a]	84[a]	23	94.49 ± 0.81	36.79 ± 0.52	25.98 ± 0.18	68.15 ± 0.58	38.33 ± 0.27	24.28 ± 0.25	40.10 ± 0.38	69.16 ± 0.66
		(590–672)	(150–200)	(77–90)		(84.81–102.15)	(28.64–40.63)	(23.74–28.33)	(62.27–74.23)	(34.93–40.96)	(21.65–27.68)	(36.51–45.32)	(62.84–77.04)
						4.68	7.76	3.89	4.66	3.82	5.74	5.24	5.22
♂	26	796[b]	218[b]	96[b]	29[c]	96.95 ± 0.87	38.29 ± 0.53	26.08 ± 0.25	70.14 ± 0.62	38.77 ± 0.32	24.42 ± 0.23	40.46 ± 0.68	72.51 ± 0.54
		(705–962)	(182–240)	(83–112)	(28–30)	(88.53–104.85)	(33.07–46.10)	(24.01–28.29)	(63.55–75.72)	(35.80–42.17)	(22.60–26.65)	(30.15–44.20)	(65.44–77.24)
						4.59	7.02	4.90	4.48	4.27	4.86	8.52	3.77

[a]Sample size reduced by 28. [b]Sample size reduced by 21. [c]Sample size reduced by 23.

vegetation during the remainder of the year (Gabrielson, 1928; Smith, 1979, 1982). In Jackson Co., the characteristic evidence of porcupine feeding by girdling or completely stripping the bark from the top 2–3 m of conifers was most commonly observed on ponderosa pine, less on lodgepole pine and sugar pine (*Pinus lambertiana*), and rarely or not at all on Engelmann spruce and Douglas-fir (Gabrielson, 1928). In regions where ponderosa pine does not occur, Douglas-fir and western hemlock (*Tsuga heterophylla*) are eaten by porcupines. In some regions, deciduous trees, especially maples (*Acer*), oaks (*Quercus*), and birch (*Betula*), also are fed upon (Dodge, 1982). Woody shrubs, especially blackberry and raspberry (*Rubus*), elderberry (*Sambucus*), and buckbrush may be included in the winter diet (Dodge, 1982; Gabrielson, 1928). Mass of the stomach contents of 18 porcupines caught in late summer–autumn averaged 0.78 kg (range, 0.11–1.38 kg—Gabrielson, 1928).

In spring and summer, the diet shifts primarily to succulent, low-growing terrestrial plants or to aquatic species. Skunk cabbage (*Lysichitum americanum*), sedge, violet (*Viola*), dandelion (*Taraxacum*), water-lily (*Nymphaea*), arrowhead (*Sagittaria*), pondweed (*Potamogeton*), and liverwort (*Riccia*) commonly are sought by porcupines. Also, porcupines often feed on forage crops, especially alfalfa and clover, and on truck and garden crops, including potatoes, carrots, lettuce, cabbage, and sweet corn (Dodge, 1982; Gabrielson, 1928).

Porcupines seem to have an insatiable craving for salt.

Reproduction, Ontogeny, and Mortality.—Both males (Shadle, 1952) and females (Dodge, 1982) attain sexual maturity at about 16–18 months. The testes descend in June–July and remain in the scrotum throughout the breeding season. The breeding season extends from September to November (Table 11-2), with females that fail to breed successfully during the first estrus possibly recycling 25–30 days later (Dodge, 1982; Woods, 1973). Estrus lasts for 8–12 h.

Precopulatory behavior is complex. In captivity, males sing (a low whine in some, but others may produce loud vocalizations), engage in a three-legged walk while holding the genitals with a forepaw, and rub the genitals on foreign objects (Shadle, 1946). Litter soiled with urine of near-estrous females is examined carefully. Males become aggressive and are prone to fight with other males (Shadle et al., 1946). Among females, the sensitivity of the genital region is reduced as estrus approaches, and the anogenital region may be rubbed on various objects. At this stage, the male approaches the female, sniffs the anogenital region, rubs noses, rears on the hind legs, and approaches the female with penis erect. The female may flee the advances of the male if not receptive, but if receptive, the female

also may rear on the hind legs and face the male. The male then wets the ventral surface of the female with urine, whereupon the female may threaten the male and flee or may exhibit sexually aggressive behavior by backing into the male with tail erect and genitals exposed (Shadle, 1946; Shadle et al., 1946). When both individuals are standing erect, the male sometimes attempts to push the female onto its back, possibly accounting for occasional, albeit seemingly erroneous, reports of venter-to-venter copulation. During observed copulations, the female held the tail erect and the male mounted from the rear by standing erect on hind legs and tail; the male held its forelegs well above the body of the female and did not grasp the female. The female actively pushed against the pelvic thrusts of the male (Shadle, 1946). Coitus was repeated several times. Ovulation is almost always from the right ovary with implantation in the right uterine horn (Dodge, 1982).

Gestation ranges from 205 to 217 days (Shadle, 1951). Parturition is usually in April–May, but may be as late as September (Dodge, 1982). One young is the rule; rare reports of multiple births are unverified (Table 11-2). Body mass of the highly precocious young ranges from 402.7 to 604.5 g among females (n = 4) and from 413.9 to 542.0 g for males (n = 3; Shadle, 1951). A 595-g male delivered by Caesarean section of an adult female killed in the wild had wet, limp quills that soon became stiff and sharp (Tryon, 1947). By 12 days of age, it weighed 679.5 g; by 28 days, 849.4 g; by 47 days, 1.3 kg; by 81 days, 1.8 kg; by 357 days, 5.4 kg; by 618 days, 6.3 kg; and by 820 days, 7.1 kg. It was tractable, playful, and tended to follow anything that moved away from it for the 1st year, but thereafter became irritable and belligerent (Tryon, 1947).

Of inimical factors, mankind, with control programs and automobiles, is responsible for much porcupine mortality (Dodge, 1982). Of natural enemies, the fisher (*Martes pennanti*) is considered the most serious, but coyotes (*Canis latrans*), mountain lions (*Puma concolor*), bobcats (*Lynx rufus*), and the great horned owl (*Bubo virginianus*) probably take some porcupines in Oregon. In British Columbia, 15 (3.6%) of 414 carnivores of 11 species harvested for fur contained remains of porcupines; fishers (29.4%) and red foxes (33.3%; *Vulpes vulpes*) most commonly exhibited evidence of preying on porcupines (Quick, 1953a).

Of 168 individuals collected near the juncture of Grant, Morrow, and Wheeler counties, 49 (29.2%) were ≤12 months old, 44 (26.2%) were 12–27 months old, 30 (17.9%) were 27–44 months old, and 45 (26.8%) were >44 months old (Smith, 1977). If the sample were representative of the population, such an age distribution might be a reflection of considerable mortality among young-of-the-year, as suggested by Smith (1977), and relatively low mortality among adults. Maximum longevity recorded in the wild is a 130-month-old female (Brander, 1971).

Habits.—Porcupines do not hibernate and are active throughout the year. Activity is mostly nocturnal or crepuscular, but those feeding in trees may be observed at any time as they usually do not retire to dens during the daylight hours. Some use is made of dens in winter, but porcupines sometimes remain in trees at temperatures far below freezing (Woods, 1973). Trees selected for resting usually are large; "witches brooms" of mistletoe (*Phoradendron*) in Douglas-fir trees often are used for protection in inclement weather in north-central Oregon (Smith, 1982:239). In Jackson Co., Gabrielson (1928) reported that porcupines denned in fissures between boulders in lava fields adjacent to ponderosa pine forests or along watercourses. Although porcupines usually are solitary except during the breeding season and when females are caring for young, some large dens seem to be communal; smaller crevices serve as natal dens. Dens and primary-foraging trees commonly are defended against use by conspecifics. Dens and entrances thereto commonly are heaped with fecal pellets. Although dens on northern slopes may be used by porcupines, those on south or southwestern slopes seem to be selected more often (Gabrielson, 1928). During late spring and summer, porcupines wandered from denning areas in lava fields to feed in meadows and in agricultural areas as far as 3 km away, but at the onset of autumn rains in October, they returned to the denning area (Gabrielson, 1928).

In north-central Oregon, average distances moved between daily resting sites ranged from 34 m in January to 442 m in May, reflecting the trend of restricted movements in winter and more extensive movements in spring and summer (Smith, 1979). The greatest distance moved between recaptures during summer was 2.6 km. Minimum-area home ranges for six porcupines followed by use of radiotelemetry ranged from 3.9 to 82.1 ha; in general, home-range size tended to be least in winter and greatest in spring and summer (Smith, 1979). Smith (1979) did not record seasonal shifts in areas used as reported in Jackson Co., Oregon (Gabrielson, 1928), or elsewhere.

Porcupines make a variety of vocalizations and sounds with the teeth, some of which can be heard at considerable distance (Woods, 1973). They seemingly are intelligent and are able to learn quickly; they reportedly have good memories and especially remember being mistreated (Shadle, 1950). In captivity, porcupines become gentle and easily handled if treated and trained with patience and kindness (Shadle, 1950). Although porcupines taken by Caesarean section or captured as neonates may become interesting pets, they also may become intractable at sexual maturity (Tyron, 1947). We advocate leaving wild animals in the wild!

Because of their feeding habits and their insatiable craving for salt, porcupines sometimes are destructive. The habit of feeding on the bark and cambium of the terminal 2–3 m of trees, clipping branches, and girdling other parts of trees does not endear the porcupine to the forest industry. Also, porcupines commonly chew tool handles, boat paddles, shoes, steering wheels, and other objects impregnated with human perspiration; tires, hydraulic lines, and electrical wiring coated with salt applied to icy roads are subject to being chewed. The glue used in plywood also is sought;

thus, signboards, boats, buildings, and anything else constructed of plywood may be damaged by porcupines. The economic impact of the porcupine is controversial; Woods (1973) claimed that most authorities agreed that damage did not warrant widespread control measures, but Dodge (1982) indicated that porcupines can inflict considerable damage and require control. Then, later in the same publication, he stated that most foresters consider damage by porcupines insignificant. In western Oregon, Dodge and Canutt (1969) indicated that most damage occurred in 10–30-year-old stands that had been thinned and that dominant trees were attacked first. Control measures employed include shooting, poisoning, and attempts to reestablish the fisher (Gladson, 1979).

Remarks.—A popular misconception is that the porcupine can "throw" its quills; this folklore originated >320 years ago (Woods, 1973). When approached, porcupines present their backsides and sometimes thrash their tails from side to side. Indeed, a quill occasionally might become detached and be propelled a few centimeters, but it lacks sufficient mass and balance to travel far or to pierce arrowlike the clothing or flesh of a person. Upon finding a dead porcupine on the road, gather a few intact quills and try to throw them; the supposed danger from thrown quills should be dispelled quickly.

We do not mean to imply that quills are not superb protective devices for porcupines. One of us (BJV) allowed a museum specimen to fall a few centimeters onto the back of the hand; a half-dozen quills became embedded sufficiently deeply to require that they be extracted by use of pliers. A dog with a face full of quills requires general anesthesia to have them extracted. In addition to the pain and potential for infection, quills that break or otherwise cannot be extracted immediately tend to become buried more and more deeply by action of surrounding muscles. Shadle (1947) and Shadle and Po-Chedley (1949) reported that quills worked their way through skin and muscle at ≅1 mm/h and sometimes traveled several centimeters before emerging through the skin >4 cm from entry points.

Because they present their backsides, porcupines may be captured and transported by hand. A foot, or more safely a forked stick, placed on the tail prevents its thrashing; a hand then can be slipped beneath the tail and the quills depressed by sliding the hand posteriorly and firmly grasping the tail like a club (Shadle, 1950). The animal must be held clear of the legs and body while being carried to avoid claws, quills, and teeth.

FAMILY MYOCASTORIDAE—NUTRIA

The family is monotypic.

Myocastor coypus Kerr
Nutria

1792. *Myocastor* Kerr, p. 225.
1805. *Myopotamus bonariensis* É. Geoffroy St.-Hilaire, 6:82.
1917. *Myocastor coypus bonariensis*: Thomas, 20:100.

Description.—*Myocastor coypus* (Plate XIX) is a large rat-like semiaquatic rodent. Body dimensions (in mm) for one adult female from Oregon were: total length, 990; tail length, 440; length of hind foot, 140; and ear length, 30. The individual weighed 8.2 kg. The species has a hunched body; a round, nearly hairless tail; a valvular mouth and nose; and pentadactyl feet with naked soles. The toes of the hind feet, except for the hallux, are included in a web; the pollex is reduced. As an adaptation to the aquatic environment, the eyes, nostrils, and small ears are set high on the head and the four or five pairs of mammae are set high on the sides. The pelage consists of long, coarse guard hairs and soft, dense underfur; the underfur is thickest on the venter and, unlike that of most furbearers, is the most valuable part of the pelt (Willner, 1982). Overall, the pelage color usually ranges from dark brown to yellow-brown, but melanistic and albanistic individuals are known to occur in Oregon. The muzzle is frosted with white hairs.

The skull of the nutria is heavy and somewhat angular (Fig. 11-134); like that of the porcupine (*Erethizon dorsatum*, the only other histricomorph rodent in North

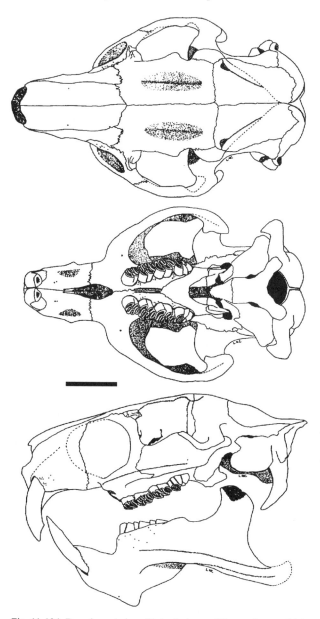

Fig. 11-134. Dorsal, ventral, and lateral views of the cranium, and lateral view of the dentary of the nutria, *Myocastor coypus* (OSUFW 3185). Scale bar equals 25 mm.

America north of Mexico), the large, round infraorbital foramen serves as passage for the deep masseter muscle. Skull dimensions (in mm) for an adult female from Oregon were: occipitonasal length, 118.43; nasal length, 43.60; length of maxillary toothrow, 28.95; zygomatic breadth, 74.18; breadth of braincase, 36.81; breadth of molariform toothrow, 24.59; skull depth, 34.99; and length of mandible, 81.07. The skull can be distinguished from that of the porcupine by the paroccipital processes extending ventrally far beyond the auditory bullae and the posterior margin of the incisive foramina set even with the anterior margin of the zygomatic plate. Also, the molariform toothrows converge anteriorly (Fig. 11-134), and dP4 is retained throughout life (Woods et al., 1992). The cheekteeth are hypsodont and the incisors are wide and have an orange surface on their anterior face.

Distribution.—The nutria, a native of the Patagonian region of South America, was introduced deliberately or inadvertently into Europe, North America, northern Asia, Japan, East Africa, and the Middle East (Woods et al., 1992). In North America, populations of the nutria are established in British Columbia; Washington, Oregon, Idaho, Nevada, New Mexico, Texas, Oklahoma, Arkansas, Louisiana, Mississippi, Alabama, Florida, Georgia, North Carolina, Virginia, Maryland, Delaware, Kentucky, and Tennessee; and in Nuevo León and Tamaulipas (Fig. 11-135; Kinler et al. 1987; Willner, 1982). Willner (1982) indicated that the species was reported from at least 31 states.

In Oregon, museum specimens are limited to areas in the southern Willamette Valley and central Coastal Region (Fig. 11-135). However, trappers reported taking large numbers of nutrias for fur in the interior valleys south to Douglas Co., in counties along the Pacific Coast south to Coos Co., and sometimes a few in the Klamath and Harney basins, and in counties that border or include the Columbia, Snake, and Malheur rivers (Table 11-27). We are skeptical of the reported taking of nutrias in Jefferson Co.

Geographic Variation.—According to Hall (1981) the race established in North America was *M. c. bonariensis.*

Habitat and Density.—Nutrias are semiaquatic, thus usually occur in or adjacent to rivers, lakes, sloughs, marshes, ponds, and temporarily flooded fields. Areas supporting both freshwater and brackish-water plant communities are used. On the William L. Finley National Wildlife Refuge, Benton Co., false loosestrife (*Ludwigia palustris*), nodding beggar-ticks (*Bidens cernua*), Oregon ash (*Fraxinus latifolia*), creeping spike-rush (*Eleocharis palustris*), simplestem bur-reed (*Sparganium emersum*), bentgrass (*Agrostis alba*), willow (*Salix*), and speedwell (*Veronica*) ranked first through eighth in order of frequency of occurrence, and all occurred on ≥30% of 106 3,010-m² plots established on Cabell Marsh and Cattail Pond combined (Wentz, 1971).

Also on the William L. Finley National Wildlife Refuge, numbers of nutrias in February–March and August–September ranged from 25 to 41–42 on Cabell Marsh and from 26 to 56 on Cattail Pond (Wentz, 1971). The size of the two water areas differed seasonally; thus, densities ranged from 0.65 to 20.75/ha and from 6.42 to 22.67/ha on the two areas, respectively. The lower densities during winter were attributed, in part, to nutrias moving to temporary ponds and flooded fields during winter; harvest of

Fig. 11-135. Localities in Oregon from which museum specimens of the nutria, *Myocastor coypus*, were examined (listed in APPENDIX, p. 531), and the distribution of the nutria in North America. (Range map redrawn from Willner, 1982:1061, fig. 53.2, with modifications.) The species has been recorded in several other states (Hall, 1981).

nutrias was not permitted on the refuge. Similarly, the higher densities in summer were attributed in part to nutrias returning to permanent water areas and the size of the marshes and ponds decreasing during the summer drought.

Diet.—In 1969–1970, nutrias were observed to feed on 40 of the 87 species of plants known to occur on Cabell Marsh and Cattail Pond (Wentz, 1971). Willow was fed upon in 12.3% of 438 bouts of feeding observed, false loosestrife in 9.4%, bur-reed in 8.9%, beggar-ticks in 7.5%, spike-rush in 6.2%, common smartweed (*Polygonum hydropiperoides*) in 5.7%, and panicgrass (*Panicum capillare*) in 5.5%. Except for the last two species, those observed to be eaten most frequently occurred in >30% of vegetation samples distributed throughout the area. The remaining 33 species fed upon were observed to be consumed during <5% of the bouts (Wentz, 1971). Of the 47 species not observed to be consumed by nutrias, 38 were coarse in texture, seasonal, unpalatable, or rare. Annual forage ratios (relationship between frequency that a species is fed upon and its relative abundance in the environment) were <1.0 for all species fed upon, indicating no active selection by nutrias. However, in spring, the forage ratio for broadleaf arrowhead (*Sagittaria latifolia*) was 1.70, indicating active selection; nutrias were considered responsible for essentially eliminating this species and

drastically reducing the already rare cat-tails (*Typha latifolia*—Wentz, 1971). Although one species of plant was essentially eliminated and another was drastically reduced by foraging by nutrias, Wentz (1971) concluded that overall the foraging by nutrias had little ecological impact on marshes in western Oregon.

In other areas of western Oregon, nutrias also are known to forage on some field crops, on fruit and nut trees, and on deciduous and coniferous forest trees (Kuhn and Peloquin, 1974). In other regions where the species was introduced, the nutria is considered to be a dominant force in community development in marshes (Woods et al., 1992).

Reproduction, Ontogeny, and Mortality.—All 33 Oregon males ≥6 months old when examined contained spermatozoa in both the testes and the epididymides (Peloquin, 1969). Among females, vaginas usually became perforate between 4 and 9 months of age; 71 of 72 females ≥9 months old had perforate vaginas (Peloquin, 1969).

In western Oregon, 60 pregnant females contained an average of 5.5 (range, 1–11) implanted embryos; 16 (26.7%) contained an average of 2.4 embryos undergoing resorption (Peloquin, 1969). An average of 0.6 embryo/ pregnancy was lost through resorption (Table 11-2). Counts of corpora lutea in 55 pairs of ovaries averaged 7.6 (range, 3–17) per female; thus, 27.6% of ova shed either were not fertilized or were fertilized and failed to implant (Peloquin, 1969).

Births, based on estimated age at capture for 113 individuals, occurred in all months of the year with peaks in January, March, May, and October (Peloquin, 1969:43, fig. 8). Some females gave birth to a first litter when 12–15 months old, and second litters sometimes were produced by females 18–20 months old. Based on a 120–130-day gestation period, under exceptionally favorable environmental conditions production of nearly three litters in a 1-year period may be possible (Brown, 1975; Peloquin, 1969).

Young nutrias are precocial. Neonates are covered with soft, downy fur; the tail hairs are silky, but are replaced by coarse hairs by 1 month of age (Willner, 1982). Nursing the 1st day of life is irregular, but thereafter young nurse for 20–30-min periods at 2–3-h intervals. Both captive and wild nutrias nurse while the maternal female is on land and in a prone position (Peloquin, 1969). In the wild, young nurse for about 7 weeks, but in captivity, young nurse for as long as 11 weeks. However, some individuals forceably weaned at 1 week of age survived (Peloquin, 1969).

Ten nutria young born in captivity in Oregon weighed an average (±*SE*) of 217 ± 12.6 g at birth (Peloquin, 1969). Gains in body mass were rapid; neonates doubled in mass in 2 weeks and averaged gaining 453 g/month during the first 12 months. Males and females averaged 2.8 and 1.9 kg, respectively, at sexual maturity (Peloquin, 1969).

No information regarding mortality is available for nutrias in Oregon, and we are hesitant to interpret sample sizes for age-mass data presented by Peloquin (1969:20, table 2) as representative of the age structure of a Willamette Valley population. However, in a sample of 223 animals, body masses for only 15 (6.7%) individuals >12 months of age were presented. Such suggests either a rapid turnover of the population or a sampling procedure strongly biased in favor of capturing individuals ≤12 months old. Because mortality elsewhere was 54–74%

Table 11-27.—Reported annual harvest of nutrias (Myocastor coypus) in Oregon by county, 1969–1996. Data provided by Oregon Department of Fish and Wildlife.

County	Trapping season																										
	1969–1970	1970–1971	1971–1972	1972–1973	1973–1974	1974–1975	1975–1976	1976–1977	1977–1978	1978–1979	1979–1980	1980–1981	1981–1982	1982–1983	1983–1984	1984–1985	1985–1986	1986–1987	1987–1988	1988–1989	1989–1990	1990–1991	1991–1992	1992–1993	1993–1994	1994–1995	1995–1996
Baker																											59
Benton	332	485	1,268	635	400	543	689	890	1,676	1,723	1,474	2,444	970	1,631	1,617	1,257	1,547	1,388	1,409	910	481	792	774	714	68	1,212	
Clackamas	124	191	261	83	328	100	241	541	438	478	359	546	428	227	442	292	300	479	356	84	115	45	115	578	84	175	
Clatsop	47	147	381	403	797	3,178	2,881	3,640	3,549	1,408	1,117	2,263	1,701	1,027	793	814	202	957	1,033	332	224	163	168	115		174	
Columbia	70	176	171	136	524	1,116	859	1,941	1,347	806	455	402	621	545	382	450	466	324	1,058	619	211	156	163	52	39	382	230
Coos							21	3		5				7		5	5	5	2		18		16	1	11	3	
Crook																											
Curry																											
Deschutes																											
Douglas		150	49	12	62	138	84	157	335	276	366	367	168	234	381	368	326	302	435	162	88	131	113	250	191	276	19
Gilliam													1						3								
Grant																											
Harney														2													
Hood River																											
Jackson																											
Jefferson														40													
Josephine																											
Klamath																	2	6									
Lake	4																										
Lane	641	1,051	867	879	1,838	1,350	2,137	1,949	1,429	2,151	1,985	941	1,285	1,091	2,198	1,370	1,710	2,008	1,470	873	823	1,340	723	1,187	1,723		10
Lincoln	220	9	70	142	186	248	497	559	671	570	564	559	251	199	485	556	412	446	491	593	217	77	110	130	15	4	
Linn	684	289	1,226	979	1,224	2,120	2,459	3,024	2,856	2,213	3,053	3,333	2,272	2,018	2,345	4,304	2,370	4,132	4,413	2,932	2,150	1,633	1,870	2,056	1,882	2,300	47
Malheur	1												4														
Marion	611	437	345	439	854	1,079	1,487	1,330	1,448	874	1,558	1,248	1,517	1,133	1,349	925	1,467	1,654	800	763	1,010	1,289	1,297	851	873		
Morrow																											
Multnomah	10	6	119	5	107	339	408	329	91	60	150	130	160	59	110	36	146	141	178	249	140	101	205	190	77		5
Polk	260	194	134	94	423	544	61	674	323	665	1,206	210	771	611	663	721	681	796	267	228	196	219	77	122	221		6
Sherman																											
Tillamook	3	1	17	6	28	28	57	50	41	40	43	108	60	65	101	44	68	129	26	12	35	32	54	70	82		2
Umatilla																											
Union														5													
Wallowa																											
Wasco																			1								
Washington	270	109	165	76	46	447	450	516	392	156	394	184	303	294	354	122	186	377	61	89	67	69	62	179	109		
Wheeler													5						30								
Yamhill	474	234	181	131	151	146	292	579	628	611	275	584	262	613	562	690	216	617	685	164	851	88	173	110	150	133	11
Totals	3,753	3,479	5,950	4,687	5,141	11,286	11,812	15,997	16,272	11,814	11,611	15,834	9,534	10,612	10,262	13,511	9,057	12,916	15,021	8,599	6,569	5,356	6,555	6,424	6,394	7,744	389
Trappers reporting[a]	721	773	731	847	1,102	1,350	1,377	1,755	2,746	2,899	3,216	3,373	3,069	3,980	3,293	3,104	3,575	3,397	3,833	1,985	1,536	1,228	1,390	1,454	1,293	1,464	687
Average price/pelt	$1.51	$1.50	$2.70	$2.75	$3.96	$4.49	$4.64	$5.68	$6.80	$5.59	$9.97	$11.75	$7.59	$5.44	$4.12	$5.17	$5.20	$6.30	$2.98	$2.76	$4.03	$2.52	$2.30	$2.33	$4.74	$2.12	$6.04

aNot all sought nutrias and not all were successful.

annually (Woods et al., 1992), we suspect that sampling bias played a significant role.

Nutrias reportedly are sensitive to low temperatures, with mortality attaining 80–90% following a few days of freezing temperatures (Newson, 1966; Norris, 1967). Kuhn and Peloquin (1974) reported that fur trappers claimed that almost no nutrias were captured following record-setting low temperatures (–22°C) for nearly 2 weeks in mid-December 1972. However, the reported harvest that year was only 21.2% less than that in the previous year (Table 11-27), suggesting that the decline was not as severe as it seemingly was thought to be. The reported harvest during the trapping season following a period of –15°C temperatures in January 1969 (U.S. Department of Commerce, 1969) increased 21.8% from the 3,082 nutrias taken during the 1968–1969 season (Table 11-27). Although animals that have lost tails, ears, and feet following frostbite commonly are observed following periods of unusually low temperatures, the impact of low temperatures on nutria populations may not be as great as previously supposed.

We have no information specific to Oregon regarding predators taking nutrias, but elsewhere a wide variety of predatory mammals, birds, and reptiles reportedly prey on the species (Woods et al., 1992). Harvest for fur likely is a significant inimical factor for nutrias in Oregon, especially during years when fur prices are high (Table 11-27).

Habits.—Nutrias are active mostly at night (Gosling, 1979), although individuals occasionally may be observed swimming, feeding, or walking along a pond bank during the daylight hours, especially when nighttime temperatures are below freezing (Gosling, 1979). Feeding, grooming, and swimming are primary activities; nutrias may bask in the sun when temperatures are low.

Nutrias construct burrows in banks of rivers, sloughs, and ponds, sometimes causing considerable erosion. In western Oregon, burrows usually were in acutely sloping banks (45–90°) and entrances were partly above water (Peloquin, 1969). Burrow length ($n = 20$) averaged 2.4 m (range, 0.9–6.1 m) and burrow diameter averaged 22.1 cm (no range provided—Peloquin, 1969). Nest cavities commonly were among tree roots. When tunnels were flooded during high water, nutrias sought refuge in thickets of blackberry (*Rubus procerus*) brambles (Peloquin, 1969). Nutrias also construct resting and feeding platforms of matted vegetation connected by trails through vegetation. In spring and summer, nesting platforms consisting of matted vegetation partly covered by surrounding plants are used by maternal females with litters.

Nutrias are coprophagous; feces are taken directly from the anus into the mouth and reingested (Gosling, 1979).

Investigators of movements reported that most recaptured nutrias are within 400 m of their original point of capture; the greatest distance moved was 3.2 km (Willner, 1982). Until 1 month old, nutrias were not observed beyond 27.4 m from the entrances to burrows (Peloquin, 1969).

Nutrias are gregarious, commonly forming groups of 2–13 individuals consisting of a male and a female dominant over other individuals. Commonly the subordinate individuals are related. Adult males sometimes are solitary (Willner, 1982).

Remarks.—Some authorities consider the nutria to be a member of the family Capromyidae, whereas others classify both the nutria and other capromyids as members of the family Echimyidae (Woods et al., 1992).

The nutria was imported into North America for a fur-farming operation at Elizabeth Lake, California, in 1899, but the venture was unsuccessful because of reproductive failure (Evans, 1970). The first fur-farming enterprise to breed nutrias in captivity successfully was in Quebec in 1927. During the 1930s, many individuals attempted to make their fortunes by raising nutrias for fur; many nutrias escaped or, when it became obvious that the venture was unprofitable, were released (Evans, 1970). Stanley G. Jewett related to Larrison (1943) that an unknown number of nutrias were liberated from a fur farm in Tillamook Co. in 1937 during a flood. A colony was established at or near Garrison Lake from this release. F. B. Wire, Oregon State Game Supervisor, also related to Larrison (1943) that feral nutrias occurred along the Nestucca River in Tillamook Co. and were doing well. By 1946, colonies were established, presumably from this and other unsanctioned releases, in several coastal counties, along the major river systems in the interior valleys, and in the Umatilla and Grande Ronde rivers in northeastern Oregon (Kuhn and Peloquin, 1974). During the 1950s, the sale of nutrias to prospective fur farmers in Oregon was promoted by various associations, magazine and newspaper articles, and demonstrations at county fairs (Kuhn and Peloquin, 1974). In the mid-1950s, entrepreneurs were selling breeding stock for as much as $950 per pair or $1,550 for two females and a male, even though in Louisiana nutria pelts were bringing only $1.00 each on the fur market. As early as the 1945–1946 fur season, records of nutrias appeared in reports required annually of fur takers; 29 were reported captured in the 1957–1958 furbearer season. The harvest peaked during the 1976–1981 seasons, an interval during which average pelt prices also attained a peak (Table 11-27). Although prices declined in the late 1980s, the harvest has remained >5,000 annually (Table 11-27).

In Oregon, the nutria simultaneously is considered a valuable unregulated furbearer and a pest. In the Willamette Valley in the late 1960s, complaints concerning its depredations on a wide variety of crops and its burrowing activities were widespread (Kuhn and Peloquin, 1974). Such was responsible for stimulating research on its biology and control (Kuhn and Peloquin, 1974; Peloquin, 1969; Wentz, 1971). Control of numbers by use of toxic baits and by trapping was advocated (Kuhn and Peloquin, 1974). We recall few complaints during the late 1970s and early 1980s when fur prices were high. Although harvest of nutrias for fur peaked during those years (Table 11-27), we seriously doubt that a reduction in numbers by fur harvest was responsible for the apparent change in attitude. We did not note a resurgence of complaints during the early 1990s, when fur prices declined to levels nearly as low as those of the early 1970s (Table 11-27).

Most of the relatively few museum specimens from Oregon either are juveniles, are not accompanied by skulls, or have broken skulls. Hence, we are unable to provide a table of mensural data for skull and body dimensions. Apparently, the large number of specimens obtained for studies of aspects of the biology of the species were not preserved.

CHAPTER 12

ORDER CARNIVORA—FLESH EATERS

The carnivores, distributed worldwide, include about 271 species grouped into 12 extant families. The order is represented in Oregon by five canids, two ursids, three otariids, two phocids, two procyonids, nine mustelids, two mephitids, and three felids. Terrestrial carnivores range in size from the 35–70-g least weasel (*Mustela nivalis*) to the 780-kg grizzly bear (*Ursus arctos*), whereas some marine forms may attain 3,700 kg.

The three extant families of marine carnivores (only two of which occur in Oregon waters) possess many similar adaptations for life in the ocean. These forms have been treated as a separate order (Pinnipedia), as a separate suborder of Carnivora (with terrestrial carnivores classified as Fissipedia), and as a grouping without taxonomic status (Stains, 1984). Simpson (1945) classified the Pinnipedia as a suborder of Carnivora. Stains (1967), on the basis of opinions of several workers, considered the pinnipeds a separate order. Later, Stains (1984), on the basis of published evidence (Tedford, 1976) of a relationship of otariids with ursids and phocids with mustelids, grouped both terrestrial and marine forms in the order Carnivora without recognizing the pinnipeds as a taxonomic unit, but largely maintained the dichotomy in describing the Carnivora. Jones et al. (1986) followed the then-current trend of grouping marine and terrestrial families to illustrate the otariid-ursid and phocid-mustelid affinities, but more recently (Jones et al., 1992) grouped the marine forms together and arbitrarily placed them after the ursids within the order Carnivora in recognition of the present consensus of the marine families constituting a monophyletic group (Nowak, 1991). Although recognition of the pinnipeds as a separate order remains open to question (Nowak, 1991), we chose to follow Jones et al. (1992) by considering both terrestrial and marine forms within the Carnivora. Dragoo and Honeycutt (1997), on the basis of analysis of mitochondrial-DNA sequence data, found the marine carnivores (phocids, otariids, andobenids) formed a monophyletic clade sister to a mephitid-procyonid-mustelid clade with ursids and canids serving as a base for the caniform clade.

The carnivores are characterized by a high brain:body mass ratio, an enlarged middle-ear cavity, reduced (or absent) clavicle, strongly developed zygomata, a baculum (tiny or sometimes missing in felids—Maser and Toweill, 1984), and a transverse barrel-shaped condyloid process seated in a transverse trenchlike glenoid fossa. Among many of the largely terrestrial families, P4 and m1 often are modified as carnassials; this, combined with a jaw hinge that permits little lateral movement, produces a highly efficient scissorlike arrangement for shearing flesh. In general, carnassials are best developed in the felids and canids, less so in the procyonids and mustelids, and least in the ursids. The Recent marine forms have conical teeth as an adaptation for grasping slick prey, but some fossil forms had carnassials (Repenning, 1976). The jaw hinge in the marine forms is similar to that of terrestrial forms.

KEY TO THE FAMILIES OF CARNIVORA OF OREGON

1a. Front limbs modified as flippers; claws on first and fifth toes on hind foot rudimentary or lacking; i3 always lacking..2
1b. Front limbs never modified as flippers, although toes commonly webbed; claws present on all toes; i3 present (except in *Enhydra*)..3
2a. Hind limbs capable of being rotated forward; pinnae present; postorbital processes present; I1 and I2 with transverse groove on occlusal surface........................... ....**Otariidae** (p. 381).
2b. Hind limbs not capable of being rotated forward; pinnae absent; postorbital processes absent; I1 and I2 not grooved on occlusal surface................................. ....**Phocidae** (p. 390).
3a. Claws retractile; ≤30 teeth..................**Felidae** (p. 450).
3b. Claws not retractile; ≥32 teeth...............................4
4a. Tail <14% of total length; combination of bunodont molariform dentition and condylobasal length >200 mm..**Ursidae** (p. 373).
4b. Tail >14% of total length; condylobasal length <200 mm or if >200 mm then P4 and m1 secodont.............5
5a. Digitigrade; four toes on front feet (plus dewclaw); 42 teeth; doglike.....................................**Canidae** (p. 354).
5b. Plantigrade or semiplantigrade; five functional toes on front feet; ≤40 teeth; not doglike..................................6
6a. Black mask across eyes or eyes ringed with brown or black and partially ringed with white; tail annulated; 40 teeth.....................................**Procyonidae** (p. 397).
6b. No mask or eye rings; tail not annulated; ≤38 teeth......7
7a. Pelage black with a white patch or vertical stripe between the eyes and two continuous or four to six segmented white body stripes; margin of palate nearly even with end of toothrow......................**Mephitidae** (p. 440).
7b. Pelage not entirely black and white; margin of palate extending well beyond end of toothrow........................ ....**Mustelidae** (p. 405).

FAMILY CANIDAE—DOGLIKE MAMMALS

The family Canidae consists of 36 Recent species of mammals and, except for some insular regions, is distributed worldwide. Five species—the gray wolf (*Canis lupus*), coyote (*Canis latrans*), red fox (*Vulpes vulpes*), kit fox (*Vulpes velox*), and gray fox (*Urocyon cinereoargenteus*)—occur or have occurred in Oregon. The canids are medium-sized carnivores with long, bushy tails; pointed noses; pointed, erect ears; long slender legs; short, nonretractile claws; a grooved baculum; and well-developed carnassial teeth. Canids are digitigrade; those in Oregon have five toes on the front feet (including a dewclaw) and four toes on the hind feet. All North American species possess a "tail gland" consisting of modified sebaceous glands on the dorsal surface of the tail.

Canids are efficient, highly cursorial predators; they often exhibit extraordinary endurance in the pursuit of prey. Some species hunt in packs, others singly or in pairs. The senses of smell and hearing are amazingly acute; they can detect prey or danger at great distances. Nevertheless, in addition to vertebrate prey, much of the diet consists of insects, crustaceans, and plant materials.

All produce one litter annually. Litter size and reproductive season differ among the species (Table 12-1).

Fossil canids first appeared in the Upper Eocene of Asia; canids made their way to Europe and North America during the Oligocene and diversified rapidly and became widespread in the Miocene (Carroll, 1988; Kurtén and Anderson, 1980; Savage and Russell, 1983). By the Pliocene-Pleistocene (Blancan) relatively few groups remained. Pleistocene-age fossils of extant and extinct canids are known from throughout most of the world (Savage and Russell, 1983). Considerable controversy remains regarding relationships of Recent species at and above the generic level (Wozencraft, 1989).

KEY TO THE CANIDAE OF OREGON

1a. Hind foot >175 mm long; incisors prominently lobed; postorbital processes convex on dorsal surface; basilar length >147 mm..2
1b. Hind foot <175 mm long; incisors not prominently lobed (may be slightly lobed in *Vulpes vulpes*); postorbital processes concave on dorsal surface; basilar length <147 mm..3
2a. Size large (as much as 80 kg); nose pad >25 mm wide; anteroposterior diameter of upper canine ≥11 mm at alveolus; P4 ≥22 mm long at alveolus; tip of little-worn upper canine not extending below mental foramen in closed mandible..
................................*Canis lupus*, gray wolf (p. 360).
2b. Size medium (usually <30 kg); nose pad <25 mm wide; anteroposterior diameter of upper canine <11 mm at alveolus; P4 <22 mm long at alveolus; tip of little-worn upper canine extending below mental foramen in closed mandible..
................................*Canis latrans*, coyote (p. 355).

3a. Tail with black median-dorsal stripe slightly raised as a mane; inferior margin of mandible with prominent step near angle; temporal ridges 2–3 mm wide and forming broadly rounded, lyre-shaped pattern on roof of skull.....
.............*Urocyon cinereoargenteus*, gray fox (p. 363).
3b. Tail lacking black mane; inferior margin of mandible lacking a step; temporal ridges <1 mm wide and forming V-shaped or slightly U-shaped pattern on roof of skull..4
4a. Tail usually tipped with white; feet black; condylobasal length >135 mm..........*Vulpes vulpes*, red fox (p. 369).
4b. Tail tipped with black; feet no darker than color of body; condylobasal length <125 mm............................
......................................*Vulpes velox*, kit fox (p. 366).

Canis latrans Say
Coyote

1823a. *Canis latrans* Say, 1:168.
1897g. *Canis lestes* Merriam, 11:25.
1913b. *Canis latrans lestes* Grinnell, 3:285.
1949. *Canis latrans umpquensis* Jackson, 62:31.

Description.—The coyote (Plate XX) is a typical canid intermediate in size between the foxes (*Urocyon* and *Vulpes*) and the gray wolf (*Canis lupus*; Table 12-2). The pelage is grayish, buff, pinkish cinnamon, or brownish, or a combination of those colors, often overlain by blackish-tipped hairs on the ears, muzzle, feet, and dorsum. The lips and eyelids are black, accentuated by contrasting borders of white fur. The underparts are paler than the remainder of the body. Considerable variation in color and markings of coyotes is evident among individuals and regionally east and west of the Cascade Range. True albinos with pink eyes and pink foot pads are known to have occurred in Oregon (Green, 1947).

The skull of the coyote (Fig. 12-1) can be distinguished from those of foxes by size (basilar length usually >130 mm), convex dorsal surfaces on the postorbital processes, and prominently lobed incisors. It can be distinguished from that of the gray wolf by an anterior-posterior diameter of the canine <11 mm and a length of P4 <22 mm, and the tip of the upper canine extending below the mental foramen in the

Table 12-1.—*Reproduction in Oregon canids.*

Species	Number of litters	Litter size $\overline{X}$	Range	n	Reproductive season	Basis	Authority	State or province
Canis latrans	1[a]	6.5		159	Jan.–May	Embryos	Hamlett, 1938	Oregon
Canis lupus	1	6.5[b]	3–11	27	Feb.–Apr.	Embryos	Rausch, 1967	Alaska
	1	5.3[c]		13	Feb.–Apr.	Embryos	Rausch, 1967	Alaska
Urocyon cinereoargenteus	1	4.5	3–7	10	Jan.–May	Embryos	Layne and McKeon, 1956	New York
	1	4.4	3–7	32		Implant sites	Layne and McKeon, 1956	New York
	1	4.0	2–6	24	Jan.–Apr.	Embryos	Layne, 1958b	Illinois
	1	3.6	2–5	32		Implant sites	Layne, 1958b	Illinois
Vulpes velox	1	4.0	3–5	5	Dec.–Mar.	Young at den	Morrell, 1972	California
	1	4.5	4–5	11	Dec.–Feb.	Young at den	Egoscue, 1962b	Utah
Vulpes vulpes	1	6.2	4–8	6		Young at den[d]	Multiple authorities[e]	Oregon
	1	7.1	4–12	29	Dec.–Feb.	Implant sites	Storm et al., 1976	Illinois
	1	6.8	2–9	34	Dec.–Feb.	Embryos	Storm et al., 1976	Illinois
	1	6.7	3–12	48	Dec.–Feb.	Embryos	Storm et al., 1976	Iowa

[a]Dixon (1920) reported a simultaneously pregnant and lactating female.
[b]Adults.
[c]Females 2 years old.
[d]May be biased; see text.
[e]Livezey and Evenden (1943) and Wilcomb (1948) combined.

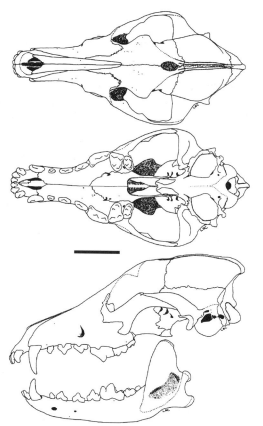

Fig. 12-1. Dorsal, ventral, and lateral views of the cranium, and lateral view of the dentary of the coyote, *Canis latrans* (OSUFW 1519). Scale bar equals 50 mm.

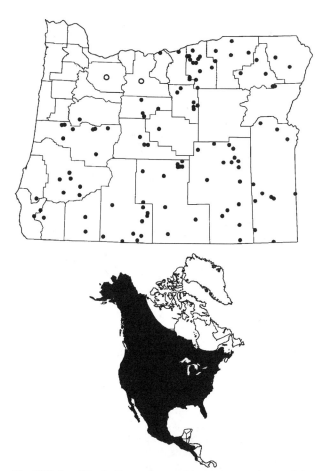

Fig. 12-2. Localities in Oregon from which museum specimens of the coyote, *Canis latrans*, were examined (listed in APPENDIX, p. 512), and the distribution of the coyote in North America. (Range map redrawn from Hall, 1981:926, map 489, with modifications.) Open symbol indicates record for county.

closed lower jaw. Ranges (estimated from graphs) of zygomatic breadth, greatest length of the skull, and bite force (distance between anterior margins of P4s at lingual alveoli/length of upper molariform toothrow X 100) of the coyote ($\cong$91–$\cong$108 mm, $\cong$171–$\cong$224 mm, and $\cong$63–$\cong$77%, respectively) rarely overlap with those of the gray wolf ($\cong$122–$\cong$153 mm, $\cong$222–$\cong$277 mm, and $\cong$78–$\cong$96%, respectively—Paradiso and Nowak, 1971). Also, the nose pad of the coyote is <25 mm wide, whereas that of the gray wolf is >25 mm.

Howard (1949) indicated that the alveolar length of the upper molariform row is ≤3.1 times greater than the palatal width (at the alveoli of p1s) in the coyote, but only ≥2.7 times greater in domestic dogs of similar size. However, Jackson (1951) suggested that the method might be more reliable with some races of coyotes than others and that skulls of collie and Russian wolfhound breeds of domestic dogs may not be separable from those of coyotes by use of the ratios.

Distribution.—Researchers have depicted the distribution of the coyote variously, with occupied areas often differing by hundreds of thousands of square kilometers (Bekoff, 1977, 1982; Hall, 1981; Hall and Kelson, 1959; Voigt and Berg, 1987; Young, 1951). Some, but certainly not all, of the discrepancies may be attributed to the relatively recent conclusion that some wild canids in New England were coyotes (Silver and Silver, 1969) and that the species had been transplanted to Georgia and Florida (Bekoff, 1982). Our homogenization of published ranges illustrates that the species occurs throughout most of North America (Fig. 12-2), but doubtlessly includes unoccupied

areas and excludes some where coyotes occur. In Oregon, collection sites for museum specimens of the coyote are fairly uniformly distributed except for the northwestern corner of the state (Fig. 12-2). Young (1951:29, fig. 4) claimed that coyotes originally were absent from northwestern Oregon but invaded the region in 1907–1945. Coyotes reportedly were harvested for fur in all counties of the state between 1969 and 1996 (Table 12-3).

The coyote and its progenitor, Johnston's coyote (*Canis* †*lepophagus*) are common as Pleistocene fossils. Fossils of *Canis latrans* are known from many Rancholabrean sites, including some in Oregon; fossils of *C.* †*lepophagus* are known from Blancan sites in Washington, Idaho, Nevada, and many other western states (Kurtén and Anderson, 1980; Nowak, 1978). Both species are believed to have evolved from the smaller mid-Pliocene (Hemphillian) species, *Canis* †*davisi* (Kurtén and Anderson, 1980), which also is known from Oregon sites (Nowak, 1978).

Geographic Variation.—Two of the 19 geographic races of *C. latrans* occur in Oregon: the larger and paler *C. l. lestes* east of the Cascade Range and the smaller and more richly colored *C. l. umpquensis* west of the Cascade Range (Hall, 1981). That coyotes that reportedly invaded northwestern Oregon after 1907 (Young, 1951) could be considered to have differentiated sufficiently in <9 years

Table 12-2.—Means (±SE), ranges (in parentheses), and CVs of measurements of skull and skin characters for female and male canids from regions in Oregon. Skin characters were recorded from specimen tags; skull characters <150 mm were measured to the nearest 0.01 mm, larger characters to the nearest 1 mm. SE and CV not provided if n <10.

Species and region	Sex	n	Total length	Tail length	Hind foot length	Ear length	Mass (kg)	Condylobasal length	Basilar length	Postglenoid length	Length of maxillary toothrow	Length of P4	Zygomatic breadth	Cranial breadth	Least interorbital breadth	Length of mandible
Canis latrans West of Cascade Range	♀	13	1,056[a] (855–1,215)	339[a] (310–380)	173[a] (130–200)	122[b] (120–123)	8.50[a] (6.40–10.00)	159.77 ± 4.00 (144.00–185.00) 9.03	144.46 ± 3.31 (129.00–163.00) 8.27	41.46[c] (40.00–43.09)	79.54 ± 0.61 (75.60–82.53) 2.76	18.83 ± 0.21[d] (17.59–19.73) 3.74	91.46 ± 1.04 (86.42–98.27) 4.11	56.27 ± 0.47 (52.62–59.08) 2.99	29.98 ± 0.59 (27.27–34.33) 7.12	130.70 ± 1.37 (123.84–139.17) 3.79
	♂	10	1,200	320	195	125	13.00	164.80 ± 4.84 (147.00–195.00) 9.29	148.00 ± 4.61 (133.00–180.00) 9.85	42.98[e] (42.84–43.12)	81.39 ± 1.46 (76.44–89.92) 5.67	19.26[f] (17.19–20.51)	94.33 ± 1.35 (88.78–98.56) 4.54	57.44 ± 0.86 (53.05–62.02) 4.75	31.05 ± 0.81 (27.86–34.33) 8.19	135.74 ± 2.73 (125.82–151.49) 6.37
East of Cascade Range	♀	30						157.70 ± 1.03 (143.00–167.00) 3.56	142.70 ± 0.95 (131.00–151.00) 3.64		82.22 ± 0.48 (76.29–86.22) 3.19	19.48 ± 0.16 (18.13–21.27) 4.46	92.07 ± 0.59 (88.00–99.24) 3.49	56.43 ± 0.29 (51.77–59.21) 2.77	30.98 ± 0.42 (26.00–35.70) 7.51	135.04 ± 0.92 (124.06–143.06) 3.71
	♂	30						165.93 ± 0.99 (156.00–176.00) 3.26	149.97 ± 0.87 (140.00–160.00) 3.16		85.72 ± 0.42 (81.53–91.62) 2.66	20.42 ± 0.17 (18.74–22.51) 4.46	97.04 ± 0.41 (92.58–102.39) 2.34	57.64 ± 0.32 (53.83–60.54) 3.03	32.89 ± 0.33 (29.38–37.76) 5.45	141.69 ± 0.82 (133.93–150.79) 3.18
Canis lupus	♀	10						215.01 ± 3.00 (202.00–230.57) 4.41	193.65 ± 2.64 (183.00–211.07) 4.31	51.59[d] (49.75–53.42)	100.26 ± 0.66 (98.02–103.39) 2.07	23.11[d] (21.23–24.12)	127.61 ± 1.79 (115.67–136.69) 4.43	66.00 ± 0.51 (63.44–68.75) 2.44	45.11 ± 0.73 (41.48–48.55) 5.14	173.74 ± 2.09 (165.00–182.13) 3.81
	♂	17					45.45	233.06 ± 2.83 (213.34–250.00) 5.00	210.16 ± 2.66 (193.00–228.00) 5.21	53.53[c] (50.00–57.09)	107.41 ± 1.19 (95.05–117.00) 4.59	25.19 ± 0.25[d] (23.45–26.58) 3.96	134.58 ± 1.72 (112.17–143.34) 5.26	69.60 ± 0.89 (65.42–78.00) 5.33	48.23 ± 0.70 (39.03–52.17) 5.99	185.52 ± 2.15 (163.92–201.00) 4.77
Urocyon cinereoargenteus	♀	9	914[a] (876–989)	381[a] (368–393)	124[a] (111–140)	75[a] (74–76)	4.54[a] (3.63–5.44)	113.62 (105.54–116.51)	104.04 (97.04–109.78)	27.31[a] (26.29–29.47)	47.94 (45.88–50.02)	10.0 (9.38–11.04)	62.93 (56.78–67.63)	44.32 (42.28–46.11)	22.84 (20.27–25.30)	87.74 (80.44–90.11)
	♂	11	911[a] (884–930)	356[a] (338–370)	132[a] (127–136)	76[a] (75–77)	4.47[a] (3.50–5.44)	116.74 ± 1.13 (110.67–121.51) 3.22	106.85 ± 1.04 (101.61–112.16) 3.22	28.79	49.69 ± 0.47 (47.22–51.86) 3.11	10.18 ± 0.18[d] (9.39–10.96) 5.44	64.92 ± 0.51 (62.04–67.05) 2.59	45.19 ± 0.36 (43.47–47.30) 2.64	23.57 ± 0.39 (21.97–25.53) 5.45	86.23 ± 0.94 (82.41–91.59) 3.59
Vulpes velox	♀	2						115.69 (107.67–123.70)	107.94 (100.68–115.19)		55.56 (50.94–60.17)	11.25 (10.26–12.23)	62.64 (57.54–67.74)	44.68 (41.82–47.54)	22.83 (20.27–25.39)	87.85 (80.07–95.62)
Vulpes vulpes East of Cascade Range	♀	2	925	349	155	75	3.86	130.40 (127.99–132.80)	121.61 (118.19–123.82)	30.70	60.29 (58.54–62.03)	13.22 (12.68–13.75)	69.25 (69.20–69.30)	46.25 (46.09–46.40)	24.91 (24.57–25.24)	100.55 (97.90–103.19)
	♂	3						134.65 (131.97–136.30)	124.62 (122.72–126.37)		63.38 (63.03–63.64)	13.30 (12.88–13.96)	70.34 (67.16–72.71)	47.26 (46.09–48.38)	25.58 (22.80–27.35)	102.06 (100.68–103.65)
Western Oregon and Cascade Range	♀	3	946 (875–987)	327 (285–393)	148 (140–165)	90 (80–100)	5.26 (4.97–5.81)	127.73 (122.71–133.04)	117.67 (112.55–121.89)	30.61 (29.89–31.43)	58.04 (52.70–62.30)	12.32 (11.50–12.96)	68.77 (66.96–70.14)	45.53 (45.23–45.73)	25.29 (23.83–26.51)	96.68 (92.93–100.20)
	♂	1						141.36	131.84	33.64	64.36	14.53	75.71	48.21	27.30	107.14

[a]Sample reduced by 10. [b]Sample size reduced by 11. [c]Sample size reduced by 9. [d]Sample size reduced by 1. [e]Sample size reduced by 1. [f]Sample size reduced by 7.

Table 12-3.—Reported annual harvest of coyotes (Canis latrans) in Oregon by county, 1969–1996. Data provided by Oregon Department of Fish and Wildlife.

County	Trapping season 1969–1970	1970–1971	1971–1972	1972–1973	1973–1974	1974–1975	1975–1976	1976–1977	1977–1978	1978–1979	1979–1980	1980–1981	1981–1982	1982–1983	1983–1984	1984–1985	1985–1986	1986–1987	1987–1988	1988–1989	1989–1990	1990–1991	1991–1992	1992–1993	1993–1994	1994–1995	1995–1996
Baker	64	4	63	91	231	291	580	420	604	445	522	661	371	518	236	279	334	481	116	357	215	211	399	650	359	340	149
Benton		12	2	6	10	36	39	123	23	41	77	32	46	60	15	27	35	23	38	33	46	10	13	10	16	50	1
Clackamas	27	4	15	52	18	43	30	16	66	129	104	130	109	95	75	60	94	67	48	23	12	20	20	21	16	4	40
Clatsop	42	1	13	18	11	40	9	12	19	35	69	33	21	70	81	55	149	129	102	124	57	26	72	34	63	56	6
Columbia	12	1	1	13	18		23	34	75	46	69	29	55	39	61	80	25	45	54	37	70	8	11	21	37	14	2
Coos	1	1				6	6	45	50	35	51	63	12	39	29	30	24	10	12	10	2	6	15	3	10	1	
Crook	39	23	72	29	2	3	68	440	539	187	387	663	468	219	297	392	352	531	221	201	86	93	223	183	107	168	24
Curry		6				5		8	58	62	66	51	8	39	20	16	11	17	5	6	1	7			4	1	
Deschutes	26	18	17	38	42	163	216	273	389	352	540	601	266	224	222	249	255	294	374	124	88	217	198	150	122	220	73
Douglas	49	22	17	30	10	49	59	55	275	251	363	279	95	284	180	179	138	149	212	126	42	46	58	39	31	104	21
Gilliam		3	30	33	11	1	3	6	14	15	75	121	99	27	54	154	60	139	81	204	85	56	31	50	75	31	6
Grant	38	47	15	21	68	129	187	307	523	264	498	654	275	356	293	299	255	571	660	204	96	194	63	113	88	143	44
Harney	32	81	56	549	447	522	195	632	1,057	297	1,166	1,854	1,214	1,669	641	1,454	599	1,454	1,111	573	369	249	419	635	501	925	207
Hood River	9	28		8	16		15	33	13	29	28	87	48	35	39	42	47	55	65	44	26	35	16	6	1	9	4
Jackson	29	36	5		60	66	51	67	132	231	242	231	197	107	113	108	117	137	63	30	23	29	24	29	41	6	
Jefferson					4	47	67	48	55	65	103	208	113	112	103	121	134	114	159	41	9	23	114	84	27	50	14
Josephine	14	22	11	32	33	51	32	64	133	105	172	121	26	76	59	42	133	32	15	12	8	3	19	26	21	24	2
Klamath	86	9	25	73	236	193	133	191	288	609	631	363	267	641	447	510	498	617	183	147	165	220	282	287	266	195	116
Lake	117	219	116	217	235	165	321	467	332	999	1,643	827	1,251	734	1,036	779	1,009	1,212	576	596	645	504	880	624	994	162	
Lane	35	21	56	86	119	159	156	229	274	187	532	176	101	199	232	133	234	191	143	55	32	30	70	142	53	58	26
Lincoln	79	10	25	23	57	63	64	75	79	82	80	96	113	169	77	114	157	143	71	74	9	57	24	31	25	25	19
Linn	17	3	8	11	36	47	58	88	87	88	102	129	93	146	127	175	140	123	52	36	22	20	83	78	36		
Malheur	183	181	165	219	282	441	475	558	711	701	2,019	1,709	1,175	1,325	965	822	1,081	1,498	1,685	637	222	500	580	833	584	978	135
Marion	4	1	13	2	1	15	11	26	40	112	62	55	39	8	60	65	71	33	30	11	16	29	18	5	5	5	18
Morrow	12	5	13	74	44	55	41	191	238	216	30	179	118	62	44	74	28	33	5	15	8	7	7	5	5	10	56
Multnomah	4	1	2	12	2	15		6	7	4	15	2	16	11	6	4	5	20		13		6	4	7	5		1
Polk	23	11	11	17	7	10	33	31	22	46	62	111	106	83	143	82	183	119	77	70	39	41	26	24	41		
Sherman							1	2	10	29	62	32	47	28	5	15	74	14	9		14	7	25	3	1		4
Tillamook	2	2	14	51	57	41	66	135	36	135	110	120	88	77	90	98	107	152	37	5	15	43	43	25	41	58	8
Umatilla	20	40	29	84	98	90	149	217	263	118	204	197	239	79	213	492	766	518	88	37	30	100	150	43	148	192	30
Union	17	9	12	22	28	103	106	98		225	305	200	137	91	130	133	157	158	42	88	31	33	33	21	33	63	16
Wallowa	69	166	130	148	103	142	88	149	142	131	342	149	140	220	268	230	314	344	94	34	97	77	129	127	152	7	
Wasco	2	3	3	5	11	22	43	109	106	78	55	219	170	107	218	173	143	168	54	103	79	107	96	79	30	27	
Washington	51	1	3	4	2	22	22	29	28	57	55	49	60	93	57	67	49	109	25	51	14	13	20	19	42	27	
Wheeler	24	27	34	74	49	92	114	131	52	45	185	292	187	229	121	121	142	115	15	30	35	39	58	58	65	57	18
Yamhill	1		4	3	11	7	5	25	16	39	45	29	82	12	21	67	127	91	42	14	18	53	26	21	21	21	24
Totals	1,128	1,110	897	1,963	2,419	3,240	3,286	5,112	6,938	5,563	10,775	11,538	7,408	9,520	6,100	7,707	7,031	10,059	10,072	4,490	2,761	3,049	3,849	4,921	3,214	5,144	1,276
Trappers reporting*	721	773	731	847	1,102	1,350	1,377	1,755	2,746	2,899	3,216	3,373	4,030	3,980	3,293	3,104	3,575	3,397	3,833	1,985	1,536	1,228	1,390	1,454	1,293	1,464	687
Average price/pelt	$7.24	$6.93	$8.69	$13.33	$19.45	$14.62	$22.88	$41.22	$29.88	$44.83	$32.40	$36.91	$42.92	$31.33	$27.94	$21.68	$28.42	$39.04	$18.74	$10.59	$10.42	$13.00	$28.85	$23.09	$23.87	$13.25	$16.10

*Not all sought coyotes and not all were successful.

(type of the latter subspecies was collected in 1916) to require a separate trinomial (Jackson, 1949) causes us to suspect the validity of the date of the range extension or that the observed differences in size and color were the result of individual variation rather than geographic variation.

Habitat and Density.—The coyote is nearly ubiquitous in Oregon; it occurs in habitats ranging from grasslands to shrub-steppe to boreal forests and from remote wilderness to highly urbanized areas (Voigt and Berg, 1987). Young (1951) considered the coyote an "edge" species in that it had extended its range into regions (including northwestern Oregon) in which logging or fires had removed broad expanses of mature forests. Thus, almost any area productive enough to support coyotes will be occupied by the species.

Estimates of density of coyote populations are extremely difficult to obtain and likely even more difficult to verify. Knowlton (1972), studying the coyote in Texas, estimated densities of 0.2–0.4/km² over a large portion of the range of the species, with densities as high as 1.9–2.3/km² in extremely favorable areas. Bekoff (1977) indicated that these estimates fit well with those of other workers. More recent estimates also seem to fall within those ranges (Babb and Kennedy, 1989; Gese et al., 1989; Pyrah, 1984; Todd and Keith, 1983). We know of no specific information on density of coyotes in Oregon.

Diet.—Between 1935 and 1992, >100 references to articles concerning foods eaten by the coyote in various of the contiguous United States were listed in *Wildlife Review*; combined, these articles were based on several hundred thousand stomach or fecal samples. Although we did not peruse all of these articles, the most common conclusions presented in those reviewed concerned the opportunistic feeding habits of the coyote, the uniqueness of the diet in comparison with that reported in articles published earlier, and the paucity of direct evidence that coyote predation affected populations of either wild or domestic animals significantly.

In Oregon, the diet of the coyote has been studied in recent years in the Coast Range (Witmer and deCalesta, 1986), on the west slope of the Cascade Range (Toweill and Anthony, 1988b), and in the *Pinus-Juniperus* zone of western Deschutes Co. (Van Vuren and Thompson, 1982); all three studies were based on examination of fecal droppings. An earlier study involved 671 stomach samples from Oregon, but the results were combined with those from numerous other states in the final report (Young, 1951).

In the Coast Range, the mountain beaver (*Aplodontia rufa*) was the food item occurring in coyote feces (*n* = 309) most frequently at all seasons (69.9–90.0%); second-most frequent remains of arvicoline rodents and deer (*Odocoileus hemionus*) in spring (8% each), fruit (huckleberry, *Vaccinium*, and blackberry, *Rubus*) in summer and autumn (20.4%), and deer in winter (21.9%—Witmer and deCalesta, 1986). Brush rabbits (*Sylvilagus bachmani*) at 12.4–13.5%, Townsend's chipmunks (*Tamias townsendii*) at 9.4–9.7%, and creeping voles (*Microtus oregoni*) at 10.6–12.5% ranked high during summer–winter. Deer in winter was considered to have been consumed as carrion, but remains of fawn deer (hooves) were considered to have been from those killed by coyotes (Witmer and deCalesta, 1986); however, evidence to support either contention was not presented.

On the west slope of the Cascade Range, fruit (huckleberry; blackberry; manzanita, *Arcostaphylos*; cherry,

Prunus; and elderberry, *Sambucus*) occurred in such a large proportion of samples (83% of 324) in summer that it ranked first in frequency of occurrence (38% of 844) among samples collected throughout the year. Rodents occurred most frequently in spring (60% of 192 samples), ungulates most frequently in winter (45% of 190) and autumn (43% of 138). Overall, of 844 samples collected throughout the year, rodents ranked second (36%) behind fruit, followed by ungulates (26%) and lagomorphs (24%—Toweill and Anthony, 1988b). Deer was the prey species identified most frequently (24%) throughout the year; it occurred in 61% of the droppings collected in December, 60% in January, and 55% in February, and a second, but lower, peak of occurrence was evident in spring and summer after the birth of fawns. Because deer ranked first or second among prey species occurring in the diet of coyotes during 9 months of the year, Toweill and Anthony (1988b) contended that their study did not support the general conclusion (Bekoff, 1982) that most deer are eaten as carrion.

In central Oregon, the seasonality in frequency of occurrence of various food remains in 308 fecal droppings deposited throughout the year by three coyotes reflected the seasonal availability of those food items on the 275-ha study area (Van Vuren and Thompson, 1982). Deer and lagomorphs (mostly *Sylvilagus nuttallii*) occurred in samples collected in all months of the year. The highest incidence of deer occurred during and immediately following the deer-hunting season; most was considered to be parts discarded by hunters when field dressing their deer. Another peak in frequency of deer was primarily of fawns and occurred during the fawning season (primarily June) and in the following 4 months. The frequency of occurrence of hibernating and nonhibernating rodents, insectivores, birds, and snakes (Serpentes) was <50% in all months and, in general, greatest in spring and summer, least in winter. Insects peaked at ≅60% in August, juniper (*Juniperus*) berries (the only fruit recorded) were eaten only in October–March, and domestic sheep was detected in only three fecal samples, all collected during months when no sheep were on the study area (Van Vuren and Thompson, 1982).

Reproduction, Ontogeny, and Mortality.—In Oregon, pregnant females were collected from late January to 1 May and unweaned young or lactating females were collected from the 1st week of March to mid-July; the peak number of pregnant females was collected from early March to mid-April (Hamlett, 1938). These dates are earlier than for any adjoining state (Hamlett, 1938). Average litter size (Table 12-1) based on counts of embryos was identical with that reported for California, slightly less than that reported for Washington, and ≅0.5 embryo greater than that reported for Idaho and Nevada (Hamlett, 1938). The average number of young found in 176 dens was 6.1, representing a loss of ≅0.4 young from the preparturition estimate of average litter size.

During their 3rd–10th week of life, coyote young are particularly vulnerable to being preyed upon by raptorial birds and coyotes residing nearby (Gier, 1975). As they travel farther from the natal den, dogs, automobiles, and other coyotes become a hazard. In Kansas, ≤50% of young coyotes survived to 1 July in the year of their birth; also, from 1 April–31 July, 10% of the adults were killed (Gier, 1975). Gier (1968, 1975) indicated that reduced natality in response to inclement climatic conditions and inadequate food supply caused a quick reduction in coyote populations.

Habits.—Their secretive nature, largely nocturnal or crepuscular activity, and wide-ranging movements have limited study of coyote behavior to captive individuals or to small numbers of free-living individuals on small areas (Bekoff, 1977; Kleiman and Brady, 1978). Nevertheless, the literature on coyote behavior is voluminous. Kleiman and Brady (1978) were critical of reports of much of the early field work on coyote behavior and of life-history information included in regional works on mammals because they commonly consisted of a combination of original research and anecdotal accounts, some of which were of dubious validity.

Although the coyote often is considered only a moderately social species, it possesses a highly developed communication system that facilitates development and maintenance of long-term social relationships (Lehner, 1978). Gier (1975) indicated that coyotes did not develop a social organization beyond familial relations. However, coyotes are now known not only to live singly and in pairs, but also to develop brief aggregations (≤22 individuals) and permanent territorial packs (three–seven individuals), especially if a stable food resource is available (Camenzind, 1978). Packs develop well-defined social hierarchies, and both pairs and packs defend territories from intrusion by nomadic singles or individuals from pairs or other groups (Camenzind, 1978).

Coyotes interact with other species of mammals variously. Domestic dogs are "mortal enemies" during most of the year, but, during their respective breeding seasons, they may become much more amicable as coyote-dog hybrids are known from throughout the range of the coyote (Gier, 1975:260). Mountain lions (*Puma concolor*), bears (*Ursus americanus*), and, formerly, the gray wolf (*Canis lupus*) are enemies at all seasons. Coyotes and badgers (*Taxidea taxus*) reportedly hunt cooperatively on occasion (Gier, 1975). Coyotes apparently exclude red foxes (*Vulpes vulpes*) to boundary areas between adjacent territories (Harrison et al., 1989; Voigt and Earle, 1983). And, of course, other mammals, either as prey or as carrion, serve as the prime source of the ≅250 kg of meat required by the average coyote annually (Gier, 1975).

Perhaps no other feature of the coyote, ecological, morphological, physiological, or behavioral, is as well known as its voice. Descriptions or phonetic representations of calls are common in earlier works on mammals (Bailey, 1936), in popular accounts of coyote life history (Dobie, 1961), and in modern scientific literature (Bekoff, 1978; Lehner, 1978; McCarley, 1975). Lehner (1978) arranged vocalizations of coyotes into 11 categories, but indicated that the types were somewhat arbitrary because several calls grade into one another. The most common types are the bark and bark-howl, often given in sequence and believed to indicate threat or alarm. On numerous occasions in summer and autumn 1992 we were treated to the coyote's bravado and "*yap, yap-yowl*" song when one approached our semiurban home between 0330 and 0430 h and challenged our kennel-housed springers, which, except when exhorted to respond, slept blissfully through the serenade.

Coyotes have been timed to run at ≅64 km/h (Nowak, 1991); even those crippled by loss of a front foot attained a speed of ≅32 km/h (Thompson, 1976). Nevertheless, slow, crippled, unwary, inattentive, and less wily coyotes are, or

have been, eliminated by humankind and its dogs, traps, poisons, guns, and automobiles. Thus, the coyote has undergone intense (natural?) selection and has become a larger (Silver and Silver, 1969), smarter, and more adaptable species (Gier, 1975). The following anecdote exemplifies the versatility and cleverness of the modern coyote.

In Morrow Co., about 0530 h on an already scorching day in August 1967, as one of us (BJV) drove westward on I-84, a coyote, loping through the sagebrush (*Artemisia*) ≅50 m south of and parallel to the eastbound lanes, suddenly made a 90° turn, jumped through the fence dashed onto the freeway, scooped a road-killed black-tailed jackrabbit (*Lepus californicus*) from the pavement, turned, and, carrying the jackrabbit, jumped back through the fence, all within the quarter minute or so that traveling at the 70-mile/h speed limit allowed the coyote to remain in view. From its obvious adroitness at detecting and retrieving the jackrabbit, we strongly suspect that this individual made a habit of scavenging road-killed animals from the freeway. Such illustrates the ever-adaptable coyote's ability to exploit environments and situations created by humans.

Remarks.—We are reluctant to add more verbiage to either side of the voluminous and nearly century-long controversy regarding the nature, extent, or economics of the predatory habits of the coyote. During the 20th century in Oregon, as in most western states, literally thousands of coyotes were killed by trappers and hunters whose salaries were paid by the federal government, state wildlife management agency, or stockmen organizations; by individuals who sought to collect bounties paid with county funds; and by hunters and trappers for income derived from the sale of pelts. In some states, coyotes were eliminated on large areas (Gier, 1975; Nowak, 1979). Nevertheless, despite hunting, trapping, poisoning, and much strong language, the coyote increased in numbers and expanded its range not only in Oregon but in numerous parts of the remainder of North America (Gier, 1975).

No doubt, the most welcomed change in control programs directed toward reducing losses to the livestock industry by coyote predation is that related to the overall objective of such programs. Initially, it was to eliminate the coyote, but when that obviously failed, the objective shifted to eliminating the offending individual(s). More recently, researchers have developed and tested far cheaper and more effective means of reducing or preventing depredations on livestock, with less hazard to nontarget species, other than by killing coyotes. Appropriate fencing (deCalesta and Cropsey, 1978; B. Thompson, 1979) and guard dogs (Lorenz and Coppinger, 1985) either singly or in combination seemingly are effective. However, herdsmen, not society as a whole (as for many lethal control programs), must bear most or all of the cost of these measures. Thus, it is the economics of coyote control that speaks loudest.

Canis lupus Linnaeus
Gray Wolf

1758. [*Canis lupus*] Linnaeus, 1:39.
1839. *Canis lupus* var. *fusca* Richardson, p. 5.
1850. *Lupus gigas* Townsend, 2:75.
1937. *Canis lupus irremotus* Goldman, 18:41.

Description.—*Canis lupus* (Plate XX) is the largest canid, not only in Oregon (Table 12-2), but in the remainder of the world as well (Nowak, 1991). The heaviest recorded wolf weighed 80 kg, but the average is only slightly more than half that for adult males, slightly less than half that for adult females (Mech, 1970). A recently collected one (OSUFW 2932) from Oregon weighed 40 kg, another (OSUFW 8727) was estimated to weigh 41–50 kg.

The wolf has relatively long legs, a narrow and deep chest, and elbows that turn inward and foot pads that turn outward. These, combined with the digitigrade feet possessed by all canids, make the wolf highly adapted for running.

The pelage of gray wolves is long over the body and tail, but relatively short on the legs and face. The predominant color usually is gray (hence the vernacular name), but the legs, flanks, and venter sometimes are yellowish or light brownish. These colors are overlain by long, black guard hairs on the dorsum, tail, and mane; some individuals are essentially entirely black. Occasionally, but more commonly in higher latitudes, all-white individuals occur.

The skull of *C. lupus* is massive (Fig. 12-3; Table 12-2); the muzzle is relatively short and heavy, the zygomata flare widely, and the dentary is stout. The greatest length of the skull is >215 mm in males, >205 mm in females; the sagittal

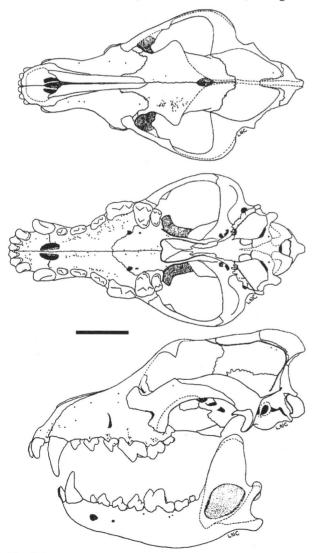

Fig. 12-3. Dorsal, ventral, and lateral views of the cranium, and lateral view of the dentary of the gray wolf, *Canis lupus* (OSUFW 2415). Scale bar equals 50 mm.

crest extends far beyond the braincase proper to support the huge temporalis muscles that close the jaw; the anteroposterior diameter of C1 is >11 mm; and the anteroposterior length of P4 is ≅25 mm. Iljin (1941) indicated that the orbital angle (between the horizontal plane across the roof of the skull and a plane across the center of the orbit) was <46° in wolves, but ≥52° in domestic dogs; hybrids were intermediate, with most 47–50°.

Other differences that may be useful in separating wolves and domestic dogs include the presence of a precaudal gland, a single annual breeding season, and large, spherically convex bullae in the wolf (Mech, 1970). However, the gray wolf is a wild dog, is capable of breeding with many domestic dogs and producing fertile offspring, and is considered by many authorities to be the progenitor of the domestic dog and possibly a conspecific of the domestic dog (Mech, 1970).

Distribution.—The geographic range of *C. lupus* includes Europe, most of Asia, northeastern Africa, and most of North America south through central Mexico to Oaxaca (Hall, 1981; Nowak, 1991). The historic range was the most extensive of any wild mammal (Nowak, 1991). Historically, in North America north of Mexico, the range of the gray wolf probably extended from Alaska, northern Canada, and northern coastal Greenland, south through all of Canada and the United States except for most of California and the southeastern states east of a line from Missouri to central Texas (Fig. 12-4; Hall, 1981). The present-day range is considered to extend only as far south as the Canada–United States border except for extensions into the northern Midwest and northern Rocky Mountains, and disjunct pockets in Idaho and the central highlands of Mexico (Carbyn, 1987).

In Oregon, the gray wolf was considered to occur mainly in the Willamette Valley and west to the Coast, at European settlement, and to continue to occur west of the Cascade Range during the first third of the 19th century (Bailey, 1936). Olterman and Verts (1972:8, fig. 1) recorded 80 museum specimens collected at 34 localities in Oregon; one locality was in the Wallowa Mountains, one each in western Deschutes and Klamath counties, two in western Lake Co., one in the Klamath Mountains, Josephine Co., and the remainder along the west slope of the Cascade Range. Our range map (Fig. 12-4) represents all museum specimens, including those taken subsequently.

Fossil remains of gray wolves are known from late Irvingtonian or early Rancholabrean sites throughout much of the United States (including Oregon) and western Canada (Kurtén and Anderson, 1980).

Geographic Variation.—Two of the 24 nominal subspecies of *C. lupus* are considered to occur in Oregon: the smaller, darker *C. l. fuscus* in the western three-fifths of the state and the larger, lighter *C. l. irremotus* in the eastern two-fifths of the state (Goldman, 1937; Hall, 1981). The geographic ranges of the races of the gray wolf depicted by Hall (1981:932, map 490) indicate that the range of *C. l. youngi* extends into the southeastern corner of Oregon, but no marginal records are shown or listed for Oregon or nearby in Nevada. We know of no museum specimens from the southeastern quarter of Oregon.

Habitat and Density.—During early European settlement, the gray wolf in Oregon was considered to be a forest mammal (Bailey, 1936). And, indeed, the last major stronghold of

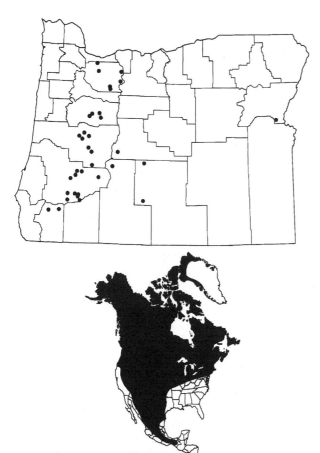

Fig. 12-4. Localities in Oregon from which museum specimens of the gray wolf, *Canis lupus*, were examined (listed in APPENDIX, p. 513), and the former distribution of the gray wolf in North America. (Range map redrawn from Hall, 1981:932, map 490, with modifications.) The species was extirpated from Oregon and the range in North America much reduced (Carbyn, 1987; Paradiso and Nowak, 1982).

the wolf in the state was the dense forests of the remote eastern parts of Clackamas, Linn, Lane, and Douglas counties. Elsewhere in historic times, the wolf was not restricted to forested areas, but its presence likely was determined by the availability of ungulate prey (Carbyn, 1987). Thus, the wolf was associated with forested areas probably because they provided some refuge from persecution by humans and they supported populations of prey species.

Diet.—The gray wolf is primarily a meat-eater. It feeds largely on ungulates and beavers (*Castor canadensis*), but in a sense is opportunistic in that it sometimes consumes insects, small mammals, fish, and occasionally even garbage (Carbyn, 1987; Mech, 1974). However, small prey species rarely contribute significantly to the overall diet of wolves. Wolves shift consumption of larger prey species with season and with changes in vulnerability of prey (Carbyn, 1987).

Wolves, because they hunt cooperatively in packs, are capable of killing the largest ungulate prey such as moose (*Alces alces*) and bison (*Bos bison*). In Oregon, the primary ungulate prey of wolves likely was deer (*Odocoileus*) and elk (*Cervus elaphus*). Prey animals killed by wolves tend to be young, old, or not in prime physical condition (Mech, 1970, 1974; Mech et al., 1971), suggesting that most wild ungulates may not be vulnerable to predation by

wolves. Elsewhere, only 7.8% of moose brought to bay and 6.7% of white-tailed deer (*Odocoileus virginianus*) chased were killed by wolves (Mech, 1970; Mech and Frenzel, 1971). When prey is plentiful, average consumption of meat usually is 10–21% of the body mass/day but may be as high as 37%/day (Carbyn, 1983). The hams and viscera (especially omental fat) of prey usually are consumed first (Mech et al., 1971). Wolves are capable of fasting or subsisting on little food for long periods, and conversely, when killing is easy, wolves often underuse the prey they kill (Carbyn, 1987). Although wolves sometimes cache surplus prey, caching is not believed to contribute significantly to their survival. Wolves are known to engage in cannibalism during times of food stress; injured members of the pack are especially vulnerable.

Reproduction, Ontogeny, and Mortality.—Although self-maintaining populations of gray wolves occurred in Oregon nearly to the mid-20th century and perhaps beyond, we found no records for reproduction in the state.

Reproduction in wolves is highly integrated with the social structure of wolf packs (Peterson, 1979). Under usual circumstances only the dominant (alpha) male and alpha female within each pack mate and produce young; thus, reproduction is suppressed in many animals and the proportion of the population involved in reproduction is, in part, a reflection of pack size. In an unexploited population, an estimated 38% of females fail to mate annually (Carbyn, 1987), but in an exploited population, essentially all females breed (Rausch, 1967). If subordinate females mate, they may be driven from the pack at least temporarily (Peterson, 1979). Pack members other than the parents also are involved in rearing the young produced by the dominant pair, thus "it is of considerable survival advantage to have but a single litter of pups, born to high-ranking members of the pack" (Peterson, 1979:219).

Females in captivity may become capable of reproducing at 10 months, but in the wild, few mate before 22 months and many delay until the 3rd year or later (Carbyn, 1987). Estrus lasts about 5–7 days, occurs about 2 weeks later in females becoming receptive for the first time, and is preceded by a bloody discharge. During coitus, the bulbous penis locks behind the vaginal sphincter; the pair is tied together and helpless for ≥30 min (Carbyn, 1987). Gestation is 63 days (Mech, 1974); litter size is variable and age related (Table 12-1). Also, production of young may decline with a decline in food resources during the previous winter (Mech, 1977).

In unexploited populations, mortality during the 1st year of life ranges from 57 to 94% and during the 2nd year it is 55%; among adults it is about 80% annually (Mech, 1970, 1974). Survival of young and population growth are dependent upon availability of food during the rearing season (Jordan et al., 1967).

Habits.—Wolves commonly function in packs consisting of two to nine (maximum, 36) individuals; packs basically are family units (Mech, 1970, 1974; Rausch, 1967). In Minnesota, only 8% of 323 wolves sighted were traveling alone (Mech et al., 1971). Pack cohesion is based on ties developed during courtship of adults and those developed during the early life of young. In newly formed packs, a dominance hierarchy develops, consisting (in order of dominance) of the adult male, adult female, and a linear order among the young (Mech, 1974). In larger packs, a separate order develops among males and among females, but the dominant individual usually is a male (Mech, 1974). A complex system of body positions and facial expressions is used by wolves to demonstrate and to communicate their hierarchical status to other members of the pack; fighting rarely occurs within packs.

Wolves may be active at any time, but tend to be more active at night; in Minnesota, wolves were resting during 62% of 171 observations in daylight hours, traveling during 28%, and feeding during 10% (Mech et al., 1971). Wolves travel extensively; 1-day trips of 72 km have been recorded (Mech, 1974). Estimates of minimum-area home ranges acquired through use of radiotelemetry in Minnesota were 110–2,552 km² (Mech and Frenzel, 1971). In Alaska, one pack of nine distinctively colored individuals relocated 13 times from 12 March to 22 April 1958 traveled >1,125 km in an area 80 by 160 km (Burkholder, 1959).

In addition to use of body position and facial expressions, communication within and among wolf packs is by vocalization and scent-marking. Howling, distinctive among individuals, is used to assemble the pack and to advertise the territory. Scent-marking (consisting of deposition of urine, feces, or both on prominent objects combined with conspicuous ground-scratching), especially at trail crossings, is believed also to be territorial advertisement (Mech, 1974). Mech et al. (1971) hypothesized that wolf packs maintained exclusive territories and that the activities of lone wolves and nonbreeding groups were restricted to the spaces between the defended areas. During prey shortages, prey within these undefended areas is attacked only as a last resort by wolf packs (Hoskinson and Mech, 1976).

Wolves locate prey by scent, tracking, or chance encounter; once located, wolves may stalk prey before attempting a chase (Mech, 1970). Large prey animals such as moose that stand their ground usually are safe, but deer and the young of larger species cannot defend themselves, so must attempt to escape by running. However, running prey seems to trigger chasing by wolves (Mech, 1970). Most chases are short; wolves quickly abandon chases of prey that commence to outdistance them. Wolves attack deer on the head, shoulder, flank, or rump, but larger ungulates are attacked on the rump. Hamstringing (severing the Achilles tendons to disable prey animals), attributed to wolves in some earlier accounts (Bailey, 1936; Mace, 1970), has not been documented in the wild (Burkholder, 1959; Mech, 1974). Large ungulates are dangerous prey; evidence of broken bones is common in wolves (Rausch, 1967).

The limiting effects of predation by wolves on populations of ungulate-prey species remain controversial (Bergerud et al., 1983; Bergerud and Snider, 1988; Hoskinson and Mech, 1976; Thompson and Peterson, 1988).

Remarks.—Bailey (1936) claimed that at European settlement wolves were common in Oregon in the foothills of the Cascade Range west to the Pacific Coast. Wolves long have been labeled the archpredator and were killed at every opportunity. Between 1913, when the newly organized Oregon Game Commission commenced to offer payment for evidence that a wolf was killed, and 1946, when the last wolf was presented, bounties were paid on only 393 individuals (Olterman and Verts, 1972). The threat obviously was more a perception than a reality.

Olterman and Verts (1972) considered wolves to have been extirpated from Oregon as none was taken in the state since the one bountied in 1946. However, a specimen (OSUFW 2932) was collected in southern Baker Co. in 1974 (Verts, 1975) and another (OSUFW 8727) was killed in eastern Douglas Co. in 1978. Sighting or howling of wolf-like canids is reported occasionally, but likely most is of coyotes or feral domestic dogs.

We cannot state unequivocally either that wolves no longer exist in the state or that a few widely scattered individuals or pairs still occur here. Confirmation of the occurrence of the species in Oregon would require examination of the skull of a specimen. To advocate, or even to condone, the killing of a wolf in the state would be unconscionable. If the wolf still occurs in Oregon, its numbers certainly are too low for it to constitute a serious threat to the livestock industry or to populations of game species. Therefore, it is the long-held public tenet that the wolf is the archpredator, not the wolf itself, that must be eliminated. Then, upon being sighted, if one of the wolf-like canids actually were a wolf, the first human reaction would not be to kill it. We can live happily without need to examine the skull and confirm the species identification of the last wolf in Oregon, if indeed we did not do so already.

Urocyon cinereoargenteus (Schreber)
Common Gray Fox

1775. *Canis cinereo argenteus* Schreber, theil 2, heft 13, pl. 92.
1899b. *Urocyon californicus townsendi* Merriam, 16:103.
1933. *Urocyon cinereoargenteus townsendi*: Grinnell, 40:110.

Description.—The gray fox (Plate XXI) is among the smaller canids in Oregon (Table 12-2). The basic color of the gray fox is grizzled gray, but the stiff middorsal hairs have long black tips that extend onto the tail as a black mane. Guard hairs are banded white, gray, and black, producing the grizzled appearance (Fritzell and Haroldson, 1982). The throat, venter, and inside of the legs are white; a cinnamon-rufous border to the white throat extends onto the flanks and underside of the tail. A blackish patch on the side of the face extends onto the lower jaw and a grayish black "exclamation point" extends upward from the interior margin of each eye. The ears are cinnamon on the exterior and lined with long whitish hairs extending from the interior margins. The feet are grayish white.

The skull of the gray fox (Fig. 12-5) is distinguishable from that of other canids by the relatively short rostrum; somewhat flaring zygomata; prominent, widely separated, and 2–3-mm-wide, lyre-shaped temporal ridges extending from the postorbital processes to the supraoccipital; and a step in the inferior margin of the mandible anterior to the angle. The depression above the postorbital process is deeper than in *Vulpes*.

The gland on the dorsal surface of the tail measures 6.5–8.0 mm wide and from ≅125 to ≅200 mm long; this is about one-third to one-half the length of the tail and the longest of any canid (Hildebrand, 1952). Usually, three pairs of mammae are present.

Distribution.—The gray fox occurs from extreme southern Manitoba, Ontario, and Quebec, south throughout the United States east of the Great Plains, and from the Columbia River, Oregon, central Nevada and Utah, and northern Colorado southward through most of Middle America into northwestern South America (Fig. 12-6; Fritzell and Haroldson, 1982; Hall, 1981). In Oregon,

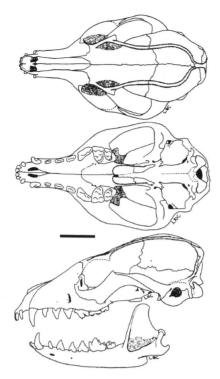

Fig. 12-5. Dorsal, ventral, and lateral views of the cranium, and lateral view of the dentary of the gray fox, *Urocyon cinereoargenteus* (OSUFW 1180). Scale bar equals 25 mm.

Fig. 12-6. Localities in Oregon from which museum specimens of the gray fox, *Urocyon cinereoargenteus*, were examined (listed in APPENDIX, p. 577), and the distribution of the gray fox in North America. (Range map redrawn from Hall, 1981:944, map 495, with modifications.)

except for a partial skeleton from Skeleton Cave, Deschutes Co. (USNM 248151), specimens of the gray fox on deposit in museums were taken only west of the Cascade Range (Fig. 12-6). However, several gray foxes reportedly were taken in Hood River, Deschutes, and Klamath counties immediately east of the Cascade Range (Bailey, 1936), and a few were reported from scattered counties farther east in the state (Table 12-4). We know of no transplants of gray foxes to eastern Oregon, and the species is not reported to occur in the neighboring states of Washington (Dalquest, 1948; Larrison, 1970) and Idaho (Davis, 1939; Larrison and Johnson, 1981) or the northern half of Nevada (Hall, 1946). Fichter and Williams (1967) suspected that occasional local reports of gray foxes in Idaho actually were red foxes (*Vulpes vulpes*) with grizzled gray on the posterior dorsum; we suspect that reports of gray foxes in Oregon counties east of, but not adjacent to, the Cascade Range have a similar basis.

Fossils of the gray fox are known from many localities outside Oregon dating from the Early Pleistocene (Irvingtonian—Kurtén and Anderson, 1980); we are not aware of any from Oregon.

Geographic Variation.—Of the 16 named subspecies only *U. c. townsendi* occurs in Oregon (Fritzell and Haroldson, 1982; Hall, 1981).

Habitat and Density.—In western United States, many authorities claimed that gray foxes tended to favor brushy vegetation in broken terrain (Fritzell and Haroldson, 1982). In Utah, gray foxes followed by use of radiotelemetry were found more frequently than expected in brushy meadows and less frequently than expected in an area in which blackbush (*Coleogyne ramosissima*) dominated (Trapp, 1978). In California, plowed fields and other bare areas were avoided, but, if available within home ranges, ripar-

ian areas, old-field habitats, and human-use areas were used by gray foxes more frequently than expected (Fuller, 1978).

No estimates of density are available for gray foxes in Oregon. In other regions, densities of 1.2–2.1/km² were reported (Fritzell and Haroldson, 1982).

Diet.—Most, if not all, samples consisting of 35 fecal passages of foxes collected March–May and 50 collected June–August ≅16 km southwest of Corvallis, Benton Co., were considered to be those deposited by gray foxes (Wilcomb, 1956). In spring, remains of lagomorphs and rodents occurred in ≅42% and ≅67% of the droppings, respectively; fruits, insects, and other groupings of vertebrates each occurred in ≅15% or fewer of the samples. In summer, fruits occurred in ≅70% of the feces, but occurrence of lagomorphs and rodents remained high (in ≅50% and ≅40% of the samples, respectively). Percent frequency of occurrence of food remains in 20 droppings collected in the vicinity of three den sites occupied by gray foxes on the same area were California ground squirrels (*Spermophilus beecheyi*), 18%; black-tailed jackrabbits (*Lepus californicus*), 7%; brush rabbits (*Sylvilagus bachmani*), 17%; small rodents (mostly *Microtus*), 12%; blackberries (*Rubus*), 23%; small birds, 5%; insects, 8%; domestic chickens, 23%; and vegetation, 20% (Wilcomb, 1948).

In Utah, remains of mammals occurred in 95% of 240 fecal droppings (60 collected during each of the four

Table 12-4.—Reported annual harvest of gray foxes (Urocyon cinereoargenteus) in Oregon by county, 1969–1996. Data provided by Oregon Department of Fish and Wildlife.

County	1969–1970	1970–1971	1971–1972	1972–1973	1973–1974	1974–1975	1975–1976	1976–1977	1977–1978	1978–1979	1979–1980	1980–1981	1981–1982	1982–1983	1983–1984	1984–1985	1985–1986	1986–1987	1987–1988	1988–1989	1989–1990	1990–1991	1991–1992	1992–1993	1993–1994	1994–1995	1995–1996
Baker																			2				1				1
Benton	2	2	8	12	11	4	8	7	11	7	4	22	3	16	29	2	11	4	3		1		1				
Clackamas	20	20	6	12	55	36	49	42	27	35	15	9	30	35	31	35	16	4	9	11							
Clatsop																				3							
Columbia				2	1		1	1									1	1									
Coos								1	3	3	3	8		6	8	5	6	10	19	4	2	3				20	
Crook																											
Curry	1							2	7	1	3	13	2	13	6	10	11	14	14	14			20	18	13		7
Deschutes		1						2					3		1												
Douglas		6	11	4	18	17	27	49	86	104	196	153	52	95	74	132	130	104	120	25	17	21	36	28	17	24	11
Gilliam																											
Grant																	3										
Harney																											
Hood River	5						5			1	1															2	
Jackson		16	7	8	16	33	28	17	48	34	43	16	43	64	34	36	22	76	56	53	10	30	40	27	14	31	28
Jefferson																											
Josephine	15	22	16	16	25	51	58	31	46	36	40	42	41	59	57	39	34	27	56	36	37	32	51	31	18	47	7
Klamath						2	2		3	1	1	1	1					3	1								
Lake																											14
Lane	18	20	8	18	61	36	38	28	28	34	83	56	38	21	11	11	24	20	14	14	2	5	8	26	15	12	
Lincoln	6				3	3	1			7	3																
Linn	34	22	14	27	43	69	43	24	22	23	22	7	13	11	8	11	15	26	17	48	4		8	5	6	3	1
Malheur										1		3	1	2								1					
Marion	12	8	3	16	13	25	17	10	6	8	6	3	5	9	9	26	13	2	1				1		1	2	
Morrow																											
Multnomah	7	1	4				6	5			2	1		3	4	2											
Polk	4	2	3		8		2	3	4	1	2	4		2	3	8	5	1	2				1				
Sherman																											
Tillamook					2											2								1			
Umatilla																											
Union																											
Wallowa																1											
Wasco																											
Washington	10	1	8	7	2	8	5	14	11	6	9	2	10	22	16	14	8	1	1		1						
Wheeler																											
Yamhill	10	16	8	2	9	6	4	8	2	13	7	1	15	12	4	6	1										
Totals	148	136	96	133	267	301	294	246	304	317	440	341	254	371	295	340	300	295	315	210	74	92	167	137	85	142	69
Trappers reporting[a]	721	773	731	847	1,102	1,350	1,377	1,755	2,746	2,899	3,216	3,373	3,069	3,980	3,292	3,104	3,575	3,397	3,833	1,985	1,536	1,228	1,390	1,454	1,293	1,464	687
Average price/pelt[b]	$2.88	$2.45	$3.26	$6.80	$16.98	$10.61	$20.50	$28.09	$27.27	$36.65	$39.04	$31.28	$26.36	$35.74	$30.97	$24.70	$18.13	$31.84	$25.83	$13.24	$6.70	$6.00	$9.00	$8.00	$11.00	$8.00	$9.00

[a]Not all sought gray foxes and not all were successful.
[b]Rounded to the nearest dollar after the 1989–1990 season.

seasons), arthropods in 51%, and fruits in 67%; mammals occurred most frequently in winter and spring samples, arthropods in spring and summer samples, and fruits in fall and winter samples (Trapp, 1978). In California, 51% of 195 stomachs contained rodents and lagomorphs and 13% contained game and poultry; mammals composed 61.4% of the volume, fruits 7.9%, and insects 2.1% (Grinnell et al., 1937). Thus, it appears that the diet of gray foxes in Oregon compares reasonably well with that of gray foxes in other western states.

Reproduction, Ontogeny, and Mortality.—The gestation period was reported as 63, 59, and 53 days by various authorities (Fritzell, 1987; Fritzell and Haroldson, 1982).

Wilcomb (1956) claimed that young in Oregon were born in April or May and that litter size ranged from two to five, with four the most common number of young. Based on studies conducted elsewhere in the range (Table 12-1; Fritzell and Haroldson, 1982), these values appear reasonable, but they were not quantified and methods used to obtain them were not described.

At birth, the average mass of gray foxes in Georgia was 85 g; at 11 days, 128 g; at 19 days, 167 g; at 23 days, 232 g; at 30 days, 247 g; at 51 days, 584 g; and at 78 days, 1,151 g (Wood, 1958). At ≅6 months of age, five young raised in captivity averaged 2,875 g (Taylor, 1943). Of 78 females 10–21 months of age at capture, 92.3 percent had produced litters or exhibited other evidence of sexual maturation during the first breeding season after their birth (Wood, 1958).

Based on age structure of populations, 1st-year mortality among gray foxes commonly is in excess of 50%, sometimes >60% (Layne, 1958b; Lord, 1961c; Wood, 1958). Fewer than 5% exceed 4 years of age (Wood, 1958).

Habits.—Although the gray fox may be seen abroad during daylight hours, most activity is nocturnal or crepuscular (Fritzell and Haroldson, 1982). Usually, the gray fox is secretive. Ground dens used by gray foxes frequently are the modified dens of other species; hollow logs, abandoned buildings, refuse piles, and rocky outcrops also may be used as den sites. Dens, especially natal dens, commonly are concealed by thick, brushy vegetation; foxes often move to other dens when disturbed (Nicholson et al., 1985).

The gray fox is unique among other North American canids in that it climbs trees to escape pursuit, to forage, and to rest. It simply runs up sloping trunks or jumps from branch to branch, and even a vertical bole without branches may be climbed as high as 18 m by grasping it with the forefeet and pushing with the hind feet (Fritzell and Haroldson, 1982). It runs down sloping trunks head first, but backs down vertical trunks.

No studies of movements of gray foxes have been conducted in Oregon, but minimum-area home ranges based on radiotelemetry data averaged 113 ha for adult females (*n* = 4) and 102 ha for adults males (*n* = 4) in Utah (Trapp, 1978). In California, home ranges of four females averaged 122 ha (Fuller, 1978). These values compare favorably with those obtained for gray foxes elsewhere in the range (Fritzell and Haroldson, 1982). Home ranges of females during the littering and nursing seasons may be reduced to ≤20% of home ranges during prereproductive seasons (Nicholson et al., 1985).

Remarks.—Some authorities have placed the gray fox in the genus *Vulpes* (Clutton-Brock et al., 1976) or the genus *Canis* (Van Gelder, 1978). We, like most others (Fritzell and Haroldson, 1982; Jones et al., 1992; Nowak, 1991; Sheldon, 1992; Wayne and O'Brien, 1987; Wozencraft, 1989), have included it in the genus *Urocyon*.

Vulpes velox (Say)
Kit Fox

1823a. *Canis velox* Say, 1:487
1888c. *Vulpes macrotis* Merriam, 4:136.
1931. *Vulpes macrotis nevadensis* Goldman, 21:250.
1990. *Vulpes velox macrotis*: Dragoo, Choate, Yates, and O'Farrell, 71:328.

Description.—*Vulpes velox* (Plate XXI) is the smallest canid that occurs in Oregon (Hall, 1946). It has the typical canid conformation; the body is slim, the legs are long and thin, and the ears are large and erect. The tail is about 40% of the total length and is tipped with black, but has no dark mane on the dorsal surface (McGrew, 1979). The feet and legs are whitish anteriorly, light rusty-brown posteriorly. The dorsum is grizzled brownish-gray medially blending to grizzled gray then to light buff laterally and finally to white on the chest and venter. The head and posterior surface of the ears are slightly lighter than the dorsum; anteriorly, the ears contain long white hairs.

The skull of the kit fox (Fig. 12-7) is narrow with a long, slender rostrum (Table 12-2); the bullae are inflated (McGrew, 1979); and the temporal ridges are not pronounced. The condylobasal length is <125 mm.

Distribution.—*Vulpes velox* occurs throughout the Great Plains, Great Basin, and southwestern deserts from southern Alberta, Saskatchewan, and Manitoba south to Jalisco and from southeastern Oregon south to Baja California Sur (Fig. 12-8; Dragoo et al., 1990; Hall, 1981). In Oregon, museum specimens are from Deschutes and Malheur counties (Fig. 12-8), and other records of *V. velox* are from near Klamath

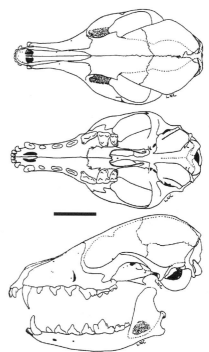

Fig. 12-7. Dorsal, ventral, and lateral views of the cranium, and lateral view of the dentary of the kit fox, *Vulpes velox* (OSUFW 6088). Scale bar equals 25 mm.

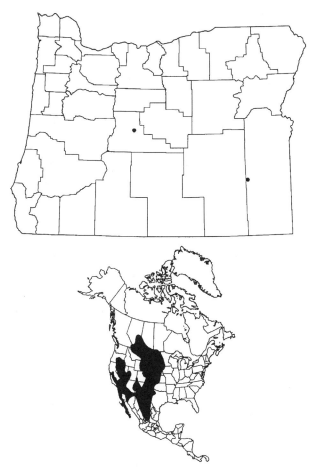

Fig. 12-8. Localities in Oregon from which museum specimens of the kit fox, *Vulpes velox*, were examined (listed in APPENDIX, p. 578), and the distribution of the kit fox in North America. (Range map redrawn from Hall, 1981:941, map 494, with modifications.)

Falls, Klamath Co. (Laughlin and Cooper, 1973) and from the southern half of Harney and Malheur counties (Benedict and Forbes, 1979; DeStefano, 1993).

In June 1992, we observed and photographed a litter of five at entrances to a natal den <8 km from the Oregon-Nevada border south of Denio, Nevada. The distance from Oregon is nearly within the known distance traveled by an individual during 1 night (Egoscue, 1962b) and less than one-fourth the recorded dispersal distance for young (Egoscue, 1956).

Geographic Variation.—Only one of the two currently recognized subspecies is reported to occur in Oregon: *V. v. macrotis* (Dragoo et al., 1990).

Habitat and Density.—Throughout its range, *V. velox* is associated with shrub-steppe and desert habitats; shadscale (*Atriplex confertifolia*), black greasewood (*Sarcobatus vermiculatus*), and big sagebrush (*Artemisia tridentata*) commonly are dominant shrubs within our region, which includes the range of the kit fox (McGrew, 1979). In Utah, ≅75% of the kit foxes observed were in areas in which ground cover was <20%, soils were light colored and loamy, and the elevation was <1,675 m (McGrew, 1979). Characteristics near a cave in which 12 kit fox skulls plus fragments of others were collected in central Malheur Co. in 1974–1976 were similar to those described in Utah (Benedict and Forbes, 1979).

In a survey for kit foxes in southern Malheur and Harney counties in 1991–1992, Keister and Immell (1994) searched along ≅1,748 km of secondary roads and trails at night by use of spotlights and used camera traps at 186 locations for 3,202 trap-nights. They observed four kit foxes and possibly three others and obtained pictures of 14 kit foxes. They also measured height, density, and proportion of various vegetative types on two 30-m transects at 39 randomly chosen sites. Height of vegetation and percent cover were not significantly different between random sites and those at which kit foxes were observed. Seemingly, vegetative composition at random sites was not compared statistically with that at sites at which kit foxes were observed or photographed. Nevertheless, Keister and Immell (1994) claimed that they had documented the occurrence of kit foxes in habitats similar to those reportedly used elsewhere by the species.

On a 64.8-km^2 study area in Utah, the density of adults was 15.4–21.6/100 km^2 in October–mid-February and that of adults and young in mid-March was 35.5–38.6/100 km^2 (Egoscue, 1962). After the study area was expanded to 103.6 km^2 and after a 3-year decline in numbers, density was 9.7/100 km^2 (Egoscue, 1975). Morrell (1972) counted six adults and eight young on a 2.59-km^2 area in the San Joaquin Valley, California, the equivalent of 231.7 adults/100 km^2 and 540.5 adult and juvenile kit foxes/100 km^2. Density of the Utah population was correlated directly with indices to populations of black-tailed jackrabbits (*Lepus californicus*). Egoscue (1975) considered that maximum density was set by the territorial requirements of kit foxes, but those in the California population did not maintain specific hunting territories (Morrell, 1972).

Keister and Immell (1994) reported their estimates of density for kit foxes on an ≅3,626-km^2 area in southern Harney and Malheur counties variously at 0.8–4.2/100 km^2 (p. 13) and 1.2–4.0/100 km^2 (p. 18). Although they admitted (p. 13) that the methods used likely produced "rough estimates at best," we consider the logic used in concocting the estimates to be totally without reasonable bases. We fear that these density values may be widely quoted and used in decisions regarding kit foxes in Oregon.

Diet.—In Utah, 94% of the prey mass as indicated by remains at dens consisted of black-tailed jackrabbits; the remainder consisted of kangaroo rats (*Dipodomys ordii* and *D. microps*), Audubon's cottontails (*Sylvilagus audubonii*), and small birds, especially horned larks (*Eremophila alpestris*—Egoscue, 1962b). Adult jackrabbits weigh approximately the same as kit foxes and were considered to be the largest prey that kit foxes can handle (Egoscue, 1962b). Reptiles were eaten infrequently. When the jackrabbit population declined, kit foxes did not shift their diet to Piute ground squirrels (*Spermophilus mollis*) that became exceedingly abundant (Egoscue, 1975). Egoscue (1975) considered the low benefit:cost ratio prevented kit foxes from preying extensively on small mammals, especially while feeding weanling young. However, in California, kangaroo rats (*D. nitratoides* and *D. ingens*) were considered staples of the diet, with Audubon's cottontails ranking second (Morrell, 1972). Considerable numbers of insects were consumed, but they did not compose much of the mass of food eaten. Some plant materials were consumed, particularly grasses and filaree (*Erodium*—Morrell, 1972). In California, weanling young were fed kangaroo rats; rarely was >5 min required for an adult to catch a kangaroo rat and return to the natal den (Morrell, 1972). Adult kit foxes require about 175 g of meat/day (Egoscue, 1962b).

Reproduction, Ontogeny, and Mortality.—In California and Utah, the breeding season extends from December to February with most young born in February or March (Egoscue, 1962b; Morrell, 1972). The gestation period is unknown, but Egoscue (1962b) speculated that it was 49–55 days, the same as for *V. vulpes*. However, litter size averages only two-thirds that of *V. vulpes* (Table 12-1).

A male neonate weighed 39.3 g and measured (in mm) as follows: total length, 162; length of tail, 51; length of hind foot, 22; length of ear, 11 (Egoscue, 1966). The eyes and ear meatuses were closed; the claws were white, well formed, and strongly recurved. None of the teeth had erupted. Except for the feet, ears, inguinal region, and distal one-half of the tail, the body was covered with hairs 1–3 mm long. The head, neck, and lower sides were buffy red, brightest on the head and neck; the back, thighs, and proximal one-half of the tail were gray-brown (Egoscue, 1966). The vibrissae were ≅3 mm long. Different color phases of neonates have been described, with most of the differences related to the ocherous buff on the legs; young in some litters are all the same color, but other litters may be of mixed colormorphs. The black tip on the tail develops at about 2 months of age (Morrell, 1972).

In Utah, the smallest individual caught weighed 507.6 g, and 1 month later it weighed 1,088.6 g (Egoscue, 1962b). Egoscue (1962b:489, fig. 9) indicated that both sexes attained the approximate adult mass by 1 August.

Benedict and Forbes (1979) suggested that presence of skulls of kit foxes in the Owyhee River Cave, Malheur Co., with evidence of use as a natal den by bobcats (*Lynx rufus*) was the result of predation on kit foxes by bobcats. Coyotes (*Canis latrans*) also may prey on kit foxes occasionally (McGrew, 1979). Raptors may take young kit foxes (Egoscue, 1975). Egoscue (1956) considered kit foxes to exhibit little wariness of traps or poisons and to be particularly vulnerable to being killed by cyanide guns used in coyote-control programs. The scarcity of kit foxes in Oregon (DeStefano, 1993; Mace, 1970) and elsewhere (McGrew, 1979) has been attributed to antipredator campaigns directed toward reduction of populations of other canids. However, some investigators indicated that numbers of kit foxes actually increased in areas subjected to intensive control programs directed toward reducing populations of coyotes by trapping and poisoning (McGrew, 1979). Keister and Immell (1994:23) did not believe coyote-control measures to be a serious threat to kit foxes in Oregon because such measures were "normally not practiced" on their ≅3,626-km² study area. Egoscue (1962b) and Morrell (1972) agreed that mortality resulting from collisions with motor vehicles was a primary inimical factor. In Utah, eight (11.3%) of 71 marked foxes were known to have been killed by vehicles during a 4-year study (Egoscue, 1962b). In Oregon, three kit foxes were killed on a ≅110-km section of U. S. 95 during a 2-year period (Keister and Immell, 1994).

Habits.—The activity of *V. velox* is largely nocturnal, although adults with young commonly emerge from dens shortly before sunset and young >1 month old may be active outside the den in late afternoon. Neither individuals nor families seem to have exclusive hunting territories, but certain dens seem to be the exclusive property of specific family groups. During a 15-month radiotelemetry study in California, only two of the 41 dens used by one family group were known to have been entered by nonfamily members (Morrell, 1972). Large brood dens are occupied until May–June, then smaller dens are used. Unmated individuals use small dens year-round. Most dens are on flat ground with sparse vegetation (Egoscue, 1962b; Morrell, 1972). Natal dens commonly have multiple (three–five or more) entrances and contain considerable amounts of prey remains and fecal droppings (Egoscue, 1962b). All young in litters commonly are moved to other dens at intervals of 2–4 weeks (Morrell, 1972).

In October–November, the male joins the female at the brood den (Morrell, 1972). Pairing may be monogamous or polygamous (Egoscue, 1962b) and the same individuals do not necessarily pair in consecutive years (Morrell, 1972). During lactation, adult females spend most of the time at natal dens; thus, the male must do most of the hunting (Egoscue, 1956). At other times pairs often hunt together, although not in an organized fashion (Morrell, 1972). Kit foxes may hunt as far as 3.2 km from the natal den (Grinnell et al., 1937).

Family units disband and young disperse by mid-September; a 32-km dispersal distance was recorded for a juvenile (Egoscue, 1956). A kit fox that inadvertently escaped from captivity returned a distance of 32 km to the area from which it was captured (Egoscue, 1956).

Kit foxes seem to be particularly unwary; they often may be approached to within a few meters, they enter live traps, and they commonly defecate and urinate on traps set for small mammals (Egoscue, 1962b).

Remarks.—The kit fox formerly was referred to as *Vulpes macrotis,* a species distinct from the swift fox (*V. velox*) and with eight geographic races, of which *V. m. nevadaensis* was reported to occur in Oregon (Hall, 1981). Based on a morphometric study, Waithman and Roest (1977) synonymized four of the five subspecies they considered, but *V. m. nevadaensis* was not among them. Several previous workers had treated *V. macrotis* as a subspecies of *V. velox* (McGrew, 1979). Dragoo et al. (1990:327) found that "two of the previously recognized subspecies of the kit fox were more similar genetically to the swift fox than to two other previously recognized subspecies of the kit fox." Because of the negligible genetic differentiation, they synonymized the two species of arid-land foxes; the name *V. velox* has priority. However, the two arid-land foxes formerly considered distinct at the species level could be distinguished by use of morphometric data. Dragoo et al. (1990) concluded that these differences justified recognition of the two former species as subspecies: *V. v. velox* and *V. v. macrotis.*

Olterman and Verts (1972) considered *V. velox* to be endangered, if indeed the species still occurred in Oregon. Although additional records and sightings have been reported (Benedict and Forbes, 1979; DeStefano, 1993; Keister and Immell, 1994; Laughlin and Cooper, 1973), we have no reason to suggest a change in status. Although listed earlier as a threatened species by the Oregon Department of Fish and Wildlife, *V. velox* did not appear on the agency's lists of sensitive species in December 1995.

Vulpes vulpes (Linnaeus)
Red Fox

1758. [*Canis*] *vulpes* Linnaeus, 1:40.
1852. *Vulpes macrourus* Baird, p. 309.
1900. *Vulpes cascadensis* Merriam, 2:665.
1936. *Vulpes fulvus cascadensis*: Bailey, 55:281.
1936. *Vulpes fulvus macrourus*: Bailey, 55:284.
1962. *Vulpes vulpes macroura*: Halloran, 43:432.
1981. *Vulpes vulpes cascadensis*: Hall, 2:937.

Description.—The red fox (Plate XXI) is a medium-sized canid (Table 12-2) that occurs in three color phases: red, silver, and cross (Voigt, 1987). The red phase is the most common and, in Oregon, is deep rusty dorsally grading to rusty grizzled with white posteriorly and to a lighter rust along the sides. The underfur is light gray. The throat and chest are white with gray underfur; the belly hair is all white. The back of the ears and front of the legs are black. The tail is sparsely overlain with black-tipped hairs and is tipped with white. The silver phase is entirely black except for an area on the posterior dorsum grizzled with white and sometimes an area grizzled with white on the shoulders, and the white tip on the tail (Voigt, 1987). A nearly all-black specimen (OSUFW 9017) was collected in Union Co.; unfortunately the skin was destroyed during preparation. One of us (BJV) collected a black-phase *V. vulpes* and saw one other in Benton Co. Other than the specimen from Wallowa Co. described by Bailey (1936), we examined no cross-phase red foxes from Oregon.

The skull of the red fox (Fig. 12-9) is distinguishable from those of other canids by a combination of a depression in the postorbital process, temporal ridges ≤1 mm wide and not lyre-shaped, and a condylobasal length >135 mm. The upper incisors may be slightly lobed.

Distribution.—Now that the natural distribution of the gray wolf (*Canis lupus*) has been reduced, the red fox has the widest distribution of any mammal other than humans (Nowak, 1991; Voigt, 1987). It occurs throughout North America north of Mexico except for some western prairie, southwestern desert, and western coastal areas (Fig. 12-10); throughout Europe and Asia except for the southeastern tropical zone; and in northern Africa. It has been introduced into Australia, where it has come to occupy much of the continent (Hall, 1981; Henry, 1986; Nowak, 1991; Voigt, 1987). Specimens of Oregon red foxes on deposit in museums are from the Willamette Valley, the Wallowa Mountains, and mountain and desert areas in Wasco, Klamath, Deschutes, Lake, Union, and Harney counties (Fig. 12-10). However, in the past 2 decades, large numbers of red foxes reportedly were taken by trappers and hunters in the Willamette Valley and other interior valleys west of the Cascade Range and in Malheur Co.; a few reportedly were taken in coastal counties and in other counties east of the Cascade Range (Table 12-5). Bailey (1936) did not consider the red fox to occur in the Willamette Valley and, indeed, no specimen from that area predates publication of his monograph. However, specimens taken

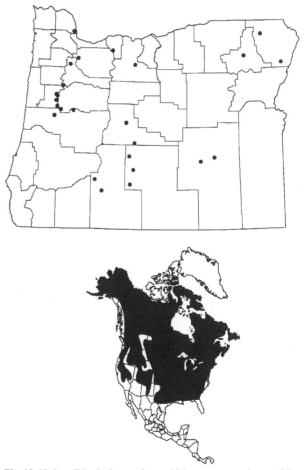

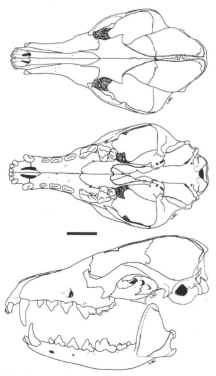

Fig. 12-9. Dorsal, ventral, and lateral views of the cranium, and lateral view of the dentary of the red fox, *Vulpes vulpes* (OSUFW 8788). Scale bar equals 25 mm.

Fig. 12-10. Localities in Oregon from which museum specimens of the red fox, *Vulpes vulpes*, were examined (listed in APPENDIX, p. 369), and the distribution of the red fox in North America. (Range map redrawn from Hall, 1981:938, map 493, with modifications.) The geographic range at settlement did not include the Northeast, the Southeast, much of the Midwest, and the southern Great Plains in the United States (Voigt, 1987).

Table 12-5.—*Reported annual harvest of red foxes (Vulpes vulpes) in Oregon by county, 1969–1996. Data provided by Oregon Department of Fish and Wildlife.*

Trapping season

County	1969–1970	1970–1971	1971–1972	1972–1973	1973–1974	1974–1975	1975–1976	1976–1977	1977–1978	1978–1979	1979–1980	1980–1981	1981–1982	1982–1983	1983–1984	1984–1985	1985–1986	1986–1987	1987–1988	1988–1989	1989–1990	1990–1991	1991–1992	1992–1993	1993–1994	1994–1995	1995–1996
Baker	56	43	40	105	78	53	74	52	24	55	17	30	30	66	38	16	15	14	5	4	6	7	15	7	12	8	1
Benton	20	11	11	22	32	29	37	12	25	13	43	31	93	76	95	63	99	35	69	55	18	4	16	10	13	25	
Clackamas																						7					
Clatsop													1						1								
Columbia	1	3	8	3	8	13	18	21	22	7					1			1	3	4		1		1		1	
Coos		3								9		5	4	1	1	3	3	4	2				2		1		
Crook																								2			
Curry													1														
Deschutes														3			2		1	2	1						
Douglas		7	4	10	21	26	44	71	138	155	136	113	82	77	123	74	48	32	11	5	12	22	20	20	56		6
Gilliam																											
Grant																		3									
Harney																					1						
Hood River		8				8	12	5			1	5		8			1	2	3	2							
Jackson					3		2						8				1										
Jefferson																									1		
Josephine					1	3				3		3		6					12		1		1				
Klamath	9				1	1	1		1								1	1									
Lake																	1										
Lane	9	35	69	105	143	162	169	83	77	127	53	95	116	36	87	73	66	126	37	30	15	54	79	33	48		
Lincoln	2					2	1							1	1	1	1	1	1								
Linn	93	30	148	298	261	221	114	50	208	116	121	65	137	147	149	112	120	195	107	53	33	16	17	81	16	33	
Malheur		5	3	7	10	20	24	28	15	47	51	20	69	33	90	86	141	137	187	80	107	93	81	120	39	8
Marion	48	49	119	102	188	59	133	75	139	49	83	66	128	107	112	67	95	155	26	25	27	13	6	1			
Morrow					6																				1		
Multnomah	9	1	1				8	5	9	3	3	2	2	1	1	2	6	2	1	1	1						
Polk	11	12	23	60	82	50	71	29	1	40	15	24	48	14	10	10	21	25	18	13					1	1	
Sherman												1															
Tillamook		4		3			1		4																		
Umatilla																						1	1		1	17	2
Union																											
Wallowa											1																
Wasco	7	2	40	2	11	8	17	9	6	7	2	5		4		1					1	1			1		
Washington													1	2	1				1	1							
Wheeler																											
Yamhill	11	16	13	4	3	7	12	6	21	14	4	11	8	7	10	15	5	4	10	3	2	1		1	1		1
Totals	276	202	254	517	707	827	691	691	436	699	630	543	537	756	584	682	562	555	782	458	239	216	245	227	221	227	18
Trappers reporting[a]	721	773	731	847	1,102	1,350	1,377	1,755	2,746	2,899	3,216	3,373	3,069	3,980	3,292	3,104	3,575	3,397	3,833	1,985	1,536	1,228	1,390	1,454	1,293	1,464	687
Average price/pelt[b]	$6.44	$5.33	$7.72	$18.55	$27.12	$24.23	$38.39	$50.35	$53.14	$64.19	$48.94	$49.73	$45.01	$33.48	$34.24	$36.29	$20.00	$28.05	$17.75	$12.30	$11.79	$10.00	$14.00	$11.00	$14.00	$18.00	$18.00

[a]Not all sought red foxes and not all were successful.
[b]Rounded to the nearest dollar after the 1989–1990 season.

in Benton Co. in 1942 (OSUFW 2507) and 1947 (OSUFW 3671) are on deposit in the Oregon State University, Department of Fisheries and Wildlife mammal collection.

Pleistocene (Wisconsinan) fossils of *V. vulpes* are abundant in cave deposits in California, Idaho, and elsewhere (Kurtén and Anderson, 1980). Remains of red foxes were identified from 9,785 ± 220-year-old and 7,875 ± 100-year-old strata at an archaeological site at The Dalles, Wasco Co. (Cressman, 1960).

Geographic Variation.—Two of the 10 currently recognized North American subspecies are considered to occur in Oregon: *V. v. cascadensis* west of a line connecting The Dalles, Hood River Co.; Bend, Deschutes Co.; and Klamath Falls, Klamath Co.; and *V. v. macroura* in the eastern fifth of the state (Hall, 1981). Descriptions of the red phase of both races invariably include mention of the yellow color of the pelt (Bailey, 1936; Fichter and Williams, 1967). Although those from the Willamette Valley that we examined were light colored, we considered them more rusty than yellow. Hall (1981) and Livezey and Evenden (1943) referred those from the Willamette Valley to *V. v. cascadensis,* but Wilcomb (1948, 1956) referred to specimens from the region as the "eastern red fox" (no trinomial provided) in reference to claimed introductions from eastern and midwestern United States.

Habitat and Density.—West of the Cascade Range, the harvest of red foxes for fur during the past 2 decades has been greatest in those counties in which the habitat is a heterogeneous mixture of croplands, grasslands (either pasture or for seed production), and woodlands (Table 12-5), the typical habitat of the species in eastern United States and Europe (Voigt, 1987). However, the red fox is highly adaptable and the species is known to occur in nearly all kinds of habitat from arctic tundra to temperate deserts (Voigt, 1987), so it might be encountered almost anywhere in the state. Bailey (1936:284) indicated that the red fox was not found in "deserts or arid valleys" in Oregon, but was confined to montane areas, "meadow valleys along streams," and areas near large lakes in eastern Oregon. Although the red fox has become more widely distributed in the 6 decades since publication of Bailey's (1936) work (Table 12-5), we have no information to refute his assessment of habitats occupied by the species.

The gradual increase in the reported harvest of red foxes for fur in Malheur Co. from none in the 1969–1970 and 1970–1971 furbearer seasons to 187 during the 1988–1989 season despite both increases and decreases in prices paid for pelts during the interval suggests that red foxes increased in density. Such an increase, if it is real, may parallel an increase that occurred a decade earlier in Idaho (Fichter and Williams, 1967).

We have no information regarding the actual density of red foxes in Oregon, but elsewhere in North America estimates range from 0.1 to 1.0/km² over broad areas, but as high as 3.0/km² in limited areas (Voigt, 1987). Zabel and Taggart (1989) reported a density of adults on Round Island, Alaska, of ≅10/km². In Great Britain, densities calculated from counts of family groups each considered to consist of 2.25 adults and 4.7 offspring were as high as ≅35 individuals/km² (Harris and Rayner, 1986).

Diet.—Voles (especially *Microtus*), lagomorphs, insects, and vegetable material (fruits and berries) constitute staples in the diet of red foxes throughout North America. How-

ever, red foxes are known to prey on birds and their eggs, opossums (*Didelphis virginiana*), insectivores, a variety of other rodents, and occasionally other carnivores (Scott and Klimstra, 1955). Red foxes are quick to take advantage of windfalls such as free-ranging poultry, concentrations of waterfowl and colonial-nesting birds (and their eggs) in nesting areas, and recently released game-farm pheasants (*Phasianus colchicus*—Sargeant, 1972; Scott and Klimstra, 1955; Wilcomb, 1956; Zabel and Taggart, 1989). In addition to being a nonspecific predator, the red fox also is an efficient scavenger; in local situations it consumes quantities of garbage and carrion (both livestock and wild animals—Voigt, 1987). In England, earthworms (*Lumbricus terrestris*) are a major prey item (MacDonald, 1980); these were identified largely by the presence of chaetae, thus may have been overlooked in food remains during studies of the diet of North American red foxes. Seemingly, the abundance of earthworms on the surface should provide a readily available winter food in western Oregon.

Food remains recovered in and at the entrance to a natal den occupied by red foxes in Benton Co. included those of California ground squirrels (*Spermophilus beecheyi*), black-tailed jackrabbits (*Lepus californicus*), brush rabbits (*Sylvilagus bachmani*), rufous-sided towhees (*Pipilo erythrophthalmus*), domestic turkeys, and domestic chickens (Livezey and Evenden, 1943). Percent frequency of occurrence of food remains in 20 fecal passages collected at entrances to each of five dens or den complexes in Benton Co. were California ground squirrels, 34%; black-tailed jackrabbits, 22%; small rodents, 11%; domestic chickens, 11%; ring-necked pheasants, 10%; small birds, 6%; livestock (carrion), 5%; insects, 4%; blackberries (*Rubus*), 3%; skunks (presumably *Mephitis mephitis*), 2%; brush rabbits, 1%; and moles (presumably *Scapanus townsendii*), 1%. Vegetation was identified in 20% of the droppings (Wilcomb, 1948).

Reproduction, Ontogeny, and Mortality.—Red foxes are monestrous with the breeding season in North America extending from December to May; date of breeding is positively related to latitude (Storm et al., 1976; Voigt, 1987). Wilcomb (1956:5) claimed that "In Oregon, the young are born in April or May." He did not provide supporting evidence for his statement, but seemingly (Wilcomb, 1948) based it on Bailey's (1936:283) report of the littering season in "eastern localities." One of us (BJV) observed behavior indicative of mating in red foxes (pairs running together—Scott, 1943) in late January in the Willamette Valley on several occasions. As the reported gestation period is 51–53 days (Voigt, 1987) and breeding may occur over a period of ≅10 weeks (Storm et al., 1976), birth of red foxes in the Willamette Valley might be expected from about mid-March to mid-May, with the peak probably about mid-April. We expect dates of breeding and littering might vary elsewhere in the state with latitude and perhaps with elevation.

Litters of seven (Livezey and Evenden, 1943), four, and eight (Wilcomb, 1948) were found in dens in Benton Co., and two litters of four and one of five were recorded by Wilcomb (1948), but he did not indicate the locality in western Oregon or whether the litters were counted before or after birth. Litter size based on counts of young at dens may be biased by neonatal mortality, by parts of litters occupying two or more dens, and, late in the period of parental care, by young bedding away from dens during daylight hours (Storm et al., 1976). Conversely, two litters occasionally are combined, producing a counterbias, but because of their rarity (nine instances at 509 dens in Iowa and Illinois) combined litters are not considered to produce a major bias in estimates of litter size (Storm et al., 1976).

Average litter size for a small sample from Oregon (Table 12-1) was 0.7 young/litter greater than that probably obtained by the same methods from other regions (Storm et al., 1976:21, table 8). However, the average is within that possible based on counts that do not include neonatal losses (Table 12-1).

Average (range in parentheses) measurements (in mm) of six neonates and one fetus of a female killed by a hunter during parturition were total length, 211 (200–219); length of tail, 67 (60–72); length of hind foot, 32 (30–34); and length of ear, 13 (12–14). Body mass averaged 110.6 g (range, 94.2–119.5 g). The pelage of the neonates was dark brown with light brown legs, whitish-brown feet, and creamy-white footpads and nails. Hairs ranged from 3 to 10 mm long. The eyes and ears were closed (Storm and Ables, 1966).

Young remain in the natal den during the 1st month of life; by 10 weeks, they commence to explore in the vicinity of the den on their own; and by 12 weeks, they begin to explore their parents' home range (Samuel and Nelson, 1982). Most females breed when ≅10 months old; the proportion failing to breed differs regionally (Layne and McKeon, 1956; Storm et al., 1976). Spermatogenesis commences in October and spermatozoa are present in the epididymides at least through March (Storm et al., 1976).

In Illinois and Iowa, >80% of 786 tagged animals recovered (of 1,987 tagged as juveniles) were shot, trapped, or killed on roadways by automobiles and >80% of these losses were during October–February (Storm et al., 1976). Despite the bias against natural mortality in assessing causes of mortality by use of tag returns, it is obvious that much mortality in red fox populations is related to activities of humans. Subadults were 1.2 times more vulnerable to hunting and trapping than adults and 1.5 times more vulnerable to all types of mortality; the average fox (all age-classes) had a 52% chance of dying during the forthcoming year (Storm et al., 1976).

Habits.—Red foxes may be annually monogamous, polygynous (Sheldon, 1992; Zabel and Taggart, 1989), or possibly polygamous (Voigt, 1987). Monogamy, with the male helping with rearing of offspring by bringing prey items to the lactating female and the young, probably is the most common mating system. However, when food resources are exceptionally abundant both males and females may improve their reproductive success by adopting a polygynous mating system, but they may shift to the less successful monogamous system when food resources decline (Zabel and Taggart, 1989). Females involved in polygynous relationships generally are kin (Voigt, 1987). Occasionally, a female will rear its litter alone (possibly with low or no success); sometimes barren females assist a pair in rearing its litter; two females that mate with the same male are known to have combined litters for rearing (Zabel and Taggart, 1989); and a litter born to one female is known to have been divided between two dens (Storm et al., 1976). Mating behavior illustrates the flexibility of the red fox that extends to social behavior, diet, denning, and many other aspects of the ecology of the species, and that likely contributes significantly to success of the species in the face of urbanization, increasingly widespread agriculture, and direct attempts to reduce density of its populations.

In spring and summer, red foxes center their activity around natal dens. A natal den excavated in Benton Co. had an entrance ≅20 cm wide and ≅38 cm high; the tunnel tapered to a subsequently constant diameter of ≅13 cm for the remainder of the ≅14 m excavated (Livezey and Evenden, 1943). Foraging activity largely is crepuscular or nocturnal, but sometimes red foxes hunt during the daylight hours, especially in winter (Ables, 1969). During the mating and rearing seasons, red foxes usually restrict their movements to ≅400–900 ha (Sargeant, 1972; Scott, 1943; Storm, 1965); much of their area is visited nightly and many of the same sites are visited for several nights consecutively (Storm, 1965). Areas used by red foxes usually are exclusive territories maintained by marking with urine and occasionally by agonistic behavior (Voigt, 1987). In early autumn, family units begin to disintegrate with the dispersal of subadults; by mid-October some subadult males averaged dispersing 19 km and subadult females averaged 8 km (Storm et al., 1976). In Illinois and Iowa, by the following March, 80% of the males and 37% of the females had moved ≥8 km; of those that survived another year, 96% of males and 58% of females had moved that distance. Males and females dispersed as far as 211 km and 108 km from their natal dens, respectively (Storm et al., 1976). A subadult male red fox marked in August in southern Wisconsin was recovered 394 km to the southeast in west-central Indiana the following May (Ables, 1965).

The red fox has been referred to as "cat-like" because of its hunting strategies, which, of course, are dictated to some extent by the morphology of the species (Henry, 1986). Hunting strategy depends largely upon the prey being stalked: insects are captured and eaten, usually after being flushed incidentally to other activities; small terrestrial mammals usually are hunted from pathways—when they are sighted, smelled, or heard, the fox crouches, then lunges upward at about 45° and descends with legs together but outstretched in an attempt to pin the prey beneath the feet; birds and arboreal mammals are hunted by stalking followed by a quick dash; lagomorphs are stalked followed by an all-out chase (Henry, 1986). In one analysis of 434 hunts by red foxes, 82% of 94 insect hunts, 23% of 257 mammal hunts, but only 2% of 83 bird hunts were successful (Henry, 1986).

Remarks.—Historically, the New World red foxes were designated *Vulpes fulva* (Desmarest, 1820), separate from the Old World *Vulpes vulpes* (Linnaeus, 1758). However, Rausch (1953), Churcher (1959), and Vogt and Arakaki (1971) provided opinion or evidence that red foxes were conspecific with a Holarctic distribution. The larger European red fox was introduced into eastern United States in colonial times, but the impact of these introductions on 20th-century red foxes in North America was not detectable because differences in morphology were sufficient to distinguish between skulls of specimens from Europe and New England. However, from Europe through Asia to Alaska then to eastern North America, certain dental characters and some skull characters exhibited a distinct cline, suggesting "an intergrading series extending across the whole Palearctic region" (Churcher, 1959:519). Thus, the North American and Eurasian forms were considered conspecific.

The origin of the red fox in the Willamette Valley seems to be as problematic as the origin of the species in eastern United States (Voigt, 1987). Bailey (1936:282, fig. 64) depicted the range of *V. v. cascadensis* as encompassing the Cascade Range and the northern half of the Coast Range, but excluding the Willamette Valley. However, he qualified his range map with the statement (p. 282) that "the limits of their range are not well defined." Graf (1947:20) recorded a hearsay report of a red fox taken on the Callapooya River, Linn Co., in 1921, but indicated that he knew of no other records in the Willamette Valley until a sudden influx in 1940 that he termed a "reappearance." He further indicated (p. 19) that the red fox "had been absent from that part of the Willamette Valley for so long that many of the farmers actually failed to recognize it." Farmers in the region became acutely aware of red foxes when their abundance increased dramatically in 1945 coincidentally with an irruption of voles (*Microtus*), and even more so the following year with the disappearance of the voles and the increase in predation on poultry by red foxes (Graf, 1947). Wilcomb (1948:9) indicated that a specimen referable to *V. v. cascadensis* was taken near Junction City, Lane Co., in 1936, but considered that the external appearance of those "taken during the course" of his investigation in Benton Co. in 1947–1948 "checked closely" with the description of the eastern red fox (no trinomial indicated) provided by Anthony (1928). He further chronicled hearsay reports of two introductions between 1910 and 1947 of red foxes from North Carolina and South Dakota and another in 1900 from an unstated source. We find it interesting that neither Bailey (1936) nor Graf (1947) commented on these introductions. Mace (1970:50), presumably referring to the red fox throughout Oregon, claimed that "The native animals disappeared following [European] settlement and the foxes present today are descendants of eastern stock transplanted in the Willamette Valley for hunting purposes." However, he presented no evidence to support his contention and we found none elsewhere.

We believe that color and markings of red foxes within an area even as small as the Willamette Valley (≅8,000 km²) are too variable to serve as criteria by which to establish the origin of the present genetic stock or to rule out swamping of introduced characters through hybridization with native animals. And, although we cannot refute claimed introductions, we suggest that a more likely explanation for the present-day presence of the red fox in the Willamette Valley, if indeed the species did not occur in the area naturally before European settlement, is the expansion of the range following an increase in density of populations in neighboring areas. A similar increase in density and expansion of the range of the red fox is believed to have occurred in eastern United States following European settlement; the increase may or may not have been related to introductions from Europe or to changes in habitat accompanying settlement (Churcher, 1959; Voigt, 1987). Also, the reported rarity or absence of red foxes in seemingly suitable habitats for long periods followed by an increase in abundance and recognition of the presence of the species by residents is a pattern recorded elsewhere in the West (Fichter and Williams, 1967). Nevertheless, establishment of the origin of red foxes occupying the Willamette Valley, like that of the red fox in eastern United States, awaits application of molecular and biochemical techniques.

In retrospect, we likely could have obtained considerable information regarding reproduction and perhaps other aspects of the biology of red foxes in Oregon by taking advantage of the relatively large numbers of red foxes harvested for fur west of the Cascade Range during the early 1980s, perhaps in conjunction with a study of reproduction in the raccoon (*Procyon lotor*—Fiero and Verts, 1986a). We implore future researchers to take advantage of such windfall sources of information.

FAMILY URSIDAE—BEARS

The family Ursidae, according to the most recent treatment, consists of six genera with nine species in Europe, Asia, northern Africa, North America, and western South America (Wozencraft, 1993); some authorities recognize only three genera with eight species (Nowak, 1991). The latter authority placed the lesser panda (*Ailurus fulgens*) with the Procyonidae rather than in this family.

Two species, *Ursus americanus* and *U. arctos*, occur or have occurred in Oregon. The bears are large, heavy-bodied, and stocky-legged carnivores. They have short, rounded, and erect ears; extremely abbreviated tails; and long, recurved claws. The skull is massive. All four premolars may be present in both jaws, but more often one or more of the small first three premolars are lost. The remaining molariform teeth are bunodont; P4 and m1 are not modified as carnassials. A baculum is present. Bears are pentadactyl and plantigrade, with the soles, except for the toe and plantar pads, covered with fur. Bears are capable of bipedal locomotion for short distances.

Bears tend to be solitary except for females with young and during the mating season. They are omnivores, with berries, fruit, nuts, roots, tubers, fish, and other vertebrate animals constituting the primary foods. Although bears commonly move with a slow, shuffling gait, they can be surprisingly agile and capable of rapid movements. Bears usually seek to avoid contact with humans, but in park-type situations they can become habituated to humans then become nuisances or even dangerous to the incautious (Storer and Tevis, 1955).

Bears require several years to attain sexual maturity, produce relatively few young per litter, produce litters at intervals of 1–4 years, and are comparatively long lived. After successful matings, implantation is delayed for several months. In winter, bears spend as long as 4 months in a dormant state; heart and respiration rates drop to about half the active rate but body temperature is depressed only slightly. Dormant bears arouse easily, and at lower latitudes during warm periods, bears may leave their winter dens for short intervals (Jonkel, 1987).

The fossil record for the bear family extends to the late Eocene in Europe and Asia and to the early Oligocene in North America but no earlier than the Pleistocene in South America and North Africa (Savage and Russell, 1983).

KEY TO THE URSIDAE OF OREGON

1a. A distinct hump on the shoulders; face concave; claws on front feet >65 mm long, longer than claws on hind feet; length of M2 >150% of length of M1; extirpated...
.................................***Ursus arctos***, grizzly bear (p. 378).
1b. Shoulders lacking a hump; face convex; claws on front

feet <65 mm long, about equal in length to claws on hind feet; length of M2 <150% of length of M1..............
..........................*Ursus americanus*, black bear (p. 374).

Ursus americanus Pallas

Black Bear

1780. *Ursus americanus* Pallas, fasc. 14:5.
1854. *Ursus americanus* var. *cinnamomum* Audubon and Bachman, 3:125.
1903*c. Ursus altifrontalis* Elliot, 3:234.
1913*b. Ursus americanus altifrontalis*: Grinnell, 3:284.
1924. *Euarctos altifrontalis* Miller, 128:90.
1924. *Euarctos cinnamomum* Miller, 128:91.
1936. *Euarctos americanus cinnamomum* Bailey, 55:319.
1936. *Euarctos americanus altifrontalis* Bailey, 55:321.

Description.—Black bears (Plate XXII) are the largest extant carnivores in Oregon (Table 12-6); they are, however, smaller than grizzly bears (*U. arctos*). Range in body mass usually is about 45–120 kg for adult females and about 70–140 kg for adult males, although occasionally a male

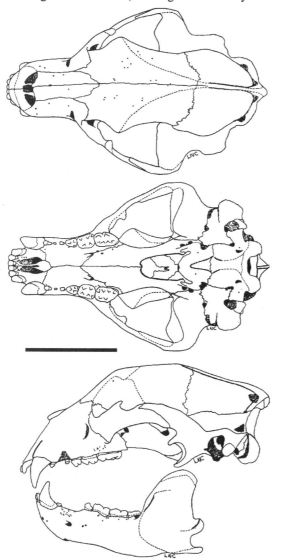

Fig. 12-11. Dorsal, ventral, and lateral views of the cranium, and lateral view of the dentary of the black bear, *Ursus americanus* (OSUFW 3175). Scale bar equals 10 cm.

weighs as much as 280 kg. The body is heavily built; the legs are stout; the feet are large; the head, eyes, and ears are relatively small; and the tail is extremely short. The claws are recurved but relatively short ($\cong$2–3 cm long), with those on the forefeet equal to or only slightly longer than those on the hind feet. The characteristic "humped" shoulders and "dished" (i.e., concave) face of the grizzly bear are absent. The pelage is black or brown; both phases may occur within the same litter produced by females of either color (Bailey, 1936). White and bluish phases are known to occur (Pelton, 1982).

The skull of the black bear (Fig. 12-11) is massive; it is characterized by prominent ridges on the frontals, strong and rugose sagittal and lambdoidal crests, and heavy zygomata. M2 is <1.5 times as long as M1.

Distribution.—Except for the Great Basin and arid Southwest regions, the black bear occurs or formerly occurred throughout much of North America as far south as Zacatecas (Fig. 12-12; Hall, 1981). Based both on localities at which museum specimens were collected (Fig. 12-12) and on the reported harvest by big-game management unit in the early 1980s (Fig. 12-13), black bears in Oregon occur in the Cascade Range and west to the Pacific Ocean, and in the Blue and Wallowa mountains; the species is absent from arid regions of central and southeastern Oregon. This is essentially

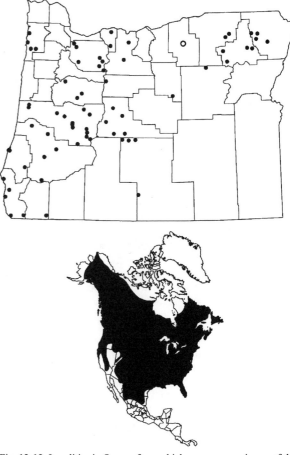

Fig. 12-12. Localities in Oregon from which museum specimens of the black bear, *Ursus americanus*, were examined (listed in APPENDIX, p. 578), and the distribution of the black bear in North America. (Range map redrawn from Hall, 1981:950, map 497, with modifications.) The black bear was extirpated from broad areas in North America (Pelton, 1982). Open symbol indicates record for county.

Table 12-6.—*Means (±SE), ranges (in parentheses) and CVs of measurements of skull characters for female and male bears* (Ursus) *from regions in Oregon. Skull characters <150 mm were measured to the nearest 0.1 mm, larger characters to the nearest 1 mm. SE and CV not provided if* n *<10.*

Species and region	Sex	n	Condylobasal length	Basilar length	Postglenoid length	Length of maxillary toothrow	Zygomatic breadth	Cranial breadth	Least inoterorbital breadth	Length of mandible
Ursus americanus										
West of Cascade Range	♀	7	220.8 (209–231)	197.7 (183–207)	69.0	82.9 (80.0–86.3)	132.7 (124.0–149.0)	78.3 (74.0–82.8)	48.7 (46.0–52.1)	152.5 (146–161)
	♂	7	251.7 (233–282)	228.7 (212–260)	84.0[a] (80.0–88.0)	92.50 (85.0–102.0)	158.9 (149.0–179.0)	86.1 (78.0–95.7)	63.6 (56.0–77.8)	175.0 (162–193)
East of Cascade Range	♀	5	230.5 (218–246)	209.7 (199–224)		84.9 (78.0–94.0)	140.4 (131.0–148.0)	76.9 (70.0–81.0)	51.3 (46.5–55.0)	156.5 (144–168)
	♂	11	250.0 ± 3.6 (229–270) 4.80	226.3 ± 3.6 (206–247) 5.20		90.2 ± 0.9 (87.0–97.0) 3.34	146.8 ± 2.78 (129.0–165.0) 6.28	81.4 ± 0.9 (77.0–85.0) 3.57	56.1 ± 1.7 (50.0–68.0) 9.81	171.7 ± 2.6 (158–189) 4.99
Ursus arctos	♀	2	292.3 (290–295)	264.5 (263–266)		113.5 (108.5–118.5)	172.8 (168.0–177.5)	92.5 (91.0–94.0)	61.5 (60.0–63.0)	207.0 (202–212)

[a]Sample size reduced by 5.

the same distribution depicted by Bailey (1936).

Bears of the genus *Ursus* arose in the Old World in the early Pliocene; a small, now extinct species, *U.* †*abstrusus*, first appeared in North America in the late Pliocene and possibly was conspecific with the small Eurasian *U.* †*minimus* (Kurtén and Anderson, 1980). This stock gave rise to North American black bears in the early Pleistocene that gradually increased in body size until, in the Wisconsinan, they were as large as modern grizzly bears. Body size has decreased markedly in the Holocene (Kurtén and Anderson, 1980).

Geographic Variation.—Two of the 16 recognized subspecies of black bears occur in Oregon: the larger *U. a. altifrontalis* west of the Cascade Range and the smaller *U. a. cinnamomum* east of the Cascade Range (Bailey, 1936; Hall, 1981). Hall's (1981:950, map 497) range map indicated that *U. a. californiensis* occurred in the southern portion of the Cascade Range, but no records for that sub-

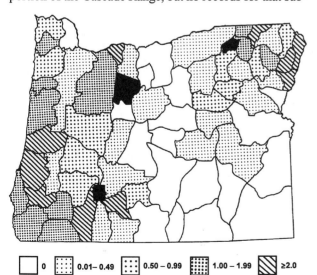

Fig. 12-13. Distribution of the harvest/100 km^2 of black bears, *Ursus americanus*, by big-game management unit, 1982–1983. Data provided by the Oregon Department of Fish and Wildlife. The two northern solid black areas are Indian Reservations; the southern black area is Crater Lake National Park.

species were listed. Bailey (1936) claimed that the proportions of individuals with black or brown pelages differed in the two regions with a ratio of about 4:1 in favor of black west of the Cascade Range and about 1.3:1 in favor of brown east of the Cascade Range.

Habitat and Density.—In general, black bears usually are associated with forested areas, particularly those in the vicinity of water, an association likely responsible for the absence of the species in much of central and southeastern Oregon (Bailey, 1936). Within this general habitat type, black bears tend to select early seral communities over more advanced developmental stages. Within home ranges of bears in southwestern Washington, all available habitat types were included in proportion to their availability, but bears used portions of their home ranges encompassing the various types disproportionately (Lindzey and Meslow, 1977a). The habitat type used most often consisted of clear-cuttings 7–12 years postharvest with vegetation dominated by brushy species, particularly salal (*Gaultheria shallon*), huckleberries (*Vaccinium*), and small western hemlock trees (*Tsuga heterophylla*). Although berries and fruits produced by some of these brushy species are particularly attractive to black bears and food availability may be responsible for the association of bears with this habitat type, Lindzey and Meslow (1977a) claimed that the attractiveness of recent clear-cuttings to bears in Washington and Oregon was not seasonal but persisted throughout the year.

In northern California, monthly assessment of habitat use from distribution of tracks, radiolocations, sightings, feeding sites, and fecal droppings in nine types revealed that bears tended to shift their activities from one type to another on the basis of seasonal availability of food resources. In May 55% and in June 18% of activity was concentrated in wet meadows that constituted only 1% of the total area; in July–September, <10% of bear activity was in this type. In wet meadows, bears foraged on new growth of grasses and forbs in spring, but in summer used this type largely for wallowing (Kelleyhouse, 1980). Mixed conifer forest covered ≅30% of the area and included ≅20–62% of bear activity; use was least in August when black bears moved to logged areas, where they foraged on insects in rotting logs, or to manzanita (*Arctostaphylos*)-coffeeberry

(*Rhamnus california*) brush, to feed on berries (Kelleyhouse, 1980).

In southwestern Oregon, 74% of 114 sightings and tracks of bears were in "coniferous and *Pseudotsuga*-sclerophyll forests" (McCollum, 1973:24); 14% were in clear-cuttings and 8% in natural brushfields. Most evidence of bears in clear-cuttings was within 91.5 m of the edge.

In southwestern Washington, estimated density ranged from 11.2 to 14.9/1,000 ha, somewhat greater than the 1.1–7.7/1,000 ha reported in several of the other conterminous states in the United States (Lindzey and Meslow, 1977*b*). Bailey (1936) suggested that the 5,504 black bears claimed to reside on national forests in Oregon probably constituted about 75% of the total population of the species in the state. According to Oregon Department of Fish and Wildlife biologists who prepared *Oregon's black bear management plan 1993–1998*, 18,000 black bears occupied 55,552 km² designated to be bear habitat in Oregon in 1980, and 25,000 occurred in 102,400 km² of bear habitat in 1993. They indicated that the population numbers were based on comparable or more conservative estimates of density obtained in adjoining states. Relative density based on harvest records is greatest in the region west of the Cascade Range, particularly in the southern portion of that region (Fig. 12-13).

Diet.—Black bears throughout their range are largely herbivorous. In spring, they consume new growth of grasses and forbs; in summer, the diet shifts to mostly berries and fruits; and in autumn, they forage voraciously on acorns, nuts, and berries (Kolenosky and Strathearn, 1987; Pelton, 1982). In spring, newborn ungulates sometimes are preyed upon, but by 2 weeks or so of age, most young ungulates are sufficiently agile to escape predatory attempts by black bears. In summer, larval and adult insects are consumed, and in summer and autumn spawning anadromous fish often are eaten. Black bears seek foods high in proteins and fats, particularly in spring when they are recovering from body attrition during dormancy and during autumn when they are storing fats on which to overwinter. Thus, black bears are quick to take advantage of windfalls such as livestock carrion, garbage dumps, and inadequately protected foods at campsites.

The only systematic study of the diet of black bears in Oregon of which we are aware was conducted March–July on two areas in the Coast Range. Where bears damaged an average of 29.4 trees/ha (98% Douglas-fir, *Pseudotsuga menziesii*) by removing bark to ingest cambium and sap, percents frequency of occurrence and volume (the latter in parentheses) of identified items in fecal droppings ($n = 61$) were graminoids, 61% (38%); forbs, 51% (32%); shrubs, 20% (12%); cambium, 12% (9%); animal material, 23% (8%); and miscellaneous, 3% (1%). Where damage to trees by bears was low ($\bar{X} = 2.7$/ha), the values for droppings ($n = 92$) were 54% (37%), 29% (20%), 50% (38%), 2% (1%), 17% (4%), and 1% (trace), respectively (Noble, 1994). Where tree damage was high, bears had significantly lower albumin:globulin ratios in their blood than where tree damage was low, suggesting that poorer nutritional condition possibly was involved when bears feed on cambium and sap (Noble, 1994).

Bears in Oregon seemingly are as much food generalists as they are elsewhere. However, being a food generalist sometimes can be a problem for a bear. Bailey (1923*a*)—based on the evidence of a broken tree from which most fruits were removed, and numerous masses of bear feces containing partially masticated fruits nearby and scattered along the trail leading from the tree—reported that a black bear in Coos Co. consumed an enormous quantity of cascara (*Rhamnus purshiana*) fruits despite their powerful cathartic effect.

Reproduction, Ontogeny, and Mortality.—In a sample from Oregon (70% from northwestern Oregon), mature follicles (≥8 mm) occurred in ovaries of females killed between 14 June and 19 July; no corpora lutea were found before 22 July, but between that date and 1 January ovaries of six of nine 3–8-year-old females contained corpora lutea (Lindzey, 1976*a*). In the same sample, females <3 years old did not breed, but 76.6% bred as 3-year-olds (Lindzey, 1976*a*). In another Oregon sample (94% from western Oregon), ≅30% of female black bears bred when 2 years old, >80% bred by 3 years of age, and essentially all bred by 4.5 years of age (C. E. Trainer, in litt., 30 June 1992). Of nine females examined in west-central Idaho, three bred for the first time at 3.5 years, five at 4.5 years, and one at 5.5 years (Reynolds and Beecham, 1980). In some other regions, age at first breeding may be even greater (Jonkel and Cowan, 1971). Potentially, after attaining sexual maturity, females can produce litters at 2-year intervals; some females in some populations extend the interval by 1–2 additional years (Lindzey, 1976*a*).

Litter size in black bears, like that of lagomorphs and some other species for which appropriate data are available, seems to be related to maternal age. In samples from Oregon (94% from western Oregon), litter size averaged 1.7, 1.8, 2.0, and 2.5 for 2-, 3-, 4–10-, and 11–16-year-old females ($n = 25$), respectively, based on counts of corpora lutea and 1.8, 1.8, and 1.7 in 3-, 4–10-, and 11–16-year-old females ($n = 48$), respectively, based on counts of implantation sites (C. E. Trainer, in litt., 30 June 1992). Lindzey (1976*a*) recorded an average of 1.3, 1.6, 2.3, 2.6, 2.0, and 1.6 implantation sites for 5-, 6-, 8-, 9-, 10-, and 11–19-year-old females ($n = 30$), most of which came from the same region as those in the previous study. In west-central Idaho, 15 of 16 females (ages not reported) produced twins (Reynolds and Beecham, 1980).

After fertilization and formation of the blastocysts, further development is retarded or ceases for ≅5 months; implantation occurs in late November or December (Wimsatt, 1963). Young are born "between the last week in January and the first 10 days of February" (Lindzey and Meslow, 1977*b*:411) while maternal females are dormant in their winter dens. Neonates weigh 0.2–0.3 kg (half that of a newborn beaver, *Castor canadensis*), are ≅200 mm long, and are naked, blind, and helpless (Pelton, 1982). Growth is rapid; young may weigh 4 kg at 6 weeks of age when the maternal female emerges and 30 kg at 9 months (Kolenosky and Strathearn, 1987). Young den with their mothers the first winter and most accompany their mothers until 16–17 months of age (Kolenosky and Strathearn, 1987; Pelton, 1982).

Habits.—During spring–autumn seasons, black bears tend to be more active during daylight and crepuscular periods, but in the month or so before and after the period of winter dormancy they are less active overall and more nocturnal (Amstrup and Beecham, 1976). In the active season, peaks of activity occurred at 1000 and 2100 h, and inactivity was observed most often between 0100 and 0400 h (Reynolds and Beecham, 1980).

In autumn, black bears, throughout most of their geographic range, fatten, become more and more lethargic, enter dens, and remain inactive throughout winter (Beecham et al., 1983; Lindzey, 1976a). Den sites commonly are under stumps and logs (Noble, 1994) or in holes in hillsides, but may include hollow trees, rock caves, drainage culverts, abandoned buildings, or, in some regions, even unsheltered depressions (Beecham et al., 1983; Lindzey and Meslow, 1976a). Dens usually are well hidden by dense vegetation and, in colder regions, often open to the north or west, possibly to take advantage of the insulative effect of snow cover for a greater portion of the dormant period (Beecham et al., 1983).

In southwestern Washington, an area with a relatively mild climate, adult females entered dens first (21 October–5 November), followed by yearlings (5–20 November), then by adult males (15–29 November—Lindzey and Meslow, 1976b). Although a few individuals shifted den sites early in the inactive period, most remained in their dens for 105–157 ($\overline{X}= 126$) days. Adult males and 2-year-olds (yearlings when they became inactive) resumed activity first, usually the 2nd week of March. Females with newborn young first emerged the 3rd week of March, but remained in the den or in the vicinity of the den for 2–3 additional weeks, presumably to care for young (Lindzey and Meslow, 1976b). In west-central Idaho, climatically more similar to areas within the geographic range of the black bear in northeastern Oregon, less difference in date of immergence was noted; during a 4-year period, adult females ($n = 22$) entered dens 9 October–24 November, whereas adult males ($n = 13$) entered dens 30 October–25 November and yearlings ($n = 8$) immerged 10 October–27 November (Beecham et al., 1983). Phenology of plants on which bears foraged seemed to play a role in determining times of immergence among years. Dates of emergence in the same region were 11–30 April, about a month later than in the coastal region (Beecham et al., 1983; Lindzey and Meslow, 1976b). In some regions, emergence occurs when temperatures of ≥10°C cause rapid snowmelt (Rogers, 1974). McCollum (1973) believed that black bears in southwestern Oregon entered winter dens from late October through December and emerged April to mid-May. Much of his evidence was hearsay.

During the period of dormancy, black bears do not feed, urinate, or defecate, and their body temperature remains near that during active seasons. In Idaho, average ($\pm SD$) rectal temperature of immobilized bears ($n = 338$) during the active season was 38 ± 1°C whereas that of bears in winter dens was 36 ± 2°C (Beecham et al., 1983). Respiration rate and heart rate of dormant bears averaged about 43 and 66%, respectively, of those of active bears (Beecham et al., 1983). Strangely, the footpads of all age-classes except young-of-the-year are shed during the period of dormancy; one worker suggested that tender soles may be responsible for restricted movements of bears upon emergence (Rogers, 1974), but others believe that other aspects of the physiology are involved (Beecham et al., 1983).

Size of home ranges of male black bears commonly is 2 or more times that of females, and size of those of both sexes seems to vary inversely with the quality of the habitat, particularly the quantity, quality, and distribution of food. On a 1,953-ha island in southwestern Washington, home ranges averaged 505 ha for males and 235 ha for females (Lindzey and Meslow, 1977a). In west-central Idaho, home ranges of adult males averaged 112.1 km² and those of females averaged 48.9 km² (Amstrup and Beecham, 1976), 20 times larger than in the lush coastal forest of Washington. These two assessments seem to represent extremes in home-range size for the species in western United States. At least part of the difference may be related to differences in delineating home ranges; investigators in the former study circumscribed home ranges by connecting outermost locations (a subjective technique), whereas investigators in the latter study used the minimum-area method. In southwestern Oregon, six marked bears were retrapped, resighted, or recovered 0–7.2 km from the site of marking (McCollum, 1973).

McCollum (1973) reported a strong tendency for nuisance black bears to "home" when transplanted to areas <55.5 km away. Four bears moved 10.5–32.0 km returned home; of 14 moved 48.3–64.2 km, eight returned home, five did not return, and one was not relocated; and of three moved 64.4–80.3 km, one returned home and two were not relocated.

Home ranges of radio-tracked bears of both sexes commonly overlap extensively (Amstrup and Beecham, 1976; Lindzey and Meslow, 1977a). Black bears commonly exhibit considerable intraspecific tolerance even at places where they congregate, such as garbage dumps or patches of huckleberries. Except for females with young, which exhibit the least tolerance of conspecifics and are agonistic most often, individuals may forage amicably within 2–8 m of each other (Herrero, 1983). At a garbage dump in Alberta, 79% of 131 dominant interactions were instigated by females with young and 61% of these were directed toward adult males or subadult bears (Herrero, 1983). Most agonistic behaviors involved distinct displays; actual contact between bears rarely was observed.

In spring, black bears commonly strip the bark from second-growth Douglas-fir and western hemlock trees ≅13–30 cm in diameter. Usually the damage is restricted to the height that the bear can reach, but sometimes bears climb trees and strip bark from trunks of trees ≥10 m above the ground. After the coarse bark is peeled away with claws and teeth, the cambium is scraped and licked to obtain nutrients (Scheffer, 1952).

Remarks.—The black bear is a game mammal in Oregon, with an open season of about 90 days statewide in autumn and a much shorter season in restricted areas in spring. The objective of the spring hunting season is largely to reduce stripping of bark in forest plantings where damage is extensive. Landowners whose property is damaged may kill bears thought to be responsible. In autumn, hunting pressure from 1975 to 1993 ranged over a 2-fold magnitude and the kill per unit effort ranged over a span of nearly 4-fold (Table 12-7). We do not know if this represents changes in numbers of black bears or is related to some social phenomenon. During the early 1990s, considerable public sentiment against hunting of bears was reported by the local news media in Oregon. On 8 December 1994, regulations derived from a referendum passed a month earlier prohibiting the use of baits or dogs in the taking of black bears became effective. We expect numbers of bears reported taken by hunters to decline and the number of complaints regarding damage by black bears to increase.

Noble (1994) suggested that planting grasses, clover (*Trifolium repens*), and trailing blackberry (*Rubus ursinus*)

Table 12-7.—*Number of hunting permits sold, number of hunters, number killed, and number killed per unit effort during the autumn hunting season for black bears* (Ursus americanus), *and other harvest of black bears in Oregon, 1975–1993. Data provided by Oregon Department of Fish and Wildlife.*

Year	Permits sold	Number hunters	Number killed	Hunter-days	Harvest/1000 hunter-days	Other harvest[a]
1975	17,924	16,247	1,841	148,092	12.4	
1976	14,660	11,043	1,074	102,557	10.5	
1977	15,847	12,883	920	133,570	6.9	
1978[b]	8,770		506			
1979	15,705	11,324	812	118,338	6.9	
1980[b]	14,762	11,072	958			
1981	15,503	10,124	783	113,722	6.9	
1982	21,586	16,756	1,313	196,713	6.8	
1983[b]	25,474	20,500	1,420			132
1984[b]	26,753		1,350[c]			98
1985[b]	25,863		1,250[c]			61
1986	25,928	20,748	1,376	239,346	5.7	136
1987	25,496	17,666	954	202,879	4.7	64
1988	20,771	15,920	803	169,335	4.7	161
1989	19,467	16,781	664	202,125	3.3	191
1990	20,375	17,080	888	217,459	4.1	158
1991	12,020[d]	9,569	1,172	117,616	10.0	203
1992	16,573	11,882	805	160,004	5.0	220
1993	17,190	13,749	1,179	183,666	6.4	134

[a]Number of bears killed during permit-only hunts in spring in areas where damage to forest plantings by bears is reported. Includes nuisance bears killed by landowners or those killed by personnel of the U.S. Department of Agriculture, Animal and Plant Health Inspection Service.
[b]No surveys conducted.
[c]Estimated from hunter reports.
[d]Deadline for tag sales advanced 36 days.

on skid roads, log landings, and other sites at which tree seedlings often do not survive well and the protection of natural stands of devil's club (*Oplopanax horridum*), salmonberry (*Rubus spectabilis*), and cow-parsnip (*Heracleum lanatum*) might be a means of reducing damage to Douglas-fir trees by bears. Oregon Department of Fish and Wildlife biologists who prepared *Oregon's black bear management plan 1993–1998* claimed that artificial feeding of bears to reduce damage to trees was effective in Washington and in Oregon. However, they indicated that feeding can be expensive, and it has the potential of attracting additional bears and increasing productivity and survival of bears in the problem area. We remain skeptical of the effectiveness of the technique in reducing damage to trees.

Ursus arctos Linnaeus
Grizzly or Brown Bear

1758. [*Ursus*] *arctos* Linnaeus, 1:47.
1815. *Ursus horribilis* Ord, 2:291.
1914. *Ursus klamathensis* Merriam, 27:185.
1918. *Ursus mirus* Merriam, 41:40.
1918. *Ursus idahoensis* Merriam, 41:54.
1963. *U*[*rsus*]. *a*[*rctos*]. *horribilis* Rausch, 41:33.

Description.—The grizzly bear (Plate XXII) is the largest terrestrially adapted carnivore among the Recent mammal fauna of North America (Table 12-6). Specimens from present-day populations average ≅135 kg (females)–≅200 kg (males); maximum weights for the two sexes are ≅280 and ≅500 kg, respectively (Jonkel, 1987). Nevertheless, there is considerable geographic variation in size (Kurtén and Anderson, 1980).

Externally, the grizzly bear is characterized by its heavy body, stout legs of about equal length, large feet, relatively small head, small eyes, short ears, and extremely short tail. A "hump" over the shoulders, the result of the mass of scapular muscles, is an identifying feature. The claws on the forefeet are exceptionally long and much less recurved

than in the black bear (*Ursus americanus*). Grizzly bears exhibit considerable variation in the color of their pelage; various shades of brown predominate, but some, or parts of some, may be nearly black, almost grayish, or somewhat yellowish (Storer and Tevis, 1955).

The skull of the grizzly bear (Fig. 12-14) is massive; the sagittal and lambdoidal crests are extensive and rugose. A broad area in the nasofrontal region is depressed, giving the grizzly bear a "dished face." M2 is >1.5 times as long as M1.

Distribution.—The historical range of *U. arctos* in North America included all of Alaska (except some islands) and Yukon, and all but the southeastern portion of Northwest Territories, then extended southward through much of the prairie provinces and the Great Plains into central Mexico and west to the Pacific Ocean (Fig. 12-15). There is some evidence that earlier the range may have extended

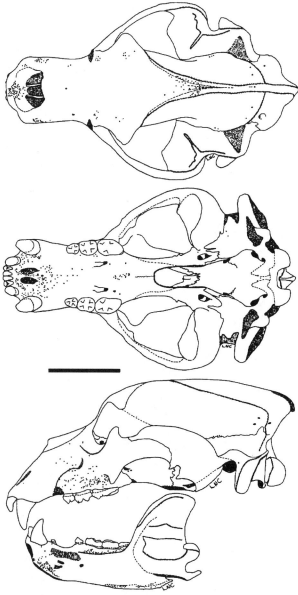

Fig. 12-14. Dorsal, ventral, and lateral views of the cranium, and lateral view of the dentary of the grizzly bear, *Ursus arctos* (OSUFW 3640). Scale bar equals 10 cm.

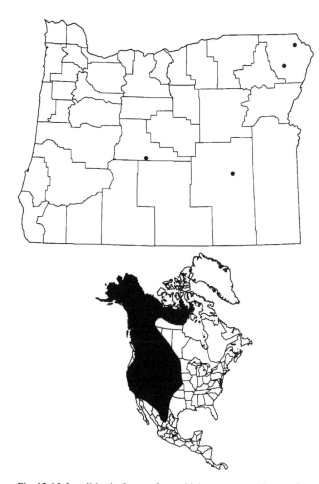

Fig. 12-15. Localities in Oregon from which museum specimens of the grizzly bear, *Ursus arctos*, were examined (listed in APPENDIX, p. 578), and the former distribution of the grizzly bear in North America. (Range map redrawn from Craighead and Mitchell, 1982:515, fig. 25.1.) The grizzly bear was extirpated from much of its original range (Craighead and Mitchell, 1982; Jonkel, 1987).

through parts of eastern North America. The present-day range in North America includes the same regions of Alaska, Yukon, and Northwest Territories, but southward includes most of British Columbia, western Alberta, northeastern Washington, extreme northern Idaho, western Montana, and northwestern Wyoming (Craighead and Mitchell, 1982; Jonkel, 1987). The historical range also covered much of Eurasia, but populations in some regions have been extirpated (Pasitschniak-Arts, 1993).

The grizzly bear has been extirpated from Oregon (Olterman and Verts, 1972); the last verified specimen was taken on Chesnimnus Creek in Wallowa Co. on 14 September 1931 (USNM 250124). Based on written and oral records, grizzly bears seemingly were widespread in Oregon before European settlement; they occurred in the Cascade Range; the Siskiyou, Blue, Steens, and Wallowa mountains; the Klamath, Rogue, Umpqua, and Willamette valleys; and some of the high desert country south and east of Bend (Bailey, 1936). Nevertheless, only a few specimens were deposited in systematic collections, and these are from Deschutes, Harney, and Wallowa counties (Fig. 12-15).

Ursus arctos probably arose from the mid-Pleistocene *Ursus †etruscus* of Europe. In the late Pleistocene *U. arctos* occurred in Alaska in sympatry with the now-extinct giant

short-faced bear (*†Arctodus simus*) that was widespread in North America. After the last glacial period (Wisconsinan), *U. arctos* extended its range into and far beyond its historic range (Kurtén and Anderson, 1980). Late Pleistocene fossils are known from West Virginia, Idaho, Wyoming, and Ohio; Holocene-age fossils are known from Idaho, Illinois, California, and Ontario (Kurtén and Anderson, 1980).

Geographic Variation.—Only one of the two currently recognized North American subspecies occurred in Oregon: *U. a. horribilis* (Rausch, 1963).

Habitat and Density.—The grizzly bear is a species of the rugged, largely inaccessible, spacious wilderness areas where either by law or by remoteness its populations are protected from overexploitation by humans (Craighead and Mitchell, 1982). Within such areas, the requirements for food, dens, and cover may be met by a variety of vegetative and physiographic types. A variety of forest types, alpine and subalpine communities, early seral communities in riparian or burned-over areas, grasslands, and shrublands appropriately intermixed and interconnected with travel lanes serve as habitat for these wide-ranging mammals. Succulent, high-protein, and high-energy foods; dry dens not subject to being disturbed by humans; and sufficient vegetation to serve as a shield from human view seem to be habitat requisites for grizzly bears (Craighead and Mitchell, 1982; Jonkel, 1987). Thus, precisely how communities provide these elements (which plant and animal species are involved) does not seem as critical as the elements themselves. Obviously, grizzly bears can persist in parklike habitats that lack some of these elements and in the face of considerable interactions with humans, but not without undesirable behavioral changes.

We found no estimates of numbers of grizzly bears in Oregon at any time before their extirpation, but from Bailey's (1936) anecdotal accounts they must have been scarce by the beginning of the present century. Within the conterminous states, estimated densities of one bear/25 km^2 and one bear/80.3 km^2 were recorded in relatively small areas during the last quarter century (Craighead and Mitchell, 1982; Jonkel, 1987). Its populations are considered threatened in this region (Wozencraft, 1993).

Diet.—Grizzly bears consume a wide variety of vegetable and animal materials: grasses, leaves, stems, shoots, roots, tubers, berries, nuts, acorns, cones, fungi, eggs, insects and their larvae, fish, small mammals, and ungulates both as prey and as carrion (Pasitschniak-Arts, 1993). Foods consumed vary both temporally and spatially, and with availability (Jonkel, 1987), but in some regions at least, about 50–60% of the diet consists of animal material (Craighead and Mitchell, 1982). Grizzly bears, although extremely powerful and capable of surprisingly rapid movements, lack the speed to be efficient predators of healthy, vigorous ungulate species. Consequently, ungulates preyed upon are largely the very young; those debilitated by age, disease, or starvation; or those handicapped by deep snow (Jonkel, 1987). Some large prey may be ambushed (Pasitschniak-Arts, 1993).

Grizzly bears are quick to take advantage of abundant sources of high-energy and high-protein foods. Ungulate carcasses on winter ranges, spawning salmonids, crops of pine (*Pinus*) nuts, berry patches, and garbage dumps often attract numbers of grizzly bears; in such situations, a social hierarchy is established to permit sharing of a food source (Jonkel, 1987).

Reproduction, Ontogeny, and Mortality.—Female grizzly bears commonly enter a short estrous period for the first time when they are 3.5 years old; although they may copulate frequently, they do not become pregnant. Older females sometimes have estrous periods lasting 3–4 weeks. The mating system is polygamous; although dominant males may attempt to defend females, females commonly mate with more than one male. First litters usually are produced when the female is ≥5.5 years old; subsequent litters are produced at intervals of 2–3 or more years, with older females exhibiting the longer intervals (Craighead and Mitchell, 1982).

Mating usually is in late spring or early summer. The fertilized ova develop to the blastocyst stage, then remain unimplanted in the lumen of the uterus for ≅5 months until the female enters winter dormancy. Implantation is in midautumn followed by an active period of embryonic development of only 6–8 weeks (Craighead and Mitchell, 1982; Pasitschniak-Arts, 1993). The one–four (usually two–three) young are born while the female is dormant in its winter den; the young weigh 340–680 g and are naked and blind (Nowak, 1991). Young weigh ≅15 kg by 3 months of age. Young are permitted to nurse for 1.5–2.5 years; they remain with the maternal female for 2–3 years, sometimes longer (Pasitschniak-Arts, 1993). Young usually are defended vigorously.

Other than humans, grizzly bears have no predators. Loss of habitat through encroachment of humans and illegal hunting are the most significant inimical factors. In unexploited populations, these bears have a potential longevity of 20–30 years (Pasitschniak-Arts, 1993).

Habits.—During spring and autumn, grizzly bears tend to be active at all hours, but during summer, they usually are inactive during the daylight hours. Grizzly bears are active for 5–9 months depending on the latitude, food supply, and snow depth; the remaining 3–7 months they are dormant. During summer and autumn, grizzly bears gorge on high-energy foods and accumulate an extensive layer of fat to serve as insulation and as a source of energy and water during dormancy. About 2–8 weeks before entering dormancy, they usually dig their own dens and line them with vegetation; occasionally they use natural cavities (Pasitschniak-Arts, 1993). In contrast to some small mammals (e.g., *Tamias*), during dormancy the bears do not feed, urinate, or defecate, and their body temperature declines by no more than ≅5°C (Craighead and Mitchell, 1982; Pasitschniak-Arts, 1993). The heart rate in these bears during sleep becomes lower in autumn, from ≅40 beats/min in October to ≅8–10 beats/min in November (Folk et al., 1972). During the period of dormancy, males lose ≅22% of their autumn body mass, whereas females lose ≅40%. Nevertheless, following arousal in spring, these bears continue to fast and to rely on body reserves for 10–14 days before returning to the usual activity mode.

Grizzly bears are capable of extensive movements through a variety of habitat types; thus, the size of their home range is dependent upon the abundance and distribution of food; age, sex, social status, and body condition of the individual; season; and topography (Craighead and Mitchell, 1982; Pasitschniak-Arts, 1993). In various parts of the geographic range, average annual home-range size (minimum-area polygon) in various ecosystems ranged from 24 to 1,398 km^2 for males and from 12 to 430 km^2 for females (LeFranc et al., 1987).

Intraspecific communication in grizzly bears largely is by olfaction although visual and auditory cues sometimes play a role. Grizzly bears may communicate by body orientation, "chuffing" (a vocal sound), marking trees by removing bark, and depositing urine and feces.

We likely would be considered to be remiss if we did not comment on the potential threat that grizzly bears pose to livestock, and occasionally to humans. At localities where the two lived in contact, grizzly bears often took livestock. Storer and Tevis (1955) provide numerous accounts of such interactions in California; we can think of no reason that they should not have occurred in Oregon during pioneer times. At the present time, fewer than 1,000 grizzly bears reside in the conterminous United States; they certainly do not constitute much of a threat to the livestock industry. Conversely, some of the remaining grizzly bears live in localities where the potential for contact with humans (often naive humans) is considerably greater. The grizzly bear is dangerous and unpredictable (especially females with young), and numerous accounts of humans being killed or maimed by them have been chronicled (Nowak, 1991; Storer and Tevis, 1955).

Remarks.—Without doubt, no other North American mammals have had such a bizarre taxonomic history as the brown and grizzly bears. Storer and Tevis (1955) provided a resumé of the sequence of taxonomic changes from the original application of a name to the species in North America by Ord in 1815, through the application of species names to a myriad of morphological variants by Merriam (1918), to the synonymizing of the New World and Old World species by Rausch (1953) under the epithet coined by Linnaeus nearly 200 years earlier.

As illustration of the unholy taxonomic mess the grizzly bear had become, Kurtén and Anderson (1980) claimed that 232 Recent and 39 fossil "species" and "subspecies" names had been applied to the taxon. Wozencraft (1993) listed 151 synonyms for *arctos*; Pasitschniak-Arts (1993) listed 99. Hall (1981) and Hall and Kelson (1959), doubtlessly the most commonly referenced taxonomic treatments of North American mammals, listed 77 species of brown and grizzly bears; 73 of the 77 were named by Merriam. The same authorities illustrated the distribution of type localities for each North American species and for the named subspecies of four species recognized by Merriam. In several instances, type localities of two or more of the taxa occur within a distance likely less than the diameter of the home range of one individual and certainly within the gene pool of one population. Simpson (1945) suggested that by following Merriam's (1918) concept of the species as it applied to bears it was possible for twin bears to be of different species!

Rausch (1963) recognized one Holarctic species, *Ursus arctos*, with two extant subspecies in North America: *U. a. middendorffi* restricted to three islands off the coast of Alaska and *U. a. horribilis* occupying the remainder of the range. Hall (1984) recognized nine subspecies in the New World. *U. a. middendorffii* commonly is referred to as the brown or Kodiak bear and *U. a. horribilis* as the grizzly bear; sometimes the name brown bear is applied to all members of the taxon. Nevertheless, the variety of vernacular names that has been applied to the species probably is responsible for some of the "taxonomic" confusion, at least among hunters and the lay public. Jonkel (1987) listed 12 vernacular names and indicated that others were in use.

FAMILY OTARIIDAE—EARED SEALS

The family Otariidae consists of seven Recent genera and 14 species distributed along coastal regions of northeastern Asia, North America, South America, southern Africa, southern Australia, New Zealand, and numerous oceanic islands (Nowak, 1991; Wozencraft, 1993). Two species, the California sea lion (*Zalophus californianus*) and the northern sea lion (*Eumetopias jubatus*) are common along the coast of Oregon. A third species, the northern fur seal (*Callorhinus ursinus*), except when on Arctic breeding grounds, is primarily pelagic, but on rare occasions comes ashore along the Oregon coast, especially if sick or injured (Scheffer, 1950a).

Otariids are characterized by hind flippers that can be turned forward beneath the body to facilitate terrestrial locomotion, small cartilaginous pinnae, nails on three digits of each rear flipper, I1 and I2 with a transverse groove on their occlusal surfaces, bullae little inflated, postorbital processes on the frontal bones, and in males, the presence of a scrotum. A tail is present; males possess a baculum. Otariids are highly adapted to the marine environment: the body is fusiform; both front and rear limbs are modified into large flippers with the digits enclosed in a web; and all teeth except the incisors are conical, pointed, and recurved to aid in the capture of the slick fish and squid that serve as their primary food. The pelage of sea lions is coarse with sparse underfur, whereas that of fur seals is thick and of fine quality; both have solid-colored pelage, without stripes or splotches. In sea lions the rostrum is rather blunt, but in fur seals it is decidedly pointed.

Otariids are exceptionally fast and maneuverable swimmers; speeds of 17 km/h for California sea lions and 26 km/h for northern fur seals were recorded (Walker et al., 1964). On land, they move ponderously and laboriously despite the reversible hind limbs. California sea lions are sometimes trained to perform in circuses and shows.

Otariids form large breeding colonies; females produce offspring in late spring shortly after returning to the breeding grounds. One young usually is produced; twinning is rare and one of the pair almost always does not survive. Both sexes attain sexual maturity at 3–9 years depending on the species (Nowak, 1991), but the success of adult males in forming and maintaining harems is related to their age and size. Rarely are males <10 years old able to maintain a breeding territory. Breeding usually occurs within a few days of parturition, but in some species at least, the blastocyst undergoes a delay in implantation of several months (Pitcher and Calkins, 1981). Of the otariids, only the northern sea lion breeds along the coast of Oregon.

The earliest otariids were small fur seals of mid-Miocene age; in the late Miocene, a line that led to *Callorhinus* (northern fur seal) diverged from the main otariid line in the North Pacific Ocean (Repenning and Tedford, 1977). The sea lions arose in the North Pacific Ocean from the *Arctocephalus* (southern fur seals) line in the late Pliocene or earliest Pleistocene (Repenning and Tedford, 1977).

KEY TO THE OTARIIDAE OF OREGON

1a. Underfur soft; nose tapering to an acute point; palate deeply notched posteriorly; combined width of nasals anteriorly 80–90% of their length.................................... ....*Callorhinus ursinus*, northern fur seal (p. 381).
1b. No underfur; nose more or less blunt; palate lacking deep notch on posterior margin; combined width of nasals anteriorly ≤75% of their length.........................2
2a. Males with prominent sagittal crest producing stepped forehead; zygomatic breadth <180 mm among males, <130 mm among females; postorbital processes rounded on anterior edge; P4 and M1 separated by space less than the length of a cheektooth.................. ....*Zalophus californianus*, California sea lion (p. 388).
2b. Males lacking a prominent crest on forehead; zygomatic breadth >220 mm among males, >145 mm among females; postorbital processes squarish; P4 and M1 separated by space more than the length of a cheektooth.. ....*Eumetopias jubatus*, northern sea lion (p. 385).

Callorhinus ursinus (Linnaeus)
Northern Fur Seal

1758. [*Phoca*] *ursina* Linnaeus, 1:37.
1859. *Callorhinus ursinus*: Gray, 1859:359.
1898. *Callorhinus alascanus* ?? *alascensis* Jordan and Clark, pt. 1, p. 45.

Description.—The northern fur seal (Plate XXIII) is the smallest member of the family in Oregon waters (Table 12-8). The species exhibits extreme sexual dimorphism; adult males weigh 181–272 kg, adult females only 43–50 kg (Nowak, 1991). The flippers are nearly twice as large as those of a comparably sized individual of any other species of otariid. The ears are cylindrical and the auditory meatus is covered with wax to prevent entry of water (Fiscus, 1978). The rostrum is short and slightly downcurved. The facial area is short and rather broad, but tapers abruptly to a pointed nose with little step between forehead and muzzle.

The pelage consists of coarse guard hairs and an exceedingly thick underfur with nearly 60,000 hairs/cm² that serves as waterproof insulation (Scheffer, 1962). Adult males are dark gray or brown dorsally and reddish brown ventrally; a grayish cape covers the neck and shoulders. Adult females are grayish brown dorsally and reddish brown ventrally (Nowak, 1991). On rookeries, the pelage quickly becomes stained yellowish brown by mud and fecal material. The vibrissae in adults are white (Kenyon and Wilke, 1953).

The skull of the northern fur seal (Fig. 12-16) is short with triangular postorbital processes, a deeply notched palate, and six postcanine teeth. The combined width of the nasals anteriorly is equal to 80–90% of their length. I3 is much larger than other incisors, about half as large as the canine.

Distribution.—The northern fur seal is restricted largely to the waters and some insular areas of the North Pacific region, where its range extends from the Pribilof and Commander islands in the Bering Sea south to the Channel Islands off the southern California coast (about 33°N latitude) and along the coasts of Japan and Russia to about 35°N latitude (Fig. 12-17; Kenyon and Wilke, 1953; Nowak, 1991). In Oregon, most museum specimens were collected along the northern coast (Fig. 12-17).

The earliest fossil record of *C. ursinus* likely is of late

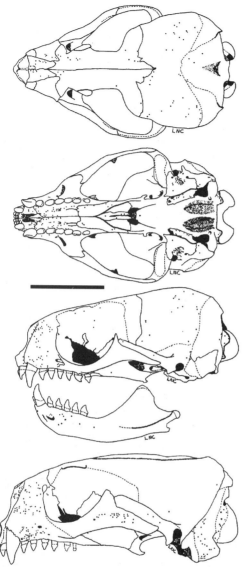

Fig. 12-16. Dorsal, ventral, and lateral views of the cranium, and lateral view of the dentary of a female northern fur seal, *Callorhinus ursinus* (OSUFW 1284) and lateral view of the cranium of a male (OSUFW 1284). Scale bar equals 60 mm.

Table 12-8.—*Means (±SE), ranges (in parentheses), and CVs of measurements of skull and skin[*] characters for female and male pinniped carnivores from Oregon. Skin characters were recorded from specimen tags; skull characters <150 mm were measured to the nearest 0.01 mm, larger characters <150 mm were measured to the nearest 1 mm. SE and CV not provided if n <10.*

Species	Sex	n	Total length	Tail length	Ear length	Mass (kg)	Condylobasal length	Basilar length	Postglenoid length	Length of maxillary toothrow	Zygomatic breadth	Cranial breadth	Least interorbital breadth	Length of mandible
Callorhinus ursinus	♀	3	1,180	57			170.0 (160–190)	152.8 (141.46–173.0)	60.48	45.43 (45.0–45.85)	98.31 (92.0–107.0)	89.95 (83.0–95.0)	21.29 (18.34–23.54)	106.19 (96.0–120.0)
	♂	1	2,050	120			247.0	224.0	74.00	71.0	146.0	98.0	45.0	173.0
Eumetopias jubatus	♀	5	2,247[b] (2,180–2,350)	125[c] (110–140)	28	206.7[a] (113.4–300.0)	306.8 (300–312)	276.8 (271–282)	88.88 (80.4–99.0)	108.56 (106–116)	169.82 (165.0–175.0)	123.52 (121.0–126.0)	62.39 (61.0–64.0)	222.64[d] (214.18–229.0)
	♂	5	3,353			499.5[c] (453.6–545.4)	361.92[d] (345–374)	331.2[d] (301–353)	104.68[d] (98.0–109.7)	129.39[d] (126.0–134.56)	217.98 (203.0–234.0)	133.54 (125.0–139.0)	89.82 (83.0–96.0)	278.06 (266.0–291.24)
Phoca vitulina	♀	5	1,464[b] (1,360–1,520)			65.15[b] (45.45–75.0)	196.9 (180–210)	178.1 (161–190)	73.13 (71.0–75.0)	46.62 (37.25–57.21)	116.67 (108.2–124.5)	91.79 (87.8–95.7)	11.36 (9.65–13.96)	128.02 (116.07–135.97)
	♂	2	1,549	279		54.4	211.5 (210–213)	185.50 (185–186)	73.00 (71.0–75.0)	59.5 (57–62)	131.00 (126.0–136.0)	97.50 (96.0–99.0)	13.64 (13.0–14.28)	138.76 (138.0–139.51)
Zalophus californianus	♂	14	2,267 ± 29.5[c] (2,159–2,440) 4.1	121[f] (115–130)	38[g] (35–40)	269.8[h] (191.7–453.6)	277.2 ± 2.08 (261–293) 2.81	249.4 ± 2.19 (232–264) 3.28	82.57 ± 0.71 (79.0–88.0) 3.21	81.86 ± 1.86 (72–95) 8.52	158.71 ± 2.65 (144.0–175.0) 6.25	105.86 ± 1.07 (101.0–114.0) 3.79	49.14 ± 0.86 (44.0–54.0) 6.57	200.79 ± 2.35 (185.0–217.0) 4.39

[*]Although measures of hindfoot length were recorded, the data were not included because of extreme variation believed to stem from differences in techniques used to measure these characters. [b]Sample size reduced by 2. [c]Sample size reduced by 3. [d]Sample size reduced by 4. [e]Sample size reduced by 10. [f]Sample size reduced by 12. [g]Sample size reduced by 5.

Miocene age (Repenning and Tedford, 1977).

Geographic Variation.—The species is considered to be monotypic.

Habitat and Density.—The northern fur seal usually is pelagic in coastal waters of Oregon, remaining 16–145 km offshore (Baker et al., 1970). Occasionally, one comes ashore, especially if sick or injured. In winter 1949–1950, after an especially severe storm with temperatures far below normal, numerous emaciated young died at sea and were washed ashore along the coasts of Oregon and Washington. Scheffer (1950a) examined 19 from Oregon and 10 from Washington; he estimated from the proportion of the sample marked before the young left the breeding grounds that ≅700 carcasses had come ashore from Elk River south of Port Orford, Curry Co., Oregon to Copalis, Grays Harbor Co., Washington.

Numbers of northern fur seals that winter off the coast of Oregon vary seasonally and with the availability of prey. No doubt, overall numbers in the population directly affect numbers that occur at times off the coast of Oregon (see Remarks). Antonelis and Perez (1984), based on a total population of 340,000 adult females in the Pribilof Islands, estimated that numbers off the coast of Oregon peaked in April at ≅45,500. Nevertheless, few Oregonians are aware of the fur seals that occur at times off the coast.

Diet.—In a summary of reports of foods eaten by 672 *C. ursinus* obtained from Washington to southeastern

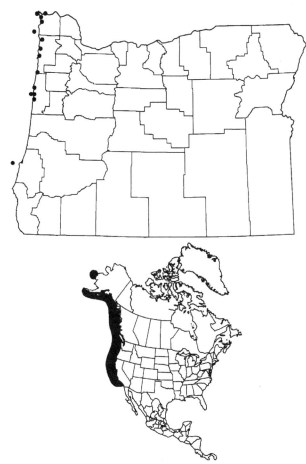

Fig. 12-17. Localities along the coast of Oregon from which museum specimens of the northern fur seal, *Callorhinus ursinus*, were examined (listed in APPENDIX, p. 512), and the distribution of the northern fur seal in North American waters. (Range map redrawn from Ronald et al., 1982:801, fig. 40.28.)

Alaska, Scheffer (1950*b*) found 72.4% of the contents consisted of herring (*Clupea pallasi*) with an additional 6.2% probably of herring. Salmon (*Oncorhynchus*) composed 6.7% and eulachon (*Thaleichthys pacificus*) 2.3%; none of the remaining identified material composed as much as 2.0%. Of 1,295 seal stomachs examined that contained food (an unknown number were empty), 307 contained squids (*Gonatus fabricii* and *Loligo opalescens*), 256 pollack (*Theragra chalcogramma*), 221 herring, 156 crustaceans, and 107 deep-sea smelts (*Bathylagus*). Only 52 contained salmon (Scheffer, 1950*b*). Only herring and squid were represented in the two specimens taken along the Oregon coast (Scheffer, 1950*b*). However, Baker et al. (1970:9) contended that "Anchovy, hake, squid, saury, and rockfish are principal foods off California and Oregon."

Of 374 fur seals taken off the coast of Washington during winter 1967–1968, 251 contained food remains in their stomachs; 60 (23.9% of those with food remains) contained salmon, which composed nearly one-third of the total volume of food in all the stomachs. Northern anchovies (*Engraulis mordax*) occurred in 13.9% of the stomachs and composed about 16% of the volume of food; capelin (*Mallotus villosus*) occurred in 13.1% and composed nearly 10%; eulachon occurred in 9.6% and composed 11.6%; and rockfish (*Sebastes*) occurred in 4.0% and composed 14.3% (National Marine Fisheries Service, 1970).

The total annual food consumption of northern fur seals while along the coasts of Oregon and Washington (December–June) was estimated to be 33.6 million kilograms, 76% of which was fish and 24% squid. The intake, in millions of kilograms, included herring, 5.9; rockfish, 5.5; anchovies, 4.0; and squid, 3.8 (Antonelis and Perez, 1984).

Reproduction, Ontogeny, and Mortality.—Adult male northern fur seals, heavy with fat after feeding at sea all winter, arrive on rookeries in May–June. The primary rookeries are on the Pribilof (United States), Commander, Robben, and Kuril (Russia) islands in the Bering Sea and Sea of Okhotsk. A new rookery was formed in 1968 on San Miguel Island in the Channel Islands off the southern California Coast (Baker et al., 1970), and another was formed on Bogoslof Island in the southern Bering Sea in 1976 (Loughlin et al., 1994). One young was born on the coast of Washington in 1959 and another along the coast of northern California in 1983 (Baker et al., 1970; Nowak, 1991). Although possibly newsworthy, the latter records likely are of little biological importance.

A portion of those males ≥10 years old are able to establish and maintain territories on the rookery sites; the remainder ultimately join younger males on hauling grounds at some distance from the rookeries if unsuccessful in attempts to displace males that have become exhausted through defense of their territories (Baker et al., 1970). Females commence to arrive on the rookeries in late June; those ≥3–5 years old that have attained sexual maturity seek a specific location on the rookery. Characteristics that determine qualities of localities sought by females are not completely clear, although proximity to the ocean is involved (Baker et al., 1970). Adult males that have prime territories thus acquire the largest harems, sometimes ≥100 females. Although territorial males guard females in their harems from interlopers and males on adjacent territories, sometimes engaging in ferocious battles, they cannot prevent a female from leaving their territory (Baker et al., 1970).

Pregnant females give birth usually within 1 day of arriving on the rookery, then mate again, usually with a territorial male, during the following 5 days (Baker et al., 1970). Within a week after parturition, the female leaves the young ashore and spends 5–14 days at sea feeding. Upon returning, the female identifies its offspring by location, sight, and smell. After nursing the young, the female returns to the sea for another long feeding bout; the sequence is repeated until the young is weaned at 3–4 months of age. About 60% of females >4 years old give birth each year, but the pregacy rate of those in their prime (8–13 years old) is ≅90% (Baker et al., 1970).

Following fertilization, embryonic development proceeds to the blastocyst stage then ceases; the blastocyst remains free in the lumen of the uterus until November, when it implants and development proceeds. Thus, the gestation period, as measured from copulation to parturition, is a few days shy of 1 year. Twinning is rare, and of those recorded only one of the pair lived to weaning (Fiscus, 1978). During embryonic development the deciduous teeth erupt, then are shed before birth; the shed teeth with partly resorbed roots often can be found in the amniotic fluid (Kubota and Matsumoto, 1963).

At birth, females average ≅4.5 kg and males average ≅5.5 kg (Fiscus, 1978). The young have open eyes, are precocious, and, except for the flippers and nose pad, are covered with a black pelage. They can swim at birth, but usually do not commence to enter the water until about 4 weeks old; they become accomplished swimmers by the time that they acquire a new gray pelage and leave the rookery in November. Despite the long intervals between feedings, the young are dependent essentially entirely on milk of the maternal female; to make this possible, the milk is composed of ≅46% fat and the juvenile stomach occupies much of the body cavity (Wilke, 1958). Weaning is abrupt and young must learn to catch fish and squid quickly; often young do not gain, and sometimes they lose, body mass for several months after leaving the rookery.

Malnutrition (including starvation related to rejection by maternal females), internal parasites, inadvertent injuries by adults involved in territorial defense and mating, and bacterial infections are primary causes of death of young on the rookeries (Baker et al., 1970). Killer whales (Orcinus orca), sharks (Lamniformes), and northern sea lions (Eumetopias jubatus) probably take a toll at sea, but severe winter storms probably are the greatest hazard to juveniles. Contamination of the fur with oil causes loss of its insulating properties, leading to hypothermia and death. Some fur seals become entangled in nets, leading to their demise. Finally, former harvests for fur, oil, and seal meal (Riley, 1967) reduced numbers and perhaps reproductive potential (see Remarks).

Habits.—Between leaving the rookeries in November and arriving thereon the following May–June, northern fur seals remain in the open sea (Kenyon and Wilke, 1953). Although the seals usually occur alone or in small groups, loose groups of as many as 100 have been observed (Baker et al., 1970). By late November, females and subadult males appear off the panhandle of Alaska; in December, many often occur off the coast of Washington; and by January, they are scattered from southeastern Alaska to southern California. Young-of-the-year occur from southeastern Alaska to Washington in January, and some continue southward and occur off the coast of California in March–May (Fiscus, 1978; Kenyon and Wilke, 1953). Most adult males remain in the northern latitudes (Antonelis and Perez, 1984). Fur seals that breed on the subarctic rookeries commence the northward migration from California waters in March, pass Oregon and Washington in March–June, and many occur in the Gulf of Alaska in April–June. Most arrive on the rookeries in June–July. A similar pattern of migration occurs in the western Pacific region, with fur seals wintering in the Pacific Ocean along Honshu, Japan, and in the Sea of Japan as far south as the South Korean island of Cheju Do (Panin and Panina, 1968). Some intermixing of herds occurs as individuals originating from both Pribilof and Commander island rookeries occur in wintering herds on both sides of the North Pacific rim (Baker et al., 1970). Fur seals that breed on the Channel Islands, California, are not believed to migrate (Fiscus, 1978).

Northern fur seals feed mostly at night and during crepuscular hours because their prey migrates toward the surface at dark. They mostly rest and sleep on the surface during the day, although they may feed if prey is available. They commonly sleep on their backs with both hind flippers and one front flipper touching to form an arch over the supine body, a position commonly referred to as the "jug handle." Only the nose and the three flippers are above water (Fiscus, 1978). Some individuals make relatively shallow dives (50–60 m), others make deep dives (≅175 m, maximum 207 m), and still others make both shallow and deep dives (Nowak, 1991). Small prey is swallowed whole; larger fish are torn apart by shaking.

Remarks.—The saga of the exploitation of the fur seal resource appears as alternating periods of wantonness and remorseful conservatism combined with questionable management decisions. It was not until Gavrilo L. Pribylov discovered the rookery on the island that he named St. George (Pribilof Islands) in 1786 that the concentration of ≅2.5 million fur seals became known (Riley, 1967). During the following 30 years probably >1.6 million seals were killed, causing the Russians to restrict the kill to <10,000 annually and prohibit the killing of females commencing in 1834. From discovery to 1868, when the Pribilof Islands became a United States possession as part of the Alaska Purchase, probably >2.5 million fur seals had been killed on the rookery. During the following 2 years, >225,000 were killed before the United States provided protection for the seals. Nevertheless, commencing the following year, the right to take seals from the islands was leased to private companies by the federal government. Quotas were established in 1874, but by 1889, another 2 million had been taken. Although the kill on the islands was regulated to some degree, the wasteful pelagic sealing activities (many seals killed in the sea sink) were not. Many of those taken at sea were pregnant females or females whose young waited on the rookery to be nursed; thus, for each female killed two seals died. More than 1 million seal skins were obtained at sea between 1868 and 1909 (Riley, 1967). International arbitration to prevent pelagic sealing was unsuccessful. When the last lease expired in 1909, fewer than 150,000 fur seals remained. Fur seals in Alaskan waters were in dire trouble.

Finally, Great Britain, Japan, Russia, and the United States adopted the North Pacific Fur Seal Convention of 1911 that prohibited pelagic sealing except by aborigines with primitive weapons. Each participating country was allocated a proportion of the annual harvest taken on the rookeries. The convention remained in effect until the start of World War II in the Pacific theater; a provisional agreement protected the fur seals until 1957 when a new convention with provisions similar to those of the original was agreed upon by Canada, the USSR, Japan, and the United States (Riley, 1967). Between adoption of the original convention in 1911 and 1965, the fur seal population increased to ≅1.5 million (Riley, 1967); from 1943 to 1955, the harvest averaged ≅70,000 annually. Based on the prediction of greater productivity and survival at lower population levels, ≅300,000 females were harvested on the Pribilof Islands in the period 1956–1968, in addition to a pelagic harvest of ≅16,000 females for research purposes during 1958–1974. This also was in addition to the annual harvest of 30,000–96,000 juvenile males. The population did not respond as anticipated; overall survival fluctuated and production of young males declined.

Part of the Pribilof Islands was set aside as a research area commencing in 1973, but harvest on the remaining areas continued; 24,000–29,000 were harvested annually until 1979. All commercial harvests were terminated on the Pribilof Islands in 1984 and aboriginal harvest in recent years has been <2,000 annually (Conservation plan for the northern fur seal, Callorhinus ursinus, prepared by the National Marine Mammal Laboratory and Office of Protected Resources, National Marine Fisheries Service, 1993). The estimated number of northern fur seals in 1992 was 1,322,446, of which 982,000 were on the Pribilof Islands (Loughlin et al., 1992).

Eumetopias jubatus (Schreber)
Northern Sea Lion

1776. *Phoca jubata* Schreber, theil 3, heft 17, pl. 83B, and p. 300.
1828. *Otaria stellerii*: Lesson, 13:420.
1885. *Eumetopias Stelleri*: True, 7:607.
1902. *Eumetopias jubata*: Allen, 16:113.

Description.—The northern sea lion (Plate XXIII) is the largest member of the family Otariidae (Table 12-8). It exhibits marked sexual dimorphism; adult males weigh as much as 1,120 kg and adult females as much as 350 kg (Loughlin et al., 1987). Among adult females and subadult males the upper body is slim, whereas in adult males it is heavy and muscular. The head is bearlike with a moderate step between muzzle and forehead. The rear flippers are relatively small and can be turned beneath the body for travel on land.

The pelage is light buff to reddish brown, slightly darker on the chest and abdomen; overall the pelage appears darker when wet. However, beneath the water *E. jubatus* appears white, whereas *Zalopus* appears black. Females are lighter than males and have long, coarse hair on the chest, shoulders, and back. The nose pad and flippers are black. Neonates are dark brown to black until their first molt at 4–6 months of age (Loughlin et al., 1987).

The skull of the northern sea lion (Fig. 12-18) is massive with a long narrow palate extending far posterior to M1. The postorbital processes are squarish and the sagittal crest is low. The nasals are long triangles tapering to a point posteriorly. Among adults, a diastema (as wide as 30 mm in males and 26 mm in females) separates P4 and M1; however, the diastema is absent in young-of-the-year and only ≅10 mm wide in yearlings (Orr and Poulter, 1967). There are five cheekteeth.

Distribution.—The northern sea lion occurs around the North Pacific Ocean rim from Hokkaido, Japan, and the Kuril Islands and Sea of Okhotsk, through the Aleutian Islands and Bering Sea, and south along the west coast of North America to the Channel Islands, California (Fig. 12-19; Loughlin et al., 1987). In surveys along the Oregon coast, northern sea lions were observed ashore at 10 sites extending from the South Jetty at the mouth of the Columbia River, Clatsop Co., south to Rogue Reef, Curry Co. (R. F. Brown and S. D. Riemer, in litt., June 1992). The distribution of museum specimens is similar (Fig. 12-19).

A Pliocene-age fossil femur from California is similar to that of *Eumetopias*, but the oldest fossil otariid with single-rooted cheekteeth and assignable to *Eumetopias* is from Japan and is ≅2 million years old (Repenning, 1976). A left radius from a mid-Pleistocene deposit at Cape Blanco, Curry Co., was assigned to *Eumetopias* (Packard, 1947a).

Geographic Variation.—The species is considered to be monotypic (Loughlin et al., 1987).

Habitat and Density.—Onshore sites used by northern sea lions may be classified either as rookeries used for birthing, breeding, and rearing young or as resting areas (commonly referred to as "haulouts") used by breeders at other times and by nonbreeders whenever they come ashore. Rookeries usually are on beaches on islands or reefs largely free of disturbance by humans or other terrestrial animals. The beaches may be sand, gravel, stones, or bedrock. The primary rookeries along the coast of Oregon are at Pyramid Rock on Rogue Reef, Curry Co., and at Long Brown Rock on Orford Reef, Curry Co.; a few young are born on Three Arch Rocks, Tillamook Co. (R. F. Brown and S. D. Riemer, in litt., June 1992). Areas used as rookeries also may serve as resting areas at other seasons, but other areas may include human-made structures in addition to beaches and rocks. R. F. Brown and S. D. Riemer (in litt., June 1992) identified seven such areas in addition to the three rookeries used by northern sea lions: South Jetty, Columbia River, Clatsop Co.; Ecola Point, Clatsop Co.; Cascade Head, Tillamook Co.; Seal Rock, Lincoln Co.; Sea Lion Caves, Lane Co.; Cape Arago, Coos Co.; and Blanco Reef, Curry Co.

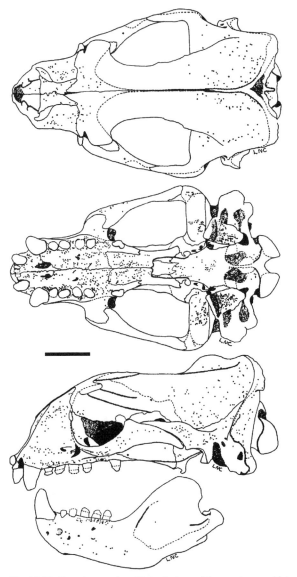

Fig. 12-18. Dorsal, ventral, and lateral views of the cranium, and lateral view of the dentary of a male northern sea lion, *Eumetopias jubatus* (OSUFW 2990). Note I2 not present in specimen. Scale bar equals 60 mm.

Along the Oregon coast during most years, northern sea lions occur on traditional rookeries from April through mid-September, with numbers peaking in May and again in August (Mate, 1975). However, during a year that relatively few California sea lions (*Zalophus californianus*) arrived on Rogue Reef in September, northern sea lions extended their stay into October. Mate (1975) suggested that either "niche pressure" or "avoidance behavior" might be responsible for *E. jubatus* leaving rookeries as the highly vocal male *Z. californianus* arrived.

No early records of abundance of the northern sea lion along the coast of Oregon are available, but Mate (1973) suggested that numbers may have been many times greater than the number of sea lions that presently use onshore areas of the state. In response to complaints of extensive depredations on salmon (*Oncorhynchus*), the Oregon Legislature authorized a $2.50 bounty on sea lions in 1900 that subsequently was increased to $5.00, then to $10.00, and finally reduced to $0.50 in 1926 (Bailey, 1936). Many hundreds were killed, likely many more than were presented for bounty (Bailey, 1936); in 1925 alone, 1,387 were presented for bounty (Pearson and Verts, 1970). The bounty remained in effect until 1933, but few were presented for bounty in 1930–1931 and no data were available for 1932–1933 (Pearson and Verts, 1970). Kenyon and Scheffer (1953) reported that ≅1,000 remained along the coast, and Kenyon and Rice (1961)

estimated 650–1,000 in 1960. Pearson and Verts (1970) counted 862 during an aerial survey of the coast in June 1968; they considered their count to represent 80% of total numbers, thus estimated a population of 1,078. Olterman and Verts (1972) reported that B. R. Mate claimed 1,500–2,500 occurred along the coast of Oregon.

Eumetopias jubatus is migratory (Mate, 1975, 1982*b*), with essentially all adult and subadult males absent south of British Columbia by mid-October and a marked decline in females and young by midwinter (Mate, 1982*b*). Thus, counts would be expected to vary seasonally. In 1989, 2,261 adults were counted on rookeries along the Oregon coast on 22 June and 593 neonates were counted on 6 July (Loughlin et al., 1992). The latter authors reported (pp. 231–232) that the Oregon population had "remained stable since about 1975 at about 2,500 nonpups and 600 pups." Other information tends to indicate that the population along the coast may be increasing slightly (Fig. 12-20), but improvement in survey techniques may account for some of the increase (R. F. Brown and S. D. Riemer, in litt., June 1992). Nevertheless, the overall population in the North Pacific Ocean declined from 240,000–300,000 in the late 1950s and early 1960s to an estimated 116,000 in 1989 (Loughlin et al., 1992). Diseases, availability of prey, commercial harvests, entanglement in debris and fishing gear, and killing by fishermen may have been contributory factors to the decline in abundance (Merrick et al., 1987). The role of commercial fisheries and other factors in altering the abundance or availability of prey, thereby possibly affecting populations of northern sea lions, is not known. However, changes in the Gulf of Alaska ecosystem and their effect on nutritional status were believed responsible for significant declines in body mass and body measurements recorded for northern sea lions from the 1970s to the mid-1980s (National Marine Fisheries Service, 1992).

Diet.—Based on 14.5 h of observation in March–July, prey brought to the surface and consumed by northern sea lions at the mouth of the Rogue River, Curry Co. (*n* = 93 observations), consisted of 82% lampreys (*Lampetra*

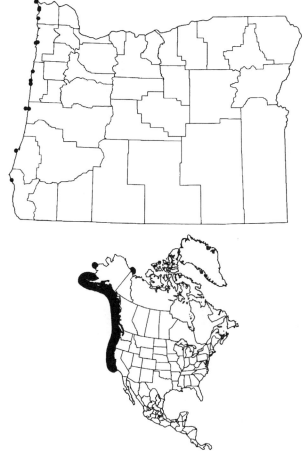

Fig. 12-19. Localities along the coast of Oregon from which museum specimens of the northern sea lion, *Eumetopias jubatus*, were examined (listed in APPENDIX, p. 518), and the distribution of the northern sea lion in North American waters. (Range map redrawn from Ronald et al., 1982:797, map 40.22.)

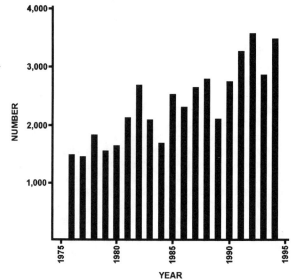

Fig. 12-20. Aerial counts of northern sea lions (exclusive of young-of-the-year) along the coast of Oregon, June–July 1976–1994. Data provided by the Oregon Department of Fish and Wildlife (R. F. Brown and S. D. Riemer, in litt., June 1992; R. F. Brown, in litt., March 1995).

tridentata), 2% salmonids, and 16% unidentified items (Jameson and Kenyon, 1977). In a subsequent study conducted at monthly intervals, most northern sea lions entered the river in June or July; of 64 observations of the species feeding in the river during those months, 18 (28.1%) prey were lampreys, 22 (34.4%) were unidentified fish other than salmonids, 23 (35.9%) were unidentified fish, and only 1 (1.6%) was identified as a salmonid (Roffe and Mate, 1984).

The stomach of one individual collected 42 km W Yaquina Head, Lincoln Co., contained three rockfish (*Sebastes*). Three individuals collected off southern California had eaten rockfish, sanddabs (*Citharichthys*), and curlfin sole (*Pleuronichthys*); two of them also had ingested stones (Fiscus and Baines, 1966). In Alaskan waters, the species consumed most frequently was the capelin (*Mallotus villosus*), followed in decreasing order by the sand lance (*Ammodytes hexapterus*), herring (*Clupea pallasi*), and rockfish. One of 18 individuals had consumed salmon (*Oncorhynchus*) and seven had ingested stones or pebbles (Fiscus and Baines, 1966). The purpose of the ingested stones is unknown (Loughlin et al., 1987).

Octopus and other cephalopods (especially Gonatidae) also contribute significantly to the diet of the northern sea lion in Alaskan waters (Loughlin et al., 1987). Pitcher (1981) indicated that walleye pollock (*Theragra chalcogramma*) was consumed in increasing amounts in the 20 years before 1980, possibly in response to an increase in abundance of that prey species. However, biomass of pollock in that region seems to have declined since the mid-1980s. Northern sea lions also are known to prey on harbor seals (*Phoca vitulina*—Pitcher and Fay, 1982) and young of northern fur seals (*Callorhinus ursinus*—Gentry and Johnson, 1981), but their overall contribution to the diet likely is not significant (Loughlin et al., 1987) and their impact is only a small fraction of the overall mortality of young of those prey species (Gentry and Johnson, 1981).

Reproduction, Ontogeny, and Mortality.—Adult males are spermatogenic from April to August (Loughlin et al., 1987). Those 6–15 years old (most 9–13 years old) establish and defend territories on rookeries by mid-May in Oregon (Mate, 1973). Upon returning to the rookeries, pregnant females give birth ≅3 days later and usually breed again 11–14 days following parturition. The peak of birthing along the coast of Oregon is about the 1st week in June, and most adult females breed by mid-July at which time breeding males abandon their territories (Mate, 1973). Territories occupied by males and birthing sites used by females commonly are occupied by the same individuals for several years. Nonpregnant females may breed with males unable to hold territories but that occupy onshore sites near the rookeries. Normally, only one young is produced; 85% of females copulate only once during a breeding season (Loughlin et al., 1987). Embryonic development is interrupted at the blastocyst stage for 3–4 months (National Marine Fisheries Service, 1992; Pitcher and Calkins, 1981). Gestation is ≅354 days (Loughlin et al., 1987).

Number of offspring produced per adult female on rookeries is difficult to estimate because adult females are not always on shore. On Orford Reef in 1969–1971, when numbers of young ranged from 296 to 413, the number of young per adult female ranged from 0.44 to 0.75 (Mate, 1973). Numbers of young counted during aerial surveys of the two major rookeries (Orford and Rogue reefs) in summer 1984–1991 ranged from 265 to 588. Surface surveys also

were conducted during 3 years; these produced counts greater than those made in the same year from the air. For example, in 1990, when 574 were counted from the air, 790 were counted from the surface (R. F. Brown and S. D. Riemer, in litt., June 1992).

At birth, neonates weigh 16–22 kg, are ≅1 m long, and have erupted teeth. Body mass and length differ little between sexes; body mass doubles and length increases ≅25% by 6–10 weeks of age (Scheffer, 1945). The stomachs of several 6–10-week-old young contained milk (a 41-kg female contained 1 l) and pebbles. Neonates are capable of swimming shortly after birth and can swim strongly in rough seas by 6–7 weeks (Scheffer, 1945). Neonates nurse at frequent intervals commencing shortly after birth, but at about 9 days, maternal females begin to spend part of the night at sea to feed at intervals of 1–3 days. Young usually are weaned before 1 year of age, but instances of females suckling a neonate and a yearling are known (Gentry and Withrow, 1978).

Growth continues to ≅8 years of age in females and to ≅10 years of age in males (Fiscus, 1961), with males attaining 3 times the body mass of females (Gentry and Withrow, 1978). Females attain sexual maturity at 3–8 years of age and most breed annually; males attain sexual maturity at 3–7 years, but usually are unable to hold territories on rookeries until 9 years old (Pitcher and Calkins, 1981; Thorsteinson and Lensink, 1962). Longevity is ≅30 years for females and ≅15 years for males (Pitcher and Calkins, 1981).

Neonates often are killed on rookeries by being crushed by adults, bitten or tossed by adult females other than the mother, or preyed upon. On some rookeries, mortality of neonates is about 10% and about half of females and three-fourths of males die before 3 years of age (Loughlin et al., 1987). However, on Orford Reef in 1969–1971, calculated mortality of young ranged from 22 to 36%, and on Simpson Reef, mortality of young was 43–83% (Mate, 1973). Predators in various regions include sharks (*Lamna ditrops*), killer whales (*Orcinus orca*), and grizzly bears (*Ursus arctos*—Loughlin et al., 1987).

Habits.—The northern sea lion is largely a mammal of the continental shelf (Fiscus and Baines, 1966; Gentry and Withrow, 1978). It is gregarious, forming large groups on traditional rookeries and resting areas; at sea, it may form rafts of several hundred individuals (Orr and Poulter, 1967). Large adult males arrive on rookeries first and commence to exhibit territorial behavior almost immediately (Orr and Poulter, 1967). Initially, territorial defense is more bluff than battle, but aggression increases as more territories are occupied. Real battles ensue when an intruder attempts to displace a territorial male; the combatants lunge at each other's head and neck, inflicting bloody wounds. Disagreements with males on adjacent territories usually are largely vocal, but if males charge each other, one usually turns sideways and lies with its head down as if asleep; the other male ceases its aggression (Orr and Poulter, 1967). Territorial males initiate the copulatory sequence by face-to-face contact with females, sometimes with noses touching and sometimes with open-mouth facial contact. Unreceptive females maintain face-to-face contact to avoid copulation. Although females often form groups within territories of males, these are not harems in the sense either of females being forceably restrained by the male (Orr and Poulter, 1967) or of social bonding among individuals forming the groups (Gentry and Withrow, 1978). Rather, it seems that groups are formed within territories to avoid boundaries contested by adjacent territorial males (Gentry and Withrow, 1978). Territorial males may attempt to herd females, but usually are unsuccessful; females cross territorial boundaries essentially at will (Gentry and Withrow, 1978). Territorial males engage in hundreds of territorial displays, fight with intruders, attempt to herd females, and sire a new generation, all without feeding for as long as 60 days (Gentry and Withrow, 1978).

Adult females on rookeries become agonistic toward other females for 3–10 days after birth of their young. They guard their young and fight over space. However, the intense aggression abates ≅2 weeks postpartum when females commence to forage nightly and young form groups of 4–12 to swim and play in tidepools (Loughlin et al., 1987; Orr and Poulter, 1967).

At the end of the breeding season in August, adult males abruptly leave the rookery, whereas females and young remain for a month or more. During winter, northern sea lions may move >450 km offshore for periods as long as 4 months and forage at depths as great as 273 m (National Marine Fisheries Service, 1992).

Remarks.—Because of range-wide declines in abundance during the previous 10–15 years, on 10 April 1990, the northern sea lion was classified as a threatened species (Nowak, 1991). It is considered "vulnerable" by the Oregon Department of Fish and Wildlife as of December 1995.

Zalophus californianus (Lesson)
California Sea Lion

1828. *Otaria californiana* Lesson, 13:420.
1866. *Zalophus* Gill, 5:7.
1880. *Zalophus californianus*: Allen, 12:276.

Description.—The California sea lion (Plate XXIII) is the midsized Oregon otariid (Table 12-8). The species exhibits considerable sexual dimorphism: adult males weigh 200–400 kg, whereas females weigh 50–110 kg (Nowak, 1991). The body is slender and tapering, the flippers are relatively small, and the rear flippers can be rotated beneath the body for moving on land. The short, coarse pelage lacks dense underfur; when dry, it usually is chocolate brown, but individuals with light-tan pelage are known to occur. The pelage appears black when wet. In older males, the pate is tan.

The skull of the California sea lion (Fig. 12-21) is relatively narrow; the nasals are slender and do not taper markedly; the postorbital processes are rounded (not squarish) anteriorly; zygomatic breadth is <180 mm in males, <130 mm in females; and the gap between P4 and M1 is no greater than the length of a cheektooth. Among adult males, the sagittal crest is raised enormously, producing a markedly stepped facial profile and a doglike appearance. There are five or six cheekteeth.

Distribution.—*Zalophus californianus* occurs as three disjunct populations: one off the coast of Japan (possibly extirpated), one in the Galápagos Islands, and one along the coast of North America from Vancouver Island, British Columbia, south to the tip of Baja California Sur and in the Gulf of California (Fig. 12-22; Mate, 1978). Its occurrence along the coast of Oregon is seasonal. Most museum specimens have been collected along the northern portion

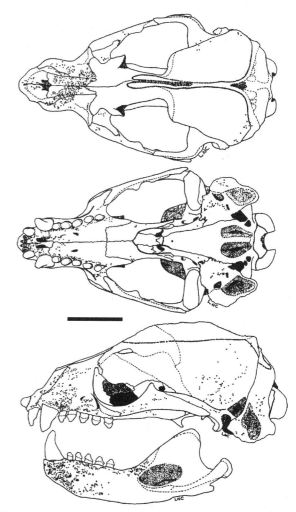

Fig. 12-21. Dorsal, ventral, and lateral views of the cranium, and lateral view of the dentary of a male California sea lion, *Zalophus californianus* (OSUFW 2211). Scale bar equals 60 mm.

of the Oregon coast (Fig. 12-22).

Fossils are known only from Pleistocene deposits of southern California (Kurtén and Anderson, 1980). Bones of *Z. californianus* were recovered from kitchen middens of prehistoric aboriginals along the coast of Oregon (Bailey, 1936).

Geographic Variation.—Only one of the three subspecies occurs along the coast of North America: *Z. c. californianus* (Mate, 1978).

Habitat and Density.—The North American population of *Z. californianus* breeds in June–July in the Gulf of California and in coastal areas from the southern tip of Baja California Sur north to the Farallon Islands off the coast near San Francisco, California (Mate, 1978; Roffe and Mate, 1984). They usually are nearshore animals, often coming ashore on sandy beaches or rocky areas. Off southern California, *Z. californianus* was reported to occur as far as ≅63 km from shore (Fiscus and Baines, 1966).

Commencing in mid-August, subadult and adult males migrate northward; some may reach British Columbia, but many stop and spend the winter at sites along the way. The primary areas where *Z. californianus* comes ashore in Oregon are at Cascade Head, Tillamook Co.; Cape Arago, Coos Co.; Rogue Reef, Curry Co.; and Orford Reef, Curry Co. (R. F. Brown, in litt., April 1988), but it also occurs in

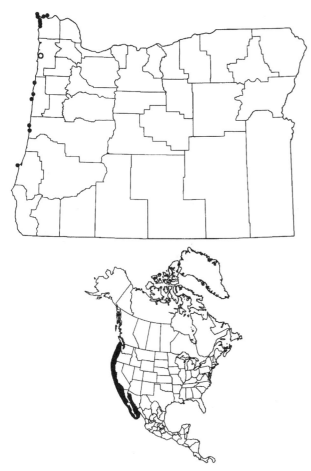

Fig. 12-22. Localities along the coast of Oregon from which museum specimens of the California sea lion, *Zalophus californianus*, were examined (listed in APPENDIX, p. 578), and the distribution of the California sea lion in North American waters. (Range map redrawn from Ronald et al., 1982:798.) Open symbol indicates record for county.

Yaquina Bay (Bayer, 1981), at Sea Lion Caves, Lane Co., and on Simpson Reef, Coos Co. (Mate, 1975). Females usually remain within the breeding range (Mate, 1978) or some may migrate southward (Nowak, 1991). Thus, numbers that occur along the Oregon coast vary with season and from year to year; ≅5,500 pass through waters off the coast of Oregon, but a count of 1,938 made in September 1984 was the maximum observed at any one time (R. F. Brown, in litt., April 1988). Many enter rivers to feed, and occasionally some individuals migrate far up the larger rivers. Mate (1978) reported that one moved 235 km up the Columbia River, upstream of Bonneville Dam, Multnomah Co. Whether the individual circumvented the dam by climbing the fish ladder or by going through the locks during upstream movements is unknown, but in going downstream it hopped on a barge and rode through the locks.

Roffe and Mate (1984) observed California sea lions in the Rogue River, Curry Co., from November to June with peak numbers (60–95/survey) in March or April. None was observed in July–September. On the Rogue Reef, west of the mouth of the Rogue River, numbers peaked in October at ≅1,100, but numbers in the river peaked after some had migrated farther northward and only ≅400 remained on the reef. Their occurrence in the river coincided with the upriver

migration of winter steelhead (*Oncorhynchus mykiss*), and their numbers there increased to the end of that run and peaked as smolts and spawned steelhead migrated back to the ocean (Roffe and Mate, 1984).

North American populations of California sea lions were decimated in the 19th century for skins and oil, and into the early 20th century for dogfood and skins (Mate, 1982*a*). The breeding population may have been as low as 600 in the mid-1920s, but by 1940 it was ≅2,000 and by 1958 it was ≅13,000 (Bartholomew and Boolootian, 1960). Estimates during the 1960s indicated that the population was increasing rapidly (Mate, 1982*a*). Mate (1978) estimated that numbers in this population were between 100,000 and 125,000 animals, but later estimated a minimum of 75,000–100,000 (Mate, 1982*a*).

Diet.—At the mouth of the Rogue River, Curry Co., Jameson and Kenyon (1977) observed California sea lions with prey on nine occasions during March–June. Five items being consumed were identified; all were lampreys (*Lampetra tridentata*). Later, during January–May in the same area, Roffe and Mate (1984) observed *Z. californianus* on the surface with prey on 83 occasions; 19 (22.9%) were salmonids, 8 (9.6%) were lampreys, 9 (10.8%) were unidentified fishes other than salmonids, and 47 (56.6%) were unidentified fishes. Fecal droppings of *Z. californianus* collected at sites along the Rogue River contained otoliths of eulachon (*Thaleichthys pacificus*), chinook salmon (*Oncorhynchus tshawytscha*), and steelhead. Gastrointestinal tracts of 26 (92.9%) of 28 California sea lions collected in the river contained remains of one–51 lampreys, 15 (53.6%) contained remains of one–12 steelhead, and 11 (39.3%) contained remains of one–21 individuals among at least eight other taxa of fish. Steelhead composed 19% of the diet, one-third of which consisted of fish that had spawned (Roffe and Mate, 1984).

Off California, prey in stomachs of a small sample of *Z. californianus* consisted exclusively of squid (mostly *Loligo opalescens*), northern anchovies (*Engraulis mordax*), and Pacific hake (*Merluccius productus*). One individual consumed 125 squid and another ate 133 anchovies; a third individual ingested 18 hake, 2 unidentified squid, and a rock, a total of >5 l of material (Fiscus and Baines, 1966).

Reproduction, Ontogeny, and Mortality.—Pregnant females arrive on California rookeries in June and usually give birth within a day or two. Most young are born at night (Peterson and Bartholomew, 1967). Adult females enter the estrous condition within 2 weeks after parturition, as evidenced by a swollen and pinkish vulva. Although males sometimes court females by barking, shaking the head, and nuzzling the female's genital area, it is usually the female that solicits the male to mount. Characteristic postures and movements are used by the female to invite sexual attention (Peterson and Bartholomew, 1967). The pair may copulate on land or in the water. Females are no longer receptive after copulating. The gestation period is ≅11 months, including a 3-month delay in implantation (Nowak, 1991).

Neonate California sea lions weigh ≅6 kg at birth. They are born with their eyes open; within 15 min they can scratch, shake, and groom their pelage, and within 30 min they can walk. They begin to suckle during the 1st day. Maternal females are extremely attentive for 2–4 days; they commonly pick up their offspring and move it closer, and they bark and make agonistic displays toward nearby females. After ≅4 days, the maternal female commences to spend increasingly longer periods at sea. When 2–3 weeks old, the young form groups of as many as 200. Maternal females locate their young by vocalizing, then make certain of their identification by visual and olfactory inspection of young that respond by bleating. Young commence to forage on their own at ≅5 months but may not be weaned until 11–12 months of age; occasionally, a female may nurse a neonate and a yearling simultaneously (Nowak, 1991).

Females attain sexual maturity at 4–5 years and most give birth to one offspring each year (Mate, 1978; Peterson and Bartholomew, 1967). Males may attain sexual maturity at about the same age, but they usually are not able to acquire and defend territories on the rookeries until ≅9 years of age. Longevity in the wild probably is ≅14 years, although one is known to have lived 31 years in captivity (Mate, 1978).

Sharks (Lamniformes) and killer whales (*Orcinus orca*) frequently prey on California sea lions (Mate, 1978). Leptospirosis (see Chapter 5) possibly was responsible for fewer than one-fourth as many *Z. californicus* wintering on Rogue Reef, Curry Co., in 1971 as during the previous 3 years (Mate, 1975).

Habits.—California sea lions are superb swimmers and divers. They can attain speeds of 24–32 km/h and depths of 140 m; they can stay under water for as long as 20 min. However, most dives are much shallower and last only 2–5 min (Mate, 1978). When pressed, they "porpoise" on the surface. They body surf on swells, toss kelp into the air, and leap completely out of the water in play.

Adult males commence to arrive on rookeries and to establish territories in May; peak numbers occur on rookeries in late June–July with some remaining into August. Few males occupy territories for >2 weeks; the average is ≅9 days. However, a territory may be occupied by a succession of males (Peterson and Bartholomew, 1967). Territorial males bark almost continuously, but intruders bent on displacing territorial males are silent. Fighting for territorial rights includes a series of lunges at the chest, flank, or flipper by participants; the well-padded chest receives most of the blows. Wounding occurs sufficiently frequently that most individuals have a pattern of scars and wounds by which they can be identified. A territory is relinquished by a territorial male only after it has been forced from the territory by another male. Most territories are adjacent to the ocean, presumably a characteristic related to the requirement that these mammals be wet while in the sun. Females ignore territorial boundaries and are not herded by territorial males, thus, are relatively independent. Males maintain territories only where females aggregate; thus territorial activities of males shift with localities occupied by females. While territorial males sleep or are otherwise inattentive, nonterritorial males often sneak into their territories and attempt to mount females. These intruders sometimes remain undetected for extended periods, but usually flee at the first bark when discovered by a territorial male (Peterson and Bartholomew, 1967).

Females are highly gregarious, but maintain individual space through threats and barks. Females with neonates vigorously defend a tiny territory around themselves and their offspring; they rout any young but their own. Thus, mother-offspring bonds develop quickly (Peterson and Bartholomew, 1967).

Remarks.—*Zalophus californianus* is the "trained seal" of circuses and other animal acts. Such training often requires a year or more, but once trained these sea lions remember their acts well (Nowak, 1991).

California sea lions have become much more visible along the coast of Oregon in recent years. Increasing numbers and reduced harassment in response to the Marine Mammal Protection Act of 1972 have lead to greater use of estuaries by the species (Bayer, 1981), which, in turn, has made it more visible to the public. Such possibly was responsible for *Z. californianus* being sighted with large fish on the surface and becoming entangled in fishing gear more frequently, leading to the assumption that the species contributed significantly to declines in catches (Roffe and Mate, 1984). Nevertheless, studies of the diet of the species conducted in areas where these sea lions, salmonids, and other sport and commercial fishes concentrate tend to indicate that the role of predation by California sea lions in the declines is of relatively little consequence. For example, Roffe and Mate (1984) reported that predation by *Z. californianus* accounted for only 0–5.1% of various runs of salmonids in the Rogue River.

FAMILY PHOCIDAE—HAIR SEALS

The family Phocidae consists of 10 Recent genera and 19 species distributed in polar and temperate seas of the world plus some tropical seas and some inland bodies of water (Nowak, 1991; Wozencraft, 1993). The phocids are represented in Oregon by the harbor seal (*Phoca vitulina*) and the northern elephant seal (*Mirounga angustirostris*). The ringed seal (*Phoca hispida*), ribbon seal (*Phoca fasciata*), and hooded seal (*Cystophora cristata*) have been reported as having occurred as extremely rare visitors in California waters (Mate, 1981; B. R. Mate, pers. comm., 18 October 1994). As these individuals must have passed along the coast of Oregon, their occurrence along the Oregon shore is remotely possible (see Chapter 14).

The phocids are characterized by the inability to rotate the hind limbs beneath the body, absence of pinnae, absence of postorbital processes on the frontal bones, absence of a groove on the occlusal surface of I1 and I2, nasals extending posteriorly between the frontals, and in males, no scrotum but abdominal testes. A stubby tail is present; males possess a baculum. Phocids are even more highly adapted than otariids for living in the marine environment: the body is fusiform; the postcanine teeth are simple pegs or have three or more cusps to aid in capturing the fish, lampreys, squid, and shellfish that constitute the primary foods; both front and hind legs are modified into flippers; the hind flippers extend posteriorly and move in an up-and-down motion much like the caudal appendage of a cetacean; the front flippers are much smaller than the hind flippers and are set far forward on the body; and the pelage is much reduced, consisting largely of stiff hairs with little or no underfur. Despite the inability to use the hind legs to aid locomotion on land, phocids can move surprisingly fast with hunching and wiggling motions. The pelage of *Phoca* is striped or splotched, but *Mirounga* is more or less uniform in color, although scarring in the neck region acquired during fights or mating may appear as a dark-colored yoke (Nowak, 1991).

Males establish and defend territories; *Mirounga* males

commonly form harems and mate with several females (Nowak, 1991). Until recently, only *Phoca vitulina* was known to breed along the Oregon coast, but since 1993 a breeding colony of *Mirounga* has been observed on Shell Island, Coos Co. Gestation is extended because of a long delay in implantation.

The oldest fossil remains are from the middle Miocene (Carroll, 1988; Tedford, 1976). In contrast with the present worldwide distribution, phocids are believed to have evolved along Atlantic Ocean shores in areas adjacent to the Mediterranean Sea (Carroll, 1988). Phocids reached the North Pacific in the Pleistocene (Carroll, 1988).

KEY TO THE PHOCIDAE OF OREGON

1a. Size large (males to 2,300 kg, females to 800 kg); males with tubular inflatable proboscis; cheekteeth simple pegs; incisors 2/1..............***Mirounga angustirostris***, northern elephant seal (p. 391).

1b. Size medium (<170 kg); males lacking tubular inflatable proboscis; cheekteeth with three or more cusps; incisors 3/2............***Phoca vitulina***, harbor seal (p. 394).

Mirounga angustirostris (Gill)

Northern Elephant Seal

1866. *Macrorhinus angustirostris* Gill, 5(1):13.
1904. [*Mirounga*] *angustirostris*: Elliot, 4:545.

Description.—The northern elephant seal (Plate XXIV) is the largest pinniped carnivore that occurs along the North Pacific coast. It exhibits considerable sexual dimorphism, with adult males attaining a body mass of as much as 2,300 kg and females as much as 800 kg (Stewart and Huber, 1993). The flippers are relatively small; the rear ones are directed posteriorly and are without nails. The most distinctive external feature is the elongated tubular proboscis with a deep transverse cleft among adult (5+ years old) males. The proboscis can be inflated and directed into the opened mouth to produce vocalizations. Among females and subadult males, the rostrum is pointed with a small, gradually stepped forehead. Also among males, the neck and chest lose most of their hair and the skin thereon becomes rough and thickened at puberty (Stewart and Huber, 1993). The eyes are directed forward.

The skull of adult males is massive (Fig. 12-23); that of females is relatively small and delicate. It lacks postorbital processes, but the interorbital area is broad and flattened. The zygomata are broad and heavy. The three to seven (usually five) cheekteeth in both jaws are small, peglike, evenly spaced, and single-rooted; the canines are large; and the two upper and one lower incisors are small.

The adult pelage consists of coarse grayish or brownish hairs without underfur; individuals that have been to sea for long periods may be slightly greenish from algae or copepods on the dorsum (Baldridge, 1977; Mate, 1981). During the molt, not only is the pelage shed, but the entire cornified epidermis sloughs off in large patches, creating a mottled appearance. Neonates are black, but the natal pelage is replaced at 3–4 weeks with silvery-gray hair that changes to tan on the venter and brown on the dorsum by 11–16 weeks (Stewart and Huber, 1993).

Distribution.—Northern elephant seals occur from the southern tip of Baja California Sur northward along the

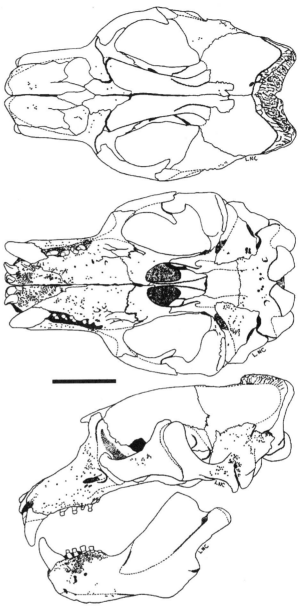

Fig. 12-23. Dorsal, ventral, and lateral views of the cranium, and lateral view of the dentary of a male northern elephant seal, *Mirounga angustirostris* (OSMNH PCD-40). Scale bar equals 10 cm.

coast of California, Oregon, Washington, and British Columbia. They also range widely in the Pacific Ocean except during breeding and molting seasons (Stewart and Huber, 1993). In Oregon, individuals reportedly were sighted ashore 6.4 km south of Bandon, Coos Co. (Freiburg and Dumas, 1954), and at Simpson Reef, Coos Co. (Mate, 1969, 1970). In addition, specimens were obtained at Seal Rock, Depoe Bay, and Newport, Lincoln Co., and at the mouth of the Columbia River, Clatsop Co. (Fig. 12-24). Although Shell Island, Cape Arago, Coos Co., seems to be the only place in Oregon used regularly by elephant seals, occasional sighting of juveniles and subadults might be expected almost anywhere along the coast of Oregon (J. Hodder et al., in litt.).

Although the fossil record of elephant seal-like mammals extends to the Miocene, the only fossils of *Mirounga* are from Pleistocene deposits of southern

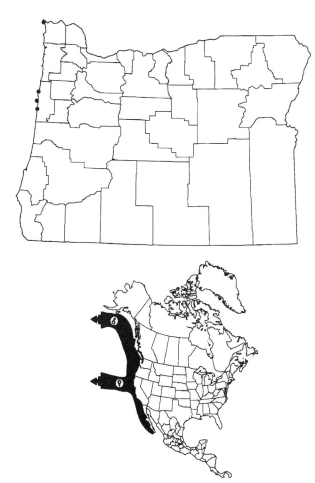

Fig. 12-24. Localities along the coast of Oregon from which museum specimens of the northern elephant seal, *Mirounga angustirostris*, were examined (listed in APPENDIX, p. 529), and the distribution of the northern elephant seal in North American waters (range map redrawn from Stewart and Huber, 1993:3, fig. 4, with modifications.)

California (Ray, 1976).

Geographic Variation.—The species is monotypic.

Habitat and Density.—Except for rookeries on sandy beaches, where they give birth and breed, then return later in the year to molt, northern elephant seals spend the remainder of the year at sea. There are few beaches suitable for establishment of rookeries in Oregon as most are subject to being wave swept at high tide (J. Hodder et al., in litt.). Adult females forage west of northern California and Oregon as far as Midway Island in the Hawaiian Chain, whereas adult males move farther north to forage in the Gulf of Alaska and along the eastern Aleutian Islands (Stewart and Huber, 1993). Juveniles of both sexes and subadult males forage in nearshore waters along the coasts of northern California, Oregon, Washington, and British Columbia (Stewart and Huber, 1993); they are reported onshore with increasing frequency (Cowan and Carl, 1945; J. Hodder et al., in litt.; Mate, 1969, 1970; Scheffer and Kenyon, 1963; Stewart and Huber, 1993).

Although nothing is known regarding abundance of northern elephant seals before 1840, populations likely already were decimated by that time as sealing operations conducted along the coasts of California and Baja California during the last half of the 19th century met with poor success (Stewart et al., 1994). Bartholomew and Hubbs

(1960) estimated from published accounts that by 1890 fewer than 100, perhaps as few as 20, remained alive. On several occasions, the species was thought to be extinct, but each time a few were found, whereupon collectors often killed most of them to preserve for posterity (Le Boeuf, 1985). The only active rookery in 1890–1920 was on Isle de Guadalupe 247 km west of the coast of Baja California. Although collecting and poaching continued, numbers on the rookery increased; 264 were counted in 1922 just before the Isle de Guadalupe became a reserve and elephant seals were given protection by the Mexican government (Stewart et al., 1994). Subsequently, northern elephant seals were observed on other islands in the region; rookeries later developed at those sites. By 1960, the population was estimated to be ≅15,000, of which 90% still were on Isle de Guadalupe (Bartholomew and Hubbs, 1960). Since 1965, populations have increased at a rate of 6.3% annually; the population in 1991 was estimated to number 127,000 (Stewart et al., 1994).

Northern elephant seals were observed in increasing numbers at Shell Island, Cape Arago, after 1982, with as many as 60 counted during 1 month. Of 22 tagged individuals observed at this locality between 1979 and 1991, all had been marked in California: 18 (82%) at Año Nuevo, 3 (14%) at the Farallon Islands, and 1 (5%) at San Miguel (J. Hodder et al., in litt.). Before 1993, most were observed in April–June and in September–November; during the birthing season (January–February) usually no more than one individual was observed at a time (J. Hodder et al., in litt.). However, 8 individuals were seen during the reproductive season in 1993, 15 in 1994, 32 in 1995, and 20 in 1996 (J. Hodder et al., in litt.).

Diet.—The major components of the diet of northern elephant seals are cephalapods and fish (Antonelis et al., 1994; Condit and Le Boeuf, 1984). Most (70%) prey species consumed by northern elephant seals older than juveniles occur at depths below the level of light penetration; only 30% are from shallower zones. Of the 53 taxa identified in materials recovered by lavage from stomachs of 193 elephant seals on San Miguel Island, California, 32 were cephalopods, 7 were teleost fishes, 4 were crustaceans, 6 were elasmobranch fishes, 3 were cyclostomes, and 1 was a tunicate (Antonelis et al., 1994). Four of the five most frequently occurring taxa (all in ≥38.9% of the stomachs) were cephalopods, three of which are highly luminescent, possibly contributing to their being detected by elephant seals at great depths. The fifth major item was Pacific hake (*Merluccius productus*); no other fish occurred in more than 14% of the samples (Antonelis et al., 1994). All five taxa make daily vertical migrations to well within depths of average dives made by northern elephant seals (see Habits).

No information regarding the diet of the northern elephant seal in United States territorial waters off the coast of Oregon is available.

Reproduction, Ontogeny, and Mortality.—Fifteen rookeries are established on islands and mainland sites along the coast of California and Baja California (Stewart et al., 1994), but until 1993, northern elephant seals were not known to reproduce at any site along the coast of Oregon (J. Hodder et al., in litt.). Three young were born at Shell Island, Coos Co., in January–February 1993, five in 1994, and an undetermined number in 1995 and 1996.

Because of high tides and wave-swept beaches during winter storms, none was known to have survived for >6 weeks (J. Hodder et al., in litt.). However, in 1997, five of seven born on Shell Island were believed to have survived and gone to sea (J. Hodder, pers. comm., 15 September 1997).

In California, adult males commence to arrive on the rookeries in early December and most that will reproduce arrive during that month. Males become sexually mature at 6–7 years of age, but usually do not breed until 9–10 years old. Females commence to arrive on rookeries by mid-December and all are gathered in harems by the end of January. Females as young as 2 years may breed, but most give birth for the first time when 4 years old. Upon arrival on the rookeries, most females are pregnant and usually give birth within 6 days.

Males engage in violent and bloody fights to establish a dominance hierarchy that functions to ensure that the highest-ranking among them are afforded a position closest to the harems (Le Boeuf, 1985). Fewer than one-third of males copulate, and the few dominant males on rookeries are responsible for siring most offspring. Some males may sire 50 offspring per year; one male, dominant in a harem for 4 consecutive years, was known to inseminate a minimum of 225 females (Le Boeuf, 1985). Males fast during the ≅3 months they are on a rookery and lose about one-third of their body mass. They become gaunt and listless; mortality within the group often is ≅50% after the long period without food (Le Boeuf, 1985). High-ranking breeding males usually die within 1–2 years of attaining peak reproductive success (Le Boeuf, 1974).

Parturient females also are subjected to great stress. They also fast for the ≅34 days they are on rookeries. In addition to producing an offspring they must nurse it for an average of 28 days and remate before going back to sea to feed (LeBoeuf, 1985). Not only must females provide energy for their own metabolism and that of their offspring, they must provide energy for growth of the offspring and supply them with sufficient reserves to remain on the rookeries, learn to swim, and commence to forage on their own. Milk produced by northern elephant seals contains an average of 54.4% fat, 9.0% protein, and only 32.8% water (Le Boeuf and Ortiz, 1977). Females lose an average of 42% of their body mass while on the rookeries. They remate about 4 days before weaning their offspring and returning to sea (Le Boeuf and Ortiz, 1977). The gestation period is ≅11 months and probably includes a 2–3 month delay in implantation (Stewart and Huber, 1993).

Weanlings also fast for 8–12 weeks, and they lose ≅25% of their body mass (Le Boeuf, 1985). Neonates weigh ≅34 kg at birth, but at weaning they weigh an average of 136 kg (Le Boeuf and Ortiz, 1977). A few male neonates, either by deception or through fostering, nurse successively more than one lactating female and become "super-weaners" that weigh as much as 238 kg (Le Boeuf, 1985). Commencing 2–3 weeks after weaning, young M. angustirostris make hesitant attempts to enter the water of tidepools. Despite rather pathetic first attempts to swim, young elephant seals are accomplished swimmers within a week and expert divers a month later (Le Boeuf, 1985).

In California, the major cause of death of young on rookeries is starvation upon becoming separated from the maternal female during the 1st week of life (Stewart and Huber, 1993). Another common mortality factor among young on rookeries is trampling by adult males during male-male interactions (Le Boeuf, 1974). Killer whales (Orcinus orca) and great white sharks (Carcharodon carcharias) occasionally prey on northern elephant seals of any cohort that has gone to sea (Stewart and Huber, 1993).

Habits.—Except for the time spent on the rookery during the breeding season and a month or so ashore while undergoing the molt, the northern elephant seal is truly a pelagic mammal. After learning to swim and dive, juveniles leave the rookery in late April and move northward, foraging along the coasts of Oregon, Washington, and British Columbia; the ≅50% that remain alive return to the rookery or nearby sites in September–October (Le Boeuf and Laws, 1994). Adult males, exhausted after the 3-month breeding season during which they have lost one-third of their body mass, commonly sleep for several days after females have left the rookery (Le Boeuf, 1985). They leave rookeries in the Channel Islands, California, in late February–March, travel to foraging grounds in the Gulf of Alaska and Aleutian Islands, and return to the Channel Islands by July. They spend ≅1 month undergoing the annual molt, then again travel to the Alaskan foraging grounds to recover body reserves before returning to the rookeries in December–January (Stewart and Huber, 1993). Adult females, after weaning their young and remating, leave the rookeries in January–February, move north and west to their foraging grounds, and return to the rookery area in April–May to molt. They leave the area in May–June and travel to their foraging area a second time, then return to the rookeries to give birth to their young and remate in early January (Stewart and Huber, 1993).

Adult females equipped with time-depth recorders just before they left the rookery after weaning their young spent the next 2.5 months at sea; 83–90% of that time was spent underwater (Le Boeuf et al., 1986, 1989). They made an average of three dives per hour, with dives averaging 17–22.5 min with an average of only 3.5 min between dives. The average dive was to a depth of 500–700 m; three individuals made dives in excess of 1,000 m and one dive was estimated to be 1,250 m deep (Le Boeuf et al., 1989).

Although elephant seals rarely were on the surface >5 min at a time, on rare occasions an animal rested on the surface for about an hour. After underwater activity recommenced, initial dives were about half the average depth, but depth of each successive dive increased in a stairstep pattern (Le Boeuf et al., 1989). Subsequent studies indicated that all sex and age-classes made deep and lengthy dives; even young 3.5 months old on their first trip to sea made dives as deep as 553 m and as long as 22 min. However, both depth and duration of dives increased with age (Le Boeuf, 1994). Later studies produced depth records of 1,585 m for males and 1,561 m for females, and a maximum duration of 77 min by an adult male (Stewart and DeLong, 1990); these are the deepest dives recorded for any mammal.

During dives, elephant seals develop bradycardia, with heart rates becoming progressively slower to ≅4 beats/min and with blood flow restricted or blocked to some areas of the body (Stewart and Huber, 1993). While on land, elephant seals undergo periods of apnea (as long as 21 min) and bradycardia (≅15% less than normal heart rate—Le Boeuf et al., 1986) to reduce metabolism and to conserve stored energy and water.

The heavy layer of blubber provides a high-quality insulation that, combined with regulation of blood flow to the flippers, provides effective thermoregulation while elephant seals are in the water. When on land and in the sun, elephant seals flip sand on their backs to aid in thermoregulation.

Remarks.—Although specimens were known from Alaska (Willett, 1943), British Columbia (Cowan and Carl, 1945), and Washington (*Seattle Times*, 15 April 1947:3), the northern elephant seal was not considered a part of the fauna of Oregon until Freiburg and Dumas (1954) reported collecting the skull of a dead adult male on the beach 6.4 km south of Bandon, Coos Co., in 1952 (Olterman and Verts, 1972). Mate (1969) reported observing two individuals ashore on Shell Island in Simpson Reef, Coos Co., during a 26-day period in September–October 1968. He also indicated that others had observed a northward progression in use of beaches by elephant seals during that decade. Subsequent observations revealed progressively greater use of this area (J. Hodder et al., in litt.).

Phoca vitulina Linnaeus
Harbor Seal

1758. *Phoca* Linnaeus, 1:37.
1864. *Halicyon richardii* Gray, p. 28.
1873. *Halicyon richardsi*: Sclater, in Clark, 1873:556.
1897. *Phoca vitulina richardsi*: Trouessart, 1897:385.

Description.—The harbor seal (Plate XXIV) exhibits less sexual dimorphism than other pinniped carnivores that occur in Oregon waters (Table 12-8); range in body mass is 70–170 kg for males and 50–150 kg for females (Nowak, 1991). The body is plump but tapers to small rear flippers permanently extended posteriorly. The head is large and rounded, the eyes are large, the limbs are short, and the nostrils sit dorsally on the muzzle. The nostrils can be closed when the harbor seal dives. The tongue is notched at the tip.

The pelage consists of overhairs as long as 9 cm that overlay short underhairs. Markings are extremely variable, but in general the pelage is gray or brownish gray with numerous small spots of black that may coalesce to form splotches. The hair has little insulative value. The flippers and nose pad are naked.

The skull of the harbor seal (Fig. 12-25) is rounded; the interorbital area is extremely narrow and without postorbital processes; the posterior margin of the palate is V-shaped, revealing the nasal septum; and the cheekteeth are three cusped. There are three upper and two lower incisors; I3 is larger than I1 or I2.

Distribution.—Harbor seals occur in nearshore regions of both North Atlantic and North Pacific oceans; they occur as far south as central Florida and northern Portugal in the Atlantic and to Baja California Norte and Hokkaido, Japan, in the Pacific (Fig. 12-26; Bigg, 1981). Harbor seals have been observed to come ashore at 32 sites along the Oregon coast and in estuaries from the Columbia River to Hunter's Island, Curry Co. (Harvey et al., 1990:66, fig. 1). Most collection sites of museum specimens from Oregon are along the northern coast (Fig. 12-26).

The earliest fossil record of *P. vitulina* is of early Pleistocene age (≅2 million years ago) from Cape Blanco, Oregon. Other Pleistocene records are from California and Alaska (Kurtén and Anderson, 1980).

Geographic Variation.—Only one of the four nominal

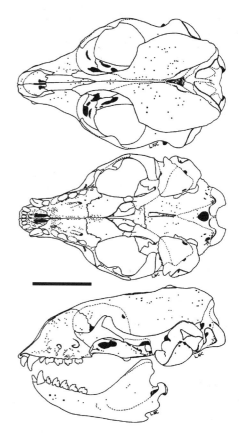

Fig. 12-25. Dorsal, ventral, and lateral views of the cranium, and lateral view of the dentary of the harbor seal, *Phoca vitulina* (OSUFW 2991). Scale bar equals 60 mm.

subspecies occurs along the coast of Oregon: *P. v. richardsi* (Bigg, 1981).

Habitat and Density.—Harbor seals commonly come ashore along the Oregon coast on estuarine mud flats, sandy beaches, rocky headlands, and offshore rocks and islands. They occasionally use seemingly inappropriate sites such as a rarely used boat dock, a grassy area accessible only at high tide, and an area separated by only 20 m of water from a viewpoint used by humans (Harvey et al., 1990). Nearshore or estuarine waters are used for foraging (Spalding, 1964).

Harbor seals were believed to be abundant when the region was settled by Europeans, but bounties and hired seal hunters were responsible for reducing numbers. Scheffer (1928) reported that harbor seals killed along the Oregon coast and submitted for bounty payments numbered 1,375 in 1925, 350 the following year. Following these losses, Scheffer (1928) considered the status of the population to be "precarious." The statewide bounty was discontinued in 1933, but a new bounty system was established in 1935 for seals killed in the Columbia River, with payment increasing from $5.00 in 1935 to $25.00 in 1955 (Pearson and Verts, 1970); it remained at $25.00 until enactment of the Marine Mammal Protection Act of 1972. On the Columbia River alone, ≅300 were presented for bounty in 1938–1942, but as ≅40% of those shot sink, probably ≅500 were killed. However, numbers presented for bounty declined rapidly; only 14 were submitted from the Columbia River in 1962 (Pearson and Verts, 1970). By 1967–1968, harbor seals were uncommon in the river;

Fig. 12-26. Localities along the coast of Oregon from which museum specimens of the harbor seal, *Phoca vitulina*, were examined (listed in Appendix, p. 547), and the distribution of the harbor seal in North American waters. (Range map redrawn from Bigg, 1981:6, fig. 2.)

they were sighted in the river on only 10 of 31 days and <10 were sighted on 7 of the 10 days (Pearson and Verts, 1970). Reduction in numbers, either through killing or through continual harassment, had nearly excluded harbor seals from the Columbia River. Under protection, harbor seals again commenced to use the Columbia River; ≅250 were in the river in spring 1984 and ≅2,100 in winter 1985 (Harvey, 1988).

Pearson and Verts (1970) reported observing 307 harbor seals along the Oregon coast and in estuaries on 17–19 May 1968 from shore points. They believed that this count probably represented 60% of numbers remaining in Oregon waters and that the species might be endangered. However, they likely underestimated the population, as Harvey (1988) reported that actual numbers are 2–14 times counts of harbor seals on land. Olterman and Verts (1972) reported that B. R. Mate observed 800 at two sites combined and estimated a minimum of 1,200 along the coast. Harvey (1988) reported that Mate counted 2,244 harbor seals during an aerial survey conducted in July 1975. Based on annual aerial censuses between that survey and one in June 1983 when he counted 3,733 harbor seals, Harvey et al. (1990) considered the population to have increased by an average of 8.1%/year. R. F. Brown (in litt., April 1988) reported a maximum count of 5,325 in February 1985, of

which 2,100 were in the Columbia River, and a continued increase in abundance of 6–8%/year.

Diet.—Harbor seals are opportunistic feeders that prey primarily on benthic and schooling fishes (Harvey, 1988), but may include some cephalopods and crustaceans in their diet (Scheffer and Sperry, 1931). Consequently, foods consumed by harbor seals may differ with seasonal and local abundances of various prey species (Harvey, 1988). In addition, knowledge of prey species and quantities of each consumed by harbor seals depends greatly upon the method used to evaluate their diet. For example, small prey species—such as lampreys (*Lampetra tridentata*), herring (*Clupea pallasi*), shiner surfperch (*Cymatogaster aggregata*), and eulachon (*Thaleichthys pacificus*) — probably are consumed under water, whereas large fishes such as salmonids are brought to the surface, torn apart, and consumed. Thus, direct observation biases records in favor of larger fishes and those likely of greatest concern to commercial and sport fishermen.

Because most large fish are brought to the surface at least once while being consumed by harbor seals, Brown and Mate (1983) were able to use direct observation to evaluate the impact of seal predation on an experimental run of chum salmon (*Oncorhynchus keta*) in a tributary (Whiskey Creek) to Netarts Bay, Tillamook Co., in 1978–1980. Harbor seals rarely were observed to take chum salmon in the bay other than as the salmon approached the narrow mouth of the creek, which they could navigate only at high tide. Although seals obviously responded to the concentrations of salmon, the impact of their predation on them was negligible (1.5–7.2% of returning fish annually). Fecal samples collected where harbor seals came ashore in Netarts Bay indicated that the most frequently consumed species was the Pacific sand lance (*Ammodytes hexapterus*); other frequently consumed species were Pacific sanddab (*Citharichthys sordidus*), rex sole (*Glyptocephalus zachirus*), slender sole (*Lyopsetta exilis*), dover sole (*Microstomus pacificus*), English sole (*Parophrys vetulus*), and staghorn sculpin (*Leptocottus armatus*). At least 24 other taxa of fishes were represented among the 2,048 individuals for which otoliths or teeth were recovered from 95 fecal samples (Brown and Mate, 1983).

Roffe and Mate (1984) combined direct observations with information obtained from examination of fecal droppings at sites along the Rogue River, Curry Co., at which harbor seals had come ashore and of stomach contents of harbor seals collected in the river. As expected, salmonids were observed most frequently being eaten by harbor seals (30.0% of 60 observations of prey being consumed), but lampreys were identified nearly as frequently (26.7%), and unidentified nonsalmonid fishes contributed 16.7% of the total. Of 89 fecal samples with recognizable prey, lampreys occurred in 81% and contributed 69% of the biomass consumed (Roffe and Mate, 1984). Remains of lampreys contributed 56.1% of the diet based on examination of stomach contents of 14 seals; salmonids contributed only 10% (Roffe and Mate, 1984).

Harvey (1988) estimated numbers and biomass of fish consumed by harbor seals along the Oregon coast for 1980 based on a population of 5,034 seals, twice the number actually observed on shore sites during aerial surveys that year. He accounted for differences related to body mass and basal metabolic rate of seals of various sex and age groups. Taxa of fish and numbers of each consumed were obtained from identification and numbers of otoliths reported in two published studies (Brown and Mate, 1983; Roffe and Mate, 1984), a thesis (Graybill, 1981), and an agency report (Beach et al., 1985) on diets of harbor seals in four regions of the Oregon coast. Size of fish was obtained from length of otoliths by considering otolith length–body length relationships of the various taxa of fish and loss of otolith length in the digestive tracts of seals. Harvey (1988) identified 63 taxa of fish of which harbor seals along the coast of Oregon consumed nearly 113 million individuals totalling 5,667,400 kg in 1 year. Of these, 12 species contributed 80.5% of the individuals and 73.3% of the biomass consumed (Table 12-9). Four taxa of salmonids contributed 593,400 kg, or 10.5% of the annual biomass consumed by harbor seals. For all taxa taken in the commercial fishery, harbor seals consumed ≅3 million kg, 24.6% of the commercial take (Harvey, 1988). Only 1.0% of the annual natural mortality of the English sole was attributable to harbor seals; this was the only species of fish for which appropriate information was available for estimating the contribution of harbor seal predation to its natural mortality (Harvey, 1988).

Harbor seals consume 5–6% of their body mass daily (Bigg, 1981). However, digestion is extremely rapid; thus, stomachs of a considerable proportion of those collected are empty or nearly so.

Reproduction, Ontogeny, and Mortality.—In Tillamook and Netarts bays, Tillamook Co., the birthing season commences during the first half of May and peaks about 1 month later. Adjusted counts at shore sites indicated that young composed 22.4–24.3% of herds in Tillamook Bay at peak of birthing. No young were seen in the Rogue River and few were seen in the Columbia River (Brown and Mate, 1983); none was seen in Yaquina Estuary (Lincoln Co.—Bayer, 1985). In Coos Bay, Coos Co., birthing commences in April and peaks in May (Graybill, 1981).

Following weaning of young 2–6 weeks after parturition, adult females enter an estrous period that may last 1–9 weeks (Bigg, 1981). Courtship and mating are rarely observed as they probably occur in the water. Mating likely is promiscuous; adult males apparently are fertile for ≅9 months of the year (Bigg, 1981). Ovulation in at least some females seems to be induced. Following fertilization and development to the blastocyst stage, implantation is delayed 1.5 to 3 months (Bigg, 1981).

The natal pelage (lanugo) of harbor seals usually is shed in utero (Shaughnessy and Fay, 1977), and young are born with adult-type, spotted pelage (Bigg, 1981). An occasional individual (possibly born prematurely) may retain lanugo at birth (Allen, 1980; Shaughnessy and Fay, 1977). In Tillamook and Netarts bays, the adult-type pelage is molted in July–early September (Brown and Mate, 1983); loss and replacement require 1–2 months (Bigg, 1981).

Young are able to swim at birth; at 2–3 days they can dive for 2 min and at 10 days for 8 min (Bigg, 1981). They are capable of producing a cry much like that of a domestic sheep. Young nurse for ≅1 min at 3–4-h intervals. At the approach of danger, maternal females grab their young and dive into the water for safety. Young are abandoned by the maternal female at weaning. Newly weaned young feed mostly on crustaceans, especially shrimps (Crangon), for 1.5–3 months. Males mature at 3–6 years, females at 2–5 years; 85–92% of mature females produce young annually (Bigg, 1981).

Mortality among young may be 21% during the 1st year of life. Among adults mortality is greater among males than females (Bigg, 1981). Harbor seals are preyed upon by northern sea lions (Eumetopias jubatus—Pitcher and Fay, 1982), killer whales (Orcinus orca), and sharks (Lamniformes—Mate, 1981). Maximum longevity is ≅32 years (Bigg, 1981).

Habits.—Harbor seals are solitary in the water, but commonly assemble in small groups (sometimes >500) of mixed sex and age when they come ashore. There is no social hierarchy, but both males and females usually maintain 0.5 m or more of individual space; the space is maintained by snorts, growls, butts, and bites (Bigg, 1981).

From marked individuals and those followed by radio-telemetry, some harbor seals move great distances whereas others move little or not at all. For example, one individual was sighted 27 times in 9 months, always in Netarts Bay, Tillamook Co., whereas another made at least three trips between Netarts and Tillamook bays (≅25 km by sea), still another made a round trip between Netarts Bay and Whale Cove, Lincoln Co. (≅75 km), and a third marked in Netarts Bay was sighted in Winchester Bay, Douglas Co. (220 km). Also, the tag attached to a harbor seal was recovered in a net in Humboldt Bay, California, 550 km south of the tagging site in Netarts Bay (Brown and Mate, 1983). Dissimilarities in movements of individual harbor seals also were reported to occur elsewhere (Pitcher and McAllister, 1981). Movements seemingly permitted use of specific estuaries for feeding or for birthing and caring for young (Brown and Mate, 1983).

Remarks.—Shaughnessy and Fay (1977) recounted events that led to the original spelling of the subspecific name richardii and its justified emendation to richardsi. We followed their counsel in use of richardsi.

Conflicts between harbor seals and commercial and sport fishermen likely will remain as long as fish (and those who wish to catch them) and seals exist. As formerly avid sport anglers, we can attest to the frustration of trolling a herring for most of an incoming tide without a strike only to have a harbor seal surface 50 m away with a bright coho salmon (Oncorhynchus kisutch) in its jaws. For those whose livelihood depends upon a catch, such a sight must transcend frustration to pure anger. On the other side of the coin, for the city dweller who catches sight of a pod of

Table 12-9.—*Estimated annual consumption of fish species most commonly eaten by harbor seals* (Phoca vitulina) *along the Oregon coast, 1980. Data from Harvey (1988:136, table 12).*

Vernacular name	Scientific name	Number eaten	Biomass consumed (kg)
Staghorn sculpin	Leptocottus armatus	11,730,700	732,000
Pacific herring	Clupea pallasi	5,909,000	458,200
Shiner surfperch	Cymatogaster aggregata	10,481,300	447,300
English sole	Parophrys vetulus	9,463,600	434,100
Rex sole	Glyptocephalus zachirus	3,127,900	337,500
Chinook salmon	Oncorhynchus tshawytscha	48,700	309,000
Sablefish	Anoplopoma fimbria	509,200	280,300
Pacific sand lance	Ammodytes hexapterus	41,771,100	254,300
Northern anchovy	Engraulis mordax	7,766,100	244,200
Coho salmon	Oncorhynchus kisutch	40,200	223,100
Starry flounder	Platichthys stellatus	589,300	222,100
Pacific lamprey	Lampetra tridentata	792,600	214,000
51 other taxa combined		21,963,200	1,947,500

seals ashore, or an individual seal that surfaced with a salmon, a day at the coast is a memorable event. Consequently, we maintain that no species of mammal, no matter the apparent severity of its negative economic impact, is totally without redeeming value.

FAMILY PROCYONIDAE—RINGTAIL AND RACCOON

The procyonids are represented in Oregon by the raccoon (*Procyon lotor*) and the ringtail (*Bassariscus astutus*). Both of these species have distinctive facial markings, annulated tails, 40 teeth, and rounded, erect ears, and are pentadactyl, omnivorous, largely nocturnal, partly arboreal, and commonly associated with habitats near water. Raccoons in Oregon are classified as furbearers, and an annual season for trapping them or hunting them with dogs is provided; ringtails are protected year-round.

As a group, the Procyonidae is the subject of some debate. Stains (1984) and Nowak (1991) indicated that the family includes seven Recent genera and 19 species; both included the red panda (*Ailurus fulgens*), thereby extending the range of the family to include parts of southeast Asia. Wozencraft (1989) did not include the red panda, thus reduced- the diversity to six genera and 18 species, and limited the geographic distribution of the family to the New World. However, he indicated that the five species of *Bassaricyon* (olingos) probably were conspecific, that the five insular species of *Procyon* probably were conspecific with *P. lotor,* and that the North American and South American forms of *Nasua* (coatimundis) probably were distinct species. Thus, by this classification, the family would still include six genera but only 10 species.

The earliest-known fossil procyonids occurred in Early Oligocene deposits in Europe and North America; mammals recognized as *Procyon* first occurred in deposits of Miocene age (Hemphillian) of Texas and *Bassariscus* first occurred in Miocene-age (Barstovian) rocks of southern California (Savage and Russell, 1983). Fossils of *P. lotor* first appeared in Pleistocene (late Irvingtonian) deposits in Florida, whereas those of *B. astutus* have been found in Wisconsinan-age deposits in California, Nevada, New Mexico, and Texas (Kurtén and Anderson, 1980).

KEY TO THE PROCYONIDAE OF OREGON

1a. Face mask black; five–seven dark bands usually completely encircling tail; palate extends far beyond posterior margin of M3; baculum recurved distally with bilobed tip.................***Procyon lotor***, raccoon (p. 400).
1b. Face mask white; eight dark bands on tail broken on ventral surface; palate extends no farther than posterior margin of M3; baculum straight or slightly curved often with spatulate tip..
............................***Bassariscus astutus***, ringtail (p. 397).

Bassariscus astutus (Lichtenstein)
Ringtail

1830. *B*[*assaris*]. *astuta* Lichtenstein, 1827:119.
1859. *Bassaris raptor* Baird, 2(2):19.
1894*b. Bassariscus astutus flavus* Rhoads, 45:416.
1913*b. Bassariscus astutus raptor*. Grinnell, 3:289.

Description.—The ringtail (Fig. 12-27) is shaped somewhat like a marten (*Martes americana*) with an elongate

Fig. 12-27. Photograph of the ringtail, *Bassariscus astutus*. (From Oregon Historical Society negative no. G IV C-31b.)

body and tail of approximately equal length (Table 12-10); the tail has eight dark bands alternating with seven buffy bands, but the dark bands are broken on the ventral surface of the tail. The bands become progressively longer posteriorly. The dorsum is buff overlain with blackish or dark-brownish guard hairs; the venter is whitish or light buff. The face mask is whitish, but the eyes are ringed with a narrow band of black. The feet are pentadactyl and the toes are equipped with semiretractile claws and, except for the pads, furred on the ventral surface. The ears are long and set wide apart on a nearly horizontal plane. Eight males captured near Myrtle Creek, Douglas Co., in 1992 averaged 925 g (range, 747–1,097 g); one female weighed 734 g (T. P. Farrell, in litt., September 1993).

The skull of the ringtail (Fig. 12-28) is moderately heavy with a relatively narrow rostrum, a large braincase, well-developed postorbital processes, and a palate that does not extend posteriorly to M2. P4 is sectorial. A sagittal crest usually is lacking, but if present it is restricted to the posterior part of the cranium as the temporal ridges tend to be set wide apart. The hind foot can be rotated 180° so that the ringtail can descend trees or rocks head first (Trapp, 1972).

Distribution.—The ringtail occurs from southwestern Oregon south to the tip of Baja California Sur and eastward through Nevada, Utah, Colorado, and Kansas then south to Oaxaca (Fig. 12-29; Hall, 1981). In Oregon, museum specimens of *B. astutus* have been collected in Lane, Douglas, Klamath, Jackson, Josephine, and Curry counties (Fig. 12-29). Some sight records compiled by R. E. Richards, University of Colorado at Colorado Springs, are beyond the range based on museum specimens (Fig. 12-30); some of the sight records were accompanied by photographs, but the veracity of others is open to question.

Although representatives of the genus *Bassariscus* are recorded from deposits dating to the Miocene, only Recent material is known for *B. astutus* and it is from within the present geographic range of the species (Poglayen-Neuwall and Toweill, 1988); none is from Oregon.

Geographic Variation.—Only one of the 14 subspecies currently recognized occurs in Oregon: *B. a. raptor* (Hall, 1981).

Habitat and Density.—In the Myrtle Creek drainage in Douglas Co., one female and four male ringtails were radio-located 157 times while they were in dens. Occupied

Table 12-10.—*Means (±SE), ranges (in parentheses) and CVs of measurements of skull and skin characters for female and male ringtails (Bassariscus astutus) and raccoons (Procyon lotor) from regions in Oregon. Skin characters were recorded from specimen tags; skull characters were measured to the nearest 0.01 mm. SE and CV not provided if* n <10.

Species and region	Sex	n	Total length	Tail length	Hind foot length	Ear length	Mass (kg)	Condylobasal length	Basilar length	Postglenoid length	Length of maxillary toothrow
Bassariscus astutus	♀	3	744 (724–763)	362 (356–368)	72				75.89 (74.84–76.47)	24.24[a] (23.81–24.66)	30.53 (29.82–31.16)
	♂	3	742[a] (724–760)	359[a] (358–360)	69[a] (67–70)	49[a] (46–53)	0.90[a] (0.83–0.97)	77.96 (75.80–79.05)	70.24 (68.14–71.45)	24.85[a] (24.16–25.53)	31.09 (30.29–31.73)
Procyon lotor											
Western and Central Oregon	♀	30	780 ± 33.8[b] (714–826) 7.5	299 ± 30.6[b] (260–360) 17.7	124 ± 3.5[c] (120–127) 4.0	54 ± 6.0[c] (48–60) 15.7	9.39	109.36 ± 0.58[a] (102.59–114.15) 2.84	99.09 ± 0.53 (92.61–103.26) 2.91	31.41 ± 0.27[a] (29.05–34.81) 4.55	35.42 ± 0.26 (31.29–37.85) 4.05
	♂	30	1,100[c]	360[c]	125[c]	65[c]	10.00[c]	114.71 ± 0.44 (109.40–119.49) 2.11	104.14 ± 0.46 (96.93–108.84) 2.41	33.08 ± 0.29 (28.10–36.06) 4.93	36.54 ± 0.29 (34.85–43.86) 4.38
Eastern Oregon	♀	1						108.05	98.57		43.07
	♂	8						114.81 (108.08–120.31)	103.59 (96.25–108.30)		45.14 (43.48–46.58)

[a]Sample size reduced by 1. [b]Sample size reduced by 27. [c]Sample size reduced by 28.

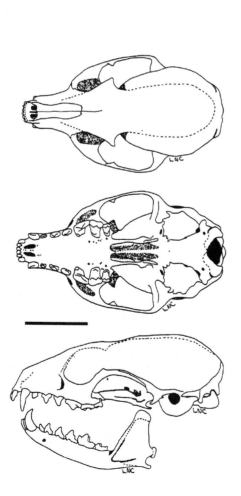

Fig. 12-28. Dorsal, ventral, and lateral views of the cranium, and lateral view of the dentary of the ringtail, *Bassariscus astutus* (OSUFW 1109). Scale bar equals 25 mm.

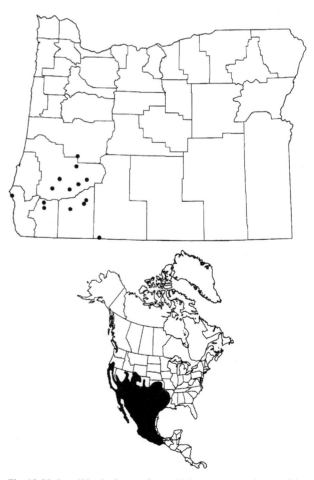

Fig. 12-29. Localities in Oregon from which museum specimens of the ringtail, *Bassariscus astutus*, were examined (listed in APPENDIX, p. 512), and the distribution of the ringtail in North America. (Range map redrawn from Hall, 1981:964, map 500, with modifications.)

Table 12-10.—*Extended.*

Zygomatic breadth	Cranial breadth	Least interorbital breadth	Length of mandible
48.19 (44.47–50.70)	33.63 (33.53–33.71)	15.82 (15.00–16.24)	52.18 (51.57–53.29)
47.79 (45.97–50.40)	34.45 (33.18–35.27)	15.30 (14.49–15.74)	54.54 (53.13–56.46)
76.89 ± 1.12 (48.03–82.42) 7.98	52.96 ± 0.73 (33.42–56.53) 7.55	25.14 ± 0.43 (15.57–28.87) 9.42	81.67 ± 1.07 (52.47–86.17) 7.20
82.79 ± 0.64 (74.48–90.66) 4.26	56.31 ± 0.29 (52.53–58.85) 2.90	25.95 ± 0.26 (22.75–28.28) 5.46	87.20 ± 0.41 (82.24–91.33) 2.54
80.39	51.36	27.42	81.13
79.49 (68.13–88.63)	53.83 (52.13–56.01)	25.95 (23.27–30.54)	86.08 (79.74–90.63)

dens were in old-growth forest (Douglas-fir, *Pseudotsuga menziesii*; incense cedar, *Calocedrus decurrens*; western red cedar, *Thuja plicata*; and sugar pine, *Pinus lambertiana*) 104 (66.2%) times, in a 3–4-year-old clear-cutting in a previously burned-over area (madrone, *Arbutus menziesii*; big-leaf maple, *Acer macrophyllum*; and Douglas-fir seedlings) 34 (21.7%) times, and in 20–35-year-old clear-cuttings (Douglas-fir) 19 (12.1%) times. Understory plants prominent near den sites included salal (*Gaultheria shallon*), sword-fern (*Polystichum munitum*), Oregongrape (*Berberis nervosa*), poison oak (*Rhus diversiloba*), and trillium (*Trillium ovatum* and *T. petiolatum*).

Grinnell et al. (1937) indicated that ringtails never occurred >0.8 km from water.

Diet.—Examination of 56 fecal droppings collected near entrances to dens of radio-tracked ringtails northeast of Myrtle Creek, Douglas Co., revealed that the species in Oregon, as elsewhere (Grinnell et al., 1937; Taylor, 1954;

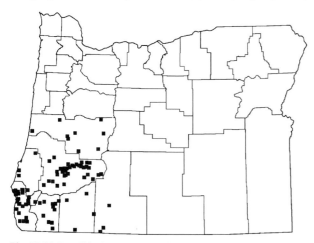

Fig. 12-30. Localities in Oregon at which ringtails, *Bassariscus astutus*, reportedly were sighted. Data provided by R. E. Richards, University of Colorado at Colorado Springs.

Toweill and Teer, 1977; Trapp, 1978), is decidedly omnivorous (Alexander et al., 1995). More than 90% of the Oregon samples contained remains of foods of both plant and animal origin. Some plant foods, such as berries (*Rubus*, *Gaultheria*) and seeds (*Acer*), occurred in the diet seasonally, whereas others, such as leaves, scales, and bracts of conifers (*Pseudotsuga*, *Calocedrus*, *Abies*), occurred year-round. Remains of mammals occurred in 66% of the samples; murid rodents were identified most frequently, with the red tree vole (*Phenacomys longicaudus*) occurring in 23% of the samples. Such suggests both nocturnal and arboreal foraging habits (Alexander et al., 1995). Insects also occurred in 66% of the samples and contributed most to the diversity of organisms consumed; carabid beetles (Coleoptera) occurred most frequently in spring, Orthoptera in summer, and Lepidoptera and Orthoptera in autumn (Alexander et al., 1995).

Reproduction, Ontogeny, and Mortality.—A female followed by radiotelemetry 31 March–6 November 1992 in Douglas Co. was located eight times in the same den in a 1-m-diameter, 9-m-tall Douglas-fir snag during the period 15 June–20 July (T. P. Farrell, in litt., September 1993). The female had cohabited with a radio-equipped male in another den 66 days earlier. Subsequent observations confirmed that the female gave birth and cared for a litter in the snag during that period. From 29 July to 8 September, the female was radio-located nine times, each time at a different den site, but all dens were located in the same 16.1-ha block as the den occupied 15 June–20 July. Ringtails with young in other regions change dens frequently, often daily after young are 20 days old (Poglayen-Neuwall and Toweill, 1988). We know of no other information regarding these topics for ringtails in Oregon.

Elsewhere, most ringtails mate in March–April and produce litters of one–four, rarely five, young in May–June after a gestation of 51–54 days (Poglayen-Neuwall and Toweill, 1988).

Neonates have cream-colored skin with pigmented areas marking dark tail rings, eyelids, and nose pad; the animals have sparse whitish hairs ≅3 mm long on the dorsum and ≅1 mm long on the venter. The eyelids and auditory meatuses are closed. The dried umbilical cord remains attached for 2–3 days postpartum (Richardson, 1942). White eye rings are evident by 10 days and the eyes open at 31–34 days (Richardson, 1942) or 22 days (E. Bailey, 1974). The auditory meatuses open at 29 days and the ears droop until ≅2.5 months (Richardson, 1942). Teeth appear at 5 weeks and young are fully furred and commence to eat meat at 7 weeks (E. Bailey, 1974). At 37 days, young were able to walk a few steps rather than crawl. Adult conformation is attained by 134 days (Richardson, 1942).

Poglayen-Neuwall and Toweill (1988) believed that great horned owls (*Bubo virginianus*) were the primary predator of ringtails, although coyotes (*Canis latrans*), bobcats (*Lynx rufus*), and raccoons (*Procyon lotor*) were known to prey on the species.

Habits.—Ringtails are strongly nocturnal (Grinnell et al., 1937), but in captivity, at least, may be active in daylight hours during the reproductive season (E. Bailey, 1974). Captive ringtails also are aggressive toward each other except during the reproductive period. Ringtails are superb climbers, often making use of a variety of rock-climbing techniques employed by humans who enjoy the sport

(Trapp, 1972). Neonates squeak and older young may make vocalizations described as a "chuck," "chirp," "growl," or "explosive bark" (E. Bailey, 1974; Richardson, 1942).

Some observers think that ringtails den in pairs, others think that individuals den alone; anecdotal accounts of their behavior in captivity indicate that pairs show mutual tolerance except perhaps for a few days after parturition (Grinnell et al., 1937). In Douglas Co., on 4 April, a female to which a beacon radio was attached simultaneously occupied a den with a radio-equipped male on at least one occasion. Of 151 denning sites used by ringtails in Douglas Co., 71 (47.0%) were inside standing cedar, Douglas-fir, or big-leaf maple snags; 41 (27.2%) were in live cedar, Douglas-fir, big-leaf maple, or sugar pine trees; and 39 (25.8%) were under shed bark or log piles often covered with blackberry (*Rubus*) brambles. The aspect at 144 den sites was significantly different from expected on the basis of random directions (Fig. 12-31). We had no measure of slopes available. Trapp (1972) indicated that captive ringtails were claustrophilic, commonly wedging themselves into tight crevices or several often denning together in a narrow cardboard box.

Home-range areas of one female and four male ringtails in the Myrtle Creek drainage, Douglas Co., were estimated from convex polygons constructed by connecting 16.1-ha (40-acre) blocks in which the animals were located in dens by radiotelemetry (Fig. 12-32). Average home-range area for the males was 332.3 ha (range, 161.9–550.6 ha); the area for the female was 267.2 ha. The overlap of the home range of the female with that of one of the males (Fig. 12-32) was temporal as well as spatial. In Utah, home-range areas of nine adult males averaged 139 ha and those of four females averaged 129 ha (Trapp, 1978).

Ringtails are noted for their lack of wariness, an attribute long recognized (Audubon and Bachman, 1849). We reviewed letters assembled by R. E. Richards, University of Colorado at Colorado Springs, from several persons who live in remote cabins in southwestern Oregon and who described ringtails taking food from their hands and other equally incautious behaviors.

Remarks.—Information regarding reproductive activities, habitat, home range, and dens in Douglas Co. was derived from data provided by T. P. Farrell, Oregon

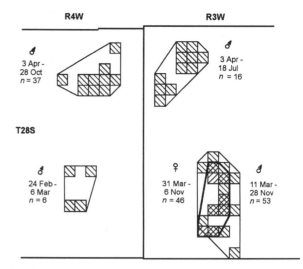

Fig. 12-32. Convex-polygon home ranges of one female and four male ringtails in Douglas Co. constructed around 16.1-ha (40-acre) blocks in which an individual was radio-located in a den one or more times during the interval indicated. Sample size (*n*) is the number of times ringtails were located in dens. Home range of the female, indicated by the wide-line polygon, is entirely within the home range of a male. Data provided by T. P. Farrell, Oregon Department of Fish and Wildlife.

Department of Fish and Wildlife; analyses and interpretations are ours.

The ringtail in Oregon is classified as a "sensitive species–undetermined status" by the Oregon Department of Fish and Wildlife "based on infrequent sightings or reports" and is afforded protection against taking by administrative rule (Marshall, 1992:RING 1).

Bassariscus astutus sometimes is referred to as the "ringtailed cat" or "civet cat," especially in popular literature. Such misnomers should be avoided as this species is not a cat (felid) but a procyonid, even though it may be somewhat cat-like in its climbing ability and perhaps in other behaviors. Also, "civet cat" is a vernacular name sometimes applied to the western spotted skunk (*Spilogale gracilis*), creating the possibility of confusing two very different Oregon mammals. The less commonly used Mexican vernacular name "cacomistle" is not used much here and should be reserved for the Central American congener *B. sumichrasti*.

Procyon lotor (Linnaeus)

Common Raccoon

1758. [*Ursus*] *lotor* Linnaeus, p. 148.
1899b. *Procyon psora pacifica* Merriam, 16:107.
1923. *Procyon lotor pacifica*: Grinnell, 21:316.
1930. *Procyon lotor excelsus* Nelson and Goldman, 11:458.

Description.—The raccoon (Plate XXV) is a moderately large (Table 12-10), heavily furred mammal sufficiently bear-like in overall conformation to have caused Linnaeus to describe the species originally as a member of the genus *Ursus*. However, the raccoon is best characterized by its distinctive markings. The face has a dark brownish-black eye mask sharply separated from whitish nose patches, eyebrow lines, and exclamation points between the eyes. The inside of the ears is ringed with white, adding to the facial contrasts. The tail is annulated with five–seven (usually six) dark rings separated by light gray or tan rings. The five digits on both manus and pes are

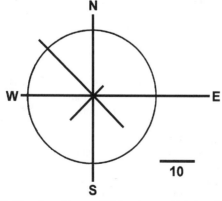

Fig. 12-31. Direction of slope at 144 den sites at which ringtails in Douglas Co. were radio-located. Length of line (note scale) indicates number of den sites on slope; circle indicates expected number based on random selection. Differences between numbers observed and numbers expected were significant ($\chi^2 = 32.889$, *d.f.* = 7, *P* < 0.01). Data provided by T. P. Farrell, Oregon Department of Fish and Wildlife.

unwebbed, the soles naked, and the manus well adapted for manipulating small objects (Lotze and Anderson, 1979). The hind legs are longer than the front; thus, in a walking gait, the rump is higher than the head.

The skull of the raccoon (Fig. 12-33) is heavy with a broad rostrum and braincase; heavy, moderately high-crowned molars with prominent cusps; and a mandible with a high, posteriorly curved coronoid process (Goldman, 1950). M1–M3 are squarish or subtriangular. The palate extends far posterior to M3.

The baculum in adults is a heavy, strongly recurved bone bilobed distally. The average mass of the baculum increases significantly with each year class; maximum mass was 5.7 g in a sample of 514 from northwestern Oregon (Fiero and Verts, 1986b). Average length increases significantly to age-class 3 (≅2.5 years old), but not beyond; the greatest length recorded was 119.5 mm in a 3.5-year-old male (Fiero and

Verts, 1986a).

The largest raccoon in Oregon that we recorded weighed 10.9 kg; however, we are confident that larger individuals occur here. Stains (1956) reported that 15 of 559 raccoons in Kansas weighed ≥13.6 kg, with one weighing 19.1 kg. Scott (1951) reported a Wisconsin male that weighed 28.4 kg.

Distribution.—Except for parts of the Rocky Mountain and Great Basin areas, the raccoon occurs from central Canada south to Panama (Fig. 12-34; Hall, 1981). The raccoon occurs in suitable habitats throughout Oregon; it does not occur in high mountain regions or in desert regions except near permanent water sources (Fig. 12-34).

Geographic Variation.—Two of the 25 currently recognized subspecies of the raccoon occur in Oregon: the relatively small and extremely dark *P. l. pacificus* in the western two-thirds of the state and the large and light-colored *P. l. excelsus* in the eastern one-third of the state (Goldman, 1950; Hall, 1981). The latter is considered the largest race of raccoons (Goldman, 1950).

Habitat and Density.—The characteristics that contribute most to habitats being suitable for raccoons are water and trees, although in certain situations habitats lacking either or both may be used by raccoons (Stains, 1956). The unmistakable tracks of raccoons can almost always be found by walking 100 m or so along the muddy banks of a stream, slough, lake, or pond. A trap with pan covered with aluminum foil

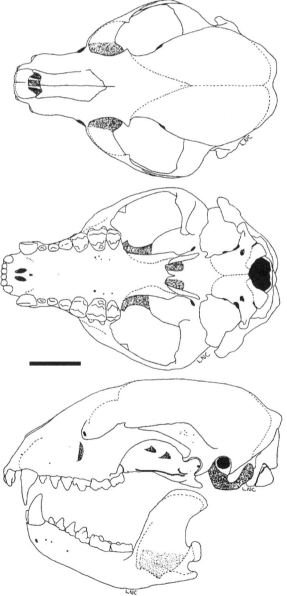

Fig. 12-33. Dorsal, ventral, and lateral views of the cranium, and lateral view of the dentary of the raccoon, *Procyon lotor* (OSUFW 3187). Scale bar equals 25 mm.

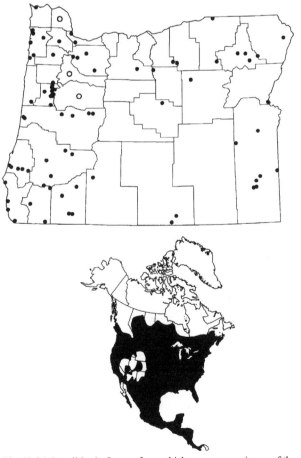

Fig. 12-34. Localities in Oregon from which museum specimens of the raccoon, *Procyon lotor*, were examined (listed in APPENDIX, p. 547), and the distribution of the raccoon in North America. (Range map redrawn from Hall, 1981:971, map 502, with modifications.) Open symbol indicates record for county.

placed 1–2 cm beneath the surface of clear water in these situations almost always proves irresistible to fondling by any raccoon that passes. A trap set on the bank and baited with a "worm" of sardine mixture squeezed from a catsup squeeze-bottle onto a nearby log or rock (Verts, 1961) usually proves equally irresistible to passing raccoons.

Hollow trees provide dens that serve to insulate the occupying raccoon from inclement weather (Stains, 1961). However, in areas where trees are scarce or absent, raccoons use dens constructed and abandoned by beavers (*Castor canadensis*), badgers (*Taxidea taxus*), or other species. They may den in natural cavities in rocky areas.

We know of no estimates of density for Oregon. However, from the reported harvest of raccoons by county (Table 12-11), it is evident that counties west of the Cascade Range, especially those in which agriculture is extensive, produce the most raccoons. Not only are the requisite water and trees more abundant, but agricultural commodities that can be used by raccoons for food are more widespread west of the Cascade Range.

Diet.—Raccoons can and do eat almost anything organic. We know of no reports listing items ingested by raccoons in Oregon, but results of studies conducted elsewhere (Baker et al., 1945; Hamilton, 1936b, 1940; Llewellyn and Uhler, 1952; Stains, 1956; Yeager and Rennels, 1943) indicate that plant materials (fruits, seeds, berries, acorns) usually compose more than half of the diet of raccoons. Vertebrates and invertebrates, especially crayfish (Decapoda) and clams (Pelecypoda), interchangeably rank second and third depending upon season and locality. That raccoons exploit windfall sources is exemplified by a study conducted on a refuge before and after the waterfowl hunting season (Yeager and Elder, 1945); plant materials composed 64% and 23% of the bulk of the diet of raccoons before and after the hunting season, respectively, whereas crippled waterfowl composed 3% and 65% of the diet during the same periods. Raccoons also commonly take advantage of the abundance and ease of foraging on sweet corn, melons, and other truck crops, often eliciting the wrath of the farmer. Thus, the raccoon is best classified as a dietary generalist.

Reproduction, Ontogeny, and Mortality.—Ordinarily, adult and some yearling female raccoons enter estrus and breed once each year; some females delay until nearly 2 years old to breed. Females that fail to become pregnant, abort or resorb their litters, or lose their young at or near birth may enter estrus and ovulate a second time 80–140 days later (Sanderson and Nalbandov, 1973). Ovulation is spontaneous, several reports to the contrary notwithstanding (Sanderson and Nalbandov, 1973), and females that do not become pregnant become pseudopregnant. Those that become pregnant or pseudopregnant have corpora lutea that persist 14–16 days beyond the normal 63-day gestation period. Embryos that implant and survive for ≥1 month produce pigmented sites of implantation that persist for ≥10 months in captives, longer in wild raccoons (Sanderson and Nalbandov, 1973); counts of pigmented sites of implantation are useful in estimating litter size and potential productivity (Fiero and Verts, 1986a; Sanderson and Nalbandov, 1973).

Although individual males commonly have 3–4-month periods during which they are incapable of breeding, at least a portion of the males in populations of raccoons seem to be capable of breeding at all seasons (Sanderson and Nalbandov, 1973). Mass of the testes in Illinois raccoons was maximal in December, minimal in June–August. Except for a few accounts of female raccoons breeding at odd seasons, raccoons usually are born in March–June; in Illinois, the peak of births was in April (Sanderson and Nalbandov, 1973). Inclement weather may delay breeding (Dorney, 1953; Sanderson and Nalbandov, 1973). We know of no information regarding dates of breeding in the raccoon in Oregon, but we would not be surprised if the peak was a bit later than the 10 February peak in Illinois (Sanderson and Nalbandov, 1973:44, fig. 5).

Based on the presence and number of pigmented sites of implantation in female raccoons obtained from hunters and trappers during the November–February furbearer seasons in northwestern Oregon, only 31.8% (41 of 129) of ≅1.5-year-olds (yearlings) had produced young, whereas 91.8% (169 of 184) of older females had produced young; litter size averaged ($\pm SE$) 2.2 ± 0.2 (range, 1–4; $n = 41$) in yearlings and 3.2 ± 0.1 (range, 1–6; $n = 169$) in older females (Fiero and Verts, 1986a). Nearly half of parous females ≥2.5 years old possessed pigmented sites in two or three discernible sets based on the intensity of the pigment. Counts of darker sets always averaged greater than those of the next lightest sets in the subsequent age-class although counts of the darkest set did not differ significantly among older age-classes of raccoons. Such suggests that pigmented sites of implantation disappeared with time or that newly formed sites in older females had a greater probably of obscuring implantation sites formed earlier (Fiero and Verts, 1986a).

Female raccoons ≅2.5 years old contribute the most young to the population because of their slightly larger litter size and because they form the greatest proportion of parous females. Nevertheless, yearling females have the greatest potential to affect recruitment because they form the greatest proportion of the female cohort. In a 2-year study of reproduction of Oregon raccoons, the proportion of yearlings that had produced offspring was amazingly constant at 31.2–32.8% (Fiero and Verts, 1986a). However, in longer studies conducted elsewhere the proportion of parous yearlings varied greatly (Fritzell et al., 1985).

At birth, the back and sides are of uniform color, but the color of individuals may range from black to dark gray to dark brown to light tan (Montgomery, 1968) to yellow mixed with gray (Hamilton, 1936b). Hairs of some individuals are of uniform color, but those of others may be tipped with black. The face mask and tail rings are visible as pigmented skin; both areas are sparsely haired, but the color of the hairs does not correspond to that of the skin on the tail, causing the rings not to be readily distinguishable on the tails of some individuals (Montgomery, 1968). By 1 week of age about half have visible tail rings and by 2 weeks most have fully haired tail rings. Montgomery (1968) reported that no loose hairs were found by combing young <7 weeks old, but that a progressively larger percentage of the surface of young raccoons was covered by agouti-type guard hairs, so by 7 weeks of age their pelage resembled that of adults.

For 51 captive young, the average age at which the eyes opened was 23.4 days (range, 18–30 days); the first solid food was consumed at 10 weeks; and by 15 weeks, little milk was being obtained through nursing (Montgomery,

Table 12-11.—Reported annual harvest of raccoons (Procyon lotor) in Oregon by county, 1969–1996. Data provided by Oregon Department of Fish and Wildlife.

County	Trapping season																										
	1969–1970	1970–1971	1971–1972	1972–1973	1973–1974	1974–1975	1975–1976	1976–1977	1977–1978	1978–1979	1979–1980	1980–1981	1981–1982	1982–1983	1983–1984	1984–1985	1985–1986	1986–1987	1987–1988	1988–1989	1989–1990	1990–1991	1991–1992	1992–1993	1993–1994	1994–1995	1995–1996
Baker	46	33	19	53	59	36	67	46	115	69	133	71	88	96	85	84	114	173	119	39	33	24	225	39	75	85	65
Benton	24	38	88	129	169	246	231	185	293	390	242	265	283	543	498	687	525	556	237	100	145	98	147	75	149	205	95
Clackamas	53	41	48	71	163	119	220	130	438	307	314	452	667	623	612	877	384	222	149	141	62	111	119	154	184	124	68
Clatsop	204	87	87	128	302	334	302	344	528	388	328	321	264	206	283	387	294	370	404	193	96	137	283	161	182	224	53
Columbia	85	31	53	78	115	209	249	283	382	352	162	196	404	272	227	389	414	351	230	148	98	127	120	48	63	71	40
Coos	42	18	46	78	140	260	434	257	357	440	405	460	276	478	406	540	663	749	613	275	177	221	347	240	249	208	198
Crook	9	5	9	20	10	10	17	11	16	9	7	27	13	16	13	26	24	42	52	4	3	2	12	15	14	13	5
Curry	16	10	4	20	34	51	73	115	210	212	184	352	183	349	263	368	220	230	150	81	19	52	52	47	36	30	37
Deschutes	41	16	11	4	20	10	13	17	24	37	12	37	26	25	35	57	89	70	60	64	27	24	46	35	44	38	4
Douglas	168	100	78	146	266	452	403	629	1,470	1,154	915	1,056	968	1,233	1,067	1,477	1,037	1,699	1,197	497	421	420	566	363	366	362	230
Gilliam		5	3	7	1	3	3	6	3	1	1	1	23	10	10	12	11	20	18	18	9	9	22	27	15	4	9
Grant	25	30	44	63	101	47	18	51	43	18	57	34	41	62	106	94	175	162	141	47	37	31	7	27	6	15	2
Harney	8	5	9	37	52	18	51	24	16	20	19	17	17	26	30	42	34	40	34	14	20	5	5	7	3	15	13
Hood River	23	10	27	7	9	21	93	64	22	47	32	98	42	71	67	82	51	69	61	52	5	122	89	92	76	37	29
Jackson	70	48	45	87	78	133	149	258	421	460	416	633	604	1,411	882	999	692	1,534	1,721	592	189	245	518	117	66	163	5
Jefferson			2		48	23	23	10	7	2	16	15	32	38	60	64	98	48	52	26		11	45	12	28	11	25
Josephine	86	32	77	59	87	93	63	82	211	244	289	406	502	776	454	534	387	511	614	175	121	153	246	72	48	95	25
Klamath	141	89	65	186	192	138	144	122	117	147	118	160	140	144	115	129	224	292	321	67	115	90	88	56	74	78	8
Lake	30	7	2	26	66	66	25	27	31	5	26	9	9	16	29	32	40	45	38	21	5	7	7	5	6	11	6
Lane	308	221	202	386	593	840	948	918	1,073	1,050	1,229	1,183	877	1,265	1,017	1,086	1,034	1,169	1,125	435	225	178	258	199	328	221	159
Lincoln	207	70	77	252	284	398	409	281	299	387	400	348	385	215	367	570	703	519	407	360	149	63	124	190	113	160	145
Linn	139	57	111	379	347	583	477	380	494	588	602	506	423	637	712	932	1,073	1,205	558	337	303	135	353	260	231	374	113
Malheur	37	27	43	25	69	76	70	104	46	37	86	80	39	49	53	108	108	211	114	86	14	20	27	22	57	41	2
Marion	104	72	82	206	161	384	284	274	656	735	474	523	510	742	838	769	889	531	334	200	121	107	241	210	453	257	134
Morrow	6	4	14	17	5	5	7	4	28	72	41	19	18	20	35	16	72	65	43	45	27	19	34	13	10	4	5
Multnomah	28	31	18	26	6	34	70	111	108	126	124	77	123	106	164	182	122	121	69	63	61	61	41	15	41	39	3
Polk	29	27	16	94	86	141	148	246	343	274	250	310	150	272	339	385	458	638	251	95	107	119	136	166	134	87	46
Sherman			5					4			15	48	17	19	14	3	13	15	6	3		4	4	10	2	5	
Tillamook	101	97	70	170	226	284	266	331	372	363	416	499	240	265	266	322	329	314	264	223	136	76	156	131	113	178	92
Umatilla	28	19	21	38	33	48	46	29	75	40	99	121	201	188	142	191	298	300	199	78	96	42	43	77	38	57	15
Union	37	13	15	5	22	20	13	20	26	19	32	53	37	48	34	57	44	70	39	32	24	23	17	17	16	14	3
Wallowa	142	85	77	93	55	70	66	62	63	80	75	62	58	132	77	138	82	113	97	50	25	10	33	55	40	56	14
Wasco	8	15	9		3	27	58	79	82	54	51	113	80	109	72	65	150	89	76	35	61	115	131	130	56	61	77
Washington	101	22	30	16	40	118	109	165	268	235	172	326	220	558	503	510	219	360	305	65	90	74	94	91	197	138	39
Wheeler	2	4	5	9	1	1	6	8	9	13	9	128	19	16	13	15	21	18	75	29	6	6	4	10	6	9	
Yamhill	108	62	38	97	94	121	78	479	366	499	330	641	436	703	639	730	455	369	243	87	79	88	117	168	264	99	103
Totals	2,456	1,431	1,550	2,995	3,897	5,036	5,694	6,156	9,012	8,874	8,091	9,556	8,382	11,804	10,527	12,967	11,546	13,288	10,418	4,798	3,232	3,014	4,754	3,316	3,778	3,584	1,842
Trappers reporting[a]	721	773	731	847	1,102	1,350	1,377	1,755	2,746	2,899	3,216	3,373	3,069	3,980	3,292	3,104	3,575	3,397	3,833	1,985	1,536	1,228	1,390	1,454	1,293	1,464	687
Average price/pelt[b]	$3.06	$1.98	$3.77	$6.85	$10.15	$10.13	$15.80	$20.69	$20.72	$24.43	$21.00	$23.88	$23.00	$16.46	$16.83	$16.89	$17.17	$19.81	$10.98	$4.40	$5.54	$3.00	$8.00	$5.00	$8.00	$6.00	$11.00

[a] Not all sought raccoons and not all were successful.
[b] Rounded to the nearest dollar after the 1989–1990 season.

1969). Growth accelerated through the first 9 weeks, slumped during weeks 10 and 11 as young commenced to eat solid food, increased again during the 12th week, then declined markedly to the 16th week. Functional weaning commenced when young were about 8 weeks old and was essentially complete by the 12th week of age; however, some individuals continued to have small amounts of milk in their stomachs to the 16th week of age when examinations ceased (Montgomery, 1969).

Average (±SE) age (in days) at eruption of the deciduous teeth (d) was dI1, 34.0 ± 2.8; dI2, 25.4 ± 2.1; dI3, 26.2 ± 1.2; dC1, 29.3 ± 0.9; dPM1, 64.5 ± 1.9; dPM2, 46.2 ± 1.0; dPM3, 49.2 ± 1.2; dPM4 48.7 ± 1.0; and di1, 28.5 ± 2.2; di2, 37.3 ± 2.1; di3, 33.0 ± 2.4; dc1, 29.3 ± 0.8; dpm1, 60.7 ± 1.8; dpm2, 43.4 ± 1.6; dpm3, 48.4 ± 1.2; dpm4, 48.7 ± 1.0. Similar data for permanent teeth were I1, 65.6 ± 1.4; I2, 73.3 ± 1.3; I3, 96.6 ± 1.7; C1, 111.7 ± 3.9; M1, 81.0 ± 1.2; and i1, 65.9 ± 1.0; i2, 72.6 ± 1.2; i3, 85.5 ± 1.9; c1, 105 ± 3.6; m1, 78.1 ± 1.5. A large proportion of dI1 and di2, and a small proportion of di1, dI2, and di3, either did not erupt or erupted and were shed before the animals were first examined (Montgomery, 1964). Thus, to ≅3.5 months of age, the sequence of eruption of the teeth can be used as a reasonably reliable criterion of age (Montgomery, 1964). The congruence among several criteria for estimating ages of raccoons older than ≅3.5 months was investigated by Fiero and Verts (1986b); several commonly used methods seemed to produce equally reliable results with large samples.

In North Dakota, survival rates for radio-equipped raccoons in the March–August period were 0.93 and 0.81 for adults and yearlings, respectively; similar data during the same season for Minnesota were 0.97 for adults and 0.83 for yearlings (Fritzell and Greenwood, 1984). Causes of death mostly were related to automobile traffic and being shot or trapped. Undoubtedly, the greatest mortality factors for raccoons in Oregon also involve encounters with automobiles and harvest for sport and fur (Table 12-11). The decline in number of hunters and trappers harvesting raccoons and the decline in reported harvest in recent years tend to suggest that economics (fur prices) plays more of a role in the extent of the harvest than do changes in relative abundance of raccoons. In addition to the harvest for fur, there are records of raccoons being preyed upon by bobcats (*Lynx rufus*), red foxes (*Vulpes vulpes*), coyotes (*Canis latrans*), and several species of raptorial birds (Stains, 1956). We presume that raccoons in Oregon occasionally fall prey to these species, but we suspect that, compared with the reported harvest or losses to accidents, losses to predators have relatively minor effects on density.

Habits.—Few, if any, behavior patterns of wild mammals are more widely recognized, and equally widely misinterpreted, than the so-called "food-washing" behavior of the raccoon. Wild raccoons, in foraging along streams, commonly sit on their hind legs (sometimes in the water), stare vacantly into space, and search for food by "dabbling" with splayed-fingered forepaws. Captive raccoons commonly carry food items to water-filled vessels in their pens and "douse" the items in the water. In the past, some observers have interpreted the "dousing" behavior as a means of moistening the food, whereas others considered the activity a means of cleansing the

food. Lyall-Watson (1963) showed that "dousing" by raccoons was related to the object's shape, odor, size, and distance from water, and to the water depth, nature of the substratum beneath the water, and hunger of the raccoon, but not to the texture, cleanliness, or wetness of the object or to the water temperature. He also found that foraging by "dabbling" was exhibited by 16-week-old raccoons at their first encounter with free water, thus was an unlearned behavior. However, "dousing" was not exhibited until young raccoons had first found food by "dabbling." Lyall-Watson (1963) concluded that "dousing" by captive raccoons, a behavior not observed in the wild, was to satisfy the appetitive fixed-action pattern for "dabbling" for food not provided for in cages.

Raccoons taken from the wild in King Co., Washington, during the breeding season, acclimated to captivity, then paired in a neutral arena for 1 h usually commenced to interact by hissing and lashing their tails. Within 10–60 s, an aggressive posture and attitude consisting of an elevated tail, laid-back ears, raised hair on the shoulders, and constant hissing were assumed (Barash, 1974). In five of six pairings of individuals captured <5 km apart, one individual exhibited subordinateness by retreating to a corner of the arena and lowering the chin, neck, and venter to the floor; in the other pairing, the raccoons arched their backs, bared their teeth, growled loudly, and fought viciously, tearing clumps of fur from each other (Barash, 1974). However, of seven pairings of raccoons captured >5 km apart, subordinate-dominant polarity was exhibited by only one pair; the remaining six engaged in fighting. Barash (1974) suggested that neighboring raccoons recognized each other and exhibited a degree of tolerance not evident among individuals from more remote localities.

In North Dakota, movements of raccoons followed by radiotracking indicated that adult males maintained home ranges averaging 2,560 ha (range, 670–4,946 ha) that did not overlap with adjacent home ranges of adult males by >10% (Fritzell, 1978). Males rarely approached each other closer than 3 km; however, in one instance, two adult males met and traveled together for ≥1 h, whereas, in another instance, two adult males met, and one fled the scene at high speed. Home ranges of adult females averaged 806 ha (range, 229–1,632 ha) and overlapped each other and those of adult males extensively; however, parous females were never located with yearlings or adults of either sex (Fritzell, 1978). Based on this study, Fritzell (1978:260) suggested that adult males exhibited territoriality "probably in response to competition for access to females."

Remarks.—When female raccoons become pseudopregnant commonly there is a dramatic change in behavior; what was a tractable pet often becomes a ferocious, snarling biter. During late winter or early spring, one of us commonly has received a telephone call from a distraught parent whose pet female raccoon has just bitten, often severely, a member of the family. The sudden and drastic change in behavior usually elicits a fear of rabies. Although rabies in raccoons is fairly common in some regions of the United States, it is relatively rare in the Pacific Northwest. Despite the low incidence of rabies in raccoons here, and the low probability of a wild-caught mammal developing rabies months after capture as an infant, the almost 100% mortality among humans that develop symptoms of rabies demands that the offending pet be tested for rabies (which requires that it be sacrificed) and that postexposure inoculation be initiated to protect the bitten person. Although we suspect that pseudopregnancy, not rabies, is responsible for eliciting most bites in late winter or early spring by female raccoons maintained as pets, we always advise that the callers consult with their veterinarian and physician. We also strongly advocate leaving wild mammals in the wild!

FAMILY MUSTELIDAE—MUSTELIDS

Family Mustelidae includes 54 species in 21 genera distributed worldwide except for Australia. Although commonly referred to as the weasel family, it includes not only the weasels, but also the otters, badger, wolverine, mink, fisher, and marten. Nine species occur in Oregon. All are small to medium-sized, pentadactyl carnivores with short muzzles, relatively short legs, highly developed anal scent glands, and small, rounded ears. Except in the sea otter (*Enhydra lutris*), the carnassials are moderately to well developed. The condyloid process of the mandible is locked into the glenoid fossa by the anteriorly curved postglenoid process; in the sea otter it is impossible to separate the mandible from the dorsal skull without breaking bone. The baculum is well developed.

Most mustelids are efficient predators on terrestrial vertebrates, but minks and otters commonly feed on fish and crayfish (Decapoda), and sea otters feed almost exclusively on sea urchins (Echinoidea) and molluscs. Vertebrate prey is killed by a bite to the head or neck.

Mustelids have one litter annually. Gestation is extended in most species because development is interrupted between fertilization and implantation. For example, gestation averages 352 days (range, 338–358 days) in the fisher (*Mustela pennanti*—Hall, 1942), because development is interrupted for ≅270 days (Enders and Pearson, 1943). Litter size ranges from one in the sea otter to a maximum of 17 in the mink (*Mustela vison*—Enders, 1952). Young of some species such as the sea otter require long periods of maternal care.

Many mustelids are valued for their fur. Minks are raised commercially for their pelts.

Although all may be observed abroad in daylight hours occasionally, mustelids, except the sea otter and weasels, are largely nocturnal. Most mustelids are active at all seasons.

Mustelids first appeared in the fossil record in the Early Oligocene of Europe and North America. The first fossils representing extant genera that occur in Oregon were those of *Martes* in the Miocene (Orleanian—Savage and Russell, 1983).

KEY TO THE MUSTELIDAE OF OREGON

1a. With 38 teeth...2
1b. With ≤36 teeth...4
  2a. Body heavy (≥10 kg); pelage on dorsum blackish with pale brown or tan bands across the forehead, and from shoulders to base of tail; tail <25% of length of head and body; condylobasal length >121 mm; sagittal crest extending ≥10 mm beyond lambdoidal crest..............
  ....***Gulo gulo***, wolverine (p. 426).

  2b. Body light (≤5.5 kg); pelage on dorsum pale golden brown to black, lacking bands of pale brown or tan; tail >25% of length of head and body; condylobasal length <121 mm; sagittal crest extending <10 mm beyond lambdoidal crest 3
3a. Pelage on dorsum pale golden brown; glandular area on abdomen of both sexes ≅40 mm long; throat and chest splotched with yellowish orange; tail <290 mm long; condylobasal length <95 mm; P4 <9.5 mm long and without an exposed external median rootlet..............
  ....***Martes americana***, marten (p. 406).
3b. Pelage on dorsum dark brown to black with light frosting of white hairs; no glandular area on abdomen; chest and genital area splotched with white; tail >290 mm long; condylobasal length >95 mm; P4 >9.5 mm long and with an exposed external median rootlet...............
  ....***Martes pennanti***, fisher (p. 411).
  4a. Toes not strongly webbed; tail not thickened at base to merge gradually with body; 34 teeth; terrestrial..........5
  4b. Toes on hind feet strongly webbed; tail thickened at base to merge gradually with body; 32 or 36 teeth (never 34); aquatic...8
5a. Pelage yellowish gray with black and white markings on face, and thin white line extending from nose pad to shoulders; claws on front feet >25 mm long; M1 triangular; condylobasal length ≥85 mm..............
  ....***Taxidea taxus***, badger (p. 428).
5b. Pelage color and markings differ, but lacking black and white markings on face and white line from nose pad to shoulders; claws on front feet <25 mm long; M1 dumbbell-shaped; condylobasal length <85 mm...................6
  6a. Pelage more or less uniformly dark brownish-black (occasionally with a few small white spots on the venter); basilar length >50 mm; length of upper toothrows >20 mm in males and >17.8 mm in females.................
  ....***Mustela vison***, mink (p. 422).
  6b. Pelage golden brown to brown on dorsum and white or yellowish on venter, white or yellow on venter usually continuous with white on legs (winter pelage may be pure white with black tip on tail in some areas of the state); basilar length <50 mm; length of upper toothrows <20 mm in males and <17.8 mm in females.................7
7a. Tail <44% of length of head and body, <110 mm long; postglenoid length >46% of condylobasal length in males, >48% in females..
  ....***Mustela erminea***, ermine (p. 415).
7b. Tail >44% of length of head and body, >110 mm long; postglenoid length <46% of condylobasal length in males, <48% in females..
  ....***Mustela frenata***, long-tailed weasel (p. 418).
8a. Pelage dark brown with whitish lower jaw and chest; tail long and tapering; fifth toe on hind foot not the longest; carnassials moderately developed; anterior edge of coronoid process straight and essentially at a right angle to toothrow; 36 teeth....................
  ....***Lutra canadensis***, river otter (p. 433).
8b. Pelage usually dark brown with grizzled areas on head; tail short and thick; fifth toe on hind foot the longest; molariform teeth strongly modified for crushing; anterior edge of coronoid process curved posteriorly and usually angled posteriorly ≥15°; 32 teeth....................
  ....***Enhydra lutris***, sea otter (p. 437).

Martes americana (Turton)
American Marten

1806. [*Mustela*] *americanus* Turton, 1:60.
1819. *Mustela vulpina* Rafinesque, p. 82.
1890c. *Mustela caurina* Merriam, 4:27.
1902b. *Mustela caurina origenes* Rhoads, 54:458.
1912. *Martes caurina caurina* Miller, 79:93.
1912. *Martes caurina origenes* Miller, 79:93.
1953. *Martes americana caurina*: Wright, 34:84.
1959. *Martes americana vulpina*: Hall and Kelson, 2:901.

Description.—The marten (Plate XXV) is weasel-like with a long body and pointed face. The legs are short; the toes, including the pads, are completely furred; and the nails are semiretractile. The tail is bushy and <290 mm long. Range-wide, sexual dimorphism is marked (Clark et al., 1987), with males weighing 470–1,300 g and females 280–850 g (Nowak, 1991).

The skull of the marten (Fig. 12-35) is similar to that of its congener, the fisher (*M. pennanti*), except it averages slightly smaller (Table 12-12): condylobasal length <95 mm and P4 <9.5 mm long (Hall, 1981). The carnassials are well developed, M1 has an enormously expanded medial lobe, and P4 is without an exposed external median rootlet.

The pelage in winter is luxuriant, with a dense underfur and a sparse covering of guard hairs. The color varies, but usually is a golden brown shading to dark brown on the feet and tail; the head is lighter. The throat and chest are splotched with orange or yellow. A small dark stripe extends upward from the medial corner of each eye, giving the appearance of a vertical eyebrow. In summer during the molt, the pelage becomes bleached and unkempt. Both sexes possess a glandular area ≅40 mm long in the midline of the venter (Hall, 1926).

Distribution.—Martens occur or have occurred through most of Alaska, much of Canada south through Massachusetts, southern New York, and northern Pennsylvania into Ohio; the range also includes most of Michigan, Wisconsin, and Minnesota, and montane regions of California, Oregon, Washington, Idaho, Montana, Wyoming, Colorado, Utah, and northern New Mexico. The species does not occur in the prairie regions of southern Canada (Fig. 12-36; Clark et al., 1987; Hall, 1981). Martens were extirpated in much of the southern part of their original range, but have reinvaded northern Minnesota in recent times (Clark et al., 1987) and were reintroduced with varying success by agencies in several states and provinces (Berg, 1982). In Oregon, museum specimens of martens were taken in the Cascade Range, southern Coast Range, and Blue and Wallowa mountains (Fig. 12-36).

Harvest records (Table 12-13) indicate that martens occur in the Blue and Wallowa mountains, in the Cascade Range, and to a limited extent in the Coast Range. The species seems to be absent from the northern Coast Range,

Table 12-12.—*Means (±SE), ranges (in parentheses) and CVs of measurements of skull and skin characters for female and male mustelids other than* Mustela *from regions in Oregon. Skin characters were recorded from specimen tags; skull characters were measured to the nearest 0.01 mm.* SE *and* CV *not provided if* n <10.

Species and region	Sex	n	Total length	Tail length	Hind foot length	Ear length	Mass (g)	Condylobasal length	Basilar length	Postglenoid length	Length of maxillary toothrow
Enhydra lutris	?	4						128.92 (126.05–131.48)	111.65 (109.27–114.26)		46.55 (45.88–47.55)
Lutra canadensis	♀	3	1,292[a] (1,059–1,524)	401	129	20		113.18 (111.79–114.68)	100.58 (99.96–101.75)	48.88[a] (48.17–49.58)	36.83 (35.80–37.86)
	♂	2						113.11 (112.21–114.01)	100.19 (99.53–100.86)	47.42 (47.23–47.61)	37.10 (36.25–37.95)
Martes americana Cascade Range to coast	♀	21	597[b] (570–652)	189[b] (165–206)	77[b] (63–83)			71.86 ± 0.52 (66.51–78.68) 3.35	64.18 ± 0.45 (59.80–69.03) 3.19	27.41[c] (26.89–27.90)	26.87 ± 0.21 (24.83–29.79) 3.56
	♂	28	621[d] (600–660)	189[d] (152–207)	87[d] (70–95)			78.23 ± 0.46 (72.22–81.92) 3.11	70.18 ± 0.44 (64.59–73.98) 3.29	27.65[e] (27.09–28.07)	29.60 ± 0.19 (27.33–31.20) 3.56
Wallowa and Blue mountains	♀	6	568[f] (563–572)	207[f] (195–218)	66[f] (59–73)			73.91 (71.86–75.63)	66.70 (64.38–68.51)	27.78[a] (27.06–28.34)	27.78 (27.34–28.44)
	♂	14	646[g] (585–677)	212[g] (185–230)	90[g] (75–96)	46	1,033	80.25 ± 0.58 (73.87–82.55) 2.71	72.60 ± 0.51 (66.88–74.28) 2.63	28.91 ± 0.29[h] (26.53–30.06) 3.43	30.15 ± 0.27 (27.27–31.46) 3.36
Martes pennanti	♀	3						106.19 (98.74–114.77)	96.86 (89.49–104.11)	34.92[a] (33.31–36.67)	40.51 (38.09–43.86)
Taxidea taxus	♀	30	760 ± 20.0[i] (740–780) 3.7	152 ± 2.0[i] (150–154) 1.86	109 ± 1.0[i] (108–110) 1.3			124.70 ± 0.82 (113.86–136.66) 3.58	112.18 ± 0.59 (107.08–122.24) 2.92		41.63 ± 0.22 (38.98–44.31) 2.92
	♂	30	746[d] (691–800)	149[d] (103–199)	116[d] (106–130)			128.29 ± 0.93 (115.87–138.60) 3.97	115.09 ± 0.87 (106.89–124.42) 4.12		42.58 ± 0.27 (39.95–45.41) 3.43

[a]Sample size reduced by 1. [b]Sample size reduced by 15. [c]Sample size reduced by 18. [d]Sample size reduced by 23. [e]Sample size reduced by 25. [f]Sample size reduced by 4. [g]Sample size reduced by 10. [h]Sample size reduced by 3. [i]Sample size reduced by 28.

the Columbia Basin, the southeastern high desert, and the Willamette Valley.

Ancestral martens probably invaded North America across Beringia during the Pliocene (Dillon, 1961). The first of these invasions was responsible for the eastern group of subspecies and the second for the western group. Hagmeier (1961) considered the latter group to exhibit greater similarities with the European sable (*M. zibellina*) than with the eastern group of subspecies, but Youngman (1975) disagreed except for the shape of M1. Wright (1953) considered the two North American groups to intergrade in western Montana. Youngman and Schueler (1991) synonymized the slightly larger Pleistocene and Holocene fossil species *M. †nobilis,* known from numerous western cave deposits, with *M. americana.* However, Grayson (1993) indicated that others do not consider this extinct species and the present-day marten to be conspecific. Indeed, Graham and Graham (1993:31) indicated that they had "retained the name *M. †nobilis* because of the morphological information that it conveys, [realizing] that it may be relegated to a lower taxonomic status later," but in the same volume, Anderson (1993:21) indicated that she would "call the large late Pleistocene-Holocene 'caurina' marten *M. americana †nobilis* to distinguish it from the smaller types of the same age (specimens the size of the extant marten from the same area)." Although

Anderson (1970) earlier had designated †*nobilis* a distinct species, neither she nor Graham and Graham (1993), more recently, seemed particularly inclined to defend separation of the two taxa at the species level.

Geographic Variation.—Two of the 14 subspecies recognized by Hall (1981) are recorded as occurring in Oregon: *M. a. caurina* and *M. a. vulpina.* Hall (1981:983, map 507) depicted ranges of two other subspecies (*M. a. humboldtensis* and *M. a. sierrae*) as extending into Oregon, but listed no marginal records for either race from the state. Hagmeier (1958:1) considered that in *M. americana* "the partitioning of the subspecies was completely arbitrary," but Dillon (1961) reanalyzed Hagmeier's (1958) data and concluded that recognition of at least some geographic races was warranted. Clark et al. (1987) tentatively recognized eight subspecies in two groups, an eastern group with five subspecies and a western group with three. They synonymized *M. a. vulpina* with *M. a. caurina*; thus, by their classification all martens that occur in Oregon are of the latter race.

Habitat and Density.—The marten is a forest species capable of tolerating a variety of habitat types if food and cover are adequate (Strickland and Douglas, 1987). In California, in an area dominated by lodgepole pine (*Pinus contorta*) with western juniper (*Juniperus occidentalis*) and hemlock (*Tsuga*) in some parts, martens selected areas where cover was <3 m above the snow and used mixed-age stands over even-age stands because the former supported more prey (Hargis and McCullough, 1984). Farther north in California, martens below 2,050-m elevation used riparian areas with lodgepole pine in much greater frequency than its availability, but did not use areas with other pines, mixed conifers, or brush in proportion to their availability. Above 2,050 m, martens used red fir (*Abies magnifica*) much more frequently than it was available,

Table 12-12. —*Extended.*

Length of P4	Zygomatic breadth	Cranial breadth	Least interorbital breadth	Length of mandible
11.49 (11.04–11.86)	95.58 (92.50–99.05)	78.24 (73.59–81.05)	38.38 (37.01–40.25)	85.56 (79.39–88.85)
11.92 (11.84–12.08)	70.07 (66.57–73.62)	54.54 (53.36–55.57)	23.91 (22.43–25.74)	71.52 (70.39–72.55)
12.26 (12.02–12.50)	74.14 (73.98–74.29)	56.89 (56.17–57.61)	24.56 (24.52–24.60)	71.91 (71.75–72.06)
7.27 ± 0.08 (6.56–7.93) 5.27	40.25 ± 0.31 (37.78–42.95) 3.49	33.34 ± 0.26 (30.76–35.15) 3.53	16.33 ± 0.15 (15.19–17.80) 4.27	46.35 ± 0.39 (42.93–50.89) 3.82
7.87 ± 0.06 (7.24–8.60) 4.22	45.39 ± 0.45 (38.73–49.30) 5.21	34.95 ± 0.25 (32.65–38.19) 3.85	17.96 ± 0.17 (16.03–19.56) 4.94	51.57 ± 0.35 (45.84–55.24) 3.59
7.72 (7.17–8.02)	43.22 (42.20–44.98)	34.34 (32.43–35.88)	17.13 (16.78–17.64)	48.47 (46.71–49.62)
8.54 ± 0.11 (7.88–9.10) 4.66	47.75 ± 0.68 (42.73–50.18) 5.34	35.62 ± 0.27 (33.94–37.18) 2.82	18.39 ± 0.29 (16.55–20.05) 5.98	53.19 ± 0.41 (49.36–55.55) 2.86
11.0 (10.71–11.57)	64.77 (60.18–67.14)	42.74 (41.63–44.47)	25.45 (23.92–27.0)	73.19 (66.78–80.46)
10.77 ± 0.09 (9.57–11.78) 4.53	77.89 ± 0.59 (73.00–86.56) 4.19	57.12 ± 0.32 (53.92–61.04) 3.04	27.79 ± 0.34 (24.27–32.01) 5.82	85.23 ± 0.53 (80.21–93.91) 3.38
10.73 ± 0.11 (9.67–12.19) 5.53	83.09 ± 0.65 (76.38–89.17) 4.27	58.41 ± 0.29 (55.93–61.67) 2.76	29.62 ± 0.31 (26.64–33.92) 5.82	88.65 ± 0.66 (81.99–96.16) 4.05

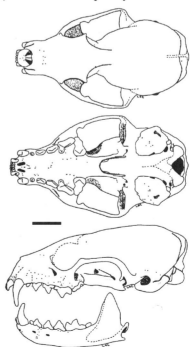

Fig. 12-35. Dorsal, ventral, and lateral views of the cranium, and lateral view of the dentary of the marten, *Martes americana* (OSUFW 1251). Scale bar equals 15 mm.

Fig. 12-36. Localities in Oregon from which museum specimens of the marten, *Martes americana*, were examined (listed in APPENDIX, p. 523), and the distribution of the marten in North America. (Range map are drawn from Hall, 1981:983, map 507, with modifications.) The species was extirpated from much of the southern part of its range east of the Rocky Mountains (Clark et al., 1987).

but used riparian areas, mixed conifers, pines, and brush much less than their availability (Spencer et al., 1983).

Extensive logging and forest fires reduce the value of areas to martens, sometimes for many years (Strickland and Douglas, 1987). In addition to these areas supporting fewer individuals, martens in these areas have shorter life spans, are less productive, and suffer higher natural and trapping mortality than those in undisturbed forests (Thompson, 1994). In addition, martens captured significantly less mass of food per kilometer of foraging travel in logged forests (Thompson and Colgan, 1994). We know of no published quantitative information regarding habitats used by martens in Oregon, but studies that should produce this type of information are in progress (E. L. Bull, pers. comm., 30 March 1994).

Although Hawley and Newby (1957:176, fig. 1) did not indicate the size of their study area in Montana, it appears that the area censused encompassed ≅28 km² of the 41-km² area illustrated; based on this estimation, densities of martens on their study area ranged from ≅0.5 to ≅1.0/km². Subsequently, Weckwerth and Hawley (1962) used the same study area; they indicated that it covered 41 km², but the permanent sampling was conducted in approximately the same area as used by Hawley and Newby (1957). Thus, den-

sities of resident animals were ≅0.4–1.0/km² and, when combined with values for resident juveniles were ≅0.6–1.4/km². These estimates are similar to those provided in several other studies summarized by Strickland and Douglas (1987). In Montana, density of marten populations was related directly to the abundance of small mammals (Hawley and Newby, 1957; Weckwerth and Hawley, 1962).

We know of no estimates of density for Oregon. Based on records of harvest for 1969–1995, martens in Oregon are most abundant in the Cascade Range, particularly the southern portion, followed by the Blue and Wallowa mountains, and with a few in the Coast Range (Table 12-13). Based on reports of personnel of the U.S. Forest Service and National Park Service, Yocom (1974*b*) reported that most records of martens were in Klamath Co. and in the Blue and Wallowa mountains.

Diet.—In Montana, remains of mammals occurred in 93.3% of 1,758 fecal droppings of martens; birds occurred in 12.0%, insects in 19.0%, and fruits in 29.2%. The occurrence of mammals tended to be high at all seasons but was highest in winter; the occurrence of other items was more seasonal (Weckwerth and Hawley, 1962). Of the mammals, muroid rodents, especially *Clethrionomys*, contributed most to the diet. A minimum of 19 species were identified and included murids, zapodids, sciurids, geomyids, leporids, soricids, mustelids, cervids, and chiropterans. Insects were largely Hymenoptera, but Coleoptera, Hemiptera, Diptera, and Orthoptera were represented. Birds were eaten mostly during the nesting season. Among fruits, huckleberries (*Vaccinium*) ranked highest and strawberries (*Fragaria*) second in summer; in autumn and winter, rose (*Rosa*) and hawthorn (*Crataegus*) pomes were used. Berries that grow on tall bushes were present but were little used, likely because they were unavailable to martens (Weckwerth and Hawley, 1962). The occasional occurrence of cervids in the diet indicates that martens eat carrion (Murie, 1961).

In California (Zielinski et al., 1983) and in Wyoming (Murie, 1961) the diet of martens is much the same as that in Montana. Mammals compose most of the diet, but a different subset of species contribute the most seasonally. For example, among sciurids, *Tamiasciurus* and *Glaucomys* tend to be preyed upon mostly in winter and *Spermophilus* and *Tamias* are preyed upon mostly in summer. Such also reflects the seasonal availability of prey (Zielinski et al., 1983). Other nearly exclusively winter prey in California were snowshoe hares (*Lepus americanus*) and deer mice (*Peromyscus maniculatus*), but ground-dwelling voles (*Microtus*, *Phenacomys*, *Clethrionomys*) became largely inaccessible when snow covered the ground (Zielinski et al., 1983). In addition, in areas with different physiographic and vegetative features, the proportions of each type of food item and the species composing each of those types were slightly different (Murie, 1961). Also, in California, the diet of martens contained prey species in different proportions between samples collected in different years (Hargis and McCullough, 1984). Such supports the conclusion that the marten is an opportunistic feeder (Strickland and Douglas, 1987); Zielinski et al. (1983) indicated that martens adjust their activity seasonally to best take advantage of prey availability.

On Vancouver Island, British Columbia, small mammals, birds, deer (*Odocoileus*), and salmonid fish com-

Table 12-13.—Reported annual harvest of martens (Martes americana) in Oregon by county, 1969–1995. Data provided by Oregon Department of Fish and Wildlife.

County	1969–1970	1970–1971	1971–1972	1972–1973	1973–1974	1974–1975	1975–1976	1976–1977	1977–1978	1978–1979	1979–1980	1980–1981	1981–1982	1982–1983	1983–1984	1984–1985	1985–1986	1986–1987	1987–1988	1988–1989	1989–1990	1990–1991	1991–1992	1992–1993	1993–1994	1994–1995
Baker															2	2		1	2		4	8	13	10	14	
Benton															1											
Clackamas											2															
Clatsop											2															
Columbia																										
Coos	4								1									2								
Crook														2												
Curry	1																			1						
Deschutes	10	19	7		19	4	32	7	22	4		64	8	28	32	10	4	3	27	64	70	37	5	3	7	9
Douglas			17	4		7	2		3		47	19	20	21			8	2	6		5		3			
Gilliam													1													
Grant							1					3			2	2	3		6	19	26	37	12			
Harney																										
Hood River		1																	4	4	4	1	2			
Jackson					5	4	1													9	9	7	2	5	2	
Jefferson			1					4							1						1					
Josephine																										
Klamath	21	16	4	1	20	23	12	10	41	54	2	66	6	17	22	30	23	1	38	25	44	24	9	6	10	
Lake												4														
Lane				2		3			7		34	31	2		11	9	17	48	36	26	33	18	5	15		6
Lincoln								1	3								3									
Linn									2		1		1	6					4	3	3	6	1	2		
Malheur																										
Marion							2					2		5												
Morrow																										
Multnomah																										
Polk																										
Sherman																										
Tillamook								4					1													
Umatilla													1													
Union	2	3	1												10	18	8	1		1		9			9	
Wallowa				2		2	18	4		4		1	1			10	10	29	6	10	1	2			2	1
Wasco																										
Washington																										
Wheeler																										
Yamhill																										
Totals	38	39	30	9	50	34	27	74	66	90	93	194	39	84	88	82	80	94	147	160	161	122	52	41	44	16
Trappers reporting[a]	721	773	731	847	1,102	1,350	1,377	1,755	2,746	2,899	3,216	3,373	3,069	3,980	3,292	3,104	3,575	3,397	3,833	1,985	1,536	1,228	1,390	1,454	1,293	1,464
Average price/pelt[b]	$6.61	$5.77	$5.08	$7.93	$5.10	$8.36	$14.85	$15.85	$14.29	$16.97	$14.87	$11.73	$11.57	$20.46	$30.31	$31.73	$22.78	$35.47	$39.02	$40.51	$24.14	$26.00	$31.00	$17.00	$15.00	$16.00

[a] Not all sought martens and not all were successful.
[b] Rounded to the nearest dollar after the 1989–1990 season.

pose most of the food eaten by martens caught by trappers (Nagorsen et al., 1989). *Peromyscus* composed >50% of the small-mammal prey. Some of the deer and salmonid fish possibly were trap bait, but Nagorsen et al. (1989) believed that much of the deer was consumed as carrion and the fish were spawners. Guillermo and Nagorsen (1989) hypothesized that because of the great size differences males and females possibly occupied different feeding niches, thereby reducing competition between the sexes. However, in a companion article, Nagorsen et al. (1989) indicated that the small differences found in diet between the sexes did not support the hypothesis.

Reproduction, Ontogeny, and Mortality.—A large proportion of female martens commonly become sexually active as yearlings ($\cong$15 months old) and produce their first litters as 2-year-olds. In Ontario, 78% of 376 yearling females collected in November-December had bred, but in one district during 1 year, none of 47 bred, possibly because of a decline in prey (Strickland and Douglas, 1987). Although abundance of prey was not mentioned, this and a small sample may have been responsible for Jonkel and Weckwerth (1963) indicating that females in Montana did not breed as yearlings. In Ontario, 93% of 504 older females collected in November–December had bred (Strickland and Douglas, 1987).

Both male and female martens may mate with more than one individual of the opposite sex, but temporary parings may occur and the two may travel together for several days. During breeding, the male grasps the female by the nape; considerable vocalization and struggling may ensue before the female permits copulation. In captivity, copulation required as much as 90 min (Markley and Bassett, 1942), but pairs observed in the wild coupled for much shorter periods (Henry and Raphael, 1989). Although of brief duration, the activity left the male exhausted. Also, in the wild, martens were observed to copulate in trees as well as on the ground (Henry and Raphael, 1989).

In Montana, the breeding season is in July–August (Jonkel and Weckwerth, 1963); in some other areas it may commence as early as late June (Strickland et al., 1982*b*). Induced ovulation is probable (Clark et al. (1987). Development progresses to the blastocyst stage, the blastocyst remains free in the lumen of the uterus for 7–8.5 months, and postimplantation development requires $\cong$27 days (Jonkel and Weckwerth, 1963). Gestation periods range from 220 to 276 days (Strickland and Douglas, 1987). The 56-day difference may be related to date of fertilization as the interval to implantation is correlated with photoperiod in captives (Clark et al., 1987). Young usually are born in March–April.

In males, spermatogenesis commences when they are yearlings (Jonkel and Weckwerth, 1963). However, yearling males may not be effective breeders because baculum changes that occur with age may not be sufficient for yearlings to induce ovulation in females (Strickland and Douglas, 1987). In adults, the testes are scrotal continuously and are enlarged from late June to September (Strickland and Douglas, 1987).

Counts of corpora lutea in sectioned ovaries of bred females are considered to produce the most reliable estimates of litter size in wild martens (Strickland and Douglas, 1987). Although counts of corpora lutea may produce slight overestimates because all ova may not be fertilized and all blastocysts may not implant, corpora lutea are present from shortly after ovulation to parturition, permitting collection of specimens throughout autumn and winter.

In Ontario, during a 12-year period, the number of corpora lutea per pregnant wild-caught female averaged 3.5 (range of annual averages, 2.8–3.7—Strickland and Douglas, 1987). Adults averaged slightly more than yearlings (Table 12-14). Fecundity in marten populations may be influenced by the availability of suitable prey, by changes in number of young per litter, or proportion of pregnant females, or a combination thereof (Strickland and Douglas, 1987).

Table 12-14.—*Reproduction in Oregon mustelids.*

Species	Number litters	Litter size $\bar{X}$	Range	n	Reproductive season	Basis	Authority	State or province
Gulo gulo	1	2.6		5	Early summer	Embryos	Liskop et al., 1981	British Columbia
	1	3.5	1–6	54	Spring–summer	Embryos	Rausch and Pearson, 1972	Alaska-Yukon
Lutra canadensis	1	3.0	2–4	43		Corpora lutea	Tabor and Wight, 1977	Oregon
		2.8	2–4	35		Blastocysts	Tabor and Wight, 1977	Oregon
		2.8	2–3	4		Embryos	Tabor and Wight, 1977	Oregon
	1	2.1	1–3	9	Mar.–Apr.	Embryos	Hamilton and Eadie, 1964	New York
		2.4	1–5	16		Corpora lutea	Hamilton and Eadie, 1964	New York
Martes americana	1	3.3[a]	—[b]	295	Jul.–Aug.	Corpora lutea	Strickland and Douglas, 1987	Ontario
	1	3.6[c]	—[b]	504	Jul.–Aug.	Corpora lutea	Strickland and Douglas, 1987	Ontario
Martes pennanti	1	2.7	1–4	26	Mar.–Apr.	Live births	Hall, 1942	British Columbia[d]
	1	3.4	2–5	54	Mar.–Apr.	Corpora lutea	Wright and Coulter, 1967	Maine
	1	2.7	2–4	22		Corpora lutea	Eadie and Hamilton, 1958	New York
Mustela vison	1	4.0	2–8	8	Mar.	Family groups	Mitchell, 1961	Montana
Taxidea taxus	1	2.9	1–4	58	Jun.–Jul.	Combination[e]	Messick and Hornocker, 1981	Idaho
	1	3.0	2–4	4		Embryos	Hall, 1946	Nevada

[a]Yearling females.
[b]Captives produce one to five offspring.
[c]Adult females.
[d]Captive animals at a fur farm.
[e]Combination of counts of implant sites, embryos, and corpora lutea.

Habits.—Martens are active year-round, although they may remain in their dens for a day or two during inclement weather (Strickland and Douglas, 1987) and, during periods of extreme cold, use subnivean dens (Zielinski et al., 1983). However, daily activity is not consistent among seasons, but is usually considered to be greater in daylight hours in summer than winter. In northern California, although activity was poorly correlated with temperature, martens tended not to be active at both high and low temperatures. Daytime activity was least and night time activity was greatest in winter (Zielinski et al., 1983).

In northern California, 74% of 155 daytime resting sites of nine radio-tracked martens were in snags, logs, stumps, and tree canopies; the average size of snags, stumps, and logs used by martens was significantly greater than the average size of those available (Martin and Barrett, 1991). These findings tended to corroborate those made earlier in the same area (Spencer et al., 1983). Stumps were used more in summer, whereas snags and logs were used more in winter; 16% of the resting sites were reused. Martin and Barrett (1991) emphasized the contribution of suitable resting sites to martens and recommended that silvicultural practices that produced snags ≥4 m tall, stumps ≥80 cm high, and logs ≥10 m long be adopted.

In Montana, home-range areas of six males captured 13–73 times during 188–788-day periods averaged 2.4 km² (range, 0.9–4.4 km²) and those of five females captured 7–17 times during 54–660-day periods averaged 0.7 km² (range, <0.1–1.8 km²). Except for two individuals that occupied nearly completely overlapping areas, home ranges of males overlapped little. Home ranges of one or more females often were included within the home range of a male (Hawley and Newby, 1957).

From an analysis of published home-range sizes from nine sites throughout the range of the species, Buskirk and McDonald (1989) reported that home ranges of males averaged 1.93 times larger than those of females and were more variable than those of females. When differences related to sex were removed, home-range size differed significantly among sites. Also, they found no geographic pattern in home-range size, and no relationship between home-range size and mean annual temperature, latitude, longitude, or elevation. Home-range size was correlated significantly with body mass in females, but not in males. From these relationships, Buskirk and McDonald (1989) concluded that studies of home-range size in relation to resource abundance would provide an indication of habitat conditions for martens.

Martens commonly use elevated perches from which to pounce on terrestrial prey; they also may follow tracks of prey in snow, excavate burrows, enlarge openings to tree dens, and rob birds' nests (Spencer and Zielinski, 1983). Martens cache prey or parts thereof and return to consume them later, sometimes within minutes or after a day or so at other times (Henry et al., 1990; Murie, 1961).

Remarks.—We know little firsthand of the marten in Oregon, but we suspect that populations here likely will not increase greatly if short-rotation timber harvest and single-species replanting continue as recommended forest-management practices. Other practices, more of the past than of the present—such as burning or otherwise removing slash, snags, and downed logs, and large clear-cuttings—likely are detrimental to marten populations.

Martes pennanti (Erxleben)
Fisher

1777. [*Mustela*] *pennanti* Erxleben, p. 470.
1898c. *Mustela canadensis pacifica* Rhoads, 19:435.
1912. *Martes pennanti pacifica*: Miller, 79:94.

Description.—The fisher (Fig. 12-37) is considerably larger than the marten (*Martes americana*; Table 12-12), and males are considerably larger than females (Strickland et al., 1982a). It is more stockily built than the weasels (*Mustela*), but is similar in that the head is somewhat pointed, the body elongate, and the legs short. The feet are plantigrade and pentadactyl, the claws are semiretractile, and, except for the pads, the feet are covered with short fur. As in the tree squirrels (*Sciurus*, *Tamiasciurus*, *Glaucomys*), the hind legs can be rotated nearly 180°, permitting decent of trees head first.

The pelage is long except on the face; dorsal guard hairs may be 70 mm long, whereas those on the venter usually are <30 mm (Powell, 1981). The fur is dark brown grading to black on the rump and legs; the tail is black. The face, neck, and shoulders are grizzled. The venter is brown and irregularly splotched with white or cream in the genital region and on the chest.

Fig. 12-37. Photograph of the fisher, *Martes pennanti*. (From Oregon Historical Society negative no. G IV C-1b.)

The skull of the fisher (Fig. 12-38) is characterized by a large braincase, a narrow rostrum, weak but wide zygomata to allow passage of large temporalis muscles, well-developed carnassials, and in males, well-developed sagittal and lambdoidal crests. The sagittal crest in females is small in comparison with that of males; in old males the crest extends beyond the braincase by as much as 15 mm. The upper carnassial possesses an exposed median rootlet, a character that distinguishes the skull of the fisher from that of the marten. Hall (1926) claimed that fishers had no abdominal glands, but Powell (1981:2) indicated that they "probably are present." The distal end of the baculum is expanded and pierced by a small, round or oval foramen.

Distribution.—Presently, the geographic range of the fisher extends across the boreal forest region in the southern half of Canada with extensions into the United States in the New England and Great Lakes states and in the Rocky Mountains and Cascade, Coast, and Sierra Nevada ranges. Formerly, the range extended farther south in Alberta, Saskatchewan, and Manitoba, and through most of the Appalachian Mountains and Midwest (Fig. 12-39; Powell, 1981). Only three specimens of fishers from Oregon are on deposit in systematic collections, two from Lane Co. and one from Douglas Co. (Fig. 12-39).

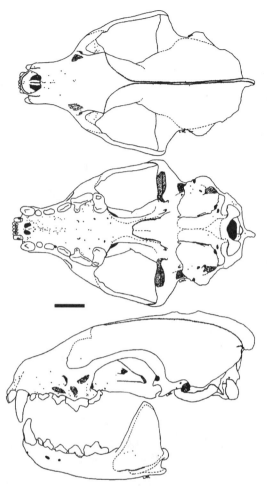

Fig. 12-38. Dorsal, ventral, and lateral views of the cranium, and lateral view of the dentary of the fisher, *Martes pennanti* (OSUFW 1236). Scale bar equals 15 mm.

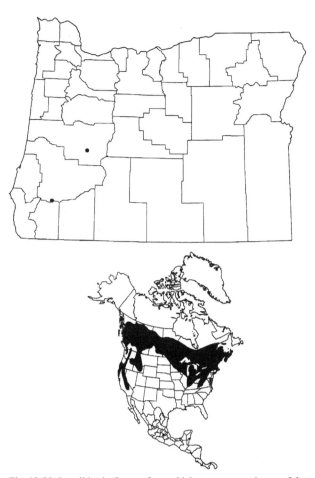

Fig. 12-39. Localities in Oregon from which museum specimens of the fisher, *Martes pennanti*, were examined (listed in APPENDIX, p. 523), and the distribution of the fisher in North America. (Range map redrawn from Powell, 1981:2, fig. 2, with modifications.) The fisher no longer occurs in much of eastern United States (Powell, 1981).

Extinct mustelids believed ancestral to fishers are known from the Pliocene of China, and an extinct fisher (*Martes †divuliana*) from the mid-Pleistocene of North America probably was related to them. *M. pennanti* first appeared in the late Pleistocene of North America, but is not believed to be a direct descendant of *M. †divuliana* although closely related to it (Kurtén and Anderson, 1980). Extralimital late Pleistocene and Holocene records of *M. pennanti* are known from the Midwest and Southeast (Kurtén and Anderson, 1980).

Geographic Variation.—Hall (1981) recognized three subspecies, of which one (*M. p. pacifica*) was indicated to occur in Oregon. Hagmeier (1959) concluded that subspecies should not be recognized in the fisher.

Habitat and Density.—Fisher habitat from a variety of localities within its geographic range commonly is described as widespread, continuous-canopy forests at relatively low elevations (Powell, 1981). In Maine, radio-tracked fishers were found exclusively in forested habitats, but deciduous stands were avoided, mixed stands were used in proportion to their availability, and coniferous stands were used more than expected (Arthur et al., 1989a). From trapping records in northeastern British Columbia, Quick (1953b) concluded that fishers were most abundant in mountainous regions, less abundant in

foothill regions, and least abundant in muskeg regions. The abundance and distribution of appropriate prey species and suitable den sites likely contribute to the ability of habitats to support fisher populations (Douglas and Strickland, 1987).

We know of no analysis of fisher habitat based on the few records in Oregon. However, based on 88 records obtained from 1955 to 1991, Aubry and Houston (1992:75) concluded that fishers in Washington, like those elsewhere, occupied a "wide variety of densely forested habitats at low to mid-elevations." West of the Cascade Range, all records were for sites at elevations of 100–1,800 m, whereas east of the range where snow depths are less, records were for sites at 600–2,200 m. West of the Cascade Range, 26.1% of 46 records were in the subalpine Pacific fir (*Abies amabilis*) zone, 54.3% in the western hemlock (*Tsuga heterophylla*) zone, and 19.6% in the Sitka spruce (*Picea sitchensis*) zone; east of the range, 53% of 19 records were from the subalpine fir (*Abies lasiocarpa*) zone, 37% from the grand fir (*Abies grandis*)–Douglas-fir (*Pseudotsuga menziesii*) zone, and 10% from the timberline-alpine zone (Aubry and Houston, 1992). Based on their survey and on research of others, Aubry and Houston (1992:76) suggested that habitat for fishers would be enhanced by "minimizing forest fragmentation, both in remaining old-growth and in second-growth forests; maintaining a high degree of forest-floor structural diversity in intensively managed plantations; preserving large snags and live trees with dead tops; maintaining continuous canopies in riparian zones; and protecting swamps and other forest wetlands." Fishers were considered "very rare" in Washington (Aubry and Houston, 1992:75).

Even where numbers of fishers are sufficient to permit an annual harvest, densities are low. In Maine, in a 500-km^2 area considered to have the highest density of fishers in the state, estimates of density ranged from 9.5 to 35.7/100 km^2 in summer and from 5.0 to 12.0/100 km^2 in winter (Arthur et al., 1989b). The maximum recorded density of fishers was ≅38.5/100 km^2 (Hamilton and Cook, 1955).

Diet.—Fishers are primarily carnivorous. Small and medium-sized forest mammals are the primary prey; porcupines (*Erethizon dorsatum*), snowshoe hares (*Lepus americanus*), tree squirrels, mice (*Peromyscus*), and voles (*Microtus, Clethrionomys*) are among the most common species preyed upon (Powell, 1981). In addition, fishers eat considerable amounts of carrion, particularly deer (*Odocoileus*). Seasonally, birds, bird eggs, amphibians, fish, and insects may be added to the diet. In some regions, fruits and mast are eaten.

False truffles (*Rhizopogon*) occurred in 50% and composed the largest volume (28%) in the diet of eight fishers collected in northwestern California during winter 1977–1978. Other major items were domestic cattle, brush rabbits (*Sylvilagus bachmani*), black-tailed deer (*Odocoileus hemionus*), broad-footed moles (*Scapanus latimanus*), and western gray squirrels (*Sciurus griseus*). The deer and cattle were thought to have been obtained as carrion; no porcupine was detected, but quills were found in pelts of some of the fishers (Grenfell and Fasenfest, 1979).

Because snowshoe hares contribute significantly to the diet of fishers, their population cycles have been suggested to be responsible for fluctuations in fisher populations (Powell, 1979a). However, when snowshoe hare populations declined in Minnesota, fishers shifted their diet to small mammals, and their body mass and fecundity did not decline, suggesting that they were not dependent upon snowshoe hares (Kuehn, 1989). In Oregon, snowshoe hare populations do not cycle (Wolff, 1982), indicating that the species should provide a relatively stable, although not particularly abundant, food source for fishers.

Reproduction, Ontogeny, and Mortality.—Both males and females attain reproductive maturity as yearlings, and essentially all females breed (Eadie and Hamilton, 1958; Wright and Coulter, 1967). Because the baculum may not be sufficiently developed to induce ovulation during copulation, yearling males are believed not to be effective breeders (Douglas and Strickland, 1987).

The breeding season extends from March through May and is considered to vary geographically (Powell, 1981); in both British Columbia and Maine breeding is in April–May (Hall, 1942; Wright and Coulter, 1967). Embryonic development proceeds to the ≅800-cell stage (Enders and Pearson, 1943) then ceases; the blastocysts remain free in the lumen of the uterus until implantation in January–April. The postimplantation period is about 30 days (Wright and Coulter, 1967). Most young are born in April–May. In 15 captives in British Columbia, the average gestation period was 352 days (range, 338–358 days—Hall, 1942).

Females become estrous 6–8 days after parturition and remain receptive for 2–3 days (Hall, 1942); those that fail to mate may become estrous again ≅14 days later (Douglas and Strickland, 1987).

Average litter size is about three (Table 12-14). Wright and Coulter (1967) found fewer blastocysts or embryos than corpora lutea in only three of 21 pregnant females, indicating that intrauterine losses were low. Shea et al. (1985) found that counts of corpora lutea increased with age: 2.2 for age-class 1 ($n = 86$), 2.8 for age-class 2 ($n = 22$), 3.1 for age-class 3 ($n = 21$), and 3.8 for age-classes 4 – 8 ($n = 12$). Thus, reported average litter sizes may be dependent upon the proportion of the several age-classes of females represented in the sample.

Neonates are partly covered with fine gray fur and have closed eyes and ears. By the 3rd day, they weigh ≅40 g and are covered with fur. The fur becomes brown in <3 months and remains so during the remainder of the summer and autumn. Weaning commences at 2–2.5 months, but young may nurse occasionally until 4 months old, when they first are able to kill prey. Young begin to become independent at ≅5 months (Powell, 1981).

Although a few records of fishers being preyed upon exist, they usually are not subject to predation by species other than humans (Powell, 1981). Trapping for fur is believed to have been a major cause of reduction of populations in several regions (Powell, 1979a, 1981).

Habits.—Fishers are active year-round, but more active in summer than in winter (Arthur and Krohn, 1991). Various authorities have reported them to be active at night, during the day, and both day and night, and with crepuscular peaks of activity (Powell, 1981). More recent investigations suggest that all may be correct, depending on season, snow depth, and denning behavior (Arthur and Krohn, 1991). Fishers are capable of making extensive movements in relatively short periods; Hamilton and Cook (1955) reported a fisher moving 45 km in 2 days and another moving 10–11 km in a few hours. Nevertheless, home-range areas commonly average ≅15 km^2 for adult females and ≅20 km^2 for adult males (Powell, 1981). In Maine, home ranges of adult males averaged 30.9 km^2 (range, 10.6–78.3 km^2, $n = 7$) and those of adult females averaged 16.3 km^2 (range, 8.1–39.1 km^2, $n = 6$—Arthur et al., 1989b). Also in Maine, juvenile fishers dispersed from natal home ranges commencing in autumn (males) and continuing to April (females). Distances between natal home ranges and home ranges established after dispersal averaged 10.8 km (range, 4.1–19.5 km) for males and 11.2 km (range, 5.0–18.9 km) for females. The relatively short dispersal distances were considered sufficient to repopulate an exploited population, but inadequate to recolonize large areas from which fishers had been extirpated (Arthur et al., 1993). Home ranges of males may overlap those of several females, but both males and females tend to exclude conspecifics of the same sex (Arthur et al., 1989b).

Fishers have the reputation of being fleet and agile arboreal mammals, but recent observers suggest that arboreal abilities of the species are overrated. Powell (1980) reported an instance of tree-to-tree travel by a fisher, but indicated that he previously had followed fishers for >240 km without detecting other such evidence and claimed that other workers also had similar experiences.

The ability of fishers to prey on porcupines is unique and renowned, but porcupines must be on the ground for fishers to kill them. Contrary to legend, fishers do not flip

Table 12-15.—*Means (±SE), ranges (in parentheses), and CVs of measurements of skull and skin characters for female and male weasels and the mink* (Mustela) *from regions in Oregon. Skin characters were recorded from specimen tags; skull characters were measured to the nearest 0.01 mm. SE and CV not provided if* n <10.

Species and region	Sex	n	Total length	Tail length	Hind foot length	Ear length	Mass (kg)	Condylobasal length	Basilar length	Postglenoid length	Length of maxillary toothrow
Mustela erminea											
West of Cascade Range	♀	26	217 ± 1.8 (204–240) 4.2	56 ± 0.9 (47–66) 8.3	26 ± 0.3 (24–29) 5.3	15 ± 0.4 (10–20) 14.8	43.6 ± 1.55[a] (31–57.7) 17.78	32.03 ± 0.12 (30.92–33.69) 1.92	28.82 ± 0.13 (27.71–30.76) 2.29	16.82 ± 0.06[b] (16.18–17.39) 1.79	8.88 ± 0.06 (8.36–9.65) 3.21
	♂	26	244 ± 2.9 (186–267) 6.1	69 ± 1.1 (59–78) 7.7	31 ± 0.2 (28–33) 3.8	17 ± 0.4[c] (13–19) 10.9	65.7 ± 1.78[c] (45–86.5) 13.00	35.68 ± 0.16 (34.25–37.42) 2.34	31.99 ± 0.15 (30.96–33.54) 2.43	18.38 ± 0.09[d] (17.69–19.32) 2.49	10.04 ± 0.08 (9.30–10.78) 3.87
East of Cascade Range	♀	1	209	51	26	15	56.0	32.30	29.17	16.71	8.93
	♂	1	234	55	29	12	84.2	35.82	31.93	17.95	10.47
Mustela frenata											
Coast Range	♀	5	344 (320–370)	124 (114–128)	40 (36–44)	23[e] (22–23)	127.0	43.43 (41.06–46.00)	39.19 (37.67–41.71)	19.99[e] (19.35–20.63)	13.24 (12.84–13.53)
	♂	15	425 ± 4.5 (373–445) 4.1	160 ± 2.4 (145–176) 5.8	48 ± 0.7 (44–52) 5.1	24[f] (23–25)	261.8[f] (214–292.3)	49.67 ± 0.49 (44.74–52.69) 3.83	45.10 ± 0.48 (39.99–48.09) 4.08	22.13 ± 0.23 (20.73–24.03) 4.10	15.39 ± 0.19 (12.96–16.19) 4.96
From southern coast north through the interior valleys	♀	3	332 (294–362)	122 (121–123)	39 (38–42)			44.12 (41.24–45.91)	39.86 (37.68–41.45)		13.58 (12.56–14.22)
	♂	4	393 (255–448)	168 (162–176)	45 (32–50)	24 (15–28)	326.4 (248–378.4)	48.34 (37.11–53.28)	43.52 (32.82–48.63)	23.12[a] (22.11–24.00)	14.94 (10.03–17.10)
Northern Cascade Range	♂	2	426 (410–442)	165 (160–170)	47 (46–48)			48.78 (47.82–49.74)	44.09 (43.11–45.06)	20.78	15.31 (15.19–15.42)
Northern half of Oregon east of Cascade Range	♀	9	316 (286–348)	112 (99–125)	35 (27–45)	19[g] (18–20)	93.1	40.21 (36.63–44.59)	36.54 (33.55–39.99)	19.32[h] (19.24–19.40)	12.17 (10.09–13.98)
	♂	20	363 ± 6.5 (310–410) 8.0	135 ± 3.0 (112–159) 10.1	42 ± 0.8 (37–49) 8.1	19[i] (13–23)	226.8[j] (203.6–250)	45.13 ± 0.44 (40.84–47.20) 4.39	41.18 ± 0.39 (37.32–43.60) 4.32	20.62[k] (20.10–20.91)	14.03 ± 0.15 (12.71–15.03) 4.75
Southern half of Oregon east of Cascade Range	♀	8	337 (310–379)	122 (108–139)	37 (35–38)			41.59 (39.48–46.42)	37.89 (36.02–42.50)		12.69 (12.07–14.27)
	♂	15	379 ± 8.9 (281–450) 9.1	143 ± 1.7[a] (132–154) 4.6	43 ± 0.7[a] (39–47) 6.1	21[l] (19–22)	257.2[l] (176.9–337.4)	47.35 ± 0.44 (43.89–50.16) 3.59	43.14 ± 0.39 (39.77–45.56) 3.49	21.88[f] (20.90–22.51)	14.57 ± 0.18 (13.39–16.19) 4.75
Jackson and Josephine counties	♂	1	445	174	52			49.43	45.04		14.37
Mustela vison	♀	19	518 ± 16.7[m] (430–661) 10.7	171 ± 6.9[m] (150–233) 13.4	62 ± 2.0[m] (52–76) 10.8	21[n] (15–27)	826.1[n] (511–1,476)	63.04 ± 0.89 (58.05–73.02) 6.14	56.81 ± 0.83 (51.73–65.69) 6.35	26.93 ± 0.40[m] (24.51–29.77) 4.93	20.57 ± 0.36 (16.15–23.63) 7.55
	♂	31	609 ± 16.4[o] (515–795) 10.4	203 ± 8.9 (150–290) 16.9	68 ± 1.1 (57–74) 6.1	23[p] (15–29)	1,200.0[q] (798–1,533)	68.32 ± 0.61 (62.10–76.41) 4.98	61.90 ± 0.58 (55.54–69.94) 5.26	29.00 ± 0.39[r] (26.95–30.84) 4.51	22.21 ± 0.23 (19.67–25.31) 5.79

[a]Sample size reduced by 1. [b]Sample size reduced by 4. [c]Sample size reduced by 3. [d]Sample size reduced by 5. [e]Sample size reduced by 2. [f]Sample size reduced by 11. [g]Sample size reduced by 6. [h]Sample size reduced by 7. [i]Sample size reduced by 17. [j]Sample size reduced by 18. [k]Sample size reduced by 14. [l]Sample size reduced by 13. [m]Sample size reduced by 8. [n]Sample size reduced by 12. [o]Sample size reduced by 16. [p]Sample size reduced by 25. [q]Sample size reduced by 22. [r]Sample size reduced by 20.

porcupines upside down to attack their undersides (Powell, 1982b). Fishers attack porcupines by biting at their faces where they are unprotected by quills. A porcupine under attack by a fisher frequently places its face against a tree, whereupon the fisher climbs the tree and descends head first to reattack the porcupine. Often 30 min or more are required for a fisher to kill a porcupine; the liver, heart, and lungs are eaten first (Powell, 1981). The remainder of the porcupine is consumed as it is skinned to avoid the quills; 2–3 days may be required for the porcupine to be eaten. Nevertheless, porcupine quills commonly are found in fishers, especially in males (Arthur et al., 1989a). No other major predator competes with fishers for porcupines.

Fishers commonly use ground burrows, tree cavities, and tree nests as resting sites; tree nests are witches'-brooms or other clumped growth and occasionally nests of birds or other mammals (Arthur et al., 1989a). Tree cavities are used by most maternal females with young and ground burrows are used mostly in winter. Fishers often burrow through >1 m of snow to reach a ground burrow (Arthur et

Table 12-15. —*Extended.*

Length of P4	Zygomatic breadth	Cranial breadth	Least interorbital breadth	Length of mandible
3.31 ± 0.03	16.27 ± 0.11	14.88 ± 0.15	7.18 ± 0.05	16.15 ± 0.08
(2.95–3.72)	(15.22–17.63)	(13.85–16.66)	(6.73–7.70)	(15.22–17.12)
5.23	3.33	5.14	3.71	2.52
3.68 ± 0.04	18.67 ± 0.12	16.35 ± 0.09	8.31 ± 0.06	18.34 ± 0.12
(3.34–4.20)	(17.21–19.88)	(15.13–17.61)	(7.87–8.93)	(17.24–19.95)
5.72	3.28	3.07	3.99	3.46
3.50	16.65	14.17	7.30	16.30
3.84	19.12	18.45	7.98	18.81
4.87	23.89	20.22	9.49	24.53
(4.60–5.10)	(23.16–25.30)	(19.11–21.13)	(8.61–10.14)	(23.22–25.8)
5.52 ± 0.08	29.13 ± 0.44	22.29 ± 0.18	11.47 ± 0.25	29.31 ± 0.44
(4.64–6.02)	(24.66–31.66)	(21.47–23.62)	(9.51–13.00)	(24.35–31.85)
5.56	5.85	3.07	8.41	5.83
4.82	23.48	19.61	9.62	24.66
(4.54–4.97)	(22.02–24.72)	(18.70–20.18)	(8.63–10.34)	(22.54–25.92)
5.14	28.10	21.76	11.33	27.01
(3.67–5.92)	(20.27–32.92)	(17.09–25.18)	(8.63–12.90)	(19.20–31.74)
5.79	27.76	21.29	10.98	28.57
(5.78–5.80)	(26.91–28.61)	(21.28–21.30)	(10.97–10.98)	(28.33–28.81)
4.47	21.99	19.69	8.51	22.34
(3.99–5.00)	(20.20–25.34)	(18.37–22.14)	(7.66–9.40)	(19.87–25.68)
5.10 ± 0.07	25.39 ± 0.35	21.56 ± 0.18	9.80 ± 0.15	25.83 ± 0.33
(4.53–5.76)	(22.03–28.11)	(19.28–22.57)	(8.55–10.72)	(22.83–27.39)
5.97	6.11	3.80	6.67	5.69
4.61	22.53	19.62	8.85	23.11
(4.20–5.06)	(20.96–25.90)	(18.67–21.24)	(8.02–9.61)	(22.07–26.82)
5.24 ± 0.08	26.57 ± 0.30	21.26 ± 0.17	10.46 ± 0.18	27.29 ± 0.32
(4.76–6.09)	(24.26–28.38)	(20.28–22.29)	(9.29–11.89)	(25.09–29.37)
5.87	4.40	3.07	6.65	4.49
5.19	30.78	23.0	10.96	29.01
7.16 ± 0.09	36.64 ± 0.63	27.94 ± 0.31	14.33 ± 0.26	38.33 ± 0.70
(6.53–8.12)	(32.06–43.87)	(25.00–31.37)	(12.09–17.52)	(34.23–45.96)
5.25	7.55	4.79	7.91	7.99
7.49 ± 0.06	40.27 ± 0.55	29.50 ± 0.28	15.74 ± 0.24	42.06 ± 0.47
(6.82–8.19)	(33.95–45.99)	(26.45–32.81)	(13.06–18.56)	(37.51–47.30)
4.60	7.66	5.26	8.58	6.16

al., 1989*a*).

Remarks.—Numbers of fishers throughout their range declined dramatically during the first half of the 20th century largely from overtrapping and destruction of habitats (Berg, 1982; Powell, 1979*a*). In Oregon, a trapping season on fishers remained open until 1937 (Powell, 1982*b*), but records for the previous decade indicate that fewer than a dozen were taken annually (Olterman and Verts, 1972). Fishers were never extirpated from the state, as reports of sightings and the occasional capture of an individual continued through the 1950s (Olterman and Verts, 1972). In 1961, 11 fishers from British Columbia were released on Buck Lake, Klamath Co., and 13 others from the same source were released at two sites in Wallowa Co. (Kebbe, 1961). The claimed justification for transplanting fishers into Oregon was the control of porcupines (Kebbe, 1961). From a questionnaire survey of management agencies in states that had reintroduced fishers, Berg (1982) concluded that the success of the transplants into Oregon was questionable. Powell (1982*b*:77) cited an agency report authored by R. Ingram that indicated that the Oregon introductions "barely supplemented the existing population." Yocom and McCollum (1973) reported 29 sight records of fishers in Oregon in the late 1960s and early 1970s by personnel of the U.S. Forest Service and National Park Service. Most of the records were in the vicinity of Crater Lake National Park and northward in the Cascade Range; three records were from the Wallowa Mountains. In 1981, 13 fishers from Minnesota were transplanted into Oregon (Berg, 1982).

Relatively little information is available regarding the biology of the fisher in the western part of its range. From recent summaries of information about the species (Douglas and Strickland, 1987; Strickland et al., 1982*a*; Powell, 1981), it seems that much of the primary research information is available only in agency reports and unpublished theses and dissertations.

Mustela erminea Linnaeus
Ermine

1758. *Mustela* Linnaeus, 1:45.
1896*b. Putorius streatori* Merriam, 11:13.
1899. *Putorius (Arctogale) muricus* Bangs, 1:71.
1929. *Mustela cicognanii streatori* Taylor and Shaw, 2:11.
1929. *Mustela cicognanii lepta* Taylor and Shaw, 2:11.
1936. *Mustela cicognanii muricus* Bailey, 55:292.
1945. *Mustela erminea streatori* Hall, 26:76.
1945. *Mustela erminea murica* Hall, 26:77.

Description.—The ermine (Plate XXVI) is the smallest member of the genus in Oregon (Table 12-15). Like its congeners, the ermine has a long, cylindrical neck and body and short legs, but the tail is relatively short (<44% of the length of the head and body). The head is flattened and somewhat triangular, the eyes are small and slightly protruding, and the ears are rounded. The usual vertebral formula is 7 C, 14 T, 6 L, 3 S, 15–19 Ca, total 45–49 (Hall, 1951*a*). The 10 mammae are unevenly spaced in two rows of five each.

The summer pelage is brown dorsally, usually white or yellowish ventrally; in some individuals the throat and belly are splotched with brown or are entirely brown. The toes and feet usually are brown, but some individuals may have white on the toes and feet at any season. The soles are covered with fur in winter, but in summer the small bare foot pads are surrounded by fur. West of the Cascade Range, ermines remain in brown pelage year-round. In some populations east of the Cascade Range, ermines replace the brown summer pelage with a white winter pelage (Plate XXVI) following the second annual molt; in other populations, some individuals remain brown, others are mottled (piebald), and others turn white. White and piebald individuals usually are females (Hall, 1951*a*). The tip of the tail remains black at all seasons, an adaptation to trick raptors into striking short, especially when in the white winter coat (Powell, 1982a). Timing of the molts is controlled primarily by photoperiod but may be affected by temperature (Rust, 1962).

The skull of the ermine (Fig. 12-40) is long and flattened, and the thin zygomata are barely wider than the cranium. The postglenoid length is >46% of the condylobasal length in males, >48% in females. Ermines exhibit strong sexual dimorphism in size; males have skulls with basilar lengths 9–24% greater than those of females (Hall, 1951a) and males may weigh 40–80% more than females (King, 1983). The carnassial teeth are well developed, reflecting the carnivorous diet. Irregular openings in inflated areas of the dorsal frontals of some individuals are the result of infestations by a nematode (*Skrjabingylus nasicola*) in the frontal sinuses.

Distribution.—The ermine occurs throughout boreal regions of the Holarctic. In North America, except for interior Greenland, the ermine occurs south to central California, central Nevada, southern Utah, and northern New Mexico in the west; it is absent from the Great Plains; and it extends south to northern Iowa, southern Wisconsin, southern Michigan, and Maryland in the east (Fig. 12-41; Hall, 1981). The ermine has been introduced into New Zealand from Great Britain (King, 1983). Most museum specimens of *M. erminea* collected in Oregon were from west of the Cascade Range, but we found a few scattered records for the species in eastern Oregon (Fig. 12-41).

Fossils of *M. erminea* are known from the late Pliocene of Europe; the species likely reached North America in the early Pleistocene as fossils of that age are known from Arkansas and Kansas. Late Pleistocene fossils are known from Texas, Missouri, Colorado, Idaho, and Alaska (Kurtén and Anderson, 1980).

Geographic Variation.—Two of the 20 North American subspecies are reported to occur in Oregon: the larger, longer-tailed *M. e. streatori* with brown lower hind legs west of the Cascade Range and the smaller, shorter-tailed *M. e. muricus* with white hind legs below the knee east of the Cascade Range (Hall, 1951a, 1981).

Habitat and Density.—*Mustela erminea* was captured

during each of 3 years on an untreated control plot established in 125-year-old Douglas-fir (*Pseudotsuga menziesii*) forest in Lane Co., a plot in a nearby clear-cutting (commencing immediately after timber harvest), and in a nearby plot in a 4-year-old clear-cutting in which debris remaining from harvest of timber had been burned (Hooven and Black, 1976). Although numbers of ermines captured were small, these findings suggest that ermines in western Oregon may not avoid forests and shrub forests as they are reported to do in the eastern part of their range in North America (Simms, 1979a). Nevertheless, the species occurs in early seral communities both in Oregon and elsewhere (Simms, 1979b).

We know of no estimates of density of *M. erminea* in Oregon, but we suspect that, in comparison with common species of small rodents, the density is low. For example, in three habitats in Lane Co., Hooven and Black (1976) recorded 0.5–1.0 captures/ha (number of individuals captured not available) annually for *M. erminea*, but annual captures of 4.0–56.8/ ha for *Peromyscus maniculatus*, 2.5–47.3/ha for *Microtus oregoni*, and 7.5–39.0/ha for *Tamias townsendii*. In Ontario, density of *M. erminea* averaged 6.0/100 ha on a 95-ha study area; in prime habitat on the same area, density averaged 10.5/100 ha (Simms, 1979b). The density of the population on the 95-ha area remained relatively stable throughout the 22-month study, but most

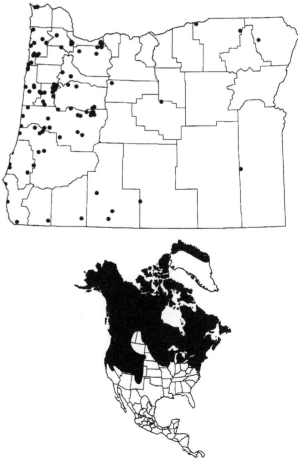

Fig. 12-41. Localities in Oregon from which museum specimens of the ermine, *Mustela erminea*, were examined (listed in APPENDIX, p. 530), and the distribution of the ermine in North America. (Range map redrawn from Hall, 1981:990, map 509, with modifications.)

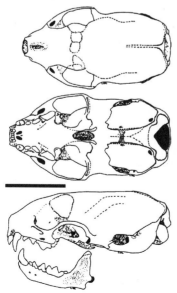

Fig. 12-40. Dorsal, ventral, and lateral views of the cranium, and lateral view of the dentary of the ermine, *Mustela erminea* (OSUFW 6160). Scale bar equals 15 mm.

individuals resided on the area for <1 year (Simms, 1979b). Elsewhere, populations may be less stable and tend to reflect changes in density of primary prey species (Fagerstone, 1987a; King, 1983).

Diet.—No report of foods eaten by the ermine in Oregon is available. Elsewhere in North America, the principal foods of the ermine in winter are mice and voles (mostly *Microtus* with some *Peromyscus*) and shrews (Soricidae—Aldous and Manweiler, 1942; Hamilton, 1933b). Other small mammals such as *Clethrionomys*, *Tamias*, and *Ochotona*, with a rare larger mammal such as *Sylvilagus* (juveniles) and an occasional bird, compose most of the remainder of the ermine's diet (Criddle and Criddle, 1925; Dixon, 1931; Seton, 1929c). When mice and voles become scarce, ermines are known to turn to earthworms (Oligochaeta) as a source of emergency food (Osgood, 1936).

Ermines kill mammalian prey in a manner similar to that employed by the long-tailed weasel (*Mustela frenata*); the nape of the prey is bitten, its body grasped with the forefeet and partially entwined by the ermine's body, and the lower body of the prey kicked and scratched with the hind feet. Blood that flows from wounds is lapped, but ermines, like long-tailed weasels, do not "suck" blood from their prey, a common misconception. Ermines also kill more than they can consume when the opportunity arises, a trait that has earned them the title of "wanton killers" (Bailey, 1936). However, the title is inappropriate because killing and caching surplus prey to be consumed later is the method by which the species stores energy while avoiding becoming obese. Ermines too fat to follow prey into their burrows would be at a disadvantage (Simms, 1979a).

The diet of the ermine is somewhat more specialized than that of the long-tailed weasel. For the ermine, most prey is mouse-sized, whereas its larger congener consumes "substantial quantities" of rat-sized prey and even an occasional rabbit-sized prey item (Rosenzweig, 1966:606). Differences in degree of specialization on mouse-sized prey possibly permit coexistence of the two species (Rosenzweig, 1966). The prey of females is smaller than that of males. Simms (1979a) believed that ermines were better able to prey on voles beneath the snow than were long-tailed weasels.

Fitzgerald (1977) considered ermines and the much less abundant long-tailed weasels to have been responsible for essentially all winter mortality in subnivean populations of *Microtus montanus* on the University of California Sagehen Creek Field Station in the Sierra Nevada Range near Truckee, California. Predation on voles by ermines was highest when vole populations were lowest, but light during the period that voles increased to peak density. These findings were suggested to support Pearson's (1966) hypothesis that predation during and following the population low was responsible for the timing and the amplitude of the vole cycle in that region (Fitzgerald, 1977). In an experiment attempted in British Columbia, ermines were introduced onto trapping grids to ascertain if they could control numbers of mice and voles thereon. However, some ermines returned to their original sites of capture far from the grids, consequently, to a large extent, invalidating the experiment (Sullivan and Sullivan, 1980).

Reproduction, Ontogeny, and Mortality.—Female ermines mate for the first time often before they are 3 months old (Hamilton, 1958a); in Europe, females are reported to

sometimes mate while still in the nest and before their eyes open (Fagerstone, 1987a; King, 1983). Adult females usually mate soon after parturition (Wright, 1963); females that fail to mate remain in estrus until autumn, thus most females become pregnant. Young females possibly mate with the same male with which their mother mates, but rapid turnover in males prevents (or at least greatly reduces the probability thereof) young females from mating with their fathers (Fagerstone, 1987a). Males become sexually mature at ≅11 months of age; both males and females are sexually inactive in winter. Copulation lasts 2–20 min and usually is repeated several times (Fagerstone, 1987a).

Ermines are believed to be induced ovulators as corpora lutea do not occur in unmated females and attempts to artificially inseminate females, even after injections of gonadotropins, were unsuccessful (King, 1983). Following successful mating, development proceeds for ≅14 days; development ceases for 9–10 months with blastocysts free in the lumen of the uterus (Wright, 1942). Development of the fetuses after implantation requires 4 weeks, but gestation requires an average of ≅280 days (King, 1983). Young usually are born in April or May (Hamilton, 1933b).

We know of no extensive study of the productivity of ermine populations in Oregon or elsewhere based on counts of corpora lutea, blastocysts, embryos, or pigmented sites of implantation. In New York, litters born in captivity averaged about six and ranged from four to 10 (Hamilton, 1933b). In Europe, productivity is related to the abundance of food resources during development of the fetuses (King, 1983).

In New York, neonates had 2-mm-long white hair over the neck and shoulders; the ears and eyes were closed, the latter appearing as dark spots beneath the skin of the lids; and a 4-mm-long piece of umbilical cord remained attached. The body mass of six averaged 1.7 g. At 7 days of age, the skin from ears to shoulders was marked with dark pigment; body mass of one male was 4.7 g and there was no sexual dimorphism. At 14 days, a brown mane extended from forehead to shoulders; the dorsum was darker; and the venter remained a clear pink and was nearly naked. Five individuals averaged 10 g; there was no detectable sexual dimorphism in size. At 21 days, the mane was prominent with longest hairs 11 mm. Fine white hairs covered the venter. The canines and premolars were through the gums, but the incisors had not erupted. Five individuals averaged 16.1 g; one female weighed 16.7 g, indicating the absence of sexual dimorphism at this age. At 35 days, the eyes and ears were open; eyes were blue but turned brown 1–2 days after opening. The female weighed 28.2 g; four males averaged 32.6 g (range, 30.0–34.7 g). The young averaged consuming >50% of their body mass per day. At 45 days, the males averaged 55.0 g, and the female weighed 37 g. At this age, the young ate the viscera of mice, but were unable to strip muscle tissue from bones and skin (Hamilton, 1933b). Considering that adult ermines in Oregon are 5–15% smaller than those in New York (Hall, 1951a), body mass of young in Oregon might be expected to be slightly smaller at the same age.

Ermines are preyed upon by a variety of raptors, carnivores, and snakes (Serpentes—Fagerstone, 1987a). Included among the predators of the ermine is its congener, the long-tailed weasel. Simms (1979a:504) believed that the southward distribution of the ermine in eastern North America was limited by "interference interactions" with the larger species. Even if this is true in the East, the ≅2,000-

km overlap in geographic ranges of the two species in the West (Figs. 12-41 and 12-43) causes us to suspect that factors other than interspecific interactions are involved in limiting the southward distribution of ermines.

Habits.—Ermines are active throughout the year. Based on telemetry studies conducted in Europe, ermines are active for short bouts at 3–5-h intervals throughout the diel cycle; bouts usually lasted 10–45 min, but occasionally extended to 4 h (King, 1983).

In Ontario, home ranges averaged 20–25 ha for males and 10–15 ha for females; home ranges overlapped considerably (Simms, 1979b). Home ranges of males are nonoverlapping, but overlap those of females; females use only a portion of the home range of a male (Fagerstone, 1987a). Nevertheless, females tend to avoid the male whose home range overlaps their own (Powell, 1979b); males tend to be dominant except immediately after the breeding season. An individual may move as far as 6 km during one hunting period. In Alaska, a male moved an airline distance of 35.4 km between August and March (Burns, 1964); the age of the individual was not estimated, but juveniles are known to move long distances in search of breeding territories (Fagerstone, 1987a). Territories are marked with products of the anal glands. Larger males tend to be dominant over smaller males.

Male and female ermines do not associate except during the breeding season, a period when juveniles remain in nests under maternal care. At this time, males may bring prey as appeasement to the then-dominant female and remain in attendance for 1–2 weeks (Fagerstone, 1987a). Such behavior may be responsible for reports of males being involved in caring for young.

Remarks.—In Britain, *M. erminea* is referred to as the "stoat," and considerable scientific literature about the species has accumulated there. The species in Europe is considerably larger than the North American ermine. In Britain, body mass averages 320 g; in Ireland, 233–365 g; and on the continent, 208–283 g (King, 1983). In North America, average body mass ranges from 56 to 206 g depending on latitude (larger in the north—Hall, 1951a).

Mustela frenata Lichtenstein
Long-tailed Weasel

1831. *Mustela frenata* Lichtenstein, pl. 42 and corresponding text.
1896b. *Putorius washingtoni* Merriam, 11:18.
1896b. *Putorius saturatus* Merriam, 11:21.
1929. *Mustela washingtoni* Taylor and Shaw, 2:11.
1936. *Mustela longicauda arizonensis* Bailey, 55:287.
1936. *Mustela longicauda saturata* Bailey, 55:289.
1936. *Mustela xanthogenys oregonensis* Bailey, 55:291.
1936. *Mustela frenata nevadensis* Hall, 473:91.
1936. *Mustela frenata effera* Hall, 473:93.
1936. *Mustela frenata altifrontalis* Hall, 473:94.
1936. *Mustela frenata saturata*: Hall, 473:106.
1936. *Mustela frenata oregonensis* Hall, 473:107.

Description.—The long-tailed weasel (Plate XXVI) is the larger of the two weasels in Oregon, but smaller (Table 12-15) than the other member of the genus, the mink (*Mustela vison*). Males are considerably larger than females (Table 12-15). The head is flattened and somewhat triangular, the body and neck are elongate and almost cylindrical, the legs are short, and the tail is long (>44% of the length of head and body). The usual vertebral formula is 7 C, 14 T, 6 L, 3 S, 19–23 Ca, total 49–53; most individuals

have ≥20 caudual vertebrae (Hall, 1951a). The 10 mammae often are distributed unevenly in two rows of five each.

The summer pelage is brown dorsally; the venter is yellowish to orangish except for a white chin and sometimes one or more scattered brown spots. The tip of the tail is black. Some individuals have white or orange toes. The pelage is molted twice annually. In the Cascade Range and eastward through the remainder of the state, the winter pelage often is white although in parts of the region some individuals may remain brown in winter (Hall, 1951a). Where some individuals remain brown and some molt to white, all brown individuals are males (Hall, 1951a). Individuals that develop white pelage in winter retain the black tip on the tail. In the Coast Range and western interior valleys, the winter pelage is indistinguishable from the summer pelage. Captive individuals moved to where the normal winter pelage differs from that in their native region retain the winter pelage color of their native region (Hall, 1951a).

The skull of *M. frenata* (Fig. 12-42) is elongate and flattened; the thin zygomata are only slightly wider than the cranium. The auditory bullae are inflated. The portion of the skull anterior to the cranium is relatively long (postglenoid length <46% and <48% of condylobasal length in males and females, respectively). P4 and m1 exhibit strongly sectorial modifications for shearing meat. Some specimens have inflated areas with irregularly shaped openings in the dorsal frontals as a result of infestations by a nematode (*Skrjabingylus nasicola*) in the frontal sinuses.

Distribution.—The long-tailed weasel occurs south of a line connecting central British Columbia, Lake Superior, and the northern boundary of Maine through most of the conterminous United States, Mexico, and Central America, and into South America as far south as Bolivia (Fig. 12-43; Hall, 1981; Nowak, 1991). In Oregon, the species occurs throughout the state (Fig. 12-43).

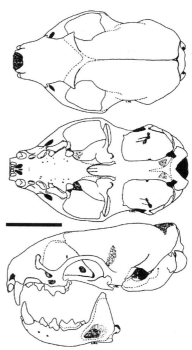

Fig. 12-42. Dorsal, ventral, and lateral views of the cranium, and lateral view of the dentary of the long-tailed weasel, *Mustela frenata* (OSUFW 1122). Scale bar equals 15 mm.

Mustela frenata has a long fossil history, with records extending from the early Pleistocene; fossils are known from >30 widely scattered sites in the late Pleistocene of the United States and northern Mexico. Some workers believe that *M. frenata* may have descended from the late Pliocene fossil species *M. †rexroadensis,* known from sites in Idaho and Kansas (Kurtén and Anderson, 1980).

Geographic Variation.—Six of the 35 named geographic races are proclaimed to occur in Oregon: the darker *M. f. altifrontalis* along most of the Coast Range, the smaller *M. f. effera* with a light-colored underside of the tail in the northern half of the state east of the Cascade Range, the larger *M. f. nevadensis* with a light-colored underside of the tail in the southern half of the state east of the Cascade Range, *M. f. oregonensis* with a frontonasal white patch along the west slope of the Cascade Range and the extreme southern portion of the Coast Range, *M. f. saturata* without a frontonasal white patch and a light-colored underside of the tail in the Siskiyou Mountains, and the lighter-colored *M. f. washingtoni* without a frontonasal white patch and a light-colored underside of the tail in the Cascade Range as far south as Mt. Jefferson (Hall, 1951a, 1981).

Habitat and Density.—Except in desert areas, where it may be excluded by the absence of free water, *M. frenata* may occur in any habitat: mature forests, second-growth woodlands, clear-cuttings, prairies, meadows, and alpine

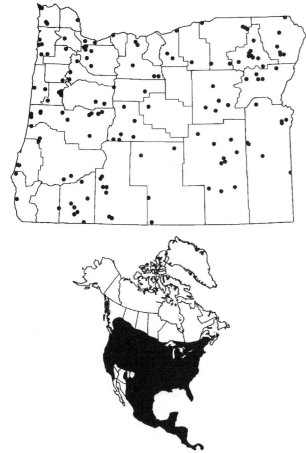

Fig. 12-43. Localities in Oregon from which museum specimens of the long-tailed weasel, *Mustela frenata*, were examined (listed in APPENDIX, p. 530), and the distribution of the long-tailed weasel in North America. (Range map redrawn from Hall, 1981:996, map 511, with modifications.)

tundra (Fagerstone, 1987a; Hall 1951a; Svendsen, 1982). However, where water is available, the abundance of suitable prey may determine which habitats are used most frequently (Hall, 1951a). In Harney Co., *M. frenata* was captured in greasewood (*Sarcobatus vermiculatus*) and marsh habitats, but not in big sagebrush (*Artemisia tridentata*) or grassland habitats (Feldhamer, 1977b). Greatest numbers of *Peromyscus maniculatus* were captured in greasewood habitats and greatest numbers of *Microtus montanus* were captured in marsh habitats, suggesting that these habitats were selected on the basis of abundance of prey. In Lane Co., one *M. frenata* was captured on a plot in a 4-year-old clear-cutting in which slash had been burned but none was caught on plots in a 125-year-old Douglas-fir (*Pseudotsuga menziesii*) forest and in a 1–3-year-old unburned clear-cutting (Hooven and Black, 1976). During the year that the weasel was caught, numbers of both *Peromyscus maniculatus* and *Microtus* (two species) were most abundant on the burned clear-cutting. Pfeiffer (1953) speculated that weasels that he caught in mountain beaver (*Aplodontia rufa*) runways were there to prey on juveniles, as most were captured during the breeding season of that species. However, *Peromyscus, Microtus,* and other genera of small mammals that weasels prey upon also are known to use mountain beaver runways.

We know of no estimates of density for *M. frenata* in Oregon or elsewhere in the Pacific Northwest. Reviewers of the biology of the species (Fagerstone, 1987a; Svendsen, 1982) mostly referenced estimates of 30–50 years ago (Craighead and Craighead, 1956; Glover, 1943; Quick, 1951) that range from 0.4 to 38/km². Polderboer et al. (1941) found evidence of four long-tailed weasels on their 65-ha study area in Iowa (62/km²). No doubt, the paucity of information is related to the relatively low density and extensive movements of long-tailed weasels (Fagerstone, 1987a); the wide range of estimates likely is related to the great differences in habitats in which *M. frenata* was censused.

Diet.—Long-tailed weasels are primarily carnivorous although occasionally a fiber, seeds of a berry (*Ribes* or *Vaccinium*), or other plant residue may be found in a fecal dropping. In Colorado, 23 (29.9%) of 77 droppings contained plant residues, many of which were believed to have been stomach contents of herbivorous vertebrate prey (Quick, 1951).

The only information regarding diet of *M. frenata* in Oregon is an anecdotal account of one preying on a larval salamander (*Dicamptodon*—Sturges, 1955). Although amphibians, reptiles, birds, and insects are reported to be among animals consumed by long-tailed weasels, small mammals, particularly *Peromyscus, Microtus, Reithrodontomys,* and *Tamias,* with an occasional *Clethrionomys, Sylvilagus,* or *Spermophilus* and a rare *Ochotona, Thomomys,* or *Zapus,* constitute the principal animals eaten (Hamilton, 1933b; Polderboer et al., 1941; Quick, 1944, 1951). Males, because of their larger size, take the larger prey, whereas females tend to be limited to the smaller species. Records exist of *M. frenata* preying on big brown bats (*Eptesicus fuscus*—Mumford, 1969), eggs of blue-winged teal (*Anas discors*—Teer, 1964), broad-footed moles (*Scapanus latimanus*—Ingles, 1939), Belding's ground squirrels (*Spermophilus beldingi*—Grinnell et al., 1937), and dusky-footed woodrats (*Neotoma fuscipes*—Vestal, 1938), all species that occur in Oregon. Weasels doubtlessly prey on many other species as the opportunity arises.

Long-tailed weasels commonly kill prey by biting at the base of the neck after grasping it with all four feet and entwining it with their body (Allen, 1938; Leopold, 1937; Miller, 1931). Usually the spinal cord or brain stem of prey is severed. Within the burrow system of prey, weasels use a throat-bite and kill by suffocation (Fagerstone, 1987a). With larger prey such as cottontail rabbits (*Sylvilagus floridanus*), both captive and free-living *M. frenata* lapped blood from wounds made in the head and neck (Allen, 1938; Leopold, 1937). Captive weasels usually ate only the cranial region of smaller prey immediately after killing it, but usually returned later to consume most of the remainder (Quick, 1951). Weasels are noted for killing surplus prey when the opportunity arises; most of the surplus is stored and eaten at a later time.

Because of the elongated bodies that enable weasels to enter confined spaces in search of prey, they have metabolic rates 50–100% greater than mammals of the same mass that have more conventionally shaped bodies (Brown and Lasiewski, 1972). Thus, the food requirement of long-tailed weasels also is greater than that of most mammals. Sanderson (1949) reported that captive juvenile long-tailed weasels consumed 22% of their body mass daily when 5–7 weeks old and 24% when 8–10 weeks old; adults consumed 18%. Wild-caught *M. frenata* may consume more than twice as much when first made captive (Fagerstone, 1987a). Because long-tailed weasels are slightly larger than ermines, they are less able to hunt voles in subnivean tunnels; thus, Simms (1979a) believed that the northward distribution of the long-tailed weasel was limited by snow cover.

Reproduction, Ontogeny, and Mortality.—Based on studies of captives, long-tailed weasels breed in July–August; zygotes develop for ≅8 days, whereupon development ceases and the blastocysts remain free-floating in the lumen of the uterus for 170–300 days. After implantation, further development for 23–27 days is required before parturition in April–May. Wright (1948) reported that gestation periods ranged from 205 to 337 days and that differences in gestation periods for successive litters produced by the same female ranged from 15 to 107 days.

Only one litter is produced annually (Wright, 1947). Litter size in captives often is six to nine; Wright (1948) was reluctant to provide an average as maternal females sometimes consume neonates. We know of no studies of natality in wild-caught *M. frenata* in Oregon or elsewhere based on counts of blastocysts, embryos, or pigmented sites of implantation.

Neonates in a litter of six born in captivity weighed an average of 3.1 g; the dorsum was covered with fine white hair, but the pink flesh was not obscured. The legs, tail, and venter were naked. The eyes and ears were closed. At 7 days, the fine white hair extended onto the legs, but the venter remained bare; the nose pad and claws had turned a light brown. Twelve individuals 7 days old weighed an average of 7.6 g; the largest weighed 9.2 g. At 14 days, the hair was longer and more dense; the ears were slightly brown. Males were separable by size; also, males could crawl, but females could not. No teeth had erupted. Five males averaged 17.0 g and six females averaged 13.5 g. At 21 days, the white natal fur was changing to gray; the tip of the tail was black. The venter was still bare. Both sexes could crawl and eat solid food. The canines and premolars had erupted, but the incisors had not. Males averaged 27 g and females averaged 21 g. At 28 days, males averaged

38.5 g and females averaged 30.5 g. At 36 days, some had both eyes open and some had only one eye open. At 42 days, males averaged 81 g and females averaged 62 g. At 49 days, males averaged 101 g and females averaged 72.5 g (Hamilton, 1933b).

In a litter of seven ≅35-day-old young caught in Manitoba the six males averaged 107.6 g and the one female weighed 86.2 g. When ≅74 days old, the males had averaged gaining ≅240 g, whereas the female gained only ≅150 g (Sanderson, 1949). Adults of both sexes in the Manitoba race (*M. f. longicauda*) average slightly larger than any of the races that occur in Oregon (Hall, 1951a); thus growth might be expected to be slightly less rapid in Oregon weasels.

Wright (1948) indicated that four females (littermates) attained the estrous condition at 96 days of age; they bred but produced no young, likely because the males were not fertile. Another captive exhibited a swollen vulva indicative of estrus at 84 days of age. Sanderson (1949) reported a wild-caught female maintained in captivity from about 5 weeks of age attained the estrous condition at 9 weeks of age. Females that produced young and nursed them entered estrus 65–104 days after parturition, whereas females that produced young but failed to nurse them entered estrus 36–71 days after parturition (Wright, 1948). Females prevented from mating seemed to remain in estrus for 2–3 weeks.

Although body mass of juvenile males (young-of-the-year) often equals that of adult males (sexually mature) by late summer or early autumn, the mass of the baculum of juveniles usually is not more than half that of adults until they commence to attain sexual maturity at about 11 months of age (Wright, 1947). Therefore, males are not capable of breeding until ≅15 months of age (Wright, 1947). In sexually mature males, mass of the testis and epididymis (combined) reaches a peak in July, commences to decline in August, and reaches a nadir by December, when the mass is nearly equal to the mass of the testis and epididymis of juveniles (Wright, 1947).

Although *M. frenata* is a predator, it often falls victim to a variety of raptorial birds, carnivorous mammals, and snakes (Serpentes—Fagerstone, 1987a). In Pennsylvania, the number of long-tailed weasels submitted for bounty was correlated negatively with the number of red foxes (*Vulpes vulpes*) and gray foxes (*Urocyon cinereoargenteus*) submitted for bounty; this observation and the frequency with which remains of weasels are found at fox dens caused Latham (1952) to suggest that the decline in weasel populations was a result of predation by foxes. We found positive correlations between numbers of weasels and numbers of gray foxes ($r^2 = 54.6$), red foxes ($r^2 = 31.6$), and both species combined ($r^2 = 42.3$) reported taken by trappers during the 1969–1970 to 1991–1992 trapping seasons in Oregon (Tables 12-4, 12-5, and 12-16). Number of weasels harvested was correlated positively with number of trappers ($r^2 = 66.0$) and price per pelt ($r^2 = 45.3$; Table 12-16), also suggesting that the small number of weasels harvested in Oregon did not reflect the effects of fox predation strongly. In western Washington, of 25 *M. frenata* killed on highways by automobiles, 24 were males, 16 of which were killed in June and July; the strong sex and seasonal biases suggest that mate-seeking likely was involved in the deaths (Buchanan, 1987).

Habits.—Long-tailed weasels are active throughout the year. Although they sometimes are considered to be

Table 12-16.—Reported annual harvest of weasels[a] (Mustela) in Oregon by county, 1969–1996. Data provided by Oregon Department of Fish and Wildlife.

Trapping season

County	1969–1970	1970–1971	1971–1972	1972–1973	1973–1974	1974–1975	1975–1976	1976–1977	1977–1978	1978–1979	1979–1980	1980–1981	1981–1982	1982–1983	1983–1984	1984–1985	1985–1986	1986–1987	1987–1988	1988–1989	1989–1990	1990–1991	1991–1992	1992–1993	1993–1994	1994–1995	1995–1996
Baker	1	3		3	1	3	5	17	4		13	2	8	8	10	5	1	7	6	1	1	2	4	10	9	25	2
Benton		1		1		2	1	1	1		2	2	1	1	1	2	2	1	1		2	2		4			
Clackamas	2	2	4		1			2		1	3	4	4	3	2	2	4										
Clatsop	11	6	7	1	2	3	3	8	9	1	2				5		6	6	3		2	1					
Columbia	1		1	2	1	3	3	1	3	1	6	2	2		5	3	4	5	1	1							
Coos		1		3				1	1		6	1	1	2			3	4					2		1		
Crook						1	7	7	1																1	4		
Curry														3									1	1			
Deschutes		3				3	2	1	3	3		3	2		1	2		3	1			1	1	1			
Douglas		10		1			1	2	2	12		9			1	1				1	1	1					
Gilliam											1																
Grant	1			3	3		4	4	1		2	2	1	1	1	4	7		45	15	10	6					
Harney	1		2	2	1	1		7	6	1	1	7	2	1	1	2	1	1	11	11	1		1		2	2	
Hood River	1	2		3			2	4	1	2	3	5	6				6		2	1	1					5	
Jackson	1		1		1				2			5		5	5	1			3	1	1						
Jefferson							2							3					1								
Josephine	1						1	1					6			1	2		2						2	2	
Klamath	1	1	1	6	4	2	14	3	2	8	4	7		5	1	3	19	8	1	3	1	3	2	1	2	3	
Lake	3	3			1	1	4	4		2	2			1		1	1	10	4				1	1			
Lane	1	2	1	6	7	7	3	7	11	21	11	10	5	16	2	6	6	3	2	3	2	1					
Lincoln	1			2	5		3	2	2	4	2	1	1	4	5	2	1	6	3	2	2	1	1		1	1	
Linn	2		4	2	1	3	3	1	6	12	6	6	3	5	1	5	3	8		7	1						
Malheur	2	3					6	7	1	6	3	2	3	2	3	1	5	3	1	1	1			1		2	
Marion		5	3	4	1			4		3			6		2	1			23		1						
Morrow	3							1	1										1								
Multnomah	3	1	1	1		1	3	3	1						4												
Polk	2	1					3	3			1			2		2	2	1	1	1							
Sherman																											
Tillamook	4			3	8	4	11	6	4	6	7	2	2		3	3	3	3	2	1	1	1					
Umatilla	1		1	1	1		4	4	2	2	2	4	1		5	2	2		1	2			1				
Union	5	2	3	4	2	3	3	3	2	12	11	10	3	9	3	3	1	3	1	1	3	3	3	1			
Wallowa	2	3	12	4	1		1	4	8		10	20	2	2	9	2	6	10			1	1	1	2			
Wasco	2						1			1																	
Washington	2	1	1		2	2		2	2		4	1	1	22	8	5	4	1	1	1			1		1		
Wheeler																						1					
Yamhill									2		3	3	3		10		2	2									
Totals	44	45	17	40	54	41	36	90	91	59	115	111	91	89	84	71	67	106	146	52	30	17	17	24	14	49	2
Trappers reporting[b]	721	773	731	847	1,102	1,350	1,377	1,755	2,746	2,899	3,216	3,373	3,069	3,980	3,292	3,104	3,575	3,397	3,833	1,985	1,536	1,228	1,390	1,454	1,293	1,464	687
Average price/pelt[c]	$0.42	$0.32	$0.19	$0.44	$0.37	$0.46	$0.75	$0.87	$0.63	$0.50	$2.16	$1.02	$0.58	$0.50	$0.50												

[a] Species not separated; we suspect that most were M. frenata.
[b] Not sought weasels and not all were successful.
[c] Average price/pelt not available for trapping seasons after 1983–1984.

nocturnal, 98% of those livetrapped during a study in Colorado were caught during the daylight hours (Svendsen, 1982). Conversely, radio-tracked weasels left their dens ≅1 h after sundown and were active 1–3 h before returning to a den (Fagerstone, 1987*a*).

Weasels can swim, but seem to avoid water; they also can climb, and have been observed to pursue arboreal prey, but they mostly hunt on the ground. When snow is on the ground they often seek prey by burrowing beneath the snow. Dens used by long-tailed weasels commonly are those appropriated from prey and modified to suit the size and requirements of the weasel; the nest is enlarged, often is lined with the fur of prey, and sometimes contains fecal droppings of the new owner. Feces commonly are deposited at entries to dens or sometimes in latrines randomly established along trails (Quick, 1944).

In Iowa, tracks in snow indicated that long-tailed weasels hunted within 226 m of their dens (Polderboer et al., 1941). A circle with that dimension as a radius would encompass 16 ha, about midway in the range of home-range sizes of long-tailed weasels where prey was abundant (Fagerstone, 1987*a*). In Michigan, home-range areas for four individuals, calculated as circles with radii equal to the maximum distance each were snow-tracked from their dens, were 32.5, 114.4, 136.8, and 164.8 ha (Quick, 1944). One individual traveled 5.5 km during 1 night. Males commonly travel twice as far as females during nightly excursions except when the female is foraging to supply offspring with prey.

Males locate females in estrus by olfaction. Upon locating an estrous female, a male grasps the female by the nape and resists the female's struggles (Fagerstone, 1987*a*). The male then grasps the female's abdomen with the front legs and commences copulatory thrusts. Copulation may last 2–3 h and may be repeated (Wright, 1948).

Intraspecific interactions, even those related to mating and rearing young, do not seem to be reasonably well understood. Some authorities consider long-tailed weasels to be solitary except during the mating season, whereas others provide evidence that the male remains with the female and aids in rearing offspring (Fagerstone, 1987*a*). Productivity of habitats may be involved in type of social structure; where resources are abundant, a linear social grouping may be exhibited, but in less productive areas weasels may be solitary (Fagerstone, 1987*a*).

Remarks.—We were surprised by the paucity of information regarding the biology of weasels in the Pacific Northwest. Because we prepared only a few long-tailed weasels as museum specimens, and because of the small numbers harvested in Oregon, we suspect that the species may not be particularly abundant in much of the region.

Mustela vison Schreber
Mink

1777. *Mustela vison* Schreber, pl. 127B.
1896. *Putorius vison energumenos* Bangs, 27:5.
1912. *Mustela vison energumenos* Miller, 79:101.
1936. *Lutreola vison energumenos* Bailey, 55:293.

Description.—The mink (Plate XXVI) is a semiaquatic species. The body is elongate and cylindrical, the legs are short and stout, and the tail is bushy and about half as long as the head and body (Table 12-15). The head is flattened dorsoventrally, the ears are small and rounded, and the nose is pointed. The feet are furred except for the toe pads and the toes are semiwebbed. In Idaho, males and females of the same race that occurs in Oregon (Hall, 1981) averaged 0.78 and 0.53 kg, respectively, considerably less than some other races (Eagle and Whitman, 1987).

The pelage consists of dense grayish underfur and long, lustrous guard hairs; dorsally, it is dark brown to blackish, but somewhat lighter on the venter. The color darkens toward the tip of the tail. Many individuals, but not all (Hall, 1946, 1981), have white splotches on the chin and sometimes elsewhere on the venter; the configuration of the splotches is unique and may be used to identify individuals (McCabe, 1949).

The skull of the mink (Fig. 12-44) is somewhat flattened and, among members of the genus in Oregon, the largest with basilar length >50 mm. The rostrum is more elongate and squarish, and the bullae are less inflated, than in congeners. The carnassials are well developed, M1 is flattened and dumbbell shaped, and the palate extends well beyond the toothrows. Sagittal and lambdoidal crests become increasingly developed with age, especially in males.

Distribution.—The mink occurs throughout Alaska; Canada, except for northeastern Northwest Territories and northwestern Quebec; and the United States, except for the desert Southwest (Fig. 12-45; Hall 1981). The mink was introduced into northern Europe and northern Asia (Eagle and Whitman, 1987). In Oregon, museum specimens were collected throughout the state, but collection localities reflect the association of minks with wetlands, especially east of the Cascade Range (Fig. 12-45). During the 1969–1995 trapping seasons for furbearers, minks reportedly were taken in all 36 counties in Oregon (Table 12-17).

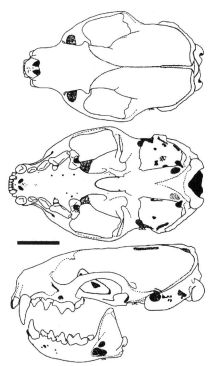

Fig. 12-44. Dorsal, ventral, and lateral views of the cranium, and lateral view of the dentary of the mink, *Mustela vison* (OSUFW 1102). Scale bar equals 15 mm.

Table 12-17.—Reported annual harvest of minks (Mustela vison) in Oregon by county, 1969–1995. Data provided by Oregon Department of Fish and Wildlife.

County	1969–1970	1970–1971	1971–1972	1972–1973	1973–1974	1974–1975	1975–1976	1976–1977	1977–1978	1978–1979	1979–1980	1980–1981	1981–1982	1982–1983	1983–1984	1984–1985	1985–1986	1986–1987	1987–1988	1988–1989	1989–1990	1990–1991	1991–1992	1992–1993	1993–1994	1994–1995
Baker	45	3	10	16	18	18	26	39	35	3	33	45	89	29	32	39	16	29	56	20	20	10	49	6	6	3
Benton	1	7	6	11	1	9	9	12	9	11	28	30	9	14	59	22	24	15	7		5	1	8	1	7	1
Clackamas	12	11	4	16	18	39	11	12	14	14	31	30	12	22	9	24	23	17	14	9	5		8	3	4	
Clatsop	78	86	57	54	51	59	108	95	111	50	92	95	56	77	72	76	40	44	72	20	11	19	3	5	11	12
Columbia	28	6	3	16	8	21	33	38	49	31	42	35	55	36	34	46	40	39	30	10	13	11	10	2		4
Coos	7	4	9	13	13	26	27	30	18	19	28	30	21	19	25	25	17	20	32	23	61	13	2		3	29
Crook	4	7	16	25	9	36	42	41	51	7	40	90	20	9	12	31	62	44	111	45	13	8	32	43	36	17
Curry		1		2	2	2	1	3	1	1	6	3	3	2	3	7	2	4	3		1					
Deschutes	52	15	17	15	26	44	36	34	22	22	48	82	32	48	35	52	101	87	68	34	41	44	41	30	27	13
Douglas	20	17	39	18	38	68	27	52	49	52	63	48	42	67	40	48	24	21	39	18	21	32	19	11	17	16
Gilliam		1	1								1	7	4	1					1	4						
Grant	46	29	33	21	55	26	32	37	68	19	84	61	37	16	23	9	35	84	93	97	49	11	17	1	1	5
Harney	4	4	8	4	20	3	5	7	16	8	4	3	15	6	5	9	10	9	19	34	97	17	1	55	2	4
Hood River	6	5	16	2	8	5	9	15	19	7	9	5	11	11	11	12	7	9	29	5	16	13	1	3	4	1
Jackson	14	12	19	38	4	38	13	28	30	23	43	123	56	64	70	48	46	58	102	53	58	53	31	2	3	2
Jefferson	2	1	4	7		12	12	17	17	14	30	19	41	73	27	30	63	25	28	39	12	5				1
Josephine	8	3	6		6		2	4	2	2	10	28	12	27	41	15	22	33	34	21	18	15	29	4	1	
Klamath	189	145	202	215	259	175	189	254	198	265	180	324	297	260	244	413	260	419	284	212	201	174	224	67	125	38
Lake	13	2	13	8	34	30	14	17	6	3	24	8	3	10	23	19	23	54	37	42	25	36	10	9	25	36
Lane	90	43	31	77	87	97	141	99	112	112	172	160	104	167	118	124	98	94	101	61	18	25	36	22	58	25
Lincoln	20	13	9	9	27	34	40	20	20	32	48	46	22	32	18	20	16	13	21	13	2	6	2	28		1
Linn	41	5	13	27	54	49	27	59	59	27	72	32	63	65	67	127	125	90	42	28	36	22	43	28	10	19
Malheur	12	4	9	12	43	30	18	27	27	14	30	135	31	19	38	51	18	75	61	39	73	15	20	14	17	15
Marion	29	17	6	20	25	53	58	45	47	62	61	60	35	77	41	91	48	39	30	22	31	5	27	20	14	12
Morrow			1	3	3	1	10	10	7	20	11	19	7	9	12	10	2	6	3	12	14	6	9		1	7
Multnomah	5	7	3	7	6		4	4	10	2	6	5	11	10	8	4	6	6	6	2	1	3	6	1		3
Polk	5	1	4	9	2	18	17	17	18	1	5	14	14	6	9	9	7	7	1	1	2	2	2	4	2	1
Sherman			2	1		5	6				1	7		2	1	1	5		6	2	2	1	1			
Tillamook	63	30	20	22	32	31	40	37	27	27	35	33	15	6	24	17	11	22	13	8	3	1	2	3	3	4
Umatilla	21	63	7	29	49	45	31	24	40	40	69	147	91	33	76	84	44	127	147	21	44	6	27	27	21	18
Union	21	20	1	5	26	25	22	25	19	19	33	40	32	14	19	23	23	29	40	48	30	6	9	9	15	4
Wallowa	105	59	44	133	81	93	82	72	68	68	95	84	70	105	138	117	79	175	141	115	123	69	46	78	71	71
Wasco	21	2	3			17	10	42	36	8	8	5	6	15	7	7	9	7	25	16	3	5	4	3	1	1
Washington	5	4	5	1	2	12	11	22	7	15	10	40	10	18	46	56	31	17	11	6	10	4	7	6	7	6
Wheeler	1	3	3	1		1					2	3	4	1		4	4	21	2	2	32					
Yamhill	10	7	8	6	9	8	9	24	4	15	12	40	19	26	32	40	15	9	3	4	2	1	4			6
Totals	978	635	633	838	1,017	1,114	1,058	1,328	1,257	1,013	1,466	1,936	1,349	1,396	1,416	1,703	1,352	1,747	1,723	1,117	1,060	638	741	460	492	370
Trappers reporting[a]	721	773	731	847	1,102	1,350	1,377	1,755	2,746	2,899	3,216	3,373	3,069	3,980	3,292	3,104	3,575	3,397	3,833	1,985	1,536	1,228	1,390	1,454	1,293	1,464
Average price/pelt[b]	$5.31	$3.29	$4.96	$9.08	$9.37	$7.11	$7.29	$10.20	$8.24	$13.91	$15.64	$17.31	$12.49	$10.43	$12.48	$12.68	$10.40	$17.10	$18.55	$16.49	$13.29	$12.00	$10.00	$11.00	$7.00	

[a]Not all sought minks and not all were successful.
[b]Rounded to the nearest dollar after the 1989–1990 season.

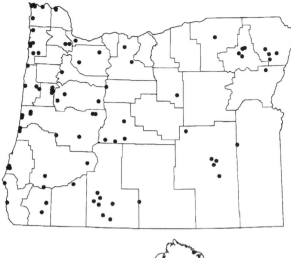

Fig. 12-45. Localities in Oregon from which museum specimens of the mink, *Mustela vison*, were examined (listed in APPENDIX, p. 531), and the distribution of the mink in North America. (Range map redrawn from Hall, 1981:1002, map 513 and Banfield, 1974:331, map 145, with modifications.)

Fossil minks are known from cave deposits in Maryland, Arkansas, and Kansas that date to the early to mid-Pleistocene; nevertheless, fossils of the species tend to be relatively uncommon (Kurtén and Anderson, 1980).

Geographic Variation.—Of the 15 recognized subspecies, only *M. v. energumenos* occurs in Oregon (Hall, 1981).

Habitat and Density.—Throughout its geographic range, the mink is associated with river, lake, pond, or marsh environments, and the extent and density of mink populations are regulated largely by the distribution and abundance of such wetlands (Arnold and Fritzell, 1990; Linscombe et al., 1982). In Oregon, such is reflected in harvest statistics; counties with the most extensive wetlands are those from which the greatest numbers of minks reportedly are taken (Table 12-17). The mink's wide distribution in its native lands and its ability to exploit new environments in other parts of the world where it was introduced suggest that, other than the requirements for some of the wide variety of vertebrates and invertebrates that it can use for food, elements of the physical environment are the critical components of mink habitat. In Manitoba, most activity of radio-tracked male minks was "on large, well-flooded, semipermanent and permanent wetlands with irregular shorelines and large areas of open water" (Arnold and Fritzell, 1990:2205). Nevertheless, in a multivariate analysis, habitat (i.e.,

vegetative substrate) explained only 26% of the variation in wetland use, only slightly more than explained by variation in abundances of birds (24%) on which the minks fed. From other studies, appropriate den sites seemed to rank high among habitat components for minks. In Idaho, of 30 den sites, 53.5% were in log jams, 19% in brush and debris, 17.5% in rock crevices, 9% in stream banks, and 1% in beaver (*Castor canadensis*) lodges (Melquist et al., 1981). A juvenile female mink fitted with a beacon radio transmitter in Minnesota used six different dens during an 11-day period and 11 different dens during the following 16 days, all in a 31-ha area; the same den was used on consecutive nights only once, but five dens were used more than one time. Most of the dens were believed to have been burrows excavated by muskrats (*Ondatra zibethicus*—Schladweiler and Storm, 1969).

Along the Madison River in southwestern Montana, estimated density of minks on a 33-km² study area was 6.3 and 3.3/km² during years that 115 and 40 minks, respectively, were removed from the area by commercial trappers (Mitchell, 1961). We know of no estimates of density for any area in Oregon. However, numbers of minks harvested annually in Oregon during a recent 26-year period never reached 2,000 (Table 12-17); numbers harvested in Oregon were ≅5% of those in some of the top mink-producing states during the same interval (Linscombe et al., 1982).

Diet.—As minks are adapted to forage in both terrestrial and aquatic environments, a wide variety of vertebrate and invertebrate prey is consumed. Prey species differ by season and region, but in an analysis of studies of diet of minks in their native range based on examination of both fecal droppings and contents of digestive tracts, mammals ranked first at all seasons, followed, in order, by invertebrates, birds, fish, amphibians, and reptiles in summer, and by fish, invertebrates, birds, amphibians, and reptiles in fall and winter (Eagle and Whitman, 1987). Among mammals eaten, small rodents, muskrats, and leporids are most common, but other mammal prey species, even chiropterans (Goodpaster and Hoffmeister, 1950), occasionally fall victim to minks. Minks are able to detect ultrasounds of the frequency emitted by small mammals on which they commonly feed (Powell and Zielinski, 1989). In summer, crayfish (Decapoda), aquatic beetles (Coleoptera), and waterfowl (Anatidae) and other marsh birds commonly are preyed upon. In winter, after many birds migrate and invertebrates become less available, fish, made less active by the cold, contribute more to the diet of minks. Amphibians and reptiles usually are minor items in the diet of minks, but occasionally minks in a local area may prey extensively on frogs (Anura) and salamanders (Urodela—Eagle and Whitman, 1987).

Species of rough fish are eaten more commonly by minks than salmonids, except during spawning, when salmonids become especially vulnerable (Eagle and Whitman, 1987). Nevertheless, in a unique experiment conducted on two 100-m-long channels of a forked stream in Quebec, habitat-improvement practices to favor eastern brook trout (*Salvelinus fontinalis*) on one channel not only doubled numbers and biomass of trout and crayfish, but also resulted in a >50% increase in mink activity. Minks avoided the more wary trout and concentrated their aquatic feeding activity on the easier-to-catch crayfish (Burgess and Bider, 1980).

In central Idaho, the nearest locale to Oregon at which a

study of the diet of the mink was conducted, of 659 fecal droppings examined, 387 (58.7%) contained remains of fishes, 280 (42.5%) mammals, 127 (19.3%) birds and eggs, 155 (23.5%) invertebrates, and 13 (2.0%) reptiles. Most of the fishes were ≤15 cm long. Of the mammals, voles (*Microtus*), deer mice (*Peromyscus*), and muskrats occurred in greatest frequency, but jumping mice (*Zapus*), leporids, chipmunks (*Tamias*), and shrews (*Sorex*) also were represented (Melquist et al., 1981). Of the avian material, 44.1% was waterfowl. Of the invertebrates, most were beetles (Coleoptera); all occurrences of reptiles were of garter snakes (*Thamnophis*). Although niches of river otters (*Lutra canadensis*) and minks overlap, minks tend to take a greater diversity of prey species than river otters, to feed more on terrestrial prey, not to feed on bottom fish such as sculpins (*Cottus*), to be active more at night, and to change foraging areas less frequently (Melquist et al., 1981).

We were surprised that no studies of the diet of minks had been conducted in Oregon, especially in view that minks commonly deposit their feces in latrines at prominent sites in their home ranges where they can be collected easily. Nevertheless, given the cosmopolitan nature of the diet of minks as revealed by investigations conducted elsewhere, we seriously doubt that such studies conducted in Oregon would reveal differences in the diet other than might be expected from differences in faunal composition or from exploitation of a locally abundant food resource.

Reproduction, Ontogeny, and Mortality.—Relatively little information regarding reproduction in wild *M. vison* is available (Table 12-14); most studies of this aspect of the biology of the species have been conducted with captive minks raised for their fur (Eagle and Whitman, 1987). In addition, based on a study of captives, counts of pigmented sites of implantation may not be a useful technique for estimating fecundity in *M. vison* (Elder, 1952). This, combined with the relatively short period that embryos can be counted and the need to collect females at a season that population levels may be depressed by the annual harvest for fur, makes estimation of productivity in minks particularly difficult.

Based on a study of captive minks in Pennsylvania, females attain reproductive competence and breed for the first time at ≅10 months of age, then each year thereafter. One captive produced 58 young in 11 litters (Enders, 1952), a feat unlikely to be duplicated in the wild. Females may be in estrus in late February–early April, but individuals are receptive within a period of ≅3 weeks. Photoperiod plays a major role in stimulating onset of estrus. Ovulation is induced and occurs about 48 h after copulation. Copulation, ovulation, or pregnancy do not shorten estrus, and females may copulate (sometimes with different males) and ovulate more than once. Many fur ranchers believe that more young are produced from the last mating, but empirical evidence does not support the belief (Enders, 1952). Thus, both superfetation and superfecundation are known to occur.

Copulation usually begins, and often proceeds, as a rough-and-tumble fight. Ultimately, the male is able to grasp the nape of the female's neck and with the hind feet move the female's body into position and attain intromission. After a bout of thrusting and presumably ejaculation, the male rolls the pair onto their sides for a few moments respite. The male maintains the hold on the female's neck and bouts of copulatory thrusting with pauses for rest are repeated again

and again. Finally the male releases the female's neck and copulation ends; intromission may last 3 h.

Females mating early during the breeding season undergo a short delay in implantation; gestation may last as long as 70 days. Females that mate later undergo a shorter delay; gestation may be as short as 40 days, but postimplantation development requires only 28–30 days (Enders, 1952). Most females give birth in May (Enders, 1952).

Neonates are pink and naked; sexual dimorphism in size is evident at birth (Enders, 1952). Young attain adult size by November.

The deciduous teeth commence to erupt at 16–21 days of age and the full complement of 28 teeth erupts by 35–49 days. The sequence of eruption is dI1, dp1, dp2, dP1, dc1, dC1, dP2, dp3 dP3, dI2, dI1, di1, and di3. The permanent teeth commence to erupt when young are 44–47 days old and the full complement of 34 teeth erupts by 64–71 days of age. The sequence of eruption is I1, I2, i1, i2, I3, C1, m1, M1, c1, P2, p2, P4, m2, P3, p3, and p4 with the sequence in M1–c1 and p3–p4 pairs sometimes reversed (Aulerich and Swindler, 1968).

Some of the larger carnivores, and sometimes raptors such as great horned owls (*Bubo virginianus*), occasionally prey on minks. Parasites and diseases, environmental contaminants (especially mercury), and harvests for fur probably are responsible for much of the annual losses (Linscombe et al., 1982).

Habits.—In Idaho, minks were active during 60% of 108 radiotelemetry observations at night and during 46% of 308 daytime observations (Melquist et al., 1981). In Manitoba, six males followed by radiotelemetry were active when located 9% of the time during daylight hours, 27% during crepuscular hours, and 71% during nighttime hours. Overall activity was greatest in April, least in May, and intermediate during June–July (Arnold and Fritzell, 1987). Although Marshall (1936) used tracks in snow to infer much about activity of minks, he observed that minks were not active on nights following snowfalls when temperatures fell below −12°C.

In Idaho, minks foraged on shore along overhanging banks, and in holes and crevices; they also peered into the water, and when they sighted prey quickly dived after it (Melquist et al., 1981). Although dives are of short duration, they are characterized by rapid bradycardia (within 1 s) as an oxygen-conserving measure (West and Van Vliet, 1986). Logjams were particularly attractive sites from which to hunt aquatic prey, possibly because aquatic prey animals concentrated at such sites. Of time spent foraging, 55% was from logjams, 20% in riparian vegetation, and 25% along stream banks (Melquist et al., 1981).

In southwestern Montana, minimum-area home ranges for two females recaptured seven and 10 times and not known to have left the 33-km² study area during a 2-year study were 7.8 and 20.4 ha, respectively. Some individuals that left the area and were captured by trappers had traveled long distances. An adult female moved 38.6 km, six young males averaged moving 18.0 km (range, 3.2–45.1 km), and two juvenile males moved 23.3 and 29.8 km (Mitchell, 1961). In Wisconsin, seven females recaptured 2–65 days between captures averaged moving 93 m (range, 23–366 m); a male caught nearly a year after marking had moved only 805 m (McCabe, 1949). In Michigan in winter, the average extent of nightly movements from dens for eight females was 123 m (range, 40–161 m); two males averaged 1,006 m when females were present and one male moved 4,829 m when no female was present (Marshall, 1936). In Manitoba, nightly movements of male minks followed by radiotelemetry averaged 2.3 km in April, 1.0 km in May, 0.8 km in June, and 0.3 km in July; monthly home-range areas ranged from 294 to 970 ha and April–July home ranges averaged 1,141 ha. The extensive movements in April were attributed to the breeding season being in March–April in the region (Arnold and Fritzell, 1987). Minks are solitary and, except during the breeding season, tend to be asocial. At onset of the breeding season, males move widely in search of receptive females, explaining in part, the greater movements recorded for males.

Remarks.—Linscombe et al. (1982) suggested that the infrequency of research projects on minks was the result of populations being relatively secure, thereby reducing their priority in research planning. In addition, minks are not particularly easy to live-capture, handle, or mark by conventional methods (Marshall, 1936; McCabe, 1949). Finally, changes in fashion, the growing sentiment against harvest of furbearers, and the shift of mink ranching from North America to Europe and Asia have combined to reduce the value of the wild mink as a furbearer. Thus, harvests of wild minks, and with them the power of the voice of the fur trapper, have declined. Not only is the need for precise information less critical for management of mink populations (at least, at the moment), but funds for research are budgeted for species and items requiring immediate management decisions. Nevertheless, the mink remains a challenging species from which an ingenious and industrious student might wrest new and interesting information through application of modern research techniques.

Gulo gulo (Linnaeus)
Wolverine

1758. [*Ursus*] *luscus* Linnaeus, 1:47.
1823. *Gulo Luscus* Sabine, p. 650.
1903*d. Gulo luteus* Elliot, 3:260.
1913*b. Gulo luscus luteus*: Grinnell, 3:291.
1935. *Gulo gulo luscus* Degerbøl, 2(4):35.

Description.—The wolverine (Plate XXVII) is the largest terrestrial mustelid in Oregon that, to some degree, resembles a small bear (*Ursus*—Hash, 1987; Pasitschniak-Arts and Larivière, 1995). It is powerfully built with a broad, dog-like head; short, rounded ears; small eyes; a slightly humped back; relatively short legs; plantigrade and pentadactyl feet; and a bushy, somewhat drooping tail. Males are ≅10% longer in all dimensions and 30% heavier than females (Hall, 1981). In Montana, males averaged 12.7 kg and females averaged 8.3 kg; the largest individual, a male, weighed 15.9 kg (Hornocker and Hash, 1981).

The pelage consists of a dense, woolly, crimped underfur 2–3 cm long overlain by coarse, stiff, and somewhat shaggy guard hairs ≅10 cm long (Hash, 1987). Fur on the tail is about twice as long as on the body. In winter, stiff bristles grow between the toes and foot pads. The base color is blackish brown with a pale-brown stripe extending along the sides from the head or shoulders to the base of the tail. Lighter markings often produce a face mask. The throat and chest are splotched with yellowish white and a ventral gland is marked with a narrow streak of white. The

2–3-cm-long, recurved claws are pale. The fur commonly is used as an attractive trim for parkas in the Arctic and is unique in that frost forming from a person's breath may be brushed away easily, thus avoiding the fur becoming wetted (Quick, 1952).

The skull of the wolverine (Fig. 12-46) is massive with heavy zygomata diverging posteriorly; small orbits; and relatively large, nearly rectangular postorbital fossae to permit passage of the large temporalis muscles. In lateral view, the dorsal margin of the skull forms a nearly smooth arc with the high point immediately posterior to the postorbital processes (Wilson, 1982). The sagittal and lambdoidal crests are well developed; the former extends

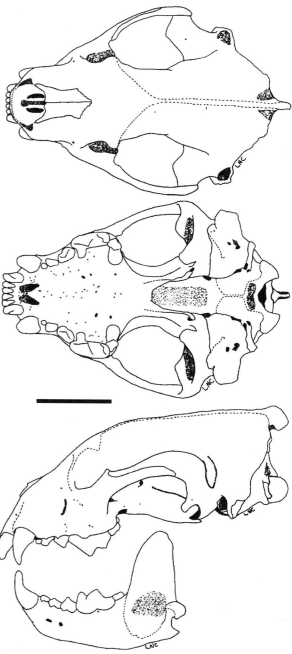

Fig. 12-46. Dorsal, ventral, and lateral views of the cranium, and lateral view of the dentary of the wolverine, *Gulo gulo* (OSUFW 3174). Scale bar equals 5 cm.

beyond the braincase as a rounded projection (Fig. 12-46). The teeth are exceptionally large and the carnassials are well developed; wolverines are capable of crushing all but the heaviest bones of the largest herbivores (Hash, 1987). The dumbbell-shaped M1, typical among mustelids, is set medially to the line of the toothrow. The rostrum is squarish and the braincase trapezoidal, but not so pronounced as in the badger (*Taxidea taxus*; Fig. 12-46).

Distribution.—The wolverine has a Holarctic distribution; in North America the historical geographic range included all of Alaska and Canada, and extended southward through montane regions of the West to southern California, Utah, and Colorado, and in the Midwest and East to southern Indiana and Pennsylvania (Fig. 12-47; Hall, 1981). The wolverine no longer occurs in the midwestern and eastern areas of the United States and Canada and is restricted to isolated wilderness areas in the remaining conterminous United States (Hash, 1987; Pasitschniak-Arts and Larivière, 1995; van Zyll de Jong, 1975; Wilson, 1982).

In Oregon, the wolverine was long thought to have been extirpated (Bailey, 1936), but in 1965 a large male was killed on Three-Fingered Jack in Linn Co. (Kebbe, 1966). Subsequently, a series of sightings of wolverines or their

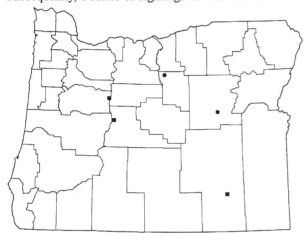

Fig. 12-47. Localities in Oregon from which museum specimens (circles) of the wolverine, *Gulo gulo*, were examined (listed in APPENDIX, p. 519) and at which wolverines reportedly were captured (squares) since 1965 (Kebbe, 1966; Oregon State Game Commission, 1970; Trainer et al., 1981), and the distribution of the wolverine in North America. (Range map redrawn from Hall, 1981:1009, map 516, with modifications.) The wolverine has been extirpated over much of the southern portion of its range (Hash, 1987).

tracks in the 1960s and early 1970s were reported secondhand (Nowak, 1973; Yocom, 1974a), and one each reportedly was caught on Steens Mountain, Harney Co., in 1973 (Trainer et al., 1981) and Broken Top Mountain, Deschutes Co., in 1969 (Oregon State Game Commission, 1970). In addition, we preserved the skull and most of the skeleton of a male taken by a county trapper in Wheeler Co. in 1986 (OSUFW 7878) and a partial skeleton and tufts of fur found in Grant Co. in 1992 (OSUFW 9471).

Gulo gulo is known in North America from cave deposits in Pennsylvania and Maryland of mid-Pleistocene age and from Holocene deposits in Alaska, Yukon Territory, Idaho, Wyoming, and Colorado (Kurtén and Anderson, 1980). The genus *Gulo* is known from the Pliocene in Europe (Savage and Russell, 1983).

Geographic Variation.—All North American wolverines are regarded as *G. g. luscus* (Wilson, 1982).

Habitat and Density.—The wolverine occurs in a broad range of wilderness habitats. Coastal forest, inland forest of coastal composition, boreal forest, and even tundra areas are occupied depending on elevation and latitude (Hash, 1987). Within these types, ecotonal areas, especially those near marshes, seem to be significant components of habitats used by wolverines. Open areas seem to be avoided, skirted, or crossed rapidly (Hornocker and Hash, 1981). Nevertheless, the critical component seems to be the absence of human activity or development (Hash, 1987).

Even under near-optimal conditions, densities of wolverine populations are extremely low, possibly accounting, in part, for the paucity of estimates. Quick (1953b), based on harvest of wolverines on registered traplines in northeastern British Columbia, estimated a density of 0.48/100 km² and an annual harvest in 1947 of ≅10% of the population. Hornocker and Hash (1981), from livetrapping records, estimated a minimum population density of 1.54/100 km² in northwestern Montana; they considered the density to be stable during their 5-year study (1972–1977).

Records of numbers of wolverines harvested annually in British Columbia revealed, despite wide fluctuations at 2–4-year intervals, a general decline from an average of ≅300 in 1919 to ≅100 in 1962. Then, with continued biennial fluctuations, the annual harvest increased to >500 animals by 1975 (van Zyll de Jong, 1975) and to >600 was responsible for the sudden appearance of wolverines where essentially none was seen for decades (Blus et al., 1994; Johnson, 1977; Kebbe, 1966; Newby and McDougal, 1964; Patterson and Bowhay, 1968). A similar increase in numbers in Montana a decade earlier was attributed to dispersal of wolverines from Canada (Newby and Wright, 1955), but Hornocker and Hash (1981) believed that wolverines had survived in Montana because of the vast expanse of official wildernesses. Because one individual taken in Oregon was a female (Oregon State Game Commission, 1970), the possibility of a selfmaintaining population of wolverines in Oregon cannot be discounted, but it seems more likely that those occasionally seen or killed in the state were dispersers from populations farther north.

Diet.—Wolverines are largely meat-eaters, but are known to consume berries in season (Rausch and Pearson, 1972). Although wolverines are capable of killing large herbivores (especially in deep snow), much of the food eaten by wolverines likely is obtained as carrion (Hash, 1987; Wilson, 1982). One of three suggested reasons for the decline in numbers of wolverines in Canada was the decline in the availability of carrion because numbers of wolves (*Canis lupus*) had declined (van Zyll de Jong, 1975). Surplus prey and scavenged carrion are cached beneath ice or snow or placed in tree branches; the wolverine's powerful jaws permit it to consume frozen meat. In Montana, 27% of 56 fecal droppings contained deer (*Odocoileus*) and elk (*Cervus*) carrion (Hornocker and Hash, 1981). Also, galliform birds, lagomorphs, sciurids, small murids, porcupines (*Erethizon dorsatum*), muskrats (*Ondatra zibethicus*), and even fish are preyed upon (Hash, 1987).

Reproduction, Ontogeny, and Mortality.—Female wolverines do not breed as young-of-the-year; 7–50% breed when 1 year old, 50–85% when 2 years old, and 62–92% when ≥3 years old (Banci and Harestad, 1988; Liskop et al., 1981; Rausch and Pearson, 1972). Some females ≥6 years of age may not breed (Liskop et al., 1981). Breeding usually is in late spring or early summer; development proceeds to the blastocyst, which lies dormant in the lumen of the uterus until late autumn to midwinter. Implantation can occur as early as November and as late as February, but unimplanted blastocysts can be found in some females during all of the same months (Banci and Harestad, 1988). Postimplantation development is believed to require 30–40 days; thus parturition occurs in December–March (Banci and Harestad, 1988; Rausch and Pearson, 1972). Gestation in a female bred in captivity lasted 272 days (Mehrer, 1976).

Some males attain sexual maturity at 14–15 months of age (Rausch and Pearson, 1972), but most apparently do not until >2 years of age (Banci and Harestad, 1988). Average mass of the testes commences to increase in February–March and reaches a peak in May–June (Banci and Harestad, 1988; Rausch and Pearson, 1972). Liskop et al. (1981) reported evidence of onset of spermatogenesis as early as January.

Productivity in Yukon Territory, based on counts of corpora lutea, increases with maternal age; females ≥6 years old (*n* = 5) had an average (±*SE*) of 4.4 ± 0.49 corpora lutea. The average number of embryos (3.2 ± 0.2) was significantly less (*P* < 0.01) than the average number of corpora lutea (3.9 ± 0.2) for 23 postimplantation females, but did not differ significantly (*P* > 0.05) from the average number of pigmented sites of implantation (3.3 ± 0.3) for 18 postpartum females (Banci and Harestad, 1988). Averages of counts of embryos, pigmented implantation sites, and corpora lutea for larger samples from Alaska and Yukon Territory were similar, but were considerably less in British Columbia (Table 12-14).

Habits.—Wolverines do not hibernate, but may be inactive during inclement weather. Activity is greatest at night, but where wolverines are relatively common, sighting a wolverine during daylight hours might be expected (Wilson, 1982).

The wolverine has a reputation for having a mean and savage disposition, but according to Hash (1987) the reputation likely was derived from observations of individuals caught in traps or in cages. Trapped or caged individuals can exhibit defensive aggression unmatched by most other species in similar circumstances (Hash, 1987). Wolverines are powerful animals that often can escape from traps, can tear into buildings or food caches, and can kill even the largest cervids.

Wolverines travel widely; in northwestern Montana, average home-range areas for animals followed by radiotelemetry were 422 km² for males and 388 km² for females. Individuals of both sexes made extensive exploratory movements. The size and shape of home ranges were not affected by physical features of the landscape. Home ranges of individuals of the same and opposite sex overlapped and no area was defended (Hornocker and Hash, 1981).

Male wolverines mark sites within their home ranges with visual or olfactory signs; females do not mark. Prominent trees, especially isolated ones, are marked by climbing and depositing scent (product of anal glands) on the tree or ground; 70% of 157 marks found by Koehler et al. (1980) were of this type. Males also gnaw or bite a limb or root, sometimes in combination with scent-marking; scratch the ground in a fashion similar to that used by a domestic dog (scent may or may not be deposited); and deposit feces or scent without use of a visual sign. One male marked 20 sites in 2.5 km of travel, whereas others may travel several kilometers without marking. Koehler et al. (1980) believed that marking was not a behavior to maintain exclusive territories, but may function to maintain the solitary nature of the wolverine in time, but not in space. Instead of being a threat, a mark may indicate to other wolverines that it may be more profitable for them to hunt elsewhere (Koehler et al., 1980). Some European workers claim that males are territorial, excluding males but not females (Wilson, 1982).

Remarks.—The wolverine in North America long was considered a species separate from the Old World wolverine and was known by the name *Gulo luscus*. Some authorities, notably Hall (1981), continued to distinguish between North American and Eurasian wolverines at the species level even though others believed them conspecific (Degerbøl, 1935; Kurtén and Rausch, 1959a). Authorities who recently have provided systematic treatments of the Carnivora of the world have recognized only one species (Ewer, 1973; Stains, 1984; Wozencraft, 1993).

Taxidea taxus (Schreber)
American Badger

1778. *Ursus taxus* Schreber, 3:520.
1825. *Meles jeffersonii* Harlan, p. 309.
1878. *Taxidea sulcata* Cope, 17:227.
1891. *Taxidea americana neglecta* Mearns, 3:250.
1901. *Taxidea neglecta*: Miller and Rehn, 30:218.
1950. *Taxidea taxus montana* Schantz, 31:90.
1972. *Taxidea taxus jeffersonii* Long, 53:732.

Description.—The badger (Plate XXVII) is a medium-sized (Table 12-12), but powerfully built carnivore strongly adapted for digging. The body is flattened, ovoid in cross section; the legs are short but stout; the toes of the forefeet are partly webbed and equipped with long, curved claws; the hind feet are shaped like miniature shovels; and each eye is equipped with a nictitating membrane that can be extended to cover it. The ears are rounded and densely covered with fur, but seem large in comparison with those of many digging mammals. The tail is short and brushlike. Males average larger than females (Table 12-12; Long, 1972).

Dorsally, the long, shaggy pelage is mottled grayish blending on the venter to light tannish or whitish. A white stripe extends from the nose pad to the shoulders. In some individuals in the southern part of the species range the stripe may extend to the rump; Long (1973) recorded one such individual from Idaho, thus some Oregon badgers may be so marked. The face is black with white splotches surrounding a black "badge" on either cheek; the feet are blackish (Long, 1973). Badgers molt annually commencing on the head and progressing in the middorsal region to the tail, thence laterally; the molt commences later at higher latitudes (Long, 1975).

When the skull of the badger (Fig. 12-48) is viewed dorsally, the rostrum and braincase have an outline similar to that of a splitting wedge; the zygomata flare as wide as or wider than the braincase. The postorbital processes are short and form a nearly equilateral triangle. The palate extends far beyond M1. The occlusal surfaces of M1 and P4 are triangular with the hypotenuse labial in the former,

lingual in the latter (Long, 1973). The carnassials are moderately developed.

Distribution.—The badger occurs from northern Alberta south through southeastern British Columbia, Washington, and Oregon east of the Cascade Range; through most of California to the tip of Baja California Sur; and eastward through most of Saskatchewan, southern Manitoba, and extreme southwestern Ontario south to Texas and to Puebla, with an eastward extension through the Prairie Peninsula to Ohio (Fig. 12-49; Long, 1972). Records of badgers in Connecticut and New York are thought to be from colonies established through introductions by humans (Nugent and Choate, 1970). In Oregon, the badger occurs throughout the region east of the Cascade Range and in eastern Jackson Co. (Fig. 12-49; see Remarks).

Fossil badgers of late Pliocene and Pleistocene ages are known from numerous localities within the range and from extralimital localities in Alaska, Kentucky, Maryland, and Pennsylvania (Kurtén and Anderson, 1980; Long, 1972). *Taxidea* probably was derived from the larger †*Pliotaxidea nevadensis*, an extinct Pliocene species known from Nevada and Oregon (Kurtén and Anderson, 1980).

Geographic Variation.—Four subspecies are recognized currently, one of which, *T. t. jeffersonii*, occurs in Oregon (Long, 1972).

Habitat and Density.—Except for forested areas, bad-

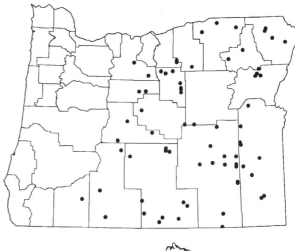

Fig. 12-48. Dorsal, ventral, and lateral views of the cranium, and lateral view of the dentary of the badger, *Taxidea taxus* (OSUFW 1504). Scale bar equals 5 cm.

Fig. 12-49. Localities in Oregon from which museum specimens of the badger, *Taxidea taxus*, were examined (listed in APPENDIX, p. 574), and the distribution of the badger in North America. (Range map redrawn from Long, 1972:735–749, figs. 3, 4, 5, and 7, with modifications.)

gers occur wherever prey is abundant. In Oregon, such locations may include steppe and shrub-steppe areas dominated by sagebrush (*Artemisia tridentata*), some of which may have been converted to irrigated hay fields and other agricultural uses where lagomorphs and sciurids are abundant. Also included may be high desert country dominated by sagebrush, greasewood (*Sarcobatus vermiculatus*), and saltbush (*Atriplex confertifolia*) populated by sciurids and geomyids; meadows partly surrounded by dryland habitat dominated by rabbit-brush (*Chrysothamnus nauseosus* and *C. viscidiflorus*), bitter-brush (*Purshia tridentata*), and scattered junipers (*Juniperus occidentalis*) inhabited by lagomorphs, sciurids, and arvicolines; and even edges of marshes and riparian zones with nesting waterfowl (Anatidae) and aquatic furbearing rodents. Areas with standing water and areas with dense vegetation usually are avoided. Light sandy soils, or even gravelly soils, are ideal for the extensive burrowing activity engaged in by badgers; heavy clay soils, although sometimes occupied, are less attractive habitats (Hall, 1946; Seton, 1929c).

In southwestern Idaho, estimated density of badger populations on a 50-km² study area declined from 5.3 to 3.2/km² during a 3-year investigation (Messick and Hornocker, 1981). We likely would consider these estimates exaggerated had not 215 individuals been captured on the 50-km² area during the 3-year period. The high density of badgers, like the unusual concentration of raptors in the region, likely is linked to the abundance of the Piute ground squirrel (*Spermophilus mollis*) on the area. Lindzey (1982) reported that he earlier had estimated 0.4 badgers/km² on a study area on the Utah-Idaho border; Seton (1929c) estimated a similar density for dry prairies in Manitoba, but about 0.1/km² in areas with heavy clay soils. We know of no estimates of density for badger populations in Oregon.

Diet.—Although the badger often is categorized justly as an opportunistic feeder, small mammals, arthropods, reptiles, and birds usually form the largest components of the diet. In some studies, plant remains occur in 10–20% of the samples, but compose a relatively insignificant part of the diet (Messick and Hornocker, 1981; Snead and Hendrickson, 1942). However, Hart and Trumbo (1983) indicated that plant material (especially corn and sunflower seeds) composed 35% of the stomach contents of badgers taken during the furbearer season in South Dakota.

On the Snake River Birds of Prey Area in southwestern Idaho, Piute ground squirrels were the major prey species except during a year of reproductive failure, when badgers shifted to other rodents and to lagomorphs (Messick and Hornocker, 1981). In south-central Idaho, the major prey was Belding's ground squirrels (*Spermophilus beldingi*); otherwise the diets of badgers on the two areas were similar (Messick et al., 1981). In Minnesota, radio-tagged badgers preyed on 14 species of small mammals, of which remains of the plains pocket gopher (*Geomys bursarius*) occurred most frequently in fecal droppings (Lampe, 1982). In Alberta, the northern pocket gopher (*Thomomys talpoides*) was a major prey item. In Iowa, ground squirrels (*Spermophilus tridecemlineatus* and *S. franklinii*) were the primary prey of badgers (Snead and Hendrickson, 1942). In central Oregon, a female badger became an extremely effective predator on young marmots (*Marmota flaviventris*—S. Thompson, 1979). Although to our knowledge no quantitative study of the diet of the badger

has been conducted in Oregon, the frequency with which we observed excavations in hay fields leads us to suspect that, over much of its range in Oregon, Belding's ground squirrels form a major component of its diet. The high ranking of these fossorial and semifossorial prey species attests to the digging ability of the badger, as many are consumed at seasons when badgers must excavate them from their hibernacula. Nevertheless, largely terrestrial mammals such as *Microtus*, *Clethrionomys*, *Peromyscus*, *Reithrodontomys*, *Mus*, *Sylvilagus*, *Zapus*, *Sorex*, *Tamias*, and even the arboreal *Tamiasciurus* and aquatic *Ondatra zibethicus* sometimes are preyed upon by badgers (Hart and Trumbo, 1983; Lampe, 1982; Messick and Hornocker, 1981; Snead and Hendrickson, 1942).

Some authorities claim that badgers cache food and eat carrion, but others find little evidence of these dietary behaviors. In Nevada, carrion (jackrabbits, *Lepus*, killed during a control program) was consumed frequently and many badgers were captured in traps baited with meat or fish (Hall, 1946), but in Idaho, badgers did not feed on abundant road-killed jackrabbits and could not be baited to traps with carcasses of Piute ground squirrels (Messick and Hornocker, 1981).

Reproduction, Ontogeny, and Mortality.—Badgers breed in summer; the three recorded observations of copulation were in late June–July although at widely separated sites latitudinally (Messick, 1987). Males are spermatogenic May–late August, with mass of testes attaining a peak in July (Wright, 1969). Males do not attain reproductive competence until >1 year of age, but some females become pubertal as juveniles; in southwestern Idaho, >30% of females bred when 4–5 months of age (Messick and Hornocker, 1981). In a sample from Idaho and Utah, 38% of yearling females did not breed, but all older females bred (Lindzey, 1982). Ovulation is induced by copulation, implantation is delayed until February (≅6.5 months) with blastocysts free in the lumen of the uterus, and young are born in March–April (Wright, 1966).

Litter size ranges from one to five (Lindzey, 1982); in southwestern Idaho, average litter size for 20 females ≤1 year of age was 2.45, whereas for 38 females >1 year old the average was 3.11 (Messick and Hornocker, 1981). No information regarding productivity in Oregon is available, but we doubt that it differs greatly from that available for nearby states (Table 12-14).

Most mortality is human-caused (Messick and Hornocker, 1981; Messick et al., 1981); some badgers are trapped for their fur (Table 12-18), some are killed by automobiles or farm machinery, and some are shot because of their potential (but rare) depredations on poultry or because of the potential (but rarely actual) hazard of their excavations to livestock. Removal of rodent prey is considered the most effective means of controlling numbers of badgers (Minta and Marsh, 1988). A few badgers are preyed upon by coyotes (*Canis latrans*), golden eagles (*Aquila chrysaetos*), and possibly ravens (*Corvus corax*—Messick and Hornocker, 1981).

Habits.—Badgers are largely nocturnal, but occasionally may be seen abroad during daylight hours (Lindzey, 1978). During the warmer months, badgers are active for most of the hours of darkness (Harlow, 1979). When ambient temperatures are below –17°C, badgers may stay below ground to avoid the added metabolic costs of ther

Table 12-18.—Reported annual harvest of badgers (Taxidea taxus) in Oregon by county, 1969–1996. Data provided by Oregon Department of Fish and Wildlife.

County	1969–1970	1970–1971	1971–1972	1972–1973	1973–1974	1974–1975	1975–1976	1976–1977	1977–1978	1978–1979	1979–1980	1980–1981	1981–1982	1982–1983	1983–1984	1984–1985	1985–1986	1986–1987	1987–1988	1988–1989	1989–1990	1990–1991	1991–1992	1992–1993	1993–1994	1994–1995	1995–1996
Baker	18	2	2	9	55	50	32	86	65	51	85	35	47	92	34	28	47	82	62	39	9	23	6	11	21	16	1
Benton																											
Clackamas									9ᵃ				21ᵃ	10ᵃ													
Clatsop														15ᵃ		1ᵃ											
Columbia																											
Coos														3ᵃ		8ᵃ			4ᵃ								
Crook		6		15		3	4	130	72	17	54	18	54	32	23	34	50	45	41	13	6	2	4	3	2	2	6
Curry															7ᵃ												3
Deschutes	1		2	7		10	26	28	21	16	30	19	14	19	11	15	24	23	27	3	13	8	3	13	1	6	1
Douglas					1ᵃ					2ᵃ			3ᵃ	4ᵃ	6ᵃ												
Gilliam							3			4	9	13	7	7	13	3	6	11	16	23	6	18	13	6	2	2	
Grant	15	33	14	8	28	21	23	58	71	26	44	33	29	19	12	5	14	57	42	23	18	10	6	2	15	2	
Harney	6	1	32	26	45	14	49	44	17	69	46	70	99	48	81	46	66	100	33	16	17	15	12	3	53		
Hood River											3																
Jackson											1																
Jefferson						4		9	1	1	15	7	8	2		3	3	5	3	2	2	1	1		1	1	
Josephine			1ᵃ				1ᵃ	1ᵃ	2ᵃ																		
Klamath	5	2	5	10	11	3	23	27	33	30	10	5	32	16	29	55	59	46	6	22	27	6	2	4	4	4	4
Lake	3	8	13	29	26	11	25	23	28	73	50	53	82	37	41	49	79	81	31	21	38	17	39	26	34	34	6
Lane													1ᵃ														
Lincoln													3ᵃ		1ᵃ												
Linn													7ᵃ	7ᵃ	4ᵃ	4ᵃ			18ᵃ								
Malheur	42	28	20	36	74	64	160	154	74	74	235	127	115	163	30	103	119	87	130	16	4	15	36	91	26	41	2
Marion								1ᵃ					2ᵃ	10ᵃ				7ᵃ	7ᵃ								
Morrow		1		22		18	38	86	54	38	19	24	53	53	7	2	5	1	2	1	11		1			1	
Multnomah													1ᵃ														
Polk													5ᵃ	5ᵃ													
Sherman						1			4		1		4	4					1		2						
Tillamook																	1					2					
Umatilla	4	4		9	15	22	24	30	20	20	20	5	44	17	8	3	31	37	12	4	6	1	2	38	4	4	
Union	6		7	8	11	24	13	10	11	11	15	5	7	21	1	2	5	4	6	2	2	1	2		1	1	
Wallowa	12	11	13	7	10	6	23	23	22	34	35	57	10	34	31	40	19	63	55	2	3	6	9	8	1	8	
Wasco				2	2		13	13	1	6	4	1	21	26	3	4	4	10	4	4	5	2	4	1	8		
Washington															1ᵃ												
Wheeler	1	2	5		4	6	7	7	3	18	5	7	7	21	2	4	4	4	7								
Yamhill														2ᵃ											2		2
Totals	113	97	98	156	242	306	277	687	653	404	776	450	554	784	291	406	481	638	687	201	112	173	147	215	108	180	23
Trappers reporting[b]	721	773	731	847	1,102	1,350	1,377	1,755	2,746	2,899	3,216	3,373	3,069	3,980	3,292	3,104	3,575	3,397	3,833	1,985	1,536	1,228	1,390	1,454	1,293	1,464	687
Average price/pelt	$2.90	$3.25	$4.35	$5.81	$8.80	$6.69	$15.93	$20.20	$12.98	$24.00	$13.45	$9.74	$12.20	$8.28	$9.36	$6.06	$4.92	$7.74	$5.04	$5.22	$5.28	$5.65	$7.52	$5.82	$10.03	$8.55	$10.15

ᵃCounty records obviously erroneous; trappers likely provided county of residence rather than county in which animals were taken.

ᵇNot all sought badgers and not all were successful.

moregulation, as den temperatures may be as much as 20°C above ambient temperatures. In Minnesota, a radio-tagged female badger was active above ground on several nights when minimum temperatures ranged from –18 to –29°C (Sargeant and Warner, 1972). In Wyoming, one badger did not surface for 21 days during winter (Harlow, 1979) and two badgers in an enclosure did not surface for 70 days (Harlow, 1981a). One individual was known to enter torpor on 30 occasions during which heart rate was reduced from 55 to 25 beats/min, body temperature fell by 9°C, and energy expenditure during each bout was reduced by 27% (Harlow, 1981a). Bouts of torpor averaged 29 h of which 15 h were required to enter torpor and 6 h were required for arousal. Energy conserved amounted to 339 kJ/bout; Harlow (1981a) estimated that the 30 bouts of torpor would extend winter survival without food by ≅10 days. Upon emergence after a long fast, badgers did not gorge but reduced the rate of food passage through the digestive tract by 15% and increased assimilation efficiency by 11% (Harlow, 1981b). Body mass of male badgers increases from spring to fall, but that of females may decrease during lactation then increase rapidly after weaning of young (Messick and Hornocker, 1981). In Wyoming, by November, 31% of the body mass was composed of fat; by March, fat stores were reduced by 37% (Harlow, 1981a).

Badgers exhibit nearly unbelievable prowess as diggers. Hall (1946) recounted several anecdotes concerning attempts to capture badgers by excavating their dens; in all instances the diggers, sometimes numbering as many as 10 men, were outdistanced by the badger even in stony soil. Because badgers dig at the face of their tunnel with the forefeet and push the excavated soil to the rear with their hind feet, they literally "swim" through the earth. The tunnel behind the badger is packed so tightly with excavated soil that it often is difficult to discern the route the badger followed.

In Utah, only 15% of 127 dens used by badgers as daytime retreats were dug by badgers on the day they were used; the abundance of old dens (1.6/ha) probably contributed to their frequent use. Of dens used, 84% were used for only 1 day, 44% were on the periphery of the badger's home range, and three of the dens were used sequentially by more than one badger (Lindzey, 1978). Natal dens were used more consistently; these dens were characterized by abundance of soil at the entrance and the abundance of badger hair and tracks in the soil (Lindzey, 1976b). Also, most natal dens had only one entrance, a main tunnel that branched to permit badgers to pass, one or two chambers 46–55 cm in diameter, and many side pockets packed with feces and earth. Maximum depth of natal dens was 2.3 m (Lindzey, 1976b). Also in Utah, badgers opened tunnels of Uinta ground squirrels (Spermophilus armatus) and partially blocked the entrance, then waited inside for a ground squirrel to seek refuge therein (Balph, 1961).

The home-range area of a radio-tagged female badger in Minnesota was 761 ha in summer, 53 ha in autumn, and only 2 ha in winter (Sargeant and Warner, 1972). In Utah, during the nonbreeding season, home-range areas of three radio-tagged females ranged from 137 to 304 ha, and those of two males were 538 and 627 ha (Lindzey, 1978). Home ranges of males overlapped those of several females and were suspected of being larger during the breeding season (Lindzey, 1978). Among badgers in Idaho, juveniles had the largest average home-range areas (males, 90–3,430 ha; females, 340–1,490 ha) followed by adults (males, 80–340 ha; females, 40–380 ha) and yearlings (males, 50–80 ha, females, 20–120 ha—Messick and Hornocker, 1981). Nevertheless, despite also having overlapping home ranges and sometimes being caught in the same trap within hours of each other, badgers were considered asocial. Family groups and the short-duration mating bond were the only consistent social groupings (Messick and Hornocker, 1981). However, intraspecific aggressive interactions increase as the breeding season approaches. In south-central Idaho, wounds were evident in 11% of 101 badgers caught in June, but 37% of 41 captured in August had wounds (Messick et al., 1981).

Campbell and Clark (1983) observed a pair of badgers in copula. The male grasped the female by the neck and the pair fell upon their right sides; the female kicked dirt between its legs. A smaller male with penis extended harassed the pair and nipped at the male's back. The pair turned on their left sides and the female again kicked dirt between its legs. The female finally broke free 21 min after first being observed. The two males fought for 4 min before the smaller of the two left the scene. Campbell and Clark (1983) were unsure whether the male had ejaculated.

Several authors reported badgers hunting in association with coyotes (Canis latrans) and interpreted the association as one that provides some benefit to one or both species involved. However, Lehner (1981) disputed applying the terms "commensalism" and "social mutualism" as the association is neither obligatory nor prolonged, so is not symbiotic. He suggested the relationship was more likely a type of phoresy whereby an individual of one species is carried along by (i.e., follows) an individual of the other species.

Remarks.—Olterman and Verts (1972) indicated that two museum specimens, one each from Benton and Lane counties, probably were mislabeled. We concur that the badgers were not collected west of the Cascade Range, but we offer a more cogent explanation for the apparent discrepancy. The label of neither contains the name of the county in which the specimen was collected. One (USNM 81809) was collected on 5 August 1896 by E. A. Preble at "Pengra"; the other (USNM 249523) was collected on 11 April 1930 by J. C. Winters at "Cottonwood, below Flynn."

According to McArthur (1992), two communities named "Pengra" existed in Oregon at one time, one each in Lane and Deschutes counties. Because several other specimens collected by Preble the day before and the day after the badger was collected at Pengra (Fig. 2-2) were labeled with localities in Deschutes Co., we are certain the "Pengra" referred to was in Deschutes Co.

McArthur (1992) listed "Cottonwood" as a former community in Lake Co., but he did not list "Flynn" as a geographic name in Oregon. However, on the General Highway Map of Benton Co. (1975), "Flynn" is shown as a locality on the Southern Pacific Railroad at the west edge of Philomath, Benton Co. We found neither a "Flynn" in Lake Co., nor a "Cottonwood" in Benton Co. Accepting the "Cottonwood" in Lake Co. as the locality for the other specimen seems equally compelling as "Flynn" in Benton Co., especially considering that the "Cottonwood" locality is well within the region from which numerous other specimens of badgers were collected (Fig. 12-49). Thus, the evidence to support the claim of badgers occurring west of the Cascade Range in Oregon, at best, is scant.

Some authors claim that badgers make interesting and tractable pets (Perry, 1939), whereas others claim quite the opposite (Fry, 1928). We strongly believe that the proper place for a badger is at the top of a food pyramid in a natural ecosystem, not as a member of a household!

Lutra canadensis (Schreber)
River Otter

1776. *Mustela lutra canadensis* Schreber, theil 3, heft 18, pl. 126B.
1898c. *Lutra hudsonica pacifica* Rhoads, 19:429.
1898a. *Lutra canadensis pacifica*: Allen, 10:460.

Description.—The river otter (Plate XXVIII) is adapted for both terrestrial and aquatic environments (Melquist and Dronkert, 1987). The heavily muscled, somewhat cylindrical body is thickest at the thorax and tapers posteriorly to a thick, dorsoventrally flattened tail composing ≅40% of the total length (Table 12-12). Anteriorly, the body tapers to a blunt and slightly flattened head. The legs are short and powerful; the feet are pentadactyl and plantigrade, and the toes are webbed. The ears are small and rounded; the meatuses are closed during submersion. The eyes are small, forwardly directed, and set high on the head. The vibrissae are long, stiff, and highly sensitive for detecting prey in murky water.

In western Oregon, body mass for male otters with pelts removed averaged 6.0 kg (range, 4.0–8.2 kg) for 44 young-of-the-year and 8.0 kg (range, 5.3–11.6 kg) for 96 older animals. For females, averages were 5.2 kg (range, 3.6–7.0 kg; $n = 40$) and 6.7 kg (range, 4.8–9.1 kg; $n = 73$) for the two age-classes, respectively (Tabor, 1974). Tabor (1974) believed that body mass before removal of the pelt averaged ≅20% greater.

The underfur is grayish, short, and dense, and overlain by longer, stiff and shiny guard hairs. The dorsum is brown, the venter a lighter brown or tan; the lower jaw and throat are whitish.

The skull of the river otter (Fig. 12-50) is flattened with a short and broad rostrum, a strongly pinched interorbital region, and a somewhat flaring braincase. The bullae are flattened, the palate extends beyond M1 by more than the length of M1, and the anterior edge of the coronoid process is straight and essentially at a right angle to the lower toothrow. The carnassial teeth are moderately developed, but M1, the posterior portion of m1, and all of m2 are adapted for crushing. River otters possess 14 pairs of ribs.

Distribution.—Except for the area north of the Brooks Range in Alaska, northeastern Northwest Territories, and parts of the desert Southwest, the river otter occurred throughout most of North America north of Mexico (Fig. 12-51; Hall, 1981). Toweill and Tabor (1982:688, fig. 36-1) indicated that the species was extirpated from much of the Midwest, Great Plains, and Southwest. Widely scattered reintroductions have been attempted (Melquist and Dronkert, 1987), some of which were with otters from Oregon (Berg, 1982).

In Oregon, most museum specimens were collected west of the Cascade Range, but three were collected in eastern Klamath Co. and one each was collected in Deschutes, Wallowa, and Malheur counties (Fig. 12-51). In the 1969–1996 fur seasons, river otters reportedly were taken in 35 of the 36 counties in Oregon; otters were not taken in Lake Co. (Table 12-19).

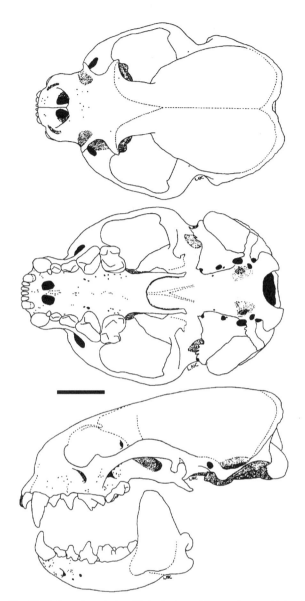

Fig. 12-50. Dorsal, ventral, and lateral views of the cranium, and lateral view of the dentary of the river otter, *Lutra canadensis* (OSUFW 3186). Scale bar equals 25 mm.

Lutra canadensis first appears in the fossil record in the late Pleistocene in cave deposits in Maryland and Pennsylvania ≅800,000 years old. Fossils from later deposits are known from several midwestern and southeastern states (Kurtén and Anderson, 1980). The species is believed to have an ancestry in China; otters crossed the Beringian Land Bridge in the middle to late Pleistocene.

Geographic Variation.—Only one of the seven subspecies recognized currently occurs in Oregon: *L. c. pacifica* (Hall, 1981). Tabor (1974) indicated that average body mass for all sex and age-classes of skinned otters from Klamath Co. was greater than that for otters from counties along the Pacific Coast or from other counties in Oregon. Unfortunately, no measure of variance was provided so we could not test if the differences were significant.

Habitat and Density.—The river otter everywhere is associated with river, lake, pond, or marsh habitats, but may make extensive overland excursions from one such habitat to

Table 12-19.—Reported annual harvest of river otters (Lutra canadensis) in Oregon by county, 1969–1996. Data provided by Oregon Department of Fish and Wildlife.

County	Trapping season																										
	1969–1970	1970–1971	1971–1972	1972–1973	1973–1974	1974–1975	1975–1976	1976–1977	1977–1978	1978–1979	1979–1980	1980–1981	1981–1982	1982–1983	1983–1984	1984–1985	1985–1986	1986–1987	1987–1988	1988–1989	1989–1990	1990–1991	1991–1992	1992–1993	1993–1994	1994–1995	1995–1996
Baker	2								7	2	4	7	7	6		6	10	1	13	11	11		8	20	22	10	
Benton		13	4	5	2	2	9	6	2	2	2	11	3	5	19	16	8	13	6	1	1	14	8	8	15	18	
Clackamas	10	3	5	29	17	13	17	17	4	23	33	12	14	29	18	20	18	22	20	26	18	24	8	8	40	8	
Clatsop	14		20	13	17	13	14	14	7	7	13	13	14	17	18	16	7	20	10	26	16	24	24	8	12	68	
Columbia	1			3	2	5	7	7	7	8	8	2	4	8	7	24	28	30	22	10	16	24	11	3	21	19	1
Coos	12	15	14	32	33	26	27	27	16	21	35	32	24	30	8	30	1	44	22	37	2	11	4	4	21	2	
Crook					5	1	1									3	3			1		4			2	8	
Curry	9	4	5	9	5	1	4	4	6	16	31	14	3	25	25	15	8	8	4		1	7			9	4	
Deschutes		10	8	5	5	9	21	21	14	8	3	3	18	5	5	17	10	17	13	11	12	12	8	8	9	3	1
Douglas	26	34	37	22	32	37	70	70	41	62	64	59	36	60	25	91	38	64	56	39	24	26	32	42	50		
Gilliam							1	1			6	1	1						1			1					
Grant		1									2			4		1											
Harney																											
Hood River	1		3	1	2	2	4	4	3	8	4	2	2	4	4	3	2	1	1	2		2	2	2	1		
Jackson			1	2	5	2	7	7	8	5	13	3	1	3	11	11	5	13	10	6	8	14	21	17	17	12	
Jefferson		1	2	4	4		7	7	5	10	10	6	8	3	2	3	23	1	1	8		3	3	16	16	5	
Josephine	4	7	13	5	5	4	17	17	10	13	13	8	4	6	8	12	13	22	9	19	29	16	15	34	34	34	
Klamath	17	19	14	23	8	12	19	19	21	31	31	42	13	8	15	13	18	30	12	26	27	22	27	40	40	25	
Lake				1																					1		
Lane	28	30	19	31	70	52	87	87	38	85	98	55	57	24	30	31	56	49	22	41	30	30	46	40	40	48	1
Lincoln	26	15	34	17	49	34	16	16	18	17	53	17	10	22	26	36	27	33	16	34	17	17	15	14	14	12	
Linn	5	9	5	33	10	10	5	5	5	6	18	4	6	4	4	23	7	10	7	14	10	11	20	13	13	13	
Malheur	3		1	9	1	1	3	3	5	1	1	12	1	1		4	3	4	12	2	1	6	6	5	9	9	
Marion	5	3	6	1	5	5	5	5	12	31	13	28	22	6	9	11	23	11	5	4	8	9	8	8	8	11	1
Morrow																						4	4		1		
Multnomah	8	4	10	2	2	1			3	13	10	2	6	6	12	6	2	2	2	5	4	1	4	4	13	6	
Polk	16	2	23	2	6	5	26	26	28	8	9	7	3	5	5	5	10	6	14	5	7	3	10	11	11	33	
Sherman												1		1	2		1	2	3	1				5	11	15	
Tillamook	20	20	51	41	36	18	54	54	25	40	43	48	27	19	31	28	36	44	15	8	9	22	19	34	34	353	1
Umatilla				6								2	2	1		1	2	1	1			5	4		7	7	
Union														1		1				1		2				5	
Wallowa	5	2	2		1	1			1	1	1	8	3	3	1	1	6	5	8	16	2	14	9	6	6	6	
Wasco		2	5		9	4	8	8	5	6	7	1	4	3		16	7	7	4	2	1				10	4	
Washington	5	2	1	2	16	7	8	8	4	4	8	2	7	3	2	1	9	16	3	1	3	5	6	6	17	9	1
Wheeler																1	3	2	1	1							
Yamhill	9	2	3	7	7	4	12	12	5	5	13	4	1	6	3	5	3	11	4	4	5	3	6	5	5	7	
Totals	226	198	265	331	298	339	276	439	291	413	558	409	295	315	299	416	383	454	474	290	337	252	320	308	466	486	8
Trappers reporting[a]	721	773	731	847	1,102	1,350	1,377	1,755	2,746	2,899	3,216	3,373	3,069	3,980	3,292	3,104	3,575	3,397	3,833	1,985	1,536	1,228	1,390	1,454	1,293	1,464	687
Average price/pelt[b]	$26.17	$23.60	$31.84	$46.80	$35.77	$34.44	$41.97	$56.90	$44.89	$57.34	$51.09	$41.38	$21.60	$33.38	$34.66	$18.38	$23.05	$29.39	$26.67	$27.11	$27.22	$21.00	$36.00	$40.00	$65.00	$48.00	$48.00

[a] Not all sought river otters and not all were successful.
[b] Rounded to the nearest dollar after the 1989–1990 season.

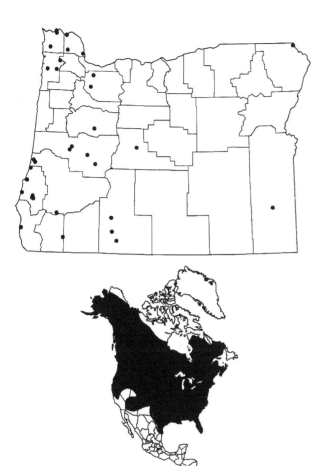

Fig. 12-51. Localities in Oregon from which museum specimens of the river otter, *Lutra canadensis*, were examined (listed in APPENDIX, p. 521), and the distribution of the river otter in North America. (Range map redrawn from Hall, 1981:1032, map 523.) The river otter no longer occurs in much of the Midwest, Great Plains, and Southwest (Melquist and Dronkert, 1987).

another. We saw river otters or their sign in upper reaches of bays, along large rivers, and on lake shores. In western Oregon, 67% of the harvest of otters is concentrated in areas west of the Coast Range (Tabor, 1974:5, fig. 2).

In central Idaho, river otters used valley habitats more than mountain habitats; also, they were associated with streams more frequently than with lakes, reservoirs, and ponds (Melquist and Hornocker, 1983). Estimates of density were 1.7–3.7 otters/10 km of waterway for different drainages that totaled 158 km of waterway that excluded ponds, lakes, and small streams that radio-tracked otters did not use. For all drainages combined during the 1 year that all were censused, the estimated density was 2.7 otters/10 km of waterway. We found no comparable estimates of density for any area in Oregon.

In Maine, the occurrence of river otters was correlated positively with presence of beaver (*Castor canadensis*) sign, shape and length of watersheds, and an index to shoreline complexity (especially the amount of shallow foraging habitat) and negatively with the proportion of mixed hardwood-softwood (the latter reflecting steep gradients and low productivity—Dubuc et al., 1990). In winter after the surfaces of beaver ponds are frozen, river otters often rift beaver dams, thereby lowering the water level in the

ponds (Reid et al., 1988). Beavers do not repair the dams in winter because they do not receive requisite visual and acoustic stimuli while swimming under the ice with ears closed. Such behavior by otters may increase prey density in partly drained ponds and concentrate certain prey species in the fast-moving water below the rift, but the primary value of dam-rifting may be simply to obtain access to water bodies covered with ice (Reid et al., 1988).

River otters establish latrines to which they return to defecate, urinate, scent-mark, and groom. In Massachusetts, such sites were significantly more likely to be on points of land or isthmuses, near mouths of streams, and near beaver bank dens, and to have coniferous trees >30 cm in diameter at breast height than occurred at random points along the same lake shores or river banks (Newman and Griffin, 1994).

Diet.—In western Oregon, of 75 river otters captured during the furbearer trapping season (15 November–15 February), fish occurred in digestive tracts of 80%, crustaceans in 33%, amphibians in 12%, birds in 8%, and molluscs in 11% (Toweill, 1974). Of the 60 otters containing fish, sculpins (Cottidae) occurred in 38%; salmon, trout, or chars (Salmonidae) in 30%; chubs, squawfish, carp, or dace (Cyprinidae) in 30%; bullheads (Ictaluridae) in 8%; and sunfishes (Centrarchidae) in 8%, and 33% contained other species or fishes unidentified to family. Of the 25 otters that had eaten crustaceans, crayfish (*Pacifasticus*) occurred in 76%, pillbugs (*Armadillidium vulgare*) in 20%, and crabs (*Hemigrapsus nudus*) in 4%. Of amphibians eaten, frogs (Anura) were consumed most frequently. Liers (1951) claimed that his tame otters when given a choice selected crayfish over frogs. Of special interest in the Oregon study was the consumption of a rough-skinned newt (*Taricha granulosa*) by one individual, apparently without being affected by its skin toxin (Toweill, 1974). The great dependence upon fishes as a major source of food by otters is reflected in studies of the diet based on examination of fecal droppings collected at all seasons in a variety of wetland habitats at scattered localities throughout much of the range of *L. canadensis* (Greer, 1955; Hamilton, 1961; Melquist and Hornocker, 1983; Sheldon and Toll, 1964). Nevertheless, Liers (1951) reported that tame river otters did not thrive on a diet of live fish, but did well on horse meat with some vegetable material and nutritional supplements. In central Idaho, birds (principally waterfowl, Anatidae) and mammals (mostly muskrats, *Ondatra zibethicus*) occurred in <1–12% and 1–4% of seasonal samples, respectively, but amphibians were not recorded (Melquist and Hornocker, 1983). We suspect that the diet of river otters in northeastern Oregon may be more similar to that of the species in central Idaho than in western Oregon.

Reproduction, Ontogeny, and Mortality.—Both in Oregon (Tabor and Wight, 1977) and elsewhere (Hamilton and Eadie, 1964), female river otters in the wild breed for the first time when ≅2 years old and produce their first litter when ≅3 years old. Embryonic development proceeds to the blastocyst stage then ceases for ≅10 months; blastocysts are free in the lumen of the uterus during this period. In western Oregon, implantation with resumption of development occurs in early February, and as implantation to birth requires ≅60 days (Hamilton and Eadie, 1964), parturition is believed to occur in April (Tabor and Wight, 1977). Thus, females give birth about 12 months after breeding. Females seemingly remate shortly after giving birth (Hamilton and Eadie, 1964; Liers, 1951). In Oregon, 112 (99.1%) of 113 adult females had bred (Tabor and Wight, 1977).

Average mass of testes and epididymides for 42 males born during the previous littering season and taken in western Oregon during the 1970–1971 and 1971–1972 trapping seasons for furbearers was 1.8 g (range, 1.0–3.2 g); the average mass was 10.1 g (range, 3.5–18.2 g; $n = 38$) for yearlings and 12.5 g (range, 6.0–20.0 g; $n = 53$) for older males (Tabor, 1974). Cauda epididymides of males born during the previous littering season contained no spermatozoa, but those of 32% of yearlings and 55% of older animals contained spermatozoa (Tabor, 1974). In New York, mass of testes of males <21 months old was ≤ 7 g and the cauda epididymides of males this age or younger did not contain spermatozoa. Males ≥ 23 months old had testes that weighed 9–30 g (most ≥ 12 g) and spermatozoa in their cauda epididymides (Hamilton and Eadie, 1964). Males taken in November–December at an age of 30–32 months had testes that weighed less than those of males 23–24 months old taken in March–April. Although mass of testes seemingly regresses during the nonbreeding season, whether or not males are fertile throughout the year is unknown (Hamilton and Eadie, 1964).

Although average litter size increases slightly with maternal age, differences between age-classes in Oregon were not significant (Tabor and Wight, 1977). Either some ova are not fertilized or there is some loss of blastocysts (or some blastocysts were not recovered) during the quiescent period as average counts of corpora lutea commonly exceed counts of blastocysts or embryos (Table 12-14). In Idaho, estimated productivity based on age structure of populations averaged 2.3 (recalculated from data provided, but reported as 2.4) offspring per breeding female ($n = 32$; Melquist and Hornocker, 1983:19), similar to estimates of natality based on counts of embryos, blastocysts, and corpora lutea (Table 12-14).

Neonates have not been described, but Hamilton and Eadie (1964) described two 132-g fetuses that they considered to be near term. Total length, tail length, and length of hind foot (in mm) of one of the similar-sized fetuses were 275, 64, and 28, respectively. The fetuses were fully furred, brownish black on the dorsum grading to grayish on the venter; the lips, cheeks, chin, and throat were somewhat lighter as in the adult. Hair on the dorsum was 6–7 mm long, somewhat shorter on the venter. The eyes were closed, but the ears were unfolded and free and no teeth were erupted (Hamilton and Eadie, 1964). Liers (1951:6) indicated that young river otters in captivity were "quite helpless" for 5–6 weeks, but their eyes opened at $\cong 5$ weeks and at $\cong 7$ weeks they moved to one corner of their kennels to defecate and urinate. At 10–12 weeks they commenced to play outside their nests.

Most mortality in river otters is related to the activities of humans (Melquist and Hornocker, 1983). Legal and illegal hunting and trapping, and encounters with automobiles, trains, and domestic dogs probably are responsible for most otter deaths. In the water, otters probably are safe from predation (Melquist and Hornocker, 1983), but when making trips overland, coyotes (*Canis latrans*) are known to prey on them (Grinnell et al., 1937).

Habits.—In Idaho in spring and summer, river otters were active $\cong 30\%$ of the time during most daylight hours (1000–1900 h) and 60–90% of the time during most hours of darkness (2200–0700 h); during crepuscular hours, activity decreased in the morning and increased during the evening. In autumn, otters were active $\cong 30\%$ of the time from 0900 to 2000 h, $\cong 45\%$ of the time from 2000 to 0100 h, and $\cong 60\%$ of the time from 0100 to 0900 h. In winter, otters were active 50–80% of the time during daylight and early evening hours (0700–2000 h), after which activity declined to $\cong 40\%$ of the time until midnight; no records for early morning hours (2400–0700 h) were available (Melquist and Hornocker, 1983:35, fig. 13).

River otters are considered among the more social members of the family Mustelidae. In addition to the adult male–adult female association during the breeding season and the maternal female–young association after parturition, a variety of groupings of otters in different sex and age-classes have been observed (Melquist and Hornocker, 1983:51, table 18).

In Idaho, home-range length for solitary otters at all seasons averaged 32 km (range, 26–42 km; $n = 4$) for juveniles (both sexes), 49 km (range, 10–81 km; $n = 4$) for yearlings (both sexes), and 44 km (range, 31–58; $n = 3$) for adult females. Insufficient data were available to estimate home-range length for an adult male. For family groups, average home-range length was 34 km (range, 25–39 km; $n = 5$; Melquist and Hornocker, 1983). Although home ranges overlapped extensively, intraspecific confrontations seemed to be resolved by mutual avoidance. Personal space was defended, but defense of specific bits of landscape as recorded for many other species was not observed (Melquist and Dronkert, 1987). In Idaho, wounds or scars that might indicate agonistic interactions were not found among 60 otters captured for study (Melquist and Hornocker, 1983).

River otters commonly are thought to be exceedingly playful (Liers, 1951), but most of this reputation seems to have been derived from observations of tame or captive individuals. Among wild otters, play was observed during only 17 (5.8%) of 294 observation periods; in all but one instance juveniles were involved (Melquist and Dronkert, 1987). Slides in snow or mud produced by otters using the same route repeatedly are rare, but otters commonly engage in sliding over snow or ice as a rapid and energetically efficient means of travel (Melquist and Dronkert, 1987).

Remarks.—Tabor and Wight (1977) applied a model (Henny et al., 1970) to estimates of fecundity and age distribution derived from a sample of river otters obtained from trappers in an attempt to ascertain the status (rate of increase) of the otter population in western Oregon. Their conclusion that the population density was stable was based on the invalid premise that rate of increase could be derived from a fecundity schedule and a life table constructed from an age distribution assumed to be stable (Caughley, 1977; Caughley and Birch 1971). We call attention to this not to deride the work of a deceased colleague and his former graduate student, but, in part, to explain our failure to refer to this aspect of their paper. Also, and more importantly, we hope to deter others from falling victim to the same tautology, as one of us (BJV) would have done had it not been for an astute colleague.

Enhydra lutris (Linnaeus)
Sea Otter

1758. [*Mustela*] *lutris* Linnaeus, 1:45.
1843. *Enhydra lutris*: Gray, p. 72.
1991. *Enhydra lutris kenyoni* Wilson, 72:23.

Description.—The sea otter (Plate XXVIII) is considered to be the largest of the living Mustelidae by some (Estes, 1980), a characteristic attributed to the wolverine (*Gulo gulo*) by others (Hall, 1981). Nonetheless, the sea otter is large (Table 12-12; males weigh as much as 45 kg, females as much as 32.5 kg—Estes, 1980); several characteristics other than its size distinguish it from other mustelids and, in some instances, from all other mammals.

The head is large and somewhat rounded, the neck thick, and the body robust. The eyes are small and the ears can be rolled somewhat as in the otariid seals; the ears point downward when the animal dives, upward when it is on the surface (Kenyon, 1969). The front legs are short; the forepaws are rounded, mittenlike, and covered on the ventral surface by a friction pad; and the toes are equipped with short, retractile claws. The vertebrae are modified to permit greater spinal flexibility (Estes, 1980). The pelvis is long and narrow, and is loosely connected to the sacrum. Most of the hind legs are enclosed in the skin of the body. The hind legs are oriented posteriorly and the hind feet are flipperlike; a small rounded spot at the terminal end of each digit is an unfurred pad. The fifth toe is the longest. The third and fourth digits are appressed, thereby providing rigidity to the flipper for propulsion. The tail is less than one-half the length of the head and body, is dorsoventrally flattened, and does not taper appreciably.

Except for the foot pads, eyes, inside of the pinnae, and nose pad, the entire body is furred, although the fur on the hind feet is short and sparse. The pelage is composed of bundles consisting of one guard hair ($\cong$64 micra diameter) and $\cong$70 underfur hairs ($\bar{X}$ = 7.2 micra diameter) arising from each of numerous oval pores; a sweat gland and two sebaceous glands are associated with each bundle. Hair density is $\cong$100,000/cm^2 in $\cong$1,300 bundles/cm^2 (Kenyon, 1969); hair density may be somewhat less in summer than in winter. The pelage usually is dark brown with the head, neck, and shoulders lighter colored and somewhat grizzled in older individuals.

The skull of the sea otter (Fig. 12-52) is massive, often slightly asymmetrical with the left side slightly larger; the rostrum is squarish, the narial opening large, and the sagittal and lambdoidal crests low. The skull and other bones of individuals that have fed on sea urchins (Echinoidea) often are purple from the polyhydroxynaphthoquinone in their tests (shells—Fox, 1953). The two lower incisors in each half of the jaw are scoop-shaped and protrude. The postcanine teeth are bunodont and adapted for crushing. Malocclusion is common. The molariform teeth often contain carieslike pits believed to be formed when a grain of sand is pressed repeatedly into a break in the enamel (Kenyon, 1969). The anterior edge of the coronoid process is curved posteriorly and usually angled $\geq$15° posteriorly.

Distribution.—Originally, sea otters occurred along the Pacific Coast from northern Baja California Sur to Alaska, then along the Aleutian and Pribilof islands, Alaska; the Commander Islands, southeast coast of Kamchatka Peninsula, Kuril Islands and southeastern Sakhalin, Russia; and

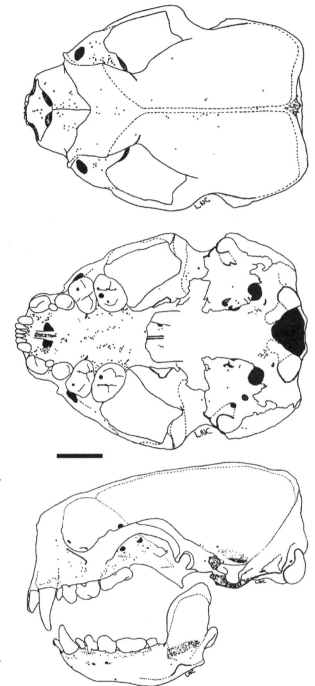

Fig. 12-52. Dorsal, ventral, and lateral views of the cranium, and lateral view of the dentary of the sea otter, *Enhydra lutris* (OSUFW 2421). Scale bar equals 25 mm.

northern Hokkaido, Japan (Kenyon, 1969). Presently, much-reduced populations occur along the coast of California between Point Conception and Monterey Bay; off Washington, British Columbia, and southeastern Alaska (all reintroduced); southwest from Prince William Sound through the Aleutian Islands, Alaska; and along the Commander Islands, Kamchatka Peninsula, and Kuril Islands, Russia (Jameson et al., 1982; Kenyon, 1969; Wilson et al., 1991).

The sea otter, apparently once abundant in coastal waters of Oregon, was extirpated here by fur hunters during the 18th and 19th centuries (Bailey, 1936; Wilson et al., 1991). No recently killed specimen was preserved after 1875 (Fig. 12-53; Bailey, 1936). One was reported seen near Neahkahnie, Tillamook Co., in August and November 1961 and February 1962 (Pedersen and Stout, 1963), but was unverified by other observers. Kenyon (1969) speculated that sea otters from Prince William Sound, Alaska, occasionally may wander south along the coast as far as Oregon. Reintroductions in 1970–1971 were not successful (see Remarks).

Remains of *E. lutris* are known from Pleistocene deposits in California and Alaska. The ancestry of *E. lutris* "can be traced back to late Miocene †*Enhydriodon* in California" (Kurtén and Anderson, 1980:160). †*Enhydriodon* consisted of two groups of extinct sea otter-like mammals, one of Pliocene-Pleistocene age in Asia and Africa and the other of Miocene-Pliocene age in Europe and North America that gave rise to the modern sea otter (Kurtén and Anderson, 1980). Recovery of bones of sea otters from middens of presettlement cultures along the Oregon coast is common (Bailey, 1936).

Geographic Variation.—A newly named race is considered to have been the sea otter that occurred along the coast of Oregon: *E. l. kenyoni* (Wilson et al., 1991). Its

Fig. 12-53. Locality along the coast of Oregon from which museum specimens of the sea otter, *Enhydra lutris*, were examined (listed in APPENDIX, p. 517), and the former distribution of the sea otter in North American waters (range map redrawn from Garshelis, 1987:644, fig. 1, with modifications.) The sea otter no longer occurs along the coast of Oregon and the present distribution of the species throughout North American waters is much reduced (Wilson et al., 1991).

skull is slightly shorter with longer nasals and mandibles than the nominate race that occurs off the coast of Asia, and the skull is slightly longer with shorter nasals and longer mandibles than the race that occurs along the coast of California (*E. l. nereis*).

Habitat and Density.—Ocean areas near reefs, islets, or points of land that provide feeding and resting areas sheltered from waves are especially attractive to sea otters. Sea otters may forage in areas with either rocky or soft bottoms. In California, foraging dives usually are ≤20 m, but in Alaska foraging dives commonly are ≥40 m (Estes, 1980); the maximum known depth of a dive by a sea otter was 97 m (Newby, 1975). Although commonly associated with kelp (Phaeophyta) beds in some areas, habitats that contain kelp are not requisite for sea otters (Kenyon, 1969).

Jameson (1975), from an evaluation of coastal regions of Oregon as potential sea otter habitat, considered the northern 145 km of the Oregon coast to be of little value and 83% of the 185-km central coast to be poor or unsuitable habitat, but 85% of the 160-km southern coast to be good to excellent habitat. This evaluation was based on the presence of rocky substrates, reefs, and kelp beds; availability of food resources; protection from storms; water depth; threat of pollution; and likelihood of human encroachment.

Sea otters in southwestern Alaska, Aleutian Islands, and Kuril Islands number ≅100,000, and ≤1,000 in southeastern Alaska, 100 at Vancouver Island, British Columbia, <50 in Washington, and about 2,000 in California (Estes, 1980; Jameson et al., 1982).

Diet.—In 1971–1973, Jameson (1975) conducted observations at Simpson Reef, near Cape Arago, Coos Co., to ascertain foods eaten by translocated sea otters. Of 425 food items observed when otters brought them to the surface, 271 (63.8%) were purple or red sea urchins (*Strongylocentrotus purpuratus* and *S. franciscanus*), 9 (2.1%) were other echinoderms, 61 (14.4%) were molluscs, 22 (5.2%) were crabs (*Cancer* and *Pugettia*), and 62 (14.6%) were unidentified. Urchins were taken approximately in proportion to their availability. Jameson (1975) believed that sea otters, if they became established as a self-sustaining population, would have a profound effect on flora and invertebrate fauna of the reefs, as detectable changes were evident from the small number of otters that had occupied the area for 1–2 years.

In other regions, sea otters are known to forage extensively on fish, most of which are sluggish bottom dwellers (Kenyon, 1969). Nevertheless, where fish are eaten, nearly half of the diet consists of invertebrates.

Reproduction, Ontogeny, and Mortality.—Females become sexually mature at ≅4 years and males attain reproductive competence at 5–6 years (Kenyon, 1969). Copulation and parturition may occur at any season (Sinha et al., 1966), although may be greater at some seasons. Most young are born in May–June in Alaska, but in December–February in California.

A gestation period of 7.5–9.0 months was estimated for sea otters in the wild, and 6.5–7.0 months for a captive female (Estes, 1980). Ovulation may be induced by copulation; implantation is delayed, but the duration of delay is unknown (Sinha et al., 1966). Although production of a young annually is possible, adult females mostly give birth at 2-year intervals (Estes, 1980; Kenyon, 1969). There is one record of twins, but likely they cannot be reared to independence; none has been reported in the wild (Estes, 1980). Females enter the estrous condition ≤1 month after parturition if the young does not survive, but otherwise usually do not enter estrus until after the young becomes independent (Estes, 1980; Kenyon, 1969). Although individual males may not be reproductively active at all times, populations contain some reproductively active males at all times (Estes, 1980).

Young are born in the water (Sandegren et al., 1973); neonates weigh 1.4–2.3 kg (Kenyon, 1969). Young are dependent upon maternal care for 6–8 months; young are nursed and given solid food commencing shortly after birth, and learn to swim, dive, groom, and forage while under maternal care. Much of early life is spent lying on the female's chest, or when larger, with the head resting on the female. In California, females with small young in November observed during daylight hours spent an average of 10% of the time nursing young, 25% grooming young, 11% grooming themselves, 2% feeding, 14% swimming, and 48% resting (Sandegren et al., 1973). Females with larger young in February and April spent an average of 8% of the time nursing young, 16% grooming young, 9% grooming themselves, 26% feeding, 12% swimming, and 37% resting. Sandegren et al. (1973) emphasized that the percentages did not total 100% because nursing was in conjunction with grooming of young, resting, or swimming. Females averaged nursing their young for 9 min each bout an average of six times daily.

Maternal females tend to be solitary and aggressive (Sandgren et al., 1973). However, if they lose their young they may adopt one that has been orphaned (Kenyon, 1969).

In California, a minimum of 9% and perhaps as many as 15% of 657 dead sea otters examined between 1968 and 1979 were killed by white sharks (*Carcharodon carcharias*—Ames and Morejohn, 1980). One individual found dead along the coast of Oregon after the 1971–1972 reintroductions was believed to have been killed by a shark (Jameson et al., 1982). Killer whales (*Orcinus orca*), bald eagles (*Haliaeetus leucocephalus*), and rough seas also contribute to mortality in sea otters (Kenyon, 1969).

Habits.—Sea otters are nearshore marine mammals that rarely come on land except in extremely remote areas or when sick or injured. When not foraging beneath the surface, sea otters spend most of their lives on their backs, either resting, currying their fur, or consuming prey brought from the sea floor. When resting, the forepaws are held together on the chest and the hind feet are held above the surface of the water to avoid heat loss through the unfurred footpads. Often, sea otters wrap a piece of kelp around their bodies when resting to avoid drifting with the tides.

Beneath the surface, sea otters swim by undulatory movements of the rear portion of the body, including the tail and hind feet (flippers). While foraging, the flippers may be used as paddles for propulsion and maneuvering. On the surface, sea otters swim on their backs with alternate strokes of the flippers. The forefeet are not used in swimming, but are folded across the chest both on and beneath the surface (Kenyon, 1969).

Sea otters tend to forage most during morning and evening hours, but may dive for food at any time of the day or night (Shimek and Monk, 1977). Length and frequency of dives depend upon the type of prey (Estes, 1980); maximum dive time is believed to be ≅6 min (Kenyon, 1969), but most dives are much shorter. Some prey may require several dives to loosen from the substrate. Food is located largely by touch, captured largely between the forepaws, and brought to the surface in a loose flap of skin of the armpit. More in some regions than in others, a stone tool may be used to remove prey from the substrate; the tool often is brought to the surface, placed on the chest, and used as an anvil on which to pound open prey. In California, one otter became adept at detecting inhabited "pop-top" beverage cans on the bottom, bringing them to the surface, ripping them open with its teeth, and consuming the octopuses (*Octopus*) within (McCleneghan and Ames, 1976).

Unlike the pinniped carnivores, sea otters have no insulative blubber; therefore, they must depend on air trapped in their fur for insulation. Consequently, the fur is curried frequently, vigorously, and meticulously, to maintain the protective barrier against hypothermia (Estes, 1980; Kenyon, 1969). They often squeeze water from the fur and blow air into the fur. After eating, sea otters commonly roll onto their sides to wash scraps of food from their fur (Estes, 1980; Kenyon, 1969). The nemesis of sea otters is spilled oil, which causes the fur to lose its insulative qualities.

Sex and age cohorts within populations of sea otters are segregated. Males of all ages (except young under maternal care) tend to occupy small areas with shallow and relatively rough seas; populations may be dense. Females rarely enter those areas. Females occur in much broader and less discrete areas between those occupied by males (most of the occupiable habitat) and at much lower densities. Adult males enter areas occupied by females to mate with them, but subadult males do not (Estes, 1980). Males are known to defend territories that include a female against intrusion by other males (Calkins and Lent, 1975).

Sea otters produce a variety of sounds and essentially all individuals are "right handed" (Kenyon, 1969).

Remarks.—In 1970, 29 sea otters were released at Port Orford, Curry Co.; in 1971, 24 more were released at the same site and 40 were released at Cape Arago, Coos Co. All were from Amchitka Island, Alaska (Jameson et al., 1982). In 1970, sea otters were observed at nine localities 5–80 km from the release site; all but one was north of the release site. In 1972, four or five sea otters were observed 204 km north of Cape Arago and three were seen 290 km north of the same site (Jameson et al., 1982). Jameson et al. (1982) believed the northward movements were attempts by some of the otters to return to their home areas; some were believed to have been recruited into other translocated populations north of Oregon. Eleven otters were known to have died by August 1973, but 10 young were known to have been born after the transplant. Twenty-one were counted during censuses in 1972, 23 in 1973, and 21 again in 1974; however, from 1975 to 1981, numbers observed declined from 13 to one. Only one young was seen after 1976. Jameson et al. (1982) suggested that the failure of sea otters to be reestablished along the coast of Oregon was attributable to emigration and mortality that resulted in populations too small to be self-sustaining. They also indicated that habitats in Oregon might not have been as suitable or possibly were of poorer quality than those farther north where populations were reestablished through similar translocations.

FAMILY MEPHITIDAE—SKUNKS

The mephitids consist of nine species of New World skunks in three genera, two of which occur in Oregon (*Spilogale* and *Mephitis*) and two species of stink-badgers (*Mydaus*) restricted to Indonesia and the Philippine Islands. Traditionally, these species were included in the Mustelidae (Stains, 1967, 1984; Wozencraft, 1989, 1993), but Dragoo and Honeycutt (1997), on analyzing mitochondrial-DNA sequences, found that the skunks and the stink-badgers formed a clade separate from the remaining genera of mustelids. They placed the skunks and stink-badgers in a separate family.

Both species of skunks that occur in Oregon are black with white stripes or stripes broken into linear series of spots, and the margin of the palate is nearly even with the end of the toothrow. The upper molar is set transversely and is squarish with an anteroposterior constriction near the labial edge. Both species are equipped with muscle-encapsulated anal scent glands capable of propelling musk toward approaching enemies. Not only is the musk highly odoriferous and persistent, it is temporarily blinding and, if ingested, causes nausea and possibly more serious reactions (Jackson, 1961). Embryonic development is interrupted between fertilization and implantation for ≅6 months in the western spotted skunk (*Spilogale gracilis*—Mead, 1968a), but only briefly (≤19 days) in the striped skunk (*Mephitis mephitis*—Wade-Smith and Richmond, 1978).

KEY TO THE MEPHITIDAE OF OREGON

1a. Pelage black with a thin white stripe from nose pad to forehead, a round white spot on pate, and two white stripes extending posteriorly on dorsum (sometimes abbreviated or absent); rostrum lower than plane of frontals; mastoid region not inflated, making sides of posterior margin of skull slightly concave when viewed in dorsal aspect; inferior margin of mandible with distinct step at the angle of the jaw.................................
.......................***Mephitis mephitis***, striped skunk (p. 445).

1b. Pelage black with three white spots on head, one on forehead and one in front of each ear (the latter two often confluent with lateral body stripes), and four to six white body stripes (usually broken into a series of spots); rostrum nearly level with plane of frontals; mastoid region inflated, making sides of posterior margin of skull slightly convex when viewed in dorsal aspect; inferior margin of mandible without a step at the angle of the jaw..
........***Spilogale gracilis***, western spotted skunk (p. 441).

Table 12-20.—*Means (±SE), ranges (in parentheses) and CVs of measurements of skull and skin characters for female and male mephitids from regions in Oregon. Skin characters were recorded from specimen tags; skull characters were measured to the nearest 0.01 mm. SE and CV not provided if* n <10.

Species and region	Sex	n	Total length	Tail length	Hind foot length	Ear length	Mass (g)	Condylobasal length	Basilar length	Postglenoid length	Length of maxillary toothrow
Mephitis mephitis											
Northwestern Oregon	♂	2	632 (613–650)	217 (210–224)	78 (73–82)			76.24 (73.12–79.36)	67.03 (63.62–70.43)	25.78	23.61 (22.67–24.54)
Southwestern Oregon	♀	1	550	250	62	20	1,500	69.73	60.55	25.38	21.45
	♂	8	672 (613–740)	272 (235–305)	79[a] (65–88)			72.99 (68.63–78.76)	63.97 (59.89–69.16)	25.09[b] (24.53–25.65)	23.19 (22.04–24.38)
Columbia Basin	♀	5	655 (625–681)	274 (247–307)	75 (66–82)			73.04 (71.51–75.11)	64.30 (62.80–65.93)		22.68 (21.46–23.18)
	♂	2	666 (642–690)	293 (288–298)	70 (69–71)			74.74 (74.20–75.28)	65.43 (64.80–66.05)		23.72 (23.58–23.86)
Remainder of Oregon	♀	6	668 (580–730)	279 (235–315)	76[a] (67–81)	18		74.04 (69.61–77.72)	64.31 (60.06–67.95)	26.59[c] (24.48–27.92)	24.43 (23.35–25.76)
	♂	10	681 ± 12.7 (620–758) 5.9	289 ± 12.2 (260–390) 13.3	79[c] (74–87)			78.75 ± 1.21 (73.27–83.41) 4.88	69.10 ± 1.04 (64.52–73.03) 4.78	27.31[b] (25.53–28.52)	24.32 ± 0.32 (22.87–26.15) 4.17
Spilogale gracilis											
West of Cascade Range	♀	27	386 ± 4.8 (330–457) 6.5	125 ± 2.5[a] (96–150) 10.2	45 ± 0.6[c] (40–52) 6.5	27 ± 0.5[d] (22–30) 8.3	503 ± 47.1[e] (266–965) 33.78	54.10 ± 0.39 (50.46–59.03) 3.78	47.22 ± 0.37 (43.81–51.53) 4.08	20.83 ± 0.19[f] (19.48–22.31) 3.79	17.55 ± 0.12 (16.46–18.78) 3.47
	♂	26	411 ± 4.4 (356–453) 5.5	132 ± 2.8 (102–162) 10.7	49 ± 0.6 (42–54) 7.5	28 ± 1.2 (17–32) 15.4	709 ± 51.7[g] (446–1,200) 27.28	57.70 ± 0.44 (52.32–62.62) 3.85	50.56 ± 0.44 (45.44–55.71) 4.40	22.58 ± 0.21[h] (20.22–24.65) 4.26	18.45 ± 0.15 (17.09–20.61) 4.06
East of Cascade Range	♀	8	373 (356–385)	138 (125–155)	40 (38–43)	27		48.79 (47.73–49.68)	43.06 (42.01–44.30)		15.58 (14.88–16.17)
	♂	9	409 (380–455)	144 (115–163)	46 (43–50)			54.51 (50.49–56.91)	48.09 (45.04–50.10)	22.02	17.05 (15.88–18.24)

[a]Sample size reduced by 1. [b]Sample size reduced by 6. [c]Sample size reduced by 2. [d]Sample size reduced by 9. [e]Sample size reduced by 14. [f]Sample size reduced by 11. [g]Sample size reduced by 12. [h]Sample size reduced by 5.

Spilogale gracilis Merriam
Western Spotted Skunk

1890a. *Spilogale gracilis* Merriam, 3:83.
1890b. *Spilogale saxatilis* Merriam, 4:13.
1890b. *Spilogale phenax latifrons* Merriam, 4:15.
1899b. *Spilogale olympica* Elliot, 1:270.
1906. *Spilogale gracilis saxatilis* Howell, 26:23.
1933. *Spilogale gracilis latifrons*: Grinnell, 40:106.

Description.—*Spilogale gracilis* (Plate XXVIII) is smaller (Table 12-20) and more weasel-like than the striped skunk (*Mephitis mephitis*). The feet are pentadactyl and plantigrade; the toes on the forefeet are equipped with ≅7-mm-long, recurved claws, whereas those on the hind feet are slightly more than half that length and less recurved. There are four pads on the soles. The pelage is black with a somewhat pentagonal white patch between the eyes, a round white subauricular patch on each side of the head (often confluent with the body stripes), and four or six segmented white body stripes. The median pair of stripes usually is narrower than the next pair of stripes and the lateralmost pair of stripes frequently is abbreviated or missing. Thus, the color pattern of each individual is unique. The distal one-third to one-half of the tail is white, but unlike the tail hairs of the striped skunk, the individual hairs are either all white or all black. The distal tail hairs can be spread to form a huge white plume (Bailey, 1936:plate 44a).

Table 12-20.— *Extended.*

Length of P4	Zygomatic breadth	Cranial breadth	Least interorbital breadth	Length of mandible
7.71	48.01	30.21	22.45	50.94
(6.70–8.71)	(47.42–48.60)	(29.59–30.82)	(21.90–23.00)	(49.33–52.54)
7.37	44.78	28.75	20.16	47.20
7.65	46.09	28.85	21.38	48.80
(7.24–7.96)	(42.65–49.55)	(26.97–30.30)	(20.23–22.97)	(46.48–52.54)
7.75	45.84	28.66	21.76	48.69
(7.20–8.54)	(43.87–47.84)	(27.54–29.80)	(21.42–22.08)	(47.44–50.00)
7.59	47.01	28.41	22.95	49.68
(7.50–7.67)	(45.45–48.56)	(28.33–28.49)	(22.91–22.98)	(49.29–50.07)
7.48	46.22	29.00	21.62	49.52
(6.94–8.14)	(42.53–49.24)	(28.45–29.58)	(19.95–23.98)	(46.48–51.14)
7.79 ± 0.17	49.48 ± 0.74	30.00 ± 0.47	23.08 ± 0.36	52.37 ± 0.97
(6.58–8.67)	(46.83–54.08)	(27.32–32.35)	(21.45–24.66)	(48.14–56.23)
7.20	4.73	5.00	4.87	5.88
6.67 ± 0.06	34.42 ± 0.27	25.53 ± 0.24	15.54 ± 0.09	34.63 ± 0.29
(6.05–7.21)	(32.49–38.45)	(23.79–28.98)	(14.40–16.63)	(32.34–39.26)
4.76	4.05	4.97	3.28	4.30
6.83 ± 0.08	37.24 ± 0.39	26.76 ± 0.23	16.53 ± 0.21	37.26 ± 0.33
(5.45–7.74)	(32.87–41.78)	(24.51–28.91)	(14.77–19.57)	(32.88–40.48)
6.28	5.29	4.38	6.44	4.51
5.98	31.19	23.67	13.34	30.56[a]
(5.48–6.26)	(29.77–33.20)	(22.31–24.37)	(12.70–14.26)	(29.67–31.67)
6.37	35.27	24.48	14.61	34.66
(5.71–7.31)	(32.93–37.84)	(23.17–26.40)	(13.49–15.82)	(33.02–36.48)

The skull of the western spotted skunk (Fig. 12-54) is flattened with the rostrum nearly level with the plane of the frontals. The mastoid region is greatly inflated, producing a flared appearance in the posterior margin of the skull. As in *Mephitis,* there are 34 teeth (Table 1-1), and the frontal bones often are partly eroded by a nematode parasite (*Skrjabingylus*). There are 14 pairs of ribs in contrast to 16 in *Mephitis.*

Spilogale gracilis is endowed with muscle-encapsulated musk glands similar to those of *M. mephitis* and, similarly, can eject musk from two papillae located immediately inside the anal sphincter. Three components of the musk compose 89–98% of the volatile compounds: (*E*)-2-butene-1-thiol, 3-methyl-1-butanethiol, and 2-phenylethanethiol. Some differences in proportions of the three compounds were found between sexes (Wood et al., 1991). Six minor compounds were phenylmethanethiol, 2-methyl-quinoline, 2-quinoline methanethiol, bis[(*E*)-2-butenyl] disulfide, (*E*)-2-butenyl 3-methylbutyl disulfide, and bis(3-methylbutyl) disulfide (Wood et al., 1991). Except for 2-phenylethanethiol, all occur in the musk of the striped skunk, but the three thioacetate derivatives found in the musk of the striped skunk are missing in the musk of the spotted skunk (Wood et al., 1991). These differences and different proportions of the compounds probably account for the odor of the musk being considered somewhat more pungent or acrid, but spreading less widely than that of *M. mephitis* (Dalquest, 1948; Hall, 1946).

Distribution.—*Spilogale gracilis* occurs west of the continental divide from southwestern British Columbia, southeastern Washington, southwestern Montana, and north-central Wyoming south through western United States to the southern tip of Baja California Sur and through central Mexico to Costa Rica (Hall, 1981; Hall and Kelson, 1959). In Oregon, museum specimens have been collected throughout most of the state; we do not believe the species occurs in most of the Willamette Valley (Fig. 12-55). Records of reported harvest of spotted skunks for fur for 1969–1996 indicate that the species was taken in all counties except Gilliam (Table 12-21).

The earliest fossil records of *Spilogale* are from late Pliocene deposits in Kansas and Texas of the small *S. †rexroadi,* believed to be directly ancestral to Recent species of *Spilogale* (Kurtén and Anderson, 1980). The larger Recent species are known from early Pleistocene deposits in Maryland and Arizona and late Pleistocene deposits throughout the present ranges of *S. gracilis* and *S. putorius* (Kurtén and Anderson, 1980).

Geographic Variation.—Two races of *S. gracilis* occur in Oregon: the larger *S. g. saxatilis* east of the Cascade Range and the smaller *S. g. latifrons* west of the Cascade Range (Hall and Kelson, 1959; Merriam, 1890b).

Habitat and Density.—Bailey (1936) described habitats commonly used by *S. gracilis* in eastern Oregon as canyons, cliffs, rimrocks, lava fields, and arid valleys. Maser et al. (1981b) indicated that in coastal Oregon *S. gracilis* was common in alder (*Alnus rubra*)-salmonberry (*Rubus spectabilis*), riparian alder, riparian hardwood, and tanoak (*Lithocarpus densiflorus*) habitats.

In sampling "only closed-canopy, upland forest (without riparian vegetation)" in three seral stages in the southern Oregon Coast Range for northern flying squirrels (*Glaucomys sabrinus*), Carey and Kershner

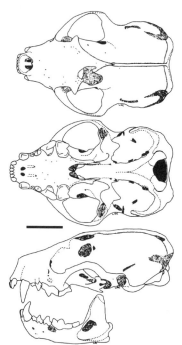

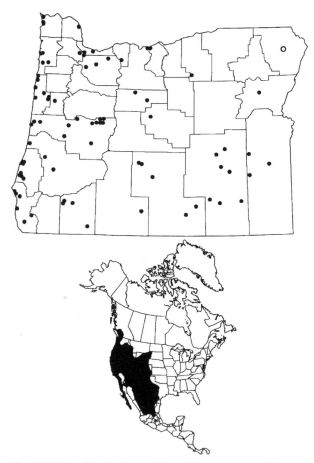

Fig. 12-54. Dorsal, ventral, and lateral views of the cranium, and lateral view of the dentary of the western spotted skunk, *Spilogale gracilis* (OSUFW 2640). Scale bar equals 15 mm.

Fig. 12-55. Localities in Oregon from which museum specimens of the western spotted skunk, *Spilogale gracilis*, were examined (listed in APPENDIX, p. 564), and the distribution of the western spotted skunk in North America. (Range map redrawn from Hall and Kelson, 1959:930, map 472.) Open symbol indicates record for county.

(1997:30) captured 144 *S. gracilis*. Rates of capture were 1.2/1,000 trap-nights in stands >200 years old, 0.4/1,000 trap-nights in 90–120-year-old naturally regenerated stands with (p. 30) "well-developed understories and substantial legacies of live and dead old-growth trees" from earlier stands, and 0.2/1,000 trap-nights in 40–70-year-old stands with little understory. Carey and Kershner (1997:33) claimed that despite significant variation in rates of capture and "in vegetation structure among seral stages" that they were unable "to develop meaningful predictive models from general structural variables that are proximate measures of habitat."

In western Oregon, where the species are sympatric, *S. gracilis* commonly uses tunnel systems constructed by the mountain beaver (*Aplodontia rufa*—Hooven et al., 1979; Lovejoy, 1972; Pfeiffer, 1953). Thus, descriptions of habitat for this syntopic species should serve for *S. gracilis* in western Oregon as well.

Both in Oregon and elsewhere, spotted skunks commonly reside in barns, sheds, and houses or under the floors thereof (Bailey, 1936; Crabb, 1948; Hall, 1946; Maser et al., 1981*b*). Piles of hay, logs, and refuse also may be used by spotted skunks. Such sites may be attractive to spotted skunks because of the protection they afford and because rats (*Rattus*) and mice (*Mus*) living therein may serve as a food source. Spotted skunks frequently are associated with the runways and stick-houses of woodrats (*Neotoma*—Bailey, 1936; Grinnell et al., 1937).

Although no estimates of density are available for *S. gracilis* in Oregon, from records of spotted skunks harvested for fur (Table 12-21), it appears that counties with a section adjacent to the Pacific Ocean may support a greater abundance of the species. On a largely agricultural area in Iowa, the estimated density of the formerly conspecific *S. putorius* in late winter and early spring was 8.1/km² (Crabb, 1948).

Diet.—No investigation of the diet of *S. gracilis* has

been conducted in Oregon, and, indeed, none seems to have been conducted for the species anywhere in its range. In Iowa, based on the examination of fecal droppings, the congener *S. putorius* consumed primarily small mammals in winter (frequency of occurrence = 90.4%, *n* = 75) and spring (86.6%, *n* = 330), but in summer and autumn, use of this food category declined (35.5%, *n* = 254; and 58.3%, *n* = 185, respectively— Crabb, 1941). A considerable portion of the small mammals consumed in winter was carrion consisting of cottontail rabbits (*Sylvilagus floridanus*) shot because of their depredations on shrubs. Seasonal frequency of occurrence of arthropods was the converse of that of small mammals; they appeared in 14.6% of the droppings in winter, 47.6% in spring, 92.4% in summer, and 80.5% in autumn. Plant material composed 10–35% of the diet and birds composed 5–10% depending upon the season (Crabb, 1941). In autumn, Selko (1937) found mammal remains in 47% of 59 fecal droppings that composed 30.7% of the total food remains; arthropods and birds occurred in 44% and 27% of the samples and composed 30.8% and 25.1% of the food remains, respectively. Although vegetable material occurred in 27% of the samples, it composed only 1.6% of the food remains (Selko, 1937).

In Washington, the stomach of one *S. gracilis* contained remains of three red-backed voles (*Clethrionomys gapperi*—Dalquest, 1948). *S. gracilis* often is accused of preying on small mammals caught in live traps (Hooven et al., 1979; Maser et al., 1981*b*).

Because *S. putorius* and *S. gracilis* were considered conspecific (see Remarks), we suspect that most state and regional works on mammals published since 1941 have relied heavily on the works of Crabb (1941) and Selko (1937) in Iowa as the source of information concerning the diet of spotted skunks regardless of the locality. Likewise, we were forced to depend largely on these publications for dietary information on a congener of the western spotted skunk, explaining our reluctance to provide more specific details regarding the identification and quantification of food items consumed. From the meager information published before 1941 (Bailey, 1936) and from the nine stomachs examined by Maser et al. (1981*b*), the information for *S. putorius* from Iowa, in general, seems to reflect reasonably well the diet of *S. gracilis* in Oregon.

Reproduction, Ontogeny, and Mortality.—Most, but not all, juvenile males become sexually mature in September of the year of their birth when only 4–5 months old. Although known to be capable of siring offspring, the contribution of juvenile males to the reproductive effort is unknown (Mead, 1968*a*). Older males become fertile in May. Regression of the testes commences in October after the breeding season and continues until March (Mead, 1968*a*).

Females commenced to breed about 20 September on the Olympic Peninsula, Washington, and most had bred by the end of the 1st week of October. Presumably all age-classes of females, including young-of-the-year, breed, as 41 of 45 (91%) *S. g. latifrons* (the race that occurs in western Oregon) collected after 25 September had bred (Mead, 1968*a*). After development, the blastocysts remain free and randomly distributed in the lumen of the uterus from October to March; the inner cell mass ceases to develop but the trophoblast continues to develop slowly (Mead, 1968*a*). Ovaries of females with unimplanted blastocysts contain "small, relatively inactive-appearing corpora lutea and large amounts of type I [granular] interstitial tissue" (Mead, 1968*a*:380). In California, implantation may occur as early as late March (Grinnell et al., 1937), but the earliest implantation in *S. g. latifrons* was on 8 April in a female held captive from 22 January (Mead, 1968*a*). Parturition in wild-caught captive *S. g. latifrons* occurred from 17 April to 26 May; dates of parturition for six specimens from western Oregon were from 17 April to 20 May (Mead, 1968*a*:384, table 8). Mead (1968*a*) considered the gestation period to be about 210–230 days, with the blastocysts unimplanted and developing slowly for 180–200 days. Adult females are anestrous from June to early September. Partly in contradiction to this timing sequence, Walker (1930*a*:229) reported that a wild-caught female *S. g. latifrons* from western Oregon that had given birth in captivity in 1924 was placed with a male during "the following winter" and gave birth to a litter on 12 April 1925. Thus, gestation periods somewhat less than that reported by Mead (1968*a*) seem possible.

Average litter size based on implanted embryos for 11 pregnant females presented by Mead (1968*a*:384, table 8) was 3.8, not 3.9 as stated in text (p. 385). Average litter size based on counts of implantation sites was slightly higher (Table 12-22).

We know of no information regarding ontogeny of *S. gracilis*. In Iowa, a female neonate *S. putorius* weighed 9.5 g; at 1 week a male weighed 22.5 g; at 2 weeks a female weighed 45.5 g; a 3-week-old male weighed 73.5 g; at 4 weeks, three individuals (two males, one female) averaged 120.2 g; at 5 weeks, the same three averaged 154.3 g; at 6.5 weeks, 176.3 g; at 8.5 weeks, 295.0 g; at 10 weeks, 346.0 g; at 11.5 weeks, 357.3 g; and at 15 weeks, 543.3 g (Crabb, 1944). The mass of the female exceeded that of either male from 5 to 6.5 weeks, but by 15 weeks it weighed only 71.0–81.8% as much as the males.

At birth, the black and white markings were visible on the nearly hairless skin, the eyes and ears were closed, and no teeth were visible; by 3 weeks of age, the young were covered with 1-cm-long fur and were able to crawl weakly, but the eyes remained closed and no teeth had erupted; the eyes opened on day 32; and by day 36, the teeth could be detected by touch and the young could walk, stamp their feet, and spread their tail hairs (Crabb, 1944). A male emitted musk at 46 days of age. Seasonally in Iowa, adult males averaged 651–736 g and adult females averaged 481–538 g; males averaged heaviest in spring and lightest in summer, whereas females averaged heaviest in autumn and the same during each of the other three seasons (Crabb, 1944).

The annual harvest of spotted skunks for fur in Oregon since 1969–1970 ranged from 87 to 1,544 animals (Table 12-21); in comparison with the harvest of several other furbearers, the harvest of spotted skunks is of little consequence and probably largely incidental to the harvest of other more valuable furbearers. Because of their proclivity for den sites in the proximity of human habitation, encounters with humans and domestic pets probably are responsible for the demise of many spotted skunks in Oregon as elsewhere (Crabb, 1948).

Habits.—*Spilogale gracilis* does not hibernate, but tends to be less active during periods of unusually cold weather (Dalquest, 1948). It is primarily nocturnal (McCullough and Fritzell, 1984) and highly secretive. Despite its secretive nature, several naturalists who have encountered it or have spent the night in a building it occupied have remarked about the amount of noise *S. gracilis* makes as it goes about its nighttime activities (Dalquest, 1948; Grinnell et al., 1937; Maser et al., 1981*b*). Dalquest (1948) remarked that the species was rarely killed on roads in Washington, but Crabb (1948) indicated that *S. putorius* was a frequent victim of automobile traffic in Iowa and Maser et al. (1981*b*) indicated that some were killed by automobiles in Oregon. Nevertheless, we do not recall ever seeing a spotted skunk killed on roads here, but such may be a reflection of infrequent trips through areas inhabited by the species.

Although equipped with a superb defense system, the spotted skunk commonly tolerates considerable disturbance before resorting to the use of musk (Bailey, 1936; Maser et al., 1981*b*). However, when approached too closely or too rapidly, *S. gracilis* shows displeasure by stamping its forefeet, then doing a "handstand" for ≤5 s (Walker, 1930*a*) with tail hairs flared over the back and sometimes with anus everted. Sometimes, if pressed, a spotted skunk will expel musk from the handstand position (Maser et al., 1981*b*; Seton, 1929*c*), but more often, a spotted skunk, with all four feet on the ground,

Table 12-21.—Reported annual harvest of spotted skunks (Spilogale gracilis) in Oregon by county, 1969–1996. Data provided by Oregon Department of Fish and Wildife.

County	1969–1970	1970–1971	1971–1972	1972–1973	1973–1974	1974–1975	1975–1976	1976–1977	1977–1978	1978–1979	1979–1980	1980–1981	1981–1982	1982–1983	1983–1984	1984–1985	1985–1986	1986–1987	1987–1988	1988–1989	1989–1990	1990–1991	1991–1992	1992–1993	1993–1994	1994–1995	1995–1996
Baker	1																		3		1					1	
Benton			15		2											3	15	17	16	2	1		6		5	17	
Clackamas		2			9												17	14	3	18	11	20	9	3		4	
Clatsop	23			21		37	34	35	4	3	27	32	5	31	53	17	17	14	51	18	11		5	3			
Columbia	37			18	2	11	9	8	16	2	11	6	19	30	56	15	51	22	32	3	2			5			
Coos		4	8	10	6	47	30	60	66	70	66	4	113	25	24	17	13	13	26	1	3		2			5	
Crook		3		2			12									13	2	9	3	9	1		1				
Curry	6		9			4	40	39	44	71	105		69	144	36	161	90	73	42	25		5		1			
Deschutes		5		3	5	5	15	1		1	6		3	1		3	1	1	5			1	1		1	1	
Douglas	2	106	2	19	14	76	95	264	285	411	304	91	224	66	168	54	105	84	104	82	57	70	40	12		97	
Gilliam																											2
Grant	2	6	1	12	2	11		1	1		2	5	5	2	3	1	3	3	2			2	2				
Harney		15			2				4	1	1	4	17	4	5	12	3	1	4		1	1	2				
Hood River			1				1	1				1			1	6	2					106	6				
Jackson		1		5	3	1	2	8			18	7	34	42	25	24	62	39	65	22	9	6		13		15	
Jefferson					2		1	1					2	2					2								
Josephine		1		4	7	24	8	15	20	36		27	25	40	44	42	38	25	45		10	7		3	3	3	
Klamath	7	4	6	18	3	7	17	3		24	25	25	24	56	14	7	13	35	12	1	6	3	2		8	3	
Lake		5	1	7			3		26	12	12	5	1	6	9	3	54	12	23	23	3				2	4	
Lane	64	126	147		230	125	267	196	502	59	610	412	476	402	250	290	313	316	417	164	217	108	141	179	116	159	
Lincoln	6	1	35	72	23	13	49	5	62	41	12	30	25	34	62	52	82	7	6	9	2					2	
Linn	6		8	9		10	32	28	36	11			20	24	28	33	58	39	44	27	3	2	14		3	3	
Malheur	10	2			2							2		2	1	5	5	4		3		2					
Marion	7								2										6	13			5				
Morrow																						1				5	
Multnomah		2		1					21										18	9		1	15				
Polk	1						20		21	9	4	8	5		4	2	12	12	18			2					
Sherman																				1							
Tillamook	2	29		45		43	91	20	51	56	88	46	30	36	37	12	49	55	27	15	11	2	11		1		
Umatilla	1	1							2									8				1				1	
Union			12																1								
Wallowa		1		1						5	6			3	2	5	1	1	8	5		6					
Wasco					2	9	8		3	3		6			2	4	4		1	5	2	2		1			
Washington	3				1	3	5								23	4		5	17	2		2	1		3		
Wheeler												3		3		2	2								1		
Yamhill	1			7	1		12	5	18	22	7	18	4	2	2	12	29	29	7	8		1					
Totals	169	305	87	254	516	271	628	595	1,104	672	1,544	1,141	978	1,255	792	931	976	931	1,032	463	377	227	384	279	168	321	2
Trappers reporting[a]	721	773	731	847	1,102	1,350	1,377	1,755	2,746	2,899	3,216	3,373	3,069	3,980	3,292	3,104	3,575	3,397	3,833	1,985	1,536	1,228	1,390	1,454	1,293	1,464	687
Average price/pelt[b]	$1.15	$0.61	$1.59	$2.26	$2.01	$2.21	$2.28	$4.80	$7.25	$9.84	$7.14	$6.44	$4.85	$1.51	$2.69	$2.95	$3.07	$3.37	$4.85	$1.83	$3.50	$6.12	$4.93				

[a] Not all sought spotted skunks and not all were successful.

[b] Price/pelt not available during some years.

Table 12-22.—*Reproduction in Oregon mephitids.*

Species	Number of litters	Litter size $\overline{X}$	Range	n	Reproductive season	Basis	Authority	State or province
Mephitis mephitis	1	7.3[a]	5–9	21	Feb.–Jun.[b]	Embryos	Verts, 1967b	Illinois
	1	7.2	4–11	39		Implant sites	Verts, 1967b	Illinois
Spilogale gracilis	1	3.8	3–5	5	Sep.–May	Embryos	Mead, 1968a	Oregon[c]
	1	4.1	3–6	23		Implant sites	Mead, 1968a	Three states[d]

[a]Includes embryos undergoing resorption; $\overline{X} = 6.3$ for living embryos.
[b]Young thought to have been born in mid-July, may have resulted from second estrus period after failure to conceive during first estrous or loss of all embryos during pregnancy.
[c]Six additional records for *S.g.latifrons* from Washington and possibly Oregon did not alter the mean but extended the range from 2 to 5.
[d]Oregon, Washington, and Nevada.

bends its body in a U-shape with both head and rear directed toward the intruder before expelling musk (Walker, 1930a). Thus, the characteristic "handstand" likely is largely a response to some perceived threat, but the pose at times seems to be elicited in other contexts (Walker, 1930a).

Spilogale, unlike *Mephitis*, is an expert climber, and occasionally may expel musk on a baying dog or approaching person from a vantage point in a tree (Cuyler, 1924). Also, *S. putorius* sometimes dens in cavities in trees, but more often in hollow logs and crevices in rocky outcrops (Crabb, 1948; McCullough and Fritzell, 1984). In Missouri, most dens occupied by *S. putorius* opened in a north-facing direction in summer, but in winter 80% faced southward (McCullough and Fritzell, 1984).

Spilogale uses a technique similar to that employed by *M. mephitis* to open eggs too large to bite; in addition to forcefully propelling the egg between the hind legs with the forefeet, the skunk kicks the egg with a hind foot as it passes (Van Gelder, 1953). The procedure is repeated until the egg strikes an object sufficiently hard to crack the shell, whereupon the contents are eaten.

In Iowa, Crabb (1948) snow-tracked individual *S. putorius* on eight occasions; distances moved during foraging activities for 1 night ranged from 100 to 4,482 m. In Missouri, minimum-area home ranges obtained by radiotracking four males for 3 weeks to 13 months ranged from 55 to 2,907 ha; home-range size for individuals for which ≥28 radio-positions were obtained during a season was greatest in spring, the reproductive season for *S. putorius* (see Remarks). Thus, movements of *S. gracilis* in Oregon might be expected to be greatest in late summer or early autumn.

Little is known regarding social interactions of spotted skunks. Walker (1930a) mentioned that one of three littermates caged together when about a year old performed "handstands" apparently in response to the others scampering about the pen. He also mentioned (p. 228) that two individuals housed together made "threatening gestures" toward each other as they tugged on the same piece of food. In such situations, the odor of musk sometimes was detected, but Walker (1930a:228) thought such release of musk was accidental rather than "a well-aimed discharge."

Remarks.—*Spilogale gracilis* commonly is referred to as the "civet cat," but it is neither a civet (family Viverridae) nor a cat (family Felidae). It also commonly is referred to as the "hydrophobia skunk" or "phoby cat," but there is no evidence that the incidence of rabies in the species approaches that reported in striped skunks in skunk-rabies

areas (see Remarks in *M. mephitis* account).

The taxonomic status of spotted skunks is much debated. Although separated from the earlier-named eastern spotted skunk (*S. putorius*) on the basis of color markings and skull morphology (Merriam, 1890a), *S. gracilis* was synonymized with *S. putorius* by Van Gelder (1959) on the basis of overlap in morphological characters. Mead (1968a, 1968b) provided evidence that *S. gracilis* and *S. putorius* were temporally isolated reproductively, therefore should be recognized as specifically distinct. The former breeds in September, undergoes a delay in implantation of 180–200 days, and gives birth in late April–May, whereas the latter breeds in March–April and gives birth in late May–early June. In addition, the baculum in *S. gracilis* may be longer and more curved, and may have a less bulbous proximal end than in *S. putorius* (Mead, 1967:612, fig. 5). Finally, although not absolutely definitive at present, the chromosome complement of western and eastern spotted skunks may be different (Hsu and Mead, 1969; Lee and Modi, 1983). Nevertheless, Hall (1981) and Wozencraft (1989, 1993) considered the western and eastern spotted skunks to be conspecific. We suspect that continued recognition only of *S. putorius* by some workers is a result of the absence of a published taxonomic treatment in which *gracilis* is reelevated to species level. However, Jones et al. (1992:5) concluded that "available data on reproductive isolation are sufficient to warrant separation" of *putorius* and *gracilis* at the species level.

A modern, quantitative, life-history study of the western spotted skunk not only would add immeasurably to information regarding the fauna of the West, but if conducted with sufficient depth and breadth, likely would establish a student as a mammalogist of considerable report.

Mephitis mephitis (Schreber)
Striped Skunk

1776. *Viverra mephitis* Schreber, theil 3, heft 17, pl. 121.
1858. *Mephitis occidentalis* Baird, 8(pt. 1):194.
1898a. *Mephitis spissigrada* Bangs, 12:31.
1899b. *Mephitis foetulenta* Elliot, 1:269.
1901. *Chincha occidentalis notata* Howell, 20:36.
1901. *Chincha occidentalis major* Howell, 20:37.
1901. *Chincha platyrhina* Howell, 20:39.
1931b. *Mephitis mephitis major*: Hall, 37:1.
1933. *Mephitis mephitis occidentalis*: Grinnell, 40:106.
1936. *Mephitis mephitis notata*: Hall, 473:67.
1936. *Mephitis mephitis spissigrada*: Hall, 473:67.

Description.—The striped skunk (Plate XXVIII) is a house cat-sized mammal (Table 12-20) with a pointed head; slightly upturned, nearly spherical nose pad; beady black eyes; pentadactyl, plantigrade feet with long, curved claws on the forefeet and short, straight claws on the hind feet; and usually 12 (range, 10–15) mammae (Verts, 1967*b*). Usually the pelage is entirely black except for a narrow white stripe on the forehead and nose, a white pate leading to two diverging white stripes on the back that extend partly or completely to the rump, white hairs or hairs with white bases in the tail, and sometimes a pencil-like tuft of white hairs extending beyond the black hairs of the tail. Entirely white skunks, or skunks with the black replaced by seal brown or the white replaced by yellow, are known to occur (Detlefsen and Holbrook, 1921).

The skull of the striped skunk (Fig. 12-56) is heavy and angular; it is widest near the posterior attachment of the zygomata (Wade-Smith and Verts, 1982). The postorbital processes are not prominent. Because the frontal region is inflated, the forehead is convex. The bullae are not inflated; thus, the margins of the mastoid regions are slightly concave. The posterior margin of the skull is squarish. P4 and m1 are moderately trenchant, but M1 and m2 are modified for crushing. M1 is squarish and its posterior margin is even with the posterior margin of the palate. The coronoid process is conical, the inferior margin of the mandible has a distinct step near the angle of the jaw, and sagittal and lambdoidal crests are prominent in older individuals.

Without doubt, the most widely recognized characteristic of the striped skunk is the pungent odor of the musk that it produces. A muscle-encapsulated gland ($\cong$25 mm in diameter in adults) positioned on each side of the anus produces a yellowish, sometimes curdy, liquid that can be expelled 2–3 m through papillae extended through the partially everted anus (Blackman, 1911). Seven compounds, each contributing >1% of the total, compose 96–98% of the volatile compounds (Wood, 1990); these are (*E*)-2-butene-1-thiol (38–44%), 3-methyl-1-butane-thiol (18–26%), (*S*)-(*E*)-2-butenyl thioacetate (12–18%), (*S*)-3-methylbutanyl thioacetate (2–3%), 2-methyl-quinoline 4–11%), 2-quinolinemethanethiol (3–12%), and (*S*)-2-quinolinemethyl thioacetate (1–4%).

Distribution.—The striped skunk occurs from southern Northwest Territories and central Quebec, south through essentially all of the United States to extreme northern Baja California Norte, northern Durango, and central Tamaulipas (Fig. 12-57; Hall, 1981). Although reported from 4,200 m (Nelson, 1930), *M. mephitis* usually occurs below 1,800 m (Grinnell et al., 1937). *M. mephitis* occurs in suitable habitats throughout Oregon except it is absent at higher elevations (Fig. 12-57).

Fossil remains of Pleistocene-age *M. mephitis* are widespread throughout the United States; the earliest fossils were from Late Blancan-age deposits (Kurtén and Anderson, 1980).

Geographic Variation.—Four of the 13 nominal subspecies occur in Oregon: *M. m. major* in most of the state east of the Cascade Range except for the Deschutes-Columbia Plateau Province, *M. m. notata* in the Columbia Basin, *M. m. occidentalis* in northwestern Oregon, and *M. m. spissigrada* in southwestern Oregon (Hall, 1981). We know of no quantitative treatment of variation among groups purported to occur in the state. However, in a large sample of striped skunks obtained on a 1,336-km² area in Illinois, Verts (1967*b*) recorded variation in color pattern and skull and body dimensions among adults of a magnitude approximately equal to that used to separate various races (Howell, 1901).

Habitat and Density.—Although occasionally occurring in forested areas, *M. mephitis* occurs commonly in more open habitats. In western Oregon, striped skunks occur in areas with dunes, prairies, meadows, shrubs, and seral forest communities in coastal regions (Maser et al., 1981*b*) and in hay and grass fields intermixed with croplands and brushlands in the interior valleys. East of the Cascade Range, striped skunks usually do not occur far out into the broad desert areas vegetated by sagebrush (*Artemisia*), but are common in meadows, hay fields, shrublands, tules, and weedy areas near streams, lakes, and marshes (Bailey, 1936). Elsewhere, several authorities described habitats of striped skunks as composed of "woodlands, brushy corners, and open fields broken by wooded ravines and rocky outcrops" (Wade-Smith and Verts, 1982:4), but Verts (1967*b*) found that skunks in Illinois were more abundant in intensively cultivated lands rather than in a mixture of woodlands, brushlands, and cultivated lands.

We know of no estimates of density for striped skunks in Oregon, but in other regions estimates ranged from 0.7 to 18.5/km², but most were 1.8–4.8/km² (Wade-Smith and Verts, 1982). We suspect that most populations in Oregon fall near the low ends of these ranges.

Diet.—Striped skunks are primarily insectivorous, with grasshoppers (Locustidae), beetles (Coleoptera, especially Carabidae and Scarabaeidae), and moth larvae (Lepidoptera) composing a large proportion of the diet. Mam-

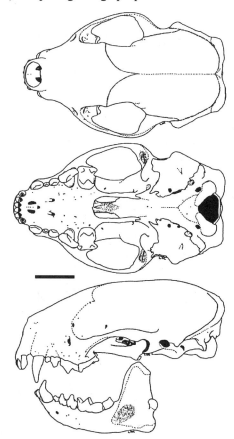

Fig. 12-56. Dorsal, ventral, and lateral views of the cranium, and lateral view of the dentary of the striped skunk, *Mephitis mephitis* (OSUFW 5554). Scale bar equals 15 mm.

mals, especially rodents, contribute significantly to the diet of striped skunks when insects are not available; rodents most likely are obtained as nestlings or as carrion (Verts, 1967b). In Oregon, striped skunks are known to eat large beetles (Cicindelidae—Maser, 1973).

Materials of plant origin also are ingested by striped skunks; however, some controversy exists regarding the quantity ingested as food and that ingested incidental to attempts to escape from traps. Verts (1967b) believed that corn (*Zea mays*), black cherries (*Prunus serotina*), nightshades (*Solanum*), and ground-cherries (*Physalis heterophylla*) were the only plants ingested as food by striped skunks in Illinois. Hamilton (1936a) considered "fruit" to be a major component of the diet of striped skunks in New York especially in autumn and winter, but investigators in other regions provided evidence to support the premise that 80–90% of the diet of striped skunks was of animal origin (Verts, 1967b).

Striped skunks may prey heavily on eggs and hatchlings of ground-nesting birds. In areas with concentrations of nests, some authorities claimed that depredations by striped skunks accounted for a large proportion of losses to predators (Verts, 1967b), whereas others reported that depredations by striped skunks composed only a small proportion of total losses (Bailey, 1971).

Reproduction, Ontogeny, and Mortality.—Other than

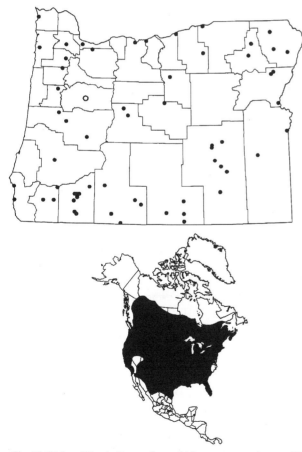

Fig. 12-57. Localities in Oregon from which museum specimens of the striped skunk, *Mephitis mephitis*, were examined (listed in APPENDIX, p. 523), and the distribution of the striped skunk in North America. (Range map redrawn from Hall, 1981:1021, map 519.) Open symbol indicates record for county.

anecdotal accounts (Bailey, 1936; Maser et al., 1981b; Wight, 1931), essentially nothing is known regarding reproduction in *M. mephitis* in Oregon. Wight (1931) believed that the mating period occurred 1–15 March in Oregon, a period about midway in the 2-month period recorded throughout the range of the species (Wade-Smith and Verts, 1982). He further commented on the synchronization of breeding and parturition in skunks in Oregon; however, Hamilton (1963) found *M. mephitis* in New York had a wide range of breeding dates and Verts (1967b) found some variation in averages and ranges of breeding dates among years in Illinois.

In Illinois, some males in samples collected throughout the year had spermatozoa in their epididymides (Verts, 1967b). Size and mass of testes of both adults and juveniles increased from August to March then declined; thus, year-round breeding capability of males is problematic.

Ovulation is induced and occurs 40–50 h after copulation. Implantation occurs by 19 days after breeding (Wade-Smith and Richmond, 1978), but commonly is sooner (Verts, 1967b). The gestation period commonly is reported as 63 days for *M. mephitis*, but a range of 59–77 days is more likely (Wade-Smith and Verts, 1982). Because the gestation period averages significantly longer (Wade-Smith and Richmond, 1978) and blood-progesterone levels increase more slowly (Wade-Smith et al., 1980) among females bred earlier in the breeding season, a short period of delayed implantation (development interrupted at blastocyst stage) may be responsible for the observed variation in length of gestation.

Although litter sizes ranging from six to ≅12 (Bailey, 1936) and two to 10 (Maser et al., 1981b) were offered for *M. mephitis* in Oregon, no quantitative evidence was presented in support of these contentions. In Illinois (Table 12-22) and Ohio (Bailey, 1971), litter sizes averaged near the minimum of Bailey's (1936) range and near the midpoint of that provided by Maser et al. (1981b).

Neonates weigh 32–35 g, and the skin is wrinkled, pinkish, and sparsely covered with 2–3-mm-long hair; the future color pattern is discernible from pigments in the skin even before birth (Verts, 1967b). In Illinois, captive skunks weighed ≅210 g by 40 days of age (Verts, 1967b), but attained that body mass by 26 days of age in New York (Shaw, 1928; Wade-Smith and Richmond, 1975). The sexes are distinguishable at birth. The eyes opened at 17–35 days ($X = 22$ days), the pinnae unfolded and the auditory meatuses opened by 24–27 days, tooth eruption occurred at 34–52 days, and weaning was complete by 56 days (Verts, 1967b). Only the permanent dentition is functional. Musk is present at birth and may be emitted by day 8 (Verts, 1967b).

In northwestern Illinois, survival from birth to 1 year of age was estimated at <35% (Verts, 1967b). Longevity may be as long as 10 years in captivity (Schwartz and Schwartz, 1981), but striped skunks probably do not live longer than 5–6 years in the wild (Casey and Webster, 1975). Various authorities reported that great horned owls (*Bubo virginianus*), mountain lions (*Puma concolor*), eagles (*Aquila chrysaetos* and *Haliaeetus leucocephalus*), coyotes (*Canis latrans*), badgers (*Taxidea taxus*), foxes (*Vulpes vulpes* and *Urocyon cinereoargenteus*), and bobcats (*Lynx rufus*) prey on striped skunks (Verts, 1967b). Although raptors are not deterred by skunk musk, most mammalian predators likely do not prey on skunks unless near starvation (Hamilton, 1943). Probably the greatest inimical factors for striped skunks are diseases and encounters with automobiles and farm machinery, although some are trapped for fur (Table 12-23).

Habits.—Striped skunks tend to be docile and unobtrusive; except during the breeding season or when infected with rabies, they are not aggressive or belligerent even toward conspecifics (Wade-Smith and Verts, 1982). Pregnant and pseudopregnant females are aggressive toward males and fight viciously if the male attempts to mount. Threat postures and defensive behaviors, including discharge of musk, usually are in response to the approach of an intruder such as a human or dog. When approached slowly, striped skunks usually arch their backs, elevate their tails, spread the hairs on the tail, and stamp their front feet while shuffling backwards; sometimes the latter behavior is so vigorous that the skunk assumes a handstand temporarily (Seton, 1920). If the approach is rapid or if the intruder approaches too closely, the skunk, while facing the intruder, bends its body into a U, everts the anus, and discharges droplets of musk over a 30–45° arc in the direction of the intruder (Verts, 1967b). Contrary to popular belief, striped skunks do not normally smell of musk; sometimes the odor of musk is weak on one that has just discharged musk. Striped skunks usually are silent, but may make a variety of sounds depending upon the situation; in captivity, probably the most frequent sound made by *M. mephitis* is a loud hiss produced by pregnant and lactating females when disturbed (Verts, 1967b).

Foraging behavior used by striped skunks depends upon prey sought: bees disturbed by scratching on the hive are caught as they exit; eggs are propelled between the hind legs (as a center "centers" a football) until the shell is broken by contact with a hard object, and grasshoppers are pounced upon with the front feet (Verts, 1967b). Aural, visual, olfactory, and tactile cues, in that order, are used to detect prey (Langley, 1979). Most prey is consumed immediately upon capture, but prey may be rolled on the ground to remove chitinous spines (caterpillars, Lepidoptera) or skin toxins (toads, Bufonidae—Wade-Smith and Verts, 1982). Nams (1991) found that striped skunks reacted to the odor of a specific food from increasingly greater distances when bits of that food were offered sequentially, but the reaction distance decreased when the skunks were offered a different food and the reaction distance to the original food declined to its original level. Thus, striped skunks are thought to develop olfactory "search images" that result in more efficient foraging by cueing on a specific type of food.

In Illinois, female striped skunks with their offspring may den together until the young are 3–5 months old, but some family units are maintained for ≤2 months (Verts, 1967b). In spring and summer, skunks that used dens commonly shifted dens frequently; they commonly bedded above ground, often in grass waterways or fencerows, and such bedding sites rarely were reused (Storm, 1972). In winter, striped skunks often den communally; such communes usually consist of a male with several females, but singles and groups consisting solely of males or solely of females are known denning arrangements (Verts, 1967b). Dens or aboveground retreats used in daytime commonly are on or near the periphery of home ranges; seemingly a foraging area is connected to a denning area by pathways across an otherwise unused area (Storm, 1972; Verts, 1967b).

In Illinois, striped skunks usually become active shortly after sunset, remain active throughout most of the night, and return to their den or denning area shortly before dawn (Storm, 1972; Verts, 1967b). Typically, upon leaving the den or aboveground retreat, they move rapidly along a frequently used route to a habitually used foraging area; those that use two or more retreats alternately or in sequence commonly use the same foraging area (Storm, 1972; Verts, 1967b). Minimum-area home ranges based on radiotelemetry positions in Illinois averaged 512 ha for adult males (*n* = 2), 378 ha for adult females (*n* = 5), 284 ha for juvenile males (*n* = 6), and 234 ha for juvenile females (*n* = 13); home ranges averaged 2.6 times longer than wide (Storm, 1972).

Remarks.—Skunk musk probably is the subject of nearly as many myths and misconceptions as opossum (*Didelphis virginiana*) reproduction. Contrary to popular belief in some regions, musk is not urine and not derived from urine, but is the product of greatly enlarged anal glands. Musk is a powerful irritant to the eyes and is a depressant to the central nervous system; ingestion may cause severe distress or death in other mammals (Wade-Smith and Verts, 1982). We know of no scientific evidence that skunk musk rubbed on the chest serves as a cure for the common cold, but we would not be surprised if symptoms were prolonged by the added irritation to the respiratory system caused by breathing the musk.

Methods of removing the odor of musk from hands, clothes, or pets are a perennial topic and include long lists of items and procedures (Cuyler, 1924; Jackson, 1961). Years ago, a skunk trapper told one of us (BJV) that a certain brand of shaving lotion will remove the odor of musk from the hands. We found that ethyl alcohol, the major component of shaving lotion and sans the perfume, aids in reducing the odor, especially if applied immediately after musk is deposited on the skin. Clothes and pets require several washings and banishment to the out-of-doors for days or weeks. We give a standard response to those who ask about the effectiveness of the frequently prescribed tomato juice for removing the odor of musk from the skin, clothes, or pets: "Don't waste good tomato juice—add a little vodka and drink it; it won't reduce the odor, but the odor won't bother you so much!"

The annual reported incidence of rabies in striped skunks commonly exceeds that in any other species of domestic or wild mammal in the United States. However, most cases are recorded in the Midwest, Texas, and California, regions frequently designated skunk-rabies areas. In Oregon, relatively few skunks or other terrestrial mammals are reported to be infected with rabies (see CHAPTER 5). However, during epizootics, the actual prevalence of the disease among skunks may be many times greater than that indicated by the reported incidence (Verts and Storm, 1966). Peak numbers of infected skunks usually are reported during the second quarter of the year, the period of gestation and lactation. Thus, contact during breeding; reactivation of latent infections by stresses related to overwintering, pregnancy, and lactation; or possibly aerosol transmission of rabies virus in winter dens may be involved in the epizoology of skunk rabies (Wade-Smith and Verts, 1982). Attempts to reduce the prevalence of rabies by re-

Table 12-23.—*Reported annual harvest of striped skunks (Mephitis mephitis) in Oregon by county, 1969–1996. Data provided by Oregon Department of Fish and Wildlife.*

County	Trapping season																										
	1969–1970	1970–1971	1971–1972	1972–1973	1973–1974	1974–1975	1975–1976	1976–1977	1977–1978	1978–1979	1979–1980	1980–1981	1981–1982	1982–1983	1983–1984	1984–1985	1985–1986	1986–1987	1987–1988	1988–1989	1989–1990	1990–1991	1991–1992	1992–1993	1993–1994	1994–1995	1995–1996
Baker	11	6		13	9	1	18	40	59	40	2	44	13	45	20	15	9	41	50	15	16	4	6	10	24	16	2
Benton		6	2	33	27	14	42	32	41	29	30	48	17	38	50	25	25	36	31	13	1	27	24	9	15	36	
Clackamas	1	2	8	34	4	4	13	16	62	27	41	45	82	37	92	42	33	14	48	24	2	9	15	10	18	21	
Clatsop	3	9		15	15		5	2	12		11	10			2	13	3	3	11						1	1	
Columbia	12	1		2	2	17	10	11	14	20	11	10	6	6	6	14	13	3	11	1	2		4			3	
Coos					4	6	10	3	18		20	8	11	12	19	10	2	3	5	5	14	14		1	1	2	
Crook			13	3											3	3	25	13	11	11	1	1			6	1	
Curry	25						7	6	1								6	1	2	5		2			6	3	
Deschutes		6			7		2	2	19	17	23	16	15	28	6	14	7	17	10	5	2	3	3		7	2	
Douglas	25	8	6	2	24	40	66	79	279	259	246	303	229	177	162	162	155	88	136	42	24	40	32	92	46	11	8
Gilliam								2	2	2				2				6	13	14	5	4	4	2	1		
Grant	5		1	1	3	1	1	1	5		4	12	1		1	6		2	5	13	3			3	1		
Harney												1		7			1		1	1	1			3		1	
Hood River	12	1	5	5			36	59	18	8	6	40	17	7	2	2	6	7	13	13	1	4	2	20	40	200	
Jackson	12	22	2	2	25	48	24	50	122	75	271	98	216	316	276	66	94	114	129	29	24	31	17	37	17	30	
Jefferson					15				1	2	5	2	9	18	13	6	10	119	5	4	2			1	12	3	
Josephine	1	6	5		14	30	54	53	93	144	89	146	119	120	101	141	87	83	99	28	49	31	25	10	20	26	9
Klamath	23	8	14	26	89	33	15	66	19	30	28	60	34	68	50	45	83	80	64	38	34	68	119	5	27	72	
Lake					1		2	3	26		8	10	35	22	5	16	21	42	53	14	14	14	31		9	6	
Lincoln	11	40	150	12	12	35	44	44	53	669	120	51	49	44	42	11	19	17	50	7	10	5	8	13	21	4	2
Linn	15	6	7	25	58	68	77	74	79	24	15	38	1	11	6	1	1	1	1	34	22	17	11	14	28	33	4
Malheur	1	1	1	1	6	1	1	9	10	72	39	23	31	40	73	65	61	38	71	39	4	7	1	3	2	7	
Marion	25	41	4	51	49	26	24	59	52	47	19	48	78	2	2	10	16	5	10	10	18	25	17	11	46	29	5
Morrow	3				1	1		3	3	3	51	2		82	97	110	43	25	66		2		2	1			
Multnomah			1	2							11	3	8	2	2	2	6	1						1			
Polk		1	1	3	22	4	4	6	42	7	27	28	10	3	23	94	62	65	37	4	12	8	4	1	3	6	
Sherman								1						44									2			1	
Tillamook	1	4	7	7		7	17	17	3	7	45	38	6	6	1	1	13	49	6	3	3	1	5	2	5	8	1
Umatilla	2	3	2	2	1	1	1		3		5	6	13	3	29	7	3	5	32	1	1	5	2	6	4		
Union	1				17				9	2	15	21		33	5	5	7	42	17		1	2	13	4	14		
Wallowa	34	39	5	10	17	4	4	4	7	7	18	13	4	8	11	11	7	54		10	12	8	4	25	8	8	4
Wasco	52	5			4	4		3	7	1	1	16		10	8	8	45	19	25		9	17	9	1	1	10	
Washington		19	5	4	19	19	36	28	20	36	48	48	25	63	24	23	28				1		2				
Wheeler										2				3				1								1	
Yamhill	11	15	3	6	14	10	5	7	30	54	35	50	16	17	20	44	27	46	23	13	6	3	8	8	13	8	
Totals	280	219	268	218	472	377	501	679	1,094	1,604	1,234	1,212	1,043	1,272	1,149	960	908	1,046	1,063	393	279	345	375	293	382	547	35
Trappers reporting[a]	721	773	731	847	1,102	1,350	1,377	1,755	2,746	2,899	3,216	3,373	3,069	3,980	3,292	3,104	3,575	3,397	3,833	1,985	1,536	1,228	1,390	1,454	1,293	1,464	687
Average price/pelt	$1.09	$1.12	$1.24	$1.49	$2.33	$2.07	$2.63	$3.26	$2.52	$4.82	$3.40	$3.07	$3.22	$1.50	$1.70	$2.26	$2.24	$2.19	$2.98	$3.60	$3.97	$4.04	$4.78	$4.88	$6.01	$4.37	$6.20

[a]Not all sought striped skunks and not all were successful.

ducing skunk populations through poisoning or trapping campaigns probably have little effect because of the usual rapid turnover of skunk populations. Because of their potential as reservoirs of rabies, we strongly recommend against maintaining striped skunks as pets.

FAMILY FELIDAE—CATS

Felids are distributed worldwide except for the Australian region, Madagascar, and most oceanic islands. The most recent treatment of the world's mammals listed 18 genera (up from five genera—Honacki et al., 1982) and 36 species of cats (Wilson and Reeder, 1993*a*). However, the classification of the felids is beset with the problem of no "convincing phylogenetic analysis on which to base decisions as to recognition of genera" (Jones et al., 1992:6), possibly accounting for the recent increase in number of genera recognized. Although the cats exhibit the greatest range in size among the carnivores, they are the least variable morphologically and karyologically (Wozencraft, 1989). In addition, convergence of characters is common in cats. These features no doubt are responsible for disagreements that have led to the sequential or simultaneous use of *Felis* and *Lynx* as generic names for the bobcat and lynx at various times in the recent past. The same features likely are responsible for other disagreements concerning the systematic relationships and nomenclature of some other cats.

The felids are represented in Oregon by the mountain lion (*Puma concolor*), the bobcat (*Lynx rufus*), and the lynx (*L. canadensis*). All felids are digitigrade with five toes on the forefeet and four on the hind feet; each toe is equipped with a sharp, recurved, and retractile claw. The incisors are small, chisel-like, and set in a straight line; the canines are long, sharp, and recurved; and P4 and m1 are strongly modified as carnassial teeth for shearing flesh and bones. The vibrissae are long and sensitive; the eyes are relatively large; and the tongue is covered with horny, recurved papillae. The baculum is rudimentary. In Oregon, bobcats are considered furbearers and an annual season for trapping them is provided; mountain lions may be taken during an open season; and the lynx records (Coggins, 1969; Verts, 1975) may be of animals that dispersed from populations farther north in the United States and Canada.

The cats have a fossil history extending from the Upper Eocene to Recent in Europe, Asia, and North America; middle Miocene to Recent in Africa; and late Pliocene to Recent in South America. Exceptionally large cats, some exceeding the size of any modern species, were common in North America to as recent as 8,000 years ago (Kurtén and Anderson, 1980). At least three of these large cats, the sabertooth (†*Smilodon fatalis*), the scimitar cat (†*Homotherium serum*), and the jaguar (*Panthera onca*) are known from the Pleistocene of Oregon (Kurtén and Anderson, 1980).

KEY TO THE FELIDAE OF OREGON

1a. Tips of ears lacking prominent tufts of hairs; tail >50% of length of head and body; P2 present......................
......................***Puma concolor***, mountain lion (p. 450).
1b. Tips of ears with prominent tufts of hairs; tail <30% of length of head and body; P2 absent..........................2
   2a. Tail >50% of length of hind foot; tail tipped with black

only on dorsal surface; foramina between auditory bulla and occipital condyle converging to form single triangular opening; posterior margin of heel pad trilobed...........................***Lynx rufus***, bobcat (p. 458).
   2b. Tail <50% of length of hind foot; tail tipped with black on all surfaces; foramina between auditory bulla and occipital condyle with distinctly separate openings; posterior margin of heel pad smooth..........................
......................................***Lynx canadensis***, lynx (p. 455).

Puma concolor (Linnaeus)
Mountain Lion

1771. *Felis concolor* Linnaeus, p. 522.
1834. *Puma concolor* Jardine, 2:266.
1832. *Felix* [sic] *oregonensis* Rafinesque, 1:62.
1896. *Felis californica* May, p. 22.
1904. [*Puma*] *concolor oregonensis* Elliot, 4:454.
1929. *Felis concolor californica*: Nelson and Goldman, 10:347.
1943*b*. *Felis concolor missoulensis* Goldman, 24:229.

Description.—*Puma concolor* (Plate XXIX) is the largest felid in Oregon (Table 12-24), and except for the jaguar (*Panthera onca*), the largest felid in the Western Hemisphere. In a sample mostly from northeastern Oregon, 12 males averaged 54.9 kg (range, 39.5–75.8 kg) and 12 females averaged 39.6 kg (range, 28.1–48.0 kg—Toweill and Meslow, 1977). On the west slope of the Cascade Range, 23 adult males averaged 54.6 kg (range, 38.5–72.6 kg) and 19 adult females averaged 37.7 kg (range, 29.5–62.6 kg—Toweill et al., 1984). Elsewhere, exceptionally large males are reported to exceed 120 kg (Robinette et al., 1961).

The skull of the mountain lion (Fig. 12-58) is massive and possesses large sagittal and lambdoidal crests. The canines are large and slightly recurved; P4 and m1 are highly modified for shearing flesh (carnassials). As in the house cat, P2 is missing. The heel pads of both fore- and hind feet have three lobes on the posterior margin; lobed heel pads are useful in separating the tracks of mountain lions from those of other felids, as lynxes (*Lynx canadensis*) have unlobed heel pads and those of bobcats (*L. rufus*), although lobed, typically do not overlap those of mountain lions in size (Lindzey, 1987). A somewhat conical and slightly dorsoventrally flattened baculum 3.9–10.2 mm long and 1.7–7.2 mm wide was present in 10 (76.9%) of 13 males from Oregon (Maser and Toweill, 1984). Females have eight mammae, but only six are functional (Lechleitner, 1969).

The dense and soft dorsal fur typically is tawny, but slate gray and reddish brown individuals are known. We examined one reddish brown skin from Oregon; however, skins of these large mammals often are not preserved. The venter is whitish. The mystacial region, back of the ears, and tip of the tail are brownish black. The upper lip is white. Young mountain lions are light tan spotted with brownish black.

Distribution.—Originally, *P. concolor* had the broadest geographic range of any mammal in the Western Hemisphere, ranging from northernmost British Columbia and Alberta; central Alberta; southern Manitoba, Ontario, and Quebec; and most of New Brunswick south through the United States and Central and South America to the southern tip of Chile (Fig. 12-59; Hall, 1981; Lindzey, 1987). Presently, the only populations that occur east of the Great Plains are in Florida and possibly western Arkansas and eastern Oklahoma (Dixon, 1982; Lindzey,

1987). The mountain lion occurs throughout western Oregon, but east of the Cascade Range the species probably is limited largely to the Ochoco, Blue, and Wallowa mountains (Fig. 12-59).

Fossil mountain lions are known from ≅30 sites of late-Pleistocene age (<500,000 years old) in western United States and Mexico (Kurtén and Anderson, 1980).

Geographic Variation.—Three of the 15 nominal sub-

species of *P. concolor* occur in Oregon: the smaller *P. c. californica* in the Siskiyou Mountains and southern interior valleys of southwestern Oregon; the lighter-colored *P. c. missoulensis* in the Wallowa Mountains; and the darker *P. c. oregonensis* throughout the remainder of the state (Goldman, 1943b; Hall, 1946, 1981). Hall's (1981:1042, map 525) range map shows the range of *P. c. kaibabensis* extending into southeastern Oregon, but no marginal records for Oregon or northwestern Nevada are shown or listed.

Habitat and Density.—Mountain lions range over broad areas and dispersers move long distances; thus, they are highly eurytopic (Kurtén and Anderson, 1980). Although it is possible to observe mountain lions in almost any habitat type, they usually are found in remote forested areas and often in dense vegetation, especially in winter. However, they sometimes cross open country; we know of one observed in a largely residential area in the Willamette Valley. Residents of suburban areas of several cities in the interior valleys have reported encounters with mountain lions; in some instances agency personnel found it necessary to dispatch mountain lions that repeatedly invaded residential areas.

In Wyoming, mountain lions did not use habitats in relation to their availability. Mixed conifer and curlyleaf mountain-mahogany (*Cercocarpus ledifolius*) vegetation

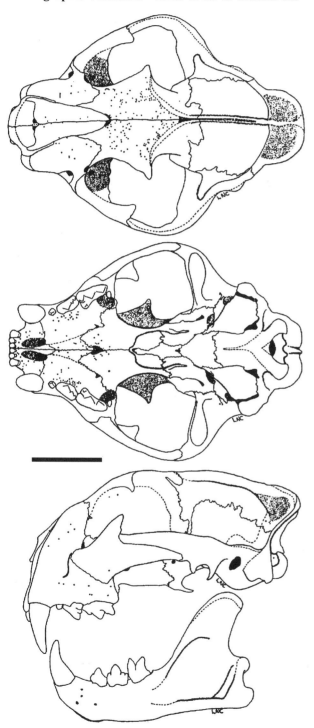

Fig. 12-58. Dorsal, ventral, and lateral views of the cranium, and lateral view of the dentary of the mountain lion, *Puma concolor* (OSUFW 3221). Scale bar equals 50 mm.

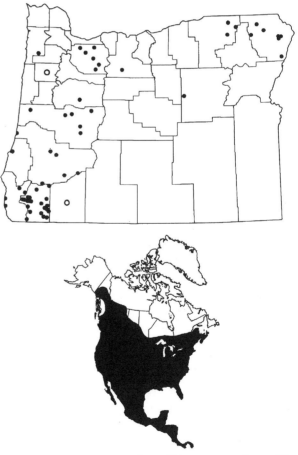

Fig. 12-59. Localities in Oregon from which museum specimens of the mountain lion, *Puma concolor*, were examined (listed in APPENDIX, p. 547), and the former distribution of the mountain lion in North America. (Range map redrawn from Hall, 1981:1042, map 525.) The mountain lion was extirpated from most of its range east of the Rocky Mountains (Lindzey, 1987). Open symbol indicates record for county.

Table 12-24.—*Means (±SE), ranges (in parentheses), and CVs of measurements of skull and skin characters for female and male Felidae from regions in Oregon. Skin characters were recorded from specimen tags; skull characters <150 mm were measured to the nearest 0.01 mm, larger characters to the nearest 1 mm. SE and CV not provided if* n <10.

Species and region	Sex	n	Total length	Tail length	Hind foot length	Ear length	Mass (kg)	Condylobasal length	Basilar length	Postglenoid length	Length of maxillary toothrow
Lynx canadensis	♀	4						115.79 (113.19–119.96)	102.61 (99.49–105.66)		40.46 (39.19–42.90)
	♂	1						118.54	106.62		40.27
Lynx rufus West of Cascade Range	♀	32						107.44 ± 0.47 (103.07–114.64) 2.41	96.04 ± 0.42 (90.84–101.6) 2.38	37.01 ± 0.39[a] (34.33–39.12) 3.32	36.59 ± 0.24 (34.17–39.10) 3.53
	♂	30	891[b] (870–912)	154[b] (141–160)	177[b] (170–185)	75[b] (70–80)	8.44[b] (7.10–11.05)	115.56 ± 0.75 (107.29–127.24) 4.29	103.39 ± 0.73 (95.28–115.08) 4.02	39.87[c] (38.72–41.50)	38.87 ± 0.30 (35.95–42.92) 4.38
East of Cascade Range	♀	15	800	150	163	64	6.26	115.02 ± 2.16 (105.97–130.13) 5.64	103.02 ± 2.05 (93.66–117.07) 5.98	41.29[d] (38.41–43.65)	38.29 ± 0.77 (34.09–42.66) 6.04
	♂	9	889	159	167	70	9.53	120.11 (110.25–131.49)	107.71 (97.13–118.96)	41.19[e] (40.26–42.13)	39.66 (35.91–42.36)
Puma concolor Western and central Oregon	♀	13	1,841	736	227			156.33 ± 4.00 (138.0–191.0) 9.23	138.18 ± 3.66 (120.0–168.0) 9.55	55.21[f] (53.77–5.65)	58.51 ± 0.66 (54.20–63.04) 4.09
	♂	13						174.46 ± 2.95 (155.0–202.46) 6.09	155.04 ± 2.95 (136.0–182.6) 6.86	62.2	63.30 ± 0.53 (59.24–66.36) 3.01
Southwestern Oregon	♀	9						148.50 (140.0–154.0)	130.00 (124.0–136.5)		57.48 (54.43–62.17)
	♂	15						173.89 ± 1.73 (168.0–182.0) 2.98	154.50 ± 1.36 (149.0–159.5) 2.64		64.24 ± 0.56 (61.83–66.81) 2.60
Northeastern Oregon	♀	1	1,953	702	276	92	14.69	196.0	175.0	58.1	67.34
	♂	1						179.0	160.0		66.40

[a]Sample size reduced by 22. [b]Sample size reduced by 27. [c]Sample size reduced by 25. [d]Sample size reduced by 12. [e]Sample size reduced by 7. [f]Sample size reduced by 11.

and steep (>50% slope), rugged terrain were used significantly more than expected; Douglas-fir (*Pseudotsuga menziesii*), juniper (*Juniperus*)-sagebrush (*Artemisia*)-grass, and lodgepole pine (*Pinus contorta*) vegetation, and riparian zones and moderate (20–40% slope) topography were used randomly; and sagebrush-grasslands and gentle (<20% slope) topography were avoided (Logan and Irwin, 1985). Habitats used proportionately greater than their availability tended to be attractive to prey species and to provide concealment that conferred advantages to the mountain lions that hunted them (Logan and Irwin, 1985); mountain lions hunt by stealth (Hornocker, 1970). In Utah and Arizona, areas selected as home areas by mountain lions were characterized by absence of ongoing or recent (≤6 years) timber harvests, below-average densities of roads, and few or no sites permanently occupied by humans (Van Dyke et al., 1986*a*, 1986*b*). Van Dyke et al. (1986*b*) believed that areas in which habitats were temporarily or permanently altered by humans constituted lower-quality habitats for mountain lions and that areas occupied by humans, even in the absence of significant impact on the habitat, become unsuitable to mountain lions.

On a 520-km² study area on the Idaho Primitive Area, Seidensticker et al. (1973) captured nine-18 mountain lions (residents, transients, and juveniles) each winter during an 8-year period. However, they indicated that they were aware of additional individuals present on the study area during some winters. During 7 years for which data were complete, the number of resident males remained constant at three individuals; during the first 3 years of the study, the number of resident females remained constant at six, but later in the study losses among resident females were not replaced. Currier (1983) indicated that intraspecific relationships were responsible for setting maximum limits for mountain lion populations of about one per 25–50 km². Van Dyke et al. (1986*a*) indicated that 61% of the variation in rates that tracks of mountain lions crossed roads in Utah and Arizona was explained by density of populations. However, insofar as we were able to determine, they presented no estimates of density. We know of no published estimates of density for mountain lion populations in Oregon.

Diet.—In Oregon, the flesh of deer (*Odocoileus hemionus*) was found in stomachs of 13 (52%) of 25 mountain lions collected in December in Wallowa and

Table 12-24.—*Extended.*

Zygomatic breadth	Cranial breadth	Least interorbital breadth	Length of mandible
87.54 (85.23–91.29)	57.70 (56.83–58.76)	28.14 (26.42–29.32)	83.78 (81.16–87.14)
88.06	59.27	30.62	84.71
83.13 ± 0.44 (76.34–87.8) 2.92	53.01 ± 0.28 (49.90–55.48) 2.88	23.62 ± 0.24 (20.95–26.54) 5.61	76.85 ± 0.36 (72.58–82.19) 2.58
88.99 ± 0.65 (81.70–97.3) 4.12	54.32 ± 0.27 (50.83–57.33) 2.79	26.07 ± 0.33 (21.55–29.50) 7.08	82.37 ± 0.65 (74.93–92.06) 4.47
87.68 ± 1.87 (79.24–100.11) 6.41	54.33 ± 0.58 (52.34–57.30) 3.18	25.65 ± 0.67 (23.19–29.92) 7.83	82.15 ± 1.85 (74.20–94.92) 7.75
92.64 (84.41–101.05)	56.31 (53.97–58.40)	27.66 (26.42–29.73)	85.85 (80.13–93.38)
128.37 ± 2.44 (114.87–149.95) 6.85	68.75 ± 1.34 (57.34–79.0) 7.03	37.84 ± 0.78 (33.85–44.06) 7.41	125.62 ± 2.14 (113.98–144.51) 6.14
142.93 ± 1.56 (131.81–149.85) 3.93	71.95 ± 0.68 (66.63–77.06) 3.39	43.26 ± 0.77 (39.01–47.10) 6.43	139.79 ± 1.38 (128.58–147.91) 3.56
123.21 (116.30–131.90)	68.71 (66.12–72.84)	36.87 (33.74–40.38)	121.70 (116.90–126.69)
146.06 ± 2.36 (134.43–160.0) 4.85	71.92 ± 0.73 (67.91–75.63) 3.05	44.38 ± 1.03 (38.51–47.32) 6.96	142.08 ± 1.49 (136.55–150.75) 3.16
153.17	76.0	45.38	151.0
149.44	75.78	43.47	148.9

eastern Lane counties. Elk (*Cervus elaphus*) occurred in two (8%), porcupine (*Erethizon dorsatum*) in three (12%), vegetation in two (8%), and unidentified material in two (8%); two (8%) of the stomachs were empty (Toweill and Meslow, 1977). No livestock was identified; Toweill and Meslow (1977) believed that its absence was related to the common practice of confining livestock close to human habitation in winter.

In Idaho, elk or deer composed 70% of 235 food items of mountain lions found in 198 droppings collected in winter; 5.5% of the items were remains of snowshoe hares (*Lepus americanus*), and the remaining items were various species of small mammals and grass (Hornocker, 1970). Of 53 elk and 40 deer killed by mountain lions, 40 (75.5%) and 25 (62.5%), respectively, were either <1.5 years old or >8.5–9.5 years old (Hornocker, 1970), suggesting that naive or senescent individuals were easier prey.

In Utah and Nevada, deer, porcupines, and domestic sheep occurred most frequently and contributed most to the diet of mountain lions (Robinette et al., 1959). In winter, mountain lions took a significantly greater proportion of male deer than occurred in the population. This difference possibly was a reflection of greater activity by male deer during the rut and the greater tendency for male deer to be associated with ledges and broken terrain that seem to be used more frequently by mountain lions.

Reproduction, Ontogeny, and Mortality.—Within mountain lion populations, those individuals that establish and maintain home areas (residents) constitute the productive element; transient females do not breed, and transient males rarely breed (Seidensticker et al., 1973). Females attain sexual maturity when they are about 2.5 years old and weigh about 36 kg; because of the land-tenure requirement (see Habitats), females that weigh as much as 41–45 kg occasionally remain virginal (Robinette et al., 1961). Robinette et al. (1961) considered that males that attained 57 kg were mature, although Toweill et al. (1988) found a reproductively competent male that weighed 47.2 kg and an aspermatic male that weighed 66.2 kg. Mountain lions seem to be polygynous in that several females may breed with the resident male whose home area overlaps their home areas. However, the mating system may be promiscuous, as pair bonds are of short duration and an individual might have several partners during its life (Sleidensticker et al., 1973). Gestation periods ranging from 82 to 98 days have been reported (Allen, 1950; Rabb, 1959).

In northeastern Oregon, 24 (52%) of 46 males (all age-classes) killed in December–January 1976–1982 contained spermatozoa in their epididymides (Toweill et al., 1988). On the west slope of the Cascade Range, eight (44%) of 18 males >24 months old were spermatogenic.

The number of young per litter (Table 12-25) seems to be related to the method used to estimate litter size; prepartum and early postpartum averages usually exceed those acquired from counts of older young (Robinette et al., 1961). Obviously, counts of older young reflect some mortality. In Utah and Nevada, of 66 prenatal litters 48 (72.7%) contained either three or four embryos; one litter each contained one and six embryos (Robinette et al. 1961). In northeastern Oregon, 21 (41.2%) of 51 females (all age-classes) were considered reproductively active (Toweill et al., 1988). On the west slope of the Cascade Range, eight (36.4%) of 22 females >24 months old exhibited evidence of previous reproductive activity.

Although young may be produced at any season, more seem to be born in summer and early autumn. Of 145 litters from Utah and Nevada with known dates of birth or dates of birth estimated on the basis of body mass of young, 98 (67.6%) were born in June–October (Robinette et al., 1961). Nevertheless, among the 145 litters at least four were born in each month.

Body mass of neonates probably is about 450 g (Robinette et al., 1961; Young and Goldman, 1946). Body masses (some estimated) of several growing young at different ages were approximately 0.8–1.0 kg at 1 week, 2.7 kg at 6 weeks, 3.6–4.1 kg at 8 weeks, 6.8–8.6 kg at 13 weeks, 15.9–22.7 kg at 6 months, and 29.5–36.4 kg at 1 year (Robinette et al., 1961). Two young from northeastern Oregon estimated to be 4 days old weighed 768.7 and 796.5 g (Toweill, 1986b). Gay and Best (1996) concluded that growth of the skull continued throughout most of the life of a mountain lion; females grow until 5–6 years old, whereas males grow until 7–9 years old.

Eruption of deciduous teeth in one Oregon specimen followed largely the same general sequential pattern, but was somewhat earlier than that reported by other workers

Table 12-25.—*Reproduction in Oregon felids.*

Species	Number of litters	Litter size $\overline{X}$	Range	n	Reproductive season	Basis	Authority	State or province
Lynx canadensis	1	4.6[a]		78		Implant sites	Brand and Keith, 1979	Alberta
		3.4[a]		100		Implant sites	Brand and Keith, 1979	Alberta
Lynx rufus	1	2.3		51[b]		Implant sites	Toweill, 1986a	Oregon[c]
	1	3.0		27[d]		Implant sites	Toweill, 1980	Oregon[e]
	1	3.2	1–8	356	Jan.–Aug.	Embryos	Gashwiler et al., 1961	Utah
	1	2.8		16		Young at side[f]	T. Bailey, 1974	Idaho
Puma concolor		2.4	1–4	38		Embryos	C. E. Trainer[g], in litt.	Oregon
		2.7		163		Implant sites	C. E. Trainer[g], in litt.	Oregon
		2.4	1–4	6		Implant sites	Toweill et al., 1988	Oregon
		2.8	1–4	11		Implant sites	Toweill et al., 1984	Oregon
	½[h]	2.6	2–3	9		Young at side	Hornocker, 1970	Idaho
	½[h]	3.4	1–6	66	All year	Embryos	Robinette et al., 1961	Utah-Nevada

[a] Adults only; larger value during period of abundance of snowshoe hares (*Lepus americanus*); smaller value during period of hare scarcity. Yearlings produced young only during years of abundance, but average litter size was significantly less ($\overline{X} = 3.9$, $n = 129$).
[b] Females =3.5 years old; litter size was 2.0 for 1.5-year olds ($n = 5$) and 2.1 for 2.5-year-olds ($n = 14$).
[c] West slope of the central Cascade Mountains.
[d] All age classes combined.
[e] Eastern Oregon.
[f] Most young =3 months old; thus, some mortality may have occurred.
[g] Biologist, Oregon Department of Fish and Wildlife.
[h] Most produced litters alternate years.

(Toweill, 1986b). The eyelids were separated by 8 days, but the animal did not open its eyes until 19 days of age; the auditory meatuses commenced to open on day 5, the pinnae commenced to become erect on day 7, and the young animal commenced to orient toward sounds by day 10. The young commenced to mew and hiss at 5 days of age and to purr while nursing from a bottle at 10 days of age. The individual could walk at 7 days, although the hind legs did not contribute significantly until 10 days of age; by 25 days, the animal could climb and pounce (Toweill, 1986b).

Seemingly, the two greatest natural causes of mortality among mountain lions are being killed by other (especially adult male) mountain lions while young and being killed while attacking a prey animal (Gashwiler and Robinette, 1957; Hornocker, 1970).

Habits.—In Idaho, mountain lions were active as much during the day as at night (Hornocker, 1970), but in Utah, mountain lion activity was largely crepuscular with peaks ≤1 h before sunrise and ≤1 h after sunset. However, mountain lions in the latter study, when disturbed by humans, shifted their activity largely to nighttime (Van Dyke et al., 1986b). Mountain lions commonly use caves as retreats (Kurtén and Anderson, 1980).

Mountain lions are largely solitary mammals; the only appreciable associations are of the female and its young and of the female and male during estrus. In a study involving radiotelemetry in Idaho, adult males were never located in association with other adult males, but were associated with lone females and females with large young most commonly and about equally; they were rarely associated with females with small young (Seidensticker et al., 1973). Despite the infrequency of associations, mountain lions maintain a somewhat loose social organization based on a system of land tenure (Seidensticker et al., 1973; Van Dyke et al., 1986b). Mountain lions are classified as residents or transients based on whether they restrict their activities to a specific area and are part of the reproductive component of the population or wander through areas maintained by residents and do not contribute to the population reproductively. Hornocker (1969b) believed that most transients were 2-year-olds that had not yet dispersed from the area; as dispersers they moved as far as 161 km. Originally referred to as "territories" (Hornocker, 1969b), ranges used by mountain lions later became known as home areas (Seidensticker et al., 1973). In Idaho, home areas of females ranged from 13 to 52 km²; females with litters of large young had larger home areas, presumably because of the requirement of greater resources for the growing young (Hornocker, 1970). Home areas of males were larger than those of females and overlapped those of several females, but not those of other males (Hornocker, 1970; Seidensticker et al., 1973). Size of home areas varies seasonally and from year to year (Currier, 1983). Although resident females seemingly use scrapes, urine, feces, anal-gland scent, and other means of communication to advertise their presence, the areas they use are not defended against conspecifics as home areas occupied by females commonly overlap (Hornocker, 1969b; Seidensticker et al., 1973). Thus, the land-tenure system tends to limit population size as population size does not follow resource abundance (Hornocker, 1970; Seidensticker et al., 1973).

Mountain lions pursued by hounds usually tree, enter caves, or seek refuge on cliffs. Mountain lions tranquilized and handled for research purposes usually moved away from the site at which they were captured (Seidensticker et al., 1973).

Remarks.—Before 1961, there was a bounty on mountain lions in Oregon; in 1968–1969, the mountain lion, except when involved in damage complaints, was given complete protection. Commencing in 1970, a limited-entry hunting season was established in parts of northeastern Oregon, the Cascade Range, and the southern Coast Range. During the past 3 decades, both the number of mountain lion hunters and annual kill by hunters have increased >15-fold (Table 12-26). During the past decade, the number of mountain lions killed illegally, killed by automobiles, or killed to reduce depredations on livestock also has increased ≅4-fold (Table 12-26). During 1986–1990, complaints of damage to livestock attributed to mountain lions increased moderately in most regions of the state except for the southeast and southwest. In the former, only one complaint was registered during the 5-year period; in the latter, the number of complaints increased ≅3-fold; approximately one-half of the complaints were from the southwest region during each of the 5 years. On 8 December 1994, regulations derived from a referendum passed a month earlier prohibiting taking of mountain lions with the aid of dogs became effective. We expect that numbers of mountain lions reported taken by hunters will decline and that complaints regarding damage to livestock will increase.

The mountain lion, long known by the epithet *Felis concolor,* was returned to the genus *Puma* by Wozencraft (1993) based on skull morphology, karyology, and studies of pelage patterns in young conducted by others (Kratochvíl, 1982; Pocock, 1917; Weigel, 1961). Also, numerous vernacular names have been applied to *P. concolor* that, in addition to mountain lion, include puma, cougar, catamount, panther, and painter.

Lynx canadensis Kerr
Lynx
1792. *Lynx canadensis* Kerr, p. 157.

Description.—The lynx (Plate XXIX) is only slightly larger than the bobcat (*L. rufus*; Table 12-24); for body mass, average values for females were greater than those of bobcats, but those of males were less (Quinn and Parker, 1987; Rolley, 1987). Nevertheless, long legs and long fur produce the illusion that the lynx is considerably larger than it actually is (Quinn and Parker, 1987).

The skull of the lynx (Fig. 12-60) is large with large lambdoidal and sagittal crests, widely separated orbits, and an abbreviated rostrum; the foramen between each auditory bulla and occipital condyle has two distinctly separate openings. As in the bobcat, P2 is missing and P4 and m1 are strongly modified as carnassials. There are five toes on the forefeet, four on the hind feet; the heel pad is unlobed. The feet are broad and the load/unit surface area is low, thereby permitting the lynx to traverse deep snow easily.

Lynxes are grizzled grayish brown in winter, but more reddish brown in summer. The undersides and legs are buffy white; the ears are brown with a white spot and long (4–5 cm) black tufts; the face is marked with white and the throat

Table 12-26.—*Number hunting permits authorized, number of hunters, and annual harvest by region and other losses of mountain lions* (Puma concolor) *in Oregon, 1970–1993. Data provided by Oregon Department of Fish and Wildlife.*

Year	Eastern Oregon			Western Oregon			Totals			
	Permits authorized	Number hunters	Number killed	Permits authorized	Number hunters	Number killed	Permits authorized	Number hunters	Number killed	Other losses[a]
1970	25	16	10				25	16	10	
1971		15				3		68	18	
1972	75	46	22				75	46	22	
1973	83	55	16				83	55	16	
1974	75	34	16				75	34	16	
1975	95	52	15				95	52	15	
1976	115	52	14	10	8	2	125	60	16	
1977	115	54	25	25	19	2	140	73	27	
1978	105	64	24	25	16	10	130	81	34	
1979	115	54	19	25	17	4	140	71	23	
1980	120	56	16	40	33	11	160	89	27[b]	17
1981	115	52	25	46[c]	31	8	161	83	33	20
1982	122	69	43	46	29	14	168	98	57	12
1983	132	51	41	56	34	13	188	85	54	10
1984[d]	167		42	96		37	263		79	22
1985[d]	207		36	155		26	362		62	27
1986	232	161	62	230	146	70	462	307	132	22
1987	227	157	76	230	180	90	457	337	166	43
1988	237	163	63	205	162	69	442	325	132	36
1989	226	153	65	225	203	80	451	356	145	38
1990	241	178	78	230	185	77	471	363	155	49
1991	252	173	86	230	192	69	482	365	155	
1992	267	189	93	250	202	94	517	391	187	22
1993	285	201	82	275	212	78	560	413	160	31

[a]Illegal kills, road kills, and kills to reduce depredations on livestock.
[b]Original total reported as 32, but data set contained arithmetic error.
[c]Reported later in long-term summaries as 43.
[d]No hunter surveys conducted.

is white; and a black tip completely encircles the tail (Tumlison, 1987).

Distribution.—The geographic range of *L. canadensis* includes all of Alaska and Canada (except the northeastern part of Northwest Territories) and the United States south to a line from southern Oregon to southern Colorado, southern Iowa, southern Indiana, and southern Maryland (Fig. 12-61; Hall, 1981). Either the species has been extirpated from much of its range in the United States or much of the range in the United States is based on specimens that dispersed southward when prey populations farther north crashed periodically (Mech, 1973, 1980; Quinn and Parker, 1987:684, fig. 1).

The lynx in Oregon is known from only 12 specimens: one from the Willamette Valley, two from the Cascade Range, one from Steens Mountain, one from the Stinkingwater Mountains, five from the Blue Mountains (three at one locality), one from the Wallowa Mountains (not seen—Coggins, 1969), and one locality from the Blue Mountians that could not be found (Fig. 12-61).

Lynxes are known from 200,000-year-old deposits in Alberta and Utah and 80,000-year-old deposits in Alaska, Idaho, Wyoming, and Yukon (Kurtén and Anderson, 1980).

Geographic Variation.—*Lynx c. canadensis* is considered the subspecies of lynxes collected in Oregon; the only other named subspecies (*L. c. subsolanus*) is restricted to Newfoundland (Hall, 1981).

Habitat and Density.—Habitats used by lynxes often are defined in terms of habitats used by their primary prey species; thus, good snowshoe hare (*Lepus americanus*) habitat usually is considered to be good lynx habitat (Quinn and Parker, 1987). Dense thickets of young conifers interspersed with small patches of grasses, forbs, and ferns seem to be prime habitat for snowshoe hares in Oregon (Black, 1965; Orr, 1934, 1940). Elsewhere, dense tangles of red raspberry (*Rubus idaeus*) canes, quaking aspen (*Populus tremuloides*), and willows (*Salix*) or of black spruce (*Picea mariana*), alder (*Alnus*), and hazelnut (*Corylus cornuta*) are considered excellent habitats for snowshoe hares (Keith, 1966).

Lynxes commonly occur at altitudes and latitudes at which snow cover is deep in winter. Because of their relatively smaller feet, bobcats cannot travel in habitats covered by deep snow. Thus, the geographic ranges of the lynx and the bobcat exhibit little sympatry (Figs. 12-61 and 12-63) and where sympatric the two species rarely are syntopic

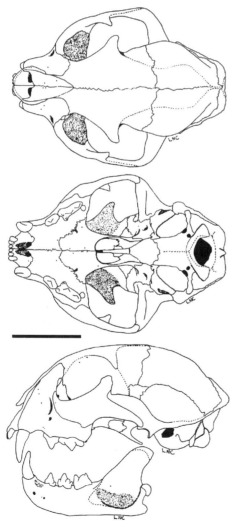

Fig. 12-60. Dorsal, ventral, and lateral views of the cranium, and lateral view of the dentary of the lynx, *Lynx canadensis* (OSUFW 1153). Scale bar equals 50 mm.

Fig. 12-61. Localities in Oregon from which museum specimens (circles) of the lynx, *Lynx canadensis*, were examined (listed in Appendix, p. 522) and were reported (square—Coggins, 1969), and the distribution of the lynx in North America. (Range map redrawn from Hall, 1981:1050, map 530.) The occurrence of the lynx in most of the contiguous United States likely is the result of dispersal during declines in population density of their primary prey, snowshoe hares (*Lepus americanus*—Quinn and Parker, 1987).

(Quinn and Parker, 1987).

In Oregon, two lynxes were taken in atypical habitats: one taken "five miles north of Imnaha, Wallowa County, Oregon" in 1964 (Coggins, 1969) was collected in bunch-grass-rimrock habitat (V. L. Coggins, pers. comm., 26 April 1993), and one taken "6 miles N., 5 miles E. Corvallis, Benton County, Oregon" in 1974 (Verts, 1975:22) was in a suburban residential area in North Albany. Another was caught near Drewsey, Harney Co., by Russell Clark on 8 January 1993 (KU 151986).

Among northern populations, a 10-year cycle of density long has been recognized to occur in the lynx and to follow, slightly out of phase, a similar cycle in populations of the snowshoe hare (Elton and Nicholson, 1942b; Keith, 1963; MacLulich, 1937). Such has been referred to as "a classic case of cyclic oscillation in population density" (Odum, 1971:191, fig. 7-16). During a cycle in Alberta, the density of lynxes in winter was 8.5/100 km² in 1964–1965, 2.3/100 km² in 1966–1967, 6.9/100 km² in 1967–1968, 10.0/100 km² in 1971–1972, 7.7/100 km² in 1972–1973, 3.8/100 km² in 1973–1974, and 3.1/100 km² in 1974–1975 (Brand et al., 1976). Although the precise mechanism responsible for initiating the cycles is debated (Keith et al., 1984; Krebs et al., 1986a, 1986b; Wolff, 1982), the role of the snowshoe hare in contributing to cycles of density in populations of the lynx, its primary predator, seems evident. Although populations of snowshoe hares in Oregon likely do not cycle (Wolff, 1982), cycles in numbers of *L. americanus* in more northern populations probably contributed directly to the occurrence of the lynx in Oregon. Lynxes are known to disperse exceptionally long distances (as far as 300 km south of the known breeding range) as prey populations decline (Mech, 1973, 1980).

Of the 12 specimens known from Oregon, one each was collected in 1897, 1964, 1974, and 1993; two in 1920; and three each in 1916 and 1927. We were unable to confirm a specimen collected near Granite, Grant Co., in 1935 referred to by Coggins (1969), although all three collected in 1927 that we examined were from that locality. Peaks in density of lynx populations in Alaska reportedly occurred in 1916–1918, 1926–1928, 1963–1966, and 1974–1975 (Quinn and Parker, 1987). A peak in density also was reported to occur in Canada in 1896 (MacLulich, 1937). Thus, collection dates of most lynxes in Oregon follow closely peaks in populations farther north; even the collection of lynxes in Oregon in 1920 may be related to an exceptionally high peak in 1914–1916 (Quinn and Parker, 1987). Thus, self-maintaining populations of lynxes likely have not existed in historic times in Oregon, but records of their occurrence here likely are of dispersers from within currently occupied areas farther north that immigrate and persist for a short time.

Diet.—No information regarding diet is available for the specimens collected in Oregon. Elsewhere, lynxes feed largely on snowshoe hares. In Alberta, when snowshoe hares were abundant, 90% of lynx carcasses obtained from trappers contained remains of snowshoe hares in their stomachs and snowshoe hares composed 97% of the biomass consumed by lynxes. When snowshoe hares were scarce, 35% of the lynxes contained remains of hares in their stomachs and hares composed 65% of the biomass consumed by lynxes (Brand and Keith, 1979). Other items consumed

by lynxes included sciurids, mice and voles, carnivores, ungulates, and birds; most occurred in food remains relatively infrequently and contributed only a small amount of the biomass consumed (Brand and Keith, 1979).

Reproduction, Ontogeny, and Mortality.—No records of lynxes reproducing in Oregon are known. Although immigrating lynxes occasionally are reported to reproduce (Mech, 1973), we suspect that these cats rarely, if ever, reproduce in Oregon.

Quinn and Parker (1987) list the breeding season as extending from mid-March to early April. In Alberta, during years that the density of snowshoe hare populations is high, some female lynxes commence to breed during the winter following their birth, but during years that hares are scarce, only those in their second winter or older breed. Among the latter group, 73% of 78 bred when hares were abundant, but only 33% of 100 bred when hares were scarce (Brand and Keith, 1979). Average litter size also declined when hares were scarce (Table 12-25). Mortality rates among young increased as population density of hares declined, but those of adults decreased. During peaks of lynx abundance, mortality resulting from harvest was greatest (≅30% annually), but when population density was low, the harvest declined (17%—Brand and Keith, 1979).

Young lynxes remain with their mothers until near onset of the breeding season following their birth.

Habits.—Although adult lynxes usually are considered to be solitary, Parker (1981) found, by tracking in the snow, that two to four lynxes traveled together on 25% of the length of the trails. Usually the lynxes traveled single file, but in good snowshoe hare habitat, the lynxes fanned out and seemed to hunt cooperatively. Multiple-lynx hunting units may be females with older young (Quinn and Parker, 1987).

Lynxes are more active at night than during the day. They usually bed in the snow at the site of a kill, use another bed for midday resting, and sometimes use another bed either at a second kill or for another midday rest (Parker, 1981). On clear days, lynxes often bed at the base of large trees on south slopes to take advantage of solar heat, but on stormy days, they seek shelter (Parker, 1981).

In open habitats, lynxes tend to travel in straight lines; chases seemingly follow coincidental sighting of hares. In seral communities in which hares are abundant, lynxes usually travel in a zig-zag pattern in search of hares. From interpretation of sign, when hares are sighted, lynxes immediately give chase. In 198 chases of snowshoe hares, lynxes were successful 34 times; in two chases of ruffed grouse (*Bonasa umbellus*), lynxes were successful both times (Parker, 1981). Distances between kills and percentage of successful chases increase with the number of lynxes traveling together. Lynxes kill an average of one hare per day (Parker, 1981).

Based on several studies, home ranges of lynxes usually are 16–20 km², but may range from 12 to 243 km² (Quinn and Parker, 1987). Home ranges of lynxes of different sexes overlap, but same-sex lynxes maintain exclusive home ranges. Urine marking especially, but possibly in combination with feces and products of anal glands, is used to delineate exclusive-use areas (Quinn and Parker, 1987).

Remarks.—Some authorities (Kurtén and Rausch, 1959b) consider *L. canadensis* Kerr, 1792, to be conspecific with the much larger Palearctic *L. lynx* (Linnaeus, 1758) that resembles the bobcat (*Lynx rufus*), whereas others consider differences worthy of specific recognition (Tumlison, 1987).

Published reports (*Hells Canyon Falcon*, 6[1]:10, 1996) of the need to preserve certain regions in Oregon for lynxes notwithstanding, no evidence of self-maintaining populations of lynxes in the state exists. We consider use of such tactics to enlist public support for conservation efforts to constitute fraud and potentially to detract from support of legitimate causes.

Lynx rufus (Schreber)
Bobcat

1777. *Felis rufa* Schreber, theil 3, heft 95, pl. 109B.
1817a. *Lynx fasciatus* Rafinesque, 2:46.
1899b. *Lynx fasciatus pallescens* Merriam, 16:104.
1902. *Lynx unita* Merriam, 15:71.
1936. *Lynx rufus unita* Bailey, 55:267.
1936. *Lynx rufus pallescens* Bailey, 55:268.
1936. *Lynx rufus fasciatus* Bailey, 55:269.

Description.—The bobcat (Plate XXIX) is the smallest wild felid in Oregon, with females being considerably smaller than males (Table 12-24). *L. rufus* is about twice the size of the common house cat, but its legs are longer, its tail is shorter (except for the Manx and similar domestic breeds), and its body is more muscular and compact (Kelson, 1946). The lambdoidal and sagittal crests on the bobcat skull (Fig. 12-62) are relatively larger; the face is longer, more massive, and more abrupt; the cranium is smaller; the jaw articulation is modified to permit a wider gape; the vertebral column is modified to accommodate relatively heavier musculature; and the clavical is much smaller. Also, the foramina between the auditory bulla and occipital condyle converge to form a single, triangular opening. In general, the bobcat is more adapted for leaping than the house cat (Kelson, 1946). The pupils are vertically elliptical. An arrowhead-shaped baculum 2.7–6.3 mm long and 1.8–3.9 mm wide was present in all six males examined in Oregon (Maser and Toweill, 1984). The feet are relatively small and the bobcat is not well adapted to negotiate deep snow. P2 is present in the bobcat and lynx (*Lynx canadensis*), but not in the house cat or mountain lion (*Puma concolor*).

In general, the variously spotted pelage is yellowish with grayish overtones in winter and with reddish overtones in summer, reflecting the two annual molts. The ears are black with a large white spot on their dorsal surfaces and are equipped with short black tufts. The tail is black-tipped dorsally, whitish ventrally, and there may be several blackish bars proximate to the tip. The venter is white with dark spots and the legs and feet are whitish with dark spots or bars. The sides of the face are extended by a ruff of fur. Bobcats in western Oregon possess more distinct markings than those in eastern Oregon.

Distribution.—The bobcat occurs from slightly north of the United States-Canada border south through the United States and the central highlands of Mexico to southern Oaxaca (Fig. 12-63; Hall, 1981). In Oregon, *L. rufus* occurs statewide (Fig. 12-63).

The fossil record for the bobcat extends from the late Blancan (≅2,000,000 years ago) of Texas. More recent fossils have been found throughout much of the present-day range of the species (Kurtén and Anderson, 1980).

Geographic Variation.—Two of the 12 named subspecies of bobcats occur in Oregon: the darker-colored *L. r. fasciatus* west of the Cascade Range and the lighter-colored *L. r. pallescens* east of the Cascade Range (Hall, 1981).

Habitat and Density.—The bobcat is among the more eurytopic carnivores in North America; in Oregon, it inhabits all habitats except intensively cultivated lands and areas at high altitudes. Overall, it tends to be found more commonly in early successional stages where the understory is dense and prey abundance is greatest. Several authorities agree that the most critical feature of the habitat is the presence of ledges; only bogs are a satisfactory replacement for ledges (McCord and Cardoza, 1982).

On the west slope of the Cascade Range, the proportion that various habitats were included in home ranges of bobcats did not differ from that available. However, on a seasonal basis, bobcats selected for sawtimber and against dense shrubs in autumn, and for sparse vegetation and against a mixture of sapling, pole, and sawtimber forest in winter (Toweill, 1986a). Toweill (1986a) suggested that habitats were selected to avoid deep snow. In Idaho, bobcats tended to concentrate their activities at lower elevations, on south-southwest slopes, in rocky terrain, and in open areas in winter, and used higher elevations and a

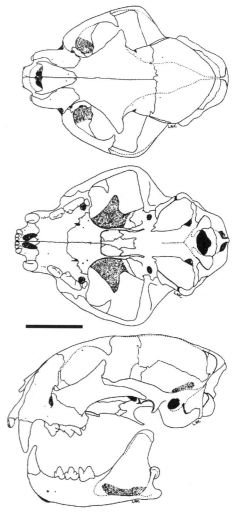

Fig. 12-62. Dorsal, ventral, and lateral views of the cranium, and lateral view of the dentary of the bobcat, *Lynx rufus* (OSUFW 8402). Scale bar equals 50 mm.

variety of vegetative and topographic types in summer (Koehler and Hornocker, 1989). Again, avoidance of deep snow was offered as a partial explanation for seasonal differences in use of habitats; proximity of escape cover and availability of prey also were believed to influence habitat use (Koehler and Hornocker, 1989).

Toweill (1986a) estimated minimum and maximum known densities for bobcats on a 279-km² study area in coniferous forest on the west slope of the central Cascade Range at 2.9–3.6/100 km² (2.2–2.5/100 km² for adults). He also referenced an unpublished report of estimated densities of 12–43/100 km² in the Coast Range of Oregon and unpublished reports of even higher densities in other states. T. Bailey (1974) estimated 5.4 adults/100 km² on a 648-km² area in Idaho, and Jones and Smith (1979) reported a maximum density of 27.8/100 km² on a 28.8-km² desert area in Arizona.

Diet.—Although perhaps not quite as numerous as for the coyote (*Canis latrans*), published reports of foods eaten seemingly have consumed an inordinate share of research time and resources directed toward furthering knowledge of the bobcat. In Oregon alone, at least four such reports (Nussbaum and Maser, 1975; Toweill, 1982; Toweill and Anthony, 1988a; Witmer and deCalesta, 1986) have been published in the past 2 decades despite an earlier (Bailey,

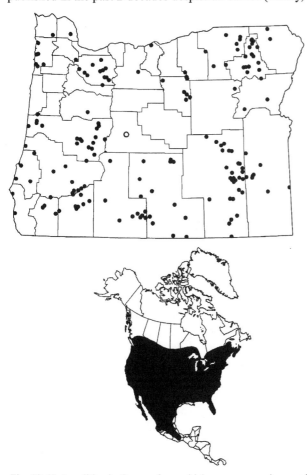

Fig. 12-63. Localities in Oregon from which museum specimens of the bobcat, *Lynx rufus*, were examined (listed in Appendix, p. 522), and the distribution of the bobcat in North America. (Range map redrawn from Hall, 1981:1053, map 531.) Open symbol indicates record for county.

1936), albeit possibly biased (Toweill, 1979), account and other information regarding diets of Oregon bobcats (Young, 1958).

In eastern Oregon, remains of lagomorphs were identified in contents of 47.5% of 200 stomachs examined; domestic sheep occurred in 13.5%, sage grouse (*Centrocercus urophasianus*) in 11.5%, various mice in 6.5%, ground squirrels (probably mostly *Spermophilus*) in 6.0%, "pine" squirrels (*Tamiasciurus douglasii* and *T. hudsonicus*) in 4.0%, deer (*Odocoileus*) in 1.5%, and a variety of small mammals and birds in ≤1.5% (Bailey, 1936:268). More recently, in stomach and intestinal contents of a sample of 117 bobcats (87% contained food remains) collected in January, arvicoline rodents were represented in 35%, lagomorphs in 33%, sigmodontine rodents in 29%, birds in 13%, sciurids in 4%, porcupines (*Erethizon dorsatum*) in 6%, deer in 3%, and a variety of other food items each in ≤3% (Toweill, 1982). Both the sage grouse and domestic sheep were absent, likely a reflection of the decline in numbers of sage grouse (Crawford and Lutz, 1985) and, at least in part, of the decline in the sheep industry in eastern Oregon. Toweill (1982) claimed that bobcats <1 year old fed more heavily on mice, woodrats (*Neotoma*), and birds, and less heavily on lagomorphs than older bobcats; however, our analysis of data that he presented (p. 314, table 3) indicated no significant differences ($\chi^2 = 2.22$, $d.f. = 4$) in proportions of the various food types eaten by the two age-classes.

Nussbaum and Maser (1975:262) reported that remains of lagomorphs occurred in 52.5% and those of rodents in 70.0% of 143 fecal droppings collected throughout the Coast Range. However, later in the same report (p. 263, table 1) they presented data that indicate that these values were obtained by summing percent frequencies of occurrence for each species within the two groups. We have no reason to doubt that 44.1% of the samples contained remains of brush rabbits (*Sylvilagus bachmani*) and 8.4% contained remains of snowshoe hares (*Lepus americanus*) or that deer mice (*Peromyscus maniculatus*) occurred in 12.6%, creeping voles (*Microtus oregoni*) in 7.0%, Townsend's voles (*M. townsendii*) in 8.4%, dusky-footed woodrats (*Neotoma fuscipes*) in 11.2%, mountain beavers (*Aplodontia rufa*) in 8.4%, Townsend's chipmunks (*Tamias townsendii*) in 7.0%, Pacific jumping mice (*Zapus trinotatus*) in 5.6%, Douglas' squirrels (*Tamiasciurus douglasii*) in 4.9%, California ground squirrels (*Spermophilus beecheyi*) in 2.8%, unidentified *Microtus* in 1.4%, and unidentified *Phenacomys* in 0.7%. However, we find it difficult to accept that neither of the two species of lagomorphs and none of the 11 categories of rodents occurred in the same fecal sample. Although the values for percent frequency of occurrence for lagomorph and rodent groupings seemingly were inflated by faulty logic, it is evident that the two groups contributed most to the diet of bobcats in the Coast Range. Deer occurred in only 2.1% and domestic sheep in only 0.7% of the samples.

In a more restricted area in the Coast Range, Witmer and deCalesta (1986) found that mountain beavers were the predominant food item of bobcats; their remains occurred in 84% of 100 fecal samples in spring, 73.7% of 57 in summer and autumn, and 62.2% of 90 in winter. Percent frequency of occurrence in each of the three seasonal groupings (in the same order) for other major food items were brush rabbits,

5.0, 3.5, 10.0; creeping voles, 6.0, 10.5, 10.0; bushy-tailed woodrats (*Neotoma cinerea*), 0.0, 8.8, 5.6; Douglas' squirrel, 2.0, 7.0, 0.0; Townsend's chipmunk, 5.0, 3.5, 1.1; deer, 6.0, 5.3, 22.2; and unidentified *Phenacomys*, 3.0, 0.0, 7.8. Although hooves of fawn deer in the summer and autumn sample were considered evidence of predation, the much greater percentage frequency of occurrence of deer remains in winter was believed to have been the result of bobcats eating carrion (Witmer and deCalesta, 1986). Domestic livestock was not listed among items identified.

In the central Cascade Range, lagomorphs, represented solely by the snowshoe hare, occurred in 70.6% of 34 fecal samples. Northern flying squirrels (*Glaucomys sabrinus*), Douglas' squirrels, and creeping voles each occurred in 11.8%; Townsend's chipmunks and western pocket gophers (*Thomomys mazama*) each in 8.8%; and deer mice, bushy-tailed woodrats, and unidentified *Microtus* each in 5.9% (Nussbaum and Maser, 1975). Four other species of small mammals each occurred in only one (2.9%) sample; neither deer nor livestock was listed. In a later study, remains of snowshoe hares occurred in 30.4% of 494 fecal samples collected throughout the year, more frequently than any other food item on an annual basis (Toweill and Anthony, 1988a). Remains of deer occurred in 22.1%, mountain beavers in 12%, and Townsend's chipmunks and western pocket gophers each in 5%; no other species occurred in >4% of the samples on an annual basis. Deer and snowshoe hares were fairly evenly distributed among the four seasons, but mountain beavers were twice as frequent in spring and summer samples than in winter samples; mountain beavers were not identified in autumn samples (Toweill and Anthony, 1988a). Again, livestock was not reported to occur in the diet.

Despite the apparent discrepancies and contradictions evident among the five studies abstracted herein, a pattern regarding dietary selections by bobcats is evident. Bobcats in Oregon, as elsewhere, feed largely on mammals, especially those in the 0.5–3.0-kg range, but sometimes take birds and occasionally other vertebrates. They are capable of taking deer and livestock, but usually select the younger and smaller individuals. Thus, adults, at least, are capable of exploiting a variety of food items and in doing so seem to take advantage of local or seasonal abundances. However, young-of-the-year do not seem capable of shifting to alternative prey when lagomorph populations crash (Bailey, 1972).

Reproduction, Ontogeny, and Mortality.—Some female bobcats attain reproductive competence during the reproductive season following their birth, whereas males are thought to delay becoming fertile until their 2nd year (Crowe, 1975b). Males are thought to be fertile continuously, but spermatogenic activity may decline in summer and autumn with a decline in volume of the testes (Duke, 1954). Females are polyestrous, with as many as three ≅44-day cycles (Crowe, 1975b). Ovulation is considered spontaneous and corpora lutea are persistent, perhaps for years (Crowe, 1975b; Duke, 1949).

The length of the gestation period of the bobcat seems to have been particularly elusive. Hamilton (1943) reported a 50-day gestation period for the bobcat, whereas Young (1958) reported 50–60 days. Ewer (1973) challenged 50–60 days as being too short based on comparisons with gestation periods of other felids. Crowe (1975b) erroneously

attributed to Ewer (1973) the gestation period of 70 days that he used in estimating dates of birth; however, the "*about 70 days*" seems to have originated with Seton (1929a:226). Gashwiler et al. (1961) seemingly used a 63-day gestation period similarly; they (p. 78) attributed a gestation period of "63 days or more" to Palmer (1954:146), who actually indicated that young were "born after a gestation period of probably more than 63 days." Several regional works on mammals list the gestation period as about 62 days (Barbour and Davis, 1974; Jones and Birney, 1988; Lowery, 1974; Mumford and Whitaker, 1982; Peterson, 1966; Sealander, 1979; Webster et al., 1985), but others list it as "about 50 days" (Davis, 1960:116; Lechleitner, 1969:209), "about 60 days" (Merritt, 1987:308), "50 to 60 days" (Hoffmeister, 1986:528), "50 to 62 days" (Hazard, 1982:153), "60–62 days" (Larrison, 1970:169), "50–70 days" (Clark and Stromberg, 1987:248), "60–70 days" (Nowak, 1991:1198), and "about 10 weeks" (Jones et al., 1983:315). Seemingly, the gestation period of the bobcat is either variable or not known precisely.

On the western slope of the central Cascade Range, none of 26 females obtained from trappers in the season following their birth exhibited evidence of having bred, whereas 5 (27.8%) of 18 yearlings, 14 (51.9%) of 27 2.5-year-olds, and 51 (82.3%) of 62 ≥3.5-year-olds had bred (Toweill, 1986a). Numbers of implantation sites also increased with age in western Oregon (Table 12-25), but comparable data are not available for eastern Oregon. In Washington, both the proportion of females breeding and the average (±*SE*) number of implantation sites were greater east of the Cascade Range for yearlings (48%, 2.80 ± 0.18, *n* = 31) and for 2.5-year-olds (100%, 2.60 ± 0.21, *n* = 15) than west of the range (40%, 1.97 ± 0.12, *n* = 75 and 94%, 2.26 ± 0.15, *n* = 72, for the two age-classes, respectively). Adults ≥3.5 years old averaged 2.71 ± 0.11 (*n* = 53) implantation sites in eastern Washington and 2.54 ± 0.09 (*n* = 157) in western Washington (Knick et al., 1985).

In both eastern (*n* = 39) and western (*n* = 92) Oregon, estimated dates of birth for bobcats <1 year old indicated that most (76.9% and 78.3%, respectively) were born in May or earlier; however, 7.7% and 2.2%, respectively, were born as late as August (Toweill, 1980). Gashwiler et al. (1961) reported that months of collection for a sample of 356 pregnant female bobcats from Utah extended from January to September, but 69.1% of the pregnant females were taken in April and May. Duke (1954) reported two pregnant females were taken in October. Thus, the breeding season may extend from January at least through late August (Gashwiler et al., 1961). The long littering season possibly is the result of some polycyclic females failing to breed during the first or second (or both) estrous periods (Crowe, 1975b).

A few records of bobcats hybridizing with house cats have been recorded; the bobcat is always the sire (Gashwiler et al., 1961; Young, 1958).

At birth, the eyes of neonates are closed; most report them to open at 9–11 days (Grinnell et al., 1937; Palmer, 1954; Pollack, 1950), but Young (1958) reported the eyes remained closed for only 3–9 days after birth. A body mass of about 283–340 g at birth seems to be the consensus, with female neonates nearer the lower value (Gashwiler et al., 1961; Young, 1958); deviation from this range seems explainable on the basis of premature birth or weighing

Table 12-27.—Reported annual harvest of bobcats (Lynx rufus) in Oregon by county, 1969–1996. Data provided by Oregon Department of Fish and Wildlife.

County	Trapping season[a] 1969–1970	1970–1971	1971–1972	1972–1973	1973–1974	1974–1975	1975–1976	1976–1977	1977–1978	1978–1979	1979–1980	1980–1981	1981–1982	1982–1983	1983–1984	1984–1985	1985–1986	1986–1987	1987–1988	1988–1989	1989–1990	1990–1991	1991–1992	1992–1993	1993–1994	1994–1995	1995–1996
Baker	133		61	70	56	50	56	28	27	42	43	36	55	50	31	54	77	54	54	38	16	13	21	53	31	39	14
Benton	3	5	8	38	34	40	69	59		48	50	66	23	33	56	25	39	74	47	59	13	13	37	18	20	26	
Clackamas	19	13	15	24	35	32	23	31	40	48	53	96	75	72	132	163	115	147	153	64	42	28	32	25	38	32	15
Clatsop	71	2	13	23	73	89	61	71	64	52	82	83	33	64	118	147	198	192	227	185	98	43	91	21	59	77	2
Columbia	25	4			36	27	29	25	57	19	42	31	40	27	31	50	26	47	103	53	39	30	5	15	12	14	1
Coos	6	11	7	28	20	49	59	85	169	173	154	198	106	160	203	218	237	123	142	69	34	30	49	25	58	52	27
Crook	59	34	46	25	16	16	52	58			53	123	65	56	46	70	87	117	112	57	35	36	53	66	44	76	59
Curry	1			10	43	47	61	62	161	187	182	279	67	165	151	154	169	173	105	62	23	36	58	16	25	50	20
Deschutes	64	15	11	43	24	39	61	30	1		57	40	40	24	25	70	44	41	41	45	36	29	58	50	30	42	21
Douglas	45	30	34	86	106	166	213		376	459	638	727	407	543	444	632	529	626	480	307	161	175	204	153	144	196	88
Gilliam		20	13		5						5	19	9	8	13	15	18	3	5	3	1	1		7	11	6	
Grant	88	80	52	76	99	40	68				61	80	72	67	76	96	115	127	130	85	42	33	24	49	23	35	64
Harney	44	27	99	95	59		31	51			88	136	103	65	69	88	116	128	154	104	48	51	130	122	170	43	
Hood River		4	13	10	10	20	27	9			7	24	23	17	13	9	11	11	13	12	20	11	10	6	6	14	5
Jackson	9	10	5		23	22	40	130	177	178	237	141	155	121	128	243	258	296	178	101	140	108	86	107	143	46	
Jefferson				10	16	16	44	10			35	26	49	20	18	37	30	44	36	22	13	13	19	32	14	31	2
Josephine	12	17	18		45	59	27	29	71	154	149	87	27	45	78	65	75	100	87	58	50	32	71	46	53	68	25
Klamath	43	19	31	79	58	42	56	14			42	68	91	79	101	141	163	136	185	94	66	39	59	51	74	117	39
Lake	77	166	111	127	67	75	63				102	104	93	191	95	114	192	169	211	156	142	89	80	151	126	222	29
Lane	90	88	106	270	240	228	349	386	484	523	536	243	492	369	371	409	387	443	224	124	114	110	146	113	189	85	
Lincoln	112	53	82	103	87	166	127	190	311	220	217	180	199	202	244	288	288	244	185	99	40	62	50	39	24	18	
Linn	9	18	16	47	41		93	161	52	156	129	138	106	116	181	218	153	163	118	49	58	61	45	84	51	36	
Malheur	146	194	181	40	21	21	20		40	86	139	80	54	40	77	115	111	104	107	101	58	72	135	95	98	8	
Marion	2	6	12	3	16	22	19	93			53	83	90	34	34	109	80	94	89	15	17	3	19	21	26	28	23
Morrow	15	8	7	12	3	10	6	1	14	8	8	6	7	12	14	8	16	8	15	5	6	3	4	1	7	8	
Multnomah	4		6	2	2	12	10	15	16	16	5	1	8	18	13	26	11	23	13	10		8	1	20	6	2	2
Polk	10	17	14	18	18	20	42	37	45	71	52	29	17	68	28	50	144	52	46	17	11	35	12	5	9	5	
Sherman					5	1				9	6	8	2		3	3	5		8	8			1			1	
Tillamook	36	35	4		83	151	149	226	211	238	232	202	140	135	170	234	280	312	171	45	26	62	38	42	45	11	
Umatilla	19	22	11	23	26	18	15		25	15	41	19	13	13	37	41	44	31	24	31	31	46	20	27	8		
Union	24	9	12	4	2	1	1		36	45	42	16	34	34	40	49	53	20	22	29	24	39	20	35	13		
Wallowa	151	156	118	26	18	29	18	11	43	23	47	39	55	60	61	93	89	55	34	22	45	77	46	50	30		
Wasco		2	8	8	32	46	79	2	24	34	28	25	23	37	60	40	73	34	21	21	17	38	21	24	18		
Washington	6	1	4		14	3	21	13	43	50	5	22	24	24	10	29	39	8	6	11	20	5	8	10	2		
Wheeler	33	59	21	18	21	26	23		29	35	25	20	25	24	60	45	62	25	14	14	12	22	19	32	5		
Yamhill	10	2	7	14	21	13	38	16	93	28	17	18	28	13	33	54	52	71	30	14	4	34	14	25	8		
Totals	1,366	1,110	1,123	1,399	1,645	1,473	1,720	1,998	2,276	2,553	3,694	4,095	2,677	3,059	3,010	3,661	4,207	4,410	4,451	2,851	1,767	1,267	1,590	1,733	1,557	2,066	781
Trappers reporting[b]	721	773	731	847	1,102	1,350	1,377	1,755	2,746	2,899	3,216	3,373	3,069	3,980	3,292	3,104	3,575	3,397	3,833	1,985	1,536	1,228	1,390	1,454	1,293	1,464	687
Average price/pelt[c]	$14.75	$13.66	$20.96	$39.43	$53.55	$36.97	$102.94	$103.21	$71.74	$115.06	$93.15	$119.27	$97.22	$115.37	$119.63	$151.79	$148.00	$233.02	$157.08	$98.67	$51.60	$50.00	$79.00	$49.00	$66.00	$24.00	$44.00

[a] 1969–1970 to 1976–1977, no closed season, no bag limit. 1977–1978, season closed east of the Cascade Range; 92-day season west of the Cascade Range and no bag limit. 1978–1979, season closed east of the Cascade Range and no bag limit. 1979–1980 to 1985–1986, 62-day season west of the Cascade Range and no bag limit. 1986–1987 to 1992–1993, 62-day season west of the Cascade Range and no bag limit and bag limit of three east of the Cascade Range. 1993–1994 to 1995–1996, 62-day season and bag limit of five east of the Cascade Range and 46-day season and bag limit of three east of the Cascade Range and 77-day season and no bag limit west of the Cascade Range.

[b] Not all sought bobcats and not all were successful.

[c] Rounded to the nearest dollar after the 1989–1990 season.

several days after birth (Gashwiler et al., 1961). Eruption of the deciduous teeth commences by 2 weeks of age and is complete by 2 months of age (Crowe, 1975*b*). Permanent incisors are fully erupted by 5.3 months and permanent canines by 6.3 months; the permanent dentition is complete by 8 months (Crowe, 1975*b*). The young usually are weaned by 2 months of age, but commonly remain with the female until autumn or early winter when they are two-thirds to three-fourths adult size (Gashwiler et al., 1961).

In Idaho, Bailey (1972) reported <3% annual mortality among adults in an unexploited bobcat population during a 3-year period. However, he considered young-of-the-year to be extremely vulnerable to low densities of prey; in a year in which lagomorph populations crashed, he found no young surviving to autumn.

There are few wild predators other than the mountain lion capable of preying on adult bobcats, but foxes (*Vulpes vulpes*), coyotes, and great horned owls (*Bubo virginianus*) are known to take young bobcats. Without doubt, the greatest inimical factor in Oregon is trapping for fur (Table 12-27). Toweill (1986*a*:61, fig. 5) implied that harvest of bobcats for fur in four western Oregon counties in 1971–1984 was related to price per pelt received; however, he did not present a coefficient of determination for the relationship.

Habits.—In Oregon, bobcats were active for periods of 4–8 h, then inactive for 1–8 h; bouts of activity and inactivity seemed more related to temperatures than to intervals of light and darkness. In winter, bobcats tended to avoid activity during periods of low temperatures, but in summer, activity seemed to be initiated as temperatures commenced to fall (Toweill, 1986*a*). Activity of the two sexes was similar, although females tended to exhibit lower rates of travel and a more bimodal activity pattern (Toweill, 1986*a*). Bobcats spend periods of inactivity at den sites consisting of natural cavities, hollow logs, or protected areas under logs.

Bobcats scent-mark with feces, urine, and products of anal glands; they also scrape the ground with their hind feet, often in combination with marking with urine or feces. In late summer and early autumn, females with young commonly mark "special places" within their home ranges rather than their home-range boundaries (T. Bailey, 1974:442). Feces are covered 52–60% of the time (T.

Bailey, 1974).

Minimum-area home ranges for bobcats on a shrub-desert area in Idaho ranged from 9.1 to 45.3 km² for females and from 6.5 to 107.9 km² for males; the average size of home ranges of four males was 2.2 times greater than for eight females (T. Bailey, 1974). In Oregon, on the west slope of the Cascade Range, minimum-area home ranges were 6.8–39.3 km² for females and 7.8–37.6 km² for males; the means (25.2 km² for females and 24.9 km² for males) were nearly identical (Toweill, 1986*a*). In Idaho, home ranges were used more intensively by females than by males; females were located by radiotelemetry within 1.6 km of a protected resting place for several days before they moved to another part of their home range, whereas males did not exhibit a similar pattern. Males rarely returned to a resting place used previously. Home-range areas of females overlapped only 0.1% with those of other females, whereas they overlapped 23% with those of males. In addition to the small overlap, the overlapping areas were seldom used by bobcats (T. Bailey, 1974). Although direct defense of areas by occupants is not known to occur, the exclusive use of home ranges by individuals, the extent and nature of scent-marking, and the failure of juveniles raised on the area or transients to occupy established home ranges except upon removal of the occupant suggest that the social organization of bobcats is a based on a land-tenure system (T. Bailey, 1974). The smaller, more intensely used home ranges of females likely are a reflection of feeding and protecting young being the exclusive duty of the female, whereas the larger home ranges of males that overlap those of several females likely are a reflection of selection for males that successfully breed the largest number of females (T. Bailey, 1974). In Oregon, the social organization seemed to be similar, except that overlap of home ranges of females was greater, topography influenced size of home ranges, and the arrangement of home ranges may have been influenced by harvest pressure (Toweill, 1986*a*).

Remarks.—The redundancy of research conducted on some aspects of the biology of bobcats exemplifies the need for field biologists to commence to shift from observational-type, hypothesis-formation research to hypothesis-testing research, by either deliberate or fortuitous manipulation of attributes of populations or environments. We are not alone in our contention (Romesburg, 1981, 1991).

CHAPTER 13

ORDER ARTIODACTYLA—EVEN-TOED UNGULATES

The artiodactyls, represented in Oregon by three families and seven species, include 219 species in 10 families distributed worldwide (Grubb, 1993). These hooved mammals were introduced into Australia and New Zealand, and some species have been extirpated from much of their original ranges throughout the world (Nowak, 1991). Fossil artiodactyls are known since the Eocene in many parts of the world, including Oregon (Savage and Russell, 1983).

The artiodactyls are characterized by the main axis of the foot aligned between the third and fourth digits. In all forms that occur in Oregon (but not in all artiodactyls), the digits terminate in hooves, the stomach is four chambered and digestion involves rumination, a caecum is present, upper incisors are absent, lower canines are incisiform, a broad diastema separates canines and premolars, and antlers or horns are present (except in some females and younger cohorts). All are herbivorous. The ruminant digestion process that involves cud chewing, chemical treatment, and microbial action permits rapid ingestion of large quantities of relatively low-quality forage and later mastication, thereby reducing exposure to potential predation (Nowak, 1991).

In Oregon as elsewhere, the artiodactyls include most of the big-game species, and many hundreds of some species are killed by sport hunters each year.

KEY TO THE ARTIODACTYLA OF OREGON

1a. White on chin and throat divided by two or three darker bands; males with laterally flattened, branched, deciduous horn sheaths over bony cores permanently attached to frontal bones; two toes present; metatarsal glands absent...(Antilocapridae) *Antilocapra americana*, pronghorn (p. 484).
1b. Chin and throat, if white, undivided by dark bands; males either with deciduous antlers or with horns with permanent horn sheaths; four toes present; metatarsal glands present or absent...2
2a. Males and females with permanent horn sheaths over bony cores attached to frontal bones; nasal bones and lacrimal bones joined by a suture...............3 (Bovidae)
2b. Males with deciduous antlers shed annually from bony pedicels on frontal bones, and females lacking antlers; nasal bones and lacrimal bones separated by prelacrimal vacuity....................................5 (Cervidae)
3a. Head massive, with short, upturned horns; heavy shaggy cape over head and shoulders; cattle-like in appearance...........................*Bos bison*, bison (p. 491).
3b. Head in proportion to body, with horns recurved or spiral, not upturned; no shaggy cape over head and shoulders; not cattle-like in appearance.......................4
4a. Pelage entirely white; horns black, smooth, recurved, and slightly divergent; horns similar in both sexes; tail >150 mm long...
........*Oreamnos americanus*, mountain goat (p. 493).
4b. Pelage grayish or brownish with white venter and rump patch; horns grayish, ridged, and often frayed on tips; horns among males massive and spiraled, among

females slender and falcate; tail <150 mm long......
................*Ovis canadensis*, mountain sheep (p. 497).
5a. Long brown hairs on neck forming dark mane; conspicuous tawny rump patch; tail inconspicuous; upper canine present; adult males with massive antlers...
............................*Cervus elaphus*, elk or wapiti (p. 464).
5b. No mane; rump patch white if present; tail conspicuous; upper canine absent; males with antlers..............6
6a. Metatarsal glands ≅25 mm long; openings of metatarsal and interdigital glands surrounded by white hairs; tail light brown on dorsal surface, white on ventral surface; lacrimal pits broad and shallow; antlers with all tines originating from a main beam...
...*Odocoileus virginianus*, white-tailed deer (p. 479).
6b. Metatarsal glands ≥75 mm long; openings of metatarsal and interdigital glands surrounded by brown or tan hairs; tail black on dorsal surface or white with a black tip; lacrimal pits narrow and deep; antlers dichotomously branched..............................*Odocoileus hemionus*, mule deer—black-tailed deer (p. 469).

FAMILY CERVIDAE—DEER, ELK, AND MOOSE

The cervids, consisting of 43 species in 16 genera (Wilson and Reeder, 1993a), are native to North America, South America, Eurasia, and northern Africa and, through introductions, are established essentially worldwide (Nowak, 1991). In Oregon, the cervids include two species of deer (*Odocoileus virginianus* and *O. hemionus*), the elk or wapiti (*Cervus elaphus*), and possibly the moose (*Alces alces*; see CHAPTER 14). These are among the larger land mammals in the state. The deer and elk are the most sought big-game species; many hundreds are killed each year for sport and food. *O. virginianus* west of the Cascade Range was afforded protection under the Endangered Species Act in 1968 (Gavin, 1984).

All cervids possess four digits on each foot, each equipped with a hoof; the third and fourth are largest, are set anteriorly, and are the primary weight-bearers, whereas the second and fifth are set slightly to the rear and, in a normal pace on hard surfaces, do not contact the substrate. Metatarsal and metacarpal bones of the third and fourth digits on the front and hind limbs, respectively, are fused to form cannon bones that lengthen and lighten the distal portion of the legs. The long, slim legs are heavily muscled at their bases, thus designed for speed. Metatarsal glands are large and surrounded by a tuft of longer hairs.

Upper incisors are absent, upper canines are present or absent depending on the species, and the three premolars are molariform. In the lower jaw, the incisors are present, the canine is incisiform (sometimes erroneously referred to as the fourth incisor), and a broad diastema separates the canine from the premolars. The incisiform teeth occlude against a callous pad in the upper jaw. Molariform teeth are selenodont and brachydont. The lacrimal is separated from the maxilla and nasal by a vacuity. The lacrimal canal has two openings, one inside and one outside the orbit. An indentation (pit) in the lacrimal holds a facial gland that opens at the inner corner of the eye.

Unique among the cervids is the presence of antlers; in Oregon species, antlers normally are limited to males, but occasionally an antlered female is recorded. Antlers are bony projections that grow from special pedicels on the frontal bones during late winter to early summer. They are soft, tender, and covered with velvetlike skin while growing, but when mature, the blood supply is cut off; the velvet dries and is rubbed off on vegetation, leaving hard, bony projections. Usually, males grow their first antlers as they attain sexual maturity; the first set of antlers commonly is small and usually unbranched. Antlers are shed after the breeding season, usually in January–February; regrowth of antlers commences shortly thereafter. Subsequent sets of antlers often are larger and more complex, but because diet and general health of the animal affect antler growth, size and complexity of antlers are not good criteria of age in cervids. Very old animals may have stunted or malformed antlers. Although day length, contributions of the anterior pituitary, and production and withdrawal of androgens are known to be involved in growth, hardening, and casting of antlers (Anderson and Wallmo, 1984), precise physiological mechanisms that control antlerogenesis are not understood (G. Bubenik, 1983).

The stomach is four chambered and digestion involves rumination (regurgitation and rechewing of food). A gallbladder is absent. The placenta is cotyledonous, the cotyledons forming opposite specialized areas on the uterine wall.

Cervids first appeared in the fossil record in the early Oligocene in Asia, late Oligocene in Europe, early Miocene in North America, and Pleistocene in South America (Nowak, 1991). A Pleistocene moose-sized deer in Ireland had antlers that extended 3.5 m from tip to tip.

Cervus elaphus Linnaeus
Elk or Wapiti

1758.   *Cervus elaphus* Linnaeus, 1:67.
1897b. *Cervus roosevelti* Merriam, 11:272.
1935.   *Cervus canadensis nelsoni* Bailey, 38:188.
1936.   *Cervus canadensis roosevelti*: Bailey, 55:81.
1969.   *Cervus elaphus nelsoni*: McCullough, 88:3.
1969.   *Cervus elaphus roosevelti*: McCullough, 88:3.

Description.—The elk (Plate XXXI) is the largest cervid in Oregon (Table 13-1), with males sometimes attaining a body mass of 500 kg (Hall, 1981). In southwestern Oregon, 110 females ≥2 years old averaged 264 kg, with highest averages among 7–9 year age-classes (≅282 kg—Harper et al., 1987).

These heavy-bodied, deer-like mammals have narrow faces tapering to a naked nose pad; relatively small, pointed ears; a heavily maned neck; a back slightly humped at the shoulders; a contrasting rump patch; and a small tail. Pelage color is grayish brown to reddish brown, somewhat lighter among males in winter; females remain dark year-round. The mane is dark brown and the rump patch and tail are cream colored. The underparts (except for a whitish patch between the hind legs) and legs are dark brown to almost blackish (Hall, 1946). Neonates are rusty colored with cream-colored spots. The metatarsal glands are about 75 mm long.

In dorsal view, the skull of the elk (Fig. 13-1) is shaped somewhat like a spear point. The orbits are large and pro-

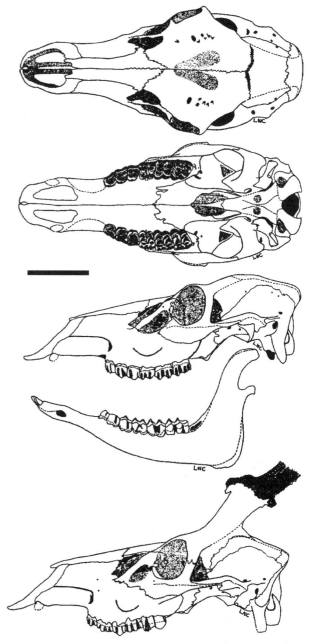

Fig. 13-1. Dorsal, ventral, and lateral views of the cranium, and lateral view of the dentary of a female elk, *Cervus elaphus* (OSUFW study); and lateral view of a male (OSUFW study), illustrating the antler pedicel. Scale bar equals 10 cm.

truding; the postorbital bar is complete. The lower canine is incisiform, but is somewhat larger than the incisors. Upper canines are present in both sexes, a characteristic unique to elk among Oregon cervids. Males possess antlers that among adults commonly are massive.

Distribution.—Originally, elk occurred in much of the United States except for New England, parts of the Southeast, Texas, and the desert Southwest. Also, the species occurred in most of the southern tier of western Canadian provinces and southern parts of those in eastern Canada (Fig. 13-2; Hall, 1981). Elk were extirpated from much of the original range; given protection and with translocations, the species has reoccupied much of the western mountains

Table 13-1.—*Means (±SE), ranges (in parentheses), and CVs of measurements of skull characters for female and male Artiodactyla from regions in Oregon. Skull characters were measured to the nearest 1 mm. SE and CV not provided if* n *<10.*

Species and region	Sex	n	Greatest length of skull	Length of maxillary toothrow	Zygomatic breadth	Length of mandibular toothrow	Length of mandibular diastema
Antilocapra americana	♀	10	262.0 ± 3.24 (245–279) 3.91	64.55 ± 1.02 (60.5–72.0) 4.99	122.7 ± 2.17 (115–137) 5.59	69.4 ± 1.12 (64–77) 5.10	61.7 ± 1.25 (56–70) 6.43
	♂	7	265.1 (255–286.5)	62.0 (59.5–64.0)	132.8 (115.5–149)	66.1 (63.5–68.5)	65.5 (60–75)
Bos bison	♀	10	471.5 ± 4.67 (455–496) 3.14	129.5 ± 3.86 (120–162) 9.44	252.7 ± 2.54 (244–270) 3.17	139.5 ± 2.5 (137–142) 2.53	79.0 ± 1.00 (78–80) 1.79
	♂	5	469.4 (279–532)	123.1 (92.32–135)	282.7 (161–370)	133.1 (92.3–163)	76.5 (58.5–103)
Cervus elaphus	♀	1	453.0	136.0	183.0	143.0	115.0
Odocoileus hemionus East of Cascade Range	♀	4	255.6 (244.5–272)	69.1 (62–75)	113.0 (108–117)	79.0 (72–86)	55.9 (54.5–59)
	♂	5	289.2 (273–305)	78.2 (73–85)	126.8 (120–139)	88.4 (83–94)	65.1 (58–72.9)
Odocoileus virginianus West of Cascade Range	♀	31	234.9 ± 1.59 (216–261) 3.78	63.7 ± 0.64 (52–70) 5.61	94.8 ± 0.75 (85–100) 4.42	72.1 ± 0.54 (67–78) 4.19	51.3 ± 0.70 (45–62) 7.64
	♂	20	253.5 ± 2.90 (237–280) 5.12	68.2 ± 1.03 (62–80) 6.79	102.6 ± 1.27 (96–113) 5.54	75.8 ± 1.17 (68–90) 6.93	57.6 ± 1.49 (49–70) 11.63
East of Cascade Range	♀	2	253.8 (245.5–262)	76.3 (74–78.6)	101.7 (100–103)	84.9 (83.5–86.2)	58.2 (55–61.4)
	♂	4	287.0 (265–303)	79.9 (73–88)	114.3 (101.9–122)	87.4 (79–95)	72.2 (62.8–85.6)

and also occurs at scattered localities within and beyond the original range (Peek, 1982:851, fig. 43.1). The species is considered conspecific with the red deer that occurs in much of Europe and Asia. Introduced populations are established in Australia, New Zealand, and Argentina (Nowak, 1991).

In Oregon, elk occur throughout the state, but are most abundant in the Blue and Wallowa mountains and in the northern Coast Range and least abundant in the southeastern high-desert region (Fig. 13-3). Museum specimens are much less widely distributed (Fig. 13-2).

Geographic Variation.—Two of the six recognized races of *C. elaphus* occur in Oregon: *C. e. nelsoni* east of the Cascade Range and *C. e. roosevelti* west of the Cascade Range. The former is slightly smaller and lighter colored; it has more slender but longer, less webbed, and more spreading antlers than the latter. Although the two races in Oregon are managed separately and often are referred to by distinctive vernacular names, even typical individuals of races are not as readily distinguishable morphologically as the two races of *Odocoileus hemionus* with similar distribution patterns in the state. At least some of the elk transplanted to reestablish herds in parts of Oregon were from regions not included within the range of the original race that occupied that area. The effect of introgression on characters used to distinguish the races, if any, is unknown.

Habitat and Density.—Elk require a mosaic of early,

forage-producing stages and later, cover-forming stages of the forest sere in close proximity (Harper et al., 1987). In western Oregon, clear-cuttings compose the primary foraging areas, attaining peak production and use 5–8 years after logging. Production of prime forage is related positively to the degree of soil disturbance, whereas use by elk is related negatively to distance from cover. Continued use of cut-over lands for foraging by elk tends to retard secondary succession (Edgerton, 1987) and to maintain production of quality forage for longer periods (Harper et al., 1987). Slash burning on ridgetops and south-facing areas may enhance production of quality forage for elk, but not immediately; burning debris in areas that already produce good forage does little to improve production (Harper et al., 1987). In some areas of western Oregon, clear-cuttings are seeded to grasses and legumes, fertilized, and grazed by domestic sheep, with one of the objectives being to enhance forage quality for elk. However, Stussy (1994) found that such practices had no apparent effect on the diets, movements, distribution, or productivity of elk, thus questioned the justification for these habitat-management practices.

Approximately 90% of use of foraging areas by elk occurs within ≅120 m of cover sufficient to hide 90% of a standing elk at ≅60 m. Such cover may be provided by later stages of the shrub and open sapling-pole stand stages and later stages of the forest sere. Hiding cover provides security for elk. Elk also require "thermal cover" that

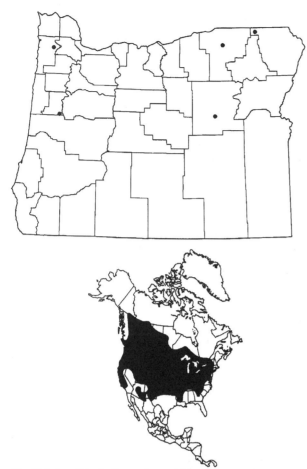

Fig. 13-2. Localities in Oregon from which museum specimens of the elk, *Cervus elaphus*, were examined (listed in APPENDIX, p. 513), and the distribution of the elk in North America. (Range map redrawn from Hall, 1981:1087, map 535, with modifications.) The species was extirpated from most of the range east of the Rocky Mountains; only a few scattered herds (mostly reintroduced) remain in the region (Peek, 1982).

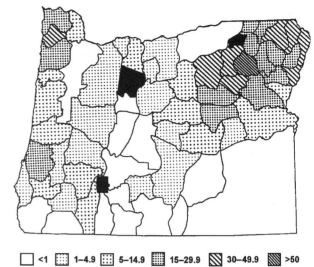

| □ <1 | ▦ 1–4.9 | ▦ 5–14.9 | ▦ 15–29.9 | ◩ 30–49.9 | ◩ >50 |

Fig. 13-3. Average annual harvest of elk/100 km² by big-game management unit in Oregon, 1991–1993. Harvest includes both sexes and all harvest methods. Data provided by the Oregon Department of Fish and Wildlife. The two northern solid black areas are Indian Reservations; the southern black area is Crater Lake National Park.

consists of forested areas with trees ≥12 m tall and with a canopy closure of ≥70%. Deciduous trees may provide thermal cover in summer, but not in winter. Such cover is sought by elk in summer when temperatures are ≥18°C and in winter when temperatures fall and winds increase (Harper et al., 1987). Nearly 100% of thermal cover used by elk in both summer and winter is ≤365 m from foraging areas; thus, elk seem to avoid traversing large, open areas (Leckenby, 1984). Such emphasizes the association of elk with "edge" created by juxtaposition of small blocks in different stages of the sere. In northeastern Oregon, of ≅800 observations of elk on summer range, the animals were in foraging areas 60% of the time, in thermal cover 34%, and in hiding cover 4% (Leckenby, 1984).

At exploration, elk reportedly were abundant in Oregon. When the Lewis and Clark Expedition overwintered at Fort Clatsop in 1805–1806, 146 elk were killed for food in <4 months. Early settlements and forts depended upon elk as a source of meat; market hunting became a respected occupation. Hides, antlers, and upper canine teeth were items of value so also served as incentives to kill elk. The area was so vast, the settlers so few, and the elk so abundant that the supply seemed endless. It was not! In 1872, the Oregon legislature passed a law prohibiting the taking or selling of elk from 1 February to 1 June each year, but failed to provide for en-

forcement of the act. By 1880, elk reportedly were scarce in some regions (Harper et al., 1987). In 1899, the state legislature provided complete protection to elk as only a few small herds remained in the Coast and Cascade ranges and in the Blue and Wallowa mountains. Some provisions for enforcement were made, and increasing populations were reported in some regions by 1916. Nevertheless, officials remained pessimistic into the mid-1920s regarding the possibility of reestablishing herds to the point of supporting hunting seasons on the species (Harper, 1982). However, a 7-day hunting season on adult males was established in Clatsop Co. in 1938; 1,243 hunters harvested 294 elk. Hunting of antlerless elk was permitted commencing in 1939 in eastern Oregon and in 1950 in parts of western Oregon (*Oregon's elk management plan*—Oregon Department of Fish and Wildlife, 1992). The harvest of elk on both sides of the Cascade Range has continued an upward trend through the past 4.5 decades (Fig. 13-4), but concomitantly hunter-success rates have declined and the proportion of adult males in populations has declined. Pedersen (1979:3) concluded that "The elk population increased mainly because of changes in the structure of the forest plant communities. Logging created openings in the forest allowing forage plants to grow."

By 1985, elk herds in the Cascade and Coast ranges were estimated to number 2,600 adult males, 30,000 adult females, and 11,000 young-of-the-year (Harper et al., 1987). Estimated densities of elk in northeastern Oregon in the early 1980s ranged from 5.5 to 3.2 to 0.8/100 ha in a northeast-southwest direction extending from the Oregon-Washington border to just south of John Day, Grant Co. (Leckenby, 1984).

Diet.—Based on epidermal fragments in feces, elk in northeastern Oregon consumed 63 species of plants during summer 1972 and 71 species in 1973 (Korfhage et al., 1980). The proportions of grasses in the diet during the 2 years, respectively, were 27% and 23%; sedges, 12% and 10%; forbs, 29% and 38%; and woody plants, 32% and 29%. Graminoids—especially sedges (*Carex geyeri* and *C. rossii*), bluebunch wheatgrass (*Agropyron spicatum*),

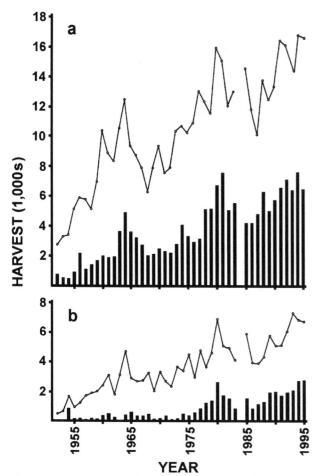

Fig. 13-4. Total number (line) and number of antlerless (bar) elk harvested annually in Oregon, 1952–1995: **a**. east of the Cascade Range (commonly referred to as Rocky Mountain elk) and **b**. west of the Cascade Range (commonly referred to as Roosevelt elk). No data were gathered in 1984. Data provided by the Oregon Department of Fish and Wildlife.

intermediate wheatgrass (*Elytrigia intermedia*), and bluegrass (*Poa*)—were used more heavily in early summer, whereas forbs—especially trail-plant (*Adenocaulon bicolor*), monkshood (*Aconitum columbianum*), and false bugbane (*Trautvetteria carolinensis*)—were used extensively in late summer. Use of browse species also was greater in late summer; woody plants commonly consumed were huckleberry (*Vaccinium membranaceum*), common snowberry (*Symphoricarpos albus*), ninebark (*Physocarpus malvaceus*), oceanspray (*Holodiscus discolor*), Pacific yew (*Taxus brevifolia*), and elderberry (*Sambucus*). Changes in diet seasonally were related to differences in phenological development caused by differences in precipitation between years; grasses and forbs developed earlier and peak use of them occurred earlier during a dry year (Korfhage et al., 1980).

Diet compositions of elk obtained by similar methods on five winter ranges also in northeastern Oregon differed greatly among ranges and during early and late parts of winter (Skovlin and Vavra, 1979). For example, on one range ponderosa pine (*Pinus ponderosa*) composed 65% of the diet in early winter and 25% in late winter, but in none of the other four ranges did the species compose >22% of the diet at either season, and it did not compose >10% in six of the seven remaining samples. Graminoids—especially sedges,

Idaho fescue (*Festuca idahoensis*), and Sandberg bluegrass (*Poa sandbergii*)—composed as much as 90% of the diet in early winter and 95% in late winter on one winter range. The least that graminoids contributed was 11% in early winter on one range and 61% in late winter on another range. Shrubs contributed relatively little to elk diets on winter ranges; the maximum that shrubs contributed to the diet was 16% in early winter and 8% in late winter. Seven of the eight remaining samples contained ≤4% shrubs (Skovlin and Vavra, 1979). Skovlin and Vavra (1979:15), in comparing diets of elk with those of mule deer (*Odocoileus hemionus*), concluded that competition between the species for forage in winter was "normally not high."

In southwestern Oregon, 90 of 93 species of plants occurring in a plant-sampling study also were included in the diet of elk. Based on the percentage of forage consumed and in order of proportion consumed, Pacific blackberry (*Rubus ursinus*), graminoids, false dandelion (*Hypochaeris radicata*), vine maple (*Acer circinatum*), salmonberry (*Rubus spectabilis*), huckleberry (*Vaccinium parvifolium*), salal (*Gaultheria shallon*), and thimbleberry (*Rubus parvifolius*) contributed most to the diet of elk. These eight plants composed 70% of the diet (Harper et al., 1987). Some seasonal differences in diet were noted; for example, vine maple was not eaten in winter and composed only 3% of the diet in autumn and 8.5% in spring, but composed 14.6% of the summer diet, more than any other species. The contribution of some other major dietary species varied less; for example, the seasonal contribution of Pacific blackberries ranged from 12.3 to 17.7% and that of graminoids from 11.7 to 21.1% (Harper et al., 1987).

In western Lane Co., young elk were strongly selective of forbs (Marquez, 1988). Himalayan blackberry (*Rubus discolor*), pearly-everlasting (*Anaphalis margaritacea*), buckhorn plantain (*Plantago lanceolata*), and white clover (*Trifolium repens*) composed 87.2% of the diet of young during July, whereas adult females consumed grasses and forbs in equal proportions. Perennial ryegrass (*Lolium perenne*) was strongly selected by adult females; it composed ≅30% of their diet (Marquez, 1988). As young matured, grasses were consumed in greater amounts.

Elk are strongly attracted to salt and can be baited into traps with salt. In northeastern Oregon, elk removed 68 kg of salt from an area in 5 days and commonly removed 23 kg from artificial licks in 2 days (Pedersen and Adams, 1977).

Reproduction, Ontogeny, and Mortality.—Although the breeding season for elk in Oregon may commence as early as mid-August and extend as late as mid-November, the peak of breeding is the last week of September and the 1st week of October, with ≅85% of females bred by mid-October (Harper et al., 1987). Development of a corpus luteum follows the usual follicular development, ovulation, copulation, and fertilization, but then in some females a second follicle ruptures and a smaller, functional secondary corpus luteum forms; apparently, the second ovulation occurs before placental development progresses sufficiently to suppress further follicular development (Peek, 1982).

Female elk in western Oregon have fewer young than their counterparts in eastern Oregon. In classifying 107,900 elk by sex and age in western Oregon and 164,426 in eastern Oregon in 1946–1983, offspring:adult female ratios were 39:100 and 47:100 in the two areas, respectively (Harper et al., 1987). Although differences in rates of juvenile mortality may contribute to differences in the ratios, the proportion of pregnant females in samples corroborates evidence of differences in productivity from the ratios. In 1964–1982, uteri of 51% of 906 yearling and adult females killed in western Oregon contained developing fetuses, whereas 67% of 1,222 females killed in eastern Oregon were pregnant. The relationship was similar in all age groups of females, but females in age groups 4–10 were most productive (Harper et al., 1987). Nutrition, lactation, and overall physical condition were major factors involved in fertility; in samples of individuals ≥3 years old, 25 (48.1%) of 52 lactating females were pregnant, whereas 37 (74.0%) of 50 nonlactating females were pregnant (Harper et al., 1987). This effect may be sufficient that many females breed no more frequently than in alternate years. The frequency of twinning in both eastern and western Oregon is low; only 0.23% of a sample of 1,281 pregnant females carried twins (Harper et al., 1987).

Of 45 yearling (1-year-old) males killed by hunters during an October hunting season in southwestern Oregon, 43 (95.6%) contained spermatozoa in their testes. However, in 13 (30.2%) of the 43, density of spermatozoa was less than that observed in adult males (Harper, 1966).

In a 187-ha enclosure in southwestern Oregon, the average number of young produced and that survived until observed with the maternal female was greater when females were bred by 2- and 3-year-old males than when bred by yearling males (Hines and Lemos, 1979). Young sired by yearling males were born mid-June to mid-September, whereas those sired by older males were born mid-May to mid-July. Hines and Lemos (1979) believed that the first estrous period occurred before yearling males attained puberty, thus explaining the lag in time of parturition. Further studies conducted on free-ranging herds in adjacent areas tended to corroborate these findings, but the magnitude of the lag was somewhat less (Hines et al., 1985). Also, fertility in females was related to nutrition as fertility was reduced by lactation and advanced age but enhanced by significant harvests of antlerless elk, thereby providing more forage per female (Hines et al., 1985). From these studies, Hines et al. (1985) suggested that depending largely on yearling males as breeders caused delays in conception that could reduce productivity by suppressing growth and survival of young, subsequent fertility of females, and increasing attainment of reproductive competence in both sexes.

Rosemary Stussy (in litt., 16 June 1997), a biologist with the Oregon Department of Fish and Wildlife, found selenium levels in livers of elk that in cattle would be considered to range from deficient to toxic, but levels were not related to reproductive performance of elk. Low levels of selenium in blood are known to reduce reproductive output in some artiodactyls (Fleuck, 1994). Stussy (in litt.) believed that selenium stored in livers of elk might not be available for use depending upon other nutritional factors and may not have been representative of selenium levels in the blood. She did not dismiss the possibility of a relationship between selenium deficiency and low productivity in elk.

Neonate elk weigh an average of ≅14 kg (range, 9–20 kg, $n = 23$), but gain body mass rapidly (Johnson, 1951). At 5–7 days, 47 elk averaged 20 kg (range, 15–27 kg—

Johnson, 1951). In Oregon, at 4 months, three of mixed sex averaged 107 kg (range, 99–114 kg); at 6–7 months, 14 males averaged 127 kg (range, 106–140 kg) and eight females averaged 119 kg (range, 108–132 kg); and at 8–9 months, 12 males averaged 124.5 kg (range, 99–141 kg) and six females averaged 110 kg (range, 95–122 kg—Hines and Lemos, 1979).

Neonate elk are equipped with deciduous lower incisors, canines, and premolars, and m1 may be erupted or covered by a thin membrane, a dental battery that distinguishes elk <4 months old (Johnson, 1951). Yearlings may be distinguished by deciduous lower canines and premolars, and the presence of m1 and m2; those ≥28 months old are distinguished by permanent canines and premolars, and the presence of partially erupted m3 (Quimby and Gaab, 1957).

Antlerogenesis, with rare exceptions, is limited to males and is strongly influenced by nutrition; it commences with the formation of 2–3-cm-long "buttons" in the first summer and autumn of life. Antlers in yearlings commonly are "spikes"; however, they may include as many as five tines, but rarely a "brow" tine. Males 2 years old commonly have antlers with three to five tines that may include a "brow" tine. Adult males in prime condition usually have six–eight-point antlers (Peek, 1982) that combined weigh as much as a neonate (Blood and Lovass, 1966). Antlers are shed in March–April, regrowth commences in May, and the velvet is shed in July; thus, as much as 13.5 kg of bone is produced in ≅3 months.

Of 54 adult female elk resident in the Cascade Range that died while being radiomonitored, 26 were poached (illegal hunting), 15 died of unknown causes, 10 were killed legally, and one each was killed by a predator, died in an accident, and died from a gunshot wound (Stussy et al., 1994). Average (± SE) annual survival rate for adult females translocated in attempts to reduce conflict with landowners or to reestablish herds (0.77 ± 0.08) was significantly less than for resident adult females (0.92 ± 0.05). Deaths among translocated elk were recorded in November–May, months in which mortality also was highest in resident elk. Of translocated elk that died, most deaths were ≤7 months after being moved; 42% exhibited evidence of malnutrition (Stussy et al., 1994).

Habits.—Adult females, their current offspring, and their female offspring of the previous year form herds that tend to remain within relatively small and distinct areas (Harper et al., 1987; Marquez, 1988; Witmer, 1982). Nevertheless, there is considerable overlap in areas used by adjacent herds and there is considerable exchange of individuals among adjacent herds (Harper et al., 1987; Marquez, 1988). Leadership of these herds usually is provided by an older female with an offspring, but other females with offspring assume leadership duties at times. A change in reproductive status may be responsible for failure of a female to reassume a leadership role in following years (Marquez, 1988). Some females, not necessarily the matriarch, spend much more time alert than other females (Marquez, 1988).

Male elk, especially the larger ones, tend to be solitary during most of the year; however, during May and June, when antler growth is rapid, males, including larger ones, sometimes form herds (Harper et al., 1987). The antlers become polished in July, at which time activity increases as males commence to search for untended females or those tended by less formidable males. Movements become extensive and males frequently are observed in areas where they otherwise rarely occur. During the rut, male elk wallow in urine-soaked grass or in muddy pools into which they have urinated; they become so foul smelling that when they are downwind their presence can be detected by humans (Harper et al., 1987). At this time, herd composition is relatively stable; adult males attempt to exclude yearling males, but when one is chased others may attempt to breed with harem females (Harper et al., 1987).

From observations by Harper (1966) in southwestern Oregon, parturient female elk separate from the herd to seek a favorable birthing site nearby. After parturition, the maternal female remains away from the herd, rarely traveling more than a few hundred meters from the young during foraging, until the young is 2–3 weeks old and able to travel. The female with its offspring then returns to the herd. While adult females forage, several young remain under the watchful eye of one adult female. When ready to nurse, an adult female calls its young; when nursing is completed, the female chases the young back to the group (Harper, 1966). Some females do not seem disturbed when nursed by young other than their own, as Harper (1966) observed young nursing for short intervals on several females in quick succession and even two young nursing the same female at the same time.

In northeastern Oregon, home-range areas of female elk in summer ranged from <150 to >6,500 ha ($n = 23$; Leckenby, 1984:15, fig. 10). Home-range areas were related negatively to the percent thermal cover available; for each 1% increase in the proportion of the area occupied by thermal cover, the size of home ranges of female elk decreased by $\cong$60 ha (Leckenby, 1984). In the Coast Range, minimum-area home ranges of individual female elk followed by radiotelemetry ranged from 58.7 to 285.5 ha; home ranges were largest in summer, smallest in winter (Witmer, 1982).

In the Blue Mountains and on the east slope of the Cascade Range, elk move to winter ranges at lower elevations, but the migrations are not as spectacular or as long as in some regions (Craighead et al., 1972). Elk usually commence to move downslope in October–November when snow depth exceeds $\cong$40 cm and upslope in March–April with regrowth of green vegetation. In northeastern Oregon, migratory movements may be $\geq$65 km and as much as 1,900 m in elevation. Many elk in the region, however, move 8–16 km and only 600–900 m in elevation (Skovlin and Vavra, 1979). On the west slope of the Cascade Range, the winter range may extend to $\cong$1,100 m during mild winters and migration distances usually are short. The longest recorded movement was $\cong$64 km (Harper et al., 1987). Adult males tend to winter at higher elevations than females and their offspring. In the Coast Range, snow depths rarely are sufficient to cause movements, and even when they are, the change in locality is of short duration.

Remarks.—In northeastern Oregon, most summer ranges for elk are on public lands, whereas winter ranges largely are on private lands (Skovlin and Vavra, 1979). Herein lies the source of most complaints of damage to crops and property.

Odocoileus hemionus (Rafinesque)
Mule Deer–Black-tailed Deer
(Synonymy under subspecies)

Description.—*Odocoileus hemionus* (Plate XXX) can be distinguished from its only congener, *O. virginianus,* by longer ears (two-thirds rather than one-half the length of the head), dichotomously branching antlers, no white hairs around interdigital and metatarsal glands, metatarsal glands $\geq$75 mm long, a deep and narrow prelacrimal pit (Fig. 13-5), and a tail brown or black dorsally or white tipped with black (Anderson and Wallmo, 1984).

Distribution.—*Odocoileus hemionus* occurs from southern Yukon Territory south through most of the region west of the 100th meridian to the southern tip of Baja California Sur and San Luis Potosi (Fig. 13-6; Anderson and Wallmo, 1984; Hall, 1981). Introduced populations are established in the Prince William Sound region and on Kodiak and Afognak islands, Alaska (Mackie et al., 1982). Scattered individuals or temporary populations occur as far east as Minnesota and Iowa (Anderson and Wallmo, 1984; Mackie et al., 1982).

Geographic variation.—Ten or 11 subspecies of *O. hemionus* are recognized by various authorities (Anderson and Wallmo, 1984; Hall, 1981; Wallmo, 1981). Of these, two, *O. h. hemionus* (mule deer) and *O. h. columbianus* (Columbian black-tailed deer), occur in Oregon. The former is larger, lighter in color, and often associated with more open habitats, whereas the latter is smaller and darker, and frequents dense, early seral forest communities. Because of the differences in size, color, marking, ecology, and behavior, and because of the voluminous information available for the two races, we decided to treat them separately. Nevertheless, we emphasize that despite these differences, the races readily intergrade and produce offspring of varying degrees of intermediacy. In summer, these intergrades are common in a $\cong$65-km-wide zone (Cronin, 1991a) in the Cascade Range. In winter, the races are separated by deep snow at higher elevations; however, intermediates do not necessarily associate with phenotypes that they resemble most closely.

Odocoileus hemionus hemionus (Rafinesque)
Mule Deer

1817b. *Cervus hemionus* Rafinesque, 1:436.
1885. *Cariacus macrotis* True, 7:592.
1898b. *Odocoileus hemionus* Merriam, 12:100.
1936. *Odocoileus hemionus macrotis* Bailey, 55:83.

Description.—The mule deer (Plate XXX) is the larger member of the genus in Oregon (Table 13-1). The ears are long (usually >225 mm—Cowan, 1936a), the metatarsal gland usually is $\geq$105 mm long and surrounded by hairs similar in color but considerably longer than those on the remainder of the metatarsus, and the tail is short and constricted basally. Basilar length is usually >263 mm in adult males and >238 mm in adult females (Cowan, 1936a). Among adult males, the antlers are dichotomously branched with tines of approximately equal length. The first set of antlers usually is unbranched but lacks an "eyeguard" (subbasal snag) irrespective of the number of tines; the third and subsequent sets of antlers often possess an eyeguard. Among adult males the angle of the main fork in the ant-

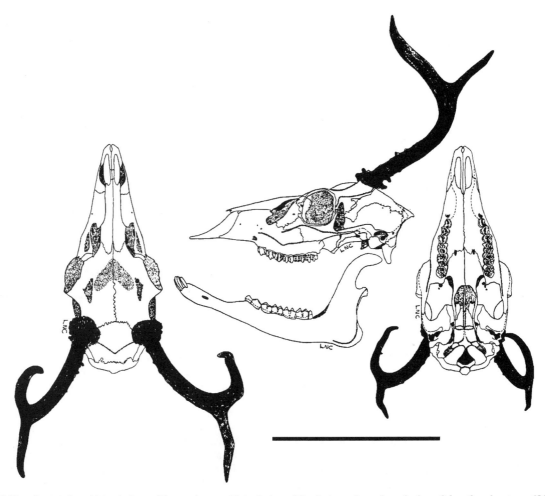

Fig. 13-5. Dorsal, ventral, and lateral views of the cranium, and lateral view of the dentary of a male mule deer, *Odocoileus hemionus* (OSUFW 1142). Scale bar equals 20 cm.

lers ranges from 78 to 98° (Cowan, 1936*a*).

In winter, the basic pelage color is cinnamon buff, darker dorsally and on the pate to slightly anterior to the eyes. The chest is nearly black at the midline grading to grayish on the sides. The nose, sides of the face, chin, and throat are whitish to pale buff. Spots of black occur immediately posterior to the nose pad and on the chin midway along the lower lip. The belly and inside of the legs are whitish to tannish. A white rump patch encircles a white tail with a black tip; the underside of the tail is bare proximally. In summer, the pelage is more reddish brown and the markings are less contrasting and more subdued. Many mule deer exhibit sufficient differences in color and markings to be recognized individually both by humans and by other deer (Geist, 1981). Neonates are spotted with white; the spotted pelage usually is molted by September (Einarsen, 1956).

Distribution.—Mule deer occur throughout Oregon east of the Cascade Range, and in summer, range into the Cascades. Museum specimens in Oregon are less widely distributed (Fig. 13-6).

Odocoileus h. hemionus arose in the Pliocene, probably from *O. h. columbianus* (Geist, 1981). Fossils are known from sites as far east as Arkansas and Texas (Kurtén and Anderson, 1980).

Habitat and Density.—Mule deer occupy a wide range of habitat types; some live in desert shrubs, some in woodlands, and some in conifer forests (Einarsen, 1956). In general, however, mule deer occupy the more open, but more rugged areas. For example, near Summer Lake, Lake Co., mule deer summer at ≅1,830 m in ponderosa pine (*Pinus ponderosa*) with an understory of bitter-brush (*Purshia tridentata*), tobacco-brush (*Ceanothus velutinus*), and manzanita (*Arctostaphylos parryana*) and winter at ≅1,370 m in big sagebrush (*Artemisia tridentata*), curlyleaf mountain-mahogany (*Cercocarpus ledifolius*), bitter-brush, and some ponderosa pine (Zalunardo, 1965).

On Steens Mountain, Harney Co., mule deer use lower elevations of summer range for birthing and early concealment of young; the extent of the area depends on the progress of snowmelt. Sites where the percent cover provided by shrubs was ≥23% and shrub height was ≥67 cm or where cover was ≥40% and height was ≥50 cm were considered the best for this purpose (Sheehy, 1978). Of six classes of shrub habitat on the mountain, tobacco-brush, big sagebrush-snowberry (*Symphoricarpos oreophilus*), and big sagebrush types were used in greater proportion than their occurrence, and low sagebrush (*A. arbuscula*) and shrubs treated with herbicide as part of range-improvement programs 17 years and 4 years earlier were used in lower proportion than their occurrence (Sheehy, 1978).

On Hart Mountain, Lake Co., male mule deer tended to

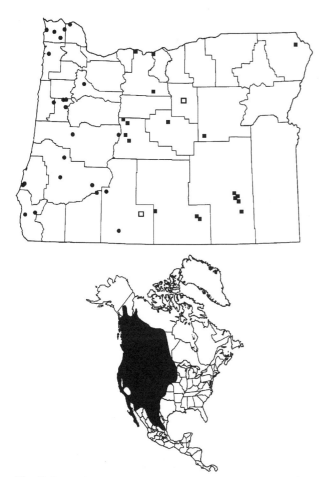

Fig. 13-6. Localities in Oregon from which museum specimens of the mule deer, *Odocoileus hemionus hemionus*, (squares) and black-tailed deer, *Odocoileus hemionus columbianus* (circles) were examined (listed in APPENDIX, p. 538), and the distribution of the mule and black-tailed deer in North America. (Range map redrawn from Hall, 1981:1090, map 536, with modifications.) Open symbol indicates record for county.

occupy stands of big sagebrush, whereas females and their offspring (especially those <8 weeks old) occupied mesic and bitter-brush communities closer to sources of water (Main, 1994). Vegetation on areas used by females and young tended to have greater species richness and denser woody cover than areas used by males. Also, females and young used areas with ≥10° slopes more than males; coyotes (*Canis latrans*), considered the primary predator of young mule deer in the region, used ≤10° slopes almost exclusively. Main (1994) believed that differential use of habitats by the sexes enhanced their respective reproductive fitnesses, females by obtaining greater security for their offspring, and males, because of their larger home ranges, by obtaining a diet of comparable or higher quality than that of females and acquiring adequate body condition to maintain them during the forthcoming rut.

Fur traders and explorers such as Ogden who traveled through eastern Oregon during the early and mid-19th century recorded numerous comments in their journals regarding the paucity of deer (Binns, 1967). Miners and early settlers in the region indicated that deer were abundant, but by the beginning of the 20th century, they again were scarce. Estimates of numbers of mule deer on national forests in Oregon in 1926–1933 ranged from 28,654 to 55,570

(Bailey, 1936). An era of protectionism commenced in 1901 with a 3.5-month open season and a five-deer either-sex bag limit; season lengths and bag limits declined, and a male-only law became effective in 1923. Refuges were established commencing in 1913, but taking of deer was permitted on some of the refuges after a series of disastrous die-offs resulting from depletion of forage on them (Edwards, 1942; Mitchell, 1944); refuges for deer were abandoned in 1955.

Mule deer were becoming abundant on nonrefuge areas also. For example, the estimated population on the winter range along the North Fork of the John Day River south of Ukiah was 3,000 in 1921, but by 1931, the population had increased to 19,500; ≅10,000 deer died during winter 1931–1932 (Cliff, 1939). By 1938, the estimated population was 20,000 animals. Mule deer remained abundant during the late 1940s and 1950s. Population estimates (although speculative) for mule deer in Oregon increased from 200,000 in 1950 to 530,000 in 1964, declined to 510,000 the next year, then increased to 570,000 in 1967 (Connolly, 1981); >75,000 (with 25–40% antlerless) were harvested annually during 12 of the 15 years between 1954 and 1968 (Fig. 13-7). In 1962, populations throughout the Great Basin region began to decline (Trainer et al., 1981). McKean and Luman (1964) claimed the population of mule deer in Oregon had declined. This precipitated an era of conservative harvests, followed by a few years of more liberal harvests then another period of conservative harvests (Fig. 13-7a). The reported harvest in 1993 was the lowest in the 43 years for which data are available (Fig. 13-7a). Population estimates released by the Oregon Department of Fish and Wildlife for 1990 were about 256,000 mule deer. Present management strategies differ in the various "management units" in eastern Oregon based on objectives of producing different buck:doe (male:female) ratios (*Mule deer management plan*—Oregon Department of Fish and Wildlife, 5 December 1990).

Connolly (1981) believed that the decline in harvest of mule deer throughout the region was not caused by a decline in hunting pressure or restrictions in hunting opportunity. In speculating about the cause of the region-wide decline in abundance of mule deer, Connolly (1981:238) proclaimed that "every identified trend in land use and plant succession on the deer ranges is detrimental to deer." He believed that quantity and quality of habitat were factors that limited abundance of mule deer.

Diet.—Although mule deer commonly are considered to be "browsers," they consume a wide variety of plant materials and, at some seasons, graze extensively. Forbs may compose 50–75% of the summer diet, but only 10% in winter. Cured grasses are used little, but green grasses are consumed when available (Hill, 1956). The cosmopolitan nature of the diet was revealed by a listing of 770 species of plants known to be consumed by *O. h. hemionus* (Mackie et al., 1982).

In winter, the critical period in the life of the mule deer, new growth of twigs of shrubs and trees is browsed, especially that of species high in fat content. Sagebrush (*Artemisia*), bitter-brush, rabbit-brush (*Chrysothamnus*), juniper (*Juniperus*), mountain-mahogany (*Cercocarpus*), and winterfat (*Eurotia lanata*) are among those commonly browsed. Where these plant species are depleted by

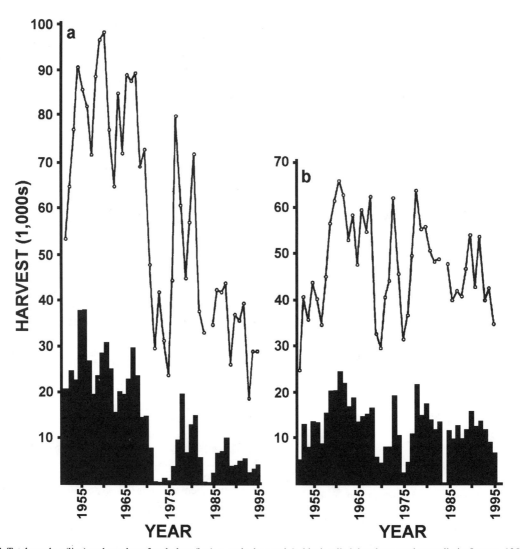

Fig. 13-7. Total number (line) and number of antlerless (bar) **a**. mule deer and **b**. black-tailed deer harvested annually in Oregon, 1952–1995. No data were gathered in 1984. Data from Oregon Department of Fish and Wildlife.

overbrowsing, deer resort to eating needles of ponderosa pine and Douglas-fir (*Pseudotsuga menziesii*—Cliff, 1939).

Even within a season, wide year-to-year differences in amounts and types of forage consumed may be related to environmental and other factors. Also, the timing of collection of samples within a season may affect results of dietary analyses. For example, analysis of stomach samples collected during 3 winters from the Interstate Herd that summers largely in the Fremont National Forest, Lake Co., Oregon, and winters in the Modoc National Forest, Modoc Co., California, indicated that green grass composed 40–50% of the diet in a mild, open winter, but <10% during a dry, cold winter (Leach, 1956). Bitter-brush was used heavily in autumn and early winter each year; sagebrush was eaten in midwinter; and forbs were consumed in spring. Dry grass was consumed when green grass was not available. Also, during an extremely cold winter with heavy snowfall, juniper composed ≥50% of the diet during December and January, whereas in open winters, juniper composed ≤15% of the diet in all monthly samples (Leach, 1956). In winter, in undisturbed portions of the Lava Beds National Monument, California, 77% of observations of feeding mule deer involved consumption of bitter-brush.

However, on burned areas, mule deer fed heavily on cheat grass (*Bromus tectorum*) and tumble mustard (*Sisymbrium altissimum*—Schones, 1978).

Einarsen (1956) claimed that the new growth of cheat grass in early spring was avidly sought by mule deer. He also indicated that consumption of large quantities of the green material commonly caused the deer to scour, that in turn could lead to death for those already weakened by malnutrition. Fungi, lichens, and mistletoe (*Phoradendron*) also seem to be relished by mule deer.

In summer, stomach contents of 29 mule deer from the Interstate Herd collected in Lake Co. consisted of 54.5% browse species, 34.9% forbs, and 10.6% grass and grass-like plants. Seventeen browse species and 26 species of forbs were identified. Tobacco-brush was the predominant browse species; it contributed 28.8% of the volume of the stomach contents and was identified in 62.1% of the stomachs (Leach, 1956). Unidentified forbs contributed 22.6% of the volume and occurred in 93.1% of the samples. Wild strawberry (*Fragaria*) was the forb identified most often; it composed 3.7% of the volume and occurred in 55.2% of the samples (Leach, 1956).

The range in daily forage consumption in captivity for

wild-caught mule deer in Colorado was 16.7–43.4 g/kg body mass, with adult males consuming least and 1–5-month-old young consuming most. Per unit of body mass, males consumed 1.4 times as much forage in summer as in winter and females consumed 1.3 times as much; young deer consumed 1.6–2.6 times as much as adults of the same sex (Alldredge et al., 1974).

Reproduction, Ontogeny, and Mortality.—Among adult males, the rut or annual period of sexual capability and excitement commences in midautumn. The antlers are polished, the neck commences to enlarge, the testes and epididymides become distended with spermatozoa, and food intake drops abruptly. A male tends a female in estrus until permitted to mate or until displaced by another male; males with the largest antlers usually are the most successful breeders. After the breeding season, the testes regress, serum-testosterone levels decline, and antlers are shed. Based largely on hearsay, Einarsen (1956) reported that the neck of adult males commenced to enlarge as early as 20 October, that mating occurred 15 November–15 December, that antlers were shed 28 December–15 March, and that young were born 5–15 June in Oregon. However, he also recorded evidence of mating as early as 1 October with young born as early as the 1st week of May; these records were considered exceptional. Estimated dates of conception for 43 females from the Interstate Herd ranged from the 2nd week of November to the end of the 1st week of January, with a peak at the end of the 1st week of December (Bischoff, 1957).

Seemingly, information regarding the ovarian and estrous cycles similar to that provided by Thomas and Cowan (1975) for *O. h. columbianus* is not available for *O. h. hemionus*. Anderson (1981) suggested that the values for *O. h. columbianus* may approximate those for the other races of *O. hemionus*. Thus, female mule deer that fail to mate probably undergo repeated estrous periods until as late as March or until bred successfully.

Parturient females separate themselves from the matriarchal association and commonly seek solitude at the margin of a meadow or open glade for birthing (Einarsen, 1956). Presumably, sites are selected that will provide hiding places for neonates. Einarsen (1956) suggested without quantification that birthing commonly occurs in the early morning hours; however, Golley (1957) reported that in a penned herd of 11 females, six (54.5%) births were in the morning, one (9.1%) at noon, and four (36.4%) in the early evening. The female leaves the immediate vicinity of the hiding place for the young and returns at intervals to nurse them. Neonates remain passive when disturbed, but within a few days flee when approached.

On Steens Mountain in 1968–1979, all of 60 young-of-the-year females 6–9 months old exhibited no evidence of reproductive activity; 58% of 26 females ≅21 months old were pregnant, with an average of 0.69 viable fetuses/female examined (1.2/pregnant female); and 97% of 190 females ≥33 months old were pregnant, with an average of 1.75 viable fetuses/female examined (1.79/pregnant female). Among the 15 pregnant females in the ≅21-month age-class, 80% contained one fetus and 20% contained two; among 185 pregnant females in older age-classes, 20.5% contained one fetus, 73.0% had two, and 6.5% had three (Trainer et al., 1981). After ≅4 months gestation, 8% of the fetuses were dead or otherwise judged not to be viable (e.g., weighed 25–68% less than a littermate). Among 345 fetuses, the sex ratio was 1:0.86 in favor of females, not significantly different from 1:1 (Trainer et al., 1981). For 249 young captured 1–16 days after birth, the sex ratio was 1:1.26 in favor of males, a significant departure from the prenatal ratio, but still not significantly different from 1:1. However, for 303 young collected or live-captured at 3–9 months of age, the sex ratio was 1:0.97 in favor of females.

Of 49 adult females collected from the Interstate Herd (south-central Oregon), 48 (98%) were pregnant and carried an average of 1.75 fetuses; 14 (29.2%) had only one fetus, 32 (66.7%) had twins, and 2 (4.2%) had triplets (Chattin, 1948). Of three females approaching 2 years of age from the same herd, one carried a fetus (Bischoff, 1958).

Average body mass at birth reported by various authors from various localities ranged from 2.7 – 4.0 kg. Average mass was less if litter size was greater than one or for females among different-sexed twins (Anderson and Wallmo, 1984). On Steens Mountain, neonates (*n* = 57) with undried umbilical cords (hence <24 h old) averaged 3.6 kg (range, 2.5 – 4.5 kg); there were no significant differences between sexes or between young born singly or as a twin (Trainer et al., 1981). Body mass of male young captured in December and March averaged ≅1 kg more than in September, but among females body mass in December was ≅2.5 kg greater than in September and ≅1 kg greater than in March (Trainer et al., 1981). However, average body condition of young, as estimated from kidney fat and marrow fat in the femur, declined as a function of age through the first 9 months of life on Steens Mountain (Trainer et al., 1979). Trainer et al. (1979) concluded that malnourishment was not a contributory factor to losses of young to predators during the first 6 months of life, but may have been during the next 3 months of life (winter). In Union Co., body condition was maintained through the first 6 months of life, then dropped precipitously between 6 and 9 months (winter period) to a level lower than that observed on Steens Mountain (Trainer et al., 1979). The pattern was believed to have differed from that on Steens Mountain as a result of competition with elk (*Cervus elpahus*) for food resources on winter range in Union Co.

In 1969–1979 on Steens Mountain, estimated mortality among young based on differences between birth rate and subsequent young:parous-female ratios averaged 32% (range, 21–48%) between birth (in June) and August and 75% (range, 53–87%) between birth and March (Trainer et al., 1981). Estimated mortality during the first 3 months was comparable with that reported for the same age-class elsewhere, but much higher during the next 6 months (autumn–winter) than reported elsewhere (Trainer et al., 1981). Although estimated mortality for young during summer was similar between the sexes, young born as singles suffered significantly higher mortality than those born as twins. Trainer et al. (1981) were unable to offer an explanation for the observed difference. We suggest, as younger, reproductive-age females have more single births than older females, that maternal experience might be involved.

Predation by coyotes (*Canis latrans*) was considered the major cause of loss of young on a 194-km² winter range on Steens Mountain; of 163 radio-tagged young, coyotes preyed on 80 (49.1%). The cause of death of 39 (23.9%) could not be determined because of paucity of remains; thus the contribution of coyote predation was believed to have been even greater (Trainer et al., 1981). However, of 65 radio-tagged young-of-the-year mule deer that died in 1971–1975, death of 31 (48%) was attributed to predation by coyotes; in 1976–1979, when 536 coyotes were removed from the winter range, 49 (50%) of 98 deaths still were attributed to coyotes (Trainer et al., 1981). The subsequent conclusion that "increased mule deer fawn survival followed aerial gunning of coyotes on mule deer winter range" (*Mule deer management plan*—Oregon Department of Fish and Wildlife, 5 December 1990:16) seemingly was based largely on probability levels of ≥0.10 (Trainer et al., 1981).

Mule deer are the most highly sought big-game species in Oregon, with annual harvest since 1952 ranging from slightly more than 16,000 to nearly 98,000 (Fig. 13-7a). Most of the harvest was of antlered males. Much of the variation ($r^2 = 0.808$) in the total annual harvest is attributable to the number of antlerless deer taken, which, in turn, doubtlessly is related directly to regulations concerning taking of these deer.

Habits.—Mule deer rest by lying on their chests and bellies with legs tucked under. Occasionally, an individual lies with its neck extended and its head on the ground and appears to sleep, but normally deer at rest are alert (Geist, 1981). When resting, or even when foraging, deer orient themselves so they do not look at the face of other individuals in the group. Deer ruminate while at rest; boluses of food are regurgitated, rechewed, and reswallowed, and are easily observed as they pass up and down the esophagus. Upon arising without being disturbed, the deer stretches its body, arches then depresses its back, and extends the tail (Geist, 1981). Then, the deer usually urinates and commonly defecates.

Mule deer are able to detect danger at long range, and when danger is detected they may hide, move away stealthily, or flee, initially by stotting, then by trotting. Stotting, a bounding gait, is used commonly by mule deer when alarmed, but the gallop, a gait employed by white-tailed deer (*O. virginianus*), is used in most other situations. Stotting is an effective escape mechanism when combined with release of odoriferous compounds from the metatarsal glands to alarm conspecifics, which also bound away, thereby distracting potential predators (Geist, 1981).

Mule deer are gregarious; they form groups of as many as 24, but >60% of groups consist of fewer than five individuals (Kucera, 1978). Except during the reproductive season, most groups consist either of males or of adult females, their young-of-the-year, and female young of previous years. In one study conducted elsewhere, groups of mixed sexes composed only 26% of those observed before the rut, but 67% of groups observed during the rut (Kucera, 1978). Females tend to remain near natal areas, but males, upon becoming independent, often disperse. These behavioral differences combined with hunting restricted largely to males have the potential of restricting gene flow, reducing effective population size with subsequent loss of genetic variation, and possibly reducing fitness (Ellsworth et al., 1994).

Among males, dominant individuals usually are those that possess the largest antlers. Most aggressive interspe-cific interactions take the form of threats, usually involving subtle postures to intimidate adversaries. Female mule deer rush an opponent and use the front legs and hooves to jab or slash if threat displays fail to accomplish the intended dominance. Males use antlers in such situations (Geist, 1981). Because mule deer are easily observed, much of their complex social organization and interactions, and the significance of them, have been described (Geist, 1981).

In many regions of Oregon, mule deer summer on ranges at higher elevations, then move to lower elevations to spend the winter. In central Oregon near Summer Lake, deer moved 0.4–128.7 km between sites at which they were marked on winter ranges and points at which they were observed later (Zalunardo, 1965). Deer that wintered in close proximity did not use the same sections of the summer range. However, of 188 individuals recaptured on the winter range, 122 (64.9%) were caught in the same trap and the remaining 66 were recaptured within 4.0 km of their original points of capture (Zalunardo, 1965).

On the Lava Beds National Monument, a few kilometers south of the Oregon-California border, minimum-area home ranges of female mule deer obtained from telemetry data ranged from 287 to 683 ha in an area with greatest cover and forage plants, and from 922 to 5,022 ha in areas with less cover and food resources (Purcell, 1980). Schones (1978) earlier found that home ranges of mule deer in winter were smaller in tall, dense vegetation on the same area.

Remarks.—*Odocoileus h. hemionus* occasionally hybridizes with *O. virginianus ochrourus* in northeastern Oregon. A discussion of interspecific hybridization in this genus is provided in the account of *O. virginianus*.

Odocoileus hemionus columbianus (Richardson)
Columbian Black-tailed Deer

1829a. *Cervus macrotis* var. ß. *Columbiana* Richardson, 1:257.
1885. *Cariacus columbianus* True, 7:592.
1858. *Cervus columbianus* Baird, 8(pt. 1):659.
1866. *Eucervus columbianus* Gray, 18:338.
1898b. *Odocoileus columbianus* Merriam, 12:100.
1912. *Odocoileus columbianus columbianus* Swarth, 10:85.
1936a. *Odocoileus hemionus columbianus* Cowan, 22:215.

Description.—Black-tailed deer (Plate XXX) usually are smaller than mule deer of the same sex and age (Table 13-1), although individuals as large as 143 kg field dressed have been recorded (Einarsen, 1946). The ears of black-tailed deer usually are <225 mm long, the metatarsal glands are ≅75 mm long, and basilar length usually is <263 mm in adult males and <243 mm in adult females (Cowan, 1936a). The pelage is dark reddish-brown, the face is brownish rather than grayish or white, the rump patch is small and does not extend much beyond the tail, and the tail is brownish or black dorsally, white ventrally, and not constricted proximally (Cowan, 1936a). The tail is naked on the underside for less than one-third its length.

The skull is similar to that of mule deer except that it usually is smaller for the same sex and age-class of *O. h. columbianus*. Cowan (1936a) claimed that the most reliable cranial characters among adults for distinguishing *O. h. columbianus* were length of the upper molariform row <75 mm in males and <70 mm in females, and palatal breadth at M3–M3 <53 mm in males and <46 mm in females. Eyeguards occur on <40% of antlers of adult males.

None of these characters is definitive for the subspecies.

The hooves of males are significantly longer than those of females among adults and significantly broader in the anterior one-third among both 1.5-year-olds and adults. Nevertheless, distinguishing sexes from tracks remains problematical (McCullough, 1965).

Distribution.—Columbian black-tailed deer occur throughout Oregon west of the Cascade Range. Museum specimens are less widely distributed (Fig. 13-6).

Odocoileus h. columbianus is believed to be ancestral to *O. h. hemionus* (Geist, 1981).

Habitat and Density.—Black-tailed deer tend to be associated with dense communities early in the forest sere. Einarsen (1946) indicated that black-tailed deer populations not only increased rapidly on the area affected by the catastrophic wildfire in northwestern Oregon that became known as the Tillamook Burn, but some individuals there quickly attained exceptionally large size. Because of the association with early seral communities and the ability to increase in numbers quickly, black-tailed deer often are responsible for damage to tree seedlings planted for reforestation. Black et al. (1969) claimed that roughly one-third of trees remaining in plantings were damaged by wild animals and 56% of all damage was by deer. Browsing by deer retards growth of trees. Strangely, Borrecco et al. (1972) indicated that treating forest plantings with certain herbicides to remove grasses and forbs tended to improve the habitat for deer without increasing the browsing of seedlings. In view of the nature of the diet of black-tailed deer and the negative impact that herbicide treatment had on mule deer habitats, we remain skeptical.

Of six plant communities on a ≅138-ha enclosure near Cedar Creek on the Tillamook Burn, Tillamook Co., black-tailed deer used the red huckleberry (*Vaccinium parvifolium*)–salal (*Gaultheria shallon*) community as much as or more than expected on the basis of its occurrence in all months, the red alder (*Alnus rubra*)–thimbleberry (*Rubus parviflorus*) community in 10 of 12 months, the big-leafed maple (*Acer macrophyllum*)–snowberry (*Symphoricarpos mollis*) community in 8 of 12 months, and the vine maple (*Acer circinatum*)–sword-fern (*Polystichum munitum*) community in 7 of 12 months. The vine maple–sword-fern community occupied 52% of the area; the other three communities occupied ≤10.5% of the area. Black-tailed deer avoided the bracken fern (*Pteridium aquilinum*)-deervetch (*Lotus crassifolius*) and thimbleberry-starflower (*Trientalis latifolia*) communities (Miller, 1968). These communities occupied 14 and 13% of the area, respectively.

Like mule deer, black-tailed deer probably were fairly scarce when early explorers came to Oregon. Members of the Lewis and Clark Expedition were hard pressed to find sufficient game to survive during the winter that they spent at Fort Clatsop, Clatsop Co. "Deer and beaver were rarely found" (Bakeless, 1947:297). Bailey (1936:87) indicated that these and other early explorers made "no mention of excessive abundance of deer in western Oregon." He attributed the low numbers to "greater abundance of wolves, mountain lions, and other predatory animals"; however, it seems more likely that the vast tracks of mature forests did not provide sufficient resources to support large populations of black-tailed deer. Bailey (1936:87) believed that

with settlement, deer were "crowded from much of their old range." More likely, habitats for deer actually improved with onset of opening of the forests for agriculture and lumber, but unrestricted harvest of deer kept populations low during the era when the philosophy of unlimited wildlife resources was prevalent. An era of protectionism in the early 20th century followed by increasingly larger portions of forest lands being returned to early stages of the sere resulted in expanding numbers of deer. Estimates of numbers (largely speculative) of black-tailed deer on national forests in Oregon increased from 39,360 in 1926 to 44,825 in 1931, then declined to 35,675 by 1933 (Bailey, 1936). In 1950, estimates of numbers in those areas remained at ≅36,000, but in only 13 years increased by >5 times (Connolly, 1981). Annual estimates of abundance (although also speculative) for black-tailed deer in all of western Oregon, with small perturbations, increased from 200,000 in 1950 to 650,000 in 1961, declined to 480,000 in the mid-1960s, then again increased to 686,000 in the early 1970s before declining to 650,000 in 1976 (Connolly, 1981). We know of no more recent estimates. As vegetation on land now in early stages of the sere develops into more mature forest communities, and if timber harvests continue to be reduced and fire-suppression activities are maintained, it seems likely that populations of black-tailed deer in western Oregon will decline.

Annual sport harvests of black-tailed deer (Fig. 13-7b) since 1952 ranged from slightly less than 25,000 to nearly 66,000. Harvests of black-tailed deer underwent perturbations similar to those that mule deer exhibited during the late 1960s and 1970s. We suspect that some of the parallelism was the result of similar regulations as well as possible changes in abundance. In recent years, annual harvests have fluctuated between 38,000 and 54,000; they have not shown the continued decline evident in harvest records of mule deer (Fig. 13-7).

From the large number of herd-composition counts and population-trend counts (number observed/unit distance traveled) reported in annual *Big game statistics* by the Oregon Department of Fish and Wildlife, we suspect that statistics of this type frequently are used in management decisions regarding harvest regulations for black-tailed deer and several other ungulates in Oregon. Application of such counts for this purpose has received considerable criticism from both theoretical and practical perspectives (Caughley, 1974; McCullough, 1993, 1994; McCullough and Hirth, 1988; McCullough et al., 1994).

Diet.—On the Cedar Creek enclosure the most-sought food of black-tailed deer in both summer and winter was Pacific blackberry (*Rubus ursinus*—Crouch, 1966, 1968b). In 1962–1963, leaves of this plant provided 556.9 kg/ha (19.3% of the total forage available) of forage for deer in September, but 195.0 kg/ha in December and only 16.8 kg/ha in March. Much of the decline in forage provided by blackberry was the result of weathering of leaves during autumn and winter (Crouch, 1968b). Thimbleberry was the primary food in summer and early autumn (Miller, 1968). Other highly selected foods were salal, red huckleberry, and cascara (*Rhamnus purshiana*). Red alder, hazel (*Corylus cornuta*), and vine maple were browsed only when green herbage was not available (Crouch, 1968b). Hines (1973:12) reported that "Thimbleberry and vine maple were the two most important browse species in summer. Each supplied relatively large amounts of leafage which was preferred by deer." However, in winter, "dormant twigs on vine maple, California hazel, creambush [*Holodiscus discolor*], and Oregongrape [*Berberis nervosa*] were rarely browsed." Hines (1973) indicated that willow (*Salix*), currant (*Ribes lacustre*), and cascara were selected strongly by black-tailed deer but that these plants were scattered and did not provide large amounts of browse for deer on the Cedar Creek enclosure.

An index derived from intensity of use of a plant species and frequency of occurrence of that species in permanent transects sampled repeatedly in 1940–1941 on the Tillamook Burn indicated that thimbleberry, salmonberry, elderberry (*Sambucus racemosa*), fireweed (*Epilobium angustifolium*), vine maple, figwort (*Scrophularia californica*), sword-fern, and Pacific blackberry were the most sought (in declining order) forage items on an annual basis. Seasonally, however, thimbleberry ranked first except during winter, when sword-fern was the most sought forage; salmonberry ranked second in autumn and spring, but salal ranked second in winter and vine maple ranked second in summer; and elderberry, huckleberry, fireweed, and salmonberry ranked third during autumn, winter, spring, and summer, respectively (Chatelain, 1947).

Planted Douglas-fir (*Pseudotsuga menziesii*) seedlings were browsed when green forage became unavailable commencing with the first snowfall. Crouch (1968b) considered browsing on the planted Douglas-fir by deer to cause little mortality and to result only in loss of the growth removed by the deer. Hines and Land (1974) reported that the proportion of Douglas-fir seedlings browsed upon by black-tailed deer on the same area differed by plant community and by density of deer; the ranges were 14.7–37.8% when densities were 11.2–23.9 deer/km² and 32.9–69.4% when densities were 27.8–44.4 deer/km². Douglas-fir in the bigleaf maple–snowberry community was browsed most heavily at both densities. Forty-five percent of the Douglas-fir trees died during the study, but some mortality was attributable to factors other than browsing.

In suburban areas, black-tailed deer often forage on landscape plantings and gardens; some homeowners plant shrubs and trees less palatable to deer thereby continue to enjoy the presence of deer without loss of valuable plantings (Happe, 1983).

Reproduction, Ontogeny, and Mortality.—Black-tailed deer are "polytocous, multiovular, spontaneous ovulators that breed during the autumn" (Thomas and Cowan, 1975:261). Mating behavior is promiscuous; copulation probably is mostly at night (Dasmann and Taber, 1956). Females of reproductive age exhibit cyclic development and atresia of follicles several weeks before first ovulation. First ovulation never results in an implanted embryo even when spermatozoa are present and ova are penetrated. Females ovulate a second time 8–9 days later and most individuals conceive at this estrus. However, if conception does not occur, 8–9-day follicular cycles are repeated, with estrus occurring at each second or third cycle (intervals of 23–29 days—Thomas and Cowan, 1975). Four captive females prevented from mating exhibited behavioral estrus five or six times at intervals of 18–30 days during periods of 107–124 days (Wong and

Parker, 1988). Corpora lutea of pregnancy become brownish corpora albicantia that persist through the life of the female; counts of these ovarian bodies commonly are used to estimate productivity.

On the McDonald-Dunn Forest (formerly McDonald State Forest), Benton Co., female black-tailed deer of reproductive age in 1968–1969 commenced to ovulate in mid-October and by 10 November 46 (93.9%) of 49 had ovulated. Based on rate of fetal growth, 14 November and 4 November were estimated dates of conception for the 2 years, respectively (Jordan and Vohs, 1976). However, females collected in spring 1970 were in better physical condition than those collected in spring 1969, despite similar dates of ovulation the previous autumns. Jordan and Vohs (1976) suggested that fetuses grew faster in 1969–1970 than in 1968–1969 and that actual dates of breeding for the 2 years probably were similar.

During a study in the Cedar Creek enclosure in 1964–1970, no female <23 months old was known to have given birth. In the wild, females usually bred for the first time when 17–18 months old and usually gave birth to one young when 24–25 months of age; older females gave birth to twins 64% of the time (Hines, 1975). However, in captivity, females commonly breed when ≅6 months old and give birth when ≅12 months of age (Mueller and Sadleir, 1975), sometimes to twins (Mueller and Sadleir, 1977). Eleven ≤3-year-old females in the Cedar Creek enclosure whose productivity was followed for 5–7 years (67 potential pregnancies) produced an average of 1.58 young/year each. Five other females with histories of producing young with morphological anomalies produced an average of 1.1 young/ year. No decline in fecundity with age of maternal females was noted; in 1970, five females ≥9 years old produced an average of 1.8 young (Hines, 1975). Hines (1975) considered the gestation period in black-tailed deer to be ≅198 days; Golley (1957) reported an average gestation of 203 days (range, 199–207; $n = 5$).

None of 20 young females of the year obtained in November on the McDonald-Dunn Forest contained corpora lutea (Jordan and Vohs, 1976). Among adult females only five (25.0%) of 20 collected in October, but 62 (91.2%) of 68 collected in November, contained corpora lutea. In November, ovaries of females ≅18 months old contained an average of 0.79 (reported incorrectly as 0.77—Jordan and Vohs, 1976:110, table 1) corpora lutea/female ($n = 24$), whereas those of adults contained an average of 1.76 ($n = 76$; summed incorrectly in original). Average number of fetuses/female collected in spring was 0.88 ($n = 8$) for females approaching 2 years of age and 1.61 ($n = 28$) for older females. Jordan (1976) indicated that only two (5.5%) of the 36 females of reproductive age were not pregnant. He also claimed both were adults (one each in age-classes 4 and 7), but to obtain an average of 0.88 fetuses/female for eight yearling females, at least one of the yearling females must be nulliparous. However, Jordan (1973) listed 0.67 fetuses/female for four yearlings in 1969 and 1.50 fetuses/female for the two yearlings collected in 1970. Thus, one yearling female must have been pregnant with twins in 1970 and two yearling females must have been nulliparous in 1969.

In Siskiyou Co., California, near the Oregon-California border, 13 adult female black-tailed deer bred from the 1st week in November to the 1st week in January with peaks in the 1st and 4th weeks of December (Bischoff, 1957). Bischoff (1958) reported 20 (91%) of 22 adult females were pregnant with an average of 1.7 fetuses. He claimed that 50% had singleton fetuses, 45% had twins, and 5% had triplets, an obvious miscalculation ($\bar{X} = 1.55$ fetuses).

In an area in northern California where levels of selenium in blood of black-tailed deer are considered deficient, supplementing the diet with two boluses (designed for domestic sheep) containing 9.5 g iron and 0.5 g selenium resulted in recruitment of 2.6 times more young into the population (Flueck, 1994). Selenium levels in mule deer and other wild ungulates in Washington were considered deficient (Fielder, 1986; Hein et al., 1994). Flueck (1994) believed that selenium available to animals has undergone a recent decline by acidification of soils, use of plant fertilizers, and intensive foraging by ruminants. White-muscle disease, a selenium-deficiency response, is common in livestock in the geographic region occupied by black-tailed deer in Oregon (Muth and Allaway, 1963:1380, fig. 1).

In the Cedar Creek enclosure, 42 (42.9%) of 98 births occurred during the first 7 days of June; 34 (34.7%) occurred in the previous 2 weeks and 17 (17.3%) were in the following 2 weeks. Only five (5.1%) young were born later (Hines, 1975). Females 2 years old were last to give birth. Birthing sites were more common on south or west slopes in vine maple–sword-fern communities with numerous trails created by earlier logging operations (Hines, 1975).

Miller (1965) described the activity and behavior of a pregnant female for ≅1 h before to nearly 2 h after giving birth to a singleton male neonate. The female fed for 15 min, bedded and ruminated for 41 min, and labored for only 7 min before the young was born. The female began to lick the neonate within 1 min of its birth and continued to do so for nearly 1 h before the neonate stood for the first time. The placenta was consumed by the female and the neonate commenced to nurse within 1 h of birth. Miller (1970a) reported four females all produced healthy offspring after experiencing severe falls within 1 month of parturition.

Within a few hours of birth, neonates exhibit no fear of humans or dogs and after being handled commonly follow handlers. Undisturbed neonates usually are attended by the maternal female with reoccurring bouts of grooming, nursing, and resting (Hines, 1975). Although the maternal female commonly stands nearby, young are not attended until they demand attention by becoming restless or by bleating. Following disturbance, maternal females lead neonates to new bedding sites by withholding nursing as encouragement to move (Hines, 1975). Except immediately following birth, twins rarely bed together; separation usually occurs as a result of differences in stamina of twin neonates. Although the maternal female was responsible for selecting the birthing area, young seemed to select their own bedding sites; sometimes after the maternal female left, the young moved a short distance to a new bedding site (Hines, 1975). Hines (1975) indicated that his observations regarding initiation of maternal female:young relations and the young selecting its bedding site were in contradiction to reports by Einarsen (1956) and Cowan (1956), respectively.

As young grow older they become bolder and more active; maternal females move greater distances from them. By 3 weeks of age, bedded young are alert and prepared to flee. They also engage in vigorous bouts of exercise while unguarded by the maternal female, likely increasing their vulnerability to predators (Hines, 1975). Association of young with conspecifics is limited exclusively to the maternal female, but after 4–6 weeks the females exhibit less agonistic behavior toward offspring of earlier years and a matriarchal hierarchy is re-formed that includes the new young (Hines, 1975). Hines (1975) observed little agonistic behavior directed toward young before parturition the following year, and these young are last to leave the parturient female.

Body mass of 141 neonates captured in 1965–1969 for marking in the Cedar Creek enclosure averaged 2.8 kg; annual averages ranged from 2.6 to 3.2 kg, the lowest following a severe winter during which pregnant females were subjected to unusual stress (Hines, 1975). Although somewhat variable, protein and fat content of milk tends to increase and lactose content tends to decrease during lactation; some of the variation may be related to the age of the maternal female (Mueller and Sadleir, 1977).

Mortality in the Cedar Creek enclosure was biased heavily toward young individuals; 65.2% of the losses in winter were among deer <1 year of age (Hines, 1975). About 75% of fawns died before 1 year of age; losses among young were not biased toward males as reported among black-tailed deer in California (Taber and Dasmann, 1954). However, mortality in the enclosure among yearlings and adults was strongly biased toward males. An average of 25% of young died during the summer following their birth and 44% died during winter. Losses of young in summer were believed to be related largely to nutrition of maternal females during pregnancy; winter losses were strongly related to the duration of snow cover (Hines, 1975).

Hines (1975) suggested that to maintain the herd on the Cedar Creek enclosure at "carrying capacity" it would have been necessary to reduce the population to ≅29 animals/km², which would have required harvest of nearly one-third of the population. He further emphasized that on the ≅52-km² McDonald-Dunn Forest, harvest of an average of 5.4 deer/km² for 17 years had not reduced subsequent harvests. The 19-year average harvest of 34/ha annually for the same area reported by Jordan and Vohs (1976) obviously was a lapsus. Although the black-tailed deer is a popular game species in Oregon, annual harvests (Fig. 13-7b) are only about one-half that possible *if* deer populations west of the Cascade Range average one-fourth as dense as in the McDonald-Dunn Forest.

Habits.—Black-tailed deer tend to be secretive and often rely on stealth or concealment rather than speed as a means of escape (Lindzey, 1943). Nevertheless, black-tailed deer commonly flush when the observer remains immobile for a few minutes. The thick tangle of shrubs, vines, and small trees in early seral communities that support populations of these deer is both supportive of and conducive to concealment rather than flight. The stotting gait, less adapted to dense vegetation, is observed less frequently in black-tailed deer than in mule deer.

Activity periods reportedly were influenced strongly by temperature (McCullough, 1960); black-tailed deer were inactive when temperatures were below -7°C, they sought shade when temperatures were >18°C, and they did not become active until dusk when daytime temperatures were ≥32°C (Hines, 1975; Miller, 1970b). In summer, black-tailed deer are more active on nights following high daytime temperatures (McCullough, 1960). Activity also is affected by phase of the lunar cycle, precipitation, and wind, with greater activity with increasing moonlight, immediately after a brief rainstorm, and on leeward slopes (Hines, 1975; Miller, 1970b). Adult and 2-year-old males tend to spend a greater part of the year at higher elevations than other deer (Lindzey, 1943; Miller, 1970b). McCullough (1960) reported that black-tailed deer favored north-facing slopes in summer, but did not exhibit differential use of north or south slopes at other times.

Black-tailed deer that occupy higher elevations during summer migrate to lower elevations, where they commonly spend the winter in logged or burned areas on exposed south slopes or in bottomlands (McCullough, 1960). On the H. J. Andrews Experimental Forest in eastern Lane and Linn counties, autumn migrations commenced in September before onset of inclement weather (McCullough, 1960). Autumn migrations in some regions are spectacular, with large numbers of deer moving en masse. Although a major function of the migrations is to avoid deep snow and to reside near accessible forage, neither snowfall, rainfall, nor temperature seems to be the stimulus that initiates autumn migrations. Onset of both spring and autumn migrations seems tied most closely to minimal levels of relative humidity (McCullough, 1964). Spring migrations in the Cascade Range commence in late March and are completed by May. Many individuals return to the home ranges they occupied the previous summer (McCullough, 1960).

Miller (1970b) calculated average annual home-range areas for various sex and age-classes of black-tailed deer in the ≅138-ha Cedar Creek enclosure as follows: adult females, 69.6 ha ($n = 8$); adult males, 100.4 ha ($n = 2$); 2-year-old females, 76.9 ha ($n = 4$); 2-year-old males, 98.8 ha ($n = 3$); yearling females, 38.9 ha ($n = 2$); yearling males, 59.9 ha ($n = 2$); young-of-the-year females, 39.3 ha ($n = 3$); and young-of-the-year males, 55.5 ha ($n = 3$). Average monthly home-range areas ranged from 12.8 to 39.6% of average annual home-range areas. Females with male offspring had larger home ranges than those with female offspring. Miller (1970b) claimed that the enclosure fence did not influence the size of home ranges as individuals rarely traveled along it. Black-tailed deer elsewhere also tend to restrict their movements to relatively small areas except during migrations (Dasmann and Taber, 1956; Zwickel et al., 1953). Hunting and other disturbances do not seem to cause these deer to extend or leave their home ranges, but actually may cause them to restrict their movements (Dasmann and Taber, 1956). Based on only one or two individuals, minimum-area home ranges of deer in suburban habitats were smaller than in forested areas or in the fringe between residential areas and forest (Happe, 1983). Deer in suburban areas tended to use areas near houses where dogs were not permitted to roam free.

Chinn (1972) found significant differences in proportions of three transferrin genotypes in samples of black-tailed deer from the McDonald-Dunn Forest and the William L. Finley National Wildlife Refuge only 21 km to the south, suggesting that genetically isolated populations may exist within short distances of each other. The relatively sedentary nature of black-tailed deer doubtlessly contributes to the differentiation.

Many polygynous species of ungulates are known to exhibit sexual segregation (Main and Coblentz, 1990). Female groups consist of a maternal female, the offspring of the year, female offspring of the previous year (Dasmann and Taber, 1956), and occasionally some female offspring produced earlier. Adult females may leave groups temporarily during the breeding season, and the groups tend to disband during the birthing period. Males of all ages except those born in the previous breeding season associate in small groups; most such groups disband during the autumn breeding season. Movements of the groups tend to be limited. The fundamental cause of sexual segregation remains much debated. Weckerly (1993) tested the oft-cited body-size hypothesis, in which intersexual resource partitioning is considered to be related to differences in body size between the sexes. He found that black-tailed deer in California indeed segregated by sex and that the spatial segregation was not a means by which intersexual competition for resources was reduced, but was (p. 491) "a consequence of each sex achieving reproductive success in a different fashion."

Miller (1971) observed mutual grooming between both sexes and among all age-classes of black-tailed deer. Grooming commences with the maternal female licking the neonate to remove the birth membranes and to dry the young. This serves to develop bonding that fortifies ties later in life. Miller (1971) found that among pairs of adult deer the dominant individual initiated the grooming session. In addition to the mechanical and comfort functions of removing debris and loose hair, mutual grooming likely reinforces social bonds that, in turn, strengthen unity of groups (Miller, 1971).

Most agonistic social interactions consist of no more than a threat, but occasionally these deer engage in combat involving striking with the front feet. Miller (1974) reported seven agonistic acts, including chases, bites, and kicks. Males, during the rut, engage in combat with the antlers (Dasmann and Taber, 1956). Dominance conflicts are common among groups of males but not among groups of females. Dominant females exhibit leadership of their groups, but dominant males do not. Adult males tend to be dominant over females (Dasmann and Tabler, 1956). Miller (1974) described a few instances in which groups of deer defended portions of their home ranges against intrusion by other groups or individuals.

Pheromones (information-bearing odoriferous compounds) produced by glands on the tarsus, metatarsus, and forehead, and those contained in urine, play significant roles in communication among black-tailed deer (Müller-Schwarze, 1971). Products of the tarsal glands function in mutual recognition. Deer sniff the tarsal glands of other individuals occasionally, seemingly to reinforce sociality of groups. Most hunters are familiar with the garliclike odor of products of the metatarsal glands produced to communicate alarm when deer are flushed or chased. Deer rub the forehead on tips of dry vegetation to deposit compounds to delineate occupied portions of the environment. Urine, either alone or combined with products of the tarsal glands, is involved in familiarization with the environment and may be used to attract or repel other individuals (Müller-Schwarze, 1971). Compounds also are produced by the interdigital glands, and deer sometimes sniff footprints of other deer, but do not seem to react in a specific manner (Müller-Schwarze, 1971).

Remarks.—*Odocoileus h. columbianus* occasionally hybridizes with *O. virginianus leucurus* in western Oregon. A discussion of interspecific hybridization in this genus is provided in the account of *O. virginianus*.

Odocoileus virginianus (Zimmermann)
White-tailed Deer

1780. *Dama virginiana* Zimmermann, 2:129.
1829. *Cervus leucurus* Douglas, 4:330.
1898. *Odocoileus leucurus* Seton-Thompson, 51:286.
1915. *Odocoileus virginianus leucurus* Lydekker, 4:162.
1929. *Odocoileus virginianus macrourus* Taylor and Shaw, 2:30.
1932a. *Odocoileus virginianus ochrourus* Bailey, 45:43.

Description.—The white-tailed deer (Plate XXX) is the smallest cervid in Oregon (Table 13-1). In addition to the characteristics shared with all members of the family, white-tailed deer possess metatarsal glands ≤25 mm long; a relatively long tail, brown dorsally with a white fringe, white ventrally, and without a basal constriction; and in males, antlers with tines arising from a main beam (Cowan, 1936a). The skull of the white-tailed deer (Fig. 13-8) is long and narrow, with complete orbits; the lacrimal pits are broad and shallow.

In winter, the pelage is a dark buffy-gray and consists of relatively long, thick, and somewhat brittle hairs; in summer, the pelage is lighter with more tawny tones and is shorter and thinner (Cowan, 1936a; Smith, 1991). The midline of the dorsum is darker and the face lighter; in males, the patch between the antlers is darker (Cowan, 1936a). Pelage markings include white on the venter, throat, muzzle, and lower lip, and around the openings of the metatarsal and interdigital glands; a black spot on the lower lip; and a grayish-white eye ring. Young white-tailed deer have two rows of 0.6–1.3-cm white spots totalling 272–342 on each side but acquire the adult-type pelage in late summer or early autumn (Smith, 1991).

Livingston (1987) was able to classify correctly 84–92% of white-tailed deer and mule deer (*O. hemionus*) by use of discriminant functions derived from lengths or widths (or both) of mandibular molariform teeth. Widths of m2 and m3, and length of p4 contributed greatly to the functions; mandibles lacking both m2 and m3 could not be classified reliably.

Distribution.—White-tailed deer occur from southwestern Northwest Territories and the southern half of the remaining southern tier of Canadian provinces south through most of the United States except for parts of the arid Southwest. The species also occurs through Central America into northern South America (Fig. 13-9; Hall, 1981; Smith, 1991).

From investigation of archaeological sites in Washington and Oregon, Livingston (1987) claimed that white-tailed deer were more abundant and more widely distributed in the region prehistorically than at settlement. However, she identified white-tailed deer at only three of 10 sites in Oregon from which she examined deer remains; all three were

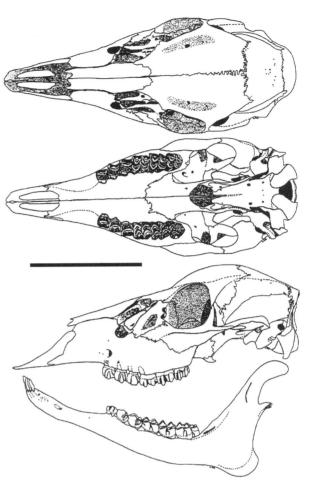

Fig. 13-8. Dorsal, ventral, and lateral views of the cranium, and lateral view of the dentary of the white-tailed deer, *Odocoileus virginianus* (OSUFW 1140). Scale bar equals 10 cm.

along the Columbia River and all were depicted as occurring within the historic range (p. 650, fig. 1).

According to records obtained from journals of early explorers and travelers through the region, white-tailed deer in western Oregon at and shortly after settlement occurred along the Columbia River from The Dalles, Wasco Co., to Astoria, Clatsop Co., and south through the fire-maintained prairies of the Willamette Valley to the Umpqua River valley near Roseburg, Douglas Co. (Cowan, 1936a; Gavin, 1978, 1979, 1984). We believe that Bailey (1936:90, fig. 12), Crews (1939:plate I), Smith (1985:245, fig. 1), and Livingston (1987:650, fig. 1) were in error in depicting the range to include the Coast Range north of the Umpqua River. Seemingly, Bailey (1936) included this region because of the statement in T. C. Peale's journal that white-tailed deer did not occur farther south than the Umpqua River and thirdhand hearsay via Ned Hollister from a hunter (p. 91) who "thought there were still a few. . . over toward the coast in Lincoln County." The other authors presumably followed Bailey (1936), as we found no other evidence for the species occurring west of the eastern foothills of the Coast Range except along the Columbia River. W. P. Smith (pers. comm., 19 July 1994), who studied white-tailed deer in Douglas Co. (Smith, 1982, 1985), agreed that pristine habitats of relatively narrow river valleys in the Coast Range north of the Umpqua River likely

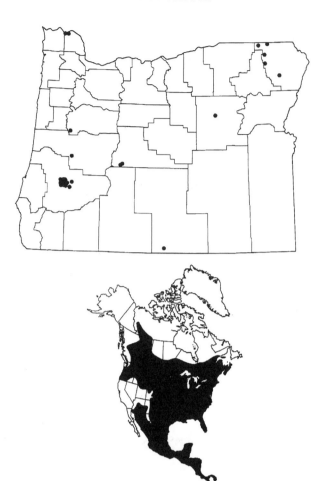

Fig. 13-9. Localities in Oregon from which museum specimens of the white-tailed deer, *Odocoileus virginianus*, were examined (listed in Appendix, p. 538), and the distribution of the white-tailed deer in North America. (Range map redrawn from Hall, 1981:1095, map 537, with modifications.)

were unsuited to the species.

At present, white-tailed deer in western Oregon occur as remnant populations in bottomlands in Clatsop and Columbia counties near Westport; on Tenasillahe Island, Clatsop Co.; on Wallace Island, Columbia Co.; and along the Umpqua River in Douglas Co. (Fig. 13-9; Gavin, 1978, 1984; Smith, 1985). Smith (1985) considered the geographic range to encompass 1,199 km² in the Douglas Co. region.

Bailey (1936:90, fig. 12) described and depicted the original distribution of white-tailed deer in eastern Oregon to include all of that portion of the state except the southern two-thirds of Lake, Harney, and Malheur counties. However, his records included sight-record accounts of early travelers in addition to identifications based solely on antlers. We consider the present range based on skulls collected nearly a century ago no less reliable (Fig. 13-9), although probably not including all of the area occupied by the species at settlement. Rohweder (1984:10) depicted the geographic range of the species in northeastern Oregon based on "recent whitetail sitings [*sic*]" as encompassing Baker, Grant, Morrow, Umatilla, Union, and Wallowa counties. He indicated that white-tailed deer were expanding their range to the south. Larrison (1970:182) claimed that the white-tailed deer was introduced "into the Blue Mts. [of southeastern Washington] in 1938 and has multiplied

and spread out through the northern part of that region." Bailey (1936:92) indicated that in "the valley and stream bottoms of the Blue Mountain section [of Oregon] . . . there are evidently a few remaining." Thus, the impact of the nearby introduction in Washington on the population in northeastern Oregon is problematic.

The genus *Odocoileus* first appeared in the Pliocene fossil record 4.2–3.2 million years ago (Savage and Russell, 1983), and *O. virginianus* appeared in later Pliocene fossil deposits over much of the historic range of the species (Kurtén and Anderson, 1980).

Geographic Variation.—Two of the 38 recognized subspecies (Smith, 1991) of white-tailed deer occur in Oregon: *O. v. leucurus* as two disjunct populations west of the Cascade Range and *O. v. ochrourus* in areas east of the Cascade Range (Hall, 1981). Gavin and May (1988:8), based on an analysis of allozymes, indicated that the population in Douglas Co. was not subspecifically distinct from *O. v. ochrourus*, but they were unsure whether those along the Columbia River were sufficiently distinct from *O. v. ochrourus* to warrant subspecific designation. On the basis of their data and reference to specimens (Cowan, 1936a) collected by S. J. Jewett (actually one was collected by O. J. Murie) at Davis Creek, Deschutes Co., Gavin and May (1988:7) suggested that the white-tailed deer in Douglas Co. possibly "were encompassed by the original distribution of white-tailed deer farther east." To us, these three specimens (MVZ 121231–121233) appear atypical of either race: the metatarsal glands are small, but neither they nor the interdigital glands are surrounded by white hairs; the lacrimal pits are broad and shallow; and the tails are relatively short, dark colored, and not surrounded by a fringe of white hairs.

Gavin and May (1988:9) further claimed that including the Douglas Co. white-tailed deer with those east of the Cascade Range (*O. v. ochrourus*) would necessitate "a new taxonomic description . . . using specimens from the lower Columbia River because the type specimen for the current designation [*O. v. leucurus*] was collected from Douglas Co." Although localities provided by Douglas (1829:331–332) with the description of the taxon are not precise by modern standards ("districts adjoining the river Columbia, more especially in the fertile Prairies of the *Cowalidske* and *Multnomah* River, within one hundred miles of the Western Ocean" and "from the confluence of the river Columbia with the Pacific"), they clearly do not include Douglas Co. Other workers have taken somewhat lesser liberties by listing the type locality as "Lower Columbia River [=Willamette River Valley]" (Hall, 1981:1093; Miller and Kellogg, 1955:802) and "Falls of the Willamette and mouth of the Columbia River, Oreg. (description based on specimens from both places)" (Bailey, 1936:89). Although we remain to be convinced of the need to do so, including the white-tailed deer in Douglas Co. with the race east of the Cascade Range would not require publication of a description of the white-tailed deer occupying habitats in the vicinity of the Columbia River.

Habitat and Density.—The Columbian White-tailed Deer National Wildlife Refuge was established on 2,105 ha of prime habitat on islands in the Columbia River and on the mainland near Cathlamet, Washington, in 1972 (Gavin, 1984). The area is only 3 m above sea level, but was diked and ditched to avoid tidal flooding and to permit cropping before the land became a refuge. Much of the information regarding the biology of the species west of the Cascade Range was derived from studies conducted on this refuge and on private lands along the North Umpqua River near Roseburg, Douglas Co.

On the mainland portion of the Columbian White-tailed Deer National Wildlife Refuge, these deer most often use habitats that provide both food and cover; they tend to be grazers rather than browsers. Vegetation ≥70 cm tall adjacent to forage species tends to attract white-tailed deer. White-tailed deer exhibit seasonality in habitat use, with sitka spruce (*Picea sitchensis*) with a grass understory selected most frequently except in summer, when open stands of western red cedar (*Thuja plicata*), red alder (*Alnus rubra*), and sitka spruce with a grass-shrub understory are used (Suring and Vohs, 1979). Areas used as pastures were invaded by Canada thistle (*Cirsium arvense*) and common rush (*Juncus effusus*) after the area became a refuge; these areas were used relatively extensively by deer (Suring and Vohs, 1979). Use of pasture lands was low despite the availability of forage there throughout the year; paucity of cover was considered the factor responsible for deer avoiding much of this habitat type except where adjacent to types that provided cover. Other communities that provided cover but little forage and those that did not provide adequate cover also were avoided (Suring and Vohs, 1979).

Gavin (1979) conducted tests of the effect of grazing and haying on use of pastures by Columbian white-tailed deer on the refuge. During summer, pastures previously not selected differentially by deer tended to be avoided when cattle were present and for 2–3 months after cattle were removed. In autumn and winter, however, pastures subjected to grazing by cattle during the previous summer were selected by deer over those from which forage had been removed by haying. Gavin (1979) observed that plant diversity tended to be reduced by haying and that the vegetation on hayed fields tended to become rank in autumn and likely less palatable and less nutritious for deer. However, he later indicated (Gavin, 1984) that both numbers of cattle and area grazed had been reduced and that a combination of heavy grazing and mowing of pastures seemed to provide the greatest diversity of palatable forage.

White-tailed deer along the Umpqua River in Douglas Co. occupy predominantly oak (*Quercus*) woodland and are almost always associated with riparian systems (Smith, 1985). They use grass-shrub, oak-savanna, open oak, closed oak, riparian, and conifer associations more frequently than expected and grassland, oak-conifer, madrone (*Arbutus menziesii*), and oak-madrone associations less frequently than expected (Smith, 1982). Use of grassland is correlated positively with production of biomass in spring and with precipitation (and "green-up") in autumn. Young-of-the-year use woodland and brushland habitats more frequently than yearlings and adults; yearlings use grassland more often but grass-shrub less often than adults. Adult males occur less frequently in grass-shrub and more frequently in conifer habitats than adult females (Smith, 1982).

Comparatively little is known regarding habitats used by white-tailed deer in northeastern Oregon. V. Coggins (pers. comm., 31 May 1994), a long-time state wildlife biologist in that region, indicated that he had seen white-tailed deer in most habitats except those on steep slopes or at high elevations. He indicated these deer made much use

of grasslands, particularly those taken out of production through conservation-reserve programs. In northern Idaho and northeastern Washington, a region similar to northeastern Oregon, white-tailed deer in winter used areas with pole-sized timber (mostly lodgepole pine, *Pinus contorta*, and Douglas-fir, *Pseudotsuga menziesii*) when snow depths were <30 cm, but moved to old-growth or mature stands of western red cedar and western hemlock (*Tsuga heterophylla*) with depauperate understories when snow depth was ≥40 cm (Pauley et al., 1993). V. Coggins (pers. comm., 31 May 1994) indicated that he had observed white-tailed deer in northeastern Oregon also to move to the protection of mature stands when snow became deep.

Scheffer (1940), in describing the newly rediscovered population of white-tailed deer along the Columbia River, estimated 100–200 (later set at 150) white-tailed deer occupied diked lands in Columbia and Clatsop counties. More recent population estimates are eight–12 white-tailed deer on Karlson Island, Clatsop Co.; 70–80 in the Wallace Island, Columbia Co.-Westport, Clatsop Co., region; and 30–50 on Tenasillahe Island, Clatsop Co. (Smith, 1985). Scheffer (1940) recorded only two–five individuals on Tenasillahe Island, Clatsop Co., an area that he listed as being in Washington.

On the 7.9-km² mainland portion of the Columbian White-tailed Deer National Wildlife Refuge, Washington, annual estimates of density in November were 27.1, 22.8, 20.8, and 25.6/km² in 1974–1977, respectively (Gavin, 1979). Gavin (1984) reported others had estimated 26.8, 24.2, and 20.1/km² during 1978–1980, respectively. Gavin et al. (1984) considered the population moderately dense and at stable equilibrium. They attributed the stability to interactions of the relatively stable habitat, absence of dispersal, low production of offspring, a nearly unlimited supply of high-quality food resulting from management practices, foot-rot infections, and predation of young by coyotes (*Canis latrans*). However, Gavin and May (1988) reported that density estimated by the same methods in 1985 was 60.8/km², more than twice earlier estimates.

Jewett (1914:5) provided hearsay evidence to support his contention that white-tailed deer formerly were common in the "Willamette Valley foothills." He indicated that a few remained along the North Umpqua River in Douglas Co. and in the Davis Lake region of Crook Co. (*sic; Deschutes Co. was created from part of Crook Co. 13 December 1916*—McArthur, 1992). He strongly advocated protection to prevent extirpation. In 1927, the Oregon Legislature established a ≅78.9-km² refuge on private land for the species along the North Umpqua River in Douglas Co. (Crews, 1939). Crews (1939) estimated 200–300 (2.5–3.8/km²) white-tailed deer on the refuge in 1938; he indicated additional white-tailed deer resided on land adjacent to the refuge. Crews (1939) also indicated that the population had not increased significantly since the refuge was established and concluded that protection alone was not sufficient to cause an increase in numbers. The refuge was dissolved in 1952 and hunting of white-tailed deer in the area again was permitted (Gavin, 1984). With inclusion of the Columbian white-tailed deer under the provisions of the Endangered Species Act, establishment of the systematic relationships of deer in the two major populations became paramount. However, in 1978, a decision was made without further study to afford the same protection to the Douglas Co. population as afforded the population along the Columbia River (Gavin, 1984). Smith (1982) estimated a density of 22.9–27.0/km² white-tailed deer (628–740 animals) on his 27.45-km² study area centered on the former refuge in Douglas Co. His expanded estimate, which included the entire range of the species in that area (1,199 km²), was ≅3,000 white-tailed deer (Smith, 1985). He attributed the 15-fold increase in area occupied and the 10-fold increase in numbers since the work of Crews (1939) to the refuge in preventing extirpation of the species during 1928–1952, to improvement of habitat resulting from agricultural practices that improved pastures, to clear-cutting of adjacent coniferous forest that created openings, and to intense predator control, initiated to enhance livestock production, that essentially eliminated loss of neonates to coyotes.

Comparable data for white-tailed deer populations in eastern Oregon are not available. Crews (1939) indicated that recent U.S. Forest Service annual game censuses included estimates of 50 white-tailed deer near Davis Lake, Deschutes Co. (Deschutes National Forest), 10 near Rudio and Cottonwood creeks, Grant Co. (Malheur National Forest), and 20 near Clackamas Lake, Roaring River, and Linney Creek, Clackamas Co., and Happy Ridge and White River, Wasco Co. (Mt. Hood National Forest). Rohweder (1984) indicated that the species was common in northeastern Oregon and that the range was expanding. V. Coggins (pers. comm., 31 May 1994) counted 301 white-tailed deer on his census routes in northeastern Oregon in 1992 and 223 in 1993. He believed that sighting of fewer animals in 1993 did not represent a decline in density, but more likely was caused by deer being less visible when he conducted his censuses in 1993.

Diet.—Scheffer (1940:276), under the subheading "Food Supply," listed several species each of trees, shrubs, and nonwoody plants known to occur in the "Columbia bottomlands" in which Columbian white-tailed deer had been found to occur. He indicated (p. 277) that "annuals and other succulent plants" were not recorded because his surveys were conducted in winter. He further stated (p. 277) that much of the land was cleared for agriculture and that "The grass of farm pastures is now an important item in the diet of the deer." Scheffer (1940:277) examined stomach contents of four deer in December–January from the same area and claimed that "all were at least half-full of soft grass and contained no traces of woody plants."

Similarly, Columbian white-tailed deer are considered by some recent researchers to feed largely by grazing on grasses and forbs rather than by browsing on woody plants (Gavin, 1979; Smith, 1982; Suring and Vohs, 1979). Gavin (1979) reported results similar to those of Scheffer (1940) for contents of stomachs of 33 deer collected throughout the year, although he provided no supporting data. Suring and Vohs (1979) reported that during 99% of 18,077 observations of feeding deer near Cathlamet, Washington, the deer were grazing. However, identification of plants by epidermal fragments in feces revealed that ≥25% of the diet of white-tailed deer on the same area consisted of browse species during September–February and as much as 50% in December (Dublin, 1980:33, fig. 6). These deer ate 28 of 42 species of trees and shrubs that occurred on the area, 17 of 29 species of grasses and grasslike plants,

and 55 of 73 species of forbs (Dublin, 1980). Smith (1982) reported that of feeding white-tailed deer observed along the Umpqua River, Douglas Co., ≥90% were grazing on grasses and forbs during November–May and >75% during the remainder of the year. We know of no analysis of foraging activity or materials ingested for white-tailed deer in northeastern Oregon. The comments of V. Coggins regarding use of grasslands by white-tailed deer in northeastern Oregon suggest that they may forage largely by grazing in that region as well. However, in northern Idaho, evergreen shrubs, including barberry (*Berberis repens*), are considered "important" winter-forage species (Pauley et al., 1993).

Reproduction, Ontogeny, and Mortality.—Based on small samples obtained in 1975–1976 on the Columbian White-tailed Deer National Wildlife Refuge in southwestern Washington, 100% of adult females, 70% of yearling females, and no young-of-the-year were reproductively active. Parturition on the refuge was largely in early to mid-June; a few females gave birth later in the summer, two as late as October. All adults for which reliable information was obtained (*n* = 5) produced twins (Gavin, 1979); comparable data for yearlings were not provided. Pregnant females were observed after 1 June, but young were captured on 9 June during 2 years and on 12 June during another year (Gavin, 1979). On the basis of the average gestation period (202 days—Smith, 1991), the rut probably occurs in late November–early December. In Douglas Co., Crews (1939) indicated that white-tailed deer bred mostly in November with a few breeding in October and December. Most young were born in June. Additional information regarding reproduction for populations in Douglas Co. is not available; no information regarding reproduction is available for the northeastern Oregon population.

Recruitment, measured on the refuge in November, was 0.75–0.85 young/female ≥3.5 years old, 0.36 young/female 2.5 years old, and 0 young/female 1.5 years old. There was a significant inverse relationship between recruitment and the estimate of density for the previous November (Gavin, 1979). However, mortality among young-of-the-year, not fecundity, was believed to regulate recruitment (Gavin, 1979). In Douglas Co., survival of marked young to the following spring was 27% in 1978, 47% in 1979, and ≥73% in 1980 (Smith, 1982). The increase in survival was attributed to a density-dependent relationship following loss of 24% of the herd during winter 1978–1979.

Based on studies conducted elsewhere, body mass, survival, and growth of neonate white-tailed deer are related to adequacy of nutrition; nursing commences immediately and neonates average gaining 0.2 kg/day (Smith, 1991). By 1 month they have tripled their body mass. Neonates have four lower incisors, commence to nibble solid food shortly, and are "functional ruminants" by 2 months of age (Smith, 1991:5). Although young commonly are weaned by 10 weeks of age, some nursing may continue into autumn. In some regions, some females may breed at 6–7 months of age, but breeding by young-of-the-year has not been recorded in Oregon populations studied. Information regarding development of young is not available for Oregon populations.

Young deer in Douglas Co. hide in the grassy glades and move only at the approach or call of the maternal female (Crews, 1939). Crews (1939) claimed that neonates were odorless and that domestic dogs could not detect them by olfaction. However, Smith (1982) used a dog with good success to locate and aid in capture of young deer for marking.

Of 137 deer that died of natural causes or were killed by automobiles ($n = 9$) on the Columbian White-tailed Deer National Wildlife Refuge in 1974–1977, 53 were young-of-the-year, 36 were adult males, 30 were adult females, 14 were yearling males, and 4 were yearling females (Gavin, 1979). Of the 53 young-of-the-year, 36 died in June–July, 8 in August–October, 6 in November–January, and 3 in February–April. Predation by coyotes was considered responsible for loss of a mimimum of 32% of 22 carcasses of young examined for puncture wounds on the refuge. Many of the deer on the refuge were affected by foot rot (necrobacillosis), a bacterial infection resulting from wet, muddy conditions in winter. Of 101 deer ≥1 year of age examined, 46.5% exhibited symptoms of the disease. Although the disease organism can invade internal organs and cause death directly, chronic infections reduce physical condition thereby contribute to death from other causes (Gavin, 1979).

Of 14 white-tailed deer carcasses examined by Crews (1939) on the former Douglas Co. refuge, six had died of gunshot wounds, three were heavily infested with botfly larvae (*Cuterebra*), one was entangled in a fence, and four died of unknown causes. In the same region, of 206 dead white-tailed deer, death of 25.7% was attributed to accidents involving motor vehicles, 25.3% to malnutrition, 2.9% to predation, 2.4% to diseases, 1.9% to entanglement in fences, and 41.8% to unknown causes (Smith, 1982). Severe winter weather, infestation by parasites, selenium deficiency in forage, and abrupt shifts in the diet may have been involved with malnutrition as mortality factors. The largest proportion of losses involving vehicular accidents occurred in July–September and possibly was related to an increase in movements by the deer resulting from a decline in availability of palatable forage during the summer drought (Smith, 1982). Of 196 carcasses for which sex and age were obtained, 51 were young-of-the-year, 7 were yearling females, 16 were yearling males, 78 were adult females, and 44 were adult males (Smith, 1982).

Habits.—White-tailed deer, when flushed, travel by graceful strides interspersed with leaps, apparently to survey surroundings for the most advantageous escape route. During flight, the tail is held erect and wagged gracefully from side to side; often the tail is visible after the remainder of the fleeing deer can no longer be seen (Crews, 1939). Olfactory and auditory capabilities are excellent; these deer distinguish their young by smelling the rump patch, and locate and differentiate food plants by odor. The ability to discriminate by sight is less developed for immobile subjects. Often, if an observer is downwind and motionless, a deer will pass close by without detecting the observer. Similarly, a bedded deer often will lie perfectly still and permit a potential observer to pass (Crews, 1939). In the relatively mild climate of Douglas Co., white-tailed deer do not yard and do not migrate (Crews, 1939).

In general, white-tailed deer tend to be active crepuscularly, but activity is affected by humidity, barometric pressure, human disturbance, and perhaps other environmental variables (Smith, 1982). During 17,872 observations on the Columbian White-tailed Deer National Wildlife Refuge, deer were feeding 79% of the time, resting 12%, moving 8%, and involved in social behaviors 1%. The deer fed more in summer (91% of the time), rested more in winter (20% of the time), moved more in spring (11% of the time), and were involved in social interactions most in summer and fall (2–3% of the time—Suring, 1975). A larger proportion (87%) of deer observed before sunrise and after sunset were feeding than after sunrise and before sunset (78%), whereas 6% were resting before sunrise and after sunset and 13% were resting after sunrise and before sunset (Suring, 1975).

White-tailed deer tend to be gregarious, with groupings usually being either matriarchal (composed of an adult female, its young, and its female offspring of earlier generations and their young) or fraternal (composed of adult and sometimes yearling males—Smith, 1991). On the refuge in Washington, overall group size for 8,130 groups averaged 2.2 (no measure of dispersion provided), but average group size ranged from 1.6 in autumn to 2.9 in summer. Overall, only 32% of males were observed in groups of two or more, only 18% in autumn (Suring, 1975). Seasonal changes in group size likely were in response to mating behavior of males in autumn and separation of females and their female offspring in anticipation of parturition.

The paucity of intraspecific interactions among males observed and the absence of copulatory behavior caused Suring (1975) to conclude that such behavior occurred largely at night. Suring (1975) also claimed that coyotes and elk (*Cervus elaphus*) had little effect on the activity of white-tailed deer, but that the deer avoided close association with domestic cattle. He observed (p. 36) "little aggression" between *O. virginianus* and *O. hemionus*, but noted "mating behavior . . . involving both sexes of both species." However, Smith (1982) indicated that in Douglas Co. only once in >13,000 observations were individuals of the two species ≤25 m apart. In the latter region, the two species were separated ecologically, with *O. virginianus* associated with woodland habitats and *O. hemionus* with grassland habitats. Smith (1982) concluded that the presence of *O. virginianus* precluded extensive use of that portion of his study area by *O. hemionus*.

On the Columbian White-tailed Deer National Wildlife Refuge, average (±*SE,* range in parentheses) size of bivariate normal home ranges among males was 65.4 ± 21.2 ha (18.6–184.7 ha) for 7 young-of-the-year, 187.2 ± 45.5 ha (54.1–316.2 ha) for 7 yearlings, and 208 ± 24.6 ha (92.4–302.4 ha) for 11 adults. For females, the same data were 154.5 ± 41.6 ha (49.4–293.6 ha) for 5 young-of-the-year, 113.6 ± 24.6 ha (37.2–274.3 ha) for 10 yearlings, and 103.6 ± 16.5 ha (35.7–316.8 ha) for 19 adults (Gavin et al., 1984). Bivariate normal home ranges for white-tailed deer on the North Umpqua River near Roseburg, Douglas Co., averaged 58.1 ha for adult females, 93.7 ha for adult males, 29.4 ha for female young-of-the-year, and 16.0 ha for male young-of-the-year (Smith, 1982). Smith (1982) claimed that home ranges of white-tailed deer in Douglas Co. were significantly smaller than those reported by Gavin (1979) for the species on the refuge. He also reported that home ranges in Douglas Co. were smaller than those reported in the literature, some smaller by an order of magnitude.

Usually, home ranges of white-tailed deer on the Columbian White-tailed Deer National Wildlife Refuge overlapped greatly. However, Gavin et al. (1984) observed some female white-tailed deer to defend against trespass by conspecific females small portions of their home ranges that contained dry bed sites. Such behavior was considered in contrast to the nonterritorial behavior usually attributed to the species.

Remarks.—Deer with characters intermediate between those of *O. virginianus* and those of *O. hemionus* reportedly are observed regularly in areas of sympatry, suggesting hybridization between the species. Because numbers of *O. virginianus* in Oregon, as elsewhere in the West, tend to be only a tiny fraction of those of *O. hemionus,* the presumed hybridization has led to concern regarding introgression and potential swamping of *O. virginianus* genetic material. The concern seemingly has increased since *O. v. leucurus* was afforded protection under the Endangered Species Act and within the Columbian White-tailed Deer National Wildlife Refuge along the lower Columbia River in Washington and Oregon.

Gavin (1984) reported that he had observed apparent hybrid deer on the mainland portion of the Columbian White-tailed Deer National Wildlife Refuge during his 1974–1976 study. He cited an unpublished report authored by M. A. Davison of the Washington Game Department, who classified 56 of 179 deer on the same area as exhibiting characteristics of both *O. v. leucurus* and *O. h. columbianus,* but considered all 85 deer observed on islands in the Columbia River to be typical *O. virginianus.* Suring (1975) observed interspecific mating behavior but no copulations between the two species on the mainland portion of the refuge. Smith (1982) showed that the same two races were segregated ecologically along the Umpqua River in Douglas Co. Krämer (1973:298) believed that hybridization of sympatric *O. virginianus* and *O. h. hemionus* in Alberta usually was prevented by a "behavioral mechanism operating at an early stage of interspecific pair formation." V. Coggins (pers. comm., 31 May 1994) indicated that individuals exhibiting all stages of intermediacy between those of *O. virginianus* and those of *O. h. hemionus* were common in northeastern Oregon. Specimens classified as hybrids in Alberta exhibited intermediate characteristics (Wishart, 1980).

Occasionally, individuals claimed to possess characteristics intermediate between those of *O. virginianus* and *O. h. columbianus* are reportedly taken in areas from which *O. virginianus* was extirpated early in the 20th century or before. Three of these so-called hybrids taken in the McDonald-Dunn Forest, Benton Co., during special hunts in 1970–1974 and examined by one of us (BJV) possessed all characteristics of *O. h. columbianus* except black bands were missing from all hairs. We believe that mutation is a plausible explanation.

Crosses of *O. h. columbianus* from Oregon with *O. virginianus* from Tennessee in captivity resulted in fertile offspring that resembled the former species (Whitehead, 1972), whereas crosses of a male *O. virginianus* from the lower Columbia River with *O. h. columbianus* from Oregon and backcrosses with its F1 hybrids produced offspring that resembled *O. virginianus* (Gavin, 1984). Cowan (1962) indicated that several offspring of a cross between a male *O. h. columbianus* and a female *O. v. ochrourus* were sterile. He also provided evidence of reduced viability of offspring produced by backcrossing a male *O.*

virginianus with its hybrid *O. virginianus × O. h. hemionus* female offspring. Whitehead (1972:65), based on small samples, found that in backcrosses of yearling *O. h. columbianus × O. virginianus* hybrids with both parental stocks "50 percent of the hybrids produced young of which 66 percent were stillborn." Cronin (1991a) referenced a thesis in support of the premise that *O. h. hemionus × O. virginianus* hybrids do not possess a running gait or antipredator behavior as effective as that possessed by either parental species.

Results of studies of mitochondrial-DNA sequences and allele frequencies at polymorphic loci (some in combination, others singly) at various localities (including Oregon and Washington) have served to produce somewhat conflicting conclusions regarding the extent of hybridization, the mechanism by which it occurs, the direction and extent of introgresion, and the time elapsed since divergence of the linages involved (Carr and Hughes, 1993; Carr et al., 1986; Cronin, 1991a, 1991b; Gavin and May, 1988). Based on allozyme data, Gavin and May (1988) considered some white-tailed deer on the refuge in southwestern Washington to possess genetic material from black-tailed deer, whereas white-tailed deer in Douglas Co. exhibited no evidence of introgression. Cronin (1991a), based on mitochrondrial-DNA genotypes, suggested that populations both on the Columbia River and in Douglas Co. probably contained genetic material introgressed from black-tailed deer.

Commencing in 1992, surveys conducted by the Oregon Department of Fish and Wildlife of those who hunted deer in northeastern Oregon were designed to separate harvest of white-tailed deer from that of mule deer. In 1992, 422 white-tailed deer reportedly were killed and in 1993 the number killed increased to 594.

FAMILY ANTILOCAPRIDAE—PRONGHORN

This endemic North American family arose in the Miocene and diversified in the Pliocene and most of the Pleistocene. Numerous species in 12 genera occurred in the grasslands and deserts of western North America. In the late Pleistocene all but one species became extinct (Kurtén and Anderson, 1980).

Antilocapra americana (Ord)
Pronghorn
1815. *Antilope Americanus* Ord, 2:292.
1818. *Antilocapra americana* Ord, 87:149.
1855. *Antilocapra anteflexa* Gray, p. 10.
1932b. *Antilocapra americana oregona* Bailey, 45:45.

Description.—The pronghorn (Plate XXXII) is a deer-sized artiodactyl (Table 13-1) with relatively long and thin legs and feet, only two digits on each foot, a fused radius and ulna, a relatively small tail, and unique horns consisting of deciduous keratin sheaths set on bony cores arising from the frontal bones. Although the horn sheaths contain hairlike striations and may have hairs embedded in their bases, they do not consist of amalgamated hairs as claimed by Hall (1981; see Remarks). Among males, the laterally flattened sheaths are 33–50 cm long, branched, and recurved, but in those females that possess horns, they are <12 cm long and simple (O'Gara, 1978). The horns and hooves are black. A pad of connective tissue beneath the posterior part of the hoof absorbs the shock of running over hard substrates. The front feet are larger and carry most of the weight when the pronghorn runs. Four mammae usually are present, but six have been observed (O'Gara, 1978). Pronghorns possess a gallbladder, a four-chambered stomach, and, in females, a bicornuate uterus. Mason (1952) reported that body mass among 19 mature males collected on Hart Mountain, Lake Co., throughout the year ranged from 45.5 to 63.6 kg.

The pelage is coarse; the hairs have numerous hexagonal air-filled cells that provide insulation. The dorsum is a light buff and is separated sharply from the white of the venter that extends high on the sides. Other markings include contrasting black and white throat patches; a black patch posterior to the nose pad; a white rump patch with erectile hairs; and a short, dark-brown mane on the nape. The rump patch can be expanded "into a great white rosette or closed down [to become] small and inconspicuous" (Bailey, 1936:70).

From dorsal and ventral perspectives, the skull of the pronghorn (Fig. 13-10) is shaped somewhat like an arrow point. A vacuity separates the nasal from the lacrimal bone; the orbit is large and complete, and projects laterally; and the lacrimal has a single orifice. The cheekteeth are hypsodont and selenodont. The lower canine is incisiform and is separated from the second premolar by a broad diastema; no teeth are present anterior to the second premolar in the upper jaw.

Distribution.—Pronghorns are known to have occurred from southwestern Manitoba, western Minnesota, eastern Texas, and Hidalgo west to southwestern Alberta, eastern Washington, southwestern Oregon, California except coastal regions, and Baja California Sur (Fig. 13-11; Hall, 1981). Present-day populations occupy disjunct regions within the former range (O'Gara, 1978). In Oregon, pronghorns originally occurred throughout the area east of the Cascade Range except for the Blue and Wallowa mountains; they also occurred west of the Cascade Range in the Rogue River valley and at the headwaters of the North Umpqua River (Bailey, 1936). We examined specimens from Harney and Lake counties (Fig. 13-11), but populations of the species are established in much of Oregon east of the Cascade Range.

An ancestral pronghorn, *A. †garcia*, is known from the Pliocene of Florida. Fossils of *A. americana* are known from numerous Pleistocene sites both within and outside the present range (Kurtén and Anderson, 1980).

Geographic Variation.—Traditionally, of the five nominal subspecies, only one, *A. a. oregona*, was considered to occur in Oregon (O'Gara, 1978). Because differences in nominal races are slight (Goldman, 1945), and because of numerous transplants without regard to race at source or release points, O'Gara (1978) considered the question of the validity of the subspecies moot. Lee et al. (1994), based on analyses of mitochondrial DNA and allozymes, did not consider populations in Oregon and northern California genetically distinct from those in the Great Plains. Therefore, they designated Oregon pronghorns to the subspecies *A. a. americana*.

Habitat and Density.—Pronghorns usually are considered denizens of open plains, but in Oregon broad areas dominated by big sagebrush (*Artemisia tridentata*) and playas (intermittent lakes) seem to form the primary habi-

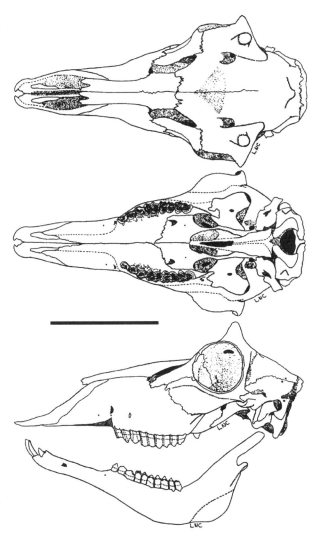

Fig. 13-10. Dorsal, ventral, and lateral views of the cranium, and lateral view of the dentary of a female pronghorn, *Antilocapra americana* (OSUFW 1605). Scale bar equals 10 cm.

tats used (Einarsen, 1948; Good and Crawford, 1978; Herrig, 1974); occasionally, herds of pronghorns even occur in areas with widely spaced junipers (*Juniperus occidentalis*) or ponderosa pines (*Pinus ponderosa*—Mace, 1954; Polenz, 1976). Within these habitats, water, either as the free compound or incorporated into succulent plants, seems to be the predominant factor influencing seasonal use of sites by pronghorns in Oregon (Good and Crawford, 1978; Herrig, 1974). In summer and autumn during years of above-average precipitation, pronghorns may use xeric sites, but during drought years, they use mesic, lowland sites and rarely move >3.2 km from sources of water (Herrig, 1974). Habitats avoided by pronghorns are those with vegetation >75 cm tall, rough terrain, and arid, flat desert. Groups composed of different sex and age-classes do not differentially use various types of sites (Herrig, 1974). On playas, pronghorns use areas vegetated by leafy arnica (*Arnica foliosa*) more frequently than other types (Good and Crawford, 1978).

Luman (1964) claimed that nearly all habitats for pronghorns in Oregon were decidedly marginal. He considered short bunchgrasses to provide the best pronghorn

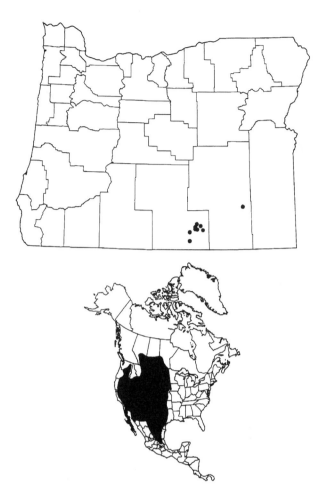

Fig. 13-11. Localities in Oregon from which museum specimens of the pronghorn, *Antilocapra americana*, were examined (listed in Appendix, p. 511), and the distribution of the pronghorn in North America. (Range map redrawn from Hall, 1981:1108, map 541, with modifications.) Currently, populations of the pronghorn occur in small disjunct regions within the range (O'Gara, 1978).

habitat, but indicated that in Oregon the original habitat of this type was replaced by shrubs through mismanagement of ranges commencing more than a century ago. Now, sagebrush with some bitter-brush (*Purshia tridentata*) and western juniper, or saltbush (*Atriplex*) and winterfat (*Eurotia lanata*) at lower elevations, form most of the habitat occupied by pronghorns. In spring, areas with cheat grass (*Bromus tectorum*) are used (Luman, 1964).

Research conducted simultaneously on study areas in southern Grant Co., and southeast of Malheur Lake, Harney Co. revealed "substantially different" habitats available on the two areas (Trainer et al., 1983:79). At the former, portions converted to range grasses were scattered through sagebrush and meadow vegetation types, whereas at the latter, only types typical of shrub-steppe were available. At both sites, pronghorns of different sex and age-classes used various types of habitat in proportions different from that available (Trainer et al., 1983).

Population densities ranging from 0.6 to 3.3 pronghorns/ km² have been reported at various times in other states (Kitchen and O'Gara, 1982). On a ≅185-km² study area in southern Grant Co., Polenz (1976) described an isolated herd that increased from 10 in 1957 to an average of ≅90 in 1963–1967, to ≅150 in 1970–1971, to ≅225 in 1972–

1976. Another herd ≅16 km away in another valley remained at ≅70 individuals during the entire period. By 1979, the estimated population on the Grant Co. area had reached 586 (3.2/km²—Trainer et al., 1983). Harvest of 21% of the herd in 1977 and 18% in 1978 plus removal of 122 animals for transplanting elsewhere reduced the herd, but by 1982 the population again had increased to an estimated 327 pronghorns (1.8/km²).

Diet.—Although Bailey (1936:75) indicated that "Supposedly, they are grass feeders," pronghorns in Oregon are primarily browsers of desert plants, thus feed largely on shrubs. Einarsen (1948) listed 12 species of shrubs, 29 species of forbs, and 22 species of grasses and grasslike plants that pronghorns were observed to consume in Oregon. However, he reported that 66.2% of the material contained in stomachs of 26 pronghorns (15 from Oregon and Nevada; 11 from Wyoming) was shrubs, 30.0% was forbs, and only 2.6% was grasses; 60.2% of the total volume was sagebrush.

On Hart Mountain, the primary foods eaten by 26 pronghorns collected throughout the year were sagebrush (60.7% of volume), *Phlox* (13.7%), other forbs (6.2%; including collinsia, *Collinsia*; lupine, *Lupinus*; lungwort, *Mertensia*; violet, *Viola*; and balsamroot, *Balsamorphiza*), and bitter-brush (4.4%). Other genera, each composing ≤3.1% of the volume of the stomach contents (in decending order), were wormwood (*Iva*), plantain (*Plantago*), fleabane (*Erigeron*), rabbit-brush (*Chrysothamnus*), clover (*Trifolium*), biscuit-root (*Lomatium*), milk-vetch (*Astragalus*), owl-clover (*Orthocarpus*), fiddleneck (*Amsinckia*), groundsel (*Senecio*), and saltbush (Mason, 1952). Forbs, including wild sweet-william (*Phlox*), were eaten most in April–July, a period during which sagebrush tended to be eaten least. Strangely, this is the season that sagebrush contains the highest levels of protein (Einarsen, 1948). Grasses, both green and dry combined, composed only 1.5% of the diet (Mason, 1952). Good (1977) indicated that pronghorns occupying playas in summer ate prostrate knotweed (*Polygonum aviculare*), poverty sumpweed (*Iva axillaris*), and evening-primrose (*Oenothera flava*).

Pronghorns commonly are described as "dainty foragers" (Einarsen, 1948:68) or "dainty feeders" (Bailey, 1936:75) because of their habit of nibbling a flower bud or whorl of tender leaves then moving on to feed a meter or two to half a kilometer away. Because pronghorns do not forage heavily in any spot and because they tend to clip stems cleanly rather than pull up plants as cattle do, rarely is evidence of them having fed at a locale detectable (Einarsen, 1948).

Reproduction, Ontogeny, and Mortality.—Female pronghorns usually attain sexual maturity at ≅16 months, but some individuals may become sexually active at ≅5 months (Mitchell, 1967; Wright and Dow, 1962). In Oregon, most females reportedly breed in a 15–20-day period commencing about 20 August; rarely do females breed after 20 September (Einarsen, 1948). The gestation period is ≅250 days; in Wyoming, gestation among captives averaged 252 days (O'Gara, 1978). Although females reportedly produce twins more frequently than one young (O'Gara, 1969a, 1978), in Oregon, of 975 births recorded in 1936–1940, 454 (46.6%) were of twins and 521 (53.4%) were of one young (Einarsen, 1948). In 1953–1954, on or near Hart Mountain National Wildlife Refuge, Lake Co., 75 females nursed 113 young; 38 nursed one young, 36 nursed twins, and 1 nursed triplets (Hansen, 1955).

Although twins frequently are produced, three–seven ova are shed and fertilized, and develop to the blastocyst stage. Through a complex two-stage process unique to pronghorns, the number of blastocysts that implant and develop to term is reduced usually to one or two (O'Gara, 1969a). O'Gara (1969a:222) speculated that production of litters of several young was advantageous earlier in the evolutionary history of the pronghorn, but that "predation or other factors have made two precocious young of greater survival value than several smaller and less well-developed young in recent times."

Based on material collected in Montana, Wyoming, and Colorado, mass of testes, size of interstitial cells, and degree of spermatogensis decreased in October–December and were at a nadir in January and February. Spermatozoa were present in the tubules of adult and yearling males from April to December and a few were present even in January and February, but mass of the testes did not attain a peak until August (O'Gara et al., 1971). The rut occurred in mid-September. Among males, horn sheaths were cast in late November or early December, and the bony cores were covered with thick, black skin containing long gray hairs on the proximal half; ≅20 mm of hard horn covered the tips of the bony cores (O'Gara et al., 1971). New hard horn commenced to form at the tip of bony cores beneath horn sheaths in place at the time of rut and maximum development of the testes; by mid-October the new horn beneath the old sheaths was ≅10 mm long. After old sheaths are cast, horn develops proximally from these tips, incorporating some of the hairs on the skin covering the base of the bony core (Fig. 13-12). In Oregon, horn sheaths reportedly are cast 15 October–15 November, and among

adult males the horn is 6.3–7.6 cm long by January (Einarsen, 1948). Among females, casting of horn sheaths is aseasonal. Occasionally second, and sometimes even third, horn sheaths ride atop most recently formed sheaths of females (O'Gara, 1969b). These older sheaths are removed easily and sometimes are shed when animals are handled. Horn casting in females is similar to that described for castrated males (O'Gara, 1969b).

In Oregon, most pronghorns are born during the 3rd week of May (Einarsen, 1948). Parturient females isolate themselves, lie down, drop their young after a short period of labor, lick the neonates dry, and move 400–800 m away to watch for possible danger. During the 1st day of life, neonates spend considerable time asleep, but can rise, wobbling, to their feet and exhibit interest in anyone who comes near. They are capable of making a bleating sound to which adult pronghorns of both sexes, but especially the maternal female, may respond by dashing close by the potential threat (Einarsen, 1948). By the 3rd day, young pronghorns are difficult to catch and by the 4th day they are impossible for a man to run down. Young nurse frequently but for brief periods (usually <2 min). By 1 week, they can run smoothly and effortlessly, and they commence to sample solid foods (Einarsen, 1948). Young are weaned by the end of August when they are ≅3 months old and weigh 20–25 kg (Einarsen, 1948).

Because most adult females and possibly some 5-month-old females breed each year, and because twins are produced at about half of the births, pronghorn populations have considerable potential to increase. Some observers believe that mortality among neonates must be high, as many herds on high-desert ranges in Oregon in summer have young:adult-female ratios of 30–40:100 (Polenz,

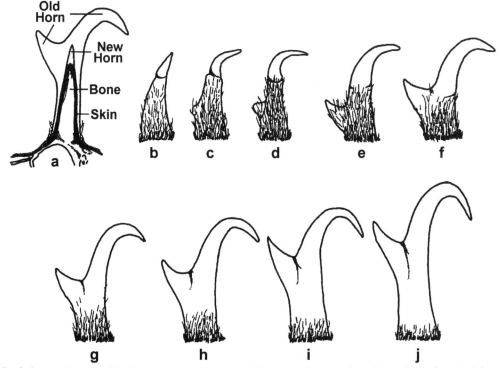

Fig. 13-12. Morphology and growth of the horn in pronghorns: **a**. lateral, longitudinal section through horn just before sheath is cast; **b–j**. horn of male pronghorn immediately after sheath is cast and stage of horn at approximatly monthly intervals thereafter. Redrawn after O'Gara and Matson (1975).

1976). During a period of increase in numbers of pronghorns on and near Hart Mountain National Wildlife Refuge, estimated mortality among young during summer 1953–1954 ranged from 8.0 to 23.3% and young:adult-female ratios were 75–99:100 (Hansen, 1955). Nevertheless, our analysis of data obtained from the Oregon Department of Fish and Wildlife indicated that a commonly obtained index to relative abundance (number observed/kilometer along ≅6,400 km of aerial transect) was not correlated positively with young:adult-female ratios during the previous summer ($r^2 = 19.58$, $F = 2.678$; $P = 0.229$), as expected if production and survival of young affect pronghorn population density. We suspect that the low reliability of the indices in reflecting either production and survival of young or abundance of pronghorns may be responsible, at least in part, for our failure to find a positive relationship.

Of 42 carcasses of young found on and near Hart Mountain National Wildlife Refuge, 69% died when <2 months of age, 17% when 3–4 months old, 10% when 5–6 months old, and 4% when 7–8 months old (Hansen, 1955). Of 225 pronghorn carcasses examined, including the 42 reported on by Hansen (1955), 38% were <2 months old when they died, 11% were 3–6 months old, 8% were 7–11 months old, and 43% were ≥12 months old (Yoakum, 1957). Yoakum (1957) believed that mortality in pronghorns could not be attributed to only one factor. He considered diseases, parasites, predation, severe winter weather, drought, natural accidents, and illegal hunting to be contributory factors to mortality among pronghorns. For the 15 years previous to his study, an estimated average of 3.7% of the population was killed annually by hunters (Yoakum, 1957).

In 1981–1982 on a 337-km² study area southeast of Malheur Lake in Harney Co., 55 (98.2%) of 56 radio-tracked neonates died at <40 days of age; the other individual died before it was 120 days old. However, ≅120 km north on the Grant Co. study area, 39 (52%) of 75 radio-tracked neonates were still alive at 40 days and 15 (20%) were alive at 120 days of age (Trainer et al., 1983). Cause of death of 83 (91.2%) of 91 that died at ≤40 days of age was attributed directly to predation, with coyotes (*Canis latrans*) believed responsible for 55 (60.4%) of those lost. Felids, raptors, badgers (*Taxidea taxus*), and unidentified predators were responsible for loss of 28 (30.8%) of the young pronghorns (Trainer et al., 1983). Only eight were believed to have died of starvation, disease, or unidentified nonpredator causes. Based on differences between estimated birth rate (1.85 fetuses/adult female) and young/adult female in late-July–August, neonatal mortality was 67% in 1981 and 43% in 1982 on the Grant Co. area and 95% and 99% during the same years on the Harney Co. area (Trainer et al., 1983). Greater heterogeneity of habitat, less distance to water, and shorter distances moved by young were believed to enhance survival (Trainer et al., 1983).

Early information regarding age in pronghorns based on erosion of the teeth indicated that most individuals were short lived, with few attaining an age of >7–9 years (Einersen, 1948; Kitchen and O'Gara, 1982). However, more recent information based on counts of cementum annuli, a technique that seems more reliable (McCutchen, 1969), indicates that pronghorns may attain an age of ≥15.5 years (Kerwin and Mitchell, 1971).

Habits.—Because pronghorns occur in open areas with relatively short vegetation, they rely on their excellent eye-sight and great speed for protection from potential enemies. The ability of pronghorns to discern something new in their environment or to detect movement is truly phenomenal; moving objects 5–6 km away attract their attention. Pronghorns can attain a speed of 80 km/h easily and may be able to reach 100 km/h or even more under favorable conditions (Einarsen, 1948). They commonly race with vehicles that pass nearby (Einarsen, 1948). They are particularly graceful at high speeds, often leaping 4–6 m at each bound. Despite the leaping ability, they negotiate fences by passing through or beneath the wires rather than by jumping over them (Einarsen, 1948).

Pronghorns may be active throughout the 24-h period, but sleep in catnaps at any time (Einarsen, 1948). They frequently arouse for a few seconds to scan their surroundings, and they seem capable of detecting and responding to strange odors while asleep. They are intensely curious; they commonly scrutinize any new activity in their area (Einarsen, 1948). Hunters sometimes take advantage of the pronghorn's curiosity by luring them to within range with a flag or other device.

In winter, pronghorns associate in bands or herds (Bruns, 1977; Einarsen, 1948) sometimes numbering 50 or more individuals. Disbanding was noted after 25 March on an Alberta-Montana study area (Bruns, 1977). In 1982 on a study area in Grant Co., average herd size increased from 4.4 to 22.4 and was positively related ($r = 0.95$) to season from June to September, whereas ≅120 km south on a study area in Harney Co., average herd size ranged from 2.2 to 7.7 and was not as strongly related ($r = 0.44$) to season. On the Grant Co. area, the entire population was grouped into three herds by the 3rd week of October and into only two herds by the 3rd week of November (Trainer et al., 1983). The herd is the basic social unit and tends to maintain the same membership (Bruns, 1977; Kitchen, 1974). A linear social hierarchy based on age and body mass develops within herds with adult males at the top, then adult females, and finally young (Bruns, 1977; Kitchen, 1974). A variety of vocalizations and specific behavior patterns are used to establish and maintain hierarchical status. Among females, the hierarchy serves to reduce competitive interactions for resources; among males, its ultimate function is to provide a mechanism by which breeders are selected (Kitchen, 1974).

Adult males when 3–4 years old attempt to establish territories by marking areas not included in established territories or in relatively unused portions of existing territories. Marking includes deposit of materials from the subauricular (cheekpatch) glands on freshly clipped vegetation and a combination of pawing the ground then urinating and defecating on the spot (Kitchen, 1974). During establishment of territories, marking may be as frequent as 23 cheek-patch rubs/h and three paw-urinate-defecate acts/h; males with established territories mark much less frequently, but tend to remark the same sites. In Montana, territories ranged from 0.23 to 4.34 km² (Kitchen, 1974). Large dominant males commence to defend territories in late winter or early spring; bachelor males may be tolerated for as much as a month longer, but not afterward until midautumn except for the 1–2 days that the territorial male is busy rutting. Territories tend to remain relatively unchanged from year to year. Defense of territories involves a varied behavioral repertoire that includes watching, chasing, vocalization (snort-wheeze), approaching, a combination feed-thrash-mark-walk-threaten display, cheek-patch display, and sparing (Kitchen, 1974). Territorial males usually do not exhibit the same sequence of behavioral acts. Many, perhaps most, males fail to establish territories and are destined to use home ranges between established territories or in peripheral, and often marginal, habitats available to the herd.

Groups of females use home ranges that include parts of territories of several adult males. Yearling females tend to join female groups to which their maternal parent belongs (Kitchen, 1974). During the rut, territorial males attempt to hold groups of females on their areas. Females are more receptive to sexual advances of territorial males than to those of bachelor males. Frequently, females solicit courtship by territorial males but almost never by bachelor males; they often reject bachelor males even when they are in estrus (Kitchen, 1974). Following the rutting season, pronghorns form large aggregations composed of all sex and age groups.

During extremely cold periods, pronghorns select microhabitats with wind velocities 63% lower, 24% less snow, and 87% softer snow than average for the region. Bedded animals orient with the anterior portions of their bodies away from the wind, but may place the head alongside the body when cold is extreme (Bruns, 1977). Pronghorns have an average body temperature of 38.5°C (Lonsdale et al., 1971).

All but ≅100 pronghorns left a study area in Grant Co. (Bear Valley north of Seneca) in late November or early December (coincidental with 20–30-cm snowfalls) and moved along a circuitous route to winter range ≅113 km southeastward. In spring, many, but not all, pronghorns returned to Bear Valley (Trainer et al., 1983).

Pronghorns sometimes harass potential predators to avoid being preyed upon; group size seems critical to the success of such strategy. For example, a group of three feeding pronghorns on Hart Mountain, Lake Co., being stalked, then chased by a coyote (Canis latrans) was observed to move toward, then join, a group of nine individuals foraging 200 m away. The group of 12, led by two individuals, ran toward the coyote; when the "stamping and noisy" herd was ≅20 m away, the coyote fled as the herd continued to chase for 10 s and to approach within 10 m (Berger, 1979:198). Because the probability of an individual being preyed upon likely is related inversely to the size of the group of which it is a member, it is selectively advantageous for an individual to cooperate in harassing a predator regardless of its relatedness to other members of the group. Failure to join such a cooperative venture likely would put the individual at greater risk of falling victim to the predator (Berger, 1979). During flight from a predator, the expanded rump patch serves to alert other members of the herd and, with the white side markings, is believed to contribute to confusing the predator, especially when combined with the characteristic erratic changes in direction herds make while fleeing (Kitchen, 1974).

Remarks.—Because shedding of horns in pronghorns is similar to the exfoliation process of horns observed occasionally in some Old World antelopes, O'Gara and Matson (1975) classified the pronghorn as a member of the family Bovidae. However, Solounias (1988) demonstrated that the horn cores in pronghorns were not epiphy-

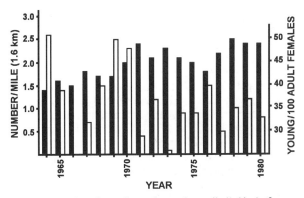

Fig. 13-13. Number of pronghorns observed per mile (1.6 km) of census route flown (solid bars) and number of young per 100 adult females observed the previous summer (open bars), 1964–1980 ($r^2 = 19.58$, $F = 2.678$; $P = 0.229$). Data provided by Oregon Department of Fish and Wildlife in "Oregon Antelope Report" published in *Transactions Interstate Antelope Conference*, 1964–1980.

seal (derived from dermal ossifications [ossa cornua] that fuse onto the frontals early in life) as occurs in the Bovidae, but were apophyseal (elongations of the periosteum of the frontals) as in Cervidae and other families within the Cervoidea. Thus, the horn cores are homologous with the antler bases of the Cervidae, not with the horn cores of the Bovidae (A. Bubenik, 1983a, 1983b). Also, Solounias (1988) indicated that growth of the keratinous sheath (horn) in the pronghorn commenced at the tip, then at the prong, and proceeded proximally toward the base, incorporating hairs in the integument covering the bony core. In Bovidae, the keratin sheath grows from the base and moves distally as the bony core grows. Thus, the pronghorn was returned to the family Antilocapridae in the Cervoidea.

According to reports of early explorers, trappers, and other travelers who kept journals, pronghorns originally were abundant and widely distributed in Oregon as elsewhere in the West (Bailey, 1936). After settlement commenced in the mid-19th century, pronghorns rapidly became fewer in number and less widely distributed. As early as 1875, restrictions were imposed on the hunting of pronghorns (Mace, 1954). In 1901, the state legislature restricted the taking of pronghorns to 15 July–1 November, but failed to establish a bag limit; the regulation was feebly enforced. In 1913, the taking of pronghorns in Oregon was prohibited by legislative act (Einarsen, 1938). By 1915, the most reliable estimate of numbers was ≅2,000 and a range limited to small portions of Malheur, Harney, Lake, Crook, and Klamath counties (Bailey, 1936). During summer in 1936–1937, after more than 2 decades of protection, pronghorn numbers were estimated to have increased to ≅17,000, with most occurring in Malheur, Harney, and Lake counties; ≅2,000 of those wintered in Nevada (Einarsen, 1938). A 4-day hunting season was established in the southern portions of Malheur, Harney, and Lake counties in 1938; 300 either-sex permits were authorized, 242 were issued, 175 pronghorns were killed, and 23 other pronghorns were reported to have died of gunshot wounds (Einarsen, 1939). Because some groups of hunters shot indiscriminately into fleeing herds (an average of nine shots was required to kill a pronghorn), Einarsen (1939) believed other pronghorns were shot and ultimately died but were unretrieved and unrecorded. Subsequently, permit seasons were limited to pronghorns with horns longer than the ears, then later a few permits for the taking of "hornless" pronghorns were issued in restricted areas. Such regulations seemingly have contributed significantly to the reduction of crippling losses. Estimated numbers of pronghorns remained about the same as estimates for 1938 until the early 1940s, but when the estimated population declined to ≅8,000 in 1946, hunting was not permitted for 3 years. Estimated numbers fluctuated between 10,000 and 14,000 during the late 1940s and early 1950s, and limited hunting was permitted commencing in 1949 (Yoakum, 1957).

Numbers of pronghorns observed per kilometer of census transect flown annually (Fig. 13-13) and numbers of pronghorns harvested annually (Fig. 13-14) indicate that populations in Oregon continued to increase since the 1950s. Doubtlessly, some of this increase is the result of

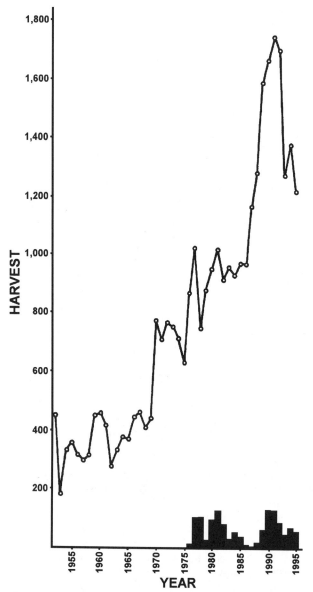

Fig. 13-14. Number of pronghorns with horns longer than the ears (adult males) harvested annually (line) and number of pronghorns with horns less than the length of the ears (females and young males) harvested annually (bars) by hunters, 1952–1995. Data from Oregon Department of Fish and Wildlife annual reports.

the establishment of the Sheldon-Hart Mountain National Wildlife refuges in Oregon and adjacent Nevada, and management for the species on those areas. Protection from indiscriminate hunting and efforts to improve range conditions also contributed to these increases.

Pronghorns commonly are referred to as "antelope," but the latter term should be limited to reference to the Old World antelopes of the family Bovidae. "Pronghorn" should be used, not only to emphasize the uniqueness of Antilocapridae, but also in recognition that it is the only extant member of the only family of mammals endemic to the Nearctic (Anderson and Jones, 1984).

FAMILY BOVIDAE—SHEEP, GOAT, AND BISON

The bovids, represented by 137 species (Wilson and Reeder, 1993*a*), originally occurred throughout Africa, most of North America and Eurasia, and western Greenland, but through some combination of domestication, introduction, and escape, now occur in essentially all habitable lands (Grubb, 1993; Nowak, 1991; Simpson, 1984). Of the 45 genera currently recognized, most are endemic to Africa; only four occurred naturally in North America: *Bos* (bison), *Ovis* (mountain sheep), *Oreamnos* (mountain goat), and *Ovibos* (muskox). Of the three species that occur in Oregon, the mountain goat was introduced, the bison was extirpated about the time of exploration and settlement by Europeans and now occurs only in semidomestic herds, and the mountain sheep was extirpated and successfully reintroduced. Fossil bovids are unknown in North America from deposits earlier than the Pleistocene (Savage and Russell, 1983; Simpson, 1984).

Bovids are large mammals. Bison may attain a height at the shoulder of nearly 2 m and a body mass of >900 kg; a semidomestic male weighed 1,724 kg (Meagher, 1986). Mountain sheep and mountain goats are much smaller; neither attains a body mass of >137 kg (Rideout and Hoffmann, 1975; Shackleton, 1985). Bovids in Oregon (but not all bovids) possess four digits on each foot; the third and fourth are largest and bear the weight of the animal, whereas the second and fifth are much smaller, are set higher and to the rear, and normally do not touch the substrate. Other characteristics of members of the family that occur in Oregon (but not all members of the family) include a lacrimal canal with an orifice restricted to inside the rim of the orbit and absence of a vacuity between lacrimal and nasal bones.

Upper incisors and upper canines are absent, premolars are molariform, and lower canines are incisiform, and a broad diastema separates lower canines and premolars. Molariform teeth are selenodont and hypsodont. The stomach is four chambered and digestion involves rumination (regurgitation and rechewing of food). A gallbladder is present.

The uniqueness of the family among artiodactyls lies largely in the unbranched horns consisting of bony extensions of the frontal bones covered by keratinized sheaths. Males of all wild species possess horns, and females of many species are similarly adorned, although their horns frequently are smaller than those of males. Most species possess two horns, but one wild species in India, *Tetracerus quadricornis,* and a domestic breed of *Ovis* possess two pairs (Simpson, 1984; Vaughan, 1986). Koopman (1967) and Simpson (1984), in summaries of characteristics of the family, indicated that the horns of bovids were nondeciduous, but O'Gara and Matson (1975) reported that at least some individuals may occasionally shed the keratin sheath by exfoliation.

Bos bison Linnaeus
Bison

1758.   [*Bos bison*] Linnaeus, 1:72.
1897 [1898]*b. Bison bison athabascae* Rhoads, 49:498.
1932*c. Bison bison oregonus* Bailey, 45:48.

Description.—The bison (Plate XXXI) is the largest artiodactyl in Oregon, sometimes attaining a body mass of >1,700 kg among males (Meagher, 1986). The elongation of the neural processes of the thoracic vertebrae and accompanying musculature compose a shoulder "hump" that produces the illusion that the bison is even larger than its actual size. In addition, the dark-brown pelage of the neck, hump, and shoulders is long and shaggy, further accentuating the size. The hindquarters are relatively slim and the pelage thereon is relatively short. The legs are short and heavy, and equipped with black hooves on the four toes. The tail is short and tufted. Short, curved, black horns arise from frontals of both sexes (Fig. 13-15). Both horns and body features are smaller and slimmer in females (Meagher, 1986).

Distribution.—At time of settlement by Europeans, bison herds were distributed throughout most of North America from Alaska to northern Mexico and from the Cascade and Sierra Nevada ranges east to near the Atlantic Coast (Fig. 13-16; Hall, 1981; Meagher, 1986). The primary range was through the grassland areas of central United States and Canada. The present range of free-ranging bison is restricted largely to parks, preserves, and other public lands (Meagher, 1986).

Records of *B. bison* in North America do not extend earlier than the late Pleistocene (Meagher, 1986).

Although "native bison . . . were never observed in Oregon historically" (Van Vuren and Bray, 1985:56), we examined specimens from eight localities and found published records for three others (Fig. 13-16). Bison in Oregon formerly were believed to be largely isolated from herds farther to the east by the "desert scrub" (Haines, 1967:12). However, records in Idaho, Oregon (Van Vuren and Bray, 1985), and Nevada (Van Vuren and Deitz, 1993) tend to indicate that bison were widespread, although likely not abundant, in the northern Great Basin. Three verbal accounts by native Americans place the extirpation of bison from Oregon in the early 1800s, just before explorers reached the region (Bailey, 1923*b*; Van Vuren and Bray, 1985). The only bison in Oregon at present are in semidomestic herds maintained on several ranches.

Geographic Variation.—Two subspecies are recognized currently, one of which, *B. b. athabascae*, occurred in Oregon (Hall, 1981; Meagher, 1986).

Habitat and Density.—Bison primarily are animals of grasslands, but occurred where vegetation suitable for grazing was present in biomes as diverse as deciduous forest, boreal forest, and desert (Meagher, 1986). In Oregon, the few records of occurrence (Van Vuren and Bray, 1985) were

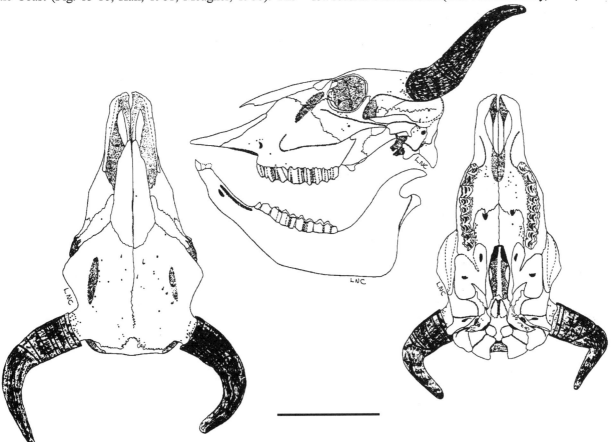

Fig. 13-15. Dorsal, ventral, and lateral views of the cranium, and lateral view of the dentary of a female bison, *Bos bison* (OSUFW 1573). Scale bar equals 20 cm.

Fig. 13-16. Localities in Oregon from which museum specimens (circles) of the bison, *Bos bison*, were examined (listed in APPENDIX, p. 512) and reported additional records (squares—Van Vuren and Bray, 1985), and the former distribution of the bison in North America. (Range map redrawn from Hall, 1981:1109, map 542.) Presently, the species exists as scattered, small populations maintained in parks, preserves, and other public lands, and on private ranches (Meagher, 1986).

in the region dominated by steppe and shrub-steppe vegetation consisting largely of bunchgrass and sagebrush (*Artemisia*—Franklin and Dyrness, 1969).

At time of settlement, 30 million bison were estimated to inhabit North America, but by 1903 only 1,644 were known to exist; the decline largely was a result of overhunting (Meagher, 1986). Numerous writers have recorded narratives of the slaughter and of the near demise of the species (Meagher, 1986). At the present time, bison probably number fewer than 100,000.

Diet.—In southern Utah, on both renovated rangeland and shrub-steppe communities, bison consumed mostly grasses and sedges with forbes and shrubs each composing <10% of the diet (Van Vuren, 1980; Van Vuren and Bray, 1983). The distribution of foraging groups of bison was not related significantly to elevation, distance to water, slope, or aspect, but was related to the distribution of food plants selected by bison most often (Van Vuren, 1980).

Reproduction, Ontogeny, and Mortality.—Although both sexes occasionally attain reproductive competence as yearlings, most females do not breed until 2–4 years old, and, although most males are fertile by 3 years of age, most do not breed until 6 years old unless older males are absent (Meagher, 1986). The breeding season is July–September,

with most copulations concentrated within a 10–14-day period. Gestation is ≅285 days, with most young born in April–May. One young is usual, two are rare.

Neonates are precocious, are born without a hump, are usually colored bright reddish-tan, and weigh 15–25 kg; young begin to darken at ≅2.5 months and by 8–9 months weigh 135–180 kg (Meagher, 1986). As yearlings they weigh 225–315 kg. Although young may not be weaned until 7–8 months old, they commence to graze as early as 5 days of age and have survived when orphaned at 7–8 weeks of age. Maximum age is >20 years, possibly as much as 40 years (Meagher, 1986).

Habits.—Bison are gregarious and seasonally migratory; in mountainous regions they follow the greening of forage upward in spring and the onset of snows to lower elevations in autumn. There is considerable segregation of the sexes; most adult males occur as singles, pairs, or small bands whereas adult females, young, yearlings, 2–3-year-old males, and sometimes a few adult males form separate (cow) groups. Cow groups are small (half are <15, most are <30 animals), highly mobile (most remain on a range <2 days), and dynamic (stable membership persists only 1–4 days) units (Van Vuren, 1980). Any contact between groups leads to exchange of members; the only stable groups are females and their offspring.

Although bison are grazers and usually associated with grasslands, they enter woodlands at times to seek protection from the sun, insects, or storms. Severe winter weather and spring storms are the greatest cause of mortality; such mortality is differential among sex and age groups (Meagher, 1986).

Males form linear dominance hierarchies that confer breeding status; these hierarchies are based on aggressive interactions rather than age or body mass. Much of the interaction occurs during rut; most interactions tend to be as stylized threats and signals of subordinance, but some may include fights that occasionally result in injury or death (Meagher, 1986). Females also exhibit linear dominance; status is established early and rarely contested.

Most activity by bison is during the daylight hours; bouts of grazing are interspersed with periods of loafing and ruminating (Meagher, 1986). Movements are mostly at a walking pace, but bison are capable of trotting, galloping, and bounding, and can attain 60 km/h (Meagher, 1986).

Remarks.—Bison were long known by the scientific epithet *Bison* Hamilton-Smith, 1827, a generic name that still commonly is used in print (e.g., Grubb, 1993; Nowak, 1991; Van Vuren and Deitz, 1993). Because fertile hybrids between bison and domestic cattle are possible, Van Gelder (1977) considered the species to be congeners and that the older generic name *Bos* should apply to the bison. Gentry (1978) and Groves (1981) concurred on paleontological and morphological grounds, respectively, and Miyamoto et al. (1989), based on similarities within two sets of mitochondrial-DNA sequences, agreed that bison and cattle were congeners. Finally, although somewhat reluctantly, Jones et al. (1992) applied the name *Bos* as the generic name of the bison in their checklist of North American mammals north of Mexico.

Periodically, some old-timer claims that a grandfather saw bison roaming the Willamette Valley in the mid-19th century and showed them depressions where "buffalo wallowed" (e.g., *Gazette-Times*, Corvallis, Oregon, 26 May

1993). Dirk Van Vuren, an authority on bison, especially on the distribution of bison in Oregon, indicated (in litt., 16 August 1993) that buffalo wallows were so variable that the only absolute means of identifying them was to see bison in them. He found claims of modern bison in the Willamette Valley incredible in view of the availability of detailed histories of the region. Bison meat was among the favorite foods of the early explorers of the region, but none mentioned bison in the Willamette Valley, although several provided accounts of bison distribution in their journals. Also, the valley was settled by pioneers who crossed the Great Plains and were familiar with bison, yet the presence of bison in the valley was not chronicled. Finally, for those who claim to have found bison skulls, those of modern bison are strikingly similar to those of some breeds of domestic cattle and those of Pleistocene and early Holocene forms of bison; thus, a detailed knowledge of bison morphology is requisite for certain identification of bison skulls. Although we found no remains of modern bison from the Willamette Valley on deposit in museums, several skulls of *B. b. †antiquus*, a large Wisconsinan-age subspecies that persisted in some regions as late as 10,000 years ago (Kurtén and Anderson, 1980) were examined. Whether these individuals lived in the Willamette Valley or their remains, in addition to erratic boulders and other debris, were transported into the valley during the cataclysmic Bretz flood of ≅13,000 years ago that originated in Montana and swept across Idaho and eastern Washington (Allen et al., 1986), is unknown.

Oreamnos americanus (Blainville)
Mountain Goat

1816. *R[upicapra]. americana* Blainville, p. 80.
1817a. *Oreamnos* Rafinesque, 2:44.
1895b. *Oreamnos montanus* Merriam, 1:19.
1912. *Oreamnos americanus americanus* Hollister, 25:185–186.

Description.—The mountain goat (Plate XXXII) is a stockily built bovid with black scimitar-shaped horns as long as 305 mm (Cowan and Guiguet, 1965), large black hooves (toes 3 and 4) and prominent dewclaws (toes 2 and 5), and an entirely white, woolly pelage. Sometimes the pelage contains scattered brown hairs on the dorsum and rump (Rideout and Hoffmann, 1975). A ≅30-mm-long beard, pointed ears, and a squarish muzzle also are characteristic. A black, curved horn gland occurs posterior to the horns and exudes products onto the horns. Males are larger (more in northern populations than in southern) and have longer, larger-diameter, and more evenly curved horns than females (Cowan and McCrory, 1970).

The skull of the mountain goat (Fig. 13-17) is much more lightly constructed than that of the mountain sheep (*Ovis canadensis*); the horns and frontals that support them are smaller, the parietals are larger, and the frontal and cornual sinuses are smaller and without additional septa (Rideout and Hoffmann, 1975). There are no upper incisors or canine, and the lower canine is incisiform. Cowan and McCrory (1970) reported that 32% of 167 skulls from throughout the range of the species exhibited dental anomalies. Horn growth is retarded in winter and, commencing the 2nd winter, an annulus is formed; thus, counts of annuli plus one provide a measure of age in years.

Distribution.—Mountain goats originally occurred in montane regions of southeastern Alaska, eastern Yukon Territory, and extreme western Northwest Territories, much of British Columbia, and extreme western Alberta south into Montana, Idaho, and Washington (Fig. 13-18). Introduced populations are established in areas beyond the original range in Alaska, British Columbia, Montana, Washington, Oregon, Colorado, South Dakota, and Utah (Rideout and Hoffmann, 1975).

Bailey (1936) argued that mountain goats did not occur naturally in Oregon at settlement. Hall and Kelson (1959) cited an article by Grant (1905) in the Ninth Annual Report of the New York Zoological Society in support of the occurrence of the species in the Wallowa Mountains and on Mt. Jefferson in the Cascade Range. Lyman (1988:19), in attempting to use biogeographic evidence to refute claims that *O. americanus* was not native to the Olympic Mountains, Washington, cited Grant (1905) as "documented historic evidence" of the species in the Cascade Range in Oregon and in the Wallowa Mountains. However, Grant (1905) provided no sources for his records (Matthews and Coggins, 1995). Lyman (1988) also referenced an unpublished report of remains of mountain goats at Rattlesnake Creek, Malheur Co. On the basis of several anecdotal accounts, published reports of archaeological evidence of mountain goats on the Idaho side of Hells Canyon, and an unpublished report of similar evidence at Camp Creek on the Oregon side of Hells Canyon, Matthews and Coggins (1995:70) concluded that mountain goats were "indigenous to the northeast corner of Oregon and most likely portions of the Oregon Cascades." Mountain goat populations have been established at several sites in the 20th century through introductions (see Remarks for a history of introductions in Oregon).

Pleistocene fossils of mountain goats are recorded from Washington, Nevada, California, Arizona, and Mexico (Harington, 1971); thus, it seems likely that the species possibly occurred in Oregon. However, if mountain goats actually occupied part of the area now considered to be part of Oregon, it was long before settlement (Lyman, 1988). Hoffmann and Taber (1967) believed that mountain goats were eliminated from southern parts of the range during a warm period at the end of the last glacial advance.

Geographic Variation.—Some authorities recognize four subspecies of mountain goats (Hall, 1981), whereas others believe that recognition of subspecies within this taxon is unwarranted (Cowan and McCrory, 1970). The original stock introduced into the Wallowa Mountains was from within the region in which mountain goats are considered to be *O. a. americanus*, but additional introductions in 1985–1989 were from ranges of two other races (Matthews and Coggins, 1995). The stock introduced into Baker Co. may have come from ranges occupied by three of the four subspecies. The conclusion of Cowan and McCrory (1970) and the possible homogenization of genetic material from several regions cause us to regard the question of subspecies of mountain goats in Oregon as moot.

Habitat and Density.—In the Wallowa Mountains, Wallowa Co., a ≅7,100-ha area intensively used by mountain goats had less timber and more slide rock and cliff rock than the total ≅22,440-ha area available to goats. Other habitats, all of which covered <5% of the area, occupied approximately equal proportions of the total area and the

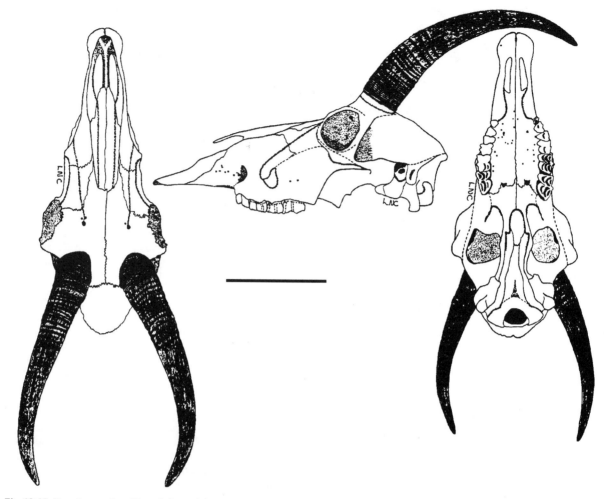

Fig. 13-17. Dorsal, ventral, and lateral views of the cranium of the mountain goat, *Oreamnos americanus* (OSUFW study). Scale bar equals 10 cm.

area used intensively by goats (Vaughan, 1975). Use of various habitats differed seasonally: ridgetops, consisting of <1% of the area available, were used extensively (74% of the time) in summer; alpine meadows covered 4.2% of the area but were used 8–13% of the time in summer–autumn; slide rock covered 17.7% of the area and was used 64% of the time in autumn, but only 8–13% of the time the remainder of the year. Cliff rock covered 19.8% of the area and was used 60% of the time in spring, 24% of the time in winter, and 3–9% of the time during summer–autumn. Timber covered 42.0% of the area, but was used only 19% of the time in winter and completely avoided at other seasons; and open timber covered 13.9% of the area and was used 35% of the time in winter, 19% each in spring and autumn, and was avoided in summer (Vaughan, 1975:50–53, tables 12 and 13). Seasonal differences in elevation of areas occupied also were noted: average elevation used was 2,719 m in summer, 2,468 m in autumn, 2,372 m in winter, and 2,323 m in spring (Vaughan, 1975).

In the Wallowa Mountains, five goats released in 1950 increased to ≅26 by 1964 (Vaughan, 1975). A permit-only hunting season was established in 1965; five goats were killed by hunters each year, from 1965 to 1968. The estimated population declined to ≅10 individuals by 1969, the hunting season was closed, and by 1971, the estimate of numbers increased to ≅22; by 1973, the minimum number

was 29 (Vaughan, 1975). In 1994, numbers were estimated at 55 individuals (Matthews and Coggins, 1995). Based on Vaughan's (1975:48) "intensive use area" of ≅7,100 ha, the highest densities probably were about 0.77/100 ha.

In 20 years, populations established through introductions increased from 10 to >200 in South Dakota (Harmon, 1944) and from 14 to 250–300 in Colorado (Hibbs et al., 1969). The population in South Dakota stabilized at 300–400 about 7 years later and remained so for a minimum of 20 years; the area available was ≅13,000 ha with a primary-use area of <850 ha (Richardson, 1971). In the Olympic Mountains, Washington, 11–12 goats introduced in the late 1920s (Moorhead and Stevens, 1982) increased to an estimated (±*SE*) 1,175 ± 171 by 1983 on a 50,063-ha area (2.3/100 ha—Houston et al., 1986), then declined to 389 ± 106 by 1990 (0.8/100 ha—Houston et al., 1991). By comparison, the population in the Wallowa Mountains seems to have done rather poorly. Bartels (1976) considered the paucity of south-facing slopes to be limiting in that region; Vaughan (1975) believed that prenatal and neonatal mortality, poor nutrition during winter, and poor quality of winter range were interrelated and acted together to limit recruitment. In Montana, reproductive output was density dependent; "goats were better able to increase from lower densities" (Swenson, 1985:841).

Diet.—In the Wallowa Mountains, mountain goats fed

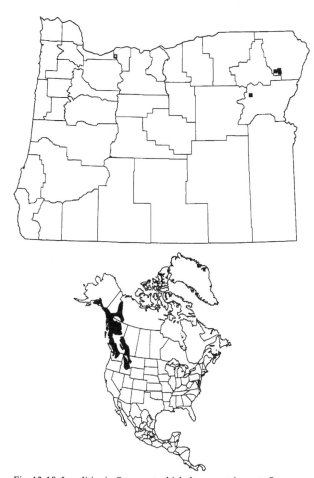

Fig. 13-18. Localities in Oregon at which the mountain goat, *Oreamnos americanus*, was introduced or became established naturally from introduced herds (Matthews and Coggins, 1995), and the distribution of the mountain goat in North America. (Range map redrawn from Rideout and Hoffmann, 1975:2, fig. 3, with modifications.) The open symbol indicates a locality at which introduced mountain goats subsequently disappeared (Matthews and Coggins, 1995).

along ridgetops in July, in meadows in August and September, and in slide-rock areas in October. In December and January, goats fed along ridges blown free of snow, but during storms goats were restricted to timbered areas, where they fed on conifers and mountain-mahogany (*Cercocarpus ledifolius*—Vaughan, 1975). By March, the primary feeding area was "overgrazed to the point that practically all vegetative material was removed" (Vaughan, 1975:63–64). With snowmelt and concomitant growth of plants in spring, mountain goats foraged in open areas. To our knowledge, no study of the dietary habits of mountain goats in Oregon has been conducted. Studies conducted elsewhere, however, indicate that the dietary habits of mountain goats are extremely variable (Hibbs et al., 1969).

In Montana, 124 species of plants were observed to have been eaten by mountain goats (Saunders, 1955). Grasses and grasslike plants (hairgrass, *Deschampsia*; bluegrass, *Poa*; fescue, *Festuca*; sedges, *Carex*; rushes, *Juncus*; woodrush, *Luzula*) were eaten more frequently than other plants at all seasons; forbs (lupine, *Lupinus*; Jacob's-ladder, *Polemonium*; lungwort, *Mertensia*; sweetvetch, *Hedysarum*; smartweed, *Polygonum*; cinquefoil, *Potentilla*) were next most commonly consumed in summer and autumn, but third in winter and spring. Coni-

fers (juniper, *Juniperus*; pines, *Pinus*; Douglas-fir, *Pseudotsuga*; fir, *Abies*) were second in winter and spring, but were eaten little at other seasons. Shrubs (except huckleberry, *Vaccinium*, and willow, *Salix*), lichens, and ferns contributed relatively little to the diet of mountain goats (Saunders, 1955).

In South Dakota, mosses and lichens composed 60% of the diet in winter; bearberry (*Arctostaphylos uvi-ursi*), 20%; pine twigs and needles, 10%; and miscellaneous items (including currant, *Ribes*; serviceberry, *Amelanchier*; rose, *Rosa*; willow; fleabane, *Erigeron*; and grasses), 10%. Aspen (*Populus tremuloides*), gooseberry (*Ribes*), and mint (*Monarda*) were avoided (Harmon, 1944). In Colorado, bunchgrasses composed 90% of the diet in winter (Hibbs et al., 1969). In another Colorado study, grasses composed 60–68% of material ingested, forbs 25–29%, and woody plants 7% (Johnson et al., 1978). In a study comparing diets of mountain goats and mountain sheep in Colorado, "goats ate more dicots, particularly forbs, whereas sheep ate more grasses and sedges" (Dailey et al., 1984:803). In one area in British Columbia, alpine fir (*Abies lasiocarpa*) was consumed during 21.6% of 1,561 observations of feeding and during 29.4% of 533 observations in another area (Geist, 1971).

In British Columbia (Hebert and Cowan, 1971) and possibly elsewhere, forage consumed by mountain goats is low in sodium. Although loss of sodium in their feces increases in spring, mountain goats do not exhibit evidence of blood-sodium deficiency, because at that season they seek and use salt licks. In Olympic National Park, they reportedly acquire salt by eating earth where hikers have voided urine (Scheffer, 1993a). Salt blocks commonly are used to lure mountain goats into traps (Rideout, 1974).

Reproduction, Ontogeny, and Mortality.—In British Columbia, the rut is most intense during November; in the week before 2 November, 11 of 13 courtship attempts were repulsed; in the remainder of the month of November only eight of 26 were repulsed and in December only two of 18 were repulsed (Geist, 1964). In South Dakota, mating activity was confined to November and early December (Richardson, 1971). The mating season in other areas was similar, with young born in May–June and gestation reported variously as 147, 180, 185.8 ± 1.3 ($\pm SD$), and 192 days (Hayssen et al., 1993). We know of no quantitative evaluation of the time and duration of the reproductive season or length of gestation in Oregon.

In South Dakota, no young-of-the-year participated in reproductive activities, but yearling males were observed to participate and about half of 2-year-old females were accompanied by young (Richardson, 1971). In Montana, yearling males collected in October–November possessed secondary spermatocytes and spermatids in seminiferous tubules, or later, spermatozoa in epididymides (Henderson and O'Gara, 1978). Houston et al. (1989) believed that sexual maturity was related to attainment of a minimum body mass ($\bar{X} = 70$ kg) irrespective of age. They also believed that attainment of minimum mass was related to availability of resources.

Vaughan (1975) reported that three young were produced in the Wallowa Mountains each year, 1972 and 1973. All six young were produced by different females, causing him to suggest that production of offspring was no more frequent than during alternate years. No twins were observed, but elsewhere twins are common in some populations, and triplets have been observed (Lentfer, 1955; Vaughan, 1975). At least 10 sexually mature females were in the Wallowa Mountain population during each of the 2 years, and no mortality was observed, suggesting that low productivity was responsible for the slow increase in numbers observed. Stevens (1982) indicated that forage quality, body condition, and number of offspring/100 adult females in Olympic National Park, Washington, were related positively.

Nursing time commences to decline early, and neonates often commence to forage when <1 week old and may be "nutritionally weaned" by 4 weeks of age. They may not be dependent upon milk as early as 1–2 weeks of age. However, occasionally, a maternal female may permit its yearling or even its 2-year-old to suckle, although remain outwardly antagonistic toward unrelated goats (Hutchins and Hansen, 1982). These "retained young" bed close to and "bleat" when they become separated from the maternal female; maternal females do not defend retained young against threats from other goats as they commonly do for younger offspring (Hutchins and Hansen, 1982). Suckling by older offspring is believed to enhance social bonds.

Growth of mountain goats maintained in captivity was seasonal, with body mass stable (young) or declining (≥1-year-olds) during winter (Houston et al., 1989). Losses in mass among males commenced about 1 month before onset of breeding and became more pronounced with age. Body mass increased in summer except among lactating females. Maximum body mass was attained at 5 years for males and 3–4 years for females; mass among younger cohorts varied considerably, but was greater when population density was lower (Houston et al., 1989).

A variety of large carnivores reportedly prey on mountain goats at least occasionally; the cougar (*Puma concolor*) probably is the most serious predator (Rideout and Hoffmann, 1975). However, avalanches and rock slides probably constitute the foremost danger to mountain goats, as they have evolved a stereotyped behavior pattern of running to cliffs with overhanging rock and pressing themselves against the rock wall in response to loud noises (Geist, 1971).

Habits.—Mountain goats are denizens of high-altitude, remote, and barren montane regions where they are capable of moving through exceedingly rugged and precipitous terrain with speed and agility. Harmon (1944) reported a mountain goat to move down a >9-m, nearly perpendicular cliff in three bounds at a dead run and another to clear a log 1.2 m above the ground in a 2.4-m-long leap while running through 4.7 cm of snow. Nevertheless, mountain goats are known to fall occasionally, whereupon they spread their legs and slide to a stop; by doing so, they often survive bad falls (Geist, 1971).

Mountain goats shift altitude seasonally and seek shelter of timbered areas to avoid deep snow; nevertheless, windswept ridges, blown free of snow, often are used in winter (Vaughan, 1975). Mountain goats do not seem to seek shelter from strong winds, however (Geist, 1971).

Unlike mountain sheep, mountain goats bed at dusk and do not move at night (Geist, 1971). In Colorado, mountain goats fed approximately 1.5 times as much in 3-h periods before 0900 h and after 1500 h as during either of the two 3-h periods between 0900 h and 1500 h (Hibbs et al., 1969).

The stiletto-like horns of mountain goats can be lethal weapons. Geist (1967) described results of an altercation between a 3-year-old male and an old female caught in a corral trap during which the male received 32 stab wounds in the head, shoulder, chest, flank, and haunches. Three wounds had penetrated the chest, punctured both lungs, and pierced the heart muscle, and another wound had entered the rumen. Nevertheless, no evidence of wounds were visible until the unconscious goat was euthanized and skinned. Certainly, capture in a corral trap created an unusual circumstance for the wild goats; thus, the severity of the combative interaction certainly was much greater than might be expected among free-living individuals where fleeing the aggressor was possible. Most agonistic social interactions develop no further than "intense threat displays" (Geist, 1964:551). Even when displays fail to elicit the intended submissive response and actual fights ensue, most blows are directed toward the haunches because interacting goats align parallel, head to opponent's rear. As partial protection from stab wounds, a "rump shield" of thick skin covers the posterior, flanks, and brisket of males. The skin may be as much as 22 mm thick near the anus and taper to 5–6 mm ventrally and anteriorly (Geist, 1967). Nevertheless, in "all-out battles, old males still get severely punctured and lacerated" (Geist, 1967:194).

Mountain goats are gregarious, but less so than some other ungulates. Groups composed of an adult female and offspring of the previous two or three breeding seasons often coalesce with other such groups to form temporary bands (Geist, 1964). Males live alone or in pairs; occasionally they join one of the female-based groups for short periods. Males may court females falteringly at any season, but advances are soundly repulsed by females except just before the rut when females gradually become less agonistic (Geist, 1964).

At the onset of rut, males, especially older ones, soon acquire large, dark patches on the rump, belly, and flanks where they have thrown dirt in the process of digging "rutting pits" (Geist, 1964). The males sit on their haunches with back arched and head directed toward the ground; they paw the ground with one front leg. Sometimes the front legs are used alternately to throw dirt toward the belly and flanks in the construction of pits that may be 45 cm long by 30 cm wide. Males rest in the pits and may urinate in them; sometimes more than one male uses a pit (Geist, 1964). As the rut progresses, males may join the female-based groups and their courtship advances commence to be tolerated. Males near females remain largely inactive; they sit in rutting pits, stand, and occasionally and cautiously court a female. When females enter estrus, they may become aggressive, charging or rushing at courting males or other females. Nearly a month after onset of the rut, females commence to permit copulation (Geist, 1964).

Courtship consists of the male approaching the female (usually from the rear) in a crouched position with tail elevated, ears directed anteriorly, nostrils flared, tongue flicking in and out, and head jerking from side to side. The male licks the female's flank or rear and taps the female on the flank or between the hind legs with a front leg. The female turns and threatens the male with its horns, whereupon the male turns away and shows the broadside of the face while continuing to flick the tongue in and out and jerk the head from side to side (Geist, 1964). Near the peak of the rut, males may approach females rapidly and kick them solidly with a front leg; sometimes a second kick is delivered with the other leg, which may lead to mounting and copulation. Upon being courted by a male, females sometimes squat and urinate; the male tests the urine and lipcurls (flehmen). In one study, 81% of courtship advances were directed toward adult females and 19% toward yearlings and young; the latter were rejected aggressively (Geist, 1964).

Adult males, although most heavily armed and armored, are least predisposed to fight, resulting in a social system in which they are subordinate to females and yearlings (Geist, 1964).

Remarks.—The mountain goat actually is not a goat (*Capra*), but a mountain antelope more closely related to the chamois (*Rupricapra rupricapra*) of Europe and southwest Asia than to the goats (Nowak, 1991).

In March 1950, three adult males, two adult females, and one female young from Chopaka Mountain, Washington, were introduced into the Wallowa Mountains at the base of Chief Joseph Mountain. The goats, sans one adult female that died shortly after release, moved a few kilometers south onto Hurwal Divide (Bartels, 1976). By the early 1960s, mountain goats had become established farther west on Hurricane Divide. In 1965–1968, 20 individuals were removed by hunting and the population declined to ≅10 individuals. By the mid-1970s, the number of goats had risen to ≅30 (Bartels, 1976; Vaughan, 1975). Because inbreeding was suspected of being responsible for poor reproductive performance, 25 individuals from Olympic National Park and eight individuals from Misty Fiord, Alaska, were introduced on Hurricane Divide in 1985–1989.

In the Cascade Range, eight goats were released on Tanner Butte, Hood River Co., in 1969–1971, and seven others were added in 1975–1976; all came from Olympic National Park. Regular surveys of this population were not conducted, but one to four goats were reported to occur in the area until the mid-1980s. In the early 1980s, we commonly saw one of these goats as we drove eastward on Interstate 84. None has been reported in the area since 1990 (Matthews and Coggins, 1995).

In 1983–1986, six mountain goats from Clearwater, Idaho; eight from Olympic National Park; and seven from Misty Fiord were introduced at Pine Creek in the Elkhorn Mountains, Baker Co. (Matthews and Coggins, 1995; W. R. Humphrey, pers. comm., 24 August 1993). In 1992, 29 individuals including nine young were counted during aerial surveys. In 1993, despite poor visibility, 14 adults and nine young were seen (Matthews and Coggins, 1995; W. R. Humphrey, pers. comm., 24 August 1993).

As a matter of principle, we oppose introduction of exotic wild organisms irrespective of the potential they may offer for sport or commerce. Mountain goats, in addition to numerous other organisms, have had a negative impact on some ecosystems to which they were added. They were introduced (or reintroduced—Anunsen and Anunsen, 1993; Lyman, 1988) onto the Olympic Peninsula in 1925–1929 (Houston et al., 1986), where their increase was followed by measures to control the population (Houston et al., 1991) and finally a call for their elimination, which, in turn, precipitated a long, intense, and verbose controversy (Anunsen and Anunsen, 1993; Hutchins, 1995; Lyman, 1994; Scheffer, 1993a, 1993b). Likely, it is to the benefit of Oregon that mountain goat populations here have remained

small, if not to its environment, then certainly to its tranquility. Nevertheless, alpine ecosystems are fragile, and the impact that mountain goats have had on them here has not been assessed, if indeed often considered. Perhaps the effort to establish the historic distribution of the mountain goat to include parts of Oregon is an attempt to avoid the criticism of the introduction of an exotic if one of the recently established populations should increase to the point that damage to the alpine ecosystem is evident. Considering that claimed specimen-based evidence of mountain goats in Oregon remains unpublished, we are strongly skeptical of the proclaimed historic occurrence of the species in the state.

Ovis canadensis Shaw
Mountain or Bighorn Sheep

1804. *Ovis canadensis* Shaw, 51:text to pl. 610.
1829. *Ovis californianus* Douglas, 4:332.
1900. *Nemorhoedus palmeri* Cragin, 11:611.
1912. *Ovis canadensis californiana*: Miller, 79:396.

Description.—The mountain sheep (Plate XXXII) is a medium-sized, largely brownish bovid with a white rump patch, muzzle, venter, and rear portion of the legs. The tail is blackish brown on the exposed surface (Shackleton, 1985). The hooves (toes 3 and 4) are equipped with a rubberlike pad that facilitates negotiating rocky terrain; toes 2 and 5 are reduced to small dewclaws. The ears are relatively small and somewhat pointed. Both sexes are equipped with horns; those of males are massive and spiral outward, whereas those of females are relatively thin, recurved, and mostly directed upward and posteriorly. Horn growth reflects nutritional status, thus is largely seasonal with distinct annuli formed in winter. As the horns of adult males form nearly complete circles (Fig. 13-19), the tips face anteriorly and often are damaged ("brooming"), with as much as 30 cm removed (Shackleton, 1985), thereby lessening the value of counts of annuli in estimating age. Age in females >5 years old is difficult to estimate from annuli in horns (Shackleton, 1985). Mountain sheep possess lacrimal, inguinal, and interdigital glands, the products of which may be involved in a social context.

The skull of the mountain sheep (Fig. 13-19) is acutely triangular. In males, modifications to support the massive horns and to permit their use in butting are striking: the cornual and frontal sinuses are greatly expanded and contain bony septa, and the blunt-end horn cores cover most of the braincase and curve laterally and posteriorly (Shackleton, 1985). In females, the smaller horns require less extensive modifications of the skull for support. There are no upper incisors and only rarely upper canines; lower incisors are appressed and lower canines are incisiform. The six cheekteeth in both upper and lower jaws are semihypsodont and selenodont; sometimes p2, or more rarely P2, is absent (Shackleton, 1985).

Distribution.—The geographic range of mountain sheep extended from southeastern British Columbia and southwestern Alberta south along the Cascade and Sierra Nevada ranges into Baja California Sur and Sonora and eastward through Montana to western North Dakota, South Dakota, and Nebraska; central Colorado and New Mexico; extreme western Texas; and eastern Coahuila (Fig. 13-20; Hall, 1981).

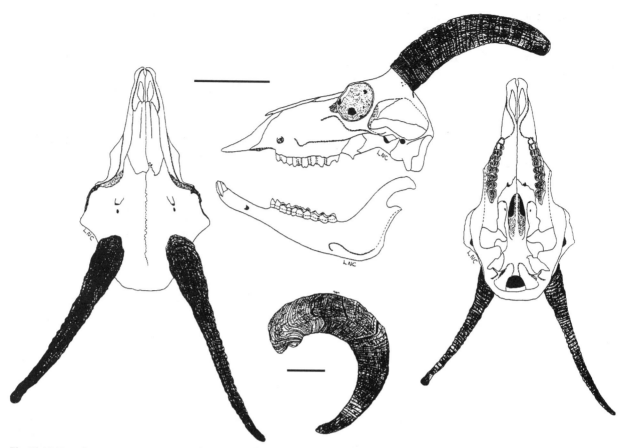

Fig. 13-19. Dorsal, ventral, and lateral views of the cranium, and lateral view of the dentary of a female mountain sheep, *Ovis canadensis* (OSUFW study); and lateral view of the horn sheath of a male (OSUFW study). Scale bar equals 10 cm.

Mountain sheep once occurred throughout much of the nonforested region of Oregon east of the Cascade Range, but were extirpated before the middle of the 20th century (see Remarks for the history of extirpation and reintroduction). In 1954–1994, the Oregon Department of Fish and Wildlife released mountain sheep at 38 sites in Baker, Gilliam, Grant, Harney, Lake, Malheur, Sherman, and Wallowa counties. The agency considered populations established at 26 sites, including all counties in which animals were released; three herds were established through natural dispersal. The range map (Fig. 13-20) includes records based on museum specimens from original herds and sites of reestablished herds.

Pleistocene records of *O. canadensis* are known from numerous sites in western North America; many of these specimens are exceptionally large (Kurtén and Anderson, 1980). Shackleton (1985), in a review of the biology of the species, indicated that Pleistocene records for Oregon existed. We found no records for Oregon mentioned in the three references he cited in support of his contention.

Geographic Variation.—Two of the seven recognized subspecies occurred in Oregon: *O. c. canadensis* in the northeastern corner of the state and the somewhat smaller *O. c. californiana* in the remainder of the range (Hall, 1981). Mountain sheep reintroduced into Wallowa and Baker counties originated from stock obtained in Alberta, Washington, Idaho, and Colorado; those introduced elsewhere originated from stock obtained in British Columbia. Thus, stock of races appropriate to the respective regions

seems to have been used for reintroductions.

Habitat and Density.—"Mountain sheep" is an appropriate name for members of this species as they "are never far from precipitous terrain" (Geist, 1971:4). Requisite components of mountain sheep habitat are "visibility, escape terrain, and abundant continuous forage" (Risenhoover et al., 1988:347). Open areas on rocky slopes, ridges, rimrocks, cliffs, and canyon walls with adjacent grasslands or meadows, but few trees, provide those requisites and form the primary habitat of this species. Within these areas, mountain sheep use various physiognomic and vegetative aspects seasonally and for specific activities. Shannon et al. (1975) were able to explain 65% of the variation in seasonal distribution of mountain sheep by use of 11 habitat variables that included slope, distance to escape terrain, elevation, aspect, biomass and nitrogen content of palatable grasses, forest cover, and snow depth. An occasional tree may be used as a thermal retreat, but vegetation that obstructs long-distance vision (the primary sense used to detect potential danger) is avoided (Kornet, 1978; Risenhoover and Bailey, 1985). Invasion of conifers resulting from suppression of wildfires has rendered some of the original habitats occupied by mountain sheep unsuitable for reintroduction. However, management of these habitats by controlled burning likely will render them capable of supporting herds of mountain sheep. In some areas occupied by mountain sheep, controlled burning is employed as a management tool already (Coggins and Matthews, 1992).

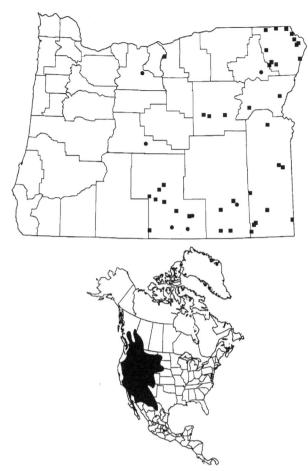

Fig. 13-20. Localities in Oregon from which museum specimens (circles) obtained from original herds of the mountain sheep, *Ovis canadensis*, were examined (listed in APPENDIX, p. 539), localities of reintroduced herds, and herds naturally established from reintroduced populations (squares—*Oregon's bighorn sheep management plan 1992–1997*, Oregon Department of Fish and Wildlife), and the distribution of the mountain sheep in North America. (Range map redrawn from Hall, 1981:1117, map 545, with modifications.)

On Hart Mountain, Lake Co., mountain sheep bedded in four areas, all classed as cliff-shrub habitat (Kornet, 1978). On arising, the sheep moved to adjacent bunchgrass (Idaho fescue, *Festuca idahoensis*) and big sagebrush (*Artemisia tridentata*)-bunchgrass habitats; they foraged above the cliffs 42% of the time, into the sagebrush basin (big sagebrush; low sagebrush, *A. arbuscula*; Thurber needlegrass, *Stipa thurberiana*; Idaho fescue; milk-vetch, *Astragalus*) 28% of the time, and onto the lower slopes 20% of the time. Boulder and rock outcrop habitats were used for resting and standing (Kornet, 1978). In autumn, sheep fed largely on top of Hart Mountain in Sandberg bluegrass (*Poa sandbergii*) and low sagebrush habitats; in spring these areas were seldom used. In spring, the cliff-shrub habitat (big sagebrush; rockspirea, *Holodiscus dumosus*; spike-fescue, *Hesperochloa kingii*) was used extensively by females and immatures (Kornet, 1978). Payer (1992:80) considered that water was "a limiting resource . . . and its availability affected the distribution" of sheep on Hart Mountain.

On Steens Mountain, Harney Co., nine of 20 described habitats were used by mountain sheep. In spring and summer, cliffrock (currant, *Ribes*; bristleweed, *Haplopappus*; bottlebrush, *Sitanion*; bluegrass, *Poa*) and cliffrock-shrub (big sagebrush; needlegrass; lupine, *Lupinus*; violet, *Viola*) habitats were used extensively by mountain sheep; in autumn, shrub-bluebunch wheatgrass (*Agropyron spicatum*), cliffrock-shrub, meadow (cinquefoil, *Potentilla*; sedges, *Carex*; licorice-root, *Ligusticum*; clover, *Trifolium*), and cliffrock (in order) were used most frequently. In winter, shrub-bluebunch wheatgrass habitat was used 68% of the time. In all seasons, habitats used most for foraging also were used most for resting (Van Dyke, 1978).

Distinct, although sometimes partly overlapping, areas are used by groups consisting mostly of adult males and those composed of females and their offspring (Kornet, 1978; Van Dyke, 1978). On Hart Mountain in 1990–1991, only small differences in use of habitats were noted between years; males were observed ≅50% of the time in big sagebrush-bunchgrass habitats and at frequencies ≤20% in the four remaining habitats. Females and immatures were observed ≅40% of the time in cliff-talus shrub habitats, ≅30% in big sagebrush-bunchgrass habitats, and ≤20% in the remaining three habitat types (Payer, 1992). A few kilometers to the northeast on Poker Jim Ridge (a northeastern extension of the escarpment along the northwestern boundary of Hart Mountain National Antelope Refuge, Lake Co.), much greater differences between years were noted. In 1990 males used low sagebrush plateau nearly 40% of the time, low sagebrush-bunchgrass and cliff-shrub habitats each ≅20% of the time, and the remaining three habitat types ≤12% of the time. In 1991, males used juniper-low sagebrush ≅35% of the time, low sagebrush plateau ≅20% of the time, low sagebrush-bunchgrass and big sagebrush-bunchgrass each slightly >15% of the time, and the remaining two types <10% of the time (Payer, 1992). Comparable data for females and immatures on Poker Jim Ridge were not presented.

Numbers of mountain sheep in many of the reestablished populations in Oregon have increased rapidly. For example, four individuals were released on Steens Mountain in November 1960 and seven more were added the following March (*Oregon's bighorn sheep management plan 1992–1997*— Oregon Department of Fish and Wildlife). By 1965, the estimated population was 35; by 1970, it was 70; by 1975, it was 100; and by 1977, the low estimate in December was 133 (Van Dyke, 1978). By 1992, the estimate was 250. Hunting of older (horns with three-quarters curl) males was initiated on Steens Mountain in 1968, and by 1975, 22 had been taken; to 1993, 178 were killed on the mountain by permit holders. In 1992, all populations of mountain sheep in the state were estimated to total 2,410. During 1965–1991, 765 hunters harvested 591 legal males (*Oregon's bighorn sheep management plan 1992–1997*— Oregon Department of Fish and Wildlife). Van Dyke (1990) indicated that limiting harvest of males to older cohorts by imposing the three-quarters curl restriction was without biological basis, was enforced inconsistently, and possibly was responsible for hunters abandoning kills that did not meet the requirement, then shooting another male. Likely, harvest of adult males and capture and removal of both sexes and all age-classes for transplanting have been responsible for many of the reestablished populations continuing to be productive.

On Hart Mountain National Antelope Refuge, density of mountain sheep on male-occupied ranges in summer was

0.6–0.9/km² in 1990 and 1.1–2.1/km² in 1991. Densities on female-occupied ranges were 5.3–24.5/km² and 8.9–34.7/km² in the 2 years, respectively (Payer, 1992).

Diet.—Although forbs and shrubs are consumed and may contribute significantly to the diet seasonally, the staple of the diet of mountain sheep is grasses. Nevertheless, there is "no consistent species-specific pattern" in seasonal changes in the diet of mountain sheep (Shackleton, 1985:4). On areas managed by controlled burning, mountain sheep are able to improve the quality of their winter diet by being better able to select the most nutritious forage; quality of the forage itself was not improved significantly by fire (Hobbs and Spowart, 1984).

Kornet (1978) observed mountain sheep to feed on bluebunch wheatgrass and giant wildrye (*Elymus cinereus*) on Hart Mountain. She also observed sheep to forage on mountain snowberry (*Symphoricarpos oreophilus*), common chokecherry (*Prunus virginiana*), rockspirea, Wood's rose (*Rosa woodsii*), and greasewood (*Sarcobatus vermiculatus*).

In British Columbia, spring–fall rumen samples (*n* = 6) contained 85% grasses, 2% forbs, and 13% shrubs, whereas the range composition during that period was 62–78% grasses, 12–20% forbs, and 5–8% shrubs. In winter, the diet as evaluated by counts of grazed stems consisted of 43–62% grasses, 3–4% forbs, and 21–49% shrubs (Blood, 1967). In Yellowstone National Park, Wyoming, microhistological examination of feces indicated that 54–59% of the winter diet was grasses and sedges, 6–7% was forbs, and 35–45% was shrubs and trees (Keating et al., 1985). In Alberta, examination of feces by the same method revealed that the winter diet of adult males differed significantly from that of females and immatures. Shank (1982) believed that spatial segregation was responsible for the differences, not selection for specific forage species by the different cohorts. Because sex and age-classes of mountain sheep are segregated spatially and forage available to the cohorts differs (Kornet, 1978; Payer, 1992; Van Dyke, 1978), we have little doubt that the diet of adult males differs from that of females and immatures in Oregon.

Reproduction, Ontogeny, and Mortality.—Female mountain sheep in most populations do not produce their first offspring until ≅3 years old, whereas in some populations at least some females produce young when ≅2 years of age (Shackleton, 1985). On Steens Mountain, young:female ratios approached 1:1, a finding consistent with production of twins, production of young by 2-year-olds, or both, and with annual continued production of young by older females. One set of twins was observed in 1976, three sets in 1977 (Van Dyke, 1978). The relatively small number of twins, but the relatively large proportion of the population composed of 1-year-olds is circumstantial evidence of breeding by 2-year-olds (Van Dyke, 1978). Kornet (1978) observed no evidence of twins on Hart Mountain in 1976 or 1977.

On Steens Mountain, the first young were born in mid-April in 1977 and the peak of birthing was in the last week of April and the 1st week of May (Van Dyke, 1978). On Hart Mountain during the same year, young were born from mid-April to late May with the peak of births in the 1st week of May (Kornet, 1978). The gestation period of *O. c. californiana* in captivity averaged 174.2 days (range, 171–178 days; *n* = 20—Shackleton et al., 1984). Therefore, the rut for these Oregon populations probably extends from about mid-October to the end of November. In Alberta, the rut lasts about 2–3 weeks and estrous females were noted as early as 19 November and as late as 11 December (Geist, 1971).

Neonates born to captive females weighed an average of 3.3 kg (range, 2.8–3.7 kg; *n* = 9—Geist, 1971). Geist (1971) believed that nutrition, size of the maternal female, and severity of the winter during pregnancy affected size at birth. He speculated that under optimum conditions neonates might weigh as much as 5.0–5.5 kg. Neonates are precocial and follow the maternal female shortly after birth. They suckle in short bouts as frequently as 3 times/h (Shackleton, 1985). Some commence to graze as early as 14 days of age, but many commence later. Young usually are weaned by 4–6 months of age (Geist, 1971). Behavioral maturity is attained by 2–3 years by females and by 6–7 years by males (Geist, 1971).

Although mortality among young may be high, especially in stable or declining populations, most populations in Oregon are increasing (*Oregon's bighorn sheep management plan 1992–1997*—Oregon Department of Fish and Wildlife). Van Dyke (1978) reported finding only one dead young and observing three unthrifty young and six with diarrhea during a 2-year study on Steens Mountain. However, young:maternal-female ratios among sheep observed each month declined from as high as 96:100 at 1 month of age to as low as 18:100 by 20 months of age (Van Dyke, 1978). From repeated censuses of young, Cottam (1985) concluded that some mortality occurred during the first 3 months of life on Hart Mountain. He suggested that losses likely were from lungworm (*Protostrongylus*)-induced pneumonia; lungworms can be transmitted to fetuses transplacentally (Hibler et al., 1972; Kistner and Wise, 1979). In Alberta, where the birthing period extended from 17 May to 21 July, survival of young to 1 year of age averaged 44% for those born in May but only 11% for those born in June–July. Festa-Bianchet (1988) suggested that inadequate nutrition of maternal females during lactation was responsible for the greater mortality in late-born offspring as quality of forage declined as the summer progressed. He further indicated that more late-born young were produced where the density of females was highest. Whether populations in Oregon have increased to the point that this phenomenon is operational is unknown.

In 1984, scabies (*Psoroptes ovis*) was introduced into mountain sheep near the Wenaha River, Wallowa Co., when 27 individuals from Idaho were released into the herd. Nearby herds in Washington and Oregon also became infested, with >50% of the individuals therein affected (Foreyt et al., 1990). Mortality in *O. c. canadensis* in Oregon, was not observed, but in Washington, 50 in a herd of 80 *O. c. californiana* succumbed (Foreyt et al., 1990).

In 1986, a herd near the Lostine River, Wallowa Co., became infected with *Pasturella* pneumonia, seemingly from contact with domestic sheep (Coggins, 1988). The population declined from ≅100 to ≅34 individuals and survival of neonates was low for 2 years, but improved subsequently. By 1992, the population was estimated to have increased to 55 (Coggins and Matthews, 1992).

Wolves (*Canis lupus*) were believed to have been a major predator of mountain sheep (Shackleton, 1985), but in Oregon, their present threat is essentially nil, if indeed they ever constituted much of a threat here. Coyotes (*Canis latrans*), bobcats (*Lynx rufus*), cougars (*Puma concolor*), and wolverines (*Gulo gulo*) are known to take mountain sheep occasionally (Shackleton, 1985). In addition to the ability to negotiate precipitous slopes rapidly, groups of mountain sheep sometimes exhibit defensive behavior toward predators (Hornocker, 1969*a*; Shank, 1977). Sheep in large groups forage significantly farther from escape terrain and spend less time alert than those in small groups (Risenhoover and Bailey, 1985).

Regulated hunting of mountain sheep in reestablished herds commenced in 1965 on Hart Mountain. Permits for taking sheep were issued for Steens Mountain in 1968; the Owyhee River Canyon, Malheur Co., in 1973; and both Strawberry Mountain, Grant Co., and Hurricane Divide in 1978. From 1965 to 1993, permits were issued for taking one or more mountain sheep in 25 reestablished herds and 685 were taken by 866 permit-holders (Fig. 13-21).

Habits.—On Steens Mountain, most sheep were active early in the morning and late in the evening but only about half were active at midday, the typical pattern observed in most populations (Van Dyke, 1978). However, on Hart Mountain, the sheep "did not exhibit the pattern of morning and afternoon feeding periods separated by a distinct midday rest" (Kornet, 1978:20). Van Dyke (1978) found some evidence of nocturnal activity in that occasionally sheep bedded at one locale at dark were observed 400–800 m away at first light.

Kornet (1978) observed mountain sheep to flee from approaching mule deer (*Odocoileus h. hemionus*) on three occasions, yet feed among the deer on four other occasions. She attributed the greater boldness to groups containing adult males and to the deer first being visible at a distance. On five occasions sheep were observed to move away from domestic cattle. Sheep fled to the nearest cliff terrain from human disturbance and in nine instances did not return to the area of the disturbance for >36 h (Kornet, 1978).

Minimum-area home-range size for radio-tracked adult males averaged 25.2 km^2 (range, 13.6–41.1 km^2; n = 11) on Poker Jim Ridge and 15.5 km^2 (range, 11.8–18.1 km^2; n = 3) on Hart Mountain (Payer, 1992). Comparable data for females were not presented.

Mountain sheep are capable of moving with speed and agility through the precipitous terrain in which they live. They sometimes use the stiff-legged "stotting" gait similar to that used by mule deer in rough terrain.

Except immediately before and during the rut, mountain sheep associate in groups consisting either of ≥3-year-old males or of adult females and immatures of both sexes. Strong dominance relationships are maintained in groups of males, but are weaker in groups of females and immatures (Shackleton, 1985). Among males, horn size, body size, and fighting ability determine social status (Geist, 1971), but among females characteristics that determine within-group status are less obvious (Shackleton, 1985).

Among males, establishing and maintaining social position is a year-round activity (Geist, 1971). Within groups, the dominant male acts toward other sheep as if they were females, thus does not behave aggressively toward subordinates. It is attempts at social climbing by subordinates that precipitate dominance fights; these are the most

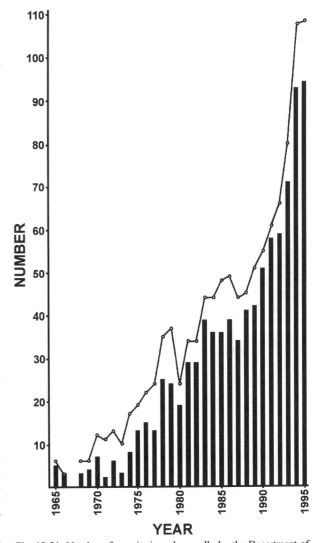

Fig. 13-21. Number of permits issued annually by the Department of Fish and Wildlife for the taking of mountain sheep (line) and number of mountain sheep taken annually by hunters, 1965–1995 (bars). Data from Oregon Department of Fish and Wildlife annual reports.

spectacular social interactions among mountain sheep (Geist, 1971). Males with gross differences in horn size do not fight; horn displays and sexual behavior usually are sufficient stimulus for the male with the smaller horns to behave subordinately by horning, nuzzling, and rubbing on the individual with the larger horns. However, males with similar-sized horns engage in fights that involve sledge-hammerlike blows with the horns. Thus, not only are the horns exhibited as a symbol of rank, they also serve as the "weapons" with which disagreements regarding rank are settled. Rarely does the dominant male overpower a challenging smaller male with heavy blows; instead it displays or courts the subordinate and uses its heavy horns to parry the butts delivered by the subordinate (Geist, 1971). Sometimes fighting males attract other males that may join the melee in support of one of the males. Among mountain sheep, rank has its privileges, as dominant males do most of the breeding; thus, selection is strong for large horns by which dominance is acquired and maintained (Geist, 1971). Aggressive interactions also occur among members of groups of females and immatures, but they usually involve physical contact, rarely displays.

At onset of the reproductive season, males join groups containing adult females. Males examine females for evidence of estrus by approaching and sniffing, and when females urinate in response to being sniffed, males sniff their genital areas and lipcurl (flehmen) as a test of the estrous state. As estrus approaches, a large dominant male commences to tend the female, driving competing males away. Fights at this time involve physical contact, but do not include displays that occur during encounters concerning dominance. Subordinates sometimes are able to mount estrous females, but the females avoid mating with smaller males (Geist, 1971). When a large male becomes exhausted from dealing with attempts by subordinates to mate with an estrous female, the female sometimes courts the large male by behaving like an aggressive subordinate. This behavior stimulates the male to mount (Geist, 1971).

Remarks.—"Originally, mountain sheep inhabited every canyon, cliff, and lava butte as well as many of the rough lava beds of Oregon east of the Cascade Mountains" (Bailey, 1936:65). By 1915, sheep were extirpated in the southern part of the state (Bailey, 1936) and by 1945 those in northeastern Oregon were gone (*Oregon's bighorn sheep management plan 1992–1997*— Oregon Department of Fish and Wildlife). Excessive hunting, parasites and diseases introduced with domestic sheep, and unregulated grazing by livestock are believed to have contributed to the extirpation of mountain sheep in Oregon.

In 1939, 23 mountain sheep from Montana were released on Hart Mountain; a remnant population persisted until 1947. In 1954, 20 mountain sheep from British Columbia were released in a large pen erected on the west slope of Hart Mountain, and were maintained there until 1957, when 18 were released on Hart Mountain (Lemos and Willis, 1990; Mace, 1969). From this population during the following 35 years, 591 individuals were captured and transplanted to 29 other sites in Oregon, 4 in Nevada, and 3 in Washington. In addition, a population was established on the Owyhee River, Malheur Co., from dispersal of animals from Idaho (*Oregon's bighorn sheep management plan 1992–1997*— Oregon Department of Fish and Wildlife).

In 1971, 20 mountain sheep from Alberta were released in Hells Canyon, Wallowa Co.; all disappeared in time. Also in 1971, 20 additional sheep from Alberta were released on the Silver Creek burn on the Lostine River, Wallowa Co. The latter introduction was successful, and surplus animals from this population and animals from Colorado, Idaho, and Washington were used to stock other areas in northeastern Oregon, resulting in the increase in number of populations in the region. Other populations were established in Wallowa Co. through dispersal of animals from these introduced populations (*Oregon's bighorn sheep management plan 1992–1997*— Oregon Department of Fish and Wildlife).

CHAPTER 14

SPECIES OF POSSIBLE OCCURRENCE IN OREGON

Micoureus alstoni (J. A. Allen, 1900)
Murine Opossum

Description.—This pouchless, long-tailed, woodrat (*Neotoma*)-sized, didelphid opossum is native to Central America. It is grayish brown on the dorsum and cream colored on the venter.

Remarks.—Two specimens in the San Diego Natural History Museum (SDNHM 17595 and SDNHM 17596) prepared by S. G. Jewett and labeled *Marmosa alstoni* were collected in stalks of bananas at Powell's Market in Portland, Multnomah Co., one in June 1938, the other in January 1939. Doubtlessly, these specimens arrived in shipments of fruit from Central America. No further evidence of the species in Oregon is available, possibly because bananas are no longer shipped as entire stalks but are separated into "hands," thereby eliminating suitable hiding places for mammals of this size.

Oryctolagus cuniculus (Linnaeus, 1758)
Old World Rabbit

Description.—In wild populations, this rabbit is similar to the eastern cottontail (*Sylvilagus floridanus*), except it averages slightly larger and lighter in color. However, individuals maintained as pets and for food commonly are white, black, brown, or splotched combinations thereof, and may weigh as much as 7.3 kg (Nowak, 1991).

Remarks.—We observed uncaged individuals of this species on several occasions in the Willamette Valley. Most were in a suburban area or in the vicinity of a farmstead, but several were far afield. Although we know of no self-maintaining wild population of the species in Oregon, the species was introduced successfully (and to the detriment of the ecosystem) on the San Juan Islands, Washington (Larrison, 1970; Nowak, 1991). We seriously doubt that one of the "domestic strains" could become established as a wild population because rabbits whose pelage contrasts markedly with the environment likely are subject to high rates of predation. However, the destruction wrought by introduced populations elsewhere (especially on islands) should be sufficient for management agencies to view release of unwanted pet rabbits with some trepidation.

Dipodomys merriami Mearns, 1890
Merriam's Kangaroo Rat

Description.—Merriam's kangaroo rat is sufficiently similar in size and color that it might be confused with *D. ordii* and *D. microps*, but it differs from those species by lacking a hallux on the hind foot.

Remarks.—*Dipodomys merriami* has been recorded in northern Nevada <48 km from the Oregon border (Hall, 1946), where it commonly is associated with alkali flats and sandy areas in which common shrub dominants are saltbush (*Atriplex*) and greasewood (*Sarcobatus*). We found no four-toed kangaroo rats among those examined from southern Harney and Malheur counties, but the possibility of the species occurring in some unsampled desert area in that region of Oregon remains.

Peromyscus boylii (Baird, 1855)
Brush Mouse

Description.—The brush mouse is a large peromyscine rodent with a tail longer than head and body, moderately long ears (shorter than hind feet), and accessory cusps on M1.

Remarks.—Specimens of the brush mouse were captured at Beswick, Siskiyou Co., California, within 3.2 km of the Oregon border (Grinnell, 1933). In 1971, one of us (BJV) sampled the small-mammal community along the east bank of the Klamath River about 10 km west of Keno, Klamath Co., south to near the California border. Several typical *P. maniculatus gambelii* were captured, but no *P. boylii*. The late M. L. Johnson, Tacoma, Washington, related to us that he had had a similar experience in the same region. Nevertheless, the occurrence of the brush mouse in extreme southern Oregon remains a possibility.

Chinchilla sp. Bennett, 1829
Chinchilla

Description.—Chinchillas are grayish, pygmy rabbit (*Brachylagus idahoensis*)-sized rodents native to South America; two species currently are recognized. They have large ears, a long bushy tail, four toes on each foot, and enormous auditory bullae.

Remarks.—Yoakum (1959) reported three individuals resided in Horse Cave, ≅3.2 km NE Bend, Deschutes Co., in November 1958; the source of the animals was unknown. Two were captured and not returned to the wild; no evidence of the third was present the following April. We know of no other evidence of the species in the state.

Phoca hispida (Schreber, 1775)
Ringed Seal

Description.—The ringed seal is much like the harbor seal (*P. vitulina*) except that it is smaller and has a long, rough dorsal pelage streaked with black and sprinkled with narrow rings of white. The venter is somewhat lighter than the dorsum and also is dotted with rings of white.

Remarks.—The ringed seal normally occurs on or under landfast ice in a circumpolar distribution. Mate (1981) reported that an individual was recorded from the California coast, but provided no particulars or reference. An individual, to get to California, must have passed the Oregon coast; thus the possibility exists for the occurrence of the species in the state.

Phoca fasciata (Zimmermann, 1783)
Ribbon Seal

Description.—The ribbon seal is much like other members of the subfamily Phocinae except that the pelage is brownish with yellowish rings encircling the neck, flippers, and torso. The yellowish bands are less prominent in females.

Remarks.—These seals normally are denizens of the edge of drift ice of the northwestern Pacific Ocean. Except for a specimen taken alive on 16 November 1962 near Morro Bay, California, the only North American specimens are from the Arctic coast of Alaska (Roest, 1964). The presence of a specimen in California suggests the possible occurrence of the species along the Oregon coast.

Cystophora cristata (Erxleben, 1777)
Hooded Seal

Description.—The hooded seal is about twice the size of the harbor seal (*Phoca vitulina*) with a black head, a silvery-gray body splotched with black, and, among adult males, an enlarged nasal cavity inflatable to form a crest or hood. Males also are able to inflate the mucous nasal septum into a red, coconut-sized, balloonlike structure usually from the left nostril.

Remarks.—The hooded seal is a species usually associated with ice floes in the North Atlantic Ocean. Individuals are known to wander as far west as northern Alaska and, in recent years in the Atlantic Ocean, were recorded as far south as Florida and Spain (Nowak, 1991). B. Mate, Oregon State University (pers. comm., 18 October 1994), advised us that a hooded seal had been identified recently in San Diego, California. It seems likely that the California specimen wandered southward after first reaching northern Alaska. Thus, the possibility of a wanderer coming ashore along the Oregon coast exists.

Alces alces (Linnaeus, 1758)
Moose

Description.—The moose is the largest member of the family Cervidae; males may attain a height at the shoulder of >2 m and weigh >800 kg. The pelage is blackish or dark brownish grading to dark gray or grayish brown on the venter and legs. The muzzle is broad and overhanging, the palmate antlers of adult males are massive, and a "bell" (waddlelike flap of skin on the throat) is present.

Remarks.—Bailey (1936) included an account of the moose in his *The Mammals and Life Zones of Oregon* on the basis of a hearsay report published by Dice (1919) that the species formerly occurred in the Blue Mountains of southeastern Washington. He also included an abbreviated account of the ill-fated introduction of five semidomestic moose at Tahkenitch Lake, Douglas Co., in October 1922. Shay (1976) further chronicled Oregon's moose adventure, which ended in 1939 with the humane dispatching of the last survivor, blinded by birdshot and unable to extract itself from between two logs. In 1974, a moose swam the Snake River and resided in Oregon for ≅2 weeks before returning to Idaho (Verts and Carraway, 1984). P. E. Matthews, Oregon Department of Fish and Wildlife (pers. comm., 5 May 1993), indicated that reports by hunters and others of the occurrence of moose in northeastern Oregon were becoming more frequent. We were unable to acquire physical evidence of the species in the state, thus, chose not to include it in our list of mammals for the state.

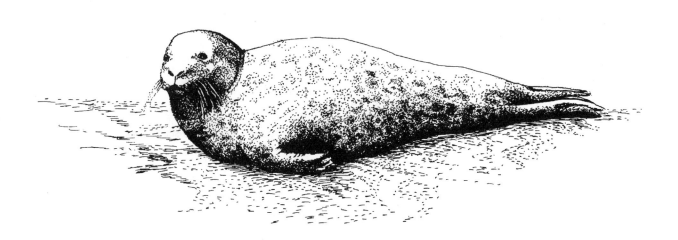

Glossary

accessory cusp—small protuberence located on labial side of occlusal surface of first and second upper molars in some *Peromyscus*.

acrocentric—type of chromosome in which centromere is subterminal.

alisphenoid bone—one of pair of bones anterior to squamosal bones that form part of braincase (Fig. 1-1).

allopatric—occupying geographically nonoverlapping ranges.

altricial—of, or pertaining to, newborn animals dependent upon maternal care; usually blind, naked, and helpless.

alveolus (pl. **alveoli**)—socket for a tooth.

amalgamated hairs—united mass of hairs.

anestrus—period of sexual inactivity between two periods of sexual activity in cyclically breeding mammals.

angle of the jaw (**angular process**)—posteriormost projection of dentary bone ventral to condyle (Fig. 1-1).

anterior—of, pertaining to, or toward front end.

anteromedial tine—in shrews, projection between and at front edge of first upper incisors.

anteroposterior—oriented from head toward tail.

antitragus—prominence of posteroventral margin of pinnae opposite tragus.

antlers—branched, bony deciduous projections from frontal bones in cervids.

apex—highest point.

appressed—squeezed together.

aquatic—living wholly or chiefly in or on water.

arboreal—inhabiting or frequenting trees.

arid—excessively dry.

articulation—joint or juncture between bones or cartilages in skeleton.

atresia—involution of ovarian follicle not destined to produce functional ovum.

atretic—having condition of atresia.

auditory bulla (pl. **auditory bullae**)—bony capsule enclosing middle ear, opening externally through external auditory meatus (Fig. 1-1).

auricular—of, or pertaining to, ear.

autosome—chromosome other than sex chromosome.

backcross—result of breeding an individual to one of the individual's parents.

baculum (pl. **bacula**)—accessory bone in penis of males of some species of mammals, homologous to os clitoris in females.

bajada—broad, alluvial slope extending from base of mountain range into basin.

basal length—distance on skull from anteriormost lower border of foramen magnum to line connecting anteriormost parts of premaxilla (Fig. 1-2).

basilar length—distance on skull from anteriormost lower border of foramen magnum to line connecting posteriormost margins of alveoli of first upper incisors (Fig. 1-2); in Lagomorpha, distance is from anteriormost upper border of foramen magnum to line connecting posteriormost margins of alveoli of second upper incisors.

basioccipital bone—single bone forming part of ventral surface of braincase and extending from anterior lip of foramen magnum to basisphenoid bone (Fig. 1-1).

basisphenoid bone—single bone forming part of ventral surface of braincase and extending from basioccipital bone anterior to pterygoids and presphenoid bones (Fig. 1-1).

bicolored—of two colors.

bicuspid—having two cusps on occlusal surface of tooth.

bifurcate—forked; two-pronged.

bilobed—divided into two lobes.

bipedal—pertaining to locomotion on only two legs.

blastocyst—single-layered, ball-shaped stage in development of embryo.

bout—spell of activity or period of action with distinct beginning and end.

brachydont—of, or pertaining to, teeth with short crowns, well-developed roots, and only narrow canals in roots.

bradycardia—slower than normal heart rate.

braincase—posterior portion of skull that encloses and protects brain.

breadth of braincase—distance across skull just posterior to distal base of zygomata in Chiroptera, Rodentia, and Carnivora; distance across skull just posterior to distal edges of external auditory meatuses in Lagomorpha; distance across skull at broadest point of cranium in Insectivora.

breadth of molariform toothrow—distance between outermost points of first upper molars.

brow tine—first tine above base of antler extending over forehead.

bunodont—of, or pertaining to, squarish teeth with low cusps, adapted for crushing.

cache—items safely hidden or concealed, intended for later retrieval; or process of hiding or concealing such items.

caecum—blind pouch in which large intestine begins and into which ileum of small intestine opens from one side.

calcar (pl. **calcaria**)—in bats, bony or cartilaginous spur attached to calcaneum (heelbone) that supports trailing edge of uropatagium (Fig. 9-1).

canines—unicuspid, single-rooted, usually elongated teeth anteriormost in maxilla, and their counterparts in dentary bone (Fig. 1-1).

carnassial teeth—in some carnivores, fourth upper premolar and first lower molar modified bladelike as an adaptation for shearing flesh and bones.

carnivorous—subsisting or feeding on animal tissues.

character displacement—situation in which differences between two species are accentuated where their geographic ranges overlap and are weakened or lost in other parts of their ranges (Brown and Wilson, 1956).

characteristic—feature typical or distinctive of organism.

cheekteeth—collectively premolars and molars, especially when premolars and molars cannot be readily differentiated; set in maxilla and dentary bone when present.

chiropatagium—flight membrane of a bat situated between third and fifth digits of forelimb; functions primarily in providing thrust in flight (Fig. 9-1).

chromosome—a rodlike body in each cell of living organisms that contains genetic material that transmits hereditary characters.

cingulum—ridge around base of tooth.

clade—group of biological taxa (as species) of organisms descended from common ancestor.

cloaca—common chamber for digestive, urinary, and reproductive tracts.

colonial—of, or pertaining to, mode of life in which several to many members of a population live in close proximity.

commensals—different species of organisms living together in which one obtains benefits from the other without damaging or benefiting it.

condyle—articulating surface.

condylobasal length—least distance between posterior margin of occipital condyles and anterior margin of premaxilla (Fig. 1-2).

condyloid process—rounded process by which dentary bone articulates with squamosal bone (Fig. 1-1).

congeners—different species of organisms being members of same genus.

conspecifics—members of the same species.

convergence—development or possession of similar characteristics by organisms of different groups caused by similarity in habits or environment.

convergent—tending to move toward or approach each other.

coprophagous—of, or pertaining to, ingestion of fecal material for purpose of extracting additional nutrients; usually normal behavior among many species of mammals.

cornoid process—process on uppermost edge of dorsal projection of dentary bone (Fig. 1-1).

corpus albicans (pl. **corpora albicantia**)—contrastingly colored fibrous scar remaining in ovary after resorption of corpus luteum.

corpus luteum (pl. **corpora lutea**)—glandular body that fills cavity of Graafian follicle after discharge of ovum; it produces progesterone, a steroid that maintains pregnancy.

cotyledonous—of, or pertaining to, type of placenta in which villi are arranged in discrete lobules, typical of ruminants.

crepuscular—of, or pertaining to, activity at twilight.

cryptogam—a plant reproducing by means of spores, e.g., ferns, mosses, algae, or fungi.

cusp—protuberance on occlusal surface of tooth.

dactylopatagium—flight membrane of bats extending between index and third, or middle, fingers (Fig. 9-1).

deciduous—of, or pertaining to, periodic shedding of teeth, antlers, hair, or other parts.

dental formula—fractionlike arrangement of numbers of each type of teeth in half of upper jaw written above numbers of each type of teeth in half of lower jaw, plus total number of teeth in all four quadrants (see Table 1-1 for dental formulas of Oregon mammals).

dentary bone—one of a pair of bones that compose the mandible (lower jaw) in mammals (Fig. 1-1).

dentine—material almost identical with bone in chemical composition, but harder and denser; composing principal mass of mammal teeth and sometimes exposed on their occlusal surfaces.

dentition—collectively, teeth.

dewclaw—vestigial claw or hoof on foot of mammal that in ordinary gait does not reach the ground.

diastema—space between adjacent teeth, usually considered greater in width than that of one tooth (Fig. 1-1).

dichotomous—divided into two similar or equivalent parts.

diel—of, or pertaining to, 24-h period.

digit—any finger or toe.

diphyodont—having two sets of teeth, milk or deciduous set replaced by permanent set.

digitigrade—walking on digits or toes, with posterior portion of foot raised off ground.

disjunct—marked by separation.

distal—located most distant from center of body.

distichous—arranged in two rows, as in barbs of feather.

diurnal—performed in, or belonging to, daytime.

divergent—tending to move away from each other.

dorsal—on, or belonging to, back of animal.

dorsum—back, especially entire back surface of animal.

drey—nest constructed by squirrels of sticks, leaves, moss, cambium, and other materials.

duff—partly decayed organic matter on forest floor.

ecotone—transition area between two adjacent ecological communities.

electrophoresis—method of determining similarities and differences among protein fractions by action of electromotive force.

elytra—hard anterior wings in beetles that serve to protect posterior pair.

emergence—act of exiting place of hibernation.

enamel—intensely hard, calcareous substance that forms thin layer capping, or partly covering, teeth of most mammals.

endemic—restricted to, or native to, particular area or region.

endogenous—arising from internal structural or functional causes.

enterohepatic—of, or involving, intestine and liver.

enzootic—disease particular to, or constantly present in, a locality.

epidermis—outer epithelial layer of skin.

epididymis (pl. **epididymides**)—elongated mass at back of testis composed chiefly of greatly convoluted efferent tubes of that organ; stores sperm while they mature.

epiphyseal—of, or belonging to, end of bone that ossifies separately and later becomes fused to shaft of bone.

epipubic bones—paired bones that project anteriorly from front of pubis into abdominal wall of marsupials.

estivation— state of dormancy not induced by low temperatures in which normal body functions are maintained; occurs during summer.

estrous—of, relating to, or characteristic of estrus.

estrus—state of sexual excitability during which female will accept male and is capable of conceiving.

eurytopic—tolerant of wide range of environmental factors.

eutherians—mammals that belong to Infraclass Eutheria, composing all extant mammals except monotremes and marsupials.

exoccipital bone—one of pair of bones lying on either side of foramen magnum and forming all or part of occipital condyles in mammals, ventral to supraoccipital bones and posterior to basioccipital bones (Fig. 1-1); often with ventral projections (paroccipital processes).

external auditory meatus—passage leading from external-ear opening to eardrum or tympanic membrane (Fig. 1-1).

extinct—no longer in existence.

extirpate—to eliminate from broad geographic areas, but not necessarily to cause to become extinct.

falcate—hooked or curved like sickle.

febrile—marked by presence of fever.

fecundity—power of producing, or having produced, offspring.

femur—proximal long bone in upper portion of hind leg.

fenestrated maxilla—maxilla in which anterior-superior portion is provided with one to several small openings, as in lagomorphs (Fig. 10-6).

fibula—smaller of two long bones in lower hind leg immediately distal to femur.

fimbriated—equipped with fringe of stiff hairs.

fissure—narrow opening, chasm, or crack of some length and considerable depth.

flank—fleshy part of side between ribs and hip.

flehmen—behavioral response, usually of a male, to odor of urine of estrous female; "typically the head is raised, the lips turned back and the nose wrinkled and breathing is stopped for a moment" (Ewer, 1968:202).

flippers—broadened and flattened front limbs adapted for swimming.

foramen (pl. **foramina**)—opening, orifice, or perforation, especially through a bone.

foramen magnum—opening in base of braincase through which spinal cord passes (Fig. 1-1).

fossorial—adapted to digging; living beneath surface of ground.

friction pads—paired digital callosities on hind toes of some soricids.

frontal bones—anteriormost pair of bones that form roof of braincase, posterior to nasal bones and maxillae and anterior to parietal bones (Fig. 1-1).

frugivorous—feeding on fruit.

fusiform—shaped like spindle; tapering toward each end.

gestation period—interval between copulation and parturition.

glean—to pick up, or gather, scattered grain or other produce.

glenoid fossa—depression in squamosal bones in which dentary bones articulate (Fig. 1-1).

Graafian follicle—a vesicle in the ovary of a mammal enclosing a developing ovum.

granivore—animal that feeds on seeds or grain.

gravid—pregnant.

greatest length of skull—distance between anteriormost point of nasals and posteriormost point of skull (for all orders except Chiroptera); for Chiroptera, anteriormost point is at anterior face of incisors.

gregarious—marked by inclination to associate with others of same kind.

hallux—big toe; first medial digit of pes.

hemoglobin—iron-bearing protein pigment in red blood cells, functioning primarily in transport of oxygen.

herbivore—animal that consumes plant material as primary component of its diet.

heterochromatin—densely staining chromatin appearing as nodules in or along chromosomes.

heterodont dentition—teeth differentiated into incisors, canines, premolars, and molars.

heterothermic—characterized by variable body temperature, usually slightly higher than temperature of environment.

hibernaculum (pl. **hibernacula**)—shelter occupied during winter by dormant animal.

hibernation—period of hypothermia and dormancy during colder parts of year; metabolic activity is reduced to nearly zero.

home range—"that area traversed by the individual in its normal activities of food gathering, mating, and caring for young. Occasional sallies outside the area, perhaps exploratory in nature, should not be considered as in part of the home range" (Burt, 1943:351).

homoiothermic—having relatively uniform body temperature maintained nearly independent of environmental temperature.

homology—similarity attributable to common origin.

horns—keratin sheaths growing over permanent bony cores on frontal bones of bovids and antilocaprids; used for offense, defense, and social interaction; also, pair of projections, such as horns of uterus.

humerus—proximal long bone in upper portion of foreleg.

hybrid—progeny of two animals of different species or varieties.

hydric—characterized by abundance of moisture.

hypotonic—having lower osmotic pressure than normal.

hypsodont—of, or pertaining to, teeth having high or deep crowns and short roots.

immergence—act of entering place of hibernation.

incisiform—shaped like an incisor.

incisive foramen (pl. **incisive foramina**)—hole in bony roof of mouth at juncture of premaxilla and maxilla through which passes nasal branches of palatine arteries and nasopalatine ducts of Jacobson (Fig. 1-1).

incisors—anteriormost kind of teeth found in mammals, usually unicuspid and chisel-shaped; usually functioning as nipping or pinching teeth; if present, set in premaxillae and dentary bones (Fig. 1-1).

infraorbital foramen (pl. **infraorbital foramina**)—opening on side of rostrum (maxilla) leading from facial area into orbit.

inguinal—of, or pertaining to, or in region of groin.

insectivore—animal that feeds on insects; member of order Insectivora.

insectivorous—feeding on insects.

intergrade—to blend one into another through intermediates; animal that possesses characteristics of two definable and adjacent geographic races.

interorbital breadth—least distance across frontal bones between orbits (eye sockets; Fig. 1-2).

interparietal bone—unpaired bone on posterior roof of braincase immediately anterior to supraoccipital bone and between two parietal bones (Fig. 1-1); not present in some mammals.

intramuscular—going into muscle.

intraperitoneal—going into abdominal cavity.

introgression—spread of genes of one species into gene complex of another.

intromission—introduction of penis into, or its maintenance within, vagina during coitus.

jugal bones—paired bones that form part of zygomata between maxillae and squamosal bones (Fig. 1-1).

karyotype—chromosomal complement of cell.

keel—flap of skin that occurs along posterior edge of calcar of some bats (Fig. 9-1).

Krummholz—stunted forest, characteristic of most alpine regions.

labial—on cheek side; toward lips; buccal.

lacrimal bones—paired bones in front of orbits through which tear ducts pass (Fig. 1-1).

lacrimal pits—depression in lacrimal bones, especially in cervids, in which preorbital glands are situated (Fig. 1-1).

lambdoidal crest—ridge extending laterally along posterior-superior margin of braincase and near juncture of supraoccipital and parietal bones.

lateral—at, near, or toward sides.

length of auditory bulla—distance from anteriormost to posteriormost points of auditory bulla (Fig. 1-1) parallel to long axis of skull.

length of diastema—distance from posteriormost edge of alveolus of lower incisors to anteriormost edge of alveolus of molariform toothrow (Fig. 1-1).

length of ear from notch—distance from notch at bottom of ear to distal tip of pinna (Fig. 1-3).

length of hind foot—distance from back of heel to end of longest toenail (Fig. 1-3).

length of mandible—distance from anteriormost point of dentary bone, in front of incisors to distalmost point of condyloid process (Fig. 1-1).

length of tail—distance from base of tail where it attaches to pelvis to distal end, exclusive of hair on tip of tail; always measured along top of tail (Fig. 1-3).

lingual—on tongue side.

loops of enamel—ringlike or angular configurations of enamel surrounding puddles of dentine on occlusal surfaces of teeth and separated from each other by reentrant angles, especially in Arvicolinae (Fig. 11-94).

mammae—milk-producing glands and accessory parts unique to mammals and usually located on ventral surface.

mandible—lower jaw, composed of dentary bones.

manus—forefoot or hand.

marine—of, or pertaining to, the sea.

marsupium—fold of skin forming pouch and enclosing mammary glands in some marsupial mammals.

mast—nuts and acorns on forest floor used as food by animals.

mastoid breadth—greatest width of skull across mastoid processes (Fig. 1-2).

maxilla (pl. **maxillae**)—one of a pair of large bones that form much of skull anterior to braincase and that bear all upper teeth except incisors (Fig. 1-1).

melanism—dense black pigmentation of pelage.

mental foramen (pl. **mental foramina**)—small opening, sometimes paired, near anterior end of each ramus of mandible.

mesic—having, or characterized by, moderate amount of moisture.

metacarpal—long bone in forefoot that separates proximal bone of a digit from a carpal (bone of wrist); region of forefoot containing these bones.

metatarsal—long bone in hind foot that separates proximal bone of a digit from a tarsal (bone of ankle); region of hind foot containing these bones.

metatarsal gland—skin gland in metatarsal region of hind foot that produces odoriferous compounds used in communication; often surrounded by tuft of long hairs, as among cervids.

microcosm—miniature universe; ecosystem.

micron (pl. **micra**)—one thousandth of a millimeter.

molariform—molarlike.

molariform teeth—collectively premolars and molars (see **cheekteeth**).

molars—posteriormost kind of teeth in mammals, commonly complex and many cusped (except simple in phocids and otariids), and lacking deciduous precursors (Fig. 1-1); if present, set in maxillae and dentary bones.

monotypic—including only one kind or type.

morphology—study of features that compose form and structure of animals.

morphometry—measurements of form and structure of animals.

multiparous—having had more than one litter.

nadir—lowest point.

nape—back of neck.

naris (pl. **nares**)—opening of nose, either external or internal.

nasal bone—paired bones that form roof of nasal cavity, usually situated dorsal to premaxillae and maxillae.

nasal length—distance from most anterior to most posterior points of nasal bones.

nasal septum—bony and cartilaginous partition between nasal passages.

natality—rate at which animals are added to a population through birth.

natatorial—adapted for swimming.

nectarivorous—feeding on nectar.

neonate—newborn animal.

neopallium—nonolfactory region of cortex of brain.

niche—role of organism in ecological community, involving especially its way of life and its effect on the environment.

nictitating membrane—thin membrane beneath lower lid of eye and capable of extending across eye in some animals.

nocturnal—of, or pertaining to, the night; especially active at night.

nulliparous—never having been pregnant.

occipital condyle—smooth and sometimes raised surface on exoccipital bone (and sometimes part of basioccipital bone) by which skull articulates with first cervical vertebra (atlas; Fig. 1-1).

occipitonasal length—least distance between posteriormost part of occiput of skull and anteriormost part of nasal bones (Fig. 1-2).

occiput—in general, posterior part of skull.

occlusal surface—grinding or biting surface of tooth.

ocherous—colored a reddish yellow.

omnivorous—feeding on both plant and animal material.

orbit—bony socket in skull that surrounds and protects eyeball (Fig. 1-1).

os clitoris (pl. **os clitores**)—small accessory bone in clitoris of females of some species of mammals, homologous to baculum in males.

ovulation—process of discharging ova from ovary.

palatal length—distance between anterior edge of premaxillae and anteriormost point on edge of palate (Fig. 1-2).

palate—roof of mouth; structures that separate mouth from nasal cavity; formed by parts of premaxillae, maxillae, and palatine bones.

palatilar length—distance from posterior edge of alveoli of first incisors and anteriormost point on posterior edge of palate (Fig. 1-2).

palatine bones—paired bones situated anterior to pterygoids and posterior and medial to maxillae and forming posterior edge of hard palate (Fig. 1-1).

papilla (pl. **papillae**)—small nipplelike projection of skin, often containing tactile corpuscles.

parapatry—of, or pertaining to, having adjacent, but nonoverlapping, geographic ranges; special case of allopatry.

pararhinal—alongside nose.

parietal bones—paired bones roofing braincase, posterior to frontal bones and anterior to supraoccipital bones (Fig. 1-1).

paroccipital process—ventral, spinelike extension of exoccipital bones, immediately posterior to auditory bullae (Fig. 1-1).

parturition—process of giving birth.

patagium—fold of skin connecting forelimbs and hind limbs.

pate—crown of head.

pectoral—of, on, or pertaining to chest.

pedicel—stalk or stem supporting another structure, such as projections on frontal bones that support antlers in male cervids.

pelage—collectively, hair of a mammal.

pentadactyl—having five fingers and five toes.

perineal—of, or pertaining to, area between anus and external genitalia.

periosteal bone—bone that lies immediately beneath the periosteum, a membrane of connective tissue that covers all bones except articular surfaces; adhesion lines used in determining age in some mammals.

periphery—edge.

pes—hind foot.

phylogenetic—based on natural evolutionary relationships.

pinna (pl. **pinnae**)—fleshy external ear.

piscivorous—feeding on fishes.

pitfall—trap constructed of container (e.g., can or bucket) buried flush with ground, into which animals fall and are captured.

plagiopatagium—flight membrane of bat supported by side of body, humerus, radius and ulna, fifth digit, and hind legs (Fig. 9-1); functions primarily in providing lift during flight.

plantar tubercle—small prominence on sole of foot.

plantigrade—walking on sole of foot in such a way that heel touches ground.

playa—dry desert basin formed by evaporation of water in a shallow lake.

plumose—featherlike, like tail of squirrel.

pneumonic—of, or pertaining to, lungs.

pod—number of animals closely clustered together, especially phocids and otariids.

pollex—thumb; first medial digit on manus.

polytocous—producing more than one young at a time.

polytypic species—species represented by several types or subdivisions.

posterior—of, pertaining to, or toward rear end.

postglenoid length—distance from posterior edge of interior of glenoid fossa to posteriormost point on occipital condyles (Fig. 1-2).

postorbital breadth—least distance across top of skull posterior to postorbital processes (Fig. 1-2).

postpartum—following parturition.

precocial—of, or pertaining to, newborn animals capable of much independent activity from birth.

prehensile—adapted for seizing or grasping by wrapping around.

prelacrimal vacuity—open space in bony covering of each side of rostrum in cervids and antilocaprids; anterior to lacrimal bones, ventral to nasal bones, and posterior and superior to premaxillae (Fig. 1-1).

premaxilla (pl. premaxillae)—one of a pair of bones at anterior end of rostrum that contribute both to rostrum and to bony palate; always containing upper incisors, if present (Fig. 1-1).

premolars—cheekteeth anterior to molars and posterior to canines; simple to complex in structure with deciduous precursors; if present, set in maxillae and dentary bones (Fig. 1-1).

presphenoid bone—single, complex bone that lies anterior to basisphenoid bone and posterior to vomer bone; forms part of floor and lateral walls of braincase (Fig. 1-1).

primaparous—having given birth one time.

prism—angular loop of enamel on occlusal surface of cheekteeth.

prismatic—formed of prisms or shaped like prism.

proboscis—long, more or less, flexible snout.

procumbent—slanting forward, like teeth of some species of mammals.

propatagium—flight membrane of bat anterior to and supported by humerus and radius and ulna; functions as flap to reduce turbulence and increase lift during low-speed flight (Fig. 9-1).

proximal—located nearest to center of body.

protuberance—bump, swelling, or bulge that projects above or beyond surrounding surface.

pterygoid bones—paired bones posterior to palatines and projecting ventrally from floor of braincase (Fig. 1-1).

quadrupedal—pertaining to locomotion on four legs.

radius—larger of two long bones in forearm immediately distal to humerus.

ramus (pl. rami)—either left or right half of jaw.

recurved—curved backward or posteriorly.

reentrant angle—infolding of enamel layer of cheekteeth.

relict—surviving as remnant of vanishing race, type, or species.

retractile—able to be drawn back and in.

ricochetal gait—mode of bipedal saltatorial locomotion whereby individual travels in series of hops in zigzag pattern.

riparian—of, relating to, living on, or located on bank of or near watercourse.

rookery—breeding ground of gregarious animals.

roost—resting place.

rooted—as of tooth, having contracted root nearly closing pulp cavity and preventing further growth.

rostrum—facial portion of skull anterior to orbits; snout.

rudimentary—having characters imperfectly developed or represented only by vestige.

ruminant—animal having three- or four-chambered stomach and regurgitating and rechewing ingested food.

rump—hindquarters or buttocks.

rut—annual recurrent state of sexual excitement in some mammals.

sagittal crest—median dorsal ridge on skull extending posteriorly from near region of orbits; formed by union of temporal ridges (Fig. 7-1); enlarges with age.

saltatorial—of, or pertaining to, locomotion by leaping.

sanguivorous—feeding on blood.

scour—to suffer from diarrhea.

scrotum—external pouch that contains testes in most male mammals.

sebaceous—relating to, secreting, or composed of fatty matter.

secodont—of, or pertaining to, teeth adapted for shearing, e.g., carnassial teeth.

selenodont—of, or pertaining to, teeth with cresent-shaped cusps in artiodactyls that function in grinding rough, herbaceous material.

senescence—quality, or state, of growing old.

sere—series of developmental communities leading to and including climax ecological community in a region.

serrate—notched or toothed on edge; sawtoothed.

skull depth—distance from highest point of cranium to most ventral point of skull in a line perpendicular to long axis of skull (Lagomorpha, Rodentia: Sciuridae, skull usually set on 1-mm slide, then 1 mm subtracted from measurement); for all other Rodentia, ventral point is portion of palate located between molar toothrows.

spatulate—shaped like spatula; flat, thin.

species—group of populations of organisms, usually of similar morphology and potentially freely interbreeding, reproductively isolated from all other such groups of populations.

sphenoidal fissure—longitudinal opening on inside of orbit separating anterior edge of alisphenoid bones from frontal bones (Fig. 11-45); present in some geomyids.

sphenopalatine vacuity—paired longitudinal slits lateral to basisphenoid and presphenoid bones; present in some woodrats.

squamosal bones—paired bones that form major portion of lateral wall of braincase and posterior part of zygoma; location of articulation with dentary bones in mammals (Fig. 1-1).

stotting—highly developed form of locomotion characterized by bounding gait (modified gallop) in mule and black-tailed deer; may be faster means of locomotion than galloping in broken terrain and dense vegetation.

sublachrimal ridge—lateral extension of maxilla anterior and ventral to lacrimal bone; prominent in some talpids.

subnivean—situated, or occurring, under snow.

subspecies—geographical race.

sudoriferous—producing or conveying sweat.

superfecundation—successive fertilization of two or more ova, from same ovulation, by different males.

superfetation—successive fertilization of two or more ova, of different ovulations, resulting in embryos of unlike ages in same pregnancy.

supraoccipital bone—single bone posterior to parietal bones

(and interparietal if present) and forming posterior and dorsal portion of braincase (Fig. 1-1); dorsal to exoccipital bones.

supraoccipital process—shelflike lateral extension of frontal bones dorsal to orbit; present in leporids.

suture—line of union between bones of skull.

sympatric—occupying same area, as of distributions of two species.

synchrony—simultaneous occurrence.

synonymize—to combine two or more different names that have been applied to the same taxon.

synonymy—list of scientific names, by date and author, used in different publications to designate a species or other taxonomic group.

syntopy—state of sympatric species occupying same habitat.

taxon (pl. **taxa**)—any group of organisms sufficiently unique to be distinguished from other groups, given a distinctive name, and ranked in a taxonomic category.

temporal ridges—paired ridges extending from near orbit to supraoccipital bone along lateral margins of dorsum of braincase (Fig. 11-90) or converging posteriorly to form middorsal sagittal crest; serving as origin for temporalis muscle.

terrestrial—living wholly or chiefly on land.

territory—any defended area; not to be confused with home range.

tibia—larger of two long bones in lower hind leg immediately distal to femur.

tine—slender projecting part or point from tooth (Fig. 8-7) or antler (Fig. 13-5).

torpid (**torpor**)—dormant; characterized by suspended animation.

total length—distance from anterior tip of animal's nose to distal tip of its tail, exclusive of hair on tip of tail; always measured with animal on its back (Fig. 1-3).

tragus (pl. **tragi**)—prominence on lower medial margin of pinna anterior to notch (Fig. 9-1).

transverse—lying or passing across.

truncate—cut off or squared off abruptly; not tapering.

tubercle (**tuberculate**)—small, knobby prominence, as rounded cusp on occlusal surface of cheektooth.

turbinate bone—convoluted or scroll-shaped bone in nasal passages of skull (Fig. 1-1).

ulna—smaller of two long bones in forearm immediately distal to humerus.

unguiculate—having nails or claws.

unguligrade—walking on toenails or hooves.

unicuspid—of, or pertaining to, simplex tooth, tooth possessing occlusal surface composed of single cusp (Fig. 1-1).

unthrifty—lacking vigor.

uropatagium (**interfemoral membrane**)—flight membrane of bat situated between hind limbs; in vespertilionid and mollossid bats incorporating all or part of tail (Fig. 9-1); functions in flight as control surface and is used to catch insect prey in flight.

venter—belly and abdominal region, especially entire surface thereof.

ventral—on, or belonging to, venter of animal.

vernacular name—common name of organism as distinguished from Latin scientific name.

vestigial—trace of organ or structure remaining from earlier stage of development of individual, or from earlier evolutionary form.

vibrissa (pl. **vibrissae**)—stiff, tactile hair (whisker) growing on face or elsewhere on mammal.

volant—capable of flying.

vomer bone—single bone that forms part of floor of rostrum and bony nasal septum (Fig. 1-1).

voucher specimen—necessary parts (for mammals, usually skin and skull) preserved for confirming identity of taxon and used in study of aspects of its biology.

wash—to overspread with tint of another color; usually associated with pelage color.

webbed—possessing membrane uniting all or part of fingers or toes.

xeric—deficient in moisture.

yard—to congregate in winter quarters, often in relatively small areas; a place where mammals congregate, usually cervids.

zygoma (pl. **zygomata**)—arch of bone extending from rostrum to braincase; formed by jugal bone and parts of maxilla and squamosal bones (Fig. 1-1); serves to protect eye.

zygomatic breadth—greatest distance between outer margins of zygomata (Fig. 1-2).

Appendix. Specimens Examined

Ammospermophilus leucurus, n = 80
(Plotted in Fig. 11-16, p. 179)

Harney Co.—(1 EOSC, 1 PSM); Alvord Desert, T33S, R35E, Sec. 21 (2 OSUFW); 2 mi. N Fields (2 OSUFW); 0.25 mi. E Fields (1 OSUFW); 3 mi. S, 1.5 mi. E Fields (1 PSU); 6 mi. N Frenchglen (2 SOSC); S side Harney Lake, T27S, R29E, Sec. 36 (1 OSUFW); Tumtum Lake (6 USNM). **Lake Co.**—Adel (2 CM, 1 SDNHM); Adel, Warner Valley (6 OSUFW); 10 mi. N Adel (1 PSU); 2 mi. N, 4 mi. W Adel (2 OSUFW); 3.7 mi. S, 2.3 mi. E Adel, Soda Lake (2 PSU); 6 mi. SE Adel, T40S, R24E, Sec. 2 (1 PSU); 10 mi. NE Adel, Calderwood Hot Springs (2 PSU, 4 USNM); SW base Mt. Warner, Warner Valley, 4,500 ft. (1 USNM); South Warner Lake (1 CM, 1 USNM); Warner Valley* (3 OSUFW, 2 SDNHM); Warner Valley, Fisher Hot Springs (4 SDNHM); Warner Valley, Twentymile Creek (2 SDNHM); E side Warner Valley from Plush (2 SDNHM). **Malheur Co.**—16 mi. S Adrain (1 SDNHM); Dry Creek (3 PSM); 16 mi. NNE McDermitt (1 PSM); 4 mi. N Rome (4 CRCM); 4 mi. SW Rome, 3,850 ft. (1 USNM); near Vale (2 AMNH, 1 USNM); 30 mi. SSW Vale, T23S, R43E, Sec. 22 (1 PSM); 31 mi. SSW Vale, T23S, R43E, Sec. 28 (2 PSM); Watson (10 USNM).

Antilocapra americana, n = 23
(Plotted in Fig. 13-11, p. 486)

Harney Co.—Mann Lake area (1 OSUFW). **Lake Co.**—Adel (1 UO); 20 mi. NE Adel, Jacob's Ranch (3 OSUFW); Desert Lake (2 CM, 3 USNM); Hart Mt. (2 CM, 1 PSM, 1 USNM); Hart Mt. Refuge (1 USNM); E of Hart Mt. (2 CM); Mt. Warner, Hart Mt., 5,500 ft. (5 USNM); 5 mi. S Plush, Drake Spring (1 MVZ).

Antrozous pallidus, n = 81
(Plotted in Fig. 9-39, p. 117)

Baker Co.—Home, 2,000 ft. (3 USNM). **Crook Co.**—Twelve Mile Creek (1 USNM). **Douglas Co.**—T28S, R7W, Sec. 31 (1 PSU). **Gilliam Co.**—Thirtymile Bridge (1 PSU). **Grant Co.**—0.4 mi. N Sherburn Ranch off White Horse Rd., turn E, go 4.6 mi. to reservoir* (1 SOSC). **Harney Co.**—7 mi. S Andrews, 4,300 ft. (1 MVZ); Catlow Cave, Catlow Valley* (4 UMMZ, 1 LACM, 2 SDNHM); 20 mi. N Folly Farm (1 UMMZ); Pike Creek, 25 mi. N Fields (1 SOSC); 3.3 mi. S Princeton (4 PSM); Road Springs, T37S, R29E, NW¼ Sec. 10 (1 SOSC). **Jackson Co.**—East Pilot Rock Springs, Mt. Ashland, 3.1 mi. from HWY 99 (2 PSM, 1 SOSC); Pilot Rock Springs, T41S, R2E, NW¼ Sec. 3 (1 SOSC); 4 mi. S Ruch, T39S, R3W, Sec. 15 (2 PSM); Talent (1 SOSC); first bridge pass junction Little Applegate and Applegate rivers (1 SOSC); 1.5 mi. on Sams Valley Rd. from HWY 234 (1 SOSC). **Jefferson Co.**—0.5 mi. N Willowdale (1 PSU). **Josephine Co.**—Peavine Mt. (2 SDNHM); T39S, R5W, Sec. 14 (1 SOSC). **Klamath Co.**—Ft. Klamath (3 USNM); Klamath Falls (1 SOSC); 35.1 mi. W Lakeview (1 SOSC). **Lake Co.**—1.8 mi. W HWY 395 and 17.1 mi. N Lakeview, Woodchopper Springs (3 SOSC). **Lane Co.**—E Eugene (1 OSUFW); E of Eugene, near Fall Creek (1 OSUFW); 11 mi. W Junction City (17 PSM). **Malheur Co.**—Butch Lake, S end Jordan Crater lava flow, T29S, R44E, NE¼ Sec. 6 (2 PSM); Dry Creek, T24S, R41E, NE¼ Sec. 4 (1 PSM); 2 mi. NW Riverside (2 USNM); Rome Hilton Reservoir (1 SOSC). **Umatilla Co.**—17 mi. ENE Umatilla (4 KU). **Wasco Co.**—Cow Canyon* (1 SDNHM); mouth Deschutes River, Columbia River, 300 ft. (1 MVZ); Fort Dalles (1 USNM); Memaloose State Park (1 PSU); Memaloose State Park, 2.5 mi. E Mosier, T2N, R12E, SE¼ Sec. 31 (4 PSM). **Wheeler Co.**—Kimberly (1 PSU).

Aplodontia rufa, n = 464
(Plotted in Fig. 11-2, p. 155)

Benton Co.—3.75 mi. S, 5.5 mi. E Alsea (4 CM); 0.5 mi. S Blodgett (1 OSUFW); 2 mi. S, 4 mi. W Blodgett (2 OSUFW); 6 mi. SW Blodgett,

T12S, R7W, SW¼ Sec. 8 (6 OSUFW); 1 mi. N, 6 mi. E Grass Mt. (1 OSUFW); 1 mi. from K.P. Cabin, north trail, Marys Peak (1 OSUFW); 2.2 mi. S, 9.5 mi. W Philomath, Marys Peak (1 OSUFW); South Fork Alsea River (1 PSM). **Clackamas Co.**—(1 SDNHM, 1 USNM); Bissell (1 USNM); Boring (2 PSU); Bull Run (1 PSM); Carver (2 PSM); Damascus (1 PSU); Eagle Creek (2 USNM); 20 mi. S Estacada (1 SDNHM); 4 mi. S, 6 mi. W Estacada (1 OSUFW); 5 mi. E Estacada (1 PSU); 5 mi. SE Estacada (1 PSU); Jennings Lodge (2 UMMZ); Marmot (6 USNM); 8 mi. S Molalla (1 CRCM); Mt. Scott (1 PSU); Mulino (2 OSUFW); Oregon City, T2S, R2E, Sec. 21 (1 PSU, 1 UW[LB]); 0.5 mi. N, 5 mi. E Oregon City (1 PSU); 4 mi. SE Oregon City, 240 ft. (2 MVZ); Roslyn Lake (1 PSM); T1S, R2E, Sec. 25 (1 PSU); T2S, R1E, Sec. 17 (1 PSU). **Clatsop Co.**—(1 PSM); Astoria (2 USNM); 1 mi. N Gearhart (6 MVZ); Mishawaka (1 USNM); site of Old Ft. Clatsop (11 MVZ). **Columbia Co.**—1 mi. N Deer Island (1 PSU); 7 mi. SE Rainier (6 MVZ). **Coos Co.**—(1 USNM); 9 mi. N Bandon (1 PSM); 0.75 mi. S Bandon (1 PSM); 4 mi. SE Bandon (12 PSM); 8 mi. NNE Bandon (1 PSM); Coos Bay (1 OSUFW); Coos Bay, south slough (3 LACM); Coos Head (4 LACM); Coquille City (3 USNM); Delmar, 9 mi. S Marshfield (2 MVZ); Marshfield (2 AMNH). **Curry Co.**—Agness (1 FMNH); Brookings (2 SOSC); 10 mi. E Brookings (3 PSM); Port Orford (1 USNM); 2 mi. S Port Orford (7 PSM); 10 mi. ESE Port Orford, Elk River (2 PSM); 5 mi. N, 2 mi. E Wedderburn (1 OSUFW). **Douglas Co.**—Gardiner (4 FMNH, 2 MVZ); 0.5 mi. N Gardiner (2 PSM); 13 mi. N, 8.5 mi. E Glide, Disappearing Creek, T24S, R2W, Sec. 3 (16 KU, 1 TTU); 13.5 mi. N, 8.5 mi. E Glide, T24S, R2W, Sec. 3 (5 KU); Reedsport (2 PSU). **Jackson Co.**—Ashland Butte (2 UO); 5 mi. S Ashland (1 SOSC); 6 mi. S Ashland (2 SOSC); 11 mi. S, 3 mi. E Ashland (1 SOSC); N base Ashland Peak, 5,200 ft. (2 USNM); Siskiyou (20 USNM); 4.4 mi. S Anderson Creek Rd. off Wagner Creek Rd. S of Talent (1 SOSC); 0.3 mi. W Colstine Rd. on Mt. Ashland Rd. (2 SOSC); Tolman Creek Rd., T39S, R1E, NW¼SE¼ Sec. 34 (1 SOSC). **Josephine Co.**—13 mi. SW Galice, Briggs Creek, 3,000 ft. (1 SOSC). **Klamath Co.**—Anna Creek, Mt. Mazama (3 USNM); 1 mi. S Crater Lake (1 MVZ); Crater Lake National Park, Munson Valley (1 MVZ, 1 USNM); Ft. Klamath (3 USNM). **Lane Co.**—7.5 mi. S Cottage Grove (1 OSUFW); Eugene (1 USNM); Florence (2 AMNH, 14 FMNH); Heceta (1 OSUFW); 3 mi. S Heceta Head (2 UO); McKenzie Bridge (3 UO); 10 mi. S McKenzie Bridge, O'Leary Mt., 5,000 ft. (2 UO, 3 USNM); 7 mi. W McKenzie Bridge (1 KU); Mercer (3 UO); 1 mi. N, 6 mi. W Reed (1 OSUFW); Scio (1 UO); Seaton [= Mapleton] (6 USNM); Spencer Butte (2 UO); 3 mi. S Spencer Butte (1 MVZ); Springfield (1 MVZ); 6 mi. E Springfield (2 MVZ); Springfield Butte (6 MVZ); near Swisshome (1 KU); 3 mi. W Veneta (8 MVZ); Vida (1 OSUFW, 4 UO, 2 USNM). **Lincoln Co.**—1.4 mi. S Burnt Woods (2 OSUFW); 0.125 mi. N, 1 mi. W Burnt Woods, T11S, R8W, Sec. 22 (14 KU, 1 KU); 3 mi. S, 2 mi. W Burnt Woods (1 OSUFW); 3.125 mi. N, 6.67 mi. E Burnt Woods, T11S, R8W, Sec. 3/4 (5 KU); 2.5 mi. S, 1 mi. E Burnt Woods, T11S, R8W, Sec. 36 (7 KU); 3 mi. S, 1 mi. E Burnt Woods (1 OSUFW); Cascade Head Experimental Forest (10 PSM); Delake (2 KU); 2.5 mi. N, 0.5 mi. E Eddyville, T10S, R9W, Sec. 33 (1 FHSU, 7 KU, 4 LACM, 22 OSUFW, 2 TTU, 3 UCONN, 1 UNSW); 2 mi. S, 1 mi. E Eddyville, Salmon Creek Rd., T11S, R9W, Sec. 13 (1 DWU, 4 KU, 2 MSUMZ, 3 OSUFW, 1 TTU); 2.25 mi. S, 2 mi. E Eddyville (1 OSUFW); 1.5 mi. NW Eddyville (1 PSM); 2 mi. N, 6 mi. W Fisher (2 OSUFW); 1.5 mi. N, 0.5 mi. E Harlan (1 OSUFW); 1.5 mi. N, 1 mi. E Harlan (1 OSUFW); 2.75 mi. N, 1 mi. E Harlan (1 OSUFW); 1 mi. N, 2.5 mi. W Nashville, T10S, R8W, W½ Sec. 27/E½ Sec. 33 (1 UNSW, 1 FHSU, 2 JFBM, 4 OSUFW, 1 SUVM, 2 UCONN); Newport (1 KU); Newport, mouth of Yaquina Bay (12 USNM); 6 mi. W Nortons, T10S, R9W, Sec. 33 (1 KU); 0.5 mi. W Siletz (1 OSUFW); 2 mi. S, 6 mi. E Tidewater, along Alsea River (1 OSUFW); 5 mi. S, 2 mi. W Toledo (3 OSUFW); Wecoma Beach (1 MVZ). **Linn Co.**—3 mi. S Gates (1 OSUFW). **Marion Co.**—Silver Creek Falls (1 MSUMZ); 1 mi. N, 1 mi. W Stayton (2 OSUFW); 1 mi. N, 1 mi. E Stayton (1 OSUFW). **Multnomah Co.**—(2 LACM); Columbia Gorge near Corbett (1 EOSC); Forest Park, T1N, R1W, SE¼ Sec. 23 (1 PSU); Gresham (2 LACM, 2

---

*Indicates locality not plotted on range map.

PSU); Portland (2 AMNH); Sandy (1 UW[WWC]); T1N, R1W, Sec. 1 (1 PSU). **Tillamook Co.**—Blaine (2 KU, 5 UMMZ, 1 USNM); Cascade Head Experimental Forest (5 PSM); Hebo Lake (2 PSM); 3 mi. N Manzanita (3 OSUFW); Oceanview (1 PSM); Pacific City (1 PSM); Tillamook (12 AMNH, 1 OSUFW, 1 UMMZ); Wilson River (1 USNM). **Washington Co.**—11 mi. NW Linton (6 MVZ); near North Fork Gales Creek (1 PSU); 0.75 mi. N HWY 6 near North Fork Gales Creek (1 PSU).

Bassariscus astutus, n = 17
(Plotted in Fig. 12-29, p. 398)

Curry Co.—Port Orford (1 USNM). **Douglas Co.**—25 mi. SE Glide (1 SDNHM); Riddle (1 SDNHM); 7 mi. S, 10 mi. E Roseburg, T28S, R4W, NE¼SW1/16 Sec. 26 (1 KU); Tiller (1 OSUFW); between North Umpqua and Diamond Lake near Rock Creek (1 SOSC); 2 mi. W Steamboat (2 PSU). **Jackson Co.**—Beagle, T35S, R2W, S½ Sec. 11 (2 OSUFW); China Gulch, Rogue River* (1 PSM); Prospect (1 USNM); T33S, R2E, SE¼ (1 SOSC). **Josephine Co.**—Galice (1 UO); T33S, R8W, Sec. 26 (1 SOSC). **Klamath Co.**—200 m downstream from Salt Caves, S side Klamath River (1 SOSC). **Lane Co.**—9 mi. S Disston, T23S, R1W, Sec. 13 (1 OSUFW).

Bos bison, n = 30
(Plotted in Fig. 13-16, p. 492)

Crook Co.—along Crooked River, near Post (1 HM). **Grant Co.**—Flagtail Mt. (1 HM); Izee (1 USNM). **Harney Co.**—Malheur Lake (2 KU, 2 LACM, 4 MVZ, 3 PSM, 11 USNM); 20 mi. E Malheur Lake (2 KU, 1 USNM). **Malheur Co.**—Cow Creek (1 USNM). **Wallowa Co.**—near Joseph (1 HM).

Brachylagus idahoensis, n = 164
(Plotted in Fig. 10-5, p. 129)

Baker Co.—(1 SDNHM); Baker [City] (1 UMMZ); 10 mi. N Baker [City] (14 UMMZ). **Crook Co.**—20 mi. S Paulina, 12 Mile Creek (1 USNM). **Deschutes Co.**—Redmond (1 USNM). **Harney Co.**—(1 LACM); Beaties Butte (1 USNM); Beckley (1 UMMZ); Burns (4 OSUFW, 3 UO, 2 USNM); 10 mi. W Burns (12 PSM); 40 mi. W Burns (1 UO); 12 mi. SE Burns (1 OSUFW); Catlow Valley (1 OSUFW, 2 USNM); Rock Creek Ranch, Catlow Valley, T33S, R30E, Sec. 28 (1 OSUFW); Crane (1 SDNHM, 36 USNM); Drewsey (1 USNM); Malheur Cave (4 USNM); 2 mi. S Mud Lake, S Harney Lake Road (1 MNWR); Narrows (1 SDNHM, 1 USNM); Sageview (1 FMNH, 1 SDNHM, 1 USNM); Voltage (1 USNM); Waverly (1 FMNH, 1 SDNHM, 2 USNM). **Klamath Co.**—Klamath Falls (1 UMMZ). **Lake Co.**—20 mi. NE Adel, Jacobs Ranch (5 OSUFW, 1 USNM); Desert Lake* (6 CM); base of Doherety's Rim, T40S, R27E, Sec. 36 (1 OSUFW, 1 SDNHM); Fort Rock (2 MVZ, 1 SDNHM); Fremont (7 USNM); Guano Creek, Jacobs Ranch (1 PSM); Guano Creek, T38S, R27E, Sec. 34 (1 OSUFW); Guano Valley (1 OSUFW); 20 mi. NW Plush, Rabbit Creek Basin (1 OSUFW); Silver Lake (2 USNM). **Malheur Co.**—Bone Springs, Sheep Head Mountain, 5,200 ft. (3 USNM); Cold Springs (1 USNM); Cow Creek (1 USNM); Ironside (19 AMNH, 1 SDNHM); 1 mi. W Ironside, 3,800 ft. (2 MVZ); Malheur (1 USNM); Mahogany Mountain, S base near Cow Creek (1 USNM); McDermitt (1 USNM); 6 mi. S Riverside (1 USNM); Rome, Owyhee River (2 USNM).

Callorhinus ursinus, n = 22
(Plotted in Fig. 12-17, p. 383)

Clatsop Co.—(3 UW); Clatsop Beach (1 CRCM); 1 mi. SW Columbia River south jetty (1 UW); ca. 6 mi. N Gearhart (1 UW); 3 mi. S Peter Iredale wreck (2 CRCM); Seaside (1 USNM, 1 CRCM). **Coos Co.**—ca. 20 mi. W Coos Bay, 43°21'N, 124°40'W (1 UW). **Lincoln Co.**—4 mi. SW Depoe Bay (1 USNM); Newport (1 PSM); Newport beach (1 USNM); 2 mi. off coast near Newport (1 OSUFW); 6 mi. S Newport (1 SDNHM); Roads End (1 SDNHM). **Tillamook Co.**—13 mi. W-NW Cape Falcon (1 USNM); Nehalem Bay (1 PSM, 1 USNM);

Netarts Bay (1 SDNHM); Tillamook Bay (1 SDNHM).

Canis latrans, n = 223
(Plotted in Fig. 12-2, p. 356)

Baker Co.—5 mi. SW Powder River canyon, Medical Springs (1 USNM). **Clackamas Co.**—(1 OSUFW). **Coos Co.**—Eden Ridge (1 OSUFW). **Crook Co.**—6 mi. SE Prineville (3 MVZ). **Curry Co.**—11 mi. E Agness (1 USNM); Stockyards Camp, [42°30'N, 124°W] (1 USNM). **Deschutes Co.**—1 mi. SE Arnold Ice Cave (1 MVZ); Bend (2 USNM); Horse Ridge (1 USNM). **Douglas Co.**—Canyonville (11 OSUFW); Drew (1 USNM); 12 mi. N Drew (1 USNM); Glide (1 OSUFW); Lane Mt. (2 OSUFW); Round Prairie (1 PSM). **Gilliam Co.**—Arlington (2 USNM); 8 mi. N Lone Rock (2 USNM); 9 mi. N Lone Rock (1 USNM). **Harney Co.**—Alvord Lake (1 USNM); Anderson Valley, T27S, R34E, S½ (1 OSUFW); 20 mi. N Andrews (6 USNM); 20 mi. W Andrews (1 USNM); Andrews Valley* (1 OSUFW); Blitzen Creek* (1 USNM); Blitzen Valley* (1 SDNHM); Burns (3 USNM); 34 mi. SW Burns, near Iron Mt. (2 OSUFW); Cottonwood Creek, S of Alvord Lake (3 USNM); 3 mi. N Harriman (1 USNM); 10 mi. N Harriman (2 USNM); 10 mi. NW Harriman (1 USNM); 20 mi. N Harney (3 USNM); Pine Creek (2 USNM); 20 mi. NW Riley, Silver Creek (1 OSUFW); Rock Creek Sink (3 USNM); Shirk (1 USNM); 2 mi. W Sugarloaf Lookout, Malheur National Forest (1 MVZ). **Jackson Co.**—5 mi. E Eagle Point (1 MVZ); 1 mi. N of milepost #11 on HWY 66 east of Ashland (2 SOSC). **Jefferson Co.**—NW Culver (1 PSM); 7 mi. E Hay Creek, Foley Ranch (5 USNM); 7 mi. E Hay Creek, Trout Creek (1 USNM); 5 mi. N Terrebonne (1 PSM); Warm Springs (1 OSUFW). **Josephine Co.**—15 mi. SW Galice, Briggs Creek (1 UO). **Klamath Co.**—8 mi. N Bly (1 USNM); 10 mi. N Bly (1 USNM); 4 mi. NW Bly (1 USNM); Bonanza (2 OSUFW); Hunt Ranch* (1 OSUFW); Ft. Klamath (3 USNM); Lost River, Klamath (1 USNM); Merrill (1 MVZ); D. O'Conner Ranch* (1 OSUFW); Weyerhauser Camp #4, T40S, R5E, NW¼ Sec. 2 (1 SOSC); T33S, R14E, Sec. 32 (1 PSM); T39S, R12E, Sec. 11 (1 PSM). **Lake Co.**—Christmas Lake (1 USNM); Fremont (2 USNM); 12 mi. S Hampton (2 USNM); 14 mi. S Hampton (2 USNM); Hart Mt. Wildlife Refuge* (7 OSUFW); Lakeview (2 USNM); Mt. Warner, 5,500 ft. (6 USNM); Paisley (1 USNM); Stauffer (1 USNM); 7 mi. N Stauffer (1 USNM); 3 mi. W Stauffer (2 USNM); Sycan Marsh (1 USNM); Sycan Valley (2 PSM). **Lane Co.**—Coburg Hills (1 OSUFW); 3 mi. S Crow (1 PSM); 8 mi. W Junction City, T15S, R6W, Sec. 36 (1 OSUFW); 11 mi. W Junction City (1 PSM); McKenzie Bridge (1 UO); S fork McKenzie River (2 KU); Oakridge (1 OSUFW); 1 mi. W Vida (1 PSM); Wallehille* (1 OSUFW). **Linn Co.**—2.5 mi. SW Crawfordsville, T14S, R2W, Sec. 22 (1 OSUFW). **Malheur Co.**—(2 PSM); Clover Creek Ranch* (1 PSM); Cord, Palemeno Butte (6 USNM); Crooked Creek (1 USNM); E base Disaster Peak, 5,500 ft. (2 USNM); Duch Mts.* (1 USNM); Ironside (2 AMNH); 4 mi. W Jamieson (1 USNM); Jordan Valley (1 PSM); McDermitt (1 USNM); 15 mi. S Mooreville (2 USNM); Rome (2 PSM); Rome, Owyhee River (1 MVZ, 1 USNM); Sheepshead Mts. (2 USNM); Vale (1 USNM). **Morrow Co.**—Big Butler Creek (1 USNM); Cecil (7 USNM); 2 mi. E Cecil (1 USNM); 6 mi. SE Cecil (2 PSM); 10 mi. S Hardman (1 USNM); 4 mi. W Hardman (1 USNM); Heppner (1 USNM); Ione (2 USNM); Juniper Canyon, 16 mi. N Ione (1 USNM); mouth of Juniper Canyon (1 USNM); 3 mi. E mouth of Juniper Canyon (1 USNM); Lena (6 USNM); Rhea Creek (1 MVZ); 25 mi. SW Umatilla (1 MVZ). **Sherman Co.**—Gilore Ranch* (1 USNM); 10 mi. N Sink (1 USNM). **Umatilla Co.**—Echo, Butler Creek (4 USNM); 20 mi. S Echo (1 USNM); 13 mi. N North Fork (1 USNM); near Pendleton (1 PSU); 2.5 mi. SW Van Sycle (1 USNM). **Union Co.**—(2 EOSC); Cove (1 EOSC); Starkey Experimental Forest* (1 PSM); Starkey Experimental Forest, T4S, R34E, NW¼ Sec. 11 (1 PSM); 4 mi. NNW Summerville (1 EOSC); Thorn Creek, near Medical Springs (2 USNM). **Wallowa Co.**—Wallowa (1 OSUFW); T1S, R45E, Sec. 35 (1 PSM). **Wasco Co.**—(2 SDNHM). **Wheeler Co.**—3 mi. E Fossil, Tupper Ranger Station, Umatilla National Forest (1 USNM); 10 mi. E head of Pine Creek (1 USNM); 7 mi. S Spray (2 USNM); 12 mi. S Spray (1 USNM); 14 mi. S Spray (2 USNM); 7 mi. SE Spray (1 USNM).

Additional locality: Deer Creek Ridge* (1 USNM).

Canis lupus, *n* = 88
(Plotted in Fig. 12-4, p. 362)

Baker Co.—Huntington, near Durbin Creek (1 OSUFW). **Clackamas Co.**—(1 OSUFW); 28 mi. above Cazadero (1 OSUFW); Clackamas Lake (1 SDNHM); Estacada, Buckeye Creek [T7S, R6E, Sec. 14] (1 USNM); Estacada, Cafin Creek (1 USNM); 35 mi. E Estacada (1 OSUFW); source of Molalla River (1 OSUFW). **Deschutes Co.**—Crane Prairie (1 USNM). **Douglas Co.**—Anchor (1 MVZ); head of Days Creek (1 MVZ); Devil's Knobs, 4 mi. from Drew (3 MVZ); 9 mi. N, 12 mi. W Diamond Lake (1 OSUFW); Drew (2 OSUFW); Glide (4 OSUFW, 4 USNM); Rock Creek, 14 mi. above Glide (1 PSM); 20 mi. NE Glide (1 OSUFW); Hoaglin (=Idleyld Park) (1 SDNHM); Tiller (2 PSM); 15 mi. SE Tiller on Elk Creek (2 USNM). **Josephine Co.**—near head Dog Creek, T34S, R7W, Sec. 3 (1 MVZ); Mt. Peavine, 40 mi. N Rogue River (8 MVZ, 2 PSM, 12 SDNHM, 2 USNM). **Klamath Co.**—Cresent Lake (1 MVZ, 1 SDNHM). **Lake Co.**—30 mi. S Silver Lake, Sycan Marsh (1 USNM); South Ice Cave, 40 mi. S Bend [T23S, R14E, Sec. 18] (1 USNM). **Lane Co.**—head of Big Fall Creek (4 MVZ); Buck Mt. (1 MVZ); in mts. back of Fall Creek (1 UO); Leaburg (1 MVZ); Little Fall Creek (2 OSUFW); Oakridge (2 MVZ, 1 SDNHM); 6 mi. above Oakridge, Willamette River (1 MVZ); 20 mi. S Oakridge (1 SDNHM). **Linn Co.**—Cascadia, Canyon Creek (1 SDNHM, 2 USNM); Cascadia, Moose Creek (2 USNM); Foster (2 MVZ, 1 PSM, 1 USNM); 6 mi. NE Foster (2 OSUFW).

Additional localities: Umpqua National Forest* (1 SDNHM); Cascade Mts.* (1 MVZ).

Castor canadensis, *n* = 238
(Plotted in Fig. 11-69, p. 257)

Baker Co.—Baker, vicinity Powder River (2 MVZ); Boylen Ranch, Pine City, T8S, R46E (1 USNM). **Benton Co.**—2 mi. S, 3.5 mi. E Alsea, Trout Creek (1 OSUFW); Berry Creek, Dunn Forest (5 OSUFW). **Clatsop Co.**—(2 UW); near Astoria (5 USNM); near Astoria, Lower Columbia River (11 USNM); Big Creek or upper waters of Klaskanie River (2 KU); Big Creek or upper waters of Klaskanie River (2 KU); 12 mi. N Nehalem, Soapstone Creek, T4N, R9W, Sec. 22 (1 OSUFW); Westport (1 USNM). **Columbia Co.**—(1 MVZ, 9 USNM); Clatskanie (1 USNM); 5--6 mi. E Clatskanie (1 USNM); 1 mi. N Deer Island (3 PSU); Rainier drainage (1 USNM). **Coos Co.**—(1 PSM); Bandon (1 PSM); 2 mi. S Bandon (2 PSM); 1.75 mi. E Bandon (2 PSM); 1 mi. NE Bandon (1 PSM). **Deschutes Co.**—Crooked River (1 SDNHM); Squaw Creek, Metolius River, T14S, R11E, Sec. 19 (2 USNM); Tumalo (3 USNM). **Douglas Co.**—Reedsport (1 PSU); 8 mi. E HWY 101 on Smith River Rd. (2 PSU); 4 mi. NE junction HWY 101 and Smith River Rd. (7 PSU). **Gilliam Co.**—T4S, R21E, Sec. 2 (1 PSU). **Grant Co.**—6 mi. E Seneca, Bear Creek (2 USNM). **Harney Co.**—Burns (1 USNM); Malheur Lake (1 SDNHM, 1 USNM); Malheur National Wildlife Refuge (12 USNM); Malheur National Wildlife Refuge, Bridge Creek, Blitzen Valley (5 USNM); Malheur National Wildlife Refuge, Martha Lake (1 EOSC); Malheur National Wildlife Refuge, Witzel Lane bridge (1 USNM); Princeton, Malheur National Wildlife Refuge (1 UW); Silver Creek (12 USNM); Steens Mt. (1 USNM); Steens Mt., Big Fish Creek, 7,500 ft. (2 USNM); Suntex (1 USNM); Suntex, Silver Creek, T22S, R26E, Sec. 30 (5 USNM). **Jackson Co.**—Howard Prairie Dam at Hyatt Lake Rd. (1 SOSC); Rogue River (1 USNM). **Jefferson Co.**—Metolius River, T13S, R9E, Sec. 10 (2 USNM). **Josephine Co.**—Bolan Lake (2 PSM); Galice at Rogue River (1 UO). **Klamath Co.**—Bly (1 USNM); Bly, T36S, R14E, Sec. 2 (7 PSM); Bly, T36S, R14E, Sec. 35 (2 PSM); Bly, T37S, R15E, Sec. 6 (6 PSM); 6 mi. above Chiloquin, Williamson River (1 MVZ); Sprague River (2 MVZ); Wood River (1 MVZ); T36S, R15E, Sec. 6 (6 PSM). **Lake Co.**—(1 MVZ); Bly, T31S, R14E, Sec. 31 (1 PSM); Dog Lake (2 PSM); Fishhole Mt. (11 PSM); Thomas Creek (1 MVZ); 0.5 mi. above Thomas Creek Ranger Station (2 USNM). **Lane Co.**—Creswell (1 USNM); 3 mi. S, 5 mi. W Creswell (1 OSUFW); McKenzie Bridge (2 UO). **Lincoln Co.**—Fisher (2 USNM). **Linn Co.**—Fish Lake (1 PSM). **Malheur Co.**—mouth Jordan Creek, Rome, Owyhee River (1 USNM); Jordan Creek, 25 mi. W Jordan Valley (2 USNM); Ontario (1 SDNHM);

Snake River near Ontario (4 USNM). **Marion Co.**—Pudding River near Woodburn (1 OSUFW). **Morrow Co.**—Willow Creek, 3 mi. S Ione (1 MVZ). **Multnomah Co.**—(1 USNM); Beaver Creek (1 PSU); Bonneville (4 USNM); Bull Run watershed (1 PSU); near Fairview, Mud Lake (1 USNM); Portland (1 PSM, 1 PSU); near Portland (6 USNM); Troutdale (1 USNM); 0.5 mi. N Troutdale on Sandy River (1 USNM). **Sherman Co.**—mouth of Deschutes River (1 OSUFW); 1 mi. N, 10 mi. W Kent (2 OSUFW). **Tillamook Co.**—(1 PSM); Blaine (1 OSUFW, 1 USNM); Cloverdale (1 OSUFW); Cloverdale, Nestucca River (2 OSUFW); Foley Creek above Nehalem (3 MVZ); Hebo (2 PSM); Manhattan Beach (3 OSUFW); Nehalem (1 OSUFW); 5 mi. E Nehalem (1 PSM); 7 mi. E Nehalem, North Fork Nehalem River (1 OSUFW); 10 mi. E Nehalem River, Mohler (2 USNM); 12 mi. E Nehalem (1 PSM); Rockaway, Clear Lake (1 OSUFW). **Umatilla Co.**—(1 LACM); Walla Walla River valley (1 LACM). **Union Co.**—Dry Creek, N end Grand Ronde Valley (1 MVZ); Mill Creek, N end Grand Ronde Valley (1 MVZ); North Powder (1 USNM); Starkey Experimental Forest, Blue Mts., Meadow Creek, T3S, R34E, NE¼NW¼ Sec. 35 (2 PSM). **Wasco Co.**—2.5 mi. S, 2.5 mi. W Shaniko (1 UW). **Wheeler Co.**—Clarno (1 MVZ).

Cervus elpahus, *n* = 5
(Plotted in Fig. 13-2, p. 466)

Benton Co.—4.5 mi. S, 13 mi. W Alsea, Five Rivers (1 OSUFW). **Grant Co.**—50 mi. N Burns, Aldrich Mts., Malheur National Forest (1 OSUFW). **Tillamook Co.**—Lees Camp (1 PSU). **Umatilla Co.**—above Pendleton (1 PSM). **Wallowa Co.**—that part of Wenaha Unit lying N. of Wenaha River and E. and North Fork Wenaha River (1 KU).

Clethrionomys californicus, *n* = 2,662
(Plotted in Fig. 11-97, p. 298)

Benton Co.—1 mi. S, 6 mi. W Alpine (1 OSUFW); 4.4 km N, 7.6 km W Alsea (13 USNM); 8 mi. N, 1.5 mi. E Alsea, T12S, R7W, SW¼ Sec. 29 (1 MSUMZ, 1 OSUFW); 3 mi. S, 4 mi. E Alsea, T14S, R7W, Sec. 22 (3 OSUFW); 1 km S, 4.4 km E Alsea, T14S, R7W, Sec. 9 (4 USNM); 1.2 km S, 4.2 km E Alsea, T14S, R7W, Sec. 9 (11 USNM); 8 km S, 2.8 km E Alsea, T14S, R7W, Sec. 32 (8 USNM); 2.5 mi. N, 1.5 mi. W Corvallis (1 OSUFW); 3 mi. S, 10 mi. W Corvallis, T12S, R7W, NE¼NE¼ Sec. 21 (2 OSUFW); 4 mi. S, 6 mi. E Harlan (1 OSUFW); Marys Peak (1 OSUFW, 1 PSM); Marys Peak, T12S, R7W, Sec. 17 (2 OSUFW); Marys Peak campground (1 PSM); McDonald Forest, Soap Creek burn (1 OSUFW); Parker Creek near Marys Peak, T12S, R7W, Sec. 20 (1 OSUFW); 1 mi. S, 6 mi. W Philomath, Marys Peak, T12S, R7W, Sec. 13 (1 OSUFW); 2 mi. S, 8 mi. W Philomath (4 OSUFW); 2 mi. S, 9 mi. W Philomath, top Marys Peak, T12S, R7W, SW¼ Sec. 21 (1 OSUFW); 2 mi. S, 10 mi. W Philomath, Marys Peak, T12S, R7W, NE¼NE¼ Sec. 21 (4 OSUFW); 3 mi. S, 4 mi. W Philomath (4 OSUFW); 3 mi. S, 7 mi. W Philomath, South Fork Rock Creek (1 OSUFW); 0.4 km S, 12.5 km W Philomath, T12S, R7W, Sec. 10 (1 USNM); 1 km S, 11.3 km W Philomath, T12S, R7W, Sec. 11/14 (3 USNM); 1.5 km S, 10.1 km W Philomath, T12S, R7W, Sec. 13 (3 USNM); 1.5 km S, 11.3 km W Philomath, T12S, R7W, Sec. 14 (4 USNM); 2.4 km S, 15.5 km W Philomath, T12S, R7W, Sec. 17/20 (7 USNM); 4 km S, 10.1 km W Philomath, T12S, R7W, Sec. 24/25 (2 USNM); 9 mi. SW Philomath, Marys Peak (5 PSM); 1 mi. E Summit, T11S, R7W, Sec. 8 (1 KU); Wells (1 USNM). **Clackamas Co.**—Aschoff Buttes (2 PSU); Barlow campground, near Zigzag River (1 PSU); 0.5 mi. S Bull Run Lake (1 PSU); 0.5 mi. SE Bull Run Lake (1 PSU); 0.5 mi. N, 2.75 mi. W Government Camp (1 PSU); 0.75 mi. N, 2.75 mi. W Government Camp, Barlow campground (3 PSU); Oregon City (1 USNM); 4.2 mi. N, 8 mi. E Sandy, T1S, R6E, Sec. 29 (10 CRCM); 6 mi. S, 7 mi. E Sandy, T3S, R6E, Sec. 18 (3 CRCM); Wilhoit Springs, SE Molalla (1 PSM); 9 mi. N, 4 mi. W Zigzag (9 CRCM); 5 mi. S, 6 mi. E Zigzag (18 CRCM). **Clatsop Co.**—Astoria (3 USNM); Nehalem Summit (1 PSM); site of Old Ft. Clatsop (2 MVZ). **Coos Co.**—1.75 mi. E Bandon (1 PSM); 9 mi. NE Bandon (11 PSM); 4 mi. SE Bandon (8 PSM); Point Arago State Park (1 LACM); 3.2 km N, 0.3 km W Remote, T29S, R10W, Sec. 21 (19 USNM); 7.3 km N, 4 km W Sitkum, T27S, R10W, Sec. 17/20 (1

USNM); 7.4 km N, 4.2 km W Sitkum, T27S, R10W, Sec. 17 (3 USNM); 7.5 km N, 12.5 km W Sitkum, T27S, R11W, Sec. 21 (14 USNM); 7.5 km N, 12.7 km W Sitkum, T27S, R11W, Sec. 21 (1 USNM); 7.6 km N, 5.2 km W Sitkum, T27S, R10W, Sec. 18 (6 USNM); 8.4 km N, 4.3 km W Sitkum, T27S, R10W, Sec. 17 (12 USNM); 2.8 km S, 5 km E Sitkum, T28S, R9W, Sec. 19 (3 USNM); 2.9 km S, 4.7 km E Sitkum, T28S, R9W, Sec. 19 (7 USNM); 3 km S, 10.8 km E Sitkum, T28S, R9W, Sec. 23 (6 USNM); 3 km S, 11 km E Sitkum, T28S, R9W, Sec. 23 (7 USNM); 4.8 km S, 9.7 km E Sitkum, T28S, R9W, Sec. 27 (16 USNM). **Curry Co.**—10 mi. E Brookings (5 PSM); Little Redwood Forest (2 PSM); 8.6 mi. E HWY 101 on Winchuck River Rd. (1 SOSC); T41S, R12W, Sec. 10 (1 SOSC). **Deschutes Co.**—15 mi. W Bend, Tumalo Creek, 6,100 ft. (2 MVZ); 1 mi. SE Broken Top, 7,100 ft. (2 MVZ); Sparks Lake (2 OSUFW). **Douglas Co.**—Anchor (1 USNM); Broken Arrow campground (1 PSU); 5 mi. S Canyonville on HWY 99 near Roseburg (2 CM); 3.2 mi. S, 3.6 mi. E Canyonville, T31S, R4W, Sec. 17, 2,000 ft. (21 UW[CR]); Diamond Lake (13 SDNHM, 3 SOSC, 2 UMMZ); N end Diamond Lake (1 OSUFW); Diamond Lake campground (1 PSU); 0.6 km N, 2.2 km W Drain, T22S, R5W, Sec. 7 (2 USNM); 0.8 km N, 2.2 km W Drain, T21S, R5W, Sec. 7 (5 USNM); 1.4 km N, 1.6 km W Drain, T22S, R5W, Sec. 7 (5 USNM); 7.6 km N, 6.4 km W Drain, T21S, R6W, Sec. 22 (6 USNM); 10 km N, 10.6 km W Drain, T21S, R6W, Sec. 7 (5 USNM); 11.3 km N, 10.5 km W Drain, T21S, R6W, Sec. 5 (2 USNM); 11.5 km N, 10.5 km W Drain, T21S, R6W, Sec. 5 (1 USNM); 4.5 mi. N, 8 mi. W Drain, T21S, R4W, Sec. 27, 1,200 ft. (1 UW[CR]); 5.5 mi. N, 4 mi. W Drain, T21S, R6W, Sec. 15, 1,280 ft. (52 UW[CR]); 6 mi. N, 4 mi. W Drain, T21S, R6W, Sec. 15, 1,200 ft. (48 UW[CR]); 6.75 mi. N, 5.5 mi. E Drain, T21S, R4W, Sec. 7 (2 UW[CR]); 9.7 km S, 8.4 km W Elkton, T23S, R8W, Sec. 20 (13 USNM); 9.7 km S, 16.2 km W Elkton, T23S, R9W, Sec. 22 (1 USNM); 11.1 km S, 7.4 km W Elkton, T23S, R8W, Sec. 28 (27 USNM); 14.9 km S, 6.2 km W Elkton, T24S, R8W, Sec. 9 (10 USNM); 15.4 km S, 6 km W Elkton, T24S, R8W, Sec. 10 (3 USNM); 15.5 km S, 5.8 km W Elkton, T24S, R8W, Sec. 10 (3 USNM); 15.7 km S, 5.6 km W Elkton, T24S, R8W, Sec. 10 (8 USNM); 25.2 km S, 5.4 km W Elkton, T25S, R8W, Sec. 10 (3 USNM); 25.6 km S, 5.8 km W Elkton, T25S, R8W, Sec. 10 (2 USNM); Francis Creek, T24S, R1W, SE¼SE¼ Sec. 12 (5 SOSC); Louis Creek, T28S, R3W, NW¼SW¼ Sec. 29 (4 SOSC); Pass Creek, T24S, R1W, SE¼ (4 SOSC); 4 mi. S, 6.4 mi. W Riddle, Iron Mt., T31S, R7W, Sec. 11, 2,000 ft. (3 UW[CR]); 2 km N, 25 km W Sutherlin, T25S, R8W, Sec. 10 (6 USNM); 2.1 km N, 24.8 km W Sutherlin, T25S, R8W, Sec. 10 (1 USNM); 2.2 km N, 26.2 km W Sutherlin, T25S, R9W, Sec. 9 (5 USNM); 2.4 km N, 24.6 km W Sutherlin, T25S, R8W, Sec. 10 (3 USNM); 2.8 km N, 26 km W Sutherlin, T25S, R9W, Sec. 9 (9 USNM); 3.8 km N, 12.1 km W Sutherlin, T25S, R7W, Sec. 1 (5 USNM); 4 km N, 12.1 km W Sutherlin, T25S, R7W, Sec. 1 (9 USNM); 4.2 km N, 12.1 km W Sutherlin, T25S, R7W, Sec. 1 (5 USNM); 6.5 km N, 6.7 km W Tenmile, T28S, R8W, Sec. 9 (13 USNM); 4.5 mi. N, 8 mi. E Tiller, T30S, R1W, Sec. 2 (36 CRCM); 4.5 mi. N, 9 mi. E Tiller (32 CRCM); 5 mi. N, 5 mi. E Tiller, Straight Creek (2 OSUFW); 7.5 mi. N, 8 mi. E Tiller (9 CRCM); 9 mi. N, 8.5 mi. E Tiller, T29S, R1W, Sec. 13 (4 CRCM); 10 mi. N, 5.5 mi. E Tiller, T29S, R1W, Sec. 12 (29 CRCM); 11 mi. N, 5 mi. E Tiller (26 CRCM); 11.5 mi. N, 5 mi. E Tiller, T28S, R1W, Sec. 26 (9 CRCM); 12 mi. N, 4 mi. E Tiller (26 CRCM); 12 mi. N, 5.5 mi. E Tiller (35 CRCM); 12 mi. N, 17 mi. E Tiller (33 CRCM); 12 mi. N, 18.5 mi. E Tiller (28 CRCM); 12.5 mi. N, 5 mi. E Tiller, T28S, R1W, Sec. 12 (5 CRCM); 12.5 mi. N, 5.5 mi. E Tiller, T28S, R1W, Sec. 31 (6 CRCM); 0.2 mi. N, 4.1 mi. W Umpqua P.O., T25S, R7W, Sec. 15, 800 ft. (8 UW[CR]); 2.2 mi. N, 9 mi. W Umpqua P.O., T25S, R8W, Sec. 15, 850 ft. (7 UW[CR]); 2.7 mi. N, 5 mi. W Umpqua, Riverview, T25S, R7W, Sec. 7, 950 ft. (23 UW[CR]); 4 mi. N, 1 mi. W Umpqua P.O., Lost Creek, T25S, R7W, Sec. 1, 1,500 ft. (20 UW[CR]); 6 mi. N, 10.5 mi. W Umpqua, Miner Creek, T24S, R8W, Sec. 29, 900 ft. (22 UW[CR]); 4.8 km S, 4.5 km W Yoncalla, T23S, R5W, Sec. 19 (2 USNM); 4.8 km S, 4.7 km W Yoncalla, T23S, R5W, Sec. 19 (7 USNM); 4.9 km S, 4.5 km W Yoncalla, T25S, R5W, Sec. 19 (8 USNM); T24S, R1W, SE¼ Sec. 26 (2 SOSC). **Hood River Co.**—Brooks Meadows, 8 mi. S, 4.5 mi. E Parkdale (1 PSU); Hood River meadows (2 PSM); N slope Mt. Hood, 2,800 ft. (3 AMNH); Mt. Hood, Government Camp, 3,600 ft. (2 AMNH); W base Mt. Hood, Yokums (1 AMNH); 2 mi. W Parkdale, 1,500 ft. (2 USNM). **Jackson Co.**—Ashland, T38S, R3E, Sec. 30 (15 SOSC); 3 mi. N, 11 mi. E Ashland (2 UMMZ); 33 mi. NE Ashland, 2.5 mi. E USFS Rd. 3706 on USFS Rd. 3705 (2 SOSC); Ashland Natural Research Area, T40S, R1E, SE¼ Sec. 4 (3 SOSC); N base Ashland Peak, 5,000 ft. (2 USNM); N base Ashland Peak, 5,200 ft. (1 USNM); Buck Peak, T39S, R4E, SW¼ Sec. 12 (1 SOSC); halfway between Drew and Crater Lake (1 USNM); Hyatt Reservoir (1 PSU); Imnaha Creek, T33S, R4E, W½ Sec. 15 (1 SOSC); 2 mi. S, 7 mi. W Lincoln (1 OSUFW); 5 mi. S Lithia Park (1 SOSC); Mt. Ashland, T40S, R1E, Sec. 21 (1 SOSC); Mt. Ashland, T40S, R1E, Sec. 9 (1 SOSC); Prospect (4 MVZ, 4 USNM); 11 mi. N Prospect (13 CRCM); 11.5 mi. N, 0.5 mi. W Prospect (18 CRCM); 13.5 mi. N, 2.5 mi. W Prospect (32 CRCM); 14 mi. N, 0.5 mi. W Prospect, T30S, R2E, Sec. 24/25 (34 CRCM); 2.5 mi. N, 8.5 mi. E Prospect (15 CRCM); 5 mi. N, 8 mi. E Prospect (20 CRCM); 5.5 mi. N, 4 mi. E Prospect (5 CRCM); Siskiyou (1 USNM); Skeeter Swamp, T36S, R4E, SW¼ Sec. 6 (1 SOSC); Soda Creek, T37S, R3E, NE¼SE¼ Sec. 18 (1 SOSC); Spring Creek, T33S, R4E, NE¼ Sec. 22 (3 SOSC); 4.8 mi. up Tolman Creek Rd. on Bull Gap Rd. (1 SOSC); 2 mi. N, 1 mi. E Union Creek (8 CRCM); Upper E. Fork Ashland Creek, T40S, R1E, SE¼SE¼ Sec. 4 (1 SOSC); 300 yds. upstream from Store Creek campgrounds (1 SOSC); Rd. 3706, 1 mi. W HWY 140 (1 SOSC); 4 mi. S Siskiyou Blvd. off Tolman Creek Rd. (1 SOSC); 1.1 mi. E USFS 3706 on USFS Rd. 3705 (2 SOSC); 0.25 mi. E junction Crane Prairie Rd. and Weyerhauser Rd. 107-63 (1 SOSC); 1.4 mi. E junction Crane Prairie Rd. and Weyerhauser Rd. 170-63, T39S, R4E, SE¼ Sec. 17 (1 SOSC); 1.6 mi. W HWY 99 on Mt. Ashland Rd. (1 SOSC); T37S, R3E, SW¼ Sec. 20 (4 SOSC); T39S, R1W, SE¼ Sec. 14 (1 SOSC); T39S, R4E, NW¼W½SW¼ Sec. 17 (2 SOSC). **Jefferson Co.**—3 mi. S, 9 mi. W Camp Sherman, HWY 20 (1 OSUFW). **Josephine Co.**—Bolan Lake (3 PSM); 1 mi. N, 10.5 mi. E Cave Junction (1 HSU); 2 mi. N, 11.5 mi. E Cave Junction (4 HSU); 2.5 mi. S, 9.5 mi. E Cave Junction (1 HSU); 3 mi. S, 10 mi. E Cave Junction (1 SOSC); 4 mi. S, 11.5 mi. E Cave Junction (2 HSU); Grants Pass (1 CM, 1 FMNH); 20 mi. SW Grants Pass, N fork Slate Creek, 1,400 ft. (2 USNM); 7 mi. E Selma (1 SOSC); 0.5 mi. S, 3.5 mi. E Takilma (1 HSU); 2 mi. S, 2 mi. E Takilma (1 HSU); 3.5 mi. S, 5.5 mi. E Takilma, T41S, R7W, Sec. 15 (2 HSU); Gray back campground, T39S, R6W, Sec. 30 (1 SOSC); drainage east of Cherry Creek near J.J. Creek (1 SOSC). **Klamath Co.**—Anna Creek, 4 mi. NW Ft. Klamath, 4,500 ft. (2 MVZ); Anna Creek, Mt. Mazama, 6,000 ft. (1 USNM); S side Como Lake (1 SOSC); Crater Lake National Park (12 localities within 10 mi. Crater Lake) (2 AMNH, 13 KU, 5 MVZ, 2 OSUFW, 6 PSU, 20 USNM); 0.5 mi. S, 3 mi. W Cresent Lake (1 OSUFW); 0.13 mi. E Heavenly Twin Lake (1 OSUFW); 9 mi. N, 3 mi. E Heavenly Twin Lakes (2 SOSC); S end Lake of the Woods (1 MVZ); Odell Creek, 4,700 ft. (1 LACM); ca. 1.5 mi. S Odell Lake, T23S, R6½E, Sec. 36 (2 PSU); Seven Lakes basin (1 SOSC); Sky Lakes basin (1 SOSC); T38S, R5E, SE¼NE¼ Sec. 21 (2 SOSC). **Lane Co.**—(2 PSM); 12.1 km S, 2.4 km W Alsea, T15S, R8W, Sec. 11 (4 USNM); 6 mi. above Belknap Springs, McKenzie River (1 OSUFW); Blue River (2 USNM); 4.4 km N, 6.8 km E Blue River, T16S, R5E, Sec. 7 (3 USNM); 6.8 km N, 7.6 km E Blue River, T15S, R5E, Sec. 32 (3 USNM); 9 mi. N Blue River Lookout Creek, T15S, R6E, Sec. 30 (2 OSUFW); 8.5 mi. E Blue River, T16S, R5E, Sec. 25/26 (1 CRCM); 8.75 mi. E Blue River (2 CRCM); 9 mi. E Blue River (427 CRCM); 1.5 mi. N, 2 mi. E Blue River (5 CRCM); 4 mi. N, 4.5 mi. E Blue River, T16S, R5E, Sec. 32 (17 CRCM); 5 mi. N, 9 mi. E Blue River (23 CRCM); 5 mi. N, 10.5 mi. E Blue River, T15S, R5E, Sec. 30 (4 CRCM); 5.5 mi. N, 5 mi. E Blue River, T15S, R5E, Sec. 28 (1 HSU); 5.5 mi. N, 7 mi. E Blue River, T17S, R5E, Sec. 7 (1 HSU); 5.5 mi. N, 13.5 mi. E Blue River, T16S, R6E, Sec. 26 (4 CRCM); 7 mi. N, 9 mi. E Blue River, T15S, R5E, Sec. 13 (8 CRCM); 2.5 mi. S, 7 mi. E Blue River (8 CRCM); 3.5 mi. S, 10 mi. E Blue River (16 CRCM); 3.5 mi. S, 10.25 mi. E Blue River (2 CRCM); 3.5 mi. S, 10.5 mi. E Blue River (1 CRCM); 5.5 mi. NE Blue River (5 PSM); 10 mi. NE Blue River, T15S, R5E, Sec. 35 (3 OSUFW); 10 mi. SE Blue River, East Fork Creek, T17S, R6E, Sec. 7 (3 OSUFW); 1.5 mi. S Cushman, 0.5 mi. E Canary, T18S, R11W, Sec. 33 (1 OSUFW); ca. 40 mi. SW Eugene (1 PSM); 9.5 mi. N, 1 mi. W Gardiner (5 PSM); Gold Lake, 4,800 ft. (6 LACM); H. J. Andrews Experimental Forest, T15S, R5E, Sec. 23 (2 OSUFW); H. J. Andrews Experimental

Forest, T15S, R5E, NW¼ Sec. 24, 3,500 ft. (24 PSM); H. J. Andrews Experimental Forest, T15S, R5E, NW¼ Sec. 28, 2,000 ft. (3 PSM); H. J. Andrews Experimental Forest, T15S, R5E, NE¼ Sec. 32, 2,000 ft. (26 PSM); H. J. Andrews Experimental Forest, T15S, R5E, SW¼ Sec. 33, 2,500 ft. (23 PSM); H. J. Andrews Experimental Forest, T15S, R5E, NW¼ Sec. 34, 2,700 ft. (12 PSM); H. J. Andrews Experimental Forest, T16S, R4E, NE¼ Sec. 19, 1,800 ft. (5 PSM); Horsepasture Mt. (1 OSUFW); 3 mi. S Huckleberry Mt., 2,300 ft. (1 LSU); Loon Creek (1 PSM); 0.4 km N, 18.5 km W Lorane, T20S, R7W, Sec. 12 (9 USNM); 0.8 km N, 19.2 km W Lorane, T20S, R7W, Sec. 12 (8 USNM); 2 km N, 6.4 km W Lorane, T20S, R5W, Sec. 5 (6 USNM); 2.6 km N, 6.2 km W Lorane, T20S, R5W, Sec. 5 (2 USNM); 2.8 km N, 6.2 km W Lorane, T20S, R5W, Sec. 5 (6 USNM); 3 km N, 19.5 km W Lorane, T20S, R7W, Sec. 1 (7 USNM); 3.2 km N, 26 km W Lorane, T19/20S, R7W, Sec. 32/5 (4 USNM); 4.8 km N, 17.5 km W Lorane, T19S, R6W, Sec. 31 (8 USNM); 6.8 km N, 11.1 km W Lorane, T19S, R6W, Sec. 23 (7 USNM); Mazama Ridge, 300 ft. above Horse Lake shelter, T18S, R7½E (1 OSUFW); McCredie Hot Springs (1 SDNHM); McKenzie Bridge (3 OSUFW, 2 SDNHM, 2 USNM); 2.6 km S McKenzie Bridge, T16S, R5E, Sec. 26 (10 USNM); 3.1 km N, 3 km W McKenzie Bridge, T16S, R5E, Sec. 3 (13 USNM); 3.3 km N, 2.4 km W McKenzie Bridge, T16S, R5E, Sec. 3 (12 USNM); 6.6 km N, 0.3 km W McKenzie Bridge, T15S, R5E, Sec. 25 (7 USNM); 6 mi. N, 17 mi. E McKenzie Bridge (1 OSUFW); 6 km N, 0.6 km E McKenzie Bridge, T15S, R5E, Sec. 25 (1 USNM); 6 km N, 7.5 km E McKenzie Bridge, T15S, R6E, Sec. 26 (5 USNM); 6.4 km N, 0.6 km E McKenzie Bridge, T15S, R5E, Sec. 25 (4 USNM); 1.6 km S, 5.2 km W McKenzie Bridge, T16S, R5E, Sec. 20 (4 USNM); 3.8 km S, 14.8 km E McKenzie Bridge, T16S, R7E, Sec. 33 (1 USNM); 4.4 km S, 14.3 km E McKenzie Bridge, T16S, R7E, Sec. 32 (45 USNM); 5.4 km S, 8.75 km E McKenzie Bridge, T16S, R6E, Sec. 35 (4 USNM); 4 mi. N McKenzie Bridge, 2,600–2,700 ft. (5 UW[OG]); 5 mi. S McKenzie Bridge, Spring Creek, T16S, R6E, Sec. 26 (1 OSUFW); 2 mi. N, 2 mi. W McKenzie Bridge, T16S, R5E, Sec. 3/4, 3,800 ft. (5 UW[OG]); 6 mi. N, 5 mi. W McKenzie Bridge, T15S, R5E, NW¼ Sec. 19 (1 KU); 2 mi. S, 0.5 mi. E McKenzie Bridge, T16S, R5E, Sec. 25, 2,800 ft. (4 UW[OG]); 3 mi. S, 5 mi. E McKenzie Bridge, T15S, R6E, Sec. 35, 2,000 ft. (4 UW[OG]); 3.5 mi. S, 9 mi. E McKenzie Bridge, T16S, R7E, Sec. 32, 3,500 ft. (12 UW[OG]); McKenzie Pass, Frog Camp (2 SDNHM); 4.5 mi. N, 0.25 mi. E Nimrod (2 CRCM); 1 mi. N, 4 mi. E Oakridge, Salmon Creek (3 OSUFW); 4 mi. N Rainbow, T15S, R5E, Sec. 32, 1,700 ft. (4 UW[OG]); Scott Lake (1 KU); N slope Three Sisters (1 OSUFW, 1 SDNHM, 10 USNM); NW slope Middle Sister, Three Sisters (1 OSUFW); N base Three Sisters (4 USNM); W base Three Sisters (1 OSUFW); Three Sisters, Alder Springs, 4,300 ft. (1 USNM); Three Sisters, 5,000 ft. (2 USNM); Three Sisters, 6,000 ft. (6 USNM); 23 mi. SE Vida (3 PSM); 4.5 mi. S Wolf Rock, T15S, R5E, Sec. 24, 3,400 ft. (1 UW[OG]). **Lincoln Co.**—N side Alsea Bay (4 PSM); 15 mi. W Haralan, Peterson Ridge, T12S, R11W, Sec. 12 (2 KU); 2 mi. S, 12 mi. W Harlan, Beaver Creek, T12S, R10W, Sec. 20 (2 KU); 5 mi. S, 10 mi. W Harlan, Gold Creek, T13S, R10W, Sec. 2 (12 KU); 7 mi. S, 5 mi. W Harlan, West Creek, T13S, R9W, Sec. 16 (25 KU); 7 mi. S, 10 mi. W Harlan, Death Ridge, T13S, R10W, Sec. 14 (10 KU); 7 mi. S, 11 mi. W Harlan, Boulder Creek, T13S, T10W, Sec. 15 (15 KU); Marys Peak, T12S, R7W, Sec. 16 (1 OSUFW); Otis (4 USNM); Taft (1 USNM); 0.6 km N, 16.9 km E Waldport, T13S, R10W, Sec. 13/14 (10 USNM); 0.9 km N, 16.9 km E Waldport, T13S, R10W, Sec. 13 (4 USNM); 4 km S, 3.2 km E Yachats, T15S, R12W, Sec. 12 (17 USNM); Yaquina Bay (1 USNM). **Linn Co.**—Big Lake (9 PSM); 1.5 mi. N Big Lake (5 PSM); 6.75 mi. N, 8 mi. E Blue River (3 CRCM); 7 mi. N, 8 mi. E Blue River, T15S, R5E, Sec. 15 (2 CRCM); 7.5 mi. N, 7.5 mi. E Blue River, T15S, R5E, Sec. 15 (17 CRCM); 8 mi. N, 8 mi. E Blue River (19 CRCM); 9 mi. N, 8 mi. E Blue River (7 CRCM); 9.5 mi. N, 9 mi. E Blue River, T15S, R5E, Sec. 2 (8 CRCM); 1 mi. N, 28 mi. E Cascadia (4 CM); 2 mi. S, 17 mi. E Cascadia (1 OSUFW); H. J. Andrews Experimental Forest, T14S, R6E, NE¼ Sec. 20, 4,400 ft. (22 PSM); H. J. Andrews Experimental Forest, T14S, R6E, NE¼ Sec. 28, 4,600 ft. (12 PSM); H. J. Andrews Experimental Forest, T15S, R5E, SE¼ Sec. 11, 3,200 ft. (3 PSM); H. J. Andrews Experimental Forest, T15S, R5E, SW¼ Sec. 13, 3,000 ft. (27 PSM); H. J. Andrews Experimental Forest, T15S, R6E,

NW¼ Sec. 7, 5,000 ft. (9 PSM); H. J. Andrews Experimental Forest, T15S, R6E, NE¼ Sec. 7, 4,300 ft. (32 PSM); Ice Cap Camp, Willamette Forest (1 PSM); 9.2 km N, 1.2 km W McKenzie Bridge, T15S, R5E, Sec. 14 (12 USNM); 10.25 km N, 0.8 km W McKenzie Bridge, T15S, R5E, Sec. 14 (12 USNM); 13.2 km N, 0.8 km E McKenzie Bridge, T15S, R5E, Sec. 1 (4 USNM); North Santiam River, 3,400 ft.* (1 MVZ); Permilia Lake, W base Mt. Jefferson (7 USNM); 32.5 mi. E Sweethome, Lost Prairie campground (2 PSU); 8 mi. E Upper Soda, Tombstone Prairie (1 OSUFW); 9 mi. E Upper Soda, Hackleman Creek (2 OSUFW); 13 mi. E Upper Soda, Toad Creek Rd. (4 OSUFW); 13.7 mi. E Upper Soda (1 OSUFW); 19 mi. E Upper Soda (1 OSUFW); 1 mi. S, 6.5 mi. E Upper Soda (1 OSUFW); 1 mi. S Wolf Rock, T15S, R5E, Sec. 1, 3,300 ft. (4 UW[OG]); 1 mi. S, 1 mi. W Wolf Rock, T15S, R5E, Sec. 2, 3,000 ft. (4 UW[OG]); 4 mi. S, 1 mi. W Wolf Rock, T15S, R5E, Sec. 14, 2,600 ft. (1 UW[OG]). **Marion Co.**—4.5 mi. S, 1.5 mi. E Bagby Hot Springs, 4,000 ft. (7 LSU). **Multnomah Co.**—Big Bend Mt. (6 PSU); Larch Mt. (2 PSU); Portland (3 AMNH); 8 mi. N, 11 mi. E Sandy (1 CRCM); 9 mi. N, 10 mi. E Sandy (3 CRCM); 10 mi. N, 10 mi. E Sandy (15 CRCM); 10 mi. N, 11 mi. E Sandy (19 CRCM); 11 mi. N, 10 mi. E Sandy (23 CRCM); 11 mi. N Zigzag (1 CRCM); 8 mi. N, 3 mi. E Zigzag (10 CRCM). **Polk Co.**—3.5 mi. S, 2 mi. W Gold Creek (1 KU); 3.75 mi. S, 2 mi. W Gold Creek (1 KU); Salmon River summit (3 PSM). **Tillamook Co.**—Cascade Head Experimental Forest (2 PSM); Netarts (1 OSUFW); Tillamook (2 OSUFW); S of Tillamook Bay near Netarts Bay (1 PSM). **Wasco Co.**—Clear Lake (5 UMMZ); 4 mi. W Pine Grove (1 OSUFW); 6 mi. W Pine Grove (1 OSFUW); 8 mi. W Pine Grove (1 OSUFW); 1 mi. N, 5 mi. W Pine Grove (1 OSUFW). **Washington Co.**—18.5 mi. NW Portland (1 MVZ).

Clethrionomys gapperi, n = 312
(Plotted in Fig. 11-99, p. 303)

Baker Co.—Anthony (22 AMNH, 1 MVZ); Bourne (7 USNM); Cornucopia (1 USNM). **Crook Co.**—Canyon Creek, Lookout Mt., Ochoco National Forest (3 SOSC); Ochoco Ranger Station, 4,000 ft. (1 MVZ). **Grant Co.**—Austin (3 USNM); 8 mi. SE Austin, Cold Spring, 4,900 ft. (4 MVZ); Beech Creek (2 USNM); Blue Mt. Springs (1 PSM); 12 mi. S Canyon City, 5,500 ft. (2 MVZ); 9 mi. S, 5 mi. W Canyon City, Windfall Creek (1 PSU); John Day, Strawberry Mt. (1 USNM); 1 mi. N, 10 mi. E Long Creek, T10S, R32E, Sec. 5 (1 OSUFW); 8 mi. S, 5 mi. W Long Creek (1 OSUFW); 3 mi. N, 6 mi. E Prairie City, T12S, R34E, NW¼ Sec. 23 (4 KU); 21 mi. SE Prairie City, N fork Malheur River, 5,000 ft. (6 MVZ); 4 mi. N, 14.5 mi. E Seneca, T16S, R33½E, NE¼NW¼ Sec. 17 (5 KU); 5 mi. N, 15.5 mi. E Seneca, T16S, E33½E, NE¼SE¼ Sec. 8 (2 KU); 6 mi. N, 15.5 mi. E Seneca, T16S, R33½E, NE¼NE¼ Sec. 5 (1 KU); Strawberry Butte (3 USNM); Strawberry Lake (1 SDNHM, 2 USNM); Strawberry Mts. (16 USNM); N slope Strawberry Mt., 8,400 ft. (1 CRCM); E slope Strawberry Mt., 8,600 ft. (1 CRCM); E slope Strawberry Mt., 8,700 ft. (1 CRCM). **Harney Co.**—16 mi. N, 20 mi. E Burns (1 OSUFW); 10 mi. N Harney (10 USNM). **Umatilla Co.**—(3 PSU); Langdon Lake (3 MSUMZ, 1 UW[EB], 5 UW[WWC], 1 CRCM); 8.4 mi. N Meacham, T1N, R36E, NW¼ Sec. 33, 4,390 ft. (2 PSU); 10 mi. W Meacham (1 USNM); Mill Creek, T1N, R35E, Sec. 24 (8 PSU); Rock Creek, 10 mi. W Tollgate (3 UW[EB]); Tollgate (2 MSUMZ, 9 UW[WWC]); 3 mi. N Tollgate (2 MVZ, 1 UW[WWC]); 3 mi. W Tollgate (4 UW[WWC]); 4 mi. W Tollgate (2 UW[EB], 3 UW[WWC]); 5 mi. W Tollgate (1 UW[WWC]); 0.5 mi. N, 2 mi. W Tollgate (1 UW[WWC]); 0.5 mi. N, 2.5 mi. W Tollgate (2 UW[WWC]); 1 mi. N, 3 mi. W Tollgate (5 UW[WWC]); 0.5 mi. S, 1 mi. E Tollgate (2 PSU); 2 mi. NW Tollgate (1 UW[WWC]); 3 mi. NW Tollgate (1 UW[WWC]); 16 mi. E Weston (10 UW[WWC]); 20 mi. SE Weston (3 UW[WWC]). **Union Co.**—Anthony Lake (8 PSM); Clark Creek guard station, Elgin (2 CRCM); Elgin (9 USNM); Grande Ronde Lake, 7,100 ft. (5 MVZ); Jubilee Forest Camp, Umatilla National Forest (2 PSM); Kamela (4 USNM); 10 mi. E Kamela, 4,900 ft. (2 USNM); Sprout Springs lookout (1 MSUMZ); Starkey Experimental Forest, Battle Creek, T4S, R34E, SE¼ Sec. 15 (1 PSM); Starkey Experimental Forest, Battle Creek, T4S, R34E, SW¼ Sec. 22 (3 PSM); Starkey Experimental Forest, Little Bear Creek, T4S, R34E, W½ Sec. 2 (5 PSM); Starkey Experimental Forest, Tin Trough

Pit, T4S, R34E, NE¼ Sec. 34 (4 PSM); 8 mi. S, 18 mi. E Union (1 OSUFW); West Eagle Meadows, [West Eagle Creek, Wallowa Mts.] (1 EOSC). **Wallowa Co.**—Aneroid (1 USNM); Aneroid Lake (3 SDNHM); head of Beaver Creek (2 EOSC); Cold Spring Creek (1 UW[WWC]); 25 mi. N Enterprise at Sled Springs (12 USNM); Hot Point* (1 EOSC); Hurricane Creek, Wallowa Mts., 5,500 ft. (1 UW[WWC]); Joseph (1 OSUFW); Joseph, Lick Creek (1 UMMZ); 12 mi. S, 9 mi. E Joseph (1 PSU); 20 mi. SE Joseph, Lick Creek (17 UMMZ); Lick Creek (2 EOSC); 10 mi. S, 1 mi. E Lostine, T2S, R43E, Sec. 35 (4 OSUFW); 13 mi. S, 2 mi. E Lostine, T3S, R43E, Sec. 14 (1 OSUFW); 16 mi. S, 3 mi. E Lostine, T3S, R43E, Sec. 36 (5 MVZ, 1 OSUFW); 19 mi. S, 4 mi. E Lostine, T4S, R44E, Sec. 17 (1 OSUFW); 17 mi. SE Lostine, Lostine River, Wallowa Mts., T2S, R43E, Sec. 35 (3 OSUFW); Moffet Camp, Umatilla National Forest (1 PSM); Sled Springs (1 EOSC); Stanley Ranger Station (1 USNM); Two Pan campground, Lostine River (1 EOSC); Wallowa Lake, 4,000 ft. (17 USNM, 4 UW[WWC]). **Wheeler Co.**—7 mi. S, 11 mi. W Mitchell (12 MVZ); 12 mi. SW Mitchell, Wildwood camp, Ochoco National Forest (4 OSUFW, 1 UMMZ).

Corynorhinus townsendii, n = 213
(Plotted in Fig. 9-37, p. 114)

Benton Co.—2.5 mi. S, 6 mi. W Alsea (1 OSUFW); 3.75 mi. S, 5.5 mi. E Alsea (2 CM); Corvallis (1 OSUFW); 3 mi. N Corvallis (1 OSUFW). **Clackamas Co.**—Carver (2 MVZ, 10 SDNHM, 3 USNM); Carver, Grants Cave (4 FMNH, 12 PSM, 9 UMMZ). **Clatsop Co.**—Cannon Beach (1 OSUFW). **Crook Co.**—Paulina Cave* (2 OSUFW). **Curry Co.**—Gold Beach (5 FMNH, 1 USNM); 1 mi. S Ophir (1 MVZ); mouth of Pistol River (1 MVZ); 10 mi. S Port Orford (2 SOSC); 1 mi. E Wedderburn, 50 ft. (1 MVZ). **Deschutes Co.**—6 mi. SE Bend (2 PSM); 14 mi. SE Bend (1 SDNHM); Boyd Cave (1 UW[WWC]); 5 mi. SE Deschutes River, near Bend, Lava Cave (2 OSUFW); Skeleton Cave, 8 mi. S Bend (1 PSU). **Douglas Co.**—10 mi. E Roseburg, T27S, R4W, Sec. 18 (1 OSUFW). **Harney Co.**—Catlow Valley (2 SDNHM); Malheur Cave (1 USNM); 11.6 mi. S, 2.3 mi. E Malheur National Wildlife Refuge Environmental Field Station (1 SOSC); 3.3 mi. S Princeton (1 PSM); Roaring Spring Cave, E rim Catlow Valley (2 MVZ). **Jackson Co.**—Ashland (1 SOSC); 8 mi. SW Prospect (3 OSUFW, 1 SOSC); Willow Creek, 5 mi. WSW Central Point (1 MVZ); 11 mi. SE Lakecreek, T37S, R3E, Sec. 16, 2,600 ft. (2 PSM); Little Butte Creek, SE HWY 140 (13 PSM); 7 mi. SW Medford, T38S, R2W, SE¼ Sec. 33 (1 PSM); 11.5 mi. from HWY 140 on Methodist Camp Rd. by way of Lake Creek Rd. (2 SOSC); 11.8 mi. E junction Butte Falls HWY and Crater Lake HWY (8 SOSC); T35S, R1E, Sec. 31 (2 SOSC); T37S, R3E, center Sec. 5 (2 SOSC); T37S, R2E, SE¼ Sec. 13 (3 SOSC); T37S, R3E, Sec. 16 (1 SOSC). **Josephine Co.**—Oregon Caves [National Monument] (4 OSUFW, 1 PSU, 2 USNM); on Applegate Rd. from Williams HWY* (1 SOSC). **Lake Co.**—T37S, R23E, Sec. 1 (1 SOSC). **Lane Co.**—Creswell (1 USNM); Eugene (7 UO); 11 mi. W Junction City (2 PSM); McKenzie Bridge (1 UO, 3 USNM); Vida (1 USNM). **Linn Co.**—Big Lake (1 PSU). **Malheur Co.**—15 mi. SW Cord (6 USNM); Horse Camp Canyon, T34S, R44E, Sec. 1 (1 SOSC); Jordan Crater (6 PSM); Owyhee Spring, 3.5 mi. SE Rome (1 PSM); Twin Pit Cave* (2 PSU); Westfall (1 PSM). **Marion Co.**—Silver Creek Cave (1 PSM). **Multnomah Co.**—Portland (2 SDNHM). **Tillamook Co.**—Nehalem (1 OSUFW); Tillamook (1 PSM); Lees Camp, Wilson River (1 OSUFW). **Umatilla Co.**—Milton-Freewater (21 UW[LB]); 5 mi. E Milton (1 PSM, 2 UW[WWC]). **Union Co.**—(1 EOSC); Catherine Creek, 9 mi. SE Union (1 EOSC); Catherine Creek Ice Caves (1 EOSC); Elgin (2 UW[WWC]); 6 mi. N Elgin (3 UW[WWC]); Ice Caves near Catherine Creek guard station (1 EOSC); Ice Caves (1 EOSC); Mt. Harris (1 EOSC); Union Ice Caves (1 EOSC). **Wallowa Co.**—ca. 1 mi. N Dug Bar (1 PSU); Hells Canyon* (2 PSU); 20 mi. S Imnaha (1 OSUFW). **Wasco Co.**—Cow Canyon* (1 SDNHM); Maupin (1 SDNHM); Memaloose State Park (1 PSM); 8 mi. S, 8 mi. W The Dalles (1 OSUFW). **Washington Co.**—13 mi. E Cedar Mill (2 PSU); 4 mi. SE Gaston (2 OSUFW); Tigard (1 PSM). **Wheeler Co.**—16 mi. E Antelope (2 OSUFW).

Didelphis virginiana, n = 84
(Plotted in Fig. 7-2, p. 41)

Benton Co.—Corvallis (5 OSUFW); 1 mi. N Corvallis (1 OSUFW); 2 mi. N Corvallis (2 OSUFW); 3 mi. N Corvallis (1 OSUFW); 7 mi. N Corvallis (1 OSUFW); 3 mi. W Corvallis (1 PSM); 4 mi. N, 0.5 mi. W Corvallis (1 OSUFW); 0.8 mi. N, 1.3 mi. E Corvallis (1 OSUFW); 0.9 mi. N, 1.3 mi. E Corvallis (1 OSUFW); 3.5 mi. N, 0.75 mi. E Corvallis (1 OSUFW); 9 mi. N, 3 mi. E Corvallis (1 OSUFW); 0.6 mi. S, 0.1 mi. W Corvallis (1 OSUFW). **Clackamas Co.**—(1 UW[EB]); Advance (1 MVZ); Lake Grove, Child Rd. near Tualatin River (11 MVZ); Lake Oswego (2 MVZ); Milwaukie (1 OSUFW); 3 mi. E Milwaukie (2 PSU); Mt. Scott (1 UW); Oregon City (1 MVZ, 1 PSU, 5 UW[EB]); 3 mi. W Sandy (1 UW); 2.5 mi. NE Wilsonville, Kruse Loop (2 MVZ); 82nd St. and Lindy (2 PSU); 82nd St. and Johnson (1 PSU). **Clatsop Co.**—Cannon Beach junction, T5N, R10W, Sec. 9 (1 OSUFW); Seaside (1 PSM). **Columbia Co.**—St. Helens (1 PSU). **Douglas Co.**—Roseburg area (1 OSUFW). **Lane Co.**—4.5 mi. SW Alpine, T15S, R6W, Sec. 10, 200 ft. S Benton-Lane Co. line (1 OSUFW). **Linn Co.**—2 mi. E Corvallis (1 OSUFW); 1 mi. N, 1 mi. E Lebanon (1 OSUFW); city limits of Scio on HWY 226, E of city center (1 OSUFW). **Marion Co.**—0.5 mi. S, 1 mi. E Independence (1 OSUFW); Keizer (2 CRCM). **Multnomah Co.**—Gresham (3 PSU); Portland (11 PSU); Portland, T1S, R1E, Sec. 29 (1 PSU); Rocky Butte (1 PSU); Sauvie Island (1 UW[WWC]); Washington Park Zoo (2 PSM). **Tillamook Co.**—1 mi. W Nehalem (1 PSM). **Umatilla Co.**—Pilot Rock, McKay Creek (1 USNM). **Washington Co.**—Aloha (1 PSU); Banks (1 EOSC); 3 mi. S Tigard (1 PSU).

Dipodomys californicus, n = 102
(Plotted in Fig. 11-63, p. 250)

Jackson Co.—Ashland (1 USNM); 11 mi. S Ashland, T41S, R7W, Sec. 4, 4,400 ft. (1 SOSC); 9 mi. S, 5 mi. E Ashland (1 SOSC); 10 mi. S, 2 mi. E Ashland (1 SOSC); 0.1 mi. from Bald Mt. R. toward Anderson Butte on Anderson Creek Rd. (1 SOSC); 0.5 mi. from Bald Mt. R. toward Anderson Butte on Anderson Creek Rd. (2 SOSC); 12 mi. S, 0.5 mi. E Ashland on interstate highway 5 (1 OSUFW); Brownsboro (5 SDNHM, 22 USNM); 2 mi. E Brownsboro (14 SDNHM); 3 mi. E Brownsboro (9 SDNHM); 0.3 mi. W Colestine Rd. off Mt. Ashland Rd. (1 SOSC); 0.75 mi. W Pinehurst, 3,500 ft. (1 MVZ); N end Upper Table Rock (1 SOSC); 3.5 mi. S junction HWY I-5 and Mt. Ashland Rd. (1 SOSC); 1.6 mi. W junction HWY 99 and Mt. Ashland Rd. (1 SOSC). **Klamath Co.**—NE of Bly (1 SOSC); NE Bly on east side of Campbell Res. (1 SOSC); 5 mi. NE Chiloquin (3 PSM); 4.75 mi. S, 2.5 mi. E Dairy (5 PSU); Fort Spring (1 PSM); Klamath Falls (2 USNM); 3 mi. E Lorella (1 HSU); 5 mi. E Lorella, T39S, R14E, Sec. 17 (2 PSU); 2 mi. N, 5 mi. E Lorella (1 OSUFW); 6 mi. E Malin, Malone campground (4 OSUFW); 7 mi. E Malin, Malone campground (1 OSUFW); Swan Lake Valley (9 PSM, 1 USNM); 3 mi. W Swan Lake, 4,300 ft. (3 MVZ); Tule Lake (1 USNM). **Lake Co.**—2 mi. NE Valley Falls (1 SOSC).

Additional locality: middle Barnes Valley* (2 SOSC).

Dipodomys microps, n = 62
(Plotted in Fig. 11-65 , p. 252)

Harney Co.—bluff above SW corner Alvord Desert (1 EOSC); Alvord Lake (2 MVZ); 2 mi. S Borax Spring, S end Alvord Lake, 4,300 ft. (6 MVZ); Buena Vista (1 PSM); 31 mi. S Burns, Malheur National Wildlife Refuge (1 OSUFW); 18 mi. S, 1 mi. W Burns (1 OSUFW); 23 mi. S, 2.3 mi. W Burns, NE Harney Lake, Malheur National Wildlife Refuge (1 OSUFW); 23 mi. S, 9.8 mi. W Burns, NW Harney Lake, Malheur National Wildlife Refuge (1 OSUFW); 1.5 mi. E Denio, 4,200 ft. (2 MVZ); 0.5 mi. N, 2 mi. E Denio, T41S, R35E, Sec. 15 (1 OSUFW); 3.6 mi. N Fields (1 SOSC); 0.75 mi. S Fields (1 OSUFW); 1 mi. N, 0.25 mi. E Fields, T37S, R33E, center Sec. 31 (1 KU); 1 mi. S, 1 mi. E Fields, T38S, R33E, S½NW¼ Sec. 30 (1 KU); 17 mi. N Frenchglen, Jack Mountain, T30S, R30E, Sec. 34 (2 OSUFW); 17.5 mi. N, 1.5 mi. E Frenchglen, T29S, R31E, SE¼SE¼ Sec. 4 (1 OSUFW); Malheur Environmental Field Station, 4,200 ft. (5 CRCM); Malheur National Wildlife Refuge, N edge Harney Lake, T26S, R30E, Sec. 28 (3 HSU); Malheur National Wildlife Refuge, SE edge Harney Lake, T27S, R30E, Sec. 19 (1 HSU); Harney Lake dunes, Malheur National Wildlife Refuge, T26S, R30E, Sec. 27 (1 CM); Narrows (7 USNM); 3 mi. NE Narrows (1 CRCM); South Coyote Butte (1 CRCM); Tumtum Lake (2 USNM);

Whitehorse Creek (1 USNM); White Horse Sink (1 USNM). **Lake Co.**—Abert Lake, 24.6 mi. N, 6.4 mi. E Lakeview (2 PSU); 8 mi. S Adel (1 MVZ); NE edge Alkali Lake (3 MVZ); 24.5 mi. N, 6 mi. E Lakeview, SE shore Abert Lake (3 PSU); Summer Lake (1 USNM); 5 mi. N Valley Falls (1 SOSC); 1.5 mi. W Valley Falls (1 SOSC). **Malheur Co.**—Rome (1 USNM); Sponge Springs, T20S, R43E, NE¼ Sec. 34 (1 PSM); Watson (1 USNM).

Dipodomys ordii, n = 745
(Plotted in Fig. 11-67, p. 255)

Baker Co.—Huntington (2 USNM); 12 mi. E Huntington (5 SDNHM). **Crook Co.**—mouth of Bear Creek, Crooked River, 3,400 ft. (8 MVZ); Powell Butte (2 USNM); Prineville (11 USNM); 20 mi. S Prineville, Crooked River (1 OSUFW); 7 mi. W Prineville (8 MVZ); 17 mi. S, 3 mi. E Prineville (1 OSUFW); 2 mi. NE Prineville (2 MVZ); 4 mi. SW Prineville (13 MVZ); 2 mi. N, 4 mi. E Redmond (1 OSUFW). **Deschutes Co.**—Alfalfa (1 PSU); Arnold Ice Cave, 16 mi. SE Bend (3 USNM); 4.5 mi. SE Bend (1 KU); Lower Bridge (5 OSUFW); 1 mi. N, 1.5 mi. E Lower Bridge (1 OSUFW); 1 mi. N, 7 mi. W Redmond (4 KU); 2 mi. S, 1 mi. E Redmond (1 OSUFW); 1 mi. N, 18 mi. E Sisters (1 PSU); 6 mi. N, 4 mi. E Sisters (1 OSUFW); 6 mi. N, 5 mi. E Sisters (1 OSUFW); 2 mi. S, 5 mi. E Sisters (1 OSUFW); 12 mi. NE Sisters (1 PSM); 2 mi. W Terrebonne (1 OSUFW); 3 mi. W Terrebonne, T14S, R13E, Sec. 18 (16 OSUFW); 4 mi. W Terrebonne (7 OSUFW); 6 mi. W Terrebonne, Lower Bridge (1 OSUFW); 7 mi. W Terrebonne, Lower Bridge (1 OSUFW); 0.2 mi. N, 4 mi. W Terrebonne (2 OSUFW); 0.5 mi. N, 3 mi. W Terrebonne (2 OSUFW); 0.5 mi. N, 3.5 mi. W Terrebonne (1 OSUFW); 0.5 mi. S, 4 mi. W Terrebonne (2 OSUFW); 1 mi. N, 5 mi. W Terrebonne, Lower Bridge (2 OSUFW); 1 mi. N, 6 mi. W Terrebonne, Lower Bridge (1 OSUFW); 1 mi. N, 7 mi. W Terrebonne, Lower Bridge (1 OSUFW); 0.5 mi. S, 3 mi. W Terrebonne (1 OSUFW); 1 mi. S, 3 mi. W Terrebonne (1 JFBM, 6 OSUFW); 2 mi. S, 8 mi. W Terrebonne (1 OSUFW). **Gilliam Co.**—Arlington (1 MVZ); 2.5 mi. S Heppner Junction (1 KU); Willows (1 OSUFW, 15 USNM); Willows Junction (2 USNM). **Harney Co.**—E edge of Alvord Basin, Lake Rd. (1 EOSC); Alvord Desert (1 EOSC); Alvord Lake (1 MVZ, 5 USNM); Alvord Valley* (4 USNM); Beatys Butte (2 SDNHM); Blitzen (1 USNM); Blitzen Valley* (8 SDNHM); Burns (5 USNM); 22 mi. S Burns, T27S, R31E, Sec. 35 (1 SOSC); 14 mi. S, 13 mi. W Burns (4 OSUFW); 21 mi. S, 1.5 mi. W Burns (1 OSUFW); 23 mi. S, 2.3 mi. W Burns, NE Harney Lake, Malheur National Wildlife Refuge (3 OSUFW); 11 mi. S, 3 mi. E Burns (2 SOSC); 15 mi. S, 4.5 mi. E Burns (1 OSUFW); 20 mi. S, 5 mi. E Burns, Malheur National Wildlife Refuge (1 OSUFW); Coyote Buttes, T27S, R30E, Sec. 13 (1 HSU); 7 mi. S Crane, Windy Point, T26S, R33E, Sec. 11 (3 OSUFW); 1.5 mi. E Denio (4 MVZ); Diamond (1 USNM); Diamond Craters (1 SOSC); Fields (1 SDNHM); 1.5 mi. N Fields (1 OSUFW); 3 mi. N Fields (3 SOSC); 5 mi. N Fields (1 SOSC); 6 mi. N Fields (1 SOSC); 45 mi. N, 17 mi. E Fields, T30/31S, R35E, Sec. 35/2 (1 KU); 1 mi. S, 1 mi. E Fields, T38S, R33E, S½NW¼ Sec. 30 (2 KU); Frenchglen (1 FMNH); 20 mi. N Frenchglen, T29S, R31E, Sec. 9 (3 OSUFW); 5 mi. N, 1 mi. E Frenchglen, Steens Mt. (2 USNM); 18 mi. N, 1 mi. E Frenchglen, T29S, R31E, SW¼NW¼ Sec. 4 (1 OSUFW); 0.5 mi. S, 2.5 mi. W Frenchglen, T32S, R32E, NW¼ Sec. 6 (3 KU); 9 mi. S, 1 mi. W Frenchglen (1 OSUFW); S shore Harney Lake, T27S, R29E, Sec. 36 (4 OSUFW); Harney Sand Dunes, W end Malheur Lake (5 PSM); 7.8 mi. E Lawen (4 CM); Malheur (2 CM); Malheur Lake, Springer Ranch (9 LACM); Malheur National Wildlife Refuge (1 EOSC); Malheur National Wildlife Refuge, SE edge Harney Lake (4 HSU); Malheur National Wildlife Refuge, SW corner Martha Lake (1 OSUFW); Malheur National Wildlife Refuge, T26S, R31E, Sec. 27 (1 CM); Malheur National Wildlife Refuge, T26S, R30E, Sec. 27 (5 CM); Malheur National Wildlife Refuge, T26S, R30E, Sec. 28 (6 CM); Malheur National Wildlife Refuge, T26S, R30E, Sec. 35 (1 CM); Malheur National Wildlife Refuge headquarters, T26S, R31E, Sec. 35 (3 CM); 4 mi. E Malheur National Wildlife Refuge headquarters (1 SOSC); 5.4 mi. E Malheur National Wildlife Refuge headquarters (1 SOSC); Narrows (7 SDNHM, 28 USNM); 1 mi. S Narrows, 4,200 ft. (3 MVZ); 2 mi. due SW Narrows, T27S, R31E, Sec. 3 (1 OSUFW); 5 mi. SW Narrows (18 MVZ); Princeton (2 SDNHM);

Shirk (3 USNM); Smyth Creek, 10 mi. E Diamond (1 MVZ); South Coyote Butte (2 CRCM); Steens Mt., T30S, R32E, Sec. 31 (3 OSUFW); Steens Mt., T31S, R32½E, Sec. 3 (6 OSUFW); Steens Mt., T31S, R32½E, Sec. 4 (8 OSUFW); Steens Mt., T31S, R32½E, Sec. 5 (3 OSUFW); Trout Creek (2 SDNHM); Voltage (3 USNM); 1 mi. N, 6 mi. W Voltage (1 OSUFW); 3 mi. S Whitehorse Ranch headquarters (1 UW[WWC]); Wildhorse Creek (1 EOSC); Wrigh's Well between Princeton and Malheur Cave (1 OSUFW). **Jefferson Co.**—4 mi. NNW Culver (2 PSM); 6 mi. S Culver (2 SOSC); 7 mi. E Culver (2 PSM); 7–12 mi. E Culver (5 PSM); Gateway (3 OSUFW, 3 USNM); Madras (4 SDNHM); 10 mi. S Madras (8 PSU); 5 mi. N, 1 mi. E Madras (1 OSUFW); 10 mi. S, 2 mi. E Madras (5 PSU); 12 mi. S, 2.75 mi. E Madras (2 PSU). **Lake Co.**—Adel (6 USNM); Adel, Twenty Mile Creek (8 OSUFW); 10 mi. N Adel (1 PSU); 9 mi. S Adel, mouth of 20 Mile Creek (1 MVZ); 7 mi. E Adel (2 PSU); NE edge Alkali Lake, 4,200 ft. (10 MVZ); Fort Rock (1 LACM, 4 SDNHM); Fort Rock campground (1 SOSC); Fort Rock State Park (2 OSUFW, 2 PSM); Fremont (10 USNM); Guano Valley (1 SDNHM); 4 mi. E Hart Mt. National Antelope Refuge (1 SOSC); 5.4 mi. E Hart Mt. National Antelope Refuge (1 SOSC); 2 mi. N Hot Springs campground (1 SOSC); Paisley (2 USNM); 5 mi. NE Paisley (1 OSUFW); Plush (1 CRCM); Silver Lake (12 USNM); 1 mi. N Valley Falls (3 SOSC); 2 mi. N Valley Falls (1 SOSC); S Warner Lake (1 CM); Warner Valley (1 USNM). **Malheur Co.**—S end Butch Lake (5 PSM); Bone Springs, Sheepshead Mts., 5,200 ft. (1 USNM); Brogan (1 OSUFW, 1 PSU); Cord, Barren Valley, 3,950 ft. (1 USNM); Cottonwood Mt., T17S, R42E, SE¼ Sec. 26, 3,400 ft. (1 PSM); head of Crooked Creek, Owyhee Desert (1 USNM); Crowley guard station, T27S, R39E, N½ Sec. 17 (2 PSM); 0.5 mi. N Crows Nest Reservoir (2 PSM); Jordan Craters (2 PSM); 8 mi. W Jordan Valley (1 MVZ); 7 mi. S, 2 mi. E Juntura (1 OSUFW); 3 mi. N McDermitt [Nevada] (2 PSM); Ontario (1 PSM); Ontario, 5 mi. S Wood Ranch (4 USNM); Owyhee (2 USNM); Riverside (15 USNM); Rome (7 USNM); Rome, mouth of Jordan Creek (4 USNM); 4 mi. N Rome (2 CRCM); 0.5 mi. S Rome, 3,600 ft. (5 MVZ); 1 mi. S Rome (5 MVZ, 1 OSUFW); 1 mi. S Rome via U.S. HWY 95 (10 OSUFW); 0.5 mi. SE Rome (2 PSM); Ryegrass Creek, Owyhee Desert, 4,000 ft. (7 USNM); Sand Hollow (16 AMNH); Vale (6 AMNH); near Vale (2 SDNHM); 3 mi. N Vale (2 MVZ); 2 mi. S Vale, Herrett Ranch (3 USNM); 5 mi. S Vale (5 MSUMZ); 5.75 mi. S Vale (11 PSM); 5 mi. S, 2 mi. W Vale (2 OSUFW); 6 mi. S, 2 mi. W Vale (2 OSUFW); Watson (3 USNM). **Morrow Co.**—Boardman (1 FMNH, 2 PSM, 15 SDNHM); 14.3 mi. Boardman* (1 PSU); 2 mi. N, 1 mi. E Boardman, Umatilla National Wildlife Refuge (1 OSUFW); 0.25 mi. S, 5 mi. W Boardman (18 PSU); 1 mi. S, 5.5 mi. W Boardman (4 PSU); 1.5 mi. S, 5.5 mi. W Boardman (14 PSU); 3 mi. S, 6 mi. W Boardman (1 PSU); 4 mi. S, 6 mi. E Boardman, E of U.S. Navy bombing range, gate 6106 (2 OSUFW); 6 mi. S, 10 mi. E Boardman (2 OSUFW); Cecil (1 MVZ, 6 PSM, 19 SDNHM); Heppner (1 USNM); 3 mi. W Irrigon (2 OSUFW); T5N, R26E, NW¼ Sec. 31 (2 PSM). **Sherman Co.**—Biggs Junction (2 PSM). **Umatilla Co.**—Hermiston (3 USNM); 1 mi. N, 3 mi. E McNary, Columbia River (1 OSUFW); Umatilla (10 MVZ, 19 USNM); Wallula (1 UW[WWC]). **Wasco Co.**—Bake Oven (1 USNM). **Wheeler Co.**—(1 UW); Twickenham (11 USNM).

Enhydra lutris, n = 5
(Plotted in Fig. 12-53, p. 438)

Curry Co.—Port Orford (1 OSUFW, 3 USNM).
Additional locality: Oregon* (1 USNM).

Eptesicus fuscus, n = 413
(Plotted in Fig. 9-33, p. 108)

Baker Co.—Anthony (1 AMNH); 8 mi. NE Unity, Middle Fork Burnt River (46 PSM); T7S, R43E, Sec. 23 (1 PSU). **Benton Co.**—Alsea (1 PSU); 3.75 mi. S, 5.5 mi. E Alsea (1 CM); Corvallis (9 OSUFW); 4 mi. N Corvallis (23 OSUFW); 7 mi. N Corvallis, Peavy Arboretum (1 OSUFW); 5 mi. S, 1 mi. W Corvallis (1 OSUFW); 10 mi. SE Corvallis, Irish Bend, T14S, R5W, Sec. 1 (1 OSUFW); W.L. Finley National Wildlife Refuge, T13S, R5W, SW¼ Sec. 20 (7 PSM). **Clackamas Co.**—Jennings

Lodge (1 SDNHM); Wilsonville (1 PSU). **Columbia Co.**—7 mi. E Wauna (1 MVZ). **Coos Co.**—Charleston (1 LACM). **Crook Co.**—mouth Bear Creek, Crooked River, 3,400 ft. (1 MVZ); T17S, R23E, Sec. 19 (1 PSU). **Curry Co.**—Marial, T33S, R10W, Sec. 9 (1 SOSC); Squaw Valley, T35S, R14E, Sec. 33 (1 SOSC). **Deschutes Co.**—Arnold Ice Cave, 16 mi. SE Bend (3 USNM); Bend (3 USNM); Farewell Bend, Deschutes River (4 USNM); Indian Ford campground, T14S, R9E, Sec. 13 (1 OSUFW, 1 PSM); Paulina Lake (2 USNM); Sand Springs* (1 PSU); Sisters (4 SDNHM, 1 UO, 7 USNM); 12 mi. NW Sisters (22 UW[EB]); 3 mi. W Terrebonne (1 OSUFW). **Douglas Co.**—Drew (1 USNM); Reston (2 USNM). **Grant Co.**—Camp Creek, junction FR36 and 1195* (1 PSM); 6 mi. S Fox (2 PSM); 4 mi. S, 9.5 mi. E Fox (1 PSM); 21 mi. SE Prairie City, North Fork Malheur River (4 MVZ); T11S, R29E, SW¼ Sec. 35 (3 PSM). **Harney Co.**—Blitzen Valley (4 SDNHM); ca. 15 mi. N Burns (1 PSU); Catlow Valley, T35S, R32E, NE¼ Sec. 26 (8 PSM); Malheur National Wildlife Refuge headquarters (5 PSM); Pike Creek, 25 mi. N Fields (1 SOSC); Steens Mt., Kiger Gorge, 6,000 ft. (1 USNM); Steens Mt., T33S, R32¾E, NE¼ Sec. 7 (1 PSM); Steens Mt., T33S, R34E, Sec. 3 (1 OSUFW); Steens Mt., T33S, R34E, Sec. 9 (1 OSUFW); T33S, R32½E, NW¼ Sec. 36, 4,960 ft. (1 EOSC). **Hood River Co.**—Brooks Meadows (1 PSU). **Jackson Co.**—Ashland (5 SOSC); W of Ashland (1 SOSC); 25 mi. E Ashland (1 SOSC); Buck Rock, off Emigrant Creek Rd. (1 SOSC); SE Grizzly Mt. (1 SOSC); McKee Bridge, T40S, R3W, Sec. 5 (1 SOSC); 7 mi. SW Medford, T38S, R1W, Sec. 35, 3,600 ft. (2 PSM); East Pilot Rock Spring, Mt. Ashland (3 PSM); 4 mi. S Ruch, T39S, R3W, Sec. 15 (1 MVZ, 3 PSM, 1 SOSC); Shoat Springs (1 SOSC); T39S, R3W, NE¼ Sec. 28 (1 SOSC). **Josephine Co.**—10 mi. S, 10 mi. E Cave Junction (1 OSUFW); Galice Creek, Rogue River Valley (4 UO); Grants Pass (2 CM, 1 SOSC); 13 mi. SW Grants Pass (2 UO); Jumpoff Joe Creek, 1 mi. E HWY I-5 (2 PSM); Merlin (1 SOSC); 3.3 mi. [unknown direction] Waldo Hill (1 SOSC); 1 mi. N of USFS 355 on USFS Rd. 3794, 2 mi. W HWY 199 (2 SOSC). **Klamath Co.**—Anna Creek, Mt. Mazama (3 USNM); 20 mi. W Cresent Lake (2 UO); Ft. Klamath (1 USNM); Good Luck Pond, T36S, R9E, Sec. 21 (1 SOSC); Hog Creek (1 SDNHM); 1 mi. S, 6 mi. W Keno (2 OSUFW); 35.1 mi. W Lakeview (1 SOSC); Onion Springs (1 SOSC); Sheep Pond, T33S, R10E, Sec. 31 (3 SOSC). **Lake Co.**—Adel, Warner Valley (1 OSUFW); Fremont (5 USNM); 10 mi. SW Silver Lake, West Silver Creek, 4,650 ft. (1 USNM); S of Warner Lake (1 CM); Woodchopper Spring, NW Chandler State Park (7 SOSC). **Lane Co.**—Eugene (2 UO, 3 USNM); E of Eugene, T18S, R1W (1 SOSC); H. J. Andrews Experimental Forest (1 PSM); 10 mi. W Junction City (1 PSM); 11 mi. W Junction City (1 PSM); Mercer, Dowell Ranch (1 USNM); McKenzie Bridge (1 OSUFW, 3 UO); 10 mi. E McKenzie Bridge, Lost Creek ranger station (2 OSUFW); 11 mi. E McKenzie Bridge, Lost Creek ranger station (1 OSUFW); 3 mi. SE Oakridge (1 UO); N base Three Sisters, 5,000 ft. (1 UO); Triangle Lake (2 UO); Vida (3 SDNHM, 4 UO, 9 USNM). **Lincoln Co.**—Cascade Head Experimental Forest (10 PSM); 2 mi. SE Harlan, T12S, R8W, NE¼ Sec. 16 (25 PSM). **Linn Co.**—Holly (1 SDNHM); 3.5 mi. NW Shedd (2 PSM); T13S, R1W, Sec. 21 (1 PSU). **Malheur Co.**—Castle Rock (1 PSU); canyon located 8 mi. NE McDermitt, 5,800 ft. (1 USNM); Succor Creek State Park (1 PSM). **Morrow Co.**—0.8 mi. E Boardman (1 PSM). **Multnomah Co.**—Portland (6 AMNH, 1 MVZ, 6 PSM, 1 PSU, 2 SDNHM, 1 UW[WWC]); Troutdale (1 PSU); T1S, R1E (1 PSU). **Polk Co.**—Monmouth (1 PSM). **Tillamook Co.**—Nehalem (1 OSUFW); Tillamook (1 OSUFW). **Umatilla Co.**—Meacham (1 USNM). **Union Co.**—0.5 mi. NE Elgin (1 PSU); Palmer Junction (1 SDNHM); Starkey Experimental Forest, T4S, R34E, Sec. 14 (2 PSM); Starkey Experimental Forest, Bear Creek, T3S, R34E, NW¼ Sec. 2 (1 PSM); Starkey Experimental Forest, Meadow Creek, T3S, R34E, SE¼ Sec. 27 (1 PSM); T2S, R40E, Sec. 3 (1 PSU). **Wallowa Co.**—15 mi. S, 2 mi. E Lostine, T3S, R43E, Sec. 25 (1 OSUFW); 16 mi. S, 3 mi. E Lostine, T3S, R43E, Sec. 36 (2 MVZ, 2 OSUFW); 19 mi. S, 4 mi. E Lostine, T4S, R44E, Sec. 17 (1 OSUFW); 21 mi. S, 6 mi. E Lostine, T4S, R44E, Sec. 26 (1 OSUFW); Powwatka Ridge (8 PSM); Rockwall Springs, 11.5 mi. NNE Wallowa, T3N, R43E, SW¼NW¼ Sec. 14 (8 PSM); T3S, R43E, NW¼SE¼ Sec. 36 (6 PSM). **Wasco Co.**—mouth Deschutes River (2 MVZ); The Dalles (1 PSM, 1 PSU); 3 mi. S The Dalles (1 USNM); 25 mi. SW The Dalles, near Five Mile Creek (1 USNM); Memaloose State Park, 2.5 mi. E Mosier, T2N, R12E, SE¼ Sec. 31 (4 PSM). **Washington Co.**—Cedar Mill (1 PSM); Hillsboro (1 PSU). **Wheeler Co.**—T12S, R21E, Sec. 5 (1 PSU). **Yamhill Co.**—2.5 mi. S, 0.5 mi. E Dayton (1 KU). **Additional localities:** N slope Three Sisters, 5,000 ft.* (2 USNM); Canyon Rd.* (1 PSU).

Erethizon dorsatum, $n = 177$
(Plotted in Fig. 11-133, p. 346)

Baker Co.—Anthony (1 AMNH); Bulger Flat* (1 EOSC); Cornucopia (1 UMMZ); Eagle Creek area (3 KU); Haines (1 EOSC); halfway between Lily White and Medical Springs (1 EOSC); Unity, East Camp Creek (1 OSUFW); Whitney (2 USNM). **Benton Co.**—Corvallis, 3520 Country Club Dr. SW (1 OSUFW). **Clackamas Co.**—25 mi. SE Estacada on Roaring River (1 USNM). **Curry Co.**—just S of Brookings on HWY 101 (1 SDNHM). **Deschutes Co.**—near Bend (1 PSM); 11 mi. W Bend, Tumalo Creek (1 MVZ); 15 mi. W Bend, Tumalo Creek (1 MVZ); 16 mi. E Bend (1 KU); 12 mi. SE Bend, Gosney Rd. (1 OSUFW); 12 mi. S Hampton (1 USNM); 14 mi. S Hampton (5 USNM); Paulina Creek, Deschutes River (1 USNM). **Douglas Co.**—6 mi. S Canyonville (1 SOSC). **Harney Co.**—3 mi. N Alberson (2 USNM); 5 mi. N Alberson (1 USNM); 7 mi. N Alberson (1 USNM); 8 mi. N Alberson (2 USNM); 9 mi. N Alberson (1 USNM); 10 mi. N Alberson (1 USNM); 5 mi. W Alberson (1 USNM); 6 mi. SW Alberson (1 USNM); Alvord Ranch (2 USNM); Burns (1 USNM); 8 mi. NE Burns (2 USNM); Burns, Grant Kesterson Ranch (1 USNM); 16 mi. NE Burns, Willians Rd. (1 USNM); Coffee Pot Creek, 3 mi. N Harney City (1 USNM); vicinity of Denio (1 MVZ); Fields (4 USNM); SE Fields (2 USNM); Malheur National Wildlife Refuge (1 PSU); 3 mi. N Mule (1 USNM); 7 mi. N Mule (1 USNM); 10 mi. N Mule (1 USNM); 10 mi. W Mule (1 USNM); 5 mi. SW Mule (1 USNM); Shirk P.O. (1 USNM); Princeton, Malheur National Wildlife Refuge (2 UW); 10 mi. S Smith (1 USNM); Steens Mt. (1 USNM); Venator (2 USNM); 5 mi. S Venator (1 USNM); 10 mi. S Venator (3 USNM); 12 mi. S Venator (5 USNM); 10 mi. SE Venator (1 USNM). **Jackson Co.**—Ashland (1 SOSC); 10 mi. NE Butte Falls (1 USNM); Emigrant Lake (1 SOSC); Gold Hill (1 OSUFW); 6 mi. S Jacksonville (1 SOSC). **Jefferson Co.**—Hay Creek (7 USNM); 10 mi. E Hay Creek (2 USNM); 11 mi. E Hay Creek (1 USNM); 12 mi. E Hay Creek (1 USNM); 16 mi. E Hay Creek (1 USNM); 4 mi. N Santiam Pass (1 OSUFW). **Klamath Co.**—Bonanza (1 PSM); Ft. Klamath (1 UMMZ, 1 USNM); 2.5 mi. S, 3 mi. E Ft. Klamath (1 KU); Klamath Falls (3 CM, 1 USNM). **Lake Co.**—Finley Corral (1 PSM); 12 mi. NE Fort Rock (1 USNM); Gearhart Mt. (1 PSM); Hart Mt. (1 PSM, 1 SDNHM); 10 mi. SW Silver Lake, 4,650 ft. (1 USNM). **Malheur Co.**—Cord (1 USNM); 1 mi. W Ironside (1 MVZ). **Morrow Co.**—Lena (39 USNM). **Sherman Co.**—Miller, mouth of Deschutes River (1 USNM); 2 mi. WSW Rufus (1 MVZ). **Umatilla Co.**—2 mi. W Bells Ranch, Juniper Canyon (1 USNM); 3.5 mi. W Ring (1 USNM). **Union Co.**—Jim Creek, Kilridge, probably near Medical Springs (1 USNM); Starkey (1 EOSC); Sugar Loaf near Medical Springs (1 USNM); 0.75 mi. W Taylor, probably near Medical Springs (1 USNM); 3 mi. N Union (1 EOSC); 1 mi. SW Union (1 EOSC). **Wallowa Co.**—16 mi. N, 4 mi. E Enterprise (1 OSUFW). **Wasco Co.**—Maupin (1 SDNHM); 2.5 mi. S, 1.75 mi. W Shaniko (1 UW); 5 mi. S, 5 mi. W Shaniko (2 UW); 15 mi. S, 8 mi. E The Dalles (1 OSUFW). **Wheeler Co.**—3 mi. S Fossil (1 USNM); 6 mi. S Fossil (1 USNM); 6 mi. W Fossil (1 USNM); 12 mi. W Fossil (1 USNM); 4 mi. E Fossil (1 USNM); 8 mi. SW Fossil (1 USNM); 4 mi. N, 1 mi. E Mitchell (1 KU); 12 mi. S Spray (2 USNM); 18 mi. S Spray (1 USNM); 14 mi. W Spray (1 USNM); 14 mi. SW Spray (1 USNM).

Euderma maculatum, $n = 2$
(Plotted in Fig. 9-35, p. 112)

Harney Co.—Mickey Hot Springs, T33S, R35E, Sec. 13 (1 SOSC). **Wheeler Co.**—0.2 km S, 0.15 km E Clarno (1 PSU).

Eumetopias jubatus, $n = 20$
(Plotted in Fig. 12-19, p. 386)

Clatsop Co.—Cannon Beach (1 PSM, 1 CRCM); Columbia River south jetty (4 UW); 4.5 mi. N Gearhart (1 PSU); 6 mi. N Gearhart (1

PSU); 2 mi. W Warrenton (1 MVZ). **Coos Co.**—Cape Arago, Simpson Reef (1 PSM). **Curry Co.**—Port Orford (1 FMNH). **Lane Co.**—Heceta Beach (1 PSM); S of Newport, 45 fathoms off Heceta Head (1 PSM). **Lincoln Co.**—Agate Beach (1 PSM); Fogerty Creek, T8S, R11W, Sec. 28 (1 OSUFW); north jetty in Newport (1 USNM). **Tillamook Co.**—(1 UW); Netarts Bay (1 SDNHM); Netarts spit (1 PSM); Short Beach (1 UW).

Glaucomys sabrinus, n = 414
(Plotted in Fig. 11-44, p. 219)

Baker Co.—Anthony (8 AMNH, 2 MVZ); Bourne (1 USNM); Cornucopia (3 USNM); Lily White, Wallowa Mt. field station, Wallowa Mts. (1 EOSC). **Benton Co.**—Corvallis (3 OSUFW, 1 USNM); Corvallis, T11S, R5W, Sec. 17 (1 OSUFW); 2 mi. N Corvallis, T11S, R5W, Sec. 21 (1 OSUFW); 4 mi. N Corvallis (1 SOSC); 7 mi. N Corvallis, T11S, R5W, Sec. 1 (1 OSUFW); 0.5 mi. N, 4 mi. W Corvallis (1 OSUFW); 2 mi. N, 5 mi. W Corvallis (1 OSUFW); 3 mi. N, 3 mi. W Corvallis (1 OSUFW); Marys Peak (1 PSM); 4.5 mi. SW Monroe (1 PSM); near HWY 34, 1.5 mi. from Oregon State University campus, Corvallis (1 OSUFW); 2 mi. S, 9 mi. W Philomath, Marys Peak (1 OSUFW); Sulphur Springs near Albany (1 USNM). **Clackamas Co.**—Estacada (1 SDNHM, 1 USNM); near Lake Oswego (1 PSU); 2 mi. N Lake Oswego (1 PSU); 2 mi. E Liberal (2 PSM); Marmot (1 USNM); Molalla (1 PSM); 3.5 mi. N, 8 mi. E Sandy (2 CRCM); 3.8 mi. N, 8 mi. E Sandy (1 CRCM); 4.2 mi. N, 8 mi. E Sandy, T1S, R6E, Sec. 29 (1 CRCM); 8 mi. N, 3 mi. E Zigzag (1 CRCM); 9 mi. N, 4 mi. W Zigzag (2 CRCM); 5 mi. S, 6 mi. E Zigzag (2 CRCM). **Clatsop Co.**—1 mi. W Hammond (1 OSUFW); Nehalem summit HWY 53 (1 PSM). **Columbia Co.**—2 mi. W Deer Island (1 PSU). **Coos Co.**—7.6 km N, 5.2 km W Drain, T28S, R9W, Sec. 19 (1 USNM); 8.4 km N, 4.3 km W Sitkum, T27S, R10W, Sec. 17 (1 USNM); 2.8 km S, 5 km E Sitkum, T28S, R9W, Sec. 19 (1 USNM); 2.9 km S, 4.7 km E Sitkum, T28S, R9W, Sec. 19 (2 USNM); 3 km S, 10.8 km E Sitkum, T28S, R9W, Sec. 23 (1 USNM); 4.8 km S, 9.7 km E Sitkum, T28S, R9W, Sec. 27 (1 USNM); 1.8 mi. S, 2.9 mi. E Sitkum, T28S, R9W, Sec. 19, 1,600 ft. (43 UW). **Curry Co.**—Gold Beach (1 FMNH, 3 SDNHM, 1 USNM); Port Orford (4 USNM). **Deschutes Co.**—Davis Creek (1 OSUFW); Davis Mt. (6 OSUFW, 1 PSM); Paulina Lake (1 USNM). **Douglas Co.**—3.2 mi. S, 3.6 mi. E Canyonville, T31S, R4W, Sec. 17, 2,000 ft. (1 UW[CR]); 7.6 km N, 6.4 km W Drain, T21S, R6W, Sec. 22 (1 USNM); 3 mi. W Drain (1 AMNH); 4.5 mi. N, 8 mi. W Drain, T21S, R4W, Sec. 27, 1,200 ft. (8 UW); 5.5 mi. N, 4 mi. W Drain, T21S, R6W, Sec. 15, 1,280 ft. (2 UW, 1 UW[CR]); 6.75 mi. N, 5.5 mi. E Drain, T21S, R4W, Sec. 7, 1,000 ft. (6 UW); Elkhead (1 USNM); 14.9 km S, 6.2 km W Elkton, T24S, R8W, Sec. 9 (1 USNM); 6 mi. S Gunter (1 MVZ); 2 km N, 25 km W Sutherlin, T25S, R8W, Sec. 10 (3 USNM); 2.7 mi. N, 5 mi. W Umpqua, T25S, R7W, Sec. 7, 950 ft. (3 UW[CR]); 4 mi. N, 1 mi. W Umpqua, T25S, R7W, Sec. 1, 1,500 ft. (2 UW, 4 UW[CR]); 6 mi. N, 10.5 mi. W Umpqua, T24S, R8W, Sec. 29, 900 ft. (4 UW, 4 UW[CR]). **Grant Co.**—15 mi. NW Austin, 5,400 ft. (1 USNM); Beech Creek (1 USNM); 4.5 mi. S, 4 mi. E Long Creek, T10S, R31E, Sec. 31 (1 OSUFW); T12S, R31E, Sec. 12 (1 OSUFW). **Hood River Co.**—Brooks Meadows, 9 mi. ENE Mt. Hood, 4,300 ft. (2 MVZ); N slope Mt. Hood, 2,800 ft. (2 USNM); 2 mi. W Parkdale, 1,500 ft. (1 USNM). **Jackson Co.**—Fish Lake (1 UO); 4 mi. E Gold Hill (1 SOSC); 2 mi. N, 2 mi. W Gold Hill (3 OSUFW); Hyatt Lake, T39S, R3E, Sec. 15 (1 SOSC); 9 mi. N Rogue River, T35S, R4W, Sec. 3 (2 SOSC); Spring Creek, T33S, R4E, NE¼ Sec. 22 (2 SOSC); Little Chinquipin #3W* (1 SOSC). **Josephine Co.**—Cave Creek campground (1 KU); Cave Junction, T39S, R6W, Sec. 33 (2 HSU); 1.5 mi. S, 9 mi. E Cave Junction (1 KU); Grants Pass (1 SOSC). **Klamath Co.**—Crater Lake (4 MVZ, 1 USNM); 4 mi. S Crater Lake, Crater Mt. (1 USNM); Crater Lake National Park, Red Blanket Creek (2 USNM); 20 mi. W Cresent Lake (5 UO); Ft. Klamath (2 OSUFW); Ft. Klamath, Sun Creek Canyon, 4,200 ft. (17 USNM); Lake of the Woods (1 SOSC); S of Snow Lakes (1 SOSC); W of Sonya Lake (1 SOSC); Upper Klamath (1 FMNH). **Lake Co.**—Fishhole Creek (4 PSM); Gearhart Mt. (11 PSM); 16 mi. NW Lakeview (2 PSM). **Lane Co.**—12.1 km S, 2.4 km W Alsea (3 USNM); Belknap Springs (2 OSUFW); near Blue River, T15S, R5E, Sec. 14, 800 m (1 KU); 4.5 mi. N Blue River, T15S, R4E, Sec. 32 (1 KU); 4 mi. W Blue

River (1 KU); 6 mi. E Blue River, Swamp Creek (3 OSUFW); 0.5 mi. N, 2 mi. W Blue River, Elk Creek, T16S, R4E, Sec. 19 (1 OSUFW); 1.25 mi. N, 5 mi. W Blue River, T16S, R3E, Sec. 15 (1 HSU); 5.5 mi. N, 7 mi. E Blue River, T15S, R5E, Sec. 26 (4 HSU); 3.5 mi. S, 8 mi. W Blue River, T17S, R3E, Sec. 7 (1 HSU); 1.5 mi. S Canary, 0.5 mi. E Cushman, T18S, R11W, Sec. 33 (1 OSUFW); 1.5 mi. S Cushman, 0.5 mi. E Canary, T18S, R11W, Sec. 33 (1 OSUFW); Elmira (1 UO); Eugene (3 UO); Gold Lake, 4,800 ft. (1 LACM); H. J. Andrews Experimental Forest* (4 PSM); 0.4 km N, 18.5 km W Lorane, T20S, R7W, Sec. 12 (3 USNM); 3 km N, 19.5 km W Lorane, T20S, R7W, Sec. 1 (1 USNM); 6.8 km N, 11.1 km W Lorane, T19S, R6W, Sec. 23 (2 USNM); McKenzie Bridge (2 UO); 6 km N, 0.6 km E McKenzie Bridge, T15S, R5E, Sec. 25 (2 USNM); 6.4 km N, 0.6 km E McKenzie Bridge, T15S, R5E, Sec. 25 (1 USNM); 11 mi. E McKenzie Bridge (1 UO); 1 mi. S, 1 mi. W McKenzie Bridge, King Creek, T16S, R5E, Sec. 26 (1 OSUFW); 10 mi. SE McKenzie Bridge, Lost Creek Ranger Station (2 USNM); Mercer (1 LACM, 1 PSM); Vida (1 OSUFW, 1 SDNHM, 1 UO, 2 USNM); Vida, McKenzie River hatchery (1 OSUFW); Willamette National Forest, near Blue River (2 OSUFW); 4 km S, 3.2 km E Yachats, T15S, R12W, Sec. 16 (1 USNM); T15S, R5E, Sec. 32 (6 PSM); T16S, R4E, Sec. 19 (2 PSM); T16S, R4E, Sec. 24 (1 PSM). **Lincoln Co.**—1.25 mi. W Harlan (1 KU); Otis (1 USNM); 4.8 km S, 3.2 km E Yachats, T15S, R12W, Sec. 12 (6 USNM); Yaquina Bay (1 USNM). **Linn Co.**—5 mi. S, 6.5 mi. W Albany, T12S, R4W, Sec. 6 (1 KU); H. J. Andrews Experimental Forest* (7 PSM); Lost Prairie camp, 32.5 mi. E Sweet Home (1 PSU); 8 mi. E Upper Soda, Tombstone Prairie (2 OSUFW); 13 mi. E Upper Soda, Toad Creek Rd. (1 OSUFW); confluence of Santiam Rivers (1 OSUFW); T15S, R5E, Sec. 11 (2 PSM). **Marion Co.**—Salem (1 OSUFW, 1 SDNHM, 2 UMMZ). **Morrow Co.**—10 mi. SE Hardman (1 PSM). **Multnomah Co.**—(1 PSU); Latourelle (1 OSUFW); Multnomah Falls (1 PSU); Portland (1 PSM, 1 PSU, 4 USNM); 9 mi. N, 10 mi. E Sandy (2 CRCM); 10 mi. N, 10 mi. E Sandy, T1N, R6E, Sec. 27 (4 CRCM); 11 mi. N, 10 mi. E Sandy (3 CRCM); 10 mi. N, 1.5 mi. W Zigzag (1 CRCM). **Tillamook Co.**—Beaver (1 FMNH, 1 KU); Blaine (1 OSUFW); Cascade Head Experimental Forest (2 PSM); Nehalem (2 KU, 1 OSUFW, 1 PSM); Netarts (1 FMNH, 1 LACM, 2 OSUFW, 2 SDNHM); 1 mi. S Oceanview (1 OSUFW); 0.5 mi. E Oceanview (4 OSUFW); 1 mi. E Oceanview (1 OSUFW, 1 PSM); Sandlake (1 OSUFW); Tillamook (1 KU, 1 MVZ, 4 OSUFW, 1 SDNHM, 1 UMMZ); 10 mi. S Tillamook (2 SDNHM). **Umatilla Co.**—Lehman (1 SDNHM); Lookingglass Creek (1 CRCM); Pikes Peak (1 UW[WWC]); Tollgate (1 UW[WWC]); 1 mi. E Tollgate (1 CRCM); 16 mi. E Weston (2 UW[WWC]). **Wallowa Co.**—Enterprise (1 SDNHM); 25 mi. N Enterprise at Sled Springs, 4,600 ft. (2 USNM); Little Sheep Creek* (1 PSM); 16 mi. S, 3 mi. E Lostine, 5,500 ft. (3 MVZ); Wallowa Lake, 4,000 ft. (1 USNM); S end Wallowa Lake (2 UO); 1 mi. S Wallowa Lake on west fork Wallowa River, 5,000 ft. (1 UW[EB]). **Washington Co.**—4 mi. E Gaston (1 AMNH); Helvitia Sta. (1 SDNHM); Hillsboro (1 USNM). **Wheeler Co.**—Derr R.S., Ochoco National Forest (1 SDNHM); Spray, Kahler Basin (1 UW[WWC]). **Yamhill Co.**—Dayton (1 OSUFW); McMinnville (1 PSM); 4 mi. N Newberg (1 AMNH).

Gulo gulo, n = 2
(Plotted in Fig. 12-47, p. 427)

Grant Co.—9 mi. S, 4 mi. E John Day, T15S, R32E, Sec. 4 (1 OSUFW). **Wheeler Co.**—5 mi. W Fossil, T6S, R20E, SE¼ Sec. 34 (1 OSUFW).

Lasionycteris noctivagans, n = 193
(Plotted in Fig. 9-26, p. 102)

Baker Co.—Anthony (5 AMNH, 5 USNM); Cornucopia (1 UMMZ); Home (1 USNM); 8 mi. NE Unity (16 PSM). **Benton Co.**—Sulphur Springs, 7 mi. N Corvallis (1 PSM); 1 mi. N, 2 mi. E Wrens, T11S, R6W, NW¼NE¼ Sec. 23, 920 ft. (1 KU). **Clackamas Co.**—0.3 mi. N, 1 mi. W Wemme (1 OSUFW); T5S, R6E, Sec. 1 (1 PSU). **Clatsop Co.**—7 mi. E Wauna (1 MVZ). **Coos Co.**—Remote (1 SDNHM). **Crook Co.**—20 mi. SE Prineville, Crooked River (1 USNM); Whiskey Spring (1

PSU). **Deschutes Co.**—Bend (1 USNM); Indian Ford campround (1 PSM); Sisters (1 SDNHM, 1 UO, 5 USNM). **Douglas Co.**—Anchor (3 USNM); Reston (1 USNM); T24S, R1W, Sec. 29 (1 PSU). **Grant Co.**—Camp Creek (1 PSM); Crane Prairie (1 PSU); Keeney Meadow (3 PSM); Kingbolt Spring (1 PSU); 21 mi. SE Prairie City, North Fork Malheur River, 5,000 ft. (2 MVZ); T11S, R29E, SW¼ Sec. 35 (2 PSM); T11S, R30E, SE¼ Sec. 32 (3 PSM). **Harney Co.**—10 mi. N Harney (1 USNM); 11 mi. N Harney, Blue Mts. (1 USNM); Malheur Lake (2 SDNHM); Steens Mt., Kiger Gorge, 6,000 ft. (1 USNM); 1.9 mi. on Long Hollow Rd. from Alvord Rd. (1 SOSC); T33S, R32½E, NE¼ Sec. 4, 4,420 ft. (1 EOSC); T33S, R32½E, NW¼ Sec. 36 (1 EOSC). **Hood River Co.**—Brooks Meadows, 9 mi. ENE Mt. Hood, 4,300 ft. (3 MVZ); T2S, R10E, Sec. 2 (1 PSU). **Jackson Co.**—Ashland (1 SOSC); 6 mi. S Ashland (1 SOSC); 3 mi. N, 10 mi. E Ashland, T38S, R3E, Sec. 30 (1 SOSC); 7 mi. S, 13 mi. W Ashland, 3,200 ft. (2 UW); East Pilot Rock Spring, Mt. Ashland (1 PSM); 7 mi. S Medford, T38S, R1W, Sec. 35, 3,600 ft. (5 PSM, 1 SOSC); Pilot Rock Rd., Mt. Ashland (1 SOSC); Shoat Springs (4 SOSC); 2.6 mi. N Dead Indian Rd. on Conde Creek Rd. (1 SOSC); 3.1 mi. S HWY 238 on West Forest Creek Rd., 1.9 mi. on Oregon Belle Loop Rd. (1 SOSC); T36S, R4E, SW¼ Sec. 32 (1 SOSC). **Josephine Co.**—Little Sugarloaf Mt. (1 SOSC); 3.8 mi. E O'Brian on 4007 to Waldo Hill (2 SOSC); 1 mi. E HWY I-5, Jumpoff Joe Creek (3 PSM); 3.5 mi. E HWY I-5 on Jumpoff Joe Creek Rd. (1 SOSC); Swede Basin Rd. (1 SOSC). **Klamath Co.**—11 mi. S, 3.5 mi. W Crater Lake (1 SOSC); Davis Lake (1 SDNHM); Dice Springs, T34S, R13E, Sec. 28 (1 SOSC); Dice Crane Springs, T32S, R8E, Sec. 24 (1 SOSC); Dixie (1 SOSC); 2.3 mi. S Keno, Onion Springs (2 SOSC); 35.1 mi. W Lakeview (1 SOSC); West Sink Creek, east of Mt. Thielson (1 USNM); 3 mi. W Swan Lake, 4,300 ft. (1 MVZ); T41S, R15E, Sec. 13 (1 SOSC). **Lake Co.**—9 mi. S, 3.5 mi. W Adel, Twelve Mile Creek (2 OSUFW); 3 mi. W Lakeview (1 SOSC); Woodchopper Springs (1 SOSC). **Lane Co.**—Belknap Springs (1 OSUFW); Mercer (1 SDNHM); Triangle Lake (2 UO); T15S, R5E, SW¼ Sec. 27, 2,000 ft. (1 PSM); T15S, R5E, SE¼ Sec. 26, 2,500 ft. (1 PSM); T15S, R5E, SE¼ Sec. 28, 1,750 ft. (1 PSM). **Linn Co.**—15.2 mi. NE Foster, junction Moose and Quartsville creeks (1 FMNH). **Malheur Co.**—Cottonwood Creek, T26S, R38E, NW¼ Sec. 29 (1 PSM); Succor Creek State Park, T24S, R46E, Sec. 20 (2 PSM); Three-Forks (1 PSU). **Marion Co.**—(1 PSU). **Multnomah Co.**—T1N, R1E, Sec. 34 (1 PSU). **Polk Co.**—Pedee (1 UCONN). **Tillamook Co.**—Beaver (1 OSUFW); Tillamook (1 OSUFW). **Umatilla Co.**—Meachem (2 SDNHM, 1 USNM); Pendleton (1 UW[WWC]). **Union Co.**—Catherine Creek State Park (1 PSU); Elgin (3 USNM). **Wallowa Co.**—3 mi. below Frazier Lake (1 SDNHM); Lick Creek R.S. (1 SDNHM, 1 UMMZ); 16 mi. S, 3 mi. E Lostine, 5,500 ft. (4 MVZ); 17 mi. SE Lostine, Lostine River, Wallowa Mts. (3 OSUFW); Minam State Park (1 PSU); Powwatka Ridge (1 PSM, 1 PSU); Rockwall Springs, 11.5 mi. NNE Wallowa, T3N, R43E, SW¼NW¼ Sec. 14 (14 PSM); Sled Springs (1 USNM); 9.5 mi. NNE Wallowa, T3N, R43E, center Sec. 26 (5 PSM). **Wasco Co.**—Memaloose State Park, 2.5 mi. E Mosier (1 PSM). **Washington Co.**—Beaverton (1 USNM); 18.5 mi. NW Portland (2 MVZ). **Wheeler Co.**—Bridge Creek (1 MVZ); 12 mi. SW Mitchell, Wildwood camp, Ochoco National Forest (1 OSUFW); T7S, R25E, Sec. 7 (2 PSU); T11S, R21E, Sec. 9 (1 PSU).

Lasiurus cinereus, n = 50
(Plotted in Fig. 9-23, p. 99)

Baker Co.—Anthony (2 AMNH); Baker City (1 EOSC); 8 mi. NE Unity (3 PSM); T7S, R43E, Sec. 23 (1 PSU). **Clackamas Co.**—Oregon City (1 SDNHM). **Clatsop Co.**—site of Old Ft. Clatsop (1 MVZ). **Crook Co.**—Whiskey Spring (1 PSU). **Curry Co.**—10 mi. E Brookings (1 PSM). **Douglas Co.**—0.5 mi. N Gardiner (1 PSM). **Grant Co.**—Canyon City (1 PSM); Coxie Meadow (1 PSM); Crane Prairie (1 PSU); Keeney Meadow (1 PSM); T11S, R29E, SW¼ Sec. 35 (1 PSM). **Harney Co.**—Pike Creek, W of Alvord Desert (1 SOSC); Voltage (1 USNM); T33S, R32½E, NW¼ Sec. 36, 4,960 ft. (1 EOSC). **Jackson Co.**—Ashland (1 OSUFW); East Pilot Rock Spring, T41S, R2E, NW¼SE¼ Sec. 3 (1 PSM, 1 SOSC); Medford (1 SOSC); Shaw Reservoir, T39S, R4E, NW¼ Sec. 13 (1 SOSC); 3.1 mi. S HWY 238 on West Forest Creek Rd., then 1.9 mi. on Oregon Belle Loop Rd. (1 SOSC); 5 mi. off HWY 238 on West Forest Creek Rd. (2 SOSC). **Josephine Co.**—Ft. Vannoy (1 SOSC);

Merlin [T35S, R6W] (1 SOSC); 1 mi. E USFS Rd. 3696 from junction with USFS Rd. 355 and HWY 199 (1 SOSC). **Lane Co.**—McKenzie Bridge (1 USNM). **Lincoln Co.**—Cascade Head Experimental Forest (1 PSM). **Malheur Co.**—Disaster Peak, 6,500 ft. (1 USNM); 22 mi. SW Rome, Crooked Creek, Owyhee River (1 USNM). **Marion Co.**—Salem (1 USNM). **Multnomah Co.**—Portland (1 PSM, 2 SDNHM). **Tillamook Co.**—(1 UMMZ); Tillamook (1 OSUFW). **Umatilla Co.**—Pearson Creek, T2S, R33E, Sec. 19 (1 PSU); 2–3 mi. E Stanfield, T4N, R29E, S edge (1 KU). **Union Co.**—La Grande (1 EOSC); 5 mi. SSE LaGrande, T4S, R38E, NW¼ Sec. 3 (1 PSM); Palmer Junction (1 SDNHM). **Wheeler Co.**—12 mi. SW Mitchell, Wildwood camp, Ochoco National Forest (1 OSUFW).

Additional locality: Sandy River Area* (1 PSU).

Lemmiscus curtatus, n = 342
(Plotted in Fig. 11-125, p. 333)

Crook Co.—2 mi. SW Barnes (1 MVZ). **Deschutes Co.**—8 mi. S, 17 mi. E Bend (1 CM); 5 mi. SW Cline Falls State Park, T16S, R12E, Sec. 4 (1 OSUFW). **Grant Co.**—5.3 mi. N, 1.3 mi. E Long Creek (1 OSUFW); 2 mi. S, 1 mi. E Long Creek (1 OSUFW); T9S, R30E, Sec. 15 (1 OSUFW); T9S, R30E, Sec. 22 (1 OSUFW). **Harney Co.**—Blitzen (1 USNM); 16 mi. N Burns, Cold Springs (1 SDNHM); 35 mi. W Burns (31 USNM); 36 mi. W Burns, Squaw Butte Range (19 MVZ); 12 mi. S, 14 mi. W Burns, T25S, R28E, Sec. 10 (6 OSUFW); Cottonwood Creek (5 PSM); Coyote Buttes (5 HSU); 7.5 mi. N, 3 mi. E Diamond (1 OSUFW); 10 mi. S, 18 mi. E Frenchglen, slope Steens Mt., T33S, R34E, Sec. 7 (2 OSUFW); Malheur National Wildlife Refuge* (1 EOSC); Squaw Butte (10 USNM); Steens Mt. (1 SDNHM); Steens Mt., T32S, R32_E, Sec. 36 (1 OSUFW); Steens Mt., T32S, R33E, Sec. 31 (1 SOSC); Steens Mt., T32S, R33E, Sec. 32 (1 OSUFW); 0.4 mi. N Steens Mt. Rd., 350 m E McCoy Ridge Rd., T32S, R33E, S½ Sec. 32 (2 OSUFW); Steens Mt., E rim, T33S, R33E, Sec. 12, 9,500 ft. (7 PSM, 1 SOSC); Steens Mt., T33S, R33E, Sec. 13 (1 OSUFW); Steens Mt., Wildhorse Lake Cirque, T33S, R33E, Sec. 36, 9,600 ft. (1 PSM); T32S, R32¾E, Sec. 20 (1 KU, 1 OSUFW); T32S, R32¾E, Sec. 33 (7 OSUFW); T32S, R32¾E, Sec. 36 (1 OSUFW); T32S, R33E, Sec. 31 (6 OSUFW); T32S, R33E, Sec. 32 (3 OSUFW); T32½S, R33E, Sec. 35 (1 OSUFW); T33S, R33E, Sec. 1 (3 OSUFW); T33S, R33E, Sec. 2 (1 OSUFW); T33S, R33E, Sec. 13 (1 OSUFW). **Jefferson Co.**—(26 PSM); 0.5 mi. E Boyce Corral* (5 PSU); 6 mi. N Culver (1 PSM); 5 mi. E Culver (31 PSM); 6 mi. E Culver (5 PSM); 11 mi. S, 1 mi. E Madras (1 PSU); 12 mi. S, 2.75 mi. E Madras (3 PSU); T11S, R13E, Sec. 35 (10 PSM); T11S, R18E, Sec. 7 (2 PSM); T11E, R11E, Sec. 35 (6 PSM); T12S, R13E, NW¼ Sec. 13 (6 PSM); T12S, R13E, Sec. 24 (3 PSM); T12S, R14E, Sec. 17 (2 PSM); T12S, R14E, Sec. 19 (1 PSM); T13S, R11E, Sec. 35 (1 PSM); T13S, R11E, Sec. 35/36 (17 PSM); T13S, R12E, Sec. 12 (32 PSM). **Klamath Co.**—(1 PSM). **Lake Co.**—(1 PSM, 1 UCONN); Buck Pasture, Hart Mt. Antelope Refuge, T35S, R26E, Sec. 17 (1 KU, 1 SOSC); Fort Rock (3 SDNHM); Hot Springs camp, Hart Mt. Antelope Refuge, T35S, R26E, NE¼NW¼ Sec. 32 (2 SOSC); 6 mi. W Lakeview (1 MVZ); Rock Creek, Warner Mts. (2 USNM); Warren Mts. (3 SDNHM). **Malheur Co.**—(3 PSM); Cedar Mts. (4 USNM); Cottonwood Creek (1 PSM); Coyote Wells, 2.5 mi. N Skull Springs (1 SDNHM); Creston, 12 mi. W Skull Springs (1 USNM); Crown Creek Burn* (1 PSM); 8 mi. N Disaster Peak (1 USNM); McDermitt (1 USNM); Ironside (3 AMNH); Jordan Crater, T28S, R43E, NW¼ Sec. 9 (2 PSM); Mud Springs, T38S, R40E, SW¼ Sec. 29 (3 PSM); 3 mi. NW Road Reservoir, T27S, R43E, NW¼ Sec. 21 (5 PSM); 25 mi. W Rome on Bone Springs Rd., T31S, R37E, Sec. 24 (1 PSU); 6.5 mi. SE Whitehorse Ranch, T38S, R39E, E½ Sec. 1 (5 PSM); T30S, R46E, SW¼ Sec. 2 (3 PSM); T32S, R43E, SW¼ Sec. 33 (1 PSM); T34S, R44E, NW¼ Sec. 15 (2 PSM). **Wasco Co.**—Antelope (5 USNM); 8 mi. NW Bake Oven (1 USNM).

Lepus americanus, n = 180
(Plotted in Fig. 10-15, p. 141)

Baker Co.—Forshey Meadows [T7S, R44E, Sec. 31] (1 EOSC); Lily White field station (1 EOSC); 6 mi. W Sumpter (1 USNM). **Benton Co.**—Chintimini Mt. [= Marys Peak], 4,000 ft. (1 USNM). **Clackamas**

Co.—36 mi SE Estacada on Colowash River* (1 USNM); 0.5 mi N, 1.5 mi E Government Camp (2 PSU). **Clatsop Co.**—Hammond (1 PSM); site of Old Fort Clatsop (3 MVZ); Olney (3 PSM); 0.8 mi. E Young River Falls, T7N, R9W, Sec. 22 (1 PSU). **Columbia Co.**—Scappoose (1 PSU). **Crook Co.**—(1 OSUFW). **Deschutes Co.**—(1 OSUFW); Bend (1 OSUFW); Black Butte on back road, T14S, R9E, Sec. 3 (1 OSUFW); mouth of Davis Creek (1 OSUFW); W slope Three Sisters (1 OSUFW); Deschutes River, mouth of Davis Creek (10 OSUFW); Davis Creek (1 OSUFW); 3 mi W Paulina Lake (1 MVZ); 7 mi NW Sisters (1 KU); W slope Three Sisters (1 OSUFW). **Grant Co.**—12 mi S Canyon City, 5,500 ft. (4 MVZ); 15 mi NW Austin (1 USNM); Strawberry Mt. (1 USNM). **Harney Co.**—15 mi N Burns (1 SDNHM). **Hood River Co.**—Brooks Meadows, 9 mi ENE Mt. Hood (4 MVZ); N slope Mt. Hood, 2,800 ft. (2 USNM); Parkdale (1 UW); 2 mi W Parkdale (1 USNM). **Jackson Co.**—5 mi S Ashland (1 SOSC); Parker Mt. (1 SOSC); half way between Drew and Crater Lake (1 USNM); 11.5 mi. from HWY 66 on Dead Indian Rd. (1 SOSC); T38S, R3E, Sec 30 (1 SOSC). **Klamath Co.**—Anna Springs, Crater Lake (1 MVZ); 20 mi. W Cresent Lake (1 UO); Ft. Klamath (1 PSM, 3 UMMZ, 19 USNM); Ft. Klamath, head Wood River (1 OSUFW); 10 mi SW Ft. Klamath (7 UMMZ); Horsefly Mt. (2 PSM); S end Lake of the Woods (1 MVZ); Odell Creek, 4,700 ft. (1 LACM); W side Upper Klamath Lake (1 PSM); Upper Klamath Lake (1 AMNH, 2 UMMZ); 4 mi. NW Merrill (1 MSUMZ). **Lake Co.**—Gearhart Mt. (8 PSM). **Lane Co.**—Alder Springs, 20 mi. E McKenzie Bridge, Three Sisters, 4,300 ft. (2 UO, 1 USNM); Belknap Springs, T16S, R6E, SW¼ Sec. 9 (1 KU, 2 OSUFW); Blue River (1 USNM); 5.5 mi. NE Blue River (1 PSM); 3 mi. S Florence (1 SDNHM); Huckleberry Lake Rd. (1 KU); McKenzie Bridge (1 OSUFW, 1 UO); Mercer (1 OSUFW, 1 SDNHM); Mercer Lake (1 OSUFW); 3 mi. SE Oakridge (2 UO). **Lincoln Co.**—Cascade Head Experimental Forest (8 PSM). **Linn Co.**—E of Cascadia, T13S, R5E, NW¼ Sec. 31 (1 KU); 4 mi. E Upper Soda on HWY 20, T13S, R5E, SW¼ Sec. 33 (1 KU). **Multnomah Co.**—14 mi. E Corbett, T1N, R6E, Sec. 29 (1 HSU); near McCord Creek, off of HWY I-84 (1 SOSC). **Polk Co.**—Valsetz (1 PSM). **Tillamook Co.**—Blaine (4 UMMZ); Cascade Head Experimental Forest (2 PSM); Nehalem (1 KU); Netarts (1 SDNHM, 1 USNM); Tillamook (5 AMNH, 1 KU, 1 MVZ, 1 OSUFW, 2 SDNHM, 3 UMMZ, 1 UO); 9 mi S Tillamook (1 MVZ); Wilson River (1 PSM). **Umatilla Co.**—5 mi. W Tollgate (1 UW[WWC]). **Union Co.**—Catherine Creek, 7 mi. E Telocaset, 3,500 ft. (1 MVZ); Kamelia (1 USNM); Starkey Experimental Forest, Barn Draw (1 PSM); Starkey Experimental Forest, Tin Trough Pit (1 PSM). **Wallowa Co.**—Bear Creek Ranger Station (1 SDNHM); 15.2 mi N Enterprise (1 OSUFW); 32 mi N Enterprise (1 SDNHM); 16 mi S, 3 mi E Lostine (4 MVZ); 18 mi S, 2 mi E Lostine (1 MVZ); Wallowa Lake (4 USNM). **Wheeler Co.**—Derr Ranger Station (1 SDNHM).

Additional locality: head of Willamette River* (1 USNM).

Lepus californicus, n = 217
(Plotted in Fig. 10-17, p. 146)

Baker Co.—Baker [City] (24 UMMZ); 10 mi. N Baker (2 UMMZ); 20 mi. NE Baker [City] (1 KU); Burkemont (1 AMNH); Keating (1 KU); 3 mi. N, 3 mi. W Keating (1 OSUFW); 2 mi. N, 0.2 mi. E Keating (1 OSUFW); T14S, R36E, Sec. 23 (1 PSU). **Benton Co.**—Corvallis (2 OSUFW); 7 mi. N Corvallis, E. E. Wilson Game Management Area (4 OSUFW); 6 mi. N, 8 mi. E Corvallis (1 USNM); 5 mi. SW Philomath (3 USNM). **Clackamas Co.**—8 mi. S Oregon City (1 PSU). **Crook Co.**—8 mi. N, 0.5 mi. E Millican (1 OSUFW); Prineville (3 USNM); 4 mi. SW Prineville (1 MVZ). **Deschutes Co.**—Bend (5 USNM); 4 mi. S, 3 mi. E Bend (1 OSUFW); Laidlaw [Tumalo] (1 USNM); 7 mi. N, 6 mi. W Redmond, Lower Bridge (1 OSUFW); 3 mi. W Terrebonne (1 OSUFW). **Douglas Co.**—Comstock (1 USNM); Drain (1 USNM); Looking Glass (1 USNM); Mrytle Creek (1 UMMZ); Roseburg (2 OSUFW, 1 UO). **Gilliam Co.**—Willows (1 USNM); Willows Junction (2 USNM). **Grant Co.**—2 mi. E Dayville (1 PSU). **Harney Co.**—Drewsey (1 USNM); Dry Lake (2 PSM); Malheur Cave, 4,000 ft. (1 USNM); 0.5 mi. N Princeton (1 EOSC); Wagontire (1 CRCM). **Jackson Co.**—Asland (1 USNM); 3 mi. W Phoenix (1 MVZ); 4 mi. S, 3 mi. E Rogue River (1 OSUFW); Shoat Springs, 5.8 mi. S off Highway 66 on

Cop-Co Road (2 SOSC). **Jefferson Co.**—4 mi. NNW Culver (1 PSM); Hay Creek (2 USNM); 1 mi. E Hay Creek (1 USNM); 2 mi. E Hay Creek (2 USNM); 5 mi. W Hay Creek (1 USNM); 8 mi. S Hay Creek (1 USNM); Gateway (1 USNM); 2 mi. S Madras (1 OSUFW); 5 mi. N, 1 mi. E Madras (1 USNM). **Josephine Co.**—Grants Pass, junction Rogue and Applegate rivers, 900 ft. (4 USNM); Grants Pass (4 CM); 1 mi. N Grants Pass (1 SOSC). **Klamath Co.**—Bly (1 PSM); 4 mi. NE Bly (3 PSM); 5 mi. E Bonanza (1 PSM); Dairy (2 OSUFW); Ft. Klamath (2 USNM, 1 OSUFW); Klamath Falls (1 USNM, 4 OSUFW, 2 CM); Klamath Spring (1 UMMZ); 3 mi. W MacDole (1 SOSC); 3 mi. N, 2 mi. E Merrill (2 OSUFW); Upper Klamath Lake (3 FMNH); Worden (2 OSUFW, 1 PSM). **Lake Co.**—(1 USNM); Adel, Warner Valley (1 USNM); 20 mi. NE Adel, Jacobs Ranch (2 USNM); Alkali Lake, Alkali Ranch (1 MVZ); Christmas Lake (2 PSU); Fremont (5 USNM); Lakeview (1 OSUFW); Mt. Warner (1 USNM); Plush (1 USNM); 7 mi. N Valley Falls (1 SOSC); 6 mi. [N] Nevada border, T41S, R25E, Sec. 5 (1 PSU). **Lane Co.**—Eugene, Spencer's Butte (1 USNM). **Malheur Co.**—(4 PSM); Beulah (2 USNM); 5 mi. S, 1 mi. E Huntington (1 OSUFW); Ironside (29 AMNH); Jordan Valley (1 USNM); Juntura (1 USNM); Ontario (2 USNM); 5 mi. SW Ontario, Wood Ranch, 2,000 ft. (5 USNM); 2 mi. NW Riverside (2 USNM); Rome, Owyhee River (1 USNM); 1 mi. S Rome (1 MVZ); Skull Spring (1 USNM); Vale (1 USNM); near Vale (1 AMNH); 1.5 mi. S Vale (2 USNM); 2 mi. E Vale, T18S, R45E, Sec. 21 (2 OSUFW); 1 mi. N, 4 mi. W Vale (1 OSUFW); Watson (1 USNM); 1 mi. SW Westfall, Branson Ranch (1 USNM). **Morrow Co.**—1 mi. N, 2 mi. "past" Boardman (1 PSU); 4 mi. SW Irrigon (1 MVZ). **Multnomah Co.**—Portland (1 SDNHM). **Umatilla Co.**—Hermiston (1 PSM); Umatilla (6 USNM); 1 mi. N, 2.5 mi. E Umatilla (1 OSUFW). **Wasco Co.**—Antelope (1 USNM). **Yamhill Co.**—1 mi. E Amity, T5S, R4W, N½ Sec. 28 (1 OSUFW).

Additional locality: Rogue River Valley* (1 USNM).

Lepus townsendii, n = 45
(Plotted in Fig. 10-19, p. 150)

Baker Co.—Keating (1 PSM). **Grant Co.**—5.5 mi. NNE Long Creek (1 MVZ). **Klamath Co.**—Upper Klamath Lake (4 FMNH). **Lake Co.**—(1 OSUFW); 20 mi. NE Adel (2 USNM); 20 mi. NE Adel, Jacobs Ranch (2 OSUFW); Lakeview (2 OSUFW, 2 CM); 2 mi. NE Lakeview (1 OSUFW); 3 mi. N Lakeview (1 PSM); Mt. Warner (3 USNM); Shirk's Ranch (1 OSUFW); T27S, R16E, Sec. 19 (1 OSUFW); Plush (1 CM). **Malheur Co.**—Cow Creek Lake (1 USNM); Ironside (7 AMNH); 5.5 mi. E McDermitt (1 PSM). **Morrow Co.**—Heppner (1 USNM). **Sherman Co.**—2.125 mi. N, 1.25 mi. E Kent (1 OSUFW); 4 mi. S, 5 mi. E Rufus, Klondike Road (1 OSUFW). **Umatilla Co.**—1.5 mi. W HWY 395, 15 mi. S Pilot Rock (1 EOSC); Umatilla (1 USNM). **Union Co.**—3 mi. W Medical Springs (1 OSUFW); Telocaset (1 EOSC, 1 SOSC). **Wallowa Co.**—3 mi. N Enterprise (1 PSU). **Wasco Co.**—Antelope (1 USNM); 3 mi. S, 14 mi. E Dufur (1 OSUFW); Shaniko (2 UW).

Lutra canadensis, n = 63
(Plotted in Fig. 12-51, p. 435)

Clackamas Co.—near Molalla (2 PSM); Upper Clackamas River (1 PSM). **Clatsop Co.**—Big Creek (1 PSM); Columbia River* (2 PSM); Gnat Creek (1 PSM); Little Joe Creek off of Necanicum River (1 PSM). **Columbia Co.**—Rock Creek, Nehalem drainage (1 PSM); N tip Sauvies Island (1 PSM); Westport Slough (1 PSM). **Coos Co.**—1.75 mi. E Bandon (1 PSM); 1 mi. NE Broadbent, T29S, R12W, NW¼ Sec. 33, 50 ft. (1 PSM); 1.75 mi. NE Broadbent, T29S, R12W, center Sec. 28 (1 PSM); 2.5 mi. NNE Broadbent, T29S, R12W, center Sec. 21 (1 PSM); Lakeside, Tenmile Creek (1 OSUFW); Willanch Slough, 0.5 mi. above Tidewater near Coos Bay (2 OSUFW). **Curry Co.**—Gold Beach (1 SDNHM). **Deschutes Co.**—(1 OSUFW); Bend (1 OSUFW). **Douglas Co.**—Glendale (1 USNM); Smith River, near Gardiner (1 USNM); junction of Scholfield Rd. and HWY 38, 100 ft. (1 PSU). **Josephine Co.**—13 mi. SW Grants Pass (2 UO). **Klamath Co.**—(2 MVZ); Ft. Klamath (1 USNM); Klamath Falls (1 CM); Klamath Lake (2 PSM); Klamath River* (3 PSM). **Lane Co.**—(3 KU); near Eugene (2 PSM); 4 mi. SW Eugene

(3 UO); McKenzie River* (1 KU); S fork McKenzie River* (1 PSM); 3 mi. SE Oakridge (1 UO); 0.5 mi. from Reserve, Goodman Creek (1 OSUFW). **Linn Co.**—Cascadia (1 SDNHM). **Malheur Co.**—12 mi. W Fairland* (1 USNM); Jordan Creek* (2 PSM); Rome, Owyhee River (4 PSM, 1 USNM). **Tillamook Co.**—(1 UMMZ); Devils Lake Fork at Wilson River (2 PSM); North Fork Trask River (2 PSM); Tillamook (1 OSUFW). **Wallowa Co.**—Coon Hollow Creek, Snake River, near Washington, T6N, R47E (1 OSUFW).

Lynx canadensis, n = 11
(Plotted in Fig. 12-61, p. 456)

Benton Co.—6 mi. N, 5 mi. E Corvallis, T10S, R4W, Sec. 34 [North Albany] (1 OSUFW). **Deschutes Co.**—35 mi. W Bend (1 USNM). **Grant Co.**—Granite (1 PSM, 2 USNM). **Harney Co.**—Diamond, Kiger Creek, 4,300 ft. (1 USNM); Stinkingwater Mountains, near Drewsey (1 KU). **Klamath Co.**—Ft. Klamath (1 USNM). **Umatilla Co.**—Lehman, Blue Mts., N. fork John Day River (1 USNM); 2 mi. W Vansycle (1 USNM).
Additional locality: Trout Valley, Blue Mts.* (1 USNM).

Lynx rufus, n = 539
(Plotted in Fig. 12-63, p. 459)

Baker Co.—12 mi. NE Bridgeport (1 USNM); Unity (1 PSM). **Clackamas Co.**—Estacada (10 USNM); Baty Butte [T6S, R3E, NW¼] (2 USNM); Big Bottoms [T6S, R7E, Sec. 36] (1 USNM); Cold Springs [T4S, R5E, Sec. 23] (8 USNM); High Ford* (4 USNM); Hillock Burn* (4 USNM); Hot Springs [T7S, R6E, NW¼] (4 USNM); Oak Grove Butte [T6S, R7E, Sec. 8] (6 USNM); North Fork* (6 USNM); Roaring River (5 USNM); Roaring River Hill* (4 USNM); Rocky Ridge* (1 USNM); Squaw Lakes [T4S, R6E, Sec. 13] (1 USNM); near Turrells Ranch* (1 USNM); Collywash [=Collawash] Burn [T7S, R6E] (3 USNM); Marmot (5 USNM); SE Molalla (1 PSM); 8 mi. NW Molalla, 14 mi. SE Oregon City (2 OSUFW); Molalla River* (5 OSUFW); Oak Grove R.S. [T6S, R7E, Sec. 8] (1 PSM). **Clatsop Co.**—7 mi. E Elsie (1 USNM); Floten* (1 PSM). **Columbia Co.**—Trenholm [T5N, R2W, Sec. 29] (1 OSUFW). **Coos Co.**—(32 OSUFW); 4 mi. SE Bandon (1 PSM); Bridge (2 USNM); Marshfield (1 USNM). **Curry Co.**—Gold Beach (1 USNM); Port Orford (2 USNM); Port Orford, 2 mi. SE Knapp's Sheep Range (1 USNM). **Deschutes Co.**—15 mi. E Paulina* (1 UW[WWC]). **Douglas Co.**—(61 OSUFW); 12 mi. N Drew (1 USNM); 4 mi. E Drew (1 USNM); 10 mi. E Drew (4 USNM); 15 mi. E Drew (1 USNM); 22 mi. E Drew (1 USNM); 5 mi. SE Drew (1 USNM); 7 mi. SE Drew (1 USNM); 7.5 mi. SE Drew (1 USNM); 8 mi. SE Drew (1 USNM); Ft. Umpqua (2 USNM); Glendale (2 USNM); near Hoaglin, Umpqua River (1 PSM); Roseburg (1 MVZ); T23S, R10W, Sec. 16 (1 OSUFW). **Gilliam Co.**—9 mi. NE Lonerock (2 USNM). **Grant Co.**—Granite (1 USNM). **Harney Co.**—Alberson, Stonehouse Canyon (1 USNM); Alberson, Warm Springs (2 USNM); Alvord Lake, Alvord Ranch (2 USNM); 20 mi. N Andrews (1 USNM); Mickey Ranch, Nude Valley (1 MVZ); 1 mi. NE Buchanan (1 USNM); 16 mi. N Burns (5 USNM); 16 mi. NE Burns (1 USNM); vicinity of Denio (2 MVZ); 15 mi. SW Drewsey (1 USNM); Fields (1 USNM); 35 mi. SE Frenchglen (1 OSUFW); 3 mi. N Harney (3 USNM); 12 mi. N Harney, Pine Creek (2 USNM); 12 mi. E Harriman (1 USNM); 15 mi. E Harriman (1 USNM); 2 mi. S Mule (1 USNM); 8 mi. SW Mule, Riddle Mts. (2 USNM); Princeton (3 USNM); Sageview (1 USNM); Smith (1 USNM); 8 mi. E Smith, Armstrong Creek (5 USNM); Steens Mt., Kiger Gorge, 8,000 ft. (1 USNM); Steens Mt., Riddle Creek near Diamond (2 MVZ); 2 mi. W Sugarloaf Lookout, Malheur National Forest (1 MVZ); 6 mi. S Venator (1 USNM); 10 mi. S Venator (2 USNM); 18 mi. S Venator (1 USNM); 22 mi. W Venator (1 USNM); 22 mi. SW Venator (4 USNM). **Jackson Co.**—3 mi. N Butte Falls (1 USNM); 10 mi. NE Butte Falls (4 USNM); 3 mi. S Derby (2 USNM); 1 mi. W Derby (2 USNM); 3 mi. E Derby (2 USNM); Evans Creek, 3 mi. N Rogue River (2 USNM); 43 mi. NE Grants Pass (4 USNM). **Josephine Co.**—near Galice (1 UO); Galice Creek, Rogue River Valley (1 UO); Grants Pass (1 USNM). **Klamath Co.**—3 mi. S Beatty (1 PSM); near Bly (5 PSM); near Bly, 4 mi. NE Deming Creek (1 PSM); 10 mi. E Bly, Sprague River Valley (1 USNM); 6 mi. NW Bly (1 USNM);

3 mi. NE Bly (1 USNM); 9 mi. NE Bly (1 USNM); Bonanza (1 PSM); 0.5 mi. N, 6 mi. W Cresent (1 OSUFW); Ft. Klamath (1 USNM); Longbranch Creek* (1 PSM); Swan Lake Valley (2 PSM). **Lake Co.**—Abert Lake (8 PSM); 1 mi. N HWY 140, mile post 83 between Bly and Lakeview (1 OSUFW); 12 mi. NE Fort Rock (1 USNM); 12 mi. S Hampton (2 USNM); 14 mi. S Hampton (8 USNM); Mts. E of Goose Lake, Sherlock Range (1 MVZ); Langell Valley (1 PSM); Plush (1 USNM); Silver Lake (1 USNM); 3 mi. W Stauffer (3 USNM); Summer Lake Game Refuge (1 OSUFW); Sycan Valley (5 PSM); Whitworth Creek (2 PSM). **Lane Co.**—(2 OSUFW); Belle (7 USNM); 37 mi. S Blue River (13 USNM); 37 mi. SE Blue River (4 USNM); 2 mi. S, 3 mi. W Greenleaf (1 OSUFW); Lickety Creek (1 PSM); Lowell (1 USNM); McKenzie Bridge (1 UO); 10 mi. SW McKenzie Bridge, Fort McKenzie River (1 USNM); South Fork McKenzie River (2 OSUFW); Mercer (1 KU, 1 OSUFW); Oakridge (5 USNM); 40 mi. S Oakridge (1 USNM); 20 mi. SE Oakridge (2 USNM); 12 mi. N, 10 mi. E Oakridge, T19S, R5E, Sec. 18 (1 OSUFW); Reed (6 USNM); Reserve (1 PSM); 0.25 mi. N Reserve (1 OSUFW); 6 mi. SW Reserve, Eagle's Nest (1 OSUFW); 1 mi. SE Reserve (1 OSUFW); Wall Creek (1 USNM); T16S, R5E (3 KU). **Lincoln Co.**—(5 OSUFW); Fisher, Five River Valley (1 USNM). **Linn Co.**—Cascadia (1 USNM). **Malheur Co.**—(4 OSUFW); 3 mi. S Beatty* (1 PSM); Cord, Duck Mt. (5 USNM); Fangollano (2 SDNHM); Jordan Valley (1 PSM); Mooreville (2 USNM); 2 mi. SW Mooreville (1 USNM); 4.5 mi. SW Mooreville (1 USNM); Owyhee River near Rome (3 USNM); 2 mi. NW Riverside (1 USNM); Rome (3 PSM). **Morrow Co.**—9 mi. SW Hardman (1 USNM); 10 mi. SW Hardman (1 USNM). **Sherman Co.**—2 mi. E Rufus (1 PSM). **Tillamook Co.**—Blaine (1 UMMZ); Foley Creek (1 USNM); Hebo (1 USNM); Netarts (1 OSUFW); Sandlake (1 OSUFW, 1 PSM); Tillamook (1 OSUFW); Trask River* (5 OSUFW). **Umatilla Co.**—Battle Mt. (1 PSM); Duncan (2 MVZ); 20 mi. S Echo (2 USNM); Gibbon (1 MVZ); 10 mi. SE Milton (1 UW[WWC]); 3 mi. E Ring near Helix (2 USNM); Weston Mt. (1 PSM). **Union Co.**—(1 PSM); Catherine Creek (2 PSM); 2 mi. E Cove (1 PSM); Deep Creek* (1 PSM); 7 mi. S, 4 mi. E Elgin, Indian Creek, T1S, R40E, Sec. 28 (1 OSUFW); Elk Mt. (1 PSM); Glass Hill (2 PSM); Gordon Creek N of Elgin (2 PSM); Ladd Canyon (2 PSM); mouth Looking Glass Creek (3 PSM); 6 mi. W Medical Springs (1 EOSC); Minam (2 PSM); Minam River* (3 PSM); Mt. Emily (1 PSM); Mt. Harris, 3 mi. up Indian Creek (1 PSM); mouth North Fork Looking Glass Creek (1 PSM); 3 mi. above Palmer Junction (1 PSM); Shaw Mt. (1 PSM); South Fork Cabin Creek N of Elgin (1 PSM); Summit Rd. between Meacham and Ukiah (1 PSM). **Wallowa Co.**—Deur Creek* (1 PSM); near Troy (1 PSM); 16 mi. NE Wallowa (1 USNM). **Wasco Co.**—Maupin (2 USNM); 7 mi. from Maupin* (1 USNM). **Washington Co.**—Hermann (1 USNM); Yamhill River* (1 OSUFW). **Wheeler Co.**—10 mi. W Fossil near Condon (1 USNM); 8 mi. S Spray (1 USNM); 14 mi. S Spray (1 USNM); 7 mi. SW Spray (1 USNM); 10 mi. SE Spray (1 USNM).
Additional localities: Sink Lake, about 5 mi. N Bunchgras Butte* (1 USNM); western Oregon* (40 OSUFW); eastern Oregon* (5 OSUFW).

Marmota flaviventris, n = 126
(Plotted in Fig. 11-14, p. 174)

Baker Co.—19 mi. ENE Baker [City] (1 KU); 3 mi. NE Huntington, 2,100 ft. (5 USNM); 2 mi. W Unity (1 PSU). **Crook Co.**—Prineville (1 USNM); 6 mi. W Prineville (1 PSM); 9.5 mi. SE Prineville, Crooked River (1 MVZ); 3 mi. N, 6 mi. E Wildcat Mt. (1 MVZ). **Deschutes Co.**—15 mi. W Bend, Happy Valley Ridge (1 MVZ); 16 mi. W Bend, Tumalo Creek, 6,500 ft. (3 MVZ); Paulina Lake (1 USNM). **Douglas Co.**—Diamond Lake (1 SDNHM). **Grant Co.**—Barry Ranch* (3 EOSC); Dale (1 USNM); 12 mi. S, 3 mi. E Dayville (1 OSUFW). **Harney Co.**—Blitzen (1 USNM); Blitzen Valley (3 SDNHM); Buchanan (4 USNM); 18 mi. S Burns, T25S, R31E, Sec. 3 (1 OSUFW); 27 mi. S Burns (1 OSUFW); Crane (1 USNM); 6 mi. S Frenchglen, T32S, R32E, Sec. 29 (1 OSUFW); Harney (2 USNM); Malheur Cave, 4,000 ft. (1 USNM); Smyth Creek, 10 mi. E Diamond, 5,500 ft. (1 MVZ). **Hood River Co.**—Mt. Hood, Cloud Cap Inn (3 USNM). **Jackson Co.**—Greensprings HWY 100 yds. W Jenny Creek (1 SOSC); Mt. Pitt (1 UO). **Jefferson Co.**—Camp Sherman (1 PSU); W slope Haystack Butte (1 OSUFW). **Klamath**

Co.—Algoma (2 MVZ); 7 mi. S Bly (1 USNM); 7 mi. E Bly (1 CM); 1 mi. NW Bly (1 PSM); 6 mi. NE Bly (1 USNM); 5 mi. E Bonanza (1 PSM); Crater Lake (2 USNM); Crater Lake National Park, T30S, R5E, E½ Sec. 24 (2 PSU); Crater Lake National Park, head Munson Valley (2 MVZ); Ft. Klamath (1 USNM); Fourmile Lake (1 UO); Klamath Falls (1 CM); Linkville [=Ft. Klamath], Klamath Basin (4 USNM); Upper Klamath Lake (5 FMNH). **Lake Co.**—Adel (1 USNM); 5 mi. N, 1 mi. E Adel, T38S, R24E, Sec. 23 (1 SOSC); Fishhole Mt. (1 PSM); 15 mi. SE Fort Rock (1 SDNHM); Fremont National Forest, Fishhole Creek, Scaler Cabin (1 PSM); Gearhart Mt., Horseglades (2 PSM); Gearhart Mt., Finley Corral (1 PSM); Guano Creek, Shirks Ranch, T38S, R27E, Sec. 34 (1 OSUFW); Hart Mt. (1 PSM); 10 mi. NE Lakeview, Camas Valley, T38S, R20E, Sec. 32 (1 PSU); Summer Lake (1 USNM); Sycan marsh (3 USNM). **Lane Co.**—Huckleberry Lake forest camp (1 KU). **Malheur Co.**—2 mi. SW Arock (1 MVZ); Cedar Mts. (1 USNM); Cord, Barren Valley, 3,950 ft. (1 USNM); Cord, Barren Valley, 4,000 ft. (1 USNM); Disaster Peak (1 USNM); Ironside (1 AMNH); Mahogony Mts., 6,500 ft. (1 USNM); canyon located 8 mi. NE McDermitt, 6,300 ft. (1 USNM); Owyhee River cave (1 PSU); Riverside (1 USNM); 8 mi. W Vale, Little Valley Canyon (1 EOSC); Westfall (1 USNM); 1 mi. SW Westfall (2 USNM). **Marion Co.**—Breitenbush Lake, 5,500 ft. (1 MVZ). **Umatilla Co.**—Cayuse (1 SDNHM); 10 mi. SE Milton (1 UW[WWC]). **Union Co.**—Elgin (2 USNM); 5 mi. N Elgin (1 EOSC); 1.5 mi. SE Ladd Canyon, T4S, R39E, SE¼ Sec. 24 (1 PSM); 1 mi. S La Grande (1 EOSC); 8 mi. SE La Grande, Foothill Rd. (2 EOSC); Thief Valley Reservoir (1 EOSC). **Wallowa Co.**—Chesnimus Rd. (1 OSUFW); Wallowa Lake (1 UW[WWC]). **Wasco Co.**—Ft. Dalles (1 USNM); Shaniko (1 UW); Waupinita Canyon, 3.8 mi. S, 4.5 mi. W Maupin (1 PSU). **Wheeler Co.**—2 mi. E Rock Creek bridge, several mi. W Picture Gorge HWY 26 (2 AMNH).

Additional localities: N base Three Sisters, 5,000 ft. (2 USNM); N slope Three Sisters (1 USNM); Cow Creek Lake* (5 USNM).

Martes americana, n = 134
(Plotted in Fig. 12-36, p. 408)

Baker Co.—(13 KU); Anthony (1 AMNH); Cornucopia (1 UO). **Clackamas Co.**—1.5 mi. N, 3.125 mi. E Government Camp (1 PSU). **Coos Co.**—Remote (1 USNM). **Curry Co.**—Port Orford (8 USNM). **Deschutes Co.**—(3 KU, 2 OSUFW); Davis Creek (1 OSUFW); Summit McKenzie Pass (1 OSUFW); N base Three Sisters (1 USNM). **Douglas Co.**—1.25 mi. W Loon Lake, T23S, R10W, NE¼ Sec. 10 (1 PSM); 2.5 mi. SW Loon Lake, T23S, R10W, NW¼ Sec. 22 (3 PSM); 4 mi. SW Loon Lake, T23S, R10W, SW¼ Sec. 33 (3 PSM); Smith River [at coast] (2 USNM). **Grant Co.**—Granite (1 SDNHM); 21 mi. SE Prairie City, North Fork Malheur River, 5,000 ft. (1 MVZ); Strawberry Butte (1 USNM); Strawberry Mts. (4 USNM). **Hood River Co.**—Mt. Hood (4 USNM). **Jackson Co.**—(1 KU); NE base Mt. Pitt [=Mt. McLoughlin] (4 USNM). **Jefferson Co.**—Mt. Jefferson (4 USNM). **Klamath Co.**—Crater Lake (1 SOSC, 3 USNM); 3 mi. S entrance Crater Lake (1 CL); Crater Lake National Park (2 CL, 1 OSUFW); Crater Lake National Park, 1.25 mi. SE Kerr Notch, Kerr Valley, 6,400 ft. (1 CL); Crater Lake National Park, Mt. Shasta overlook* (1 CL); Crater Lake National Park, Palasade viewpoint (1 CL); Ft. Klamath (7 OSUFW, 1 USNM); Mt. Mazama (1 USNM); Odell Lake (1 OSUFW). **Lake Co.**—Gearhart Mt. (5 PSM). **Lane Co.**—Belle (9 USNM); 12 mi. N, 2 mi. W Florence, Heceta Head (1 OSUFW); Oakridge (9 OSUFW, 2 PSM, 2 USNM); Reed (14 USNM). **Lincoln Co.**—Newport (2 USNM). **Umatilla Co.**—near Tollgate (1 KU). **Wallowa Co.**—Aneroid Lake (1 SDNHM); Big Sheep Creek, Imnaha River Tributary, T4S, R46E, Sec. 27 (2 OSUFW); Eagle River, Wallowa Mts. (1 UMMZ); 12 mi. S, 9 mi. E Joseph, Upper Big Sheep Breek, T4S, R46E, SW¼ Sec. 27 (1 OSUFW); Lostine River (1 USNM).

Martes pennanti, n = 3
(Plotted in Fig. 12-39, p. 412)

Douglas Co.—Glendale (1 USNM). **Lane Co.**—Oakridge (2 OSUFW).

Mephitis mephitis, n = 121
(Plotted in Fig. 12-57, p. 447)

Baker Co.—Anthony (2 AMNH, 2 MVZ); 3 mi. NE Huntington, 2,100 ft. (2 USNM); S of Lily White (1 EOSC). **Benton Co.**—4 mi. N Corvallis, T11S, R5W, Sec. 12 (1 OSUFW). **Clackamas Co.**—Boring (1 SDNHM). **Clatsop Co.**—Gearhart (1 MVZ). **Crook Co.**—Crane Prairie* (1 OSUFW); Lookout Mt., Howard (1 USNM). **Curry Co.**—Gold Beach (1 FMNH); Port Orford (2 USNM). **Deschutes Co.**—Sisters (2 USNM); 3 mi. W Tumalo (1 SDNHM). **Douglas Co.**—Roseburg (1 OSUFW, 1 USNM). **Gilliam Co.**—Willows (2 USNM). **Harney Co.**—Blitzen Valley (2 SDNHM); Burns (1 USNM); 1.5 mi. S Burns (1 USNM); Diamond, 4,300 ft. (2 USNM); Harney (2 USNM); Narrows (2 SDNHM, 1 USNM); Shirk (1 USNM). **Jackson Co.**—Ashland (1 SOSC); Derby (1 USNM); 2 mi. S Derby (1 USNM); 4 mi. S Derby (1 USNM); 3 mi. W Derby (2 USNM); 3 mi. E Derby (2 USNM); 1 mi. N Talent (1 SOSC); White City (1 SOSC); 7 mi. from junction Crater Lake HWY on Butte Falls HWY (1 SOSC). **Josephine Co.**—Grants Pass (1 USNM); Grants Pass, Rogue River Valley (5 USNM); 14 mi. W Grants Pass (1 SDNHM); 40 mi. NE Grants Pass (1 USNM). **Klamath Co.**—(2 MVZ); 8 mi. W Bly, Sprague River (1 OSUFW); 2 mi. NE Bly (2 USNM); Ft. Klamath (1 OSUFW, 6 USNM); Klamath Falls (2 CM); Tule Lake (2 USNM). **Lake Co.**—Adel (1 OSUFW); 8 mi. S Adel (2 SDNHM); 1 mi. N Lakeview, Hunters Hot Spring (1 MVZ); Plush (1 USNM); S Warner Valley* (1 SDNHM). **Lane Co.**—E edge Cheshire (1 OSUFW); Eugene (1 USNM); Oakridge (2 OSUFW). **Linn Co.**—(1 OSUFW). **Malheur Co.**—Fangollano (1 USNM); Owyhee (1 USNM). **Morrow Co.**—Lena (8 USNM); Lena, Butler Creek (6 USNM). **Multnomah Co.**—Portland (1 SDNHM). **Polk Co.**—McCoy (1 USNM). **Sherman Co.**—mouth of Deschutes River, Columbia River (1 MVZ); Miller, mouth of Deschutes River (1 USNM). **Tillamook Co.**—Tillamook, Simmons Creek (1 OSUFW). **Umatilla Co.**—Umatilla (1 USNM). **Union Co.**—Elgin (2 USNM); La Grande (2 EOSC). **Wallowa Co.**—Enterprise (1 SDNHM); Sheep Creek, 15 mi. E Joseph (1 UW[WWC]); 16 mi. NE Wallowa (1 USNM). **Wasco Co.**—The Dalles (8 USNM). **Washington Co.**—Forest Grove (1 AMNH). **Wheeler Co.**—6 mi. S Fossil (1 USNM). **Yamhill Co.**—Lafayette (2 UW[WWC]).

Microdipodops megacephalus, n = 175
(Plotted in Fig. 11-61, p. 248)

Crook Co.—Powell Butte (1 SDNHM). **Harney Co.**—(2 OSUFW); Alvord Desert, T33S, R35E, Sec. 21 (1 OSUFW, 1 USNM); Alvord Lake (11 MVZ, 5 USNM); 7 mi. S Andrews, 4,300 ft. (4 MVZ); Beatys Butte (1 USNM); Big Sand Gap Springs (2 EOSC, 1 CRCM); 0.9 mi. SE Big Sand Gap Springs, Alvord Basin, 4,100 ft. (1 CRCM); 1.1 mi. S Big Sand Gap Springs (2 EOSC); 1.3 [mi.] E along Borax Lake Rd., Alvord Basin (1 CRCM); 2 mi. S Borax Spring, S end Alvord Lake, 4,300 ft. (20 MVZ); 31 mi. S Burns, Harney Lake (2 OSUFW); 18 mi. S, 1 mi. W Burns, Harney Lake (2 OSUFW); 20 mi. S, 5 mi. W Burns (1 OSUFW); 21 mi. S, 1.5 mi. W Burns, dunes N of Harney Lake (1 OSUFW); 23 mi. S, 5.5 mi. W Burns (1 OSUFW); 1.5 mi. E Denio, 4,200 ft. (6 MVZ); 1.4 mi. E along Serrano Rd., Alvord Basin (3 CRCM); 8.6 mi. E along Serrano Rd., Alvord Basin (2 CRCM); 10 mi. S, 9 mi. E Fields (1 PSU); 6 mi. due SE Frenchglen, T32S, R32½E, Sec. 14 (1 OSUFW); Horse Springs, 1.8 mi. SE Big Sand Gap, Alvord Basin (2 CRCM); 0.8 mi. S Horse Springs, Alvord Basin (2 CRCM); 1.2 mi. S Horse Springs (1 CRCM); 2 mi. S Horse Springs (1 CRCM); Little Sand Gap, Alvord Basin (1 CRCM); near Malheur Environmental Field Station (1 PSU); Malheur National Wildife Refuge, N edge Harney Lake, T26S, R30E, Sec. 28 (1 HSU); Malheur National Wildlife Refuge, Harney Lake Dunes, T26S, R30E, Sec. 28 (3 CM); 2.5 mi. N junction Mickey Hot Springs Rd. and Windmill Rd., Alvord Basin (1 CRCM); Narrows (8 USNM); 3 mi. E Narrows (10 CRCM); 1 mi. S Narrows, 4,200 ft. (3 MVZ); 5 mi. SW Narrows, 4,000 ft. (24 MVZ); 2.8 mi. E along Poweline Rd. (2 CRCM); Tumtum Lake (1 USNM); Whitehorse Creek (1 EOSC); Whitehorse Creek, Alvord Basin (1 EOSC); Whitehorse sink (1 USNM); Wildhorse Creek, Alvord Basin (1 CRCM); 5.3 mi. WSW HWY 205 on Double-O Ranch Rd., 1 mi. S to dunes (6 SOSC); E edge of Alvord

Basin along Coyote Lake Rd. (1 EOSC). **Lake Co.**—(1 UCONN); NE edge Alkali Lake, 4,200 ft. (20 MVZ); 2.5 mi. NE Alkali Lake ranch (2 MVZ); 10 mi. N, 1.5 mi. E Fort Rock, 1.5 mi. E Cabin Lake campgrounds (1 OSUFW). **Malheur Co.**—head of Crook Creek, Owyhee Desert, 4,000 ft. (6 USNM).

Microtus californicus, n = 245
(Plotted in Fig. 11-108, p. 312)

Douglas Co.—Drain (10 USNM); 4 km W Drain, T22S, R6W, Sec. 13 (3 USNM); 0.8 km N, 2.2 km W Drain (1 USNM); Francis Creek, T24S, R1W, SE¼ Sec. 12 (1 SOSC); Roseburg (2 USNM); 7 mi. NW Roseburg (1 OSUFW); Sutherlin (1 OSUFW); 6 mi. W Sutherlin, Tyee Mt. (1 OSUFW); 1 mi. S, 10 mi. W Sutherlin (1 OSUFW); Winston (1 KU); Ben Irvine Reservoir, ca. 5–7 mi. SW Winston (1 KU). **Jackson Co.**—(1 SOSC); Ashland (1 SDNHM, 41 SOSC, 7 USNM); Ashland, W slope of valley (1 USNM); SE Ashland, Senic Hills Cemetary Rd. near Bear Creek (1 SOSC); Ashland, Scenic Hills Memorial Park (1 SOSC); Ashland, 0.25 mi. NE Scenic Hills Memorial Park (2 SOSC); 2 mi. E Ashland (9 PSM); 4 mi. E Ashland, Dead Indian Rd. (1 MVZ, 3 SOSC); 6.2 mi. E Ashland (1 SOSC); 5.5 mi. N, 11 mi. E Ashland (1 OSUFW); 0.25 mi. S, 0.125 mi. W Ashland (3 OSUFW); Brownsboro (3 USNM); 2 mi. E Dark Hollow (2 MVZ); Denman Wildlife Management Area, 1 mi. W Agate Rd. [near White City] (9 SOSC); 3 mi. N, 0.25 mi. W Eagle Point (3 OSUFW); 1 mi. N, 8 mi. E Eagle Point, Lick Creek, T36S, R1E, Sec. 3 (1 OSUFW); Emigrant Lake (3 SOSC); NW side Emigrant Lake (1 SOSC); 2.1 mi. E Emigrant Lake (1 SOSC); 1 mi. S Emigrant Lake, T39S, R2E, NW¼ Sec. 46 (1 SOSC); W slope Grizzly Peak, Ashland (1 USNM); Ice House Lake, 9 mi. E HWY 66 on Dead Indian Rd. (2 SOSC); Medford (5 OSUFW); 6 mi. S Medford (12 MVZ); 4 mi. E Medford (24 LACM); 1 mi. N, 5 mi. W Medford (2 OSUFW); 7 mi. S, 15 mi. W Medford (1 OSUFW); 5 mi. SW Medford (2 PSM); Pinehurst (2 SOSC); 6.5 mi. S Prospect (1 SOSC); Rogue River (1 SOSC); 9 mi. N Rogue River, T35S, R4W, Sec. 3, 1,200 ft. (2 SOSC); Rogue River valley (7 USNM); Shoat Springs (1 PSM); Siskiyou (4 USNM); Talent (14 SOSC); 3 mi. N Talent (1 SOSC); Lower Table Rock (2 SOSC); Upper Table Rock, T36S, R2W, NE¼ Sec. 2 (1 SOSC); White City (3 SOSC); 1 mi. W HWY 99 on Mt. Ashland Rd. (1 SOSC); 14.8 mi. from HWY66 on Dean Indian Rd. (2 SOSC); 0.3 mi. W Colestine Rd. on Mt. Ashland Rd. (1 SOSC); junction Dead Indian Rd. and Emigrant Creek Rd. (2 SOSC); 0.9 mi. W old HWY99 on Mt. Ashland Rd. (3 SOSC); T36S, R3W, SW¼ (2 SOSC); T38S, R2E, Sec. 23/24 (1 SOSC). **Josephine Co.**—13 mi. SW Galice, Briggs and Horse creeks, Ferren Ranger Station (1 USNM); Grants Pass (5 FMNH, 1 OSUFW, 1 SDNHM, 3 USNM); 20 mi. SW Grants Pass, N Fork Slate Creek (1 USNM); 4 mi. S Selma (6 MVZ); 7 mi. E Selma (1 SOSC); T39S, R6W, Sec. 30 (3 SOSC). **Lane Co.**—6 mi. E Cottage Grove (1 OSUFW); 2 mi. N, 2 mi. E Cottage Grove (1 OSUFW); 3 mi. S, 4 mi. E Eugene, T18S, R3W, Sec. 3 (1 KU).

Microtus canicaudus, n = 342
(Plotted in Fig. 11-111, p. 318)

Benton Co.—Corvallis (2 KU, 27 OSUFW, 9 PSM, 3 USNM); 0.5 mi. N Corvallis, T11S, R5W, Sec. 22 (5 OSUFW); 1 mi. N Corvallis (1 OSUFW); 2 mi. N Corvallis (11 PSM); 2.5 mi. N Corvallis (1 OSUFW); 3 mi. N Corvallis (1 KU); 3 mi. N Corvallis, T11S, R5W, Sec. 14 (1 OSUFW); 4 mi. N Corvallis (2 KU); 4 mi. N Corvallis, T11S, R5W, Sec. 12 (1 OSUFW); 4.5 mi. N Corvallis (1 OSUFW); 5 mi. N Corvallis (2 OSUFW, 3 PSM); 6 mi. N Corvallis, T11S, R4W, Sec. 6 (2 OSUFW); 7 mi. N Corvallis, T11S, R4W, Sec. 6 (2 OSUFW); 9 mi. N Corvallis (2 OSUFW); 10 mi. N Corvallis (1 OSUFW); 0.5 mi. S Corvallis, T11S, R5W, Sec. 22 (1 OSUFW); 1 mi. S Corvallis, T11S, R5W, Sec. 15 (5 OSUFW); 1 mi. S Corvallis, T12S, R5W, Sec. 3 (1 OSUFW); 2 mi. S Corvallis (1 OSUFW); 3 mi. S Corvallis (11 PSM); 10 mi. S Corvallis (1 OSUFW); 12 mi. S Corvallis (1 KU); 0.5 mi. W Corvallis (1 OSUFW); 0.65 mi. W Corvallis (2 OSUFW); 1 mi. W Corvallis (4 OSUFW); 1.25 mi. W Corvallis (1 OSUFW); 1.5 mi. W Corvallis (2 OSUFW); 2 mi. W Corvallis (1 OSUFW); 4.5 mi. W Corvallis (1 OSUFW); 1 mi. N, 2 mi.

W Corvallis (2 OSUFW); 1 mi. N, 1 mi. E Corvallis (1 OSUFW); 4 mi. N, 2 mi. E Corvallis (2 SOSC); 5 mi. N, 3 mi. E Corvallis (1 OSUFW); 5 mi. N, 5 mi. E Corvallis, 200 ft. (5 CM); 6 mi. N, 3 mi. E Corvallis (1 OSUFW); 7 mi. N, 1 mi. E Corvallis (1 OSUFW); 7 mi. N, 3 mi. E Corvallis, E. E. Wilson Wildlife Refuge (2 OSUFW); 0.5 mi. S, 3 mi. W Corvallis (2 OSUFW); 1 mi. S, 0.25 mi. W Corvallis (2 OSUFW); Finley Wildlife Refuge, T12S, R5W, NE¼ (5 KU); 2 mi. S, 1 mi. W Greenberry (1 OSUFW); 3 mi. S, 1 mi. W Greenberry (1 OSUFW); 1.5 mi. S Philomath (2 OSUFW); 1.5 mi. E Philomath (3 PSM); 2 mi. S, 3 mi. E Philomath (1 OSUFW); 5 mi. S, 5 mi. W Philomath (1 OSUFW); S of Soap Creek, T10S, R5W, Sec. 26 (4 OSUFW); T10S, R4W, Sec. 8 (1 OSUFW); T11S, R4W, Sec. 5 (2 OSUFW); T11S, R4W, Sec. 8 (1 OSUFW); T11S, R5W, Sec. 19 (1 OSUFW); T12S, R5W, Sec. 3 (2 OSUFW); T12S, R5W, Sec. 7 (1 OSUFW); T12S, R5W, Sec. 9 (1 OSUFW). **Clackamas Co.**—(5 PSM); Canby (4 USNM); Jennings Lodge (1 SDNHM); 7.5 mi. SE Molalla (1 PSM); 8 mi. SE Molalla (2 PSM). **Columbia Co.**—Scapoose airport (1 PSM). **Lane Co.**—5 mi. from Junction City on River Rd. (2 OSUFW); 8 mi. W Junction City (10 PSM); 8–10 mi. W Junction City (9 PSM); 10 mi. W Junction City (3 PSM); 11 mi. W Junction City (1 PSM). **Linn Co.**—Albany (1 USNM); 2 mi. W Albany, T11S, R4W, Sec. 2 (7 OSUFW); 5 mi. S, 1 mi. W Albany, T11S, R4W, SW¼ Sec. 35 (1 KU); 5 mi. S, 1.75 mi. W Albany, T11S, R4W, SW¼ Sec. 35 (1 MSUMZ, 3 OSUFW); 2.5 mi. SW Albany (1 PSM); 5 mi. SE Albany, T11S, R3W, Sec. 36 (1 OSUFW); 1 mi. N, 1 mi. E Albany (1 OSUFW); 0.5 mi. S, 6 mi. W Halsey (1 OSUFW); 13 mi. S, 8 mi. E Salem (1 OSUFW); 3.5 mi. S, 6.5 mi. W Shedd (1 OSUFW). **Marion Co.**—1 mi. N, 4 mi. E Macleay (1 OSUFW); 2 mi. N, 4 mi. E Macleay (2 OSUFW); Salem (2 MVZ, 5 SDNHM, 2 UMMZ); 3 mi. E Salem (1 USNM); 9 mi. E Salem (1 OSUFW); Silverton (1 SOSC); 1 mi. N, 2 mi. E Talbot (1 OSUFW); 3 mi. S, 2 mi. W Turner (4 OSUFW); Woodburn (16 PSM). **Multnomah Co.**—Gresham (1 USNM); Portland (22 AMNH, 5 KU, 2 LACM, 1 OSUFW, 1 PSM, 2 SDNHM); St. Johns, Portland (1 PSM). **Polk Co.**—McCoy (6 FMNH, 9 USNM); Monmouth (2 OSUFW); 0.25 mi. N, 1.5 mi. E Monmouth (1 OSUFW). **Washington Co.**—Banks (1 USNM); Beaverton (1 FMNH, 1 SDNHM); 2 mi. S Cornelius (1 OSUFW); Forest Grove (1 UMMZ); ca. 3 mi. E Gaston (1 UW); 4 mi. SE Gaston (1 OSUFW); Hillsboro (3 USNM); Scholls (2 PSU). **Yamhill Co.**—4 mi. E Carlton (1 AMNH); Dayton (14 PSM); 3.5 mi. S Dayton (4 OSUFW); 5 mi. S Dayton (1 OSUFW); 7.5 mi. W McMinnville (1 UW[WWC]); 2 mi. S, 3 mi. E Newberg (1 OSUFW); Sheridan (2 USNM); 3 mi. W Yamhill (1 OSUFW).

Microtus longicaudus, n = 824
(Plotted in Fig. 11-113, p. 321)

Baker Co.—Anthony (64 AMNH, 8 MVZ, 1 SDNHM); Bourne (4 USNM); Cornucopia (4 UMMZ, 3 USNM); 2.5 mi. NE Cornucopia, East Pine Creek (1 USNM); 0.3 mi. SE Rock Creek Butte, Elkhorn Mts., 8,500 ft. (1 CRCM); 0.7 mi. down Empire Gulch Rd.* (1 EOSC). **Benton Co.**—1 mi. S, 6 mi. W Alpine (1 OSUFW); 3 mi. S, 4 mi. E Alsea (1 OSUFW); 3.75 mi. S, 5.5 mi. E Alsea (5 CM); 10 mi. SE Alsea (1 HSU); Marys Peak (1 PSM); Parker Creek Falls, Marys Peak (1 OSUFW); 2 mi. S, 9 mi. W Philomath, head of Fork Rock Creek, Marys Peak (1 OSUFW); 3 mi. S, 6 mi. W Philomath, Marys Peak (1 OSUFW); 1.5 km S, 11.3 km W Philomath (1 USNM); 1 mi. S Summit (1 OSUFW). **Clackamas Co.**—Estacada (1 PSM); Jennings Lodge (2 SDNHM); 8 mi. SE Molalla (1 PSM). **Clatsop Co.**—1 mi. N, 2 mi. W Hammond, Fort Stevens State Park (1 OSUFW). **Columbia Co.**—7 mi. SE Rainier (1 MVZ); 4 mi. E Vernonia (2 PSM); T4N, R1W, NE¼ Sec. 34 (1 PSM); T6N, R1W, SW¼ Sec. 7 (1 PSM). **Crook Co.**—Lookout Mt., Ochoco National Forest, Canyon Creek (1 SOSC); Marks Creek, 12 mi. N Howard (9 USNM); Maury Mts. (5 USNM); Ochoco Ranger Station, 4,000 ft. (3 MVZ); Paulina (1 SDNHM); 13 mi. N, 2 mi. W Paulina, T14S, R23E, SE¼NW¼ Sec. 29 (1 KU); Walton Lake, Ochoco National Forest* (1 SOSC). **Curry Co.**—Brookings, T40S, R14W, Sec. 36 (1 PSM, 1 SOSC); Gold Beach (19 FMNH, 2 SDNHM, 3 USNM); S edge Gold Beach (14 PSM); Harbor (1 USNM); 0.25 mi. N Harris Beach State Park (1 SOSC); Little Redwood Forest Camp (5 PSM); 1.2 mi. S Pistol River (4 SOSC); Wedderburn (16 USNM). **Deschutes Co.**—near Beaver Marsh, near head of Little Meadows, Deschutes River* (2 USNM); Bend (1 USNM); 8 mi. W Bend (1 OSUFW); 15 mi. W Bend, Tumalo Creek, 6,100 ft. (1 MVZ); 5 mi. E Bend (1 SDNHM); Davis Lake (1 SDNHM); E fork Deschutes River* (2 USNM); Farewell Bend, Deschutes River (1 USNM); Indian Ford campground, T14S, R9E, Sec. 13 (1 OSUFW); La Pine (3 USNM); Paulina Lake (10 SDNHM, 5 USNM); 3 mi. W Paulina Lake (2 MVZ); Sand Spring, T21S, R16E, Sec. 13 (1 OSUFW); Squaw Creek Canyon (1 CM); 8 mi. W Terrebonne, T14S, R12E (1 OSUFW). **Douglas Co.**—Diamond Lake (1 SDNHM, 4 USNM); 24.4 km S, 6 km W Elkton (1 USNM); Oxbow Burn, Bum Creek, T20S, R7W, Sec. 19 (50 PSM). **Gilliam Co.**—6 mi. NE Lone Rock, NW corner Blue Mts. (1 USNM). **Grant Co.**—Austin (4 USNM); 7 mi. E Austin, Summit Creek, 4,700 ft. (4 MVZ); 8 mi. E Austin, Cold Spring, 4,900 ft. (6 MVZ); Beech Creek (1 USNM); John Day, Strawberry Mt. (1 USNM); 20 mi. N John Day (1 UW[EB], 1 UW[WWC]); 14 mi. N Mt. Vernon (1 UW[WWC]); 3 mi. N, 6 mi. E Prairie City, T12S, R34E, NW¼ Sec. 23 (2 KU); 21 mi. SE Prairie City, N fork Malheur River, 5,000 ft. (17 MVZ); Strawberry Mts. (1 USNM). **Harney Co.**—Buena Vista (1 SDNHM); Burns (4 USNM); 35 mi. W Burns, Squaw Butte Experimental Range (1 USNM); 50 mi. W Burns, Squaw Butte (9 USNM); 20 mi. N, 32 mi. W Burns, T19S, R25E, SE¼ Sec. 26 (1 KU); 30 mi. NW Burns, Emigrant Creek (4 USNM); 2 mi. E Crane (1 USNM); 10 mi. E Diamond, Smyth Creek, 5,500 ft. (17 MVZ); Diamond Crater, T29S, R32E, Sec. 3 (1 HSU); Fish Lake, Steens Mt., T32½S, Sec. 31 (1 OSUFW); 2.4 mi. W Fish Lake, T32S, R32¾E, Sec. 36 (1 OSUFW); 6 mi. E Frenchglen, T32S, R32½E, Sec. 14 (1 OSUFW); 10 mi. N Harney (10 USNM); Horse Pasture Field, Malheur National Wildlife Refuge, T27S, R31E, Sec. 3 (1 CM); Princeton (1 SDNHM); Steens Mt., 8,000 ft. (2 USNM); Steens Mt., 8,800 ft. (1 USNM); Steens Mt., Fish Lake, 8,000 ft. (1 USNM); Steens Mt., Kiger Gorge, 7,500 ft. (4 USNM); Steens Mt., Kiger Gorge, 6,000 ft. (1 USNM); Steens Mt., Kiger Gorge, 6,900 ft. (5 USNM); Steens Mt., Leeburg Basin* (3 USNM); Steens Mt., Little Blitzen Gorge, T33S, R33E, Sec. 10 (1 OSUFW); Steens Mt., head of Little Fir Creek, 7,500 ft.* (1 USNM); Steens Mt., T32S, R32½E, Sec. 10 (1 KU, 2 OSUFW); Steens Mt., T32S, R32½E, Sec. 25 (1 OSUFW); Steens Mt., T32S, R32¾E, Sec. 20 (3 OSUFW); Steens Mt., T32S, R32¾E, Sec. 27 (2 OSUFW); Steens Mt., T32S, R32¾E, Sec. 33 (5 OSUFW); Steens Mt., T32S, R32¾E, Sec. 35 (3 OSUFW); Steens Mt., T32S, R33E, Sec. 31 (1 OSUFW); Steens Mt., T32S, R33E, Sec. 32 (1 KU, 6 OSUFW); Steens Mt., T33S, R33E, Sec. 2 (1 OSUFW); Steens Mt., T33S, R33E, Sec. 3 (1 OSUFW); Steens Mt., T33S, R33E, Sec. 5 (1 OSUFW); Steens Mt., T37S, R33E, Sec. 32 (1 OSUFW); T33S, R32½E, NE¼ Sec. 4 (1 EOSC); T34S, R32¾E, NW¼ Sec. 33, 5,680 ft. (1 EOSC). **Hood River Co.**—Brooks Meadows, 9 mi. ENE Mt. Hood, 4,300 ft. (2 MVZ); 1 mi. E Cascade Locks (3 PSM); 6 mi. S Hood River, Odell (1 OSUFW); 8 mi. W Hood River, Viento State Park (3 OSUFW); 1 mi. W Viento (1 OSUFW); T3N, R9E, NW¼ Sec. 35 (1 PSM). **Jackson Co.**—Ashland (4 SOSC); Ashland Peak, N base Long's Camp, 3,300 ft. (1 USNM); W of Ashland (1 PSM); 3 mi. N, 11 mi. E Ashland, T38S, R4E, Sec. 18 (1 OSUFW); 5 mi. N, 21.5 mi. E Ashland (3 OSUFW); 7 mi. N, 13 mi. E Ashland, T38S, R3E, Sec. 4 (1 OSUFW); Bolan Lake (1 PSM); 2 mi. E Butte Falls (1 LACM); Conde Creek (1 PSM); Dead Indian area* (1 PSM); Hooper Springs, 10.8 mi. from HWY66 on Dean Indian Rd. (2 SOSC); E. of Hyatt Lake (1 SOSC); head of Lost Creek, 4,800 ft. (1 MVZ); Mt. Ashland, T40S, R1E, Sec. 9 (1 SOSC); Siskiyou (5 USNM); Soda Creek, T39S, R4E, NW¼ Sec. 8 (1 SOSC); 1.2 mi. N Dead Indian Rd. on Conde Creek Rd. (1 SOSC); T38S, R3E, NW¼ Sec. 15 (5 SOSC); T39S, R4E, SE¼ Sec. 8 (12 SOSC); T39S, R4E, NW¼SW¼ Sec. 9 (2 SOSC); T39S, R4E, Sec. 16 (1 SOSC); T39S, R4E, NW¼W½SW¼ Sec. 17 (4 SOSC); T39S, R4E, Sec. 18 (1 SOSC); T39S, R4E, Sec. 23 (2 SOSC); T40S, R1E, Sec. 21 (1 SOSC). **Jefferson Co.**—4 mi. NNW Culver (4 PSM); Mill Creek, 20 mi. W Warm Springs (2 USNM); 9 mi. NE Sisters (1 PSM). **Josephine Co.**—Bolan Lake (14 PSM). **Klamath Co.**—Beaver Marsh (1 SDNHM); Bly (1 SDNHM); Bly, Sprague Forest Camp (2 PSM); 15 mi. W Bly (1 USNM); 5 mi. E Bly (1 SOSC); Buck Peak* (1 SOSC); E slope Cascade Divide, Crater Lake National Park, ca. 6,400 ft.* (1 KU); Crater Lake* (1 MVZ, 7 USNM); Ft. Klamath (19 USNM); Klamath Falls (1 USNM);

2 mi. S Lake of the Woods, Rainbow Creek (1 MVZ); Marshy Spring area, E slope Cascade Divide, ca. 6,300 ft.* (1 KU); 12 mi. E Mt. Thielson, West Sink Creek (1 USNM); site of Old Ft. Klamath (1 MVZ); Swan Lake valley (2 USNM); Upper Klamath Lake (1 FMNH); Upper Munson Meadows, Crater Lake National Park, 6,300 ft. (1 KU); T38S, R5E, SE¼NE¼ Sec. 21 (1 SOSC). **Lake Co.**—Adel (1 SDNHM); 14 mi. SW Adel, Barley Camp, Warner Mts., T41S, R21E, E edge Sec. 8 (1 MVZ, 3 OSUFW); Buck Pasture, Hart Mt. National Antelope Refuge, T35S, R26E, Sec. 17 (6 KU); Dairy Creek Ranger Station (2 SDNHM); Drews Creek (1 USNM); site of Old Ft. Warner (2 SDNHM); Gearhart Mt. (21 PSM); Hart Mt. (1 SDNHM, 2 UMMZ); Hart Mt., Guano Creek, site of Old Camp Warner, T36S, R25E, Sec. 25 (5 OSUFW); Hart Mt. Antelope Refuge, T36S, R25E, NW¼ Sec. 24 (1 PSM, 1 SOSC); Hart Mt. Antelope Refuge, 2.5 mi. W headquarters (2 SOSC); Hart Mt. Antelope Refuge, Hill top Reservoir (1 SOSC); Hart Mt. Antelope Refuge, Hot Springs campground (1 SOSC); Hart Mt. Antelope Refuge, 1.4 mi. SE Hot Springs (4 SOSC); 17 mi. N Lakeview, Chandler State Park (1 SOSC); 2 mi. E Lakeview (3 MVZ); 15 mi. E Lakeview (1 USNM); 25 mi. NW Lakeview, mountain meadows at Finley Corrals (3 USNM); near Laughton Lake* (1 SOSC); Rock Creek, Warner Mt. (2 SDNHM); Silver Lake, W Silver Creek, 4,650 ft. (1 USNM); Thomas Creek Ranger Station, Fremont National Forest (1 SDNHM); Warner Mts. (3 USNM); Willow Creek forest camp (2 SOSC); Yamsay Mts., W of Silver Creek, 7,000 ft. (7 USNM); T35S, R17E, Sec. 17 (2 PSM). **Lane Co.**—Eugene (3 UW-X); 3 mi. W Eugene (1 UW-X). **Lincoln Co.**—Cape Perpetua (1 SDNHM); Cascade Head Experimental Forest (6 PSM); Depoe Bay (4 PSM); Newport (1 SDNHM); Thelake (1 SDNHM); 5 mi. W Waldport (1 PSM); Yachats (1 SDNHM). **Malheur Co.**—(3 AMNH); Cedar Mts., T26S, R41E (2 USNM); Cottonwood Creek, T26S, R38E, NW¼ Sec. 29 (3 PSM); Cow Creek Lake (2 USNM); 10 mi. NE Crowley Creek, Summit (1 USNM); Crowley guard station, T27S, R39E, N½ Sec. 17 (1 PSM); Disaster Peak, 5,600 ft. (1 USNM); Disaster Peak, 7,000 ft. (1 USNM); Jordan Valley, 4,200 ft. (6 USNM); 0.25 mi. W Jordan Valley (2 PSM); Mahogony Mts., 6,500 ft. (4 USNM); McDermitt (1 USNM); 8 mi. NE McDermitt, 6,300 ft. (1 USNM); 2 mi. NW Riverside (4 USNM); Rockville (1 USNM); Skull Spring (1 USNM); Succor Creek State Park, T23S, R46E, NW¼ Sec. 15, 2,460 ft. (4 PSM). **Morrow Co.**—3 mi. S, 2 mi. E Hardman (4 KU). **Multnomah Co.**—2.5 mi. W Bonneville (5 PSM). **Polk Co.**—2.2 mi. SW Airlie (2 OSUFW). **Sherman Co.**—5 mi. N, 19 mi. E The Dalles (1 OSUFW). **Tillamook Co.**—(1 PSM); Blaine (1 UMMZ); Hebo Lake (2 PSM); Miami River* (1 PSM); 2 mi. up Miami River (7 PSM); Netarts (5 SDNHM); 2 mi. E Netarts (1 PSM); 2 mi. S Twin Rocks (10 PSM). **Umatilla Co.**—Langdon Lake (1 MVZ); MacDougal camp, 7 mi. W Tollgate (3 UW[EB]); Meacham (2 USNM); Rock Creek, 10 mi. W Tollgate (1 UW[EB]); Tollgate (1 UW[WWC]); 4 mi. W Tollgate (2 UW[WWC]); Tollgate Ranger Station (1 SDNHM). **Union Co.**—10 mi. W Elgin (1 UW[EB]); Starkey Experimental Forest, Battle Creek Spring, T4S, R34E, SW¼ Sec. 22 (1 PSM); Starkey Experimental Forest, Bear Creek, T3S, R34E, NW¼ Sec. 2 (1 PSM); Starkey Experimental Forest, Meadow Creek, T3S, R34E, SE¼ Sec. 27 (1 PSM); Starkey Experimental Forest, Meadow Creek camp, T3S, R34E, NE¼ Sec. 35 (6 PSM); 7 mi. E Telocaset, Catherine Creek, 3,500 ft. (1 MVZ). **Wallowa Co.**—Lick Creek Ranger Station (2 UMMZ); 6 mi. S Lostine, T2S, R43E, Sec. 10 (1 OSUFW); 16 mi. S Lostine, T3S, R43E, Sec. 25 (5 OSUFW); 10 mi. S, 1 mi. E Lostine, T2S, R43E, Sec. 35 (2 OSUFW); 16 mi. S, 2 mi. E Lostine, T3S, R43E, Sec. 25 (2 OSUFW); 16 mi. S, 3 mi. E Lostine, T3S, R43E, Sec. 36 (9 MVZ, 1 OSUFW); 19 mi. S, 4 mi. E Lostine, T4S, R44E, Sec. 17 (1 OSUFW); 17 mi. SE Lostine, Lostine River, Wallowa Mts. (3 OSUFW); 15 mi. N Paradise, Horse Creek, 2,000 ft. (1 USNM); Troy, 1,700 ft. (1 USNM); 8 mi. W Wallowa (1 UW[EB]); Wallowa Lake, 4,000 ft. (9 USNM); S of Wallowa Lake, 8,000 ft. (1 USNM); S of Wallowa Lake, 8,500 ft. (1 USNM); 2 mi. S Wallowa Lake on west fork Wallowa River, 5,000 ft. (1 UW[EB]). **Wasco Co.**—Clear Lake (3 UMMZ). **Washington Co.**—2 mi. W Carpenter Creek (1 OSUFW). **Wheeler Co.**—Bridge Creek (1 MVZ); 7 mi. S, 11 mi. W Mitchell (12 MVZ); Wildwood Camp, Ochoco National Forest (2 KU, 1 OSUFW).

Microtus montanus, n = 1,220
(Plotted in Fig. 11-115, p. 323)

Baker Co.—(1 MSUMZ); Anthony Creek game area (1 EOSC); 11 mi. N Baker [City], Radium Hot Springs (8 CRCM); Brownlee (1 MSUMZ); Home, 1,850 ft. (5 USNM); Homestead, 1,800 ft. (4 USNM); Keating (2 USNM); Oxbow (1 MSUMZ); S of Unity, West Camp Creek (2 PSM). **Crook Co.**—2 mi. SW Barnes (10 MVZ); Crooked River, mouth of Bear Creek, 3,400 ft. (1 MVZ); Crooked River, 3 mi. E Bear Creek (1 MVZ); Marks Creek, Ochoco National Forest (2 USNM); Prineville (1 USNM); 13 mi. S Prineville (6 PSM); Ragh Ranger Station, Ochoco National Forest, T15S, R25E, Sec. 9 (1 SOSC). **Deschutes Co.**—Bend (9 USNM); 19 mi. SW Bend, Deschutes River (2 OSUFW); Cold Spring* (1 KU); mouth of Davis Creek (1 OSUFW); La Pine (1 USNM); Little Lava Lake (6 PSM); Indian Ford Campground (1 PSM); Lower Bridge (6 OSUFW, 1 PSM); Sisters (2 OSUFW, 1 SDNHM, 7 USNM); 6 mi. E Sisters (1 OSUFW); 4 mi. N, 6 mi. E Sisters (1 OSUFW); Sparks Lake (1 UMMZ); Squaw Creek Canyon (2 CM); 1 mi. N Three Creek Lake (19 PSM). **Douglas Co.**—Diamond Lake (2 SDNHM); N end of Diamond Lake (1 OSUFW); Diamond Lake near outlet (1 OSUFW). **Gilliam Co.**—1.5 mi. S Heppner Junction (3 KU). **Grant Co.**—7 mi. E Austin, Summit Creek, 4,700 ft. (3 MVZ); 8 mi. E Austin, Cold Spring, 4,900 ft. (29 MVZ); Camp Creek, T12S, R32E, NW¼ Sec. 2 (1 PSM, 23 SOSC); 2 mi. SE Canyon City (2 LSU); 15 mi. S John Day (1 UW[EB]); 2.5 mi. E John Day (2 LSU); John Day River near Moon Creek (1 EOSC); Lake Creek guard station, Malheur National Forest, T16S, R33½E Sec. 10 (1 OSUFW); 1 mi. N, 10.5 mi. S Long Creek, T10S, R32E, Sec. 5 (1 OSUFW); 5.3 mi. N, 1.3 mi. E Long Creek (2 OSUFW); N fork Malheur River, Malheur National Forest* (1 KU); 5 mi. W Mt. Vernon (3 MVZ); Prairie City (1 USNM); 3 mi. N, 6 mi. E Prairie City, T12S, R34E, NW¼ Sec. 23 (1 KU); 21 mi. SE Prairie City, N fork Malheur River (2 MVZ); 3 mi. S, 7 mi. W Silvies, T18S, R30E, SW¼ Sec. 25 (2 KU); Strawberry Mts. (1 USNM); Strawberry Mts., 1 mi. W Wickiup campground (10 HSU). **Harney Co.**—marsh below Alvord Ranch (2 EOSC); 0.25 mi. S, 0.25 mi. W Ant Hill, T29S, R33E, SW¼NE¼ Sec. 36, 5,000 ft. (1 KU); Benson Boat Landing, Malheur National Wildlife Refuge, T26S, R31E, Sec. 25 (7 CM); Blitzen Valley* (33 SDNHM); Buchanan (1 USNM); Buena Vista (1 SDNHM); Buena Vista Substation, Malheur National Wildlife Refuge (1 OSUFW); Burns (3 USNM); 4 mi. E Burns, 4,150 ft. (14 MVZ); 22 mi. S, 8 mi. E Burns, S Malheur National Wildlife Refuge (1 OSUFW); 24 mi. S, 7 mi. E Burns, Malheur National Wildlife Refuge (2 OSUFW); 25 mi. S, 7 mi. E Burns, N of headquarters, Malheur National Wildlife Refuge (2 OSUFW); 26 mi. S, 3 mi. E Burns, Malheur National Wildlife Refuge (1 OSUFW); 26 mi. S, 14 mi. E Burns, Malheur National Wildlife Refuge (1 OSUFW); 28 mi. S, 4 mi. E Burns, N Wright's Pond, Malheur National Wildlife Refuge (2 OSUFW); 40 mi. S, 12 mi. E Burns, Malheur National Wildlife Refuge (3 OSUFW); 28 mi. NW Burns (2 OSUFW); 30 mi. NW Burns (25 USNM); 5 mi. NNE Burns, 4,200 ft. (1 MVZ); 2 mi. E Crane (1 USNM); 1 mi. NW Crane, Crane Hot Springs (2 PSU); Diamond, 4,300 ft. (19 USNM); 7.5 mi. N, 3 mi. E Diamond (1 OSUFW); 1 mi. W Drewsey (1 OSUFW); Frenchglen (1 PSM, 1 UMMZ); 0.5 mi. E Frenchglen, Warm Springs #1, Malheur National Wildlife Refuge (1 OSUFW); 4.5 mi. N, 0.5 mi. E Frenchglen, Malheur National Wildlife Refuge (2 OSUFW); Horse Pasture Field, Malheur National Wildlife Refuge, T27S, R31E, Sec. 3 (20 CM); Jackman Park, 2 mi. SE Fish Lake, 7,500 ft. (3 UW[WWC]); S Little Juniper Field, Malheur National Wildlife Refuge, T31S, R32½E, Sec.29–30 (1 CM); 5 mi. N of P Ranch, Malheur National Wildlife Refuge, T31S, R31E, Sec. 7 (2 OSUFW); Malheur Lake (1 USNM, 5 UW[EB]); Malheur National Wildlife Refuge (4 OSUFW, 1 USNM); Malheur National Wildlife Refuge, North Swamp Field (4 OSUFW); 1 mi. N Malheur National Wildlife Refuge headquarters (8 HSU); Narrows (2 LACM, 1 SDNHM, 22 USNM); 2.5 mi. SW Narrows (1 PSM); Page Springs campground, 3 mi. SE Frenchglen, 4,200 ft. (7 UW[WWC]); Page Springs Dam, 2 mi. ESE Frenchglen, 4,200 ft. (3 UW[WWC]); Princeton (1 SDNHM); 24 mi. N, 12 mi. W Riley, T19S, R25E, NE¼ SW¼ Sec. 26 (1 KU); Roaring Springs (2 UW[WWC]); Shirk P.O. (2 USNM); Sod House Ranch, Blitzen Valley (1 PSM); Steens Mt. (2 OSUFW); Steens Mt., Big Fish Creek, 7,500 ft. (3 USNM); Steens Mt., Blitzen Canyon (1 SDNHM); Steens Mt., Fish Lake, 8,000 ft. (2 FMNH, 2 KU, 16 UMMZ, 1 USNM); Steens Mt., Fish Lake, T32½S, R33E, Sec. 31 (1 OSUFW); Steens Mt., Kiger Gorge, 6,900 ft. (1 USNM); Steens Mt., Kiger Gorge, 7,500 ft. (1 USNM); Steens Mt., N end, 6,400 ft. (1 FMNH); Steens Mt., Wild Horse Gulch (1 USNM); Steens Mt., T31S, R32½E, Sec. 3 (1 KU); Steens Mt., T31S, R32½E, Sec. 13 (2 OSUFW); Steens Mt., T32S, R32½E, Sec. 25 (3 OSUFW); Steens Mt., T32S, R32½E, Sec. 10 (2 OSUFW); Steens Mt., T32S, R32¾E, Sec. 33 (1 OSUFW); Steens Mt., T32S, R32¾E, Sec. 36 (6 OSUFW); Steens Mt., T32S, R33E, Sec. 32 (1 KU, 1 OSUFW); Steens Mt., T32½S, R33E, Sec. 29 (14 OSUFW); Steens Mt., T32½S, R33E, Sec. 33 (1 OSUFW); Steens Mt., T32½S, R33E, Sec. 35 (1 OSUFW); Steens Mt., T33S, R33E, Sec. 1 (1 KU, 8 OSUFW); Steens Mt., T33S, R33E, Sec. 2 (8 OSUFW); Steens Mt., T33S, R33E, Sec. 3 (1 KU, 5 OSUFW); Steens Mt., Little Blitzen Gorge, T33S, R33E, Sec. 10 (1 OSUFW); Steens Mt., T33S, R34E, Sec. 9 (1 OSUFW); Summit, N end Alvord Valley (2 USNM); Voltage (12 USNM); 3 mi. S Whitehorse Ranch (1 UW[WWC]); Willow Spring* (1 SOSC); T32S, R32½E, NW¼ Sec. 17 (1 EOSC). **Hood River Co.**—6 mi. E Cascade Locks (6 PSM); Hood River (8 SDNHM, 4 USNM); Hood River, Van Horn Butte (10 USNM). **Jackson Co.**—(1 SOSC); Brownsboro (7 USNM); Conde Creek (1 PSM); Ice House Lake, 9 mi. up Dead Indian Rd. from HWY66 (1 SOSC); 1.5 mi. S Siskiyou summit W of I-5 (1 SOSC); 1.6 mi. W old HWY99 on Mt. Ashland Rd. (2 SOSC); T38S, R3E, Sec. 10 (2 SOSC); T39S, R3E, SE¼NE¼ Sec. 33 (1 SOSC). **Jefferson Co.**—(4 PSM); Camp Sherman (1 PSM); 4 mi. from Camp Sherman, Deschutes National Forest* (1 KU); 4 mi. NNW Culver (89 PSM); 6 mi. SE Culver (6 PSM); 4 mi. ESE Culver (1 PSM); Hay Creek (1 SDNHM, 14 USNM); 12 mi. E Hay Creek, Foley Creek (6 USNM); Trout Creek, Ochoco National Forest (1 USNM). **Klamath Co.**—(2 OSUFW); 1.5 mi. SW Bonanza (1 PSM); 0.75 mi. SE Bonanza (1 PSM); 5 mi. NE Chiloquin (1 PSM); Crater Lake National Park* (1 OSUFW); near S entrance Crater Lake National Park (1 OSUFW); Dairy (1 SDNHM); Ft. Klamath (1 AMNH, 18 OSUFW, 1 PSM, 11 USNM); Klamath Basin, Lost River (7 USNM); Klamath Falls (12 CM, 4 OSUFW, 17 USNM); 8 mi. NW Klamath Falls (2 OSUFW); Munson Creek, Crater Lake Headquarters (4 OSUFW); Ross Reservoir* (2 SOSC); Spring Creek (1 SDNHM); Swan Lake Valley (5 USNM); Teddy Powers Meadow (3 PSM); Tule Lake (1 USNM); Upper Klamath marsh (4 USNM); Yamsay River, Yamsay Mts., 4,800 ft. (5 USNM); 4 mi. NW Weyerhauser Camp 4, junction HWY101 and W38* (3 SOSC); 4.6 road mi. NW Weyerhauser Camp 4 on HWY66* (3 SOSC). **Lake Co.**—Adel (1 CM, 1 OSUFW, 1 SDNHM); 9 mi. S Adel, Twenty Mile Creek (2 OSUFW); Buck Creek, Hart Mt. National Antelope Refuge (12 KU); site of Old Ft. Warner, Guano Creek, Hart Mt., T36S, R25E, Sec. 25 (1 OSUFW, 6 SDNHM); Fremont National Forest* (7 UMMZ); Gearhart Mt. (31 PSM); Goose Lake Mts. (1 USNM); Goose Lake Valley (3 PSM); Hart Mt. (5 PSM); Indian Springs, Hart Mt. Antelope Refuge* (1 SOSC); near Hart Mt. Antelope Refuge headquarters (1 SOSC); Hart Mt. Antelope Refuge, 1 mi. from headquarters (2 SOSC); 2 mi. W hdqrs Hart Mt. Antelope Refuge (28 KU); 3.2 mi. W hdqrs Hart Mt. Antelope Refuge (23 KU); Hart Mt. Antelope Refuge, Hot Springs camp, T35S, R26E, NE¼NW¼ Sec. 32 (1 SOSC); Hart Mt. Antelope Refuge, 1.8 mi. S Hot Springs camp (1 SOSC); 6 mi. W Lakeview (6 MVZ); 15 mi. E Lakeview (1 SDNHM); 10 mi. SE Lakeview, 6,600 ft. (3 USNM); 25 mi. SE La Pine, T24S, R14E, Sec. 17 (1 OSUFW); Loufton Lake* (1 SOSC); N slope Rock Creek, Warner Mt. (1 SDNHM); Plush (9 USNM); Quartz Mt., Fishhole guard station (3 SOSC); Quartz Mt., Beaver Dam Lake (2 SOSC); 10 mi. SW Silver Lake, W Silver Creek, 4,650 ft. (8 USNM); Sycan Marsh (4 USNM); 2 mi. N Valley Falls off HWY 395 (1 SOSC); 6 mi. S Valley Falls along Crooked Creek (2 SOSC); 0.5 mi. S West Side School* (1 MVZ); Willow Creek, Hart Mt. National Antelope Refuge (7 KU); Winter Rim Burn (1 PSM). **Malheur Co.**—(4 AMNH); 4 mi. S, 1 mi. E Adrian (1 OSUFW); Antelope Reservoir, 10 mi. SW Jordan Valley, T31S, R45E, SE¼SE¼ Sec. 33 (2 PSM); Battle Mt. (1 SDNHM); S end Butch Lake, T29S, R44E, NE¼ Sec. 6 (1 PSM); Bogus Lake, T29S, R41E, SE¼ Sec. 9 (1 PSM); Cottonwood Creek, T26S, R38E, NW¼ Sec. 29 (2 PSM); head of Crooked Creek, Owyhee Desert, 4,000 ft. (6 USNM); George Harper Ranch* (1 USNM); Ironside, 4,000 ft. (16 AMNH); Jordan Valley,

4,200 ft. (6 USNM); 7.2 mi. N Jordan Valley (1 PSU); 0.25 mi. W Jordan Valley (1 PSM); 0.5 mi. W Jordan Valley (2 PSM); Kane Springs, T21S, R43E, Sec. 4 (3 PSM); McDermitt (7 USNM); Mud Springs, T38S, R40E, SW¼ Sec. 29 (7 PSM); 20 mi. from Owyhee River, Bowden Ranch on Rattlesnake Creek (1 USNM); Pharmacy Butte, 0.5 mi. W Jordan Valley, T30S, R46E, SW¼ Sec. 2 (1 PSM); Riverside (2 OSUFW); Sheaville (2 USNM); Soldier Creek exclosure, 8.5 mi. WSW Jordan Valley, T30S, R45E, SW¼ Sec. 15, 4,420 ft. (1 PSM); Vale (2 USNM); near Vale (1 AMNH); 2 mi. N Vale (1 PSM); 3 mi. S Vale (1 PSM); 3.5 mi. S Vale (2 PSM); Watson (2 USNM); 6 mi. SE Whitehorse Ranch, T37S, R37E, SW¼ Sec. 27 (2 PSM). **Morrow Co.**—3 mi. S, 2 mi. E Hardman (3 KU); 2 mi. SE Heppner (1 UW[WWC]); T5N, R26E, NE¼ Sec. 24 (2 PSM). **Sherman Co.**—Biggs (2 PSM); 1 mi. S Biggs Junction (2 PSM). **Umatilla Co.**—Juniper Canyon (1 UW[WWC]); Langdon Lake (3 UW[WWC]); 7 mi. W Langdon lake (2 UW[WWC]); Meachem (1 USNM); Meacham Lake, 3,800 ft. (4 MVZ); 1 mi. E Milton-Freewater (1 UW[WWC]); Pendleton (1 SDNHM); Tollgate (1 AMNH); Tollgate, biology lodge (5 MSUMZ); 4 mi. W Tollgate (1 AMNH); 0.5 mi. N, 2.5 mi. W Tollgate (1 UW[WWC]); 1 mi. N, 3 mi. W Tollgate (2 UW[WWC]); 3 mi. NW Tollgate (3 UW[WWC]); Weston (5 UW[WWC]); 10 mi. E Weston (1 UW[WWC]); 16 mi. E Weston (1 UW[WWC]); 4 mi. SE Weston (1 UW[WWC]); 300 yds. NE junction I-80N and Barnhart Rd.* (2 SOSC). **Union Co.**—(1 EOSC); head of Beaver Creek (1 EOSC); Beaver Meadows, 18 mi. SW North Powder, 5,300 ft. (1 USNM); Cove (1 EOSC); Elgin (1 EOSC, 2 USNM); La Grande (5 EOSC); 8 mi. E La Grande (1 EOSC); 30–36 mi. SE Pendleton, Starkey Range (8 USNM); Starkey Experimental Forest, T4S, R34E, NE¼ Sec. 16 (1 PSM); Starkey Experimental Forest, Meadow Creek camp, T3S, R34E, NE¼ Sec. 35 (2 PSM); Starkey Experimental Forest, Warm Spring, T3S, R34E, SW¼ Sec. 27 (1 PSM); Telocaset (1 USNM); Union (1 EOSC); T6S, R41E, Sec. 2 (1 EOSC). **Wallowa Co.**—(2 PSM); College Creek Ranger Station, Imnaha River, Wallowa National Forest (1 USNM); Dug Bar, Snake River (2 SDNHM); 3 mi. SW Imnaha, Sheep Creek (2 PSU); 5 mi. SE Imnaha, Little Sheep Creek (1 PSU); between Minam and Wallowa (1 PSM); 15 mi. NE Paradise, Horse Creek (1 USNM); 3 mi. N Troy (1 PSM); Wallowa (1 SDNHM); Wallowa Lake (6 USNM). **Wasco Co.**—Antelope (1 PSM); 2 mi. S, 2 mi. E The Dalles (1 OSUFW); Wapinitia (4 USNM). **Wheeler Co.**—Derr Ranger Station, Ochoco National Forest (2 SDNHM); Fossil (1 OSUFW, 7 UW[WWC]); 1 mi. N Fossil (5 UW[WWC]); 10 mi. E Fossil (1 UW[WWC]); 6 mi. N, 5 mi. E Mitchell (1 OSUFW); 7 mi. S, 11 mi. W Mitchell (11 MVZ); Ochoco National Forest* (1 UMMZ); 11 mi. N Service Creek (1 UW[WWC]); 4 mi. E Service Creek (4 UW[WWC]); Squaw Butte (5 USNM).

Additional localities: head of Crooked Creek* (5 USNM); N slope Rock Creek, Warner Mt.* (1 SDNHM).

Microtus oregoni, n = 690
(Plotted in Fig. 11-118, p. 326)

Benton Co.—(1 OSUFW, 2 PSM); 3.75 mi. S, 5.5 mi. E Alsea (1 CM); 3.5 mi. SSW Blodgett on Woods Creek (1 OSUFW); Camp Adair (2 OSUFW); Corvallis (2 UCONN, 1 USNM); 6 mi. N Corvallis (1 PSM); 7 mi. N Corvallis (13 PSM); 8 mi. N Corvallis (1 PSM); 2 mi. S Corvallis (2 PSM); 10 mi. W Corvallis, McDonald Forest (1 OSUFW); 1 mi. N, 2 mi. W Corvallis (2 OSUFW); 1 mi. N, 3 mi. W Corvallis, McDonald Forest (1 OSUFW); 2 mi. N, 4 mi. W Corvallis, McDonald Forest (1 OSUFW); 3 mi. N, 3 mi. W Corvallis (1 OSUFW); 3 mi. N, 4 mi. W Corvallis (2 OSUFW); 8 mi. N, 5 mi. E Corvallis, E. E. Wilson Wildlife Area (2 OSUFW); 8 mi. S, 21 mi. W Corvallis, Marys Peak (1 OSUFW); 2 mi. SSW Corvallis (2 PSM); 4.5 mi. SW Corvallis, T11S, R5W, Sec. 8 (1 OSUFW); top Marys Peak (1 OSUFW); McDonald Forest (2 OSUFW, 2 PSM); Missouri Bend (1 SOSC); 0.4 km S, 12.5 km W Philomath, T12S, R7W, Sec. 10 (1 USNM); 1.5 km S, 11.3 km W Philomath, T12S, R7W, Sec. 14 (1 USNM); 1 mi. S, 6 mi. W Philomath, 1.25 mi. W Stillson Creek, T12S, R7W, Sec. 13 (1 OSUFW); 1 mi. S, 9 mi. W Philomath (1 KU); 2 mi. S, 8 mi. W Philomath, Marys Peak (2 OSUFW); 2 mi. S, 9 mi. W Philomath, top Marys Peak, T12S, R7W, SW¼ Sec. 21 (2 OSUFW); 2 mi. S, 10 mi. W Philomath, T12S, R7W, Sec. 21, top

Marys Peak (1 OSUFW); 4 mi. SW Philomath (1 PSM); 5 mi. SW Philomath (4 USNM); 9 mi. SW Philomath, Marys Peak (3 PSM); 0.25 mi. W Stillson Creek, Marys Peak, T12S, R7W, Sec. 2 (2 OSUFW); 1.25 mi. SE Summit (2 PSM); Wells (3 USNM); T11S, R5W, Sec. 3 (1 KU, 1 OSUFW), T11S, R5W, Sec. 32 (2 KU). **Clackamas Co.**—Clackamas (1 UW[EB]); Jennings Lodge (1 SDNHM); 2 mi. E Liberal (5 PSM); 8 mi. SE Molalla (3 PSM); Mulino (1 OSUFW, 1 SDNHM); Oregon City (5 USNM, 3 UW[EB]); 2 mi. N, 4 mi. E Sandy (3 CRCM); 3.5 mi. N, 8 mi. E Sandy (1 CRCM); T5S, R8½E, Sec. 26 (1 PSU). **Clatsop Co.**—(1 UMMZ); Astoria (2 MVZ, 12 USNM); 7.5 mi. S Cannon Beach (1 MVZ); Gearhart (1 PSM); Hurricane Ridge* (1 UMMZ); site of old Ft. Clatsop (44 MVZ); 1 mi. N site of Old Ft. Clatsop (1 MVZ); Warrenton (2 MVZ); T8N, R10W, NW¼ Sec. 6 (6 PSM). **Columbia Co.**—3 mi. SW Rainier (2 MVZ); 7 mi. SE Rainier (9 MVZ). **Coos Co.**—2 mi. SE Bandon (2 PSM); 4 mi. SE Bandon (1 PSM); Bastendorf Beach (1 LACM); Cape Arago State Park (5 LACM); Charleston (1 LACM); 5 mi. SW Charleston, Sunset Bay (3 LACM); Coos Bay, South Slough (1 LACM); Coos Head (1 LACM); 7.3 km N, 4 km W Sitkum, T27S, R10W, Sec. 17/20 (2 USNM); 7.6 km N, 5.2 km W Sitkum, T27S, R10W, Sec. 18 (1 USNM). **Curry Co.**—10 mi. E Brookings (7 PSM); Gold Beach (4 FMNH); Long Ridge Forest Camp (2 PSM); 1 mi. S Long Ridge Forest Camp (1 PSM). **Douglas Co.**—Bradley Creek (3 SDNHM); Diamond Lake (3 SDNHM); 0.8 km N, 2.2 km W Drain, T21S, R5W, Sec. 7 (1 USNM); 6.2 km N, 12.7 km E Drain, T24S, R4W, Sec. 27 (1 USNM); 25.6 km S, 5.6 km W Elkton, T25S, R8W, Sec. 10 (1 USNM); Francis Creek, T24S, R1W, SE¼SE¼SE¼ Sec. 12 (6 SOSC); Lookingglass (1 USNM); Louis Creek, T28S, R3W, N½SW¼ Sec. 29 (5 SOSC); Oxbow Burn, Bum Creek, T20S, R7W, Sec. 19 (52 PSM); Oxbow Burn, 20 mi. SE Crow (3 PSM); Scottsburg (2 USNM); Sutherlin (1 OSUFW); 2.2 km N, 26.2 km W Sutherlin, T25S, R9W, Sec. 9 (1 USNM); 3.4 km N, 11.9 km W Sutherlin, T25S, R7W, Sec. 1 (1 USNM); 3.6 km N, 11.9 km W Sutherlin, T25S, R7W, Sec. 1 (5 USNM); 12 mi. N, 17 mi. E Tiller (1 CRCM); 12.5 mi. N, 5 mi. E Tiller, T28S, R1W, Sec. 32 (1 CRCM); 2.2 mi. N, 9 mi. W Umpqua P.O., T25S, R8W, Sec. 15, 850 ft. (1 UW[CR]). **Hood River Co.**—Brooks Meadows, 9 mi. ENE Mt. Hood, 4,300 ft. (9 MVZ); Hood River (1 SDNHM). **Jackson Co.**—Ashland (1 SOSC); Ashland Research Natural Area, T39S, R1E, SW¼ Sec. 28 (3 SOSC); 1 mi. S Emigrant Lake, T39S, R2E, NW¼ Sec. 29 (2 SOSC); W slope Grizzly Peak, Ashland, 3,500 ft. (1 USNM); W slope Grizzly Peak, Ashland, 5,500 ft. (1 USNM); 6 mi. S Medford (1 MVZ); 7 mi. S Medford (2 PSM, 1 SOSC); 2.5 mi. N, 8.5 mi. E Prospect (1 CRCM); 5 mi. N, 8 mi. E Prospect (1 CRCM); 0.3 mi. W Colestene Rd. on Mt. Ashland (1 SOSC); T37S, R4E, Sec. 23 (1 SOSC). **Josephine Co.**—1 mi. S, 0.5 mi. W Cave Junction (1 LACM); 3 mi. NE Grants Pass, W fork Jones Creek, 1,600 ft. (1 USNM); S slope Grants Pass Peak, 2,300 ft. (1 USNM); 8.5 mi. W Selma (1 LACM); Sourgrass Creek, T35S, R9W, SW¼NE¼ Sec. 3 (2 SOSC). **Klamath Co.**—Crater Lake, N slope Glacier Lake, 7,800 ft. (2 USNM); Sky Lakes Basin (1 SOSC). **Lane Co.**—5 mi. N Alpha, T15S, R9W, Sec. 31 (1 OSUFW); 5 mi. N, 6 mi. W Alpha, T15S, R9W, Sec. 31 (5 OSUFW); 2 mi. S, 1 mi. E Austa (6 CM); 4 mi. E mouth Big Creek (2 KU); Blue River (1 USNM); 1.25 mi. N, 5 mi. W Blue River, T16S, R3E, Sec. 15 (1 HSU); 1.5 mi. N, 2 mi. E Blue River (1 CRCM); 5 mi. N, 7 mi. E Blue River, Upper Lookout Creek Tributary, T15S, R5E, Sec. 35 (1 JFBM); 5 mi. N, 9 mi. E Blue River (1 CRCM); 5.5 mi. N, 13.5 mi. E Blue River, T16S, R6E, Sec. 26 (2 CRCM); 7 mi. N, 9 mi. E Blue River, T15S, R5E, Sec. 13 (1 CRCM); 5 km N, 7.6 km E Blue River, T16S, R5E, Sec. 6 (2 USNM); 6.8 km N, 7.6 km E Blue River, T15S, R5E, Sec. 32 (1 USNM); 2.5 mi. S, 7 mi. E Blue River (1 CRCM); 3.5 mi. S, 8 mi. W Blue River, T17S, R3E, Sec. 7 (1 HSU); 3.5 mi. S, 10 mi. E Blue River (1 CRCM); 5.5 mi. NE Blue River (1 PSM); 10 mi. NE Blue River, Upper Lookout Creek Tributary, T15S, R5E, Sec. 25 (1 JFBM); Eugene (2 USNM); H. J. Andrews Experimental Forest, T15S, R5E, NW¼ Sec. 24 (1 PSM); H. J. Andrews Experimental Forest, T15S, R5E, NW¼ Sec. 28 (4 PSM); H. J. Andrews Experimental Forest, T15S, R5E, NW¼ Sec. 34 (1 PSM); H. J. Andrews Experimental Forest, T16S, R4E, NE¼ Sec. 19 (11 PSM); H. J. Andrews Experimental Forest, T16S, R4E, NE¼ Sec. 24 (26 PSM); H. J. Andrews Experimental Forest, T16S, R5E, SW¼ Sec. 6 (14 PSM); 3 mi. N, 7 mi. W Horton (1 OSUFW);

Mapleton (1 USNM); McKenzie Bridge (1 OSUFW, 1 USNM); 2.6 km S McKenzie Bridge, T16S, R5E, Sec. 26 (1 USNM); 6.6 km N, 0.3 km W McKenzie Bridge, T15S, R5E, Sec. 25 (2 USNM); 6 km N, 7.5 km E McKenzie Bridge, T15S, R6E, Sec. 26 (1 USNM); 6.4 km N, 0.6 km E McKenzie Bridge, T15S, R5E, Sec. 25 (1 USNM); 1.6 km S, 5.2 km W McKenzie Bridge, T16S, R5E, Sec. 20 (3 USNM); 2.6 km S, 1.4 km E McKenzie Bridge, T16S, R5E, Sec. 25 (4 USNM); 3.8 km N, 14.8 km E McKenzie Bridge, T16S, R7E, Sec. 33 (16 USNM); 4.4 km N, 14.3 km E McKenzie Bridge, T16S, R7E, Sec. 32 (2 USNM); 4 mi. N McKenzie Bridge, T15S, R5E, Sec. 25, 2,600–2,700 ft. (3 UW[OG]); 10 mi. E McKenzie Bridge (14 USNM); 3 mi. S, 8.5 mi. E McKenzie Bridge, T16S, R7E, Sec. 32, 3,500 ft. (7 UW[OG]); 3.5 mi. S, 9 mi. E McKenzie Bridge, T16S, R7E, Sec. 32, 3,500 ft. (2 UW[OG]); 10 mi. SE McKenzie Bridge, Lost Creek Ranger Station (6 USNM); Olallie Creek Camp, McKenzie River (1 PSM); Siltcoos Forest Camp (1 OSUFW); N base Three Sisters, 5,000 ft. (2 USNM); Vida (2 SDNHM, 3 USNM); 23 mi. SE Vida (5 PSM). **Lincoln Co.**—Alsea Bay (2 PSM); Cascade Head Experimental Forest (2 PSM); Chintimini Mt., 4,000 ft. (1 USNM); Delake (2 UMMZ); Eddyville (2 USNM); 2 mi. S, 12 mi. W Harlan, Beaver Creek, T12S, R10W, Sec. 20 (1 KU); 5 mi. S, 10 mi. W Harlan, Gold Creek, T13S, R10W, Sec. 2 (4 KU); 7 mi. S, 5 mi. W Harlan, West Creek, T13S, R9W, Sec. 16 (1 KU); Newport (3 SDNHM); 7 mi. S Newport (1 UMMZ); 0.5 mi. E Newport (1 UMMZ); 2 mi. E Newport (1 UMMZ); Otis (1 USNM); Siletz River (1 USNM); 5 mi. NE Waldport (1 PSM); 4 km S, 3.2 km E Yachats, T15S, R12W, Sec. 12 (1 USNM); Yaquina Bay (2 USNM). **Linn Co.**—1.5 mi. N Big Lake (1 PSM); 8 mi. N, 8 mi. E Blue River (1 CRCM); 9 mi. N, 8 mi. E Blue River (1 CRCM); 6 mi. E Cascadia, Canyon Creek (1 OSUFW); N of Corvallis water works, T11S, R4W, Sec. 31/32 (2 OSUFW); H. J. Andrews Experimental Forest, T14S, R6E, NE¼ Sec. 20, 4,400 ft. (2 PSM); H. J. Andrews Experimental Forest, T14S R6E, NE¼ Sec. 28, 4,600 ft. (3 PSM); H. J. Andrews Experimental Forest, T15S, R5E, SE¼ Sec. 11, 3,200 ft. (3 PSM); H. J. Andrews Experimental Forest, T15S, R6E, NE¼ Sec. 7, 4,300 ft. (2 PSM); H. J. Andrews Experimental Forest, T15S, R6E, NW¼ Sec. 7, 5,000 ft. (17 PSM); 3.5 mi. N, 3.5 mi. W Sweet Home, T13S, R1W, Sec. 15 (2 OSUFW); 4 mi. E Upper Soda (1 OSUFW); 8 mi. E Upper Soda, Tombstone Prairie, T13S, R5E, Sec. 35 (6 OSUFW); 9 mi. E Upper Soda (1 OSUFW); T15S, R1W, SE¼NE¼ Sec. 13 (1 SOSC); T15S, R2W, SE¼NE¼ Sec. 13 (1 SOSC). **Marion Co.**—Aumsville (1 USNM); Detroit (1 OSUFW); Salem (1 OSUFW); 1 mi. N, 1 mi. E Talbot (1 OSUFW). **Multnomah Co.**—Latourell (1 OSUFW); Portland (17 AMNH, 4 LACM, 1 PSU, 5 SDNHM, 2 USNM); 10 mi. N, 10 mi. E Sandy, T1N, R6E, Sec. 27 (1 CRCM). **Polk Co.**—3.5 mi. S, 2 mi. W Gold Creek (1 OSUFW). **Tillamook Co.**—(1 UMMZ); Beaver (1 KU); Blaine (1 AMNH, 2 MVZ, 1 UMMZ); Cloverdale (1 UMMZ); 1.7 mi. NW Dolph (2 UMMZ); Hebo Lake (5 PSM); 2 mi. up Miami River (1 PSM); Nehalem (1 PSM); Netarts (1 OSUFW, 1 PSM, 25 SDNHM, 5 UMMZ); swamp near Netarts Bay (1 OSUFW); Tillamook (1 KU, 1 LACM, 1 PSM, 1 SDNHM, 2 UMMZ, 2 USNM); 2 mi. S Twin Rocks (1 PSM). **Wasco Co.**—Clear Lake (5 PSM). **Washington Co.**—Beaverton (5 FMNH); Blodgett Arboretum on HWY 6* (1 PSU); 5 mi. SE Hillsboro (1 USNM); 18.5 mi. W Portland, 1,300 ft. (4 MVZ); 21.5 mi. NW Portland (1 MVZ).

Microtus richardsoni, n = 249
(Plotted in Fig. 11-120, p. 329)

Baker Co.—Anthony (4 AMNH); Bourne (5 USNM); Cornucopia (1 UMMZ, 5 USNM). **Clackamas Co.**—0.5 mi. N, 2.75 mi. W Government Camp (1 PSU); Mt. Hood, Government Camp, 3,600 ft. (2 USNM); Marmot (1 USNM). **Deschutes Co.**—(1 PSM); 15 mi. W Bend, Tumalo Creek, 6,100 ft. (17 MVZ); head of Falls Creek, SE of South Sister, 6,450 ft. (2 CRCM); Sparks Lake (2 OSUFW, 2 UMMZ); 1 mi. N Three Creek Lake (8 PSM). **Douglas Co.**—Bradley Creek (3 SDNHM); Diamond Lake (1 PSM, 1 SDNHM); Louis Creek, T28S, R3W, SE¼SE¼ Sec. 30 (1 SOSC). **Grant Co.**—15 mi. NW Austin (2 USNM); 21 mi. SE Prairie City, N fork Malheur River, 5,000 ft. (7 MVZ); Strawberry Butte (2 USNM); Strawberry Mts. (3 USNM). **Hood River Co.**—Brooks Meadows guard station, T2S, R10E, Sec. 2 (1 PSU); Elk Cove, N side

Mt. Hood, 5,600 ft. (3 CRCM); W base Mt. Hood, Yokums (2 USNM); Mt. Hood, near timberline (7 USNM). **Jackson Co.**—Spring Creek, T33S, R4E, NE¼ Sec. 22 (1 SOSC). **Klamath Co.**—Anna Creek, Mt. Mazama, 6,000 ft. (12 USNM); Castle Crest Gardens near Crater Lake National Park headquarters (2 OSUFW); Crater Lake Park (2 CM); Crater Lake (1 MVZ, 21 USNM); S side Crater Lake (2 UMMZ); 0.5 mi. S Crater Lake (3 UMMZ); 1 mi. S Crater Lake (1 MVZ); 0.5 mi. N Government Camp, Crater Lake National Park (3 MVZ); Munson Valley, Crater Lake (1 USNM); Odell Creek, 4,700 ft. (1 LACM); Red Blanket Cabin (2 MVZ); Upper Munson Meadows, Crater Lake National Park, ca. 6,300 ft. (1 KU, 2 CRCM). **Lane Co.**—2 mi. W Blue River, Elk Creek, T16S, R4E, Sec. 19 (1 OSUFW); 6 mi. W Blue River, Hagen Creek, T16S, R3E, Sec. 22 (2 OSUFW); 5.5 mi. N, 5 mi. E Blue River, T15S, R5E, Sec. 28 (1 HSU); 10 mi. NE Blue River, Upper Lookout Creek Tributary, T15S, R5E, Sec. 25 (6 OSUFW); Gold Lake, 4,800 ft. (2 LACM); H. J. Andrews Experimental Forest, T15S, R5E, Sec. 32 (2 PSM); McKenzie Bridge (1 USNM); 6.6 km N, 0.3 km W McKenzie Bridge (2 USNM); 6 km N, 7.5 km E McKenzie Bridge (1 USNM); 1.6 km S, 5.2 km W McKenzie Bridge (1 USNM); 4.4 km S, 14.3 km E McKenzie Bridge (1 USNM); 6 mi. E McKenzie Bridge (1 OSUFW); Olallie Creek Camp, McKenzie River (1 PSM); 13 mi. S, 1 mi. E Rainbow, Dutch Oven, T18S, R5E, Sec. 22 (1 OSUFW); N base Three Sisters, 5,000 ft. (2 USNM); 23 mi. SE Vida (5 PSM). **Linn Co.**—1.5 mi. N Big Lake (3 PSM); 7 mi. N, 8 mi. E Blue River, T15S, R5E, Sec. 15 (1 CRCM); W base Mt. Jefferson, Permilia Lake (2 USNM); 13 mi. E Upper Soda (2 OSUFW); 20.5 mi. E Upper Soda, Lost Lake Creek, T13S, R7½E, Sec. 21 (1 OSUFW); 2 mi. N, 18 mi. E Upper Soda, stream off Lost Lake on HWY 20 (2 OSUFW). **Marion Co.**—Detroit (1 USNM). **Umatilla Co.**—Langdon Lake (1 FMNH). **Union Co.**—9.9 mi. N, 5.5 mi. W Elgin (1 PSU); Grande Ronde Lake (2 MVZ); Moffet Camp (1 PSM). **Wallowa Co.**—Aneroid Lake (1 LACM, 3 SDNHM); Anthony Lake (9 PSM); Eagle Cap Wilderness Area* (1 SOSC); SE side Jewett Lake, 8,350 ft. (1 CRCM); E side John Henry Lake, Wallowa Mts., 7,200 ft. (1 CRCM); S of Joseph, Wallowa Mts., 8,000 ft. (1 USNM); S of Joseph, Wallowa Mts., 8,500 ft. (1 USNM); Lick Creek Ranger Station (1 UMMZ); 16 mi. S Lostine, T3S, R43E, Sec. 25 (2 OSUFW); 1 mi. S, 3 mi. E Lostine, T3S, R43E, Sec. 36 (1 OSUFW); 16 mi. S, 2 mi. E Lostine, T3S, R43E, Sec. 25 (1 OSUFW); 16 mi. S, 3 mi. E Lostine, T3S, R43E, Sec. 36, 5,500 ft. (7 MVZ, 2 OSUFW); 19 mi. S, 4 mi. E Lostine, T4S, R44E, Sec. 17 (1 OSUFW); 21 mi. S, 5 mi. E Lostine, T4S, R44E, Sec. 27 (2 OSUFW); 21 mi. S, 6 mi. E Lostine, T4S, R44E, Sec. 26 (2 OSUFW); 17 mi. SE Lostine, Lostine River, Wallowa Mts. (1 OSUFW); Minam Lake (1 USNM); Stanley Ranger Station, Wallowa National Forest* (2 SDNHM); Wallowa Lake, 6,500 ft. (12 USNM); S of Wallowa Lake, Blue Mts., 8,100–8,300 ft. (9 USNM); 8 mi. S Wallowa Lake, Blue Mts., ca. 10,000 ft. (1 USNM).

Microtus townsendii, n = 560
(Plotted in Fig. 11-123, p. 331)

Benton Co.—(3 OSUFW, 6 PSM); Adair Village (1 KU); 3 mi. S, 2 mi. E Alsea, N Fork Alsea River, T14S, R7W, Sec. 20 (1 OSUFW); 3 mi. S, 4 mi. E Alsea, T14S, R7W, Sec. 22 (1 OSUFW); 3.25 mi. S, 4.25 mi. E Alsea (2 CM); 3.5 mi. SSW Blodgett (1 PSM); Corvallis (20 OSUFW, 5 PSM); Corvallis, T10S, R4W, Sec. 7 (1 OSUFW); 1 mi. N Corvallis (1 OSUFW); 2 mi. N Corvallis (8 PSM); 2 mi. N Corvallis, T11S, R5W, Sec. 21 (1 OSUFW); 4.5 mi. N Corvallis (1 OSUFW); 6 mi. N Corvallis, T11S, R5W, Sec. 11 (1 OSUFW); 7 mi. N Corvallis (1 OSUFW); 8 mi. N Corvallis (1 OSUFW, 5 PSM); 9 mi. N Corvallis, E. E. Wildlife Area (2 OSUFW); 2 mi. S Corvallis (1 PSM); 3 mi. S Corvallis, near airport, T11S, R5W, Sec. 22 (2 OSUFW, 1 PSM); 0.5 mi. W Corvallis (2 PSM); 1 mi. W Corvallis (2 OSUFW); 3 mi. W Corvallis (1 OSUFW); 0.1 mi. E Corvallis (1 OSUFW); 1 mi. N, 1 mi. W Corvallis (2 OSUFW); 1 mi. N, 2 mi. W Corvallis (1 OSUFW); 5 mi. N, 5 mi. E Corvallis, 200 ft. (2 CM); 7 mi. N, 3 mi. E Corvallis, E. E. Wilson Game Management Area (1 OSUFW); 1 mi. S, 1 mi. W Corvallis (1 OSUFW); 10.5 mi. S, 1 mi. W Corvallis, Finley National Wildlife Refuge, T13S, R5W, Sec. 29 (1 HSU, 13 OSUFW); 6 mi. NE Corvallis (1 OSUFW); 2 mi. SSW Corvallis (1 PSM); W.L. Finley National Wildlife Refuge (1

HSU); 2 mi. S, 1 mi. W Greenberry (1 OSUFW); 4 mi. S, 2 mi. W Greenberry (1 OSUFW); 7 mi. N, 2 mi. W Monroe, Finley National Wildlife Refuge (1 OSUFW); Philomath (3 OSUFW); 0.5 mi. E Philomath (1 OSUFW); 2 mi. E Philomath on HWY 20 (1 OSUFW); 1 mi. N, 7 mi. W Philomath, Old Peak Rd., Marys Peak, T12S, R7W, Sec. 11 (3 OSUFW); 6 mi. SW Philomath (1 OSUFW); Prairie Mt., Bull Prairie (1 PSM); 0.25 mi. SE Summit (1 PSM); Wells (1 USNM); near Wrens (1 PSM); T10S, R4W, Sec. 8 (1 OSUFW); T11S, R5W, Sec. 33 (2 KU); T12S, R5W, Sec. 4 (1 OSUFW); T12S, R5W, Sec. 7 (1 OSUFW). **Clackamas Co.**—(1 PSU); Clackamas (1 UW[EB]); Jennings Lodge (2 PSM); Milwaukie (4 CM); 8 mi. SE Molalla (3 PSM); Mulino (1 OSUFW); Oregon City (1 USNM, 7 UW[EB]); 1 mi. N Oregon City (1 MVZ); 2 mi. S, 2 mi. E Oregon City (1 OSUFW); 2 mi. N, 3 mi. W Wilsonville (2 OSUFW). **Clatsop Co.**—site of Old Ft. Clatsop (7 MVZ); 1 mi. N site of Old Ft. Clatsop (3 MVZ); Puget Sound* (1 UMMZ). **Columbia Co.**—(1 PSM); 1 mi. S Goble (1 OSUFW); 7 mi. SE Rainier (14 MVZ). **Coos Co.**—Bandon (1 USNM); 2 mi. N Bandon (3 PSM); 2 mi. S Bandon (1 PSM); 4 mi. S Bandon (6 PSM); 7.5 mi. S Bandon, T30S, T15W, Sec. 3 (1 OSUFW); 10.5 mi. S Bandon, T30S, R15W, Sec. 10 (2 OSUFW); 1.75 mi. E Bandon, T28S, R14W, SE¼ Sec. 2 (1 SOSC); 1.75 mi. SE Bandon (9 PSM); 4 mi. SE Bandon (4 PSM); Cape Arago State Park (7 LACM); Charleston (1 LACM); 3 mi. S Charleston, Sunset Bay (17 LACM); Coos Bay, South Bay marsh (1 LACM); Coos Bay, South Slough (3 LACM); 9 mi. NE Coquille (1 PSM); Delmar, 9 mi. S Marshfield (3 MVZ); Empire (16 USNM); Marshfield (7 FMNH); 2.5 mi. N Myrtle Point (1 OSUFW); 3.2 km N, 0.3 km W Remote, T29S, R10W, Sec. 21 (1 USNM); 1 mi. W Twomile (1 PSM); T29S, R14W, Sec. 9 (1 PSM). **Douglas Co.**—Charlotte Creek, Lower Umpqua River (1 PSM); Curtin (1 OSUFW); Drew (1 USNM); Gardiner (1 FMNH, 1 MVZ); Oxbow Burn, Bum Creek (3 PSM); Oxbow Burn, Sisters Creek (1 PSM); Scottsburg (1 FMNH, 1 USNM); Sutherlin (13 OSUFW); 11.5 mi. N, 5 mi. E Tiller, T28S, R1W, Sec. 26 (1 CRCM); 12.5 mi. N, 5.5 mi. E Tiller, T28S, R1W, Sec. 31 (1 CRCM). **Hood River Co.**—T3N, R9E, NW¼ Sec. 35 (1 PSM); T3N, R9E, NE¼ Sec. 34 (3 PSM). **Jackson Co.**—Mosquite Ranger Station* (1 USNM); Prospect (1 MVZ, 1 USNM); 8 mi. SW Prospect (1 SOSC); halfway between Drew and Crater Lake (3 USNM). **Josephine Co.**—6 mi. N, 10 mi. W Selma (1 SOSC). **Lane Co.**—6.8 km N, 7.6 km E Blue River (1 USNM); Cushman, N bank Siuslaw River (2 CM); Eugene (1 USNM, 2 UW-X); W side Eugene along Willow Creek, T18S, R4W, NE¼ Sec. 4 (1 OSUFW); W side Eugene along Willow Creek, T18S, R4W, NE¼ Sec. 9 (2 OSUFW); Florence (2 FMNH); 5 mi. S Florence (1 LACM); 2 mi. SW Florence (5 PSM); 9 mi. N, 1 mi. W Gardiner (4 PSM); 10 mi. W Junction City (8 PSM); 1 mi. N, 4 mi. W Marcola (1 OSUFW); 3.1 km N, 3 km W McKenzie Bridge, T16S, R5E, Sec. 3 (2 USNM); 6.4 km N, 0.6 km E McKenzie Bridge, T15S, R5E, Sec. 25 (1 USNM); 2.6 km S, 1.4 km E McKenzie Bridge, T16S, R5E, Sec. 25 (1 USNM); Mercer (1 UMMZ); Mercer, Dowell Ranch (1 UMMZ); Smith River SW of Eugene (1 PSM). **Lincoln Co.**—0.5 mi. W Agate Beach (2 UMMZ); Cascade Head Experimental Forest (7 PSM); 15 mi. W Harlan, Peterson Ridge, T12S, R11W, Sec. 12 (1 KU); Newport (1 SDNHM, 6 USNM); 6.5 mi. S Toledo, Beaver Creek, T12S, R10W, SE¼ Sec. 16 (2 KU); Yaquina Bay (5 USNM); Yaquina Head, Agate Beach (3 UMMZ). **Linn Co.**—Albany (2 OSUFW, 3 USNM); 5 mi. S, 1.75 mi. W Albany, T11S, R4W, SW¼ Sec. 35 (2 MUSMZ, 6 OSUFW); 5 mi. S, 3 mi. W Albany (1 OSUFW); 3 mi. S, 4 mi. E Corvallis near railroad tracks (1 OSUFW); E of Corvallis water works, T12S, R4W, Sec. 5 (1 OSUFW); 4 mi. S, 1.5 mi. W Halsey (1 OSUFW); 3.5 mi. S, 7 mi. W Shedd (1 OSUFW); Shelburn (1 USNM); 1 mi. E Sweet Home (1 OSUFW); 1 mi. S, 1 mi. E Sweet Home (1 OSUFW); 1 mi. N Tangent (1 OSUFW); T12S, R4W, Sec. 5 (1 KU, 1 OSUFW). **Marion Co.**—Aumsville (2 USNM); 1 mi. S, 2.5 mi. W Pratum (1 OSUFW); Salem (1 MVZ, 1 OSUFW, 1 USNM); 6 mi. S Salem (1 PSM); Silverton (2 SOSC); 5 mi. N Talbot (1 OSUFW); 1 mi. N, 2 mi. E Talbot (2 OSUFW); Woodburn (1 PSM). **Multnomah Co.**—Latourelle (2 OSUFW); near Linnton (1 AMNH); Government Island (1 PSM); Portland (6 AMNH, 2 KU, 4 LACM, 1 MVZ, 4 PSM, 16 SDNHM, 12 USNM); Sauvies Island (1 CM, 27 PSM); T1N, R1E, SW¼ Sec. 21 (2 PSM). **Polk Co.**—McCoy (4 CM, 4 FMNH). **Tillamook Co.**—1 mi. N, 1 mi. W Barview (1 OSUFW); 1 mi. N Bay City (3 AMNH); Beaver (2 OSUFW, 1 UMMZ); Blaine (3 MVZ, 1 OSUFW, 3 UMMZ); 5 mi. S, 1 mi. W Dayton (1 OSUFW); 2 mi. S Hebo (1 PSM); 2.66 mi. S Hebo (1 PSM); Hebo Lake Forest camp (1 LACM); 2 mi. up Miami River (4 PSM); Nehalem (1 UW-X); 1 mi. W Nehalem (1 OSUFW); Netarts (2 AMNH, 1 LACM, 4 SDNHM, 1 USNM); Netarts divide (3 PSM); 4 mi. N Netarts (1 OSUFW); Oceanside (2 LACM); Pacific City (1 PSM); Tillamook (2 AMNH, 4 KU, 2 LACM, 3 MVZ, 5 OSUFW, 1 PSM, 1 SDNHM, 1 UMMZ); 1 mi. S Tillamook (1 OSUFW); 6.5 mi. SW Tillamook, 10 ft. (2 AMNH); 9 mi. SE Tillamook (1 OSUFW). **Washington Co.**—Forest Grove (1 AMNH); 18.5 mi. NW Portland (4 MVZ); 21.5 mi. NW Portland (1 MVZ). **Yamhill Co.**—2 mi. S, 1.5 mi. W Amity (1 UW); 4 mi. E Carlton, 170 ft. (4 AMNH); Dayton (1 PSM); 2 mi. S, 1.5 mi. E Dayton (1 OSUFW); 3 mi. NW McMinnville (1 UW[EB]); Sunset Tunnel, HWY 26 (1 UW).

Mirounga angustirostris, $n = 4$
(Plotted in Fig. 12-24, p. 392)

Clatsop Co.—Fort Stevens State Park (1 PSM). **Lincoln Co.**—Depoe Bay (1 PSM); 0.5 mi. S Newport jetty (1 PSM); Seal Rock (1 OSUFW).

Mus musculus, $n = 183$
(Plotted in Fig. 11-93, p. 294)

Baker Co.—20 mi. NE Baker [City] (1 KU); Burkemont (1 AMNH). **Benton Co.**—Corvallis (25 OSUFW); 4 mi. N Corvallis (1 OSUFW); 0.5 mi. S Corvallis (1 OSUFW); 10 mi. S Corvallis (1 OSUFW); 1 mi. W Corvallis (4 OSUFW); 2 mi. W Corvallis (3 OSUFW); 5 mi. W Corvallis (1 OSUFW); 1 mi. N, 2 mi. W Corvallis (2 OSUFW); 1 mi. N, 3 mi. W Corvallis (2 OSUFW); 2 mi. N, 4 mi. W Corvallis (1 OSUFW); 6 mi. N, 1 mi. W Corvallis (1 OSUFW); 3 mi. N, 2 mi. E Corvallis (1 OSUFW); 5 mi. N, 5 mi. E Corvallis, 200 ft. (3 CM); 7 mi. N, 4 mi. W Corvallis (1 OSUFW); 0.5 mi. S, 0.5 mi. W Corvallis (1 OSUFW); 1 mi. S, 1 mi. W Corvallis (2 OSUFW); 3 mi. S, 2 mi. W Corvallis (1 OSUFW); 2 mi. SWW Corvallis (1 PSM). **Clackamas Co.**—3 mi. W Estacada, Springwter Rd. (2 OSUFW). **Coos Co.**—0.75 mi. S Bandon (1 PSM); 1.75 mi. E Bandon (1 PSM); Coquille (2 USNM). **Curry Co.**—Gold Beach (1 USNM). **Douglas Co.**—Elkhead (1 USNM); 6 mi. N Roseburg (1 SOSC). **Gilliam Co.**—2.5 mi. S Heppner Junction (1 KU). **Grant Co.**—5.3 mi. N, 1.3 mi. E Long Creek, Morgrass-Pass Creek (1 OSUFW); 5 mi. W Mt. Vernon, Schauffer Ranch (3 MVZ); 21 mi. SE Prairie City, North Fork Malheur River, 5,000 ft. (1 MVZ). **Harney Co.**—4 mi. E Burns (1 MVZ); 26 mi. S, 6 mi. E Burns, Malheur Environemental Field Station (1 OSUFW); Malheur Lake (1 SDNHM). **Jackson Co.**—Ashland (7 SOSC); 1 mi. E Ashland (1 PSM); 2 mi. E Ashland (1 PSM); 0.25 mi. S junction HWY I-5 and HWY 66 in Ashland (1 SOSC); Gold Hill (1 SOSC); Medford (1 SOSC, 9 UW[WWC]); Senic Hills Cenetary Rd. near Bear Creek SE of Ashland (1 SOSC); Talent (1 SOSC); 3 mi. S Talent (1 SOSC). **Klamath Co.**—Ft. Klamath (7 USNM); 7.5 mi. S Klamath Falls (4 SOSC); 2 mi. E Klamath Falls at Embree Place (1 USNM); Klamath Branch Experiment Station (1 OSUFW); Muck Experiment Station (1 OSUFW); 0.5 mi. S Township Rd. near Ady Canal, Worden (2 SOSC); Worden (1 OSUFW). **Linn Co.**—Albany (1 OSUFW); 4.5 mi. S, 4 mi. E Albany (1 OSUFW); 3.5 mi. S, 4 mi. E Corvallis (1 OSUFW); 2 mi. N, 1 mi. E Scio (3 OSUFW); T11S, R5W, Sec. 36 (1 SOSC). **Malheur Co.**—1 mi. S Jamison (1 EOSC); Ontario (2 USNM); 5 mi. SW Ontario, Wood Ranch, 2,000 ft. (3 USNM); 2 mi. NW Riverside (1 USNM); near Vale (1 AMNH). **Marion Co.**—1 mi. N, 3 mi. E Salem (2 OSUFW). **Morrow Co.**—T5N, R26E, NW¼ Sec. 22 (1 PSM). **Multnomah Co.**—Portland (1 AMNH, 3 PSM, 1 PSU, 2 SDNHM, 18 USNM). **Polk Co.**—0.5 mi. S, 3 mi. E Rickreall (1 OSUFW); 1 mi. S, 3 mi. E Rickreall (2 OSUFW). **Sherman Co.**—mouth of Deschutes River, Columbia River (1 MVZ); Miller, mouth of Deschutes River (1 OSUFW). **Umatilla Co.**—0.5 mi. NE Freewater (1 UW[WWC]). **Union Co.**—(2 EOSC); Cove (1 EOSC); La Grande (6 EOSC); 10 mi. N La Grande (1 EOSC). **Wasco Co.**—mouth of Deschutes River, Columbia River (2 MVZ). **Washington Co.**—18.5 mi. NW Portland (8 MVZ); Tigard (1 PSM).

Mustela erminea, n = 171
(Plotted in Fig. 12-41, p. 416)

Benton Co.—0.5 mi. N Adair Village (1 KU); 1.2 km S, 4.2 km E Alsea, T14S, R7W, Sec. 9 (1 USNM); 3 mi. W Alsea Falls, T14S, R7W, Sec. 35 (1 OSUFW); 3 mi. N Corvallis, T11S, R5W, Sec. 10 (1 OSUFW); 7 mi. N Corvallis (1 OSUFW); 10 mi. S Corvallis (3 OSUFW); 2.5 mi. N, 1.5 mi. W Corvallis (2 OSUFW); 3 mi. N, 4 mi. W Corvallis (1 OSUFW); 10 mi. N, 2 mi. E Corvallis (1 OSUFW); Lewisburg (1 OSUFW); near top Marys Peak (1 OSUFW); McDonald Forest (4 OSUFW, 1 PSM); 4 mi. W Philomath, T12S, R6W, Sec. 17 (1 OSUFW); 9 mi. SW Philomath, Marys Peak (1 PSM). **Clackamas Co.**—Oregon City (1 PSM); 3.8 mi. N, 8 mi. E Sandy (1 CRCM); 6 mi. S, 7 mi. E Sandy, T3S, R6E, Sec. 18 (1 CRCM); 9 mi. N, 4 mi. W Zigzag (2 CRCM). **Clatsop Co.**—Astoria (1 USNM); 0.5 mi. W site of Old Ft. Clatsop Junction, T8W, R10W, Sec. 35 (1 PSU). **Coos Co.**—4 mi. SE Bandon (6 PSM); Bullards Beach State Park (1 PSU); Remote (1 OSUFW); 3 km S, 10.8 km E Sitkum (1 USNM). **Crook Co.**—Walton Lake, Ochoco National Forest (1 SOSC). **Curry Co.**—10 mi. E Brookings (2 PSM); Gold Beach (2 FMNH); Port Orford (1 USNM). **Douglas Co.**—8 mi. S, 2.5 mi. W Alma (1 OSUFW); 20 mi. S Crow, Oxbow Burn (5 PSM); Gardiner (1 FMNH); 12 mi. N, 18 mi. E Reedsport (1 OSUFW); 2.4 km N, 24.6 km W Sutherlin, T25S, R8W, Sec. 10 (1 USNM); 3.8 km N, 12.1 km W Sutherlin, T25S, R7W, Sec. 1 (1 USNM). **Jackson Co.**—adjacent to Howard Prairie Dam on Hyatt Lake (1 SOSC). **Jefferson Co.**—20 mi. W Warm Springs, Mill Creek (1 USNM). **Josephine Co.**—north North Fork Deer Creek (1 SOSC). **Klamath Co.**—Ft. Klamath (1 USNM); Klamath Falls (1 USNM); 8 mi. NW Klamath Falls (1 OSUFW). **Lake Co.**—Gearhart Mt. (1 PSM). **Lane Co.**—4 mi. N Blue River (1 OSUFW); 3.5 mi. S, 8 mi. W Blue River, T17S, R3E, Sec. 7 (2 HSU); 5.5 mi. N, 5 mi. E Blue River, T15S, R5E, Sec. 28 (1 HSU); 5.5 mi. N, 7 mi. E Blue River, T15S, R5E, Sec. 26 (2 HSU); 7 mi. S Cottage Grove (1 PSM); 14 mi. W Eugene (1 OSUFW); ca. 40 mi. SW Eugene (4 PSM); 9 mi. N, 1 mi. W Gardiner (1 PSM); H. J. Andrews Experimental Forest, T15S, R5E, Sec. 32 (2 PSM); H. J. Andrews Experimental Forest, T15S, R5E, Sec. 33 (1 PSM); 0.4 km N, 18.5 km W Lorane, T20S, R7W, Sec. 12 (1 USNM); 0.8 km N, 19.2 km W Lorane, T20S, R7W, Sec. 12 (2 USNM); 2 km N, 6.4 km W Lorane, T20S, R5W, Sec. 5 (1 USNM); 2.8 km N, 6.2 km W Lorane, T20S, R5W, Sec. 5 (2 USNM); McKenzie Bridge (1 UO); 1.6 km S, 5.2 km W McKenzie Bridge, T16S, R5E, Sec. 20 (1 USNM); Mercer (1 SDNHM, 1 UO); 1 mi. N Mohawk River, T15S, R1W, SE¼NE¼ Sec. 13 (1 SOSC); Oakridge (1 OSUFW); 12 mi. N, 4 mi. W Rainrock (1 OSUFW); Vida (2 UO); 23 mi. S Vida (8 PSM). **Lincoln Co.**—(1 PSM); 5 mi. S Burnt Woods (5 PSM); 2.5 mi. N, 0.5 mi. E Eddyville, T10S, R9W, Sec. 33 (1 OSUFW); 1.5 mi. NW Eddyville (1 PSM); 15 mi. W Harlan, Peterson Ridge, T12S, R11W, Sec. 12 (1 KU); Lincoln City, T10S, R9W (1 UNSW); 2 mi. N Lincoln City (1 OSUFW); Newport (1 USNM). **Linn Co.**—9 mi. N, 8 mi. E Blue River (1 CRCM); 6 mi. N, 5 mi. E Lacomb (1 OSUFW); 12 mi. N, 2 mi. W Eugene (1 OSUFW); 13.2 km N, 0.8 km E McKenzie Bridge, T15S, R5E, Sec. 1 (1 USNM); Scio (1 PSM). **Malheur Co.**—Cord, Barren Valley (1 USNM). **Marion Co.**—Salem (1 OSUFW); Silver Creek Falls State Park, T8S, R1E, Sec. 19 (1 PSU). **Multnomah Co.**—9 mi. N, 10 mi. E Sandy (2 CRCM); 10 mi. N, 11 mi. E Sandy, T1N, R6E, Sec. 27 (1 CRCM); 11 mi. N Zigzag (1 CRCM). **Tillamook Co.**—Beaver (1 OSUFW, 1 UMMZ); head of Ben Smith Creek Rd., halfway between Forest Grove and Tillamook (4 OSUFW); Blaine (1 KU, 2 OSUFW, 2 UMMZ); Cascade Head (3 USNM); 2.66 mi. S Hebo, Mt. Hebo (1 PSM); 2 mi. SE Hebo P.O. (1 OSUFW); Nehalem (1 OSUFW); 5 mi. ± Nehalem at Magnolia Junction* (1 PSM); E side Netarts Bay (1 PSU); Tillamook (3 KU, 1 LACM, 1 MVZ, 7 OSUFW, 3 PSM, 1 USNM); between Dolph and HWY 101 (1 UMMZ); T1N, R7W, SW¼ Sec. 18 (1 OSUFW). **Umatilla Co.**—0.5 mi. S, 1 mi. E Tollgate, T3N, R38E, Sec. 4 (1 PSU); Umatilla (1 USNM). **Wallowa Co.**—3 mi. SW Enterprise (1 OSUFW). **Washington Co.**—Beaverton (1 FMNH); Blodgett Arboretum near HWY 6 (1 PSU); Cedar Mill (1 PSU); 4 mi. SE Gaston (1 OSUFW); Sylvan (1 PSU).

Mustela frenata, n = 250
(Plotted in Fig. 12-43, p. 419)

Baker Co.—(1 EOSC); Anthony (3 AMNH); Baker (1 SDNHM); 5 mi. E Baker City (1 EOSC); Bourne (2 USNM); Haines (1 USNM). **Benton Co.**—1 mi. W Alsea (1 OSUFW); Corvallis (2 OSUFW); 4 mi. S Corvallis (1 OSUFW); Soap Creek, T12S, R5E, Sec. 3 (1 OSUFW). **Clackamas Co.**—4 mi. W Eagle Creek (1 OSUFW); Jennings Lodge (1 UMMZ); Milwalkie (3 PSM); 8 mi. SE Molalla (1 PSM); Mulino (1 OSUFW, 1 PSM); Oswego (1 SDNHM). **Clatsop Co.**—1 mi. W Hamlet (1 PSM); 15 mi. NE Nehalem (2 PSM); site of Old Ft. Clatsop (1 MVZ); T8N, R10W, NW¼ Sec. 6 (1 PSM). **Columbia Co.**—Deer Island (5 PSM). **Coos Co.**—1 mi. SE Bandon (1 PSM); 2.5 mi. SSE Bandon (1 PSM). **Crook Co.**—Prineville (4 USNM). **Curry Co.**—Gold Beach (3 FMNH); Langlois (1 SDNHM). **Deschutes Co.**—(1 EOSC); Bend (1 USNM); 18 mi. S, 12 mi. W Bend (1 OSUFW); head of Davis Creek (1 PSM); mouth of Davis Creek (1 PSM); Davis Mt. (1 SDNHM); Paulina Lake (1 SDNHM); Sisters (2 USNM). **Douglas Co.**—Anchor (1 USNM); Loon Lake (1 PSM); 1.25 mi. SE Scottsburg (1 PSM); Tiller (1 SOSC); Tyler* (1 SOSC). **Gilliam Co.**—Condon (1 SDNHM); Eightmile Creek, fork of Thirtymile Creek* (1 USNM); Tenmile Creek (1 USNM); Willows (1 USNM). **Grant Co.**—12 mi. S, 5 mi. W Canyon City (1 PSU); John Day (1 EOSC); Long Creek (1 SDNHM); 10 mi. S, 11 mi. W Mt. Vernon, Aldrich Mts., Murders Creek, T15S, R28E, SW¼ Sec. 12 (1 KU); Strawberry Mts. (2 USNM). **Harney Co.**—(1 OSUFW); Buchanan (1 USNM); Burns (1 SDNHM, 1 UO, 1 USNM); 29 mi. S, 6 mi. E Burns, Malheur National Wildlife Refuge (1 OSUFW); 25 mi. NW Burns (1 USNM); 1 mi. S, 4 mi. E Crane (1 OSUFW); Malheur Lake (1 SDNHM); Sageview, Poker Jim Ridge, Rock Creek (1 USNM); Shirk P.O. (1 USNM); Steens Mt. Rd., 8,600 ft.* (1 SOSC); Steens Mt., Kiger Gorge, 6,000 ft. (2 USNM); Voltage (1 USNM). **Hood River Co.**—1 mi. E Cascade Locks on HWY 80N (1 PSM). **Jackson Co.**—Ashland (1 SOSC); Denman Wildlife Refuge (1 SOSC); 43 mi. NE Grants Pass (1 USNM); Medford (2 USNM); 6 mi. S Medford (1 MVZ); W of Medford, old stage rd.-orchard (1 SOSC); Mt. Pitt (1 USNM); Siskiyou (3 USNM). **Jefferson Co.**—Hay Creek (1 SDNHM). **Josephine Co.**—13 mi. SW Grants Pass (1 UO). **Klamath Co.**—Anna Creek, Mt. Mazama, 6,000 ft. (2 USNM); Crater Lake National Forest* (1 USNM); Crater Lake National Park (1 MVZ); Ft. Klamath (4 USNM); 8 mi. S Klamath Falls, 660 yds E Spring Lake Rd. (1 SOSC); 6 mi. S, 2 mi. W Klamath Falls (1 OSUFW); Wood River Mts. (2 USNM); mountain W of Ft. Klamath, 5,500 ft. (3 USNM). **Lake Co.**—Fort Rock (1 OSUFW, 1 SDNHM); Gearhart Mt. (4 PSM); 30 mi. SW Lakeview, Dog Lake Ranger Station (1 USNM); 3 mi. W Stauffer (1 USNM). **Lane Co.**—near Ada (1 OSUFW); Belle (2 USNM); 1.25 mi. N, 5 mi. W Blue River, T16S, R3E, Sec. 15 (1 HSU); Cow Creek Lake* (1 USNM); Eugene (1 SDNHM, 1 UO); 10 mi. N Florence (1 OSUFW); H. J. Andrews Experimental Forest, T16S, R4E, Sec. 24 (1 PSM); 5 mi. S Junction City (1 OSUFW); McKenzie Bridge (1 UO); Mercer (1 SDNHM); Oakridge (1 OSUFW, 1 PSM); Reed (2 USNM); Vida (1 UO). **Lincoln Co.**—3.5 mi. S, 2 mi. E Elk City, T11S, R9W, Sec. 36 (1 KU); 3.5 mi. E Newport (1 OSUFW). **Linn Co.**—7 mi. E Cascadia (1 OSUFW); Permilia Lake, W base Mt. Jefferson (1 USNM); 4 mi. S, 8 mi. E Upper Soda (1 OSUFW). **Malheur Co.**—(1 EOSC); Cedar Mts. (2 USNM); Ironside, 4,000 ft. (3 AMNH); Jordan Valley (1 USNM); Riverside (1 USNM); 2 mi. NW Riverside (1 USNM); 1.5 mi. S Vale at Herrett Ranch (2 USNM); 3.5 mi. S Vale (4 PSM); 2 mi. N, 4 mi. E Vale (1 OSUFW). **Marion Co.**—Salem (2 USNM); 10 mi. S Salem (1 SOSC); Woodburn (1 PSM). **Morrow Co.**—Hardman (1 USNM). **Multnomah Co.**—Gresham (1 PSM); Portland (1 PSM, 1 PSU). **Polk Co.**—15 mi. N Corvallis (1 OSUFW). **Tillamook Co.**—Beaver (4 OSUFW); Blaine (1 KU, 2 OSUFW, 2 UMMZ); 2 mi. S Garibaldi (1 OSUFW); Netarts (1 SDNHM); Tillamook (2 AMNH, 4 KU, 1 LACM, 2 MVZ, 7 OSUFW, 1 PSM, 12 SDNHM). **Umatilla Co.**—near Tollgate (3 UW[WWC]); 0.5 mi. N, 2 mi. W Tollgate (1 UW[WWC]); 4 mi. NW Tollgate, 4,300 ft. (1 UW[WWC]); 10 mi. W Ukiah (1 OSUFW); 15 mi. E Ukiah, 4,000 ft. (1 USNM); Umatilla (1 USNM). **Union Co.**—(2 EOSC); Elgin (1 USNM); Glass Hill, T3S,

R38E, Sec. 32 (1 PSM); Hot Lake (1 EOSC); Island City (1 EOSC); La Grande (1 UW[WWC]); 20 mi. E Lehman (1 SDNHM); 7.5 mi. WNW Medical Springs (1 PSM); 0.5 mi. E Union, T4S, R40E, NW¼ Sec. 17 (2 PSM); 3.7 mi. N, 6.2 mi. W Union, T3S, R39E, Sec. 31 (1 PSU); HWY 82, near Gateway Auto* (1 EOSC). **Wallowa Co.**— Coverdale Rd., T3S, R46E, Sec. 20 (1 PSU); mouth of Crow Creek, T3N, R45E, SE¼ Sec. 26 (1 OSUFW); Enterprise (1 SDNHM); 4 mi. E Lostine (1 EOSC); 15 mi. N Paradise at Horse Creek, 2,000 ft. (1 USNM); Wallowa lake (1 USNM); south Wallowa Lake, Wallowa Mts. (1 USNM). **Wasco Co.**—Antelope (3 USNM); 7 mi. E Antelope (4 USNM); 4 mi. S The Dalles (1 MVZ); 7 mi. S The Dalles, near Endersby (1 OSUFW); Wapinitia (1 USNM). **Washington Co.**—Gales Creek (1 USNM). **Yamhill Co.**—3 mi. S Gaston (1 PSU).

Additional localities: Fish Lake* (1 USNM); Ochoco R.S., Ochoco National Forest on Coyle Creek* (1 SDNHM).

Mustela vison, n = 196
(Plotted in Fig. 12-45, p. 424)

Baker Co.—Anthony (3 AMNH). **Benton Co.**—Corvallis (2 OSUFW, 1 USNM); 3 mi. N Corvallis, Jackson Creek (1 OSUFW); 5 mi. N, 0.25 mi. W Corvallis (1 OSUFW); Willamette River near Corvallis (1 OSUFW). **Clackamas Co.**—near Molalla (1 PSM); Wemme (1 PSM). **Clatsop Co.**—Astoria (1 UMMZ); Big Creek (1 PSM); Burke Lake (1 PSM); Meacoxie Creek* (1 PSM); Necanicum River (1 PSM); Perkins Creek* (1 PSM); Youngs Bay (1 PSM). **Columbia Co.**— Anundie Island, Clatskanie River (1 PSM); Clatskanie River (1 PSM); Columbia River* (1 PSM). **Coos Co.**—1.75 mi. E Bandon (2 PSM); 1.75 mi. SE Bandon (1 PSM); 4 mi. SE Bandon (1 PSM). **Curry Co.**— Gold Beach (3 FMNH); Port Orford (2 USNM). **Deschutes Co.**—(4 OSUFW); Bend (1 OSUFW); Davis Creek (1 OSUFW); La Pine (1 USNM); Paulina Lake (4 USNM). **Douglas Co.**—5 mi. N, 1 mi. W Gardiner (1 PSM); 9 mi. N, 1 mi. W Gardiner (1 PSM); Glendale (5 USNM); Riddle (1 OSUFW); 2 mi. S, 8.5 mi. E Toketee Falls, T27S, R5E, Sec. 6 (1 OSUFW). **Harney Co.**—Blitzen Valley, Malheur National Wildlife Refuge (1 UW); 30 mi. NW Burns (1 USNM); Diamond (1 USNM); Narrows (1 USNM); Voltage (1 USNM). **Jackson Co.**—43 mi. NE Grants Pass (1 USNM). **Jefferson Co.**—Mt. Jefferson, 44°40', 124°05' (2 USNM). **Josephine Co.**—Grants Pass, Rogue River (3 OSUFW); 13 mi. SW Grants Pass (1 UO). **Klamath Co.**—(1 MVZ); 1 mi. W Chiloquin (1 PSM); Ft. Klamath (1 AMNH, 1 OSUFW); Klamath Falls (2 OSUFW); Klamath River* (25 PSM); Pelican Bay, Klamath Lake (1 MVZ); Sprague River (1 PSM); on Sprague River near town of Sprague River (7 PSM); Upper Klamath Lake (3 PSM). **Lake Co.**—Gearhart Mts. (1 PSM). **Lane Co.**—Belknap Springs, McKenzie River (1 OSUFW); Belle (15 USNM); McKenzie Bridge (2 UO); 3 mi. S Mercer, Clear Lake (1 KU); Mercer Lake (2 OSUFW); Oakridge (4 OSUFW); 40 mi. W Oakridge (1 USNM); Reed (3 USNM). **Lincoln Co.**—NW of Burnt Woods, T11S, R9W (4 KU); 1 mi. S, 1 mi. W Burnt Woods (1 OSUFW); Cascade Head Experimental Forest (1 PSM); Newport (2 USNM). **Linn Co.**—Halsey (1 SOSC); 1 mi. W Lebanon (1 OSUFW); 0.5 mi. E Lost Lake (1 PSM); 3 mi. N, 8 mi. W Tangent (1 OSUFW). **Malheur Co.**—2 mi. NW Riverside (2 USNM). **Marion Co.**—Salem (1 OSUFW). **Multnomah Co.**—Portland (1 SDNHM, 1 USNM). **Tillamook Co.**—Beaver (1 OSUFW); Blaine (1 SDNHM); Nehalem (1 OSUFW, 1 SDNHM); Netarts (1 OSUFW, 3 SDNHM); Netarts Bay (1 OSUFW); Tillamook (1 MVZ, 3 OSUFW, 2 SDNHM). **Umatilla Co.**—Pendleton (1 SDNHM). **Union Co.**—Alicel, Grande Ronde River (1 EOSC); 3 mi. S, 1 mi. E Imbler (1 OSUFW); La Grande (1 EOSC); road between La Grande and Hot Lake (1 EOSC). **Wallowa Co.**—(1 EOSC); Aneroid Lake (5 UO); 3 mi. SW Enterprise, T2S, R44E, Sec. 9 (1 PSM); Lostine River* (3 PSM); Sheep Creek, 10 mi. E Joseph (1 UW[WWC]); Wallowa Lake (1 USNM); Wallowa River* (5 PSM). **Wasco Co.**—6 mi. W Dufur (1 SDNHM); near Maupin (1 MVZ). **Washington Co.**—(1 SDNHM); Beaverton (1 USNM); Tualatin River (1 PSM). **Wheeler Co.**—18 mi. E Mitchell (1 SOSC).

Additional locality: Wallowa Island, Columbia River* (1 PSM).

Myocastor coypus, n = 15
(Plotted in Fig. 11-135, p. 350)

Benton Co.—3 mi. N Corvallis (1 OSUFW); 6 mi. N Corvallis (1 OSUFW); 1 mi. E Corvallis (1 OSUFW); 2 mi. S, 3 mi. W Corvallis (1 OSUFW); 7 mi. S, 1 mi. E Corvallis (1 OSUFW); Long Tom River near Willamette River (1 OSUFW); slough off Willamette River, Black Lake area, 8.8 mi. S, 1.5 mi. E Corvallis (1 OSUFW); near Wrens (1 OSUFW). **Lane Co.**—(1 PSM); 11 mi. W Junction City, T15S, R6W, Sec. 23, 500 ft. (1 PSM). **Lincoln Co.**—Old Slice Slough, Yaquina River, near Toledo (1 OSUFW). **Linn Co.**—3 mi. NW Lebanon (1 UW); 2 mi. S, 4 mi. W Tangent (1 OSUFW). **Marion Co.**—3 mi. W Woodburn (1 OSUFW). **Multnomah Co.**—0.25 mi. N park entrance on Oxbow Rd. (1 SOSC).

Myotis californicus, n = 235
(Plotted in Fig. 9-4, p. 80)

Baker Co.—Anthony (1 USNM). **Benton Co.**—1 mi. N, 1 mi. E Bellfountain (1 OSUFW); Corvallis (1 MVZ, 3 OSUFW); 6 mi. N, 4 mi. E Corvallis (1 OSUFW); Philomath (1 OSUFW); 5 mi. SW Philomath (2 USNM); W.L. Finley Wildlife Refuge, 11 mi. S Corvallis, T13S, R5W, SW¼ Sec. 20 (1 PSM). **Clackamas Co.**—Estacada (1 UMMZ); Gladstone (5 UW[LB]); Marmot (1 USNM). **Clatsop Co.**—Jewell, Fishhawk Creek (1 PSM); site of Old Ft. Clatsop (5 MVZ). **Columbia Co.**—(1 UW); Goble (1 MVZ); 7 mi. SE Rainier (1 MVZ). **Coos Co.**— 1 mi. S Bandon (3 PSM); 1.75 mi. E Bandon (1 PSM); 2.5 mi. SE Bandon (2 MVZ); 4 mi. SE Bandon (6 PSM); Charleston (1 LACM); Remote (1 SDNHM). **Crook Co.**—mouth of Bear Creek, Crooked River, 3,400 ft. (1 MVZ). **Curry Co.**—10 mi. E Brookings (4 PSM); Gold Beach (1 PSU); 20 mi. above Gold Beach (1 SDNHM); mouth of Powder River* (1 KU); 9 mi. SE Port Orford (1 MVZ). **Deschutes Co.**—14 mi. SE Bend (1 SDNHM); Sisters (3 USNM); 6 mi. NW Sisters (1 PSM). **Douglas Co.**—Reston (1 USNM); 6 mi. S, 1 mi. E Scottsburg (1 OSUFW); 4.6 km N, 12 km W Sutherlin (1 USNM). **Grant Co.**—ca. 6 mi. S Fox (2 PSM); Kingbolt Spring (1 PSU); T8S, R27E, Sec. 36 (1 PSU); T11S, R29E, SW¼ Sec. 35 (2 PSM); T11S, R30E, SE¼ Sec. 32 (1 PSM). **Harney Co.**—4 mi. N Princeton, Windy Point (1 OSUFW); 3.3 mi. S Princeton (2 PSM). **Hood River Co.**—N slope Mt. Hood, 2,800 ft. (2 USNM). **Jackson Co.**—Ashland (2 SOSC); 2.5 mi. S Ashland, T39S, R1E, Sec. 28, 3,650 ft. (3 PSM); 3 mi. S Ashland, T39S, R1E, Sec. 28 (2 SOSC); 3 mi. N, 10 mi. E Ashland, T38S, R3E, Sec. 30 (2 SOSC); Brownsboro (1 USNM); 6.7 mi. NE Butte Falls, T34S, R2E, NE¼ Sec. 26 (1 SOSC); Grizzly Mt. (1 SOSC); S slope Grizzly Peak (1 SOSC); 7 mi. S Medford, T38S, R1W, Sec. 35 (1 PSM); 5 mi. SW Medford (1 SOSC); 7 mi. SW Medford, T38S, R2W, SE¼ Sec. 33 (1 PSM); 2.5 mi. NE Office, T38S, R3W, SE¼ Sec. 13 (1 PSM); East Pilot Rock Spring, Mt. Ashland, T41S, R2E, center Sec. 3 (1 PSM); Prospect (1 MVZ); Shoat Springs (1 SOSC); Talent (1 SOSC); 1.1 mi. SE Thompson Creek Rd. 3 mi. S Applegate (1 SOSC). **Josephine Co.**— 6.125 mi. E Cave Junction (1 SOSC); 2 mi. N, 3 mi. W Kerby (2 SOSC); 4 mi. S, 9 mi. W O'Brien (1 SOSC); 0.5 mi. up Shorthorn Gulch from Jumpoff Joe Creek (1 SOSC); 3.5 mi. E HWY I-5 on Jump-off-Joe Creek Rd. (1 SOSC); 1 mi. N USFS Rd. 355 on USFS Rd. 3794, 2 mi. W HWY 199 (1 SOSC). **Klamath Co.**—Gerber (2 SOSC); 2 mi. S Hayden Mt., Klamath River, T40S, R6E, Sec. 10 (1 OSUFW); 2.3 mi. S Keno, Onion Springs (1 SOSC); Mosquito Ranger Station* (1 USNM). **Lake Co.**—Silver Lake (1 SDNHM); South Warner Lakes (1 SDNHM); Woodchopper Springs (1 SOSC). **Lane Co.**—Eugene (1 SDNHM, 1 USNM); 0.5 mi. S Florence (1 MVZ); 9 mi. N, 1 mi. W Gardiner (17 PSM); 9.5 mi. N, 1 mi. W Gardiner (15 PSM); Gettings Creek bridge at HWY I-5 crossing (1 PSU); McKenzie Bridge (2 USNM); T15S, R7W, Sec. 10 (3 PSM). **Lincoln Co.**—Cascade Head Experimental Forest (1 PSM); Delake (12 SDNHM, 1 UMMZ); Driftwood Beach State Park (1 PSU); 1 mi. W Otis, mouth Salmon River (1 OSUFW); 1 mi. NE Otis (5 PSM); Waldport (1 SDNHM). **Linn Co.**—Holly (1 SDNHM). **Malheur Co.**—Bogus Lake (1 PSM); Cottonwood Creek, T26S, R38E, NW¼ Sec. 29 (12 PSM). **Marion Co.**—Salem (1 CM); Woodburn (1 PSM). **Tillamook Co.**—Blaine (1 SDNHM); near Nehalem (1 PSM); Netarts (1 MVZ, 1 OSUFW, 1 SDNHM); Tillamook (1 AMNH, 3 KU, 1 LACM,

1 MVZ, 3 OSUFW, 1 UMMZ); 7 mi. SE Tillamook (1 OSUFW); Wheeler Hospital (1 PSM). **Union Co.**—Elgin (1 USNM); Lookingglass (1 USNM); Starkey Experimental Forest, T3S, R34E, NW¼ Sec. 2 (1 PSM). **Wallowa Co.**—Rockwall Springs, 11.5 mi. NNE Wallowa, T3N, R43E, SW¼NW¼ Sec. 14 (9 PSM); 9.5 mi. NNE Wallowa, T3N, R43E, center Sec. 26 (1 PSM); Wallowa Lake (1 USNM). **Wasco Co.**—head Little Badger Creek trail, Mt. Hood National Forest (1 PSU); Memaloose State Park, 2.5 mi. E Mosier, T2N, R12E, SE¼ Sec. 31 (5 PSM, 1 PSU); junction USFS Rd. 27 and Jordan Creek (1 PSU). **Washington Co.**—Beaverton (1 PSM); 4 mi. E Gaston (1 OSUFW); 18.5 mi. NW Portland (3 MVZ). **Yamhill Co.**—Dayton (3 PSM).

Myotis ciliolabrum, n = 45
(Plotted in Fig. 9-6, p. 83)

Baker Co.—Homestead, 1,800 ft. (1 USNM). **Crook Co.**—Paulina Cave* (1 OSUFW); Twelvemile Creek (2 USNM); T18S, R22E, Sec. 9 (1 PSU). **Deschutes Co.**—Sisters (1 USNM); Skeleton Cave (1 UW). **Grant Co.**—(1 EOSC). **Harney Co.**—Frenchglen (1 PSM); Mickey Hot Springs (2 SOSC); 3.3 mi. E Princeton (1 PSM); T41S, R30E, SW¼ Sec. 15 (1 SOSC). **Hood River Co.**—N slope Mt. Hood (1 USNM). **Jefferson Co.**—0.5 mi. N Willowdale (1 PSU). **Klamath Co.**—ca. 25 mi. E Chiloquin, T33S, R10E, Sec. 31 (1 SOSC); T32S, R8E, Sec. 24 (1 SOSC). **Lake Co.**—0.25 mi. SW Desert Lake, Hart Mt. National Wildlife Refuge (2 SOSC); SW base Mt. Warner, Warner Valley, 4,500 ft. (1 USNM). **Malheur Co.**—Butch Lake, T29S, R44E, NE¼ Sec. 6 (1 PSM); Cow Lakes (1 SDNHM); canyon located 8 mi. NE McDermitt, 5,600 ft. (2 USNM); 2 mi. NW Riverside (3 USNM); Rockville (1 USNM); Rome (1 USNM); Sheaville (1 USNM); Skull Spring (1 USNM); Vale (1 USNM). **Morrow Co.**—Rock Creek bridge on HWY 207 (1 PSU). **Sherman Co.**—Miller, mouth of Deschutes River (4 USNM). **Union Co.**—La Grande (1 EOSC). **Wallowa Co.**—24 mi. S, 8 mi. E Imnaha, Hell's Canyon, Snake River (1 OSUFW); Wenaha, T6N, R43E, Sec. 30 (1 EOSC, 1 PSU). **Wasco Co.**—Memaloose State Park, 2.5 mi. E Mosier, T2N, R12E, SE¼ Sec. 31 (1 PSM). **Wheeler Co.**—Crown Rock, John Day River (3 USNM).

Myotis evotis, n = 324
(Plotted in Fig. 9-9, p. 85)

Baker Co.—Anthony (2 AMNH); Cornucopia (1 UMMZ); 2.5 mi. NE Cornucopia, East Pine Creek (1 USNM); field station at Lily White, Wallowa Mts., T7S, R44E, Sec. 7 (2 EOSC). **Benton Co.**—3.75 mi. S, 5.5 mi. E Alsea (1 CM); Corvallis (1 OSUFW); 4.5 mi. N Corvallis, 250 ft. (9 OSUFW); W.L. Finley Wildlife Refuge, 11 mi. S Corvallis (1 PSM); just N Lewisburg (1 KU); 1.5 mi. NW Wrens (1 PSM). **Clackamas Co.**—Gladstone (4 UW[LB]); Jennings Lodge (1 SDNHM); Molalla (1 PSM); T6S, R5E, Sec. 2 (1 PSM). **Clatsop Co.**—1 mi. W Hamlet (4 PSM). **Coos Co.**—4 mi. SE Bandon (1 PSM). **Crook Co.**—Cave Basin, T17S, R23E, Sec. 19 (1 PSU); Twelvemile Creek (2 USNM). **Deschutes Co.**—Davis Lake (1 UMMZ); Indian Ford campground, T14S, R9E, Sec. 13 (1 OSUFW); Sisters (1 USNM); 12 mi. NW Sisters (13 UW[EB]); Sparks Lake, 5,426 ft. (1 MVZ). **Douglas Co.**—T25S, R1W, SW¼ Sec. 7 (1 PSU); T29S, R1W, Sec. 13 (1 PSU). **Grant Co.**—Cold Spring, 8 mi. E Austin (2 MVZ); T11S, R30E, SE¼ Sec. 32 (1 PSM). **Harney Co.**—Burns (1 SDNHM); 6.1 mi. W Domingo Pass* (2 SOSC); 13.1 mi. S, 13.6 mi. W Fields (1 SOSC); Malheur National Wildlife Refuge headquarters, T26S, R31E, Sec. 35 (1 CM); Road Springs, T37S, R29E, NW¼ Sec. 10 (1 SOSC); Steens Mt., T33S, R32¾E, NE¼ Sec. 7 (2 PSM); Steens Mt., T32S, R32¾E, Sec. 28 (1 OSUFW); Willow Springs, Beaty's Butte R.S. (1 SOSC); T21S, R27E, Sec. 34 (1 PSU); T41S, R30E, SW¼ Sec. 15 (1 SOSC). **Hood River Co.**—T2S, R10E, Sec. 2 (1 PSU). **Jackson Co.**—Dixie (1 SOSC); 2 mi. SW Green Springs Summit (1 SOSC); S slope Grizzly Peak (1 SOSC); 3.5 mi. S Jacksonville, T38S, R2W, Sec. 19 (1 PSM); 7 mi. SW Medford, T38S, R2W, SE¼ Sec. 33 (2 PSM); East Pilot Rock Spring, Mt. Ashland, T41S, R2E, center Sec. 3 (7 PSM, 1 SOSC); 0.25 mi. pass Shale City Rd. on Dead Indian HWY, 0.8 mi. on Grizzly Peak Rd. (2 SOSC); Shoat Springs (3 SOSC); 1.4 mi. S HWY 66 off

Copco Rd. (1 SOSC); 2.5 mi. off Mt. Ashland Rd. on Pilot Rock Rd. (1 SOSC). **Josephine Co.**—3 mi. S, 10 mi. E Cave Junction (1 SOSC); Grants Pass (1 SOSC); Oregon Caves [National Park] (2 USNM); 1 mi. N on USFS Rd. 3794 from junction of USFS Rd. 355 and HWY 199 (1 SOSC); 1.9 mi. S Greyback campground (1 SOSC); Jumpoff Joe Creek, 1 mi. E HWY I-5 (1 PSM). **Klamath Co.**—(1 PSM); 6 mi. S, 8 mi. W Bly (1 PSM); Crater Lake National Park headquarters (1 MVZ); Dice Spring, T34S, R13E, Sec. 23 (1 SOSC); 2 mi. SE Lake of the Woods, Rainbow Creek (1 MVZ); Royston Spring, T37S, R13E, SW¼ Sec. 33 (1 PSM); Tsuga Lake, Sky Lakes Basin (1 SOSC). **Lake Co.**—1 mi. N Chandler Wayside (1 SOSC); Desert Lake, Hart Mt. (1 SOSC); 9 mi. N Fort Rock (1 SOSC); Fremont (1 USNM); N of Lakeview (1 USNM); Woodchopper Springs (4 SOSC); T41S, R16E, Sec. 18 (1 SOSC). **Lane Co.**—Eugene (1 SOSC); 5 mi. W Nimrod (1 KU); T21S, R2W, Sec. 24 (1 PSU). **Lincoln Co.**—Cascade Head Experimental Forest (20 PSM); Delake (6 SDNHM, 2 UMMZ); Yaquina (1 OSUFW); T12S, R11W, Sec. 36 (1 PSU). **Linn Co.**—1.7 mi. NE Peoria, T13S, R4W, NE¼ Sec. 5, 240 ft. (2 PSM); 3.5 mi. NW Shedd, T12S, R4W, NE¼ Sec. 27, 250 ft. (2 PSM). **Malheur Co.**—(1 EOSC); Cedar Mt. (1 USNM); Cottonwood Creek, T18S, R42E, center Sec. 3, 3,160 ft. (7 PSM); Disaster Peak, 7,000 ft. (1 USNM); Eagle Creek Canyon, T39S, R43E, Sec. 13 (1 SOSC); Horse Camp Canyon, T34S, R44E, Sec. 1 (1 SOSC). **Marion Co.**—Salem (2 SDNHM). **Morrow Co.**—Cold Springs camp* (1 SDNHM). **Multnomah Co.**—Multnomah Falls (1 SDNHM). **Tillamook Co.**—Tillamook (2 OSUFW, 1 PSM); Tillamook, Cape Meares (2 OSUFW); 9 mi. SE Tillamook, Pleasant Valley, T2S, R9W, Sec. 29 (2 OSUFW). **Umatilla Co.**—Lehman (1 SDNHM). **Union Co.**—LaGrande (1 PSM); 9 mi. NE LaGrande, T1S, R37E, SE¼ Sec. 31 (2 PSM); Mt. Emily, T1S, R37E, Sec. 15 (2 EOSC); Starkey Experimental Forest, Dinnerbucket Pond, T4S, R34E, SE¼ Sec. 14 (3 PSM); Starkey Experimental Forest, T4S, R34E, Sec. 14 (6 PSM); Starkey Experimental Forest, T3S, T34E, Sec. 15 (4 PSM); Starkey Experimental Forest, T3S, R34E, SW¼ Sec. 35 (1 PSM); Starkey Experimental Forest, Bear Creek, T3S, R34E, NW¼ Sec. 2 (1 PSM); West Eagle Meadow, Eagle Creek, Wallowa Mts. (1 EOSC). **Wallowa Co.**—Joseph (1 SDNHM); Powwatka Ridge (19 PSM); Rockwall Springs, 11.5 mi. NNE Wallowa, T3N, R43E, SW¼NW¼ Sec. 14 (83 PSM); 9.5 mi. NNE Wallowa, T3N, R43E, center Sec. 26 (16 PSM); Wallowa Lake (1 USNM); T6N, R43E, Sec. 30 (1 PSU). **Wasco Co.**—Antelope (1 SDNHM); Clear Lake (1 UMMZ); Forest Creek campground, T4S, R10E, Sec. 34 (1 PSU). **Wheeler Co.**—Fossil (1 SDNHM); 7 mi. S, 11 mi. W Mitchell (1 MVZ); T12S, R21E, Sec. 5 (1 PSU).

Myotis lucifugus, n = 386
(Plotted in Fig. 9-12, p. 88)

Baker Co.—2.5 mi. NE Cornucopia, East Pine Creek (1 USNM); Keating (4 KU); 8 mi. NE Unity, Middle Fork Burnt River (16 PSM). **Benton Co.**—3.75 mi. S, 5.5 mi. E Alsea (7 CM); Corvallis (5 OSUFW); 3.5 mi. N Corvallis, T11S, R5W, Sec. 28 (1 OSUFW); 11 mi. S Corvallis, W.L. Finley National Wildlife Refuge (3 PSM). **Clackamas Co.**—Carver, Grants Cave (1 FMNH); Estacada (1 PSM); Jennings Lodge (2 SDNHM); Oswego (2 SDNHM); West Linn (1 PSU). **Clatsop Co.**—Gearhart (1 OSUFW); Olney (1 PSM). **Columbia Co.**—0.8 mi. W Goble (3 PSM); 2.5 mi. W Rainier, T7N, R2W, NW¼ Sec. 7 (3 PSM); 3 mi. E St. Helens, T4N, R1W, SE¼ Sec. 1 (3 PSM). **Coos Co.**—1.75 mi. E Bandon, T28S, R14W, SE¼ Sec. 29, 100 ft. (9 PSM); Dora (1 PSU); 4 mi. S Millington (1 MVZ). **Crook Co.**—T17S, R20E, Sec. 1 (1 PSU). **Curry Co.**—3 mi. E Wedderburn (1 SOSC). **Deschutes Co.**—Bend (3 USNM); La Pine (1 USNM); Paulina Lake (1 USNM); Sparks Lake (1 OSUFW). **Douglas Co.**—NW of Tyee off HWY 138 (2 PSU); T28S, R6E, NW¼ Sec. 20 (2 SOSC). **Gilliam Co.**—Thirtymile Creek bridge (1 PSU). **Harney Co.**—26 mi. S, 7 mi. E Burns, Malheur National Wildlife Refuge (3 OSUFW); Fish Lake (1 PSM); Malheur Lake (13 SDNHM, 5 UMMZ, 1 USNM); Malheur National Wildlife Refuge headquarters (1 MVZ, 9 OSUFW); Narrows (2 SDNHM, 1 USNM); 4 mi. N Princeton, Windy Point (1 MVZ, 3 OSUFW); 3.3 mi. S Princeton (2 PSM); Steens Mt. (3 USNM); Steens Mt., Little Blitzen Canyon, T33S, R32¾E, Sec. 26 (2 OSUFW);

Steens Mt., T31S, R32½E, Sec. 3 (2 OSUFW); Steens Mt., T32S, R33E, Sec. 32 (1 OSUFW); Steens Mt., T33S, R33E, Sec. 5 (2 OSUFW); Trout Creek (1 USNM); Voltage (3 USNM); T33S, R32¾E, central Sec. 7 (2 EOSC); T41S, R30E, SW¼ Sec. 5 (1 SOSC). **Hood River Co.**—Brooks Meadows (1 PSU). **Jackson Co.**—Ashland (1 PSM, 2 SOSC); Indian Creek (1 SOSC); Shoat Spring, 6.8 mi. S of HWY 66 on Copco Rd. (4 SOSC); 0.6 mi. N HWY 140 on Butte Falls Rd., 28.2 mi. E Crater Lake HWY (1 SOSC); Shoat Spring, T40S, R4E, SE¼ Sec. 34 (1 PSM); T33S, R1W, SW¼NE¼ Sec. 14 (1 SOSC). **Jefferson Co.**—16.5 mi. S, 3 mi. W Madras, near Crooked River (1 OSUFW); Suttle Lake (1 OSUFW). **Josephine Co.**—Bolan Lake (2 PSM); Grants Pass (1 SOSC); T33S, R9W, Sec. 35 (1 SOSC). **Klamath Co.**—Ft. Klamath (24 MVZ); 1 mi. S Ft. Klamath (7 MVZ); Hog Creek (1 SDNHM); 3 mi. E Keno (4 MVZ, 1 SOSC); Klamath Falls (1 FMNH, 10 MVZ); Lake of the Woods (1 OSUFW); 4 mi. N Modoc Point (2 OSUFW); Malin (2 SDNHM); Mt. Lakai church* (1 OSUFW); 2 mi. N Olene (3 MVZ); Sheep Pond, T33S, R10E, Sec. 31 (1 SOSC); Upper Klamath Lake (17 SDNHM, 1 UMMZ); Upper Klamath Lake, White Pelican Resort (1 MVZ); Upper Klamath Lake, Rocky Point (1 MVZ, 9 OSUFW, 4 PSM); Worden (1 OSUFW). **Lake Co.**—9 mi. S, 3.5 mi. W Adel, Twentymile Creek (1 OSUFW); 10 mi. SE Lakeview, 6,600 ft. (1 USNM); South Warner Valley* (1 SDNHM); T37S, R21E, Sec. 11 (2 SOSC). **Lane Co.**—Blue River (1 USNM); mouth of Cummins Creek (3 MVZ); Junction City (1 PSU); McKenzie Bridge (3 USNM); 8 mi. E McKenzie Bridge (1 KU); Mercer (1 SDNHM); T21S, R1W, Sec. 24 (1 PSU). **Lincoln Co.**—Cascade Head Experimental Forest (14 PSM); Delake (1 PSM, 14 SDNHM); Taft (6 PSM). **Linn Co.**—5 mi. N, 2 mi. E Lebanon (2 OSUFW). **Malheur Co.**—Cottonwood Creek, T18S, R42E, center Sec. 3, 3,160 ft. (1 PSM); Cow Creek bridge (2 PSM); Dry Creek (2 PSM); Riverside (2 USNM); 3.5 mi. SE Rome, Owyhee Spring (2 PSM); Sheaville (1 USNM); Succor Creek State Park, T24S, R46E, Sec. 20 (1 PSM); Watson (1 USNM). **Marion Co.**—(1 PSU); 1 mi. W Gervais (2 PSU); Salem (1 SDNHM); 3 mi. S St. Paul (1 PSU). **Morrow Co.**—0.8 mi. E Boardman (1 PSM). **Multnomah Co.**—Portland (4 PSM, 6 SDNHM); Sauvie Island, T3N, R1W, Sec. 14 (1 OSUFW, 3 PSM). **Tillamook Co.**—Blaine (1 MVZ); Nehalem (1 OSUFW); Netarts (1 SDNHM). **Umatilla Co.**—Hat Rock State Park (2 PSU); Mecham (1 SDNHM); Pendleton (2 SDNHM). **Union Co.**—Hot Lake hotel (1 EOSC); Ladd Canyon Pond (1 PSU); Park Manor apts. [La Grande] (1 EOSC). **Wallowa Co.**—Lick Creek R.S. (1 PSM); 10 mi. S, 1 mi. E Lostine, T2S, R43E, Sec. 35 (3 OSUFW); 21 mi. S, 5 mi. E Lostine, T4S, R44E, Sec. 27 (3 OSUFW); Wallowa Lake (17 SDNHM). **Wasco Co.**—Clear Lake (2 PSM); Memaloose State Park (1 PSU); Memaloose State Park, 2.5 mi. E Mosier, T2N, R12E, SE¼ Sec. 31 (11 PSM); 0.5 mi. N, 0.5 mi. W South Junction, T8S, R14E, Sec. 20 (1 OSUFW); Tygh Valley (1 PSU). **Wheeler Co.**—T12S, R21E, Sec. 5 (1 PSU). **Yamhill Co.**—Willamina (1 PSU).

Myotis thysanodes, *n* = 32
(Plotted in Fig. 9-16, p. 92)

Baker Co.—Balm Creek mine (1 PSU). **Benton Co.**—3.75 mi. S, 5.5 mi. E Alsea (2 CM). **Clackamas Co.**—Carver (1 USNM). **Clatsop Co.**—site of Old Ft. Clatsop (1 MVZ). **Columbia Co.**—3 mi. ENE Clatskanie (1 PSU). **Douglas Co.**—T24S, R1W, SE¼ Sec. 29 (1 PSU); T29S, R1W, Sec. 13 (1 PSU). **Grant Co.**—South Fork of Lon Creek at Keeney Meadows (2 PSM). **Jackson Co.**—Ashland (2 SOSC); 2.5 mi. S Ashland, T39S, R1E, Sec. 28, 650 ft. (1 PSM); 3 mi. S Ashland, T39S, R1E, Sec. 28 (1 SOSC); E of Pilot Rock Spring, Mt. Ashland (1 PSM); 7 mi. S Medford (1 SOSC); 7 mi. SW Medford, T38S, R2W, SE¼ Sec. 33 (1 PSM); Pompadour Bluff Cave (1 SOSC); Shoat Springs (1 SOSC). **Josephine Co.**—1 mi. E on USFS Rd. 3696 from USFS Rd. 355, T36S, R6W, Sec. 17 (1 OSUFW). **Klamath Co.**—Upper Klamath Lake, Rocky Point (1 OSUFW); 4.6 mi. SE Dead Indian Rd., 2.9 mi. S Moon Prairie Rd. (1 SOSC). **Lane Co.**—T23S, R2E, Sec. 11 (1 PSU). **Lincoln Co.**—Cascade Head Experimental Forest (3 PSM). **Tillamook Co.**—Tillamook (1 MVZ, 1 PSM). **Union Co.**—Rainbow Ponds, FSRD 5125* (1 EOSC); T5S, R41E, Sec. 21 (1 PSU). **Wallowa Co.**—Rockwall Springs, 11.5 mi. NNE Wallowa, T3N, R43E, SW¼NW¼ Sec. 14 (1 PSM); Wenaha, T6N, R43E, Sec. 30 (1 EOSC).

Myotis volans, *n* = 211
(Plotted in Fig. 9-19, p. 94)

Baker Co.—Anthony (1 AMNH, 1 USNM); Sparta (1 USNM); 8 mi. NE Unity (1 PSM); T7S, R43E, Sec. 23 (1 PSU). **Benton Co.**—3.75 mi. S, 5.5 mi. E Alsea (3 CM); 1 mi. N, 2 mi. E Wrens, T11S, R6W, NW¼NE¼ Sec. 23, 920 ft. (1 KU). **Clackamas Co.**—Carver, Grants Cave (1 PSM); Eagle Creek (1 PSM); Estacada (1 SDNHM). **Clatsop Co.**—site of Old Ft. Clatsop (6 MVZ). **Crook Co.**—Whiskey Spring (1 PSU); YCC camp along Canyon Creek, Ochoco National Forest (1 SOSC). **Curry Co.**—10 mi. E Brookings (1 PSM); 9 mi. SSE Port Orford (1 MVZ). **Deschutes Co.**—Paulina Lake (5 USNM); 12 mi. NW Sisters (17 UW[EB]); Sparks Lake, 5,426 ft. (1 MVZ). **Grant Co.**—(1 PSM); 12 mi. S Canyon City, 5,500 ft. (1 MVZ); 10.3 mi. S Keeney Camp (1 PSM); 21 mi. SE Prairie City, North Fork Malheur River, 5,000 ft. (1 MVZ); T11S, R29E, SW¼ Sec. 35 (3 PSM); T11S, R30E, SE¼ Sec. 32 (3 PSM). **Harney Co.**—Grey Springs reservoir (1 SOSC); Sageview (1 SDNHM); Smyth Creek, 10 mi. E Diamond, 5,500 ft. (1 MVZ); Steens Mt., Little Blitzen Canyon, T33S, R32¾E, Sec. 26 (1 OSUFW); Steens Mt., Little Blitzen Gorge, T33S, R33E, Sec. 10 (1 OSUFW); T21S, R27E, Sec. 34 (1 PSU); T40S, R29E, Sec. 6 (1 SOSC). **Jackson Co.**—Chappal Springs, T40S, R2W, Sec. 34 (2 SOSC); Grizzly Mt. (1 SOSC); Shoat Springs (3 SOSC); 1.4 mi. S Hwy 66 off Copco Rd. (1 SOSC); 0.6 mi. N on Butte Falls Rd. from HWY 140, 28.2 mi. on 140 from Crater Lake HWY (1 SOSC). **Josephine Co.**—Little Sugarloaf Mt. (1 SOSC); Oregon Cave [National Park] (3 USNM); 3.5 mi. E of HWY I-5 on Jumpoff Joe Creek Rd. (1 SOSC); 0.25 mi. up Shorthorn Gulch from Jumpoff Joe Creek Rd. (1 SOSC). **Klamath Co.**—Dice Springs, T34S, R13E, Sec. 28 (1 SOSC); Good Luck Pond, T36S, R9E, Sec. 21 (2 SOSC); Onion Springs (2 SOSC); West Sink Creek, E of Mt. Thielson (1 USNM). **Lake Co.**—Desert Lake, 15.7 mi. S of Hart Mt. (1 SOSC); Foster Flat R.S. (1 SDNHM); Fremont (6 USNM); 10 mi. SW Silver Lake, W Silver Creek, 4,650 ft. (1 USNM); Woodchopper Springs (4 SOSC); T40S, R21E, Sec. 11 (1 SOSC). **Lane Co.**—McKenzie Bridge (2 USNM); 10 mi. SE McKenzie Bridge, Lost Creek Ranger Station (1 USNM); T16S, R2W, SW¼ Sec. 17 (1 PSU). **Lincoln Co.**—Cascade Head Experimental Forest (7 PSM). **Malheur Co.**—Castle Rock (1 PSU). **Tillamook Co.**—Netarts (1 SDNHM). **Union Co.**—Starkey Experimental Forest, Bear Creek (4 PSM); 9.5 mi. NNE Wallowa, T3N, R43E, center Sec. 26 (10 PSM); T3S, R34E, SW¼SW¼ Sec. 35 (4 PSM); T5S, R41E, Sec. 21 (1 PSU). **Wallowa Co.**—Kirkland Lookout (1 PSM); Powwatka Ridge (14 PSM, 1 PSU); Rockwall Springs, 11.5 mi. NNE Wallowa, T3N, R43E, SW¼NW¼ Sec. 14 (45 PSM); Starkey Experimental Forest (1 PSM); 9.5 to 11.5 mi. NNE Wallowa (10 PSM); Wenaha, T6N, R43E, Sec. 30 (1 EOSC); T2S, R46E, Sec. 3 (1 PSU). **Wasco Co.**—Clear Lake (1 PSM); head of Little Badge Creek trail (1 PSU); Memaloose State Park (1 PSU). **Washington Co.**—Lake Oswego (1 LACM); 18.5 mi. NW Portland (3 MVZ). **Wheeler Co.**—T7S, R25E, Sec. 6 (1 PSU); T12S, R21E, Sec. 5 (1 PSU).

Myotis yumanensis, *n* = 514
(Plotted in Fig. 9-21, p. 96)

Baker Co.—(1 USNM); Homestead (1 USNM); 8 mi. NE Unity, Middle Fork of Burnt River (3 PSM); 1–2 mi. N oxbow along Snake River* (2 EOSC). **Benton Co.**—6 mi. (by road) W Alpine (1 HSU); 3.75 mi. S, 5.5 mi. E Alsea (1 CM); 10 mi. SE Alsea (1 HSU); Corvallis (1 UMMZ); Corvallis, Kiger Island (1 OSUFW); 3 mi. S, 2 mi. W Corvallis (1 OSUFW); 8 mi. NNE Corvallis, T10S, R5W, NE¼ Sec. 36 (1 PSM). **Clackamas Co.**—Eagle Creek (1 PSM); Estacada (1 FMNH); Molalla (2 FMNH, 2 PSM); Rhododendron (10 PSM). **Clatsop Co.**—Gearhart (1 OSUFW); Hamlet (1 OSUFW); site of Old Ft. Clatsop (3 MVZ); Olney (5 PSM); 7 mi. E Wauna (1 KU). **Columbia Co.**—3 mi. E St. Helens, T3N, R1W, NW¼ Sec. 27 (6 PSM). **Coos Co.**—1.75 mi. E Bandon (9 PSM); Charleston (1 LACM). **Crook Co.**—20 mi. SE Prineville, Crooked River (1 USNM); Twelve Mile Creek (1 USNM); T17S, R21E, Sec. 7 (1 PSU). **Curry Co.**—10 mi. E Brookings (11 PSM); 1 mi. N, 2 mi. E Gold Beach (1 OSUFW); 9 mi. SSE Port Orford (3 KU, 7 MVZ); Squaw Valley, T35S, R14W, Sec. 33 (1 SOSC); Wedderburn (1 USNM); 3 mi. E Wedderburn (3 SOSC). **Deschutes Co.**—12 mi.

NW Sisters (2 UW[EB]). **Douglas Co.**—19 mi. E Canyonville, 1 mi. S Milo (1 LACM); Cow Creek, T32S, R7W, SW¼SW¼ Sec. 20 (1 SOSC); Roseburg (1 USNM); 10 mi. SE Steamboat (2 LACM); T29S, R1W, Sec. 13 (1 PSU). **Gilliam Co.**—Lonerock (2 USNM). **Harney Co.**—Dolman Ranch* (2 PSU); Frenchglen (5 CM, 3 UMMZ); Malheur Environmental Field Station headquarters (3 SOSC); Malheur Lake (5 UMMZ); Malheur National Wildlife Refuge headquarters (9 CM, 3 HSU, 3 SOSC). **Hood River Co.**—T2S, R10E, Sec. 2 (1 PSU). **Jackson Co.**—Copco Pond, 1.4 mi. S HWY 66, T40S, R4E, NW¼ Sec. 10 (2 PSM); East Pilot Rock Springs (1 SOSC); inlet to Reeder Reservoir, Ashland Research Natural Area (2 SOSC); Shoat Springs, T40S, R4E, SE¼ Sec. 34 (1 PSM, 2 SOSC); 0.6 mi. on Butte Falls Rd. from HWY 140 (1 SOSC); 2.8 mi. S Butte Falls (1 SOSC). **Jefferson Co.**—Opal Springs (1 PSU). **Josephine Co.**—(1 SOSC); 1 mi. E on USFS Rd. 3696 from USFS Rd. 355 (1 SOSC). **Klamath Co.**—Bly (1 PSM); NW shore Gerber Reservoir (1 HSU); Ft. Klamath (28 MVZ, 1 UMMZ); 1 mi. S Ft. Klamath (9 MVZ); 3 mi. E Keno (91 MVZ); Klamath Agency (4 OSUFW); Klamath Falls (13 MVZ); 35.1 mi. E Lakeview (1 SOSC); 2 mi. N Olene (28 MVZ); Upper Klamath Lake (1 KU, 9 PSM, 32 UMMZ); Upper Klamath Lake, White Pelican Resort (24 MVZ); Upper Klamath Lake, Rocky Point (1 KU, 14 OSUFW, 29 PSM); T36S, R9E, Sec. 3 (1 SOSC); T37S, R12E, Sec. 11 (1 SOSC); T41S, R15E, Sec. 13 (1 SOSC). **Lake Co.**—Plush, Honey Creek (1 UMMZ). **Lane Co.**—mouth of Cummins Creek (1 MVZ); 9 mi. N, 1 mi. W Gardiner (6 PSM); McKenzie Bridge (5 USNM); Mercer (1 SDNHM); Vida (1 USNM). **Lincoln Co.**—Cascade Head Experimental Forest (4 PSM); Delake (2 PSM, 1 UMMZ). **Malheur Co.**—Cow Creek Bridge (2 PSM); Hardpan Pond, T33S, R44E, Sec. 9 (1 SOSC); Horse Camp Canyon, T34S, R44E, Sec. 1 (1 SOSC); Ontario (1 EOSC). **Marion Co.**—4 mi. S Salem, Willamette Slough (1 CM). **Morrow Co.**—Anson Wright Park (1 PSU). **Multnomah Co.**—Portland (1 PSM, 3 USNM); Sauvies Island (1 FMNH, 1 UMMZ). **Tillamook Co.**—(2 KU); Blaine (2 KU, 5 MVZ, 1 SDNHM, 2 UMMZ); Mohler (2 MVZ); Tillamook (2 KU, 6 OSUFW); Wilson River, McNamer's Camp* (1 USNM). **Umatilla Co.**—17 mi. ENE Umatilla (5 KU). **Union Co.**—0.5 mi. NE Elgin (1 PSU); La Grande (2 EOSC). **Wallowa Co.**—Hells Canyon* (1 PSU); Minam State Park (6 PSU); 10 mi. S, 1 mi. E Lostine, T2S, R43E, Sec. 35 (1 OSUFW). **Wasco Co.**—Memaloose State Park (1 PSU); Memaloose State Park, 2.5 mi. E Mosier, T2N, R12E, SE¼ Sec. 31 (1 PSM); The Dalles (1 PSM). **Wheeler Co.**—3 mi. E Mitchell (1 PSU).

Neotoma cinerea, n = 671
(Plotted in Fig. 11-83, p. 281)

Baker Co.—Anthony (17 AMNH, 1 MVZ); 10 mi. N Baker (1 UMMZ); 20 mi. NE Baker [City] (1 KU); 21 mi. NE Baker [City] (1 KU); Bourne (4 USNM); Cornucopia (5 UMMZ); Homestead, 1,800 ft. (3 USNM); Huntington (1 USNM). **Benton Co.**—3.5 mi. E Alsea (1 OSUFW); 7 mi. N, 4 mi. E Alsea, near Marys Peak (2 OSUFW); 3.75 mi. S, 5.5 mi. E Alsea (2 CM); 6 mi. from Corvallis, T11S, R5W, SEC. 1 (1 OSUFW); 0.5 mi. N, 2.5 mi. W Corvallis (1 OSUFW); 4 mi. N, 3 mi. W Corvallis (1 OSUFW); 3 mi. S, 5 mi. E Kings Valley, Berry Creek (2 OSUFW); Marys Peak (1 SOSC); 0.25 mi. N, 3.25 mi. W Philomath (1 OSUFW); 2 mi. S, 8 mi. W Philomath, Marys Peak (1 OSUFW); 4 mi. S, 5 mi. W Philomath, Westwood Tree Farm, Starker (1 OSUFW); 5 mi. SW Philomath (5 USNM); 15 mi. W SW Philomath (1 OSUFW); Wells (3 USNM); 5 mi. N, 1 mi. W Wrens (1 OSUFW). **Clackamas Co.**—Bissell (1 USNM); Cherryville (1 PSU); Eagle Creek, 8 mi. SE Bissell (1 PSU); 0.5 mi. N, 1.5 mi. E Government Camp (1 PSU); 9.5 mi. NE Molalla, 14.5 mi. SE Oregon City (1 OSUFW); Rhododendron (1 UMMZ); near Second Creek, T5S, R5E, Sec. 24 (1 PSU). **Clatsop Co.**—Astoria (1 UMMZ); 3 mi. S Astoria (1 OSUFW); Cannon Beach (1 PSM); 1 mi. W Hamlet (11 PSM); 15 mi. N Nehalem (1 PSM). **Columbia Co.**—2 mi. N Deer Island (1 PSU); 0.5 mi. W Deer Island (1 PSU); Keasey (1 UW); 7 mi. SE Rainier (4 MVZ). **Coos Co.**—Empire (1 USNM); 16 mi. E North Bend, East Fork Millicoma (1 OSUFW). **Crook Co.**—4 mi. W mouth Bear Creek, Crooked River (1 MVZ); Ochoco Ranger Station, Ochoco National Forest (1 SOSC); Prineville (5 USNM); 1 mi. N, 5 mi. E Prineville (1 USNM); 4 mi. SW Prineville,

3,300 ft. (3 MVZ); Twelvemile Creek (1 USNM). **Curry Co.**—Port Orford (2 USNM). **Deschutes Co.**—Arnold Ice Cave, 16 mi. SE Bend (1 USNM); Bend, Tumalo Creek (1 OSUFW); 9 mi. E Bend (1 OSUFW); 10 mi. S, 14 mi. E Bend (1 OSUFW); head of Davis Creek (6 OSUFW); Lower Bridge (1 OSUFW); Paulina Lake (3 USNM); 1 mi. E Redmond, Redmond air center (1 OSUFW); 3 mi. N, 1 mi. W Redmond (1 OSUFW); Sisters (1 OSUFW, 4 USNM); 3 mi. W Terrebonne (9 OSUFW); 4 mi. W Terrebonne (10 OSUFW); 1 mi. N, 2 mi. W Terrebonne, Upper Deschutes Canyon (2 OSUFW); 1 mi. S, 2 mi. W Terrebonne (1 OSUFW); 1 mi. S, 3 mi. W Terrebonne (2 OSUFW); 1 mi. S, 4 mi. W Terrebonne, T14S, R13E, SEC. 18 (1 OSUFW); 1 mi. S, 6 mi. W Terrebonne, 3,100 ft. (1 OSUFW). **Douglas Co.**—Drew (2 USNM); Ft. Umpqua (2 USNM); Gardiner (7 FMNH, 1 USNM); Loon Lake Ridge (1 PSM); 14 mi. E Reedsport (1 PSU); near Tiller (1 CRCM); Winchester Bay (1 PSU); 2.7 mi. N, 5 mi. W Umpqua, T25S, R7W, Sec. 7, 950 ft. (1 UW[CR]). **Gilliam Co.**—Condon (1 SDNHM); 2.5 mi. S Heppner Junction (3 KU); 4 mi. E John Day River* (1 PSM); exit to Phillipi Canyon (1 PSM); Willows (4 USNM). **Grant Co.**—Austin (2 USNM); Beech Creek (6 USNM); Blue Mt. Spring (2 PSM); 12 mi. S Canyon City (1 MVZ); 1 mi. N, 10.5 mi. E Long Creek, T10S, R32S, SEC. 4 (1 OSUFW); Mt. Vernon Hot Springs, Warm Springs Creek (1 EOSC); 14 mi. NW Ritter (1 PSU); Strawberry Lake, Strawberry Mts. (1 USNM). **Harney Co.**—Buchanan (13 USNM); Burns (11 USNM); 9 mi. S, 2 mi. E Burns (1 OSUFW); Catlow Valley, Home Creek Ranch (1 OSUFW); Cow Creek Lake (3 USNM); Diamond, 4,300 ft. (4 USNM); 3.75 mi. N, 3 mi. E Diamond (2 KU); Diamond Crater, T29S, R32E, Sec. 3 (2 HSU); Drewsey (1 USNM); 12 mi. S, 5 mi. E Fields, Red Point, T40S, R35E, SEC. 23 (1 OSUFW); 3.25 mi. S, 8.5 mi. E Frenchglen, T32S, R32½E (1 OSUFW); Harney (12 USNM); Malheur Cave, 4,000 ft. (2 USNM); Malheur National Wildlife Refuge headquarters, T26S, R31E, Sec. 35 (2 CM, 1 HSU, 1 SOSC); Narrows (1 USNM); Shirk P.O. (3 USNM); Steens Mt., Kiger Gorge, 6,000 ft. (3 USNM); Steens Mt., Kiger Gorge, 6,800 ft. (3 USNM); Steens Mt., T33S, R33E, SEC. 2 (1 OSUFW); Steens Mt., T33S, R33E, SEC. 3 (1 OSUFW); Trout Creek, T39S, R36E, NE¼ Sec. 26 (1 PSM); Voltage (1 USNM); Wright's Point (2 UO); T32S, R32½E, NW¼ Sec. 17 (1 EOSC). **Hood River Co.**—Brooks Meadow, 9 mi. ENE Mt. Hood, 4,300 ft. (2 MVZ); Eagle Creek, 0.4 mi. SE Cascade Fish Hatchery (1 UW); NW slope Mt. Hood, 2,800 ft. (1 USNM); Mt. Hood, Cooper Spur (1 OSUFW); 11 mi. S Mt. Hood, Sherwood camp (2 UW[WWC]); 2 mi. W Parkdale, 1,500 ft. (2 USNM). **Jackson Co.**—Ashland Research Natural Area (1 SOSC); Hyatt Reservoir (1 PSU); Prospect (4 USNM); Siskiyou (7 USNM); 1.6 mi. W HWY 99 on Mt. Ashland Rd. (1 SOSC); 1.7 mi. W Old HWY 99 off Mt. Ashland Rd. (1 SOSC); 2.5 mi. on SW slope Mt. Ashland Rd. (1 SOSC); 1 mi. W HWY99 on Mt. Ashland Rd. (1 SOSC); T38S, R3E, Sec. 30 (1 SOSC). **Jefferson Co.**—near Camp Sherman, Deschutes National Forest (1 KU); 1 mi. N, 3 mi. W Culver (2 OSUFW); Hay Creek (2 OSUFW; 1 SDNHM, 4 USNM); 12 mi. SE Hay Creek (1 OSUFW); 10 mi. S Madras (1 PSU); 10 mi. S, 0.75 mi. W Madras (1 PSU); 2 mi. S, 5 mi. E Madras (2 CM); Ollalie Forest camp (1 PSU); Suttle Lake (4 KU); Warm Springs (4 USNM); Warm Springs, Mill Creek, 20 mi. W Agency (1 USNM). **Josephine Co.**—Bolan Lake (3 PSM); 17 mi. E Cave Junction (1 KU); 1 mi. N, 2 mi. W Merlin, Quartz Creek (1 OSUFW); Oregon Caves game refuge (1 OSUFW). **Klamath Co.**—Crater Lake National Park* (1 OSUFW; 1 USNM); 2 mi. W Cresent Lake, 5,000 ft. (1 UO); Ft. Klamath (1 OSUFW; 6 USNM); Fort Spring, Fremont National Forest (1 PSM); Fourmile Lake (1 UO, 2 USNM); NW shore Gerber Reservoir (1 HSU); 3 mi. E Horsefly Mt. (2 PSM); Klamath Falls (1 USNM); Mosquito Ranger Station (1 USNM); Naylox (2 USNM); Pine Flats, 16 mi. E on HWY140 at Mitchell-Hawkins county Rd.* (1 SOSC); head of Red Blanket Creek, Crater Lake National Park (1 USNM); W Silver Creek, Yamsay Mts., 4,700 ft. (1 USNM); Spencer Creek, 4,100 ft. (4 MVZ); T37S, R14E, Sec. 1 (1 PSU); T40S, R15E, Sec. 9 (1 PSM). **Lake Co.**—14 mi. SW Adel, Warner Mts., Barley Camp, T41S, R21E, E edge Sec. 8 (2 OSUFW); Christmas Lake (1 USNM); Fremont (2 USNM); Gearhart Mt. (1 PSM); Lakeview, Warner Canyon (1 UO); Plush (7 USNM); "Scaler Cabin"* (1 PSM); Silver Lake (2 USNM); Sycan Marsh (1 USNM); S Warner Lake (1 CM). **Lane Co.**—Belknap Springs (1 OSUFW); 5 mi. N, 9 mi. E Blue River, Mack Creek, T15S, R5E, Sec. 36 (1 KU, 1 OSUFW); 0.5 mi. E Canary (1

OSUFW); 7 mi. S Cottage Grove (1 PSM); Dutch Oven campground, South Fork McKenzie River (1 OSUFW); Eugene (3 USNM); Florence (9 FMNH, 1 USNM); 9 mi. N, 1 mi. W Gardiner (1 PSM); Heceta, Smith Ranch (1 OSUFW); Lost Creek Ranger Station, 10 mi. SE McKenzie Bridge (1 USNM); Mapleton (1 USNM); McCredie Springs (1 OSUFW); McKenzie Bridge (1 UO); 2 mi. N, 2 mi. W McKenzie Bridge, T16S, R5E, Sec. 3/4, 3,800 ft. (1 UW[OG]); O'Leary Mt., 10 mi. S McKenzie Bridge (2 USNM); Mercer (1 OSUFW); 3 mi. SE Oakridge (5 UO); Seaton [=Mapleton] (1 USNM); Vida (1 UO). **Lincoln Co.**—1.4 mi. S Burnt Woods (1 OSUFW); Cascade Head Experimental Forest (9 PSM); Delake (Ocean Lake) (1 KU); Drift Creek (1 OSUFW); 2 mi. N, 6 mi. W Fisher (1 OSUFW); Newport (1 USNM); 0.1 mi. N, 1 mi. E Newport (1 OSUFW); 1 mi. NE Otis (3 PSM); Siletz (2 UW[WWC]); Yachats (1 SDNHM); Yaquina Bay (2 LACM, 1 USNM). **Linn Co.**—2 mi. S, 14 mi. E Corvallis (1 OSUFW); Fish Lake (4 PSM); 9 mi. from Lacomb Gate, Green Peter Mt. (2 OSUFW); Permilia Lake, W Mt. Jefferson (1 USNM). **Malheur Co.**—Beulah (2 USNM); Bogus Lake, T29S, R41E, SE¼ Sec. 9 (3 PSM); near Bowden Crater, Bowden Ranch, T34S, R41E, Sec. 27 (1 USNM); Butch Lake, S end Jordan Crater lava flow, T29S, R44E, NE¼ Sec. 6 (1 PSM); Cedar Mt. (3 USNM); Cord, Barren Valley, 3,950 ft. (4 USNM); Cottonwood Creek, T18S, R42E, center Sec. 3, 3,160 ft. (1 PSM); Cottonwood Mt. lookout (1 PSM); Crooked Creek, Owyhee Desert (3 USNM); 3 mi. N Danner (5 HSU); Disaster Peak (1 USNM); Dry Creek, T24S, R41E, NE¼ Sec. 4 (2 PSM); Indian Reservoir (1 PSM); Ironside, 4,000 ft. (10 AMNH); 1 mi. N Jamieson, T15S, R43E, NE¼ Sec. 10, 2,526 ft. (6 PSM); Jordan Crater, T28S, R43E, SW¼ Sec. 4 (1 PSM); 20 mi. S, 3 mi. W Jordan Valley (1 OSUFW); 21 mi. N McDermitt, 5,300 ft. (1 MVZ); Owyhee River, T31S, R42E, NE¼NW¼ Sec. 31 (1 PSM); Riverside (1 USNM); 2 mi. NW Riverside (3 USNM); 4 mi. N Rome (1 CRCM); Sheaville (1 USNM); Vale (5 USNM); Watson (1 USNM); Westfall, Branson Ranch (10 USNM). **Multnomah Co.**—0.5 mi. S Bull Run Reservoir #1 (3 PSU); Larch Mt. (1 PSU); Latourelle (1 OSUFW); Portland (1 PSM, 1 PSU, 2 SDNHM, 1 UMMZ); 4 mi. S, 3 mi. E Scappoose (1 PSU). **Polk Co.**—(1 MVZ). **Sherman Co.**—Millers (1 USNM); Miller, mouth Deschutes River, east side (5 OSUFW). **Tillamook Co.**—Beaver (2 KU, 1 PSM, 1 UO); Blaine (1 CM, 1 KU, 1 MVZ, 1 OSUFW, 3 UMMZ, 1 USNM); Dolph, Little Nestucca River (8 UMMZ); between Dolph and HWY 101 (6 UMMZ); East Beaver Creek (1 PSM); Lees Camp (1 PSU); 0.75 mi. N Neskowin P.O. (1 OSUFW); Netarts (6 OSUFW, 2 SDNHM, 1 UO); Ocean view (1 OSUFW); Tillamook (1 LACM, 1 MVZ, 3 OSUFW, 1 UMMZ). **Umatilla Co.**—(1 UW[WWC]); Blalock Mt. (2 UW[WWC]); W of Langdon Lake (1 KU); Pendleton (1 USNM); South Fork* (1 UW[WWC]); Tollgate (2 MSUMZ, 1 UW[WWC]); 3 km W Tollgate (3 CRCM); 4 mi. W Tollgate (1 AMNH, 9 UW[WWC]); 6 mi. S Tollgate (1 UW[WWC]); 0.5 mi. N, 2 mi. W Tollgate (1 UW[WWC]); 3 mi. NW Tollgate (1 UW[WWC]); Umatilla (2 USNM); 20 mi. SE Weston (1 UW[WWC]). **Union Co.**—Clarks Creek guard station, Elgin (1 CRCM); Fox Hill (1 EOSC); La Grande (2 EOSC); 20 airmiles NW La Grande, 5,000 ft. (2 EOSC); Cricket Flats near Elgin (1 EOSC); Palmer Junction (1 CRCM); Starkey Experimental Forest, T4S, R34E, SW¼ Sec. 9 (1 PSM); Starkey Experimental Forest, T4S, R34E, NE¼ Sec. 16 (1 PSM); Starkey Experimental Forest, Battle Creek, T4S, R34E, NW¼ Sec. 23 (1 PSM); Starkey Experimental Forest, Bear Creek, T3S, R34E, NW¼ Sec. 2 (1 PSM). **Wallowa Co.**—Aneroid Lake (1 UO); Billy Meadows, T4N, R46E, Sec. 15 (1 OSUFW); Coverdale Rd. E of Joseph, T3S, R46E, Sec. 20 (1 PSU); 35 mi. N Joseph (1 OSUFW); Imnaha River* (3 CRCM); mi. post #1, Minam River* (1 OSUFW); Troy (4 PSM); Wallowa Lake, 4,000 ft. (2 USNM). **Wasco Co.**—Columbia River, mouth Deschutes River (2 MVZ); 8 mi. SE Dufur (1 PSM); Shaniko (1 UW); The Dalles (2 USNM); 2.5 mi. S The Dalles (1 PSU); 7 mi. W The Dalles (1 PSU); 6 mi. W Wamic, T4S, R11E, Sec. 11, Rock Creek (2 OSUFW); 1 mi. N, 6 mi. W Wamic (3 OSUFW); 30 mi. NW Warm Springs (1 UMMZ); T3N, R12E, SW¼ Sec. 33 (4 PSM); T3N, R12E, SW¼ Sec. 34 (1 PSM). **Washington Co.**—Beaverton (1 USNM); 1 mi. N, 1 mi. E Helvetia (1 UW); Hillsboro (1 PSU); 21.5 mi. NW Portland (8 MVZ). **Wheeler Co.**—7 mi. S, 11 mi. W Mitchell (3 MVZ); Shelton State Park, SE of Fossil (1 PSU); Ochoco National Forest, Ochoco Springs, 100 yds. S USFS Rd. 127 (1 SOSC).

Additional localities: Little Butte Creek* (1 USNM); Lone Creek* (5 USNM); John Day River* (1 USNM).

Neotoma fuscipes, n = 275
(Plotted in Fig. 11-85, p. 284)

Benton Co.—Adair Village, T10S, R4W, NE¼ Sec. 23 (3 KU, 1 OSUFW); 1 mi. N, 1 mi. E Adair Village, T10S, R4W, NW¼ Sec. 19 (1 OSUFW); 2 mi. N Bellfountain (1 PSM); Corvallis (1 KU, 15 OSUFW, 2 USNM); Corvallis, T11S, R5W, Sec. 23 (1 OSUFW); Corvallis, T12S, R5W, Sec. 28 (1 OSUFW); N. Corvallis (1 KU); 0.5 mi. N Corvallis (2 OSUFW); 1 mi. N Corvallis (3 OSUFW); 2 mi. N Corvallis (3 OSUFW); 3 mi. N Corvallis (1 OSUFW); 4 mi. N Corvallis (1 OSUFW); 5 mi. N Corvallis (2 OSUFW); 7 mi. N Corvallis, E. E. Wilson Game Managment Area (4 KU, 2 OSUFW); 8 mi. N Corvallis (1 OSUFW); 0.5 mi. S Corvallis (1 OSUFW); 1 mi. S Corvallis (1 KU); 0.5 mi. W Corvallis (1 OSUFW); 2 mi. W Corvallis (1 OSUFW); 1 mi. N, 0.5 mi. W Corvallis (1 OSUFW); 1 mi. N, 1 mi. E Corvallis (3 OSUFW); 5 mi. N, 5 mi. W Corvallis, Berry Creek experiment station (2 OSUFW); 8 mi. N, 1 mi. E Corvallis, Soap Creek (2 OSUFW); 8.5 mi. N, 2 mi. E Corvallis (1 OSUFW); 9 mi. N, 2 mi. E Corvallis, T10S, R5W, NW¼ Sec. 1 (5 KU); 9 mi. N, 2.5 mi. E Corvallis, T10S, R5W, NW¼ Sec. 1 (7 KU, 1 OSUFW); 8 mi. S, 8 mi. W Corvallis (1 OSUFW); 10 mi. S, 0.5 mi. W Corvallis (1 OSUFW); 7.5 mi. S, 3 mi. E Corvallis (1 OSUFW); 4 mi. NE Corvallis, T11S, R5W, Sec. 5 (1 OSUFW); Corvallis airport (1 KU); 2 mi. SE Greasy Creek (1 PSM); 1 mi. S, 3 mi. W Greenberry (1 OSUFW); 2 mi. S, 3 mi. W Greenberry (1 OSUFW); 1 mi. S, 1.5 mi. E Hoskins (1 OSUFW); 1.75 mi. S Lewisburg, T11S, R5W, N¼ Sec. 24 (1 KU); McDonald Forest, junction Schneder and Nettleton rds. (1 OSUFW); Monroe (1 OSUFW); 1 mi. S, 1 mi. W Monroe (1 OSUFW); 1 mi. S, 1 mi. W Philomath (1 OSUFW); 5 mi. W Wellsdale (1 OSUFW); T11S, R5W, SE¼ Sec. 13 (2 KU). **Clackamas Co.**—8 mi. S, 1 mi. W Molalla (1 OSUFW); 8 mi. SE Molalla (2 PSM); 0.25 mi. S, 0.5 mi. E Sunnyside (1 OSUFW). **Coos Co.**—3 mi. SSE Bandon (5 PSM); 4 mi. SE Bandon (5 PSM); 3 mi. SW Broadbent (2 PSM). **Curry Co.**—Gold Beach (9 FMNH); Long Ridge Forest Camp (2 PSM); Port Orford (1 USNM); T38S, R12W, Sec. 23 (1 PSM). **Douglas Co.**—Drain (2 USNM); Reston (2 USNM); 4 mi. S, 6.4 mi. W Riddle, Iron Mt., T31S, R7W, Sec. 11, 2,000 ft. (5 UW[CR]); 7 mi. NE Roseburg, Short's Ranch (4 OSUFW); 4.5 mi. N, 8 mi. E Tiller, T30S, R1W, Sec. 2 (1 CRCM). **Jackson Co.**—(1 SOSC); Ashland (1 SOSC); Ashland, W slope Grizzly Pass 3,500 ft. (2 USNM); 3 mi. S Ashland (1 SOSC); 1.9 mi. W Ashland (1 SOSC); 1 mi. N, 4 mi. E Ashland (1 OSUFW); 3 mi. S, 10 mi. E Ashland (1 SOSC); 12 mi. S, 0.5 mi. E Ashland on HWY interstate 5 (2 OSUFW); Brownsboro (2 USNM); 4 mi. up Dead Indian Rd. from HWY 66 [T38S, R2E, Sec. 32] (1 SOSC); 3.5 mi. up Emigrant Creek Rd. on N. facing slope (1 SOSC); Emigrant Lake (1 SOSC); W of Emigrant Lake, T39S, R2E, Sec. 8 (2 SOSC); 0.5 mi. S of SW arm of Emigrant Lake, T39S, R2E, Sec. 4 (1 SOSC); S portion of Emigrant Lake (1 SOSC); 1 mi. S Emigrant Lake, T39S, R2E (1 SOSC); Gold Hill (4 OSUFW, 1 PSM, 2 SDNHM); between Jacksonville and Ruch (1 MVZ); 6 mi. S Medford (3 MVZ); 5 mi. W Medford (1 SOSC); Mt. Ashland (1 SOSC); Prospect (1 MVZ); Prospect, 8 mi. SW Cold Spring Rd. (2 SOSC); 9 mi. N Rogue River, T35S, R4W, Sec. 3, 1,200 ft. (2 SOSC); Soda Mt., T41S, R3E, Sec. 10 (1 SOSC); inlet of Sampson Creek into Emigrant Lake (1 SOSC); 70 yds. SE Copco & Agate Flat Rd. [T41S, R4E, Sec. 3] (1 SOSC); Copco Rd. 7 mi. off HWY 66 [T40S, R4E, Sec. 34] (1 SOSC); N Valley View Rd. [T38S, R1E, Sec. 17/18/19] (1 SOSC); 0.9 mi. W Old Hwy 99 on Mt. Ashland Rd. (1 SOSC); 3 mi. off old HWY99 along Mt. Ashland Rd. (1 SOSC); roadside by Rogue River (1 SOSC); Wagner Creek Rd. 1 mi. up Yank Gluch (1 SOSC). **Josephine Co.**—10 mi. E Cave Junction (1 KU); 3 mi. S, 10 mi. E Cave Junction (1 SOSC); 3 mi. S, 11 mi. E Cave Junction (1 SOSC); Galice, 500 ft. (1 USNM); Farren Ranger Station, 13 mi. SW Galice (3 USNM); Grants Pass (1 CM, 2 FMNH, 7 USNM); 1 mi. S Grants Pass (1 MVZ); 5 mi. W Grants Pass (1 MVZ); 13 mi. SW Grants Pass (3 UO); S slope Grants Pass Peak, 3,600 ft. (3 USNM); Oregon Caves National Monument (1 OSUFW); 0.25 mi. S Jump-off-Joe Creek Rd. 1 mi. E I-5, T34S, R5W, SW¼ Sec. 31 (1 SOSC). **Klamath Co.**—5 mi. E Bonanza (3 PSM); 1.5 mi. SW Bonanza (10

PSM); 3.7 mi. S, 1.9 mi. E Keno (1 PSU); 17 mi. N, 15 mi. W Klamath Falls (1 OSUFW); 8 mi. NW Klamath Falls, Geary Ranch (1 OSUFW); 2 mi. N, 4 mi. E Lorella, 5,000 ft. (1 OSUFW); 7 mi. E Malin, Malone campground (1 OSUFW); Pankey Lake (6 PSM); adjacent of Willow Valley Reservoir, T41S, R14½E Sec. 19 (2 PSU); T37S, R12E, Sec. 9 (1 PSM); T39S, R11½E, Sec. 14 (1 PSM); T39S, R12E, Sec. 14 (2 PSM); T40S, R11½E, Sec. 4 (1 PSM); T40S, R12E, Sec. 2 (1 PSM); T41S, R12E, Sec. 2 (1 PSM); T41S, R14E, Sec. 12 (1 PSM); T41S, R14½E, Sec. 24 (1 PSM). **Lake Co.**—3 mi. from Valley Falls (1 SOSC). **Lane Co.**—2 mi. N, 0.5 mi. E London Springs (1 OSUFW); Mabel (1 UO); 3 mi. SE Oakridge (9 UO). **Linn Co.**—5.5 mi. W Albany (1 OSUFW); 0.6 mi. E Crabtree (1 PSU); 6 mi. E Crabtree (1 PSU); 2 mi. N, 1 mi. E Scio (1 OSUFW); T11S, R2W, Sec. 10 (1 PSU). **Marion Co.**—Salem (2 USNM); 6 mi. S Salem, Sunnyside (1 OSUFW); 5 mi. N, 2.5 mi. E Stayton (1 OSUFW). **Polk Co.**—4 mi. N, 2 mi. W Dalles, Pioneer Loop Rd. (1 OSUFW); Monmouth (2 PSM); ca. 2 mi. S Monmouth (2 KU); ca. 3 mi. S Monmouth (3 KU); 2 mi. S, 1 mi. W Monmouth (1 OSUFW). **Yamhill Co.**—1.5 mi. S, 1.5 mi. W Amity (1 UW).

Additional localities: Rogue River Valley* (2 USNM); Sheep Creek* (1 UW[CR]).

Neotoma lepida, n = 96
(Plotted in Fig. 11-87, p. 287)

Harney Co.—Alvord Desert, T33S, R35E, Sec. 21 (1 OSUFW); Big Sand Gap Springs (2 EOSC); Borax Spring, S end Alvord Lake, 4,300 ft. (3 MVZ); Buena Vista (8 SDMNH, 5 UMMZ); 31 mi. S, 7 mi. E Burns, Malheur National Wildlife Refuge (2 OSUFW); Coyote Butte, Malheur Environmental Field Station (1 SOSC); Diamond (3 USNM); 4 mi. N, 6 mi. W Diamond (1 UW[WWC]); Diamond Craters, 5 mi. N, 5 mi. W Diamond, 4,200 ft. (3 UW[WWC]); 13 mi. S Fields (1 OSUFW); 45 mi. N, 17 mi. E Fields, T30S, R35E, Sec. 35 (3 KU); 6.75 mi. S, 12 mi. E Fields, T39S, R36E, Sec. 25, Trout Creek (2 OSUFW); 7 mi. S, 3 mi. E Fields (11 PSU); 16 mi. N, 3 mi. E Frenchglen (2 OSUFW); 18 mi. N, 1 mi. E Frenchglen, T29S, R31E, SW¼NW¼ Sec. 4 (1 OSUFW); S shore Harney Lake, T27S, R30E, Sec. 32 (1 OSUFW); Howard Camp, Ochoco National Forest, T19S, R26E, NW¼ Sec. 22 (1 SOSC); Malheur Environmental Field Station (1 EOSC, 1 CRCM); 3 mi. E Narrows (1 CRCM); 1 mi. S Narrows (3 MVZ); 2 mi. S, 11 mi. W Riley (1 OSUFW); 4 mi. S, 12 mi. W Riley (1 OSUFW); Steens Mt., T30S, R31E, Sec. 14 (3 OSUFW); Steens Mt., T31S, R32½E, Sec. 23 (2 OSUFW); Trout Creek, T39S, R36E, NE¼ Sec. 26 (1 PSM); Voltage (1 USNM); Whitehorse Creek (1 USNM); 3 mi. S Whitehorse Ranch headquarters (1 UW[WWC]). **Lake Co.**—S Warner Lake, T40S, R23E, Sec. 24 (1 CM). **Malheur Co.**—4 mi. S Adrian (1 SDNHM); 16 mi. S Adrian (4 SDNHM); Bogus Lake (1 PSM); Dry Creek, T24S, R43E, Sec. 34 (1 PSM); Owyhee Reservoir (1 SOSC); 4 mi. N Rome (7 CRCM); 1 mi. S Rome (1 MVZ); 3 mi. S Rome (2 OSUFW); Succor Creek, T23S, R46E, NW¼ Sec. 15, 2,460 ft. (4 PSM); Vale (2 USNM); Watson (3 USNM).

Neurotrichus gibbsii, n = 570
(Plotted in Fig. 8-22, p. 67)

Benton Co.—8 mi. W Albany, T11S, R5W, Sec. 17 (1 OSUFW); 4.4 km N, 7.6 km W Alsea, T13S, R8W, Sec. 20 (1 USNM); 1 km S, 4.4 km E Alsea (2 USNM); 1.2 km S, 4.2 km E Alsea, T14S, R7W, Sec. 9 (1 USNM); 3 mi. N Blodgett (1 PSM); 2 mi. S, 0.5 mi. E Blodgett (1 OSUFW); Corvallis (2 KU, 6 OSUFW, 5 UCONN); 1 mi. N Corvallis (1 KU, 1 OSUFW); 2 mi. N Corvallis (1 OSUFW); 6 mi. N Corvallis (1 PSM, 1 UMMZ); 1 mi. S Corvallis, T11S, R5W, Sec. 3 (3 OSUFW); 4 mi. S Corvallis (1 OSUFW); 11 mi. S Corvallis (1 PSM); 2 mi. W Corvallis, T11S, R5W, Sec. 17 (1 OSUFW); 5 mi. W Corvallis (1 UMMZ); 2 mi. N, 4 mi. W Corvallis (3 OSUFW); 3.5 mi. NW Corvallis, T11S, R5W, Sec. 17 (1 OSUFW); 1 mi. SW Corvallis, T12S, R5W, Sec. 3 (1 OSUFW); Lewisburg (1 OSUFW); 3 mi. W Lewisburg (1 OSUFW); Marys Peak (1 OSUFW); 1 mi. W Philomath (1 PSM); 4 mi. W Philomath (1 PSM); 1.5 km N, 10.1 km W Philomath, T12S, R7W, Sec. 13 (1 USNM); 1.5 km S, 11.3 km W Philomath, T12S, R7W, Sec. 14 (3 USNM); 2.4 km S, 15.5 km W Philomath, T12S, R7W, Sec. 17/20 (5

USNM); 4 km S, 10.1 km W Philomath, T12S, R7W, Sec. 24/25 (4 USNM); 4 mi. SW Philomath, T12S, R6W, NE¼ Sec. 29 (4 PSM); 1.25 mi. SE Summit, T11S, R7W, SW¼ Sec. 8, 650 ft. (11 PSM); 1 mi. S, 3 mi. E Wrens (1 OSUFW). **Clackamas Co.**—Boring (1 PSU); Estacada (1 LACM); Government Camp (1 PSU); Jennings Lodge (2 LACM, 1 OSUFW, 1 SDNHM); Lake Oswego (1 OSUFW; 2 PSU); Milwaukie (1 PSM); SE Molalla (1 PSM); 7 mi. S Molalla (3 PSM); Oregon City (1 UW[EB]); 2 mi. N, 4 mi. E Sandy (1 CRCM); 6 mi. S, 7 mi. E Sandy, T3S, R6E, Sec. 18 (1 CRCM); Still Creek Forest Camp, Mt. Hood National Forest (1 PSM); Wemme (1 SDNHM); 0.33 mi. N, 1 mi. W Wemme, T2S, R7E, Sec. 31 (1 OSUFW); Wilhoit Springs (1 PSM); 9 mi. N, 4 mi. W Zigzag (1 CRCM); 5 mi. S, 6 mi. E Zigzag (2 CRCM). **Clatsop Co.**—(1 PSM); Astoria (1 MVZ, 1 USNM); site of Old Ft. Clatsop (23 MVZ); Seaside (1 USNM); T8N, R10W, NW¼ Sec. 6 (1 PSM); T8N, R10W, W½ Sec. 6 (1 PSM). **Columbia Co.**—2 mi. W Deer Island (1 PSU); 2 mi. W Rainier (1 PSM); 7 mi. SE Rainier (3 MVZ). **Coos Co.**—1.75 mi. S Bandon (1 PSM); 1.75 mi. E Bandon (1 PSM); 1.75 mi. NNE Bandon (1 PSM); 4 mi. SE Bandon (3 PSM); 3.2 km N, 0.3 km W Remote, T29S, R10W, Sec. 21 (1 USNM); 7.4 km N, 4.2 km W Sitkum, T27S, R10W, Sec. 17 (6 USNM); 7.5 km N, 12.5 km W Sitkum, T27S, R11W, Sec. 21 (3 USNM); 8.4 km N, 4.3 km W Sitkum, T27S, R10W, Sec. 17 (2 USNM); 2.9 km S, 4.7 km E Sitkum, T28S, R9W, Sec. 19 (4 USNM); 3 km S, 10.8 km E Sitkum, T28S, R9W, Sec. 23 (2 USNM); 4.8 km S, 9.7 km E Sitkum, T28S, R9W, Sec. 27 (3 USNM). **Curry Co.**—10 mi. E Brookings (3 PSM); Gold Beach (1 FMNH); 0.5 mi. above mouth N side Hunter Creek (1 MVZ). **Deschutes Co.**—Indian Ford Camp (1 PSM). **Douglas Co.**—3.2 mi. S, 3.6 mi. E Canyonville, T31S, R4W, Sec. 17, 2,000 ft. (1 UW[CR]); 20 mi. W Cresent Lake (3 UO); Diamond Lake (1 SOSC, 1 USNM); 4 km W Drain, T22S, R6W, Sec. 13 (1 USNM); 10 km N, 10.5 km W Drain, T21S, R6W, Sec. 7 (1 USNM); 5.5 mi. N, 4 mi. W Drain, T21S, R6W, Sec. 15, 1,280 ft. (1 UW[CR]); 6 mi. N, 4 mi. W Drain, T21S, R6E, Sec. 15, 1,200 ft. (1 UW[CR]); Elkhead (1 USNM); 9.7 km S, 16.2 km W Elkton, T23S, R9W, Sec. 22 (1 USNM); 11.1 km S, 7.4 km W Elkton, T23S, R8W, Sec. 28 (7 USNM); 14.9 km S, 6.2 km W Elkton, T24S, R8W, Sec. 9 (1 USNM); 15.5 km S, 5.8 km W Elkton, T24S, R8W, Sec. 10 (2 USNM); 15.7 km S, 5.6 km W Elkton, T24S, R8W, Sec. 10 (4 USNM); 2 km N, 25 km W Sutherlin, T25S, R8W, Sec. 10 (1 USNM); 2.2 km N, 26.2 km W Sutherlin, T25S, R8W, Sec. 9 (1 USNM); 2.8 km N, 26 km W Sutherlin, T25S, R8W, Sec. 9 (2 USNM); 3.6 km N, 11.9 km W Sutherlin, T25S, R7W, Sec. 7 (1 USNM); 4 km N, 12.1 km W Sutherlin, T25S, R7W, Sec. 1 (1 USNM); 6.5 km N, 6.7 km W Tenmile (2 USNM); 7.5 mi. N, 8 mi. E Tiller (7 CRCM); 10 mi. N, 9 mi. E Tiller, T29S, R1W, Sec. 12 (2 CRCM); 12 mi. N, 4 mi. E Tiller (1 CRCM); 12.5 mi. N, 5.5 mi. E Tiller, T28S, R1W, Sec. 31 (1 CRCM); 0.2 mi. N, 4.1 mi. W Umpqua P.O., T25S, R7W, Sec. 15, 800 ft. (8 UW[CR]); 2.7 mi. N, 5 mi. W Umpqua, Riverview, T25S, R7W, Sec. 7, 950 ft. (5 UW[CR]); 6 mi. N, 10.5 mi. W Umpqua, Miner Creek, T24S, R8W, Sec. 29, 900 ft. (5 UW[CR]). **Hood River Co.**—Brooks Meadow, 9 mi. ENE Mt. Hood (1 MVZ). **Jackson Co.**—(1 SOSC); Ashland (3 SOSC); 5 mi. S Ashland (2 SOSC); 6 mi. S Ashland (1 SOSC); 2 mi. E Dark Hollow (1 MVZ); Emigrant Lake (1 SOSC); Imnaha Creek, T33S, R4E, W½ Sec. 15 (3 SOSC); 5 mi. N Lorane (1 OSUFW); McKenzie Bridge, mouth of South River (1 OSUFW); Prospect near Rogue River bridge (1 SOSC); 5 mi. N, 6 mi. E Prospect (1 OSUFW); 5 mi. N, 8 mi. E Prospect (1 CRCM); 5.5 mi. N, 4 mi. E Prospect (1 CRCM); 13.5 mi. N, 2.5 mi. W Prospect (4 CRCM); S end Reeder Reservoir, Ashland Research Natural Area, T39S, R1E, SW¼ Sec. 28 (4 SOSC); Siskiyou (1 USNM); Soda Creek, T39S, R4E, SE¼ Sec. 8 (1 SOSC); 1.4 mi. S South Fork Little Butte Creek on Soda Creek, T37S, R3E, SW¼ Sec. 20 (1 SOSC); 3 mi. S Talent (2 SOSC); 2 mi. N, 1 mi. E Union Creek (1 CRCM); 1.4 mi. N Dead Indian Rd. on Conde Creek Rd. (1 SOSC); T37S, R3E, NE¼ Sec. 13 (1 SOSC). **Josephine Co.**—4 mi. S, 3 mi. E Cave Junction, T40S, R5E, Sec. 3 (1 HSU); Galice (1 UO). **Klamath Co.**—Anna Creek, Mt. Mazama (1 USNM); Crater Lake* (1 USNM); Ft. Klamath (4 USNM). **Lane Co.**—2 mi. S, 1 mi. E Austa (1 CM); 8.75 mi. E Blue River, 2,400 ft. (1 CRCM); 9 mi. E Blue River, T16S, R5E, Sec. 25/26 (7 CRCM); 1.25 mi. N, 5 mi. W Blue River, T16S, R3E, Sec. 15 (4 HSU); 4.4 km N, 6.8 km E Blue River, T16S, R5E, Sec. 7 (2 USNM);

6.8 km N, 7.6 km E Blue River, T15S, R5E, Sec. 32 (3 USNM); 1.5 mi. N, 2 mi. E Blue River (5 CRCM); 4 mi. N, 4.5 mi. E Blue River, T16S, R5E, Sec. 32 (3 CRCM); 5 mi. N, 9 mi. E Blue River (17 CRCM); 5.5 mi. N, 13.5 mi. E Blue River, T16S, R6E, Sec. 26 (8 CRCM); 7 mi. N, 9 mi. E Blue River, T15S, R5E, Sec. 13 (2 CRCM); 7.25 mi. N, 9.25 mi. E Blue River, 3,600 ft. (1 CRCM); 3.5 mi. S, 8 mi. W Blue River, T17S, R3E, Sec. 7 (1 HSU); 2.5 mi. S, 7 mi. E Blue River (3 CRCM); 3.5 mi. S, 10 mi. E Blue River (8 CRCM); Eugene (1 UO, 1 USNM); Gold Lake, 4,800 ft. (1 LACM); 3 mi. S Huckleberry Mt., 2,300 ft. (1 LSU); 11 mi. W Junction City (1 PSM); 5 mi. N Lorane (1 OSUFW); 2 km N, 6.4 km W Lorane, T20S, R5W, Sec. 5 (4 USNM); 2.8 km N, 6.2 km W Lorane, T20S, R5W, Sec. 5 (1 USNM); 3 km N, 19.5 km W Lorane, T20S, R7W, Sec. 1 (1 USNM); 6.8 km N, 11.1 km W Lorane, T19S, R6W, Sec. 23 (2 USNM); McCredie Hot Springs (1 SDNHM); McKenzie Bridge (1 OSUFW, 2 PSM, 1 UO, 1 USNM); 2.6 km S McKenzie Bridge, T16S, R5E, Sec. 26 (2 USNM); 3.3 km N, 2.4 km W McKenzie Bridge, T16S, R5E, Sec. 3 (3 USNM); 6 km N, 0.6 km E McKenzie Bridge, T15S, R5E, Sec. 25 (2 USNM); 6 km N, 7.5 km E McKenzie Bridge, T15S, R6E, Sec. 26 (1 USNM); 6.6 km N, 0.3 km W McKenzie Bridge, T15S, R5E, Sec. 25 (1 USNM); 6.4 km N, 0.6 km E McKenzie Bridge, T15S, R5E, Sec. 25 (2 USNM); 1.6 km N, 5.2 km W McKenzie Bridge, T16S, R5E, Sec. 20 (4 USNM); 3.8 mi. S, 14.8 km E McKenzie Bridge, T16S, R7E, Sec. 33 (1 USNM); 4.4 mi. S, 14.3 km E McKenzie Bridge, T16S, R7E, Sec.32 (8 USNM); 5.4 mi. S, 8.75 km E McKenzie Bridge, T16S, R6E, Sec. 35 (4 USNM); 3.5 mi. S, 9 mi. E McKenzie Bridge, T16S, R7E, Sec. 32, 3,500 ft. (1 UW[OG]); 4.5 mi. N Nimrod (2 CRCM); 18.6 mi. E Oakridge (2 PSM); 3 mi. SE Oakridge (15 UO); 3.9 mi. W Salt Creek Falls (1 PSM); Vida (1 UO, 2 USNM); Vida, McKenzie River (1 MVZ); 23 mi. SE Vida (5 PSM); Waldo Lake, 1.6 mi. N HWY 58 (1 PSM); T15S, R5E, Sec. 11 (3 PSM); T15S, R5E, Sec. 32 (3 PSM); T15S, R5E, Sec. 33 (4 PSM); T15S, R5E, Sec. 34 (1 PSM); T16S, R4E, Sec. 24 (5 PSM). **Lincoln Co.**—NE Alsea Bay (1 PSM); Cascade Head Experimental Forest (3 PSM); Delake (2 KU; 1 OSUFW); 5 mi. S, 10 mi. W Harlan, Gold Creek, T13S, R10W, Sec. 2 (5 KU); 7 mi. S, 5 mi. W Harlan, West Creek, T13S, R9W, Sec. 16 (1 KU); 7 mi. S, 10 mi. W Harlan, Death Ridge, T13S, R10W, Sec. 14 (1 KU); Newport (1 SDNHM); Rose Lodge (1 USNM); 0.6 km N, 16.9 km E Waldport (3 USNM); Yaquina Bay (3 USNM). **Linn Co.**—Albany (1 OSUFW); 8 mi. W Albany (1 OSUFW); 6.75 mi. N, 8 mi. E Blue River, 2,500 ft. (1 CRCM); 7 mi. N, 8 mi. E Blue River, T15S, R5E, Sec. 15 (15 CRCM); 7.5 mi. N, 7.5 mi. E Blue River, T15S, R5E, Sec. 15 (5 CRCM); 8 mi. N, 8 mi. E Blue River (1 CRCM); 9 mi. N, 8 mi. E Blue River (1 CRCM); 9.5 mi. N, 9 mi. E Blue River, T15S, R5E, Sec. 2 (3 CRCM); 9.2 km N, 1.2 km W McKenzie Bridge, T15S, R5E, Sec. 14 (3 USNM); 10.25 km N, 0.8 km W McKenzie Bridge, T15S, R5E, Sec. 14 (1 USNM); 13.2 km N, 0.8 km E McKenzie Bridge, T15S, R5E, Sec. 1 (1 USNM); 8 mi. S, 4 mi. W Upper Soda, T15S, R4E, S½ Sec. 5 (1 KU). **Marion Co.**—Jefferson (1 USNM); Minto State Park, N Santiam HWY (1 PSM); Salem (1 AMNH, 1 OSUFW); 4.5 mi. E Salem, T7S, R2W, SW¼ Sec. 26 (2 PSM); T7S, R2W, SW¼ Sec. 31 (1 PSM). **Multnomah Co.**—Audubon Sanctuary [Portland] (1 PSU); Multnomah Falls (1 USNM); Portland (2 AMNH, 3 MVZ, 1 OSUFW, 2 PSM, 6 PSU, 6 SDNHM); 10 mi. N, 11 mi. E Sandy (12 CRCM); 11 mi. N, 10 mi. E Sandy (2 CRCM); St. Johns, Portland (1 PSM); 11 mi. N Zigzag (1 CRCM); 8 mi. N, 3 mi. E Zigzag (4 CRCM); T1N, R5E, NE¼ Sec. 29 (1 PSM). **Polk Co.**—4 mi. N, 6 mi. W Dalles (1 OSUFW). **Tillamook Co.**—Blaine (1 UMMZ); Cloverdale (2 UMMZ); 5 mi. SW Cloverdale (1 OSUFW); Nehalem (1 OSUFW); Netarts (1 OSUFW, 6 SDNHM); Tillamook (1 MVZ, 3 PSM, 1 SDNHM). **Washington Co.**—Cedar Mill (2 PSU); Gaston (4 OSUFW); 5 mi. S, 1 mi. E Hillsboro (1 PSU); Laurelwood (1 PSU); Oswego (1 LACM); 18.5 mi. NW Portland (3 MVZ); Tigard (1 PSM). **Yamhill Co.**—Dayton (3 PSM).

Ochotona princeps, *n* = 360
(Plotted in Fig. 10-2, p. 126)

Baker Co.—Anthony (2 MVZ, 1 SDNHM); Anthony, Bradley's Gulch (12 AMNH); Bourne (8 USNM); 4 mi. E Bourne (1 USNM); Cornucopia (2 FMNH, 22 UMMZ); Cornucopia, near head of Pine Creek (5 USNM); 2.5 NE Cornucopia at E Pine Creek (1 USNM); 0.2 mi. NW

Upper Twin Lake, 8,000 ft. (3 CRCM). **Clackamas Co.**—Estacada (1 LACM, 3 SDNHM); 1.5 mi. N, 2.25 mi. E Government Camp (5 PSU); Rhododendron (1 PSU). **Deschutes Co.**—15 mi. W Bend (6 MVZ); Tumalo Creek, 16 mi. W Bend, 6,500 ft. (1 MVZ); Broken Top, 20 mi. W Bend, 8,000 ft. (1 MVZ); Davis Creek (1 OSUFW); Davis Lake, 4,390 (2 MVZ); near Davis Lake (1 SDNHM); Elk Lake, Lava Flow (1 OSUFW); Paulina Lake (4 SDNHM, 1 USNM); Sisters (1 UO); Sparks Lake (2 SDNHM, 4 OSUFW); N Slope Three Sisters (2 OSUFW); N Slope Three Sisters, 6,000 ft. (2 USNM); N Slope Three Sisters, Alder Spring, 4,300 ft. (2 USNM); W base Three Sisters (1 OSUFW). **Grant Co.**—Austin (1 USNM); Granite (1 LACM); 8 mi. N Granite (1 SDNHM); 1 mi. E Granite (1 SDNHM); Strawberry Butte (2 USNM); Strawberry Lake (4 SDNHM); Strawberry Mts. (8 USNM); E slope Strawberry Mt., 8,700 ft. (1 CRCM). **Harney Co.**—Steens Mt. (3 FMNH, 1 OSUFW, 2 SDNHM, 6 UMMZ); Steens Mt., Kiger Gorge, 6,300 ft. (14 USNM). **Hood River Co.**—Mt. Hood, Cloud Cap Inn (3 USNM); Mt. Hood (1 PSU); Mt. Hood, Copper Spur (1 OSUFW). **Jackson Co.**—13 mi. E Ashland (2 MVZ); Fish Lake (4 UO). **Jefferson Co.**—2 mi. S, 10 mi. W Camp Sherman (1 OSUFW); Olallie Lake, 5,000 ft. (1 MVZ); 1.3 mi. W Santiam Pass (2 KU); 2.5 mi. W Suttle Lake (1 KU); 0.5 mi. E County Line, US 20 (4 OSUFW). **Klamath Co.**—Anna Creek, Mt. Mazama, 6,000 ft. (2 USNM); Crater Lake (13 USNM); 1 mi. S Crater Lake (1 MVZ); Bald Top Mountain, Crater Lake National Park (1 USNM); Fourmile Lake (3 UO); Kerr Notch, Crater Lake National Park (1 MVZ, 1 USNM); Mosquito Ranger Station, Crater Lake National Forest [Park] (2 USNM); Wizard Island, Crater Lake (6 MVZ); 15 mi. SW Ft. Klamath (4 UMMZ); Lake of the Woods (1 MVZ); 3 mi. S, 6 mi. E Lake of the Woods (1 SOSC); Mt. Thielson (2 FMNH, 2 UMMZ); Fourmile Lake (1 USNM). **Lake Co.**—20 mi. NE Adel (2 SDNHM); Cougar Peak, 7,925 ft. (1 USNM); Crane Mountain, Warner Mts. (1 OSUFW); Ft. Warner Creek (1 SDNHM); Guano Valley (1 SDNHM); Hart Mt. (1 FMNH, 1 PSM, 5 SDNHM, 4 UMMZ); Mt. Warner, 5,500 ft. (2 USNM); Thomas Creek Ranger Station, Fremont National Forest, 6,000 ft. (3 USNM). **Lane Co.**—Belknap Springs (3 OSUFW); 4 mi. NE McKenzie Bridge, H. J. Andrews Experimental Forest (2 PSM); Lost Creek Ranger Station, 10 mi. SE McKenzie Bridge (1 USNM); McKenzie Pass (2 USNM); McKenzie Pass, Summit (1 UO); Salt Creek Falls (1 PSM); 0.5 mi. S Waldo Lake (1 PSM); Benson Lake, NW base Three Sisters Mts., T15S, R7E (1 KU, 3 OSUFW); Frog Camp, Three Sisters (1 OSUFW); White Branch Creek, Middle Sister, Three Sisters Mts. (1 OSUFW); W of Middle Sister (1 UO). **Linn Co.**—1.5 mi. N Big Lake (3 PSM); 2 mi. N Big Lake (3 PSM); 0.5 mi. N, 19.5 mi. E Cascadia (9 PSU); Fish Lake, 3,190 ft. (3 PSM); H. J. Andrews Experimental Forest, T15S, R6E, Sec. 7 (4 PSM); 6.5 mi. W Hogg Pass, T13S, R7½E, Sec. 29 (2 OSUFW); Hogg Butte, T13S, R7½E, Sec. 22 (1 OSUFW); Permilia [Pamelia] Lake (2 SDNHM, 9 USNM); Pamelia Lake, T10S, R8E (1 OSUFW); 1 mi. W Pamelia Lake (1 OSUFW); Santiam Junction, T13S, R7E (1 OSUFW); 0.5 mi. W Santiam Junction, Highway 222 (1 OSUFW); 14 mi. N, 20 mi. E Sweet Home (2 OSUFW); Lost Prairie Camp Ground, 32.5 mi. E Sweet Home (5 PSU); 17 mi. E Upper Soda (2 OSUFW); 19 mi. E Upper Soda, near Sawyer's Ice Cave (1 OSUFW). **Multnomah Co.**—14 mi. E Corbett, T1N, R6E, Sec. 29 (5 HSU); Multnomah Falls (2 PSM); 0.5 mi. E Multnomah Falls (4 PSU); 0.5 mi. E Multnomah Falls, Columbia River (2 OSUFW). **Union Co.**—(1 EOSC); Hidden Lake, Wallowa Mts. (2 SDNHM). **Wallowa Co.**—(1 EOSC); Aneroid Lake (1 SDNHM, 2 USNM); Bear Creek Ranger Station, Wallowa National Forest (3 SDNHM); Eagle Cap, 9,000 ft, Wallowa Mts. (2 MVZ); talus slopes near Frances Lake, Eagles Cap Wilderness Area (1 OSUFW); forks of Imnaha [River], 7,500 ft. (1 USNM); 16 mi. S, 3 mi. E Lostine, 5,500 ft. (3 MVZ); 18 mi. S, 2 mi. E Lostine, 8,800 ft. (9 MVZ); 18 mi. S, 3 mi. E Lostine, 7,000 ft. (4 MVZ); 18 mi. S, 4 mi. E Lostine, 6,500 ft. (1 MVZ); 19 mi. S, 5 mi. E Lostine (2 MVZ); 18 mi. SE Lostine, Lostine River, Wallowa Mts. (2 OSUFW); 19 mi. SE Lostine, W Fork Lostine River, Wallowa Mts. (1 OSUFW); Minam River, Wallowa National Forest (1 SDNHM); 0.1 mi. NE Swamp Lake, Wallowa Mts., 7,950 ft. (2 CRCM); 1 mi. S Wallowa (1 UW); Wallowa Lake (8 USNM); 1.5 mi. above Wallowa Lake up west fork Wallowa River, 4,800 ft. (2 UW[EB]); 1 mi. S Wallowa Lake on west fork Wallowa River, 5,000 ft. (2 UW[EB]).

Odocoileus hemionus columbianus, n = 24
(Plotted in Fig. 13-6, p. 471)

Benton Co.—Corvallis (1 OSUFW); 8 mi. N Corvallis, McDonald Forest, T10S, R5W (1 OSUFW); McDonald Forest refuge (1 OSUFW). **Clatsop Co.**—30 mi. E Astoria, base Saddle Mt. on Nehalem River (1 USNM); ca. 4 mi. S Jewell (1 UW); 5 mi. NW Mist, Nehalem River basin (1 PSU); Seaside (1 USNM). **Columbia Co.**—near Clatskenie (1 MVZ). **Coos Co.**—4 mi. SE Bandon (1 PSM); Two Mile Creek, T29S, R14W, Sec. 19 (1 PSM). **Curry Co.**—Adams Ranch, 42°30', 124°15' (1 USNM); Agness (2 FMNH). **Douglas Co.**—Roseburg (1 OSUFW); 12 mi. N, 18.5 mi. E Tiller (1 CRCM); Yoncalla (1 USNM). **Jackson Co.**—Rogue River N.F., Prospect R.D. 6560 Rd., 3.5 mi. W junction with HWY 230 (1 SOSC). **Klamath Co.**—Crater Lake (1 USNM); Klamath Falls (1 CM). **Lane Co.**—Eugene, Hendricks Park (1 UO); T17S, R8E, Sec. 31 (1 OSUFW). **Lincoln Co.**—Nashville (1 AMNH). **Marion Co.**—hills around Silverton (1 OSUFW). **Tillamook Co.**—Kilches River (1 PSM).

Odocoileus hemionus hemionus, n = 38
(Plotted in Fig. 13-6, p. 471)

Crook Co.—Ochoco R. S., 4,000 ft. (1 MVZ). **Deschutes Co.**—7 mi. S, 11 mi. W Bend (1 MVZ); 1 mi. W Indian Ford camp (1 OSUFW); 7 mi. N, 5 mi. W Sisters (1 OSUFW); 3 mi. S Three Creeks Lake, T17S, R9E, Sec. 34 (1 OSUFW). **Grant Co.**—Izee, Blue Mts. (1 PSM). **Harney Co.**—about 5 mi. NE Alvord Ranch, 4,800 ft. (1 MVZ); Diamond (1 USNM); Steens Mt. (3 USNM); Steens Mt., Cocamonco C. H., 6,800 ft. (2 USNM); Steens Mt., Kiger Gorge, 8,000 ft. (3 USNM). **Klamath Co.**—(1 PSM). **Lake Co.**—Gearhart Mt. (3 PSM); Hart Mt. refuge (11 USNM); Warner Mt. (1 PSM). **Sherman Co.**—head of Deschutes River (1 USNM). **Wallowa Co.**—11 mi. N, 31 mi. E Wallowa (1 OSUFW). **Wasco Co.**—S. Mosier (1 UW); 5 mi. S, 6.5 mi. W Shaniko (2 UW). **Wheeler Co.**—4 mi. SE Deer Meadows* (1 PSM).

Odocoileus virginianus, n = 73
(Plotted in Fig. 13-9, p. 480)

Columbia Co.—Clatskanie (1 SDNHM); 3 mi. E Westport (1 USNM). **Deschutes Co.**—Davis Creek, T22S, R8E (3 MVZ); head of Davis Creek (2 USNM). **Douglas Co.**—(1 OSUFW); 7 mi. S Glide (1 AMNH); 4 mi. W Glide (6 AMNH); 5 mi. W Glide (7 AMNH); 6 mi. W Glide (8 AMNH); 4 mi. SW Glide (5 AMNH); 5 mi. SW Glide (4 AMNH); 6 mi. SW Glide (4 AMNH); 5 mi. WSW Glide (1 AMNH); 3 mi. SSE Glide (1 AMNH); Roseburg (2 PSM); Roseburg, Oak Creek (1 OSUFW); Roseburg, Winchester area (2 OSUFW); 1 mi. N Roseburg (1 AMNH); 2 mi. N Roseburg (1 AMNH); 1 mi. E Roseburg (2 AMNH); 2 mi. E Roseburg (1 AMNH); 4.5 mi. E Roseburg (1 OSUFW); 5 mi. NE Roseburg (2 AMNH); 7 mi. NE Roseburg (1 AMNH); 8 mi. NE Roseburg (2 MVZ); 6 mi. NNE Roseburg (1 AMNH); 6 mi. ENE Roseburg (1 AMNH). **Grant Co.**—Beech Creek (2 USNM). **Lake Co.**—Willow Creek, T39S, R18E, Sec. 27 (1 MVZ). **Lane Co.**—Cottage Grove (1 UO). **Wallowa Co.**—Bartlett Bench, Wenaha Unit (1 PSM); 3 mi. S Enterprise, T2S, R44E, NE¼ Sec. 14 (1 OSUFW); Flora Rd. grade* (1 OSUFW); Hoodoo Spring, T6N, R42E, Sec. 29 (1 PSM); Noregard area [T3N, R41E, Sec. 2] (1 KU); Wallowa River, T2N, R41E, Sec. 26 (1 PSM).

Ondatra zibethicus, n = 250
(Plotted in Fig. 11-127, p. 336)

Baker Co.—Anthony (1 AMNH); Baker [City] (1 UMMZ); 20 mi. ENE Baker [City], Powder River (1 KU); Burkemont (1 AMNH); Home, 1,850 ft. (1 USNM). **Benton Co.**—Corvallis (1 OSUFW, 1 PSM); Corvallis, T11S, R5W, Sec. 32 (1 OSUFW); 10 mi. N Corvallis (1 OSUFW); 11 mi. S Corvallis, Finley National Wildlife Refuge (1 OSUFW); 1 mi. S, 1 mi. W Corvallis (1 OSUFW); 9 mi. E Monroe, T13S, R5W, Sec. 25 (1 OSUFW). **Clackamas Co.**—Boring (1 PSU); Gladstone (1 PSM). **Clatsop Co.**—6 mi. SE Astoria (8 MVZ); Camp

Clatsop (1 MVZ); Gearhart (2 MVZ); 1 mi. N Gearhart (8 MVZ); Warrenton (1 MVZ); T7N, R10W, Sec. 27 (3 PSU). **Columbia Co.**—0.5 mi. S Columbia City (1 PSU); along Columbia River between St. Helens and Deer Island (9 PSU); Warren (1 PSM). **Coos Co.**—1.75 mi. E Bandon (6 PSM). **Crook Co.**—Prineville (1 PSM); 8 mi. below Prineville Dam on Crooked River (1 OSUFW); Twelvemile Creek (1 USNM). **Deschutes Co.**—(1 EOSC); Lower Bridge (1 PSM). **Douglas Co.**—Loon Lake (3 PSM); Reedsport (3 PSU). **Gilliam Co.**—Willows (2 USNM). **Grant Co.**—Barry Ranch* (1 EOSC); Strawberry Mts., 44°15', 118°45' (1 USNM). **Harney Co.**—Burns (6 USNM); Emigrant Creek, 30 mi. NW Burns (2 USNM); Steens Mt., Kiger Gorge, 6,000 ft. (1 USNM); 2 mi. S Malheur Environmental Field Station, Portal Rd. (1 PSU); Narrows (1 LACM, 1 OSUFW); Narrows Creek (1 OSUFW); Voltage (1 SDNHM); Voltage, Malheur Lake bird reservation (11 USNM). **Jackson Co.**—Ashland (1 SOSC); 1 mi. S Ashland (1 SOSC); 6 mi. E Ashland (1 SOSC); 3 mi. N Central Point (1 SOSC); Emigrant Lake (1 SOSC); Jacksonville pond (1 SOSC). **Klamath Co.**—Bonanza (10 PSM); 0.75 mi. SE Bonanza (2 PSM); Klamath Falls, Big A canal, 5608 Villa Dr. (1 SOSC); 5 mi. S Klamath Falls (1 OSUFW); 7 mi. S, 1 mi. W Klamath Falls, Klamath River (1 OSUFW); Klamath Marsh near Crater Lake on HWY 97 (1 LACM); Miller Island Refuge (1 OSUFW); 1 mi. N, 4 mi. E Rocky Point (5 OSUFW); near town of Sprague River on Sprague River (11 PSM); 1 mi. N Spring Lake, T40S, R9E, NW¼ Sec. 11 (1 SOSC). **Lake Co.**—Fishhole Creek (12 PSM); HWY 93* (1 PSU). **Lane Co.**—Elmira (1 UO); Florence (4 FMNH); 2 mi. N, 1 mi. E Junction City (1 OSUFW); Mercer (3 OSUFW); Dowell Ranch, E arm Mercer Lake (1 UO). **Lincoln Co.**—Cape Foulweather (1 UO); Devils Lake (4 PSM); Salmon River, tidewater (2 PSM); Siletz River, tidewater (3 PSM); 4 mi. S Toledo, T11S, R10W, Sec. 26 (1 OSUFW); 1 mi. E Toledo (1 EOSC). **Malheur Co.**—Butch Lake, S end Jordan Crater lava flow, T29S, R44E, NE¼ Sec. 6 (4 PSM); Cows Lake (1 SDNHM); Ironside (1 AMNH); Ontario (1 UW[WWC]); Rome, Crooked River (2 PSM). **Marion Co.**—Salem (2 OSUFW); T4S, R2E, N½ Sec. 34 (1 PSU). **Morrow Co.**—Heppner (1 USNM); Willow Creek, 3 mi. S Ione (2 MVZ). **Multnomah Co.**—Crystal Springs at Reed College (1 PSU); Gresham (1 PSU); Portland (6 AMNH, 1 LACM, 2 MVZ, 7 PSM, 3 SDNHM, 2 UMMZ, 6 USNM); Portland, Switzer Lake (2 USNM); 5 mi. E Portland (3 UW); 4 mi. N, 6 mi. E Portland (1 OSUFW). **Sherman Co.**—mouth of Deschutes River (2 OSUFW). **Tillamook Co.**—Tillamook (3 OSUFW); 4 mi. S Tillamook (2 KU); 10 mi. S Tillamook (1 SDNHM). **Umatilla Co.**—(2 LACM); Milton-Freewater (2 UW[WWC]); 2 mi. N Milton-Freewater (2 UW[WWC]); 5 mi. E Umatilla (1 UW[WWC]). **Union Co.**—3 mi. E La Grande (1 EOSC); stream near La Grande (1 EOSC); 1.75 mi. N, 1.5 mi. E Starkey (1 OSUFW); Summerville (1 EOSC). **Wallowa Co.**—(1 KU). **Wasco Co.**—The Dalles (1 USNM); T3N, R12E, SE¼ Sec. 33 (2 PSM). **Washington Co.**—Beaverton (1 FMNH, 1 PSU); 1 mi. N Tualatin (1 PSU); T2S, R1W, NW¼ Sec. 23 (1 PSU); T2S, R2W, Sec. 21 (3 PSU). **Yamhill Co.**—McMinnville (1 UW[WWC]).

Onychomys leucogaster, n = 142
(Plotted in Fig. 11-81, p. 277)

Crook Co.—Buck Creek (1 USNM). **Deschutes Co.**—Bend, back of Pilot Butte (5 UO). **Gilliam Co.**—Arlington (1 MVZ); Willows (2 USNM); Willows Junction (1 USNM). **Harney Co.**—(1 EOSC); E edge of Alvord Basin along Coyote Creek, Lake Rd. (1 EOSC); Alvord Lake, Alvord Desert (1 MVZ); N of Alvord Ranch (1 EOSC); Alvord Valley* (1 USNM); Blitzen Valley* (1 SDNHM); Burns (1 USNM); 31 mi. S Burns, Malheur National Wildlife Refuge (1 OSUFW); 23 mi. S, 2.3 mi. W Burns, NE Harney Lake, Malheur National Wildlife Refuge (3 OSUFW); 5 mi. NNE Burns (1 MVZ); Coyote Butte, T27S, R30E, Sec. 13 (2 HSU); 1.5 mi. E Denio (1 MVZ); 1 mi. S, 1 mi. E Fields, T38S, R33E, NW¼ Sec. 30 (1 KU); Harney Lake dunes, Malheur National Wildlife Refuge, T26S, R30E, Sec. 28 (1 CM); Harney (2 USNM); Malheur Lake, Springer Ranch (2 LACM); Malheur National Wildlife Refuge headquarters, T26S, R32E, Sec. 35 (1 CM); Narrows (2 SDNHM, 1 UO, 2 USNM); 1 mi. S Narrows (6 MVZ); 7 mi. E Narrows, Sod House Springs, T27S, R31E, Sec. 3 (2 OSUFW); Steens Mt., T31S,

R31½E, Sec. 4 (9 OSUFW); Steens Mt., T31S, R32½E, Sec. 3 (4 OSUFW); Steens Mt., T31S, R32½E, Sec. 5 (1 OSUFW); Steens Mt., T33S, R35E, Sec. 21 (1 OSUFW); Voltage (1 USNM); Whitehorse Sink (1 USNM). **Jefferson Co.**—Gateway (2 USNM). **Klamath Co.**—Klamath Basin* (2 USNM); Klamath Falls (3 USNM); Swan Lake Valley (2 USNM); Tule Lake (1 USNM). **Lake Co.**—NE edge Alkali Lake (1 MVZ); Fort Rock State Park (2 OSUFW); Fremont (1 USNM); Glass Butte (1 USNM); Goose Lake Valley (1 USNM); Silver Lake (1 USNM); Plush (1 USNM). **Malheur Co.**—Brogan (2 OSUFW); head of Crooked Creek, Owyhee Desert, 4,000 ft. (1 USNM); Ironside, 4,000 ft. (9 AMNH); 5 mi. W Jordan Valley (1 PSM); 8 mi. W Jordan Valley, 4,500 ft. (1 MVZ); 21 mi. N McDermitt, 5,300 ft. (2 MVZ); Rockville (1 USNM); Rome, Owyhee River (4 USNM); 4 mi. N Rome (1 CRCM); 0.5 mi. S Rome, 3,600 ft. (4 MVZ); near Vale (10 AMNH); 3 mi. N Vale, 2,600 ft. (3 MVZ); 6.5 mi. SE Whitehorse Ranch (2 PSM). **Morrow Co.**—14.3 mi. W Boardman (1 PSU); 1.5 mi. S, 5.5 mi. W Boardman (2 PSU); 4 mi. S, 1 mi. E Boardman, T3N, R25E, Sec. 33 (1 OSUFW); Heppner (2 USNM); 2.5 mi. SW Irrigon (4 MVZ). **Sherman Co.**—Columbia River, mouth of John Day River (1 MVZ). **Umatilla Co.**—Umatilla (2 MVZ, 5 USNM). **Wheeler Co.**—Squaw Butte* (1 USNM); Twickenham (1 USNM).

Additional locality: Rock Creek* (1 USNM).

Ovis canadensis, n = 7
(Plotted in Fig. 13-20, p. 499)

Deschutes Co.—Pine Mt. (1 PSM). **Harney Co.**—Steens Mt. (1 USNM). **Lake Co.**—Adel (1 USNM); Hart Mt. (1 SDNHM); near Lakeview (1 MVZ). **Union Co.**—Catherine Creek (1 SDNHM). **Wasco Co.**—Maupin (1 SDNHM).

Perognathus longimembris, n = 20
(Plotted in Fig. 11-57, p. 241)

Harney Co.—NE corner Alvord Basin (3 EOSC); 7 mi. S Andrews, 4,300 ft. (4 MVZ); 2 mi. N Fields (1 OSUFW); 3 mi. N Fields (3 SOSC); Horse Springs, Alvord Basin, 4,100 ft. (2 EOSC); Little Sand Gap, Alvord Basin (1 EOSC); Tumtum Lake (3 USNM). **Malheur Co.**—Rome (1 USNM); 0.5 mi. S Rome, 3,600 ft. (2 MVZ).

Perognathus parvus, n = 823
(Plotted in Fig. 11-59, p. 245)

Baker Co.—Baker [City] (2 UMMZ); W side of road across from Forshey Meadow [T7S, R44E, Sec. 31] between Sparta and Lily White (2 EOSC); 12 mi. SE Huntington, Snake River (2 SDNHM). **Crook Co.**—2 mi. SW Barnes (3 MVZ); 2 mi. SE Barnes (1 MVZ); mouth Bear Creek, Crooked River, 3,400 ft. (20 MVZ); Prineville (7 USNM); 2 mi. E Prineville (2 MVZ); 3 mi. S, 3 mi. E Prineville (1 SOSC); 13.5 mi. S, 2.5 mi. E Prineville (1 OSUFW); Twelvemile Creek (1 USNM). **Deschutes Co.**—(6 PSM); Bend (1 MVZ); 14.2 mi. SE Bend (1 KU); 16 mi. SE Bend, Arnold Ice Cave (2 USNM); 5 mi. SW Cline Falls State Park, T16S, R12E, Sec. 4 (1 OSUFW); Lower Bridge, T14S, R12E, Sec. 9 (1 OSUFW); Lower Bridge, T14S, R12E, Sec. 16 (3 OSUFW); 1 mi. N, 7 mi. W Redmond (1 KU); 1 mi. N, 18 mi. E Sisters (1 PSU); 3 mi. W Terrebonne (1 OSUFW); 4 mi. W Terrebonne (2 OSUFW); 6 mi. W Terrebonne, T14S, R12E, Sec. 16 (2 OSUFW); 1 mi. N, 5 mi. W Terrebonne, Lower Bridge (1 OSUFW); 1 mi. N, 7 mi. W Terrebonne (1 OSUFW); 1 mi. S, 3 mi. W Terrebonne (1 OSUFW). **Gilliam Co.**—Arlington (8 MVZ); 5.3 mi. N Condon, T3S, R21E, SW¼ Sec. 13 (1 PSU); 2.5 mi. S Heppner Junction (9 KU); mouth John Day River, Columbia River (1 MVZ); Willows (9 OSUFW, 1 PSU, 1 SOSC, 16 USNM); Willows Junction (2 USNM). **Grant Co.**—8 mi. E Austin, Cold Spring, 4,900 ft. (2 MVZ); 5.3 mi. N, 1 mi. E Long Creek, Morgrass-Pass Creek (1 OSUFW); 20 mi. S, 2 mi. W Long Creek (1 OSUFW); T9S, R30E, Sec. 15 (1 OSUFW). **Harney Co.**—Alvord Ranch (1 EOSC); N of Alvord Ranch (1 EOSC); 7 mi. S Andrews (1 MVZ); 0.25 mi. S, 0.25 mi. W Ant Hill, T29S, R33E, SW¼NE¼ Sec. 36, 5,000 ft. (2 KU); Burns (1 OSUFW, 5 USNM); 22 mi. S Burns, T27S, R31E, Sec. 35 (1

OSUFW); 12 mi. S, 15 mi. W Burns (4 OSUFW); 9 mi. S, 2 mi. E Burns (1 OSUFW); 23 mi. S, 5.5 mi. W Burns, NW shore Harney Lake (1 OSUFW); 11 mi. S, 3 mi. E Burns, S side of Wright's Point (2 SOSC); 24 mi. S, 2 mi. E Burns, Malheur National Wildlife Refuge, Sec. 21, S of Mud Lake (2 OSUFW); 42 mi. S, 12 mi. E Burns, Malheur National Wildlife Refuge (2 OSUFW); 40 mi. S, 12 mi. E Burns, near Buena Vista substation, Malheur National Wildlife Refuge (1 OSUFW); 30 mi. NW Burns, Campbell Ranch (1 USNM); Coyote Buttes, T27S, R30E, Sec. 13 (7 HSU); Crane (5 USNM); 7 mi. S Crane, Windy Point, T26S, R33E, Sec. 11 (2 OSUFW); 5 mi. N Denio (2 USNM); 1.5 mi. E Denio, 4,200 ft. (3 MVZ); Diamond (1 SDNHM, 5 USNM); 10 mi. E Diamond, Smyth Creek, 5,500 ft. (6 MVZ); 4 mi. N, 6 mi. W Diamond (2 UW[WWC]); 5 mi. N, 5 mi. W Diamond, Diamond Craters (6 UW[WWC]); 1.5 mi. N Fields (2 OSUFW); 1 mi. N, 0.25 mi. E Fields, T37S, R33E, center Sec. 31 (1 KU); 22 mi. N, 9 mi. E Fields, Pike Creek, 4,400 ft. (1 UW[WWC]); 1 mi. S, 1 mi. E Fields, T38S, R33E, S½NW¼ Sec. 30 (1 KU); 6.75 mi. S, 12 mi. E Fields, Trout Creek, T39S, R36E, Sec. 25 (2 OSUFW); 7 mi. S, 3 mi. E Fields (4 PSU); 1 mi. E Fish Lake, Steens Mt. (6 UO); 19.5 mi. N Frenchglen, T29S, R31E, NW¼SW¼ Sec. 29, 4,200 ft. (1 SUVM); 20 mi. N Frenchglen, T29S, R31E, Sec. 9 (1 OSUFW); 2 mi. N, 4 mi. E Frenchglen, Malheur National Wildlife Refuge (1 OSUFW); 16.5 mi. N, 1.5 mi. E Frenchglen, T29S, R31E, SE¼SE¼ Sec. 9 (1 OSUFW); 17.5 mi. N, 1.5 mi. E Frenchglen, T29S, R31E, SE¼SE¼ Sec. 4 (1 JFBM, 2 OSUFW); 0.5 mi. S, 2.5 mi. W Frenchglen, T32S, R32E, NW¼ Sec. 9 (1 KU); 4 mi. S, 10 mi. E Frenchglen, 5,500 ft. (4 UW[WWC]); 5 mi. S 10 mi. E Frenchglen, 6,500 ft. (1 UW[WWC]); 8 mi. SE Frenchglen, 5,500 ft. (3 UW[WWC]); 5 mi. S, 11 mi. E Frenchglen, T32S, R32¾E, SW¼NE¼ Sec. 34, 6,600 ft. (2 KU, 2 OSUFW); Harney (1 USNM); Harney Lake dunes, Malheur National Wildlife Refuge, T26S, R30E, Sec. 28 (1 CM); Jackman Park, 2 mi. SE Fish Lake, 7,500 ft. (5 UW[WWC]); head of Kiger Gorge, Steens Mt., 9,000 ft. (1 UW[WWC]); 7.8 mi. E Lawen (3 CM); Malheur Environmental Field Station (2 PSU, 1 CRCM); Malheur Lake, Springer Ranch (1 LACM); Malheur National Wildlife Refuge, Main Rd. (1 OSUFW); headquarters Malheur National Wildlife Refuge, T26S, R30E, Sec. 27 (1 CM); headquarters Malheur National Wildlife Refuge, T26S, R31E, Sec. 35 (9 CM); SW corner Martha Lake (1 OSUFW); Narrows (1 SDNHM, 5 UO, 12 USNM); 1 mi. S Narrows, 4,200 ft. (3 MVZ); 6 mi. E Narrows (1 SDNHM); 5 mi. SW Narrows (3 MVZ); Rock Creek* (2 USNM); Shirk P.O. (5 USNM); Skull Creek Rd. (3 SOSC); Steens Mt., 1 mi. N Fish Lake (1 HSU); Steens Mt., 14 mi. N "P" Ranch (1 SDNHM); N end Steens Mt., 6,400 ft. (1 FMNH); Steens Mt., T31S, R32½E, Sec. 3 (4 OSUFW); Steens Mt., T31S, R32½E, Sec. 23 (1 OSUFW); Steens Mt., T32S, R32½E, Sec. 10 (3 OSUFW); Steens Mt., 6 mi. E Frenchglen, T32S, R32½E, Sec. 14 (2 OSUFW); Steens Mt., 2.4 mi. W Fish Lake, T32S, R32¾E, Sec. 36 (3 OSUFW); Steens Mt., T32S, R33E, Sec. 31 (1 OSUFW); Steens Mt., T32S, R33¾E, Sec. 20 (1 OSUFW); Steens Mt., T33S, R33E, Sec. 2 (1 OSUFW); Steens Mt., Little Blitzen Gorge, T33S, R33E, Sec. 10 (2 OSUFW); Tumtum Lake (7 USNM); Voltage (1 SDNHM, 6 USNM); Whitehorse Sink (2 USNM); Wright's Point (2 UO); T40S, R36E, Sec. 4, Cotton Creek Rd. (1 PSM); 5.3 mi. SW HWY 205 on Double-O Ranch Rd., 1 mi. S to dunes (1 SOSC); 2 mi. above P Ranch, rock cliff east side Blitzen River (4 UO). **Jackson Co.**—1 mi. N Howard Prairie Reservoir (2 PSM); T38S, R3E, Sec. 26, 0.3 mi. N Dead Indian Rd. off Shale City Rd. (2 SOSC). **Jefferson Co.**—4 mi. NNW Culver (3 PSM); Gateway (5 OSUFW, 3 USNM); 3 mi. S Madras, T11S, R13E, SE¼ Sec. 35 (2 OSUFW); 3 mi. S Madras, T11S, R13E, SW¼ Sec. 36, Crooked River National Grassland (1 OSUFW); 10 mi. S Madras (10 PSU); 10 mi. S, 0.75 mi. W Madras (1 PSU); 2 mi. S, 5 mi. E Madras (1 CM); 1 mi. N, 1.5 mi. W Opal City, T13S, R12E, NW¼ Sec. 12 (4 OSUFW). **Klamath Co.**—Anna Creek, 4 mi. NW Ft. Klamath (4 MVZ); 4 mi. N, 2 mi. E Chiloquin (1 OSUFW); 1 mi. S, 12 mi. E Chiloquin (3 CM); 5 mi. NE Chiloquin (3 PSM); Crater Lake National Park, east entrance above Wheeler Creek (1 OSUFW); 4 mi. off HWY 66 to Devil's Lake (1 SOSC); Klamath Basin, Lost River* (7 USNM); Klamath Falls (6 CM); 1.5 mi. S Summit Stage Station, T25S, R11E, 4,700 ft. (9 MVZ); Swan Lake (1 PSM); Swan Lake Valley (2 USNM); Tule Lake (5 USNM); Williamson River* (3 USNM). **Lake Co.**—(1 UCONN); Adel, Twenty Mile Creek (5 OSUFW);

9 mi. S Adel, mouth of Twentymile Creek (3 MVZ); 7.2 mi. E Adel on HWY 140 (1 PSU); 20 mi. NE Adel, Hart Mt., T36S, R25E, Sec. 25, site of Old Camp Warner (2 OSUFW); NE edge Alkali Lake, 4,200 ft. (2 MVZ); Buck Pasture, Willow Creek, Hart Mt. Antelope Refuge, T35S, R26E, Sec. 17 (2 KU); Cabin Lake (2 UCONN); "Camp Creek," Fremont National Forest* (2 PSM); Fremont (6 USNM); Guano Valley (5 SDNHM); Hart Mt. (1 SDNHM); Hart Mt. National Wildlife Refuge, Hot Springs campground (2 SOSC); Hart Mt. National Wildlife Refuge, 5 mi. E headquarters on Frenchglen Rd. (1 SOSC); 3 mi. N Lakeview (2 MVZ); Silver Lake (1 SDNHM, 4 USNM); Summer Lake (3 USNM). **Malheur Co.**—Brogan (2 OSUFW); Cord, Barren Valley, 3,950 ft. (1 USNM); Cow Creek Lakes (1 USNM); Ironside (33 AMNH); 1 mi. W Ironside (11 MVZ); 5 mi. W Jordan Valley (1 PSM); 8 mi. W Jordan Valley, 4,500 ft. (3 MVZ); Malheur [City] (1 USNM); McDermitt (1 USNM); 3 mi. N McDermitt (2 PSM); 21 mi. N McDermitt (2 MVZ); 5 mi. SW Ontario, Wood Ranch, 2,000 ft. (1 USNM); Owyhee (1 USNM); 2 mi. NW Riverside (4 USNM); Rockville (3 USNM); Rome, Owyhee River (2 USNM); 4 mi. N Rome (1 CRCM); 1 mi. S Rome (6 MVZ); 3 mi. S Rome (2 OSUFW); 3 mi. S, 3 mi. W Rome (1 OSUFW); 3 mi. S, 1 mi. E Rome (1 OSUFW); Sand Hollow (9 AMNH); Skull Spring (1 USNM); Three Forks Crossing, S of Jordan Valley (3 PSU); Vale (6 UO, 2 USNM); near Vale (2 AMNH); 3 mi. N Vale, 2,600 ft. (6 MVZ); Watson (3 USNM); 3 mi. SE Folly Farm Rd. (1 PSM). **Morrow Co.**—Boardman (6 SDNHM); 4 mi. W Boardman, T4N, R24E, Sec. 10 (1 PSU); 4.5 mi. W Boardman on Columbia River (1 KU); 4.6 mi. W Boardman on Columbia River (1 KU); 14.3 mi. W Boardman (1 PSU); 5 mi. S Boardman (2 OSUFW); 0.25 mi. S, 5 mi. W Boardman (11 PSU); 0.5 mi. S, 5 mi. W Boardman (12 PSU); 0.5 mi. S, 5.5 mi. W Boardman (4 PSU); 1 mi. S, 5 mi. W Boardman (1 PSU); 1.5 mi. S, 5.5 mi. W Boardman (29 PSU); 3 mi. S, 6 mi. W Boardman (1 PSU); 5 mi. S, 5 mi. W Boardman, T4N, R24E, Sec. 33 (1 HSU, 2 OSUFW); 5 mi. S, 5 mi. E Boardman (1 OSUFW); 3.75 mi. S, 2.5 mi. E Boardman (1 HSU, 35 OSUFW); Cecil (5 SDNHM); Cecil, McIntyre Camp (5 PSM); Cecil, 6 mi. SE McIntyre Camp (1 PSM); Crown Rock, John Day River (4 USNM); Heppner (2 USNM); 3.5 mi. W Irrigon, McCormick Ranch, Umatilla National Wildlife Refuge, T5N, R26E, NW¼NW ¼ Sec. 22 (23 OSUFW); 2.5 mi. SW Irrigon, 300 ft. (11 MVZ). **Sherman Co.**—Miller, mouth of Deschutes River (5 OSUFW, 8 USNM); mouth of John Day River, Columbia River (1 MVZ). **Umatilla Co.**—Umatilla (13 MVZ, 1 PSU, 7 USNM); 1 mi. S Umatilla (1 PSU). **Wallowa Co.**—15 mi. N Paradise at Horse Creek, 2,000 ft. (1 USNM). **Wasco Co.**—(1 PSM); Antelope (1 USNM); The Dalles (24 MVZ, 14 USNM); Maupin (1 USNM); mouth of Deschutes River, Columbia River (6 MVZ). **Wheeler Co.**—4 mi. E Service Creek (1 UW[WWC]); 5 mi. E Service Creek (2 UW[WWC]); Spray (2 UW[WWC]); Twickenham (6 USNM).

Peromyscus crinitus, n = 222
(Plotted in Fig. 11-75, p. 269)

Baker Co.—Huntington (5 USNM); 3 mi. NE Huntington (1 USNM). **Crook Co.**—Prineville (2 SDNHM, 16 USNM); 6 mi. S Prineville, 3,100 ft. (1 LSU); 20 mi. SE Prineville, Crooked River (2 USNM). **Harney Co.**—E rim of Alvord Basin* (2 EOSC); Buchanan (1 USNM); Buena Vista (4 PSM, 2 SDNHM); Burns (4 USNM); 9 mi. S, 2 mi. E Burns (1 OSUFW); 10 mi. S, 3 mi. E Burns, Wright's Point (1 OSUFW); Diamond (10 USNM); 10 mi. E Diamond, Smyth Creek, 5,000 ft. (2 MVZ); Diamond Crater, T29S, R32E, Sec. 3 (11 HSU); Diamond Craters (1 UW[WWC]); Diamond Craters, 5 mi. N, 5 mi. W Diamond, 4,200 ft. (2 UW[WWC]); Drewsey (1 USNM); 13 mi. S Fields (5 OSUFW); 2 mi. S, 6 mi. W Fields (1 OSUFW); 6.75 mi. S, 12 mi. E Fields, Trout Creek, T39S, R36E, Sec. 25 (2 OSUFW); 12 mi. S, 5 mi. E Fields, Red Point schoolhouse, T40S, R35E, Sec. 23 (3 OSUFW); 15.5 mi. N, 2.5 mi. E Frenchglen (1 OSUFW); 17.5 mi. N, 1.5 mi. E Frenchglen, T29S, R31E, Sec. 4 (2 OSUFW); 18 mi. N, 1 mi. E Frenchglen, T29S, R31E, SW¼NW¼ Sec. 4 (3 OSUFW); 0.5 mi. S, 2.5 mi. W Frenchglen, T32S, R32E, NW¼ Sec. 9 (2 KU); 2 mi. S, 3 mi. E Frenchglen, 4,500 ft. (1 OSUFW); Harney (4 USNM); Malheur Cave (4 USNM); 11.6 mi. S, 2.3 mi. E Malheur Environmental Field Station, Malheur National Wildlife Refuge (1 SOSC); 11.6 mi. S, 1.6 mi. E

Malheur Environmental Field Station, Malheur National Wildlife Refuge (2 SOSC); 11.7 mi. S Malheur Environmental Field Station and 2.8 mi. E by center patrol Rd. (1 SOSC); Narrows (1 USNM); top of Serrano Point (7 EOSC); Steens Mt., T34S, R32¾E, SW¼ Sec. 7 (2 PSM); Trout Creek, T39S, R36E, Sec. 26 (3 PSM); T32S, R32½E, center Sec. 28 (3 EOSC). **Jefferson Co.**—Hay Creek (1 USNM); Warm Springs (8 USNM). **Lake Co.**—Buck Pasture, Willow Creek, Hart Mt. National Antelope Refuge (1 KU); Hot Springs camp, Hart Mt. National Antelope Refuge (1 SOSC); 9 mi. S Silver Lake (1 PSU). **Malheur Co.**—Bogus Lake, T29S, R41E, SE¼ Sec. 9 (1 PSM); Disaster Peak (1 USNM); head of Crooked Creek, Owyhee Desert (5 USNM); Dry Creek, T24S, R41E, NE¼ Sec. 4 (6 PSM); Jordan Valley, Rocky Butte, 4,600 ft. (4 USNM); 0.5 mi. W Jordan Valley (14 PSM); 21 mi. N McDermitt, 5,300 ft. (1 MVZ); Pharmacy Butte, 0.5 mi. W Jordan Valley, T30S, R46E, SW¼ Sec. 2 (2 PSM); 2 mi. NW Riverside (10 USNM); 1 mi. S Rome, 3,600 ft. (1 MVZ); Sheaville (1 USNM); Succor Creek State Park, T23S, R46E, NW¼ Sec. 15, 2,460 ft. (19 PSM); Succor Creek State Park, T24S, R46E, Sec. 20 (1 PSM, 1 PSU); Vale (16 USNM); Watson (1 USNM). **Wasco Co.**—Maupin (8 USNM). **Wheeler Co.**—2 mi. S, 2 mi. E Clarno (1 OSUFW).

Peromyscus maniculatus, n = 7,178
(Plotted in Fig. 11-77, p. 271)

Baker Co.—Anthony (10 MVZ, 1 SDNHM); 10 mi. N Baker (6 UMMZ); 20 mi. NE Baker [City] (8 KU); Camp Creek, 6 mi. from Unity (1 OSUFW); Cornucopia (11 PSM, 1 MSUMZ, 6 UMMZ); Eagle Creek area (2 KU); W slope N ridge of Elkhorn Peak, Elkhorn Mts., 8,200 ft. (1 CRCM); Haines (1 EOSC); 3 mi. E Halfway (4 MSUMZ); 15 mi. S, 11 mi. W Hereford, T15S, R37E, Sec. 7 (1 KU); Long Creek, SE of Unity (1 PSM); Oxbow (2 MSUMZ); 0.25 mi. S Oxbow (1 MSUMZ); 3 mi. SW Oxbow (1 MSUMZ); 0.3 mi. SE Rock Creek Butte, Elkhorn Mts., 8,500 ft. (3 CRCM); 0.4 mi. SE Rock Creek Butte, Elkhorn Mts., 8,400 ft. (1 CRCM); 0.2 mi. SSE Rock Creek Butte, Elkhorn Mts., 8,600 ft. (2 CRCM); above Twin Lakes, Elkhorn Mts., 7,900 ft. (1 CRCM); 3 mi. SE Unity (1 PSM); 0.1 mi. N Upper Twin Lake, Elkhorn Mts., 7,700 ft. (3 CRCM); 0.3 Upper Twin Lake, Elkhorn Mts., 8,500 ft. (3 CRCM); 0.4 Upper Twin Lake, Elkhorn Mts., 8,200 ft. (1 CRCM); 0.3 mi. NW Upper Twin Lake, Elkhorn Mts., 8,300 ft. (1 CRCM); 0.2 mi. NNW Upper Twin Lake, Elkhorn Mts., 8,100 ft. (5 CRCM); 0.3 mi. NNW Upper Twin Lake, Elkhorn Mts., 8,300 ft. (2 CRCM); 0.3 mi. WNW Upper Twin Lake, Elkhorn Mts., 8,500 ft. (1 CRCM); Wallowa Mt. field station (1 EOSC). **Benton Co.**—(4 PSM); 1 mi. N, 1 mi. W Adair, 200 ft. (1 OSUFW); 6 mi. (by road) W Alpine (4 HSU); 3 mi. N, 3 mi. E Alsea, Crooked Creek, T13S, R7W, Sec. 21 (1 KU); 5 mi. N, 3 mi. E Alsea, Marys Peak, 2,597 ft. (2 OSUFW); 8 mi. N, 1.5 mi. E Alsea, T12S, R7W, SW¼ Sec. 29 (1 OSUFW); 8 mi. N, 2 mi. E Alsea, Marys Peak (2 OSUFW); 3 mi. S, 2 mi. E Alsea (1 OSUFW); 3 mi. S, 2 mi. E Alsea, T14S, R7W, Sec. 20 (5 OSUFW); 3 mi. S, 4 mi. E Alsea, T14S, R7W, Sec. 22 (5 OSUFW); 3.25 mi. S, 4.25 mi. E Alsea (14 CM); 3.75 mi. S, 5.5 mi. E Alsea (36 CM); 10 mi. SE Alsea (10 HSU); 4 mi. S Blodgett (2 OSUFW); 1.5 mi. W Blodgett (1 OSUFW); 6 mi. S, 2 mi. W Blodgett, Marys Peak (1 OSUFW); Corvallis (64 localities within 12 mi. Corvallis) (18 CM, 4 KU, 158 OSUFW, 2 PSM, 3 SOSC); 13 mi. W Corvallis (2 PSM); 5 mi. S, 15 mi. W Corvallis, T12S, R7W, SW¼ Sec. 21 (1 OSUFW); 2 mi. W Hoskins (2 OSUFW); 1 mi. S, 1.5 mi. E Hoskins (2 CM); 2 mi. S, 2 mi. E Hoskins (1 OSUFW); 7 mi. E Kings Valley (1 OSUFW); Marys Peak (11 localities on Marys Peak) (22 OSUFW, 1 PSM); McDonald Forest (2 OSUFW); 3 mi. W Monroe, 288 ft. (1 OSUFW); Philomath (13 localities within 7 mi. Philomath) (2 KU, 17 OSUFW, 13 USNM); South Fork Alsea River* (7 PSM); 2 mi. N Wrens, 1,000 ft. (1 OSUFW); 0.5 mi. S, 1.5 mi. E Wrens (1 CM, 6 OSUFW); 0.5 mi. S, 3 mi. E Wrens, T11S, R6W (1 OSUFW). **Clackamas Co.**—Alpine campground, T3S, R9E, Sec. 7 (2 PSU); Canby (1 PSU); Carver (2 PSU); 2 mi. N Clackamas (1 PSU); 0.5 mi. S Clackamas (4 PSU); 1 mi. N, 5 mi. E Clackamas, T2S, R3E, Sec. 6 (1 PSU); 3 mi. E Damascus (2 PSU); Estacada (1 OSUFW); 1 mi. N, 3.5 mi. W Government Camp (3 PSU); Jennings Lodge (7 PSM); 2 mi. N, 1.5 mi. W Ladd Hill (4 PSU); 2 mi. E Liberal (4 PSM); Milwaukie (1 PSU); E of Milwaukie,

Linnwood Swamp (3 PSU); 2 mi. E Milwaukie (1 PSU); 1 mi. NE Milwaukie (2 OSUFW); SE Molalla (1 PSM); 7 mi. SE Molalla (2 PSM); 8 mi. SE Molalla (1 PSM); Mulino (2 OSUFW); Oak Grove (1 PSM); 0.5 mi. S, 3 mi. E Oregon City (1 PSU); 1 mi. S, 0.5 mi. E Oregon City (1 PSU); 19 mi. S Portland (1 OSUFW); Rhododendron (1 KU, 1 OSUFW); 2 mi. N, 4 mi. E Sandy (1 CRCM); 3.8 mi. N, 8 mi. E Sandy (1 CRCM); 4.2 mi. N, 8 mi. E Sandy, T1S, R6E, Sec. 29 (1 CRCM); 2 mi. S, 2 mi. E Sandy (1 UW); 6 mi. S, 7 mi. E Sandy, T3S, R6E, Sec. 18 (3 CRCM); West Linn (1 PSM); 9 mi. N, 4 mi. W Zigzag (1 CRCM); Dale Ickes Junior High School (2 PSU). **Clatsop Co.**—Astoria (10 MVZ, 3 OSUFW, 7 UMMZ); 6 mi. S, 1 mi. E Astoria, Tucker Creek, 100 ft. (1 OSUFW); 7.5 mi. S Cannon Beach (4 MVZ); site of Old Ft. Clatsop (10 MVZ); W of Old Ft. Clatsop Junction, T8N, R10W, Sec. 35, 250 ft. (1 PSU); Elsie (1 CM); Klootchy Creek campground (8 LACM); Quartz Creek Rd., T4N, R7W, Sec. 10, 1,000 ft. (1 PSU); Seaside (3 PSM); Warrenton (3 MVZ); 0.8 mi. E Youngs River Falls, T7N, R9W, Sec. 22 (1 PSU); T7N, R9W, Sec. 6 (1 OSUFW); T8N, R7W, NE¼ Sec. 18 (7 PSM); T8N, R8W, SW¼ Sec. 12 (1 PSM); T8N, R8W, NE¼ Sec. 20/SW¼ Sec. 16 (2 PSM); T8N, R10W, W½ Sec. 6 (3 PSM); T8N, R11W, E½ Sec. 35 (3 PSM). **Columbia Co.**—(2 PSU); Deer Island (7 PSM); 2 mi. N Deer Island (4 PSU); 2 mi. W Deer Island (8 PSU); 5 mi. S Goble (1 PSM); 0.5 mi. N, 0.5 mi. E Keasey (1 UW); 7 mi. SE Rainier (19 MVZ); Tankton* (1 PSU); 1 mi. N, 0.5 mi. W Vernonia (15 UW); T4N, R1W, W½ Sec. 22 (8 PSM); T4N, R1W, N½ Sec. 34 (2 PSM); T4N, R1W, NE¼ Sec. 34 (12 PSM); T6N, R1W, SW¼ Sec. 7 (1 PSM); T6N, R2W, NW¼ Sec. 18 (2 PSM); T6N, R2W, E½ Sec. 13 (1 PSM); T7N, R2W, NE¼ Sec. 7 (3 PSM); T7N, R2W, SE¼ Sec. 26 (6 PSM). **Coos Co.**—1.75 mi. S Bandon (1 PSM); 2 mi. S Bandon (1 PSM); 4 mi. S Bandon (1 PSM); 7.5 mi. S Bandon, T30S, R15W, Sec. 3 (1 OSUFW); 9 mi. S Bandon (3 PSM); 10.5 mi. S Bandon, T30S, R15W, Sec. 10 (2 OSUFW); 1 mi. E Bandon (6 MVZ); 1.75 mi. E Bandon (13 PSM); 3 mi. SE Bandon (1 OSUFW, 8 PSM); 4 mi. SE Bandon (39 PSM); 9 mi. SE Bandon (2 PSM); 1.75 mi. NNE Bandon (1 PSM); Bastendorf Beach (7 LACM); Bullard Beach State Park (1 PSU); Camp Terramar* (1 SOSC); Cape Arago State Park (10 LACM); Charleston (24 LACM, 1 PSU); 1 mi. S Charleston (2 LACM); 3 mi. S Charleston, Sunset Bay (2 LACM); 5 mi. SW Charleston, Sunset Bay (14 LACM); Coos Bay, South Bay marsh (1 LACM); Coos Bay, South Slough (8 LACM); Coos Head (16 LACM, 60 UO); Coos Spit (29 UO); Empire (2 UO); Marshfield (3 FMNH, 1 UO); 2.5 mi. N McCalough Bridge (1 SOSC); 3.2 km N, 0.3 km W Remote, T29S, R10W, Sec. 21 (1 USNM); 7.3 km N, 4 km W Sitkum, T27S, R10W, Sec. 17/20 (2 USNM); 7.4 km N, 4.2 km W Sitkum, T27S, R10W, Sec. 17 (1 USNM); 7.5 km N, 12.5 km W Sitkum, T27S, R11W, Sec. 21 (1 USNM); 7.6 km N, 5.2 km W Sitkum, T27S, R10W, Sec. 18 (2 USNM); 8.4 km N, 4.3 km W Sitkum, T27S, R10W, Sec. 17 (1 USNM); 3 km S, 10.8 km E Sitkum, T28S, R9W, Sec. 23 (1 USNM); 3 km S, 11 km E Sitkum, T28S, R9W, Sec. 23 (1 USNM); 4.8 km S, 9.7 km E Sitkum, T28S, R9W, Sec. 27 (3 USNM); Sunset Bay (14 LACM, 2 UO); Sunset Bay State Park (1 MVZ). **Crook Co.**—2 mi. SW Barnes (2 MVZ); mouth of Bear Creek, Crooked River (1 MVZ); Big Summit R.S., Ochoco National Forest (1 SOSC); Ochoco R.S., 4,000 ft. (1 MVZ); W side Ochoco summit (2 PSM); 6 mi. S, 5 mi. W Paulina (1 OSUFW); Powell Butte (1 SDNHM); Prineville (1 MVZ); 0.8 mi. S Prineville (3 LSU); 6 mi. S Prineville, 3,100 ft. (1 LSU); 13 mi. S Prineville (2 PSM); 20 mi. S Prineville, Crooked River (3 OSUFW); 2 mi. W Prineville (1 MVZ); 3 mi. W Prineville (1 MVZ); 7 mi. W Prineville (1 MVZ); 4 mi. SW Prineville (19 MVZ); 2 mi. NE Prineville (6 MVZ); Prineville Dam, T17S, R16E, Sec. 10 (1 OSUFW). **Curry Co.**—Agness (2 FMNH); 1.5 mi. ENE Agness, N bank Rogue River (2 MVZ); Brookings (1 PSU, 2 SOSC); 10 mi. E Brookings (15 PSM); 1.5 mi. N, 5 mi. E Brookings (2 OSUFW); Gold Beach (16 FMNH, 4 PSM, 2 SDNHM, 1 SOSC); 3 mi. above Gold Beach, S side Rogue River (3 MVZ); Harris Beach State Park (1 MVZ, 4 SOSC); 0.5 mi. N Harris Beach State Park (1 SOSC); Little Redwood Forest camp (16 PSM); 3.4 mi. W Loeb campground (2 SOSC); Long Ridge Forest camp (16 PSM); Port Orford (44 UO); 8 mi. S Port Orford, Humbug State Park (1 OSUFW); 2 mi. S Port Orford (4 PSM); 10 mi. E Port Orford (4 PSM); 1.75 mi. SE Port Orford (6 PSM); 10 mi. ESE Port Orford, Elk River (16 PSM); S side Rogue River up from HWY 101 (1 OSUFW); stateline*

(1 SDNHM); Winchuck campground, T41S, R12W, NW¼ Sec. 10 (2 SOSC); junction North Bank Rd. and North Fork of Chetco River (1 SOSC). **Deschutes Co.**—Bend (44 UO); Bend, T17S, R11E, Sec. 15 (3 OSUFW, 1 SDNHM); Bend, Deschutes River bank (6 UO); 3 mi. W Bend (2 KU); 11 mi. W Bend, Tumalo Creek, 4,700 ft. (4 MVZ); 15 mi. W Bend, Tumalo Creek, 6,100 ft. (3 MVZ); 16 mi. W Bend, 6,500 ft. (2 MVZ); 5 mi. S, 3 mi. E Bend (1 PSU); 8 mi. S, 17 mi. E Bend (5 CM); 4 mi. SE Bend (3 KU); 14 mi. SE Bend (9 KU); Big Springs campground (1 PSU); Broken Top Mt. (3 SDNHM); W side Broken Top, 7,100 ft. (2 CRCM); W side Broken Top, 7,300 ft. (3 CRCM); 5 mi. SW Cline Falls State Park, T16S, R12E, Sec. 4 (1 OSUFW); 0.25 mi. N, 2.5 mi. E Cloverdale, T15S, R11E, Sec. 3 (9 OSUFW); Cold Spring (2 KU); head of Davis Creek (9 OSUFW); mouth of Davis Creek, Deschutes River (15 OSUFW); W side Green Lakes Basin, E side South Sister, 5,500 ft. (2 CRCM); Indian Ford Camp (13 PSM); Little Lava Lake (5 PSM); Lower Bridge (3 different locality descriptors) (5 OSUFW, 1 SOSC); Paulina Lake (2 SDNHM); 3 mi. W Paulina Lake, 5,700 ft. (5 MVZ); Redmond (7 UO); 1 mi. N, 7 mi. W Redmond (5 KU); Sisters (8 OSUFW, 3 PSM); 4 mi. S Sisters (1 PSM); 4.5 mi. W Sisters, Cold Springs campground (2 HSU); 0.25 mi. E Sisters (1 PSU); 7 mi. E Sisters (2 CM); 1.25 mi. N, 4 mi. W Sisters, Cold Spring campground (11 PSU); 5.5 mi. N, 3 mi. W Sisters, Black Butte, T14S, R9E, Sec. 3 (7 OSUFW); 4 mi. N, 6 mi. E Sisters, Holmes Rd., T14S, R11E, Sec. 21, 3,100 ft. (2 OSUFW); Sparks Lake, 5,426 ft. (1 MVZ, 2 OSUFW); Squaw Creek Canyon (1 CM); Terrebonne (6 localities within 5 mi. Terrebonne) (24 OSUFW, 1 PSU); 1 mi. N Three Creek Lake (2 PSM); NW slope Three Sisters (1 OSUFW); W base Three Sisters (3 OSUFW); Tumalo State Park (1 PSM); T17S, R11E, Sec. 15 (2 OSUFW). **Douglas Co.**—Bum Creek, Oxbow Burn (6 PSM); 5 mi. S Canyonville on HWY 99, near Roseburg (2 CM); 13 mi. E Canyonville, S fork Umpqua River (1 LACM); 19 mi. E Canyonville, 1 mi. S Milo (1 LACM); 3.2 mi. S, 3.6 mi. E Canyonville, T31S, R4W, Sec. 17, 2,000 ft. (1 UW[CR]); Diamond Lake (1 PSM, 9 SDNHM); N end Diamond Lake (1 OSUFW); 13 mi. N, 5 mi. E Diamond Lake, 0.5 mi. W Windingo Pass (2 PSU); 1.4 km N, 1.6 km W Drain, T22S, R5W, Sec. 7 (1 USNM); 10 km N, 10.6 km W Drain, T21S, R6W, Sec. 7 (1 USNM); 4.5 mi. N, 8 mi. W Drain, T21S, R4W, Sec. 27, 1,200 ft. (11 UW[CR]); 5.5 mi. N, 4 mi. W Drain, T21S, R6W, Sec. 15, 1,280 ft. (10 UW[CR]); 6 mi. N, 4 mi. W Drain, T21S, R6W, Sec. 15, 1,200 ft. (5 UW[CR]); 6.75 mi. N, 5.5 mi. E Drain, T21S, R4W, Sec. 7 (2 UW[CR]); 9.7 km S, 8.4 km W Elkton, T23S, R8W, Sec. 20 (2 USNM); 9.7 km S, 16.2 km W Elkton, T23S, R9W, Sec. 22 (1 USNM); 11.1 km S, 7.4 km W Elkton, T23S, R8W, Sec. 28 (7 USNM); 14.9 km S, 6.2 km W Elkton, T24S, R8W, Sec. 9 (2 USNM); 15.4 km S, 6 km W Elkton, T24S, R8W, Sec. 10 (2 USNM); 15.5 km S, 5.8 km W Elkton, T24S, R8W, Sec. 10 (6 USNM); 15.7 km S, 5.6 km W Elkton, T24S, R8W, Sec. 10 (4 USNM); 24.4 km S, 6 km W Elkton, T25S, R8W, Sec. 10 (3 USNM); 25.2 km S, 5.4 km W Elkton, T25S, R8W, Sec. 10 (2 USNM); Gardiner (20 FMNH, 9 MVZ); 0.5 mi. N Gardiner (3 PSM); Hoaglin (7 UO); Louis Cr., T28S, R3W, NW¼SW¼ Sec. 29 (1 SOSC); Lower Umpqua River* (5 PSM); Middle Creek, T31S, R6W, SW¼ Sec. 30 (1 SOSC); 5 mi. S Mt. Thielsen (3 MVZ); SW Reedsport (1 PSM); 3 mi. N Reedsport, T21S, R12W, Sec. 10 (1 PSU); 4 mi. S, 6.4 mi. W Riddle, Iron Mt., T31S, R7W, Sec. 11, 2,000 ft. (7 UW[CR]); Roseburg, North Umpqua River (157 UO); 7 mi. NW Roseburg (1 OSUFW); Scottsburg (5 FMNH); 7 mi. E Skeleton Mt., 4,360 ft.* (1 LACM); 2 km N, 25 km W Sutherlin, T25S, R8W, Sec. 10 (8 USNM); 2.2 km N, 26.2 km W Sutherlin, T25S, R9W, Sec. 9 (2 USNM); 2.4 km N, 24.6 km W Sutherlin, T25S, R8W, Sec. 10 (1 USNM); 2.8 km N, 26 km W Sutherlin, T25S, R8W, Sec. 9 (1 USNM); 3.4 km N, 11.9 km W Sutherlin, T25S, R7W, Sec. 1 (6 USNM); 3.6 km N, 11.9 km W Sutherlin, T25S, R7W, Sec. 1 (5 USNM); 3.8 km N, 12.1 km W Sutherlin, T25S, R7W, Sec. 1 (1 USNM); 4 km N, 12.1 km W Sutherlin, T25S, R7W, Sec. 1 (1 USNM); 4.2 km N, 12.1 km W Sutherlin, T25S, R7W, Sec. 1 (1 USNM); 4.3 km N, 12.1 km W Sutherlin, T25S, R7W, Sec. 1 (6 USNM); 4.4 km N, 12.1 km W Sutherlin, T25S, R7W, Sec. 1 (10 USNM); 0.75 mi. S Tenmile (1 OSUFW); 4 mi. N, 6 mi. W Tenmile (1 PSU); 6.5 km N, 6.7 km W Tenmile, T28S, R8W, Sec. 9 (3 USNM); 4.5 mi. N, 8 mi. E Tiller, T30S, R1W, Sec. 2 (1 CRCM); 4.5 mi. N, 5 mi. E Tiller (1 CRCM); 9 mi. N, 8.5 mi. E Tiller, T29S, R1W, Sec. 13 (5 CRCM); 10

mi. N, 9 mi. E Tiller, T29S, R1W, Sec. 12 (9 CRCM); 11 mi. N, 5 mi. E Tiller (2 CRCM); 11.5 mi. N, 5 mi. E Tiller, T28S, R1W, Sec. 26 (1 CRCM); 12 mi. N, 17 mi. E Tiller (7 CRCM); 12 mi. N, 18 mi. E Tiller, Castle Rock, T28S, R2E, Sec. 27 (1 OSUFW); 12 mi. N, 19 mi. E Tiller (3 CRCM); 12.5 mi. N,5 mi. E Tiller, T28S, R1W, Sec. 32 (3 CRCM); 12.5 mi. N, 5.5 mi. E Tiller, T28S, R1W, Sec. 31 (1 CRCM); 0.2 mi. N, 4.1 mi. W Umpqua P.O., T25S, R7W, Sec. 15, 800 ft. (20 UW[CR]); 2.2 mi. N, 9 mi. W Umpqua P.O., T25S, R8W, Sec. 15, 850 ft. (7 UW[CR]); 2.7 mi. N, 5 mi. W Umpqua, Riverview, T25S, R7W, Sec. 7, 950 ft. (11 UW[CR]); 4 mi. N, 1 mi. W Umpqua P.O., Lost Creek, T25S, R7W, Sec. 1, 1,500 ft. (3 UW[CR]); 6 mi. N, 10.5 mi. W Umpqua P.O., T24S, R8W, Sec. 29, 900 ft. (2 UW[CR]); Winchester Bay (2 PSM); 1 mi. S Winchester Bay (3 OSUFW); 1 mi. S, 1 mi. W Winchester Bay (1 PSU); 4.8 km S, 4.5 km W Yoncalla, T23S, R5W, Sec. 19 (1 USNM); 0.25 mi. on Smith River Rd. from junction with HWY 101 (1 PSU). **Gilliam Co.**—2.5 mi. S Heppner Junction (5 KU); 4 mi. E John Day River* (17 PSM); exit to Phillipi Canyon (42 PSM); Willows (3 OSUFW). **Grant Co.**—Blue Mt. Springs (2 PSM); Bull Prairie Recreation Area* (1 PSU); 12 mi. S, 5 mi. W Canyon City, Windy Point Rd. (2 PSU); 1.5 mi. SE Canyon City (2 LSU); Cold Spring, 8 mi. E Austin (10 MVZ); 20 mi. N John Day (1 UW[EB]); 15 mi. S John Day (1 UW[LB]); 1 mi. N, 10 mi. E Long Creek, Kahler Butte, T10S, R32E, Sec. 4 (1 OSUFW); 5.3 mi. N, 1.3 mi. E Long Creek (3 OSUFW); 2 mi. S, 1 mi. E Long Creek (1 OSUFW); Malheur National Forest, Lake Creek guard station, T16S, R33½E, Sec. 10 (1 OSUFW); Malheur National Forest, T16S, R34E, Sec. 10 (1 OSUFW); Malheur National Forest, T18S, R32E, Sec. 23 (2 OSUFW); North Fork Malheur River camp, Malheur National Forest 5,000 ft. (3 KU, 16 MVZ); 12 mi. N Mt. Vernon, Beech Creek (2 UW[WWC]); 5 mi. W Mt. Vernon (5 MVZ); 11 mi. S, 6 mi. E Prairie City, T15S, R36E, Sec. 25 (1 KU); N slope Strawberry Mt. (21 CRCM); E slope Strawberry Mt. (17 CRCM); Turtle Cove, John Day River (4 MVZ); Warm Springs Creek (2 EOSC). **Harney Co.**—E rim Alvord Basin* (1 EOSC); W edge Alvord Desert, 4,400 ft. (2 UW[WWC]); N of Alvord Ranch (1 EOSC); Andrews, Serrano Point (3 EOSC); 7 mi. S Andrews (3 MVZ); 0.25 mi. S, 0.25 mi. W Ant Hill, T29S, R33E, SW¼NE¼ Sec. 36, 5,000 ft. (6 KU); Blitzen Valley* (8 SDNHM); 2 mi. S Borax Spring, S end Alvord Lake (2 MVZ); Buena Vista (2 OSUFW, 23 SDNHM); 15 mi. N Burns (1 UW[EB]); 20 mi. N Burns (1 OSUFW); 31 mi. S Burns, Malheur National Wildlife Refuge (1 OSUFW); 50 mi. S Burns (1 OSUFW); 7 mi. N, 2 mi. E Burns (1 OSUFW); 12 mi. S, 15 mi. W Burns (4 OSUFW); 10 mi. S, 3 mi. E Burns, Wright's Point (1 OSUFW); 11 mi. S, 3 mi. E Burns, S side Wright's Point (2 SOSC); 12 mi. S, 2.5 mi. E Burns, Wright's Point (1 OSUFW); 22 mi. S, 8 mi. E Burns (1 OSUFW); 5 mi. NNE Burns (5 MVZ); Catlow Valley, Rock Creek Ranch (3 OSUFW); Coyote Butte, T27S, R30E, Sec. 13 (3 HSU); W edge Coyote Butte (2 CRCM); S of Coyote Butte (1 CRCM); 7 mi. S Crane, T26S, R33E, Sec. 11 (1 OSUFW); 1 mi. NW Crane, Crane Hot Springs (1 PSU); 1.5 mi. E Denio (1 MVZ); 3.75 mi. N, 3 mi. E Diamond, T29S, R33E, center E¼ Sec. 11 (4 KU); 4 mi. N, 6 mi. W Diamond (2 UW[WWC]); 5 mi. NW Diamond (6 UW[WWC]); Diamond Crater, T29S, R32E, Sec. 3 (2 HSU); Diamond Craters, 5 mi. N, 5 mi. W Diamond, 4,200 ft. (15 UW[WWC]); 5 mi. N Fields (2 SOSC); 13 mi. S Fields (1 OSUFW); 45 mi. N, 17 mi. E Fields, T30/31S, R35E, Sec. 35/2 (2 KU); 7 mi. S, 3 mi. E Fields (4 PSU); 8 mi. S, 4 mi. E Fields, 4,400 ft. (1 OSUFW); Fir Canyon, Steens Mt.* (1 UO); 2.4 mi. W Fish Lake, T32S, R32¾E, Sec. 36 (1 OSUFW); Frenchglen (4 PSM); 19.5 mi. N Frenchglen, T29S, R31E, NW¼SW¼ Sec. 9 (1 OSUFW); 20 mi. N Frenchglen, Sage Canyon, T29S, R31E, Sec. 9 (2 OSUFW); 6 mi. due E Frenchglen, T32S, R32½E, Sec. 14 (1 OSUFW); 17.5 mi. N, 1.5 mi. E Frenchglen, T29S, R31E (2 OSUFW); 18 mi. N, 1 mi. E Frenchglen, 5,000 ft. (1 OSUFW); 0.5 mi. S, 2.5 mi. W Frenchglen, T32S, R32E, NW¼ Sec. 9 (9 KU); 2 mi. S, 2.5 mi. E Frenchglen, T32S, R31E, W½ Sec. 17 (1 OSUFW); 4 mi. S, 10 mi. E Frenchglen, 6,500 ft. (10 UW[WWC]); 5 mi. S, 10 mi. E Frenchglen, 500 ft. (1 UW[WWC]); 5 mi. S, 11 mi. E Frenchglen, T32S, R32¾E, SW¼NE¼ Sec. 34, 6,600 ft. (1 KU); 2 mi. SE Frenchglen (2 UW[WWC]); 6 mi. SE Frenchglen, T32S, R32½E, Sec. 14 (3 OSUFW); 8 mi. SE Frenchglen, 5,500 ft. (2 UW[WWC]); N edge Harney Lake (1 UW[WWC]); Harney Sand Dunes, W end Malheur Lake (5 PSM); 5.25 mi. S, 10.5 mi. W Hines, T24S,

R28E, NW¼ Sec. 24 (1 KU); 7 mi. S, 12 mi. W Hines, T24S, R28E, center Sec. 26 (1 KU); Jackman Park, 2 mi. SE Fish Lake, 7,500 ft. (12 UW[WWC]); head of Kiger Gorge, Steens Mt. Rd., 9,000 ft. (7 UW[WWC]); 7.8 mi. E Lawen (7 CM); Malheur Lake (4 LACM, 8 UW[EB]); Malheur Environmental Field Station (5 EOSC, 3 PSU); 0.25 mi. S Malheur Environmental Field Station (1 PSU); Malheur National Wildlife Refuge (132 different locality descriptors) (120 CM, 6 HSU, 5 OSUFW, 2 SOSC, 1 UCONN, 1 UW[WWC]); Narrows (2 LACM, 1 OSUFW, 60 UO); 1 mi. S Narrows, 4,200 ft. (5 MVZ); 2 mi. SW Narrows, T27S, R31E, Sec. 3 (1 OSUFW); 5 mi. SW Narrows (1 MVZ); Page Springs campground, 4,200 ft. (5 UW[WWC]); 6 mi. SE Page Springs, Steens Mt. Rd., 5,500 ft. (1 UW[WWC]); 6 mi. up the 70 mi. loop road from Page Springs campground toward Fish Lake, 5,500 ft. (3 UW[WWC]); Pike Creek, 22 mi. N, 9 mi. E Fields, 4,400 ft. (1 UW[WWC]); 1.5 mi. S, 10 mi. W Riley, T24S, R25E, N½ Sec. 3 (1 KU); Roaring Springs* (3 UW[WWC]); top Serrano Point (4 EOSC); 1 mi. up Skull Creek Rd. (1 SOSC); Smyth Creek, 10 mi. E Diamond (9 MVZ); Squaw Butte Range, 36 mi. W Burns (1 MVZ); Steens Mt. (27 localities on Steens Mt.) (6 CM, 8 FMNH, 101 OSUFW, 2 PSM); Trout Creek, T39S, R36E, NE¼ Sec. 26 (4 PSM); Wright's Point (353 UO); Voltage (1 SDNHM); 2 mi. above P Ranch house, cliff above Blitzen River (5 UO); 3 mi. above P Ranch house, cliff above Blitzen River (52 UO). **Hood River Co.**—1 mi. S, 1.5 mi. E Badger Lake, 5,400 ft. (1 LSU); Brooks Meadows, 9 mi. ENE Mt. Hood (14 MVZ); 1 mi. E Cascade Locks, Oxbow Fish Hatchery (1 PSM); Cloud Cap Inn, N side Mt. Hood (1 PSU); Cooper Spur, NE side Mt. Hood, 6,600 ft. (1 CRCM); Eagle Creek, 3.9 mi. S, 3.5 mi. E Cascade Fish Hatchery (1 UW); Eagle Creek National Forest campground (10 LACM); Elk Cove, N side Mt. Hood, 5,500 ft. (1 CRCM); head east fork Herman Creek, 3,600 ft. (1 UW); Hood River Meadows (1 PSM); Lamberson Spur, E side Mt. Hood, 7,200 ft. (1 CRCM); Mt. Hood, Cooper Spur (2 OSUFW); Mt. Hood Meadows (1 PSU); 3 mi. W Mt. Hood summit, 5,400 ft. (2 LSU); 2 mi. S Odell (3 PSU); Snowshoe Lodge, N side Mt. Hood (1 PSU); N side Wahtum Lake, 3,700 ft. (3 UW); T3N, R9E, NW¼ Sec. 35 (9 PSM); T3N, R9E, SW¼ Sec. 35 (8 PSM). **Jackson Co.**—(1 SOSC); Ashland (24 SOSC); 4 mi. E Ashland (1 SOSC); 24 km E Ashland (1 SOSC); 3 mi. N, 10 mi. E Ashland (2 SOSC); 4.3 mi. NW Ashland (1 SOSC); 4 mi. SE Ashland (1 SOSC); 13 mi. SE Ashland (1 SOSC); Ashland Research Natural Area, T40S, R1E, Sec. 4 (13 SOSC); Brownsboro (1 SOSC); Butte Falls (3 OSUFW); 2 mi. E Butte Falls (2 LACM); Cedar Guard Station (1 SOSC); 1.4 mi. E junction Crane Prairie Rd. on Weyerhauser Rd. 107-63 (1 SOSC); 22 mi. W Crater Lake (4 UW[EB], 3 UW[WWC]); Dead Indian Rd.* (1 SOSC); 1.7 mi. E Dead Indian Rd. on Emigrant Creek Rd. (8 SOSC); 2 mi. E Dead Indian Rd. on Emigrant Creek Rd. (1 SOSC); junction Dead Indian Rd. and Conde Cr. Rd. (2 SOSC); 10 mi. E HWY 66 on Dead Indian Rd. (1 SOSC); Emigrant Lake (2 SOSC); 2.1 mi. E Emigrant Lake (1 SOSC); NW side Emigrant Lake (1 SOSC); SW tip Emigrant Lake (1 SOSC); Fish Lake (1 SOSC); Griffin Creek, Cemetary Hill (1 PSM); 10.8 mi. from HWY 66 on Dead Indian Rd. at Hooper Springs (7 SOSC); Jacksonville (1 SOSC); 2 mi. SW Jacksonville (4 LACM); Imnaha Creek, T33S, R4E, W½ Sec. 15 (1 SOSC); 5 mi. N, 2 mi. W Lincoln, Hyatt Lake (1 OSUFW); Medford (2 SOSC); 7 mi. S Medford (3 PSM); 6 mi. N, 5 mi. E Medford (1 OSUFW); Mt. Ashland, T40S, R1E, Sec. 9 (3 SOSC); Mt. Ashland Rd. (3 SOSC); 3 mi. S Mt. Ashland turnoff on old HWY 99 (1 SOSC); Mt. Ashland Rd. 0.3 mi. W Colestine Rd. (2 SOSC); 0.9 mi. W HWY 99 on Mt. Ashland Rd. (1 SOSC); 1.6 mi. W HWY 99 on Mt. Ashland Rd. (13 SOSC); 1.7 mi. W HWY 99 on Mt. Ashland Rd. (1 SOSC); 1.5 mi. W on Mt. Ashland access rd. (1 SOSC); 1.7 mi. S HWY 66 on Neil Creek Rd. (6 SOSC); Pilot Rock Rd., 3 mi. E HWY 99 (2 SOSC); 11 mi. N Prospect (2 CRCM); 11.5 mi. N, 0.5 mi. W Prospect (2 CRCM); 13.5 mi. N, 2.5 mi. W Prospect (1 CRCM); 14 mi. N, 0.5 mi. W Prospect, T30S, R2E, Sec. 24/25 (7 CRCM); 2.5 mi. N, 8.5 mi. E Prospect (12 CRCM); 5.5 mi. N, 4 mi. E Prospect (6 CRCM); Rogue River (1 SOSC); 9 mi. N Rogue River (4 SOSC); Sams Valley (1 SDNHM); Shoat Springs, 6.8 mi. S HWY 66 on Copco Rd. (2 SOSC); 2 mi. S HWY on Soda Mt. Rd. (1 SOSC); Spring Creek, T33S, R4E, NE¼ Sec. 22 (3 SOSC); Talent (1 SOSC); S of Talent (1 SOSC); 4 mi. S Talent (1 SOSC); 2 mi. N, 1 mi. E Union Creek (5 CRCM); 4 mi. up Tolman Creek Rd. from HWY 99 (1 SOSC); 6.1 mi.

Up Toleman Cr. Rd. from junction 392 and 3901 (1 SOSC); 2.5 mi. E USFS Rd. 3706 on USFS Rd. 3705 (1 SOSC); 7.1 mi. S HWY 66 on Lepco Rd. (1 SOSC); T36S, R2W, NE¼ Sec. 2 (1 SOSC); T38S, R2E, NE¼NE¼ Sec. 25 (2 SOSC); T39S, R3E, SE¼NE¼ Sec. 35 (2 SOSC); T40S, R1E, Sec. 10 (1 SOSC). **Jefferson Co.**—(3 PSM); Ashwood (1 PSU); Camp Sherman (9 PSM); Culver (4 PSM); 3 mi. E Culver (3 PSM); 5 mi. E Culver (4 PSM); 4 mi. NNW Culver (6 PSM); 6 mi. SE Culver (1 PSM); 4 mi. SSW Culver (1 PSM); Gateway (5 OSUFW); Hay Creek (1 OSUFW); 12 mi. SE Hay Creek (1 OSUFW); 3 mi. S Madras, T11S, R13E, SW¼ Sec. 36 (1 OSUFW); 10 mi. S Madras (2 PSU); 7 mi. E Madras (1 OSUFW); 10 mi. S, 0.75 mi. W Madras (4 PSU); 2 mi. S, 5 mi. E Madras (14 CM); 8 mi. S, 3 mi. E Madras (2 PSU); 15 mi. S, 2.5 mi. E Madras (1 OSUFW); Metolius (52 UO); 1 mi. N, 1.5 mi. W Opal City, T13S, R12E, NW¼ Sec. 12 (2 OSUFW); Santiam Pass (1 KU); Suttle Lake (6 KU, 2 OSUFW); 1 mi. E Suttle Lake (1 OSUFW); Warm Springs (2 OSUFW); 20 mi. W Warm Springs, Mill Creek (2 OSUFW); Willow Creek* (1 PSU); T13S, R11E, Sec. 35 (3 PSM); T13S, R12E, Sec. 12 (11 PSM). **Josephine Co.**—Bald Mt. (3 PSU); Bolan Lake (23 PSM, 6 SDNHM); 10 mi. E Cave Junction (2 KU); 17 mi. E Cave Junction (2 KU); 3 mi. S, 10 mi. E Cave Junction (1 SOSC); 1 mi. S, 0.5 mi. W Cave Junction, 1,400 ft. (1 LACM); 10 mi. S, 8 mi. E Cave Junction (1 OSUFW); 10 mi. S, 10 mi. E Cave Junction, Bolan Lookout, T41S, R6W, Sec. 7 (5 OSUFW); Cedar guard station, T39S, R6W, Sec. 30, 1,900 ft. (1 PSU); Grants Pass (5 CM, 5 FMNH, 2 PSM); Lava Camp Lake (1 KU); Oregon Caves National Monument (3 OSUFW, 1 CRCM); 5 mi. W O'Brian (1 SOSC); 8.5 mi. W Selma (1 LACM); Sourgrass Cr., T35S, R9W, SE¼NE¼ Sec. 2 (1 SOSC). **Klamath Co.**—Anna Creek, 4 mi. NW Ft. Klamath (1 MVZ); 1 mi. S, 12 mi. E Chiloquin (7 CM); 5 mi. NE Chiloquin (1 PSM); Crater Lake (25 localities within 10 mi. Crater Lake) (14 MVZ, 4 OSUFW, 7 PSU, 94 UO); 4.7 mi. S, 2.5 mi. E Dairy (2 PSU); Ft. Klamath (5 OSUFW); 10 mi. SW Ft. Klamath (4 UMMZ); Heavenly Twin Lake (1 OSUFW); 3 mi. S, 6 mi. W Keno (1 OSUFW); Klamath Falls (13 CM); 16 mi. E Klamath Falls, 4,200 ft. (2 OSUFW); 9 mi. S, 13 mi. W Klamath Falls (1 OSUFW); 2 mi. N, 4 mi. E Lorella (1 OSUFW); 4 mi. SW Merrill (2 OSUFW); Miller Hill, T39S, R9E, Sec. 28 (1 OSUFW); Odell Creek, 4,700 ft. (1 LACM); 3 mi. W Swan Lake, 4,300 ft. (2 MVZ); Upper Klamath Lake (4 FMNH); Upper Klamath Lake, Rocky Point (7 PSM); 4 mi. N HWY 97 on HWY 62 (1 OSUFW); 3 mi. W HWY 39 on Old Midland Rd. (1 OSUFW). **Lake Co.**—Adel (1 SDNHM); 9 mi. S Adel, Twentymile Creek, Warner Mts. (3 MVZ, 3 OSUFW); 9 mi. N, 8 mi. E Adel (1 PSU); 10 mi. NE Adel (1 PSU); 14 mi. SW Adel, Twentymile Creek, Warner Mts., T41S, R21E, E edge Sec. 8 (6 OSUFW); Alkali Lake (4 MVZ); 2.5 mi. NE Alkali Lake Ranch (2 MVZ); 0.5 mi. S Bobs Lake (1 SOSC); Buck Pasture, Hart Mt. Antelope Refuge (1 KU); Cabin Lake Ranger Station (1 SOSC); 6.5 mi. N Chandler State Park (1 SOSC); 11.6 mi. N Chandler State Park (1 SOSC); Fort Rock (2 SDNHM); Fort Rock State Park (2 OSUFW, 1 PSM); 3.5 mi. N, 4 mi. E Fort Rock (3 HSU); Gearhart Mt. (30 PSM); Guano Creek, site Old Camp Warner, T36S, R25E, Sec. 25 (6 OSUFW); Hart Mt. (2 PSM, 2 SDNHM); 2 mi. W hdqrs. Hart Mt. Antelope Refuge (2 KU); 3.2 mi. W hdqrs. Hart Mt. Antelope Refuge (1 KU); Hot Springs camp, Hart Mt. Antelope Refuge (3 localities within 0.5 mi. Hot Springs camp) (5 SOSC); 3 mi. W Lakeview (1 MVZ); 2 mi. E Lakeview (6 MVZ); 14 mi. E Lakeview, Vernon Spring (5 CM); 25 mi. SE La Pine, T24S, R14E, Sec. 17 (1 OSUFW); Quartz Mt., 6 mi. S HWY 66 toward Laughton Lake (1 SOSC); Silver Lake (1 OSUFW, 80 UO); 9 mi. S Silver Lake (1 PSU); 7 mi. N, 1 mi. W Silver Lake (4 PSU); 16 mi. S, 8 mi. W Silver Lake, T31S, R13E, Sec. 20 (1 OSUFW); 19 mi. S, 7 mi. W Silver Lake, T31S, R13E, Sec. 20 (1 OSUFW); Summer Lake (1 OSUFW); 1 mi. S Summer Lake (1 OSUFW); Valley Falls (1 SOSC); 2 mi. N Valley Falls (3 SOSC); 2 mi. N Valleytoll* (1 SOSC); 1.66 mi. S, 6.125 mi. W Walton* (7 PSU); Warner Canyon (48 UO); Willow Creek, Hart Mt. Antelope Refuge (21 KU); Willow Creek Forest camp (1 SOSC); 5.4 mi. E headquarters on Frenchglen Rd. (1 SOSC). **Lane Co.**—5 mi. N, 6 mi. W Alpha, T15S, R9W, Sec. 31 (1 OSUFW); 0.5 mi. S, 1 mi. W Alvadore, 430 ft. (2 OSUFW); 2 mi. S, 1 mi. E Austa (12 CM); Belknap Springs (3 OSUFW); 4 mi. E mouth Big Creek (6 KU); 4.4 km N, 6.8 km E Blue River, T16S, R5E, Sec. 7 (1 USNM); 5 km N, 7.6 km E Blue River, T16S, R5E, Sec.

6 (1 USNM); 6.8 km N, 7.6 km E Blue River, T15S, R5E, Sec. 32 (2 USNM); 2.5 mi. S, 7 mi. E Blue River (2 CRCM); 2 mi. W Blue River, Elk Creek, T16S, R4E, Sec. 19 (3 OSUFW); 6 mi. E Blue River, T16S, R5E, Sec. 20 (2 OSUFW); 9 mi. E Blue River, T16S, R3E, Sec. 25/26 (4 CRCM); 1.5 mi. N, 2 mi. E Blue River, 1,600 ft. (7 CRCM); 4 mi. N, 4.5 mi. E Blue River, T16S, R5E, Sec. 32 (1 CRCM); 4 mi. N, 9 mi. E Blue River, Mack Creek, T15S, R5E, Sec. 35 (17 OSUFW); 5 mi. N, 9 mi. E Blue River (2 CRCM); 5 mi. N, 10.5 mi. E Blue River, T15S, R5E, Sec. 30 (1 CRCM); 5.5 mi. N, 13.5 mi. E Blue River, T16S, R6E, Sec. 26 (3 CRCM); 7.25 mi. N, 9.25 mi. E Blue River, T15S, R5E, Sec. 13 (1 CRCM); 15 mi. SSE Blue River (8 UW-X); 5 mi. N Cougar Dam, T16S, R5E, Sec. 6 (5 UW[OG]); 1 mi. N, 10.5 mi. E Cougar Dam, T16S, R6E, Sec. 25, 2,900 ft. (8 UW[OG]); 5 mi. SE Crow, 400 ft. (1 OSUFW); Cushman (6 CM); Eugene (2 OSUFW, 83 UO, 5 UW-X); 5 mi. S Eugene, Spencer Butte (1 OSUFW); 5 mi. W Eugene (4 LACM); 4 mi. N, 7 mi. W Eugene, near Fern Ridge reservoir (2 OSUFW); Florence (6 FMNH); 9 mi. N Florence (8 LSU); 12 mi. N Florence (1 MVZ); 5 mi. S Florence, W side Siltcoos Lake (1 LACM); 3 mi. N, 1 mi. W Florence (2 OSUFW); 2 mi. SW Florence (3 PSM); 9 mi. N, 1 mi. W Gardiner (8 PSM); Gibson Island* (23 UW-X); Gold Lake, 4,800 ft. (5 LACM); Goshen (6 UW-X); 0.5 mi. N, 3 mi. E Greenleaf, 400 ft. (1 OSUFW); H. J. Andrews Experimental Forest, T15S, R5E, NW¼ Sec. 24, 3,500 ft. (58 PSM); H. J. Andrews Experimental Forest, T15S, R5E, NW¼ Sec. 28, 2,000 ft. (53 PSM); H. J. Andrews Experimental Forest, T15S, R5E, NE¼ Sec. 32, 1,650 ft. (52 PSM); H. J. Andrews Experimental Forest, T15S, R5E, NE¼ Sec. 32, 2,000 ft. (94 PSM); H. J. Andrews Experimental Forest, T15S, R5E, SW¼ Sec. 33, 2,500 ft. (103 PSM); H. J. Andrews Experimental Forest, T15S, R5E, NW¼ Sec. 34, 2,700 ft. (20 PSM); H. J. Andrews Experimental Forest, T16S, R4E, NE¼ Sec. 19, 1,800 ft. (73 PSM); H. J. Andrews Experimental Forest, T16S, R4E, NE¼ Sec. 24, 1,400 ft. (127 PSM); H. J. Andrews Experimental Forest, T16S, R5E, NW¼ Sec. 6, 1,600 ft. (92 PSM); H. J. Andrews Experimental Forest, T16S, R5E, SW¼ Sec. 6, 1,600 ft. (58 PSM); Hand Lake, Willamette National Forest, T15S, R7½E (1 OSUFW); Heceta, Smith Ranch (1 OSUFW); 4 mi. NE Heceta Head, Big Creek (1 OSUFW); Horsepasture Mt. (1 OSUFW); 2 mi. SE Jasper (1 KU); 12 mi. W Junction City (1 PSM); 0.4 km N, 18.5 km W Lorane, T20S, R7W, Sec. 12 (3 USNM); 0.8 km N, 19.2 km W Lorane, T20S, R7W, Sec. 12 (1 USNM); 2 km N, 6.4 km W Lorane, T20S, R5W, Sec. 5 (2 USNM); 2.8 km N, 6.2 km W Lorane, T20S, R5W, Sec. 5 (1 USNM); 3 km N, 19.5 km W Lorane, T20S, R7W, Sec. 1 (1 USNM); 3.2 km N, 26 km W Lorane, T19/20S, R7W, Sec. 32/5 (1 USNM); 6.8 km N, 11.1 km W Lorane, T19S, R6W, Sec. 23 (1 USNM); 4.8 km N, 17.5 km W Lorane, T19S, R6W, Sec. 31 (1 USNM); McKenzie Bridge (4 OSUFW, 2 PSM); 2.6 km S McKenzie Bridge, T16S, R5E, Sec. 26 (1 USNM); 2.5 km N, 4.5 km W McKenzie Bridge, T15S, R5E, Sec. 31 (1 USNM); 3.3 km N, 2.4 km E McKenzie Bridge, T16S, R5E, Sec. 3 (1 USNM); 6 km N, 0.6 km E McKenzie Bridge, T15S, R5E, Sec. 25 (3 USNM); 6 km N, 7.5 km E McKenzie Bridge, T15S, R6E, Sec. 26 (2 USNM); 1.6 km S, 5.2 km W McKenzie Bridge, T16S, R5E, Sec. 20 (4 USNM); 2.6 km S, 1.4 km E McKenzie Bridge, T16S, R5E, Sec. 25 (2 USNM); 3.8 km S, 14.8 km E McKenzie Bridge, T16S, R7E, Sec. 33 (15 USNM); 4.4 km S, 14.3 km E McKenzie Bridge, T16S, R7E, Sec. 32 (2 USNM); 5.4 km S, 8.75 km E McKenzie Bridge, T16S, R6E, Sec. 35 (3 USNM); 4 mi. N McKenzie Bridge, T15S, R5E, Sec. 25, 2,600–2,700 ft. (41 UW[OG]); 2.5 mi. E McKenzie Bridge, 1,500 ft. (1 OSUFW); 10 mi. E McKenzie Bridge, Lost Creek R.S. (1 OSUFW); 2 mi. N, 2 mi. W McKenzie Bridge, T16S, R5E, Sec. 3/4, 3,800 ft. (10 UW[OG]); 2 mi. S, 0.5 mi. E McKenzie Bridge, T16S, R5E, Sec. 26, 2,800 ft. (5 UW[OG]); 3 mi. S, 8.5 mi. E McKenzie Bridge, T16S, R7E, Sec. 32, 3,500 ft. (85 UW[OG]); 3.5 mi. S, 9 mi. E McKenzie Bridge, T16S, R7E, Sec. 32, 3,500 ft. (10 UW[OG]); Mercer (2 OSUFW); 6 mi. SW Monroe (2 PSM); 4 mi. N Nimrod, T16S, R3E, Sec. 14 (2 CRCM); Oakridge (1 UO); 5 mi. W Oakridge (6 UW-X); 2.5 mi. N Rainbow, T16S, R5E, Sec. 5 (1 UW-X); 4 mi. N Rainbow, T15S, R5E, Sec. 32, 1,700 ft. (22 UW[OG]); Siltcoos Forest camp (2 OSUFW); 10 mi. N, 1 mi. W Swisshome, T15S, R9W, Sec. 30/31 (4 OSUFW); 4 mi. W Terrebonne (1 OSUFW); W base Three Sisters (4 OSUFW); NW Base Three Sisters, Frog camp (3 OSUFW); Vida (3 OSUFW, 1 SOSC); 4.5 mi. S Wolf Rock, T15S, R5E, Sec. 24, 3,400 ft.

(6 UW[OG]); 4 km S, 3.2 km E Yachats, T15S, R12W, Sec. 16 (1 USNM). **Lincoln Co.**—(2 PSM); Alsea Bay (13 PSM); 0.5 mi. N, 1 mi. E Burnt Woods, 500 ft. (3 OSUFW); Cape Perpetua Recreation Area (1 MVZ); Cascade Head (2 PSM); Cascade Head Experimental Forest (14 PSM); 7 mi. S, 10 mi. W Harlan, Death Ridge, T13S, R10W, Sec. 14 (3 KU); 1 mi. N, 5 mi. E Kernville (1 OSUFW); 6 mi. S, 0.5 mi. E Lost Creek State Park (1 OSUFW); Newport (2 FMNH, 1 MVZ, 1 PSM, 2 SDNHM); 2 mi. N Newport (2 UMMZ); 2 mi. S Newport, Yaquina Bay (2 UMMZ); 5 mi. S Newport (1 PSU); 7 mi. S Newport, Beaver Creek (24 UMMZ); 0.5 mi. E Newport (2 UMMZ); 1 mi. E Newport (1 OSUFW, 4 UMMZ); 2 mi. E Newport, Beaver Creek (4 LACM, 11 UMMZ); 5 mi. E Newport (2 OSUFW); 0.5 mi. NE Newport (2 UMMZ); South Beach (2 OSUFW); Taft (2 KU, 1 OSUFW); Waldport (1 PSU); 0.6 km N, 16.9 km E Waldport, T13S, R10W, Sec. 13/14 (1 USNM); 0.9 km N, 16.9 km E Waldport, T13S, R10W, Sec. 13 (1 USNM); Yachats (3 PSM); 1.6 km N, 5.8 km E Yachats, T14S, R11W, Sec. 17 (1 USNM); 4.8 km S, 3.2 km E Yachats, T15S, R12W, Sec. 12 (4 USNM); Yaquina (4 LACM). **Linn Co.**—(1 OSUFW, 4 PSM); Albany (3 OSUFW, 1 SOSC); 2 mi. S Albany, 400 ft. (1 OSUFW); 5 mi. S, 7 mi. W Albany (1 OSUFW); 6.5 mi. S, 4.5 mi. W Albany (1 OSUFW); 3 mi. S, 2 mi. E Albany, 400 ft. (1 OSUFW); 7 mi. SW Albany (1 OSUFW); 1.5 mi. N Big Lake (10 PSM); 7 mi. N, 8 mi. E Blue River, T15S, R5E, Sec. 15 (1 CRCM); 7.5 mi. N, 7.5 mi. E Blue River, T15S, R5E, Sec. 15 (4 CRCM); 7.75 mi. N, 8 mi. E Blue River (1 CRCM); 8 mi. N, 8 mi. E Blue River, 3,600 ft. (2 CRCM); 9.5 mi. N, 9 mi. E Blue River, T15S, R5E, Sec. 2 (1 CRCM); 9 mi. N, 8 mi. E Blue River (1 CRCM); 2 mi. E Cascadia, Canyon Creek Rd., T13S, R3E, Sec. 33, 1,000 ft. (17 OSUFW); 5 mi. E Cascadia, 1,200 ft. (2 OSUFW); 7 mi. E Cascadia, 1,200 ft. (3 OSUFW); 8 mi. E Cascadia, 1,200 ft. (1 OSUFW); 18 mi. E Cascadia, Hackleman Creek (7 OSUFW); 21 mi. E Cascadia, Hackleman Creek (1 OSUFW); 1 mi. N, 28 mi. E Cascadia (12 CM); 1.2 mi. N, 8.5 mi. E Cascadia (3 OSUFW); 1.3 mi. N, 15 mi. E Cascadia (1 OSUFW); 1.75 mi. N, 29 mi. E Cascadia (6 OSUFW); 1.8 mi. N, 23 mi. E Cascadia (1 OSUFW); 0.3 mi. S, 2 mi. E Cascadia (1 OSUFW); 1 mi. S, 2.5 mi. E Cascadia (3 OSUFW); 1 mi. S, 6 mi. E Cascadia, Falls Creek (1 OSUFW); 0.5 mi. E Corvallis (2 OSUFW); 1 mi. E Corvallis, 250 ft. (1 OSUFW); 3 mi. S, 4 mi. E Corvallis (2 OSUFW); 0.6 mi. E Crabtree (2 PSU); Fish Lake (6 PSM); H. J. Andrews Experimental Forest, T14S, R6E, NE¼ Sec. 20, 4,400 ft. (92 PSM); H. J. Andrews Experimental Forest, T14S, R6E, NE¼ Sec. 28, 4,600 ft. (76 PSM); H. J. Andrews Experimental Forest, T15S, R5E, SE¼ Sec. 11, 3,200 ft. (35 PSM); H. J. Andrews Experimental Forest, T15S, R5E, SW¼ Sec. 13, 3,000 ft. (64 PSM); H. J. Andrews Experimental Forest, T15S, R6E, NW¼ Sec. 7, 5,000 ft. (132 PSM); H. J. Andrews Experimental Forest, T15S, R6E, NE¼ Sec. 7, 4,300 ft. (98 PSM); Lebanon (1 OSUFW); Lost Prairie, Santiam Pass, T13S, R6E, Sec. 27 (1 OSUFW, 14 PSU); 2 mi. W Lyons, 700 ft. (2 OSUFW); Marion Creek, North Santiam River, 2,400 ft. (4 MVZ); 9.2 km N, 1.2 km W McKenzie Bridge, T15S, R5E, Sec. 14 (1 USNM); 10.25 km N, 0.8 km W McKenzie Bridge, T15S, R5E, Sec. 14 (3 USNM); 13.2 km N, 0.8 km E McKenzie Bridge, T15S, R5E, Sec. 1 (1 USNM); 0.3 mi. N Peoria, 237 ft. (1 OSUFW); 1.5 mi. N Peoria, T13S, R4W, Sec. 11 (1 OSUFW); 3 mi. N Peoria (1 OSUFW); 1.5 mi. N, 1 mi. W Peoria, sea level (2 OSUFW); 8 mi. E Sweet Home, 1,000 ft. (1 OSUFW); 8.9 mi. E Sweet Home, 1,000 ft. (1 OSUFW); 32.5 mi. E Sweet Home (2 PSU); 4 mi. W Tangent (1 OSUFW); 5 mi. W Tangent, Colorado Lake, 250 ft. (2 OSUFW); 6 mi. W Tangent (1 OSUFW); 2 mi. N, 7 mi. W Tangent, 25 ft. (3 OSUFW); 1 mi. S, 2 mi. W Tangent, Calapooya River, T12S, R4W, Sec. 27 (36 OSUFW); 8 mi. E Upper Soda, Tombstone Prairie, T13S, R5E, Sec. 35, 4,000 ft. (9 OSUFW, 1 SUVM); 9 mi. E Upper Soda, Hackleman Creek, 4,000 ft. (3 OSUFW); 13 mi. E Upper Soda, Toad Creek Rd., 3,500 ft. (4 OSUFW, 1 SUVM); 13.7 mi. E Upper Soda, Toad Creek (1 OSUFW); 17 mi. E Upper Soda, 3,000 ft. (2 OSUFW); 0.5 mi. S, 7 mi. E Upper Soda, Ram Creek, 3,800 ft. (2 OSUFW); 1 mi. S Wolf Rock, T15S, R5E, Sec. 1, 3,300 ft. (19 UW[OG]); 1 mi. S, 1 mi. W Wolf Rock, T15S, R5E, Sec. 2, 3,000 ft. (8 UW[OG]); 4 mi. S, 1 mi. W Wolf Rock, T15S, R5E, Sec. 14, 2,600 ft. (3 UW[OG]). **Malheur Co.**—(1 PSM); Butch Lake, S end Jordan Crater lava flow, T29S, R44E, NE¼ Sec. 6 (12 PSM); Bogus Lake (1 PSM); Brogan (5 OSUFW); Cottonwood Creek, T18S, R42E, center Sec. 3 (2 PSM); Cottonwood

Creek, T26S, R38E, NW¼ Sec. 29 (5 PSM); Crowley guard station, T27S, R39E, N½ Sec. 17 (3 PSM); 3 mi. N Danner (2 HSU); 6.4 mi. W Harper Junction, T20S, R41E, Sec. 34 (1 PSU); 1 mi. W Ironside (14 MVZ); Jordan Valley (2 PSM); 3 mi. N McDermit (9 PSM); Pharmacy Butte, 0.5 mi. W Jordan Valley (2 PSM); Pole Creek, T19S, R39E, NW¼NE¼ Sec. 31 (1 PSM); 25 mi. W Rome, T31S, R37E, Sec. 24 (2 PSU); 0.5 mi. S Rome (1 MVZ); 1 mi. S Rome (9 MVZ); 40 mi. S, 0.5 mi. E Rome, T38S, R43/44E, SE¼ Sec. 25 (1 OSUFW); Soldier Creek exclosure, 8.5 mi. WSW Jordan Valley, T30S, R45E, SW¼ Sec. 15, 4,420 ft. (2 PSM); Succor Creek Rd. (2 PSU); Succor Creek State Park, T23S, R46E, NW¼ Sec. 15, 2,460 ft. (8 PSM); Vale (4 UO); 2 mi. N Vale (3 PSM); 3 mi. N Vale (4 MVZ); 3 mi. S Vale (6 PSM); 5.75 mi. S Vale, T19S, R45E, NE¼SW¼ Sec. 20, 2,700 ft. (3 PSM); 7 mi. SSE Vale, T19S, R45E, NW¼NW¼ Sec. 26, 2,700 ft. (1 PSM); T29S, R44E, NE¼ Sec. 6 (2 PSM). **Marion Co.**—4.5 mi. S, 1.5 mi. E Bagby Hot Springs, 4,000 ft. (7 LSU); 1 mi. N, 3 mi. E Brooks (1 OSUFW); Detroit (3 OSUFW); 10 mi. E Detroit (2 PSU); 2 mi. E Jefferson, 400 ft. (1 OSUFW); Salem (2 KU, 1 OSUFW); Silver Creek Recreation area (1 UW[EB]); 1 mi. N, 1 mi. E Talbot (14 OSUFW); Woodburn (2 PSM). **Morrow Co.**—4 mi. W Boardman, T4N, R24E, Sec. 10 (1 PSU); 4.6 mi. W Boardman (2 KU); 0.5 mi. S, 5 mi. W Boardman (7 PSU); 6 mi. S, 10 mi. E Boardman (1 OSUFW); Cecil, 6 mi. SE McIntyre Camp (1 PSM); Coyote Spring Wildlife Refuge (1 PSU); 3 mi. S, 2 mi. E Hardman (4 KU); Heppner (2 UW[WWC]); 2 mi. SE Heppner (5 UW[WWC]); 3 mi. W Irrigon (1 OSUFW); T5N, R26E, SE¼ Sec. 30 (6 PSM); T5N, R26E, NW¼ Sec. 31 (1 PSM). **Multnomah Co.**—(1 PSU); 8 mi. SE Bonneville (2 PSM); Bridal Veil Falls Creek (2 PSU); Gresham, Sandy River (1 PSU); Hayden Island (1 PSU); Horsetail Falls, 39 mi. E Portland (3 PSU); Johnson Creek (1 PSU); Larch Mt. (1 PSU); Latourelle (8 OSUFW); 1 mi. N, 1 mi. E Latourell Falls, 500 ft. (1 OSUFW); Portland (1 CM, 3 KU, 7 LACM, 18 MVZ, 3 OSUFW, 19 PSM, 43 PSU, 41 SDNHM, 3 UMMZ); Portland, Kelly Butte (1 OSUFW); Portland, Reed Canyon (3 PSU); Portland, Rocky Butte (5 PSU); 2 mi. N, 10 mi. W Portland, 900 ft. (1 OSUFW); 5 mi. S, 3 mi. W Portland (2 PSU); 6 mi. N, 11 mi. E Sandy (1 CRCM); 8 mi. N, 11 mi. E Sandy (5 CRCM); 9 mi. N, 10 mi. E Sandy (1 CRCM); 10 mi. N, 10 mi. E Sandy, T1N, R6E, Sec. 27 (3 CRCM); 10 mi. N, 11 mi. E Sandy (1 CRCM); 11 mi. N, 10 mi. E Sandy (3 CRCM); 5 mi. E Sandy River, 5 mi. N HWY 30 (1 PSU); 1 mi. S, 0.2 mi. W Sellwood Bridge (1 PSU); 10 mi. S, 0.2 mi. W Sellwood Bridge (1 PSU); Springdale (1 PSU); 11 mi. N Zigzag (1 CRCM); 10 mi. N, 1.5 mi. W Zigzag (2 CRCM); junction old and new German Town rds. (1 PSU); T1N, R1W, Sec. 3 (1 PSU); T1N, R1E, S½ Sec. 20 (1 PSM); T1N, R1E, SW¼ Sec. 21 (6 PSM); T1S, R1E, Sec. 5 (34 PSU); T1S, R1E, Sec. 13 (3 PSU); T1S, R1E, Sec. 32 (3 PSU); T1N, R2E, Sec. 5 (1 PSU); T1N, R2E, Sec. 21 (1 PSU); T1N, R2E, Sec. 24 (1 PSU); T1N, R2E, Sec. 29 (6 PSU); T1N, R2E, Sec. 30 (3 PSU); T1N, R4E, NW¼ Sec. 20 (17 PSM); T1N, R5E, NE¼ Sec. 29 (1 PSM); T1N, R6E, Sec. 29 (2 PSU). **Polk Co.**—1 mi. S, 2 mi. W Buena Vista (1 OSUFW); 3 mi. S, 2 mi. W Dalles (1 OSUFW); 3.5 mi. S, 2 mi. W Gold Creek, Gooseneck Creek, T7S, R7W, SE¼ Sec. 5 (9 OSUFW); 3.75 mi. S, 2 mi. W Gold Creek, Gooseneck Creek, T7S, R7W, Sec. 5 (3 OSUFW); 4 mi. S, 1.5 mi. W Gold Creek, T7S, R7W, center Sec. 5 (1 OSUFW); McCoy (3 CM, 2 FMNH). **Sherman Co.**—1 mi. S, 0.5 mi. W Biggs (10 CM); mouth Deschutes River, Columbia River (9 MVZ); Miller, head Deschutes River (1 OSUFW); Miller, mouth Deschutes River (6 OSUFW); T2N, R16E, SW¼ Sec. 1 (3 PSM); T2N, R16E, SE¼ Sec. 2 (11 PSM); T2N, R16E, N½ Sec. 9 (8 PSM). **Tillamook Co.**—0.5 mi. N Bay City, 50 ft. (1 OSUFW); Blaine (2 KU, 2 UMMZ); Cascade Head Experimental Forest (7 PSM); Drift Creek Rd. 400 yds N HWY 6 (1 PSU); Neahkahnie Beach (1 PSU); Hebo Lake (2 PSM); Mt. Hebo (1 OSUFW); Nehalem (1 PSU); 1 mi. W Nehalem (1 OSUFW); Netarts (2 LACM, 19 OSUFW, 4 PSM, 2 SDNHM); 1 mi. W Netarts (1 OSUFW); 3 mi. S, 1.5 mi. E Netarts, T2S, R10W (1 OSUFW); Oceanside (1 OSUFW); Oceanview (2 OSUFW); 2.75 mi. N, 3.5 mi. W Sand Lake, Cape Lookout (1 PSU); Tillamook (8 KU, 2 LACM, 6 MVZ, 3 OSUFW, 3 SDNHM); 5 mi. S Tillamook, Cape Lookout (1 PSU); 7 mi. SE Tillamook (2 OSUFW); between Tillamook and Netarts (3 KU). **Umatilla Co.**—Bear Creek, 10 mi. W Tollgate (2 UW[EB]); Echo (2 SDNHM); Langdon Lake (3 KU, 2 MSUMZ, 1 MVZ, 1 UW[EB], 7 UW[WWC]);

MacDougal camp, 6 mi. W Tollgate (1 UW[EB]); 2.5 mi. S MacDougal camp, 7 mi. W Tollgate (6 UW[EB]); near Meacham (1 CM); 6.8 mi. N Meacham, T1N, R36E, SW¼ Sec. 20, 4,290 ft. (3 PSU); 8.4 mi. N Meacham, T1N, R36E, NW¼ Sec. 33, 4,390 ft. (3 PSU); 2 mi. NW Meacham (1 MVZ); Meacham Lake (3 MVZ); Milton-Freewater (1 UW[WWC]); Pendleton (1 OSUFW, 3 SDNHM); 5 mi. E Pendleton (1 OSUFW); Rock Creek, 10 mi. W Tollgate (3 UW[EB]); Tollgate (1 MSUMZ, 8 UW[WWC]); 2 mi. W Tollgate (1 UW[WWC]); 4 mi. W Tollgate (4 UW[EB], 2 UW[WWC]); 5 mi. W Tollgate (2 UW[EB], 2 UW[WWC]); 2 mi. E Tollgate (1 UW[WWC]); 1 mi. N, 3 mi. W Tollgate (1 UW[WWC]); 3 mi. NW Tollgate (1 UW[WWC]); 1 mi. S Umatilla (1 PSU); 15 mi. E Weston (1 UW[WWC]); 16 mi. E Weston (8 UW[WWC]); 20 mi. E Weston, Tollgate (1 UW[WWC]); 4 mi. SE Weston (2 UW[WWC]); 20 mi. SE Weston, Tollgate (2 UW[WWC]); 300 yds NE junction I-80N and Barnhart Rd.* (1 SOSC); 1 mi. S stateline, Cottonwood Creek (2 UW[WWC]); T1N, R36E, Sec. 33 (1 PSU). **Union Co.**—Case slough (1 EOSC); Catherine Creek, 7 mi. E Telocaset, 3,500 ft. (2 MVZ); Clarks Creek guard station, Elgin (4 CRCM); Grande Ronde valley (1 EOSC); Ladd Marsh (2 EOSC); La Grande (3 EOSC); 7 mi. SE La Grande on Foothill Rd. (1 EOSC); 8 mi. SE La Grande (1 EOSC); Spout Springs lookout* (1 UW[WWC]); Starkey Experimental Forest, Bear Creek, T3S, R34E, NW¼ Sec. 2 (5 PSM); Starkey Experimental Forest, Little Bear Creek, T4S, R34E, W½ Sec. 2 (3 PSM); Starkey Experimental Forest, Meadow Creek, T3S, R34E, SE¼ Sec. 27 (2 PSM); Summerville (1 EOSC); 6 mi. E Tollgate (1 UW[EB]); T3S, R38E, SE¼NW¼ Sec. 27 (1 EOSC); T6S, R41E, Sec. 2 (1 EOSC). **Wallowa Co.**—Aneroid Lake (1 SDNHM); Hurricane Creek, S of Joseph, 6,000 ft. (1 CRCM); Ice Lake, Wallowa Mts. (1 MVZ); 0.3 mi. E Jewett Lake, Wallowa Mts., 8,350 ft. (1 CRCM); 0.4 mi. NW Jewett Lake, Wallowa Mts., 8,450 ft. (1 CRCM); 0.5 mi. S, 0.5 mi. W Joseph (1 OSUFW); Lick Creek R.S. (1 PSM); 2.5 mi. SW Imhaha, Little Sheep Creek (2 PSU); 6 mi. S Lostine, T2S, R43E, Sec. 10 (2 OSUFW); 7.5 mi. S Lostine, Pole Bridge Forest (1 OSUFW); 13 mi. S, 2 mi. E Lostine, T3S, R43E, Sec. 14 (1 OSUFW); 16 mi. S, 3 mi. E Lostine (7 MVZ); 21 mi. S, 5 mi. E Lostine, T4S, R44E, Sec. 27 (1 OSUFW); 21 mi. S, 6 mi. E Lostine, T4S, R44E, Sec. 26 (2 OSUFW); 17 mi. SE Lostine, Lostine River, Wallowa Mts. (5 OSUFW); 2 mi. N Minam, Minam State Park (1 PSU); ENE ridge Pete's Point, 8,900 ft. (1 CRCM); Thorn Creek R.S. (2 SDNHM); Troy (4 PSM); 8 mi. W Wallowa (1 UW[EB]); Wallowa Lake (1 OSUFW, 54 UO, 6 UW[WWC]); Wallowa Lake, 3 mi. S Joseph (3 PSU); Wenaha Forks (1 EOSC); Williamson camp, Lostine River (2 UW). **Wasco Co.**—6 mi. S, 12 mi. W Clarno (1 OSUFW); Clear Creek, T5S, R9E, Sec. 4 (1 PSU); mouth of Deschutes River, Columbia River (15 MVZ); Frog Lake, T4S, R9E, Sec. 17 (1 OSUFW); Memaloose State Park, T3N, R12E, SW¼ Sec. 32 (3 PSM); Memaloose State Park, T3N, R12E, SW¼ Sec. 33 (11 PSM); Memaloose State Park, T3N, R12E, SE¼ Sec. 33 (2 PSM); Memaloose State Park, T3N, R12E, SW¼ Sec. 34 (5 PSM); Miller Creek Canyon, 0.25 mi. N HWY 26 (1 UW); Miller Island, The Dalles (1 PSM); Moody, mouth of Deschutes River (2 OSUFW); 2.5 mi. S, 0.75 mi. W Shaniko (4 UW); Shaniko (4 PSM); The Dalles (1 MVZ); 2 mi. W The Dalles (3 PSU); 14 mi. SW The Dalles, Fivemile Creek (5 OSUFW); 20 mi. NW Warm Springs (2 UMMZ); 30 mi. NW Warm Springs (14 UMMZ); T2N, R14E, NE¼ Sec. 24 (13 PSM); T2N, R15E, NW¼ Sec. 22 (11 PSM). **Washington Co.**—Beaverton (2 PSU, 1 SDNHM); near Beaverton (3 PSM); 0.5 mi. N, 2 mi. W Beaverton (4 PSU); 0.5 mi. S, 1 mi. E Beaverton (4 PSU); Cornelius Pass (1 PSU); 4 mi. SE Gaston (1 OSUFW); 5 mi. S, 1 mi. E Hillsboro (20 PSU); 2 mi. S Laurel (4 OSUFW); North Plains (1 PSU); 18.5 mi. NW Portland (11 MVZ); Tigard (1 PSU); 0.3 mi. S, 1.5 mi. W Tigard (2 PSU); 2 mi. S, 0.7 mi. E Tigard (1 PSU); Trolley Park, near HWY 6 (1 PSU); 1 mi. N Tualatin (1 PSU); Tualatin River (1 UO); 3 mi. N junction 190th and Farmington Rd. (3 PSU); T1N, R2W, Sec. 8 (1 PSU); T1N, R4W, Sec. 28 (2 PSU). **Wheeler Co.**—Bridge Creek (5 MVZ); Butte Creek Summit (1 UW[WWC]); 2 mi. S, 2 mi. E Clarno (1 OSUFW); 7 mi. S, 11 mi. W Mitchell (12 MVZ); Shelton State Park, 10 mi. S Fossil (8 UW[WWC]); Wildwood Camp, Ochoco National Forest, T12S, R20E, Sec. 33 (1 OSUFW, 1 SDNHM). **Yamhill Co.**—(1 KU); 5 mi. NW Grande Ronde (1 UMMZ); Lafayette (2 UW[EB], 3 UW[WWC]); 2.8 mi. S McMinnville, T5S, R4W, Sec. 8 (2 PSU).

Additional localities: Dixie Butte* (1 USNM); Fish Creek* (2 UO); Fish Lake* (49 UO); Homedale Rd.* (2 SOSC); Canyon Creek, Ochoco National Forest* (54 UO).

Peromyscus truei, n = 200
(Plotted in Fig. 11-79, p. 274)

Crook Co.—Crooked River, mouth of Bear Creek, 3,400 ft. (5 MVZ); Crooked pRiver, 4 mi. W mouth Bear Creek, 3,100 ft. (1 MVZ); 6 mi. S Prineville, 3,100 ft. (1 LSU); 13.5 mi. S, 2.5 mi. E Prineville (1 OSUFW); 20 mi. SE Prineville, Crooked River (2 USNM); 2 mi. N, 4 mi. E Redmond (1 OSUFW). **Deschutes Co.**—10 mi. S, 14 mi. E Bend (1 OSUFW); 5 mi. N, 5 mi. W Hampton, T21S, R20E, S½ Sec. 15 (2 KU); Lower Bridge (2 OSUFW); 3 mi. N, 1 mi. W Redmond (1 OSUFW); 2 mi. S, 1 mi. E Redmond (1 OSUFW); 3 mi. N, 3 mi. E Sisters (1 OSUFW); 3 mi. N, 4 mi. E Sisters (1 OSUFW); 5 mi. N, 5 mi. E Sisters (1 OSUFW); 2 mi. S, 5 mi. E Sisters (1 OSUFW); Squaw Creek Canyon (2 CM); 3 mi. W Terrebonne (15 OSUFW); 4 mi. W Terrebonne (8 OSUFW); 0.5 mi. N, 4 mi. W Terrebonne (2 OSUFW); 1 mi. N, 1 mi. W Terrebonne (1 OSUFW); 1 mi. N, 3 mi. W Terrebonne, Steamboat Rock (2 OSUFW); 1 mi. N, 5 mi. W Terrebonne, Lower Bridge (2 OSUFW); 0.5 mi. S, 3 mi. W Terrebonne (1 OSUFW); 1 mi. S, 3 mi. W Terrebonne, Cottontail Ln., 3,100 ft. (9 OSUFW); 2 mi. S, 8 mi. W Terrebonne (1 OSUFW). **Grant Co.**—Falls Creek, 3 mi. N Mt. Vernon (1 EOSC). **Harney Co.**—Blitzen River canyon, T33S, R32½E, NW¼ Sec. 36 (2 EOSC); 3.75 mi. N, 3 mi. E Diamond (2 KU); 45 mi. N, 17 mi. E Fields (4 KU); 0.25 mi. N, 18 mi. W Riley, T23S, R24E, SE¼ Sec. 29 (1 KU); 1.5 mi. S, 10 mi. W Riley, T24S, R25E, N½ Sec. 3 (2 KU). **Jackson Co.**—Ashland (4 SOSC); NE of Ashland (1 SOSC); 3 mi. E Ashland (1 SOSC); 4 mi. E Ashland (2 SOSC); 4 mi. SE Ashland (8 PSM); 0.25 mi. past Ashland city dump (1 SOSC); 0.5 mi. S Belmont* (1 MVZ); Brownsboro (1 USNM); Emigrant Creek Rd., Cascade Riding Stables (1 SOSC); inlet to Emigrant Lake (4 SOSC); 1 mi. S Emigrant Lake (6 SOSC); 2.1 mi. E Emigrant Lake (1 SOSC); 2 mi. W Jacksonville (3 LACM); Kinney Creek (1 SOSC); S of Medford (1 PSM); 6 mi. S Medford (4 MVZ); 7 mi. S, 15 mi. W Medford (1 OSUFW); Mt. Ashland (1 SOSC); 3 mi. S, 4 mi. W Shady Cove (1 HSU); Dead Indian Rd. (1 SOSC); 4 mi. up Dead Indian Rd. (3 SOSC); 7 mi. up Dead Indian Rd. (3 SOSC); 21 mi. E Emigrant Lake (2 SOSC); 22 mi. SE Dead Indian Rd. on Emigrant Creek Rd. (1 SOSC); 2 mi. W Emigrant Creek Rd. from junction with Dead Indian Rd. (1 SOSC); 1.5 mi. up Emigrant Creek Rd. from Dead Indian Rd. (1 SOSC); Emigrant Creek Rd. 1.8 mi. E Dead Indian Rd. (5 SOSC); 2.3 mi. ESE Dead Indian Rd. on Emigrant Creek Rd. (5 SOSC); 3.2 mi. E Dead Indian Rd. on Emigrant Creek Rd. (1 SOSC); 3 mi. up Emigrant Creek Rd. E of Dead Indian Rd. (7 SOSC); 1.7 mi. up Mt. Ashland Rd. (1 SOSC); Shale City Rd., 0.9 mi. N Dead Indian Rd., T38S, R2E, Sec. 22 (1 PSU); Shoat Springs, 6.8 mi. S HWY 66 oh Copco Rd. (1 SOSC); White City (1 SOSC). **Jefferson Co.**—10 mi. S Madras (1 PSU); 2 mi. S, 5 mi. E Madras (3 CM); 10 mi. W Metolius (3 OSUFW); 10 mi. N, 4 mi. W Sisters (1 OSUFW); 9 mi. N, 6.5 mi. E Sisters (1 OSUFW); Warm Springs (1 USNM). **Josephine Co.**—Briggs Creek, 3,000 ft. (4 USNM); Galice (1 USNM); Grants Pass (1 CM, 2 FMNH, 1 OSUFW, 4 USNM); 2 mi. N, 1 mi. W Grants Pass (4 OSUFW); Grants Pass Peak (1 USNM); Gulch Creek at Illinois River* (1 SOSC); Little Grayback Rd., T39S, R7W, Sec. 34, 1,700 ft. (3 PSU); 4.6 mi. SW O'Brien (1 SOSC); 4 mi. S Selma, 1,350 ft. (1 MVZ); 2 mi. S, 6 mi. E Takilma, west branch camp (1 OSUFW); 0.25 mi. up [N] Shorthorn Gulch Rd. from Jump-off-Joe Creek Rd. (2 PSM); Black Bar on Rogue River, T33S, R9W, Sec. 35 (1 SOSC); T39S, R6W, Sec. 30 (1 SOSC). **Klamath Co.**—SW of Keno, ca. 0.5 mi. E Klamath Rd. (1 PSU); 0.7 mi. N, 0.7 mi. E Keno (1 PSU). **Lake Co.**—Oatman Flat, 7 mi. N, 1 mi. W Silver Lake (1 PSU); Oatman Flat, 8 mi. NW Silver Lake (1 PSU); Silver Lake (1 OSUFW); 4.5 mi. N, 0.5 mi. W Summer Lake (1 KU).

Phenacomys albipes, n = 102
(Plotted in Fig. 11-102, p. 306)

Benton Co.—1 mi. S, 6 mi. W Alpine, Alsea OSUFW); 4.4 km N, 7.6 km W Alsea, T13S, R8W, Sec. 20 (1 USNM); 3 mi. S, 3 mi. E Alsea,

Alsea Falls, Fall Creek (1 OSUFW); 3 mi. S, 4 mi. E Alsea (1 OSUFW); Alsea Falls Park (1 OSUFW); 3.75 mi. S, 5.5 mi. E Alsea (3 CM); 5 mi. S, 1.5 mi. W Blodgett (1 OSUFW); 5 mi. S, 5 mi. W Corvallis (1 OSUFW); Marys Peak, Conners Camp, 3,000 ft. (1 PSM); 1 mi. S, 7 mi. W Philomath, Marys Peak, T12S, R6W, Sec. 14 (1 OSUFW); 3 mi. S, 8 mi. W Philomath, Marys Peak, T12S, R6W, Sec. 30 (1 OSUFW); 6 mi. S, 8 mi. W Philomath, Dinner Creek, T13S, R7W, Sec. 1 (1 KU); 6 mi. SW Philomath (1 OSUFW); Summit (1 SOSC); 1 mi. S Summit (2 KU); 1.25 mi. SE Summit, T11S, R7W, SW¼ Sec. 8, 650 ft. (5 PSM); 3 mi. NW Wren, T11S, R6W, SW¼ Sec. 18 (1 PSM, 1 UW). **Clatsop Co.**—Nehalem summit (1 PSM); site of Old Ft. Clatsop, 100 ft. (2 MVZ); T8N, R10W, W½ Sec. 6 (1 PSM). **Columbia Co.**—7 mi. SE Rainier, 100 ft. (2 MVZ); Scappoose (1 PSU). **Coos Co.**—1.75 mi. E Bandon, T28S, R14W, SE¼ Sec. 29, 100 ft. (4 PSM); 3 mi. SE Bandon (1 PSM); 4 mi. SE Bandon, T29S, R14W, SW¼ Sec. 9, 250 ft. (5 PSM). **Curry Co.**—3 mi. above Gold Beach, S side Rogue River (1 MVZ). **Douglas Co.**—11.5 km N, 10.5 km W Drain, T21S, R6W, Sec. 5 (1 USNM); 9.7 km S, 16.2 km W Elkton, T23S, R9W, Sec. 22 (1 USNM); Gardiner (1 MVZ); Oxbow Burn, T20S, R8W (1 OSUFW); SW Reedsport (1 PSM); 3.6 km N, 11.9 km W Sutherlin, T25S, R7W, Sec. 1 (1 USNM); Threemile Creek (1 PSM); 4 km S, 3.2 km E Yachats, T15S, R12W, Sec. 16 (1 USNM). **Josephine Co.**—T38S, R6W, SW¼ Sec. 5 (1 SOSC). **Lane Co.**—Big Fall Creek (1 OSUFW); 6 mi. W Blue River, Hagen Creek, T16S, R3E, Sec. 22 (1 OSUFW); 14 mi. S, 8 mi. E Blue River, T18S, R5E, Sec. 23 (1 OSUFW); H. J. Andrews Experimental Forest, T15S, R5E, NE¼ Sec. 28, 1,850 ft. (1 PSM); 3.2 km N, 26 km W Lorane, T19/20S, R7W, Sec. 32/5 (1 USNM); 2 mi. W Vida (1 SDNHM); 23 mi. SE Vida, T18S, R5E, Sec. 23 (1 PSM); 4 mi. E Yachats, T15S, R12W, NE¼ Sec. 15, 160 ft. (1 PSM). **Lincoln Co.**—head of Alsea Bay up Shad Creek (1 PSM); Cascade Head Experimental Forest, T6S, R10W, SW¼ Sec. 21 (10 PSM); 15 mi. W Harlan, Peterson Ridge, T12S, R11W, Sec. 12 (1 KU); 2 mi. S, 8 mi. W Harlan, T12S, R10W, Sec. 13 (1 KU); 5 mi. S, 10 mi. W Harlan, Gold Creek, T13S, R10W, Sec. 2 (1 KU); 7 mi. S, 10 mi. W Harlan, Death Ridge, T13S, R10W, Sec. 14 (2 KU); 7 mi. S, 11 mi. W Harlan, Boulder Creek, T13S, R10W, Sec. 15 (2 KU); 1 mi. E Newport (2 UMMZ); Rose Lodge (1 USNM). **Tillamook Co.**—1 mi. N Bay City (1 KU); Blaine (1 MVZ); 2 mi. up Miami River (5 PSM); Netarts (4 SDNHM); Netarts, Whiskey Creek (1 PSM); Tillamook (1 SDNHM); 6.5 mi. SW Tillamook, 20 ft. (2 AMNH); 7 mi. SW Tillamook along Netarts Bay, 25 ft. (1 AMNH).

Phenacomys intermedius, n = 91
(Plotted in Fig. 11-104, p. 308)

Baker Co.—Anthony Lake (1 PSM); 0.2 mi. NNW Upper Twin Lake, Elkhorn Mts., 8,000 ft. (1 CRCM). **Clackamas Co.**—6 mi. S, 7 mi. E Sandy, T3S, R6E, Sec. 18 (1 CRCM). **Deschutes Co.**—15 mi. W Bend, Tumalo Creek, 6,100 ft. (1 MVZ); mouth of Davis Creek, Deschutes River (1 SDNHM); 9 km N, 6 km W Sisters, bank of Lost Lake Creek (1 OSUFW); E side South Sister, 7,800 ft. (1 CRCM); Three Sisters, 5,000 ft. (2 USNM); Three Sisters, 7,000 ft. (1 USNM); T16S, R9E, Sec. 26 (1 OSUFW). **Douglas Co.**—Diamond Lake (1 PSM, 1 USNM). **Grant Co.**—21 mi. SE Prairie City, North fork Malheur River, 5,000 ft. (1 MVZ). **Harney Co.**—18 mi. N, 5 mi. E Riley, T20S, R26E, SE¼ Sec. 26 (1 KU). **Hood River Co.**—Mt. Hood (1 USNM); Tilly Jane Creek, NE side Mt. Hood, 6,450 ft. (1 CRCM). **Klamath Co.**—Anna Creek, 4 mi. NW Ft. Klamath, 4,500 ft. (3 MVZ); Crater Lake near Glacier Peak* (2 USNM); 1.8 mi. S Crater Lake National Park headquarters (1 SOSC); 6 mi. N Geary Ranch on Lake of the Woods Rd.* (1 OSUFW). **Lake Co.**—Bald Butte (1 PSM); Gearhart Mt. (3 PSM). **Lane Co.**—2.6 km S McKenzie Bridge, T16S, R5E, Sec. 26 (1 USNM); 3.8 km S, 14.8 km E McKenzie Bridge (7 USNM); Three Sisters (1 SDNHM). **Linn Co.**—Big Lake (1 OSUFW); 1.5 mi. N Big Lake, T13S, R7½E, NE¼ Sec. 35, 4,800 ft. (12 PSM); 1.5 mi. N Big Lake (5 PSM); 1.5 mi. S Big Lake (8 PSM); 10 mi. E Upper Soda, Tombstone Prairie (1 OSUFW). **Multnomah Co.**—8 mi. N, 3 mi. E Zigzag (1 CRCM). **Wallowa Co.**—Aneroid Lake (1 SDNHM, 3 USNM); 0.4 mi. NW Jewett Lake, Wallowa Mts., 8,450 ft. (2 CRCM); E side John Henry Lake, Wallowa Mts., 7,200 ft. (2 CRCM); 16 mi. S, 2 mi. E Lostine, T3S, R43E, Sec. 25 (1 USNM); 16

mi. S, 3 mi. E Lostine (1 MVZ); 19 mi. S, 4 mi. E Lostine, T4S, R44E, Sec. 17 (2 OSUFW); 21 mi. S, 5 mi. E Lostine, T4S, R44E, Sec. 27 (5 OSUFW); 21 mi. S, 6 mi. E Lostine, T4S, R44E, Sec. 26 (1 OSUFW); Wallowa Lake, 4,000 ft. (1 USNM). **Wasco Co.**—Clear Lake (4 PSM); 6 mi. W Pine Grove (1 OSUFW); 0.5 mi. N, 5 mi. W Pine Grove (1 OSUFW).
    Additional locality: Blue Mts.* (1 USNM).

Phenacomys longicaudus, n = 461
(Plotted in Fig. 11-106, p. 310)

Benton Co.—(8 PSM); 3.5 mi. W Adair (2 PSM); 3.75 mi. W Adair (3 PSM); 4 mi. W Adair (2 OSUFW); 3.75 mi. WNW Adair (2 PSM); 7 mi. W Alpine, Alsea Falls Park (1 OSUFW); 4 mi. S Alpine (1 PSM); 5 mi. S Alpine (1 PSM); 1 km S, 4.4 km E Alsea, T14S, R7W, Sec. 9 (1 USNM); 1 mi. N Bellfountain (1 OSUFW); near Blodgett (1 FMNH); Corvallis (1 UMMZ); SW Corvallis (2 PSM); 6 mi. N Corvallis, McDonald Forest, Nettleton Rd. (11 OSUFW, 1 PSM); 7 mi. N Corvallis (1 PSM, 7 UMMZ); 5 mi. N, 2 mi. W Corvallis (1 OSUFW); 7 mi. N, 2 mi. W Corvallis (1 OSUFW); 8 mi. N, 0.5 mi. W Corvallis (1 OSUFW); McDonald Forest (2 OSUFW, 4 PSM, 2 UW); 3 mi. S Monroe (1 PSM); 4.5 mi. W Monroe (1 OSUFW); 5.75 mi. W Monroe, Glenbrook (2 OSUFW); 4 mi. SW Monroe, 600 ft. (1 OSUFW); 4.5 mi. SW Monroe, T15S, R6W, Sec. 2 (26 OSUFW, 17 PSM); 1.5 mi. S, 1.5 mi. W Philomath (1 OSUFW); 1 km S, 11.3 km W Philomath, T12S, R7W, Sec. 11/14 (2 USNM); 5 mi. W Summit on Sealy Creek Rd. (1 PSM). **Clackamas Co.**—near Molalla, Schoenborns Ranch (2 OSUFW); 8 mi. S Molalla (3 CRCM); 7 mi. SE Molalla (3 PSM); 7.5 mi. SE Molalla (21 PSM); 8 mi. SE Molalla (12 PSM); 8.5 mi. SE Molalla (2 PSM). **Clatsop Co.**—T5N, R8W, Sec. 6 (1 PSM). **Coos Co.**—8 mi. SE Bandon (1 PSM); 1 mi. NW Bill Peak near road, 1,200 ft. (1 PSM); 3 mi. SW Broadbent (2 PSM); Head Light* (1 LACM); Marshfield (2 SDNHM, 1 USNM); 3 km S, 10.8 km E Sitkum, T28S, R9W, Sec. 23 (1 USNM); 4.8 km S, 9.7 km E Sitkum, T28S, R9W, Sec. 27 (1 USNM). **Curry Co.**—Agness (8 SDNHM); 10 mi. E Brookings (2 PSM); 20 mi. E Gold Beach (1 SDNHM); 20 mi. E Gold Beach, mouth of Lobster Creek (2 USNM); 23 mi. E Gold Beach (1 SDNHM); 20 mi. up Rogue River, Gold Beach (5 SDNHM); Port Orford (1 SDNHM); Rogue River, 2 mi. E HWY 101 on N side, in owl pellet (1 PSM); Winchuck River at Wheeler Creek (1 UW). **Douglas Co.**—7.6 km N, 6.4 km W Drain, T21S, R6W, Sec. 22 (1 USNM); 3 mi. E Elkton near HWY 38 (5 PSM); 15.7 km S, 5.6 km W Elkton, T24S, R8W, Sec. 10 (1 USNM); near Roseburg (1 AMNH); 3.8 km N, 12.1 km W Sutherlin, T25S, R7W, Sec. 1 (1 USNM); 4 km N, 12.1 km W Sutherlin, T25S, R7W, Sec. 1 (1 USNM); 6.5 km N, 6.7 km W Tenmile, T28S, R8W, Sec. 9 (1 USNM); 9 mi. N, 8.5 mi. E Tiller, T29S, R1W, Sec. 13 (3 CRCM); 11.5 mi. N, 5 mi. E Tiller, T28S, R1W, Sec. 26 (1 CRCM); 12.5 mi. N, 5.5 mi. E Tiller, T28S, R1W, Sec. 31 (1 CRCM); 4.9 km S, 4.5 km W Yoncalla, T25S, R5W, Sec. 19 (2 USNM). **Hood River Co.**—1 mi. E Cascade Locks at Oxbow hatchery (2 PSM). **Jackson Co.**—14 mi. N, 5 mi. E Prospect, T30S, R2E, Sec. 24/25 (1 CRCM). **Josephine Co.**—Oregon Caves National Monument, guide shack (1 OSUFW). **Lane Co.**—5 mi. N, 9 mi. E Blue River (1 CRCM); 6.8 km N, 7.6 km E Blue River, T15S, R5E, Sec. 32 (7 USNM); 2.5 mi. S, 7 mi. E Blue River (1 CRCM); 3.5 mi. S, 10 mi. E Blue River (1 CRCM); 2.75 mi. W Cheshire near Jones Creek (3 OSUFW); 2.8 mi. W Cheshire, Jones Creek (2 OSUFW); 3 mi. W Cheshire (1 PSM); 5.5 mi. NW Cheshire (2 PSM); 6 mi. NW Cheshire (1 PSM); 6.75 mi. WNW Cheshire (1 OSUFW); 6 mi. NNE Coburg (6 PSM); 17 mi. SE Cottage Grove (1 AMNH); 2.5 mi. SW Donna (1 PSM); 7 mi. NW Elmira (1 PSM); Eugene (1 SOSC, 1 USNM); Eugene, Spencer Butte (3 OSUFW); 1 mi. W Junction City (1 PSM); 11 mi. W Junction City (5 PSM); 5 mi. N Lorane (6 SDNHM); 2 mi. E Lorane (1 PSM); 0.8 km N, 19.2 km W Lorane, T20S, R7W, Sec. 12 (1 USNM); 2 km N, 6.4 km W Lorane, T20S, R5W, Sec. 5 (1 USNM); 2.8 km N, 6.2 km W Lorane, T20S, R5W, Sec. 5 (1 USNM); 6.6 km N, 0.3 km W McKenzie Bridge, T16S, R5E, Sec. 25 (1 USNM); 6 km N, 0.6 km E McKenzie Bridge, T15S, R5E, Sec. 25 (3 USNM); 6.4 km N, 0.6 km E McKenzie Bridge, T15S, R5E, Sec. 25 (3 USNM); 2.6 km S, 1.4 km W McKenzie Bridge, T16S, R5E, Sec. 25 (1 USNM); 5.4 km S, 8.75 km E McKenzie Bridge, T16S, R6E, Sec. 35 (1 USNM); Meadows (1 USNM); 4.5 mi. SW Monroe (1

PSM); 6 mi. SW Monroe near Merle Hewitt Ranch (1 OSUFW); 4.5 mi. N Nimrod (1 CRCM); 1.5 mi. NW Noti (1 PSM); 4.5 mi. W Vida (1 PSM); 0.25 mi. NE Walterville (1 PSM). **Lincoln Co.**—10 mi. SW Grande Ronde (2 PSM); 15 mi. W Harlan, Peterson Ridge, T12S, R11W, Sec. 12 (2 KU); 2 mi. S, 12 mi. W Harlan, Beaver Creek, T12S, R10W, Sec. 20 (1 KU); 7 mi. S, 10 mi. W Harlan, Death Ridge, T13S, R10W, Sec. 14 (1 KU); 7 mi. S, 11 mi. W Harlan, Boulder Creek, T13S, R10W, Sec. 15 (6 KU); near Nashville (1 FMNH); Stott Mt., head of Slick Creek, between Grand Ronde and Valsetz, 2,800 ft. (1 PSM); 4.8 km S, 3.2 km E Yachats, T15S, R12W, Sec. 12 (1 USNM); headwaters of Siletz & Salmon rivers (1 PSM). **Linn Co.**—7.5 mi. N, 7.5 mi. E Blue River, T15S, R5E, Sec. 15 (1 CRCM); 9.2 km N, 1.2 km W McKenzie Bridge, T15S, R5E, Sec. 14 (1 USNM); 10.25 km N, 0.8 km W McKenzie Bridge, T15S, R5E, Sec. 14 (1 USNM); 13.2 km N, 0.8 km E McKenzie Bridge, T15S, R5E, Sec. 1 (1 USNM); 4.5 mi. S Sodaville (1 OSUFW). **Multnomah Co.**—10 mi. N, 10 mi. E Sandy, T1N, R6E, Sec. 27 (1 CRCM); 11 mi. N, 10 mi. E Sandy (2 CRCM); 10 mi. N, 1.5 mi. W Zigzag (1 CRCM). **Polk Co.**—2.2 mi. SW Airlie (2 OSUFW); near Falls City (1 FMNH); near Grande Ronde (5 PSM). **Tillamook Co.**—Cape Lookout (1 SDNHM); 3 mi. E Cape Lookout, headwaters Tillamook River (8 PSM); Cape Meares (1 PSM); W of Cape Meares (2 PSM); E of Cape Meares (2 PSM); between Cape Meares and Oceanside (1 PSM); Cascade Head, T6S, R10W, NE¼ Sec. 17 (1 PSM); Fall Creek, Netarts Bay (1 MVZ); Foley Creek (1 UW); 8 mi. SE Hebo (4 KU); N Nehalem (1 OSUFW); Neskowin (1 MVZ, 1 OSUFW); Netarts (1 MVZ, 4 OSUFW, 1 PSM, 1 SDNHM, 1 UMMZ); Netarts watershed (6 PSM); Netarts, Whiskey Creek (5 OSUFW); Netarts Bay (1 PSM); 0.75 mi. SE Netarts Summit (1 UW); ca. 1 mi. SE Netarts Summit (1 UW); 1.5 mi. SE Netarts Summit (1 UW); Oceanside (7 PSM); NW side Oceanside (1 PSM); 1 mi. NE Oceanside (2 PSM); 2 mi. NE Oceanside (2 PSM); 5 mi. NE Pleasant Valley (2 PSM); Tillamook (1 LACM, 12 OSUFW, 9 PSM, 1 PSU, 1 SDNHM); vicinity of Tillamook (1 UW); Tillamook, Cape Lookout (2 OSUFW); 4 mi. W Tillamook (1 OSUFW); 0.75 mi. E Tillamook (1 PSM); 3 mi. SW Tillamook (2 PSM, 2 UW); 4 mi. SW Tillamook (1 LACM); 5 mi. SE Tillamook (1 SDNHM); S Tillamook Bay (8 PSM); S Tillamook Bay, T1S, R10W, Sec. 21 (23 PSM); Bay Ocean Rd. close to Tillamook Bay (1 UW); Tillamook Head (1 PSM); Elkoff Rd., SW Tillamook Co. (1 UW); above Wheeler (1 UW). **Washington Co.**—3 mi. E Gaston (3 AMNH); 4 mi. E Gaston (1 AMNH, 1 MVZ, 1 CRCM); 5 mi. E Gaston (2 AMNH); 8 mi. E Gaston (1 AMNH, 1 KU); 1.5 mi. SE Laurelwood (2 PSM). **Yamhill Co.**—9 mi. W Carlton (2 MVZ); 4 mi. N Newberg (11 AMNH); 7 mi. NW Newberg (1 AMNH).

Phoca vitulina, n = 43
(Plotted in Fig. 12-26, p. 395)

Clatsop Co.—Arcadia wayside near Arch Cape (1 CRCM); Astoria (1 UW); 2.5 mi. N Astoria [=1.9 mi. N Peter Iredale wreck] (1 CRCM); Chute Drift, river mile 20 (1 UW); N side south jetty Clatsop spit (1 CRCM); mouth Columbia River (1 PSM, 1 UW); beach between Gearhart and wreck of Peter Iredale (1 UW); 3 mi. N Gearhart (1 UW); Hammond (1 CRCM); Harbor Drift, river mile 36 (2 UW); 1 mi. S Peter Iredale wreck (1 CRCM); Prairie Channel, river mile 24 (1 UW); Red Slough, river mile 37 (1 UW); Seaside (1 CRCM); Sunset Beach (1 UW); Tenasillahe Island (3 UW); Tongue Point, 2 mi. ENE Astoria (1 CRCM); 0.5 mi. W Tongue Point (1 CRCM); Woody Island drift (12 UW). **Columbia Co.**—Lower Channel Drift, river mile 47 (4 UW); Wallace [Island], river mile 50 (1 UW). **Curry Co.**—south end Agate Beach, Port Orford (1 PSM). **Lincoln Co.**—7 mi. N Newport, Beverly Beach (1 OSUFW); 1 mi. S Yachats (1 PSM). **Additional locality:** coast* (1 PSM).

Pipistrellus hesperus, n = 69
(Plotted in Fig. 9-30, p. 106)

Grant Co.—T8S, R28E, Sec. 28 (1 PSU). **Harney Co.**—1.6 mi. S Big Sand Gap Springs (1 SOSC); Mickey Hot Springs (1 SOSC); 4 mi. N Princeton, Windy Point (1 MNWR, 1 OSUFW); 3.3 mi. S Princeton (2 PSM); 3.3 mi. NE Princeton (2 SOSC); Trout Creek (2 SDNHM); 6.9

mi. E Fields Rd., 3.9 mi. S Pike Creek at Sand Gap (1 SOSC); T33S, R32½E, NE¼ Sec. 45, 4,420 ft. (1 EOSC). **Malheur Co.**—Cottonwood Creek, T18S, R42E, center, Sec. 3, 3,160 ft. (6 PSM); Cow Creek Lake (1 USNM); Owyhee Canyon, T22S, R45E, Sec. 3 (4 PSM); Owyhee Spring, 3.5 mi. SE Rome (2 PSM); 2 mi. NW Riverside (1 USNM); 15 mi. N, 1 mi. W Rockville, Succor Creek State Park (1 OSUFW); Succor Creek State Park, T24S, R46E, Sec. 20 (21 PSM); Twin Springs, T22S, R43E, NE¼NW¼ Sec. 35 (4 PSM); Watson (13 USNM). **Sherman Co.**—Miller, mouth of Deschutes River (1 USNM). **Wasco Co.**—Moody (1 UO). **Wheeler Co.**—T11S, R21E, Sec. 9 (1 PSU).

Procyon lotor, n = 1,261
(Plotted in Fig. 12-34, p. 401)

Baker Co.—Huntington (1 USNM). **Benton Co.**—(50 KU); 5 mi. SE Blodget (1 PSM); Corvallis (2 OSUFW); 4.5 mi. N Corvallis, T11S, R5W, Sec. 14 (1 OSUFW); 7 mi. N Corvallis (1 OSUFW); 4 mi. S Corvallis (1 OSUFW); 5 mi. S Corvallis (1 OSUFW); 4 mi. E Corvallis (1 OSUFW); 6 mi. S, 0.5 mi. W Corvallis (1 OSUFW); 5 mi. S, 3 mi. E Corvallis (1 OSUFW); 1 mi. S Monroe (1 PSM). **Clackamas Co.**—(253 KU); Aurora (1 UW[LB]); Collywash [Collawash] Burn, Clackamas River, 50 mi. above Estacada, [T7S, R6E] (2 USNM); Estacada (2 USNM); Mulino (1 UO). **Clatsop Co.**—(2 KU); Astoria (2 UMMZ). **Columbia Co.**—(2 KU). **Coos Co.**—1.75 mi. E Bandon (3 PSM); 4 mi. SE Bandon (4 PSM); Bridge (1 USNM); 2.5 mi. NNE Broadbent (2 PSM); Remote (1 USNM). **Crook Co.**—Big Summit Prairie, Ochoco National Forest [T14S, R21E] (1 USNM). **Curry Co.**—5 mi. ENE Brookings (1 PSM); Gold Beach (1 FMNH); Pistol River (1 USNM); Port Orford (5 USNM). **Douglas Co.**—(6 KU); Gazley, Morgan Gulch (1 MVZ); Glendale (3 USNM); 24 mi. E Glide (1 USNM); 14 mi. NE Glide (1 USNM); Roseburg (1 OSUFW). **Gilliam Co.**—10 mi. W Hardman (1 USNM); 4 mi. SW Hardman (1 USNM). **Grant Co.**—(1 EOSC); Monument (1 PSM). **Jackson Co.**—Ashland (1 SOSC); Beagle (5 OSUFW); 8 mi. SW Prospect (1 SOSC); T38S, R1W, SW¼ Sec. 33 (1 SOSC). **Josephine Co.**—32 mi. S Grants Pass (4 USNM); 43 mi. NE Grants Pass (3 USNM); Illinois River (1 OSUFW). **Klamath Co.**—2 mi. S Casey State Park, HWY 62 (1 SOSC). **Lake Co.**—Adel (1 USNM); Adel, 20-mile Ranch (2 USNM); 9 mi. S Adel, mouth Twentymile Creek (1 MVZ); Twomile Creek* (1 PSM). **Lane Co.**—(129 KU, 1 PSM); 37 mi. SE Blue River* (1 USNM); Leaburg (1 UO); McKenzie Bridge (1 UO); 5 mi. W McKenzie Bridge (2 KU); McKenzie River* (2 KU); main part of McKenzie River* (3 KU); S fork McKenzie River* (1 KU); Mercer (1 OSUFW); Mercer Lake (1 OSUFW); 5 mi. N Springfield (1 OSUFW). **Lincoln Co.**—(38 KU); 5 mi. N, 0.5 mi. W Fisher (1 OSUFW). **Linn Co.**—(39 KU). **Malheur Co.**—Arock (1 PSM); Dry Creek (1 USNM); 10 mi. W Fairyland* (2 USNM); 12 mi. W Fairyland* (1 USNM); 8 mi. SE Harper (1 USNM); Jordan Valley (1 PSM); 2 mi. NW Riverside (2 USNM); Rome (2 PSM); 7 mi. S Rome (1 USNM). **Marion Co.**—(133 KU). **Multnomah Co.**—(121 KU); Portland (1 SDNHM, 1 USNM); Sauvies Island (1 PSM). **Polk Co.**—(92 KU); 11 mi. N Corvallis (1 OSUFW). **Tillamook Co.**—(34 KU, 1 UMMZ); Beaver (1 KU); Lees Camp, Wilson River (1 PSM); Nehalem (1 PSM); 1 mi. W Pacific City (1 UW[WWC]); Tillamook (1 KU); Wheeler (1 OSUFW). **Umatilla Co.**—Lehman Springs (1 EOSC); Vansycle (1 USNM). **Union Co.**—2 mi. N, 4.5 mi. E Catherine guard station, T5S, R41E, Sec. 7 (1 OSUFW); Cove (1 EOSC); Starr Lane (1 EOSC). **Wallowa Co.**—15 mi. SW Enterprise (1 USNM); Imnaha [Walla walla Co. sic] (3 USNM). **Wasco Co.**—(1 KU); 25 mi. N, 10 mi. E Madras (1 OSUFW). **Washington Co.**—(112 KU); Forest Grove (1 PSU). **Wheeler Co.**—18 mi. W Fossil (1 MVZ); 2 mi. E Tupper R.R. Station* (1 USNM). **Yamhill Co.**—(123 KU); 6 mi. NW McMinnville (1 USNM).

Puma concolor, n = 149
(Plotted in Fig. 12-59, p. 451)

Clackamas Co.—0.5 mi. from Carver near Clackamas River, 4–5 mi. from Oregon City (1 OSUFW); Estacada (3 USNM); Cold Springs (1 USNM); Collywash Burn [=Collawash] [T7S, R6E] (1 USNM); Oak Grove Butte [T6S, R7E, Sec. 8] (3 USNM); Roaring River Mt.* (5

USNM); Marmot (1 USNM). **Coos Co.**—Marshfield (1 USNM). **Curry Co.**—(7 USNM); 10 mi. E Agness (1 USNM); 12 mi. E Agness (3 USNM); Gold Beach (1 USNM); NE of Illahe (2 USNM); source of Pistol River (14 USNM). **Douglas Co.**—Drew (2 USNM); 19 mi. E Drew (1 USNM); Ft. Umpqua (1 USNM); Glendale (2 USNM); Glide (1 USNM); 2 mi. S, 4 mi. E Sutherlin, T25S, R5W, Sec. 26 (1 OSUFW); Umpqua National Forest* (4 SDNHM). **Grant Co.**—Dayville, Murder Creek (8 USNM). **Jackson Co.**—14 mi. NW Rock Creek* (1 USNM). **Josephine Co.**—(1 SDNHM); Bald Mt., 40 mi. W Selma (3 USNM); Centennial Creek (1 USNM); Diamond Peak* (3 USNM); Florence Creek (1 USNM); Galice (1 UO); Grants Pass (1 UO, 9 USNM); 25 mi. W Grants Pass (1 UO); 43 mi. NE Grants Pass (8 USNM); 10 mi. S Grants Pass (2 USNM); 15 mi. S Grants Pass (1 USNM); 25 mi. S Grants Pass (2 USNM); Kerby Peak (4 USNM); Josephine Creek (1 USNM); Marble Mt. (2 USNM); Murphy Peak (2 USNM); Pine Creek, 29 mi. NW Selma (1 USNM); 35 mi. W Selma (1 USNM); 10 mi. NW Selma, Horse Mt. (2 USNM); 15 mi. NW Selma (1 USNM); 10 mi. SE Selma (2 USNM); 15 mi. SE Selma (1 USNM); head Soldier Creek, 15 mi. NW Selma (1 USNM); 15 mi. W Waldo (1 USNM); Williams (1 USNM). **Lane Co.**—Hills Creek Reservoir (1 KU); Little Fall Creek (1 SDNHM); McKenzie Bridge (1 KU, 2 UO); McKenzie Ranger District* (2 KU); Oakridge (1 PSM); Reed (1 USNM); Sourgrass Mt., North Fork Willamette [River] (1 PSM). **Linn Co.**—Santiam River above Cascadia (1 MVZ). **Polk Co.**—(1 USNM). **Tillamook Co.**—Niagara Creek (1 MVZ). **Umatilla Co.**—24 mi. NNW La Grande (1 PSM); Tollgate (1 OSUFW); T4N, R36E, Sec. 4 (1 KU). **Wallowa Co.**—Bear Gulch (3 PSM); Imnaha (1 PSM); 3 mi. W Imnaha, 2 mi. from Sheep Creek up Bear Gulch (1 OSUFW); Olson Meadows (1 PSM); Upper Imnaha River (1 USNM); 1 mi. N, 10 mi. W Wallowa, Minam River (1 OSUFW); Wildcat Creek (2 PSM). **Wasco Co.**—Quartz Creek (1 USNM).

Additional localities: Chetco unit* (1 EOSC); Micky Burn* (1 USNM); N fork Battle Creek* (1 USNM).

Rattus norvegicus, n = 99
(Plotted in Fig. 11-89, p. 289)

Benton Co.—Corvallis (20 OSUFW, 1 SDNHM, 1 SOSC); 0.25 mi. N Corvallis (1 OSUFW); 1 mi. N Corvallis (4 OSUFW); 2 mi. N Corvallis (1 OSUFW); 7 mi. N Corvallis, E. E. Wilson Game Management Area (2 OSUFW); 9 mi. N Corvallis (1 OSUFW); 3 mi. S Corvallis (4 OSUFW); 6 mi. S Corvallis (1 OSUFW); 3 mi. N, 2 mi. W Corvallis (1 OSUFW); 5 mi. N, 1 mi. W Corvallis (1 OSUFW); 7 mi. N, 1 mi. E Corvallis (1 OSUFW); 1.5 mi. S, 0.5 mi. E Corvallis (2 OSUFW); 2 mi. S, 1 mi. E Corvallis (1 OSUFW); 3 mi. S, 3 mi. E Corvallis (1 OSUFW); 8 mi. N, 4 mi. E Corvallis, North Albany (1 OSUFW). **Clackamas Co.**—8 mi. S, 2 mi. W Molalla (2 OSUFW). **Clatsop Co.**—Westport (1 OSUFW); T8N, R8W, NE¼ Sec. 20/SW¼ Sec. 16 (1 PSM). **Columbia Co.**—1 mi. N, 7 mi. W Clatskanie (2 OSUFW). **Coos Co.**—Coquille (1 OSUFW). **Gilliam Co.**—2.5 mi. S Heppner Junction (1 KU). **Jackson Co.**—Ashland (3 SOSC); Medford (1 OSUFW, 2 SOSC); Talent (1 SOSC). **Lane Co.**—Cottage Grove (6 OSUFW). **Linn Co.**—2 mi. N, 4 mi. E Albany (1 OSUFW); 1 mi. S, 1 mi. E Albany (1 OSUFW); 5 mi. W Lebanon (1 OSUFW); Scio (1 SDNHM); 0.5 mi. N, 1 mi. E Sweet Home (1 OSUFW); 4 mi. W Tangent (2 OSUFW); Truax Slough, NE of Corvallis (1 OSUFW). **Marion Co.**—1 mi. N, 3 mi. E Salem (1 OSUFW); 4 mi. W Woodburn (1 OSUFW). **Morrow Co.**—3 mi. W Irrigon (1 OSUFW); T5N, R26E, NW¼ Sec. 22 (4 PSM). **Multnomah Co.**—Portland (5 OSUFW, 5 PSU, 1 SDNHM); 4 mi. N, 6 mi. E Portland, Willamette Slough (2 OSUFW). **Polk Co.**—1 mi. N, 2 mi. E Dalles (1 OSUFW). **Umatilla Co.**—Milton-Freewater (1 EOSC); Pendleton (1 SDNHM). **Union Co.**—La Grande (1 EOSC). **Washington Co.**—Raleigh Hills (1 PSU). **Yamhill Co.**—1.5 mi. E Yamhill (1 OSUFW).

Rattus rattus, n = 20
(Plotted in Fig. 11-91, p. 291)

Clatsop Co.—3 mi. S Astoria (1 OSUFW); Hammond (1 OSUFW). **Coos Co.**—Bastendorf Beach (2 LACM); Coos Head (2 LACM); Coquille, T28S, R13W, Sec. 1 (1 OSUFW); 1 mi. S, 7 mi. E Eastside (1

OSUFW). **Douglas Co.**—3 mi. N Reedsport, T21S, R12W, Sec. 10 (1 PSU). **Lane Co.**—12 mi. N Florence (1 OSUFW); Siltcoos Lake (1 OSUFW). **Marion Co.**—Salem (2 PSU). **Multnomah Co.**—Portland (2 SDNHM). **Tillamook Co.**—Netarts (3 OSUFW, 2 SDNHM).

Reithrodontomys megalotis, n = 238
(Plotted in Fig. 11-73, p. 264)

Baker Co.—3 mi. NE Huntington, 2,100 ft. (5 USNM); Oxbow (2 MSUMZ). **Crook Co.**—4 mi. SW Prineville (1 MVZ). **Deschutes Co.**—Lower Bridge (1 OSUFW); 1 mi. N, 7 mi. W Redmond (2 KU). **Gilliam Co.**—Columbia River, mouth of John Day River (10 MVZ); 1.5 mi. S Heppner Junction (3 KU); 2.5 mi. S Heppner Junction (3 KU); Willows (1 SDNHM, 9 USNM); Willows, 2 mi. S Columbia River (1 SDNHM). **Grant Co.**—20 mi. S, 2 mi. W Long Creek (2 OSUFW). **Harney Co.**—marsh, Alvord Ranch (1 EOSC); Blitzen Valley (2 SDNHM); Burns (1 USNM); 40 mi. S, 12 mi. E Burns, near Buena Vista substation, Malheur National Wildlife Refuge (4 OSUFW); 9 mi. S Crane, shore of Malheur Lake, T26S, R33E, Sec. 10 (1 OSUFW); Diamond, 4,300 ft. (2 USNM); Folly Farm, Sunrise Valley, 3,950 ft. (1 USNM); 4.5 mi. N, 0.5 mi. E Frenchglen, S Knox Pond, Malheur National Wildlife Refuge (2 OSUFW); 2 mi. S, 2.5 mi. E Frenchglen (1 OSUFW); Malheur Environmental Field Station, T27S, R31E, Sec. 5 (1 OSUFW); 0.25 mi. NE Malheur Environmental Field Station (1 SOSC); Malheur Lake (1 PSM); Malheur Lake Refuge (1 OSUFW); Malheur National Wildlife Refuge (3 OSUFW); Narrows (6 UO, 20 USNM); 5 mi. SW Narrows (2 MVZ); 2.5 mi. S, 15 mi. W Princeton (1 PSU); 10 mi. SE Princeton (1 OSUFW); Springer Ranch* (1 USNM); Steens Mts., T31S, R32½E, Sec. 3 (1 OSUFW); Steens Mts., T31S, R32½E, Sec. 13 (1 OSUFW); Voltage (1 LACM, 2 SDNHM). **Hood River Co.**—Hood River (1 USNM). **Jackson Co.**—Ashland (6 SOSC, 3 USNM); Ashland airport (2 SOSC); 1 mi. E Ashland (1 SOSC); 2 mi. N, 4 mi. E Ashland (1 SOSC); 2 mi. E Dark Hollow, 1,600 ft. (1 MVZ); 6 mi. S Medford (2 MVZ); 4 mi. E Medford (4 LACM); near Talent (1 SOSC); 4 mi. from HWY 66 on Dead Indian Rd. (1 SOSC); 1.8 mi. E Dead Indian Rd. on Emigrant Creek Rd. (1 SOSC). **Jefferson Co.**—(1 PSM); 3 mi. NNW Culver (1 PSM); 4 mi. NNW Culver (2 PSM); T13S, R12E, Sec. 12 (1 PSM). **Josephine Co.**—Grants Pass (1 FMNH); 4 mi. S Selma (2 MVZ). **Klamath Co.**—Klamath Falls (2 CM, 4 USNM); Klamath Experiment Station (2 OSUFW); Poe Valley, Rajnus Ranch, T39S, R11½E (1 OSUFW). **Lake Co.**—NE edge Alkali Lake (1 MVZ); Buck Pasture, Willow Creek, Hart Mt. Antelope Refuge (1 KU); 6 mi. W Lakeview (2 MVZ); S Warner Lake (1 CM). **Malheur Co.**—(3 PSM); Brogan (1 PSU); 10 mi. N McDermitt (1 PSM); 7.2 mi. N Jordan Valley (3 PSU); 0.25 mi. W Jordan Valley (2 PSM); 0.5 mi. W Jordan Valley (6 PSU); 1 mi. W Ironside (1 MVZ); Kane Springs, T21S, R43E, Sec. 4 (1 PSM); Sheaville (1 USNM); Soldier Creek Exclosure, 8.5 mi. WSW Jordan Valley, T30S, R45E, SW¼ Sec. 15, 4,420 ft. (1 PSM); Vale (1 UO, 2 USNM); near Vale (6 AMNH); 2 mi. N Vale (1 PSM); 3 mi. N Vale (1 MVZ); Watson (3 USNM); 6.5 mi. SE Whitehorse Ranch (3 PSM); 0.7 mi. S, 0.7 mi. W Worden* (1 PSU). **Morrow Co.**—Boardman (1 SDNHM); 0.25 mi. S, 5 mi. W Boardman (3 PSU); 0.5 mi. S, 5 mi. W Boardman (3 PSU); 1.5 mi. S, 5.5 mi. W Boardman (3 PSU); Cecil (1 PSM, 1 SDNHM); 6 mi. SE Cecil (1 PSM); Irrigon (1 MVZ); 2.5 mi. SW Irrigon (2 MVZ); T5N, R26E, NE¼ Sec. 29 (1 PSM). **Sherman Co.**—Columbia River, mouth Deschutes River (10 MVZ); Millers (2 SDNHM, 2 USNM); 5 mi. N, 19 mi. E The Dalles (1 OSUFW). **Umatilla Co.**—Umatilla (2 USNM). **Union Co.**—Hot Lake (2 EOSC); 2.1 mi. S, 3.2 mi. W North Powder (1 PSU). **Wallowa Co.**—Dug Bar, Snake River (1 SDNHM); 14 mi. up Wenaha River above Troy (1 PSM). **Wasco Co.**—Columbia River, mouth Deschutes River (10 MVZ); Maupin (1 PSM); W end of The Dalles, T1N, R13E, Sec. 3 (1 KU); 3 mi. E Tygh Valley (2 USNM). **Wheeler Co.**—4 mi. E Service Creek (1 UW[WWC]); Sheldon State Park (1 UW[WWC]); Twickenham (1 USNM).

Scapanus latimanus, n = 52
(Plotted in Fig. 8-24, p. 68)

Jackson Co.—(1 SOSC); Ashland, 1,975 ft. (4 SOSC, 1 USNM); Ashland, S slope Grizzly Peak, 2,300 ft. (1 USNM); Ashland, W slope

Grizzly Peak, 3,500 ft. (1 USNM); 6 mi. S Ashland (1 SDNHM, 1 USNM); Brownsboro (1 USNM); Gold Hill (1 SDNHM); Gold Hill near Medford (1 USNM); Goose Lake* (1 SDNHM); 2 mi. N Jacksonville (1 SOSC); Klamath Falls (1 SOSC); 6 mi. S Medford, Belmont Orchard, 1,600 ft. (4 MVZ); 5 mi. W Medford (5 PSM); 8 mi. SW Prospect (2 SOSC); 1.5 mi. N Rogue River (1 OSUFW); Talent (1 SOSC); near Talent, Southern Oregon Experimental Station (1 MVZ); Claton Creek* (1 SOSC). **Josephine Co.**—Hugo (1 USNM); 7.2 mi. E Selma (1 SOSC). **Klamath Co.**—0.75 mi. S Bonanza (1 PSM); 1 mi. E Bonanza (1 PSM); 3 mi. E Bonanza (1 PSM); Crater Lake National Park, administrative area (1 MVZ); Crater Lake National Park, Mt. Mazama, 7,000 ft. (1 USNM); Ft. Klamath (3 USNM); Klamath Falls (1 CM, 1 SOSC); 8 mi. S Klamath Falls, T39S, R8E, Sec. 33 (1 OSUFW); 2 mi. E Klamath Falls (1 SOSC); S end Lake of the Woods, campground (1 MVZ); Pelican Bay (1 USNM). **Lake Co.**—Fremont (1 USNM); Fort Rock (1 USNM); Gearhart Mt., Finley Corral (2 PSM); Goose Lake Valley (1 SDNHM).

Scapanus orarius, n = 402
(Plotted in Fig. 8-26, p. 70)

Baker Co.—Baker [City] (1 USNM); Cornucopia (1 USNM); Halfway (1 USNM); Home, Snake River, 2,000 ft. (1 USNM). **Benton Co.**—1 mi. S, 6 mi. W Alpine (1 OSUFW); Alsea (1 OSUFW); 4 mi. N, 4 mi. E Alsea, T13S, R7W, Sec. 15 (6 OSUFW, 1 KU); 4.4 km N, 7.6 km W Alsea, T13S, R8W, Sec. 20 (1 USNM); 1 km S, 4.4 km E Alsea, T14S, R7W, Sec. 9 (1 USNM); 1.2 km S, 4.2 km E Alsea, T14S, R7W, Sec. 9 (1 USNM); 3.75 mi. S, 5.5 mi. E Alsea (2 CM); Corvallis (1 OSUFW); 6 mi. N Corvallis, McDonald Forest (1 OSUFW); 5 mi. N, 5 mi. E Corvallis, 200 ft. (1 CM); 1 km S, 11.3 km W Philomath, T12S, R7W, Sec. 11/14 (1 USNM); 1.5 km S, 10.1 km W Philomath, T12S, R7W, Sec. 13 (2 USNM); 1.5 km S, 11.3 km W Philomath, T12S, R7W, Sec. 14 (2 USNM); 2.4 km S, 15.5 km W Philomath, T12S, R7W, Sec. 17/20 (2 USNM); 4 km S, 10.1 km W Philomath, T12S, R7W, Sec. 24/25 (2 USNM); 3 mi. S, 1 mi. W Philomath, T12S, R6W, Sec. 26 (1 KU); 5 mi. SW Philomath (1 USNM); 1.25 mi. SE Summit, T11S, R7W, SW¼ Sec. 8, 650 ft. (1 PSM). **Clackamas Co.**—Aurora (2 UW[EB], 2 UW[LB], 1 UW[LaB]); Clackamas (2 UW[EB]); Howard (2 USNM); Jennings Lodge (1 FMNH, 2 KU, 12 PSM, 6 SDNHM); Milwaukie (7 PSM); Oregon City (1 UW[DG], 12 UW[EB], 4 UW[LB]); 6 mi. S Portland (1 OSUFW); Ripplebrook Ranger District, T6S, R6E, Sec. 3 (1 PSU); 2 mi. N, 4 mi. E Sandy (3 CRCM); 3.8 mi. N, 8 mi. E Sandy (2 CRCM); 0.33 mi. N, 1 mi. W Wemme, T2S, R7E, Sec. 31 (1 OSUFW); 9 mi. N, 4 mi. W Zigzag (2 CRCM); 3.5 mi. N, 8 mi. E Zigzag (1 CRCM). **Clatsop Co.**—Astoria (1 UMMZ); site of Old Ft. Clatsop (1 MVZ); 1 mi. N site of Old Ft. Clatsop (1 MVZ). **Columbia Co.**—Rainier, 5 mi. up Fern Hill Rd. (1 SOSC); 10 mi. NE Vernonia (1 PSM). **Coos Co.**—0.75 mi. S Bandon (1 PSM); 2 mi. S Bandon (1 PSM); 1.75 mi. E Bandon (9 PSM); 3 mi. SE Bandon (2 PSM); 3 mi. SSE Bandon (1 PSM); 4 mi. SE Bandon (3 PSM); Charleston (2 LACM); Myrtle Point (1 USNM); 4.8 km S Sitkum (1 USNM); 7.4 km N, 4.2 km W Sitkum, T27S, R10W, Sec. 17 (2 USNM); 7.5 km N, 12.5 km W Sitkum, T27S, R11W, Sec. 21 (2 USNM); 7.6 km N, 5.2 km W Sitkum, T27S, R10W, Sec. 18 (2 USNM); 8.4 km N, 4.3 km W Sitkum, T27S, R10W, Sec. 17 (1 USNM); 2.9 km S, 4.7 km E Sitkum, T28S, R9W, Sec. 19 (1 USNM); 3 km S, 10.8 km E Sitkum, T28S, R9W, Sec. 23 (2 USNM); 3 km S, 11 km E Sitkum, T28S, R9W, Sec. 23 (1 USNM); 4.8 km S, 9.7 km E Sitkum, T28S, R9W, Sec. 27 (1 USNM). **Crook Co.**—Prineville (1 USNM). **Curry Co.**—Brookings (1 SOSC); 10 mi. E Brookings (4 PSM); Little Redwood Forest Camp (1 PSM). **Douglas Co.**—9 mi. N, 8.5 mi. E Blue River, T29S, R1W, Sec. 13 (1 CRCM); 3.2 mi. S, 3.6 mi. E Canyonville, T31S, R4W, Sec. 17, 2,000 ft. (3 UW[CR]); 7.6 km N, 6.4 km W Drain, T21S, R5W, Sec. 22 (1 USNM); 11.3 km N, 10.5 km W Drain, T21S, R6W, Sec. 5 (2 USNM); 5.5 mi. N, 4 mi. W Drain, T21S, R6W, Sec. 15, 1,280 ft. (1 UW[CR]); 6 mi. N, 4 mi. W Drian, T21S, R6W, Sec. 15, 1,200 ft. (2 UW[CR]); 9.7 km S, 16.2 km W Elkton, T23S, R9W, Sec. 22 (1 USNM); 11.1 km S, 7.4 km W Elkton, T23S, R8W, Sec. 28 (5 USNM); 14.9 km S, 6.2 km W Elkton, T24S, R8W, Sec. 9 (5 USNM); 4.5 mi. N Gardiner (1 PSM); 9 mi. N Gardiner (1 PSM); 3.2 km N, 26 km W Lorane, T19/20S, R7W, Sec. 32/5 (1 USNM); Reston (1 USNM);

4 mi. S, 6.4 mi. W Riddle, Iron Mt., T31S, R7W, Sec. 11, 2,000 ft. (2 UW[CR]); 2.4 km N, 24.6 km W Sutherlin, T25S, R8W, Sec. 10 (2 USNM); 2.8 km N, 26 km W Sutherlin, T25S, R8W, Sec. 9 (1 USNM); 3.6 km N, 11.9 km W Sutherlin, T25S, R7W, Sec. 1 (1 USNM); 3.8 km N, 12.1 km W Sutherlin, T25S, R7W, Sec. 1 (1 USNM); 6.5 km N, 6.7 km W Sutherlin, T28S, R8W, Sec. 9 (1 USNM); 4.5 mi. N, 8 mi. E Tiller, T30S, R1W, Sec. 2 (2 CRCM); 11 mi. N, 5 mi. E Tiller (1 CRCM); 12 mi. N, 4 mi. E Tiller (2 CRCM); 12 mi. N, 17 mi. E Tiller (1 CRCM); 12.5 mi. N, 5.5 mi. E Tiller, T28S, R1W, Sec. 31 (2 CRCM); 0.2 mi. N, 4.1 mi. W Umpqua P.O., T25S, R7W, Sec. 15, 800 ft. (1 UW[CR]); 2.2 mi. N, 9 mi. W Umpqua P.O., T25S, R8W, Sec. 15, 850 ft. (1 UW[CR]); 4 mi. N, 10.5 mi. W Umpqua, Miner Creek, T24S, R8W, Sec. 29, 900 ft. (4 UW[CR]); 4 mi. N, 1 mi. W Umpqua P.O., Lost Creek, T25S, R7W, Sec. 1, 1,500 ft. (3 UW[CR]). **Grant Co.**—6 mi. E Dayville (1 SDNHM). **Jackson Co.**—Ashland (2 SOSC); Emigrant Lake (1 SOSC); 11 mi. N Prospect (2 CRCM); 1.5 mi. N Rogue River, off East Evans Creek Rd. (2 OSUFW). **Lane Co.**—(2 PSM); Ada, near Siltcoos Lake, T19S, R11W, Sec. 30 (1 OSUFW); 2 mi. S, 1 mi. E Austa (1 CM); 8.75 mi. E Blue River (1 CRCM); 9 mi. E Blue River, T16S, R5E, Sec. 25/26 (1 CRCM); 4.4 km N, 6.8 km E Blue River, T16S, R5E, Sec. 20 (8 USNM); 5 km N, 7.6 km E Blue River, T16S, R5E, Sec. 6 (1 USNM); 6.8 km N, 7.6 km E Blue River, T15S, R5E, Sec. 32 (1 PSM, 5 USNM); 1.5 mi. N, 2 mi. E Blue River (1 CRCM); 4 mi. N, 4.5 mi. E Blue River, T16S, R5E, Sec. 32 (2 CRCM); 5 mi. N, 9 mi. E Blue River (4 CRCM); 5.5 mi. N, 13.5 mi. E Blue River, T16S, R6E, Sec. 26 (1 CRCM); 7 mi. N, 9 mi. E Blue River, T15S, R5E, Sec. 13 (4 CRCM); 2.5 mi. S, 7 mi. E Blue River (1 CRCM); 3.5 mi. S, 10 mi. E Blue River (1 CRCM); Cushman (1 LACM); Florence (1 OSUFW); 0.4 km N, 18.5 km W Lorane, T20S, R7W, Sec. 12 (2 USNM); 2 km N, 6.4 km W Lorane, T20S, R5W, Sec. 5 (3 USNM); 2.8 km N, 6.2 km W Lorane, T20S, R5W, Sec. 5 (3 USNM); 3 km N, 19.5 km W Lorane, T20S, R7W, Sec. 1 (1 USNM); 4.8 km N, 17.5 km W Lorane, T19S, R6W, Sec. 31 (1 USNM); McKenzie Bridge (2 USNM); 2.6 km S McKenzie Bridge, T16S, R5E, Sec. 26 (1 USNM); 3.1 km N, 3 km W McKenzie Bridge, T16S, R5E, Sec. 3 (1 USNM); 3.3 km N, 2.4 km W McKenzie Bridge, T16S, R5E, Sec. 3 (7 USNM); 6.6 km N, 0.3 km W McKenzie Bridge, T15S, R5E, Sec. 25 (5 USNM); 6 km N, 0.6 km E McKenzie Bridge, T15S, R5E, Sec. 25 (2 USNM); 6 km N, 7.5 km E McKenzie Bridge, T15S, R6E, Sec. 26 (3 USNM); 6.4 km N, 0.6 km E McKenzie Bridge, T15S, R5E, Sec. 25 (4 USNM); 1.6 km S, 5.2 km W McKenzie Bridge, T16S, R5E, Sec. 20 (3 USNM); 2.6 km S, 1.4 km E McKenzie Bridge, T16S, R5E, Sec. 25 (2 USNM); 3.8 km S, 14.8 km E McKenzie Bridge, T16S, R7E, Sec. 33 (1 USNM); 5.4 km S, 8.7 km E McKenzie Bridge, T16S, R6E, Sec. 35 (2 USNM); 6 mi. NE Mercer (1 LACM); 4 mi. N Nimrod, T16S, R3E, Sec. 14 (1 CRCM); 4.5 mi. N Nimrod (2 CRCM); just W of Oakridge Rigdon R.S., T21S, R2E, Sec. 11 (1 USNM); N base Three Sisters (2 USNM); Vida (4 USNM). **Lincoln Co.**— Cascade Head Experimental Forest, T16S, R10W, Sec. 21 (4 PSM); 2 mi. S, 12 mi. W Harlan, Beaver Creek, T12S, R10W, Sec. 20 (1 KU); 7 mi. S, 5 mi. W Harlan, West Creek, T13S, R9W, Sec. 16 (1 KU); Newport (1 USNM); 0.6 km N, 16.9 km E Waldport, T13S, R10W, Sec. 13/24 (2 USNM); 1.6 km N, 5.8 km E Yachats, T14S, R11W, Sec. 17 (1 USNM); 4 km S, 3.2 km E Yachats, T15S, R12W, Sec. 12 (4 USNM); Yaquina Bay (4 USNM). **Linn Co.**—7 mi. N, 8 mi. E Blue River, T15S, R5E, Sec. 15 (4 CRCM); 7.5 mi. N, 7.5 mi. E Blue River, T15S, R5E, Sec. 15 (5 CRCM); 8 mi. N, 8 mi. E Blue River (8 CRCM); 9 mi. N, 8 mi. E Blue River (1 CRCM); 9.2 km N, 1.2 km W McKenzie Bridge, T15S, R5E, Sec. 14 (1 USNM); 10.25 km N, 0.8 km W McKenzie Bridge, T15S, R5E, Sec. 14 (1 USNM); 13.2 km N, 0.8 km E McKenzie Bridge, T15S, R5E, Sec. 1 (1 USNM). **Multnomah Co.**—Portland (3 AMNH, 3 MVZ, 3 OSUFW, 4 PSM, 2 PSU, 9 SDNHM, 1 UMMZ, 2 USNM); 9 mi. N, 10 mi. E Sandy (1 CRCM); 10 mi. N, 11 mi. E Sandy (1 CRCM); 11 mi. N Zigzag (2 CRCM); 8 mi. N, 3 mi. E Zigzag (1 CRCM); 10 mi. N, 1.5 mi. W Zigzag (1 CRCM). **Polk Co.**—6 mi. N, 2 mi. E Kings Valley (1 OSUFW). **Sherman Co.**—2 mi. S Columbia River, HWY 206 (1 EOSC); 3 mi. S Columbia River, HWY 206 (1 EOSC). **Tillamook Co.**—Blaine (1 KU); Nehalem (2 KU, 1 OSUFW, 1 PSM); Netarts (1 SDNHM); Oceanside (1 OSUFW); Tillamook (1 LACM, 1 MVZ, 9 OSUFW, 2 SDNHM, 1 UMMZ, 2 USNM); 2 mi. up Miami River (1 PSM). **Umatilla Co.**—2 mi. W Pendleton (1 SDNHM). **Union Co.**—Cove (1 EOSC). **Wasco Co.**—

Dufur (2 SDNHM); Maupin (1 SDNHM); 7 mi. SE Maupin (1 SDNHM). **Washington Co.**—18.5 mi. NW Portland (1 MVZ); HWY 26 and Cornell Rd. (1 PSU). **Yamhill Co.**—0.75 mi. S Rex (1 MVZ).

Scapanus townsendii, n = 528
(Plotted in Fig. 8-29, p. 72)

Benton Co.—Adair Village, T10S, R4W, Sec. 30 (2 OSUFW, 2 PSM); 7 mi. W Alsea, Fall Creek (1 OSUFW); 4 mi. SW Blodett, T12S, R7W, NW¼ Sec. 9 (1 PSM); Corvallis (23 OSUFW, 2 UCONN, 2 UMMZ); Corvallis, T11S, R5E, Sec. 3 (5 OSUFW); Corvallis, T11S, R5W, Sec. 15 (1 OSUFW); Corvallis, T11S, R5E, Sec. 30 (1 OSUFW); Corvallis, T11S, R5W, Sec. 31 (1 OSUFW); Corvallis, T12S, R5W, Sec. 4 (1 OSUFW); 1.5 mi. N Corvallis (1 OSUFW); 3 mi. N Corvallis, T11S, R5W, Sec. 29 (2 OSUFW); 5 mi. N Corvallis, T11S, R5W, Sec. 1 (2 OSUFW); 1 mi. S Corvallis, T12S, R5W, Sec. 3 (4 OSUFW); 0.5 mi. W Corvallis, T11S, R5W, Sec. 35 (1 OSUFW); 1 mi. W Corvallis, T11S, R5W, Sec. 34 (2 KU, 11 OSUFW); 2 mi. W Corvallis, T11S, R5W, Sec. 34 (3 OSUFW); 6 mi. N, 3 mi. W Corvallis (1 OSUFW); 6 mi. N, 2 mi. E Corvallis, T10S, R4W, Sec. 31 (1 OSUFW, 1 KU); 11 mi. N, 3 mi. E Corvallis (1 OSUFW); 1 mi. S, 1 mi. W Corvallis (1 OSUFW); 1 mi. NW Corvallis, T11S, R5W, Sec. 28 (1 OSUFW); 1 mi. SW Corvallis, T12S, R5W, Sec. 3 (2 OSUFW); 3 mi. W Monroe (1 OSUFW); 0.25 mi. S, 0.25 mi. W Monroe (1 OSUFW); 1 mi. W Philomath (1 PSM); 2 mi. E Philomath on HWY 20 (1 OSUFW); 1 mi. S, 6 mi. W Philomath near Stillson Creek, Marys Peak (1 OSUFW); 10 mi. S, 8 mi. W Philomath (1 OSUFW); 1.5 km S, 11.3 km W Philomath (1 USNM); 5 mi. SW Philomath (1 PSM, 2 USNM); 1.25 mi. SE Summit, T11S, R7W, SW¼ Sec. 8, 650 ft. (1 PSM); Wells (1 USNM); T13S, R5W, Sec. 28 (2 PSM). **Clackamas Co.**—Aurora (23 UW[LB], 11 UW[LaB]); Boring (1 PSU); Clackamas (1 PSM, 13 UW[EB]); Gladstone (2 UW[RM]); Jennings Lodge (1 KU, 1 LACM, 12 PSM, 1 OSUFW, 14 SDNHM, 1 UMMZ); Milwaukie (5 PSM, 2 PSU); 8 mi. S Molalla (1 CRCM); Mulino (8 OSUFW); Oak Grove (1 PSM); Oregon City (3 USNM, 20 UW[EB]); Oswego (1 MVZ, 1 USNM); Wilsonville (1 OSUFW, 8 SDNHM); 1 mi. N, 1.5 mi. E Wilsonville, T3S, R1E, NW¼ Sec. 18 (1 KU). **Clatsop Co.**—7.5 mi. S Cannon Beach (1 MVZ); site of Old Ft. Clatsop (2 MVZ). **Columbia Co.**—Deer Island (2 PSM); Rainier, 5 mi. up Fern Hill Rd. (6 SOSC); 7 mi. SE Rainier (2 MVZ); Sauvie Island (3 CRCM); T4N, R1W, W½ Sec. 22 (1 PSM); T6N, R2W, NW¼ Sec. 12 (1 PSM). **Coos Co.**—1.75 mi. S Bandon (7 PSM); 2 mi. S Bandon (4 PSM); 1.75 mi. E Bandon (1 PSM); 3 mi. SE Bandon (1 PSM); 4 mi. SE Bandon (3 PSM); Charleston (1 LACM); Coquille (1 OSUFW, 1 USNM); 9 mi. NE Coquille (2 PSM). **Curry Co.**—Brookings (1 PSM); Gold Beach (1 FMNH, 2 OSUFW, 1 USNM); 1 mi. S Pistol River (2 MVZ); mouth of Joe Hall Creek and Chetco River (1 SOSC). **Douglas Co.**—13 mi. E Canyonville, S fork Umpqua River (1 LACM); 3 mi. S, 0.5 mi. E Cleveland (1 OSUFW); Drain (1 USNM); 10 km N, 10.6 km W Drain (1 USNM); 3 mi. NE Grants Pass, James Creek, 1,600 ft. (1 USNM); Louis Creek, T28S, R3W, NW¼SW¼ Sec. 29 (1 SOSC); 1.5 mi. W Roseburg (1 OSUFW); Scottsburg (1 OSUFW); 0.5 mi. W Sutherlin (1 OSUFW). **Jackson Co.**—Ashland (1 SOSC); Butte Falls R.D., T33S, R4E, NW¼NE¼ Sec. 26 (1 SOSC); Exposition Park [in Central Point] (1 SOSC); Medford (1 LACM, 1 SOSC); 1.5 mi. N Rogue River off E Evans Creek Rd. (3 OSUFW). **Josephine Co.**—0.5 mi. W Grants Pass, T36S, R5W, Sec. 18 (2-OSUFW); 1 mi. W Grants Pass along Rogue River (1 OSUFW); 3.5 mi. W Grants Pass, Lower River Rd. (1 OSUFW); 4 mi. S Williams (1 SOSC). **Lane Co.**—(5 PSM); near Ada (1 OSUFW); Coburg, Reed's Pasture (1 MVZ); Eugene (1 OSUFW, 7 UO); 1 mi. S, 2 mi. E Junction City (1 OSUFW); 2 mi. S, 1 mi. E Junction City (2 OSUFW); 0.5 mi. S, 3 mi. E Mapleton (1 LSU); 1.2 mi. N, 2.6 mi. E Mapleton (1 LSU); Mercer (1 OSUFW); Mercer Lake, Dowell Ranch (3 OSUFW); Noti (1 OSUFW). **Lincoln Co.**—3.5 mi. S Burnt Woods (1 OSUFW); Cascade Head Experimental Forest (10 PSM); Alsea HWY near Fall Creek (1 OSUFW); 1.25 mi. W Harlan (1 KU); 3 mi. E Tidewater (1 EOSC); 1 mi. W Toledo (1 UMMZ); Waldport (1 OSUFW). **Linn Co.**—4.5 mi. S, 6.5 mi. W Albany (7 OSUFW); 5 mi. S, 5.25 mi. W Albany (2 HSU, 1 JFBM, 1 OSUFW); 6 mi. S, 5.25 mi. W Albany (3 OSUFW); 9 mi. SW Albany, T11S, R5W, Sec. 36 (1 OSUFW); 0.6 mi. E

Corvallis (1 OSUFW); 2.5 mi. S, 2 mi. E Halsey (1 OSUFW); 5 mi. W Lebanon (1 OSUFW); North Albany, Gibson Hill (1 OSUFW); 1.5 mi. S, 1 mi. E Shedd (1 OSUFW); 3.5 mi. S, 6.75 mi. W Shedd (1 OSUFW); 3.5 mi. S, 7 mi. W Shedd (2 OSUFW); 0.5 mi. E Sweet Home (1 OSUFW); 5 mi. NW Sweethome, T13S, R1W, Sec. 4 (1 PSM); 33117 White Pine Rd., T12S, R4W (1 OSUFW). **Marion Co.**—1 mi. N, 3 mi. E Brooks (1 OSUFW); 0.5 mi. N, 0.25 mi. W Mt. Angel, T6S, R1W, Sec. 3 (1 KU); Salem (2 AMNH, 1 MVZ, 1 OSUFW, 1 USNM); 10 mi. N Salem (1 OSUFW); 0.5 mi. N, 2.5 mi. E Stayton (1 OSUFW); 6 mi. N, 3 mi. E Stayton (1 OSUFW); Woodburn (4 PSM). **Multnomah Co.**—Government Island (2 OSUFW); Gresham (1 OSUFW, 1 PSU); 3 mi. SE Gresham, T1S, R3E, Sec. 11 (1 OSUFW); Portland (17 AMNH, 1 KU, 5 MVZ, 2 OSUFW, 10 PSM, 4 PSU, 7 SDNHM, 1 USNM); 1 mi. W Rosewood* (1 OSUFW); Sauvies Island (1 PSU). **Polk Co.**—Falls City, Perry Rd. (1 OSUFW); 2 mi. S, 0.5 mi. W Hopewell (1 OSUFW); 5 mi. S, 1 mi. E Independence (2 OSUFW); 2 mi. S, 1 mi. E Monmouth (1 OSUFW); 0.5 mi. S, 3 mi. E Rickreall (1 OSUFW). **Tillamook Co.**—(2 LACM); Beaver (4 KU, 2 OSUFW, 1 UMMZ); Blaine (1 KU, 2 MVZ, 2 OSUFW, 2 UMMZ); Cloverdale (3 OSUFW); Hebo Lake (1 PSM); Mt. Hebo (1 OSUFW); Nehalem (1 OSUFW, 1 PSM, 1 UW[WWC]); 1 mi. N Nehalem (1 PSM); 1 mi. SW Nehalem (1 OSUFW); Netarts (1 OSUFW, 2 USNM); Oceanside (1 LACM); Pleasant Valley (3 UMMZ); Tillamook (2 AMNH, 6 KU, 2 LACM, 2 MVZ, 11 OSUFW, 1 SDNHM, 1 UMMZ, 1 UO, 1 USNM); 1 mi. S Tillamook (2 OSUFW); 9 mi. S Tillamook (2 UMMZ); 2 mi. W Tillamook (1 PSM); 7 mi. E Tillamook, Wilson River (2 PSM); 30 mi. E Tillamook, King's Mt. (1 PSU); 2 mi. SE Tillamook (1 OSUFW); 2 mi. up Miami River (1 PSM). **Washington Co.**—Beaverton (3 USNM); Cedar Mill (1 EOSC, 7 PSM); 2 mi. W Hillsboro, Rock Creek (1 PSU); 0.5 mi. N Metzger (1 OSUFW); Portland (2 PSU); 18.5 mi. NW Portland (1 MVZ); Tigard (1 OSUFW, 1 PSU). **Yamhill Co.**—McMinnville (1 OSUFW); 3 mi. E Newberg (5 MVZ).

Sciurus carolinensis, n = 1
(Plotted in Fig. 11-34, p. 204)

Marion Co.—Salem (1 OSUFW).

Sciurus griseus, n = 289
(Plotted in Fig. 11-36, p. 208)

Benton Co.—Corvallis (1 OSUFW, 1 USNM); 4 mi. N Corvallis (1 KU); 6 mi. N Corvallis, Coffin Butte (1 OSUFW); 2 mi. W Corvallis (1 OSUFW); 1 mi. N, 2 mi. W Corvallis (1 OSUFW); 1.7 mi. N, 0.7 mi. W Corvallis (1 OSUFW); 5 mi. N, 4 mi. W Corvallis (2 OSUFW); 9 mi. N, 1 mi. W Corvallis (1 OSUFW); 9 mi. S, 2 mi. W Corvallis (1 OSUFW); 6 mi. S, 2 mi. W Philomath (4 OSUFW); 5 mi. S, 1.3 mi. E Philomath (1 OSUFW). **Clackamas Co.**—Alspaugh (2 USNM); Marmot (1 USNM). **Coos Co.**—Coos Head Municipal Park (2 LACM); 1 mi. N Powers (1 OSUFW); 1.5 mi. N, 0.5 mi. W Powers (1 OSUFW); 3 mi. N, 1 mi. W Powers (1 OSUFW). **Curry Co.**—Brookings (1 SOSC); 10 mi. E Brookings (1 PSM); Gold Beach (1 PSU); Little Redwood Forest Camp (4 PSM); Long Ridge Forest Camp (2 PSM); 17 mi. up river from mouth of Rogue River (1 USNM). **Deschutes Co.**—Bend (1 OSUFW); Indian Ford Camp (3 PSM); 5 mi. E Sisters (1 SDNHM). **Douglas Co.**—Anchor (1 USNM); Elkhead (1 USNM); Glendale (1 USNM); 7 km W Glide (1 USNM); Reston (1 USNM); Riddle (2 USNM, 1 CRCM); Roseburg (1 USNM); 15 mi. E Roseburg (2 OSUFW); 5 mi. NW Roseburg (3 UO); 3 mi. W Sutherlin (3 OSUFW); 5 mi. N, 3 mi. E Sutherlin (1 OSUFW). **Hood River Co.**—Hood River (1 PSU, 4 USNM); 2 mi. S Odell (1 PSU). **Jackson Co.**—Ashland (5 SOSC); Ashland, canyon adjacent to Clayton Creek (2 SOSC); Ashland, SW Perozzi Acres (2 SOSC); Bald Mt., T37S, R1E, Sec. 31 (13 SOSC); SW Bald Mt. (1 SOSC); S slope Bald Mt. (5 SOSC); S Bald Mt., summit near Wagner Creek Rd. (13 SOSC); Emigrant Lake (5 SOSC); S of Emigrant Lake (2 SOSC); 1.5 to 2 mi. up old HWY 99 from Emigrant Lake (2 SOSC); Gold Hill (1 OSUFW); Prospect (1 KU, 6 OSUFW, 1 SOSC); 8 mi. SW Prospect (3 AMNH, 1 SOSC); 3 mi. S Ruch (1 SOSC); S Shale City (4 SOSC); 0.5 mi. S Shale City off Dean Indian Rd. (1 SOSC); 0.25 to 1 mi. S Shale City (1 SOSC); 0.5 to 0.75 mi. S Shale City (2 SOSC); 0.5 to 1 mi. S

Shale City (2 SOSC); 1 to 2 mi. S Shale City (1 SOSC); 1 mi. S Shale City (5 SOSC); 1.5 mi. S Shale City (2 SOSC); 2 mi. S Shale City (2 SOSC); 5 mi. S Talent (1 SOSC); 5 mi. NE Trail (1 LACM); Old Craggy Mt. off Dead Indian Rd.* (5 SOSC); 1 mi. S Mt. Ashland Rd. and Colestine Rd. (1 SOSC); end of Neil Creek Rd. (1 SOSC); 1.5 mi. up old HWY 99 from HWY 66 (2 SOSC); 2 mi. S on HWY 99 from HWY 66 (1 SOSC); 1.6 mi. S USFS Rd. 3925 (1 SOSC). **Jefferson Co.**— Camp Sherman (1 PSU); 20 mi. W Warmsprings, Mill Creek (1 USNM). **Josephine Co.**—15 mi. E Cave Junction (2 KU); 2.5 mi. S, 9.5 mi. E Cave Junction (1 KU); 7 mi. S, 9 mi. E Cave Junction (1 SOSC); Grants Pass (4 CM); 13 mi. SW Grants Pass (5 UO). **Klamath Co.**—Bonanza (1 PSM); Ft. Klamath (1 MVZ); 3 mi. N Ft. Klamath (1 LACM); near Johnston Prairie* (1 CM); Modoc Point (1 USNM); Pokegama (20 USNM). **Lane Co.**—Coast Range* (1 PSU); Elmira (1 UO); 4 mi. N Elmira (6 OSUFW); 9 mi. W Eugene (1 CM); 10 mi. W Junction City (1 PSM); Mabel (1 UO); McKenzie Bridge (1 PSM); 12 mi. SSW Monroe (1 OSUFW); 3 mi. SE Oakridge (4 UO). **Linn Co.**—Lebanon (1 UW[WWC]); 5 mi. S Lebanon (1 UW[WWC]); Shelburn (1 USNM). **Marion Co.**—Salem (1 SDNHM, 1 UMMZ, 1 USNM); 8 mi. N, 2.5 mi. E Stayton (1 OSUFW). **Multnomah Co.**—Bonneville (1 OSUFW); Cooper Mt. near Beaverton-Vora Rd. (2 PSU); Portland (2 LACM); Sandy River* (1 USNM). **Wasco Co.**—Friend (1 USNM); Ft. Dalles (2 USNM); 5 mi. S, 1 mi. E Mosier (1 OSUFW); Quartz Butte (1 PSU); W of Rowena (1 PSM); The Dalles (1 LACM, 1 OSUFW); 14 mi. SW The Dalles, Mill Creek (2 SDNHM); 14 mi. SW The Dalles, Fivemile Creek (3 OSUFW, 1 PSU); Wamic (1 USNM); Wapinitia (3 USNM); 6 mi. W Wapinitia (1 PSM). **Washington Co.**—Beaverton (1 OSUFW); Cedar Mills (1 PSU); 4 mi. E Gaston (1 UW[WWC]); 2 mi. S Scholls (9 OSUFW); 1 mi. S, 1 mi. W Scholls (7 OSUFW); Tigard (1 PSU); 1 mi. N Tualatin (1 PSU); T1N, R4W, Sec. 28 (1 PSU). **Yamhill Co.**—2 mi. S, 1 mi. W Carlton (1 PSU); Dayton (2 KU, 7 OSUFW, 6 PSM); 3.5 mi. E Newberg (1 KU); 1 mi. N, 2 mi. W Yamhill (1 OSUFW).

Sciurus niger, n = 44
(Plotted in Fig. 11-38, p. 210)

Baker Co.—Baker City (1 PSM). **Clackamas Co.**—3 mi. W Estacada (1 PSU); 3 mi. N Lake Oswego (1 PSU); Milwaukie (3 PSU); Oak Grove (1 PSU). **Lane Co.**—E Eugene (3 OSUFW). **Marion Co.**— 9 mi. N Salem (4 OSUFW); 2 mi. N, 1 mi. W Waconda (2 OSUFW); 2 mi. N, 1.3 mi. W Waconda (2 OSUFW); 2 mi. N, 1.5 mi. W Waconda, near bank Willamette River (1 OSUFW). **Multnomah Co.**—Portland (1 PSU); Pier Park, Portland (1 PSU). **Union Co.**—(2 EOSC); La Grande (3 EOSC); 0.5 mi. W Union (1 OSUFW). **Washington Co.**—Cedar Mill (2 PSU); 2 mi. E Gaston (1 USNM); 2 mi. S Scholls (7 OSUFW); 1 mi. S, 1 mi. W Scholls (3 OSUFW); Tualatin (1 PSU). **Yamhill Co.**—Dayton (2 RA); Grand Island (1 PSU).

Sorex bairdi, n = 400
(Plotted in Fig. 8-2, p. 49)

Benton Co.—6 mi. N Alder Creek (1 OSUFW); 2 mi. W Corvallis (1 PSM); 6 mi. N, 3 mi. W Corvallis (1 OSUFW); Marys Peak, T12S, R6W, NW¼ Sec. 30 (2 OSUFW); T11S, R7W, SW¼ Sec. 8, 650 ft. (1 PSM). **Clackamas Co.**—4.5 mi. N Alder Creek, S. Fork Bull Run River, T1S, R6E, Sec. 29 & 32 (2 OSUFW); 6 mi. N Alder Creek, Cedar Creek, T1S, R6E, Sec. 29 (1 OSUFW); Estacada (1 KU); 8 mi. SE Molalla (1 PSM); Oregon City (1 USNM); 2 mi. N, 4 mi. E Sandy (2 CRCM); 3.5 mi. N, 8 mi. E Sandy (1 CRCM); 3.8 mi. N, 8 mi. E Sandy (1 CRCM); 6 mi. S, 7 mi. E Sandy, T3S, R6E, Sec. 18 (8 CRCM); Wemme (1 SDNHM); Wilhoit Springs, T6S, R2E, Sec. 16 (1 PSM); 9 mi. N, 4 mi. W Zigzag (8 CRCM); 5 mi. S, 6 mi. E Zigzag (4 CRCM). **Clatsop Co.**—Astoria (6 USNM). **Columbia Co.**—7 mi. SE Rainier (1 MVZ); 1 mi. N, 0.5 mi. W Vernonia (1 UW); T4N, R1W, N½ Sec. 34 (1 PSM). **Coos Co.**—3 mi. S Charleston, Sunset Bay (1 LACM); 5 mi. SW Charleston, Sunset Bay (2 LACM); Delmar, 9 mi. S Marshfield [= Coos Bay] (1 MVZ); Point Arago State Park (1 LACM). **Lane Co.**—4.4 km N, 6.8 km E Blue River, T16S, R5W, Sec. 7 (4 USNM); 6.8 km N, 7.6 km E Blue River, T15S, R5E, Sec. 32 (16 USNM); 1.5 mi. N, 2 mi. E

Blue River (1 CRCM); 7 mi. N, 8 mi. E Blue River, T15S, R5E, Sec. 15 (1 CRCM); 10 mi. NE Blue River, Lookout Creek Upper Tributary, T15S, R5E, Sec. 25 (1 OSUFW); 5 mi. N Cougar Dam, T16S, R5E, Sec. 6 (3 UW[OG]); McKenzie Bridge (3 OSUFW, 1 PSM); 2.6 km S McKenzie Bridge, T16S, R5E, Sec. 26 (4 USNM); 3.1 km N, 3 km W McKenzie Bridge, T16S, R5E, Sec. 3 (7 USNM); 3.3 km N, 2.4 km W McKenzie Bridge, T16S, R5E, Sec. 3 (5 USNM); 6 km N, 0.6 km E McKenzie Bridge, T15S, R5E, Sec. 25 (13 USNM); 6 km N, 7.5 km E McKenzie Bridge, T15S, R6E, Sec. 26 (2 USNM); 6.4 km N, 0.6 km E McKenzie Bridge, T15S, R5E, Sec. 25 (3 USNM); 6.6 km N, 0.3 km E McKenzie Bridge, T15S, R5E, Sec. 25 (9 USNM); 1.6 km S, 5.2 km W McKenzie Bridge, T16S, R5E, Sec. 20 (6 USNM); 3.8 km S, 14.8 km E McKenzie Bridge, T16S, R7E, Sec. 33, 1,070 m (5 USNM); 4.4 km S, 14.3 km E McKenzie Bridge, T16S, R7E, Sec. 32 (8 USNM); 5.4 km S, 8.75 km E McKenzie Bridge, T16S, R6E, Sec. 35 (3 USNM); 4 mi. N McKenzie Bridge, T15S, R5E, Sec. 25, 2,600 ft. (1 UW[OG]); 1 mi. N, 8 mi. W McKenzie Bridge, T16S, R4E, SW¼ Sec. 9 (1 KU); 3.5 mi. N, 9 mi. E McKenzie Bridge, T16S, R7E, Sec. 32, 3,500 ft. (4 UW[OG]); 3 mi. S, 8.5 mi. E McKenzie Bridge, T16S, R7E, Sec. 32, 3,500 ft. (1 UW[OG]); 1 mi. N Mohawk River, T15S, R1W, SE¼NE¼ Sec. 13 (1 SOSC); 4 mi. N Nimrod, T16S, R3E, Sec. 14 (1 CRCM); 4 mi. N Rainbow, T15S, R5E, Sec. 32, 1,700 ft. (3 UW[OG]); Upper tributary Lookout Creek, T15S, R5E, Sec. 25 (19 USNM); Vida (1 SDNHM, 1 USNM); T15S, R5E, NW¼ Sec. 24, 2,700 ft. (1 PSM); 5 mi. S Wolf Rock, T15S, R5E, Sec. 25, 2,800 ft. (1 UW[OG]); T15S, R5E, NW¼ Sec. 24, 3,500 ft. (6 PSM); T15S, R5E, Sec. 26 (1 USNM); T15S, R5E, NW¼ Sec. 28, 2,000 ft. (2 PSM); T15S, R5E, NE¼ Sec. 32, 1,650 ft. (8 PSM); T15S, R5E, NE¼ Sec. 32, 2,000 ft. (6 PSM); T15S, R5E, SW¼ Sec. 33, 2,500 ft. (3 PSM); T16S, R4E, NE¼ Sec. 24, 1,400 ft. (4 PSM); T16S, R5E, NW¼ Sec. 6 (1 PSM); T16S, R5E, SW¼ Sec. 6, 1,600 ft. (5 PSM); T18S, R5E, Sec. 23, 3,000 ft. (1 PSM). **Lincoln Co.**—Cascade Head, T6S, R10W, SW¼ Sec. 21, 240 ft. (17 PSM); Delake (1 KU); 7 mi. S, 10 mi. W Harlan, Death Ridge, T13S, R10W, Sec. 14 (2 KU); Newport (1 MVZ, 1 OSUFW, 2 SDNHM); Otis (2 USNM); Taft (3 MVZ). **Linn Co.**—7.5 mi. N, 7.5 mi. E Blue River, T15S, R5E, Sec. 15 (1 CRCM); 8 mi. N, 8 mi. E Blue River (5 CRCM); 9 mi. N, 8 mi. E Blue River (7 CRCM); 2 mi. S, 17 mi. E Cascadia, T14S, R5E, S½ Sec. 1 (1 KU); 9.2 km N, 1.2 km W McKenzie Bridge, T15S, R5E, Sec. 14, 810 m (12 USNM); 10.25 km N, 0.8 km W McKenzie Bridge, T15S, R5E, Sec. 14 (5 USNM); 13.2 km N, 0.8 km E McKenzie Bridge, T15S, R5E, Sec. 1 (2 USNM); Permilia [= Pamelia] Lake, W. base Mt. Jefferson (1 PSM, 12 USNM); Trout Creek F., T13S, R3E, NE¼NW¼ Sec. 32, 1,200 ft. (2 PSM); 9 mi. E Upper Soda, T13S, R4E, Sec. 27 (1 OSUFW); 1 mi. S Wolf Rock, T15S, R5E, Sec. 1, 3,300 ft. (2 UW[OG]); 4 mi. S, 1 mi. W Wolf Rock, T15S, R5E, Sec. 14, 2,600 ft. (5 UW[OG]); T14S, R6E, NE¼ Sec. 20, 4,400 ft. (3 PSM); T14S, R6E, NE¼ Sec. 28, 4,600 ft. (1 PSM); T15S, R5E, SE¼ Sec. 11, 3,200 ft. (4 PSM); T15S, R5E, SW¼ Sec. 13, 3,000 ft. (5 PSM); T15S, R6E, NE¼ Sec. 7, 4,300 ft. (4 PSM); T15S, R6E, NW¼ Sec. 7, 5,000 ft. (4 PSM). **Marion Co.**—14 mi. N Detroit at Elk Lake (1 UCD). **Multnomah Co.**—2 mi. S, 2.5 mi. E Multnomah Falls, Oneata Creek, T1N, R6E, Sec. 22 (4 OSUFW); Portland (1 OSUFW, 2 PSU, 1 USNM); 6 mi. N, 11 mi. E Sandy (2 CRCM); 8 mi. N, 3 mi. E Zigzag (3 CRCM). **Polk Co.**—3.75 mi. S, 2 mi. W Gold Creek, T7S, R7W, Sec. 5 (2 KU, 1 OSUFW). **Tillamook Co.**—Arra Wanna* (1 LACM); 14 mi. SE Beaver (1 UCD); Blaine (1 OSUFW); 5 mi. SW Cloverdale (1 CMNH, 1 OSUFW); 5 mi. E Hebo Lake, 1600 ft. (1 UCD); Nestucca River* (6 MVZ); Netarts (1 LACM, 1 OSUFW, 1 PSM, 19 SDNHM); 1 mi. E Netarts (2 PSM); Netarts Bay (1 SDNHM); Netarts summit (1 PSM); Sandlake (1 LACM); Tillamook (3 KU, 1 JFBM, 1 OSUFW, 1 PSM, 3 SDNHM, 2 USNM); 5 mi. S Tillamook (2 PSU); 8 mi. S Tillamook (1 KU); 7 mi. SE Tillamook (2 OSUFW). **Washington Co.**—Tigard (2 PSU). **Yamhill Co.**—3 mi. NW McMinnville (1 UW[EB]).

Sorex bendirii, n = 351
(Plotted in Fig. 8-4, p. 51)

Benton Co.—1 mi. S, 6 mi. W Alpine (4 OSUFW); 4.4 km N, 7.6 km W Alsea, T13S, R8W, Sec. 20 (1 USNM); 3.75 mi. S, 5.5 mi. E Alsea (1 CM); 1 km S, 4.4 km E Alsea, T14S, R7W, Sec. 9 (1 USNM);

1.2 km S, 4.2 km E Alsea, T14S, R7W, Sec. 9 (2 USNM); 10 mi. SE Alsea (1 HSU); 5.5 mi. N Corvallis (1 PSM); 3 mi. W Corvallis (1 OSUFW); Marys Peak, 9 mi. SW Philomath (1 PSM); 1 km S, 11.3 km W Philomath, T12S, R7W, Sec. 11/14 (1 USNM); 1.5 km S, 11.3 km W Philomath, T12S, R7W, Sec. 14 (2 USNM); 4 km S, 10.1 km W Philomath, T12S, R7W, Sec. 24/25 (1 USNM); 3 mi. N Summit, T10S, R7W, NE¼ Sec. 27 (4 PSM); Wrens, Marsh Mill Pond, T11S, R6W, Sec. 36 (1 OSUFW). **Clackamas Co.**—8 mi. SE Molalla (1 PSM); Mulino (1 PSM, 1 SDNHM); Oregon City (1 USNM); 3.5 mi. N, 8 mi. E Sandy (2 CRCM); 4.2 mi. N, 8 mi. E Sandy (3 CRCM); 9 mi. N, 4 mi. W Zigzag (9 CRCM); 5 mi. S, 6 mi. E Zigzag (1 CRCM). **Clatsop Co.**—Astoria (1 OSUFW, 2 USNM); 7.5 mi. S Cannon Beach (1 MVZ); site of Old Fort Clatsop (8 MVZ). **Columbia Co.**—3 mi. SW Rainier (2 MVZ); 7 mi. SE Rainier (12 MVZ); T5N, R2W, Sec. 19 (1 PSU). **Coos Co.**—(1 PSM); 1.75 mi. S Bandon, T29S, R14W, NE¼ Sec. 6 (1 PSM); 1.75 mi. E Bandon, T28S, R14W, SE¼ Sec. 29, 50 ft. (4 PSM); 4 mi. SE Bandon, T29S, R14W, SW¼ Sec. 9 (8 PSM); 3 mi. S Charleston (1 LACM); 7.6 km N, 5.2 km W Sitkum, T27S, R10W, Sec. 18 (1 USNM); 8.4 km N, 4.3 km W Sitkum, T27S, R10W, Sec. 17 (1 USNM); 2.8 km S, 5 km E Sitkum, T28S, R9W, Sec. 19 (1 USNM); 4.8 km S, 9.7 km E Sitkum, T28S, R9W, Sec. 27 (1 USNM). **Curry Co.**—Gold Beach (2 FMNH, 1 PSM). **Douglas Co.**—ca. 9 mi. E Days Creek (1 PSM); 7.2 km S, 16.2 km W Elkton, T23S, R9W, Sec. 22 (1 USNM); 15.7 km S, 5.6 km W Elkton, T24S, R8W, Sec. 10 (1 USNM); Gardiner (1 AMNH, 2 MVZ); 3.2 km N, 26 km W Lorane, T19/20S, R7W, Sec. 32/5 (1 USNM); Louis Creek, T28S, R3W, SE¼SE¼ Sec. 30 (1 SOSC); 3.8 km N, 12.1 km W Sutherlin, T25S, R7W, Sec. 1 (1 USNM); 6.5 km N, 6.7 km W Tenmile, T28S, R8W, Sec. 9 (1 USNM); 4.5 mi. N, 9 mi. E Tiller (1 CRCM); 11 mi. N, 5 mi. E Tiller (1 CRCM); 12 mi. N, 4 mi. E Tiller (1 CRCM); 12 mi. N, 17 mi. E Tiller (2 CRCM); T20S, R7W, NE¼ Sec. 19 (2 PSM). **Hood River Co.**—Camas Prairie, 12 mi. SW Mt. Hood (1 USNM); Oxbow Hatchery [Cascade Locks] (1 PSM). **Jackson Co.**—Crater Lake National Forest, Rustler Peak [T34S, R4E, Sec. 16] (1 USNM); 2 mi. N, 1 mi. E Union Creek (1 CRCM). **Klamath Co.**—Ft. Klamath (1 USNM); 18 mi. SE Ft. Klamath, near Williamson River (1 USNM); Klamath Falls (1 CM). **Lane Co.**—12.1 km S, 2.4 km W Alsea, T15S, R8W, Sec. 11 (2 USNM); ca. 6.8 km N, 7.6 km E Blue River, T15S, R5E, Sec. 32 (1 USNM); 6 mi. E Blue River, Swamp Creek, T16S, R5E, Sec. 20 (1 OSUFW); 9 mi. E Blue River, T16S, R5E, Sec. 25/26 (1 CRCM); 0.5 mi. N, 2 mi. W Blue River, Elk Creek, T16S, R4E, Sec. 19 (1 OSUFW); 1.25 mi. N, 5 mi. W Blue River, T16S, R3E, Sec. 15 (2 HSU); 1.5 mi. N, 2 mi. E Blue River (3 CRCM); 4 mi. N, 4.5 mi. E Blue River, T16S, R5E, Sec. 32 (1 CRCM); 4 mi. N, 9 mi. E Blue River, Mack Creek Tributary, T15S, R5E, Sec. 35 (1 OSUFW); 5 mi. N, 9 mi. E Blue River (2 CRCM); 5 mi. N, 10.5 mi. E Blue River, T15S, R5E, Sec. 30 (1 OSUFW); 5.5 mi. N, 5 mi. E Blue River, T15S, R5E, Sec. 28 (4 HSU); 5.5 mi. N, 13.5 mi. E Blue River, T16S, R6E, Sec. 26 (2 CRCM); 2.5 mi. S, 7 mi. E Blue River (3 CRCM); 3.5 mi. S, 10 mi. E Blue River (4 CRCM); 10 mi. NE Blue River, Upper Lookout Creek Tributary, T15S, R5E, Sec. 25 (3 OSUFW); Eugene (3 SDNHM, 1 USNM); H. J. Andrews Experimental Forest (4 PSM); H. J. Andrews Experimental Forest, T15S, R5E, SW¼ Sec. 33 (2 PSM); 0.4 km N, 18.5 km W Lorane, T20S, R7W, Sec. 12 (1 USNM); 3 km N, 19.5 km W Lorane, T20S, R7W, Sec. 1 (1 USNM); Mabel (2 UO); Marshfield [Florence] (1 FMNH); McCredie Hot Springs (1 SDNHM); McKenzie Bridge (1 OSUFW, 1 PSM, 1 PSU, 1 UO); 2.6 km S McKenzie Bridge, T16S, R5E, Sec. 26 (1 USNM); 6 km N, 0.6 km E McKenzie Bridge, T15S, R5E, Sec. 25 (8 USNM); 6.4 km N, 0.6 km E McKenzie Bridge, T15S, R5E, Sec. 25 (14 USNM); 4.4 km N, 6.8 km E Blue River, T16S, R5E, Sec. 7 (1 USNM); 3.3 km N, 2.4 km W McKenzie Bridge, T16S, R5E, Sec. 3 (2 USNM); 6 km N, 7.5 km E McKenzie Bridge, T15S, R6E, Sec. 26 (1 USNM); 1.6 km S, 5.2 km W McKenzie Bridge, T16S, R5E, Sec. 20 (13 USNM); 5.4 km S, 8.75 km E McKenzie Bridge, T16S, R6E, Sec. 35 (6 USNM); 4.4 km S, 14.3 km E McKenzie Bridge, T16S, R7E, Sec. 32 (2 USNM); 3.8 km S, 14.8 km E McKenzie Bridge, T16S, R7E, Sec. 33 (1 USNM); 14 mi. S McKenzie Bridge (1 OSUFW); 12 mi. S, 1 mi. W McKenzie Bridge (1 USNM); 3 mi. S, 5 mi. E McKenzie Bridge, T15S, R6E, Sec. 35, 2,000 ft. (1 UW[OG]); 3.5 mi. S, 9 mi. E McKenzie Bridge, T16S, R7E, Sec. 32, 3,500 ft. (2 UW[OG]); Mercer

(1 UO); 1 mi. N Mohawk River, T15S, R2W, SE¼NE¼ Sec. 13 (2 SOSC); 4 mi. N Nimrod, T16S, R3E, Sec. 14 (1 CRCM); 1 mi. N, 4 mi. E Oakridge, Salmon Creek Rd. (1 OSUFW); 3 mi. SE Oakridge (1 UO); Spencer Butte (1 UO); Vida (1 SDNHM, 1 USNM); 23 mi. SE Vida (2 PSM); 4.5 mi. S Wolf Rock, T15S, R5E, Sec. 24, 3,400 ft. (1 UW[OG]); 4 mi. S, 1 mi. W Wolf Rock, T15S, R5E, Sec. 14, 2,600 ft. (2 UW[OG]); T18S, R5E, NE¼ Sec. 22 (2 PSM); T18S, R5E, SE¼ Sec. 22 (2 PSM). **Lincoln Co.**—Otis (1 USNM); N side Alsea Bay (1 PSM); 5 mi. S Burnt Woods (1 PSM); Cascade Head Experimental Forest, T6S, R10W, SW¼ Sec. 21, 240 ft. (10 PSM); 1.5 mi. NW Eddyville, T11S, R9W, SE¼ Sec. 5 (1 PSM); 15 mi. W Harlan, Peterson Ridge, T12S, R11W, Sec. 12 (1 KU); 7 mi. S, 10 mi. W Harlan, Death Ridge, T13S, R10W, Sec. 14 (1 KU); 7 mi. S, 11 mi. W Harlan, Boulder Creek, T13S, R10W, Sec. 15 (2 KU); Newport (2 MVZ); 0.6 km N, 16.9 km E Waldport, T13S, R10W, Sec. 13/24 (3 USNM); 4 km S, 3.2 km E Yachats, T15S, R12W, Sec. 12 (3 USNM). **Linn Co.**—7 mi. N, 8 mi. E Blue River, T15S, R5E, Sec. 15 (2 CRCM); 7.5 mi. N, 7.5 mi. E Blue River, T15S, R5E, Sec. 15 (5 CRCM); 8 mi. N, 8 mi. E Blue River (1 CRCM); 9 mi. N, 8 mi. E Blue River (2 CRCM); 5 mi. S, 1 mi. E Crawfordsville (1 OSUFW); 9.2 km N, 1.2 km W McKenzie Bridge, T15S, R5E, Sec. 14 (10 USNM); 10.25 km N, 0.8 km W McKenzie Bridge, T15S, R5E, Sec. 14 (9 USNM); 13.2 km N, 0.8 km E McKenzie Bridge, T15S, R5E, Sec. 1 (4 USNM). **Multnomah Co.**—5 mi. S, 7.5 mi. E Multnomah Falls, Columbia Gorge, T1S, R7E, Sec. 4 (1 OSUFW); Portland (1 USNM); Sherwood District, Portland (1 PSU); 8 mi. N, 11 mi. E Sandy (3 CRCM); 10 mi. N, 11 mi. E Sandy (1 CRCM); 10 mi. N, 1.5 mi. W Zigzag (7 CRCM). **Tillamook Co.**—1 mi. N Bay City (1 AMNH); Blaine (1 UMMZ); Foley Creek Road (3 PSM); 2 mi. S Hebo, T4S, R10W, Sec. 25 (2 PSM); 2.66 mi. S Hebo (2 PSM); Hebo Lake Forest camp (4 LACM); E of Nehalem on Sweethome Creek (1 PSM); Netarts (3 SDNHM); Ridge Summit between Netarts and Tillamook (1 PSM); Pacific City (1 OSUFW, 1 UMMZ); Smith Lake (1 PSM); Tillamook (1 KU, 1 MVZ, 1 OSUFW, 1 SDNHM, 1 UMMZ); 7 mi. S Tillamook (1 AMNH); 6.5 mi. SW Tillamook (3 AMNH); 2 mi. up Miami River (5 PSM); Gauldy 1A* (2 PSM). **Washington Co.**—Beaverton (1 USNM); 18.5 mi. NW Portland (1 MVZ).

Sorex merriami, n = 25
(Plotted in Fig. 8-6, p. 52)

Grant Co.—3 mi. S, 7 mi. W Silvies, T18S, R30E, SW¼ Sec. 25 (1 KU). **Harney Co.**—28 mi. S, 4 mi. W Burns, Harney Lake dunes (1 OSUFW); 5.5 mi. W Fish Lake on Steens Mt. Rd., then N 250 m, T32S, R32¾E, Sec. 33 (1 OSUFW); 5.5 mi. W Fish Lake on Fish Lake Rd., T32S, R33¾E, Sec. 33 (2 UW[WWC]); 3 mi. S, 8 mi. E Frenchglen, Steens Mt. (6 OSUFW); 3 mi. S, 8.5 mi. E Frenchglen, Steens Mt., T32S, R33¾E, Sec. 36 (2 OSUFW); 5 mi. S, 14 mi. E Frenchglen (1 OSUFW); 5.5 mi. S, 14 mi. E Frenchglen, Steens Mt., T32S, R33E, Sec. 31 (2 OSUFW); 2 mi. S Narrows (1 UW[WWC]); Ochoco National Forest, T19S, R25E, NE¼SW¼ Sec. 26 (1 KU); 26 mi. N, 9 mi. W Riley, T19S, R25E, SE¼NE¼ Sec. 13 (1 KU); Snow Mt. district, Ochoco National Forest (1 OSUFW); Steens Mt. (2 OSUFW); Steens Mt., T32S, R32¾E, Sec. 29 (1 OSUFW). **Lake Co.**—Hart Mt. National Wildlife Refuge, T36S, R26E, Sec. 17 (1 KU). **Wasco Co.**—Antelope (1 USNM).

Sorex monticolus, n = 340
(Plotted in Fig. 8-8, p. 54)

Baker Co.—Anthony (4 MVZ, 1 USNM); Bourne (3 USNM); 15 mi. S, 11 mi. W Hereford, T15S, R37E, Sec. 7 (3 KU); 0.3 mi. SE Rock Creek Butte, Elkhorn Mts. (1 CRCM); 12 mi. S, 2 mi. W Unity, T15S, R37E, Sec. 18 (3 KU). **Clackamas Co.**—4.5 mi. N Alder Creek, T1S, R6E, Sec. 29/32 (1 OSUFW); 6 mi. N Alder Creek, T1S, R6E, Sec. 29 (1 OSUFW); 2 mi. N, 4 mi. E Sandy (14 CRCM); 3.5 mi. N, 8 mi. E Sandy (4 CRCM); 3.8 mi. N, 8 mi. E Sandy (6 CRCM); 4.2 mi. N, 8 mi. E Sandy, T1S, R6E, Sec. 29 (1 CRCM); 6 mi. S, 7 mi. E Sandy, T3S, R6E, Sec. 18 (10 CRCM); 9 mi. N, 4 mi. W Zigzag (16 CRCM); 5 mi. S, 6 mi. E Zigzag (36 CRCM). **Clatsop Co.**—3 mi. N, 1 mi. W Knappa, mouth of Big Creek (1 OSUFW); Seaside (1 PSM); T8N, R7W, NE¼ Sec. 18 (1 PSM); T8N, R10W, W½ Sec. 6 (3 PSM); T9N, R6W, N½

Sec. 19 (1 PSM). **Columbia Co.**—1 mi. W Rainier (1 PSM); 7 mi. NE Vernonia (1 PSM); T6N, R1W, SW¼ Sec. 7 (2 PSM). **Crook Co.**—Paulina Ranger district (1 OSUFW); 13 mi. N, 2 mi. W Paulina (1 OSUFW); 17 mi. N, 3 mi. W Paulina (1 OSUFW). **Grant Co.**—2 mi. W Logdell, T16S, R30E, Sec. 5 (1 KU); 4 mi. W Logdell, T16S, R29E, Sec. 1 (1 KU); 3 mi. E Logdell, T16S, R30E, Sec. 2 (1 KU); 6 mi. S, 2 mi. W Logdell, T16S, R30E, Sec. 32 (1 KU); 8 mi. S, 5 mi. W Long Creek, T11S, R31E, Sec. 19 (7 OSUFW); N Fork Malheur River (1 MVZ); 3 mi. N, 6 mi. E Prairie City, T12S, R34E, NW¼ Sec. 23 (7 KU); 11 mi. S, 6 mi. E Prairie City, T15S, R36E, Sec. 25 (3 KU); 11 mi. S, 7 mi. E Prairie City, T15S, R36E, Sec. 26 (2 KU); 26 mi. S, 10 mi. E Prairie City, T17S, R35E, Sec. 17 (2 KU); 26 mi. S, 12 mi. E Prairie City, T17S, R35E, Sec. 15 (2 KU); 26 mi. S, 13 mi. E Prairie City, T17S, R35E, Sec. 14 (1 KU); 26 mi. S, 15 mi. E Prairie City, T17S, R35E, Sec. 19 (3 KU); 27 mi. S, 17 mi. E Prairie City, T17S, R35E, Sec. 27 (2 KU); 27 mi. S, 18 mi. E Prairie City, T17S, R35E, Sec. 26 (1 KU); 3 mi. S, 7 mi. W Silvies, T18S, R30E, SE¼ Sec. 26 (1 KU); 4 mi. N, 14.5 mi. E Seneca, T16S, R33½E, NE¼NW¼ Sec. 17 (8 KU); 5 mi. N, 15.5 mi. E Seneca, T16S, E33½E, NE¼SE¼ Sec. 8 (1 KU); 6 mi. N, 15.5 mi. E Seneca, T16S, R33½E, NW¼NW¼ Sec. 4 (6 KU); 4 mi. S, 3 mi. W Seneca, T17S, R31E, Sec. 20 (2 KU); 3 mi. S, 4 mi. W Seneca, T17S, R31E, Sec. 18 (2 KU); 4 mi. S, 4 mi. W Seneca, T17S, R30E, Sec. 24 (2 KU); 2 mi. S, 4 mi. W Seneca, T17S, R30E, Sec. 13 (1 KU); 15 mi. N, 14 mi. W Seneca, T14S, R28E, Sec. 14 (1 KU); 23 mi. N, 21 mi. W Seneca, T14S, R28E, Sec. 10 (3 KU); Strawberry Butte (1 USNM). **Harney Co.**—Diamond (1 USNM); 5 mi. S, 11 mi. E Frenchglen, T32S, R32¾E, SW¼NE¼ Sec. 34, 6,600 ft. (3 KU); 5 mi. S, 14 mi. E Frenchglen, T32S, R33E, Sec. 31 (1 OSUFW); Steens Mt., Kiger Gorge (2 USNM); Little Blitzen Gorge, Steens Mt., T33S, R33E, Sec. 10 (3 OSUFW); Malheur Refuge (2 BC/FC); Steens Mt., McCoy Creek (1 USNM); 18 mi. N, 4 mi. W Riley, T20S, R26E, SE¼ Sec. 26 (9 KU); 26 mi. N, 11 mi. W Riley, T19S, R25E, Sec. 13 (1 KU); Steens Mt. (1 MVZ, 1 OSUFW); Steens Mt., T32S, R32¾E, Sec. 36 (1 OSUFW); Steens Mt., T32S, R33E, Sec. 32 (1 OSUFW); Steens Mt., T33S, R33E, Sec. 2 (1 OSUFW); Steens Mt., T33S, R33E, Sec. 3 (1 OSUFW); T19S, R25E, NE¼SW¼ Sec. 26 (1 KU); T32S, R32½E, NW¼ Sec. 17 (2 EOSC); T32S, R33E, Sec. 31 (11 OSUFW). **Hood River Co.**—1 mi. E Cascade Locks, Oxbow Fish Hatchery (1 PSM); 2 mi. W Parkdale, 1,500 ft. (2 USNM). **Multnomah Co.**—2 mi. S, 2.5 mi. E Multnomah Falls (3 OSUFW); 6 mi. N, 11 mi. E Sandy (4 CRCM); 8 mi. N, 11 mi. E Sandy (4 CRCM); 9 mi. N, 10 mi. E Sandy (9 CRCM); 10 mi. N, 11 mi. E Sandy, T1N, R6E, Sec. 27 (40 CRCM); 11 mi. N, 10 mi. E Sandy (6 CRCM); 11 mi. N Zigzag (3 CRCM); 10 mi. N, 1.5 mi. W Zigzag (9 CRCM); 8 mi. N, 3 mi. E Zigzag (7 CRCM). **Tillamook Co.**—2 mi. upstream Miami River (1 PSM). **Union Co.**—La Grande (1 EOSC); 0.5 mi. E Otis Spring (1 PSM); Starkey Experimental Forest, Tin Trough Pit, T4S, R34E, NE¼ Sec. 34 (3 PSM). **Wallowa Co.**—6 mi. S Lostine, T2S, R43E, Sec. 10 (1 OSUFW); 10 mi. S, 1 mi. E Lostine, T2S, R43E, Sec. 25 (1 OSUFW); 13 mi. S, 2 mi. E Lostine (1 OSUFW); 15 mi. S, 2 mi. E Lostine (2 OSUFW); 19 mi. S, 4 mi. E Lostine, T4S, R44E, Sec. 17 (2 OSUFW); 21 mi. S, 6 mi. E Lostine, T4S, R44E, Sec. 26 (1 OSUFW); 7 mi. S, 11 mi. W Mitchell (1 MVZ); Wallowa Lake (2 MVZ, 5 USNM); Wallowa Mts. (1 USNM); Wallowa Mts., S of Wallowa Lake (1 USNM). **Wheeler Co.**—8 mi. S, 5 mi. W Mitchell (3 OSUFW); T14S, R25E, SW¼ Sec. 34 (1 KU).

Sorex pacificus, n = 589
(Plotted in Fig. 8-10, p. 55)

Benton Co.—1 mi. S, 6 mi. W Alpine (2 OSUFW); Alsea (1 OSUFW); 10 mi. W Alsea (1 OSUFW); 4.4 km N, 7.6 km W Alsea, T13S, R8W, Sec. 20 (1 USNM); Corvallis (2 UCONN); 3 mi. N, 4 mi. W Corvallis, T11S, R5W, Sec. 20 (1 KU); Marys Peak (3 OSUFW, 1 PSM); 2 mi. S, 9 mi. W Philomath, T12S, R7W, NW¼ Sec. 21 (1 OSUFW); 3 mi. S, 6 mi. W Philomath, T12S, R7W, SE¼ Sec. 14 (1 OSUFW); 3 mi. S, 8 mi. W Philomath, T12S, R6W, NW¼ Sec. 30 (1 OSUFW); 1.5 km S, 11.3 km W Philomath, T12S, R7W, Sec. 14 (1 USNM); 2.4 km S, 15.5 km W Philomath, T12S, R7W, Sec. 17/20 (1 USNM); 4 km S, 10.1 km W Philomath, T12S, R7W, Sec. 24/25 (1

USNM); 5 mi. SW Philomath (1 USNM). **Coos Co.**—Coos Bay, South Slough (1 LACM); Delmar, 9 mi. S Marshfield (1 MVZ); Marshfield [= Coos Bay] (1 FMNH); Point Arago State Park (1 LACM); Sunset Bay (1 LACM). **Douglas Co.**—Ft. Umpqua (1 USNM); Oxbow Burn, 300 ft. (3 PSM); 4 km N, 12.1 km W Sutherlin, T25S, R7W, Sec. 1 (1 USNM); 4.5 mi. N, 8 mi. E Tiller, T30S, R1W, Sec. 2 (1 CRCM); 11 mi. N, 5 mi. E Tiller (1 CRCM); 11.5 mi. N, 5 mi. E Tiller, T28S, R1W, Sec. 26 (2 CRCM); 12 mi. N, 4 mi. E Tiller (1 CRCM); 12 mi. N, 5.5 mi. E Tiller, T28S, R1W, Sec. 32 (2 CRCM). **Jackson Co.**—Upper Ashland Research Natural Area (2 SOSC); 0.6 mi. N junction FSR 3903 and E Fork Ashland Creek (1 SOSC); 6 mi. S Ashland (1 SOSC); Imnaha Creek, T33S, R4E, W½ Sec. 15 (2 KU, 1 SOSC); 11 mi. N Prospect (1 CRCM); 11.5 mi. N, 0.5 mi. W Prospect (3 CRCM); 13.5 mi. N, 2.5 mi. W Prospect (8 CRCM); 14 mi. N, 0.5 mi. W Prospect (2 CRCM); 5.5 mi. N, 4 mi. E Prospect (2 CRCM); Spring Creek, T33S, R4E, NE¼ Sec. 22 (1 OSUFW, 4 KU, 2 SOSC); T37S, R3E, SW¼ Sec. 20 (1 SOSC). **Klamath Co.**—0.125 mi. E Heavenly Twin Lake (1 OSUFW); Sphagnum Bog, Crater Lake National Park (2 SOSC). **Lane Co.**—5 mi. Alpha, T15S, R9W, Sec. 31 (2 OSUFW); 12.1 km S, 2.4 km W Alsea, T15S, R8W, Sec. 11 (2 USNM); 6 mi. W Blue River, Hagan Creek, T16S, R3E, Sec. 22 (4 OSUFW); 6 mi. E Blue River, Cougar Creek (1 OSUFW); 6 mi. E Blue River, Swamp Creek, T16S, R5E, Sec. 20 (4 OSUFW, 6 USNM); 9 mi. E Blue River, T16S, R5E, Sec. 32 (14 CRCM); 0.5 mi. N, 2 mi. W Blue River, Elk Creek, T16S, R4E, Sec. 19 (2 OSUFW); 1.5 mi. N, 2 mi. E Blue River (13 CRCM); 4 mi. N, 4.5 mi. E Blue River (3 CRCM); 4 mi. N, 9 mi. E Blue River, Mack Creek Tributary, T15S, R5E, Sec. 35 (2 OSUFW); 5 mi. N, 9 mi. E Blue River (12 CRCM); 5 mi. N, 10.5 mi. E Blue River, T15S, R5E, Sec. 30 (6 CRCM); 5.5 mi. N, 13.5 mi. E Blue River, T16S, R6E, Sec. 26 (2 CRCM); 7 mi. N, 9 mi. E Blue River, T15S, R5E, Sec. 13 (30 CRCM); 4.4 km N, 6.8 km E Blue River, T16S, R5E, Sec. 7 (1 USNM); 5 km N, 7.6 km E Blue River, T16S, R5E, Sec. 6 (1 USNM); 6.8 km N, 7.6 km E Blue River, T15S, R5E, Sec. 32 (5 USNM); 2.5 mi. S, 5.5 mi. E Blue River, Cougar Creek, T16S, R5E, Sec. 32 (2 OSUFW); 2.5 mi. S, 7 mi. E Blue River (23 CRCM); 3.5 mi. S, 10 mi. E Blue River (25 CRCM); 4 mi. S, 10 mi. E Blue River, East Fork Creek, T17S, R6E, Sec. 7 (2 OSUFW); 4 mi. S, 10.5 mi. E Blue River, East Fork Creek, T17S, R6E, Sec. 7 (3 OSUFW, 1 USNM); 5.5 mi. NE Blue River, 1,500 ft. (1 PSM); 10 mi. NE Blue River, Lookout Creek, T15S, R6E, Sec. 30 (2 OSUFW); 10 mi. NE Blue River, Upper Lookout Creek tributary, T15S, R5E, Sec. 24 (3 OSUFW); 5 mi. N Cougar Dam, T16S, R5E, Sec. 6 (3 UW[OG]); Elk Creek, McKenzie Bridge (1 USNM); 8 mi. N Florence (2 OSUFW); 5 mi. S Florence, W side Siltcoos Lake (2 LACM); Gold Lake, 4,800 ft. (5 LACM); Hagan Cr., T16S, R3E, Sec. 22 (2 OSUFW); Mapleton (1 USNM); McKenzie Bridge (2 OSUFW); 2.6 km S McKenzie Bridge, T16S, R5E, Sec. 26 (5 USNM); 3.1 km N, 3 km W McKenzie Bridge, T16S, R5E, Sec. 3 (9 USNM); 3.3 km N, 2.4 km W McKenzie Bridge, T16S, R5E, Sec. 3 (4 USNM); 6.6 km N, 0.3 km W McKenzie Bridge, T16S, R5E, Sec. 25 (2 USNM); 6 km N, 0.6 km E McKenzie Bridge, T15S, R5E, Sec. 25 (7 USNM); 6.4 km N, 0.6 km E McKenzie Bridge, T15S, R5E, Sec. 25 (4 USNM); 1.6 km S, 5.2 km W McKenzie Bridge, T16S, R5E, Sec. 20 (1 USNM); 2.6 km S, 1.4 km E McKenzie Bridge, T16S, R5E, Sec. 25 (3 USNM); 3.8 km S, 14.8 km E McKenzie Bridge, T16S, R7E, Sec. 33 (6 USNM); 4.4 km S, 14.3 km E McKenzie Bridge, T16S, R7E, Sec. 32 (8 USNM); 5.4 km S, 8.75 km E McKenzie Bridge, T16S, R6E, Sec. 35 (7 USNM); 4 mi. N McKenzie Bridge, T15S, R5E, Sec. 25, 2,600 ft. (9 UW[OG]); 2 mi. N, 2 mi. W McKenzie Bridge, T16S, R5E, Sec. 3/4, 3,800 ft. (1 UW[OG]); 1 mi. N, 8 mi. E McKenzie Bridge, T16S, R4E, SW¼ Sec. 9 (1 OSUFW); 6 mi. N, 0.5 mi. E McKenzie Bridge (1 OSUFW); 2 mi. S, 0.5 mi. E McKenzie Bridge, T16S, R5E, Sec. 25, 2,800 ft. (2 UW[OG]); 3 mi. S, 5 mi. E McKenzie Bridge, T15S, R6E, Sec. 35, 2,000 ft. (1 UW[OG]); 3 mi. S, 8.5 mi. E McKenzie Bridge, T16S, R7E, Sec. 32, 3,500 ft. (1 UW[OG]); Mercer, Grass Mt. refuge (1 OSUFW); 4 mi. N Nimrod, T16S, R3E, Sec. 14 (24 CRCM); 4.5 mi. N Nimrod (15 CRCM); 1 mi. N, 4 mi. E Oakridge (1 OSUFW); 4 mi. N Rainbow, T15S, R5E, Sec. 32, 1,700 ft. (2 UW[OG]); Swamp Creek, T16S, R5E, Sec. 20 (3 OSUFW, 6 USNM); 4.8 km S, 3.2 km E Yachats, T15S, R12W, Sec. 12 (1 USNM); T15S, R5E, NW¼ Sec. 28, 2,000 ft. (1 PSM); T15S, R5E, NE¼ Sec. 32, 1,650 ft. (3 PSM); T15S, R5E, NE¼ Sec. 32, 2,000 ft. (1

PSM); T15S, R5E, SW¼ Sec. 33, 2,500 ft. (3 PSM); T16S, R4E, NE¼ Sec. 19, 1,800 ft. (6 PSM); T16S, R4E, NE¼ Sec. 24, 1,400 ft. (6 PSM); T18S, R5E, Sec. 23, 3,000 ft. (1 PSM). **Lincoln Co.**—Alsea Bay (1 PSM); 1.25 mi. N (4 KU); 8 mi. W Harlan, Nettle Creek, T12S, R9W, Sec. 28 (11 USNM); 10 mi. W Harlan, Flynn Creek, T12S, R10W, Sec. 12 (9 USNM); 15 mi. W Harlan, Peterson Ridge, T12S, R11W, Sec. 12 (6 KU); 2 mi. S, 8 mi. W Harlan, Flynn Creek, T12S, R10W, Sec. 13 (1 TTU); 15 mi. W Harlan, Peterson Ridge, T12S, R11W, Sec. 12 (2 KU); 2 mi. S, 12 mi. W Harlan, T12S, R10W, Sec. 20 (2 OSUFW, 5 KU, 2 TTU); 5 mi. S, 10 mi. W Harlan, Gold Creek, T13S, R10W, Sec. 2 (1 OSUFW, 2 KU, 2 TTU); 7 mi. S, 5 mi. W Harlan, West Creek, T13S, R9W, Sec. 16 (1 OSUFW, 1 KU, 1 TTU); 7 mi. S, 10 mi. W Harlan, T13S, R10W, Sec. 14 (5 KU); 7 mi. S, 11 mi. W Harlan, Boulder Creek, T13S, R10W, Sec. 15 (5 KU); Newport (4 USNM); 9 mi. N Newport (1 SDNHM); Taft (1 SDNHM); 10 mi. N Tidewater, S side Death Ridge (1 USNM); 10 mi. N Tidewater, S fork Drift Creek (2 USNM); 10 mi. N Tidewater, Gold Creek (1 USNM); 10 mi. N Tidewater at Murphy (3 USNM); Yaquina Bay (1 USNM); T12S, R10W, Sec. 13/24 (2 USNM). **Linn Co.**—7 mi. N, 8 mi. E Blue River, T15S, R5E, Sec. 15 (22 CRCM); 7.5 mi. N, 7.5 mi. E Blue River, T15S, R5E, Sec. 15 (19 CRCM); 8 mi. N, 8 mi. E Blue River (16 CRCM); 9 mi. N, 8 mi. E Blue River (7 CRCM); 9.5 mi. N, 9 mi. E Blue River, T15S, R5E, Sec. 2 (11 CRCM); 7 mi. E Cascadia (1 UCD); 9.2 km N, 1.2 km W McKenzie Bridge, T15S, R5E, Sec. 14 (4 USNM); 10.25 km N, 0.8 km W McKenzie Bridge, T15S, R5E, Sec. 14 (3 USNM); 13.2 km N, 0.8 km E McKenzie Bridge, T15S, R5E, Sec. 1 (7 USNM); 1 mi. S Wolf Rock, T15S, R5E, Sec. 1, 3,300 ft. (1 UW[OG]); 1 mi. S, 1 mi. W Wolf Rock, T15S, R5E, Sec. 2, 3,000 ft. (2 UW[OG]); 4 mi. S, 1 mi. W Wolf Rock, T15S, R5E, Sec. 14, 2,600 ft. (5 UW[OG]); T12S, R2E, Sec. 19 (1 OSUFW); T14S, R6E, NE¼ Sec. 20, 4,400 ft. (1 PSM); T14S, R6E, NE¼ Sec. 28, 4,600 ft. (1 PSM); T15S, R5E, SE¼ Sec. 11 (1 PSM); T15S, R5E, SW¼ Sec. 13, 3,000 ft. (1 PSM); T15S, R6E, NE¼ Sec. 7, 4,300 ft. (1 PSM); T15S, R6E, NW¼ Sec. 7, 5,000 ft. (1 PSM). **Tillamook Co.**—Cascade Head (1 PSM).

Sorex palustris, *n* = 188
(Plotted in Fig. 8-12, p. 57)

Baker Co.—Anthony (16 AMNH, 1 MVZ); Bourne (1 SDNHM, 16 USNM); Cornucopia (1 UMMZ, 7 USNM). **Clackamas Co.**—Barlow campground, near Zigzag River (3 PSU); 0.75 mi. N, 2.75 mi. W Government Camp (6 PSU); Barlow campground, Government Camp (3 PSU); Twin Bridges campground, Government Camp (7 PSU). **Crook Co.**—Howard (1 USNM); Ochoco Spring, Ochoco National Forest (2 SOSC); Peaslee Creek, Ochoco National Forest (1 SOSC). **Deschutes Co.**—11 mi. W Bend, Tumalo Creek (1 MVZ); 15 mi. W Bend, Tumalo Creek, 6,100 ft. (2 MVZ); mouth of Davis Creek, Deschutes River (1 OSUFW); SE of South Sister, head of Falls Creek, 6,450 ft. (1 CRCM); Sparks Lake (2 OSUFW). **Douglas Co.**—Bradley Creek (1 SDNHM); Diamond Lake (1 SDNHM); Diamond Lake near outlet (1 OSUFW). **Grant Co.**—5 mi. E Austin, 4,200 ft. (1 USNM); 3 mi. S, 9 mi. E Austin, T12S, R35½E, NW¼NE¼ Sec. 1 (1 KU); Beech Creek [T13S, R30E, Sec. 28] (2 USNM); John Day, Alpine Creek, Strawberry Mt.* (1 USNM); 21 mi. SE Prairie City, North Fork Malheur River, 5,000 ft. (14 MVZ); 6 mi. N, 15.5 mi. E Seneca, T16S, R33½E, NE¼NE¼ Sec. 5 (1 KU); Strawberry Lake (1 SDNHM); Strawberry Butte (3 USNM); Strawberry Mt. (6 USNM). **Harney Co.**—26 mi. N, 11 mi. W Riley, T19S, R25E, Sec. 13 (1 KU); 18 mi. N, 5 mi. E Riley, T20S, R26E, SE¼ Sec. 26 (1 KU); Steens Mt. (1 USNM); Steens Mt., Kiger Gorge, 6,900 ft. (1 USNM); Steens Mt., T32S, R32¾E, Sec. 20 (3 OSUFW); Steens Mt., T32S, R32¾E, Sec. 27 (1 OSUFW); Steens Mt., T32S, R32¾E, Sec. 35 (1 OSUFW); Steens Mt., T33S, R32½E, Sec. 10 (1 OSUFW); Steens Mt., Little Blitzen Gorge, T33S, R32¾E, Sec. 26 (1 OSUFW); Steens Mt., Little Blitzen Gorge, T33S, R33E, Sec. 10 (1 OSUFW). **Hood River Co.**—Brooks Meadows, 9 mi. E Mt. Hood (1 MVZ); Elk Cove, N side Mt. Hood, 5,600 ft. (1 CRCM); Mt. Hood (2 UW[WWC]); Tilly Jane Creek, NE side Mt. Hood, 6,450 ft. (1 CRCM). **Jackson Co.**—6.4 mi. N Dead Indian Rd. on Conde Creek Rd. (1 SOSC). **Jefferson Co.**—Hay Creek (1 USNM); Warm Springs, 20 mi. W Mill Creek (2 USNM).

Klamath Co.—Crater Lake National Park (1 OSUFW, 5 USNM); Crater Lake National Park, Upper Munson Meadow, 6,400 ft. (1 CRCM); Crater National Forest Camp 76 (1 USNM); 6 mi. E Cresent Lake (1 UO); Ft. Klamath (1 USNM); Klamath Falls, Link River Dam (1 SDNHM); Munson Valley (1 MVZ). **Lake Co.**—near Drew Creek (1 USNM); Gearhart Mt. (1 PSM). **Lane Co.**—4 mi. N, 9 mi. E Blue River, Mack Creek Tributary, T15S, R5E, Sec. 35 (2 OSUFW); Gold Lake, 4,800 ft. (1 LACM); McKenzie Bridge (2 OSUFW, 1 UO, 1 USNM); 2 mi. S McKenzie Bridge, King's Creek, T16S, R6E, Sec. 26 (1 OSUFW); 4.4 km S, 14.3 km E McKenzie Bridge, T16S, R7E, Sec. 32 (3 USNM); 3 mi. SE Oakridge (1 UO); Three Sisters, NW base North Side Peak (1 USNM); Upper Horse Creek, 200 yds. above Skyline Trail (1 OSUFW). **Linn Co.**—Permilia Lake, W base Mt. Jefferson (1 USNM). **Malheur Co.**—Disaster Peak, 7,000 ft. (1 USNM). **Multnomah Co.**—9 mi. N, 10 mi. E Sandy (1 CRCM). **Umatilla Co.**—Rock Creek, 10 mi. W Tollgate (1 UW[EB]). **Union Co.**—Grande Ronde Lake, 7,100 ft. (1 MVZ). **Wallowa Co.**—Aneroid Lake (1 UO); 10 mi. S, 1 mi. E Lostine, T2S, R43E, Sec. 35 (1 OSUFW); 16 mi. S, 3 mi. E Lostine, T3S, R43E, Sec. 36 (1 OSUFW); 21 mi. S, 5 mi. E Lostine, T4S, R44E, Sec. 27 (1 OSUFW); 17 mi. SE Lostine, Lostine River (1 OSUFW); 16 mi. S, 3 mi. E Lostine (1 MVZ); Wallowa Lake, 4,000 ft. (3 USNM); Wallowa Lake, 5,000 ft. (1 USNM); Wallowa Lake, 8,000 ft. (1 USNM); Wallowa Lake (15 MVZ); near Wallowa Lake (1 UW[EB]). **Wheeler Co.**—7 mi. S, 11 mi. W Mitchell (2 MVZ).

Sorex preblei, *n* = 48
(Plotted in Fig. 8-14, p. 58)

Deschutes Co.—East Lake, Paulina Mts., 6,300–6,400 ft. (1 USNM). **Grant Co.**—12.5 mi. S, 6 mi. E John Day, T15S, R32E, SE¼ Sec. 26 (1 KU, 1 OSUFW); 3 mi. S, 4 mi. Seneca, T17S, R31E, Sec. 18 (1 KU); 1 mi. N, 9 mi. W Silvies, T18S, R30E, Sec. 3 (1 KU). **Harney Co.**—20 mi. N, 32 mi. W Burns, T19S, R25E, NE¼SE¼ Sec. 26 (1 OSUFW); 22 mi. S, 6 mi. E Burns (1 OSUFW); 31 mi. SE Burns (1 OSUFW); Diamond, 4,300 ft. (1 USNM); 3 mi. S, 8 mi. E Frenchglen, 5,900 ft. (1 OSUFW); 5 mi. S, 11 mi. E Frenchglen, T32S, R32¾E, SW¼NE¼ Sec. 34, 6,600 ft. (4 KU); 5 mi. S, 14 mi. E Frenchglen, Steens Mt., T32S, R33E, Sec. 31, 7,300 ft. (14 OSUFW); 5.5 mi. S, 14 mi. E Frenchglen, Steens Mt., T32S, R33E, Sec. 31, 7,400 ft. (6 OSUFW); Horse Pasture Field, Malheur National Wildlife Refuge, T27S, R31E, Sec. 3 (1 CM); 0.25 mi. NE Malheur Environmental Field Station (1 SOSC); S Malheur Environmental Field Station (1 PSU); Steens Mt., T32S, R32E, Sec. 33, 6,400 ft. (1 OSUFW); Steens Mt., T32S, R32¾E, Sec. 33, 6,400 ft. (3 OSUFW); Steens Mt., T32S, R32¾E, Sec. 35 (1 OSUFW); Steens Mt., T32S, R33E, Sec. 32, 7,400 ft. (2 OSUFW). **Klamath Co.**—Beaver Marsh, T28S, R8E, SE¼NW¼ Sec. 16. **Lake Co.**—Hart Mt. National Wildlife Refuge, 0.2 mi. NNE Lookout, T36S, R26E, SW¼ Sec. 2 (1 KU, 1 OSUFW). **Malheur Co.**—Jordan Valley, 4,200 ft. (1 USNM). **Wallowa Co.**—23 mi. N Enterprise at Sled Springs, 4,600 ft. (1 USNM).

Sorex sonomae, *n* = 1,069
(Plotted in Fig. 8-16, p. 60)

Benton Co.—1 mi. S, 6 mi. W Alpine (1 OSUFW); 4.4 km N, 7 mi. W Alsea, T13S, R8W, Sec. 20 (2 USNM); 5 mi. N, 4 mi. E Alsea (1 OSUFW); 2 mi. S, 3 mi. W Alsea (1 OSUFW); 2 mi. S, 1 mi. E Alsea (1 CM); 3.75 mi. S, 5.5 mi. E Alsea (1 CM); 1.2 km S, 4.2 km E Alsea, T14S, R7W, Sec. 9 (1 USNM); 8 km S, 2.8 km E Alsea, T14S, R7W, Sec. 32 (1 USNM); 6 mi. N Corvallis (1 PSM); 2 mi. N, 4 mi. W Corvallis, Oak Creek (1 OSUFW); Marys Peak, T12S, R7W, Sec. 28 (1 PSM); Missouri Bend (1 OSUFW); 2 mi. S, 9 mi. W Philomath, T12S, R7W, SW¼ Sec. 16 (1 OSUFW); 3 mi. S, 9 mi. W Philomath (1 OSUFW); 3 mi. S, 9 mi. W Philomath, Marys Peak (2 OSUFW); 5 mi. S, 5 mi. W Philomath, T13S, R7W, Sec. 1 (1 KU); 1 km S, 11.3 km W Philomath, T12S, R7W, Sec. 11/14 (3 USNM); 1.5 km S, 10.1 km W Philomath, T12S, R7W, Sec. 13 (5 USNM); 4 km S, 10.1 km W Philomath, T12S, R7W, Sec. 24/25 (2 USNM); 5 mi. SW Philomath (1 USNM); 6 mi. SW Philomath, 550 ft (1 OSUFW); 8.5 mi. SW Philomath (1 OSUFW); Prairie Mt. Rd., T14S, R6W, Sec. 31 (1 PSM); T11S, R7W, SW¼ Sec. 8, 650 ft.

(2 PSM); T12S, R7W, Sec. 14 (4 USNM). **Coos Co.**—1.75 mi. E Bandon, T28S, R14W, SE¼ Sec. 29, 100 ft. (11 PSM); 9 mi. NE Bandon, 250 ft. (1 PSM); 3 mi. SE Bandon (1 PSM); 4 mi. SE Bandon, T29S, R14W, SW¼ Sec. 9, 250 ft. (3 PSM); 4 mi. SE Bandon, T29S, R14W, NW¼ Sec. 16, 400 ft. (14 PSM); Bullards Beach State Park (1 PSU); Cape Arago State Park (2 LACM); Bastendorf Beach (1 LACM); Charleston (2 LACM); 3 mi. S Charleston, Sunset Bay (1 LACM); 5 mi. SW Charleston, Sunset Bay (4 LACM); Coos Head (6 LACM); Marshfield [= Coos Bay] (1 FMNH, 1 USNM); Myrtle Point (1 USNM); 3.2 km N, 0.3 km W Remote (6 USNM); 7.3 km N, 4 km W Sitkum, T27S, R10W, Sec. 17/20 (2 USNM); 7.4 km N, 4.2 km W Sitkum, T27S, R10W, Sec. 17 (7 USNM); 7.5 km N, 12.5 km W Sitkum, T27S, R11W, Sec. 21 (5 USNM); 7.6 km N, 5.2 km W Sitkum, T27S, R10W, Sec. 18 (2 USNM); 8.4 km N, 4.3 km W Sitkum, T27S, R9W, Sec. 19 (1 USNM); 2.9 km S, 4.7 km E Sitkum, T28S, R9W, Sec. 27 (4 USNM); 3 km S, 10.8 km E Sitkum, T28S, R9W, Sec. 23 (1 USNM); 3 km S, 11 km E Sitkum, T28S, R9W (2 USNM); 4.8 km S, 9.7 km E Sitkum, T28S, R9W, Sec. 27 (3 USNM); Sunset Bay (5 LACM); T27S, R13W, NW¼ Sec. 18, 250 ft. (1 PSM); T28S, R14W, SE¼ Sec. 29, 50 ft. (1 PSM); T29S, R14W, NE¼ Sec. 6, 100 ft. (1 PSM). **Curry Co.**—10 mi. E Brookings, T41S, R12W, NW¼ Sec. 3, 240 ft. (2 PSM); Gold Beach (1 PSM, 1 SDNHM); 3 mi. N Gold Beach (3 MVZ); 0.5 mi. N Harris Beach St. Park (1 SOSC); Port Orford (1 CMNH). **Douglas Co.**—3.2 mi. S, 3.6 mi. E Canyonville, O'Shea Creek, T31S, R4W, Sec. 17, 2,000 ft. (2 OSUFW, 2 UW[CR]); 1 mi. S, 33 mi. E Days Creek at Abbott Cr., T30S, R2E, Sec. 24/25 (2 OSUFW); 4 mi. N, 17 mi. E Diamond Lake (1 SOSC); 0.8 km N, 6.2 km W Drain, T22S, R6W, Sec. 7 (1 USNM); 1.4 km N, 1.6 km W Drain, T22S, R5W, Sec. 7 (1 USNM); 10 km N, 10.5 km W Drain, T21S, R6W, Sec. 7 (3 USNM); 10 km N, 10.6 km W Drain, T21S, R6W, Sec. 7 (1 USNM); 11.3 km N, 10.5 km W Drain, T21S, R6W, Sec. 5 (2 USNM); 4.5 mi. N, 8 mi. W Drain, T21S, R4W, Sec. 27, 1,200 ft. (12 UW[CR]); 5.5 mi. N, 4 mi. W Drain, T21S, R6W, Sec. 15, 1,280 ft. (1 OSUFW, 6 UW[CR]); 6 mi. N, 4 mi. W Drain, T21S, R6W, Sec. 15, 1,200 ft. (5 OSUFW, 20 UW[CR]); 6 mi. N, 5 mi. W Drain (2 OSUFW); 9.7 km S, 8.4 km W Elkton, T23S, R8W, Sec. 20 (4 USNM); 11.1 km S, 7.4 km W Elkton, T23S, R8W, Sec. 28 (3 USNM); 14.9 km S, 6.2 km W Elkton, T24S, R8W, Sec. 9 (3 USNM); 15.5 km S, 5.8 km W Elkton, T24S, R8W, Sec. 10 (3 USNM); 24.4 km S, 6 km W Elkton, T25S, R8W, Sec. 10 (3 USNM); 25.2 km S, 5.4 km W Elkton, T25S, R8W, Sec. 10 (4 USNM); Francis Creek, T24S, R1W, SE¼SE¼ Sec. 12 (1 SOSC); Francis Creek, upstream, Rd. 12.1, T24S, R1W, SE¼SE¼ Sec. 12 (3 SOSC); Francis Creek, downstream Rd. 12.1, T24S, R1W, SE¼SE¼ Sec. 12 (1 SOSC); Gardiner (2 FMNH, 3 MVZ, 1 USNM); 1.3 mi. E Gardiner (3 MVZ); 9 mi. S Idleyld Park, T27S, R3W, SE¼ Sec. 35 (13 OSUFW); Louis Creek, T28S, R3W, NW¼SW¼ Sec. 29 (7 SOSC); Louis Creek, T28S, R3W, SE¼SE¼ Sec. 30 (1 SOSC); Middle Creek, T31S, R6W, SW¼ Sec. 30 (1 SOSC); Oxbow Burn, 300 ft, T20S, R7W, Sec. 19 (12 PSM); Pass Creek, 0.4 mi. from Caton Creek, T24S, R1W, SE¼ Sec. 26 (1 SOSC); 2 km N, 25 km W Sutherlin, T25S, R8W (2 USNM); 2.2 km N, 26.2 km W Sutherlin, T25S, R8W, Sec. 9 (2 USNM); 2.4 km N, 24.6 km W Sutherlin, T25S, R8W, Sec. 10 (1 USNM); 2.8 km N, 26 km W Sutherlin, T25S, R8W, Sec. 9 (1 USNM); 3.6 km N, 11.9 km W Sutherlin, T25S, R7W, Sec. 1 (2 USNM); 3.8 km N, 12.1 km W Sutherlin, T25S, R7W, Sec. 1 (4 USNM); 4.2 km N, 12.1 km W Sutherlin, T25S, R7W, Sec. 1 (1 USNM); 4.3 km N, 12.1 km W Sutherlin, T25S, R7W, Sec. 1 (6 USNM); 6.5 km N, 6.7 km W Tenmile (2 USNM); 4.5 mi. N, 8 mi. E Tiller, T30S, R1W, Sec. 2 (3 CRCM); 4.5 mi. N, 9 mi. E Tiller (9 CRCM); 7.5 mi. N, 8 mi. E Tiller (9 CRCM); 9 mi. N, 8.5 mi. E Tiller, T29S, R1W, Sec. 13 (1 CRCM); 10 mi. N, 9 mi. E Tiller, T29S, R1W, Sec. 12 (3 CRCM); 11 mi. N, 5 mi. E Tiller (2 CRCM); 11.5 mi. N, 5 mi. E Tiller, T28S, R1W, Sec. 26 (3 CRCM); 12 mi. N, 4 mi. E Tiller (10 CRCM); 12 mi. N, 5.5 mi. E Tiller, T28S, R1W, Sec. 32 (9 CRCM); 12 mi. N, 17 mi. E Tiller (1 CRCM); 12 mi. N, 19 mi. E Tiller (15 CRCM); 12.5 mi. N, 5.5 mi. E Tiller, T28S, R1W, Sec. 31 (5 CRCM); 0.2 mi. N, 4.1 mi. W Umpqua P.O., T25S, R7W, Sec. 15, 800 ft. (4 UW[CR]); 2.2 mi. N, 9 mi. W Umpqua P.O., T25S, R8W, Sec. 15, 850 ft. (1 DWU, 4 OSUFW, 11 UW[CR]); 2.7 mi. N, 5 mi. W Umpqua P.O., Riverview, T25S, R7W, Sec. 7, 950 ft. (5 OSUFW, 5 UW[CR]); 4 mi. N, 1 mi. W Umpqua P.O., Lost Creek, T25S, R7W, Sec. 1, 1,500 ft. (2 OSUFW, 3 UW[CR]); 6 mi. N, 10.5 mi. W Umpqua P.O., Miner Creek, T24S, R8W, Sec. 29, 900 ft. (7 OSUFW, 13 UW[CR]); 4.8 km S, 4.7 km W Yoncolla, T23S, R5W (3 USNM); 4.9 km S, 4.5 km W Yoncolla, T25S, R5W, Sec. 19 (1 USNM); T20S, R7W, Sec. 19 (1 PSM); T21S, R12W, SE¼ Sec. 9 (1 PSM). **Jackson Co.**—6 mi. S Ashland (3 SOSC); 3 mi. S, 10 mi. E Ashland (1 SOSC); Lower Ashland Natural Research Area, T39S, R1E, SW¼ Sec. 28 (1 SOSC); Upper Ashland Natural Research Area, T40S, R1E, SE¼SE¼ Sec. 4 (2 SOSC); Ashland Research Natural Area, T40S, R1E, SE¼ Sec. 4 (1 SOSC); Imnaha Creek, T33S, R4E, W½ Sec. 15 (1 OSUFW, 4 KU); 11 mi. N Prospect (2 CRCM); 11.5 mi. N, 0.5 mi. W Prospect (2 CRCM); 13.5 mi. N, 2.5 mi. W Prospect (7 CRCM); 14 mi. N, 0.5 mi. W Prospect, T30S, R2E, Sec. 24/25 (15 CRCM); 2.5 mi. N, 8.5 mi. E Prospect (9 CRCM); 5.5 mi. N, 4 mi. E Prospect (2 CRCM); Spring Creek, T33S, R4E, NE¼ Sec. 22 (3 OSUFW, 4 KU); 2 mi. N, 1 mi. E Union Creek (1 CRCM); T37S, R4E, Sec. 23 (1 OSUFW); 0.5 mi. SW of T40S, R1E, Sec. 3/10 sign (1 SOSC). **Josephine Co.**—Bolan Lake, 5,551 ft, T41S, R6W, Sec. 7 (2 PSM); Grants Pass (1 CM); 5 mi. E Holland, T39S, R6W, Sec. 33 (4 OSUFW); 0.5 mi. S, 11.5 mi. E Kerby, T39S, R6W, Sec. 8 (1 OSUFW); 2 mi. S, 12 mi. E Kerby, T39S, R7W, Sec. 9 (1 OSUFW); Sourgrass Creek, T35S, R9W, Sec. 2 (1 SOSC); Sourgrass Creek, T35S, R9W, Sec. 3 (2 SOSC); Sourgrass Creek, T35S, R9W, SW¼ NE¼ Sec. 3 (1 SOSC); Sourgrass Creek, T35S, R9W, SE¼NE¼ Sec. 3 (3 SOSC); 2 mi. S, 4 mi. E Takilma, Skag Hope, T41S, R7W, Sec. 7 (3 OSUFW, 1 TTU); 3 mi. S, 4 mi. E Takilma, T41S, R7W, Sec. 8 (12 OSUFW); 3.5 mi. S, 4 mi. E Takilma, Page Mt., T41S, R7W, Sec. 8/9/16/17 (3 OSUFW); 0.5 mi. S, 11.5 mi. E Kerby, T39S, R6W, Sec. 8 (1 OSUFW); T39S, R6W, Sec. 33 (2 OSUFW); T41S, R7W, Sec. 7 (1 OSUFW); T41S, R7W, Sec. 8/9/16/17 (2 OSUFW). **Klamath Co.**—1.8 mi. S park headquarters, Crater Lake National Park (1 SOSC). **Lane Co.**—5 mi. N Alpha, T15S, R9W, Sec. 31 (4 OSUFW); 5 mi. N, 6 mi. W Alpha (2 OSUFW); 12.1 km S, 2.4 km W Alsea, T15S, R8W, Sec. 11 (1 USNM); 4 mi. E Big Creek (1 KU); 4.4 km N, 6.8 km E Blue River, T16S, R5E, Sec. 7 (1 USNM); 6.8 km N, 7.6 km E Blue River, T15S, R5E, Sec. 32 (5 USNM); 1 mi. N Blue River, T16S, R4E, NE¼ Sec. 20 (4 KU); 4 mi. N Blue River, T15S, R4E, W½ Sec. 32 (2 KU); 6 mi. S Blue River, T17S, R4E, SW¼ Sec. 20 (1 KU); 6 mi. W Blue River (1 OSUFW); 6 mi. E Blue River, T16S, R5E, Sec. 20 (2 OSUFW); 9 mi. E Blue River, T16S, R5E, Sec. 25/26 (10 CRCM); 1.5 mi. N, 2 mi. E Blue River (6 CRCM); 2 mi. N, 2 mi. E Blue River, T16S, R4E, S½ Sec. 10 (1 KU, 1 TTU); 4 mi. N, 4.5 mi. E Blue River, T16S, R5E, Sec. 32 (4 CRCM); 5 mi. N, 1 mi. E Blue River, T15S, R4E, S½ Sec. 27 (1 KU); 5 mi. N, 9 mi. E Blue River (17 CRCM); 5.5 mi. N, 13.5 mi. E Blue River, T16S, R6E, Sec. 26 (3 CRCM); 7 mi. N, 9 mi. E Blue River, T15S, R5E, Sec. 13 (1 CRCM); 2.5 mi. S, 7 mi. E Blue River (2 CRCM); 3.5 mi. S, 10 mi. E Blue River (13 CRCM); 4 mi. S, 10 mi. E Blue River, Fork Creek, T17S, R6E, Sec. 7 (1 USNM); 5.5 mi. NE Blue River, 1,500 ft. (2 PSM); 5 mi. N Cougar Dam, T16S, R5E, Sec. 6, 1,600–1,800 ft. (4 UW[OG]); East Fork Creek, T16S, R6E, Sec. 7 (4 OSUFW); Elk Creek, T16S, R4E, Sec. 19 (5 OSUFW); Eugene, 5 mi. S Spencer Butte (2 SDNHM); Eugene, 3 mi. W Spencer Butte (2 SDNHM); Foley Ridge, T16S, R6E, Sec. 20/21 (1 OSUFW); H. J. Andrews Experimental Forest, 1,700 ft (1 OSUFW); Hagan Creek, T16S, R3E, Sec. 22 (7 OSUFW); King Creek, T16S, R5E, Sec. 26 (1 USNM, 2 USNM); Lookout Creek, T15S, R6E, Sec. 30 (6 OSUFW); 0.4 km N, 18.5 km W Lorane, T20S, R7W, Sec. 12 (7 USNM); 2 km N, 6.4 km W Lorane, T20S, R5W, Sec. 5 (2 USNM); 2.8 km N, 6.2 km W Lorane, T20S, R5W, Sec. 5 (2 USNM); 3 km N, 19.5 km W Lorane, T20S, R7W, Sec. 1 (6 USNM); 3.2 km N, 26 km W Lorane, T19/20S, R7W, Sec. 32/5 (1 USNM); 4.8 km N, 17.5 km W Lorane, T19S, R6W, Sec. 31 (1 USNM); 5 mi. E Lowell, Fall Creek Reservoir (3 OSUFW); Mapleton (1 USNM); McKenzie Bridge (12 OSUFW); 2.6 km S McKenzie Bridge (5 USNM); 3.1 km N, 3 km W McKenzie Bridge, T16S, R5E, Sec. 3 (2 USNM); 3.3 km N, 2.4 km W McKenzie Bridge, T16S, R5E, Sec. 3 (2 USNM); 6.6 km N, 0.3 km W McKenzie Bridge, T16S, R5E, Sec. 25 (2 USNM); 6 km N, 0.6 km E McKenzie Bridge, T15S, R5E, Sec. 25 (6 USNM); 6 km N, 7.5 km E McKenzie Bridge, T15S, R6E, Sec. 26 (2 USNM); 6.4 km N, 0.6 km E McKenzie Bridge, T15S, R5E, Sec. 25 (5 USNM); 1.6 km S, 5.2 km W McKenzie Bridge, T16S, R5E, Sec. 20, 460 m (5 USNM); 2.6 km S, 1.4 km E McKenzie Bridge, T16S, R5E, Sec. 25 (1 USNM); 3.8 km S, 14.8

km E McKenzie Bridge, T16S, R7E, Sec. 33 (1 USNM); 4.4 km S, 14.3 km E McKenzie Bridge, T16S, R7E, Sec. 32 (4 USNM); 4 mi. N McKenzie Bridge, T15S, R5E, Sec. 25, 2,600 ft. (3 UW[OG]); 1 mi. N, 8 mi. W McKenzie Bridge, T16S, R4E, SW¼ Sec. 9 (4 KU); 2 mi. N, 2 mi. W McKenzie Bridge, T16S, R5E, Sec. 3/4, 3,800 ft. (2 UW[OG]); 5 mi. N, 1 mi. W McKenzie Bridge, T15S, R5E, S½ Sec. 14 (9 KU, 4 OSUFW); 6 mi. N, 5 mi. W McKenzie Bridge, T15S, R5E, NW¼ Sec. 19 (1 KU); 2 mi. S, 0.5 mi. E McKenzie Bridge, T16S, R5E, Sec. 25, 2,800 ft. (4 UW[OG]); 3 mi. S, 5 mi. E McKenzie Bridge, T15S, R6E, Sec. 35, 2,000 ft. (4 UW[OG]); McRae Lookout, T15S, R5E, SW¼SW¼ Sec. 26 (3 OSUFW); Mercer [= Florence] (1 MVZ); 1 mi. W Mohawk River, T15S, R1W, SE¼NE¼ Sec. 13 (1 SOSC); 4 mi. N Nimrod, T16S, R3E, Sec. 14 (2 CRCM); 4.5 mi. N Nimrod (18 CRCM); 5 mi. W Oakridge, Shady Del campground, T21S, R2E, Sec. 15 (2 UCD); 1 mi. N O'Leary Mt., T16S, R7E, Sec. 32 (1 UW-X); 4 mi. N Rainbow, T15S, R5E, Sec. 32, 1,700 ft. (1 UW[OG]); 2.5 mi. N, 2.5 mi. E Rainbow, T16S, R5E, Sec. 5 (1 UW-X); Spring Creek, T16S, R6E, Sec. 26 (1 OSUFW); Swamp Creek, T16S, R5E, Sec. 20 (7 OSUFW); Upper Lookout Creek Tributary, T15S, R5E, Sec. 24 (3 OSUFW); Vida (4 OSUFW, 1 SDNHM, 3 USNM); 11 mi. S Yachats, 6.5 mi. E HWY 101 on Big Creek (1 UCD); 4.8 km S, 3.2 km E Yachats, T15S, R12W, Sec. 12 (1 USNM); T15S, R5E, NW¼ Sec. 24, 3,500 ft. (4 PSM); T15S, R5E, Sec. 26 (1 USNM); T15S, R5E, NW¼ Sec. 28, 2,000 ft. (1 PSM); T15S, R5E, NE¼ Sec. 32, 1,600 ft. (2 PSM, 4 USNM); T15S, R5E, NE¼ Sec. 32, 2,00 ft. (4 PSM); T15S, R5E, SW¼ Sec. 33, 2,500 ft. (6 PSM); T15S, R5E, NW¼ Sec. 34, 2,700 ft. (2 PSM); T15S, R8W, Sec. 11 (1 USNM); T16S, R4E, NE¼ Sec. 19, 1,800 ft. (10 PSM); T16S, R4E, NE¼ Sec. 24, 1,400 ft. (13 PSM); T16S, R5E, SW¼ Sec. 6, 1,600 ft. (1 PSM); T16S, R5E, Sec. 26 (1 USNM); T18S, R5E, Sec. 23, 3,000 ft. (4 PSM); T20S, R12W, NE¼ Sec. 4 (1 PSM). **Lincoln Co.**—4 mi. S, 8 mi. W Alsea, T14S, R9W, Sec. 14/23 (1 OSUFW); 8 mi. W Harlan, Nettle Creek, T12S, R9W, Sec. 28 (13 USNM); 10 mi. W Harlan, Flynn Creek, T12S, R10W, Sec. 12 (6 USNM); 15 mi. W Harlan, T12S, R11W, Sec. 11, Peterson Ridge (30 KU); 2 mi. S, 8 mi. W Harlan, Flynn Creek, T12S, R10W, Sec. 13 (6 OSUFW, 1 TTU); 2 mi. S, 12 mi. W Harlan, Beaver Creek, T12S, R10W, Sec. 20 (15 KU); 5 mi. S, 10 mi. W Harlan, Gold Creek, T13S, R10W, Sec. 2 (10 KU, 9 OSUFW); 7 mi. S, 5 mi. W Harlan, West Creek, T13S, R9W, Sec. 16 (4 KU, 2 TTU); 7 mi. S, 10 mi. W Harlan, Death Ridge, T13S, R10W, Sec. 14 (1 OSUFW, 5 KU); 7 mi. S, 11 mi. W Harlan, Boulder Creek, T13S, R10W, Sec. 13 (19 KU, 6 OSUFW); 7 mi. S, 11 mi. W Harlan, Death Ridge, T13S, R10W, Sec. 14 (6 OSUFW); Newport (3 MVZ, 1 OSUFW, 2 SDNHM); 2 mi. E Newport (1 LACM); Taft (1 LACM); 10 mi. N Tidewater, S side Death Ridge (5 USNM); 10 mi. N Tidewater, S fork Drift Creek (3 USNM); 10 mi. N Tidewater at Murphy (15 USNM); 0.6 km N, 16.9 km E Waldport, T13S, R10W, Sec. 13/24 (1 USNM). **Linn Co.**—7 mi. N, 8 mi. E Blue River, T15S, R5E, Sec. 15 (8 CRCM); 7.5 mi. N, 7.5 mi. E Blue River, T15S, R5E, Sec. 15 (3 CRCM); 8 mi. N, 8 mi. E Blue River (3 CRCM); 9 mi. N, 8 mi. E Blue River (1 CRCM); 9.5 mi. N, 9 mi. E Blue River, T15S, R5E, Sec. 2 (1 CRCM); about 10 mi. SE Cascadia (1 SUVM); 10.25 km N, 0.8 km W McKenzie Bridge, T15S, R5E, Sec. 14 (4 USNM); 13.2 km N, 0.8 km E McKenzie Bridge, T15S, R5E, Sec. 1 (2 USNM); 1 mi. S Wolf Rock, T15S, R5E, Sec. 1, 3,300 ft. (2 UW[OG]); 4 mi. S, 1 mi. W Wolf Rock, T15S, R5E, Sec. 14, 2,600 ft. (3 UW[OG]); T14S, R6E, NE¼ Sec. 20, 4,400 ft. (2 PSM); T14S, R6E, NE¼ Sec. 28, 4,600 ft. (2 PSM); T15S, R5E, SE¼ Sec. 11, 3,200 ft. (1 PSM); T15S, R5E, SW¼ Sec. 13, 3,000 ft. (5 PSM); T15S, R6E, NE¼ Sec. 7, 4,300 ft. (2 PSM). **Tillamook Co.**—Sand Lake (1 LACM).

Sorex trowbridgii, n = 4,450
(Plotted in Fig. 8-18, p. 61)

Benton Co.—(7 PSM); Alsea (4 PSM); 7.5 mi. W Alsea (1 PSM); 4.4 km N, 7.6 km W Alsea, T13S, R8W, Sec. 20 (13 USNM); 4 mi. N, 4 mi. E Alsea (1 OSUFW); 3.75 mi. S, 5.5 mi. E Alsea (31 CM); 1 km S, 4.4 km E Alsea, T14S, R7W, Sec. 9 (1 USNM); 1.2 km S, 4.2 km E Alsea, T14S, R7W, Sec. 9 (15 USNM); 8 km S, 2.8 km E Alsea, T14S, R7W, Sec. 32 (9 USNM); 10 mi. SE Alsea (3 HSU); 3 mi. N Blodgett (1 PSM); Corvallis (1 OSUFW, 1 PSM, 5 UCONN); 3 mi. N Corvallis,

T11S, R5W, Sec. 15 (1 OSUFW); 7 mi. N Corvallis, McDonald Forest (12 PSM); 8 mi. N Corvallis (2 PSM); 1 mi. N, 3 mi. W Corvallis (1 OSUFW); 2 mi. N, 4 mi. W Corvallis (2 OSUFW); 0.25 mi. N, 0.25 mi. E Corvallis (1 KU); 5 mi. S, 14 mi. W Corvallis (1 OSUFW); 4 mi. from Corvallis (1 OSUFW); 2 mi. SW Hoskins (1 PSM); McDonald Forest, Peavy Arboritum pond, T10S, R4W, Sec. 31 (1 OSUFW); Marys Peak, 4,000 ft. (1 PSM); 1 mi. W Philomath, T12S, R6W, NE¼ Sec. 10, 250 ft. (1 PSM); 1 mi. S, 6 mi. W Philomath, Marys Peak, T12S, R7W, Sec. 13 (1 OSUFW); 2 mi. S, 7 mi. W Philomath, Marys Peak (2 OSUFW); 2 mi. S, 9 mi. W Philomath, Marys Peak (2 OSUFW); 3 mi. S, 9 mi. W Philomath, T12S, R7W, Sec. 28 (2 OSUFW); 5 mi. S, 5 mi. W Philomath, T13S, R7W, Sec. 1 (1 KU); 1 km S, 11.3 km W Philomath, T12S, R7W, Sec. 11/14 (8 USNM); 1.5 km S, 10.1 km W Philomath, T12S, R7W, Sec. 13 (14 USNM); 1.5 km S, 11.3 km W Philomath, T12S, R7W, Sec. 14 (22 USNM); 2.4 km S, 15.5 km W Philomath, T12S, R7W, Sec. 17/20 (12 USNM); 4 km S, 10.1 km W Philomath, T12S, R7W, Sec. 24/25 (12 USNM); 5 mi. SW Philomath (4 USNM); 6 mi. SW Philomath on HWY 34 (1 OSUFW); 8 mi. SW Philomath (1 OSUFW); 8.5 mi. SW Philomath (1 OSUFW); 9 mi. SW Philomath, Marys Peak (2 PSM); 1.5 mi. N Summit (1 PSM); 5.7 mi. NE Summit (2 PSM); 0.5 mi. S, 3 mi. E Wrens, McDonald Game Preserve, T11S, R6W (2 OSUFW); 2.4 mi. NW Wrens (1 PSM). **Clackamas Co.**—4.5 mi. N Alder Creek, S. Fork Bull Run River, T1S, R6E, Sec. 29/32 (6 OSUFW); 6 mi. N Alder Creek, Cedar Creek, T1S, R6E, Sec. 29 (8 OSUFW); Arra[h] Wanna (1 MVZ, 1 UMMZ); Estacada (1 MVZ); [Lake] Oswego (2 KU, 1 MVZ); 2 mi. E Liberal, Camp Onahnee (2 PSM); 8 mi. SE Mollala (2 PSM); Mulino (2 OSUFW); Oregon City (3 UW[EB]); 4 mi. SE Oregon City (1 MVZ); 2 mi. N, 4 mi. E Sandy (2 CRCM); 3.5 mi. N, 8 mi. E Sandy (9 CRCM); 4.2 mi. N, 8 mi. E Sandy, T1S, R6E, Sec. 29 (2 CRCM); 6 mi. S, 7 mi. E Sandy, T3S, R6E, Sec. 18 (4 CRCM); Still Creek Forest Camp, Mt. Hood National Forest (1 PSM); 5 mi. S, 6 mi. E Zigzag (8 CRCM). **Clatsop Co.**—Astoria (1 PSM, 5 MVZ, 12 USNM); Empire (1 USNM); Klootchy Creek campground (1 LACM); site of Old Fort Clatsop (23 MVZ); 1 mi. W site of Old Fort Clatsop (3 MVZ); T8N, R10W, W½ Sec. 6 (3 PSM). **Columbia Co.**—2 mi. W Deer Island (2 PSU); Longview Bridge (1 PSM); Rainier, Goble Creek (1 PSM); 1 mi. W Rainier (1 PSM); 4 mi. W Rainier (1 PSM); 7 mi. SE Rainier (14 MVZ); 0.5 mi. W St. Helens (1 PSU); 7 mi. N Vernonia (1 PSM); 7 mi. NE Vernonia (2 PSM). **Coos Co.**—1.75 mi. S Bandon (2 PSM); 9 mi. S Bandon (4 PSM); 1.75 mi. E Bandon (7 PSM); 9 mi. NE Bandon (1 PSM); 3 mi. SE Bandon (1 PSM); 4 mi. SE Bandon (21 PSM); Marshfield [=Coos Bay] (2 AMNH, 1 USNM); 3.2 km N, 0.3 km W Remote, T29S, R10W, Sec. 21 (16 USNM); 7.3 km N, 4 km W Sitkum, T27S, R10W, Sec. 17/20 (3 USNM); 7.4 km N, 4.2 km W Sitkum, T27S, R10W, Sec. 17 (16 USNM); 7.5 km N, 12.5 km W Sitkum, T27S, R11W, Sec. 21 (12 USNM); 7.5 km N, 12.7 km W Sitkum, T27S, R11W, Sec. 21 (7 USNM); 7.6 km N, 5.2 km W Sitkum, T27S, R10W, Sec. 18 (14 USNM); 8.4 km N, 4.3 km W Sitkum, T27S, R10W, Sec. 17 (19 USNM); 2.8 km S, 5 km E Sitkum, T28S, R9W, Sec. 19 (3 USNM); 2.9 km S, 4.7 km E Sitkum, T28S, R9W, Sec. 19 (9 USNM); 3 km S, 11 km E Sitkum, T28S, R9W, Sec. 23 (1 USNM); 4.8 km S, 9.7 km E Sitkum, T28S, R9W, Sec. 27 (4 USNM); 5.8 km S, 10.8 km E Sitkum (2 USNM). **Curry Co.**—5 mi. N Agness (4 HSU); 2 mi. W Agness, N side Rogue River (1 MVZ); 10 mi. E Brookings (12 PSM); Gold Beach (4 FMNH, 4 PSM); Little Redwood Forest Camp (1 PSM); Long Ridge Forest Camp (1 PSM); 1 mi. S Long Ridge Forest Camp (2 PSM); Port Orford (1 USNM); 2 mi. S Port Orford (1 PSM). **Douglas Co.**—(13 PSM); 8 mi. S, 2.5 mi. W Alma, Oxbow Burn (1 OSUFW); 3.2 mi. S, 3.6 mi. E Canyonville, T31S, R4W, Sec. 17, 2,000 ft. (15 UW[CR]); Charlotte Creek, Lower Umpqua River (1 PSM); 20 mi. S Crow, Oxbow Burn (28 PSM); 1 mi. S, 33 mi. E Days Creek, Abbott Creek, T30S, R2E, Sec. 24/25 (3 OSUFW); Diamond Lake (1 SDNHM); 0.6 km N, 2.2 km W Drain, T22S, R5W, Sec. 7 (4 USNM); 0.8 km N, 2.2 km W Drain, T21S, R5W, Sec. 7 (20 USNM); 1.4 km N, 1.6 km W Drain, T22S, R5W, Sec. 7 (18 USNM); 4 km N, 15 mi. from Drain, Elk Creek, T22S, R6W, Sec. 13 (7 USNM); 7.6 km N, 6.4 km W Drain, T21S, R6W, Sec. 22 (25 USNM); 10 km N, 10.5 km W Drain, T21S, R6W, Sec. 7 (5 USNM); 10 km N, 10.6 km W Drain, T21S, R6W, Sec. 7 (16 USNM); 11.3 km N, 10.5 km W Drain, T21S, R6W, Sec. 5 (10 USNM); 11.5 km N, 10.5 km W Drain, T21S, R6W, Sec. 5 (19

USNM); 6.2 km N, 12.7 km E Drain, T24S, R4W, Sec. 27 (5 USNM); 4.5 mi. N, 8 mi. W Drain, T21S, R4W, Sec. 27, 1,200 ft. (34 UW[CR]); 5.5 mi. N, 4 mi. W Drain, T21S, R6W, Sec. 15, 1,280 ft. (38 UW[CR]); 6 mi. N, 4 mi. W Drain, T21S, R6W, Sec. 15, 1,200 ft. (50 UW[CR]); Drew (1 USNM); Eel Creek, Lower Umpqua River (2 PSM); 9.7 km S, 8.4 km W Elkton, T23S, R8W, Sec. 20 (31 USNM); 9.7 km S, 16.2 km W Elkton, T23S, R9W, Sec. 22 (17 USNM); 11.1 km S, 7.4 km W Elkton, T23S, R8W, Sec. 28 (29 USNM); 14.9 km S, 6.2 km W Elkton, T24S, R8W, Sec. 9 (9 USNM); 15.4 km S, 6 km W Elkton, T24S, R8W, Sec. 10 (8 USNM); 15.5 km S, 5.8 km W Elkton, T24S, R8W, Sec. 10 (24 USNM); 15.7 km S, 5.6 km W Elkton, T24S, R8W, Sec. 10 (19 USNM); 24.4 km S, 6 km W Elkton, T25S, R8W, Sec. 10 (4 USNM); 25.2 km S, 5.8 km W Elkton, T25S, R8W, Sec. 10 (10 USNM); 25.6 km S, 5.8 km W Elkton, T25S, R8W, Sec. 10 (11 USNM); Fairview, T25S, R2W, Sec. 25 (2 OSUFW); Gardiner (6 MVZ); Myrtle Creek (1 SDNHM); Reedsport (1 PSM); SW Reedsport (3 PSM); 3.2 km N, 0.3 km W Remote, T22S, R5W, Sec. 21 (1 USNM); Reston (1 USNM); 4 mi. S, 6.4 mi. W Riddle, Iron Mt., T31S, R7W, Sec. 11, 2,000 ft. (7 UW[CR]); 3 km S, 10.8 km E Sitkum, T28S, R9W, Sec. 23 (10 USNM); 2 km N, 25 km W Sutherlin, T25S, R8W, Sec. 10 (10 USNM); 2.1 km N, 24.8 km W Sutherlin, T25S, R8W, Sec. 10 (3 USNM); 2.2 km N, 26.2 km W Sutherlin, T25S, R9W, Sec. 9 (12 USNM); 2.4 km N, 24.6 km W Sutherlin, T25S, R8W, Sec. 10 (6 USNM); 2.8 km N, 26 km W Sutherlin, T25S, R8W, Sec. 9 (9 USNM); 3.4 km N, 11.9 km W Sutherlin, T25S, R7W, Sec. 1 (5 USNM); 3.6 km N, 11.9 km W Sutherlin, T25S, R7W, Sec. 1 (3 USNM); 3.8 km N, 12.1 km W Sutherlin, T25S, R7W, Sec. 1 (13 USNM); 4 km N, 12.1 km W Sutherlin, T25S, R7W, Sec. 1 (13 USNM); 4.2 km N, 12.1 km W Sutherlin, T25S, R7W, Sec. 1 (5 USNM); 4.3 km N, 12.1 km W Sutherlin, T25S, R7W, Sec. 1 (2 USNM); 4.4 km N, 12.1 km W Sutherlin, T25S, R7W, Sec. 1 (1 USNM); 6.5 km N, 6.7 km W Tenmile, T28S, R8W, Sec. 9 (22 USNM); 1 mi. N, 21 mi. E Tiller, Abbot Creek, T30S, R2E, Sec. 25 (1 OSUFW); 4 mi. N, 9 mi. E Tiller, Upper Freezout Creek, T30S, R1W, Sec. 1/2 (1 OSUFW); 4.5 mi. N, 8 mi. E Tiller, T30S, R1W, Sec. 2 (10 CRCM); 4.5 mi. N, 9 mi. E Tiller (9 CRCM); 7.5 mi. N, 8 mi. E Tiller (5 CRCM); 9 mi. N, 8.5 mi. E Tiller, T29S, R1W, Sec. 13 (5 CRCM); 10 mi. N, 9 mi. E Tiller, T29S, R1W, Sec. 12 (8 CRCM); 11 mi. N, 4 mi. E Tiller, Smith Ridge, T28S, R1W, Sec. 32 (1 OSUFW); 11.5 mi. N, 5 mi. E Tiller, T28S, R1W, Sec. 26 (9 CRCM); 12 mi. N, 4 mi. E Tiller (4 CRCM); 12 mi. N, 5 mi. E Tiller, T28S, R1W, Sec. 31 (17 CRCM); 12 mi. N, 5.5 mmi. E Tiller (11 CRCM); 12 mi. N, 17 mi. E Tiller (6 CRCM); 12 mi. N, 18 mi. E Tiller, Castle Rock, T28S, R2E, Sec. 27 (3 OSUFW); 12 mi. N, 19 mi. E Tiller (6 CRCM); 0.2 mi. N, 4.1 mi. W Umpqua P.O., T25S, R7W, Sec. 15, 800 ft. (65 UW[CR]); 2.2 mi. N, 9 mi. W, Umpqua P.O., T25S, R8W, Sec. 15, 850 ft. (45 UW[CR]); 2.7 mi. N, 5 mi. W Umpqua, Riverview, T25S, R7W, Sec. 7, 950 ft. (61 UW[CR]); 4 mi. N, 1 mi. W Umpqua P.O., Lost Creek, T25S, R7W, Sec. 1, 1,500 ft. (47 UW[CR]); 6 mi. N, 10.5 mi. W Umpqua, Miner Creek, T24S, R8W, Sec. 29, 900 ft. (87 UW[CR]); 1 mi. S Winchester Bay (1 MVZ); 4.8 km S, 4.5 km W Yoncalla, T23S, R5W, Sec. 19 (5 USNM); 4.8 km S, 4.7 km W Yoncalla, T23S, R5W, Sec. 19 (11 USNM); 4.9 km S, 4.5 km W Yoncalla, T23S, R5W, Sec. 19 (39 USNM). **Hood River Co.**—2 mi. W Parkdale, 1,500 ft. (3 USNM); T3N, R9E, NW¼ Sec. 35 (1 PSM); T3N, R9E, SW¼ Sec. 35 (1 PSM). **Jackson Co.**—S Ashland, end of Neil Creek Road (1 SOSC); 3 mi. S Ashland (1 PSM); 4 mi. SE Ashland (1 PSM); Ashland Research Natural Area, T40S, R1E, SE¼ Sec. 4 (20 SOSC); Ashland Research Natural Area, E fork Ashland Creek, inlet to Reeder Reservoir (5 SOSC); 2 mi. E Butte Falls (1 LACM); Imnaha Creek, T33S, R4E, W½ Sec. 15 (2 SOSC); S Medford, 4,700 ft. (1 PSM); 7 mi. S Medford (3 PSM); 11 mi. N Prospect (15 CRCM); 11.5 mi. N, 0.5 mi. W Prospect (4 CRCM); 13.5 mi. N, 2.5 mi. W Prospect (13 CRCM); 14 mi. N, 0.5 mi. W Prospect, T30S, R2E, Sec. 24/25 (4 CRCM); 2.5 mi. N, 8.5 mi. E Prospect (3 CRCM); 5 mi. N, 6 mi. E Prospect, Dead Soldier Camp, T32S, R4E, Sec. 5 (5 OSUFW); 5 mi. N, 8 mi. E Prospect (2 CRCM); 5.5 mi. N, 4 mi. E Prospect (3 CRCM); Siskiyou (3 USNM); Spring Creek, T33S, R4E, NE¼ Sec. 22 (1 KU); Talent (1 SOSC); 4 mi. W Talent (1 SOSC); 2 mi. N, 1 mi. E Union Creek (5 CRCM); 4.7 mi. S Anderson Creek Road on Wagner Creek Road (1 SOSC). **Josephine Co.**—Bolan Lake (3 PSM); 13 mi. SW Grants Pass (1 UO); Oregon Caves National Monument (1 OSUFW). **Klamath**

Co.—[Cr]esent Lake National Forest campground (1 LACM); Swan Lake Valley (2 USNM). **Lake Co.**—Gearhart Mountain (6 PSM); Lakeview (1 USNM). **Lane Co.**—(27 PSM); 5 mi. N, 6 mi. W Alpha, T15S, R9W, Sec. 31 (2 OSUFW); 12.1 km S, 2.4 km W Alsea, T15S, R8W, Sec. 11 (11 USNM); 2 mi. S, 1 mi. E Austa (4 CM); Belknap Springs (1 OSUFW); 1 mi. N Belknap Springs, McKenzie River (1 OSUFW); Blue River (1 USNM); 6 mi. W Blue River, Hagan Creek, T16S, R3E, Sec. 22 (6 OSUFW); ca. 6 mi. E Blue River, Swamp Creek, T16S, R5E, Sec. 20 (2 OSUFW); 9 mi. E Blue River, T16S, R5E, Sec. 25/26 (49 CRCM); 0.5 mi. N, 2 mi. W Blue River, Elk Creek, T16S, R4E, Sec. 19 (5 OSUFW); 7 mi. N, 8 mi. E Blue River (29 CRCM); 4.4 km N, 6.8 km E Blue River, T16S, R5E, Sec. 7 (57 USNM); 5 km N, 7.6 km E Blue River, T16S, R5E, Sec. 6 (16 USNM); 6.8 km N, 7.6 km E Blue River, T15S, R5E, Sec. 32 (87 USNM); 2 mi. N, 2 mi. W Blue River, Elk Creek, T16S, R4E, Sec. 19 (1 OSUFW); 3.5 mi. S, 8 mi. W Blue River, T17S, R3E, Sec. 7 (2 HSU); 1.5 mi. N, 2 mi. E Blue River (37 CRCM); 4 mi. N, 9 mi. E Blue River, Mack Creek Tributary, T15S, R5E, Sec. 35 (2 OSUFW); 5 mi. N, 9 mi. E Blue River (58 CRCM); 5 mi. N, 10.5 mi. E Blue River, T15S, R5E, Sec. 30 (6 CRCM); 5.5 mi. N, 7 mi. E Blue River, T15S, R5E, Sec. 26 (1 HSU); 5.5 mi. N, 13.5 mi. E Blue River, T16S, R6E, Sec. 26 (17 CRCM); 6 mi. N, 8 mi. E Blue River, McRae Lookout, T15S, R5E, SW¼SW¼ Sec. 26 (1 OSUFW); 7 mi. N, 9 mi. E Blue River, T15S, R5E, Sec. 13 (22 CRCM); 7.25 mi. N, 9.25 mi. E Blue River (10 CRCM); 2.5 mi. S, 5.5 mi. E Blue River, T16S, R5E, Sec. 32 (1 OSUFW, 38 CRCM); 2.5 mi. S, 7 mi. E Blue River (4 CRCM); 3.5 mi. S, 10 mi. E Blue River (28 CRCM); 3.5 mi. S, 10.25 mi. E Blue River, 2,800 ft. (1 CRCM); 4 mi. S, 10.5 mi. E Blue River, East Fork Creek, T17S, R6E, Sec. 7 (2 OSUFW); 5.5 mi. NE Blue River (9 PSM); ca. 10 mi. NE Blue River, Lookout Creek, T15S, R6E, Sec. 30 (2 OSUFW); 5 mi. N Cougar Dam, T16S, R5E, Sec. 6 (2 UW[CR]); 1 mi. N, 10.5 mi. E Cougar Dam, T16S, R6E, Sec. 25, 2,900 ft. (1 UW[OG]); Eugene (1 OSUFW, 2 USNM); W Eugene on HWY 126 (2 OSUFW); 9 mi. N, 1 mi. W Gardiner (2 PSM); H. J. Andrews Experimental Forest, T15S, R5E, SE¼ Sec. 15, 700 ft. (1 PSM); H. J. Andrews Experimental Forest, T15S, R5E, NW¼ Sec. 24, 3,500 ft. (28 PSM); H. J. Andrews Experimental Forest, T15S, R5E, NW¼ Sec. 28, 2,000 ft. (39 PSM); H. J. Andrews Experimental Forest, T15S, R5E, NW¼ Sec. 32, 1,500 ft. (2 PSM); H. J. Andrews Experimental Forest, T15S, R5E, NE¼ Sec. 32, 1,600 ft. (2 PSM); H. J. Andrews Experimental Forest, T15S, R5E, NE¼ Sec. 32, 1,650 ft. (42 PSM); H. J. Andrews Experimental Forest, T15S, R5E, NE¼ Sec. 32, 2,000 ft. (38 PSM); H. J. Andrews Experimental Forest, T15S, R5E, SW¼ Sec. 33, 2,500 ft. (43 PSM); H. J. Andrews Experimental Forest, T15S, R5E, NW¼ Sec. 34, 2,700 ft. (20 PSM); H. J. Andrews Experimental Forest, T16S, R4E, NE¼ Sec. 19, 1,800 ft. (27 PSM); H. J. Andrews Experimental Forest, T16S, R4E, NE¼ Sec. 24, 1,400 ft. (78 PSM); H. J. Andrews Experimental Forest, T16S, R5E, NW¼ Sec. 6, 1,600 ft. (15 PSM); H. J. Andrews Experimental Forest, T16S, R5E, SW¼ Sec. 6, 1,600 ft. (35 PSM); 10 mi. W Junction City (2 PSM); 0.4 km N, 18.5 km W Lorane, T20S, R7W, Sec. 12 (17 USNM); 0.8 km N, 19.2 km W Lorane, T20S, R7W, Sec. 12 (17 USNM); 2 km N, 6.4 km W Lorane, T20S, R5W, Sec. 5 (28 USNM); 2.8 km N, 6.2 km W Lorane, T20S, R5W, Sec. 5 (38 USNM); 3 km N, 19.5 km W Lorane, T20S, R7W, Sec. 1 (22 USNM); 3.2 km N, 26 km W Lorane, T19/20S, R7W, Sec. 32/5 (11 USNM); 4.8 km N, 17.5 km W Lorane, T19S, R6W, Sec. 31 (22 USNM); 6.8 km N, 11.1 km W Lorane, T19S, R6W, Sec. 23 (18 USNM); 5 mi. E Lowell, Fall Creek Reservoir (1 OSUFW); Mabel (6 UO); Marshfield (2 AMNH); McKenzie Bridge (4 OSUFW, 4 UO); 4 mi. N McKenzie Bridge, T15S, R5E, Sec. 25, 2,600–2,700 ft. (5 UW[OG]); 5 mi. S McKenzie Bridge, T16S, R6E, Sec. 26 (2 OSUFW); 2.6 km S McKenzie Bridge, T16S, R5E, Sec. 26 (49 USNM); 1 mi. N, 8 mi. W McKenzie Bridge, T16S, R4E, SW ¼ Sec. 9 (1 OSUFW, 2 KU); 5 mi. N, 1 mi. W McKenzie Bridge, T15S, R5E, S½ Sec. 14 (5 KU); 6 mi. N, 5 mi. W McKenzie Bridge, T15S, R5E, NW¼ Sec. 19 (1 KU); 3.1 km N, 3 km W McKenzie Bridge, T16S, R5E, Sec. 3 (50 USNM); 3.3 km N, 2.4 km W McKenzie Bridge, T16S, R5E, Sec. 3 (73 USNM); 6.6 km N, 0.3 km W McKenzie Bridge, T15S, R5E, Sec. 25 (21 USNM); 6 km N, 0.6 km E McKenzie Bridge, T15S, R5E, Sec. 25 (28 USNM); 6 km N, 7.5 km E McKenzie Bridge, T15S, R6E, Sec. 26 (31 USNM); 6.4 km N, 0.6 km E McKenzie Bridge, T15S, R5E, Sec. 25 (27 USNM); 1

mi. S, 1 mi. W McKenzie Bridge, King Creek, T16S, R5E, Sec. 26 (4 OSUFW); 1.6 km S, 5.2 km W McKenzie Bridge, T16S, R5E, Sec. 20 (48 USNM); 2.6 km S, 1.4 km W McKenzie Bridge, T16S, R5E, Sec. 25 (22 USNM); 3.3 km S, 0.8 km W McKenzie Bridge, T16S, R5E, Sec. 3 (1 USNM); 2 mi. S, 0.5 mi. E McKenzie Bridge, T16S, R5E, Sec. 25, 2,800 ft. (10 UW[OG]); 3 mi. N, 5 mi. E McKenzie Bridge, T15S, R6E, Sec. 35, 2,000 ft. (6 UW[OG]); 3.5 mi. S, 9 mi. E McKenzie Bridge, T16S, R7E, Sec. 32, 3,500 ft. (2 UW[OG]); 3.8 km S, 14.8 km E McKenzie Bridge, T16S, R7E, Sec. 33 (10 USNM); 4.4 km S, 14.3 km E McKenzie Bridge, T16S, R7E, Sec. 32 (55 USNM); 5.4 km S, 8.75 km E McKenzie Bridge, T16S, R6E, Sec. 35 (36 USNM); Mercer (2 SDNHM); Mercer, Grass Mt. Refuge (1 OSUFW); 4 mi. N Nimrod, T16S, R3E, Sec. 14 (6 CRCM); 4.5 mi. N Nimrod (5 CRCM); 4 mi. N, 1 mi. E Nimrod, T16S, R3E, Sec. 15 (8 OSUFW); 4 mi. N, 2 mi. E Nimrod, T16S, R3E, Sec. 14 (7 OSUFW); 4.6 mi. E Oakridge (1 PSM); 7.4 mi. E Oakridge (1 PSM); 11 mi. E Oakridge (1 PSM); 14.1 mi. E Oakridge (2 PSM); 18.6 mi. E Oakridge (1 PSM); 1 mi. N, 4 mi. E Oakridge, Salmon Creek Rd. (6 OSUFW); 3 mi. SE Oakridge (4 UO); 4 mi. N Rainbow, T15S, R5E, Sec. 32, 1,700 ft. (8 UW[OG]); 2 mi. S, 0.5 mi. E Rainbow, Blue River, T16S, R5E, Sec. 32 (8 OSUFW); Spencer Butte (3 UO); N slope Three Sisters, 6,000 ft. (2 USNM); Vida (3 OSUFW, 1 PSM, 1 SDNHM, 8 UO, 7 USNM); 2 mi. W Vida, Fish Hatchery (4 USNM); 23 mi. SE Vida (27 PSM); 1.66 mi. S, 6.125 mi. W Walton (1 PSU). **Lincoln Co.**—Alsea (4 PSM); Cascade Head Experimental Forest (20 PSM); Delake (2 KU, 3 UMMZ); 15 mi. N Harlan, Peterson Ridge, T12S, R11W, Sec. 11 (7 KU); 2 mi. S, 8 mi. W Harlan, T12S, R10W, Sec. 13 (1 KU); 2 mi. S, 12 mi. W Harlan, T12S, R10W, Sec. 20 (3 KU); 5 mi. S, 10 mi. W Harlan, Gold Creek, T13S, R10W, Sec. 2 (11 KU); 7 mi. S, 5 mi. W Harlan, West Creek, T13S, R9W, Sec. 16 (112 KU); 7 mi. S, 10 mi. W Harlan, Death Ridge, T13S, R10W, Sec. 14 (112 KU); 7 mi. S, 11 mi. W Harlan, Boulder Creek, T13S, R10W, Sec. 15 (4 KU); Newport (1 MVZ, 4 SDNHM); 2 mi. N Newport (7 UMMZ); 7 mi. S Newport, Beaver Creek (3 UMMZ); 1 mi. E Newport (11 UMMZ); 0.5 mi. NE Newport (1 UMMZ); Otis (12 USNM); 2 mi. S, 13 mi. W Philomath on HWY 34 (2 OSUFW); Rose Lodge (1 USNM); Taft (1 KU, 3 LACM, 3 MVZ, 1 UMMZ); 10 mi. N Tidewater at Murphy, T11S, R9W, Sec. 7 (1 USNM); 0.6 km N, 16.9 km E Waldport, T13S, R10W, Sec. 13/14 (10 USNM); 0.9 km N, 16.9 km E Waldport, T13S, R10W, Sec. 13 (3 USNM); 1.6 km N, 5.8 km E Waldport, T14S, R11W, Sec. 17 (4 USNM); 4 km S, 3.2 km E Yachats, T15S, R12W, Sec. 16 (16 USNM); 4.8 km S, 3.2 km E Yachats, T15S, R12W, Sec. 12 (31 USNM); Yaquina Bay (2 USNM). **Linn Co.**—10 mi. N, 2 mi. E Albany (1 OSUFW); 2 mi. N Big Lake (1 PSM); 6.75 mi. N, 8 mi. E Blue River, T15S, R5E, Sec. 15 (24 CRCM); 7.5 mi. N, 7.5 mi. E Blue River, T15S, R5E, Sec.15 (22 CRCM); 8 mi. N, 8 mi. E Blue River (30 CRCM); 9 mi. N, 8 mi. E Blue River (28 CRCM); 9.5 mi. N, 9 mi. E Blue River, T15S, R5E, Sec. 2 (50 CRCM); 2 mi. S, 17 mi. E Cascadia, T14S, R5E, S½ Sec. 1 (11 KU); 9.75 mi. S, 3.5 mi. E Cascadia (1 OSUFW); Fish Lake (1 PSM); H. J. Andrews Expermental Forest, T14S, R6E, NE¼ Sec. 28, 4,600 ft. (9 PSM); H. J. Andrews Experimental Forest, T14S, R6E, NE¼ Sec. 20, 4,400 ft. (8 PSM); H. J. Andrews Experimental Forest, T15S, R6E, NE¼ Sec. 7, 4,300 ft. (17 PSM); H. J. Andrews Experimental Forest, T15S, R6E, NW¼ Sec. 7, 5,000 ft. (14 PSM); H. J. Andrews Experimental Forest, T15S, R5E, SW¼ Sec. 13, 3,000 ft. (23 PSM); H. J. Andrews Experimental Forest, T15S, R5E, SE¼ Sec. 11, 3,200 ft. (36 PSM); 1.5 mi. N Lacomb (1 PSM); 5 mi. N Lacomb (2 PSM); 3.3 mi. N Lacomb (3 PSM); 1.7 mi. S Lyons (4 PSM); 9.2 km N, 1.2 km W McKenzie Bridge, T15S, R5E, Sec. 14 (23 USNM); 10.25 km N, 0.8 km W McKenzie Bridge, T15S, R5E, Sec. 14 (46 USNM); 13.2 km N, 0.8 km E McKenzie Bridge, T15S, R5E, Sec. 1 (52 USNM); 5.9 mi. E Scio (4 PSM); 1 mi. S, 2 mi. W Tangent, Calapooya River, T12S, R4W, Sec. 27 (1 OSUFW); Trout Creek Forest Camp, T13S, R3E, NE¼ Sec. 32 (1 PSM); 8 mi. S, 4 mi. W Upper Soda, T15S, R4E, S½ Sec. 5 (13 KU); 3.2 mi. N Waterloo (1 PSM); 1 mi. S Wolf Rock, T15S, R5E, Sec. 1, 3,300 ft. (1 UW[OG], 1 UW-X). **Marion Co.**—4.5 mi. S, 1.5 mi. E Bagby Hot Springs, 4,000 ft. (6 LSU). **Multnomah Co.**—5 mi. S Bonneville, Tanner Creek (1 MVZ); 1.5 mi. W Dodson (1 PSU); Latourelle (2 OSUFW); 4 mi. S, 2.5 mi. E Multnomah Falls, N. Fork Bull Run River, T1N, R6E, Sec. 34 (2 OSUFW); 6 mi. S, 6.5 mi. E

Multnomah Falls, Big Band Creek, T1S, R7E, Sec. 8 (1 OSUFW); Portland (1 KU, 3 AMNH, 1 USNM, 2 PSM, 1 MVZ); Portland, Council Crest (1 MVZ); Portland, 1 mi. S St. Johns Bridge (2 PSU); Portland, N Marine Drive, S Columbia Blvd, E of NE 148th (2 PSU); 6 mi. N, 11 mi. E Sandy (4 CRCM); 8 mi. N, 11 mi. E Sandy (6 CRCM); 9 mi. N, 10 mi. E Sandy (2 CRCM); 10 mi. N, 11 mi. E Sandy, T1N, R6E, Sec. 27 (14 CRCM); 11 mi. N, 10 mi. E Sandy (7 CRCM); Springdale (1 MVZ); 11 mi. N Zigzag (2 CRCM); 8 mi. N, 3 mi. E Zigzag (3 CRCM); 9 mi. N, 4 mi. W Zigzag (15 CRCM); 10 mi. N, 1.5 mi. W Zigzag (3 CRCM). **Polk Co.**—3.5 mi. S, 2 mi. W Gold Creek, Gooseneck Creek, T7S, R7W, SE¼ Sec. 5 (1 OSUFW); Salmon River Summit (2 PSM). **Tillamook Co.**—Blaine (2 UMMZ, 1 UO); Cascade Head (3 PSM); Cascade Head Experimental Forest (24 PSM); Hebo Lake (2 PSM); Nehalem (2 OSUFW); 1 mi. W Nehalem (1 OSUFW); Netarts (5 LACM, 5 MVZ, 13 OSUFW, 4 PSM, 17 SDNHM, 6 UMMZ, 6 UO, 2 USNM); Netarts Bay (2 SDNHM); Netarts Summit, Beaver Swamp (1 PSM); Oceanside (1 OSUFW); Sand Lake (2 MVZ); Tillamook (2 AMNH, 2 KU, 1 MVZ); 5 mi. S Tillamook (1 PSU). **Wasco Co.**—1 mi. N, 5 mi. W Pine Grove, T5S, R11E, Sec. 17 (9 OSUFW); 1 mi. N, 6 mi. W Pine Grove, T5S, R10E, Sec. 24 (3 OSUFW); 1 mi. N, 7 mi. W Pine Grove (1 OSUFW); 2 mi. N, 6 mi. W Pine Grove (1 OSUFW). **Washington Co.**—Cedar Mill (1 PSU); Cornell Rd., [Cedar Mill] (2 PSU); 5 mi. S, 1 mi. E Hillsboro (2 PSU); 18.5 mi. NE Portland (1 MVZ).

Additional locality: Three Sisters, Alder Spring, 4,300 ft.* (1 USNM).

Sorex vagrans, n = 1,909
(Plotted in Fig. 8-20, p. 63)

Baker Co.—Anthony (4 AMNH); 4 mi. S Baker [City] (1 EOSC); 15 mi. S, 11 mi. W Hereford, T15S, R37E, Sec. 7 (1 KU); 3 mi. E Halfway (1 MSUMZ); Oxbow (1 MSUMZ). **Benton Co.**—(1 PSM); Alsea (2 PSM); 9 mi. E Alsea (1 MSB); 6 mi. S, 6 mi. E Alsea, Williams Creek (1 OSUFW); 4.4 km N, 7.6 km W Alsea, T13S, R8W, Sec. 20 (1 USNM); 3.25 mi. S, 4.25 mi. E Alsea (2 CM); 3.75 mi. S, 5.5 mi. E Alsea (2 CM); 8 km S, 2.8 km E Alsea, T14S, R7W, Sec. 32 (1 USNM); 10 mi. SE Alsea (1 HSU); 4 mi. S Blodgett (1 OSUFW); Corvallis (1 KU, 1 MVZ, 22 OSUFW, 2 PSM, 2 UCONN); 0.25 mi. N Corvallis (1 OSUFW); 1 mi. N Corvallis (3 OSUFW); 2 mi. N Corvallis, 250 ft. (1 OSUFW, 2 PSM); 2.5 mi. N Corvallis, T11S, R4W, Sec. 3 (1 OSUFW); 5 mi. N Corvallis (1 OSUFW); 5.5 mi. N Corvallis (1 OSUFW); 7 mi. N Corvallis, 800 ft. (2 PSM); 8 mi. N Corvallis (1 PSM); 0.5 mi. S Corvallis (1 OSUFW); 1 mi. S Corvallis (1 KU); 2 mi. S Corvallis (8 PSM); 3 mi. S Corvallis (1 PSM); 10 mi. S Corvallis, W. L. Finley National Wildlife Refuge, [T13S, R5W, Sec. 21/27/28] (35 OSUFW); 12 mi. S Corvallis, 200 ft. (3 PSM); 0.5 mi. W Corvallis, T11S, R5W, Sec. 34 (1 KU); 1 mi. W Corvallis (5 OSUFW); 1.25 mi. W Corvallis (1 OSUFW); 2 mi. W Corvallis (4 OSUFW, 1 PSM); 3 mi. W Corvallis (1 OSUFW); 1 mi. N, 2 mi. W Corvallis (1 OSUFW); 1 mi. N, 3 mi. W Corvallis (1 OSUFW); 1 mi. N, 4 mi. W Corvallis (1 OSUFW); 2 mi. N, 1 mi. W Corvallis (2 OSUFW); 2 mi. N, 4 mi. W Corvallis (3 OSUFW); 2.5 mi. N, 1.5 mi. W Corvallis (1 OSUFW); 3 mi. N, 4 mi. W Corvallis, T11S, R5W, Sec. 20 (2 KU); 4 mi. N, 1 mi. W Corvallis (1 OSUFW); 6 mi. N, 3 mi. W Corvallis (1 OSUFW); 5 mi. N, 5 mi. E Corvallis, 200 ft. (3 CM); 9 mi. N, 3 mi. E Corvallis, T10S, R4W, Sec. 17 (3 KU); 8.5 mi. N, 2 mi. E Corvallis (11 OSUFW); 0.5 mi. S, 2.25 mi. W Corvallis (1 OSUFW); 0.5 mi. S, 3 mi. W Corvallis (1 OSUFW); 10.5 mi. S, 1 mi. W Corvallis (283 OSUFW); 3 mi. NW Corvallis (1 OSUFW); 5 mi. NW Corvallis (1 OSUFW); 4 mi. NE Corvallis (1 OSUFW); 4.5 mi. SW Corvallis (1 OSUFW); Greasy Creek on Beaver Creek Rd. (1 OSUFW); 2 mi. S, 1 mi. W Greenberry (1 OSUFW); 12 mi. S, 3 mi. W Harlan, T14S, R9W, Sec. 13 (1 OSUFW); Lewisburg (1 OSUFW); Missouri Bend, [T14S, R8W, Sec. 8] (1 OSUFW); 1.5 mi. E Philomath (1 OSUFW); 2 mi. S, 9 mi. W Philomath (1 OSUFW); 2 mi. S, 13 mi. W Philomath (1 OSUFW); 3 mi. S, 9 mi. W Philomath, Marys Peak, T12S, R7W, SW¼ Sec. 21 (3 OSUFW); 0.4 km S, 12.5 km W Philomath, T12S, R7W, Sec. 10 (50 USNM); 1.5 km S, 11.3 km W Philomath, T12S, R7W, Sec. 14 (1 USNM); 2.4 km S, 15.5 km W Philomath, T12S, R7W, Sec. 17/20 (8 USNM); 5 mi. SW Philomath (4 USNM); 1.25 mi. SE Summit, T11S,

R7W, SW¼ Sec. 8, 650 ft. (1 PSM); 0.5 mi. S, 3 mi. E Wrens (2 OSUFW); T10S, R4W, SE¼ Sec. 30 (1 PSM); T11S, R5W, Sec. 20 (1 OSUFW); T11S, R7W, SW¼ Sec. 8, 650 ft. (1 PSM); T12S, R5W, SE¼ Sec. 3, 200 ft. (2 PSM); T12S, R7W, NW¼ Sec. 21 (2 OSUFW). **Clackamas Co.**—Camp Millard, [T2S, R4E, Sec. 31] (1 KU); Clackamas (6 UW[EB]); Estacada, Camp Millard (1 KU); 3 mi. NE Estacada (2 OSUFW); Jennings Lodge (1 MVZ, 1 SDNHM); Milwaukie (1 PSM); Molalla (1 PSM); 8 mi. SE Mollala (6 PSM); Mulino (1 OSUFW); Oak Grove (2 PSM); Oregon City (1 USNM, 2 UW[EB], 3 UW[LB]); Oswego (1 PSM); Lake Oswego (1 OSUFW); 6 mi. S, 7 mi. E Sandy, T3S, R6E, Sec. 18 (1 CRCM); Wilhoit Springs (1 PSM); 5 mi. S, 6 mi. E Zigzag (4 CRCM). **Clatsop Co.**—site of Old Ft. Clatsop, 100 ft. (18 MVZ); Gearhart (1 PSM); Seaside (1 PSM); Welsh Island, T9N, R6W, N½ Sec. 19 (2 PSM); T8N, R8W, SW¼ Sec. 16/NE¼ Sec. 20 (1 PSM); T8N, R10W, NW¼ Sec. 6 (10 PSM); T8N, R11W, E½ Sec. 35 (5 PSM). **Columbia Co.**—2 mi. W Deer Island, 250 ft., [T5N, R2W, Sec. 12] (2 PSU); 0.5 mi. E Deer Island, 300 ft., [T5N, R1W, Sec. 8] (1 PSU); 7 mi. SE Rainier, 100 ft. (12 MVZ); Reed Island, T6N, R1W, NW¼ Sec. 18 (5 PSM); 7 mi. SE Vernonia (1 PSM); T4N, R1W, N½ Sec. 34 (8 PSM); T4N, R1W, NE¼ Sec. 34 (5 PSM); T6N, R1W, SW¼ Sec. 7 (8 PSM); T6N, R2W, E½ Sec. 13 (8 PSM); T6N, R2W, NW¼ Sec. 12 (5 PSM); T7N, R2W, SE¼ Sec. 26 (2 PSM). **Coos Co.**—7.5 mi. S Bandon, T30S, R15W, Sec. 15 (2 OSUFW); 10.5 mi. S Bandon, T30S, R15W, Sec. 10 (1 OSUFW); 1.75 mi. E Bandon (5 PSM); 2 mi. SE Bandon (1 PSM); Cape Arago State Park (4 LACM); 3 mi. S Charleston, Sunset Bay (7 LACM); 5 mi. SW Charleston, Sunset Bay (5 LACM); Coos Bay, South Bay marsh (5 LACM); Coos Bay, South Slough (6 LACM); Delmar, 9 mi. S Marshfield [= Coos Bay] (3 MVZ); Marshfield [= Coos Bay] (1 FMNH); Point Arago State Park (1 LACM); 7.3 km N, 4 km W Sitkum, T27S, R10W, Sec. 17/20 (15 USNM); 7.4 km N, 4.2 km W Sitkum, T27S, R10W, Sec. 17 (1 USNM); Sunset Bay (2 LACM); 1 mi. N Two Mile (1 PSM). **Crook Co.**—Blue Mts., T13S, R22E, SW¼ Sec. 25 (1 OSUFW); Blue Mts., T15S, R24E, SE¼ Sec. 26 (2 OSUFW); Blue Mts., T15S, R24E, NW¼NE¼ Sec. 36 (2 OSUFW); Bridge* (1 SDNHM); 13 mi. N, 2 mi. W Paulina, T14S, R23E, SE¼NW¼ Sec. 29 (1 OSUFW); 13 mi. S Prineville (3 PSM). **Curry Co.**—Brookings (1 PSM); Gold Beach (3 FMNH, 2 PSM, 2 USNM); 0.25 mi. N Harris Beach St. Park, [T40S, R14W, Sec. 35] (1 SOSC); Port Orford (1 SDNHM). **Deschutes Co.**—(12 PSM); 5 mi. W Bend, Tumalo Creek (1 SDNHM); mouth of Davis Creek, Deschutes River (9 OSUFW); Indian Ford camp, T14S, R9E, Sec. 13, 3,200 ft. (2 PSM); Lower Bridge (1 OSUFW); Paulina Lake (1 SDNHM); Pringle Falls, T21S, R9E, SE¼SE¼ Sec. 14 (4 PSM); 8.5 mi. N, 8.6 mi. W Sisters (1 OSUFW); Sparks Lake, T18S, R8E, Sec. 11 (4 OSUFW); 1 mi. N Three Creek Lake (6 PSM). **Douglas Co.**—Bradley Creek (1 SDNHM); Diamond Lake, [T27S, R6E, Sec. 32] (1 PSM, 2 SDNHM); Diamond Lake, Diamond Lake campground (3 PSU); north end Diamond Lake near outlet, [T27S, R5½E, Sec. 31] (4 OSUFW); 13 mi. N, 5 mi. E Diamond Lake, 1.5 mi. W Windigo Pass (1 PSU); Drain (1 USNM); 4 km W Drain (2 USNM); 0.8 km N, 2.2 km W Drain, T22S, R5W, Sec. 7 (2 USNM); 10 km N, 10.5 km W Drain, T21S, R6W, Sec. 7 (1 USNM); 15.7 km S, 5.6 km W Drain, T24S, R8W, Sec. 10 (1 USNM); 9.7 km S, 16.2 km W Elkton, T23S, R9W, Sec. 22 (1 USNM); Gardiner (2 MVZ); 6 mi. N Gardiner (1 MVZ); 1.3 mi. E Gardiner (5 MVZ); Louis Creek, T28S, R3W, SE¼SE¼ Sec. 30 (1 SOSC); Oxbow Burn, 300 ft., T20S, R7W, Sec. 19 (2 PSM); Reedsport (1 PSM); 2 km N, 25 km W Sutherlin, T25S, R8W, Sec. 1 (1 USNM); 3.6 km N, 11.9 km W Sutherlin, T25S, R7W, Sec. 1 (1 USNM); 4.3 km N, 12.1 km W Sutherlin, T25S, R7W, Sec. 1 (1 USNM); 10 mi. N, 9 mi. E Tiller, T29S, R1W, Sec. 12 (1 CRCM); 11 mi. N, 5 mi. E Tiller (1 CRCM); 11.5 mi. N, 5 mi. E Tiller (1 CRCM); 2.2 mi. N, 9 mi. W Umpqua P.O., T25S, R8W, Sec. 15, 850 ft. (4 UW[CR]); Winchester Bay (1 PSM); T29S, R3E, NE¼SE¼ Sec. 33 (1 SOSC); T29S, R3E, SE¼NE¼ Sec. 35 (1 SOSC). **Grant Co.**—Blue Mt. springs (1 PSM); Camp Creek (59 PSM); 2 mi. W Logdell, T16S, R30E, Sec. 5 (1 KU); 8.5 mi. W Logdell, T16S, R29E, Sec. 6 (2 KU); 6 mi. S, 2 mi. W Logdell, T16S, R30E, Sec. 32 (5 KU); 1 mi. N, 10.5 mi. E Long Creek, T10S, R32E, Sec. 5 (1 OSUFW); 3 mi. N, 6 mi. E Prairie City, T12S, R34E, NW¼ Sec. 23 (32 KU); 11 mi. S, 6 mi. E Prairie City, T15S, R36E, Sec. 25 (2 KU); 26 mi. S, 15 mi. E Prairie City, T17S, R35E, Sec. 19 (1 KU); 27 mi. S, 17 mi. E

Prairie City, T17S, R35E, Sec. 27 (3 KU); 5 mi. W Seneca, T17S, R30E, Sec. 2 (1 KU); 4 mi. N, 14.5 mi. E Seneca, T16S, R33½E, NE¼NW¼ Sec. 17 (4 KU); 6 mi. N, 14 mi. W Seneca, T17S, R29E, Sec. 10 (1 KU); 6 mi. N, 15.5 mi. E Seneca, T16S, R33½E, NW¼NW¼ Sec. 4 (4 KU); 6 mi. N, 15.5 mi. E Seneca, T16S, R33½E, NE¼NE¼ Sec. 5 (4 KU); 6 mi. N, 21 mi. W Seneca, T16S, R28E, Sec. 26 (1 KU); 15 mi. N, 15 mi. E Seneca, T14S, R28E, Sec. 15 (1 KU); 1 mi. S, 11 mi. W Seneca, T17S, R29E, Sec. 12 (1 KU); 2 mi. S, 5 mi. W Seneca, T17S, R30E, Sec. 14 (2 KU); 2 mi. S, 10 mi. W Seneca, T17S, R29E, Sec. 13 (4 KU); 2 mi. S, 12 mi. W Seneca, T17S, R29E, Sec. 11 (1 KU); 3 mi. S, 4 mi. W Seneca, T17S, R31E, Sec. 18 (3 KU); 4 mi. S, 3 mi. W Seneca, T17S, R31E, Sec. 20 (3 KU); 4 mi. S, 6 mi. W Seneca, T17S, R30E, Sec. 22 (1 KU); 6 mi. S, 10 mi. W Seneca, T17S, R30E, Sec. 31 (1 KU); 6 mi. S, 11 mi. W Seneca, T17S, R29E, Sec. 36 (2 KU); 3 mi. S, 6 mi. W Silvies, T18S, R30E, SW¼ Sec. 25 (1 KU); Strawberry Mts., 1 mi. W Wickiup campground (1 HSU); T17S, R35E, Sec. 9/10 (1 KU). **Harney Co.**—(3 EOSC); Borax Lake (3 EOSC); 20 mi. N, 32 mi. W Burns, T19S, R25E, SE¼ Sec. 26 (2 OSUFW); 26 mi. S, 8.5 mi. E Burns, Malheur National Wildlife Refuge (5 OSUFW); 28 mi. S, 4 mi. W Burns, Harney Lake dunes (1 OSUFW); 7 mi. N, 19 mi. W Drewsey, T19S, R33E, NW¼ Sec. 15 (1 KU); 6 mi. N, 0.5 mi. E Frenchglen (2 OSUFW); 14 mi. N, 3 mi. E Harney, T20S, R33E, NE¼ Sec. 9 (3 KU); Horse Pasture Field, Malheur National Wildlife Refuge, T27S, R31E, Sec. 3 (16 CM); 1 mi. N Narrows (1 OSUFW); 26 mi. N, 11 mi. W Riley, T19S, R25E, Sec. 13 (2 KU); Snow Mt. district, Ochoco National Forest (1 OSUFW); Steens Mt. (1 OSUFW); Steens Mt., Fish Lake region (1 OSUFW); Steens Mt., T32S, R32¾E, Sec. 33, 6,400 ft. (1 OSUFW); Steens Mt., T32S, R32¾E, Sec. 35 (3 OSUFW); Steens Mt., T32S, R32¾E, Sec. 36 (1 OSUFW); Steens Mt., T32S, R33E, Sec. 32 (10 OSUFW); Steens Mt., T32½S, R33E, Sec. 33, 7,400 ft. (2 OSUFW); Steens Mt., T33S, R33E, Sec. 2 (1 OSUFW); Steens Mt., T33S, R33E, Sec. 3 (1 OSUFW); Steens Mt., Little Blitzen Gorge, T33S, R33E, Sec. 10 (4 OSUFW); Steens Mt., T33S, R34E, Sec. 9, 5,200 ft. (1 OSUFW); Tule Springs* (7 EOSC); T19S, R25E, NE¼SW¼ Sec. 26 (1 KU); T33S, R32½E, NE¼ Sec. 4 (1 EOSC). **Hood River Co.**—9 mi. ENE Mt. Hood, Brooks Meadow, 4300 ft., [T2S, R10E, Sec. 11] (4 MVZ); 2 mi. W Parkdale, 1500 ft. (1 USNM); T3N, R9E, NW¼ Sec. 35 (1 PSM). **Jackson Co.**—Ashland (5 SDNHM, 5 SOSC, 1 USNM); near Ashland (1 SOSC); 1 mi. E Ashland (1 PSM); 2 mi. E Ashland (11 PSM); 4.5 mi. N, 12.8 mi. E Ashland (1 SOSC); 5 mi. N, 13 mi. E Ashland (1 SOSC); 4.5 mi. S, 2.5 mi. W Ashland (1 OSUFW); Lower Ashland Natural Research Area, T39S, R1E, SW¼ Sec. 28 (1 SOSC); Ashland, Senic Hills Memorial Park, [T39S, R1E, Sec. 11] (4 SOSC); Senic Hills cemetary (1 SOSC); Bear Creek at junction with Mt. Ave., Ashland (1 SOSC); Denman Game Management Area, [T36S, R1W, Sec. 30] (2 SOSC); 3 mi. N, 0.25 mi. W Eagle Point (1 OSUFW); Emigrant Lake, Neil Creek Rd., [T39S, R2E, SW¼] (1 SOSC); Adjacent to Howard Prairie Dam Rd. on Hyatt Lake Rd., [T39S, R3E, Sec. 2] (1 SOSC); Imnaha Creek, T33S, R4E, W½ Sec. 15 (1 KU); Medford (1 SOSC); 6 mi. S Medford (1 KU); 4 mi. E Medford (3 LACM); 11.5 mi. N, 0.5 mi. W Prospect (2 CRCM); 13.5 mi. N, 2.5 mi. W Prospect (7 CRCM); 14 mi. N, 0.5 mi. W Prospect, T30S, R2E, Sec. 24/25 (2 CRCM); 5 mi. N, 8 mi. E Prospect (2 CRCM); Spring Creek, T33S, R4E, NE¼ Sec. 22 (1 KU); T36S, R2W, NW¼ Sec. 14 (1 SOSC). **Jefferson Co.**—(4 PSM); 4 mi. NNW Culver (1 PSM); Suttle Lake (7 KU). **Josephine Co.**—T39S, R7W, SW¼ Sec. 18 (2 SOSC). **Klamath Co.**—Sphagnum Bog, Crater Lake National Park, [T30S, R5E, Sec. 4/5] (13 SOSC); panhandle Crater Lake National Park (1 SOSC); Crater Lake headquarters, [T31S, R6E] (1 OSUFW); 2 mi. S park headquarters, Crater Lake National Park, [T31S, R6E, Sec. 20] (9 SOSC); Crater Lake National Park, Castle Crest gardens (1 OSUFW); Crater Lake National Park, Munson Creek (2 SOSC, 1 CRCM); Crater Lake National Park, Upper Munson Meadow, 6,400 ft. (1 CRCM); Crater Lake National Park, Wizard Island (1 OSUFW); Ft. Klamath (8 OSUFW); 8 mi. NW Klamath (1 OSUFW); Seven Lakes Basin, North Lake, [T33S, R5E] (2 SOSC). **Lake Co.**—Gearhart Mt., Finley Corral (29 PSM); Hart Mt. National Antelope Refuge, Buck Pasture (3 KU); Hart Mt. National Antelope Refuge, 2 mi. W headquarters (2 KU); Hart Mt. National Antelope Refuge, 3.2 km W headquarters, T35S, R26E, Sec. 17 (8 KU); Hart Mt. National Antelope Refuge, 3.2 km W headquarters, Willow Creek (1

KU). **Lane Co.**—Belknap Springs (1 OSUFW); 9 mi. E Blue River (1 CRCM); 7 mi. N, 9 mi. E Blue River, T15S, R5E, Sec. 13 (6 CRCM); 5 km N, 7.6 km E Blue River, T16S, R5E, Sec. 6 (1 USNM); 6.8 km N, 7.6 km E Blue River, T15S, R5E, Sec. 32 (1 USNM); 3.5 mi. S, 10 mi. E Blue River (3 CRCM); 0.5 mi. S, 5 mi. E Cottage Grove (1 OSUFW); 15 mi. S Cottage Grove (3 PSM); 5 mi. N Cougar Dam, T16S, R5E, Sec. 6 (1 UW[OG]); 1 mi. N, 10.5 mi. E Cougar Dam, T16S, R6E, Sec. 25, 2,900 ft. (1 UW[OG]); Cushman, north bank Siuslaw River (6 CM); Eugene (2 OSUFW, 2 USNM); 9 mi. N Florence (3 LSU); 5 mi. S Florence, W side Siltcoos Lake (1 LACM); 9 mi. N, 1 mi. W Gardiner (4 PSM); Gold Lake, 4,800 ft. (2 LACM); H. J. Andrews Experimental Forest, T15S, R5E, SW¼ Sec. 13, 3,000 ft. (1 PSM); H. J. Andrews Experimental Forest, T16S, R4E, NE¼ Sec. 24, 1,400 ft. (6 PSM); Heceta, Cape Creek, [T16S, R12W, Sec. 22] (2 OSUFW); 8 mi. W Junction City (3 PSM); 10 mi. W Junction City, 400 ft. (5 PSM); 10 mi. W Junction City, 450 ft. (1 PSM); Lookout Creek (1 PSM); 1 mi. S, 4 mi. E Lowell (5 OSUFW); McKenzie Bridge (2 OSUFW, 1 USNM); 2.6 km S McKenzie Bridge, T16S, R5E, Sec. 26 (4 USNM); 3.1 km N, 3 km W McKenzie Bridge, T16S, R5E, Sec. 3 (2 USNM); 3.3 km N, 2.4 km W McKenzie Bridge, T16S, R5E, Sec. 3 (3 USNM); 6 km N, 0.6 km E McKenzie Bridge, T15S, R5E, Sec. 25 (1 USNM); 6 km N, 7.5 km E McKenzie Bridge, 1,400 ft., T15S, R6E, Sec. 26 (1 USNM); 6.4 km N, 0.6 km E McKenzie Bridge, T15S, R5E, Sec. 25 (2 USNM); 1.6 km S, 5.2 km W McKenzie Bridge, T16S, R5E, Sec. 20 (6 USNM); 2.6 km S, 1.4 km E McKenzie Bridge, T16S, R5E, Sec. 25 (1 USNM); 3.8 km S, 14.8 km E McKenzie Bridge, T16S, R7E, Sec. 33 (56 USNM); 4.4 km S, 14.3 km E McKenzie Bridge, T16S, R7E, Sec. 32 (14 USNM); 5.4 km S, 8.75 km E McKenzie Bridge, T16S, R6E, Sec. 35 (2 USNM); 2 mi. S, 0.5 mi. E McKenzie Bridge, T16S, R5E, Sec. 25, 2,800 ft. (2 UW[OG]); 3 mi. S, 8.5 mi. E McKenzie Bridge, T16S, R7E, Sec. 32 (11 UW-X); 10 mi. SE McKenzie Bridge (2 USNM); Mercer [= Florence], Dowell Ranch, [T17S, R12W, Sec. 25] (1 SDNHM); Mercer [= Florence], Grass Mt. Refuge (1 OSUFW); Mercer Lake, T17S, R12W, NE¼NW¼ Sec. 31 (1 PSM); 1 mi. N O'Leary Mt., T16S, R6E, Sec. 32 (2 UW-X); Scott Lake (1 KU); N base Three Sisters, 5000 ft. (1 USNM); N slope Three Sisters Mts., 7,000 ft. (2 SDNHM); NW slope Three Sisters (1 OSUFW); Upper Horse Lake meadow, [T18S, R7½E, Sec. 22] (1 OSUFW); Vida (1 USNM); 1.5 mi. [no direction given] Lalone Rd.* (1 CRCM). **Lincoln Co.**—Ayers Lake, T13S, R9W, SE¼ Sec. 4 (3 OSUFW); 5 mi. S Burnt Woods, 900 ft. (7 PSM); Cape Horncreek, T12S, R9W, N½ Sec. 21 (2 OSUFW); Cascade Head Experimental Forest, 240 ft. (2 PSM); 5 mi. S, 1 mi. E Elk City, Deer Creek, T12S, R10W, W½ Sec. 11 (3 KU); 1 mi. W Harlan (1 KU); 15 mi. W Harlan, Peterson Ridge, T12S, R11W, Sec. 12 (2 KU); 5 mi. S, 10 mi. W Harlan, Gold Creek, T13S, R10W, Sec. 2 (3 KU); 7 mi. S, 10 mi. W Harlan, T13S, R10W, Sec. 14 (1 KU); Newport (2 OSUFW, 1 PSM, 6 SDNHM); 3 mi. N Siletz (1 OSUFW); 6 mi. S Toledo, Horse Creek, T12S, R10W, Sec. 9/16 (5 KU); 0.6 km N, 16.9 km E Waldport, T13S, R10W, Sec. 13/24 (1 USNM); 4 km S, 3.2 km E Yachats, T15S, R12W, Sec. 12 (2 USNM). **Linn Co.**—Albany (2 OSUFW); 5 mi. S, 7 mi. W Albany (1 OSUFW); Big Lake, T13S, R7½E, SW¼ Sec. 2, 4,640 ft. (1 PSM); Big Lake (5 PSM); 1.5 mi. N Big Lake (22 PSM); 2 mi. N Big Lake (1 PSM); 8 mi. N, 8 mi. E Blue River (4 CRCM); 9 mi. N, 8 mi. E Blue River (1 CRCM); 9.5 mi. N, 9 mi. E Blue River, T15S, R5E, Sec. 2 (3 CRCM); 1 mi. N, 28 mi. E Cascadia (2 CM); 0.6 mi. S, 3 mi. E Corvallis (1 OSUFW); 2 mi. S, 5 mi. E Corvallis, Peoria Rd. (1 OSUFW); 1 mi. S, 1 mi. E Foster (1 OSUFW); 6 mi. S, 3 mi. E Halsey, T14S, R4W, Sec. 20 (6 KU); 0.5 mi. W Lebanon (1 OSUFW); 2 mi. N, 1 mi. E Lebanon (1 OSUFW); Lost Prairie, Santiam Pass, [T13S, R6E, Sec. 27] (1 OSUFW); 9.2 km N, 1.2 km W McKenzie Bridge, T15S, R5E, Sec. 14 (10 USNM); 10.25 km N, 0.8 km W McKenzie Bridge, T15S, R5E, Sec. 14 (7 USNM); 13.2 km N, 0.8 km E McKenzie Bridge, 1,020 ft., T15S, R5E, Sec. 1 (6 USNM); 0.5 mi. N, 1.5 mi. W Sweet Home (1 OSUFW); 2 mi. N, 3 mi. W Tangent (1 OSUFW); 1 mi. S, 2 mi. E Tangent (1 OSUFW); 9 mi. W Upper Soda, Tombstone Prairie (1 OSUFW); 3.3 mi. N Waterloo (1 PSM); 1 mi. S Wolf Rock, T15S, R5E, Sec. 1, 3,300 ft. (1 UW[OG], 1 UW-X); T14S, R6E, NE¼ Sec. 20, 4,400 ft. (18 PSM); T14S, R6E, NE¼ Sec. 28, 4,600 ft. (21 PSM); T15S, R6E, NW¼ Sec. 7, 5,000 ft. (37 PSM). **Malheur Co.**—Cottonwood Creek (3 PSM). **Marion Co.**—14 mi. N

Detroit at Elk Lake (1 UCD); Salem (1 MVZ, 1 OSUFW, 2 SDNHM, 4 USNM); 6 mi. S, 2 mi. W St. Paul, T5S, R3W, Sec. 13 (18 KU, 4 OSUFW, 4 TTU); Silver Creek Recreation camp (1 UW[EB]); 1 mi. N, 1 mi. E Talbot (2 OSUFW); Woodburn (6 PSM). **Multnomah Co.**—Government Island, [T1N, R2/3E] (14 PSU); McGuire Island, [T1N, R3E, Sec. 18/19] (25 SOSC); Multnomah Falls (1 PSM); Oregon shore, [T1N, R3E, Sec.28] (7 SOSC); Portland (1 AMNH, 1 KU, 2 LACM, 9 MVZ, 6 OSUFW, 5 PSM, 1 PSU, 8 SDNHM, 1 SOSC, 5 USNM); 10 mi. N, 11 mi. E Sandy (1 CRCM); 1 mi. SE St. John's Bridge, [T1N, R1W, Sec. 12] (1 PSU); Sauvie Island (1 OSUFW); 8 mi. N, 3 mi. E Zigzag (14 CRCM); T1N, R1E, S½ Sec. 20 (2 PSM); T1N, R1E, SW¼ Sec. 21 (1 PSM); T1N, R4E, NW¼ Sec. 20 (1 PSM); T3N, R1W, NW¼ Sec. 26 (1 PSM). **Polk Co.**—McCoy (1 FMNH); Independence (1 MVZ). **Tillamook Co.**—0.25 mi. W Barview (1 OSUFW); Bayocean (1 SDNHM); Beaver (1 OSUFW); Blaine (1 MVZ); Hebo Lake (7 PSM); Hebo Lake, 5 mi. E Hebo (1 UCD); 2 mi. up Miami River (2 PSM); Nehalem (1 MSUMZ, 3 OSUFW, 2 PSM); 1 mi. W Nehalem (4 OSUFW); Netarts (6 LACM, 1 MVZ, 1 OSUFW, 17 SDNHM); Oceanside (1 OSUFW); Ocean View (1 OSUFW); Tillamook (3 KU, 2 LACM, 2 OSUFW, 10 PSM); 9 mi. S Tillamook (7 PSM). **Umatilla Co.**—(1 PSM); Tollgate (1 MSUMZ); 4 mi. W Tollgate (1 UW[EB]); 4 mi. SE Weston (2 UW[WWC]); 16 mi. SE Weston (2 UW[WWC]). **Union Co.**—(2 EOSC); Anthony Lake, 7,100 ft. (2 PSM); La Grande (2 EOSC); La Grande, Grande Ronde River (1 EOSC); Mt. Emily, W of Summerville (1 EOSC); Starkey Experimental Forest (6 PSM); Starkey Experimental Forest, Bear Creek (6 PSM); 2 mi. S, 5 mi. W Starkey (1 OSUFW); Starkey Experimental Forest, Meadow Creek (1 PSM). **Wallowa Co.**—(2 EOSC); Aneroid Lake (1 SDNHM); Beaver Ridge, Wenache area (1 EOSC); 6 mi. S Lostine, T2S, R43E, Sec. 10 (2 OSUFW); 10 mi. S, 1 mi. E Lostine, T2S, R43E, Sec. 35 (3 OSUFW); 13 mi. S, 2 mi. E Lostine, T3S, R43E, Sec. 14 (1 OSUFW); 15 mi. S, 2 mi. E Lostine, T3S, R43E, Sec. 25 (3 OSUFW); 16 mi. S, 2 mi. E Lostine, T3S, R43E, Sec. 25 (2 OSUFW); 16 mi. S, 3 mi. E Lostine, T3S, R43E, Sec. 36 (1 OSUFW); 19 mi. S, 4 mi. E Lostine, T4S, R44E, Sec. 17 (1 OSUFW); 21 mi. S, 6 mi. E Lostine, T4S, R44E, Sec. 26 (2 OSUFW); 2 mi. below Minam Lake (1 SDNHM). **Wasco Co.**—1 mi. N, 5 mi. W Pine Grove (2 OSUFW); 1 mi. N, 6 mi. W Pine Grove (3 OSUFW); 1 mi. N, 8 mi. W Pine Grove (2 OSUFW); 2 mi. N, 6 mi. W Pine Grove (1 OSUFW); 2 mi. N, 7 mi. W Pine Grove (1 OSUFW). **Washington Co.**—Beaverton (1 PSM, 2 PSU); Cedar Mill (1 PSU); Hillsboro (2 MVZ); Forest Grove (1 PSU); 18.5 mi. NW Portland (1 MVZ); Scholls (4 PSU); 1.25 mi. W Sylvan (2 UW); Tigard (2 PSU). **Wheeler Co.**—Wildwood Camp, Ochoco National Forest (1 OSUFW). **Yamhill Co.**—7.5 mi. W McMinnville (1 UW[WWC]); 2 mi. S, 1 mi. W McMinnville (1 OSUFW); Sheridan (2 USNM); 0.5 mi. E Sheridan (2 OSUFW); 1.5 mi. E Yamhill (1 OSUFW).

Additional locality: Willamette National Forest* (1 OSUFW).

Spermophilus beecheyi, n = 396
(Plotted in Fig. 11-18, p. 181)

Benton Co.—(1 PSM); 3.75 mi. S, 5.25 mi. E Alsea (1 CM); Corvallis (3 OSUFW, 1 USNM); 9 mi. N Corvallis, T10S, R4W, Sec. 17 (2 OSUFW); 1 mi. S Corvallis (1 KU); 5 mi. N, 5 mi. E Corvallis, 200 ft. (4 CM); 7 mi. NE Corvallis, T11S, R4W, Sec. 3 (1 OSUFW); 8.5 mi. S Philomath (1 OSUFW); 2 mi. S, 9 mi. W Philomath, T12S, R2W, Sec. 21 (1 OSUFW); 5 mi. SW Philomath (3 USNM). **Clackamas Co.**—Aurora (2 UW[LB]); 5 mi. S Canby (1 PSU); 1 mi. N, 5 mi. E Clackamas, T2S, R3E, Sec. 6 (2 PSU); Gladstone (1 PSM); Jennings Lodge (2 KU, 3 PSM, 3 SDNHM); Mulino (7 OSUFW); Oak Grove (1 USNM); Oregon City (1 PSU, 1 USNM); Rock Creek, Clackamas River (1 PSU); 8 mi. E Scott Mills (1 OSUFW). **Clatsop Co.**—site of Old Ft. Clatsop (1 MVZ); T8N, R11W, E½ Sec. 35 (3 PSM). **Crook Co.**—4 mi. N, 4 mi. E Prineville (1 OSUFW). **Curry Co.**—5 mi. N Agness (1 HSU); 10 mi. E Brookings (2 PSM); Gold Beach (4 FMNH); Little Redwood Forest Camp (3 PSM); Long Ridge Forest Camp (2 PSM). **Douglas Co.**—(1 EOSC); 22 mi. E Drew near Abbott Butte (1 USNM); Elkhead (4 USNM); Reston (1 USNM); Roseburg (6 USNM); Scottsburg (1 USNM); Winston, Wildlife Safari (4 CRCM); Smith River Rd. 0.25 mi. from HWY 101 junction (1

PSU). **Hood River Co.**—(1 UW); Cascade Locks, HWY 30 (4 PSU); Hood River (1 PSM, 4 USNM); Mt. Hood, 2,800 ft. (1 USNM); Parkdale (1 UMMZ); White River, T4S, R11E, Sec. 26 (1 PSU); T2S, R10E, Sec. 14 (2 PSU); T3N, R9E, SW¼ Sec. 35 (1 PSM). **Jackson Co.**—Ashland (3 SOSC); 0.5 mi. N Ashland (1 SOSC); 1 mi. N Ashland (1 SOSC); 1.5 mi. N Ashland (1 SOSC); 4 mi. S Ashland (1 SOSC); 7 mi. S Ashland (1 PSU); 14 mi. S Ashland (1 MVZ); 17 km E Ashland (1 SOSC); 3 mi. NE Ashland, Emigrant Creek (1 SOSC); Brownsboro (1 SOSC); Emigrant Lake (1 SOSC); NW shore Emigrant Lake (1 SOSC); 1 mi. off Fern Valley Rd. at Hilldale Orchard Rd. (2 SOSC); Greensprings Rd. (1 SOSC); Hooper Springs, Dead Indian Rd. (1 SOSC); Medford (1 SOSC); 6 mi. S Medford (2 MVZ); Prospect (7 USNM); 5 mi. N Rogue River (2 CRCM); 2.8 mi. from Shoat Springs on Copco Rd. (1 SOSC); Siskiyou (1 USNM); 7.5 mi. S Hopkins Ranch on Copco Rd. (1 SOSC); 90 yds. S Copco Rd. and Agate Flat Rd. (1 SOSC); Copco Rd., 75 yds. S of bridge at Fall Creek (1 SOSC); 0.7 mi. N Dead Indian Rd. off Conde Rd. (1 SOSC); 1 mi. off Fern Valley Rd. at Hilldale Orchard Rd. (2 SOSC); 30 yds. from Fern Valley Rd. (1 SOSC); 0.5 mi. E Mountain Ave. on Pilot View Rd. (1 SOSC). **Jefferson Co.**—Warm Springs (2 USNM). **Josephine Co.**—Bolan Lake (2 PSM); 13 mi. SW Galice, Briggs and Horse creeks, Farren Ranger Station (1 USNM); Grants Pass (20 CM, 12 USNM); near Grants Pass, Rogue River Valley (4 USNM); 13 mi. SW Grants Pass (4 UO). **Klamath Co.**—Bonanza (1 PSM); S boundary Crater Lake National Park (1 MVZ); 3 mi. N Ft. Klamath (1 LACM); 10 mi. SW Ft. Klamath (1 UMMZ); 9.1 mi. N, 3.4 mi. E Keno (1 PSU); Klamath Falls (3 CM); 1 mi. S, 8 mi. W Klamath Falls (1 OSUFW); 8 mi. NW Klamath Falls, Geary Ranch (1 PSU); Klamath Lake* (3 USNM); Naylox, Klamath Lake (2 USNM); Upper Klamath Lake* (1 FMNH). **Lake Co.**—Lakeview (1 SOSC); 30 mi. SW Lakeview, Dog Lake (1 USNM). **Lane Co.**—Belknap Springs (1 PSU); 1 mi. E mouth Big Creek (1 KU); 2.5 mi. E mouth Big Creek (1 KU); Eugene (1 CRCM, 3 USNM); W of Eugene (3 UO); 2 mi. N Elmira (1 UO); 0.5 mi. E Franklin (1 OSUFW); Hand Lake, Willamette National Forest, T15S, R7½E, NE¼ Sec. 34 (1 OSUFW); 1 mi. N Junction City (1 OSUFW); Mapleton (1 USNM); 10 mi. W McKenzie Bridge (5 USNM); 3 mi. SE Oakridge (6 UO); Roosevelt Beach (1 UMMZ); 1.66 mi. S, 6.125 mi. W Walton (1 PSU). **Lincoln Co.**—3.5 mi. S, 2 mi. E Elk City, T11S, R9W, Sec. 36 (2 KU); 1 mi. N, 0.5 mi. W Nortons, T10S, R8W, Sec. 29 (2 KU). **Linn Co.**—1 mi. S Albany (1 OSUFW); 1 mi. N, 2 mi. W Crabtree (1 OSUFW); Hand Lake, Willamette National Forest (1 OSUFW); 3 mi. NW Lebanon (3 UW). **Marion Co.**—Salem (6 AMNH, 3 OSUFW, 1 PSM, 2 UMMZ); 2 mi. S, 1 mi. E Salem (1 OSUFW); 9 mi. S Silverton, T8S, R1W, Sec. 23 (1 OSUFW); 6 mi. S, 5 mi. E Silverton (1 OSUFW). **Multnomah Co.**—Corbett (1 PSU); 2.5 mi. SE Gresham (2 PSU); Linton (1 PSU); Portland (3 AMNH, 2 LACM, 1 PSM, 3 PSU, 3 USNM); 10 mi. E Portland (1 PSU); Sauvies Island (1 SDNHM); Sellwood [=Portland] (1 USNM); Pier Park, Portland (1 PSU); St. Johns (1 PSM). **Polk Co.**—Dallas (1 USNM); 5 mi. N, 3 mi. W Hoskins (1 OSUFW); McCoy (4 USNM). **Sherman Co.**—Miller (3 USNM); Miller, mouth Deschutes River (4 OSUFW); 1 mi. W Rufus (1 OSUFW). **Tillamook Co.**—(1 KU); Beaver (1 OSUFW); Blaine (1 OSUFW, 8 UMMZ); Mohler, Cook Creek (1 OSUFW); 10 mi. N Nehalem (1 PSM); Neskowin (1 MVZ); 2 mi. S Oretown (1 KU); Tillamook (1 AMNH, 1 KU, 1 OSUFW, 1 UMMZ); 7 mi. SE Tillamook (2 OSUFW). **Wasco Co.**—1 mi. SW Celilo (1 MVZ); mouth of Deschutes River, Columbia River (3 MVZ); 5 mi. W Dufur (2 PSU); 6.2 mi. W Dufur (1 PSU); Ft. Dalles (2 USNM); Moody (1 OSUFW); 8 mi. S Pine Grove (1 PSU); 3 mi. W Rowena (1 UW); The Dalles (1 OSUFW, 1 SDNHM, 20 USNM); 7 mi. W The Dalles (1 PSU); 12 mi. W The Dalles (4 PSU); 8 mi. S, 4 mi. W The Dalles (1 OSUFW); 8.5 mi. S, 5 mi. W The Dalles (1 OSUFW); 4 mi. S, 7 mi. E The Dalles (1 PSU); 14 mi. SW The Dalles, Fivemile Creek (1 OSUFW); 19 mi. SW The Dalles (1 PSU); Maupin (1 SDNHM, 8 USNM); Wamic (3 USNM); Wapinitia (1 USNM); Warm Springs River (2 USNM). **Washington Co.**—Beaverton (1 USNM); Forest Grove (12 USNM); Hillsboro (1 PSM); 11 mi. NW Linton (1 MVZ); 18.5 mi. NW Portland (4 MVZ); NW Miller Rd. [Portland] (1 PSU). **Yamhill Co.**—(1 PSU, 1 UW); Yamhill (2 PSU).

    Additional localities: Wetherbee Rd. 0.5 mi. from fish hatchery Rd.* (1 SOSC); H. J. Andrews Experimental Forest* (3 PSM).

Spermophilus beldingi, n = 598
(Plotted in Fig. 11-20, p. 185)

    Baker Co.—(1 EOSC); Baker (6 UMMZ); 10 mi. N Baker (3 UMMZ); 20 mi. NE Baker [City] (2 KU); 9.5 mi. SE Baker [City] (1 KU); Burkemont mine (4 AMNH); Forshey Meadow [T7S, R44E, Sec.31] (1 EOSC); Haines (1 USNM, 1 UW); Home, 1,850 ft. (10 USNM); Keating, Balm Creek (1 OSUFW); 4 mi. after Keating* (1 EOSC); 3 mi. N Pine Creek (1 CRCM); Sparta (2 AMNH, 2 SDNHM); Sumpter, Black Mt. (1 OSUFW); along road between Sparta turnoff and Forshey Meadow (1 EOSC). **Crook Co.**—2 mi. SW Barnes (3 MVZ); 3 mi. E mouth Bear River, Crooked River (2 MVZ); 40–50 mi. E Bend (1 OSUFW); Buck Creek (1 USNM); Howard (4 USNM); N slope Maury Mts. (1 USNM); Ochoco R.S., 4,000 ft. (3 MVZ); Paulina Butte lookout, 60 mi. NE Prineville (1 PSU); Prineville (7 mi. W Prineville (1 MVZ); 15 mi. N, 16 mi. E Prineville (1 OSUFW); 3 mi. N, 6 mi. E Wildcat Mt. (1 MVZ). **Deschutes Co.**—2 mi. S, 20 mi. W Bend, near Sparks Lake, 5,600 ft. (1 UW); Broken Top, 20 mi. W Bend, 8,000 ft. (2 MVZ); Davis Lake (1 UMMZ); Lower Bridge (1 OSUFW); Redmond (1 SDNHM); Sisters (1 OSUFW); 7 mi. S Sisters (1 MVZ); Sunriver (3 OSUFW); Swampy Lakes, 6,500 ft. (1 MVZ); 5 mi. N, 1 mi. W Terrebonne (2 OSUFW); 1 mi. N, 2 mi. E Terrebonne (3 OSUFW); 1 mi. N Three Creek Lake (2 PSM); Tumalo Creek, 16 mi. W Bend, 6,500 ft. (1 MVZ). **Douglas Co.**—Diamond Lake (5 PSM); S end Diamond Lake (1 SOSC); Diamond Lake, Diamond Lake campground, 5,183 ft. (1 PSU). **Gilliam Co.**—6 mi. E Lonerock (1 USNM). **Grant Co.**—Austin (2 USNM); Beech Creek (2 USNM); Camp Creek, Aldrich Mts., 4,900 ft. (1 CRCM); 12 mi. S, 5 mi. E Canyon City, Windy Point Rd. (2 PSU); Cold Spring, 8 mi. E Austin (7 MVZ); 10 mi. N Dayville (1 SDNHM); 3 mi. S, 4 mi. E Izee (7 OSUFW); 14 mi. SE Izee (1 OSUFW); John Day (1 OSUFW); 10 mi. S John Day (1 UW[EB]); Kimberly (1 OSUFW); Long Creek, Morgrass-Pass Creek (1 OSUFW); 30 mi. N Long Creek (2 UW[EB]); 4 km E Monument (9 UW[WWC]); 6 mi. W Monument (4 OSUFW); 3 mi. S, 4 mi. E Monument (1 OSUFW); 6 mi. S, 2 mi. E Monument (1 OSUFW); 6 mi. S, 3 mi. E Monument (1 OSUFW); Mt. Vernon (2 USNM); Mt. Vernon, 3.5 mi. S Ingle Butte Ranch (5 KU); 3.5 mi. S Mt. Vernon (1 KU); 21 mi. SE Prairie City, North fork Malheur River (1 MVZ); 1 mi. N Seneca, 4,700 ft. (22 CRCM); 3 mi. S Seneca (2 UW[WWC]); meadow SE junction FS1409 and 1409C, Aldrich Mts., 4,900 ft. (4 CRCM); T15S, R30E, Sec. 23 (1 LSU). **Harney Co.**—Blitzen Valley (3 SDNHM); Buchanan (6 USNM); Burns (2 KU, 2 OSUFW, 2 USNM); 1 mi. N Burns (1 OSUFW); 1 mi. S Burns (1 OSUFW); 26 mi. E Burns (1 MVZ); 13 mi. N, 3 mi. E Burns (1 OSUFW); 20 mi. S, 7 mi. E Burns, Malheur Wildlife Refuge headquarters (2 OSUFW); 26 mi. S, 6 mi. E Burns (1 OSUFW); 28 mi. NW Burns, Emigrant Creek (1 OSUFW, 1 PSU); Crane (1 USNM); 4.75 mi. N, 3 mi. E Diamond (1 PSU); Diamond, head of Cocoamonga Creek, 6,300 ft. (5 USNM); Drewsey (1 USNM); 2–3 mi. NW Fish Lake, 7,100 ft. (1 UW[WWC]); 0.6 mi. S, 18 mi. E Frenchglen, S side Fish Lake, Steens Mt. (1 OSUFW); 6 mi. S, 14 mi. E Frenchglen (1 OSUFW); 10 mi. N Harney (4 USNM); Home Creek (1 SDNHM); Idlewild Forest campground (1 OSUFW); Malheur Environmental Field Station (1 SOSC); Malheur Lake (2 UW[EB]); Malheur National Wildlife Refuge, T25S, R31E, Sec. 16 (1 OSUFW); Malheur National Wildlife Refuge headquarters, T26S, R31E, Sec. 35 (4 CM, 1 OSUFW, 1 CRCM); Narrows (5 USNM); 16.5 mi. SE New Cuyama* (1 KU); 25 mi. SW Sageview (1 SDNHM); Shirk (1 SDNHM); Silvies Valley (1 SDNHM); Steens Mt. (2 OSUFW, 3 UMMZ, 2 USNM); N end Steens Mt. (2 FMNH); Steens Mt., Big Fish Lake, 8,000 ft. (1 USNM); Steens Mt., Kiger Gorge, 6,000 ft. (9 USNM); Steens Mt., Wild Horse Canyon (1 USNM); Steens Mt., T32S, R32¾E, Sec. 36 (2 OSUFW); Steens Mt., T33S, R33E, Sec. 2 (1 OSUFW); Voltage (1 SDNHM); White Horse Creek (1 USNM). **Jackson Co.**—Fish Lake (5 PSU); 14 mi. SE Pilot Rock (1 SDNHM). **Jefferson Co.**—5 mi. E Culver (1 PSM); Hay Creek (3 OSUFW, 10 USNM); 12 mi. E Hay Creek at Foley Creek (3 USNM); Haystack Reservoir (1 PSM); 10 mi. S Madras (3 PSU); T12S, R10E, Sec. 13 (2 PSM). **Klamath Co.**—(3 LACM, 6 PSM); Aspen Lake (1 OSUFW); Bly, Sprague Forest Camp (10 PSM); 9.5 mi. NW Bonanza (3 PSM); 0.75 mi. SE Bonanza (5 PSM); 1 mi. N, 8 mi. E Chinchalo,

Klamath Forest (1 OSUFW); 6 mi. E Cresent Lake, 4,500 ft. (8 UO); 7 mi. SE Cresent Lake, 5,000 ft. (1 UO); Gerber Reservoir (2 SOSC); Ft. Klamath (1 OSUFW, 33 USNM); Klamath Falls (3 CM, 2 SDNHM, 4 USNM); near Klamath Falls, Spring Lake Rd. (1 OSUFW); 7.5 mi. S Klamath Falls (1 PSM); 9 mi. S Klamath Falls (1 SOSC); 7 mi. S, 0.25 mi. W Klamath Falls (1 SOSC); 7.5 mi. S, 0.25 mi. W Klamath Falls (1 SOSC); 8 mi. S, 0.25 mi. W Klamath Falls (6 SOSC); Lost River, Klamath Basin* (1 USNM); Merrill (1 SDNHM); 5 mi. NE Merrill (1 SOSC); 3 mi. W Swan Lake, 4,300 ft. (40 MVZ); Swan Lake Valley, Klamath Basin (2 USNM); Upper Klamath Lake (6 FMNH); 0.5 mi. from HWY 66 on road to Keno Recreation Area (1 SOSC); T33S, R14E, Sec. 13 (1 PSM). **Lake Co.**—14 mi. W Adel, Blue Creek, Warner Mts. (1 OSUFW); Fort Rock (1 OSUFW); Gearhart Mt., Camp Creek (1 PSM); Hart Mt. (3 PSM); Lakeview (4 CM, 1 USNM); E of Lakeview, Camas Prairie (1 CM); 10 mi. NE Lakeview, Camas Valley, T38S, R20E, Sec. 32 (1 PSU); Mt. Warner (2 USNM); 10 mi. S Plush (1 CRCM); 5 mi. E Quartz Pass, 5,200 ft. (1 MVZ); 10 mi. S Summer Lake (1 OSUFW). **Malheur Co.**—5 mi. W Arock (5 OSUFW); Battle Mt. (1 SDNHM); Beulah (14 USNM); 10 mi. N Beulah (1 UW[WWC]); Cedar Mts. (1 USNM); Cow Lake (1 SDNHM); Disaster Peak, 7,800 ft. (1 USNM); Ironside, 4,000 ft. (10 AMNH); 7 mi. from mouth of Jordan Creek (1 USNM); Jordan Valley, 4,200 ft. (4 USNM); 2 mi. W Jordan Valley (4 MVZ); Mahogany Mts., 6,500 ft. (3 USNM); near McDermitt (2 USNM); canyon located 8 mi. NE McDermitt, 8,000 ft. (1 USNM); Ontario (1 SDNHM); Po Valley on Lost River* (1 USNM); head of Rattlesnake Creek (1 USNM); Rock Reservoir* (3 PSM); Rockville (1 USNM); 25 mi. W Rome on Bone Springs Rd., T31S, R37E, Sec. 24 (1 PSU); Soldier Creek ranch, near Stone House (1 PSU); 3.5 mi. S Vale (1 PSM); Westfall (1 SDNHM). **Morrow Co.**—Heppner (1 SDNHM, 11 USNM, 3 UW[WWC]); 16 mi. E Heppner (6 USNM); 3 mi. S, 3 mi. E Heppner (3 OSUFW); 2 mi. SE Heppner (4 UW[WWC]). **Umatilla Co.**—10 mi. S Meacham (1 UMMZ); 4.2 mi. S, 7.5 mi. W Pilot Rock, 2,900 ft. (1 CRCM); 10.8 mi. S, 8 mi. W Pilot Rock, 3,400 ft. (2 CRCM). **Union Co.**—Alicel (1 EOSC); Hot Lake (2 EOSC); Mt. Harris Rd. out of Imbler (1 EOSC); La Grande (4 EOSC, 1 SOSC); 1.5 mi. E La Grande (1 EOSC); 8 mi. NW La Grande (1 UW[EB]); Ladd marsh (2 EOSC); North Powder (1 UW[EB]); 2 mi. N, 3 mi. E Starkey (1 OSUFW); Telocaset (2 USNM); Thief Valley Reservoir (2 EOSC); 3 mi. N Union along HWY (1 EOSC); Chuncha Ranch, 14 mi. up Grande Ronde River from La Grande (1 EOSC); Foot Hill Rd. (3 EOSC). **Wallowa Co.**—Enterprise (1 SDNHM); 1 mi. N, 8 mi. E Enterprise (1 PSM); Joseph (4 UMMZ, 4 USNM); Zumwalt (1 OSUFW). **Wasco Co.**—1.7 mi. S, 3.7 mi. E Antelope, 3,400 ft. (4 CRCM); 6 mi. SE Maupin, T5S, R14E, Sec. 22 (1 OSUFW); Shaniko (3 UW). **Wheeler Co.**—Bridge Creek (1 MVZ); 7 mi. S, 11 mi. W Mitchell (2 MVZ); Shelton State Park, SE of Fossil (1 PSU, 1 UW[WWC]); Twickenham (2 USNM); 1 mi. S, 5 mi. W Winlock (1 PSU).

Spermophilus canus, n = 286
(Plotted in Fig. 11-28, p. 197)

Baker Co.—(1 AMNH); 9 mi. N Baker (1 CM); 10 mi. N Baker (3 UMMZ); 3 mi. NE Huntington (3 USNM). **Crook Co.**—2 mi. SW Barnes (1 MVZ); Prineville (8 USNM). **Deschutes Co.**—Bend, W bank Deschutes River (11 USNM); Redmond (1 USNM); 1 mi. S, 4 mi. W Terrebonne (1 OSUFW). **Harney Co.**—16 mi. E Alvord Lake, Whitehorse Sink (2 USNM); Burns (2 USNM); 23 mi. S, 9.8 mi. W Burns, NW Harney Lake (2 OSUFW); Crane (5 USNM); Drewsey (2 USNM); Fish Lake, Steens Mt. (1 UO); Malheur National Wildlife Refuge, near headquarters, T26S, R31E, Sec. 35 (1 OSUFW); Narrows (18 USNM); Rock Creek (4 USNM). **Jefferson Co.**—5 mi. N Culver (1 PSM); 6 mi. S, 4 mi. E Culver (2 OSUFW); Gateway (1 OSUFW, 2 PSM, 10 USNM); Hay Creek (20 USNM); 2.5 mi. N Madras (2 OSUFW); 3 mi. S Madras, T11S, R13E, SW¼ Sec. 36 (1 OSUFW); Warm Springs (3 USNM). **Lake Co.**—10 mi. N Christmas Lake (1 USNM); 4 mi. S, 26 mi. E Christmas Valley, T27S, R21E, Sec. 31 (1 OSUFW); Fort Rock (2 SDNHM); 20 mi. E Fort Rock (1 SDNHM); Fremont (1 USNM); 4 mi. E Hart Mt. Antelope Refuge headquarters on Frenchglen Rd. (1 SOSC); 6 mi. E Hart Mt. Antelope Refuge headquarters on Frenchglen Rd. (2 SOSC); Plush (1 USNM); Summer Lake (2 USNM). **Malheur Co.**—(6 AMNH, 2 EOSC, 1 SOSC); Cedar Mts. (1 USNM); Ironside,

4,000 ft. (5 AMNH); 1 mi. N Jamieson, T15S, R43E, SE¼SE¼ Sec. 3, 2,600 ft. (2 PSM); 30 mi. S Jordan Valley, T35S, R46E, Sec. 10 (1 PSU); 5 mi. W Jordan Valley (1 PSM); Ontario (16 KU, 1 SDNHM); Ontario, W bank Snake River (5 MVZ, 1 OSUFW); 0.25 mi. N Ontario (3 MVZ); 5 mi. SW Ontario (23 USNM); 2 mi. NW Riverside (1 USNM); 41 mi. S, 37 mi. W Rome, T36S, R37E, Sec. 23 (1 OSUFW); Vale (40 KU, 12 USNM); 2 mi. N Vale (1 PSM); 2.5 mi. N Vale (2 MVZ); 1.5 mi. S Vale (13 USNM); 3.5 mi. S Vale (3 PSM); 4.5 mi. S Vale (2 PSM); 5.75 mi. S Vale (1 PSM); 7 mi. S Vale (1 PSM); 1 mi. SE Vale (3 PSU); 6.5 mi. SSE Vale (1 PSM); 7 mi. SSE Vale (1 PSM); Westfall (1 SDNHM, 1 USNM). **Union Co.**—(1 EOSC); 2 mi. N, 2 mi. W North Powder (2 CM). **Wasco Co.**—7 mi. E Antelope (7 USNM); 1 mi. S, 1 mi. E Maupin (1 PSU); Shaniko (1 SDNHM). **Wheeler Co.**—Squaw Butte (1 USNM).

Spermophilus columbianus, n = 198
(Plotted in Fig. 11-22, p. 188)

Baker Co.—SW slope Antelope Hill, 10 mi. W Hanes (1 EOSC); Anthony (16 AMNH, 1 SDNHM); Baker (1 UMMZ); Baker, Pine Creek, Old Elkhorn mine (1 OSUFW); Bourne (3 USNM); Cornucopia (4 PSM, 11 UMMZ, 2 USNM); 3 mi. N Cornucopia (1 UMMZ); Eagle Creek area (7 KU); 10 mi. NW Unity, South Sister Creek, 5,000 ft. (1 MVZ); 8 mi. NE Unity (1 PSM). **Grant Co.**—Austin (2 USNM); 12 mi. S Canyon City, 5,500 ft. (3 MVZ); head Canyon Creek (1 PSM); Cold Spring, 8 mi. E Austin, 4,900 ft. (17 MVZ); Dixie Butte (6 USNM); Dixie Forest Service Camp, Hwy 20 (2 PSU); 6 mi. S, 9 mi. W Izee (1 OSUFW); 16 mi. S, 17 mi. E John Day (1 USNM); 12.5 mi. S, 2 mi. E Mt. Vernon, Bear Valley, T15S, R30E, Sec. 29 (1 KU); 21 mi. SE Prairie City, North Fork Malheur River (6 MVZ); Stark Spring, 11 mi. S, 3 mi. W John Day, 6,100 ft. (1 CRCM); Strawberry Butte (2 USNM); Strawberry Mts. (1 USNM); 7 mi. S, 5 mi. E Suplee, Harris Ranch (1 OSUFW); 0.3 mi. S, Windfall Spring, 5,200 ft. (1 CRCM); 1.5 mi. S Windfall Spring, 5,000 ft. (3 CRCM); meadow SE junction FS1542 and FS1409C, Aldrich Mts., 4,850 ft. (3 CRCM). **Harney Co.**—10 mi. N Harney (4 USNM); 10 mi. S, 6 mi. E Suplee (2 OSUFW). **Malheur Co.**—Ironside, 4,500 ft. (3 AMNH). **Morrow Co.**—25 mi. S Heppner (1 USNM); 20 mi. SSE Heppner (1 USNM). **Umatilla Co.**—Adams (2 PSU); 5 mi. NW Albee (1 MVZ); Battle Mt. State Park, 13.6 mi. S, 6.3 mi. W Pilot Rock, 4,000 ft. (1 CRCM); Cayuse (1 SDNHM); Meacham (1 UMMZ); 1.5 mi. N Milton (1 KU); Spofford (3 UW[WWC]); 4 mi. W Tollgate (1 FMNH); Ukiah (1 SDNHM); 8 mi. SE Walla Walla, Cottonwood Canyon (3 FMNH, 2 UW). **Union Co.**—Elgin (5 USNM); 6 mi. N Elgin (1 EOSC); 9.9 mi. N, 5.5 mi. W Elgin, T3N, R38E, Sec. 26 (1 PSU); 13.3 mi. N, 7.2 mi. W Elgin, T3N, R38E, Sec. 9 (1 PSU); Glass Hill Rd. (1 EOSC); Grande Ronde Lake, 7,100 ft. (2 MVZ); Kamela, 20 mi. From La Grande (1 OSUFW); La Grande (1 EOSC); La Grande, Bowman Hicks Mill (1 EOSC); 1.5 mi. SE Ladd Canyon, T4S, R39E, SE¼ Sec. 24 (2 PSM); Meadow Creek, T3S, R34E, NW¼NE¼ Sec. 35 (1 PSM); Morgan Lake watertower (2 EOSC); Mt. Harris, T1S, R39E, Sec. 35 (2 EOSC, 1 OSUFW); Pelican Creek, N of Hilgard (1 EOSC); Starkey (1 EOSC); 2 mi. N, 5 mi. W Starkey (1 OSUFW); Summerville (1 EOSC); Summerville, Pumpkin Ridge (1 EOSC); 1 mi. E Red Bridge Park, between HWY and river (1 EOSC); T6S, R41E, Sec. 2 (1 EOSC). **Wallowa Co.**—(1 EOSC); Aneroid Lake, T4S, R45E (1 OSUFW, 1 UO); Davis Creek, T3N, R45E (1 OSUFW); Eagle Cap Wilderness Area (1 SOSC); 25 mi. N Enterprise at Sled Springs, 4,600 ft. (2 USNM); 0.4 mi. NW Jewett Lake (4 CRCM); 0.6 mi. ENE Jewett Lake (2 CRCM); Joseph (2 UMMZ); Little Sheep Creek (2 USNM); 17 mi. S Lostine, 5,300 ft. (1 UO); 10 mi. S, 1 mi. E Lostine, T2S, R43E, Sec. 35 (1 OSUFW); 16 mi. S, 3 mi. E Lostine (3 MVZ); 17 mi. SE Lostine, Lostine River, Wallowa Mts. (3 OSUFW); Minam Lake (1 SDNHM); NNE face Petes Point, 8,450 ft. (1 CRCM); Wallowa (1 USNM); Wallowa Lake, Wallowa Mts., 4,000 ft. (8 USNM); S end Wallowa Lake (3 UO).

Spermophilus elegans, n = 7
(Plotted in Fig. 11-24, p. 191)

Malheur Co.—near McDermitt (4 USNM); head Quinn River (3 USNM).

Spermophilus lateralis, n = 752
(Plotted in Fig. 11-26, p. 194)

Baker Co.—Anthony (16 AMNH, 1 SDNHM); Cornucopia (1 USNM); 2 mi. SW Eagle Creek, FSRD S679 (1 EOSC); Home, 1,850 ft. (2 USNM); Homestead, 1,800 ft. (7 USNM); Huntington (2 USNM); 3 mi. NE Huntington, 2,100 ft. (3 USNM); Lily White (1 EOSC); Rock Creek (1 USNM); 8 mi. NE Unity (2 PSM); 0.5 mi. N Upper Twin Lake, Elkhorn Mts., 8,500 ft. (1 CRCM); between Lily White and Sparta (1 EOSC); 0.25 [no label] NW Wallowa Mt. field station* (1 EOSC). **Crook Co.**—Canyon Creek, Ochoco National Forest (1 SOSC); Howard (3 USNM); NW of Howard, Ochoco National Forest (2 USNM); 8 mi. N, 35 mi. E Prineville, T13S, R21E, NE¼ Sec. 25 (1 KU); 18 mi. N, 20 mi. E Prineville (1 OSUFW); 20 mi. SE Prineville, Crooked River (1 USNM); 2 mi. E Steins Piller (1 OSUFW); T15S, R25E, NW¼ Sec. 17 (1 SOSC). **Deschutes Co.**—(1 PSM); Arnold Ice Cave (1 USNM); Bend (3 SDNHM, 1 UO, 7 USNM); Bend, Tumalo Creek (2 OSUFW); 3 mi. W Bend (4 KU); 16 mi. W Bend (2 MVZ); 2 mi. N, 4 mi. W Bend, T17S, R11E, Sec. 15 (1 OSUFW); 18 mi. S, 12 mi. W Bend, Fall River hatchery (1 OSUFW); 3 mi. S, 1 mi. E Bend (1 OSUFW); 14 mi. S, 22 mi. E Bend (1 OSUFW); Black Butte, T14S, R9E, Sec. 3 (25 different locality descriptors) (4 CM, 2 HSU, 111 OSUFW); Cold Spring (8 KU); 13 mi. N, 18 mi. W La Pine (1 OSUFW); 3 mi. N, 3 mi. E La Pine (1 OSUFW); 10 mi. N, 3 mi. E La Pine (1 OSUFW); McKenzie Pass summit (1 OSUFW); Paulina Lake (1 OSUFW, 2 SDNHM, 4 USNM); Sisters (3 OSUFW, 7 UO, 4 USNM); Sisters, Indian Ford (1 OSUFW); 2 mi. S Sisters (1 OSUFW); 4.5 mi. W Sisters, Cold Springs campground (2 HSU); 14 mi. E Sisters (1 KU); 5 mi. N, 3 mi. E Sisters (1 OSUFW); 1 mi. S, 6 mi. W Sisters (1 OSUFW); 15 mi. S, 8 mi. W Sisters, Little Three Creek Lake (1 OSUFW); E side South Sister, 7,600 ft. (4 CRCM); Sparks Lake (1 OSUFW); Swampy Lake, 6,500 ft. (1 MVZ). **Douglas Co.**—Diamond Lake (5 USNM); 6 mi. N Diamond Lake (1 SOSC); 12 mi. NW Diamond Lake (1 OSUFW). **Gilliam Co.**—Lonerock (10 USNM); Willows Junction (1 USNM). **Grant Co.**—Austin (1 USNM); Beech Creek (1 USNM); Canyon City (1 SDNHM); Dale (2 USNM); 10 mi. N Dayville (2 UW[WWC]); Elk Flat, Malheur National Forest* (1 KU); 1 mi. N Indian Springs campground, Malheur National Forest, 7,600 ft. (1 CRCM); 8 mi. S John Day (1 UW[EB]); 10 mi. S John Day (4 UW[EB]); 20 mi. S, 11 mi. E John Day, T17S, R33E, NW¼ Sec. 2 (1 KU); 1 mi. N, 10 mi. E Long Creek, Kahler Butte, T10S, R32E, Sec. 5 (1 OSUFW); Malheur National Forest, T18S, R32E, Sec. 23 (1 OSUFW); North Fork Malheur River camp, Malheur National Forest (1 KU); 8 mi. S, 1.5 mi. E Prairie City, Malheur National Forest (1 OSUFW); 10 mi. S, 2 mi. E Prairie City, Strawberry Mt. campground (1 OSUFW); Strawberry Butte, 7,000 ft. (3 USNM); Strawberry Mts. (7 USNM); N slope Strawberry Mt., 8,200 ft. (2 CRCM); N slope Strawberry Mt., 8,300 ft. (1 CRCM); N slope Strawberry Mt., 8,500 ft. (1 CRCM); N slope Strawberry Mt., 8,700 ft. (1 CRCM); E slope Strawberry Mt., 8,600 ft. (2 CRCM); Strawberry Mts., Wickiup campground (1 HSU). **Harney Co.**—Anderson Valley, T27S, R34E, S½ (1 OSUFW); Buchanan (7 USNM); Burns (2 USNM); 23 mi. S Burns, T27S, R31E, Sec. 35 (1 OSUFW); 28 mi. NE Burns (1 OSUFW); Drewsey (3 USNM); 45 mi. N, 17 mi. E Fields, T30/31S, R35E, Sec. 35/2 (1 KU); 8.6 mi. SE Fish Lake, 9,100 ft. (1 UW[WWC]); 9 mi. SE Fish Lake, 9,100 ft. (1 UW[WWC]); 3 mi. S, 6 mi. E Frenchglen, 5,200 ft. (1 UW[WWC]); Harney (2 USNM); Hunter Spring, Malheur National Forest (1 KU); Idlewild Forest (1 OSUFW); Pike Creek, 4,100 ft. (1 EOSC); Shirk P.O. (2 USNM); Steens Mt., Barren Valley (1 USNM); Steens Mt., Kiger Gorge, 6,000 ft. (2 USNM); Steens Mt., Wildhorse Lake Cirque (2 PSM); Steens Mt., T32S, R32½E, Sec. 25 (2 OSUFW); Steens Mt., T32S, R33E, Sec. 27 (1 OSUFW); Steens Mt., T33S, R33E, Sec. 2 (1 OSUFW); T33S, R32¾E, central Sec. 7 (1 EOSC). **Hood River Co.**—Cloud Cap Inn, Mt. Hood, 6,800 ft. (4 USNM); Cooper Spur, NE side Mt. Hood, 6,600 ft. (6 CRCM); 3 mi. S, 15 mi. W Friend (1 OSUFW); Lamberson Spur, 7,200 ft. (1 CRCM); W base Mt. Hood near Tollgate, 1,500 ft. (1 USNM); W slope Mt. Hood near timberline (2 USNM); Mt. Hood, above timberline, 6,800 ft. (1 USNM); 15 mi. S Mt. Hood, Sahalie Falls (1 KU); Tilly Jane Creek, 6,450 ft. (1 CRCM). **Jackson Co.**—Ashland (1 SOSC); 6 mi. S Ashland (1 SOSC); 14 mi. E Ashland (1 SOSC); 0.75

mi. S, 13 mi. E Butte Falls (1 SOSC); 2 mi. NE Four Corners (1 PSM); near Howard Prairie off Dead Indian Rd. (1 SOSC); 2.5 mi. E Shale City (1 PSM); Siskiyou (1 USNM); 1.5 mi. S junction Hyatt Lake and Dead Indian Rd. (1 SOSC); 2 mi. before Jackson/Klamath county line* (1 SOSC); 2.5 mi. E USFS Rd. 3706 on USFS Rd. 3705 (1 SOSC). **Jefferson Co.**—12 mi. N, 2 mi. E Camp Sherman (2 OSUFW); Gateway (2 USNM); 12 mi. E Hay Creek on Foley Creek (2 USNM); 12 mi. SE Hay Creek (2 OSUFW); 10 mi. NW Sisters, T13S, R9E, Sec. 32 (1 UW[WWC]); 3 mi. E Suttle Lake (1 KU); Warm Springs (1 CM, 1 SDNHM); 20 mi. W Warm Springs, Mill Creek (4 OSUFW, 6 USNM). **Josephine Co.**—Bolan Mt. (1 SDNHM); Cave Creek campground (3 KU); 17 mi. E Cave Junction (4 KU); 10 mi. S, 8 mi. E Cave Junction (1 OSUFW); 13 mi. SW Galice, Briggs Creek, Farren Ranger Station (1 USNM); 15 mi. SW Galice (6 UO); Oregon Caves [National Monument] (3 OSUFW). **Klamath Co.**—Anderson Spring (1 USNM); Anna Creek, Mt. Mazama, 6,000 ft. (1 KU, 4 USNM); W of Buck Peak, T38S, R6E, Sec. 7/8 (1 SOSC); W of Buck Peak, T38S, R6E, Sec. 8/17 (1 SOSC); SW of Buck Peak, T38S, R6E, Sec. 8 (1 SOSC); SW of Buck Peak, T38S, R6E, NW¼ Sec. 16 (1 SOSC); Chiloquin (1 USNM); 4 mi. N, 2 mi. E Chiloquin (1 OSUFW); 2 mi. N Cottonwood Creek* (1 SOSC); Crater Lake* (2 LACM, 29 USNM); Crater Lake, 1 mi. S rim village (1 PSU); Crater Lake National Park headquarters (1 OSUFW); S. entrance Crater Lake National Park (2 USNM); 1 mi. S entrance Crater Lake National Park (1 KU); Cresent Lake, 4,800 ft. (5 LACM); 20 mi. W Cresent Lake (23 UO); 6 mi. E Cresent Lake, 4,500 ft. (6 UO); 12 mi. E Diamond Lake (1 SOSC); Fourmile Lake (1 UO, 9 USNM); Ft. Klamath (49 USNM); 3 mi. N Ft. Klamath (3 LACM); near Keno (1 SOSC); 10 mi. W Keno (1 SOSC); Klamath Falls (1 CRCM); within 8 mi. Klamath Falls (1 OSUFW); Naylox, Klamath Lake (3 USNM); Odell Lake, 4,700 ft. (3 LACM); Upper Klamath Lake, Rocky Point (3 OSUFW); T38S, R6E, Sec. 8 (1 SOSC); T38S, R6E, Sec. 8/15 (1 SOSC). **Lake Co.**—Adel, Warner Valley (1 OSUFW); 14 mi. SW Adel, Barley Camp, Warner Mts., T41S, R21E, E edge Sec. 8 (3 OSUFW); Chandler State Park (2 SOSC); head of Drews Creek Valley (2 USNM); Fort Rock (1 OSUFW); Fremont (4 USNM); Hart Mt. (1 SDWNHM); Hole in the Ground (1 OSUFW); Lakeview (2 CM, 1 UO); 5.6 mi. S, 6.8 mi. E Lakeview, Vernon Springs (1 PSU); 7.2 mi. S, 7.6 mi. E Lakeview, Willow Creek F.C. (1 PSU); 10 mi. SE Lakeview (2 USNM); Mt. Warner (1 USNM); 10 mi. S Plush (3 CRCM); Rogger Meadow (1 PSU); 9 mi. S Silver Lake (1 PSU); 16 mi. S, 7 mi. W Silver Lake, T31S, R13E (1 OSUFW); 18 mi. S, 6 mi. W Silver Lake (1 OSUFW); 10 mi. SW Silver Lake, 4,650 ft. (2 USNM); South Warner Lake (1 CM); 12 mi. S, 3 mi. W Summer Lake (2 OSUFW). **Lane Co.**—Gold Lake, 4,800 ft. (2 LACM); Huckleberry Lake Rd. (1 KU); 10 mi. S McKenzie Bridge, O'Leary Mt. (2 USNM); 17 mi. E McKenzie Bridge (6 UO); McKenzie Pass (2 USNM); Scott Mt. summit (1 OSUFW); Three Sisters (1 USNM); N base Three Sisters, 5,000 ft. (5 UO, 2 USNM); W base Three Sisters (4 OSUFW); NW base Three Sisters, Benson Lake (1 OSUFW); Three Sisters, Frog Camp, 5,000 ft. (1 UO). **Malheur Co.**—(10 PSM); Beulah (1 USNM); Brogan (7 OSUFW); Cedar Mts. (3 USNM); Disaster Peak (1 USNM); Ironside, 4,000 ft. (1 AMNH); Ironside, 4,500 ft. (8 AMNH); Ironside, 5,500 ft. (1 AMNH); 8 mi. NE McDermitt, McDermitt Creek, 6,000 ft. (3 USNM); 2 mi. NW Riverside (7 USNM); Westfall (5 USNM). **Morrow Co.**—Heppner (2 USNM); 15 mi. S Heppner (2 USNM); 2 mi. SE Heppner (2 UW[WWC]). **Sherman Co.**—Miller (1 OSUFW); Miller, mouth of Deschutes River (5 OSUFW, 6 USNM). **Umatilla Co.**—12 mi. N Langdon Lake, 6,000 ft. (1 UW[ERB]); Meacham (1 USNM). **Union Co.**—Clarks Creek guard station, Elgin (1 CRCM); Kamela (1 USNM); 10 mi. E Kamela, 4,900 ft. (1 USNM); 3 mi. N, 4 mi. E Starkey (1 OSUFW). **Wallowa Co.**—(1 EOSC); Aneroid Lake, Lookout Mt. (1 UO); 0.1 mi. SE Jewett Lake, 8,350 ft. (1 CRCM); Joseph (1 CM); 21 mi. S, 5 mi. E Lostine, T4S, R44E, Sec. 27 (1 OSUFW); 17 mi. SE Lostine, Lostine River, Wallowa Mts. (1 OSUFW); Lostine River, 5 mi. up from Lostine (1 UW[WWC]); 15 mi. N Paradise at Horse Creek, 2,000 ft. (1 USNM); Troy, 1,700 ft. (4 USNM); Wallowa (1 USNM); Wallowa Canyon* (2 USNM); Wallowa Lake (3 USNM); near Wallowa Lake (2 UW[EB]); 1 mi. S Wallowa Lake (5 UW[EB]); Wenaha Forks (1 EOSC). **Wasco Co.**—7 mi. E Antelope (2 USNM); Friend (2 USNM); Maupin (2 USNM); 2.5 mi. S, 2 mi. W Shaniko (2

UW); Wapinitia (4 USNM); 8 mi. W Wapinitia, 2,500 ft. (5 USNM); mts. N of Warm Springs River, 2,800 ft. (1 USNM); 20 mi. N Warm Springs at Mill Creek, T8S, R9E (1 OSUFW); White River, T4S, R11E, Sec. 26 (1 PSU). **Wheeler Co.**—Crown Rock, John Day River (3 USNM); 13 mi. S Fossil (1 UW[WWC]); Francisville, near Crown Rock (1 USNM); 12 mi. SW Mitchell, Wildwood Camp (1 OSUFW).

Additional localities: Goose Lake Mts.* (1 USNM); Cow Creek Lake* (1 USNM); N slope Mt. Jefferson (1 SDNHM).

Spermophilus mollis, n = 8
(Plotted in Fig. 11-30, p. 199)

Malheur Co.—2 mi. S Burns Junction (5 UU); E base Disaster Peak, 5,500 ft. (1 USNM); Rome, Owyhee River (2 USNM).

Spermophilus washingtoni, n = 66
(Plotted in Fig. 11-32, p. 202)

Gilliam Co.—Willows (6 USNM); Willows Junction (1 USNM). **Morrow Co.**—(3 EOSC); Boardman (1 SDNHM); Boardman bombing range (5 PSM); 4 mi. S, 6 mi. E Boardman (1 OSUFW); Heppner (9 USNM); 18 mi. S Irrigon (5 SDNHM); Pine City, T1N, R27E, Sec. 10 (8 OSUFW). **Umatilla Co.**—Cold Springs (1 USNM); Pendleton (1 PSM, 16 USNM); E edge Pendleton (2 SDNHM); 2 mi. W Pendleton (1 SDNHM); Pilot Rock (2 SDNHM, 1 USNM); Umatilla (1 USNM); Vinson, 30 mi. E Heppner (2 USNM).

Spilogale gracilis, n = 185
(Plotted in Fig. 12-55, p. 442)

Baker Co.—Baker [City] (1 USNM). **Benton Co.**—4 mi. SW Blodgett (1 PSM). **Clackamas Co.**—Estacada (4 USNM); Marmot (2 USNM); 8 mi. NW Molalla, 14 mi. SE Oregon City (1 OSUFW); Oak Grove Butte, [T6S, R7E, Sec. 8] (1 USNM). **Clatsop Co.**—Gearhart (2 MVZ); 1 mi. N Gearhart (1 MVZ); site of Old Ft. Clatsop (2 MVZ). **Columbia Co.**—7 mi. SE Rainier (1 MVZ). **Coos Co.**—3 mi. SE Bandon (1 PSM); 4 mi. SE Bandon (9 PSM); 9 mi. NE Bandon (1 PSM); 1.75 mi. NNE Bandon (1 PSM); Charleston (4 LACM); Coos Head (1 LACM); Empire (1 USNM); Marshfield (3 FMNH). **Crook Co.**—Prineville (1 USNM). **Curry Co.**—Gold Beach (3 FMNH); Hubbard Creek, 8 mi. SE Port Orford (1 USNM); Little Redwood Forest camp (1 PSM); Long Ridge Forest camp (1 PSM); Port Orford (4 USNM); 2 mi. S Port Orford (3 PSM). **Douglas Co.**—Gardiner (5 FMNH); Roseburg (1 USNM); 6 mi. N, 10.5 mi. W Umpqua P.O., T24S, R8W, Sec. 29, 900 ft. (1 UW-CR). **Harney Co.**—Alvord Valley, 15 mi. NE Andrews (1 USNM); Burns (2 USNM); Harney (2 USNM); Narrows (6 USNM, 1 UW); Sageview (2 USNM); Shirk P.O. (1 USNM); Steens Mt., 12 mi. N Diamond (1 SDNHM). **Hood River Co.**—Mt. Hood, Mt. Laurel Creek (1 USNM). **Jackson Co.**—Beagle (1 OSUFW, 1 PSM); 43 mi. NE Grants Pass (2 USNM); Evans Creek (1 SDNHM); 3 mi. off HWY 66 on Dead Indian Rd. (1 SOSC). **Jefferson Co.**—Hay Creek (1 USNM); Warm Springs, 9 mi. NW Agency (1 USNM). **Klamath Co.**—4 mi. N Bly (1 PSM). **Lake Co.**—Fort Rock (1 SDNHM); 21 mi. SE Fort Rock (1 SDNHM); Fremont (2 USNM); Plush (4 USNM); Warner Valley from Rock Creek (1 USNM). **Lane Co.**—(3 CRCM); 4.4 km N, 6.8 km E Blue River (1 USNM); 5.5 mi. NE Blue River (1 PSM); 6 mi. N, 7 mi. W Cottage Grove, T19S, R4E, Sec. 28 (1 OSUFW); Eugene (1 UO, 2 USNM); 13 mi. N, 4.33 mi. E Florence (1 OSUFW); Hermann (10 USNM); Mabel (1 UO); McKenzie Bridge (6 UO); McKenzie Bridge, Horsepasture Trail (1 PSM); 7 mi. W McKenzie Bridge (1 KU); S fork McKenzie River (2 KU); Mercer (1 UO); Oakridge (3 OSUFW); Oakridge, Hall Creek (1 USNM); Reed (3 USNM). **Lincoln Co.**—1.4 mi. S Burnt Woods (1 OSUFW); Cascade Head Experimental Forest camp (5 PSM); Delake (1 SDNHM); Depoe Bay (1 UMMZ); 2.75 mi. N, 1 mi. E Harlan (1 OSUFW); 3 mi. N, 3 mi. W Nortons, T10S, R9W, Sec. 13 (1 KU); Yaquina (2 USNM). **Malheur Co.**—Cedar Mts. (3 USNM); Cord, Barren Valley, 3,950 ft. (3 USNM); 2 mi. NW Riverside (1 USNM). **Morrow Co.**—10 mi. SW Hardman (2 USNM). **Multnomah Co.**—Fairview (1 SDNHM); Fulton (1 USNM); Portland (1 PSM). **Sherman Co.**—Miller, mouth Deschutes River (1

OSUFW, 1 PSM, 1 USNM). **Tillamook Co.**—Beaver (1 KU); Blaine (1 KU, 1 UMMZ); Netarts (2 OSUFW, 4 SDNHM, 2 USNM); Tillamook (3 KU, 2 LACM, 1 MVZ, 1 OSUFW). **Wallowa Co.**—(1 KU). **Wasco Co.**—mouth of Deschutes River (3 OSUFW); Moody (1 OSUFW). **Washington Co.**—Beaverton (3 FMNH).

Sylvilagus bachmani, n = 185
(Plotted in Fig. 10-8, p. 133)

Benton Co.—(1 OSUFW); Corvallis (1 KU, 1 USNM, 2 OSUFW); near Corvallis (3 PSM); 4 mi. N Corvallis, T11S, R5W, Sec. 5 (1 OSUFW); 7 mi. N Corvallis, E. E. Wilson Game Management Area (22 OSUFW); 8 mi. N Corvallis (2 CRCM); 9 mi. N Corvallis (1 KU); 9 mi. N Corvallis, E. E. Wilson Game Management Area, T10S, R4W, Sec. 17 (1 OSUFW); 7 mi. S, 3 mi. W Corvallis, W side Finley Refuge (1 OSUFW); E. E. Wilson Area, T10S, R5W (1 SOSC); Kiger Island, SE Corvallis (2 OSUFW); 5 mi. SW Philomath (2 USNM). **Clackamas Co.**—Jennings Lodge (1 PSM); Marmot (1 SDNHM); Molino (1 OSUFW); Mollala (1 PSM); Mt. Scott (1 KU). **Columbia Co.**—7 mi. SE Rainier (1 MVZ); 4 mi. E Scapoose (1 PSM); near Veronia (1 UW). **Coos Co.**—Bandon (1 USNM); 1.75 mi. S Bandon (1 PSM); 3 mi. SE Bandon (4 PSM); 3 mi. SSE Bandon (2 PSM); 4 mi. SE Bandon (14 PSM); Coos Bay, south slough Rd. (1 LACM); Norway (2 USNM). **Curry Co.**—1.5 mi. N Brookings, Harris State Park (1 SOSC); 6 mi. S Carpenterville, T39S, R14W, Sec. 25 (1 OSUFW); Gold Beach (5 FMNH); Little Redwood Forest Camp (5 PSM); Long Ridge Forest Camp (1 PSM); Port Orford (1 OSUFW); Winchuck campground, T41S, R12W, NW¼ Sec. 10 (1 SOSC). **Douglas Co.**—Anchor (1 USNM); Reston (1 USNM); Roseburg (1 USNM). **Josephine Co.**—Grants Pass (1 SOSC); Jump-off Joe Creek Road and Bonney Ranch Road [T34S, R5W, SW¼SW¼ Sec. 32] (1 SOSC). **Lane Co.**—18 mi. S Blue River, T19S, R4E, Sec. 21 (1 OSUFW); 8 mi. W Creswell (1 PSU); Eugene (1 USNM); Florence (8 FMNH); 0.5 mi. N, 4 mi. E Lorane (1 OSUFW); McKenzie Bridge (1 UO); Mercer, Grass Mt. Refuge (1 OSUFW); Vida (1 UO). **Lincoln Co.**—1.5 mi. N, 1 mi. E Harlan (1 OSUFW); 1 mi. S, 3 mi. W Harlan (1 OSUFW); 5 mi. NE Yachats (1 PSM). **Linn Co.**—2 mi. S, 4 mi. E Albany (1 OSUFW); 2 mi. E Corvallis (1 OSUFW); 10 mi. S, 7 mi. E Corvallis (1 OSUFW); Peoria (1 OSUFW); 0.5 mi. N Peoria (1 OSUFW). **Marion Co.**—Salem (2 AMNH, 2 USNM); 8 mi. N Salem, Highway 99W and Quinaby Road (1 OSUFW); Woodburn (1 PSM). **Multnomah Co.**—W slope Larch Mt. (1 PSM); Portland (2 SDNHM, 2 AMNH, 2 USNM, 1 PSM). **Polk Co.**—Grand Ronde (1 USNM); Independence (1 USNM); 2 mi. N Independence (1 OSUFW); Monmouth (1 PSM); 3 mi. S, 1 mi. W Monmouth (1 OSUFW). **Tillamook Co.**—Blaine (19 UMMZ, 1 PSM); Netarts (1 CM, 6 OSUFW, 1 PSU, 1 UO); Tillamook (1 AMNH, 1 LACM, 1 OSUFW, 1 SDNHM, 1 UMMZ, 1 UO); 9 mi. SE Tillamook (1 LACM). **Washington Co.**—Beaverton (4 FMNH, 2 USNM); 1 mi. S Cornelius (1 OSUFW); Forest Grove (1 OSUFW); 18.5 mi. NW Portland (1 MVZ); Route 4, Box 338, Sherwood (1 PSU). **Yamhill Co.**—4 mi. E Dayton on Willamette River (1 PSU); 4 mi. N McMinnville (1 OSUFW).

Sylvilagus floridanus, n = 546
(Plotted in Fig. 10-10, p. 135)

Benton Co.—1 mi. S ODF&W Regional Office, on Highway 99W [Adair Village] (1 OSUFW); Corvallis (2 KU, 15 OSUFW, 1 SOSC); 0.25 mi. N Corvallis (2 OSUFW); 1 mi. N Corvallis (1 OSUFW); 2 mi. N Corvallis (3 OSUFW); 3 mi. N Corvallis (1 OSUFW); 3.5 mi. N Corvallis (1 OSUFW); 7 mi. N Corvallis, E. E. Wilson Game Management Area (1 KU, 392 OSUFW); 1 mi. S Corvallis (1 OSUFW); 4 mi. S Corvallis, Airport Road (1 KU, 1 OSUFW); 0.25 mi. W Corvallis (1 OSUFW); 3 mi. N, 3 mi. W Corvallis (1 OSUFW); 3 mi. N, 4 mi. W Corvallis (1 OSUFW); 5 mi. N, 5 mi. E Corvallis, 200 ft. (1 CM); 6 mi. N, 1.5 mi. E Corvallis (1 OSUFW); 7 mi. N, 6 mi. E Corvallis (1 OSUFW); 8 mi. N, 2 mi. E Corvallis (1 OSUFW); 9 mi. N, 3 mi. E Corvallis, Wellsdale, T10S, R4W, Sec. 17 (55 KU); 10 mi. N, 0.5 E Corvallis (2 OSUFW); 2 mi. S, 2 mi. W Corvallis (1 OSUFW); 2.5 mi. S, 1 mi. W Corvallis (2 OSUFW); 5 mi. S, 1.5 mi. W Corvallis (1 OSUFW); 1.5 mi.

S, 1 mi. E Corvallis (1 OSUFW); 8.8 mi. S, 1.5 mi. E Corvallis, Black Lake (7 OSUFW); 0.25 mi. W Greenberry (1 OSUFW); 0.25 mi. N Lewisburg, Highway 99N [99W] (1 OSUFW); 7 mi. N Lewisburg, junction Highway 99W (2 OSUFW); Highway 99W, just S Marys River (1 OSUFW). **Clackamas Co.**—Mt. Scott (1 KU). **Columbia Co.**—2 mi. W Deer Island (1 PSU). **Lane Co.**—25 mi. S, 1 mi. W Corvallis (2 OSUFW). **Linn Co.**—(1 OSUFW); 3 mi. E Corvallis from OSU (1 OSUFW); Fayetteville, T13S, R4W, Sec. 3 (1 OSUFW); 0.8 mi. N, 2 mi. E Halsey (1 OSUFW); 3.25 mi. S, 3.5 mi. E Lebanon, T12S, R1W, Sec. 32 (1 OSUFW); 1 mi. N Harrisburg (1 KU); 1 mi. N Peoria (1 OSUFW); 4 mi. S Peoria on Peoria Road (1 OSUFW); 2 mi. E Peoria (1 OSUFW). **Marion Co.**—1 mi. N Marion (1 OSUFW). **Multnomah Co.**—Delta Park (2 PSU); Gresham (1 PSU); Portland (1 PSM, 9 PSU, 3 OSUFW); 8 mi. S, 10 mi. E Portland (1 OSUFW); Smith Lake (1 PSU); St. Johns (1 PSU); Swan Island (1 PSU). **Umatilla Co.**—Milton-Freewater (1 EOSC); 1 mi. S Umapine (1 OSUFW); 2 mi. S Umapine (1 OSUFW), 1 mi. S, 1 mi. W Umapine (4 OSUFW); 1 mi. S, 1.5 mi. W Umapine (1 OSUFW).

Sylvilagus nuttallii, n = 451
(Plotted in Fig. 10-12, p. 139)

Baker Co.—Baker [City] (2 UMMZ); 10 mi. N Baker [City] (11 UMMZ); Burkemont (3 AMNH); Homestead, 1,800 ft. (1 USNM); Huntington (1 USNM); 11 mi. E Unity (1 KU); junction Powder and Snake rivers, Braunlee Dam backwater (1 EOSC). **Crook Co.**—2 mi. SW Barnes (1 MVZ); Prineville (2 USNM); 20 mi. S Prineville (2 USNM); 4 mi. SW Prineville (2 MVZ); 1 mi. N, 5 mi. E Prineville (2 OSUFW); 4 mi. N, 4 mi. E Prineville (1 OSUFW); 3 mi. S, 1 mi. E Prineville (1 OSUFW); Twelvemile Creek (3 USNM). **Deschutes Co.**—Arnold Ice Cave (1 USNM); Bend (1 USNM); 3 mi. N Bend (1 OSUFW); 16 mi. E Bend, Horse Ridge (9 OSUFW); 1 mi. N, 2.5 mi. E Brothers (1 OSUFW); Buckhorn Canyon, 2–3 mi. N Highway 126 (2 OSUFW); 3 mi. N, 4 mi. W Cline Falls (1 OSUFW); 1–4 mi. E Deschutes River, N of Lower Bridge Road, Deschutes Canyon (23 OSUFW); S of Lower Bridge, N Oden Falls, Upper Deschutes Canyon (9 OSUFW); 1 mi. N, 1.5 mi. E Lower Bridge, W of Steamboat Rock (1 OSUFW); 2 mi. N Redmond (2 OSUFW); 5 mi. S Redmond (1 OSUFW); 7 mi. N, 6 mi. W Redmond, Lower Bridge (3 OSUFW); Sisters (1 OSUFW, 3 USNM); 2 mi. E Sisters on Highway to Redmond (1 OSUFW); 8 mi. NE Sisters (1 KU); 10 mi. NE Sisters (1 KU); 17 mi. NE Sisters, Lower Bridge (19 OSUFW); 1.5 mi. W Terrebonne (3 OSUFW); 2 mi. W Terrebonne (6 OSUFW); 2.25 mi. W Terrebonne (1 OSUFW); 2.5 mi. W Terrebonne (2 OSUFW); 2.75 mi. W Terrebonne (1 OSUFW); 3 mi. W Terrebonne, T14S, R13E, Sec. 18 (62 OSUFW); 3.5 mi. W Terrebonne, T14S, R13E, Sec. 18 (3 OSUFW); 3.75 mi. W Terrebonne (3 OSUFW); 4 mi. W Terrebonne (3 OSUFW); 5 mi. W Terrebonne (7 OSUFW); 0.25 mi. N, 3.5 mi. W Terrebonne (2 OSUFW); 0.5 mi. N, 2.75 mi. W Terrebonne (1 OSUFW); 0.5 mi. N, 3 mi. W Terrebonne (4 OSUFW); 0.5 mi. N, 3.25 mi. W Terrebonne (1 OSUFW); 0.5 mi. N, 3.5 mi. W Terrebonne (1 OSUFW); 0.5 mi. N, 3.75 mi. W Terrebonne (3 OSUFW); 0.5 mi. N, 4 mi. W Terrebonne (3 OSUFW); 0.75 mi. N, 2.5 mi. W Terrebonne (2 OSUFW); 1 mi. N, 3 mi. W Terrebonne (1 OSUFW); 1 mi. N, 4 mi. W Terrebonne (1 OSUFW); 1 mi. N, 5 mi. W Terrebonne (3 OSUFW); 1 mi. N, 6 mi. W Terrebonne, Lower Bridge (1 OSUFW); 1.5 mi. N, 5 mi. W Terrebonne (6 OSUFW). **Gilliam Co.**—3 mi. S Olex (1 OSUFW); Willows (3 USNM). **Grant Co.**—4 mi. W Monument, N fork John Day River (1 EOSC); Mt. Vernon (1 USNM). **Harney Co.**—Blitzen River (1 UO); Burns (2 USNM, 1 OSUFW); Diamond (1 USNM); Diamond Craters, T29S, R32E, Sec. 3 (3 HSU); Drewsey (2 USNM); 15.5 mi. N, 2.5 mi. E Frenchglen (1 OSUFW); Lake Alvord (1 USNM); Malheur Cave, 4,000 ft. (1 USNM); Narrows (2 USNM); Shirk P.O. (1 USNM); Steens Mt. (1 SDNHM, 1 USNM). **Jackson Co.**—Ashland, W slope Grizzly Peak, 3,500 ft. (1 USNM). **Jefferson Co.**—Warm Springs, 9 mi. W Agency (1 USNM); Cove Palisades State Park Pennisula (1 PSU); 7 mi. SE Culver, Haystack Butte (15 OSUFW); 8 mi. S, 8 mi. W Culver, Squaw Creek (2 OSUFW); 4 mi. S, 3 mi. E Culver, Haystack Butte (2 OSUFW); Hay Creek (2 USNM, 1 PSM); 8.5 mi. S, 2.5 E Madras (10 PSU); 10 mi. S Madras (1 PSU); 15 mi. S, 2.5 mi. E Madras (1 OSUFW); 14 mi. NE

Sisters, Squaw Creek Road (53 OSUFW). **Josephine Co.**—Grants Pass, junction of Rogue and Applegate rivers (1 USNM). **Klamath Co.**—Keno (1 PSM); Klamath Falls (3 CM); near Merrill Dump (1 OSUFW); 0.5 mi. S Spring Lake (1 SOSC). **Lake Co.**—Abert Lake (1 USNM); Adel (1 USNM); 20 mi. NE Adel (2 USNM); Fremont (5 USNM); Lakeview (1 OSUFW, 1 UO, 1 USNM); 10 mi. W Plush (1 USNM); Silver Lake (1 USNM). **Malheur Co.**—Brogan (3 OSUFW); Cedar Mountains (1 USNM); Crooked Creek, Owyhee Desert, 4,000 ft. (1 USNM); E base Disaster [Peak], 5,500 ft. (1 USNM); Ironside, 4,000 ft. (6 AMNH); 1 mi. W Ironside (2 MVZ); 7 mi. from mouth of Jordan Creek (1 USNM); 3 mi. W Juntura (1 USNM); 8 mi. NE McDermitt [Nevada] (2 USNM); 5 mi. SW Ontario, Wood Ranch (2 USNM); 13 mi. SW Owyhee, Owyhee River, T22S, R45E, Sec. 3 (1 PSU); 2 mi. NW Riverside (4 USNM); Rome, Owyhee River (1 USNM); Rome (1 MVZ); 3 mi. S, 1 mi. E Rome (1 OSUFW); near Vale (2 AMNH); Vale (10 USNM); Westfall (2 USNM). **Morrow Co.**—14.25 mi. W Boardman (3 PSU); Heppner (1 USNM); 1.5 mi. SW Irrigon (1 MVZ). **Sherman Co.**—Columbia River, mouth of Deschutes River (1 MVZ); Miller, mouth of Deschutes River (1 USNM, 1 OSUFW). **Umatilla Co.**—Freewater (1 UW[WWC]); Pendleton (1 SDNHM); Umatilla (1 USNM); 1 mi. N, 2.5 mi. E Umatilla (1 OSUFW); 16 mi. SE Weston (2 UW[WWC]). **Union Co.**—(1 EOSC); Cove (1 EOSC); 0.5 mi. S, 1 mi. W LaGrande (1 OSUFW); 2 mi. W College Place on Whitman Dr. [La Grande] (1 EOSC); 1 mi. N, 10.5 mi. W North Powder (1 OSUFW). **Wallowa Co.**—Enterprise (1 USNM). **Wasco Co.**—Columbia River, mouth of Deschutes River (1 MVZ); The Dalles, 133 ft. (3 USNM); 7 mi. S, 4.5 mi. W The Dalles (1 OSUFW); 7 mi. S, 9 mi. E The Dalles (1 OSUFW); 14 mi. SW The Dalles, Five Mile Creek (1 OSUFW); Maupin (5 USNM). **Wheeler Co.**—(1 UMMZ); Fossil (1 MVZ, 1 UW[WWC]); 1 mi. N Fossil (1 UW[WWC]); Twickenham (1 USNM).

Additional localities: Ft. Dallis, 135 ft.* (1 USNM); Willow Springs, Deschutes Valley* (1 USNM).

Tadarida brasiliensis, n = 50
(Plotted in Fig. 9-41, p. 121)

Douglas Co.—Roseburg, T27S, R5W, NW¼ Sec. 19 (3 SOSC). **Jackson Co.**—Ashland (1 OSUFW, 4 PSM, 1 PSU, 24 SOSC); Central Point (1 SOSC); Medford (1 PSM, 9 SOSC); 0.25 mi. W Medford (1 SOSC); White City (2 SOSC); T39S, R4W (1 SOSC). **Klamath Co.**—Klamath Falls, T38S, R9W, Sec. 28 (2 SOSC).

Tamias amoenus, n = 1,377
(Plotted in Fig. 11-4, p. 160)

Baker Co.—W slope Antelope Hill, 10 mi. W Haines (1 EOSC); Anthony (32 AMNH, 10 MVZ, 2 SDNHM); 6 mi. NE Baker City (1 AMNH); Bourne (3 USNM); Burkemont (2 AMNH); 3 mi. E Burkemont (2 AMNH); Cornucopia, 4,740 ft. (1 PSM, 1 MSUMZ, 2 UMMZ, 9 USNM); Eagle Creek area (16 KU); Lily White field station (1 EOSC); McEwen (1 USNM); Sumpter (1 SDNHM); Unity, 0.5 mi. SE Reservoir (1 PSM); Wallowa Mt. field station* (1 EOSC); SE ridge of Point 8569, WSW slope Elkhorn Mts., 8,300 ft. (1 CRCM). **Crook Co.**—Grizzly Mt., 5,000 ft. (1 MVZ); Howard, Lookout Mt., 6,600 ft. (6 USNM); Maury Mts. (4 USNM); Ochoco Forest* (1 USNM); Ochoco Ranger Station, 4,000 ft. (4 MVZ); Prineville (8 USNM, 1 UW); 20 mi. S Prineville (1 OSUFW); 3 mi. W Prineville (1 MVZ); 7 mi. W Prineville (2 MVZ); 18 mi. N, 12.5 mi. E Prineville (1 OSUFW); 2 mi. NE Prineville (2 MVZ); 20 mi. SE Prineville, Crooked River (2 USNM); 3 mi. N, 6 mi. E Wildcat Mt., 4,700 ft. (1 MVZ). **Deschutes Co.**—(1 LACM, 1 PSM); Bend (1 MVZ, 1 OSUFW, 8 USNM); Bend, Tumalo Creek (2 OSUFW); 3 mi. W Bend (7 KU); 11 mi. W Bend, Tumalo Creek (1 MVZ); 15 mi. W Bend, Tumalo Creek (2 MVZ); 16 mi. W Bend, Tumalo Creek (9 MVZ); 1 mi. S, 1 mi. W Bend (1 OSUFW); 3 mi. S, 1 mi. E Bend (1 OSUFW); 5 mi. S, 8 mi. E Bend (2 OSUFW); 7 mi. NW Bend, T17S, R11E, Sec. 9 (1 OSUFW); 4 mi. NE Bend (1 OSUFW); 14 mi. SE Bend (1 KU); Black Butte, T14S, R9E, Sec. 3 (15 different locality descriptors) (3 CM, 54 OSUFW); Broken Top Mt. (1 SDNHM); Cold Spring campground, 4 mi. W Sisters (2 PSU); Cold Springs campground,

4.5 mi. W Sisters (7 HSU); Cold Spring campground, 0.25 mi. N, 4 mi. W Sisters (4 PSU); Cold Spring campground, 1.25 mi. N, 4 mi. W Sisters (7 PSU); Cougar Rock, Grand View Loop, T14S, R10E, Sec. 10 (1 OSUFW); head Davis Creek, T22S, R8E, Sec. 18 (4 OSUFW); Deep Canyon, T14S, R11E, Sec. 34/35/26/25 (1 OSUFW); East Lake, Paulina Mts. (1 OSUFW); 1 mi. S East Lake, Paulina Mt. (1 MVZ); Elk Lake (1 PSU); head of Fall Creek (1 CRCM); Fourmile Spring (1 PSM); 5 mi. N, 5 mi. W Hampton, T21S, R20E, S½ Sec. 15 (1 KU); Indian Ford, Sisters (1 OSUFW); Indian Ford camp (1 PSM); La Pine (4 USNM); 3 mi. N, 3 mi. E La Pine (2 OSUFW); 10 mi. N, 3 mi. E La Pine (1 OSUFW); Lower Bridge (1 OSUFW); Lower Bridge, Deschutes River (1 OSUFW); Paulina Butte Lookout (2 PSU); Paulina Lake (3 SDNHM, 4 USNM); 3 mi. W Paulina Lake (4 MVZ); Redmond (1 PSM, 2 SDNHM); 3 mi. N, 1 mi. W Redmond (1 OSUFW); 1.5 mi. S, 2.5 mi. W Redmond (1 OSUFW); Sisters (2 LACM, 7 OSUFW, 5 PSM, 3 USNM); 20 mi. N Sisters (2 SOSC); 1.7 mi. N, 5.8 mi. E Sisters (1 OSUFW); 15 mi. S, 8 mi. W Sisters, Little Three Creek Lake (1 OSUFW); Skeleton Cave (7 PSU); E side South Sister, 7,000 ft. (1 CRCM); Sparks Lake (3 OSUFW); 3 mi. W Terrebonne, 3,000 ft. (1 CM, 2 OSUFW); 4 mi. W Terrebonne (10 OSUFW); 1 mi. N, 5 mi. W Terrebonne, Lower Bridge (2 OSUFW); 0.5 mi. S, 4 mi. W Terrebonne (3 OSUFW); 1 mi. S, 2 mi. W Terrebonne (1 OSUFW); 1 mi. S, 3 mi. W Terrebonne (2 HSU, 16 OSUFW); 1 mi. N Threecreek Lake (3 PSM); Three Sisters (1 USNM); Three Sisters, 5,000 ft. (1 USNM); Tumalo (1 LACM); Tumalo State Park (1 PSM); T21S, R10E, Sec. 35 (1 PSU). **Douglas Co.**—Diamond Lake (6 SDNHM, 9 USNM); S end Diamond Lake (1 SOSC); 5 mi. S Mt. Thielson (2 MVZ). **Gilliam Co.**—Lonerock (1 USNM). **Grant Co.**—(1 LACM); Beech Creek (1 SDNHM, 7 USNM); Canyon City (1 SDNHM); 12 mi. S Canyon City (1 MVZ); 9 mi. S, 3 mi. E Canyon City (1 PSU); Cold Spring, 8 mi. E Austin (27 MVZ); 14 mi. N Dayville (2 UW[WWC]); 1 mi. S, 5 mi. W Granite (1 OSUFW); 9 mi. N Kimberly, Tamarack Springs (1 OSUFW); 1 mi. N Indian Springs campground, Malheur National Forest (1 CRCM); 10 mi. E Long Creek, T10S, R32E, Sec. 5 (5 UW); 10.5 mi. E Long Creek, Kahler Butte, T10S, R32E, Sec. 5 (1 OSUFW); 1 mi. N, 10 mi. E Long Creek, T10S, R32E, Sec. 5 (2 OSUFW, 3 UW); 4.5 mi. S, 4 mi. E Long Creek, Jordan Creek, T10S, R31E, Sec. 31 (1 OSUFW); Magone Lake, T12S, R31E, Sec. 12 (1 OSUFW); 21 mi. SE Prairie City, North Fork Malheur River (4 MVZ); ca. 14 mi. NW Ritter (2 PSU); Strawberry Butte (2 USNM); Strawberry Mts. (10 USNM); Strawberry Mts., Wickiup campground (1 HSU); N slope Strawberry Mts., 8,200 ft. (1 CRCM); N slope Strawberry Mts., 8,300 ft. (1 CRCM); N slope Strawberry Mts., 8,500 ft. (1 CRCM); E slope Strawberry Mts., 8,600 ft. (1 CRCM); E slope Strawberry Mts., 8,700 ft. (1 CRCM); Summit Creek, 7 mi. E Austin (3 MVZ); Trout Meadows, Whitman National Forest* (1 LACM). **Harney Co.**—Anderson Valley (1 PSU); Buchanan (5 USNM); 25 mi. NW Burns, T22S, R26E, Sec. 11 (1 OSUFW); 28 mi. NW Burns (4 OSUFW); 2 mi. E Crane (3 USNM); Delintment Lake (1 OSUFW); Drewsey (1 USNM); 6 mi. E Frenchglen, T32S, R32½E, Sec. 14 (1 OSUFW); 4 mi. S, 9 mi. E Frenchglen, 6,000 ft. (1 UW[WWC]); 4 mi. S, 10 mi. E Frenchglen, 500 ft. (3 UW[WWC]); Idlewild Park, T21S, R30E, Sec. 34 (2 OSUFW); Jackman Park, 2 mi. E Fish Lake, 20 mi. ESE Frenchglen, 7,500 ft. (3 UW[WWC]); Multon Creek, T20S, R26E, Sec. 10 (1 OSUFW); Riddle Mt. near Mahon Ranch, T28S, R34E (2 OSUFW); 0.25 mi. N, 18 mi. W Riley, T23S, R24E, SE¼ Sec. 29 (1 KU); 1.5 mi. S, 10 mi. W Riley, T24S, R25E, N½ Sec. 3 (1 KU); Smyth Creek, 10 mi. E Diamond (6 MVZ); Steens Mt. (2 USNM); Steens Mt., Big Fish Creek, 7,500 ft. (1 USNM); Steens Mt., head of Blitzen Gorge, 9,000 ft. (1 USNM); Steens Mt., Fish Lake, T32½S, R33E, Sec. 31, 8,000 ft. (1 OSUFW, 2 PSM, 1 PSU, 2 USNM); Steens Mt., 1 mi. N Fish Lake (4 HSU); 4 mi. W Fish Lake, T32S, R33E, Sec. 36 (2 OSUFW); Steens Mt., Kiger Gorge, 6,000 ft. (14 USNM); Steens Mt., Kiger Gorge, 6,800 ft. (2 USNM); Steens Mt., Kiger Gorge, 8,000 ft. (1 USNM); Steens Mt., Leiburg Basin* (3 USNM); Steens Mt., Little Blitzen Gorge, T33S, R33E, Sec. 10 (1 OSUFW); Steens Mt., Little Blitzen River, T33S, R33E, Sec. 24 (1 OSUFW); Steens Mt., T32S, R32E, Sec. 33 (1 OSUFW); Steens Mt., T32S, R32½E, Sec. 25 (1 OSUFW); Steens Mt., T32S, R32¾E, Sec. 27 (1 KU); Steens Mt., T32S, R32¾E, Sec. 28 (1 OSUFW); Steens Mt., T32S, R32¾E, Sec. 32 (1 KU); Steens Mt., T32S, R32¾E, Sec. 33

(1 OSUFW); Steens Mt., T32S, R33E, Sec. 27 (1 OSUFW); Steens Mt., T32S, R33E, Sec. 31 (2 OSUFW); Steens Mt., T32S, R33E, Sec. 32 (2 OSUFW); Steens Mt., T32S, R33E, Sec. 33 (3 OSUFW); Steens Mt., T32½S, R33E, Sec. 25 (1 OSUFW); Steens Mt., T32¾S, R32E, Sec. 26 (1 OSUFW); Steens Mt., T33S, R33E, Sec. 2 (2 OSUFW); Steens Mt., T33S, R33E, Sec. 5 (2 OSUFW); T32S, R31E, W½ Sec. 17 (1 OSUFW); T33S, R32½E, NW¼ Sec. 26 (1 EOSC). **Hood River Co.**—11 mi. E Bear Springs (2 SDNHM); Brooks Meadows, 9 mi. ENE Mt. Hood (5 MVZ); 2 mi. N Brooks Meadows guard station (1 PSU); Cloud Cap Summit, Mt. Hood (2 USNM); Mt. Hood, Cooper Spur (1 OSUFW); N slope Mt. Hood, 2,800 ft. (3 USNM); W slope near timberline Mt. Hood (3 USNM); Tilly Jane Creek, NE side Mt. Hood, 6,450 ft. (1 CRCM); T2S, R9E, Sec. 28 (4 PSU). **Jackson Co.**—14 mi. E Ashland (1 SOSC); 24 mi. E Ashland (1 SOSC); N base Ashland Peak, 6,000 ft. (1 USNM); Benson Gulch Rd. (1 PSU); Conde Creek (1 SOSC); Jackson Lake* (1 SOSC); Mt. Ashland Rd. (1 SOSC); Siskiyou (12 USNM); 4 mi. W Siskiyou Pass (1 MVZ); 4 mi. NE Union Creek (1 MVZ); Shale City Rd. 0.3 mi. N junction with Dead Indian Rd. (1 SOSC); Conde Creek Rd. 1.3 mi. N junction with Dead Indian Rd. (2 SOSC); Conde Creek Rd. 1.2 mi. up from junction with Dead Indian Rd. (1 SOSC); 2.5 mi. E USFS Rd. 3706 on USFS Rd. 3705 (3 SOSC); 0.5 mi. E junction Yew Spring Hookup Rd. and Crane Prairie Rd. (2 SOSC); 0.25 mi. E junction Crane Prairie Rd. and Weyerhauser Rd. 107-63 (2 SOSC); 1.4 mi. E junction Crane Prairie Rd. and Weyerhauser Rd. 107-63 (3 SOSC); 2 mi. E junction Crane Prairie Rd. and Weyerhauser Rd. 107-63 (1 SOSC); T39S, R4W, SE¼ Sec. 8 (1 SOSC). **Jefferson Co.**—Booth Lake, SE base Three-Fingered Jack, T13S, R8E, Sec. 18 (1 OSUFW); 10 mi. NW Cougar Rock, Grand View Loop (1 OSUFW); Camp Sherman (4 PSM); 4 mi. from Camp Sherman* (3 KU); 12 mi. E Foley Creek (2 USNM); 4 mi. S Geneva, Squaw Creek Canyon (1 OSUFW); Hay Creek (4 USNM); 12 mi. SE Hay Creek (4 OSUFW, 1 PSU); Haystack Rock (2 PSU); 10 mi. S Madras (2 PSU); 3 mi. S, 2.5 mi. E Madras (1 OSUFW); 5 mi. W Sisters, T13S, R9E, Sec. 32 (2 OSUFW); 9 mi. N, 7 mi. E Sisters (2 OSUFW); 20 mi. W Warm Springs, Mill Creek (8 OSUFW); Suttle Lake (1 OSUFW); Warm Springs (1 SDNHM). **Josephine Co.**—Bolan Lake (6 PSM); Bolan Mt. (7 SDNHM); 10 mi. S, 8 mi. E Cave Junction (1 OSUFW); Lave Caves Park* (2 SDNHM); near Oregon Caves [National Monument], Lake Mt., Siskiyou Mts., 6,600 ft. (2 USNM); Oregon Caves State Game Refuge (1 OSUFW). **Klamath Co.**—(1 SOSC); Anna Creek, 4 mi. NW Ft. Klamath, Mt. Mazama, 6,000 ft. (34 MVZ, 6 USNM); Aspen Lake (2 PSM); Bald Top Mt., Crater Lake National Park (1 MVZ); 15 mi. E Bly (1 USNM); 4.5 mi. SE Bly (1 SOSC); 9 mi. SE Bly (1 PSM); 4 mi. S, 1 mi. W Bonanza (2 OSUFW); W of Buck Peak, T38S, R6E, Sec. 7/8 (2 SOSC); 4 mi. N, 2 mi. E Chiloquin (1 OSUFW); 1 mi. S, 12 mi. E Chiloquin (1 CM); 1 mi. N Cottonwood Creek (1 SOSC); Crater Lake, Godfrey Glen Wayside (1 PSU); Crater Lake National Park (3 MVZ, 6 PSU, 15 USNM); Crater Lake National Park, Red Blanket Creek (2 USNM); Crater Lake National Park, Sun Creek (1 USNM); Crater Lake National Park, Union Peak (1 USNM); Cresent Lake, 4,800 ft. (1 LACM); 12 mi. E Diamond Lake (1 SOSC); 18 mi. S, 12 mi. W Fall River Trout hatchery, T23S, R11E, Sec. 24, 4,200 ft. (1 OSUFW); Ft. Klamath (1 AMNH, 69 USNM); 10 mi. SW Ft. Klamath (1 UMMZ); Fort Spring, Fremont National Forest (5 PSM); 1 mi. Government Camp, 7,100 ft.* (1 MVZ); Hooper Springs (1 SOSC); Horseglade (2 PSM); W of Keno, near Klamath River (1 SOSC); 10 mi. W Keno (1 SDNHM); 9.1 mi. N, 3.4 mi. E Keno (1 PSU); 10.1 mi. N, 2.7 mi. E Keno (2 PSU); Klamath Falls (7 CM, 3 USNM); 6 mi. E Klamath Falls (1 SOSC); 8 mi. E Klamath Falls (1 OSUFW); Klamath Lake, Naylox (3 USNM); S end Lake of the Woods (1 MVZ); 5 mi. N, 2 mi. W Merrill (1 OSUFW); Odell Creek, 4,700 ft. (1 LACM); Old Ft. Klamath (1 MVZ); Swan Lake Valley (3 USNM); South Buck Peak, T38S, R6E (1 SOSC); Upper Klamath Lake, Rocky Point Resort (1 SOSC); Upper Klamath Marsh (1 USNM); West Sink Creek, 12 mi. E Mt. Thielson (1 USNM); Yainax Butte (1 PSM); 4 mi. N HWY 97 on HWY 62 (1 OSUFW); T38S, R6E, Sec. 8/9 (1 SOSC); T38S, R6E, E½ Sec. 8/15 (2 SOSC). **Lake Co.**—14 mi. W Adel, Blue Creek, Warner Mts. (2 OSUFW); 14 mi. SW Adel, Barley Camp, Warner Mts. (3 OSUFW); Chandler State Park (1 SOSC); 15 mi. N Christmas Lake (1 USNM); 30 mi. N Christmas Lake (1 USNM); Drews Reservoir (1 PSM); site of Old Ft. Warner, Hart Mt. (1

SDNHM); Fremont (5 USNM); Gearhart Mt. (37 PSM); Hart Mt. (2 PSM); Hot Spring camp, Hart Mt. Antelope Refuge (2 SOSC); 0.5 mi. N Hot Springs camp, Hart Mt. National Antelope Refuge (1 SOSC); 1 mi. S Hot Spring camp, Hart Mt. National Antelope Refuge (1 SOSC); Lakeview (4 PSM, 2 SDNHM, 1 USNM); 10 mi. N Lakeview, Chandler State Park (1 SOSC); 3 mi. E Lakeview (1 MVZ); 14 mi. E Lakeview, Vernon Spring (1 CM); 15 mi. E Lakeview (1 SDNHM); 20 mi. NW Lakeview (2 OSUFW); 25 mi. NW Lakeview, Finley Corrals (2 USNM); 15 mi. N, 27 mi. W Paisley, Silver Creek (1 OSUFW); 9 mi. S Silver Lake (2 PSU); 17 mi. S, 7 mi. W Silver Lake, Silver Creek (1 OSUFW); 18 mi. S, 6 mi. W Silver Lake (1 OSUFW); 17 mi. S, 6 mi. E Silver Lake (1 OSUFW); Sycan Marsh (1 USNM); Warner Mts., Hart Mt. (7 USNM); West Silver Creek, Silver Lake, 4,650 ft. (3 SDNHM); West Silver Creek, Yamsay Mts., 7,000 ft. (2 USNM); Willow Creek Forest camp (1 SOSC). **Lane Co.**—Hand Lake, T15S, R7½E (1 OSUFW); 10 mi. S McKenzie Bridge, O'Leary Mt. (2 USNM); NW base Middle Sister, Frog Camp, Three Sisters (1 OSUFW); NW slope Middle Sister, Three Sisters (1 OSUFW); N slope Three Sisters (1 OSUFW); W base Three Sisters (2 OSUFW). **Linn Co.**—(1 PSM); Big Lake (2 PSM); 1.5 mi. N Big Lake (3 PSM); McKenzie Pass summit (1 OSUFW); 13.7 mi. E Upper Soda, Toad Creek Rd. (1 OSUFW); 20 mi. E Upper Soda (1 OSUFW). **Malheur Co.**—Cedar Mts. (13 USNM); Ironside (2 AMNH); Ironside, 4,000 ft. (3 USNM); Ironside, 4,500 ft. (9 AMNH); Ironside, 5,500 ft. (7 AMNH). **Morrow Co.**—20 mi. SSE Heppner (1 USNM). **Umatilla Co.**—(1 MSUMZ); Blalock Mt. (1 KU, 1 UW[WWC]); 13 mi. E Dale (1 SOSC); Langdon Lake (3 KU, 1 MSUMZ, 4 UW[WWC], 2 CRCM); 4 mi. W Langdon Lake (2 MSUMZ, 6 UW[WWC]); Meachem (3 USNM); 8.4 mi. N Meacham (1 PSU); Meacham Lake (1 MVZ); 2 mi. NW Meacham Lake (1 MVZ); 10 mi. E Milton (2 AMNH); 10 mi. SE Milton (1 UW[WWC]); 16 mi. SE Pendleton (1 USNM); Pikes Peak (1 UW[WWC]); Tollgate (3 KU, 19 UW[WWC]); near Tollgate (1 UW[WWC]); 3 mi. W Tollgate (3 UW[WWC]); 4 mi. W Tollgate (11 UW[WWC]); 8 mi. E Tollgate (1 UW[WWC]); 0.5 mi. N, 2.5 mi. W Tollgate (1 UW[WWC]); 1 mi. N, 3 mi. W Tollgate (4 UW[WWC]); 3 mi. NW Tollgate (3 UW[WWC]); 3.25 mi. NW Tollgate (1 UW[WWC]); 4 mi. NW Tollgate (1 UW[WWC]); 15 mi. E Ukiah (1 USNM); 16 mi. E Weston (25 UW[WWC]). **Union Co.**—(2 EOSC, 3 UW); Anthony Lake (5 PSM); Catherine Creek, 7 mi. E Telocaset (3 MVZ); Clarks Creek guard station, Elgin (5 CRCM); Elgin (8 USNM); 4 mi. W Elgin (1 PSM); Grande Ronde Lake, 7,100 ft. (2 MVZ); Kamela (6 USNM); La Grande (4 EOSC); 4 mi. SW EOSC campus [La Grande] in Morgan Lake drainage (1 EOSC); Glasshill Rd. (1 EOSC); Fox Hill Rd. (1 EOSC); 1 mi. E Red Bridge Park between HWY and river (1 EOSC); Lick Creek R.S. (1 PSM); Meadow Creek, Starkey Experimental Forest (1 PSM); 1 mi. N, 4 mi. W Medical Springs (1 OSUFW); Morgan Lake, La Grande (2 EOSC); Sheep Creek* (1 PSM); Starkey Experimental Forest (4 PSM); Telocaset (1 USNM); 8 mi. S, 18 mi. E Union (1 OSUFW). **Wallowa Co.**—(1 OSUFW, 1 PSM); Eagle Cap Wilderness Area* (2 SOSC); Echo Lake, Hurricane Creek, S of Joseph, 8,400 ft. (1 CRCM); 25 mi. N Enterprise at Sled Sprinps, 4,600 ft. (6 USNM); Fall Creek campground, Wallowa Mts. (1 UW[WWC]); Hurricane Creek, Wallowa Mts. (1 UW[WWC]); Joseph (2 CM); S of Joseph, 8,300 ft. (2 USNM); 50 mi. NE Joseph (1 OSUFW); 20 mi. SE Joseph, Lick Creek (16 UMMZ); Joseph Creek near Horse Creek, 5,000 ft. (5 USNM); 10 mi. N, 3 mi. E Lewis (1 OSUFW); 10.5 mi. N, 4 mi. E Lewis (2 OSUFW); 11 mi. N, 3 mi. E Lewis (1 OSUFW); 6 mi. S Lostine, T2S, R43E, Sec. 10 (1 OSUFW); 16 mi. S, 3 mi. E Lostine, T3S, R43E, Sec. 36 (8 MVZ, 1 OSUFW); 18 mi. S, 2 mi. E Lostine, 8,800 ft. (1 MVZ); 19 mi. S, 4 mi. E Lostine, T4S, R44E, Sec. 17 (5 OSUFW); 17 mi. SE Lostine, Lostine River, Wallowa Mts. (2 OSUFW); Marr Flat, Wallowa National Forest (6 SDNHM); 2 mi. N Minam, Minam State park (1 PSU); 15 mi. N Paradise at Horse Creek, 2,000 ft. (4 USNM); NNE face Pete's Point, 9,000 ft. (1 CRCM); ENE ridge Pete's Point, 8,900 ft. (1 CRCM); 8 mi. W Pine Grove* (1 PSU); Sheep Creek (1 OSUFW); Swamp Lake, Wallowa Mts., 7,800 ft. (1 CRCM); Wade Springs, Kuhn Ridge, Wallowa National Forest (1 SDNHM); Wallowa (1 UMMZ, 1 USNM); Wallowa Lake, 4,000 ft. (1 SDNHM, 16 USNM, 2 UW[WWC]); S of Wallowa Lake, Wallowa Mts. (2 USNM); Wallowa Lake, 3 mi. S Joseph (1 PSU); Whiskey Creek near Wallowa Canyon (11 USNM). **Wasco Co.**—(4 UW);

Antelope (1 USNM); 2 mi. N, 3 mi. E Antelope (1 OSUFW); 15 mi. E Eukiah* (1 UW); Mt. Hood HWY* (1 PSM); Shaniko (1 SDNHM, 3 UW); 14 mi. SW The Dalles, Fivemile Creek (3 OSUFW); 15 mi. SW The Dalles (1 USNM); 25 mi. SW The Dalles (1 USNM); 6 mi. SW Wamic (1 LACM); Wapinitia (9 USNM); 6 mi. W Wapinitia (3 PSM, 2 USNM); 20 mi. W Warm Springs, Mill Creek (6 USNM); Warm Springs River, 2,300 ft.* (2 USNM); Warm Springs River, 2,400 ft.* (1 USNM); 30 mi. NW Warm Springs (1 UMMZ). **Wheeler Co.**—Mitchell (1 PSU); 7 mi. S, 11 mi. W Mitchell (10 MVZ); 12 mi. SW Mitchell, Wildwood Camp, Ochoco National Forest, T12S, R20E, Sec. 33 (4 OSUFW); Shelton State Park (3 PSU); 8 mi. N, 4 mi. W Spray (1 OSUFW).

Additional localities: China Hat Butte, Deschutes National Forest, 5,000 ft.* (1 SDNHM); Deschutes National Forest* (1 SDNHM); Three Sisters Mts.* (2 UMMZ); Turpin Bridge, Fremont National Forest* (1 SDNHM).

Tamias minimus, n = 429
(Plotted in Fig. 11-6, p. 164)

Baker Co.—10 mi. N Baker [City] (7 UMMZ); 6 mi. NE Baker City (1 AMNH); Burkemont (1 AMNH); Sparta (1 USNM). **Crook Co.**—2 mi. SW Barnes (4 MVZ); Camp Meeker at Camp Creek (2 USNM); Prineville (6 USNM); 2 mi. NE Prineville (13 MVZ); Scissors Springs, Big Summit Rd., Ochoco National Forest (1 SOSC); West Camp Creek (1 USNM). **Deschutes Co.**—Arnold Ice Cave, 16 mi. SE Bend (2 USNM); west bank Deschutes River, Bend (1 USNM); 3 mi. W Bend (1 KU); 10 mi. S, 8 mi. W Bend (1 OSUFW); 4 mi. SE Bend (2 KU); 14 mi. SE Bend (2 KU); 7 mi. SE Brothers (4 KU); head of Davis Creek (1 OSUFW); 18 mi. S, 12 mi. W Fall River hatchery (1 OSUFW); La Pine (1 USNM); Lower Bridge (1 SOSC); Skeleton Cave, Deschutes National Forest (12 PSU); Sisters (1 USNM); 3 mi. W Terrebonne (1 OSUFW). **Douglas Co.**—5 mi. S Mt. Thielson (1 MVZ). **Grant Co.**—12 mi. S, 5 mi. E Canyon City (1 PSU); Strawberry Mts. (1 USNM). **Harney Co.**—(1 PSM); Alvord Lake (1 MVZ, 4 USNM); N of Alvord Ranch, 4,100 ft. (1 EOSC); Alvord Valley* (1 USNM); Beatys Butte (1 SDNHM, 1 USNM); Blitzen Valley* (4 SDNHM); 2 mi. S Borax Spring, S end Alvord Lake (1 MVZ); Buchanan (2 USNM); Buena Vista (1 OSUFW); Burns (6 USNM); 22 mi. S Burns, Malheur National Wildlife Refuge, T27S, R31E, Sec. 35 (1 OSUFW); 12 mi. S, 14 mi. W Burns, T25S, R28E, Sec. 10 (3 OSUFW); 23 mi. S, 2.3 mi. W Burns, NW Harney Lake (2 OSUFW); 23 mi. S, 9.8 mi. W Burns, NW Harney Lake, Malheur National Wildlife Refuge (3 OSUFW); 24 mi. S, 7 mi. E Burns, Malheur National Wildlife Refuge (1 OSUFW); 25 mi. NW Burns (4 USNM); Cottonwood Creek Rd. (1 PSM); Coyote Butte, T27S, R30E, Sec. 13 (2 HSU); Crane (3 USNM); Delintment Lake, T19S, R26E, Sec. 29 (1 OSUFW); 0.5 mi. N, 2 mi. E Denio, T41S, R35E, Sec. 15 (1 OSUFW); Diamond (2 USNM); 13 mi. S Fields (1 OSUFW); 15 mi. E Frenchglen, 1 mi. N Fish Lake, Steens Mt. (1 OSUFW); 15.5 mi. N, 3 mi. E Frenchglen (2 OSUFW); 2 mi. S, 2.5 mi. E Frenchglen, T32S, R31E, W½ Sec. 17 (1 OSUFW); 7.8 mi. E Lawen (1 CM); Malheur Lake, T26S, R31E, Sec. 35 (5 OSUFW); Malheur Lake, Springer Ranch* (1 LACM); Malheur Environmental Field Station, 4,200 ft. (1 CRCM); Malheur National Wildlife Refuge (1 EOSC); Malheur National Wildlife Refuge, T26S, R31E, Sec. 35 (2 OSUFW); Malheur National Wildlife Refuge, S edge Harney Lake, T27S, R29¾E, Sec. 35 (3 HSU); Malheur National Wildlife Refuge, SE edge Harney Lake, T28S, R29¾E, Sec. 35 (4 HSU); Malheur National Wildlife Refuge, SE edge Harney Lake, T27S, R30E, Sec. 19 (1 HSU); Malheur National Wildlife Refuge, Harney Lake dunes, T26S, R30E, Sec. 27 (4 CM); Malheur National Wildlife Refuge, Harney Lake dunes, T26S, R30E, Sec. 28 (1 CM); Malheur National Wildlife Refuge headquarters, T26S, R31E, Sec. 35 (1 CM); Narrows (1 LACM, 3 SDNHM, 10 USNM); 1 mi. S Narrows (11 MVZ); 5 mi. S Narrows (1 MVZ); 3 mi. E Narrows (2 CRCM); 2 mi. SW Narrows, Steens Mt., T27S, R31E, Sec. 3 (1 KU); Rock Creek Sink (5 USNM); Silver Lake (1 PSM); Smyth Creek, 10 mi. E Diamond (12 MVZ); Steens Mt.* (2 USNM); Steens Mt., Fish Lake, 8,000 ft. (1 USNM); Steens Mt., dry fork of Home Creek (4 SDNHM); N end Steens Mt., 6,400 ft. (7 FMNH); Steens Mt., T31S, R32½E, Sec. 23 (2 OSUFW); Steens Mt., T32S, R32 ¾E, Sec. 36 (1 OSUFW); Steens Mt., T32S, R32¾E, Sec. 32 (1 OSUFW);

Steens Mt., T32S, R33E, Sec. 31 (2 OSUFW); Steens Mt., T32S, R33E, Sec. 32 (4 OSUFW); Steens Mt., T32S, R33E, Sec. 36 (1 OSUFW); Steens Mt., T33S, R33E, Sec. 2 (1 KU); Voltage (2 SDNHM, 6 USNM). **Jackson Co.**—2.5 mi. E of USFS Rd. 3706 on USFS Rd. 3705 (1 SOSC). **Jefferson Co.**—3 mi. S, 2.5 mi. E Madras (1 OSUFW); 12 mi. S, 2.75 mi. E Madras (8 PSU). **Klamath Co.**—Anna Creek, 4 mi. NW Ft. Klamath, Mt. Mazama, 6,000 ft. (1 MVZ); Dutchman Flat NE of Bly (1 SOSC); Lost River, Klamath Basin (5 USNM); 4 mi. SW Merrill, Mitchell Ranch (1 OSUFW); Tule Lake (1 USNM). **Lake Co.**—Adel, Warner Valley (1 OSUFW); 20 mi. N Adel (4 OSUFW, 2 SDNHM); 9 mi. S Adel, Twentymile Creek, Warner Valley (5 OSUFW); 10 mi. NE Adel (3 PSU); NE edge Alkali Lake (1 MVZ); Cabin Lake (1 UCONN); Chandler State Park (1 SOSC); 15 mi. N Christmas Lake (1 USNM); Dutchman Flat NE of Bly* (1 SOSC); Fort Rock (1 LACM); Fort Rock State Park (4 OSUFW, 1 PSM); Fremont (4 USNM); near Fremont, Silver Lake (2 USNM); Guano Creek, 9 mi. S Hart Mt. (3 OSUFW); Guano Lake (1 PSM); Guano Valley (2 SDNHM); Hart Mt.* (1 SDNHM); Hart Mt. National Wildlife Refuge, near headquarters (3 SOSC); Hot Springs camp, Hart Mt. National Wildlife Refuge (1 SOSC); 14 mi. S Hot Springs camp, Hart Mt. National Wildlife Refuge (1 SOSC); Lakeview (1 USNM); 13 mi. N Lakeview, Chandler State Park (2 SOSC); 17 mi. N Lakeview (2 SOSC); 10 mi. S Plush (2 CRCM); Silver Lake, Silver Creek, 4,650 ft. (1 USNM); Sycan Marsh (1 USNM); 50 yds N Rock Creek campground, Hart Mt. National Wildlife Refuge (1 SOSC); T39S, R21E, Sec. 2 (1 PSU). **Malheur Co.**—(8 AMNH, 1 USNM); Battle Mt. (1 SDNHM); Beulah (2 USNM); Brogan (2 OSUFW); Butte Creek (1 PSM); Cedar Mts. (4 USNM); Cold Springs (4 USNM); 6 mi. W Cord, Duck Creek Pass, 5,000 ft. (1 USNM); Cow Creek Lake (1 USNM); Cow Lake (2 SDNHM); 0.5 mi. N Crows Nest Reservoir (1 PSM); Ironside (20 AMNH); 1 mi. W Ironside (23 MVZ); 8 mi. N Jordan Valley, 4,500 ft. (1 USNM); 5 mi. W Jordan Valley (1 PSM); 8 mi. W Jordan Valley (2 MVZ); Mahogony Mts., 6,500 ft. (3 USNM); McDermitt (4 USNM); 21 mi. N McDermitt (1 MVZ); canyon located 8 mi. NE McDermitt, 6,300 ft. (1 USNM); 5 mi. SW Ontario, 2,000 ft. (9 USNM); Rome, Owyhee River (2 USNM); 2 mi. NW Riverside (8 USNM); 25 mi. W Rome on Bone Springs Rd., T31S, R37E, Sec. 24 (1 PSU); Ryegrass, Owyhee desert, 4,400 ft. (1 USNM); Ryegrass, Owyhee desert, 4,500 ft. (3 USNM); Scale Reservoir (1 PSM); Sheep Head Mts., Bone Springs, 5,200 ft. (3 USNM); Skull Spring (8 USNM); Soldier Creek exposure (1 PSM); Succor Creek Rd.* (1 PSU); Vale (1 AMNH, 2 USNM). **Umatilla Co.**—Pikes Peak (1 UW[WWC]). **Wheeler Co.**—Mitchell (1 PSU).

Tamias senex, n = 267
(Plotted in Fig. 11-8, p. 167)

Deschutes Co.—Arnold Ice Cave (2 USNM); Bend (3 USNM); 4 mi. W Bend (1 OSUFW); 11 mi. W Bend, Tumalo Creek (6 MVZ); 16 mi. W Bend, 6,500 ft. (3 MVZ); 1 mi. S, 3 mi. W Bend (1 OSUFW); 3 mi. S, 2 mi. E Bend (1 OSUFW); Black Butte, T14S, R9E, Sec. 3 (23 different locality descriptors) (1 HSU, 58 OSUFW); head of Davis Creek, T22S, R8E, Sec. 18 (1 OSUFW); Fourmile Spring, Deschutes National Forest* (2 PSM); Indian Ford (1 OSUFW); Indian Ford campground (5 PSM); 10 mi. N, 3 mi. E La Pine (1 OSUFW); Lava Lake (1 OSUFW); Little Lava Lake (3 PSM); 3 mi. W Paulina Lake, 5,700 ft. (3 MVZ); 5 mi. W Sisters, T13S, R9E, Sec. 32 (1 OSUFW); 6 mi. W Sisters (1 OSUFW); 0.25 mi. N, 4 mi. W Sisters (1 PSU); 1.25 mi. N, 4 mi. W Sisters, Cold Spring camp (6 PSU); 6.5 mi. N, 5 mi. E Sisters (1 OSUFW); 0.75 mi. S, 4 mi. W Sisters (1 OSUFW); 6 mi. S, 4 mi. W Sisters, 4,000 ft. (1 CM, 1 OSUFW); 6 mi. S, 4 mi. E Sisters, 4,000 ft. (1 CM); Sparks Lake (2 OSUFW); 3 mi. W Terrebonne, 3,000 ft. (1 CM); 1 mi. S, 3 mi. W Terrebonne (1 OSUFW); Tumalo Creek, Bend (2 OSUFW). **Douglas Co.**—Diamond Lake (2 PSM, 12 SDNHM, 2 SOSC). **Jackson Co.**—14 mi. E Ashland, Green Spring Rd. (1 CM); 1 mi. W Crater Lake National Park (1 UMMZ); halfway between Drew and Crater Lake (4 USNM). **Jefferson Co.**—Camp Sherman (6 PSM); 12 mi. N, 2 mi. E Camp Sherman (2 OSUFW); Ollalie Forest camp (1 PSU); 1 mi. W Santiam Pass (1 KU); Suttle Lake, 3,435 ft. (22 KU, 6 OSUFW, 1 UO); 2.5 mi. W Suttle Lake (2 KU); 1 mi. E Suttle Lake (1 OSUFW); Warm Springs

(1 SDNHM); 20 mi. W Warm Springs, Mill Creek (4 OSUFW, 6 USNM); T13S, R8E, Sec. 18 (1 PSU). **Klamath Co.**—Anna Creek, Mt. Mazama, 6,000 ft. (2 USNM); Crater Lake [10 localities within 5 mi. of Crater Lake] (10 USNM); Cresent Lake, 4,800 ft. (2 LACM, 2 SDNHM, 1 USNM); Fourmile Lake (2 USNM); Ft. Klamath (1 CM, 2 PSM, 1 UMMZ, 10 USNM); 7 mi. NW Ft. Klamath (4 OSUFW); 12 mi. SW Ft. Klamath (4 UMMZ); Klamath Falls (2 CM, 1 USNM); 11 mi. SE La Pine (1 OSUFW); Naylox (2 USNM); Red Blanket Creek, Crater Lake National Park (1 MVZ); West Silver Creek, Yamsay Mts., 7,000 ft. (1 USNM); Upper Klamath Lake, Rocky Point (4 OSUFW); Upper Klamath Marsh (1 USNM); Yamsay River, Yamsay Mts., 4,800 ft. (1 USNM). **Lake Co.**—1 mi. E Bear Flat (1 SOSC); 10 mi. N Fort Rock (1 SOSC); Gearhart Mt. (3 PSM); Lakeview (1 USNM); West Silver Creek, Silver Lake, 4,650 ft. (1 USNM). **Lane Co.**—Gold Lake, 4,800 ft. (6 LACM); W base Three Sisters, 5,000 ft. (1 OSUFW, 1 PSM); NW base Three Sisters, Scott Lake, T15S, R7½E, SW¼ (1 OSUFW). **Linn Co.**—1.5 mi. N Big Lake, T13S, R7½E, NE¼ Sec. 35, 4,800 ft. (3 PSM); Brownsboro* (2 USNM); 1 mi. N, 28 mi. E Cascadia (1 CM).

Additional locality: Turpin Ridge, Fremont National Forest* (1 SDNHM).

Tamias siskiyou, n = 426
(Plotted in Fig. 11-10, p. 169)

Curry Co.—13 mi. E Agness (7 MVZ); 1 mi. N, 8 mi. E Agness (4 HSU); Brookings (1 SDNHM); 10 mi. E Brookings (1 PSM); Gold Beach (9 FMNH, 2 USNM); 8 mi. NE Gold Beach (2 MVZ); Lobster Creek (1 SDNHM); 5 mi. E Pistol River (6 MVZ); Port Orford (1 USNM); 10 mi. E of coast, 200 yds S junction Rds. 333 and 3506* (8 MVZ); stateline* (2 SDNHM). **Douglas Co.**—13 mi. S, 2 mi. E Clearwater, Umpqua National Forest (1 OSUFW); 5 mi. S Mt. Thielson (1 MVZ). **Jackson Co.**—Ashland (2 SOSC); 3 mi. S Ashland (1 SOSC); 5 mi. S Ashland (1 SOSC); 8.5 mi. W Ashland (1 SOSC); Lower Ashland Research Natural Area, T39S, R1E, SW¼ Sec. 28 (1 SOSC); Upper Ashland Research Natural Area, T40S, R1E (3 SOSC); 12 mi. E Greensprings Rd. summit, Ashland (2 SOSC); 14 mi. E Ashland, Dead Indian Rd. (1 MVZ); 3 mi. N, 10 mi. E Ashland (1 SOSC); N base Ashland Peak, 5,000 ft. (5 USNM); N base Ashland Peak, 5,200 ft. (3 USNM); N base Ashland Peak, Longs Camp, 3,300 ft. (5 USNM); Butte Falls, NE of Medford (6 OSUFW); Conde Creek Research Area on Dead Indian Rd. (2 SOSC); East Fork Ashland Creek, T40S, R1E, SE¼SE¼ Sec. 4 (3 SOSC); 5 mi. N, 2 mi. W Lincoln, Hyatt Lake (1 OSUFW); 6 mi. SW Medford (1 PSM); Mt. Ashland summit (1 SOSC); Mt. Ashland, 5,000 ft. (1 SOSC); 1 mi. E Mt. Ashland (1 SOSC); Mt. Ashland Rd. (3 SOSC); Prospect (1 MVZ, 3 OSUFW, 10 SOSC, 7 USNM); 10 mi. E Prospect, Bessie Creek (1 MVZ); 2.5 mi. N, 8.5 mi. E Prospect (1 CRCM); 5 mi. N, 8 mi. E Prospect (2 CRCM); 8 mi. SW Prospect, 1,800 ft. (3 CM, 1 SOSC); 12 mi. E Shady Cove (10 MVZ); Siskiyou (40 USNM); 7 mi. W Siskiyou Pass (2 MVZ); Talent (1 SOSC); Union Creek (8 MVZ); Union Creek at Natural Bridge (1 MVZ); 5 mi. W Union Creek (4 MVZ); 2 mi. N, 1 mi. E Union Creek (1 CRCM); 4 mi. NE Union Creek (10 MVZ); 5 mi. NE Union Creek (1 MVZ); 3 mi. SW Union Creek (1 MVZ); junction of Lake in the Woods and Howard Prairie rds. (1 SOSC); 0.7 mi. N Dead Indian Rd. on Conde Rd. (2 SOSC); 1.4 mi. N Dead Indian Rd. (2 SOSC); 1 mi. off HWY 99 on Mt. Ashland Rd. (1 SOSC); Neil Creek Rd. 1.6 mi. SW HWY 66 (1 SOSC); 0.8 mi. E USFS Rd. 3706 on USFS Rd. 3705 (2 SOSC); 2 mi. N Dead Indian Rd. on USFS Rd. 3706 (3 SOSC); 5.6 mi. S HWY 99 on Tolman Creek Rd. (1 SOSC); 1.1 mi. E USFS Rd. 3706 (4 SOSC); junction Weyerhauser Rd. and and Crane Prairie Rd. (6 localities within 2 mi.) (11 SOSC); 2.8 mi. from HWY 66 on HWY 99 (1 SOSC); 11.5 mi. from HWY 66 on Dead Indian Rd. (1 SOSC); T39S, R4W, SE¼ Sec. 8 (1 SOSC); T39S, R4E, Sec. 16/17 (1 SOSC); T40S, R1E, Sec. 3 (1 SOSC); T40S, R1E, Sec. 9 (1 SOSC). **Josephine Co.**—Bolan Lake (30 PSM, 6 SDNHM); Brooking* (1 MVZ); Cave Creek campground, T40S, R6W, SE¼ Sec. 5 (8 KU); 10 mi. E Cave Junction (3 KU); 15 mi. E Cave Junction (1 KU); 17 mi. E Cave Junction (3 KU); 1 mi. S, 9 mi. E Cave Junction (1 SOSC); 1.5 mi. S, 9 mi. E Cave Junction, T39S, R6W, SW¼ Sec. 30 (17 KU); 10 mi. S, 8 mi. E Cave Junction (1 OSUFW); 10 mi. S, 10 mi. E Cave Junction, Bolan Lake,

T41S, R6W, Sec. 7 (1 OSUFW); 3 mi. W Galice (10 MVZ); 13 mi. SW Galice, Briggs and Horse creeks, Farren Ranger Station, 3,700 ft. (8 USNM); 15 mi. SW Galice (6 UO); 18 mi. SW Galice, Briggs Valley (6 MVZ); 25 mi. W Grants Pass (4 UO); Gray Back Rd. 0.5 mi. E Cedar guard station, T39S, R6W, Sec. 30, 1,900 ft. (2 PSU); Grayback campground (1 SOSC); Oregon Caves National Monument (4 MVZ, 3 OSUFW, 5 SDNHM). **Klamath Co.**—Anna Creek, 4 mi. NW Ft. Klamath, 4,500 ft. (6 MVZ); Big Marsh Creek, 8 mi. S Cresent Junction, T25S, R7E, Sec. 18 (3 PSU); N of Buck Peak, T38S, R6E, Sec. 8 (1 SOSC); W of Buck Peak, T38S, R6E, Sec. 7/8 (3 SOSC); SE of Buck Peak, T38S, R6E, Sec. 8 (1 SOSC); Crater Lake National Park (2 PSU, 1 SOSC); S entrance Crater Lake National Park (1 MVZ); Crater Lake National Park, Goodby Picnic area, T31S, R6E, Sec. 19/20 (1 PSU); Crater Lake National Park, T31S, R7½E (1 OSUFW); Crater Lake National Park, Upper Munson Meadows (1 OSUFW); 20 mi. W Cresent Lake (27 UO); 6 mi. E Cresent Lake (1 UO); 7 mi. E Cresent Lake (1 UO); S base Hammond Peak, Cresent Lake (1 UO); S end Lake of the Woods (1 MVZ); ca. 1.5 mi. S Odell Lake, T23S, R6½E, Sec. 36 (1 PSU); Sun Creek, Crater Lake National Park (1 MVZ); E base Union Peak (1 MVZ); T38S, R5E, NE¼SE¼ Sec. 21 (2 SOSC); T38S, R6E, Sec. 8/9 (1 SOSC); T38S, R6E, Sec. 8/15 (2 SOSC). **Lane Co.**—Gold Lake (1 LACM); 2.5 mi. W McKenzie Bridge (1 KU); 6 mi. W McKenzie Pass (1 KU); Middle Sister, Three Sisters, Camp Riley (1 OSUFW); 27.3 mi. SE Oakridge (1 KU); N base Three Sisters, 5,000 ft. (1 USNM); NW base Three Sisters, Scott Lake, T15S, R7½E, SW¼ (4 OSUFW); Upper Horse Creek, near Horse Lake (1 OSUFW). **Linn Co.**—19 mi. E Upper Soda (1 OSUFW).

Tamias townsendii, *n* = 1,871
(Plotted in Fig. 11-12, p. 170)

Benton Co.—(1 OSUFW); Alsea (1 SDNHM); 5 mi. N, 3 mi. E Alsea (1 OSUFW); 6 mi. N, 5 mi. E Alsea (1 OSUFW); 7 mi. N, 3 mi. E Alsea, Marys Peak (1 OSUFW); 2 mi. S, 3 mi. W Alsea (2 OSUFW); 3 mi. S, 2 mi. E Alsea, T14S, R7W, Sec. 20 (2 OSUFW); 3 mi. S, 3 mi. E Alsea, T14S, R7W, Sec. 20 (1 OSUFW); 3.75 mi. S, 5.5 mi. E Alsea (3 CM); 10 mi. SE Alsea (12 HSU); 5 mi. E Bellfountain (2 OSUFW); 1 mi. S, 0.5 mi. W Blodgett (1 OSUFW); 4 mi. S, 0.75 mi. W Blodgett (1 OSUFW); 6 mi. S, 1.5 mi. W Blodgett (2 OSUFW); 6 mi. S, 2 mi. W Blodgett (1 OSUFW); Corvallis (3 KU, 4 OSUFW, 1 USNM); 1.5 mi. N Corvallis, T11S, R5W, Sec. 15 (1 OSUFW); 2.5 mi. N Corvallis, T11S, R5W, Sec. 14 (1 OSUFW); 4 mi. N Corvallis, T11S, R5W, Sec. 15 (1 OSUFW); 5 mi. N Corvallis (1 OSUFW); 7 mi. N Corvallis (1 PSM); 8 mi. N Corvallis, T11S, R5W, Sec. 25 (1 OSUFW); 9 mi. N Corvallis, T10S, R5W, Sec. 21 (1 OSUFW); 10 mi. N Corvallis, 250 ft. (1 CM); 10.5 mi. N Corvallis (2 MVZ); 1 mi. S Corvallis, T12S, R5W, Sec. 12 (4 OSUFW); 3 mi. S Corvallis, T11S, R4W, Sec. 20 (1 OSUFW); 10 mi. S Corvallis, T14S, R6W, Sec. 4 (2 OSUFW); 5 mi. W Corvallis (1 SOSC); 0.25 mi. N, 0.25 mi. W Corvallis (1 OSUFW); 1 mi. N, 3 mi. W Corvallis (1 OSUFW); 2 mi. N, 9 mi. W Corvallis (1 OSUFW); 2.5 mi. N, 1.5 mi. W Corvallis (1 OSUFW); 0.4 mi. S, 0.1 mi. W Corvallis (1 OSUFW); 1 mi. S, 1 mi. W Corvallis (1 OSUFW); 2 mi. S, 17 mi. W Corvallis (1 OSUFW); 5 mi. S, 3 mi. E Corvallis (1 OSUFW); 4 mi. NW Corvallis, T11S, R5W, Sec. 5 (1 OSUFW); 5 mi. NW Corvallis, Oak Creek, T11S, R5W, Sec. 19 (1 OSUFW); 7 mi. NE Corvallis (1 OSUFW); 23 mi. SW Corvallis, Grass Mt., T13S, R8W, Sec. 21 (5 OSUFW); 4 mi. S, 2 mi. W Greenberry (1 OSUFW); 1 mi. S, 1.5 mi. E Hoskins (1 OSUFW); Marys Peak (3 OSUFW, 1 PSM); Philomath, T13S, R6W, Sec. 15 (1 OSUFW); 3.5 mi. W Philomath (1 OSUFW); 5 mi. N, 4.5 mi. W Philomath (1 OSUFW); 2 mi. S, 8 mi. W Philomath, Marys Peak (3 OSUFW); 2 mi. S, 9 mi. W Philomath, Marys Peak, North Fork Rock Creek (4 OSUFW); 4 mi. S, 5 mi. W Philomath, Starker Forest (1 OSUFW); 4 mi. S, 7 mi. E Philomath, Marys Peak (1 OSUFW); 5 mi. SW Philomath (5 USNM); near summit Prairie Mt. (1 PSM); T10S, R5W, Sec. 10 (1 OSUFW). **Clackamas Co.**—(3 AMNH, 1 OSUFW, 1 USNM); Barlow campground, Zigzag River (1 PSU); Bissell (1 USNM); Boring (3 PSU); N side Bull Run Lake (1 PSU); SE side Bull Run Lake (1 PSU); 0.5 mi. S Bull Run Lake (1 PSU); Cherryville (1 PSU); 1 mi. N Clackamas (1 PSU); 5 mi. E Clackamas (1 PSU); Clackamas River* (1 PSU); Government Camp (4

MVZ, 1 PSU); 0.25 mi. N, 2.75 mi. W Government Camp (2 PSU); 0.5 mi. N, 1.75 mi. W Government Camp (1 PSU); 0.75 mi. N, 2.75 mi. W Government Camp (2 PSU); 1 mi. S, 2.25 mi. E Government Camp (1 PSU); 2 mi. N, 1.5 mi. W Ladd Hill (2 PSU); Lake Oswego (2 PSU); 2 mi. N Lake Oswego (1 PSU); Logan (3 AMNH); 4 mi. NE Aurora (1 PSU); Minthorne, 2 mi. S Portland (1 OSUFW); 5 mi. E Molalla, T5S, R2E, Sec. 22 (1 OSUFW); 7 mi. SE Molalla (1 PSM); Mulino (5 OSUFW, 1 PSU); Oregon City (1 USNM); 1 mi. S, 0.5 mi. E Oregon City (1 PSU); Rhododendron (1 LACM, 1 PSU, 4 MVZ, 1 SDNHM); Still Creek Forest camp, Mt. Hood National Forest (1 PSM); Timothy Lake Rd. at Oak Grove Fork Crossing, T5S, R8½E, Sec. 26 (2 PSU); Trillium Lake, S slope Mt. Hood, T3S, R8½E, Sec. 36 (1 OSUFW); Tryon Creek State Park (1 PSU); Wilsonville (1 PSU); 9 mi. N, 4 mi. W Zigzag (1 CRCM); 5 mi. S, 6 mi. E Zigzag (3 CRCM); junction of Oak Grove and Timothy Lake Rd. (3 PSU); T2S, R1E, Sec. 15 (2 PSU); T2N, R2E, Sec. 12 (1 PSU); T5S, R8E, Sec. 26 (13 PSU). **Clatsop Co.**—(1 USNM); Astoria (1 PSM, 1 UMMZ, 9 USNM); Green Mt., 10 mi. E Jewell (2 KU); 3 mi. S, 1 mi. W Hammond (4 OSUFW); Klootchy Creek campground (1 LACM); site of Old Ft. Clatsop (18 MVZ); Saddle Mt. (1 SDNHM); Saddle Mt. State Park access road, 3.7 mi. off HWY 26 (1 CRCM); Seaside (1 KU, 1 MVZ, 1 SOSC, 11 USNM); 6 mi. S, 4 mi. E Seaside (1 OSUFW); T4N, R7W, Sec. 10, 1,000 ft. (1 PSU); T8N, R10W, W½ Sec. 6 (1 PSM). **Columbia Co.**—(1 PSU, 1 UW); Deer Island (1 PSU); Rainier (1 SDNHM); 5 mi. W Rainier (1 UMMZ); 7 mi. SE Rainier (5 MVZ); 13 mi. E Scappoose, between St. Helens and Vernonia (1 PSU). **Coos Co.**—8 mi. S Bandon (1 OSUFW); 10 mi. S Bandon (1 PSM); 1.75 mi. E Bandon (2 PSM); 4 mi. SE Bandon (5 PSM); 1 mi. NE Bridge E of Indian Mary Park (1 MVZ); Charleston (2 LACM); 5 mi. SW Charleston, Sunset Bay (2 LACM); Coos Head (5 LACM); Delmar, 9 mi. S Marshfield (1 MVZ); Empire (4 USNM); 4 mi. W Glasgow (1 OSUFW); 11 mi. NE Illahe (4 MVZ); Marshfield (9 FMNH, 1 PSM, 17 USNM); Myrtle Point (8 USNM); Remote, 5 mi. up Sandy Creek (1 SDNHM); 7.5 km N, 12.7 km W Sitkum, T27S, R11W, Sec. 21 (1 USNM); 2.9 km S, 4.7 km E Sitkum, T28S, R9W, Sec. 19 (1 USNM); 4.8 km S, 9.7 km E Sitkum, T28S, R9W, Sec. 27 (1 USNM); 4.6 mi. N, 2.6 mi. W Sitkum, T27S, R10W, Sec. 17, 1,100 ft. (6 UW); 4.7 mi. N, 3.2 mi. W Sitkum, T27S, R10W, Sec. 18, 1,200 ft. (28 UW); 4.7 mi. N, 7.8 mi. W Sitkum, T27S, R11W, Sec. 21, 260 ft. (12 UW); 1.8 mi. S, 2.9 mi. E Sitkum, T28S, R9W, Sec. 19, 1,600 ft. (27 UW). **Curry Co.**—Agness (1 FMNH); 5 mi. N Agness (3 HSU); 16 mi. NE Gold Beach, N side Rogue River (2 MVZ); Little Redwood Forest camp (11 PSM); 13 mi. NE of bridge at Lobster Creek, Rd. 343 (10 MVZ); Port Orford (2 USNM); 2 mi. S Port Orford (1 PSM); 3 mi. E Port Orford (1 MVZ); 6 mi. E Port Orford, Elk Creek (1 MVZ); 18 mi. E Port Orford (1 MVZ); 10 mi. ESE Port Orford (2 PSM). **Douglas Co.**—Anchor (7 USNM); 3.2 mi. S, 3.6 mi. E Canyonville, T31S, R4W, Sec. 17, 2,000 ft. (1 UW[CR]); Drain (2 USNM); 4.5 mi. N, 8 mi. W Drain, T21S, R4W, Sec. 27, 1,200 ft. (9 UW); 5.5 mi. N, 4 mi. W Drain, Peterson Point, T21S, R6W, Sec. 15, 1,280 ft. (8 UW, 6 UW[CR]); 6 mi. N, 4 mi. W Drain, T21S, R6W, Sec. 15, 1,200 ft. (6 UW, 2 UW[CR]); 6.75 mi. N, 5.5 mi. E Drain, T21S, R4W, Sec. 7, 1,000 ft. (6 UW); Elkhead (1 USNM); 11.1 km S, 7.4 km W Elkton, T23S, R8W, Sec. 28 (1 USNM); 15.4 km S, 6 km W Elkton, T24S, R8W, Sec. 10 (1 USNM); 7.2 km S, 0.8 km E Elkton, T23S, R7W, Sec. 17 (1 USNM); Gardiner (3 FMNH); Glendale (33 USNM); Louis Creek, T28S, R3W, NW¼ Sec. 29 (2 SOSC); Middle Creek, T31S, R6W, SW¼ Sec. 30 (3 SOSC); Middle Creek, T31S, R7W, SE¼ Sec. 25 (2 SOSC); Oxbow Burn, ca. 40 mi. SW Eugene (21 PSM); Reston (2 USNM); 4 mi. W Riddle, Iron Mt., T31S, R7W, Sec. 11, 2,000 ft. (10 UW[CR]); 5 mi. SE Roseburg, T27S, R5W, Sec. 36 (1 OSUFW); Scottsburg (1 FMNH, 2 USNM); Smith River near Gardiner (1 USNM); 1.2 mi. N, 15 mi. W Sutherlin, T25S, R8W, Sec. 10, 1,020 ft. (1 UW); 1.2 mi. N, 15.5 mi. W Sutherlin, T25S, R8W, Sec. 10, 1,000 ft. (3 UW); 14 mi. NW Sutherlin (1 MVZ); 7.5 mi. N, 8 mi. E Tiller (2 CRCM); 11.5 mi. N, 5 mi. E Tiller, T28S, R1W, Sec. 26 (1 CRCM); 12.5 mi. N, 5 mi. E Tiller, T28S, R1W, Sec. 32 (1 CRCM); 12.5 mi. N, 5.5 mi. E Tiller, T28S, R1W, Sec. 31 (4 CRCM); 0.2 mi. N, 4.1 mi. W Umpqua P.O., T25S, R7W, Sec. 15, 800 ft. (3 UW[CR]); 2.2 mi. N, 9 mi. W Umpqua P.O., T25S, R8W, Sec. 15, 850 ft. (1 UW, 4 UW[CR]); 2.7 mi. N, 5 mi. W Umpqua, T25S, R7W, Sec. 7, 950 ft. (17 UW, 8

UW[CR]); 4 mi. N, 1 mi. W Umpqua P.O., T25S, R7W, Sec. 1, 1,500 ft. (1 UW, 5 UW[CR]); 6 mi. N, 10.5 mi. W Umpqua, T24S, R8W, Sec. 29, 900 ft. (18 UW, 9 UW[CR]); 4.8 km S, 4.5 km W Yoncalla, T23S, R5W, Sec. 19 (1 USNM). **Hood River Co.**—(3 UW); Bear Springs, Mt. Hood (2 SDNHM); 1 mi. E Bear Springs, Mt. Hood (1 SDNHM); Brooks Meadows, 9 mi. ENE Mt. Hood (2 KU); Eagle Creek National Forest campground (5 LACM); 10 mi. E Government Camp (3 SDNHM); Hood River (1 USNM); 2 mi. S, 8 mi. W Hood River (1 OSUFW); Hood River Meadows (3 PSM); Government Camp, Mt. Hood, 3,600 ft. (2 USNM); N slope Mt. Hood, 2,800 ft. (6 USNM); W base Mt. Hood (1 USNM); SE Mt. Hood, E base Cascade Mts. (2 USNM); Mt. Hood Loop HWY, near Cloud Cap cutoff (1 PSM); Parkdale (2 UMMZ, 2 UW); 2 mi. W Parkdale, 1,500 ft. (2 USNM); T3N, R9E, SW¼ Sec. 35 (5 PSM). **Jackson Co.**—Imnaha Creek, T33S, R4E, W½ Sec. 15 (1 SOSC). **Josephine Co.**—1 mi. NE of bridge E of Indian Mary Peak (1 MVZ); 1 mi. NE Indian Mary campground on Stratton Creek Rd. (2 MVZ); 4 mi. NE Indian Mary campground on Hog Creek Rd. (7 MVZ). **Lane Co.**—near Ada, T19S, R11W, Sec. 30 (1 OSUFW); 2 mi. S, 1 mi. E Austa (1 CM); Belknap Springs (2 OSUFW); 4 mi. N Belknap Springs (1 UO); Blue River (2 USNM); near Blue River, T15S, R5E, Sec. 14, 800 m (3 KU); near Blue River, T17S, R4E, SW¼ Sec. 20, 375 m (1 KU); 2 mi. W Blue River, Elk Creek, T16S, R4E, Sec. 19 (1 OSUFW); 4 mi. W Blue River (1 KU); 6 mi. E Blue River, Swamp Creek, T16S, R5E, Sec. 20 (4 OSUFW); 9 mi. E Blue River (1 CRCM); 4 mi. N, 9 mi. E Blue River, Mack Creek Tributary, T15S, R5E, Sec. 35 (2 OSUFW); 5 mi. N, 9 mi. E Blue River (3 CRCM); 5 mi. N, 10.5 mi. E Blue River, T15S, R5E, Sec. 30 (1 CRCM); 5.5 mi. N, 7 mi. E Blue River, T15S, R5E, Sec. 26 (5 HSU); 7 mi. N, 9 mi. E Blue River, T15S, R5E, Sec. 13 (2 CRCM); 4.4 km N, 6.8 km E Blue River, T16S, R5E, Sec. 7 (1 USNM); 5 km N, 7.6 km E Blue River, T16S, R5E, Sec. 6 (1 USNM); 2.5 mi. S, 7 mi. E Blue River (1 CRCM); 10 mi. NW Blue River, Lookout Creek Tributary, T15S, R5E, Sec. 35 (1 OSUFW); 10 mi. NE Blue River, Upper Tributary Lookout Creek, T15S, R5E, Sec. 25 (2 OSUFW); 10 mi. SE Blue River, East Fork Creek, T17S, R6E, Sec. 7 (1 OSUFW); 5 mi. N Cougar Dam, T16S, R5E, Sec. 6 (3 UW[OG]); Elmira (1 UO); Empire (1 UO); Eugene (2 OSUFW, 2 USNM); Florence (4 FMNH, 8 USNM); 6 mi. N Florence, Sutton Lake (1 UO); 4 mi. S Florence, W side Woahink Lake (1 LACM); 5 mi. S Florence, W side Siltcoos Lake (2 LACM); 5.5 mi. S Florence (1 PSM); Heceta Head (2 UO); 4 mi. NE Heceta Head, Big Creek (1 OSUFW); H. J. Andrews Experimental Forest* (1 CRCM); H. J. Andrews Experimental Forest, T15S, R5E, SW¼ Sec. 13, 3,000 ft. (24 PSM); H. J. Andrews Experimental Forest, T15S, R5E, NW¼ Sec. 24, 3,500 ft. (24 PSM); H. J. Andrews Experimental Forest, T15S, R5E, NW¼ Sec. 28, 2,000 ft. (32 PSM); H. J. Andrews Experimental Forest, T15S, R5E, NE¼ Sec. 32, 2,000 ft. (65 PSM); H. J. Andrews Experimental Forest, T16S, R4E, NE¼ Sec. 19, 1,800 ft. (17 PSM); H. J. Andrews Experimental Forest, T16S, R4E, NE¼ Sec. 24, 1,400 ft. (28 PSM); H. J. Andrews Experimental Forest, T16S, R5E, NW¼ Sec. 6, 1,600 ft. (73 PSM); H. J. Andrews Experimental Forest, T15S, R5E, SW¼ Sec. 33, 2,500 ft. (27 PSM); H. J. Andrews Experimental Forest, T15S, R5E, NW¼ Sec. 34, 2,700 ft. (14 PSM); Horsepasture Mt. (1 OSUFW); 3 mi. S Huckleberry Mt., 2,300 ft. (1 LSU); Mapleton (3 USNM); McKenzie Bridge (6 OSUFW, 1 SDNHM, 1 UMMZ, 3 UO, 6 USNM); McKenzie Bridge, Ollalie Creek camp (2 PSM); 0.5 mi. W McKenzie Bridge (1 USNM); 2.5 mi. W McKenzie Bridge (1 KU); 10 mi. E McKenzie Bridge (1 USNM); 5 mi. S McKenzie Bridge, Spring Creek, T16S, R6E, Sec. 26 (1 OSUFW); 10 mi. S McKenzie Bridge, Horse Pasture Mt. (2 UO); 10 mi. S McKenzie Bridge, O'Leary Mt. (7 USNM); 10 mi. W McKenzie Bridge (1 UO); 0.5 mi. S, 8 mi. E McKenzie Bridge (2 OSUFW); 3 mi. S, 8.5 mi. E McKenzie Bridge, T16S, R7E, Sec. 32, 3,500 ft. (5 UW[OG]); 3.5 mi. S, 9 mi. E McKenzie Bridge, T16S, R7E, Sec. 32, 3,500 ft. (1 UW[OG]); 10 mi. SE McKenzie Bridge, Lost Creek Ranger Station (1 OSUFW, 2 USNM); Mercer (2 OSUFW); 4.5 mi. N Nimrod (1 CRCM); 2 mi. SE Oakridge (1 UO); 3 mi. SE Oakridge (16 UO); 3 mi. N Reed (1 OSUFW); mouth of Salt Creek* (3 UO); Seaton (5 USNM); Siltcoos camp (3 PSM); Spencer Butte (1 UO); Triangle Lake (1 PSM); 1 mi. S, 3 mi. W Veneta (1 MVZ); Vida (2 UO, 3 USNM); 1.66 mi. S, 6.125 mi. W Walton (3 PSU); T16S, R6E, Sec. 17 (1 USNM). **Lincoln Co.**—0.25 mi. W Agate Beach, Yaquina Head

(1 OSUFW); 6 mi. N, 1 mi. W Burnt Woods, 1,000 ft. (1 CM); Cascade Head Experimental Forest (12 PSM); Delake (1 PSM); Depoe Bay (1 LACM); 1.25 mi. W Harlan (2 KU); Newport (1 MVZ, 9 USNM); 1 mi. E Newport (1 UMMZ); Otis (2 USNM); 1.5 mi. S Salado (1 OSUFW); Waldport (1 OSUFW, 1 PSU, 1 UMMZ); Yachats (1 OSUFW); Yaquina Bay (6 USNM). **Linn Co.**—(1 PSM); 7 mi. N, 8 mi. E Blue River, T15S, R5E, Sec. 15 (2 CRCM); 7.5 mi. N, 7.5 mi. E Blue River, T15S, R5E, Sec. 15 (2 CRCM); 9 mi. N, 8 mi. E Blue River (5 CRCM); 9.5 mi. N, 9 mi. E Blue River, T15S, R5E, Sec. 2 (3 CRCM); 12 mi. S, 4 mi. E Brownsville (2 OSUFW); 2 mi. E Cascadia (1 OSUFW); 7 mi. E Cascadia (2 OSUFW); 8 mi. E Cascadia (3 OSUFW); 0.5 mi. N, 4 mi. E Cascadia (1 OSUFW); 1 mi. S, 6 mi. E Cascadia, Falls Creek (1 OSUFW); 17 mi. E Cascadia, 1 mi. E Tombstone Prairie (3 OSUFW); 2.75 mi. S, 3.5 mi. E Corvallis (1 OSUFW); 1.5 mi. E South Shore camp Detroit Reservoir (2 PSU); Fish Lake (18 PSM); Green Peter, Candy Rd. (2 OSUFW); H. J. Andrews Experimental Forest, T14S, R6E, NE¼ Sec. 20, 4,400 ft. (29 PSM); H. J. Andrews Experimental Forest, T14S, R6E, NE¼ Sec. 28, 4,600 ft. (46 PSM); H. J. Andrews Experimental Forest, T15S, R5E, SE¼ Sec. 11, 3,200 ft. (43 PSM); H. J. Andrews Experimental Forest, T15S, R6E, NE¼ Sec. 7, 4,300 ft. (107 PSM); H. J. Andrews Experimental Forest, T15S, R6E, NW¼ Sec. 7, 5,000 ft. (33 PSM); Ice Cap camp, Willamette National Forest (1 PSM); 5 mi. S Lebanon (1 MSUMZ); Lost Prairie campground (22 PSU); summit McKenzie Pass (1 OSUFW); Permilia Lake, east base Mt. Jefferson (11 USNM); 2 mi. W Tangent, T12S, R5W, Sec. 22 (2 OSUFW); Trout Creek Forest camp, T13S, R4E, Sec. 32 (1 OSUFW); 8 mi. E Upper Soda, Tombstone Prairie (5 OSUFW); 9 mi. E Upper Soda, Hackleman Creek, T13S, R6E, Sec. 27 (1 HSU, 1 OSUFW); 13 mi. E Upper Soda, Toad Creek Rd. (1 HSU, 8 OSUFW); 17 mi. E Upper Soda (1 HSU, 1 OSUFW). **Marion Co.**—4.5 mi. S, 1.5 mi. E Bagby Hot Springs, 4,000 ft. (8 LSU); Bald Mt., head Clackamas River (1 USNM); Bald Peak State Park (1 PSU); Detroit (1 OSUFW, 7 USNM); Salem (7 AMNH, 1 OSUFW, 1 UMMZ, 2 USNM); 12 mi. N Salem (1 OSUFW); 5 mi. E Salem (1 OSUFW); 6 mi. S, 5 mi. E Silverton (2 OSUFW); 9 mi. SE Silverton, T8S, R1E, Sec. 14 (1 OSUFW); 4 mi. N, 2.5 mi. E Stayton (1 OSUFW); Woodburn (2 PSM). **Multnomah Co.**—1 mi. N, 5 mi. E Clackamas, T2S, R3E, Sec. 6 (4 PSU); 13 mi. E Corbett, T1N, R6E, Sec. 29 (1 HSU); 14 mi. E Corbett, T1N, R6E, Sec. 29 (2 HSU); Gresham, 15 mi. E Portland (1 USNM); Larch Mt. (1 PSU); Latourelle (1 OSUFW); 0.5 mi. E Multnomah Falls, Columbia River (1 OSUFW); Portland (10 AMNH, 1 LACM, 2 MVZ, 7 PSM, 33 PSU, 1 UMMZ, 22 USNM); E. Portland (2 USNM); Ramapo, 10 mi E Portland (1 USNM); 8 mi. N, 11 mi. E Sandy (1 CRCM); 9 mi. N, 10 mi. E Sandy (1 CRCM); Sellwood [= Portland] (2 USNM); 11 mi. N Zigzag (1 CRCM); 8 mi. N, 3 mi. E Zigzag (3 CRCM); T1N, R2E, Sec. 22 (1 PSU); T1S, R1E, Sec. 31 (1 PSU); T1S, R2E, Sec. 24 (1 PSU). **Polk Co.**—1 mi. S, 2 mi. W Buena Vista (1 OSUFW); Grande Ronde (3 USNM); 3 mi. S, 3 mi. E Monmouth (1 OSUFW); 4 mi. S, 3 mi. E Willamina, Mill Creek (1 OSUFW). **Tillamook Co.**—(1 UW); Beaver (2 KU, 1 MVZ, 3 OSUFW); Blaine (2 UMMZ); Elk Creek Rd. near HWY 6 (1 PSU); Hebo Lake (2 PSM); 2 mi. up Miami River, Tillamook (2 PSM); Mt. Hebo (3 PSM); Nehalem (2 KU, 1 OSUFW, 1 SDNHM); Netarts (3 LACM, 8 OSUFW, 2 PSM, 2 PSU, 10 SDNHM, 3 USNM); Netarts Bay (7 UO); Tillamook (12 AMNH, 1 MVZ, 3 OSUFW, 1 SDNHM, 1 SOSC, 1 UO, 10 USNM); 7 mi. SE Tillamook (3 OSUFW); Wilson River, Tillamook (1 USNM). **Wasco Co.**—(3 UW); Clear Lake, T4S, R9E, Sec. 32 (1 OSUFW); Frog Lake, T4S, R9E, Sec. 17 (1 OSUFW); 20 mi. W Maupin, HWY 216 (1 PSU); Mt. Hood HWY, Mt. Hood National Forest* (5 PSM); The Dalles (1 OSUFW); 14 mi. SW The Dalles, Fivemile Creek (2 OSUFW); 22 mi. SW The Dalles (1 USNM); 25 mi. SW The Dalles (2 USNM); Wapinitia (1 USNM); 20 mi. NW Warm Springs (1 OSUFW); T5S, R9E, Sec. 4 (1 PSU). **Washington Co.**—(3 PSU); Beaverton (3 AMNH, 1 PSM, 1 PSU, 1 SDNHM, 1 USNM); 1.75 mi. NW Beaverton (1 PSU); Forest Grove (7 USNM); Gales Creek (1 UW); 5 mi. S, 1 mi. E Hillsboro (4 PSU); 18.5 mi. NW Portland (9 MVZ); 21.5 mi. NW Portland (2 MVZ); Scholls (1 PSU); Sherwood (2 PSU); Sherwood Junction (2 PSU); 20 mi. S, 0.7 mi. E Tigard (1 PSU); 5 mi. SW Tigard (2 PSU); Tualatin (1 AMNH); 5 mi. S, 1 mi. W Tualatin (1 PSU); T1N, R1W, Sec. 7 (1 PSU). **Yamhill Co.**—Dundee, T4S, R3W, Sec. 13 (2 PSU); Grand Ranch* (1 UMMZ);

McMinnville (1 PSU); 7.5 mi. W McMinnville (1 UW[WWC]); Trask Mt. (1 KU); 2 mi. N, 10 mi. E Yamhill (1 PSU).

Additional localities: Columbia River* (1 USNM); Rogue River Mts., 3,000 ft.* (1 USNM).

Tamiasciurus douglasii, n = 1,065
(Plotted in Fig. 11-40, p. 213)

Baker Co.—12 mi. S, 15 mi. E Prairie City, North Fork Malheur River (2 MVZ); 14 mi. S, 15 mi. E Prairie City, T15S, R35½E, Sec. 22 (1 KU); 21 mi. SE Prairie City, North Fork Malheur River (3 MVZ); 7.5 mi. N, 7 mi. W Unity, T12S, R36E, Sec. 7 (1 USNM); 5 mi. S, 8 mi. W Unity, T14S, R35½E, Sec. 12 (1 USNM); 8 mi. S, 1 mi. E Unity, T14S, R37E, Sec. 27 (3 KU). **Benton Co.**—6 mi. W Albany, T11S, R4W, Sec. 17 (1 OSUFW); 1 mi. S, 3 mi. E Alpine (1 OSUFW); 5 mi. N, 4 mi. E Alsea (1 OSUFW); 3.75 mi. S, 5.5 mi. E Alsea (1 CM); 10 mi. SE Alsea (1 HSU); 4 mi. S, 1 mi. W Blodgett (1 OSUFW); Chintimini Mt., 4,000 ft. (1 USNM); Chintimini Mt., 5,000 [sic] ft. (1 USNM); Corvallis (8 OSUFW, 3 USNM); 4 mi. N Corvallis, T11S, R4W, Sec. 17 (1 OSUFW); 5 mi. N Corvallis (1 KU); 6 mi. N Corvallis (1 PSM); 9 mi. N Corvallis (1 KU); 0.5 mi. S Corvallis (1 KU); 3 mi. S Corvallis (1 OSUFW); 15 mi. W Corvallis (1 OSUFW); 1.7 mi. N, 0.7 mi. W Corvallis (1 OSUFW); 2.5 mi. N, 1.5 mi. W Corvallis (1 OSUFW); 5 mi. N, 4 mi. W Corvallis (1 OSUFW); 5 mi. N, 5 mi. W Corvallis (1 OSUFW); 7 mi. N, 1 mi. W Corvallis, E. E. Wilson Game Management Area (1 OSUFW); 2 mi. S, 1.7 mi. W Corvallis (1 OSUFW); 2 mi. S, 17 mi. W Corvallis (1 OSUFW); 4 mi. S, 3 mi. W Corvallis (2 OSUFW); 8 mi. S, 8 mi. W Corvallis (1 OSUFW); 10 mi. S, 3.5 mi. W Corvallis (1 OSUFW); 5 mi. S, 1 mi. E Corvallis (1 OSUFW); 5 mi. S, 3 mi. E Corvallis (2 OSUFW); 2 mi. N, 6 mi. W Fisher (1 OSUFW); Marys Peak campground (1 OSUFW); 7 mi. W Philomath (1 OSUFW); 7.5 mi. W Philomath (2 OSUFW); 1 mi. S Philomath (1 OSUFW); 0.5 mi. N, 1 mi. W Philomath (2 OSUFW); 0.25 mi. S, 10 mi. W Philomath (1 OSUFW); 1.5 mi. S, 7 mi. W Philomath (1 OSUFW); 2 mi. S, 5 mi. W Philomath (2 OSUFW); 2.2 mi. S, 9.5 mi. W Philomath, Marys Peak (1 OSUFW); 4 mi. S, 5 mi. W Philomath, Starker Forest (1 OSUFW); 5 mi. S, 6 mi. W Philomath (1 OSUFW); 5 mi. SW Philomath (6 USNM); N base Prairie Mt. (2 PSM); 1 mi. N, 2 mi. E Summit, 250 ft. (1 CM); Wells (2 USNM); 3 mi. N Wrens, 600 ft. (1 CM); 3 mi. W Wrens, above Marys River (1 OSUFW); T11S, R5W, Sec. 17 (1 OSUFW). **Clackamas Co.**—0.5 mi. S, 4 mi. E Beaver Creek (1 PSU); 1 mi. N, 5 mi. E Clackamas, T2S, R3E, Sec. 6 (5 PSU); North Fork Clackamas River (1 PSM); Eagle Creek (1 USNM); Eagle Creek, Clackamas River [T2S, R4E] (1 PSU); Jennings Lodge (1 OSUFW, 2 PSM, 5 SDNHM); 2 mi. N Lake Oswego (1 PSU); Logan (1 AMNH, 4 FMNH, 1 UMMZ); Milwaukie (1 PSM); 3 mi. N Molalla (1 PSU); 5 mi. E Molalla, T5S, R2E, Sec. 22 (1 OSUFW); 8 mi. S, 1 mi. W Molalla (1 OSUFW); 7.5 mi. SE Molalla (2 PSM); Mulino (4 OSUFW); Oak Grove, T2S, R1E, Sec. 11 (1 PSU); West Linn (1 PSU); T2S, R1E, Sec. 15 (1 PSU). **Clatsop Co.**—Astoria (1 USNM); Hammond (1 PSM); Knappa (1 OSUFW); site of Old Ft. Clatsop (3 MVZ); Saddle Mt. (1 OSUFW); Seaside (3 USNM); 2 mi. W Warrenton (1 OSUFW); 3 mi. S Warrenton High School on HWY 101 (1 USNM); T7N, R10W, Sec. 27 (1 PSU). **Columbia Co.**—(1 UW); Birkenfeld (1 OSUFW); Deer Island, T5N, R2W, Sec. 14 (2 PSU); 0.25 mi. E Keasey (1 UW); 7 mi. SE Rainier (8 MVZ); 0.5 mi. N, 2 mi. W St. Helens (1 OSUFW). **Coos Co.**—4 mi. SE Bandon (15 PSM); Bridge (1 USNM); Coquille City (1 USNM); Empire (11 USNM); Marshfield (9 USNM); Myrtle Point (8 USNM); Remote (2 SDNHM); 4.7 mi. N, 7.8 mi. W Sitkum, T27S, R11W, Sec. 21, 260 ft. (1 UW[CR]). **Crook Co.**—Allen Creek campground, Ochoco National Forest (1 SOSC); Buck Creek, 43°20', 120° (1 USNM); Howard (4 USNM); Maury Mts. (1 USNM); 20 mi. W Mitchell (1 OSUFW); 10.5 mi. S, 4 mi. W Mitchell, T13S, R21E, SW¼ Sec. 28 (1 KU); 11 mi. S, 6 mi. W Mitchell, T13S, R21E, NW¼ Sec. 21 (1 KU); Ochoco R. S., 4,000 ft. (1 MVZ); 10.5 mi. N, 1.5 mi. W Prineville (2 OSUFW); 5 mi. N, 46 mi. E Prineville, T14S, R23E, NW¼ Sec. 1 (1 USNM); 8 mi. N, 35 mi. E Prineville (6 KU, 5 USNM); 1 mi. S, 59 mi. E Prineville, T15S, R25E, SE¼ Sec. 1 (3 USNM); 3 mi. S, 23 mi. E Prineville, T15S, R20E, NE¼ Sec. 18 (1 USNM). **Curry Co.**—1 mi. N, 8 mi. E Agness (1 HSU); 10 mi. E Brookings (2 PSM); Gold Beach (8 FMNH); 13 mi. S Gold Beach (1 PSM); Chetco (1 FMNH); Little Redwood Forest camp (3 PSM); Long Ridge Forest camp (3 PSM); Port Orford (1 UO, 6 USNM); 15 mi. from mouth Rogue River (1 USNM). **Deschutes Co.**—Arnold Ice Cave, 16 mi. SE Bend (3 USNM); Bend (2 USNM); 5 mi. W Bend (1 SDNHM); 11 mi. W Bend, Tumalo Creek, 4,700 ft. (3 MVZ); 15 mi. W Bend, Tumalo Creek, 6,100 ft. (6 MVZ); 16 mi. W Bend, Tumalo Creek, 6,500 ft. (4 MVZ); 3 mi. S, 13 mi. W Bend, Swampy Lakes (2 MVZ); 15 mi. S, 25 mi. W Bend, Cultus Lake, T20S, R7E, SE¼ Sec. 13 (7 KU); 19 mi. S, 25 mi. W Bend, Crane Prairie, T21S, R7E, Sec. 12 (7 KU); Black Butte, T14S, R9E, Sec. 3 (6 different locality descriptors) (3 CM, 13 OSUFW); Cold Spring (1 KU); Davis Creek, near Maston's Camp* (1 OSUFW); head Davis Creek (2 OSUFW); mouth Davis Creek, Deschutes River (4 OSUFW); Davis Mt. (6 OSUFW); Fourmile Spring* (1 PSM); Indian Ford camp (11 PSM); La Pine (2 OSUFW, 1 PSM, 7 USNM); 5 mi. N La Pine (1 UW); 3 mi. N, 3 mi. E La Pine (1 OSUFW); 7 mi. E McKenzie Pass (1 KU); Paulina Lake (2 USNM); Paulina Mts. (2 USNM); Sisters (6 OSUFW, 1 PSM, 1 UO, 2 USNM); 1.5 mi. N, 4 mi. W Sisters, Cold Springs camp (2 PSU); 4 mi. N, 9 mi. W Sisters (3 MVZ); 5 mi. N, 3 mi. W Sisters, T14S, R9E, NW¼ Sec. 13 (1 USNM); 9 mi. N, 3 mi. W Sisters, T13S, R9E, Sec. 26 (2 KU, 2 USNM); Sparks Lake (1 KU, 1 OSUFW); 3 mi. W Wickiup Rd. from HWY 97, T22S, R10E (1 OSUFW). **Douglas Co.**—Anchor (6 USNM); Diamond Lake (2 SDNHM, 1 UMMZ, 1 USNM); 13 mi. N, 5 mi. E Diamond Lake, 0.5 mi. W Windigo Pass (1 PSU); Drain (1 USNM); 3 mi. W Drain (1 AMNH); 4.5 mi. N, 8 mi. W Drain, T21S, R4W, Sec. 27, 1,200 ft. (1 UW[CR]); 5.5 mi. N, 4 mi. W Drain, T21S, R6W, Sec. 15, 1,280 ft. (4 UW[CR]); 6.75 mi. N, 5.5 mi. E Drain, T21S, R4W, Sec. 7, 950 ft. (1 UW[CR]); 11.1 km S, 7.4 km W Elkton, T23S, R8W, Sec. 28 (1 USNM); Gardiner (6 FMNH, 1 USNM); Glendale (3 USNM); 3.2 km N, 26 k/m W Lorane, T19/20S, R7W, Sec. 32/5 (1 USNM); 5 mi. S Mt. Thielson, 6,200 ft. (3 MVZ); Reston (2 USNM); 4 mi. S, 6.4 mi. W Riddle, Iron Mt., T31S, R7W, Sec. 11, 2,000 ft. (8 UW[CR]); Scottsburg (1 USNM); 1.2 mi. N, 15.5 mi. W Sutherlin, T25S, R8W, Sec. 10, 1,100 ft. (2 UW[CR]); 1.5 mi. N, 15.5 mi. W Sutherlin, T25S, R8W, Sec. 15, 1,000 ft. (1 UW[CR]); 0.2 mi. N, 4.1 mi. W Umpqua P.O., T25S, R7W, Sec. 15, 800 ft. (1 UW[CR]); 2.2 mi. N, 9 mi. W Umpqua P.O., T25S, R8W, Sec. 15, 850 ft. (1 UW[CR]); 2.7 mi. N, 5 mi. W Umpqua P.O., T25S, R7W, Sec. 7, 950 ft. (2 UW[CR]); 4 mi. N, 1 mi. W Umpqua P.O., T25S, R7W, Sec. 1, 1,500 ft. (2 UW[CR]); 6 mi. N, 10.5 mi. W Umpqua, T24S, R8W, Sec. 29, 900 ft. (3 UW[CR]). **Grant Co.**—Dixie Mt. (3 PSM); 8 mi. E Austin, Cold Spring, 4,900 ft. (2 MVZ); 8 mi. S, 2 mi. E John Day, T15S, R32E, NW¼ Sec. 8 (1 KU); 9 mi. S, 14 mi. E John Day, T14S, R34E, Sec. 30 (1 KU, 1 USNM); 11 mi. S, 3 mi. W John Day, T15S, R31E, Sec. 20 (2 KU, 1 USNM); 11 mi. S, 16 mi. W John Day (2 KU); 13 mi. S, 30 mi. E John Day, T15S, R36E, NE¼ Sec. 30 (1 KU); 13 mi. S, 31 mi. E John Day, T15S, R36E, NW¼ Sec. 32 (2 USNM); 14 mi. S, 5 mi. E John Day, T16S, R32E, Sec. 3 (2 USNM); 14 mi. S, 28 mi. E John Day, T16S, R35E, Sec. 2 (2 KU, 1 USNM); 15 mi. S, 4 mi. E John Day, T16S, R32E, NW¼ Sec. 10 (1 USNM); 15 mi. S, 8 mi. E John Day, T16S, R33E, Sec. 17 (15 KU); 16 mi. S, 7 mi. E John Day, T16S, R32E, Sec. 13 (1 KU); 16 mi. S, 17 mi. E John Day, T16S, R33½E, Sec. 14 (1 KU); 17 mi. S, 10 mi. E John Day, T16S, R33E, Sec. 21 (1 KU); 18 mi. S, 11 mi. E John Day, T16S, R33E, Sec. 27 (2 KU); 18 mi. S, 24 mi. E John Day, T16S, R34E, SW¼ Sec. 24 (2 USNM); 18 mi. S, 29 mi. E John Day, T16S, R35E, SE¼ Sec. 24 (2 USNM); 20 mi. S, 11 mi. E John Day, T17S, R33E, Sec. 2 (1 KU, 1 USNM); 20 mi. S, 16 mi. E John Day, T17S, R33½E, NE¼ Sec. 3 (1 USNM); 20 mi. S, 24 mi. E John Day, T16S, R34E, Sec. 36 (3 KU); 23 mi. S, 25 mi. E John Day, T17S, R35E, Sec. 17 (2 KU); 24 mi. S, 11 mi. E John Day, T17S, R33E, Sec. 22 (2 KU); 25 mi. S, 28 mi. E John Day, T17S, R35E, Sec. 28 (5 KU, 1 USNM); 26 mi. S, 12 mi. E John Day, T17S, R33E, Sec. 35 (2 KU); 10 mi. S, 11 mi. W Mt. Vernon, T15S, R28E, SW¼ Sec. 11 (1 KU); T16S, R35E, Sec. 23 (1 KU). **Harney Co.**—1 mi. NW Burns (1 USNM); 10 mi. N Harney (1 USNM). **Hood River Co.**—(3 UW); Bear Spring, Mt. Hood (2 MVZ, 1 SDNHM); Brooks Meadows, 9 mi. ENE Mt. Hood (4 MVZ); 4 mi camp, Herman Creek trail [NW corner of county] (1 UW); Mt. Hood (4 USNM); Mt. Hood, Cooper Spur (2 OSUFW); Mt. Hood, Cloud Cap Inn, 6,800 ft. (1 USNM); Mt. Hood, N slope, 2,800 ft. (1 USNM); Oak Grove (2 PSU); Parkdale (1 AMNH, 2

UMMZ); 2 mi. W Parkdale, 1,500 ft. (6 USNM); T2S, R9E, Sec. 28 (4 PSU). **Jackson Co.**—6 mi. S Ashland (1 SOSC); 14 mi. E Ashland, Dead Indian summit (1 CM, 2 MVZ); Bald Mt.* (2 SOSC); Howard Prairie Lake resort (1 SOSC); 10 mi. E Medford, Lost Creek, 5,000 ft. (1 MVZ); Perrozzi Acres* (2 SOSC); Pilot Rock, 14 mi. SE California Gulch (1 SDNHM); Prospect (1 OSUFW, 1 USNM); 2 mi. N Prospect (1 SOSC); 6 mi. N Prospect (1 KU); 8 mi. N Prospect (1 KU); 8 mi. SW Prospect (3 CM, 1 KU, 1 OSUFW, 2 SOSC); Siskiyou (8 USNM); halfway between Drew and Crater Lake (4 USNM); Rogue River (1 USNM); Upper Tolman Creek Rd. 1 mi. E Ashland Loop Rd. (1 SOSC); 1.4 mi. S junction USFS Rd. 3706 and USFS Rd. 3705 (1 SOSC); junction Dead Indian Rd. and Fish Lake Rd. (1 PSU); T37S, R4E, Sec. 26 (1 SOSC); T38S, R3E, Sec. 30 (1 SOSC). **Jefferson Co.**—Booth Lake, SE base Three-Fingered Jack, T13S, R8E, Sec. 18 (1 OSUFW); S Candle Creek camp, T11S, R9E, SE¼ Sec. 35 (5 KU); 12 mi. E Hay Creek on Foley Creek (3 USNM); 12 mi. SE Hay Creek (1 OSUFW, 1 PSU, 2 USNM); 13 mi. N, 9 mi. W Sisters, T13S, R8E, NE¼ Sec. 1 (2 KU); 20 mi. N, 4 mi. W Sisters, T11S, R9E, NE¼ Sec. 35 (9 KU); Suttle Lake, 3,45 ft. (2 KU, 4 OSUFW, 1 UO); 20 mi. W Warmsprings, Mill Creek (1 PSM, 3 USNM). **Josephine Co.**—Bolan Lake (1 PSM, 1 SDNHM); Grants Pass (1 LACM); 13 mi. SW Grants Pass (2 UO); Oregon Caves State Game Refuge (1 OSUFW). **Klamath Co.**—Anna Creek, 4 mi. NW Ft. Klamath, 4,500 ft. (1 MVZ); Bly (1 CM); 15 mi. E Bly (1 USNM); Chiloquin (1 USNM); Crater Lake National Park* (3 MVZ, 4 USNM); Crater Lake National Park, Lost Creek ranger station, T31S, R7½E, Sec. 1 (1 OSUFW); Crater Lake National Park, Pumice Flat, SW Bear Bluff, 6,200 ft. (2 KU); Crater Lake National Park, 0.25 mi. S Red Cone Crater, 6,300 ft. (2 KU); Crater Lake National Park, east entrance (1 USNM); Crater Lake National Park, south entrance (2 USNM); Crater Lake National Park, Sun Creek (1 USNM); Crater Lake National Park, N side Timber Crater (1 MVZ); Cresent Lake, 4,800 ft. (4 LACM); Cresent Lake, Diamond Peak, 6,800 ft. (1 UO); 20 mi. W Cresent Lake (20 UO); 12 mi. E Diamond Lake (1 CM); 20 mi. W Fremont near Lake Co. line (2 OSUFW); Ft. Klamath (6 OSUFW, 3 UMMZ, 34 USNM, 2 CRCM); Klamath Falls (3 CM, 1 OSUFW, 1 CRCM); 12 mi. N, 30 mi. W Klamath Falls (2 OSUFW); Klamath Lake (2 FMNH); 3 mi. E Lake of the Woods (1 SOSC); Naylox (1 USNM); Swan Lake Valley (2 USNM); Upper Klamath Lake (3 FMNH, 2 UMMZ); West Sink Creek, 12 mi. E Mt. Thielsen (2 USNM); Yamsay Mts., W Silver Creek, 7,000 ft. (1 USNM). **Lake Co.**—14 mi. SW Adel, Barley Camp (2 MVZ, 2 OSUFW); Camas Springs (1 SDNHM); N base Crook Peak, Warner Mts., T37S, R22E, Sec. 17 (2 OSUFW); Finley Corrals (2 CM); Fremont (1 USNM); Lakeview (1 OSUFW, 4 UO); Lakeview, Camas Spring (1 SDNHM); 8 mi. NE Lakeview, Salt Creek ranger station (1 OSUFW); 10 mi. SE Lakeview, 6,600 ft. (1 USNM); Quartz Mt., Fishhole Guard Station (2 SOSC); Rogger Meadow, 7.2 mi. N, 5 mi. W summit Grane Mt., Warner Mt., 6,500 ft. (1 CRCM); 12 mi. S, 3 mi. W Summer Lake (2 OSUFW); Turpin Ridge, Fremont National Forest (1 SDNHM); 1.66 mi. S, 6.125 mi. W Walton* (1 PSU); Warner Mts., 5,700 ft. (2 USNM); E Warner Mts., 6,500 ft. (2 USNM); Warner Mts., E of Goose Lake (4 USNM); 10 mi. SW West Silver Creek, Silver Lake, 4,650 ft. (1 USNM); 2.2 mi. E HWY 395 on HWY 140 (1 SOSC). **Lane Co.**—Belknap Springs (2 OSUFW); 4 mi. N Belknap Springs (1 UO); 2 mi. N, 3 mi. E Cottage Grove (1 OSUFW); Cresent* (1 SDNHM); Cushman, 200 ft. (2 CM); 1 mi. S, 2 mi. W Donna (1 OSUFW); 4 mi. N Elmira (1 OSUFW); Eugene (4 OSUFW, 1 UO, 1 USNM); Florence (1 USNM); 9 mi. N, 1 mi. W Gardiner, Siltcoos camp (3 PSM); Heceta Head (1 CM); H. J. Andrews Experimental Forest, T15S, R5E, Sec. 22 (1 PSM); Huckleberry Lake Rd. (3 KU); 10 mi. W Junction City (1 PSM); McKenzie Bridge (2 OSUFW, 1 UMMZ, 8 UO, 4 USNM); McKenzie Bridge, mouth Smith River (1 OSUFW); McKenzie Pass, 5,300 ft. (1 USNM); Mercer (1 UO); Mercer, Dowell Ranch (1 OSUFW); 11 mi. N, 15 mi. E Oakridge, T19S, R5E, Sec. 25 (1 OSUFW); 3 mi. SE Oakridge (5 UO); Pengra (1 USNM); 2 mi. S Prairie Mt. (2 PSM); Scott Lake, NW base Middle Sister, Three Sisters, T15S, R7½E (4 OSUFW, 1 PSM); Seaton (2 USNM); 10 mi. N, 43 mi. E Springfield, T16S, R5½E, SW¼ Sec. 34 (5 USNM); 12 mi. N, 50 mi. E Springfield, T15S, R7½E, SW¼ Sec. 34 (2 USNM); N base Three Sisters (1 UO); Vida (1 OSUFW, 1 UO); T16S, R9W, Sec. 29 (1 PSU). **Lincoln Co.**—2 mi. S, 1 mi. W

Burnt Woods, T11S, R8W, Sec. 31 (1 KU); 2 mi. S, 17 mi. W Corvallis (1 OSUFW); 5 mi. S, 20 mi. W Corvallis (1 OSUFW); 6 mi. W Harlan (2 OSUFW); Newport (2 MVZ, 6 USNM); 1 mi. N, 0.5 mi. W Nortons, T10S, R8W, Sec. 29 (1 KU); 3 mi. N, 3 mi. W Nortons, T10S, R9W, Sec. 13 (1 KU); Yaquina (1 USNM); Yaquina Bay (4 USNM); T11S, R9W, Sec. 25 (1 KU). **Linn Co.**—3 mi. E Albany, Knox Butte, T10S, R3W, Sec. 35 (1 OSUFW); Big Lake (1 PSM); 1.5 mi. N Big Lake (4 PSM); 16 mi. E Cascadia, Hackleman Creek (1 OSUFW); 14 mi. from Lacomb, Green Peter Mt. (1 USNM); Lost Prairie, Santiam Pass, T13S, R6E, Sec. 27 (1 OSUFW, 7 PSU); 2 mi. E Lost Prairie (1 OSUFW); North Santiam River, 3,400 ft. (3 MVZ); Permilia [sic] Lake, Mt. Jefferson (4 USNM); Sweet Home (2 USNM); 1.5 mi. E Sweet Home (1 OSUFW). **Marion Co.**—Bald Mt., head of Clackamas River (1 USNM); Detroit (1 PSU, 1 USNM); Salem (5 AMNH, 1 KU, 1 OSUFW, 2 UMMZ, 4 USNM); Silver Falls State Park (1 KU); 5 mi. S, 6 mi. E Silverton (1 OSUFW); Woodburn (1 PSM). **Multnomah Co.**—Crown Point (1 SDNHM); 0.25 mi. SE Foster* (1 PSU); Larch Mt. (1 PSM); Latourelle (1 OSUFW); Portland (9 AMNH, 2 CM, 1 MVZ, 8 PSU, 12 USNM); Sellwood [=Portland] (1 USNM); Troutdale (1 PSU); T1N, R1W, Sec. 7 (1 PSU). **Polk Co.**—2 mi. W Dallas (1 OSUFW); McCoy (3 UO, 2 USNM); Grande Ronde (1 USNM); 5 mi. S Willamina (2 OSUFW); T10S, R5W, Sec. 2 (1 PSM). **Tillamook Co.**—Beaver (2 OSUFW); Blaine (3 KU, 4 OSUFW, 1 SOSC, 1 UMMZ); Cascade Head Experimental Forest (2 PSM); Dolph (1 OSUFW); Hebo (1 PSM); 1 mi. E Kelso* (1 PSU); Mt. Hebo (2 OSUFW); Nehalem (2 PSM); Netarts (5 OSUFW, 1 PSM, 1 PSU, 1 SDNHM, 1 UMMZ, 1 UO); Ocean View (2 OSUFW); 4 mi. N Pacific City (1 PSU); Tillamook (1 MVZ, 2 OSUFW, 1 SDNHM, 2 USNM); 5 mi. SE Tillamook (1 OSUFW). **Wasco Co.**—(1 PSU, 1 UW); Clear Lake (1 PSM); Pine Grove, T5S, R11E, Sec. 30 (1 OSUFW); 8 mi. W Maupin (1 PSM); 20 mi. W Maupin on HWY 216 (1 PSU); The Dalles (1 OSUFW); 14 mi. SW The Dalles, Fivemile Creek (2 OSUFW, 1 PSU); 25 mi. SW The Dalles, 2,500 ft. (1 USNM); Wapinitia (3 USNM); 6 mi. W Wapinitia (2 PSM); 10 mi. W Wapinitia, 2,500 ft. (1 USNM). **Washington Co.**—Beaverton (2 OSUFW, 4 USNM); Cooper Mt., Aloha (1 PSU); Dilley (1 USNM); Farmington (1 PSU); Forest Grove (1 AMNH, 8 USNM); Forest Park [T1S, R1W, Sec. 11] (1 PSU); 0.1 mi. N Scoggins Valley Rd., Gaston (1 EOSC); Oak Grove* (1 PSU); 6 mi. W Portland (1 MVZ); 18.5 mi. NW Portland (5 MVZ); Sherwood (1 PSU); Sherwood Junction (1 PSU); 3 mi. S Tigard (3 PSU); Wildwood Trail [near Germantown Rd., Portland] (1 PSU); T1S, R5W, Sec. 3 (1 PSU); T2S, R2W, Sec. 21 (1 PSU). **Wheeler Co.**—7 mi. S, 11 mi. W Mitchell, 4,850 ft. (5 MVZ); 8 mi. S, 8 mi. W Mitchell, T13S, R20E, SE¼ Sec. 15 (1 KU); 6 mi. S, 10 mi. E Mitchell, T13S, R23E, NW¼ Sec. 2 (3 KU); 11 mi. S, 7 mi. W Mitchell, T13S, R20E, NW¼ Sec. 3 (1 KU); 12 mi. SW Mitchell, Wildwood Camp, Ochoco National Forest, [44.52°N, 120.33°W] (3 OSUFW); 8 mi. N, 45 mi. E Prineville, T13S, R23E, SW¼ Sec. 23 (3 USNM); 12 mi. N, 26 mi. E Prineville, T12S, R20E, NE¼ Sec. 34 (4 USNM); Spray, Kahler Basin (2 UW[WWC]). **Yamhill Co.**—1.5 mi. N, 2 mi. E Dayton (1 OSUFW); Newberg (1 SDNHM); 3.5 mi. E Newberg (2 MVZ); 2 mi. N, 1 mi. E Yamhill (1 PSU).

Additional localities: Clackamas Lake* (1 USNM); Bear Springs* (1 USNM); Camp Harney* (1 USNM).

Tamiasciurus hudsonicus, n = 374
(Plotted in Fig. 11-42, p. 217)

Baker Co.—Anthony (13 AMNH, 6 MVZ, 2 SDNHM); 11 mi. N, 22 mi. E Baker City, T7S, R43E, SE¼ Sec. 24 (2 USNM); 18 mi. N, 19 mi. E Baker [City], T6S, R43E, Sec. 10 (1 KU, 7 USNM); 18 mi. N, 22 mi. E Baker City, T7S, R43E, SE¼ Sec. 24 (1 USNM); 19 mi. N, 36 mi. E Baker [City] (13 KU, 8 USNM); Cornucopia (2 USNM); Eagle Creek area (3 KU); 4.5 mi. N, 8.25 mi. W Haines (2 PSU); 4 mi. N, 5 mi. W Halfway, T7S, R45E, NW¼ Sec. 28 (4 KU); 9 mi. N, 2 mi. E Halfway, T6S, R46E, NW¼ Sec. 34 (6 KU); 8 mi. NE Halfway (2 MSUMZ); 6 mi. S, 26 mi. E John Day, T14S, R35½E, Sec. 26 (1 USNM); 6 mi. S, 29 mi. E John Day (2 USNM); Lily White field station, 0.25 mi. up canyon (1 EOSC); 0.25 mi. N junction S700 and S780, 1.5 mi. S Lily White (1 EOSC); 2 mi. N, 10 mi. E Ponderosa, T6S, R43E, SE¼ Sec. 16 (6 KU);

14 mi. S, 15 mi. E Prairie City, T15S, R35½E, Sec. 22 (4 KU); 21 mi. SE Prairie City, North Fork Malheur River (1 MVZ); Rock Creek (1 USNM); Unity (2 OSUFW); 6 mi. N, 5.5 mi. W Unity, T12S, R36E, NE¼ Sec. 16 (2 USNM); 7.5 mi. N, 7 mi. W Unity, T12S, R36E, Sec. 7 (3 USNM); 5 mi. S, 8 mi. W Unity, T14S, R35½E, Sec. 12 (1 KU, 1 USNM); 8 mi. S, 1 mi. E Unity, T14S, R37E, Sec. 27 (9 KU). **Grant Co.**—10 mi. E Austin, 4,000 ft. (1 USNM); 2 mi. S, 23 mi. W Baker [City], T9S, R36E, Sec. 30 (5 KU, 1 USNM); 10 mi. S, 3 mi. E Bates, T13S, R35E, NE¼ Sec. 12 (2 KU); Beech Creek (2 USNM); Dixie Mt. (1 PSM); 12 mi. N, 5 mi. W Granite, T7S, R35E, Sec. 5 (2 KU); 8 mi. S, 20 mi. E John Day, T15S, R35E, Sec. 31 (1 USNM); 9 mi. S, 14 mi. E John Day, T14S, R34E, Sec. 30 (1 KU); 10 mi. S, 13 mi. E John Day, T15S, R33E, NE¼ Sec. 24 (2 KU); 11 mi. S, 3 mi. W John Day, T15S, R31E, Sec. 20 (1 USNM); 11 mi. S, 16 mi. W John Day (1 KU); 12 mi. S, 27 mi. E John Day (2 USNM); 13 mi. S, 8 mi. E John Day, T15S, R33E, SE¼ Sec. 29 (1 KU, 1 USNM); 13 mi. S, 30 mi. E John Day, T15S, R36E, NE¼ Sec. 30 (1 USNM); 14 mi. S, 22 mi. E John Day, T16S, R34E, Sec. 3 (4 KU, 2 USNM); 14 mi. S, 28 mi. E John Day, T16S, R35E, Sec. 2 (4 KU, 2 USNM); 15 mi. S, 4 mi. E John Day, T16S, R32E, NW¼ Sec. 10 (1 USNM); 15 mi. S, 16 mi. E John Day, T16S, R33½E, SW¼ Sec. 3 (1 USNM); 16 mi. S, 7 mi. E John Day, T16S, R32E, Sec. 13 (1 KU); 16 mi. S, 17 mi. E John Day, T16S, R33½E, Sec. 14 (2 KU, 3 USNM); 17 mi. S, 10 mi. E John Day, T16S, R33E, Sec. 21 (7 KU, 3 USNM); 18 mi. S, 11 mi. E John Day, T16S, R33E, Sec. 27 (3 KU); 18 mi. S, 24 mi. E John Day, T16S, R34E, SW¼ Sec. 24 (2 KU, 1 USNM); 20 mi. S, 11 mi. E John Day, T17S, R33E, Sec. 2 (2 KU); 20 mi. S, 24 mi. E John Day, T16S, R34E, Sec. 36 (5 KU); 24 mi. S, 11 mi. E John Day, T17S, R33E, Sec. 22 (2 KU); 25 mi. S, 28 mi. E John Day, T17S, R35E, Sec. 28 (1 KU); 9.5 mi. S Mt. Vernon, T15S, R29E, Sec. 15 (1 KU); 10 mi. S, 12 mi. W Mt. Vernon, T15S, R28E, Sec. 15 (1 KU); 10 mi. S, 2 mi. E Prairie City, T14S, R34E, NW¼ Sec. 31 (5 KU); 16 mi. S, 15 mi. E Prairie City, T15S, R35½E, Sec. 27 (3 KU); 21 mi. S, 11 mi. E Prairie City, T16S, R34E, NE¼ Sec. 23 (1 KU); 22 mi. S, 13 mi. E Prairie City, T16S, R35E, NW¼ Sec. 20 (1 KU); 24 mi. S, 12 mi. E Prairie City, T17S, R34E, NE¼ Sec. 1 (1 KU); 25 mi. S, 12 mi. E Prairie City, T17S, R35E, SW¼ Sec. 7 (1 KU); 26 mi. S, 12 mi. E Prairie City, T17S, R35E, NE¼ Sec. 19 (1 KU); 2 mi. N, 10 mi. W Unity, T13S, R35½E, Sec. 3 (4 KU, 1 USNM); 8.5 mi. N, 10 mi. W Unity, T11S, R35½E, NE¼ Sec. 34 (1 USNM); T16S, R34E, Sec. 12 (6 KU). **Malheur Co.**—Ironside, 4,500 ft. (4 AMNH). **Morrow Co.**—20 mi. SSE Heppner (1 USNM). **Umatilla Co.**—(1 MSUMZ, 1 UW[WWC]); Blalock Mt. (3 UW[WWC]); 12 mi. E Dale (1 SOSC); Langdon Lake (1 KU, 1 UW[WWC]); 3 mi. W Langdon Lake (1 UW[WWC]); 4 mi. NW Langdon Lake (1 UW[WWC]); Meacham (1 UMMZ, 3 USNM); 10 mi. W Meacham (1 USNM); 2 mi. NW Meacham, 3,600 ft. (1 MVZ); 10 mi. E Milton (1 AMNH); 15 mi. E Milton (2 AMNH); 8 mi. S, 15 mi. E Pendleton, Blue Mts. (2 OSUFW); 18 mi. SE Pendleton (1 USNM); Tollgate (1 MVZ, 1 UMMZ, 2 UW[WWC]); 3 mi. W Tollgate (2 UW[WWC]); 4 mi. W Tollgate (1 UW[EB], 1 UW[WWC]); 1 mi. N, 3 mi. W Tollgate (1 UW[WWC]); 2 mi. N, 0.5 mi. W Tollgate Springs (1 UW[WWC]); Ukiah, North Fork John Day River (1 OSUFW); Ukiah, 23 mi. E Blue Mts. (1 KU); 15 mi. E Ukiah (1 OSUFW); 23 mi. E Ukiah, Blue Mts., 4,800 ft. (1 KU); 26 mi. E Ukiah (2 SOSC); 10 mi. SE Ukiah (2 USNM); 17 mi. SE Ukiah, Tower Mt. (1 OSUFW); 16 mi. E Weston (7 UW[WWC]); 17 mi. E Weston (1 UW[WWC]); T5N, R39E, Sec. 6 (3 UW[WWC]). **Union Co.**—(1 EOSC); Elgin (2 USNM); Fox hill Rd. (1 EOSC); Glasshill Rd. (1 EOSC); 4 mi. N, 14 mi. W Haines, T7S, R37E, NW¼ Sec. 18 (13 KU); 6 mi. N, 1 mi. W Ponderosa, T5S, R41E, SE¼ Sec. 27 (5 KU); Starky Experimental Forest (10 PSM, 1 PSU); 7 mi. E Telocaset, Catherine Creek (2 MVZ); 6 mi. E Tollgate (1 MVZ); 1 mi. W junction of Meadow Creek and Smith Creek (4 UW); T3S, R38E, SE¼NW¼ Sec. 27 (1 EOSC). **Wallowa Co.**—Aneroid Lake (3 UO); Eagle Cap Wilderness area (1 SOSC); 15.2 mi. N Enterprise (1 OSUFW); 25 mi. N Enterprise at Sled Springs (6 USNM); 5 mi. E Enterprise, T2S, R43E, Sec. 1 (3 KU); 1 mi. N, 5 mi. W Enterprise, T1S, R43E, Sec. 36 (2 KU); 26 mi. N, 7.6 mi. W Enterprise (1 OSUFW); 3 mi. S, 2.5 mi. W Enterprise (2 OSUFW); Fall Creek campground, Wallowa Mts. (3 UW[WWC]); Hurricane Creek, Wallowa Mts., 5,500 ft. (2 UW[WWC]); 4 mi. up Hurricane Creek (2 UW[WWC]); 8 mi. up Hurricane Creek (1 UW[WWC]); Joseph (1 CM, 1 OSUFW); 10 mi. E Joseph (2 PSM); 5 mi. S, 4 mi. E Joseph (2 OSUFW); Lick Creek R.S. (2 FMNH, 1 PSM); 16 mi. S Lostine, T3S, R43E, Sec. 25 (1 OSUFW); 16 mi. S, 3 mi. E Lostine (6 MVZ); 17 mi. SE Lostine, Lostine River, Wallowa Mts. (4 OSUFW); Marr Flat, Wallowa National Forest (1 SDNHM); 2 mi. N Minam, Minam State Park (1 PSU); 15 mi. N Paradise at Horse Creek, 2,000 ft. (3 USNM); Peavine Creek, T3N, R46E, Sec. 29 (1 OSUFW); Sheep Creek (2 OSUFW); Wallowa Lake, 4,000 ft. (13 USNM); S end Wallowa Lake (1 UO); 2 mi. E Wallowa Lake, 4,000 ft. (1 CRCM).

Tamiasciurus unclassified, $n = 1,032$
(Not plotted)

Baker Co.—13 mi. N, 20 mi. W Baker City, T7S, R37E, SW¼ Sec. 7 (2 USNM); Bourne (5 USNM); 6 mi. S, 26 mi. E John Day, T14S, R35½E, Sec. 26 (11 KU, 5 USNM); 6 mi. S, 29 mi. E John Day (12 USNM); Lookout Mt. (2 UMMZ); 12 mi. S, 15 mi. E Prairie City, North Fork Malheur River (1 MVZ); 14 mi. S, 15 mi. E Prairie City, T15S, R35½E, Sec. 22 (2 KU); 21 mi. SE Prairie City, North Fork Malheur River (9 MVZ); 5 mi. NW Sumpter (2 OSUFW); Unity (1 KU); Unity, East Camp Creek (2 OSUFW); 1.5 mi. N, 8 mi. W Unity, T13S, R35½E, SE¼ Sec. 1 (5 USNM); 6 mi. N, 5.5 mi. W Unity, T12S, R36E, NE¼ Sec. 16 (2 USNM); 7.5 mi. N, 7 mi. W Unity, T12S, R36E, Sec. 7 (1 KU, 14 USNM); 2.5 mi. S, 7 mi. W Unity, T13S, R36E, NW¼ Sec. 33 (2 USNM); 5 mi. S, 8 mi. W Unity, T14S, R35½E, Sec. 12 (1 KU, 8 USNM); 4 mi. S, 5 mi. W Unity, T14S, R36E, Sec. 9 (4 USNM); 8 mi. S, 1 mi. E Unity, T14S, R37E, Sec. 27 (12 KU). **Grant Co.**—Aldrich Mts., meadow SE junction FS1542 and 1409C, 4,850 ft. (1 CRCM); Austin (1 USNM); 8 mi. E Austin, Cold Spring, 4,900 ft. (16 MVZ); 2 mi. S, 23 mi. W Baker [City], T9S, R36E, Sec. 30 (18 KU, 10 USNM); 10 mi. S, 3 mi. E Bates, T13S, R35E, NE¼ Sec. 12 (2 KU); Beech Creek (2 SDNHM, 2 USNM); Blue Mt. Spring (1 PSM); 12 mi. S Canyon City, 5,300 ft. (1 MVZ); 9 mi. S, 3 mi. W Canyon City (1 PSU); 9 mi. S, 5 mi. W Canyon City, T15S, R31E, Sec. 19 (1 PSU); 3 mi. S Dale, T9S, R36E, Sec. 30 (1 KU); 14 mi. N Dayville (1 UW[WWC]); Dixie Mt. (6 PSM); Elkhorn Springs, Malheur National Forest (1 KU); 2 mi. W Fall Mt. lookout, Aldrich Mts., 8,400 ft. (1 CRCM); 4 mi. S Granite (1 OSUFW); 12 mi. N, 2 mi. W Granite, T7S, R35E, SE¼ Sec. 2 (2 KU); 12 mi. N, 5 mi. W Granite, T7S, R35E, Sec. 5 (2 KU); 2 mi. N, 26 mi. E John Day, T13S, R35E, NE¼ Sec. 12 (10 USNM); 9 mi. N, 18 mi. E John Day, T12S, R34E, SE¼ Sec. 11 (11 USNM); 11 mi. S, 16 mi. W John Day, T15S, R29E, Sec. 19 (27 KU, 5 USNM); 7 mi. S, 14 mi. E John Day, T14S, R34E, SW¼ Sec. 30 (1 USNM); 8 mi. S, 14 mi. E John Day, T14S, R34E, SW¼ Sec. 31 (2 USNM); 8 mi. S, 20 mi. E John Day, T15S, R35E, Sec. 31 (3 KU, 12 USNM); 9 mi. S, 1.5 mi. E John Day, T15S, R32E, SE¼ Sec. 7 (1 KU); 9 mi. S, 14 mi. E John Day, T14S, R34E, Sec. 30 (11 KU, 12 USNM); 10 mi. S, 12 mi. E John Day, T15S, R33E, Sec. 12 (6 KU); 10 mi. S, 13 mi. E John Day, T15S, R33E, NE¼ Sec. 24 (11 KU, 3 USNM); 11 mi. S, 3 mi. W John Day, T15S, R31E, Sec. 20 (7 KU, 10 USNM); 11 mi. S, 16 mi. W John Day (2 USNM); 12 mi. S, 27 mi. E John Day (18 KU, 10 USNM); 13 mi. S, 8 mi. E John Day, T15S, R33E, SE¼ Sec. 29 (18 KU, 3 USNM); 13 mi. S, 30 mi. E John Day, T15S, R36E, NE¼ Sec. 30 (12 KU, 7 USNM); 13 mi. S, 31 mi. E John Day, T15S, R36E, NW¼ Sec. 32 (2 KU, 14 USNM); 14 mi. S, 5 mi. E John Day, T16S, R32E, Sec. 3 (11 KU, 6 USNM); 14 mi. S, 22 mi. E John Day, T16S, R34E, Sec. 3 (10 KU, 204 USNM); 14 mi. S, 28 mi. E John Day, T16S, R35E, Sec. 2 (16 KU, 18 USNM); 15 mi. S, 4 mi. E John Day, T16S, R32E, NW¼ Sec. 10 (9 USNM); 15 mi. S, 8 mi. E John Day, T16S, R33E, Sec. 17 (15 KU); 15 mi. S, 16 mi. E John Day, T16S, R33½E, SW¼ Sec. 3 (4 USNM); 16 mi. S, 7 mi. E John Day, T16S, R32E, Sec. 13 (6 KU); 16 mi. S, 17 mi. E John Day, T16S, R33½E, Sec. 14 (16 KU, 3 USNM); 16 mi. S, 25 mi. E John Day, T16S, R35E, Sec. 17 (2 USNM); 17 mi. S, 10 mi. E John Day, T16S, R33E, Sec. 21 (21 KU, 15 USNM); 17 mi. S, 12 mi. E John Day, T16S, R33E, Sec. 13 (23 KU); 17 mi. S, 22 mi. E John Day, T16S, R34E, NE¼ Sec. 14 (17 KU, 2 USNM); 18 mi. S, 11 mi. E John Day, T16S, R33E, Sec. 27 (14 KU, 6 USNM); 18 mi. S, 24 mi. E John Day, T16S, R34E, SW¼ Sec. 24 (16 KU, 15 USNM); 18 mi. S, 29 mi. E John Day, T16S, R35E, SE¼

Sec. 24 (8 KU, 1 USNM); 20 mi. S, 11 mi. E John Day, T17S, R33E, Sec. 2 (5 KU, 2 USNM); 20 mi. S, 16 mi. E John Day, T33½E, NE¼ Sec. 3 (5 USNM); 20 mi. S, 24 mi. E John Day, T16S, R34E, Sec. 36 (11 KU); 22 mi. S, 10 mi. E John Day, T17S, R33E, Sec. 10 (22 KU); 23 mi. S, 25 mi. E John Day, T17S, R35E, Sec. 17 (3 KU, 2 USNM); 24 mi. S, 11 mi. E John Day, T17S, R33E, Sec. 22 (6 KU, 3 USNM); 25 mi. S, 28 mi. E John Day, T17S, R35E, Sec. 28 (12 KU, 8 USNM); 26 mi. S, 12 mi. E John Day, T17S, R33E, Sec. 35 (12 KU, 3 USNM); 6 mi. N Kimberly, Tamarack Springs (2 OSUFW); 9 mi. N Kimberly, Tamasack Springs (1 OSUFW); Mt. Vernon, Beech Creek forest camp (1 UW[WWC]); 8 mi. S, 11 mi. W Mt. Vernon, T15S, R28E, Sec. 3 (1 KU); 10 mi. S, 11 mi. W Mt. Vernon, T15S, R28E, SW¼ Sec. 11 (6 KU); 10 mi. S, 12 mi. W Mt. Vernon, T15S, R28E, Sec. 15 (1 KU); 10.5 mi. S, 12 mi. W Mt. Vernon, T15S, R28E, SE¼ Sec. 15 (1 KU); 11.5 mi. S, 2 mi. E Mt. Vernon, T15S, R30E, SW¼ Sec. 20 (1 KU); 10 mi. S, 2 mi. E Prairie City, T14S, R34E, NW¼ Sec. 31 (3 KU); 14 mi. S, 15 mi. E Prairie City, T15S, R35½E, Sec. 22 (5 KU); 16 mi. S, 15 mi. E Prairie City, T15S, R35½E, Sec. 27 (2 KU); 18 mi. S, 15 mi. E Prairie City, T16S, R35E, NE¼ Sec. 9 (1 KU); 21 mi. S, 11 mi. E Prairie City, T16S, R34E, NE¼ Sec. 23 (1 KU); 22 mi. S, 13 mi. E Prairie City, T16S, R35E, SE¼ Sec. 19 (1 KU); 23 mi. S, 12 mi. E Prairie City, T16S, R34E, SW¼ Sec. 36 (6 KU); 25 mi. S, 12 mi. E Prairie City, T17S, R35E, SW¼ Sec. 7 (1 KU); Seneca, Bear Creek, T16S, R31E, SE¼ Sec. 36 (1 USNM); 3 mi. E Seneca, T16S, R32E, Sec. 31 (2 KU); Strawberry Mts. (3 USNM); Strawberry Mts., Wickiup campground (2 HSU); 2 mi. N, 10 mi. W Unity, T13S, R35½E, Sec. 3 (15 KU, 12 USNM); 8.5 mi. N, 10 mi. W Unity, T11S, R35½E, NE¼ Sec. 34 (2 USNM); T16S, R34E, Sec. 12 (16 KU); T16S, R35E, Sec. 23 (1 KU).

Taxidea taxus, n = 154
(Plotted in Fig. 12-49, p. 429)

Baker Co.—10 mi. NE Baker [City] (1 KU); 13 mi. NE Baker [City] (1 KU); Keating (1 KU); 5 mi. N on road from HWY 51 toward Lily White* (1 EOSC). **Crook Co.**—4 mi. SW Prineville, 3,300 ft. (1 MVZ); 8 mi. S Suplee (1 USNM). **Deschutes Co.**—(1 EOSC); Bend, Bear Creek (2 USNM); Bend, Deschutes River (2 USNM); 1 mi. N, 2.5 mi. E Brothers (1 OSUFW); La Pine (1 SDNHM); Pengra [≅La Pine] (1 USNM). **Gilliam Co.**—8 mi. N Lone Rock (2 USNM); 9 mi. N Lone Rock (1 USNM). **Grant Co.**—10 mi. S Izee (1 USNM). **Harney Co.**—6 mi. N Alberson (1 USNM); 5 mi. W Alberson (1 USNM); south Blitzen Valley, Malheur National Wildlife Refuge* (1 UW); 1 mi. NW Burns P.O. (2 USNM); Denio (1 MVZ); Fields (13 USNM); 20 mi. N Harney City (1 USNM); Harriman (2 USNM); Narrows (1 USNM); Princeton (3 USNM); Steens Mt.* (1 USNM); Venator (1 USNM); 10 mi. S Venator (1 USNM); 12 mi. S Venator (5 USNM); near Harney Lake dunes on Double-O Ranch Rd. (1 SOSC). **Jackson Co.**—3 mi. E Butte Falls (2 USNM). **Jefferson Co.**—Hay Creek (2 MVZ, 5 USNM); 7 mi. E Hay Creek, Trout Creek (4 USNM); 7 mi. E Hay Creek, Foley Ranch (6 USNM). **Klamath Co.**—Ft. Flamath (3 USNM); Klamath Falls (1 CM); T24S, R11E, Sec. 20 (1 PSU). **Lake Co.**—Adel (1 USNM); 20 mi. NE Adel (2 OSUFW); Drews Valley (1 MVZ); Fishhole Meadows (1 PSM); Fishhole Mt. (1 PSM); above Flynn at Cottonwood (1 USNM); Fremont (2 USNM); Gearhart Mts., near Finley Corral (1 PSM); 12 mi. S Hampton (1 USNM); 14 mi. S Hampton (4 USNM); Lakeview (2 MVZ); Mt. Warner, 5.500 ft. (1 USNM); 3 mi. S Stauffer (1 USNM); 11 mi. N on R3615 from HWY 140* (1 SOSC). **Malheur Co.**—(1 PSM); Beulah (1 USNM); Cedar Mts. (2 USNM); Cord, Barren Valley, 3,950 ft. (1 USNM); 16 mi. SE Cord (1 USNM); Ironside (1 AMNH); Rattlesnake Creek (1 USNM); Riverside (1 USNM); Riverside Wildlife Management Area (1 PSU); 20 mi. S Rome, Sheephead Mt. (1 USNM). **Morrow Co.**—(10 USNM); Tupper R. S. [T6S, R27E, Sec. 8], 2 mi. E Tupper (1 USNM). **Umatilla Co.**—Echo, Butte Creek (1 USNM); Juniper Canyon, T5N, R32E, Sec. 13/14 (6 USNM); 10 mi. SE Milton (1 UW[WWC]). **Union Co.**—2 mi. N Island City (1 PSM); Starkey Experimental Forest, T4S, R34E, SW¼ Sec. 12 (1 PSM); Thorn Creek, near Medical Springs (1 USNM). **Wallowa Co.**—Charolais Rd., T2N, R45E, NW¼NW¼ Sec. 29 (1 PSM); Chico, Imnaha National Forest (1 USNM); 16 mi. NE Wallowa (2 USNM); T1N, R48E, SE¼SE¼ Sec. 20 (1 PSM). **Wasco Co.**—(2 UW);

Antelope (1 USNM); Maupin (1 SDNHM). **Wheeler Co.**—7 mi. W Fossil (1 USNM); 8 mi. E Fossil (1 USNM); 5 mi. SW Fossil (2 USNM); 7 mi. S Spray (1 USNM); 13 mi. S Spray (8 USNM); 14 mi. S Spray (2 USNM); 16 mi. S Spray (1 USNM).

Thomomys bottae, n = 238
(Plotted in Fig. 11-46, p. 225)

Curry Co.—Brookings (3 PSM, 2 SDNHM, 2 USNM); Brookings, 8.5 mi. up Gardner Ridge Rd. (1 SOSC); 3 mi. S Brookings (1 SOSC); 1.5 mi. N, 1.5 mi. W Brookings, Harris Beach State Park (1 OSUFW); 6 mi. below Loeb State Park (1 SOSC); Pistol River (1 KU, 1 USNM); mouth of Pistol River (1 USNM); Pistol River, T38S, R14W, Sec. 19 (6 USNM); 1.5 mi. S Pistol River, 300 ft. (14 MVZ); 3 mi. S Pistol River (2 USNM); stateline (4 SDNHM). **Douglas Co.**—Riddle (4 USNM); Roseburg (5 USNM); Yoncalla (1 USNM). **Jackson Co.**—(1 SOSC); Applegate River, 1 mi. S Beaver Creek (5 MVZ); Ashland (2 OSUFW, 21 SOSC, 9 USNM); 1 mi. W Ashland (2 SOSC); Brownsboro (1 USNM); Central Point (1 SOSC); 1 mi. N, 1 mi. W Jacksonville (5 OSUFW); Medford (1 LACM); vivinity of Medford (1 PSM); 10 mi. N Medford (1 SDNHM, 1 USNM); 6 mi. S Medford (18 MVZ); 7 mi. S Medford (8 MVZ); 4 mi. E Medford (1 LACM); Mt. Ashland Rd. (1 SOSC); Phoenix (1 SOSC); Pompadour Butte, T39S, R1E, Sec. 5 (1 SOSC); 5 mi. N Rogue River (1 CRCM); 9 mi. N Rogue River (1 SOSC); 3 mi. W Shady Cove (1 SOSC); 0.25 mi. NE Scenic Hills Memorial Park (1 SOSC); Shoat Springs (1 SOSC); Soda Mt. (1 SOSC); Steinman, Siskiyou Mts., 3,000 ft. (1 MVZ); Talent (1 SOSC); 3 mi. S Talent (3 SOSC); Wagner Gap, 4,600 ft. (5 MVZ); White City (1 SOSC); 3 mi. W White City (2 OSUFW); 4 mi. on Dead Indian Rd. from HWY 66 (1 SOSC); 11.5 mi. from HWY 66 on Dead Indian Rd. (1 SOSC); 2.5 mi. E USFS Rd. 3706 on USFS Rd. 3705 (1 SOSC). **Josephine Co.**—2 mi. S Cave Junction (3 USNM); 13 mi. SW Galice, Briggs and Horse creeks, Ferren R.S., 2,000 ft. (3 USNM); Grants Pass (6 FMNH, 21 USNM); 2 mi. S Grants Pass (1 OSUFW); 1 mi. W Grants Pass (2 USNM); 2 mi. W Grants Pass, N side Rogue River (18 MVZ); 3.5 mi. W Grants Pass, River Road (4 MVZ); 5 mi. W Grants Pass, N side Rogue River (2 MVZ); Waldo (1 USNM); Jump off Joe Creek Rd. and Shorthorn Gulch (1 SOSC). **Klamath Co.**—Keno (1 USNM); Klamath Falls (8 CM); 2 mi. N, 7 mi. W Klamath Falls (2 OSUFW); 3 mi. S, 7 mi. W Klamath Falls (1 OSUFW); 4 mi. S, 14 mi. W Klamath Falls (1 OSUFW); 5 mi. S, 13 mi. W Klamath Falls (1 OSUFW); 6 mi. S, 11 mi. W Klamath Falls (1 OSUFW); 7 mi. S, 23 mi. W Klamath Falls (1 OSUFW); Geary Ranch headquarters, W Klamath Lake* (1 USNM); 2 mi. N, 7 mi. W Midland (2 USNM); W. Klamath Lake (1 USNM). **Lane Co.**—Cottage Grove (3 USNM).

Thomomys bulbivorus, n = 635
(Plotted in Fig. 11-48, p. 229)

Benton Co.—(5 PSM); Corvallis (1 HSU, 3 KU, 1 MVZ, 20 OSUFW, 2 USNM); probably near Corvallis (1 KU); 10 km S Corvallis (1 UW); 2.5 mi. N Corvallis (1 OSUFW); 4 mi. N Corvallis, T11S, R5W, SW¼NE¼ Sec. 13 (10 LSU); 8 mi. N Corvallis (1 PSM); 1.5 mi. S Corvallis, T12S, R5W, Sec. 3 (1 OSUFW); 2 mi. S Corvallis (1 OSUFW); 1 mi. W Corvallis, T12S, R5W, Sec. 4 (1 OSUFW); 1 mi. W Corvallis, T12S, R5W, Sec. 10 (2 OSUFW); 2 mi. W Corvallis (3 OSUFW); 1 mi. E Corvallis (4 SOSC); 2 mi. E Corvallis (1 OSUFW); 1 mi. N, 2 mi. W Corvallis (1 HSU, 1 OSUFW); 1.5 mi. N, 0.25 mi. W Corvallis (1 OSUFW); 3 mi. N, 4 mi. W Corvallis (1 OSUFW); 4 mi. N, 0.5 mi. W Corvallis (28 KU, 1 MUSMZ, 1 OSUFW); 0.25 mi. N, 2 mi. E Corvallis (1 OSUFW); 4 mi. N, 0.5 mi. E Corvallis (1 OSUFW); 4 mi. N, 0.75 mi. E Corvallis, T11S, R5W, NE¼SW¼ Sec. 12 (2 LSU); 4.5 mi. N, 0.5 mi. E Corvallis (10 KU); 5 mi. N, 0.25 mi. E Corvallis (1 OSUFW); 5 mi. N, 2 mi. E Corvallis (1 OSUFW); 5 mi. N, 5 mi. E Corvallis, 200 ft. (8 CM); 4 mi. NW Corvallis (4 CM, 4 PSM); 17 mi. N NW Corvallis, 6 mi. W Adair (1 OSUFW); 4 mi. NE Corvallis (3 CM); 2.5 mi. SE Corvallis, Kiger Island (7 MVZ); Granger (1 KU); 2 mi. N, 16 mi. W Lebanon (1 OSUFW); 0.25 mi. S Monroe, T14S, R5W, NW¼ Sec. 35 (6 KU); 0.25 mi. S, 1 mi. E Monroe, T14S, R5W, NW¼ Sec. 35 (13 KU, 2 OSUFW); North Albany, Gibson Hill (1 OSUFW); Philomath (1 OSUFW); 0.25

mi. S Philomath Rd. (1 KU); 5 mi. SW Philomath (2 USNM); T11S, R5W, Sec. 3 (1 OSUFW). **Clackamas Co.**—Auora (5 UW[EB], 13 UW[LB]); 2 mi. E Colton (4 MVZ); 7 mi. SE Molalla (1 MVZ); 7.5 mi. SE Molalla (23 PSM); 8 mi. SE Molalla (1 PSM, 1 UMMZ); Oregon City (3 UW[EB]); 3 mi. S Oregon City (1 PSU); Oswego (1 MVZ); Mulino (4 OSUFW). **Lane Co.**—Eugene (1 UO, 8 USNM); Eugene, Spencers Butte (1 USNM); 7 mi. W Eugene (1 UW[WWC]); 3 mi. S Junction City (1 OSUFW); 2 mi. S, 1 mi. E Junction City (1 OSUFW). **Linn Co.**—4 mi. S, 6 mi. W Albany (1 OSUFW); 4.5 mi. S, 6.5 mi. W Albany (2 MSUMZ, 12 OSUFW); 5 mi. S, 4.25 mi. W Albany (1 OSUFW); 5 mi. S, 4.5 mi. W Albany, T11S, R4W, SE¼ Sec. 32 (7 KU); 5.5 mi. S, 4.5 mi. W Albany (1 OSUFW); 0.5 mi. E Corvallis (1 OSUFW); 0.6 mi. E Corvallis (1 OSUFW); 1 mi. E Corvallis (4 SOSC) 3 mi. S, 2 mi. E Lebanon (2 OSUFW). **Marion Co.**—1 mi. N, 3 mi. E Brooks (1 OSUFW); Salem (2 AMNH, 12 MVZ, 1 OSUFW, 14 USNM); 8 mi. N Salem (1 OSUFW); 4 mi. S Salem (2 KU); 0.5 mi. N, 2.5 mi. E Stayton (1 OSUFW); 1 mi. N, 5 mi. E Stayton (1 OSUFW); Woodburn (16 PSM). **Multnomah Co.**—Portland (6 AMNH, 1 KU, 14 PSM, 4 PSU, 7 SDNHM, 6 USNM); T1N, R1W, Sec. 1 (2 PSU). **Polk Co.**—Grande Ronde (3 USNM); Independence (1 USNM); McCoy (5 FMNH, 1 USNM); Rickreall, T7S, R4W, center Sec. 30 (23 KU); 1 mi. S, 2 mi. W Rickreall (1 OSUFW); 1 mi. S, 3 mi. E Rickreall (1 OSUFW). **Washington Co.**—Banks (4 USNM); Beaverton (7 FMNH, 1 PSU, 5 USNM); Cedar Mill (1 PSM); Dilley (2 PSU); Forest Grove (1 CM, 1 UO); Gaston (1 OSUFW, 1 PSM, 1 PSU, 1 SDNHM); Hillsboro (2 KU, 4 LACM, 2 LSU, 17 MVZ, 1 PSU, 4 USNM); 5.5 mi. S, 1.5 mi. W Hillsboro (30 KU); Laurel (1 PSU); Merriweather golf course [Hillsboro] (2 PSU); 2 mi. S Mountaindale (1 OSUFW); North Plains (1 PSU); 1.5 mi. S, 1 mi. W Scholls, T2S, R2W, SE¼ Sec. 20 (15 KU, 3 OSUFW, 1 SM); Tigard (1 PSM, 2 PSU); 1 mi. W Skyline School on Cornelius Rd. (5 MVZ); field near OSBH laboratory (2 PSU); T1N, R4W, Sec. 36 (1 PSU); T1S, R2W, Sec. 3 (1 PSU); T2S, R1W, Sec. 18 (1 PSU); T2S, R1W, Sec. 28 (1 PSU). **Yamhill Co.**—Dayton (1 OSUFW, 3 PSM, 1 PSU); 3.5 mi. S Dayton (51 KU); 2.5 mi. S, 0.5 mi. E Dayton (5 OSUFW); Lafayette (1 UW[WWC]); McMinnville (3 PSU); 7.5 mi. W McMinnville (1 UW[WWC]); 2 mi. S, 3 mi. E Newberg (18 KU, 1 OSUFW, 1 PSM); Parrott Mt. (1 EOSC); Sheridan (2 USNM); 0.5 mi. E Sheridan, T5S, R6W, Sec. 25 (4 OSUFW); 0.25 mi. NE Sheridan (4 OSUFW); Yamhill River (1 USNM); 1 mi. N Yamhill (16 KU, 4 PSM).

Thomomys mazama, *n* = 1,394
(Plotted in Fig. 11-51, p. 232)

Benton Co.—Alsea (8 USNM); 3 mi. W Alsea, T14S, R8W, Sec. 9 (6 USNM); 8 mi. SW Alsea, Lobster Creek, T14S, R9W (4 USNM); 10 mi. SW Alsea (1 USNM); 5 mi. S SW Blodgett (1 OSUFW); Chintimini Mts., 4,000 ft. (1 USNM); Corvallis (1 CRCM); 3 mi. N Corvallis (1 CRCM); Marys Peak, T12S, R7W, Sec. 29 (23 USNM); 2 mi. S, 9 mi. W Philomath, Marys Peak (1 OSUFW); 0.4 km S, 12.5 km W Philomath (3 USNM); 5 mi. SW Philomath (1 USNM); Summit (1 USNM); 2 mi. NE Summit Marys Peak, T11S, R7W, Sec. 5 (1 USNM). **Clackamas Co.**—(1 SDNHM); 1 mi. N Bull Run (2 PSU); Canby (10 USNM); 3 mi. SE Canby (2 SDNHM, 3 USNM); 1 mi. N Government Camp (3 MVZ); 2 mi. SE Government Camp, T3S, R8½E (5 USNM); Molalla (1 CRCM); 8 mi. S Molalla (3 CRCM); 20 mi. E Molalla (1 SDNHM); 7.5 mi. SE Molalla (18 PSM); Oregon City (29 USNM); 2 mi. W Rhododendron, T3S, R7E, Sec. 9 (13 USNM); Wilsonville (1 FMNH). **Clatsop Co.**—Elsie (1 OSUFW, 12 PSM, 5 USNM); 1 mi. S Elsie, Nehalem River, T4N, R7W, Sec. 5 (4 USNM); 2 mi. S Elsie, Nehalem River, T4N, R7E, Sec. 7 (5 USNM); 5 mi. E Elsie (1 USNM); 12 mi. NE Elsie, Green Mt. area (3 OSUFW); 1 mi. W Hamlet (1 PSM); Jewell (3 SDNHM); 2.5 mi. S Jewell on Nehalem River (18 USNM); 3 mi. S Jewell, T5N, R7W, Sec. 23 (9 USNM); 6 mi. S Jewell (3 PSM); 1.5 mi. NW Jewell (2 USNM); 2 mi. SE Jewell (3 USNM). **Columbia Co.**—Scappoose (2 KU, 1 OSUFW, 21 PSM, 14 SDNHM, 4 UMMZ, 71 USNM); 1 mi. NE Scappoose (3 USNM). **Curry Co.**—Brookings (2 USNM); 7 mi. N Brookings (1 USNM); 11 mi. N Brookings (3 USNM); Oregon-California state line, Oregon coast (1 USNM); Gold Beach (10 FMNH, 1 KU, 4 SDNHM, 10 USNM); mouth Pistol River (4 USNM);

Wedderburn (8 USNM); Wedderburn, T36S, R14E, Sec. 30 (10 USNM). **Deschutes Co.**—11 mi. W Bend, Tumalo Creek (10 MVZ); 15 mi. W Bend, 6,100 ft. (7 MVZ); 16 mi. W Bend, 6,500 ft. (6 MVZ); W side Broken Top, 7,000 ft. (2 CRCM); 1 mi. SE Broken Top, 7,100 ft. (1 MVZ); Davis Creek (1 OSUFW); mouth of Davis Creek (5 OSUFW); Devil's Lake, 1 mi. W Sparks Lake, T18S, R8E, Sec. 10 (6 USNM); W side Green Lakes basin, E of South Sister, 6,500 ft. (3 CRCM); Indian Ford campground (1 OSUFW, 3 PSM); Indian Ford Creek (3 PSM); La Pine (12 USNM); 6 mi. N La Pine, T21S, R11W (5 USNM); 2.5 mi. N, 9 mi. E La Pine, 5,600 ft. (1 KU); 3.5 mi. N, 5 mi. E La Pine (1 KU); 5 mi. N, 2.5 mi. E La Pine, 4,200 ft. (1 KU); 6 mi. N, 3 mi. E La Pine, T21S, R11E, SE¼ Sec. 7 (2 MVZ); 9 mi. N, 3 mi. E La Pine, T20S, R11E, Sec. 32 (2 MVZ); Little Lava Lake (1 PSM); Paulina Lake, T21S, R13E, Sec. 31 (1 SDNHM, 11 USNM); Paulina Lake, Paulina Mts. (3 OSUFW); 0.25 mi. S Paulina Lake, 6,400 ft. (2 KU); Paulina Prairie, T21S, R11E, Sec. 19 (5 MVZ); Pengra (2 USNM); Sisters (4 USNM); 12 mi. N Sisters (1 SDNHM); 4 mi. N, 9 mi. W Sisters (2 MVZ); 4 mi. N, 6 mi. E Sisters, Black Butte, T14S, R9E (2 OSUFW); Sparks Lake, T18S, R8E, Sec. 11 (1 UMMZ, 1 UOMNH, 16 USNM); 1 mi. N Three Creek Lake (11 PSM). **Douglas Co.**—Anchor (4 USNM); Diamond Lake, T28S, R6E, Sec. 7 (4 PSM, 14 SDNHM, 3 UMMZ, 11 USNM); Ft. Umpqua (1 USNM); 2 mi. E Scottsburg, T22S, R9W, Sec. 16 (11 USNM). **Hood River Co.**—Brooks Meadows, 9 mi. ENE Mt. Hood (9 MVZ); Cascade Locks, Oxbow Fish Hatchery (3 MVZ, 16 PSM, 4 SDNHM); 6 mi. E Cascade Locks (14 PSM); Cloud Cap Inn, Mt. Hood (5 USNM); above Elkcove, N side Mt. Hood (1 CRCM); 4.5 mi. E Oxbow (3 PSM); Cooper Spur, Mt. Hood (2 UMMZ, 4 USNM); Dee (2 USNM); 0.5 mi. E Hood River P.O., 70 ft. (3 KU); 1.5 mi. S Hood River P.O., 4,000 ft. (5 KU); 3 mi. S Hood River bridge (5 PSM); 2 mi. E Hood River (1 PSM); 2 mi. SW Hood River (2 KU, 1 MVZ); 2 mi. SE Hood River (1 PSM); Hood River Meadows (2 PSM); Mt. Hood near timberline (4 USNM); W base Mt. Hood (1 USNM); 2 mi. S Odell (2 PSU); Parkdale, 4,200 ft. (6 USNM, 2 UW); 1 mi. W Parkdale (1 KU); 2 mi. W Parkdale, 1,500 ft. (6 USNM); 3 mi. S Parkdale, T1S, R10E, Sec. 17 (8 USNM); 2 mi. SW Parkdale, 2,800 ft. (2 KU). **Jackson Co.**—Ashland (1 SOSC, 1 UOMNH); E of Ashland (1 SOSC); 14 mi. E Ashland (1 SOSC); 5.5 mi. N, 12 mi. E Ashland, Dead Indian area (1 OSUFW); Ashland Research Natural Area (1 SOSC); N base Ashland Peak, Longs Camp, 3,300 ft. (2 USNM); N base Ashland Peak, 4,600 ft. (1 USNM); Butte Falls R.D., T33S, R4E, SE¼ Sec. 18 (2 SOSC); halfway between Drew and Crater Lake (5 USNM); Mt. Ashland summit (1 SOSC); Mt. Ashland Rd. (7 SOSC); Prospect, Rogue River Valley (5 MVZ, 1 USNM); 10 mi. E Prospect, Bessie Creek (4 MVZ); 8 mi. SW Prospect, 1,800 ft. (3 CM, 6 USNM); inlet to Reeder Reservoir (1 SOSC); Siskiyou, 4,131 ft. (2 USNM); Soda Creek, T39S, R4E, NW¼ Sec. 8 (1 SOSC); 2 mi. N, 1 mi. E Union Creek (1 CRCM); Union Creek Forest camp, Crater Lake HWY, 4,720 ft. (2 MVZ); Wagner Gap, 4,600 ft. (1 MVZ); 0.6 mi. N junction East Fork Ashland Creek off FS Rd 3903B (2 SOSC); 10.3 mi. NE HWY 66 on Dead Indian Rd. (1 SOSC); 11.5 mi. from HWY 66 on Dead Indian Rd. (2 SOSC). **Jefferson Co.**—Camp Sherman, Pioneer Ford camp (9 PSM); Jack Lake, NE base Three Fingered Jack (1 USNM). **Josephine Co.**—Bolan Lake (9 PSM); 13 mi. SW Grant Pass (4 UOMNH); Winona (1 USNM). **Klamath Co.**—1 mi. N Altamont, Klamath Branch Experiment Station (2 OSUFW); Anna Creek, 4 mi. NW Ft. Klamath, Mt. Mazama (2 MVZ, 3 USNM); Buck Peak, T38S, R6E, NW¼ Sec. 16 (1 SOSC); Crater Lake* (11 MVZ, 2 OSUFW, 5 USNM); 1 mi. S Crater Lake (1 MVZ); Crater Lake National Park (2 USNM); Crater Lake, Annie Springs (1 USNM); Crater Lake National Park, Bald Top Mt. (2 USNM); Crater Lake National Park, Munson Meadows (1 OSUFW); Crater Lake National Park headquarters (5 MVZ); Crater National Forest, camp 76 (3 USNM); Davis Lake, T23S, R7E, Sec. 11 (3 USNM); Ft. Klamath (16 USNM); 2 mi. N, 18 mi. W Kirk (1 OSUFW); Klamath Agency (2 SDNHM); 2 mi. N, 0.5 mi. W Klamath Agency (1 OSUFW); 2 mi. N, 1 mi. W Klamath Agency (1 OSUFW); Klamath Falls (16 CM, 1 USNM); 6 mi. S Klamath Falls (1 OSUFW); 9 mi. N, 14 mi. W Klamath Falls, Aspen Lake (1 OSUFW); 3 mi. S, 22 mi. W Klamath Falls (1 OSUFW); 7 mi. S, 23 mi. W Klamath Falls (1 OSUFW); W. Klamath Lake, Geary Ranch (1 OSUFW); S end Lake of the Woods (1 MVZ); 2 mi. S Lake of the Woods, Rainbow Creek (1

MVZ); Merrill (2 MVZ); 4 mi. W Midland (3 OSUFW); 4 mi. N, 2 mi. W Midland (1 OSUFW); 4 mi. N, 11 mi. W Midland, near Round Lake (1 OSUFW); Pelican Bay, 4,200 ft. (1 USNM); Rustler Peak, Crater National Forest (5 USNM); 3 mi. W Swan Lake (2 MVZ); Upper Klamath Lake, N Modoc Point (2 PSM); Upper Klamath Lake, Rocky Point (2 OSUFW). **Lane Co.**—Belknap Springs (2 OSUFW); 4 mi. E mouth Big Creek (1 KU); 7 mi. N, 9 mi. E Blue River (1 CRCM); 4.4 km N, 6.8 km E Blue River, T16S, R5E, Sec. 7 (1 USNM); Deadwood (2 USNM); 6 mi. N Florence, Mercer Lake (16 USNM); 13 mi. N Florence, China Creek (1 USNM); 14 mi. N Florence, Big Creek (6 USNM); H. J. Andrews Experimental Forest* (2 PSM); 4 mi. NE Heceta Head, Big Creek (11 USNM); Huckleberry Lake (2 KU); Mable (1 SDNHM); Mapleton (5 OSUFW, 9 USNM); 12 mi. N Mapleton, Indian Creek, T16S, R10W, Sec. 4 (3 USNM); 3 mi. NW Mapleton, T17S, R10W, Sec. 28 (6 USNM); McKenzie Bridge (2 OSUFW, 1 SDNHM, 2 UOMNH, 7 USNM); 3.1 km N, 3 km W McKenzie Bridge (2 USNM); 3.8 km S, 14.8 km E McKenzie Bridge, T16S, R7E, Sec. 33 (15 USNM); 4.4 km S, 14.3 km E McKenzie Bridge, T16S, R7E, Sec. 32 (1 USNM); 10 mi. S McKenzie Bridge, Horse Pasture Mt. (2 UOMNH); 10 mi. S McKenzie Bridge, O'Leary Mts. (4 USNM); 10 mi. E McKenzie Bridge (3 USNM); 2 mi. N, 2 mi. W McKenzie Bridge, T16S, R5E, Sec. 3/4, 3,800 ft. (1 UW[OG]); 0.5 mi. N, 3 mi. E McKenzie Bridge (4 LSU); 0.9 mi. S, 7.0 mi. E McKenzie Bridge (10 LSU); 10 mi. SE McKenzie Bridge, Lost Creek Ranger Station (2 USNM); Mercer (6 OSUFW, 1 PSU, 2 MVZ, 7 SDNHM, 2 UOMNH, 3 USNM); NW base Middle Sister, Three Sisters (5 MVZ, 1 SDNHM); Mobile* (1 USNM); 5 mi. N Noti, T16S, R6W, Sec. 31 (7 USNM); Scott Lake, NW base Three Sisters, 4,800 ft. [T15S, R7½E, S½] (6 KU, 5 USNM); Seaton (6 USNM); 14 mi. N Swisshome, Paris (4 USNM); 4 mi. NE Swisshome, T16S, R8W (1 USNM); Tenmile Creek, 5.5 mi. E HWY 101 (4 MVZ); Tenmile Creek, T15S, R11W, Sec. 33 (1 USNM); W base Three Sisters (5 OSUFW, 1 SDNHM); 7 mi. SE Yachats, Tenmile Creek (17 USNM). **Lincoln Co.**—Devil's Lake, T7S, R11W, Sec. 2 (12 USNM); 5 mi. E Eddyville, T11S, R8W, Sec. 16 (2 USNM); 9 mi. E Fisher (1 OSUFW); Oceanlake (1 OSUFW); 7 mi. E Yachats, T14S, R11W (7 USNM); 18 mi. E Yachats, Wilkinson Creek (2 USNM); 5 mi. up Yachats River (3 SDNHM). **Linn Co.**—Big Lake (7 PSM); 20 mi. E Cascadia, Tombstone Prairie (1 OSUFW); 25 mi. E Cascadia, Lost Prairie campground (1 OSUFW); Lost Prairie, Santiam Pass, T13S, R6E, SEC. 27 (1 OSUFW, 9 USNM); 10.25 km N, 0.8 km W McKenzie Bridge, T15S, R5E, Sec. 14 (1 USNM); Santiam Pass, Big Lake, T14S, R7½E (1 OSUFW); 8 mi. E Upper Soda, Tombstone Prairie (8 OSUFW); 1 mi. S, 0.25 mi. E Upper Soda (1 OSUFW); 1 mi. S, 6 mi. E Upper Soda, Lost Prairie (1 OSUFW). **Marion Co.**—Silver Creek recreation area (1 UW[EB]). **Multnomah Co.**—Tigard (1 PSM). **Polk Co.**—Black Rock (2 USNM); Pedee (3 USNM). **Tillamook Co.**—Beaver (3 KU, 3 PSM, 5 UMMZ, 1 SDNHM, 1 UOMNH, 24 USNM); Blaine (1 CM, 3 KU, 3 MVZ, 3 SDNHM, 5 UMMZ, 7 USNM); 14 mi. NE Cedar Butte, T1N, R8W, Sec. 10 (7 USNM); Hebo Lake (2 PSM); Lees camp (1 PSU); 15 mi. E Mohlar, Cook Creek, junction to Blue Lake (1 OSUFW); Mt. Hebo (2 SDNHM, 3 USNM); 6 mi. E Mt. Hebo, Hebo, T4S, R9W, Sec. 14 (13 USNM); 7 mi. E Mt. Hebo, Hebo, T4S, R9W, Sec. 13 (3 USNM); Neskowin (2 OSUFW); 13 mi. N Summit, Cedar Butte (2 USNM); Tillamook (1 KU, 3 LACM, 10 MVZ, 1 SDNHM, 2 UO, 12 USNM); 4 mi. S Tillamook (1 PSM); 5 mi. W Tillamook (6 USNM); 3 mi. E Tillamook, T1S, R9W, Sec. 27 (1 USNM); 4 mi. E Tillamook (4 PSM); 7 mi. E Tillamook, T1S, R8W, Sec. 8 (1 PSM, 3 USNM); 9 mi. E Tillamook, Wilson River area, T1S, R8W, Sec. 10 (1 OSUFW, 5 USNM); 10 mi. E Tillamook, Wilson River, T1S, R8W, Sec. 27 (2 OSUFW, 6 USNM); 6 mi. S, 3 mi. E Tillamook (1 OSUFW); 18 mi. NE Tillamook, Wilson River (3 USNM); SE Tillamook, Camp Murphy, T2S, R7W, SEC. 24 (3 OSUFW); 4 mi. SE Tillamook, T2S, R9W, Sec. 3 (4 USNM); 5 mi. SE Tillamook, T2S, R9W, Sec. 10 (3 USNM); 16 mi. SE Tillamook (3 UMMZ); 21 mi. SE Tillamook (1 OSUFW); 9 mi. ENE Tillamook, T1S, R8W, Sec. 8 (1 USNM). **Wasco Co.**—Mosier, Rowena Crest (2 USNM). **Washington Co.**—Bull Mt. (1 SDNHM); Forest Grove, David's Hill (1 USNM); Gales Creek (1 USNM); Hillsboro (3 USNM, 4 CRCM); 6 mi. SW Hillsboro (2 USNM); 8 mi. SE Hillsboro (2 USNM); 11 mi. SE Hillsboro (2 USNM); 12 mi. SE Hillsboro (1 USNM); Laurel Hill (1 SDNHM); 3 mi.

N, 2.5 mi. W Manning, T3N, R4W, NE¼ Sec. 35 (1 KU); Sherwood (1 SDNHM); Sherwood, Parrot Mt. (5 USNM); Tigard (1 USNM); 4 mi. W Tigard, T2S, R1W, Sec. 4 (7 USNM); 2 mi. S Timber, T3N, R5W, Sec. 34 (8 USNM). **Yamhill Co.**—Bald Peak State Park, [T2S, R3W, Sec. 16] (2 PSU); Lafayette (3 UW[WWC]); Mt. Hebo (2 OSUFW); 10 mi. NE Hebo, Mt. Hebo (1 OSUFW); 7.5 mi. W McMinnville (2 UW[WWC]); 3 mi. E Newberg (3 MVZ)

Additional localities: N base Three Sisters, 5,000 ft. (15 USNM); N slope Three Sisters, 7,000 ft. (2 USNM); Three Sisters* (1 SDNHM).

Thomomys talpoides, $n = 1,179$
(Plotted in Fig. 11-53, p. 235)

Baker Co.—Anthony [just S of 45° line N of Sparta] (26 AMNH, 4 MVZ, 3 SDNHM, 1 USNM); Baker [City] (2 USNM); Bourne (6 USNM); Burkemont (2 AMNH); 2.5 mi. NE Cornucopia, East Pine Creek (1 USNM); Haines (1 EOSC); 1.5 mi. NW Haines, 3,535 ft. (1 EOSC); Home (7 USNM); Homestead, 1,800 ft. (6 USNM); Homestead, 3,500 ft. (2 USNM); Huntington (1 USNM); 3 mi. NE Huntington, 2,100 ft. (4 USNM); McEwen (2 USNM); Rock Creek (2 USNM); 0.5 mi. SE Rock Creek Butte, Elkhorn Mts., 8,300 ft. (1 CRCM); 3 mi. SE Rock Creek Butte, Elkhorn Mts., 8,500 ft. (1 CRCM). **Crook Co.**—(1 OSUFW); 3 mi. E Bear Creek, Crooked River, 3,600 ft. (6 MVZ); Big Canyon Creek, mouth of Davis Creek (1 PSU); Howard (9 USNM); Ochoco R. S., 4,000 ft. (4 MVZ); Paulina (1 SDNHM); 20 mi. S Paulina (1 SDNHM); 25 mi. E Prineville (4 USNM); 26 mi. NE Prineville, Marks Creek (1 OSUFW, 1 SDNHM); 4 mi. SW Prineville (1 MVZ); Big Summit Prairie (1 SDNHM); Trout Creek guard station, 20 mi. NE Prineville (20 MVZ); Walton Lake (1 SOSC); T17S, R19E, Sec. 35 (2 OSUFW). **Deschutes Co.**—(1 PSM); Bend (1 OSUFW, 2 PSM, 10 USNM); Bend, west bank Deschutes River (2 USNM); 1 mi. N Bend (6 MVZ); 3 mi. N Bend, 3,400 ft. (1 KU); 6 mi. N Bend, 3,613 ft. (19 MVZ); 0.5 mi. S Bend (1 OSUFW); 1 mi. S Bend (2 KU); 2.2 mi. S Bend (2 MVZ); 3.4 mi. S Bend (1 MVZ); 6 mi. N, 1 mi. E Bend (2 MVZ); 4 mi. S, 2 mi. W Bend (6 MVZ); Deschutes (4 KU); Farewell Bend, Deschutes River (2 USNM); Redmond (2 SDNHM); 1 mi. E Redmond (1 OSUFW); 1.5 mi. N Sans Spring* (1 OSUFW); Sisters (12 OSUFW); 7 mi. E Sisters (2 CM); 1 mi. N, 7 mi. E Sisters (1 OSUFW); 1 mi. N Terrebonne, 2,800 ft. (1 KU); 2.5 mi. W Terrebonne, along Deschutes River (1 OSUFW); Tumalo, E side Deschutes River, 3,100 ft. (6 MVZ); 4 mi. W Tumalo (2 KU). **Gilliam Co.**—2.5 mi. S Heppner Junction (1 KU); mouth John Day River, Columbia River (2 MVZ); Willows (14 USNM); Willows Junction (1 USNM). **Grant Co.**—(1 EOSC); Austin (7 USNM); 8 mi. E Austin, Cold Spring, 4,900 ft. (16 MVZ); Bull Prairie (1 CRCM); Camp Creek (1 SOSC); Dayville (2 SDNHM); Dixie Butte (1 USNM); 6 mi. N Kimberly, Tamarack Springs (1 OSUFW); 5.1 mi. N, 1.7 mi. W Long Creek, Morgrass-Pass Creek (1 OSUFW); Magone Lake, T12S, R31E, Sec.12 (2 OSUFW); Mt. Vernon (1 USNM); 21 mi. SE Prairie City, North Fork Malheur River, 5,000 ft. (12 MVZ); 5 mi. S Ritler Butte, T9S, R30E, Sec. 15 (1 OSUFW); Strawberry Mts. (4 USNM); E slope Strawberry Mt., 8,700 ft. (1 CRCM); T15S, R26E, NW¼ Sec. 33 (1 SOSC). **Harney Co.**—Alvord Lake (1 USNM); Blitzen Valley* (10 SDNHM); Burns (3 OSUFW, 10 USNM); 47 mi. S, 10 mi. E Burns, N Baca Lake, Malheur National Wildlife Refuge (1 OSUFW); 30 mi. NW Burns (6 USNM); Buchanan (2 USNM); Camp Harney (1 USNM); 2 mi. E Crane (5 USNM); Diamond, 4,300 ft. (10 USNM); 10 mi. E Diamond, Smyth Creek, 5,000 ft. (7 MVZ); Drewsey (3 USNM); 0.75 mi. W Fish Lake, T32½S, R33E, Sec. 5 (1 OSUFW); 1.5 mi. S, 2.5 mi. E Frenchglen, T32S, R32½E, NW¼ Sec. 17 (4 KU); 5 mi. S, 10 mi. E Frenchglen, 6,500 ft. (1 UW[WWC]); 5 mi. S, 11 mi. E Frenchglen, T32S, R32¾E, SW¼NE¼ Sec. 34, 6,600 ft. (3 KU); 8.5 mi. SE Frenchglen, T32S, R32½E, Sec. 24 (1 OSUFW); Harney (1 USNM, 1 UW); Jackman Park, 2 mi. SE Fish Lake, 7,500 ft. (5 UW[WWC]); Malheur Lake (2 LACM, 1 USNM); ca. 8 km S Malheur Lake, Malheur Field Station (1 UW); Narrows (1 USNM); 24 mi. N, 12 mi. W Riley, T19S, R25E, NE¼SW¼ Sec. 26 (1 KU); Shirk P.O. (15 USNM); N end Steens Mt. (7 FMNH); Steens Mt., Big Fish Creek, 7,500 ft. (4 USNM); Steens Mt., Fish Lake (2 UOMNH); Steens Mt., Fish Lake, 8,000 ft. (1 CM, 2 FMNH, 16 UMMZ, 2 USNM); Steens Mt., 4.2

mi. W Fish Lake (1 SOSC); Steens Mt., head of Kiger Gorge, 9,000 ft. (1 UW[WWC]); Steens Mt., Kiger Gorge, 6,000 ft. (8 USNM); Steens Mt., Kiger Gorge, 6,800 ft. (1 USNM); Steens Mt., Wildhorse Lake (1 SOSC); Steens Mt., T30S, R32E, Sec. 31 (1 OSUFW); Steens Mt., T32S, R32½E, Sec. 25 (2 OSUFW); Tumtum Lake (1 USNM); Voltage (1 SDNHM, 1 USNM). **Hood River Co.**—Cascade Locks, Oxbow Fish Hatchery (4 SDNHM); 2 mi. S Hood River (1 SOSC); 0.5 mi. E Hood River (3 KU). **Jefferson Co.**—(5 PSM); Culver (1 PSM); 2 mi. E Culver (1 KU); 19 mi. E Culver (8 PSM); 4 mi. NNW Culver (11 PSM); Foley Farm, Sunrise Valley, 3,950 ft. (4 USNM); Gateway (1 USNM); Hay Creek (6 OSUFW, 15 USNM); 12 mi. E Hay Creek, Foley Creek (1 USNM); 2 mi. S Madras, 2,400 ft. (3 KU); Metolius (2 USNM); Trout Creek, Ochoco National Forest (2 USNM); Warm Springs (3 USNM); 0.5 mi. S, 3 mi. E Warm Springs (1 OSUFW). **Klamath Co.**—3 mi. S, 2 mi. E Beatty (1 OSUFW); 1 mi. S, 4 mi. W Bly (1 OSUFW); 1 mi. S, 7 mi. W Bly (1 OSUFW); 2 mi. N, 6 mi. W Bly (1 OSUFW); 2 mi. N, 8 mi. E Bly (1 OSUFW); 9 mi. W Bonanza (7 PSM); 0.75 mi. E Bonanza (1 PSM); 1 mi. E Bonanza (1 PSM); 3 mi. E Bonanza (6 PSM); 9.5 mi. NW Bonanza (1 PSM); 9.5 mi. SW Bonanza (1 PSM); 2 mi. N, 21 mi. E Kirk (1 OSUFW); 14 mi. N, 1.5 mi. W Sprague River (1 OSUFW); Teddy Power's Meadow* (4 PSM); Yamsay River, 4,800 ft. (6 USNM); Yamsay Mts., West Silver Creek, 7,000 ft. (3 USNM); T39S, R11½E, Sec. 3 (5 PSM); T39S, R12E, Sec. 2 (2 PSM); T39S, R12E, Sec. 8 (6 PSM). **Lake Co.**—Adel (1 OSUFW, 4 USNM); 20 mi. NE Adel, Hart Mt., site of Old Ft. Warner (4 OSUFW, 2 SDNHM); Fort Rock (2 OSUFW); 9 mi. N Fort Rock (1 SOSC); Foster Flat R.S. (2 SDNHM); Fremont (4 USNM); Gearhart Mt., Finley Corral (17 PSM); Guano Creek, Hart Mt. National Antelope Refuge (1 SOSC); Hart Mt. (1 LACM, 1 SDNHM); 0.25 mi. S Hot Spring camp, Hart Mt. National Wildlife Refuge (2 SOSC); Ingram R.S., Fremont National Forest (2 PSM); Lakeview (4 USNM); 2 mi. E Lakeview, 5,200 ft. (1 MVZ); 14 mi. E Lakeview, Vernon Spring (1 CM); 25 mi. NW Lakeview at Finley Corrals (2 USNM); Mt. Warner, 5,500 ft. (1 USNM); 2 mi. N, 6 mi. W Paisley (3 KU); 10 mi. S Plush (2 CRCM); Robertson Draw, Hart Mt. National Wildlife Refuge (1 SOSC); Silver Lake (5 USNM); 1 mi. N Silver Lake, 4,600 ft. (1 KU); 1 mi. NW Silver Lake P.O., 4,400 ft. (2 KU); 10 mi. W Silver Lake, West Silver Creek, 4,650 ft. (1 USNM); 1 mi. N Summer Lake (1 KU); Thomas Creek R.S. (1 LACM); 23 mi. W Valley Falls, Deadhorse Creek (9 OSUFW). **Malheur Co.**—Battle Mt. (1 SDNHM); Beulah (4 USNM); Cedar Mts. (2 USNM); Cord, Barren Valley, 3,950 ft. (2 USNM); Cow Creek Lake (3 USNM); 10 mi. above Harper (1 SDNHM); Ironside, 4,000 ft. (11 AMNH); 1 mi. W Ironside (3 MVZ); Jordan Valley, 4,200 ft. (8 USNM); 2 mi. S Jordan Valley (5 OSUFW); 30 mi. S, 11 mi. W Jordan Valley (5 PSM); 3 mi. W Juntura (2 USNM); Mahogany Mts., 6,500 ft. (2 USNM); canyon located 8 mi. NE McDermitt, 5,800 ft. (3 USNM); 14 mi. S Riverside (1 SDNHM); 2 mi. NW Riverside (8 USNM); Rockville (14 USNM); Rome, Owyhee River, 3,850 ft. (3 USNM); Skull Spring (3 USNM); Vale (1 AMNH); near Vale (1 AMNH); Watson (7 USNM). **Morrow Co.**—0.25 mi. S, 5 mi. W Boardman (1 PSU); 6 mi. E Cecil (1 SDNHM); Irrigon (3 MVZ). **Sherman Co.**—Columbia River, at mouth of Deschutes River (1 KU); Biggs Junction (1 PSM); Miller, mouth Deschutes River (2 KU, 18 MVZ, 1 OSUFW, 8 USNM); Rufus (2 USNM); 2 mi. WSW Rufus (4 MVZ). **Umatilla Co.**—Echo (1 SDNHM); 1 mi. E Freewater (3 UW[WWC]); 2 mi. E Freewater (1 UW[WWC]); Hermiston (1 SDNHM); Langdon Lake, ca. T3N, R38E (1 MVZ, 4 UW[WWC]); 3 mi. W Langdon Lake (1 UW[WWC]); 4 mi. W Langdon Lake (1 KU, 2 UW[WWC]); 4 mi. NW Langdon Lake (1 MVZ); Meacham (7 USNM); 6.8 mi. N Meacham, T1N, R36E, NW¼ Sec. 33, 4,390 ft. (5 PSU); 2 mi. NW Meacham (5 MVZ); Meacham Lake, 3,800 ft. (2 MVZ); Milton (3 UW[WWC]); 1.6 mi. S Milton (1 UW[WWC]); 1 mi. E Milton (3 UW[WWC]); 2 mi. E Milton (2 UW[WWC]); 4 mi. E Milton (1 UW[WWC]); 6 mi. SW Milton (1 UW[WWC]); 7 mi. SW Milton (2 UW[WWC]); 7.5 mi. SW Milton (1 UW[WWC]); 8 mi. SE Milton (1 UW[WWC]); Milton-Freewater (2 EOSC); 2 mi. N, 1 mi. E Milton-Freewater (2 UW[WWC]); Pendleton (6 OSUFW, 2 PSM, 5 SDNHM, 6 USNM); 4 mi. W Summit, Tollgate (1 UW[WWC]); 7 mi. W Tollgate (1 UW[WWC]); Ukiah (2 LACM, 1 SDNHM); 10 mi. SE Ukiah (2 LACM, 3 USNM); Umapine (2 PSM); Umatilla (2 MVZ, 12 USNM); 16 mi. E Weston (1 UW[WWC]). **Union**

Co.—(7 EOSC); Anthony Lake (1 PSM); W side Anthony Lake (2 PSM); Elgin (1 PSM, 1 USNM); 9.9 mi. N, 5.5 mi. W Elgin, T3N, R38E, Sec. 26 (2 PSU); 13.3 mi. N, 7.2 mi. W Elgin (1 PSU); 13.25 mi. N, 7.25 mi. W Elgin (2 PSU); Island City (1 EOSC); 0.2 mi. E Island City (1 OSUFW); La Grande, T3S, R38E, NE¼SE¼ Sec. 8 (1 EOSC); 9 mi. S, 2 mi. W La Grande (4 OSUFW); Summerville (1 EOSC); Telocaset (2 USNM); 7 mi. E Telocaset, Catherine Creek (2 MVZ); Airport rd. (1 EOSC); Carmen Milles farm, 0.5 mi. W Cove Rd. (2 EOSC). **Wallowa Co.**—(2 EOSC); Bear Creek, Snake River (1 SDNHM); Billy Meadows R.S. (1 SDNHM); Chico (1 SDNHM); Dug Bar, Snake River (1 SDNHM); 25 mi. N Enterprise at Sled Springs, 4,600 ft. (1 SDNHM, 6 USNM); 3 mi. S, 2 mi. W Enterprise (1 OSUFW); Imnaha (1 SDNHM); 10 mi. S, 1 mi. E Lostine (1 MVZ); 16 mi. S, 2 mi. E Lostine, T3S, R43E, Sec. 25 (2 OSUFW); 16 mi. S, 3 mi. E Lostine (4 MVZ); 21 mi. S, 6 mi. E Lostine, T4S, R44E, SEC. 26 (1 OSUFW); 17 mi. SE Lostine, Lostine River (3 OSUFW); 15 mi. N Paradise at Horse Creek, 2,000 ft. (4 USNM); Pittsburg Landing, Snake River (1 SDNHM); Roland Bar Landing (1 SDNHM); Temperance Creek (1 SDNHM); Troy, 1,600 ft. (7 USNM); Troy, 1,700 ft. (1 USNM); Wallowa Lake, 4,000 ft. (9 USNM); S. Wallowa Lake, Wallowa Mts. (7 USNM); Wenaha Forks (1 EOSC); T4N, R47E, Sec. 35 (64 UW). **Wasco Co.**—7 mi. E Antelope (2 USNM); Friend (1 LACM, 4 USNM); Ft. Dalles (2 USNM); Mosier (2 UMMZ); Rowena Crest, Mosier (3 OSUFW); The Dalles (14 MVZ, 3 PSU, 19 USNM); 2 mi. W [The] Dalles (2 PSU); 0.5 mi. NW The Dalles, 90 ft. (4 KU); Wamic (2 USNM); Wapinitia (1 USNM). **Wheeler Co.**—15 mi. S Fossil, Twickenham, on the John Day River (4 OSUFW); 7 mi. S, 11 mi. W Mitchell (88 MVZ); 13 mi. SW Mitchell, Corral Flat, T13S, R20E, SEC. 6 (2 OSUFW); 11 mi. N Service Creek (1 UW[WWC]); 8 mi. N, 4 mi. W Spray (1 OSUFW); Wildwood campground, Ochoco National Forest, T12S, R20E (1 OSUFW).

Thomomys townsendii, $n = 171$
(Plotted in Fig. 11-55, p. 239)

Harney Co.—Lake Alvord (1 MVZ, 7 USNM); S end Lake Alvord (4 MVZ); Borax Spring, S end lake Alvord, 4,300 ft. (2 MVZ); 20 mi. S, 10.1 mi. W Burns, N Harney Lake (1 OSUFW); 22.5 mi. S, 14.5 mi. E Burns (5 OSUFW); 26 mi. S, 7 mi. E Burns, Sod House School (2 OSUFW); 28 mi. S, 3 mi. W Burns, Harney Lake bed (1 OSUFW); Fields (2 SDNHM); Malheur Environmental Field Station (4 localities within 2 mi. of Station) (2 OSUFW, 5 PSU, 1 SOSC); 2 mi. W Malheur National Wildlife Refuge headquarters (3 HSU); 1 mi. E Malheur National Wildlife Refuge headquarters (2 OSUFW); Harney Lake dunes, Malheur National Wildlife Refuge, T26S, R30E, Sec. 28 (6 CM); Malheur National Wildlife Refuge, T26S, R31E, Sec. 35 (2 CM); Malheur National Wildlife Refuge, Redhouse field, Double-O Ranch, T26S, R28E, Sec. 13 (1 CM); 4 mi. SW Narrows (5 MVZ); 5 mi. SW Narrows (3 MVZ); North Coyote, Malheur Environmental Field Station (2 SOSC); 15 mi. W, 2.5 mi. S Princeton, Malheur Environmental Field Station (1 PSU); Stergen Meadows, Pueblo Mts. (1 KU); Trout Creek (1 SDNHM); Tum Tum Lake (6 USNM). **Malheur Co.**—Brogan Valley, 1 mi. E of city (1 EOSC); near N end Diseaster Peak, near head Whitehorse Creek (2 USNM); 1 mi. W Malheur Butte (2 UMMZ); 3 mi. N McDermitt (2 PSM); Ontario (1 USNM); Ontario, W side Snake River (2 MVZ); 2.5 mi. N Ontario (20 MVZ); 5 mi. SW Ontario, 2,000 ft. (21 USNM); Owyhee (7 USNM); Vale (6 USNM); Sand Hollow near Vale (1 AMNH); 2 mi. N Vale (3 PSM); 1.5 mi. S Vale (32 USNM); 5 mi. S Vale (1 PSM); 1 mi. N, 5 mi. E Vale (1 OSUFW); 5 mi. S, 2 mi. W Vale (1 OSUFW); 1 mi. SW Westfall (2 USNM).

Urocyon cinereoargenteus, $n = 49$
(Plotted in Fig. 12-6, p. 364)

Benton Co.—Corvallis (1 PSM); 9 mi. S, 2 mi. W Corvallis (1 OSUFW); 7 mi. SE Corvallis (1 CRCM); Sulphur Springs (2 USNM). **Clackamas Co.**—0.5 mi. S, 2 mi. E Carver (1 OSUFW); 7 mi. SE Oregon City (2 PSM); SE Molalla (14 PSM). **Coos Co.**—Coquille (1 USNM). **Curry Co.**—Stockyards Camp, 42°30′N, 124°W (2 USNM). **Deschutes Co.**—Skeleton Cave, 14 mi. S Bend (1 USNM). **Douglas Co.**—Glendale

(4 USNM). **Jackson Co.**—2 mi. S Gold Hill (1 SOSC). **Josephine Co.**—Galice Creek, Rogue River Valley (1 UO); 40 mi. NE Grants Pass (1 USNM); 43 mi. NE Grants Pass (9 USNM). **Lane Co.**—T16S, R5W, Sec. 29 (2 PSM). **Linn Co.**—1 mi. N Harrisburg (1 OSUFW); 2 mi. E Scio (1 UO). **Polk Co.**—2 mi. S Buena Vista (1 CRCM); 1 mi. S Suver (1 MVZ). **Washington Co.**—Gaston (1 OSUFW).

Ursus americanus, n = 111
(Plotted in Fig. 12-12, p. 374)

Clackamas Co.—Clackamas Lake (1 USNM); Clackamas River, Mt. Hood National Forest (2 USNM); Baty Butte [T6S, R3E, NW¼] (1 USNM); N. Fork Cold Springs (6 USNM); Granite Peak [T7S, R7E, Sec. 5/6] (1 USNM); Oak Grove [T2S, R1E, Sec. 2/1/11/12] (2 USNM); Roaring River Mt. (1 USNM); Shell Rock, [T6S, R8E, Sec. 3] (1 USNM); Oregon City (1 USNM). **Coos Co.**—Marshfield (2 USNM); Remote (1 USNM). **Curry Co.**—Gold Beach (2 USNM); 27 mi. S Gold Beach (1 USNM); 25 mi. E Harbor (1 USNM); Lobster Creek (1 MVZ); Port Orford (2 USNM). **Deschutes Co.**—Bend (1 USNM); 35 mi. W Bend, Lava Lake (1 USNM); La Pine (1 USNM); SE of Paulina Mts. (1 USNM); 1 mi. E Sisters (1 OSUFW). **Douglas Co.**—Drain (1 USNM); Glendale (2 USNM); Glide (1 USNM); Rockcrest* (1 OSUFW); Roseburg (1 KU); Smith River (3 USNM). **Grant Co.**—Dale (1 USNM). **Jefferson Co.**—N slope Mt. Jefferson, timberline (1 MVZ). **Josephine Co.**—21 mi. SE Williams (1 USNM). **Lake Co.**—Fort Rock (2 USNM); 12 mi. N Fort Rock (1 USNM); 12 mi. NE Fort Rock (1 USNM); Gearhart Mt. (1 PSM); 30 mi. NW Silver Lake (6 USNM). **Lane Co.**—Belle (2 USNM); near Blue River, Lucky Boy Rd. (1 KU); Eugene (1 MVZ); Eugene, Collison Ranch, Fall Creek (1 USNM); 8 mi. W Gilchrest Butte, Little Deschutes River, Lapine (1 PSM); South Fork McKenzie River (2 OSUFW); 40 mi. S Oakridge (3 USNM); 10 mi. E Oakridge, Wall Creek (1 USNM); 10 mi. SE Oakridge, Wall Creek (1 USNM); 20 mi. SE Oakridge (1 USNM); Reed (2 USNM); Reserve (2 PSM); 0.5 mi. S Reserve, T20S, R1E, Sec. 2 (1 OSUFW); 2 mi. S Reserve (1 OSUFW); 3 mi. S Reserve (1 OSUFW); 5 mi. S Reserve (1 OSUFW). **Linn Co.**—Cascadia, Rock Creek (1 USNM); 6 mi. from summit of mts. above Cascadia (1 MVZ); 1 mi. S Scio, 1 mi. S Thomas Creek (1 OSUFW). **Morrow Co.**—(1 USNM). **Tillamook Co.**—Blaine (1 OSUFW); Garibaldi (1 USNM); Mt. Hebo (1 KU); Netarts Bay (1 PSM). **Umatilla Co.**—Duncan (1 SDNHM, 1 USNM); Meacham (1 USNM). **Union Co.**—Boulder Creek, Wallowa National Forest (1 USNM); Chicken Creek, Blue Mts., T5S, R35E, Sec. 35 (1 USNM); Meadow Creek, Gray's Ranch, Blue Mts. (1 USNM); Mud Springs, near Cove, Wallowa National Forest (2 USNM); Sheep Creek, Blue Mts., T5S, R35E, Sec. 35 (3 USNM). **Wallowa Co.**—head of Bear Creek (1 USNM); Enterprise (1 USNM); Wallowa (1 USNM); 12 mi. W Wallowa (1 USNM); Wallowa Mts. (1 USNM). **Wasco Co.**—Dufur (3 USNM); Wapinita (1 SDNHM). **Wheeler Co.**—15 mi. SE Mitchell, P C Touchets Range (2 USNM).

Additional localities: Jack Creek* (1 USNM); Blue River* (3 USNM); Blue Mts., Mathews Range* (1 USNM).

Ursus arctos, n = 6
(Plotted in Fig. 12-15, p. 378)

Deschutes Co.—South Ice Caves (1 USNM). **Harney Co.**—Malheur Lake, found on dry lake bed (1 USNM). **Wallowa Co.**—Enterprise, 45°20', 117°20' (1 USNM); Wallowa Mts.* (1 USNM); Wallowa National Forest, near Billy Meadows, T4N, R46E (1 USNM).

Additional locality: Cascade Mts.* (1 AMNH).

Vulpes velox, n = 7
(Plotted in Fig. 12-8, p. 367)

Deschutes Co.—Bend (2 USNM). **Klamath Co.**—Ft. Klamath, Wood River (1 USNM). **Malheur Co.**—Pollick (1 SDNHM, 1 USNM); 20 mi. S Pollick between Rome and Pollick* (1 USNM). **Wallowa Co.**—Wallowa (1 USNM).

Vulpes vulpes, n = 36
(Plotted in Fig. 12-10, p. 369)

Benton Co.—Corvallis (2 OSUFW); 10 mi. S Corvallis (1 OSUFW); 1 mi. S, 1 mi. E Corvallis (1 OSUFW); 5 mi. W Monroe, T14S, R5W, Sec. 24 (2 OSUFW). **Clackamas Co.**—Canby (1 PSU). **Columbia Co.**—Sauvie Island (1 CRCM). **Deschutes Co.**—Bend (2 USNM); South Ice Caves, [T22S, R14E, Sec. 20] (1 USNM). **Harney Co.**—18 mi. S, 14 mi. W Burns (1 OSUFW); 10 mi. S, 6 mi. E Burns, S of Wrights Point, T25S, R32E, NW¼NW¼ Sec. 6 (1 KU). **Hood River Co.**—Mt. Hood (1 USNM). **Klamath Co.**—Crater Lake National Park (1 MVZ); Ft. Klamath (1 USNM). **Lake Co.**—Fremont (1 USNM); Silver Lake (3 SDNHM); Thompson Valley (1 USNM). **Lane Co.**—0.5 mi. SW Franklin, T16S, R5W, Sec. 29 (2 PSM); T15S, R1W, SW¼ Sec. 10 (1 PSM). **Linn Co.**—(3 OSUFW); 8 mi. N, 1.5 mi. E Albany (1 OSUFW); near Harrisburg (1 KU). **Marion Co.**—3.5 mi. W Woodburn (1 OSUFW). **Union Co.**—1 mi. SE La Grande (1 OSUFW). **Wallowa Co.**—18 mi. SE Joseph, Big Sheep Basin (2 USNM); Wallowa (1 USNM); Wallowa Mts.* (1 SDNHM). **Wasco Co.**—Maupin (1 MVZ).

Zalophus californianus, n = 15
(Plotted in Fig. 12-22, p. 389)

Clatsop Co.—Astoria (1 UW); Chute Drift, river mile 20 (1 UW); jetty sands on Clatsop Spit (1 CRCM); Gearhart approach (1 CRCM); 0.5 mi. S Peter Iredale wreck, Ft. Stevens State Park (1 CRCM); Seaside (1 PSM); 1 mi. N Seaside (1 CRCM); Sunset Beach (1 PSM). **Coos Co.**—about 5 mi. S Coos Bay, Simpson Reef (1 LACM). **Lane Co.**—near Florence (1 PSM); Sutton Creek (1 PSM). **Lincoln Co.**—Agate Beach wayside (1 PSM); Lost Creek picnic area (1 PSM); Salishan Beach (1 PSM). **Tillamook Co.**—(1 PSM).

Zapus princeps, n = 286
(Plotted in Fig. 11-129, p. 341)

Baker Co.—Anthony (8 AMNH, 2 SDNHM); Bourne (14 USNM); Cornucopia (9 USNM); 3 mi. N Cornucopia (2 UMMZ); 2.5 mi. NE Cornucopia, East Pine Creek (2 USNM); 2.5 mi. N, 14.5 mi. W Haines, Anthony Lake (1 OSUFW); 4.5 mi. N, 8.25 mi. W Haines (1 PSU); Lily White, Wallow Mt. field station (1 EOSC); McEwen (3 USNM). **Crook Co.**—Chuckwagon Spring, Round Mt. (1 SOSC); Howard (3 USNM); 12 mi. N Howard, Marks Creek (4 USNM); Ochoco R.S., 4,000 ft. (4 MVZ). **Grant Co.**—Austin (5 USNM); 8 mi. E Austin, Cold Spring, 4,900 ft. (4 MVZ); 15 mi. NW Austin (2 USNM); Beech Creek (2 USNM); Camp Creek (4 SOSC); 1 mi. N, 10 mi. E Long Creek, Kahler Butte, T10S, R32E, Sec. 4 (2 OSUFW); 21 mi. SE Prairie City, North Fork Malheur River (21 MVZ); Strawberry Butte (3 USNM); Strawberry Mts. (13 USNM); Strawberry Mt., John Day (1 USNM). **Harney Co.**—Diamond, 4,300 ft. (4 USNM); Harney (2 USNM); Steens Mt., 0.5 mi. N Fish Lake (1 HSU); Steens Mt., 1 mi. N Fish Lake (1 HSU); Steens Mt., Kiger Gorge, 6,000 ft. (2 USNM); Steens Mt., Kiger Gorge, 6,900 ft. (8 USNM); Steens Mt., Kiger Gorge, 7,500 ft. (1 USNM); Steens Mt., Kiger Gorge, 8,000 ft. (1 USNM); Steens Mt., Little Blitzen, T33S, R33E, Sec. 10 (1 OSUFW); Steens Mt., T32S, R32¾E, Sec. 20 (1 OSUFW); Steens Mt., T32S, R32¾E, Sec. 26 (1 OSUFW); Steens Mt., T32½S, R33E, Sec. 33 (2 OSUFW); Steens Mt., T33S, R33E, Sec. 4 (1 OSUFW). **Jackson Co.**—Ashland (2 SOSC); Ashland Natural Research Area, T40S, R1E, Sec. 4 (2 SOSC); Ashland Research Natural Area, T39S, R1E, SW¼ Sec. 28 (1 SOSC); N base Ashland Peak, 5,000 ft. (2 USNM); N base Grizzly Peak, 3,300 ft. (1 USNM); W slope Grizzly Peak, 4,600 ft. (1 USNM); Howard Prairie Dam Rd. off Hyatt Lake Rd. (1 SOSC); T37S, R3E, SW¼ Sec. 20 (1 SOSC); T40S, R4W, Sec. 18 (1 SOSC). **Jefferson Co.**—12 mi. E Hay Creek at Foley Creek (2 USNM). **Klamath Co.**—West Silver Creek, Yamsay Mts., 7,000 ft. (1 USNM); T35S, R15E, Sec. 25 (1 PSM). **Lake Co.**—14 mi. SW Adel, Barley Camp, Warner Mts., T41S, R21E, E edge Sec. 8 (2 OSUFW, 1 SDNHM); Dairy Creek (1 SDNHM); Gearhart Mt. (14 PSM); 3 mi. N Finley Corrals (1 PSM); 0.5 mi. S Hot Spring camp, Hart Mt. Antelope Refuge, T35S, R26E, NE¼NW¼ Sec. 32 (2 SOSC); Hart Mt. Antelope Refuge,

Warmsprings campground* (1 SOSC); 2 mi. E Lakeview, 5,200 ft. (3 MVZ); Warner Mts. (1 USNM); T36S, R16E, Sec. 12 (1 PSM). **Malheur Co.**—E base Disaster Peak, 5,600 ft. (1 USNM); Ironside, 4,000 ft. (9 AMNH); Jordan Valley, 4,200 ft. (4 USNM). **Umatilla Co.**—Bear Canyon, Umatilla National Forest (1 SDNHM); Meacham (1 USNM); 4 mi. W Tollgate (1 UW). **Union Co.**—Anthony Lake (6 PSM); Elgin (3 USNM); Summerville (1 EOSC); Umatilla National Forest, Jubilee Forest camp (1 PSM); West Eagle meadow, West Eagle Creek (2 EOSC). **Wallowa Co.**—(1 EOSC); Aneroid Lake (7 SDNHM, 7 UO); Cloverdale Rd., T3S, R46E, Sec. 29 (2 PSU); 24 mi. SE Joseph, Lick Creek campground (1 PSU); 17 mi. S Lostine, 5,300 ft. (1 UO); 10 mi. S, 1 mi. E Lostine, T2S, R43E, Sec. 35 (4 OSUFW); 13 mi. S, 2 mi. E Lostine, T3S, R43E, Sec. 14 (1 OSUFW); 16 mi. S, 3 mi. E Lostine (10 MVZ); 17 mi. SE Lostine, Lostine River, Wallowa Mts. (6 OSUFW); Minam Lake (1 USNM); 15 mi. N Paradise at Horse Creek, 2,000 ft. (2 USNM); Sled Springs (1 SDNHM); Umatilla National Forest, Mottet camp (3 PSM); Umatilla National Forest, Timothy Guard Station (1 PSM); 8 mi. W Wallowa (1 UW); near Wallowa Lake, 4,500 ft. (3 FMNH); North Fork Wallowa River, 5,000 ft.* (1 FMNH). **Wheeler Co.**—7 mi. S, 11 mi. W Mitchell (20 MVZ); 12 mi. SW Mitchell, Wildwood Camp, Ochoco National Forest (1 OSUFW); T13S, R20E, center Sec. 2 (1 SOSC).

Zapus trinotatus, n = 702
(Plotted in Fig. 11-131, p. 344)

Benton Co.—(1 PSM); 1 mi. N, 4 mi. W Albany, E. E. Wilson Game Management Area (1 OSUFW); 10 mi. W Alsea (1 OSUFW); 5 mi. N, 5 mi. E Alsea (2 CM); 3.25 mi. S, 4.25 mi. E Alsea (3 CM); 3.25 mi. S, 5.5 mi. E Alsea (2 CM); 3.75 mi. S, 5.5 mi. E Alsea (5 CM); Corvallis (8 OSUFW, 2 UCONN); 2 mi. N Corvallis (1 KU); 3 mi. N Corvallis (1 KU); 5 mi. N Corvallis (1 OSUFW); 7 mi. N, 3 mi. E Corvallis, E. E. Wilson Game Management Area (1 OSUFW); 10.5 mi. S, 1 mi. W Corvallis (2 OSUFW); Marys Peak (2 PSM); McDonald Forest (6 OSUFW, 2 PSM); 1 mi. S, 6 mi. W Philomath, 0.25 mi. W Stillson Creek, Marys Peak (1 OSUFW); 4 km S, 10.1 km W Philomath, T12S, R7W, Sec. 24/25 (1 USNM). **Clackamas Co.**—Barlow campground, near Zigzag River (1 PSU); Collins Lake, Government Camp (1 PSU); Government Camp (2 PSU); 2.75 mi. W Government Camp, Twin Bridges campground (1 PSU); 0.25 mi. N, 2.75 mi. W Government Camp (2 PSU); 0.75 mi. N, 2.75 mi. W Government Camp (9 PSU); 1 mi. N, 2.75 mi. W Government Camp (4 PSU); 0.5 mi. S, 1.5 mi. E Government Camp, Still Creek (5 PSU); 8 mi. SE Molalla (6 PSM); Mulino (1 OSUFW); Oregon City (1 OSUFW, 1 UW[LB]); 2 mi. N, 4 mi. E Sandy (4 CRCM); 3.5 mi. N, 8 mi. E Sandy (3 CRCM); 3.8 mi. N, 8 mi. E Sandy (9 CRCM); 4.2 mi. N, 8 mi. E Sandy, T1S, R6E, Sec. 29 (7 CRCM); 6 mi. S, 7 mi. E Sandy, T3S, R6E, Sec. 18 (10 CRCM); Still Creek [campground near Government Camp] (1 PSU); 9 mi. N, 4 mi. W Zigzag (3 CRCM); 5 mi. S, 6 mi. E Zigzag (5 CRCM); T5S, R8½E, Sec. 26 (2 PSU). **Clatsop Co.**—Arch Cape (1 SDNHM); Astoria (2 USNM); 7.5 mi. S Cannon Beach (1 MVZ); site of Old Ft. Clatsop (11 MVZ); Seaside (2 USNM). **Columbia Co.**—Quincy (1 CRCM); 7 mi. E Rainier (11 MVZ). **Coos Co.**—1.75 mi. E Bandon (18 PSM); 1.75 mi. SE Bandon (1 PSM); 3 mi. SE Bandon (1 OSUFW); 4 mi. SE Bandon (11 PSM); Charleston (3 LACM); 3 mi. S Charleston, Sunset Bay (1 LACM); 5 mi. SW Charleston, Sunset Bay (4 LACM); 5 mi. SE Charleston, Sunset Bay (4 LACM); Coos Bay, South Slough (1 LACM); Coos Head (2 LACM); Cape Arago State Park (6 LACM); Sunset Bay (1 LACM); Marshfield (1 USNM); 7.5 km N, 12.5 km W Sitkum, T27S, R11W, Sec. 21 (2 USNM). **Curry Co.**—Gold Beach (4 FMNH, 1 PSM). **Deschutes Co.**—11 mi. W Bend, Tumalo Creek (2 MVZ); 15 mi. W Bend, Tumalo Creek, 6,100 ft. (1 MVZ); head of Falls Creek, SE of South Sister, 6,450 ft. (1 CRCM). **Douglas Co.**—Anchor (6 USNM); Crystal Springs, 1 mi. E Umpqua inlet to Lake Lemolo (2 SOSC); Diamond Lake near outlet (2 OSUFW); 11.5 km N, 10.5 km W Drain (1 USNM); 4.5 mi. N, 8 mi. E Drain, T21S, R4W, Sec. 27, 1,200 ft. (3 UW[CR]); Elkhead (1 USNM); Fish Creek (1 PSM); Ft. Umpqua (1 USNM); Francis Creek, T24S, R1W, SE¼SE¼ Sec. 12 (1 SOSC); Gardiner (2 FMNH, 5 MVZ); Glendale (1 USNM); Lake Creek, 0.25 mi. below Diamond Lake (3 OSUFW); Louis Creek, T28S, R3W,

SE¼SE¼ Sec. 30 (4 SOSC); Lower Umpqua, West Eel Creek (1 PSM); Middle Creek, T31S, R6W, SW¼ Sec. 30 (6 SOSC); Reston (1 USNM); 4.4 km N, 12.1 km W Sutherlin, T25S, R7W, Sec. 1 (4 USNM); 4.5 mi. N, 9 mi. E Tiller (1 CRCM). **Hood River Co.**—Brooks Meadows, 9 mi. ENE Mt. Hood, 4,300 ft. (10 MVZ); Cascade Locks, Oxbow Hatchery (1 PSM); Elk Cove, N side Mt. Hood (9 CRCM); above Elk Cove, N side Mt. Hood, 5,800 ft. (1 CRCM); Hood River Meadow (1 PSU); W slope Mt. Hood near timberline (1 USNM); W base Mt. Hood (1 USNM); north side Wahtum Lake, 3,700 ft. (2 UW). **Jackson Co.**—Ashland Butte (1 UO); halfway between Drew and Crater Lake (2 USNM); Prospect (2 USNM); 13.5 mi. N, 2.5 mi. W Prospect (1 CRCM); 14 mi. N, 0.5 mi. W Prospect, T30S, R2E, Sec. 24/25 (1 CRCM); 2.5 mi. N, 8.5 mi. E Prospect (1 CRCM); 5 mi. N, 8 mi. E Prospect (1 CRCM); Spring Creek, T33S, R4E, NE¼ Sec. 22 (1 SOSC); T31S, R2E, NE¼ Sec. 8 (1 SOSC). **Josephine Co.**—1 mi. N, 10.5 mi. E Cave Junction, T39S, R6W, Sec. 8 (1 HSU); Grants Pass, junction of Rogue and Applegate rivers (1 USNM). **Klamath Co.**—Anna Creek, Mt. Mazama, 6,000 ft. (2 USNM); Crater Lake (5 MVZ, 9 USNM); 0.5 mi. S Crater Lake (3 UMMZ); 1 mi. S Crater Lake (1 MVZ); Crater Lake National Park, E slope Cascade divide, 6,300 ft. (1 KU); 20 mi. W Cresent Lake, 5,000 ft. (13 UO); 0.5 mi. N Government Camp, Crater Lake National Park (2 MVZ); Munson Creek, Crater Lake headquarters (1 OSUFW); Ft. Klamath (2 USNM); Odell Creek, 4,700 ft. (3 LACM); Upper Munson Meadow, Crater Lake [National Park] (1 KU, 1 USNM). **Lane Co.**—Ada, near Siltcoos Lake, T19S, R11W, Sec. 30 (2 OSUFW); 2 mi. S, 1 mi. E Austa (1 CM); Belknap Springs, McKenzie Bridge (1 OSUFW); 4 mi. E mouth Big Creek (3 KU); 0.5 mi. N, 2 mi. W Blue River, Elk Creek (1 CRCM); 1.5 mi. N, 2 mi. E Blue River (1 CRCM); 4 mi. N, 4.5 mi. E Blue River, T16S, R5E, Sec. 32 (1 CRCM); 5 mi. N, 10.5 mi. E Blue River, T15S, R5E, Sec. 30 (5 CRCM); 5.5 mi. N, 5 mi. E Blue River, T15S, R5E, Sec. 28 (1 HSU); 5.5 mi. N, 13.5 mi. E Blue River, T16S, R6E, Sec. 26 (1 CRCM); 1 mi. S, 11 mi. E Blue River (1 OSUFW); 2.5 mi. S, 7 mi. E Blue River (1 CRCM); 10 mi. NE Blue River, Upper Lookout Creek, T15S, R5E, Sec. 25 (1 JFBM); 1 mi. N, 10.5 mi. E Cougar Dam, T16S, R6E, Sec. 25, 2,900 ft. (1 UW[OG]); Eugene (3 USNM); 5 mi. S Eugene (1 OSUFW); 14 mi. W Eugene (1 OSUFW); 8 mi. N Florence (1 OSUFW); 5 mi. S Florence, W side Siltcoos Lake (1 LACM); 2 km N, 6.4 km W Lorane, T20S, R5W, Sec. 5 (1 USNM); Mapleton (1 USNM); McKenzie Bridge (6 OSUFW, 5 UO, 5 USNM); 3.3 km N, 2.4 km W McKenzie Bridge, T16S, R5E, Sec. 3 (1 USNM); 6 km N, 0.6 km E McKenzie Bridge, T15S, R5E, Sec. 25 (1 USNM); 6 km N, 7.5 km E McKenzie Bridge, T15S, R6E, Sec. 26 (2 USNM); 6.4 km N, 0.6 km E McKenzie Bridge, T15S, R5E, Sec. 25 (1 USNM); 1.6 km S, 5.2 km W McKenzie Bridge, T16S, R5E, Sec. 20 (14 USNM); 3.8 km S, 14.8 km E McKenzie Bridge, T16S, R7E, Sec. 13 (2 USNM); 4.4 km S, 14.3 km E McKenzie Bridge, T16S, R7E, Sec. 32 (2 USNM); 5 mi. S McKenzie Bridge, Spring Creek, T16S, R6E, Sec. 26 (1 JFBM); 10 mi. S McKenzie Bridge, Horse Pasture Mt. (4 UO); 10 mi. SE McKenzie Bridge (6 USNM); McKenzie River* (1 OSUFW); Mercer (1 UO); 1 mi. N Mohawk River, T15S, R1W, SE¼NE¼ Sec. 13 (4 SOSC); 4 mi. N Nimrod, T16S, R3E, Sec. 14 (1 CRCM); 4.5 mi. N Nimrod (1 CRCM); 3 mi. SE Oakridge (5 UO); Pot Holes Meadow, T17S, R6E (1 OSUFW); Spencer Butte (2 UO); Sutton Lake, 6 mi. N Florence (1 MVZ); N base Three Sisters, 5,000 ft. (1 USNM); Three Sisters, Alder Springs, 4,300 ft. (2 USNM); Upper Horse Lake, T18S, R7E, Sec. 22 (1 OSUFW); Vida (1 OSUFW, 14 UO, 3 USNM). **Lincoln Co.**—(1 KU, 1 USNM); Cascade Head Experimental Forest (10 PSM); Depoe Bay (1 SDNHM); Delake (1 KU, 1 MVZ, 2 PSM); 1.5 mi. NW Eddyville (13 PSM); 15 mi. W Harlan, Peterson Ridge, T12S, R11W, Sec. 12 (8 KU); 2 mi. S, 8 mi. W Harlan, T12S, R10W, Sec. 13 (1 KU); 2 mi. S, 12 mi. W Harlan, Beaver Creek, T12S, R10W, Sec. 20 (1 KU); 5 mi. S, 10 mi. W Harlan, Gold Creek, T13S, R10W, Sec. 2 (16 KU); 7 mi. S, 11 mi. W Harlan, Boulder Creek, T13S, R10W, Sec. 15 (1 KU); Lincoln Beach (1 PSM); Newport (2 MVZ, 1 OSUFW, 5 SDNHM, 1 UO, 3 USNM); 2 mi. E Newport (2 LACM); Otis, Cascade Head (1 USNM); Waldport (2 USNM); Yaquina Bay (6 USNM). **Linn Co.**—Big Lake (1 OSUFW); 6.75 mi. N, 8 mi. E Blue River (2 CRCM); 7 mi. N, 8 mi. E Blue River, T15S, R5E, Sec. 15 (2 CRCM); 9.5 mi. N, 9 mi. E Blue River, T15S, R5E, Sec. 2 (2 CRCM); 2 mi. E Cascadia (1 OSUFW); 2.5 mi. N, 29 mi. E Cascadia, 1 mi. E

Lost Lake campground (1 OSUFW); Fish Lake (5 PSM); Lost Prairie, 32.5 mi. E Sweet Home (2 PSU); 10.25 km N, 0.8 km W McKenzie Bridge, T15S, R5E, Sec. 14 (1 USNM); North Santiam River, 3,400 ft.* (3 MVZ); 32.5 mi. E Sweet Home, Lost Prairie campground (7 PSU); 8 mi. E Upper Soda, Tombstone Prairie, 4,000 ft. (2 SUVM); 13 mi. E Upper Soda, Toad Creek Rd., 3,500 ft. (1 SUVM). **Marion Co.**—Salem (1 USNM). **Multnomah Co.**—near Linnton (1 AMNH); Portland (2 AMNH, 3 MVZ, 1 PSM); 6 mi. N, 11 mi. E Sandy (4 CRCM); 8 mi. N, 11 mi. E Sandy (9 CRCM); 9 mi. N, 10 mi. E Sandy (1 CRCM); 10 mi. N, 10 mi. E Sandy (3 CRCM); 10 mi. N, 11 mi. E Sandy (7 CRCM); 11 mi. N, 10 mi. E Sandy (1 CRCM); 11 mi. N Zigzag (7 CRCM); 10 mi. N, 1.5 mi. W Zigzag (3 CRCM). **Polk Co.**—2 mi. S Buena Vista (3 OSUFW). **Tillamook Co.**—Beaver (1 OSUFW); Blaine (3 MVZ, 2 UMMZ); 8.5 mi. E Hebo (1 OSUFW); Hebo Lake (7 PSM); Hebo Lake Forest camp (4 LACM); Mt. Hebo (1 SDNHM); Nehalem (1 OSUFW); Netarts (3 OSUFW, 1 LACM, 2 PSM, 17 SDNHM, 6 UO, 1 USNM); Oceanside (2 LACM); Tillamook (4 AMNH, 1 KU, 1 MVZ, 3 OSUFW, 2 PSM, 1 UO); 9 mi. S Tillamook (1 MVZ); head of Ben Smith Rd., halfway between Forest Grove and Tillamook (1 OSUFW). **Wasco Co.**—Clear Creek, T5S, R9E, Sec. 4 (5 PSU); Clear Lake (1 PSM). **Washington Co.**—Beaverton (2 USNM); North Plains (1 PSU); 18.5 mi. NW Portland (6 MVZ). **Yamhill Co.**—Dayton (2 PSM).

LITERATURE CITED

Ables, E. D. 1965. An exceptional fox movement. Journal of Mammalogy, 46:102.

———. 1969. Activity studies of red foxes in southern Wisconsin. The Journal of Wildlife Management, 33:145–153.

Adams, L. 1959. An analysis of a population of snowshoe hares in northwestern Montana. Ecological Monographs, 29:141–170.

Albright, R. 1959. Bat banding at Oregon Caves. The Murrelet, 40:26–27.

Alcorn, J. R. 1940. Life history notes on the Piute ground squirrel. Journal of Mammalogy, 21:160–170.

———. 1941. Counts of embryos in Nevadan kangaroo rats (genus Dipodomys). Journal of Mammalogy, 22:88–89.

———. 1944. Notes on the winter occurrence of bats in Nevada. Journal of Mammalogy, 25:308–310.

Alden, R. H., and J. D. Ifft. 1943. Early naturalists in the far West. Occasional Papers of the California Academy of Sciences, 20:1–59.

Aldous, C. M. 1951. The feeding habits of pocket gophers (*Thomomys talpoides moorei*) in the high mountain ranges of central Utah. Journal of Mammalogy, 32:84–87.

Aldous, S. E., and J. Manweiler. 1942. The winter food habits of the short-tailed weasel in northern Minnesota. Journal of Mammalogy, 23:250–255.

Aleksiuk, M. 1968. Scent-mound communication, territoriality, and population regulation in beaver (*Castor canadensis* Kuhl). Journal of Mammalogy, 49:759–762.

Alexander, L. F. 1996. A morphometric analysis of geographic variation within *Sorex monticolus*. Miscellaneous Publications, Museum of Natural History, University of Kansas, 88:1–54.

Alexander, L. F., and B. J. Verts. 1992. Clethrionomys californicus. Mammalian Species, 406:1–6.

Alexander, L. F., B. J. Verts, and T. P. Farrell. 1994 [1995]. Diet of ringtails (*Bassariscus astutus*) in Oregon. Northwestern Naturalist, 75:97–101.

Allan, P. F. 1944. Mating behavior of Dipodomys ordii richardsoni. Journal of Mammalogy, 25:403–404.

———. 1946. Notes on Dipodomys ordii richardsoni. Journal of Mammalogy, 27:271–273.

Alldredge, A. W., J. F. Lipscomb, and F. W. Whicker. 1974. Forage intake rates of mule deer estimated with fallout cesium-137. The Journal of Wildlife Management, 38:508–516.

Allen, D. L. 1938. Notes on the killing technique of the New York weasel. Journal of Mammalogy, 19:225–229.

———. 1942. Populations and habitats of the fox squirrel in Allegan County, Michigan. The American Midland Naturalist, 27:338–379.

Allen, H. 1864. Monograph of the bats of North America. Smithsonian Miscellaneous Collection, 7(publ. 165):1–85.

———. 1865. On a new genus of Vespertilionidae. Proceedings of the Academy of Natural Sciences of Philadelphia, 17:173–175.

———. 1866. Notes on the Vespertilionidae of tropical America. Proceedings of the Academy of Natural Sciences of Philadelphia, 18:279–288.

———. 1894. A monograph of the bats of North America. Bulletin of the United States National Museum, 43:1–198.

Allen, J. A. 1875. Synopsis of the American Leporidae. Proceedings of the Boston Society of Natural History, 17:430–436.

———. 1880. History of North American pinnipeds: a monograph of the walruses, sea-lions, sea-bears and seals of North America. United States Geological and Geographical Survey of the Territories, Miscellaneous Publications, 12:1–785.

———. 1890a. A review of some of the North American ground squirrels of the genus Tamias. Bulletin of the American Museum of Natural History, 3:45–116.

———. 1890b. List of mammals collected by Mr. Clark P. Streator in British Columbia, with descriptions of two new subspecies of Sciurus. Bulletin of the American Museum of Natural History, 3:161–168.

———. 1890c. Descriptions of a new species and a new subspecies of the genus Lepus. Bulletin of the American Museum of Natural History, 3:159–160.

———. 1891. Description of a new species of big-eared bat, of the genus Histiotus, from southern California. Bulletin of the American Museum of Natural History, 3:195–198.

———. 1893a. List of mammals collected by Mr. Charles P. Rowley in the San Juan Region of Colorado, New Mexico and Utah, with descriptions of new species. Bulletin of the American Museum of Natural History, 5:69–84.

———. 1893b. On a collection of mammals from the San Pedro Martir Region of Lower California, with notes on other species, particularly of the genus Sitomys. Bulletin of the American Museum of Natural History, 5:181–202.

———. 1894. On the mammals of Aransas County, Texas, with descriptions of new forms of Lepus and Oryzomys. Bulletin of the American Museum of Natural History, 6:165–198.

———. 1895. List of mammals collected in the Black Hills Region of South Dakota and in western Kansas by Mr. Walter W. Granger, with field notes by the collector. Bulletin of the American Museum of Natural History, 7:259–274.

———. 1896. Descriptions of new North American mammals. Bulletin of the American Museum of Natural History, 8:233–240 + 2 pl.

———. 1898a. Nomenclatorial notes on certain North American mammals. Bulletin of the American Museum of Natural History, 10:449–461.

———. 1898b. Revision of the chickarees, or North American red squirrels (subgenus Tamiasciurus). Bulletin of the American Museum of Natural History, 10:249–298.

———. 1900. Note on the generic names Didelphis and Philander. Bulletin of the American Museum of Natural History, 13:185–190.

———. 1902. The generic and specific names of some of the Otariidae. Bulletin of the American Museum of Natural History, 16:111–118.

Allen, J. E., M. Burns, and S. C. Sargent. 1986. Cataclysms on the Columbia: a layman's guide to the features produced by the catastrophic Bretz floods in the Pacific Northwest. Timber

Press, Portland, Oregon, 211 pp.

Allen, R. 1950. Notes on the Florida panther, *Felis concolor coryi* Bangs. Journal of Mammalogy, 31:279–280.

Allen, S. 1980. Notes on the births and deaths of harbor seal pups at Double Point, California. The Murrelet, 61:41–43.

Allison, E. C., J. W. Durham, and V. A. Zullo. 1962. Cold-water late Cenozoic faunas of northern California and Oregon. Geological Society of America, Special Paper, 68:2 (abstract only).

Allison, I. S. 1966. Fossil Lake, Oregon: its geology and fossil faunas. Oregon State University, Studies in Geology, 9:1–48.

———. 1979. Pluvial Fort Rock Lake, Lake County, Oregon. State of Oregon Department of Geology and Mineral Industries, Special Paper, 7:1–72.

———. 1982. Geology of pluvial Lake Chewaucan, Lake County, Oregon. Oregon State University Press, Studies in Geology, 11:1–79.

Allred, D. M. 1973. Small mammals of the National Reactor Testing Station, Idaho. The Great Basin Naturalist, 33:246–250.

Allred, D. M., and D. E. Beck. 1963a. Ecological distribution of some rodents at the Nevada Atomic Test Site. Ecology, 44:211–214.

———. 1963b. Range of movement and dispersal of some rodents at the Nevada Atomic Test Site. Journal of Mammalogy, 44:190–200.

Allred, D. M., D. E. Beck, and C. D. Jorgensen. 1963. Biotic communities of the Nevada Test Site. Brigham Young University Science Bulletin, Biological Series, 2(2):1–52.

Alt, D. D., and D. W. Hyndman. 1978. Roadside geology of Oregon. Mountain Press Publishing Co., Missoula, Montana, 278 pp.

Ambrose, R. F., and T. E. Meehan. 1977. Aggressive behavior of *Perognathus parvus* and *Peromyscus maniculatus.* Journal of Mammalogy, 58:665–668.

Ames, J. A., and G. V. Morejohn. 1980. Evidence of white shark, *Carcharodon carcharias,* attacks on sea otters, *Enhydra lutris.* California Fish and Game, 66:196–209.

Amstrup, S. C., and J. [J.] Beecham. 1976. Activity patterns of radio-collared black bears in Idaho. The Journal of Wildlife Management, 40:340–348.

Andersen, D. C. 1978. Observations on reproduction, growth, and behavior of the northern pocket gopher (*Thomomys talpoides*). Journal of Mammalogy, 59:418–422.

Andersen, D. C., and J. A. MacMahon. 1981. Population dynamics and bioenergetics of a fossorial herbivore, *Thomomys talpoides* (Rodentia: Geomyidae), in a spruce-fir sere. Ecological Monographs, 51:179–202.

———. 1985a. The effects of catastrophic ecosystem disturbance: the residual mammals at Mount St. Helens. Journal of Mammalogy, 66:581–589.

———. 1985b. Plant succession following the Mount St. Helens volcanic eruption: facilitation by a burrowing rodent, Thomomys talpoides. The American Midland Naturalist, 114:62–69.

Andersen, D. C., J. A. MacMahon, and M. L. Wolfe. 1980. Herbivorous mammals along a montane sere: community structure and energetics. Journal of Mammalogy, 61:500–519.

Anderson, A. E. 1981. Morphological and physiological characteristics. Pp. 27–97, *in* Mule and black-tailed deer of North America (O. C. Wallmo, ed.). University of Nebraska Press, Lincoln, 605 pp.

Anderson, A. E., and O. C. Wallmo. 1984. Odocoileus hemionus. Mammalian Species, 219:1–9.

Anderson, A. O., and D. M. Allred. 1964. Kangaroo rat burrows at the Nevada Test Site. The Great Basin Naturalist, 24:93–101.

Anderson, E. 1970. Quaternary evolution of the genus *Martes* (Carnivora, Mustelidae). Acta Zoologica Fennici, 130:1–133.

———. 1993. Evolution, prehistoric distribution, and systematics of *Martes*. Pp. 13–25, *in* Martens, sables, and fishers: biology and conservation (S. W. Buskirk, A. S. Harestad, M. G. Raphael, and R. A. Powell, eds.). Cornell University Press, Ithaca, New York, 484 pp.

Anderson, J. L., and R. Boonstra. 1979. Some aspects of reproduction in the vole *Microtus townsendii.* Canadian Journal of Zoology, 57:18–24.

Anderson, P. K., P. H. Whitney, and J-P. Huang. 1976. Arvicola richardsoni: ecology and biochemical polymorphism in the Front Ranges of southern Alberta. Acta Theriologica, 21:425–468.

Anderson, R. J. 1977. Relation of the northern pocket gopher to forest habitats in south-central Oregon. M.S. thesis, Oregon State University, Corvallis, 46 pp.

Anderson, R. M. 1942. Canadian voles of the genus Phenacomys with description of two new Canadian subspecies. The Canadian Field-Naturalist, 56:56–60.

———. 1965. Methods of collecting and preserving vertebrate animals. National Museum of Canada Bulletin, 69:1–199.

Anderson, S. 1955. Small mammals from Gilliam County, Oregon. The Murrelet, 36:26–27.

———. 1959. Distribution, variation, and relationships of the montane vole, Microtus montanus. University of Kansas Publications, Museum of Natural History, 9:415–511.

Anderson, S., and J. K. Jones, Jr. 1984. Orders and families of Recent mammals of the World. John Wiley & Sons, New York, 686 pp.

Anderson, S. E. 1974. *Eutamias minimus* and *E.amoenus*: morphological cluster analysis. M.S. thesis, Portland State University, Portland, Oregon, 48 pp.

Anthony, E. L. P., and T. H. Kunz. 1977. Feeding strategies of the little brown bat, *Myotis lucifugus,* in southern New Hampshire. Ecology, 58:775–786.

Anthony, H. E. 1913. Mammals of northern Malheur County, Oregon. Bulletin of the American Museum of Natural History, 32:1–27 + 2 pls.

———. 1928. Field book of North American mammals. Descriptions of every mammal known north of the Rio Grande, together with brief accounts of habits, geographical ranges, etc. G. P. Putnam's Sons, New York, 674 pp.

Anthony, R. G., and M. L. Morrison. 1985. Influence of glyphosate herbicide on small-mammal populations in western Oregon. Northwest Science, 59:159–168.

Anthony, R. G., E. D. Forsman, G. A. Green, G. Witmer, and S. K. Nelson. 1987. Small mammal populations in riparian zones of different-aged coniferous forests. The Murrelet, 68:94–102.

Antonelis, G. A., Jr., and M. A. Perez. 1984. Estimated annual food consumption by northern fur seals in the California Current. California Cooperative Oceanic Fisheries Investigations (CalCOFI) Report, 25:135–145.

Antonelis, G. A., M. S. Lowry, C. H. Fiscus, B. S. Stewart, and R. L. DeLong. 1994. Diet of the northern elephant seal. Pp. 211–223, *in* Elephant seals: population ecology, behavior, and physiology (B. J. Le Boeuf and R. M. Laws, eds.). University of California Press, Berkeley, 414 pp.

Anunsen, C. S., and R. Anunsen. 1993. Response to Scheffer. Conservation Biology, 7:954–957.

Arata, A. A. 1967. Muroid, gliroid, and dipodoid rodents. Pp. 226–253, *in* Recent mammals of the world: a synopsis of families (S. Anderson and J. K. Jones, Jr., eds.). The Ronald Press Company, New York, 453 pp.

Arbogast, B. S. 1992. Intraspecific phylogeography of the New World flying squirrels (*Glaucomys*). M.S. thesis, Louisiana State University, [Baton Rouge], 100 pp.

Armitage, K. B. 1962. Social behaviour of a colony of the yellow-bellied marmot (*Marmota flaviventris*). Animal Behaviour, 10:319–331.

———. 1979. Food selectivity by yellow-bellied marmots. Journal of Mammalogy, 60:628–629.

———. 1982. Marmots and coyotes: behavior of prey and predator. Journal of Mammalogy, 63:503–505.

Armitage, K. B., and J. F. Downhower. 1974. Demography of yellow–bellied marmot populations. Ecology, 55:1233–1245.

Armstrong, D. M. 1979. Ecological distribution of rodents in Canyonlands National Park, Utah. The Great Basin Naturalist, 39:199–205.

Armstrong, D. M., and J. K. Jones, Jr. 1971. Sorex merriami. Mammalian Species, 2:1–2.

Armstrong, F. H. 1957. Notes on *Sorex preblei* in Washington state. The Murrelet, 38:6.

Arnold, T. W., and E. K. Fritzell. 1987. Activity patterns, movements, and home ranges of prairie mink. Prairie Naturalist, 19:25–32.

———. 1990. Habitat use by male mink in relation to wetland characteristics and avian prey abundances. Canadian Journal of Zoology, 68:2205–2208.

Arthur, S. M., and W. B. Krohn. 1991. Activity patterns, movements, and reproductive ecology of fishers in southcentral Maine. Journal of Mammalogy, 72:379–385.

Arthur, S. M., W. B. Krohn, and J. R. Gilbert. 1989a. Habitat use and diet of fishers. The Journal of Wildlife Management, 53:680–688.

———. 1989b. Home range characteristics of adult fishers. The Journal of Wildlife Mangement, 53:674–679.

Arthur, S. M., T. F. Paragi, and W. B. Krohn. 1993. Dispersal of juvenile fishers in Maine. The Journal of Wildlife Management, 57:868–874.

Atsatt, P. R., and T. Ingram. 1983. Adaptation to oak and other fibrous, phenolic-rich foliage by a small mammal, *Neotoma fuscipes*. Oecologia, 60:135–142.

Aubry, K. B., and D. B. Houston. 1992. Distribution and status of the fisher (*Martes pennanti*) in Washington. Northwestern Naturalist, 73:69–79.

Audubon, J. J., and J. Bachman. 1841. Descriptions of new species of quadrupeds inhabiting North America. Proceedings of the Academy of Natural Sciences of Philadelphia, 1:92–103.

———. 1842. Descriptions of new species of quadrupeds inhabiting North America. Journal of the Academy of Natural Sciences of Philadelphia, series 1, part 2, 8:285.

———. 1849. The quadrupeds of North America. G. R. Lookwood, New York, 2:1–334.

———. 1854. The quadrupeds of North America. G. R. Lookwood, New York, 3:1–348.

Aulerich, R. J., and D. R. Swindler. 1968. The dentition of the mink (*Mustela vison*). Journal of Mammalogy, 49:488–494.

Axelrod, D. I. 1950. Classification of the Madro-Tertiary flora. Carnegie Institution of Washington, Contributions to Palaeontology, 590:1–22.

Babb, J. G., and M. L. Kennedy. 1989. An estimate of minimum density for coyotes in western Tennessee. The Journal of Wildlife Management, 53:186–188.

Bachman, J. J. 1837. Observations, on the different species of hares (genus *Lepus*) inhabiting the United States and Canada. Journal of the Academy of Natural Sciences of Philadelphia, 7:282–361.

———. 1839a. Additional remarks on the genus Lepus, with corrections of a former paper, and descriptions of other species of quadrupeds found in North America. Journal of the Academy of Natural Sciences of Philadelphia, 8:75–101.

———. 1839b. Description of several new species of American quadrupeds. Journal of the Academy of Natural Sciences of Philadelphia, part 1, 8:57–74.

———. 1839c. The following species must be added to the list of Mr. Townsend's quadrupeds. Journal of the Academy of Natural Sciences of Philadelphia, 8:101–105.

———. 1839d. Monograph of the species of squirrel inhabiting North America. Proceedings of the Zoological Society of London, 1838:85–103.

———. 1842. Observations on the genus Scalops, (shrew moles), with descriptions of the species found in North America. Boston Journal of Natural History, 4:26–35.

Bailey, B. 1929. Mammals of Sherburne County, Minnesota. Journal of Mammalogy, 10:153–164.

Bailey, E. P. 1974. Notes on the development, mating behavior, and vocalization of captive ringtails. The Southwestern Naturalist, 19:117–119.

Bailey, J. A. 1969a. Exploratory study of nutrition of young cottontails. The Journal of Wildlife Management, 33:346–353.

———. 1969b. Quantity of soft pellets produced by caged cottontails. The Journal of Wildlife Management, 33:421.

Bailey, J. A., and R. J. Siglin. 1966. Some food preferences of young cottontails. Journal of Mammalogy, 47:129–130.

Bailey, T. N. 1971. Biology of striped skunks on a southwestern Lake Erie marsh. The American Midland Naturalist, 85:196–207.

———. 1972. Ecology of bobcats with special reference to social organization. Ph.D. dissert., University of Idaho, Moscow, 82 pp. (not seen, cited in Crowe, 1975a).

———. 1974. Social organization in a bobcat population. The Journal of Wildlife Management, 38:435–446.

Bailey, V. 1898. Descriptions of eleven new species and subspecies of voles. Proceedings of the Biological Society of Washington, 12:85–90.

———. 1900. Revision of American voles of the genus Microtus. North American Fauna, 17:1–88.

————. 1914. Eleven new species and subspecies of pocket gophers of the genus Thomomys. Proceedings of the Biological Society of Washington, 27:115–118.

————. 1915. Revision of the pocket gophers of the genus Thomomys. North American Fauna, 39:1–136.

————. 1923a. Bears eat cascara berries. Journal of Mammalogy, 4:53–54.

————. 1923b. Buffalo in Oregon. Journal of Mammalogy, 4:254–255.

————. 1926 [1927]. A biological survey of North Dakota. North American Fauna, 49:1–226.

————. 1930. Animal life of Yellowstone National Park. Charles C Thomas, Publisher, Springfield, Illinois, 241 pp.

————. 1931. Mammals of New Mexico. North American Fauna, 53:1–412.

————. 1932a. The northwestern white-tail deer. Proceedings of the Biological Society of Washington, 45:43–44.

————. 1932b. The Oregon antelope. Proceedings of the Biological Society of Washington, 45:45–46.

————. 1932c. Buffalo of the Malheur Valley, Oregon. Proceedings of the Biological Society of Washington, 45:47–48.

————. 1935. A new name of the Rocky Mountain elk. Proceedings of the Biological Society of Washington, 38:187–190.

————. 1936. The mammals and life zones of Oregon. North American Fauna, 55:1–416.

Bailey, V., and C. C. Sperry. 1929. Life history and habits of grasshopper mice, genus Onychomys. United States Department of Agriculture Technical Bulletin, 145:1–19.

Baird, S. F. 1852. [Zoology of the valley of the Great Salt Lake of Utah]. In Exploration and Survey of the Valley of the Great Salt Lake of Utah, including reconnaissance of a new route through the Rocky Mountains (H. Standbury). Special Session, U.S. Senate, exec. no. 3. Philadelphia, Lippincott, Grambo & Co., Appendix C (not seen, cited in Coues and Allen, 1877).

————. 1855. Characteristics of some new species of North American Mammalia, collected chiefly in connection with the United States Surveys of a railroad route to the Pacific. Proceedings of the National Academy of Sciences of Philadelphia, 7:333–336.

————. 1857 [1858]. Mammals. In Reports of explorations and surveys, to ascertain the most practicable and economical route for a railroad from the Mississippi River to the Pacific Ocean. Beverley Tucker, Printer, Washington, D.C., 8(part 1):1–757 + pls. 17–60.

————. 1859. Mammals of the boundary. United States and Mexican Boundary Survey, 2(2):1–62 + 27 pls.

Bakeless, J. 1947. Lewis & Clark: partners in discovery. William Morrow & Company, New York, 498 pp.

Baker, A. E. M. 1974. Interspecific aggressive behavior of pocket glophers Thomomys bottae and T. talpoides (Geomyidae: Rodentia). Ecology, 55:671–673.

Baker, J. K. 1962. Notes on the Myotis of the Carlsbad Caverns. Journal of Mammalogy, 43:427–428.

Baker, R. C., F. Wilke, and C. H. Baltzo. 1970. The northern fur seal. United States Fish and Wildlife Service Circular, 336:1–21.

Baker, R. H., C. C. Newman, and F. Wilke. 1945. Food habits of the raccoon in eastern Texas. The Journal of Wildlife Management, 9:45–48.

Baker, R. J., and S. L. Williams. 1972. A live trap for pocket gophers. The Journal of Wildlife Management, 36:1320–1322.

Baldridge, A. 1977. The barnacle Lepas pacifica and the alga Navicula grevillei on northern elephant seals, Mirounga angustirostris. Journal of Mammalogy, 58:428–429.

Baldwin, E. M. 1981. Geology of Oregon. Third ed. Kendall/Hunt Publishing Company, Dubuque, Iowa, 170 pp.

Balph, D. F. 1961. Underground concealment as method of predation. Journal of Mammalogy, 42:423–424.

Banci, V., and A. [S.] Harestad. 1988. Reproduction and natality of wolverine (Gulo gulo) in Yukon. Annales Zoologica Fennici, 25:265–270.

Bancroft, W. L. 1967. Record fecundity for Reithrodontomys megalotis. Journal of Mammalogy, 48:306–308.

Banfield, A. W. F. 1974. The mammals of Canada. The University of Toronto Press, Toronto, 438 pp.

Bangs, O. 1896. Notes on the synonymy of the North American mink with description of the new species. Proceedings of the Boston Society of Natural History, 27:1–6 + 2 pls.

————. 1898a. Descriptions of two new skunks of the genus Mephitis. Proceedings of the Biological Society of Washington, 12:31–33.

————. 1898b. A list of the mammals of Labrador. The American Naturalist, 32:489–507.

————. 1899. Descriptions of some new mammals from western North America. Proceedings of the New England Zoölogical Club, 1:65–72.

Barash, D. P. 1974. Neighbor recognition in two "solitary" carnivores: the raccoon (Procyon lotor) and the red fox (Vulpes fulva). Science, 185:794–796.

Barbour, R. W., and W. H. Davis. 1969. Bats of America. The University Press of Kentucky, Lexington, 286 pp.

————. 1974. Mammals of Kentucky. The University Press of Kentucky, Lexington, 322 pp.

Barclay, R. M. R. 1984. Observations on the ecology and behaviour of bats at Delta Marsh, Manitoba. The Canadian Field-Naturalist, 98:331–336.

————. 1985. Long- versus short-range foraging strategies of hoary (Lasiurus cinereus) and silver-haired (Lasionycteris noctivagans) bats and the consequences for prey selection. Canadian Journal of Zoology, 63:2507–2515.

————. 1986. The echolocation calls of hoary (Lasiurus cinereus) and silver-haired (Lasionycteris noctivagans) bats as adaptations for long- versus short-range foraging strategies and the consequences for prey selection. Canadian Journal of Zoology, 64:2700–2705.

Barclay, R. M. R., P. A. Faure, and D. R. Farr. 1988. Roosting behavior and roost selection by migrating silver-haired bats (Lasionycteris noctivagans). Journal of Mammalogy, 69:821–825.

Barclay, R. M. R., D. W. Thomas, and M. B. Fenton. 1980. Comparison of methods used for controlling bats in buildings. The Journal of Wildlife Management, 44:502–506.

Barkalow, F. S., Jr. 1967. A record gray squirrel litter. Journal of Mammalogy, 48:141.

Barnes, L. G., and E. D. Mitchell. 1975. Late Cenozoic northeast Pacific Phocidae. Conseil International pour l'Exploration de la Mer, Rapports et Procès–verbaux des Réunions, 169:34–42.

Barnett, S. A. 1975. The rat: a study in behavior. Revised ed. The University of Chicago Press, Chicago, 318 pp.

Barnett, S. A., and M. M. Spencer. 1951. Feeding, social behaviour and interspecific competition in rats. Behaviour, 3:229–242.

Barr, T. R. B. 1963. Infectious diseases in the opossum: a review. The Journal of Wildlife Management, 27:53–71.

Barry, R. E., Jr., A. A. Fressola, and J. A. Bruseo. 1989. Determining the time of capture for small mammals. Journal of Mammalogy, 70:660–662.

Barss, J. M., and R. B. Forbes. 1984. A spotted bat (*Euderma maculatum*) from north–central Oregon. The Murrelet, 65:80.

Bartels, R. 1976. Mountain goats of the Wallowas. Oregon Wildlife, 31(12):3–4.

Bartholomew, G. A., and R. Boolootian. 1960. Numbers and population structure of the pinnipeds on the California Channel Islands. Journal of Mammalogy, 41:366–375.

Bartholomew, G. A., and T. J. Cade. 1957. Temperature regulation, hibernation, and aestivation in the little pocket mouse, *Perognathus longimembris*. Journal of Mammalogy, 38:60–72.

Bartholomew, G. A., and C. L. Hubbs. 1960. Population growth and seasonal movements of the northern elephant seal, *Mirounga angustirostris*. Mammalia, 24:313–324.

Batzli, G. O. 1974. Influence of habitat structure on a population of voles. Bulletin of the Southern California Academy of Sciences, 73:83–85.

Batzli, G. O., and F. A. Pitelka. 1971. Condition and diet of cycling populations of the California vole, *Microtus californicus*. Journal of Mammalogy, 52:141–163.

Batzli, G. O., L. L. Getz, and S. S. Hurley. 1977. Suppression of growth and reproduction of microtine rodents by social factors. Journal of Mammalogy, 58:583–591.

Bayer, R. D. 1981. California sea lions in the Yaquina River estuary, Oregon. The Murrelet, 62:56–59.

———. 1985. Six years of harbor seal censusing at Yaquina estuary, Oregon. The Murrelet, 66:44–49.

Beach, R. J., A. C. Geiger, S. J. Jeffries, S. D. Treacy, and B. L. Troutman. 1985. Marine mammals and their interactions with fisheries of the Columbia River and adjacent waters, 1980–1982. National Marine Fisheries Service, Northwest and Alaska Fisheries Center, Seattle, Washington, Processed Report, 85-04:1–316.

Beacham, T. D. 1979*a*. Dispersal tendency and duration of life of littermates during population fluctuations of the vole, *Microtus townsendii*. Oecologia, 42:11–21.

———. 1979*b*. Survival in fluctuating populations of the vole *Microtus townsendii*. Canadian Journal of Zoology, 57:2375–2384.

———. 1980. Growth rates of the vole Microtus townsendii during a population cycle. Oikos, 35:99–106.

———. 1982. Townsend's vole (bias). Pp. 186–188, *in* CRC handbook of census methods for terrestrial vertebrates (D. E. Davis, ed.). CRC Press, Inc., Boca Raton, Florida, 397 pp.

Bear, G. D., and R. M. Hansen. [1966]. Food habits, growth and reproduction of white-tailed jackrabbits in southern Colorado. Colorado State University, Agricultural Experiment Station Technical Bulletin, 90:1–59.

Beatley, J. C. 1976. Environments of kangaroo rats (*Dipodomys*) and effects of environmental change on populations in southern Nevada. Journal of Mammalogy, 57:67–93.

Bee, J. W., and E. R. Hall. 1956. Mammals of northern Alaska on the Arctic Slope. Miscellaneous Publications, Museum of Natural History, University of Kansas, 8:1–309.

Bee, J. W., D. Murariu, and R. S. Hoffmann. 1980. Histology and histochemistry of specialized integumentary glands in eight species of North American shrews (Mammalia, Insectivora). Travaux de Museum d'Histoire Naturelle "Grigore Antipa," 22:547–569.

Beecham, J. J., D. G. Reynolds, and M. G. Hornocker. 1983. Black bear denning activities and den characteristics in west-central Idaho. International Conference on Bear Research and Management, 5:79–86.

Beer, J. [R.] 1942. Notes on the winter food of beavers in the Palouse Prairies, Washington. Journal of Mammalogy, 23:444–445.

———. 1954. A record of the hoary bat from a cave. Journal of Mammalogy, 35:116.

———. 1955. Survival and movements of banded big brown bats. Journal of Mammalogy, 36:242–248.

———. 1956. A record of a silver-haired bat in a cave. Journal of Mammalogy, 37:282.

———. 1959. A collection of deer mice from Otter Rock, Oregon. The Murrelet, 40:28–29.

Beer, J. R., and R. K. Meyer. 1951. Seasonal changes in the endocrine organs and behavior patterns of the muskrat. Journal of Mammalogy, 32:173–191.

Beer, J. R., and W. Truax. 1950. Sex and age ratios in Wisconsin muskrats. The Journal of Wildlife Management, 14:323–331.

Bekoff, M. 1977. Canis latrans. Mammalian Species, 79:1–9.

———. 1978. Behavioral development in coyotes and eastern coyotes. Pp. 97–126, *in* Coyotes: biology, behavior, and management (M. Bekoff, ed.). Academic Press, New York, 384 pp.

———. 1982. Coyote: *Canis latrans*. Pp. 447–459, *in* Wild mammals of North America: biology, management, and economics (J. A. Chapman and G. A. Feldhamer, eds.). The Johns Hopkins University Press, Baltimore, 1147 pp.

Belden, L. M. 1965. Agonistic behavior in montane voles, *Microtus montanus,* from different parent population densities. M.S. thesis, Oregon State University, Corvallis, 97 pp.

Belk, M. C., and H. D. Smith. 1991. Ammospermophilus leucurus. Mammalian Species, 368:1–8.

Belk, M. C., C. L. Pritchett, and H. D. Smith. 1990. Patterns of microhabitat use by *Sorex monticolus* in summer. The Great Basin Naturalist, 50:387–389.

Bell, G. P. 1980. Habitat use and response to patches of prey by desert insectivorous bats. Canadian Journal of Zoology, 58:1876–1883.

———. 1982. Behavioral and ecological aspects of gleaning by a desert insectivorous bat, *Antrozous pallidus* (Chiroptera: Vespertilionidae). Behavioral Ecology and Sociobiology, 10:217–223.

Belwood, J. J., and M. B. Fenton. 1976. Variation in the diet of *Myotis lucifugus* (Chiroptera: Vespertilionidae). Canadian Journal of Zoology, 54:1674–1678.

Benedict, E. M., and R. B. Forbes. 1979. Kit fox skulls in a southeastern Oregon cave. The Murrelet, 60:25–27.

Beneski, J. T., Jr., and D. W. Stinson. 1987. Sorex palustris. Mammalian Species, 296:1–6.

Bennett, E. T. 1829. The gardens and menagerie of the Zoological Society delineated. J. Sharpe, London (not seen, cited in the National Union Catalog, 46:681).

Berg, W. E. 1982. Reintroduction of fisher, pine marten, and river otter. Pp. 159–173, in Midwest furbearer management (G. C. Sanderson, ed.). Proceedings of a Symposium held at the 43rd Midwest Fish and Wildlife Conference, Wichita, Kansas, 7–8 December 1981, 195 pp.

Berger, J. 1979. "Predator harrassment" as a defensive strategy in ungulates. The American Midland Naturalist, 102:197–199.

Berger, P. J., and N. C. Negus. 1982. Stud male maintenance of pregnancy in Microtus montanus. Journal of Mammalogy, 63:148–151.

Berger, P. J., E. H. Sanders, P. D. Gardner, and N. C. Negus. 1977. Phenolic compounds functioning as reproductive inhibitors in Microtus montanus. Science, 195:575–577.

Bergerud, A. T., and J. B. Snider. 1988. Predation in the dynamics of moose populations: a reply. The Journal of Wildlife Management, 52:559–564.

Bergerud, A. T., W. Wyett, and B. Snider. 1983. The role of wolf predation in limiting a moose population. The Journal of Wildlife Management, 47:977–988.

Berkenhout, J. 1769. Outlines of the natural history of Great Britain and Ireland. London, 1:5.

Berry, R. J. 1981a. Preface. Pp. xv-xvii, in Biology of the house mouse (R. J. Berry, ed.). Symposia of the Zoological Society of London, 47:1–715.

———. 1981b. Town mouse, country mouse: adaptation and adaptability in Mus domesticus (M.musculus domesticus). Mammal Review, 11:91–136.

Berry, R. J., W. N. Bonner, and J. Peters. 1979. Natural selection in house mice (Mus musculus) from South Georgia (South Atlantic Ocean). Journal of Zoology (London), 189:385–398.

Best, T. L. 1988. Morphologic variation in the spotted bat Euderma maculatum. The American Midland Naturalist, 119:244–252.

Betts, B. J. 1976. Behaviour in a population of Columbian ground squirrels, Spermophilus columbianus columbianus. Animal Behaviour, 24:652–680.

———. 1990. Geographic distribution and habitat preferences of Washington ground squirrels (Spermophilus washingtoni). Northwestern Naturalist, 71:27–37.

———. 1991. Spatial behavior as a predictor of infanticidal individuals in Columbian ground squirrels. Northwestern Naturalist, 72:85–91.

Beule, J. D., and A. T. Studholme. 1942. Cottontail rabbit nests and nestlings. The Journal of Wildlife Management, 6:133–140.

Bickford, C. E. 1980. Aspects of the social structure of the California ground squirrel (Spermophilus beecheyi) in western Oregon. M.S. thesis, Oregon State University, Corvallis, 56 pp.

Bigg, M. A. 1981. Harbor seal: Phoca vitulina Linnaeus, 1758 and Phoca largha Pallas, 1811. Pp. 1–27, in Handbook of marine mammals. Vol. 2. Seals (S. H. Ridgway and R. J. Harrison, eds.). Academic Press, London, 359 pp.

Binns, A. 1967. Peter Skene Ogden: fur trader. Binfords & Mort, Publishers, Portland, Oregon, 358 pp.

Bischoff, A. I. 1957. The breeding season of some California deer herds. California Fish and Game, 43:91–96.

———. 1958. Productivity of some California deer herds. California Fish and Game, 44:253–259.

Bishop, S. C. 1947. Curious behavior of a hoary bat. Journal of Mammalogy, 28:293–294, 409.

Bittner, S. L., and O. J. Rongstad. 1982. Snowshoe hare and allies. Pp. 146–163, in Wild mammals of North America: biology, management, and economics (J. A. Chapman and G. A. Feldhamer, eds.). The Johns Hopkins University Press, Baltimore, 1147 pp.

Black, C. C. 1963. A review of the North American Tertiary Sciuridae. Bulletin of the Museum of Comparative Zoology, 130:111–248 + 22 pls.

Black, H. C. 1965. An analysis of a population of snowshoe hares, Lepus americanus washingtoni Baird, in western Oregon. Ph.D. dissert., Oregon State University, Corvallis, 285 pp.

Black, H. C., and E. H[F]. Hooven. 1974. Response of small-mammal communities to habitat changes in western Oregon. Pp. 177–186, in Wildlife and forest management in the Pacific Northwest (H. C. Black, ed.). School of Forestry, Oregon State University, Corvallis, 236 pp.

Black, H. C., E. J. Dimock II, W. E. Dodge, and W. H. Lawrence. 1969. Survey of animal damage on forest plantations in Oregon and Washington. Transactions of the North American Wildlife and Natural Resources Conference, 34:388–408.

Black, H. L. 1972. Differential exploitation of moths by the bats Eptesicus fuscus and Lasiurus cinereus. Journal of Mammalogy, 53:598–601.

———. 1974. A north temperate bat community: structure and prey populations. Journal of Mammalogy, 55:138–157.

———. 1976. American kestrel predation on the bats Eptesicus fuscus, Euderma maculatum, and Tadarida brasiliensis. The Southwestern Naturalist, 21:250–251.

Blackburn, D. F. 1973. Courtship behavior among white-tailed jackrabbits. The Great Basin Naturalist, 33:203–204.

Blackman, M. W. 1911. The anal glands of Mephitis mephitica. Anatomical Record, 5:491–515.

Blainville, H. M. D. de. 1816. Sur plusieurs espèces d'animaux Mammifères de l'ordre des Ruminants. Bulletin des Sciences, Société Philomathique de Paris, 1816:73–82 (not seen, cited in Royal Society Catalogue of Scientific Papers [1800–1863], 1:406).

Blair, W. F. 1943. Populations of the deer-mouse and associated small mammals in the mesquite association of southern New Mexico. Contributions from the Laboratory of Vertebrate Biology, University of Michigan, 21:1–40.

———. 1953. Population dynamics of rodents and other small mammals. Advances in Genetics, 5:1–41.

Blake, B. H. 1972. The annual cycle and fat storage in two populations of golden-mantled ground squirrels. Journal of Mammalogy, 53:157–167.

Blaustein, A. R. 1972. Notes on the behavior of the dark kangaroo mouse, Microdipodops megacephalus Merriam (Rodentia, Heteromyidae). Occasional Papers of the Biological Society of Nevada, 33:1–2.

———. 1980. Behavioral aspects of competition in a three-species rodent guild of coastal southern California. Behavioral Ecology and Sociobiology, 6:247–255.

Blaustein, A. R., and A. C. Risser, Jr. 1974a. Dominance relationships of the dark kangaroo mouse (Microdipodops megacephalus) and the little pocket mouse (Perognathus longimembris) in captivity. The Great Basin Naturalist, 34:312–316.

———. 1974b. Interspecific interactions between the Great Basin pocket mouse (Perognathus parvus) and the little pocket mouse (P. longimembris). The Murrelet, 55:23–25.

Bleich, V. C., and O. A. Schwartz. 1975. Observations on the home range of the desert woodrat, Neotoma lepida intermedia. Journal of Mammalogy, 56:518–519.

Blood, D. A. 1967. Food habits of the Ashnola bighorn sheep herd. The Canadian Field-Naturalist, 81:23–29.

Blood, D. A., and A. L. Lovass. 1966. Measurements and weight relationships in Manitoba elk. The Journal of Wildlife Management, 30:135–140.

Blus, L. J. 1966. Relationship between litter size and latitude in the golden mouse. Journal of Mammalogy, 47:546–547.

Blus, L. J., L. Fitzner, and R. E. Fitzner. 1993 [1994]. Wolverine specimen from south–central Washington state. Northwestern Naturalist, 74:22.

Boag, D. A., and J. O. Murie. 1981. Population ecology of Columbian ground squirrels in southwestern Alberta. Canadian Journal of Zoology, 59:2230–2240.

Bogan, M. A. 1972. Observations on parturition and development in the hoary bat, Lasiurus cinereus. Journal of Mammalogy, 53:611–614.

———. 1974. Identification of Myotis californicus and M. leibii in southwestern North America. Proceedings of the Biological Society of Washington, 87:49–56.

Bonaccorso, F. J., and J. H. Brown. 1972. House construction of the desert woodrat, Neotoma lepida lepida. Journal of Mammalogy, 53:283–288.

Bond, R. M. 1945. Range rodents and plant succession. Transactions of the North American Wildlife Conference, 10:229–234.

Bookhout, T. A. 1964. Prenatal development of snowshoe hares. The Journal of Wildlife Management, 28:338–345.

———. 1965. Breeding biology of snowshoe hares in Michigan's Upper Peninsula. The Journal of Wildlife Management, 29:296–303.

Boonstra, R. 1977. Predation on Microtus townsendii populations: impact and vulnerability. Canadian Journal of Zoology, 55:1631–1643.

———. 1978. Effect of adult Townsend voles (Microtus townsendii) on survival of young. Ecology, 59:242–265.

———. 1980. Infanticide in microtines: importance in natural populations. Oecologia, 46:262–265.

Boonstra, R., and C. J. Krebs. 1978. Pitfall trapping of Microtus townsendii. Journal of Mammalogy, 59:136–148.

———. 1979. Viability of large- and small-sized adults in fluctuating vole populations. Ecology, 60:567–573.

Booth, E. S. 1942. Observations on the young of Eutamias. The Murrelet, 23:84.

———. 1943. Observations on the kangaroo rat in captivity. The Murrelet, 24:10–11.

———. 1946. Notes on the life history of the flying squirrel. Journal of Mammalogy, 27:28–30.

———. 1947. A new chipmunk from Washington state. The Murrelet, 28:7–8.

Borell, A. E., and R. Ellis. 1934. Mammals of the Ruby Mountains region of northeastern Nevada. Journal of Mammalogy, 15:12–44.

Borrecco, J. E., and E. F. Hooven. 1972. Northern distribution record of the California meadow mouse (Microtus californicus) in Oregon. The Murrelet, 53:32–33.

Borrecco, J. E., H. C. Black, and E. F. Hooven. 1972. Response of black-tailed deer to herbicide-induced habitat changes. Proceedings of the Western Association of Game and Fish Commissioners, 52:437–451.

———. 1979. Response of small mammals to herbicide-induced habitat changes. Northwest Science, 53:97–106.

Bourlière, F. 1954. The natural history of mammals. A. A. Knopf, New York, 363 pp.

Boutin, S., Z. Tooze, and K. Price. 1993. Post-breeding dispersal by female red squirrels (Tamiasciurus hudsonicus): the effect of local vacancies. Behavioral Ecology, 4:151–155.

Bowers, M. A. 1982. Foraging behavior of heteromyid rodents: field evidence of resource partitioning. Journal of Mammalogy, 63:361–367.

———. 1986. Geographic comparison of microhabitats used by three heteromyids in response to rarefaction. Journal of Mammalogy, 67:46–52.

Bowers, M. A., and H. D. Smith. 1979. Differential habitat utilization by sexes of the deermouse, Peromyscus maniculatus. Ecology, 60:869–875.

Bowles, J. H. 1921. Notes on the California gray squirrel (Sciurus griseus griseus) in Pierce County, Washington. The Murrelet, 2(3):12–13.

Bowyer, R. T., and K. D. Curry. 1983. Use of a roller press to obtain cuticular impressions of guard hairs on acetate strips. Journal of Mammalogy, 64:531–532.

Boyd, S. K., and A. R. Blaustein. 1985. Familiarity and inbreeding avoidance in the gray-tailed vole (Microtus canicaudus). Journal of Mammalogy, 66:348–352.

Bradbury, J. W. 1977. Social organization and communication. Pp. 1–72, in Biology of bats (W. A. Wimsatt, ed.). Academic Press, New York, 3:1–651.

Bradford, D. F. 1974. Water stress of free-living Peromyscus truei. Ecology, 55:1407–1414.

Bradley, W. G. 1967. Home range, activity patterns and ecology of the antelope ground squirrel in southern Nevada. The Southwestern Naturalist, 12:231–252.

———. 1968. Food habits of the antelope ground squirrel in southern Nevada. Journal of Mammalogy, 49:14–21.

Brand, C. J., and L. B. Keith. 1979. Lynx demography during a snowshoe hare decline in Alberta. The Journal of Wildlife Management, 43:837–849.

Brand, C. J., L. B. Keith, and C. A. Fischer. 1976. Lynx responses to changing snowshoe hare densities in central Alberta. The Journal of Wildlife Management, 40:416–428.

Brand, C. J., R. H. Vowles, and L. B. Keith. 1975. Snowshoe hare mortality monitored by telemetry. The Journal of Wildlife Management, 39:741–747.

Brander, R. B. 1971. Longevity of wild porcupines. Journal of Mammalogy, 52:835.

Brandt, C. A. 1989. Mate choice and reproductive success of pikas. Animal Behaviour, 37:118–132.

Brandt, J. F. 1835. Mammalium Rodentium exoticorum novorum vel minus rite cognitorum. Musei Academici Zoologici descriptiones et icones. Mémoires de l'Académie Impériale des Sciences de Saint-Pétersbourg, series 6, 3:357–442 (not seen, cited in Royal Society Catalogue of Scientific Papers [1800–1863], 1:573).

———. 1855. Beiträge zur nähern Kenntniss der Säugethiere Russland's. Kaiserliche Akademie der Wissenschaften, Saint Pétersburg Academie Science Mémoirs, 9(part 2):2–336.

Brass, D. A. 1994. Rabies in bats: natural history and public health implications. Livia Press, Ridgefield, Connecticut, 335 pp.

Braun, C. E., and R. G. Streeter. 1968. Observations on the occurrence of white-tailed jackrabbits in the alpine zone. Journal of Mammalogy, 49:160–161.

Brenner, F. J. 1968. A three-year study of two breeding colonies of the big brown bat, *Eptesicus fuscus*. Journal of Mammalogy, 49:775–778.

Brigham, R. M., and M. B. Fenton. 1986. The influence of roost closure on the roosting and foraging behaviour of *Eptesicus fuscus* (Chiroptera: Vespertilionidae). Canadian Journal of Zoology, 64:1128–1133.

Brigham, R. M., J. E. Cebek, and B. C. Hickey. 1989. Intraspecific variation in the echolocation calls of two species of insectivorous bats. Journal of Mammalogy, 70:426–428.

Broadbooks, H. E. 1939. Food habits of the vagrant shrew. The Murrelet, 20:62–66.

———. 1958. Life history and ecology of the chipmunk, *Eutamias amoenus*, in eastern Washington. University of Michigan, Museum of Zoology, Miscellaneous Publications, 103:1–42.

———. 1961. Homing behavior of deer mice and pocket mice. Journal of Mammalogy, 42:416–417.

———. 1965. Ecology and distribution of the pikas of Washington and Alaska. The American Midland Naturalist, 73:299–335.

———. 1968. Molts of the yellow-pine chipmunk, Eutamias amoenus. The American Midland Naturalist, 79:364–387.

———. 1970. Home ranges and territorial behavior of the yellow-pine chipmunk, *Eutamias amoenus*. Journal of Mammalogy, 51:310–326.

Brodie, E. D., Jr., and C. Maser. 1967. Analysis of great horned owl pellets from Deschutes County, Oregon. The Murrelet, 48:11–12.

Bronson, F. H. 1979. The reproductive ecology of the house mouse. The Quarterly Review of Biology, 54:265–299.

Bronson, M. T. 1979. Altitudinal variation in the life history of the golden-mantled ground squirrel (*Spermophilus lateralis*). Ecology, 60:272–279.

Brothwell, D. 1981. The Pleistocene and Holocene archaeology of the house mouse and related species. Pp. 1–13, *in* Biology of the house mouse (R. J. Berry, ed.). Symposia of the Zoological Society of London, 47:1–715.

Brown, B. A., J. O. Whitaker, [Jr.], T. P. Frence, and C. Maser. 1986. Note on food habits of the screech owl and the burrowing owl of southeastern Oregon. The Great Basin Naturalist, 46:421–426.

Brown, J. H., and R. C. Lasiewski. 1972. Metabolism of weasels: the cost of being long and thin. Ecology, 53:939–943.

Brown, J. H., and G. A. Lieberman. 1973. Resource utilization and coexistence of seed-eating desert rodents in sand dune habitats. Ecology, 54:788–797.

Brown, J. H., G. A. Lieberman, and W. F. Dengler. 1972. Woodrats and cholla: dependence of a small mammal population on the density of cacti. Ecology, 53:310–313.

Brown, L. G., and L. E. Yeager. 1945. Fox squirrels and gray squirrels in Illinois. Bulletin of the Illinois Natural History Survey, 23:449–549.

Brown, L. N. 1964. Breeding records and notes on *Phenacomys silvicola* in Oregon. Journal of Mammalogy, 45:647–648.

———. 1967a. Ecological distribution of mice in the Medicine Bow Mountains of Wyoming. Ecology, 48:677–679.

———. 1967b. Ecological distribution of six species of shrews and comparison of sampling methods in the central Rocky Mountains. Journal of Mammalogy, 48:617–622.

———. 1967c. Seasonal activity patterns and breeding of the western jumping mouse (Zapus princeps) in Wyoming. The American Midland Naturalist, 78:460–470.

———. 1970. Population dynamics of the western jumping mouse (*Zapus princeps*) during a four-year study. Journal of Mammalogy, 51:651–658.

———. 1975. Ecological relationships and breeding biology of the nutria (*Myocastor coypus*) in the Tampa, Florida, area. Journal of Mammalogy, 56:928–930.

———. 1977. Litter size and notes on reproduction in the giant water vole (*Arvicola richardsoni*). The Southwestern Naturalist, 22:281–282.

Brown, R. F., and B. R. Mate. 1983. Abundance, movements, and feeding habits of harbor seals, *Phoca vitulina*, at Netarts and Tillamook bays, Oregon. Fishery Bulletin, 81:291–301.

Brown, R. N., C. H. Southwick, and S. C. Golian. 1989. Male-female spacing, territorial replacement, and the mating system of pikas (*Ochotona princeps*). Journal of Mammalogy, 70:622–627.

Brown, W. L., Jr., and E. O. Wilson. 1956. Character displacement. Systematic Zoology, 5:49–64.

Bruns, E. H. 1977. Winter behavior of pronghorns in relation to habitat. The Journal of Wildlife Management, 41:560–571.

Brunton, D. F. 1981. Nocturnal aggregations of white-tailed jack-rabbits at Rimbey, Alberta. Blue Jay, 39:120–121.

Bryant, M. D. 1945. Phylogeny of Nearctic Sciuridae. The American Midland Naturalist, 33:257–390.

Bubenik, A. B. 1983a. Taxonomy of *Pecora* [sic] in relation to morpho-physiology of their cranial appendices. Pp. 163–185, *in* Antler development in Cervidae (R. D. Brown, ed.). Caesar Kleberg Wildlife Research Institute, Kingsville, Texas, 480 pp.

———. 1983b. Proposals for standardized nomenclature for bony appendices in *Pecora* [sic]. Pp. 187–194, *in* Antler development in Cervidae (R. D. Brown, ed.). Caesar Kleberg Wildlife Research Institute, Kingsville, Texas, 480 pp.

Bubenik, G. A. 1983. The endocrine regulation of the antler cycle. Pp. 73–107, *in* Antler development in Cervidae (R. D. Brown, ed.). Caesar Kleberg Wildlife Research Institute, Kingsville, Texas, 480 pp.

Buchanan, J. B. 1987. Seasonality in the occurrence of long-tailed weasel road-kills. The Murrelet, 68:67–68.

Buchanan, J. B., R. W. Lundquist, and K. B. Aubry. 1990. Winter populations of Douglas' squirrels in different-aged Douglas-fir forests. The Journal of Wildlife Management, 54:577–

581.

Buchler, E. R. 1975. Food transit time in *Myotis lucifugus* (Chiroptera: Vespertilionidae). Journal of Mammalogy, 56:252–255.

———. 1976. The use of echolocation by the wandering shrew (*Sorex vagrans*). Animal Behaviour, 24:858–873.

Buchler, E. R., and S. B. Childs. 1981. Orientation to distant sounds by foraging big brown bats (*Eptesicus fuscus*). Animal Behaviour, 29:428–432.

———. 1982. Use of the post-sunset glow as an orientation cue by big brown bats (*Eptesicus fuscus*). Journal of Mammalogy, 63:243–247.

Buckner, C. H. 1970. Direct observation of shrew predation of insects and fish. Blue Jay, 28:171–172.

Budeau, D. A., and B. J. Verts. 1986. Relative brain size and structural complexity of habitats of chipmunks. Journal of Mammalogy, 67:579–581.

Buech, R. R. 1984. Ontogeny and diurnal cycle of fecal reingestion in the North American beaver (*Castor canadensis*). Journal of Mammalogy, 65:347–350.

Bull, E. L., M. G. Henjum, and R. S. Rohweder. 1989. Diet and optimal foraging of great gray owls. The Journal of Wildlife Management, 53:47–50.

Burgess, S. A., and J. R. Bider. 1980. Effects of stream habitat improvements on invertebrates, trout populations, and mink activity. The Journal of Wildlife Management, 44:871–880.

Burkhardt, J. W., and E. W. Tisdale. 1969. Nature and successional status of western juniper vegetation in Idaho. Journal of Range Management, 22:264–270.

Burkholder, B. L. 1959. Movements and behavior of a wolf pack in Alaska. The Journal of Wildlife Management, 23:1–11.

Burnett, C. D., and T. H. Kunz. 1982. Growth rates and age estimation in *Eptesicus fuscus* and comparison with *Myotis lucifugus*. Journal of Mammalogy, 63:3–41.

Burns, J. A. 1982. Water vole *Microtus richardsoni* (Mammalia, Rodentia) from the late Pleistocene of Alberta. Canadian Journal of Earth Science, 19:628–631.

Burns, J. J. 1964. Movements of a tagged weasel in Alaska. The Murrelet, 45:10.

Burroughs, R. D. 1961. The natural history of the Lewis and Clark Expedition. Michigan State University Press, [East Lansing], 340 pp.

Burt, W. H. 1934. The mammals of southern Nevada. Transactions of the San Diego Society of Natural History, 7:375–427.

———. 1943. Territoriality and home range concepts as applied to mammals. Journal of Mammalogy, 24:346–352.

———. 1948. The mammals of Michigan. The University of Michigan Press, Ann Arbor, 288 pp.

———. 1960. Bacula of North American mammals. Miscellaneous Publications of the Museum of Zoology, University of Michigan, 113:1–75.

Burton, D. H., and H. C. Black. 1978. Feeding habits of Mazama pocket gophers in south-central Oregon. The Journal of Wildlife Management, 42:383–390.

Busher, P. E., R. J. Warner, and S. H. Jenkins. 1983. Population density, colony composition, and local movements in two Sierra Nevadan beaver populations. Journal of Mammalogy, 64:314–318.

Buskirk, S. W., and L. L. McDonald. 1989. Analysis of variability in home-range size of the American marten. The Journal of Wildlife Management, 53:997–1004.

Butler, L. 1962. Periodicities in the annual muskrat population figures for the province of Saskatchewan. Canadian Journal of Zoology, 40:1277–1286.

Buwalda, J. P., and B. N. Moore. 1930. The Dalles and Hood River formations, and the Columbia River Gorge. Carnegie Institution of Washington, Contributions to Palaeontology, 404:11–26.

Byre, V. J. 1990. A group of young peregrine falcons prey on migrating bats. The Wilson Bulletin, 102:728–730.

Cabrera, A. 1932. Los mamíferos de Marruecos. Trabajos del museo nacional de ciencias naturales, Madrid, sec. zool., 57:264 (not seen, cited in The National Union Catalog, 88:525).

Cahalane, V. H. 1947. Mammals of North America. The Macmillan Company, New York, 682 pp.

Calhoun, J. B. 1963. The ecology and sociology of the Norway rat. United States Department of Health, Education, and Welfare, Public Health Service Publication, 1008:1–288.

Calkins, D. [G.], and P. C. Lent. 1975. Territoriality and mating behavior in Prince William Sound sea otters. Journal of Mammalogy, 56:528–529.

Camenzind, F. J. 1978. Behavioral ecology of coyotes on the National Elk Refuge, Jackson, Wyoming. Pp. 267–294, *in* Coyotes: biology, behavior, and management (M. Bekoff, ed.). Academic Press, New York, 384 pp.

Cameron, D. M., Jr. 1967. Gestation period of the golden-mantled ground squirrel (*Citellus lateralis*). Journal of Mammalogy, 48:492–493.

Cameron, G. N. 1971. Niche overlap and competition in woodrats. Journal of Mammalogy, 52:288–296.

———. 1973. Effect of litter size on postnatal growth and survival in the desert woodrat. Journal of Mammalogy, 54:489–493.

Cameron, G. N., and D. G. Rainey. 1972. Habitat utilization by *Neotoma lepida* in the Mohave Desert. Journal of Mammalogy, 53:251–266.

Camp, C. L. 1918. Excavations of burrows of the rodent *Aplodontia*, with observations on the habits of the animal. University of California Publications in Zoology, 17:517–536.

Campbell, C. A. R. 1925. Bats, mosquitoes and dollars. Stratford Company, Boston, 262 pp. (not seen, cited in Storer, 1926).

Campbell, T. M. III, and T. W. Clark. 1983. Observation of badger copulatory and agonistic behavior. The Southwestern Naturalist, 28:107–108.

Campbell, W. 1968. Alexandrian rat predation on ancient murrelet eggs. The Murrelet, 49:38.

Cantor, L. F., and T. G. Whitman. 1989. Importance of belowground herbivory: pocket gophers may limit aspen to rock outcrop refugia. Ecology, 70:962–970.

Canutt, P. R. 1969. Relative damage by small mammals to reforestation in Washington and Oregon. Pp. 55–59, *in* Wildlife and reforestation in the Pacific Northwest (H. C. Black, ed.). School of Forestry, Oregon State University, Corvallis, 91 pp.

Carbyn, L. G. 1983. Wolf predation on elk in Riding Mountain National Park, Manitoba. The Journal of Wildlife

Management, 47:963–976.

———. 1987. Gray wolf and red wolf. Pp. 359–376, in Wild furbearer management and conservation in North America (M. Novak, J. A. Baker, M. E. Obbard, and B. Malloch, eds.). Ontario Ministry of Natural Resources, Toronto, 1150 pp.

Carey, A. B. 1978. Sciurid interspecific aggression. Journal of Mammalogy, 59:206–207.

———. 1995. Spotted owl ecology: theory and methodology—a reply to Rosenberg et al. Ecology, 76:648–652.

Carey, A. B., and J. E. Kershner. 1996 [1997]. Spilogale gracilis in upland forests of western Washington and Oregon. Northwestern Naturalist, 77:29–34.

Carey, A. B., and J. W. Witt. 1991. Track counts as indices to abun-dances of arboreal rodents. Journal of Mammalogy, 72:192–194.

Carey, A. B., S. P. Horton, and B. L. Biswell. 1992. Northern spotted owls: influence of prey base and landscape character. Ecological Monographs, 62:223–250.

Carleton, M. D. 1984. Introduction to rodents. Pp. 255–265, in Orders and families of Recent mammals of the world (S. Anderson and J. K., Jones, Jr., eds.). John Wiley & Sons, New York, 686 pp.

Carleton, M. D., and G. G. Musser. 1984. Muroid rodents. Pp. 289–379, in Orders and families of Recent mammals of the world (S. Anderson and J. K. Jones, Jr., eds.). John Wiley & Sons, New York, 686 pp.

Carleton, W. M. 1966. Food habits of two sympatric Colorado sciurids. Journal of Mammalogy, 47:91–103.

Carr, S. M., and G. A. Hughes. 1993. Direction of introgressive hybridization between species of North American deer (Odocoileus) as inferred from mitochondrial-cytochrome-b sequences. Journal of Mammalogy, 74:331–342.

Carr, S. M., S. W. Ballinger, J. N. Derr, L. H. Blankenship, and J. W. Bickham. 1986. Mitochondiral DNA analysis of hybridization between white-tailed deer and mule deer in west Texas. Proceedings of the National Academy of Science, 83:9576–9580.

Carraway, L. N. 1987. Analysis of characters for distinguishing Sorex trowbridgii from sympatric S. vagrans. The Murrelet, 68:29–30.

———. 1988. Records of reproduction in Sorex pacificus. The Southwestern Naturalist, 33:479–501.

———. 1990. A morphologic and morphometric analysis of the "Sorex vagrans species complex" in the Pacific Coast region. Special Publications, The Museum, Texas Tech University, 32:1–76.

———. 1995. A key to Recent Soricidae of the western United States and Canada based primarily on dentaries. Occasional Papers of the Museum of Natural History, The University of Kansas, 174:1–49.

Carraway, L. N., and P. K. Kennedy. 1993. Genetic variation in Thomomys bulbivorus, an endemic to the Willamette Valley, Oregon. Journal of Mammalogy, 74:952–962.

Carraway, L. N., and B. J. Verts. 1982. A bibliography of Oregon mammalogy. Oregon Agricultural Experiment Station, Special Report, 644:1–47.

———. 1985. Microtus oregoni. Mammalian Species, 233:1–6.

———. 1988. Relative brain size in some western Sorex. The Southwestern Naturalist, 33:386–388.

———. 1991a. Pattern and color aberrations in pelages of Scapanus townsendii. Northwest Science, 65:16–21.

———. 1991b. Neurotrichus gibbsii. Mammalian Species, 387:1–7.

———. 1991c. Neotoma fuscipes. Mammalian Species, 386:1–10.

———. 1993. Aplodontia rufa. Mammalian Species, 431:1–10.

———. 1994a. Relationship of mandibular morphology to relative bite force in some Sorex from western North America. Pp. 201–210, in Advances in the biology of shrews (J. F. Merritt, G. L. Kirkland, Jr., and R. K. Rose, eds.). Carnegie Museum of Natural History Special Publication, 18:1–458.

———. 1994b. Sciurus griseus. Mammalian Species, 474:1–7.

———. 1995 [1996]. Distinguishing skulls of Oregon Tamias amoenus and Tamias minimus. Northwestern Naturalist, 76:144–145.

Carraway, L. N., L. F. Alexander, and B. J. Verts. 1993. Scapanus townsendii. Mammalian Species, 434:1–5.

Carraway, L. N., B. J. Verts, M. L. Jones, and J. O. Whitaker, Jr. 1996. A search for age-related changes in bite force and diet in shrews. The American Midland Naturalist, 135:231–240.

Carraway, L. N., E. Yensen, B. J. Verts, and L. F. Alexander. 1993 [1994]. Range extension and habitat of Peromyscus truei in eastern Oregon. Northwestern Naturalist, 74:81–84.

Carroll, L. E., and H. H. Genoways. 1980. Lagurus curtatus. Mammalian Species, 124:1–6.

Carroll, R. L. 1988. Vertebrate paleontology and evolution. W. H. Freeman and Company, New York, 698 pp.

Cary, M. 1917. Life zone investigations in Wyoming. North American Fauna, 42:1–95.

Casey, G. A., and W. A. Webster. 1975. Age and sex determination of striped skunks (Mephitis mephitis) from Ontario, Manitoba, and Quebec. Canadian Journal of Zoology, 53:223–231.

Caslick, J. W. 1956. Color phases of the roof rat, Rattus rattus. Journal of Mammalogy, 37:255–257.

Cassin, J. 1853. Description of a new mole of the genus Scalops, from Oregon; a specimen of which is in the collection of the Exploring Expedition made by the U.S. ships Vincennes and Peacock, under the command of Captain Charles Wilkes, of the United States Navy. Proceedings of the Academy of Natural Sciences of Philadelphia, 6:299.

———. 1858. United States exploring expedition during the years 1838, 1839, 1840, 1841, 1842 under the command of Charles Wilkes U.S.N. Mammalogy and ornithology, 8:1–466.

Caughley, G. 1974. Interpretation of age ratios. The Journal of Wildlife Management, 38:557–562.

———. 1977. Analysis of vertebrate populations. John Wiley & Sons, New York, 234 pp.

Caughley, G., and L. C. Birch. 1971. Rate of increase. The Journal of Wildlife Management, 35:658–663.

Chaney, R. W. 1925. The Mascall flora—its distribution and climatic relation. Carnegie Institution of Washington, Contributions to Palaeontology, 349:23–48.

Chapman, J. A. 1971a. Organ weights and sexual dimorphism of the brush rabbit. Journal of Mammalogy, 52:453–455.

———. 1971b. Orientation and homing of the brush rabbit

(*Sylvilagus bachmani*). Journal of Mammalogy, 52:686–699.

———. 1974. Sylvilagus bachmani. Mammalian Species, 34:1–4.

Chapman, J. A., and A. L. Harman. 1972. The breeding biology of a brush rabbit population. The Journal of Wildlife Management, 36:816–823.

Chapman, J. A., and G. S. Lind. 1973. Latitude and litter size of the California ground squirrel, *Spermophilus beecheyi*. Bulletin of the Southern California Academy of Science, 72:101–105.

Chapman, J. A., and D. E. C. Trethewey. 1972a. Movements within a population of introduced eastern cottontail rabbits. The Journal of Wildlife Management, 36:155–158.

———. 1972b. Factors affecting trap responses of introduced eastern cottontail rabbits. The Journal of Wildife Management, 36:1221–1226.

Chapman, J. A., and B. J. Verts. 1969. Interspecific aggressive behavior in rabbits. The Murrelet, 50:17–18.

Chapman, J. A., J. G. Hockman, and M. M. Ojeda C. 1980. Sylvilagus floridanus. Mammalian Species, 136:1–8.

Chappell, M. A., and G. A. Bartholomew. 1981. Activity and thermoregulation of the antelope ground squirrel Ammospermophilus leucurus in winter and summer. Physiological Zoology, 54:215–223.

Chatelain, E. F. 1947. Food preferences of the Columbian black-tailed deer *Odocoileus hemionus columbianus* (Richardson) on the Tillamook Burn, Oregon. M.S. thesis, Oregon State College, Corvallis, 64 pp.

Chattin, J. E. 1948. Breeding season and productivity in the Interstate deer herd. California Fish and Game, 34:25–31.

Chew, R. M., and B. B. Butterworth. 1964. Ecology of rodents in Indian Cove (Mojave Desert), Joshua Tree National Monument, California. Journal of Mammalogy, 45:203–225.

Chew, R. M., B. B. Butterworth, and R. Grechman. 1959. The effects of fire on the small mammal population of chaparral. Journal of Mammalogy, 40:253.

Childs, J. E. 1995. Special feature: zoonoses. Journal of Mammalogy, 76:663.

Childs, J. E., J. N. Mills, and G. E. Glass. 1995. Rodent-borne hemorragic fever viruses: a special risk for mammalogists? Journal of Mammalogy, 76:664–680.

Chinn, E. K. K. 1972. Transferrin variation and population differentiation in black-tailed deer. M.S. thesis, Oregon State University, Corvallis, 50 pp.

Christian, J. J. 1956. The natural history of a summer aggregation of the big brown bat, Eptesicus fuscus fuscus. The American Midland Naturalist, 55:66–95.

Churcher, C. S. 1959. The specific status of the New World red fox. Journal of Mammalogy, 40:513–520.

Clark, D. R., Jr., C. O. Martin, and D. M. Swineford. 1975. Organochlorine insecticide residues in the free-tailed bat (*Tadarida brasiliensis*) at Bracken Cave, Texas. Journal of Mammalogy, 56:429–443.

Clark, F. H. 1938. Age of sexual maturity in mice of the genus Peromyscus. Journal of Mammalogy, 19:230–234.

Clark, F. W. 1972. Influence of jackrabbit density on coyote population change. The Journal of Wildlife Management, 36:343–356.

Clark, J. W. 1873. [On the skull of a seal]. Proceedings of the Zoological Society of London, 1873:556–557.

Clark, L. D. 1962. Experimental studies of the behavior of an aggressive, predatory mouse, *Onychomys leucogaster*. Pp. 179–186, *in* Roots of behavior: genetics, instinct, and socialization in animal behavior (E. L. Bliss, ed.). Hoeber Medical Division—Harper & Brothers, New York, 339 pp.

Clark, T. W. 1968. Food uses of the Richardson ground squirrel (*Spermophilus richardsonii elegans*) in the Laramie Basin of Wyoming. The Southwestern Naturalist, 13:248–249.

———. 1970a. Richardson's ground squirrel (*Spermophilus richardsonii*) in the Laramie Basin, Wyoming. The Great Basin Naturalist, 30:55–70.

———. 1970b. Early growth, development, and behavior of the Richardson ground squirrel (Spermophilus richardsoni elegans). The American Midland Naturalist, 83:197–205.

———. 1971. Ecology of the western jumping mouse in Grand Teton National Park, Wyoming. Northwest Science, 45:229–238.

———. 1973. Local distribution and interspecies interactions in microtines, Grand Teton National Park, Wyoming. The Great Basin Naturalist, 33:205–217.

Clark, T. W., and R. H. Denniston [II]. 1970. On the descriptive ethology of Richardson's ground squirrel. The Southwestern Naturalist, 15:193–200.

Clark, T. W., and M. R. Stromberg. 1987. Mammals in Wyoming. University of Kansas, Museum of Natural History, Public Education Series, 10:1–314.

Clark, T. W., E. Anderson, C. Douglas, and M. Strickland. 1987. Martes americana. Mammalian Species, 289:1–8.

Clarkson, D. A., and L. S. Mills. 1994. Hypogeous sporocarps in forest remnants and clearcuts in southwest Oregon. Northwest Science, 68:259–265.

Cliff, E. P. 1939. Relationship between elk and mule deer in the Blue Mountains of Oregon. Transactions of the North American Wildlife Conference, 4:560–569.

Clutton-Brock, J., G. B. Corbet, and M. Hills. 1976. A review of the family Canidae, with a classification by numerical methods. Bulletin of the British Museum of Natural History (Zoology), 29:117–199 (not seen, cited in Wilson and Reeder, 1993b).

Cockrum, E. L. 1955. Reproduction in North American bats. Transactions of the Kansas Academy of Science, 58:487–511.

———. 1973. Additional longevity records for American bats. Journal of the Arizona Academy of Science, 8:108–110.

Cockrum, E. L., and S. P. Cross. 1964–1965. Time of bat activity over water holes. Journal of Mammalogy, 45:635–636 and 46:356.

Coggins, V. L. 1969. Lynx taken in northeastern Oregon. The Murrelet, 50:16.

———. 1988. The Lostine Rocky Mountain bighorn sheep die-off and domestic sheep. Proceedings of the Biennial Symposium, Northern Wild Sheep and Goat Council, 6:57–64.

Coggins, V. L., and P. E. Matthews. 1992. Lamb survival and herd status of the lostine bighorn herd following a *Pasturella* die-off. Proceedings of the Biennial Symposium, Northern Wild Sheep and Goat Council, 8:147–154.

Colbert, E. H. 1938. Pliocene peccaries from the Pacific Coast region of North America. Pp. 241–269 + 6 pls. *in* Studies on Cenozoic vertebrates of western North America. Carnegie

Institution of Washington, Contributions to Palaeontology, 487:1–281.

Cole, R. W. 1970. Pharyngeal and lingual adaptations in the beaver. Journal of Mammalogy, 51:424–425.

Colvin, D. V. 1973. Agonistic behaviour in males of five species of voles Microtus. Animal Behaviour, 21:471–480.

Commissaris, L. R. 1959. Notes on the yuma myotis in New Mexico. Journal of Mammalogy, 40:441–442.

Conaway, C. H. 1952. Life history of the water shrew (Sorex palustris navigator). The American Midland Naturalist, 48:219–248.

Conaway, C. H., and D. W. Pfitzer. 1952. Sorex palustris and Sorex dispar from the Great Smoky Mountains National Park. Journal of Mammalogy, 33:106–108.

Condit, R., and B. J. Le Boeuf. 1984. Feeding habits and feeding grounds of the northern elephant seal. Journal of Mammalogy, 65:281–290.

Conley, W. 1976. Competition between Microtus: a behavioral hypothesis. Ecology, 57:224–237.

Connell, J. H. 1954. Home range and mobility of brush rabbits in California chaparral. Journal of Mammalogy, 35:392–405.

Connolly, G. E. 1981. Trends in populations and harvests. Pp. 225–243, in Mule and black-tailed deer of North America (O. C. Wallmo, ed.). University of Nebraska Press, Lincoln, 605 pp.

Constantine, D. G. 1961a. Locality records and notes on western bats. Journal of Mammalogy, 42:404–405.

———. 1961b. Spotted bat and big free-tailed bat in northern New Mexico. The Southwestern Naturalist, 6:92–97.

———. 1967. Rabies transmission by air in bat caves. Public Health Service Publication, 1617:1–51.

Cooper, W. 1837. On two species of Plecotus inhabiting the United States Territory. Annals of the Lyceum of Natural History of New-York, 4:71–75.

———. 1868. The fauna of Montana Territory. The American Naturalist, 2:528–538.

Cope, E. D. 1868. [untitled]. Proceedings of the Academy of Natural Sciences of Philadelphia, 1868:2.

———. 1878. Descriptions of new vertebrata from the Upper Tertiary formations of the West. Proceedings of the American Philosophical Society, 17:219–232.

———. 1883. The extinct Rodentia of North America. The American Naturalist, 17:370–381.

Corbet, G. B. 1978. The mammals of the Palaearctic Region: a taxonomic review. Cornell University Press, Ithaca, New York, 314 pp.

Corn, P. S., and R. B. Bury. 1986. Habitat use and terrestrial activity by red tree voles (Arborimus longicaudus) in Oregon. Journal of Mammalogy, 67:404–406.

———. 1988. Distribution of the voles Arborimus longicaudus and Phenacomys intermedius in the central Oregon Cascades. Journal of Mammalogy, 69:427–429.

Cornely, J. E., and B. J. Verts. 1988. Microtus townsendii. Mammalian Species, 325:1–9.

Cornely, J. E., L. N. Carraway, and B. J. Verts. 1992. Sorex preblei. Mammalian Species, 416:1–3.

Coss, R. G., and D. H. Owings. 1978. Snake-directed behavior by snake naive and experienced California ground squirrels in a simulated burrow. Zeitschrift für Tierpsychologie,

48:421–435.

Costain, D. B. 1978. Dynamics of a population of Belding's ground squirrels in Oregon. M.S. thesis, Oregon State University, Corvallis, 66 pp.

Costain, D. B., and B. J. Verts. 1982. Age determination and age-specific reproduction in Belding's ground squirrels. Northwest Science, 56:230–235.

Cottam, C., and C. S. Williams. 1943. Speed of some wild mammals. Journal of Mammalogy, 24:262–263.

Cottam, D. F. 1985. Lamb production differences of bighorn sheep on Hart Mountain, Oregon. M.S. thesis, Oregon State University, Corvallis, 60 pp.

Couch, L. K. 1925. Storing habits of Microtus townsendii. Journal of Mammalogy, 6:200–201.

———. 1927. Migrations of the Washington black-tailed jack rabbit. Journal of Mammalogy, 8:313–314.

———. 1930. Notes on the pallid yellow-bellied marmot. The Murrelet, 11(2):3–6.

Coues, E. 1875. A critical review of the North American Saccomyidae. Proceedings of the Academy of Natural Sciences of Philadelphia, 27:272–327.

———. 1877. Precursory notes on American insectivorous mammals, with descriptions of new species. Bulletin of the United States Geologic and Geographical Survey of the Territories, 3(3):631–653.

Coues, E., and J. A. Allen. 1877. Monographs of North American Rodentia. United States Geological Survey of the Territories, 11:1–1091.

Coulombe, H. N. 1970. The role of succulent halophytes in the water balance of salt marsh rodents. Oecologia, 4:223–247.

Coutts, R. A., M. B. Fenton, and E. Glen. 1973. Food intake by captive Myotis lucifugus and Eptesicus fuscus (Chiroptera: Vespertilionidae). Journal of Mammalogy, 54:985–990.

Cowan, I. McT. 1933. Some notes on the hibernation of Lasionycteris noctivagans. The Canadian Field-Naturalist, 47:74–75.

———. 1936a. Distribution and variation in deer (genus Odocoileus) of the Pacific Coastal Region of North America. California Fish and Game, 22:155–246.

———. 1936b. Nesting habits of the flying squirrel Glaucomys sabrinus. Journal of Mammalogy, 17:58–60.

———. 1942. Food habits of the barn owl in British Columbia. The Murrelet, 23:48–53.

———. 1956. Life and times of the Coast black-tailed deer. Pp. 523–618, in The deer of North America: the white-tailed, mule and black-tailed deer, genus Odocoileus. Their history and management (W. P. Taylor, ed.). The Stackpole Company, Harrisburg, Pennsylvania and The Wildlife Management Institute, Washington, D.C., 668 pp.

———. 1962. Hybridization between the black-tail deer and the white-tail deer. Journal of Mammalogy, 43:539–541.

Cowan, I. McT., and M. G. Arsenault. 1954. Reproduction and growth in the creeping vole, Microtus oregoni serpens Merriam. Canadian Journal of Zoology, 32:198–208.

Cowan, I. McT., and S. G. Carl. 1945. The northern elephant seal (Mirounga angustirostris) in British Columbia waters and vicinity. The Canadian Field-Naturalist, 59:170–171.

Cowan, I. McT., and C. J. Guiguet. 1965. The mammals of British Columbia. British Columbia Provincial Museum, Handbook, 11:1–414.

Cowan, I. McT., and W. McCrory. 1970. Variation in the mountain goat, *Oreamnos americanus* (Blainville). Journal of Mammalogy, 51:60–73.

Cox, G. W. 1989. Early summer diet and food preferences of northern pocket gophers in north central Oregon. Northwest Science, 63:77–82.

Cox, G. W., and D. W. Allen. 1987. Soil translocation by pocket gophers in a Mima moundfield. Oecologia, 72:207–210.

Cox, G. W., and J. Hunt. 1990. Form of Mima mounds in relation to occupancy by pocket gophers. Journal of Mammalogy, 71:90–94.

Crabb, W. D. 1941. Food habits of the prairie spotted skunk in southeastern Iowa. Journal of Mammalogy, 22:349–364.

———. 1944. Growth, development and seasonal weights of spotted skunks. Journal of Mammalogy, 25:213–221.

———. 1948. The ecology and management of the prairie spotted skunk in Iowa. Ecological Monographs, 18:201–232.

Cragin, F. W. 1900. Goat-antelope from the cave fauna of Pikes Peak Region. Bulletin of the Geological Society of America, 11:610–612.

Craighead, J. J. [1959]. Predation by hawks, owls, and gulls. Pp. 35–42, *in* The Oregon meadow mouse irruption of 1957–1958. Federal Cooperative Extension Service, Oregon State College, Corvallis, 88 pp.

Craighead, J. J., and F. C. Craighead, Jr. 1956. Hawks, owls and wildlife. The Stackpole Company, Harrisburg, Pennsylvania and The Wildlife Management Institute, Washington, D.C., 443 pp.

Craighead, J. J., and J. A. Mitchell. 1982. Grizzly bear: *Ursus arctos*. Pp. 515–556, *in* Wild mammals of North America: biology, management, and economics (J. A. Chapman and G. A. Feldhamer, eds.). The Johns Hopkins University Press, Baltimore, 1147 pp.

Craighead, J. J., G. Atwell, and B. W. O'Gara. 1972. Elk migrations in and near Yellowstone National Park. Wildlife Monographs, 29:1–48.

Cramblet, H. M., and R. L. Ridenhour. 1956. Parturition in *Aplodontia*. Journal of Mammalogy, 37:87–90.

Cranford, J. A. 1977. Home range and habitat utilization by *Neotoma fuscipes* as determined by radiotelemetry. Journal of Mammalogy, 58:165–172.

———. 1978. Hibernation in the western jumping mouse (*Zapus princeps*). Journal of Mammalogy, 59:496–509.

Crawford, J. A., and R. S. Lutz. 1985. Sage grouse population trends in Oregon, 1941–1983. The Murrelet, 66:69–74.

Cressman, L. S. 1960. Cultural sequences at The Dalles, Oregon: a contribution to Pacific Northwest prehistory. Transactions of the American Philosophical Society, 50(10):1–108.

Crews, A. K. 1939. A study of the Oregon white-tailed deer, *Odocoileus virginianus leucurus* (Douglas). M.S. thesis, Oregon State College, Corvallis, 46 pp. + 14 pls.

Criddle, N., and S. Criddle. 1925. The weasels of southern Manitoba. The Canadian Field-Naturalist, 39:142–148.

Crompton, A. W., and F. A. Jenkins, Jr. 1979. Origin of mammals. Pp. 59–73, *in* Mesozoic mammals: the first two-thirds of mammalian history (J. A. Lillegraven, Z. Kielan-Jaworowska, and W. A. Clemens, eds.). University of California Press, Berkeley, 311 pp.

Cronin, M. A. 1991*a*. Mitochondrial and nuclear genetic relationships of deer (*Odocoileus* spp.) in western North America. Canadian Journal of Zoology, 69:1270–1279.

———. 1991*b*. Mitochondrial-DNA phylogeny of deer (Cervidae). Journal of Mammalogy, 72:553–566.

Cross, S. P. 1965. Roosting habits of *Pipistrellus hesperus*. Journal of Mammalogy, 46:270–279.

———. 1969. Behavioral aspects of western gray squirrel ecology. Ph.D. dissert., University of Arizona, Tucson, 168 pp.

———. 1979. Oregon bats. Oregon Wildlife, 34(5):3–8.

Crouch, G. L. 1966. Preferences of black-tailed deer for native forage and Douglas-fir seedlings. The Journal of Wildlife Management, 30:471–475.

———. 1968*a*. Clipping of woody plants by mountain beaver. Journal of Mammalogy, 49:151–152.

———. 1968*b*. Forage availability in relation to browsing of Douglas-fir seedlings by black-tailed deer. The Journal of Wildlife Management, 32:542–553.

Crowcroft, P. 1955. Territoriality in wild house mice, *Mus musculus* L. Journal of Mammalogy, 36:299–301.

Crowe, D. M. 1975*a*. A model for exploited bobcat populations in Wyoming. The Journal of Wildlife Management, 39:408–415.

———. 1975*b*. Aspects of ageing, growth, and reproduction of bobcats from Wyoming. Journal of Mammalogy, 56:177–198.

Csuti, B. A. 1979. Patterns of adaptation and variation in the Great Basin kangaroo rat (*Dipodomys microps*). University of California Publications in Zoology, 111:1–69.

Currie, P. O., and D. L. Goodwin. 1966. Consumption of forage by black-tailed jackrabbits on salt-desert ranges of Utah. The Journal of Wildlife Management, 30:304–311.

Currier, M. J. P. 1983. Felis concolor. Mammalian Species, 200:1–7.

Cuvier, F. 1825. Des dents des mammifères, considérées comme caractères zoologiques. F. G. Levrault, Strasbourg. 258 pp. (not seen, cited in Coues and Allen, 1877).

———. 1831. Supplément à l'histoire naturelle générale et particulière de Buffon. Mammifères. 1:335 (not seen, cited in Gill and Coues, 1877).

Cuyler, W. K. 1924. Observations on the habits of the striped skunk (Mephitis mesomelas varians). Journal of Mammalogy, 4:180–189.

Czaplewski, N. J., J. P. Farney, J. K. Jones, Jr., and J. D. Druecker. 1979. Synopsis of bats of Nebraska. Occasional Papers, The Museum, Texas Tech University, 61:1–24.

Dailey, T. V., N. T. Hobbs, and T. N. Woodard. 1984. Experimental comparisons of diet selection by mountain goats and mountain sheep in Colorado. The Journal of Wildlife Management, 48:799–806.

Dale, F. H. 1939. Variability and environmental responses of the kangaroo rat, Dipodomys heermanni saxatilis. The American Midland Naturalist, 22:703–731.

Dalke, P. D. 1942. The cottontails rabbits in Connecticut. State Geological and Natural History Survey Bulletin, 45:1–97.

Dalquest, W. W. 1941*a*. Distribution of cottontail rabbits in Washington state. The Journal of Wildlife Management, 5:408–411.

———. 1941*b*. Ecologic relationships of four small mammals in western Washington. Journal of Mammalogy, 22:170–173.

―――. 1942. Geographic variation in northwestern snowshoe hares. Journal of Mammalogy, 23:166–183.

―――. 1943a. Seasonal distribution of the hoary bat along the Pacific coast. The Murrelet, 24:20–24.

―――. 1943b. The systematic status of the races of the little big-eared bat Myotis evotis H. Allen. Proceedings of the Biological Society of Washington, 56:1–2.

―――. 1944. The molting of the wandering shrew. Journal of Mammalogy, 25:146–148.

―――. 1947a. Notes on the natural history of the bat Corynorhinus rafinesquii in California. Journal of Mammalogy, 28:17–30.

―――. 1947b. Notes on the natural history of the bat, Myotis yumanensis, in California, with a description of a new race. The American Midland Naturalist, 38:224–247.

―――. 1948. Mammals of Washington. University of Kansas Publication, Museum of Natural History, 2:1–444.

Dalquest, W. W., and D. R. Orcutt. 1942. The biology of the least shrew-mole, Neurotrichus gibbsii minor. The American Midland Naturalist, 27:387–401.

Dalquest, W. W., and M. C. Ramage. 1946. Notes on the long-legged bat (Myotis volans) at Old Fort Tejon and vicinity, California. Journal of Mammalogy, 27:60–63.

Dalquest, W. W., and V. B. Scheffer. 1945. The systematic status of the races of the mountain beaver (Aplodontia rufa) in Washington. The Murrelet, 26:34–37.

Daly, J. C., and J. L. Patton. 1986. Growth, reproduction, and sexual dimorphism in Thomomys bottae pocket gophers. Journal of Mammalogy, 67:256–265.

Daly, M., M. I. Wilson, and P. Behrends. 1984. Breeding of captive kangaroo rats, Dipodomys merriami and D. microps. Journal of Mammalogy, 65:338–341.

Dasmann, R. F., and R. D. Taber. 1956. Behavior of Columbian black-tailed deer with reference to population ecology. Journal of Mammalogy, 37:143–164.

Davis, D. E. 1948. The survival of wild brown rats on a Maryland farm. Ecology, 29:437–448.

―――. 1951a. The relation between level of population and pregnancy in Norway rats. Ecology, 32:459–461.

―――. 1951b. A comparison of reproductive potential of two rat populations. Ecology, 32:469–475.

―――. 1953. The characteristics of rat populations. The Quarterly Review cf Biology, 28:373–401.

Davis, R. 1969. Growth and development of young pallid bats, Antrozous pallidus. Journal of Mammalogy, 50:729–736.

Davis, R. B., C. F. Herreid II, and H. L. Short. 1962. Mexican free-tailed bats in Texas. Ecological Monographs, 32:311–346.

Davis, W. B. 1937. Variations in Townsend pocket gophers. Journal of Mammalogy, 18:145–158.

―――. 1938. Relation of size of pocket gophers to soil and altitude. Journal of Mammalogy, 19:338–342.

―――. 1939. The Recent mammals of Idaho. The Caxton Printers, Ltd., Caldwell, Idaho, 400 pp.

―――. 1960. The mammals of Texas. Texas Game and Fish Commission Bulletin, 41:1–252.

Davis, W. B., and G. H. Lowery, Jr. 1940. The systematic status of the Louisiana muskrat. Journal of Mammalogy, 21:212–213.

Davis, W. H., and R. W. Barbour. 1970. Life history notes on some southwestern Myotis. The Southwestern Naturalist, 15:261–263.

Davis, W. H., and H. B. Hitchcock. 1965. Biology and migration of the bat, Myotis lucifigus, in New England. Journal of Mammalogy, 146:296–313.

Davis, W. H., R. W. Barbour, and M. D. Hassell. 1968. Colonial behavior of Eptesicus fuscus. Journal of Mammalogy, 49:44–50.

Dawson, M. R. 1958. Later Tertiary Leporidae of North America. University of Kansas, Paleontological Contributions to Vertebrates, 6:1–75 + 2 pls.

Dawson, M. R., and L. Krishtalka. 1984. Fossil history of the families of Recent mammals. Pp. 11–57, in Orders and families of Recent mammals of the world (S. Anderson and J. K. Jones, Jr., eds.). John Wiley & Sons, New York, 686 pp.

deCalesta, D. S., and M. G. Cropsey. 1978. Field test of a coyote-proof fence. Wildlife Society Bulletin, 6:256–259.

Degerbøl, M. 1935. Report of the mammals collected by the fifth Thule expedition to Arctic North America: Zoology I. Mammals, 2(4):1–67.

DeKay, J. E. 1842. Zoology of New-York, or the New-York fauna. Part I. Mammalia. W. & A. White & J. Visscher, Albany, New York, 146 pp. + 33 pls.

DeLong, K. T. 1966. Population ecology of feral house mice: interference by Microtus. Ecology, 47:481–487.

Denniston, R. H., II. 1957. Notes on breeding and size of young in the Richardson ground squirrel. Journal of Mammalogy, 38:414–416.

Desmarest, A. G. 1820. Mammalogie. In Tableau Encyclopédique et Méthodique. Paris, 1:203.

DeStefano, S. 1992 [1993]. Observations on kit foxes in southeastern Oregon. Northwestern Naturalist, 73:54–56.

Detlefsen, J. A., and F. M. Holbrook. 1921. Skunk breeding. Journal of Heredity, 12:242–254.

Dewsberry, D. A., D. Q. Estep, and D. L. Lanier. 1977. Estrous cycles of nine species of muroid rodents. Journal of Mammalogy, 58:89–92.

Dice, L. R. 1917. Systematic position of several American Tertiary lagomorphs. University of California Publications, Bulletin of the Department of Geology, 10:179–183.

―――. 1919. The mammals of southeastern Washington. Journal of Mammalogy, 1:10–22 + 2 pls.

―――. 1925. Wood rat damage to fruit trees in eastern Oregon. Journal of Mammalogy, 6:282.

―――. 1926. Notes on Pacific Coast rabbits and pikas. Occasional Papers of the Museum of Zology, University of Michigan, 166:1–28.

Diersing, V. E. 1984. Lagomorphs. Pp. 241–254, in Orders and families of Recent mammals of the world (S. Anderson and J. K. Jones, Jr., eds.). John Wiley & Sons, New York, 686 pp.

Dietz, C. L. 1973. Bat walking behavior. Journal of Mammalogy, 54:790–792.

Dillon, L. S. 1961. Historical subspeciation in the North American marten. Systematic Zoology, 10:49–64.

Dills, G. G., and T. Manganiello. 1973. Diel temperature fluctuations of the Virginia opossum (Didelphis virginiana virginiana). Journal of Mammalogy, 54:763–765.

Di Salvo, A. F., J. Palmer, and L. Ajello. 1969. Multiple

pregnancy in *Tadarida brasiliensis cynocephala*. Journal of Mammalogy, 50:152.

Dixon, J. [S.] 1920. Control of the coyote in California. California Agricultural Experiment Station Bulletin, 320:379–397.

———. 1931. Pika versus weasel. Journal of Mammalogy, 12:72.

Dixon, K. R. 1982. Mountain lion: *Felis concolor*. Pp. 711–727, *in* Wild mammals of North America: biology, management, and economics (J. A. Chapman and G. A. Feldhamer, eds.). The Johns Hopkins University Press, Baltimore, 1147 pp.

Dobie, J. F. 1961. The voice of the coyote. University of Nebraska Press, Lincoln, 386 pp.

Dobson, F. S. 1983. Agonism and territoriality in the California ground squirrel. Journal of Mammalogy, 64:218–225.

Dobson, G. E. 1890. A monograph of the Insectivora systematic and anatomical. Part 3, fasciculus 1, London.

Dodds, D. G. 1965. Reproduction and productivity of snowshoe hares in Newfoundland. The Journal of Wildlife Management, 29:303–315.

Dodge, W. E. 1982. Porcupine: *Erethizon dorsatum*. Pp. 355–366, *in* Wild mammals of North America: biology, management, and economics (J. A. Chapman and G. A. Feldhamer, eds.). The Johns Hopkins University Press, Baltimore, 1147 pp.

Dodge, W. E., and P. R. Canutt. 1969. A review of the status of the porcupine (*Erethizon dorsatum epixanthum*) in western Oregon. United States Department of the Interior, Fish and Wildlife Service and United States Department of Agriculture, Forest Service, 25 pp.

Donat, F. 1933. Notes on the life history and behavior of *Neotoma fuscipes*. Journal of Mammalogy, 14:19–26.

Dorney, R. S. 1953. Some unusual juvenile raccoon weights. Journal of Mammalogy, 34:122–123.

Dougherty, J. F. 1940. Skull and skeletal remains of the camel Paratyilopus cameloides (Wortman) from the John Day deposits, Oregon. Pp. 49–58 + 1 pl., *in* Studies of Cenozoic vertebrates and stratigraphy of western North America. Carnegie Institution of Washington, Contributions to Palaeontology, 514:1–194.

Douglas, C. L. 1967. New records of mammals from Mesa Verde National Park, Colorado. Journal of Mammalogy, 48:322–323.

———. 1969. Comparative ecology of pinyon mice and deer mice in Mesa Verde National Park, Colorado. University of Kansas Publication, Museum of Natural History, 18:421–504.

Douglas, C. W., and M. A. Strickland. 1987. Fisher. Pp. 511–529, *in* Wild furbearer management and conservation in North America (M. Novak, J. A. Baker, M. E. Obbard, and B. Malloch, eds.). Ontario Ministry of Natural Resources, Toronto, 1150 pp.

Douglas, D. 1829. Observations on two undescribed species of North American Mammalia. The Zoological Journal, 4:330–332.

Dowding, E. S. 1955. Endogone in Canadian rodents. Mycologia, 47:51–57.

Downs, T. 1956. The Mascall fauna from the Miocene of Oregon. University of California Publications in Geological Sciences, 31:199–354 + pls 5–12.

Doyle, A. T. 1987. Microhabitat separation among sympatric microtines, Clethrionomys californicus, Microtus oregoni, and M. richardsoni. The American Midland Naturalist, 118:258–265.

———. 1990. Use of riparian and upland habitats by small mammals. Journal of Mammalogy, 71:14–23.

Drabek, C. M. 1994. Summer and autumn temporal activity of the montane vole (*Microtus montanus*) in the field. Northwest Science, 68:178–184.

Dragoo, J. W., and R. L. Honeycutt. 1997. Systematics of mustelid-like carnivores. Journal of Mammalogy, 78:426–443.

Dragoo, J. W., J. R. Choate, T. L. Yates, and T. P. O'Farrell. 1990. Evolutionary and taxonomic relationships among North American arid-land foxes. Journal of Mammalogy, 71:318–332.

Drickamer, L. C., and B. M. Vestal. 1973. Patterns of reproduction in a laboratory colony of *Peromyscus*. Journal of Mammalogy, 54:523–528.

Druecker, J. D. 1972. Aspects of reproduction in *Myotis volans*, *Lasionycteris noctivagans*, and *Lasiurus cinereus*. Ph.D. dissert., University of New Mexico, Albuquerque, 68 pp.

Drummond, P. C., and D. R. Jones. 1979. The initiation and maintenance of bradycardia in a diving mammal, the muskrat, *Ondatra zibethica*. The Journal of Physiology, 290:253–271.

Dublin, H. T. 1980. Relating deer diets to forage quality and quantity: the Columbian white-tailed deer (*Odocoileus virginianus leucurus*). M.S. thesis, University of Washington, [Seattle], 135 pp.

Dubuc, L. J., W. B. Krohn, and R. B. Owen. 1990. Predicting occurrence of river otters by habitat on Mount Desert Island, Maine. The Journal of Wildlife Management, 54:594–599.

Duke, K. L. 1944. The breeding season in two species of Dipodomys. Journal of Mammalogy, 25:155–160.

———. 1949. Some notes on the histology of the ovary of the bobcat (Lynx) with special reference to the corpora lutea. Anatomical Record, 103:111–122.

———. 1954. Reproduction in the bobcat (Lynx rufus). Anatomical Record, 120:816–817 (abstract only).

Duke, S. D., G. C. Bateman, and M. M. Bateman. 1979. Longevity record for *Myotis californicus*. The Southwestern Naturalist, 24:693.

Dunmire, W. W. 1961. Breeding season of three rodents on White Mountain, California. Journal of Mammalogy, 42:489–493.

Dunstan, T. C., and J. F. Harper. 1975. Food habits of bald eagles in north-central Minnesota. The Journal of Wildlife Management, 39:140–143.

Durrant, S. D. 1935. Occurrence of the spotted bat in Utah. Journal of Mammalogy, 16:226.

Durrant, S. D., and R. M. Hansen. 1954. Distribution patterns and phylogeny of some western ground squirrels. Systematic Zoology, 3:82–85.

Eadie, W. R. 1951. A comparative study of the male accessory genital glands of *Neürotrichus*. Journal of Mammalogy, 32:36–43.

Eadie, W. R., and W. J. Hamilton, Jr. 1958. Reproduction in the fisher in New York. New York Fish and Game Journal, 5:77–83.

Eagle, T. C., and J. S. Whitman. 1987. Mink. Pp. 615–624, *in* Wild furbearer management and conservation in North America (M. Novak, J. A. Baker, M. E. Obbard, and B. Malloch, eds.). Ontario Ministry of Natural Resources, Toronto, 1150 pp.

Easterla, D. A. 1965. The spotted bat in Utah. Journal of Mammalogy, 46:665–668.

———. 1966. Yuma myotis and fringed myotis in southern Utah. Journal of Mammalogy, 47:350–351.

———. 1970. First records of the spotted bat in Texas and notes on its natural history. The American Midland Naturalist, 83:306–308.

———. 1971. Notes on young and adults of the spotted bat, *Euderma maculatum*. Journal of Mammalogy, 52:475–476.

———. 1973. Ecology of the 18 species of Chiroptera at Big Bend National Park, Texas. Part 2. The Northwest Missouri State   University Studies, 34:54–165.

———. 1976. Notes on the second and third newborn of the spotted bat, Euderma maculatum, and comments on the species in Texas. The American Midland Naturalist, 96:499–501.

Easterla, D. A., and J. O. Whitaker, Jr. 1972. Food habits of some bats from Big Bend National Park, Texas. Journal of Mammalogy, 53:887–890.

Ecke, D. H. 1954. An invasion of Norway rats in southwest Georgia. Journal of Mammalogy, 35:521–525.

Edge, E. R. 1931. Seasonal activity and growth in the Douglas ground squirrel. Journal of Mammalogy, 12:194–200.

———. 1934. Burrows and burrowing habits of the Douglas ground squirrel. Journal of Mammalogy, 15:189–296.

Edge, W. D., J. O. Wolff, and R. L. Carey. 1995. Density-dependent responses of gray-tailed voles to mowing. The Journal of Wildlife Management, 59:245–251.

Edge, W. D., R. L. Carey, J. O. Wolff, L. M. Ganio, and T. Manning. 1996. Effects of Guthion 2S^R on *Microtus canicaudus*: a risk assessment validation. Journal of Applied Ecology, 33:269–278.

Edgerton, P. J. 1987. Influence of ungulates on the development of the shrub understory of an upper slope mixed conifer forest. Pp. 162–167, *in* Proceedings—symposium on plant-herbivore interactions (F. D. Provenza, J. T. Flinders, and E. D. McArthur, compilers). United States Department of Agriculture, Forest Service, General Technical Report, INT-222:1–179.

Edson, J. M. 1932. Hibernation in the northern jumping mouse. The Murrelet, 13:55–56.

Edwards, O. T. 1942. Survey of winter deer range, Malheur National Forest, Oregon. The Journal of Wildlife Management, 6:210–220.

Edwards, R. Y. 1955. The habitat preferences of the boreal Phenacomys. The Murrelet, 36:35–38.

Egoscue, H. J. 1956. Preliminary studies of the kit fox in Utah. Journal of Mammalogy, 37:351–357.

———. 1957. The desert woodrat: a laboratory colony. Journal of Mammalogy, 38:472–481.

———. 1960. Laboratory and field studies of the northern grasshopper mouse. Journal of Mammalogy, 41:99–110.

———. 1962*a*. The bushy-tailed wood rat: a captive colony. Journal of Mammalogy, 43:328–337.

———. 1962*b*. Ecology and life history of the kit fox in Toole County, Utah. Ecology, 43:481–497.

———. 1964. Ecological notes and laboratory life history of the canyon mouse. Journal of Mammalogy, 45:387–396.

———. 1966. Description of a newborn kit fox. The Southwestern Naturalist, 11:501–502.

———. 1975. Population dynamics of the kit fox in western Utah. Bulletin of the Southern California Academy of Sciences, 74:122–127.

Egoscue, H. J., J. G. Bittmenn, and J. A. Petrovich. 1970. Some fecundity and longevity records for captive small mammals. Journal of Mammalogy, 51:622–623.

Einarsen, A. S. 1938. Life history and management of antelope in Oregon. Transactions of the North American Wildlife Conference, 3:381–387.

———. 1939. Oregon's open season on antelope in 1938. Transactions of the North American Wildlife Conference, 4:216–220.

———. 1946. Management of black-tailed deer. The Journal of Wildlife Management, 10:54–59.

———. 1948. The pronghorn antelope and its management. The Wildlife Management Institute, Washington, D.C., 238 pp.

———. 1956. Life of the mule deer. Pp. 363–414, *in* The deer of North America: the white-tailed, mule and black-tailed deer, genus *Odocoileus*. Their history and management (W. P. Taylor, ed.). The Stackpole Company, Harrisburg, Pennsylvania and The Wildlife Management Institute, Washington, D.C., 668 pp.

Eisenberg, J. F. 1963. The intraspecific social behavior of some cricetine rodents of the genus Peromyscus. The American Midland Naturalist, 69:240–246.

———. 1964. Studies on the behavior of Sorex vagrans. The American Midland Naturalist, 72:417–425.

———. 1981. The mammalian radiations: an analysis of trends in evolution, adaptation, and behavior. The University of Chicago Press, Chicago, 610 pp.

Eisenberg, J. F., and D. E. Wilson. 1978. Relative brain size and   feeding strategies in the Chiroptera. Evolution, 32:740–751.

———. 1981. Relative brain size and demographic strategies in didelphid marsupials. The American Naturalist, 118:1–15.

Elder, W. H. 1952. Failure of placental scars to reveal breeding history in mink. The Journal of Wildlife Management, 16:110.

Elftman, H. O. 1931. Pleistocene mammals of Fossil Lake, Oregon. American Museum Novitates, 481:1–21.

Ellerman, J. R. 1940. The families and genera of living rodents. I. Rodents other than Muridae. British Museum of Natural History, London, 1:1–689.

Elliot, D. G. 1899*a*. Preliminary descriptions of new rodents from the Olympic Mountains. Field Columbian Museum, Zoological Series, 1:225–228.

———. 1899*b*. Catalogue of mammals from the Olympic Mountains Washington with descriptions of new species. Field Columbian Museum, Zoological Series, 1:241–276 + pls. 41–62.

———. 1903*a*. A list of mammals obtained by Edmund Heller, collector for the museum, from the coast region of northern California and Oregon. Field Columbian Museum, Zoology Series, 3:175–197.

———. 1903*b*. Descriptions of apparently new species and subspecies of mammals from California, Oregon, the Kenai Peninsula, Alaska, and Lower California, Mexico. Field Columbian Museum, Zoology Series, 3:153–173.

———. 1903*c*. Descriptions of apparently new species of mammals of the genera Heteromys and Ursus from Mexico and Washington. Field Columbian Museum, Zoology Series, 3:233–237.

———. 1903*d*. Descriptions of twenty-seven apparently new species and subspecies of mammals. Field Columbian Museum, Zoology Series, 3:239–261.

———. 1904. The land and sea mammals of Middle America and the West Indies. Field Columbian Museum, Zoology Series, 4:441–850.

Elliott, C. E., and J. T. Flinders. 1980. Postemergence development and interyear residence of juvenile Columbian ground squirels in the Idaho Primitive Area. The Great Basin Naturalist, 40:362–364.

———. 1984. Cranial measurements of the Columbian ground squirrel (*Spermophilus columbianus columbianus*), with special reference to subspecies taxonomy and juvenile skull development. The Great Basin Naturalist, 44:505–508.

———. 1991. Spermophilus columbianus. Mammalian Species, 372:1–9.

Ellis, L. S., and L. R. Maxson. 1979. Evolution of the chipmunk genera *Eutamias* and *Tamias*. Journal of Mammalogy, 60:331–334.

Ellison, L. 1946. The pocket gopher in relation to soil erosion on mountain range. Ecology, 27:101–114.

Ellison, L., and C. M. Aldous. 1952. Influence of pocket gophers on vegetation of subalpine grassland in central Utah. Ecology, 33:177–186.

Ellsworth, D. L., R. L. Honeycutt, N. J. Silvy, M. H. Smith, J. W. Bickham, and W. D. Klimstra. 1994. White-tailed deer restoration to the southeastern United States: evaluating genetic variation. The Journal of Wildlife Management, 58:686–697.

Elton, C. 1936. House mice (*Mus musculus*) in a coal-mine in Ayrshire. The Annals and Magazine of Natural History, including Zoology, Botany, and Geology, series 10, 17:553–558.

Elton, C., and M. Nicholson. 1942*a*. Fluctuations in numbers of the muskrat (*Ondatra zibethica*) in Canada. The Journal of Animal Ecology, 11:96–126.

———. 1942*b*. The 10-year cycle in numbers of the lynx in Canada. The Journal of Animal Ecology, 11:215–244.

Enders, R. K. 1952. Reproduction in the mink (Mustela vison). Proceedings of the American Philosophical Society, 96:691–755.

Enders, R. K., and O. P. Pearson. 1943. The blastocyst of the fisher. Anatomical Record, 85:285–287.

Engler, C. H. 1943. Carnivorous activities of big brown and pallid bats. Journal of Mammalogy, 24:96–97.

English, P. F. 1923. The dusky-footed wood rat (Neotoma fuscipes). Journal of Mammalogy, 4:1–9.

Engstrom, M. D., and J. R. Choate. 1979. Systematics of the northern grasshopper mouse (*Onychomys leucogaster*) on the central Great Plains. Journal of Mammalogy, 60:723–739.

Erickson, H. R. 1966. Muskrat burrowing damage and control

procedures in New York, Pennsylvania and Maryland. New York Fish and Game Journal, 13:176–187.

Erlien, D. A., and J. R. Tester. 1984. Population ecology of sciurids in northwestern Minnesota. The Canadian Field-Naturalist, 98:1–6.

Errington, P. L. 1937. Drowning as a cause of mortality in muskrats. Journal of Mammalogy, 18:497–500.

———. 1961. Muskrats and marsh management. The Stackpole Company, Harrisburg, Pennsylvania, 183 pp.

———. 1962. Disease cycles in nature—epizootiology of a disease of muskrats. Pp. 7–25, *in* The problems of laboratory animal disease (R. J. C. Harris, ed.). Academic Press, New York, 265 pp.

———. 1963. Muskrat populations. Iowa State University Press, Ames, 665 pp.

Erxleben, J. C. P. 1777. Systema regni animalis per classes ordines, genera, species, varietates, cum synonymia et historia animalivm. Class I. Mammalia. Lipsiae, impensis Weygandianis, 1:1–636.

Escherich, P. C. 1981. Social biology of the bushy-tailed woodrat, *Neotoma cinerea*. University of California Publications in Zoology, 110:1–132.

Estep, D. Q., and D. A. Dewsbury. 1976. Copulatory behavior of *Neotoma lepida* and *Baiomys taylori*: relationships between penile morphology and behavior. Journal of Mammalogy, 57:570–573.

Estes, J. A. 1980. Enhydra lutris. Mammalian Species, 133:1–8.

Evans, F. C., and R. Holdenried. 1943. A population study of the Beechey ground squirrel in central California. Journal of Mammalogy, 24:231–260.

Evans, J. 1970. About nutria and their control. United States Department of the Interior, Bureau of Sport Fisheries and Wildlife, Resource Publication, 86:1–65.

Ewer, R. F. 1968. Ethology of mammals. Plenum Press, New York, 418 pp.

———. 1971. The biology and behaviour of a free-living population of black rats (*Rattus rattus*). Animal Behaviour Monographs, 4:127–174.

———. 1973. The carnivores. Cornell University Press, Ithaca, New York, 494 pp.

Eydoux, J. F. T., and F. L. P. Gervais. 1836. Mammifères observés pendant l'expédition de la Favorite. *in* Voyage autour du monde sur la corvette de l'état La Favorite pendant les années 1830–1832, sous le commandement de M. Laplace, capitaine de frégate. Magasin de Zoologie, Paris, 6:1–200 (not seen, cited in Baird, 1858).

Fafarman, K. R., and R. J. Whyte. 1979. Factors influencing nighttime roadside counts of cottontail rabbits. The Journal of Wildlife Management, 43:765–767.

Fagerstone, K. A. 1987*a*. Black-footed ferret, long-tailed weasel, short-tailed weasel, and least weasel. Pp. 549–573, *in* Wild furbearer management and conservation in North America (M. Novak, J. A. Baker, M. E. Obbard, and B. Malloch, eds.). Ontario Ministry of Natural Resources, Toronto, 1150 pp.

———. 1987*b*. Comparison of vocalizations between and within *Spermophilus elegans elegans* and *S. richardsonii*. Journal of Mammalogy, 68:853–857.

———. 1988. The annual cycle of Wyoming ground squirrels

in Colorado. Journal of Mammalogy, 69:678–687.

Fagerstone, K. A., G. K. LaVoie, and R. E. Griffith, Jr. 1980. Black-tailed jackrabbit diet and density on rangeland and near agricultural crops. Journal of Range Management, 33:229–233.

Falls, J. B. 1968. Activity. Pp. 543–570, in Biology of Peromyscus (Rodentia) (J. A. King, ed.). The American Society of Mammalogists, Special Publication, 2:1–593.

Fashing, N. J. 1973. Implications of karyotypic variation in the kangaroo rat, Dipodomys heermanni. Journal of Mammalogy, 54:1018–1020.

Fautin, R. W. 1946. Biotic communities of the northern desert shrub biome of western Utah. Ecological Monographs, 16:251–310.

Feldhamer, G. A. 1977a. Double captures of four rodent species in Oregon. Northwest Science, 51:91–93.

———. 1977b. Factors affecting the ecology of small mammals on Malheur National Wildlife Refuge. Ph.D. dissert., Oregon State University, Corvallis, 84 pp.

———. 1979a. Age, sex ratios and reproductive potential in black-tailed jackrabbits. Mammalia, 43:473–478.

———. 1979b. Vegetative and edaphic factors affecting abundance and distribution of small mammals in southeast Oregon. The Great Basin Naturalist, 39:207–218.

———. 1979c. Home range relationships of three rodent species in southeast Oregon. The Murrelet, 60:50–57.

Fenton, M. B. 1969. The carrying of young by females of three species of bats. Canadian Journal of Zoology, 47:158–159.

———. 1970. The deciduous dentition and its replacement in Myotis lucifugus (Chiroptera: Vespertilionidae). Canadian Journal of Zoology, 48:817–820.

Fenton, M. B., and R. M. R. Barclay. 1980. Myotis lucifugus. Mammalian Species, 12:1–8.

Fenton, M. B., and G. P. Bell. 1979. Echolocation and feeding behaviour in four species of Myotis (Chiroptera). Canadian Journal of Zoology, 57:1271–1277.

———. 1981. Recognition of species of insectivorous bats by their echolocation calls. Journal of Mammalogy, 62:233–243.

Fenton, M. B., D. C. Tennant, and J. Wyszecki. 1987. Using echolocation calls to measure the distribution of bats: the case of Euderma maculatum. Journal of Mammalogy, 68:142–144.

Fenton, M. B., C. G. van Zyll de Jong, G. P. Bell, D. B. Campbell, and M. Laplante. 1980. Distribution, parturition dates, and feeding of bats in south-central British Columbia. The Canadian Field-Naturalist, 94:416–420.

Ferrell, C. S. 1995. Systematics and biogeography of the Great Basin pocket mouse, Perognathus parvus. M.S. thesis, University of Nevada-Las Vegas, 68 pp.

Ferron, J. 1981. Comparative ontogeny of behaviour in four s pecies of squirrels (Sciuridae). Zeitschrift für Tierphysiologie, 55:193–216.

Festa-Bianchet, M. 1981. Reproduction in yearling female Columbian ground squirrels (Spermophilus columbianus). Canadian Journal of Zoology, 59:1032–1035.

———. 1988. Birth date and survival in bighorn lambs (Ovis canadensis). Journal of Zoology (London), 214:653–661.

Festa-Bianchet, M., and D. A. Boag. 1982. Territoriality in adult female Columbian ground squirrels. Canadian Journal of Zoology, 60:1060–1066.

Fichter, E., and R. Williams. 1967. Distribution and status of the red fox in Idaho. Journal of Mammalogy, 48:219–230.

Field, R. J., and G. Feltner. 1974. Wolverine. Colorado Outdoors, 23:1–6.

Fielder, P. C. 1986. Implications of selenium levels in Washington mountain goats, mule deer and Rocky Mountain elk. Northwest Science, 60:15–20.

Fiero, B. C., and B. J. Verts. 1986a. Age-specific reproduction in raccoons in northwestern Oregon. Journal of Mammalogy, 67:169–172.

———. 1986b. Comparison of techniques for estimating age in raccoons. Journal of Mammalogy, 67:392–395.

Findley, J. S. 1945. The interesting fate of a flying squirrel. Journal of Mammalogy, 26:437.

———. 1954. Reproduction in two species of Myotis in Jackson Hole, Wyoming. Journal of Mammalogy, 35:434.

———. 1955. Speciation of the wandering shrew. University of Kansas Publications, Museum of Natural History, 9:1–68.

———. 1972. Phenetic relationships among bats of the genus Myotis. Systematic Zoology, 21:31–52.

———. 1987. The natural history of New Mexico mammals. University of New Mexico Press, Albuquerque, 164 pp.

———. 1993. Bats: a community perspective. Cambridge University Press, New York, 167 pp.

Findley, J. S., and C. Jones. 1964. Seasonal distribution of the hoary bat. Journal of Mammalogy, 45:461–470.

———. 1965. Comments on spotted bats. Journal of Mammalogy, 46:679–680.

Findley, J. S., and G. L. Traut. 1970. Geographic variation in Pipistrellus hesperus. Journal of Mammalogy, 51:741–765.

Findley, J. S., A. H. Harris, D. E. Wilson, and C. Jones. 1975. Mammals of New Mexico. University of New Mexico Press, Albuquerque, 360 pp.

Finley, R. B., Jr. 1958. The wood rats of Colorado: distribution and ecology. University of Kansas Publications, Museum of Natural History, 10:213–552.

Fisch, G. G., and E. J. Dimock II. 1978. Shoot clipping by Douglas squirrels in regenerating Douglas fir. The Journal of Wildlife Management, 42:415–418.

Fiscus, C. H. 1961. Growth in the Steller sea lion. Journal of Mammalogy, 42:218–223.

———. 1978. Northern fur seal. Pp. 152–159, in Marine mammals of eastern North Pacific and Arctic waters (D. Haley, ed.). Pacific Search Press, Seattle, Washington, 256 pp.

Fiscus, C. H., and G. A. Baines. 1966. Food and feeding behavior of Steller and California sea lions. Journal of Mammalogy, 47:195–200.

Fish, F. E. 1982. Function of the compressed tail of surface swimming muskrats (Ondatra zibethicus). Journal of Mammalogy, 63:591–597.

Fisher, J. L. 1976. An autecological study of the pinyon mouse, Peromyscus truei, in southwestern Oregon. M.S. thesis, Southern Oregon State College, Ashland, 100 pp.

Fisler, G. F. 1965. Adaptations and speciation in harvest mice of the San Francisco Bay region. University of California Publications in Zoology, 71:1–108.

———. 1971. Age structure and sex ratio in populations of

Reithrodontomys. Journal of Mammalogy, 52:653–662.

———. 1976. Agonistic signals and hierarchy changes of antelope squirrels. Journal of Mammalogy, 57:94–102.

Fitch, H. S. 1947. Predation by owls in the Sierran foothills of California. The Condor, 49:137–151.

Fitch, H. S., and L. L. Sandidge. 1953. Ecology of the opossum on a natural area in northeastern Kansas. University of Kansas Publications, Museum of Natural History, 7:305–338.

Fitch, H. S., and H. W. Shirer. 1970. A radiotelemetric study of spatial relationships in the opossum. The American Midland Naturalist, 84:170–186.

Fitch, H. S., F. Swenson, and D. F. Tillotson. 1946. Behavior and food habits of the red-tailed hawk. The Condor, 48:205–237.

Fitzgerald, B. M. 1977. Weasel predation on a cyclic population of the montane vole (*Microtus montanus*) in California. The Journal of Animal Ecology, 46:367–397.

FitzGerald, S. M., and L. B. Keith. 1990. Intra- and inter-specific dominance relationships among arctic and snowshoe hares. Canadian Journal of Zoology, 68:457–464.

Fitzner, R. E., and J. N. Fitzner. 1975. Winter food habits of short-eared owls in the Palouse Prairie. The Murrelet, 56(2):2–4.

Flahaut, M. R. 1939. Unusual location of hibernating jumping mice. The Murrelet, 20:17–18.

Fleming, A. S., P. Chee, and F. Vaccarino. 1981. Sexual behaviour and its olfactory control in the desert woodrat (*Neotoma lepida lepida*). Animal Behaviour, 29:727–745.

Fleming, T. H. 1972. Aspects of the population dynamics of three species of opossums in the Panama Canal Zone. Journal of Mammalogy, 53:619–623.

Fletcher, R. A. 1963. The ovarian cycle of the gray squirrel, *Sciurus griseus nigripes*. M.A. thesis, University of California, Berkeley, 30 pp.

Flinders, J. T., and R. M. Hansen. 1972. Diets and habitats of jackrabbits in northeastern Colorado. Colorado State University, Range Science Department, Science Series, 12:1–29.

———. 1973. Abundance and dispersion of leporids within a shortgrass ecosystem. Journal of Mammalogy, 54:287–291.

———. 1975. Spring population responses of cottontails and jackrabbits to cattle grazing shortgrass prairie. Journal of Range Management, 28:290–293.

Flueck, W. T. 1994. Effect of trace elements on population dynamics: selenium deficiency in free-ranging black-tailed deer. Ecology, 75:807–812.

Flyger, V., and J. E. Gates. 1982. Fox and gray squirrels. Pp. 209–229, *in* Wild mammals of North America: biology, management, and economics (J. A. Chapman and G. A. Feldhamer, eds.). The Johns Hopkins University Press, Baltimore, 1147 pp.

Folk, G. E., Jr., M. A. Folk, and J. J. Minor. 1972. Physiological condition of three species of bears in winter dens. Pp. 107–124, *in* Bears—their biology and management (S. Herrero, ed.). International Union for Conservation of Nature and Natural Resources, Morges, Switzerland, new series, 23:1–371.

Forbes, R. B. 1966. Notes on a litter of least chipmunks. Journal of Mammalogy, 47:159–161.

Forbes, R. B., and L. W. Turner. 1972. Notes on two litters of Townsend's chipmunks. Journal of Mammalogy, 53:355–359.

Forcum, D. L. 1966. Postpartum behavior and vocalizations of snowshoe hares. Journal of Mammalogy, 47:543.

Foreyt, W. J., V. [L.] Coggins, and P. Fowler. 1990. Psoroptic scabies in bighorn sheep in Washington and Oregon. Proceedings of the Biennial Symposium, Northern Wild Sheep and Goat Council, 7:135–142.

Forsman, E. [D.] 1975. A preliminary investigation of the spotted owl in Oregon. M.S. thesis, Oregon State University, Corvallis, 127 pp.

———. 1980. Habitat utilization by spotted owls in the west-central Cascades of Oregon. Ph.D. dissert., Oregon State University, Corvallis, 95 pp.

Forsman, E. [D.], and C. Maser. 1970. Saw-whet owl preys on red tree mice. The Murrelet, 51:10.

Forsman, E. D., E. C. Meslow, and H. M. Wight. 1984. Distribution and biology of the spotted owl in Oregon. Wildlife Monographs, 87:1–64.

Foster, J. B. 1961. Life history of the phenacomys vole. Journal of Mammalogy, 42:181–198.

Fox, D. L. 1953. Animal biochromes and structural colours: physical, chemical, distributional & physiological features of coloured bodies in the animal world. Cambridge University Press, Cambridge, United Kingdom, 379 pp.

Francq, E. N. 1969. Behavioral aspects of feigned death in the opossum, Didelphis marsupialis. The American Midland Naturalist, 81:556–568.

———. 1970. Electrocardiograms of the opossum, *Didelphis marsupialis*, during feigned death. Journal of Mammalogy, 51:395.

Franklin, J. F., and C. T. Dyrness. 1969. Vegetation of Oregon and Washington. United States Department of Agriculture Forest Service Research Paper, PNW-80:1–216.

———. [1988]. Natural vegetation of Oregon and Washington. [Reprint of 1973 edition with bibliographic supplement]. Oregon State University Press, [Corvallis], 452 pp.

Frase, B. A., and R. S. Hoffmann. 1980. Marmota flaviventris. Mammalian Species, 135:1–8.

Freeman, P. W. 1981. Correspondence of food habits and morphology in insectivorous bats. Journal of Mammalogy, 62:166–173.

Freiburg, R. E., and P. C. Dumas. 1954. The elephant seal (*Mirounga angustirostris*) in Oregon. Journal of Mammalogy, 35:129.

French, A. R. 1976. Selection of high temperatures for hibernation by the pocket mouse, *Perognathus longimembris*: ecological advantages and energetic consequences. Ecology, 57:185–191.

———. 1977. Circannual rhythmicity and entrainment of surface activity in the hibernator, *Perognathus longimembris*. Journal of Mammalogy, 58:37–43.

French, N. R., B. G. Maza, and A. P. Aschwanden. 1967. Life spans of *Dipodomys* and *Perognathus* in the Mojave Desert. Journal of Mammalogy, 48:537–548.

French, N. R., R. McBride, and J. Detmer. 1965. Fertility and population density of the black-tailed jackrabbit. The Journal of Wildlife Management, 29:14–26.

French, R. L. 1956. Eating, drinking, and activity patterns in *Peromyscus maniculatus sonoriensis*. Journal of

Mammalogy, 37:74–79.

Fritzell, E. K. 1978. Aspects of raccoon (*Procyon lotor*) social organization. Canadian Journal of Zoology, 56:260–271.

———. 1987. Gray fox and island gray fox. Pp. 408–420, *in* Wild furbearer management and conservation in North America (M. Novak, J. A. Baker, M. E. Obbard, and B. Malloch, eds.). Ontario Ministry of Natural Resources, Toronto, 1150 pp.

Fritzell, E. K., and R. J. Greenwood. 1984. Mortality of raccoons in North Dakota. The Prairie Naturalist, 16:1–4.

Fritzell, E. K., and K. J. Haroldson. 1982. Urocyon cinereoargenteus. Mammalian Species, 189:1–8.

Fritzell, E. K., G. F. Hubert, Jr., B. E. Meyen, and G. C. Sanderson. 1985. Age specific reproduction in Illinois and Missouri raccoons. The Journal of Wildlife Management, 49:901–905.

Frost, D. R., and R. M. Timm. 1992. Phylogeny of plecotine bats (Chiroptera: "Vespertilionidae"): summary of the evidence and proposal of a logically consistent taxonomy. American Museum Novitates, 3034:1–16.

Frum, W. G. 1953. Silver-haired bat, *Lasionycteris noctivagans*, in West Virginia. Journal of Mammalogy, 34:499–500.

Fry, W. 1928. The California badger. California Fish and Game, 14:204–210.

Fuller, C. A., and A. R. Blaustein. 1990. An investigation of sibling recognition in a solitary sciurid, Townsend's chipmunk, Tamias townsendii. Behaviour, 112:36–52.

Fuller, T. K. 1978. Variable home-range sizes of female gray foxes. Journal of Mammalogy, 59:446–449.

Furlong, E. L. 1932. Distribution and description of skull remains of the Pliocene antelope Sphenophalos from the northern Great Basin Province. Pp. 27–36 + 5 pls., *in* Papers concerning the palaeontology of California, Oregon and the northern Great Basin Province. Carnegie Institution of Washington, Contributions to Palaeontology, 418:1–113.

Gabrielson, I. N. 1928. Notes on the habits and behavior of the porcupine in Oregon. Journal of Mammalogy, 9:33–38 + 1 pl.

———. 1931. The birds of the Rogue River Valley, Oregon. Condor, 33:110–121.

Gage, K. L., R. S. Ostfeld, and J. G. Olson. 1995. Nonviral vector-borne zoonoses associated with mammals in the United States. Journal of Mammalogy, 76:695–715.

Gander, F. F. 1929. Experiences with wood rats, Neotoma fuscipes macrotis. Journal of Mammalogy, 10:52–58.

Gannon, W. L. 1988. Zapus trinotatus. Mammalian Species, 315:1–5.

Gannon, W. L., and T. E. Lawlor. 1989. Variation of the chip vocalization of three species of Townsend chipmunks (genus *Eutamias*). Journal of Mammalogy, 70:740–753.

Gano, K., and W. H. Rickard. 1982. Small mammals of a bitterbrush-cheatgrass community. Northwest Science, 56:1–7.

Ganskopp, D., B. Myers, and S. Lambert. 1993. Black-tailed jackrabbit preferences for eight forages used for reclamation of Great Basin rangelands. Northwest Science, 67:246–250.

Gardner, A. L. 1973. The systematics of the genus Didelphis (Marsupialia: Didelphidae) in North and Middle America. Special Publications, The Museum, Texas Tech University, 4:1–81.

Garrison, T. E., and T. L. Best. 1990. Dipodomys ordii. Mammalian Species, 353:1–10.

Garshelis, D. L. 1987. Sea otter. Pp. 643–655, *in* Wild furbearer management and conservation in North America (M. Novak, J. A. Baker, M. E. Obbard, and B. Malloch, eds.). Ontario Ministry of Natural Resources, Toronto, 1150 pp.

Gashwiler, J. S. 1959. Small mammal study in west-central Oregon. Journal of Mammalogy, 40:128–139.

———. 1963. Pouch capacity of Cooper's chipmunks. The Murrelet, 44:7–8.

———. 1965. Longevity and home range of a Townsend chipmunk. Journal of Mammalogy, 46:693.

———. 1970a. Plant and mammal changes on a clearcut in west-central Oregon. Ecology, 51:1018–1026.

———. 1970b. Further study of conifer seed survival in a western Oregon clearcut. Ecology, 51:849–854.

———. 1971. Deer mouse movement in forest habitat. Northwest Science, 45:163–170.

———. 1972. Life history notes on the Oregon vole, *Microtus oregoni*. Journal of Mammalogy, 53:558–569.

———. 1976a. Notes on the reproduction of Trowbridge shrews in western Oregon. The Murrelet, 57:58–62.

———. 1976b. A new distribution record of Merriam's shrew in Oregon. The Murrelet, 57:13–14.

———. 1976c. Biology of Townsend's chipmunks in western Oregon. The Murrelet, 57:26–31.

———. 1977. Reproduction of the California red-backed vole in western Oregon. Northwest Science, 51:56–59.

———. 1979. Deer mouse reproduction and its relationship to the tree seed crop. The American Midland Naturalist, 102:95–104.

Gashwiler, J. S., and W. L. Robinette. 1957. Accidental fatalities of the Utah cougar. Journal of Mammalogy, 38:123–126.

Gashwiler, J. S., W. L. Robinette, and O. W. Morris. 1960. Foods of bobcats in Utah and eastern Nevada. The Journal of Wildlife Management, 24:226–229.

———. 1961. Breeding habits of bobcats in Utah. Journal of Mammalogy, 42:76–84.

Gates, W. H. 1937. Notes on the big brown bat. Journal of Mammalogy, 18:97–98.

Gavin, T. A. 1978. Status of Columbian white-tailed deer *Odocoileus virginianus leucurus*: some quantitative uses of biogeographic data. Pp. 185–202, *in* Threatened deer. International Union for Conservation of Nature and Natural Resources, Morges, Switzerland, 434 pp.

———. 1979. Population ecology of the Columbian white-tailed deer. Ph.D. dissert., Oregon State University, Corvallis, 149 pp.

———. 1984. Whitetail populations and habitats: Pacific Northwest. Pp. 487–496, *in* White-tailed deer: ecology and management (L. K. Halls, ed.). Stackpole Books, Harrisburg, Pennsylvania, 870 pp.

Gavin, T. A., and B. May. 1988. Taxonomic status and genetic purity of Columbian white-tailed deer. The Journal of Wildlife Management, 52:1–10.

Gavin, T. A., L. H. Suring, P. A. Vohs, Jr., and E. C. Meslow. 1984. Population characteristics, spatial organization, and natural mortality in the Columbian white-tailed deer. Wildlife Monographs, 91:1–41.

Gay, S. W., and T. L. Best. 1996. Age-related variation in skulls of the puma (*Puma concolor*). Journal of Mammalogy, 77:191–198.

Gazin, C. L. 1932. A Miocene mammalian fauna from southeastern Oregon. Pp. 37–86 + 6 pls., *in* Papers concerning the palaeontology of California, Oregon and the northern Great Basin Province. Carnegie Institution of Washington, Contributions to Palaeontology, 418:1–113.

Gehman, S. D. 1984. Activity patterns and behavior of free-living Nuttall's cottontails. M.S. thesis, Oregon State University, Corvallis, 82 pp.

Geiser, F., and G. J. Kenagy. 1987. Polyunsaturated lipid diet lengthens torpor and reduces body temperature in a hibernator. American Journal of Physiology, 21:R897–R901.

Geist, V. 1964. On the rutting behavior of the mountain goat. Journal of Mammalogy, 45:551–568.

———. 1967. On fighting injuries and dermal shields of mountain goats. The Journal of Wildlife Management, 31:192–194.

———. 1971. Mountain sheep: a study in behavior and evolution. The University of Chicago Press, Chicago, 383 pp.

———. 1981. Behavior: adaptive strategies in mule deer. Pp. 157–223, *in* Mule and black-tailed deer of North America (O. C. Wallmo, ed.). University of Nebraska Press, Lincoln, 605 pp.

Geluso, K. N. 1978. Urine concentrating ability and renal structure of insectivorous bats. Journal of Mammalogy, 59:312–323.

Geluso, K. N., J. S. Altenbach, and D. E. Wilson. 1976. Bat mortality: pesticide poisoning and migratory stress. Science, 194:184–186.

———. 1981. Organochlorine residues in young Mexican free-tailed bats from several roosts. The American Midland Naturalist, 105:249–257.

Gennaro, A. L. 1968. Northern geographic limits of four rodent genera Peromyscus, Perognathus, Dipodomys, and Onychomys in the Rio Grande Valley. The American Midland Naturalist, 80:477–493.

Genoways, H. H., and J. K. Jones, Jr. 1969. Taxonomic status of certain long-eared bats (genus *Myotis*) from southwestern United States and Mexico. The Southwestern Naturalist, 14:1–13.

Gentry, A. W. 1978. Bovidae. Pp. 540–572, *in* Evolution of African mammals (V. J. Maglio and H. B. S. Cooke, eds.). Harvard University Press, Cambridge, Massachusetts, 641 pp.

Gentry, R. L., and J. H. Johnson. 1981. Predation by sea lions on northern fur seal neonates. Mammalia, 45:423–430.

Gentry, R. L., and D. E. Withrow. 1978. Steller sea lion. Pp. 166–171, *in* Marine mammals of eastern North Pacific and Arctic waters (D. Haley, ed.). Pacific Search Press, Seattle, Washington, 256 pp.

Geoffroy St.-Hilaire, É. 1803. Catalogue des mammifères du Muséum National d'Histoire Naturelle. Muséum National d'Histoire Naturelle, Paris, 272 pp. (not seen, cited in Wilson and Reeder, 1993*b*).

———. 1805. Mémoire sur un nouveau genre de Mammifères nommé *Hydromis*. Annales du Muséum d'Histoire Naturelle, Paris, 6:81–92.

Geoffroy St.-Hilaire, I. 1824. Mémoire sur une Chauve-Souris

Américaine, formant une nouvelle espèce dans le genre Nyctinome. Annals de Sciencias Naturaes, 1:337–347 (not seen, cited in Royal Society of London Catalogue of Scientific Papers [1800–1863], 2:832).

George, S. B., and J. D. Smith. 1991. Inter- and intraspecific variation among coastal and island populations of *Sorex monticolus* and *Sorex vagrans* in the Pacific Northwest. Pp. 75–91, *in* The biology of the Soricidae (J. S. Findley and T. L. Yates, eds.). The Museum of Southwestern Biology, Special Publication, 1:1–91.

Gerard, A. S., and G. A. Feldhamer. 1990. A comparison of two survey methods for shrews: pitfalls and discarded bottles. The American Midland Naturalist, 124:191–194.

Gese, E. M., O. J. Rongstad, and W. R. Mytton. 1989. Population dynamics of coyotes in southeastern Colorado. The Journal of Wildlife Management, 53:174–181.

Gettinger, R. D. 1984. A field study of activity patterns of *Thomomys bottae*. Journal of Mammalogy, 65:76–84.

Ghiselin, J. 1965. *Thomomys bottae* in Granite Springs Valley, Pershing County, Nevada. Journal of Mammalogy, 46:525.

———. 1970. Edaphic control of habitat selection by kangaroo mice (*Microdipodops*) in three Nevadan populations. Oecologia, 4:248–261.

Gier, H. T. 1968. Coyotes in Kansas. Kansas State College, Agricultural Experiment Station Bulletin, 393:1–118 (revised).

———. 1975. Ecology and behavior of the coyote (*Canis latrans*). Pp. 247–262, *in* The wild canids: their systematics, behavioral ecology and evolution (M. W. Fox, ed.). Van Nostrand Reinhold Company, New York, 508 pp.

Giger, R. D. 1965. Surface activity of moles as indicated by remains in barn owl pellets. The Murrelet, 46:32–36.

———. 1973. Movements and homing in Townsend's mole near Tillamook, Oregon. Journal of Mammalogy, 54:648–659.

Gile, J. D., and J. W. Gillett. 1979. Fate of ^{14}C dieldrin in a simulated terrestrial ecosystem. Archives of Environmental Contamination and Toxicology, 8:107–124.

Gill, A. E. 1977. Food preferences of the California vole, *Microtus californicus*. Journal of Mammalogy, 58:229–233.

Gill, T. 1866. Prodrome of a monograph of the pinnipedes. Proceedings of the Commission of the Essex Institute, 5(1):1–13 (not seen, cited in Gill and Coues, 1877).

Gill, T., and E. Coues. 1877. Material for a bibliography of North American mammals. Pp. 951–1081, *in* Report of the United States Geological Survey of the Territories (F. V. Hayden). Government Printing Office, Washington, D.C., 9:1–1081.

Gillesberg, A-M., and A. B. Carey. 1991. Arboreal nests of *Phenacomys longicaudus* in Oregon. Journal of Mammalogy, 72:784–787.

Gillette, L. N. 1980. Movement patterns of radio-tagged opossums in Wisconsin. The American Midland Naturalist, 104:1–12.

Gilston, R. R. 1976. Population dynamics of *Microtus canicaudus*. M.S. thesis, Oregon State University, Corvallis, 75 pp.

Giuliano, W. M., J. A. Litvaitis, and C. L. Stevens. 1989. Prey selection in relation to sexual dimorphism of fishers (*Martes pennanti*) in New Hampshire. Journal of Mammalogy, 70:639–641.

Gladson, J. 1979. The porcupine: its life, loves and distribution.

Oregon Wildlife, 34(11):10–12.

Glass, B. P. 1982. Seasonal movements of Mexican free-tailed bats *Tadarida brasiliensis mexicana* banded in the Great Plains. The Southwestern Naturalist, 27:127–133.

Glass, B. P., and R. J. Baker. 1965. *Vespertilio subulatus* Say, 1823: proposed supression under the plenary powers (Mammalia, Chiroptera). Bulletin of Zoological Nomenclature, 22:204–205.

———. 1968. The status of the name *Myotis subulatus* Say. Proceedings of the Biological Society of Washington, 81:257–260.

Glass, G. E., G. W. Korch, and J. E. Childs. 1988. Seasonal and habitat differences in growth rates of wild *Rattus norvegicus.* Journal of Mammalogy, 69:587–592.

Gleason, R. S., and D. R. Johnson. 1985. Factors influencing nesting success of burrowing owls in southeastern Idaho. The Great Basin Naturalist, 45:81–84.

Glendenning, R. 1959. Biology and control of the coast mole, *Scapanus orarius orarius* True, in British Columbia. Canadian Journal of Animal Science, 39:34–44.

Glover, F. A. 1943. A study of the winter activities of the New York weasel. Pennsylvania Game News, 14(6):8–9.

Gmelin, J. F. 1788. Caroli a Linné Systema naturae per regna tria naturae, secundum classes, ordines, genera, species cum characteribus, differentiis, synonymis, locis. 13th ed., 1:1–500.

Goehring, H. H. 1972. Twenty-year study of *Eptesicus fuscus* in Minnesota. Journal of Mammalogy, 53:201–207.

Goertz, J. W. 1964. Habitats of three Oregon voles. Ecology, 45:846–848.

Goldenberg, D. L. 1980. The effect of maternal condition on sex ratio of offspring of grey-tailed voles. M.S. thesis, Oregon State University, Corvallis, 47 pp.

Goldman, E. A. 1921. Two new rodents from Oregon and Nevada. Journal of Mammalogy, 2:232–233.

———. 1925. A new kangaroo rat of the genus Dipodomys from Oregon. Proceedings of the Biological Society of Washington, 38:33–34.

———. 1931. Two new desert foxes. Journal of the Washington Academy of Sciences, 21:249–251.

———. 1937. The wolves of North America. Journal of Mammalogy, 18:37–45.

———. 1938. Notes on the voles of the Microtus longicaudus group. Journal of Mammalogy, 19:491–492.

———. 1939. Remarks on pocket gophers, with special reference to Thomomys talpoides. Journal of Mammalogy, 20:231–244.

———. 1943*a.* The systematic status of certain pocket gophers, with special reference to *Thomomys monticola.* Journal of the Washington Academy of Science, 33:146–147.

———. 1943*b.* Two new races of the puma. Journal of Mammalogy, 24:228–231.

———. 1945. A new pronghorn antelope from Sonora. Proceedings of the Biological Society of Washington, 58:3–4.

———. 1950. Raccoons of North and Middle America. North American Fauna, 60:1–153.

Golley, F. B. 1957. Gestation period, breeding and fawning behavior of Columbian black-tailed deer. Journal of Mammalogy, 38:116–120.

Gomez, D. M. 1992. Small-mammal and herpetofauna abundance in riparian and upslope areas of five forest conditions. M.S. thesis, Oregon State University, Corvallis, 118 pp.

Good, J. R. 1977. Habitat factors affecting pronghorn use of playas in south central Oregon. M.S. thesis, Oregon State University, Corvallis, 64 pp.

Good, J. R., and J. A. Crawford. 1978. Factors influencing pronghorn use of playas in south central Oregon. Proceedings of the Eighth Biennial Pronghorn Antelope Workshop, pp. 182–205.

Goodpaster, W. [W.], and D. F. Hoffmeister. 1950. Bats as prey for mink in Kentucky cave. Journal of Mammalogy, 31:457.

Goodwin, D. L., and P. O. Currie. 1965. Growth and development of black-tailed jack rabbits. Journal of Mammalogy, 46:96–98.

Gordon, B. L. 1977. Monterey Bay area: natural history and cultural imprints. Second ed. Boxwood Press, Pacific Grove, California, 321 pp.

Gordon, K. 1936. Territorial behavior and social dominance among Sciuridae. Journal of Mammalogy, 17:171–172.

———. 1938. Observations on the behavior of Callospermophilus and Eutamias. Journal of Mammalogy, 19:78–84.

———. 1943. The natural history and behavior of the western chipmunk and the mantled ground squirrel. Oregon State Monographs, Studies in Zoology, 5:1–104.

Gosling, L. M. 1979. The twenty-four hour activity cycle of captive coypus (*Myocastor coypus*). Journal of Zoology (London), 187:341–367.

Gosling, N. M. 1977. Winter record of the silver-haired bat, *Lasionycteris noctivagans* Le Conte, in Michigan. Journal of Mammalogy, 58:657.

Gould, E. 1955. The feeding efficiency of insectivorous bats. Journal of Mammalogy, 36:399–407.

Graf, R. P. 1985. Social organization of snowshoe hares. Canadian Journal of Zoology, 63:468–474.

Graf, W. 1947. Mouse populations in relation to predation by foxes and hawks. The Murrelet, 28:18–21.

———. 1955. Cottontail rabbit introductions and distribution in western Oregon. The Journal of Wildlife Management, 19:184–188.

Graham, R. W., and M. A. Graham. 1993. The Quaternary biogeography and paleoecology of *Martes* in North America. Pp. 26–58, *in* Martens, sables, and fishers: biology and conservation (S. W. Buskirk, A. S. Harestad, M. G. Raphael, and R. A. Powell, eds.). Cornell University Press, Ithaca, New York, 484 pp.

Grange, W. B. 1932. Observations on the snowshoe hare, Lepus americanus phaeonotus Allen. Journal of Mammalogy, 13:1–19.

Grant, M. 1905. The Rocky Mountain goat. New York Zoological Society, Ninth Annual Report, pp. 230–261 (not seen, cited in Hall and Kelson, 1959).

Gray, J. A., Jr. 1943. Rodent populations in the sagebrush desert of the Yakima Valley, Washington. Journal of Mammàlogy, 24:191–193.

Gray, J. E. 1837. Description of some new or little known Mammalia, principally in the British Museum collection. The Magazine of Natural History, and Journal of Zoology, Botany,

Mineralogy, Geology, and Meteorology, new series, 1:577–587.

———. 1842. Descriptions of some new genera and fifty unrecorded species of Mammalia. The Annals and Magazine of Natural History, including Zoology, Botany, and Geology, series 1, 10:255–267.

———. 1843. List of the specimens of Mammalia in the collection of the British Museum. British Museum (Natural History) Publications, London, 216 pp. (not seen, cited in Gill and Coues, 1877).

———. 1855. Notice of the horns of an unrecorded species of the prong horn (Antilocapra), in the collection of the Derby Museum, Liverpool. Proceedings of the Zoological Society of London, 1855:9–11.

———. 1859. On the sea-lions, or lobos marinos of the Spaniards, on the coast of California. Proceedings of the Zoological Society of London, 1859:357–361.

———. 1864. Notes on seals (Phocidae), including the description of a new seal (Halicyon richardii) from the west coast of North America. Proceedings of the Zoological Society of London, 1864:27–34.

———. 1866. On the long-eared or mule deer of North America (Eucervus). The Annals and Magazine of Natural History, including Zoology, Botany, and Geology, series 3, 18:338–339.

———. 1868. Synopsis of the species of *Saccomyine,* or pouched mice, in the collection of the British Museum. Proceedings of the Zoological Society of London, 1868:199–206.

———. 1869. On the white-toothed American beaver. The Annals and Magazine of Natural History, including Zoology, Botany, and Geology, series 4, 4:293.

Graybill, M. R. 1981. Haul out patterns and diet of harbor seals, *Phoca vitulina*, in Coos County, Oregon. M.S. thesis, University of Oregon, Eugene, 55 pp. (not seen, cited in Harvey, 1988).

Grayson, D. K. 1977. On the Holocene history of some northern Great Basin lagomorphs. Journal of Mammalogy, 58:507–513.

———. 1979. Mount Mazama, climatic change, and Fort Rock Basin archaeofaunas. Pp. 427–457, *in* Volcanic activity and human ecology (P. D. Sheets and D. K. Grayson, eds.). Academic Press, New York, 644 pp.

———. 1993. The desert's past: a natural prehistory of the Great Basin. Smithsonian Institution Press, Washington, D.C., 356 pp.

Grayson, D. K., and S. D. Livingston. 1989. High-elevation records for *Neotoma cinerea* in the White Mountains, California. The Great Basin Naturalist, 49:392–395.

Green, D. D. 1947. Albino coyotes are rare. Journal of Mammalogy, 28:63.

Green, G. A. 1983. Ecology of breeding burrowing owls in the Columbia Basin, Oregon. M.S. thesis, Oregon State University, Corvallis, 51 pp.

Green, G. A., R. E. Fitzner, R. G. Anthony, and L. E. Rogers. 1993. Comparative diets of burrowing owls in Oregon and Washington. Northwest Science, 67:88–93.

Green, J. S., and J. T. Flinders. 1979a. Techniques for capturing pygmy rabbits. The Murrelet, 60:112–113.

———. 1979b. Homing by a pygmy rabbit. The Great Basin Naturalist, 39:88.

———. 1980a. Brachylagus idahoensis. Mammalian Species, 125:1–4.

———. 1980b. Habitat and dietary relationships of the pygmy rabbit. Journal of Range Management, 33:136–142.

Green, R. G., and C. A. Evans. 1940a. Studies on a population cycle of snowshoe hares on the Lake Alexander area I. Gross annual censuses, 1932–1939. The Journal of Wildlife Management, 4:220–238.

———. 1940b. Studies on a population cycle of snowshoe hares on the Lake Alexander area. III. Effect of reproduction and mortality of young hares on the cycle. The Journal of Wildlife Management, 4:347–358.

Greenhall, A. M. 1982. House bat management. United States Department of the Interior, Resource Publication, 143:1–33.

Greenwald, G. S. 1956. The reproductive cycle of the field mouse, *Microtus californicus.* Journal of Mammalogy, 37:213–222.

———. 1957. Reproduction in a coastal California population of the field mouse Microtus californicus. University of California Publications in Zoology, 54:421–446.

Greer, K. R. 1955. Yearly food habits of the river otter in the Thompson Lakes region, northwestern Montana, as indicated by scat analysis. The American Midland Naturalist, 54:299–313.

Grenfell, W. E., and M. Fasenfest. 1979. Winter food habits of fishers, *Martes pennanti*, in northwestern California. California Fish and Game, 65:186–189.

Griffin, D. R., F. A. Webster, and C. R. Michael. 1960. The echolocation of flying insects by bats. Animal Behaviour, 8:141–154.

Grim, J. N. 1958. Feeding habits of the southern California mole. Journal of Mammalogy, 39:265–268.

Grinnell, H. W. 1914. Three new races of vespertilionid bats from California. University of California Publications in Zoology, 12:317–320.

———. 1918. A synopsis of the bats of California. University of California Publications in Zoology, 17:223–404.

Grinnell, J. 1912. The Warner Mountain cony. Proceedings of the Biological Society of Washington, 25:129–130.

———. 1913a. The species of the mammalian genus *Sorex* of west-central California. University of California Publications in Zoology, 10:179–195.

———. 1913b. A distributional list of the mammals of California. Proceedings of the California Academy of Sciences, series 4, 3:265–390.

———. 1921. Revised list of the species in the genus Dipodomys. Journal of Mammalogy, 2:94–97.

———. 1922. A geographical study of the kangaroo rats of California. University of California Publications in Zoology, 24:1–124.

———. 1923. A systematic list of the mammals of California. University of California Publications in Zoology, 21:313–324.

———. 1933. Review of the Recent mammal fauna of California. University of California Publications in Zoology, 40:71–234.

———. 1935. Differentiation in pocket gophers of the Thomomys bottae group in northern California and southern Oregon. University of California Publications in Zoology,

40:403–416.

Grinnell, J., and J. [S.] Dixon. 1918. Natural history of the ground squirrels of California. California State Commission of Horticulture, Monthly Bulletin, 7:597–708.

Grinnell, J., and J. M. Linsdale. 1929. A new kangaroo rat from the upper Sacramento Valley, California. University of California Publications in Zoology, 30:453–459.

Grinnell, J., and T. I. Storer. 1924. Animal life in the Yosemite: an account of the mammals, birds, reptiles, and amphibians in a cross-section of the Sierra Nevada. University of California Press, Berkeley, 752 pp.

Grinnell, J., and H. S. Swarth. 1912. The mole of southern California. University of California Publications in Zoology, 10:131–136.

Grinnell, J., J. [S.] Dixon, and J. M. Linsdale. 1930. Vertebrate natural history of a section of northern California through the Lassen Peak region. University of California Publications in Zoology, 35:1–594.

———. 1937. Furbearing mammals of California. Their natural history, systematic status, and relations to man. University of California Press, Berkeley, 1:1–375.

Gross, J. E., L. C. Stoddart, and F. H. Wagner. 1974. Demographic analysis of a northern Utah jackrabbit population. Wildlife Monographs, 40:1–68.

Groves, C. P. 1981. Systematic relationships in the Bovini (Artiodacytla, Bovidae). Zeitschrift für Zoologische Systematik und Evolutionsforschung, 19:264–278.

Grubb, P. 1993. Order Artiodactyla. Pp. 377–414, *in* Mammal species of the world: a taxonomic and geographic reference. Second ed. (D. E. Wilson and D. M. Reeder, eds.). Smithsonian Institution Press, Washington, D.C., 1206 pp.

Gufar, Z., D. Hoyle, J. Keech, B. Roberts, L. R. Powers, and E. Yensen. 1980. Records of the pinyon mouse, *Peromyscus truei*, from southwestern Idaho. Journal of the Idaho Academy of Science, 16:1–2.

Guillermo, R. G., and D. W. Nagorsen. 1989. Geographic and sexual variation in the skull of Pacific Coast marten (*Martes americana*). Canadian Journal of Zoology, 67:1386–1393.

Gunderson, H. L. 1976. Mammalogy. McGraw-Hill Book Company, New York, 483 pp.

Gunderson, H. L., and J. R. Beer. 1953. The mammals of Minnesota. Minnesota Museum of Natural History, University of Minnesota, Occasional Papers, 6:1–190.

Günther, F. R. S. 1880. Notes on some Japanese Mammalia. Proceedings of the Scientific Meetings of the Zoological Society of London, 1880:440–443.

Gunther, P. M., B. S. Horn, and G. D. Babb. 1983. Small mammal populations and food selection in relation to timber harvest practices in the western Cascade Mountains. Northwest Science, 57:32–44.

Hacker, A. L. 1991. Population attributes and habitat selection of recolonizing mountain beaver. M.S. thesis, Oregon State University, Corvallis, 64 pp.

Hacker, A. L., and B. E. Coblentz. 1993. Habitat selection by mountain beavers recolonizing Oregon Coast Range clearcuts. The Journal of Wildlife Management, 57:847–853.

Hafner, M. S., and D. J. Hafner. 1979. Vocalizations of grasshopper mice (genus *Onychomys*). Journal of Mammalogy, 60:85–94.

Hagen, J. B., and L. G. Forslund. 1979. Comparative fertility of four age classes of female gray-tailed voles, *Microtus canicaudus*, in the laboratory. Journal of Mammalogy, 60:834–837.

Hagmeier, E. M. 1958. The inapplicability of the subspecies concept in North American marten. Systematic Zoology, 7:1–7.

———. 1959. A re-evaluation of the subspecies of fisher. The Canadian Field-Naturalist, 73:185–197.

———. 1961. Variation and relationships in North American marten. The Canadian Field-Naturalist, 75:150–168.

Haines, F. 1967. Western limits of the buffalo range. The American West, 4(4):2–12, 66–67.

Hall, E. R. 1926. The abdominal skin gland of Martes. Journal of Mammalogy, 7:227–229.

———. 1927. An outbreak of house mice in Kern County, California. University of California Publications in Zoology, 30:189–203.

———. 1928. Notes on the life history of the sage-brush meadow mouse (*Lagurus*). Journal of Mammalogy, 9:201–204.

———. 1931a. Critical comments on mammals from Utah, with descriptions of new forms from Utah, Nevada and Washington. University of California Publications in Zoology, 37:1–13.

———. 1931b. Description of a new mustelid from the later Tertiary of Oregon, with assignment of Parictis primaevus to the Canidae. Journal of Mammalogy, 12:156–158.

———. 1933a. Critical comments on mammals from Utah, with descriptions of new forms from Utah, Nevada and Washington. University of California Publications in Zoology, 37:1–13.

———. 1933b. Sorex leucogenys in Arizona. Journal of Mammalogy 14:153–154.

———. 1936. Mustelid mammals from the Pleistocene of North America with systematic notes on some Recent members of the genera Mustela, Taxidea, and Mephitis. Carnegie Institute of Washington Publication, 473:41–119.

———. 1938. Notes on the meadow mice Microtus montanus and M. nanus with description of a new subspecies from Colorado. Proceedings of the Biological Society of Washington, 51:131–134.

———. 1940. A new race of Belding ground squirrel from Nevada. The Murrelet, 21:59–61.

———. 1942. Gestation period in the fisher with recommendations for the animal's protection in California. California Fish and Game, 28:143–147.

———. 1945. Four new ermines from the Pacific Northwest. Journal of Mammalogy, 26:75–85.

———. 1946. Mammals of Nevada. University of California Press, Berkeley, 710 pp.

———. 1951a. American weasels. University of Kansas Publications, Museum of Natural History, 4:1–466.

———. 1951b. A synopsis of the North American Lagomorpha. University of Kansas Publication, Museum of Natural History, 5:119–202.

———. 1962. Collecting and preparing study specimens of vertebrates. Miscellaneous Publications, University of Kansas, Museum of Natural History, 30:1–46.

———. 1981. The mammals of North America. Second ed. John Wiley & Sons, New York, 1:1–600 + *90*, 2:601–1181 + *90*.

————. 1984. Geographic variation among brown and grizzly bears (*Ursus arctos*) in North America. Special Publication, Museum of Natural History, University of Kansas, 13:1–16.

Hall, E. R., and E. L. Cockrum. 1953. A synopsis of the North American microtine rodents. University of Kansas Publications, Museum of Natural History, 5:373–498.

Hall, E. R., and F. H. Dale. 1939. Geographic races of the kangaroo rat, Dipodomys microps. Occasional Papers of the Museum of Zoology, Louisiana State University, 4:47–63.

Hall, E. R., and D. M. Hatfield. 1934. A new race of chipmunk from the Great Basin of western United States. University of California Publications in Zoology, 40:321–326.

Hall, E. R., and K. R. Kelson. 1959. The mammals of North America. Ronald Press Company, New York, 1:1–546 + *79*, 2:547–1083 + *79.*

Hall, E. R., and R. T. Orr. 1933. A new race of pocket gopher found in Oregon and Washington. Proceedings of the Biological Society of Washington, 46:41–44.

Hall, E. R., and W. B. Witlow. 1932. A new black-tailed jackrabbit from Idaho. Proceedings of the Biological Society of Washington, 45:71–72.

Hall, J. G. 1960. Willow and aspen in the ecology of beaver on Sagehen Creek, California. Ecology, 41:484–494.

Halloran, A. F. 1962. An Arizona specimen of the red fox. Journal of Mammalogy, 43:432.

Hamilton, W. J., Jr. 1930. The food of the Soricidae. Journal of Mammalogy, 11:26–39.

————. 1933*a*. The insect food of the big brown bat. Journal of Mammalogy, 14:155–156.

————. 1933*b*. The weasels of New York. The American Midland Naturalist, 14:289–344.

————. 1936*a*. Seasonal food of skunks in New York. Journal of Mammalogy, 17:240–246.

————. 1936*b*. The food and breeding habits of the raccoon. Ohio Journal of Science, 36:131–140.

————. 1939. Observations on the life history of the red squirrel in New York. The American Midland Naturalist, 22:732–745.

————. 1940. The summer food of minks and raccoons on the Montezuma Marsh, New York. The Journal of Wildlife Management, 4:80–84.

————. 1943. The mammals of eastern United States. An account of Recent land mammals occurring east of the Mississippi. Handbooks of American Natural History, Comstock Publishing Co., Inc., Ithaca, New York, 2:1–432.

————. 1949. The bacula of some North American vespertilionid bats. Journal of Mammalogy, 30:97–102.

————. 1955. Mammalogy in North America. Pp. 661–688, *in* A century of progress in the natural sciences 1853–1953. California Academy of Sciences, San Francisco, 807 pp.

————. 1958*a*. Early sexual maturity in the female short-tailed weasel. Science, 127:1057.

————. 1958*b*. Life history and economic relations of the opossum (*Didelphis marsupialis virginiana*) in New York state. Cornell University Agricultural Experiment Station Memoir, 354:1–48.

————. 1961. Late fall, winter, and early spring foods of 141 otters from New York. New York Fish and Game Journal, 8:106–109.

————. 1963. Reproduction of the striped skunk in New York.

Journal of Mammalogy, 44:123–124.

Hamilton, W. J., Jr., and A. H. Cook. 1955. The biology and management of the fisher in New York. New York Fish and Game Journal, 2:13–35.

Hamilton, W. J., Jr., and W. R. Eadie. 1964. Reproduction in the otter, *Lutra canadensis.* Journal of Mammalogy, 45:242–252.

Hamlett, G. W. D. 1938. The reproductive cycle of the coyote. United States Department of Agriculture Technical Bulletin, 616:1–11.

Hammer, E. W. 1966. Opossum range extended south into Lane County, Oregon. The Murrelet, 47:8.

Hammer, E. W., and C. Maser. 1969. The use of cow chips by the sagebrush vole (*Lagurus curtatus*). The Murrelet, 50:35–36.

————. 1973. Distribution of the dusky-footed woodrat, *Neotoma fuscipes* Baird, in Klamath and Lake counties, Oregon. Northwest Science, 47:123–127.

Handley, C. O., Jr. 1959. A revision of American bats of the genera Euderma and Plecotus. Proceedings of the United States National Museum, 110:95–246.

Hanken, J., and P. W. Sherman. 1981. Multiple paternity in Belding's ground squirrel litters. Science, 212:351–353.

Hansen, C. G. 1956. An ecological survey of the vertebrate animals on Steen's Mountain, Harney County, Oregon. Ph.D. dissert., Oregon State College, Corvallis, 199 pp.

Hansen, E. L. 1955. Survival of the pronghorn antelope in south central Oregon during 1953 and 1954. M.S. thesis, Oregon State College, Corvallis, 117 pp.

————. 1965. Muskrat distribution in south-central Oregon. Journal of Mammalogy, 46:669–671.

Hansen, R. M. 1956. Decline in Townsend ground squirrels in Utah. Journal of Mammalogy, 37:123–124.

————. 1960. Age and reproductive characteristics of mountain pocket gophers in Colorado. Journal of Mammalogy, 41:323–335.

————. 1962. Movements and survival of *Thomomys talpoides* in a Mima-mound habitat. Ecology, 43:151–154.

Hansen, R. M., and G. D. Bear. 1963. Winter coats of white-tailed jackrabbits in southwestern Colorado. Journal of Mammalogy, 44:420–422.

————. 1964. Comparison of pocket gophers from alpine, subalpine and shrub-grassland habitats. Journal of Mammalogy, 45:638–640.

Hansen, R. M., and J. T. Flinders. 1969. Food habits of North American hares. Colorado State University, Range Science Department, Science Series, 1:1–18.

Hansen, R. M., and R. S. Miller. 1959. Observations on the plural occupancy of pocket gopher burrow systems. Journal of Mammalogy, 40:577–584.

Hansen, R. M., and M. J. Morris. 1968. Movement of rocks by northern pocket gophers. Journal of Mammalogy, 49:391–399.

Happe, P. J. 1983. The use of suburban habitats by Columbian black-tailed deer. M.S. thesis, Oregon State University, Corvallis, 95 pp.

Hardy, R. 1945. The influence of types of soil upon the local distribution of some mammals in southwestern Utah. Ecological Monographs, 15:71–108.

————. 1949. Notes on mammals from Arizona, Nevada, and

Utah. Journal of Mammalogy, 30:434–435.

Hargis, C. D., and D. R. McCullough. 1984. Winter diet and habitat selection of marten in Yosemite National Park. The Journal of Wildlife Management, 48:140–146.

Harington, C. R. 1971. A Pleistocene mountain goat from British Columbia and comments on the dispersal history of *Oreamnos*. Canadian Journal of Earth Sciences, 8:1081–1093.

Harlan, R. 1825. Fauna Americana: being a description of the mammiferous animals inhabiting North America. J. Harding, Printer, Philadelphia. 318 pp.

Harlow, H. J. 1979. A photocell to measure winter activity of confined badgers. The Journal of Wildlife Management, 43:997–1001.

———. 1981*a*. Torpor and other physiological adaptations of the badger (Taxidea taxus) to cold environments. Physiological Zoology, 54:267–275.

———. 1981*b*. Effect of fasting on rate of food passage and assimilation efficiency in badgers. Journal of Mammalogy, 62:173–177.

Harmon, W. H. 1944. Notes on mountain goats in the Black Hills. Journal of Mammalogy, 25:149–151.

Harper, F. 1929. Mammal notes from Randolph County, Georgia. Journal of Mammalogy, 10:84–85.

Harper, J. A. 1966. Ecological study of Roosevelt elk. Oregon State Game Commission, Game Report, 1:1–29.

———. 1982. Roosevelt elk—an historical outlook. Oregon Wildlife, 37(4):3–6.

Harper, J. A., et al. 1987. Ecology and management of Roosevelt elk in Oregon. Revised ed. Oregon Department of Fish and Wildlife, [Portland], 70 pp.

Harper, P. A., and A. S. Harestad. 1986. Vole damage to coniferous trees on Texada Island. Forestry Chronicles, 62:429–432.

Harriman, A. E. 1973. Self-selection of diet in northern grasshopper mice (Onychomys leucogaster). The American Midland Naturalist, 90:97–106.

Harris, A. H. 1974. *Myotis yumanensis* in interior southwestern North America, with comments on *Myotis lucifugus*. Journal of Mammalogy, 55:589–607.

Harris, A. H., and L. N. Carraway. 1993. *Sorex preblei* from the Late Pleistocene of New Mexico. The Southwestern Naturalist, 38:56–58.

Harris, A. H., and J. S. Findley. 1962. Status of *Myotis lucifugus phasma* and comments on variation in *Myotis yumanensis*. Journal of Mammalogy, 43:192–199.

Harris, J. H. 1984. An experimental analysis of desert rodent foraging ecology. Ecology, 65:1579–1584.

———. 1986. Microhabitat segregation in two desert rodent species: the relation of prey availability to diet. Oecologia, 68:417–421.

———. 1987. Variation in the caudal fat deposit of *Microdipodops megacephalus*. Journal of Mammalogy, 68:58–63.

Harris, M. A., and J. O. Murie. 1984. Inheritance of nest sites in female Columbian ground squirrels. Behavioral Ecology and Sociobiology, 15:97–102.

Harris, S., and J. M. V. Rayner. 1986. Urban fox (*Vulpes vulpes*) population estimates and habitat requirements in several British cities. The Journal of Animal Ecology, 55:575–591.

Harrison, D. J., J. A. Bissonette, and J. A. Sherburne. 1989. Spatial relationships between coyotes and red foxes in eastern Maine. The Journal of Wildlife Management, 53:181–185.

Harrison, J. L. 1958. Range of movement of some Malayan rats. Journal of Mammalogy, 39:190–206.

Hart, E. B., and M. Trumbo. 1983. Winter stomach contents of South Dakota badgers. The Great Basin Naturalist, 43:492–493.

Hartman, C. G. 1928. The breeding season of the opossum (*Didelphis virginiana*) and the rate of intra-uterine and post natal development. Journal of Morphology and Physiology, 46:143–215.

———. 1952. Possums. The University of Texas Press, Austin, 174 pp.

Hartman, G. D., and T. L. Yates. 1985. Scapanus orarius. Mammalian Species, 253:1–5.

Harvey, J. T. 1988. Population dynamics, annual food consumption, movements, and dive behaviors of harbor seals, *Phoca vitulina richardsi*, in Oregon. Ph.D. dissert., Oregon State University, Corvallis, 177 pp.

Harvey, J. T., R. F. Brown, and B. R. Mate. 1990. Abundance and distribution of harbor seals (*Phoca vitulina*) in Oregon, 1975–1983. Northwestern Naturalist, 71:65–71.

Hash, H. S. 1987. Wolverine. Pp. 575–585, *in* Wild furbearer management and conservation in North America (M. Novak, J. A. Baker, M. E. Obbard, and B. Malloch, eds.). Ontario Ministry of Natural Resources, Toronto, 1150 pp.

Haskell, H. S., and H. G. Reynolds. 1947. Growth, developmental food requirements, and breeding activity of the California jack rabbit. Journal of Mammalogy, 28:129–136.

Hatfield, D. M. 1935. A natural history study of Microtus californicus. Journal of Mammalogy, 16:261–271.

Hatt, R. T. 1927. Notes on the ground squirrel, Callospermophilus. Occasional Papers of the Museum of Zoology, University of Michigan, 185:1–22.

———. 1932. The vertebral column of ricochetal rodents. Bulletin of the American Museum of Natural History, 63:599–745.

———. 1944. A large beaver-felled tree. Journal of Mammalogy, 25:313.

Hattan, G. 1976. Oregon's gray squirrels. Oregon Wildlife, 31(11):6–7.

Hatton, L. E., Jr., and R. S. Hoffmann. 1979. The distribution of red squirrels (*Tamiasciurus*) in eastern Oregon. The Murrelet, 60:23–25.

Haufler, J. B., and J. G. Nagy. 1984. Summer food habits of a small mammal community in the pinyon-juniper ecosystem. The Great Basin Naturalist, 44:145–150.

Hawes, D. B. 1975. Experimental studies of competition among four species of voles. Ph.D. dissert., University of British Columbia, Vancouver, 107 pp.

Hawes, M. L. 1975. Ecological adaptations in two species of shrews. Ph.D. dissert., University of British Columbia, Vancouver, 211 pp.

———. 1976. Odor as a possible isolating mechanism in sympatric species of shrews (*Sorex vagrans* and *Sorex obscurus*). Journal of Mammalogy, 57:404–406.

———. 1977. Home range, territoriality, and ecological separation in sympatric shrews, *Sorex vagrans* and *Sorex obscurus.* Journal of Mammalogy, 58:354–367.

Hawley, V. D., and F. E. Newby. 1957. Marten home ranges and population fluctuations. Journal of Mammalogy, 38:174–184.

Hayden, F. V. 1869. A new species of hare from the summit of Wind River Mountains. The American Naturalist, 3:113–116.

Hayden, P., and J. J. Gambino. 1966. Growth and development of the little pocket mouse, *Perognathus longimembris.* Growth, 30:187–197.

Hayden, P., and R. G. Lindberg. 1976. Survival of laboratory-reared pocket mice, *Perognathus longimembris.* Journal of Mammalogy, 57:266–272.

Hayes, J. P., and S. P. Cross. 1987. Characteristics of logs used by western red-backed voles, *Clethrionomys californicus,* and deer mice, *Peromyscus maniculatus.* The Canadian Field-Naturalist, 101:543–546.

Hayes, J. P., S. P. Cross, and P. W. McIntire. 1986. Seasonal variation in mycophagy by the western red-backed vole, *Clethrionomys californicus,* in southwestern Oregon. Northwest Science, 60:250–257.

Hayes, J. P., E. G. Horvath, and P. Hounihan. 1995. Townsend's chipmunk populations in Douglas-fir plantations and mature forests in the Oregon Coast Range. Canadian Journal of Zoology, 73:67–73.

Hayman, R. W., and G. W. C. Holt. 1940. *In* Ellerman, J. R., The families and genera of living rodents, British Museum, London, 1:1–689.

———. 1941. *In* Ellerman, J. R., The families and genera of living rodents, British Museum, London, 2:1–690.

Hayssen, V. 1991. Dipodomys microps. Mammalian Species, 389:1–9.

Hayssen, V., A. van Tienhoven, and A. van Tienhoven. 1993. Asdell's patterns of mammalian reproduction: a compendium of species-specific data. Cornell University Press, Ithaca, New York, 1023 pp.

Hayward, B., and R. Davis. 1964. Flight speeds in western bats. Journal of Mammalogy, 45:236–242.

Hayward, C. L. 1949. The short-tailed weasel in Utah and Colorado. Journal of Mammalogy, 30:436–437.

Hazard, E. B. 1982. The mammals of Minnesota. University of Minnesota Press, Minneapolis, 280 pp.

Heaton, T. H. 1985. Quaternary paleontology and paleoecology of Crystal Ball Cave, Millard County, Utah: with emphasis on mammals and description of a new species of fossil skunk. The Great Basin Naturalist, 45:337–390.

Hebert, D., and I. McT. Cowan. 1971. Natural salt licks as a part of the ecology of the mountain goat. Canadian Journal of Zoology, 49:605–610.

Hein, R. G., P. A. Talcott, J. L. Smith, and W. L. Myers. 1994. Blood selenium values of selected wildlife populations in Washington. Northwest Science, 68:185–188.

Heisinger, J. F. 1962. Periodicity of reingestion in the cottontail. The American Midland Naturalist, 67:441–448.

Heller, H. C. 1971. Altitudinal zonation of chipmunks (*Eutamias*): interspecific aggression. Ecology, 52:312–319.

Henderson, C. B. 1990. The influence of seed apparency, nutrient content and chemical defenses on dietary preference in *Dipodomys ordii.* Oecologia, 82:333–341.

Henderson, R. E., and B. W. O'Gara. 1978. Testicular development of the mountain goat. The Journal of Wildlife Management, 42:921–922.

Hendrickson, G. O. 1938. Winter food and cover of Mearns cottontail. Transactions of the North American Wildlife Conference, 23:787–793.

Hennessy, D. F., D. H. Owings, M. P. Rowe, R. G. Coss, and D. W. Leger. 1981. The information afforded by a variable signal: constraints on snake-elicited tail flagging by California ground squirrels. Behaviour, 78:188–226.

Hennings, D., and R. S. Hoffmann. 1977. A review of the taxonomy of the *Sorex vagrans* species complex from western North America. Occasional Papers of the Museum of Natural History, The University of Kansas, 68:1–35.

Henny, C. J., W. S. Overton, and H. M. Wight. 1970. Determining parameters for populations by using structural models. The Journal of Wildlife Management, 34:690–703.

Henny, C. J., C. Maser, J. O. Whitaker, Jr., and T. E. Kaiser. 1982. Organochlorine residues in bats after a forest spraying with DDT. Northwest Science, 56:329–337.

Henry, D. B., and T. A. Bookhout. 1969. Productivity of beavers in northeastern Ohio. The Journal of Wildlife Management, 33:927–932.

Henry, J. D. 1986. Red fox: the catlike canine. Smithsonian Institution Press, Washington, D.C., 174 pp.

Henry, S. E., and M. G. Raphael. 1989. Observations of copulation of free-ranging American marten. Northwest Naturalist, 70:32–33.

Henry, S. E., M. G. Raphael, and L. F. Ruggiero. 1990. Food caching and handling by marten. The Great Basin Naturalist, 50:381–383.

Herd, R. M., and M. B. Fenton. 1983. An electrophoretic, morphological, and ecological investigation of a putative hybrid zone between *Myotis lucifugus* and *Myotis yumanensis* (Chiroptera: Vespertilionidae). Canadian Journal of Zoology, 61:2029–2050.

Hermanson, J. W., and T. J. O'Shea. 1983. Antrozous pallidus. Mammalian Species, 213:1–8.

Herreid, C. F., II. 1961. Notes on the pallid bat in Texas. The Southwestern Naturalist, 6:13–20.

Herreid, C. F., II, and R. B. Davis. 1966. Flight patterns of bats. Journal of Mammalogy, 47:78–86.

Herrero, S. [M.] 1983. Social behaviour of black bears at a garbage dump in Jasper National Park. International Conference on Bear Research and Management, 5:54–70.

Herrig, D. M. 1974. Use of range sites by pronghorns in south central Oregon. M.S. thesis, Oregon State University, Corvallis, 71 pp.

Hershkovitz, P. 1949. Status of names credited to Oken, 1816. Journal of Mammalogy, 30:289–301.

Heske, E. J. 1987a. Responses of a population of California voles, *Microtus californicus,* to odor-baited traps. Journal of Mammalogy, 68:64–72.

———. 1987b. Pregnancy interruption by strange males in the California vole. Journal of Mammalogy, 68:406–410.

Heske, E. J., and J. M. Repp. 1986. Laboratory and field evidence for avoidance of California voles (*Microtus californicus*) by western harvest mice (*Reithrodontomys megalotis*). Canadian Journal of Zoology, 64:1530–1534.

Hibbs, D., F. A. Glover, and D. L. Gilbert. 1969. The mountain goat in Colorado. Transactions of the North American Wildlife and Natural Resources Conference, 34:409–418.

Hibler, C. P., R. E. Lange, and C. J. Metzer. 1972. Transplacental transmission of *Protostrongylus* spp. in bighorn sheep. Journal of Wildlife Diseases, 8:389.

Hickling, G. J., J. S. Millar, and R. A. Moses. 1991. Reproduction and nutrient reserves of bushy-tailed wood rats (*Neotoma cinerea*). Canadian Journal of Zoology, 69:3088–3092.

Hickman, J. C., ed. 1993. The Jepson manual: higher plants of California. University of California Press, Berkeley, 1400 pp.

Hight, M. E., M. Goodman, and W. Prychodko. 1974. Immunological studies of the Sciuridae. Systematic Zoology, 23:12–25.

Hildebrand, M. 1952. The integument in Canidae. Journal of Mammalogy, 33:319–428.

———. 1961. Voice of the grasshopper mouse. Journal of Mammalogy, 42:263.

Hill, E. P. 1982. Beaver: *Castor canadensis*. Pp. 256–281, *in* Wild mammals of North America: biology, management, and economics (J. A. Chapman and G. A. Feldhamer, eds.). The Johns Hopkins University Press, Baltimore, 1147 pp.

Hill, J. E. 1934. External characters of newborn pocket gophers. Journal of Mammalogy, 15:244–245.

Hill, J. E., and J. D. Smith. 1984. Bats: a natural history. University of Texas Press, Austin, 243 pp.

Hill, R. R. 1956. Forage, food habits, and range of mule deer. Pp. 415–429, *in* The deer of North America: the white-tailed, mule and black-tailed deer, genus *Odocoileus*. Their history and management (W. P. Taylor, ed.). The Stackpole Company, Harrisburg, Pennsylvania and The Wildlife Management Institute, Washington, D.C., 668 pp.

Hill, R. W. 1983. Thermal physiology and energetics of *Peromyscus*; ontogeny, body temperature, metabolism, insulation, and microclimatology. Journal of Mammalogy, 64:19–37.

Hillborn, R. 1982. Townsend's vole (method). P. 185, *in* CRC handbook of census methods for terrestrial vertebrates (D. E. Davis, ed.). CRC Press, Inc., Boca Raton, Florida, 397 pp.

Hines, W. W. 1973. Black-tailed deer populations and Douglas-fir reforestation on the Tillamook Burn, Oregon. Oregon State Game Commission Game Research Report, 3:1–59.

———. 1975. Black-tailed deer behavior and population dynamics in the Tillamook Burn, Oregon. Oregon Wildlife Commission, Wildlife Research Report, 5:1–31.

Hines, W. W., and C. E. Land. 1974. Black-tailed deer and Douglas-fir regeneration in the Coast Range of Oregon. Pp. 121–132, *in* Wildlife and forest management in the Pacific Northwest (H. C. Black, ed.). School of Forestry, Oregon State University, Corvallis, 236 pp.

Hines, W. W., and J. C. Lemos. [1979]. Reproductive performance by two age-classes of male Roosevelt elk in southwestern Oregon. Oregon Department of Fish and Wildlife, Wildlife Research Report, 8:1–54.

Hines, W. W., J. C. Lemos, and N. A. Hartmann, Jr. 1985. Male breeding efficiency in Roosevelt elk of southwestern Oregon. Oregon Department of Fish and Wildlife, Wildlife Research Report, 15:1–25.

Hinschberger, M. S. 1978. Occurrence and relative abundance of small mammals associated with riparian and upland habitats along the Columbia River. M.S. thesis, Oregon State University, Corvallis, 78 pp.

Hirata, D. N., and R. D. Nass. 1974. Growth and sexual maturation of laboratory-reared, wild *Rattus norvegicus*, *R.rattus*, *R.exulans* in Hawaii. Journal of Mammalogy, 55:472–474.

Hirshfeld, J. R., Z. C. Nelson and W. G. Bradley. 1977. Night roosting behavior in four species of desert bats. The Southwestern Naturalist, 22:427–433.

Hitchcock, C. L., and A. Cronquist. 1973. Flora of the Pacific Northwest: an illustrated manual. University of Washington Press, Seattle, 730 pp.

Hobbs, N. T., and R. A. Spowart. 1984. Effects of prescribed fire on nutrition of mountain sheep and mule deer during winter and spring. The Journal of Wildlife Management, 48:551–560.

Hobbs, R. J., S. L. Gumon, V. J. Hobbs, and H. A. Mooney. 1988. Effects of fertiliser addition and subsequent gopher disturbance on a serpentine annual grassland community. Oecologia, 75:291–295.

Hodgson, J. R. 1972. Local distribution of *Microtus montanus* and *M. pennsylvanicus* in southwestern Montana. Journal of Mammalogy, 53:487–499.

Hoffman, G. R. 1960. The small mammal components of six climax plan associations in eastern Washington and northern Idaho. Ecology, 41:571–572.

Hoffmann, R. S. 1955. Merriam shrew in California. Journal of Mammalogy, 36:561.

———. 1958. The role of reproduction and mortality in population fluctuations of voles (*Microtus*). Ecological Monographs, 28:79–109.

———. 1993. Order Lagomorpha. Pp. 807–827, *in* Mammal species of the world: a taxonomic and geographic reference (D. E. Wilson and D. M. Reeder, eds.). Smithsonian Institution Press, Washington, D.C., 1206 pp.

Hoffmann, R. S., and R. D. Fisher. 1978. Additional distributional records of Preble's shrew (*Sorex preblei*). Journal of Mammalogy, 59:883–884.

Hoffmann, R. S., and D. L. Pattie. 1968. A guide to Montana mammals: identification, habitat, distribution and abundance. University of Montana, Missoula, 133 pp.

Hoffmann, R. S., and R. D. Taber. 1967. Origin and history of Holarctic tundra ecosystems, with special reference to their vertebrate faunas. Pp. 143–170, *in* Arctic and alpine environments (H. E. Wright, Jr. and W. H. Osburn, eds.). Indiana University Press, Bloomington, 308 pp.

Hoffmann, R. S., P. L. Wright, and F. E. Newby. 1969. The distribution of some mammals in Montana. I. Mammals other than bats. Journal of Mammalogy, 50:579–604.

Hoffmann, R. S., C. G. Anderson, R. W. Thorington, Jr., and L. R. Heaney. 1993. Family Sciuridae. Pp. 419–465, *in* Mammal species of the world: a taxonomic and geographic reference (D. E. Wilson and D. M. Reeder, eds.). Smithsonian Institution Press, Washington, D.C., 1206 pp.

Hoffmeister, D. F. 1941. Two new subspecies of the piñon mouse, *Peromyscus truei*, from California. Proceedings of the Biological Society of Washington, 54:129–132.

———. 1951. A taxonomic and evolutionary study of the piñon mouse, *Peromyscus truei*. Illinois Biological Monograph, 21(4):1–104.

———. 1955. Mammals new to Grand Canyon National Park. Plateau, 28:1–7.

———. 1956. A record of *Sorex merriami* from northeastern Colorado. Journal of Mammalogy, 37:276.

———. 1969. The species problem in the Thomomys bottae-Thomomys umbrinus complex of pocket gophers in Arizona. Miscellaneous Publications, University of Kansas, Museum of Natural History, 51:75–91.

———. 1970. The seasonal distribution of bats in Arizona: a case for improving mammalian range maps. The Southwestern Naturalist, 15:11–22.

———. 1981. Peromyscus truei. Mammalian Species, 161:1–5.

———. 1986. Mammals of Arizona. The University of Arizona Press and the Arizona Game and Fish Department, [Tucson], 602 pp.

Holden, M. E. 1993. Family Dipodidae. Pp. 487–499, *in* Mammal   species of the world: a taxonomic and geographic reference. Second ed. (D. E. Wilson and D. M. Reeder, eds.). Smithsonian Institution Press, Washington, D.C., 1206 pp.

Holdenreid, R., and H. B. Morlan. 1956. A field study of wild mammals and fleas of Santa Fe County, New Mexico. The American Midland Naturalist, 55:369–381.

Holechek, J. L. 1982. Sample preparation techniques for microhistological analysis. Journal of Range Management, 35:267–268.

Holekamp, K. E. 1984. Natal dispersal in Belding's ground squirrels (*Spermophilus beldingi*). Behavioral Ecology and Sociobiology, 16:21–30.

Holler, N. R., and H. M. Marsden. 1970. Onset of evening activity of swamp rabbits and cottontails in relation to sunset. The Journal of Wildlife Management, 34:349–353.

Hollister, N. 1910. Descriptions of two new muskrats. Proceedings of the Biological Society of Washington, 23:1–2.

———. 1911. Four new mammals from the Canadian Rockies. Smithsonian Miscellaneous Collection, 56(26):1–4.

———. 1912. The names of the Rocky Mountain goats. Proceedings of the Biological Society of Washington, 25:185–186.

———. 1913. Three new subspecies of grasshopper mice. Proceedings of the Biological Society of Washington, 26:215–216.

———. 1916. The generic names Epimys and Rattus. Proceedings of the Biological Society of Washington, 29:126.

Holmes, A. C. V., and G. C. Sanderson. 1965. Populations and movements of opossums in east-central Illinois. The Journal of Wildlife Management, 29:287–295.

Honacki, J. H., K. E. Kinman, and J. W. Koeppl (eds.). 1982. Mammal species of the world: a taxonomic and geographic reference. Allen Press, Inc., and The Association of Systematics Collections, Lawrence, Kansas, 694 pp.

Hooper, E. T. 1938. Geographical variation in wood rats of the species Neotoma fuscipes. University of California Publications in Zoology, 42:213–246.

———. 1940. Geographic variation in bushy-tailed wood rats. University of California Publications in Zoology, 42:407–424.

———. 1944. Additional records of the Merriam shrew in Montana. Journal of Mammalogy, 25:92.

Hooven, E. F. 1959. Dusky-footed woodrat in young Douglas fir. Oregon Forest Research Center, Research Note, 41:1–24.

———. 1966. A test of thiram on two rabbit-infested areas of Oregon. Tree Planters' Notes, 79:1–3.

———. 1969. The influence of forest succession on populations of small animals in western Oregon. Pp. 30–38, *in* Wildlife and reforestation in the Pacific Northwest (H. C. Black, ed.). School of Forestry, Oregon State University, Corvallis, 92 pp.

———. 1971*a*. Effects of clearcutting a Douglas-fir stand upon small animal populations in western Oregon. Ph.D. dissert., Oregon State University, Corvallis, 170 pp.

———. 1971*b*. Pocket gopher damage on ponderosa pine plantations in southwestern Oregon. The Journal of Wildlife Management, 35:346–353.

———. 1973*a*. Notes on the water vole in Oregon. Journal of Mammalogy, 54:751–753.

———. 1973*b*. Response of the Oregon creeping vole to the clearcutting of a Douglas-fir forest. Northwest Science, 47:256–264.

———. 1976. Changes in small-mammal populations related to   abundance of Douglas-fir seed. Oregon State University, Forest Research Laboratory, Research Note, 57:1–2.

———. 1977. The mountain beaver in Oregon: its life history and control. Oregon State University, Forest Research Laboratory,   Research Paper, 30:1–20.

Hooven, E. F., and H. C. Black. 1976. Effects of some clearcutting practices on small-mammal populations in western Oregon. Northwest Science, 50:189–208.

Hooven, E. F., H. C. Black, and J. C. Lowrie. 1979. Disturbance of small mammal live traps by spotted skunks. Northwest Science, 53:79–81.

Hooven, E. F., R. F. Hoyer, and R. M. Storm. 1975. Notes on the vagrant shrew, *Sorex vagrans*, in the Willamette Valley of western Oregon. Northwest Science 49:163–173.

Hoover, R. L. 1954. Seven fetuses in western fox squirrels (*Sciurus niger rufiventer*). Journal of Mammalogy, 35:447–448.

Hopkins, D. D., and R. B. Forbes. 1979. Size and reproductive patterns of the Virginia opossum in northwestern Oregon. The Murrelet, 60:95–98.

———. 1980. Dietary patterns of the Virginia opossum in an urban environment. The Murrelet, 61:20–30.

Horn, E. E. 1923. Some notes concerning the breeding habits of Thomomys townsendi [*sic*], observed near Vale, Malheur County, Oregon, during the spring of 1921. Journal of Mammalogy, 4:37–39.

Horner, B. E. 1968. Gestation period and early development in *Onychomys leucogaster brevicuadus*. Journal of Mammalogy, 49:513–515.

Hornocker, M. G. 1969*a*. Defensive behavior in female bighorn sheep. Journal of Mammalogy, 50:128.

———. 1969*b*. Winter territoriality in mountain lions. The Journal of Wildlife Management, 33:457–464.

———. 1970. An analysis of mountain lion predation upon mule deer and elk in the Idaho Primitive Area. Wildlife Monographs,   21:1–39.

Hornocker, M. G., and H. S. Hash. 1981. Ecology of the wolverine in northwest Montana. Canadian Journal of Zoology, 56:1286–1301.

Hoskinson, R. L., and L. D. Mech. 1976. White-tailed deer

migration and its role in wolf predation. The Journal of Wildlife Management, 40:429–441.

Houston, D. B., B. B. Moorhead, and R. W. Olson. 1986. An aerial census of mountain goats in the Olympic Mountain Range, Washington. Northwest Science, 60:131–136.

———. 1991. Mountain goat population trends in the Olympic Mountain Range, Washington. Northwest Science, 65:212–216.

Houston, D. B., C. T. Robbins, and V. Stevens. 1989. Growth in wild and captive mountain goats. Journal of Mammalogy, 70:412–416.

Hováth, O. 1966. Observation of parturition and maternal care of the bushy-tailed wood rat (*Neotoma cinerea occidentalis Baird* [sic]). The Murrelet, 47:6–8.

Howard, W. E. 1949. A means to distinguish skulls of coyotes and domestic dogs. Journal of Mammalogy, 30:169–171.

———. 1952. A live trap for pocket gophers. Journal of Mammalogy, 33:61–65.

Howard, W. E., and H. E. Childs, Jr. 1959. Ecology of pocket gophers with emphasis on Thomomys bottae mewa. Hilgardia, 29:277–358.

Howard, W. E., and R. E. Cole. 1967. Olfaction in seed detection by deer mice. Journal of Mammalogy, 48:147–150.

Howard, W. E., and M. E. Smith. 1952. Rate of extrusive growth of incisors of pocket gophers. Journal of Mammalogy, 33:485–487.

Howell, A. B. 1920a. A second record of Phenacomys albipes in California, with a description of the species. Journal of Mammalogy, 1:242–243.

———. 1920b. Microtus townsendii in the Cascade Mountains of Oregon. Journal of Mammalogy, 1:141–142.

———. 1920c. A study of the California jumping mice of the genus Zapus. University of California Publications in Zoology, 21:225–238.

———. 1920d. Some Californian experiences with bat roosts. Journal of Mammalogy, 1:169–177.

———. 1921. Description of a new species of Phenacomys from Oregon. Journal of Mammalogy, 2:98–100.

———. 1923. Descriptions of a new microtine rodent from Oregon, with remarks on some contiguous forms. Journal of Mammalogy, 4:33–37.

———. 1924. The mammals of Mammoth, Mono County, California. Journal of Mammalogy, 5:25–36.

———. 1926a. Voles of the genus Phenacomys. II. Life history of the red tree mouse Phenacomys longicaudus True. North American Fauna, 48:39–66.

———. 1926b. Voles of the genus Phenacomys. I. Revision of the genus Phenacomys. North American Fauna, 48:1–37.

———. 1928. The food and habitat preferences of Phenacomys albipes. Journal of Mammalogy, 9:153–154.

Howell, A. H. 1901. Revision of the skunks of the genus Chincha. North American Fauna, 20:1–63.

———. 1906. Revision of the skunks of the genus Spilogale. North American Fauna, 26:1–55.

———. 1915. Revision of the American marmots. North American Fauna, 37:1–80.

———. 1918. Revision of the American flying squirrels. North American Fauna, 44:1–64.

———. 1919. Descriptions of nine new North American pikas. Proceedings of the Biological Society of Washington, 32:105–110.

———. 1922. Diagnosis of seven new chipmunks of the genus Eutamias, with a list of the American species. Journal of Mammalogy, 3:178–185.

———. 1924. Revision of the American pikas (genus Ochotona). North American Fauna, 47:1–57.

———. 1925. Preliminary descriptions of five new chipmunks from North America. Journal of Mammalogy, 6:51–54.

———. 1928. Descriptions of six new North American ground squirrels. Proceedings of the Biological Society of Washington, 41:211–214.

———. 1929. Revision of the American chipmunks (genera Tamias and Eutamias). North American Fauna, 52:1–157.

———. 1931. Preliminary descriptions of four new North American ground squirrels. Journal of Mammalogy, 12:160–162.

———. 1938. Revision of the North American ground squirrels, with a classification of the North American Sciuridae. North American Fauna, 56:1–256.

Hsu, T. C., and K. Benirschke. 1969. *Microtus oregoni* (creeping vole). An atlas of mammalian chromosomes. Vol. 3, Folio 121, Springer-Verlag, New York, unpaged.

Hsu, T. C., and R. A. Mead. 1969. Mechanisms of chromosomal changes in mammalian speciation. Pp. 8–17, *in* Comparative mammalian cytogenetics (K. Benirschke, ed.). Springer-Verlag, New York, 473 pp.

Hubbard, C. A. 1922. Some data upon the rodent *Aplodontia*. The Murrelet, 3(1):14–18.

Hudson, G. E., and M. Bacon. 1956. New records of *Sorex merriami* for eastern Washington. Journal of Mammalogy, 37:436–438.

Hudson, J. W. 1962. The role of water in the biology of the antelope ground squirrel Citellus leucurus. University of California Publications in Zoology, 64:1–51.

Huestis, R. R. 1931. Seasonal pelage differences in Peromyscus. Journal of Mammalogy, 12:372–375.

Huestis, R. R., and E. Barto. 1936. Flexed-tailed Peromyscus. Journal of Heredity, 27:73–75.

Huey, L. M. 1959. Longevity notes on captive *Perognathus*. Journal of Mammalogy, 40:412–415.

Humphrey, S. R. 1975. Nursery roosts and community diversity of Nearctic bats. Journal of Mammalogy, 56:321–346.

———. 1982. Bats: Vespertilionidae and Molossidae. Pp. 52–70, *in* Wild mammals of North America: biology, management, and economics (J. A. Chapman and G. A. Feldhamer, eds.). The Johns Hopkins University Press, Baltimore, 1147 pp.

Humphrey, S. R., and T. H. Kunz. 1976. Ecology of a Pleistocene relict, the western big-eared bat (*Plecotus townsendii*), in the southern Great Plains. Journal of Mammalogy, 57:470–494.

Hundertmark, K. J. 1982. Food selection and juvenile survival in Nuttall's cottontail in central Oregon. M.S. thesis, Oregon State University, Corvallis, 50 pp.

Hunsaker, D., II. 1977. Ecology of New World marsupials. Pp. 95–156, *in* The biology of marsupials (D. Hunsaker II, ed.). Academic Press, New York, 537 pp.

Huntly, N. J., A. T. Smith, and B. L. Ivins. 1986. Foraging behavior of the pika (*Ochotona princeps*), with comparisons of grazing versus haying. Journal of Mammalogy, 67:139–

148.

Husar, S. L. 1976. Behavioral character displacement: evidence of food partitioning in insectivorous bats. Journal of Mammalogy, 57:331–338.

Hutchins, M. 1995. Olympic Mountain goat controversy continues. Conservation Biology, 9:1324–1326.

Hutchins, M., and C. Hansen. 1982. Mother-infant interactions among free-ranging, non-native, mountain goats (*Oreamnos americanus*) in Olympic National Park. Pp. 58–67, *in* Ecological research in National Parks of the Pacific Northwest (E. E. Starkey, J. F. Franklin, and J. W. Matthews, eds.). Oregon State University, Forest Research Laboratory Publication, Corvallis, 142 pp.

Hutchinson, J. H. 1966. Notes on some upper Miocene shrews from Oregon. University of Oregon, Museum of Natural History Bulletin, 2:1–23.

———. 1968. Fossil Talpidae (Insectivora, Mammalia) from the later Tertiary of Oregon. University of Oregon, Museum of Natural History Bulletin, 11:1–117.

Iljin, N. A. 1941. Wolf-dog genetics. Journal of Genetics, 42:359–414.

Illiger, D. C. 1811. Prodromus systematis mammalium et avium additis terminis zoographicis utriusque classis, eorumque versione Germanica. C. Salfeld, Berolini, 302 pp.

Ingles, L. G. 1939. Observations on a nest of the long-tailed weasel. Journal of Mammalogy, 20:253–254.

———. 1947. Ecology and life history of the California gray squirrel. California Fish and Game, 33:139–158.

———. 1949. Hunting habits of the bat, *Myotis evotis*. Journal of Mammalogy, 30:197–198.

———. 1959. A quantitative study of mountain beaver activity. The American Midland Naturalist, 61:419–423.

———. 1960. A quantitative study of the activity of the dusky shrew (*Sorex vagrans obscurus*). Ecology, 41:656–660.

———. 1961a. Home range and habitats of the wandering shrew. Journal of Mammalogy, 42:455–462.

———. 1961b. Reingestion in the mountain beaver. Journal of Mammalogy, 42:411–412.

———. 1965. Mammals of the Pacific States: California, Oregon and Washington. Stanford University Press, Stanford, California, 506 pp.

Innes, D. G. L., and J. S. Millar. 1982. Life-history notes on the heather vole, *Phenacomys intermedius levis*, in the Canadian Rocky Mountains. The Canadian Field Naturalist, 96:307–311.

Iverson, S. L. 1967. Adaptations to arid environments in *Perognathus parvus* (Peale). Ph.D. dissert., University of British Columbia, Vancouver, 130 pp.

Izor, R. J. 1979. Winter range of the silver-haired bat. Journal of Mammalogy, 60:641–643.

Jackson, H. H. T. 1915. A review of the American moles. North American Fauna, 38:1–100.

———. 1918. Two new shrews from Oregon. Proceedings of the Biological Society of Washington, 31:127–130.

———. 1921. Two unrecognized shrews from California. Journal of Mammalogy, 2:161–162.

———. 1922. New species and subspecies of *Sorex* from western America. Journal of the Washington Academy of Science, 12:262–264.

———. 1928. A taxonomic review of the American long-tailed shrews (genera *Sorex* and *Microsorex*). North American Fauna, 51:1–238.

———. 1949. Two new coyotes from the United States. Proceedings of the Biological Society of Washington, 62:31–32.

———. 1951. Part II. Classification of the races of the coyote. Pp. 229–342, *in* The clever coyote. The Stackpole Company, Harrisburg, Pennsylvania and the Wildlife Management Institute, Washington, D.C., 411 pp.

———. 1961. Mammals of Wisconsin. University of Wisconsin Press, Madison, 504 pp.

Jahoda, J. C. 1970. The effects of rainfall on the activity of (*Onychomys leucogaster breviauritus*). American Zoologist, 10:326 (abstract only).

James, T. R., and R. W. Seabloom. 1969. Reproductive biology of the white-tailed jack rabbit in North Dakota. The Journal of Wildlife Mangement, 33:558–568.

James, W. B. 1953. The Merriam shrew in Washington state. Journal of Mammalogy, 34:121.

James, W. B., and E. S. Booth. 1954. Biology and life history of the sagebrush vole. Walla Walla College Publications of the Department of Biological Sciences and the Biological Station, 4:1–20 (first published 20 July 1952; revised 15 January 1954).

Jameson, E. W., Jr. 1952. Food of deer mice, *Peromyscus maniculatus* and *P. boylei*, in the northern Sierra Nevada, California. Journal of Mammalogy, 33:50–60.

———. 1954. Insects in the diet of pocket mice, *Perognathus parvus*. Journal of Mammalogy, 35:592–593.

———. 1955. Observations on the biology of *Sorex trowbridgei* [*sic*] in the Sierra Nevada, California. Journal of Mammalogy, 36:339–345.

———. 1965. Food consumption of hibernating and nonhibernating *Citellus lateralis*. Journal of Mammalogy, 46:634–640.

Jameson, E. W., Jr., and H. J. Peeters. 1988. California mammals. University of California Press, Berkeley, 403 pp.

Jameson, R. J. 1975. An evaluation of attempts to reestablish the sea otter in Oregon. M.S. thesis, Oregon State University, Corvallis, 66 pp.

Jameson, R. J., and K. W. Kenyon. 1977. Prey of sea lions in the Rogue River, Oregon. Journal of Mammalogy, 58:672.

Jameson, R. J., K. W. Kenyon, A. M. Johnson, and H. M. Wight. 1982. History and status of translocated sea otter populations in North America. Wildlife Society Bulletin, 10:100–107.

Janes, S. W., and J. M. Barss. 1985. Predation by three owl species on northern pocket gophers of different body mass. Oecologia, 67:76–81.

Jannett, F. J., Jr., and J. Z. Jannett. 1974. Drum-marking by Arvicola richardsoni and its taxonomic significance. The American Midland Naturalist, 92:230–234.

Jannett, F. J., Jr., J. A. Jannett, and M. E. Richmond. 1979. Notes on reproduction in captive *Arvicola richardsoni*. Journal of Mammalogy, 60:837–838.

Janson, R. G. 1946. A survey of the native rabbits of Utah with reference to their classification, distribution, life histories and ecology. M.S. thesis, Utah State Agricultural College, [Logan], 103 pp.

Jardine, W. 1834. The natural history of the Felinae. The naturalist's library, W. H. Lizars, Edinburgh, 2:266 (not seen, cited in Hall, 1981 and The National Union Catalog, 278:166).

Jenkins, H. O. 1948. A population study of the meadow mice (*Microtus*) in three Sierra Nevada meadows. Proceedings of the California Academy of Sciences, 26:43–67.

Jenkins, S. H. 1979. Seasonal and year-to-year differences in food selection by beavers. Oecologia, 44:112–116.

———. 1980. A size-distance relation in food selection by beavers. Ecology, 61:740–746.

Jenkins, S. H., and P. E. Busher. 1979. Castor canadensis. Mammalian Species, 120:1–8.

Jenkins, S. H., and B. D. Eshelman. 1984. Spermophilus beldingi. Mammalian Species, 221:1–8.

Jewett, S. G. 1914. The white-tailed and other deer in Oregon. The Oregon Sportsman, 2(8):5–9.

———. 1920. Notes on two species of Phenacomys in Oregon. Journal of Mammalogy, 1:165–168.

———. 1931. Nest and young of Sorex vagrans vagrans. Journal of Mammalogy, 12:163.

———. 1955. Free-tailed bats and melanistic mice in Oregon. Journal of Mammalogy, 36:458–459.

Jewett, S. G., and H. W. Dobyns. 1929. The Virginia opossum in Oregon. Journal of Mammalogy, 10:351.

Jewett, S. G., and E. R. Hall. 1940. A new race of beaver from Oregon. Journal of Mammalogy, 21:87–89.

Johannessen, C. L., W. A. Davenport, A. Millet, and S. McWilliams. 1971. The vegetation of the Willamette Valley. Annals of the Association of American Geographers, 61:286–302.

Johansen, K. 1962. Buoyancy and insulation in the muskrat. Journal of Mammalogy, 43:64–68.

Johnson, B. K., R. D. Schultz, and J. A. Bailey. 1978. Summer forages of mountain goats in the Sawatch Range, Colorado. The Journal of Wildlife Management, 42:636–639.

Johnson, D. E. 1951. Biology of the elk calf, *Cervus elaphus nelsoni*. The Journal of Wildlife Management, 15:396–410.

Johnson, D. R. 1961. The food habits of rodents on rangelands of southern Idaho. Ecology, 42:407–410.

Johnson, D. R., and M. H. Maxell. 1966. Energy dynamics of Colorado pikas. Ecology, 47:1059–1061.

Johnson, D. R., N. C. Nydegger, and G. W. Smith. 1987. Comparison of movement-based density estimates for Townsend ground squirrels in southwestern Idaho. Journal of Mammalogy, 68:689–691.

Johnson, D. W., and D. M. Armstrong. 1987. Peromyscus crinitus. Mammalian Species, 287:1–8.

Johnson, L. A., J. P. Flook, and H. W. Hawk. 1989. Sex preselection in rabbits: live births from X and Y sperm separated by DNA and cell sorting. Biology of Reproduction, 41:199–203.

Johnson, M. K. 1975. Tent building in mountain beavers (*Aplodontia rufa*). Journal of Mammalogy, 56:715–716.

Johnson, M. K., and R. M. Hansen. 1979a. Coyote food habits on the Idaho National Engineering Laboratory. The Journal of Wildlife Management, 43:951–956.

———. 1979b. Foods of cottontails and woodrats in south-central Idaho. Journal of Mammalogy, 60:213–215.

Johnson, M. L. 1967. An external identification character of the heather vole, *Phenacomys intermedius*—the orange aural hairs. The Murrelet, 48:17.

———. 1968. Application of blood protein electrophoretic studies to problems in mammalian taxonomy. Systematic Zoology, 17:23–30.

———. 1973. Characters of the heather vole *Phenacomys* and the red tree vole, *Arborimus*. Journal of Mammalogy, 54:239–244.

Johnson, M. L., and S. B. Benson. 1960. Relationship of the pocket gophers of the Thomomys mazama-Thomomys talpoides complex in the Pacific Northwest. The Murrelet, 41:17–22.

Johnson, M. L., and C. W. Clanton. 1954. Natural history of *Sorex merriami* in Washington State. The Murrelet, 35:1–4.

Johnson, M. L., and S. B. George. 1991. Species limits with the *Arborimus longicaudus* species-complex (Mammalia: Rodentia) with a description of a new species from California. Contributions in Science, Natural History Museum of Los Angles County, 429:1–16.

Johnson, M. L., and C. Maser. 1982. Generic relationships of *Phenacomys albipes*. Northwest Science, 56:17–19.

Johnson, M. L., and B. T. Ostenson. 1959. Comments on the nomenclature of some mammals of the Pacific Northwest. Journal of Mammalogy, 40:571–577.

Johnson, M. L., C. W. Clanton, and J. Girard. 1948. The sagebrush vole in Washington state. The Murrelet, 29:44–47.

Johnson, R. D., and J. E. Anderson. 1984. Diets of black-tailed jack rabbits in relation to population density and vegetation. Journal of Range Management, 37:79–83.

Johnson, R. E. 1977. An historical analysis of wolverine abundance and distribution in Washington. The Murrelet, 58:13–16.

Johnson, S. R. 1971. The thermal regulation, microclimate, and distribution of the mountain beaver, *Aplodontia rufa pacifica* Merriam. Ph.D. dissert., Oregon State University, Corvallis, 164 pp.

Johnstone, R. 1979. An unusual nest site of the Pacific jumping mouse with notes on its natural history. The Murrelet, 60:72.

Jones, C. 1961. Additional records of bats in New Mexico. Journal of Mammalogy, 42:538–539.

———. 1965. Ecological distribution and activity periods of bats of the Mogollon Mountains area of New Mexico and adjacent Arizona. Tulane Studies in Zoology, 12:93–100.

Jones, F. L. 1950. Recent records of the wolverine (*Gulo luscus luteus*) in California. California Fish and Game, 36:320–322.

———. 1955. Records of southern wolverine, *Gulo luscus luteus*, in California. Journal of Mammalogy, 36:569.

Jones, G. S. 1978. *Microtus longicaudus* with grooved incisors. The Murrelet, 59:104–105.

Jones, G. S., J. O. Whitaker, Jr., and C. Maser. 1978. Food habits of jumping mice (*Zapus trinotatus* and *Z. princeps*) in western North America. Northwest Science, 52:57–60.

Jones, J. H., and N. S. Smith. 1979. Bobcat density and prey selection in central Arizona. The Journal of Wildlife Management, 43:666–672.

Jones, J. K., Jr., and E. C. Birney. 1988. Handbook of mammals of the north-central states. University of Minnesota Press, Minneapolis, 346 pp.

Jones, J. K., Jr., and J. R. Choate. 1978. Distribution of two species of long-eared bats of the genus *Myotis* on the northern Great Plains. The Prairie Naturalist, 10:49–52.

Jones, J. K., Jr., D. C. Carter, and H. H. Genoways. 1973a. Checklist of North American mammals north of Mexico. Occasional Papers, The Museum, Texas Tech University, 12:1–14.

———. 1975. Revised checklist of North American mammals north of Mexico. Occasional Papers, The Museum, Texas Tech University, 28:1–14.

———. 1979. Revised checklist of North American mammals north of Mexico, 1979. Occasional Papers, The Museum, Texas Tech University, 62:1–17.

Jones, J. K., Jr., D. M. Armstrong, R. S. Hoffmann, and C. Jones. 1983. Mammals of the northern Great Plains. University of Nebraska Press, Lincoln, 379 pp.

Jones, J. K., Jr., R. P. Lampe, C. A. Spenrath, and T. H. Kunz. 1973b. Notes on the distribution and natural history of bats in south-eastern Montana. Occasional Papers, The Museum, Texas Tech University, 15:1–12.

Jones, J. K., Jr., D. C. Carter, H. H. Genoways, R. S. Hoffmann, and D. W. Rice. 1982. Revised checklist of North American mammals north of Mexico, 1982. Occasional Papers, The Museum, Texas Tech University, 80:1–22.

Jones, J. K., Jr., D. C. Carter, H. H. Genoways, R. S. Hoffmann, D. W. Rice, and C. Jones. 1986. Revised checklist of North American mammals north of Mexico, 1986. Occasional Papers, The Museum, Texas Tech University, 107:1–22.

Jones, J. K., Jr., R. S. Hoffmann, D. W. Rice, C. Jones, R. J. Baker, and M. D. Engstrom. 1992. Revised checklist of North American mammals north of Mexico, 1991. Occasional Papers, The Museum, Texas Tech University, 146:1–23.

Jonkel, C. J. 1987. Brown bear. Pp. 457–473, in Wild furbearer management and conservation in North America (M. Novak, J. A. Baker, M. E. Obbard, and B. Malloch, eds.). Ontario Ministry of Natural Resources, Toronto, 1150 pp.

Jonkel, C. J., and I. McT. Cowan. 1971. The black bear in the spruce-fir forest. Wildlife Monographs, 27:1–57.

Jonkel, C. J., and R. P. Weckwerth. 1963. Sexual maturity and implantation of blastocysts in the wild pine marten. The Journal of Wildlife Management, 27:93–98.

Jordan, D. S., and G. A. Clark. 1898. The history, condition, and needs of the herd of fur seals resorting to the Pribilof Islands. In The fur seals and fur-seal islands of the North Pacific Ocean (D. S. Jordan). United States Government Printing Office, Washington, D.C., United States Treasury Document no. 207(part 1):1–249.

Jordan, J. W. 1973. Natality of black-tailed deer in McDonald State Forest. M.S. thesis, Oregon State University, Corvallis, 36 pp.

Jordan, J. W., and P. A. Vohs, Jr. 1976. Natality of black-tailed deer in McDonald State Forest, Oregon. Northwest Science, 50:108–113.

Jordan, P. A., P. C. Shelton, and D. L. Allen. 1967. Numbers, turnover, and social structure of the Isle Royale wolf population. American Zoologist, 7:233–252.

Jorgensen, C. D., and C. L. Hayward. 1965. Mammals of the Nevada Test Site. Brigham Young University Science Bulletin, Biological Series, 6(3):1–81.

Junge, J. A., and R. S. Hoffmann. 1981. An annotated key to the long-tailed shrews (genus Sorex) of the United States and Canada, with notes on Middle America Sorex. Occasional Papers of The Museum of Natural History, The University of Kansas, 94:1–48.

Kalinowski, S. A., and D. S. deCalesta. 1981. Baiting regimes for reducing ground squirrel damage to alfalfa. Wildlife Society Bulletin, 9:268–272.

Kallen, F. C. 1977. The cardiovascular systems of bats: structure and function. Pp. 289–483, in Biology of bats (W. A. Wimsatt, ed.). Academic Press, New York, 3:1–651.

Kartman, L., F. M. Prince, and S. F. Quan. [1959]. Epizootiologic aspects. Pp. 43–54, in The Oregon meadow mouse irruption of 1957–1958. Federal Cooperative Extension Service, Oregon State College, Corvallis, 88 pp.

Kaufman, D. W., and G. A. Kaufman. 1982. Effect of moonlight on activity and microhabitat use by Ord's kangaroo rat (Dipodomys ordii). Journal of Mammalogy, 63:309–312.

Kaufman, G. A., and D. W. Kaufman. 1989. An artificial burrow for the study of natural populations of small mammals. Journal of Mammalogy, 70:656–659.

Kaufmann, A. F. 1976. Epidemiologic trends of leptospirosis in the United States, 1965–1974. Pp. 177–189, in The biology of parasitic spirochetes (R. C. Johnson, ed.). Academic Press, New York, 402 pp.

Kavanau, J. L., and R. M. Havenhill. 1976. Compulsory regime and control of environment in animal behavior, part 3. Light level   preferences of small nocturnal mammals. Behaviour, 59:203–225.

Kawamichi, T. 1976. Hay territory and dominance rank of pikas (Ochotona princeps). Journal of Mammalogy, 57:133–148.

Keating, K. A., L. R. Irby, and W. F. Kasworm. 1985. Mountain sheep winter food habits in the upper Yellowstone Valley. The Journal of Wildlife Management, 49:156–161.

Kebbe, C. E. 1955a. Muskrat migration from Klamath Marsh. The Murrelet, 36:26.

———. 1955b. Status of the opossum in Oregon. The Murrelet, 38:42.

———. 1956. Record of white muskrat from Baker County, Oregon. The Murrelet, 37:11.

———. 1960. Oregon's beaver story. Oregon State Game Commission Bulletin, 15(2):3–6.

———. 1961. Return of the fisher. Oregon State Game Commission Bulletin, 16(12):3–7.

———. 1966. Wolverine killed in Oregon. The Murrelet, 47:65.

Keen, R., and H. B. Hitchcock. 1980. Survival and longevity of the little brown bat (Myotis lucifugus) in southeastern Ontario. Journal of Mammalogy, 61:1–7.

Keister, G. P., Jr., and D. Immell. 1994. Continued investigations of kit fox in southeastern Oregon and evaluation of status. Oregon Department of Fish and Wildlife, Technical Report, 94–5–01:1–33.

Keith, J. O., R. M. Hansen, and A. L. Ward. 1959. Effect of 2,4-D on abundance and foods of pocket gophers. The Journal of Wildlife Management, 23: 137–145.

Keith, L. B. 1963. Wildlife's ten-year cycle. University of Wisconsin Press, Madison, 201 pp.

———. 1964. Daily activity pattern of snowshoe hares. Journal of Mammalogy, 45: 626–627.

———. 1966. Habitat vacancy during a snowshoe hare decline. The Journal of Wildlife Management, 30:828–832.

———. 1974. Some features of population dynamics in mammals. Proceedings of the International Congress of Game Biologists, 11:17–58.

Keith, L. B., and L. A. Windberg. 1978. A demographic analysis of the snowshoe hare cycle. Wildlife Monographs, 58:1–70.

Keith, L. B., E. C. Meslow, and O. J. Rongstad. 1968. Techniques for snowshoe hare population studies. The Journal of Wildlife Management, 32:801–812.

Keith, L. B., J. R. Cary, O. J. Rongstad, and M. C. Brittingham. 1984. Demography and ecology of a declining snowshoe hare population. Wildlife Monographs, 90:1–43.

Kelleyhouse, D. G. 1980. Habitat utilization by black bears in northern California. International Conference on Bear Research and Management, 4:221–227.

Kellogg, L. 1912. Pleistocene rodents of California. University of California Publications, Bulletin of the Department of Geology, 7:151–168.

———. 1914. Aplodontia chryseola, a new mountain beaver from the Trinity Region of northern California. University of California Publications in Zoology, 12:295–296.

Kellogg, R. 1918. A revision of the Microtus californicus group of meadow mice. University of California Publications in Zoology, 21:1–42.

Kelson, K. R. 1946. Notes on the comparative osteology of the bobcat and the house cat. Journal of Mammalogy, 27:255–264.

Kelt, D. A. 1988. Dipodomys californicus. Mammalian Species, 324:1–4.

Kemp, G. A., and L. B. Keith. 1970. Dynamics and regulation of red squirrel (Tamiasciurus hudsonicus) populations. Ecology, 51:763–779.

Kenagy, G. J. 1972. Saltbush leaves: excision of hypersaline tissue by a kangaroo rat. Science, 178:1094–1096.

———. 1973. Daily and seasonal patterns of activity and energetics in a heteromyid rodent community. Ecology, 54:1201–1219.

———. 1976. The period of daily activity and its seasonal changes in free-ranging and captive kangaroo rats. Oecologia, 24:105–140.

———. 1981. Effects of day length, temperature, and edogenous control on annual rhythms of reproduction and hibernation in chipmunks (Eutamias ssp.). Journal of Comparative Physiology, 141A:369–378.

Kenagy, G. J., and B. M. Barnes. 1988. Seasonal reproductive patterns in four coexisting rodent species from the Cascade Mountains, Washington. Journal of Mammalogy, 69:274–292.

Kenagy, G. J., and G. A. Bartholomew. 1985. Seasonal reproductive patterns in five coexisting California desert rodent species. Ecological Monographs, 55:371–397.

Kennicott, R. 1863. Descriptions of four new species of Spermophilus, in the collections of the Smithsonian Institution. Proceedings of the Academy of Natural Sciences of Philadelphia, 15:157–158.

Kenward, R. E. 1983. The causes of damage by red and grey squirrels. Mammal Review, 13:159–166.

Kenyon, K. W. 1969. The sea otter in the eastern Pacific Ocean. North American Fauna, 68:1–352.

Kenyon, K. W., and D. W. Rice. 1961. Abundance and distribution of the Steller sea lion. Journal of Mammalogy, 42:223–234.

Kenyon, K. W., and V. B. Scheffer. 1953. The seals, sea lions, and sea otter of the Pacific coast. United States Fish and Wildlife Service Wildlife Leaflet, 344:1–28.

Kenyon, K. W., and F. Wilke. 1953. Migration of the northern fur seal, Callorhinus ursinus. Journal of Mammalogy, 34:86–98.

Kerr, R. 1792. The animal kingdom, or zoological system, of the celebrated Sir Charles Linneaus. Class I. Mammalia. London, J. Murray and R. Faulder, 644 pp. (not seen, cited in Baird, 1858).

Kerwin, M. L., and G. J. Mitchell. 1971. The validity of the wear-age technique for Alberta pronghorns. The Journal of Wildlife Management, 35:743–747.

Kindschy, R. R. 1985. Response of red willow to beaver use in southeastern Oregon. The Journal of Wildlife Management, 49:26–28.

King, C. M. 1983. Mustela erminea. Mammalian Species, 195:1–8.

King, J. A. (ed.). 1968. Biology of Peromyscus (Rodentia). The American Society of Mammalogists, Special Publication, 2:1–593.

King, J. A., E. O. Price, and P. L. Weber. 1968. Behavioral comparisons within the genus Peromyscus. Papers of the Michigan Academy of Science, Arts and Letters, 53:113–136.

Kingdon, J. 1974. East African mammals: an atlas of evolution in Africa. Volume II, Part B. Hares and rodents. The University of Chicago Press, Chicago, 343–704 + i-lvii.

Kinler, N. W., G. Linscombe, and P. R. Ramsey. 1987. Nutria. Pp. 327–343, in Wild furbearer management and conservation in North America (M. Novak, J. A. Baker, M. E. Obbard, and B. Malloch, eds.). Ontario Ministry of Natural Resources, Toronto, 1150 pp.

Kirkland, G. L., Jr., and J. S. Findley. 1996 [1997]. First Holocene record for Preble's shrew (Sorex preblei) in New Mexico. The Southwestern Naturalist, 41:320–322.

Kirkland, G. L., Jr., and J. N. Layne (eds.). 1989. Advances in the study of Peromyscus (Rodentia). Texas Tech University Press, Lubbock, 366 pp.

Kistner, T. P., and D. Wise. 1979. Transplacental transmission of Protostrongylus sp. in California bighorn sheep (Ovis canadensis californiana) in Oregon. Journal of Wildlife Diseases, 15:561–562.

Kitchen, D. W. 1974. Social behavior and ecology of the pronghorn. Wildlife Monographs, 38:1–96.

Kitchen, D. W., and B. W. O'Gara. 1982. Pronghorn: Antilocapra americana. Pp. 960–971, in Wild mammals of North America: biology, management, and economics (J. A. Chapman and G. A. Feldhamer, eds.). The Johns Hopkins University Press, Baltimore, 1147 pp.

Kleiman, D. G., and C. A. Brady. 1978. Coyote behavior in the context of recent canid research: problems and perspectives. Pp. 163–188, in Coyotes: biology, behavior, and management (M. Bekoff, ed.). Academic Press, New York, 384 pp.

Klimstra, W. D., and E. L. Corder. 1957. Food of the cottontail in southern Illinois. Transactions of the Illinois State Academy of Science, 50:247–256.

Kline, P. D. 1963. Notes on the biology of the jackrabbit in Iowa. Iowa Academy of Science, 70:196–204.

———. 1965. Factors influencing roadside counts of cottontails. The Journal of Wildlife Management, 29:665–671.

Klingener, D. 1971. The question of cheek pouches in Zapus.

Journal of Mammalogy, 52:463–464.

———. 1984. Gliroid and dipodoid rodents. Pp. 381–388, *in* Orders and families of Recent mammals of the world (S. Anderson and J. K. Jones, Jr., eds.). John Wiley & Sons, New York, 686 pp.

Knick, S. T., J. D. Brittell, and S. J. Sweeney. 1985. Population characteristics of bobcats in Washington state. The Journal of Wildlife Management, 49:721–728.

Knowlton, F. F. 1972. Preliminary interpretations of coyote population mechanics with some management implications. The Journal of Wildlife Management, 36:369–382.

Koehler, G. M. 1990. Snowshoe hare, *Lepus americanus*, use of forest successional stages and population changes during 1985–1989 in north-central Washington. The Canadian Field-Naturalist, 105:291–293.

Koehler, G. M., and M. G. Hornocker. 1989. Influences of seasons on bobcats in Idaho. The Journal of Wildlife Management, 53:197–202.

Koehler, G. M., M. G. Hornocker, and H. S. Hash. 1980. Wolverine marking behavior. The Canadian Field Naturalist, 94:339–341.

Koeppl, J. W., and R. S. Hoffmann. 1981. Comparative postnatal growth of four ground squirrel species. Journal of Mammalogy, 62:41–57.

Koeppl, J. W., R. S. Hoffmann, and C. F. Nadler. 1978. Pattern analysis of acoustical behavior in four species of ground squirrels. Journal of Mammalogy, 59:677–696.

Koeppl, J. W., N. A. Slade, and R. S. Hoffmann. 1975. A bivariate home range model and its application to ethological data. Journal of Mammalogy, 56:81–90.

Koford, C. B., and M. R. Koford. 1948. Breeding colonies of bats, *Pipistrellus hesperus* and *Myotis subulatus melanorhinus*. Journal of Mammalogy, 29:417–418.

Koford, R. R. 1982. Mating system of a territorial tree squirrel (*Tamiasciurus douglasii*) in California. Journal of Mammalogy, 63:274–283.

———. 1992. Does supplemental feeding of red squirrels change population density, movements, or both? Journal of Mammalogy, 73:930–932.

Kolenosky, G. B., and S. M. Strathearn. 1987. Black bear. Pp. 443–454, *in* Wild furbearer management and conservation in North America (M. Novak, J. A. Baker, M. E. Obbard, and B. Malloch, eds.). Ontario Ministry of Natural Resources, Toronto, 1150 pp.

Koopman, K. F. 1967. Artiodactyls. Pp. 385–406, *in* Recent mammals of the world: a synopsis of families (S. Anderson and J. K. Jones, Jr., eds.). The Ronald Press Company, New York, 453 pp.

———. 1984. Bats. Pp. 145–186, *in* Orders and families of Recent mammals of the world (S. Anderson and J. K. Jones, Jr., eds.). John Wiley & Sons, New York, 686 pp.

———. 1993. Order Chiroptera. Pp. 139–241, *in* Mammal species of the world: a taxonomic and geographic reference. Second ed. (D. E. Wilson and D. M. Reeder, eds.). Smithsonian Institution Press, Washington, D.C., 1206 pp.

Koprowski, J. L. 1991a. Response of fox squirrels and gray squirrels to a late spring-early summer food shortage. Journal of Mammalogy, 72:367–372.

———. 1991b. Damage due to scent marking by eastern gray and fox squirrels. Proceedings Great Plains Wildlife Damage

Conference, 10:101–105.

———. 1993a. Sex and species biases in scent-marking by fox squirrels and eastern grey squirrels. Journal of Zoology (London), 230:319–323.

———. 1993b. Behavioral tactic, dominance, and copulatory success among male fox squirrels. Ethology Ecology & Evolution, 5:169–176.

———. 1993c. The role of kinship in field interactions among juvenile gray squirrels (*Sciurus carolinensis*). Canadian Journal of Zoology, 71:224–226.

———. 1993d. Alternative reproductive tactics in male eastern gray squirrels: "making the best of a bad job." Behavioral Ecology, 4:165–171.

Korfhage, R. C., J. R. Nelson, and J. M. Skovlin. 1980. Summer diets of Rocky Mountain elk in northeastern Oregon. The Journal of Wildlife Management, 44:746–750.

Kornet, C. A. 1978. Status and habitat use of California bighorn sheep on Hart Mountain, Oregon. M.S. thesis, Oregon State University, Corvallis, 49 pp.

Krämer, A. 1973. Interspecific behavior and dispersion of two sympatric deer species. The Journal of Wildlife Management, 37:288–300.

Kramm, K. R. 1972. Body temperature regulation and torpor in the antelope ground squirrel, *Ammospermophilus leucurus*. Journal of Mammalogy, 53:609–611.

Kratochvíl, J. 1982. Karyotyp und System der Familie Felidae (Carnivora, Mammalia). Folia Zoologica, 31:289–304.

Krebs, C. J. 1966. Demographic changes in fluctuating populations of *Microtus californicus*. Ecological Monographs, 36:239–273.

Krebs, C. J., and I. Wingate. 1976. Small mammal communities of the Kluane region, Yukon Territory. The Canadian Field-Naturalist, 90:379–389.

———. 1985. Population fluctuations in the small mammals of the Kluane region, Yukon Territory. The Canadian Field Naturalist, 99:51–61.

Krebs, C. J., S. Boutin, and B. S. Gilbert. 1986a. A natural feeding experiment on a declining snowshoe hare population. Oecologia, 70:194–197.

Krebs, C. J., J. A. Redfield, and M. J. Taitt. 1978. A pulsed-removal experiment on the vole *Microtus townsendii*. Canadian Journal of Zoology, 54:79–95.

Krebs, C. J., M. S. Gaines, B. L. Keller, J. H. Myers, and R. H. Tamarin. 1973. Population cycles in small rodents. Science, 179:35–41.

Krebs, C. J., B. S. Gilbert, S. Boutin, A. R. E. Sinclair, and J. N. M. Smith. 1986b. Population biology of snowshoe hares. I. Demography of food-supplemented populations in the southern Yukon, 1976–84. Journal of Animal Ecology, 55:963–982.

Krebs, J. W., M. L. Wilson, and J. E. Childs. 1995. Rabies—epidemiology, prevention, and future research. Journal of Mammalogy, 76:681–694.

Kritzman, E. B. 1970. Niche fit and overlap of *Peromyscus maniculatus* and *Perognathus parvus* in eastern Washington. M.S. thesis, University of Washington, Seattle, 86 pp.

———. 1972. A captive shrew-mole and her litter. The Murrelet, 53:47–49.

———. 1974. Ecological relationships of *Peromyscus maniculatus* and *Perognathus parvus* in eastern Washington.

Journal of Mammalogy, 55:172–188.

Krutzsch, P. H. 1946. Some observations on the big brown bat in San Diego County, California. Journal of Mammalogy, 27:240–242.

———. 1954a. North American jumping mice (genus Zapus). University of Kansas Publications, Museum of Natural History, 7:349–472.

———. 1954b. Notes on the habits of the bat, *Myotis californicus*. Journal of Mammalogy, 35:539–545.

———. 1955. Observations on the Mexican free-tailed bat, *Tadarida mexicana*. Journal of Mammalogy, 36:236–242.

Krutzsch, P. H., and T. A. Vaughan. 1955. Additional data on the bacula of North American bats. Journal of Mammalogy, 36:96–100.

Kubota, K., and K. Matsumoto. 1963. On the deciduous teeth shed into the amnion of the fur seal embryo. Tokyo Medical and Dental University Bulletin, 10:89–93.

Kucera, T. E. 1978. Social behavior and breeding system of the desert mule deer. Journal of Mammalogy, 59:463–476.

Kuehn, D. W. 1989. Winter foods of fishers during a snowshoe hare decline. The Journal of Wildlife Management, 53:688–692.

Kuhl, H. 1820. Beiträge zur Zoologie und vergleichenden Anatomie. Frankfurt am Main, abth. 1 (not seen, cited in Gill and Coues, 1877).

Kuhn, L. W., and W. D. Edge. 1990. Controlling moles. Oregon State University, Extension Service, EC 987:1–4.

Kuhn, L. W., and E. P. Peloquin. 1974. Oregon's nutria problem. Proceedings Vertebrate Pest Conference, 6:101–105.

Kuhn, L. W., W. Q. Wick, and R. J. Pedersen. 1966. Breeding nests of Townsend's mole in Oregon. Journal of Mammalogy, 47:239–249.

Kulwich, R., L. Struglia, and P. B. Pearson. 1953. The effect of coprophagy on the excretion of B vitamins by the rabbit. The Journal of Nutrition, 49:639–645.

Kunz, T. H. 1971. Reproduction of some vespertilionid bats in central Iowa. The American Midland Naturalist, 846:477–486.

———. 1973. Resource utilization: temporal and spatial components of bat activity in central Iowa. Journal of Mammalogy, 54:14–32.

———. 1974. Reproduction, growth, and mortality of the vespertilionid bat, *Eptesicus fuscus*, in Kansas. Journal of Mammalogy, 55:1–13.

———. 1982. Lasionycteris noctivagans. Mammalian Species, 172:1–5.

Kunz, T. H., and R. A. Martin. 1982. Plecotus townsendii. Mammalian Species, 175:1–6.

Kunz, T. H., E. L. P. Anthony, and W. T. Rumage III. 1977. Mortality of little brown bats following multiple pesticide applications. The Journal of Wildlife Management, 41:476–483.

Kurta, A. 1982. Flight patterns of *Eptesicus fuscus* and *Myotis lucifugus* over a stream. Journal of Mammalogy, 63:335–337.

Kurta, A., and R. H. Baker. 1990. Eptesicus fuscus. Mammalian Species, 356:1–10.

Kurta, A., and T. H. Kunz. 1988. Roosting metabolic rate and body temperature of male little brown bats (*Myotis lucifugus*) in summer. Journal of Mammalogy, 69:645–651.

Kurta, A., and M. E. Stewart. 1990. Parturition in the silver-haired bat, *Lasionycteris noctivagans*, with a description of neonates. The Canadian Field-Naturalist, 104:598–600.

Kurtén, B. 1968. Pleistocene mammals of Europe. Aldine Publishing Company, Chicago, 317 pp.

———. 1972. The age of mammals. Columbia University Press, New York, 250 pp.

Kurtén, B., and E. Anderson. 1980. Pleistocene mammals of North America. Columbia University Press, New York, 442 pp.

Kurtén, B., and R. Rausch. 1959a. Biometric comparisons between North American and European mammals. I. A comparison between Alaskan and Fennoscandian wolverine (*Gulo gulo* Linnaeus). Acta Arctica, 11:5–20.

———. 1959b. Biometric comparisons between North American and European mammals. II. A comparison between the northern lynxes of Fennoscandia and Alaska. Acta Arctica, 11:21–44.

Laban, C. 1966. A study of behavior in the deer mouse *Peromyscus maniculatus rubidus* Osgood during its 24-hour cycle of activity in a simulated natural habitat. Ph.D. dissert., Oregon State University, Corvallis, 94 pp.

Lahrman, F. W. 1980. A concentration of white-tailed jack rabbits. Blue Jay, 38:130.

Lair, H. 1985. Length of gestation in the red squirrel, *Tamiasciurus hudsonicus*. Journal of Mammalogy, 66:809–810.

Lambin, X. 1994. Natal philopatry, competition for resources, and inbreeding avoidance in Townsend's voles (*Microtus townsendii*). Ecology, 75:224–235.

Lampe, R. P. 1982. Food habits of badgers in east central Minnesota. The Journal of Wildlife Management, 46:790–795.

Lampman, B. H. 1947. A note on the predaceous habit of the water shrew. Journal of Mammalogy, 28:181.

Langer, W. L. 1964. The black death. Scientific American, 210(2):114–121.

Langley, W. M. 1979. Preference of the striped skunk and opossum for auditory over visual prey stimuli. Carnivore, 2:31–34.

Larrison, E. J. 1943. Feral coypus in the Pacific Northwest. The Murrelet, 24:3–9.

———. 1970. Washington mammals: their habits, identification, and distribution. The Seattle Audubon Society, [Seattle], 243 pp.

Larrison, E. J., and D. R. Johnson. 1973. Density changes and habitat affinities of rodents of shadscale and sagebrush associations. The Great Basin Naturalist, 33:255–264.

———. 1981. Mammals of Idaho. University Press of Idaho, Moscow, 166 pp.

Latham, R. M. 1952. The fox as a factor in the control of weasel populations. The Journal of Wildlife Management, 16:516–517.

Laughlin, J. M., and A. L. Cooper. 1973. A range extension of the kit fox in Oregon. The Murrelet, 54:23.

Laundré, J. W. 1989. Horizontal and vertical diameter of burrows of five small mammal species in southeastern Idaho. The Great Basin Naturalist, 49:646–649.

———. 1993. Effects of small mammal burrows on water infiltration in a cool desert environment. Oecologia, 94:43–

48.

Laurance, W. F., and T. D. Reynolds. 1984. Winter food preferences of captive-reared northern flying squirrels. The Murrelet, 65:20–22.

LaVal, R. K. 1973. Observations on the biology of Tadarida brasiliensis cynocephala in southeastern Louisiana. The American Midland Naturalist, 89:112–120.

Lawhon, D. K., and M. S. Hafner. 1981. Tactile discriminatory ability and foraging strategies in kangaroo rats and pocket mice (Rodentia: Heteromyidae). Oecologia, 50:303–309.

Lawler, R. M., and K. N. Geluso. 1986. Renal structure and body size in heteromyid rodents. Journal of Mammalogy, 67:367–372.

Laycock, W. A., and B. Z. Richardson. 1975. Long-term effects of pocket gopher control on vegetation and soils of a subalpine grassland. Journal of Range Management, 28:458–462.

Layne, J. N. 1951. The use of the tail by an opossum. Journal of Mammalogy, 32:464–465.

———. 1952. The os genitale of the red squirrel, Tamiasciurus. Journal of Mammalogy, 33:457–459.

———. 1954. The biology of the red squirrel, Tamiasciurus hudsonicus loquax (Bangs), in central New York. Ecological Monographs, 24:227–267.

———. 1958a. Notes on mammals of southern Illinois. The American Midland Naturalist, 60:219–254.

———. 1958b. Reproductive characteristics of the gray fox in southern Illinois. The Journal of Wildlife Management, 22:157–163.

———. 1968. Ontogeny. Pp. 148–253, in Biology of Peromyscus (Rodentia) (J. A. King, ed.). The American Society of Mammalogists, Special Publication, 2:1–593.

———. 1972. Tail autotomy in the Florida mouse, Peromyscus floridanus. Journal of Mammalogy, 53:62–71.

Layne, J. N., and W. H. McKeon. 1956. Some aspects of red fox and gray fox reproduction in New York. New York Fish and Game Journal, 3:44–74.

Leach, H. R. 1956. Food habits of the Great Basin deer herds of California. California Fish and Game, 42:243–308.

Le Boeuf, B. J. 1974. Male-male competition and reproductive success in elephant seals. American Zoologist, 14:163–176.

———. 1985. Elephant seals. The Boxwood Press, Pacific Grove, California, 48 pp.

———. 1994. Variation in the diving pattern of northern elephant seals with age, mass, sex, and reproductive condition. Pp. 237–252, in Elephant seals: population ecology, behavior, and physiology (B. J. Le Boeuf and R. M. Laws, eds.). University of California Press, Berkeley, 414 pp.

Le Boeuf, B. J., and R. M. Laws. 1994. Elephant seals: an introduction to the genus. Pp. 1–26, in Elephant seals: population ecology, behavior, and physiology (B. J. Le Boeuf and R. M. Laws, eds.). University of California Press, Berkeley, 414 pp.

Le Boeuf, B. J., and C. L. Ortiz. 1977. Composition of elephant seal milk. Journal of Mammalogy, 58:683–685.

Le Boeuf, B. J., D. P. Costa, and A. C. Huntley. 1986. Pattern and depth of dives in northern elephant seals, Mirounga angustirostris. Journal of Zoology (London), 208:1–7.

Le Boeuf, B. J., Y. Naito, A. C. Huntley, and T. Asaga. 1989.

Prolonged, continuous, deep diving by northern elephant seals. Canadian Journal of Zoology, 67:2514–2519.

Lechleitner, R. R. 1957. Reingestion in the black-tailed jack rabbit. Journal of Mammalogy, 38:481–485.

———. 1958a. Certain aspects of behavior of the black-tailed jack rabbit. The American Midland Naturalist, 60:145–155.

———. 1958b. Movements, density, and mortality in a black-tailed jack rabbit population. The Journal of Wildlife Management, 22:371–384.

———. 1959. Sex ratio, age classes and reproduction of the black-tailed jack rabbit. Journal of Mammalogy, 40:63–81.

———. 1969. Wild mammals of Colorado: their appearance, habits, distribution, and abundance. Pruett Publishing Company, Boulder, Colorado, 254 pp.

Leckenby, D. A. 1984. Elk use and availability of cover and forage habitat components in the Blue Mountains, northeast Oregon, 1976–1982. Oregon Department of Fish and Wildlife, Wildlife Research Report, 14:1–40.

Le Conte, J. 1831. Appendix of the American editor. Pp. 431–439, in The animal kingdom arranged in conformity with its organization. By the Baron Cuvier, and translated from the French, with notes and additions by H. M'Murtrie. 1:1–448 + 4 pls.

———. 1853. Descriptions of three new species of American Arvicolae, with remarks upon some other American rodents. Proceedings of the Academy of Natural Sciences of Philadelphia, 6:404–415.

———. 1856. Observations on the North American species of bats. Proceedings of the Academy of Natural Sciences of Philadelphia, 7:431–438.

Lee, M. R., and W. S. Modi. 1983. Chromosomes of Spilogale pygmaea and S. putorius leucoparia. Journal of Mammalogy, 64:493–495.

Lee, T. E., Jr., J. W. Bickham, and M. D. Scott. 1994. Mitochondrial DNA and allozyme analysis of North American pronghorn populations. The Journal of Wildlife Management, 58:307–318.

Leege, T. A., and R. M. Williams. 1967. Beaver productivity in Idaho. The Journal of Wildlife Management, 31:326–332.

Leffler, S. R. 1964. Fossil mammals from the Elk River Formation, Cape Blanco, Oregon. Journal of Mammalogy, 45:53–61.

LeFranc, M. N., Jr., M. B. Moss, K. A. Patnode, and W. C. Sugg III (eds.). 1987. Grizzly bear compendium. Interagency Grizzly Bear Committee, Washington, D.C., 540 pp.

Lehner, P. N. 1978. Coyote communication. Pp. 127–162, in Coyotes: biology, behavior, and management (M. Bekoff, ed.). Academic Press, New York, 384 pp.

———. 1981. Coyote-badger associations. The Great Basin Naturalist, 41:347–348.

Lemen, C. A. 1978. Seed size selection in heteromyids. Oecologia, 35:13–19.

Lemen, C. A., and P. W. Freeman. 1987. Competition for food and space in a heteromyid community in the Great Basin Desert. The Great Basin Naturalist, 47:1–6.

Lemos, J. C., and M. J. Willis. 1990. Population status and transplanting history of California bighorn sheep in Oregon. Proceedings of the Biennial Symposium, Northern Wild Sheep and Goat Council, 8:3–11.

Lentfer, J. W. 1955. A two-year study of the Rocky Mountain

goat in the Crazy Mountains, Montana. The Journal of Wildlife Management, 19:417–429.

Leonard, M. L., and M. B. Fenton. 1983. Habitat use by spotted bats (Euderma maculatum, Chiroptera: Vespertilionidae): roosting and foraging behaviour. Canadian Journal of Zoology, 61:1487–1491.

———. 1984. Echolocation calls of Euderma maculatum (Vespertilionidae): use in orientation and communication. Journal of Mammalogy, 65:122–126.

Leopold, A. 1937. Killing technique of the weasel. Journal of Mammalogy, 18:98–99.

Lesson, R. P. 1828. Dictionnaire classique d'histoire naturelle, 13:420 (not seen, cited in Miller and Kellogg, 1955).

Levenson, H., and R. S. Hoffmann. 1984. Systematic relationships among taxa of the Townsend chipmunk group. The Southwestern Naturalist, 29:157–168.

Levenson, H., R. S. Hoffmann, C. F. Nadler, L. Deutsch, and S. D. Freeman. 1985. Systematics of the Holarctic chipmunks (Tamias). Journal of Mammalogy, 66:219–242.

Levin, E. Y., and V. Flyger. 1971. Uroporphyrinogen III cosynthetase activity in the fox squirrel (Sciurus niger). Science, 174:59–60.

Lewis, S. E. 1993. Effect of climatic variation on reproduction by pallid bats (Antrozous pallidus). Canadian Journal of Zoology, 71:1429–1433.

———. 1994. Night roosting ecology of pallid bats (Antrozous pallidus) in Oregon. The American Midland Naturalist, 132:219–226.

Lewis, T. H. 1983. The anatomy and histology of the rudimentary eye of Neurotrichus. Northwest Science, 57:8–15.

Li, C. Y., C. Maser, Z. Maser, and B. A. Caldwell. 1986. Role of three rodents in nitrogen fixation in western Oregon: another aspect of mammal-mycorrhizal fungus-tree mutualism. The Great Basin Naturalist, 46:411–414.

Licht, P., and P. Leitner. 1967. Behavioral responses to high temperatures in three species of California bats. Journal of Mammalogy, 48:52–61.

Lichtenstein, M. H. C. 1830. Erläuterungen der Nachrichten des Franc. Hernandez von den vierfüssigen Thieren Neuspaniens. Koeniglich-preussische Akademie der Wissenschaften, Berlin, 1827:89–127 (not seen, cited in The National Union Catalog, 332:46).

———. 1831. Darstellung neuer oder wenig bekannter Säugethiere in Abbildungen und Beschreibungen von Fünf und Sechzig Arten auf Fünfzig Colorirten Steindrucktafeln nach den Originalen des Zoologischen Museums der Universität zu Berlin. C. Güderitz, Berlin, pl. 42 and corresponding text, unpaged.

Lidicker, W. Z., Jr. 1966. Ecological observations on a feral house mouse population declining to extinction. Ecological Monographs, 36:27–50.

———. 1985. Dispersal. Pp. 420–454, in Biology of New World Microtus (R. H. Tamarin, ed.). Special Publication, The American Society of Mammalogists, 8:1–893.

———. 1988. Solving the enigma of microtine "cycles." Journal of Mammalogy, 69:225–235.

Liers, E. E. 1951. Notes on the river otter (Lutra canadensis). Journal of Mammalogy, 32:1–9.

Lim, B. K. 1987. Lepus townsendii. Mammalian Species, 288:1–6.

Lindsay, S. L. 1982. Systematic relationship of parapatric tree squirrel species (Tamiasciurus) in the Pacific Northwest. Canadian Journal of Zoology, 60:2149–2156.

Lindsey, G. D. 1983. Rhodamine B: a systemic fluorescent marker for studying mountain beavers (Aplodontia rufa) and other animals. Northwest Science, 57:16–21.

Lindzey, F. 1987. Mountain lion. Pp. 657–668, in Wild furbearer management and conservation in North America (M. Novak, J. T. Baker, M. E. Obbard, and B. Malloch, eds.). Ontario Ministry of Natural Resources, Toronto, 1150 pp.

Lindzey, F. G. 1976a. Black bear population ecology. Ph.D. dissert., Oregon State University, Corvallis, 105 pp.

———. 1976b. Characteristics of the natal den of the badger. Northwest Science, 50:178–180.

———. 1978. Movement patterns of badgers in northwestern Utah. The Journal of Wildlife Management, 42:418–422.

———. 1982. Badger: Taxidea taxus. Pp. 653–663, in Wild mammals of North America: biology, management, and economics (J. A. Chapman and G. A. Feldhamer, eds.). The Johns Hopkins University Press, Baltimore, 1147 pp.

Lindzey, F. G., and E. C. Meslow. 1976a. Characteristics of black bear dens on Long Island, Washington. Northwest Science, 50:236–242.

———. 1976b. Winter dormancy in black bears in southwestern Washington. The Journal of Wildlife Management, 40:408–415.

———. 1977a. Home range and habitat use by black bears in southwestern Washington. The Journal of Wildlife Mangement, 41:413–425.

———. 1977b. Population characteristics of black bears on an island in Washington. The Journal of Wildlife Management, 41:408–412.

Lindzey, J. S. 1943. A study of Columbian black-tailed deer Odocoileus hemionus columbianus (Richardson). M.S. thesis, Oregon State College, Corvallis, 62 pp.

Linnaeus, C. 1758. Systema naturae per regna tria naturae, secundum classes, ordines, genera, species, cum characteribus differentiis, synonymis, locis. 10th ed. L. Salvii, Stockholm, 1:1–824.

———. 1766. Systema naturae per regna tria naturae, secundum classes, ordines, genera, species, cum characteribus, differentiis, synonymis, locis. Regnum Animale. 12th ed. Laurentii Salvii, Holmiae, 1(part 1):1–532.

———. 1771. Regni animalis. Pp. 521–552, in Appendix, Mantissa Plantarum altera. Uppsala, 588 pp.

Linscombe, G., N. Kinler, and R. J. Aulerich. 1982. Mink: Mustela vison. Pp. 629–643, in Wild mammals of North America: biology, management, and economics (J. A. Chapman and G. A. Feldhamer, eds.). The Johns Hopkins University Press, Baltimore, 1147 pp.

Linsdale, J. M. 1938. Environmental responses of vertebrates in the Great Basin. The American Midland Naturalist, 19:1–206.

———. 1946. The California ground squirrel: a record of observations made on the Hastings Natural History Reservation. University of California Press, Berkeley, 475 pp.

Linsdale, J. M., and L. P. Tevis, Jr. 1956. A five-year change in an assemblage of wood rat houses. Journal of Mammalogy, 37:371–374.

Liskop, K. S., R. M. F. S. Sadleir, and B. P. Saunders. 1981. Reproduction and harvest of wolverine (*Gulo gulo*) in British Columbia. Worldwide Furbearer Conference Proceedings, 1:469–477.

Litvaitis, J. A. 1990. Differential habitat use by sexes of snowshoe hares (*Lepus americanus*). Journal of Mammalogy, 71:520–523.

Litvaitis, J. A., A. G. Clark, and J. H. Hunt. 1986. Prey selection and fat deposits of bobcats (*Felis rufus*) during autumn and winter in Maine. Journal of Mammalogy, 67:389–392.

Livezey, B. C., and B. J. Verts. 1979. Estimates of age and age structure in Mazama pocket gophers, *Thomomys mazama*. The Murrelet, 60:38–41.

Livezey, R., and F. Evenden, Jr. 1943. Notes on the western red fox. Journal of Mammalogy, 24:500–501.

Livingston, S. D. 1987. Prehistoric biogeography of white-tailed deer in Washington and Oregon. The Journal of Wildlife Management, 51:649–654.

Llewellyn, L. M., and F. H. Dale. 1964. Notes on the ecology of the opossum in Maryland. Journal of Mammalogy, 45:113–126.

Llewellyn, L. M., and F. M. Uhler. 1952. The foods of fur animals of the Patuxent Research Refuge, Maryland. The American Midland Naturalist, 48:193–203.

Loehr, K. A., and A. C. Risser, Jr. 1977. Daily and seasonal activity patterns of the Belding ground squirrel in the Sierra Nevada. Journal of Mammalogy, 58:445–448.

Logan, K. A., and L. L. Irwin. 1985. Mountain lion habitats in the Big Horn Mountains, Wyoming. Wildlife Society Bulletin, 13:257–262.

Lokemoen, J. T., and H. F. Duebbert. 1976. Ferruginous hawk nesting ecology and raptor populations in northern South Dakota. Condor, 78:464–470.

Long, C. A. 1972. Taxonomic revision of the North American badger, *Taxidea taxus*. Journal of Mammalogy, 53:725–759.

———. 1973. Taxidea taxus. Mammalian Species, 26:1–4.

———. 1975. Molt in the North American badger, *Taxidea taxus*. Journal of Mammalogy, 56:921–924.

Long, C. A., and W. C. Kerfoot. 1963. Mammalian remains from owl-pellets in eastern Wyoming. Journal of Mammalogy, 44:129–131.

Lonsdale, E. M., B. Bradach, and E. T. Thorne. 1971. A telemetry system to determine body temperature in pronghorn antelope. The Journal of Wildlife Management, 35:747–751

Lord, G. H., A. C. Todd, and C. Kabat. 1956a. Studies on Errington's disease in muskrats. I. Pathological changes. American Journal of Veterinary Research, 17:303–306.

Lord, G. H., A. C. Todd, C. Kabat, and H. Mathiak. 1956b. Studies on Errington's disease in muskrats. II. Etiology. American Journal of Veterinary Research, 17:307–310.

Lord, J. K. 1863. Notes on two species of mammals. Proceedings of the Scientific Meetings of the Zoological Society of London, 1863:95–98.

Lord, R. D., Jr. 1960. Litter size and latitude in North American mammals. The American Midland Naturalist, 64:488–499.

———. 1961a. Potential life span of cottontails. Journal of Mammalogy, 42:99.

———. 1961b. Seasonal changes in roadside activity of cottontails. The Journal of Wildlife Management, 25:206–209.

———. 1961c. A population study of the gray fox. The American Midland Naturalist, 66:87–109.

Lorenz, J. R., and L. Coppinger. 1985. Introducing livestock-guarding dogs. Oregon State University Extension Service Circular, 1224:1–3.

Lotze, J-H., and S. Anderson. 1979. Procyon lotor. Mammalian Species, 119:1–8.

Loughlin, T. R., M. A. Perez, and R. L. Merrick. 1987. Eumetopias jubatus. Mammalian Species, 283:1–7.

Loughlin, T. R., A. S. Perlov, and V. A. Vladimirov. 1992. Range-wide survey and estimation of total number of Steller sea lions in 1989. Marine Mammal Science, 8:220–239.

Loughlin, T. R., et al. 1994. Status of the northern fur seal population in the United States during 1992. Pp. 9–28, *in* Fur seal inves- tigations, 1992 (E. H. Sinclair, ed.). United States Department of Commerce, National Oceanic and Atmospheric Administration Technical Memorandum, NMFS-AFSC-45:1–190.

Lovejoy, B. P. 1972. A capture-recapture analysis of a mountain beaver population in western Oregon. Ph.D. dissert., Oregon State University, Corvallis, 105 pp.

Lovejoy, B. P., and H. C. Black. 1979a. Movements and home range of the Pacific mountain beaver, Aplodontia rufa pacifica. The American Midland Naturalist, 101:393–402.

———. 1979b. Population analysis of the mountain beaver, *Aplodontia rufa pacifica*, in western Oregon. Northwest Science, 53:82–89.

Lovejoy, B. P., H. C. Black, and E. F. Hooven. 1978. Reproduction, growth, and development of the mountain beaver (*Aplodontia rufa pacifica*). Northwest Science, 52:323–328.

Lowery, G. H., Jr. 1974. The mammals of Louisiana and its adjacent waters. Lousiana State University Press, Baton Rouge, 565 pp.

Loy, W. G. 1976. Atlas of Oregon. University of Oregon Books, Eugene, 215 pp.

Luckens, M. M., and W. H. Davis. 1964. Bats: sensitivity to DDT. Science, 146:948.

Ludwig, D. R. 1984a. Microtus richardsoni. Mammalian Species, 223:1–6.

———. 1984b. *Microtus richardsoni* microhabitat and life history. Pp. 319–331, *in* Winter ecology of small mammals (J. F. Merritt, ed.). Special Publication of Carnegie Museum of Natural History, 10:1–380.

———. 1988. Reproduction and population dynamics of the water vole, *Microtus richardsoni*. Journal of Mammalogy, 69:532–541.

Luman, I. D. 1964. Oregon antelope report, 1964. Transactions 1964 Interstate Antelope Conference, 9 unnumbered pp.

Lyall-Watson, M. 1963. A critical re-examination of food "washing" behaviour in the raccoon (*Procyon lotor* Linn.). Proceedings of the Zoological Society of London, 141:371–393.

Lydekker, R. 1915. Catalogue of the ungulate mammals in the British Museum (Natural History), 4:1–438.

Lyman, C. P. 1943. Control of coat color in the varying hare *Lepus americanus* Erxleben. Bulletin of the Museum of Comparative Zoology, 93:391–461.

Lyman, R. L. 1988. Significance for wildlife management of the Late Quaternary biogeography of mountain goats

(*Oreamnos americanus*) in the Pacific Northwest, U.S.A. Arctic and Alpine Research, 20:13–23.

———. 1994. The Olympic mountain goat controversy: a different perspective. Conservation Biology, 8:898–901.

Lyon, M. W., Jr. 1904. Classification of the hares and their allies. Smithsonian Miscellaneous Collection, 45:321–447.

MacArthur, R. A. 1978. Winter movements and home range of the muskrat. The Canadian Field-Naturalist, 92:345–349.

MacArthur, R. A., and M. Aleksiuk. 1979. Seasonal microenvironments of the muskrat (*Ondatra zibethicus*) in a northern marsh. Journal of Mammalogy, 60:146–154.

MacDonald, D. W. 1980. The red fox, *Vulpes vulpes*, as a predator upon earthworms, *Lumbricus terrestris*. Zeitschrift für Tierphysiologie, 52:171–200 (not seen, cited in Voigt, 1987).

Mace, R. U. 1954. Oregon's pronghorn antelope. Oregon State Game Commission, Wildlife Bulletin, 1:1–26.

———. 1969. Bighorns in Oregon. Oregon State Game Commission Bulletin, 24(8):3–5.

———. 1970. Oregon's furbearing animals. Oregon State Game Commission. Wildlife Bulletin, 6:1–82.

MacFarlane, J. D. 1977. Some aspects of reproduction in female *Microtus townsendii*. M.A. thesis, University of British Columbia, Vancouver, 97 pp.

MacFarlane, J. D., and J. M. Taylor. 1982*a*. Nature of estrus and ovulation in *Microtus townsendii* (Bachman). Journal of Mammalogy, 63:104–109.

———. 1982*b*. Pregnancy and reproductive performance in the Townsend's vole, *Microtus townsendii* (Bachman). Journal of Mammalogy, 63:165–168.

MacIsaac, G. L. 1977. Reproductive correlates of the hip gland in voles (*Microtus townsendii*). Canadian Journal of Zoology, 55:939–941.

Mackie, R. J., K. L. Hamlin, and D. F. Pac. 1982. Mule deer: *Odocoileus hemionus*. Pp. 862–877, *in* Wild mammals of North America: biology, management, and economics (J. A. Chapman and G. A. Feldhamer, eds.). The Johns Hopkins University Press, Baltimore, 1147 pp.

MacLulich, D. A. 1937. Fluctuations in the numbers of the varying hare (*Lepus americanus*). University of Toronto Studies, Biology Series, 43:1–136 (not seen, cited in Odum, 1971).

MacMillen, R. E. 1964. Population ecology, water relations, and social behavior of a southern California rodent fauna. University of California Publications in Zoology, 71:1–59.

———. 1972. Water economy of nocturnal desert rodents. Symposia of the Zoological Society of London, 31:147–174.

———. 1983. Water regulation in *Peromyscus*. Journal of Mammalogy, 64:38–47.

MacMillen, R. E., and E. A. Christopher. 1974. The water relations of two populations of noncaptive desert rodents. Pp. 117–137, *in* Environmental physiology of desert organisms (N. F. Hadley, ed.). Halstead Press, New York, 283 pp.

MacNab, J. A., and J. C. Dirks. 1941. The California red-backed mouse in the Oregon Coast range. Journal of Mammalogy, 22:174–180.

Madison, D. M. 1985. Activity rhythms and spacing. Pp. 373–419, *in* Biology of New World *Microtus*. (R. H. Tamarin, ed.). Special Publication, The American Society of Mammalogists, 8:1–893.

Main, M. B. 1994. Advantages of habitat selection and sexual segregation in mule and white-tailed deer. Ph.D. dissert., Oregon State University, Corvallis, 121 pp.

Main, M. B., and B. E. Coblentz. 1990. Sexual segregation among ungulates: a critique. Wildlife Society Bulletin, 18:204–210.

Manning, R. W., and J. K. Jones, Jr. 1989. Myotis evotis. Mammalian Species, 329:1–5.

Manning, T., W. D. Edge, and J. O. Wolff. 1995. Evaluating population-size estimators: an empirical approach. Journal of Mammalogy, 76:1149–1158.

Marcot, B. G. 1984. Winter use of some northwestern California caves by western big-eared bats and long-eared myotis. The Murrelet, 65:46.

Markham, O. D., and F. W. Whicker. 1973. Seasonal data on reproduction and body weights of pikas (*Ochotona princeps*). Journal of Mammalogy, 54:496–498.

Markley, M. H., and C. F. Bassett. 1942. Habits of captive marten. The American Midland Naturalist, 28:605–616.

Marquez, M. 1988. Leadership behavior and foraging strategies of a herd of Roosevelt elk inhabiting the Oregon Coast Range. M.S. thesis, Oregon State University, Corvallis, 72 pp.

Marsden, H. M., and N. R. Holler. 1964. Social behavior in confined populations of the cottontail and the swamp rabbit. Wildlife Monographs, 13:1–39.

Marshall, D. B. 1992. Sensitive vertebrates of Oregon. Oregon Department of Fish and Wildlife, Portland (species accounts paged separately).

Marshall, L. G. 1984. Monotremes and marsupials. Pp. 59–115, *in* Orders and families of Recent mammals of the world (S. Anderson and J. K. Jones, Jr., eds.). John Wiley & Sons, New York, 686 pp.

Marshall, L. G., J. A. Case, and M. O. Woodburne. 1990. Phylogenetic relationships of the families of marsupials. Current Mammalogy, 2:33–505

Marshall, W. H. 1936. A study of the winter activities of the mink. Journal of Mammalogy, 17:382–392.

Martell, A. M., and A. Radvanyi. 1977. Changes in small mammal populations after clearcutting of northern Ontario black spruce forest. The Canadian Field-Naturalist, 91:41–46.

Marti, C. D. 1969. Some comparisons of the feeding ecology of four owls in north-central Colorado. The Southwestern Naturalist, 14:163–170.

Marti, C. D., and C. E. Braun. 1975. Use of tundra habitats by prairie falcons in Colorado. Condor, 77:213–214.

Martin, C. O., and D. J. Schmidly. 1982. Taxonomic review of the pallid bat, Antrozous pallidus (Le Conte). Special Publications, The Museum, Texas Tech University, 18:1–48.

Martin, J. E. 1984. A survey of Tertiary species of *Perognathus* (Perognathinae) and a description of a new genus of Heteromyinae. Carnegie Museum of Natural History, Special Publication, 9:90–121.

Martin, K. J. 1994. Movements and habitat associations of northern flying squirrels in the central Oregon Cascades. M.S. thesis, Oregon State University, Corvallis, 46 pp.

Martin, P. 1971. Movements and activities of the mountain beaver (*Aplodontia rufa*). Journal of Mammalogy, 52:717–

723.

Martin, S. K., and R. H. Barrett. 1991. Resting site selection by marten at Sagehen Creek, California. Northwest Naturalist, 72:37–42.

Martinsen, D. L. 1968. Temporal patterns in the home range of chipmunks (*Eutamias*). Journal of Mammalogy, 49:83–91.

Mascarello, J. T. 1978. Chromosomal, biochemical, mensural, juvenile, and cranial variation in desert woodrats (*Neotoma lepida*). Journal of Mammalogy, 59:477–495.

Maser, C. 1965. Nest of bushy-tailed wood rat fifty feet above the ground. The Murrelet, 46:46.

———. 1966*a*. A second *Neotoma cinerea* nest fifty feet above the ground. The Murrelet, 47:72.

———. 1966*b*. Life histories and ecology of *Phenacomys albipes*, *Phenacomys longicaudus*, *Phenacomys silvicola*. M.S. thesis, Oregon State University, Corvallis, 221 pp.

———. 1966*c*. Notes on a captive *Sorex vagrans*. The Murrelet, 47:51–53.

———. 1967. A live trap for microtines. The Murrelet, 48:58.

———. 1969. Bacula of *Tamiasciurus*. The Murrelet, 50:10–11.

———. 1973. Notes on predators upon cicindelids. Cicindela, 5:21–23.

———. 1975. Characters useful in identifying certain western Oregon mammals. Northwest Science, 49:158–159.

Maser, C., and E. D. Brodie, Jr. 1966. A study of owl pellet contents from Linn, Benton and Polk counties, Oregon. The Murrelet, 47:9–14.

Maser, C., and C. Brown. 1972. Bacula of two western moles. Northwest Science, 46:319–321.

Maser, C., and J. F. Franklin. 1974. Checklist of vertebrate animals of the Cascade Head Experimental Forest. United States Forest Service Research Bulletin, PNW–51:1–32.

Maser, C., and E. W. Hammer. 1972. A note on the food habits of barn owls in Klamath County, Oregon. The Murrelet, 53:28.

Maser, C., and E. F. Hooven. 1969. A new locality record of *Phenacomys albipes*. The Murrelet, 50:22.

———. 1970*a*. Dental abnormalities in *Microtus longicaudus*. The Murrelet, 51:11.

———. 1970*b*. Low altitude records of *Microtus richardsoni* in the Oregon Cascades. The Murrelet, 51:12.

———. 1974. Notes on the behavior and food habits of captive Pacific shrews, *Sorex pacificus pacificus*. Northwest Science, 48:81–95.

Maser, C., and M. L. Johnson. 1967. Notes on the white-footed vole (*Phenacomys albipes*). The Murrelet, 48:23–27.

Maser, C., and Z. Maser. 1988. Mycophagy of red-backed voles, *Clethrionomys californicus* and *C. gapperi*. The Great Basin Naturalist, 48:269–273.

Maser, C., and C. Shaver. 1976. Acute incisor malocclusion in a Townsend ground squirrel. The Murrelet, 57:17–18.

Maser, C., and G. S. Strickler. 1978. The sage vole, *Lagurus curtatus*, as an inhabitant of subalpine sheep fescue, *Festuca ovina*, communities on Steens Mountain—an observation and interpretation. Northwest Science, 52:276–284.

Maser, C., and R. M. Storm. 1970. A key to the Microtinae of the Pacific Northwest (Oregon, Washington, Idaho). OSU Book Stores, Inc., Corvallis, Oregon, 162 pp.

Maser, C., and D. E. Toweill. 1984. Bacula of mountain lion,

Felis concolor, and bobcat, *F. rufus*. Journal of Mammalogy, 65:496–497.

Maser, C., R. Anderson, and E. L. Bull. 1981*a*. Aggregation and sex segregation in northern flying squirrels in northeastern Oregon, an observation. The Murrelet, 62:54–55.

Maser, C., E. W. Hammer, and S. H. Anderson. 1970. Comparative food habits of three owl species in central Oregon. The Murrelet, 51:29–33.

———. 1971. Food habits of the burrowing owl in central Oregon. Northwest Science, 45:19–26.

Maser, C., E. W. Hammer, and M. L. Johnson. 1969. Abnormal coloration in *Microtus montanus*. The Murrelet, 50:39.

Maser, C., Z. Maser, and R. Molina. 1988. Small-mammal mycophagy in rangelands of central and southeastern Oregon. Journal of Range Management, 41:309–312.

Maser, C., J. M. Trappe, and R. A. Nussbaum. 1978*a*. Fungal-small mammal interrelationships with emphasis on Oregon coniferous forests. Ecology, 59:799–809.

Maser, C., J. M. Trappe, and D. C. Ure. 1978*b*. Implications of small mammal mycophagy to the management of western coniferous forests. Transactions of the North American Wildlife and Natural Resources Conference, 43:78–88.

Maser, C., Z. Maser, J. W. Witt, and G. Hunt. 1986. The northern flying squirrel: a mycopophagist in southwestern Oregon. Canadian Journal of Zoology, 64:2086–2089.

Maser, C., B. R. Mate, J. F. Franklin, and C. T. Dyrness. 1981*b*. Natural history of Oregon Coast mammals. United States Department of Agriculture, Forest Service, General Technical Report, PNW–133:1–496.

Maser, C., S. Shaver, C. Shaver, and B. Price. 1980. A note on the food habits of the barn owl in Malheur County, Oregon. The Murrelet, 61:78–80.

Maser, C., E. W. Hammer, C. Brown, R. E. Lewis, R. L. Rausch, and M. L. Johnson. 1974. The sage vole, *Lagurus curtatus* (Cope, 1868), in the Crooked River National Grassland, Jefferson County, Oregon. A contribution to its life history and ecology. Säugetierkundliche Mitteilungen, 22:193–222.

Maser, Z., and C. Maser. 1987. Notes on mycophagy of the yellow-pine chipmunk (*Eutamias amoenus*) in northeastern Oregon. The Murrelet, 68:24–27.

Maser, Z., C. Maser, and J. M. Trappe. 1985. Food habits of the northern flying squirrel (*Glaucomys sabrinus*) in Oregon. Canadian Journal of Zoology, 63:1084–1088.

Mason, E. 1952. Food habits and measurements of Hart Mountain antelope. The Journal of Wildlife Management, 16:387–389.

Mate, B. R. 1969. Northern extension of range of shore occupation by *Mirounga angustirostris*. Journal of Mammalogy, 50:639.

———. 1970. Oldest tagged northern elephant seal recovered in Oregon. California Fish and Game, 56:137.

———. 1973. Population kinetics and related ecology of the northern sea lion (*Eumetopias jubatus*) and the California sea lion (*Zalophus californianus*), along the Oregon coast. Ph.D. dissert., University of Oregon, Eugene, 94 pp.

———. 1975. Annual migrations of the sea lions *Eumetopias jubatus* and *Zalophus californianus* along the Oregon coast. Rapports et Procès-verbaux des Réunions. Conseil International pour l'Exploration de la Mer, 169:455–461.

————. 1978. California sea lion. Pp. 172–177, *in* Marine mammals of eastern North Pacific and Arctic waters (D. Haley, ed.). Pacific Search Press, Seattle, Washington, 256 pp.

————. 1981. Marine mammals. Pp. 372–458, *in* Natural history of Oregon coast mammals (C. Maser, B. R. Mate, J. F. Franklin, and C. T. Dyrness). United States Department of Agriculture, Forest Service, General Technical Report, PNW-133:1–496.

————. 1982*a*. History and present status of the California sea lion, *Zalophus californianus*. Pp. 303–309, *in* Mammals in the seas. Food and Agriculture Organization of the United Nations, Fisheries Series 5, 4:1–531.

————. 1982*b*. History and present status of the northern (Steller) sea lion, *Eumetopias jubatus*. Pp. 311–317, *in* Mammals in the seas. Food and Agriculture Organization of the United Nations, Fisheries Series 5, 4:1–531.

Matthew, W. D. 1902. List of the Pleistocene fauna from Hay Springs, Nebraska. Bulletin of the American Museum of Natural History, 16:317–322.

Matthews, P. E., and V. L. Coggins. 1994 [1995]. Status and history of mountain goats in Oregon. Biennial Symposium of the Northern Wild Sheep and Goat Council, 9:69–74.

Matthey, R. 1958. Un nouveau type de détermination chromosomique du sexe chez les mammifères *Ellobius lutgescens* Th. et *Microtus* (*Chilotus*) *oregoni* Bachm. (Muridés-Microtinés). Experimentia, 14:240–241.

Mawby, J. E. 1960. A new American occurrence of *Heterosorex* Gaillard. Journal of Paleontology, 34:950–956.

Maxson, J. H. 1928. Merychippus isonesus (Cope) from the later Tertiary of the Crooked River Basin, Oregon. Pp. 55–58, *in* Papers concerning the Palaeontology of the Cretaceous and late Tertiary of Oregon, of the Pliocene of northwestern Nevada, and of the late Miocene and Pleistocene of California. Carnegie Institution of Washington, Contributions to Palaeontology, 393:1–58.

Maxwell, C. S., and M. L. Morton. 1975. Comparative thermo-regulatory capabilities of neonatal ground squirrels. Journal of Mammalogy, 56:821–828.

May, W. B. 1896. California game "marked down." Southern Pacific R. R. Co., San Francisco, California, 64 pp. (not seen, cited in Miller and Kellogg, 1955).

Mayr, E. 1965. Animal species and evolution. The Belknap Press of Harvard University Press, Cambridge, Massachusetts, 797 pp.

McAllister, J. A., and R. S. Hoffmann. 1988. Phenacomys intermedius. Mammalian Species, 305:1–8.

McAllister, K. R. 1995. Washington State recovery plan for the pygmy rabbit. Washington Department of Fish and Wildlife, Olympia, 73 pp.

McArthur, L. A. 1992. Oregon geographic names. Sixth ed. revised by L. L. McArthur. Oregon Historical Society Press, Portland, 957 pp.

McCabe, R. A. 1949. Notes on live-trapping mink. Journal of Mammalogy, 30:416–423.

McCabe, T. T., and B. D. Blanchard. 1950. Three species of Peromyscus. Rood Associates, Santa Barbara, California, 136 pp.

McCarley, H. 1975. Long-distance vocalizations of coyotes (*Canis latrans*). Journal of Mammalogy, 56:847–856.

McCarty, R. 1978. Onychomys leucogaster. Mammalian Species, 87:1–6.

McCleneghan, K., and J. A. Ames. 1976. A unique method of prey capture by a sea otter, *Enhydra lutris*. Journal of Mammalogy, 57:410–412.

McCollum, M. T. 1973. Habitat utilization and movements of black bears in southwest Oregon. M.S. thesis, California State University, Humboldt, [Arcata], 67 pp.

McComb, W. C., K. McGarigal, and R. G. Anthony. 1993. Small mammal and amphibian abundance in streamside and upslope habitats of mature Douglas-fir stands, western Oregon. Northwest Science, 67:7–15.

McCord, C. M., and J. E. Cardoza. 1982. Bobcat and lynx: *Felis rufus* and *F. lynx*. Pp. 728–766, *in* Wild mammals of North America: biology, management, and economics (J. A. Chapman and G. A. Feldhamer, eds.). The Johns Hopkins University Press, Baltimore, 1147 pp.

McCornack, E. C. 1914. A study of Oregon Pleistocene. University of Oregon Bulletin, new series, 12(2):1–16.

————. 1920. Contributions to the Pleistocene history of Oregon. University of Oregon Leaflet Series, Geological Bulletin, 6(3, part 2): 1–23.

McCracken, G. F. 1984. Communal nursing in Mexican free-tailed bat maternity colonies. Science, 223:1090–1091.

McCrady, E., Jr. 1938. The embryology of the opossum. American Anatomical Memoirs, 16:1–233.

McCulloch, C. Y., and J. M. Inglis. 1961. Breeding periods of the Ord kangaroo rat. Journal of Mammalogy, 42:337–344.

McCullough, C. R., and E. K. Fritzell. 1984. Ecological observations of eastern spotted skunks on the Ozark Plateau. Transactions, Missouri Academy of Sciences, 18:25–32.

McCullough, D. R. 1960. An ecological study of the Columbian black-tailed deer in a logged environment. M.S. thesis, Oregon State College, Corvallis, 63 pp.

————. 1964. Relationship of weather to migratory movements of black-tailed deer. Ecology, 45:249–256.

————. 1965. Sex characteristics of black-tailed deer hooves. The Journal of Wildlife Management, 29:210–212.

————. 1969. The tule elk: its history, behavior, and ecology. University of California Publications in Zoology, 88:1–209.

————. 1993. Variation in black-tailed deer herd composition counts. The Journal of Wildlife Management, 57:890–897.

————. 1994. What do herd composition counts tell us? Wildlife Society Bulletin, 22:295–300.

McCullough, D. R., and D. H. Hirth. 1988. Evaluation of the Petersen-Lincoln estimator for a white-tailed deer population. The Journal of Wildlife Management, 52:534–544.

McCullough, D. R., F. W. Weckerly, P. I. Garcia, and R. R. Evett. 1994. Sources of inaccuracy in black-tailed deer herd composition counts. The Journal of Wildlife Management, 58:319–329.

McCully, H. 1967. The broad-handed mole, *Scapanus latimanus*, in a marine littoral environment. Journal of Mammalogy, 48:480–481.

McCutchen, H. E. 1969. Age determination of pronghorns by the incisor cementum. The Journal of Wildlife Management, 33:172–175.

McDaniel, L. L. 1967. Merriam's shrew in Nebraska. Journal of Mammalogy, 48:493.

McDonald, D. L., and L. G. Forslund. 1978. The development

of social preferences in the voles *Microtus montanus* and *Microtus canicaudus*: effects of cross-fostering. Behavioral Biology, 22:497–508.

McEwan, E. H., N. Aitchison, and P. E. Whitehead. 1974. Energy metabolism of oiled muskrats. Canadian Journal of Zoology, 52:1057–1062.

McGrew, J. C. 1979. Vulpes macrotis. Mammalian Species, 123:1–6.

McGrew, P. O. 1941. The Aplodontoidea. Field Museum of Natural History, Geology Series Publication, 510, 9:1–30.

McIntire, P. W. 1980. Population structure and microhabitat utilization of *Eutamias siskiyou* within a variously treated shelterwood cut. M.A. thesis, Southern Oregon State College, Ashland, 112 pp.

McKay, D. O. 1975. Dynamics of a population of mountain cottontail rabbits in central Oregon. M.S. thesis, Oregon State University, Corvallis, 63 pp.

McKay, D. O., and B. J. Verts. 1978a. Habitat preference and dispersion of Nuttall's cottontails. Northwest Science, 52:363–368.

———. 1978b. Estimates of some attributes of a population of Nuttall's cottontails. The Journal of Wildlife Management, 42:159–168.

McKean, J. W., and I. D. Luman. 1964. Oregon's 1962 decline in mule deer harvest. Proceedings of the Annual Conference of the Western Association of Game and Fish Commissioners, 44:177–180.

McKeever, S. 1963. Seasonal changes in body weight, reproductive organs, pituitary, adrenal glands, thyroid gland, and spleen of the Belding ground squirrel (*Citellus beldingi*). American Journal of Anatomy, 113:153–173.

———. 1964a. The biology of the golden-mantled ground squirrel, *Citellus lateralis*. Ecological Monographs, 34:383–401.

———. 1964b. Food habits of the pine squirrel in northeastern California. The Journal of Wildlife Management, 28:402–404.

———. 1965. Reproduction in *Citellus beldingi* and *C. lateralis* in north-eastern California. Journal of Reproduction and Fertility, 9:384–385.

———. 1966. Reproduction in *Citellus beldingi* and *Citellus lateralis* in northeastern California. Symposia of the Zoological Society of London, 15:365–385.

McLaughlin, C. A. 1984. Protogomorph, sciuromorph, castorimorph, myomorph (geomyoid, anomaluroid, pedetoid, and ctenodactyloid) rodents. Pp. 267–288, *in* Orders and families of Recent mammals of the world (S. Anderson and J. K. Jones, Jr., eds.). John Wiley & Sons, New York, 686 pp.

McLeod, J. A. 1948. Preliminary studies on muskrat biology in Manitoba. Transactions of the Royal Society of Canada, series 3, 42:81–95 (not seen, cited in Errington, 1963).

———. 1950. A consideration of muskrat populations and population trends in Manitoba. Transactions of the Royal Society of Canada, 44:69–79 (not seen, cited in Errington, 1963).

M'Closkey, R. T. 1972. Temporal changes in populations and species diversity in a California rodent community. Journal of Mammalogy, 53:657–676.

McMahon, E. E., C. C. Oakley, and S. P. Cross. 1981. First record of the spotted bat (*Euderma maculatum*) from Oregon. The Great Basin Naturalist, 41:270.

McManus, J. J. 1969. Temperature regulation in the opossum, *Didelphis marsupialis virginiana*. Journal of Mammalogy, 50:550–558.

———. 1970. Behavior of captive opossums, Didelphis marsupialis virginia-na. The American Midland Naturalist, 84:144–169.

———. 1974. Didelphis virginiana. Mammalian Species, 40:1–6.

McMillin, J. H., and D. A. Sutton. 1972. Additional information on chromosomes of *Aplodontia rufa* (Sciuridae). The Southwestern Naturalist, 17:307–308.

McNeil, W. L. 1956. The pallid bat discovered near Spokane, Washington. The Murrelet, 37:34–35.

Mead, J. I. 1987. Quaternary records of pika, *Ochotona*, in North America. Boreas, 16:165–171.

Mead, R. A. 1967. Age determination in the spotted skunk. Journal of Mammalogy, 48:606–616.

———. 1968a. Reproduction in western forms of the spotted skunk (genus *Spilogale*). Journal of Mammalogy, 49:373–390.

———. 1968b. Reproduction in eastern forms of the spotted skunk (genus *Spilogale*). Journal of Zoology, 156:119–136.

Meagher, M. 1986. Bison bison. Mammalian Species, 266:1–8.

Mearns, E. A. 1890. Description of supposed new species and subspecies of mammals, from Arizona. Bulletin of the American Museum of Natural History, 2:277–307.

———. 1891. Observations on the North American badgers, with especial reference to the forms found in Arizona, with descriptions of a new subspecies from southern California. Bulletin of the American Museum of Natural History, 3:239–251.

———. 1896. Perliminary description of a new subgenus and six new species and subspecies of hares, from the Mexican Border of the United States. Proceedings of the United States National Museum, 18:551–565.

———. 1907. Mammals of the Mexican boundary of the United States: a descriptive catalogue of the species of mammals occurring in that region; with a general summary of the natural history, and a list of trees. Bulletin of the United States National Museum, 56:1–501.

Mech, L. D. 1970. The wolf: the ecology and behavior of an endangered species. The Natural History Press, Garden City, New York, 384 pp.

———. 1973. Canada lynx invasion of Minnesota. Biological Conservation, 5:151–152.

———. 1974. Canis lupus. Mammalian Species, 37:1–6.

———. 1977. Productivity, mortality, and population trends of wolves in northeastern Minnesota. Journal of Mammalogy, 58:559–574.

———. 1980. Age, sex, reproduction, and spatial organization of lynxes colonizing northeastern Minnesota. Journal of Mammalogy, 61:261–267.

Mech, L. D., and L. D. Frenzel, Jr. 1971. An analysis of the age, sex, and condition of deer killed by wolves in northeastern Minnesota. United States Department of Agriculture, Forest Service, Research Paper, NC–52:35–59.

Mech, L. D., K. L. Heezen, and D. B. Siniff. 1966. Onset and cessation of activity in cottontails rabbits and snowshoe hares in relation to sunset and sunrise. Animal Behaviour, 14:410–413.

Mech, L. D., L. D. Frenzel, Jr., R. R. Ream, and J. W. Winship. 1971. Movements, behavior, and ecology of timber wolves in northeastern Minnesota. United States Department of Agriculture, Forest Service, Research Paper, NC–52:1–35.

Meehan, T. E. 1978. The occurrence, energetic significance and initiation of spontaneous torpor in the Great Basin pocket mouse, *Perognathus parvus*. Dissertation Abstracts International, 38B:5169.

Méhely, L. 1900. Magyarország denevéreinek monographiája [Monographia Chiropterorum Hungariae]. V. Hornyánsky, Budapest, 372 pp. + 22 pls. (not seen, cited in The National Union Catalog, 373:457).

Mehrer, C. F. 1976. Gestation period in the wolverine, *Gulo gulo*. Journal of Mammalogy, 57:570.

Meier, K. E. 1983. Habitat use by opossums in an urban environment. M.S. thesis, Oregon State University, Corvallis, 69 pp.

Melquist, W. E., and A. E. Dronkert. 1987. River otter. Pp. 627–641, *in* Wild furbearer management and conservation in North America (M. Novak, J. A. Baker, M. E. Obbard, and B. Malloch, eds.). Ontario Ministry of Natural Resources, Toronto, 1150 pp.

Melquist, W. E., and M. G. Hornocker. 1983. Ecology of river otters in west central Idaho. Wildlife Monographs, 83:1–60.

Melquist, W. E., J. S. Whitman, and M. G. Hornocker. 1981. Resource partitioning and coexistence of sympatric mink and river otter populations. Worldwide Furbearer Conference Proceedings, 1:187–220.

Meredith, D. H. 1972. Subalpine cover associations of Eutamias amoenus and Eutamias townsendii in the Washington Cascades. The American Midland Naturalist, 88:348–357.

Merriam, C. H. 1884a. Descriptions of a new genus and species of the Soricidae. Atophyrax bendirii Merriam. Transactions of the Linnean Society of New York, 2:217–225.

———. 1884b. The mammals of the Adirondack region, northeastern New York. 1974 Reprint Edition by Arno Press, Inc. of the L. S. Foster, New York edition, 316 pp.

———. 1886. Description of a new species of bat from the western United States (*Vespertilio ciliolabrum* sp. nov.). Proceedings of the Biological Society of Washington, 4:1–4.

———. 1888a. Description of a new Spermophile from California. Annals of the New York Academy of Sciences, 4:317–321.

———. 1888b. Description of a new species of meadow mouse from the Black Hills of Dakota. The American Naturalist, 22:934–935.

———. 1888c. Description of a new fox from southern California. Proceedings of the Biological Society of Washington, 4:135–138.

———. 1889a. Preliminary revision of the North American pocket mice (genera Perognathus et Cricetodipus auct.) with descriptions of new species and subspecies and a key to the known forms. North American Fauna, 1:1–37.

———. 1889b. Description of a new genus (Phenacomys) and four new species of Arvicolinae. Pp. 27–35, *in* Descriptions of fourteen new species and one new genus of North American mammals. North American Fauna, 2:1–52.

———. 1889c. Description of a new species of ground squirrel from the arid lands of the Southwest. Pp. 19–21, *in* Descriptions of fourteen new species and one new genus of North American mammals. North American Fauna, 2:1–52.

———. 1890a. Annotated list of mammals of the San Francisco Mountain plateau and desert of the little Colorado in Arizona, with notes on their vertical distribution, and descriptions of new species. North American Fauna, 3:43–86.

———. 1890b. Contribution toward a revision of the little striped skunks of the genus Spilogale, with descriptions of seven new species. Pp. 1–16, *in* Descriptions of twenty-six new species of North American mammals. North American Fauna, 4:1–60.

———. 1890c. Description of a new marten (Mustela caurina) from the Northwest Coast Region of the United States. Pp. 27–29, *in* Descriptions of twenty-six new species of North American mammals. North American Fauna, 4:1–60.

———. 1890d. Descriptions of three new kangaroo rats, with remarks on the identity of Dipodomys ordii of Woodhouse. Pp. 41–49, *in* Descriptions of twenty-six new species of North American mammals. North American Fauna, 4:1–60.

———. 1890e. Descriptions of two new species of Evotomys from the Pacific Coast Region of the United States. Pp. 25–26, *in* Descriptions of twenty-six new species of North American mammals. North American Fauna, 4:1–60.

———. 1891. Results of a biological reconnaissance of south-central Idaho. North American Fauna, 5:1–416.

———. 1892. The geographic distribution of life in North America with special reference to the Mammalia. Proceedings of the Biological Society of Washington, 7:1–64.

———. 1894a. Descriptions of eight new pocket mice (genus Perognathus). Proceedings of the Academy of Natural Sciences of Philadelphia, 46:262–268.

———. 1894b. Preliminary descriptions of eleven new kangaroo rats of the genera Dipodomys and Perodipus. Proceedings of the Biological Society of Washington, 9:109–116.

———. 1895a. Synopsis of the American shrews of the genus Sorex. North American Fauna, 10:57–98.

———. 1895b. The earliest available name for the mountain goat. Science, new series, 1:19.

———. 1896a. The mammals of Mount Mazama, Oregon. Mazama, 1:204–251.

———. 1896b. Synopsis of the weasels of North America. North American Fauna, 11:1–45.

———. 1897a. A new bat of the genus *Antrozous* from California. Proceedings of the Biological Society of Washington, 11:179–180.

———. 1897b. *Cervus roosevelti*, a new elk from the Olympics. Proceedings of the Biological Society of Washington, 11:271–275.

———. 1897c. Description of a new flying squirrel from Ft. Klamath, Oregon. Proceedings of the Biological Society of Washington, 11:225.

———. 1897d. Descriptions of eight new pocket gophers of the genus *Thomomys*, from Oregon, California, and Nevada. Proceedings of the Biological Society of Washington, 11:213–216.

———. 1897e. Descriptions of two new red backed mice (*Evotomys*) from Oregon. Proceedings of the Biological

Society of Washington, 11:71–72.

———. 1897*f*. Notes on the chipmunks of the genus *Eutamias* occurring west of the east base of the Cascade-Sierra system, with descriptions of new forms. Proceedings of the Biological Society of Washington, 11:189–212.

———. 1897*g*. Revision of the coyotes or prairie wolves, with descriptions of new forms. Proceedings of the Biological Society of Washington, 11:19–33.

———. 1897*h*. Three new jumping mice (*Zapus*) from the northwest. Proceedings of the Biological Society of Washington, 11:103–104.

———. 1897*i*. Two new moles from California and Oregon. Proceedings of the Biological Society of Washington, 11:101–102.

———. 1897*j*. The voles of the subgenus *Chilotus*, with descriptions of new species. Proceedings of the Biological Society of Washington, 11:73–75.

———. 1898*a*. Descriptions of six new ground squirrels from the western United States. Proceedings of the Biological Society of Washington, 12:69–71.

———. 1898*b*. The earliest generic name for the North American deer, with descriptions of five new species and subspecies. Proceedings of the Biological Society of Washington, 12:99–104.

———. 1899*a*. Descriptions of six new rodents of the genera *Aplodontia* and *Thomomys*. Proceedings of the Biological Society of Washington, 13:19–21.

———. 1899*b*. Results of a biological survey of Mount Shasta California. North American Fauna, 16:1–179.

———. 1900. Preliminary revision of the North American red foxes. Proceedings of the Washington Academy of Science, 2:661–676.

———. 1901*a*. Descriptions of three new kangaroo mice of the genus *Microdipodops*. Proceedings of the Biological Society of Washington, 14:127–128.

———. 1901*b*. Descriptions of twenty-three new pocket gophers of the genus *Thomomys*. Porceedings of the Biological Society of Washington, 14:107–117.

———. 1901*c*. Two new rodents from northwestern California. Proceedings of the Biological Society of Washington, 14:125–126.

———. 1902. A new bobcat (Lynx unita) from the Rocky Mountains. Proceedings of the Biological Society of Washington, 15:71–72.

———. 1904*a*. Unrecognized jack rabbits of the *Lepus texianus* group. Proceedings of the Biological Society of Washington, 17:135–138.

———. 1904*b*. New and little known kangaroo rats of the genus *Perodipus*. Proceedings of the Biological Society of Washington, 17:139–146.

———. 1904*c*. Jack rabbits of the *Lepus campestris* group. Proceedings of the Biological Society of Washington, 17:131–134.

———. 1908. Four new rodents from California. Proceedings of the Biological Society of Washington, 21:145–148.

———. 1913. Six new ground squirrels of the *Citellus mollis* group from Idaho, Oregon, and Nevada. Proceedings of the Biological Society of Washington, 26:137.

———. 1914. Descriptions of thirty apparently new grizzly and brown bears from North America. Proceedings of the

Biological Society of Washington, 27:173–196.

———. 1918. Review of the grizzly and big brown bears of North America. North American Fauna, 41:1–136.

———. 1930. A nest of the California gray squirrel (Sciuirus griseus). Journal of Mammalogy, 11:494.

Merriam, J. C. 1913. Tapir remains from late Cenozoic beds of the Pacific Coast region. University of California Publications, Bulletin of the Department of Geology, 7:169–175.

———. 1916. Mammalian remains from a late Tertiary formation at Ironside, Oregon. University of California Publications, Bulletin of the Department of Geology, 10:129–135.

———. 1917. Relationships of Pliocene mammalian faunas from the Pacific Coast and Great Basin provinces of North America. University of California Publications, Bulletin of the Department of Geology, 10:421–443.

Merriam, J. C., C. Stock, and C. L. Moody. 1916. An American Pliocene bear. University of California Publications, Bulletin of the Department of Geology, 10:87–109.

———. 1925. The Pliocene Rattlesnake Formation and fauna of eastern Oregon, with notes on the geology of the Rattlesnake and Mascall deposits. Pp. 43–92, *in* Papers concerning the palaeontology of the Pleistocene of California and the Tertiary of Oregon, Carnegie Institution of Washington, Contributions to Palaeontology, 347:1–92.

Merrick, R. L., T. R. Loughlin, and D. G. Calkins. 1987. Decline in abundance of the northern sea lion, *Eumetopias jubatus*, in Alaska, 1956–86. Fishery Bulletin, 85:351–365.

Merritt, J. F. 1981. Clethrionomys gapperi. Mammalian Species, 146:1–9.

———. 1987. Guide to the mammals of Pennsylvania. University of Pittsburg Press, Pittsburg, 408 pp.

Merritt, J. F., and J. M. Merritt. 1978*a*. Population ecology and energy relationships of *Clethrionomys gapperi* in a Colorado subalpine forest. Journal of Mammalogy, 59:576–598.

———. 1978*b*. Seasonal home ranges and activity of small mammals of a Colorado subalpine forest. Acta Theriologica, 23:195–202.

Meserve, P. L. 1974. Ecological relationships of two sympatric woodrats in a California coastal sage scrub community. Journal of Mammalogy, 55:442–447.

———. 1976. Food relationships of a rodent fauna in a California coastal sage scrub community. Journal of Mammalogy, 57:300–317.

———. 1977. Three-dimensional home ranges of cricetid rodents. Journal of Mammalogy, 58: 549–558.

Messick, J. P. 1987. North American badger. Pp. 587–597, *in* Wild furbearer management and conservation in North America (M. Novak, J. A. Baker, M. E. Obbard, and B. Malloch, eds.). Ontario Ministry of Natural Resources, Toronto, 1150 pp.

Messick, J. P., and M. G. Hornocker. 1981. Ecology of the badger in southwestern Idaho. Wildlife Monographs, 76:1–53.

Messick, J. P., M. C. Todd, and M. G. Hornocker. 1981. Comparative ecology of two badger populations. Worldwide Furbearer Conference Proceedings, 2:1290–1304.

Meyer, N. J. 1947. The histology and linkage relations of flexed-tailed, and inherited skeletal anomaly in Peromyscus, the deer mouse. M.A. thesis, University of Oregon, Eugene, 53 pp.

Meyers, A. V. 1946. Beaver management in Oregon. Oregon State Game Commission Bulletin, 1(2):1,3–4.

Meyers, S. M., and J. O. Wolff. 1994. Comparative toxicity of azinphos-methyl to house mice, laboratory mice, deer mice, and gray-tailed voles. Archives of Environmental Contamination and Toxicology, 26:478–482.

Michener, G. R. 1976. Tail autotomy as an escape mechanism in *Rattus rattus.* Journal of Mammalogy, 57:600–603.

———. 1977. Effects of climatic conditions on the annual activity and hibernation cycle of Richardson's ground squirrels and Columbian ground squirrels. Canadian Journal of Zoology, 55:693–703.

———. 1983. Kin identification, matriarchies, and the evolution of sociality in ground-dwelling sciurids. Pp. 528–572, *in* Advances in the study of mammalian behavior (J. F. Eisenberg and D. G. Kleiman, eds.). Special Publication, The American Society of Mammalogists, 7:1–753.

———. 1984. Age, sex, and species differences in the annual cycles of ground-dwelling sciurids: implications for sociality. Pp. 81–107, *in* The biology of ground-dwelling squirrels: annual cycles, behavioral ecology, and sociality (J. O. Murie and G. R. Michener, eds.). University of Nebraska Press, Lincoln, 459 pp.

———. 1992. Sexual differences in over-winter torpor patterns of Richardson's ground squirrels in natural hibernacula. Oecologia, 89:397–406.

———. 1993. Sexual differences in hibernaculum contents of Richardson's ground squirrels: males store food. Pp. 109–118, *in* Life in the cold: ecological, physiological, and molecular mechanisms (C. Carey, G. L. Florant, B. A. Wunder, and B. Horwitz, eds.). Westview Press, Boulder, [Colorado], 575 pp.

Mickey, A. B. 1961. Record of the spotted bat from Wyoming. Journal of Mammalogy, 42:401–402.

Mihok, S. 1976. Behaviour of subarctic red-backed voles (*Clethrionomys gapperi athabascae*). Canadian Journal of Zoology, 54:1932–1945.

———. 1979. Behavioural structure and demography of subarctic *Clethrionomys gapperi* and *Peromyscus maniculatus.* Canadian Journal of Zoology, 57:1520–1535.

Millar, J. S. 1970a. The breeding season and reproductive cycle of the western red squirrel. Canadian Journal of Zoology, 48:471–473.

———. 1970b. Variations in fecundity of the red squirrel, *Tamiasciurus hudsonicus* (Erxleben). Canadian Journal of Zoology, 48:1055–1058.

———. 1972. Timing of breeding of pikas in southwestern Alberta. Canadian Journal of Zoology, 50:665–669.

———. 1973. Evolution of litter-size in the pika, *Ochotona princeps* (Richardson). Evolution, 27:134–143.

———. 1974. Success of reproduction in pikas, *Ochotona princeps* (Richardson). Journal of Mammalogy, 55:527–542.

———. 1989. Reproduction and development. Pp. 169–232, *in* Advances in the study of *Peromyscus* (Rodentia) (G. L. Kirkland, Jr., and J. N. Layne, eds.). Texas Tech University Press, Lubbock, 366 pp.

Millar, J. S., and S. C. Tapper. 1973. Growth rates of pikas in Alberta. The Canadian Field-Naturalist, 87:457–459.

Millar, J. S., and F. C. Zwickel. 1972. Characteristics and

ecological significance of hay piles of pikas. Mammalia, 36:657–667.

Millar, J. S., D. G. L. Innes, and V. A. Loewen. 1985. Habitat use by non-hibernating small mammals of the Kananaskis Valley, Alberta. The Canadian Field-Naturalist, 99:196–204.

Miller, F. L. 1965. Behavior associated with parturition in black-tailed deer. The Journal of Wildlife Mangement, 29:629–631.

———. 1968. Observed use of forage and plant communities by black-tailed deer. The Journal of Wildlife Management, 32:142–148.

———. 1970a. Accidents to parturient black-tailed deer. The American Midland Naturalist, 83:303–304.

———. 1970b. Distribution patterns of black-tailed deer (*Odocoileus hemionus columbianus*) in relation to environment. Journal of Mammalogy, 51:248–260.

———. 1971. Mutual grooming by black-tailed deer in northwestern Oregon. The Canadian Field-Naturalist, 85:295–301.

———. 1974. Four types of territoriality observed in a herd of black-tailed deer. Pp. 644–660, *in* The behaviour of ungulates and its relation to management (V. Geist and F. Walther, eds.). International Union for Conservation of Nature and Natural Resources, Morges, Switzerland, new series, 24(pt. 2):513–940.

Miller, F. W. 1931. A feeding habit of the long-tailed weasel. Journal of Mammalogy, 12:164.

Miller, G. S., Jr. 1896. Genera and subgenera of voles and lemmings. North American Fauna, 12:1–84.

———. 1897a. Revision of the North American bats of the family Vespertilionidae. North American Fauna, 13:1–135 + 3 pls.

———. 1897b. Descriptions of a new vole from Oregon. Proceedings of the Biological Society of Washington, 11:67–68.

———. 1899. Descriptions of six new American rabbits. Proceedings of the Academy of Natural Sciences of Philadelphia, 51:380–390.

———. 1899 [1900]. Preliminary list of New York mammals. Bulletin of the New York Museum, 6:271–390.

———. 1912. List of North American land mammals in the United States National Museum, 1911. Bulletin of the United States National Museum, 79:1–401.

———. 1914. Two new North American bats. Proceedings of the Biological Society of Washington, 27:211–212.

———. 1924. List of North American Recent mammals, 1923. Bulletin of the United States National Museum, 128:1–673.

Miller, G. S., Jr., and G. M. Allen. 1928. The American bats of the genera Myotis and Pizonyx. Bulletin of the United States National Museum, 144:1–218.

Miller, G. S., Jr., and R. Kellogg. 1955. List of North American Recent mammals. United States National Museum Bulletin, 205:1–954.

Miller, G. S., Jr., and J. A. G. Rehn. 1901. Systematic results of the study of North American land mammals to the close of the year 1900. Proceedings of the Boston Society of Natural History, 30:1–352.

Miller, R. S. 1964. Ecology and distribution of pocket gophers (Geomyidae) in Colorado. Ecology, 45:256–272.

Miller, S. J. 1979. The archaeological fauna of four sites in Smith

Creek Canyon. Pp. 272–329, *in* The archaeology of Smith Creek canyon, eastern Nevada (D. R. Tuohy and D. L. Rendall, eds.). Nevada State Museum Anthropological Papers, 17:1–394.

Mills, J. N., et al. 1995. Guidelines for working with rodents potentially infected with hantavirus. Journal of Mammalogy, 76:716–722.

Mills, L. S. 1995. Edge effects and isolation: red-backed voles on forest remnants. Conservation Biology, 9:395–403.

Mills, R. S., G. W. Barrett, and M. P. Farrell. 1975. Population dynamics of the big brown bat (*Eptesicus fuscus*) in southwestern Ohio. Journal of Mammalogy, 56:591–604.

Minta, S. C., and R. E. Marsh. 1988. Badgers (*Taxidea taxus*) as occasional pests in agriculture. Proceedings Vertebrate Pest Conference, 13:199–208.

Mitchell, G. E. 1944. The Murderers Creek deer herd. Transactions of the North American Wildlife Conference, 9:167–172.

———. 1950. Wildlife-forest relationships in the Pacific Northwest region. Journal of Forestry, 48:26–30.

Mitchell, G. J. 1967. Mimimum breeding age of female pronghorn antelope. Journal of Mammalogy, 48:489–490.

Mitchell, J. L. 1961. Mink movements and populations on a Montana river. The Journal of Wildlife Management, 25:48–54.

Miyamoto, M. M., S. M. Tanhauser, and P. J. Laipis. 1989. Systematic relationships in the artiodactyl tribe Bovini (family Bovidae), as determined from mitochondrial DNA sequences. Systematic Zoology, 38:342–349.

Modi, W. S. 1985. Chromosomes of six species of New World microtine rodents. Mammalia, 49:357–363.

———. 1986. Karyotypic differentiation among two sibling species pairs of New World microtine rodents. Journal of Mammalogy, 67:159–165.

Mogart, J. R. 1985. Carnivorous behavior by a white-tailed antelope ground squirrel, *Ammospermophilus leucurus*. The Southwestern Naturalist, 30:304–305.

Mohr, W. P., and C. O. Mohr. 1936. Recent jack rabbit populations at Rapidan, Minnesota. Journal of Mammalogy, 17:112–114.

Montgomery, G. G. 1964. Tooth eruption in preweaned raccoons. The Journal of Wildlife Management, 28:582–584.

———. 1968. Pelage development of young raccoons. Journal of Mammalogy, 49:142–145.

———. 1969. Weaning of captive raccoons. The Journal of Wildlife Management, 33:154–159.

Moore, A. W. 1929a. Extra-uterine pregnancy in Peromyscus. Journal of Mammalogy, 10:81.

———. 1929b. Some notes upon Utah mammals. Journal of Mammalogy, 10:259–260.

———. 1933. Food habits of Townsend and coast moles. Journal of Mammalogy, 14:36–40.

———. 1937. Some effects of altitude and latitude on the Columbian ground squirrel. Journal of Mammalogy, 18:368–369.

———. 1939. Notes on the Townsend mole. Journal of Mammalogy, 20:499–501.

———. 1940. Wild animal damage to seed and seedlings on cut-over Douglas fir lands of Oregon and Washington. United States Department of Agriculture Technical Bulletin, 706:1–27.

———. 1942. Shrews as a check on Douglas-fir regeneration. Journal of Mammalogy, 23:37–41.

———. 1943a. Notes on the sage mouse in eastern Oregon. Journal of Mammalogy, 24:188–191.

———. 1943b. The pocket gopher in relation to yellow pine reproduction. Journal of Mammalogy, 24:271–272.

Moore, A. W., and E. H. Reid. 1951. The Dalles pocket gopher and its influence on forage production of Oregon mountain meadows. United States Department of Agriculture Circular, 884:1–36.

Moorhead, B. B., and V. Stevens. 1982. Introduction and dispersal of mountain goats in Olympic National Park. Pp. 46–50, *in* Ecological research in National Parks of the Pacific Northwest (E. E. Starkey, J. F. Franklin, and J. W. Matthews, eds.). Oregon State University, Forest Research Laboratory Publication, Corvallis, 142 pp.

Morhardt, J. E., and J. W. Hudson. 1966. Daily torpor induced in white-footed mice (*Peromyscus* spp.) by starvation. Nature, 212:1046–1047.

Morrell, S. 1972. Life history of the San Joaquin kit fox. California Fish and Game, 58:162–174.

Morrison, M. L., and R. G. Anthony. 1989. Habitat use by small mammals on early-growth clear-cuttings in western Oregon. Canadian Journal of Zoology, 67:805–811.

Morse, T. E., and H. M. Wight. 1969. Dump nesting and its effect on production in wood ducks. The Journal of Wildlife Management, 33:284–293.

Morton, M. L. 1975. Seasonal cycles of body weights and lipids in Belding ground squirrels. Bulletin of the Southern California Academy of Sciences, 74:128–143.

Morton, M. L., and J. S. Gallup. 1975. Reproductive cycle of the Belding ground squirrel (*Spermophilus beldingi beldingi*): seasonal and age differences. The Great Basin Naturalist, 35:427–433.

Morton, M. L., and P. W. Sherman. 1978. Effects of a spring snowstorm on behavior, reproduction, and survival of Belding's ground squirrels. Canadian Journal of Zoology, 56:2578–2590.

Morton, M. L., and H. L. Tung. 1971. Growth and development in the Belding ground squirrel (*Spermophilus beldingi beldingi*). Journal of Mammalogy, 52:611–616.

Morton, S. R. 1979. Diversity of desert dwelling mammals: a comparison of Australia and North America. Journal of Mammalogy, 60:253–264.

Moses, R. A., and J. S. Millar. 1992. Behavioural asymmetries and cohesive mother-offspring sociality in bushy-tailed wood rats. Canadian Journal of Zoology, 70:597–604.

Mossman, A. S. 1955. Reproduction of the brush rabbit in California. The Journal of Wildlife Management, 19:177–184.

Muchlinski, A. E., and A. L. Carlisle. 1982. Urine concentration by an undisturbed, naturally arousing hibernator (*Spermophilus lateralis*): water balance implications. Journal of Mammalogy, 63:510–512.

Mueller, C. C., and R. M. F. S. Sadleir. 1975. Attainment of early puberty in female black-tailed deer (*Odocoileus hemionus columbianus*). Theriogenology, 3:101–105.

———. 1977. Changes in the nutrient composition of milk of black-tailed deer during lactation. Journal of Mammalogy,

58:421–423.

Mueller, H. C. 1968. The role of vision in vespertilionid bats. The American Midland Naturalist, 79:524–525.

Mullen, R. K. 1971. Energy metabolism and body water turnover rates of two species of free-living kangaroo rats, *Dipodomys merriami* and *Dipodomys microps*. Comparative Biochemistry and Physiology, 39A:379–390.

Müller-Schwarze, D. 1971. Pheromones in black-tailed deer (*Odocoileus hemionus columbianus*). Animal Behaviour, 19:141–152.

Mumford, R. E. 1958. Population turnover in wintering bats in Indiana. Journal of Mammalogy, 39:253–261.

———. 1969. Long-tailed weasel preys on big brown bats. Journal of Mammalogy, 50:360.

Mumford, R. E., and J. O. Whitaker, Jr. 1982. Mammals of Indiana. Indiana University Press, Bloomington, 537 pp.

Murie, A. 1961. Some food habits of the marten. Journal of Mammalogy, 42:516–521.

Murie, J. O., and M. A. Harris. 1978. Territoriality and dominance in male Columbian ground squirrels (*Spermophilus columbianus*).Canadian Journal of Zoology, 56:2402–2412.

———. 1982. Annual variation of spring emergence and breeding in Columbian ground squirrels (*Spermophilus columbianus*). Journal of Mammalogy, 63:431–439.

Murie, J. O., D. A. Boag, and V. K. Kivett. 1980. Litter size in Columbian ground squirrels (*Spermophilus columbianus*). Journal of Mammalogy, 61:237–244.

Murray, K. F., and A. M. Barnes. 1969. Distribution and habitat of the woodrat, *Neotoma fuscipes*, in northeastern California. Journal of Mammalogy, 50:43–48.

Musser, G. G., and M. D. Carleton. 1993. Family Muridae. Pp. 501–755, *in* Mammal species of the world: a taxonomic and geographic reference. Second ed. (D. E. Wilson and D. M. Reeder, eds.). Smithsonian Institution Press, Washington, D.C., 1206 pp.

Musser, G. G., and S. D. Durrant. 1960. Notes on *Myotis thysanodes* in Utah. Journal of Mammalogy, 41:393–394.

Muth, O. H., and W. H. Allaway. 1963. The relationship of white muscle disease to the distribution of naturally occurring selenium. Journal of the American Veterinary Medical Association, 142:1379–1384.

Muul, I. 1969. Mating behavior, gestation period, and development of *Glaucomys sabrinus*. Journal of Mammalogy, 50:121.

Myers, R. F. 1960. *Lasiurus* from Missouri caves. Journal of Mammalogy, 41:114–117.

Nadler, C. F. 1966. Chromosomes and systematics of American ground squirrels of the subgenus *Spermophilus*. Journal of Mammalogy, 47:579–596.

———. 1968. The chromosomes of *Spermophilus townsendii* (Rodentia: Sciuridae) and report of a new subspecies. Cytogenetics, 7:144–157.

Nadler, C. F., R. S. Hoffmann, and K. R. Greer. 1971. Chromosomal divergence during evolution of ground squirrel populations (Rodentia: *Spermophilus*). Systematic Zoology, 20:298–305.

Nadler, C. F., R. S. Hoffmann, J. H. Honacki, and D. Pozin. 1977. Chromosomal evolution in chipmunks, with special emphasis on A and B karyotypes of the subgenus Neotamias.

The American Midland Naturalist, 98:343–353.

Nadler, C. F., R. I. Sukernik, R. S. Hoffmann, N. N. Vorontsov, C. F. Nadler, Jr., and I. I. Fomichova. 1974. Evolution in ground squirrels—I. transferrins in Holarctic populations of *Spermophilus*. Comparative Biochemistry and Physiology, 47A:663–681.

Nagorsen, D. W. 1983. Winter pelage colour in snowshoe hares (*Lepus americanus*) from the Pacific Northwest. Canadian Journal of Zoology, 61:2313–2318.

———. 1985. A morphometric study of geographic variation in the snowshoe hare (*Lepus americanus*). Canadian Journal of Zoology, 63:567–579.

———. 1990. The mammals of British Columbia: a taxonomic catalogue. Royal British Columbia Museum Memoir, 4:1–140.

Nagorsen, D. W., K. F. Morrison, and J. E. Forsberg. 1989. Winter diet of Vancouver Island marten (*Martes americana*). Canadian Journal of Zoology, 67:1394–1400.

Nams, V. O. 1991. Olfactory search images in striped skunks. Behaviour, 119:267–284.

National Marine Fisheries Service. 1970. Fur seal investigations, 1968. National Marine Fisheries Service, Special Scientific Report, 617:1–125.

———. 1992. Recovery plan for the Steller sea lion (*Eumetopias jubatus*). Prepared by the Steller Sea Lion Recovery Team for the National Marine Fisheries Service, Silver Spring, Maryland, 92 pp.

Naylor, B. J., J. F. Bendell, and S. Spires. 1985. High density of heather voles, *Phenacomys intermedius* in jack pine, *Pinus banksiana*. The Canadian Field-Naturalist, 99:494–497.

Neal, F. D., and J. E. Borrecco. 1981. Distribution and relation of mountain beaver to opening in sampling stands. Northwest Science, 55:79–86.

Nee, J. A. 1969. Reproduction in a population of yellow-bellied marmots (*Marmota flaviventris*). Journal of Mammalogy, 50:756–765.

Negus, N. C., and P. J. Berger. 1977. Experimental triggering of reproduction in a natural population of *Microtus montanus*. Science, 196:1230–1231.

Negus, N. C., and J. S. Findley. 1959. Mammals of Jackson Hole, Wyoming. Journal of Mammalogy, 40:371–381.

Negus, N. C., and A. J. Pinter. 1965. Litter sizes of *Microtus montanus* in the laboratory. Journal of Mammalogy, 46:434–437.

Neider, D. S. 1963. Population dynamics of enclosed *Microtus montanus* (Peale) in relation to cover and density. M.S. thesis, Oregon State University, Corvallis, 71 pp.

Nelson, E. W. 1909. The rabbits of North America. North American Fauna, 29:1–314.

———. 1927. Description of a new subspecies of beaver. Proceedings of the Biological Society of Washington, 40:125–126.

———. 1930. Wild animals of North America. Intimate studies of big and little creatures of the mammal kingdom. National Geographic Society, Washington, D.C., 254 pp.

Nelson, E. W., and E. A. Goldman. 1929. List of the pumas, with three described as new. Journal of Mammalogy, 10:345–350.

———. 1930. Six new raccoons of the Procyon lotor group. Journal of Mammalogy, 11:453–459.

Nelson, Z. C., and M. K. Yousef. 1979. Thermoregulatory responses of desert woodrats, *Neotoma lepida*. Comparative Biochemistry and Physiology, A. Comparative Physiology, 63:109–113.

Newby, F. E., and J. J. McDougal. 1964. Range extension of the wolverine in Montana. Journal of Mammalogy, 45:485–487.

Newby, F. E., and P. L. Wright. 1955. Distribution and status of the wolverine in Montana. Journal of Mammalogy, 36:248–253.

Newby, T. C. 1975. A sea otter (*Enhydra lutris*) food dive record. The Murrelet, 56:7.

Newman, D. G., and C. R. Griffin. 1994. Wetland use by river otters in Massachusetts. The Journal of Wildlife Management, 58:18–23.

Newman, J. R. 1976. Population dynamics of the wandering shrew *Sorex vagrans*. The Wasmann Journal of Biology, 34:235–250.

Newson, R. M. 1966. Reproduction in the feral coypus (*Myocastor coypus*). Symposia of the Zoological Society of London, 15:323–334.

Nice, M. M., C. Nice, and D. Ewers. 1956. Comparison of behavior development in snowshoe hares and red squirrels. Journal of Mammalogy, 37:64–74.

Nicholson, W. S., E. P. Hill, and D. Briggs. 1985. Denning, pup-rearing, and dispersal in the gray fox in east-central Alabama. The Journal of Wildlife Management, 49:33–37.

Nikolai, J. C., and D. M. Bramble. 1983. Morphological structure and function in desert heteromyid rodents. Great Basin Naturalist Memoirs, 7:44–64.

Nistler, D. L., B. J. Verts, and L. N. Carraway. 1993. Aging techniques, juvenile breeding, and body mass in *Thomomys bulbivorus*. Northwestern Naturalist, 74:25–28.

Nixon, C. M., L. P. Hansen, and S. P. Havera. 1986. Demographic characteristics of an unexploited population of fox squirrels *(Sciurus niger)*. Canadian Journal of Zoology, 64:512–521.

Nixon, C. M., D. M. Worley, and M. W. McClain. 1968. Food habits of squirrels in southeast Ohio. The Journal of Wildlife Management, 32:294–305.

Noback, C. R. 1951. Morphology and phylogeny of hair. Annals New York Academy of Sciences, 53:476–492.

Noble, W. O. 1994. Characteristics of spring foraging ecology among black bears in the central Coast Ranges [*sic*] of Oregon. M.S. thesis, Oregon State University, Corvallis, 96 pp.

Nolte, D. L., G. Epple, D. L. Campbell, and J. R. Mason. 1993. Response of mountain beaver (*Aplodontia rufa*) to conspecifics in their burrow system. Northwest Science, 67:251–255.

Norris, J. D. 1967. A campaign against feral coypus (*Myocastor coypus*) in Great Britain. The Journal of Applied Ecology, 4:191–199.

Norton, A. C., A. V. Beran, and G. A. Misrahy. 1964. Electroencephalograph during "feigned sleep" in the opossum. Nature, 204:162–163.

Novak, M. 1987. Beaver. Pp. 282–312, *in* Wild furbearer management and conservation in North America (M. Novak, J. A. Baker, M. E. Obbard, and B. Malloch, eds.). Ontario Ministry of Natural Resources, Toronto, 1150 pp.

Nowak, R. M. 1973. Return of the wolverine. National Parks Conservation, 47:20–23.

———. 1978. Evolution and taxonomy of coyotes and related *Canis*. Pp. 3–16, *in* Coyotes: biology, behavior, and management (M. Bekoff, ed.). Academic Press, New York, 384 pp.

———. 1979. North American Quaternary *Canis*. Monograph of the Museum of Natural History, University of Kansas, 6:1–154.

———. 1991. Walker's mammals of the world. Fifth ed. The Johns Hopkins University Press, Baltimore, 1:1–642, 2:643–1629.

Nowak, R. M., and J. L. Paradiso. 1983. Walker's mammals of the world. Fourth ed. The Johns Hopkins University Press, Baltimore, 1:1–568.

Nugent, R. F., and J. R. Choate. 1970. Eastward dispersal of the badger, *Taxidea taxus*, into the northeastern United States. Journal of Mammalogy, 51:626–627.

Nungesser, W. C., and E. W. Pfeiffer. 1965. Water balance and maximum concentrating capacity in the primitive rodent *Aplodontia rufa*. Comparative Biochemistry and Physiology, 14:289–297.

Nungesser, W. C., E. W. Pfeiffer, D. A. Iverson, and J. F. Wallerius. 1960. Evaluation of renal countercurrent hypothesis in Aplodontia. Federation Proceedings, 19:362.

Nussbaum, R. A., and C. Maser. 1969. Observations of *Sorex palustris* preying on *Dicamptodon ensatus*. The Murrelet, 50:23–24.

———. 1975. Food habits of the bobcat, *Lynx rufus*, in the Coast and Cascade ranges of western Oregon in relation to present management policies. Northwest Science, 49:261–266.

Nydegger, N. C., and D. R. Johnson. 1989. Size and overlap of Townsend ground squirrel home ranges. The Great Basin Naturalist, 49:108–112.

Odum, E. P. 1971. Fundamentals of ecology. Third ed. W. B. Saunders Company, Philadelphia, 574 pp.

O'Farrell, M. J. 1974. Seasonal activity patterns of rodents in a sagebrush community. Journal of Mammalogy, 55:809–823.

O'Farrell, M. J., and A. R. Blaustein. 1972. Ecological distribution of sagebrush voles, *Lagurus curtatus*, in south-central Washington. Journal of Mammalogy, 53:632–636.

———. 1974. Microdipodops megacephalus. Mammalian Species, 46:1–3.

O'Farrell, M. J., and W. G. Bradley. 1970. Activity patterns of bats over a desert spring. Journal of Mammalogy, 51:18–26.

O'Farrell, M. J., and W. A. Clark. 1984. Notes on the white-tailed antelope squirrel, *Ammospermophilus leucurus*, and the pinyon mouse, *Peromyscus truei*, in north central Nevada. The Great Basin Naturalist, 44:428–430.

O'Farrell, M. J., and E. H. Studier. 1973. Reproduction, growth, and development in *Myotis thysanodes* and *M. lucifugus* (Chiroptera: Vespertilionidae). Ecology, 54:18–30.

———. 1975. Population structure and emergence activity patterns in Myotis thysanodes and M. lucifugus (Chiroptera: Vespertilionidae) in northeastern New Mexico. The American Midland Naturalist, 93:368–376.

———. 1980. Myotis thysanodes. Mammalian Species, 137:1–5.

O'Farrell, M. J., W. G. Bradley, and G. W. Jones. 1967. Fall and winter bat activity at a desert spring in southern Nevada. The Southwestern Naturalist, 12:163–171.

O'Farrell, T. P. 1965. Home range and ecology of snowshoe hares in interior Alaska. Journal of Mammalogy, 46:406–418.

O'Farrell, T. P., R. J. Olson, R. O. Gilbert, and J. D. Hedlund. 1975. A population of Great Basin pocket mice, *Perognathus parvus*, in the shrub-steppe of south-central Washington. Ecological Monographs, 45:1–28.

O'Gara, B. W. 1969a. Unique aspects of reproduction in the female pronghorn (*Antilocapra americana* Ord). The American Journal of Anatomy, 125:217–232.

———. 1969b. Horn casting by female pronghorns. Journal of Mammalogy, 50:373–375.

———. 1978. Antilocapra americana. Mammalian Species, 90:1–7.

O'Gara, B. W., and G. Matson. 1975. Growth and casting of horns by pronghorns and exfoliation of horns by bovids. Journal of Mammalogy, 56:829–846.

O'Gara, B. W., R. F. Moy, and G. D. Bear. 1971. The annual testicular cycle and horn casting in the pronghorn (*Antilocapra americana*). Journal of Mammalogy, 52:537–544.

Ohno, S. [C.] 1967. Sex chromosomes and sex-linked genes. Springer-Verlag, New York, 192 pp.

Ohno, S. [C.], J. Jainchill, and C. Stenius. 1963. The creeping vole (*Microtus oregoni*) as a gonosomic mosaic I. The OY/XY constitution of the male. Cytogenetics, 2:232–239.

Ohno, S. C., C. Stenius, and L. Christian. 1966. The XO as the normal female of the creeping vole (*Microtus oregoni*). Pp. 182–187, *in* Chromosomes today (C. D. Darlington and K. R. Lewis, eds.). Oliver and Boyd, London, 274 pp.

Oken, L. 1816. Lehrbuch des Naturgeschichte. Jena (not seen, cited in Hershkovitz, 1949)

Olterman, J. H., and B. J. Verts. 1972. Endangered plants and animals of Oregon *IV. Mammals*. Oregon State University, Agricultural Experiment Station, Special Report, 364:1–47.

Ord, G. 1815. [Remarks on the natural history of North American mammals.] Pp. 290–361, *in* A new geographical, historical, and commercial grammar; and present state of the several kingdoms of the world (W. Guthrie). Second ed. Johnson and Warner, Philadelphia, 2:1–603 (not seen, cited in Gill and Coues, 1877).

———. 1818. Sur plusieurs animaux de l'Amérique septentrionale et entre autres sur le Rupicapra americana, l'Antilope americana, le Cervus major ou Wapiti. etc. Jour. de Phys., Chim., Hist. Nat., et des Arts, 87:146–155 (not seen, cited in Gill and Coues, 1877).

Oregon Department of Fish and Wildlife. 1993–94. Oregon wildlife and commercial fishing codes. Oregon Department of Fish and Wildlife, Portland, 146 pp.

Oregon State Game Commission. 1946a. Beaver live-trapping. Oregon State Game Commission Bulletin, 1(6):4.

———. 1946b. Oregon's marsupial. Oregon State Game Commission Bulletin, 1(5):5.

———. 1970. Rare wolverine to be placed on exhibit. Oregon State Game Commission Bulletin, 25(1):7.

———. 1973. Squirrel transplant. Oregon State Game Commission Bulletin, 28(4):11.

Orr, E. L., W. N. Orr, and E. M. Baldwin. 1992. Geology of Oregon. Fourth ed. Kendall/Hunt Publishing Company, Dubuque, Iowa, 254 pp.

Orr, R. T. 1934. Description of a new snowshoe rabbit from eastern Oregon, with notes on its life history. Journal of Mammalogy, 15:152–154.

———. 1935. Descriptions of three new races of brush rabbit from California. Proceedings of the Biological Society of Washington, 48:27–30.

———. 1939. Longevity in Perognathus longimembris. Journal of Mammalogy, 20:505.

———. 1940. The rabbits of California. Occasional Papers, California Academy of Science, 19:1–227.

———. 1942. Observations on the growth of young brush rabbits. Journal of Mammalogy, 23:298–302.

———. 1950. Unusual behavior and occurrence of a hoary bat. Journal of Mammalogy, 31:456–457.

———. 1954. Natural history of the pallid bat, *Antrozous pallidus* (LeConte). Proceedings of the California Academy of Sciences, series 4, 28:165–246.

Orr, R. T., and T. C. Poulter. 1967. Some observations on reproduction, growth, and social behavior in the Steller sea lion. Proceedings of the California Academy of Sciences, series 4, 35:193–226.

Orr, W. N., and E. L. Orr. 1981. Handbook of Oregon plant and animal fossils. Privately published, Eugene, Oregon, 285 pp.

Osborne, T. O., and W. A. Sheppe. 1971. Food habits of *Peromyscus maniculatus* on a California beach. Journal of Mammalogy, 52:844–845.

Osborne, W. A. 1932. Mice plague in Australia. Nature, 129:755.

Osgood, F. L. 1935. Apparent segregation of sexes in flying squirrels. Journal of Mammalogy, 16:231.

———. 1936. Earthworms as a supplementary food of weasels. Journal of Mammalogy, 17:64.

Osgood, W. H. 1900. Revision of the pocket mice of the genus Perognathus. North American Fauna, 18:1–72.

———. 1901. A new white-footed mouse from California. Proceedings of the Biological Society of Washington, 14:193–194.

———. 1909a. Revision of the mice of the American genus Peromyscus. North American Fauna, 28:1–285 + foldout map.

———. 1909b. The status of *Sorex merriami*, with description of an allied new species from Utah. Proceedings of the Biological Society of Washington, 22:51–53.

O'Shea, T. J., and T. A. Vaughan. 1977. Nocturnal and seasonal activities of the pallid bat, *Antrozous pallidus*. Journal of Mammalogy, 58:269–284.

Ostfeld, R. S. 1986. Territoriality and mating systems of California voles. The Journal of Animal Ecology, 55:691–706.

Ostfeld, R. S., and L. L. Klosterman. 1986. Demographic substructure in a California vole population inhabiting a patchy environment. Journal of Mammalogy, 67:693–704.

Owings, D. H., M. Borchert, and R. Virginia. 1977. The behaviour of California ground squirrels. Animal Behaviour, 25:221–230.

Packard, E. L. 1947a. A fossil sea lion from Cape Blanco, Oregon. Oregon State Monographs, Studies in Geology, 6:13–22.

———. 1947b. A pinniped humerus from the Astoria Miocene

of Oregon. Oregon State Monographs, Studies in Geology, 7:23–32.

———. 1952. Fossil edentates of Oregon. Oregon State Monographs, Studies in Geology, 8:1–16 + 2 pls.

Pagels, J. F., and T. W. French. 1987. Discarded bottles as a source of small mammal distribution data. The American Midland Naturalist, 118:217–219.

Pagels, J. F., and C. Jones. 1974. Growth and development of the free-tailed bat, *Tadarida brasiliensis cynocephala* (Le Conte). The Southwestern Naturalist, 19:267–276.

Palisot de Beauvois, A. M. F. J. 1796. Catalogue raisonné du muséum de Mr. C. W. Peale. S. H. Smith, Philadelphia, 42 pp. (not seen, cited in Miller and Kellogg, 1955).

Pallas, P. S. 1780. Spicilegia Zoologica: quibus novae imprimis et obscurae animalium species iconibus, descriptionibus atque commentariis illustrantur. fasc. 14:5 (not seen, cited in Baird, 1857).

Palmer, F. G. 1937. Geographic variation in the mole Scapanus latimanus. Journal of Mammalogy, 18:280–314.

Palmer, R. S. 1954. The mammal guide: mammals of North America north of Mexico. Doubleday and Company, Garden City, New York, 384 pp.

Palmer, T. S. 1897. Jack rabbits of the United States. United States Department of Agriculture, Division of Ornithology and Mammalogy Bulletin, 8:1–88.

Panin, K. I., and G. K. Panina. 1968. Fur seal ecology and migration to the Sea of Japan during winter and spring. Pp. 66–76, *in* Pinnipeds of the North Pacific (V. A. Arsen'ev and K. I. Panin, eds.). Keter Press, Jerusalem, 274 pp. Translated from Russian by the Israel Program for Scientific Translations, Jerusalem, 1971.

Paradiso, J. L., and A. M. Greenhall. 1967. Longevity records for American bats. The American Midland Naturalist, 78:251–252.

Paradiso, J. L., and R. M. Nowak. 1971. A report on the taxonomic status and distribution of the red wolf. United States Fish and Wildlife Service, Special Scientific Report, 145:1–36.

———. 1982. Wolves: *Canis lupus* and allies. Pp. 460–474, *in* Wild mammals of North America: biology, management, and economics (J. A. Chapman and G. A. Feldhamer, eds.). The Johns Hopkins University Press, Baltimore, 1147 pp.

Parker, G. R. 1981. Winter habitat use and hunting activities of lynx (*Lynx canadensis*) on Cape Breton Island, Nova Scotia. Worldwide Furbearer Conference Proceedings, 1:221–248.

Parkinson, A. 1979. Morphologic variation and hybridization in *Myotis yumanensis sociabilis* and *Myotis lucifugus carissima*. Journal of Mammalogy, 60:489–504.

Parsons, H. J., D. A. Smith, and R. F. Whittam. 1986. Maternity colonies of silver-haired bats, *Lasionycteris noctivagans*, in Ontario and Saskatchewan. Journal of Mammalogy, 67:598–600.

Pasitschniak-Arts, M. 1993. Ursus arctos. Mammalian Species, 439:1–10.

Pasitschniak-Arts, M., and S. Larivière. 1995. Gulo gulo. Mammalian Species, 499:1–10.

Patten, B. C., Jr., and M. A. Patten. 1956. Swimming ability of the little brown bat. Journal of Mammalogy, 37:440–441.

Patterson, B. D., and L. R. Heaney. 1987. Preliminary analysis of geographic variation in red-tailed chipmunks (*Eutamias*

ruficaudus). Journal of Mammalogy, 68:782–791.

Patterson, J. R., and E. L. Bowhay. 1968. Four recent records of wolverine in Washington. The Murrelet, 49:12–13.

Pattie, D. L. 1969. Behavior of captive marsh shrews (*Sorex bendirii*). The Murrelet, 50:27–32.

———. 1973. Sorex bendirii. Mammalian Species, 27:1–2.

Patton, J. L. 1993. Family Geomyidae. Pp. 469–476, *in* Mammal species of the world: a taxonomic and geographic reference. Second ed. (D. E. Wilson and D. M. Reeder, eds.). Smithsonian Institution Press, Washington, D.C., 1206 pp.

Patton, J. L., and M. F. Smith. 1990. The evolutionary dynamics of the pocket gopher *Thomomys bottae*, with emphasis on California populations. University of California Publications in Zoology, 123:1–161.

Patton, J. L., H. MacArthur, and Suh Y. Yang. 1976. Systematic relationships of the four-toed populations of *Dipodomys heermanni*. Journal of Mammalogy, 57:159–163.

Patton, J. L., M. F. Smith, R. D. Price, and R. A. Hellenthal. 1984. Genetics of hybridization between the pocket gophers *Thomomys bottae* and *Thomomys townsendii* in northeastern California. The Great Basin Naturalist, 44:431–440.

Pauley, G. R., J. M. Peek, and P. Zager. 1993. Predicting white-tailed deer habitat use in northern Idaho. The Journal of Wildlife Management, 57:904–913.

Payer, D. C. 1992. Habitat use and population characteristics of bighorn sheep on Hart Mountain National Antelope Refuge, Oregon. M.S. thesis, Oregon State University, Corvallis, 105 pp.

Peale, T. R. 1848. Mammalia and ornithology. United States exploring expedition during the years 1838, 1839, 1840, 1841, 1842 under the command of Charles Wilkes, U.S.N.C. Sherman, Philadelphia, 8:1–338 (reprint by Arno Press, New York, 1978).

Pearson, E. W. 1962. Bats hibernating in silica mines in southern Illinois. Journal of Mammalogy, 43:27–33.

Pearson, J. P. 1972. The influence of behavior and water requirements on the distribution and habitat selection of the gray-tailed vole (*Microtus canicaudus*) with notes on *Microtus townsendii*. Ph.D. dissert., Oregon State University, Corvallis, 56 pp.

Pearson, J. P., and B. J. Verts. 1970. Abundance and distribution of harbor seals and northern sea lions in Oregon. The Murrelet, 51:1–5.

Pearson, O. P. 1942. On the cause and nature of a poisonous action produced by the bite of a shrew (*Blarina brevicauda*). Journal of Mammalogy, 23:159–166.

———. 1960a. Habits of harvest mice revealed by automatic photographic recorders. Journal of Mammalogy, 41:58–74.

———. 1960b. Habits of *Microtus californicus* revealed by automatic photographic records. Ecological Monographs, 30:231–249.

———. 1964. Carnivore-mouse predation: an example of its intensity and bioenergetics. Journal of Mammalogy, 45:177–188.

———. 1966. The prey of carnivores during one cycle of mouse abundance. The Journal of Animal Ecology, 35:217–233.

———. 1971. Additional measurements of the impact of carnivores on California voles (*Microtus californicus*). Journal of Mammalogy, 52:41–49.

———. 1985. Predation. Pp. 535–566, *in* Biology of New World

Microtus (R. H. Tamarin, ed.). Special Publication, The American Society of Mammalogists, 8:1–893.

Pearson, O. P., M. R. Koford, and A. K. Pearson. 1952. Reproduction of the lump-nosed bat (*Corynorhinus rafinesquei*) in California. Journal of Mammalogy, 33:273–320.

Pease, J. L., R. H. Vowles, and L. B. Keith. 1979. Interaction of snowshoe hares and woody vegetation. The Journal of Wildlife Management, 43:43–60.

Pedersen, D. [R. J.] 1979. Northeast Oregon elk research. Oregon Wildlife, 34(11):3–5.

Pedersen, R. J. 1963. The life history and ecology of Townsend's mole *Scapanus townsendii* (Bachman) in Tillamook County Oregon. M.S. thesis, Oregon State University, Corvallis, 60 pp.

———. 1966. Nesting behavior of Townsend's mole. The Murrelet, 47:47–48.

Pedersen, R. J., and A. W. Adams. 1977. Summer elk trapping with salt. Wildlife Society Bulletin, 5:72–73.

Pedersen, R. J., and J. Stout. 1963. Oregon sea otter sighting. Journal of Mammalogy, 44:415.

Peek, J. M. 1982. Elk: *Cervus elaphus*. Pp. 851–861, *in* Wild mammals of North America: biology, management, and economics (J. A. Chapman and G. A. Feldhamer, eds.). The Johns Hopkins University Press, Baltimore, 1147 pp.

Pelikán, J. 1981. Patterns of reproduction in the house mouse. Pp. 205–229, *in* Biology of the house mouse (R. J. Berry, ed.). Symposia of the Zoological Society of London, 47:1–715.

Peloquin, E. P. 1969. Growth and reproduction in the feral nutria *Myocastor coypus* (Molina) near Corvallis, Oregon. M.S. thesis, Oregon State University, Corvallis, 55 pp.

Pelton, M. R. 1982. Black bear: *Ursus americanus*. Pp. 504–514, *in* Wild mammals of North America: biology, management, and economics (J. A. Chapman and G. A. Feldhamer, eds.). The Johns Hopkins University Press, Baltimore, 1147 pp.

Pengelley, E. T., and K. C. Fisher. 1961. Rhythmical arousal from hibernation in the golden-mantled ground squirrel, Citellus lateralis tescorum. Canadian Journal of Zoology, 39:105–120.

———. 1963. The effect of temperature and photoperiod on the yearly hibernating behavior of captive golden-mantled ground squirrels (Citellus lateralis tescorum). Canadian Journal of Zoology, 41:1103–1120.

Perkins, J. M. 1977. Bat homing. M.S. thesis, Portland State University, Portland, Oregon, 97 pp.

———. 1987. Distribution, status, and habitat affinities of Townsend's big-eared bat (*Plecotus townsendii*) in Oregon. Oregon Department of Fish and Wildlife, Nongame Wildlife Program, Technical Report, 86-5-01:1–49.

Perkins, J. M., and S. P. Cross. 1988. Differential use of some coniferous forest habitats by hoary and silver-haired bats. The Murrelet, 69:21–24.

Perkins, J. M., J. M. Barss, and J. Peterson. 1990. Winter records of bats in Oregon and Washington. Northwestern Naturalist, 71:59–62.

Perry, M. L. 1939. Notes on a captive badger. The Murrelet, 20:49–53.

Peters, W. K. H. 1866. Ueber einige weniger bekannte fledertiere.

Monatsberichte der Königlich Preussischen Akademie der Wissenschaften, Berlin, 1865:641–674 (not seen, cited in The National Union Catalog, 452:607).

Peterson, A. W. 1950. Backward swimming of the muskrat. Journal of Mammalogy, 31:453.

Peterson, R. L. 1956. *Phenacomys* eaten by speckled trout. Journal of Mammalogy, 37:121.

———. 1966. The mammals of eastern Canada. Oxford University Press, Toronto, 465 pp.

Peterson, R. O. 1979. Social rejection following mating of a subordinate wolf. Journal of Mammalogy, 60:219–221.

Peterson, R. S., and G. A. Bartholomew. 1967. The natural history and behavior of the California sea lion. Special Publication, The American Society of Mammalogists, 1:1–79.

Petrides, G. A. 1949. Sex and age determination in the opossum. Journal of Mammalogy, 30:364–378.

Pfeifer, S. R. 1982. Variability in reproductive output and success of *Spermophilus elegans* ground squirrels. Journal of Mammalogy, 63:284–289.

Pfeiffer, E. W. 1953. Animals trapped in mountain beaver (*Aplodontia rufa*) runways, and the mountain beaver in captivity. Journal of Mammalogy, 34:396.

———. 1956a. Notes on reproduction in the kangaroo rat, *Dipodomys*. Journal of Mammalogy, 37:449–450.

———. 1956b. The male reproductive tract of a primitive rodent, *Aplodontia rufa*. The Anatomical Record, 124:629–635.

———. 1958. The reproductive cycle of the female mountain beaver. Journal of Mammalogy, 39:223–235.

———. 1960. Cyclic changes in the morphology of the vulva and clitoris of *Dipodomys*. Journal of Mammalogy, 41:43–48.

Phillips, G. L. 1966. Ecology of the big brown bat (Chiroptera: Vespertilionidae) in northeastern Kansas. The American Midland Naturalist, 75:168–198.

Phillips, J. A. 1981. Growth and its relationship to the initial annual cycle of the golden-mantled ground squirrel, *Spermophilus lateralis*. Canadian Journal of Zoology, 59:865–871.

Pimentel, R. A. 1949. Black rat and roof rat taken in the central Oregon Coast strip. The Murrelet, 30:52.

Pinkel, D., B. L. Gledhill, S. Lake, D. Stephenson, and M. A. Van Dilla. 1982. Sex preselection in mammals? Separation of sperm bearing Y and "O" chromosomes in the vole *Microtus oregoni*. Science, 218:904–906.

Pinter, A. J. 1970. Reproduction and growth for two species of grasshopper mice (*Onychomys*) in the laboratory. Journal of Mammalogy, 51:236–243.

Pinter, A. J., and N. C. Negus. 1965. Effects of nutrition and photopheriod on reproductive physiology of *Microtus montanus*. American Journal of Physiology, 208:633–638.

Pisano, R. G., and T. I. Storer. 1948. Burrows and feeding of the Norway rat. Journal of Mammalogy, 29:374–383.

Pitcher, K. W. 1981. Prey of the Steller sea lion, *Eumetopias jubatus*, in the Gulf of Alaska. Fishery Bulletin, 79:467–472.

Pitcher, K. W., and D. G. Calkins. 1981. Reproductive biology of Steller sea lions in the Gulf of Alaska. Journal of Mammalogy, 62:599–605.

Pitcher, K. W., and F. H. Fay. 1982. Feeding by Steller sea lions on harbor seals. The Murrelet, 63:70–71.

Pitcher, K. W., and D. C. McAllister. 1981. Movements and haulout behavior of radio-tagged harbor seals. The Canadian Field-Naturalist, 95:292–297.

Pocock, R. I. 1917. The classification of existing Felidae. The Annals and Magazine of Natural History, including Zoology, Botany, and Geology, series 8, 20:329–351.

Poglayen-Neuwall, I., and D. E. Toweill. 1988. Bassariscus astutus. Mammalian Species, 327:1–8.

Polderboer, E. B., L. W. Kuhn, and G. O. Hendrickson. 1941. Winter and spring habits of weasels in central Iowa. The Journal of Wildlife Management, 5:115–119.

Polenz, A. 1976. A mountain valley antelope herd. Oregon Wildlife, 31(7):3–5.

Pollack, E. M. 1950. Breeding habits of the bobcat in northeastern United States. Journal of Mammalogy, 31:327–330.

Pomel, [N.] A. 1848. Études sur les carnassiers insectivores (Extrait). Archives de Sciences Physiques et Naturelles, 2me partie, Classification des insectivores, 9:244–251 (not seen, cited in Gill and Coues, 1877).

Ports, M. A., and S. B. George. 1990. Sorex preblei in the northern Great Basin. The Great Basin Naturalist, 50:93–95.

Pouché, R. M., and G. L. Bailie. 1974. Notes on the spotted bat (Euderma maculatum) from southwest Utah. The Great Basin Naturalist, 34:254–256.

Powell, R. A. 1979a. Fishers, population models, and trapping. Wildlife Society Bulletin, 7:149–154.

———. 1979b. Mustelid spacing patterns: variation on a theme by Mustela. Zeitschrift für Tierpsychologie, 50:153–165.

———. 1980. Fisher arboreal activity. The Canadian Field-Naturalist, 94:90–91.

———. 1981. Martes pennanti. Mammalian Species, 156:1–6.

———. 1982a. Evolution of black-tipped tails in weasels: predator confusion. The American Naturalist, 119:126–131.

———. 1982b. The fisher: life history, ecology, and behavior. University of Minnesota Press, Minneapolis, 217 pp.

Powell, R. A., and W. J. Zielinski. 1989. Mink response to ultrasound in the range emitted by prey. Journal of Mammalogy, 70:637–638.

Powers, R. A., and B. J. Verts. 1971. Reproduction in the mountain cottontail rabbit in Oregon. The Journal of Wildlife Management, 35:605–613.

Pray, L. 1921. Opossum carries leaves with its tail. Journal of Mammalogy, 2:109–110.

Preble, E. A. 1899. Revision of the jumping mice of the genus Zapus. North American Fauna, 15:1–43.

Price, K., and S. Boutin. 1993. Territorial bequeathal by red squirrel mothers. Behavioral Ecology, 4:144–150.

Price, M. V. 1978. Seed dispersion preferences of coexisting desert rodent species. Journal of Mammalogy, 59:624–626.

Proctor, J., and K. Whitten. 1971. A population of the valley pocket gopher (Thomomys bottae) on a serpentine soil. The American Midland Naturalist, 85:517–521.

Provost, E. E., and C. M. Kirkpatrick. 1952. Observations of the hoary bat in Indiana and Illinois. Journal of Mammalogy, 33:110–113.

Pucek, M. 1959. The effect of the venom of the European water shrew (Neomys fodiens fodiens Pennant). Acta Theriologica,

3:93–104.

Purcell, A. 1980. Seasonal movements and home ranges of mule deer at Lava Beds National Monument. M.S. thesis, Oregon State University, Corvallis, 69 pp.

Pyrah, D. 1984. Social distribution and population estimates of coyotes in north-central Montana. The Journal of Wildlife Management, 48:679–690.

Quadagno, D. M. 1968. Home range size in feral house mice. Journal of Mammalogy, 49:149–151.

Quay, W. B. 1953. Seasonal and sexual differences in the dorsal skin gland of the kangaroo rat (Dipodomys). Journal of Mammalogy, 34:1–14.

———. 1965. Variation and taxonomic significance in the sebaceous caudal glands of pocket mice (Rodentia: Heteromyidae). The Southwestern Naturalist, 10:282–287.

———. 1970. Integument and derivatives. Pp. 1–56, in Biology of bats (W. A. Wimsatt, ed.). Academic Press, New York, 2:1–477.

Quick, H. F. 1944. Habits and economics of the New York weasel in Michigan. The Journal of Wildlife Management, 8:71–78.

———. 1951. Notes on the ecology of weasels in Gunnison County, Colorado. Journal of Mammalogy, 32:281–290.

———. 1952. Some characteristics of wolverine fur. Journal of Mammalogy, 33:492–493.

———. 1953a. Occurrence of porcupine quills in carnivorous mammals. Journal of Mammalogy, 34:256–259.

———. 1953b. Wolverine, fisher, and marten studies in a wilderness region. Transactions of the North American Wildlife Conference, 18:513–533.

Quimby, D. C., and J. E. Gaab. 1957. Mandibular dentition as an age indicator in Rocky Mountain elk. The Journal of Wildlife Management, 21:435–451.

Quinn, N. W. S., and G. Parker. 1987. Lynx. Pp. 683–694, in Wild furbearer management and conservation in North America (M. Novak, J. A. Baker, M. E. Obbard, and B. Malloch, eds.). Ontario Ministry of Natural Resources, Toronto, 1150 pp.

Rabb, G. B. 1959. Reproductive and vocal behavior in captive pumas. Journal of Mammalogy, 40:616–617.

Racey, K. 1928. Notes on Phenacomys intermedius. The Murrelet, 9:54–56.

———. 1929. Observations on Neurotrichus gibbsii gibbsii. The Murrelet, 10:61–62.

Racey, P. A. 1973. Environmental factors affecting the length of gestation in heterothermic bats. Journal of Reproduction and Fertility, suppl., 19:175–189.

Rafinesque, C. S. 1817a. Descriptions of seven new genera of North American quadrupeds. American Monthly Magazine and Critical Review, 2(1):44–46.

———. 1817b. Museum of Natural Sciences. American Monthly Magazine and Critical Review, 1:431–442.

———. 1819. Description of a new species of marten (Mustela vulpina). American Journal of Science, 1819:82–84.

———. 1832. Cougars of Oregon. Atlantic Journal, 1:62–63.

Ramirez, P., Jr., and M. [G.] Hornocker. 1981. Small mammal populations in different-aged clearcuts in northwestern Montana. Journal of Mammalogy, 62:400–403.

Randall, J. A. 1978. Behavioral mechanisms of habitat segregation between sympatric species of Microtus: habitat preference and interspecific dominance. Behavioral Ecology and Sociobiology, 3:187–202.

Randall, J. A., and R. E. Johnson. 1979. Population densities and habitat occupancy by *Microtus longicaudus* and *M. montanus*. Journal of Mammalogy, 60:217–219.

Raphael, M. G. 1984. Late fall breeding of the northern flying squirrel, *Glaucomys sabrinus*. Journal of Mammalogy, 65:138–139.

Rasmussen, D. L. 1974. New Quaternary fossil localities in the Upper Clark Fork River Valley, western Montana. Northwest Geology, 3:62–70.

Rasweiler, J. J., IV. 1977. The care and management of bats as laboratory animals. Pp. 519–618, *in* Biology of bats (W. A. Wimsatt, ed.). Academic Press, New York, 3:1–651.

Rausch, R. A. 1967. Some aspects of the population ecology of wolves, Alaska. American Zoologist, 7:253–265.

Rausch, R. A., and A. M. Pearson. 1972. Notes on the wolverine in Alaska and the Yukon Territory. The Journal of Wildlife Management, 36:249–268.

Rausch, R. L. 1953. On the status of some Arctic mammals. Arctic, 6:91–148 (not seen, cited in Voigt, 1987).

———. 1963. Geographic variation in size of North American brown bears, Ursus arctos L., as indicated by condylobasal length. Canadian Journal of Zoology, 41:33–45.

Ray, C. E. 1976. Geography of phocid evolution. Systematic Zoology, 25:391–419.

Redfield, J. A., C. J. Krebs, and M. J. Taitt. 1977. Competition between *Peromyscus maniculatus* and *Microtus townsendii* in grasslands of coastal British Columbia. The Journal of Animal Ecology, 46:607–616.

Reduker, D. W. 1983. Functional analysis of the masticatory apparatus in two species of *Myotis*. Journal of Mammalogy, 64:277–286.

Reduker, D. W., T. L. Yates, and I. F. Greenbaum. 1983. Evolutionary affinities among southwestern long-eared *Myotis* (Chiroptera: Vespertilionidae). Journal of Mammalogy, 64:666–677.

Reeves, H. M., and R. M. Williams. 1956. Reproduction, size, and mortality in the Rocky Mountain muskrat. Journal of Mammalogy, 37:494–500.

Reichman, O. J. 1975. Relation of desert rodent diets to available resources. Journal of Mammalogy, 56:731–751.

Reichman, O. J., T. G. Whitman, and G. A. Ruffner. 1982. Adaptive geometry of burrow spacing in two pocket gopher populations. Ecology, 63:687–695.

Reid, D. G., S. M. Herrero, and T. E. Code. 1988. River otters as agents of water loss from beaver ponds. Journal of Mammalogy, 69:100–107.

Reid, V. H., R. M. Hansen, and A. L. Ward. 1966. Counting mounds and earth plugs to census mountain pocket gophers. The Journal of Wildlife Management, 30:327–334.

Reimer, J. D., and M. L. Petras. 1967. Breeding structure of the house mouse, *Mus musculus*, in a population cage. Journal of Mammalogy, 48:88–99.

———. 1968. Some aspects of commensal populations of *Mus musculus* in southwestern Ontario. The Canadian Field-Naturalist, 82:32–42.

Reith, C. C. 1980. Shifts in time of activity by *Lasionycteris noctivagans*. Journal of Mammalogy, 61:104–108.

Rensberger, J. M. 1975. *Haplomys* and its bearing on the origin of the aplodontoid rodents. Journal of Mammalogy, 56:1–14.

Repenning, C. A. 1976. Adaptive evolution of sea lions and walruses. Systematic Zoology, 25:375–390.

Repenning, C. A., and F. Grady. 1988. The microtine rodents of the Cheetah Room fauna, Hamilton Cave, West Virginia, and the spontaneous origin of *Synaptomys*. United States Geological Survey Bulletin, 1853:1–32.

Repenning, C. A., and R. H. Tedford. 1977. Otarioid seals of the Neogene. United States Geological Survey Professional Paper, 992:1–93 + 24 pls.

Reynolds, D. G., and J. J. Beecham. 1980. Home range activities and reproduction of black bears in west-central Idaho. International Conference on Bear Research and Management, 4:181–190.

Reynolds, H. C. 1945. Some aspects of the life history and ecology of the opossum in central Missouri. Journal of Mammalogy, 26:361–379.

Reynolds, R. P. 1981. Elevational record for *Euderma maculatum* (Chiroptera: Vespertilionidae). The Southwestern Naturalist, 26:91–92.

Reynolds, R. T. 1970. Nest observations of the long-eared owl (*Asio otus*) in Benton County, Oregon, with notes on their food habits. The Murrelet, 51:8–9.

Reynolds, T. D., and W. L. Wakkinen. 1987. Characteristics of the burrows of four species of rodents in undisturbed soils in southeastern Idaho. The American Midland Naturalist, 118:245–250.

Rhoades, F. 1986. Small mammal mycophagy near woody debris accumulations in the Stehekin River Valley, Washington. Northwest Science, 60:150–153.

Rhoads, S. N. 1894a. Descriptions of three new rodents from California and Oregon. The American Naturalist, 28:67–71.

———. 1893 [1894b]. Geographic variation in Bassariscus astutus, with description of a new species. Proceedings of the Academy of Natural Sciences of Philadelphia, 45:413–418.

———. 1894c. Descriptions of four new species and two subspecies of white-footed mice from the United States and British Columbia. Proceedings of the Academy of Natural Sciences of Philadelphia, 46:253–261.

———. 1894d. Description of a new genus and species of arvicoline rodent from the United States. The American Naturalist, 28:182–185.

———. 1894e. Some proposed changes in the nomenclature of the American mammals. *In* Reprint of the North American Zoology of 1890 (G. Ord). Haddonfield, New Jersey. Appendix, pp. 1–52 (not seen, cited in Zoological Records, 31:18).

———. 1894 [1895a]. A new jumping mouse from the Pacific slope. Proceedings of the Academy of Natural Sciences of Philadelphia, 47:421–422.

———. 1895b. Additions to the mammal fauna of British Columbia. The American Naturalist, 29:940–942.

———. 1897 [1898a]. A revision of the west American flying squirrels. Proceedings of the Academy of Natural Sciences of Philadelphia, 49:314–327.

———. 1897 [1898b]. Notes on living and extinct species of North American Bovidae. Proceedings of the Academy of Natural Sciences of Philadelphia, 49:483–502.

————. 1898c. Contributions to a revision of the North American beavers, otters and fishers. Transactions of the American Philosophical Society, new series, 19:417–439 + 5 pls.

————. 1901 [1902]a. On the common brown bats of peninsular Florida and southern California. Proceedings of the Academy of Natural Sciences of Philadelphia, 53:618–619.

————. 1902b. Synopsis of the American martens. Proceedings of the Academy of Natural Sciences of Philadelphia, 54:443–460.

Rhodes, M. N. 1982. Morphological changes in introduced eastern cottontails in Oregon. M.S. thesis, Oregon State University, Corvallis, 45 pp.

Richardson, A. H. 1971. The Rocky Mountain goat in the Black Hills. South Dakota Game, Fish and Parks Bulletin, 2:1–24.

Richardson, J. 1828. Short characters of a few quadrupeds procured on Capt. Franklin's late expedition. The Zoological Journal, 3:516–520.

————. 1829a. Fauna boreali-americana; or the zoology of the northern parts of British America: containing descriptions of the objects of natural history collected on the late Northern Land Expeditions, under command of Captain Sir John Franklin, R.N. Part first, containing the quadrupeds. John Murray, London, 1:1–300.

————. 1829b. On Aplodontia, a new genus of the Order Rodentia, constituted from the reception of the sewellel, a burrowing animal which inhabits the north western coast of America. The Zoological Journal, 4:333–337.

————. 1839. The zoölogy of Captain Beechey's voyage. Compiled from collections and notes made, &c., during a voyage to the Pacific and Behring's Straits, performed in his Majesty's ship Blossom, under the command of Capt. F. W. Beechey, R.N., &c. Mammalia, Henry G. Bohn, London.

Richardson, W. B. 1942. Ring-tailed cats (Bassariscus astutus): their growth and development. Journal of Mammalogy, 23:17–26.

Richens, V. B. 1965. An evaluation of control on the Wasatch pocket gopher. The Journal of Wildlife Management, 29:413–425.

————. 1966. Notes on the digging activity of a northern pocket gopher. Journal of Mammalogy, 47:531–533.

Rickard, W. H. 1960. The distribution of small mammals in relation to the climax vegetation mosaic in eastern Washington and northern Idaho. Ecology, 41:99–106.

Rickart, E. A. 1972. An analysis of barn owl and great horned owl pellets from western Nebraska. The Prairie Naturalist, 4:35–38.

————. 1982. Annual cycles of activity and body composition in Spermophilus townsendii mollis. Canadian Journal of Zoology, 60L:3298–3306.

————. 1986. Postnatal growth of the Piute ground squirrel (Spermophilus mollis). Journal of Mammalogy, 67:412–416.

————. 1987. Spermophilus townsendii. Mammalian Species, 268:1–6.

————. 1988. Population structure of the Piute ground squirrel (Spermophilus mollis). The Southwestern Naturalist, 33:91–96.

————. 1989. Variation in renal structure and urine concentrating capacity among ground squirrels of the Spermophilus townsendii complex (Rodentia: Sciuridae).

Comparative Biochemistry and Physiology, A. Comparative Physiology, 92:531–534.

Rickart, E. A., and E. Yensen. 1991. Spermophilus washingtoni. Mammalian Species, 371:1–5.

Rickart, E. A., R. S. Hoffmann, and M. Rosenfeld. 1985. Karyotype of Spermophilus townsendii artemesiae (Rodentia: Sciuridae) and chromosomal variation in the Spermophilus townsendii complex. Mammalian Chromosome Newsletter, 26:94–102.

Riddle, B. R., and J. R. Choate. 1986. Systematics and biogeography of Onychomys leucogaster in western North America. Journal of Mammalogy, 67:233–255.

Rideout, C. B. 1974. Comparison of techniques for capturing mountain goats. The Journal of Wildlife Management, 38:573–575.

Rideout, C. B., and R. S. Hoffmann. 1975. Oreamnos americanus. Mammalian Species, 63:1–6.

Riley, F. 1967. Fur seal industry of the Pribilof Islands, 1786–1965. United States Fish and Wildlife Circular, 275:1–12.

Risenhoover, K. L., and J. A. Bailey. 1985. Foraging ecology of mountain sheep: implications for habitat management. The Journal of Wildlife Management, 49:797–804.

Risenhoover, K. L., J. A. Bailey, and L. A. Wakelyn. 1988. Assessing the Rocky Mountain bighorn sheep management problem. Wildlife Society Bulletin, 16:346–352.

Rissman, E. F., and R. E. Johnston. 1985. Female reproductive development is not activated by male California voles exposed to family cues. Biology of Reproduction, 32:352–360.

Robey, E. H., Jr., H. D. Smith, and M. C. Belk. 1987. Niche pattern in Great Basin rodent fauna. The Great Basin Naturalist, 47:488–496.

Robinette, W. L., J. S. Gashwiler, and O. W. Morris. 1959. Food habits of the cougar in Utah and Nevada. The Journal of Wildlife Management, 23:261–273.

————. 1961. Notes on cougar productivity and life history. Journal of Mammalogy, 42:204–217.

Robinson, D. J., and I. McT. Cowan. 1954. An introduced population of the gray squirrel (Sciurus carolinensis Gmelin) in British Columbia. Canadian Journal of Zoology, 32:261–282.

Robinson, J. W., and R. S. Hoffmann. 1975. Geographical and interspecific cranial variation in big-eared ground squirrels (Spermophilus): a multivariate study. Systematic Zoology, 24:79–88.

Robinson, S. R. 1980. Antipredator behavior and predator recognition in Beldings ground squirrels. Animal Behaviour, 28:840–852.

Roest, A. I. 1951. Mammals of the Oregon Caves area, Josephine County. Journal of Mammalogy, 32:345–351.

————. 1953. Notes on pikas from the Oregon Cascades. Journal of Mammalogy, 34:132–133.

————. 1964. A ribbon seal from California. Journal of Mammalogy, 45:416–420.

Roffe, T. J., and B. R. Mate. 1984. Abundances and feeding habits of pinnipeds in the Rogue River, Oregon. The Journal of Wildlife Management, 48:1262–1274.

Rogers, L. E., and J. D. Hedlund. 1980. A comparison of small mammal populations occupying three distinct shrub-steppe communities in eastern Oregon. Northwest Science, 54:183–

186.

Rogers, L. L. 1974. Shedding of foot pads by black bears during denning. Journal of Mammalogy, 55:672–674.

Rogers, M. A. 1991a. Evolutionary differentiation with the northern Great Basin pocket gopher, *Thomomys townsendii.* II. Genetic variation and biogeographic considerations. The Great Basin Naturalist, 51:127–152.

———. 1991b. Evolutionary differentiation with the northern Great Basin pocket gopher, *Thomomys townsendii.* I. Morphological variation. The Great Basin Naturalist, 51:109–126.

Rogowitz, G. L. 1990. Seasonal energetics of the white-tailed jackrabbit (*Lepus townsendii*). Journal of Mammalogy, 71:277–285.

———. 1992. Reproduction of white-tailed jackrabbits on semi-arid range. The Journal of Wildlife Management, 56:676–684.

Rogowitz, G. L., and J. A. Gessaman. 1990. Influence of air temperature, wind and irradiance on metabolism of white-tailed jackrabbits. Journal of Thermal Biology, 15:125–131.

Rogowitz, G. L., and M. L. Wolfe. 1991. Intraspecific variation in life-history traits of the white-tailed jackrabbit (*Lepus townsendii*). Journal of Mammalogy, 72:796–806.

Rohweder, R. [S.] 1984. Oregon's Idaho deer. Oregon Wildlife, 39(2):9–11.

Rohweder, R. [S.], J. Melland, and C. Maser. 1979. A new record of Washington ground squirrels in Oregon. The Murrelet, 60:28–29.

Rolley, R. E. 1987. Bobcat. Pp. 671–681, *in* Wild Furbearer Management and Conservation in North America (M. Novak, J. A. Baker, M. E. Obbard, and B. Malloch, eds.). Ontario Ministry of Natural Resources, Toronto, 1150 pp.

Romesburg, H. C. 1981. Wildlife science: gaining reliable knowledge. The Journal of Wildlife Management, 45:293–313.

———. 1991. On improving the natural resources and environmental sciences. The Journal of Wildlife Management, 55:744–756.

Ronald, K., J. Selley, and P. Healey. 1982. Seals: Phocidae, Otariidae, and Odobenidae. Pp. 769–827, *in* Wild mammals of North America: biology, management, and economics (J. A. Chapman and G. A. Feldhamer, eds.). The Johns Hopkins University Press, Baltimore, 1147 pp.

Rongstad, O. J. 1969. Gross prenatal development of cottontail rabbits. The Journal of Wildlife Management, 33:164–168.

Rongstad, O. J., and J. R. Tester. 1971. Behavior and maternal relations of young snowshoe hares. The Journal of Wildlife Management, 35:338–346.

Rood, J. P. 1966. Observations on the reproduction of Peromyscus in captivity. The American Midland Naturalist, 76:496–503.

Rosatte, R. C. 1987. Advances in rabies research and control: applications for the wildlife profession. Wildlife Society Bulletin, 15:504–511.

Rosenberg, D. K. 1991. Characteristics of northern flying squirrel and Townsend's chipmunk populations in second- and old-growth forests. M.S. thesis, Oregon State University, Corvallis, 61 pp.

Rosenberg, D. K., and R. G. Anthony. 1992. Characteristics of northern flying squirrel populations in young second- and old-growth forests in western Oregon. Canadian Journal of Zoology, 70:161–166.

———. 1993a. Differences in Townsend's chipmunk populations between second- and old-growth forests in western Oregon. The Journal of Wildlife Management, 57:365–373.

———. 1993b. Differences in trapping mortality rates of northern flying squirrels. Canadian Journal of Zoology, 71:660–663.

Rosenberg, D. K., K. A. Swindle, and R. G. Anthony. 1994. Habitat associations of California red-backed voles in young and old-growth forests in western Oregon. Northwest Science, 68:266–272.

Rosenberg, D. K., C. J. Zabel, B. R. Noon, and E. C. Meslow. 1994. Northern spotted owls: influence of prey base—a comment. Ecology, 75:1512–1515.

Rosenzweig, M. L. 1966. Community structure in sympatric Carnivora. Journal of Mammalogy, 47:602–612.

Ross, A. 1961. Notes on food habits of bats. Journal of Mammalogy, 42:66–71.

———. 1967. Ecological aspects of the food habits of insectivorous bats. Proceedings of the Western Foundation of Vertebrate Zoology, 1:204–263.

Roth, D., and L. R. Powers. 1979. Comparative feeding and roosting habits of three sympatric owls in southwestern Idaho. The Murrelet, 60:12–15.

Roth, E. L. 1972. Late Pleistocene mammals from Klein Cave, Kerr County, Texas. Texas Journal of Science, 24:75–84.

Rowan, W., and L. B. Keith. 1956. Reproductive potential and sex ratios of snowshoe hares in northern Alberta. Canadian Journal of Zoology, 34:273–281.

Rowe, F. P. 1970. The response of wild house mice (*Mus musculus*) to live-traps marked by their own and by a foreign mouse odour. Journal of Zoology (London), 162:517–520.

Rubin, R. D. 1979. A model for the relationship between water and energy metabolism and body size in pocket mice. Dissertation Abstracts International, 40B:1074–1075.

Ruffer, D. G. 1965a. Burrows and burrowing behavior of *Onychomys leucogaster.* Journal of Mammalogy, 46:241–247.

———. 1965b. Sexual behaviour of the northern grasshopper mouse (*Onychomys leucogaster*). Animal Behaviour, 13:447–452.

———. 1966. Observations on the calls of the grasshopper mouse (*Onychomys leucogaster*). Ohio Journal of Science, 66:219–220.

———. 1968. Agonistic behavior of the northern grasshopper mouse (*Onychomys leucogaster breviauritus*). Journal of Mammalogy, 49:481–487.

Rusch, D. A., and W. G. Reeder. 1978. Population ecology of Alberta red squirrels. Ecology, 59:400–420.

Rust, A. K. 1978. Activity rhythms in the shrews, Sorex sinuosus Greinnell and Sorex trowbridgii Baird. The American Midland Naturalist, 99:369–382.

Rust, C. C. 1962. Temperature as a modifying factor in the spring pelage change of short-tailed weasels. Journal of Mammalogy, 43:323–328.

Rust, H. J. 1946. Mammals of northern Idaho. Journal of Mammalogy, 27:308–327.

Ruth, F. S. 1954. Wolverine seen in Squaw Valley, California.

Journal of Mammalogy, 35:594–595.

Sabine, J. 1823. Zoological appendix. Pp. 647–668, *in* Narrative of a journey to the shores of the polar sea in 1819, 20, 21, and 22 (J. Franklin). John Murray, London, 768 pp. Reprinted 1969 by Greenwood Press, Publisher, New York.

Sadleir, R. M. F. S. 1969. The ecology of reproduction in wild and domestic mammals. Methuen and Company, Ltd., London, 321 pp.

Sakai, H. F., and B. R. Noon. 1993. Dusky-footed woodrat abundance in different-aged forests in northwestern Califonia. The Journal of Wildlife Management, 57:373–382.

Salmon, T. P., and R. E. Marsh. 1989. California ground-squirrel trapping influenced by anal-gland odors. Journal of Mammalogy, 70:428–431.

Samuel, D. E., and B. B. Nelson. 1982. Foxes: *Vulpes vulpes* and allies. Pp. 475–490, *in* Wild mammals of North America: biology, management, and economics (J. A. Chapman and G. A. Feldhamer, eds.). The Johns Hopkins University Press, Baltimore, 1147 pp.

Sandegren, F. E., E. W. Chu, and J. E. Vandevere. 1973. Maternal behavior in the California sea otter. Journal of Mammalogy, 54:668–679.

Sanderson, G. C. 1949. Growth and behavior of a litter of captive long-tailed weasels. Journal of Mammalogy, 30:412–415.

———. 1961. Estimating opossum populations by marking young. The Journal of Wildlife Management, 25:20–27.

Sanderson, G. C., and A. V. Nalbandov. 1973. The reproductive cycle of the raccoon in Illinois. Illinois Natural History Survey Bulletin, 31(2):29–85.

Sandidge, L. L. 1953. Food and dens of the opossum (*Didelphis virginiana*) in northeastern Kansas. Transactions of the Kansas Academy of Science, 56:97–106.

Sargeant, A. B. 1972. Red fox spatial characteristics in relation to waterfowl predation. The Journal of Wildlife Management, 36:225–236.

Sargeant, A. B., and D. W. Warner. 1972. Movements and denning habits of a badger. Journal of Mammalogy, 53:207–210.

Sather, J. H. 1958. Biology of the Great Plains muskrat in Nebraska. Wildlife Monographs, 2:1–35.

Saunders, J. K., Jr. 1955. Food habits and range use of the Rocky Mountain goat in the Crazy Mountains, Montana. The Journal of Wildlife Management, 19:429–437.

Saussure, M. H. de. 1860. Note sur quelques mammifères du Mexique. Revue et Magasin de Zoologie, Paris, series 2, 12:241–293.

Savage, D. E., and D. E. Russell. 1983. Mammalian paleofaunas of the world. Addison-Wesley Publishing Company, Reading, Massachusetts, 432 pp.

Say, T. 1823*a*. *In* James, E., Account of an expedition from Pittsburg to the Rocky Mountains, performed in the years 1819 and '20. H. C. Carey and I. Lea, Philadelphia, 1:1–503.

———. 1823*b*. *In* James, E., Account of an expedition from Pittsburg to the Rocky Mountains, performed in the years 1819 and '20. H. C. Carey and I. Lea, Philadelphia, 2:1–442.

Schaefer, V. H. 1981. A test of the possible reduction of the digging activity of moles in pastures by increasing soil nitrogen. Acta Theriologica, 26:118–123.

———. 1982. Movements and diel activity of the coast mole *Scapanus orarius* True. Canadian Journal of Zoology, 60:480–482.

Schaefer, V. H., and R. M. F. S. Sadleir. 1981. Factors influencing molehill construction by the coast mole (*Scapanus orarius* True). Mammalia, 45:31–38.

Schantz, V. S. 1950. A new badger from Montana. Journal of Mammalogy, 31:90–93.

Scharf, D. W. 1935. A Miocene mammalian fauna from Sucker Creek, southeastern Oregon. Pp. 97–118 + 2 pls., *in* Papers concerning the palaeontology of California, Nevada and Oregon, CarnegieInstitution of Washington, Contributions to Palaeontology, 453:1–125.

Scheffer, T. H. 1925. Notes on the breeding of beavers. Journal of Mammalogy, 5:129–130.

———. 1928. Precarious status of the seal and sea-lion on our Northwest coast. Journal of Mammalogy, 9:10–16.

———. 1929. Mountain beavers in the Pacific Northwest: their habits, economic status, and control. United States Department of Agriculture Farmers' Bulletin, 1598:1–18.

———. 1930*a*. Bat matters. The Murrelet, 11:11–13.

———. 1930*b*. Determining the rate of replacement in a species. Journal of Mammalogy, 11:466–469.

———. 1938*a*. Breeding records of Pacific Coast pocket gophers. Journal of Mammalogy, 19:220–224.

———. 1938*b*. Pocket mice of Washington and Oregon in relation to agriculture. United States Department of Agriculture Technical Bulletin, 608:1–15.

———. 1941. Ground squirrel studies in the Four-Rivers Country, Washington. Journal of Mammalogy, 22:270–279.

———. 1946. Re-allocation of the Townsend ground squirrel. Journal of Mammalogy, 27:395–396.

———. 1952. Spring incidence of damage to forest trees by certain mammals. The Murrelet, 33:38–41.

Scheffer, T. H., and C. C. Sperry. 1931. Food habits of the Pacific harbor seal, Phoca richardii. Journal of Mammalogy, 12:214–226.

Scheffer, V. B. 1940. A newly located herd of Pacific white-tailed deer. Journal of Mammalogy, 21:271–282.

———. 1941. Management studies of transplanted beavers in the Pacific Northwest. Transactions of the North American Wildlife Conference, 6:320–326.

———. 1945. Growth and behavior of young sea lions. Journal of Mammalogy, 26:390–392.

———. 1950*a*. Winter injury to young fur seals on the Northwest coast. California Fish and Game, 36:378–379.

———. 1950*b*. The food of the Alaska fur seal. United States Fish and Wildlife Service Wildlife Leaflet, 329:1–16.

———. 1962. Pelage and surface topography of the northern fur seal. North American Fauna, 64:1–206.

———. 1993*a*. The Olympic goat controversy: a perspective. Conservation Biology, 7:916–919.

———. 1993*b*. Reply to the Anunsens. Conservation Biology, 7:958.

Scheffer, V. B., and K. W. Kenyon. 1963. Elephant seal in Puget Sound, Washington. The Murrelet, 44:23–24.

Schladweiler, J. L., and G. L. Storm. 1969. Den-use by mink. The Journal of Wildlife Management, 33:1025–1026.

Schones, R. S. 1978. The effects of prescribed burning on mule deer wintering at Lava Beds National Monument. M.S. thesis,

Oregon State University, Corvallis, 70 pp.

Schowalter, D. B. 1980. Swarming, reproduction, and early hibernation of *Myotis lucifugus* and *M. volans* in Alberta, Canada. Journal of Mammalogy, 61:350–354.

Schowalter, D. B., and J. R. Gunson. 1979. Reproductive biology of the big brown bat (*Eptesicus fuscus*) in Alberta. The Canadian Field-Naturalist, 93:48–54.

Schowalter, D. B., W. J. Dorward, and J. R. Gunson. 1978. Seasonal occurrence of silver-haired bats (*Lasionycteris noctivagans*) in Alberta and British Columbia. The Canadian Field-Naturalist, 92:288–291.

Schramm, P. 1961. Copulation and gestation in the pocket gopher. Journal of Mammalogy, 42:167–170.

Schreber, J. C. D. von. 1775. Die Säugthiere in Abbildungen nach der Natur mit Beschreibungen. Verlegts Wolfgang Walther, Erlangen, theil 2, heft 13, pl. 92.

———. 1776. Die Säugthiere in Abbildungen nach der Natur, mit Beschreibungen, theil 3, heft 17, pl. 83B and p. 300; theil 3, heft 17, pl. 121; theil 3, heft 18, pl. 126B.

———. 1777. Die Säugthiere in Abbildungen nach der Natur, mit Beschreibungen, theil 3, heft 95, pl. 109B; pl. 127B.

———. 1778. Die Säugthiere in Abbildungen nach der Natur, mit Beschreibungen, theil 3, p. 520.

Schreiber, R. K. 1978. Bioenergetics of the Great Basin pocket mouse, Perognathus parvus. Acta Theriologica, 23:469–487.

———. 1979. Coefficients of digestibility and caloric diet of rodents in the northern Great Basin Desert. Journal of Mammalogy, 60:416–420.

Schwartz, A. 1955. The status of the species of the *Brasiliensis* group of the genus *Tadarida*. Journal of Mammalogy, 36:106–109.

Schwartz, C. W., and E. R. Schwartz. 1981. The wild mammals of Missouri. Revised ed. University of Missouri Press and Missouri Department of Conservation, Columbia, 356 pp.

Schwartz, O. A., and V. C. Bleich. 1975. Comparative growth in two species of woodrats, *Neotoma lepida intermedia* and *Neotoma albigula venusta*. Journal of Mammalogy, 56:653–666.

Scott, T. G. 1943. Some food coactions of the northern plains red fox. Ecological Monographs, 13:427–479.

Scott, T. G., and W. D. Klimstra. 1955. Red foxes and a declining prey population. Southern Illinois University Publications, Monograph Series, 1:1–123.

Scott, W. B. 1937. A history of land mammals in the Western Hemisphere. The Macmillan Company, New York, 786 pp.

Scott, W. E. 1951. Wisconsin's first prairie spotted skunk, and other notes. Journal of Mammalogy, 32:363.

Scrivner, J. H., and H. D. Smith. 1981. Pocket gophers (*Thomomys talpoides*) in successional stages of spruce-fir forest in Idaho. The Great Basin Naturalist, 41:362–367.

Sealander, J. A. 1979. A guide to Arkansas mammals. River Road Press, Conway, Arkansas, 313 pp.

Seidel, D. R., and E. S. Booth. 1960. Biology and breeding habits of the meadow mouse, *Microtus montanus*, in eastern Washington. Walla Walla College Publications of the Department of Biological Sciences and the Biological Station, 29:1–14.

Seidensticker, J. C., IV. 1968. Notes on the food habits of the great-horned owl in Montana. The Murrelet, 49:1–3.

Seidensticker, J. C., IV, M. G. Hornocker, W. V. Wiles, and J. P.

Messick. 1973. Mountain lion social organization in the Idaho Primitive Area. Wildlife Monographs, 35:1–60.

Selko, L. F. 1937. Food habits of Iowa skunks in the fall of 1936. The Journal of Wildlife Management, 1:70–76.

Selleck, D. M., and B. Glading. 1943. Food habits of nesting barn owls and marsh hawks at Dune Lakes, California, as determined by the "cage nest" method. California Fish and Game, 29:122–131.

Seton, E. T. 1920. Acrobatic skunks. Journal of Mammalogy, 1:140.

———. 1929a. Order Carnivora or flesh-eaters: cats, wolves, and foxes. *In* Lives of game animals: an account of those land animals in America, north of the Mexican border, which are considered "game," either because they have held the attention of sportsmen, or received the protection of law. Doubleday, Doran & Company, Inc., Garden City, New York, 1(1):1–337.

———. 1929b. Order Rodentia, etc., squirrels, rabbits, armadillo, and opposum. *In* Lives of game animals: an account of those land animals in America, north of the Mexican border, which are considered "game," either because they have held the attention of sportsmen, or received the protection of law. Doubleday, Doran & Company, Inc., Garden City, New York, 4(2):441–949.

———. 1929c. Order Carnivora or flesh-eaters: bears, coons, badgers, skunks, and weasels. *In* Lives of game animals: an account of those land animals in America, north of the Mexican border, which are considered "game," either because they have held the attention of sportsmen, or received the protection of law. Doubleday, Doran & Company, Inc., Garden City, New York, 2(2):369–746.

Seton-Thompson, E. 1898. A list of big game of North America. Forest and Stream, 51:285–287 (not seen, cited in Smith, 1991).

Severaid, J. H. 1945. Breeding potential and artificial propagation of the snowshoe hare. The Journal of Wildlife Management, 9:290–295.

———. 1950. The gestation period of the pika (*Ochotona princeps*). Journal of Mammalogy, 31:356–357.

Shackleton, D. M. 1985. Ovis canadensis. Mammalian Species, 230:1–9.

Shackleton, D. M., R. G. Peterson, J. Haywood, and A. Blottrell. 1984. Gestation period in *Ovis canadensis*. Journal of Mammalogy, 65:337–338.

Shadle, A. R. 1946. Copulation in the porcupine. The Journal of Wildlife Management, 10:159–162 + 1 pl.

———. 1947. Porcupine spine penetration. Journal of Mammalogy, 28:180–181.

———. 1950. Feeding, care, and handling of captive porcupines (*Erethizon*). Journal of Mammalogy, 31:411–416.

———. 1951. Laboratory copulations and gestations of porcupine, *Erethizon dorsatum*. Journal of Mammalogy, 32:219–221.

———. 1952. Sexual maturity and first recorded compulation of a 16-month male porcupine, *Erethizon dorsatum*. Journal of Mammalogy, 33:239–241.

Shadle, A. R., and D. Po-Chedley. 1949. Rate of penetration of a porcupine spine. Journal of Mammalogy, 30:172–173.

Shadle, A. R., M. Smelzer, and M. Metz. 1946. The sex reactions of porcupines (Erethizon d. dorsatum) before and after copulation. Journal of Mammalogy, 27:116–121.

Shamel, H. H. 1931. Notes on the American bats of the genus Tadarida. Proceedings of the United States National Museum, 78(19):1–27.

Shank, C. C. 1977. Cooperative defense by bighorn sheep. Journal of Mammalogy, 58:243–244.

———. 1982. Age-sex differences in the diets of wintering Rocky Mountain bighorn sheep. Ecology, 63:627–633.

Shannon, N. H., R. J. Hudson, V. C. Brink, and W. D. Kitts. 1975. Determinants of spatial distribution of Rocky Mountain bighorn sheep. The Journal of Wildlife Management, 39:387–401.

Shaughnessy, P. D., and F. H. Fay. 1977. A review of the taxonomy and nomenclature of North Pacific harbour seals. Journal of Zoology (London), 182:385–419.

Shaw, G. 1801. General zoology or systematic natural history. Thomas Davison, London, 2:1–560.

———. 1804. Naturalist's miscellany. E. Nodder, London, 51:text to pl. 610.

Shaw, W. T. 1920. The cost of a squirrel and squirrel control. Washington Agricultural Experiment Station Popular Bulletin, 118:1–19.

———. 1921. Moisture and altitude as factors in determining the seasonal activities of the Townsend ground squirrel in Washington. Ecology, 2:189–192.

———. 1924. Alpine life of the heather vole (Phenacomys olympicus). Journal of Mammalogy, 5:12–15.

———. 1925a. The food of ground squirrels. The American Naturalist, 59:250–264.

———. 1925b. Breeding and development of the Columbian ground squirrel. Journal of Mammalogy, 6:106–113.

———. 1926. The storing habit of the Columbian ground squirrel. The American Naturalist, 60:367–373.

———. 1928. The spring and summer activities of the dusky skunk in captivity. New York State Museum Handbook, 4:1–92.

———. 1936. Moisture and its relation to the cone-storing habit of the western pine squirrel. Journal of Mammalogy, 17:337–349.

———. 1944. Brood nests and young of two western chipmunks in the Olympic Mountains of Washington. Journal of Mammalogy, 25:274–284.

Shay, R. 1976. The great moose adventure. Oregon Wildlife, 31(2):3–5.

Shea, M. E., N. L. Rollins, R. T. Bowyer, and A. G. Clark. 1985. Corpora lutea number as related to fisher age and distribution in Maine. The Journal of Wildlife Management, 49:37–40.

Sheehy, D. P. 1978. Characteristics of shrubland habitat associated with mule deer fawns at birth and during early life in southeastern Oregon. Oregon Department of Fish and Wildlife, Information Report Series, Wildlife, 78–1:1–31.

Sheldon, J. W. 1992. Wild dogs: the natural history of the nondomestic Canidae. Academic Press, Inc., San Diego, 248 pp.

Sheldon, W. G., and W. G. Toll. 1964. Feeding habits of the river otter in a reservoir in central Massachusetts. Journal of Mammalogy, 45:449–455.

Sheppard, D. H. 1969. A comparison of reproduction in two chipmunk species (Eutamias). Canadian Journal of Zoology, 47:603–608.

———. 1971. Competition between two chipmunk species (Eutamias). Ecology, 52:320–329.

Sherman, H. B. 1937. Breeding habits of the free-tailed bat. Journal of Mammalogy, 18:176–187.

Sherman, P. E. 1973. Agonistic behavior and dominance in Townsend's chipmunks (Eutamias townsendii). M.S. thesis, Portland State University, Portland, Oregon, 52 pp.

Sherman, P. W. 1977. Nepotism and the evolution of alarm calls. Science, 197:1246–1253.

———. 1981. Reproductive competition and infanticide in Belding's ground squirrels and other animals. Pp. 311–331, in Natural selection and social behavior: recent research and new theory (R. D. Alexander and D. W. Tinkle, eds.). Westview Press, Inc. Boulder, Colorado, 627 pp.

Sherman, P. W., and M. L. Morton. 1984. Demography of Belding's ground squirrels. Ecology, 65:1617–1628.

Sherrell, P. E. 1969. A new locality record for Phenacomys albipes. The Murrelet, 50:39.

Shields, P. W. 1960. Movement patterns of brush rabbits in northwestern California. The Journal of Wildlife Management, 24:381–386.

Shimek, S. J., and A. Monk. 1977. Daily activity of sea otter off the Monterey Peninsula, California. The Journal of Wildlife Management, 41:277–283.

Short, H. L. 1978. Analysis of cuticular scales on hairs using the scanning electron microscope. Journal of Mammalogy, 59:261–268.

Shotwell, J. A. 1951. A fossil sea-lion from Fossil Point, Oregon. Proceedings of the Oregon Academy of Sciences, 2:97 (abstract only).

———. 1955. Review of the Pleistocene beaver Dipoides. Journal of Paleontology, 29:129–144.

———. 1956. Hemphillian mammalian assemblage from northeastern Oregon. Bulletin of the Geological Society of America, 67:717–738.

———. 1958a. Evolution and biogeography of the aplodontid and mylagaulid rodents. Evolution, 12:451–484.

———. 1958b. Inter-community relationships in Hemphillian (Mid-Pliocene) mammals. Ecology, 39:271–282.

———. 1963. Mammalian fauna of the Drewsey Formation, Bartlett Mountain, Drinkwater and Otis Basin local faunas. Pp. 70–77, in The Juntura Basin: studies in Earth history and paleoecology (J. A. Shotwell, ed.). Transactions of the American Philosophical Society, new series, 53(1):1–77.

———. 1967a. Late Tertiary geomyoid rodents of Oregon. University of Oregon, Museum of Natural History Bulletin, 9:1–51.

———. 1967b. Peromyscus of the late Tertiary in Oregon. University of Oregon, Museum of Natural History Bulletin, 5:1–35.

———. 1968. Miocene mammals of southeast Oregon. University of Oregon, Museum of Natural History Bulletin, 14:1–67.

———. 1970. Pliocene mammals of southeast Oregon and adjacent Idaho. University of Oregon, Museum of Natural History Bulletin, 17:1–103.

Shotwell, J. A., and D. E. Russell. 1963. Mammalian fauna of the Upper Juntura formation, and the Black Butte local fauna. Pp. 42–69, in The Juntura Basin: studies in Earth history and paleoecology (J. A. Shotwell, ed.). Transactions of the

American Philosophical Society, new series, 53(1):1–77.

Shufeldt, R. W. 1885. Description of Hesperomys truei, a new species belonging to the subfamily Murinae. Proceedings of the United States National Museum, 8:403–408.

Shump, K. A., Jr., and A. U. Shump. 1982. Lasiurus cinereus. Mammalian Species, 185:1–5.

Silver, H., and W. T. Silver. 1969. Growth and behavior of the coyote-like canid of northern New England with observations on canid hybrids. Wildlife Monographs, 17:1–41.

Silver, J. 1933. Mole control. United States Department of Agriculture, Farmer's Bulletin, 1716:1–17.

Simms, D. A. 1979a. North American weasels: resource utilization and distribution. Canadian Journal of Zoology, 57:504–520.

———. 1979b. Studies of an ermine population in southern Ontario. Canadian Journal of Zoology, 57:824–832.

Simons, E. L., and T. M. Bown. 1984. A new species of Peratherium (Didelphidae; Polyprotodonta): the first African marsupial. Journal of Mammalogy, 65:539–548.

Simpson, C. D. 1984. Artiodactyls. Pp. 563–587, in Orders and families of Recent mammals of the world (S. Anderson and J. K. Jones, Jr., eds). John Wiley & Sons, New York, 686 pp.

Simpson, G. G. 1945. The principles of classification and a classification of mammals. Bulletin of the American Museum of Natural History, 85:1–350.

Simpson, M. R. 1993. Myotis californicus. Mammalian Species, 428:1–4.

Sinha, A. A., C. H. Conaway, and K. W. Kenyon. 1966. Reproduction in the female sea otter. The Journal of Wildlife Management, 30:121–130.

Skalski, J. R., and B. J. Verts. 1981. Dynamics of a transferrin polymorphism in a population of Sylvilagus nuttallii. Oecologia, 48:329–332.

Skovlin, J. [M.], and M. Vavra. 1979. Winter diets of elk and deer in the Blue Mountains, Oregon. United States Department of Agriculture, Forest Service, Research Paper, PNW-260:1–21.

Skryja, D. D. 1974. Reproductive biology of the least chipmunk (Eutamias minimus operarius) in southeastern Wyoming. Journal of Mammalogy, 55:221–224.

Slipp, J. W. 1942. Nest and young of the Olympic dusky shrew. Journal of Mammalogy, 23:211–212.

Slobodchikoff, C. N., T. A. Vaughan, and R. M. Warner. 1987. How prey defenses affect a predator's net energetic profit. Journal of Mammalogy, 68:668–671.

Small, R. J., and B. J. Verts. 1983. Responses of a population of Perognathus parvus to removal trapping. Journal of Mammalogy, 64:139–143.

Smartt, R. A. 1978. A comparison of ecological and morphological overlap in a Peromyscus community. Ecology, 59:216–220.

Smith, A. T. 1974. The distribution and dispersal of pikas: influences of behavior and climate. Ecology, 55:1368–1376.

Smith, A. T., and M. L. Weston. 1990. Ochotona princeps. Mammalian Species, 352:1–8.

Smith, C. C. 1968. The adaptive nature of social organization in the genus of three [sic] squirrels Tamiasciurus. Ecological Monographs, 38:31–63.

———. 1970. The coevolution of pine squirrels (Tamiasciurus) and conifers. Ecological Monographs, 40:349–371.

———. 1978. Structure and function of the vocalizations of tree squirrels (Tamiasciurus). Journal of Mammalogy, 59:793–808.

———. 1981. The indivisible niche of Tamiasciurus: an example of nonpartitioning of resources. Ecological Monographs, 51:343–363.

Smith, C. F. 1936. Notes on the habiats of the long-tailed harvest mouse. Journal of Mammalogy, 17:274–278.

———. 1942. The fall food of the brushfield pocket mouse. Journal of Mammalogy, 23:337–339.

Smith, G. W. 1977. Population characteristics of the porcupine in northeastern Oregon. Journal of Mammalogy, 58:674–676.

———. 1979. Movements and home range of the porcupine in northeastern Oregon. Northwest Science, 53:277–282.

———. 1982. Habitat use by porcupines in a ponderosa pine/Douglas-fir forest in northeastern Oregon. Northwest Science, 56:236–240.

Smith, G. W., and D. R. Johnson. 1985. Demography of a Townsend ground squirrel population in southwestern Idaho. Ecology, 66:171–178.

Smith, G. W., and N. C. Nydegger. 1985. A spotlight, line-transect method for surveying jack rabbits. The Journal of Wildlife Management, 49:699–702.

Smith, H. C. 1961. Early biological surveys of Oregon. Oregon State Game Commission Bulletin, 16(8):3–6.

———. 1988. The wandering shrew, Sorex vagrans, in Alberta. The Canadian Field-Naturalist, 102:254–256.

Smith, H. D., G. H. Richins, and C. D. Jorgensen. 1978. Growth of Dipodomys ordii (Rodentia: Heteromyidae). The Great Basin Naturalist, 38:215–221.

Smith, L. 1941. An observation on the nest-building behavior of the opossum. Journal of Mammalogy, 22:201–202.

Smith, M. C. 1968. Red squirrel responses to spruce cone failure in interior Alaska. The Journal of Wildlife Management, 32:305–317.

Smith, W. P. 1982. Status and habitat use of Columbian white-tailed deer in Douglas County, Oregon. Ph.D. dissert., Oregon State University, Corvallis, 273 pp.

———. 1985. Current geographic distribution and abundance of Columbian white-tailed deer, Odocoileus virginianus leucurus (Douglas). Northwest Science, 59:243–251.

———. 1991. Odocoileus virginianus. Mammalian Species, 388:1–13.

Smolen, M. J., and B. L. Keller. 1987. Microtus longicaudus. Mammalian Species, 271:1–7.

Snead, E., and G. O. Hendrickson. 1942. Food habits of the badger in Iowa. Journal of Mammalogy, 23:380–391.

Solounias, N. 1988. Evidence from horn morphology on the phylogenetic relationships of the pronghorn (Antilocapra americana). Journal of Mammalogy, 69:140–143.

Soper, J. D. 1946. Mammals of the northern Great Plains along the international boundary in Canada. Journal of Mammalogy, 27:127–153.

Sorenson, M. W. 1962. Some aspects of water shrew behavior. The American Midland Naturalist, 68:445–462.

Spalding, D. J. 1964. Comparative feeding habits of the fur seal, sea lion and harbour seal on the British Columbia coast. Fisheries Research Board of Canada Bulletin, 146:1–52.

Sparks, D. R. 1968a. Diet of black-tailed jackrabbits on sandhill rangeland in Colorado. Journal of Range Management,

21:203–208.

———. 1968*b*. Occurrence of milk in stomachs of young jackrabbits. Journal of Mammalogy, 49:324–325.

Spencer, A. W., and D. Pettus. 1966. Habitat preferences of five sympatric species of long-tailed shrews. Ecology, 47:677–683.

Spencer, A. W., and H. W. Steinhoff. 1968. An explanation of geographic variation in litter size. Journal of Mammalogy, 49:281–286.

Spencer, D. A. [1959]. Biological and control aspects. Pp. 15–25, *in* The Oregon meadow mouse irruption of 1957–1958. Federal Cooperative Extension Service, Oregon State College, Corvallis, 88 pp.

Spencer, J. L. 1955. Reingestion in three species of lagomorphs. Lloydia, 18:197–199.

Spencer, W. D., and W. J. Zielinski. 1983. Predatory behavior of pine martens. Journal of Mammalogy, 64:715–717.

Spencer, W. D., R. H. Barrett, and W. J. Zielinski. 1983. Marten habitat preferences in the northern Sierra Nevada. The Journal of Wildlife Management, 47:1181–1186.

Speth, R. L. 1969. Patterns and sequences of molts in the Great Basin pocket mouse, *Perognathus parvus.* Journal of Mammalogy, 50:284–290.

Speth, R. L., C. L. Pritchett, and C. D. Jorgensen. 1968. Reproductive activity of *Perognathus parvus.* Journal of Mammalogy, 49:336–337.

Stager, K. E. 1945. Tadarida mexicana in Oregon. Journal of Mammalogy, 26:196.

Stains, H. J. 1956. The raccoon in Kansas. University of Kansas, Museum of Natural History and State Biological Survey of Kansas, Miscellaneous Publication, 10:1–76.

———. 1961. Comparison of temperatures inside and outside two tree dens used by raccoons. Ecology, 42:410–413.

———. 1967. Carnivores and pinnipeds. Pp. 325–354, *in* Recent mammals of the world: a synopsis of families (S. Anderson and J. K. Jones, Jr., eds.). The Ronald Press Company, New York, 453 pp.

———. 1984. Carnivores. Pp. 491–521, *in* Orders and families of Recent mammals of the world (S. Anderson and J. K. Jones, Jr., eds.). John Wiley & Sons, New York, 686 pp.

Stanton, F. W. 1944. Douglas ground squirrel as a predator on nests of upland game birds in Oregon. The Journal of Wildlife Management, 8:153–161.

Stanton, N. L., L. M. Shults, and M. Parker. 1992. Variation in reproductive potential of Wyoming ground squirrels in southeastern Wyoming. The Prairie Naturalist, 24:261–271.

Starrett, A., and P. Starrett. 1956. Merriam shrew, *Sorex merriami*, in Colorado. Journal of Mammalogy, 37:276–277.

States, J. B. 1976. Local adaptations in chipmunk (*Eutamias amoenus*) populations and evolutionary potential at species' borders. Ecological Monographs, 46:221–256.

Stehn, R. A., and M. E. Richmond. 1975. Male-induced pregnancy termination in the prairie vole, Microtus ochrogaster. Science, 187:1211–1213.

Stephens, F. 1906. California mammals. The West Coast Publishing Company, San Diego, California, 351 pp.

Sterling, K. B. 1974. Last of the naturalists: the career of C. Hart Merriam. Arno Press, New York, 478 pp.

Stevens, V. 1982. Factors reflecting mountain goat condition and habitat quality: a comparison of sub-populations in Olympic National Park. Pp. 51–57, *in* Ecological research in National Parks of the Pacific Northwest (E. E. Starkey, J. F. Franklin, and J. W. Matthews, eds.). Oregon State University, Forest Research Laboratory Publication, Corvallis, 142 pp.

Stewart, B. S., and R. L. DeLong. 1990. Sexual differences in migrations and foraging behavior of northern elephant seals. American Zoologist, 30:44A (abstract only).

Stewart, B. S., and H. R. Huber. 1993. Mirounga angustirostris. Mammalian Species, 449:1–10.

Stewart, B. S., et al. 1994. History and present status of the northern elephant seal population. Pp. 29–48, *in* Elephant seals: population ecology, behavior, and physiology (B. J. Le Boeuf and R. M. Laws, eds.). University of California Press, Berkeley, 414 pp.

Stewart, R. W., and J. R. Bider. 1977. Summer activity of muskrats in relation to weather. The Journal of Wildlife Management, 41:487–499.

Stewart, T. W. 1979. Epizootiology of leptospirosis in some wild mammals in western Oregon. M.S. thesis, Oregon State University, Corvallis, 29 pp.

Stickel, L. F. 1954. A comparison of certain methods of measuring ranges of small mammals. Journal of Mammalogy, 35:1–15.

Stienecker, W. E. 1977. Supplemental data on the food habits of the western gray squirrel. California Fish and Game, 63:11–21.

Stienecker, W. [E.], and B. M. Browning. 1970. Food habits of the western gray squirrel. California Fish and Game, 56:36–48.

Stirton, R. A. 1935. A review of the Tertiary beavers. University of California Publications in Geological Sciences, 23:391–457 + 1 chart.

Stock, A. D. 1972. Swimming ability in kangaroo rats. The Southwestern Naturalist, 17:98–99.

Stock, C. 1930. Carnivora new to the Mascall Miocene fauna of eastern Oregon. Carnegie Institution of Washington, Contributions to Paleontology, 404:43–48 + 1 pl.

Stoddart, L. C. 1984. Site fidelity and grouping of neonatal jack rabbits, *Lepus californicus.* Journal of Mammalogy, 65:136–137.

———. 1985. Severe weather related mortality of black-tailed jack rabbits. The Journal of Wildlife Management, 49:696–698.

Stones, R. C., and C. L. Hayward. 1968. Natural history of the desert woodrat, *Neotoma lepida.* The American Midland Naturalist, 80:458–476.

Storch, G., and Z. Qui. 1983. The Neogene mammalian faunas of Ertemte and Harr Obo in Inner Mongolia (Nei Mongol), China. 2. Moles—Insectivora: Talpidae. Senckenbergiana Lethaea, 64:89–127.

Storer, T. I. 1922. The young of the California gray squirrel. Journal of Mammalogy, 3:188–189.

———. 1926. Bats, bat towers, and mosquitoes. Journal of Mammalogy, 7:85–90 + 1 pl.

———. 1930. Summer and autumn breeding of the California ground squirrel. Journal of Mammalogy, 11:235–237.

———. 1931. A colony of Pacific pallid bats. Journal of Mammalogy, 12:244–247.

———. 1969. Mammalogy and the American Society of

Mammalogists, 1919–1969. Journal of Mammalogy, 50:785–793.

Storer, T. I., and D. E. Davis. 1953. Studies on rat reproduction in San Francisco. Journal of Mammalogy, 34:365–373.

Storer, T. I., and L. P. Tevis, Jr. 1955. California grizzly. University of California Press, Berkeley, 335 pp.

Storm, G. L. 1965. Movements and activities of foxes as determined by radio-tracking. The Journal of Wildlife Management, 29:1–13.

———. 1972. Daytime retreats and movements of skunks on farmlands in Illinois. The Journal of Wildlife Management, 36:31–45.

Storm, G. L., and E. D. Ables. 1966. Notes on newborn and full-term wild red foxes. Journal of Mammalogy, 47:116–118.

Storm, G. L., R. D. Andrews, R. L. Phillips, R. A. Bishop, D. B. Siniff, and J. R. Tester. 1976. Morphology, reproduction, dispersal, and mortality of Midwestern red fox populations. Wildlife Monographs, 49:1–82.

Strickland, M. A., and C. W. Douglas. 1987. Marten. Pp. 531–546, in Wild furbearer management and conservation in North America (M. Novak, J. A. Baker, M. E. Obbard, and B. Malloch, eds.). Ontario Ministry of Natural Resources, Toronto, 1150 pp.

Strickland, M. A., C. W. Douglas, M. Novak, and N. P. Hunziger. 1982a. Fisher: Martes pennanti. Pp. 586–598, in Wild mammals of North America: biology, management, and economics (J. A. Chapman and G. A. Feldhamer, eds.). The Johns Hopkins University Press, Baltimore, 1147 pp.

———. 1982b. Marten: Martes americana. Pp. 599–612, in Wild mammals of North America: biology, management, and economics (J. A. Chapman and G. A. Feldhamer, eds.). The Johns Hopkins University Press, Baltimore, 1147 pp.

Stroud, D. C. 1982. Population dynamics of Rattus rattus and R. norvegicus in a riparian habitat. Journal of Mammalogy, 63:151–154.

Studier, E. H., and M. J. O'Farrell. 1972. Biology of Myotis thysanodes and M. lucifugus (Chiroptera: Vespertilionidae). 1 Thermoregulation. Comparative Biochemistry and Physiology, 41A:467–595.

Stuewer, F. W. 1948. A record of red bats mating. Journal of Mammalogy, 29:180–181.

Sturges, F. W. 1955. Weasel preys on larval salamander. Journal of Mammalogy, 36:567–568.

———. 1957. Habitat distributions of birds and mammals in Lostine Canyon, Wallowa Mountains, northeast Oregon. Ph.D. dissert., Oregon State University, Corvallis, 130 pp.

Stussy, R. J. 1994. The effects of forage improvement practices on Roosevelt elk in the Oregon Coast Range. M.S. thesis, Oregon State University, Corvallis, 77 pp.

Stussy, R. J., W. D. Edge, and T. A. O'Neil. 1994. Survival of resident and translocated female elk in the Cascade Mountains of Oregon. Wildlife Society Bulletin, 22:242–247.

Sullins, G. L. 1976. Evaluation of baits and baiting techniques for Belding's ground squirrels. M.S. thesis, Oregon State University, Corvallis, 32 pp.

Sullins, G. L., and B. J. Verts. 1978. Baits and baiting techniques for control of Belding's ground squirrels. The Journal of Wildlife Management, 42:890–896.

Sullins, G. L., D. O. McKay, and B. J. Verts. 1976. Estimating ages of cottontails by periosteal zonations. Northwest Science, 50:17–22.

Sullivan, R. M. 1985. Phyletic, biogeographic, and ecologic relationships among montane populations of least chipmunks (Eutamias minimus) in the Southwest. Systematic Zoology, 34:419–448.

Sullivan, T. P. 1980. Comparative demography of Peromyscus maniculatus and Microtus oregoni populations after logging and burning of coastal forest habitats. Canadian Journal of Zoology, 58:2252–2259.

———. 1990. Responses of red squirrel (Tamiasciurus hudsonicus) populations to supplemental food. Journal of Mammalogy, 71:579–590.

Sullivan, T. P., and W. Klenner. 1992. Response to Koford: red squirrels and supplemental feeding. Journal of Mammalogy, 73:933–936.

Sullivan, T. P., and C. J. Krebs. 1981. Microtus population biology: demography of M. oregoni in southwestern British Columbia. Canadian Journal of Zoology, 59:2092–2102.

Sullivan, T. P., and D. S. Sullivan. 1980. The use of weasels for natural control of mouse and vole populations in a coastal coniferous forest. Oecologia, 47:125–129.

———. 1982. Population dynamics and regulation of the Douglas squirrel (Tamiasciurus douglasii) with supplemental food. Oecologia, 53:264–270.

Sullivan, T. P., B. Jones, and D. S. Sullivan. 1989. Population ecology and conservation of the mountain cottontail, Sylvilagus nuttallii nuttallii, in southern British Columbia. The Canadian Field-Naturalist, 103:335–340.

Sullivan, T. P., D. S. Sullivan, and C. J. Krebs. 1983. Demographic responses of a chipmunk (Eutamias townsendii) population with supplemental food. The Journal of Animal Ecology, 52:743–755.

Sullivan, T. P., D. R. Crump, H. Wieser, and E. A. Dixon. 1990. Response of pocket gophers (Thomomys talpoides) to an operational application of synthetic semiochemicals of stoat (Mustela erminea). Journal of Chemical Ecology, 16:941–949.

Sumner, L., and J. S. Dixon. 1953. Birds and mammals of the Sierra Nevada with records from Sequoia and Kings Canyon national parks. University of California Press, Berkeley, 483 pp.

Suring, L. H. 1975. Habitat use and activity patterns of the Columbian white-tailed deer along the lower Columbia River. M.S. thesis, Oregon State University, Corvallis, 59 pp.

Suring, L. H., and P. A. Vohs, Jr. 1979. Habitat use by Columbian white-tailed deer. The Journal of Wildlife Management, 43:610–619.

Sutton, D. A. 1987. Analysis of Pacific Coast Townsend chipmunks (Rodentia: Sciuridae). The Southwestern Naturalist, 32:371–376.

———. 1993. Tamias townsendii. Mammalian Species, 435:1–6.

Sutton, D. A., and C. F. Nadler. 1969. Chromosomes of the North American chipmunk genus Eutamias. Journal of Mammalogy, 50:524–535.

———. 1974. Systematic revision of three Townsend chipmunks (Eutamias townsendii). The Southwestern Naturalist, 19:199–211.

Svendsen, G. E. 1978. Castor and anal glands of the beaver

(*Castor canadensis*). Journal of Mammalogy, 59:618–620.

———. 1980. Seasonal change in feeding patterns of beaver in southeastern Ohio. The Journal of Wildlife Management, 44:285–290.

———. 1982. Weasels: *Mustela* species. Pp. 613–628, *in* Wild mammals of North America: biology, management, and economics (J. A. Chapman and G. A. Feldhamer, eds.). The Johns Hopkins University Press, Baltimore, 1147 pp.

Svihla, A. 1929 [1930]. Breeding habits and young of the red-backed mouse *Evotomys*. Papers of the Michigan Academy of Science, Arts and Letters, 11:485–490.

———. 1931. The Olympic red-backed mouse. The Murrelet, 12:54.

———. 1932. A comparative life history study of the mice of the genus *Peromyscus*. University of Michigan, Museum of Zoology, Miscellaneous Publication, 24:1–39.

———. 1934. The mountain water shrew. The Murrelet, 15:44–45.

———. 1936. Notes on the hibernation of a western chipmunk. Journal of Mammalogy, 17:289–290.

Svihla, A., and R. D. Svihla. 1931. The Louisiana muskrat. Journal of Mammalogy, 12:12–28.

———. 1933. Notes on the jumping mouse Zapus trinotatus trinotatus Rhoads. Journal of Mammalogy, 14:131–133.

Svihla, R. D. 1931. Mammals of the Uinta Mountains region. Journal of Mammalogy, 12:256–266.

———. 1936. Breeding and young of the grasshopper mouse (Onychomys leucogaster fuscogriseus). Journal of Mammalogy, 17:172–173.

Svoboda, P. L., and J. R. Choate. 1987. Natural history of the Brazilian free-tailed bat in the San Luis Valley of Colorado. Journal of Mammalogy, 68:224–234.

Swarth, H. S. 1912. Report on a collection of birds and mammals from Vancouver Island. University of California Publications in Zoology, 10:1–124.

Swenson, J. E. 1985. Compensatory reproduction in an introduced mountain goat population in the Absaroka Mountains, Montana. The Journal of Wildlife Management, 49:837–843.

Swift, R. J. 1977. The reproductive cycle of the western gray squirrel in Butte County, California. M.S. thesis, California State University, Chico, 78 pp.

Taber, R. D. 1969. Criteria of sex and age. Pp. 325–401, *in* Wildlife management techniques. Third ed. (R. H. Giles, Jr., ed.). The Wildlife Society, Washington, D.C., 623 pp.

Taber, R. D., and R. F. Dasmann. 1954. The black-tailed deer of the chaparral: its life history and management in the north Coast Range of California. California Fish and Game, Game Bulletin, 8:1–163.

Tabor, J. E. 1974. Productivity, survival, and population status of river otter in western Oregon. M.S. thesis, Oregon State University, Corvallis, 62 pp.

Tabor, J. E., and H. M. Wight. 1977. Population status of river otter in western Oregon. The Journal of Wildlife Management, 41:692–699.

Taitt, M. J., and C. J. Krebs. 1981. The effect of extra food on small rodent populations: II. Voles (*Microtus townsendii*). The Journal of Animal Ecology, 50:125–137.

———. 1983. Predation, cover, and food manipulations during a spring decline of *Microtus townsendii*. The Journal of Animal Ecology, 52:837–848.

———. 1985. Population dynamics and cycles. Pp. 567–620, *in* Biology of New World *Microtus* (R. H. Tamarin, ed.). Special Publication, The American Society of Mammalogists, 8:1–893.

Taitt, M. J., J. H. W. Gipps, C. J. Krebs, and Z. Dundjerski. 1981. The effect of food and cover on declining populations of *Microtus townsendii*. Canadian Journal of Zoology, 59:1593–1599.

Tallmon, D., and L. S. Mills. 1994. Use of logs within home ranges of California red-backed voles on the remnant of forest. Journal of Mammalogy, 75:97–101.

Tamarin, R. H., and S. R. Malecha. 1971. The population biology of Hawaiian rodents: demographic parameters. Ecology, 52:383–394.

———. 1972. Reproductive parameters in *Rattus rattus* and *Rattus exulans* of Hawaii, 1968 to 1970. Journal of Mammalogy, 53:513–528.

Tanner, V. M. 1927. Some of the smaller mammals of Mount Timpanogos, Utah. Journal of Mammalogy, 8:250–251.

Taylor, W. P. 1910. Two new rodents from Nevada. University of California Publications in Zoology, 5:283–302.

———. 1915. Description of a new subgenus (*Arborimus*) of *Phenacomys*, with a contribution to knowledge of the habits and distribution of *Phenacomys longicaudus* True. Proceedings of the California Academy of Sciences, series 4, 5:111–161.

———. 1916. The status of the beavers of western North America, with a consideration of the factors in their speciation. University of California Publications in Zoology, 12:413–495.

———. 1918. Revision of the rodent genus *Aplodontia*. University of California Publications in Zoology, 17:435–504.

———. 1919. Notes on mammals collected principally in Washington and California between the years 1853 and 1874 by Dr. James Graham Cooper. Proceedings of the California Academy of Science, series 4, 9:69–121.

———. 1943. The gray fox in captivity. Texas Game and Fish, 1 (9): 12–13, 19.

———. 1954. Food habits and notes on life history of the ring-tailed cat in Texas. Journal of Mammalogy, 35:55–63.

Taylor, W. P., and W. T. Shaw. 1929. Provisional list of land mammals of the state of Washington. Occasional Papers of the Charles R. Conner Museum, 2:1–32.

Tedford, R. H. 1976. Relationship of pinnipeds to other carnivores (Mammalia). Systematic Zoology, 25:363–374.

Teer, J. G. 1964. Predation by long-tailed weasels on eggs of blue-winged teal. The Journal of Wildlife Management, 28:404–406.

Terry, C. J. 1978. Food habits of three sympatric species of Insectivora in western Washington. The Canadian Field-Naturalist, 92:38–44.

———. 1981. Habitat differentiation among three species of Sorex and Neurotrichus gibbsii in Washington. The American Midland Naturalist, 106:119–125.

Tevis, L. [P.], Jr. 1950. Summer behavior of a family of beavers in New York state. Journal of Mammalogy, 31:40–65.

———. 1952. Autumn foods of chipmunks and golden-mantled ground squirrels in the northern Sierra Nevada. Journal of

Mammalogy, 33:198–205.

———. 1953. Stomach contents of chipmunks and mantled squirrels in northeastern California. Journal of Mammalogy, 34:316–324.

———. 1955. Observations on chipmunks and mantled squirrels in northeastern California. The American Midland Naturalist, 53:71–78.

———. 1956. Responses of small mammal populations to logging of Douglas-fir. Journal of Mammalogy, 37:189–196.

Thaeler, C. S., Jr. 1968a. An analysis of the distribution of pocket gophers in northeastern California (genus *Thomomys*). University of California Publications in Zoology, 86:1–46.

———. 1968b. An analysis of three hybrid populations of pocket gophers (genus *Thomomys*). Evolution, 22:543–555.

———. 1980. Chromosome numbers and systematic relations in the genus *Thomomys* (Rodentia: Geomyidae). Journal of Mammalogy, 61:414–422.

Thomas, D. C., and I. McT. Cowan. 1975. The pattern of reproduction in female Columbian black-tailed deer, *Odocoileus hemionus columbianus.* Journal of Reproduction and Fertility, 44:261–272.

Thomas, D. W. 1988. The distribution of bats in different ages of Douglas-fir forests. The Journal of Wildlife Management, 52:619–626.

Thomas, D. W., G. P. Bell, and M. B. Fenton. 1987. Variation in echolocation call frequencies recorded from North American vespertilionid bats: a cautionary note. Journal of Mammalogy, 68:842–847.

Thomas, O. 1893. On two new members of the genus *Heteromys* and two of *Neotoma.* The Annals and Magazine of Natural History, including Zoology, Botany, and Geology, series 6, 12:233–235.

———. 1894. Descriptions of some new neotropical Muridae. The Annals and Magazine of Natural History, including Zoology, Botany, and Geology, series 6, 14:346–366.

———. 1904. A new bat from the United States, representing the European *Myotis* (*Leuconoe*) *Daubentoni.* The Annals and Magazine of Natural History, including Zoology, Botany, and Geology, series 7, 13:382–384.

———. 1912. On mammals from central Asia, collected by Mr. Douglas Carruthers. The Annals and Magazine of Natural History, including Zoology, Botany, and Geology, series 8, 9:391–408.

———. 1917. On small mammals from the delta of the Parana. The Annals and Magazine of Natural History, including Zoology, Botany, and Geology, series 8, 20:95–100.

Thompson, B. C. 1976. Running speeds of crippled coyotes. Northwest Science, 50:181–182.

———. 1979. Evaluation of wire fences for coyote control. Journal of Range Management, 32:457–461.

Thompson, I. D. 1994. Marten populations in uncut and logged boreal forests in Ontario. The Journal of Wildlife Management, 58:272–280.

Thompson, I. D., and P. W. Colgan. 1994. Marten activity in uncut and logged boreal forests in Ontario. The Journal of Wildlife Mangement, 58:280–288.

Thompson, I. D., and R. O. Peterson. 1988. Does wolf predation alone limit the moose population in Pukaskwa Park?: a comment. The Journal of Wildlife Management, 52:556–559.

Thompson, S. D. 1982a. Microhabitat utilization and foraging behavior of bipedal and quadrapedal heteromyid rodents. Ecology, 63:1303–1312.

———. 1982b. Structure and species composition of desert heteromyid rodent species assemblages: effect of a simple habitat manipulation. Ecology, 63:1313–1321.

———. 1982c. Spatial utilization and foraging behavior of the desert woodrat, *Neotoma lepida.* Journal of Mammalogy, 63:570–581.

Thompson, S. E., Jr. 1979. Socioecology of the yellow-bellied marmot (*Marmota flaviventris*) in central Oregon. Ph.D. dissert., University of California, Berkeley, 223 pp.

Thorington, R. W., Jr., and L. R. Heaney. 1981. Body proportions and gliding adaptations of flying squirrels (Petauristinae). Journal of Mammalogy, 62:101–114.

Thorsteinson, F. V., and C. J. Lensink. 1962. Biological observations of Steller sea lions taken during an experimental harvest. The Journal of Wildlife Management, 26:353–359.

Tiemeier, O. W. 1965. Study area, reproduction, growth and development, age distribution, life span, censusing, live trapping and tagging, crop damage, predation, and habits. Pp. 5–39, in The black-tailed jack rabbit in Kansas. Kansas State University, Agricultural Experiment Station, Technical Bulletin, 140:1–75.

Tiemeier, O. W., and M. L. Plenert. 1964. A comparison of three methods for determining the age of black-tailed jackrabbits. Journal of Mammalogy, 45:409–416.

Todd, A. W., and L. B. Keith. 1983. Coyote demography during a snowshoe hare decline in Alberta. The Journal of Wildlife Management, 47:394–404.

Tomasi, T. E., and R. S. Hoffmann. 1984. *Sorex preblei* in Utah and Wyoming. Journal of Mammalogy, 65:708.

Tomich, P. Q. 1962. The annual cycle of the California ground squirrel Citellus beecheyi. University California Publications in Zoology, 65:213–282.

———. 1986. Mammals in Hawaii: a synopsis and notational bibliography. Second ed. Bishop Museum Special Publication, 76:1–375.

Torke, K. G., and J. W. Twente. 1977. Behavior of *Spermophilus lateralis* between periods of hibernation. Journal of Mammalogy, 58:385–390.

Toweill, D. E. 1974. Winter food habits of river otters in western Oregon. The Journal of Wildlife Management, 38:107–111.

———. 1979. Bobcat populations, a review of available literature. Oregon Department of Fish and Wildlife, Information Report Series, Wildlife, 79–2:1–28.

———. 1980. Sex and age structure in Oregon bobcat populations. Oregon Department of Fish and Wildlife, Information Report Series, Wildlife, 80–1:1–32.

———. 1982. Winter foods of eastern Oregon bobcats. Northwest Science, 56:310–315.

———. 1986a. Resource partitioning by bobcats and coyotes in a coniferous forest. Ph.D dissert., Oregon State University, Corvallis, 155 pp.

———. 1986b. Notes on the development of a cougar kitten. The Murrelet, 67:20–23.

Toweill, D. E., and R. G. Anthony. 1988a. Annual diet of bobcats in Oregon's Cascade Range. Northwest Science, 62:99–103.

———. 1988b. Coyote foods in a coniferous forest in Oregon. The Journal of Wildlife Management, 52:507–512.

Toweill, D. E., and C. Maser. 1985. Food of cougars in the

Cascade Range of Oregon. The Great Basin Naturalist, 45:77–80.

Toweill, D. E., and E. C. Meslow. 1977. Food habits of cougars in Oregon. The Journal of Wildlife Management, 41:576–578.

Toweill, D. E., and J. E. Tabor. 1982. River otter: *Lutra canadensis*. Pp. 688–703, *in* Wild mammals of North America: biology, management, and economics (J. A. Chapman and G. A. Feldhamer, eds.). The Johns Hopkins University Press, Baltimore, 1147 pp.

Toweill, D. E., and J. G. Teer. 1977. Food habits of ringtails in the Edwards Plateau region of Texas. Journal of Mammalogy, 58:660–663.

Toweill, D. E., C. Maser, L. D. Bryant, and M. L. Johnson. 1988. Reproductive characteristics of eastern Oregon cougars. Northwest Science, 62:147–150.

Toweill, D. E., C. Maser, M. L. Johnson, and L. D. Bryant. 1984. Size and reproductive characteristics of western Oregon cougars. Proceedings of the Mountain Lion Workshop, 2:176–184.

Townsend, J. K. 1839. Narrative of a journey across the Rocky Mountains to the Columbia River, and a visit to the Sandwich Islands, Chili &c. with a scientific appendix. H. Perkins, Philadelphia, 352 pp.

———. 1850. On the giant wolf of North America (Lupus gigas). Journal Academy of Sciences of Philadelphia, 2:75–79.

Trainer, C. E., N. A. Hartmann, and T. P. Kistner. 1979. Measurements, weights, and physical condition of fawns and adult female mule deer in eastern Oregon. Oregon Department of Fish and Wildlife, Information Report Series, Wildlife, 79–1:1–61.

Trainer, C. E., M. J. Willis, G. P. Keister, Jr., and D. P. Sheehy. 1983. Fawn mortality and habitat use among pronghorn during spring and summer in southeastern Oregon, 1981–82. Oregon Department of Fish and Wildlife, Wildlife Research Report, 12:1–117.

Trainer, C. E., J. C. Lemos, T. P. Kistner, W. C. Lightfoot, and D. E. Toweill. 1981. Mortality of mule deer fawns in southeastern Oregon, 1968–1979. Oregon Department of Fish and Wildlife, Wildlife Research Report, 10:1–113.

Trapp, G. R. 1972. Some anatomical and behavioral adaptations of ringtails, *Bassariscus astutus*. Journal of Mammalogy, 53:549–557.

———. 1978. Comparative behavioral ecology of the ringtail and gray fox in southwestern Utah. Carnivore, 1 (part 2):3–32.

Trapp, G. R., and C. Trapp. 1965. Another vocal sound made by snowshoe hares. Journal of Mammalogy, 46:705.

Trappe, J. M., and R. W. Harris. 1958. Lodgepole pine in the Blue Mountains of northeastern Oregon. Pacific Northwest Forest and Range Experiment Station Research Paper, 30:1–22.

Trethewey, D. E. C., and B. J. Verts. 1971. Reproduction in eastern cottontail rabbits in western Oregon. The American Midland Naturalist, 86:463–476.

Trivers, R. L., and D. E. Willard. 1973. Natural selection of parental ability to vary the sex ratio of offspring. Science, 179:90–92.

Trombulak, S. C. 1985. The influence of interspecific competition on home range size in chipmunks (*Eutamias*).

Journal of Mammalogy, 66:329–337.

Trouessart, E.-L. 1897. Catalogus mammalium tam viventium quam fossilium. R. Friedländer & Sohn, Berolini, fasc. 3, 664 pp.

———. 1904. Catalogus mammalium tam viventium quam fossilium. R. Friedländer & Sohn, Berolini., suppl., 929 pp.

True, F. W. 1884 [1885]. A provisional list of the mammals of North and Central America, and the West Indian islands. Proceedings of the United States National Museum 7:587–611.

———. 1886. A new bat from Puget Sound. Science, 8:588.

———. 1890. Description of a new species of mouse, Phenacomys longicaudus, from Oregon. Proceedings of the United States National Museum, 13:303–304.

———. 1894. Diagnoses of new North American mammals, p. 2 (preprint of Proc. USNM, 17:354).

———. 1895. Diagnosis of some undescribed wood rats (genus Neotoma) in the National Museum. Proceedings of the United States National Museum, 17:353–355.

———. 1896. A revision of the American moles. Proceedings of the United States National Museum, 19:1–111 + 4 pls.

———. 1905. Diagnosis of a new genus and species of fossil sea-lion from the Miocene of Oregon. Smithsonian Miscellaneous Collection, 438:47–49.

———. 1909. A further account of the fossil sea lion Pontolis magnus, from the Miocene of Oregon. Pp. 143–148, *in* Contributions to the Tertiary paleontology of the Pacific Coast. I. The Miocene of Astoria and Coos Bay, Oregon (W. H. Dall, ed.). United States Geological Survey Professional Paper, 59:1–278.

Trulio, L. A., W. J. Loughry, D. F. Hennessy, and D. H. Owings. 1986. Infanticide in California ground squirrels. Animal Behaviour, 34:291–294.

Tryon, C. A., Jr. 1947. Behavior and post-natal development of a porcupine. The Journal of Wildlife Management, 11:282–283.

Tryon, C. A., [Jr.], and H. N. Cunningham. 1968. Characteristics of pocket gophers along an altitudinal transect. Journal of Mammalogy, 49:699–705.

Tryon, C. A., [Jr.], and D. P. Snyder. 1973. Biology of the eastern chipmunk, *Tamias striatus*: life tables, age distributions, and trends in population numbers. Journal of Mammalogy, 54:145–168.

Tumlison, R. 1987. Felis lynx. Mammalian Species, 269:1–8.

Tunberg, A. D., W. E. Howard, and R. E. Marsh. 1984. A new concept in pocket gopher control. Proceedings Vertebrate Pest Conference, 11:7–16.

Turner, L. W. 1972a. Autecology of the Belding ground squirrel in Oregon. Ph.D. dissert., University of Arizona, [Tucson], 149 pp.

———. 1972b. Habitat differences between Spermophilus beldingi and Spermophilus columbianus in Oregon. Journal of Mammalogy, 53:914–917.

———. 1973. Vocal and escape responses of Spermophilus beldingi to predators. Journal of Mammalogy, 54:990–993.

Turner, W. J. 1937. Studies on porphyria. I. Observations on the fox-squirrel, Sciurus niger. Journal of Biological Chemistry, 118:519–530.

Turton, W. 1806. A general system of nature, through the three grand kingdoms of animals, vegetables, and minerals, systematically divided into their several classes, orders, genera, species, and varieties, with their habitations, manners, economy, structure, and pecularities. Lackington, Allen and Co., London, 1:1–940.

Tuttle, M. D., and L. R. Heaney. 1974. Maternity habits of *Myotis leibii* in South Dakota. Bulletin of the Southern California Academy of Science, 73:80–83.

Tuttle, M. D., and D. Stevenson. 1982. Growth and survival of bats. Pp. 105–150, *in* Ecology of bats (T. H. Kunz, ed.). Plenum Press, New York, 425 pp.

Twente, J. W., Jr. 1955. Some aspects of habitat selection and other behavior of cavern-dwelling bats. Ecology, 36:706–732.

Twining, H., and A. L. Hensley. 1943. The distribution of muskrats in California. California Fish and Game, 29:64–78.

Tyser, R. W. 1975. Taxonomy and reproduction of *Microtus canicaudus*. M.S. thesis, Oregon State University, Corvallis, 67 pp.

Uhlig, H. G. 1955. The determination of age of nestling and sub-adult gray squirrels in West Virginia. The Journal of Wildlife Management, 19:479–483.

Ure, D. C., and C. Maser. 1982. Mycophagy of red-backed voles in Oregon and Washington. Canadian Journal of Zoology, 60:3307–3315.

U.S. Department of Commerce. 1969. Climatological data, Oregon: January 1969. Environmental Science Service Administration, 75(1):1–19.

Van Dyke, F. G., R. H. Brocke, and H. G. Shaw. 1986*a*. Use of road track counts as indices of mountain lion presence. The Journal of Wildlife Management, 50:102–109.

Van Dyke, F. G., R. H. Brocke, H. G. Shaw, B. B. Ackerman, T. P. Hemker, and F. G. Lindzey. 1986*b*. Reactions of mountain lions to logging and human activity. The Journal of Wildlife Management, 50:95–102.

Van Dyke, W. A. 1978. Population characteristics and habitat utilization of bighorn sheep, Steens Mountain, Oregon. M.S. thesis, Oregon State University, Corvallis, 87 pp.

————. 1990. Bighorn sheep harvest regulations in Oregon: management considerations. Proceedings of the Biennial Symposium, Northern Wild Sheep and Goat Council, 7:252–255.

Van Gelder, R. G. 1953. The egg-opening technique of a spotted skunk. Journal of Mammalogy, 34:255–256.

————. 1959. A taxonomic revision of the spotted skunks (genus *Spilogale*). Bulletin of the American Museum of Natural History, 117:229–392.

————. 1977. Mammalian hybrids and generic limits. American Museum Novitates, 2635:1–25.

————. 1978. A review of canid classification. American Museum Novitates, 2646:1–25.

Van Gelder, R. G., and W. W. Goodpaster. 1952. Bats and birds competing for food. Journal of Mammalogy, 33:491.

Van Horne, B. 1982. Demography of the longtail vole *Microtus longicaudus* in seral stages of coastal coniferous forest, southeast Alaska. Canadian Journal of Zoology, 60:1690–1709.

Van Vuren, D. 1980. Ecology and behavior of bison in the Henry Mountains, Utah. M.S. thesis, Oregon State University, Corvallis, 39 pp.

————. 1991. Yellow-bellied marmots as prey of coyotes. The American Midland Naturalist, 125:135–139.

Van Vuren, D., and K. B. Armitage. 1991. Duration of snow cover and its influence on life-history variation in yellow-bellied marmots. Canadian Journal of Zoology, 69:1755–1758.

Van Vuren, D., and M. P. Bray. 1983. Diets of bison and cattle on a seeded range in southern Utah. Journal of Range Management, 36:499–500.

————. 1985. The Recent geographic distribution of *Bison bison* in Oregon. The Murrelet, 66:56–58.

Van Vuren, D., and F. C. Deitz. 1993. Evidence of *Bison bison* in the Great Basin. The Great Basin Naturalist, 53:318–319.

Van Vuren, D., and S. E. Thompson, Jr. 1982. Opportunistic feeding by coyotes. Northwest Science, 56:131–135.

van Zyll de Jong, C. G. 1975. The distribution and abundance of the wolverine (*Gulo gulo*) in Canada. The Canadian Field Naturalist, 89:431–437.

————. 1982. An additional morphological character useful in distinguishing two similar shrews *Sorex monticolus* and *Sorex vagrans*. The Canadian Field-Naturalist, 96:349–350.

————. 1984. Taxonomic relationships of Nearctic small-footed bats of the *Myotis leibii* group (Chiroptera: Vespertilionidae). Canadian Journal of Zoology, 62:2519–2526.

————. 1985. Handbook of Canadian mammals. 2. Bats. National Museums of Canada, Ottawa, 212 pp.

Vaughan, M. R. 1975. Aspects of mountain goat ecology, Wallowa Mountains, Oregon. M.S. thesis, Oregon State University, Corvallis, 113 pp.

Vaughan, T. A. 1954. Mammals of the San Gabriel Mountains of California. University of Kansas Publications, Museum of Natural History, 7:513–582.

————. 1961. Vertebrates inhabiting pocket gopher burrows in Colorado. Journal of Mammalogy, 42:171–174.

————. 1963. Movements made by two species of pocket gophers. The American Midland Naturalist, 69:367–372.

————. 1969. Reproduction and population densities in a montane small mammal fauna. Miscellaneous Publications, University of Kansas, Museum of Natural History, 51:51–74.

————. 1974. Resource allocation in some sympatric, subalpine rodents. Journal of Mammalogy, 55:764–795.

————. 1982. Stephen's woodrat: a dietary specialist. Journal of Mammalogy, 63:53–62.

————. 1986. Mammalogy. Third ed. CBS College Publishing, Philadelphia, 576 pp.

Vaughan, T. A., and P. H. Krutzsch. 1954. Seasonal distribution of the hoary bat in southern California. Journal of Mammalogy, 35:431–432.

Vaughan, T. A., and T. J. O'Shea. 1976. Roosting ecology of the pallid bat, *Antrozous pallidus*. Journal of Mammalogy, 57:19–42.

Vaughan, T. A., and W. P. Weil. 1980. The importance of arthropods in the diet of *Zapus princeps* in a subalpine habitat. Journal of Mammalogy, 61:122–124.

Vertrees, J. D. [1959]. Economic and control aspects. Pp. 3–13, *in* The Oregon meadow mouse irruption of 1957–1958. Federal Cooperative Extension Service, Oregon State College, Corvallis, 88 pp.

Verts, B. J. 1961. A convenient method of carrying and

dispensing baits. Journal of Mammalogy, 42:283.

———. 1963. Movements and populations of opossums in a cultivated area. The Journal of Wildlife Management, 27:127–129.

———. 1967a. Summer breeding of brush rabbits. The Murrelet, 48:19.

———. 1967b. The biology of the striped skunk. University of Illinois Press, Urbana, 218 pp.

———. 1975. New records for three uncommon mammals in Oregon. The Murrelet, 56:22–23.

———. 1988. Two bats caught on a plant. The Murrelet, 69:36–38.

Verts, B. J., and L. N. Carraway. 1980. Natural hybridization of Sylvilagus bachmani and introduced S. floridanus in Oregon. The Murrelet, 61:95–98.

———. 1981. Dispersal and dispersion of an introduced population of Sylvilagus floridanus. The Great Basin Naturalist, 41:167–175.

———. 1984. Keys to the mammals of Oregon. Third ed. O.S.U. Book Stores, Inc., Corvallis, Oregon, 178 pp.

———. 1986. Replacement in a population of Perognathus parvus subjected to removal trapping. Journal of Mammalogy, 67:201–205.

———. 1987a. Microtus canicaudus. Mammalian Species, 267:1–4.

———. 1987b. Thomomys bulbivorus. Mammalian Species, 273:1–4.

———. 1991. Summer breeding and fecundity in the camas pocket gopher, Thomomys bulbivorus. Northwestern Naturalist, 72:61–65.

———. 1995. Phenacomys albipes. Mammalian Species, 494:1–5.

Verts, B. J., and D. B. Costain. 1988. Changes in sex ratios of Spermophilus beldingi in Oregon. Journal of Mammalogy, 69:186–190.

Verts, B. J., and S. D. Gehman. 1991. Activity and behavior of free-living Sylvilagus nuttallii. Northwest Science, 65:231–237.

Verts, B. J., and G. L. Kirkland, Jr. 1988. Perognathus parvus. Mammalian Species, 318:1–8.

Verts, B. J., and G. L. Storm. 1966. A local study of prevalence of rabies among foxes and striped skunks. The Journal of Wildlife Management, 30:419–421.

Verts, B. J., L. N. Carraway, and R. L. Green. 1997. Sex-bias in prenatal parental investment in the eastern cottontail (Sylvilagus floridanus). Journal of Mammalogy, 78:1164–1171.

Verts, B. J., S. D. Gehman, and K. J. Hundertmark. 1984. Sylvilagus nuttallii: a semiarboreal lagomorph. Journal of Mammalogy, 65:131–135.

Verts, B. J., S. A. Ludwig, and W. Graf. 1972. Distribution of eastern cottontail rabbits in Linn County, Oregon, 1953–1970. The Murrelet, 53:25–26.

Vestal, E. H. 1938. Biotic relations of the wood rat (Neotoma fuscipes) in the Berkeley Hills. Journal of Mammalogy, 19:1–36.

Vigors, N. A. 1830. In Observations on the quadrupeds found in the district of upper Canada extending between York and Lake Simcoe, with the view of illustrating their geographical distribution, as well as of describing some species hitherto

unnoticed (Dr. Gapper). The Zoological Journal, 5:201–204.

Villa-R., B. 1966 [1967]. Los murcielagos de Mexico. Instituto de Biología, Universidad Nacional Autónoma de México, Mexico, D. F., 491 pp.

Virchow, D. 1977. Use of telemetry in the study of pocket gopher (Geomys bursarius) movements and activity patterns. Proceedings the Nebraska Academy of Sciences and Affiliated Societies, 87:23 (abstract only).

Vleck, D. 1981. Burrow structure and foraging costs in the fossorial rodent, Thomomys bottae. Oecologia, 49:391–396.

Vogel, S. 1994. Nature's pumps. American Scientist, 82:464–471.

Vogt, D. W., and D. T. Arakaki. 1971. Karyotype of the American red fox (Vulpes fulva). Journal of Heredity, 62:318–319.

Voigt, D. R. 1987. Red fox. Pp. 379–392, in Wild furbearer management and conservation in North America (M. Novak, J. A. Baker, M. E. Obbard, and B. Malloch, eds.). Ontario Ministry of Natural Resources, Toronto, 1150 pp.

Voigt, D. R., and W. E. Berg. 1987. Coyote. Pp. 345–357, in Wild furbearer management and conservation in North America (M. Novak, J. A. Baker, M. E. Obbard, and B. Malloch, eds.). Ontario Ministry of Natural Resources, Toronto, 1150 pp.

Voigt, D. R., and B. D. Earle. 1983. Avoidance of coyotes by red fox families. The Journal of Wildlife Management, 47:852–857.

Vollmer, A. T., B. G. Maza, P. A. Medica, F. B. Turner, and S. A. Bamberg. 1976. The impact of off-road vehicles on a desert ecosystem. Environmental Management, 1:115–129.

von Bloeker, J. C., Jr. 1937. Mammal remains from detritus of raptorial birds in California. Journal of Mammalogy, 18:360–361.

Vorhies, C. T. 1935. The Arizona specimen of Euderma maculatum. Journal of Mammalogy, 16:224–226.

Voth, E. H. 1968. Food habits of the Pacific mountain beaver, Aplodontia rufa pacifica Merriam. Ph.D. dissert., Oregon State University, Corvallis, 263 pp.

Voth, E. H., C. Maser, and M. L. Johnson. 1983. Food habits of Arborimus albipes, the white-footed vole, in Oregon. Northwest Science, 57:1–7.

Wade-Smith, J., and M. E. Richmond. 1975. Care, management, and biology of captive striped skunks (Mephitis mephitis). Laboratory Animal Science, 25:575–584.

———. 1978. Reproduction in captive striped skunks (Mephitis mephitis). The American Midland Naturalist, 100:452–455.

Wade-Smith, J., and B. J. Verts. 1982. Mephitis mephitis. Mammalian Species, 173:1–7.

Wade-Smith, J., M. E. Richmond, R. A. Mead, and H. Taylor. 1980. Hormonal and gestational evidence for delayed implantation in the striped skunk, Mephitis mephitis. General and Comparative Endocrinology, 42:509–515.

Wagner, A. 1845. Berichte über die Säugethiere, &c. Archiv für Naturgeschichte gegrundet von Wiegmann, 1845:148 (not seen, cited in Baird, 1858).

Wagner, F. H., and L. C. Stoddart. 1972. Influence of coyote predation on black-tailed jackrabbit populations in Utah. The Journal of Wildlife Management, 36:329–342.

Wai-Ping, V., and M. B. Fenton. 1989. Ecology of spotted bat (Euderma maculatum) roosting and foraging behavior. Journal of Mammalogy, 70:617–622.

Waithman, J., and A. Roest. 1977. A taxonomic study of the kit fox, *Vulpes macrotis*. Journal of Mammalogy, 58:157–164.

Walker, A. 1930*a*. The "handstand" and some other habits of the Oregon spotted skunk. Journal of Mammalogy, 11:227–229.

———. 1930*b*. Notes on the forest Phenacomys. Journal of Mammalogy, 11:233–235.

———. 1942. The fringed bat in Oregon. The Murrelet, 23:62.

Walker, E. P. et al. 1964. Mammals of the world. The Johns Hopkins University Press, Baltimore, 2:647–1500.

Walker, K. M. 1949. Distribution and life history of the black pocket gopher, *Thomomys niger* Merriam. M.S. thesis, Oregon State College, Corvallis, 94 pp.

———. 1955. Distribution and taxonomy of the small pocket gophers of northwestern Oregon. Ph.D. dissert., Oregon State College, Corvallis, 200 pp.

Wallace, R. E. 1946. A Miocene mammalian fauna from Beatty Buttes, Oregon. Pp. 113–134 + 6 pls., *in* Fossil vertebrates from western North America and Mexico. Carnegie Institution of Washington, Contributions to Paleontology, 551:1–195.

Wallen, K. 1982. Social organization in the dusky-footed woodrat (*Neotoma fuscipes*): a field and laboratory study. Animal Behaviour, 30:1171–1182.

Wallmo, O. C. 1981. Mule and black-tailed deer distribution and habitats. Pp. 1–25, *in* Mule and black-tailed deer of North America (O. C. Wallmo, ed.). University of Nebraska Press, Lincoln, 605 pp.

Walters, R. D., and V. D. Roth. 1950. Faunal nest study of the woodrat, *Neotoma fuscipes monochroura* Rhoads. Journal of Mammalogy, 31:290–292.

Walton, D. W., and N. J. Siegel. 1966. The histology of the pararhinal glands of the pallid bat, *Antrozous pallidus*. Journal of Mammalogy, 47:357–360.

Waring, G. H. 1966. Sounds and communications of the yellow-bellied marmot (*Marmota flaviventris*). Animal Behaviour, 14:177–183.

Warner, G. M. 1971. Vocalizations of the Townsend chipmunk (*Eutamias townsendii*). M.S. thesis, Portland State University, Portland, Oregon, 37 pp.

Warner, R. M. 1985. Interspecific and temporal dietary variation in an Arizona bat community. Journal of Mammalogy, 66:45–51.

Warner, R. M., and N. J. Czaplewski. 1984. Myotis volans. Mammalian Species, 224:1–4.

Warnock, R. G., and A. W. Grundmann. 1963. Food habits of some mammals of the Bonneville Basin as determined by the contents of the stomach and intestine. Proceedings of the Utah Academy of Sciences, Arts and Letters, 40:66–73.

Warren, E. R. 1926. Notes on the breeding of wood rats of the genus Neotoma. Journal of Mammalogy, 7:97–101.

———. 1927. The beaver. The Williams & Wilkins Company, Baltimore, 177 pp.

Waterhouse, G. R. 1838 [1839]. [On a new species of hare from North America]. Proceedings of the Zoological Society of London, 1838:103–105.

Watkins, L. C. 1977. Euderma maculatum. Mammalian Species, 77:1–4.

Wayne, R. K., and S. J. O'Brien. 1987. Allozyme divergence within the Canidae. Systematic Zoology, 36:339–355.

Webster, D. B., and M. Webster. 1971. Adaptive value of hearing and vision in kangaroo rat predator avoidance. Brain, Behavior and Evolution, 4:310–322.

———. 1972. Kangaroo rat auditory thresholds before and after middle ear reduction. Brain, Behavior and Evolution, 5:41–53.

———. 1975. Auditory systems of the Heteromyidae: functional morphology and evolution of the middle ear. Journal of Morphology, 152:153–170.

Webster, F. A., and D. R. Griffin. 1962. The role of the flight membranes in insect capture by bats. Animal Behaviour, 10:332–340.

Webster, W. D., and J. K. Jones, Jr. 1982. Reithrodontomys megalotis. Mammalian Species, 167:1–5.

Webster, W. D., J. F. Parnell, and W. C. Biggs, Jr. 1985. Mammals of the Carolinas, Virginia, and Maryland. The University of North Carolina Press, Chapel Hill, 255 pp.

Weckerly, F. W. 1993. Intersexual resource partitioning in black-tailed deer: a test of the body size hypothesis. The Journal of Wildlife Management, 57:475–494.

Weckwerth, R. P., and V. D. Hawley. 1962. Marten food habits and population fluctuations in Montana. The Journal of Wildlife Management, 26:55–74.

Weigel, I. 1961. Das Fellmuster der wildebenden Katzenarten und Hauskatze in vergleichender und stammesgeschichtlicher. Saüetierkundliche Mitteilungen, 9:1–20.

Weigl, P. D., and D. W. Osgood. 1974. Study of the northern flying squirrel, Glaucomys sabrinus, by temperature telemetry. The American Midland Naturalist, 92:482–486.

Weil, J. W. 1975. Agonistic behavior in three species of *Microtus* (*M. canicaudus*, *M. oregoni*, and *M. townsendii*). M.S. thesis, Oregon State University, Corvallis, 121 pp.

Weiss, N. T., and B. J. Verts. 1984. Habitat and distribution of pygmy rabbits (*Sylvilagus idahoensis*) in Oregon. The Great Basin Naturalist, 44:563–571.

Wells-Gosling, N., and L. R. Heaney. 1984. Glaucomys sabrinus. Mammalian Species, 229:1–8.

Wentz, W. A. 1971. The impact of nutria (*Myocastor coypus*) on marsh vegetation in the Willamette Valley, Oregon. M.S. thesis, Oregon State University, Corvallis, 41 pp.

West, N. E. 1968. Rodent-influenced establishment of ponderosa pine and bitterbrush seedlings in central Oregon. Ecology, 49:1009–1011.

West, N. H., and B. N. Van Vliet. 1986. Factors influencing the onset and maintenance of bradycardia in mink. Physiological Zoology, 9:451–463.

Westoby, M. 1980. Black-tailed jack rabbit diets in Curlew Valley, northern Utah. The Journal of Wildlife Management, 44:942–948.

Whitaker, J. O., Jr. 1962. Endogone, Hymenogaster, and Melanogaster as small mammal foods. The American Midland Naturalist, 67:152–156.

———. 1972. Food habits of bats from Indiana. Canadian Journal of Zoology, 50:877–883.

Whitaker, J. O., Jr., and T. W. French. 1984. Foods of six species of sympatric shrews from New Brunswick. Canadian Journal of Zoology, 62:622–626.

Whitaker, J. O., Jr., and C. Maser. 1976. Food habits of five western Oregon shrews. Northwest Science, 50:102–107.

Whitaker, J. O., Jr., and L. L. Schmeltz. 1973. Food and external parasites of *Sorex palustris* and food of *Sorex cinereus* from St. Louis County, Minnesota. Journal of Mammalogy, 54:283–285.

Whitaker, J. O., Jr., S. P. Cross, and C. Maser. 1983. Food of vagrant shrews (*Sorex vagrans*) from Grant County, Oregon, as related to livestock grazing pressures. Northwest Science, 57:107–111.

Whitaker, J. O., Jr., C. Maser, and S. P. Cross. 1981*a*. Food habits of eastern Oregon bats, based on stomach and scat analysis. Northwest Science, 55:281–292.

———. 1981*b*. Foods of Oregon silver-haired bats, *Lasionycteris noctivagans*. Northwest Science, 55:75–77.

Whitaker, J. O., Jr., C. Maser, and L. E. Keller. 1977. Food habits of bats of western Oregon. Northwest Science, 51:46–55.

Whitaker, J. O., Jr., C. Maser, and R. J. Pedersen. 1979. Food and ectoparasitic mites of Oregon moles. Northwest Science, 53:268–273.

White, J. A. 1953. Genera and subgenera of chipmunks. University of Kansas Publication, Museum of Natural History, 5:543–561.

White, N. R., and A. S. Fleming. 1987. Auditory regulation of woodrat (*Neotoma lepida*) sexual behaviour. Animal Behaviour, 35:1281–1297.

White, T. D. 1984. Locomotion and the role of the epipubic bones in the brush-tail possum. The Australian Mammal Society Bulletin, 8:179 (abstract only).

Whitehead, C. J., Jr. 1972. A preliminary report on white-tailed and black-tailed deer crossbreeding studies in Tennessee. Proceedings of Annual Conference of the Southeastern Association of Game and Fish Commissioners, 25:65–69.

Whitford, W. G. 1976. Temporal fluctuations in density and diversity of desert rodent populations. Journal of Mammalogy, 57:351–369.

Whitlow, W. B., and E. R. Hall. 1933. Mammals of the Pocatello Region of southeastern Idaho. University of California Publications in Zoology, 40:235–275.

Whorton, B. J. 1989. Kernel methods for estimating the utilizations distribution in home range studies. Ecology, 70:164–168.

Wick, W. Q., and A. S. Landforce. 1962. Mole and gopher control. Cooperative Extension Service, Oregon State University, Extension Bulletin, 804:1–16.

Wied-Neuwied, M. A. P., zu Prinz. 1841. Reise in das innere Nord-America in den jahren 1832 bis 1834. J. Holscher, Colbenz, 2:99 (not seen, cited in The National Union Catalog, 662:153).

Wight, H. M. 1918. The life-history and control of the pocket gopher in the Willamette Valley. Oregon Agricultural College Experiment Station Bulletin, 153:1–55.

———. 1922. The Willamette Valley gopher. The Murrelet, 3(3):6–8.

———. 1925. Notes on the tree mouse, Phenacomys silvicola. Journal of Mammalogy, 6:282–283.

———. 1928. Food habits of the Townsend's mole Scapanus townsendii (Bachman). Journal of Mammalogy, 9:19–33.

———. 1930. Breeding habits and economic relations of The Dalles pocket gopher. Journal of Mammalogy, 11:40–48.

———. 1931. Reproduction in the eastern skunk (*Mephitis mephitis nigra*). Journal of Mammalogy, 12:42–47.

Wilcomb, M. J. 1948. Fox populations and food habits in relation to game birds in the Willamette Valley, Oregon. M.S. thesis, Oregon State College, Corvallis, 95 pp.

Wilcomb, M. S[J]. 1956. Studies in wildlife management. I. Fox populations and food habits in relation to game bird survival, Willamette Valley, Oregon. Oregon Agricultural Experiment Station, Technical Bulletin, 38:1–16.

Wilke, F. 1958. Fat content of fur-seal milk. The Murrelet, 39:40.

Wilkins, K. T. 1989. Tadarida brasiliensis. Mammalian Species, 331:1–10.

Wilks, B. J. 1962. Reingestion in geomyid rodents. Journal of Mammalogy, 43:267.

Willett, G. 1943. Elephant seal in southeastern Alaska. Journal of Mammalogy, 24:500.

Williams, D. F. 1968. A new record of *Myotis thysanodes* from Washington. The Murrelet, 49:26–27.

———. 1978. Karyological affinities of the species groups of silky pocket mice (Rodentia, Heteromyidae). Journal of Mammalogy, 59:599–612.

———. 1984. Habitat associations of some rare shrews (*Sorex*) from California. Journal of Mammalogy, 65:325–328.

———. 1991. Habitats of shrews (genus *Sorex*) in forest communities of the western Sierra Nevada, California. Special Publication, The Museum of Southwestern Biology, 1:1–14.

Williams, D. F., and J. S. Findley. 1979. Sexual size dimorphism in vespertilionid bats. The American Midland Naturalist, 102:113–126.

Williams, J. E., and D. C. Cavahaugh. 1983. Differential signs of plague in young and old California ground squirrels (*Spermophilus beecheyi*). Journal of Wildlife Diseases, 19:154–155.

Williams, O., and B. A. Finney. 1964. *Endogone*—food for mice. Journal of Mammalogy, 45:265–271.

Williams, T. C., L. C. Ireland, and J. M. Williams. 1973. High altitude flights of the free-tailed bat, *Tadarida brasiliensis*, observed with radar. Journal of Mammalogy, 54:807–821.

Willner, G. R. 1982. Nutria: *Myocastor coypus*. Pp. 1059–1076, *in* Wild mammals of North America: biology, management, and economics (J. A. Chapman, and G. A. Feldhamer, eds.). The Johns Hopkins University Press, Baltimore, 1147 pp.

Willner, G. R., G. A. Feldhamer, D. E. Zucker, and J. A. Chapman. 1980. Ondatra zibethicus. Mammalian Species, 141:1–8.

Wilson, D. E. 1982. Wolverine: *Gulo gulo*. Pp. 644–652, *in* Wild mammals of North America: biology, management, and economics (J. A. Chapman and G. A. Feldhamer, eds.). The Johns Hopkins University Press, Baltimore, 1147 pp.

———. 1993. Family Castoridae. P. 467, *in* Mammal species of the world: a taxonomic and geographic reference. Second ed. (D. E. Wilson and D. M. Reeder, eds.). Smithsonian Institution Press, Washington, D.C., 1206 pp.

Wilson, D. E., and D. M. Reeder. 1993*a*. Introduction. Pp. 1–12, *in* Mammal species of the world: a taxonomic and geographic reference. Second ed. (D. E. Wilson and D. M. Reeder, eds.). Smithsonian Institution Press, Washington, D.C., 1206 pp.

———. 1993*b*. Mammal species of the world: a taxonomic and geographic reference. Second ed. Smithsonian Institution Press, Washington, D.C., 1206 pp.

Wilson, D. E., M. A. Bogan, R. L. Brownell, Jr., A. M. Burdin, and M. K. Maminov. 1991. Geographic variation in sea otters, *Enhydra lutris.* Journal of Mammalogy, 72:22–36.

Wilson, R. W. 1935. A new species of Dipoides from the Pliocene of eastern Oregon. Pp. 19–28 + 1 pl., *in* Papers concerning the palaeontology of California, Nevada and Oregon. Carnegie Institution of Washington, Contributions to Palaeontology, 453:1–195.

———. 1938*a*. New middle Pliocene rodent and lagomorph faunas from Oregon and California. Pp. 1–19 + 3 pls., *in* Studies on Cenozoic vertebrates of western North America. Carnegie Institution of Washington, Contribution to Palaeontology, 487:1–281.

———. 1938*b*. Pliocene rodents of western North America. Pp. 21–73, *in* Studies on Cenozoic vertebrates of western North America. Carnegie Institution of Washington, Contributions to Palaeontology, 487:1–281.

Wimsatt, W. A. 1945. Notes on breeding behavior, pregnancy, and parturition in some vespertilionid bats of the eastern United States. Journal of Mammalogy, 26:23–33.

———. 1963. Delayed implantation in the Ursidae, with particular reference to the black bear. Pp. 49–76, *in* Delayed implantation (A. C. Enders, ed.). University of Chicago Press, Chicago, 318 pp.

Wiseman, G. L., and G. O. Hendrickson. 1950. Notes on the life history and ecology of the opossum in southeast Iowa. Journal of Mammalogy, 31:331–337.

Wishart, W. D. 1980. Hybrids of white-tailed and mule deer in Alberta. Journal of Mammalogy, 61:716–720.

Witmer, G. W. 1982. Roosevelt elk habitat use in the Oregon Coast Range. Ph.D. dissert., Oregon State University, Corvallis, 104 pp.

Witmer, G. W., and D. S. deCalesta. 1986. Resource use by unexploited sympatric bobcats and coyotes in Oregon. Canadian Journal of Zoology, 64:2333–2338.

Witt, J. W. 1991. Fluctuations in the weight and trapping response for *Glaucomys sabrinus* in western Oregon. Journal of Mammalogy, 72:612–615.

———. 1992. Home range and density estimates for the northern flying squirrel, *Glaucomys sabrinus*, in western Oregon. Journal of Mammalogy, 73:921–929.

Wobeser, G., D. B. Hunter, and P.-Y. Daoust. 1978. Tyzzer's disease in muskrats: occurrence in free-living animals. Journal of Wildlife Diseases, 14:325–328.

Wobeser, G. A., T. R. Spraker, and V. L. Harms. 1980. Collection and field preservation of biological materials. Pp. 537–551, *in* Wildlife Management Techniques Manual. Fourth ed. (S. D. Schemnitz, ed.). The Wildlife Society, Washington, D.C., 686 pp.

Wolf, T. F., and A. I. Roest. 1971. The fox squirrel (*Sciurus niger*) in Ventura County. California Fish and Game, 57:219–220.

Wolff, J. O. 1980. The role of habitat patchiness in population dynamics of snowshoe hare populations. Ecological Monographs, 50:111–130.

———. 1982. Refugia, dispersal, predation, and geographic variation in snowshoe hare cycles. Pp. 441–449, *in* Proceedings of the world lagomorph conference (K. Myers and C. D. MacInnes, eds.). University of Guelph, Guelph, Ontario, 983 pp.

———. 1989. Social behavior. Pp. 271–291, *in* Advances in the study of *Peromyscus* (Rodentia) (G. L. Kirkland, Jr., and J. N. Layne, eds.). Texas Tech University Press, Lubbock, 366 pp.

Wolff, J. O., W. D. Edge, and R. Bentley. 1994. Reproductive and behavioral biology of the gray-tailed vole. Journal of Mammalogy, 75:873–879.

Wolff, J. O., E. M. Schauber, and W. D. Edge. 1997. Effects of habitat loss and fragmentation on the behavior and demography of gray-tailed voles. Conservation Biology, 11:945–956.

Wong, B., and K. L. Parker. 1988. Estrus in black-tailed deer. Journal of Mammalogy, 69:168–171.

Wood, A. E. 1936. Geomyid rodents from the middle Tertiary. American Museum Novitates, 866:1–31.

Wood, F. D. 1935. Notes on the breeding behavior and fertility of Neotoma fuscipes macrotis in captivity. Journal of Mammalogy, 16:107–109.

Wood, J. E. 1958. Age structure and productivity of a gray fox populations. Journal of Mammalogy, 39:74–86.

Wood, W. 1974. Muskrat origin, distribution, and range extension through the coastal areas of Del Norte Co. Calif. & Curry Co. Oregon. The Murrelet, 55:1–4.

Wood, W. F. 1990. New components in defensive secretion of the striped skunk, *Mephitis mephitis.* Journal of Chemical Ecology, 16:2057–2065.

Wood, W. F., C. G. Morgan, and A. Miller. 1991. Volatile components in defensive spray of the spotted skunk, *Spilogale putorius.* Journal of Chemical Ecology, 17:1415–1420.

Woodhouse, [S. W.] 1853. [A new species of pouched rat, of the genus Dipodomys, Gray]. Proceedings of the Academy of Natural Sciences of Philadelphia, 6:224.

Woods, C. A. 1973. Erethizon dorsatum. Mammalian Species, 29:1–6.

Woods, C. A., L. Contreras, G. Willner-Chapman, and H. P. Whidden. 1992. Myocastor coypus. Mammalian Species, 398:1–8.

Woodsworth, G. C., G. P. Bell, and M. B. Fenton. 1981. Observations of the echolocation, feeding behaviour, and habitat use of *Euderma maculatum* (Chiroptera: Vespertilionidae) in southcentral British Columbia. Canadian Journal of Zoology, 59:1099–1102.

Wozencraft, W. C. 1989. Classification of Recent Carnivora. Pp. 569–593, *in* Carnivore behavior, ecology, and evolution (J. L. Gittleman, ed.). Cornell University Press, Ithaca, New York, 620 pp.

———. 1993. Order Carnivora. Pp. 279–348, *in* Mammal species of the world: a taxonomic and geographic reference. Second ed. (D. E. Wilson and D. M. Reeder, eds.). Smithsonian Institution Press, Washington, D.C., 1206 pp.

Wright, P. L. 1942. Delayed implantation in the long-tailed weasel (*Mustela frenata*), the short-tailed weasel (*Mustela cicognanii*) and the marten (*Martes americana*). Anatomical Record, 83:341–353.

———. 1947. The sexual cycle of the male long-tailed weasel (*Mustela frenata*). Journal of Mammalogy, 28:343–352.

———. 1948. Breeding habits of captive long-tailed weasels (Mustela frenata). The American Midland Naturalist, 39:338–344.

———. 1953. Intergradation between *Martes americana* and *Martes caurina* in western Montana. Journal of Mammalogy, 34:74–86.

———. 1963. Variations in reproductive cycles in North American mustelids. Pp. 77–97, *in* Delayed implantation (A. C. Enders, ed.). University of Chicago Press, Chicago, 318 pp.

———. 1966. Observations on the reproductive cycle of the American badger (*Taxidea taxus*). Symposia of the Zoological Society of London, 15:27–45.

———. 1969. The reproductive cycle of the male American badger. Journal of Reproduction and Fertility, suppl., 6:435–445.

Wright, P. L., and M. W. Coulter. 1967. Reproduction and growth in Maine fishers. The Journal of Wildlife Management, 31:70–87.

Wright, P. L., and S. A. Dow, Jr. 1962. Minimum breeding age in pronghorn antelope. The Journal of Wildlife Management, 26:100–101.

Wrigley, R. E., J. E. Dubois, and H. W. R. Copland. 1979. Habitat, abundance, and distribution of six species of shrews in Manitoba Canada. Journal of Mammalogy, 60:505–520.

Wroot, A. J., S. A. Wroot, and J. O. Murie. 1987. Intraspecific variation in postnatal growth of Columbian ground squirrels (*Spermophilus columbianus*). Journal of Mammalogy, 68:395–398.

Yalden, D. W., and P. A. Morris. 1975. The lives of bats. Quadrangle/The New York Times Book Co., New York, 247 pp.

Yates, T. L. 1984. Insectivores, elephant shrews, tree shrews, and dermopterans. Pp. 117–144, *in* Orders and families of Recent mammals of the world (S. Anderson and J. K. Jones, Jr., eds.). John Wiley & Sons, New York, 686 pp.

Yates, T. L., and R. J. Pedersen. 1982. Moles. Pp. 37–51, *in* Wild mammals of North America: biology, management, and economics (J. A. Chapman and G. A. Feldhamer, eds.). The Johns Hopkins University Press, Baltimore, 1147 pp.

Yates, T. L., W. R. Barber, and D. M. Armstrong. 1987. Survey of North American collections of Recent mammals. Journal of Mammalogy, 68(2), suppl.:1–76.

Yeager, L. E. 1936. Winter daytime dens of opossums. Journal of Mammalogy, 17:410–411.

Yeager, L. E., and W. H. Elder. 1945. Pre- and post-hunting season foods of raccoon on an Illinois goose refuge. The Journal of Wildlife Management, 9:48–56.

Yeager, L. E., and R. G. Rennels. 1943. Fur yield and autumn foods of the raccoon in Illinois River bottom lands. The Journal of Wildlife Management, 7:45–60.

Yeatter, R. E., and D. H. Thompson. 1952. Tularemia, weather, and rabbit populations. Illinois Natural History Survey Bulletin, 26:351–382.

Yensen, E., and D. L. Quinney. 1992. Can Townsend's ground squirrels survive on a diet of exotic annuals? The Great Basin Naturalist, 52:269–277.

Yensen, E., D. L. Quinney, K. Johnson, K. Timmerman, and K. Steenhof. 1992. Fire, vegetation changes, and population fluctuations of Townsend's ground squirrels. The American Midland Naturalist, 128:299–312.

Yoakum, J. D. 1957. Factors affecting the mortality of pronghorn

antelope in Oregon. M.S. thesis, Oregon State College, Corvallis, 112 pp.

———. 1959. Recent chinchilla release in Oregon. The Murrelet, 40:19–20.

Yocom, C. F. 1970. Status of muskrats in northwestern California. The Murrelet, 51:21–22.

———. 1973. Wolverine records in the Pacific coastal states and new records for northern California. California Fish and Game, 59:207–209.

———. 1974*a*. Recent wolverine records in Washington and Oregon. The Murrelet, 55:15–18.

———. 1974*b*. Status of marten in northern California, Oregon and Washington. California Fish and Game, 60:54–57.

Yocom, C. F., and M. T. McCollum. 1973. Status of the fisher in northern California, Oregon, and Washington. California Fish and Game, 59:305–309.

Young, H., R. L. Strecker, and J. T. Emlen, Jr. 1950. Localization of activity in two indoor populations of house mice, *Mus musculus*. Journal of Mammalogy, 31:403–410.

Young, J. A., and R. A. Evans. 1981. Demography and fire history of a western juniper stand. Journal of Range Management, 34:501–506.

Young, S. P. 1951. Part I. Its history, life habits, economic status, and control. Pp. 1–226, *in* The clever coyote. The Stackpole Company, Harrisburg, Pennsylvania and the Wildlife Management Institute, Washington, D.C., 411 pp.

———. 1958. The bobcat of North America: its history, life habits, economic status and control, with list of currently recognized subspecies. The Stackpole Company, Harrisburg, Pennsylvania and The Wildlife Management Institute, Washington, D.C., 193 pp.

Young, S. P., and E. A. Goldman. 1946. The puma, mysterious American cat. The American Wildlife Institute, Washington, D.C., 358 pp.

Young, W. A. 1962. Range extension of dusky-footed wood rat. The Murrelet, 42:6.

Youngman, P. M. 1975. Mammals of the Yukon Territory. National Museums of Canada, Publications in Zoology, 10:1–192.

Youngman, P. M., and F. W. Schueler. 1991. *Martes nobilis* is a synonym of *Martes americana*, not an extinct Pleistocene-Holocene species. Journal of Mammalogy, 72:567–577.

Zabel, C. J., and S. J. Taggart. 1989. Shift in red fox, *Vulpes vulpes*, mating system associated with El Niño in the Bering Sea. Animal Behaviour, 38:830–838.

Zakrzewski, R. J. 1985. The fossil record. Pp. 1–51, *in* Biology of New World *Microtus* (R. H. Tamarin, ed.). Special Publication, The American Society of Mammalogists, 8:1–893.

Zalunardo, R. A. 1965. The seasonal distribution of a migratory mule deer herd. The Journal of Wildlife Management, 29:345–351.

Zegers, D. A. 1981. Verification of censusing techniques for the Wyoming ground squirrel. Acta Theriologica, 26:123–125.

———. 1984. Spermophilus elegans. Mammalian Species, 214:1–7.

Zegers, D. A., and O. Williams. 1979. Energy flow through a population of Richardson's ground squirrels. Acta Theriologica, 24:221–235.

Zielinski, W. J., W. D. Spencer, and R. H. Barrett. 1983. Relationship between food habits and activity patterns of pine martens. Journal of Mammalogy, 64:387–396.

Zimmermann, E. A. W. 1780. Geographische Geschichte des Menschen, und der allgemein verbreiteten vierfüssigen Thiere, 2:129 (not seen, cited in Gill and Coues, 1877).

———. 1783. Geographische Geschichte des Menschen und der allgemein verbreiteten vierfüssigen. Weygandschen, Leipzig, Thiere 3, p. 277.

Zinn, T. L., and W. W. Baker. 1979. Seasonal migration of the hoary bat, *Lasiurus cinereus*, through Florida. Journal of Mammalogy, 60:634–635.

Zoloth, S. R. 1969. Observations of the population of brush rabbits on Año Nuevo Island, California. The Wasmann Journal of Biology, 27:149–161.

Zwickel, F. [C.], G. Jones, and H. Brent. 1953. Movement of Columbian black-tailed deer in the Willapa Hills area, Washington. The Murrelet, 34:41–46.

INDEX

Abromys lordi 245
Accipiter
    cooperi 133
    gentilis 145, 163, 186, 189, 208, 214
Accipitridae 118
Accipitrinae 330
Acheta domesticus 278
†*Achyoscapter* 20
Acrididae 117
†*Adjidaumo* 20
Aegolius acadicus 67, 272, 302, 310, 319, 327
†*Aelurodon* 21
Ailurus fulgens 373, 397
Aix sponsa 220
Alces alces 362, 463, 504
†*Allomys* 19
†*Alluvisorex* 20
†*Alticamelus* 20, 21
†*Amebelodon* 21
Ammodytes hexapterus 387, 395
Ammospermophilus 3, 20
    leucurus 31, **178–180**
        description 178; diet 179; distribution 178, 179; fossils 178; geographic variation 178; habitat and density 178–179; habits 180; home range 180; morphometrics 175; Plate VIII; reproduction, ontogeny, and mortality 177, 179–180; specimens examined 511; torpor 180
†*Amphicyon* 20, 21
†*Amynodon* 19
†Anagalida 153
Anas discors 419
Anatidae 424, 430, 435
Anisonyx
    brachiura 187
    rufa 154
antelope, pronghorn
    (see *Antilocapra americana*)
Antilocapra 3
    americana 7, 17, **484–490**
        as prey 488; description 484–485; diet 486; distribution 485, 486; exploitation 489–490; fossils 485; geographic variation 485; habitat and density 485–486, 489; habits 488–489; harvest 490; horns 487; morphometrics 465; Plate XXXII; reproduction, ontogeny, and mortality 486–488; specimens examined 511
    anteflexa 484
    †*garcia* 485
    oregona 484, 485
Antilocapridae **484–490**
    distribution 484; fossils 484; species diversity 484
Antilope Americanus 484
Antrozous 3
    cantwelli 119
    pacificus 119

    pallidus **116–119**
        as prey 118; description 116; diet 117–118; distribution 116, 117; flight speed 75; fossils 116; geographic variation 116; habitat and density 116–117; habits 118–119; morphometrics 77; Plate IV; reproduction, ontogeny, and mortality 101, 118; roosts 118–119; specimens examined 511; torpor 119
Anura 424, 435
Aphelocoma coerulescens 133, 148, 209
†*Aphelops* 20, 21
Aphis 208
Aplodontia 3
    chryseola 154
    pacifica 154, 155
    rufa 6, 8, 22, **154–158**, 285, 305, 442
        as prey 156, 359, 419, 459; description 154; diet 155–156; distribution 154–155; fossils 154–155; geographic variation 155; habitat and density 155; habits 156–157; home range 157; morphometrics 154–155; Plate VII; reproduction, ontogeny, and mortality 156; specimens examined 511–512; vocalizations 157
Aplodontidae **154–158**
    species diversity 154
Aquila chrysaetos 140, 148, 151, 156, 189, 203, 208, 430, 447
Araneae 51, 52, 57, 61, 63, 86, 91, 94, 102, 249
Arborimus 311
    albipes 311
    longicaudus 311
    pomo 311
†*Archaeohippus* 20
†*Archaeolagus* 19, 20
Arctocephalus 381
†*Arctodus simus* 379
Arctogale 415
†*Arctomyoides* 20
Arctomys
    avarus 173
    Columbianus 187
    Douglasii 181
    flaviventer 173
Ardea herodias 319, 332
Arenivaga erratica 248
Armadillidium vulgare 435
Artiodactyla **463–502**
    characteristics 463; distribution 463; key to species 463; natural history 463; species diversity 463
Arvicola
    californica 311
    curtata 332
    gapperi 303
    longicaudus 320
    macropus 328
    montana 322

    mordax 320
    nanus 322
    oregoni 325
    pauperrima 332
    richardsoni 328
    townsendii 330
Arvicolinae 262, 263, **296–339**
    characteristics 296; distribution 296; key to species 296–297; natural history 296; species diversity 296
Asio
    flammeus 203, 247, 266, 272, 282, 322, 325, 332, 334, 342
    otus 67, 129, 230, 237, 256, 266, 272, 282, 310, 319, 322, 325, 327, 334, 342, 345
Astacidea 51
Athene cunicularia 186, 203, 247, 255, 266, 272, 278, 325, 334, 342
Atophyrax 49
    bendirii 49
Aulacomys arvicoloides 328
Autographa californica 86
badger, American (see *Taxidea taxus*)
Balanosphyra formicivora 209
Bassaricyon 397
Bassaris
    astuta 397
    raptor 397
Bassariscus 3, 20
    astutus 28, 30, **397–400**
        as prey 399; description 397, 398; diet 399; distribution 397, 398; Fig. 12-27; fossils 397; geographic variation 397; habitat and density 397, 398; habits 399–400; home range 400; morphometrics 398–399; reproduction, ontogeny, and mortality 399; specimens examined 512; status 400; vocalizations 400
    flavus 397
    raptor 397
    sumichrasti 400
bat
    big brown (see *Eptesicus fuscus*)
    Brazilian free-tailed
        (see *Tadarida brasiliensis*)
    hoary (see *Lasiurus cinereus*)
    myotis (see *Myotis*)
    pallid (see *Antrozous pallidus*)
    pipistrelle (see *Pipistrellus hesperus*)
    silver-haired
        (see *Lasionycteris noctivagans*)
    spotted (see *Euderma maculatum*)
    Townsend's big-eared
        (see *Corynorhinus townsendii*)
Bathylagus 383
bear
    black (see *Ursus americanus*)
    brown (see *Ursus arctos*)
    grizzly (see *Ursus arctos*)
beaver
    American (see *Castor canadensis*)

mountain (see *Aplodontia rufa*)
bison (see *Bison bison*)
Bison 492
   athabascae 491
   bison 491
   oregonus 491
bobcat (see *Lynx rufus*)
Bonasa umbellus 457
boomer (see *Aplodontia rufa*)
Bos 3, 22, 490
   †*antiquus* 493
   athabascae 491
   bison 17, **491–493**
      as prey 362; description 491; diet
      492; distribution 491; folklore 492–
      493; fossils 493; geographic varia-
      tion 491; habitat and density 491–
      492; habits 492; morphometrics
      465; Plate XXXI; reproduction, on-
      togeny, and mortality 492; speci-
      mens examined 512; taxonomy 492
Bovidae 489, **490–502**
   characteristics 490–491; distribution
   490; species diversity 490
Brachylagus 3, 13, 123, 128, 134, 138
   idahoensis 17, **128–131**, 503
      as prey 129; description 128; diet
      129; distribution 128, 129; geo-
      graphic variation 128; habitat and
      density 128–129; habits 129–130;
      home range 130; morphometrics
      130–131; Plate V; reproduction, on-
      togeny, and mortality 127, 129;
      specimens examined 512; status
      130; taxonomy 130
†*Brachypsalis* 20
Bradypus 2
†*Brontotherium* 19
Bubo virginianus 67, 129, 133, 140, 145,
   148, 176, 182, 186, 198, 208, 225, 230,
   237, 256, 266, 272, 275, 282, 285, 295,
   316, 322, 325, 327, 332, 334, 338, 342,
   345, 347, 399, 425, 447, 462
Bufonidae 448
Buteo
   jamaicensis 67, 133, 140, 176, 186,
      189, 200, 203, 208, 225, 316, 319,
      332
   lagopus 148, 200, 203, 332
   regalis 200, 203, 338
   swainsoni 148, 176, 186, 200, 203
Callorhinus 3, 381
   alascanus 381
   alascensis 381
   ursinus **381–385**
      as predator 382–383; as prey 384,
      387; description 381, 382; diet 382–
      383; distribution 381–382, 383; ex-
      ploitation 384–385; fossils 381–
      382; geographic variation 382;
      habitat and density 382; habits 384;
      harvest 384–385; morphometrics
      382; Plate XXIII; reproduction, on-
      togeny, and mortality 383–384;
      specimens examined 512; status
      384–385

Callospermophilus 158
   chrysodeirus 193
   connectens 193
   trepidus 193
   trinitatis 193
†*Camelops vitakerianus* 22
Cancer 438
Canidae **354–373**
   characteristics 354; fossils 355; distribution 354; key to species 355; natural
   history 355; species diversity 354
Canis 3, 20, 21, 366
   argenteus 363
   cinereo 363
   †*davisi* 356
   fusca 360
   fuscus 361
   irremotus 360, 361
   latrans 2, 8, 35, 67, 129, 133, 140, 145,
      147, 148, 151, 156, 163, 176, 182,
      186, 189, 200, 203, 208, 225, 247,
      256, 259, 272, 275, 282, 302, 327,
      332, 338, 347, 354, **355–360**, 432,
      459, 489
      as predator 360, 368, 399, 404, 430,
      447, 471, 473, 481, 488, 500, 501;
      as prey 360; description 355–356;
      diet 359; distribution 356; fossils
      356; geographic variation 356, 359;
      habitat and density 359; habits 360;
      harvest 358; morphometrics 357;
      persecution 360; Plate XX; repro-
      duction, ontogeny, and mortality
      355, 359–356; specimens examined
      512; vocalizations 360
   †*lepophagus* 356
   lestes 355, 356
   lupus 8, 17, 30, 259, 279, 354, 355,
      360–363, 369, 428
      as predator 360; description 360–
      361; diet 362; distribution 361, 362;
      fossils 361; geographic variation
      361; habitat and density 361–362;
      habits 362–363; morphometrics
      357; persecution 363; Plate XX; re-
      production, ontogeny, and mortal-
      ity 355, 362; specimens examined
      513; status 363; vocalizations 363
   umpquensis 355, 356
   velox 366
   vulpes 369
   youngi 361
Capra 497
Capromyidae 352
Carabidae 117, 446
Carcharodon carcharias 393, 439
Cariacus
   columbianus 474
   macrotis 469
Carnivora **354–462**
   characteristics 354; distribution 354;
   key to families 354; species diversity
   354; systematics 354
Castor 3, 21
   baileyi 256, 257
   canadensis 8, 17, 36, 56, 153, **256–**

262, 376, 402, 424, 435
   as pest 261; as prey 259, 362; benefits 261; dams and lodges 259–
   261; description 256, 257; diet 258–
   259; distribution 256–257; exploitation 261–262; folklore 261; fossils 257; geographic variation 257;
   habitat and density 257–258; habits 259–261; harvest 260; morphometrics 258–259; Plate XIV; reproduction, ontogeny, and mortality
   156, 259; specimens examined 513;
   taxonomy 261
   fiber 261
   idoneus 256, 257
   leucodonta 256
   leucodontus 257
   pacificus 256
   shastensis 256, 257
   subauratus 256
   zibethicus 335
Castoridae **256–262**
   distribution 256; species diversity 256
†*Castorides ohioensis* 257
Centrarchidae 435
Centrocercus urophasianus 459
Cerambycidae 117
Cervoidea 489
Cervidae **463–484**, 489
   characteristics 463–464; distribution
   463; fossils 464; species diversity 463
Cervus 3, 428
   canadensis 464
   Columbiana 474
   columbianus 474
   elaphus 2, 8, 463, **464–469**, 473, 483
      antlerogenesis 468; as prey 462,
      453; description 464; diet 466–467;
      distribution 464–465, 466; exploi-
      tation 466; geographic variation
      465; habitat and density 465–466;
      habits 468–469; harvest 466, 467;
      home range 469; migrations 469;
      morphometrics 465; Plate XXXI;
      reproduction, ontogeny, and mortal-
      ity 467–468; sexual segregation
      469; specimens examined 513
   hemionus 469
   leucurus 479
   macrotis 474
   nelsoni 464, 465
   roosevelti 464, 465
Ceutophilus 52
†*Chalicotherium* 20
chickaree (see *Tamiasciurus douglasi*)
Chilopoda 51, 60, 61, 66, 69, 71, 86
Chincha
   major 445
   notata 445
   occidentalis 445
   platyrhina 445
chinchilla (see *Chinchilla*)
Chinchilla 503
chipmunk
   Allen's (see *Tamias senex*)
   eastern (see *Tamias striatus*)

least (see *Tamias minimus*)
Siskiyou (see *Tamias siskiyou*)
Townsend's (see *Tamias townsendii*)
yellow-pine (see *Tamias amoenus*)
Chironomidae 81, 97
Chiroptera 19, **74–122**
 as pests 75; characteristics 74; diet 74–75; distribution 74; flight speed 75; fossils 74; key to species 75–76; population declines 75; species diversity 74
Choloepus 2
Chrysochloridae 45
Chrysopidae 117
Cicadidae 117
Cicindelidae 55, 60, 447
Circus cyaneus 129, 148, 203, 316, 332
Citellus 158
 artemisia 196
 beecheyi 181
 beldingi 184
 canus 196
 chrysodeirus 193
 connectens 193
 columbianus 187
 crebrus 184
 douglasi 181
 elegans 190
 idahoensis 196
 lateralis 193
 leucurus 178
 mollis 196
 nevadensis 190
 oregonus 184
 richardsonii 190
 ruficaudus 187
 townsendii 196, 203
 trepidus 193
 trinitatis 193
 vigilis 196
 vinnulus 178
 washingtoni 201
Citharichthys 387
 sordidus 395
Clethrionomys 3, 296, 320, 322, 408, 413, 417, 419, 430
 californicus 30, 49, **298–303**, 304, 327
  as prey 302; description 297, 298; diet 299; distribution 298; Fig. 11-95; geographic variation 298–199; habitat and density 299; habits 302; morphometrics 300; reproduction, ontogeny, and mortality 299, 302; specimens examined 513–515; taxonomy 303
 gapperi 33, 298, **303–304**, 309
  as prey 304, 442–443; description 297, 303; diet 303; distribution 303, 304; fossils 303; geographic variation 303; habitat and density 303; habits 304; home range 304; morphometrics 300; Plate XVII; reproduction, ontogeny, and mortality 302, 303–304; specimens examined 515–516
 idahoensis 303
 mazama 298, 299, 302

obscurus 298, 299
 occidentalis 303
Clupea pallasi 383, 387, 395
Colaptes auratus 103
Coleoptera 51, 52, 55, 60, 61, 63, 66, 68, 80, 84, 86, 89, 91, 94, 103, 105, 108, 111, 117, 120, 249, 399, 424, 425, 446
†*Colodon* 19
Corvus
 brachyrhynchos 313
 corax 200, 313, 430
Corynorhinus 3
 intermedius 113
 macrotis 113
 pallescens 113
 rafinesquii 113
 townsendi 113
 townsendii 17, 83, 94, 98, 102, **113–116**
  description 113; diet 114; distribution 113, 114; flight speed 75; fossils 113; geographic variation 113; habitat and density 113–114; habits 115–116; morphometrics 77; Plate IV; reproduction, ontogeny, and mortality 101, 114–115; roosts 115; specimens examined 516; taxonomy 116
Cottidae 435
cottontail
 desert (see *Sylvilagus auduboni*)
 eastern (see *Sylvilagus floridanus*)
 mountain (see *Sylvilagus nuttallii*)
Cottus 57, 425
cougar (see *Puma concolor*)
coyote (see *Canis latrans*)
Crangon 396
Cricetidae 262
Cricetinae 262
Cricetodipus parvus 245
Cricetus talpoides 234
Crotalus 163
 virdis 133, 148, 186, 200, 203, 225, 247
Crotaphytus wislizenii 249
Chrysopidae 117
†*Cupidinomys* 21
Culicidae 128
Cuterebra 483
Cyanocitta stelleri 209
Cymatogaster aggregata 395
†*Cynodontia* 3
Cyprinidae 435
Cystophora cristata 390, 504
Dama virginiana 479
Decapoda 402, 424
deer
 black-tailed
  (see *Odocoileus h. columbianus*)
 mule (see *Odocoileus h. hemionus*)
 white-tailed
  (see *Odocoileus virginianus*)
dental formulas 2, 3
Dermacentor andersoni 37
†*Desmatophoca* 19
Dicamptodon 419

ensatus 57, 62
†*Diceratherium* 20
Didelphidae 20, **40**
 fossils 40; species diversity 40
Didelphimorphia **40**
 characteristics 40; distribution 40; systematics 40
Didelphis virginiana 3, 37, **40–44**, 295, 448
 as prey 371; description 40; diet 43; distribution 41; folklore 44; geographic variation 41; habitat and density 41, 43; habits 43–44; harvest 42; home range 44; introduction 41; morphometrics 41; Plate I; reproduction, ontogeny, and mortality 43; specimens examined 516
†*Dimetrodon* 2
Diplostoma 6
 bulbivorum 229
Dipodidae **339–345**
 characteristics 340; distribution 339–340; fossils 340; key to species 340; natural history 340; species diversity 339
Dipodomys 3, 21, 242, 243, 248, 249
 californicus 28, 31, **249–251**
  description 249–250; diet 250–251; distribution 250; geographic variation 250; habitat and density 250; habits 251; morphometrics 242–243; Plate XIII; reproduction, ontogeny, and mortality 244, 251; specimens examined 516; taxonomy 251
 columbianus 254, 255
 gabrielsoni 249
 heermanni 249, 251
 idahoensis 252
 ingens 367
 merriami 503
 microps 31, **251–254**, 367, 503
  description 251–252; diet 253; distribution 252; fossils 252; geographic variation 252; habitat and density 252–253; habits 253–254; morphometrics 242–243 Plate XIII; reproduction, ontogeny, and mortality 244, 253; specimens examined 516–517
 nitratoides 367
 ordii **254–256**, 277, 367, 503
  as prey 255–256; description 254; diet 255; distribution 254, 255; fossils 254; geographic variation 254; habitat and density 254–255; habits 256; morphometrics 242–243; Plate XIII; reproduction, ontogeny, and mortality 244, 255–256; specimens examined 517
 preblei 251, 252
†*Dipoides* 21
†*Diprionomys* 20, 21
Diptera 57, 61, 63, 66, 80, 84, 86, 88, 92, 96, 99, 103, 105, 120
disease-causing organisms
 Bacillus piliformis 339

Borrelia
  burgdorferi 37
  hermsii 37
Clostridium 339
Coxiella burnetti 38
Francisella tularensis 36
hantavirus 35
Leptospira 37
  alexi 37
  autumnalis 37
  ballum 37
  canicola 37
  grippotyphosa 37
  hardjo 37
  icterohemorrhagiae 37
  pomona 37
  szwajizak 37
  tarassovi 37
Lyssavirus 35
Pasturella 500
Rickettsia rickettsii 37
Yersinia pestis 36
†Docodonta 3
domestic dog 72, 133, 348, 356, 360, 361,
  363, 436, 448, 477
†*Dromomeryx* 19, 20
Drosophila 89
Echimyidae 352
Echinoidea 405
Elanus caeruleus 313
elk (see *Cervus elaphus*)
Engraulis mordax 383, 389
Enhydra 3, 22
  kenyoni 437, 438
  lutris 8, 17, **437–439**
    as predator 438; as prey 439; de-
    scription 437; diet 438; distribution
    437–438; exploitation 437–438;
    fossils 438; geographic variation
    438X; habitat and density 438; hab-
    its 439; morphometrics 406–407;
    Plate XXVIII; reproduction, ontog-
    eny, and mortality 438–439; rein-
    troduction 439; specimens exam-
    ined 517; status 437–438, 439
  nereis 438
†*Enhydriodon* 438
†*Entoptychus* 19
†*Eomellivora* 21
†Eomyidae 20
Eozapus 339
Ephemeroptera 50, 57, 97
Epicauta 277
†*Epigaulus* 19, 20
Eptesicus 3, 117
  bernardinus 107
  fuscus 35, 89, 91, 94, 99, 102, 104,
  107–110
    as pest 110; as prey 109, 419; de-
    scription 107, 108; diet 104, 107–
    108; distribution 107, 108; Fig. 9-
    31; flight speed 75; fossils 107; geo-
    graphic variation 107; habitat and
    density 107; habits 109–110; mor-
    phometrics 77; reproduction, on-
    togeny, and mortality 101, 108–

    109; roosts 110; specimens exam-
    ined 517–518
  pallidus 107
equine encephalomyelitis 43
Equus 22
Eremophila alpestris 367
Erethizon 3
  dorsatum 345–348, 349
    as pest 348; as prey 347, 413, 428,
    453, 459; capture 348; description
    345, 346; diet 346–347; distribution
    345, 346; folklore 348; fossils 345;
    geographic variation 345; habitat
    and density 345–346; habits 348;
    home range 348; morphometrics
    347; Plate XIX; reproduction, on-
    togeny, and mortality 156, 347–
    348; specimens examined 518; vo-
    calizations 348
  dorsatus 345
  epixanthum 345
  epixanthus 345
Erethizontidae **345–348**
  distribution 345; species diversity 345
Erinaceidae 20, 45
ermine (see *Mustela erminea*)
Errington's disease 339
Euarctos
  altifrontalis 374
  americanus 374
  cinnamomum 374
†*Eucastor* 21
Eucervus columbianus 474
Euderma 3, 116
  maculata 110
  maculatum 10, 76, 105, **110–113**
    as prey 112; description 110; diet
    111; distribution 110–111, 112;
    geographic variation 111; habitat
    and density 111; habits 112–113;
    morphometrics 77; reproduction,
    ontogeny, and mortality 101, 111–
    112; Plate III; roosts 112; specimens
    examined 518; status 113; vocaliza-
    tions 112
†*Eumegamys* 153
Eumetopias 3
  jubata 385
  jubatus 381, 384, **385–388**
    as predator 396; as prey 387; de-
    scription 385; diet 386–387; distri-
    bution 385, 386; fossils 385; geo-
    graphic variation 385; habitat and
    density 385–386; habits 387–388;
    morphometrics 382; Plate XXIII;
    reproduction, ontogeny, and mortal-
    ity 387; specimens examined 518–
    519; status 386
  Stelleri 385
†*Euplocyon* 20
Eusattus 248
Eutamias 158
  albiventris 159
  amoenus 159
  cooperi 169
  ludibundus 159

  minimus 163
  ochraceus 159
  propinquus 159
  scrutator 163
  senex 166
  siskyiou 168
  townsendi 169
Evotomys
  californicus 298
  idahoensis 303
  mazama 298
  obscurus 298
Falco
  mexicanus 200, 203, 322
  peregrinus 103, 148
  sparverius 222
Falconidae 319
Felidae 445, **450–462**
  characteristics 450; distribution 450;
  fossils 450; key to species 450; species diversity 450; systematics 450
Felis 21
  californica 450
  concolor 450, 455
  missoulensis 450
  rufa 458
Felix oregonensis 450
ferret (see *Mustela putorius*)
Fiber
  mergens 335
  occipitalis 335
  osoyoosensis 335
  zibethicus 335
fisher (see *Martes pennanti*)
Fissipedia 354
formation of Oregon 16–17
Formicidae 69, 277
fox
  common gray
    (see *Urocyon cinereoargenteus*)
  kit (see *Vulpes velox*)
  red (see *Vulpes vulpes*)
  swift (see *Vulpes velox*)
Gastropoda 43, 50, 55, 57, 59, 63, 66, 68,
  69, 71, 277
Geomyidae **233–239**
  as pests 223; benefits 223; characteristics 223; control 223; distribution 223;
  key to species 223–224; natural history
  223; species diversity 223
Geomyoidea 223, 239
Geomys 6
  bursarius 430
  Townsendii 238
Glaucomys 3, 212, 408, 411
  bangsi 219
  fuliginosus 219
  klamathensis 219
  oregonensis 219
  sabrinus 216, 218, **219–223**, 441
    aerodynamics 222; as prey 222,
    460; description 219; diet 221; dis-
    tribution 219; fossils 219; geo-
    graphic variation 219; habitat and
    density 219–221; habits 222; home
    range 222; morphometrics 220;

Plate XI; reproduction, ontogeny, and mortality 162, 221–222; specimens examined 519; taxonomy 223; torpor 222

Glyptocephalus zachirus 395

†*Gomptotherium* 21

Gonatus fabricii 383

gopher (see pocket gopher)

Gopherus polyphemus 6

†*Grangerimus* 19

grasshopper mouse
    northern (see *Onychomys leucogaster*)
    southern (see *Onychomys torridus*)

gray digger (see *Spermophilus beecheyi*)

ground squirrel (also see squirrel)
    Belding's (see *Spermophilus beldingi*)
    California
        (see *Spermophilus beecheyi*)
    Columbian
        (see *Spermophilus columbianus*)
    golden-mantled
        (see *Spermophilus lateralis*)
    Merriam's (see *Spermophilus canus*)
    Piute (see *Spermophilus mollis*)
    Washington
        (see *Spermophilus washingtoni*)
    Wyoming (see *Spermophilus elegans*)

Gryllacrididae 117

Gryllidae 57

Gulo 3
    gulo **426–428**, 437
        as predator 427–428, 501; description 426–427; diet 427–428; distribution 427; fossils 427; geographic variation 427; habitat and density 427; habits 428; home range 428; Plate XXVII; reproduction, ontogeny, and mortality 428; specimens examined 519; status 427; taxonomy 428
    Luscus 426
    luscus 426, 428
    luteus 426

Haliaeetus leucocephalus 151, 338, 439, 447

Halicyon
    richardii 394
    richardsi 394

Hantavirus pulmonary syndrome 35

†*Haptodus* 2

hare
    snowshoe (see *Lepus americanus*)
    (also see jackrabbit)

†*Hemicyon* 20

Hemigrapsus nudus 435

†*Hemipsalodon grandis* 19

Hemiptera 61, 84, 86, 92, 103, 105, 114, 120

hemorrhagic disease 339

Hesperomys
    crinitus 267
    gambelii 270
    maniculatus 270
    sonoriensis 270
    truei 273

†*Hesperosorex* 21

Heteromyidae 223, **239–256**
    characteristics 239–240; key to species 240; natural history 239–240; species diversity 239; systematics 239; torpor 240

†*Heterosorex* 20

†*Hipparion* 21
    anthonyi 21

Hirundinea 57

Histiotus maculatus 110

histoplasmosis 43

†*Histricops* 21

Homoptera 80, 84, 86, 92, 94, 103, 105, 120

†*Homotherium serum* 450

house cat 35, 62, 72, 97, 222, 302, 313, 332, 458, 460

†*Hyaenodon* 19

†*Hydrosacpheus* 21

Hyla 63

Hymenoptera 61, 84, 103, 105, 107, 120, 227

†*Hypohippus* 19, 20

†*Hypolagus* 20, 21

Hypudaeus leucogaster 276

†*Hyrachyus* 19

Hystrix dorsata 345

Ichneumonidae 52

Ictaluridae 435

†*Ilingoseros* 21

†*Indarctos* 21

†*Ingentisorex* 20

Insectivora **45–73**
    distribution 45; fossils 45; key to species 45–46; species diversity 45

introductions (exotics) 41, 134–135, 206, 211–212, 289, 372–373, 497

†Ischyromyidae 153

Isopoda 51

Isoptera 80, 89, 94, 96

Ixodes 37

jackrabbit
    black-tailed (see *Lepus californicus*)
    white-tailed (see *Lepus townsendii*)

jerboa 339

kangaroo rat
    California (see *Dipodomys californicus*)
    chisel-toothed (see *Dipodomys microps*)
    Merriam's (see *Dipodomys merriami*)
    Ord's (see *Dipodomys ordii*)

key to the orders 39

Lagomorpha **123–152**
    characteristics 123; distribution 123; key to species 123; species diversity 123; systematics 123

Lagomys 124

Lagurus 335
    curtatus 332
    pauperrimus 332

Lamna ditrops 387

Lamniformes 384, 396

Lampetra tridentata 386–387, 389, 395

Lanius 325
    excubitor 332

†*Lantanotherium* 20

Larinae 313

Larus 325

Lasionycteris noctivagans 3, 35, 88, 94, 98, **100–105**, 114
    as prey 103; description 100, 102; diet 102–104; distribution 100–102; Fig. 9-24; flight speed 75; fossils 101; geographic variation 101; habitat and density 101–102; habits 103–104; morphometrics 77; reproduction, ontogeny, and mortality 101, 103; roosts 104; specimens examined 519–520; torpor 104; vocalizations 104

Lasiurus 75
    cinereus 3, 35, 76, **98–100**, 102, 104, 114
        as prey 100; description 98, 99; diet 99; distribution 98, 99; flight speed 75; fossils 98; geographic variation 98; habitat and density 98–99; habits 100; Plate III; morphometrics 77; reproduction, ontogeny, and mortality 99–101; specimens examined 520; vocalizations 100
    †*fossilis* 98

Lemmiscus 3, 296, 303, 307, 309, 311, 320, 322, 332
    curtatus 52, **332–335**
        as prey 334; description 297, 332–333; diet 333–334; distribution 333; fossils 333; geographic variation 333; habitat and density 333; habits 334–335; morphometrics 300; Plate XVIII; reproduction, ontogeny, and mortality 302, 334; specimens examined 520; taxonomy 335
    intermedius 333
    pauperrimus 332, 333

Lepidoptera 52, 63, 76, 84, 88, 91, 94, 96, 99, 102, 105, 107, 111, 114, 117, 120, 249, 277, 399, 446, 448

Leporidae **128–152**
    characteristics 128; species diversity 128; status 128

†*Leptarctus* 20

Leptocottus armatus 395

†*Leptodontomys* 21

leptospirosis 37, 43, 339

Lepus 3, 20, 21, 123, 128, 146, 430
    americanus 30, **141–145**, 147, 336, 408, 413, 453, 456, 457, 459
        as pest 145; as prey 145; cycles 143–144; description 141; diet 144; distribution 141–142; fossils 142; geographic variation 142; habitat and density 142–144; habits 145; morphometrics 142–143; Plate VI; reproduction, ontogeny, and mortality 127, 144–145; specimens examined 520–521
    artemisia 138
    bachmani 131
    bairdii 141
    californicus 22, **145–149**, 150, 360, 367
        as pest 149; as prey 148, 364, 371;

description 145–146; diet 147; distribution 146; fossils 146; geographic variation 146–147; habitat and density 147; habits 148–149; home range 149; morphometrics 142–143; Plate VI; reproduction, ontogeny, and mortality 127, 147–148; specimens examined 521

campanius 152
campestris 149
depressus 145
deserticola 145
floridanus 134
idahoensis 128
klamathensis 141, 142
nuttallii 138
oregonus 141, 142
princeps 124
sierrae 149
sylvaticus 134
texianus 145
townsendii 9, 22, 145–147, **149–152**
   as prey 151; description 149–150; diet 151; distribution 150; fossils 150; geographic variation 150; habitat and density 150–151; habits 151–152; morphometrics 142–143; Plate VI; reproduction, ontogeny, and mortality 127, 151; specimens examined 521
ubericolor 131
vigilax 145
wallawalla 145, 147
washingtoni 141, 142
Libellulidae 117
Limacidae 277
†*Limnoecus* 20
†*Liodontia* 20, 21
Locustidae 446
Loligo opalescens 383, 389
Lumbricus terrestris 371
lungworm 500
Lupus gigas 360
Lutra 3, 21
   canadensis 8, 259, 425, **433–436**
      as predator 435; as prey 436; description 433; diet 435; distribution 433, 435; fossils 433; geographic variation 433, 435; habitat and density 433, 435; habits 436; harvest 434; home range 436; morphometrics 406–407; Plate XXVIII; reproduction, ontogeny, and mortality 410, 435–436; specimens examined 521–522
   hudsonica 433
   pacifica 433
†*Lutravus* 21
Lutreola
   energumenos 422
   vison 422
†*Lutrictis* 20
Lygaeidae 249
Lyme disease 37
lynx (see *Lynx lynx*)
Lynx 3

canadensis 145, 189, 450, **455–458**
   cycles 457; description 455–456; diet 457; distribution 456; fossils 456; geographic variation 456; habitat and density 456–457; habits 457; home range 457; immigration 457; morphometrics 452–453; Plate XXIX; reproduction, ontogeny, and mortality 454, 457; specimens examined 522; status 458; taxonomy 457–458
fasciatus 458
lynx 457
pallescens 458
rufus 8, 129, 133, 148, 151, 156, 163, 182, 272, 278, 282, 285, 302, 327, 332, 334, 338, 345, 347, 450, 455, 457, **458–462**
   as predator 368, 369, 404, 447, 459–460, 501; as prey 462; description 458; diet 459–460; distribution 458, 459; fossils 458; geographic variation 458; habitat and density 458–459; habits 462; harvest 461; home range 462; hybrids 460; morphometrics 452–453; Plate XXIX; reproduction, ontogeny, and mortality 454, 460, 462; specimens examined 522
subsolanus 456
uinta 458
Lyopsetta exilis 395
†*Machairodus* 21
†*Macrognathomys* 20, 21
Macrorhinus angustirostris 391
Macroscelidea 45
Macroscelididae 45
Mallotus villosus 383, 387
mammal
   biogeographical affinities 33–34; characteristics 1–2; collections of Oregon species 8–12; evolution of faunas 19–22; nomenclature 6–7; origin 2–3, 6; present-day faunas 22–33; publications 11–13; skull bones 4; teeth 1–2
†*Mammut* 19–21
Mantidae 117
Marmosa alstoni 503
marmot, yellow-bellied
   (see *Marmota flaviventris*)
Marmota 3, 21
   avara 174
   avarus 173
   flaviventer 173
   flaviventris 9, **173–178**
      as prey 176, 430; description 173–174; diet 174, 176; distribution 174; fossils 174; geographic variation 174; habitat and density 174; habits 176–178; morphometrics 175; Plate VIII; reproduction, ontogeny, and mortality 176–177; specimens examined 522–523; vocalizations 178
Marsupialia 40
marten, American (see *Martes americana*)

Martes 3, 20
   americana 9, 189, 208, 222, 302, 310, 322, 330, **406–411**
      description 406, 407; diet 408, 410; distribution 406–407, 408; fossils 407; geographic variation 407; habitat and density 407–408; habits 410–411; harvest 409; home range 411; morphometrics 406–407; Plate XXV; reproduction, ontogeny, and mortality 410; specimens examined 523
   caurina 406, 407
   †*divuliana* 412
   humboldtensis 407
   †*nobilis* 407
   origenes 406
   pacifica 412
   pennanti 9, 17, 208, 338, 347, 406, **411–415**
      description 411, 412; diet 413; distribution 411–412; Fig. 12-37; fossils 412; geographic variation 412; habitat and density 412–413; habits 413–415; home range 413; morphometrics 406–407; reproduction, ontogeny, and mortality 410, 413; specimens examined 523; status 415
   sierrae 407
   vulpina 406, 407
   zibellina 407
Megachiroptera 74
†*Megalonychidae* 21
Megascapheus 224, 229, 338
Meles jeffersonii 428
†*Meniscomys* 19
Mephitidae **440–450**
   characteristics 440; distribution 440; key to species 440; species diversity 440; systematics 440
Mephitis 3, 445
   foetulenta 445
   major 445, 446
   mephitis 2, 8, 35, 37, 247, 313, 319, 332, 431, **445–450**
      as predator 446–447; as prey 371, 447; description 445–446; diet 446–447; distribution 446, 447; folklore 448; fossils 446; geographic variation 446; habitat and density 446; habits 448; harvest 449; home range 448; morphometrics 440–441; musk 448; Plate XXVIII; reproduction, ontogeny, and mortality 445, 447–448; specimens examined 523; vocalizations 448
   notata 445, 446
   occidentalis 445, 446
   spissigrada 445, 446
Merluccius productus 389, 392
†*Merychippus* 19, 20
†*Merycodus* 19, 20
†*Merycochocerus* 20
†*Merycoidodon* 20

Micoureus alstoni 503
Microchiroptera 74
Microdipodops 3
    megacephalus 31, 244, **248–249**
        description 248; diet 248–249; distribution 248; geographic variation 248; habitat and density 248; habits 249; morphometrics 242–243; Plate XIII; reproduction, ontogeny, and mortality 244, 249; specimens examined 523–524
    oregonus 248
Microlagus 131
Microstomus pacificus 395
Microtinae 262
†*Microtoscoptes* 21
Microtus 3, 133, 296, 267, 294, 298, 303, 309, 321, 325, 331, 335, 364, 371, 373, 408, 413, 417, 419, 425, 430, 459, 460
    abditus 320, 321
    adocetus 325, 326
    angusticeps 320, 321
    angustus 320, 321
    artemisiae 332
    arvicoloides 328
    bairdii 325, 326
    californicus 28, 36, 267, 294, 295, **311–317**, 320, 322, 330
        as prey 313, 316; description 297, 311, 312; diet 312–313; distribution 311–312; fossils 312; geographic variation 312; habitat and density 312; habits 316–317; home range 317; morphometrics 314; Plate XVII; reproduction, ontogeny, and mortality 313, 316; specimens examined 524
    canescens 320
    canicaudus 27, 35, 296, **317–320**, 322, 325, 331, 332
        as prey 319; capture 319; description 297, 317, 318; diet 318; distribution 317–320; Fig. 11-109; geographic variation 318; habitat and density 318; habits 319; home range 319; morphometrics 314; reproduction, ontogeny, and mortality 316, 318–319; specimens examined 524
    curtatus 332
    eximius 311, 312
    halli 320, 321
    longicaudus 305, **320–322**, 323
        as prey 322; description 297, 320; diet 322; distribution 320–321; fossils 320–321; geographic variation 321; habitat and density 321; habits 322; home range 322; morphometrics 314; Plate XVII; reproduction, ontogeny, and mortality 316, 322; specimens examined 524–525; taxonomy 322
    macropus 328
    micropus 323
    montanus 9, 264, 271, 296, 307, 313, 319, 320, **322–325**, 335, 419
        as prey 325, 417; description 297, 322–323; diet 324; distribution 323; fossils 323; geographic variation 323; habitat and density 323–324; habits 325; irruptions 322–325; morphometrics 314–315; Plate XVIII; reproduction, ontogeny, and mortality 316, 324–325; specimens examined 526–527
    mordax 320–322
    nanus 320, 322, 323
    oregoni 22, 69, 298, 299, 303, 307, 309, 311, 319, 320, 322, **325–328**, 330, 332, 344, 416
        as prey 327, 359, 459; description 297, 325–326; diet 327; distribution 326; Fig. 11-116; genetics 327–328; geographic variation 326; habitat and density 326–327; habits 327; home range 327; morphometrics 315; reproduction, ontogeny, and mortality 316, 327; specimens examined 527–528
    richardsoni 30, 320, 322, **328–330**, 344
        as prey 330; description 297, 328; diet 329; distribution 328, 329; fossils 328; geographic variation 328; habitat and density 328–329; habits 330; home range 330; morphometrics 315; Plate XVIII; reproduction, ontogeny, and mortality 316, 329–330; specimens examined 528
    townsendii 8, 22, 271, 319, 320, 322, 327, **330–332**
        as prey 332, 459; description 297, 330, 331; diet 331; distribution 330, 331; Fig. 11-121; geographic variation 330; habitat and density 330–331; habits 332; home range 332; morphometrics 315; reproduction, ontogeny, and mortality 316, 331–332; specimens examined 528–529
mink (see *Mustela vison*)
†*Miolabis* 20
†*Miosciurus* 19
Mirounga 3, 390
    angustirostris 390, **391–394**
        as prey 393; description 391; diet 392; distribution 391–392; diving 393; exploitation 392; fossils 391–392; geographic variation 392; habitat and density 392; habits 392–394; Plate XXIV; reproduction, ontogeny, and mortality 392–393; specimens examined 529; status 392, 394
†*Mixodontia* 123
mole
    broad-footed (see *Scapanus latimanus*)
    coast (see *Scapanus orarius*)
    Townsend's (see *Scapanus townsendii*)
Molossidae 74, **119–122**
    characteristics 119; distribution 119; natural history 119; species diversity 119; torpor 119
Molossus mexicanus 119
†*Monosaulax* 20
moose (see *Alces alces*)
†*Moropus* 19
mountain
    cottontail (see *Sylvilagus nuttallii*)
    goat (see *Oreamnos americanus*)
    lion (see *Puma concolor*)
    sheep (see *Ovis canadensis*)
mouse
    canyon (see *Peromyscus crinitus*)
    dark kangaroo (see *Microdipodops megacephalus*)
    deer (see *Peromyscus maniculatus*)
    house (see *Mus musculus*)
    northern grasshopper (see *Onychomys leucogaster*)
    Pacific jumping (see *Zapus trinotatus*)
    piñon (see *Peromyscus truei*)
    western harvest (see *Reithrodontomys megalotis*)
    western jumping (see *Zapus princeps*)
†Multituberculata 3
Muridae 57, 76, **262–339**
    distribution 262; diversity 262; key to subfamilies 262; systematics 262
Murinae 262, **288–296**
    as pests 288; characteristics 288; control 288; key to species 289; natural history 288; species diversity 288
murine typhus 43
Mus 63, 430, 442
    alexandrinus 291
    cinereus 280
    musculus 35, 291, **293–295**, 317
        as commensal 295; as prey 295; description 293, 294; diet 294; distribution 293, 294; geographic variation 293; habitat and density 293–294; habits 295; morphometrics 289; Plate XVI; reproduction, ontogeny, and mortality 294–295; specimens examined 529
    norvegicus 289
    †*petteri* 295
    rattus 291
muskrat, common (see *Ondatra zibethicus*)
Mustela 3, 20, 129, 145, 189, 222, 332, 411
    altifrontalis 418, 419
    americanus 406
    arizonensis 418
    canadensis 411, 433
    caurina 406
    cicognanii 415
    effera 418, 419
    energumenos 422, 424
    erminea 22, 237, 247, 272, 302, 322, 327, 330, **415–418**
        as predator 417; as prey 417; description 415–416; diet 417; distribution 416; fossils 416; geographic variation 416; habitat and density 416–417; habits 418; home range 418; morphometrics 414–415; Plate XXVI; reproduction, ontogeny, and mortality 417–418; specimens examined 530

frenata 133, 163, 172, 186, 200, 203, 247, 302, 322, 334, 417, **418–422**
as predator 419–420; as prey 420; description 418; diet 419–420; distribution 418–419; fossils 419; geographic variation 419; habitat and density 419; habits 420, 422; harvest 421; home range 422; morphometrics 414–415; Plate XXVI; reproduction, ontogeny, and mortality 420; specimens examined 530–531

lepta 415
longicauda 418, 420
lutra 433
lutris 437
murica 415
muricus 415, 416
nevadensis 418, 419
nivalis 354
oregonensis 418, 419
origenes 406
pacifica 411
pennanti 411
putorius 237
†*rexroadensis* 419
saturata 418, 419
streatori 415, 416
vison 8, 172, 259, 332, 338, **422–426**
as predator 424–425; as prey 425; description 422; diet 424–425; distribution 422, 424; fossils 424; geographic variation 424; habitat and density 424; habits 425–426; harvest 423; home range 425–426; morphometrics 414–415; Plate XXVII; reproduction, ontogeny, and mortality 410, 425; specimens examined 531

vulpina 406
washingtoni 418, 419
xanthogenys 418
zibellina 407
Mustelidae **405–439**
characteristics 405; distribution 405; fossils 405; key to species 405; natural history 405; species diversity 405

Mydaus 440
†*Mylagaulodon* 19
†*Mylagaulus* 19–21
Mynomes 320, 322, 328
†*Mylodon harlani* 22
Myocastor 3
bonariensis 348, 349
coypus 345, **348–352**
as pest 352; description 348–349; diet 350; distribution 349, 350; geographic variation 349; habitat and density 349–350; habits 352; harvest 351; introduction 352; Plate XIX; reproduction, ontogeny, and mortality 156, 350, 352; specimens examined 531

Myocastoridae **348–352**
species diversity 348
Myopotamus bonariensis 348

myotis
California (see *Myotis californicus*)
fringed (see *Myotis thysanodes*)
little brown (see *Myotis lucifugus*)
long-eared (see *Myotis evotis*)
long-legged (see *Myotis volans*)
western small-footed (see *Myotis ciliolabrum*
Yuma (see *Myotis yumanensis*)
Myotis 3, 100, 117
alascensis 87
altifrons 93
auriculus 86
californicus **79–82**, 84, 102
description 79; diet 80–81; distribution 79, 80; flight speed 75; fossils 79; geographic variation 79; habitat and density 79–80; habits 81–82; morphometrics 77; Plate III; reproduction, ontogeny, and mortality 81; roosts 81; specimens examined 531–532; taxonomy 82

carissima 87
caurinus 79
ciliolabrum 75, 79, 81, **82–84**, 94, 102
description 82, 83; diet 84; distribution 82–83; fossils 83; geographic variation 83; habitat and density 83–84; habits 84; morphometrics 77; Plate III; reproduction, ontogeny, and mortality 84; roosts 83–84; specimens examined 532; taxonomy 84

evotis **84–87**, 94, 102
description 84, 85; diet 81, 86; distribution 84, 85; Fig. 9-7; fossils 84; geographic variation 84; habitat and density 84–86; habits 86; morphometrics 77–78; reproduction, ontogeny, and mortality 86; roosts 86; specimens examined 532

interior 93
leibii 80, 82, 84
Leuconoe 87
longicrus 93
lucifugus **87–90**, 94–98, 102
as prey 89; description 87; diet 81, 88–89; distribution 87, 88; Fig. 9-10; fossils 87; geographic variation 87; habitat and density 87–88; habits 89–90; morphometrics 78; reproduction, ontogeny, and mortality 89; roosts 89; specimens examined 532–533; taxonomy 90; torpor 90; vocalizations 90

melanorhinus 82, 84
pacificus 84
saturatus 95
sociabilis 95
subulatus 82, 84
thysanodes 84, **90–93**, 94, 99, 102, 107
description 90, 91; diet 91–92; distribution 91, 92; Fig. 9-13; flight spped 75; geographic variation 91; habitat and density 91; habits 92–93; morphometrics 78; reproduc-

tion, ontogeny, and mortality 92; roosts 93; specimens examined 533

volans 86, 87, **93–95**, 99, 102
description 93, 94; diet 81,94; distribution 93, 94; Fig. 9-17; flight speed 75; fossils 93; geographic variation 93; habitat and density 93–94; habits 95; morphometrics 78; reproduction, ontogeny, and mortality 94–95; specimens examined 533; vocalizations 95

yumanensis 84, 88, 90, 94, **95–98**, 102, 118
as prey 97; description 95–96; diet 81, 96–97; distribution 96; flight speed 75; geographic variation 96; habitat and density 96; habits 97–98; morphometrics 78; Plate III; reproduction, ontogeny, and mortality 97; roosts 97; specimens examined 533–534; torpor 98

Myriapoda 277
†*Mystipterus* 20
Napaeozapus 339
Nasua 397
necrobacillosis 483
Nemorhoedus palmeri 497
†*Neohipparion* 21
Neotoma 3, 36, 262, 273, 442, 459, 503
alticola 280
apicalis 280
cinerea 9, **280–283**, 288
as prey 282, 460; description 280; diet 281; distribution 280, 281; fossils 280; geographic variation 280; habitat and density 280–281; habits 282–283; houses 282; morphometrics 265; Plate XV; reproduction, ontogeny, and mortality 266, 281–282; sign 283; specimens examined 534–535

fusca 280
fuscus 280
fuscipes 37, 263, 282, **283–286**, 288
as prey 285, 419, 459; description 283, 284; diet 284–285; distribution 283, 284; fossils 283; geographic variation 284; habitat and density 284; habits 285; home range 285; houses 285; morphometrics 265; Plate XV; reproduction, ontogeny, and mortality 266, 285; specimens examined 535–536

lepida 33, 268, 282, **286–288**
description 286; diet 286; distribution 286, 287; fossils 286; geographic variation 286; habitat and density 286; habits 287–288; home range 287; houses 287–288; morphometrics 265; Plate XV; reproduction, ontogeny, and mortality 266, 286–287; specimens examined 536

macrotis 283
monochroura 283, 284
nevadensis 286

occidentalis 280
pulla 280
splendens 283
stephensi 282
Neosorex 55
navigator 55
Neuroptera 80, 84, 89, 92, 103, 105
Neurotrichus 3, 21
gibbsii **65–68**
as prey 67; description 65, 66; diet 66; distribution 65–66; fossils 65; geographic variation 66; habitat and density 66; habits 67; morphometrics 67; Plate II; reproduction, ontogeny, and mortality 66–67; specimens examined 536–537
major 65
Noctuidae 86, 117
Notoedres 208
nutria (see *Myocastor coypus*)
Nyctea scandia 332
Nycteris cinerea 98
Nyctinomus brasiliensis 119
Nyctinomops macrotis 119
Nymphalis californica 194
Ochotona 3, 21, 417, 419
brunnescens 124, 125
fenisex 124
fumosa 124, 125
jewetti 124, 125
princeps 22, 30, 123, **124–128**
description 124, 126; diet 126; distribution 124–126; fossils 125; geographic variation 125; habitat and density 125; habits 127–128; morphometrics 124–125; Plate V; reproduction, ontogeny, and mortality 126–127; specimens examined 537; territories 127; vocalizations 127
shisticeps 124
taylori 124, 125
Ochotonidae **124–128**
characteristics 124; distribution 124; fossils 124; species diversity 124
Octopus 387, 439
Odocoileus 3, 37, 362, 410, 413, 428, 459
Columbianus 474
hemionus 17, 413, 463, 465, 467, **469–479**, 483, 484
as prey 359, 452; description 469; distribution 469; geographic variation 469; Plate XXXI
h. columbianus 6–7, 469, 470, 473, **474–479**, 484
description 474–475; diet 475–476; distribution 471, 475; genetics 478; habitat and density 475; habits 477–479; harvest 472, 475; home range 478; hybrids 479; intergradation 469; Plate XXXI; reproduction, ontogeny, and mortality 476–477; sexual segregation 478; specimens examined 538
h. hemionus 6–7, **469–474**, 475, 484, 501

as prey 471, 473–474; description 469–470; diet 471–473; distribution 470; exploitation 471, 474; habitat and density 470–471; habits 474; harvest 471–472; home range 474; hybrids 474; intergradation 469; morphometrics 465; Plate XXX; reproduction, ontogeny, and mortality 473–474; sexual segregation 474; specimens examined 538
leucurus 8, 479, 480, 484
macrotis 469
macrourus 479
ochrourus 474, 479, 480, 484
virginianus 28, 463, 474, **479–484**
as prey 362, 481; description 479; diet 482; distribution 479–480; fossils 480; geographic variation 480; habitat and density 480–482; habits 483–484; home range 483–484; hybrids 484; morphometrics 465; Plate XXX; reproduction, ontogeny, and mortality 482–483; specimens examined 538; taxonomy 480
Oligochaeta 43, 51, 57, 63, 66, 69, 71
Omus
audouini 55
dejeani 60
Oncorhynchus 383, 386, 387
keta 395
kisutch 396
mykiss 389
tshawytscha 389
Ondatra 3, 296, 303, 311, 320, 322
occipitalis 335, 336
osoyoosensis 335, 336, 338
zibethica 335, 339
zibethicus 36, 154, 328, **335–339**, 424
as prey 338, 428, 430, 435; cycles 336; description 335–336; diet 336, 338; diseases 339; distribution 336; fossils 336; geographic variation 336; habitat and density 336; habits 338–339; harvest 337; home range 339; lodges 338–339, Plate XVII; morphometrics 300–310; Plate XVIII; reproduction, ontogeny, and mortality 302, 338; specimens examined 538; taxonomy 339
Onychomys 3
brevicaudus 276–278
durranti 276, 277
†*fossilis* 277
fuscogriseus 277
leucogaster 247, **276–280**
as predator 277–278; as prey 278; description 276; diet 277–278; distribution 276–277; fossils 277; geographic variation 277; habitat and density 277; habits 278–280; home range 279; morphometrics 265; Plate XIV; reproduction, ontogeny, and mortality 266, 278; specimens examined 538–539; vocalizations 279
fuliginosus 276

torridus 279
Orchestoidea californiana 68, 272
Orcinus orca 384, 387, 390, 393, 396, 439
Oreamnos 3, 490
montanus 493
americanus **493–497**
as prey 496; description 493; diet 494–495; distribution 493, 495; fossils 493; geographic variation 493; habitat and density 493–494; habits 496–497; introduction 497; Plate XXXII; reproduction, ontogeny, and mortality 495–496; taxonomy 497
Orgyia pseudotsugata 104
Ornithodoros hermsi 37
Orthroptera 63, 111, 399
Oryctolagus cuniculus 123, 503
Oryctomys bottae 224
†*Osteoborus* 21
Otaria
californiana 388
stellerii 385
Otariidae **381–391**
characteristics 381; distribution 381; fossils 381; key to species 381; natural history 381; species diversity 381
Otognosis longimembris 240
Otospermophilus 158
otter
river (see *Lutra canadensis*)
sea (see *Enhydra lutris*)
Otus asio 342
Ovibos 490
Ovis 3, 490
californiana 497, 498, 500
californianus 497
canadensis 17, 493, **497–502**
as prey 500–501; description 497, 498; diet 500; distribution 497–498, 499; fossils 498; extirpation 502; geographic variation 498; habitat and density 498–500; habits 501–502; harvest 501; home range 501; Plate XXXII; reintroduction 502; reproduction, ontogeny, and mortality 500–501; specimens examined 539
Pacifasticus 435
†*Paleocastor* 19
†*Paleolagus* 19, 20
Panthera onca 450
†*Paracamelus* 21
†*Paradomnina* 20
†*Parahippus* 19, 20
†*Paraphenocomys* 311
Parapliosaccomys 1
†*Paratylops* 19
Parophrys vetulus 395
Pelecypoda 402
†*Pelycosauria* 2
Pentatomidae 117
†*Peridionomys* 20
†*Peridipodomys* 20
Perodipus
columbianus 254

microps 251
ordii 254
preblei 251
†*Peridomys* 20
Perognathus 3, 20, 21
  longimembris 240–245, 249
    description 240; diet 241; distribution 240, 241; geographic variation 240; habitat and density 240–241; habits 2443–245; morphometrics 242–243; reproduction, ontogeny, and mortality 241–244; specimens examined 539; torpor 243–244
  lordi 245, 246
  monticola 245
  mollipilosus 245, 246
  nevadensis 240
  olivaceous 245, 246
  parvus 244, 245–248, 277
    as prey 247; description 245; diet 246; distribution 245–246; fossils 245–246 geographic variation 246; habitat and density 246; habits 247; home range 247; morphometrics 242–243; Plate XIII; reproduction, ontogeny, and mortality 244, 246–247; specimens examined 539–540; taxonomy 246; torpor 247
  †*stevei* 245
Peromyscus 3, 20, 21, 63, 410, 413, 417, 419, 425, 430
  aremisiae 270
  boylii 503
  crinitus 33, 267–270, 274
    description 267; diet 268–269; distribution 267; fossils 267; geographic variation 267; habitat and density 267–268; habits 269; morphometrics 268 reproduction, ontogeny, and mortality 266, 269; specimens examined 540; torpor 269
  gambeli 270
  gambelii 270
  gilberti 273, 274
  maniculatus 7, 9, 35, 49, 90, 236, 247, 256, 262, 268, 269, 270–273, 275, 276, 280, 334, 344, 408, 416, 419, 503
    as prey 272, 459; description 270; diet 271–272; distribution 270, 271; fossils 270; geographic variation 270; habitat and density 270–271; habits 272–273; home range 273; morphometrics 268; Plate XIV; reproduction, ontogeny, and mortality 266, 272; specimens examined 540–545; torpor 273
  oreas 270
  perimekurus 270
  preblei 273, 274
  rubidus 270
  sequoiensis 273, 274
  sonoriensis 270
  Truei 273
  truei 13, 28, 273–276, 280

    as prey 275; description 273–274; diet 275; distribution 274; fossils 274; geographic variation 274; habitat and density 274–276; habits 275–276; home range 276; morphometrics 268; Plate XIV; reproduction, ontogeny, and mortality 266, 275; specimens examined 545
Phalangida 91
Phasianus colchicus 135, 371
Phenacomys 3, 296, 298, 303, 320, 322, 408, 459, 460
  albipes 22, 304–306, 309
    description 297, 304–305; diet 306; distribution 305, 306; Fig. 11-100; geographic variation 305; habitat and density 305–306; habits 306; morphometrics 301; reproduction, ontogeny, and mortality 302, 306; specimens examined 545–546; status 306; taxonomy 311
  †*brachyodus* 311
  †*deeringensis* 311
  intermedius 30, 307–309
    as prey 308–309; description 297, 307; diet 308; distribution 307, 308; fossils 307; geographic variation 307; habitat and density 307–308; habits 309; morphometrics 301; reproduction, ontogeny, and mortality 302, 308–309; specimens examined 546
  †*gryci* 311
  longicaudus 22, 62, 306, 309–311
    as prey 310; description 297, 309; diet 310; distribution 309–310; fossils 310; geographic variation 310; habitat and density 310; habits 310–311; morphometrics 301; Plate XVII; reproduction, ontogeny, and mortality 302, 310; specimens examined 546–547; taxonomy 311
  olympicus 307
  oramontis 307
  pomo 311
  silvicola 309, 310
Phoca 3, 390
  fasciata 390, 504
  hispida 390, 503
  jubata 385
  richardii 396
  richardsi 396
  ursina 381
  vitulina 21, 390, 391, 394–397, 503, 504
    as predator 395–396; as prey 387, 396; description 394; diet 395–396; distribution 394, 395; exploitation 394–395; fossils 394; geographic variation 394; habitat and density 394–395; habits 396; morphometrics 382; Plate XXIV; reproduction, ontogeny, and mortality 396; specimens examined 547; taxonomy 396

Phocidae 390–397
  characteristics 390; distribution 390; fossils 391; key to species 391; natural history 390–391; species diversity 390
Phocinae 504
physiographic provinces 22–33
  Basin and Range 22, 24–26, 30–32, 33, 56, 113
  Blue Mountains 22, 24–26, 32, 33, 113, 116
  Cascades 22, 24–26, 28–30
  Coast Range 22, 24–25, 26
  Deschutes-Columbia Plateau 22, 24–26, 29, 30, 56, 102, 201, 446
  High Lava Plains 22, 24–26, 30, 31, 56
  Klamath Mountains 22, 24–26, 27–28
  Owyhee Uplands 22, 24–25, 26, 32–33
  Willamette Valley 22, 24–25, 26–27
Pica 325
Picidae 218
pika, American
  (see *Ochotona princeps*)
Pinnipedia 354
Pipilo erythrophthalmus 371
Pipistrellus hesperus 3, 33, 105–107
  description 105, 106; diet 104, 105; distribution 105, 106; Fig. 9-28; flight speed 75; fossils 105; geographic variation 105; habitat and density 105; habits 106–107; morphometrics 78; reproduction, ontogeny, and mortality 101, 105–106; roosts 106; specimens examined 547; torpor 106
†*Pithanotharia starri* 22
Pituophis
  catenifer 133, 225
  melanoleucus 148, 200, 203
plague 36, 183–184, 293, 325
†*Platybelodon* 21
†*Platygonus* 22
Plecoptera 57
Plecotus 116
  pallescens 113
  townsendii 113
†*Plesiogulo* 21
Plethodon 63
Pleuronichthys 387
†*Pleurolicus* 19
†*Pliauchenia* 21
†*Pliocyon* 20
†*Pliohippus* 21
†*Plionictus* 21
†*Pliosaccomys* 21
†*Pliotaxidea* 21
  nevadensis 429
pneumonia 500
pocket gopher
  Botta's (see *Thomomys bottae*)
  camas (see *Thomomys bulbivorus*)
  northern (see *Thomomys talpoides*)
  western (see *Thomomys mazama*)
  Townsend's (see *Thomomys townsendii*)
pocket mouse
  little (see *Perognathus longimembris*)
  Great Basin (see *Perognathus parvus*)
Polyphylla 249

crinta 117
†*Pontolias* 19
   magnus 21, 22
porcupine, common
   (see *Erethizon dorsatum*)
†*Probainognathus* 33
†*Procamelus* 21
Procyon 3, 8, 397
   excelsus 400, 401
   lotor 35, 67, 313, 332, 338, 373, 397,
      400–405
         as predator 399; as prey 404; de-
         scription 400–401; diet 402; distri-
         bution 401; geographic variation
         401; habitat and density 401–402;
         habits 404; harvest 403; home range
         404; morphometrics 398–399; Plate
         XXV; reproduction, ontogeny, and
         mortality 402, 404; specimens ex-
         amined 547
   pacifica 400
   pacificus 401
   psora 400
Procyonidae 373, **397–405**
   characteristics 397; distribution 397;
   fossils 397; key to species 397; spe-
   cies diversity 397
†*Prodipodomys* 20
†*Prolagus* 124
pronghorn (see *Antilocapra americana*)
†*Prosomys* 21
†*Prosthennops* 19, 20, 21
   cf. *rex* 21
†*Protosciurus* 19
†*Protospermophilus* 19, 20
Protostrongylus 500
†*Pseudaeluras* 20, 21
†*Pseudotheridomys* 20
pseudotuberculosis 339
Psoroptes ovis 500
Pteromys oregonensis 219
Pugettia 438
puma (see *Puma concolor*)
Puma 3
   californica 451
   concolor 28, 156, 189, 347, **450–455**,
      458
         as predator 360, 447, 450–453, 496,
         501; description 450, 451; diet 452–
         453; distribution 450–451; fossils
         451; geographic variation 451;
         habitat and density 451–452; hab-
         its 454; harvest 455; morphome-
         trics 452–453; Plate XXIX; repro-
         duction, ontogeny, and mortality
         453–454; specimens examined
         547–548
   kaibabensis 451
   missoulensis 451
   oregonensis 450, 451
Putorius
   energumenos 422
   muricus 415
   streatori 415
   saturatus 418
   vison 422

washingtoni 418
Q fever 38
rabbit
   brush (see *Sylvilagus bachmani*)
   cottontail (see cottontail)
   jack (see jackrabbit)
   Old World (see *Oryctolagus cuniculus*)
   pygmy (see *Brachylagus idahoensis*)
rabies 35–36, 43, 79, 109, 404–405, 445,
   448, 450
raccoon, common (see *Procyon lotor*)
†*Rakomeryx* 20
rat
   black (see *Rattus rattus*)
   California kangaroo
      (see *Dipodomys californicus*)
   chisel-toothed kangaroo
      (see *Dipodomys microps*)
   Merriam's kangaroo
      (see *Dipodomys merriami*)
   Norway (see *Rattus norvegicus*)
   Ord's kangaroo (see *Dipodomys ordii*)
   (also see woodrat)
Rattus 3, 9, 36, 37, 63, 442
   alexandrinus 292
   frugivorus 292
   norvegicus **289–291**, 293, 295
         as commensal 289; description 289;
         diet 290; distribution 289; geo-
         graphic variation 289; habitat and
         density 289–290; habits 290–291;
         introduction 289; morphometrics
         290; Plate XVI; reproduction, on-
         togeny, and mortality 290; speci-
         mens examined 548
   rattus **291–293**
         as commensal 292; as pest 293;
         description 291; diet 292; distribu-
         tion 291–292; geographic variation
         292; habitat and density 292; hab-
         its 293; introduction 292; morpho-
         metrics 290; reproduction, ontog-
         eny, and mortality 292–293; speci-
         mens examined 548
red-digger (see *Spermophilus columbianus*)
Reithrodon
   longicauda 263
   megalotis 263
Reithrodontomys 3, 419, 430
   klamathensis 263
   longicauda 263
   longicaudus 264
   megalotis **263–267**, 317
         as prey 266; description 263, 264;
         diet 264; distribution 263, 264; Fig.
         11-71; fossils 263; geographic
         variation 264; habitat and density
         264; habits 266–267; home range
         267; morphometrics 265; reproduc-
         tion, ontogeny, and mortality 264–
         266; specimens examined 548; tor-
         por 267
ringtail (see *Bassariscus astutus*)
ringworm 208, 339
Rocky Mountain spotted fever 37–38, 43
Rodentia 74, **153–353**

characteristics 153; distribution 153;
key to families 153; natural history 153;
species diversity 153
Rupicapra
   americana 493
   rupicapra 497
Saccophorus 224
Salmonidae 435
Salvelinus fontinalis 309, 424
Saturniidae 117
scabies 500
†*Scalopoides* 20, 21
Scalops aeneus 71
Scalopus
   latimanus 68
   Townsendii 71
Scandentia 45
†*Scapanoscapter* 20
Scapanus 2, 3, 20, 21, 65
   alpinus 68
   dilatus 68
   latimanus 21, 28, **68–69**, 71, 183
         as prey 413, 419; description 68;
         diet 68; distribution 68; fossils 68;
         geographic variation 68; morpho-
         metrics 67; specimens examined
         548–549
   orarius 22, **69–71**
         as prey 70; capture 70; description
         69; diet 69–70; distribution 69–70;
         geographic variation 69; habitat and
         density 69; habits 70; home range
         70; morphometrics 67; Plate II; re-
         production, ontogeny, and mortal-
         ity 66, 70; specimens examined
         549–550
   schefferi 69
   Townsendii 71
   townsendii 9, 27, 69, 70, **71–73**
         as pest 73; as prey 72, 371; ben-
         efits of 73; description 71; diet 71–
         72; distribution 71, 72; Fig. 8-27;
         geographic variation 71; habitat and
         density 71; habits 72–73; morpho-
         metrics 67; reproduction, ontogeny,
         and mortality 66, 72; specimens ex-
         amined 550
Scarabaeidae 111, 446
Sciuridae **157–223**, 441
   characteristics 157; distribution 157;
   fossils 157; key to species 158–159;
   natural history 157–158; species diversity 158; torpor 158
Sciuropterus
   alpinus 219
   bangsi 219
   fuliginosus 219
   klamathensis 219
Sciurus 3, 20
   albolimbatus 212
   Belcheri 212
   californicus 212
   carolinensis 3, **203–207**, 209–211
         as pest 206–207; description 203–
         204; diet 206; distribution 204;
         geographic variation 204; habitat

and density 204, 206; habits 206; home range 206; introduction 206; morphometrics 205; Plate X; reproduction, ontogeny, and mortality 206; specimens examined 550

cascadensis 212

Douglasii 212

douglasii 212

fossor 207

griseus 3, 8, 204, **207–209**, 210–212
    as prey 208, 413; description 207; diet 208; distribution 207, 208; geographic variation 207; habitat and density 207–208; habits 209; harvest 209; morphometrics 205; Plate X; reproduction, ontogeny, and mortality 162, 208–209; specimens examined 550–551

hudsonicus 212, 216

lateralis 193

leporinus 207

molli-pilosus 212

niger 3, 204, 206, 209, **210–212**, 292
    as pest 212; description 210; diet 211; distribution 210; geographic variation 210; habitat and density 210–211; habits 211; introduction 211–212; morphometrics 205; Plate X; reproduction, ontogeny, and mortality 211; specimens examined 551

Pennsylvanica 203

pennsylvanicus 203, 204

Richardsoni 216

Sabrinus 219

vulgaris 216

Scoliidae 117

Scorpiones 117, 266, 278

Scotophilus hesperus 105

seal
    harbor (see *Phoca vitulina*)
    hooded (see *Cystophora cristata*)
    northern elephant
        (see *Mirounga angustirostris*)
    northern fur (see *Callorhinus ursinus*)
    ribbon (see *Phoca fasciata*)
    ringed (see *Phoca hispida*)

sea lion
    California (see *Zalophus californianus*)
    northern (see *Eumetopias jubatus*)

Sebastes 383, 387

Serpentes 62, 64, 70, 72, 118, 121, 183, 266, 272, 275, 332, 417

†*Sewelleladon* 19

sheep, bighorn (see *Ovis canadensis*)

shrew
    Baird's (see *Sorex bairdi*)
    dusky (see *Sorex monticolus*)
    fog (see *Sorex sonomae*)
    marsh (see *Sorex bendirii*)
    Merriam's (see *Sorex merriami*)
    montane (see *Sorex monticolus*)
    Pacific (see *Sorex pacificus*)
    Pacific water (see *Sorex bendirii*)
    Preble's (see *Sorex prebeli*)
    Trowbridge's (see *Sorex trowbridgei*)

vagrant (see *Sorex vagrans*)
    water (see *Sorex palustris*)

shrew-mole (see *Neurotrichus gibbsii*)

Sialidae 117

Sicista 339

Sigmodontinae **262–288**
    characteristics 262; key to species 263; natural history 262–263; species diversity 262

Silphidae 117

Siphonaptera 36

Sitomys
    americanus 270
    artemisiae 270
    gilberti 273

Skjrabingylus 441
    nasicola 416, 418

skunk
    striped (see *Mephitis mephitis*)
    western spotted (see *Spilogale gracilis*)

†*Smilodon fatalis* 450

Solenodontidae 45

Solifugae 249

Solpugida 117

Sorex 3, 425, 430
    amoenus 62
    bairdi 30, **46–49**, 53
        description 46; diet 48–49; distribution 46, 49; geographic variation 46, 48; habitat and density 48; habits 49; morphometrics 47; reproduction, ontogeny, and mortality 49; specimens examined 551; taxonomy 49
    bairdii 46
    bendirii 22, **49–51**, 55, 344
        description 49–50; diet 50–51; distribution 50; geographic variation 50; habitat and density 50; habits 51; morphometrics 47; Plate I; reproduction, ontogeny, and mortality 49, 51; specimens examined 551–552
    cascadensis 54
    cinereus 56
    dobsoni 62
    mariposae 60, 61
    merriami 31, **51–52**, 58
        as prey 52; description 51, 52; diet 52; distribution 51–52; geographic variation 51; habitat and density 51–52; habits 52; morphometrics 47; reproduction, ontogeny, and mortality 49, 52; specimens examined 552
    montereyensis 60
    monticolus 33, 46, 49, **52–54**, 61
        description 52–53; diet 54; distribution 53; geographic variation 53; habitat and density 53–54; habits 54; morphometrics 57; reproduction, ontogeny, and mortality 49, 54; specimens examined 552–553
    navigator 55
    nevadensi 62
    obscurus 46, 52, 65

ornatus 62

pacificus 22, 30, **54–55**, 58, 60
    description 54, 55; diet 51, 55; distribution 54, 55; geographic variation 54; habitat and density 54–55; habits 55; morphometrics 47; reproduction, ontogeny, and mortality 49, 55; specimens examined 553–554; taxonomy 55

palmeri 49

permiliensis 46

palustris 30, **55–57**
    as prey 57; description 55–56; diet 57; distribution 56; geographic variation 56; habitat and density 56; habits 51; morphometrics 48; Plate I; reproduction, ontogeny, and mortality 49, 57; specimens examined 554

preblei 31, **57–58**
    description 57, 58; distribution 57, 58; geographic variation 57, 58; habitat and density 58; morphometrics 48; specimens examined 554

setosus 52

shastensis 62

similis 52

sonomae 22, 55, **58–60**
    description 58–59; diet 51, 59–60; distribution 59; geographic variation 59; habitat and density 59; habits 60; morphometrics 48; Plate I; reproduction, ontogeny, and mortality 49, 60; specimens examined 554–556; taxonomy 60

suckleyi 62

tenelliodus 58, 59

trigonirostris 62, 65

trowbridgii 9, 22, 55, **60–62**, 344
    as prey 62; description 60; diet 51, 61; distribution 60–61; geographic variation 61; habitat and density 61; habits 62; morphometrics 48; Plate II; reproduction, ontogeny, and mortality 49, 61–62; specimens examined 556–558

vagrans xiv, 9, 46, 52, 54, 58, 61, **62–65**
    as prey 64; description 62, 63; diet 51, 63–64; distribution 62, 63; fossils 62; geographic variation 62; habitat and density 62–63; habits 64–65; home range 65; morphometrics 48; Plate II; reproduction, ontogeny, and mortality 49, 64; specimens examined 558–560; taxonomy 65

yaquinae 54, 55, 60

Soricidae 2, **46–65**, 417
    capture 46; characteristics 46; species diversity 46

†Sparassocynidae 40

Spermophilus 3, 20, 21, 36, 158, 178, 408, 419, 459
    armatus 432
    artemisiae 196

beecheyi 9, **181–184**, 209
    as predator 184; as prey 182, 364,
    371, 459; description 181; diet 182;
    distribution 181; fossils 181; geo-
    graphic variation 181; habitat and
    density 181–182; habits 182–183;
    morphometrics 175; Plate VIII; re-
    production, ontogeny, and mortal-
    ity 176, 182; specimens examined
    560–561; torpor 182–183
beldingi 33, 180, **184–187**, 188, 193,
    201
    as prey 186, 419, 430; baits 185,
    186; description 184; diet 185; dis-
    tribution 184, 185; geographic
    variation 184; habitat and density
    184–185; habits 186–187; morpho-
    metrics 175; Plate VIII; reproduc-
    tion, ontogeny, and mortality 177,
    185–186; specimens examined
    561–562; torpor 186; vocalizations
    187
canus 33, 184, **196–198**, 199, 201
    as prey 198; description 196–197;
    diet 197–198; distribution 197; geo-
    graphic variation 197; habitat and
    density 197; habits 198; morpho-
    metrics 175; Plate IX; reproduction,
    ontogeny, and mortality 198; speci-
    mens examined 562; taxonomy 198
chrysodeirus 194
columbianus 33, 176, 184, **187–190**,
    218
    as prey 189; description 187; diet
    188–189; distribution 187–188;
    fossils 188; geographic variation
    188; habitat and density 188; hab-
    its 189–190; home range 190;
    morphometrics 175; Plate IX; pos-
    tures 188–190; reproduction, on-
    togeny, and mortality 177, 189;
    specimens examined 562; torpor
    189; vocalizations 190
connectens 194
creber 184
douglasii 181
elegans 17, 184, **190–193**, 256
    description 190; diet 191; distribu-
    tion 190–191; fossils 190–191;
    geographic variation 191; habitat
    and density 191; habits 192–193;
    home range 193; morphometrics
    175; Plate IX; reproduction, ontog-
    eny, and mortality 177, 191–192;
    specimens examined 562; status
    193; taxonomy 193; torpor 192
franklini 430
idahoensis 196
lateralis 165, 167, 184, **193–196**, 218
    description 193; diet 194; distribu-
    tion 193–194; fossils 194; geo-
    graphic variation 194; habitat and
    density 194; habits 195–196; home
    range 195; morphometrics 175–
    176; Plate IX; reproduction, ontog-
    eny, and mortality 177, 194–195;

specimens examined 563–564
loringi 202
mollis 184, 196, 197, **198–201**, 256,
    367, 430
    as prey 200; description 198–199;
    diet 199–200; distribution 199; geo-
    graphic variation 199; habitat and
    density 199; habits 200; home
    range 201; morphometrics 176; re-
    production, ontogeny, and mortal-
    ity 177, 200; specimens examined
    564; taxonomy 200–201; torpor
    200
nancyae 196
nevadensis 190, 193
oregonus 194
richardsonii 190–193
ruficaudus 188
stephensi 198
taxonomic considerations 158
townsendii 196, 198, 203
    taxonomic considerations 196
trepidus 194
tridecemlineatus 6, 430
trinitatus 194
vigilis 196–198
washingtoni 30, 184, 196, **201–203**
    as prey 203; description 201; diet
    202; distribution 201–202; geo-
    graphic variation 202; habitat and
    density 202; habits 203; morphome-
    trics 176; Plate IX; reproduction,
    ontogeny, and mortality 177, 202–
    203; specimens examined 564; sta-
    tus 203; taxonomy 196, 203
†*Sphenophalos* 21
Sphingidae 117
Spilogale 3
    gracilis 272, 302, 313, 332, 400, **441–**
    445
        as predator 442–443; description
        441, 442; diet 442–443; distribution
        441, 442; fossils 441; geographic
        variation 441; habitat and density
        441–442; habits 443, 445; harvest
        444; morphometrics 440–441;
        musk 443, 445; Plate XXVIII; re-
        production, ontogeny, and mortal-
        ity 443, 445; specimens examined
        564; taxonomy 445
    latifrons 441, 443
    olympica 441
    phenax 441
    putorius 441–443, 445
    rexroadi 441
    saxatilis 441
Squamata 118, 183
squirrel (also see ground squirrel)
    Douglas' (see *Tamiasciurus douglasii*)
    eastern fox (see *Sciurus niger*)
    eastern gray (see *Sciurus carolensis*)
    northern flying (see *Glaucomys sabrinus*)
    red (see *Tamiasciurus hudsonicus*)
    western gray (see *Sciurus griseus*)
    white-tailed antelope
        (see *Ammospermophilus leucurus*)

Stenopelmatus 117
    fuscus 245, 249
†*Sthenictis* 21
stoat (see *Mustela erminea*)
Strigidae 118
Strix
    nebulosa 304, 308, 342
    occidentalis 202, 285, 302, 308–310
    varia 62
Strongylocentrotus
    franciscanus 438
    purpuratus 438
Sylvilagus 3, 123, 128, 129, 134, 417, 419,
    430
    alacer 135
    audubonii 367
    bachmani 27, **131–134**, 137
        as pest 134; as prey 133, 359, 364,
        371, 413, 459; description 131,
        132; diet 132; distribution 131–
        133; geographic variation 133;
        habitat and density 132; habits 133–
        134; home range 133–134; hybrids
        134; morphometrics 130–131; Plate
        V; reproduction, ontogeny, and
        mortality 127, 132–133; specimens
        examined 564
    floridanus 36, 123, 131, **134–138**, 145,
        149, 151, 285, 442, 503
        as game 137; as prey 420; descrip-
        tion 132, 134; diet 135–136; distri-
        bution 134–135; geographic varia-
        tion 135; habitat and density 135;
        habits 136–137; harvest 137; intro-
        duction 134–135; morphometrics
        130–131; Plate V; reproduction, on-
        togeny, and mortality 127, 136;
        specimens examined 564–565
    idahoensis 128
    llanensis 135
    mearnsii 134, 135
    nuttallii 9, 22, 131, 134, 137, **138–140**,
        162, 197
        as game 14; as prey 140, 359; de-
        scription 132, 138; diet 139; distri-
        bution 138, 139; fossils 139; genet-
        ics 140; geographic variation 138;
        habitat and density 138–139; hab-
        its 140; morphometrics 130–131;
        Plate V; reproduction, ontogeny,
        and mortality 127, 139–140; speci-
        mens examined 565
    similis 135
    sylvaticus 134
    tehamae 131, 132
    ubericolor 131, 132
†Symmetrodonta 3
†Synapsida 2
Synthliboramphus antiquum 293
Tachycineta thalassina 106
Tadarida 3
    brasiliensis 35, 83, 97, 118, **119–122**
        as prey 121; description 119; diet
        120; distribution 119–120, 121;
        flight speed 75; fossils 120; geo-
        graphic variation 120; habitat and

density 120; habits 121–122; morphometrics 78; Plate IV; reproduction, ontogeny, and mortality 120–121; roosts 121; specimens examined 565

mexicana 119

molossa 119

Talpidae 2, **65–73**

characteristics 65; distribution 65; species diversity 65

Tamias 3, 20, 21, 36, 158, 193, 380, 408, 417, 419, 425, 530

albiventris 160

alleni 166

amoenus 37, **159–163**, 164–167, 172

as prey 163; description 159–160; diet 162; distribution 160; fossils 160; geographic variation 160; habitat and density 160–162; habits 163; home range 163; morphometrics 161; Plate VII; reproduction, ontogeny, and mortality 162–163; specimens examined 565–567; torpor 163

chrysodeirus 193

cooperi 166, 169, 170, 173

dorsalis 166

leucurus 178

littoralis 169

ludibundus 160

merriami 166

minimus 9, 159–160, **163–166**, 170, 172

as prey 165; description 163–164; diet 165; distribution 164; fossils 164; geographic variation 164; habitat and density 164–165; habits 165–166; home range 165; morphometrics 161; Plate VII reproduction, ontogeny, and mortality 162, 165; specimens examined 567–568; torpor 165

ochraceus 160

ochrogenys 166

pictus 163

quadrimaculatus 166

scrutator 163, 164

senex 30, **166–168**, 170, 172

description 166–167; diet 167–168; distribution 166; geographic variation 166; habitat and density 167; habits 168; morphometrics 161; Plate VII; reproduction, ontogeny, and mortality 162, 168; specimens examined 568; taxonomy 166; torpor 168

siskiyou 28, 30, 167, **168–169**, 170, 173

description 168; diet 169; distribution 168, 169; geographic variation 168; habitat and density 168–169; habits 169; home range 169; morphometrics 161; Plate VII; reproduction, ontogeny, and mortality 169; specimens examined 568–569; taxonomy 166

striatus 43

taxonomic considerations 158

Townsendii 169

townsendii 8, 162, 167, 168, **169–173**, 416

as pest 173; as prey 172, 359, 459; benefits of 173; description 169–170; diet 171–172; distribution 170; geographic variation 170; habitat and density 170–171; habits 172–173; home range 173; morphometrics 161; Plate VII; reproduction, ontogeny, and mortality 162, 172; specimens examined 569–571; taxonomy 166; vocalizations 172

Tamiasciurus 3, 408, 411, 430

albolimbatus 212, 213

douglasii 8, **212–216**, 218

as prey 459; description 212; diet 213–214; distribution 212–213; fossils 213; geographic variation 213; habitat and density 213; habits 214–215; morphometrics 205; Plate XI; reproduction, ontogeny, and mortality 162, 214; specimens examined 571–572; taxonomy 215–216; vocalizations 214

hudsonicus 6, 33, 37, 43, 212–215, **216–218**

as prey 459; description 216; diet 217; distribution 216, 217; fossils 216; geographic variation 216; habitat and density 216–217; habits 218; morphometrics 205; Plate XI; reproduction, ontogeny, and mortality 217–218; specimens examined 572–573; taxonomy 215–216; vocalizations 218

mollipilosus 212, 213

richardsoni 216

unclassified, specimens examined 573–574

Tapirus

cf. †*californicus* 22

†*haysii* 22

†*Tardontia* 21

Taricha granulosa 435

Taxidea 3

americana 428

jeffersonii 428, 429

montana 428

neglecta 428

sulcata 428

taxus 8, 148, 163, 176, 186, 189, 200, 203, 247, 338, 360, 402, 427, **428–433**

as predator 430, 447, 488; as prey 430; description 428–429, 432; diet 430; distribution 429; fossils 429; geographic variation 429; habitat and density 429–430; habits 430, 432; harvest 431; home range 432; morphometrics 406–407; Plate XXVII; reproduction, ontogeny, and mortality 410, 430; specimens examined 574

†*Teleoceras* 21

Teleostei 57, 338

Tenebrionidae 110, 117, 278

Tenricidae 45

†*Tephroncyon* 20

†*Tetrabelodon* 21

Tetracerus quadricornis 490

Tettigoniidae 117

Thaleichthys pacificus 383, 389, 395

Thamnophis 425

ordinoides 67

Theragra chalcogramma 383, 387

†Therapsida 2

†Theriodonta 3

Thomomys 3, 6, 223, 231, 234, 419

atrogriseus 238

bachmani 238, 239

bottae 28, **224–228**, 237–239

as prey 225; description 224; diet 225; distribution 224, 225; fossils 224; geographic variation 224; habitat and density 224–225; habits 225, 228; Mima mounds 228; morphometrics 226; reproduction, ontogeny, and mortality 225, 228; specimens examined 574; taxonomy 228

bulbivora 229

bulbivorus 6, 22, 27, 223, 224, 228, **229–231**, 238, 239, 319

as prey 230; capture 231; description 229; diet 230; distribution 229; genetic variation 231; geographic variation 229; habitat and density 229–230; habits 230–231; morphometrics 226; Plate XII; reproduction, ontogeny, and mortality 228, 230; specimens examined 574–575; tunnel ventilation 231

columbianus 234

detumidus 224, 228

douglasi 6, 233

elkoenis 239

fuscus 234

helleri 231–233

hesperus 6, 231–233

laticeps 224, 228

leucodon 224, 228

mazama 6, 30, 223, 224, **231–234**, 235

as prey 460; description 231–232; diet 233; distribution 232; geographic variation 232; habitat and density 232–233; habits 233–234; morphometrics 226–227; reproduction, ontogeny, and mortality 228, 233; specimens examined 575–576; taxonomy 233

monticola 233

nasicus 231–233

nevadensis 238, 239

niger 6, 231–234

oregonus 231–234

owyhensis 239

quadratus 234

relictus 239
similis 239
talpoides 9, 33, 187, 224, 225, 232, 233, **234–238**, 239, 333
as pest 235; as prey 237, 430; capture 238; description 234; diet 236; distribution 234; fossils 234; geographic variation 234; habitat and density 234–236; habits 237–238; Mima mounds 235; morphometrics 227; Plate XIII; reproduction, ontogeny, and mortality 228, 236–237; specimens examined 576–577
townsendii 33, 224, **238–239**
description 238; diet 238; distribution 238, 239; fossils 238; geographic variation 238; habitat and density 238; habits 239; morphometrics 227; reproduction, ontogeny, and mortality 228, 238–239; specimens examined 577; taxonomy 239
umbrinus 228, 239
wallowa 234
†*Ticholeptus* 19, 20
tick-borne relapsing fever 37
†*Tomarctus* 20
Trichechus manatus 2
Trichophyton rubrum 208
†Triconodonta 3
Tricoptera 57, 80, 89, 96, 117
tularemia 36, 43, 325, 339
Tupaiidae 45
Tyto alba 67, 70, 72, 133, 148, 198, 225, 230, 237, 239, 255, 266, 272, 282, 295, 316, 319, 322, 327, 332, 345
Tytonidae 118
Tyzzer's disease 339
†*Uintatherium* 19
Urocyon 3, 222, 355
californicus 363
cinereoargenteus 8, 28, 133, 182, 208, 247, 272, 313, 319, 332, 354, **363–366**
as predator 420, 447; description 363, 364; diet 364, 366; distribution 363–364; fossils 364; geographic variation 364; habitat and density 364; habits 366; harvest 365; home range 366; morphometrics 357; Plate XXI; reproduction, ontogeny, and mortality 355, 366; specimens examined 577–578; taxonomy 366
townsendi 363
Urodela 424
Urotrichus gibbsii 65
Ursidae 21, **373–381**
characteristics 373; distribution 373; fossils 373; key to species 373–374; natural history 373; species diversity 373
Ursus 3, 400, 426
†*abstrusus* 375
altifrontalis 374, 375
americanus 8, 292, 373, **374–378**

as pest 373–378; as predator 360; description 374; diet 376; distribution 374–375; fossils 375; geographic variation 375; habitat and density 375–376; habits 376–377; harvest 375, 378; home range 377; morphometrics 375; Plate XXII; reproduction, ontogeny, and mortality 376; specimens examined 578; torpor 377
arctos 17, 189, 354, 373, 374, **378–381**
as predator 387; description 378; diet 379–380; distribution 378–379; fossils 379; geographic variation 379; habitat and density 379; habits 380; morphometrics 375; Plate XXII; reproduction, ontogeny, and mortality 380; specimens examined 578; status 379; taxonomy 380–381; vocalizations 380
californiensis 375
cinnamomum 374, 375
†*etruscus* 379
horribilis 378–380
idahoensis 378
klamathensis 378
lotor 400
luscus 426
middendorffii 380
†*minimus* 375
mirus 378
taxus 428
†*Ustatochoerus* 21
Velella velella 272
Vespertilio
californicus 79
chrysonotus 84
ciliolabrum 82
cinereus 98
evotis 84
fuscus 107
leibii 84
linereus 98
longicrus 93
lucifugus 87
melanorhinus 82
noctivagans 100
pallidus 116
volans 93
yumanensis 95
Vespertilionidae 74, **76–119**
characteristics 76; diet 76; distribution 76; natural history 76, 79
Viverra mephitis 445
Viverridae 445
vole
California (see *Microtus californicus*)
creeping (see *Microtus oregoni*)
gray-tailed (see *Microtus canicaudus*)
heather (see *Phenacomys intermedius*)
long-tailed (see *Microtus longicaudus*)
montane (see *Microtus montanus*)
red tree (see *Phenacomys longicaudus*)
sagebrush (see *Lemmiscus curtatus*)
southern red-backed

(see *Clethrionomys gapperi*)
Townsend's (see *Microtus townsendii*)
water (see *Microtus richardsoni*)
western red-backed
(see *Clethrionomys californicus*)
white-footed (see *Phenacomys albipes*)
Vulpes 3, 21, 222, 355, 363, 366
cascadensis 369, 371, 373
fulva 372
fulvus 369
macrotis 366, 368
macroura 369, 371
macrourus 369
nevadensis 366, 368
velox 247, 256, 354, **366–368**
as prey 368; description 366; diet 367–368; distribution 366–367; geographic variation 367; habitat and density 367; habits 368; morphometrics 357; Plate XXI; reproduction, ontogeny, and mortality 355, 368; specimens examined 578; status 368; taxonomy 368
vulpes 9, 27, 35, 37, 129, 182, 247, 259, 272, 275, 292, 319, 332, 338, 347, 354, 360, 364, 368, **369–373**
as predator 404, 420, 447, 462; description 369; diet 371; distribution 369, 371; fossils 371; geographic variation 371; habitat and density 371; habits 372; harvest 370; introduction 372–373; morphometrics 357; Plate XXI; reproduction, ontogeny, and mortality 355, 371–372; specimens examined 578; taxonomy 372
wapiti (see *Cervus elaphus*)
weasel
long-tailed (see *Mustela frenata*)
(also see ermine)
wolf, gray (see *Canis lupus*)
wolverine (see *Gulo gulo*)
woodrat
bushy-tailed (see *Neotoma cinerea*)
desert (see *Neotoma lepida*)
dusky-footed (see *Neotoma fuscipes*)
Zalophus 3, 385
californianus 381, 386, **388–390**
as predator 389; as prey 390; description 388, 389; diet 389; distribution 388, 389; exploitation 390; fossils 388; geographic variation 388; habitat and density 388–389; habits 390; morphometrics 382; Plate XXIII; reproduction, ontogeny, and mortality 389–390; specimens examined 578
Zapodidae 340
Zapodinae 340
Zapus 2, 3, 339, 340, 345, 419, 415, 430
major 340
montanus 343
oregonus 340
pacificus 340
princeps 33, **340–343**, 345
description 340; diet 341; distribu-

tion 340, 341; fossils 340; geographic variation 340; habitat and density 340–341; habits 342–343; morphometrics 342; Plate XIX; reproduction, ontogeny, and mortality 341–342; specimens examined 578–579; torpor 342–343

trinotatus 9, 340, **343–345** as prey 345; description 343; diet 344; distribution 343, 344; geographic variation 343; habitat and density 343–344; habits 345; morphometrics 342; Plate XIX; reproduction, ontogeny, and mortality 344–345; specimens examined 579–580

Zootermopsis nevadensis 54

Designers: B. J. Verts & Leslie N. Carraway
Compositor: Leslie N. Carraway
Text: 10/10.5 Times New Roman, PageMaker 6.5, Windows 95
Display: Times New Roman, PageMaker 6.5, Windows 95
Copyediting: Princeton Editorial Associates, Scottsdale, Arizona
Printer: Friesens, Manitoba, Canada
Color Insert: Cascade Printing Company, Corvallis, Oregon
Binder: Friesens, Manitoba, Canada